27th International Conference on Low Temperature Physics 2014

(LT 27)

Journal of Physics: Conference Series Volume 568

Buenos Aires, Argentina
6-13 August 2014

Editor:

E. Calzetta

ISBN: 978-1-63439-699-8
ISSN: 1742-6588

Printed from e-media with permission by:

Curran Associates, Inc.
57 Morehouse Lane
Red Hook, NY 12571

Some format issues inherent in the e-media version may also appear in this print version.

Printed by Curran Associates, Inc. (2015)

Institute of Physics
Dirac House, Temple Back
Bristol BS1 6BE UK

Phone: 44 1 17 929 7481
Fax: 44 1 17 920 0979

techtracking@iop.org

Additional copies of this publication are available from:

Curran Associates, Inc.
57 Morehouse Lane
Red Hook, NY 12571 USA
Phone: 845-758-0400
Fax: 845-758-2634
Email: curran@proceedings.com
Web: www.proceedings.com

EDITORIAL MATERIAL

PART 1

PART 3

LT27

Buenos Aires, Argentina

August 6 - 13, 2014

Foreword

Esteban Calzetta

Editor in chief, Physics Department, FCEN-UBA and IFIBA-CONICET, Argentina

E-mail: calzetta@df.uba.ar

Abstract. The Low Temperature Physics Conference is an international event held every three years, under the auspices of the IUPAP through its Commission C5 on Low Temperature Physics. These conferences aim at the exchange of information and views among the members of the international scientific community in the general field of Low Temperature Physics, promoting the excellence of invited and contributed presentations. It is a tradition that LT offers updates on the various topics, provided by the highest representatives of the field, as well as oral and poster contributions in the different areas.

LT27 was the first conference of the series that took place in South America. It was therefore a particular challenge for us and it is a source of pride to present these Proceedings as proof of the wonderful scientific atmosphere at the meeting.

We wish to acknowledge our host institutions, sponsors, and the many people whose work made this meeting possible. Our appreciation goes to all of them, as well as to speakers and attendants.

The Low Temperature Physics Conference is an international event held every three years, under the auspices of the IUPAP through its Commission C5 on Low Temperature Physics. These conferences aim at the exchange of information and views among the members of the international scientific community in the general field of Low Temperature Physics, promoting the excellence of invited and contributed presentations. It is a tradition that LT offers updates on the various topics, provided by the highest representatives of the field, as well as oral and poster contributions in the different areas.

LT27 was the first conference of the series that took place in South America. It was therefore a particular challenge for us and it is a source of pride to present these Proceedings as proof of the wonderful scientific atmosphere at the meeting.

We wish to acknowledge our host institutions, Departamento de Física (Facultad de Ciencias Exactas y Naturales -UBA) and Instituto de Física de Buenos Aires (IFIBA), as well as the work of the many people that made this conference possible. We are especially indebted to our institutional sponsors, the International Union of Pure and Applied Physics (IUPAP), Centro Latinoamericano de Física (CLAF), International Centre for Theoretical Physics (ICTP) and Universidad de Buenos Aires (UBA), Consejo Nacional de Investigaciones Científicas y Técnicas (CONICET), Ministerio de Ciencia, Tecnología e Innovación Productiva (MINCYT) and Agencia Nacional de Promoción Científica y Tecnológica (ANPCYT) from Argentina for financial support. The generous contributions of the organizers of the former Conference LT26 as well as our commercial sponsors Attocube, CamCool Research Limited, Cryogenic, Cryomech Inc., ICEOxford, IOP Publishing, Janis Research, Lake Shore Cryotronics, Leiden Cryogenics, Oxford Instruments and Quantum Design International have provided a substantial input to the Conference budget. We also acknowledge very encouraging auspices of Asociación Física Argentina (AFA), Academia Nacional de Ciencias Exactas, Físicas y Naturales, Asociación Argentina para el Progreso de las Ciencias and Gobierno de la Ciudad de Buenos Aires.

We list below the members of the organizing committees. A Conference Summary prepared by John Saunders and Kimitoshi Kono, Secretary and Chair, respectively, of the Commission C5 (Low Temperature Physics) of the IUPAP is also included.

We wish also to acknowledge the IOP personnel, who not only provided the outlet for these Proceedings but were also of invaluable help during the editorial process

Our appreciation goes to all of them, as well as to speakers and attendants.

Organizing Committees

Conference Chair:
Susana Hernández (Universidad de Buenos Aires/IFIBA)

Conference Co-Chair:
Victoria Bekeris (Universidad de Buenos Aires/IFIBA)

IUPAP C5 Chair:
Kimitoshi Kono (RIKEN, Japan)

Financial Managers :

Gustavo Lozano (Universidad de Buenos Aires/IFIBA,) Horacio Grinberg (Universidad de Buenos Aires/IFIBA) and Ana Maria Llois (Centro Atómico Constituyentes/ Universidad de Buenos Aires)

Proceedings Editor-in-Chief: Esteban Calzetta (Universidad de Buenos Aires/IFIBA)

Section 1 - Quantum Gases, Fluids and Solids Chair: Leszek Szybisz (Centro Atómico Constituyentes/ Universidad de Buenos Aires)

Co-Chair: Robert Hallock (University of Massachussets, Amherst, USA)

Editor: Pablo Capuzzi (Universidad de Buenos Aires/IFIBA)

Programme Committee: Eckhard Krotscheck (University of Linz, Austria), Michle Leduc (ENS-LKB-Paris, France), Paul Leiderer (University of Konstanz, Germany) and Yuichi Okuda (Tokio Institute of Technology, Japan).

Local Committee: Pablo Capuzzi (Universidad de Buenos Aires/IFIBA), Susana Hernández (Universidad de Buenos Aires/IFIBA), Anibal Iucci (Universidad Nacional de La Plata) and Javier Luzuriaga (Centro Atómico Bariloche)

Section 2 - Superconductivity Chair: Daniel Domínguez (Centro Atómico Bariloche)

Co-Chair: Michael Norman (Argonne National Laboratory, USA)

Editors: Gabriela Pasquini (Universidad de Buenos Aires/IFIBA) and Yanina Fasano (Centro Atómico Bariloche)

Programme Committee: Paul C. Canfield (Ames Laboratory, USA), Rosario Fazio (Scuola Normale Superiore, Italia) and Eli Zeldov (Weissmann Institute, Israel)

Local Committee: Armando Aligia (Centro Atómico Bariloche), Victor Correa (Centro Atómico Bariloche), Yanina Fasano (Centro Atómico Bariloche), Guillermo Jorge (Universidad de General Sarmiento), Gladys Nieva (Centro Atómico Bariloche), Gabriela Pasquini (Universidad de Buenos Aires/IFIBA), Adriana Serquis (Centro Atómico Bariloche) and Ruben Weht (Centro Atómico Constituyentes - CNEA)

Section 3 - Cryogenic Techniques and Applications Chair: Santiago Grigera (Universidad Nacional de La Plata)

Co-Chair: Henri Godfrin (Néel Institute, CNRS and Université Joseph Fourier, France)

Editor: Carlos Acha (Universidad de Buenos Aires/IFIBA)

Programme Committee: Blas Cabrera (Stanford University, USA), Hervé Courtois (Néel Institute, CNRS, France), Christian Enss (Kirchoff Institute for Physics, Heildelberg University, Germany), Tomiyoshi Haruyama (High Energy Accelerator Organization KEK, Tsukuba, Japan) and Ray Radebaugh (Cryogenics Technology Group, NIST, Boulder, USA)

Local Committee: Carlos Acha (Universidad de Buenos Aires/IFIBA) Julio Guimpel (Centro Atómico Bariloche) Alejandra Tonina (Instituto Nacional de Tecnología Industrial)

Section 4 - Magnetism and Quantum Phases Chair: Marcelo Rozenberg (Universidad de Buenos Aires/IFIBA–Orsay,France)

Co-Chair: Hidenori Takagi (University of Tokyo and Rikken, Japan)

Editor: Alberto Camjayi (Universidad de Buenos Aires/IFIBA)

Programme Committee: Philippe Bourges (CEA Saclay, France), Dragana Popovic (NHFML, Tallahassee, USA), Shoucheng Zhang (Standford, USA), Piers Coleman (Rutgers, USA) and Julian Sereni (Centro Atómico Bariloche)

Local Committee: Ana Maria Llois (Centro Atómico Constituyentes/Universidad de Buenos Aires), Laura Steren (Centro Atómico Constituyentes/Universidad de Buenos Aires), Gustavo Lozano (Universidad de Buenos Aires/IFIBA), Luis Manuel (Instituto de Física de Rosario) and Julián Sereni (Centro Atómico Bariloche)

Section 5 - Electronic Quantum Transport in Condensed Matter Chair: Liliana Arrachea (Universidad de Buenos Aires/IFIBA)

Co-Chair: Helene Bouchiat (LPS, Orsay, France)

Editor: H. Pastawski and Luis Foa-Torres (FAMAF, Universidad Nacional de Córdoba)

Programme Committee: Carlos Balseiro (Centro Atómico Bariloche), Li Lu (IOP, Beijing, China), Jukka Pekola (Aalto University, Finland) and Felix von Oppen (Freie Universitt Berlin Germany)

Local Committee: Carlos Balseiro (Centro Atómico Bariloche), Alberto Camjayi (Universidad de Buenos Aires/IFIBA), Pablo Cornaglia (Centro Atómico Bariloche), Luis Foa-Torres (Universidad Nacional de Córdoba), Edith Goldberg (Universidad Nacional del Litoral),

Horacio Grinberg (Universidad de Buenos Aires/IFIBA), Carlos Naón (Universidad Nacional de La Plata), Horacio Pastawski (Universidad Nacional de Córdoba), Rodolfo Romero (UNE), Pablo Roura Bas (Centro Atómico Constituyentes), María José Sánchez (Centro Atómico Bariloche) and Gonzalo Usaj (Centro Atómico Bariloche)

Conference Secretariat Ana Juan Congresos www.anajuan.com

Journal of Physics: Conference Series **568** (2014) 001001 doi:10.1088/1742-6596/568/0/001001

attocube
pioneers of precision

CamCool

CRYOGENIC
CRYOGENIC LIMITED

CRYOMECH
WORLD LEADERS IN CRYOREFRIGERATION FOR MORE THAN 50 YEARS

ICE
Innovative Cryogenic Engineering

IOP Publishing

JANIS

Lake Shore
CRYOTRONICS

Leiden Cyrogenics BV
Leader in Low Temperature Techniques

OXFORD
INSTRUMENTS

QuantumDesign

AAPC
ASOCIACION ARGENTINA PARA
EL PROGRESO DE LAS CIENCIAS

ANCEFN
Academia Nacional de
Ciencias Exactas, Físicas y Naturales

Buenos Aires Ciudad

Journal of Physics: Conference Series **568** (2014) 001002 doi:10.1088/1742-6596/568/0/001002

Report on 27th International Conference on Low Temperature Physics (LT27), Buenos Aires, Argentina, August 2014

Abstract. The following is an extract from the Activity Report to the IUPAP General Assembly, November 2014 by Commission C5 (Low Temperature Physics). It provides an overview of LT27, reflecting the most important and recent developments.

The LT series is the flagship international conference in low temperature physics, held triennially. LT27 was a landmark meeting: the first LT held in the southern hemisphere and the first in South America. Also notable was the fact that for the first time the Conference Chair was a woman, as was the Co-Chair.

Low temperature physics, the field which C5 represents, is a research area of extraordinary range, reach and impact. This was reflected in the topics covered at the meeting and the progress made, with research which cross-cut the traditional session topics of: Quantum Gases, Fluids and Solids; Superconductivity, Cryogenic Techniques and Applications; Magnetism and Quantum Phases; Electronic Quantum Transport in Condensed Matter. This breadth of activity was also reflected by the range of research reported by the IUPAP Young Scientist Prize winners, covering superconducting quantum devices, x-ray spectroscopy of high T_c superconductors, and the discovery of Hofstadter's butterfly in graphene systems.

The field of quantum mechanics in artificial, engineered systems (designed, nanofabricated and tuneable) is making dramatic progress. The remarkable developments in superconducting quantum circuits, operating quantum mechanically, formed the subject of the London Prize Lectures. The London Prize is an IUPAP sponsored award as well. The achievement of sufficiently long qubit lifetimes now seems enable the implementation of quantum error correction, and realization of large scale fault- tolerant quantum computers. At the same time advances in quantum sensors are expected to deliver new capabilities in sensing technologies.

Furthermore, and not least, this research area contributes to advances in fundamental understanding. For example, it has been demonstrated that a man-made macroscopic objects (realised by nanofabrication techniques) can behave as Schrödinger's cat, existing in a superposition of quantum states! And new quantum systems are emerging, with fresh opportunities for quantum control. In particular non-linear nano-electromechanical resonators, engineered in a variety of ways, in some cases with very high quality factors, can be cooled into the quantum regime. The area of engineered quantum mechanical-resonators is expected to develop strongly in future.

The Superconducting Quantum Interference Device (SQUID) constitutes a more mature technology and the LT27 meeting celebrated the SQUID's 50th anniversary. The conference heard a review of the widespread applications, realised and potential, of SQUIDs: geophysics and the discovery of mineral deposits; magnetoencephalography; dark matter (axion) detectors; photon detectors for cosmic microwave backround; high resolution x-ray spectroscopy; noise thermometry; MRI systems operating at magnetic fields four orders of magnitude lower than conventional MRI systems with several potential clinical applications, and many more. An important developing direction is SQUIDs for micro- and nanoscale detection. And in this context it was also interesting to hear a review of work over the past on the realisation of a SQUID in liquid 4He, the SHeQUID, of both fundamental interest and application as a sensitive detectors of rotation.

Nanophysics at low temperatures continues to be exploited to create novel devices with new functionalities (and acronyms), for example those based on hybrid metallic nanostructures such as the SQUIPT and HyQUID. In semiconductor-superconductor hybrid structures the focus is on the unambiguous detection, and potential manipulation of Majoranas. These particles, which are their own antiparticle, are yet to be discovered. Their quantum entanglement offers a route to topologically protected quantum computing. Developments in scanning tunnelling microscopy with superconducting tips have led to new insights into amorphous and homogeneous superconducting films, while scanning SQUID microscopy now provides a powerful technique to image complex oxide heterostructures, an important new class of materials.

These new systems and regimes of investigation challenge our fundamental understanding. A fascinating example concerns how thermodynamics applies on the nanoscale. Here it was reported that the investigation of work and dissipation in driven processes manipulating single electrons has led to an experimental realization of Maxwell's Demon!

A central theme of the conference was the discussion of a wide range of systems coming under the umbrella of quantum matter: topological insulators; graphene; unconventional superconductors, including their paradigm-superfluid ^{3}He; quantum spin liquids. The pervasive importance of the topological classification of quantum matter has been widely recognised recently, and its experimental ramifications are an active area of study. This conceptual unification provides a strong feature of low temperature physics and this conference. Another key concept in strongly correlated quantum systems is that of quantum phase transitions and quantum criticality, where a control parameter such as pressure or magnetic field is used to tune a system between ground states. The breakdown of the standard model of interacting Fermi systems at a quantum critical point remains a major unresolved issue, being actively pursued experimentally and theoretically, with significantly divergent views being presented. Here heavy fermion metals are important prototypes. There is evidence that quantum critical fluctuations enhance superconductivity in the iron-pnictides, which feeds into the debate on high T_c cuprates, where Fermi surface reconstruction is also believed to play a key role. In related work there has also been significant progress in establishing the rich behaviour of organic charge transfer salts. These systems are highly tuneable by composition and pressure, and are models for the Mott metal-insulator transition and candidate quantum spin liquids.

Superfluid ^{3}He currently provides our only established example of a topological superfluid or superconductor. The ability to fingerprint the order parameter by NMR has led to striking discoveries on how superfluidity can be manipulated by disorder with controlled anisotropy, through the introduction of aerogel. Recently this has led to the stabilization and identification of a new phase, the polar phase, not found in usual bulk liquid. Of relevance to the upsurge of interest in topological quantum matter, the surface of ^{3}He-B is predicted to host Majorana fermions. Surface acoustic studies already provide indirect evidence of a Majorana cone of dispersing surface excitations.

Both superfluid ^{3}He and ^{4}He continue to be the focus of investigations of turbulence, one of the major unsolved problems in physics. In a superfluid, the quantization of vorticity simplifies the problem and permits the confrontation of experiment with detailed theoretical simulation. For the past ten years the experimental claims for a form of superfluidity in solid ^{4}He has attracted the attention of a significant segment of the quantum fluids and solids community. The pendulum has swung towards impressive detailed studies of the elastic properties of this

Journal of Physics: Conference Series **568** (2014) 001002 doi:10.1088/1742-6596/568/0/001002

quantum solid and its dislocations, and this is now probably the best understood material from this point of view. However the enigma of supersolidity, as it is sometimes referred to, remains just that. The community only now is beginning to fully understand how to eliminate elastic artifacts from attempts to detect superfluidity with torsional pendular. There are a number of new results using this technique, and the problem remains open. However intriguing new work was presented which seemed to provide possible evidence for quasi-one dimensional superfluidity (probably along dislocations or grain boundaries) in the bulk solid, and evidence for a new class of quasi-condensate in two-dimensional helium. The interplay of solid order and superfluidity in bose-solids remains a subject of active interest for both the cold-atom and condensed matter communities.

Finally we address the strategic issues that have been discussed in previous reports, and recently in the literature. There have been shifts in the demand for ^{3}He gas, a key ingredient in dilution refrigerator technology, which have results in a significant lowering in price relative to the recently experienced high. On the other hand, the price of liquid ^{4}He has continued to increase, as a result of a combination of global supply and US government policy. While it is anticipated that new sources will come on stream, from newly developed natural gas reserves, further price increases are inevitable. Wastage of this non-renewable resource by sectors where helium recovery is not practiced cannot be justified. However this is beyond the control of our community. The trend is for our community to mitigate the impact of these price increases by installing efficient helium liquifier/recovery plant or by single-cryostat based recycling schemes. The alternative is to adopt cryogen-free technology. Indeed, the development of cryogen-free dilution refrigerator technology by a number of suppliers, has led to a relative explosion of demand. The number of dilution refrigerators (the 50th anniversary of which is approaching) is increasing rapidly: an important driver is research on quantum technologies, which rely on millikelvin temperatures for their operation. The push-button operation of these dilution refrigerator systems is attractive to users, and also facilitates practical applications. Perhaps the advent of the desk-top dilution refrigerator is not far away. More recently even the feasibility of achieving microkelvin temperatures, by a booster nuclear demagnetisation stage attached to a cryogen-free dilution refrigerator, has been demonstrated. Thus the ever-widening accessibility of ever-lower temperatures seems to be an inevitable trend, opening up to new fields.

In that context we note that the European Microkelvin community is working, with the support of the European Commission, to establish a European Microkelvin Platform as a "virtual laboratory", to promote accessibility, engage with interdisciplinary research directions, and promote discovery. The pooling of complementary expertise and collective approach to scientific challenges has been effective. Engagement and exchange of best practice with the wider international community is likely to contribute to the global development of low temperature physics, a field of established strategic importance.

LT27 conference summary respectfully submitted by
John Saunders, Secretary, C5
Kimitoshi Kono, Chair, C5

Peer review statement

All papers published in this volume of *Journal of Physics: Conference Series* have been peer reviewed through processes administered by the proceedings Editors. Reviews were conducted by expert referees to the professional and scientific standards expected of a proceedings journal published by IOP Publishing.

The spin kinetics of ^{3}He in contact with nanosized crystalline powders LaF_3

E M Alakshin, A M Gazizulina, R R Gazizulin, A V Klochkov, T R Safin, K R Safiullin, M S Tagirov and M Yu Zakharov
Kazan Federal University, Institute of Physics, 420008, Kremlevskaya, 18, Kazan, Russia

E-mail: alakshin@gmail.com

Abstract. The spin kinetics of ^{3}He in contact with nanosized crystalline powders LaF_3 has been studied by NMR methods at the temperature 1.5 K. The ^{3}He longitudinal relaxation time increases proportionally to the magnitude of the external magnetic field and the transverse relaxation time does not depend on the magnetic field. Relaxation of the gaseous and liquid ^{3}He in contact with nanosized crystalline powder LaF_3 takes place by the ^{3}He adsorbed layer. The nuclear magnetic relaxation of adsorbed ^{3}He layer on the surface of LaF_3 nanoparticles is due to the two-dimensional spin-diffusion motion.

Introduction

Hyperpolarized ^{3}He gas is widely used in neutron accelerators for the polarization of neutron beams [1–4], nuclear magnetic resonance tomography, and diverse fundamental scientific researches [5–10]. Therefore, obtaining the hyperpolarized spin state of ^{3}He nuclei using new methods remains a topical problem to date.

In 1995 the possibility of dynamic ^{3}He nuclear polarization using dielectric Van Vleck paramagnetic was discussed [11]. The channel of energy transfer from the PrF_3 to liquid ^{3}He via magnetic coupling between the nuclear spins was proposed. The "^{141}Pr - ^{3}He" system has been studied in detail [12-16].

The process previously interpreted [13] as a relaxation mechanism of ^{3}He on the surface of nanosized crystalline powders PrF_3 through the adsorbed layer may be a cross-relaxation effect between ^{141}Pr and ^{3}He nuclei. To confirm this idea the additional experiments with nanosized crystalline powders LaF_3 (diamagnetic analogue of PrF_3) have been required.

We report here the experimental investigation of ^{3}He spin kinetics in the system "LaF_3 nanoparticles - ^{3}He". The home built NMR spectrometer has been used and low temperature was achieved by ^{4}He vapor pumping. The complete details of the experimental setup were published earlier [17]. The typical thermalization time of ^{3}He adsorbed layer on a diamagnetic solid state substrate is less than 2 hours for our setup and temperature range [18]. The covering of sample surface by ^{3}He was carried out in the following sequence: gaseous ^{3}He was condensed in the experimental cell at temperature 1.5 K in small amounts (about 0.5 cm^3 STP). The pressure in the system has been monitored, and when it reached a value less than 10^{-2} mbar the next dose of ^{3}He was added. When the equilibrium pressure in the system rose above 10^{-1} mbar, the entire surface was assumed to be covered by ^{3}He layer. Further condensation of ^{3}He led to an increase of the pressure with most of the ^{3}He remaining in the gas phase. When the pressure approached the vapor saturation value (67 mbar at

1.5 K), the liquid phase appeared in the cell and further dosing of ^3He was controlled by the ^3He NMR signal amplitude until whole sample has been filled completely. Before all NMR measurements the sample was thermalized during 2 hours.

Results and discussion

The series of LaF_3 nanosamples were synthesized using different microwave irradiation time (sample 1 – 0 min, sample 2 – 20 min, sample 3 – 40 min, sample 4 – 60 min, sample 5 – 420 min), the procedure described in [13]. The X-ray analysis of synthesized samples is shown in figure 1, which identifies samples as crystalline powders LaF_3. Further on, the data on ^3He spin kinetics in the system "LaF_3 nanoparticles - ^3He" are presented for sample 2.

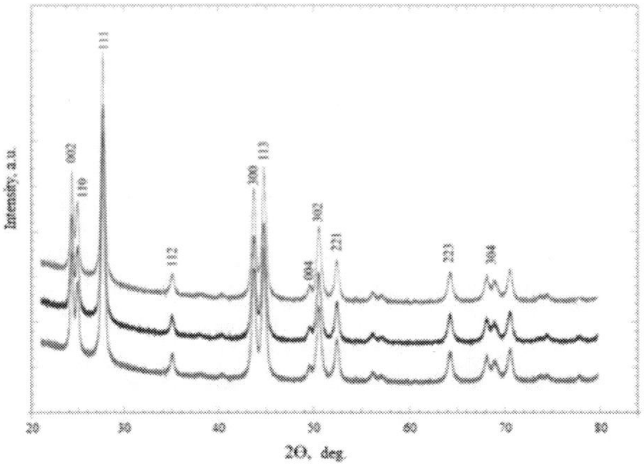

Figure 1. X-ray analysis of LaF_3 nanoparticles.

The typical image of PrF_3 (diamagnetic analog of LaF_3) nanoparticle obtained by high resolution transmission electron microscopy presented in figure 2. The size of nanoparticles is 20-40 nm.

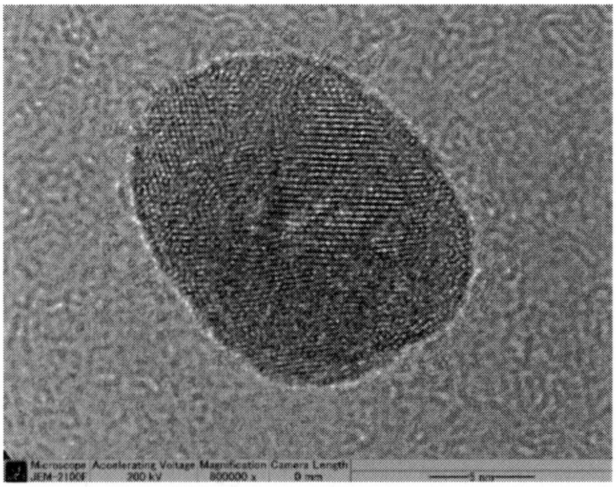

Figure 2. High-resolution transmission electron microscopy image of PrF_3 nanoparticle.

The magnetic field dependence of the ^3He nuclear magnetic relaxation times in contact with LaF_3 nanoparticles in different aggregate states of ^3He have been investigated. Dependences of the ^3He relaxation times T_1 and T_2 on the magnetic field for the three aggregate states are shown in figures 3 and 4.

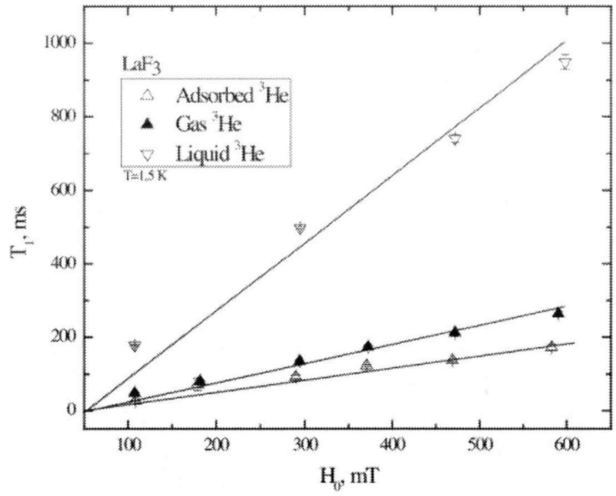

Figure 3. Longitudinal relaxation time of ^3He in contact with LaF_3 nanoparticles in different aggregate states of ^3He versus magnetic field.

It is seen in figure 3 that the longitudinal relaxation time T_1 increases proportionally to the magnitude of the external magnetic field. The transverse relaxation time T_2 is independent on the applied magnetic field (figure 4).

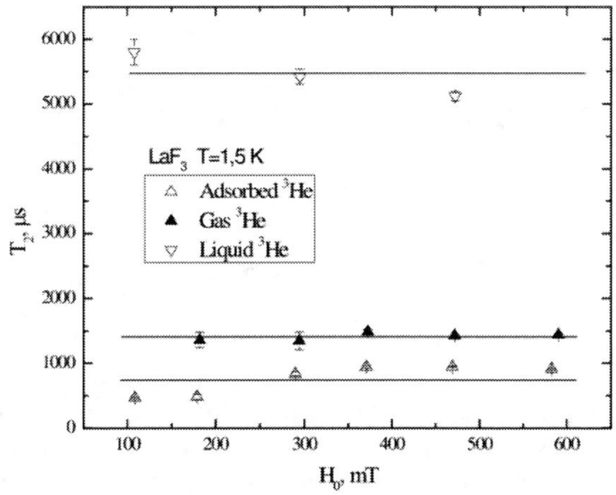

Figure 4. Transverse relaxation time of ^3He in contact with LaF_3 nanoparticles in different aggregate states of ^3He versus magnetic field.

The presence of "free" ^{3}He (gaseous or liquid) above adsorbed layer in the experimental cell significantly increase the relaxation times values, therefore the relaxation occurs through the ^{3}He adsorbed layer [19].

The ^{3}He T_1 and T_2 dependences in the gaseous and liquid phases in the cell on the total number of ^{3}He atoms (figures 3 and 4) clearly shows that the magnetic relaxation times are directly proportional to the corresponding relaxation times in the adsorbed layer and to the ratio of the number of ^{3}He spins of whole nuclear spin system to the number of ^{3}He spins in the layer, i.e.

$$T_1 = T_{1S} \cdot N_0/N_S, \quad (1)$$

where T_1, T_{1S} – longitudinal magnetization recovery times, N_0, N_S-number of spins in the whole system and the adsorbed layer, respectively;

$$T_2 = T_{2S} \cdot N_0/N_S, \quad (2)$$

where T_2, T_{2S} - times of the transverse magnetization decay, N_0, N_S-number of spins in the entire system and in the adsorbed layer, respectively.

This fact as well as the character of dependence of the relaxation times on the magnetic field plays in favor of relaxation mechanism of adsorbed ^{3}He in the two-dimensional film by the two-dimensional spin-diffusion motion, as proposed by Cowan [20, 21].

The model of 2D relaxation, proposed by Cowan, only qualitatively describes obtained data. Comparison of ^{3}He T_1 values of adsorbed layer on various diamagnetic surfaces [21-24] shows quite good similarity with presented here experimental data, which additionally supports proposed relaxation mechanism. However, there are several additional mechanisms of ^{3}He surface relaxation due to dipole-dipole interaction between ^{3}He spins and solid substrate [25-31]. The direct transfer of magnetization from ^{3}He to ^{19}F will not speed up ^{3}He nuclear transverse relaxation sufficiently to observed values, because at helium temperatures the intrinsic spin-lattice relaxation of ^{19}F in pure diamagnetic fluoride crystals is in the order of 100s of seconds [see for example 32]. The ^{3}He nuclear magnetic relaxation of adsorbed layer due to the modulation of the dipole-dipole interaction between ^{19}F and ^{3}He by the quantum zero point motion [31] is not sufficient, as well.

Conclusion

Nuclear magnetic relaxation of the gaseous and liquid ^{3}He in contact with nanosized crystalline powder LaF_3 takes place by the adsorbed layer. The nuclear magnetic relaxation of adsorbed ^{3}He layer on the surface of LaF_3 nanoparticles is due to the two-dimensional spin-diffusion motion.

Acknowledgments

The work is performed according to the Russian Government Program of Competitive Growth of Kazan Federal University, partly supported by the Russian Foundation for Basic Research and Tatarstan government (project no. 12-02-97048) and by the Ministry of Education and Science of the Russian Federation (project no. 02.G25.31.0029).

References

[1] Gentile T R, Babcock E, Borchers J A, et al., 2005 *Phyica B* **356** 96
[2] Petoukhov A K, Andersen K H, Jullien D, et al., 2006 *Phyica B* **385-386** 1146
[3] Chang L J, Mueller R, Appelt S, et al., 2004 *Phyica B* **350** e707
[4] Krimmer J, Distler M, Heil W, et al., 2009 *Nucl. Instrum. Methods Phys. Res.* **611** 18
[5] Van Beek E J, Wild J M, Kauczor H U, et al., 2004 *J. Magn. Reson. Imag.* **20** 540
[6] Schreiber W G, Morbach A E, Stavngaard T, et al., 2005 *Respir. Physiol. Neurobiol.* **148** 23
[7] Patz S, Muradian I, Hrovat M I, et al., 2008 *Acad. Radiol.* **15** 713
[8] Baudin E, Hayden M E, Tastevin G, et al., 2008 *Compt. Rend. Chim.* **11** 560
[9] Gazizulin R R, Klochkov A V, Kuzmin V V, et al., 2010 *Appl. Magn. Reson.* **38** 271
[10] Gazizulin R R, Klochkov A V, Kuzmin V V, et al., 2009 *Magn. Reson. Solids* **11** 33
[11] Tagirov M S, Tayurskii D A, 1995 *JETP Lett.* **61** 672

Journal of Physics: Conference Series **568** (2014) 012001 doi:10.1088/1742-6596/568/1/012001

[12] Alakshin E M, Gabidullin B M, Gubaidullin A T, et al., 2011 *arXiv:condmat.* **1104** 0208 - http://arxiv.org/abs/1104.0208
[13] Alakshin E M, Gazizulin R R, Egorov A V, et al., 2011 *J. Low Temp. Phys.* **162** 645
[14] Alakshin E M, Blokhin D S, Sabitova A M, et al., 2012 *JETP Lett.* **96** 181
[15] Alakshin E M, Aleksandrov A S, Egorov A V, et al., 2011 *JETP Lett.* **94** 240
[16] Alakshin E M, Gazizulin R R, Klochkov A V, et al., 2013 *JETP Lett.* **97** 579
[17] Alakshin E M, Gazizulin R R, Klochkov A V, et al., 2013 *Magnetic Resonance in Solids* **15** 1
[18] Alakshin E M, Gazizulin R R, Klochkov A V, et al., 2011 *JETP Lett.* **93** 223
[19] Hammel P C and Richardson R C, 1970 *Phys. Rev. Lett.* **51** 1441
[20] Cowan B P, 1980 *J. Phys. C* **13 (24)** 4575
[21] Cowan B P, 1983 *J. Low Temp. Phys.* **50 (1-2)** 135
[22] Alakshin E M, Gazizulin R R, Klochkov A V, et al., 2011 *J. Phys.: Conf. Ser.* **324** 012028
[23] Klochkov A V, Kuz'min V V, Safiullin K R, et. al., 2008 *JETP Lett.* **88** 823
[24] Klochkov A V, Kuzmin V V, Safiullin K R, et. al., 2009 *J. Phys.: Conf. Ser.* **150** 032019
[25] Friedman L J, Millet P J, and Richardson R C, 1981 *Phys. Rev. Lett.* **47** 1078
[26] Friedman L J, Gramila T J, and Richardson R C, 1984 *J. Low Temp. Phys.* **55** 83
[27] Singerman R W, Van Keuls F W, and Richardson R C, 1994 *Phys. Rev. Lett.* **72** 2789
[28] Van Keuls F W, Singerman R W, and Richardson R C, 1994 *J. Low Temp. Phys.* **96** 103
[29] Egorov A V, Aukhadeev F L, Tagirov M S, and Teplov M A, 1984 *JETP Lett.* **39** 584
[30] Schuhl A, Maegawa S, Meisel M W, et al., 1987 *Phys. Rev. B* **36** 6811
[31] Genio E B, Ihas G G and Sullivan N S, 1998 *J. Low Temp. Phys.* **112** 21
[32] Klochkov A V, Kurzin S P, Mukhamedshin I R, et. al., 1998 *Appl. Magn. Reson.* **14** 525

Journal of Physics: Conference Series **568** (2014) 012002 doi:10.1088/1742-6596/568/1/012002

Theoretical study of ghost imaging with cold atomic waves under the condition of partial coherence

Jun CHEN[1,2] and Yun-xian LIU[1]

[1]College of Optical and Electronic Technology, China Jiliang University, Hangzhou 310018, China

E-mail:chenjun.sun@gmail.com

Abstract. A matter wave ghost imaging mechanism is proposed and demonstrated theoretically. This mechanism is based on the Talbot-Lau effect. Periodic gratings of matter wave density, which appear as a result of interference of atoms diffracted by pulses of an optical standing wave, are utilized to produce the reference wave and the signal wave simultaneously for the ghost imaging. An advantage of this mechanism is that during the imaging process, the beam-splitter is not needed, which highly simplifies the experimental setup and makes the ghost imaging possible in the field of matter wave.

1. Introduction

Ghost imaging (GI), also known as correlated imaging has attracted a lot of interest for its great applications in quantum metrology, lithography, and holography [1]. It is a method to obtain the image of an object by correlation measurements between two optical beams [2], [3]. One optical beam passes through the object and illuminates a single-pixel (bucket) detector which provides no spatial resolution. The other optical beam does not interact with the object, but it impinges on a high-resolution point-like detector. The name "ghost" comes from the fact that the information about the spatial distribution of the object is retrieved, through the intensity correlation measurements by scanning the point-like detector in the reference path, although the beam in this path never interacts with the object to be imaged.

In recent years, the development of matter wave optics has attracted a lot of interest. Many optical effects have found their counterparts in the field of matter wave. The realization of Hanbury- Brown Twiss (HBT) effect with cold atoms opened the possibility for multi-bit information processing with matter waves [4]. It is known that, the essential physics behind the HBT effect and the GI effect is the same [5]. Both effects arise as a result of the intensity fluctuations of the stochastic wave source. The difference is that, in the GI, many bits of information of the object are treated together, whereas, in the HBT, only one bit information is required to be obtained. A natural question arises that whether it is possible to realize the GI with a matter wave. In a traditional GI experiment, a beam-splitter is an indispensable optical component to create point-to-point correlation [6]. In 2008, the computational GI was proposed, where the beam splitter disappears and a light modulator was used [7]. In 2012, an interesting GI without beam splitter was proposed, in which a periodic diffraction grating was applied [8]. In the field of matter wave, a large-area beam-splitter is quite a difficulty for experiments [9][10],

[2] To whom any correspondence should be addressed.

which limits the extension of GI from the field of optics to the field of matter wave. In this manuscript, we propose a mechanism which utilizes a periodic atomic grating to produce the point-to-point correlation of the matter wave field, and present a theoretical framework to demonstrate the GI with matter waves.

2. Matter wave Talbot-Lau interferometer

We propose a mechanism for the matter wave ghost imaging (GI) as shown in Fig.1, which can be divided into two parts. One is designed for *Talbot-Lau interferometer*, which generates periodic density patterns of the matter wave. It comprises two absorptive gratings which are standing waves of resonant optical fields (also called optical masks) [11]. The other part of the GI scheme is for *imaging*, where the correlation detection is performed at a detection plane by a bucket detector D_2 and a high-resolution detector D_1. The object to be imaged is placed behind one of the periodic density patterns on the detection plane. The matter wave transmitted from the object is the so called signal wave, and is collected by the detector D_2. Since D_2 is a bucket (single-pixel) detector, no spatial resolution of the object is gathered by D_2. The high-resolution detector D_1 is placed behind another periodic density pattern which is the so called reference wave, and collects the information of the density distribution of this pattern. For the location of D_1, it is worth noting that detector D_1 can be placed at any density pattern other than the one for D_2. According to the density correlation (the second order correlation) of the matter wave between D_1 and D_2, the image of the object would be retrieved. We would first present the theoretical derivations of the periodic density patterns in the *Talbot-Lau interferometer*, and then the *imaging* process.

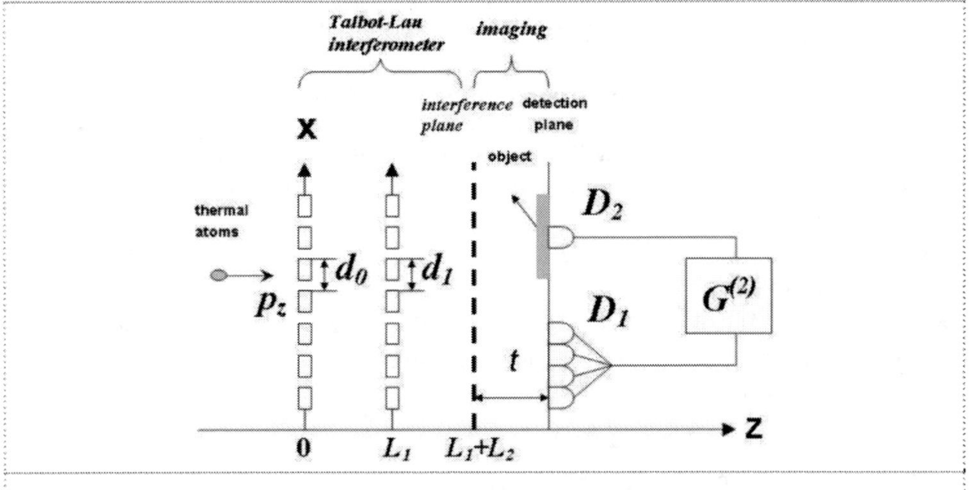

Figure 1. Scheme for the Talbot-Lau effect based ghost imaging with thermal atoms.

Talbot-Lau effect is a self-imaging of a periodic structure illuminated by partially coherent or incoherent light. In Talbot-Lau interferometer, monochromatic spatially incoherent light passes through a set of two parallel gratings separated by a known distance. At particular separations, illumination of the second grating is coherent and, subsequently, the second grating's image forms in the diffraction zone. For matter waves, the Talbot-Lau effect can be realized with atom beams manipulated by solid gratings and also optical fields [11]. Utilizing Talbot-Lau effect, point-to-point correlation of the matter wave field can be produced without beam splitters.

In the above setup of Fig.1, we assume that the matter wave source is a monochromatic incoherent atom beam. A flux of atoms enters the plane at $z=0$ with a longitudinal momentum p_z which is much greater than its transverse components. The Wigner function of the matter wave at $z=0$, is

$$W_0(x,p)=1. \tag{1}$$

The matter wave beam passes through the first atomic grating which is located at $z=0$ and is perpendicular to the propagation direction. After the first grating, the Wigner function turns out to be

$$W_1(x,p)=\int H_0(x,q)W_0(x,p-q)\mathrm{d}q=\int H_0(x,q)\mathrm{d}q=\left|h_0(x)\right|^2, \tag{2}$$

where $h_0(x)$ is the transmission function of the first grating at $z=0$, and $H_0(x, q)$ is the convolution kernel

$$H_0(x,q)=\frac{1}{(2\pi\hbar)^2}\int h_0^*(x-\Delta/2)h_0(x+\Delta/2)\exp[ip\Delta/\hbar]\,\mathrm{d}\Delta. \tag{3}$$

where Δ is a two-point spatial separation. After a free evolution by a distance L_1, followed by a passage through the second grating with a convolution kernel H_1 and another free evolution by a distance L_2, the Wigner function of the matter wave at the Interference plane, $z=L_1+L_2$, is [12]

$$W_d(x,p)=\int\left|h_0\left(x-\frac{p}{p_z}(L_1+L_2)+\frac{q}{p_z}L_1\right)\right|^2 H_1(x-\frac{p}{p_z}L_2,q)\mathrm{d}q. \tag{4}$$

We assume that the first grating has a period d_0, and its transmission function is $h_0(x)=\sum_m a_m\exp[2\pi imx/d_0]$. Likewise, the second grating has the Fourier coefficients b_m and the period d_1,

$$|h_0(x)|^2=\sum_{l\in Z}A_l\exp[2\pi il\frac{x}{d_0}],\ \text{with}\ A_l=\sum_j a_j a_{j-l}^*, \tag{5}$$

$$H_1(x,p)=\sum_{l,j\in Z}b_j b_{j-l}^*\exp[2\pi il\frac{x}{d_1}]\delta(p-\hbar\pi\frac{2j-l}{d_1}). \tag{6}$$

In the case that the two gratings are set at an equal distance $L_1=L_2=L$ and an equal period $d_1=d_2=d$, the Wigner function at the interference plane, $z=L_1+L_2$, is (see Eq.(17) in Ref. [12])

$$W_T(x,p)=\frac{1}{\hbar}\sum_{l,j,m\in Z}A_l b_j b_{j-m}\exp[2\pi i(l+m)\frac{x}{d}-2\pi i(2l+m)\frac{L}{d}\frac{p}{p_z}]\exp[\pi il(2j-m)\frac{L}{L_\lambda}] \tag{7}$$

where $L_\lambda=d^2/\lambda_d$ is the Talbot length and $\lambda_d=2\pi\hbar/p_z$ is the deBrogie wavelength.

Since the Wigner function is the Fourier transformation of the first order correlation function, $G^{(1)}(x_1,x_2)$, there is

$$\begin{aligned}G^{(1)}(x_1,x_2)&=\int W_T(x,p)\exp[-ip\frac{\Delta}{\hbar}]\,\mathrm{d}p\\&=\frac{2\pi}{\hbar}\sum_{l,j,m\in Z}A_l b_j b_{j-m}^*\exp[2\pi i(l+m)\frac{x}{d}]\exp[\pi il(2j-m)\frac{L}{L_\lambda}]\delta\left(x_2-x_1+(2l+m)\frac{\lambda L}{d}\right)\end{aligned}, \tag{8}$$

where $x=(x_1+x_2)/2$ is the mid-point position of x_1 and x_2, $\Delta=x_2-x_1$ is the two-point separation, and $G^{(1)}(x_1,x_2)\equiv\langle U_1^*(x_1)U_2(x_2)\rangle$ is the first order correlation function with $U_j(x_j)$ being the matter wave field at the position x_j ($j=1,2$).

When we pick up the $m=-2l$ components, the matter wave density at the interference plane is

$$D(x)=\left\langle U^{*}(x)U(x)\right\rangle=\frac{2\pi}{\hbar}\sum_{l,j,m\in Z}A_{l}b_{j}b^{*}_{j-m}\exp[-2\pi il\frac{x}{d}+2\pi il(j+l)\frac{L}{L_{\lambda}}]. \quad (9)$$

The first term of the above *exp* function, $\exp[-2\pi ilx/d]$, shows a periodic density pattern along the transverse direction with the period d. This density pattern is similar to the original grating and can be recognized as the image of the grating. The second term of the *exp* function, $\exp[2\pi il(j+l)L/L_{\lambda}]$, shows that the periodic density pattern appears at distances that are multiples of the Talbot length L_{λ}, which is the so called Talbot-Lau effect.

3. Matter wave ghost imaging

According to the Talbot-Lau effect, we can obtain periodic density patterns of the matter wave at the interference plane $z=L_1+L_2$. These density patterns possess the property of point-to-point correlation and can be used for ghost imaging. As shown in Fig.1, the object is placed behind one density pattern of the matter wave. The matter wave transmission of the object is collected by the bucket detector D_2, which is a single-pixel detector and can not retrieve the spatial resolution of the object. A parallel high-resolution detector D_1 is placed behind another density pattern at the reference plane. D_1 is spatial sensitive and can collect the density distribution of the matter wave field to be detected. However, no object is placed before D_1. The signal from D_1 is for the purpose of reference.

Our starting point is the first order correlation function obtained from the Talbot-Lau effect at the interference plane $z=L_1+L_2$, which has been theoretically expressed in Eq(8). When we pick up the $m=-2l$ components, the first order correlation function at the interference plane is

$$G_{T}^{(1)}(x_1,x_2)=\frac{2\pi}{\hbar}\sum_{l,j,m\in Z}A_{l}b_{j}b^{*}_{j-m}\exp[-2\pi il\frac{x}{d}]\exp[2\pi il(j+l)\frac{L}{L_{\lambda}}]\ \delta(x_2-x_1). \quad (10)$$

After a free evolution, the diffraction field-function collected by detector D_1. is $U_1(x_1)$, and there is

$$U_1(x_1)=-i\frac{M}{2\pi\hbar t_1}\int U_T(x_{01})\exp\left\{i\frac{M}{\hbar}\frac{(x_{01}-x_1)^2+(z_{01}-z_1)^2}{2t_1}\right\}\mathrm{d}x_{01}. \quad (11)$$

The field-function collected by the detector D_2. is U_2. Since D_2. is a bucket detector, U_2 does not has the spatial resolution

$$U_2=-i\frac{M}{2\pi\hbar t_2}\iint U_T(x_{02})h(x_2)\exp\left\{i\frac{M}{\hbar}\frac{(x_{02}-x_2)^2+(z_{02}-z_2)^2}{2t_2}\right\}\mathrm{d}x_{02}\mathrm{d}x_2 \quad (12)$$

where $h(x_2)$ is the transmission function of the object, t_1 and t_2 are the evolution time of the matter wave from the interference plane ($z=L_1+L_2$) to the detector D_1. and D_2., respectively. x_{0j} and x_j ($j=1,2$) are coordinates in the interference plane and detection plane, respectively. z_{0j} and z_j ($j=1,2$) are the coordinates in the matter wave propagation direction. In the above expression, M is the atom mass and $\hbar$ is the reduced Planck constant.

For an incoherent matter wave, the second order correlation function can be expressed as [13]

$$G^{(2)}(x_1,x_2)=\left\langle U_1^{*}(x_1)U_1(x_1)\right\rangle\left\langle U_2^{*}(x_2)U_2(x_2)\right\rangle+\left|G^{(1)}(x_1,x_2)\right|^2. \quad (13)$$

We set the grating functions as Eq.(5) and Eq.(6), and choose $z_{01}=z_{02}=z_0$, $z_1=z_2=z$, $t_1=t_2=t$. By substituting $U_1(x_1)$ and $U_2(x_2)$ into the expression of the first order correlation function, the correlation function at the detection plane is

$$G^{(1)}(x_1,x_2)$$
$$=\frac{M^2}{(2\pi\hbar)^2t^2}\iiint\Gamma_T(x_{01},x_{02})h(x_2)\exp\{-i\frac{M}{\hbar}[\frac{(x_{01}-x_1)^2}{2t}-\frac{(x_{02}-x_2)^2}{2t}]\}\mathrm{d}x_{01}\mathrm{d}x_{02}\mathrm{d}x_2$$

$$\propto \sum_{l,j,m} A_l b_j b^*_{j-m} \exp[2\pi i l(j+l)\frac{L}{L_\lambda}] \exp\left[\frac{iM}{\hbar t}\left(x_1 - l\frac{\pi\hbar t}{Md}\right)\right] h(x_1 - l\frac{2\pi\hbar t}{Md}) \tag{14}$$

where the integral formula $1/2\pi\int_{-\infty}^{\infty}\exp[iqx]dx = \delta(q)$ has been used. According to Eq.(14) we find that the first order correlation function between the detector D_1 and D_2, contains the image information of the object according to $G^{(1)}(x_1, x_2) \propto h(x_1 - l2\pi\hbar t/Md)$, and has a transverse periodicity according to $G^{(1)}(x_1, x_2) \propto \exp\left[iM\left(x_1 - l\pi\hbar t/Md\right)/\hbar t\right]$. The Talbot-Lau effect along the propagation direction is also shown in Eq.(14) according to $G^{(1)}(x_1, x_2) \propto \exp[2\pi i l(j+l)L/L_\lambda]$. The above theoretical derivation shows that by scanning the reference detector D_2 we can retrieve the image of the object at the bucket detector D_1.

4. Conclusion
In this manuscript, we propose a mechanism for the correlation imaging with partially coherent matter waves. According to our theoretical derivation, it is shown that a Talbot-Lau interferometer can be used to produce a point-to-point correlation field, and the correlation imaging (ghost imaging) can be realized afterwards. The advantage of this mechanism is that it can realize the matter wave correlation imaging without a beam-splitter which simplifies the experiment system and saves the setup cost.

Acknowledgments
This work is financially supported by the National Natural Science Foundation of China (No.61108005).

References
[1] B. I. Erkmen and J. H. Shapiro, "Ghost imaging: from quantum to classical to computational," Adv. Opt. Photon. **2** (4), 405-450 (2010).
[2] Y. J. Cai and S. Y. Zhu, "Ghost imaging with incoherent and partially coherent light radiation," Phys. Rev. E **71**, 056607 (2005).
[3] R. Meyers, K. S. Deacon and Y. Shih, "Ghost-imaging experiment by measuring reflected photons," Phys. Rev. A **77** (4), 041801 (2008).
[4] M. Schellekens, R. Hoppeler, A. Perrin, J. V. Gomes, D. Boiron, A. Aspect and C. I. Westbrook, "Hanbury Brown Twiss Effect for Ultracold Quantum Gases " Science **310**, 648-651 (2005).
[5] L.-G. Wang, S. Qamar, S.-Y. Zhu and M. S. Zubairy, "Hanbury Brown-Twiss effect and thermal light ghost imaging: A unified approach," Phys. Rev. A **79** (3), 033835 (2009).
[6] R. Meyers, K. S. Deacon and Y. Shih, "Ghost-imaging experiment by measuring reflected photons," Phys. Rev. A **77** (4), 041801 (2008).
[7] J. H. Shapiro, "Computational ghost imaging," Phys. Rev. A **78** (6), 061802 (2008)
[8] H. Li, Z. Chen, J. Xiong, and G. Zeng, "Periodic diffraction correlation imaging without a beam-splitter," Opt. Express **20**, 2956-2966 (2012).
[9] D. Cassettari, B. Hessmo, R. Folman, T. Maier and J. Schmiedmayer, "Beam Splitter for Guided Atoms," Phys. Rev. Lett. **85**, 5483-5487 (2000).
[10] O. Houde, D. Kadio and L. Pruvost, "Cold Atom Beam Splitter Realized with Two Crossing Dipole Guides," Phys. Rev. Lett. **85** (26), 5543 (2000).
[11] A. Turlapov, A. Tonyushkin and T. Sleator, "Talbot-Lau effect for atomic de Broglie waves manipulated with light," Phys. Rev. A **71**, 043612 (2005).
[12] K. Hornberger, J. E. Sipe, and M. Arndt, "Theory of decoherence in a matter wave Talbot-Lau interferometer," Phys. Rev. A **70**, 053608 (2004).
[13] M. Naraschewski and R. J. Glauber, "Spatial coherence and density correlations of trapped Bose gases," Phys. Rev. A **59** (6), 4595 (1999).

Dependence of dislocation damping on helium-3 concentration in helium-4 crystals

A D Fefferman,[1,*] F Souris,[1,*] A Haziot,[1,2] J R Beamish[1,3] and S Balibar[1]

[1] Laboratoire de Physique Statistique de l'ENS, associé au CNRS et aux Universités D. Diderot et P.M. Curie, 24 rue Lhomond 75231 Paris Cedex 05, France
[2] Present Address: Department of Physics, The Pennsylvania State University, University Park, Pennsylvania 16802, USA
[3] Department of Physics, University of Alberta, Edmonton, Alberta, Canada T6G 2E1
[*] These authors contributed equally.

E-mail: balibar@lps.ens.fr

Abstract. The mechanical properties of crystals are strongly affected by dislocation mobility. Impurities can bind to dislocations and interfere with their motion, causing changes in the crystal's shear modulus and mechanical dissipation. In ^{4}He crystals, the only impurities are ^{3}He atoms, and they can move through the crystal at arbitrarily low temperatures by quantum tunneling. When dislocations in ^{4}He vibrate at speeds $v < 45$ μm/sec, bound ^{3}He impurities move with the dislocations and exert a damping force $B_3 v$ on them. In order to characterize ^{4}He dislocation networks and determine the ^{3}He binding energy, it is usually assumed that B_3 is proportional to the concentration of ^{3}He bound to the dislocations. In this preliminary report, we determine B_3 in a crystal with 2.32 ppm ^{3}He and compare with our previous measurements of B_3 in natural purity ^{4}He crystals to verify the assumption of proportionality.

1. Introduction

Recent work[1, 2, 3, 4, 5] has demonstrated that the mechanical properties of hcp ^{4}He crystals can be understood in great detail by modeling their dislocations as elastic strings that vibrate between pinning points[6]. At low dislocation speeds and low temperatures, ^{3}He impurities bind to the dislocations and move with them, damping their vibrations[3]. The binding energy E_B was determined from the frequency dependence of the peak dissipation temperature under the assumption that the damping due to ^{3}He is proportional to the concentration of ^{3}He atoms bound to the dislocations [3, 5, 7]. In order to test this assumption, we are measuring the ^{3}He damping coefficient as a function of the ^{3}He concentration in the helium gas that we use to grow our crystals. In this preliminary paper, we report the ^{3}He damping coefficient for a crystal with a 2.32 ppm bulk ^{3}He concentration and compare with that of crystals made of natural purity helium. We find a linear dependence on ^{3}He concentration.

Dislocations in ^{4}He crystals are strongly pinned where they intersect, forming a network characterized by an average length L_N between network nodes and a density Λ that is given by the total dislocation length per unit volume of the crystal. In the absence of damping, the dislocations vibrate elastically between pinning points in response to an oscillating stress. (No effect of the lattice potential on dislocation vibrations has been detected in hcp ^{4}He [1].) This results in a strain ϵ_{dis} that adds to the strain ϵ_{el} for the case of immobile dislocations [8]. The additional strain ϵ_{dis} causes the shear modulus μ to decrease from the intrinsic value μ_{el}:

$$\mu = \frac{\mu_{\text{el}}}{1 + \epsilon_{\text{dis}}/\epsilon_{\text{el}}}. \tag{1}$$

The magnitude of the softening $(\mu_{\text{el}} - \mu)/\mu_{\text{el}}$ is of order 1 with mobile dislocations, but it is reduced by damping of dislocation motion by bound ^{3}He impurities or thermal phonons. We define a relaxation time $\tau = BL_N^2/\pi^2 C$, where B is the damping coefficient, $C = \mu_{\text{el}} b^2 \ln(R/r)/[4\pi(1-\nu)] = 2.3 \times 10^{-12}$ N is the tension in the dislocation line, $b = 0.367$ nm is the Burgers vector, $\nu = 0.33$ is Poisson's ratio, $R \approx L_N \approx 100$ μm is the typical spacing between dislocations, and $r \approx 1$ nm is the diameter of the dislocation core [9]. For the case of a constant L_N (i.e. no pinning by bound ^{3}He), we have[5]

$$\frac{\epsilon_{\text{dis}}}{\epsilon_{\text{el}}} = \alpha \Lambda L_N^2 \frac{1 - i\omega\tau}{1 + (\omega\tau)^2} \tag{2}$$

where $\omega/2\pi$ is the driving frequency, $\alpha = 32\,(1-\nu)\,/\pi^4 \ln(R/r) = 0.019$. Equations 1 and 2 show that as B increases, μ increases to its intrinsic value μ_{el} and the dissipation Q^{-1} =Im[μ] /Re[μ] passes through a maximum at $\omega\tau = \sqrt{1 + \alpha\Lambda L_N^2}$ [5]. We use Eqs. 1 and 2 along with measurements in the phonon damping regime to determine Λ and L_N for a given crystal [2], and then use Eqs. 1 and 2 along with measurements on the same crystal in the ^{3}He damping regime to determine the ^{3}He damping coefficient.

2. Experimental details

Our crystal growth and shear modulus measurement techniques have been explained in detail in previous publications [1, 5]. Inside the measurement cell, two piezoelectric shear plates face each other with a separation of 0.7 mm, forming a narrow gap that is filled with an oriented ^{4}He crystal (Fig. 1 inset). The orientation of the crystal is determined by photographing its growth shape through the windows of the cryostat at constant temperature and pressure. Applying a voltage V to one transducer produces a shear strain $\epsilon = Vd_{15}$ in the ^{4}He crystal, where $d_{15} = 95$ pm/V below 1 K [1]. The resulting stress σ in the ^{4}He is measured with the opposite transducer. The shear modulus is then given by $\mu = \sigma/\epsilon$. The present measurements were of a crystal with 2.32 ppm ^{3}He concentration that was oriented so that its six-fold axis of symmetry was nearly aligned with motion of the drive transducer, with spherical coordinates $\theta = 1.93$ deg and $\phi = 52.1$ deg, where θ is the polar coordinate measured relative to z and ϕ is the azimuthal coordinate measured relative to x (coordinates defined in Fig. 1 inset). As a result, the maximum shear modulus μ_{el} for this orientation was within 1 % of the intrinsic value at melting pressure of the elastic coefficient $c_{44} = 124$ bar [1, 10]. The crystal was grown at 1.4 K so that any small liquid droplets that could trap ^{3}He solidified as the crystal was cooled below 1 K.

3. Results and discussion

We used high drive, high frequency measurements to determine L_N and Λ for our crystal ($x_3 = 2.32$ ppm), as in [2]. Figure 1 shows the dissipation and shear modulus measured on cooling at a driving frequency of $\omega/2\pi = 16$ kHz and a rms driving strain $\epsilon = 2.3 \times 10^{-6}$. The shear modulus decreased monotonically while cooling because the high ϵ prevented ^{3}He atoms from binding to the dislocations, except at the very lowest temperatures where a small amount of binding occurred. At the same time, this driving strain is low enough to prevent irreversible deformation of the crystal. The overlap of the three different measurements in Fig. 1, which were made on different days, demonstrates the high degree of reproducibility of the measurement.

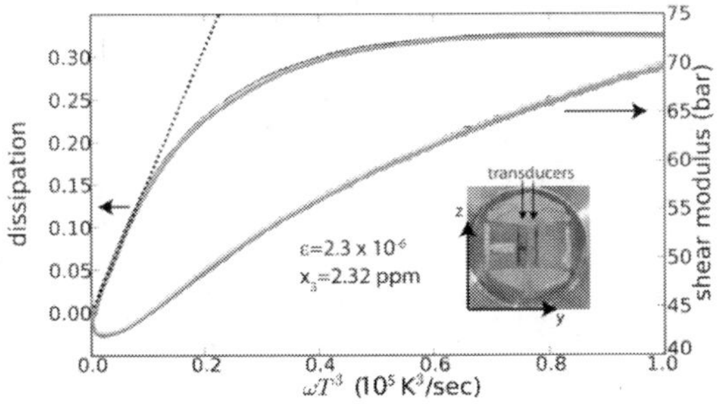

Figure 1. Solid curves: The dissipation and shear modulus measured on cooling at a driving frequency of $\omega/2\pi = 16$ kHz and a rms driving strain of 2.3×10^{-6}. The ^{3}He concentration was 2.32 ppm. The different colors correspond to measurements on different days, which demonstrate excellent reproducibility. The dashed line demonstrates the linear dependence of Q^{-1} on ωT^3 in the limit of low T, as predicted by Eq. 4, which allows us to determine the average network length $L_N = 158$ μm and dislocation density $\Lambda = 4.1 \times 10^5/\text{cm}^2$. Inset: The experimental cell used for these measurements (see text).

Since ^{3}He atoms were not binding to the dislocations in these high drive measurements, thermal phonons made the only contribution to the dislocation damping. It was shown in [2] that thermal phonons damp dislocations by the fluttering mechanism in this temperature range, so that the damping coefficient is $B_{\text{ph}} = 14.4k_B^3T^3/\pi^2\hbar^2c^3 \approx 2.1 \times 10^{-8}\ T^3$ Pa s where c is the sound speed [11, 12], and the relaxation time is $\tau_{\text{ph}} = B_{\text{ph}}L_N^2/\pi^2C$. Substituting τ_{ph} for τ in Eq. 2 and using Eq. 1 yields, at low temperatures where the phonon damping is small, a maximum softening

$$\frac{\Delta\mu}{\mu_{\text{el}}} \equiv \frac{\mu_{\text{el}} - \mu}{\mu_{\text{el}}} = \frac{\alpha\Lambda L_N^2}{1 + \alpha\Lambda L_N^2} \tag{3}$$

and a dissipation

$$Q^{-1} = \frac{\Delta\mu}{\mu_{\text{el}}}\omega\tau_{\text{ph}} = \frac{\Delta\mu}{\mu_{\text{el}}}\frac{14.4k_B^3}{\pi^4\hbar^2c^3}\frac{L_N^2}{C}\omega T^3. \tag{4}$$

The dashed line in Fig. 1 shows that Q^{-1} has a linear dependence on ωT^3, as predicted by Eq. 4. Thus we substitute the maximum softening $\Delta\mu/\mu_{\text{el}} = 0.664$ and the low temperature slope $Q^{-1}/\omega T^3 = 1.55 \times 10^{-5}$ sec/K^3 from Fig. 1 into Eqs. 3 and 4 to determine $L_N = 158$ μm and $\Lambda = 4.1 \times 10^5/\text{cm}^2$.

Having characterized the dislocation network, we can now determine the ^{3}He damping coefficient. We used measurements at low drive and low frequency, and thus low dislocation speed, for this purpose. Figure 2 shows the temperature dependence of μ and Q^{-1} in this limit, measured on cooling with the same crystal as in Fig. 1. The shear modulus increases monotonically as the crystal is cooled because ^{3}He atoms progressively bind to the dislocations and damp their motion. Because the measurements in Fig. 2 are at low frequency, damping by thermal phonons is now negligible. At the lowest temperatures, $\mu = \mu_{\text{el}}$. Near the midpoint of the temperature variation of μ at each frequency, Q^{-1} reaches a maximum at a temperature T_p.

The black circles in Fig. 3 show $1/T_p$ measured in the 2.32 ppm crystal at different driving frequencies, including the measurements shown in Fig. 2. We can derive an expression for the frequency dependence of T_p by initially assuming that the ^{3}He damping coefficient B_3 scales with some power n of x_d, the concentration of bound ^{3}He, i.e., $B_3 = B_0x_d^n$, where $x_d = x_3\exp(E_B/T)$ and x_3 is the bulk ^{3}He concentration. We will determine $B_0x_3^n$ from our fit (in previous work [7, 3, 5], it was assumed that $n = 1$). Equations 1 and 2 imply that the dissipation attains a maximum value at the temperature T_p at which $\omega\tau = \sqrt{1 + \alpha\Lambda L_N^2}$ [5].

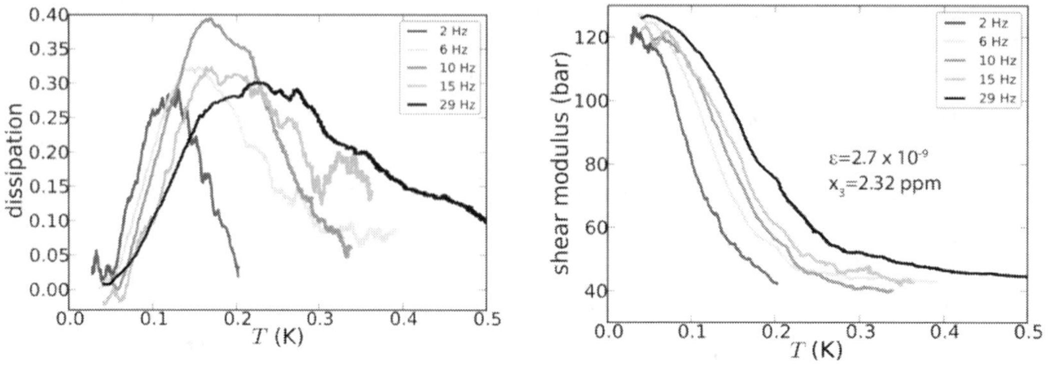

Figure 2. The dissipation and shear modulus measured on cooling at an rms driving strain of 2.7×10^{-9} with the same crystal as in Fig. 1. The ^{3}He concentration was 2.32 ppm.

Substituting $\tau = B_3 L_N^2/\pi^2 C$, since phonon damping is negligible, implies

$$f = f_\infty \exp(-nE_B/T) \tag{5}$$

where $f = \omega/2\pi$ and

$$f_\infty = \frac{\pi C\sqrt{1+\alpha\Lambda L_N^2}}{2B_0 x_3^n L_N^2} \tag{6}$$

The black solid line in Fig. 3 is a fit of Eq. 5 to the data from the 2.32 ppm crystal with best fit values nE_B =0.73 K and f_∞ = 757 Hz. Using Eq. 6 with f_∞ = 757 Hz and the values L_N = 158 μm and $\Lambda = 4.1 \times 10^5/\text{cm}^2$ determined above, we obtain $B_0 x_3^n = 3.3 \times 10^{-7}$ kg/sec m.

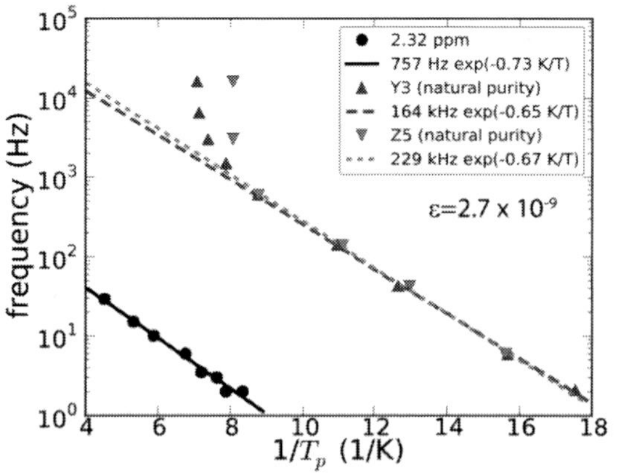

Figure 3. The black circles show $1/T_p$, the inverse temperature of the Q^{-1} peak, measured at different driving frequencies (Fig. 2). These measurements were made at low driving strain $\epsilon = 2.7 \times 10^{-9}$. The black line is a fit of Eq. 5 to the data. The red and blue data points are measurements at the same driving strain on natural purity crystals studied in [3, 5] and the red and blue lines are corresponding fits of Eq. 5.

In order to determine the value of n, we use measurements on natural purity crystals from our previous work and the analysis procedure given above. The natural purity crystal Y3 studied by Haziot *et al.* [3] had $L_N = 88$ μm and $\Lambda = 5.5 \times 10^5/\text{cm}^2$. Fefferman *et al.* [5] reported that natural purity crystal Z5 had $L_N = 96$ μm and $\Lambda = 7.9 \times 10^5/\text{cm}^2$. The frequency dependence of T_p for Y3 and Z5 and corresponding fits using Eq. 5 are shown in Figure 3. Using Eq. 6 we find that the values of $B_0 x_3^n$ for crystals Y3 and Z5 are respectively 3.8×10^{-9} and 2.6×10^{-9} kg/sec m, so that the respective ratios to the value of $B_0 x_3^n$ at 2.32 ppm are 0.012 and 7.9×10^{-3}.

The concentration of ^{3}He in the natural purity helium from our supplier was $x_3 = 3.6 \times 10^{-8}$, which is 0.016 times that of our 2.32 ppm sample. Our natural purity helium had much less ^{3}He than the natural purity helium of [13], in which $x_3 = 1.7 \times 10^{-7}$ was measured, demonstrating that x_3 is far from universal. It may vary by two orders of magnitude depending on the origin of the helium [14]. Under the assumption that x_3 of our supplier's helium has been constant over the last three years (we intend to verify this), our results are much more consistent with a linear dependence of the ^{3}He damping coefficient on ^{3}He concentration ($n = 1$) than with, e.g., a quadratic dependence. This implies that the slopes of the Arrhenius plots in [3, 5, 7] indeed give the binding energy, $E_B = 0.7$ K, as had been assumed in those works. Kim *et al.* [15] and Iwasa [16] made the estimates $E_B = 0.4$ and 0.2 K, respectively. These estimates would only be consistent with the damping measurements in the present work and in [3, 5, 7] if $n \approx 2$, but our preliminary results imply that this is not the case.

Acknowledgments
We are grateful to P. Jean-Baptiste (LSCE, CEA, Saclay, France) for the measurement of the isotopic purity of our samples. This work was supported by grants from ERC (AdG 247258-SUPERSOLID) and from NSERC Canada.

References
[1] Haziot A, Rojas X, Fefferman A D, Beamish J R and Balibar S 2013 *Phys. Rev. Lett.* **110**(3) 035301 URL `http://link.aps.org/doi/10.1103/PhysRevLett.110.035301`
[2] Haziot A, Fefferman A D, Beamish J R and Balibar S 2013 *Phys. Rev. B* **87**(6) 060509 URL `http://link.aps.org/doi/10.1103/PhysRevB.87.060509`
[3] Haziot A, Fefferman A D, Souris F, Beamish J R, Maris H J and Balibar S 2013 *Phys. Rev. B* **88**(1) 014106 URL `http://link.aps.org/doi/10.1103/PhysRevB.88.014106`
[4] Haziot A, Rojas X, Fefferman A D, Beamish J R and Balibar S 2013 *Phys. Rev. Lett.* **111**(11) 119602 URL `http://link.aps.org/doi/10.1103/PhysRevLett.111.119602`
[5] Fefferman A D, Souris F, Haziot A, Beamish J R and Balibar S 2014 *Phys. Rev. B* **89**(1) 014105 URL `http://link.aps.org/doi/10.1103/PhysRevB.89.014105`
[6] Granato A and Lucke K 1956 *J. Appl. Phys.* **27** 583
[7] Syshchenko O, Day J and Beamish J 2010 *Phys. Rev. Lett.* **104** 195301
[8] Hull D and Bacon D 2000 *Introduction to Dislocations* (Oxford, UK: Butterworth-Heinemann Elsevier)
[9] Hirth J and Lothe J 1964 *Theory of dislocations* (Malabar, Florida, USA: Krieger publishing company)
[10] Greywall D S 1977 *Phys. Rev. B* **16**(11) 5127–5128 URL `http://link.aps.org/doi/10.1103/PhysRevB.16.5127`
[11] Ninomiya T 1974 *J. Phys. Soc. Japan* **36** 399
[12] Ninomiya T 1984 *Scripta Metall. Mater.* **18** 669
[13] Vekhov Y, Mullin W and Hallock R 2013 arXiv:1311.4913v1
[14] Oxburgh E R, O'Nions R K and Hill R I 1986 *Nature* **324** 632
[15] Kim E, Xia J S, West J T, Lin X, Clark A C and Chan M H W 2008 *Phys. Rev. Lett.* **100**(6) 065301 URL `http://link.aps.org/doi/10.1103/PhysRevLett.100.065301`
[16] Iwasa I 2013 *J. Low Temp. Phys.* **171** 30

Journal of Physics: Conference Series **568** (2014) 012004 doi:10.1088/1742-6596/568/1/012004

Hetero-pairing and pseudogap phenomena in an ultracold Fermi gas with mass imbalance

Ryo Hanai and Yoji Ohashi

Department of Physics, Keio University, 3-14-1 Hiyoshi, Kohoku-ku, Yokohama 223-8522, Japan

E-mail: rhanai@rk.phys.keio.ac.jp

Abstract. We investigate strong-coupling properties of a unitary Fermi gas consisting of two different species with different masses. Including pairing fluctuations within the self-consistent T-matrix approximation, we calculate the single-particle density of states in the normal state. We show that the pseudogap phenomenon, which is characterized by a dip structure in the density of states around $\omega = 0$, appears more remarkably in the light mass component than in the heavy mass component. As a result, the pseudogap temperature, which is determined as the temperature at which the pseudogap disappears in the density of states, is higher in the former than in the latter. We also find that this different pseudogap temperatures lead to the existence of two kinds of pseudogap regions. That is, one is the ordinary pseudogap regime where the pseudogap appears in both the component, and the other case is that the light mass component only exhibits the pseudogap phenomenon. As the origin of these component-dependent pseudogap phenomena, we point out the importance of different Fermi temperatures between the two components. Since the formation of hetero-Cooper pairs and their condensation are expected in various systems, such as a ^{6}Li-^{40}K Fermi gas mixture, an exciton (polariton) gas, as well as color superconductivity, our results would be useful for the understanding of strong-coupling properties of these novel Fermi condensates.

1. Introduction

Although superfluid ^{40}K[1] and ^{6}Li[2] Fermi gases have attracted much attention in cold Fermi gas physics especially on the viewpoint of the BCS-BEC crossover[3, 4], they still belong to the same category as metallic superconductivity in the sense that Cooper pairs are formed between the same species. Thus, since the realization of these atomic Fermi superfluids, further possibilities beyond the homo-pairing have been explored in cold Fermi gas physics. At present, a tunable interaction associated with a Feshbach resonance has been realized in a ^{6}Li-^{40}K Fermi mixture[5, 6, 7, 8, 9], and ^{6}Li-^{40}K hetero-molecules have been observed[7], although their superfluid phase transition has not been reported yet. Once the superfluid phase transition is achieved in this system, we would be able to systematically study the BCS-BEC crossover physics in this hetero-pairing state by adjusting the interaction strength associated with a Feshbach resonance. Since the so-called Sarma state[10] is predicted in a mass-imbalanced Fermi gas[11], the assessment of this prediction would also become possible. In addition, since hetero-pairs have been also discussed in various fields, such as semiconductor physics (excitons [12, 13], exciton-polaritons [14, 15]), as well as high-energy physics (color superconductivity [16, 17]), the realization of a hetero-superfluid Fermi gas would widely contribute to the understanding of these novel Fermi condensates.

In the current stage of research for a superfluid Fermi gas with mass imbalance, the achievement of the superfluid phase transition temperature T_c is a crucial issue. In this regard, the so-called pseudogap phenomenon, which is characterized by a gap-like structure in the single-particle excitations above T_c, would be helpful, because this precursor phenomenon of the superfluid phase transition gradually becomes remarkable as one approaches T_c. Thus, using this, one can evaluate to what extend the system is close to the superfluid instability. Recently, a photoemission-type experiment has become possible[18], and pseudogapped single-particle excitation spectra have been observed in a mass-balanced ^{40}K Fermi gas[18]. The observed pseudogap structures agree well with strong-coupling calculations[19, 20, 21, 22, 23]. Thus, this experimental technique would also be useful for the observation of the pseudogap phenomenon in the presence of mass imbalance.

In this paper, we investigate single-particle excitations and strong-coupling effects in a unitary Fermi gas with mass imbalance. In our previous paper[24], we showed that the ordinary T-matrix approximation (TMA), which has been extensively used to successfully explain various BCS-BEC crossover physics in the mass-balanced case, breaks down in the presence of mass imbalance. To overcome this difficulty, we have extended the TMA to include higher-order pairing fluctuations[24, 25]. Although this extended T-matrix approximation (ETMA) can partially solve this difficulty, it was found to still give unphysical results in the highly mass-imbalanced regime, originating from an inconsistent treatment of the single-particle Green's function. Thus, in this paper, we further extend the ETMA to completely eliminate this inconsistency. For this purpose, we employ the self-consistent T-matrix approximation[26, 27], to examine the pseudogap phenomenon and effects of mass imbalance. From the temperature dependence of the pseudogapped density of states, we determine the pseudogap temperatures in both the light mass and the heavy mass components, each of which is defined as the temperature at which a dip structure in the density of states disappears. Throughout this paper, for simplicity, we put $\hbar = k_{\rm B} = 1$, and the system volume V is taken to be unity.

2. Formalism

We consider a two-component Fermi gas with mass imbalance, described by the Hamiltonian,

$$H = \sum_{\boldsymbol{p},\sigma={\rm L,H}} \xi_{\boldsymbol{p}\sigma} c^\dagger_{\boldsymbol{p}\sigma} c_{\boldsymbol{p}\sigma} - U \sum_{\boldsymbol{q}} \sum_{\boldsymbol{p},\boldsymbol{p}'} c^\dagger_{\boldsymbol{p}+\boldsymbol{q}/2,{\rm L}} c^\dagger_{-\boldsymbol{p}+\boldsymbol{q}/2,{\rm H}} c_{-\boldsymbol{p}'+\boldsymbol{q}/2,{\rm H}} c_{\boldsymbol{p}'+\boldsymbol{q}/2,{\rm L}}, \tag{1}$$

where $c_{\boldsymbol{p}\sigma}$ is the annihilation operator of a Fermi atom with the kinetic energy $\xi_{\boldsymbol{p}\sigma} = p^2/(2m_\sigma) - \mu_\sigma$, measured from the Fermi chemical potential μ_σ. The pseudospin σ describes the light mass component ($\sigma = {\rm L}$) with an atomic mass $m_{\rm L}$, and the heavy mass component ($\sigma = {\rm H}$) with an atomic mass $m_{\rm H}$. $-U$ is a tunable pairing interaction, which is related to the s-wave scattering length a_s as $4\pi a_s/m = -U/[1 - U\sum_{\boldsymbol{p}}(m/p^2)]$ (where $2m^{-1} = m_{\rm L}^{-1} + m_{\rm H}^{-1}$ is twice the reduced mass). In this paper, we consider the unitarity limit $(k_{\rm F}a_s)^{-1} = 0$ (where $k_{\rm F} = (3\pi^2 N)^{1/3}$ is the Fermi momentum, with N being the total number of Fermi atoms). For simplicity, we consider a uniform Fermi gas, ignoring effects of a harmonic trap.

Strong-coupling effects are conveniently described by the self-energy $\Sigma_\sigma(\boldsymbol{p}, i\omega_n)$ in the single-particle thermal Green's function,

$$G_\sigma(\boldsymbol{p}, i\omega_n) = \frac{1}{i\omega_n - \xi_{\boldsymbol{p}\sigma} - \Sigma_\sigma(\boldsymbol{p}, i\omega_n)}, \tag{2}$$

where ω_n is the fermion Matsubara frequency. In the self-consistent T-matrix approximation (SCTMA)[26, 27], the self-energy has the form

$$\Sigma_\sigma(\boldsymbol{p}, i\omega_n) = T \sum_{\boldsymbol{q},\nu_n} \Gamma(\boldsymbol{q}, i\nu_n) G_{-\sigma}(\boldsymbol{q} - \boldsymbol{p}, i\nu_n - i\omega_n). \tag{3}$$

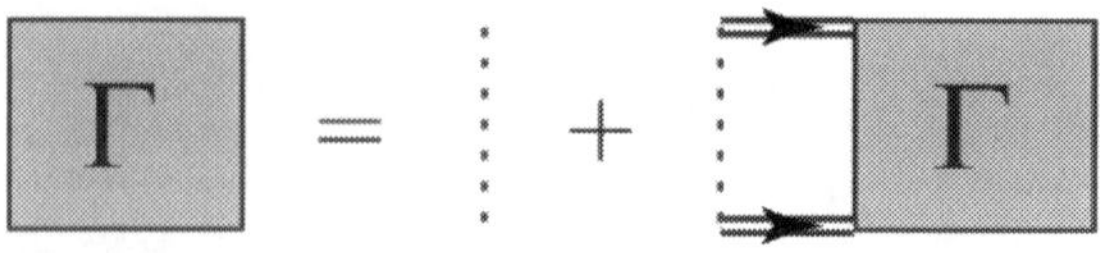

Figure 1. Particle-particle scattering matrix Γ in the SCTMA. The double solid line is the dressed Green's function G in Eq. (2). The dotted line is the pairing interaction $-U$.

Here, ν_n is the boson Matsubara frequency, and $-\sigma$ means the opposite component to σ. In Eq. (3), the particle-particle scattering matrix $\Gamma(\boldsymbol{q}, i\nu_n)$ is diagrammatically given as Fig.1, which gives $\Gamma(\boldsymbol{q}, i\nu_n) = -U/[1 - U\Pi(\boldsymbol{q}, i\nu_n)]$, where

$$\Pi(\boldsymbol{q}, i\nu_n) = T \sum_{\boldsymbol{k}, i\omega_n} G_{\rm L}(\boldsymbol{k} + \boldsymbol{q}/2, i\nu_n + i\omega_n) G_{\rm H}(-\boldsymbol{k} + \boldsymbol{q}/2, -i\omega_n) \tag{4}$$

is the pair-correlation function.

As usual, the superfluid phase transition temperature $T_{\rm c}$, as well as the chemical potentials $\mu_{\rm L}$ and $\mu_{\rm H}$, are self-consistently determined by solving the Thouless criterion [28]

$$[\Gamma(\boldsymbol{q} = 0, i\nu_n = 0)]^{-1} = 0, \tag{5}$$

together with the number equations

$$N_{\rm L} = N_{\rm H} = \frac{N}{2}, \tag{6}$$

where

$$N_\sigma = T \sum_{\boldsymbol{p}, i\omega_n} G_\sigma(\boldsymbol{p}, i\omega_n). \tag{7}$$

In the normal state above $T_{\rm c}$, we only solve the number equation (6), to determine $\mu_{\rm L}$ and $\mu_{\rm H}$. The single-particle density of states $\rho_\sigma(\omega)$ is then calculated from the analytic continued Green's function as

$$\rho_\sigma(\omega) = -\frac{1}{\pi} \sum_{\boldsymbol{p}} {\rm Im}[G_\sigma(\boldsymbol{p}, i\omega_n \to \omega + i\delta)]. \tag{8}$$

We briefly note the difference between the present SCTMA and the previous ETMA used in Refs.[24, 25]. The SCTMA is reduced to the ETMA formalism, when the dressed Green's function G in the pair-correlation function $\Pi(\mathbf{q}, i\nu_n)$ in Eq. (4) is simply replaced by the bare one,

$$G_\sigma^0(\boldsymbol{p}, i\nu_n) = \frac{1}{i\omega_n - \xi_{\boldsymbol{p}\sigma}}. \tag{9}$$

Thus, the ETMA involves the internal inconsistency that the bare Green's function G_0 in Eq. (9) is used in $\Pi(\boldsymbol{q}, i\nu_n)$, although the dressed Green's function G is used in the self-energy in Eq. (3). As shown in Ref.[24], this inconsistency becomes crucial in the highly mass imbalanced regime, leading to the unphysical vanishing $T_{\rm c}$ in the BCS region. This problem is avoided in the present SCTMA because of the fully self-consistent treatment of the dressed Green's function[26], so that we can safely discuss strong-coupling physics in the whole region $0 \le m_{\rm L}/m_{\rm H} \le 1$. For more details about the difference between the two approximations, we refer to Refs.[24, 26].

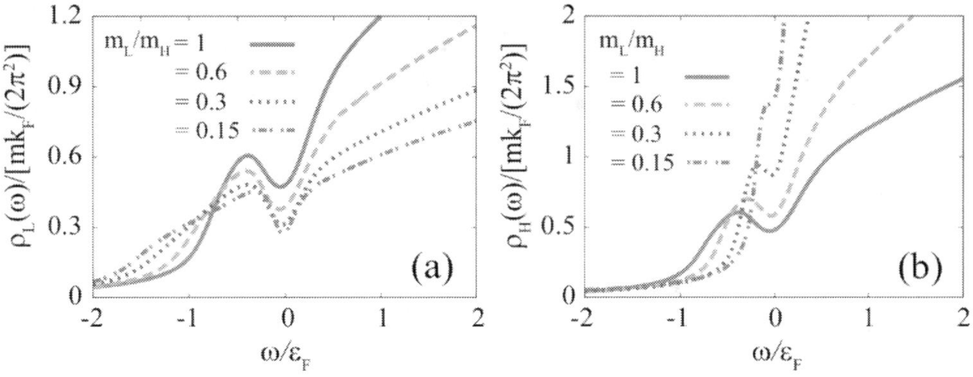

Figure 2. Calculated density of states $\rho_\sigma(\omega)$ in a mass-imbalanced unitary Fermi gas at T_c. (a) Light mass component. (b) Heavy mass component. $\varepsilon_F = k_F^2/(2m)$, where $k_F = (3\pi^2 N)^{1/3}$, and m is twice the reduced mass.

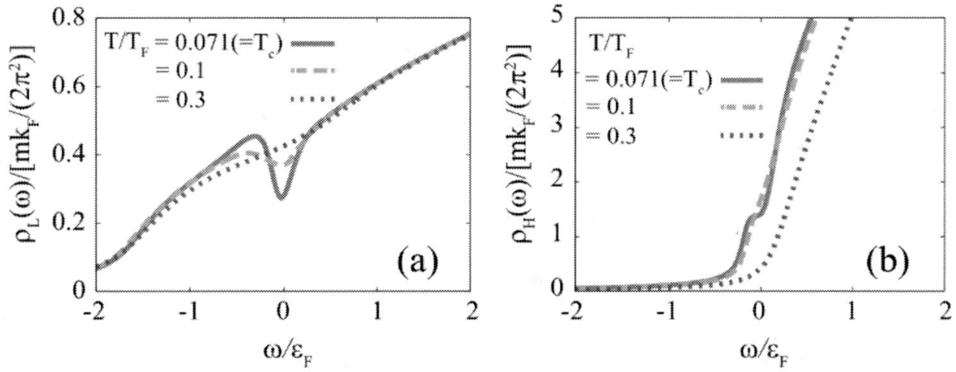

Figure 3. Calculated density of states $\rho_\sigma(\omega)$ in a unitary Fermi gas when $T \geq T_c$. (a) Light mass component. (b) Heavy mass component. We take $m_L/m_H = 0.15$. $T_F = k_F^2/(2m)$, where $k_F = (3\pi^2 N)^{1/3}$, and m is twice the reduced mass.

3. Pseudogapped density of states in a mass-imbalanced Fermi gas

Figure 2 shows the density of states in a unitary Fermi gas. In the mass-balanced case ($m_L/m_H = 1$), one sees a dip structure around $\omega = 0$ in each the component. Since the superfluid order parameter vanishes at T_c, this is the pseudogap originating from strong pairing fluctuations[22]. In the light mass component (Fig.2(a)), the pseudogap continues to exist even in the highly mass imbalanced regime ($m_L/m_H = 0.15 \ll 1$). In contrast, the dip structure in Fig. 2(b) gradually becomes obscure with decreasing the ratio m_L/m_H of the mass imbalance, to eventually disappear when $m_L/m_H = 0.15$. Thus, although the pseudogap originates from the formation of preformed Cooper pairs consisting of atoms in both the components, the appearance of this phenomenon is very different in between the two components.

This component-dependent pseudogap phenomenon is due to the fact that the Fermi temperature $T_F^L = k_F^2/(2m_L)$ in the light mass component is higher than the Fermi temperature $T_F^H = k_F^2/(2m_H)$ in the heavy mass component. The *scaled* temperature T/T_F^H in the heavy

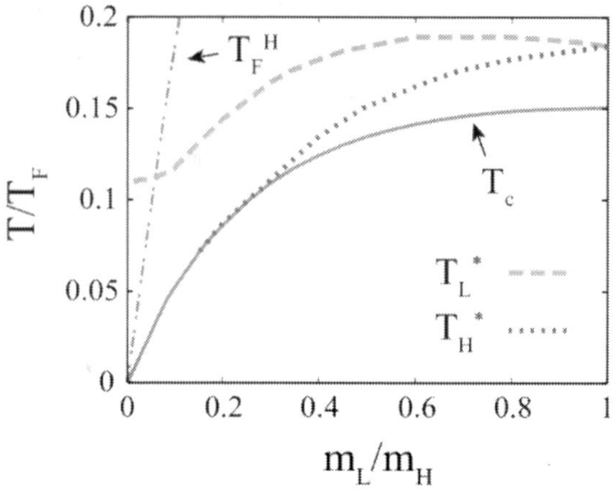

Figure 4. Pseudogap temperature in the light mass component $T_{\rm L}^*$ and that in the heavy mass component $T_{\rm H}^*$ in a unitary Fermi gas. $T_{\rm F}^{\rm H}$ is the Fermi temperature of the heavy mass component.

mass component then becomes higher than $T/T_{\rm F}^{\rm L}$ in the light mass component, so that thermal excitations smear out the pseudogapped density of states more remarkably in the former. In the highly mass imbalanced case ($m_{\rm L}/m_{\rm H} = 0.15$), when we raise the temperature from $T_{\rm c}$, while the pseudogap structure is still seen at $T/T_{\rm F} = 0.1$ (where $T_{\rm F} = k_{\rm F}^2/(2m)$) in the light mass component (Fig.3(a)), Fig.3(b) shows that even the shoulder structure seen at $T_{\rm c}$ completely disappears in the heavy mass component when $T/T_{\rm F} = 0.1$. At this temperature, one finds,

$$\frac{T}{T_{\rm F}^{\rm H}} = 0.38 > \frac{T}{T_{\rm F}^{\rm L}} = 0.06, \qquad (10)$$

which means that the heavy mass component is effectively at much higher temperature than the light mass component.

4. Component-dependent pseudogap temperatures

To identify the pseudogap regime in a quantitative manner, we conveniently introduce the pseudogap temperature T_σ^* (σ = L, H) in each component, as the temperature at which the dip structure disappears in the density of states $\rho_\sigma(\omega)$. As expected from the discussions in Sec.3, Fig. 4 shows $T_{\rm L}^* > T_{\rm H}^*$ when $m_{\rm L}/m_{\rm H} < 1$. When $m_{\rm L}/m_{\rm H} \lesssim 0.15$, the pseudogap no longer exists in the heavy mass component even at $T_{\rm c}$ (See Fig. 2(b)), so that the pseudogap temperature $T_{\rm H}^*$ is absent there.

The component-dependence on the pseudogap phenomenon naturally leads to two kinds of pseudogap regions. That is, when $T_{\rm c} \le T \le T_{\rm H}^*$, the pseudogap appears in the both components. On the other hand, the pseudogap can be seen only in the light mass component when $T_{\rm H}^* \le T \le T_{\rm L}^*$[30].

In the highly mass imbalanced regime, Fig.4 shows that the pseudogap temperature $T_{\rm L}^*$ in the light mass component becomes higher than the Fermi temperature $T_{\rm F}^{\rm H}$ in the heavy mass component. In this regard, we recall that the superfluid phenomenon is purely a quantum phenomenon. Thus, this result indicates that, although the pseudogap phenomenon is a precursor of this quantum phenomenon, the pseudogap is still seen in the light mass component even when the heavy mass component is in the classical regime ($T/T_{\rm F}^{\rm H} \gtrsim 1$).

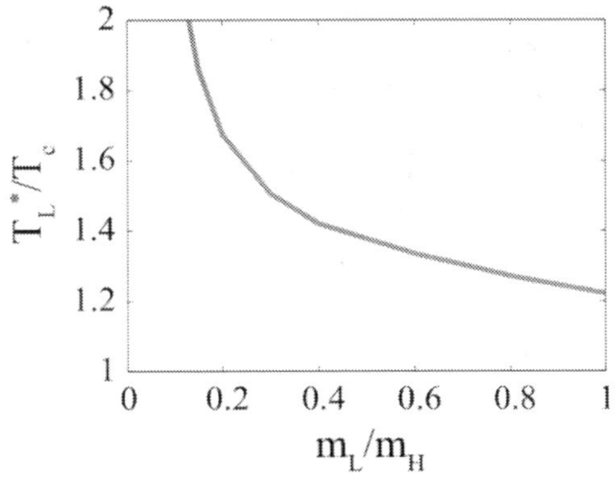

Figure 5. Calculated ratio T_L^*/T_c of the pseudogap temperature in the light mass component to the superfluid phase transition temperature, as a function of m_L/m_H.

Figure 5 shows the ratio T_L^*/T_c as a function of m_L/m_H. Since the superfluid phase transition is a quantum phenomenon, both the components should be in the Fermi degenerate regime at T_c, which leads to $T_c < T_F^H$ ($< T_F^L$). On the other hand, one finds $T_F^H < T_L^*$ in the highly mass imbalanced regime, as shown in Fig.4. Reflecting these, the ratio T_L^*/T_c becomes large when m_L/m_H becomes small, as shown in Fig. 5. This means that, even in the highly mass imbalanced regime where T_c is very low, one can observe the pseudogap in the light mass component in the relatively wide temperature region, $T_c \leq T \leq T_L^*$.

5. Summary

To summarize, we have investigated strong-coupling properties of a unitary Fermi gas consisting of two different species with different masses. Within the framework of the self-consistent T-matrix approximation, we calculated the single-particle density of states in the normal state, to examine effects of mass imbalance on the pseudogap phenomenon originating from pairing fluctuations.

We showed that the pseudogap structure is more clearly seen in the light mass component than in the heavy mass component. When we determine the pseudogap temperature in each component (which is evaluated as the temperature at which the pseudogap structure disappears in the density of states), it is higher in the light mass component (T_L^*) than in the heavy mass component (T_H^*). As a result, we obtain two kinds of pseudogap region. In the region $T_c \leq T \leq T_H^*$, the pseudogap appears in both the components. When $T_H^* \leq T \leq T_L^*$, on the other hand, the pseudogap only appears in the light mass component. In the highly mass imbalanced regime, the pseudogap phenomenon does not occur in the heavy mass component, so that the pseudogap temperature in this component is absent there.

The component-dependent pseudogap phenomena originates from the different Fermi temperatures between the two components. The Fermi temperature T_F^L in the light mass component is higher than the Fermi temperature T_F^H in the heavy mass component, so that the scaled temperature T/T_F^σ is always higher in the latter than in the former. As a result, thermal effects in the heavy mass component is more remarkable for a given temperature T, so that the pseudogap structure is easily smeared out thermally in this component, compared with the light mass component.

In the case of a ^{6}Li-^{40}K Fermi gas mixture, one finds $m_L/m_H = 6/40 = 0.15 \ll 1$. Thus, in this case, the light mass component (^{6}Li) is suitable for the observation of the pseudogap phenomenon. Since the pseudogap is a precursor of the superfluid phase transition, toward the realization of the superfluid phase transition of this Fermi gas mixture, our results would be helpful to assess to what extent the system is close to the superfluid instability using the pseudogap phenomenon. In addition, since Fermi superfluids with hetero-Cooper pairs have also been discussed in other fields, such as semiconductor physics (exciton gas and exciton-polariton condensate), as well as high-energy physics (color superconductivity), our results would also contribute to the study of these unconventional Fermi condensates.

Acknowledgments
We thank H. Tajima for fruitful discussions. R.H. was supported by KLL-PhD Program Research Grant, as well as Graduate School Doctoral Student Aid Program in Keio University. Y.O. was supported by Grant-in-Aid for Scientific research from MEXT in Japan (Grant No. 25105511 and No. 25400418).

References

[1] Regal C, Greiner M and Jin D S 2004 *Phys. Rev. Lett.* **92** 040403
[2] Zwierlein M W, Stan C A, Schunck C H, Raupach S M F, Kerman A J and Ketterle W 2004 *Phys. Rev. Lett.* **92** 120403
[3] Giorgini S, Pitaevskii L P and Stringari S 2008 *Rev. Mod. Phys.* **80** 1215
[4] Chin C, Grimm R, Julienne P S and Tiesinga E 2010 *Rev. Mod. Phys.* **82** 1225
[5] Wille E, *et. al.* 2008 *Phys. Rev. Lett.* **100** 053201
[6] Taglieber M, Voigt A C, Aoki T, Hänsch T W and Dieckmann K 2008 *Phys. Rev. Lett.* **100** 010401
[7] Voigt A C, Taglieber M, Costa L, Aoki T, Wieser W, Hänsch T W and Dieckmann K 2009 *Phys. Rev. Lett.* **102** 020405
[8] Costa L, Brachmann J, Voigt A C, Hahn C, Taglieber M, Hänsch T W and Dieckmann K 2010 *Phys. Rev. Lett.* **105** 123201
[9] Naik D, Trenkwalder A, Kohstall C, Spiegelhalder F M, Zaccanti M, Hendl G, Schreck F, Grimm R, Hanna T M and Julienne P S 2010 *Eur. Phys. J.* D **65** 55
[10] Sarma G 1963 *J. Phys. Chem. Solids* **24** 1029
[11] Baarsma J E, Gubbels K B and Stoof H T C 2010 *Phys. Rev.* A **82** 013624
[12] High A A, Leonard J R, Hammack A T, Fogler M M, Butov L V, Kavokin A V, Campman K L and Gossard A C 2012 *Nature* **483** 584
[13] Yoshioka K, Morita Y, Fukuoka K and Kuwata-Gonokami M 2013 *Phys. Rev.* B **88** 041201(R)
[14] Deng H, Weihs G, Santori C, Bloch J and Yamamoto Y 2002 *Science* **298** 199
[15] Kasprzak J, *et. al.* 2006 *Nature* **443** 409
[16] Barrois B C 1977 *Nucl. Phys.* B **129** 390
[17] Bailin D and Love A 1984 *Phys. Rep.* **107** 325
[18] Stewart J T, Gaebler J P and Jin D S 2008 *Nature* **454** 744
[19] Perali A, Pieri P, Strinati G C and Castellani C 2002 *Phys. Rev.* B **66** 024510
[20] Chen Q and Levin K 2009 *Phys. Rev. Lett.* **102** 190402
[21] Magierski P, Wlazlowski G, Bulgac A and Drut J E 2009 *Phys. Rev. Lett.* **103** 210403
[22] Tsuchiya S, Watanabe R and Ohashi Y 2009 *Phys. Rev.* A **80** 033613
[23] Haussmann R, Punk M and Zwerger W 2009 *Phys. Rev.* A **80** 063612
[24] Hanai R, Kashimura T, Watanabe R, Inotani D and Ohashi Y 2013 *Phys. Rev.* A **88** 053621
[25] Kashimura T, Watanabe R and Ohashi Y 2012 *Phys. Rev.* A **86** 043622
[26] Hanai R and Ohashi Y 2014 *J. Low Temp. Phys.* **175** 272
[27] Haussmann R 1993 *Z. Phys.* B **91** 291
[28] Thouless D J 1960 *Ann. Phys.* **10** 553
[29] Kashimura T, Watanabe R and Ohashi Y 2014 *Phys. Rev.* A **89** 013618
[30] In the highly mass imbalanced regime where the pseudogap temperature $T_{\rm H}^*$ is absent, the second pseudogap regime is given as $T_{\rm c} \le T \le T_{\rm L}^*$.

Anomalous nonlinear attenuation of ultrasound in solid ^{4}He in a torsional oscillator below 200 mK

I Iwasa[1], **J M Goodkind**[2] **and H Kojima**[3]

[1] Department of Mathematics and Physics, Kanagawa University, Kanagawa 259-1293, Japan
[2] Physics Department, University of California, San Diego, California 92093, USA
[3] Serin Physics Laboratory, Rutgers University, New Jersey 08854, USA

E-mail: `iwasa@mva.biglobe.ne.jp`

Abstract. In order to elucidate the ultra-low temperature behavior of solid ^{4}He, simultaneous measurements of longitudinal ultrasound (US) and torsional oscillation have been made. Changes in attenuation and velocity of US at 10 MHz have been measured in polycrystalline hcp ^{4}He samples (0.3 or 20 ppm of ^{3}He impurity) grown in a 1 kHz torsional oscillator (TO). In a 0.3 ppm ^{3}He sample, the US attenuation and velocity were found to depend on the US drive voltage at temperatures below 70 mK where the anomalies in the TO frequency and dissipation were also observed. The US attenuation at low T (10 mK) decreased monotonically as the drive voltage was decreased but then remained small and constant as the drive voltage was increased again. The US velocity change at low T was negative with respect to the high-T (400 mK) value, contrary to the positive sign expected from the known variation in the shear modulus. In a 20 ppm ^{3}He sample, both the US and TO anomalies shifted to 150 mK. The amplitude dependence and hysteresis of US attenuation were related to pinning of dislocations by ^{3}He impurities, and nonlinear spatial variations of the amplitude of US pulses were derived.

1. Introduction

Ultrasound (US) measurements in MHz-order frequency are suitable for studying mechanical properties of materials. The very first US measurements on solid helium [1] were probably made on polycrystalline samples to determine the phase diagram. Later US measurements on solid ^{4}He were mainly performed on single-crystalline samples to study anisotropic elasticity [2, 3] and dislocations [4, 5]. The effects of ^{3}He impurities in solid ^{4}He have been studied and explained by pinning of dislocations by ^{3}He impurity atoms [6]. Later a new anomaly has been observed below 200 mK in ^{4}He crystals containing ^{3}He impurities [7].

After the discovery of anomaly of the torsional oscillator (TO) containing solid ^{4}He [8, 9], the shear modulus of solid ^{4}He below kHz-order frequency has been extensively studied on polycrystalline and single-crystalline samples and with various ^{3}He concentrations including ultra-pure samples [10, 11]. Now the variation of the shear modulus is well understood as caused by dislocations and their interaction with ^{3}He impurities [12, 13]. Many of the TO anomalies can be accounted for by the shear modulus change of solid helium in the torsion rod [14].

We have made simultaneous measurements of ultrasound and torsional oscillation on polycrystalline ^{4}He with different ^{3}He concentrations [15]. Our motivation is to clarify the origin of the TO anomaly, to extend our knowledge on dislocations in solid ^{4}He and to shed light on

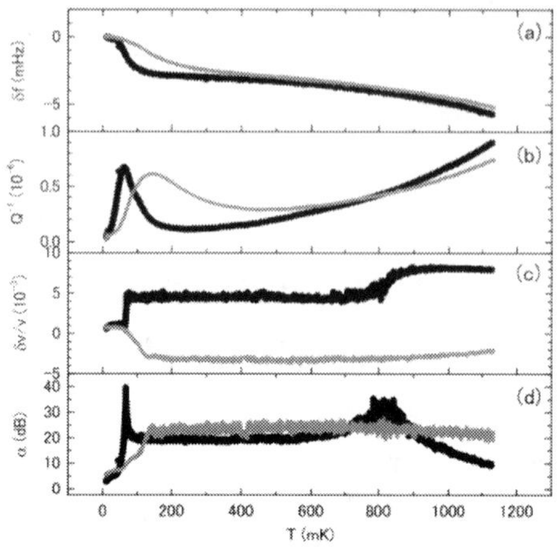

Figure 1. Simultaneously measured data on a x_3=0.3 ppm sample (black thick line) and a x_3=20 ppm sample (red thin line). (a) TO frequency shift, (b) TO dissipation, (c) US fractional velocity shift, and (d) changes in US attenuation.

the US anomaly [7]. To our knowledge, this is the first US measurement on polycrystalline ^{4}He at temperatures below 1 K.

2. Experimental apparatus and procedure

Our TO incorporating 10 MHz quartz transducers has been described in detail previously [15]. Briefly, the BeCu torsion rod of 3.6 mm diameter and 9.0 mm length is attached to a bob containing sample space. A hole of 0.8 mm diameter through the torsion rod serves to introduce helium gas into the sample space. Two fins for torsional excitation and detection are mounted on the outer wall of the bob. Two 10-MHz X-cut quartz transducers of 10 mm diameter and a cylindrical spacer (10.2 mm outer diameter, 7.9 mm inner diameter, H = 5.3 mm height) between them are inserted into the sample space and held by springs for US measurements. Polycrystalline ^{4}He is grown in the sample space by the blocked capillary method over a time interval of about two hours. After the measurements with the nominal purity ^{4}He (^{3}He concentration x_3=0.3 ppm) have been completed, helium gas is driven out of the sample space and reloaded with ^{4}He with x_3=20 ppm.

The resonance frequency of the empty TO at low temperature is 1009 Hz. The TO frequency is decreased by about 180 mHz when the sample space is filled with solid helium. For US measurements, 9.58 MHz RF pulses with 2 μs width and 1 kHz repetition frequency are applied to the transmitter transducer. The amplitude of the pulses can be varied by an attenuator. The propagated US signal through the sample is detected by the receiver transducer and measured by a super-heterodyne US measurement system. Changes in the US attenuation α and velocity $\delta v/v$ are calculated from the in-phase and out-of-phase outputs of the super-heterodyne system. The longitudinal velocity in our solid ^{4}He samples is assumed to be v = 500 m/s.

3. Results

Results of simultaneous measurements on two samples with x_3=0.3 ppm and 20 ppm are shown in Fig. 1, where the data are taken as the temperature (T) is decreased down to about 10 mK.

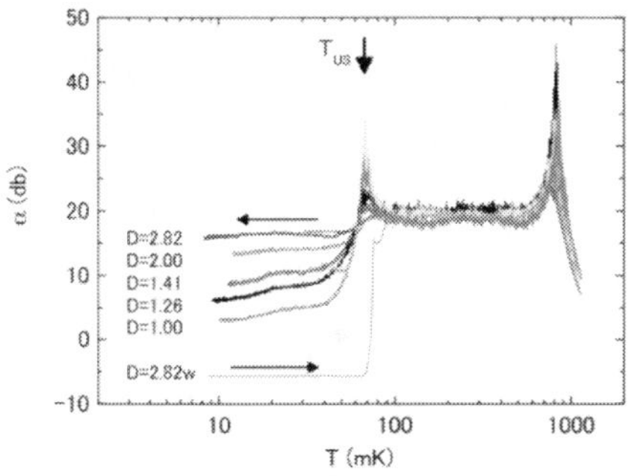

Figure 2. Temperature dependence of US attenuation on a sample with x_3=0.3 ppm. D indicates the normalized drive voltage. "w" indicates a warming run.

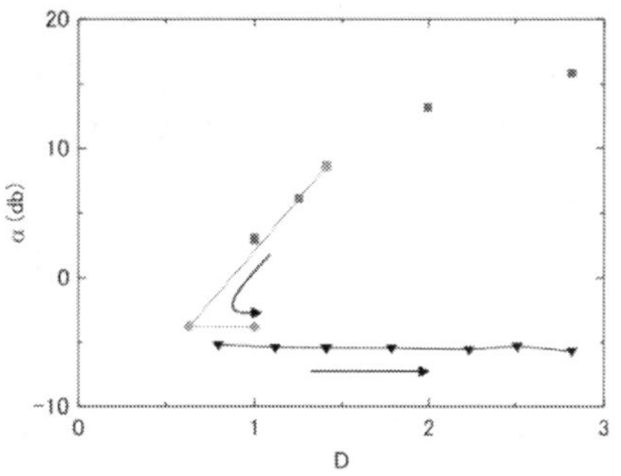

Figure 3. Attenuation at low T (about 10 mK) taken on a sample with x_3=0.3 ppm with various conditions. Blue squares: cooled from 1.1K to 10 mK at each D. Red diamonds: D was decreased and then increased at 10 mK. Black triangles: D was increased at 10 mK.

3.1. TO anomaly
Changes in TO resonant frequency, δf, and changes in TO dissipation, Q^{-1}, are shown in panel (a) and (b) of Fig. 1, respectively. Effects of temperature on "backgrounds" of frequency shift and dissipation in the empty cell have been subtracted. The TO anomaly is characterized by decreases in δf with increasing T where $T < 400$ mK and peaks in Q^{-1}. The characteristic temperature of TO anomaly, T_{TO}, is the temperature of dissipation peak; T_{TO}=70 mK for x_3=0.3 ppm, and 140 mK for 20 ppm. The total anomalous shift in frequency, $\delta f(10 \text{ mK}) - \delta f(400 \text{ mK})$, is between 2 and 3 mHz.

3.2. US anomaly
Fractional changes in US velocity, $\delta v/v$, and relative changes in attenuation, α, are shown in panel (c) and (d) of Fig. 1, respectively. US anomalies are observed around 800 mK and below 200 mK. We focus our attention to the low-T US anomaly below 200 mK. The US anomaly is characterized by decrease in $\delta v/v$ with decreasing temperature and the peak in α for x_3=0.3 ppm sample and by increase in $\delta v/v$ and decrease in α with decreasing temperature for x_3=20 ppm sample. The characteristic temperature of US anomaly, T_{US}, is the temperature of the attenuation peak and the temperature at which the velocity or attenuation changes on cooling. We find T_{US}=70 mK for x_3=0.3 ppm, and 150 mK for 20 ppm, which are close to T_{TO} at each ^{3}He concentration.

3.3. Amplitude dependent US attenuation
We have extensively studied the behavior of the sample with x_3=0.3 ppm. The attenuation strongly depends on the normalized US drive voltage, D, at $T < T_{US}$ as shown in Fig. 2. Most of the data in Fig. 2 are taken on cooling runs at different setting of D except for the data D=2.82w which is a warming run. D=1 corresponds to an excitation voltage of 89 mV peak-to-peak applied to the transmitter. The initial amplitude of US pulses excited by the transmitter should be proportional to D. Larger attenuation is observed for cooling runs at higher D.

Fig. 2 shows sharp attenuation peaks at 70 mK (low T anomaly) and 800 mK (high T anomaly). The shape of the peaks, however, is sensitive to the calibration of our US measurement system and the accurate height of the peaks cannot yet be determined.

Fig. 3 shows the attenuation as D is varied at about 10 mK. Squares are the data at the end of the cooling runs at various D shown in Fig.2. It is clear that the attenuation decreases as D is decreased. When D is changed at 10 mK, hysteretic behavior is observed. In one case (diamonds), D is first decreased from 1.4 to 0.63 and then increased to 1.0. The attenuation decreases in the initial decrease in D but does not change in the increase in D. In the other case (triangles), the attenuation remains at the same value on increasing D from 0.1 to 2.82. In this way, the behavior is nonlinear for decreasing D and linear for increasing D.

3.4. Amplitude dependence of US velocity

The US velocity also depends on D at $T < T_{US}$ as shown in Fig. 4. The velocity at 10 mK in this sample is smaller than the velocity at high T (400 mK), but in the other sample with x_3=20 ppm we find $v(10\ \mathrm{mK}) > v(400\ \mathrm{mK})$ as shown in Fig. 1(c).

3.5. Interaction between TO and US measurements

Simultaneous TO measurements while US drive voltage is changed show no change in TO characteristics although the US characteristics are changed. Similarly, simultaneous US measurements while TO excitation voltage is changed show no change in US characteristics. We, therefore, conclude that there is no measurable interaction between TO excitation and US propagation in the present TO cell.

4. Discussion

4.1. TO anomaly

Beamish et al.[14] have proposed that hardening of solid ^{4}He (i.e. increase in shear modulus) in the torsion rod is the origin of many of the observed TO anomaly. The increase in shear modulus of solid ^{4}He below about 100 mK may be explained by dislocations being pinned by ^{3}He impurity atoms and becoming immobile. As a result, the total torsional spring constant would increase and the TO frequency would increase. The shear modulus of solid ^{4}He at low T is considered to be the intrinsic property of the lattice. The shear modulus would decrease at high T because dislocations become freed from pinning by ^{3}He impurity atoms. The amount of the decrease depends on the dislocation density, distribution of network pinning length etc.

The total torsional spring constant is a sum of the contributions from the BeCu torsion rod and from solid ^{4}He in the torsion rod. The torsional spring constant of our BeCu torsion rod is calculated to be $K_{rod} = 92$ Nm and the torsional spring constant due to solid ^{4}He in the rod at low T to be $K_{He} = 6.4 \cdot 10^{-5}$ Nm. If the shear modulus of solid ^{4}He in the rod is reduced by 50% from the low T value, the resulting frequency change is estimated to be 0.17 mHz. The observed frequency change is more than 10 times larger so that it cannot be fully explained by the hardening effect of solid ^{4}He in the rod. Mechanisms other than the hardening effect of solid ^{4}He in the rod must be present. A possibility is the mechanism proposed by Iwasa [12, 17], the origin of which is pinning of dislocation by ^{3}He impurities in solid ^{4}He in the TO, but the size of the effect is not determined theoretically. Another possibility is the motion of the quartz transducers and the spacer. Although the spacer is bonded to the inner wall of the bob with vacuum grease at room temperature, the grease may become brittle at low temperature and a gap is formed between the spacer and the pressurized bob. Then the quartz transducers and the spacer are loosely connected to the bob via springs at both ends and solid helium filling the gap acts as an additional spring which contributes to the TO anomaly.

4.2. Analysis of the amplitude dependent US attenuation

Past experiments on the temperature dependence of ultrasound propagation in solid ^{4}He [4, 5] have been carried out mostly on single crystal samples. Measured US velocity, v, has been

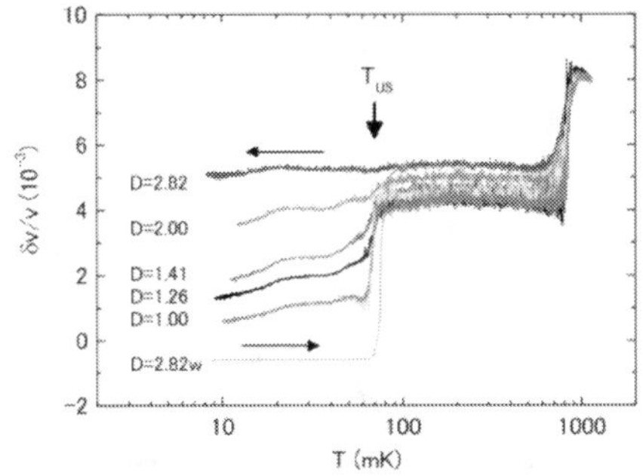

Figure 4. Temperature dependence of US fractional velocity on the same sample as in Fig. 2 with x_3=0.3 ppm. D indicates the normalized drive voltage. "w" indicates a warming run.

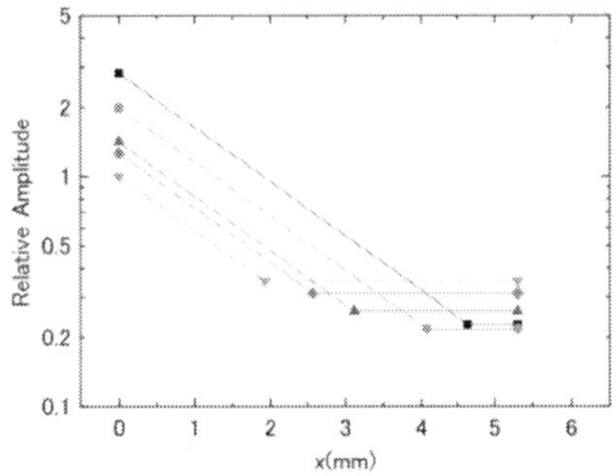

Figure 5. Nonlinear attenuation of US amplitude reconstructed from low-T data in Fig. 2. Different symbols designate different drive voltages.

analyzed by assuming the net fractional change in v is given by $\delta v/v = \delta v_d/v + \delta v_p/v$, where $\delta v_p/v = p_0 + p_4 T^4$ is introduced to account for the change in v due to phonon anharmonicity (p_0 and p_4 are fitting parameters). The dislocation line contribution $\delta v_d/v$ is written using the Granato-Lücke theory [16] which accounts for the frequency-dependent and amplitude-independent US attenuation and modulus change by treating the dislocation line as a string. The high-T US anomaly in our polycrystalline samples can be similarly analyzed by the Granato-Lücke theory as discussed in a previous report [15], where the effect of ^{3}He impurities was not taken into account. It is, however, clear that the amplitude-dependent low-T anomaly below T_{US} is not described by the Granato-Lücke model.

The observed amplitude dependence of US attenuation and velocity, the hysteresis of attenuation, and the dependence of T_{US} on x_3 are similar to the characteristics of TO anomaly and those of shear-modulus anomaly in and below the kHz region, suggesting the origin of US anomaly is also pinning of dislocations by ^{3}He atoms. None of the US measurements on single crystals of ^{4}He grown from high purity He gas (x_3=0.3 ppm) have shown such an amplitude-dependent low-T anomaly, but there is an US measurement on a single crystal of ^{4}He containing 30 ppm ^{3}He which shows amplitude dependent velocity and attenuation at temperatures below 1 K [6].

Dislocations in solid ^{4}He are strongly pinned by the network nodes and weakly pinned by ^{3}He impurity atoms. The network pinning length, L_N, is distributed over a wide range but it is independent of temperature. The resonant frequency of a dislocation segment of length L_N is given by

$$F_0 = \frac{1}{2L_N}\sqrt{\frac{C}{A}}, \tag{1}$$

where A and C are the effective mass per unit length and the line tension of the dislocation, respectively. A dislocation segment with the critical length, L_c, has the resonant frequency equal to the US frequency, F,

$$L_c = \frac{1}{2F}\sqrt{\frac{C}{A}}, \tag{2}$$

L_c slightly depends on the molar volume of solid ^{4}He and is estimated to be about 9 μm for our samples at $F = 9.58$ MHz. US attenuation at low T is mainly due to the resonant vibration of dislocation segments whose lengths are around L_c. The impurity pinning length is a function of T and x_3. The average distance between pinning ^{3}He atoms along the dislocation line is the average impurity pinning length, L_i, given by [6]

$$L_i = 3.4 \times 10^{-9} x_3^{-2/3} e^{-0.2/T} [\mathrm{m}]. \tag{3}$$

L_i decreases with decreasing temperature and becomes equal to L_c at a critical temperature, T_c. We estimate T_c=94 mK for x_3=0.3 ppm. The number of dislocation segments of length L_c decreases at $T < T_c$ due to impurity pinning so that the attenuation is expected to decrease with decreasing temperature. However, dislocations are not necessarily pinned if the amplitude of US pulse is sufficiently large.

The US amplitude is spatially varied in our US experiment as a US pulse is excited by one transducer and propagates in the solid helium sample. The US attenuation through $H = 5.3$ mm thick solid helium is estimated to be 25 dB from the attenuation difference of the run D=2.82w in Fig. 2 between 19 dB at $T > 100$ mK and -6 dB at $T < 60$ mK. Hence the amplitude of US pulse at the receiver side of the sample is about 1/18 of the incident amplitude.

The US amplitude is also temporally varied. Although the excited US pulse width is longer than the electric pulse width of 2 μs due to ringing of the transducer, it is still much shorter than the repetition interval of 1 ms.

The amplitude dependence of α at low T ($\sim$ 10 mK) is interpreted as follows. We assume that there is a critical US amplitude, A_c, above which dislocation segments are not pinned even at low T and the attenuation coefficient is given by $a = 0.54$ mm^{-1}. On the other hand, for $A \leq A_c$ dislocation segments are pinned by ^{3}He impurity atoms and the attenuation vanishes. The spatial variation of US amplitude is given by

$$A(x) = A_0 e^{-ax} \tag{4}$$

for $A_0 > A_c$, where x is the position along the propagation direction and A_0 is the initial amplitude. The US pulse is attenuated as it propagates and the amplitude becomes equal to A_c at $x = x_1$. Then attenuation becomes zero so that

$$A(x) = A_c \tag{5}$$

for $x > x_1$. According to this model of nonlinear attenuation, A_c corresponds to the amplitude of received signal, A_r. Although A_r slightly varies from run to run possibly due to ringing of the transducer, x_1 can be determined from known values of A_0 and A_r for each run and $A(x)$ can be reconstructed as shown in Fig. 5. Referring to Fig. 5, when A_0 is decreased at low T, the received amplitude should stay at $A_r = A_c$ so that x_1 becomes smaller and the net attenuation is decreased. When A_0 is further decreased to below A_c, the received amplitude should become $A_r = A_0$ because of zero attenuation. Once a dislocation segment is pinned by ^{3}He atoms, L_i at 10 mK is so small ($L_i = 3$ nm at 20 mK for x_3=0.3 ppm) that the dislocation line stays in pinned state when the amplitude is increased again. Therefore, the attenuation should stay at low value and the received amplitude should increase linearly with increasing A_0 as observed.

4.3. Velocity change

The amplitude dependent velocity at 10 mK in Fig. 4 can be analyzed similarly as the amplitude dependent attenuation. We assume the velocity is $v = v_0$ for $A < A_c$ where dislocations are pinned and $v = v_1$ for $A > A_c$ where dislocations are not pinned. Then the measured velocity can be expressed as

$$v = v_0 \left(1 + \frac{x_1}{H} \frac{v_1 - v_0}{v_0}\right). \tag{6}$$

We obtain

$$\frac{v_1 - v_0}{v_0} = 5.9 \times 10^{-3} \tag{7}$$

from the data in Fig. 4. We note v_1 is bigger than v_0 so that the US velocity decreases with decreasing temperature in the $x_3 = 0.3$ ppm sample. In contrast, the US velocity increases in the $x_3 = 20$ ppm sample with decreasing temperature as shown in Fig. 1(c). We expect the longitudinal velocity would increase from the known increase in the shear modulus at low T. Therefore, we need a further study to understand the negative velocity change in the $x_3 = 0.3$ ppm sample.

4.4. US anomaly found by Ho et al.

Ho, Bindloss and Goodkind [7] reported US anomalies in their measurements on low-density single crystals of ^{4}He containing ^{3}He impurities. Some of their observations, such as the sharp attenuation peak at a temperature, T_p, between 100 and 200 mK for samples with ^{3}He between 10 and 30 ppm and the velocity increase around T_p, are similar to our low-T anomaly. In this respect, their anomalies may possibly be related to pinning of dislocations.

5. Conclusion

We have made simultaneous measurements of longitudinal ultrasound and torsional oscillation on polycrystalline hcp ^{4}He samples containing 0.3 and 20 ppm ^{3}He impurity. Both US and TO measurements showed anomalous behavior at low temperatures which can be explained by pinning of dislocations by ^{3}He impurity atoms, but no interaction was observed between TO excitation and US propagation. The frequency shift of the TO anomaly was more than 10 times larger than the expected effect from hardening of solid helium in the torsion rod. The amplitude dependence and hysteresis of US attenuation have been analyzed by the nonlinear attenuation depending on the US amplitude which varied spatially. Relationship between our low-T US anomaly and the previously observed US anomaly by Ho et al. has been suggested.

Acknowledgments

We thank Elizabeth Eibling for the initial stage of apparatus design, Bettina Hein, Rebecca Cebulka and Michelle Goffreda for data analysis and Michael Keiderling for data acquisition. This research was supported by NSF DMR1005325.

References

[1] Vignos J H and Fairbank H A 1961 *Phys. Rev. Lett.* **6** 265
[2] Crepeau R H, Heybey O, Lee D M and Strauss S A 1971 *Phys. Rev.* **A 3** 1162
[3] Greywall D S 1971 *Phys. Rev.* **A 3** 2106
[4] Wanner R, Iwasa I and Wales S 1976 *Solid State Commun.* **18** 853
[5] Iwasa I, Araki K and Suzuki H 1979 *J. Phys. Soc. Jpn.* **46** 1119
[6] Iwasa I and Suzuki H 1980 *J. Phys. Soc. Jpn.* **49** 1722
[7] Ho P, Bindloss I P and Goodkind J M 1997 *J. Low Temp. Phys.* **109** 409
[8] Kim E and Chan M H W 2004 *Nature* **427** 225
[9] Kim E and Chan M H W 2004 *Science* **305** 1941
[10] Day J and Beamish J 2004 *Nature* **450** 853
[11] Rojas X, Haziot A, Bapst V, Balibar S and Maris H J 2010 *Phys. Rev. Lett.* **105** 145302
[12] Iwasa I 2010 *Phys. Rev.* **B 81** 104527
[13] Fefferman A D, Souris F, Haziot A, Beamish J R and Balibar S 2014 *Phys. Rev.* **B 89** 014105
[14] Beamish J R, Fefferman A D, Haziot A, Rojas X and Balibar S 2012 *Phys. Rev.* **B 85** 180501
[15] Hein B, Goodkind J M, Iwasa I and Kojima H 2013 *J. Low Temp. Phys.* **171** 322
[16] Granato A and Lücke K 1956 *J. Appl. Phys.* **27** 583
[17] Iwasa I 2013 *J. Low Temp. Phys.* **171** 30

Investigation of half-quantum vortex in superfluid ^{3}He-A phase

Y Kimura[1,2], T Kunimatsu[1], K Obara[1], H Yano[1], T Hata[1], T Takagi[3], O Ishikawa[1]

[1] Graduate School of Science, Osaka City University, Osaka 558-8585, Japan
[2] The Institute for Solid State Physics, The University of Tokyo, Chiba 277-8581, Japan
[3] Department of Applied Physics, Fukui University, Fukui 910-8507, Japan

E-mail: kimura@sci.osaka-cu.ac.jp

Abstract. In superfluid ^{3}He-A, it was theoretically predicted that the half-quantum vortex (HQV) is stable where the order parameters $\hat{\boldsymbol{d}}$ and $\hat{\boldsymbol{\ell}}$ are perpendicular to each other in the order parameter configuration. However, the existence of the HQV has not been reported experimentally. Now we are trying to detect the HQV in superfluid ^{3}He-A confined in the parallel-plates sample cell with 12 μm gap, where $\hat{\boldsymbol{d}}$ and $\hat{\boldsymbol{\ell}}$ are perpendicular to each other, by using a NMR technique under a rotation at ISSP. In order to detect the HQV, we improved the magnet system for NMR so far, but we have not observed the HQV yet. We report here the experimental details and some results for investigating the HQV.

1. Introduction

The half-quantum vortex (HQV), which has a winding number 1/2, in the superfluid ^{3}He-A phase was first proposed by Volovik and Mineev in 1976 [1] and many theoretical [2–6] and experimental [7, 8] studies have been done on this subject. However the existence of the HQV has not been confirmed so far in experiments. The superfluid ^{3}He is well known as the spin-triplet p-wave condensate, particularly the superfluid ^{3}He-A phase is the equal-spin pairing state, with the anisotropic energy gap. The order parameter for the A phase, $A_{\mu j}$, is described as $A_{\mu j} = A\hat{d}_\mu(\hat{m}_j + i\hat{n}_j)$ by using two vector fields $\hat{\boldsymbol{d}}$, which is the spin part of the order parameter, and $\hat{\boldsymbol{\ell}} = \hat{\boldsymbol{m}} \times \hat{\boldsymbol{n}}$, which is the orbital part. $\hat{\boldsymbol{\ell}}$ must be perpendicular to the sample container wall due to the pair breaking effect. In the presence of a magnetic field $\boldsymbol{H}$, $\hat{\boldsymbol{d}}$ is perpendicular to $\boldsymbol{H}$ due to the anisotropy of the magnetic energy. In the bulk liquid, $\hat{\boldsymbol{\ell}}$ is parallel to $\hat{\boldsymbol{d}}$ due to the dipole interaction. On the other hand, we can obtain the $\hat{\boldsymbol{\ell}} \perp \hat{\boldsymbol{d}}$ configuration in the parallel-plate geometry as follows. A strong magnetic field whose strength is larger than that of the dipole field, $H \gtrsim 3$ mT, is applied directed to perpendicular to the surface of the parallel-plates which has a gap as thin as the dipole coherence length $\xi_{\rm D} \sim 10$ μm [7, 8]. In the $\hat{\boldsymbol{\ell}} \perp \hat{\boldsymbol{d}}$ configuration, it is shown that the HQV can be energetically stable than the singular vortex under certain conditions, such as a magnetic field on the order of 1 T, a rotation speed larger than several rad/sec, and a temperature near $T_{\rm c}$ [6, 9]. In the A phase, the information about an angle α between two vectors $\hat{\boldsymbol{\ell}}$ and $\hat{\boldsymbol{d}}$ is provided by cw-NMR measurements. The NMR resonance frequency f is written as $f = f_{\rm L} + \dfrac{{f_{\rm A}^{\rm L}}^2}{2f_{\rm L}} \cos 2\alpha$

under the strong magnetic field [10], where $f_{\rm L}$ is the Larmor frequency and $f_{\rm A}^{\rm L}$ is the pressure- and temperature-dependent longitudinal NMR frequency of the bulk A phase. The frequency shift from the Larmor frequency, $\Delta f = f - f_{\rm L} = f_{\rm A}^{\rm L\,2}/2f_{\rm L}\ \cos 2\alpha$, is positive in the bulk liquid, whereas it is negative in the parallel-plate geometry with $\alpha = 90^\circ$ [7, 8].

2. Experiments

We are trying to detect the HQV in the superfluid A phase confined in the parallel-plate sample cell, with cw-NMR method and the rotating cryostat at ISSP [11]. The sample cell is the same as that was used in ref. [8]. It consists of 220 alternately stacked polyimide films of a 12.5 μm thickness and a 25 μm thickness. In each 12.5 μm film, there are two holes of a 3.0 mm diameter and a channel connecting between them, 0.3 mm wide and 9.5 mm long. There is also one hole of a 3.0 mm diameter in 25 μm film. By vertically sandwiching the 12.5 μm film between the 25 μm films, a space of 12.5 μm in thickness and 3.0 mm in diameter and a through hole in stacked films were obtained. The sample cell is set to the rotating cryostat at ISSP; the surface of the parallel-plate sample is perpendicular to the rotation axis, $\hat{z}$ axis.

At the ISSP rotating cryostat we could apply only a longitudinal magnetic field, $H_{\rm l}$, whose direction was parallel to $\hat{z}$ and whose maximum strength was 100 mT, by one solenoidal coil. Recently we improved the current source power and the current lead ampacity. In order to improve the homogeneity of the magnetic field, we also added a new magnet composed of two pairs of saddle-shaped coils perpendicular to each other, named the inner coil and the outer coil (see Figure 1). So we can apply a transverse magnetic field, $H_{\rm t}$, to any direction perpendicular to $\hat{z}$, and we can tilt the magnetic field direction. As a result of replacements of the current source and the cable ampacity, we are now able to apply the maximum current about 10 A corresponding to the maximum magnetic field of 1 T. In cw-NMR measurements, we set the operating frequency at 875 kHz corresponding to the NMR magnetic field $H_0 = 27.0$ mT in the normal phase. The absorption signal is observed by sweeping $H_{\rm l}$ under a liquid pressure at 3.01 MPa corresponding to $T_{\rm c} = 2.4$ mK. Additionally, in order to prevent solid ^{3}He from being formed on the surface of parallel-plates, we coated all the sample cell surface with 2.5 layers ^{4}He before introducing ^{3}He.

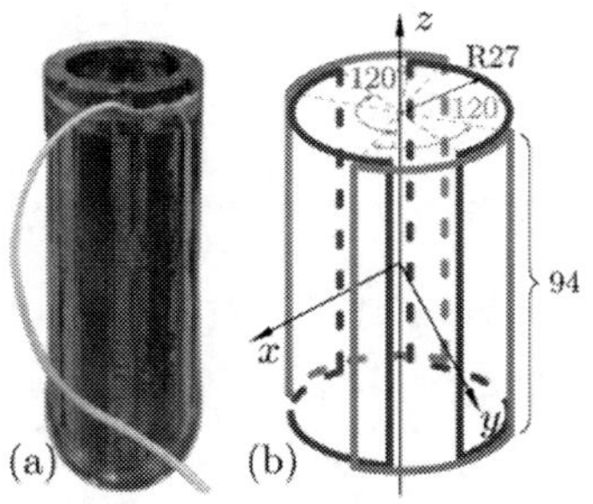

Figure 1. (a) Photograph of a bobbin of saddle-shaped coils made of copper, a part of winding superconducting wire is 27 mm in radius and 94 mm tall. (b) Schematic of two pairs of winding superconducting wire, blue line shows the inner coil of 180 turns applying a magnetic field to x direction, and red line shows the outer coil of 150 turns applying a magnetic field to y direction.

3. Results and Discussions

Figure 2(a) shows the cw-NMR absorption spectra in the normal phase with respect to a change of a current passing through the outer coil, $I_{\rm s}$, as a function of the longitudinal magnetic field $H_{\rm l}$. Applying the transverse magnetic field $H_{\rm t} = \beta I_{\rm s}$ by the saddle-shaped coils, where β is the magnetic current ratio, it was found that the longitudinal resonance field of the NMR spectrum was shifted (Figure 2(a)) and the peak height of the NMR spectrum was changed (Figure 2(b)).

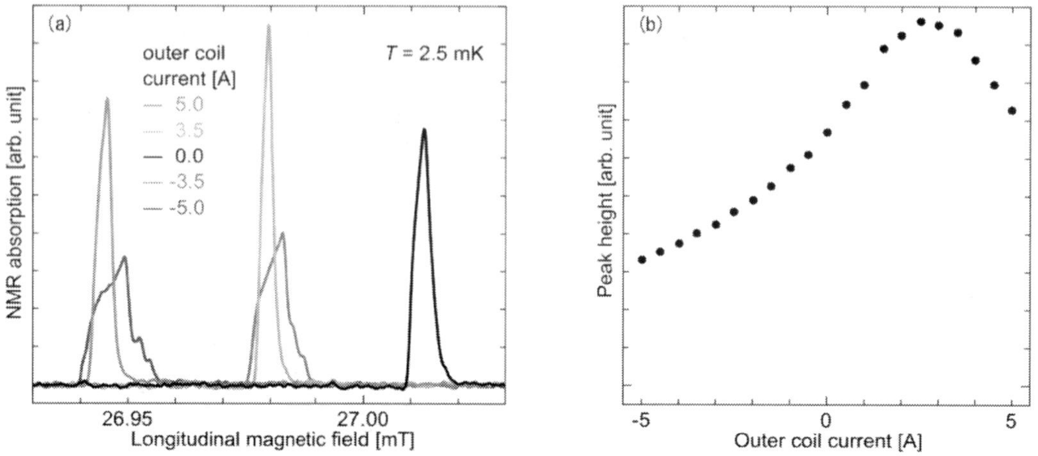

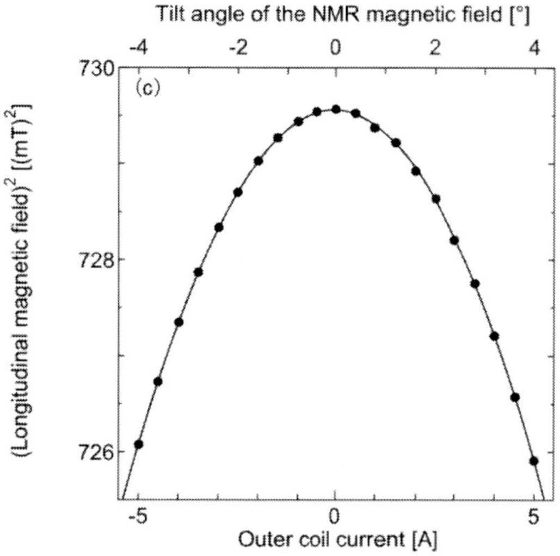

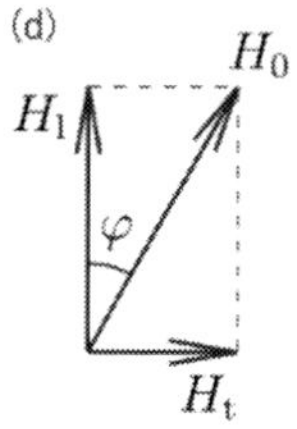

Figure 2. (a)NMR spectra at $I_s = 5.0$, 3.5, 0.0, -3.5, and -5.0 A as a function of the resonating H_l measured at 2.5 mK (normal phase). (b)Peak height of NMR spectra as a function of I_s. The vertical scale of (b) is the same as (a). (c)Square of H_l as a function of I_s. The measured values are well fitted by a quadratic function $-0.1431 I_s^2 + 729.6$ (solid line). The upper horizontal axis shows the tilt angle of the NMR magnetic field. (d)Composition of the magnetic field and the tilt angle of the field.

Maximum peak height in Figure 2(b) shows that the homogeneity of the magnetic field became better. A current which optimized the homogeneity of a magnetic field for the inner coil and the outer coil was around 2.0 A and 2.5 A, respectively.

The changes of the resonating longitudinal field are explained as follows. Since H_l, H_t, and H_0 satisfy the relation $H_0 = \sqrt{H_l^2 + H_t^2}$ at the resonant condition (Figure 2(d)), it is found that $H_l^2 = -\beta^2 I_s^2 + H_0^2$. Figure 2(c) shows the square of H_l is well fitted by a quadratic function of the outer coil current I_s. We evaluated the value of β from the fitting parameter, and we found that the value of β for the inner coil and the outer coil was 0.454 mT/A and 0.377 mT/A, respectively. The ratio of β, 0.454/0.377, is equal to the ratio of the number of turns in the saddle-shaped coils, 180/150. Using the value of β, we calculated the tilt angle φ, which is written as $\varphi = \tan^{-1}\{H_t/H_l\}$ (Figure 2(d)). The tilt angle given by the outer coil current $I_s = 2.5$ A optimizing the magnetic field was 2°.

Figure 3 shows the cw-NMR absorption spectra with respect to a change of temperature and H_t as a function of Δf. In the A phase, two peaks were observed; one small peak near the Larmor frequency and another large peak which has $\Delta f \approx -1.0$ kHz at 2.3 mK or $\Delta f \approx -3.2$ kHz at

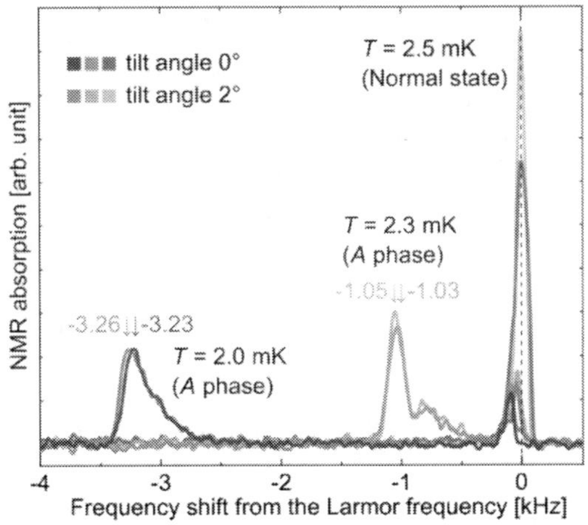

Figure 3. NMR spectra with the tilt angle $0°$ ($I_s = 0.0$ A) and $2°$ ($I_s = 2.5$ A) as a function of the frequency shift from the Larmor frequency measured at 2.5 (normal phase), and 2.3 and 2.0 mK (A phase). In the A phase, the frequency shift showed the negative shift. Numbers next to arrows show values of frequency shifts.

2.0 mK. It is thought that negatively shifted large peaks are attributed to the NMR response of the A phase in the slab geometry as described above. By tilting the magnetic field of $2°$, the amount of Δf at the large peak increased as indicated by two arrows for each temperature. This means that applying $I_s = 2.5$ A to the saddle-shaped coil produces not only the better homogeneity of the magnetic field but also the correction for α closed to $90°$. Tilting the magnetic field of φ corresponds to the changes of the angle α to $\alpha + \varphi$. Then Δf is rewritten as $\Delta f = f_{\mathrm{A}}^{\mathrm{L}^2} \big/ 2f_{\mathrm{L}} \ \cos 2(\alpha_0 + \varphi)$, where α_0 is an original angle between $\hat{\boldsymbol{\ell}}$ and $\hat{\boldsymbol{d}}$ without H_{t}. By using $f_{\mathrm{L}} = 875$ kHz, we evaluated that $f_{\mathrm{A}}^{\mathrm{L}^2} = 1.9 \times 10^9$ Hz^2 and $\alpha_0 = 81°$ at 2.3 mK, and $f_{\mathrm{A}}^{\mathrm{L}^2} = 5.7 \times 10^9$ Hz^2 and $\alpha_0 = 85°$ at 2.0 mK. According to ref. [12], $f_{\mathrm{A}}^{\mathrm{L}^2}$ is 2×10^9 Hz^2 at 2.3 mK and 6×10^9 Hz^2 at 2.0 mK. Experimental values of $f_{\mathrm{A}}^{\mathrm{L}^2}$ almost agree with literature values. The angle α_0 should not depend on the temperature, however, the obtained angle is different from each other. The reason of this discrepancy is not clear. That the negatively shifted spectrum has an asymmetry shape having the tail toward to the Larmor frequency seems to show the existence of the non-uniform texture of $\hat{\boldsymbol{\ell}}$ between parallel-plates. Both theoretical arguments on the NMR spectrum and more precise measurements are needed to explain the observed spectrum quantitatively.

Our recent theoretical calculation suggests that if the HQVs exist, the higher a rotation speed is, the smaller the amount of $|\Delta f|$ becomes. Unfortunately the experimental result does not show such a feature of the rotation dependent resonance frequency. Non-uniform texture may prevent the HQV from being formed. We have not tried to perform the cw NMR measurement a combined condition of a high magnetic field and a large rotation speed, which is suggested as the stabilized condition of the HQV theoretically. We will try to stabilize the HQV under a rotation applying a high H_{l} and a optimized H_{t}, and detect the HQV by performing the cw-NMR measurements under rotation.

4. Summary

We are trying to detect the HQV in the superfluid A phase. In order to detect the HQV, we just improved the apparatus, and fabricated the new superconducting magnet. By performing cw-NMR measurements applying H_{t} in the normal phase, we evaluated the performance of the new magnet and we found that the homogeneity of the NMR field was improved. We have not detected the HQV so far but found that $\hat{\boldsymbol{\ell}}$ seemed to be not uniform.

Acknowledgments

This work was carried out under the Visiting Researcher's Program of the Institute for Solid State Physics, the University of Tokyo, and supported by the "Topological Quantum Phenomena" (No. 22103003) Grant-in Aid for Scientific Research on Innovative Areas from the Ministry of Education, Culture, Sports, Science and Technology (MEXT) of Japan.

References

[1] Volovik G and Mineev V 1976 *JETP Lett.* **24** 561
[2] Cross M C and Brinkman W F 1977 *J. Low Temp. Phys.* **27** 683
[3] Salomaa M M and Volovik G E 1985 *Phys. Rev. Lett.* **55** 1184
[4] Vakaryuk V and Leggett A J 2009 *Phys. Rev. Lett.* **103** 057003
[5] Kawakami T, Tsutsumi Y and Machida K 2010 *J. Phys. Soc. Jpn.* **79** 044607
[6] Nakahara M and Ohmi T 2014 *Phys. Rev. B* **89** 104515
[7] Hakonen P J, Nummila K K, Simola J T, Skrbek L and Mamniashvili G 1987 *Phys. Rev. Lett.* **58** 678
[8] Yamashita M, Izumina K, Matsubara A, Sasaki Y, Ishikawa O, Takagi T, Kubota M and Mizusaki T 2010 *J. Low Temp. Phys.* **158** 353
[9] Kondo K, Ohmi T, Nakahara M, Kawakami T, Tsutsumi Y and Machida K 2012 *J. Phys. Soc. Jpn.* **81** 104603
[10] Vollhardt D and Wölfle P 1990 *The Superfluid Phases of Helium 3* 1st ed (Taylor & Francis)
[11] Kubota M, Obata T, Ishiguro R, Yamashita M, Igarashi T, Hayata E, Ishikawa O, Sasaki Y, Mikhin N, Fukuda M, Kovacik V and Mizusaki T 2003 *Physica B: Condensed Matter* **329-333** 1577
[12] Abragam A and Goldman M 1982 *Nuclear magnetism: order and disorder* (Clarendon Press)

Journal of Physics: Conference Series **568** (2014) 012007 doi:10.1088/1742-6596/568/1/012007

Extended Two-Channel Kondo Phase of a Rotational Quantum Defect in a Fermi Gas

E Fuh Chuo[1], K Ballmann[1], L Borda[2] and J Kroha[1]

[1] Physikalisches Institut, Universität Bonn, Nussallee 12, 53115 Bonn, Germany
[2] Department of Theoretical Physics, Budapest University of Technology and Economics, Budafoki ut 8, H-1111 Budapest, Hungary

E-mail: kroha@physik.uni-bonn.de

Abstract. We show by numerical renormalization group calculations that a quantum defect with a two-dimensional rotational degree of freedom, immersed in a bath of fermionic particles with angular momentum scattering, exhibits an extended 2CK phase without fine-tuning of parameters. It is stabilized by a correlation effect which causes the states with angular momentum $m=\pm1$ to be the lowest energy states of the defect. This level crossing with the non-interacting $m=0$ ground state is signaled by a plateau in the temperature-dependent impurity entropy at $S(T)=k_B\ln 2$, before the 2CK ground state value $S(0)=k_B\ln\sqrt{2}$ is reached.

1. Introduction

The two-channel Kondo (2CK) effect has intrigued condensed matter physicists ever since the problem has been formulated by Noziéres and Blandin in 1980 [1]. It arises when two identical baths or "channels" of mobile fermionic particles compete for the formation of a singlet with a quantum impurity spin S=1/2. As a result, an exotic quantum state of matter with a non-vanishing zero-point entropy of $S(0)=k_B\ln\sqrt{2}$ is formed, and the very notion of particles comprising the system breaks down [2, 3, 4]. However, the necessity of fine-tuning of system parameters to the symmetry point where the quantum impurity degree of freedom is degenerate as well as the two channels are identical (i.e., they have the same density of states and the same coupling to the impurity), has hampered the physical realization of the 2CK effect in electronic systems, see Ref. [5] for a review. For two-level systems (TLS) in a metal, proposed early-on to exhibit the 2CK effect [6], it was shown that the Kondo temperature T_K is generically lower than the TLS level splitting concatenated with the TLS tunneling transitions, so that the 2CK fixed point could not be reached [7, 8]. The realization in heavy-fermion lattice systems, albeit supported by symmetries stabilizing the 2CK fixed point [5], has remained unconfirmed due to the multitude of unclear experimental signatures [9, 10]. In an ingeneously designed semiconductor quantum dot system 2CK signatures could only be observed by fine-tuning of system parameters [11, 12].

In the present work we consider a rotational quantum defect immersed in a fermionic bath. The defect can be comprised of an atomic particle bound in a flattened or effectively two-dimensional harmonic oscillator potential. The three lowest quantum states of the isolated defect are the ground state with angular momentum $m=0$ and a doublet of degenerate angular momentum states $m=\pm1$, split off from the ground state by an excitation energy Δ_0, see

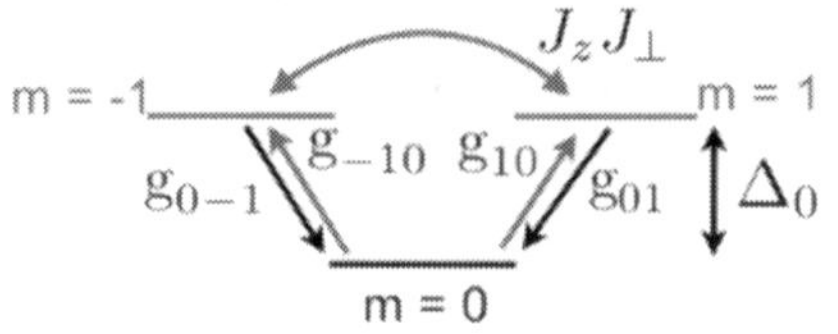

Figure 1. Level scheme of the SU(3) impurity showing the different coupling constants of the local level transitions and the level splitting Δ_0.

Fig. 1. Upon coupling to the itinerant fermion bath, transitions between these levels occur due to angular momentum scattering. The system thus obeys a partially broken SU(3) symmetry. We show by numerical renormalization group (NRG) calculations that in a wide range of parameters the excited-state doublet $m = \pm 1$ is down-renormalized by Kondo correlation effects below the ground state of the non-interacting defect. Hence, this orbital momentum doublet becomes the Kondo degree of freedom, termed pseudospin $m = \pm 1$, and the quantum spin $\sigma = \pm 1/2$ of the fermionic bath particles constitutes the channel degree of freedom, conserved by the orbital angular momentum scattering. The degeneracy of the defect's orbital doublet is stabilized by spatial parity, while the channel degeneracy is guaranteed by time reversal symmetry. The NRG calculations unambiguously show that this model exhibits an extended 2CK phase. It may explain the 2CK signatures observed in the early point contact spectroscopy experiments by Ralph and Buhrman [13], including the conductance spikes at elevated bias and their splitting in a magnetic field, as proposed by us earlier [14, 15]. We also propose that this model may be realized physically in systems of ultracold atomic gases [16].

2. Model and numerical renormalization group

The system described above is represented by the model Hamiltonian [14, 15],

$$\mathcal{H} = \sum_{\mathbf{k}\sigma m} \varepsilon_{\mathbf{k}} c^\dagger_{\mathbf{k}\sigma m} c_{\mathbf{k}\sigma m} + \Delta_0 \sum_{m=\pm 1} f^\dagger_m f_m + \sum_\sigma \left[\frac{1}{2} J_z\, S_z\, s^\sigma_z + J_\perp \left(S_{1,-1} s^\sigma_{-1,1} + S_{-1,1} s^\sigma_{1,-1} \right) \right]$$
$$+ \sum_\sigma \sum_{\substack{m,n=-1,1 \\ n-m\in\pm 1}} \left[g^{(n)}_{m0}\, S_{m,0} s^\sigma_{n-m,n} + h.c. \right] + \sum_{m\pm 1} 2 g^{(0)}_{mm} S_{m,m} s^\sigma_{0,0}\,. \qquad (1)$$

Here the first term is the kinetic energy of the fermionic bath with spin $\sigma = \pm 1/2$; the second one describes the degenerate local doublet, $m = \pm 1$, with the level spacing Δ_0 above the impurity ground state, $m = 0$. The third, fourth and fifth terms describe the interactions between the impurity and the fermionic bath, where the capital operators $S_{m,n}$ denote the generators of SU(3) in the impurity Hilbert space and the lower-case operators $s_{m,n}$ denote the generators of SU(3) in the conduction electron Hilbert space, with coupling constants $J_z, J_\perp$, and $g^{(n)}_{m_f m_i}$ as defined in Fig. 1. The transitions within the unbroken SU(2) subgroup in the degenerate subspace of excited levels, $m = \pm 1$, are written explicitly as the third term. The dynamics of the impurity operators f_m, $f^\dagger_m$ are subject to the constraint that the impurity is in one of its three quantum states at any time, $\hat{Q} = \sum_{m=0,\pm 1} f^\dagger_m f_m = 1$. In Eq. (1), the angular momentum of the mobile fermions with respect to the impurity center, $m = 0, \pm 1$, is written explicitly, while the label $\mathbf{k}$ in the first term comprises all other, continuous fermion degrees of freedom. The couplings $g^{(n)}_{m_f m_i}$ connecting the ground and the excited states are generically substantially smaller than $J_\perp$ and J_z, because the former connect angular wave functions with different number of nodes, so that their matrix elements are small, while the $J_\perp$, J_z connect states with equal number of nodes. In order to keep the number of coupling constants simple, we will show only the results for $g^{(n)}_{m_f m_i} =: g$ and for Kondo coupling $J_\perp = J_z/2 =: J$ or for $J_\perp = J_z =: J$, with g substantially smaller than J. We emphasize, however, that this does not affect the generic properties of the phase diagram and of T_K discussed in Section 3. All parameters are given in units of the unrenormalized half bandwidth D.

To analyze this model, we employed the NRG [17], adapting the algorithm outlined in Ref. [18] to the present case. Since the Hamiltonian has a large symmetry group of (partially broken) $SU(3) \times SU(2)$, the calculations are numerically demanding. Because the interaction involves transitions between three levels [the fundamental representation of SU(3)], three Wilson chains are required. Each one comes in two flavors, the conserved SU(2) channel degree of freedom, spin $\sigma = \pm 1/2$. Therefore, each site of a Wilson chain has four states, $|0\rangle$, $|\uparrow\rangle$, $|\downarrow\rangle$, and $|\uparrow\downarrow\rangle$, i.e., in each renormalization group step the dimension of the Hilbert space grows by a factor $4^3 = 64$. We exploited symmetries of the model [19] so as to decompose the total Hilbert space into it's conserved subspaces and to gain efficiency of the computer code.

3. 2CK phase of the rotational defect model

In this section we present the results of our NRG analysis, in particular the phase diagram in the plane of Kondo coupling J and level spacing Δ_0 as well as the behavior of the Kondo temperature T_K and of the entropy $S(T)$ of the model in the 2CK phase. First we discuss how the 2CK fixed point is identified and distinguished from a trivial potential scattering fixed point in the renormalization group flow: For a Fermi liquid fixed point (single-channel Kondo or potential scattering) the fixed point spectrum alternates between even and odd number of NRG iterations, corresponding to even or odd total number of (pseudo)spins and due to the different energies of the (pseudo)spin singlet or (pseudo)spin doublet ($s = 1/2$) states, see Fig 2. For the 2CK fixed point there is no alternation of the spectra [19], because a complex spin-entangled state is formed, independent of the parity of the particle number in the system, see Fig 3. Furthermore, a Fermi liquid fixed point is characterized by an equidistant fixed point spectrum for large NRG iteration numbers, while the 2CK fixed point spectrum is not equidistant [19]. The qualitatively different fixed point spectra are shown in the NRG level flow diagrams of Figs. 2 (potential scattering) and 3 (2CK), respectively. While for the pure 2CK model [1] the fixed point level spacings are known from boundary conformal field theory [19], the fixed point spectrum of

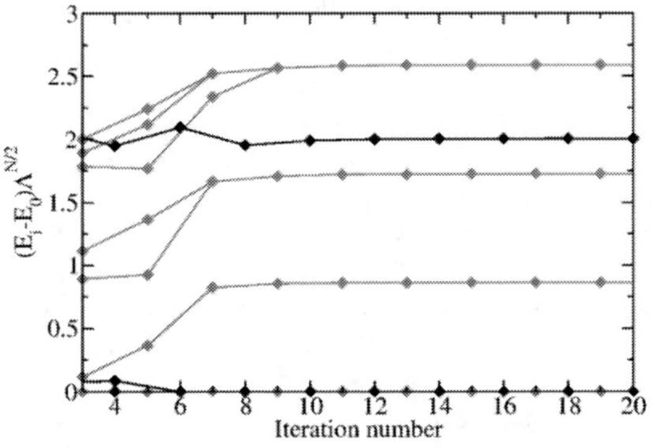

Figure 2. NRG energy level flow for the parameters $J = J_\perp = J_z/2 = 0.2$, $g = 0.002$, $\Delta_0 = 0.14$, and NRG discretization parameter $\Lambda = 4$ [18]. Potential scattering fixed point: alternation between even (black dots) and odd (red dots) number of iterations; equidistant fixed point spectra for even and odd iterations, respectively.

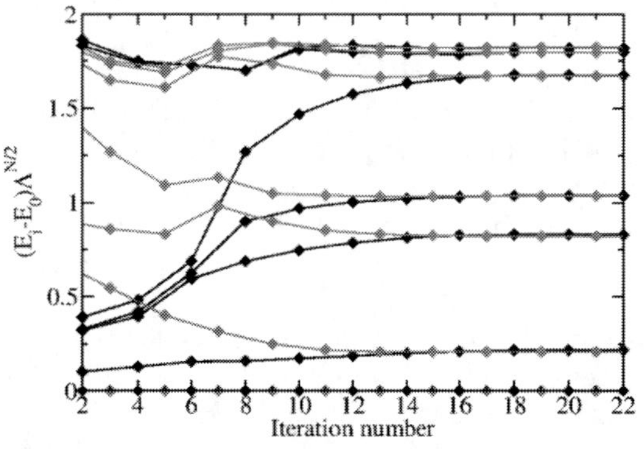

Figure 3. NRG energy level flow for the parameters $J = 0.2$, $g = 0.002$, $\Delta_0 = 0.13$, and NRG discretization parameter $\Lambda = 4$ [18]. 2CK fixed point: no alternation between even (black dots) and odd (red dots) number of iterations; non-equidistant fixed point spectrum.

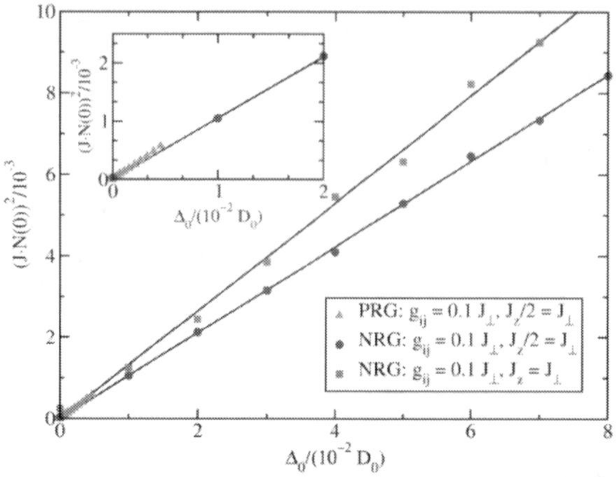

Figure 4. Phase diagram separating 2CK (above the line) and potential scattering (below the line) regimes for two different ratios of $J_z/J_\perp$, $J := J_\perp$, and $\Lambda = 3$ and iteration number $N = 25$. The green squares and blue circles are the results of NRG calculations. The red triangles are calculated with the PRG approach of Refs. [14, 15] for the parameter set of the blue data points.

our extended model, Eq. (1), shows deviations from these values due to additional potential scattering terms present in our model. Nevertheless, the 2CK state may be distinguished from the Fermi liquid fixed point by the non-equidistant level spacing. Comparison of Figs. 2 and 3 shows that the system changes, for given values of J and g, from potential scattering to 2CK upon reducing the level spacing Δ_0.

Mapping out the different fixed points in the $N(0)J - \Delta_0$ plane, with N(0) the density of states at the Fermi energy, we obtain the phase diagram of our model as shown in Fig. 4. The data points shown are unstable fixed points separating the 2CK regime (above) from the trivial potential scattering regime (below the line). It is seen that the model generically exhibits an extended 2CK phase for a wide range of system parameters. In determining the phase boundary numerically, it must be observed that a minimal coupling strength $\bar{J}$ is necessary in order for the 2CK fixed point to be detectable by the NRG within N iteration steps. This threshold $\bar{J}$ must be subtracted from the numerically determined critical value of J in order to obtain the 2CK phase boundary that would be obtained after $N \to \infty$ iterations. $\bar{J}$ may be estimated as follows. The minimal resolvable energy by NRG after N steps is $E_{\rm min} = D \cdot \Lambda^{-N/2}$, where D is the half bandwidth and Λ the discretization parameter. The minimal Kondo temperature resolvable by NRG is, therefore, $T_{K\,\rm min} = D\exp[-1/(2N(0)\bar{J}] = E_{\rm min}$. Combining the last two equations, one obtains for the threshold, $N(0)\bar{J} = 1/N \ln \Lambda$, i.e., $N(0)\bar{J} = 0.0144$ for our NRG parameters N=25, $\Lambda = 3$. This threshold has been subtracted in the phase diagram, Fig. 4. Good quantitative agreement is found with the results of a perturbative renormalization group treatment for small coupling values, as described in Refs. [14, 15].

It is known that in the pure 2CK model [1] the 2CK fixed point corresponds to an intermediate coupling strength of $J^*/D \approx 0.7$ [19]. Hence, at this coupling strength T_K should reach up to the band width, $T_K^*/D \approx 1$, and decay exponentially on both sides of J^*. The latter has been shown due to a strong/weak coupling duality of the 2CK model [1, 20]. It is different from the single-channel Kondo model, where the fixed point is at strong coupling, $J \to \infty$. Our NRG calculations show that this behavior is also true for the rotational defect model, Eq. (1): Fig. 5 shows T_K for this model as a function of J for various values of the level spacing Δ_0, calculated by NRG. It exemplifies that large values of T_K may be realized in this model.

Finally, we discuss the temperature T dependence of the impurity contribution to the entropy $S(T)$, as shown in Fig. 6. For $T > \Delta_0$, all three impurity levels are thermally excited, and $S(T)/k_B = \ln 3$. As T is reduced below Δ_0, but $T > T_K$, the entropy obtains a plateau at $S(T)/k_B \approx \ln 2$. This indicates, that for $T < \Delta_0$ not the bare excited state doublet is frozen out, but a level crossing occurs induced by Kondo correlations, so that the $m = \pm 1$ doublet becomes the interacting impurity ground state and can host Kondo fluctuations even at the

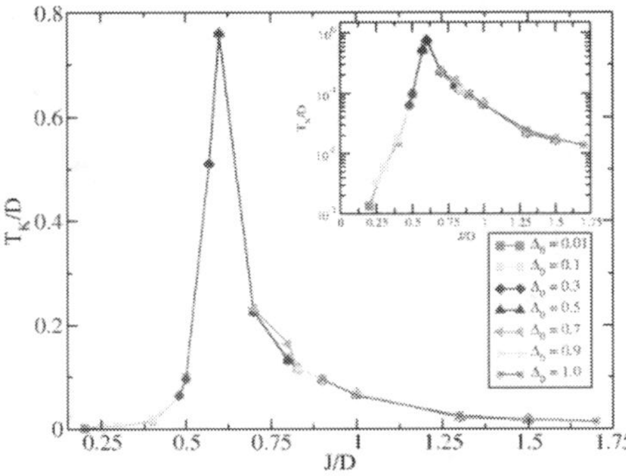

Figure 5. The Kondo temperature T_K as a function of the coupling constant $J = J_\perp = J_z/2$ is shown for various Δ_0 and for $g = 0.01J$. The inset shows the same on a logarithmic scale.

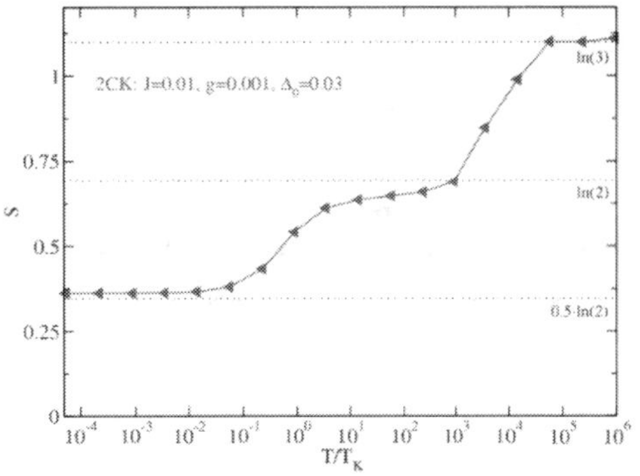

Figure 6. The impurity contribution to the entropy is shown as a function of temperature T. Three plateaus can be clearly distinguished, see text.

lowest energies. Because of the SU(2) channel symmetry, the system is then bound to flow to the 2CK fixed point for $T < T_K$. This is confirmed by the 2CK zero-point entropy of $S(0)/k_B = \ln\sqrt{2}$ [3] which is achieved in Fig. 6 at the plateau for $T \ll T_K$.

To summarize, we have demonstrated the existence of an extended 2CK phase in a rotational impurity model with partially broken SU(3) symmetry. We have also shown that at the 2CK fixed point coupling, $J^* \approx 0.7$, T_K reaches up to the band cutoff, as expected for a 2CK system, so that large T_K values can be achieved in this model.

Acknowledgments

We would like to thank Ralf Bulla, Alfred Zawadowski and Eran Sela for useful discussions. This work was supported in part by the DAAD (EFC), by the Alexander von Humboldt Foundation (LB) and by DFG through SFB 608.

References

[1] P. Nozières and A. Blandin, Journal de Physique (Paris), **41**, 193 (1980).
[2] A. M. Tsvelik and P. B. Wiegmann, Z. Phys. B **54**, 201 (1994).
[3] N. Andrei and C. Destri, Phys. Rev. Lett. **52**, 364 (1984);
[4] I. Affleck and A. W. W. Ludwig, Nucl. Phys. B, **352**, 849 (1991); Phys. Rev. Lett. **67**, 3160, (1991); Phys. Rev. B **48**, 7297 (1993).
[5] D. L. Cox and A. Zawadowski, Adv. Phys. **47**, 599 (1998).
[6] J. L. Black, K. Vladár, and A. Zawadowski, Phys. Rev. B **26**, 1559 (1982); K. Vladár and A. Zawadowski, Phys. Rev. B **28**, 1564 (1983); **28**, 1582 (1983); Phys. Rev. B **28**, 1596 (1983).
[7] I. L. Aleiner *et al.*, Phys. Rev. Lett. **86**, 2629 (2001).
[8] I. L. Aleiner and D. Controzzi, Phys. Rev. B **66**, 045107 (2002).
[9] C. L. Seaman *et al.*, Phys. Rev. Lett. **67**, 2882 (1991);
[10] T. Cichorek *et al.*, Phys. Rev. Lett. **94**, 236603 (2005).
[11] Y. Oreg, and D. Goldhaber-Gordon, Phys. Rev. Lett. **90**, 136602 (2003).
[12] R. M. Potok, I. G. Rau, H. Shtrikman, Y. Oreg, and D. Goldhaber-Gordon, Nature **446**, 167 (2007).
[13] D. C. Ralph and R. A. Buhrman, Phys. Rev. Lett. **69**, 2118 (1992); Phys. Rev. B **51** 3554 (1995).
[14] M. Arnold, T. Langenbruch, and J. Kroha, Phys. Rev. Lett. **99**, 186601 (2007).
[15] K. Ballmann and J. Kroha, Ann. Phys. (Berlin) **524**, No 3-4, 245-251 (2012).
[16] E. Fuh Chuo, K. Ballmann, L. Borda, and J. Kroha, in preparation.
[17] K. G. Wilson Rev. Mod. Phys. **47**, 775 (1975).
[18] R. Bulla, T. A. Costi, and Th. Pruschke, Rev. Mod. Phys. **80**, 395 (2008).
[19] H. B. Pang and D. L. Cox, Phys. Rev. B, **44**, 9454 (1991).
[20] Ch. Kolf and J. Kroha, Phys. Rev. B **75**, 045129 (2007).

Journal of Physics: Conference Series **568** (2014) 012008 doi:10.1088/1742-6596/568/1/012008

Pulsed transport of electrons on liquid helium confined in narrow channels

F Shaban[1], T Lorenz[1], R Rau[1], M Ashari[1], D Rees[2], K Kono[3], E Scheer[1] and P Leiderer[1]

[1] Department of Physics, University of Konstanz, Konstanz 78457, Germany
[2] NCTU, Hsinchu City, 300, Taiwan
[3] RIKEN, Hirosawa 2-1, Wako, Saitama, Tokyo, 351-0198, Japan

E-mail: paul.leiderer@uni-konstanz.de

Abstract. We have used a Source-Gate-Drain configuration with electrons on liquid helium (the Helium Field Effect Transistor "He-FET") to study the transport of "classical" electrons through narrow channels. The channels, formed by the split gate of the device, were between ten and a hundred μm long and several μm wide, and could be blocked completely by a negative bias voltage applied to the gate. In contrast to previous experiments, where the electron densities in source and drain were nearly the same and the system therefore was close to equilibrium, in the present measurement the drain was empty. The transport of the electrons through the channel was initiated by opening the gate with a short positive pulse with a duration down to nanoseconds, and the amount of electrons which passed during this time was registered. In this way, we could determine for the first time the transport properties of such a system on a nanosecond time scale and far off equilibrium. Measurements with varying gate voltage provide clear evidence for the formation of lanes of electrons in the channels.

1. Introduction

The transport of electrons through narrow constrictions has been a topic of considerable interest for decades. Most systems which have been investigated so far were either metals, where point contacts down to the size of a single atom have been realized [1], or semiconductors, e.g. MOSFETs with split gate geometry, where the effective contact size can be tuned electrically by applying a proper gate voltage [2-4]. For Surface State Electrons (SSE) on liquid helium, which form a two-dimensional Coulomb system of exceptional purity, only a small number of investigations in confined geometry exists so far. The first experiment with a Source-Gate-Drain configuration like in a Field Effect Transistor was reported by Klier et al. [5] and switching of the source-drain current I_{sd} by varying the gate voltage was demonstrated. Since then, investigations with similar set-ups have revealed interesting phenomena like the transport of strongly correlated electron systems through narrow channels [6,7,8].

Usually, experiments with SSE on liquid helium are carried out with nearly the same electron density in source and drain, the system is therefore close to equilibrium. Transport properties are derived from the response to an ac driving field, obtained by capacitive coupling to the electrodes and recorded by lock-in technique at a typical frequency in the kHz range. The behavior of the system is therefore studied on a scale of microseconds and above [8]. In contrast, we report here on pulsed dc measurements, where a temporal resolution for the electron transport down to some ten nanoseconds has been reached. Moreover, in our transport studies the system is far off equilibrium, because the electron density in the source at the channel entrance is adjusted to some finite value, whereas the drain is completely empty because all arriving electrons are removed immediately. It also should be mentioned that in our measurements here, like in nearly all other experiments with SSE on helium, the electron density was in the range of 10^{10} cm^{-2} or below, where the SSE behave like a classical system [9].

2. Experimental

The principle of our experimental set-up has been described previously [5,10]. A tiny heated filament above the source area supplies the electrons and is operated in a pulsed mode with a pulse duration of 25 ms and a repetition rate of 4 Hz to minimize the thermal load. The electrons which have passed through the gate channel to the drain are collected by a pick-up electrode. Whereas in the first experiments with a Helium FET the electrode structures consisted of gold films evaporated on glass, the substrate used in the present work was a piece of a silicon wafer, covered with 500 nm oxide, which formed the source-channel-drain area. The split-gate electrodes and also a guard electrode surrounding the whole device consisted of 100 nm gold films evaporated onto the wafer (see figure 1a). A voltage on the order of a few volts was applied to the Si substrate between the source and the drain side. Together with the voltages applied to the split gate and the guard this provided the potential landscape for the electrons on the device (figures 1b, 1c). The sample was positioned at an adjustable height in the gas space above the liquid level of a ^{4}He cryostat and was therefore covered with a superfluid He film whose thickness could be controlled between 20 and 60 nm. The temperature in all the measurements was 1.3K.

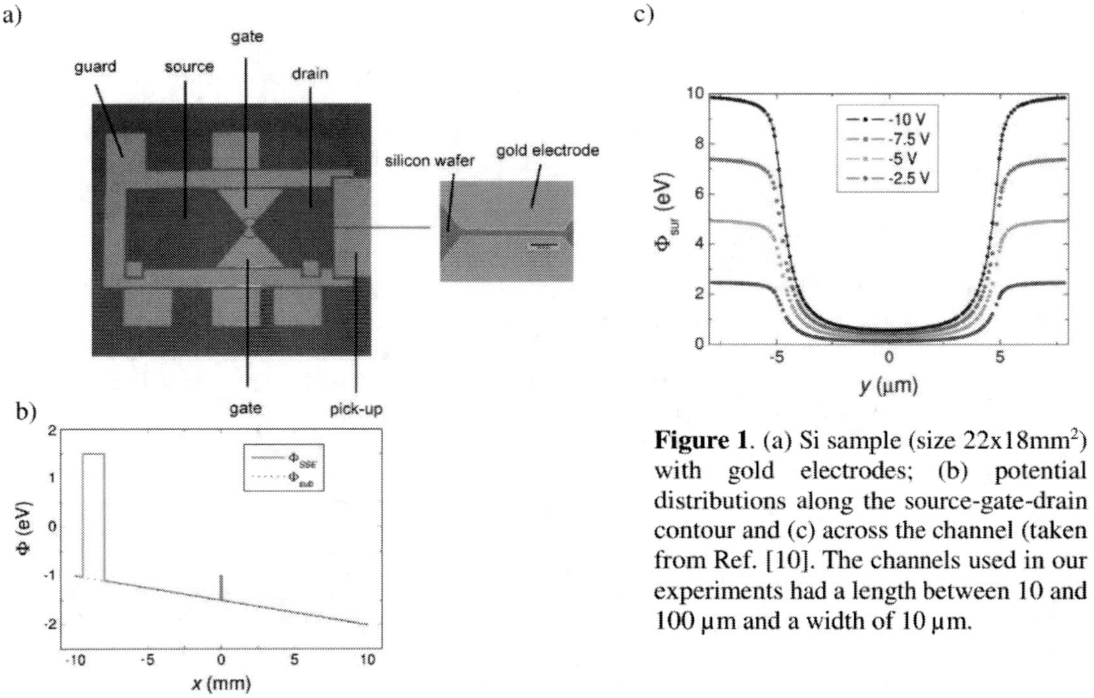

Figure 1. (a) Si sample (size 22x18mm^2) with gold electrodes; (b) potential distributions along the source-gate-drain contour and (c) across the channel (taken from Ref. [10]. The channels used in our experiments had a length between 10 and 100 μm and a width of 10 μm.

3. Results

3.1 I-V characteristics

We first describe steady state measurements of the variation of the source-drain current I_{sd} with the gate voltage V_g, as shown in figure 2. Starting at a value of V_g sufficiently negative to block the source-drain current, the gate voltage is slowly increased and the potential barrier of the gate therefore lowered. Above a certain threshold value, -0.03V in this case, the gate starts to open and a gradual increase of I_{sd} is observed. Two prominent features show up in addition: i) the current I_{sd} displays a periodic modulation (not to be mistaken as noise) which is caused by the pulse-like supply of electrons from the filament to the source and is hence an artifact of the measurement. The amplitude of this modulation increases as the gate is opened more and more, because at large I_{sd} the depletion of charge between the recharging pulses is more pronounced; ii) at small values of I_{sd} steps are observed as the gate is opening. At first sight this feature is reminiscent of the steps in *I-V* characteristics of split-gate quantum point contacts made from semiconductor heterostructures, which are a quantum phenomenon [2-4]. However, in the present case we are dealing with a classical electron system, and therefore the steps have to result from a classical effect. We relate our observation to the well-known phenomenon of particle lane formation in channels [11]: As the gate voltage is increased and the effective channel width is getting larger, the number of electron lanes which fit into the channel increases as well, and each additional lane will give rise to a step in I_{sd}. A similar feature has been observed in ac measurements of SSE transport close to equilibrium by Rees et al. [8]. It results from the strong correlations between the electrons close to Wigner crystallization and is also found in the transport of other classical systems in confined geometry [11].

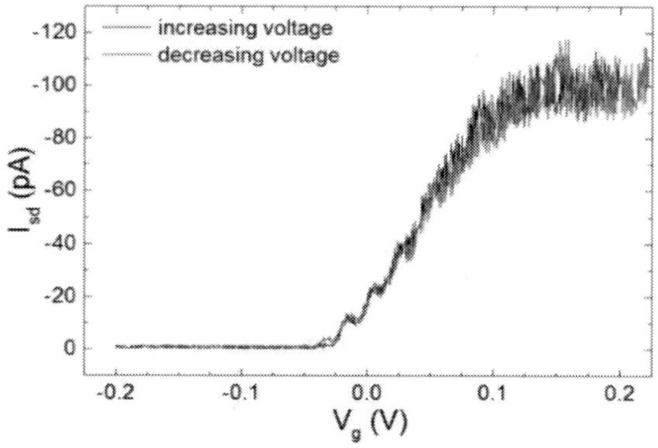

Figure 2. Source-drain current I_{sd}, collected at the pick-up electrode, as the gate voltage V_g is varied from closed to open conditions. The black line represents the opening curve, the red line the closing curve. A slight hysteresis between opening and closing voltages of 0.024V is compensated for clarity by shifting the red curve by this amount to negative values. The gate channel had in this case point contact geometry with a width of 10 μm.

3.2 Gated Transport Measurements

We now turn from the quasi-continuous measurements of the source-drain current to experiments where first electrons are added to the source for some time, with the gate being closed, and then the filament is switched off. Starting from an empty source, the chemical potential of the electron ensemble in the source increases as the density of the electrons grows, until it becomes high enough that the electrons "spill" over the gate and a pick-up current is observed. At this point the filament is switched off, and a state with a well-defined number of source electrons N_s is thus prepared. N_s can be determined by opening the gate completely and integrating over the current pulse that arrives at the drain. It was found previously that N_s grows nearly quadratically as the gate barrier is increased, due to the particular charge density profile in the source which has its maximum right at the channel entrance [10].

Such a measurement also allows one to determine the maximum number of electrons that can reversibly be stored in the source: It is known that above a critical charge density, which depends on the roughness of the substrate, SSE on the film surface locally break through the film and are then localized on the SiO_2 surface of the substrate [9]. This part of the source electrons is then immobile and does not contribute to the charge that moves to the drain when the gate is opened. However, the immobile electrons still add to the Coulomb energy when the source is charged anew. This implies that the characteristic curve which relates the number of stored (mobile) electrons N_s to the gate voltage is shifted step by step to more negative V_g values, as more and more localized electrons are present. An example is shown in figure 3. For this particular sample, the maximum charge that could safely be stored as mobile electrons was $N_s \sim 2\text{x}10^9$, corresponding to an electron density of about 10^{10} cm^{-2} at the channel entrance. In a normal experimental run the electron density was kept well below this critical value to avoid complications due to trapped surface charge.

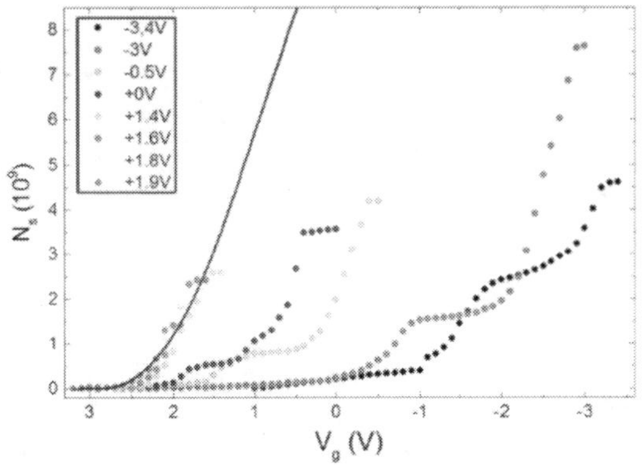

Figure 3. Number of mobile source electrons, N_s, determined as described in [10], for a sequence of runs where the source was successively charged at higher and higher potential barriers (the gate voltage where the charging of the source took place is indicated in the inset). The blue curve shows the result for the ideal case, when all the electrons collected in the source are mobile. The gradual shift of the data towards more negative gate voltage is due to an increasing number of immobile source electrons.

In the measurements of figure 3, the gate was opened completely for a long time (> 1 s) in order to collect all the mobile electrons present in the source on the pick-up electrode. In the experiments to be described now the gate was also opened completely, but then closed again after a time Δt ranging from nanoseconds to milliseconds. Figure 4 shows the collected number of electrons, N, for pulse widths in the millisecond range, normalized by the total number of electrons in the source. After an initial increase, the charge transported during the opening time of the gate saturates for $\Delta t > 4$ ms, which means that by then all mobile electrons have been removed from the source. This characteristic depletion time is caused by two contributions, namely the transport of the electrons within the source area towards the channel entrance, and the transport of the electrons through the channel. Since we are mainly interested in the latter mechanism, we now concentrate on short opening times $\Delta t < 10$ µs, where only a small fraction of the total charge N_s is removed from the source by each pulse and the redistribution of the electrons in the source area can therefore be neglected.

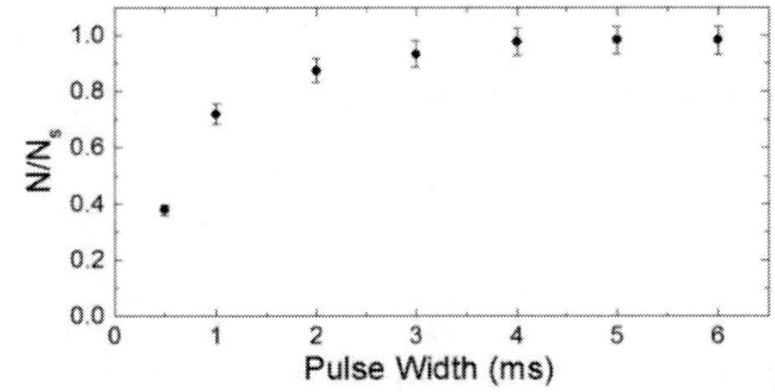

Figure 4. Normalized number of electrons moving from source to drain, as the gate is opened by positive pulses of various width. The channel in this sample was 100µm long.

One of our results – but this time for a sample with a short channel, resembling a point contact - are displayed in figure 5a, and in figure 5b on an expanded scale. In figure 5a the number of transported SSE increases linearly with the gate opening time Δt, which implies that on this time scale the charge per unit time transported through the channel (i.e. the current) is constant.

A closer look at the data on the expanded scale in figure 5b reveals that the straight line, representing the channel current, does not extrapolate through the origin, but rather suggests that a finite opening time Δt_{min} of the channel is required before the first electrons make it from the source to the drain. We identify this minimal time of about 45 ns with half the transit time of the electrons. Due to the shape of the potential barrier, those electrons which have not yet reached the middle of the channel will be pushed back into the source when the barrier is switched on again, while electrons which have travelled beyond the middle will be pushed forward and collected by the pick-up electrode. Based on this model, one can derive a rough estimate for the electron mobility in the channel from the observed Δt_{min} and the externally applied potential gradient

$$\mu = v/E = (l/2\Delta t_{min})/E$$

where l is the channel length and v the velocity of the moving electrons. The resulting value for the mobility, $\mu \sim 2.4 \times 10^4$ cm^2/Vs, is of the same order as the mobility of electrons on bulk ^{4}He in this temperature range. A more detailed analysis will be the subject of a separate paper.

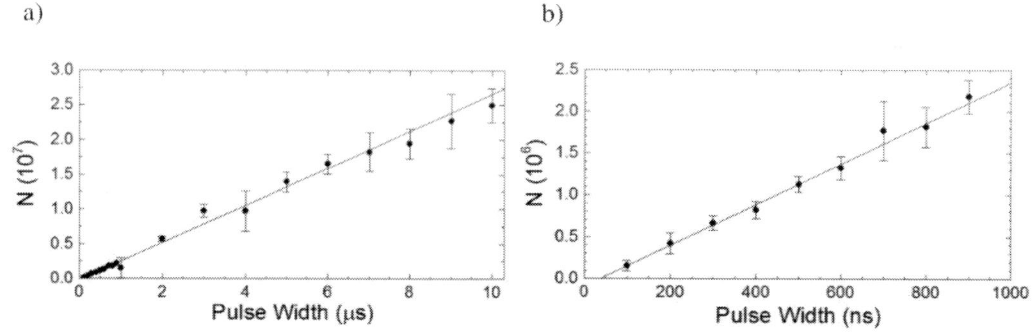

Figure 5. (a) Number of electrons moving from source to drain as the gate is opened for times short compared to the depletion time. The total number of electrons in the source was here $N_s \sim 10^9$. b) Same as in (a) for the regime of ns pulses. The shape of the gate channel was in this case like a point contact (geometrical width 10μm), therefore the channel length was not well defined, but effectively was comparable to the width. The straight lines are just guides to the eye.

4. Conclusions

The results presented here show that Surface State Electrons in confined geometry are well suited for investigating the transport of correlated electrons in the classical Coulomb regime, which is not directly accessible with electrons in solid state systems. Experiments complementary to those in MOSFETs demonstrate phenomena like the formation of particle lanes and the manipulation of SSE ensembles on the micrometer and nanosecond scale. Of particular interest will be to achieve somewhat higher electron densities, where crossover effects towards the degenerate Fermi gas are expected [9].

Acknowledgements

We thank V.B. Shikin and M. Schmid for useful discussions, and Louis Kukk for technical help. M. A. gratefully acknowledges support by DAAD, F. S. a scholarship by the University of Khartoum.

References

[1] Agraït N, Levy Yeyati A and van Ruitenbeek J M 2003 *Phys. Rep.* **377** 81
[2] van Wees B J et al. 1988 *Phys. Rev. Lett.* **60** 848
[3] Wharam D A et al. 1988 *J. Phys.* C**21** L209
[4] Levinson Y B, Lubin M I and Sukhorukov E V 1992 *Phys. Rev.* B **45** 11936
[5] Klier J, Doicescu I and Leiderer P 2000 *J. Low Temp. Phys.* **121** 603
[6] Glason P et al. 2001 *Phys. Rev. Lett.* **87** 176802
[7] Ikegami H, Akimoto H and Kono K 2010 *Phys. Rev.* **B 82** 201104(R)
[8] Rees D G et al. 2011 *Phys. Rev. Lett.* **106** 026803
[9] Günzler T, Bitnar B, Mistura G, Neser S and Leiderer P 1996 *Surf. Sci.* **361/362** 831
[10] Ashari M, Rees D G, Kono K, Scheer E and Leiderer P 2012 *J. Low Temp. Phys.* **167** 15
[11] Köppl M, Henseler P, Erbe A, Nielaba P and Leiderer P 2006 *Phys. Rev. Lett.* **97** 208302

Journal of Physics: Conference Series **568** (2014) 012009 doi:10.1088/1742-6596/568/1/012009

Application of Fibonacci oscillators in the Debye model

André A.A. Marinho[a,b], **F.A. Brito**[b], **C. Chesman**[a]

[a] Departamento de Física Teórica e Experimental, Universidade Federal do Rio Grande do Norte, Caixa Postal 1634, 59072-970 Natal, Rio Grande do Norte, Brazil

[b] Departamento de Física, Universidade Federal de Campina Grande, 58109-970 Campina Grande, Paraiba, Brazil

E-mail: andre@dfte.ufrn.br or andreofisico@hotmail.com.br

Abstract. In this paper we study the thermodynamics of a crystalline solid by applying q-deformed algebra of Fibonacci oscillators through the generalized Fibonacci sequence of two real and independent deformation parameters q_1 and q_2. We find a (q_1, q_2)-deformed Hamiltonian and consequently the q-deformed thermodynamic quantities. The results led us to interpret the deformation parameters acting as disturbance or impurities factors modifying the characteristics of a crystal structure. More specifically, we found the possibility of adjusting the Fibonacci oscillators to describe the change of thermal conductivity of a given element as one inserts impurites.

1. Introduction

The study of quantum groups and quantum algebras has attracted great interest in recent years and stimulated intense research in various fields of physics [1, 2, 3] taking into account a range of applications covering astrophysics and condensed matter, for instance, black holes and high-temperature superconductors [4].

A possible mechanism to generate a deformed version of the classical statistical mechanics consists in replacing the Gibbs-Boltzmann distribution by a deformed version. In this sense it is postulated a form of deformed entropy which implies a generalized theory of thermodynamics [5, 6, 7, 8]. In this context, it was demonstrated that a natural realization of the q-deformed thermodynamics can be done via q-calculation formalism [9, 10].

A new proposal for the q-calculation is the inclusion of two distinct deformation parameters in some physical applications. Starting from the generalization of q-algebra [11], in Ref. [12] was generalized the Fibonacci sequence. Here, the numbers are in that sequence of generalized Fibonacci oscillators, where new parameters (q_1, q_2) are introduced [13, 14, 15]. They provide a unification of quantum oscillators with quantum groups, keeping the degeneration property of the spectrum invariant under the symmetries of the quantum group. The quantum algebra with two deformation parameters may have a greater flexibility when it comes to applications in realistic phenomenological physical models [16, 17].

We know that a solid is formed by a large number of atoms bound by cohesion forces of various types. The motion of atoms in a solid is very narrow, causing each atom to move only within a small neighborhood, executing vibratory motion around its equilibrium point. In a crystalline solid, the equilibrium points of atomic vibrations form a regular spatial structure, a cubic structure, for example.

The study conducted by Anderson, Lee and Elliot [18, 19, 20] shows that the presence of defects or impurities in a crystal modifies the electrostatic potential in their neighborhood, breaking the translational symmetry of the periodic potential. This perturbation can produce electronic wave functions located near the impurity, ceasing to be propagated throughout the crystal.

The conductivity of semiconductors can be dramatically altered by the presence of impurities, i.e., different from atoms that make up the pure crystal. This property allows the production of a variety of electronic devices of the same semiconductor material. This process of placing impurities in semiconductor materials is called doping.

In the present study we follow the lines of [21, 22, 23] to address the issues of (q_1, q_2)-deformed thermodynamics through application of a generalized q-algebra to a Debye solid. We find that the deformation is related to phenomena due to impurities or disorder factors in the system. We shall mainly show that the Fibonacci oscillators may act as defects or impurities in the crystal lattice, allowing us to modify quantities such as Debye temperature, thermal and electrical conductivities.

2. Algebra of the Fibonacci oscillators

The generalization of integers usually is given by a sequence. The two well-known ways to describe a sequence are the arithmetic and geometric progressions. However, the Fibonacci sequence encompasses both. Generalizing this sequence, we get the Fibonacci oscillators, so the spectrum can now be given by the Fibonacci integer.

Defining the Fibonacci *basic number*, the q-deformed quantum oscillator is now defined by the Heisenberg algebra in terms of the annihilation and creation operators c and $c^\dagger$, respectively, and the number operator N, as follows:

$$[x_{i,q_1,q_2}] = c_i^\dagger c_i = \frac{q_2^{2x_i} - q_1^{2x_i}}{q_2^2 - q_1^2}, \tag{1}$$

$$c_i c_i^\dagger - q_1^2 c_i^\dagger c_i = q_2^{2N_i}, \qquad \text{and} \qquad c_i c_i^\dagger - q_2^2 c_i^\dagger c_i = q_1^{2N_i}, \tag{2}$$

$$[1 + x_{i,q_1,q_2}] = q_1^2[x_{i,q_1,q_2}] + q_2^{2N_i}, \quad \text{or} \quad [1 + x_{i,q_1,q_2}] = q_2^2[x_{i,q_1,q_2}] + q_1^{2N_i}. \tag{3}$$

Let us now generalize to (q_1, q_2)-deformed algebra the study we accomplished in [22, 23] for only one q-deformation parameter. Now we have the following Hamiltonian with the respective energy eigenvalues

$$\mathcal{H}_{q_1,q_2} = \frac{\hbar\omega}{2}\left(cc^\dagger + c^\dagger c\right), \tag{4}$$

$$E_{n_{i,q_1,q_2}} = \frac{\hbar\omega_E}{2}\Big([n_{i,q_1,q_2}] + [n_{i,q_1,q_2} + 1]\Big) = \frac{\hbar\omega_E}{2} + \frac{\hbar\omega_E\left(2\ln(q_2) - 2\ln(q_1)\right)n}{q_2^2 - q_1^2}. \tag{5}$$

Notice that when $q_1 = 1$ and $q_2 \to 1$ (and vice-versa) the deformation vanishes, so that we have the usual definition

$$E_n = \frac{\hbar\omega_E}{2}(2n + 1). \tag{6}$$

3. (q_1, q_2)-deformed Debye Solid

Corrections of Einstein's model are given by the Debye model, allowing us to integrate from a continuous spectrum of frequencies up to the Debye frequency ω_D, as the number of normal modes of vibration is $3N$, as is found in the literature [24, 25, 26]. Following the same mathematical development, and inserting the given parameters (q_1, q_2), we can write the specific heat for any temperature in the form:

$$c_{V_{q_1,q_2}}(T) = 3\kappa_B D(\alpha_{0_{q_1,q_2}}), \tag{7}$$

where $D(\alpha_{0_{q_1,q_2}})$ is the (q_1, q_2)-deformed Debye function, defined by

$$D(\alpha_{0_{q_1,q_2}}) = \frac{3}{(\alpha_{0_{q_1,q_2}})^3}\int_0^{\alpha_{0_{q_1,q_2}}} \frac{\alpha^4 \exp(\alpha)}{[\exp(\alpha) - 1]^2} d\alpha, \tag{8}$$

$$\alpha_{0_{q_1,q_2}} = \frac{\hbar\omega_{D_{q_1,q_2}}}{\kappa_B T} = \frac{\theta_{D_{q_1,q_2}}}{T}, \qquad \omega_{D_{q_1,q_2}} = 2\omega_D\left(\frac{\ln(q_2)-\ln(q_1)}{q_2^2-q_1^2}\right). \tag{9}$$

where $\omega_{D_{q_1,q_2}}$ and $\theta_{D_{q_1,q_2}}$, are the (q_1,q_2)-deformed Debye frequency and temperature, respectively, and ω_D is the Debye frequency characteristic. Integrating Eq.(8) by parts one finds

$$D(\alpha_{0_{q_1,q_2}}) = -\frac{3\alpha_{0_{q_1,q_2}}}{\exp(\alpha_{0_{q_1,q_2}})-1} + \frac{12}{\alpha^3_{0_{q_1,q_2}}}\Bigg\{-\frac{\pi^4}{15} - \frac{\alpha^4_{0_{q_1,q_2}}}{4} + \alpha^3_{0_{q_1,q_2}}\ln[1-\exp(\alpha_{0_{q_1,q_2}})]$$
$$+ 3\alpha^2_{0_{q_1,q_2}}Li_2[\exp(\alpha_{0_{q_1,q_2}})] - 6\alpha_{0_{q_1,q_2}}Li_3[\exp(\alpha_{0_{q_1,q_2}})] + 6Li_4[\exp(\alpha_{0_{q_1,q_2}})]\Bigg\} \tag{10}$$

where

$$Li_n(z) = \sum_{k=0}^{\infty}\frac{z^n}{k^n}, \tag{11}$$

is the polylogarithm function. We can check that, as in the usual Debye solid, the low-temperature specific heat in a q-deformed Debye solid is proportional to T^3, due to phonon excitation, a fact that is in agreement with experiments. Thus, let us express the (q_1,q_2)-deformed specific heat for low temperatures as follows:

$$c_{V_{q_1,q_2}} = \frac{12\pi^4\kappa_B}{5}\left(\frac{T}{\theta_{D_{q_1,q_2}}}\right)^3 = 1944\left(\frac{T}{\theta_{D_{q_1,q_2}}}\right)^3\frac{J}{molK}. \tag{12}$$

Through the settings for the thermal conductivity (κ) and electrical (σ) — see [27], we can rewrite the (q_1,q_2)-deformed specific heat in the useful form

$$\kappa_{q_1,q_2} = \frac{\kappa c_{V_{q_1,q_2}}}{c_V} \qquad \text{and} \qquad \sigma_{q_1,q_2} = \frac{\kappa_{q_1,q_2}\sigma}{\kappa}. \tag{13}$$

For illustration purposes, we choose iron (Fe) and chromium (Cr), two materials that can be employed in many areas of interest. In Figs.(1) we present deformed values of Fe (Fe_{q-def}) (black curve) and Cr (Cr_{q-def}) (green curve), for values $q_1 = 1$ and $q_2 = 0.1, \cdots, 1$, where for this range we assume the maximum deformation ($q_1 = 1$ and $q_2 = 0.1$) and the pure element (bulk) ($q_1 = 1$ and $q_2 = 1$). The other elements are represented by colors and indicated in the very figure.

On the left side of the Fig.(1), we can observe that before reaching their limits, black and green curves can assume the values of Debye temperatures (θ_D) of other elements. The Fe_{q-def} e.g., equates to: beryllium (Be) when $q_2 \approx 0.23$, chromium (bulk) (Cr) $q_2 \approx 0.75$ and osmium (Os) $q_2 \approx 0.94$. On the right, we have the behavior of the curves obtained for the specific heat c_V. We note that the behavior is quite different from the previous curves θ_D, i.e., the curves start at lower values (maximum deformation) until they reach their pure values. Having Fe_{q-def} as an example again, it is possible to see, as it reaches the value of specific heat capacity of all the elements, including Cr (bulk) when $q_2 \approx 0.74$.

Let us now return to Eq.(10), where we have the complete Debye function $D(\alpha_{0_{q_1,q_2}})$ to see the behavior of the thermal conductivity. Thus, in the Fig.(2), we show a comparison to the thermal conductivity κ for a pure and impure material. We have the thermal conductivity as a function of temperature (T) for Fe (bulk) and a combination of the values q_1 and q_2 for the Fe_{q-def} (impure). Therefore, we have the Fe (red) when $q_1 = q_2 = 1$ (pure), and when $q_1 = 0.2$ and $q_2 = 1.8$, we have a similar behavior to silicon (Si) (golden), whose value for the Debye temperature is $\theta_D = 645K$. Finally, for a combination of values $q_1 = 1$ and $q_2 = 1.4$ we have a curve that is similar to that of zinc (Zn) (orange), with $\theta_D = 327K$.

One should note that deformation is clearly playing the role of impurity concentration in the material sample. This is because deformation acts directly on the Debye temperature, which means that the Debye frequency is modified. Changing the Debye frequency is a clear sign of the material being modified by impurities.

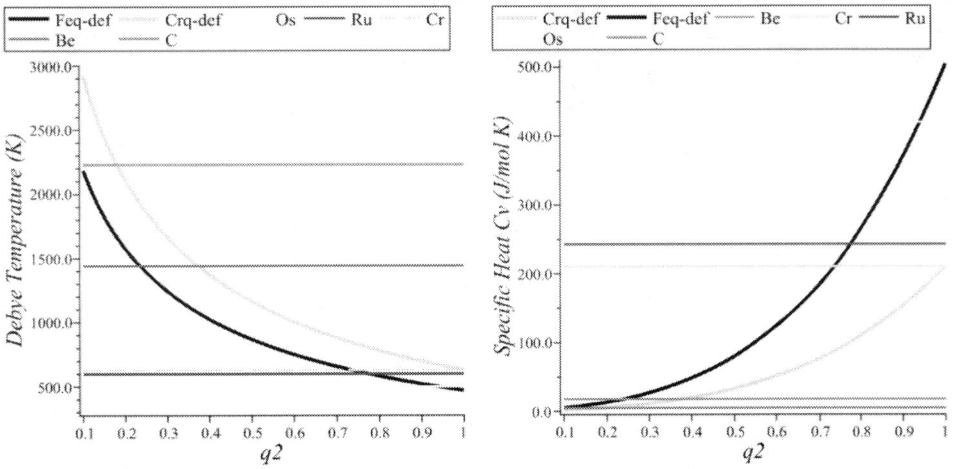

Figure 1. Debye temperature θ_D depending on the variation $q_2 = 0.1, \cdots, 1$ and $q_1 = 1$, **(left)**. Specific Heat c_V depending on the variation $q_2 = 0.1, \cdots, 1$ and $q_1 = 1$, **(right)**

4. Conclusions

The initial idea that q-algebra acts as a factor of disorder or impurity is enhanced by inserting two factors, the so-called Fibonacci oscillators. In Figs.(1,2), we note that the elements that suffer deformation may become similar to others. The existence of more degrees of freedom as in the present case of two deformation parameters, q_1 and q_2, can be well associated with different types of deformations related to two distinct phenomena of disorders or impurities such as, for instance, one due to pressure generating disorders and other due to doping, respectively.

Acknowledgments
We would like to thank CNPq, CAPES, and PNPD/PROCAD-CAPES, for partial financial support.

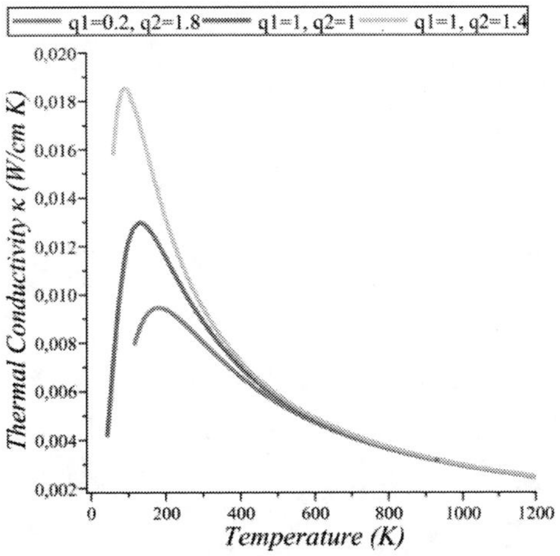

Figure 2. Thermal conductivity κ of Fe as a function of temperature in the range of $T = 0, \cdots, 1200K$, and values for q_1 and q_2 given explicitly.

References

[1] L. Biedenharn, J. Phys. A: Math. Gen. **22**, L873 (1989).
[2] A. Macfarlane, J. Phys. A: Math. Gen. **22**, 4581 (1989).
[3] A.U. Klimyk, *Spectra of Observables in the q-Oscillator and q-Analogue of the Fourier Transform*, Methods and Applications, **1**, 8, (2005).
[4] F. Wilczek, *Fractional Statistics and Anyon Superconductivity*, World Scientific, Singapore, (1990).
[5] C. Tsallis, J. Stat. Phys. **52**, 479 (1988).
[6] G. Kaniadakis, M. Lissia, A. M. Scarfone, Phys. Rev. E **71**, 046128 (2005).
[7] S. Abe, Phys. Lett. A **224**, 326 (1997).
[8] A. Lavagno, Phys. Lett. A **301**, 13-18 (2002).
[9] A. Lavagno, A. M. Scarfone e P. N. Swamy, J. Phys. A: Math. Theor. **40**, 8635-8654 (2007).
[10] Liu Hui, et al., Whuan Univ. J. Nat. Sciences **15**, 57-63 (2010).
[11] F.H. Jackson, Proc. Edin. Math. Soc. **22**, 28-39(1904).
[12] M. Arik, et al., Z. Phys. C **55**, 89-95 (1992).
[13] A. Algin, Phys. Lett. A **292**, 251-255 (2002);
A. Algin, B. Deviren, J. Phys. A: Math. Gen. **38**, 5945-5956 (2005).
A. Algin, J. Stat. Mech. Theor. Exp. **P10009**, 10 (2008).
A. Algin, E. Arslan, J. Phys. A: Math. Theor. **41**, 365006 (2008).
A. Algin, E. Arslan, Phys. Lett. A **372**, 2767-2773 (2008).
Algin A, Arik M., Kocabicakoglu D., Int. J. Theor. Phys.**47**, 1322-1332 (2008).
A. Algin, J. Stat. Mech. Theor. Exp. **P04007**, 04 (2009).
A. Algin, J. CNSNS **15**, 1372-1377 (2010).
[14] A.M. Gavrilik, A.P. Rebesh, Mod. Phys. Lett. A **22**, 949-960 (2007).
[15] A.A. Marinho, F.A. Brito, C. Chesman, Physica A **411**, 74-79 (2014).
[16] Daoud M., Kibler M., Phys. Lett A **206**, 13-17 (1995).
[17] Gong R S, Phys. Lett A **199**, 81-85 (1995).
[18] P.W. Anderson, Phys. Rev. **5**, 109 (1958).
[19] P.A. Lee, T.V. Ramakrishnan, Rev. Mod. Phys. **2**, 57 (1985).
[20] Elliott et al., Rev. Mod. Phys. **3**, 46 (1974).
[21] F.A. Brito, A.A. Marinho, Physica A **390**, 2497-2503 (2011). .
[22] A.A. Marinho, F.A. Brito, C. Chesman, Physica A **391**, 3424-3434 (2012).
[23] D. Tristant, F.A. Brito, Physica A **407**, 276-286 (2014).
[24] R.K. Patthria, *Statistical Mechanics*, Pergamon press, Oxford (1972)
[25] K. Huang, *Statistical Mechanics*, John Wiley & Sons, (1987)
[26] C. Kittel, *Introduction to Solid State Physics*, John Wiley & Sons, (1996)
[27] J.M. Ziman, *Electron and Phonons - The Theory of Transport Phenomena in Solids*, Oxford Univ. Press, (1960).

Journal of Physics: Conference Series **568** (2014) 012010 doi:10.1088/1742-6596/568/1/012010

Bubbles Containing 4, 6 and 12 Electrons in Liquid Helium-3

Wanchun Wei, Zhuolin Xie and Humphrey J. Maris[1]
Department of Physics, Brown University, Providence, Rhode Island 02912, USA

E-mail: humphrey_maris@brown.edu

Abstract. We have performed calculations of the properties of bubbles in liquid helium containing small numbers of electrons. We use an iterative approach to estimate the energy of the electrons inside a bubble of given shape, and then vary the shape of the bubble to find the minimum energy. For helium-3 we show that at zero applied pressure bubbles containing 4 electrons are unstable against breakup into single electron bubbles, but are stable at pressures between -0.23 and -0.15 bars. Bubbles with 6 electrons are stable between -0.17 and -0.05 bars and bubbles with 12 electrons are stable over the pressure range -0.1 to 0.08 bars.

1. Introduction

It was first proposed by Ferrell [1] that an electron injected into liquid helium would force open a bubble in the liquid. Since then a large number of experiments have been performed to study these so-called electron bubbles. Measurements have been made of their optical properties [2], mobility [3], and effective mass [4]. Experimental [5] and theoretical studies [6] have also been made of multi-electron bubbles (MEB) containing a large number of electrons, e.g., 10^5 or more electrons. Here we report on calculations of the properties of bubbles containing just a small number of electrons.

Electron bubbles exist because there is an energy barrier (height approximately 1 eV in helium-4) preventing an electron from entering into bulk liquid helium. Because of this barrier, when a bubble contains an electron the wave function of the electron is close to zero at the surface of the bubble. Thus the total energy of a spherical bubble containing one electron is

$$E = \frac{h^2}{8mR^2} + 4\pi R^2 \alpha + \frac{4\pi}{3} R^3 P, \tag{1}$$

where R is the bubble radius, m is the mass of the electron, α is the surface tension, and P is the applied pressure. At zero pressure the energy is minimum for a radius of

$$R_0 = \left(\frac{h^2}{32\pi m\alpha} \right)^{1/4}. \tag{2}$$

Bubbles containing 2 electrons are known to be unstable against fission [7]. In a recent paper we have considered the shape and stability of bubbles containing 4, 6, and 12 electrons in liquid helium-4 [8]. Here we use the same method to obtain results for electron bubbles in helium-3.

As already mentioned in our earlier paper [8], calculations of the structure of bubbles containing a few electrons are of interest because these bubbles may be produced when a multi-electron bubble becomes unstable and collapses. Their presence could be detected through cavitation experiments.

[1] To whom any correspondence should be addressed.

2. Calculation Method

To find the size and shape of a bubble containing Z electrons we need a method for calculating the energy of the electrons allowing for both quantum effects and the Coulomb repulsion. This method and its limitations are described in detail in ref. [8]. The first step is to consider the way to arrange Z electrons inside a sphere so as to minimize the Coulomb energy. There is an extensive literature on this problem (see [8]). For $Z = 2$ the electrons lie on the surface of the sphere at opposite ends of a diameter, and for $Z = 3$ they lie in a plane passing through the centre of the sphere. For $Z = 4$ they are at the vertices of a tetrahedron, for $Z = 6$ along Cartesian axes, and for $Z = 12$ they lie on the vertices of an icosahedron. However, for other values of Z the arrangement has lower symmetry and so we restrict attention to the special values just listed. To calculate the quantum corrections to the classical solution we place $Z - 1$ of the electrons on the surface of the sphere in the positions just mentioned, and then calculate the wave function and energy of the last electron in the potential provided by these electrons. From the wave function, we find the average distance $<\delta R>$ of the last electron from the surface of the sphere. We then move each of the $Z - 1$ inwards from the surface of the sphere by this same amount from the surface and repeat the calculation until $<\delta R>$ remains constant. Finally, the method is extended to consider a bubble shape which is non-spherical and has the symmetry of the classical solution.

3. Numerical Results

We specify the bubble shape by a radius function $R(\theta,\phi)$ which is a constant plus a term consisting of one linear combination of spherical harmonics giving a shape with the required symmetry. The linear combination is composed of a combination of $Y_{lm}(\theta,\phi)$, with different m but the same value of l. The value of l is chosen to be the smallest possible value which can give the desired symmetry. For $Z = 4$ this value of l is 3, for $Z = 6$ it is 4, and for $Z = 12$ it is 6. It is convenient to express the function $R(\theta,\phi)$ in terms of its maximum and minimum values. As an example, for $Z = 4$ this gives [8]

$$R(\theta,\phi) = \frac{R_{\max} + R_{\min}}{2} + \frac{R_{\max} - R_{\min}}{2}\left[\frac{5}{2}\cos^3\theta - \frac{3}{2}\cos\theta + \frac{1}{\sqrt{2}}\sin^3\theta\,\cos(3\phi)\right] \qquad (3)$$

In Figs. 1 and 2 we show contour plots of the total energy of bubbles at $P = 0$ as a function of $R_{\max}$ and $R_{\min}$. For the surface tension we have taken the zero temperature value of 0.156 erg cm^{-2} [9]. Just as for helium-4, we find that bubbles with $Z = 4$ or $Z = 6$ are unstable at zero pressure, i.e., there is no local minimum in the $R_{\max} - R_{\min}$ plane. Stable bubbles can be obtained at negative pressures. Bubbles with $Z = 4$ are found to be stable in the pressure range -0.23 to -0.15 bars, and for $Z = 6$ the range is -0.17 to -0.05 bars. At the lower end of the pressure range the bubble becomes unstable against explosion and at the upper end the bubble breaks up into single electron bubbles.

For $Z = 12$ the calculations predict that the bubble is stable over the pressure range -0.10 to 0.08 bars. The energy as a function of $R_{\min}$ and $R_{\max}$ is shown in Fig. 2. Figure 3 shows the variation of $R_{\min}$ and $R_{\max}$ as a function of pressure for the $Z = 12$ bubble. The shape of the $Z = 12$ bubble at three pressures is shown in Fig. 4.

It is interesting to compare these results with the calculations already performed for helium-4 [8]. For a single-electron bubble the radius varies with surface tension as $\alpha^{-1/4}$ and so a bubble in helium-3 should be larger than a helium-4 bubble by a factor of about 1.26. For multi-electron bubbles with Z large the radius varies as $\alpha^{-1/3}$. For the few-electron bubbles our results show that when the surface tension decreases the bubble size increases, but also the pressure range over which the bubble

Journal of Physics: Conference Series **568** (2014) 012010 doi:10.1088/1742-6596/568/1/012010

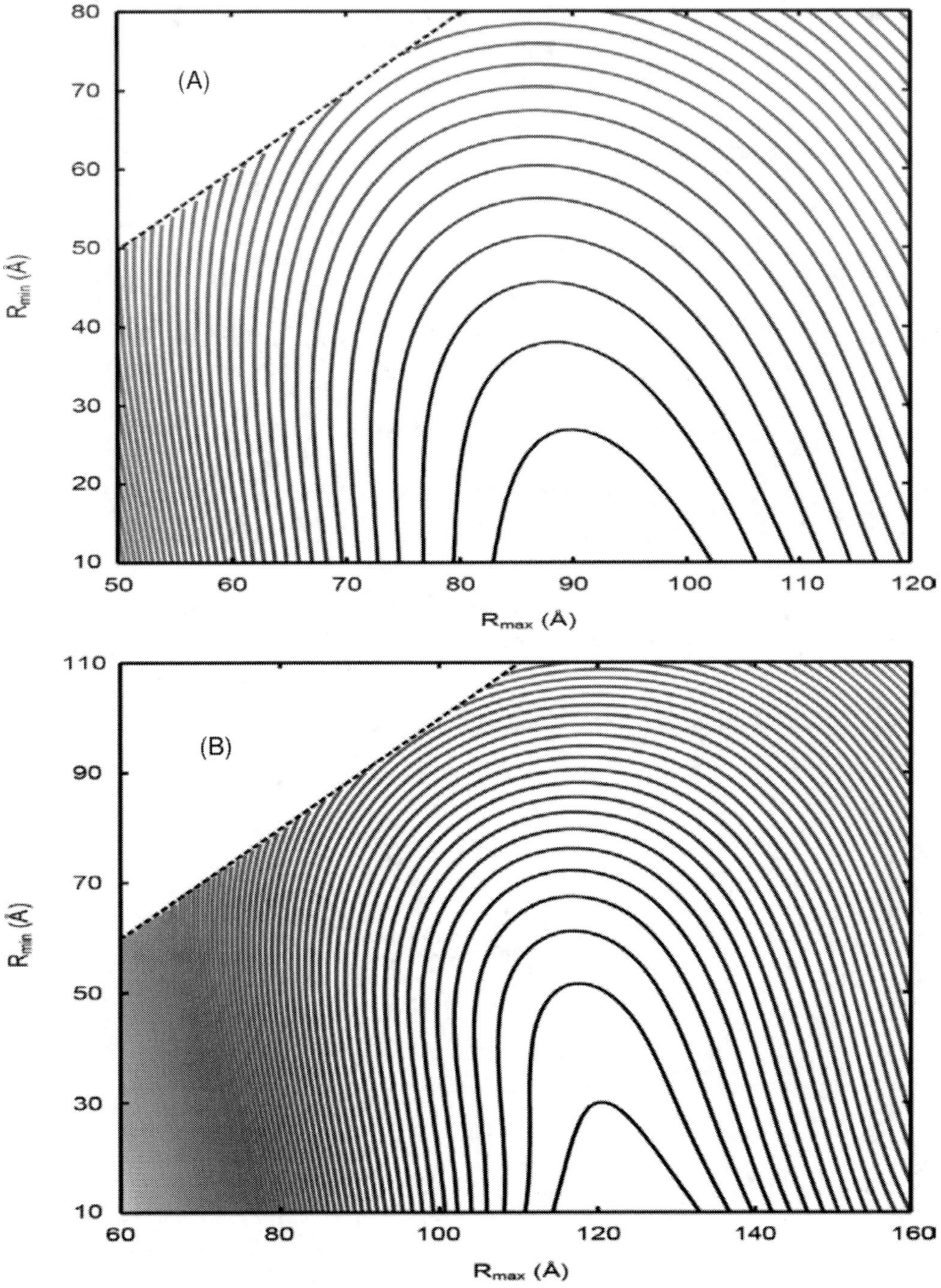

Fig. 1. Energy of bubbles at zero pressure containing (A) 4 electrons and (B) 6 electrons. The contour lines are spaced by increments of 3×10^{-14} ergs.

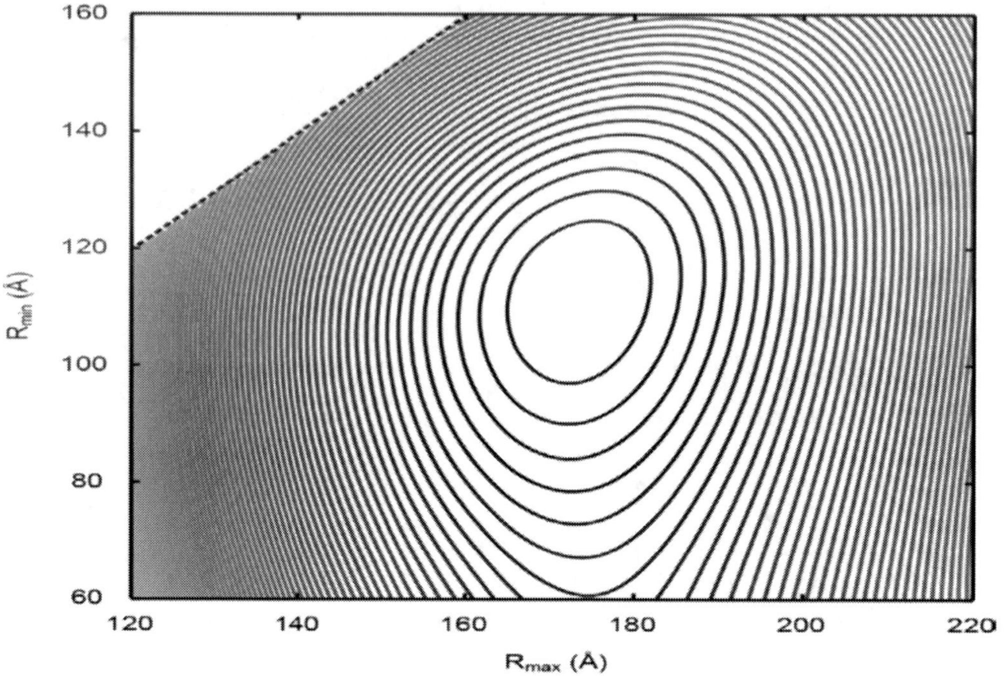

Fig. 2. Energy of bubbles at zero pressure containing 12 electrons. The contour lines are spaced by increments of 3×10^{-14} ergs.

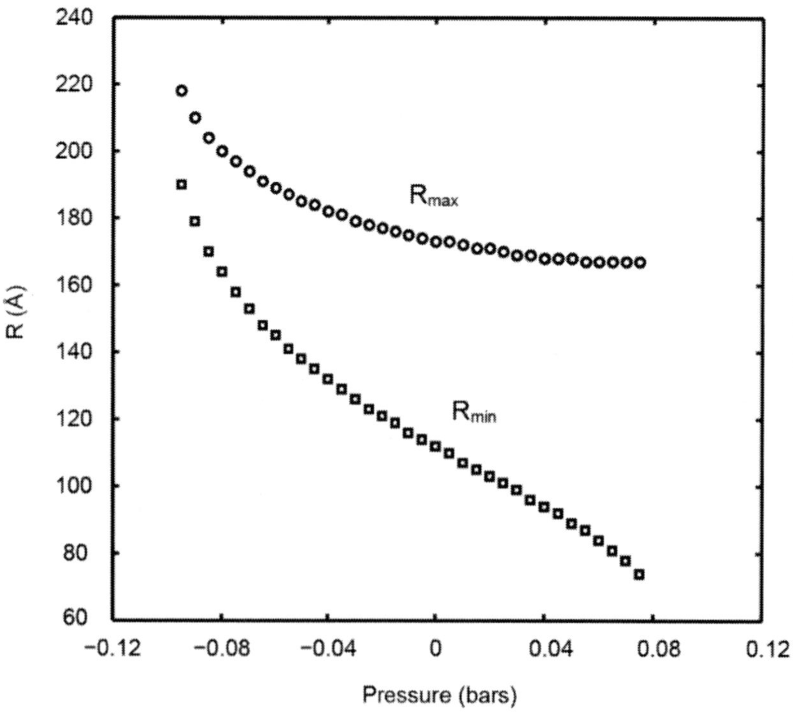

Fig. 3. Maximum and minimum radius for a bubble containing 12 electrons.

is stable decreases. This suggests that at some critical value of surface tension (different for each value of Z) there is no pressure at which the bubble is stable. Thus, since α decreases with increasing temperature and goes to zero at the critical point, there would be a critical temperature for each Z. However, it is probably not useful to calculate these temperatures with the current approach because when the temperature is raised to give a large decrease in α the vapor pressure will increase and it will be necessary to make allowance for the gas atoms inside the bubble.

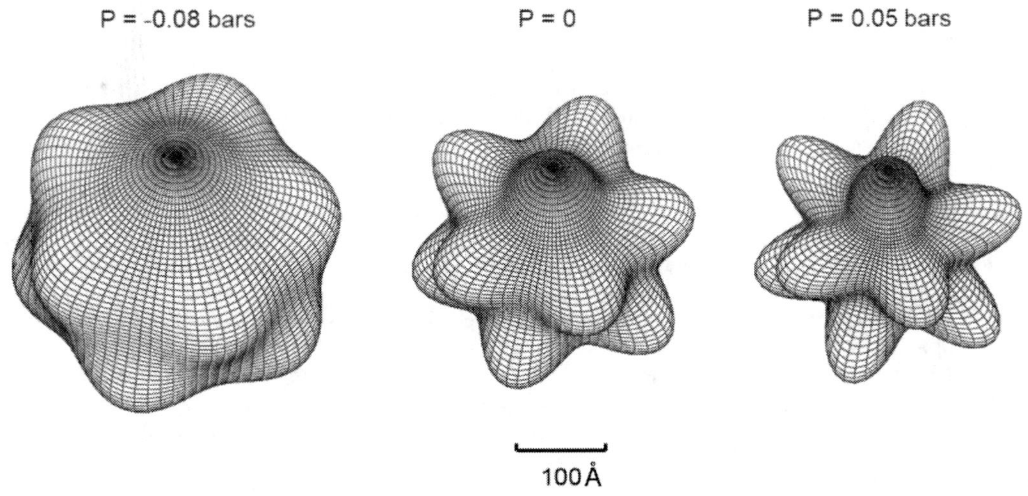

Fig. 4. Calculated shape of the $Z = 12$ bubble at pressures of -0.08, 0 and 0.05 bars.

Acknowledgments
We thank Ambarish Ghosh and Gary Williams for helpful discussions. This work was supported in part by the National Science Foundation under Grant No. DMR 0965728.

References

[1] Ferrell R A 1957 Phys. Rev. 108 167
[2] Grimes C C and Adams G 1990 Phys. Rev. B41, 6366 and 1992 Phys. Rev. B45, 2305; Parshin A Y and Pereversev S V 1990 JETP Lett. 52, 282 and 1992 Sov. Phys. JETP 74, 68
[3] Schwarz K W 1972 Phys. Rev. A6 837
[4] Poitrenaud J and Williams F I B, 1972 Phys. Rev. Lett. 29, 1230 and 1974 Phys. Rev. Lett. 32, 1213
[5] Volodin A P, Khaikin M S and Edel'man V S 1977 JETP Lett. 26 543; Albrecht U and Leiderer P 1987 Europhys. Lett. 3 705 and 1992 J. Low Temp. Phys. 86 131; Fang J, Dementyev A E, Tempere J, and Silvera I F, 2009 Rev. Sci. Inst. 80 043901; Vadakkumbatt V, Joseph E M, Pal A and Ghosh A J. 2013 J. Low Temp. Phys. 171 239; Joseph E M, Vadakkumbatt V, Pal A and Ghosh A J. 2014 J. Low Temp. Phys. 175 78
[6] Salomaa M M and Williams G A, 1981 Phys. Rev. Lett. 47, 1730 and 1983 Physics Scripta. T4, 204; Hannahs S T, Williams G A and Salomaa M M 1995 in *Proceedings of the 1995 IEEE Ultrasonics Symposium*, ed M Levy, S C Schneider, and B R McAvoy, (IEEE, Picataway, New Jersey, 1995), volume 1, p. 635; Tempere J, Silvera I F, Devreese J T 2003 Phys. Rev. B67, 035402; Guo W, Jin D and Maris H J 2008 Phys. Rev. B78 014511
[7] Dexter D L and Fowler W B, 1969 Phys. Rev. 183, 307; Maris H J 2003 J. Low Temp. Phys. 132, 77
[8] Wei W, Xie Z and Maris H J 2014 Phys. Rev. 89 064504
[9] Guo H M, Edwards D O, Sarwinski R E and Tough J T, 1971 Phys. Rev. Lett. 27 1259

Study of Exotic Ions in Superfluid Helium-4 Using a Carbon Nanotube Source

W. Wei, Z.-L. Xie, Y. Yang, G.M. Seidel and H.J. Maris[1]
Department of Physics, Brown University, Providence, Rhode Island 02912, USA

E-mail: humphrey_maris@brown.edu

Abstract. We have extended our measurements of the properties of exotic negatively-charged ions in superfluid helium-4. We measured the ion mobility using the time-of-of-flight method at temperatures in the range between 1.03 and 1.16 K. Ions were generated by an electrical discharge produced by applying a voltage to sharp tips in the helium vapor above the liquid surface. Previous studies by Ihas and Sanders, Eden and McClintock, and by our group used tungsten tips and were able to detect at least twelve exotic ions in addition to the normal electron bubble. In the present work we have experimented with tips each consisting of a stainless steel wire coated with carbon nanotubes. We have found that with these tips the strength of the exotic ion signal is substantially increased making it possible to detect several ions which previously could not be seen. The present data combined with the results of the previous studies indicate that there are at least eighteen exotic ions with different mobility.

Introduction

Studies of ion mobility in superfluid helium-4 have revealed the existence of a number of negatively charged objects in addition to the well-studied normal electron bubble (NEB). Doake and Gribbon [1] produced ions in the liquid using an α source, and discovered an ion now called the "fast ion" which has a mobility about six times larger than an NEB. Ihas and Sanders [2] and Eden and McClintock [3] generated ions by applying a voltage to a sharp tip in the vapor above the liquid to produce an electrical discharge. Ions were then drawn from the vapor into the liquid. These experiments demonstrated the existence of at least 13 so-called "exotic ions" with mobility lying between the mobility of the fast ion and the NEB. In addition, it has recently been found that as well as the ions with 13 different and discrete values of mobility, there is also a large component in the measured signal which comes from ions which have appear to have a continuous distribution of mobility [4].

Despite much effort, an understanding of the structure of these objects is still lacking:

1) One can propose that the ions are negative impurity ions. This seems unlikely because liquid helium does not usually contain any significant concentration of impurities. However, it is possible that in the electrical discharge atoms from the discharge tip are introduced into the vapor. Even if this were to happen, it seems unlikely that there could be 13 different impurities present in the vapor in roughly comparable amounts. Also each impurity species should produce a negative ion of a definite size and mobility. Thus it appears impossible to explain the continuous background in terms of impurities.

2) A second possibility is that the exotic ions are helium ions. Helium ions have been studied in a number of experiments [5] but these have a lifetime much less than the time needed to pass through

[1] To whom any correspondence should be addressed.

the mobility cell. In addition, helium ions cannot provide an explanation of all 13 different exotic ions or the continuous background.

3) Finally, it has been proposed that the exotic ions and the fast ion could be electron bubbles containing only a fraction of the wave function of an electron [6]. These bubbles could be smaller than an NEB and, as a result, have a higher mobility. The fission theory might be able to explain the existence of the continuous mobility distribution.

To investigate further it is of value to vary the experimental conditions in the mobility measurements as much as possible. In this paper we have changed the way the discharge in the vapor is produced by replacing the tungsten tips (W) used in previous experiments with carbon nanotube tips (CNT). As a result of this change, we have been able to discover several exotic ions not detected before.

2. Experiment

The mobility measurements were made with the cell as shown in Fig. 1. A discharge was formed between the tips mounted on the tip holder T and a plate P containing an array of holes. The level of liquid in the cell was between P and the grid G1. This grid was held with a dc voltage positive with respect to the dc voltage on a second grid G2; these voltages prevented any negatively charged objects entering the lower part of the cell. Application of a negative voltage pulse to G1 allowed a pulse of

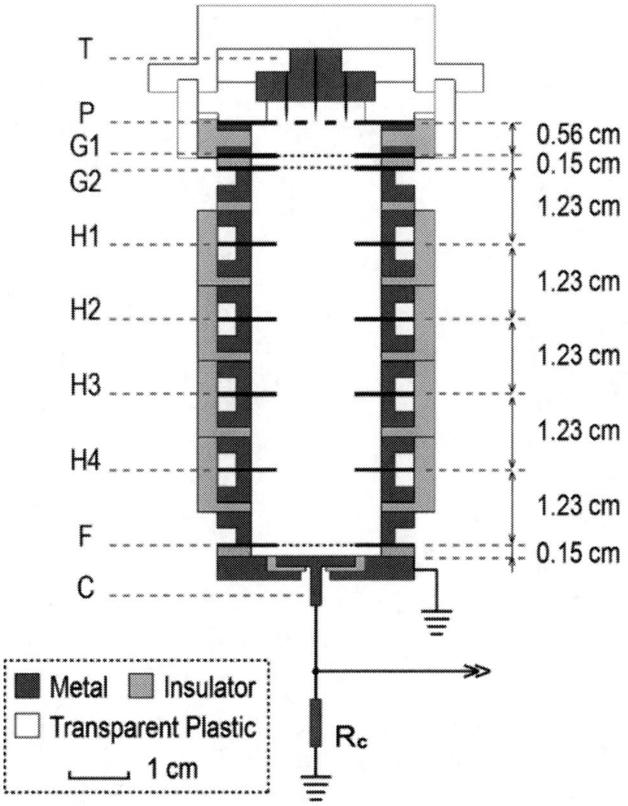

Fig. 1. Schematic diagram of the mobility cell. The function of the different components is described in the text.

negative ions to enter the drift field region below G2. The field in this drift region was made uniform by appropriate voltages applied to the field homogenizer disks H1-4. After passing through the Frisch grid F, the ions reached the collector C.

The current arriving at C passed to ground through a 2.46 MΩ load resistor R_c. The voltage at the collector was amplified and the signal averaged over 5000 traces. Measurements were made as a function of the voltages applied to produce the discharge, the temperature and the drift field.

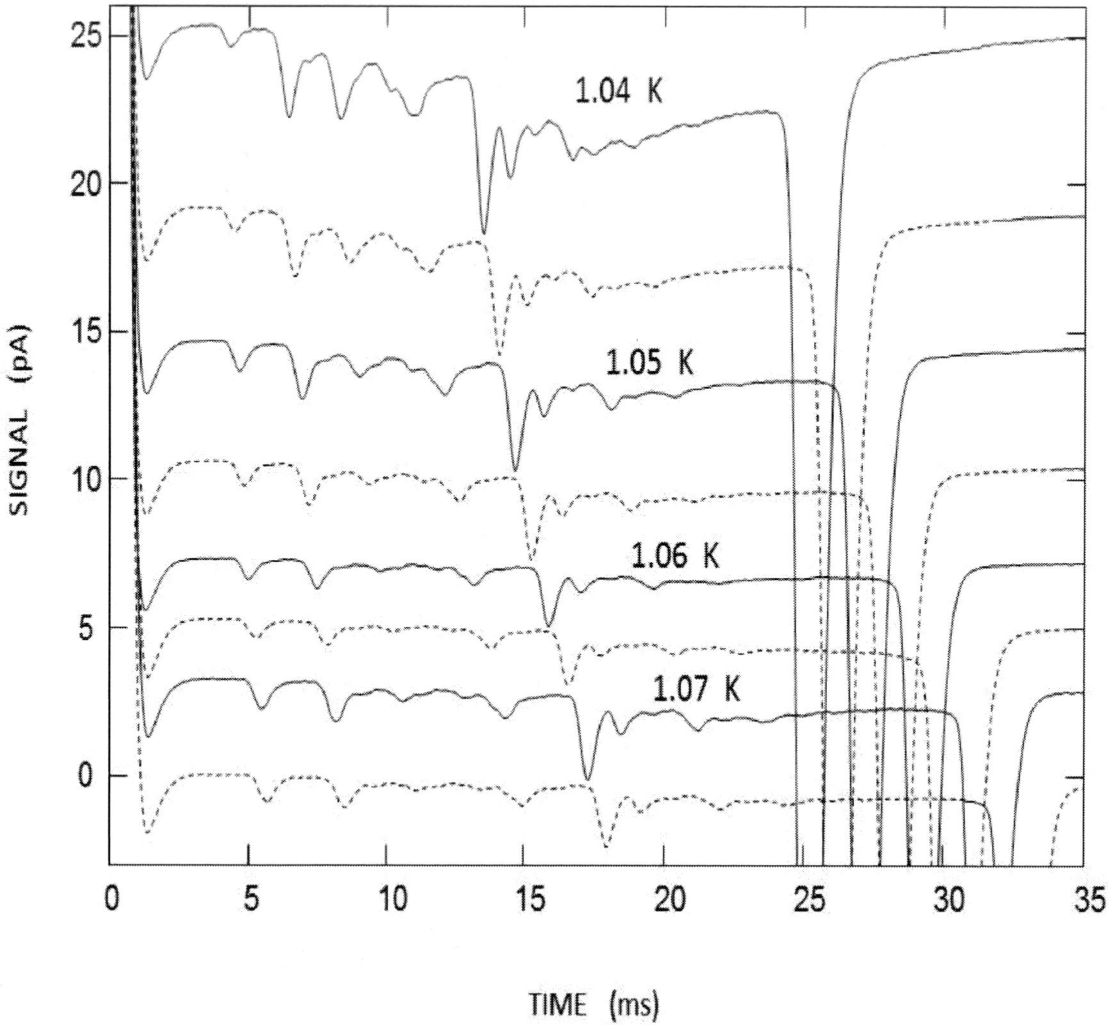

Fig. 2. Current arriving at the collector as a function of time at a series of temperatures spaced by 50 mK. The different data sets are offset for clarity. The peak at early time is the result of electronic pickup and overshoot.

To prepare CNT tips [7] we first placed one or two drops of acetone containing single-wall carbon nanotubes on a glass slide. After the drops spread and dried out, a piece of soft stainless steel capillary (0.012" OD) was coated with a thin layer of silver epoxy. This was then rolled on the slide before the epoxy dried in order to glue the CNTs onto it. The epoxy was then dried out.

Representative data obtained with the CNT tips at a series of temperatures between 1.04 K and 1.075 K are shown in Fig. 2. These data were taken with -510 V on G1, -520 V on G2 and 15.7 V on the Frisch grid F. The voltages on the tips and on the plate were adjusted slightly to optimize the discharge at each temperature. Typical values were -930 V on the tips and -530 V on the plate. We were surprised to find that the signals with the CNT were considerably stronger than with the tungsten tips.

In Fig. 3 we show in more detail the signal at 1.045 K. Fourteen different ions are detected in addition to the fast ion (arrival time 4.5 ms) and the normal electron bubbles at 26 ms.

We were interested to see if the exotic ions seen with the CNT tips were the same ions that had been seen previously with tungsten tips. It is not so simple to make a definite determination because the strength of the signal coming from a given ion varies considerably with the condition of the plasma, and this is influenced by the properties of the tips. In addition, because the CNT signal is stronger than

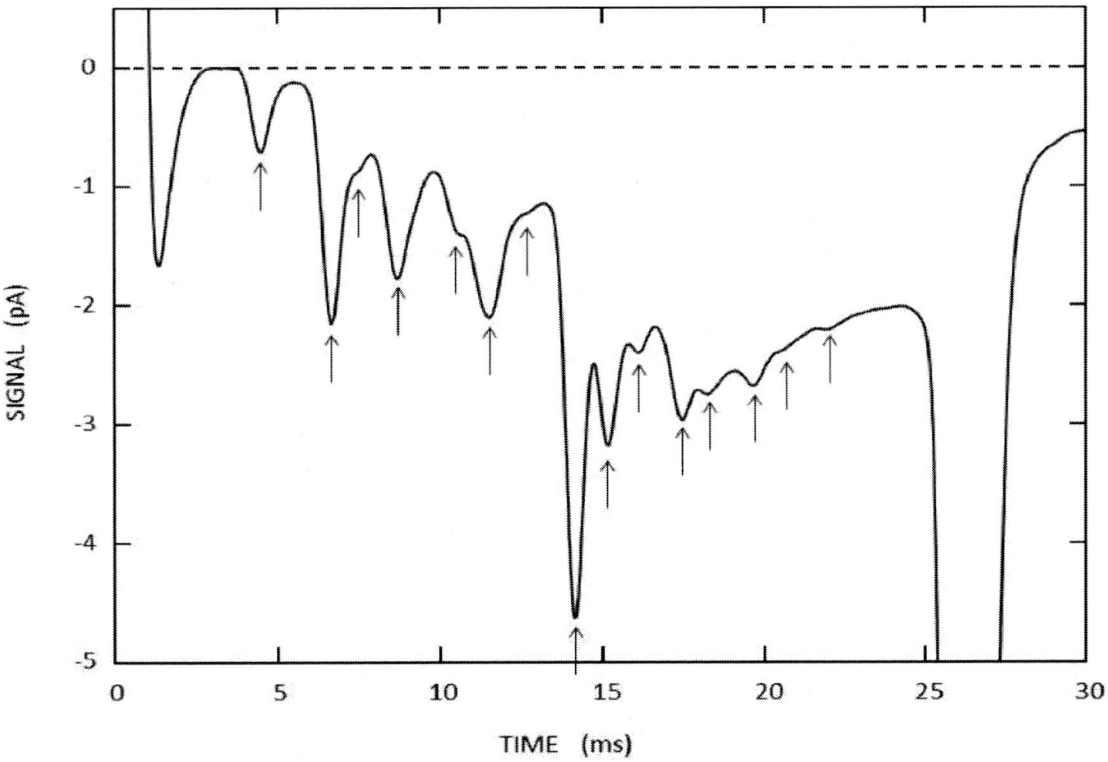

Fig. 3. Signal detected in an ion mobility experiment at 1.045 K. The voltages applied to different parts of the cell are given in the text. The normal electron bubbles arrive at 26 ms. Each arrow indicates the arrival of a particular exotic ion. The peak at 1.3 ms results from overload of the amplifier due to the electromagnetic pickup of the pulse applied to the gate grid.

the tungsten tip signal, it may not be possible to detect some of the ions seen with the CNT when the tungsten tips are used. In Fig. 4 we show a comparison of two data sets both taken at a temperature close to 1.04 K. For the CNT data there was a voltage of -974 V on the tips, -530 V on the plate, -510 V on G1, and -520 V on G2. For the tungsten data the voltage on the tips was -775 V, -550 V on the plate, -510 V on G1, and -520 V on G2. The time axis has been scaled so that the normal electron bubbles arrive at unit time, and the signal axis has been scaled and offset to make it easier to compare the two data sets. One can see from the figure that there are at least six ions that are clearly present regardless of which type of tip is used. This means that these ions cannot have a structure which incorporates the material of the tip. As far as the other ions which appear with either but not both of the sources, it seems most likely that their appearance for only one type of tip is related to the different condition of the plasma. One can see from Fig. 3 that several of the ions which do appear clearly on both traces have a large difference in their amplitude in the two traces. Combining the exotic ions which are seen with either or both the CNT and W we find that there are at least 18 different ions.

In previous work [4] we were able to show that when a fit was made to each peak coming from an exotic ion and this fit used to remove the peak, a continuous background signal was revealed. This background took on two distinct forms depending on the temperature and the voltages applied to produce the plasma. Background type #1 had a sharp cut-off at approximately half the arrival time of the normal electron bubbles. Background type #2 increased smoothly with arrival time and it was not

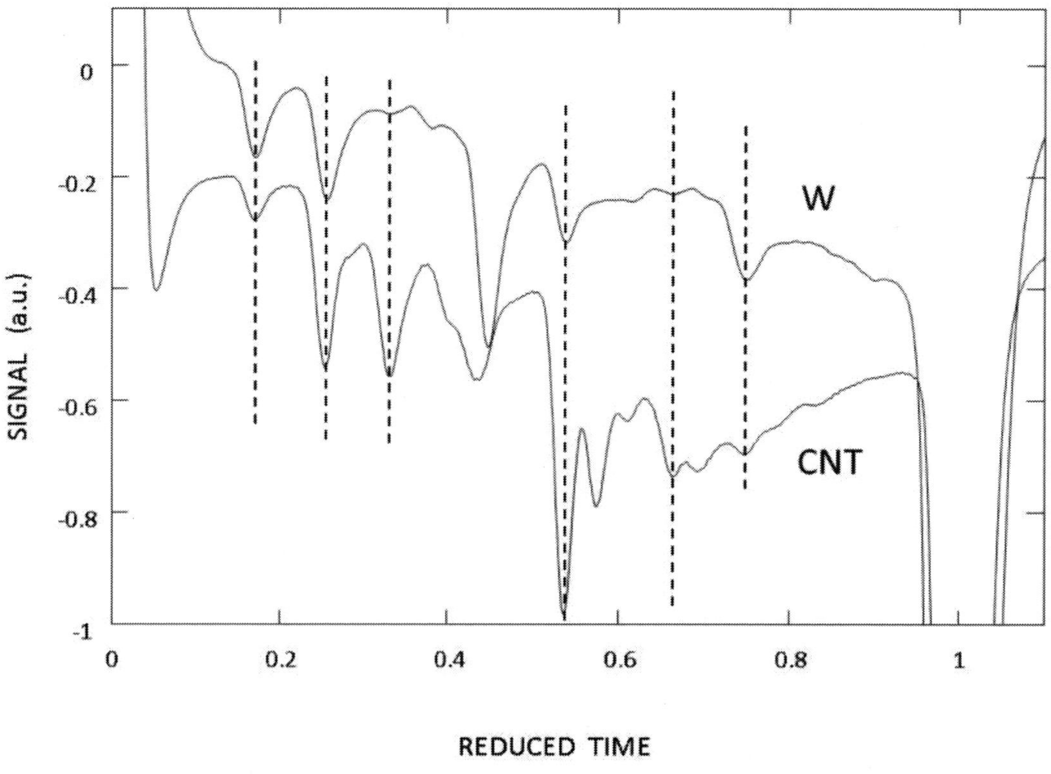

Fig. 4. Comparison of signals obtained at 1.04 K using carbon nanotube (CNT) and tungsten (W) tips. The vertical dashed lines indicate the ions which are clearly present when either type of tip s used.

clear whether there was a cut-off. We have not performed this peak subtraction procedure with the CNT data since it is made difficult by the presence of more ion peaks than were present with the tungsten tips. However, looking at the general form of the data as shown in Figs. 2-4 it appears that the type #1 background is again present.

3. Summary

As emphasized in the introduction, there is no accepted theory of the physical nature of the exotic ions. The present experiment has determined that there are at least 18 exotic ions each with a different mobility. Nearly all of the ions seen previously in experiments with tungsten tips are detected with tips of the different material used here. The experiments also confirm the existence of a signal coming from ions with a continuous distribution of mobility as seen previously[4]. These results make it very unlikely that the exotic ions can be explained in terms of impurities or negative helium ions.

We thank Leon Cooper for helpful discussions. This work was supported in part by the National Science Foundation under Grant No. DMR 0965728.

[1] Doake C S M and Gribbon P W F 1969 *Phys. Lett.* **30A** 251
[2] Ihas G G 1971 Ph.D. thesis, University of Michigan; Ihas G G and Sanders T M 1971 *Phys. Rev. Lett.* **27** 383; Ihas G G and Sanders T M 1972 *Proc. 13th. Int. Conf. on Low Temperature* vol 1ed K D Timmerhaus, W J O'Sullivan and E F Hammel, (Plenum, New York), p 477.
[3] Eden V L and McClintock P V E 1983 *Proc. 75th.Jubilee Conf. on Liquid Helium-4* (St. Andrews); Eden V L and McClintock P V E 1984 *Phys. Lett.* **102A** 197; Williams C D H, Hendry P C and McClintock P V E 1987 *Jap. J. Appl. Phys.* **26-3** 105
[4] Wei W, Xie Z-L, Seidel G M and Maris H J 2013 *J. Low Temp. Phys.* **171** 178
[5] Brehm B, Gusinow M A, and Hall J L 1967 *Phys. Rev. Lett.* **19** 737; Kristensen P, Pedersen U V, Petrunin V V, Andersen T and Chung K T 1997 *Phys. Rev.* A **55** 978; Mader D L and Novick R 1972 *Phys. Rev. Lett.* **29** 199; Blau L M, Novick R and Weinflash D 1970 *Phys. Rev. Lett.* **24** 1268; Bae Y K, Coggiola M J 1984 *Phys. Rev. Lett.* **52** 747; Kvale T J, Compton R N, Alton G D,Thompson J S and Pegg D J 1986 *Phys. Rev. Lett.* **56** 592; Andersen T, Andersen L H, Bjerre N, Hvelpund P, Posthumus J H 1994 *J. Phys*. B **27** 1135
[6] Maris H J 2000 *J. Low Temp. Phys.* **120** 173
[7] Kawasaki K, Tsukagoshi K, and Kono K 2005 *J. Low Temp. Phys.* **138** 899

Pseudogap phenomena in a two-dimensional ultracold Fermi gas near the Berezinskii-Kosterlitz-Thouless transition

M. Matsumoto, and Y. Ohashi

Department of Physics, Keio University, 3-14-1 Hiyoshi, Kohoku-ku, Yokohama, Kanagawa 223-8522, Japan

E-mail: moriom@rk.phys.keio.ac.jp

Abstract. We investigate single-particle excitations and strong-coupling effects in a two-dimensional Fermi gas. Including pairing fluctuations within a Gaussian fluctuation theory, we calculate the density of states $\rho(\omega)$ near the Berezinskii-Kosterlitz-Thouless (BKT) transition temperature $T_{\rm BKT}$. Near $T_{\rm BKT}$, we show that superfluid fluctuations induce a pseudogap in $\rho(\omega)$. The pseudogap structure is very similar to the BCS superfluid density of states, although the superfluid order parameter is absent in the present two-dimensional case. Since a two-dimensional ^{40}K Fermi gas has recently been realized, our results would contribute to the understanding of single-particle properties near the BKT instability.

1. Introduction

In ultracold Fermi gases, we can systematically study various many-body phenomena[1, 2]. For example, a tunable pairing interaction associated with a Feshbach resonance enables us to examine superfluid properties from the weak-coupling regime to the strong-coupling limit[3]. Another example is that, using an optical lattice, we can tune the dimensionality of the system[4]. Using the former advantage, the BCS-BEC crossover has been realized in ^{40}K[5] and ^{6}Li[6, 7, 8] Fermi gases. Using the latter, a two-dimensional Fermi gas has recently been realized[4, 9, 10].

In contrast to a three-dimensional Fermi gas, the BCS-type superfluid phase transition is prohibited by strong pairing fluctuations in the two-dimensional case[11, 12]. However, Berezinskii[13, 14], Kosterlitz and Thouless (BKT)[15] clarified that a two-dimensional system may have a quasi-long-range order, exhibiting superfluidity. In cold atom physics, the BKT transition has recently been realized in a ^{87}Rb Bose gas, loaded on a two-dimensional optical lattice[16]. The BKT transition in a Fermi gas is a crucial next challenge.

Since the BKT transition occurs under the situation that the BCS long-range-order is completely suppressed by strong two-dimensional pairing fluctuations, physical properties near the BKT instability would be also affected by these fluctuations. In addition, in the three-dimensional BCS state, the superfluid order parameter Δ is directly related to the single-particle excitation gap. Since the pair formation also occurs in the fermionic BKT state, it is an interesting problem how the energy gap in the BKT phase is described without the gap parameter Δ.

In this paper, we investigate the single-particle density of states $\rho(\omega)$ in a two-dimensional Fermi gas. Within a Gaussian fluctuation theory[17, 18], we examine how pairing fluctuations affect this quantity. In the three-dimensional case, they are known to cause the pseudogap phenomenon[19], which is characterized by a dip structure in $\rho(\omega)$ around $\omega = 0$. We show that this many-body phenomenon also occurs in the two-dimensional case, where the pseudogapped density of states near $T_{\rm BKT}$ is very similar to the BCS superfluid density of states. We briefly note that pseudogap phenomena enhanced by the low-dimensionality of the system have recently been discussed in Refs. [20, 21, 22].

In this paper, we set $\hbar = k_{\rm B} = 1$, and the system area is taken to be unity, for simplicity.

2. Formulation

We consider a model two-dimensional Fermi gas, consisting of two atomic hyperfine states described by pseudospin $\sigma =\uparrow, \downarrow$. We evaluate the BKT phase transition temperature by the method used in Refs. [23, 24, 25], which is based on the functional integral formalism[18] for the partition function $\mathcal{Z} = \int \mathcal{D}[\Psi, \bar{\Psi}] e^{-S[\Psi,\bar{\Psi}]}$ (which is related to the thermodynamic potential Ω as $\Omega = -T \ln \mathcal{Z}$). The action $S[\Psi, \bar{\Psi}]$ is given by

$$S[\Psi, \bar{\Psi}] = \int_0^\beta d\tau \int d\mathbf{r} \left[\sum_\sigma \bar{\Psi}_\sigma(x)[\partial_\tau + \xi_{\hat{\mathbf{p}}}]\Psi_\sigma(x) - U\bar{\Psi}_\uparrow(x)\bar{\Psi}_\downarrow(x)\Psi_\downarrow(x)\Psi_\uparrow(x) \right], \tag{1}$$

where $\beta = 1/T$, and τ is the imaginary time. Grassmann variables $\Psi_\sigma(x)$ and $\bar{\Psi}_\sigma(x)$ describe Fermi atoms with an atomic mass m, where $x = (\mathbf{r}, \tau)$. $\xi_{\hat{\mathbf{p}}} = -\nabla^2/(2m) - \mu$, where μ is the chemical potential. $-U(< 0)$ is a pairing interaction, which is related to the s-wave scattering length a_s as[26], $-1/U = (m/2\pi)\ln(2/Ck_{\rm F}a_{\rm s}) - \sum_{p \geq k_{\rm F}/\sqrt{2}} m/p^2$, where $k_{\rm F}$ is the Fermi momentum, and $C \simeq 1.78$ is the Euler constant.

We introduce the Cooper-pair field $\Delta(x)$ through the Hubbard-Stratonovich transformation[18]. Carrying out the fermion integrals, we obtain

$$\mathcal{Z} = \int \mathcal{D}[\Delta, \Delta^*] e^{-S_{\rm eff}}, \tag{2}$$

where

$$S_{\rm eff}[\Delta, \Delta^*] = \int_0^\beta \int d\mathbf{r} \frac{|\Delta(x)|^2}{U} - {\rm Tr} \ln[-\mathbf{G}^{-1}], \tag{3}$$

and the Nambu-Gorkov Green's function $\mathbf{G}$ is given by

$$\mathbf{G}^{-1}(x, x') = \begin{pmatrix} -\partial_\tau - \xi_{\hat{\mathbf{p}}} & \Delta(x) \\ \Delta^*(x) & -\partial_\tau + \xi_{\hat{\mathbf{p}}} \end{pmatrix} \delta(x - x'). \tag{4}$$

In the normal state, we include pairing fluctuations described by $\Delta(x)$ within the Gaussian fluctuation level[18]. Executing the functional integrals in terms of $\Delta(x)$ and $\Delta^*(x)$ in Eq. (2), we obtain $\Omega = \Omega_0 + T\sum_{\mathbf{q}, i\nu_n} \ln[1 - U\Pi(\mathbf{q}, i\nu_n)]$. Here, Ω_0 is the thermodynamic potential for a two-dimensional free Fermi gas, and

$$\Pi(\mathbf{q}, i\nu_n) = \sum_{\mathbf{p}} \frac{1 - f(\xi_{\mathbf{p}+\frac{\mathbf{q}}{2}}) - f(\xi_{-\mathbf{p}+\frac{\mathbf{q}}{2}})}{\xi_{\mathbf{p}+\frac{\mathbf{q}}{2}} + \xi_{-\mathbf{p}+\frac{\mathbf{q}}{2}} - i\nu_n} \tag{5}$$

is the pair-correlation function (where $\xi_{\mathbf{p}} = p^2/(2m) - \mu$, $f(\varepsilon) = 1/[e^{\beta\varepsilon} + 1]$, and ν_n is the boson Matsubara frequency). The equation for the number N of Fermi atoms is then obtained from Ω as

$$N = -\frac{\partial \Omega}{\partial \mu} = 2T \sum_{\mathbf{p}, i\omega_n} \tilde{G}(\mathbf{p}, i\omega_n). \tag{6}$$

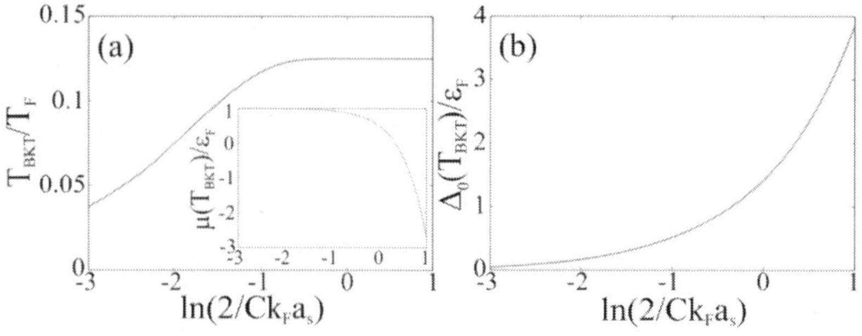

Figure 1. (a) Calculated BKT phase transition temperature $T_{\rm BKT}$. The inset shows the chemical potential $\mu(T_{\rm BKT})$. $T_{\rm F}$ is the Fermi temperature. (b) Saddle point solution (mean-field order parameter) Δ_0 at $T_{\rm BKT}$, normalized by the Fermi energy $\varepsilon_{\rm F}$.

In Eq. (6), the single-particle thermal Green's function $\tilde{G}$ has the form

$$\tilde{G}(\mathbf{p}, i\omega_n) = G_0(\mathbf{p}, i\omega_n) + G_0(\mathbf{p}, i\omega_n)\Sigma(\mathbf{p}, i\omega_n)G_0(\mathbf{p}, i\omega_n), \tag{7}$$

where $G_0(\mathbf{p}, i\omega_n) = 1/[i\omega_n - \xi_{\mathbf{p}}]$ is the bare Green's function, ω_n is the fermion Matsubara frequency, and the self-energy

$$\Sigma(\mathbf{p}, i\omega_n) = T\sum_{\mathbf{q},i\nu_n} \Gamma(\mathbf{q}, i\nu_n)G_0(\mathbf{q}-\mathbf{p}, i\nu_n - i\omega_n) \tag{8}$$

is just the same as that in the T-matrix approximation[19]. $\Gamma(\mathbf{q}, i\nu_n) = -U/[1 - U\Pi(\mathbf{q}, i\nu_n)]$ in Eq. (8) is the particle-particle scattering matrix.

We determine the chemical potential μ from Eq. (6). We then calculate the density of states $\rho(\omega)$ from the Green's function $G^{-1} = G_0^{-1} - \Sigma$ with the self-energy in Eq. (8) as[27],

$$\rho(\omega) = -\frac{2}{\pi}\sum_{\mathbf{p}} {\rm Im}\big[G(\mathbf{p}, i\omega_n \to \omega + i\delta)\big]. \tag{9}$$

While the present formalism can describe strong-coupling effects in the normal state, it does not give the BKT instability. Thus, to determine $T_{\rm BKT}$, we employ the prescription used in Refs. [23, 24, 25]. Including phase fluctuations $\theta(x)$ around the saddle point solution Δ_0 of the action $S_{\rm eff}$ in Eq. (3) ($\Delta(x) = \Delta_0 e^{i\theta(x)}$) within the Gaussian fluctuations level, we obtain

$$\mathcal{Z} = e^{-S_{\rm MF}}\int \mathcal{D}[\theta]e^{-S_{\rm FL}}, \tag{10}$$

where $S_{\rm MF}$ is given by Eq. (3) with $\Delta(x)$ being replaced by Δ_0. $S_{\rm FL} = \sum_{\mathbf{q},i\nu_n}[Jq^2 + K\nu_n^2]\theta(\mathbf{q}, i\nu_n)\theta(-\mathbf{q}, -i\nu_n)$ is the fluctuation contribution. Here,

$$J = \frac{1}{4m}\sum_{\mathbf{p}}\left[1 - \frac{\xi_{\mathbf{p}}}{E_{\mathbf{p}}}\tanh\left(\frac{E_{\mathbf{p}}}{2T}\right) - \frac{1}{2T}\frac{p^2}{2m}{\rm sech}^2\left(\frac{E_{\mathbf{p}}}{2T}\right)\right] \tag{11}$$

is the phase stiffness, and $K = \frac{1}{4}\sum_{\mathbf{p}}\left[\frac{\Delta_0^2}{E_{\mathbf{p}}^3}\tanh(\frac{E_{\mathbf{p}}}{2T}) + \frac{1}{2T}\frac{\xi_{\mathbf{p}}^2}{E_{\mathbf{p}}^2}{\rm sech}^2(\frac{E_{\mathbf{p}}}{2T})\right]$, where $E_{\mathbf{p}} = \sqrt{\xi_{\mathbf{p}}^2 + \Delta_0^2}$. Using the phase stiffness constant J in Eq. (11), we determine $T_{\rm BKT}$ from the KT-Nelson formula[28], $T_{\rm BKT} = (\pi/2)J(T_{\rm BKT})$. In this equation, Δ_0 is determined from the BCS gap equation, $1 = U\sum_{\mathbf{p}}\frac{1}{2E_{\mathbf{p}}}\tanh(\frac{E_{\mathbf{p}}}{2T})$, and μ is evaluated from the number equation $N = -\partial\Omega/\partial\mu$, where $\Omega = -T\ln\mathcal{Z}$ with $\mathcal{Z}$ being given by Eq. (10) where the θ-integration has been executed. Figure 1 shows the self-consistent solutions for $T_{\rm BKT}$, μ, and Δ_0.

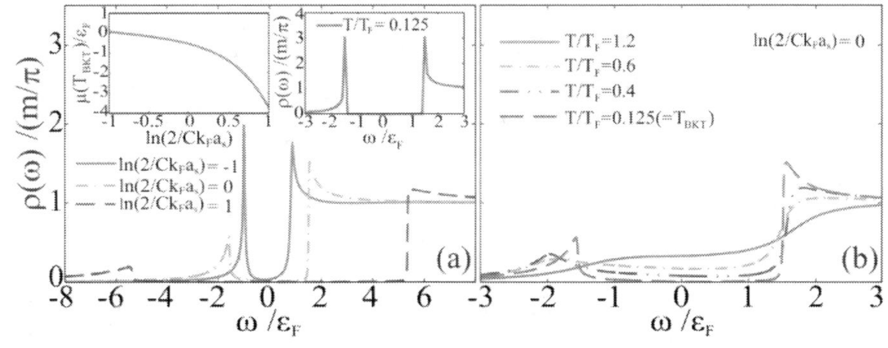

Figure 2. (a) Calculated density of states $\rho(\omega)$ at $T_{\rm BKT}$. The left inset shows the chemical potential μ used in this calculation. The right inset shows the BCS superfluid density of states, where the parameter set (Δ_0, μ) shown in Fig. 1 at $\ln(2/Ck_{\rm F}a_{\rm s}) = 0$ is used. (b) $\rho(\omega)$ at various temperatures when $\ln(2/Ck_{\rm F}a_{\rm s}) = 0$.

3. Pseudogap phenomenon near the BKT phase transition

Figure 2(a) shows the single-particle density of states $\rho(\omega)$ at $T_{\rm BKT}$. We see a clear gap-like structure even in the relatively weak-coupling case ($\ln(2/Ck_{\rm F}a_{\rm s}) = -1$), which becomes more remarkable for a stronger interaction. We recall that, although we have introduced the saddle point solution Δ_0 in calculating $T_{\rm BKT}$, this mean-field order parameter is *not* used in $\rho(\omega)$ in Eq. (9). Thus, the gap-like structure in Fig. 2(a) is the pseudogap associated with pairing fluctuations. Since the low-dimensionality enhances pairing fluctuations, the pseudogap in the present case is more remarkable than the three-dimensional case[19].

When $\ln(2/Ck_{\rm F}a_{\rm s}) \geq 0$, the pseudogap structure at $T_{\rm BKT}$ is very similar to the BCS superfluid density of states with sharp coherence peaks at the gap edges (see the right inset in Fig. 2(a)). To understand this, we note that the particle-particle scattering matrix $\Gamma(\mathbf{q}, i\nu_n)$ in Eq. (8), which describes fluctuations in the Cooper channel, would be large when $\mathbf{q} = \nu_n = 0$ near $T_{\rm BKT}$. Using this, when we approximate Eq. (8) to $\Sigma(\mathbf{p}, i\omega_n) \simeq G_0(-\mathbf{p}, -i\omega_n)T\sum_{\mathbf{q},i\nu_n}\Gamma(\mathbf{q}, i\nu_n) \equiv -G_0(-\mathbf{p}, -i\omega_n)\Delta_{\rm PG}^2$, the Green's function G in Eq. (9) is found to have the same form as the ordinary particle component of the BCS Green's function as,

$$G(\mathbf{p}, i\omega_n) = -\frac{i\omega_n + \xi_{\mathbf{p}}}{\omega_n^2 + \xi_{\mathbf{p}}^2 + \Delta_{\rm PG}^2}. \tag{12}$$

This means that the so-called pseudogap parameter $\Delta_{\rm PG}$[19, 29] plays similar roles to the BCS superfluid order parameter Δ_0, leading to the pseudogap with sharp 'coherence' peaks, as shown in Fig. 2(a). Thus, when one measures $\rho(\omega)$ near $T_{\rm BKT}$, the system would look like the mean-field BCS state, although the superfluid order parameter Δ is actually absent and strong pairing fluctuations only exist.

While the pseudogap structure is similar to the BCS density of states at $T_{\rm BKT}$, their temperature dependence is very different from each other. In the BCS state, when the temperature increases, the gap width gradually becomes small, keeping the sharp coherence peaks at the gap edges, In contrast, Fig. 2(b) shows that the pseudogap structure, as well as the sharp peaks, gradually become obscure with increasing the temperature.

4. Summary

To summarize, we have discussed strong-coupling properties of a two-dimensional ultracold Fermi gas. Including pairing fluctuations within the framework of a Gaussian fluctuation theory, we calculated the single-particle density of states $\rho(\omega)$ near the BKT phase transition. At $T_{\rm BKT}$, pairing fluctuations induce a pseudogap in $\rho(\omega)$, the structure of which is very similar to the BCS superfluid density of states with a finite excitation gap, as well as sharp coherence peaks at the gap edges. That is, the BCS-like energy gap is expected to be observed even in a two-dimensional Fermi gas near $T_{\rm BKT}$, although the superfluid order parameter is absent and the system is dominated by low-dimensional pairing fluctuations. Since the achievement of the BKT phase transition is an exciting challenge in cold Fermi gas physics, our results would be useful for the study of this superfluid state on the view point of single-particle excitations.

Acknowledgments

We thank D. Inotani, R. Hanai, and H. Tajima for discussions. Y.O. was supported by a Grant-in-Aid for Scientific Research from MEXT in Japan (Grant No. 25105511 and No. 25400418).

5. References

[1] Gurarie V and Radzihovsky L 2007 *Ann. Phys.* **332** 2
[2] Bloch I, Dalibard J and Zwerger W 2008 *Rev. Mod. Phys.* **80** 885
[3] Giorgini S, Pitaevskii L P and Stringari S 2008 *Rev. Mod. Phys.* **80** 1215
[4] Sommer A T, Cheuk L W, Ku M J H, Bakr W S and Zwierlein M W 2012 *Phys. Rev. Lett.* **108** 045302
[5] Regal C A, Greiner M and Jin D S 2004 *Phys. Rev. Lett.* **92** 040403
[6] Zwierlein M W, Stan C A, Schunck C H, Raupach S M F, Kerman A J and Ketterle W 2004 *Phys. Rev. Lett.* **92** 120403
[7] Kinast J, Hemmer S L, Gehm M E, Turlapov A and Thomas J E 2004 *Phys. Rev. Lett.* **92** 150402
[8] Bartenstein M, Altmeyer A, Riedl S, Jochim S, Chin C, Denschlag J H and Grimm R 2004 *Phys. Rev. Lett.* **92** 203201
[9] Feld M, Fröhlich B, Vogt E, Koschorreck M and Köhl M 2011 *Nature* **480** 75
[10] Fröhlich B, Feld M, Vogt E, Koschorreck M, Zwerger W and Köhl M 2011 *Phys. Rev. Lett.* **106** 105301
[11] Mermin N D and Wagner H 1966 *Phys. Rev. Lett.* **17** 1133
[12] Hohenberg P C 1967 *Phys. Rev.* **158** 383
[13] Berezinskii V L 1971 *Sov. Phys. JETP* **32** 493
[14] Berezinskii V L 1972 *Sov. Phys. JETP* **34** 610
[15] Kosterlitz J M and Thouless D J 1973 *J. Phys. C:Solid State Phys.* **6** 1181
[16] Hadzibabic Z, Krüger P, Cheneau M, Battelier B and Dalibard J 2006 *Nature* **441** 1118
[17] Nozières P and Schmitt-Rink S 1985 *J. Low Temp. Phys.* **59** 195
[18] Sá de Melo C A R, Randeria M and Engelbrecht J R 1993 *Phys. Rev. Lett.* **71** 3202
[19] Tsuchiya S, Watanabe R and Ohashi Y 2009 *Phys. Rev.* A **80** 033613
[20] Pietilä V 2012 *Phys. Rev.* A **86** 023608
[21] Watanabe R, Tsuchiya S and Ohashi Y 2013 *Phys. Rev.* A **88** 013637
[22] Klimin S N, Tempere J and Devreese J T 2012 *New J. Phys.* **14** 103044
[23] Iskin M and Sá de Melo C A R 2009 *Phys. Rev. Lett.* **103** 165301
[24] Tempere J, Klimin S N and Devreese J T 2009 *Phys. Rev.* A **79** 053637
[25] Salasnich L, Marchetti P A and Toigo F 2013 *Phys. Rev.* A **88** 053612
[26] Morgan S A, Lee M D and Burnett K 2002 *Phys. Rev.* A **65** 022706
[27] When the Green's function in Eq. (7) is used in Eq. (9), the resulting density of states is known to unphysically become negative in the strong-coupling region[19]. This difficulty is, however, avoided by using $G^{-1} = G_0^{-1} - \Sigma$. We note that the Green's function with the self-energy given by Eq. (8) is the same as that in the T-matrix approximation[19].
[28] Nelson D R and Kosterlitz J M 1977 *Phys. Rev. Lett.* **39** 1201
[29] Chen Q J and Levin K 2009 *Phys. Rev. Lett.* **102** 190402

Journal of Physics: Conference Series **568** (2014) 012013 doi:10.1088/1742-6596/568/1/012013

Equations of Motion for the Out-of-Equilibrium Dynamics of Isolated Quantum Systems from the Projection Operator Technique

N. Nessi[1] **and A. Iucci**[1]

[1]Instituto de Física La Plata (IFLP) - CONICET and Departamento de Física, Universidad Nacional de La Plata, CC 67, 1900 La Plata, Argentina

E-mail: nnessi@fisica.unlp.edu.ar

Abstract. We present a rigorous framework to obtain evolution equations for the momentum distribution and higher order correlation functions in weakly interacting systems based on the Projection Operator Technique. These equations can be numerically solved in an efficient way. We compare the solution of the equations with known results for 1D models and find an excellent agreement.

1. Introduction

The out-of-equilibrium dynamics of closed quantum systems is nowadays a very active field of research (see [1] for a review). The central question behind these efforts is to describe the dynamics of an isolated many-body system towards an eventual stationary state after it has been driven out of equilibrium. Of particular interest are the properties of the equilibrium state itself. If simple correlation functions in this equilibrium state can be reproduced by means of a canonical ensemble the system is said to thermalize, which, interestingly, has been shown not to occur in certain situations [2, 3]. These issues, that lie at the heart of statistical mechanics, have recently received a very strong impulse from the experiments with cold atomic gases. These systems are almost perfectly isolated from the environment and can remain quantum coherent for a long time compared with typical experiments duration allowing to study *in situ* the dynamics of large aggregates of interacting quantum particles [2].

From the theoretical perspective, the non-equilibrium dynamics of large, interacting many-body systems settles a formidable challenge. Among the several theoretical approaches at disposal, different kinds of evolution equations for few-body correlations are being widely employed in the contemporary literature (see, for example [4, 5, 6, 7]). In contrast to numerical methods, which are confined either to small systems or short times, as for instance exact diagonalization or time-dependent density matrix renormalization group (t-DMRG), the solution of evolution equations allows to investigate the dynamics of large systems for relatively long times and usually demands a rather modest numerical effort. However, the deduction of these kind of equations starting from first principles, i.e., the many-body Schrödinger equation or its equivalents, often involves a series of uncontrolled approximations [7, 8] that, although routinely utilized, cannot be justified in all cases.

On the other hand, the most studied type of evolution equations are kinetic equations, which are markovian, i.e. memoryless, equations for one-particle distribution functions, a typical example being the Boltzmann equation. Even though the markovian limit is well defined in some situations [9, 10] it may not be always justified. In particular, in situations in which the system is reluctant to loose memory of the initial conditions (such as integrable models) markovian kinetic equations may be loosing important features of the relaxation dynamics.

In this work we present a method for constructing evolution equations for few body correlation functions in weakly interacting systems that involves no approximations beyond a series expansion in the small parameter quantifying the strength of the interaction. The resulting equations of motion are inherently non-markovian. The approach is based on the Projection Operator Technique (POT) [9, 10, 11]. We benchmark the method by comparing the solution of the equations with known results from one dimensional (1D) systems.

2. Model and statement of the problem

We consider a system of interacting spinless fermions with Hamiltonian $H = H_0 + \alpha H_1$ with

$$H = H_0 + \alpha H_1 = \sum_{\boldsymbol{k}} \epsilon(\boldsymbol{k}) n(\boldsymbol{k}) + \alpha \sum_{\boldsymbol{k}_1,\boldsymbol{k}_2,\boldsymbol{k}_3,\boldsymbol{k}_4} V^{\boldsymbol{k}_1,\boldsymbol{k}_2}_{\boldsymbol{k}_3,\boldsymbol{k}_4} c^\dagger(\boldsymbol{k}_1) c^\dagger(\boldsymbol{k}_2) c(\boldsymbol{k}_3) c(\boldsymbol{k}_4), \tag{1}$$

where $c^\dagger(\boldsymbol{k})$ and $c(\boldsymbol{k})$ are fermionic creation and annihilation operators satisfying canonical anticommutation relations, $V^{\boldsymbol{k}_1,\boldsymbol{k}_2}_{\boldsymbol{k}_3,\boldsymbol{k}_4}$ is the momentum-space matrix element of the interaction, $\epsilon(\boldsymbol{k})$ is the dispersion relation, $n(\boldsymbol{k}) = c^\dagger(\boldsymbol{k})c(\boldsymbol{k})$ is the number operator and α is the strength of the interaction. Our results can be easily extended to the bosonic case. The hermiticity of the Hamiltonian and the symmetry in the sum indices impose $V^{\boldsymbol{k}_1,\boldsymbol{k}_2}_{\boldsymbol{k}_3,\boldsymbol{k}_4} = -V^{\boldsymbol{k}_2,\boldsymbol{k}_1}_{\boldsymbol{k}_3,\boldsymbol{k}_4} = -V^{\boldsymbol{k}_1,\boldsymbol{k}_2}_{\boldsymbol{k}_4,\boldsymbol{k}_3} = \bar{V}^{\boldsymbol{k}_4,\boldsymbol{k}_3}_{\boldsymbol{k}_2,\boldsymbol{k}_1}$, where $\bar{V}$ denotes the complex conjugate.

Furthermore, we will be interested in the special case of a translationally invariant Hamiltonian in which the particles interact via a pair potential $v(\boldsymbol{x} - \boldsymbol{y})$. In such case

$$V^{\boldsymbol{k}_1,\boldsymbol{k}_2}_{\boldsymbol{k}_3,\boldsymbol{k}_4} = \frac{1}{4V} \delta_{\boldsymbol{k}_1+\boldsymbol{k}_2,\boldsymbol{k}_3+\boldsymbol{k}_4} \left(\hat{v}(\boldsymbol{k}_1 - \boldsymbol{k}_4) - \hat{v}(\boldsymbol{k}_2 - \boldsymbol{k}_4) - \hat{v}(\boldsymbol{k}_1 - \boldsymbol{k}_3) + \hat{v}(\boldsymbol{k}_2 - \boldsymbol{k}_3) \right), \tag{2}$$

where $\hat{v}(\boldsymbol{k}) = \int d\boldsymbol{r}\, v(\boldsymbol{r}) e^{i\boldsymbol{k}\cdot\boldsymbol{r}}$ is the Fourier transform of the potential and we have written the antisymmetrized version in order to respect the symmetry conditions of the potential. We are interested in the evolution of the system starting from an arbitrary initial condition given by a density matrix $\rho(0)$ which we leave unspecified until the next section.

3. Evolution equation for the momentum distribution

Our starting point is the Liouville equation in the interaction representation ($\hbar$=1):

$$\partial_t \tilde{\rho}(t) = -i\alpha[\tilde{H}_1(t), \tilde{\rho}(t)] = \alpha L(t) \tilde{\rho}(t), \tag{3}$$

where $\tilde{O}(t) = e^{iH_0 t} O e^{-iH_0 t}$ is the interaction representation of the operator O and we have introduced the Liouville superoperator $L(t)O = -i[\tilde{H}_1(t), O]$. Our task is to find approximate solutions to the microscopic dynamics described by the Liouville equation. The POT defines a program for achieving this. We need to first identify the "slow" or "macroscopic" variables in our system and then project the dynamics into the subspace of these slow variables. In a weakly interacting system the occupation number operators emerge as natural slow variables since $[H, n(\boldsymbol{k})] = \mathcal{O}(\alpha)$. To perform the projection we first introduce the "relevant" density matrix

$$\sigma(t) = \frac{1}{Z(t)} \exp\left[-\sum_{\boldsymbol{k}} \lambda(\boldsymbol{k}, t) n(\boldsymbol{k}) \right], \tag{4}$$

where the time-dependent partition function is given by $Z(t) = \text{Tr}\left[\exp\left(-\sum_{\boldsymbol{k}} \lambda(\boldsymbol{k},t) n(\boldsymbol{k})\right)\right]$. Note that $\tilde{\sigma}(t) = \sigma(t)$. The Lagrange multipliers $\lambda(\boldsymbol{k},t)$ enforce the relation:

$$\langle n(\boldsymbol{k})\rangle_t \equiv \text{Tr}[n(\boldsymbol{k})\sigma(t)] = \text{Tr}[n(\boldsymbol{k})\rho(t)]. \tag{5}$$

In other words, $\sigma(t)$ is the density matrix that maximizes the Von Neumann entropy subject to the constraints (5). We note that although the dynamics of $\tilde{\rho}(t)$ is generated by the Liouville equation, the relevant density matrix $\sigma(t)$ is time-dependent independently of the representation. The projection of the dynamics consists in finding an equation of motion for $\sigma(t)$. To this end we introduce a projection super-operator $P(t)$ that projects the relevant density matrix $P(t)\tilde{\rho}(t) = \tilde{\sigma}(t)$. We refer the interested reader to the literature to find an explicit expression for $P(t)$ [9, 10, 11, 12, 13]. It is also useful to define the complementary projector $Q(t) = 1 - P(t)$.

Following the usual steps [9, 10, 11, 12], from the Liouville equation we obtain an equation for the slow dynamics

$$\partial_t \tilde{\sigma}(t) = \alpha P(t)L(t)\tilde{\sigma}(t) + \alpha^2 \int_0^t ds\, P(t)L(t)G(t,s)Q(s)L(s)\tilde{\sigma}(s) + \alpha P(t)L(t)G(t,0)Q(0)\tilde{\rho}(0), \tag{6}$$

where $G(s,t) = \text{T}_{\rightarrow} \exp\left[-\alpha \int_s^t ds'\, Q(s')L(s')\right]$ is an anti-chronologically ordered exponential. The first term in Eq. (6) is a mean-field term that vanishes due to momentum conservation, the second one is a memory term that can be completely expressed in terms of the past history of the $\langle n(\boldsymbol{k})\rangle_t$'s and the third term is a microscopic noise term that cannot be expressed solely as a function of the slow variables. This last term vanishes if we choose an initial condition that has the same form as the relevant density matrices, i.e., if $\rho(0) = \sigma(0)$, and we shall in the following circumscribe to this case.

Eq. (6) is equivalent to the Liouville dynamics and, in general, as difficult to solve as the original problem. It sets, however, a good starting point for approximations. To render Eq. (6) tractable we perform a perturbative expansion in the interaction strength using that $G(t,s) = I + \mathcal{O}(\alpha)$. Taking the trace $\langle n(\boldsymbol{k})\rangle_t = \text{Tr}[n(\boldsymbol{k})\sigma(t)]$ we finally obtain

$$\partial_t \langle n(\boldsymbol{k})\rangle_t = \alpha^2 \int_0^t ds\, \text{Tr}\left[n(\boldsymbol{k})L(t)L(s)\tilde{\sigma}(s)\right] + \mathcal{O}(\alpha^3). \tag{7}$$

A great simplification arises since, given the Gaussian structure of $\sigma(t)$, we can use the Wick pairing rule to evaluate the trace in (7). After a straightforward (but potentially tedious) calculation we obtain the explicit equation of motion

$$\begin{aligned} f(\boldsymbol{k},t) &= f(\boldsymbol{k},0) - 16\alpha^2 \sum_{\boldsymbol{k}_2,\boldsymbol{k}_3,\boldsymbol{k}_4} |V^{\boldsymbol{k},\boldsymbol{k}_2}_{\boldsymbol{k}_3,\boldsymbol{k}_4}|^2 \int_0^t ds\, \frac{\sin\left[(t-s)\Delta e^{\boldsymbol{k},\boldsymbol{k}_2}_{\boldsymbol{k}_3,\boldsymbol{k}_4}\right]}{\Delta e^{\boldsymbol{k},\boldsymbol{k}_2}_{\boldsymbol{k}_3,\boldsymbol{k}_4}} \\ &\times \left(f(\boldsymbol{k},s)f(\boldsymbol{k}_2,s)\bar{f}(\boldsymbol{k}_3,s)\bar{f}(\boldsymbol{k}_4,s) - f(\boldsymbol{k}_3,s)f(\boldsymbol{k}_4,s)\bar{f}(\boldsymbol{k},s)\bar{f}(\boldsymbol{k}_2,s)\right) + \mathcal{O}(\alpha^3), \end{aligned} \tag{8}$$

where $\Delta e^{\boldsymbol{k},\boldsymbol{k}_2}_{\boldsymbol{k}_3,\boldsymbol{k}_4} = \epsilon(\boldsymbol{k}) + \epsilon(\boldsymbol{k}_2) - \epsilon(\boldsymbol{k}_3) - \epsilon(\boldsymbol{k}_4)$ and, in order to ease the notation, we have defined $f(\boldsymbol{k},t) \equiv \langle n(\boldsymbol{k})\rangle_t$ and $\bar{f}(\boldsymbol{k},t) \equiv 1 - \langle n(\boldsymbol{k})\rangle_t$. This equation, in slightly different versions, has appeared many times in the literature. In Ref. [10] it was derived using the same tools that we present here but it was used only as an intermediate step to derive the Boltzmann equation whereas in Refs. [6, 14] it was heuristically derived and used to study the dynamics of infinite dimensional models and to derive a quantum version of the Boltzmann equation, respectively. We want to stress that (8) is valid also for lattice systems in which only quasi-momentum is conserved.

Although the approach is clearly perturbative, the results go beyond conventional lowest order perturbation theory because the perturbation expansion is performed inside the integro-differential equations and, therefore, the coupling is involved in a highly non-linear way in the final expressions. To asses the accuracy of the approximation is thus not a straightforward task. One possible alternative is to calculate higher order corrections to Eq. (8), which can be done systematically within the context of the projection operator technique. Then, the error could be calculated as the relative difference between the lowest order and the next-to-leading order solutions [11]. We relegate such error analysis to future work. An alternative approach is to compare results with other techniques. For instance, performing a suitable short-times approximation on (8) (details can be found in [6]) it is possible to rederive results first obtained by Moeckel and Kehrein using the flow-equation approach in lowest order perturbation theory [15]. Moreover, in [6] it was found that the solution of Eq. (8) compares well on the accessible timescales with dynamical mean-field theory results for the infinite dimensional Hubbard model. Below we provide an example for a 1D model in which the solution of the equation (8) captures highly non trivial features of the short-to-intermediate times relaxation. Regarding the validity of the approach for longer timescales than those accessible for current numerical techniques, we should mention that under certain assumptions it can be shown [14] that Eq. (8) reduces, for long enough times, to the quantum Boltzmann equation, which is expected to successfully describe the thermalization of isolated weakly interacting quantum systems after the system has effectively lost the information about the details of the initial conditions [8].

With respect to implementation details, Eq. (8) can be solved using standard techniques for systems of Volterra integral equations [19]. A straightforward algorithm for the solution using, for instance, the trapezoidal rule to perform the time integral, implies a calculation time that scales as $L^{3D} \times N^2$, where N is the number of times steps and D the space dimension. We have found an algorithm whose execution time scales as $L^{3D} \times N$ allowing us to reach large sizes and times. We finally note that the evolution equations are very suitable for parallel computing.

3.1. Results for a 1D model

Now we present results obtained using (8) in a 1D model of spinless fermions with nearest-neighbor hopping and interactions

$$H = -J \sum_j (c_j^\dagger c_{j+1} + c_{j+1}^\dagger c_j) + \alpha \sum_j n_j n_{j+1}, \tag{9}$$

where $n_j = c_j^\dagger c_j$ and sums run for $j = 0, L-1$ (L is the length of the system) and we impose periodic boundary conditions. This model can be mapped on to the XXZ model by means of a Jordan-Wigner transformation and, in equilibrium, exhibits a Luttinger liquid phase for $\alpha < 2J$. We study a system at half-filling ($k_F = \frac{\pi}{2}$) and take as initial state the ground state of the Hamiltonian with $\alpha = 0$, i.e., we consider the "quantum quench" scenario. The same situation was studied in Ref. [16] using t-DMRG where, remarkably, it was found that the short time dynamics were accurately described by the quench dynamics of the Luttinger model (LM) [17, 18], raising the question of to what extent the equilibrium low energy fixed point of a given model can describe the out of equilibrium dynamics. In Fig. 1 we show results for two different observables, the jump $Z(t)$ of the momentum distribution at k_F and the kinetic energy $e_{kin}(t)$, and compare them with the LM predictions. For $Z(t)$ we find that, after an initial Gaussian decay at very short times, the intermediate time decay is well described by a power law whose exponent is given precisely by the LM predictions. At longer times, beyond an α-dependent time scale t^*, deviations occur. These deviations will be studied in detail in a future publication. For $e_{kin}(t)$ we find that the LM prediction works well even beyond the time scale t^* associated with $Z(t)$.

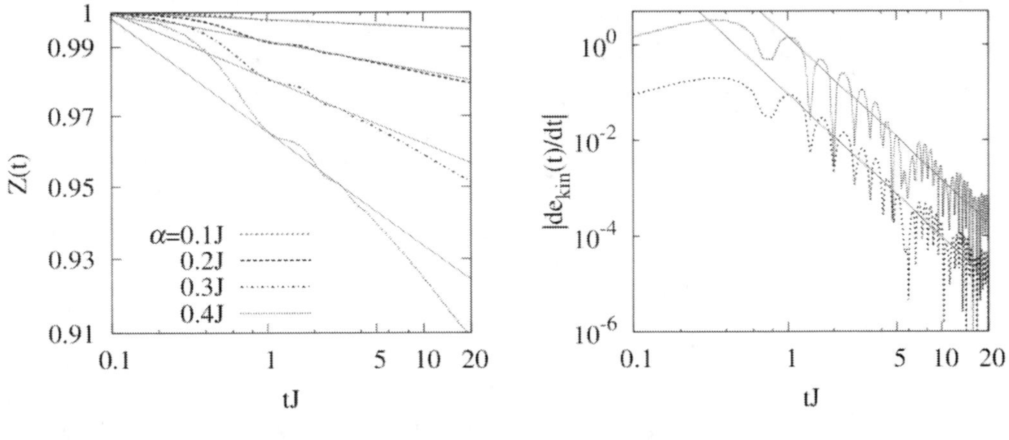

Figure 1. Right panel: Discontinuity in momentum distribution at the Fermi momentum, $Z(t)$. Solid lines are power laws with the exponent given by the LM predictions $Z(t) \sim t^{-\gamma}$, with $\gamma = \frac{1}{4}(K^2 + K^{-2} - 2)$. The Luttinger parameter is obtained from bosonization, $K = \sqrt{\frac{\pi v_F - \alpha}{\pi v_F + 3\alpha}}$. Left panel: Derivative of the kinetic energy $e_{kin}(t)$. Solid lines are power laws t^{-3} given by the LM predictions. The system size is $L = 256$.

4. Evolution equation for higher order correlations

To climb another step in the hierarchy of correlations we must switch to the Heisenberg representation:

$$\partial_t O = i[H, O] \equiv iLO. \tag{10}$$

Note the difference with the Liouvillian defined in (3). The trace operation defines a dual projection operator over the observables of the Hilbert space:

$$\mathrm{Tr}\left[OP(t)\mu\right] = \mathrm{Tr}\left[\mu \mathsf{P}(t)O\right], \tag{11}$$

where O is an observable, μ a density matrix and $\mathsf{P}(t)$ is the observable space projection operator. In the Heisenberg representation the strategy is to separate the slow and fast components of the evolution operator e^{iLt} using the projector $\mathsf{P}(t)$ and its complement $\mathsf{Q}(t)$. This allows to find an operator Langevin-like equation for the slow variables [9],

$$\partial_t \delta n(\boldsymbol{k}, t) = \int_0^t ds \sum_{\boldsymbol{k}'} \Phi_{\boldsymbol{k},\boldsymbol{k}'}(s,t)\, \delta n(\boldsymbol{k}', s) + \eta_{\boldsymbol{k}}(t), \tag{12}$$

where $\delta n(\boldsymbol{k}, t) = e^{iLt} n(\boldsymbol{k}) - \langle n(\boldsymbol{k})\rangle_t$. The memory function $\Phi_{\boldsymbol{k},\boldsymbol{k}'}(s,t)$ can be expressed in terms of the $\langle n(\boldsymbol{k})\rangle_t$'s. In particular, it is possible to show that [9]

$$\Phi_{\boldsymbol{k},\boldsymbol{k}'}(t,u) = \frac{\delta \int_0^t ds\, K_{\boldsymbol{k}}(t,s)}{\delta \langle n(\boldsymbol{k}')\rangle_u}, \tag{13}$$

where

$$K_{\boldsymbol{k}}(t,s) = -\alpha^2 \mathrm{Tr}\left\{[H_1, e^{iH_0(t-s)}[H_1, n(\boldsymbol{k})]e^{-iH_0(t-s)}]\sigma(s)\right\} + \mathcal{O}(\alpha^3). \tag{14}$$

In the last equation a perturbative expansion analogous to that performed in the interaction representation has been used. The microscopic noise term $\eta_{\boldsymbol{k}}(t)$ cannot be written entirely in terms of the $\langle n(\boldsymbol{k})\rangle_t$'s. Again, we refer the interested reader to the literature to find an explicit expression for $\eta_{\boldsymbol{k}}(t)$ [9] since for our immediate purposes we shall not need it. Using the Wick rule and taking the functional derivative Eq. (13) we find an explicit evolution equation for the time-correlation function of the slow variables $F_{\boldsymbol{k},\boldsymbol{k}'}(t) \equiv \mathrm{Tr}\left[\rho(0)\delta n(\boldsymbol{k},t)\delta n(\boldsymbol{k}',0)\right]$

$$F_{\boldsymbol{k},\boldsymbol{k}'}(t) = F_{\boldsymbol{k},\boldsymbol{k}'}(0) - 16\alpha^2 \int_0^t ds \left\{ F_{\boldsymbol{k},\boldsymbol{k}'}(s) A_{\boldsymbol{k}}(t,s) + \sum_{\boldsymbol{q}} F_{\boldsymbol{q},\boldsymbol{k}'}(s)\left[B_{\boldsymbol{k},\boldsymbol{q}}(t,s) - 2C_{\boldsymbol{k},\boldsymbol{q}}(t,s)\right] \right\}, \tag{15}$$

where

$$\begin{aligned} B_{\boldsymbol{k},\boldsymbol{q}}(t,s) &= \sum_{\boldsymbol{k}_3,\boldsymbol{k}_4} |V^{\boldsymbol{k},\boldsymbol{q}}_{\boldsymbol{k}_3,\boldsymbol{k}_4}|^2 \frac{\sin\left[(t-s)\Delta e^{\boldsymbol{k},\boldsymbol{q}}_{\boldsymbol{k}_3,\boldsymbol{k}_4}\right]}{\Delta e^{\boldsymbol{k},\boldsymbol{q}}_{\boldsymbol{k}_3,\boldsymbol{k}_4}} \left(f(\boldsymbol{k},s)\bar{f}(\boldsymbol{k}_3,s)\bar{f}(\boldsymbol{k}_4,s) + \bar{f}(\boldsymbol{k},s)f(\boldsymbol{k}_3,s)f(\boldsymbol{k}_4,s)\right), \\ C_{\boldsymbol{k},\boldsymbol{q}}(t,s) &= \sum_{\boldsymbol{k}_2,\boldsymbol{k}_4} |V^{\boldsymbol{k},\boldsymbol{k}_2}_{\boldsymbol{q},\boldsymbol{k}_4}|^2 \frac{\sin\left[(t-s)\Delta e^{\boldsymbol{k},\boldsymbol{k}_2}_{\boldsymbol{q},\boldsymbol{k}_4}\right]}{\Delta e^{\boldsymbol{k},\boldsymbol{k}_2}_{\boldsymbol{q},\boldsymbol{k}_4}} \left(f(\boldsymbol{k},s)f(\boldsymbol{k}_2,s)\bar{f}(\boldsymbol{k}_4,s) + \bar{f}(\boldsymbol{k},s)\bar{f}(\boldsymbol{k}_2,s)f(\boldsymbol{k}_4,s)\right), \end{aligned}$$

and $A_{\boldsymbol{k}}(t,s) = \sum_{\boldsymbol{k}'} B_{\boldsymbol{k}',\boldsymbol{k}}(t,s)$. To the best of our knowledge, it is the first time that Eq. (15) is derived. In order to solve Eq. (15) we need first to calculate the solution to Eq. (8) to determine the coefficients $A_{\boldsymbol{k}}(t,s)$, $B_{\boldsymbol{k},\boldsymbol{q}}(t,s)$ and $C_{\boldsymbol{k},\boldsymbol{q}}(t,s)$. Once this is done, Eq. (15) is as amenable to numerical solution as Eq. (8). Solutions of this equation for specific 1D models shall be presented elsewhere. We remark here, however, that the Heisenberg representation POT could be useful to obtain evolution equations for other correlation functions using the same scheme.

5. Conclusions

We have presented a rigorous framework to obtain evolution equations for various observables and correlations functions in weakly interacting systems. These equations can be numerically solved in an efficient way. We have put to test the accuracy of the approach comparing with known results for 1D models. Moreover, the POT formalism could allow to obtain higher order corrections to these equations. The inherent limitation of the approach is that it is valid only in the perturbative regime.

Acknowledgments

We acknowledge useful discussions with Miguel A. Cazalilla and the hospitality of the National Tsing Hua University in Taiwan, where part of the present work was realized. This work was supported by CONICET (PIP 0662), ANPCyT (PICT 2010-1907) and UNLP (PID X497), Argentina.

References

[1] A. Polkovnikov *et al* 2011 Nonequilibrium dynamics of closed interacting quantum systems *Rev. Mod. Phys.* **83**, 863.
[2] T. Kinoshita *et al* 2006 A quantum Newton's cradle *Nat.* **440**, 900.
[3] M. Rigol *et al* 2007 Relaxation in a completely integrable many-body quantum system: an ab initio study of the dynamics of the highly excited states of 1D lattice hard-core bosons *Phys. Rev. Lett.* **98**, 050405
[4] M. Tavora and A. Mitra 2013 Quench dynamics of one-dimensional bosons in a commensurate periodic potential: A quantum kinetic equation approach *Phys. Rev. B* **88**, 115144.
[5] M. Fürst *et al* 2012 Matrix-valued Boltzmann Equation for the Hubbard Chain *Phys. Rev. E* **86**, 031122.
[6] M. Stark and M. Kollar 2013 Kinetic description of thermalization dynamics in weakly interacting quantum systems *Preprint* arXiv:1308.1610v2.
[7] J. Berges 2004 Introduction to Nonequilibrium Quantum Field Theory *AIP Conf. Proc.* **3**, 739.
[8] J. Berges 2006 Range of validity of transport equations *Phys. Rev. D* **74**, 045022.

[9] H. Grabert 1982 *Projection operator techniques in nonequilibrium statistical mechanics* (Springer-Verlag).
[10] J. Rau and B. Müller 1996 From Reversible Quantum Microdynamics to Irreversible Quantum Transport *Phys. Rep.* **272**, 1.
[11] H.-P. Breuer and F. Petruccione 2002 *The theory of open quantum systems* (Oxford University Press).
[12] B. Robertson 1966 Equations of motion in non-equilibrium statistical mechanics *Phys. Rev.* **144**, 151.
[13] K. Kawasaki and J. Gunton 1973 Theory of nonlinear transport processes: Nonlinear shear viscosity and normal stress effects *Phys. Rev. A* **8**, 2048.
[14] L. Erdos *et al* 2004 On the quantum Boltzmann equation *J. Stat. Phys.* **116**, 367.
[15] M. Moeckel and S. Kehrein 2008 Interaction quench in the Hubbard model *Phys. Rev. Lett.* **100**, 175702.
[16] C. Karrasch *et al* 2012 Luttinger liquid universality in the time evolution after an interaction quench *Phys. Rev. Lett.* **109**, 126406.
[17] M. Cazalilla 2006 Effect of suddenly turning on interactions in the Luttinger model *Phys. Rev. Lett.* **97**, 156403.
[18] A. Iucci and M. Cazalilla 2009 Quantum quench dynamics of the Luttinger model *Phys. Rev. A* **80**, 063619.
[19] P. Linz 1985 *Analytical and numerical methods for Volterra equations* (Studies in Applied Mathematics).

27th International Conference on Low Temperature Physics (LT27) IOP Publishing
Journal of Physics: Conference Series **568** (2014) 012014 doi:10.1088/1742-6596/568/1/012014

Study of coupled QED-cavities using the self-consistent Mori projector method

Sumie Okubo,[1,2,*] Mikio Eto,[1] Peter Degenfeld-Schonburg,[2] and Michael J. Hartmann[3]

[1]Faculty of Science and Technology, Keio University, 3-14-1 Hiyoshi, Kohoku-ku, Yokohama 223-8522, Japan, [2]Technische Universität München, Physik Department, James Franck Straße, 85748 Garching, Germany, [3]Institute of Photonics and Quantum Science, Heriot-Watt University, Edinburgh EH14 4AS, United Kingdom

E-mail: sumieo@rk.phys.keio.ac.jp*

Abstract. We study the non-equilibrium properties of nonlinear QED-cavities coupled via photon tunneling in the presence of dissipation and coherent pumping. To illustrate the interplay between photon leakage, Rabi oscillations, and coherent photon hopping, we examine the dynamical evolution and the stationary states of finite cavity arrays in the highly nonlinear regime. Finally, we employ a cluster version of the self-consistent Mori projector method and show both its accuracy and efficiency for the determination of local quantities in the highly complex many-body situation at hand.

1. Introduction

The study of light-matter interactions in cavity electrodynamics (QED) has always been a field of high interest [1]. In recent years strong theoretical and experimental progress was made towards the goal of investigating the physics of coupled QED-cavities [2]. An array of such QED-cavities, in which photons can tunnel to neighboring sites, yields a platform to study non-equilibrium many-body problems [3]. The competition between physical processes such as Rabi oscillations driven by an external laser, coherent oscillation between the cavities due to photon tunneling, and dissipation due to the photon loss from the cavities leads to a variety of interesting many-body phenomena, with non-equilibrium phase transitions being one prominent example among them [4, 5].

In this work, we examine the non-equilibrium properties of photons in a one-dimensional array of QED-cavities. The Hamiltonian of the system is described by the Bose Hubbard Hamiltonian together with a term describing a coherent laser drive [6]. The photon loss from the cavities is described by standard Lindblad dissipators [7]. First, we investigate the time evolution of photons in a finite array of cavities to illustrate the dynamical character of this complex system. In this case, we solve the equation of motion for the density matrix numerically. Next, we employ the self-consistent Mori projection method (c-MoP), which was recently developed in Ref. [8], to study the steady state of an infinite cavity array. This method determines the reduced density matrix for a single cavity, or a cluster of a few cavities, in a self-consistent way. We show that the calculated results with two- or three-cavity clusters are in good agreement with a numerically-exact solution obtained by the time-dependent density-matrix renormalization group (t-DMRG)

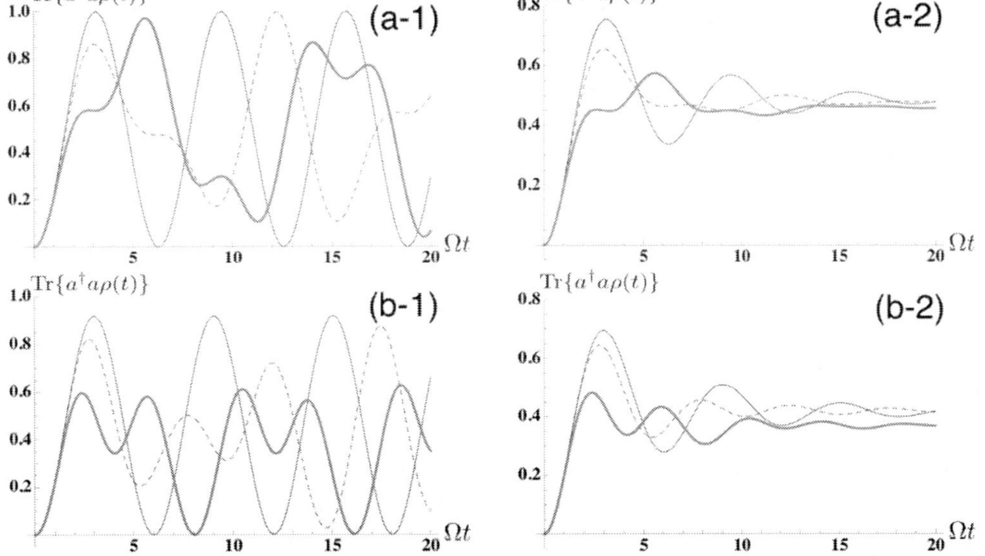

Figure 1. On-site photon numbers in an array of two cavities ($N = 2$), as a function of time. The detuning and dissipation parameters are (a-1) $\Delta = 0$ and $\gamma = 0$, (a-2) $\Delta = 0$ and $\gamma/\Omega = 0.25$, (b-1) $\Delta/\Omega = 0.3$ and $\gamma = 0$, (b-2) $\Delta/\Omega = 0.3$ and $\gamma/\Omega = 0.25$, respectively. The tunneling rate of photons is $J/\Omega = 0$ (solid line), $J/\Omega = 1$ (dashed line), and $J/\Omega = 2$ (grey solid line) in all the panels.

method [8, 9]. In contrast to t-DMRG, however, we emphasize that c-MoP is suitable for any lattice dimensions and arbitrary geometry.

2. Model

We examine an effective model for photons in a one-dimensional array of N QED-cavities, after the qubit degrees of freedom are integrated out [6]. The coherent motion of photons is described by the Bose Hubbard model in a driving laser field of strength $\Omega/2$,

$$H = \sum_{j=1}^{N} \left[\Delta a_j^\dagger a_j + \frac{U}{2} a_j^\dagger a_j^\dagger a_j a_j + \frac{\Omega}{2}(a_j^\dagger + a_j) \right] - J \sum_{j=1}^{N-1} (a_j^\dagger a_{j+1} + a_{j+1}^\dagger a_j). \tag{1}$$

Here, the Hamiltonian is written in a frame rotating with the frequency ω of the laser. $\Delta = \omega_{\text{qubit}} - \omega$ is the difference between the qubit excitation frequency and the laser frequency. Photons interact with each other in a cavity with energy U and tunnel to the neighboring cavities with rate J. We restrict ourselves to the case of $U \to \infty$, in which the number of photons is unity or zero in each cavity.

We take the photon loss from the cavities with a constant rate γ into account. The density matrix $\rho(t)$ evolves according to the Liouville equation of motion [10],

$$\dot{\rho}(t) = \mathcal{L}\rho(t) = -i\,[H, \rho(t)] + \frac{\gamma}{2} \sum_{j=1}^{N} \left[2a_j \rho(t) a_j^\dagger - a_j^\dagger a_j \rho(t) - \rho(t) a_j^\dagger a_j \right]. \tag{2}$$

We solve this equation numerically for finite arrays ($N = 2, 3, 4$) in section 3 and apply the self-consistent Mori projector approach to an infinite array in section 4.

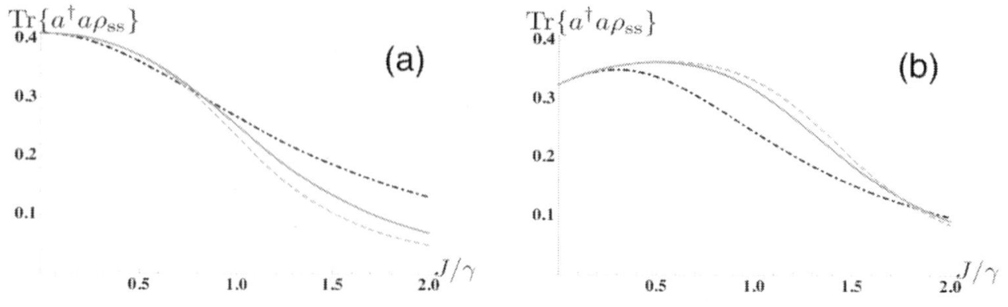

Figure 2. On-site steady-state photon numbers in a finite array of cavities as a function of the tunneling rate J with finite dissipation $[\gamma = (2/3)\Omega]$. The detuning parameter is (a) $\Delta = 0$ and (b) $\Delta/\gamma = 0.6$. The array size is $N = 2$ (dash-dotted line), $N = 3$ (red solid line), and $N = 4$ (orange dashed line) with periodic boundary conditions.

3. Time evolution and steady states of small size arrays

To elucidate the dynamical properties of the system under study, we begin with an array of two QED-cavities ($N = 2$). Figure 1 shows the time evolution of the photon number in a cavity for the initial state, $|\phi(t = 0)\rangle = |0\rangle_1 \otimes |0\rangle_2$. We tune the laser frequency to resonance, i.e. $\Delta = 0$, in panels (a-1) and (a-2) while $\Delta/\Omega = 0.3$ in (b-1) and (b-2). The photon loss from the cavities is absent in (a-1) and (b-1), whereas it is present in (a-2) and (b-2).

In the absence of dissipation due to photon loss, the photon number coherently oscillates. In panel (a-1), the Rabi oscillation changes the photon number between zero and unity for $J = 0$ (solid line). The tunneling rate J complicates the oscillation, reflecting the coexistence of the Rabi oscillation in a cavity and coherent oscillation between the cavities. In panel (b-1) with finite Δ, the Rabi frequency is given by $\sqrt{\Omega^2 + \Delta^2}$ for $J = 0$. The maximum of the photon number is less than unity in this case. In panels (a-2) and (b-2), the oscillation is damped to a steady state by the dissipation. (We also analyze the operator $\mathcal{L}$ in Eq. (1). One of its eigenvalues is zero, the eigenstate of which represents the steady state. The other eigenvalues have a negative real part, which determine the damped oscillation or overdamping.)

Next, we examine the photon numbers in the steady state with finite γ. In Fig. 2, we plot the photon numbers as a function of the tunneling rate J. The size of the array is $N = 2$, 3 or $N = 4$ with periodic boundary conditions. When the resonant condition is satisfied ($\Delta = 0$), the photon number monotonically decreases with J, as shown in Fig. 2(a). The photon number is maximal at $J = 0$ due to the Rabi oscillations in a cavity, which is disturbed by the tunneling between neighboring cavities, resulting in a decrease of the photon number with increasing J. In fact, the tunneling rate J shifts the energy levels of the two cavity system resulting in an effective detuning of the laser drive [6, 11]. Hence, the finite detuning Δ in Fig. 2(b) leads to a maximal photon number at a finite value of J. These features of the photon numbers are commonly observed for $N = 2$, 3, and 4, and also hold true for larger arrays as we show below.

4. Steady state of large size arrays

Now we investigate an infinite one-dimensional array of QED-cavities by applying a cluster version of the self-consistent Mori projection (c-MoP) method [8] to this problem.

In the c-MoP method, the entire many-body system is divided into a subsystem of interest which we refer to as the "system," and the remainder of the lattice which we refer to as the "environment." The closed equations for the reduced density matrix of the "system" are derived by imposing a self-consistent condition to that of the "environment" which holds true because

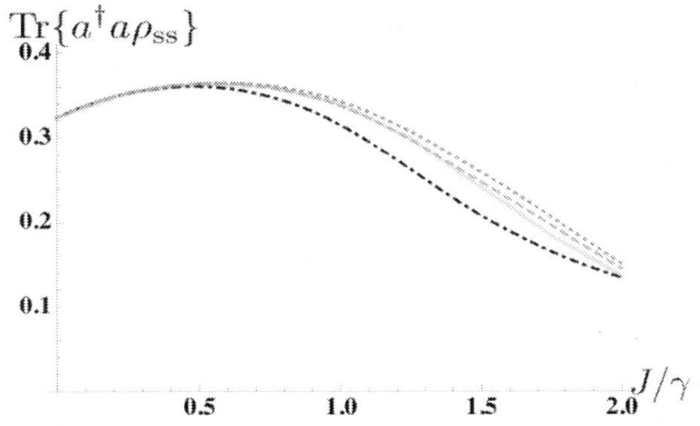

Figure 3. On-site steady-state photon numbers in an infinite array of cavities, as a function of the tunneling rate J with finite dissipation $[\gamma = (2/3)\Omega]$. The detuning parameter is $\Delta/\gamma = 0.6$. The calculation is performed by single-cavity c-MoP (black dash-dotted line), 2-cavity c-MoP cluster (red dotted line), and 3-cavity c-MoP cluster methods (orange dashed line). The data obtained by a t-DMRG integration is shown by the green solid line.

of translation invariance in the full equation of motion, see Eqs. (1) and (2). In our study, the "system" is a single cavity, or a cluster of few cavities, which couples to the "environment" via photon tunneling on the boundaries. For a detailed description, see Ref. [8] and references therein.

For the "system" of a single cavity, which we denote by the index n_0, we need to solve the following algebraic equation of motion for the steady state of the reduced density matrix ρ_{n0}^{ss},

$$0 = \mathcal{L}_{n_0}\rho_{n_0}^{ss} + iZJ\left[a_{n_0}\mathrm{Tr}_n\{a_n^\dagger\rho_n^{ss}\} + a_{n_0}^\dagger\mathrm{Tr}_n\{a_n\rho_n^{ss}\}, \rho_{n_0}^{ss}\right]$$
$$-\left(ZJ^2\sum_{j=\{+,-\}}\left[a_{n_0}^\dagger, \int_0^\infty d\tau\, e^{\tau\mathcal{L}_{n_0}}\left(d_j(\rho_n^{ss})\,\delta a_{n_0}^j\rho_{n_0}^{ss} - s_j(\rho_n^{ss})\,\rho_{n_0}^{ss}\delta a_{n_0}^j\right)\right] + \mathrm{H.c}\right), \tag{3}$$

where Z is the number of nearest neighbors, $a_n^+ = a_n^\dagger$, $a_n^- = a_n$, $\delta a_n^j = a_n^j - \mathrm{Tr}_n\{a_n^j\rho_n^{ss}\}$, and

$$d_j(\rho_n^{ss}) = \mathrm{Tr}_n\{a_n e^{\tau\mathcal{L}_n}(\delta a_n^j)^\dagger\rho_n^{ss}\}, \tag{4}$$
$$s_j(\rho_n^{ss}) = \mathrm{Tr}_n\{a_n e^{\tau\mathcal{L}_n}\rho_n^{ss}(\delta a_n^j)^\dagger\}. \tag{5}$$

In Eq. (3), ρ_n is the reduced density matrix for site n. The Liouville operator $\mathcal{L}_{n_0}(\cdot) = -i[H_{n_0},(\cdot)] + \gamma/2[2a_{n_0}(\cdot)a_{n_0}^\dagger - a_{n_0}^\dagger a_{n_0}(\cdot) - (\cdot)a_{n_0}^\dagger a_{n_0}]$ describes the dynamics of the "system" without the coupling to the "environment." As a self-consistent condition, we replace ρ_n by ρ_{n_0} in Eqs. (3)–(5), which yields closed algebraic equations for ρ_{n_0}.

We make three comments. (i) The c-MoP method up to first order in the "system"-"environment" coupling J, see the first two terms on the right hand side of Eq. (3), is equivalent to the mean-field approximation [12, 13]. The term proportional to J^2 explicitly takes system-environment correlations into account, and thus, we gain a much more accurate result than the results obtained with a mean-field Ansatz [8]. (ii) When the "system" is a cluster consisting of two or three neighboring cavities, we obtain very similar self-consistent equations for the reduced density matrix. The larger size of the "system" yields the better approximation. (iii) The numerical task to solve for ρ_{n0}^{ss} in Eq. (3) is comparable to the numerical task to solve a mean-field equation which can be done very efficiently as it only involves the dimension of the local Hilbert space. Even the integral over the dynamical map $e^{\tau \mathcal{L}_{n_0}}$ only demands the diagonalization of $\mathcal{L}_{n_0}$ which is again a local operator. In contrast to DMRG methods which work only efficient in 1D, the c-MoP equations do not change their structure by going from one-dimensional arrays to higher dimensional lattices and are therefore well suited for the treatment of problems with an arbitrary lattice geometry.

Figure 3 shows the calculated results for the photon numbers in the steady state by the c-MoP method for different values of the tunneling rate J. The "system" consists of a single cavity (black dash-dotted line), a two-cavity cluster (red solid line), and a three-cavity cluster (orange dashed line). Further we show the result obtained by a t-DMRG integration of Eq. (2) with $N = 21$ lattice sites and open boundary conditions. From this t-DMRG numerics we extract the reduced density matrix for the central site $n_0 = 11$ and the corresponding photon number (green solid line). Although we find a deviation from this numerically-exact result in the case of single-site c-MoP, the calculation with two- or three-site c-MoP clusters shows a good agreement with the t-DMRG result. This clearly indicates the suitability of the c-MoP method to examine this non-equilibrium many-body problem.

5. Conclusion

We have studied the non-equilibrium many-body problem of a one-dimensional array of QED-cavities. For a finite array, we have elucidated the complex and rich interplay between Rabi oscillations in a cavity driven by the laser, coherent oscillation induced by tunneling of photons between neighboring cavities, and the dissipative effects of photon losses from the cavities, leading to the relaxation of the array to a non-equilibrium steady state. For an infinite array, we have adopted the c-MoP method to calculate the reduced density matrix of subsystem in a self-consistent way. We have found that the two- or three-cavity c-MoP cluster method is a powerful tool to study the complex many-body behavior of coupled QED-cavities on a lattice in any dimension and arbitrary lattice geometry.

Acknowledgments

This work is supported by a Grant-in-Aid for Scientific Research (S) (No. 26220711) from Japan Society for the Promotion of Science and by the collaborative research center 631 of the German research foundation (DFG).

References

[1] H. Walther, B. T. H. Varcoe, B.-G. Englert, and T. Becker, Rep. Prog. Phys. **69**, 1325 (2006).
[2] M. J. Hartmann, F. G. S. L. Brandão, and M. B. Plenio, Laser & Photon. Rev. **2**, 527 (2008).
[3] A. A. Houck, H. E. Türeci, and J. Koch, Nat. Phys. **8**, 292 (2012).
[4] A. D. Greentree, C. Tahan, J. H. Cole, and L. C. L. Hollenberg, Nat. Phys. **2**, 856 (2006).
[5] I. Carusotto and C. Ciuti, Rev. Mod. Phys. **85**, 299 (2013).
[6] M. J. Hartmann, F. G. S. L. Brandão, and M. B. Plenio, Nat. Phys. **2**, 849 (2006).
[7] H.-P. Breuer and F. Petruccione, *The Theory of Open Quantum Systems*, Oxford University Press (2007).
[8] P. Degenfeld-Schonburg and M. J. Hartmann, Phys. Rev. B **89**, 245108 (2014).
[9] U. Schollwöck, Rev. Mod. Phys. **77**, 259 (2005).
[10] M. J. Hartmann, Phys. Rev. Lett. **104**, 113601 (2010).
[11] P. Degenfeld-Schonburg, E. del Valle, and M. J. Hartmann, Phys. Rev. A **85**, 013842 (2012).
[12] L. P. Kadanoff, J. Stat. Phys. **137**, 777 (2009).
[13] S. Diehl, A. Tomadin, A. Micheli, R. Fazio, and P. Zoller, Phys. Rev. Lett. **105**, 015702 (2010).

Superfluid helium quantum interference devices (SHeQUIDs): principles and performance

R E Packard[1] **and Y Sato**[2]
[1] Department of Physics, University of California, Berkeley, California 94720, USA
[2] Rowland Institute at Harvard, Harvard University, Cambridge, Massachusetts 02142, USA

E-mail: rpackard@berkeley.edu, sato@rowland.harvard.edu

Abstract. We describe recent progress in developing a superfluid helium analog of the superconducting dc-SQUID. The devices tested thus far are sensitive detectors of rotation as well as useful probes for studies of superfluidity. The key ingredients of the superfluid helium quantum interference device (SHeQUID) involve commercially available technology and modest cryogenic facilities.

The superfluid states of quantum liquids (^{3}He-B and ^{4}He) are described by a macroscopic order parameter of the form $\Psi = \sqrt{\rho_s} e^{i\phi}$, where ρ_s is the superfluid density and ϕ is the quantum mechanical phase [1]. When two such fluids are coupled together through a small channel, the system exhibits many surprising features. Quintessential phenomena are various Josephson effects, which occur when the channel between the two fluids is sufficiently small. Such a small channel is normally referred to as a weak link or junction. In that scenario, solving the time-dependent Schrodinger's equation applied to the coupled system results in two governing equations:

$$I = I_0 \sin \Delta\phi \tag{1}$$

and

$$\frac{d\Delta\phi}{dt} = -\frac{\Delta\mu}{\hbar}. \tag{2}$$

The first equation (dc Josephson equation [2]) describes the nonlinear relationship between the phase difference across the junction $\Delta\phi$ and the superfluid mass current I through it. The second equation (Josephson-Anderson phase evolution equation [3]) states that the chemical potential difference $\Delta\mu = m(\Delta P / \rho - s\Delta T)$ drives the phase factor to evolve in time. Here ΔP and ΔT are pressure and temperature differentials, ρ is the fluid density, s is the entropy per unit mass and m is the mass of constituent particles making up the superfluid, either ^{4}He atomic mass or twice the ^{3}He mass [1].

The dynamic behaviours predicted by the above two equations are quite counter-intuitive. For example even with the absence of any driving force ($\Delta\mu = 0$), dc mass current could appear across the junction from nonzero $\Delta\phi$. On the other hand, when a constant $\Delta\mu$ is applied across the junction, $\Delta\phi$ evolves linearly in time, resulting in oscillating mass current across the junction. This oscillation phenomenon is called the Josephson oscillation which occurs at frequency $\omega_J \equiv \Delta\mu / \hbar$.

Although electron tunnelling through a thin non-superconducting film constitutes a weak link for superconductors [2], the atomic mass of helium is too large to exhibit appreciable tunnelling. To couple superfluids weakly, one needs the equivalent of a Dayem Bridge [4], a constricted passage whose dimensions are on the order of the superfluid healing length ξ. In practice, two reservoirs of superfluid helium can be coupled weakly through an array of apertures. For reasons that are still not obvious the superfluid in an array of weak links behaves coherently with the Josephson oscillations in every aperture being in phase with all the others. An array is required to lift the miniscule mass flow signal above the experimental noise.

For superfluid ^{3}He, at zero ambient pressure, the healing length is given by [5]

$$\xi_3 = \frac{65}{(1 - T/T_c)^{1/2}} \text{ nm}, \tag{3}$$

where $T_c = 1$ mK is the superfluid ^{3}He transition temperature. This characteristic size came within the reach of nanofabrication about 30 years ago. In contrast, the healing length of ^{4}He at zero ambient pressure is given by [6]

$$\xi_4 = \frac{0.3}{(1 - T/T_\lambda)^{0.67}} \text{ nm}, \tag{4}$$

where $T_\lambda = 2.176$ K is the superfluid ^{4}He transition temperature. This sub-nanometer size requirement still remains prohibitive to this day, which explains why the investigations of Josephson effects in superfluids began with ^{3}He despite the requirement of mK cryogenics. We show below however that this issue can be circumvented by taking advantage of the diverging temperature dependence of ξ_4 near T_λ.

A flow signature consistent with the $I \propto \sin\Delta\phi$ relation was first reported in 1988 using ^{3}He [7], and the Josephson oscillation itself was directly observed a decade later [8]. Figure 1 depicts a generic apparatus similar to the one utilized for the latter experiment. A 65 X 65 array of 90 nm apertures *e*-beam lithographed in a 50 nm-thick silicon nitride membrane was used as a weak link. Fluid motion is hydrodynamically coupled to the diaphragm, which is coated with a superconducting metal. The diaphragm motion is sensed with a Paik-type displacement sensor [9], providing a direct measure of fluid motion in the weak link junction. The application of a step voltage between the diaphragm and the electrode directly above pulls up the diaphragm, creating a pressure difference between the inner and outer reservoirs. This step ΔP causes the fluid to oscillate across the aperture array, and these Josephson oscillations, whose frequency is proportional to the applied ΔP, are indeed observed.

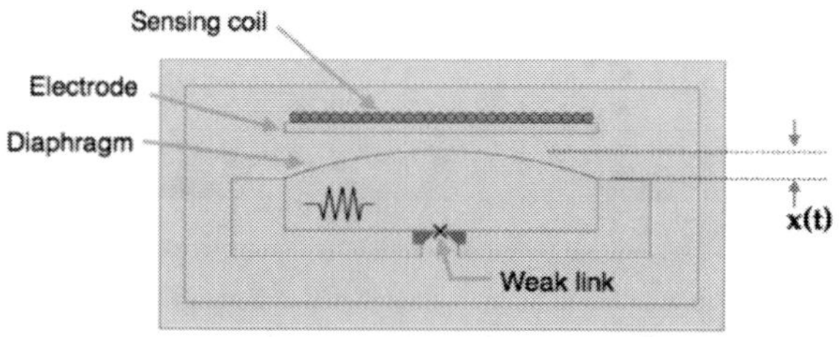

Figure 1. Depiction of generic apparatus for investigating superfluid Josephson effects. Electrode and diaphragm form a pressure pump. The SQUID-based transducer monitors the diaphragm position x(t). Weak link is formed by an array of holes. Both ΔP and ΔT (induced with the heater depicted) can be used to drive Josephson oscillations.

The superfluid helium quantum interference device (SHeQUID) exploits the Josephson oscillation phenomena discussed above. See figure 2. A torus filled with superfluid helium is interrupted by two weak links. When $\Delta\mu$ is applied across the two junctions, the current oscillates through each junction as $I_{c,1}\sin(\omega_J t)$ and $I_{c,2}\sin(\omega_J t+\Delta\phi_{ext})$, where $\Delta\phi_{ext}$ denotes the phase shift between the two oscillations. The overall amplitude of the total mass current $I_c^*\sin(\omega_J t)$ can be written as

$$I_c^* = (I_{c,1}+I_{c,2})\left\{\cos^2\left(\frac{\Delta\phi_{ext}}{2}\right)+\gamma^2\sin^2\left(\frac{\Delta\phi_{ext}}{2}\right)\right\}^{1/2}, \tag{5}$$

with $\gamma=(I_{c,1}-I_{c,2})/(I_{c,1}+I_{c,2})$. As can be seen in the above relation, the mass current amplitude becomes a function of $\Delta\phi_{ext}$, making the device an interferometer.

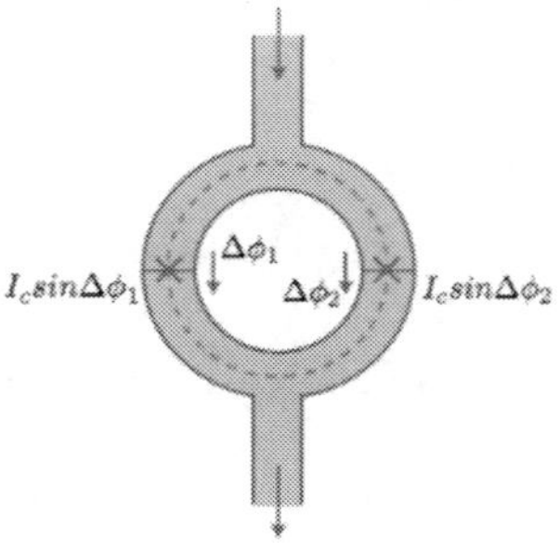

Figure 2. SHeQUID configuration. Two junctions are placed in a superfluid torus as in the case of a dc-SQUID. The phase integral condition along the closed loop depicted by a dotted line underlies the interferometry principle.

A SHeQUID is a direct analog of a dc superconducting quantum interference device (dc-SQUID [10]). In the case of a dc-SQUID, the phase shift arises from the magnetic flux through the superconducting loop, rendering the device a very sensitive magnetometer. For a SHeQUID, the role of magnetic flux can be played by various physical phenomena, and one such example is rotation. When the torus as depicted in figure 2 is rotated at speed $\vec{\Omega}$ about an axis perpendicular to its plane, the partitions containing the junctions force the fluid to flow as a solid body. Since superfluid velocity corresponds to phase gradient via $v_s=(\hbar/m)\nabla\phi$ [1], the line integral applied to a closed loop gives $\Delta\phi_{ext}=(2m/\hbar)\vec{\Omega}\cdot\vec{A}$ where $\vec{A}$ is the area vector of the torus. Substituting this into equation 5, one can see that the variation in the amplitude of the combined Josephson oscillation provides information on the rotation of the system.

This rotation-induced phase shift is equivalent to the Sagnac phase shift $\Delta\phi=(2\omega/c^2)\vec{\Omega}\cdot\vec{A}$ seen in optical interferometers where a beam of light is split and recombined while enclosing a finite area [11]. However the phase shift for a given rotation and enclosed area is very different for photons and helium. The induced phase shift is proportional to the effective mass of interfering medium. For a given change in rotation, the resultant signal per device area for a SHeQUID is larger than that for laser interferometers by ten orders of magnitude (ratio of $m_{He}c^2$ to photon energy). This suggests the development of compact sensitive rotation sensors. The first SHeQUID was constructed with ^{3}He [12], and its operation was demonstrated by inducing and detecting phase shifts from the rotating Earth.

These developments in ^{3}He preceded those in ^{4}He by almost two decades. The extremely small size of ^{4}He healing length, on the order of a few angstroms, was the primary factor for this delay. As mentioned previously, the healing length diverges near T_λ. This means that instead of attempting to fabricate sub-nanometer apertures to match the zero-temperature healing length, one could increase the healing length near T_λ to match the bigger aperture sizes available. However earlier estimates suggested that the thermally-driven phase fluctuation at 2 K would destroy the temporal coherence of fluid oscillation in a given aperture [13]. Although this thermal fluctuation is not an issue for ^{3}He at its 1mK operating temperature, at a temperature two thousand times higher this was believed to be detrimental to Josephson phenomena in ^{4}He. Temperatures so close to T_λ also meant significantly diminished superfluid density and hence even smaller signals.

Despite the aforementioned factors, early evidence for $I \propto \sin\Delta\phi$ behaviour in superfluid ^{4}He was reported near T_λ in 2001 [14], followed by the direct observation of Josephson oscillations four years later [15]. These discoveries and subsequent analysis have suggested that thermal fluctuations may actually be suppressed in an array of apertures due to coupling among them [16]. Another surprise that accompanied these discoveries is that an oscillation phenomenon at the Josephson frequency persists even when the temperature is lowered so that the aperture size is considerably larger than the superfluid healing length [15].

In the regime where the two superfluid reservoirs are strongly coupled, the analysis that leads to equation 1 is no longer valid. When a finite chemical potential difference is applied across an array of holes, superfluid is simply accelerated from one side to the other. The fluid velocity increases until it reaches a critical value, at which point a quantized vortex is stochastically nucleated. The vortex moves across the channel and removes a fixed amount of energy from the fluid flow. The velocity decreases by a fixed amount given by $\Delta v = \kappa / l_{eff}$, where $\kappa = h/m$ is the quantum of circulation and l_{eff} is the effective hydrodynamic length of the aperture. This dissipative event is known as a phase slip [17-20].

When the chemical potential difference persists as in the case of applying and holding ΔP in an apparatus depicted in figure 1, the process described above repeats, and a series of phase slip events takes place. It can be shown that the frequency for such events happens to be identical to the Josephson frequency. Both the weakly coupled Josephson regime and the strongly coupled phase slip regime have been investigated, and the current-phase relation is seen to change smoothly from linear to sinusoidal as one approaches the transition temperature [21]. It is worthwhile to note that, in the strongly coupled regime, the phase slip events occur synchronously among all the apertures but that degree of synchronicity is observed to decrease as one lowers the temperature further. The nature of the mechanism that synchronizes the array and an explaination of why that synchronization decreases at lower temperatures has been investigated both theoretically and experimentally, [22-24]. However these features are still an unresolved mystery.

Figure 3. Multi-turn and gradiometric sensing arm for a SHeQUID. The length of the arm is approximately 0.5m. From Ref [28].

Given the ease of cryogenics at 2 K, many SHeQUIDs have been constructed using ^{4}He following the first demonstration of ^{4}He SHeQUID in 2006 [25]. As a novel phase gradient meter, it has been utilized to verify the link between the wavefunction picture of superfluid and the Landau's two-fluid model [26]. It has also been used to detect the motion of a quantized vortex and the onset of quantum turbulence [27]. As a compact rotation sensor, the device has seen significant increases in sensitivity. A large area ^{4}He SHeQUID with a long multi-turn path in astatic geometry has for example been reported with an intrinsic rotational sensitivity of 1×10^{-8} rad/sec/$\sqrt{\text{Hz}}$ [28]. See figure 3. The total area enclosed is 225 cm^2, a forty-fold increase compared to the first ^{3}He SHeQUID [12]. The superfluid Josephson system has also formed an ideal test bed for studying some fascinating nonlinear phenomena such as the Fiske effect, junction size effect, and bifurcation, and those in turn have provided a set of tools to increase the sensitivity and utility of the device [29-32]. For example, flux locking with injected hydrodynamic heat flow for example allows the linearization of the device [33]. A linearized SHeQUID (figure 4) equipped with such function that can operate continuously to automatically measure time-varying rotation rates has been demonstrated in 2013 [34, 35].

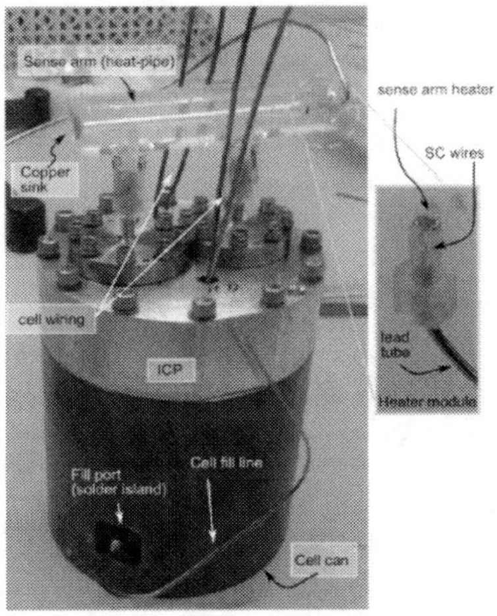

Figure 4. Continuously operating, flux-locked SHeQUID. From Ref [35].

The experimental components for the SHeQUID operation have been simplified over the years, and the requirements have become quite modest. For ^{4}He SHeQUID experiments, high-resolution thermometers with sub-nK resolution [36, 37] are used to regulate the temperature within several nano-Kelvins of the set point. A recent development involves redesigning these thermometers to utilize the temperature dependence of the magnetic susceptibility of a paramagnetic PdMn alloy [38]. Alloys with the required mixing percentage, Neodymium magnets to apply the external fields, and a dc-SQUID to read out the susceptibility change, can all be purchased commercially. Weaklink arrays are made of approximately 5000 apertures, *e*-beam lithographed in a ~50 nm-thick SiN membrane [39-41]. The aperture diameter depends on experiments, but it typically ranges from 30 nm to 70 nm. This type of aspect ratio is not demanding with the current state-of-the-art nanofabrication, and with a recipe that has been improved upon over the years, the aperture fabrication has even been successfully outsourced [42]. The temperature regime of 2 K is easily obtained, and the experiments have been carried out with simple probes dipped into a liquid ^{4}He dewar without a 1 K pot. The whole bath of liquid ^{4}He can be temperature controlled to the required precision with minimum liquid loss. With careful engineering [43], it should be feasible to load the experimental stages on cryocoolers to entirely remove the need for a cryogen bath, allowing a remote and fully automated operation.

The proof-of-concept demonstration of the SHeQUID itself as well as various techniques to illuminate and exploit its exotic properties has now been achieved to some extent. As a step towards the next phase for the SHeQUID development as an "instrument", one may explore its untapped sensitivity. The intrinsic device sensitivity is proportional to several parameters including the sensing area, number of apertures in the junctions, and superfluid density. The effects of increasing the sensing area have been studied, and although the signal increase has been reported, this approach has its limitations. Writing the fluid kinetic energy as $LI^2/2$, the superfluid hydrodynamic inductance L can be identified as $(\hbar/m_4)(dI/d\Delta\phi)^{-1}$ [2]. For a sinusoidal current-phase relation of equation 1, this leads to kinetic inductance of the junction (Josephson inductance) $L_J = \hbar/(m_4 I_c)$ at $\Delta\phi = 0$. In contrast for superfluid flowing through the rest of the interfering path of length l and cross sectional area σ, the inductance is given by $L_t = l/(\rho_s\sigma)$. The device should be designed such that $L_J >> L_t$.

In the case of dc-SQUIDs a separate pickup loop with a large flux capture area can be used while an appropriate impedance matching can be accomplished through flux transformers. In contrast a SHeQUID can only operate in a "bare" circuit because no coupling exists between two loops of neutral superfluid placed nearby. Hence as the interfering path is made significantly long, its hydrodynamic inductance starts to dominate over the Josephson inductance and the device should inevitably start to suffer signal loss. Increasing σ to decrease L_l leads to significant fluid volume increase, which could compromise the signal through increased compressibility.

A less complicated route for enhancing sensitivity may be increasing the number of apertures. With *e*-beam lithography and the several hundred micron-size membranes used thus far, a number increase by an order of magnitude is easily feasible. This numerical increase may also have an additional benefit of further suppressing the thermal fluctuations via enhanced coupling. Developing the technique to fabricate even smaller holes is also a fascinating option since that would allow the Josephson regime to extend down to lower temperatures [44]. For example, if operated near 1.5 K, superfluid fraction that contributes to the mass current can be nearly 100 % as opposed to a few % near T_λ. In principle this should lead to a much higher signal to noise ratio. This approach would require a larger number of apertures to offset the smaller signal per aperture, but with the presence of increased number, this may also have the benefit of suppressed thermal fluctuations. The ultimate phase resolution limit of SheQUIDs may be quantum fluctuations embodied in the uncertainty principle. Assuming that the vibrational noise from the environment can be controlled, it will be critical to suppress thermally excited phase fluctuations if one is to get to that ultimate limit.

Another aspect to consider is a new methodology for detecting the fluid oscillation. The current transduction mechanism relies on the fluid pushing on the diaphragm, with finite spring constant k, resulting in a competition between the Josephson energy $(\hbar / m_4)I_c$ and the diaphragm spring energy $k\Delta x^2/2$. Rewriting the latter as $k(I_c/\rho A\omega)^2/2$, where A is the diaphragm area and ω is the oscillation frequency and equating the two energies gives the maximum current amplitude that is supportable $I_{c,\max} \sim \hbar\rho^2 A^2\omega^2 / m_4 k$. To increase this upper limit in a given junction, one should use a soft and large diaphragm and operate at high frequencies. While this can be achieved to some extent with the current transducers, a new non-mechanical method to detect the fluid motion in the junctions could remove the dependence on hydrodynamic coupling, which is naturally a step-down transformer. Similarly, since the fluid oscillation occurs in the velocity domain, the signal obtained with the displacement sensor diminishes as the frequency increases. A sensitive sensor whose output is proportional to velocity instead of displacement could provide the means to take full advantage of the increased signal.

Ultrasensitive rotation sensors find applications in multidisciplinary fields including geodesy, geophysics, inertial navigation, and seismology. With the device technology rapidly developing, it is not at all inconceivable to use SHeQUIDs to complement state-of-the-art ring laser gyroscopes in their quest to test general relativistic effects on Earth [45]. Furthermore, due to their novel ability to detect quantum mechanical phase differences, SHeQUIDs may also be utilized as tools for studying various Berry's phase phenomena. For example, He-McKeller-Wilkens topological phase was recently observed with atom interferometers [46-49].

During the past decade many of the Josephson effects discovered with superfluid helium have also been detected in cold BEC gases [50-52]. Several laboratories are pursuing the development of these matter wave interferometers [53, 54]. In these projects it is a distinct advantage to be working with essentially room temperature elements rather than cryostats. However phase fluctuations limited by the finite number of atoms and the inability to easily produce large enclosed area may limit the cold atom devices. With so much effort now underway the future is bound to be interesting both for SHeQUIDs and their cold atom counterparts.

More details related to SHeQUIDs can be found in Ref [55]. For discussion on superfluid gyroscopes based on a single weaklink and a large parallel path, see Ref [56].

This work has been supported by the NSF Division of Materials Research and the Rowland Institute at Harvard.

References

[1] Tilley D R and Tilley J 1990 *Superfluidity and Superconductivity* (Bristol: Institute of Physics)
[2] Likharev K K 1986 *Dynamics of Josephson Junctions and Circuits* (New York: Academic Press)
[3] Anderson P W 1966 *Rev. Mod. Phys.* **38** 298
[4] Anderson P W and Dayem A H 1964 *Rev. Rev. Lett.* **13** 195
[5] Vollhardt D and Wolfle P 1990 *The Superfluid Phases of Helium-3* (New York: Taylor and Francis)
[6] Henkel R P, Smith E N and Reppu J D 1969 *Rev. Rev. Lett.* **23** 1276
[7] Avenel O and Varoquaux E 1988 *Phys. Rev. Lett.* **60** 416
[8] Pereverzev S V, Loshak A, Backhaus S, Davis J C and Packard R E 1997 *Nature* **388** 449
[9] Paik H 1976 *J. Appl. Phys.* **47** 1168
[10] Clarke J and Braginski A I 2004 The *SQUID Handbook: Fundamentals and Technology of SQUIDs and SQUID Systems* (Weinheim: Wiley-VCH)
[11] Post E J 1967 *Rev. Mod. Phys.* **39** 475
[12] Simmonds R, Marchenkov A, Hoskinson E, Davis J C and Packard R E 2001 *Nature* **412** 55
[13] Zimmermann W 1987 *in Proceedings of the Fifth Oregon Conference on Liquid Helium, a Report from the Department of Physics, University of Oregon*, p 118
[14] Sukhatme K, Mukharsky Y, Chui T and Pearson D 2001 *Nature* **411** 280
[15] Hoskinson E, Packard R E and Haard T 2005 *Nature* **433** 376
[16] Chui T, Holmes W and Penanen K 2003 *Phys. Rev. Lett.* **90** 085301
[17] Onsagar L 1949 *Nuovo Cimento* **6** 249
[18] Feynman R P 1955 *Progress on Low Temperature Physics* edited by Gorter CJ (Amsterdam: North-Holland) vol 1 ch 2
[19] Anderson P W 1966 *Quantum Fluids* edited by Brewer D F (Amsterdam: North-Holland) p 146
[20] Avenel O and Varoquaux E 1985 *Phys. Rev. Lett.* **55** 2704
[21] Hoskinson E, Sato Y, Hahn I and Packard R E 2005 *Nat. Phys.* **2** 23
[22] Sato Y, Hoskinson E and Packard R E 2006 *Phys. Rev. B* **74** 144502
[23] Sato Y, Hoskinson E and Packard R E 2007 *JLTP* **149** 222
[24] Pekker D, Barankov R and Goldbart P M 2007 *Phys. Rev. Lett.* **98** 175301
[25] Hoskinson E, Sato Y and Packard R E 2006 *Phys. Rev. B* **74** 100509(R)
[26] Sato Y, Joshi A and Packard R E 2007 *Phys. Rev. Lett.* **98** 195302
[27] Sato Y, Joshi A and Packard R E 2007 *Phys. Rev. B* **76** 052505
[28] Narayana S and Sato Y 2011 *Phys. Rev. Lett.* **106** 255301
[29] Narayana S and Sato Y 2010 *Phys. Rev. Lett.* **105** 205302
[30] Hoskinson E, Sato Y, Penanen K and Packard R E 2006 *AIP Conf.* **850** 117
[31] Sato Y 2010 *Phys. Rev. B* **81** 172502
[32] Narayana S and Sato Y 2011 *Phys. Rev. Lett.* **106** 055302
[33] Sato Y, Joshi A and Packard R E 2007 *Appl. Phys. Lett.* **91** 074107
[34] Joshi A and Packard R E 2013 *JLTP* **172** 162
[35] Joshi A, 2013 Ph.D. thesis. University of California at Berkeley
[36] Welander P B and Hahn I 2001 *Rev. Sci. Instr.* **72** 3600
[37] Welander P, Barmatz M and Hahn I 2000 *IEEE Trans. Instr. Meas.* **49** 253
[38] Narayana S and Sato Y 2010 *IEEE Trans. Appl. Supercond.* **20** 2402
[39] Amar A, Lozes R, Sasaki Y, Davis J C and Packard R E 1993 *J. Vac Sci. Technol. B* **11** 259
[40] Loshak A 1999 Ph.D. thesis. University of California at Berkeley 2007
[41] Joshi A, Sato Y and Packard R E 2009 *J. Phys. Conf. Ser.* **150** 012018
[42] EigenPhase Technologies, Trumansburg, NY 14886

[43] Pelliccione M, Sciambi A, Bartel J, Keller A J and Goldhaber-Gordon D 2013 *Rev. Sci. Instr.* **84** 033703
[44] Murakawa S, Chikazawa Y, Higashino R, Yoshimura K, Shibayama Y, Kuriyama K, Honda K, and Shirahama K 2010 *International Symposium on Quantum Fluids and Solids* PS1-27
[45] Schreiber K U and Wells J P R 2013 *Rev. Sci. Instr.* **84** 041101
[46] Lepoutre S, Gauguet A, Trenec G, Bucher M and Vigue J 2012 *Phys. Rev. Lett.* **109** 120404
[47] He XG and McKellar B H J 1993 *Phys. Rev. A* **47** 3424
[48] Wei H, Han R and Wei X 1995 *Phys. Rev. Lett.* **75** 2071
[49] Sato Y and Packard R E 2009 *J. Phys. Conf. Ser.* **150** 32093
[50] Albiez M, Gati R, Folling J, Hunsmann S, Cristiani M and Oberthaler M 2005 *Phys. Rev. Lett.* **95** 010402
[51] Levy S, Lahoud E, Shomroni I and Steinhauer J 2007 *Nature* **449** 579
[52] Eckel S, Lee J G, Jendrzejewski F, Murray N, Clark C W, Lobb C J, Phillips W D, Edwards M and Campbell G K 2014 *Nature* **506** 200
[53] Wright K C, Blakestad R B, Lobb C J, Phillips W D and Campbell G K 2013 *Phys. Rev. Lett.* **110** 025302
[54] Ryu C, Blackburn P W, Blinova A A and Boshier M G 2013 *Phys. Rev. Lett.* **111** 205301
[55] Sato Y and Packard R E 2012 *Rep. Prog. Phys.* **75** 016401
[56] Avenel O, Mukharsky Y and Varoquaux E 2004 *JLTP* **135** 745

Stabilization of Multi-electron Bubbles in Superfluid Helium

Isaac F Silvera[1], Jieping Fang[1], and Jacques Tempere[2]

[1]Department of Physics, Harvard University, Cambridge, MA 02138, United States
[2]TQC, departement Fysica, Univesity of Antwerpen, Universiteitsplein1, B2610, Antwerpen, Belgium

silvera@physics.harvard.edu

Abstract. Multielectron bubbles (MEBs) in liquid helium were first observed in the late 1970s, but their properties have never been explored experimentally due to their short lifetimes and the difficulty to localize them. We report the observation of long-lived MEBs in a novel cell filled with superfluid helium at static negative pressures. MEBs were extracted from the electron filled vapor sheath of a heated filament loop embedded in the superfluid helium and observed by high-speed photography. MEBs are 2D electron gases on the 3D surface of hollow helium bubbles. Diameters can range from nanometers to millimeters, depending on the number of enclosed electrons. Electrons move in angular momentum states; deformations of the surface are called spherical ripplons. The attractive electron-ripplon interaction leads to an unusual form of superconductivity. If they can be compressed, Wigner crystallization and quantum melting can be observed, as well as a new phase for localization called the ripplo-polaron lattice. MEBs are unstable to tunneling discharge when pressed against a surface. Just as Bose gases are captured in a trap for study, MEBs must also be localized away from walls. We shall discuss methods of capturing them in an electromagnetic trap embedded in the liquid helium.

1. Introduction

An equilibrium multielectron bubble in liquid helium is a spherical cavity containing electrons in a nanometer thin layer on the helium surface of the bubble so as to form a two-dimensional electron system on a spherical surface. MEBs were first observed by Volodin, Khaikin, and Edel'man [1] in the late 1970's. Free electrons were created by thermionic emission in the helium vapor phase; a planar electrode with a positive voltage was placed under the flat surface of bulk superfluid (SF) helium to create a high surface density of electrons, as a two-dimensional electron gas. Above a critical electric field (~ 4 kV/cm) corresponding to a surface electron density of ~ $2.4\times10^9\,\mathrm{cm}^{-2}$, the surface became hydrodynamically unstable and subsumed of order 10^7-10^8 electrons in the form of a single bubble. A high-speed camera recorded this. The created bubble flew, opposite to the direction of buoyant forces, to the high voltage plate with a velocity ~ 10^4 cm/s where it was annihilated. The lifetime of the bubble was about 1 ms. Shikin [2] analysed and presented a theory for the equilibrium diameter of an MEB. Diameters of nanometers to millimeters were predicted depending on the number of electrons confined in the bubble. A decade later, Albrecht and Leiderer [3, 4] used a similar geometry to produce MEBs in normal helium. MEBs with $10^5 - 10^7$ electrons were observed by small-angle light scattering using a He-Ne laser, and oscillations and fissioning of the MEBs were observed as well. The velocity of an MEB was about 10 cm/s and the lifetimes were up to 100 ms. They also proposed trapping the MEBs with an electromagnetic quadrupole cage to extend the measurement time, but no experiments were reported.

After MEBs were first observed a few decades ago, a plethora of fascinating properties were predicted including tunable surface electron density, instabilities, superconductivity, vortex states, sonoluminescence, and magnetic properties such as the quantum Hall effect, etc. reviewed in ref. [5]. In order to study these properties, stabilizing an MEB in a long-lived state is crucial. However, none of the earlier reported methods were able to create an MEB with a long enough lifetime. Silvera and Tempere proposed a new method: electrons would be created in the vapor phase of a cylindrical cell having a dome shaped. The cell, which is partially filled with SF helium, incorporated a mechanism to translate the surface of the helium up and down the cell. Single electrons require an energy of ~1 eV to penetrate the surface of the bulk helium [6]. Thus, as the helium level is moved up to the dome, the electrons, encircled by helium surfaces, are corralled into a smaller and smaller volume and eventually form an MEB at the top of the dome, localized by buoyancy. Such a system was built using a concave lens for the dome, so that the MEBs could be observed with a low temperature microscope. Large MEBs that were formed in the dome rapidly discharged or collapsed [7]. An analysis showed that the MEBs were attracted to the surface of the dome, mainly by image and buoyant forces so that the superfluid film that separated them from the surface was thinned to ~5-6 nm so that the electrons could tunnel into the surface, where they were evidently trapped. After a number of attempts to confine MEBs in this manner, the surface of the dome charged up sufficiently creating patch electric fields that repelled the MEBs so that the helium film was not thinned. In this way large MEBs were observed for a few seconds. However, they were not confined in the dome and passed out of the field of view. These bubbles were shown to be filled with electrons, as they could be steered with external electric fields. However they were challenging to control and localize, so another method was developed.

If a fine tungsten filament is heated in superfluid liquid helium a vapor sheath will form around the glowing hot regions of the wire [8]. When such a filament is formed in a loop a large bag-shaped sheath forms around the loop and this is filled with electrons due to thermionic emission from the glowing filament. It turned out that the sheath is tightly tethered to the loop, so an external electric field from a nearby electrode could be used to extract MEBs from the sheath. These MEBs were photographed by high-speed video (up to 10K frames/sec) and seen to be extracted under very turbulent conditions. They underwent large amplitude oscillations, decreasing in size, and finally disappearing after several milliseconds. These MEBs were created under the surface of bulk pumped superfluid helium at a temperature of ~1.5 K, so that there was always a positive pressure on the MEBs due to the pressure head of the helium.

Theoretical analysis has shown that for positive pressures MEBs should be dynamically unstable. Surface deformations of MEBs are called spherical ripplons; these harmonic modes can be represented by spherical harmonics. The pressure dependence of several of the modes is shown in Fig. 1. It is seen the quadrupole mode (L=1) is unstable at zero pressure and that with increasing pressure, all of the higher modes increase in frequency and then drop to zero, so that in principle they should give rise to dynamic instabilities of the MEBs. However, for low negative pressures (negative pressures below the explosion pressure) all of the modes are stable.

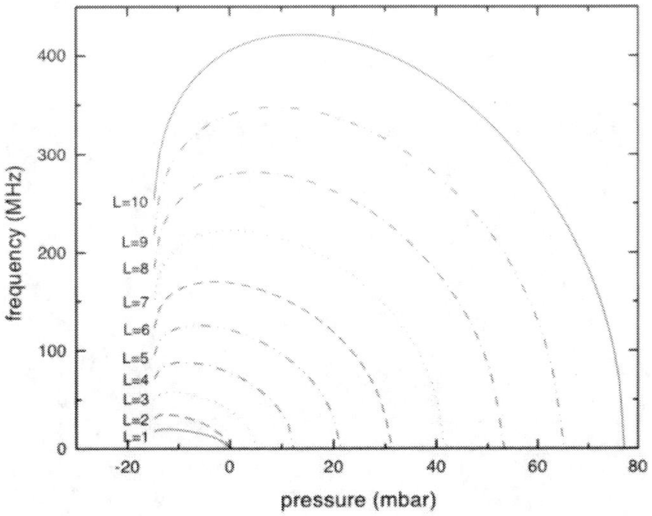

Figure 1. The frequency of ripplon modes of an MEB with N=1000 electrons as a function of pressure. L is the index of the spherical harmonics.

2. The Negative Pressure Cell.

In order to stabilize MEBs a cell was built in which MEBs could be produced under negative pressure of the helium. A hollow cubic cell was constructed out of copper with indium sealed fused silica windows on the four vertical walls. Electrical feedthroughs on the bottom flange were used to activate or monitor a tungsten filament, provide electric fields, capacitance pressure gauge, etc. On the top of the cell were mounted a superfluid tight valve made of Torlon and a bellows that is incorporated as part of the volume of the cell; this bellows could be manually expanded or contracted. The cell was submerged in a pumped bath of SF helium and then filled so that there was no vapour phase in the interior when the Torlon valve was closed. By expanding or compressing the bellows the pressure in the cell could be made negative or positive. This was detected with a capacitance gauge that also insured that the Torlon valve was SF tight: the pressure remained stable at a fixed extension of the bellows. The loop tungsten filament could be viewed through the windows and illuminated for high-speed video photograpy. Electrodes could also be mounted in the cell for extraction of MEBs.

3. Observations.

Because the cell is isochoric, compressing the bellows creates a positive pressure; in this case a sheath could not be formed when the tungsten loop filament was heated. The explanation is that it is energetically unfavorable to create a vapor sheath that will further compress the helium. On the other hand with negative pressures, below a certain threshold, vapor sheaths were easily formed as this relieves the negative stress in the fluid. At low temperatures of ~1.5 K, MEBs could be formed by several methods. With a sheath present a sudden change in the negative pressure released MEBs, observed with an external microscope and high-speed video. MEBs could also be extracted from the sheath using electrodes either below (with negative voltage) or above the loop (with positive voltage).

However, in all of these cases the MEBs had large turbulent oscillations and disappeared within a few milliseconds, just as with positive pressures in an isobaric cell.

Our idea was that in order to create a stable MEB it should be created "adiabatically", or we should find conditions to damp the oscillations. We produced shaped loops that would allow an MEB to gently slide off of the filament, but this was unsuccessful, as the sheath remained tethered to the loop. We then produced MEBs just below the lambda point of the SF helium with the idea that this might damp the oscillations. This procedure succeeded in producing long-lived MEBS. In Fig.2 we

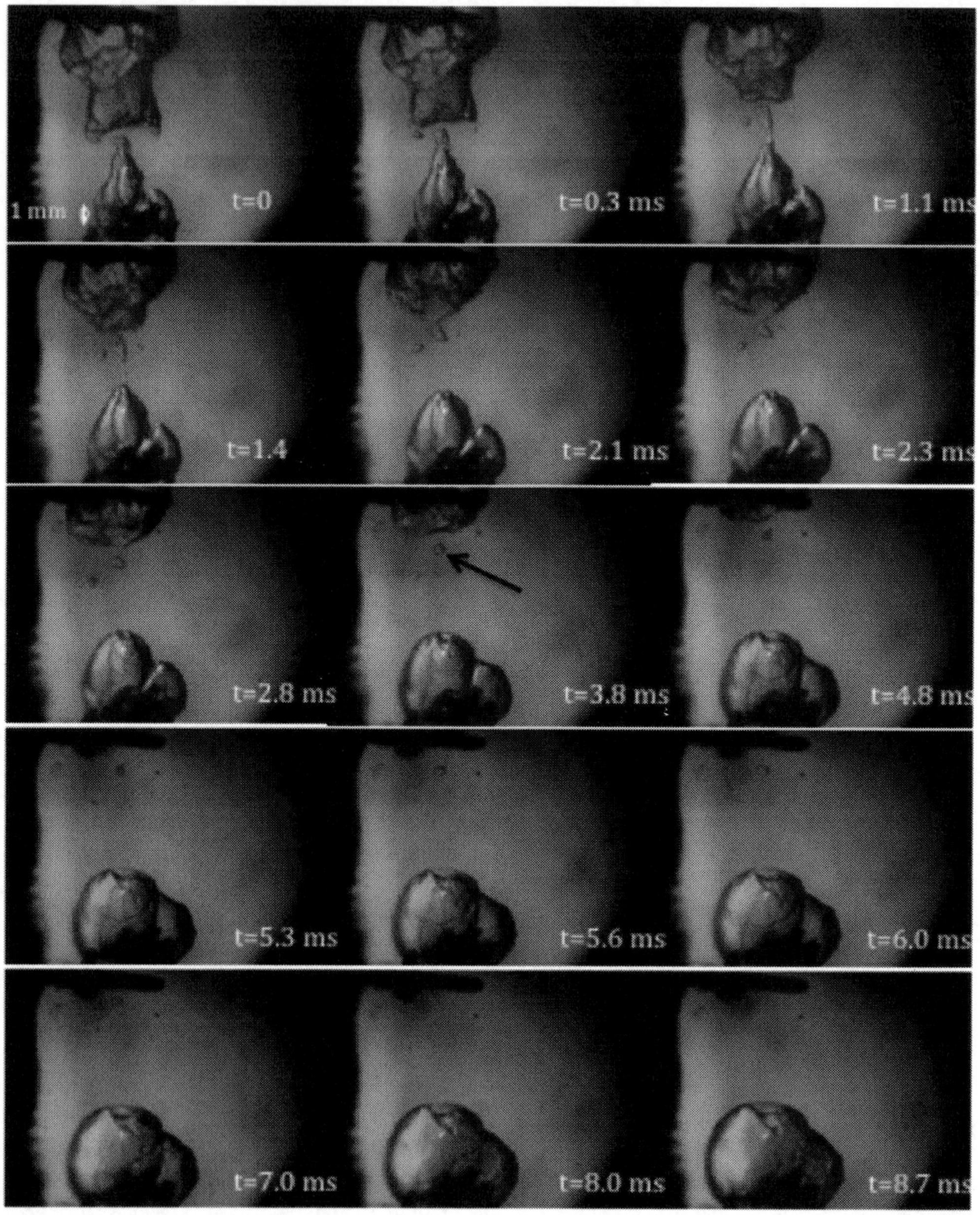

Figure 2 Selected frames from a high-speed video of stabilized MEBs. The arrow at 3.8 ms indicates one of the MEBs.

show a sequence taken at 10,000 frames/s. The filament itself was somewhat pear shaped and hottest at the top. The filament was biased at -3 KV and was below a ring electrode biased at +2 KV. We also attached metallic carbon nano tubes to the filament that produce more electrons in the sheath by field emission. In the figure one sees large hollow "cloud" that breaks loose and rises above the filament. Smaller MEBs are seen below this cloud that rapidly relax towards a spherical shape. These MEBs eventually rise out of the field of view in the superfluid helium. However, these are of a substantially different nature than earlier ones created at lower temperatures, i.e. these have smooth surfaces without oscillations and are approaching spherical shapes. This procedure was reproducible, and in one case an MEB was observed for ~30 ms, before it rose out of the field of view of our microscopy.

We have successfully produced long-lived MEBS that will rise in the SF fluid. The next step is to capture such MEBs with an radio frequency quadrupole trap that has been designed to be somewhat robust in that it can trap MEBs with a rather broad distribution of the number of electrons. To operate such a trap an electric pulse will be used to stop the MEBs as they enter the trap, to remove their kinetic energy. After capture detailed studies can be carried out on the MEBs. We note that Joseph et al [9] have reported observation of MEBs above the lambda point.

In conclusion, we have found that MEBs created around the lambda point of SF helium rapidly relax towards a spherical, and stable condition. The special properties of SF helium near the lambda point are a large heat capacity and large density of superfluid vortices that may play a role in the observed damping of large amplitude oscillations. These MEBs were produced in a negative pressure environment. It is not yet clear that this is an absolutely necessary condition, as earlier experiments showed that MEBs were produced for a few seconds of observation in a positive pressure environment.

Partial support for this research was provided by a grant from Alby and Kim Silvera. JT received support from the Fund for Scientific Research – Flanders, FWO projects G. 0115.12N and G.0122.12N. The high-speed video camera was provided by the Center for Nanoscale Systems (CNS), which is supported by the NSF award no. ECS-0335765. CNS is part of Harvard University.

[1] A. P. Volodin, M. S. Khaikin, and V. S. Edel'man, JETP Lett. **26**, 543 (1978).
[2] V. B. Shikin, JETP Lett. **27**, 39 (1978).
[3] U. Albrecht, and P. Leiderer, Europhys. Lett. **3**, 705 (1987).
[4] U. Albrecht, and P. Leiderer, J. Low Temp. Phys. **86**, 131 (1992).
[5] J. Tempere, I. F. Silvera, and J. T. Devreese, Surface Science Reports **62**, 159 (2007).
[6] W. T. Sommer, Phys. Rev. Lett. **12**, 271 (1964).
[7] J. Fang, A. E. Dementyev, J. Tempere, and I. F. Silvera, Rev. Sci. Inst. **82**, 033904 (2011).
[8] I. F. Silvera, and J. Tempere, Phys. Rev. Lett. **100**, 117602 (2008).
[9] E. M. Joseph, V. Vadakkumbatt, A. Pal, and A. Ghosh, J. Low Temp. Phys. **173** (2013).

Journal of Physics: Conference Series **568** (2014) 012017 doi:10.1088/1742-6596/568/1/012017

Lattice Relaxation of ^{4}He with ^{3}He Impurities: NMR Studies

D. Candela

Department of Physics, University of Massachusetts, Amherst, Massachusetts, MA 01003, USA

C. Huan, S. S. Kim, L. Yin, J. S. Xia, N. S. Sullivan

Department of Physics, University of Florida, Gainesville, Fl 32611, USA

E-mail: sullivan@phys.ufl.edu

Abstract.

Measurements of the ^{3}He nuclear spin relaxation times of dilute ^{3}He impurities in solid ^{4}He have been used to explore the unusual dynamics of solid 4He at low temperatures. The ^{3}He impurities move through the lattice by quantum mechanical exchange with neighboring ^{4}He atoms. Because of the larger zero point motion of the ^{3}He atoms, there is an appreciable lattice distortion that accompanies the tunneling ^{3}He atom and the tunneling motion depends on the elastic properties of the ^{4}He lattice. This motion modulates the ^{3}He-^{3}He nuclear dipole-dipole interactions and thus determines the NMR relaxation rates. We compare the observed temperature dependence of the NMR relaxation rates with that expected from the measurements of the shear modulus by Syshchenko *et al.* [Phys. Rev. Lett. 104, 195301 (2009)].

1. Introduction

Great interest has been generated by the observation of a 'giant plasticity' in solid ^{4}He by Rojas *et al.*[1, 2] that has been attributed to a remarkable mobility of dislocations in the solid at low temperatures. It was shown[3] that for very pure single crystals the dislocations can move over large distances and produce large changes in both the elastic shear modulus and the dissipation that is associated with the motion of the dislocations. This observation was preceded by the detection of a significant change in the shear modulus of solid ^{4}He[4–6]. The unusual plasticity has been invoked[3] to explain some of the previous anomalies reported for solid ^{4}He including the detection of non-classical rotational inertias in the same temperature region[7–13] and unexpected peaks in the sound attenuation[14, 15]. For recent reviews see Balibar[16] and Chan *et al.*[17].

We have carried our NMR studies of the motion of ^{3}He impurities in solid ^{4}He as a new non-invasive means of studying these unusual properties of the ^{4}He solid[18]. The ^{3}He impurities tunnel by particle-particle exchange through the lattice[19, 20] but because of the larger zero point motion of the ^{3}He atom compared to ^{4}He atoms, the lattice surrounding the impurity is distorted and the lattice must relax as these deformations move. Studies of the microscopic ^{3}He motion using NMR techniques can therefore provide a sensitive probe of any change in the elastic properties of the lattice.

2. Diffusion of Impurities in Quantum Solids

The atoms in quantum solids such as ^{4}He are weakly localized with respect to their lattice sites because of the relatively high zero point kinetic energy and the weak Van der Waals binding forces and can therefore exchange sites with nearest neighbors by quantum tunneling[21]. This effect leads to the multi-particle spin exchange in solid ^{3}He[22, 23] and the quantum diffusion of isotopic impurities such as ^{3}He in solid ^{4}He[24]. The diffusion of ^{3}He in solid ^{4}He has been studied extensively by Richards and colleagues[19, 20, 25–27] and by the team of Grigoriev[28] who interpreted the results in terms of a gas-like diffusion of the impurities with a tunneling frequency J_{34} and diffusion constant $D \sim zJ_{34}a_0^2/x_3\sigma$ where z is the lattice co-ordination number and σ the scattering cross-section for tunneling impurities. A full quantitative understanding for relatively high ^{3}He concentrations was not achieved until Landesman and others[29–31] realized that one had to account for the lattice distortion around the ^{3}He impurity that results from the larger zero point motion of the ^{3}He atom compared to the ^{4}He atoms in the lattice.

The elastic deformation around each impurity leads to a long range interaction $K_{ij} = K_0/r_{ij}^3$ between ^{3}He impurities[29]. The range of this interaction, R_K, is determined by the distance of closest approach for two impurities which occurs when the elastic energy difference is equal to the change in kinetic energy or $a_0|\nabla K(r)| = J_{34}$, for which $R_K = (\frac{3K_0}{J_{34}})^{1/4} = 8.2a_0$ where a_0 is the lattice spacing. The effective tunneling frequency $J_{eff} \sim J_{34}^2/K_0$ ~kHz. The mean separation between ^{3}He atoms is given by $R_M = x_3^{(-1/3)}a_0$, and for ^{3}He concentrations $x_3 > 180$ ppm, $R_K > R_M$, and the ^{3}He impurities are in constant (liquid-like) interaction. Experimental studies prior to 2010 were limited to concentrations $100 < x_3 < 10,000$ ppm and studied this constant interaction regime for which the the characteristic microscopic correlation time is given by $\tau_c^{-1} = 19\frac{J_{34}^2}{K_0}x_3^{-1/3}$ [30, 32]. The corresponding nuclear spin-lattice relaxation time induced by the tunneling is temperature independent and given by $T_1^{liq} = 0.86x_3^{-1/3}$.

The more interesting low concentration regime for $R_M \gg R_K$, corresponding to $x_3 \ll 100$ppm has only recently been studied in the experiments by Toda *et al.*[11, 33] and Kim *et al.*[18, 34, 35]. In this regime the separation of the ^{3}He atoms is much larger than the scattering length and collisions are rare. This is the coherent tunneling regime (originally postulated by Richards *et al.*[19]) where the group velocity of the ^{3}He impurity motion is $v_G = za_0J_{34}$. The characteristic time for a ^{3}He atom to tunnel between sites is given by $\tau^{coh} = (zJ_{34})^{-1}x_3^{1/3}$ for which the nuclear spin-lattice relaxation time is $T_1^{coh} = 2.410^{-3}x_3^{-4/3}$ for $J_{34}/(2\pi) = 0.6$ MHz in good agreement with experiments above 0.2 K. Below 0.2K we observe an anomalous peak in the nuclear spin-lattice relaxation close to where Beamish *et al.*[4, 6] observed a peak in the dissipation for the measurements of the shear modulus. Anomalous features in the attenuation of sound have also been observed[14], pointing to changes in the elasticity of the solid in this temperature region.

3. Experimental Considerations

The sample chamber consisted of a polycarbonate cylinder sealed at one end with a silver end piece that is an integral part of a cold finger extending from a nuclear demagnetization refrigerator. As shown in Fig 1. the helium sample is admitted from the opposite end of the cell which includes an *in situ* pressure gauge. A crossed-coil design was employed for NMR pulsed excitation and reception[34] with the excitation coil thermally isolated from the sample cell. The NMR spectrometer was a super-heterodyne design operating at 2.01 MHz. The latter was chosen to keep the expected nuclear spin-lattice relaxation times below 10^4 s. To enhance the signal/noise a low temperature RF preamplifier was deployed close to the sample chamber[36]. The amplifier was operated at 0.8K as determined by the bias current and was moved and thermally linked to the still of the dilution refrigerator when operating to reach the lowest temperatures of $\sim$ 10mK.

Journal of Physics: Conference Series **568** (2014) 012017 doi:10.1088/1742-6596/568/1/012017

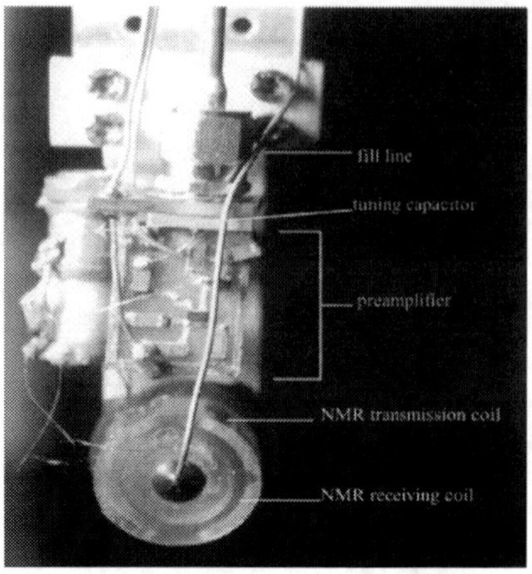

Figure 1. NMR cell and preamplifier. The NMR transmission coil overlays the receiving coil which is oriented perpendicular to the transmission coil and to the applied magnetic field.

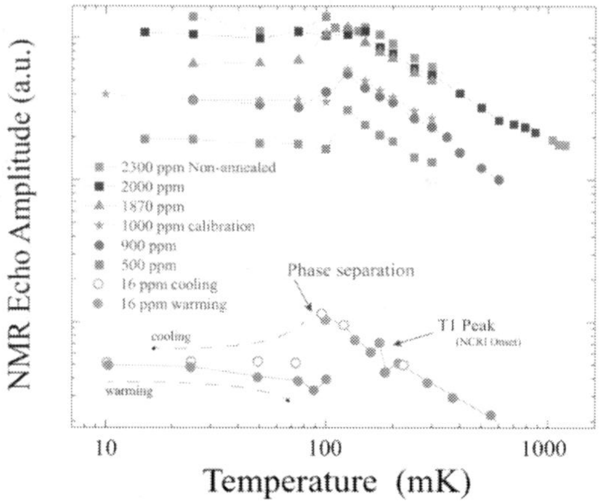

Figure 2. Temperature dependence of the NMR echo amplitude as a function of temperature for several ^{3}He concentrations. Clear phase transitions are observed with notable hysteresis for low concentrations.

Gas samples were prepared by mixing known gas volumes of ^{3}He and ^{4}He at room temperature and then condensed into the sample volume at the melting pressure corresponding to the desired molar volume. The sample cell was cooled very slowly through the melting curve as monitored by the pressure gauge, and the samples are formed by the blocked capillary method. Part of the ^{3}He composition is lost to wall adsorption during the condensation and the precise ^{3}He composition was determined from the NMR amplitudes. Fig. 2 shows typical results for the temperature dependence of the NMR amplitudes for several samples, illustrating the Curie law behavior at high temperatures followed by the formation of Fermi liquid droplets at very low temperatures[35]. Gas samples prepared with 30 ppm and 20 ppm of ^{3}He were found, for example, to have only 26 ppm and 16 ppm concentrations, respectively, in the solid sample.

4. Results

The results for measurements of both the nuclear spin-lattice times, T_1, and spin-spin relaxation times, T_2, as a function of temperature are shown in Fig. 3 for a sample with a ^{3}He concentration of 16 ppm. A sharp change in the value of T_1 by a factor of 30 is seen at the phase separation temperature of 90 mK. This transition temperature is in good agreement with the theoretical predictions of Balibar and Edwards[35, 37]. Above the transition temperature we expected a temperature independent relaxation as predicted for the coherent quantum tunneling of the isotopic impurities[29, 30] and as observed at much higher concentrations[20, 27].

The surprising result was a large peak (factor of 2 increase) in the spin-lattice relaxation time near 180 mK and accompanied by a less clear dip in the magnitude of the spin-spin lattice relaxation time in the same temperature range. We also show results for the spin-lattice relaxation time for 500 ppm for which no strong evidence of a peak is seen, consistent with the results of Schratter *et al.*[20]. The observations for the dilute sample with 16 ppm ^{3}He imply an increase at $T \sim 180$ mK in the low frequency component of the spectral densities, $J(\omega)$, of the fluctuations in the nuclear spin-spin interactions that are responsible for the relaxation. The

Journal of Physics: Conference Series **568** (2014) 012017 doi:10.1088/1742-6596/568/1/012017

increase in $J(0)$ would result in a decrease in the spectral density at the Larmor frequency as $J(\omega)$ is normalized. Since $T_2^{-1} \propto J(0)$ and $T_1^{-1} \propto J(\omega_L)$ this change in the spectral densities would lead to a peak in the value of T_1 with a concomitant decrease in the value of T_2. The value of T_2 at low temperatures is higher than the values for $T > 0.2$K, suggesting that some of the quantum tunneling that narrows the line width is reduced by scattering or other mechanisms at high temperatures. The origin of this feature is not understood.

It is worth noting that because of the very long relaxation times it takes about one day to measure each single point in Fig. 3 and for that reason the number of data points for a given sample is limited. In Fig. 4 we also show the results for a 24 ppm sample that was never cooled below the phase transition temperature. The peak in T_1 is only seen for low ^{3}He concentrations when the mean separation is larger than the range of the elastic deformation surrounding each impurity.

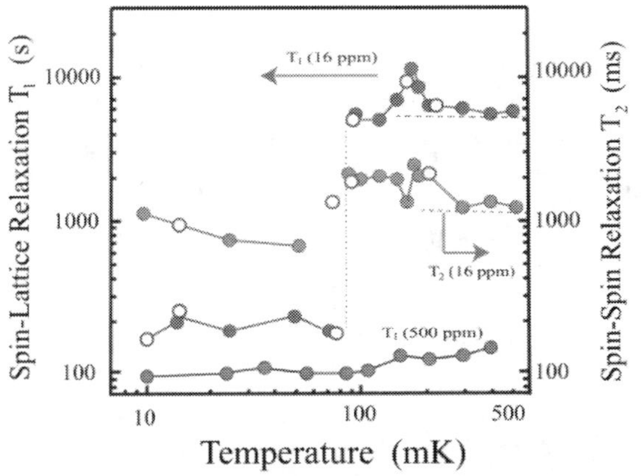

Figure 3. Temperature variation of the nuclear spin-lattice and spin-spin relaxation times for samples with helium concentrations of 16 and 500 ppm (closed circles for cooling, open circles for warming). The dashed lines show the temperature independent predictions for quantum tunneling of the ^{3}He impurities. The dotted line marks the phase separation.

5. Phenomenological model

The nuclear spin-lattice and spin-spin relaxation times are determined by the fluctuations in the ^{3}He nuclear spin dipole-dipole interactions that result from the tunneling of the ^{3}He atoms. Because of the lattice deformation surrounding a ^{3}He atom, the relaxation of the lattice deformation as the ^{3}He atom moves and thus the elasticity of the solid, play an important role in the relaxation process. The interaction between the ^{3}He atoms at sites i and j due to the elastic deformation has been given by Landesman[29] and Huang *et al.*[31] as

$$K_{ij}(r) = K_0(1 - 3\cos^2\theta_{ij})r_{ij}^{-3} \tag{1}$$

where $K_0/(2\pi) \sim$1200 MHz[29]. If we designate τ_j as an operator for the creation of a deformed site j occupied by a ^{3}He atom, we need to consider the correlation function

$$\langle \tau_j(t)\tau_k(0)\rangle = \left\langle j \left| \exp[-i\sum_p (K_{ip} - K_{pk})t] \right| k \right\rangle = \prod_p \langle j \left|\cos[(K_{jp} - K_{pk})t]\right| k\rangle \tag{2}$$

where the sum over p will only be significant for near neighbors. The frequency scale for this correlation function is set by the difference $K_{ip} - K_{pj}$ which for nearest neighbors is given by $\omega_p = 3\sqrt{\frac{8}{3}x_3^{4/3}}K_0$, the root mean square of the gradient of $K(r_{ij})$. For $K_0/2\pi = 1200$MHz we find $\omega_p/(2\pi) = 2.4$kHz.

The relevant correlation time for determining the nuclear spin-lattice relaxation times is evaluated by completing the spatial average of Eq. 2 and integrating over time. Because of the rapid oscillation of the terms $\cos[K_{ip} - K_{pj}]t$ the terms cancel except for very small t. To carry out the evaluation, the cosine terms are therefore expanded in t and the average taken over spatial co-ordinates using Anderson's method[38] before integrating. This method provides a reasonable evaluation of the high frequency component of the spectral densities but it does not account for the low frequency component such as that responsible for the dissipative term of the shear modulus[4, 6]. It is this missing component of the spectral density that we believe is responsible for the anomalous peak seen in the nuclear spin-lattice relaxation near 0.2 K.

In the absence of a microscopic theoretical model for the lattice behavior we use a phenomenological model by separating the time domain of the integration of Eq. 2 into two components: (i) the short time regime considered above, and (ii) a new long term (low frequency) additional component. We approximate this additional term by replacing the oscillatory terms by the average $\cos(\omega_p t)$ and adding a phenomenological decay term $\tau_L = \tau_0 \exp(\Delta/T)$ analogous to that invoked by Beamish *et al.*[4, 5]. The additional term in the correlation time is given by

$$\tau_{RL} = \int \langle\langle \tau_J(t)\tau_k(0)]\rangle\rangle_{avg} = \Re \int \langle j|k\rangle \exp[i\omega_p t - t/\tau_L]dt = R_0^{-1}[\frac{u}{1+u^2}] \qquad (3)$$

where $u = z\omega_p\tau_L$ and R_0 is the phonon induced transition rate that includes the lattice relaxation. We estimate R_0 from Abragam's treatment[39] of single phonon induced transitions (Eq. 83 of Ref. [39]) and find $R_0^{-1} = 2600T^{-1}$ s for $x_3 = 16$ ppm. The best fit shown in Fig. 4 for the total relaxation (adding τ_{RL} to the temperature independent coherent relaxation) is given by $R_0^{-1} = 1610T^{-1}$ s with $\tau_0 = 18$ ns. and $\Delta = 1.01$ K. This fit implies a value $K_0/(2\pi) = 1350$kHz, slightly higher than that cited by Landesman[29]. The dotted line is the calculated variation for $\Delta = 0.67$ K, the mean value for binding energy for ^{3}He atoms on dislocations[3] .

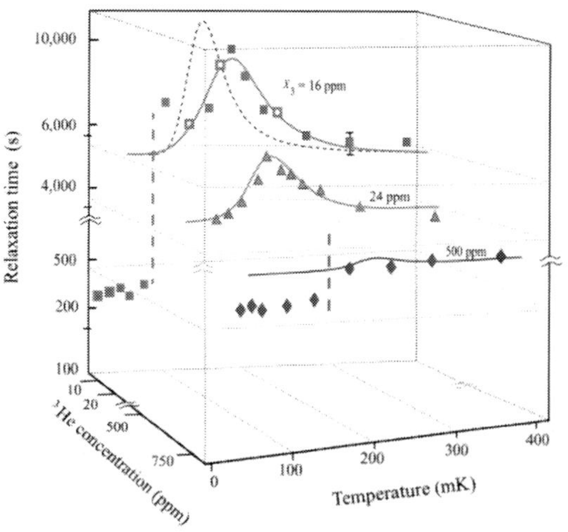

Figure 4. Comparison of the temperature dependence of the nuclear spin-lattice relaxation times for dilute ^{3}He in solid ^{4}He for concentrations $x_3 = 16$, 24 and 500 ppm ^{3}He with phenomenological model for lattice relaxation. A typical error bar is shown for the 16 ppm data. The broken vertical bars mark the phase separation into small Fermi liquid droplets. The 24 ppm sample was never cooled below the phase separation temperature.

6. Discussion

The value inferred for the activation energy of the relaxation process, $\Delta = 1.01$ K, is appreciably larger than the typical energy deduced by Fefferman *et al.*[3] for the binding of ^{3}He atoms to dislocations in very pure crystals (0.3 ppm ^{3}He) and much larger than that observed by Beamish *et al.* [4, 5] for the anomalous peaks in the shear modulus for crystals of similar quality. This is not necessarily surprising for the high ^{3}He concentration of the samples used for the NMR studies. For the expected concentrations of dislocations,[40, 41] less than 1 ppm of ^{3}He would saturate all dislocations and their effect would not be seen in the current NMR experiments.

The mechanism for the thermal activation in these samples is therefore expected to be associated with a different physical process, possibly the motion of grain boundaries in solid ^{4}He and a deeper theoretical understanding of grain boundary dynamics is needed to determine if this is the case.

7. Acknowledgments

We thank Sebastien Balibar, John Beamish, Pradeep Kumar and Alan Dorsey for many useful comments. The experiments were carried out at the High B/T Facility of the National High Magnetic Field Laboratory supported by the National Science Foundation DMR-1157490 and the State of Florida. NS also gratefully acknowledges support from the National Science Foundation through the award DMR-1303599.

References

[1] Rojas X, Haziot A, Bapst V, Balibar S and Maris H J 2010 *Phys. Rev. Lett.* **105**(14) 145302
[2] Haziot A, Rojas X, Fefferman A D, Beamish J R and Balibar S 2013 *Phys. Rev. Lett.* **110**(3) 035301
[3] Fefferman A D, Souris F, Haziot A, Beamish J R and Balibar S 2014 *Phys. Rev. B* **89**(1) 014105
[4] Day J and Beamish J 2006 *Phys. Rev. Lett.* **96** 105304–105307
[5] Syshchenko O, Day J and Beamish J 2009 *J. Phys. (Cond. Matt.)* **21** 164204–164209
[6] Syshchenko O, Day J and Beamish J 2010 *Phys. Rev. Lett.* **104** 195301
[7] Kim E and Chan M H W 2004 *Nature* **427** 225–227
[8] Kim E and Chan M H W 2004 *Science* **305** 1941–1944
[9] Clark A C, West J T and Chan M H W 2007 *Phys. Rev. Lett.* **99** 135302
[10] Reppy J D 2010 *Phys. Rev. Lett.* **104** 255301(1)–255301(4)
[11] Toda R, Onoe W, Kanemoto M, Kakuda T, Tanaka Y and Sasaki Y 2011 *J. Low Temp. Phys.* **162** 476–482
[12] Gumann P, Keiderling M C, Ruffner D and Kojima H 2011 *Phys. Rev.* **B83** 224519–224528
[13] Gumann P, Keiderling M C and Kojima H 2012 *J. Low Temp. Phys.* **112** 162–174
[14] Goodkind J M 2002 *Phys. Rev. Lett.* **89** 095301–095304
[15] Ho P C, Bindloss I P and Goodkind J M 1997 *J. Low Temp. Phys.* **109** 409–421
[16] Balibar S 2010 *Nature* **464** 176
[17] Chan M W H, Hallock R B and Reatto L 2013 *J. Low Temp. Phys.* **172** 317–363
[18] Kim S S, Huan C, Yin L, Xia J S, Candela D and Sullivan N S 2013 *Phys. Rev. B* **87**(22) 224303
[19] Richards M G, Pope J and Widom A 1972 *Phys. Rev. Lett.* **29** 708–711
[20] Schratter J, Allen A R and Richards M G 1984 *J. Low Temp. Phys.* **57** 179–187
[21] Guyer R A, Richardson R C and Zane L I 1971 *Rev. Mod. Phys.* **43** 532
[22] Roger M, Hetherington J H and Delrieu J M 1983 *Rev. Mod. Phys.* **55**(1) 1–64
[23] Bäuerle C, Chen A S, Bunkov Y M, Godfrin H and Roger M 2008 *J. Low Temp. Phys.* **113** 287–292
[24] Andreev A F and Lifshitz I M 1969 *Sov. Phys. JETP* **29** 1107–1103
[25] Greenberg A S, Thomlinson W C and Richardson R C 1972 *J. Low Temp. Phys.* **8** 3–28
[26] Richards M G, Pope J, Tofts P S and Smith J H 1976 *J. Low Temp. Phys.* **24** 1
[27] Allen A R, Richards M G and Schratter J 1982 *J. Low Temp. Phys.* **47**(3) 289–320
[28] Grigoriev V N, Esel'son B N, Mikheev V A, Slusarev V A, Strzhemechny M A and Shulman Y E 1973 *J. Low Temp. Phys.* **13** 65–79
[29] Landesman A 1975 *Physics Letters A* **54** 137–139
[30] Landesman A and Winter J M 1974 Diffusion of ^{3}He impurities in ^{4}He *Proceedings Low Temperature Physics-LT-13* (Plenum, New York) pp 73–78
[31] Huang W, Goldberg H A and Guyer R A 1975 *Phys. Rev. B* **11** 3374–3392
[32] Sullivan N S, Candela D, Kim S S, Huan C, Yin L and Xia J S 2014 *J. Low Temp. Phys.* **174** 133–139
[33] Toda R, Gumann P, Kosaka K, Kanemoto M, Onoe W and Sasaki Y 2010 *Phys. Rev. B* **81** 214515
[34] Kim S S, Huan C, Yin L, Xia J S, Candela D and Sullivan N S 2011 *Phys. Rev. Lett* **106** 185303(1) –185303(4)
[35] Huan C, Kim S S, Yin L, Xia J S, Candela D and Sullivan N S 2011 *J. Low Temp. Phys.* **162** 162–173
[36] Huan C, Kim S S, Phelps L, Xia J S, Candela D and Sullivan N S 2010 *J. Low Temp. Phys.* **158** 692–696
[37] Edwards D O and Balibar S 1989 *Phys. Rev. B* **39** 4083–4097
[38] Anderson P W 1951 *Phys. Rev.* **82** 167
[39] Abragam A 1961 *Nuclear Magnetism* (Oxford: Clarendon Press)
[40] Paalanen M A, Bishop D J and Dail H W 1981 *Phys. Rev. Lett.* **46** 664–667
[41] Iwasa I 2010 *Phys. Rev.* **B81** 104527

Journal of Physics: Conference Series **568** (2014) 012018 doi:10.1088/1742-6596/568/1/012018

NMR Studies of ^{3}He Films on Boron Nitride

Y. Tang, N. S. Sullivan

Department of Physics, University of Florida, Gainesville, Fl 32611, USA

E-mail: `sullivan@phys.ufl.edu`

Abstract. We report the results of NMR studies of the dynamics of ^{3}He adsorbed on hexagonal boron nitride. These studies can identify the phase transitions of the 2D films as a function of temperature. A thermally activated temperature dependence is observed for $2.6 < T < 8$ K compared to a linear temperature dependence for $0.7 < T < 2.6$ K. This linear dependence is consistent with that expected for thermal diffusion in a fluid for coverages of 0.4 - 0.6 of a monolayer.

1. Introduction

Helium-three monolayer and submonolayer films are of great interest because the different phases and microscopic dynamics that occur are dominated by quantum effects that lead to new phenomena in reduced dimensions at low temperatures[1, 2]. In addition the films are ideal for testing our understanding of the quantum dynamics and spin ordering in two dimensions from first principles[3]. At low temperatures it had been postulated that the large quantum mechanical zero-point kinetic energy of the ^{3}He atoms could be sufficient to prevent the system from condensing as a 2D liquid in some scenarios[4], leading to a unique quantum gas at extremely low temperatures. Experimental studies by Bhattacharyya and Gasparini [5, 6] offered evidence for such a phase but the majority of experiments [7–9] and the most recent theoretical treatments[10–12] do not support the existence of a low temperature gas phase. Recent thermodynamic measurements by Sato *et al.* [13, 14], however, for ^{3}He adsorbed on bare graphite (and on ^{4}He plated graphite), have provided new evidence in support of a self-bound state for very low coverages at very low temperatures. We have carried out NMR studies at much higher coverages of 0.41 and 0.62 of a full monolayer and at higher temperatures to determine the microscopic dynamics of the ^{3}He atoms and the mechanisms of diffusion at coverages near the fluid-solid phase transition region[12, 15]. At these coverages NMR studies by Crane and colleagues[16] for ^{3}He on BN showed evidence of a mixed coverage of fluid and solid components with the solid melting near 1.6K to form a dense fluid state with appreciable orientational order imposed by the substrate. We have carried out studies using a much higher NMR frequency which should help distinguish between the fluid and solid components.

In addition to measuring the diffusion in the fluid phase, NMR studies can be used to determine the spin exchange rates in the solid phases. Below monolayer coverage, the ^{3}He monolayer forms a commensurate $\sqrt{3} \times \sqrt{3}$ solid structure that provides a unique testing ground for the multi-particle exchange model [17, 18] invoked to explain the magnetic properties of bulk ^{3}He and ^{3}He films [15, 19, 20]. For example, coverages less than the ideal commensurate structure lead to a reduced probability of 3-particle exchange compared to the 2-particle exchange and

Figure 1. Electron micrograph of hexagonal BN sample.

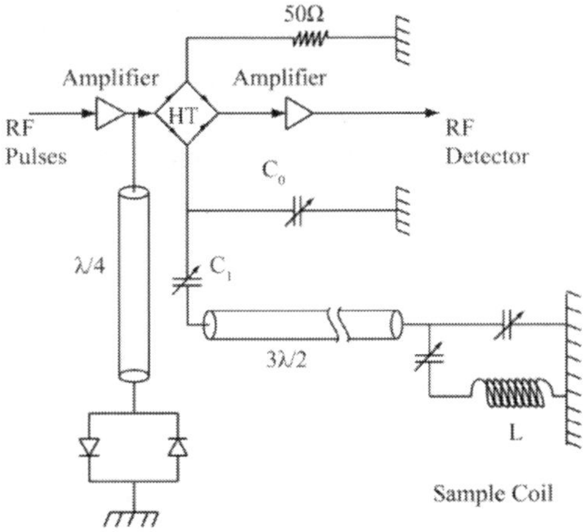

Figure 2. Sketch of NMR circuit. HT is a quadrature hybrid tee bridge.

this results in a much higher effective magnitude for the overall tunneling rate that can be seen in the ^{3}He nuclear spin lattice relaxation times[21].

2. Experimental considerations

We used powdered hexagonal boron nitride (h-BN)[22] as the substrate for the helium film[23]. h-BN has similar structure and dimensions to graphite but is an insulator rather than a semimetal and has a smaller magnetic susceptibility ($-4.8\ 10^{-7}$ emu/g)[24] than graphite ($-170\ 10^{-7}$ emu/g)[25]. The electron micrograph of Fig. 1 shows that the substrate material consists of an assembly of small platelets of sizes in the 0.5 - 2.0 μm range and thicknesses of 0.05 - 0.10 μm. The samples were prepared by baking under vacuum[26, 27] before being introduced to the sample chamber which contained a brush of fine copper wires that penetrated the powdered h-BN to ensure thermal contact with a rod extending from a dilution refrigerator. A precise amount of ^{3}He gas was introduced to the sample to produce the desired coverage which was inferred from previous adsorption isotherm studies used to characterize the same material[26].

Pulsed NMR techniques were used to measure the ^{3}He nuclear spin-spin and spin-lattice relaxation times. The NMR setup shown in Fig. 2 uses a tuned system matched at low temperatures to close to 50 Ω. A $3\lambda/2$ 50 Ω cable was used to connect the tuned circuit to a hybrid-tee bridge at room temperature that had a second fine tuning stage. The latter stage was important to correct for the small shift in tuning as the sample cell was cooled. A high magnetic field, corresponding to a Larmor frequency $\omega_L/(2\pi) = 200$ MHz, was chosen so that ω_L was much larger than the typical motional frequencies[28] of the ^{3}He atoms. For this choice the relaxation times of the solid component are expected to be much higher than the fluid component as the exchange induced relaxation time in the solid grows exponentially with Larmor frequency[28]. This choice therefore simplifies the analysis of the relaxation data. (Crane *et al.*[16] used Larmor frequencies of 2.7 and 4.5 MHz.) The relaxation times were measured using standard $90^\circ - \tau - 90^\circ$ and $90^\circ - \tau - 180^\circ$ pulse sequences followed by signal averaging (50 - 5000 times) to obtain adequate signal/noise ratios.

3. Results

The decay of the NMR echoes following repeated pulse sequences as a function of repeat time τ_r show two distinct components, a fast component with a relatively large amplitude and a slow component with a much smaller amplitude. A typical decay for the echo amplitude $E(\tau_r)$ is shown in Fig. 3 for $x = 0.62$ at $T = 1.9$K. These two components are expected from the results of Crane[16] for this temperature range if the ^{3}He forms a mixed state of fluid and solid. The surprising result from this study is the relatively small amount of solid component for temperatures $T > 0.7$K. From the flat geometry of the BN platelets we would expect an adsorbed solid component at the edges (where there are much stronger adsorption sites) to correspond to $\sim 5\%$ of the total area. This "edge" component would appear to account for an appreciable fraction of the long relaxation time component.

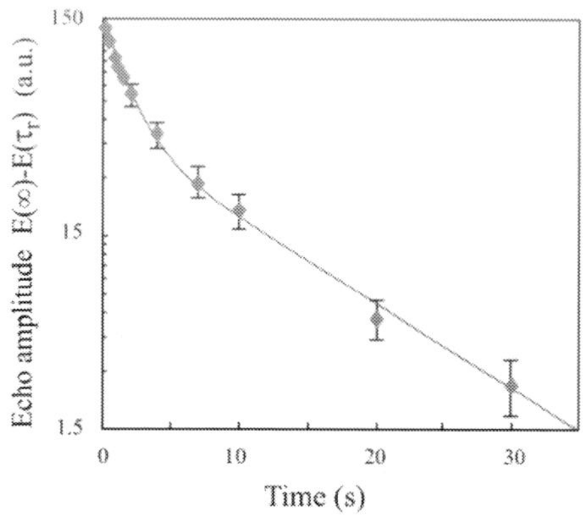

Figure 3. Typical decay observed for the NMR echo amplitudes for $x = 0.62$ at 1.9 K with $E(\infty) - E(\tau_r) = 121e^{(-\tau_r/1.57)} + 29e^{(-\tau_r/11)}$ shown by the solid line. The fast decay is attributed to a fluid component and the slower decay to a solid component.

The observed temperature dependence of the nuclear spin-lattice relaxation times for the fast decay component of the relaxation is shown in Fig. 4 for two different coverages, 0.62 and 0.41 of a full monolayer. Three different temperature regimes are observed. In the high temperature regime, $2.6 < T < 8$ K, there is a clear thermal activation with an energy $E_A = 31 \pm 2$ K, corresponding to the binding of the ^{3}He atoms to the surface and in good agreement with other studies[29]. The sharp minimum for the 0.62 coverage is a standard BPP minimum[30] given by $T_{1(min)} \approx \omega_L/M_2$ where $M_2 = 1.8 \times 10^9$ s^{-2} is the calculated NMR second moment for a full monolayer of ^{3}He in BN (and includes the ^{3}He-B and ^{3}He-N nuclear spin-spin interactions).

The most interesting and unexpected region is for intermediate temperatures, $0.7 < T < 2.6$ K, for which a linear dependence on temperature is observed, and most remarkably there is very little variation in the magnitude of T_1 for an appreciable change in the amount of ^{3}He added to the total surface coverage. These two features support a model where the surface is covered with patches of fluid of constant areal density but varying in overall size and co-existing with solid patches. Following Crane[16] the patches are believed to be nucleated at specific sites and grow continuously as the temperature decreases. If this region consisted of continuous fluid covering all the surface area of the h-BN platelets one would expect a very strong change in the absolute value of T_1 with ^{3}He coverage since the NMR second moment M_2 scales as $\sum_{pairs(ij)} R_{ij}^{-6}$. These studies show that the fluid component is close to constant in density but it is distinct from the dense self-bound liquid state observed by Sato *et al.*[13] for very low coverages. At low temperatures, $T < 0.7$ K, the solid component begins to dominate the relaxation, and we observe the onset of a stronger temperature dependence (shown by the dotted line in Fig. 4) which is attributed to the relaxation determined by the solid ^{3}He to substrate interactions.

4. Discussion

The nuclear spin-lattice relaxation time T_1 is given by[31]

$$\frac{1}{T_1} = \frac{2}{3} M_2 \sum_n n^2 J_n(n\omega_L)$$

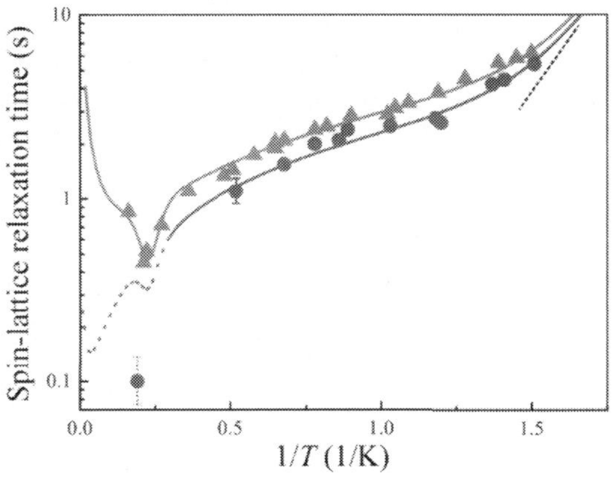

Figure 4. Temperature dependence of the nuclear spin-lattice relaxation times for ^{3}He monolayers on BN for different coverages x; red triangles $x_1 = 0.62$, blue circles $x_2 = 0.41$. Typical error bars are shown. The dotted line at high temperatures indicates a region of very poor S/N at high temperatures. The broken line at low temperatures marks the onset of a purely solid phase characteristic.

where the $J_n(\omega)$ are the normalized spectral densities of the fluctuations of the ^{3}He-^{3}He dipole-dipole interactions at frequency ω. (Note that for the spectral densities we neglect the contributions of unlike spin-spin interactions such as ^{3}He -^{14}N because, (i) they are reduced relative to the ^{3}He-^{3}He contribution by $[\gamma(^{14}\mathrm{N})/\gamma(^3\mathrm{He})]^2 \sim 10^{-2}$ and $[\gamma(^{11}\mathrm{B})/\gamma(^3\mathrm{He})]^2 \sim 10^{-1}$, and (ii) as a result of their long relaxation times in the solid platelet, their spin magnetization would be saturated and thus ineffective[32] as far as the ^{3}He relaxation is concerned. If J_n can be characterized by a simple correlation time, τ, then we have a simple Lorentzian $J(\omega) = \tau/[1 + (\omega\tau)^2]$. We have analyzed the data in terms of a sum of three relaxation rates $T_1^{-1} = T_{1H}^{-1} + T_{1I}^{-1} + T_{1L}^{-1}$ for the high, intermediate and low temperature regions, respectively. For the high temperature gas-like region we used a Lorentzian form with $\tau_H = 6.8 \times 10^{-8} \exp(31/T)$ which leads to the BPP minimum at 7.1 K. For temperatures below the minimum we can take $\omega_L\tau > 1$ and analyze the lower temperature data in terms of $T_1^{-1} = (4M_2/3\omega_L^2)[\tau_I^{-1} + \tau_L^{-1}]$. The best fit to the data for the intermediate temperature range is shown by the solid lines in Fig. 4. For this fit $\tau_I = (1.7 \pm 0.5)T^{-1}10^{-9}$ s (red line of Fig. 4) for coverage $x_1 = 0.62$, and $(1.4 \pm 0.5)T^{-1}10^{-9}$ s (blue line of Fig. 4) for coverage $x_2 = 0.41$. The linear temperature dependence is consistent with that reported by Crane (Fig. 7.2.2.14 of Ref. [16]) and interpreted in terms of a fluid component. The low temperature data had poor signal/noise and was fit empirically to $\tau_L = 2.1 \times 10^{-10}T^3$.

The most recent calculation of the spin diffusion for a fluid Fermi system in two dimensions has been given by Bruun[33] who finds a diffusion constant $D = k_BT/(4\pi^2 n\hbar I_D)$ where n is the areal density and $I_D = 2/[(\ln(E_B/k_BT) - 0.42)^2 + \pi^2] \sim 0.22$. The characteristic time $\tau_c = r_n^2/D$ where r_n is the mean separation of the ^{3}He atoms. We find from Bruun's theory $\tau_c = 2.4\,T^{-1}10^{-9}$ s in reasonable agreement with the experimental results for τ_I. The important point in this analysis is that both τ_I and M_2 are independent of the fractional coverage if the fluid forms patches (or puddles) of liquid. If the ^{3}He were to spread uniformly across the platelets in this temperature regime there would be a much stronger concentration dependence which is not observed.

5. Conclusion

The linear temperature dependence and very weak variation of the nuclear spin-lattice relaxation times at intermediate temperatures with the number of ^{3}He atoms adsorbed for submonolayer films with fractional coverages x = 0.41 and x= 0.62 on hexagonal boron nitride support the interpretation that this phase consists of patches of liquid ^{3}He at very close to constant areal density, with a small fraction of solid that becomes complete at 0.7K. Although the fluid density appears to be constant in this region there is no evidence that this is an incompressible liquid state such as reported for much lower coverages and at lower temperatures by Sato *et al.*[13].

Acknowledgments

We thank Efstratios Manousakis, Brian Cowan and Aldo Migone for many useful discussions. Support from the National Science foundation through the award DMR-1303599 is also gratefully acknowledged.

References

[1] Böhm H M, Krotschek E, Panholzer M, Godfrin H, Lauter H J and Meschke M 2010 *J. Low Temp. Phys.* **158** 194–200
[2] Nava M, Motta A, Galli D E, Vitali E and Moroni S 2012 *Phys. Rev. B* **85**(18) 184401
[3] Misguich G, Lhuillier C, Bernu B and Waldtmann C 1999 *Phys. Rev. B* **60**(2) 1064–1074
[4] Andreev A F and Kompaneets D A 1973 *Zh. Eksp. Teor. Fiz. Pis'ma Red.* **17** 376
[5] Bhattacharyya B K and Gasparini F M 1982 *Phys. Rev. Lett.* **49** 919
[6] Bhattacharyya B K and Gasparini F M 1985 *Phys. Rev. B* **31** 2719
[7] Greywall D S 1990 *Phys. Rev. B* **41** 1842–1862
[8] Schick M 1980 Theory of helium monolayers *Phase Transitions in Surface Films* ed Dash J G and Ruvalds J (New York: Plenum) p 68
[9] Godfrin H 1991 Adsorbed quantum gases *Excitations in Two-Dimensional and Three-Dimensional Quantum Fluids* ed Wyatt A G F and Lauter H J (New York: Plenum) p 445
[10] Grau V, Boronat J and Casulleras J 2002 *Phys. Rev. Lett.* **89** 045301(1)–045301(4)
[11] Um C, Kahng J, Kim Y, george T F and Pandey L N 1997 *J. Low Temp. Phys.* **107** 283–303
[12] Abraham F F and Broughton J Q 1987 *Phys. Rev. lett.* **59** 64–67
[13] Sato D, Naruse K, Matsiui T and Fukuyama H 2012 *Phys. Rev. lett.* **109** 235306(1)–235306(4)
[14] Sato D, Tsuj D, Takayoshi S, Obata K, Matsu T and Fukuyama H 2010 *J. Low Temp. Phys.* **158**(1) 201–206
[15] Fukuyama H 2008 *J. Phys. Soc. Jpn* **77** 111013(1)–111013(12)
[16] Crane T P 1998 *An NMR study of Helium-3 adsorbed on hexagonal boron nitride* Ph.D. thesis Royal Holloway University of London
[17] Roger M, Bauerle C M, Bunkov Y M, Chen A S and Godfrin H 1998 *Phys.Rev. Lett* **80** 1308–1311
[18] Godfrin H 1986 *Can. J. Phys.* **65** 1430–1434
[19] Elbs J, Winkelmann C, Bunkov Y M, Collin E and Godfrin H 2008 *J. Low Temp. Phys.* **148** 749–753
[20] Bäuerle C, Bunkov Y M, Chen A S, Cousins D J, Godfrin H, Roger M and Triqueneaux S 2000 *Physica B* **280** 95–99
[21] Parks C, Stachowiak P and Sullivan N S 2000 *J. Low Temp. Phys.* **121** 489–494
[22] Advanced Ceramics Corporation (Lot 4069) 11907 Madison Ave, Cleveland, OH 44107
[23] Shreshtha P, Alkhafaji M, Lokowitz M, Yang G and Migone A 1994 *Langmuir* **10** 3244–3246
[24] Zupan J, Komac M and Kolar D 1970 *J. Appl. Phys.* **41** 5337–5338
[25] Ganguli N and Krsihnan K S 1941 *Proc. Roy. Soc. A (Ln)* **177** 168–182
[26] Evans M D, Patel N and Sullivan N S 1992 *J. Low Temp. Phys.* **89** 653–657
[27] Crane T P and Cowan B P 2000 *Phys. Rev. B* **62** 11359–11362
[28] Guyer R A, Richardson R C and Zane L I 1971 *Rev. Mod. Phys.* **43** 532
[29] Crane T P and Cowan B P 2000 *Physica B* **284-288** 230–231
[30] Bloembergen N, Purcell E M and Pound R V 1948 *Phys. Rev.* **73** 679–716
[31] Cowan B P 1997 *Nuclear Magnetic Resonance and Relaxation* (Cambridge, UK: Cambridge University Press)
[32] Abragam A 1961 *Principles of Nuclear Magnetism* (Oxford, UK: Clarendon Press)
[33] Bruun G M 2012 *Phys. Rev. A* **85** 1–5

Spin-gap phenomenon in a strongly interacting ultracold Fermi gas

H. Tajima, R. Hanai, and Y. Ohashi

Department of Physics, Keio University, 3-14-1 Hiyoshi, Kohoku-ku, Yokohama,Kanagawa 223-8522, Japan

E-mail: `htajima@rk.phys.keio.ac.jp`

Abstract. We investigate magnetic properties of a strongly interacting ultracold Fermi gas. Within the framework of an extended T-matrix approximation, we calculate the spin susceptibility χ in the unitarity limit. We show that effects of pairing fluctuations on this magnetic quantity are quite different in between the normal state and the superfluid phase. In the normal state, pairing fluctuations cause spin-gap phenomenon near the superfluid phase transition temperature T_c, where χ is anomalously suppressed. In the superfluid phase, on the other hand, the ordinary suppression of χ by the BCS energy gap is weakened by pairing fluctuations, because they induce finite density of states inside the gap. Our results indicate that the spin susceptibility is a useful quantity for the study of pairing fluctuations in the BCS-BEC crossover regime of an ultracold Fermi gas.

1. Introduction

An ultracold Fermi gas is now widely recognized as a useful quantum many-body system[1, 2]. Indeed, using a tunable pairing interaction associated with a Feshbach resonance, we can systematically examine how a weak-coupling BCS-type Fermi superfluid continuously changes into the Bose-Einstein condensation (BEC) of tightly bound molecules[3, 4, 5, 6, 7, 8, 9, 10, 11]. In the intermediate coupling regime where strong pairing fluctuations exist, we can study strong-coupling physics beyond the mean-field level.

While the cold Fermi gas system has the above mentioned advantage, compared with the electron system, there are not so many experimental techniques to observe various physical quantities. Although this weak point is gradually being overcome by recent extensive experimental efforts[12, 13, 14, 15, 16, 17, 18], the current stage of cold Fermi gas physics still needs ideas to observe strong coupling phenomena within the current experimental techniques.

In this paper, we investigate the uniform spin susceptibility χ in a unitary Fermi gas both above and below T_c. This magnetic quantity has recently become possible to observe in this field[15, 16, 17, 18]. (Note that spins $\sigma =\uparrow,\downarrow$ in this system are actually *pseudospins* describing two atomic hyperfine states.) Since χ is sensitive to the formation of (pseudo)spin-singlet pairs, the observation of this quantity is expected to give useful information about pairing fluctuations. In this paper, including strong-coupling effects within the framework of an extended T-matrix approximation (ETMA)[19, 20, 21], we calculate χ both above and below T_c. In the normal state, we show that pairing fluctuations cause the spin-gap phenomenon, being characterized by the suppression of χ. In the superfluid phase, on the other hand, pairing fluctuations are found to enhance χ, reflecting that they smear the BCS excitation gap in the density of states. We

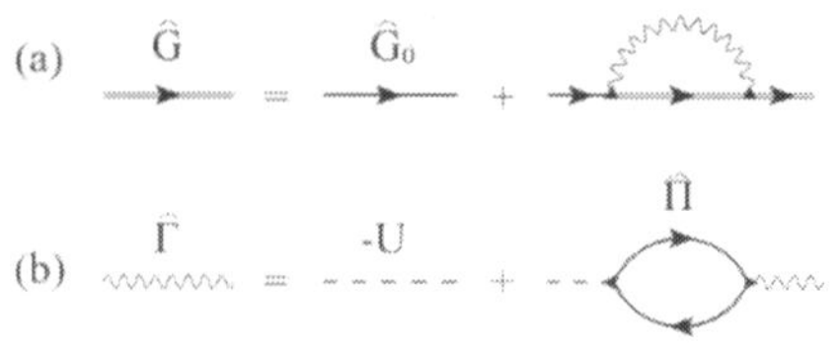

Figure 1. (a) Diagrammatic representation of the Dyson equation for the 2×2-matrix single-particle Green's function $\hat{G}$ (double-solid line). The solid line is the bare Green's function $\hat{G}^0$ with no self-energy correction. The wavy line is the particle-particle scattering matrix $\hat{\Gamma}$ which is diagrammatically given as panel (b). In panel (b), the dashed line is the pairing interaction $-U$, and $\hat{\Pi} = \{\Pi^{ij}\}$ is the pair-correlation function given in Eq. (5).

also compare our results with the recent experiment on a superfluid ^{6}Li Fermi gas. Throughout this paper, we take $\hbar = k_B = 1$, and the system volume is taken to be unity, for simplicity.

2. Formulation: Extended T-matrix approximation

We consider a two-component Fermi gas described by the ordinary BCS model. Since we deal with both the normal and superfluid phases, it is convenient to employ the Nambu representation[22]. In this formalism, the BCS Hamiltonian is written as[23, 24]

$$H = \sum_{\mathbf{p}} \Psi^\dagger_{\mathbf{p}} \left[\xi_{\mathbf{p},\uparrow} \frac{\tau_3 + 1}{2} + \xi_{\mathbf{p},\downarrow} \frac{\tau_3 - 1}{2} - \Delta\tau_1 \right] \Psi_{\mathbf{p}} - U \sum_{\mathbf{q}} \rho_+(\mathbf{q}) \rho_-(-\mathbf{q}). \tag{1}$$

Here, $\Psi^\dagger_{\mathbf{p}} = (c^\dagger_{\mathbf{p},\uparrow}, c_{-\mathbf{p},\downarrow})$ is the Nambu field, where $c_{\mathbf{p},\sigma}$ is the annihilation operator of a Fermi atom with pseudospin $\sigma =\uparrow, \downarrow$, describing two atomic hyperfine states. $\xi_{\mathbf{p},\sigma} = p^2/2m - \mu - \sigma h$ is the kinetic energy of a Fermi atom, measured from the Fermi chemical potential μ (where m is an atomic mass). In this paper, although we consider the unpolarized case, we have introduced an infinitesimally small effective magnetic field h in $\xi_{\mathbf{p},\sigma}$ to calculate the spin susceptibility. $-U$ is a pairing interaction, which is related to the s-wave scattering length a_s as $4\pi a_s/m = -U/[1 - U \sum_{\mathbf{p}}^{\omega_c} \frac{m}{p^2}]$ (where ω_c is a cut-off energy)[6]. In Eq. (1), $\rho_\pm(\mathbf{q}) = [\rho_1(\mathbf{q}) \pm i\rho_2(\mathbf{q})]/2$, where $\rho_j(\mathbf{q}) = \sum_{\mathbf{p}} \Psi^\dagger_{\mathbf{p}+\mathbf{q}/2} \tau_j \Psi_{\mathbf{p}-\mathbf{q}/2}$ ($j = 1, 2$) are the generalized density operators[23, 24] (where τ_j ($j = 1 \sim 3$) are the Pauli matrices acting on the particle-hole space). In Eq. (1), the superfluid order parameter $\Delta = U \sum_{\mathbf{p}} \langle c_{-\mathbf{p},\downarrow} c_{\mathbf{p},\uparrow} \rangle$ is taken to be real and is proportional to the τ_1 component without loss of generality. In this case, ρ_1 and ρ_2 physically describe amplitude and phase fluctuations of the order parameter Δ, respectively[23]. In this paper, we ignore effects of a harmonic trap and consider a uniform Fermi gas, for simplicity.

Strong-coupling effects on single-particle excitations are taken into account by the self-energy $\hat{\Sigma}(\mathbf{p}, i\omega_n)$ in the 2×2-matrix single-particle thermal Green's function,

$$\hat{G}(\mathbf{p}, i\omega_n) = \begin{pmatrix} G_{11}(\mathbf{p}, i\omega_n) & G_{12}(\mathbf{p}, i\omega_n) \\ G_{21}(\mathbf{p}, i\omega_n) & G_{22}(\mathbf{p}, i\omega_n) \end{pmatrix} = \frac{1}{i\omega_n - \xi_{\mathbf{p},\uparrow} \frac{\tau_3+1}{2} - \xi_{\mathbf{p},\downarrow} \frac{\tau_3-1}{2} + \Delta\tau_1 - \hat{\Sigma}(\mathbf{p}, i\omega_n)}, \tag{2}$$

where ω_n is the fermion Matsubara frequency. In the extended T-matrix approximation (ETMA), the Dyson equation for the Green's function in Eq. (2) is diagrammatically described as Fig.1. The resulting 2×2-matrix self-energy is given by

$$\hat{\Sigma}(\mathbf{p}, i\omega_n) = -T \sum_{\mathbf{q}, i\nu_l} \sum_{i,j=\pm} \Gamma^{ij}(\mathbf{q}, i\nu_l) \tau_{-i} \hat{G}(\mathbf{p} - \mathbf{q}, i\omega_n - i\nu_l) \tau_{-j}. \tag{3}$$

Here, ν_l is the boson Matsubara frequency, and

$$\hat{\Gamma}(\mathbf{q}, i\nu_l) = \begin{pmatrix} \Gamma^{-+}(\mathbf{q}, i\nu_l) & \Gamma^{--}(\mathbf{q}, i\nu_l) \\ \Gamma^{++}(\mathbf{q}, i\nu_l) & \Gamma^{+-}(\mathbf{q}, i\nu_l) \end{pmatrix} = -U \left[1 + U \begin{pmatrix} \Pi^{-+}(\mathbf{q}, i\nu_l) & \Pi^{--}(\mathbf{q}, i\nu_l) \\ \Pi^{++}(\mathbf{q}, i\nu_l) & \Pi^{+-}(\mathbf{q}, i\nu_l) \end{pmatrix} \right]^{-1}, \quad (4)$$

is the particle-particle scattering matrix, where

$$\Pi^{ij}(\mathbf{q}, i\nu_l) = -T \sum_{\mathbf{p}, i\omega_n} \mathrm{Tr} \left[\tau_i \hat{G}^0(\mathbf{p} + \mathbf{q}, i\omega_n + i\nu_l) \tau_j \hat{G}^0(\mathbf{p}, i\omega_n) \right] \quad (5)$$

is the lowest-order pair-correlation function. $\hat{G}^0(\mathbf{p}, i\omega_n)$ is the bare Green's function given by Eq.(2) with $\hat{\Sigma}$=0.

The superfluid order parameter Δ is determined from the condition, $\det[\hat{\Gamma}(\mathbf{q} = \mathbf{0}, i\nu_l = 0)]^{-1}_{h=0} = 0$[24], which gives

$$1 = U \sum_{\mathbf{p}} \frac{1}{2E_{\mathbf{p}}} \tanh \frac{E_{\mathbf{p}}}{2T}, \quad (6)$$

where $E_{\mathbf{p}} = \sqrt{(p^2/2m - \mu)^2 + \Delta^2}$. Although Eq. (6) has the same form as the ordinary BCS gap equation, the Fermi chemical potential μ in this equation is known to remarkably deviate from the Fermi energy ε_{F} in the unitarity limit[6]. This strong-coupling correction can be conveniently incorporated into the theory by solving the gap equation (6), together with the equation for the number N of Fermi atoms, $N = 2T \sum_{\mathbf{p}, i\omega_n} G_{11}(\mathbf{p}, i\omega_n)|_{h=0}$. In the normal state ($\Delta = 0$), we only solve the number equation to determine μ as a function of the temperature.

The uniform spin susceptibility $\chi(T)$ is calculated from

$$\chi(T) = \left. \frac{\partial \Delta N}{\partial h} \right|_{h \to 0} = T \sum_{\mathbf{p}, i\omega_n} \mathrm{Tr} \left. \frac{\partial \hat{G}(\mathbf{p}, i\omega_n)}{\partial h} \right|_{h \to 0}. \quad (7)$$

Here, $\Delta N = N_\uparrow - N_\downarrow$, where N_σ is the number of Fermi atoms in the σ component. In this paper, we numerically evaluate Eq. (7) by taking a small but finite value of h.

We note that the self-energy $\hat{\Sigma}_{\mathrm{TMA}}$ in the ordinary T-matrix approximation (TMA)[24, 25, 27, 28, 29, 30] is obtained by replacing the renormalized Green's function $\hat{G}$ in Eq. (2) with the bare one $\hat{G}^0$. The strong-coupling theory developed by Nozières and Schmitt-Rink (NSR)[5] is also reproduced by expanding the Green's function in Eq. (2) with the self-energy being replaced by Σ_{TMA} to $O(\hat{\Sigma}^{\mathrm{TMA}})$. While the TMA and the NSR theory have been extensively used to clarify various BCS-BEC crossover physics in ultracold Fermi gases[8, 9, 24, 25, 26, 27, 28, 29, 30], they are known to unphysically give negative spin susceptibility in the crossover region[19, 31, 32]. On the other hand, it has recently been shown that the ETMA can correctly describe the spin susceptibility in the whole BCS-BEC crossover region[19, 21, 33], and the results agree well with the recent experiment on a ^{6}Li normal Fermi gas in the BCS regime[19]. Thus, in this paper, we also employ the ETMA to examine strong-coupling corrections to χ in the unitarity limit.

3. Spin susceptibility in a unitary Fermi gas

Figure 2(a) shows the spin susceptibility χ in a unitary Fermi gas ($(k_{\mathrm{F}} a_s)^{-1} = 0$). In the normal state, the ETMA spin susceptibility is smaller than the mean-field result ('MF' in this figure), and decreases with decreasing the temperature when $T/T_{\mathrm{F}} \leq 0.37$. Since pairing fluctuations are completely ignored in the mean-field theory, this spin-gap behavior of χ originates from strong pairing fluctuations near T_{c}.

We briefly note that the present ETMA gives the first-order phase transition as shown in the inset in Fig. 2(a), which is, however, an artifact of this strong-coupling theory. This leads to the

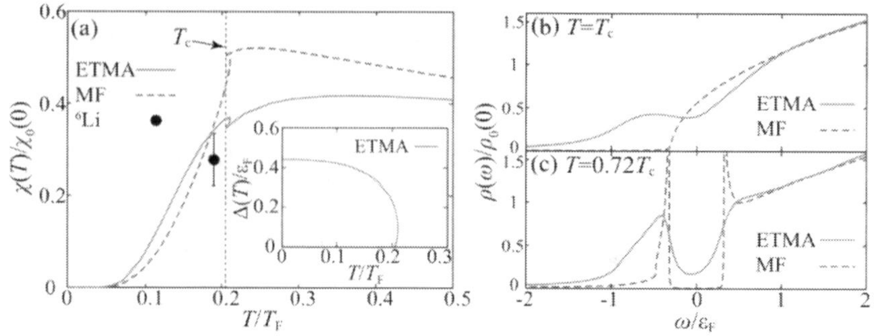

Figure 2. (a) Calculated ETMA spin susceptibility χ in a unitary Fermi gas ($(k_F a_s)^{-1} = 0$), normalized by the spin susceptibility $\chi_0(0)$ of a free Fermi gas at $T = 0$. The temperature is normalized by the Fermi temperature T_F. 'MF' is the mean-field BCS result in Eq. (8) where μ and Δ evaluated in the ETMA are used. (Note that the expression is reduced to that for a free Fermi gas above T_c, except that the ETMA result for μ is used.) The solid circle shows the recent experiment on a ^{6}Li Fermi gas[15]. The inset shows $\Delta(T)$. (b) Single-particle density of states $\rho(\omega)$ at T_c. $\rho_0(0)$ is the density of states at the Fermi level for a free Fermi gas. 'MF' is the density of states for a free Fermi gas where μ obtained in the ETMA is used. (c) $\rho(\omega)$ in the superfluid state at $T = 0.72T_c$. 'MF' is the superfluid density of states in the mean-field BCS theory where μ and Δ obtained in the ETMA are used.

singular behavior of χ near T_c seen in Fig. 2(a). In this regard, we recall that the same problem exists in the TMA[24], as well as the NSR theory[34]. Although the improvement of the ETMA so as to recover the expected second-order superfluid phase transition is a crucial issue[35, 36], we leave this as a future problem.

In the superfluid phase, apart from the above mentioned singularity just below T_c, χ decreases with decreasing the temperature, as shown in Fig. 2(a). Although this behavior is already seen in the mean-field case, the decrease is more remarkable in the latter. In the mean-field BCS theory, the spin susceptibility is given by[38]

$$\chi = \frac{1}{2T}\int_{-\infty}^{\infty} d\omega \rho(\omega)\mathrm{sech}^2\frac{\omega}{2T}, \tag{8}$$

so that the decrease of the spin susceptibility below T_c is found to be deeply related to the development of the BCS gap in the superfluid density of states $\rho(\omega)$. Evaluating ETMA density of states using the formula,

$$\rho(\omega) = -\frac{1}{\pi}\sum_{\mathbf{p}} \mathrm{Im}G_{11}(\mathbf{p}, i\omega_n \to \omega + i\delta), \tag{9}$$

we find from Fig. 2(c) that pairing fluctuations make $\rho(\omega)$ gapless even below T_c. As a result, compared with the mean-field case, the suppression of χ is less remarkable. Thus, although pairing fluctuations suppress χ above T_c, they enhance χ below T_c.

We briefly note that our result is close to the recent experiment on a ^{6}Li Fermi gas (solid circle in Fig. 2(a))[15]. Our results are also consistent with the recent quantum Monte-Carlo simulation[39], as well as the results in the self-consistent T-matrix approximation[40].

4. Summary

To summarize, we have discussed strong-coupling corrections to the spin susceptibility χ in a unitary Fermi gas. Within the framework of an extended T-matrix approximation, we showed that effects of pairing fluctuations on this magnetic quantity are different in between the normal state and the superfluid phase. In the normal state, pairing fluctuations suppress χ, leading to the spin-gap phenomenon near T_c. In the superfluid phase, pairing fluctuations enhance χ, because they induce finite density of states inside the BCS gap. Since the spin susceptibility is observable in ultracold Fermi gases, our results would be useful for the study of strong-coupling physics in the BCS-BEC crossover region using the uniform spin susceptibility.

Acknowledgments
We would like to thank T. Kashimura, R. Watanabe, and D. Inotani for discussions. H. T. was supported by KLL-PhD Program Research Grant, as well as Graduate School Doctoral Student Aid Program from Keio University. Y.O. was supported by a Grant-in-Aid for Scientific Research from MEXT in Japan (Grant No. 25105511 and No. 25400418).

References

[1] Gurarie V and Radzihovsky L 2007 *Ann. Phys.* **322** 2
[2] Bloch I, Dalibard J and Zwerger W 2008 *Rev. Mod. Phys.* **80** 885
[3] D. M. Eagles, Phys. Rev. **186**, 456 (1969).
[4] Leggett A J 1980 *Modern Trends in the Theory of Condensed Matter* ed A Pekalski and J Pryzstawa (Springer Verlag Berlin)
[5] Nozières P and Schmitt-Rink S 1985 *J. Low Temp. Phys.* **59** 195-211
[6] Sa de Melo C A R, Randeria M and Engelbrecht J R 1993 *Phys. Rev. Lett.* **71** 3202
[7] Haussmann R 1993 *Z. Phys.* B **91** 291; 1994 *Phys. Rev.* B **49** 12975
[8] Ohashi Y and Grifin A 2002 *Phys. Rev. Lett.* **89** 130402; 2003 *Phys. Rev.* A **67** 063612
[9] Chen Q J, Stajic J, Tan S N and Levin K 2005 *Phys. Rep.* **412** 1
[10] Regal C A, Greiner M and Jin D S 2004 *Phys. Rev. Lett.* **92** 040403
[11] Zwierlein M W, Stan C A, Schunck C H, Raupach S M F, Kerman A J and Ketterle W 2004 *Phys. Rev. Lett.* **92** 120403
[12] Stewart J T, Gaebler J P, Jin D S 2008 *Nature* **454** 744
[13] Nascimbène S, Navon N, Jiang K J, Chevy F, Salomon C 2010 *Nature* **463** 1057
[14] Ku M J H, Sommer A T, Cheuk L W, Zwierlein M W 2012 *Science* **335** 563
[15] Sanner C, Su E J, Keshet A, Huang W, Gillen J, Gommers R and Ketterle W 2011 *Phys. Rev. Lett.* **106** 010402
[16] Sommer A, Ku M, Roati G and Zwierlein M W 2011 *Nature (London)* **472** 201
[17] Nascimbène S, Navon N, Pilati S, Chevy F, Giorgini S, Georges A and Salomon C 2011 *Phys. Rev. Lett.* **106** 215303
[18] Lee Y R, Wang T T, Rvachov T M, Choi J H, Ketterle W and Heo M S 2013 *Phys. Rev.* A **87** 043629
[19] Kashimura T, Watanabe R and Ohashi Y *Phys. Rev.* A 2012 **86** 043622; 2014 *Phys. Rev.* A **89** 013618
[20] Hanai R, Kashimura T, Watanabe R, Inotani D and Ohashi Y 2013 *Phys. Rev.* A **88** 053621
[21] Tajima H, Kashimura T, Hanai R, Watanabe R and Ohashi Y 2014 *Phys. Rev.* A **89** 033617
[22] Schrieffer R B 1964 *Theory of Superconductivity* (Addison-Wesley)
[23] Ohashi Y, Takada S, 1997 *J. Phys. Soc. Jpn* **66** 2437
[24] Watanabe R, Tsuchiya S and Ohashi Y 2010 *Phys. Rev.* A **82** 043630; 2012 *Phys. Rev.* A **86** 063603
[25] Pieri P, Pisani L and Strinati G C 2004 *Phys. Rev.* B **70** 094508
[26] Perali A, Pieri P, Pisani L and Strinati G C 2004 *Phys. Rev. Lett.* **92** 220404
[27] Tsuchiya S, Watanabe R and Ohashi Y 2009 *Phys. Rev.* A **80** 033613; 2010 *Phys. Rev.* A **82** 033629; 2011 *Phys. Rev.* A **84** 043647
[28] Mueller E J 2011 *Phys. Rev.* A **83** 053623
[29] Palestini F, Pieri P and Strinati G C 2012 *Phys. Rev. Lett.* **108** 080401
[30] Mink M P, Jacobs V P J, Stoof H T C, Duine R A, Polini M and Vignale G 2012 *Phys. Rev.* A **86** 063631
[31] Liu X J and Hu H 2006 *Europhys. Lett.* **75** 364
[32] Parish M M, Marchetti F M, Lamacraft A and Simons B D 2007 *Nat. Phys.* **3** 124
[33] We note that the origin of the difference between the ETMA- and TMA-susceptibility can be simply understood when we approximate $\hat{\Gamma}(\mathbf{q}, i\nu_l)$ to the value at $\mathbf{q} = \nu_l = 0$ ($\equiv \Gamma$)[21]. In this case, while the dressed Green's function in the ETMA self-energy in Eq.(3) leads to the RPA (random phase approximation) type susceptibility $\chi_{\rm ETMA} \simeq \chi_{\rm DOS}/[1 + \Gamma\chi_{\rm DOS}]$ (where $\chi_{\rm DOS} = -(T/2)\sum_{\mathbf{p}, i\omega_n} {\rm Tr}\hat{G}(\mathbf{p}, i\omega_n)^2)$, the TMA (where $\hat{G}$ in Eq. (3) is replaced by $\hat{G}^0$) gives $\chi_{\rm TMA} \simeq \chi_{\rm DOS}[1 -$

Journal of Physics: Conference Series **568** (2014) 012019 doi:10.1088/1742-6596/568/1/012019

$\Gamma_0\chi_{\rm DOS}$], so that the latter becomes negative when $\Gamma\chi_{\rm DOS} > 1$. We also note that the so-called GG_0 theory[9] uses $\hat{G}^0$ for $\hat{G}$ in Eq.(3), although the dressed Green's function is partially used in $\Pi^{ij}(\mathbf{q}, i\nu_l)$. Thus, the RPA-type series is truncated as in the TMA case, so that this approach is considered to underestimate χ compared to the ETMA.
[34] Fukushima N, Ohashi Y, Taylor E, Griffin A 2007 *Phys. Rev.* A **75** 033609
[35] Ohashi Y 2005 *J. Phys. Soc. Jpn.* **74** 2659
[36] We note that Ref. [37] succeeds in avoiding the first-order phase transition by employing the combined GG_0 theory with the static approximation with respect to pairing fluctuations.
[37] Guo H, He Y, Chien C, and Levin K 2013 *Phys. Rev.* A **88** 043644
[38] Yosida K 1958 *Phys. Rev.* **110** 769
[39] Wlazlowski G, Magierski P, Drut J E, Bulgac A and Roche K J 2013 *Phys. Rev. Lett.* **110** 090401
[40] Enss T and Haussmann R 2012 *Phys. Rev. Lett.* **110** 195303

Journal of Physics: Conference Series **568** (2014) 012020 doi:10.1088/1742-6596/568/1/012020

Density functional theory investigation of two-dimensional dipolar fermions in a harmonic trap

Hande Ustunel[1], **Saeed H. Abedinpour**[2,3], **and B. Tanatar**[4]

[1] Department of Physics, Middle East Technical University, 06800, Dumlupinar Blv No: 1, Cankaya, Ankara, Turkey

[2] Department of Physics, Institute for Advanced Studies in Basic Sciences (IASBS), Zanjan 45137-66731, Iran

[3] School of Physics, Institute for Research in Fundamental Sciences (IPM), Tehran 19395-5531, Iran

[4] Department of Physics, Bilkent University, 06800, Bilkent, Ankara, Turkey

E-mail: `ustunel@metu.edu.tr`

Abstract. We investigate the behavior of polarized dipolar fermions in a two-dimensional harmonic trap in the framework of the density functional theory (DFT) formalism using the local density approximation. We treat only a few particles interacting moderately. Important results were deduced concerning key characteristics of the system such as total energy and particle density. Our results indicate that, at variance with Coulombic systems, the exchange-correlation component was found to provide a large contribution to the total energy for a large range of interaction strengths and particle numbers. In addition, the density profiles of the dipoles are shown to display important features around the origin that is not possible to capture by earlier, simpler treatments of such systems.

1. Introduction

Advances in trapping and cooling polar molecules with large permanent magnetic moments and atoms with electric moments have enabled numerous studies of dipolar quantum gases [1, 2, 3, 4, 5]. An important characteristic of these systems is that the dipole-dipole interaction is long ranged and anisotropic. Similar to the tunability of short-ranged interactions through Feshbach resonances, new techniques have been proposed [6] for controlling the strength and the sign of dipolar interactions. The interaction in three-dimensions is repulsive for dipoles aligned side-by-side and is attractive for dipoles aligned head-to-toe. Because the latter property along with collisional losses may cause instabilities in the system it is suggested that a gas trapped in two-dimensional (2D) geometry will provide a stable system.

Ground-state properties of 2D dipolar fermions have been investigated in a number of works [7, 8, 9, 10, 11, 12, 13, 14, 15, 16, 17, 18, 19]. In particular, the ground-state energy, pressure, and compressibility of a uniform gas of 2D fermions have been calculated by Lu and Shlyapnikov [17] up to second order in interaction strength. Recently, Matveeva and Giorgini [18] have performed quantum Monte Carlo (QMC) simulations to obtain the phase diagram of this system over a wide range of coupling constants. Abedinpour et al. [19] developed hypernetted-chain (HNC) approximation to calculate the pair distribution function and static structure factor which agrees very well with the QMC results.

In this work we present a theoretical study of ground-state energy and density profile of a system of 2D polarized dipolar fermions in a harmonic trap potential. Utilizing the parametrized ground-state energy [19] we develop a density-functional formulation within the local-density approximation (LDA). We solve the Kohn-Sham equations iteratively using a simple charge-mixing method using a basis of two-dimensional Hermite functions up to 10 closed shells (i.e. a total of 66 basis functions). In Section 2, we introduce the model and the resulting Kohn-Sham equations followed by a presentation of our results in Section 3. We summarize our conclusions in Section 4.

2. Method

The interaction potential between two dipoles confined to 2D is given by

$$E_{dd} = \frac{C_{dd}}{r^3}, \tag{1}$$

where r is the distance between the dipoles which are assumed to align perpendicular to the plane. In the case of an electric dipole with dipole moment d, the interaction strength is given by $C_{dd} = d^2/(4\pi\epsilon_0)$, whereas for a magnetic dipole of magnetic moment μ, it is $C_{dd} = \mu_0\mu^2/(4\pi)$. The many-particle Hamiltonian describing a system of N dipoles in a harmonic trap characterized by an angular frequency ω is

$$\hat{H} = -\frac{1}{2}\sum_{i=1}^{N}\nabla_i^2 + \frac{1}{2}\sum_{i=1}^{N}(x_i^2 + y_i^2) + \frac{1}{2}\sum_{i\neq j=1}^{N}\frac{g_{dd}}{|\vec{r}_i - \vec{r}_j|^3}, \tag{2}$$

where $\vec{r}_i = (x_i, y_i)$ is the position vector of the ith dipole. The Hamiltonian in Eq. (2) is written in harmonic units (h.u.) where lengths are measured in terms of the harmonic length $\ell = \sqrt{\hbar/m\omega}$ and energies in terms of $\hbar\omega$. In this notation, the coupling constant g_{dd} is defined as

$$g_{dd} = \frac{C_{dd}}{\hbar\omega\ell^3}. \tag{3}$$

In the LDA treatment of DFT, the total energy is written as a functional of the particle density, $n(\vec{r})$ as

$$E[n] = T_s[n] + E_{trap} + \frac{1}{2}g_{dd}\int d\vec{r}d\vec{r}'\frac{1}{|\vec{r}-\vec{r}'|^3} + \int d\vec{r}n(\vec{r})\epsilon_{XC}(n(\vec{r})), \tag{4}$$

where the first, second and third terms represent the single-particle kinetic energy, the energy associated with the harmonic trap and the classical Hartree energy respectively. The key element of the treatment that determines the accuracy of the energy is the exchange-correlation energy represented by the last term. Here $\varepsilon_{XC}(n(\vec{r}))$ is the exchange-correlation energy (per particle) of a uniform dipole gas with density $n = n(\vec{r})$. Defining the density in terms of auxiliary single-particle orbitals as

$$n(\vec{r}) = \sum_{i}^{N}|\phi_i(\vec{r})|^2, \tag{5}$$

and performing a variational minimization of the energy with respect to these orbitals yields the so-called Kohn-Sham equations:

$$\hat{H}\phi_i(\vec{r}) = \left[-\frac{1}{2}\nabla^2 + \hat{V}_{trap} + g_{dd}\int d\vec{r}'\frac{n(\vec{r}')}{|\vec{r}-\vec{r}'|^3} + \hat{V}_{xc}\right]\phi_i(\vec{r}) = \epsilon_i\phi_i(\vec{r}). \tag{6}$$

In Eq. (6), $\hat{V}_{trap} = \frac{1}{2}r^2$ and $\hat{V}_{XC}$ are the external trap and exchange-correlation potentials respectively. $\{\phi_i(\vec{r})\}$ are the Kohn-Sham orbitals and $\{\varepsilon_i\}$ are the corresponding eigenvalues.

Once the system of N nonlinear equations in Eq. (6) are solved simultaneously, the ground state density may be reconstructed from the solutions using Eq. (5).

The exchange-correlation energy $\varepsilon_{XC}(\vec{r})$ used in Eqs. (4) and (6) has been previously derived by two of the authors [19]. In order to overcome the numerical problems caused by the singularity at $r = 0$, we use a regularization scheme to treat the Hartree and exchange potentials together. The regularized interaction is given by

$$v_{\text{reg}}(r) = \frac{C_{\text{dd}}}{4\pi} \frac{1}{(r^2 + \sigma^2)^{3/2}}, \tag{7}$$

in which σ is a short range cutoff and one recovers the bare interaction in the $\sigma \to 0$ limit. With the introduction of this cutoff, the sum of the Hartree and exchange energy becomes

$$\varepsilon_{\text{HF}} = g_{dd}(4\pi n)^{3/2}\frac{32}{45} + \mathcal{O}(\sigma^1), \tag{8}$$

which is regular for small r.

3. Results and Discussion

The results presented in this paper were obtained by means of the self-consistent solution of Eq. (6) varying the two parameters of interest, namely the coupling constant g_{dd} and the particle number N. We present few-particle results, in particular $N = 3$ and $N = 21$ for small to moderate interaction strengths. The effect of varying σ is illustrated in Fig. 1(a) where the total LDA energy per particle is displayed for a fixed g_{dd} value of 1.0 for $N = 3$ and $N = 21$. Good convergence is obtained for both in a rather large range after $\sigma = 0.2$.. All results were obtained for a σ of 0.2ℓ. Fig. 1(b) displays the total LDA energy per particle for a system of 3 particles as a function of g_{dd} while Fig. 1(c) depicts the energy for 21 particles. For $g_{dd} = 0$, the energy for noninteracting, polarized particles in a harmonic trap is recovered. As g_{dd} increases, the energy is initially seen to slowly increase as expected from the repulsive nature of the interaction. However, shortly before $g_{dd} = 1.0$, the energy plunges to negative values and decreases rapidly. This behavior can be explained by considering the Hartree and exchange-correlation energies separately as also presented in Fig. 1. In both cases, the Hartree and exchange-correlation energies behave in a similar manner but with opposite signs. In the weak interaction region, the kinetic, trap and Hartree energies keep the total energy positive while as the interaction strength increases, the exchange-correlation energy dominates bringing the total energy down to large, negative values.

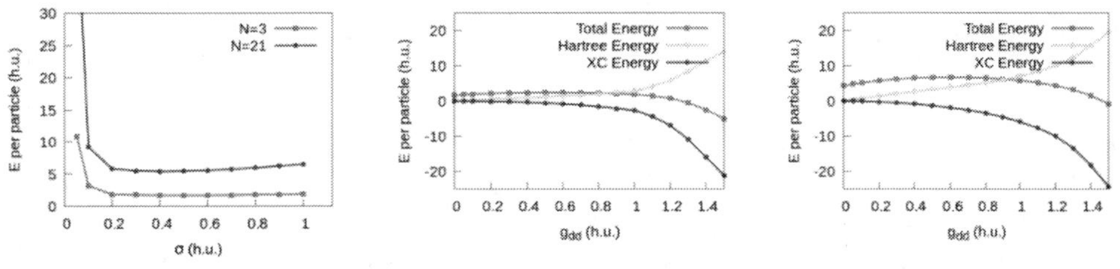

(a) Energy versus σ (b) $N = 3$ (c) $N = 21$

Figure 1. Total energy per particle as a function of σ (a); total energy (red squares), Hartree energy (green circles) and exchange-correlation energy (blue circles) per particle for $N = 3$ (b) and $N = 21$ (c) as a function of interaction strength, g_{dd}.

The transition in behavior from repulsive to attractive interaction evident from the energies, is also observed in the density profiles. Fig. 2 and Fig. 3 display density profiles for 3 and 21 particles respectively, for a series of selected g_{dd} values.

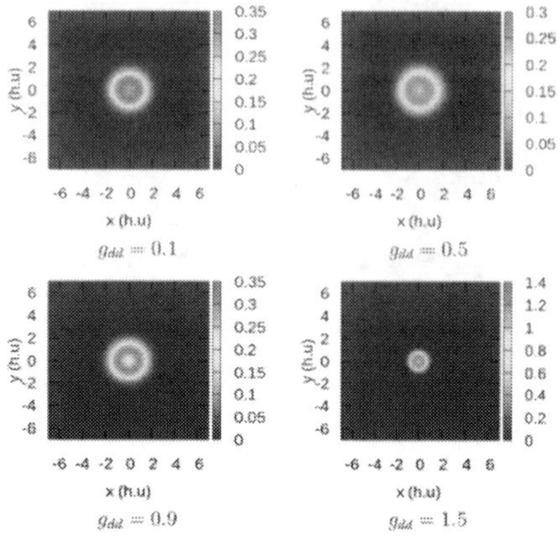

Figure 2. Density profiles for $N = 3$ and g_{dd} values varying between 0 and 1.5.

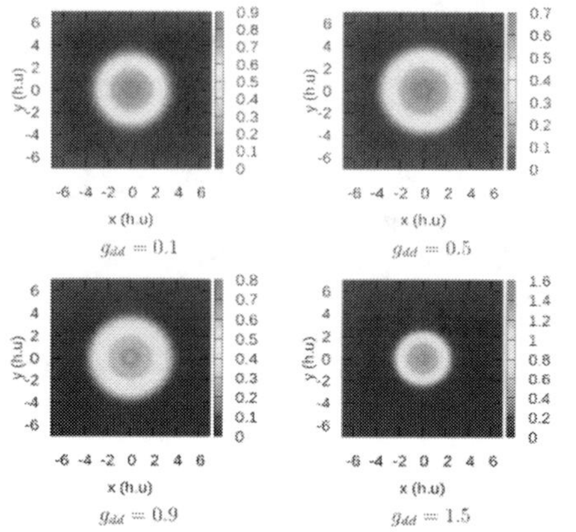

Figure 3. Density profiles for $N = 21$ and g_{dd} values varying between 0 and 1.5.

While the spatial extent of the density is mostly constant in the weak interaction region, a noticeable narrowing is observed once the exchange-correlation term begins to dominate.

In addition to the shift in behavior, the shape of the profiles demands attention. A common feature of all the density profiles obtained in this study is the central crater for $r = 0$ (see Fig. 4). This dip occurs independently of regularization cutoff and interaction strength but its existence depends upon particle number. While some particle numbers a crater as in the present case, others yield a peak. As the density decays away from the origin, well-distinguished terraces are observed as seen in Fig. 4(a) and (b), which are more pronounced for larger particle numbers. These subtle features in the density profile have not been previously identified by using simpler approximations such as variational approaches [20] or the Thomas-Fermi approximation [21, 22].

4. Conclusions

In this short manuscript, an LDA-DFT treatment of dipolar fermions, harmonically confined in two dimensions, was developed. A regularized dipolar interaction has been used to treat the Hartree and exchange-correlation terms in order to avoid singularities. Energies and density profiles for $N = 3$ and 21 particles have been presented for a range of interaction strengths ranging from weak to moderate. In the weak interaction region, the classical part of the energy (kinetic, trap and Hartree) was observed to be the deciding factor for determining the energy values as well as density profiles. For larger interaction strengths, the exchange-correlation term was seen to grow rapidly in the negative direction and dominating the behavior of the system. This result is in contrast to Coulombic systems where exchange and correlation energies are ordinarily a small portion of the total energy. It therefore underlines the importance of the full quantum mechanical treatment of the dipolar problem.

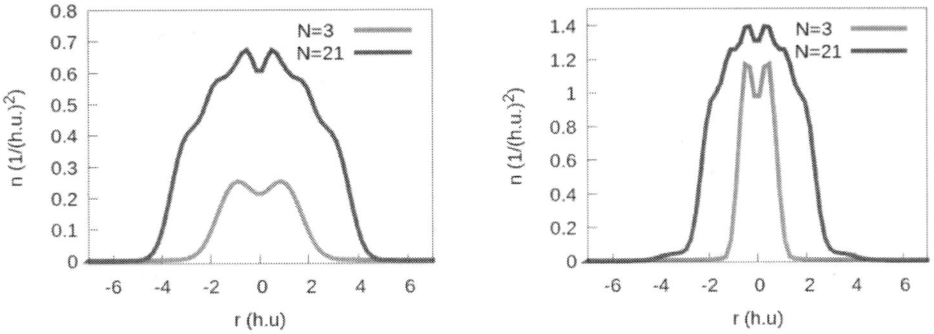

(a) $g_{dd} = 0.5$

(b) $g_{dd} = 1.5$

Figure 4. Cross-section ($y = 0$) of density profiles for $g_{dd} = 0.5$ (a) and $g_{dd} = 1.5$ (b). In each plot, radial density profiles for $N = 3$ and $N = 21$ are displayed.

References

[1] M.A. Baranov, Phys. Rep. **464**, (2008) 71.
[2] T. Lahaye, C. Menotti, L. Santos, M. Lewenstein, and T. Pfau, Rep. Prog. Phys. **72**, (2009) 126401.
[3] K. Aikawa, A. Frisch, M. Mark, S. Baier, A. Rietzler, R. Grimm, and F. Ferlaino, Phys. Rev. Lett. **108**, (2012) 210401.
[4] M. Lu, N.Q. Burdick, and B.L. Lev, Phys. Rev. Lett. **108**, (2012) 215301.
[5] M.A. Baranov, M. Dalmonte, G. Pupillo, and P. Zoller, Chemical Reviews **112** (2012) 5012.
[6] S. Giovanazzi, A. Gorlitz, and T. Pfau, Phys. Rev. Lett. **89**, (2002) 130401.
[7] G.M. Bruun and E. Taylor, Phys. Rev. Lett. **101**, (2008) 245301; **107**, (2011) 169901(E).
[8] C.-K. Chan, C. Wu, W.-C. Lee, S. Das Sarma, Phys. Rev. A **81**, (2010) 023602.
[9] Y. Yamaguchi, T. Sogo, T. Ito, and T. Miyakawa, Phys. Rev. A **82**, (2010) 013643.
[10] K. Sun, C. Wu, and S. Das Sarma, Phys. Rev. B **82**, (2010) 075105.
[11] Q. Li, E.H. Hwang, and S. Das Sarma, Phys. Rev. B **82** (2010) 235126.
[12] M. Babadi and E. Demler, Phys. Rev. A **84** (2011) 033636.
[13] M. Babadi and E. Demler, Phys. Rev. B **84**, (2011) 235124.
[14] N.T. Zinner and G.M. Bruun, Eur. Phys. J. D **65**, (2011) 133.
[15] M.M. Parish and F.M. Marchetti, Phys. Rev. Lett. **108**, (2012) 145304.
[16] F.M. Marchetti, and M.M. Parish, Phys. Rev. B **87**, (2013) 045110.
[17] Z.K. Lu and G.V. Shlyapnikov, Phys. Rev. A **85**, (2012) 023614.
[18] N. Matveeva and S. Giorgini, Phys. Rev. Lett. **109**, (2012) 200401.
[19] S.H. Abedinpour, R. Asgari, B. Tanatar, and M. Polini, Ann. Phys. (NY) **340** (2014) 25.
[20] T. Sogo, L. He, T. Miyakawa, S. Yi, H. Lu and H. Pu, New J. Phys. **11** (2009) 055017.
[21] B. Fang and B.-G. Englert, Phys. Rev. A **83**, 052517 (2011).
[22] B. P. van Zyl, E. Zaremba, and P. Pisarski, Phys. Rev. A **87**, 043614 (2013).

Drag Effect in Double-Layer Dipolar Fermi Gases

B. Tanatar, B. Renklioglu, M.O. Oktel
Bilkent University, Department of Physics, Bilkent 06800, Ankara, Turkey
E-mail: tanatar@fen.bilkent.edu.tr

Abstract. We consider two parallel layers of two-dimensional spin-polarized dipolar Fermi gas without any tunneling between the layers. The effective interactions describing screening and correlation effects between the dipoles in a single layer (intra-layer) and across the layers (inter-layer) are modeled within the Hubbard approximation. We calculate the rate of momentum transfer between the layers when the gas in one layer has a steady flow. The momentum transfer induces a steady flow in the second layer which is assumed initially at rest. This is the drag effect familiar from double-layer semiconductor and graphene structures. Our calculations show that the momentum relaxation time has temperature dependence similar to that in layers with charged particles which we think is related to the contributions from the collective modes of the system.

1. Introduction

Ultracold gases of polar atoms or molecules with their anisotropic long-range interaction are of great interest in recent years. [1, 2] Single and multi-layer structures of two-dimensional (2D) bosons or fermions are being studied from the point of view of their quantum phases, Fermi liquid properties, collective excitations, and formation of density waves. [3, 4, 5, 6, 8, 9, 10, 11, 12] Ground-state properties of 2D dipolar fermions have been investigated in a number of works [13, 14, 15]. Transport properties on the other hand are much less studied [16]. In this work, we address a transport property in dipolar fermion gases in analogy to a similar effect observed in electronic systems.

In electronic bilayer systems a transport phenomenon known as the drag effect has been studied for a long time. In the drag effect when a current is applied to one of the layers the electrons in the second will be dragged resulting in an interlayer resistivity which is related to the momentum transfer between the layers. [18, 19] Interestingly, the newly discovered 2D electronic systems such as graphene also exhibit this phenomenon [20].

In this paper, we consider the drag effect in a double-layer dipolar system of polarised fermions. It is assumed that a flow is established in one of the layers while We adapt the Hubbard approximation to describe the correlation effects between the dipoles within the same layer and use the random-phase approximation (RPA) to account for the interactions across the layers. We assume that the Femi gases in the parallel layers are in normal state so that the transfer of momentum from one layer to another gives rise to a friction-like effect. This is in contrast to the so-called dissipationless drag predicted for the same system when the fermions are in a superfluid state [21].

2. Model

We consider two infinite parallel layers of 2D spin-polarized dipolar Fermi gas, separated by a distance d. We assume that an external field aligns the dipoles perpendicularly to the to the laters, so that the intra-layer interaction is purely repulsive and isotropic and given by $V_{11}(r) = C_{dd}/(4\pi r^3)$. The inter-layer bare interaction on the other hand is $V_{12}(r) = (C_{dd}/4\pi)(r^2 - 2d^2)/(r^2 + d^2)^{5/2}$. The Fourier transform of the intra-layer interaction is not well-defined, therefore a cut-off parameter related to the width of the layer is used in various applications. [11, 12]

In this work we adapt the Hubbard approximation for the intra-layer interaction which is a familiar approach from electronic systems. [22] The Hubbard approximation allows us to obtain the intra-layer interaction in Fourier space free of any cut-off parameter and it may be regarded as an effective interaction. For the inter-layer interaction we use the bare Fourier transform as no cut-off is required in this case. Therefore, in Fourier space the intra-layer interaction $V_{11}(q)$ within a single layer and the inter-layer interaction $V_{12}(q)$ across the layers are given by [21]

$$V_{11}(q) = \frac{C_{dd}}{2}\left[\sqrt{k_F^2 + q^2} - q\right] \quad \text{and} \quad V_{12}(q) = -\frac{C_{dd}}{2} q \exp(-qd), \tag{1}$$

where the indices 1 and 2 indicate different layers and C_{dd} is the dipole-dipole coupling constant.

As a result of the momentum transfer caused by inter-layer dipole-dipole interactions, an applied current in one layer drives a current in the other one. This momentum transfer constitutes the Coulomb drag rate in the system, given by [?, 19]

$$\tau_D^{-1} = \frac{\hbar^2}{8mnk_BT\pi^2} \int_0^\infty \mathrm{d}q\ q^3 \int_0^\infty \mathrm{d}\omega\ \frac{|W_{12}(q,\omega)|^2\ \mathrm{Im}\ \chi_1(q,\omega)\ \mathrm{Im}\ \chi_2(q,\omega)}{\sinh^2(\hbar\omega/(2k_BT))}, \tag{2}$$

where k_B is the Boltzmann constant and $\chi(q,\omega)$ is the 2D polarization function. In addition, m is the mass of a dipole and n is the density of a single layer. Here, W_{12} is the dynamically screened effective interaction, defined by

$$W_{12} = \frac{V_{12}}{\epsilon(q,\omega)} \tag{3}$$

in which the dielectric function is written as

$$\epsilon(q,\omega) = [1 - V_{11}(q)\chi_1(q,\omega)]\,[1 - V_{22}(q)\chi_2(q,\omega)] - [V_{12}(q)]^2\,\chi_1(q,\omega)\chi_2(q,\omega), \tag{4}$$

in terms of the intra- and inter-layer interactions and non-interacting dynamic susceptibility of the ith layer $\chi_i(q,\omega)$ for spin-polarised fermions.

3. Results

We calculate the drag rate τ_D^{-1} between the layers of dipolar gases using Eq. (2) as a function of temperature. Note that the temperature dependence comes from various occurrences of χ's in the integrand. The dimensionless interaction strength parameter is defined as $\lambda = k_F a_0$ where a_0 denotes the characteristic length scale, obtained by $a_0 = mC_{dd}/(4\pi\hbar^2)$. Here, m is the mass of a dipole, $k_F = \sqrt{4\pi n}$ is the Fermi wave number and n is the density of a single layer.

Figures 1 and 2 show the drag rate for various values of the interaction strength λ and layer separation distance d. In general, as the distance d gets larger inter-layer interactions are weakened and the drag rate decreases. As the coupling strength λ increases, both the intra- and inter-layer interactions become stronger which results in an increased drag rate. However, we

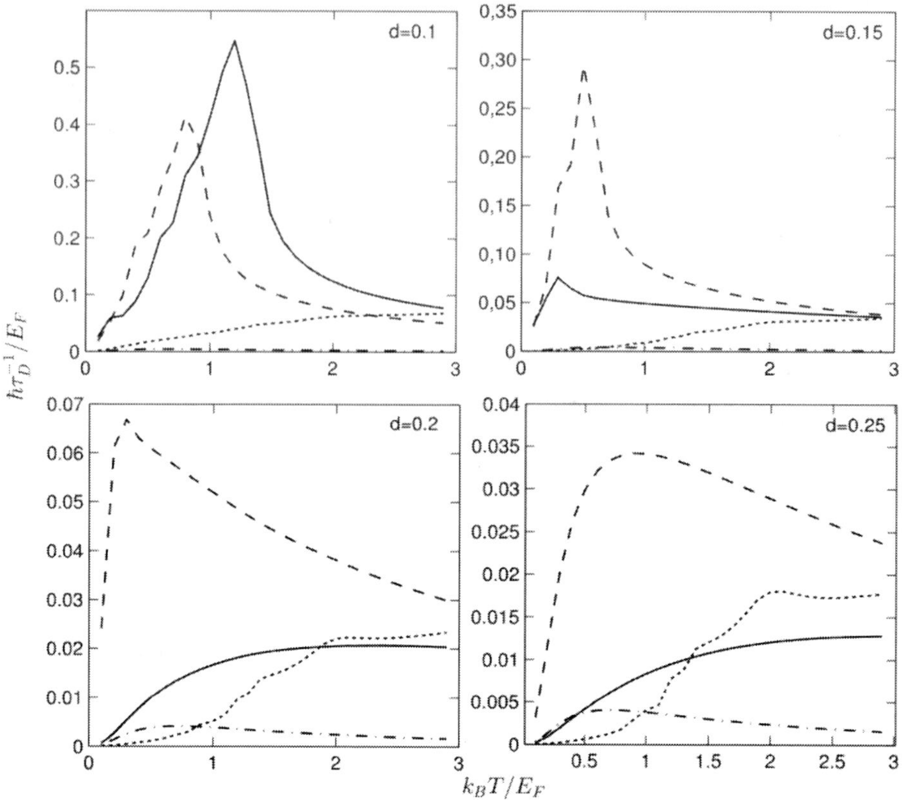

Figure 1. The dimensionless drag rate τ_D^{-1} as a function of temperature for different values of the interaction strengths, indicated by λ, as $\lambda = 0.2$ (dashed-dotted lines), $\lambda = 2.0$ (dashed lines), $\lambda = 4.0$ (solid lines) and $\lambda = 8.0$ (dotted lines). Here, the graphs are obtained for the systems with the well separation distances $d = 0.1, 0.15, 0.2$, and 0.25, respectively.

do see some anomalies especially at larger values of λ. We surmise that this is because of the Hubbard approximation model we use for the intra-layer interactions and improved interaction models are necessary for further explorations. The overall shape of the drag rate as a function of temperature (up to $3T_F$ resembles that of electronic double-layer structures. [23] Therefore, it is natural to associate the peak around T_F with the collective excitation modes of double-layer dipolar fermions in a similar manner. These are in essence collective density excitations which may be associated with in-phase and out-of-phase oscillations in the double layer system. [21] The peak position in the drag rate also depends on the separation distance d and coupling strength λ. As λ increases the collective modes become less energetic (dispersion is softened) and it becomes easier to excite the modes. This results in a decrease in the overall shape of the drag rate as a function of temperature and a shift in the peak position towards low temperature. With some irregularities we observe this general behaviour. To understand the discrepancies we need to resort to improved interaction models. It would be interesting to calculate the drag rate due to collective excitations only to understand the origin of enhancement at high temperatures.

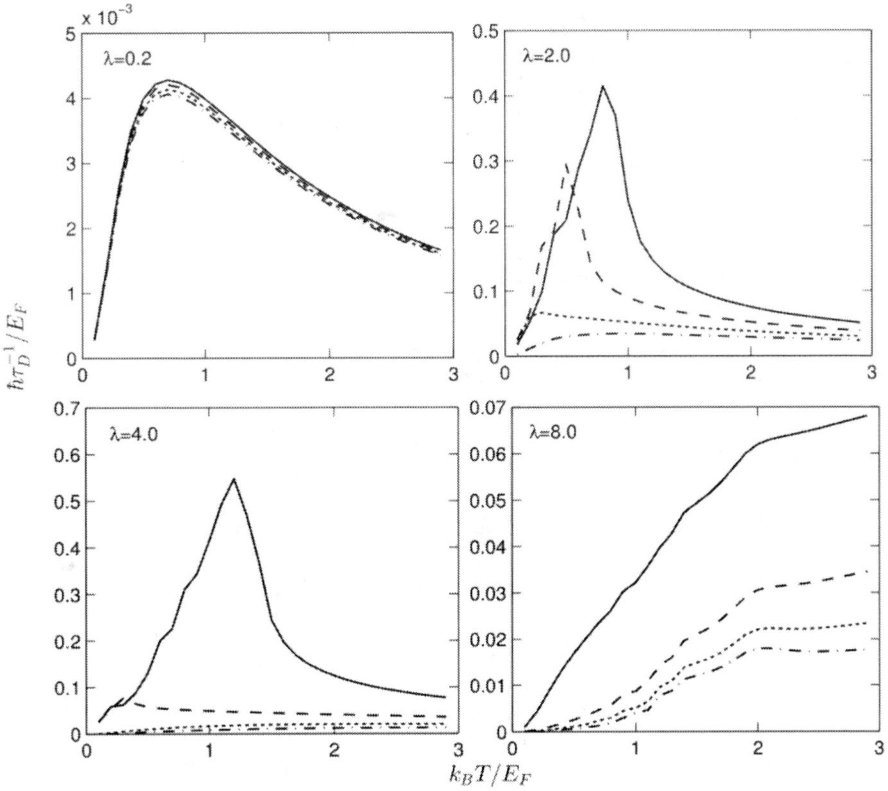

Figure 2. Temperature dependence of the dimensionless drag rate τ_D^{-1} for the corresponding values of the interaction strengths, $\lambda = 0.2, 2.0, 4.0, 8.0$. The solid, dashed, dotted and dashed-dotted lines define the well separation distances for $d = 0.1, 0.15, 0.2, 0.25$, respectively.

4. Conclusion

In conclusion we have calculated the drag rate between two layers of dipolar Fermi gas. We have found that for typical layer separation distances and moderate interaction strengths the drag rate shows a peaked behaviour as a function of temperature. We believe that this is due to the collective mode excitations in the double-layer structure similar to that encountered in electronic systems. More sophisticated screened interaction models are needed to study the strongly correlated regime (large λ). It would also be interesting to explore the effects of density imbalance in the layers.

Acknowledgments

This work is supported by TUBITAK (112T176 and 112T974) and TUBA.

References

[1] S. Ospelkaus, A. Pe'er, K.-K. Ni, J. J. Zirbel, B. Neyenhuis, S. Kotochigova, P. S. Julienne, J. Ye, and D. S. Jin, Nature Phys. **4**, 622 (2008).
[2] M. A. Baranov, Phys. Rep. **464**, 71 (2008).
[3] G. M. Bruun and E. Taylor, Phys. Rev. Lett. **101**, 245301 (2008).

[4] G. E. Astrakharchik, J. Boronat, I. L. Kurbakov, and Yu. E. Lozovik, Phys. Rev. Lett. **98**, 060405 (2007).
[5] H. P. Büchler, E. Demler, M. Lukin, A. Micheli, N.Prokof'ev, G. Pupillo, and P. Zoller, Phys. Rev. Lett. **98**, 060404 (2007).
[6]
[7] N. Matveeva and S. Giorgini, Phys. Rev. Lett. **109**, (2012) 200401.
[8] Z.-K. Lu and G. V. Shlyapnikov, Phys. Rev. A **85**, 023614 (2012).
[9] K.-K. Ni, S. Ospelkaus, M. H. G. de Miranda, A. Pe'er, B. Neyenhuis, J. J. Zirbel, S. Kotochigova, P. S. Julienne, D. S. Jin, and J. Ye, Science **322**, 231 (2008).
[10] C.-K. Chan, C. Wu, W.-C. Lee, and S. Das Sarma, Phys. Rev. A **81** 023602 (2010).
[11] Q. Li, E. H. Hwang, and S. Das Sarma, Phys. Rev. B **82**, 235126 (2010).
[12] N. T. Zinner and G. M. Bruun, Eur. Phys. J. D **65**, 133 (2011).
[13] M.M. Parish and F.M. Marchetti, Phys. Rev. Lett. **108**, (2012) 145304.
[14] F.M. Marchetti, and M.M. Parish, Phys. Rev. B **87**, (2013) 045110.
[15] S. H. Abedinpour, R. Asgari, and B. Tanatar, Ann. Phys. (NY) **340**, 25 (2014).
[16] T. Ernst, T. Paul, and Peter Schlagheck, Phys. Rev. A **81**, 013631 (2010).
[17] T. Enss, C. Kppersbusch, and L. Fritz, Phys. Rev. A **86**, 013617 (2012).
[18] A. G. Rojo, J. Phys.: Condens. Matter **11**, R31 (1999).
[19] R. Asgari, B. Tanatar, and B. Davoudi, Phys. Rev. B **77**, 115301 (2008).
[20] J. C. W. Song, D. A. Abanin, and L. S. Levitov, NANO LETTERS **13**, 3631 (2013).
[21] B. Tanatar, J. Low Temp. Phys. **171**, 632 (2013).
[22] G.F. Giuliani and G. Vignale, Quantum Theory of the Electron Liquid, Cambridge University Press, Cambridge, 2005.
[23] K. Flensberg and B.Y-K. Hu, Phys. Rev. B **52**, 14796 (1995).

Journal of Physics: Conference Series **568** (2014) 012022 doi:10.1088/1742-6596/568/1/012022

Nuclear magnetic resonance on ^{3}He confined in 2.8-nm channel of FSM16

Junko Taniguchi, Daiki Tanaka, and Masaru Suzuki
University of Electro-Communications, Chofu, Tokyo, Japan
E-mail: tany@phys.uec.ac.jp

Abstract.
We have performed continuous-wave (cw) nuclear magnetic resonance measurements for ^{3}He confined in 2.8-nm channel of FSM16. The magnetization shows a reduction from the Curie law at low temperature for the high areal density where ^{3}He in the channel consists of three portions; a solid first-layer, an amorphous-solid overlayer, and a fluid inside. By an analysis for the magnetization and the available heat capacity data, the reduction can be attributed to an interaction of spins in amorphous solid and a degenerate of fluid.

1. Introduction

One-dimensional (1D) quantum many-body systems have been attractive due to large quantum fluctuations. Trials to realize the 1D Fermi system by confining ^{3}He into nanometer-size channels are in progress.[1, 2, 3] In these cases, the attention has been paid mainly to a dilute limit, where ^{3}He atom is expected to behave like a non-interacting gas. At low areal densities, heat capacity measurements for ^{3}He in a 2.8-nm channel show that the crossover from a 2D classical gas to a 1D degenerate state occurs.[1] Very recently, a nuclear magnetic resonance (NMR) signal of 1D diffusion has been reported for a little narrower channel.[2]

On the other hand, ^{4}He confined in nanometer-size channels has been also studied to search the features of 1D system.[4] Recently, it was made clear that the superfluidity of pressurized liquid ^{4}He confined in a 2.8-nm channel has a large frequency dependence. It has attracted attention as a possible signature of the bosonic Tomonaga-Luttinger (TL) liquid.[5] In analogy with the ^{4}He system, a dense liquid ^{3}He in the channel possibly shows a feature of the TL liquid. Thus motivated, we started NMR measurements of dense liquid ^{3}He confined in the channel of the same size.

In this paper, we report the preliminary results of NMR measurements for ^{3}He confined in a 2.8-nm channel. At high areal densities, the reduction of magnetization (M) from the Curie law is observed. By taking the previous heat capacity (C) data into consideration[6, 7], it is made clear that the reduction can be attributed to the interaction of spins in the amorphous solid and the degenerate of fluid.

2. Experimental

The porous material we used is FSM (Folded Sheets Mesoporous materials) 16, whose 1D channel is 2.8 nm in diameter, and 0.2-0.5 μm in length.[8] The FSM powder was baked at 220°C for 6 hours. For a thermal contact, a mixture of FSM and 70 μm silver powders was compressed, together with a brush of fine silver wires, and sintered at 200°C for 3 hours. The sintered sample was put into a Stycast 1266 sample cell. The surface area of the sample was determined as 100±5 m^2 from a N_2 adsorption isotherm. According to the previous heat capacity measurements for ^{3}He, the areal densities of the first layer completion and the complete filling of channel are estimated to be 8.5±0.7 and 17.7±0.7 atoms/nm^2, respectively.[6, 7]

Continuous-wave NMR measurements were performed by sweeping a magnetic field. A measuring frequency was 2.089 MHz. The rf coil wound around the sample cell made a resonant circuit with a capacitance of a coaxial line. The Q factor of the circuit was as high as 85. To avoid a saturation of NMR signal, a small amplitude of the rf field was used. It was 10 mV_{pp} below 0.10 K.

The absorption line was averaged more than 20 times to improve the signal noise ratio. It was slightly broader than a Lorentzian function, due to a magnetic homogeneity of about 4×10^{-4}. M was obtained from a numerical integration of the absorption line using a convolution of a Lorentzian and a Gaussian function. The uncertainty of M was mainly due to the uncertainty of subtraction of the base line and was about 5% at 0.08 K.

The sample cell and the magnet for the static field were anchored separately to a mixing chamber of the dilution refrigerator. The temperature (T) of the sample cell was measured by means of a RuO_2 thermometer. Measurements were performed in the temperature range between 0.055 and 1.0 K.

3. Results and Discussion

The magnetization was measured for several areal densities between 10 and 18 atoms/nm^2. Above about 0.4 K, M increases inversely proportional to T, which means that it obeys the Curie law. The Curie constant obtained from the data above 0.4 K increases in proportion to the areal density.

Regarding the low temperature behavior, M at 10 atoms/nm^2 obeys the Curie law down to the lowest temperature of 0.055 K. As the areal density is increased, M becomes smaller than the Curie law. For 16 atoms/nm^2, the reduction starts at around 0.2 K (T^{-1} = 5 K^{-1}), as shown in Fig. 1.

To search the origin of the reduction, we examine the states of ^{3}He confined in the channel, on the basis of previous heat capacity measurements. In the Refs. 6 and 7, it is pointed out that ^{3}He in the channel is divided into three portions, a solid first-layer, an amorphous solid overlayer, and a fluid inside. In this paper, we discuss the temperature dependence of C for each portion. Then, we show that the heat capacity and the reduction of the magnetization can be explained by these contributions.

In the first-layer solid, the heat capacity is quite small at high temperature, and shows an upturn with decreasing temperature below about 40 mK. The temperature dependence of the upturn is T^{-2}, which is a typical one of the spin heat capacity with an exchange interaction energy (J). The interaction energy at the areal density of the first layer completion is evaluated as $J_1/k_B \sim$1.7 mK.

When the amorphous-solid layer grows (the corresponding areal density ranges from 8.5 to 13 atoms/nm^2), at high temperature, a T-linear heat capacity grows, which is characteristic to an amorphous solid.[9, 10] On the other hand, the temperature dependence at low temperature becomes weaker than T^{-2}-one, and approaches nearly a constant value, with increasing areal density. A similar temperature dependence is observed for an amorphous solid formed in a Vycor glass, which has a heterogeneity.[11] Golov et al. attributed the temperature dependence to the distribution of exchange interaction, which is caused by a range of local areal densities. Although FSM16 has a uniform structure, the local areal density of ^{3}He adsorbed on the first layer may have a distribution due to the restricted size of the channel. Therefore, in the same manner as the Vycor glass, we assume a uniform distribution of the local density. Due to the

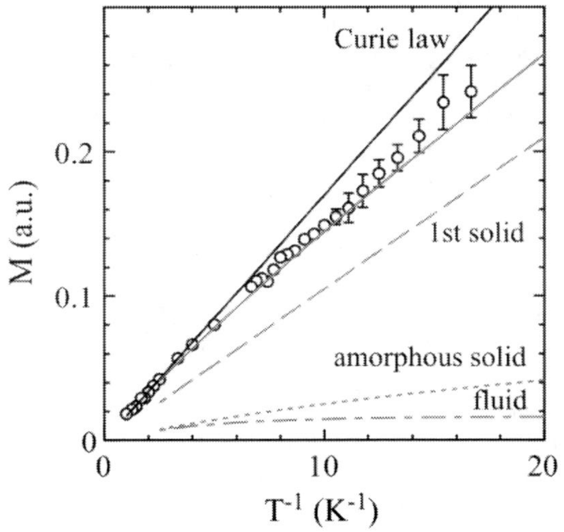

Figure 1. Magnetization as a function of the inverse of temperature. The areal density is 16 atoms/nm^2. Solid line corresponds to the Curie law. Solid curve shows the calculated magnetization, which is the sum of the contributions from the first-layer solid (dash line), the amorphous solid (dotted curve) and the fluid (dashed-dotted curve) . (see text.)

exponential dependence of J on the density, it yields a roughly uniform distribution over $\ln J$,

$$\frac{dN_{aS}(J)}{d(\ln J)} = \text{const}, \tag{1}$$

where $dN_{aS}(J)$ is the number of ^{3}He having the exchange interaction energy between J and $J + dJ$. At around the temperature of J/k_B, the heat capacity of ^{3}He shows a Schottky-like peak. Sum of the Schottky-like heat capacity for the range of J yields the weak temperature dependence.

In fitting the heat capacity of amorphous-solid layer, we use fitting parameters of a lower (J_l) and a higher (J_h) cut-off of interaction energy, and the effective areal density of amorphous solid (n_{aS}). As a lower cut-off, we adopt J_l/k_B =1.7 mK. As a higher cut-off, we use the temperature at around which the upturn due to the spin heat capacity appears. As the areal density is increased, J_h/k_B becomes high, and reaches 120 mK at 13 atoms/nm^2, where the amorphous solid is completed. We found that n_{aS} is smaller than the direct experimental values by up to 1.4 atom/nm^2. It may come from the increase in the density of the first layer, which is compressed by the overlayer.

When a fluid appears (the corresponding areal density is above 13 atoms/nm^2), there is no more growth of the upturn. On the other hand, above about 0.35 K, the increment of heat capacity by the introduction of ^{3}He is independent of temperature, and its amount is about 1.4 J·K^{-1}·mol^{-1}. With decreasing temperature, the increment of heat capacity decreases at around 150 mK, and seems to approach the T-linear dependence. The temperature dependence is quite analogous to that of 3D Fermi gas, although its value is about eighth part of the one of the Fermi gas. The quantitative inconsistency also appears in the case of bulk liquid ^{3}He, and may come from the complicated interaction peculiar to a dense liquid.[12] Here, assuming that the fluid in the channel behaves as a non-interacting 3D Fermi gas, the temperature dependence is qualitatively reproduced, when the degenerate temperature is $\sim$ 300 mK.

Journal of Physics: Conference Series **568** (2014) 012022 doi:10.1088/1742-6596/568/1/012022

Now, we calculate the magnetization assuming the same states as those for the analysis of heat capacity. J_1/k_B (1.7 mK) is smaller more than one order than the measured temperature region. Thus, the first layer contribution is expected to increase inversely proportional to temperature down to the lowest temperature, as shown as a dashed line in Fig. 1. In the amorphous-solid layer, the effective exchange interaction energy ranges from 1.7 to 120 mK. Since the magnetization of the amorphous-solid layer is smaller than the Curie law, we adopt the temperature dependence of magnetization with antiferromagnetic coupling.[13] By summing up each contribution over the distribution range of J, a weaker temperature dependence of magnetization M_{aS} than the Curie law is obtained as shown as a dotted curve in Fig. 1. For the fluid, we calculated the magnetization, assuming a non-interacting 3D Fermi gas using the degenerate temperature of 300 mK. It is shown as a dashed-dotted curve in Fig. 1. Finally, we sum these three contributions, as shown as a solid curve in Fig. 1. Within the experimental error, the observed temperature dependence is in agreement with the calculated curve.

The results show that the temperature dependences of both the heat capacity and the magnetization are explained by the contributions of the first-layer solid, the amorphous solid and the fluid. It is of significance that both data support the degenerate of fluid in the channel. In order to study the feature of the degenerate fluid in detail, it would be better to perform the measurements for the system where the first-layer and the amorphous solids are replaced by ^{4}He. In addition, the feature of 1D Tomonaga-Luttinger liquid is expected to become pronounced in the dynamical property, such as a relaxation time and a diffusion constant. We think that measurements of spin-lattice and spin-spin relaxation times (T_1 and T_2) by means of pulse NMR is useful to search the feature of degenerate fluid state in the channel.

4. Conclution

We have performed cw NMR measurements for ^{3}He confined in 2.8-nm channel of FSM16. The magnetization shows a reduction from the Curie law at low temperature above 13 atoms/nm^2. The reduction can be attributed to an interaction of spins in the amorphous solid and the degenerate of the fluid.

Acknowledgments

This work was partly supported by JSPS KAKENHI Grant Number 23740264. We also thank S. Inagaki and Y. Fukushima for providing the substrate FSM16.

References

[1] Taniguchi J, Yamaguchi A, Ishimoto H, Ikegami H, Matsushita T, Wada N, Gatica SM, Cole MW, Ancilotto F, Ikagaki S, and Fukushima Y, 2005 *Phys. Rev. Lett.* **94** 065301
[2] Yager B, Nyéki J, Casey A, Cowan BP, Lusher CP, Saunders J 2013 *Phys. Rev. Lett.* **111** 215303
[3] Matsushita T, Kuze A, Kawai R, Hieda M, Wada N, 2013 *J. Low Temp. Phys.* **171** 657
[4] Wada N, Taniguchi J, Ikegami H, Inagaki S, and Fukushima Y, 2001 *Phys. Rev. Lett.* **86** 4322
[5] Taniguchi J, Demura K, Suzuki M, 2013 *Phys. Rev.* B **88** 014502
[6] Taniguchi J, Yamaguchi A, Ishimoto H, Ikegami H, and Wada N, 2004 *J. Low Temp. Phys.* **134** 595
[7] Taniguchi J, Ph. D. thesis, Tokyo University 2004.
[8] Inagaki S, Fukushima Y, Kuroda K, 1993 *J. Chem. Soc. Chem. Commun.* **22** 680
[9] Andreev AF, 1978 *JETP Lett.* **28** 556. According to this reference, the $T-$linear heat capacity is characteristic to an amorphous solid, and ^{4}He fluid can be regarded as an amorphous solid. In this paper, we describe the state for the range of areal density between 8.5-13 atoms/nm^2 as an amorphous solid, since there is no positive signal to support a fluid state.
[10] Toda R, Hieda M, Matsushita T, and Wada N, 2009 *J. Phys.: Conf. Ser.* **150** 032112
[11] Golov A and Pobell F, 1996 *Phys. Rev.* B **53** 12647
[12] The magnetization of bulk liquid ^{3}He is quantitatively explained by the Fermi gas, differently from the heat capacity.
[13] Bonner JC and Fisher ME, 1964 *Phys. Rev.* **135** A640

Journal of Physics: Conference Series **568** (2014) 012023 doi:10.1088/1742-6596/568/1/012023

Pressure dependent attenuation peaks for quartz tuning forks in superfluid ^{4}He at mK temperatures

J T Tuoriniemi, M S Manninen and J Rysti

Low Temperature Laboratory, O.V. Lounasmaa Laboratory, Aalto University, FI-00076 AALTO, FINLAND

Abstract. Pure superfluid ^{4}He at very low temperatures is thought to be free of mechanisms producing dissipation for slowly moving objects, and the superfluid merely contributes an effective mass to the motion without any anticipated dependence on temperature below 0.1 K. To our great surprise, we found that the smallest commercially available quartz tuning fork oscillators with tines shorter than 2 mm and resonance frequency 32 kHz experienced attenuation by as much as 100 times more than the internal dissipation of the oscillator at certain values of pressure at millikelvin temperatures in superfluid ^{4}He. These features smeared out at temperatures above 20 mK, although practically no thermal excitations in ^{4}He are yet present. Another fork somewhat larger showed indication of similar anomalies as well, but at greatly lesser scale.

1. Introduction

Superfluids are called so because they can flow without any dissipation as long as certain critical flow velocity is not exceeded. Similarly, slowly moving objects can travel through such fluids without friction, but only at the limit of zero temperature. At finite temperatures, moving bodies interact with thermal quasiparticle excitations, which exchange momentum with the moving body. At sufficiently low temperatures the quasiparticles become ballistic because of very long mean free paths, which simplifies the description of their effect. In ^{4}He the low temperature excitations behave as phonons, whose contribution to the drag force vanishes in proportion to the fourth power in temperature, and below 100 mK it is negligible in most practical cases.

All sorts of oscillating bodies have traditionally been employed in studies of superfluids to indicate the temperature, for example, by measuring the resonance width of the mechanical oscillator put into the fluid. This becomes ineffective once the temperature is below about 1/10 of the superfluid transition temperature because of the reason outlined above, and then the instrument will only show its internal damping due to the nonideality of its making.

Probably the best sensors of this type in terms of low internal damping are the quartz tuning fork resonators, which have recently become popular for the superfluid research, see for example Refs. [1–5]. These devices are made from single crystal quartz and their Q-value at low temperatures in vacuum is often beyond 10^6, translating to very good sensitive range. Quartz tuning forks are mass produced as reference oscillators for various devices, such as ordinary clocks. They are available in many sizes and packing, usually tuned to the frequency 2^{15} Hz = 32768 Hz at room temperature in vacuum, and their performance just improves as the temperature is reduced. The two tines of the fork are driven into antiphase motion by virtue of the piezoelectric effect on quartz, and their oscillation can be detected with great sensitivity due to the current induced by alternating stress on the constructing material.

The tip velocity of the fork tines at typical oscillation amplitudes here is less than 1 mm/s, so that we expected practically no dissipation due to superfluid ^{4}He at millikelvin temperatures. Thus greater was our surprise, when we observed that the smallest quartz tuning forks could experience damping over 100 times larger than internal dissipation in certain ranges of pressure in such medium. This anomaly was very sensitive to pressure producing hundreds of attenuation peaks starting from the saturated vapor pressure up to the melting pressure of ^{4}He. The explanation to such behavior remains to be found. Therefore, this paper merely describes our observations and calls for suggestions about its origins.

2. Experimental setting

A versatile experimental cell with instruments for measuring the pressure, density, and temperature of helium fluids and solids over wide ranges was used to study all flavors of helium, i.e. pure ^{3}He, pure ^{4}He, and also mixtures of these. There were two quartz tuning fork resonators of different sizes placed into the experimental volume. These were meant primarily as thermometers for superfluid ^{3}He refrigerated deep below 1 mK. Work on ^{4}He was initially thought only to serve as reference to experiments on ^{3}He [6] and on helium mixtures [7].

The measurements on ^{4}He were conducted after the investigations on helium mixtures, so that the cell had to be "cleaned" from traces of ^{3}He first. For this purpose the system was warmed above the critical temperature of ^{3}He and the cell was pumped empty. The cell was then refilled with pure ^{4}He to 4 bar pressure and emptied again. Yet another partial filling cycle was performed, although the ^{3}He/^{4}He ratio of the outcoming gas, as analyzed by a mass spectrometer, was observed not to differ from what we obtain using standard ^{4}He from a gas bottle. It was concluded that the situation could not be improved any more and the remaining concentration probably did not differ from that of standard ^{4}He from natural sources.

The two quartz tuning forks employed in these experiments were deliberately of quite different design. One of these was originally housed in a parallelepiped casing with inner dimensions $0.5 \times 0.85 \times 2.75$ mm^3. To be used in helium, the top plate of the casing was removed exposing the fork to the bulk experimental volume from one of its flat sides. A photograph of this device is shown in Fig. 1. The other fork was somewhat bigger and housed in a cylindrical metal casing. The top of the cylinder was filed open in order to bring the fork into contact with helium in the main volume. The dimensions of these two quartz resonators are listed in Table 1.

The resonance parameters — frequency, width, amplitude, phase, and background (practically constant) — were first determined by measuring full spectra. Fits to the expected

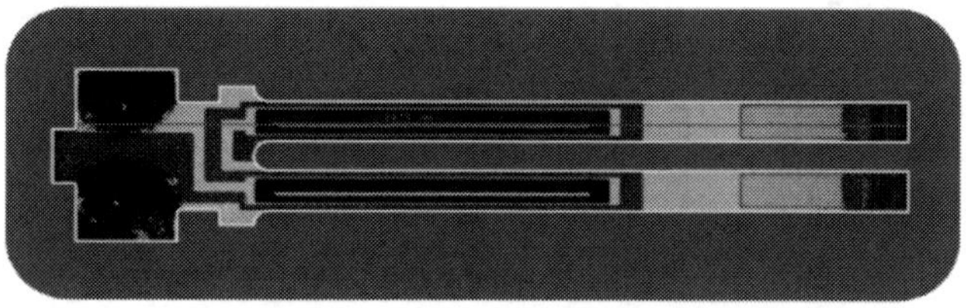

Figure 1. Small fork used in this work. The dark background oval displays the size of the casing (0.85×2.75 mm^2) of the fork. Table 1 shows the tine dimensions of this and another fork used.

Table 1. Dimensions (in mm) of the two quartz tuning forks used. "T-" stands for tine.

Type	Casing	Q-vac	T-length	T-width	T-thickness	T-spacing
ECS-.327-7-34B-TR	box	10^6	1.9	0.11	0.11	0.09
ECS-.327-8-14X	cylinder	$2 \cdot 10^6$	2.3	0.24	0.13	0.10

lineshape were excellent, the forks showed perfectly linear response, and the area under the resonance curve could be used to relate the resonance amplitude and width. Once this background information was available the forks could be driven in a single point tracking mode, which is described in detail in Ch. 3.3.1 of Ref. [8]. Resonances with rather high Q-values could then be monitored with fairly short time intervals, which were typically set at 1–10 s.

The residual resonance width of the two forks, including possible nonidealities due to the cables, electronics, etc., was about 30 mHz for the smaller and about 15 mHz for the larger one. The same value within few mHz was observed either in vacuum or in superfluid ^{3}He or ^{4}He deep below the superfluid transition temperature, that is, if the regions of the anomalous damping in superfluid ^{4}He were avoided.

3. Results

Once the unexpected attenuation peaks of the small fork were discovered, we scanned the pressure-temperature landscape through selected paths, which can be observed in Fig. 2 displaying the changes in temperature and pressure as functions of real time together with the recorded resonance width of the smaller fork. It was not possible to cover the whole phase space systematically during the time available for this experiment, so that only selected slices through the entire area can be shown.

Figure 3 shows the resonance width of the smaller fork as function of pressure over a limited range at the temperature of 10 mK. The behavior is quite dramatic, as the quality factor of the

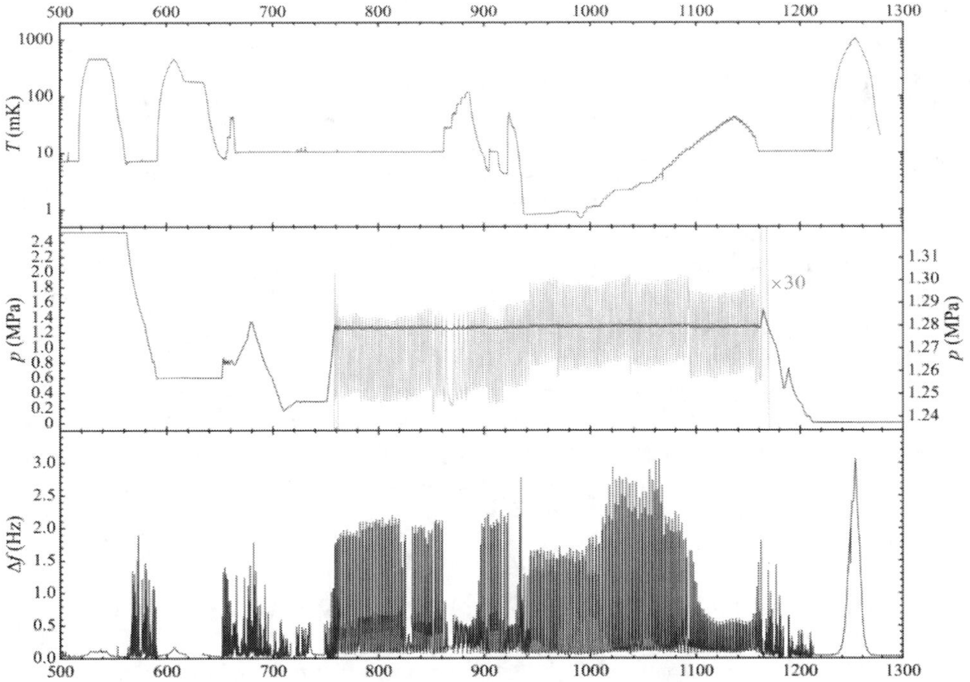

Figure 2. Temperature and pressure of ^{4}He and resonance width of the small fork as functions of time since the anomalous behavior was first encountered in the afternoon Sep. 3rd, 2013. The period of examining a narrow pressure region as the function of temperature is expanded on the pressure plot (×30, light colored trace) and the corresponding scale is on the right.

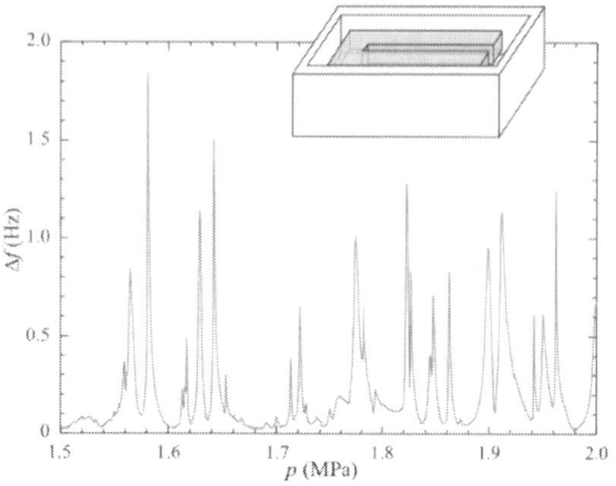

Figure 3. Some of the attenuation peaks recorded from the small fork, as the pressure was decreased between about 550—600 hours in Fig. 2. The largest peaks correspond to about factor 100 increase in the attenuation of the fork oscillation. The outline of this pattern was reproducible at later times. The box on the top schematically illustrates the geometry of the fork (not to scale).

fork changes as much as by a factor of 100 over rather narrow intervals of pressure. The pattern was reproducible as its outline remained the same over the entire span of this study, which took about 54 days and during which the temperature was cycled from over 100 mK to below 1 mK and the pressure was varied across the entire range from the saturated vapor pressure to the melting pressure of the solid phase. From time to time full resonance spectra were recorded in order to verify that the tracking algorithm functioned properly.

It is amazing that such strong coupling from the quartz resonator to superfluid ^{4}He can happen, as the medium is supposedly at the zero temperature limit. Why does this coupling exist only at certain values of pressure producing very distinct attenuation peaks? Also the temperature dependence is quite peculiar, while features as sharp as in Fig. 3 were observed only below 20 mK or so. Above that the pattern became much more rounded and remained such until the quasiparticle contribution in ^{4}He to the attenuation became appreciable causing general increase of width at any value of pressure.

A narrow interval in pressure including one of the most prominent attenuation peaks was examined more systematically as the function of temperature. The result is summarized in Fig. 4, where a 3D plot of the small fork's width is displayed as functions of temperature and pressure. The same data are shown as a contour diagram in Fig. 5. As the temperature goes down the attenuation peak suddenly becomes very sharp and begins to push towards higher pressures. It has been questioned if superfluid ^{4}He can be refrigerated to low millikelvin temperatures in the first place, while no thermal excitations are there any more to make any difference from absolute zero, but now there *is* a clear response to temperature even at this low range.

The bigger fork also showed clear indications of something similar influencing its resonance but the scale of effects was much smaller, obviously because of the larger mass involved in its motion. Typical changes in resonance width were only about a factor of two and they appeared as very sharp attenuation peaks as function of pressure, just as for the smaller fork.

4. Discussion

We can offer no proper explanation for the features just described. We can only speculate about the possible significance of traces of ^{3}He present in the system or of the role of quantized vorticity around or pinned to the quartz tuning forks. Maybe the existence of both is essential, as the ^{4}He vortices are known to capture ^{3}He in their cores possibly playing some tricks there. We even contaminated our sample deliberately by about 10 ppm of ^{3}He but this did not cause any significant change to the overall behavior. This happened at about 770 hours time in Fig. 2.

The observed pattern resembles a lot the results of our earlier studies on second sound coupled to quartz tuning forks in dilute helium mixtures [9–11]. We emphasize, however, that

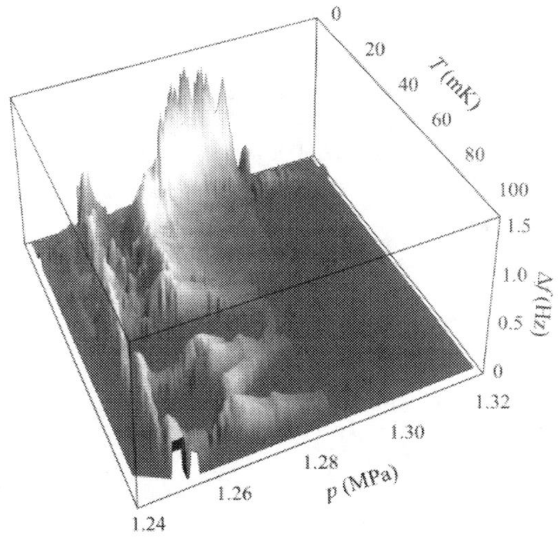

Figure 4. 3D plot of the small tuning fork resonance width as functions of pressure and temperature around one of the most prominent attenuation peaks at mK range in ^{4}He. When no data was available at a given point, the value was set to zero (the flat bottom).

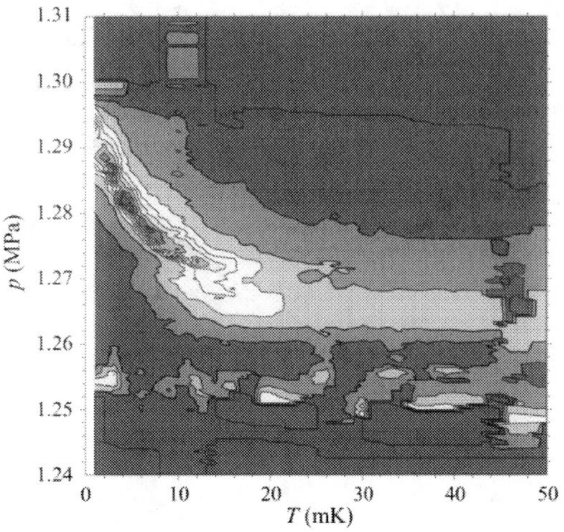

Figure 5. Same data as in Fig. 4 plotted as a contour diagram. The color encodes the resonance width: red corresponds to strong attenuation and the darkest blue corresponds to background width (or no data).

second sound is not likely to cause the present anomalies, because the resonances should appear primarily while changing temperature, not pressure. It is unlikely that sufficient amounts of ^{3}He impurities were present in order to produce pronounced second sound resonances. Support for this conclusion is also provided by the lack of such resonances, when the temperature was increased to the range of 1 K at the end of this experiment, during about 1240—1260 hours seen in Fig. 2. Only the emergence of thermal quasiparticle damping was observed.

Few cases of unexplained attenuation for quartz tuning forks in superfluid helium have been reported elsewhere [12–14]. In those cases the damping experienced by the forks varied with time in an uncontrolled manner under seemingly steady conditions. In that sense the characteristics of the earlier observations crucially differ from what we have described here. More studies using different types of tuning forks in variety of enclosures is obviously needed in order to gain better understanding of the mechanisms involved.

Acknowledgments

This work has been supported in part by European Union (FP7/2007-2013, Grant No. 228464 Microkelvin) and by Academy of Finland (LTQ CoE grant no. 250280). This research made use of the Aalto University Low Temperature Laboratory infrastructure.

References

[1] Clubb D O, Buu O V L, Bowley R M, Nyman R and Owers-Bradley J R 2004 *J. Low Temp. Phys.* **136** 1–13
[2] Blaauwgeers R et al. 2007 *J. Low Temp. Phys.* **146** 537–62
[3] Pentti E M, Tuoriniemi J T, Salmela A J and Sebedash A P 2008 *J. Low Temp. Phys.* **150** 555–60
[4] Pentti E M, Tuoriniemi J T, Salmela A J and Sebedash A P 2008 *Phys. Rev.* B **78** 064509
[5] Bradley D I, Človečko M, Fisher S N, Garg D, Guise E, Haley R P, Kolosov O, Pickett G R and Tsepelin V 2012 *Phys. Rev.* B **85** 014501

[6] Manninen M S, Kaikkonen J-P, Peri V, Rysti J, Todoshchenko I and Tuoriniemi J 2014 *J. Low Temp. Phys.* **175** 56–62
[7] Rysti J, Manninen M S and Tuoriniemi J 2014 *J. Low Temp. Phys.* **175** 739–54
[8] Pentti, E, Rysti J, Salmela, A, Sebedash, A and Tuoriniemi J 2011 *J. Low Temp. Phys.* **165** 132–65
[9] Salmela A, Tuoriniemi J, Pentti E, Sebedash A and Rysti J 2009 *J. Phys.: Conf. Ser.* **150** 012040
[10] Salmela A, Tuoriniemi J and Rysti J 2011 *J. Low Temp. Phys.* **162** 678–85
[11] Tuoriniemi J, Rysti J, Salmela A and Manninen M 2012 *J. Phys.: Conf. Ser.* **400** 012077
[12] Bradley D I, Fear M J, Fisher S N, Guénault A M, Haley R P, Lawson C R, Pickett G R, Schanen R, Tsepelin V and Wheatland L A 2014 *J. Low Temp. Phys.* **175** 379–84
[13] Garg D, Efimov V B, Giltrow M, McClintock P V E, Skrbek L and Vinen W F 2012 *Phys. Rev.* B **85** 144518
[14] Garg D, Efimov V B, Giltrow M, McClintock P V E, Skrbek L and Vinen W F 2012 *Low Temp. Phys.* **38** 1026–30

Dynamics of Fluctuactions in the BCS-BEC Crossover in Superfluid Fermi Gases

H. Vivas C.

Departamento de Física, Universidad Nacional de Colombia, Sede Manizales, A.A. 127, Colombia

E-mail: hvivasc@unal.edu.co

Abstract. The BCS-BEC crossover phenomenon in a 3D attractive Fermi gas is revisited by using finite temperature Green functions formalism derived from the quantum path integral technique and Mean Field Approximation (MFA). Collective modes for weak and strong coupling particle-particle interaction g are obtained after recovering the dynamical coefficients in the non-linear Time Dependent Ginzburg Landau (TDGL) equation, and numerically simulated under those particular states for the filling energy $\mu \sim 0$ and order parameter $\Delta(\mathbf{x}, \tau) \sim \Delta_0$. Explicit formulas for the first order average fluctuations $\langle \eta^2_{\mathbf{q},\omega} \rangle$ and the scaling law for the thermal energy associated to the non-linear collective modes are also reported.

1. Introduction

The problem of the crossover from BCS theory with cooperative Cooper pairing to the formation and condensation of composite bosons has attracted considerable attention after the discovery of high T_C superconductors. Monte Carlo simulations for the attractive (negative U) Hubbard model have shown, for instance, that the crossover region displays highly anomalous correlations in degenerate Fermi system above T_C in the intermediate coupling regime [1]. The crossover phenomenon is also of interest for other problems: excitonic condensates, superconductor-insulator transition and Cooper pair formation in resonant Feshbach channels [2],[3],[4] and more recently, in tunneling properties of bounded Fermi pairs in optical lattices [5] and superfluid Sarma phase critical behavior for unbalanced 2D chemical potentials in ultracold Fermi gases [6],[7]. By using the path integral formalism, we study a 3D continuum model of fermions with attractive interactions at finite temperature, by reconstructing the Time-Dependent Ginzburg Landau (TDGL) equation for the particular case in which the order parameter vanishes at $T = T_C$, and for arbitrary strength coupling g [8]. We show that, in the weak coupling limit, the TDGL coefficients are reduced to the classical ones obtained by Gor'kov (1959) in the framework of the microscopic approximation. This formalism is systematically extended for the broken symmetry case ($T < T_C$), s-wave scattering and the Mean-Field Approximation (MFA). In section 4 we derive the complete set of self-consistent equations for collective excitations, and solve them in two particular cases: i) for the weak-coupling interaction limit and ii) and the intermediate/strong attractive interaction potential. We show that our derivation reproduces the value for the sound velocity in the condensate in the first scenario.

Journal of Physics: Conference Series **568** (2014) 012024 doi:10.1088/1742-6596/568/1/012024

2. Time-Dependent Ginzburg Landau Equation (TDGL)

The insertion of the fluctuating (bosonic) fields $\Delta\left(\bar{\Delta}\right)$ into the effective action functional $S_{eff}\left[\bar{\Delta},\Delta\right]$ has been developed in the framework of the Randeira's prescription by starting from the Nambu propagator $(\mathbf{G}^{-1})$ [13]: $\mathbf{G}^{-1} = \mathbf{G_0}^{-1} + \Delta\sigma^{+} + \bar{\Delta}\sigma^{-}$. Here $\sigma^{\pm} = (\sigma_x \pm \sigma_y)/2$, and σ_j's are Pauli's matrices, $\mathbf{G_0}^{-1}$ represents the *non-interacting* Green's propagator given explicitly by:

$$\mathbf{G_0} = \begin{pmatrix} (i\omega_m - \xi_{\mathbf{k}})^{-1} & 0 \\ 0 & (i\omega_m + \xi_{\mathbf{k}})^{-1} \end{pmatrix}, \tag{1}$$

with $i\omega_m = 2\pi T(m+1/2)$ as the set of fermionic Matsubara frequencies, and $\xi_{\mathbf{k}} = k^2/2m - \mu$ defines the non-interacting kinetic energy for particles with momentum $\mathbf{k}$, and chemical potential μ at temperature T. The relationship with the effective action $S_{eff}\left[\bar{\Delta},\Delta\right]$ is obtained after applying the well-known Hubbard-Stratonovich transformation over Fermi fields operators [9],[10]. The remaining (bosonic) fields contribute into $S_{eff}\left[\bar{\Delta},\Delta\right]$ in the form:

$$S_{eff}\left[\bar{\Delta},\Delta\right] = \int d^3\mathbf{x}\int_0^{\beta} d\tau\left[\frac{|\Delta|^2}{g} - \mathrm{Tr}\ln\mathbf{G_0}^{-1} - \mathrm{Tr}\ln\left[1+\mathbf{G_0}\left(\Delta\sigma^{+}+\bar{\Delta}\sigma^{-}\right)\right]\right], \tag{2}$$

in which the bare interaction between Fermions is primarily encoded through the parameter g. By performing second order expansions in the last term of (2), the effective action can be reduced straightforwardly and the final result resembles the Gaussian structure discussed by Nozières and Schmitt-Rink (NSR) in the frame of the particle-particle scattering t-matrix theory [11]: $S_{eff}\left[\bar{\Delta},\Delta\right] = S_0 + \sum_q |\Delta(q)|^2 / \Gamma(q)$. Factor $\Gamma(q)$ contains the regularization condition for the *effective* $s-$wave interaction on g. Time-Dependent Ginzburg Landau equation (TDGL) becomes now apparent when $S_{eff}\left[\bar{\Delta},\Delta\right]$ includes fourth-order fluctuations in the fields $|\Delta|$ and under the minimization constraint $\delta S_{eff}/\delta\bar{\Delta} = 0$. Its final form in the space-time representation takes the form:

$$\left(a - c\frac{\nabla^2}{2m} + b\,|\Delta(\mathbf{x},t)|^2 - id\frac{\partial}{\partial t}\right)\Delta(\mathbf{x},t) = 0, \tag{3}$$

where the coefficients in (3) are obtained under specific conditions for a system close to its critical temperature T_C and slow-time variations in the fields Δ. Details in the derivation of (3) are given by the author in reference ([12]).

3. Broken Symmetry State at $\mathbf{T} < \mathbf{T_C}$

The previous analysis can be extended for the broken symmetry (superfluid) state at *low* temperatures. The first stage of our approximation comes from the condition in which the order parameter fluctuations are negligible ($\Delta(x) \approx \Delta_0$), and $\delta S_{eff}/\delta\Delta_0 = 0$. The relationship between the interaction parameter g and Δ_0 can be obtained by using the standard Gaussian integration technique. The final result can be expressed as: $1/g = \sum_{\mathbf{k}} \tanh(\beta E_{\mathbf{k}}/2)/2E_{\mathbf{k}}$, where $E_{\mathbf{k}} = \sqrt{\xi_{\mathbf{k}}^2 + \Delta_0^2}$. The regularization condition for the ultraviolet divergence in the gap equation is established by including proper terms associated to the s-wave scattering length a_s, which directly describes the two-body attraction strength g in a low density system. The renormalized gap equation then reads:

$$\frac{m}{4\pi a_s} = \sum_{\mathbf{k}}\left[\frac{1}{2\epsilon_{\mathbf{k}}} - \frac{\tanh(\beta E_{\mathbf{k}}/2)}{2E_{\mathbf{k}}}\right], \tag{4}$$

with $\varepsilon_{\mathbf{k}} = k^2/2m$. In order to investigate the nature of the fluctuating fields (denoted by $\eta(\mathbf{x},\tau)$) around the saddle point Δ_0, we write $\Delta(\mathbf{x},\tau) = \Delta_0 + \eta(\mathbf{x},\tau)$ while the generic definition for the Green's propagator in Eq. (1) recasts into: $\mathbf{G}^{-1} = \mathbf{G}_0^{-1}(\Delta_0) + \eta\sigma^+ + \bar{\eta}\sigma^-$, where $\mathbf{G}_0(\Delta_0)$ is defined as:

$$\mathbf{G}_0(\Delta_0) = \begin{pmatrix} \mathcal{G}_k(\Delta_0) & \mathcal{F}_k(\Delta_0) \\ \mathcal{F}_k^\star(\Delta_0) & -\mathcal{G}_k^\star(\Delta_0) \end{pmatrix}, \tag{5}$$

with $\mathcal{G}_k(\Delta_0) = -(i\omega_m + \xi_{\mathbf{k}})/(\omega_m^2 + E_{\mathbf{k}}^2)$, and $\mathcal{F}_k(\Delta_0) = \Delta_0/(\omega_m^2 + E_{\mathbf{k}}^2)$. The effective action functional must be rewritten in terms of the mean field values $\Delta_0, \bar{\Delta}_0$ and their corresponding fluctuating fields $\eta, \bar{\eta}$. Hence:

$$S_{eff}[\bar{\eta}, \eta] = \sum_q \left[\frac{|\eta_q|^2}{g} - \mathrm{Tr}\ln \mathbf{G}_0^{-1}(\Delta_0) + \sum_{n=1}^{\infty} \frac{(-1)^n}{n} \mathrm{Tr}\left(\mathbf{G}_0(\Delta_0)\boldsymbol{\Sigma}\right)^n \right], \tag{6}$$

where $\boldsymbol{\Sigma}$ is a off-diagonal matrix defined as $\eta_q\sigma^+ + \bar{\eta}_q\sigma^-$. The index q in Equation (6) denotes the pair-momentum and the set of Bosonic Matsubara frequencies $i\nu_n$. It is possible to relate this result with quantities of physical interest such as the generating partition functional $\mathcal{Z}$ and the Helmholtz free energy $\mathcal{F}$ through $\mathcal{F} = -kT\ln\mathcal{Z}$. Equation (6) can be rewritten in a quadratic form in the fields $\bar{\eta}, \eta$ as:

$$S_{eff}[\bar{\eta}, \eta] = S_0 + \sum_{\mathbf{q}} \eta_q^\dagger \left(K_q^{-1}\right) \eta_q, \tag{7}$$

with the spinors fields $\eta^\dagger \equiv [\bar{\eta}_q, \eta_q]$, and K_q basically defined in terms of the g, a_s and T. Calculation of $\langle \eta_q^2 \rangle$ becomes now evident from Eq. (7). For the case in which the bosonic frequencies $i\nu_n$ are different from zero, the tensor K_q^{-1} might be separated into its *static* component, which has been collected into $K_{\mathbf{q}}^{-1}$, and its *frequency-dependent* part denoted as K_ω^{-1}, where ω is the real frequency that appears after analytic continuation procedure on $i\nu$, under $i\nu \to \omega + i0^+$. Hence, $K_q^{-1} = K_{\mathbf{q}}^{-1} + K_\omega^{-1}$, with $(K_\omega^{11})^{-1} = -\sum_{\mathbf{k}}(1 - \Delta_0^2/E_{\mathbf{k}}^2)\tanh(\beta E_{\mathbf{k}}/2)/2E_{\mathbf{k}}(1 - \omega^2/4E_{\mathbf{k}}^2)$, $(K_\omega^{22})^{-1} = -\sum_{\mathbf{k}}\tanh(\beta E_{\mathbf{k}}/2)/2E_{\mathbf{k}}(1 - \omega^2/4E_{\mathbf{k}}^2)$, $(K_\omega^{12})^{-1} = i\sum_{\mathbf{k}}\omega\xi_{\mathbf{k}}\tanh(\beta E_{\mathbf{k}}/2)/4E_{\mathbf{k}}^3(1 - \omega^2/4E_{\mathbf{k}}^2)$, with $(K_\omega^{21})^{-1} = -(K_\omega^{12})^{-1}$. The expression for the partition function $\mathcal{Z} = \mathcal{Z}_0\int\mathcal{D}[\bar{\eta}, \eta]\exp\left[-\sum_{\mathbf{q}}\eta_{\mathbf{q}}^\dagger K_{\mathbf{q}}^{-1}\eta_{\mathbf{q}}\right] = \mathcal{Z}_0\prod_{\mathbf{q}}\det(K_{\mathbf{q}})$ might be modified as: $\ln\mathcal{Z} = -S_0 + \sum_{\mathbf{q}}\int d\omega\ln[\det(K_q)]$, where $\det\left(K_{\mathbf{q},\omega}^{-1}\right) =$

$$\frac{1}{4}\left[g^{-1} + \Pi_{\mathbf{q}}^{11} + \Pi_{\mathbf{q}}^{12} + \left(K_\omega^{11}\right)^{-1}\right] \times \left[g^{-1} + \Pi_{\mathbf{q}}^{11} - \Pi_{\mathbf{q}}^{12} + \left(K_\omega^{22}\right)^{-1}\right] + \frac{1}{4}\left[\left(K_\omega^{12}\right)^{-1}\right]^2, \tag{8}$$

with $2K_{\mathbf{q}}^{-1} = g^{-1} + \widehat{\mathbf{\Pi}}_{\mathbf{q}}$, and $S_0 = -\sum_q \mathrm{Tr}\ln\mathbf{G}_0^{-1}(\Delta_0)$. The expression (8) provides the core result in this section: the complete set of self-consistent equations that describes the Helmholtz free energy associated to Gaussian fluctuations in the field η, via the relationship $\mathcal{F}(\Delta_0, T, \mu) = -\beta^{-1}\ln\mathcal{Z}$, for any temperature between T and T_C, and under the regularization condition in the BSC-BEC crossover scenario (Eq. 4). A similar expression for the full dynamics in the fluctuating field $\eta(\omega, \mathbf{q})$ is also obtained: $\langle|\eta|^2\rangle_{\mathbf{q},\omega} = \mathrm{Tr}\{K_{\mathbf{q},\omega}^{-1}\}/\det\left(K_{\mathbf{q},\omega}^{-1}\right)$, where $\mathrm{Tr}\{\cdots\}$ indicates the Trace operation over the embraced quantity $\{\cdots\}$. Explicit expressions for the quantities $\Pi_{\mathbf{q}}^{11}$ ($\Pi_{\mathbf{q}}^{12}$) are given in Appendix A, while $\mathrm{Tr}\{K_{\mathbf{q},\omega}^{-1}\} = g^{-1} + \Pi_{\mathbf{q}}^{11} + (K_\omega^{11})^{-1} + (K_\omega^{22})^{-1}$. In the limit of small $\mathbf{q}$−momenta, $\Pi_{\mathbf{q}}^{11} \pm \Pi_{\mathbf{q}}^{12} \approx \bar{c}_\pm(\Delta_0, T, \mu)(q^2/2m)$.

4. Collective Modes

Collective modes are calculated in analytical form by applying the condition over the kernel $K_{\mathbf{q},\omega}^{-1}$: $\det\left(K_{\mathbf{q},\omega}^{-1}\right)=0$. For small values of the momentum and frequencies, we obtain:

$$\left[A-D\omega^2+\bar{c}_+\left(\frac{q^2}{2m}\right)\right]\left[-R\omega^2+\bar{c}_-\left(\frac{q^2}{2m}\right)\right]-B^2\omega^2=0, \tag{9}$$

with $A, B, \bar{c}_\pm, D$ and R as temperature-dependent functions [12]: $A=\sum_{\mathbf{k}}\Delta_0^2\bar{X}_{\mathbf{k}}/2E_{\mathbf{k}}^3$, $R=\sum_{\mathbf{k}}\bar{X}_{\mathbf{k}}/8E_{\mathbf{k}}^3$, $D=\sum_{\mathbf{k}}\left(1-\Delta_0^2/E_{\mathbf{k}}^2\right)\bar{X}_{\mathbf{k}}/8E_{\mathbf{k}}^3$, $B=\sum_{\mathbf{k}}\xi_{\mathbf{k}}\bar{X}_{\mathbf{k}}/4E_{\mathbf{k}}^3$. The parameters $\bar{c}_\pm$ are given explicitly in the appendix A. Results for numerical simulations are illustrated in figures (1) and (2).

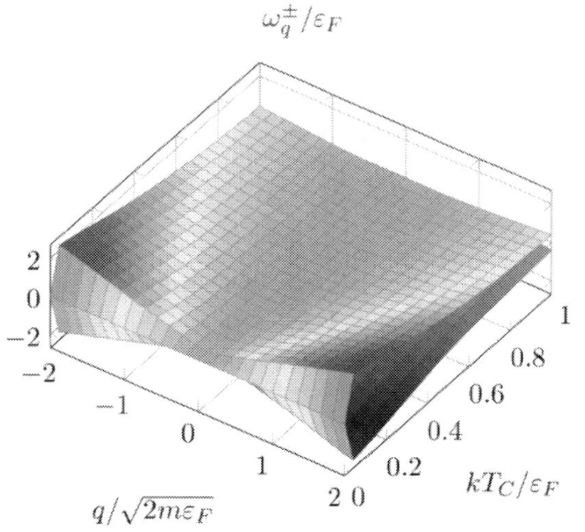

Figure 1: Low-energy topology surface for collective modes mapped through $\omega_q^\pm$ at $\mu\approx 0$, and $x=kT_C/\varepsilon_F<1$. $x\to 0$ limit corresponds to BCS regime, while $x\to\infty$ is characteristic to the pure Bose state.

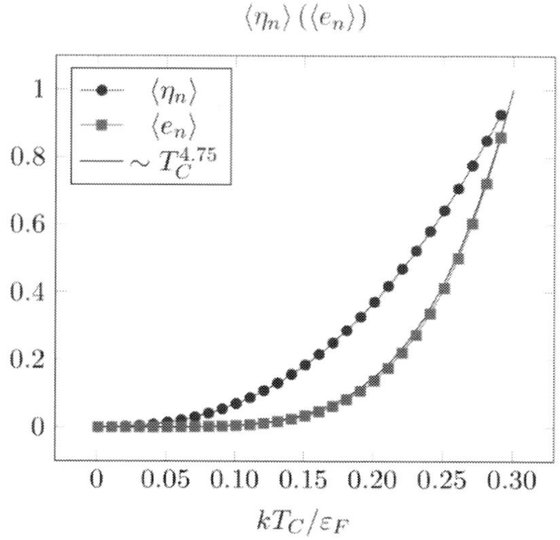

Figure 2: Order parameter and average thermal energy activated by the low frequency collective modes as a function of the critical temperature T_C. Solid line represents the scaling law for the normalized thermal energy $\langle e_n\rangle$.

$$\bar{c}_\pm=\sum_{\mathbf{k}}\left[\left(\frac{\bar{X}_{\mathbf{k}}}{4E_{\mathbf{k}}^2}-\frac{\beta\bar{Y}_{\mathbf{k}}}{8E_{\mathbf{k}}}\right)Z_{\mathbf{k}}^\pm+\frac{\beta^2\bar{X}_{\mathbf{k}}\bar{Y}_{\mathbf{k}}}{2}Z_{\mathbf{k}}'^\pm\right]. \tag{10}$$

Figure (1) illustrates collective excitations of formed pairs in two regimes: for $x\to 0$ which corresponds to the weak interaction coupling, or just the (BCS) state with ε_F playing the role of an effective cutoff energy [13]; and the strong coupling (labeled as $x\to\infty$), which corresponds to bounded states for a dilute gas of small Cooper pair molecules (Bose gas) with a condensation temperature close to $0.218\varepsilon_F$. Integrals involving in $A, B, \bar{c}_\pm ...$, peak around the Fermi surface at $T\to 0$ (BCS state) for the weak-coupling limit. In these case $A=N_n(\varepsilon_F)/2$, $B=N_n(\varepsilon_F)/4\Delta_0=\bar{c}_-$, $R=N_n(\varepsilon_F)/8\Delta_0^2$, $D=R/3$, $\bar{c}_+=2\pi N_n(\varepsilon_F)/9\Delta_0$, with $N_n(\varepsilon_F)$ as the normal state density function at zero temperature, which is related with the superfluid density function as $N_s(E_{\mathbf{k}})\,dE_{\mathbf{k}}=N_n(\xi_{\mathbf{k}})\,d\xi_{\mathbf{k}}$. In the tractable limit at $T\sim T_C$, $\Delta_0=0$; $A=0$, $D=R=7N(\varepsilon_F)\zeta(3)/16\pi^2T_C^2$, $\bar{c}_+=\bar{c}_-=4\varepsilon_F D/3$ and

$B = \left(2\sqrt{2}-1\right)\zeta(3/2)N(\varepsilon_F)/(16\pi T_C\varepsilon_F)^{1/2}$. Neglecting terms of the form ω^4, q^4 and $\omega^2 q^2$, we get a first approach for the collective excitations modes: $\omega(\mathbf{q}) = v_s(T,\Delta_0,\mu) \mid \mathbf{q} \mid$, with the *sound velocity* given by: $v_s(T,\Delta_0,\mu) = \sqrt{A\bar{c}_-/2m(B^2+AR)}$. Under this conditions, the dispersion relationship for ω_q at low critical temperature takes the linear form $\omega_q = \sqrt{\bar{c}_-/2mD}q = v_s q$, and v_s, the *sound* velocity in the condensate, takes the value $v_s = v_F/\sqrt{3}$, ($v_F = \sqrt{2\varepsilon_F/m}$ as the Fermi velocity) which corresponds to the well-known result for the *phase* mode. Figure (2) illustrates the normalized thermal energy due to the collective excitations $\omega_q^j = \pm\frac{1}{2}\left[-B/D \pm \sqrt{(B/D)^2 + 4(\bar{c}_-/D)(q^2/2m)}\right]$ (in the combination $\pm;+$) for a temperature close to the critical temperature T_C. It is defined as $\langle e(T_C)\rangle = \sum_{\mathbf{q},j}\omega_q^j n_B\left(\omega_q^j, T_C\right)$, where $n_B^{-1}\left(\omega_q^j, T_C\right) \equiv \left(\exp\left(\beta_C\omega_q^j\right)-1\right)$ is the Bose-Einstein statistical factor, $\beta_C^{-1} = kT_C$. By performing complete integration on the $\mathbf{q}$ space and physically meaningful j-modes, the thermal energy contribution associated to the collective excitations for the BEC state scales as $T_C^{19/4}$ in the strong coupling regime, in contrast with the pure linear phase mode contribution in which $\langle e\rangle \sim T_C^4$. Thermal energy and order parameter at critical temperature behavior have been normalized out of its corresponding values at $kT_C/\varepsilon_F = 0.3$, and denoted as $\langle e_n\rangle$ and $\langle \eta_n\rangle$ respectively. Further efforts might be orientated in studying high-order expansions in $\mathbf{\Sigma}$ and their *renormalization* over coefficients in (9) for different pairing mechanisms [14], as well as specific effects of artificially engineered trapping potentials in ultracold gases arrays of lattices.

5. Acknowledgements:
This work was financially supported by Universidad Nacional de Colombia, FCEN- Manizales.

6. Appendix A
The equation (10) is partially decrypted for $\bar{X}_{\mathbf{k}} = \tanh(\beta E_{\mathbf{k}}/2)$, $\bar{Y}_{\mathbf{k}} = \mathrm{sech}^2(\beta E_{\mathbf{k}}/2)$. Notice that for $\Delta_0 = 0$, $Z_{\mathbf{k}}^+ = Z_{\mathbf{k}}^-$ and $Z_{\mathbf{k}}'^+ = Z_{\mathbf{k}}'^-$, results obtained by Randeira *et-al.* in reference [8]. For $\Delta_0 \neq 0$, in simplified notation: $Z^+ = \xi/E\left(1-3\Delta_0^2/E^2\right) + (\mathbf{k}\cdot\hat{\mathbf{n}})^2/mE\left(-1+10\left(\Delta_0^2/E^2-\Delta_0^4/E^4\right)\right)$; $Z'^+ = \Delta_0^2\xi/4E^3 + (\mathbf{k}\cdot\hat{\mathbf{n}})^2\left(1-4\Delta_0^2/E^2+6\Delta_0^4/E^4\right)/4mE$; $Z^- = \xi/E - (\mathbf{k}\cdot\hat{\mathbf{n}})^2\left(1-3\Delta_0^2/E^2\right)/mE$; $Z'^- = (\mathbf{k}\cdot\hat{\mathbf{n}})^2\left(1-\Delta_0^2/E^2+2\Delta_0^4/E^4\right)/4mE$. Equation (8) for $\det\left(K_{\mathbf{q},\omega}^{-1}\right)$ is completely unveiled with $\Pi_{\mathbf{q}}^{11} \pm \Pi_{\mathbf{q}}^{12} = \sum_{\mathbf{k}} \frac{E_{\mathbf{k}+\mathbf{q}}\left(\xi_{\mathbf{k}}\xi_{\mathbf{k}+\mathbf{q}}-E_{\mathbf{k}}^2\mp\Delta_0^2\right)\tanh(\beta E_{\mathbf{k}}/2) - E_{\mathbf{k}}\left(\xi_{\mathbf{k}}\xi_{\mathbf{k}+\mathbf{q}}-E_{\mathbf{k}+\mathbf{q}}^2\mp\Delta_0^2\right)\tanh\left(\beta E_{\mathbf{k}+\mathbf{q}}/2\right)}{2E_{\mathbf{k}}E_{\mathbf{k}+\mathbf{q}}\left(E_{\mathbf{k}}^2-E_{\mathbf{k}+\mathbf{q}}^2\right)}$.

7. References
[1] Randeira M, Trivedi N, Moreo A, Scalettar R 1992 *Physical Review Letters* **69**, 2001
[2] Ohashi Y and Griffin A 2002 *Physical Review Letters* **89** 130402
[3] Bruun G and Pethick C 2004 *Physical Review Letters* **92** 140404
[4] Floerchinger S, Scherer M, Dielh S and Wetterich B 2008 *Physical Review B* **78** 174528
[5] Ohashi Y 2009 *J. Phys.: Conf. Series* **150** 032076
[6] Du J, Liang J and Liang J Q 2012 *Physical Review A* **85** 033610
[7] Salasnich L 2011 *Physical Review A* **83** 033630
[8] Engelbrecht J R, Randeira M and Sá de Melo C A R 1997 *Physical Review B* **55** 15153
[9] Auerbach A 1998 *Interacting Electrons and Quantum Magnetism* Springer Verlag (New York) pp 255
[10] Popov V N 1988 *Functional Integrals and Collective Excitations* vol II Cambridge University Press (Cambridge)
[11] Nozières P and Schmitt-Rink S 1985 *J. Low Temp. Phys.* **59** 195
[12] Vivas E H 2005 preprint http://arxiv.org/abs/cond-mat/0504600
[13] Sá de Melo C A R, Randeira M and Engelbrecht J R 1993 *Physical Review Letters* **71** 3202
[14] Baranov M A 2008 *Physics Reports* **468** 71

Journal of Physics: Conference Series **568** (2014) 012025 doi:10.1088/1742-6596/568/1/012025

Non-universal Casimir effect in saturated superfluid ^{4}He films at T$_\lambda$

John B. S. Abraham[1], **Gary A. Williams**[1], **and Konstantin Penanen**[2]

[1] Department of Physics and Astronomy, University of California, Los Angeles, CA 90095

[2] Jet Propulsion Laboratory, California Institute of Technology, Pasadena, CA 91109

E-mail: gaw@ucla.edu

Abstract. Measurements of Casimir effects in ^{4}He films in the vicinity of the bulk superfluid transition temperature T_λ have been carried out, where changes in the film thickness and the superfluid density are both monitored as a function of temperature. The Kosterlitz-Thouless superfluid onset temperature in the film is found to occur just as the Casimir dip in the film thickness from critical fluctuations becomes evident. Additionally, a new film-thickening effect is observed precisely at T_λ when the temperature is swept extremely slowly. We propose that this is a non-universal Casimir effect arising from the viscous suppression of second sound modes in the film.

Saturated liquid ^{4}He films, in contact with the bulk liquid, form an illuminating model system as a condensed matter analogue of the electromagnetic Casimir effect. Thermal fluctuations in the film are limited by the finite thickness of the film, leading to a free-energy difference with the unlimited fluctuations in the bulk [1]. This causes a change in the film thickness as the atoms move to minimize the free energy. The equilibrium thickness is determined by an energy-balance relation per atom of mass m for a film of thickness d at a height h above the bulk helium surface,

$$mgh = U_{vdW} + K_{crit} + K_G + K_2 \quad , \tag{1}$$

where the substrate van der Waals interaction U_{vdW} at the film surface [2], including retardation effects, is given by $(\gamma_0/d^3)(1+d/d_{1/2})^{-1}$. This is the main term determining the film thickness, with γ_0 the van der Waals constant equal to 3.59×10^{-13} erg Å^3 for a Cu substrate, and $d_{1/2}$ = 193 Å .

The fluctuation-induced Casimir forces K then produce small additional shifts Δd as the temperature is swept near the bulk superfluid transition temperature T_λ. The force K_{crit} comes from critical fluctuations near the transition temperature, i.e. the vortex loops in the model of Ref. [3]. Finite-size scaling theories [4, 5, 6] show that $K_{crit} = V k_B T_\lambda \theta(x)/d^3$ where $\theta(x)$ is a universal scaling function of the variable $x = t(d/\xi_0)^{1/\nu}$ with $t = (T - T_\lambda)/T_\lambda$, $\nu = 0.6717$, ξ_0 = 1.43 Å the amplitude of the coherence length above T_λ (using the normalization of Ref. [7]), and V = 45.8 Å^3 the volume per helium atom. Experiments [8, 9, 10, 11] have observed a dip in the film thickness corresponding to these critical fluctuations in agreement with simulations [12], and data collapse of films of different thickness confirmed the form of the scaling. One factor not well determined in experiments to date is the location of the Kosterlitz-Thouless (KT) superfluid transition relative to the dip in film thickness. A quartz microbalance oscillator technique at 5 MHz [10] seemed to show the transition as occurring at a temperature somewhat below the bottom of the dip, but the exact location could not be pinpointed because the high frequency greatly broadens the KT transition [13, 14, 15].

The term K_G in Eq. 1 refers to the free energy difference between bulk and film due to thermally excited Goldstone modes[16]. It was originally proposed that such modes should include the second sound mode in the case of liquid helium [11, 17, 18], but since we question this we include instead a separate term K_2 for second sound, to be discussed below. For Goldstone modes such as spin waves in the XY model the free energy difference arises because standing waves form in the direction perpendicular to the film surface, eliminating modes that can still propagate in the bulk. This gives $K_G = Vk_BT\zeta(3)/(8\pi d^3)$, a result well confirmed in computer simulations [12, 7, 19, 20] and analytic calculations [21] for XY spin waves. The experiments in helium did find a difference in film thickness between low temperatures and above T_λ similar to but larger than this prediction [9, 11]; however a later theory [18] showed that this was primarily due to energy differences in the waves at the free surface of the film and bulk, resulting in a thinning magnitude nearly double the the expression for K_G above.

We doubt very much that second sound propagates at all in saturated helium films. Second sound involves counterflow of the normal fluid of viscosity η and density ρ_n, and in a film this will halt any flow at angular frequencies ω where the viscous penetration depth $l_v = (2\eta/\rho_n\omega)^{1/2}$ is larger than the film thickness [22]. For a film with $d = 300$ Å near T_λ this is all frequencies below $\omega/2\pi = 6$ MHz. Any possible modes at higher frequencies in the film would necessarily be waveguide modes reflecting from the free surface, and because second sound involves temperature oscillations these would excite sound waves in the helium vapor, leading to very strong attenuation from the dissipative Onsager reciprocity relations for the evaporation-condensation process [22, 23]. Such waveguide modes coupled to vapor in bulk helium have been observed to be strongly attenuated (Q values of order 5) at only a few hundred Hz [23], so we believe there is almost no probability they could propagate at MHz frequencies in the film.

If there is no second sound in the film, then K_2 in Eq. 1 will just be equal to F_2V, where F_2 is the free energy per unit volume of second sound in the bulk superfluid. Second sound is well known to propagate in the bulk to within microkelvins of T_λ [24, 25], but it then ceases propagation above the superfluid transition, which will cause K_2 to drop to zero. This means from Eq. 1 that there must be a step increase at T_λ in the film thickness, which to linear order in $\Delta d/d$ is

$$\frac{\Delta d}{d} = \frac{-F_2(T_\lambda)V}{3\,U_{vdW}}\frac{\left(1+\frac{d}{d_{1/2}}\right)}{\left(1+\frac{4}{3}\frac{d}{d_{1/2}}\right)} \quad . \tag{2}$$

We report here the experimental observation of just such a step increase in the thickness at precisely T_λ, but which only becomes apparent for an extremely slow sweep near the transition. Also, in studies of the critical-fluctuation Casimir effect we have used third sound measurements on the same films to determine the KT onset temperatures, and find that the onset occurs at the very beginning of the dip in thickness.

Our experimental cell is mounted in a vacuum space inside a conventional 1 K pot cryostat. There are three intermediate thermal stages between the 1 K pot held at 1.9 K and the experimental stage, which minimizes thermal gradients. The temperature of the experimental stage is regulated by applying feedback heat to the intermediate stages using PID temperature control, while the cell itself is not directly regulated. The cell temperature is measured with a germanium thermometer attached to the outside of the cell, with a resolution of about 10 μK. The position of T_λ can be determined to within 30 μK by the change in the temperature drift of a carbon thermometer at the end of a stainless tube extending from the bottom of the cell. All thermal connections are made with gold plated copper press joints to maximize the thermal conductivity. The cell fill line capillary is connected to the bottom of the cell through

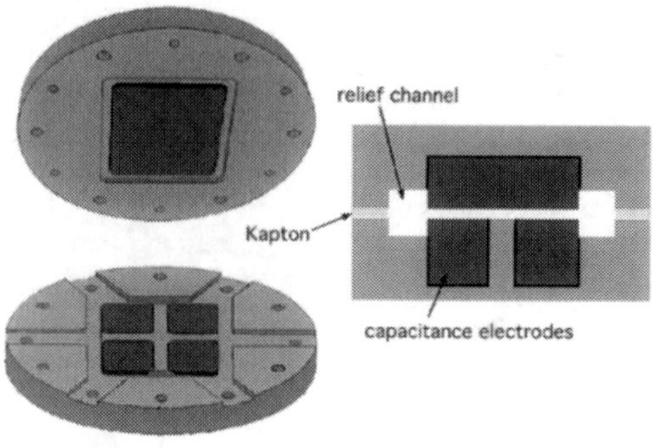

Figure 1. Schematic of the capacitor electrodes.

a pressure actuated cold valve, so that the cell can be isolated from other parts of the cryostat during measurement.

Our capacitor substrates (shown schematically in Fig. 1) are fashioned from OFHC copper, which after assembly in copper holding plates with Stycast 1266 epoxy, are diamond-turned to give a surface roughness of less than 10 nm, as checked with an atomic force microscope. The capacitor plates are cleaned and assembled in a microfabrication facility and clean room to minimize any possible dust. The lower capacitor plate is partitioned into four rectangular quadrants of size 0.9×1.2 cm, in order to probe the He films with third sound. 2 mm grounding strips minimize any cross talk between the quadrants. A relief channel 1 mm deep and 2.5 mm wide is machined out of the region surrounding both the upper and lower electrodes to prevent any capillary filling. The gap is set by placing a spacer of 60 μm thick Kapton film between the substrate holders, touching only at the outer edge of the relief channel. Six channels in the lower holder allow the helium film and vapor to enter freely.

The empty cell capacitance of a quadrant is approximately 15 pF. An identical electrode assembly is bolted to the outside of the sealed cell in the vacuum, and forms a reference capacitor for a ratio transformer bridge. The ratio transformer at room temperature is thermally insulated and temperature-regulated to 0.1 C. The thickness of the film is measured from the capacitance ratio by subtracting out the vapor contribution to the cell capacitance, using the same methods as Refs. [9, 11]. Third sound is generated in the film by applying an oscillating electric field to one of the quadrants, sweeping slowly from 3-20 Hz, and detecting resonant modes by the capacitance change of the diagonal quadrant [26]. The equilibrium film thickness well below the λ point is actually most accurately measured with the third sound speed [27].

Figure 2 shows the change in film thickness for several films near T_λ at a temperature sweep rate about 400 μK/hr. The Casimir force K_{crit} from critical fluctuations is readily apparent, causing a dip in the film thickness when the coherence length equals the film thickness, in good agreement with the previous measurements. Figure 3 shows the Casimir force scaling function, with the same good data collapse as also seen previously [9, 11]. At the dip minimum we find $x_{min} = -6.0(5)$ and $\theta_{min} = -1.2(1)$, compared with $x_{min} = -5.7(5)$ and $\theta_{min} = -1.30(3)$ found previously. This can also be compared with Monte Carlo simulations on XY spins with free boundary conditions at the top and bottom boundaries [7, 19, 20], which find $x_{min} = -5.3(1)$ or -5.43(2), and $\theta_{min} = -1.35(3)$ or -1.396(6).

Our ability to monitor third sound in these films allows us to determine the Kosterlitz-Thouless onset relative to the dip in thickness. The superfluid density averaged across the film can be extracted from the third sound speed, shown as the red symbols in Fig. 3 plotted on the right-hand axis. The signal can no longer be detected past the last points plotted, and at

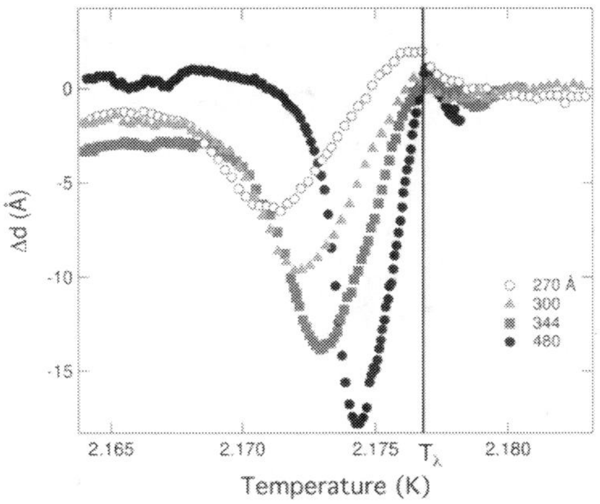

Figure 2. Change in the film thickness versus temperature.

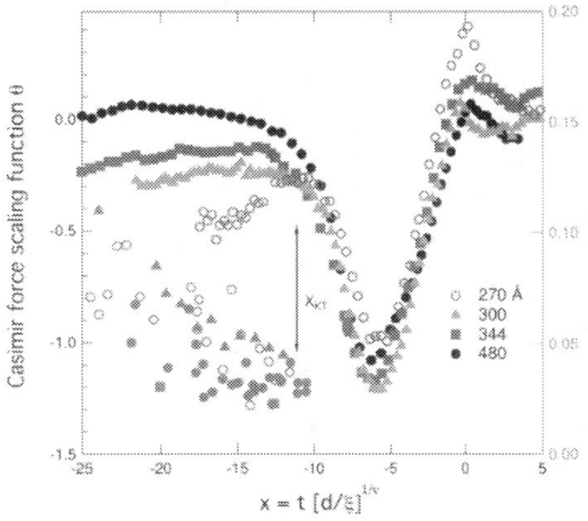

Figure 3. Casimir force scaling function from the data of Fig. 2 (left axis), and average superfluid density (right axis).

that point the value of $\bar{\rho}_s d/T_{KT}$ averaged over the different films is 4.7±1.5 ng/cm^2 K, only somewhat above the universal KT value of 3.5 ng/cm^2 K. This occurs at a value x_{KT} = -11.1(4) shown by the arrow. These very low frequency (4-5 Hz) onset values of T_{KT} are consistent with previous results in films near the lambda point [28]. The onset occurs at the very beginning of the dip in thickness, a result predicted in the vortex-loop theory [3] (though only for periodic boundary conditions). The Monte Carlo simulations [7] give the KT transition at x_{KT} = -7.6, more in the middle of the dip, but this is due to the fact that XY spins have a higher vortex core energy than found for helium films [29], and the higher core energy shifts T_{KT} to higher temperatures.

We were initially puzzled by the relative maximum in the film thickness near the transition temperature T_λ seen in Figs. 2 and 3, which was also observed in the quartz microbalance experiment [10]. We found, however, that this feature was quite dependent on the temperature sweep rate. When we increased the sweep rate to about 2 mK/hr, we found only a smooth increase through T_λ to a constant value, very similar to the behavior seen in the initial measurements of Ref. [9]. In this faster sweep we found that the dip region was essentially unchanged from the slower sweep case, even for temperatures both above and below T_{KT}. We ascribe this behavior to the fact that even above T_{KT} there is still finite-frequency superfluidity in the film [13, 14, 15], which can relatively rapidly change the film thickness and even out any free-energy gradients.

Given the unusual behavior at T_λ, we decided to sweep as slowly as we could near the transition, starting a few hundred μK below T_λ and taking upward steps of 20 μK, equilibrating for 3-5 hours at each point for the temperature and capacitance to stabilize. The results are shown in Figure 4 for three film thicknesses. Below T_λ there is little variation over this very restricted range, but right at T_λ a sudden step increase in the thickness occurs, with the magnitude of the increase a strong function of the equilibrium film thickness. We suspended the temperature stepping once the thickness started to increase; for example with the 480 Å film the increase was followed for about 15 hours before it started to level off and the temperature steps were resumed. By checking different quadrants we found the same value of the step increase, showing that the film is uniform across the cell and that this effect is not due to capillary condensation, which would tend to localize the thickness change to one quadrant. We also checked that this effect is present for temperature sweeps both up and down in temperature

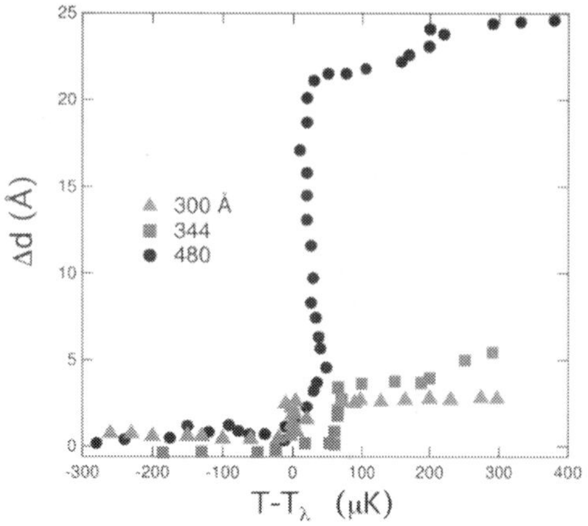

Figure 4. Change in film thickness observed at the superfluid transition for extremely slow temperature sweeps.

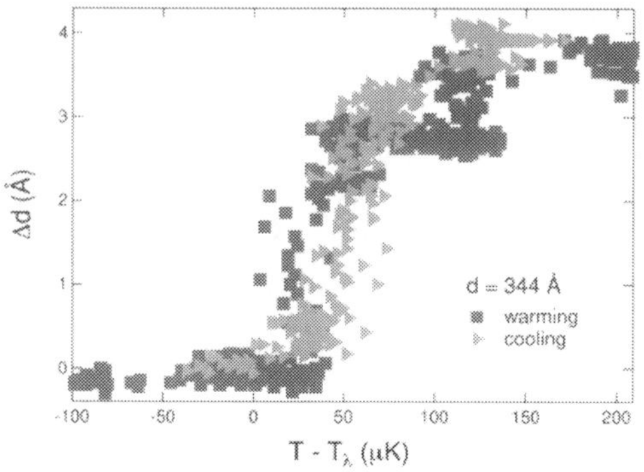

Figure 5. Change in the 344 Å film thickness for sweeps warming and cooling through the transition.

through the lambda point, shown in Fig. 5 for two sweeps of the 344 Å film, taken a week apart. To within our temperature and thickness resolution, these sweeps show that the step behavior is quite reproducible. The small offset in the exact temperature where each step occurs in Fig. 4 is not important; this is within the drift specification of our temperature bridge over the two weeks needed for each sweep.

Figure 6 shows the relative change in film thickness $\Delta d/d$ at the transition as a function of d, which increases rapidly. This is quite consistent with the rapid increase predicted by Eq. 2, since U_{vdW} is a strong function of d. The solid line in Fig. 6 is a fit to a rootfinder solution of the full Eq. 1, with the second-sound free energy F_2 as the only free parameter, yielding F_2 = -3.5 erg/cm^3. To our knowledge, there has never been a calculation or measurement of the free energy of second sound in bulk helium to compare with our value for F_2 at T_λ. This would require a detailed theory of the properties of thermally excited second sound near T_λ at MHz frequencies, a regime where the wavelength is smaller than the correlation length, well outside the low-frequency hydrodynamic regime. There has been some theoretical work in this area [30, 31], but comparison to experiment has only been at the single wavelength investigated in the light-scattering experiments [25].

An initial calculation of the free energy can be attempted using the Debye approximation commonly used for phonons in solids. We assume that the angular Debye cutoff frequency ω_D is the highest frequency where the mode propagates, e. g. where the quality factor $Q \approx 1$. In the limit $\hbar\omega/k_BT \ll 1$ (a good approximation since c_2 is relatively small), the expression for the free energy is

$$F_2 = \frac{k_BT}{2\pi^2 c_2^3}\int_0^{\omega_D} \omega^2 \ln\left(\frac{\hbar\omega}{k_BT}\right) d\omega \tag{3}$$
$$= -\frac{k_BT}{2\pi^2}\left(\frac{\omega_D}{c_2}\right)^3 \left[\frac{1}{9} - \frac{1}{3}\ln\left(\frac{\hbar\omega_D}{k_BT}\right)\right].$$

In evaluating this expression for T very close to T_λ, the value of c_2 at MHz frequencies does not go to zero, due to finite-wavenumber and finite-frequency broadening (seen in the light-scattering measurements [25] and known from theory [30, 31, 32, 33]), but becomes roughly

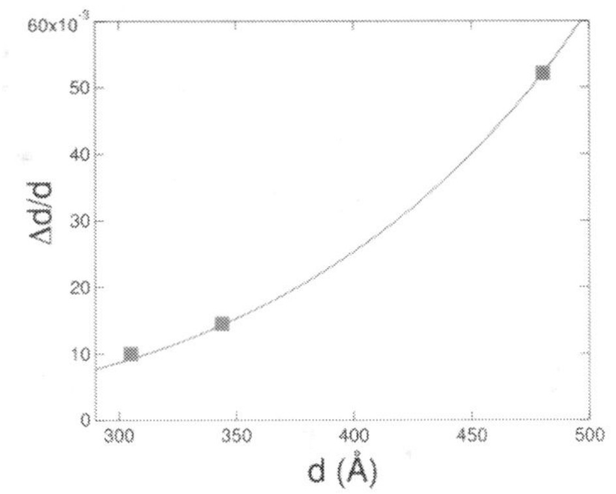

Figure 6. Relative jump in the film thickness at T_λ. The solid line is a fit to Eq. 1, giving $F_2(T_\lambda)$ = -3.5 erg/cm^3. (Color online)

constant starting about 50 μK from T_λ. The light-scattering measurements very close to T_λ are only available at high pressure (23.1 and 28.5 bars), but a rough extrapolation to the saturated vapor pressure of the films using low-frequency measurements [24] gives an estimate for c_2 at T_λ of 80 cm/s. Inserting this value and the fitting result of the step data, $F_2(T_\lambda) \approx$ -3.5 erg/cm^3 into Eq. 3 we find a value of $\omega_D/2\pi \approx 6$ MHz. This is a fairly reasonable order of magnitude estimate, since the Brillouin line in the light scattering data near T_λ at 1.93 bars [34] already shows strong attenuation, with the Q $\approx$ 3 at a center frequency of 4 MHz at 500 μK from the transition. However, the frequency dependence of the attenuation in this non-hydrodynamic regime is not well known. There is certainly a need for a more detailed theory of the free energy of second sound close to T_λ.

The long relaxation time for the appearance of the sharp step at T_λ probably arises from the complete lack of superfluidity in both the film and the bulk. The only means of mass transport to change the film thickness is then through the vapor, and this is a very slow mass diffusion process, particularly since the vapor in the 60 μm channel is also viscously constrained. It is very easy to miss the step (as we did numerous times early in this data run); the key procedural difference in observing the effect was to halt the (already very slow) temperature sweep at the first sign of even a slight change in the capacitance [35].

We note that the first sound mode very likely does not play any role in this Casimir effect in helium films. First sound is known to propagate to very high frequencies in both the bulk [36] and in films [37]. For such wavelengths smaller than the film thickness there will be little difference between the free energy of the film and the bulk, and no change at the lambda point since the modes exist both below and above T_λ.

In summary, we identify the onset of the KT transition in saturated helium films to occur right at the start of the Casimir dip in film thickness. In addition we observe a new step increase in the film thickness at T_λ that we attribute to the abrupt cessation of second sound in the bulk liquid. Since this Casimir effect is related to the unique properties of second sound in helium (the viscous suppression of the mode in films) and involves material-dependent quantities such as c_2, it will not be a universal property of XY superfluids.

Acknowledgements
We thank Joseph Rudnick for very useful discussions, Hong-Wen Jiang for the use of the AFM, and Vincent Wong for help with data analysis. A portion of this work was carried out at the Jet

Propulsion Laboratory, California Institute of Technology, under a contract with the National Aeronautics and Space Administration, and was also supported in part by the National Science Foundation, grant DMR-0906467.

References

[1] Krech M 1994 *The Casimir Effect in Critical Systems* (World Scientific)
[2] Cheng E and Cole M W 1988 *Phys. Rev. B* **38** 987
[3] Williams G A 2004 *Phys. Rev. Lett.* **92** 197003
[4] Fisher M E and DeGennes P G 1978 *C.R. Acad. Sci. (Paris)* **28** 207
[5] Danchev D 1996 *Phys. Rev. E* **53**(3) 2104 ISSN 1063-651X
[6] Krech M 1999 *J. Phys. Condens. Matter* **11** R391
[7] Vasilyev O, Gambassi A, Maciolek A and Dietrich S 2009 *Phys. Rev. E* **79** 041142
[8] Dionne R J and Hallock R B 1989 (New York: AIP) p 199 AIP Conf. Proc. No. 194
[9] Garcia R and Chan M H W 1999 *Phys. Rev. Lett.* **83** 1187
[10] Garcia R, Jordan S, Lazzaretti J and Chan M H W 2004 *J. Low Temp. Phys.* **134** 527
[11] Ganshin A, Scheidemantel S, Garcia R and Chan M H W 2006 *Phys. Rev. Lett.* **97** 075301
[12] Hucht A 2007 *Phys. Rev. Lett.* **99** 185301
[13] Bishop D and Reppy J D 1978 *Phys. Rev. Lett.* **40** 1727
[14] Hieda M, Matsuda K, Kato T, Matsushita T and Wada N 2009 *J. Phys. Soc. Jap.* **78** 033604
[15] Ambegaokar V, Halperin B, Nelson D and Siggia E 1980 *Phys. Rev. B* **21** 1806
[16] Li H and Kardar M 1992 *Phys. Rev. A* **46** 6490
[17] Kardar M and Golestanian R 1999 *Rev. Mod. Phys.* **71** 1233
[18] Zandi R, Rudnick J and Kardar M 2004 *Phys. Rev. Lett.* **93** 155302
[19] Hasenbusch M 2009 *J. Stat. Mech.* P02005
[20] Hasenbusch M 2010 *Phys. Rev. B* **81** 165412
[21] Dohm V 2013 *Phys. Rev. Lett.* **110** 107207
[22] Bergman D J 1971 *Phys. Rev. A* **3** 2058–2066
[23] Williams G A 1983 *J. Low Temp. Phys.* **50**(5-6) 455 ISSN 0022-2291
[24] Greywall D S and Ahlers G 1973 *Phys. Rev. A* **7**(6) 2145–2162
[25] Tarvin J A, Vidal F and Greytak T J 1977 *Phys. Rev. B* **15**(9) 4193–4210
[26] Hoffmann J A, Penanen K, Davis J C and Packard R E 2004 *J. Low Temp. Phys.* **135** 177
[27] Abraham J B S 2013 Ph.D. thesis UCLA
[28] Gasparini F M, Kimball M O, Mooney K P and Diaz-Avila M 2008 *Rev. Mod. Phys.* **80**(3) 1009–1059
[29] Cho H and Williams G A 1995 *Phys. Rev. Lett.* **75** 1562
[30] Halperin B I, Hohenberg P C and Siggia E D 1976 *Phys. Rev. B* **13**(3) 1299–1328
[31] Siggia E D 1976 *Phys. Rev. B* **13**(7) 3218–3221
[32] Williams G A 1992 *Phys. Rev. Lett.* **68**
[33] Williams G A 1993 *Phys. Rev. Lett.* **71** 392
[34] Winterling G, Miller J and Greytak T J 1974 *Phys. Lett.* **48A** 343
[35] Ref. [11] reported sweeping through the lambda point at rates between 10 and 40 μK/hr, finding "a small discrepancy very close to T_λ not present in earlier work". It is left unclear exactly what that discrepancy was, but from our experience 40 μK/hr is too fast to observe the step effect we found, particularly since for their relatively thin films the step would be quite small and easy to miss. Even 10 μK/hr would be too fast if the sweep is not halted and held for many hours at the first sign of a capacitance change, as in our measurements.
[36] Woods A D B and Cowley R A 1970 *Phys. Rev. Lett.* **24** 646
[37] Anderson C H and Sabisky E S 1972 *Phys. Rev. Lett.* **28** 80

Journal of Physics: Conference Series **568** (2014) 012026 doi:10.1088/1742-6596/568/1/012026

Third sound measurements of superfluid ^{4}He films on multiwall carbon nanotubes below 1 K

Emin Menachekanian, John B.S. Abraham, Bob Chen, Vito Iaia, Andrew Li, and Gary A. Williams

Department of Physics and Astronomy, University of California, Los Angeles, CA 90095

E-mail: gaw@ucla.edu

Abstract. Third sound is studied for superfluid films of 4He adsorbed on multiwall carbon nanotubes packed into an annular resonator. The third sound is generated with mechanical oscillation of the cell, and detected with carbon bolometers. A filling curve at temperatures near 250 mK shows oscillations in the third sound velocity, with maxima at the completion of the 4th and 5th atomic layers. Sharp changes in the Q factor of the third sound are found at partial layer fillings. Temperature sweeps at a number of fill points show strong broadening effects on the Kosterlitz-Thouless (KT) transition, and rapidly increasing dissipation, in qualitative agreement with the predictions of Machta and Guyer. At the 4th layer completion there is a sudden reduction of the transition temperature T_{KT}, and then a recovery back to linear variation with temperature, although the slope is considerably smaller than the KT prediction.

We have carried out third sound measurements on multiwall carbon nanotubes at temperatures below 1 K, the same tubes used in the 1.3 K measurements of Ref. [1]. The cell geometry is a little different than those measurements, the tubes (0.75 g) are now lightly packed in a single step into an annular resonator of the type we have used in the past [2], with mean diameter 2.9 cm and thickness 3.2 mm. Fig. 1 shows SEM and TEM photos of the nanotubes, which appear to range in diameter between 10 and 25 nm (larger than the 8-15 nm specified by the supplier [3]) and lengths between 1-2 μm. A superconducting coil and permanent magnet assembly is attached to the resonator using thermally insulating thin-wall stainless steel tubes to mechanically vibrate the resonator, exciting the third sound. The resonant modes are detected from the thermal oscillations by current-biased resistance bolometers, whose output goes into a LabView FFT.

Fig. 2 shows the third sound speed and Q factor as a function of the helium coverage at temperatures near 250 mK. Strong oscillations in the third sound speed are seen, due to the completion of atomic layers, quite similar to those seen in measurements on a highly oriented pyrolitic graphite (HOPG) substrate [4]. We adopt the coverage scale of those authors, taking one layer to be the relative maxima of c_3 at the completion of the fourth and fifth layers, and also using their determination of 0.076 atoms/Å^2 as the layer coverage above the first two (more dense) layers. Although oscillations in c_3 could not be detected at higher coverages, the observation of maxima in the Q factor at the 5 and 6 layer completions shows this to be a reasonable scale. We were completely unable to detect any signal (due to high dissipation) below 3.18 layers at oscillation levels up to 3.0 V applied to our drive coil, about the maximum we could sustain before heating of our dilution refrigerator became a problem. The magnitude of the third sound speed is reduced from that in Ref. [4] due to multiple scattering effects from our

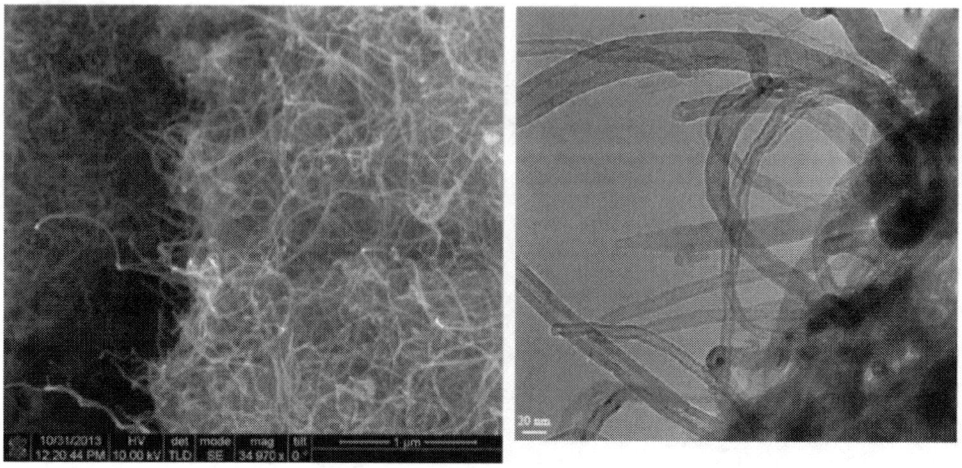

Figure 1. SEM (right) and TEM (left) pictures of the multiwall carbon nanotubes

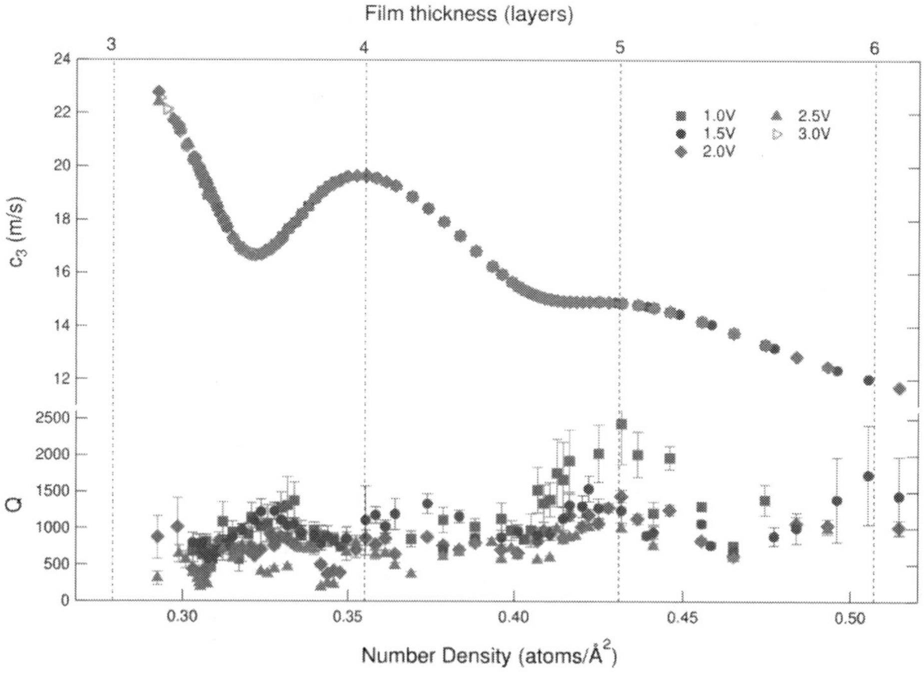

Figure 2. Third sound speed and Q near 250 mK, as a function of the helium coverage.

nanotube tangle; we deduce an index of refraction of about 1.7 comparing to their results. The speed is only very slightly shifted down with increasing drive, while the Q is more significantly lowered (and has quite a bit of scatter).

The Q is initially fairly low when the signal is first observed at 3.18 layers, and then immediately increases to a maximum before dropping rapidly to a minimum at 3.37 layers. The

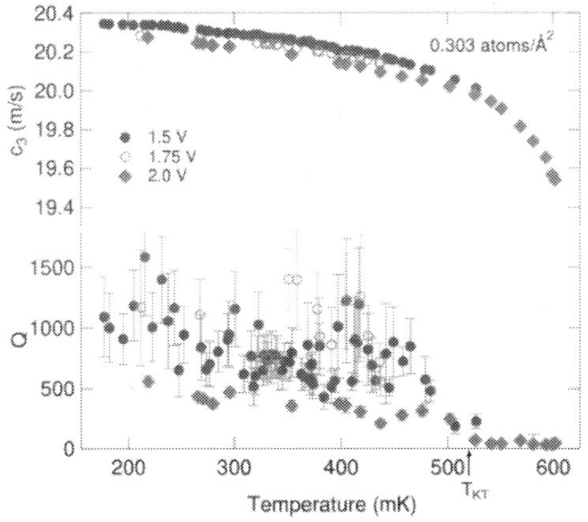

Figure 3. Temperature sweep at 3.32 layers.

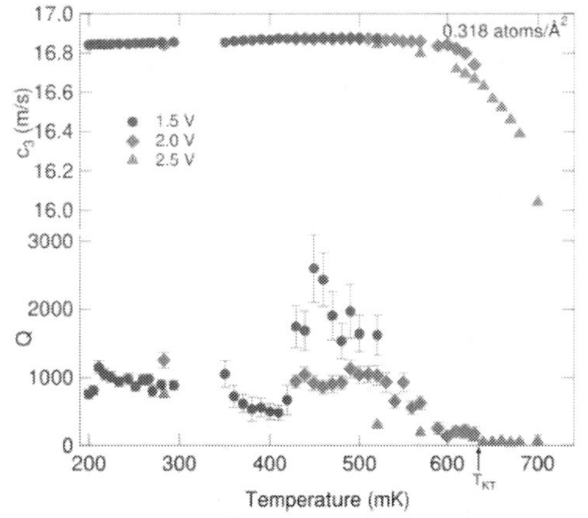

Figure 4. Temperature sweep at 3.51 layers.

origin of this dissipation is unclear, since at this low temperature there are very few thermal excitations or vortices. The formation of a gas-liquid interface at the low-density fill of the start of the fourth layer is a possible source of this attenuation, since evaporation due to the thermal component of the third sound would be dissipative. Such interfaces have been predicted for the second and higher layers of coverage on flat graphite substrates [5], though it is not entirely clear if this has been observed or not [6, 7]. The sharply curved nanotube substrate differs from the flat substrate, however, and nanotube simulations are only available for single-wall nanotubes and very low coverages [8]. With increasing coverage we find then a climb to a maximum Q near 3.5 layers, the minimum of c_3, showing that the high compressibility of the film at that point also plays a role in the Q factor.

Fig. 3 shows a temperature sweep carried out at the coverage of 3.32 layers. At low temperatures there is a slow decrease of c_3, with a polynomial fit showing primarily linear and square components in the decrease. Above 520 mK there is a sharp decrease in the Q, and the beginning of a faster drop in c_3, and we identify this as the onset of the vortex-pair

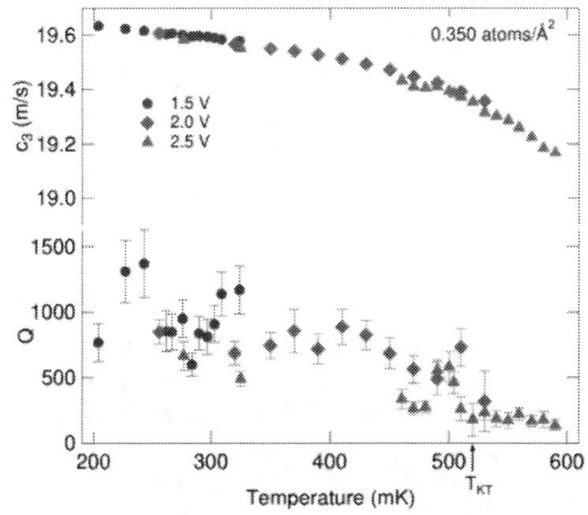

Figure 5. Temperature sweep at 3.93 layers.

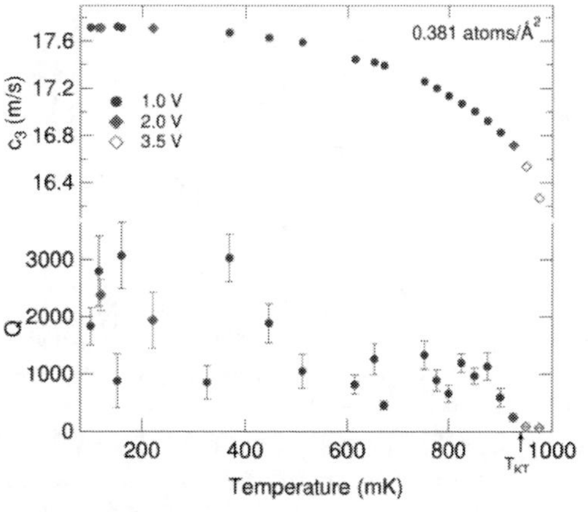

Figure 6. Temperature sweep at 4.34 layers.

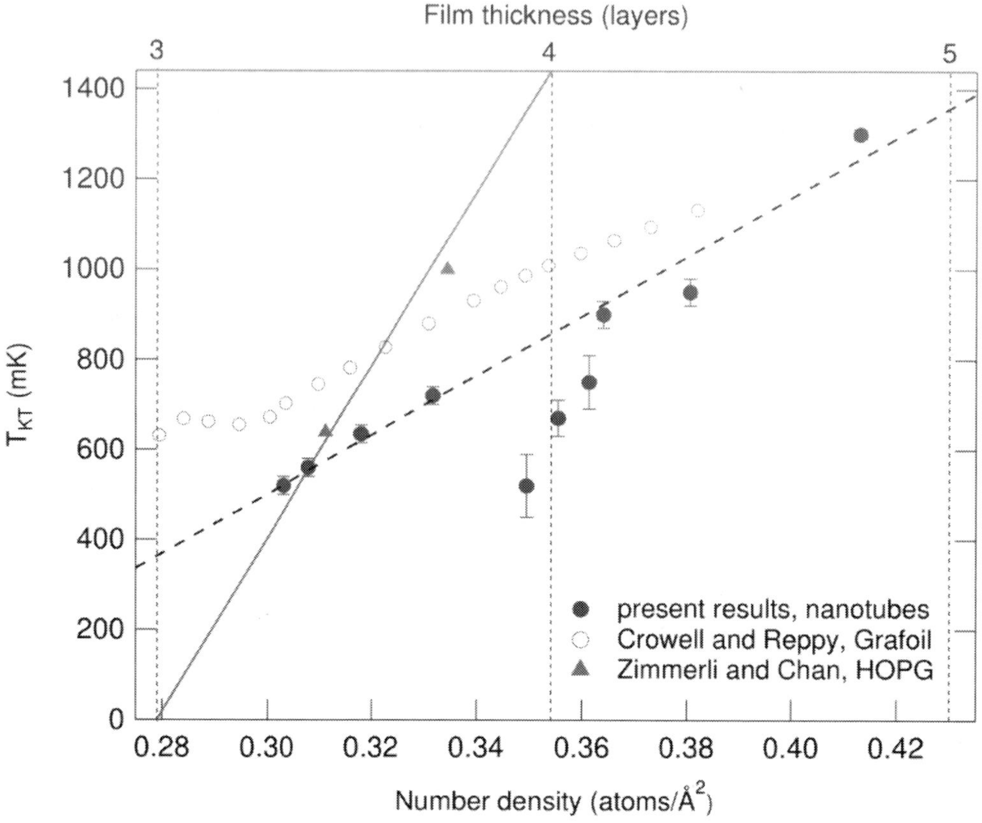

Figure 7. Superfluid onset temperatures versus coverage, compared with Refs. [4, 6].

Kosterlitz-Thouless transition. The very high dissipation prevents following c_3 to lower values, which is quite different behavior from our previous studies of third sound on alumina powder [2] where we could follow the broadened c_3 to as low as 20% of its initial value. This observation of high dissipation on the cylindrical nanotubes, however, is in qualitative agreement with the predictions of Guyer and Machta [9] for the KT transition on cylinders, where the driven vortex pairs can counter-rotate around the cylinder, giving attenuation comparable to that on flat substrates. They also predict a strong finite-size broadening of the transition due to the finite ratio of the nanotube radius to the vortex core radius, which is apparent from the relatively slow decrease of the third sound speed above T_{KT}, though due to the loss of signal this cannot be quantified.

Fig. 4 shows a temperature sweep at 3.51 layers, close to the minimum in c_3. The unusual feature here is the linear increase of c_3 with temperature. This can only occur if the superfluid density increases with temperature, or if the thickness of the film decreases with temperature. The latter possibility seems more likely, since the compressibility of the film is near its maximum here, and might further change with T. However, it is also possible that internal structural changes in the film could lead to an increase in the superfluid density. With this thicker film there is an increase in the onset temperature T_{KT}. At coverages just above the minimum in c_3 the low-temperature speed goes back to uniformly decreasing with T, but now the decrease is entirely linear at the lowest temperatures.

For coverages near the maximum in c_3 at the fourth layer completion, we find a marked depression in the superfluid onset temperatures, as shown in Fig. 5 at 3.93 layers, and then a further increase with higher coverage as in Fig. 6 at 4.34 layers. This is re-entrant superfluidity, as shown in the onset temperatures plotted in Fig. 7: adding coverage at a fixed temperature of say 600 mK would see superfluidity disappear at 3.8 layers, and then reappear at 4.1 layers. This behavior at the fourth layer completion appears to be unique to the nanotube films, as it is not observed on the Grafoil substrate [6], whose onset temperatures are also plotted on the same graph. The Graphoil data only shows a relatively constant T_{KT} near the third layer completion (visible in Fig. 7), and nothing similar at the fourth layer completion. There have been simulations [10] predicting a drop in superfluid density at layer completions, from the increasing atomic repulsion, which would translate to a drop in T_{KT} such as we observe. The effect seems to be much stronger on the nanotubes compared to flat substrates.

The dashed curve in Fig. 7 shows a linear fit to our onset temperatures (excluding the three points near the fourth layer completion), with an onset thickness extrapolating to 2.4 layers, similar to that found in [6] in the third layer. The slope of this curve, however, is only about 1/2 of the expected KT universal value, shown as the solid line in the plot for an onset thickness of three atomic layers as deduced on the HOPG substrate [4]. This is unusual, but it also seems to be the same result found on the Grafoil substrate [6], where we have plotted the reported dissipation maximum, which should track T_{KT}. It appears as if only about 1/2 of the atoms that are being added actually join the superfluid condensate. It is unclear why this occurs: possibilities are that there could be a strong localization of atoms at defects on the substrate, or if gas-liquid phase separation occurs, the gas atoms might not be superfluid, as speculated in [6]. The HOPG substrate experiment, however, appeared to find results consistent with the universal KT line, but unfortunately with only two data points. Our finding in Fig. 2 that third sound could not be observed at any temperature below about 3.2 layers brings up the question of whether the onset seen in Ref. [4] at 3.4 layers and 0.639 K was an actual KT onset, or simply the same third sound signal loss from high dissipation we observed on approaching the third-layer completion point.

We have found no evidence in our measurements of helium adsorption inside the nanotubes, though that may have occurred. The TEM pictures seem to show a few tubes with what appear to be open ends, but also some of the tubes show closed end caps. If helium did enter it would have been at the very beginning of the fill, due to the strong effects of surface tension.

Acknowledgements
We thank William Hubbard and Chris Regan for taking the electron microscope photos at the UCLA NanoSystems Institute. This work was supported in part by the U. S. National Science Foundation, Grant No. DMR 0906467.

References
[1] Menachekanian E and Williams G 2012 *J. Phys. : Conf. Series* **400** 012046
[2] Kotsubo V and Williams G A 1986 *Phys. Rev. B* **33** 6106
[3] URL `www.cheaptubes.com`
[4] Zimmerli G, Mistura G and Chan M H W 1992 *Phys. Rev. Lett.* **68** 60
[5] Clements B, Forbert H, Krotscheck E and Saarela M 1994 *J. Low Temp. Phys.* **95** 849
[6] Crowell P A and Reppy J D 1996 *Phys. Rev. B* **53** 2701
[7] Greywall D S and Busch P A 1991 *Phys. Rev. Lett.* **67** 3535
[8] Gordillo M C and Boronat J 2012 *Phys. Rev. B* **86**(16) 165409
[9] Machta J and Guyer R 1989 *J. Low Temp. Phys.* **74** 231
[10] Zimanyi G, Crowell P, Scalettar R and Batrouni G 1994 *Phys. Rev. B* **50** 6515

Vortex emissions from quantum turbulence generated by vibrating wire in superfluid ^{4}He at finite temperature

Y Wakasa, S Oda, Y Chiba, K Obara, H Yano, O Ishikawa, and T Hata
Graduate School of Science, Osaka City University, Osaka 558-8585, Japan
E-mail: `hideo@sci.osaka-cu.ac.jp`

Abstract. An oscillating object immersed in superfluid helium generates quantum turbulence, emitting quantized vortices to its surroundings. We report vortex emissions in directions parallel and perpendicular to the oscillating motion of a thin wire used as a turbulence generator. Two vibrating wires are used to detect the vortex emissions. We use superfluid ^{4}He as a medium, with the temperature set to 1.25 K, at which a small amount of normal fluid component is present. In this setup, only vortex loops with sizes larger than a certain loop diameter $D = 42$ μm can be detected. In the perpendicular direction, vortex loops are detected when the oscillation amplitude is comparable with D. In the parallel direction, however, no vortex loops are detected at the same amplitude, suggesting an anisotropic emission of vortex loops.

1. Introduction

Turbulent flow in superfluids has a simpler structure than classical turbulence in viscous fluids [1]. Since the circulation of superfluids is quantized, superfluid flows consist only of quantized vortices. These vortices carry the same quantized circulation κ and have very thin cores, e.g. the core radius of a superfluid ^{4}He vortex is $a_0 \sim 0.1$ nm. A disordered tangle of quantized vortices produces turbulence in superfluid flow, namely quantum turbulence. Quantum turbulence is a central issue in low-temperature physics [2].

In experiments, quantum turbulence can be produced by a vibrating obstacle immersed in superfluid helium, even at very low temperature [3, 4]. Since a quantized vortex cannot have an unconnected end in superfluid helium, a vortex line forms a loop or it ends at the superfluid boundary. If the boundary is moved, superfluid flows arise along the surface of the boundary, causing vortex lines remaining at the boundary to move. A strongly vibrating motion of an object stretches vortex lines attached to the object, producing vortex loops by reconnection between vortex lines. Some of the vortex loops move away from the oscillation path, and others collide with the object during the return motion, producing further vortex loops. Thus, a vortex tangle forms in the oscillation path and vortex loops are emitted from the vortex tangle.

The motions of vortices are described as being governed only by quantized circulations in superfluids at the zero temperature limit. A circular vortex loop, for instance, moves in a superfluid sea, propelled by quantized circulation with a velocity [5]

$$v = \frac{\kappa}{4\pi R}\left(\ln\frac{8R}{a_0} - \frac{1}{2}\right), \tag{1}$$

where R is the radius of the vortex loop. Despite the simplicity of the description, experimental studies on the motion of vortex loops are quite difficult because there are few methods for vortex observation.

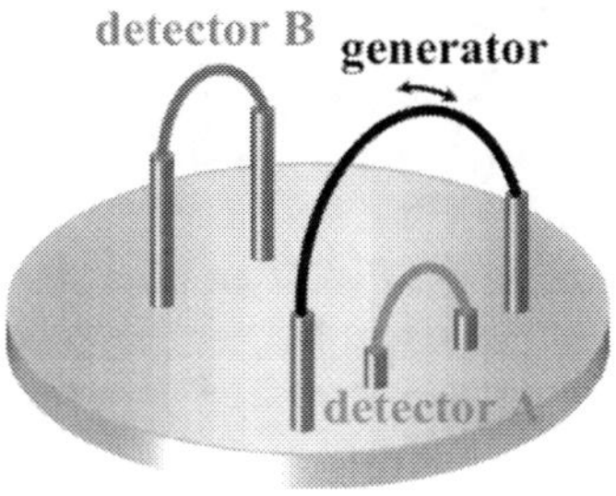

Figure 1. Schematic of vibrating wires. The detector wires are mounted around the generator wire. Detectors A and B are located in directions perpendicular and parallel, respectively, to the vibration direction of the generator indicated by the arrows. The distances between the generator and detectors A and B are 0.84 mm and 0.83 mm, respectively.

Visualization techniques for quantized vortices can trace vortex motions [6, 7], though these techniques have been limited to studies on quantized vortices at high temperatures. Also, tracer particles may affect the vortex motions.

Recently, the paths of vortices emitted from turbulence have been studied by using vibrating wires [8, 9]. A vortex-free vibrating wire, which cannot generate turbulence alone even at high vibrating velocities, is capable of detecting a vortex loop [4]. When a vortex loop collides with a vortex-free vibrating wire, the vibration of the wire is dissipated due to generation of turbulence. After the wire vibration has been stopped, the wire returns to the vortex-free state [10]. A set of two vibrating wires, a turbulence generator and a vortex detector, enables the study of the time-of-flight of a vortex emitted from the generator to the detector. In the present paper, we report vortex emissions in directions parallel and perpendicular to the vibrating directions of a generator, generated in superfluid ^{4}He at a finite temperature.

2. Experimental setup

To detect vortex flight paths, we mounted three vibrating wires made of NbTi wire with a diameter of ~2 μm in a copper chamber with a pin hole. Two detector wires, detector A and detector B, are located in directions perpendicular and parallel, respectively, to the vibration direction of a generator wire, as shown in Fig. 1. The distance between the apexes of the generator and detector A is tuned to be 0.84 mm, with the distance between the generator and detector B set to be almost the same at 0.83 mm. The bending shape of detector A is nearly identical to that of detector B. Although the wire arrangement is similar to a previous experimental design [11], the tuned distances enable us to study the direction distributions of vortex emissions from the generator more precisely. The resonance frequencies in vacuum at 4.2 K are 1145 Hz for the generator, 2906 Hz for detector A and 3282 Hz for detector B.

To study vortex emissions for large vortex loops, we performed time-of-flight measurements in superfluid ^{4}He at a temperature of 1.25 K. At a finite temperature, a normal fluid component arises from superfluid helium, dissipating superfluid vortices. Vortex loops shrink during flight and may therefore fade away. The flight distance l and the flight time τ until disappearance of a circular vortex loop with a radius R_0 are given by $l = R_0/\alpha$ and $\tau = R_0/2\alpha v_0$ [5], where α is the mutual friction coefficient between a normal fluid component and a vortex line, and v_0 is the velocity of a loop with radius R_0 given by Eq. (1). In the present setup, the detectors can detect emitted vortex loops with an initial diameter only above 42 μm, because smaller vortex loops may disappear shortly before reaching the detectors. The flight time of the fastest loop is estimated to be 0.08 s under this condition.

3. Results and discussion

3.1. Time-of-flight measurements for the perpendicular direction

Time-of-flight measurements were performed by similar methods to those used in previous studies [9, 11]. We measured the period between the start of turbulence generation by the generator and the detection of a vortex loop by the detector. The period, therefore, corresponds to the creation time of a

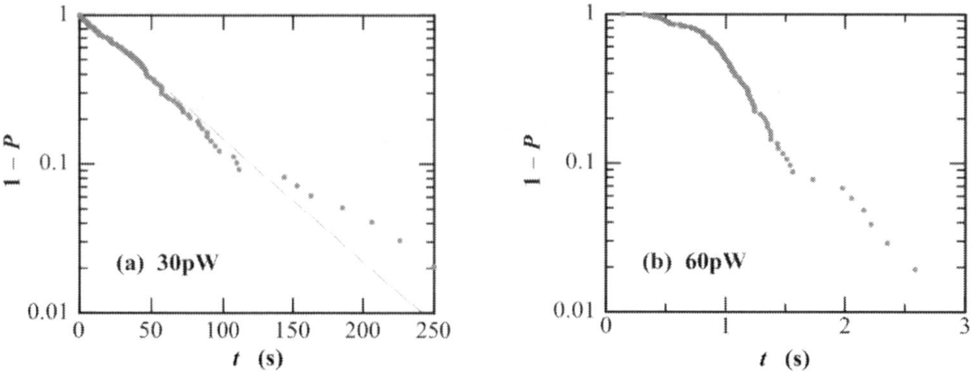

Figure 2. Distribution of vortices detected with detector A for generation powers of (a) 30 pW and (b) 60 pW. The probability P is the detection probability within a time period t. (a) The probability $1 - P$ for 30 pW follows the exponential function Eq. (2), shown by the solid line. (b) The detection periods for 60 pW are much shorter than those for 30 pW. The distribution, however, no longer shows a single exponential function (see text).

vortex loop plus the time of its flight from the generator to the detector. We measured periods repeatedly at a temperature of 1.25 K for a power of 30 pW injected in a turbulence region, and found that the observed periods were distributed widely from 0.4 s to 250 s for 100 measurements. The non-detection probability $1 - P$ is plotted as a function of time in Fig. 2(a), where P is the detection probability within a time window t. The data of the probability $1 - P$ are consistent with an exponential function

$$1 - P = \exp\left(-\frac{t - t_0}{t_1}\right), \tag{2}$$

shown as the solid line in Fig. 2(a), indicating that the vortex detection is a Poisson process. The fitting parameters t_0 and t_1 are estimated to be 0 s and 53 s, respectively. The parameter t_0 corresponds to the creation time plus the flight time of the fastest vortex loop [9]. As mentioned in the previous section, the flight time of the fastest vortex loop that can reach the detector is expected to be 0.08 s. The creation time of a vortex loop by the generator is expected to be less than 10 ms [9]. Therefore, the time t_0 is estimated to be almost zero in the present time range. The mean detection period corresponds to the fitting parameter t_1. Hence, the detector detects vortex loops at irregular intervals with a mean interval time of t_1. Since the detector can respond only to a reachable vortex loop, this result suggests that vortex loops are emitted randomly with respect to the emission direction. To study the emission rate of large vortex loops, we measured times-of-flight for different generation powers. For powers below 30 pW, we rarely detected vortex loops within a measurement time window of 500 s. For a power of 60 pW, however, the observed detection periods are much shorter than those for 30 pW, as shown in Fig. 2(b). The emission rate of large vortex loops appears to increase steeply with increasing generation power.

The data of the probability $1 - P$ for 60 pW do not fit to a single exponential function. The detection rate of vortex loops, proportional to the slope of the probability, increases at times above 0.8 s, in contrast to the 30 pW case. Increasing emission rates appear to affect the vortex emissions. It is plausible that emitted vortices reconnect between themselves at high densities, forming a tangle of vortices between the generator and the detector, as observed in the studies of superfluid ^{3}He-B [12]. Forming the tangle may reduce vortex loops reaching the detector. When a balance is achieved in the tangle, the detection rate increases, as the distribution above 0.8 s for the 60 pW case shown in Fig. 2(b).

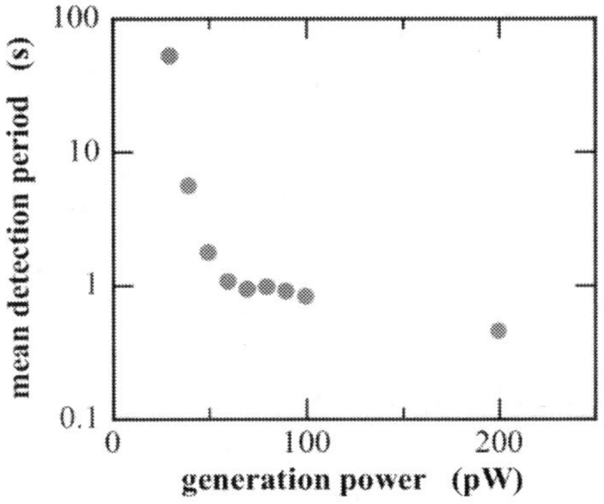

Figure 3. Mean detection period of detector A as a function of generation power. Below 50 pW, the distributions of detection period indicate a Poisson process. Above 60 pW, however, the distributions do not reveal a Poisson process. In the figure, therefore, we averaged the detection period for all powers instead of using time t_1 in Eq. (2) (see text).

3.2. Power dependence of vortex emission for the perpendicular direction

To study the emission rates, we measured the distributions of time-of-flight as a function of generation power. Below 50 pW, the data are consistent with a single exponential function, indicating a Poisson process. Above 60 pW, however, the distributions do not indicate a single exponential function. The mean detection period is equal to time t_1 of Eq. (2) in a Poisson process because the time t_0 is negligible, though we could not estimate t_1 for powers above 60 pW. Instead of using time t_1, we averaged the detection periods to estimate a mean detection period, as shown in Fig. 3. The mean detection period decreases steeply with decreasing power below 60 pW, though the value does not vary so much above 60 pW.

In the present setup, only vortex loops larger than 42 μm in diameter can reach the detector. The emission of such large vortex loops should relate to the geometry of the turbulent region. For a power of 60 pW, the generator wire produces a turbulent region of (wire thickness: 2.4 μm) × (wire length: 1.3 mm) × (vibrating peak-to-peak amplitude: 34 μm). Thus, the emitted vortex size, ~42 μm, is comparable with the vibrating amplitude. This result suggests that the vibrating wire stretches attached vortices, producing vortex loops comparable with its amplitude. The large vortex loops may propagate in the direction perpendicular to the vibration direction. In a previous paper [9], we found that a vibrating wire produces vortex loops equal to or smaller than 0.1 μm diameter. Consequently, a vibrating wire produces vortex loops with a diameter from 0.1 μm to a size comparable to its vibrating amplitude. Note that vortex loops with a larger size may not emit in this power range, because vortex loops have been rarely detected below 30 pW.

3.3. Vortex emission for the parallel direction

To study the direction distribution of vortex emissions, we measured times-of-flight for the direction parallel to the vibrating direction using detector B. Unlike in the perpendicular direction case, we did not detect vortices in a power range of the order of 10 pW to 100 pW. At 100 pW, we observed vortices within a time window of 500 s, though a few data points were still missing in the window. The detection periods are much longer than those for the perpendicular direction at the same power, indicating that large vortices are emitted anisotropically. The probability $1 - P$ of detection periods at 100 pW are well fitted to a single exponential function Eq. (2), as shown in Fig. 4, indicating a Poisson process. The fitting parameters t_0 and t_1 are estimated to be 0 s and 169 s, respectively.

This distribution suggests that vortices are emitted from the generator independently and randomly. This behavior is similar to that of the vortex emission for 30 pW in the perpendicular direction shown in Fig 2(a). The question remains, however, as to how the vibrating wire produces large vortex loops. In the parallel direction, the cross-section of a turbulent region for 100 pW has a geometry described by (wire thickness: 2.4 μm + wire bending: 0.9 μm) × (wire length: 1.3 mm). Thus, the largest vortex

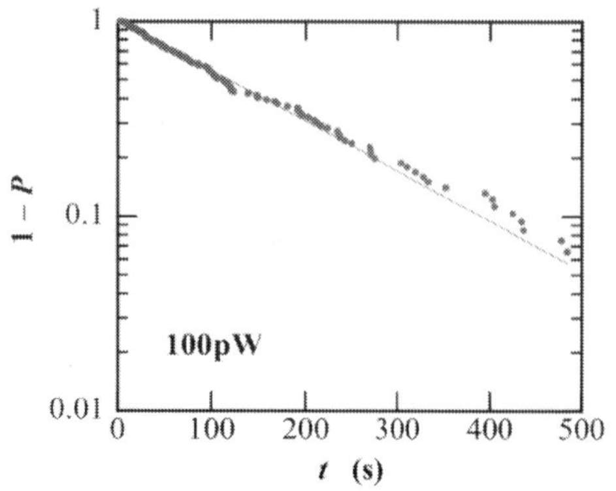

Figure 4. Distribution of vortices detected with detector B for a generation power of 100 pW. Here P is the detection probability within a time period t. The probability $1 - P$ is well fitted to the exponential function of Eq. (2), shown as the solid line.

loop produced by the generator is expected to be of the order of 1 μm, much smaller than the detected vortex loops. It is possible that large vortices may be emitted from tangles of vortices. As mentioned in the previous section, the distribution of the vortex emissions at 60 pW shown in Fig. 2(b) implies that a tangle of vortices may form around the generator. In a tangle of vortices, vortices reconnect between themselves, forming large vortex loops [12]. Therefore, it is plausible that large vortex loops may be emitted from tangles forming around the generator, though we cannot estimate the size of such tangles in this experiment.

In the present work, we observed anisotropic emissions of large vortex loops generated by a vibrating wire. In previous studies [13, 14], numerical simulations using a vibrating sphere also predict anisotropic emissions of vortex loops. Contrary to the present results, however, the emission rate of large vortex loops is higher in the direction parallel to the vibrating direction than that in the perpendicular direction. The emission rate of large vortex loops may be associated with the geometry of the vibrating object. A vibrating sphere produces a turbulence forming a column with a length equal to the vibrating amplitude and a diameter comparable with the sphere size. The turbulence produced by a vibrating wire may spread in a rather planar region defined by the vibrating amplitude and the wire length. Therefore, a vibrating wire can produce large vortex loops propagating in the perpendicular direction more easily. Further experimental and numerical studies are necessary to describe vortex emissions produced by a vibrating object.

4. Conclusions

We report the vortex emission of large vortex loops generated by a vibrating wire immersed in superfluid ^{4}He. Using vortex-free vibrating wires as a vortex detector, we measured times-of-flight in the directions perpendicular and parallel to the vibrating direction of the generator wire. In the perpendicular direction, the vibrating wire can produce vortex loops with a diameter comparable with the vibrating peak-to-peak amplitude. In the parallel direction, we also find the emission of such large vortex loops, though a higher generation power is required for large vortex loops to be detected. This anisotropy may be associated with the creation mechanism of large vortex loops.

Acknowledgments

The authors are very grateful to M. Tsubota and A. Nakatsuji for stimulating discussions. The research was supported by a Grant-in-Aid for Scientific Research (B) (Grant No. 23340108) from the Japan Society for the Promotion of Science.

References

[1] Vinen W F and Donnelly R J 2007 *Phys. Today* **60** No.4, 43

[2] Halperin W P and Tsubota M 2009 *Progress in Low Temperature Physics* vol 16 (Amsterdam: North-Holland)
[3] Fisher S N, Hale A J, Guénault A M and Pickett G R 2001 *Phys. Rev. Lett.* **86** 244
[4] Goto R, Fujiyama S, Yano H, Nago Y, Hashimoto N, Obara K, Ishikawa O, Tsubota M and Hata T 2008 *Phys. Rev. Lett.* **100** 045301
[5] Donnelly R J 1991 *Quantized Vortices in Helium II* (Cambridge, England: Cambridge University Press)
[6] Paoletti M S, Fiorito R B, Sreenivasan K R and Lathrop D P 2008 *J. Phys. Soc. Jpn.* **77** 111007
[7] Bewley G P and Sreenivasan K R 2009 *J. Low Temp. Phys.* **156** 84
[8] Kubo H, Nago Y, Nishijima A, Obara K, Yano H, Ishikawa O and Hata T 2013 *J. Low Temp. Phys.* **171** 466
[9] Nago Y, Nishijima A, Kubo H, Ogawa T, Obara K, Yano H, Ishikawa O and Hata T 2013 *Phys. Rev. B* **87** 024511
[10] Yano H, Nago Y, Goto R, Obara K, Ishikawa O and Hata T 2010 *Phys. Rev. B* **81** 220507(R)
[11] Oda S, Wakasa Y, Kubo H, Obara K, Yano H, Ishikawa O and Hata T 2014 *J. Low Temp. Phys.* **175** 317
[12] Fujiyama S, Mitani A, Tsubota M, Bradley D I, Fisher S N, Guénault A M, Haley R P, Pickett G R and Tsepelin V 2010 *Phys. Rev. B* **81** 180512
[13] Nakatsuji A, Tsubota M and Yano H 2013 *J. Low Temp. Phys.* **171** 519
[14] Nakatsuji A, Tsubota M and Yano H 2014 *Phys. Rev. B* **89** 174520

Journal of Physics: Conference Series **568** (2014) 012028 doi:10.1088/1742-6596/568/1/012028

Counterflow quantum turbulence in a square channel under the normal fluid with a Poiseuille flow

Satoshi Yui[1] **and Makoto Tsubota**[1,2]

[1] Department of Physics, Osaka City University, 3-3-138 Sugimoto, Sumiyoshi-Ku, Osaka 558-8585, Japan

[2] The OCU Advanced Research Institute for Natural Science and Technology (OCARINA), Osaka City University, 3-3-138 Sugimoto, Sumiyoshi-Ku, Osaka 558-8585, Japan

E-mail: yui@sci.osaka-cu.ac.jp

Abstract. We perform a numerical analysis of superfluid turbulence produced by thermal counterflow in He II by using the vortex filament model. Counterflow in a low aspect ratio channel is known to show the transition from laminar flow to the two turbulent states TI and TII. The present understanding is that the TI has the turbulent superfluid and the laminar normal fluid but both fluids are turbulent in the TII state. This work studies the vortex tangle in the TI state. Solid boundary condition is applied to walls of a square channel, and the velocity field of the normal fluid is prescribed to be a laminar Poiseuille profile. An inhomogeneous vortex tangle, which concentrates near the solid boundaries, is obtained as the statistically steady state. It is sustained by its characteristic space-time oscillation. The inhomogeneity of the vortex tangle shows the characteristic dependence on temperature, which is caused by two effects, namely the profile of the counterflow velocity and the mutual friction.

1. Introduction

A thermal counterflow in He II is internal convection of two fluids, namely the normal fluid and the superfluid. When the counterflow velocity exceeds a critical value, a self-sustaining tangle of quantized vortices appears to form superfluid turbulence. In low aspect ratio channels, superfluid turbulence makes the mysterious transition. The increase in the counterflow velocity is observed to change the laminar state to the first turbulent state TI, and next to the second turbulent state TII [1]. Melotte and Barenghi suggested that the transition from the TI to TII state is caused by the transition of the normal fluid from laminar to turbulent [2]. The recent developments of the visualization technique have enabled us to confirm the scenario. Guo *et al.* have followed the motion of seeded metastable He_2^* molecules by a laser-induced-fluoresence technique to observe that the normal fluid can be turbulent at relatively high velocities [3].

To understand the mysterious transition of counterflow quantum turbulence, it is necessary to address the coupled dynamics of the two fluids; the superfluid is described by the vortex filament model, the normal fluid by the Navier-Stokes equation, and they are coupled through the mutual friction [4]. However, it is difficult to solve fully the coupled dynamics. As the first essential step, we address the TI state in a square channel with prescribing the velocity field of the normal fluid to a Poiseuille profile. Our simulation obtains a statically steady state of an inhomogeneous vortex tangle.

Baggaley *et al.* [5] studied numerically a thermal counterflow between two plates. They prescribed Poiseuille and turbulent profiles for the velocity field of the normal fluid. An inhomogeneous vortex tangle was obtained, where vortices concentrated near the solid boundaries. They suggested that their results supported the scenario proposed by Melotte and Barenghi. The better understanding of the TI and TII states would be obtained by studying the flow in a low aspect ratio channel where all boundaries are solid except for the flow direction. This is because the TI and TII states are actually observed in low aspect ratio channels and another turbulence TIII state is observed in high aspect ratio channels [1].

2. Formulation

In a vortex filament model [6] a quantized vortex is represented by a filament passing through a fluid and has a definite vorticity. This approximation is very suitable in He II, since the core size of a quantized vortex is much smaller than any other characteristic length scale. At zero temperature the vortex filament moves with the superfluid velocity $\mathbf{v}_s = \mathbf{v}_{s,\omega} + \mathbf{v}_{s,b} + \mathbf{v}_{s,a}$, where $\mathbf{v}_{s,\omega}$ is the velocity field produced by vortex filaments, $\mathbf{v}_{s,b}$ by solid boundaries, and $\mathbf{v}_{s,a}$ is the applied superfluid velocity. The velocity field $\mathbf{v}_{s,\omega}$ is given by the Biot-Savart law; this work addresses the full Biot-Savart integral [7]. Since $\mathbf{v}_{s,a}$ represents the laminar flow of the superfluid, $\mathbf{v}_{s,a}$ requires irrotational condition, which is supposed to be uniform. The velocity field $\mathbf{v}_{s,b}$ is obtained by a simple procedure; it is just the field produced by an image vortex which is constructed by reflecting the filament into the surface and reversing its direction. Taking into account the mutual friction, the velocity of a point $\mathbf{s}$ on the filament is given by

$$\dot{\mathbf{s}} = \mathbf{v}_s + \alpha \mathbf{s}' \times (\mathbf{v}_n - \mathbf{v}_s) - \alpha' \mathbf{s}' \times [\mathbf{s}' \times (\mathbf{v}_n - \mathbf{v}_s)], \tag{1}$$

where α and α' are the temperature-dependent coefficients, and the prime denotes derivatives of $\mathbf{s}$ with respect to the coordinate ξ along the filament. The velocity field of the normal fluid is prescribed to be a Poiseuille profile. In a rectangular channel the Poiseuille profile is given by

$$v_n = u_0 \sum_{m=1,3,5,\cdots}^{\infty} (-1)^{\frac{m-1}{2}} \left[1 - \frac{\cosh(m\pi z/2a)}{\cosh(m\pi b/2a)}\right] \frac{\cos(m\pi y/2a)}{m^3}, \tag{2}$$

where y and z are coordinates normal to the flow direction x, and a and b are halves of the channel width along the y- and z- axes [8].

3. Numerical simulation

In this study, all simulations are performed under the following conditions. The numerical space resolution is $\Delta\xi = 8.0 \times 10^{-4}$ cm, and the time resolution is $\Delta t = 1.0 \times 10^{-4}$ s. We study thermal counterflow of He II at temperatures T=1.3 K, 1.6 K and 1.9 K. The computing box is $0.1 \times 0.1 \times 0.1$ cm^3. Periodic boundary conditions are used along the flow direction x, while solid boundary conditions are applied to the channel walls at $y = \pm 0.05$ cm and $z = \pm 0.05$ cm. All simulations start with eight randomly oriented vortex rings of radius 0.023 cm.

The vortex line density (VLD) is defined as $L = \frac{1}{\Omega}\int_{\mathcal{L}} d\xi$, where the integral is performed along all vortices in the sample volume Ω. The vortex tangle reaches the statistically steady state. Figure 1 (a) shows the time evolution of VLD. Fluctuations are larger than those in a uniform counterflow [7], which is attributable to the mechanism characteristic of this system discussed in section 4.1.

The statistically steady state is known to exhibit the characteristic relation $L^{1/2} = \gamma(v_{ns} - v_0)$ [9] with the parameters γ and v_0. We regard the counterflow velocity v_{ns} as the spatially averaged amplitude of $\mathbf{v}_{ns} = \mathbf{v}_n - \mathbf{v}_{s,a}$. Figure 1 (b) shows the VLD temporally averaged over the statistically steady state, which almost satisfies the relation. Table 1 shows the comparison

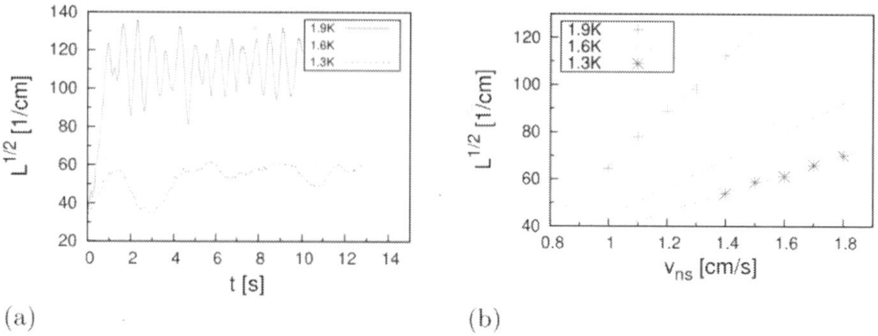

Figure 1. (Color online) (a) Vortex line density as a function of time with $v_{ns} = 1.4$ cm/s. (b) Vortex line density averaged over the statistically steady state as a function of v_{ns}.

of γ among the present work γ_{num1}, the simulation γ_{num2} under the periodic boundary condition [7] and a typical experiment γ_{exp} [10]. The values of γ_{num1} are lower than the values of γ_{num2} obtained under the uniform counterflow. The difference of γ comes from the difference of the mechanism sustaining the vortex tangle. The origin of the discrepancy between γ_{num1} and γ_{exp} is not clear, but this may be attributable to neglecting the effect of the vortex tangle on the Poiseuille flow of the normal fluid through mutual friction.

In order to estimate the inhomogeneity of the vortex tangle, we divide the computational box to 15^3 sub-volumes and define the VLD at a sub-volume as the local VLD. Figure 2 (a) shows the spatially dependence of $L'(y, z)$, which is obtained by averaging the local VLD spatially over the flow direction and temporally over the statistically steady state. We estimate the inhomogeneity of the vortex tangle by a spatial variance Δ_L of L'/L. Figure 2 (b) shows the characteristic dependence of Δ_L on v_{ns} and T. Firstly, Δ_L of 1.6K is the largest among three temperatures. Secondly, Δ_L at 1.3K and 1.6K increases with v_{ns}, while Δ_L at 1.9 K decreases with v_{ns}. The dependence on T is understood as discussed in section 4.2, but the dependence on v_{ns} is not known.

4. Discussions

Section 4.1 addresses the mechanism for sustaining the inhomogeneous vortex tangle. In section 4.2 we discuss how the vortex tangle becomes inhomogeneous depending on temperature. In section 4.3 we consider how the aspect ratio of the channel cross section affects the two fluids.

4.1. Mechanism for sustaining the inhomogeneous vortex tangle

The inhomogeneous vortex tangle is sustained by its space-time oscillation. As shown in Fig. 1 (a), the VLD shows the non-linear oscillation with large amplitude in the statistically steady state, which is much different from the case of the uniform counterflow [7]. The non-uniform mutual friction, which is produced by the non-uniform counterflow velocity, makes the dynamics

T (K)	γ_{num1} (s/cm^2)	γ_{num2} (s/cm^2)	γ_{exp} (s/cm^2)
1.3	40	53.5	59
1.6	60	109.6	93
1.9	112	140.1	133

Table 1. Line density coefficients γ. Numerical results γ_{num1} by this work, γ_{num2} by Adachi *et al.* [7] and experimental results γ_{exp} by Childers and Tough [10].

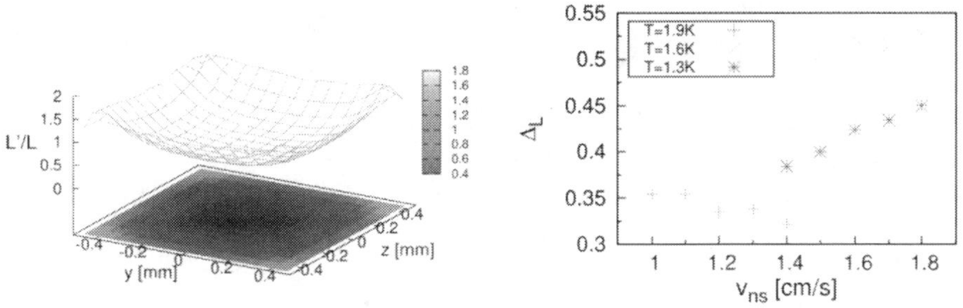

(a) (b)

Figure 2. (Color online) (a) Profiles of the vortex line density (T=1.9 K, v_{ns} = 1.4 cm/s). One can see the concentration near the solid boundaries. (b) Dependence of the spatially variance Δ_L of L'/L on v_{ns} and T.

of the vortex tangle depend on the position. This causes the space-time oscillation of the vortex tangle. In the following the detail of the oscillation is explained.

Figure 3 shows the space-time pattern of the vortex tangle at 1.9 K. The period of the non-linear oscillation is about 0.6 s, consisting of four stages (a)-(d). In Fig. 3 (a) corresponding to the minimum of the VLD, vortices are dilute, where vortices remain only near the solid walls. Then the vortices near the walls invade to the central region in Fig. 3 (b). These vortices make lots of reconnections in the central region subject to the large counterflow in Fig. 3 (c). Hence the VLD increases significantly to the maximum. Eventually in Fig. 3 (d) the Poiseuille flow excludes the vortex tangle from the central region toward the solid walls. Thus the VLD around the central region decreases, and the vortices are absorbed by the solid boundaries. Then the vortex tangle returns to the stage of Fig. 3 (a). Therefore the vortex tangle repeats the periodic motion, resulting in the non-linear oscillation of Fig. 1 (a). The vortex tangle sustained by this mechanism is more dilute than the case of the uniform counterflow for the same T and v_{ns}.

4.2. Temperature dependence of inhomogeneity
We discuss why the inhomogeneity Δ_L of the vortex tangle is maximized at some temperature, as shown Fig. 2 (b). We explain how the profile of the counterflow velocity changes depending on T, and by using this the T dependence of Δ_L is discussed.

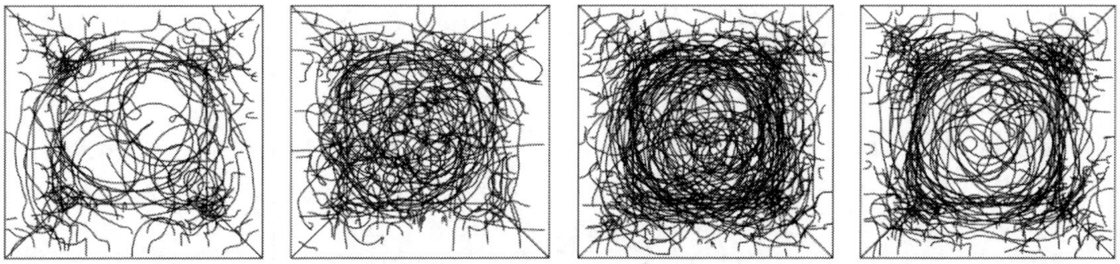

(a) Depletion (2.00 s) (b) Invasion (2.14 s) (c) Maximum (2.30 s) (d) Exclusion (2.42 s)

Figure 3. Vortex tangles in the statistically steady state viewed along the flow direction (T =1.9 K, v_{ns} = 1.4 cm), corresponding to the results of Fig. 1 (a).

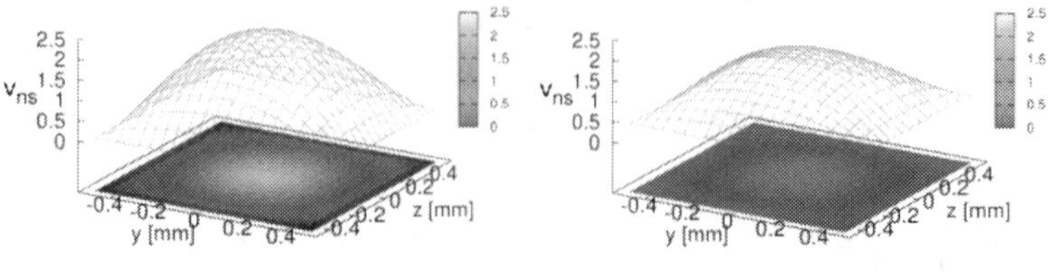

(a) T=1.3 K (ρ_{ns}=0.052) (b) T=1.9 K (ρ_{ns}=0.738)

Figure 4. (Color online) Counterflow velocity scaled by its spatial average.

The increase in T makes the counterflow velocity more uniform. The thermal counterflow requires the conservation of mass, yielding $\mathbf{v}_{s,a} = -\rho_{ns}\overline{\mathbf{v}}_n$, where $\overline{\mathbf{v}}_n$ is spatially averaged $\mathbf{v}_n$ over the cross section of the channel. The ratio $\rho_{ns} = v_{s,a}/\overline{v}_n$ increases with T. Thus the counterflow velocity $\mathbf{v}_{ns} = \mathbf{v}_n - \mathbf{v}_{s,a}$ becomes more uniform at higher T, since $\mathbf{v}_{s,a}$ is uniform and $\mathbf{v}_n$ is non-uniform. Figure 4 shows the T dependence of the counterflow velocity profile.

The non-uniform $\mathbf{v}_{ns}$ makes the vortex tangle inhomogeneous through the mutual friction term $\alpha\mathbf{s}' \times \mathbf{v}_{ns}$ in Eq. (1). The non-uniformity of $\mathbf{v}_{ns}$ decreases with T, reducing the inhomogeneity of the vortex tangle. On the other hand, the coefficient α increases with T, enhancing the inhomogeneity of the vortex tangle. Therefore the inhomogeneity Δ_L of the vortex tangle is maximized at some temperature.

4.3. Aspect ratio of the channel cross section

The aspect ratio of the channel plays an important role in the density and the profile of the vortex tangle. The increase of the aspect ratio from unity reduces the counterflow velocity gradient along the long side of the cross section. Thus the exclusion of the vortex tangle shown in Fig. 3 (d) does not work so much along the long side of the cross section. Hence the VLD should increase compared to the VLD in a low aspect ratio channel.

It would be meaningful to consider how the aspect ratio affects the normal fluid, though it is prescribed in this formulation. The linear stability analysis of the Navier-Stokes equation shows that the critical Reynolds number for turbulence transition of a viscous fluid increases significantly with decreasing the aspect ratio [11]. According to this result, the normal fluid in counterflow could remain laminar in a low aspect ratio channel even if the superfluid becomes turbulent and the vortex tangle disturbs the normal fluid, which may correspond to the TI state. In order to understand the TI state, therefore, the studies in a low aspect ratio channel like this work will be indispensable.

References

[1] Tough J T 1982 *Progress in Low Temperature Physics*, ed D F Brewer (Amsterdam: North-Holland), Vol. 8
[2] Melotte D L and Barenghi D F 1998 *Phys. Rev. Lett.* **80** 4181
[3] Guo W, Cahn S B, Nikkel J A, Vinen W F, and McKinsey D N 2010 *Phys. Rev. Lett.* **105** 045301
[4] Kivotides D 2007 *Phys. Rev.* B **76** 054503
[5] Baggaley A W and Laizet S 2013 *Phys. Fluids* **25** 115101
[6] Schwarz K W 1985 *Phys. Rev.* B **31** 5782
[7] Adachi H, Fujiyama S and Tsubota M 2010 *Phys. Rev.* B **81** 104511
[8] 1998 *The handbook of fluid dynamics*, ed R W Johnson (Boca Raton: CRC Press)
[9] Vinen W F 1957 *Proc. R. Soc. London* Ser. A **242** 493
[10] Childers R K and Tough J T 1976 *Phys. Rev.* B **13** 1040
[11] Tatsumi T and Yoshimura T 1990 *J. Fluid Mech.* **212** 437

Analysis of motion of solid hydrogen tracer particles in oscillating superfluid flows.

E Zemma[1,2], J Luzuriaga[1] and S Babuin[3]

[1]Centro Atómico Bariloche, (8400)S.C. Bariloche, CNEA, Inst. Balseiro UNC, [2]CONICET, Argentina , [3]Institute of Physics ASCR, Na Slovance 2, 182 21 Praha 8, Czech Republic

E-mail: zemma@cab.cnea.gov.ar

Abstract. We have developed a relatively simple cryostat which allows us to image turbulent flows in superfluid helium at temperatures below 2 K, using frozen H_2 particles. We analyze the statistics of the velocities of these solid tracers, which follow the turbulent flow generated by oscillating bodies. We have also studied one of the oscillators working in air at room temperature, and traced the flow with solid talcum particles for comparison. Images were recorded by a digital camera at 240 frames per second, while frequencies of the oscillators are between 20 to 45 Hz. The flow is characterized by a modified Reynolds number Re_δ based on the viscous penetration depth δ. Software in a dedicated particle tracking velocimetry code allows us to compute the trajectories and velocities of tens of thousands of particles. We have obtained the number of particles for equally spaced intervals of the velocity modulus. For the oscillators in the superfluid, the probability of finding particles at higher velocities has an exponential decay. Within our resolution the statistics in the superfluid for oscillating objects with sharp borders is largely independent of Re_δ, while the logarithmic decay at low velocities seems faster than for high velocities for rounded objects. On the other hand, for data taken in air the result is closer to a classical Gaussian distribution of velocities.

1. Introduction

Quantum Turbulence studies have a long history[1, 2] and the subject is far from being exhausted [3]. Similarities and differences have been found between classical and quantum turbulence giving rise to a continuing interest in the subject[4–9]. New experimental techniques for flow visualization[10–12] have been developed and adapted to quantum fluids.

Micron and sub-micron solid particles can trace in detail the dynamics of quantum flows. They have been used to image situations of possible dissipation mechanisms such as reconnections and Kelvin waves[13]. Other experiments image thermal counterflow and flow in channels[11, 12, 14–17]. Unsteady and oscillatory flows that have a classical analogue can also be studied by this technique and Vinen[9] has pointed out that tracer imaging of oscillatory flows in Quantum turbulence could provide valuable information.

We have been able to film H_2 solid particles flowing around oscillating objects of different shapes, which generate turbulence when submerged in superfluid helium. By analyzing video images of the particle motion, we can obtain the statistics of the particle velocities and we report the results in the present paper.

2. Experimental Details

The apparatus has been described in more detail in a previous publication[18], and here we give a brief description of the main points. We used a glass dewar with a window to allow filming from outside, images were obtained with a camera[19] at a rate of 240 frames per second. Experiments were performed between 1.7 and 2 K. Dissipation in the superfluid component in this range of temperatures is mainly produced by mutual friction between the quantized vortices and normal fluid[20].

The oscillators studied consist of a body attached to a flexible stem which moves in flexure producing a translation of the rigid body. Different shapes were tried. The first one consisted of a sphere of radius 4.5 mm with a stem in the shape of a beam of BeCu alloy 5 mm wide and 0.05 mm thick. As the wide beam modified the flow substantially, we made a second experiment with the 5 mm wide laminar beam only. A third oscillator consisting of a sphere of 5 mm radius and a cylindrical wire stem of 0.7 mm radius was also used in order to study the spherical geometry with a smaller perturbation. For comparison, we took videos of this last oscillator in air at room temperature using solid talcum particles as tracers.

Solid particles in helium are formed when a hydrogen-helium gas mixture is injected into the liquid following the procedure outlined by Bewley and co-authors[21–23]. We filmed each injection to detect the movement of particles in the fluid. We use a Hydrogen-Helium mixture with a 1/50 ratio at a pressure of 500 Torr, around a hundred cubic centimeters of gas are injected each time. We wait until the perturbations caused by the injection die out before using the video footage of particle motion.

To illuminate the tracer particles we used a green laser beam fed into the cryostat through an optical fibre. The frozen H_2 particles are not expected to absorb significant energy in the visible[24]. The illumination forms a three dimensional cone of light, which also partly diffuses in the oscillators. Other reflections where minimized by using a black screen in the background.

The time interval is 4.17 ms between frames and we use this quantity as our time reference to calculate velocities. The distances in the images are calibrated with respect to the measured dimensions of the optical fiber. The particles analyzed here occupy one or just a few pixels in the image obtained. The size of a pixel corresponds to around 70 microns in the image but the light could be scattered from a particle which is smaller than this size. It is hard to evaluate the minimum observable dimension, but we estimate our particles to be distributed from well below 70 microns to 200 microns in size.

3. Experimental Results

The images of tracer particles have to be processed and analyzed with appropriate statistical tools and we have written a particle image velocimetry code based on available MATLAB modules[25], leaving aside the non trivial question of what is really traced by the tracer particles. In each run, the oscillating body can be followed. We use this to test the program, confirming that the digital output captures the known frequency and sinusoidal motion of the oscillator. Particle detection is based on luminous contrast. In the image, spots that stand out from the background according to certain criteria (*i.e.* light intensity and diameter) are identified and their coordinates estimated in every frame of the video sequence. Digitizing the positions manually we have checked some of the data obtained from the program, both to find the best parameter settings and to assess the program performance. Trajectories and velocities of tens of thousands of particles have been computed.

At first, we must choose the region near the oscillator to be studied. The area is generally a square, approximately 10 mm in size, where illumination is higher. The edge closer to the moving body is around 1 mm away of the rest position of the oscillator to avoid including it in the images. Therefore the flow is studied up to distances of about twice the diameter of the spherical oscillators but not closer than 1 mm.

We use only one camera, therefore our image analysis can only evaluate the two dimensional velocity vector projected on the screen and is an underestimate which does not include motion parallel to the line of sight. We have evaluated the absolute value (modulus) of the velocity vector obtained and this dynamic parameter will be treated in the following. We have observed erratic velocities for our particles, and we evaluate the probability that a given one has an instantaneous velocity for a set of oscillation amplitudes in the different oscillating geometries measured.

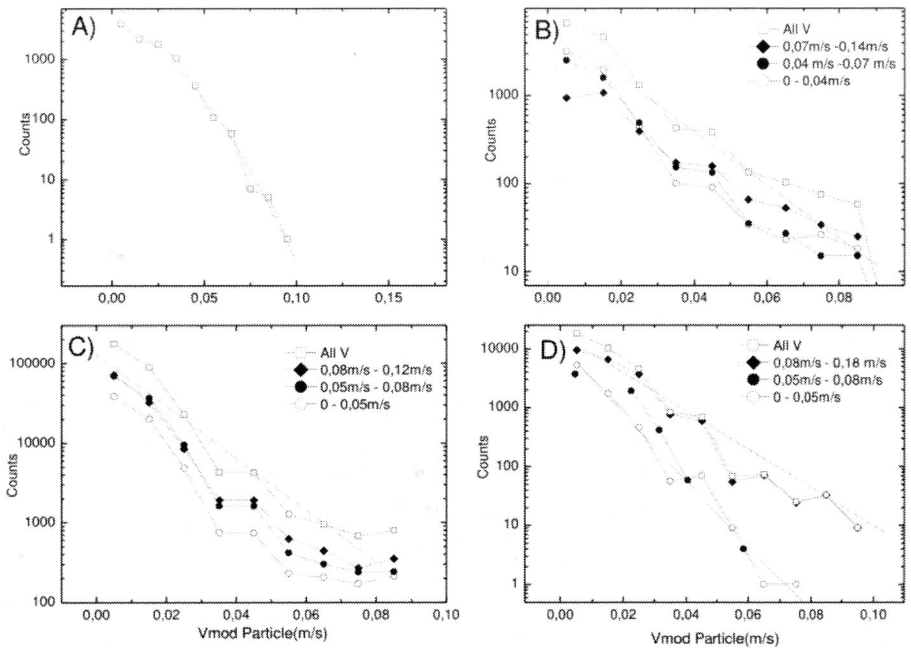

Figure 1. Un-normalized probability distribution function for the oscillating bodies; A) sphere in air with wire stem; B) sphere attached with wide flexible beam in helium; C) sphere with wire stem in helium; D) Flexible beam with no attached body, in helium. We have divided the data into groups corresponding to low, medium and high velocity of the oscillator, the intervals in which we evaluate Re_δ of the flow. A group including all of the particles observed, regardless of the oscillator velocity, is shown in open squares (labelled "All V"). It can be seen that the graphs are straight lines for the superfluid, where we have included straight lines as guides to the eye. This implies an exponentially falling probability of finding particles with higher velocities. For the sphere in air, the data can be represented by a parabola, as shown by the fitted line in the corresponding graph. The data are not normalized, so the numbers in each velocity interval are different, proportional to the number of particles captured in each measurement.

We have plotted the number of particles for equally spaced intervals of the velocity modulus, (*i.e.* an un-normalized probability distribution function (pdf)) and we show the results in Figure 1. We group our data according to the velocity amplitude of the sinusoidal motion of the oscillator as low, medium and high velocities. We have considered as well a group including all of the particles observed and calculated the statistics of these lumped data.

For oscillating flow due to vibrating objects, moving with velocity amplitude U frequency ω and with amplitude l in the limit $U/\omega << l >> \delta$, the characteristic length is not the size of the object, but the viscous penetration depth $\delta = (2\nu/\omega)^{1/2}$, where ν stands for the kinematic viscosity. The flow therefore is characterized by a modified Reynolds number[26] $Re_\delta = U\delta/\nu$. The values of Re_δ for the oscillators are shown in Table 1. In our case the limit $U/\omega << l$ is not fully realized, the oscillator is closer to $U/\omega \sim l$, however Re_δ is still more appropiate than Re because the displacement is small compared to the dimensions of the oscillators.

Oscillator	T(K)	$f = \omega/2\pi$ Hz	δ (mm)	Re_δ High	Re_δ Medium	Re_δ Low
sphere & beam	1.8	40	0.014	74	36	20
beam	1.8	37	0.014	65	43	27
sphere & wire	1.8	23	0.018	124	55	35
sphere in air	300	23	.47	4.5	-	-

Table 1. Frequency,viscous penetration depth and values for Re_δ corresponding to the maximum value for each velocity range of the different oscillators.

The pdf observed in Figure 1, for the oscillating systems studied show two slightly different types of behavior. In this graph with a logarithmic y-axis a straight line fit adjusts the results in the superfluid, implying an exponential decay in the observed velocity distribution. In the objects with sharp edges (the beam and sphere plus beam) the exponential behavior is not greatly dependent on the velocity of the oscillating object, and the straight lines corresponding to different velocities all have the same slope, within experimental error. Thus, almost independent of the velocity of the oscillating object, the number of particles whose modulus velocity acquires greater values decreases exponentially. In the sphere attached by a thin wire, the adjustment to be made is dependent on the amplitude of oscillation of the sphere. At different speeds it shows different exponential behaviors in the velocities. For values of lower Re_δ the slope is higher, meaning that the decay is faster, whereas the exponential decay is slower for high Re_δ,.

In order to compare with a classical case, with no superfluid, we have tested the sphere with thin wire oscillating in air at room temperature. We have used particles of talcum powder as tracers. The values of Re_δ and δ attainable in air are different than those of liquid helium. In this case, we have plotted the higher values of Re_δ, and we find a decay in the form of a parabola in a graph with a logarithmic y-axis, as would be expected for a Gaussian distribution of zero mean.

4. Discussion

Following the analysis found e.g. in Landau-Lifshitz[27], it can be assumed that the vorticity produced by a rounded body oscillating with small amplitude is annulled to a great extent when the body changes the direction of motion. If sharp edges are present, the cancellation is much less, so it is possible that the observed differences in the pdf may be related to the presence of sharp edges in the beam, and the sphere attached by a beam, and the more rounded geometry of the sphere attached by a wire. The fact that a (classically expected) Gaussian distribution with no exponential tails was found in the sphere in air may be significant and could be due to the absence of superfluid. Some caution is necessary in this conclusion however, because of the difference in Re_δ between liquid helium and air.

Non-Gaussian velocity distributions have been reported in other experiments in superfluid visualization[28–30] and for example in thermal counterflow a non-Gaussian tail in the pdf, proportional to v^{-3} was found[28, 29]. These measurements are not in the type of flow observed here. In this sense our data are difficult to compare with other experiments in quantum turbulence since to the best of our knowledge there are no published data on statistics of tracers in oscillating superfluid flows.

In conclusion, we present data on the visualization of turbulent flows around oscillating objects, which possibly show a different overall behavior than in classical fluids. Other observations of filamentary structures, which might be produced by quantum vortices are not numerous enough to be treated statistically and will be reported elsewhere.

Acknowledgments

This work was partially supported by 06/C432 grant from U.N. Cuyo and CONICET-Czech Academy of Sciences Scientific Cooperation agreement.

[1] Tough J 1982 (*Progress in Low Temperature Physics* vol 8) ed Brewer D (Elsevier) pp 133 – 219
[2] Donnelly R J and Swanson C E 1986 *Journal of Fluid Mechanics* **173** 387–429
[3] Barenghi C F, Skrbek L and Sreenivasan K R 2014 *Proceedings of the National Academy of Sciences* **111** 4647–4652
[4] Donnelly R J 2003 *Physica B: Condensed Matter* **329-333** 1 – 6
[5] Vinen W F 2006 *Journal of Low Temperature Physics* **145**(1) 7–24
[6] Vinen W F 2008 *Philosophical Transactions of the Royal Society* **366** 2925–2933
[7] Paoletti M and Lathrop D 2011 *Annu. Rev. Condens. Matter Phys.* **2** 213–234
[8] Tsubota M 2009 *International Journal of Emerging Multidisciplinary Fluid Sciences* **1** 229–254
[9] Vinen W 2010 *Journal of Low Temperature Physics* **161** 419–444
[10] Donnelly R, Karpetis A, Niemela J, Sreenivasan K, Vinen W and White C 2002 *Journal of low temperature physics* **126** 327–332
[11] Van Sciver S W and Barenghi C F 2009 *Progress in Low Temperature Physics* **16** 247–303
[12] Guo W, La Mantia M, Lathrop D P and Van Sciver S W 2014 *Proceedings of the National Academy of Sciences* **111** 4653–4658
[13] Fonda E, Meichle D P, Ouellette N T, Hormoz S and Lathrop D P 2014 *Proceedings of the National Academy of Sciences* **111** 4707–4710
[14] Zhang T and Van Sciver S W 2005 *Nature Physics* **1** 36–38
[15] La Mantia M, Chagovets T, Rotter M and Skrbek L 2012 *Review of Scientific Instruments* **83** 055109–055109
[16] Chung D Y and Critchlow P 1965 *Physical Review Letters* **14** 892–894
[17] Chagovets T and Van Sciver S 2011 *Physics of Fluids* **23** 107102
[18] Zemma E and Luzuriaga J 2013 *Journal of Low Temperature Physics* **173** 71–79
[19] CASIO 0 *Exilim EX ZR100* (digital camera)
[20] Paoletti M S, Fisher M E, Sreenivasan K R and Lathrop D P 2008 *Phys. Rev. Lett.* **101**(15) 154501
[21] Bewley G, Lathrop D and Sreenivasan K 2006 *Nature* **441** 588–588
[22] Bewley G P 2009 *Cryogenics* **49** 549 – 553 ISSN 0011-2275 flow Visualization at Low Temperature
[23] Bewley G, Sreenivasan K and Lathrop D 2008 *Experiments in Fluids* **44** 887–896
[24] Penney R and Hunt T K 1968 *Phys. Rev.* **169**(1) 228–228
[25] http://physics.georgetown.edu/matlab/index.html
[26] Blažková M, Schmoranzer D and Skrbek L 2007 *Phys. Rev. E* **75**(2) 025302
[27] Landau L 1987 *Fluid mechanics: volume 6 (course of theoretical physics) Author: LD Landau, EM Lifshitz, Publisher: Bu* (Butterworth-Heinemann)
[28] Paoletti M, Fisher M, Sreenivasan K and Lathrop D 2008 *Physical review letters* **101** 154501
[29] Mantia M L and Skrbek L 2014 *EPL (Europhysics Letters)* **105** 46002
[30] La Mantia M, Duda D, Rotter M and Skrbek L 2013 *Procedia IUTAM* **9** 79–85

Study of superfluid ^{3}He films with a micro-electro-mechanical device

P Zheng[1], W G Jiang[1], C S Barquist[1], H B Chan[2] and Y Lee[1]

[1] Department of Physics, University of Florida, Gainesville, Florida 32611, USA
[2] Department of Physics, The Hong Kong University of Science and Technology, Clear Water Bay, Hong Kong

E-mail: yoonslee@phys.ufl.edu

Abstract. A micro-electro-mechanical system (MEMS) device with a 1.25 μm gap between a movable plate and the substrate was immersed in superfluid ^{3}He and cooled below 500 μK at 21.2 bar. Mechanical resonances of its shear motion were monitored on warming from the base temperature. Our preliminary results demonstrate increasing damping and decreasing resonance frequencies with temperature, consistent with a thermal damping model caused by thermal quasiparticles. The quality factor (Q) of the oscillator remains surprisingly low ($Q \approx 30$) down to $0.2\,T/T_c$, about 4 orders of magnitude smaller than the value in vacuum at 4 K. The average superfluid gap was determined to be significantly suppressed from the value in bulk at the corresponding pressure.

1. Introduction

In the normal state, liquid ^{3}He – an almost ideal Fermi liquid – possesses the full symmetry that a condensed matter system can possibly have. Below around 2 mK, the system enters into unconventional *p-wave* spin-triplet superfluid phases where all of the symmetry except for translational symmetry is broken, exhibiting extraordinarily rich physical phenomena. The surface scattering in unconventional superfluids/superconductors induces quasi-particle bound states and suppresses the order parameter near the surface within the coherence length, often called Andreev surface bound states (ASBS). This generic nature of unconventional superfluid order parameter near the boundaries combined with the exotic symmetries in the superfluid phases of ^{3}He [1] lends an unfathomable source of fascinating physical phenomena [2–8] predicted to exist in confined geometry, many of which are of topological origin. Superfluid ^{3}He in confined geometry is indeed a treasure box that only a few groups have just started to probe experimentally [9, 10].

Various types of mechanical oscillators such as vibrating wires [11] and tuning forks [12] have been employed with great success to investigate bulk superfluid ^{3}He and also as reliable low temperature thermometers. In order to study quasi-two-dimensional liquid ^{3}He, we have developed comb-drive micro-electro-mechanical systems (MEMS) devices consisting of a movable center plate suspended by serpentine springs over a substrate with a well defined gap [13, 14]. When such a device is immersed in the liquid, a well defined film is formed naturally in the gap. The center plate of the device can be actuated to oscillate in its own plane (shear motion), exhibiting a resonant behavior from which one can extract physical properties of the surrounding fluid.

In superfluid ^{3}He-B with an isotropic gap, the mean free path of quasiparticles increases exponentially with decreasing temperature, and it could become much larger than the size of the experimental devices, placing the fluid-probe system in the ballistic limit. In this regime the hydrodynamic concepts cease to work, and the damping of the oscillators can be described by a simple model [15]. When an oscillator is driven at a small velocity, the damping force is proportional to the velocity. The damping coefficient decreases rapidly with temperature as $\exp(-\Delta_0/k_BT)$, where Δ_0 is the bulk energy gap and k_B is the Boltzmann constant [16]. In this work, we report our preliminary results of the resonance properties of a MEMS device submerged in superfluid ^{3}He: the temperature dependences of the resonance frequency and the damping of the oscillator in ^{3}He-B at 21.2 bar.

2. Experiment

The detailed descriptions on the design and the operation of the MEMS device employed in this study can be found in [14,17]. A differential capacitance method was used to detect the resonance motion of the MEMS device which has a movable poly-silicon plate with $200 \times 200\,\mu\text{m}^2$ size and $2\,\mu$m thickness. The gap between the plate and substrate is $1.25\,\mu$m. The device was immersed in liquid ^{3}He and cooled down to the base temperature of about $500\,\mu$K at 21.2 bar by a dilution refrigerator and a copper demagnetization stage. The system was let to warm up naturally from the base temperature during which multiple resonance spectra of the resonator were obtained by sweeping the excitation frequency through the resonance. Temperature was measured by a ^{3}He melting curve thermometer (MCT) above 1 mK and a platinum NMR thermometer below 1 mK which was calibrated against the MCT. In order to operate the Pt NMR thermometer, the ^{3}He sample was exposed to a magnetic field of 14 mT perpendicular to the plane of the film. Temperature was measured right before and after each frequency sweep to monitor possible heating during the sweep and the average of the two measurements was used to represent the temperature for the sweep. No appreciable heating was observed for all the data presented in this work.

The displacement of the oscillating plate is the solution to the equation of motion of a damped driven harmonic oscillator. Due to the phase shift of the background signal with frequency in the experiment, the amplitude of the output was used to fit to the Lorentzian, x_a, after a proper background subtraction:

$$x_a = \frac{Af_0d}{\sqrt{(f_0^2 - f^2)^2 + (df)^2}}, \tag{1}$$

where $A = F_0/(4\pi^2 m f_0 d)$ is the amplitude of the Lorentzian, F_0 the amplitude of the driving force, m the mass of the oscillator, d the full width at half maximum (FWHM), f the excitation frequency, and f_0 the resonance frequency in vacuum. Here A, f_0, d are fitting parameters and f is swept through the resonance frequency. The width d is proportional to the damping coefficient experienced by the oscillator assuming a linear velocity dependence of damping force, while the resonance frequency f_0 can be related to the fluid loading to the resonator:

$$\gamma = 2\pi d m_{total}, \tag{2}$$
$$k = 4\pi^2 m_{total} f_0^2, \tag{3}$$

where γ is the damping coefficient F_{damp}/v, m_{total} the resonator mass plus the loading mass, k the spring constant of the MEMS device which was measured at 4 K in vacuum.

3. Results and Discussion

The amplitudes of the resonance at various temperatures are shown in Fig. 1. From the base temperature to about 1.3 mK where the liquid is still in superfluid state, the spectrum evolves

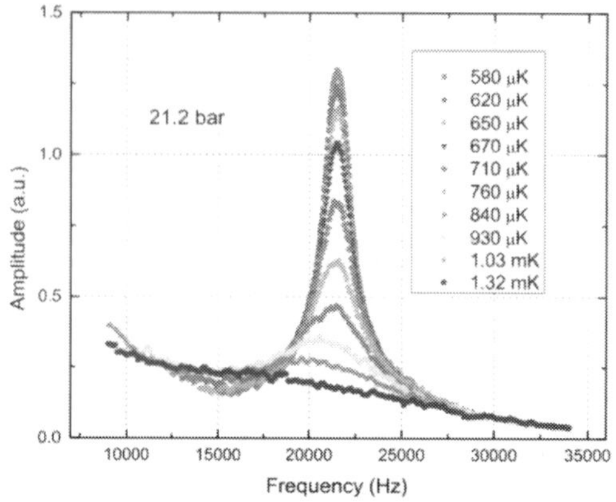

Figure 1. Temperature evolution of resonance spectra of the MEMS device at 21.2 bar with a constant excitation of 1.0 V_{pp}.

progressively from a well defined Lorentzian to a very broad hump. The width of the resonance increases rapidly as the temperature rises, indicating a fast growth of damping force. Meanwhile, the peak position shifts to lower frequencies due to the fluid mass loading to the resonator.

To understand the temperature dependence of damping force, two important length scales need to be evaluated and compared. One is the mean free path, l, of the quasiparticle which increases exponentially with decreasing temperature due to the isotropic energy gap. The other is the characteristic length, a, of the apparatus, which is related to the dimensions of the movable plate of the MEMS device in this case. The characteristic length of the MEMS device is of the order of the plate size $a \approx 200\,\mu$m. When l is much less than a, the system is in the hydrodynamic regime where the temperature dependence of the fluid viscosity, along with the geometry of the device, determines the damping force. An effort has been made to obtain the temperature dependence of the viscosity of normal ^{3}He through the damping force on a similar MEMS device, which will be published elsewhere.

On the other hand, when l is much greater than a, the system goes into the ballistic regime where the hydrodynamic concepts break down. For $T \lesssim 0.4\ T_c$, l becomes larger than a typical length scale of MEMS devices [18]. The damping force exerted on the MEMS device can be formulated through a model where scattering of thermal quasiparticles off the moving object is considered [15]. The model includes a key feature in superfluid, Andreev scattering of low-lying quasiparticles. When the object moves slowly, the velocity dependence simplifies to a linear relationship. The coefficient of proportionality, which is directly related to the FWHM through Eqn. (2), thus has a temperature dependence solely from the density of quasiparticles $\propto \exp(-\Delta_0/k_B T)$. Therefore, the width of the peak has a temperature dependence $d \propto \exp(-\Delta_0/k_B T)$ [11].

The resonance was fitted to Eqn. (1) to extract d, A, and f_0. The FWHM d are plotted against the reduced temperatures T/T_c for four excitations in Fig. 2. At the lowest temperature the peak width is still around 700 Hz, which is about 4 orders of magnitude larger than that in vacuum at 4 K. This indicates that the damping force at this temperature is still much larger than the intrinsic damping arising from the internal dissipation of the resonator. The fact that these four sets of data agree with each other demonstrates the damping coefficient is independent of driving force or resonator velocity, *i.e.*, the resonator is in the linear regime. A fit of the

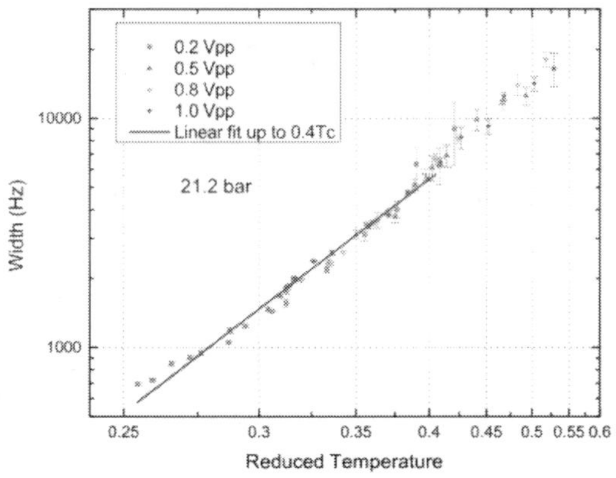

Figure 2. Temperature dependance of FWHM for various excitations. The *y-axis* scale is logarithmic while the *x-axis* scale is reciprocal. The straight line is the result of a fit of the data below 0.4 T/T_c to $d \propto \exp(-\Delta_0/k_B T)$ [11].

data below $0.4\,T/T_c$ to the exponential dependence gives $\Delta_0 = 1.57\ k_B T_c$, which is about 20 % smaller than $1.85\,k_B T_c$, the bulk gap at 21.2 bar [19]. Although it is needed to have a direct determination of T_c, the observed suppression of the gap is substantial but expected considering the suppression of the order parameters due to the boundaries of the film [20]. Reaching a lower base temperature is essential to verify the exponential temperature dependence of the damping to the intrinsic damping limit and also to extract a more accurate gap value [11].

The peak amplitude A and resonance frequency f_0 are plotted against the reduced temperatures for four excitations in Fig. 3. The left panel shows that the normalized amplitudes lie upon each other, ensuring the response of the oscillator is linear with the driving force. However, nonlinearity was observed when the resonator velocity exceeded a certain critical value, where excess damping set in and the lineshape was deformed from the Lorentzian. The amplitude decreases as temperature rises until around $0.4\,T/T_c$, where the amplitude seems to saturate. The right panel of Fig. 3 demonstrates the resonance frequency remains almost constant for $T < 0.4\ T/T_c$ before it starts to decrease at higher temperatures. This crossover occurs around the same temperature where the system enters into the hydrodynamic regime on warming. In the ballistic regime where damping is caused by direct momentum transfer between the quasiparticles and the resonator through scattering, the damping force is in phase with the velocity and no shift in resonance is expected [16]. Nevertheless, the low temperature limit resonance frequency, 21450 Hz, is about 100 Hz lower than in vacuum at 4 K. At this moment it is difficult to isolate the contributions from the fluid film only to the temperature dependences of the width, amplitude and position of the resonance peaks. A new design of the device is proposed so that the properties of films could be determined exclusively.

4. Conclusions

A MEMS device with a $1.25\,\mu$m gap has been immersed in superfluid ^{3}He and cooled down to $500\,\mu$K. Its resonance properties were measured on warming and resonance parameters such as peak width, amplitude and position were obtained as functions of temperature. The peak width is directly related to the damping coefficient and does not depend on velocity in the linear regime. Crossover from hydrodynamic to ballistic regime with decreasing temperature was observed in

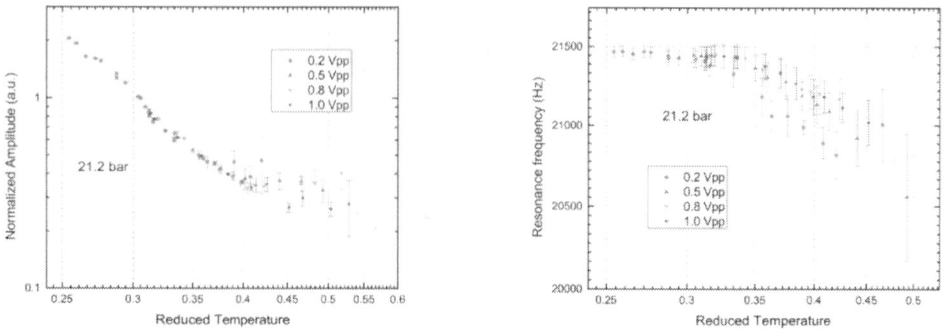

Figure 3. Log-log plots of temperature dependence of amplitude and resonance frequency for various excitations. Left panel: normalized amplitude of resonance peak against reduced temperature where fitted peak amplitudes are normalized with corresponding excitations; Right panel: resonance frequency against reduced temperature.

the temperature dependence of the resonance frequency. The temperature dependence of the width below $0.4\,T/T_c$ gives an energy gap of $1.57\,k_B T_c$ which is substantially smaller than the bulk value at the corresponding pressure. The peak amplitude also shows a significant change in its temperature dependence at a higher temperature, $0.4\,T/T_c$.

Acknowledgments

YL acknowledges stimulating discussions with the Lancaster ultra low temperature physics group. This work is supported by NSF through DMR-1205891(YL).

References

[1] Volovik G E 2003 *The Universe in a Helium Droplet* (Oxford University Press)
[2] Vorontsov A B and Sauls J A 2007 *Phys. Rev. Lett.* **98** 045301
[3] Vorontsov A B and Sauls J A 2005 *J. Low Temp. Phys.* **138** 283
[4] Chung S B and Zhang S C 2009 *Phys. Rev. Lett.* **103** 235301
[5] Mizushima T, Sato M and Machida K 2012 *Phys. Rev. Lett.* **109** 165301
[6] Wu H and Sauls J A 2013 *Phys. Rev. B* **88** 184506
[7] Tsutsumi Y and Machida K 2012 *J. Phys. Soc. Jpn.* **81** 074607
[8] Sauls J A 2011 *Phys. Rev. B* **84** 214509
[9] Levitin L V, Bennett R G, Sorovtsev E V, Parpia J M, Cowan B, Casey A J and Saunders J 2013 *Phys. Rev. Lett.* **111** 235304
[10] Casey A, Parpia J, Schanen R, Cowan B and Saunders J 2004 *Phys. Rev. Lett.* **92** 255301
[11] Guénault A M, Keith V, Kennedy C J, Mussett S G and Pickett G R 1986 *J. Low Temp. Phys.* **62** 511
[12] Bradley D I, Crookston P, Fisher S N, Ganshin A, Guénault A M, Haley R P, Jackson M J, Pickett G R, Schanen R and Tsepelin V 2009 *J. Low Temp. Phys.* **157** 476
[13] González M, Bhupathi P, Moon B H, Zheng P, Ling G, Garcell E, Chan H B and Lee Y 2011 *J. Low Temp. Phys.* **162** 661
[14] González M, Zheng P, Garcell E, Lee Y and Chan H B 2013 *Rev. Sci. Instrum.* **84** 025003
[15] Fisher S N, Guénault A M, Kennedy C J and Pickett G R 1989 *Phys. Rev. Lett.* **63** 2566
[16] Guénault A M and Pickett G R 1984 *Physica B* **126** 260
[17] Barquist C S, Bauer J, Edmonds T, Zheng P, Jiang W G, González M, Lee Y and Chan H B To be published
[18] Ono Y, Hara J and Nagai K 1982 *J. Low Temp. Phys.* **48** 167
[19] Halperin W P and Pitaevskii L P (eds) 1990 *Helium Three* (Elsevier)
[20] Vorontsov A B and Sauls J A 2003 *Phys. Rev. B* **68** 064508

Journal of Physics: Conference Series **568** (2014) 012031 doi:10.1088/1742-6596/568/1/012031

Dynamics of the forward vortex cascade in two-dimensional quantum turbulence

Andrew Forrester and Gary A. Williams
Department of Physics and Astronomy, University of California, Los Angeles, CA 90095
E-mail: gaw@ucla.edu

Abstract. The dynamics of the forward vortex cascade in 2D turbulence in a superfluid film is investigated using analytic techniques. The cascade is formed by injecting pairs with the same initial separation (the stirring scale) at a constant rate. They move to smaller scales with constant current under the action of frictional forces, finally reaching the core size separation, where they annihilate and the energy is removed by a thermal bath. On switching off the injection, the pair distribution first decays starting from the initial stirring scale, with the total vortex density decreasing linearly in time at a rate equal to the initial injection rate. As pairs at smaller scales decay, the vortex density then falls off as a power law, the same power law found in recent exact solutions of quenched 2D superfluids.

Constant-current cascades have long played a role in understanding turbulence in fluids. In classical 2D turbulence it is well known that there are two such cascades [1, 2], an inverse cascade of energy to large length scales with a $k^{-5/3}$ (Kolmogorov) energy spectrum, and a forward cascade of enstrophy (vorticity) to small length scales, with a k^{-3} (or $k^{-3}\ln(k)$) energy spectrum. The dynamics of the free decay of such turbulence has been investigated with numerical simulations [3, 4, 5, 6], where the decay is dominated by the dynamics of the inverse cascade where like-sign regions of vorticity merge, heading toward the final state that is two macroscopic counter-rotating vortices.

The nature of the cascades in 2D quantum turbulence, with its discrete quantized vortices, is considerably less well understood. A forward cascade of vortex pairs was found a number of years ago [7] for low vortex densities, comparable to or smaller than the density at the thermal Kosterlitz-Thouless transition. More recently two numerical simulations of the Gross-Pitaevskii model at high vortex densities found instead a forward energy cascade with a $k^{-5/3}$ spectrum [8, 9]. Further complicating the picture is another set of simulations which found no direct cascade, but did find an inverse energy cascade with a roughly $k^{-5/3}$ spectrum [10, 11].

In this paper we show that analytic techniques can be used to study the dynamics of the growth and decay of the forward vortex cascade. We find that the decay of the vortex density is initially linear in time, but then switches at longer times to a power-law decay with an exponent identical to that found in our recent exact solutions of quenched 2D superfluids [12]. We believe these are the first analytic solutions for the dynamics of any type of fluid turbulence.

We consider an incompressible superfluid film connected to a thermal bath at a temperature of 0.1 T_{KT}, where T_{KT} is the critical Kosterlitz-Thouless temperature where thermally excited vortex pairs drive the superfluid density to zero. The vortex dynamics are described by the same Fokker-Planck equation [13] used in describing the distribution of vortex pairs of separation r

[12, 14], with the addition of a 2D delta function to inject additional pairs of a fixed separation R into the film at a rate α,

$$\frac{\partial \Gamma}{\partial t} = \frac{1}{r}\frac{\partial}{\partial r}\left(r\frac{\partial \Gamma}{\partial r} + 2\pi K\,\Gamma\right) + \alpha\,\delta^2(\vec{r} - \vec{R}) \tag{1}$$

This equation is set in dimensionless form with lengths in units of the vortex core radius a_0, the vortex-pair distribution function Γ in units a_0^{-4}, and the time in units of the vortex diffusion time, $\tau_0 = a_0^2/2D$ with D the vortex diffusion coefficient. K is the dimensionless superfluid density, $K = \hbar^2\sigma_s/m^2k_BT$, and is determined from the Kosterlitz recursion relation [15]

$$\frac{dK}{dr} = -4\pi^3 r^3 K^2 \Gamma \tag{2}$$

with σ_s the superfluid areal density and m the ^{4}He atomic mass. The dimensionless injection rate α is given by $\alpha = a_0^2\dot{Q}\tau_0$ where $\dot{Q}$ is the number of vortex pairs injected per unit area per time at random positions and orientations across the plane.

The steady-state solutions of Eq. 1 where $\partial\Gamma/\partial t = 0$ were previously found in [7], but only for $r < R$, while for dynamics the vortices at $r > R$ also need to be included. In the limit of low injection rates the vortex densities are well below the densities at T_{KT}, and the superfluid fraction is unaffected by the vortices. The solution for $r < R$ is then a constant value

$$\Gamma = \Gamma_0 = \alpha/2\pi K_0 \tag{3}$$

where K_0 is the value at $r = 1$ where σ_s equals the "bare" superfluid density σ_s^0. For $r > R$ the solution is a quasi-thermal distribution extending from R, $\Gamma = \Gamma_0\,(r - R)^{-2\pi K_0}$, and which arises from injected pairs initially at separation R getting a thermal kick to higher separation. The steady-state flux of vortex pairs from Eq. 1 is $J = -(r\partial\Gamma/\partial r + 2\pi K\Gamma))$, and for this solution is a constant, $J = -\alpha$ for $r < R$, and zero for $r > R$.

The two curves in Fig. 1 for the lowest values of α show the steady-state distribution of pairs for the low-injection regime calculated directly from Eqs. 1 and 2 using fixed-step Runge-Kutta iteration with the injection scale $R = 400$, in agreement with the above form of Eq. 3. The delta function is approximated with a strongly peaked Gaussian form with spatial width of 2. The superfluid fraction remains unity for $r < R$, though at $\alpha = 1 \times 10^{-12}$ the vortex density is getting high enough that there is a very slight reduction for $r > R$.

The upper two curves in Fig. 1 show the distribution in the limit where the vortex density becomes comparable to that found at T_{KT}, indicated by the dashed curve in Fig. 1. The superfluid fraction is rapidly driven to zero at a finite length scale r_0 that depends on the injection rate. This effect on the superfluid density shows that this turbulent state is not just characterized by isolated dipole pairs, but in fact is a complex many-body state of smaller pairs screening the interaction of larger pairs to drive down the superfluid density. To a very good approximation the solutions in this regime can be represented as

$$\begin{aligned} \Gamma &= \Gamma_0 & (r < R) \\ &= \Gamma_0(1 + \ln(r/r_0)) & (r_0 < r < R) \\ &= \Gamma_0(1 + \ln(R/r_0)) & (r > R). \end{aligned} \tag{4}$$

The steady-state flux for this solution is also $J = -\alpha$ for $r < R$, and zero for $r > R$.

The time dependence of the cascade growth can be studied by solving Eqs. 1 and 2 as a function of both time and distance using standard numerical techniques, starting from thermal equilibrium at 0.1 T_{KT} and switching on the injection at $t = 0$. Figure 2 shows the growth of the distribution and the pair current for the very low injection rate $\alpha = 1 \times 10^{-20}$. Initially

Journal of Physics: Conference Series **568** (2014) 012031 doi:10.1088/1742-6596/568/1/012031

the distribution just broadens as the pairs get thermal kicks to larger and smaller separations, but the frictional forces on the vortex cores also give rise to a net current of pairs to smaller separations. After a few hundred diffusion times the decaying pairs reach the core size separation where they annihilate, and the energy is pulled out by the thermal bath. At longer times the distribution becomes constant at $r < R$ and increases uniformly toward the equilibrium value. The current is initially only appreciable near the injection scale, and then finally at very long times reaches the value of $-\alpha$, the cascade state where the rate of pairs being injected at R is equal to the rate of pairs being pulled out by the thermal bath at the scale a_0. The total vortex density is shown in Fig. 3, with initially a linear increase with slope α before any pairs are pulled out at a_0, and then a leveling off towards equilibrium once pairs begin annihilating.

On switching off the injection, the decay of the cascade starts at the injection scale, as shown in Fig. 4, since the current of pairs away from R is no longer being replenished by the injected pairs. Once the region of diminished pairs begins to reach the scale a_0 the distribution starts to uniformly decrease, and in fact in this regime the solution of Eqs. 3 and 4 is the exact solution found in [12] for a quenched 2D superfluid with an initial temperature equal to 0.1 T_{KT}. At long times, however, it comes to equilibrium at the bath temperature. The decaying vortex density is shown in Fig. 3, initially a linear decay with slope $-\alpha$, since this is still the rate at which pairs annihilate, but then at long times falling off as $t^{-(\pi K_0 - 1)}$, the value predicted in [12].

In the limit of higher injection rates where the superfluid density is driven to zero at finite scales ($\alpha > 1 \times 10^{-11}$) the dynamics of the cascade becomes more complicated. The gradient in the superfluid density K in Eq. 1 leads to an additional current toward large scales that subtracts from the forward current and tends to slow down both the growth and decay of the cascade. This is shown in Fig. 5 for the growth of the cascade at $\alpha = 1 \times 10^{-7}$. As the pair distribution grows the superfluid fraction is depressed to smaller and smaller length scales. The current in the cascade region $r < R$ only grows slowly since the pairs being injected are divided, with part

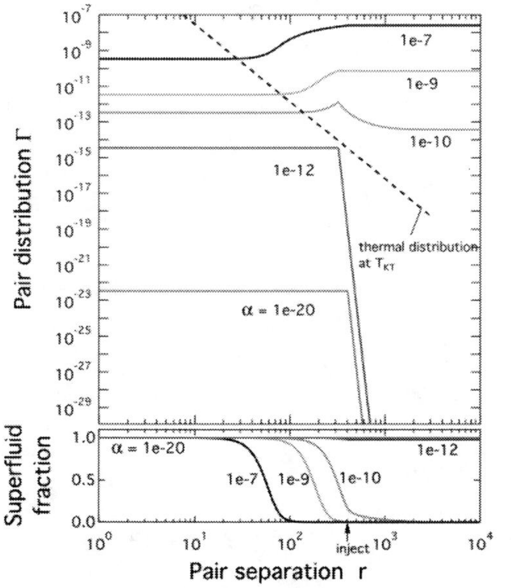

Figure 1. Steady-state solutions of Eqs. 1 and 2 for the vortex-pair distribution function at different injection rates α.

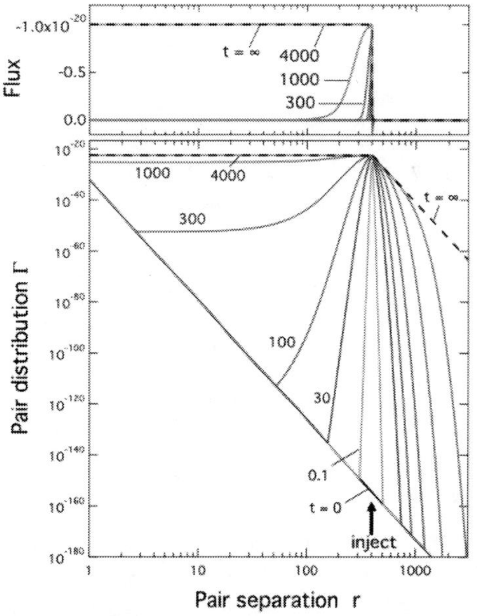

Figure 2. Growth of the pair distribution and current in time for $\alpha = 1 \times 10^{-20}$.

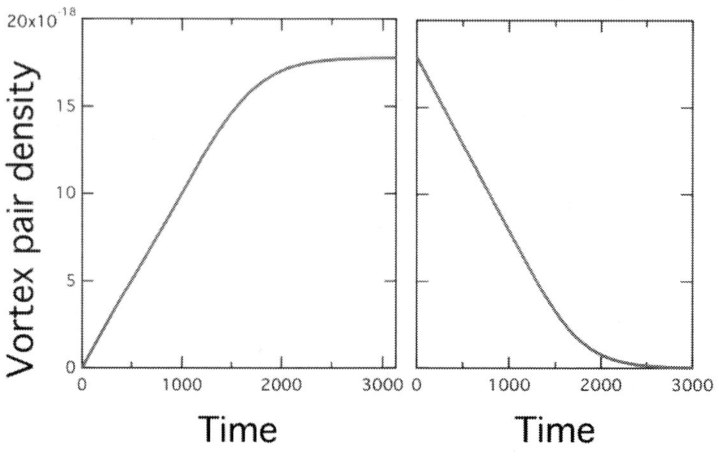

Figure 3. Growth (left) and decay (right) of the vortex-pair density, for injection rate $\alpha = 1 \times 10^{-20}$.

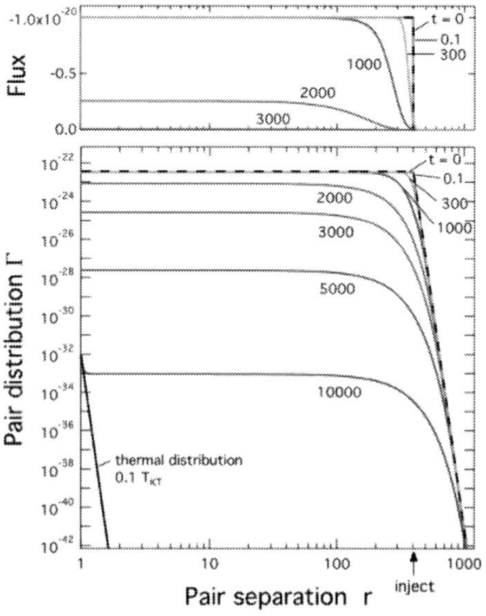

Figure 4. Decay of the pair distribution function and current after switching off the injection $\alpha = 1 \times 10^{-20}$.

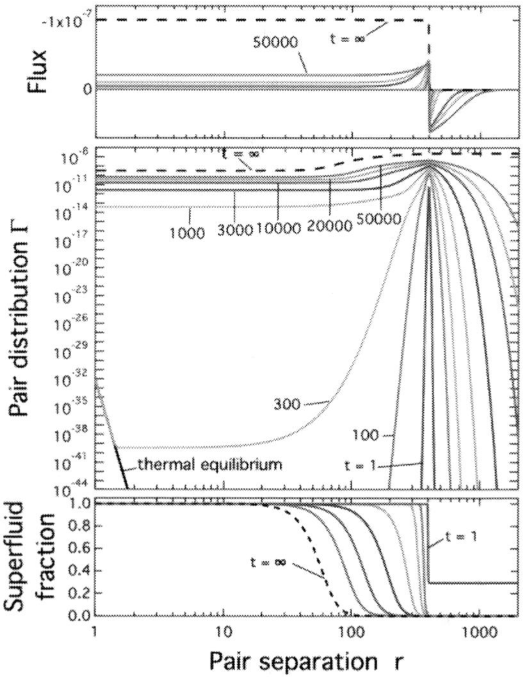

Figure 5. Growth of the pair distribution, current, and superfluid fraction for injection $\alpha = 1 \times 10^{-7}$.

going in the forward direction and others increasing their separation in the inverse direction, driven by the gradient in K. The vortex density in the cascade region grows more slowly than linear in time due to this division, with fits giving a variation as approximately $t^{2/3}$.

In this analysis we find no evidence of a forward energy cascade with spectrum $k^{-5/3}$ as found in the earlier simulations [8, 9], and this may be due to the relatively dilute vortex densities considered here, in which there will only be annihilation of same-pair vortices when their separation reaches a_0. The vortex densities in Refs. [8] and [9] were quite high, with vortices on a large fraction of the lattice sites, and in this situation there will be a much higher probability of annihilation of opposite-circulation vortices on different pairs, which might give rise to the observed forward energy cascade. The densities are so high that these systems are certainly not superfluid at length scales much larger than the vortex core size. A further difference from our system is that the Gross-Pitaevskii model that is simulated is a compressible superfluid, and the sound waves excited in the turbulence may also contribute to energy flows. And finally, these simulations had zero dissipation, whereas our forward cascade requires an arbitrarily small but finite dissipation to develop. The dissipation only enters into the dynamics through the vortex diffusion constant that sets the time scale; it does not affect the steady-state solutions of Eqs. 3 and 4. Weak dissipation still allows the possibility of "inertial" cascades, as shown by the simulations of the classical 2D Navier-Stokes equation with finite viscosity [16].

We also do not observe an inverse cascade of energy to large scales. There is an initial outward current current as pairs build up in the region $r > R$, but this falls to zero in the steady state for both the low and high injection rates. An inverse cascade requires a clustering of like-sign vortices in a negative-temperature state [10, 11], but our Kosterlitz-Thouless model assumes from the start a uniform distribution of vortices with positive and negative circulation.

Acknowledgements

This work was supported in part by the U. S. National Science Foundation, Grant No. DMR 0906467.

References

[1] Kraichnan R H and Montgomery D 1980 *Reports on Progress in Physics* **43** 547
[2] Boffetta G and Ecke R E 2012 *Annual Review of Fluid Mechanics* **44** 427–451
[3] Bracco A, McWilliams J C, Murante G, Provenzale A and Weiss J B 2000 *Physics of Fluids* **12** 2931–2941
[4] Sire C and Chavanis P H 2000 *Phys. Rev. E* **61** 6644–6653
[5] van Bokhoven L J A, Trieling R R, Clercx H J H and van Heijst G J F 2007 *Physics of Fluids* **19** 046601
[6] LaCasce J H 2008 *Physics of Fluids* **20** 085102
[7] Chu H C and Williams G A 2001 *Quantized Vortex Dynamics and Superfluid Turbulence* (*Lecture Notes in Physics* no 571) ed Barenghi C F, Donnelly R J and Vinen W F (Heidelberg: Springer-Verlag) pp 226–232
[8] Numasato R, Tsubota M and L'vov V S 2010 *Phys. Rev. A* **81**(6) 063630
[9] Chesler P M, Liu H and Adams A 2013 *Science* **341** 368–372
[10] Reeves M T, Billam T P, Anderson B P and Bradley A S 2013 *Phys. Rev. Lett.* **110**(10) 104501
[11] Billam T P, Reeves M T, Anderson B P and Bradley A S 2014 *Phys. Rev. Lett.* **112**(14) 145301
[12] Forrester A, Chu H C and Williams G A 2013 *Phys. Rev. Lett.* **110**(16) 165303
[13] Ambegaokar V, Halperin B, Nelson D and Siggia E 1980 *Phys. Rev. B* **21** 1806
[14] Chu H C and Williams G A 2001 *Phys. Rev. Lett.* **86**(12) 2585–2588
[15] Kosterlitz J M 1974 *J. Phys. C* **7** 1046
[16] Boffettal G and Musacchio S 2010 *Phys. Rev. E* **82** 016307

Intrinsic anisotropy of the superconductivity in $FeSe_{1-x}Te_x$ (x = 0.6) observed by specific heat in magnetic fields.

T Konno[1], T Adachi[2], M Imaizumi[1], T Noji[1], T Kawamata[1], Y Koike[1]

[1]Department of Applied Physics, Tohoku University, Sendai 980-8579, Japan
[2]Department of Engineering and Applied Sciences, Sophia University, Tokyo 102-8554, Japan

E-mail: t-adachi@sophia.ac.jp (T Adachi)

Abstract. Low-temperature specific heat measurements in various magnetic fields parallel and perpendicular to the c-axis for annealed single crystals of the iron-chalcogenide superconductor $FeSe_{1-x}Te_x$ have revealed that the anisotropy of the enhancement of the residual electronic-specific-heat-coefficient in the ground state by the application of magnetic field, namely, the anisotropy of the upper critical field, H_{c2}, is 2.2 for x = 0.6, which is different from the anisotropy of H_{c2} estimated from the resistivity measurements in non-annealed single crystals of x = 0.6 [Khim *et al.* 2010 *Phys. Rev.* B **81** 184511]. It appears that the anisotropy of H_{c2} strongly depends on the content of excess iron included in a crystal.

1. Introduction

In recent years, iron-based superconductors have attracted considerable attention due to their high superconducting (SC) transition temperatures, T_c's, multiple SC gaps and exotic paring mechanisms relating to orbital fluctuations and/or spin fluctuations. As-grown single crystals of the iron-chalcogenide superconductor $FeSe_{1-x}Te_x$ tend to include excess iron more or less, resulting in the marked suppression of bulk superconductivity at $x \gtrsim 0.7$ [1]. On the other hand, it has been reported that single crystals annealed in vacuum (~ 10^{-4} Pa) exhibit bulk superconductivity in a wide range of x = 0.5 – 0.9 [2,3]. In order to investigate details of the SC gap in the annealed single crystals of $FeSe_{1-x}Te_x$, we have measured the specific heat in magnetic fields for x = 0.6 – 1 [4,5]. As a result, following two points have been clarified; (i) there exist at least two kinds of SC gap and the smaller SC gap is significantly depressed by the application of magnetic field, whereas the depression of the larger SC gap is small, and (ii) the enhancement of the residual electronic-specific-heat-coefficient in the ground state, γ_0, by the application of magnetic field parallel to the c-axis is much smaller than that expected for superconductors with the typical s-wave or d-wave SC paring symmetry. This unexpected small enhancement of γ_0 by the application of magnetic field has been discussed in terms of the multi-band effect [6].

In this paper, to obtain further knowledge of the SC properties, we have investigated the magnetic-field dependence of γ_0 for $FeSe_{1-x}Te_x$ with x = 0.6, changing the direction of the applied magnetic field.

2. Experimental

Single crystals of $FeSe_{1-x}Te_x$ with x = 0.6 were grown by the Bridgman method [2,3]. As-grown crystals were annealed at 400 °C for 200 h in vacuum (~ 10^{-4} Pa). Specific heat measurements were carried out in magnetic fields of 0 – 9 T parallel and perpendicular to the c-axis on field cooling by the thermal-relaxation method, using a commercial apparatus (Quantum Design, PPMS). The Cernox thermometer used for the specific heat measurements was calibrated in each field. The heat capacity of the crystals was obtained by subtracting the heat capacity of the addenda measured in each field.

3. Results and Discussion

Figures 1(a) and (b) show the low-temperature specific heat of $x = 0.6$ in various magnetic fields parallel and perpendicular to the c-axis, respectively, plotted as C/T vs T^2. It is found that C/T increases with increasing field due to the generation of normal carriers. It is also found that the increase in C/T with increasing field is more significant in fields parallel to the c-axis than those perpendicular to the c-axis, which is a usual behavior reflecting the anisotropy of the upper critical field, H_{c2}, observed in conventional layered superconductors. The data are well fitted with the standard function, $C = \gamma_0 T + \beta T^3$, where βT^3 is the term of the phonon specific heat.

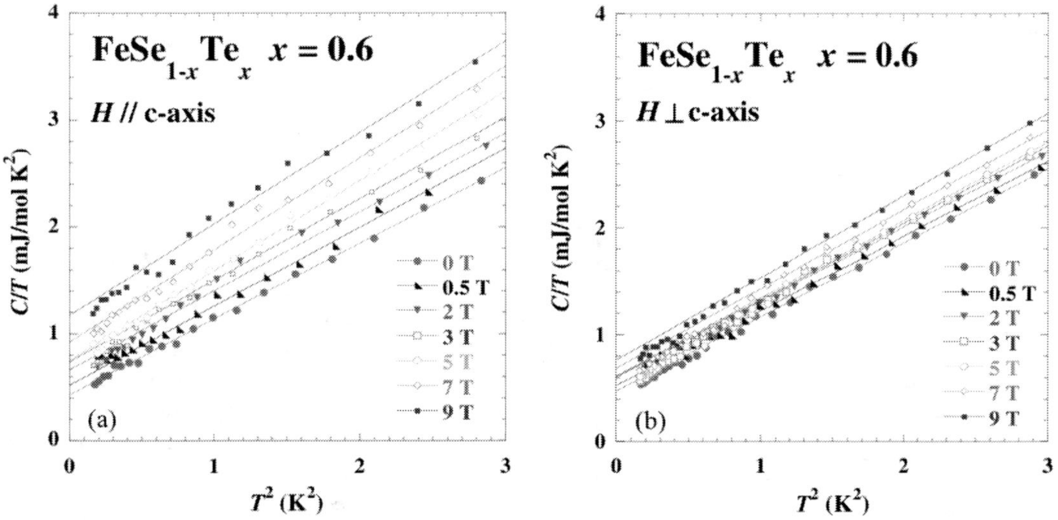

Figure 1. Temperature dependence of the specific heat, C, for $FeSe_{1-x}Te_x$ with $x = 0.6$ in various magnetic fields (a) parallel and (b) perpendicular to the c-axis on field cooling at very low temperatures down to 0.4 K, plotted as C/T vs T^2.

Figure 2 (a) displays the magnetic-field dependence of $\Delta\gamma$, defined as $\Delta\gamma = \gamma_0(H) - \gamma_0(0)$, for $FeSe_{1-x}Te_x$ with $x = 0.6$. The vertical axis is normalized by $\gamma_N - \gamma_0(0)$, where γ_N represents the electronic-specific-heat-coefficient in the normal state. It is found that the enhancement of $\Delta\gamma$ by the application of magnetic field parallel and perpendicular to the c-axis is much smaller than that expected for the typical *s*-wave or *d*-wave SC paring symmetry [4,5]. Figure 2(b) shows the magnetic-field dependence of the ratio of $\Delta\gamma$ in parallel fields to that in perpendicular fields, $\Delta\gamma_{//c}/\Delta\gamma_{\perp c}$, for $x = 0.6$. It is found that the ratio is not dependent on field and that the average value is 2.2. The $\Delta\gamma_{//c}/\Delta\gamma_{\perp c}$ is regarded as corresponding to the anisotropy of H_{c2}. According to the effective mass model [7], therefore,

$$\frac{\Delta\gamma_{//c}}{\Delta\gamma_{\perp c}} = \frac{H_{c2}(H \perp c)}{H_{c2}(H // c)} = \frac{\xi_{ab}}{\xi_c} = \sqrt{\frac{m_c^*}{m_{ab}^*}}, \tag{1}$$

where ξ_{ab} and ξ_c are SC coherence lengths in the ab-plane and along the c-axis, respectively, and m^*_{ab} and m^*_c are effective masses in the ab-plane and along the c-axis, respectively. Therefore, the anisotropy of the SC coherence length is estimated as 2.2 and the anisotropy of the effective mass is estimated as 4.8.

The present result of the anisotropy of H_{c2} is consistent with our former result obtained from the resistive SC transition just below T_c for the annealed single crystals of $x = 0.6$ that the anisotropy of H_{c2} is estimated as 2.7 [2]. However, these values are different from that obtained from the resistivity measurements using as-grown (non-annealed) single crystals of $x = 0.6$, namely, ~ 1 [8,9]. The difference might be related to excess iron. That is, since excess iron included in the as-grown single crystals is removed through the annealing in vacuum (~ 10^{-4} Pa) [10], it appears that the intrinsic two-dimensional nature of $FeSe_{1-x}Te_x$ emerges in the present annealed single crystals, while it is hidden by excess iron in the as-grown crystals.

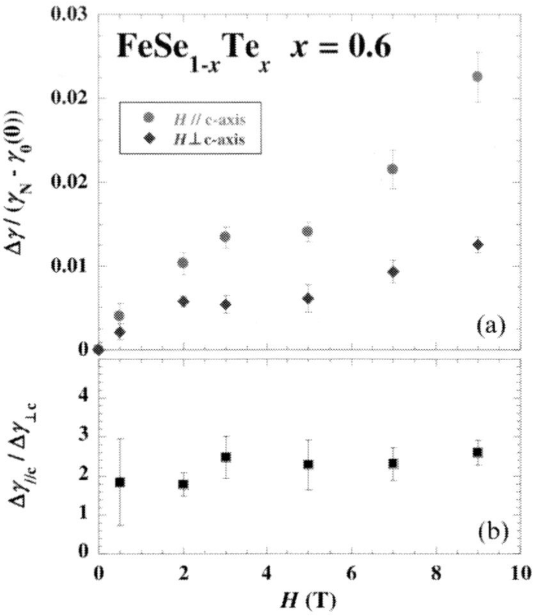

Figure 2. (a) Magnetic-field dependence of the residual electronic-specific-heat-coefficient in the ground state in magnetic fields parallel and perpendicular to the c-axis, $\Delta\gamma = \gamma_0(H) - \gamma_0(0)$, normalized by $\gamma_N - \gamma_0(0)$ for $FeSe_{1-x}Te_x$ with $x = 0.6$. (b) Magnetic-field dependence of the ratio of $\Delta\gamma$ in parallel fields to that in perpendicular fields, $\Delta\gamma_{//c}/\Delta\gamma_{\perp c}$, for $FeSe_{1-x}Te_x$ with $x = 0.6$.

4. Summary

From the low-temperature specific heat measurements in magnetic fields parallel and perpendicular to the c-axis in annealed single crystals of $FeSe_{1-x}Te_x$, the anisotropy of the enhancement of $\Delta\gamma$ by the application of magnetic field, namely, the anisotropy of H_{c2} has been found to be 2.2 for $x = 0.6$. Therefore, the anisotropy of the SC coherence length and that of the effective mass have been estimated as 2.2 and 4.8, respectively. The present result is different from the former result of the anisotropy of H_{c2} ~ 1 in non-annealed single crystals of $x = 0.6$ [8,9]. It appears that the intrinsic two-dimensional nature of $FeSe_{1-x}Te_x$ emerges by removing excess iron through the annealing.

Acknowledgements

This work was partially supported by JSPS KAKENHI Grant Number 23540399.

References

[1] Sales B C, Sefat A S, McGuire M A, Jin R Y, Mandrus D, Mozharivsky Y 2009 *Phys Rev* B **79** 094521

[2] Noji T, Suzuki T, Abe H, Adachi T, Kato M, Koike Y 2010 *J. Phys. Soc. Jpn.* **79** 084711

[3] Noji T, Imaizumi M, Suzuki T, Adachi T, Kato M, Koike Y 2012 *J. Phys. Soc. Jpn.* **81** 054708

[4] Konno T, Adachi T, Imaizumi M, Noji T, Kawamata T, Koike Y, *Physica* C, in press
[5] Konno T, Adachi T, Imaizumi M, Noji T, Kawamata T, Koike Y, submitted
[6] Bang Y 2010 *Phys. Rev. Lett.* **104** 217001
[7] Morris R C, Coleman R V, Bhandari R 1972 *Phys. Rev.* B **5** 895
[8] Khim S, Kim J-W, Choi E-S, Bang Y, Nohara M, Tagaki H, Kim K-H 2010 *Phys. Rev.* B **81** 184511
[9] Fang M, Yang J, Balakirev F F, Kohama Y, Singleton J, Qian B, Mao Z Q, Wang H, Yuan H Q 2010 *Phys. Rev.* B **81** 020509(R)
[10] Ohno M, Kawamata T, Noji T, Naruse K, Matsuoka Y, Adachi T, Nishizaki T, Sasaki T, Koike Y 2014 *J. Phys. Soc. Jpn.* **83** 044704

Journal of Physics: Conference Series **568** (2014) 022002 doi:10.1088/1742-6596/568/2/022002

Electronic Homogeneity in the Overdoped Regime of the Electron-Doped High-T_c Superconductors

M A Baqiya[1], T Adachi[2], A Takahashi[1], T Prombood[1], M Watanabe[1], K Fukumoto[1], Y Tanabe[1], and Y Koike[1]

[1]Department of Applied Physics, Tohoku University, Sendai 980-8579, Japan
[2]Department of Engineering and Applied Sciences, Sophia University, Tokyo 102-8554, Japan

Email: t-adachi@sophia.ac.jp (T Adachi)

Abstract. In order to investigate the electronic homogeneity in electron-doped high-T_c cuprates, magnetization and specific heat measurements have been performed for $Pr_{1-x}LaCe_xCuO_4$ single crystals with $x = 0.13 - 0.17$. It has been found from the magnetization curve that the hysteresis is relatively small and that the so-called second magnetization peak is absent for every x. Moreover, the Sommerfeld constant in the ground state has shown $H\ln H$ dependence on magnetic field H for every x, suggesting that the superconductivity is not pure but dirty. These results are in contrast to those observed in the overdoped regime of hole-doped $La_{2-x}Sr_xCuO_4$ where an inhomogeneous superconducting state is formed in spite of the superconductivity being pure. Therefore, it has been concluded that a rather homogeneous electronic state is realized in the electron-doped cuprates, suggesting different electronic states between hole- and electron-doped cuprates.

1. Introduction

Recently, the study of microscopic electronic inhomogeneity of the superconductivity in high-T_c cuprates has attracted interest. For hole-doped $La_{2-x}Sr_xCuO_4$ (LSCO), it has been suggested from the magnetic susceptibility measurements that both the superconducting (SC) transition temperature, T_c, and the SC volume fraction decrease markedly with increasing x in the overdoped regime [1]. The low-temperature specific heat measurements have revealed the existence of a finite value of the Sommerfeld constant in the SC ground state [2]. In addition, from the magnetization-hysteresis-loop measurements, a so-called second magnetization peak has been observed in the overdoped regime, indicating the enhancement of the vortex pinning [3]. These results suggest that the electronic state in overdoped LSCO is so inhomogeneous as to be composed of normal and SC states.

As for the electron-doped cuprates, the situation is rather complicated due to the reduction process of excess oxygen which is essential for the appearance of superconductivity. Therefore, in order to investigate whether the electronic state is inhomogeneous or not, we have carried out magnetization and specific heat measurements in electron-doped single crystals of $Pr_{1-x}LaCe_xCuO_4$ (PLCCO) with $x = 0.13 - 0.17$.

2. Experimental

Single crystals of PLCCO with $x = 0.13 - 0.17$ were grown by the travelling-solvent floating-zone method. The quality of the grown crystals was checked by the x-ray back-Laue photography to be good. The composition of the crystals was checked by the inductively coupled plasma analysis. The reduction annealing was performed in vacuum at various temperatures for various hours to obtain SC crystals. Magnetization and specific heat measurements were performed in magnetic fields up to 7 and 9 T parallel to the c-axis using a SC quantum interference device (SQUID) magnetometer and a commercial apparatus (Quantum Design, PPMS), respectively.

3. Results and Discussion

Figure 1 shows the dependence on magnetic field H of the hysteresis in the magnetization curve, ΔM, for fully reduced crystals of PLCCO with $x = 0.13 - 0.17$. The T_c's of crystals with $x = 0.13$, 0.15 and 0.17 are 20.5 K, 12.2 K and 5.1 K, respectively. It is found that, compared with ΔM of overdoped LSCO with $x = 0.198$ [3], the hysteresis is quite small and no second magnetization peak is observed in PLCCO. These results suggest that the vortex pinning is relatively weak in PLCCO, implying a rather homogeneous electronic state in contrast to the inhomogeneous one in overdoped LSCO.

Figure 2(a) shows the temperature dependence of the specific heat, C, plotted as C/T vs. T^2 for fully reduced crystals of PLCCO with $x = 0.13 - 0.17$. It is found that C at low temperatures is expressed as $C/T = \gamma_0 + \beta T^2$, where γ_0 is the residual Sommerfeld-constant in the ground state and β is the coefficient of the specific heat of phonons. For every x, γ_0 is not zero but has a finite value even in the SC state, which is consistent with the results in the former report [4]. It is noted that C/T shows no upturn due to a Schottky anomaly, suggesting that the crystals are free of any measurable magnetic impurities.

It is well known for d-wave superconductors that γ_0 exhibits $H^{1/2}$ dependence [5], which has been observed in overdoped LSCO in spite of the existence of a finite value of γ_0 in zero field. Therefore, it has been concluded that the superconductivity in overdoped LSCO is pure and that the electronic state is inhomogeneous [2]. For dirty d-wave superconductors, on the other hand, γ_0 is proportional to $H\ln H$ [6]. As shown in Fig. 2(b), the magnetic-field dependence of $\Delta\gamma_0 = \gamma_0(H) - \gamma_0(0)$ for PLCCO with $x = 0.15$ is well reproduced by $H\ln H$ rather than $H^{1/2}$. These results suggest that the observed finite value of γ_0 in zero field is not due to the inhomogeneous superconductivity but due to the

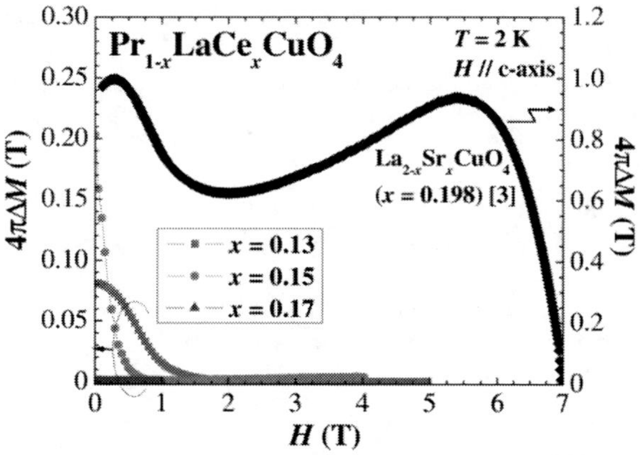

Figure 1. Magnetic-field dependence of the magnitude of the hysteresis in the magnetization curve, ΔM, for $Pr_{1-x}LaCe_xCuO_4$ with $x = 0.13 - 0.17$. ΔM for $La_{2-x}Sr_xCuO_4$ with $x = 0.198$ is also shown for comparison [3].

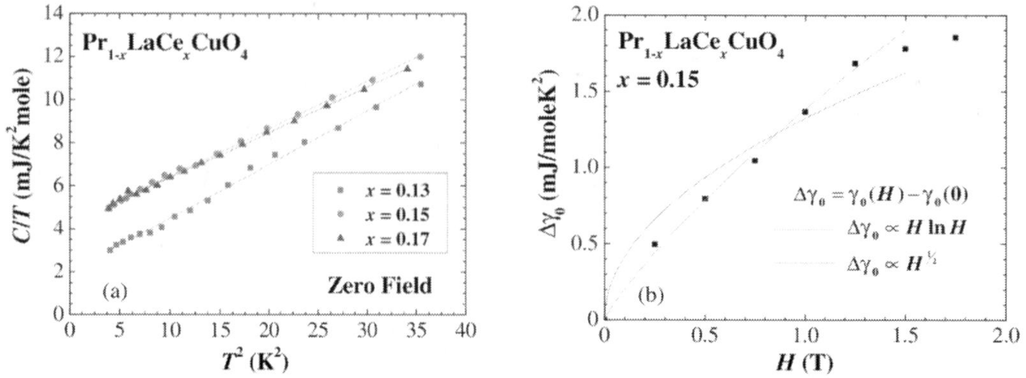

Figure 2. (a) Temperature dependence of the specific heat, C, plotted as C/T vs. T^2 in zero field for fully reduced crystals of $Pr_{1-x}LaCe_xCuO_4$ with $x = 0.13 - 0.17$. (b) Magnetic-field dependence of $\Delta\gamma_0 = \gamma_0(H) - \gamma_0(0)$ for the fully reduced crystal of $Pr_{1-x}LaCe_xCuO_4$ with $x = 0.15$.

homogeneous dirty superconductivity where probably the scattering by the residual excess oxygen is frequent.

The present results of magnetization and specific heat have revealed that a rather homogeneous SC state is formed in the electron-doped cuprates, in contrast to the inhomogeneous SC state observed in the hole-doped cuprates. One possible explanation for the different electronic states is related to the so-called non-Ce-doped superconductivity in the parent compounds of the electron-doped cuprates such as $La_{2-x}RE_xCuO_4$ (RE: rare-earth elements) with the Nd_2CuO_4–type structure [7,8]. Very recently, we have suggested that the electronic state of the electron-doped cuprates including the parent SC compounds is able to be understood in term of a band structure taking into account the strong electron correlation [9]. In order to clarify whether or not the novel electronic state leads to the homogeneous SC state in the electron-doped cuprates, further experiments to unveil the detailed electronic state of PLCCO are necessary.

4. Summary

We have investigated the electronic homogeneity in PLCCO with $x = 0.13 - 0.17$ from the magnetization and specific heat measurements. It has been found that the magnetization curve exhibits a relatively small hysteresis and that the residual Sommerfeld-constant in the ground state shows $H\ln H$ dependence. These results suggest that a rather homogeneous SC state is realized in the electron-doped cuprates, in contrast to the inhomogeneous SC state in the overdoped regime of hole-doped LSCO.

References

[1] Tanabe Y, Adachi T, Noji T and Koike Y 2005 *J. Phys. Soc. Jpn.* **74** 2893
[2] Wang Y, Yan J, Shan L, Wen H, Tanabe Y, Adachi T and Koike Y 2007 *Phys. Rev. B* **76** 064512
[3] Tanabe Y, Adachi T, Omori K, Sato H, Noji T and Koike Y 2007 *J. Phys. Soc. Jpn.* **76** 113706
[4] Balci H and Greene R L 2004 *Phys. Rev. B* **70** 140508(R)
[5] Volovik G E 1993 *JETP Lett.* **58** 469
[6] Kübert C and Hirschfeld P J 1997 *Solid State Commun.* **105** 459
[7] Tsukada A, Krockenberger Y, Noda M, Yamamoto H, Manske D, Alff L and Naito M 2005 *Solid State Commun.* **133** 427
[8] Takamatsu T, Kato M, Noji T and Koike Y 2012 *Appl. Phys. Express* **5** 073101
[9] Adachi T, Mori Y, Takahashi A, Kato M, Nishizaki T, Sasaki T, Kobayashi N and Koike Y 2013 *J. Phys. Soc. Jpn.* **82** 063713

Possible ferromagnetic phase in non-superconducting heavily overdoped cuprates of Bi-2201

K Kurashima[1], **T Adachi**[2], **K M Suzuki**[1†], **Y Fukunaga**[1], **T Kawamata**[1], **T Noji**[1], **Y Koike**[1]

[1] Department of Applied Physics, Tohoku University, Sendai, Japan
[2] Department of Engineering and Applied Sciences, Sophia University, Tokyo, Japan

E-mail: t-adachi@sophia.ac.jp (T Adachi)

Abstract. In order to investigate the possible ferromagnetic phase in the non-superconducting (non-SC) heavily overdoped (HOD) regime of high-T_c cuprates suggested by Sonier *et al.* [Proc. Natl. Acad. Sci. USA **107**, 17131 (2010)] in $La_{2-x}Sr_xCuO_4$, both the magnetization and in-plane electrical resistivity, ρ_{ab}, have been measured using single crystals of $Bi_{1.76}Pb_{0.35}Sr_{1.89}CuO_{6+\delta}$ covering a wide range of hole concentration. Magnetization curves in the non-SC HOD regime have shown a tendency to be saturated in high magnetic fields at low temperatures, which is probably a precursor of the possible ferromagnetic transition at a lower temperature. In addition, it has been found that ρ_{ab} exhibits the temperature dependence being proportional to $T^{4/3}$, which is characteristic of a metal with two-dimensional ferromagnetic fluctuations, in a wide range of hole concentration from the superconducting overdoped to the non-SC HOD regime. It has been concluded that two-dimensional ferromagnetic correlation develops at low temperatures in HOD Bi-2201 and that therefore the development of the ferromagnetic correlation is probably an intrinsic feature in the HOD cuprates.

1. Introduction

In general, electrons in the non-superconducting (non-SC) heavily overdoped (HOD) regime of high-T_c cuprates have been believed to behave as Fermi-liquid. However, phenomena contradictory to the Fermi-liquid picture have been reported, such as an upturn of the magnetic susceptibility, χ, at low temperatures in HOD $Tl_2Ba_2CuO_{6+\delta}$ [1]. Formerly, it has theoretically been suggested that ferromagnetic fluctuations suppress the *d*-wave superconductivity in the overdoped (OD) regime of hole-doped high-T_c cuprates [2]. In addition, Sonier *et al.* have reported from muon spin relaxation measurements in non-SC HOD $La_{2-x}Sr_xCuO_4$ (LSCO) with $x = 0.33$ that the spin correlation develops at low temperatures below 0.9 K and that the in-plane electrical resistivity, ρ_{ab}, is proportional to $T^{5/3}$ at high temperatures above 60 K, suggesting the existence of three-dimensional (3D) ferromagnetic fluctuations in the non-SC HOD regime [3]. However, both oxygen deficiency and disorder characteristic of non-SC HOD LSCO give us a doubt whether a ferromagnetic phase intrinsically exists in the cuprates.

The so-called Bi-2201 is one of promising high-T_c cuprates in which the partial substitution of Pb for Bi allows us to access to the non-SC HOD regime. Moreover, Bi-2201 is free from oxygen deficiency and expected to be less disordered than LSCO. Therefore, in order to investigate the possible ferromagnetic phase, we have performed both magnetization and ρ_{ab} measurements using Bi-2201 single crystals of $Bi_{1.76}Pb_{0.35}Sr_{1.89}CuO_{6+\delta}$ in the superconducting (SC) regime and non-SC OD regime.

2. Experimental

$Bi_{1.76}Pb_{0.35}Sr_{1.89}CuO_{6+\delta}$ single crystals were grown by the floating-zone technique. The quality of the crystals obtained was evaluated to be good from the observation of clear fourfold symmetric spots in the x-ray back-Laue photograph. The composition of the crystals was determined by the inductively-coupled-plasma analysis. To control the hole concentration, reduction or oxidation annealing was performed in vacuum, in argon or in oxygen of ambient or high pressure [4-6]. The magnetization was measured using a SQUID magnetometer. The ρ_{ab} was measured by the standard dc four-probe method using a commercial apparatus (Quantum Design, PPMS). The value of the hole concentration, p, in the SC crystals was determined using the empirical law of $T_c/T_c^{max} = 1 - 82.6(p[\text{per Cu}] - 0.16)^2$, where T_c^{max} is the maximum value of T_c in a system [7]. In this study, four kinds of crystals were used; HOD1 and HOD2 in the non-SC HOD regime with $p > 0.27$ per Cu, and OD1 and OD2 in the SC overdoped regime with $p = 0.26$, 0.22 per Cu and $T_c = 2.8$, 15.0 K, respectively.

3. Results and discussion

Figure 1 shows the temperature dependence of χ for crystals annealed under various conditions. The χ is given by

$$\chi(T) = \chi_0 + \chi_1(T), \quad (1)$$

where χ_0 and $\chi_1(T)$ are temperature-independent and temperature-dependent terms, respectively. The χ_0 is described as

$$\chi_0 = \chi_{Pauli} + \chi_{core} + \chi_{V.V.}, \quad (2)$$

where χ_{Pauli} is the Pauli paramagnetism of conducting electrons, χ_{core} the core diamagnetism and $\chi_{V.V.}$ the Van Vleck paramagnetism. Since only χ_{Pauli} among them depends on p and increases with increasing p, the magnitude relation of p between the crystals can be estimated roughly from the value of $\chi(T)$ at high temperatures. In fact, $\chi(T)$ of the crystal OD1 with $T_c = 2.8$ K is larger than that of the crystal OD2 with $T_c = 14.8$ K. The p value of the non-SC crystal HOD2 is estimated to be larger than that of the crystal OD1 and that of the crystal HOD1 is estimated to be larger than that of the crystal HOD2. In Fig. 1, it is also found that χ at low temperatures exhibits an upturn in crystals HOD1, HOD2 and OD1, as observed in HOD $Tl_2Ba_2CuO_{6+\delta}$ [1]. This may be due to ferromagnetic fluctuations.

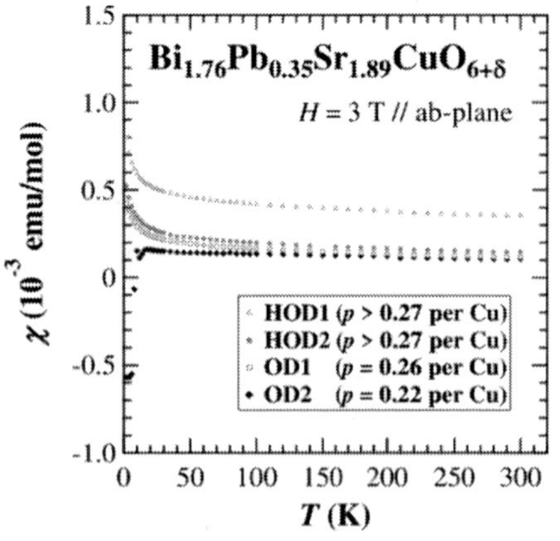

Figure 1. Temperature dependence of the magnetic susceptibility, χ, of $Bi_{1.76}Pb_{0.35}Sr_{1.89}CuO_{6+\delta}$ in the overdoped (OD) and the heavily overdoped (HOD) regimes.

Figure 2(a) shows the magnetization curves of the crystal HOD2 in the HOD regime. Surprisingly, while the magnetization is linear to magnetic field at 20 K, it tends to be saturated in high fields at low

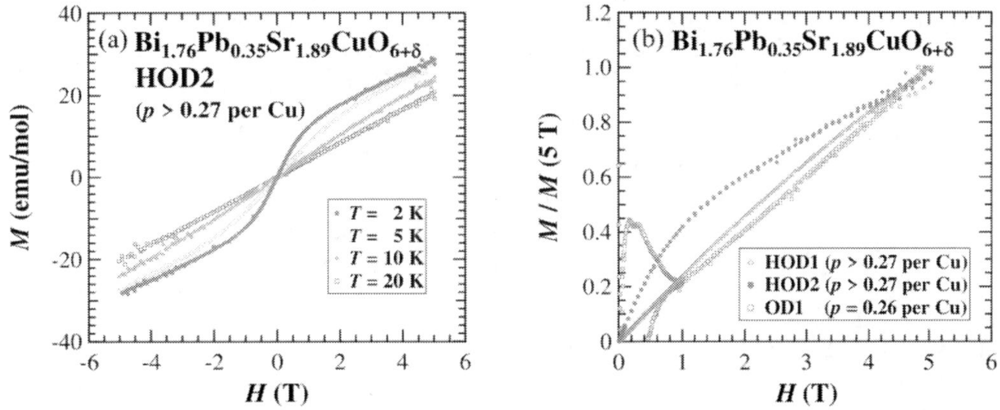

Figure 2. (a) Magnetization curves of $Bi_{1.76}Pb_{0.35}Sr_{1.89}CuO_{6+\delta}$ in the heavily overdoped (HOD) regime. (b) Magnetization curves at 2 K of $Bi_{1.76}Pb_{0.35}Sr_{1.89}CuO_{6+\delta}$ in the overdoped (OD) and HOD regimes normalized by values of magnetization at 5 T, M/M(5 T).

temperatures. This is probably a precursor of the ferromagnetic transition pointed out by Sonier *et al.* in non-SC HOD LSCO with $x = 0.33$ [3]. In Fig. 2(b), the normalized magnetization curve at 2 K is shown for crystals OD1, HOD1 and HOD2. It is clearly found that the tendency to the saturation is observed only in the HOD regime and that the saturation becomes more significant in the crystal HOD2 close to the SC regime than the crystal HOD1. Therefore, it appears that the ferromagnetic transition takes place at a low temperature below 2 K in the HOD regime and the transition temperature is the highest at the boundary between SC OD and non-SC HOD regimes.

Figure 3(a) shows the temperature dependence of ρ_{ab} of crystals OD1 and HOD1. It is found that ρ_{ab} is proportional to $T^{4/3}$ in a wide range of temperature. The p dependence of the exponent n obtained by the fitting with $\rho_{ab} = \rho_0 + AT^n$ in the normal state is shown in Fig. 3(b). It is found that the n value changes from 1 around the optimally doped regime to 2 in the underdoped regime, as pointed out by

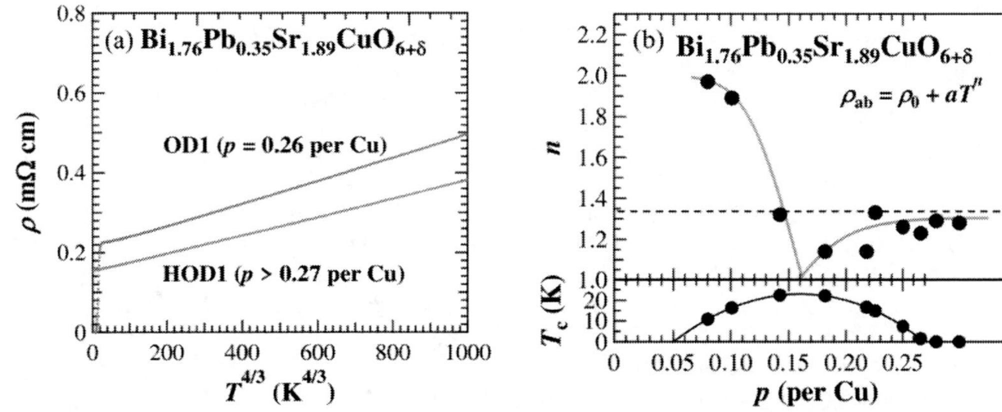

Figure 3. (a)Temperature dependence of the in-plane electrical resistivity, ρ_{ab}, of $Bi_{1.76}Pb_{0.35}Sr_{1.89}CuO_{6+\delta}$ in the overdoped (OD) and heavily overdoped (HOD) regimes. (b) Hole-concentration, p, dependences of the exponent n in $\rho_{ab} = \rho_0 + AT^n$ and T_c in $Bi_{1.76}Pb_{0.35}Sr_{1.89}CuO_{6+\delta}$. The broken line exhibits $n = 4/3$. The solid line in n vs. p is a guide line and the solid line in T_c vs. p is given by the empirical relation [7].

Ando *et al.* [8]. On the other hand, from the optimally doped regime to the HOD regime, the n value tends to increase with increasing p and to converge to 4/3 in the HOD regime. According to the self-consistent renormalization theory [9], the $T^{4/3}$ dependence of ρ_{ab} is characteristic of a metal with two-dimensional (2D) ferromagnetic fluctuations. This result is associated with the $T^{5/3}$ dependence of ρ_{ab} observed in non-SC HOD LSCO, which is characteristic of 3D ferromagnetic fluctuations [3]. The difference in dimensionality may be caused by the weakness of the magnetic coupling between CuO_2 layers due to the long interlayer distance of Bi-2201 (12.3 Å) compared with that of LSCO (6.6 Å).

From the theoretical point of view, it has been suggested that a ferromagnetic order can be formed in a metal where the electron correlation is strong enough, using the mean field approximation [10] and the Gutzwiller approximation [11], but Tahvildar-Zadeh *et al.* have reported that the ferromagnetic correlation will be collapsed by quantum fluctuations within the 3D Hubbard model [12]. Experimentally, Piriou *et al.* have recently reported that the density of states increases with increasing p in the HOD regime due to the existence of the van Hove singularity from the scanning tunneling spectroscopy in Bi-2201 single crystals [13]. If this is the case, it is possible that a ferromagnetic order appears, satisfying the Stoner's criterion. Therefore, the present results of the magnetization and ρ_{ab} suggesting the existence of ferromagnetic fluctuations may be due to the enhancement of the density of states at the Fermi level owing to the approach to the van Hove singularity in the HOD regime and a ferromagnetic transition is expected to take place at a low temperature below 2 K.

4. Summary

We have investigated the possible ferromagnetic state in the HOD regime of high-T_c cuprates using $Bi_{1.76}Pb_{0.35}Sr_{1.89}CuO_{6+\delta}$ single crystals covering a wide range of p. It has been found that χ exhibits an upturn at low temperatures in the OD and HOD regimes and that the magnetization curves exhibit a tendency to be saturated in high fields at low temperatures in the non-SC HOD regime, which is the most significant at p close to the SC regime. Moreover, $T^{4/3}$ dependence of ρ_{ab} has been observed in the OD and HOD regimes, suggesting the existence of 2D ferromagnetic fluctuations. It has been concluded that 2D ferromagnetic correlation develops at low temperatures in the HOD regime of Bi-2201, so that a ferromagnetic transition is expected to take place at a lower temperature below 2 K. Accordingly, the development of the ferromagnetic correlation might be an intrinsic feature in the HOD cuprates.

† Present address : Institute for Materials Research, Tohoku University, Sendai, Japan

References

[1] Kubo Y, Shimakawa Y, Manako T and Igarashi H 1991 *Phys. Rev. B* **43** 7875
[2] Kopp A, Ghosal A and Chakravarty S 2007 *Proc. Natl. Acad. Sci. USA* **104** 6123
[3] Sonier J E, Kaiser C V, Pacradouni V, Sabok-Sayr S A, Cochrane C, MacLaughlin D E, Komiya S and Hussey N E 2010 *Proc. Natl. Acad. Sci. USA* **107** 17131
[4] Yang W L, Wen H H, Ni Y M, Xiong J W, Chen H, Dong C, Wu F, Qin Y L and Zhao Z X 1998 *Physica C* **308** 294
[5] Kudo K, Miyoshi Y, Sasaki T and Kobayashi N 2005 *Physica C* **426** 251
[6] Murakami Y, Kambe S and Kawai M 1991 *Physica C* **178** 71
[7] Presland M R, Tallon J L, Buckley R G, Liu R S and Flower N E 1991 *Physia C* **176** 95
[8] Ando Y, Kurita Y, Komiya S, Ono S and Segawa K 2004 *Phys. Rev. Lett.* **92** 197001
[9] Hatatani M and Moriya T 1995 *J. Phys. Soc. Jpn.* **64** 3434
[10] Penn D R 1966 *Phys. Rev.* **142** 350
[11] Fazekas P and Müller-Hartmann E 1991 *Z. Phys. B* **85** 285
[12] Tahvildar-Zadeh A N, Jarrell M and Freericks J K 1997 *Phys Rev B* **55** R3332
[13] Piriou A, Jenkins N, Berthod C, Maggio-Aprile I and Fischer Ø 2011 *Nat. Commun.* **2** 221

Asymmetric Avalanche Behavior in a Zeeman-Limited Superconductor

P.W. Adams and J.C. Prestigiacomo

Department of Physics and Astronomy, Louisiana State University, Baton Rouge, Louisiana 70803, USA

E-mail: adams@phys.lsu.edu

Abstract. We have performed transport and tunneling density of states measurements of ultra-thin Al films through the first-order parallel critical field transition. The transition is intrinsically hysteretic and exhibits avalanche-like jumps in both resistivity and tunneling density states. Tunneling measurements on films with sheet resistances of a few hundred ohms show large avalanche-like collapses of the condensate on the superheating branch of the critical field hysteresis loop. In contrast, the transition back into the superconducting phase (i.e., along the supercooling branch) is always continuous.

1. Introduction

The response of a superconductor to a magnetic field that only couples to the electron's spin, sometimes referred to as a Zeeman field, remains a subject of significant interest [1, 2]. In practice, if one applies a magnetic field to a superconducing film whose thickness is much less than the coherence length, then the primary coupling between the field and the condensate is through the Zeeman splitting of the conduction electrons. The film geometry suppresses the Miessner response and the parallel field cannot induce screening currents, nor can quantized vortices enter the system. In this limit, the critical field transition is Zeeman mediated. Studies of Zeeman-limited superconductivity in low atomic mass elemental metal films have revealed several interesting effects, including a hysteretic first-order critical field transition [5, 6], incoherent Cooper pairing [7, 8], reentrance [9], and excess sub-gap states [10]. Here, we present resistivity and tunneling density of states measurements across the hysteretic Zeeman critical field transition in ultra-thin Al films. As the normal state is approached, we observe large avalanches in the density of states indicating that macroscopic regions of superconductivity suddenly and irreversibly collapse. In contrast, the transition from the normal state to the superconducting state is smooth and continuous.

In this study a magnetic field was applied parallel to the surface of superconducting Al films having a thickness that was approximately 5 times smaller than the coherence length ($\xi \sim 20$ nm). In this limit, the orbital response to the field is suppressed, and the transition occurs when the Zeeman splitting is of the order of the superconducting gap Δ_0 [11]. The conventional picture is that this Zeeman mediated transition, which is often referred to as the spin-paramagnetic (S-P) transition, occurs between a BCS ground state with a homogenous order parameter and a polarized Fermi liquid normal state. At low temperatures the Zeeman critical field is expected to be near the Clogston-Chandrasekhar value $H_c = \Delta_0/\sqrt{2}\mu_B$ [12]. As we show below, the non-equilibrium behavior of the Zeeman-limited superconducting ground state suggests that the order parameter may not be homogeneous.

2. Sample Preparation

Samples were fabricated by first preparing aluminum films from 99.999% Al targets via e-beam deposition onto fire-polished glass substrates held at 84 K. The deposition rate was typically 1 Å/s in a 0.1 μTorr vacuum. Films with thicknesses ranging from $t = 20 \rightarrow 30$ Å had normal-state sheet resistances that ranged from $R_n = 5.5$ kΩ/sq to 80 Ω/sq at 80 mK, respectively, and a disorder-independent superconducting transition temperature of $T_c \sim 2.7$ K. After the initial deposition, the films were warmed to room temperature and then exposed to air for 10-20 min in order to form a thin native oxide, which served as the tunneling barrier. A 90-Å-thick Al counter-electrode was deposited on top of the oxide thereby creating a junction with an area of about 1 x 1 mm^2. The barrier resistances ranged from 1 kΩ to 10 kΩ depending on the thickness of the electrode, exposure time, and other factors. Only junctions with barrier resistances much higher than the films' resistance were used. Transport and tunneling data were collected via a 4-probe configuration with a lock-in amplifier. The films were cooled using a dilution refrigerator equipped with a mechanical rotator allowing us to align the films to within 0.1° of parallel field.

3. Experimental Results

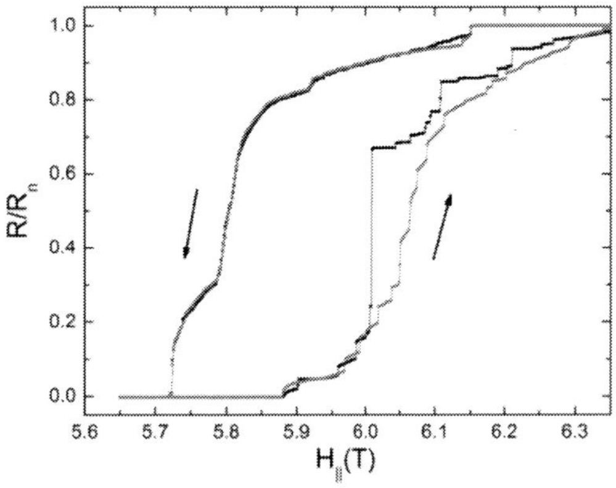

Figure 1. Parallel critical field transition in a $R_n = 570\Omega$ /sq Al film at 60 mK. The red and black lines represent two identical sweeps through the first-order transition. The arrows depict the field sweep direction.

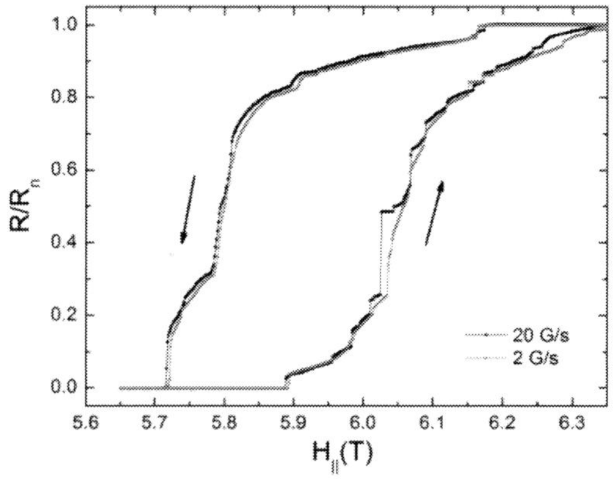

Figure 2. Parallel critical field transition for the film in Fig. 1 taken at two different sweep rates.

Shown in Fig. 1 is the resistive parallel critical field transition of a $R_n = 570\Omega$/sq Al film, taken at 60 mK. The data represent two identical sweeps through the hysteretic transition. Note that the upsweep branch (superheating branch) of the hysteresis loop is interspersed with many avalanche-like jumps in resistance, whereas the down sweep branch (supercooling branch) is somewhat smoother. These jumps are similar to what has been observed in previous transport studies of Al films having substantially more disorder than the ones used in this study [5, 6]. Since the film thicknesses is much less than the coherence length, the jumps are not due superconducting vortex motion nor to magnetic flux dynamics.

The non-equilibrium character of the hysteresis loop has both field dependent and time dependent components. In Fig. 2 we show two loops that were obtained using different field sweep rates. Note that intervals between the avalanches on the superheating branch are much flatter in the 20 G/s trace that they are in the 2 G/s trace. This clearly indicates that there is a slow temporal relaxation between avalanches. We have probed the time dependence of the transition by halting the field sweep at the midpoint of the superheating branch and then measuring the ensuing relaxation. Generally, such traces follow a stretched-exponential form with occasional spontaneous avalanches. Although transport measurements leave little doubt that avalanche-like dynamics is characteristic of the S-P transition, it has remained unclear whether or not the non-equilibrium behavior observed in transport actually represents the behavior of the condensate. For example, a sample that it almost entirely in the normal state can still have zero resistance so long as there is at least one superconducting filamentary path along its length.

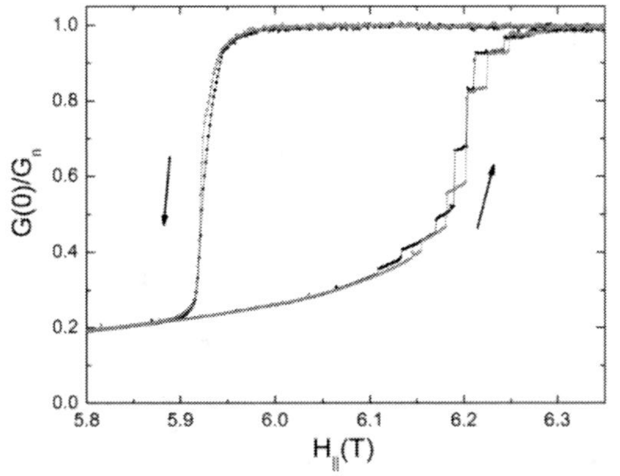

Figure 3. The zero bias tunneling conductance normalized by its normal state value. The data was taken on a $R_n = 570\ \Omega/\mathrm{sq}$ Al film as a function of parallel magnetic field at $T = 60$ mK. The red and black lines represent two separate sweeps through the hysteresis loop. The arrows depict the field sweep direction.

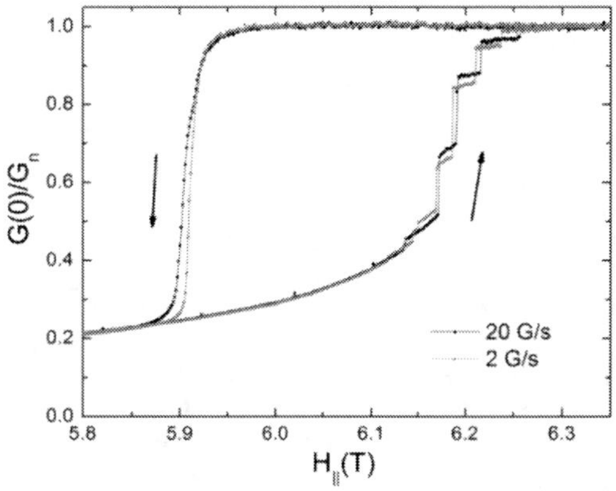

Figure 4. Normalized tunneling density of states of the film in Fig. 3 showing the parallel critical field hysteresis loop for two different sweep rates. These data were taken at 60 mK.

At low temperatures the quasiparticle tunneling conductance is proportional to the density of electronic states (DOS) of the superconducting films [13]. Since planar tunneling is an areal microscopic probe of the condensate, it is relatively insensitive to filamentary superconductivity. Indeed, previous tunneling studies of the S-P transition have shown that the order parameter exhibits a hysteresis that is very similar to what is observed in transport, but up to now avalanches have not been reported in such spectra [9]. Shown in Fig. 3 is the zero-bias tunneling conductance of the film used in Fig. 1 as a function of parallel field. Because the Al counter-electrode was relatively thick, it had a critical field (~ 3 T) that was much lower than that of the film. Thus the data in Fig. 3 represent S-I-N tunneling and reflect the quasiparticle density of states at the Fermi energy of the Al film. The precipitous drop in tunneling conductance as the field is lowered through the transition is due to the opening of the superconducting gap. Note that the superheating DOS exhibits avalanche behavior, but the supercooling branch is smooth and continuous. Similar behavior was observed in all of our moderately disordered samples.

The asymmetry of the avalanches in Fig. 3 is unusual. For instance, avalanches in the magnetization of ferromagnetic systems (Barkhausen effect) are observed on both branches of the magnetization loop [15, 16]. Similarly, avalanches are observed in Martensitic transitions when the sample is either cooled or heated through the transition [17]. In Fig. 4 we show the effects of sweep rate on the hysteresis loop obtained from the zero bias tunnel conductance. The 20 G/s hysteresis loop is slightly wider than the 2 G/s loop, as is expected. But otherwise, the two loops look very similar to each other.

4. Summary

We observe avalanche-like collapses in the condensate on the superheating branch of the S-P transition hysteresis loop of moderately disordered Al films. Our data suggest that as the magnetic field is lowered through the parallel critical field, the system can always find an appropriate ground state that can accommodate the pairing interaction, Zeeman field, and the disorder. Hence, the supercooling branch is smooth and continuous. But, for reasons that remain unclear, the system cannot find a continuous path from the superconducting phase to the normal state. One possible explanation for the asymmetric avalanche behavior is that the high-field order parameter is more complex than that of a homogeneous BCS ground state.

Acknowledgments

This work was supported by the U.S. Department of Energy, Office of Science, Basic Energy Sciences, under Award No. DE-FG02-07ER46420.

References

[1] R. Beaird, A. B. Vorontsov, and I. Vekhter, Phys. Rev. B **81**, 224501 (2010).
[2] Y.M. Xiong, S. Stadler, P.W. Adams, and G. Catelani, Phys. Rev. Lett. **106**, 247001 (2011).
[3] P. Fulde and R. A. Ferrell, Phys. Rev. **135**, A550 (1964).
[4] A. I. Larkin and Y. N. Ovchinnikov, Zh. Eksp. Teor. Fiz. **47**, 1136 (1964) [Sov. Phys. JETP **20**, 762 (1965)].
[5] Wenhao Wu and P.W. Adams, Phys. Rev. Lett. **73**, 1412 (1994).
[6] Wenhao Wu and P.W. Adams, Phys. Rev. Lett. **74**, 610 (1995).
[7] V. Yu. Butko, P.W. Adams, and I.L. Aleiner, Phys. Rev. Lett. **82**, 4284 (1999).
[8] G. Catelani, Y.M. Xiong, X.S. Wu, and P.W. Adams, Phys. Rev. B **80**, 054512 (2009).
[9] V. Yu. Butko, E.I. Meletis and P.W. Adams, Phys. Rev. Lett. **83**, 3725 (1999).
[10] Y. L. Loh, N. Trivedi, Y. M. Xiong, P. W. Adams, and G. Catelani, Phys. Rev. Lett. **107**, 067003 (2011).
[11] P. Fulde, Adv. Phys. **22**, 667 (1973)
[12] A. M. Clogston, Phys. Rev. Lett. **9**, 266 (1962); B. S. Chandrasekhar, Appl. Phys. Lett. **1**, 7 (1962).
[13] M. Tinkham, *Introduction to Superconductivity* (McGraw-Hill, New York, 1996).
[14] B. L. Altshuler, A. G. Aronov, M. E. Gershenson, and Yu. V. Sharvin, Sov. Sci. Rev., Sect. A **9**, 223 (1987).
[15] Zhu Diao, E. R. Nowak, Gen Feng, and J. M. D. Coey, Phys. Rev. Lett. **104**, 047202 (2010).
[16] S. Papanikolaou, F. Bohn, R. L. Sommer, G. Durin, S. Zapperi, and J. P. Sethna, Nat. Phys **7**, 316 (2011).
[17] E. Vives, J. Ortin, L. Manosa, I. Rafols, R. Perez-Magrane, and A. Planes, Phys. Rev. Lett. **72**, 1694 (1994).

Multiband character of β-FeSe: Angular dependence of the magnetoresistance and upper critical field

M. L. Amigó[1], **V. Ale Crivillero**[1], **D. G. Franco**[1], **and G. Nieva**[1,2]

[1] Comisión Nacional de Energía Atómica - Centro Atómico Bariloche and Instituto Balseiro, Universidad Nacional de Cuyo, S.C. de Bariloche, Argentina
[2] CICT - Consejo Nacional de Investigaciones Científicas y Técnicas, Argentina

E-mail: `gnieva@cab.cnea.gov.ar`

Abstract. We studied *ab*-plane transport properties in single crystals of the superconductor β-FeSe up to 16 T. In the normal state, below 90 K, the crystals present a strongly anisotropic positive magnetoresistance that becomes negligible above that temperature. In the superconducting state (T_c=8.87(5) K) the upper critical field anisotropy $H_{c2}\|ab/\ H_{c2}\|c$ changes with temperature and the angular dependence of the dissipation for fixed temperatures and fields reflects a strongly anisotropic behavior. Our results make evident that multiband effects are needed to describe the measured transport properties. We model the magnetoresistance and upper critical field behavior with a two-band model showing that the diffusivities ratio parameter remains unchanged going from the normal to the superconducting state.

1. Introduction

Several parent compounds of the families 1111 and 122 of the Fe based superconductors present a structural transition and spin density wave at similar temperatures. By doping this compounds the transition temperature decreases and superconductivity emerges[1]. In the case of pure β-FeSe there is a structural transition at $T_s \sim$90 K, not accompanied by spin density wave and the samples show superconductivity at low temperatures. In this compound the orthorhombic distortion and superconductivity do not compete[2]. For the Fe-based superconductors, several disconnected Fermi-surface sheets contribute to superconductivity, as revealed by angle-resolved photoemission spectroscopy. In particular for β-FeSe these experiments show a significant change in the electronic bands going through the structural transition[3]. Evidence of the multiband behavior of this material was found in the physical properties such as the magnetic penetration depth[4], the upper critical field, H_{c2}, or the Hall coefficient[5]. In this work, we grow single crystals of the β-FeSe phase and study the consequences of the structural transition in the normal and superconducting states.

2. Crystal characterization

$Fe_{1-y}Se$ single crystals were grown using KCl: $2AlCl_3$ flux in a temperature gradient (hot part of the ampule at 395°C and cold part at 385°C), for 45 days[6]. This low temperature process allows us to grow the desired phase avoiding high temperature structural transitions and decompositions. We use X Ray diffraction and Energy Dispersive Spectroscopy to characterize the structure and the composition of the single crystals. The crystals have a plate-like shape

Journal of Physics: Conference Series **568** (2014) 022005 doi:10.1088/1742-6596/568/2/022005

with the c axis oriented perpendicular to the crystal plane. The crystals have only the tetragonal β-FeSe phase present. The lattice parameters are $c = (5.52 \pm 0.01)$Å and $a = (3.77 \pm 0.01)$Å, in good agreement with the literature[7]. The composition of the samples present a slight Fe deficiency, $y = 0.04$.

3. Results and discussion

Figure 1 shows the ab-plane resistivity for a single crystal of $Fe_{0.96}Se$ with applied magnetic field parallel to the c axis ($H \| c = 0$, 8 and 16 T). The onset of the transition temperature for $H = 0$ is $T_c = (8.87 \pm 0.05)$ K. The resistivity presents a metallic-like behavior in the normal state and a change in the slope around 90 K. At this temperature there is a structural transition from a tetragonal to an orthorhombic phase[8]. Below 90 K we find a positive magnetoresistance when the field is parallel to the c axis, suggesting that its origin is related to the structural transition. A magnetic characterization of the crystals in the normal state shows no evidence of a magnetic feature at this temperature. In the literature there are similar magnetoresistance results in films[9] and policrystals[10].

In order to study the connection between the electronic properties of β-FeSe above and below the superconducting temperature we performed electrical transport measurements with an applied magnetic field up to 16 T parallel and perpendicular to the c axis, see figure 2a and b. The electrical contact configuration on the sample allows the current to flow in the ab plane, always perpendicular to the magnetic field. For $H = 0$ there is a sharp transition, and the transition width slightly increases with field. Figure 2c shows the magnetoresistance at 14 K, approximately 5 K above T_c for the field parallel to the c axis. For field parallel to the ab plane the magnetoresistance is negligible as can be seen in figure 2b. Moreover, there is a strong difference in the temperature of the superconducting transition onset (i.e. in the upper critical field, H_{c2}vs.T) for both field directions, indicating an anisotropic material.

Considering that the multiband electronic structure, characteristic of Fe-based supercon-

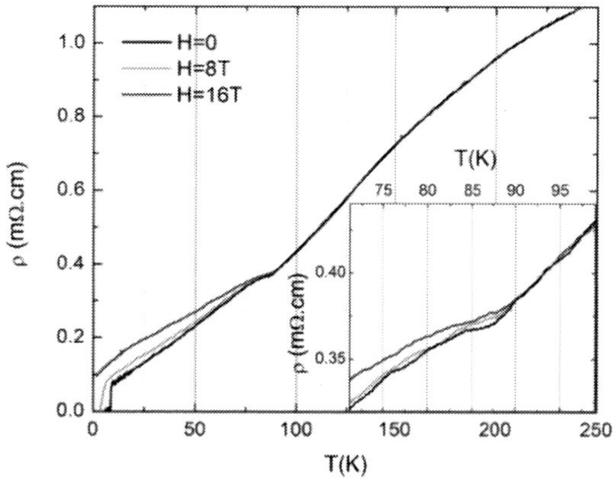

Figure 1. Temperature dependence of the ab-plane resistivity for a $Fe_{0.96}Se$ crystal with applied magnetic field parallel to the c axis ($H = 0$, 8 and 16 T). *Inset* : Detail of the resistivity near the structural transition.

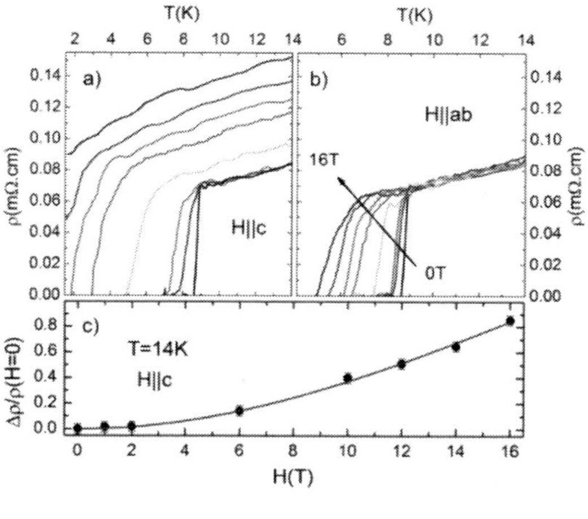

Figure 2. Temperature dependence of the ab-plane resistivity for an applied magnetic field ($H = 0$, 1, 2, 6, 10, 12, 14 and 16 T) a) parallel to the c axis and b) perpendicular to the c axis. c) Magnetoresistance as a function of the applied magnetic field parallel to the c axis for $T = 14$ K.

ductors, was found in many features of β-FeSe, we choose a two-band model to describe the magnetoresistance data in figure 2*c*. Assuming a spherical symmetry in each band[11], the magnetoresistivity for $H\|c$ is given by,

$$\begin{aligned}\frac{\Delta\rho}{\rho(H=0)} &= \frac{\rho(H)-\rho(H=0)}{\rho(H=0)} = \frac{\sigma_1\sigma_2(D_1-D_2)^2H^2}{(\sigma_1+\sigma_2)^2+H^2(D_1\sigma_1+D_2\sigma_2)^2} \\ &= \frac{\sigma_1(\rho(H=0)^{-1}-\sigma_1)D_1^2(1-\eta)^2H^2}{\rho(H=0)^{-2}+H^2D_1^2\left[\sigma_1+\eta(\rho(H=0)^{-1}-\sigma_1)\right]^2}, \end{aligned} \quad (1)$$

where σ_m and D_m are the conductivity and diffusivity in the m band, $\eta = D_2/D_1$ and $\rho(H=0)=(\sigma_1+\sigma_2)^{-1}$. For an arbitrary direction of the magnetic field, D_m must be replaced by

$$D_m(\theta)=\sqrt{(D^a_m)^2cos^2(\theta)+D^a_mD^c_msin^2(\theta)}, \quad (2)$$

where θ is the angle between the applied magnetic field and the c axis and D^a_m and D^c_m are the m band diffusivities in plane and along the c axis respectively[12].

The magnetoresistance is temperature dependent as can be observed in figure 1. For the data at 14 K, in figure 2*c* ($H\|c$) the best fitted curve using equation 1 is shown with a solid line. We used $\rho(H=0,T=14$ K$)$=0.0835 mΩ.cm obtained from figure 2*a* and the fit yields: $\eta(0°) = D^a_2/D^a_1 = 0.051$, D^a_1 =0.292 T^{-1} and σ_1 =36 mΩ^{-1}.cm^{-1}. For $H\|ab$ there is a negligible magnetoresistance, which could be explained if the diffusivities are equal in both bands, $\eta(90°)=\sqrt{\frac{D^a_2D^c_2}{D^a_1D^c_1}}=1$.

We measure the angular dependence of the resistivity at H =16 T for several temperatures. For temperatures above 90 K there is no angular dependence in accordance with the negligible magnetoresistance at high temperature. Below the structural transition temperature there is a smooth angular dependence, see figure 3. Below $T_c(H=0)$ there is a narrow angular range in which the sample is superconducting, see the data for $T=(5.12\pm0.01)$ K in figure 3. Combining equation 1 and 2 we fitted the data for $T=(29.9\pm0.1)$ K. We use the same parameters already obtained for D^a_1, D^a_2 and $\eta(90°)$. The remaining fitted parameters are σ_1=7.47$m\Omega^{-1}.cm^{-1}$ and $D^c_1=1.62.10^{-4}$T^{-1}.

According to the model proposed here, there must be a change in the diffusivities at $T\sim$90 K for the emergence of magnetoresistivity, so there must be a change in the effective masses and therefore in the band structure. Recently, some authors studied the changes in the electronic bands at different temperatures going through the structural transition [3][13][14]. There is no general agreement about a band reconstruction or variation due to symmetry change at the transition. However, such a change in the electronic band structure could be the microscopic cause of the anisotropic magnetoresistance.

Figure 4*a* shows the temperature dependence of the critical field data, obtained from the onset of the superconducting transition for both field directions. We also plotted the Werthamer Helfand and Hohenberg (WHH)[15] model (dotted line). We observe that this model departs from the experimental data. For $H\|ab$, the data curve towards lower values of H, which might indicate the presence of a finite spin paramagnetic effect contribution, that reduces the energy of the normal state, decreasing the value of H_{c2}(T)[15]. However, for the case of $H\|c$ the corresponding WHH curve is below the experimental data. This could not be explained introducing a spin paramagnetic effect contribution. A multiple-band scenario must be taken in consideration. We will describe the superconducting critical field H_{c2}(T) in a two-band scenario. In the dirty limit and including the spin paramagnetic effect, a two-band model yields[12]

$$ln(t)+\frac{1}{2}\Big(U_1(h)+U_2(h)+\frac{\lambda_0}{w}\Big)-\frac{s}{4}\sqrt{\Big(U_1(h)-U_2(h)-\frac{\lambda_-}{w^2}\Big)^2+\frac{\lambda_{12}\lambda_{21}}{w^2}}=0, \quad (3)$$

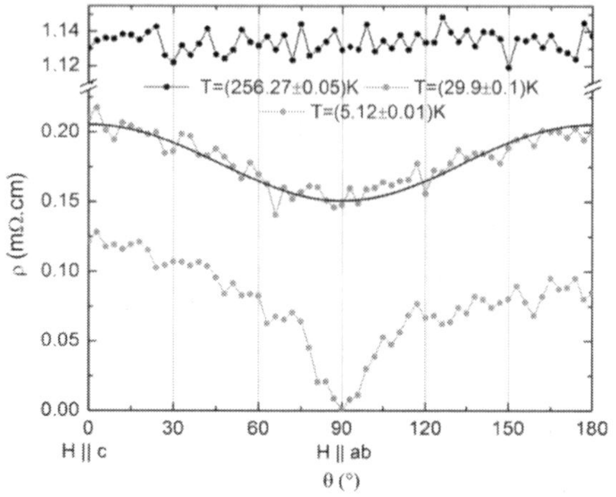

Figure 3. *ab*-plane resistivity as a function of the angle between the c axis and the applied magnetic field (H =16 T) for different temperatures. The solid line is a fit to the data using equation 1 and 2.

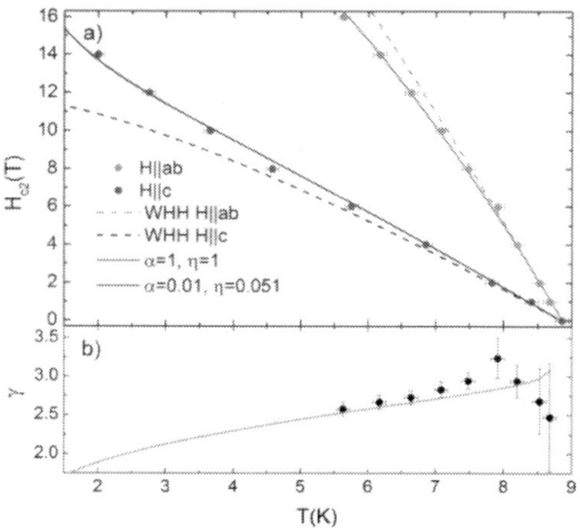

Figure 4. *a*) H_{c2}(T) for field parallel to the c axis and to the ab plane. In dotted lines is the WHH model ($\eta = 1$) with $\alpha = 0$ for both field directions. The solid lines are the fit of the equation 3. *b*) Anisotropy as a function of temperature and in continuous line is the ratio of the fit for the two directions of the magnetic field.

$$h = \frac{H_{c2}(T)}{-\frac{dH_{c2}}{dt}|_{t=1}}, \tag{4}$$

$$U_1(h) = Re\Big(\psi\big(\frac{1}{2}+(1+i\alpha)\frac{2h}{\pi^2 t}\big)-\psi\big(\frac{1}{2}\big)\Big), \qquad U_2(h) = Re\Big(\psi\big(\frac{1}{2}+(1+i\frac{\alpha}{\eta})\frac{2h\eta}{\pi^2 t}\big)-\psi\big(\frac{1}{2}\big)\Big), \tag{5}$$

where, $t = T/T_c$, $\lambda_- = \lambda_{11} - \lambda_{22}$, $\lambda_0 = \sqrt{\lambda_-^2 + 4\lambda_{12}\lambda_{21}}$, $w = \lambda_{11}\lambda_{22} - \lambda_{12}\lambda_{21}$, $s = sign(w)$ and $\psi(x)$ is the digamma function. λ_{11} and λ_{22} are the intraband coupling constant in bands 1 and 2, λ_{12} and λ_{21} quantify the interband coupling. $\alpha(\theta) = D_0/D_1(\theta)$ takes into account the spin paramagnetic effect, $D_0 = \hbar/2m$. Taking $\eta = 1$ and $\alpha = 0$ we recover the one band WHH model, without the spin orbit coupling.

To further investigate multiband characteristics in H_{c2}(T) and the relation with the parameters found in the normal state we fit equation 3 for both field directions (solid lines in figure 4*a*). We use the η values obtained from equation 1, $\eta(0°)$=0.051 and $\eta(90°)$=1. The fitted parameters are $\lambda_{11} = 0.13$, $\lambda_{22} = 0.1$, $\lambda_{12} = \lambda_{21} = 0.0087$, $\alpha(0°)$=0.01 and $\alpha(90°)$=1. Similar values of λ_{ij} are obtained for FeSe in the literature[5].

The value of $H_{c2}(0)$ strongly depends on α, so to give a reliable estimation, it is necessarily to measure at low temperatures and high fields. The values of the parameters obtained here are phenomenological and they are used to show that FeSe presents a multiple band behavior, that is necessary take into account the spin paramagnetic effect and that the different values of the diffusivities could explain the $H_{c2}(T)$ and the magnetoresistivity. Figure 4*b* present the effective anisotropy ($\gamma = H_{c2} \|ab/H_{c2} \|c$). It has a maximum near T_c and decreases with temperature as seen in others works [16][17]. This effective anisotropy behavior has also been taken as an indication of a multiband character of $H_{c2}(T)$.

4. Conclusion

The angular dependent magnetoresistance of β-FeSe crystals as well as the anisotropic upper critical field can be described by a two-band model with parameters such as the band diffusivities ratio that are valid in a wide temperature range. In the normal state, the kink in the ab plane resistivity at 90 K marks the onset of a very anisotropic magnetoresistance. Above this temperature, coincident with an structural transition, the magnetoresistance is negligible.

Acknowledgments
We thank Conicet PIP11220090100448 and Sectyp U.N.Cuyo for financial support.

References

[1] Stewart G R 2011 *Rev. Mod. Phys.* **83**(4) 1589–1652
[2] Böhmer A E, Hardy F, Eilers F, Adelmann P, Schweiss P, Wolf T and Meingast C 2013 *Phys. Rev. B* **87**(18) 180505
[3] Shimojima T, Suzuki Y, Sonobe T, Nakamura A, Sakano M, Omachi J, Yoshioka K, Kuwata-Gonokami M, Ono K, Kumigashira H, Böhmer A E, Hardy F, Wolf T, Meingast C, v Löhneysen H, Ikeda H and Ishizaka K 2014 *ArXiv e-prints* (*Preprint* 1407.1418)
[4] Khasanov R, Bendele M, Amato A, Conder K, Keller H, Klauss H-H and Luetkens H and Pomjakushina E 2010 *Phys. Rev. Lett.* **103**(8) 087004
[5] Lei H, Graf D, Hu R, Ryu H, Choi E S, Tozer S W and Petrovic C 2012 *Phys. Rev. B* **85**(9) 094515
[6] Chareev D, Osadchii E, Kuzmicheva T, Lin J Y, Kuzmichev S, Volkova O and Vasiliev A 2013 *CrystEngComm* **15**(10) 1989–1993
[7] McQueen T M, Huang Q, Ksenofontov V, Felser C, Xu Q, Zandbergen H, Hor Y S, Allred J, Williams A J, Qu D, Checkelsky J, Ong N P and Cava R J 2009 *Phys. Rev. B* **79**(1) 014522
[8] McQueen T M, Williams A J, Stephens P W, Tao J, Zhu Y, Ksenofontov V, Casper F, Felser C and Cava R J 2009 *Phys. Rev. Lett.* **103**(5) 057002
[9] Chang H H, Luo J Y, Wu C T, Hsu F C, Huang T W, Wu P M, Wu M K and Wang M J *Supercond. Sci. Tech.* **25** 035004
[10] Chu H, Li J, Li S, Zhang Y, Wang N, Wang J, Gao Y, Deng H, Sun X and Zheng D 2010 *Sci. China Phys., Mech. Astron.* **53** 1180–1186
[11] Ziman J 1965 *Principles of the Theory of Solids* (University Press)
[12] Jaroszynski J, Hunte F, Balicas L, Jo Y j, Raičević I, Gurevich A, Larbalestier D C, Balakirev F F, Fang L, Cheng P, Jia Y and Wen H H 2008 *Phys. Rev. B* **78**(17) 174523
[13] Nakayama K, Miyata Y, Phan G N, Sato T, Tanabe Y, Urata T, Tanigaki K and Takahashi T 2014 *ArXiv e-prints* (*Preprint* 1404.0857)
[14] Maletz J, Zabolotnyy V B, Evtushinsky D V, Thirupathaiah S, Wolter A U B, Harnagea L, Yaresko A N, Vasiliev A N, Chareev D A, Böhmer A E, Hardy F, Wolf T, Meingast C, Rienks E D L, Büchner B and Borisenko S V 2014 *Phys. Rev. B* **89**(22) 220506
[15] Werthamer N R, Helfand E and Hohenberg P C 1966 *Phys. Rev.* **147**(1) 295–302
[16] Vedeneev S I, Piot B A, Maude D K and Sadakov A V 2013 *Phys. Rev. B* **87**(13) 134512
[17] Lei H, Hu R, Choi E S, Warren J B and Petrovic C 2010 *Phys. Rev. B* **81**(9) 094518

Journal of Physics: Conference Series **568** (2014) 022006 doi:10.1088/1742-6596/568/2/022006

Two-dimensional macroscopic quantum tunneling in multi-gap superconductor Josephson junctions

Hidehiro Asai[1], Shiro Kawabata[1], Yukihiro Ota[2], Masahiko Machida[2]
[1]Electronics and Photonics Research Institute (ESPRIT), National Institute of Advanced Industrial Science and Technology(AIST), Tsukuba, Ibaraki 305-8568, Japan
[2]CCSE, Japan Atomic Energy Agency, Kashiwa, Chiba 277-8587, Japan

E-mail: hd-asai@aist.go.jp

Abstract. Low-temperature characters of superconducting devices yield definite probes for different superconducting phenomena. We study the macroscopic quantum tunneling (MQT) in a Josephson junction, composed of a single-gap superconductor and a two-gap superconductor. Since this junction has two kinds to the superconducting phase differences, calculating the MQT escape rate requires the analysis of quantum tunneling in a multi-dimensional configuration space. Our approach is the semi-classical approximation along a 1D curve in a 2D potential-energy landscape, connecting two adjacent potential (local) minimums through a saddle point. We find that this system has two plausible tunneling paths; an in-phase path and an out-of-phase path. The former is characterized by the Josephson-plasma frequency, whereas the latter is by the frequency of the characteristic collective mode in a two-band superconductor, Josephson-Leggett mode. Depending on external bias current and inter-band Josephson-coupling energy, one of them mainly contributes to the MQT. Our numerical calculations show that the difference between the in-phase path and the out-of-phase path is manifest, with respect to the bias-current-dependence of the MQT escape rate. This result suggests that our MQT setting be an indicator of the Josephson-Leggett mode.

1. Introduction

Macroscopic quantum tunneling (MQT) is one of the interesting phenomena of Josephson junctions (JJs) [1, 2]. A superconducting-to-resistive switching event in a current-biased JJ is related to the MQT at low temperatures (i.e., negligible contributions from thermal excitations). Different JJs show the tunneling phenomena [3–8]. The theoretical aspects are studied well, depending on types of the JJs [9–17].

Subsequent to the discovery of magnesium diboride [18] and iron-based superconductors [19], multi-band superconductivity has been extensively studied [20–24]. A lot of notable effects have been predicted in the JJs with multi-band superconductors [25–30]. In such JJs, the switching process out of the zero-voltage state is described by the dynamics of multiple gauge-invariant phase differences. Hence, the MQT may occur in a multi-dimensional configuration space, defined by the phase differences.

In this paper, we study the MQT in a hetero JJ, formed by a conventional single-gap superconductor and a two-gap superconductor. The switching events in such JJs take place in 2D configuration spaces, including the two superconducting phase differences. We calculate the MQT escape rate, taking the quantum phase dynamics in the 2D potential. This potential-energy landscape leads to the presence of two plausible tunneling paths; an in-phase path and an out-of phase path. The former is characterized by the Josephson-plasma frequency, whereas the latter is by the frequency of the characteristic collective mode in a two-band superconductor, Josephson-Leggett mode. In contrast to the in-phase path (see, e.g., [15, 17]), the contribution from the out-of-phase path to the MQT is rarely studied. Depending on external bias current and inter-band Josephson-coupling energy, one of them mainly contributes to the MQT. The in-phase path is primary for a strong inter-band coupling, while the out-of-phase path mainly contributes to the tunneling for a weak inter-band coupling. Our numerical calculations show that the difference between the in-phase path and the out-of-phase path is manifest, with respect to the bias-current-dependence of the MQT escape rate. In comparison with the MQT rate for the in-phase path, the MQT rate for the out-of-phase path drastically decreases as the bias current decreases. Hence, by systematically measuring the MQT rate, we can detect the tunneling process corresponding to the out-of-phase path, which is related to the Josephson-Leggett mode. Thus, our MQT setting would be an indicator of the Josephson-Leggett mode.

2. Model and formulation

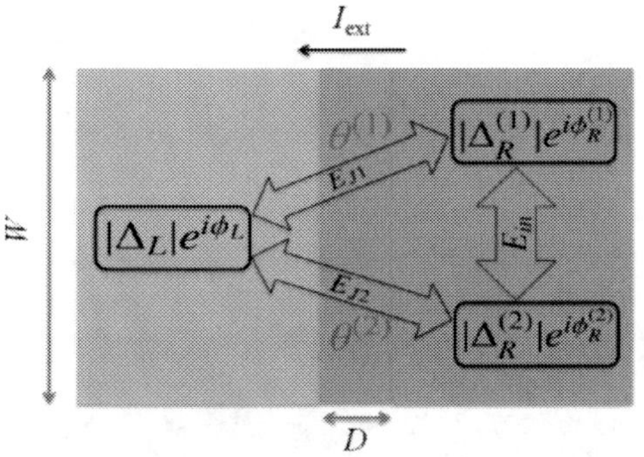

Figure 1. A schematic figure of the Josephson junction composed of a single-gap and a two-gap superconductor.

We consider a hetero Josephson junction composed of a single-gap superconductor and a two-gap superconductor, as shown in figure 1. The superconducting electrodes are divided by an insulating layer with thickness D and area W. The DC bias current $I_{\rm ext}$ is applied to the junction. The right electrode has two gaps, $|\Delta_{\rm R}^{(1)}|e^{i\phi_{\rm R}^{(1)}}$ and $|\Delta_{\rm R}^{(2)}|e^{i\phi_{\rm R}^{(2)}}$, whereas the left electrode has a single gap $|\Delta_{\rm L}|e^{i\phi_{\rm L}}$. These superconducting phases are coupled to each other via the Josephson couplings, and the two gauge-invariant phase differences, $\theta^{(1)}$ and $\theta^{(2)}$ can be defined between the electrodes. Using these phase differences, the two important variables for describing MQT, i. e. , the center-of-mass phase θ and relative phase ψ are described as,

$$\theta = \frac{\alpha_2}{\alpha_1+\alpha_2}\theta^{(1)} + \frac{\alpha_1}{\alpha_1+\alpha_2}\theta^{(2)}, \quad \psi = \theta^{(1)} - \theta^{(2)}. \tag{1}$$

A dimensionless constant α_i is related to the density of states of the ith-band electron near the interface between the right electrode and the insulator. The Josephson relation in this junction is given by $\frac{\partial\theta}{\partial t} = \frac{2e}{\hbar}V$ [29] with the electric charge e, the Planck constant $\hbar$. Hence, θ is directly coupled with the electric field.

The real-time effective Lagrangian for the JJ is [15, 17, 29]

$$\mathcal{L}(\theta,\Psi) = \frac{M}{2}\dot{\theta}^2 + \frac{M}{2}\dot{\Psi}^2 - V_{\rm 2D}(\theta,\psi), \quad M = \frac{\hbar^2}{2E_{\rm c}} \tag{2}$$

$$V_{\rm 2D}(\theta,\Psi) = -E_{\rm J,1}\cos\left(\theta + \frac{\alpha_1}{r}\Psi\right) - E_{\rm J,2}\cos\left(\theta - \frac{\alpha_2}{r}\Psi\right) - E_{\rm in}\cos\left(r\Psi\right) - E_{\rm J}\gamma\theta, \tag{3}$$

with the charging energy E_C, the Josephson energy in the ith tunneling channel $E_{J,i}$, the total Josephson energy $E_J = E_{J,1} + E_{J,2}$, the inter-band Josephson energy $E_{in} = \kappa(\hbar W|J_{in}|/2e)$, the dimensionless bias current $\gamma = I_{ext}/I_c$. Here, we use the scaled relative phase $\Psi = \psi/r$, $r = \sqrt{\alpha_1 + \alpha_2}$ in order that the mass of the relative phase equals to that of center of mass phase, M. The critical current I_c is related to E_J. The factor κ in E_{in} is the sign of the inter-band Josephson coupling constant J_{in}, which is related to the gap symmetry. We take $\kappa = +1$ in this paper.

As seen in (3), the potential for a fictitious phase particle is described in 2D (θ, Ψ) space. In this study, we calculate the MQT rate in the 2D space in the following manner. First, we find a local minimum of the potential and the saddle point which is close to the local minimum. Next, by assuming that the fictitious particle follows the path of minimal curvature [16], we numerically calculate the tunneling path passing through the saddle point. Finally, we calculate the MQT rate for tunneling barrier $V_{1D}(z)$ along the tunneling path using Wentzel-Kramers-Brillouin (WKB) formula, where z is the curvilinear abscissa along the path. In the WKB approximation, the MQT rate Γ is given by, [2]

$$\Gamma = \omega_0\left(\frac{30S_b}{\pi\hbar}\right)^{\frac{1}{2}} \exp\left(-\frac{S_b}{\hbar}\right), \quad S_b = 2\int_{\zeta_0}^{\zeta_1} d\zeta[2MV_{1D}(\zeta)]^{\frac{1}{2}}, \tag{4}$$

where ζ_0 and ζ_1 are the coordinates for the local minimum and the end point of the tunneling path, and ω_0 is the characteristic frequency along the tunneling path.

3. Numerical results

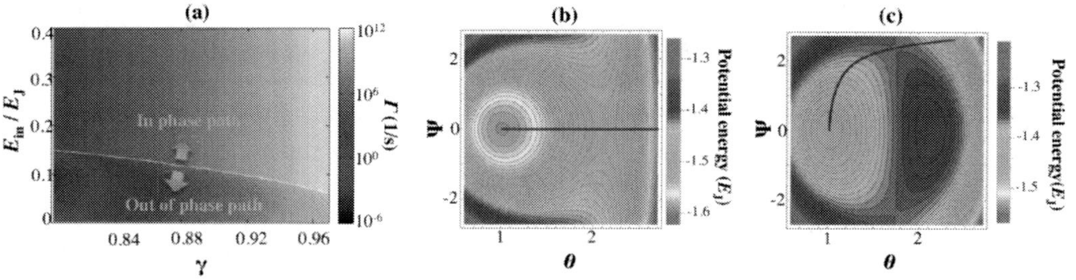

Figure 2. (a) MQT escape rate Γ as functions of external current γ and inter-band Josephson energy E_{in}. (b) and (c) are potential landscapes $V_{2D}(\theta, \Psi)$ and tunneling paths in the landscapes for $E_{in}/E_J = 0.05, \gamma = 0.85$ and $E_{in}/E_J = 0.2, \gamma = 0.85$ respectively. Black solid lines indicates the tunneling paths.

In order to investigate MQT for various tunneling path, we calculate the MQT rate Γ, varying the parameter sets (E_{in}, γ). In this study, we take $E_J = 0.01$ eV, $E_c/E_J = 0.002$, $E_{J,1} = E_{J,2}$, $\alpha_1 = \alpha_2 = 0.1$. Figure 2 (a) shows Γ as functions of E_{in} and γ. In this calculation, we find that two types of tunneling paths appear in our junction: in-phase and out-of-phase path. The white solid line in figure 2(a) indicates the boundary between the parameter region corresponding to each tunneling path. In figures 2 (b) and (c), we show the potential landscapes $V_{2D}(\theta, \Psi)$ and the tunneling paths in the landscapes for $(E_{in} = 0.05, \gamma = 0.85)$ and $(E_{in} = 0.2, \gamma = 0.85)$ respectively. The in-phase path is the straight-line path along the θ axis with $\Psi = 0$ as shown in figure 2 (b). Thus, ω_0 corresponds to the Josephson plasma frequency, $\omega_0 = \sqrt{2E_J E_c}(1-\gamma^2)^{\frac{1}{4}}/\hbar$. On the other hand, the out-of-phase path is the curved-line path with $\Psi \neq 0$. Moreover, in the vicinity of the local minimum, the path is almost parallel to the Ψ axis. Thus, ω_0 corresponds to the Josephson Leggett frequency, $\omega_0 = \sqrt{r^2 E_{\rm in} E_{\rm J}}/\hbar$. As shown in figure 2 (a), out-of-phase path only appears in weak inter-band coupling condition $E_{\rm in}/E_{\rm J} < 0.15$. Moreover, the MQT rate corresponding to the out-of-phase path drastically decreases as γ decreases compared to the in-phase path. **Figure 3 (a) and (b) show the escape rate Γ, respectively, for $E_{\rm in}/E_{\rm J} = 0.08$ and $E_{\rm in}/E_{\rm J} = 0.12$, varying γ. The peculiar kink structures in the both figures are ascribed to switching from the out-of-phase to the in-phase tunneling paths, that is, the bias-current dependence of Γ of the both paths are largely different. We check that the kink position in figure 3(a) ($\gamma = 0.947$) is consistent with the data in figure 2(a); the white solid line in figure 2(a) intersects with $E_{\rm in}/E_{\rm J} = 0.08$ at this current values. A similar analysis is done for $E_{\rm in}/E_{\rm J} = 0.12$, indicating that the kink appears at $\gamma = 0.877$. Figure 4 shows the remarkable difference between the out-of-phase path (figure 4(a)) and the in-phase path (figure 4(b)), on the potential landscape for $E_{\rm in}/E_{\rm J} = 0.08$.** The MQT along the out-of-phase path reflects the nature of the inter-band fluctuation, namely, the Josephson Leggett mode. Hence, we can detect the Josephson Leggett mode by systematically measuring the MQT rate for multi-band JJs with weak inter-band coupling.

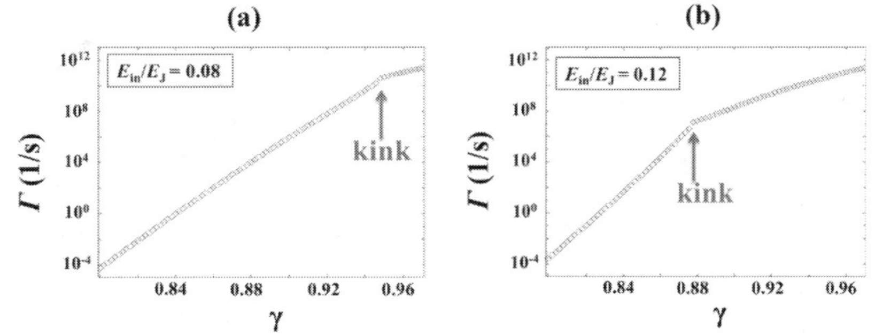

Figure 3. MQT escape rate Γ as a function external current γ for (a) $E_{\rm in}/E_{\rm J} = 0.08$, and (b) $E_{\rm in}/E_{\rm J} = 0.12$.

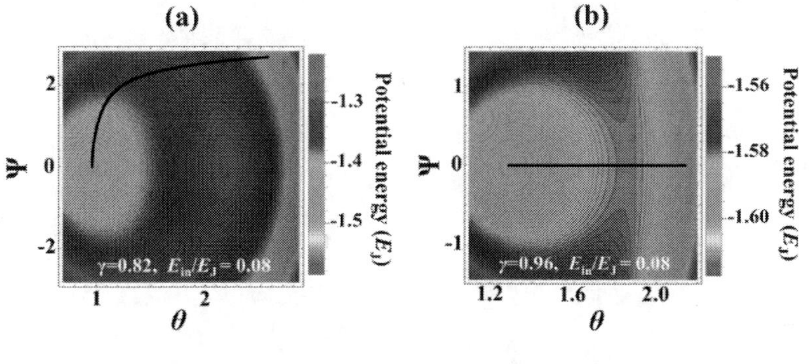

Figure 4. Potential landscapes and tunneling paths in the landscapes for (a) ($E_{\rm in}/E_{\rm J} = 0.08, \gamma = 0.82$), and (b) ($E_{\rm in}/E_{\rm J} = 0.08, \gamma = 0.96$). Black solid lines indicates the tunneling paths.

4. Conclusion

We investigated the MQT in a hetero JJ consisting of a single-gap and a two-gap superconductor. In this JJ, the MQT can be described by quantum dynamics of two phase variables: center-of-mass phase θ and relative-phase ψ. We analyzed the tunneling path in 2D potential-energy landscape for the phase dynamics, and calculated MQT rate along the path. We found two possible tunneling paths in 2D potential-energy landscape, i.e., in-phase and out-of phase path. The in-phase path is the straight-line path along the θ axis with $\psi = 0$, and related to the Josephson plasma mode. On the other hand, the out-of-phase path is the curved-line path with $\psi \neq 0$, and related to the inter-band fluctuation (i.e. Josephson Leggett mode). The relevant tunneling path changes with the external bias current and the inter-band Josephson-coupling energy. We found that the in-phase path appears when the inter-band coupling energy is strong, whereas, the out-of phase path appears when the inter-band coupling energy is weak. Moreover, in comparison with the in-phase path, the MQT rate corresponding to the out-of-phase path drastically decreases as bias current decreases. Our results indicate that we can detect the Josephson Leggett mode by systematically measuring the MQT rates for various inter-band coupling conditions.

References

[1] A. O. Caldeira and A. J. Leggett, Ann. Phys. **149**, 374 (1983).
[2] E. Simanek, *Inhomogeneous Superconductors: Granular and Quantum Effects* (Oxford University Press, New York, 1994).
[3] R. F. Voss and R. A. Webb, Phys. Rev. Lett. **47**, 265 (1981).
[4] M. H. Devoret, J. M. Martinis, and J. Clarke, Phys. Rev. Lett. **55**, 1908 (1985).
[5] T. Bauch, T. Lindström, F. Tafuri, G. Rotoli, P. Delsing, T. Claeson, and F. Lombardi, Science **311** 57 (2006).
[6] K. Inomata, S. Sato, K. Nakajima, A. Tanaka, Y. Takano, H. B. Wang, M. Nagao, H. Hatano, and S. Kawabata, Phys. Rev. Lett. **95**, 107005 (2005).
[7] X. Y. Jin, J. Lisenfeld, Y. Koval, A. Lukashenko, A. V. Ustinov, and P. Müller, Phys. Rev. Lett. **96**, 177003 (2006).
[8] Y. Kubo, A. O. Sboychakov, F. Nori, Y. Takahide, S. Ueda, I. Tanaka, A. T. M. N. Islam, and Y. Takano, Phys. Rev. B **86**, 144532 (2012).
[9] T. Kato and M. Imada, J. Phys. Soc. Jpn. **65**, 2963 (1996).
[10] S. Kawabata, S. Kashiwaya, Y. Asano, and Y. Tanaka, Phys. Rev. B **72**, 052506 (2005).
[11] M. Machida and T. Koyama, Physica C **463-465**, 84 (2007).
[12] S. Savel'ev, A. L. Rakhmanov, and F. Nori, Phys. Rev. Lett. **98**, 077002 (2007).
[13] A. O. Sboychakov, S. Savel'ev, A. L. Rakhmanov, and F. Nori, Euro. Phys. Lett. **80**, 17009 (2007).
[14] S. Kawabata, T. Bauch, and T. Kato, Phys. Rev. B **80**, 174513 (2009).
[15] Y. Ota, M. Machida, and T. Koyama, Phys. Rev. B **83**, 060503(R) (2011).
[16] N. Didier and F. W. J. Hekking Phys. Rev. B **85**, 104522 (2012).
[17] H. Asai, Y. Ota, S. Kawabata, M. Machida, and F. Nori, Phys. Rev. B **89**, 224507 (2014).
[18] J. Nagamatsu, N. Nakagawa, T. Muranaka, Y. Zenitani, and J. Akimitsu, Nature (London) **410**, 63 (2001).
[19] Y. Kamihara, T. Watanabe, M. Hirano, and H. Hosono, J. Am. Chem. Soc. **130**, 3296 (2008).
[20] K. Ishida, Y. Nakai, and H. Hosono, J. Phys. Soc. Jpn. **78**, 062001 (2009).
[21] J. Paglione and R. L. Greene, Nature Phys. **6**, 645 (2010).
[22] F. Bouquet, R. A. Fisher, N. E. Phillips, D. G. Hinks, and J. D. Jorgensen, Phys. Rev. Lett. **87**, 047001 (2001).
[23] X. X. Xi, Rep. Prog. Phys. **71**, 116501 (2008).
[24] H. Kito, H. Eisaki, and A. Iyo, J. Phys. Soc. Jpn. **77**, 063707 (2008).
[25] D. F. Agterberg, E. Demler, and B. Janko, Phys. Rev. B **66**, 214507 (2002).
[26] S. Onari and Y. Tanaka, Phys. Rev. B **79**, 174526 (2009).
[27] J. Linder, I. B. Sperstad, and A. Sudbø, Phys. Rev. B **80**, 020503(R) (2009).
[28] D. Inotani and Y. Ohashi, Phys. Rev. B **79**, 224527 (2009).
[29] Y. Ota, M. Machida, T. Koyama, and H. Matsumoto, Phys. Rev. Lett. **102**, 237003 (2009).
[30] A. A. Golubov, A. Brinkman, Y. Tanaka, I. I. Mazin, and O. V. Dolgov, Phys. Rev. Lett. **103**, 077003 (2009).
[31] A. J. Leggett, Prog. Theor. Phys. **36**, 901 (1966).

Journal of Physics: Conference Series **568** (2014) 022007 doi:10.1088/1742-6596/568/2/022007

Interference phenomena and long-range proximity effect in superconductor/ferromagnet systems

A.I. Buzdin
University of Bordeaux, LOMA UMR-CNRS 5798, F-33405,France

A.V. Samokhvalov
Institute for Physics of Microstructures, Russian Academy of Sciences, 603950, Nizhny Novgorod, Russia
Lobachevsky State University of Nizhny Novgorod, Nizhny Novgorod 603950, Russia

R.I. Shekhter
Department of Physics, University of Gothenburg, SE-412 96 Goteborg, Sweden

E-mail: a.bouzdine@loma.u-bordeaux1.fr

Abstract. We study the novel interference mechanism of the long-range proximity effect in superconductor/ferromagnet/superconductor (SFS) structures in the ballistic regime. Even a small non-collinear magnetic domain near the center of a ferromagnetic weak link is shown to restore the singlet supercurrent inherent to the normal metal. The underlying physics of the effect is the magnetic scattering of the Cooper pair by the domain, which reverses total momentum of the pair in the ferromagnet and thus compensates the phase gain before and after the spin-flip scattering. The above phenomenon opens a way to easily control the properties of SFS junctions and, inversely, to manipulate the magnetic moment via the Josephson current.

1. Introduction

Josephson junctions with a ferromagnetic (F) metal weak link is known to reveal a very strong decrease of the critical current compared to a normal metal weak link (see Refs. [1, 2] for review). This behavior is induced by the destructive effect of the exchange field h which leads to phase difference $\gamma \sim L/\xi_h$ accumulated between the electron- and hole- like parts of the total wave function while propagating along a path of the length L in the ferromagnet [3, 4]. Here $\xi_h = \hbar V_F/2h$ is a characteristic length determined by the exchange field, and V_F is the Fermi velocity. Measurable quantities should be calculated as the superpositions of fast oscillating contributions $e^{i\gamma}$ from different trajectories and, thus, rapidly vanish with the increasing distance L and displays a short-range ($\sim \xi_h$) behavior. It should be noted though, that a simple domain structure consisting of two F layers with opposite orientations of exchange field cancels the phase gain γ and suppresses the destructive effect of an exchange field [3, 5]. Another way to produce long-range proximity effect in SFS (superconductor-ferromagnet-superconductor) weak links is a composite F layers comprising three non-collinear domains, was suggested theoretically [6, 7, 8] and realized in recent experiments [9, 10]. In such a case the triplet Cooper pairs of electrons with aligned spins (with equal-spin pairs) are generated by a thin ferromagnetic domain [11, 12], located between superconducting lead and a thick central non-collinear domain. Since these triplet pairs bind electrons with exactly the same de-Broglie wave length, they do not dephase, thereby leading to long-range proximity effect.

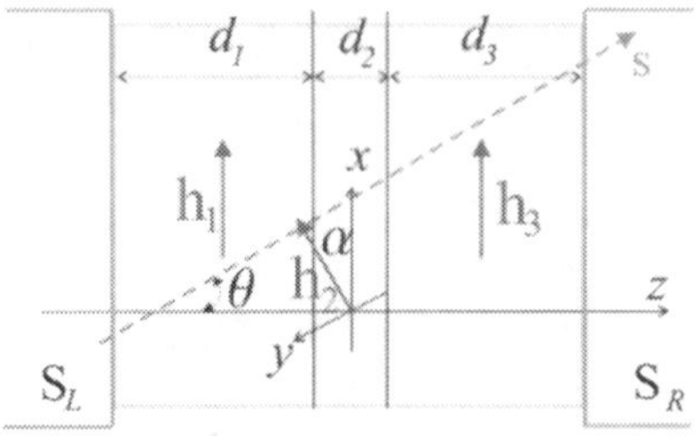

Figure 1. The schematic sketch of the SFS Josephson junction containing three ferromagnetic layers (domains) with a stepwise profile of the exchange field (1). Linear quasiparticle trajectory is shown by the red dashed line.

Here we suggest a new way of manipulating the Cooper pairs flow through a clean ferromagnet, which consists in using the composite F layer with a thin non-collinear domain $d_2 \sim \xi_h \ll d_1, d_3$, located near the center of the F barrier ($d_1 \simeq d_3$), as shown in Fig. 1. In contrast with the situation $d_1, d_3 \sim \xi_h \ll d_2$ analyzed in Refs. [6, 7, 8], this setup does not support the long-range triplet superconducting current. The thin central domain induces a special scattering of Cooper pairs which corresponds to exchange spins of two electrons forming a singlet Cooper pair (spin-flip scattering). Since modulus of the Fermi wave-vector for electrons with a spin polarized along the field is larger $|\mathbf{k}_\uparrow| > |\mathbf{k}_\downarrow|$ due to the ferromagnetic spin-splitting effect, the pair have a non-zero total momentum $\hbar\mathbf{q} = \hbar\mathbf{k}_\uparrow - \hbar\mathbf{k}_\downarrow$ ($|\mathbf{q}| \sim 1/\xi_h$). The spin-flip scattering by the domain changes spin arrangement of a pair so that the new total momentum of the Cooper pair $\hbar\mathbf{q}'$ is reversed: $\hbar\mathbf{q}' = -\hbar\mathbf{q}$. At a symmetric position of the scatterer ($d_1 \simeq d_3$) the total phase gain $\gamma \sim (d_1 - d_3)/\xi_h$ for a singlet Cooper pair should be cancelled and the long-range singlet superconducting proximity in SFS link becomes possible.

The paper is organized as follows. In Sec. 2, we briefly discuss the basic equations and methods used in calculations and calculate the long-range component of the Josephson current at the first harmonics of the current–phase relation. We discuss our results in Sec. 4.

2. Model and Methods

Let us consider the Josephson transport through a ballistic SFS junction containing three ferromagnetic layers (domains) with a stepwise profile of the exchange field

$$h(z) = \begin{cases} h\mathbf{x}_0, & \text{in domains} \quad d_1, d_3 \\ h\left(\mathbf{x}_0 \cos\alpha + \mathbf{y}_0 \sin\alpha\right), & \text{in domain} \quad d_2\,, \end{cases} \tag{1}$$

where α is the angle of the exchange field rotation in the central domain d_2 (see Fig. 1). The total length of the constriction $d = d_1 + d_2 + d_3$ is assumed to be large compared to the magnetic coherence length $\xi_h = \hbar V_F/2h$: $d \gg \xi_h$. Here we consider the limit of the short junction $d \ll \xi_n$, where $\xi_n = \hbar V_F/T_c$ is the coherence length of normal metal and T_c is the critical temperature of the S layer.

The current–phase relation of SFS Josephson junction is determined by the quasiclassical relation [4, 5]

$$I = \sum_n I_n = \sum_n a_n \sin n\varphi \frac{\langle(\mathbf{n}, \mathbf{n}_F)\cos n\gamma\rangle}{\langle(\mathbf{n}, \mathbf{n}_F)\rangle}\,, \tag{2}$$

where $\mathbf{n}$ is the unit vector normal to the junction plane, $\mathbf{n}_F$ is the unit vector along the trajectory, and a_n are the coefficients of the Fourier expansion for the current–phase relation for superconductor–normal metal junction of the same geometry. The angular brackets denote the averaging over different quasiclassical trajectories characterized by a given angle θ and a certain starting point at the superconductor surface, and for 3D constriction looks as

$$\frac{\langle(\mathbf{n},\mathbf{n}_F)\cos(n\gamma)\rangle}{\langle(\mathbf{n},\mathbf{n}_F)\rangle} = 2\int_0^{\pi/2} d\theta \sin\theta\cos\theta\,\cos(n\gamma)\,, \tag{3}$$

where $\cos\theta = (\mathbf{n},\mathbf{n}_F)$. At temperatures T close to T_c the current–phase relation (2) is sinusoidal, and the coefficient a_1 is determined by the following simple relations [5]:

$$a_1 = \frac{eT_c}{8\hbar}N\left(\frac{\Delta}{T_c}\right)^2 . \tag{4}$$

Here Δ is the temperature dependent superconducting gap, The factor N is determined by the number of transverse modes in the junction: $N = s_0^{-1}\int ds \int d\mathbf{n}_F(\mathbf{n}_F,\mathbf{n}) \sim S/s_0$, where S is the junction cross–section area, and $s_0^{-1} = (k_F/2\pi)^2$, where k_F is the Fermi momentum.

2.1. Transfer–matrix formalism for Eilenberger Equations

The phase gain $\gamma(\theta)$ along a trajectory $\mathbf{s} = s\,\mathbf{n}_F$ (see Fig. 1) is determined by the singlet part f_s of the anomalous quasiclassical Green function $f = f_s + \mathbf{f}_t\hat{\sigma}$ taken at the right superconducting electrode ($s = s_R = d/cos\theta$) [5]: $\cos\gamma = f_s(s_R)$. Here $\hat{\sigma}$ is a Pauli matrix vector in the spin space. The singlet (triplet) parts f_s, ($\mathbf{f}_t$) of the Green function f satisfy the linearized Eilenberger equations [6] written for zero Matsubara frequencies

$$-i\hbar V_F\partial_s f_s + 2\mathbf{h}\mathbf{f}_t = 0\ , \qquad -i\hbar V_F\partial_s \mathbf{f}_t + 2f_s\mathbf{h} = 0\ , \tag{5}$$

with the boundary conditions $f_s(s_L) = 1$, $\mathbf{f}_t(s_L) = 0$ at the left superconducting electrode ($s = s_L = 0$).

To consider the Josephson transport through ferromagnetic layer with an arbitrary non-collinear distribution of the magnetizations $\mathbf{M}$ and the exchange field $\mathbf{h}$ it is convenient to utilize the transfer–matrix formalism. For this, we need to solve the equations (5) for the case when the quantization axis is taken arbitrarily in the ferromagnetic layer of a thickness $d_i = z_i - z_{i-1}$. We assume that a quasiclassical trajectory $\mathbf{s}$ is characterized by a given angle θ with respect to the z-axis and exchange field $\mathbf{h} = h\,(\mathbf{x}_0\cos\alpha_i + \mathbf{y}_0\sin\alpha_i)$ lie in the plane $(x,\,y)$, as shown in Fig. 1. The triplet part $\mathbf{f}_t$ consists of two nonzero components and can be written as $\mathbf{f}_t = f_{tx}\mathbf{x}_0 + f_{ty}\mathbf{z}_0$. Defining the transfer–matrix $\hat{T}_{\alpha_i}(d_i,\theta)$ that relates the components of the Green function $\hat{f}(s) = \{f_s(s),\, f_{tx}(s),\, f_{ty}(s)\}$ at the left ($s = s_{i-1} = z_{i-1}/\cos\theta$) and right ($s = s_i = z_i/\cos\theta$) boundaries of the F layer,

$$\hat{f}(s_i) = \hat{T}_{\alpha_i}(d_i,\theta)\,\hat{f}(s_{i-1})\,. \tag{6}$$

we get the following expression:

$$\hat{T}_{\alpha_i}(d_i,\theta) = \begin{pmatrix} \cos(qs_{d_i}) & -i\cos\alpha_i\sin(qs_{d_i}) & -i\sin\alpha_i\sin(qs_{d_i}) \\ -i\cos\alpha_i\sin(qs_{d_i}) & \sin^2\alpha_i + \cos^2\alpha_i\cos(qs_{d_i}) & \sin\alpha_i\cos\alpha_i\,(\cos(qs_{d_i})-1) \\ -i\sin\alpha_i\sin(qs_{d_i}) & \sin\alpha_i\cos\alpha\,(\cos(qs_{d_i})-1) & \cos^2\alpha_i + \sin^2\alpha_i\cos(qs_{d_i}) \end{pmatrix}, \tag{7}$$

where $q \equiv 1/\xi_h = 2h/\hbar V_F$ and $s_{d_i} = d_i/\cos\theta$.

2.2. The long-range phase gain

Solving the equations (5) by the transfer matrix method for the stepwise profile of the exchange field (1), the anomalous quasiclassical Green function $\hat{f}(s_R) = \{f_s(s_R), f_{tx}(s_R), f_{ty}(s_R)\}$ at the right superconducting electrode ($s = s_R = d/\cos\theta$) can be easily expressed via the boundary conditions $\hat{f}(0) = (1, 0, 0)$ at the left superconducting electrode ($s = 0$) as follows:

$$\hat{f}(s_R) = \hat{T}_0(d_3, \theta)\, \hat{T}_\alpha(d_2, \theta)\, \hat{T}_0(d_1, \theta) \hat{f}(0)\,, \tag{8}$$

where $d = d_1 + d_2 + d_3$ is the total thickness of the ferromagnetic barrier, and the transfer–matrix $\hat{T}_\alpha(d_i, \theta)$ is determined by the expression (7). As a result, the singlet part $f_s(s_R)$ responsible for the Josephson current through the junction, can be written in the form:

$$\begin{aligned} \cos\gamma \;&=\; \cos\delta_2 \cos(\delta_1 + \delta_3) - \cos\alpha \sin\delta_2 \sin(\delta_1 + \delta_3) \\ &-\; \sin^2\alpha\, \sin\delta_1 \sin\delta_3 (1 - \cos\delta_2)\,, \end{aligned} \tag{9}$$

where $\cos\theta = (\mathbf{n}, \mathbf{n}_F)$ and $\delta_i = d_i/\xi_h \cos\theta$ $(i = 1, 2, 3)$. Introducing the shift $z_0 = (d_1 - d_3)/2$ of the central domain with respect to the weak link center it is convenient to rewrite the expression (9) in the following equivalent form:

$$\cos\gamma(\theta) = b_1\, \cos\delta_d + b_2\, \sin\delta_d - b_3\, \cos 2\delta_z\,, \tag{10}$$

where $\delta_d = d/\xi_h \cos\theta$, $\delta_z = z_0/\xi_h \cos\theta$, and the coefficients b_n $(n = 1, 2, 3)$ don't depend on the total length of the junction $d = d_1 + d_2 + d_3$:

$$\begin{aligned} b_1 &= \cos^2\delta_2 + \cos\alpha \sin^2\delta_2 + b_3 \cos\delta_2\,, \\ b_2 &= \cos\delta_2 \sin\delta_2 - \cos\alpha \cos\delta_2 \sin\delta_2 + b_3 \sin\delta_2\,, \\ b_3 &= \sin^2\alpha\, (1 - \cos\delta_2)/2\,. \end{aligned}$$

Averaging the expression (10) over the trajectory direction θ and neglecting the first two terms proportional to $\xi_h/d \ll 1$, which decrease just as for the case of homogeneous ballistic 3D SFS junction, one arrives at the following expression

$$(\cos\gamma)^{LR} = -\frac{1}{2} \sin^2\alpha(1 - \cos\delta_2)\, \cos 2\delta_z\,, \tag{11}$$

3. Results

The long-range component of the Josephson current at the first harmonic is determined by the relation:

$$I^{LR} = I_c^{LR} \sin\varphi = a_1 T_1^{LR} \sin\varphi\,, \qquad T_1^{LR} = 2 \int_0^{\pi/2} d\theta \sin\theta \cos\theta\, (\cos\gamma)^{LR}\,. \tag{12}$$

For a very thin central domain $d_2 \ll \xi_h$ in the center of the NW ($z_0 = 0$) one can easily estimate from (11, 12) the critical current of the SFS junction

$$I_c^{LR} \approx \frac{I_0}{2} \sin^2\alpha \left(\frac{d_2}{\xi_h}\right)^2 \ln\frac{\xi_h}{d_2}\,, \tag{13}$$

where $I_0 = (eT_cN/8\hbar)\,(\Delta/T_c)^2$ – is the critical current of the SNS junction for zero exchange field ($\gamma = 0$). Figure 2a shows the dependence of the maximal Josephson current $I_c^{LR} = a_1 T_1^{LR}$

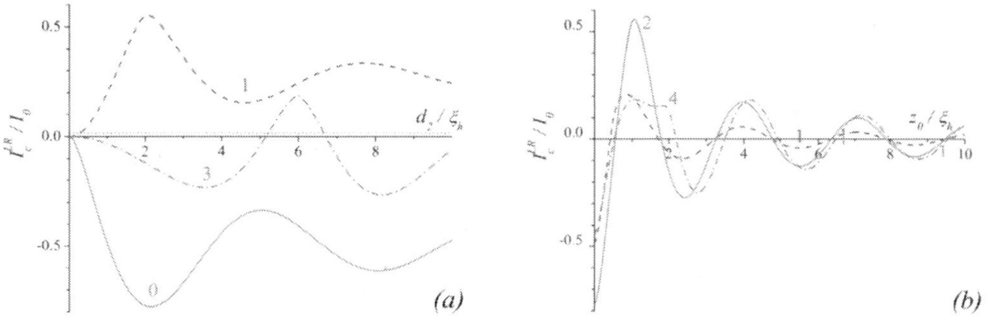

Figure 2. (a) The dependence of I_c^{LR} on the thickness d_2 of the 90° domain ($\alpha = \pi/2$) for different values of the shift of the domain z_0: $z_0 = 0$ - red solid line; $z_0 = \xi_h$ - blue dashed line; $z_0 = 3\xi_h$ - green dash-dotted line. Dotted line shows the value of $I_c = \max\{I_1\}$ in absence of domain d_2. (b) The dependence of maximal Josephson current I_c^{LR} on the shift of the central domain z_0 for different values of the d_2: $d_2 = \xi_h$ - blue dashed line; $d_2 = 2\xi_h$ - red solid line; $d_2 = 4\xi_h$ - green dash-dotted line. We have set $T = 0.9T_c$; $d = 50\xi_h$ [$I_0 = (eT_cN/8\hbar)(\Delta/T_c)^2$].

on the thickness d_2 of the 90° domain ($\alpha = \pi/2$) for different positions of the domain with respect to the weak link center. The amplitude of I^{LR} oscillates with varying the thickness of the central domain d_2 and has the first maximum at $d_2 \simeq 2.5\xi_h$. Naturally, when the central domain disappears ($d_2 \to 0$), the long-range effect vanishes. Comparison of the long-range Josephson current I^{LR} with the total supercurrent across the junction (2) shows that the difference is negligible until the outer domains are long enough: $d_1, d_3 \gg \xi_h$ [13]. Figure 2b shows the dependences of the maximal Josephson current I_c^{LR} on the position of the central domain z_0 for different values of the domain d_2 thickness. We may see that the critical current is very sensitive to the position of the central domain and the first zero of I_1 occurs already at $z_0 \simeq 0.5\xi_h$. Interestingly, that the long-range supercurrent I^{LR} is negative at a symmetric position ($z_0 = 0$) of the domain d_2. This means that the spin-flip scatterer produces the π shift effect and generates a π junction. With a displacement of the central domain the SFS junction can be switched from π to 0 state.

3.1. Arbitrary ferromagnetic barrier

The transfer–matrix formalism can be easily generalized for a layered ferromagnetic barrier with an arbitrary non-collinear distribution of the exchange field **h** which is described by the dependence $\alpha(z)$. Splitting the barrier on N thin layers of the thickness $d_i = z_i - z_{i-1}$ ($i = 1 \div N$) one consider the exchange field to be constant inside each layer i. Application of the transfer–matrix $\hat{T}_{\alpha_i}(d_i, \theta)$ (7) "layer by layer" results in the following relation between the components of the Green function $f(0)$ and $f(s_R)$ at the left and right superconducting electrodes, respectively:

$$\hat{f}(s_R) = \hat{T}_{\alpha_N}(d_N, \theta) \ldots \hat{T}_{\alpha_2}(d_2, \theta)\,\hat{T}_{\alpha_1}(d_1, \theta)\,\hat{f}(0)\,. \tag{14}$$

We apply the described transfer–matrix formalism to study the effect of smooth in the SFS constriction shown in Fig. 1a of the Letter. As the dependence $\alpha(z)$ we use a draft model of 90°–domain described by Gaussian function:

$$\alpha(z) = \frac{\pi}{2}\exp\left(-\frac{(z-z_0)^2}{2w^2}\right), \tag{15}$$

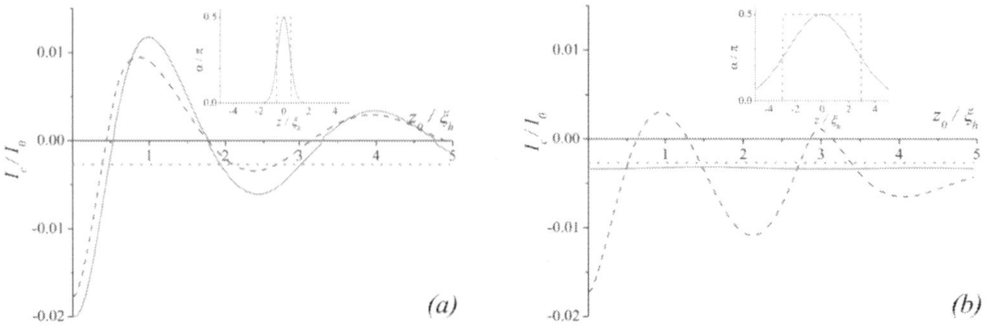

Figure 3. The dependence of I_c^{LR} on the shift z_0 of the 90°−domain (15) ($\alpha = \pi/2$) for different values of the width of the transition region w (solid red line): (a) $w = 1$; (b) $w = 5$. The blue dashed line shows the dependence of I_c^{LR} on the shift z_0 of stepwise 90°−domain for comparison. The dotted line shows the critical current I_c in the absence of the domain ($d_2 = 0$). Inset shows the coordinate dependence of the rotation angle α (15) (red solid line) and the in the relevant stepwise domain (blue dashed line). We have set $d = 20\xi_h$; $T = 0.9T_c$ [$I_0 = (4eT_cN/\hbar)(\Delta/T_c)^2$].

where z_0 is the shift of the domain with respect to the weak link center, and w describes the width of the domain. Figure 3 shows the dependences of the critical current of the SFS junction on position z_0 of the 90°−domain (15) for different values of the domain width w. We may see that the long-range effect seems to be completely disappeared if $w \gg \xi_h$ (see Fig. 3c), but it is quite robust for the domain width smaller than $2 - 3\xi_h$ and only weakly depends on the exact form of the transition region. So, the smooth (on the scale ξ_h) profile of the magnetization decreases the long-range effect and the proposed mechanism occurs to be most efficient for $d_2 \sim \xi_h$.

Certainly, the above long-range effect in the first harmonic describes the properties of the SFS constriction if the contribution of higher harmonics in the current–phase relation (2) is negligible. This approach is usually applicable everywhere, except very close to the $0 - \pi$ transition ($T_1 = 0$), because $|a_n|\,(n \geq 2) \ll |a_1|$ at $T \approx T_c$. At this $0 - \pi$ transition the contribution of the second harmonic I_2 becomes dominant. For all considered cases we obtained the positive amplitude of the second harmonic in the vicinity of these transitions, which means that they occur discontiguously by a jump between 0 and π phase states [13].

4. Discussion

In order to elucidate the peculiarities of the Cooper pairs scattering with a spin-flop transition of electrons it is convenient to introduce the new functions $f_\pm = f_s \pm f_{tx}$ which describes the pairs with zero spin projection and a reversed spin arrangement. The equations (5) can be drastically simplified if the direction of the exchange field coincides with a spin quantisation axis x:

$$\mp i\hbar V_F \partial_s f_\pm + 2hf_\pm = 0\,. \qquad (16)$$

In this case $f_\pm(s_R) = e^{\mp iqs_R} f_\pm(0)$, $f_{ty}(s_R) = f_{ty}(0) = 0$, $f_{tz}(s_R) = f_{tz}(0) = 0$. Calculating the superconducting current at the right electrode s_R we readily see that it results from the interference with the singlet component coming from the left electrode $f_s(s_R) = (f_+(s_R) + f_-(s_R))/2$ (triplet components are absent and irrelevant because the right electrode provides only the singlet component). The oscillating factors $e^{\mp iqs_R}$ in $f_\pm(s_R)$ produce, after the averaging over the trajectories directions (angle θ), a strong damping of the critical current compared to the normal metal (where these factors are absent).

Now we may easily understand the mechanism of the singlet long-range proximity effect in the presence of a small region with a non-collinear magnetization near the center of a ferromagnetic weak link. Indeed after coming through the first F layer d_1 an extra phase factor appears in $f_\pm$ functions: $f_\pm(s_{d_1}) = \mathrm{e}^{\mp iqs_{d_1}} f_s(0)$. In the absence the middle layer d_2, the $f_\pm$ at the right electrode would be $f_\pm(s_{d_1} + s_{d_3}) = \mathrm{e}^{\mp iq(s_{d_1}+s_{d_3})} f_s(0)$ and the oscillating factors will strongly damp critical current. The additional non-collinear middle layer d_2 (spin-flip scatterer) will mix up the components f_+ and f_- so that

$$f_\pm(s_{d_1} + s_{d_2}) = a_\pm \mathrm{e}^{-iqs_{d_1}} + b_\pm \mathrm{e}^{+iqs_{d_1}} .$$

Then the resulting $f_\pm$ functions at the right electrode should be

$$f_\pm(s_{d_1} + s_{d_2} + s_{d_3}) = a_\pm \mathrm{e}^{-iq(s_{d_1} \pm s_{d_3})} + b_\pm \mathrm{e}^{+iq(s_{d_1} \mp s_{d_3})}$$

and for $d_1 = d_3$ the oscillating factor at the first (second) term vanishes. This means the emergence of the long-range singlet proximity effect discussed in the present paper. Note that the additional noncollinear F layer d_2 may strongly increase the critical current, provided that it is placed at the center of the structure.

The very important property of the discussed system is that it provides a direct mechanism of the coupling between supercurrent and magnetic moment, similar to the situation discussed in Refs. [14, 15]. Since the long-range critical current I_c^{LR} depends on the profile of the magnetization, the superconducting current acts as a direct driving force on the magnetic moment and can change its orientation. Inversely, the precession of the magnetic moment shall modulate the critical current.

To summarize, we studied the interference phenomena and long-range proximity effect originated by the spin-exchange scattering in ferromagnetic ballistic weak link. In contrast to the widely discussed triplet long-range proximity effect we elucidate a new singlet long-range proximity effect. We demonstrated that this phenomenon provides an efficient way to control the Josephson current and to couple it with a magnetic moment. This offers the means for spintronic manipulation of superconducting weak links by external magnetic inhomogeneity located on a submicron length scale [13].

The authors thank A.S. Mel'nikov for stimulating discussions. This work was supported, in part, by European IRSES program SIMTECH (contract n.246937), by French ANR grant "MASH", by the Swedish VR, by the RFBR, by the program "Quantum Mesoscopic and Disordered Structures" of RAS, and by Ministry of Education and Science of RF and Lobachevsky State University (agreement 02.B.49.21.0003).

References

[1] Buzdin A I 2005 *Rev. Mod. Phys.* **77** 935
[2] Bergeret F S, Volkov A F and Efetov K B 2005 *Rev. Mod. Phys..* **77** 1321
[3] Blanter Ya M and Hekking F W J 2004 *Phys. Rev. B* **69** 024525
[4] Buzdin A I, Melnikov A S, and Pugach N G 2011 *Phys. Rev. B* **83** 144515
[5] Melnikov A S, Samokhvalov A V, Kuznetsova S M, and Buzdin A I 2012 *Phys. Rev. Lett.* **109** 237006
[6] Houzet M and Buzdin A I 2007 *Phys. Rev. B* **76** 060504(R)
[7] Alidoust M and Linder J, Rashedi G, Yokoyama T, and Sudbo A. 2010 *Phys. Rev. B* **81** 014512
[8] Volkov A F and Efetov K B 2010 *Phys. Rev. B* **81** 144522
[9] Robinson J W A, Witt J D S, and Blamire M G 2010 *Science* **329** 59
[10] Khaire T S, Khasawneh M A, Pratt W P (Jr.), and Birge N O 2010 *Phys. Rev. Lett.* **104** 137002
[11] Bergeret F S, Volkov A F, and Efetov K B 2001 *Phys. Rev. Lett.* **86** 4096
[12] Kadigrobov A , Shekhter R I and Jonson M 2001 *Europhys. Lett.* **54** 394
[13] Samokhvalov A V, Shekhter R I and Buzdin A I 2014 *Sci. Rep.* in press
[14] Buzdin A I 2008 *Phys. Rev. Lett.* **101** 107005
[15] Konschelle F and Buzdin A 2009 *Phys. Rev. Lett.* **102** 017001

Journal of Physics: Conference Series **568** (2014) 022008 doi:10.1088/1742-6596/568/2/022008

Proximity effects and vortex dynamics in superconducting heterostructures

S J Carreira[1,3], V Bekeris[1], Y J Rosen[2,4], C Monton[2] and Ivan K Schuller[2]

[1] Departamento de Física, FCEyN, Universidad de Buenos Aires and IFIBA, CONICET, Pabellón 1, Ciudad Universitaria, 1428 Buenos Aires, Argentina.

[2] Department of Physics, Center of Advance Nanoscience, University of California, San Diego, 9500 Gilman Drive, La Jolla, California 92093, USA.

E-mail: sjcarreira@gmail.com

Abstract. Vortex lattice dynamics has been studied in thin Nb superconducting films sputtered on top of a dense triangular array of V dots with and without an intermediate SiO_2 insulating layer. While the insulating layer modifies only slightly the Nb film corrugation, it reduces superconducting commensurability effects (CE) substantially. This implies that superconducting commensurability is dominated by proximity effects. Moreover, the H_{C2} (T) phase diagram of the sample without an insulating layer shows a parabolic temperature dependence near T_C and critical temperature oscillations with the periodicity of the matching field. Therefore, strong proximity effects locally suppress superconductivity leading to a superconducting mesh. When the proximity effect is decreased by an insulating layer, H_{C2}(T) follows the expected linear T dependence.

1. Introduction

In recent years, the fabrication of high quality superconducting (SC) heterostructures confined to the nm scale, gave rise to the discovery of novel phenomena. The systems under study comprise mainly SC thin films with a regular array of holes [1], blind holes [2], magnetic dots [3],[4] and insulating dots [5]. In type II superconductors with an array of pinning centers, commensurability effects (CE) take place for an integer or fractional number of flux lines per unit cell [6]. This is known as a "collective" effect as is strongly dependent on mutual repulsion between vortices. Vortex pinning in a thin film containing a non-ferromagnetic array of dots has been explained as a sum of different contributions, depending on the temperature and the geometry involved: 1) the depression of the superconducting order parameter in the surrounding of the dots [7], 2) intrinsic pinning due to the random film defects [8], 3) periodic thin film corrugation and film inhomogeneity [9] and 4) filling effects due to the quantized nature of the vortices [10]. The latter, known as the Little & Parks effect [11] has been widely studied in multiple connected superconducting strips and in thin films with a regular array of pinning centers [12],[13].

We explore pinning mechanisms and CE using ac-susceptibility, in thin SC Niobium (Nb) films, sputtered on top of a dense triangular array of Vanadium (V) dots with and without a SiO_2 insulating

[3] Present address : Dpto. de Materia Condensada. Centro Atómico Constituyentes. CNEA. Av. Gral Paz 1499 (1650), Buenos Aires, Argentina.

[4] Present address: Laboratory for Physical Sciences, College Park, MD 20740, USA.

layer used to separate the V dots from the Nb film. We show that the insulating barrier modifies weakly the film corrugation and inhibits the depression of superconductivity by proximity effects. Although the superconducting-matching effects persist, the $H_{C2}(T)$ phase diagram is strongly modified. The paper is organized as follows. In Section 2 we describe the experimental details. Results and discussions are presented in Section 3. Conclusions are drawn in Section 4.

2. Experimental

A dense submicrometric triangular array of V dots, diameter $d = (70\pm5)$ nm and thickness $h = (50\pm2)$ nm with lattice parameter $D = (110\pm10)$ nm, was deposited on a Si substrate as described elsewhere [14]. On top of the V dots, a Nb film of thickness $\tau = (100\pm5)$ nm was sputtered (sample 10KC) and for the sample 10Si/V, an additional SiO_2 film of thickness $\varepsilon = (10\pm2)$ nm was evaporated before the Nb deposition. The bulk critical temperature of V, $T_C^V(\text{bulk}) = 5.4$ K [15], is lower than the bulk critical temperature of Nb, $T_C^{Nb}(\text{bulk}) = 9$ K [16]. It is well known that critical temperature may be modified for films or heterostructures [17]. In these experiments, V dots remain normal in both samples in the field and temperature ranges studied, as we found no evidence of its SC transition [18].

3. Results and discussion

Figure 1 (a) shows the zero field temperature dependence of AC susceptibility, χ', for sample 10KC (full) and 10Si/V (open symbols). An AC magnetic field hac = 0.25 Oe has been applied perpendicular to the samples. The zero field critical temperatures (T_C), are $T_C^{10KC} = (6.60\pm0.02)$ K and $T_C^{10Si/V} = (7.17\pm0.02)$ K for sample 10KC and sample 10Si/V, respectively.

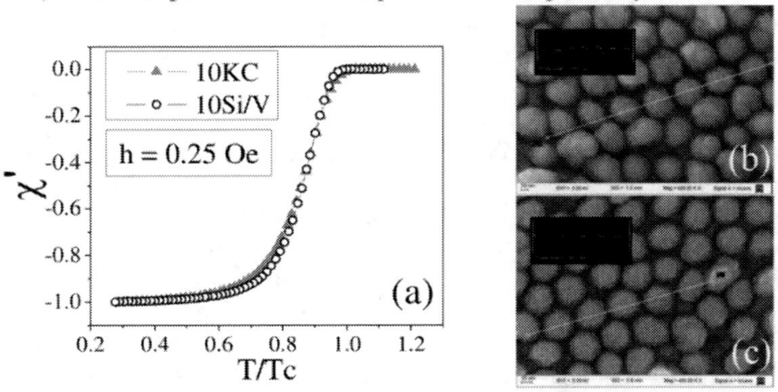

Figure 1: (a) $\chi'(T/T_C)$ (hac = 0.25 Oe, H =0) for sample 10KC (full) and 10Si/V (open symbols). (b-c) SEM images and intensity profiles of the surface of sample 10KC (b) and 10Si/V (c).

Note the similarity between both transitions *vs* t = T/Tc. In these heterostructures, periodic surface corrugation and thickness variation of the Nb thin film around the V dots can produce CE [19]. In addition, normal V dots can depress the superconducting order parameter of the Nb thin film by proximity effects, leading to vortex pinning. Since the surface Nb corrugation and the V dots lattice have the same periodicity (in fact, one produces the other), both mechanisms may lead to the same matching fields. Consequently, it is not straightforward to assess independently the strength of each pinning mechanism.

To determine the influence of the proximity effect of V dots on the Nb thin film we measured the corrugation in samples 10KC and 10Si/V by Scanning Electron Microscopy (SEM) as shown in figure 1 (b) and (c) respectively. By comparing intensity profiles from SEM images at the same magnification, we found that the 10 nm Si layer in 10Si/V reduces the surface corrugation only by ~10% compared to the surface corrugation in sample 10KC. The unexpected corrugation similarities in both samples implies that differences in pinning are not related to corrugation effects.

Figure 2 (a) shows AC susceptibility measurements *vs* DC magnetic field, applied perpendicular to the surface of both samples in zero field cooling experiments.

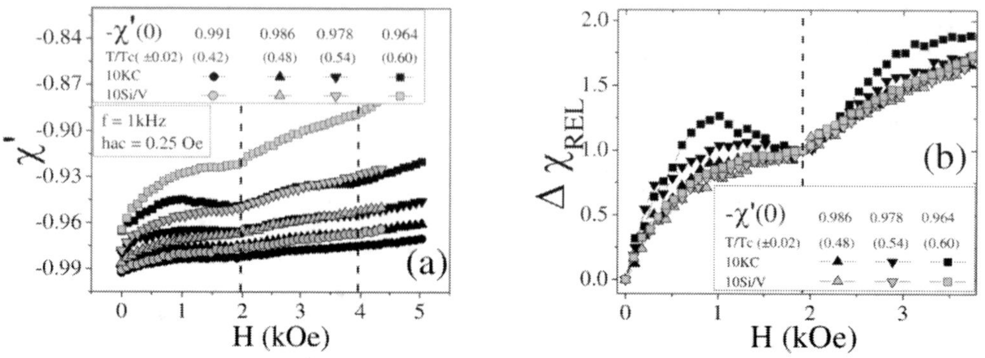

Figure 2: (a) Shielding capability *vs* DC magnetic field of the sample 10KC (full) and 10Si/V (open symbols). (b) $\Delta\chi_{REL} = (\chi(H)-\chi(0))/(\chi(H_1)-\chi(0))$ extracted from data in (a). Vertical dashed lines indicate the first and second matching fields.

We compare curves at the same reduced temperature and zero field shielding capability, for the sample 10KC (full) and 10Si/V (open symbols). Vertical lines indicate the experimental first and second matching fields H_1 = (1975±50) Oe and H_2 = (3950±50) Oe, in agreement with calculated values for a unit cell with lattice parameter d = 110 nm [6].

In figure 2 (b) we plotted $\Delta\chi_{REL} = (\chi'(H)-\chi'(0))/(\chi'(H_1)-\chi'(0))$ *vs* H, which shows clearly the reduction of vortex mobility at the first matching field. The weak temperature dependence of matching intensity in 10Si/V leads us to attribute the slightly reduced vortex mobility at matching fields to the periodic surface corrugation. This implies that the SiO_2 layer isolates the SC Nb film from the normal V dots, impeding the proximity effects and the depression of the SC order parameter thus reducing matching effects. In contrast, CE are clearly enhanced and temperature dependent for the sample 10KC. Since both samples have almost the same corrugation, this implies that proximity effects between the normal V dots and Nb film in the sample 10KC, depress the SC parameter in the film around localized regions close to each V dot. The existence of proximity effects implies that the interfaces between the V dots and the Nb film are clean. In addition, at fields other than the matching fields screening appears to be more effective in 10KC than in 10Si/V. This is expected considering that not only surface corrugation but also pinning close to the V dots play an important role in sample 10KC. At the same reduced temperature, vortex dynamics are reduced and matching effects are clearly enhanced for sample 10KC in contrast to sample 10Si/V. CE at low temperatures (below $0.5T_C$) are smeared due to the enhancement of intrinsic random pinning as temperature is reduced.

When the temperature is raised close to T_C, the coherence length ξ changes substantially, which has important implication for the behavior of these films. In figure 3 (a) we plot the phase diagram $H_{C2}(T)$ for sample 10KC in full symbols, extracted from field cooling temperature dependent susceptibility measurements at different DC fields. Critical field shows strong non linear temperature dependence near T_C, on which an oscillatory variation with periodicity of the matching field is superimposed. The parabolic dependence of $H_{C2}(T)$ for SC strips of width w is well known and follows the expression:

$$H_{C2}(T_C) = \frac{\sqrt{12}\phi_0}{2\pi w\xi(T)} = \frac{\sqrt{12}\phi_0}{2\pi w\xi(0)}\left(1-\frac{T_C(H)}{T_C}\right)^{1/2} \quad (1)$$

where $\phi_0 = 2\times10^{-7}$ Gcm^2 is the flux quantum [20].

The horizontal dashed lines indicate the first and second matching fields. As was previously discussed [21],[4], fluxoid quantization in SC wire networks gives rise to this oscillatory behavior and has been extensively studied in SC networks [12],[22], Josephson junction arrays [23], and perforated Al films [24]. These phenomena have been observed not only in SC films with an array of holes but also in films with a regular array of insulating Si dots [25], and even with magnetic dots [4],[26]. This can be understood in terms of the one-dimensional superconductivity at temperatures close to the critical temperature, when the coherence length becomes of the order or larger than the width w of the SC strips which surround the pinning centers.

In contrast, for sample 10Si/V the upper critical field, plotted in open squares in figure 3 (a), shows a linear temperature dependence and no T_C oscillations, characteristic of a continuous film with a coherence length smaller than any in-plane dimensions of the film. In this case, $\xi(0)$ was calculated using the bulk expression $H_{C2} = \phi_0/2\pi\xi^2(T)$ [20]. From the linear fit shown in dashed line in figure 3 (a), we obtain $\xi(0) = (7\pm1)$ nm.

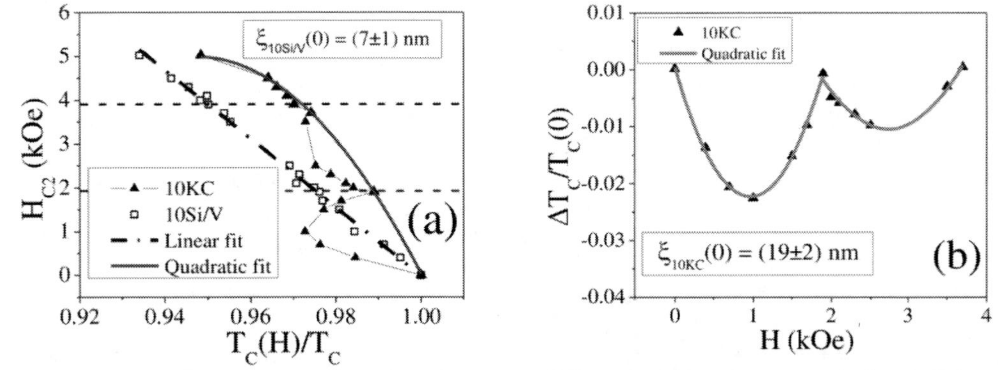

Figure 3: (a) H_{C2} *vs* $T_C(H)/T_C$ of the sample 10KC (full) and 10Si/V (open symbols). Solid and dash-dot lines are the quadratic and linear fittings for samples 10KC and 10Si/V respectively. Horizontal dashed lines indicate H_1 and H_2. (b) $\Delta T_C/T_C = (T_C-T_C(H))/T_C$ *vs* DC field for the sample 10KC and quadratic fittings of the first two oscillations after parabolic background was subtracted.

A phenomenological model [24] used to describe the SC phase diagram of wire networks determines the oscillations with a sequence of parabolas:

$$\frac{\Delta T_C}{T_C} = -\left(\frac{\xi(0)}{d}\right)^2\left[\frac{1}{4}-\left(\frac{\phi}{\phi_0}-n-\frac{1}{2}\right)^2\right] \qquad \Delta T_C = T_C - T_C(H) \qquad (2)$$

Here, n is the number of quantum flux per unit cell, $\phi = A_{u.c.}H$ is the flux per unit cell and ΔT_C is the difference between the zero field critical temperature T_C and the critical temperature in an applied field $T_C(H)$. In terms of the lattice parameter d, the area of the unit cell for a triangular array is $A_{u.c.} = (\sqrt{3}/2)d^2$. In figure 3 (b) we plot $\Delta T_C/T_C = (T_C-T_C(H))/T_C$ *vs* DC field, after subtracting the parabolic background from the data in figure 3 (a). The first and second oscillations correspond to one and two flux quanta per unit cell (n = 0 and n = 1, respectively) and from (2) we estimated the G-L coherence length $\xi(0) = (19\pm2)$ nm, which is significantly larger than the coherence length of the sample 10Si/V. The lattice parameter d = (150±40) nm, is in agreement with the expected value d = 110 nm. A parabolic fit of the data shown in figure 3 (a) to (1) together with the coherence length determined above, helps estimate the width of the SC strips, $w = (46\pm4)$ nm. This width is smaller than the lattice parameter d. This is consistent with our geometry landscape, in which the SC strips enclose regions over the normal V dots where the Nb thin film is in the normal state.

4. Conclusions

We explored the pinning mechanisms and CE using ac-susceptibility measurements, in thin Nb films containing a dense triangular array of V dots. We compared samples with similar corrugation, with and without an intermediate insulating SiO_2 layer. Stronger matching effects occur for the sample where proximity effects induce pinning (i.e. without the SiO_2 layer). Moreover, the phase diagram $H_{C2}(T)$ is strongly modified. Therefore, we conclude that the local depression of superconductivity in the Nb thin film by proximity effects with the V dots, is responsible for the observed T_C oscillations.

The research at UCSD was supported by the Office of Basic Energy Science, U.S. Department of Energy, BES-DMS funded by the Department of Energy's Office of Basic Energy Science, DMR under grant DE FG03 87ER-45332. Research at UBA was supported by ANPCyT grant PICT N° 753 and UBA grant UBACyT 661.

References

[1] Berdiyorov G R, Milošević M V and Peeters F M 2006 *Phys. Rev.* B **74** 174512.
[2] Silhanek A V, Raedts S, Van Bael M J and Moshchalkov V V 2004 *Eur. Phys. Jour.* B **37** 19.
[3] Milošević M V, Yampolskii S V and Peeters F M 2002 *Phys. Rev.* B **66** 174519.
[4] Gomez A, del Valle J, Gonzalez E M, Chiliotte C E, Carreira S J, Bekeris V, Prieto J L, Schuller Ivan K and Vicent J L 2014 *Supercond. Sci. Technol.* **27** 065017.
[5] Hoffmann A, Prieto P and Schuller Ivan K 2000 *Phys. Rev.* B **61** 6958.
[6] Vélez M, Martín J I, Villegas J E, Hoffmann A, González E M, Vicent J L and Schuller Ivan K 2008 *J. Magn. Magn. Mater.* **320** 2547.
[7] Patel U, Xiao Z L, Hua J, Xu T, Rosenmann D, Novosad V, Pearson J, Welp U, Kwok W K, and Crabtree G W 2007 *Phys. Rev.* B 76 020508(R).
[8] Chiliotte C, Pasquini G, Bekeris V, Villegas J E, Li C-P and Schuller Ivan K 2011 *Supercond. Sci. Technol.* **24** 065008.
[9] Lance Horng, Wu J C, Kang P C, Lin P H and Wu T C 2003 *Jpn. J. Appl. Phys.* **42** 2679.
[10] Zhang W J *et al.* 2012 *EPL* **99** 37006.
[11] Little W A, Parks R D 1962 *Phys. Rev. Lett.* **9** 9.
[12] Pannetier B, Chaussy J, Rammal R and Villegier J C 1984 *Phys. Rev. Lett.* **53** 1845.
[13] Behrooz A, Burns M J, Deckman H, Levine D, Whitehead B and Chaikin P M 1986 *Phys. Rev. Lett.* **57** 368.
[14] Li C-P, Roshchin I V, Batlle X, Viret M, Ott F and Schuller Ivan K 2006 *J. Appl. Phys.* **100** 074318.
[15] Sekula S T and Kernohan R H 1972 *Phys. Rev.* B **5** 904.
[16] De Blois R W and De Sorbo W 1964 *Phys. Rev. Lett.* **12** 499.
[17] Koren G and Millo O 2009 *Phys. Rev.* B **80** 054507.
[18] Carreira S J, Chiliotte C, Bekeris V, Rosen Y, Monton C and Schuller Ivan K 2014 *Supercond. Sci. Technol.* **27** 085007.
[19] Jaccard Y, Martín J I, Cyrille M -C, Vélez M, Vicent J L and Schuller Ivan K *Phys. Rev.* B **58** 8232.
[20] Tinkham M 1996 *Indroduction to Superconductivity* (New York: McGraw-Hill) 2nd. edition.
[21] Welp U, Xiao Z L, Jiang J S, Vlasko-Vlasov V K, Bader S D, Crabtree G W, Liang J, Chik H and Xu J M 2002 *Phys. Rev.* B 66 212507.
[22] Tinkham M *et al.* 1983 *Phys. Rev.* B **28** 6578; Wilks C W *et al.* 1991 *Phys. Rev.* B **43** 2721.
[23] Giovannella C and Tinkham M 1995 *Macroscopic Quantum Phenomena and Coherence in Superconducting Networks* (Singapore: World Scientific).
[24] Bezryadin A and Pannetier B 1995 *J. Low Temp. Phys.* **98** 251.
[25] Eisenmenger J, Oettinger M, Pfahler C, Plettl A, Walther P and Ziemann P 2007 *Phys. Rev.* B **75** 144514.
[26] Samokhvalov A V, Mel'nikov A S, Ader J-P and Buzdin A I 2009 *Phys. Rev.* B **79** 174502.

Magnetic irreversibility and zero resistance in granular Y358 superconductor

F T Dias[1], C P Oliveira[1], V N Vieira[1], D L Silva[1], F Mesquita[2], M L Almeida[2], J Schaf[2] and P Pureur[2]

[1]Instituto de Física e Matemática, Universidade Federal de Pelotas, Caixa Postal 354, 96010-900 Pelotas, RS, Brazil
[2]Instituto de Física, Universidade Federal do Rio Grande do Sul, Caixa Postal 15051, 91501-970 Porto Alegre, RS, Brazil

E-mail: fabio.dias@ufpel.edu.br

Abstract. We report on magnetization and magnetoresistance measurements of polycrystalline $Y_3Ba_5Cu_8O_{18}$ superconductor (Y358), prepared by solid state reaction in order to study the correlation between the magnetic irreversibility line and the zero electric resistance as a function of applied field. The magnetization measurements were performed using a MPMS-XL SQUID magnetometer and the magnetoresistance measurements were made using a PPMS, both from Quantum Design, up to 1 T. The granular microstructure was confirmed by scanning electron microscopy. In this case the grain junctions are weaker than in the Y123 system and the effect of the applied field on the grain couplings is much stronger. Consequently, in our sample the zero resistance line shifts away from the irreversibility line and is lower by more than 15 K at 1 T. We explain our results in terms of the superconducting glass model and in comparison with results obtained in other materials from the YBaCuO family.

1. Introduction

The discovery of the new member of the YBaCuO-family, the $Y_3Ba_5Cu_8O_{18}$ superconducting system [1], or simply Y358, has allowed the access to new and important results about the phenomenology of the superconducting cuprates, however, there are some differences between the family members. While the $YBa_2Cu_3O_{7-\delta}$ superconducting (Y123), one of the most studied superconducting systems, has two CuO_2 planes and one CuO chain, the $Y_3Ba_5Cu_8O_{18}$ system has five CuO_2 planes and three CuO chains. This difference reflects in some important superconducting characteristics, such as the critical temperature (T_C), for example.

The Y358 has been prepared under different growth techniques, such as conventional sintering [1,2,3,4,5,6], sol-gel [7,8], biopolymer-mediated [9] and melt-texturing [10], resulting in different specimens from polycrystalline samples to superconducting nanowires [9]. Different compounds has been derived from the Y358 system, such as $Yb_{1.8}Sm_{1.2}Ba_5Cu_8O_{18}$ [11], $RE_3Ba_5Cu_8O_{18}$ (RE = Sm and Nd) [12], $Y_3Ba_5Ca_2Cu_8O_{18}$ [13], $Y_3Ba_5Cu_8O_{y-x}F_x$ [14,15] and $Y_3Ba_5Cu_{8-x}Zn_xO_{18-\delta}$ [16]. Several studies between YBaCuO-family members have been carried out in order to compare structural, magnetic and electric properties [2,5,8]. On the other hand, there are many and important aspects and fundamental properties to be investigated in the Y358 system.

In this work we have performed magnetization and magnetoresistance measurements up to 1 T in a polycrystalline $Y_3Ba_5Cu_8O_{18}$ sample to study the magnetic irreversibility line and the zero resistance state, as well as the correlation between both behaviors. Scanning electron microscopy (SEM) and X-ray diffraction (XRD) analysis confirmed the granular microstructure and the results are explained in terms of the known models and in comparison with the Y123 system.

2. Sample preparation and characterization

Three samples of $Y_3Ba_5Cu_8O_{18}$ (Y358) were grown by solid-state reaction, starting from high purity powders (Sigma-Aldrich, 99.999%) of Y_2O_3, $BaCO_3$ and CuO. The powders were mixed in the appropriated stoichiometry ratios and reacted in air at 930 ºC for 12 h. Calcinations were repeated three times. Subsequently the powders were pressed into pellets of 10 mm in diameter and 2.5 mm of thickness under pressure. Finally the samples were sintered at 930 ºC during 24 h and annealed under oxygen flow. Previous magnetic and electric (four probe resistivity) measurements were used to verify the superconducting state of our samples. Due to the similarity between the results obtained we adopted one sample as representative of our study. The sample was cut into three pieces in order to perform the characterization and the electric and magnetic measurements.

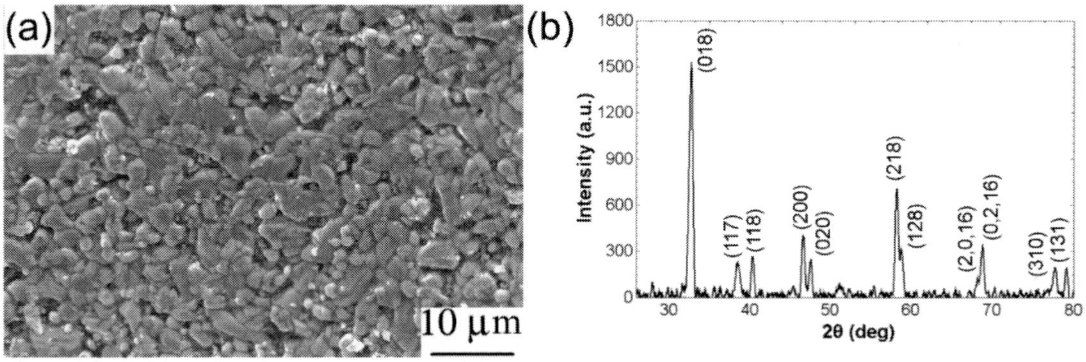

Figure 1. (a) SEM image showing the polycrystalline array with micrometric grains and (b) XRD pattern of the Y358 sample.

The characterization was made using scanning electron microscopy (JEOL JSM-6390LV microscope) and X-ray diffraction (PANalytical XPert Pro MPD diffractometer with CuKα radiation) techniques. Figure 1 shows scanning electron microscopy (SEM) image (figure a) and an XRD pattern (figure b) representative of the sample. The figure 1(a) clearly shows the polycrystalline characteristic, typical of sintered ceramic samples, with disordered grains and having an average size of few microns. The figure 1(b) shows an XRD pattern of our Y358 sample. The peaks in the diffractogram, as well as the lattice parameters obtained ($a = 3.8878$ Å, $b = 3.8185$ Å and $c = 31.116$ Å), are in agreement with previous reports [1,4], and no significant impurity was found in our sample.

3. Results and discussion

Magnetization and magnetoresistance experiments as a function of temperature were performed in the Y358 sample employing magnetic fields up to 1 T. The magnetization experiments were performed in a SQUID magnetometer (SQUID MPMS-XL from Quantum Design) and consisted of the standard method for measuring the magnetic irreversibility, based on zero-field-cooled (ZFC) and field-cooled (FC) procedures [17]. The method consisted in initially cooling down our sample below the critical temperature (T_C) in zero magnetic field (ZFC). After the magnetic field was applied and the magnetization data were collected under constant field while the temperature was increasing up to values above T_C. Thereafter the FC magnetization was measured while cooling the sample back to temperature values well below T_C in the same magnetic field. The figure 2(a) shows ZFC and FC magnetization data for a magnetic field of 0.1 T. In the figure 2(a) the temperature of the bifurcation between both curves indicates the irreversibility limit T_{IRR} (H) in the given field.

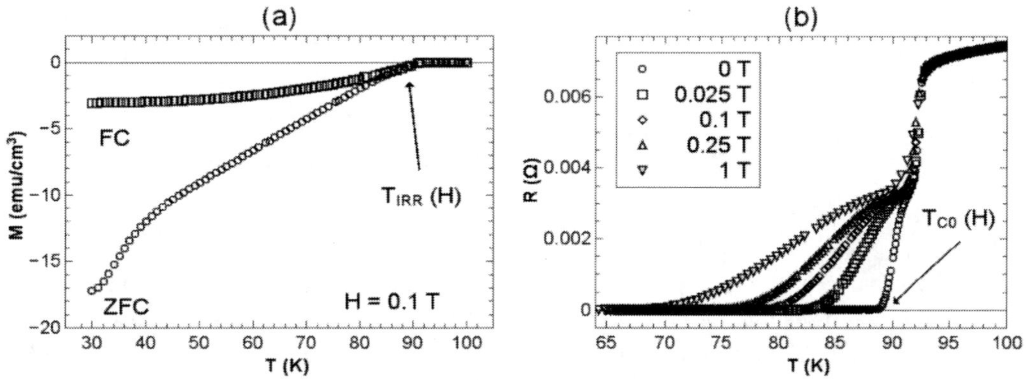

Figure 2. (a) ZFC and FC magnetization curves for H = 0.1 T showing the irreversibility limit T_{IRR} (H). (b) Magnetoresistance data measured in H = 0, 0.025, 0.1, 0.25 and 1 T. The arrow indicates the zero resistance temperature T_{C0} (H) for zero field.

The magnetoresistance experiments were made using the four-contact technique with a PPMS from Quantum Design, under magnetic fields up to 1 T and a constant measuring current of 1 mA. The figure 2(b) shows magnetoresistance results as a function of temperature for some of the magnetic fields employed. It may be observed that the magnetoresistance of our sample is strongly sensitive to the applied field. This behavior is typical of ceramic samples where superconducting granularity effects are dominant. The zero resistance temperature T_{C0} (H) was evaluated adopting a practical criterion, according to which the zero resistance temperature is obtained from the threshold of the plateau where the electric resistance falls to zero, as indicating by the arrow in the figure 2(b) for zero field.

We have determined the magnetic irreversibility limit, and the irreversibility line of our sample, for magnetic fields from 0.001 to 1 T, as shown in the figure 3(a). The continuous lines through the experimental data are fittings with Almeida-Thouless-like (AT) and Gabay-Toulouse-like (GT) [see expressions in Fig. 3a] power laws. The equation and the fitting parameters are displayed in the figure. These power laws have been largely employed to describe the irreversibility line for granular superconductors [17,18,19,20]. In the case of our sample, the results shown in figure 3(a) resemble the results obtained for pure and doped single-crystals and polycrystalline samples of $YBa_2Cu_3O_{7-\delta}$ (Y123) superconductor [17,19]. On the other hand the results shown in the figure 3(a) reveal a large regime from 0.1 T to 1 T governed by a Gabay-Toulouse power law. In the $YBa_2Cu_3O_{7-\delta}$ superconductor this power law is generally observed until magnetic fields about 0.4 T or less [17], however, similar results were found for the polycrystalline $DyBa_2Cu_3O_{7-\delta}$ system [20], where the GT power law could also be observed up to 1 T.

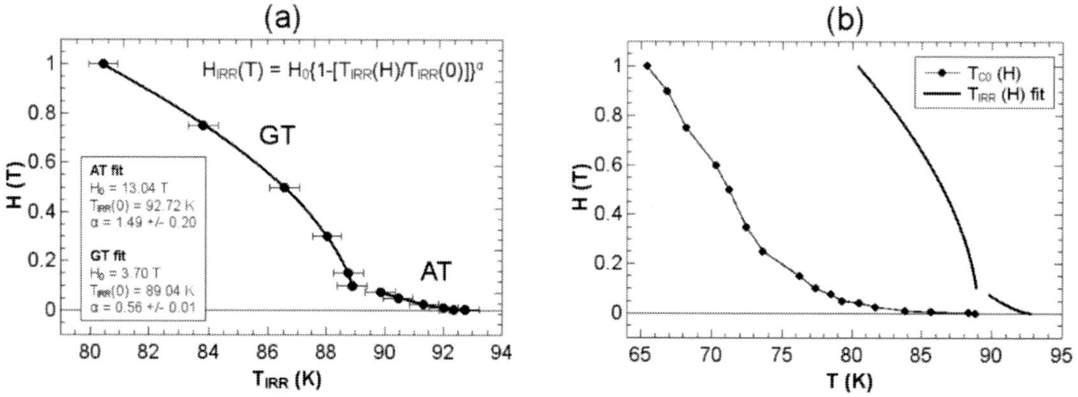

Figure 3. (a) Magnetic irreversibility data. The continuous lines are fittings with Almeida-Thouless-like (AT) and Gabay-Toulouse-like (GT) power laws. The equation and the fitting parameters are shown in the figure. (b) Zero resistance line (left) obtained from the T_{C0} (H) data together with the irreversibility line T_{IRR} (H) represented by the AT and GT fittings (right).

The figure 3(b) shows the correlation between the zero resistance line obtained from T_{C0} (H) data and the irreversibility line obtained from the T_{IRR} (H) data. In the figure the irreversibility line T_{IRR} (H) is represented by the AT and GT fittings.

The high-T_C cuprates, like the Y358 superconductor, normally are granular superconducting systems, and in this case the beginning of the magnetic irreversibility takes place when the first cluster of coupled grains is established in the sample. Consequently, the clusters of superconducting grains can retain the magnetic flux (Josephson-type), leading to magnetic irreversibility. However, the electric transport along these clusters will still be resistive because the clusters are disconnect from each other. The electric resistance falls to zero only when long-range coherence of the superconducting order parameter takes place. Therefore, in granular systems the electric resistance will be nonzero even in temperatures below the irreversibility line, vanishing only when percolation occurs and phase coherence establishes through the whole sample. The results shown in figure 3(b) clearly corroborate the aspects described above. Similar results have been found for granular Y123 samples [17], revealing the similarities between Y123 and Y358 superconducting systems.

In summary, our data show that the profiles of the magnetic irreversibility and zero resistance lines as a function of applied field of the granular Y358 superconductor, the new member of the YBaCuO-family, are similar to those of the Y123 system. However, the fact that the Gabay-Toulouse like power-law of the magnetic irreversibility limit is observed up to much higher fields than in the Y123 system apparently constitutes a significant difference between both superconducting materials. The electric resistivity shows a strong sensitivity to applied magnetic fields and is related with the considerably weaker connections between the superconducting grains in the Y358 system.

References

[1] Aliabadi A, Akhavan Farshchi Y and Akhavan M 2009 *Physica C* **469** 2012
[2] Udomsamuthirun P, Kruaehong T, Nilkamjon T and Ratreng S 2010 *J. Supercond. Nov. Magn.* **23** 1377
[3] Esmaeili A, Sedghi H, Amniat-Talab M and Talebian M 2011 *Eur. Phys. J. B* **79** 443
[4] Akduran N 2012 *J. Low Temp. Phys.* **168** 323
[5] Sujinnapram S, Udomsamuthirun P, Kruaehong T, Nilkamjon T and Ratreng S 2011 *Bull. Mater. Sci.* **34** 1053
[6] Landínez Téllez D A, Báez M C and Roa-Rojas J 2012 *Mod. Phys. Lett. B* **26** 1250067
[7] Ayas A O, Ekicibil A, Çetin S K, Coskun A, Er A O, Ufuktepe Y, Fırat T and Kıymaç K 2011 *J. Supercond. Nov. Magn.* **24** 2243
[8] Gholipour S, Daadmehr V, Rezakhani A T, Khosroabadi H, Shahbaz Tehrani F, Hosseini Akbarnejad R 2012 *J. Supercond. Nov. Magn.* **25** 2253
[9] Konne J, Davis S A, Glatzel S, Lees M R and Hall S R 2012 *Supercond. Sci. Technol.* **25** 115005
[10] Bollat S and Kutuk S 2012 *J. Supercond. Nov. Magn.* **25** 731
[11] Martinez Buitrago D, Reyes-Suarez N C, Peña J P, Ortiz-Diaz O, Otálora J and Parra Vargas C A 2013 *J. Supercond. Nov. Magn.* **26** 2269
[12] Topal U and Akdogan M 2012 J. Supercond. Nov. Magn. **25** 239
[13] Ekicibil A, Cetin S K, Ayas A O, Coskun A, Fırat T and Kıymac K 2011 *Solid State Sciences* **13** 1954
[14] Srinivasan K, Thomas G, Padaikathan P and Ashoka N V 2013 *Int. J. Eng. Res. App.* **3** 927
[15] Srinivasan K, Thomas G and Padaikathan P 2011 *J Minerals Mat. Charac. Eng.* **10** 1277
[16] Esmaeili A, Sedghi H, Golzan M M and Amniat-Talab M 2011 *J. Supercond. Nov. Magn.* **24** 2237
[17] Dias F T, Vieira V N, Rodrigues Jr. P, Obradors X, Pureur P and Schaf J 2008 *Phys. Rev. B* **77** 134503
[18] Vieira V N, da Silva J P and Schaf J 2001 *Phys. Rev. B* **64** 094516
[19] Vieira V N, Pureur P and Schaf J 2002 *Phys. Rev. B* **66** 224506
[20] Fabris F W, Roa-Rojas J and Pureur P 2001 *Physica C* **354** 304

Journal of Physics: Conference Series **568** (2014) 022010 doi:10.1088/1742-6596/568/2/022010

Detection of discretized single-shell penetration in mesoscopic vortex matter

M. I. Dolz [1], Y. Fasano [2], N. R. Cejas Bolecek [2], H. Pastoriza [2], M. Konczykowski [3], and C. J. van der Beek [3]

[1] Departamento de Física, Universidad Nacional de San Luis, CONICET, Argentina
[2] Laboratorio de Bajas Temperaturas, Centro Atómico Bariloche, CNEA, Bariloche, Argentina
[3] Laboratoire des Solides Irradiés, Ecole Polytechnique, CNRS URA-1380, Palaiseau, France

E-mail: [1] `mdolz@unsl.edu.ar`

Abstract.
We investigated configurational changes in mesoscopic vortex matter with less than thousand vortices during flux penetration in freestanding $50\,\mu$m diameter disks of $Bi_2Sr_2CaCu_2O_{8+\delta}$. High-resolution AC and DC local magnetometry data reveal oscillations in the transmittivity echoed in peaks in the third-harmonics magnetic signal fainting on increasing vortex density. By means of extra experimental evidence and a simple geometrical analysis we show that these features fingerprint the discretized entrance of single-shells of vortices having a shape that mimics the sample edge.

1. Introduction

The study of vortex configurations, flux penetration, multi-vortex and giant-vortex states stabilized by confinement have attracted much attention during the last decade [1, 2, 3, 4, 5, 6]. The vortex arrangement in superconducting disks with sizes comparable or smaller than coherence length or penetration depth is quite different to that observed in macroscopic samples with weak [7, 8] or strong [9, 10] pinning. For instance, in disks thinner than coherence length the confinement effect overwhelms the inter-vortex interaction and vortices penetrate in ring-shaped shells mimicking the sample edge [3]. However, inter-vortex interaction strongly depends on the anisotropy of the material and the resulting mesoscopic vortex configuration can significantly change with the degree of layerness of the material.

The case of mesoscopic vortex matter nucleated in the extremely-layered high-T_c $Bi_2Sr_2CaCu_2O_{8+\delta}$ studied here is rather interesting since the system presents a rich phase diagram. This is due to the dominant effect of thermal fluctuations and extreme anisotropy in samples with weak disorder. The phase diagram of macroscopic as well as mesoscopic [11] $Bi_2Sr_2CaCu_2O_{8+\delta}$ vortex matter is dominated by a first-order transition [12, 13] at T_{FOT} between a solid phase at low temperatures and a liquid [14] or decoupled gas [15, 16] of pancake vortices at high temperatures. Irrespective of the sample size, vortex penetration in $Bi_2Sr_2CaCu_2O_{8+\delta}$ is dominated by Bean-Livingston surface barriers at high temperatures and bulk pinning effects at low temperatures [17, 18]. The temperature and measuring-time ranges in which every effect dominates is determined by the thickness-to-width ratio of the samples [17, 18, 19]. This results in the solid vortex phase presenting irreversible magnetic behavior whereas the high-temperature

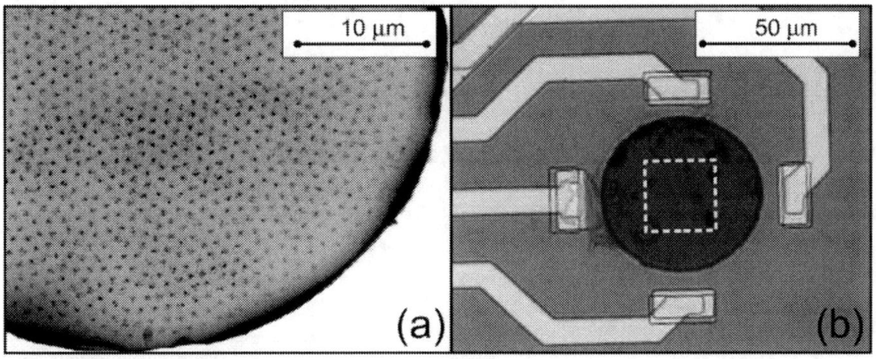

Figure 1. (a) Magnetic decoration image [24] of the vortex matter nucleated in a field-cooling process at 20 Oe in a 50 μm diameter $Bi_2Sr_2CaCu_2O_{8+\delta}$ disk. (b) Typical disk located on top of the $16 \times 16\,\mu m^2$ Hall sensor used to measure (depicted with a dotted-white frame in real size).

phase is magnetically reversible. The transition between both phases is detected as a jump in the DC local induction [12] and as a frequency-independent peak in the AC transmittivity [20, 21].

In this work we study mesoscopic vortex matter penetrating into micron-sized $Bi_2Sr_2CaCu_2O_{8+\delta}$ disks by means of local DC and AC micro-Hall-magnetometry measurements with low-noise level. We detect features in the AC transmittivity and third-harmonic signal at certain fields and ascribe them to the abrupt entrance of single vortex-shells on increasing field.

2. Experimental

We studied disks fabricated from macroscopic optimally-doped $Bi_2Sr_2CaCu_2O_{8+\delta}$ ($T_c = 90$ K) single-crystals grown by means of the traveling-solvent floating zone technique [22]. Micron-sized disks of roughly 1 μm thickness and diameter $d = 50\,\mu$m as the one shown in Fig. 1 (a) were obtained by combining optical lithography and physical ion-milling techniques [23]. The disks were mounted with micron-precision manipulators onto $2D$-electron-gas Hall-sensors with active areas of $16 \times 16\,\mu m^2$ and glued with Apiezon N grease, see Fig. 1 (b).

The magnetic properties at individual-vortex scale nucleated in field-cooling were studied by means of magnetic decoration imaging [25]. Local magnetization was measured applying DC, H, and AC , H_{ac}, fields parallel to the c-axis. We simultaneously measure the first and third harmonics of the AC induction by means of a digital-signal-processing lock-in technique [11]. We obtained the transmittivity, T', by normalizing the in-phase component of the first harmonic signal [26]. This is a magnitude extremely sensitive to discontinuities in the local induction. The modulus of the third harmonic signal is similarly converted to the magnitude $\mid T_{h3} \mid$ that has a non-negligible value for non-linear magnetic response at $H < H_{irr}$, the irreversibility field [26].

3. Results and discussion

Figure 1 (a) shows a zoom-in of a magnetic decoration image revealing the vortex arrangement in the disks for a field-cooling experiment at 20 Oe. At the edge of the sample vortices arrange in a circular-shaped shell mimicking the disk geometry. This structure is the result of the nucleation of the vortex ensemble in the high-temperature vortex liquid and the subsequent freezing of the structure at the scale of lattice spacing at the temperature at which bulk pinning sets in [7], namely $T_{irr}(20\,Oe) = 85.3\,K \sim T_{FOT}$ for the 50 μm disks [11].

For the flux-penetration at fixed temperature experiments we discuss from now on, vortices enter into the sample on increasing field above the first-penetration field, $B_p(T)$. In this experiment, a snapshot of flux-penetration would reveal a vortex arrangement certainly different

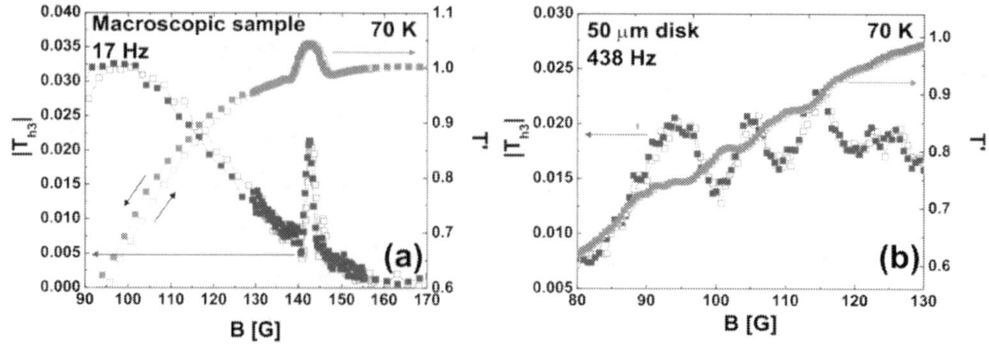

Figure 2. Field-evolution of the transmittivity T' (red points) and third-harmonic signal $| T_{h3} |$ (blue points) at 70 K. Results for (a) the macroscopic sample from which were fabricated the (b) 50 μm diameter disks of $Bi_2Sr_2CaCu_2O_{8+\delta}$. The ripple H_{ac} fields have ~ 1 Oe rms in magnitude and frequencies as indicated. Peaks in $| T_{h3} |$ echo the oscillations in T'. The open (full) points correspond to the warming (cooling) branch as indicated by the black arrows.

to the one shown in Fig. 1 (a), with an induction gradient towards the sample center. However, Fig. 1 (a) suggest that in flux-penetration experiments vortices will enter into the sample in circular-shaped shells. The entrance of these shells can be continuous or discretized, namely vortices gradually enter one by one, or a circle of vortices suddenly jump into the sample.

In macroscopic $Bi_2Sr_2CaCu_2O_{8+\delta}$ samples, the first-order phase transition is detected as a sharp peak in AC transmittivity [20, 21], a magnitude sensitive to discontinuities in the magnetic induction (proportional to the first derivative of the signal). Figure 2 (a) shows that for the macroscopic sample from which were fabricated the disks this peak develops at $H_{FOT} = 143$ G at 70 K. Peaks or oscillations in T' are echoed as peaks in $| T_{h3} |$, see Fig. 2 (a), a magnitude proportional to the third derivative of the signal. On increasing field above $H_{irr} \sim 160$ G (for 17 Hz), $| T_{h3} |$ is below the experimental resolution indicating a linear vortex magnetic response.

When vortices penetrate micron-sized disks the first-order phase transition do persist at the same H_{FOT} than in macroscopic samples [11]. In addition, if the noise level and field-resolution of the measurements are improved, additional features are detected in the T' and $| T_{h3} |$ curves for fields smaller than H_{FOT}. On increasing field at 70 K vortices penetrate and oscillations in T' are observed in the range of 80 to 130 G, see Fig. 2 (b). This contrasts with the flat T' curve measured in this field-range for the macroscopic $Bi_2Sr_2CaCu_2O_{8+\delta}$ sample. These features observed in the micron-sized disks are due to discontinuities in the local induction when vortices are entering (full points) or exiting (open points) from the disks. The amplitude of the oscillations in T' for the disks are two-to-four times smaller than the peaks measured in the macroscopic sample at H_{FOT}.

Figure 2 (b) shows that the oscillations in T' are echoed as peaks in $| T_{h3} |$. We were able to detect these peaks in the temperature-range between 70 to 90 K. For the 50 μm disks at 70 K, the $| T_{h3} |$ peaks are detected between 80 and 130 G, for a wide range of frequencies, see the examples of Fig. 3 (a). We identify the particular fields at which the peaks develop with B_B, with B the field-location of the maximum. For instance, Fig. 3(a) depicts peaks B_{90} to B_{124}. The fields at which $| T_{h3} |$ becomes negligible on increasing vortex density rises with frequency, namely H_{irr} moves upward with frequency [11].

On sweeping up field, flux-penetration in the 50 μm disks starts at $B_p = 10$ G with the first single-shell of vortices nucleating at the sample edge. The B_p field is obtained from DC local magnetization loops as indicated in the insert to Fig. 3 (b). Further increasing the field might

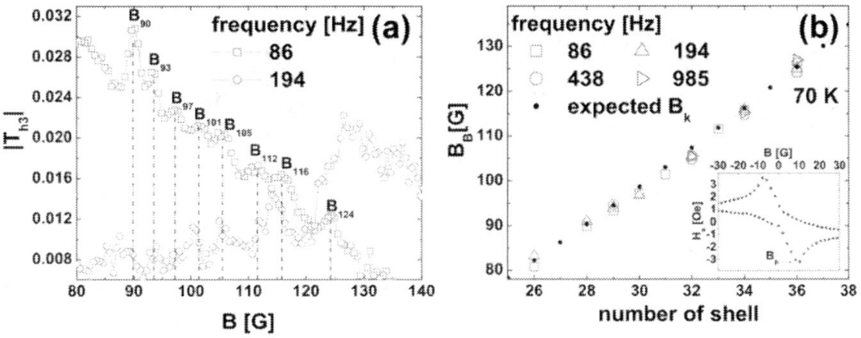

Figure 3. Third-harmonics signal of vortex matter penetrating $50\,\mu$m $Bi_2Sr_2CaCu_2O_{8+\delta}$ disks at 70 K. (a) Field-evolution of $\mid T_{h3} \mid$ for h_{ac} with different frequencies and a magnitude of 1.5 Oe rms (warming branches). The fields at which the peaks develop are indicated as $B_{\rm B}$. (b) Field-location of $\mid T_{h3} \mid$ peaks (open color points) and of the expected single-shell discretized penetration fields (full black points and dotted line), as a function of the single-shell number. Insert: Local DC magnetization loop at 70 K with the first-penetration field indicated.

entail either a continuous entrance of vortices, or a discretized process in which vortices do not enter until the increment in H is such that its vorticity allows the formation of a new single-shell of vortices. In the latter case, the new k-th single-shell of vortices jumps into the sample for a given $B_{\rm k}$ field. We propose that the peaks observed in $\mid T_{h3} \mid$ concomitant with the oscillations in T' are the fingerprint of this discretized entrance of vortices on increasing H, namely that $B_{\rm B} \sim B_{\rm k}$ for given k and B. If vortices would penetrate the sample individually in a continuous way, no such sharp features in T' and $\mid T_{h3} \mid$ are expected.

We can also consider a very simple geometrical analysis in order to relate the observed $B_{\rm B}$ features with the entrance of a k-th shell. By considering the measured $B_{\rm p}$, we estimate the number of vortices suddenly penetrating in the first single-shell ($k = 1$) as $n_1 = \pi B_{\rm p} d^2/4\Phi_0$, with Φ_0 the flux quantum. We then calculate the required increment in field in order to produce the jumping-in of the second single-shell ($k = 2$), ΔB_2. In order to do so we consider that $n_2 = (B_{\rm p} + \Delta B_2)(\pi d^2)/4\Phi_0$ and $n_2 = n_1 + \Delta n$ with the number of vortices in the second single-shell $\Delta n = (\pi d)/a_2$ with a_2 the vortex spacing. We proceed in the same fashion for the following single-shells, and obtained the B_k curve shown in Fig. 3 (b) as a function of the single-shells ordinal number k. The vortex spacing is considered in a very rough approximation as $a_k = 1.075\sqrt{\Phi_0/B_{k-1}}$ with the special case of $a_2 = 1.075\sqrt{\Phi_0/B_p}$. This is a very simple analysis but since we do not have direct imaging evidence on the penetration of vortices within a single-shell, we can not suggest a more sophisticated model.

We then identified all the observed $B_{\rm B}$ peaks from $\mid T_{h3} \mid$ measurements at several frequencies in the range of 80 to 130 G. For every $B_{\rm B}$ we then find the closest magnetic-field value for which is expected the entrance of a single-shell of vortices, $B_{\rm k}$, and construct the curves shown in Fig. 3 (b). Namely, we assume that the peaks in $\mid T_{h3} \mid$ are located at the particular field-values $B_{\rm k}$ such that its vorticity allows a single-shell of vortices with ordinal number k penetrating into the sample. In spite of the simplicity of this analysis, the agreement between the experimental points and the values expected from this geometrical model is quite remarkable.

This interpretation could be challenged by arguing that the oscillations in T' come from inhomogeneities in the sample giving different local values of $H_{\rm FOT}$ at a fixed measurement temperature. This seems not to be the case since these features in T' have a different shape than the paramagnetic peaks associated to the transition at $H_{\rm FOT}$, and they are observed in a different field-range than this transition. Nevertheless, we did an extra experimental check in order to verify our interpretation. In the case of the T' and concomitant $\mid T_{h3} \mid$ peaks associated to the first-order transition in macroscopic samples, applying an in-plane field produces a shifting of the peaks towards smaller fields [21]. When we applied such a field in our 50 μm disks we found that the field-location of the $B_{\rm B}$ peaks in $\mid T_{h3} \mid$ does not change appreciably. Therefore with these experiments we strengthen our interpretation of the peaks in $\mid T_{h3} \mid$ being the fingerprint of the discretized entrance of single-shells of vortices. Vortex imaging experiments on increasing field in films with strong periodic pinning potentials show that the entrance of a vortex shell is not performed abruptly but in steps following the terrace critical state [27]. However, in our case the samples do not have a periodic pinning potential and such a penetration state is not expected. Nevertheless, whether the penetration is performed with all vortices of a single-shell entering suddenly at the same field, or slightly gradually in the small range of fields the $\mid T_{h3} \mid$ peaks widen, can not be ascertained from our experiments.

Finally, we find that on further increasing the field above 130 Oe the oscillations in T' as well as the peaks in the third-harmonic signal progressively faint. This suggest that in this field-range the field-steps required to allow the discretized entrance of single-shell of vortices become smaller than the field-resolution of our experiments of roughly 0.1 G. Therefore this phenomenology is resolution-limited and we can not ascertain whether there is a discrete-to-continuum crossover of flux penetration in micron-sized vortex matter.

4. Conclusions

Low-noise level and high-resolution local magnetic measurements, as well as single-vortex imaging, allowed us to detect the entrance of quite-likely ring-shaped single-shells of vortices penetrating micron-sized samples. Therefore for the mesoscopic vortex matter nucleated in the extremely-layered $Bi_2Sr_2CaCu_2O_{8+\delta}$ superconductor, flux-penetration at low fields follows a discretized process in which the confinement plays a relevant role on imposing the geometry of the sample for vortex entrance. This effect is no longer detected for a system with more than 13.000 vortices indicating inter-vortex interaction becomes more relevant.

References

[1] V. V. Moshchalkov, L. Gielen, C. Strunk, R. Jonckheere, X. Qiu *et al.*, Nature **373**, 319 (1995).
[2] A. K. Geim, I. V. Grigorieva, S. V. Dubonos, J. G. S. Lok, J. C. Maan *et al.*, Nature **390**, 256 (1997).
[3] V. A. Schweigert, F. M. Peeters, and P. S. Deo, Phys. Rev. Lett. **81**, 2783 (1998).
[4] J. J. Palacios, Phys. Rev. B **58**, R5948 (1998).
[5] V. Bruyndoncx, L. Van Look, M. Verschuere, and V. V. Moshchalkov, Phys. Rev. B **60**, 10468 (1999).
[6] L. R. E. Cabral, B. J. Baelus, and F. M. Peeters, Phys. Rev. B **70**, 144523 (2004).
[7] Y. Fasano, M. De Seta, M. Menghini, H. Pastoriza and F. de la Cruz, Proc. Nat. Acad. Sci. **102**, 3898 (2005).
[8] A. P. Petrovic, Y. Fasano, R. Lortz, C. Senatore, A. Demuer *et al.*, Phys. Rev. Lett. **103**, 257001 (2009).
[9] M. Menghini, Y. Fasano, F. de la Cruz, S. S. Banerjee *et al.*, Phys. Rev. Lett. **90**, 147001 (2003).
[10] S. Demirdis, C.J. van der Beek, Y. Fasano, N. R. Cejas Bolecek *et al.*, Phys. Rev. B **84**, 094517 (2011).
[11] M. Konczykowski, Y. Fasano, M. I. Dolz, H. Pastoriza, V. Mosser, and M. Li, arXiv:1212.4564 (2012).
[12] H. Pastoriza, M. F. Goffman, A. Arribere, and F. de la Cruz, Phys. Rev. Lett. **72**, 2951 (1994).
[13] E. Zeldov, D. Majer, M. Konczykowski, V. B. Geshkenbein, V. M. Vinokur *et al.*, Nature **375**, 373 (1995).
[14] D. R. Nelson, Phys. Rev. Lett. **60**, 1973 (1988).
[15] L. I. Glazman, and A. E. Koshelev, Phys. Rev. B **43**, 2835 (1991).
[16] H. Pastoriza and P. H. Kes, Phys. Rev. Lett. **75**, 3525 (1995).
[17] N. Morozov, E. Zeldov, M. Konczykowski, and R.A. Doyle, Physica C **291**, 113 (1997).
[18] Y. M. Wang, A. Zettl, S. Ooi and T. Tamegai Phys. Rev. B, **65**, 184506(2002).
[19] N. Chikumoto, M. Konczykowski, N. Motohira, and A. P. Malozemoff, Phys. Rev . Lett. **69** 1260 (1992).
[20] N. Morozov, E. Zeldov, D. Majer and M. Konczykowski, Phys. Rev. B **54**, R3784 (1996).
[21] M. Konczykowski, C. J. van der Beek, A. E. Koshelev, V. Mosser *et al.*, Phys. Rev. Lett. **97**, 237005 (2006).
[22] T. W. Li, P.H. Kes, N.T. Hien, J. J. M. Franse and A. A. Menovsky, J. Cryst. Growth **135**, 481 (1994).
[23] M. I. Dolz, A. B. Kolton and H. Pastoriza, Phy. Rev. B **81**, 092502 (2010).
[24] Y. Fasano, M. De Seta, M. Menghini, H. Pastoriza, and F. De la Cruz, Solid State Commun. **128**, 51 (2003).

[25] Y. Fasano, J. Herbsommer, and F. de la Cruz, Phys. Stat. Sol. (b) **215**, 563 (1999).
[26] J. Gilchrist, and M. Konczykowski, Phys. C **212**, 43 (1993).
[27] A. V. Silhanek *et al.* Phys. Rev. B **83**, 024509 (2011).

Journal of Physics: Conference Series **568** (2014) 022011 doi:10.1088/1742-6596/568/2/022011

Influence of thermal gradient in vortex states of mesoscopic superconductors

E C S Duarte[1], A Presotto[1], D Okimoto[1], E Sardella[2,3], R Zadorosny[1]

[1] Departamento de Física e Química, Univ Estadual Paulista - Unesp, Ilha Solteira, SP, Brazil, CEP 15385-000

[2] UNESP-Universidade Estadual Paulista, Faculdade de Ciências, Departamento de Física, CEP 17033-360, Bauru-SP, Brazil

[3] UNESP-Universidade Estadual Paulista, IPMet-Instituto de Pesquisas Meteorológicas, CEP 17048-699, Bauru-SP, Brazil

E-mail: elwis.gapira@gmail.com

Abstract.
In general, the studies of finite size effects in mesoscopic superconductors have been carried out in such a way that the temperature parameter is constant in the entire system. However, we could have situations where a real sample is near a heater source, as an example. In such situations, gradients of temperature are present. On the other hand, mesoscopic superconductors are interesting systems due to the fact that they present confinement effects which influence all the vortex dynamics. Thus, in this work we studied the influence of thermal gradients on the vortex dynamics in mesoscopic superconductors. For this purposes, we used the time dependent Ginzburg-Landau equations. The thermal gradients produce an asymmetric distribution of the currents around the system which, in turn, yield interesting vortex configurations and difficult the formation of giant vortices.

1. Introduction

The time dependent Ginzburg-Landau (TDGL) theory has been successfully used in the last decades to describe several aspects of mesoscopic superconducting systems. Such systems present small size of the order of penetration depth $\lambda(T)$ and/or the coherence length $\xi(T)$, where T is the temperature. By using the TDGL formalism, several works have demonstrated that the confinement effects induce the formation of multi-vortex (MV) and giant vortex (GV) states [1]-[4]. This depend of the temperature and the system geometry [5]-[11]. Also, the presence of surface defects and De Gennes boundary conditions modify the vortex dynamics of mesoscopic systems [12], in such way that differs from the dynamics presented by the macroscopic ones [13]-[14]. In this work, we study the behavior of the vortex dynamics in mesoscopic systems in the presence of linear thermal gradients. Then, we analyze their magnetic properties, the vortex configurations, and the possible formation of a GV state.

2. Theoretical formalism

The formalism which we used was firstly proposed by Schimid[15], and it is a generalization of the Ginzburg-Landau (GL) equations, where a temporal evolution of the order parameter ψ

and the potential vector $\mathbf{A}$ was inserted; $|\psi|^2$ is the density of superelectrons and the induction magnet field is given by $\mathbf{B} = \nabla \times \mathbf{A}$. In dimensionless units, the TDGL equations are given by

$$\frac{\partial \psi}{\partial t} + \imath\varphi\psi = (-i\boldsymbol{\nabla} - \mathbf{A})^2\psi + \psi(1 - T(x) - |\psi|^2) , \tag{1}$$

$$\beta\frac{\partial \mathbf{A}}{\partial t} + \boldsymbol{\nabla}\varphi = \mathbf{J_S} - \kappa^2\boldsymbol{\nabla} \times \boldsymbol{\nabla} \times \mathbf{A} , \tag{2}$$

where de superconductivity current density is given by

$$\mathbf{J_S} = \mathrm{Re}(\psi^*(-\imath\nabla - \mathbf{A})\psi) . \tag{3}$$

For the thermal gradient we assumed a linear dependence of T along the x axis which is given by

$$T(x) = T_l + \frac{T_r - T_l}{a}x , \tag{4}$$

where T_r and T_l are the temperatures at the right-hand (hottest) and left-hand (coldest) side of the superconducting square. All equations have been normalized. The distances are in units of the coherence length $\xi(0)$; the magnetic field in units of the bulk upper critical field $H_{c2}(0)$ and the temperature in units of the critical temperature T_c; $\kappa = \lambda(T)/\xi(T)$ is the GL parameter. The TDGL equations are gauge invariant under the transformations: $\psi' = \psi e^{i\chi}$, $\mathbf{A}' = \mathbf{A} + \nabla\chi$, $\varphi' = \varphi - \frac{\partial\chi}{\partial t}$. We have worked in the Coulomb gauge where $\varphi' = 0$ for all times and positions. We numerically solved such equations by using the link variable method [16]. This method is preferably used because it preserves the gauge invariance of the equations once they are discretized. The TDGL equations will be used only as a relaxation method to obtain the steady state, that is, we are solely interested in the equilibrium vortex configurations.

3. Results and Discussion

We carried out the numerical simulations considering square systems with lateral size of $12\xi(0)$ and GL parameter $\kappa = 5$, which is equivalent to a Pb-In alloy[17]. We also used the following simulation parameters: $\beta = 1$; the size of the mesh was taken as $\Delta x = \Delta y = 0.125\xi(0)$ and the steps of the external magnetic field $\Delta H = 10^{-3}$. All the dynamics were analyzed under a thermal gradient which changes the value of ψ along the x axis of the system. In this work, we investigated the dynamics of two systems under two different thermal gradients, i.e, one in which the temperature varies from $0.7T_c$ to $0.9T_c$, label by S_1 and another one from $0.8T_c$ to $0.85T_c$, labeled by S_2. We also compared the results with a system at a homogeneous temperature (HT for brief) distribution by taking $T = 0.85T_c$. In Fig. 1 we present the magnetization curve as a function of the applied magnetic field of the systems under investigation. The behavior of such curves are similar to those ones exhibited by the HT systems, i.e., each discontinuity is due to one or more vortex penetration. Note that, the field corresponding to the first penetration for the S_1 system is lower than for the S_2 one. This is due to the fact that S_1 has its right-hand side hotter than S_2 which, on its turn, has a similar behavior presented by the HT system. The field sufficient for the destruction of superconductivity is higher for S_1 than for S_2 because temperature in the colder region is lower for the former system. In Fig. 2 we illustrate the profile of the modulus the current density, J/J_0, where J_0 is the despairing current, along the x axis and for three different distances w from the bottom side of the system (see inset of Fig. 1). A comparison between S_1 and S_2 is showed in Fig. 2. It is worth noticing the asymmetric distribution of J in comparison with the inset of this same figure for the HT system. However, S_2 presents a less asymmetric profile when it is compared with S_1. Also, it can be observed that J presents a higher value in the colder region and a degradation of superconductivity in the hottest one. We notice that with both perturbations S_1 and S_2 the vortices nucleate inside the superconductor one by one. As expected, the penetration always occurs at the right-hand side of the system.

We also studied the influence of non-homogeneous distribution of temperature on the giant vortex state which has been found for the HT systems [18]. In Fig. 3 we show two vortex state for S_1, S_2 and HT. In the panel 3(a) it is shown the intensity of $|\psi|$ superimposed by the streamlines $\mathbf{J}$. In such figure is clearly seen two separate vortices which is confirmed by the phase of ψ in panel 3(b). The other hand, the equivalent panels 3(c) and (d) for S_2, and 3(e) and (f) for the HT system, we clearly see the formation of giant vortex state. As a consequence, we can say that a rather non-homogeneous temperature like S_2 is not sufficient to avoid the nucleation of such state.

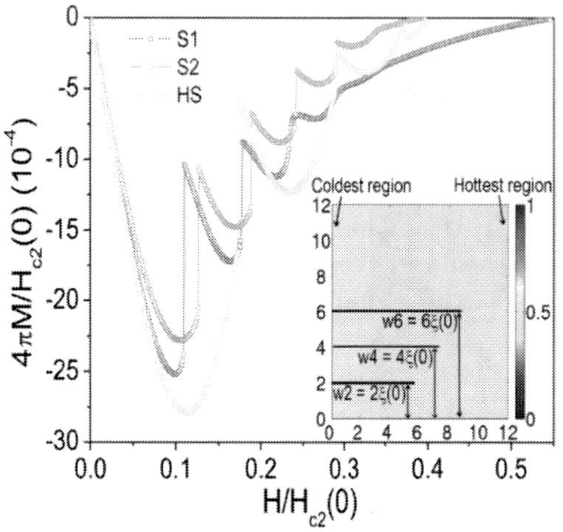

Figure 1. The magnetization curve as a function of the external applied field for the S_1, S_3 and HT systems. The inset shows the intensity of the order parameter in the initial state (without applied magnet field) under a thermal gradient.

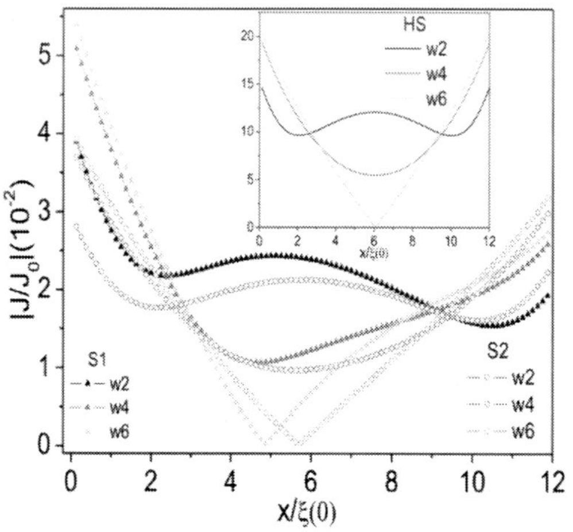

Figure 2. Profile of the modulus of the current along the x axis for three different levels of the square shown in the inset of Fig. 1 for S_1 and S_2. In the inset of the present figure it is shown the symmetric profile of the current for the HT system.

Fig. 4 illustrate the scenario associated with the third penetration. S_1 and S_2 present the same penetration dynamics with three nucleated vortices in the stationary state. However, the HT system presents four vortex state just after the third penetration (see panels 4(e) and (f)). This indicates that the thermal gradient breaks the fourfold symmetry of the multi-vortex state in those systems. On the other hand, it is interesting to note that giant vortex is created in S_1, which means that a greater variation of the temperature induces a non-conventional vortex interaction.

4. Conclusion

In this work we studied two systems with different thermal gradients by using the time dependent Ginzburg-Landau equations. It was shown that a variation of the temperature induces an asymmetric distribution of $\mathbf{J}$ along the system which is responsible for non-conventional behavior of the vortex dynamics. Depending on the conditions, such the range of variation of T and the number of penetrated vortices, giant vortex states can be avoided. Another interesting behavior is the break of the fourfold symmetric of the third penetration presented by the systems with thermal gradients in comparison with a system in a uniform temperature.

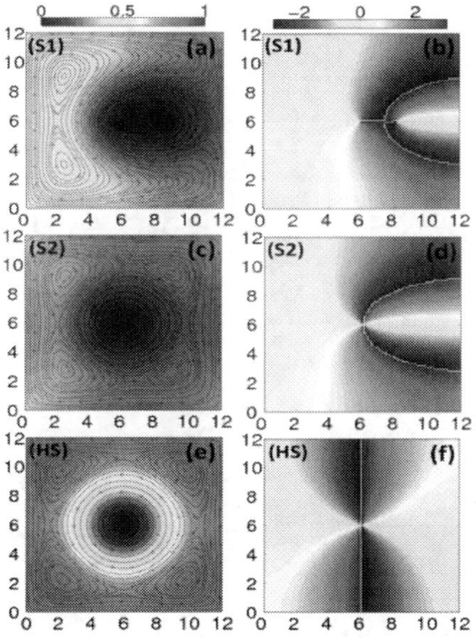

Figure 3. Left: intensity of the order parameter; right: phase of the order parameter; S_1 (a,b); S_2 (c,d); HT (e,f).

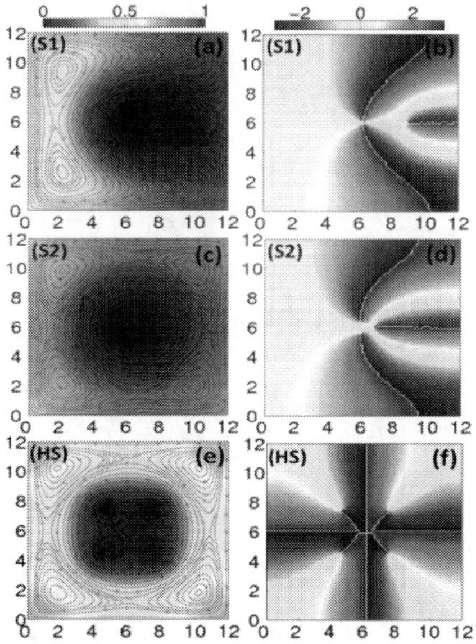

Figure 4. The same as for Fig. 3; S_1 (a,b); S_2 (c,d); HT (e,f).

5. Acknowledgments

We thank the Brazilian Agencies Fundunesp/PROPe grant 2115/002/14-PROPe/CDC and the São Paulo Research Foundation (FAPESP), grants 2013/17719-8 and 2012/04388-0 for financial support.

6. References

[1] Machida M, Kaburaki, H 1993 *Phys.Rev.Lett* **71** 3206–3209
[2] Bolech A C, Buscaglia G C, Lopez A, 1995 *Phys.Rev.B* **52** 15719–15722
[3] Deo P S, Schweigert V A, Peeters F M, Geim A K 1997 *Phys.Rev.Lett* **79** 4653–4657
[4] Schweigert V A, Peeters F M 1998 *Phys.Rev.B* **57** 13817
[5] Schweigert V A, Peeters F M 1999 *Phys.Rev.B* **60** 3084–3087
[6] Cren T, Garcia L S, Debontridder F, Roditchev D 2011 *Phys.Rev.Lett* **107** 097202
[7] Baelus B J, Kanda A, Peeters F M, Ootuka Y, Kadowaki K 2005 *Phys.Rev.B* **71** 140502
[8] Sardella E, Lisboa-Filho P N, Malvezzi A L 2008 *Phys.Rev.B* **77** 104508
[9] Presotto A, Sardella E, Zadorosny R 2013 *Physica C* **492** 75–79
[10] Palacios J J 1998 *Phys.Rev.B* **58** R5948-R5951
[11] Zadorosny R, Sardella E, Malvezzi A L, Lisboa-Filho P N, Ortiz W A 2012 *Phys.Rev.B* **85** 214511
[12] Barba-Ortega J, Gonzlez J D, Sardella E 2014 *J.Low.Temp.Phys* **174** 96–103
[13] Barba-Ortega J, Sardella E, Aguiar J A, Brant E H 2012 *Physica C* **479** 49–52
[14] Connolly M R, Milosevic M V, Bending S J, Clem J R, Tamegai T 2009 *Euro.Phys.Lett* **85** 17008
[15] Schimid A 1966 *Phys.Kondens.Materie* **5** 302–317
[16] Gropp W D, Kaper H G, Leaf G K, Levine D M, Palumbo M, Vinokur V M 1996 *J.Comput.Phys* **123** 254–266
[17] Poole Jr C P, Farach H A, Creswick R J, Prozorov Ruslan 2007 *Superconductivity* 2.ed (Amsterdam: Elsevier) p 670
[18] Sardella E, Malvezzi A L, Lisboa-Filho P N, Ortiz W A 2006 *Phys.Rev.B* **74** 014512

Journal of Physics: Conference Series **568** (2014) 022012 doi:10.1088/1742-6596/568/2/022012

On the bad metallicity and phase diagrams of $Fe_{1+\delta}X$ (X =Te, Se, S, solid solutions): an electrical resistivity study

M El Massalami[1], K Deguchi[2], T Machida[3], H Takeya[2] and Y Takano[2]

[1] Instituto de Física, Universidade Federal do Rio de Janeiro, Caixa Postal 68528, 21941-972 Rio de Janeiro RJ, Brazil

[2] National Institute for Materials Science, 1-2-1, Sengen, Tsukuba, 305-0047, Japan

[3] Department of Physics, Tokyo University of Science, 1-3 Kagurazaka, Shinjuku-ku, Tokyo 162-8601, Japan

Abstract. Based on a systematic analysis of the thermal evolution of the resistivities of Fe-based chalcogenides $Fe_{1+\delta}Te_{1-x}X_x$ (X= Se, S), it is inferred that their often observed nonmetallic resistivities are related to a presence of two resistive channels: one is a high-temperature thermally-activated process while the other is a low-temperature log-in-T process. On lowering temperature, there are often two metal-to-nonmetall crossover events: one from the high-T thermally-activated nonmetallic regime into a metal-like phase and the other from the log-in-T regime into a second metal-like phase. Based on these events, together with the magnetic and superconducting transitions, a phase diagram is constructed for each series. We discuss the origin of both processes as well as the associated crossover events. We also discuss how these resistive processes are being influenced by pressure, intercalation, disorder, doping, or sample condition and, in turn, how these modifications are shaping the associated phase diagrams.

1. Introduction

It is remarkable that the normal-state resistivities of $Fe_{1+\delta}X$ (X= Te, Se, S, or their solid solutions) [1–4] as well as those of intercalated $A_xFe_{2-y}Se_2$ (A=K, Rb, Cs, Tl, ..)[5–7] are neither truly metallic nor truly insulating. It is also remarkable that the normal-state and superconducting phase diagrams of these chalcogenides are highly irreproducible and markedly different from one another. In fact, reported resistivities of the very same stoichiometric compound do not show the same thermal/magnetic/baric/concentration evolution, $\rho(T, H, P, x)$, or the same crossover/transition events. It is then no surprise that these phase diagrams (being constructed out of such characterization) manifest a strong dependence on sample condition or history [1–4].

Although an earlier electronic structure calculations predicted a low-carrier-density metallic character [8] the isomorphous $Fe_{1+\delta}X$ compounds exhibit a variety of normal-state behavior: $Fe_{1+\delta}S$ is nonmetallic below T_{MNM} ~300 K [9] but T_{MNM} can be strongly reduced by pressure [10]. $Fe_{1+\delta}Se$ is a nonmagnetic metal, undergoes a structural phase transition at T_S ~90 K, and superconducts at T_c ~8 K [11]. Finally, $Fe_{1+\delta}Te$ is a nonmetallic paramagnet above a magnetic and structural transition at T_{MS} ~70K while a metallic antiferromagnet (AFM) below T_{MS} [12, 13].

Journal of Physics: Conference Series **568** (2014) 022012 doi:10.1088/1742-6596/568/2/022012

In this work we address the above mentioned bad metallicity of $Fe_{1+\delta}X_{1-x}Y_x$ chalcogenides. Based on the analysis of their resistivities and on the obtained phase diagrams, we identified two processes that are responsible for their bad metallic character as well as for shaping their phase diagrams: the first is a high-temperature (150 to 300 K) thermally-activated process [14] while the other is a low-temperature (< 100 K) log-in-T process [15].

2. Guidelines for analyzing $\rho(T, P, x)$

In order to rationalize the variety of functional forms of $\rho(T, P, x)$ and, in addition, so as to identify and evaluate the strength of the involved resistive channels, let us assume, based on earlier studies [16–19], that the character of their normal-state is shaped by the combined influences of crystalline electric field interactions, electronic correlations, disorder, and band filling: based on the strength of these factors, the high-temperature (T >100 K) normal state could be either metallic, Mott insulator, or an intermediate orbital-selective Mott phase (OSMP) wherein some of the Fe $3d$ orbitals are localized while the others are itinerant [14, 16–19]. The high-T phase of most of the studied intercalated $A_xFe_{2-y}Se_2$ (as well as most of $Fe_{1+\delta}X_{1-x}Y_x$, see below) compounds is reported to be an OSMP.[14, 16–19] We assumed that, within this OSMP, localized states are separated from itinerant ones by a mobility edge at E_c [20]. Then the thermal evolution of the resistivity depends on the relative strength of $|E_c - E_F|$ with respect to k_BT: if E_c is not located in the $3d$ multiplet or that $|E_c - E_F| > k_BT$, then $\rho(T)$ is governed by the Mott variable range hopping process (VRH):

$$\rho(T) = \rho_0^{vrh} \exp([T_{vrh}/T]^{\frac{1}{d+1}}), \tag{1}$$

where d=2 (3) represents a 2- (3-) dimensionality and all other terms have their usual meaning. If $|E_c - E_F| < k_BT$, as assumed for the under-study Fe-based compounds, then $\rho(T)$ is governed by the Arrhenius proces [21]:

$$\rho(T) = \rho_0^A \exp(|E_c - E_F|/T) = \rho_0^A \exp(\triangle/T). \tag{2}$$

Previous studies on these OSMPs reported that, due to the characteristic arrangement of the energy levels of the involved Fe-$3d$-orbitals as well as due to entropy arguments, a lowering of temperature often leads to a temperature-induced crossover (OSMT) from an OSMP into a metallic phase at T_X^{HT} [14, 16].

In addition to the activated high-temperature process, the low temperature resistivities of various chalcogenides[3, 22, 23] are reported to exhibit another process which, in most cases, can be approximated as a log-in-T contribution:

$$\rho(T) = \rho_o^{LT}[1 + S\ln(T_o/T)], \tag{3}$$

the logarithmic slope S is a measure of the intensity of the process while T_o and ρ_o^{LT} are characteristic, here experimentally-determined, parameters. In contrast to the activated process, the origin of such a log-in-T behavior is not well studied; as such this will be discussed below after the analysis of our results.

Depending on the relative strength of the above-mentioned two resistive channels, $\rho(T)$ would assume a variety of functional forms: $\frac{\partial\rho}{\partial T}$ would be positive for a metallic character while negative for any nonmetallic contribution [24, 25]. In addition, $\frac{\partial\rho}{\partial T}$ would be helpful in identifying transition/crossover events (e.g. a metal-to-nonmetal, MNM, crossover is manifested as a maximum in the resistivity: $\frac{\partial\rho}{\partial T}_{T_X}$=0). In general, on cooling, two crossovers may be observed: one from the high-T activated regime into a metal-like phase (metal-I) at T_X^{HT} and another from the low-T log-in-T regime into a metal-like phase (metal-II) at T_X^{LT} (see below); it is emphasized that, for $Fe_{1+\delta}(Te_{1-x}Se_x)$ $0.1 \leq x \leq 0.5$, both T_X^{LT} and T_X^{HT} events (see also [3, 26, 27]) are not accompanied by any visible symmetry-breaking process.

Journal of Physics: Conference Series **568** (2014) 022012 doi:10.1088/1742-6596/568/2/022012

3. Results

Samples preparation, annealing, and measurements (namely structural, elemental, magnetization and resistivity) of $FeTe_{1-x}Se_x$ and $FeTe_{1-x}S_x$ series were the same as the ones reported in previous works [28]. The above-mentioned resistivity analysis was applied to the measured curves of these samples: this analysis can be readily extended to other chalcogenides.

Figure 1 show $\rho(T)$ curves of two, representatives, oxygen-annealed $FeTe_{1-x}Se_x$ samples. A closer look at Fig. 1(b) reveals that $\frac{\partial\rho_n}{\partial T}(45 < T < 300K)$ exhibits, approximately, two negative values; each is taken to indicate a distinct resistive channel, stemming from a distinct origin, operating within a distinct temperature region, and has a distinct thermal evolution: the one operating within 150< T <300K is an activated process (see above) while the other, operating below 100K, is a log-in-T process.

The thermal evolution of the activated process is exhibited in Figs. 1(d, h). In spite of the polycrystalline form and the limited temperature range available for this behavior, a fit ρ(250< T <300K) to Eq. 1 [see Fig. 1(h)] indicates a simple Arrhenius expression of Eq. 2 wherein the effective single parameter $\triangle$ is taken to represent a mean localization energy separating the Fermi level from the closest mobility edge [21]. For the particular case of $FeTe_{1-x}Se_x$, this $\triangle$ ~40 K is, roughly, the same for all x [see Fig. 2(c)]: such a $\triangle$ should not be confused with that of an activated semiconductivity.

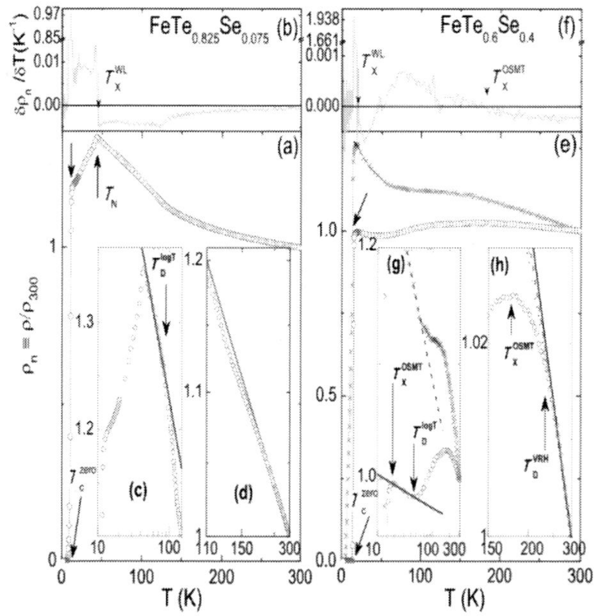

Figure 1. Normalized $\rho_n \equiv \rho/\rho_{300K}$ curves of $FeTe_{0.925}Se_{0.075}$ (a-d) and $FeTe_{0.6}Se_{0.4}$ (e-h). (a, e) ρ_n *vs* T curves. (b, f) $\frac{\partial\rho_n}{\partial T}$ *vs* T curves. (c,g) ρ_n *vs* T curves in a linear-log plot. The solid lines is a fit to Eq. 3. $S(x)$ is shown in Fig. 2(c). (d,h) ρ_n *vs* T curves in a log-reciprocal plot. The solid, dashed and dotted lines are, resp., a fit to Eq. 2, Eq. 1 (d=2) and Eq.1 (d=3). Notice that T_D^{HT} is the lower point at which $\rho_n(T)$ starts to deviate away from Eq. 2 while T_D^{LT} is the upper deviation point from Eq. 3. The cross symbols in (e) and (g) represent the resistivity of the as-prepared sample: annealing in O_2 reduces the two resistive contributions (also evident in e.g. Ref. [29]) [here $S_{\text{as-prep}}$=0.089(3) while S_{O_2}=0.013(1)].

On the other hand, Figs. 1(c, g) indicates that $\rho(T_X^{LT} < T < T_D^{LT})$ follows Eq. 3 [22, 30] wherein S, the only fit parameter, is shown in Fig. 2 (c).

In addition to the manifestation of two resistive channels, $\rho(T,x)$ of $FeTe_{1-x}Se_x$ show some other finer details: (i) $\rho(T, x < 0.1)$ curves are different from the ones with $0.1 \leq x \leq 0.5$: the dividing line, x ~0.1, coincides with the concentration beyond which magnetism is suppressed [1–3]. (ii) $\rho(T, x < 0.1)$ starts to deviate away from Eq. 2 at T_D^{HT}; on further cooling, $\rho(T, x < 0.1)$ exhibits a sharp drop at $T_{MS}(x)$ (related to the reported magnetic and structural transition[31]) followed by a metallic behavior. On cooling well below T_{MS}, $\rho(T < T_{MS}, x < 0.1)$ exhibits the transitions associated with weak and bulk superconductivity [1, 2]. (iii) $\rho(T, 0.1 \leq x \leq 0.5)$

exhibits also a deviation from Eq.2 at T_D^{HT}, followed by a crossover into a metallic state at $T_X^{HT}(x)$ [see Fig. 1(f)]. On further cooling, $\rho(T)$ once more exhibits the log-in-T behavior, a crossover into a metallic state at $T_X^{LT}(x)$, and finally, the bulk superconductivity at T_c^{zero}.

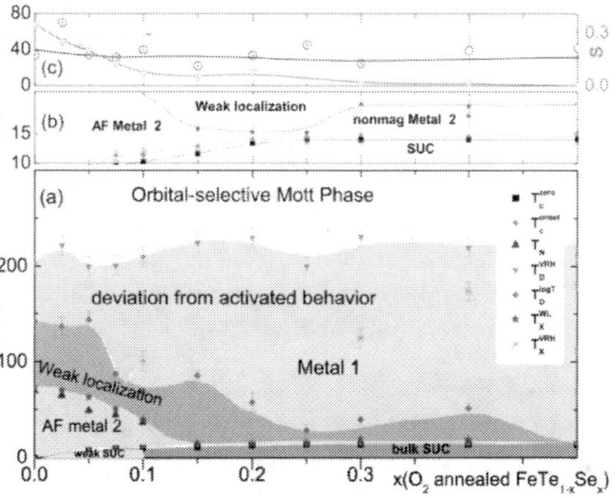

Figure 2. (a) A phase diagram of $FeTe_{1-x}Se_x$ ($0 \leq x \leq 0.5$, annealed in O_2). (b) The x-dependence of the closely-spaced T_c^{zero}, T_c^{onset} and T_X^{LT} events: in spite of their closeness, the metallic state is (re)established well before the onset of superconductivity. (c) *Left ordinate*: Δ versus x (the determination of Δ and T_{VRH} is strongly influenced by the choice among Eqs.1 or 2 and among the limits of the available T-range). *Right ordinate*: the logarithmic slope $S(x)$ as obtained from the fit of Eq. 3.

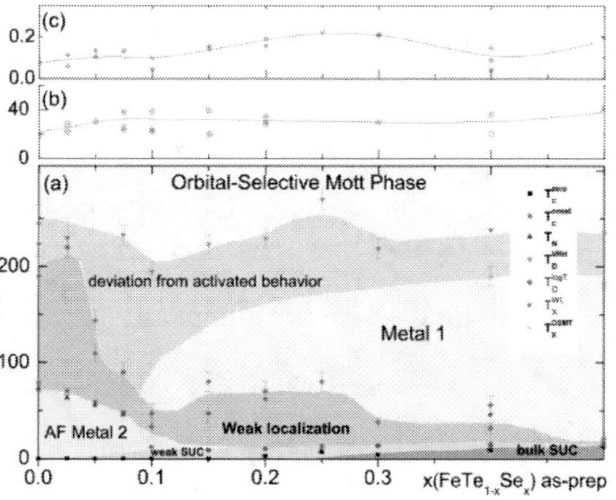

Figure 3. (a) The phase diagram of as-prepared $FeTe_{1-x}Se_x$ ($0 \leq x \leq 0.5$). The difference between this diagram and that of oxygen-annealed one (Fig. 2) emphasizes the role of annealing and oxygen intercalation in shaping the normal and superconducting states of these chalcogenides. (b) Effective Δ versus x as obtained from the fit of Eq.2 within $250 \leq T \leq 300$ K. (c) S versus x as obtained from the fit of Eq. 3.

All the above-mentioned resistivity events of oxygen-annealed $FeTe_{1-x}Se_x$ samples are collected in Fig. 2: in addition to the transitions at T_{MS} and T_c^{zero} and the crossovers at T_X^{HT} and T_X^{LT}, we also include T_D^{HT} and T_D^{LT}. Similar analyses were carried out on the resistivities of as-prepared $FeTe_{1-x}Se_x$ samples as well as those of $FeTe_{0.8}S_{0.2}$ (representative of $FeTe_{1-x}S_x$): the obtained phase diagrams are shown, respectively, in Figs. 3 and 4.

The pressure influence on the superconducting and normal-state properties of these chalcogenides was extensively studied (see e.g., $Fe_{1+\delta}Te$,[32] $Fe_{1+\delta}Se$,[33–39] $Fe_{1+\delta}S$,[10] and $Fe_{1+\delta}(Te_{1-x}Se_x)$ [40–43]). In general, $\rho(T_c < T < 300\text{K})$ is strongly reduced by P. It is noticed that, even for pressures leading to amorphization [43?] the evolution of $\rho(T_c < T < 300\text{K}, P)$ does not support an interpretation in terms of a closure of a semiconducting gap. Rather, such an influence can be envisaged as a reduction of the two resistive channels and an enhancement of the metallic state: For $Fe_{1+\delta}Te$, the metallic state (stabilized below T_{MS} at ambient pressure) is enhanced to ~240 K at 7 GPa [32], while for $Fe_{1+\delta}S$, the metallic state (stabilized above 300 K at ambient pressure) is extended downwards to ~70 K at ~3 GPa [10].

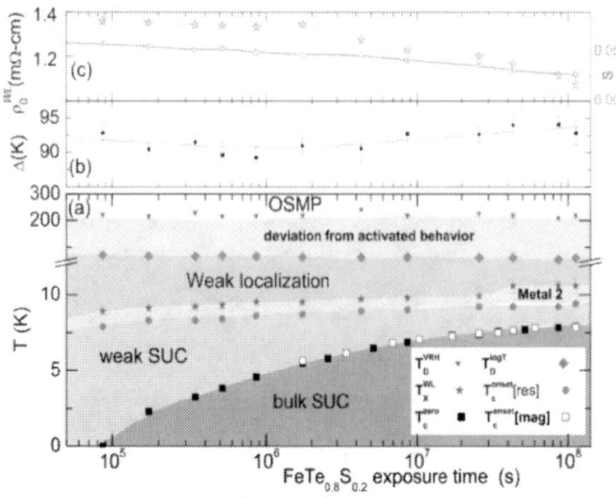

Figure 4. (a) The time evolution of the superconducting, magnetic and transport properties of as-prepared $FeTe_{0.8}S_{0.2}$. The resistivities and magnetizations (see Fig. 11 in Ref. [2]) were measured at the plotted time intervals while the sample was being continuously exposed to air at ambient temperature. Similar phase diagrams can be obtained for other intercalation processes. (b) Effective Δ *versus* x as obtained from the fit of Eq.2 within $250 \leq T \leq 300$ K. (c) the evolution of ρ_o^{LT} (left ordinate) and S (right ordinate) versus x as obtained from the fit of Eq. 3.

Finally, $T_{N,mag}$ and $T^{onset}_{c,mag}$, as determined from magnetization curves (not shown), are shown in the respective phase diagrams; evidently $T^{onset}_{c,mag}$ equals $T^{zero}_{c,\rho}$; as such $T^{onset}_{c,\rho}$ should not be taken as an onset of superconductivity, rather it is $\sim T_X^{LT}$ (see below).

4. Discussion and conclusions

We showed a consistent and a unified analysis of $\rho(x, P, T$ <300K) based on consideration of an activated and a log-in-T processes. Various x-dependent transition/crossover events (T_{MS}, T_N, T_c^{zero}, T_X^{HT}, T_X^{LT}, T_D^{HT} and T_D^{LT}) were identified and these were used to construct the corresponding x-T phase diagrams; that these diagrams are topologically equivalent to the earlier ones [1–4, 31, 44–46] emphasizes the role of these channels in shaping the phase diagrams.

The origin of the high-T activated process, as already discussed, is related to the presence of an orbital-selective Mott phase [14, 16–19]. A careful analysis of this process in $Fe_{1+\delta}X$ suggests that $Fe_{1+\delta}S$ is at the vicinity of Mott transition, $Fe_{1+\delta}Se$ is at a metallic side, while $Fe_{1+\delta}Te$ and $Fe_{1+\delta}Te_{1-x}Y_x$ solid solutions are at an intermediate orbital-selective Mott phase. A variation in any of the control parameters (crystal structure, intercalation, pressure, defects, or filling) would turn the normal-state into either a metal-like, an activated OSMP, or a Mott insulator [7, 14, 16–19]. This was elegantly illustrated in the case of $AFe_{2-y}Se_2$ [6, 14].

The origin of the low-T localization, on the other hand, can be deduced from the scanning tunneling microscopy studies of Machida *et al.*[47] which revealed (at 4.2 K, within the metallic state of $Fe_{1+\delta}Te$) the presence of an inhomogeneous distribution of local density of states (LDOS) around the randomly-distributed defect centres (suggested to be excess Fe): LDOS is increased around these centres while depleted away from them. Scattering from such nonperiodic potentials is assumed to be the cause of the low-T process [3, 22, 23, 48, 49]. Indeed, the strength of such a process is strongly controlled by the concentration of defects: this is well illustrated in the correlation of $\rho(T < T_{MS})$ with δ in $Fe_{1+\delta}Te$ [48] as well as in the correlation of $S(x)$ (see Fig. 2) with the excess Fe or chalcogen deficiency [29]. In general, the control of scattering centres can be effected by annealing (Fig. 2 and Ref. [50]), applying pressure [10, 32–43], or intercalation (Fig. 4 and Ref. [2]).

This defect-driven low-T process is manifested here as, though not limited to, a log-in-T term; three possible mechanisms can be suggested for this log-in-T behavior: (i) a typical Kondo or any of the formally analogous processes arising from scattering against nonmagnetic degenerate-state impurities [51, 52], (ii) Anderson weak localization wherein defects lead to a coherent backscattering of non-interacting electrons [53, 54], or (iii) Altshuller-Aronov[15] weak localization wherein the system of interacting electrons are subjected to a random potential. The following considerations elect the Altshuller-Aronov weak localization process: First, these chalcogenides are structurally formed from Fe_2X_2 layers that are coupled to their neighbors by weak van der Waals forces. Second, the observation of stronger correlation effects in chalcogenides[8] suggests a scenario of *interacting* electrons in a random potential which may arise from a randomly-distributed (paramagnetic and/or nonmagnetic) impurities.

The identification of the origins of these two processes makes it easy to trace down the influence of perturbations (such as pressure, intercalation, annealing, ..) on the boundary lines at T_X^{LT} and T_X^{HT} of the studied phase diagrams. While the origin of the crossover at T_X^{HT} is already discussed, that of T_X^{LT} is most probably related to the fluctuation-induced formation of incoherent Cooper pairs which is favored by disorder and lower dimensionality of chalcogenides [55, 56]: due to the onset of these fluctuations, the metallic-like character is (re)established well before the onset of superconductivity (see Figs. 2, 3, 4). This feature is reminiscent of the situation in both nonconventional underdoped HT_c, see e.g. Ref. [24], and conventional LT_c superconductors [55]. Accordingly, T_X^{LT} may be identified as the onset point, T^*, of a pseudogap phase [30]. It is emphasized that this low-temperature disorder-driven metal-nonmetal transition is different from the high-temperature Mott metal-nonmetal transition.

The competition between superconductivity and localization, observed above, may be discussed in terms of a competition between an attractive (V_p) and a repulsive (V_c) interactions. On assuming an effective coupling $\lambda_{eff} = \lambda - \mu^*$ wherein $\lambda = V_p N_{E_F}$ and $\mu^* = \frac{V_c N_{E_F}}{1+V_c N_{E_F}\ln(\frac{\omega_c}{\omega_D})}$ (all terms have their usual meaning) [57], then (i) an increase in, say, disorder would augment μ^* which, being in competition with superconductivity, would degrade and, above a critical value, eventually quench the superconductivity [55]. A further increase of disorder would eventually lead to a stronger Anderson-type localization as best illustrated in $Fe_{1.01-x}Cu_xSe$ [58]. (ii) the manifestation of T_X^{LT} and the associated metal-II phase is conditioned by the strength of the attractive term: if no or weaker superconductivity, then no T_X^{LT} event.

Finally, based on the above-mentioned remarks, it is easy to understand (i) why the resistivity behavior of most Fe-based chalcogenides are neither truly metallic nor truly insulating, (ii) why some events are not exhibited in the reported resistivities of the as-prepared or annealed/intercalated samples, and (iii) why the reported superconducting and magnetic phase diagrams are strongly influenced by sample conditions (compare Figs. 2 and 3).

Acknowledgments
We acknowledges fruitful discussions with D Edwards, N Trivedi, J Albuquerque, C M Chaves, A Troper, M A C Gusmao, L Craco, C Lewenkopf, and S Das Sarma.

References
[1] Y. Mizuguchi and Y. Takano, J. Phys. Soc. Jpn. **79**, 102001 (2010).

[2] K. Deguchi, Y. Takano, and Y. Mizuguchi, Sci. Technol. Adv. Mater. **13**, 054303 (2012).

[3] T. J. Liu *et al.*, Nat. Mater. **9**, 716 (2010) ; N. Katayama *et al.*, J. Phys. Soc. Jpn. **79**, 113702 (2010).

[4] B. C. Sales *et al.*, Phys. Rev. B **79**, 094521 (2009); M. Tropeano *et al.*, Supercond. Sci. Technol. **23**, 054001 (2010).

[5] J. Guo *et al.*, Phys. Rev. B **82**, 180520 (2010).

[6] M.-H. Fang *et al.*, Europhys. Lett. **94**, 27009 (2011) ; T. P. Ying *et al.*, Nature Sci. Rep. **2**, 426 (2012) ; D. M. Wang *et al.*, Phys. Rev. B **83**, 132502 (2011).

[7] Z. Wang *et al.*, Nature Commun **5**, 3202 (2014) ; F. Chen *et al.*, Phys. Rev. X **1**, 021020 (2011).
[8] D. J. Singh, Sci. Technol. Adv. Mater **13**, 054304 (2012).
[9] E. F. Bertaut *et al.*, Solid State Comm. **3**, 335 (1965).
[10] S. Denholme *et al.*, Sci. Technol. Adv. Mater. (2014), in press.
[11] T. M. McQueen *et al.*, Phys. Rev. Lett. **103**, 057002 (2009).
[12] W. Bao *et al.*, Phys. Rev. Lett. **102**, 247001 (2009).
[13] S. Li *et al.*, Phys. Rev. B **79**, 054503 (2009).
[14] M. Yi *et al.*, Phys. Rev. Lett. **110**, 067003 (2013) ; P. Gao *et al.*, Phys. Rev. B **89**, 094514 (2014).
[15] B. L. Altshuler and A. G. Aronov, Solid State Commun **46**, 429 (1983).
[16] R. Yu and Q. Si, Phys. Rev. Lett. **110**, 146402 (2013).
[17] L. Craco, M. S. Laad, and S. Leoni, Phys. Rev. B **84**, 224520 (2011).
[18] Z. P. Yin, K. Haule, and G. Kotliar, Phys. Rev. B **86**, 195141 (2012).
[19] E. Bascones, B. Valenzuela, and M. J. Calderón, Phys. Rev. B **86**, 174508 (2012).
[20] N. F. Mott, J Physics C: Solid State Physics **20**, 3075 (1987).
[21] D. Basko, I. Aleiner, and B. Altshuler, Annals of Physics **321**, 1126 (2006).
[22] T. J. Liu *et al.*, Phys. Rev. B **80**, 174509 (2009).
[23] H. H. Chang *et al.*, Supercond. Sci. Technol. **25**, 035004 (2012).
[24] Y. Ando *et al.*, Phys. Rev. Lett. **93**, 267001 (2004).
[25] Y. J. Song *et al.*, J. Korean Phys.Soc. **59**, 312 (2011).
[26] Y. Liu, R. K. Kremer, and C. T. Lin, Supercond. Sci. Technol. **24**, 035012 (2011).
[27] C. S. Yadav and P. L. Paulose, J. Appl. Phys. **107**, 083908 (2010).
[28] K. Deguchi *et al.*, J Supercond Nov Magn **27**, 305 (2014).
[29] Y. Sun *et al.*, Scientific Reports **4**, 4585 (2014).
[30] I. Pallecchi *et al.*, J Supercond Nov Mag **24**, 1751 (2011).
[31] A. Martinelli *et al.*, Phys. Rev. B **81**, 094115 (2010).
[32] H. Okada *et al.*, J. Phys. Soc. Jpn. **78**, 083709 (2009).
[33] Y. Mizuguchi *et al.*, Appl. Phys. Lett. **93**, 152505 (2008).
[34] S. Margadonna *et al.*, Phys. Rev. B **80**, 064506 (2009).
[35] T. Imai *et al.*, Phys. Rev. Lett. **102**, 177005 (2009).
[36] S. Medvedev *et al.*, Nat. Mater. **8**, 630 (2009).
[37] L. Li *et al.*, J Supercond Novel Mag **22**, 667 (2009).
[38] D. Braithwaite *et al.*, J. Phys: Condens. Matter **21**, 232202 (2009).
[39] G. Garbarino *et al.*, Europhys. Lett. **86**, 27001 (2009).
[40] K. Horigane *et al.*, J. Phys. Soc. Jpn. **78**, 063705 (2009).
[41] G. Tsoi *et al.*, J. Phys: Condens. Matter **21**, 232201 (2009).
[42] C.-L. Huang *et al.*, J. Phys. Soc. Jpn. **78**, 084710 (2009).
[43] A. K. Stemshorn *et al.*, J. Mater. Res. **25**, 396 (2010).
[44] M. H. Fang *et al.*, Phys. Rev. B **78**, 224503 (2008).
[45] C. Dong *et al.*, Phys. Rev. B **84**, 224506 (2011).
[46] Y. Kawasaki *et al.*, Solid State Commun. **152**, 1135 (2012).

[47] T. Machida *et al.*, Phys. Rev. B **87**, 214508 (2013).
[48] T. Machida *et al.*, Physica C **484**, 19 (2013).
[49] J. Hu, T. J. Liu, B. Qian, and Z. Q. Mao, Phys. Rev. B **88**, 094505 (2013).
[50] S. Cao *et al.*, J. Appl. Phys. **110**, 033914 (2011).
[51] A. Taraphder and P. Coleman, Phys. Rev. Lett. **66**, 2814 (1991).
[52] D. C. Ralph and R. A. Buhrman, Phys. Rev. Lett. **69**, 2118 (1992).
[53] P. W. Anderson, E. Abrahams, and T. V. Ramakrishnan, Phys. Rev. Lett. **43**, 718 (1979).
[54] G. J. Dolan and D. D. Osheroff, Phys. Rev. Lett. **43**, 721 (1979).
[55] B. Sacepe *et al.*, Nat Commun **1**, 140 (2010).
[56] P. A. Lee, N. Nagaosa, and X.-G. Wen, Rev. Mod. Phys. **78**, 17 (2006).
[57] J. B. Ketterson and S. N. Song, *Superconductivity* (Cambridge University Press, 1999).
[58] A. J. Williams *et al.*, J. Phys.: Condens. Matter **21**, 305701 (2009).

Coherent transition induced by Sr doping on the $YBa_2Cu_3O_{7-\delta}$ single crystal magnetoconductivity

L F Lopes[1], V N Vieira[1], F T Dias[1], D L Silva[1], S T Jaeckel[2], P Pureur[3] and J Schaf[3]

[1]Instituto de Física e Matemática, Universidade Federal de Pelotas, 96010-900, Pelotas, RS, Brazil
[2]Instituto Federal Farroupilha, 97555-000, Alegrete, RS, Brazil
[3]Instituto de Física, Universidade Federal do Rio Grande do Sul, 91501-970, Porto Alegre, RS, Brazil

e-mail: lutienefl@gmail.com (Lutiene Fernandes Lopes)

Abstract. We have studied by magnetoresistivity measurements the influence of the low Sr chemical doping on the conductivity thermodynamic fluctuation regimes (TFR) of a $YBa_{1.98}Sr_{0.02}Cu_3O_{7-\delta}$ single crystal. The low-current and low-frequency resistivity, $\rho(T,H)$ measurements were performed in a PPMS while DC magnetic fields up to 50kOe were simultaneously applied perpendicular to the ab plane of the single crystal and to the measurement current density. The $d\rho(T,H)/dT$ data shows that the superconducting transition of the sample occur is two stages that corresponds to the signature of a granular superconductor. At this scenario is possible to define two critical temperature transitions $T_C(H)$ and $T_{C0}(H)$ [$T_C(H) > T_{C0}(H)$]. At $T_C(H)$ the superconductivity sets in within the grains and at $T_{C0}(H)$ a coherent transition takes place and a long-range superconducting order takes place over the whole sample. Special attention is reserved to the paracoherent region [$T_{C0}(H) \leq T < T_C(H)$] of the $\rho(T,H)$ data. The field and the temperature dependence of the paracoherent TFR behavior is characterized by a power law represented by a critical exponent $S_0 \approx 3$. This exponent is an indicative that the coherence transition of the sample belongs to the 3D-XY universality class with relevant disorder that probably is associated to the low Sr doping.

1. Introduction

The contribution of the granularity to the superconductor properties of the high temperature superconductors (HTSC) has been an object of the intensity scientific investigation [1,2,3]. In particular AC magnetoresitivity and DC magnetization techniques are successful to characterize its contribution to electrical and magnetic properties of these materials [1,2,3]. The polycrystalline samples prepared by usual solid state reaction techniques are essentially classified as granular superconductors [4,5]. Otherwise the characterization of superconductor single crystals like a granular material is a controversial subject because these are idealized as homogeneous materials [3].

The chemical doping is designated for some works at the literature as an efficient way to introduce an inhomogeneous (granular) character to the superconductor magnetic and electrical properties of $YBa_2Cu_3O_{7-\delta}$ single crystals [1,2,3]. In special magnetic irreversibility studies performed in $Y_{0.98}Ca_{0.02}Ba_2Cu_3O_{7-\delta}$ [3], $YBa_{2-x}Sr_xCu_3O_{7-\delta}$ ($x \leq 0.5$) [2] and $YBa_2Cu_{2.97}Zn_{0.03}O_{7-\delta}$ [1] single crystals show that their irreversible magnetic line dynamics at low magnetic field regime is described by the Almeida-Thouless and the Gabay-Thoulouse power laws that is the magnetic signature of a frustrated and disordered material [2,3]. Otherwise the identification of a magnetoresistivity coherent transition in the electrical property of a superconductor consists in an additional way of characterizes it as a granular material [1,4,5]. As far as we know the report at the literature about magnetoresistivity coherent transition in the HTSC single crystals is scarce [1].

In this work, we report on magnetoresistivity and thermodynamic fluctuation conductivity results of a $YBa_{1.98}Sr_{0.02}Cu_3O_{7-\delta}$ single crystal with the aim of characterizes a granular scenario from the identification of a magnetoconductivity coherent transition.

2. Sample preparation and experimental methods

The single crystals of $YBa_{1.98}Sr_{0.02}Cu_3O_{7-\delta}$ were grown by self-flux method [1] and the selected ones were submitted to an extra oxygen process with the objective of improve their superconducting temperature transition T_C [1]. The single crystal of $YBa_{1.98}Sr_{0.02}Cu_3O_{7-\delta}$ [YB(Sr)CO] used at this work had its structure and superconducting transition characterized respectively by x-ray diffraction (XRD) and electrical resistivity, $\rho(T,H)$. The figure 1(a) displays the XRD result which confirms the $YBa_2Cu_3O_{7-\delta}$ orthorhombic crystalline structure along the c crystallographic axis. The c lattice parameter estimate from the figure 1a was c = (11.67 ± 0.01)Å. The superconducting transition of the YB(Sr)CO is highlight in the figure 1(b). The sample presents a sharper transition with $T_C \approx$ 92,6K. The results presented in the figures 1(a) and 1(b) to the YB(Sr)CO single crystal are in agreement with those reported from literature to samples well oxygenated of this material [6,7].

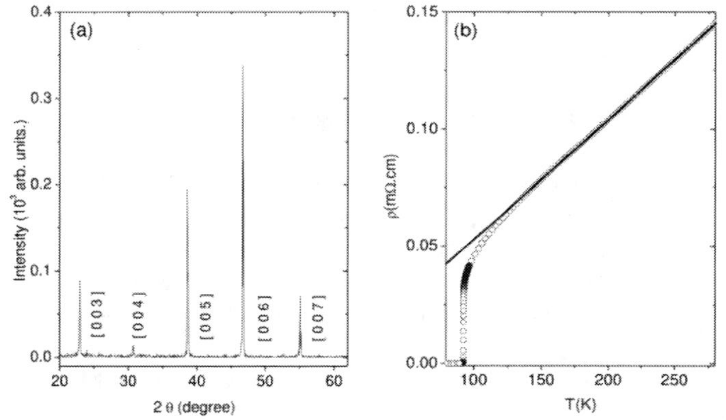

Figure 1. (a) The XRD of the YB(Sr)CO sample obtained for monochromated *CuK α* radiation oriented to its c axis. (b) The $\rho(T,H)$ superconducting transition of the YB(Sr)CO sample.

The four contact $\rho(T,H)$ measurements were performed with AC low current-low frequency PPMS inset while DC magnetic fields up to 50kOe were simultaneously applied parallel to the c axis of the single crystal ($H // c$) and to the measurement current density ($H // J$). The $\rho(T,H)$ was recorded while the temperature was decreased from $T > T_C$. During it measurements the temperature was swept very slowly (≤ 0.4 k/min) so that a high number of data points could be recorded in the temperature range of the superconducting transition and the temperature derivative of the resistivity [$d\rho(T)/dT$] could be numerically determinate.

3. Results and Discussion

The upper panels of figures 2 and 3 displays the $\rho(T,H)$ data as well as the lower panels of the figures 2 and 3 displays their correspondent $d\rho(T,H)/dT$ data to $H \leq$ 50kOe applied parallel to the c axis. The d$\rho(T,H)$/dT data of the figure 2 shows that the resistive transition of the YB(Sr)CO sample is a single step process. Otherwise The d$\rho(T,H)$/dT data of the figure 3 shows that the resistive transition of the YB(Sr)CO became a two step process to 2.5kOe < $H \leq$ 50kOe. The T_{PAR} is the paring temperature that in $\rho(T,H)$/dT plots corresponds approximately to the T_C. The observation of a two step process to the resistive superconducting transition is a clear manifestation of the granularity. At this scenario the first step corresponds to establishment of the superconductivity within the grains at T_C as well as the second step corresponds to the establishment of a long-range superconducting order over the whole sample resulting in a zero resistivity state. Contrasting the figures 2 and 3 results is reasonable to conclude that the second step resistivity transition could be induced by the magnetic field.

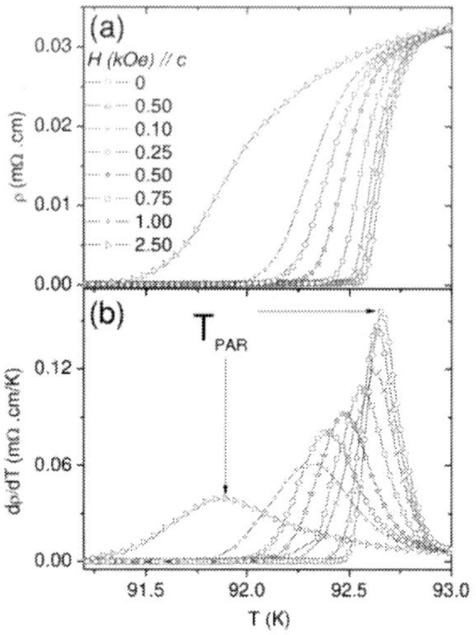

Figure 2. (a) The $\rho(T)$ and (b) the $d\rho(T)/dT$ data to the YB(Sr)CO sample to $H \leq$ 2.5kOe.

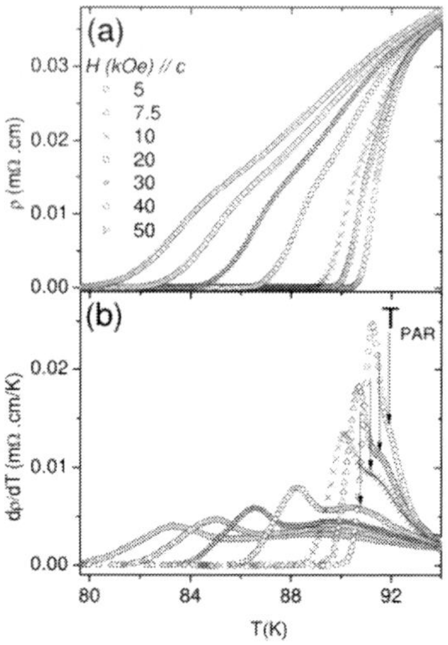

Figure 3. (a) The $\rho(T)$ and (b) the $d\rho(T)/dT$ data to the YB(Sr)CO sample to 5kOe ≤ H ≤ 50kOe.

The contribution of the TFR on the $\rho(T,H)$ data is obtained applying a method based on the Kouvel-Fischer analysis of critical phenomena [8] as highlighted in the figure 4.

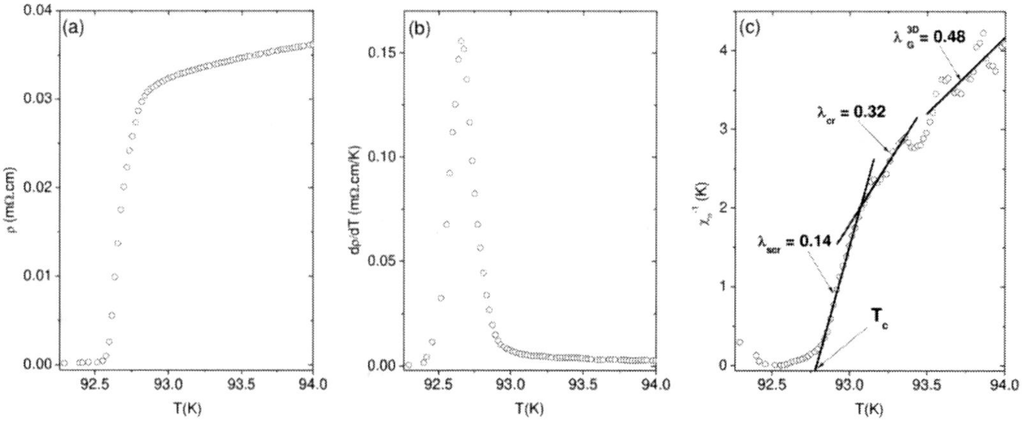

Figure 4. (a) The $\rho(T)$ data, (b) the $d\rho(T)/dT$ data and (c) the$[\chi_\sigma(T)]^{-1}$ data for $H = 0$.

According to this method the TFR on the magnetoconductivity is numerically determined from the quantity:

$$\chi_\sigma = -\frac{d}{dT}[\ln(\Delta\sigma)] \tag{1}$$

where $\Delta\sigma$ is the fluctuation conductivity extracted from the experimental data as $\Delta\sigma = \sigma - \sigma_R$ where $\sigma = \rho^{-1}$ is the measured conductivity and the subtracted regular term ($\sigma_R = \rho_R^{-1}$) is estimated by extrapolating the high temperature behavior of the $\rho(T)$ data of the figure 1(b).

When the temperature approaches T_C from above we assume that $\Delta\sigma$ diverges as $\Delta\sigma = A(t)^{-\lambda}$ where the reduced temperature is $t = (T - T_C)T_C^{-1}$ and A and λ are the criticals amplitude and exponent, respectively. Then, from equation 1 we obtain

$$\frac{1}{\chi_\sigma} = -\frac{1}{\lambda}(T - T_{c0}) \tag{2}$$

The existence of asymptotically TFR regimes becomes evident applying the equation (2) in the data of figure 4(c). The extrapolation of the solid line to the temperature axis allows the simultaneously identification of the critical temperature, T_C and the critical TFR λ_{scr}.

The λ_{scr} exponent observed close to the T_C is related to the critical fluctuations. The $[\chi_\sigma(T,H)]^{-1}$ data for $T > T_C$ identify the existence of critical, λ_{cr} and Gaussian, λ_G TFR [1,4]. However in this work we are very interesting in the paracoherent TFR behavior of the magnetoconductivity data situated at $T_{C0}(H) \leq T < T_C(H)$. In this particular temperature range, the TFR of $[\chi_\sigma(T,H)]^{-1}$ data is shown to diverge according to the power law $\Delta\sigma = A_0(t)^{-S_0}$ where A_0 is a constant, S_0 is the critical exponent related to the vortex physics [9] and $t = (T - T_{C0})T_{Co}^{-1}$ where T_{Co} is the superconducting critical temperature which marks the establishment of the zero resistance state. The figure 5 displays the paracoherent region of the $[\chi_\sigma(T,H)]^{-1}$ data to 1kOe $\leq H \leq$ 50kOe.

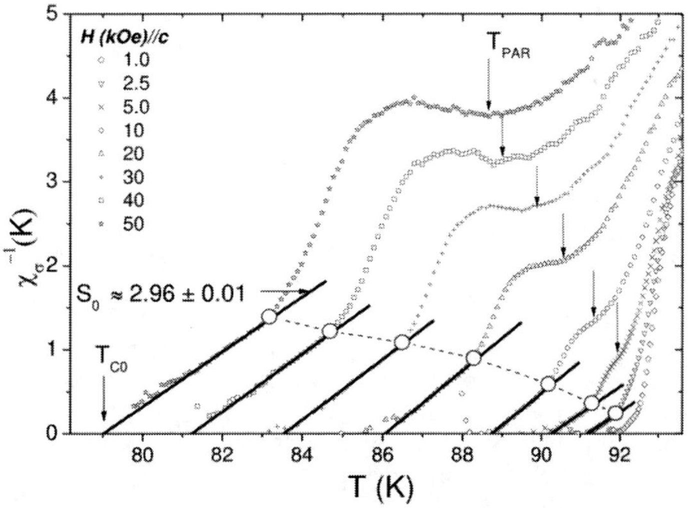

Figure 5. The $[\chi_\sigma(T)]^{-1}$ data of YB(Sr)CO single crystal in the paracoherent region.

In the figure 5 the extrapolation of the solid line of the $[\chi_\sigma(T,H)]^{-1}$ data to the temperature axis allows the simultaneously identification of the critical temperature, T_{C0} and the critical TFR S_0 to $H > 1$kOe. The T_{C0} decreases faster than T_{PAR} as the applied magnetic field is increasing to 50kOe. The open circles correspond to the temperature maximum amplitude of the lower temperature peak of the figure 3. Otherwise their temperature positions coincide with the beginning of the S_0 power law regimes which identifies it as a precursor of the coupling grain process.

In the immediate vicinity of zero resistance of the $[\chi_\sigma(T,H)]^{-1}$ data is possible identify the occurrence of a power law regime corresponding to the averaged exponent $\bar{S}_0 \approx 2.96 \pm 0.01$. In special, the $S_0 \approx 3$ was observed to a $YBa_2Cu_{2.97}Zn_{0.03}O_{7-\delta}$ single crystal to $H \le 1$kOe [1] and to the HTSC polycrystalline samples this exponent converges to 4 as $H > 1$kOe are applied [4,5].

The identification of an exponent $S_0(H) \approx 3$ in the paracoherent region of the $[\chi_\sigma(T,H)]^{-1}$ data of the figure 5 is compatible with the predictions of the Monte Carlo simulations by Young and Wengel based on the phase glass Hamiltonian presented in the equation (3) [10].

$$H = -\sum_{\langle ij \rangle} J_{ij} \cos(\theta_i - \theta_j - A_{ij}) \qquad (3)$$

where J_{ij} is the Josephson coupling energy between nearest neighbor grains i and j, θ_i and θ_j are the phases of the parameter order of the grains i and j, respectively and the vector potential A_{ij} is given to $A_{ij} = (2\pi / \phi_0) \int_i^j \vec{A}.d\vec{l}$ where ϕ_0 is the flux quantum and the line integral is evaluated between the centers of grains i and j. In this scenario the fluctuating phase of the superconductor order parameter in each grain becomes long-range ordered as a consequence of activation of weak links between the grain [5]. This result is the indicative that the coherence transition belongs to the 3D-XY universality class with relevant disorder [9].

According to the results presented in the figures 2, 3 and 5 we suggest that the partial substitution of the 1% of the Ba atoms for Sr atoms introduces a granular character to the superconductor electrical properties of the in the $YBa_2Cu_3O_{7-\delta}$ single crystal. This granular character is highlighted by the two step process of the resistive superconducting transition and the characterization of a paracoherent-coherent transition of the granular array represented by the $S_0(H) \approx 3$ critical exponent.

Acknowledgments
The authors would like to thanks CAPES scientific agency and PRONEM PROGRAM, under contract 11/2042-7, for partially financing this work.

References
[1] Vieira V N, Pureur P, Schaf J 2002 *Phys. Rev. B* **66** 224506
[2] Vieira V N and Schaf J 2002 *Phys. Rev. B* **65** 144531
[3] Vieira V N, Riegel I C and Schaf J 2007 *Phys. Rev. B* **76** 024518
[4] Pureur P, Costa R M, Rodrigues P Jr, Schaf J and Kunzler J V 2007 *Phys. Rev. B* **47** 11420
[5] Costa R M, Ferreira L M, Vieira V N, Pureur P and Schaf J 2007 *Eur. Phys. J. B* **58** 107
[6] Fietz W A and Webb W W 1969 *Phys. Rev.* **178** 657
[7] Veal B W, Kwok W K and Umezawa A 1987 *Appl. Phys. Lett.* **51** 279
[8] Kouvel J C and Fisher M E 1964 *Phys. Rev.* **136** 1626
[9] Wengel C and Young A P 1997 *Phys. Rev. B* **56** 5918
[10] Ebner C and Stroud D 1985 *Phys. Rev. B* **31** 165

Journal of Physics: Conference Series **568** (2014) 022014 doi:10.1088/1742-6596/568/2/022014

Low Sr doping effects on critical current density and pinning mechanism of $YBa_2Cu_3O_{7-\delta}$ single crystals

R F Lopes[1], V N Vieira[2], A P A Mendonça[2], F T Dias[2], D L da Silva[2], P Pureur[1], J Schaf[1] and J J Roa[3]

[1]Instituto de Física, Universidade Federal do Rio Grande do Sul, 91501-970, Porto Alegre, RS, Brazil
[2]Instituto de Física e Matemática, Universidade Federal de Pelotas, 96010-900, Pelotas, RS, Brazil
[3]Department of Materials Science and Metallurgical Engineering, University of Barcelona, 08028, Barcelona, Spain

e-mail: rovanfl@gmail.com

Abstract. We report on isotherm dc magnetization hysteresis loops, $M(H)$ to 70K $\leq T \leq$ 82.5K of $YBa_{2-x}Sr_xCu_3O_{7-\delta}$ (x = 0, 0.02 and 0.1) single crystals with the purpose to study the influence of the low Sr doping effects on the $YBa_2Cu_3O_{7-\delta}$ critical current density, $J_C(H)$ and normalized flux pinning force density, $f(h)$. The $M(H)$ measurements were performed in a commercial SQUID magnetometer with $H \leq$ 50kOe applied parallel to the c axis of the samples. The extended Bean critical state model was applied to $J_C(H)$ determination. The magnetic irreversibility fields, $H_{irr}(T)$ were obtained from $M(H)$ loops. The $f(h)$ behavior was determined from $F_P/F_{P,max}$ versus H/H_{irr} plots where F_P is the pinning force density ($\boldsymbol{F}_P = \boldsymbol{J} \times \mu_0\boldsymbol{H}$). The results show that the low Sr doped samples transport higher $J_C(H)$ than the pure one. The application of the Dew-Hughes scaling functions to the $f(h)$ plots designates the core normal punctual type as majority pinning mechanism of samples. We suggest that the core normal punctual pinning mechanism, responsible to the enhancement of the $J_C(H)$ transported by $YBa_{2-x}Sr_xCu_3O_{7-\delta}$ samples, is possibly connected to the segregation of Sr atoms precipitates at the crystalline structure of these samples.

1. Introduction

The chemical doping of the high temperature superconductors (HTSC) structure is applied successfully as a tool to investigate the effectively relation established between flux pinning mechanisms and $J_c(H,T)$ behavior in these materials. In special the enhancement of $J_c(H,T)$ is reported in the literature to $YBa_2Cu_3O_{7-\delta}$ single crystals when Y, Ba and Cu atoms are partially substituted correspondingly by Ca, Sr and Zn atoms [1-8]. However, scarce and not concluding are the results listed at the literature which investigate the interconnection between $J_c(H,T)$ and the pinning mechanisms and its chemical doping level concentration dependence at the $YBa_2Cu_3O_{7-\delta}$ single crystal structure [3-7].

Specified for the partially substitution of Ba atoms by Sr atoms at $YBa_2Cu_3O_{7-\delta}$ structure its solubility is almost 50% [9-11]. The smaller Sr atomic radius introduces lattice distortion in the $YBa_2Cu_3O_{7-\delta}$ structure [11]. Otherwise the enhancement of the $J_c(H,T)$ was confirmed to the

$YBa_2Cu_3O_{7-\delta}$ single crystals with Sr concentrations lower than 6% ($x \leq 0.12$) [2,3,5]. Some authors point the establishment of a defect structure constitutes of a high density of twin planes decorated with local Sr atom precipitates as responsible to the enhancement of the $J_c(H,T)$ [2,3] however no information about how this affects the pinning mechanism dynamics is presented.

With the purpose of collaborate to the elucidation of some aspects of the previous considerations, we report on a series of isotherm DC magnetization measurements performed in $YBa_{2-x}Sr_x Cu_3O_{7-\delta}$ ($x = 0, 0.02, 0.1$) single crystals with the aim to study the role of the low Sr doping on the enhancement of the $J_c(H,T)$ and its interconnection to the normalize pinning force density, $f(h)$ mechanism.

2. Experimental details

The $YBa_{2-x}Sr_x Cu_3O_{7-\delta}$ ($x = 0, 0.02$ and 0.1) single crystals referred in the text as ScY, ScSr002 and ScSr01 were prepared by self flux method [9,10]. The samples were analyzed by X ray diffraction (XRD) and polarized light microscopy (PLM) [9,10,13]. The XRD results for the doped samples confirm the Y123 orthorhombic structure lattice and points to the linear reduce of c crystallographic parameter as Sr doping level is enhanced. These results are in agreement with those reported in the literature for well oxygenated Sr doped single crystals [10-12]. The figure 1(a) displays the XRD obtained to ScSr01 sample. The doped samples PLM results, not showed, identify the presence of a high density of twin planes as contrasted to the pure sample [3,9,10]. The figure 1(b) displays the critical temperature transition, T_c of the samples characterized from zero field cooled DC magnetization measurements, with applied magnetic field H=10Oe. The T_c values are in agreement with those reported in the literature for $YBa_{2-x}Sr_x Cu_3O_{7-\delta}$ single crystals [10-12].

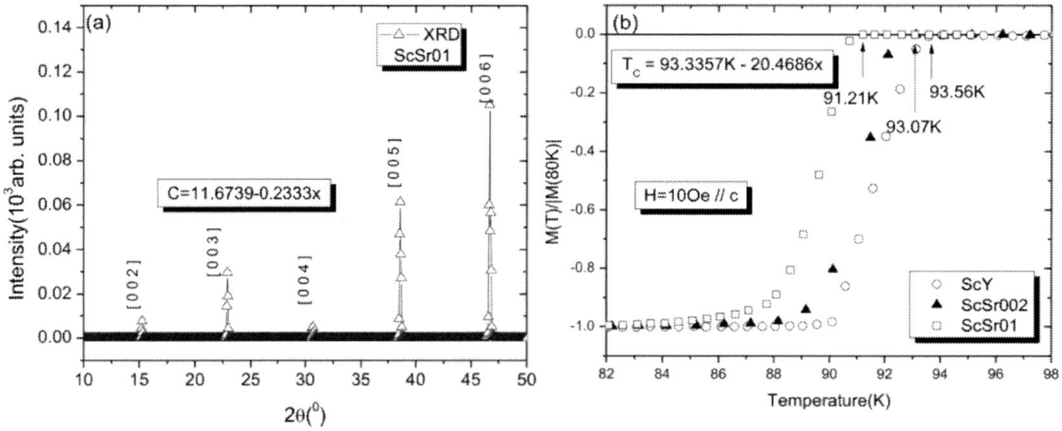

Figure 1. (a) The XRD of the sample ScSr01. (b) the $M_{ZFC}(T)$ plots where the $T_C(x)$ is indicated by the arrows.

The isothermal DC magnetization measurements were performed with a quantum design SQUID magnetometer. The usual $M(H)$ hysteresis loops measurement procedure was adopted [4] and the magnetization of the samples was recorded to $70K \leq T \leq 82.5K$ while magnetic fields, up to 50 kOe, were applied parallel to c crystallographic axis of the samples. The possible demagnetization factor contribution to the $M(H)$ data was checked.

3. Experimental results

The figure 2(a) display the magnetization hysteresis loops, $M(H,T)$. The SMP arrow identify the second magnetization peak (SMP) [2-4,8,14-16] whereas the $\mu_0H_{irr}(T)$ arrow defines the irreversible magnetic filed extracted from $M(H,T)$ data [2,8]. The figure 2(b) displays the $J_c(H,T)$ data calculated from the application of the extended Bean critical model [5-8] to the $M(H,T)$ data of the figure 2(a). According to this model the $J_c(H,T) = 20\Delta M(H)\eta^{-1}$ where η is a flat sample geometric parameter [5-8].

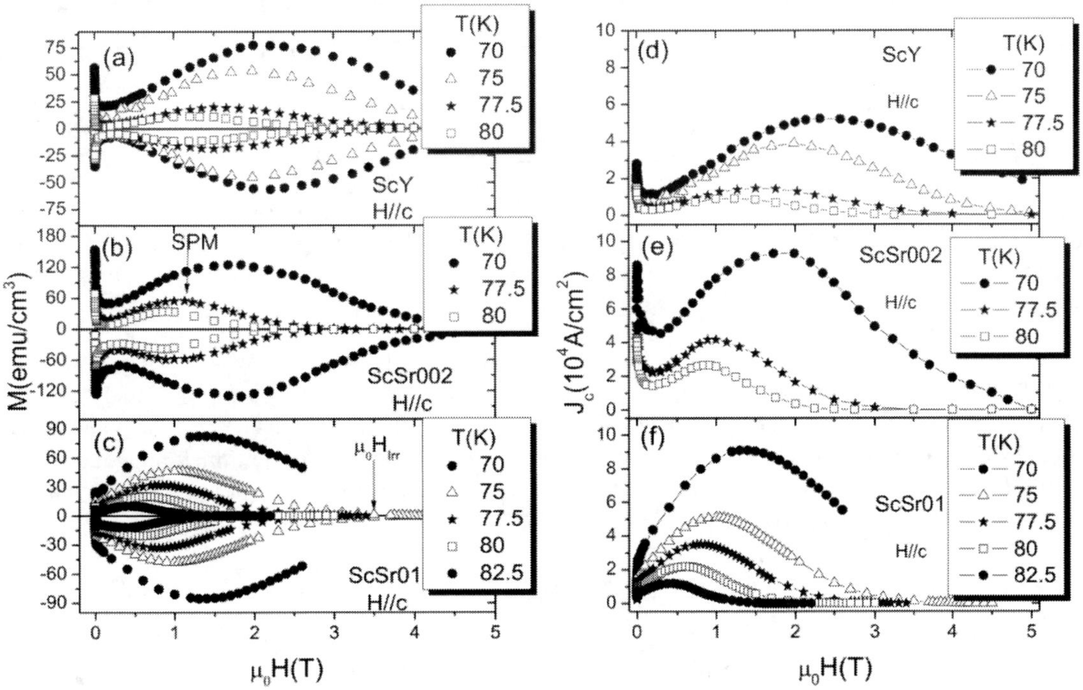

Figure 2. (a), (b) and (c) the $M(H)$ curves for the samples ScY, ScSr002 and ScSr01 respectively, the arrows indicate the irreversibility field $H_{irr}(T)$ and the second peak magnetization SPM. (d), (e) and (f) $J_C(H,T)$ for samples ScY, ScSr002 and ScSr01 respectively.

The $H_{irr}(T)$ data obtained from $M(H)$ measurements reproduce those reported by references 9 and 10 for the irreversible magnetic temperature, $T_{irr}(H)$ obtained from $M(T)$ measurements of $YBa_{2-x}Sr_xCu_3O_{7-\delta}$ (x = 0; 0.10; 0.25; 0.37 and 0.5) single crystals. The H-T plots of the $H_{irr}(T)$ data of our samples are fitted by $H_{irr}(T) \approx [1- T(H)\ T_{irr}(H=0)^{-1}]^{1.5}$ power law which is supported by giant flux creep scenario, the same behavior is reported to the $T_{irr}(H)$ data of $YBa_{2-x}Sr_xCu_3O_{7-\delta}$ ($x \leq 0.5$) single crystals [9,10]. Otherwise the identification of SPM in the $M(H)$ plots of our samples corresponds to the maximization of the flux pinning potential of samples.

The $J_c(H,T)$ of the Sr doped single crystals showed in the figure 2(b) are higher than that observed to the ScY sample. For instance to $T = 70$K and $\mu_0H \sim 2$T the $J_c(H,T)$ of ScY, ScSr002 and ScSr01 are correspondingly 5 10^4Acm^{-2}, 9.2 10^4Acm^{-2} and 8 10^4Acm^{-2}. The enhancement of the $J_c(H,T)$ observed to Sr doped samples is in agreement to those reported at the literature for $YBa_2Cu_3O_{7-\delta}$ single crystal with Sr concentrations lower than 6% ($x \leq 0.12$) [2,3,5].

The figure 3 displays the plots of the normalized pinning force density, $f = F_P/F_{P,max}$ versus the reduced field, $h = H/H_{irr}$ to ScY, ScSr002 and ScSr01 samples [1,6,8,14,17]. The $F_P(H,T)$ data, not showed here, was determinate from $\boldsymbol{F_P} = \boldsymbol{J} \times \mu_0\boldsymbol{H}$ definition, where $F_{P,max}$ corresponds to the upper $F_P(H,T)$ data.

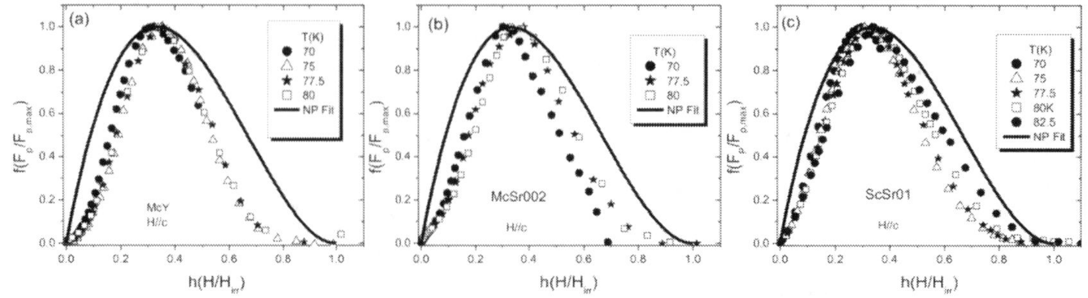

Figure 3. (a), (b) and (c) Normalized pinning force density $f = F_P/F_{P,max}$ versus reduced field $h = H/H_{irr}$ plotted with the core, normal, point pinning function of the Dew Hughes model, black line, for the samples ScY, ScSr002 and ScSr01 respectively.

The $f(h)$ data approximately scales in terms of an common behaviour to $70K \leq T \leq 82.5K$. The NP fit solid line in these figures represents a fitting associated to the Dew-Hughes (DH) model [17,18]. This model was originally conceived to classify the differently pinning mechanism of the conventional superconductors which are classified in magnetic or core types subdivided in normal or Δk modalities. In the DH model the pinning mechanisms are identified through the application of the scaling law described in the equation 1 to the $f(h)$ data performance.

$$f(h) = A(h)^p(1-h)^q, \qquad (1)$$

In the equation 1 A is a numerical parameter and p and q are fitting parameters related to the geometric characteristic of the pinning mechanisms. In the case of the HTSC the $\mu_0 H_{c2}(T)$ parameter originally applied to determine the h is substituted for $\mu_0 H_{irr}(T)$ parameter [1,6,8,14,17].

For all the previous pinning mechanisms cited in the text and associated to the equation (1) fitting, the $f(h)$ data behaviour of our samples approximately matches to the core normal point pinning mechanism, where in figures 3(a), 3(b) and 3(c) are represented by the NP fit solid line. This particular pinning mechanism is obtained adopting p = 1 and q = 2 in the equation (1) [17]. According to the DH model the other way of identify the characteristic pinning mechanism from $f(h)$ data behaviour consists in establishes the reduced field parameter value that corresponds to the maximum $f(h)$ data value [17]. This specified parameter is called h_0 and its value determined for our samples is approximately 0.34 which is in agreement with $h_0 = 0.33$ classify by DH model to the core normal point pinning mechanism [17].

The matches of the maximum $f(h)$ data range of figures 3(a), 3(b) and 3(c) to the core normal point DH fitting is a strong evidence that the majority pinning of our samples is governed by the normal point type and that the low Sr doping do not change considerably the characteristic pinning mechanism associated to the origin of the SPM in the ScY sample. Otherwise we probably associate the core normal point pinning mechanism of the ScSr002 and ScSr01 samples to the establishment of Sr precipitate clusters at its crystalline structure like remarked by Saito et al [3] to the $YBa_{1.99}Sr_{0.01}Cu_3O_{7-\delta}$ single crystal TEM analysis.

4. Conclusions

In this work we report on isotherm dc magnetization hysteresis loops, $M(H)$ to $70K \leq T \leq 82.5K$ of $YBa_{2-x}Sr_xCu_3O_{7-\delta}$ (x = 0, 0.02 and 0.1) single crystals with the purpose to study the influence of the low Sr doping effects on the $YBa_2Cu_3O_{7-\delta}$ critical current density, $J_C(H)$ and normalized flux pinning force density, $f(h)$ mechanism. The low Sr doping ($x \leq 0.1$) of the $YBa_2Cu_3O_{7-\delta}$ single crystals resulted in the enhancement of the $J_c(H,T)$ for all selected temperatures as compared to the no doped sample.

The *f*(*h*) data of studied samples approximately scales in terms of an common behaviour that according to the DH model could be associated to the manifestation of the core normal point pinning mechanism. We suggested that the establishment of Sr precipitate clusters at crystalline structure of the doped samples as responsible to the activation of the normal point pinning resulting in the enhancement of $J_c(H,T)$ of the $YBa_{2-x}Sr_xCu_3O_{7-\delta}$ single crystals.

Acknowledgments
The authors would like to thanks to CAPES and CNPQ scientific agencies and PRONEM 03/2011 (under contract: 11/2042-7) scientific program for partially financing this work.

References

[1] Rogacki K, Dabrowski B and Chmaissem O 2006 *Phys. Rev. B* **73** 224518
[2] Küpfer H, Zhukov A A, Will A, Jahn W, Meier-Hirmer R, Wolf Th, Voronkova V I, Kläser M and Saito K 1996 *Phys. Rev. B* **54** 644
[3] Saito K, Nissen H U, Beeli C, Wolf T, Schauer W and Küpfer H 1998 *Phys. Rev. B* **58** 6645
[4] Vieira V N and Schaf J 2004 *Physica C* **408-410** 533
[5] Shimoyama J, Tazaki Y, Ishii Y, Nakashima T, Horri S and Kishio K 2006 *Journal of Physics: Conference Series* **43** 235
[6] Wang Z H, Zhang H, Gao J, Yang T, Qiu L and Yao X X 2002 *Supercond. Sci. Technol.* **15** 1766
[7] Ishii Y, Tazaki Y, Nakashima T, Shimoyama J, Horri S and Kishio K 2007 *Physica C* **460-462** 1345
[8] Hussain M, Kuroda S, Takita K 1998 *Physica C* **297** 176
[9] Vieira V N and Schaf J 2002 *Phys. Rev. B* **65** 144531
[10] Vieira V N and Schaf J 2003 *Physica C* **384** 514
[11] Kakihana M, Eriksson S G, Börgesson L, Johansson L G, Ström C and Käll M 1993 *Phys. Rev. B* **47** 5359
[12] Licci F, Gauzzi A, Marezio M, Radaelli G P, Masini R and Chaillout-Bougerol C 1998 *Phys. Rev. B* **58** 15208
[13] Roa J J, Oncins G, Dias F T, Vieira V N, Schaf J and Segarra M *et al.* 2011 *Physica C* **471** 544
[14] Koblischka M R, van Dalen A J J, Higuchi T, Yoo S I and Murakami M 1998 *Phys. Rev. B* **58** 2863
[15] Küpfer H, Wolf Th, Lessing C, Zhukov, A A, Lançon X, Meier-Hirmer R, Schauer W and Wühl H 1998 *Phys. Rev. B* **58** 2886
[16] Klein L, Yacoby E R, Yeshurun Y, Erb A, Müller-Vogt G, Breit V and Wühl H 1994 *Phys. Rev. B* **49** 4403
[17] Dew-Hughes D. 1974 *Philos. Mag.* **30** 293
[18] Kramer E. 1973 J. *J. Appl. Phys.* **44** 1360

Development of $YBa_2Cu_3O_7$-Ba_2YTaO_6 nanocomposites by chemical solution deposition

P. Garcés[1], M. Coll[2], H. Castro[1], T. Puig[2] and X. Obradors[2]

[1] Universidad Nacional de Colombia, Bogotá, Colombia.
[2] Instituto de Ciencia de Materiales-Consejo Superior de Investigaciones Científicas, Barcelona, España.

E-mail: paug76@gmail.com

Abstract. In this work we report the study of chemical solution deposited (CSD) $YBa_2Cu_3O_{7-\delta}$ (YBCO) nanocomposite thin films, with diluted Ba_2YTaO_6 (BYTO) nanoparticles. We studied the influence of thermal treatment on the properties of nanocomposites, with different content of secondary phases (6%, 10% and 20%). We measured the film microstructure, nanostrain and the superconducting properties: critical temperature (T_c) and critical current density (J_c). The films were characterized using X-rays diffraction and SQUID inductive measurements. The use of two steps in the thermal treatment allowed to increase the nanostrain up to 0.30% for high nanoparticle loads (20%BYTO), decrease the nanoparticle size down to 15 nm and lead a smoother $J_c(H)$ dependence, compared with the standard (single-step) thermal process and the pristine YBCO films.

1. Introduction

The applied research on superconductors, in which transport of large currents is required, has led today to the development of tapes, based on high-T_c superconductors (HTS). These kind of tapes are called: second generation – 2G, or coated conductors, and are usually based on an YBCO layer, deposited onto a metallic substrate. These devices present a great potential for applications, given that, in addition to the competitive high values of J_c achieved, higher than 1 MA/cm^2, its manufacture can be easily scaled up to produce long lengths, by chemical methods, at low cost. A usual method of fabrication consists in depositing a superconducting layer, from a chemical solution onto a metallic substrate. To increase J_c different methods are being investigated, in order to enhance vortex pinning. An attractive and economical method is the chemical solution deposition route- CDS, in which a structure of non-superconducting or nanocomposites, is diluted inside the superconducting matrix [1].

YBCO nanocomposite materials have opened a new path to enhance vortex pinning in HTS. Fabrication of nanostructured YBCO thin films with the incorporation of nanosized oxide-secondary phases, has shown strong enhancement of YBCO performance at high magnetic fields, in a wide temperature range, ensuing the potential for use in a broad number of power applications [2-6]. Recently, it has been demonstrated that the presence of metal oxide nanoparticles in the YBCO matrix generates nano-scale defects, which act as vortex pinning centers, improving its performance. Rare-earth niobiates and tantalates (double perovskite $Ba_2RE(Ta,N)O_6$ and pyrochlore Re_3TaO_7withRE= Yb^{+3}, $Gd^{+3,}$ Er^{+3}) have gained attention as effective artificial pinning centers [7]. CSD is a cost-effective technique for preparation of large area and long length films [6]. Trifluroacetate (TFA) metal-organic decomposition (MOD) has been widely investigated as a chemical solution route [8, 9]. CSD of YBCO-Ba_2YTaO_6 (YBCO-BYTO) system is attractive, since among several nanocomposite systems, this one presents the highest percentage of random nanoparticles segregated within an epitaxial superconducting matrix [3, 5]. However, at a certain BYTO concentration (≥ 10%M), nanostrain values saturate and nanoparticles tend to agglomerate producing a current blocking effect in the YBCO matrix, degrading J_c [3, 2].Therefore, it is crucial to control nanoparticle formation process in order to engineer nanoparticle size and nanostrain to improve superconducting properties. In order to control these parameters, a two-step thermal process was developed in a previous work [2]. This process allows formation of a highly epitaxial and strained YBCO matrix, with more than 90% randomly oriented nanoparticles. It has been shown that nanoparticle size is reduced from 40-60 nm, with a one step process, down to 15-20 nm, with a two steps process. This work was made for BYTO concentration of 10% M [3].

In the present work we studied the characteristics of CSD- YBCO-BYTO nanocomposite thin films with BYTO concentrations 6%, 10% and 20% M. We found similar results to the previous work. We obtained critical current densities of ~2.6 MA/cm^{2}at 77K and a smooth $J_c(H)$ dependence.

2. Experimental details

YBCO-BYTO films were made starting with a chemical solution precursor obtained by introducing stoichiometric amounts of tantalum (V) ethoxide in a TFA-YBCO solution prepared as described elsewhere [10, 11]. Molar percentages of BYTO 6%, 10% and 20% have been investigated. The metallorganic precursor solution was spin coated on crystalline substrates (100) $LaAlO_3$ (LAO) of 5x5 mm^2. Then a thermal treatment (pyrolysis) was made, in a tubular furnace, under oxygen flow, at 300°C for 30 min. Instead of a one step thermal process used in previous works [2], the crystallization process was achieved following a two steps process, in an atmosphere of N_2, with an O_2 partial pressure of 0.2 mbar, as indicated in figure 1. A water vapor partial pressure $P(H_2O)$ of 22 mbar was added at certain stages of the process, indicated by blue boxes. It is important to note that the second heating ramp (from 575°C to 775°C) has been made under dry atmosphere, to minimize the coalescence of the already nucleated BYTO nanoparticles. After thermal process the superconducting phase was achieved by oxygen annealing at 450 °C.

The microstructure and phase analysis of the fully converted YBCO nanocomposite films were studied by two-dimensional (2D) X-ray diffraction (XRD) using a Brucker AXS GADDS diffractometer. Strain quantification (ε) was carried out based on the Williamson-Hall method [12, 13], by analyzing the symmetric (00l) 2θ Bragg diffraction integral breath, acquired in a SIEMENS D5000 diffractometer. Quantitative determination of the randomly oriented nanodots fraction (RF) was performed by 2D-XRD analysis, based on simultaneous out-of-plane measurement of both textured and random signals [2]. Both J_c and T_c were obtained by SQUID measurements.

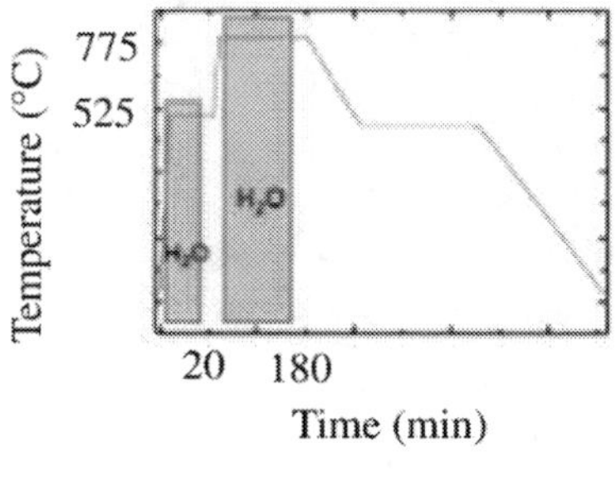

Figure 1. Schematic of the two step thermal process used to prepare YBCO nanocomposites.

3. Results and discussion

Figure 2 shows the diffractogram of YBCO films, with different molar percentages of BYTO, (6%, 10% and 20%). We observe in all films (00l) reflections, indicating c-axis orientation. The peak that appears at 2θ ≈ 43.0° corresponds to the (400) reflection of BYTO second phase. The addition of Ta in the precursor solution changed very little the **c** lattice parameter of YBCO (c ≈ 11.683 Å), indicating no Ta substitution in YBCO [2]. The CuO phase is also identified in all samples.

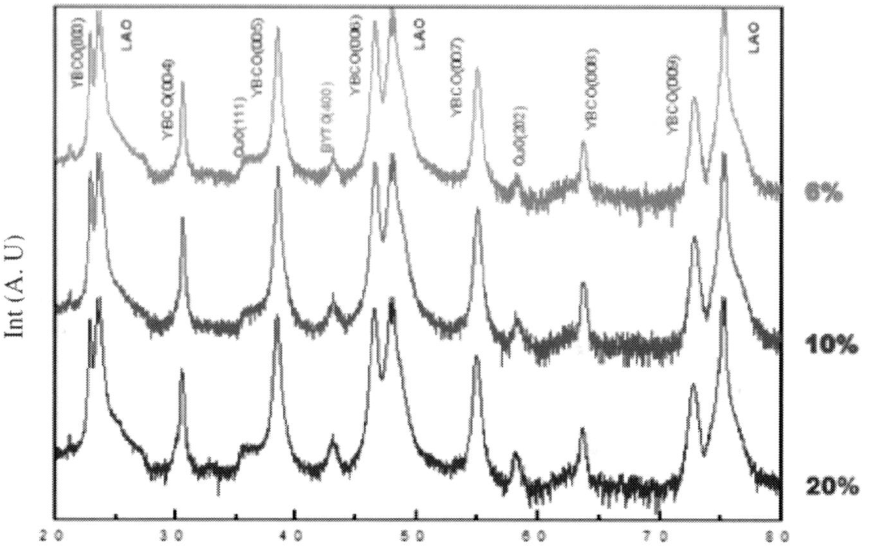

Figure 2. X-ray diffraction of YBCO-BYTO nanocomposites films with different molar percentages of BYTO (6%, 10%, and 20%).

A typical 2D-XRD pattern for YBCO with 10% mol BYTO is shown in figure 3. The (00l) diffraction spots of the YBCO phase confirm c-axis orientation, consistent with figure 2. The BYTO ring from the (220) reflection indicates that this minority phase is mostly randomly oriented, with a random fraction, RF= 95%.

The size of epitaxial BYTO nanoparticles was estimated from the Scherrer formula ($D= 0.9\lambda/\beta \cos\theta$) and ranges from 5 nm to 15 nm, as the BYTO percentage is increased. From previous TEM analysis it has been identified that epitaxial nanoparticles tend to grow at the interface with the substrate. It is important to note that the particle size of the randomly oriented nanoparticles embedded within the YBCO matrix is decreased from 60 nm to 15-20 nm when using the two step thermal process [3].

Nanostrain was quantified by means of the Williamson-Hall method [9]. In figure 4 it is observed that the two steps thermal process increases nanostrain, reaching values up to 0.30% for the 20% BYTO samples. Two steps thermal process do not degrade T_c ~ 90K, and lead to critical current densities up to 2.6 MA/cm^2 at 77K, as shown in table 1. The YBCO-20% BYTO nanocomposite show lower J_c values, compared to 6% and 10% nanocomposites, although the nanostrain values are very high. It is likely that despite decreasing nanoparticle size, such high concentration of nanoparticles start to be an obstacle to current flow percolation [9].

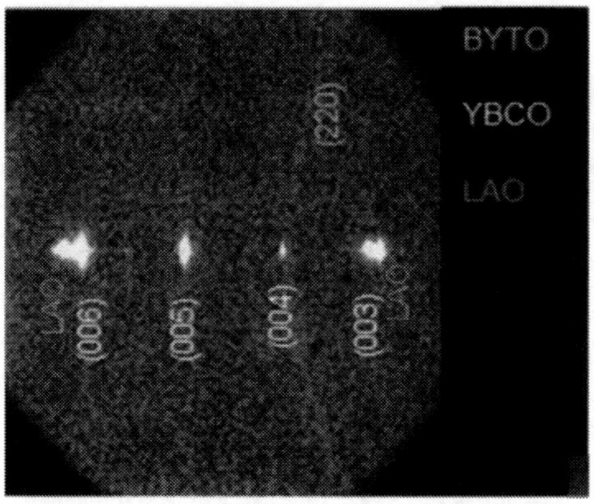

Figure 3. (a) Two-dimensional X-ray diffraction pattern for the YBCO-10% BYTO film.

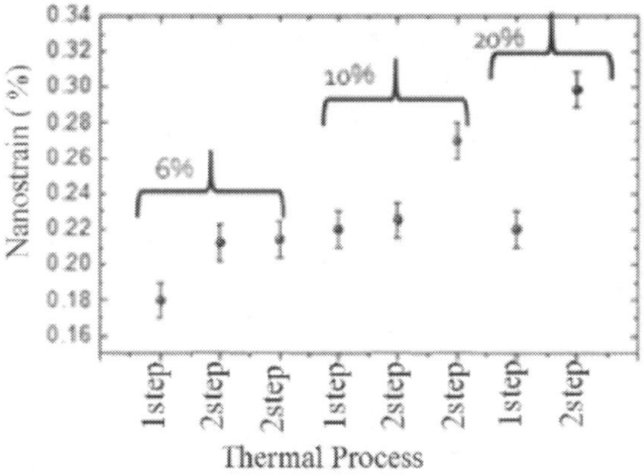

Figure 4. Nanostrain versus thermal process for YBCO nanocomposites with different percentages of BYTO. The first point in each set of data is presented for comparison and corresponds to samples prepared in a previous work with a one step thermal process [2].

Table 1. Physical properties of YBCO nanocomposite films, with different molar percentages of BYTO.

%BYTO	T_c (K)	ΔT_c(K)	J_c (77K) MA/cm^2	Particle Size of epitaxial NP (nm)
6	89	2.8	2.4	5.6
10	89	4.4	2.6	10.5
20	89	3.6	1.9	14.3

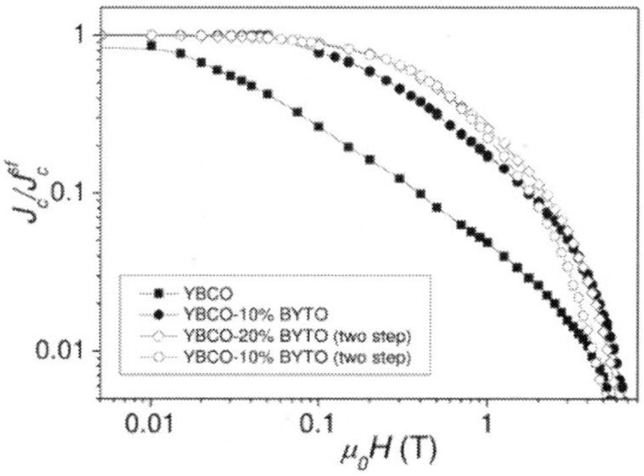

Figure 5. Magnetic field dependence of the critical current density a 77K and H ∥ c, normalized to its self-field value, for: standard YBCO sample, YBCO-10% BYTO nanocomposite made with the standard one step thermal process and two YBCO-BYTO nanocomposites (10%BYTO and 20%BYTO) made with the two step thermal process.

The pinning strength has been evaluated from the J_c (H) behavior at 77K and H ∥ c in a log-log plot, normalized at self-field, as shown in figure 5. At low fields, there is a plateau associated with the single vortex pinning regime, where vortex-vortex interactions are negligible and vortices are pinned individually. At high fields, collective effects become important and J_c starts to decrease with magnetic field [3]. All the YBCO-BYTO nanocomposites show a smooth dependence with magnetic field, compared to pristine YBCO films. Also, the two step processed YBCO-BYTO films show a smoother field dependence, compared to the single-step processed YBCO-BYTO films.

4. Conclusions

The use of a two-step heating profile to prepare YBCO nanocomposites with 6%, 10% and 20% BYTO is an effective approach to decrease nanoparticle size, increase nanostrain and improve the pinning performance while maintaining highly epitaxial YBCO films. These results demonstrate the potential for an industrially scalable chemical solution deposition approach to prepare YBCO nanocomposites.

5. References

[1] Odier P, Supardi Z, De – Barros D, Vergnières L, Ramirez – Castellanos J, Vallet – Regi M, Villard C and Weiss F 2004 *Supercon. Sci. Technol.* **17** 1303
[2] Coll M, YeS, Rouco V, Palau A, Guzman R, Gazquez J, Arbiol J, Sou H, Puig T and Obradors X2013 *Superc. Sci. Technol.* **26** 015001.
[3] Coll M, Guzman R, Garcés P, Gazquez J, Rouco V, Palau A, Ye Se, Magen C, Suo H, Castro H, Puig T and Obradors X 2014 *Superc. Sci. Technol.* **27**044008
[4] Matsumoto K and Mele P 2010*Superc. Sci. Technol.* **23**014001
[5] Obradors X, Puig T, Palau A, Pomar A, Sandiumenge F, Mele P and Matsumoto K 2010 *Comprehensive Nanoscience and Technology.* **5,** ed D Adrews G Scholes and G Wiederrecht (Amsterdam: Elsevier)
[6] Holesinger T G et al 2008 *Adv. Mater.* **20** 391
[7] Feldman D M, Holesinger T G, Maiorov B, Foltyn S R, Coulter J Y and Apodaca I 2010 *Superc. Sci. Technol.***23** 095004
[8] Obradors X, Puuig T, Ricart S, Coll M, Gazquez J, Palau A and Granados X 2012 *Superc. Sci. Technol.***25** 123001
[9] Llordes A, Palau A, Gazquez J, Coll M, Vlad R, Pomar A, Arbiol J, Guzman R, Ye S, Rouco V, Sandiumenge F, Ricart S, Puig T, Varela M, Chateigner D, Vanacken J, Gutierrez J, Moshchalkov V, Deutcher G, Magen C and Obradors X 2012 *Nat. Mater.* **11** 329
[10] Llordes A, Zalamova K, Ricart S, Palau A, Pomar A, Puig T, Hardy A, Van Bael M K and Obradors X 2010 *Chem Mater.* **22**1686
[11] Roma N, Morlens S, Ricart S, Zalamova K, Moreto J M, Pomar A, Puig T and Obradors X 2006 *Superc. Sci. Technol.* **19** 521
[12] Popa N C 1998 *Journal of Applied Crystallography.* **31** 176
[13] Williamson G K and Hall W H 1953 *Acta Metallurgia.* **1** 22

Interconnection between zero resistance and magnetic irreversibility temperatures in the hole doped $Y_{0.9}Ca_{0.1}Ba_2Cu_3O_{7-\delta}$ single crystal

D G Macedo[1], V N Vieira[1], F T Dias[1], D L da Silva[1], S Jaeckel[1], P Pureur[2] and J Schaf[2]

[1]Instituto de Física e Matemática, UFPEL, 96010-610, Pelotas, Brazil
[2]Instituto de Física, UFRGS, 91501-970, Porto Alegre, Brazil

Abstract. We report on ac magnetoresistance, $R(T,H)$ and ZFC [$M_{ZFC}(T)$] and FCC [$M_{FCC}(T)$] dc magnetizations measurements of the a hole doped $Y_{0.9}Ca_{0.1}Ba_2Cu_3O_{7-\delta}$ single crystal with the propose of to disclose the correlation between the zero resistance, $T_{R0}(H)$ and the magnetic irreversibility, $T_{irr}(H)$ temperatures. The $R(T,H)$ measurements were performed with a PPMS with the measurement current density applied parallel to the sample ab plane. The $M_{ZFC}(T)$ and $M_{FCC}(T)$ measurements were performed with a commercial SQUID magnetometer. For the both measurements $H \leq 50$kOe were applied parallel to the c axis of the sample. The contrasting of the $T_{irr}(H)$ and $T_{R0}(H)$ data profile at H-T diagram shows that for $H \leq 5$kOe the $T_{R0}(H)$ data falls systematically underneath of the $T_{irr}(H)$ data and for $H > 5$kOe the $T_{R0}(H)$ data matches to the $T_{irr}(H)$ data. We attributed to the establishment of a superconducting granular scenario provided for Ca doping as responsible for the observation of these features. At this scenario, $T_{irr}(H)$ and $T_{R0}(H)$ do not depend of the same parts of the sample. While the $T_{irr}(H)$ depends on well coupled grain clusters the $T_{R0}(H)$ depends on grain arrays traversing the whole sample. The granular aspect of this result is discussed at the light of the superconducting glass theories.

Introduction

The magnetic irreversibility of a granular superconductor is characterized by observation of Almeida-Thouless (AT) and Gabay-Toulouse (GT) power laws at low field region of its magnetic irreversibility line (MIL) [1-7]. These power laws are preceded from mean field calculations for the frustrated Ising and Heisenberg spin glass systems [1,4-6]. Otherwise the superconducting resistivity versus temperature transition of a granular superconductor is characterized as a two stage process where the lower temperature one is more affected by the magnetic field [1-3].

In a granular scenario magnetic irreversibility, $T_{irr}(H)$ and zero resistance, $T_{R0}(H)$ temperatures determined from magnetization and resistance measurements respectively are ruled by the physics of superconducting grain coupling [5,7,8]. The $T_{irr}(H)$ depends on the capacity of well coupled grain clusters transport a critical current locally given rise to the magnetic irreversibility [5,7] whereas the $T_{R0}(H)$ is achieved when a long range coherence transition takes place over the coupled grain clusters [5,7]. At the H-T diagram the $T_{irr}(H)$ and the $T_{R0}(H)$ are applied as a boundary to separate the transport of the dissipative current density from the no dissipative one [5,7]. In a homogenous (not granular) superconductor is consensual the idea that the $T_{irr}(H)$ and the $T_{R0}(H)$ temperatures match [7,8] otherwise in a granular (inhomogeneous) superconductor these temperatures are magnetic field dependent and therefore could be not coincident [5,7].

In the present work we reported precisely dc magnetization and ac magnetoresistance measurements performed in a hole doped $Y_{0.9}Ca_{0.1}Ba_2Cu_3O_{7-\delta}$ single crystal with the propose of to disclose the correlation between the zero resistance, $T_{R0}(H)$ and the magnetic irreversibility, $T_{irr}(H)$ temperatures in a inhomogeneous $YBa_2Cu_3O_{7-\delta}$ single crystal [2,5].

Sample preparation and experimental procedures

The single crystals of $Y_{0.9}Ca_{0.1}Ba_2Cu_3O_{7-\delta}$ [Y(Ca)BCO] were grown by self-flux method [5]. The selected single crystals were submitted to an extra oxygen process with the objective of improve their superconducting temperature transition, T_C [5]. The picked Y(Ca)BCO sample to the electrical and magnetic measurements had its structure and superconducting transition characterized respectively by x-ray diffraction (XRD) [2,9] and electrical resistance, $R(T,H)$ [2,10]. The figure 1A displays its XRD result. It is possible identify the characteristic YBCO-123 orthorhombic crystalline structure along its c crystallographic axis [11,12]. The c lattice parameter obtained from the XRD analysis was $c = (11.68 \pm 0.01)$Å. The $R(T)$ superconducting transition of the Y(Ca)BCO, in the absence of magnetic field, is highlight in the figure 1B. The superconducting temperature transition is defined as the temperature that corresponds to the maximum value of the resistance derivative [$dR(T,H)/dT$] plot showed in the inset of figure 1B. The single crystal shows a main transition temperature $T_C(H) \approx 89.25$K. The results presented at the figures 1A and 1B to the Y(Ca)BCO single crystal are in agreement with those reported from literature to samples well oxygenated of this material [1,2].

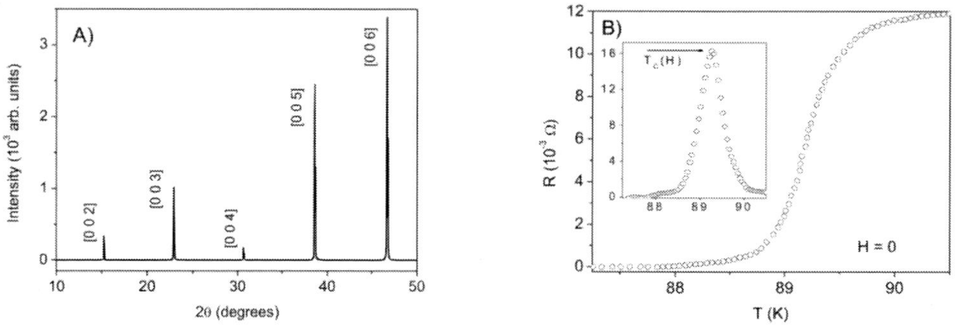

Figure 1. A) The XRD of the Y(Ca)BCO sample obtained for monochromated $CuK\ \alpha$ radiation oriented to its c axis. B) Its $R(T)$ superconducting transition to $H = 0$.

The dc magnetization and ac resistance measurements were performed at the same sample while dc magnetic fields up to 50kOe were applied parallel to its c axis (H // c).

The four contact $R(T,H)$ measurements were performed with ac low current-low frequency PPMS inset to the measurement current density applied along the ab plane of the single crystal.

In special, the contribution of density current intensity to the electrical transport properties of an inhomogeneous superconductor is very relevant [13]. At this work we applied a low excitation dc current density intensity ($J \sim 12$ Acm^{-2} that corresponds to a nominal current of $I = 0.5$mA) to all the magnetoresistance measurements with the proposition of attenuate the contribution of the electrical dissipative effects on the sample magnetoresistance response. The adoption of the last procedure allows us to focus to the majority contribution of the magnetic filed on the electrical inhomogeneous (coupling process) magnetoresistance response of our superconductor.

The $R(T,H)$ data was recorded while the sample was field cooling from $T > T_C$. During it measurements the temperature was swept very slowly ($\leq$ 0.4 K/min) so that a high number of data points could be recorded in the selected temperature range that the temperature resistance derivative [$dR(T,H)/dT$] could be numerically determinate.

The isofield dc magnetization measurements were performed with a quantum design SQUID magnetometer. The usual zero field cooled, $M_{ZFC}(T)$ and field cooled cooling, $M_{FCC}(T)$ dc magnetizations were recorded. The magnetization measurement proceeds consisted in first cooling down the sample to temperatures well below T_C in zero field (ZFC). Then the zero field cooled magnetization, $M_{ZFC}(T)$ was measured under constant magnetic field while slowly warming the sample (0.4 K/min or less) up to temperatures well above T_C. Subsequently the field cooled cooling magnetization, $M_{FCC}(T)$ was measured while cooling (0.4 K/min or less) the sample back to low temperatures in the same field (FCC).

Experimental results and discussion

The figures 2A and 2B display in their main the $\Delta M(T)$ and $dR(T,H)/dT$ plots to H = 1kOe where the $T_C(H)$, $T_{R0}(H)$ and $T_{irr}(H)$ temperatures are in eminence. Otherwise their insets highlight the correspondingly $R(T,H)$ data $M_{FCC}(T)$ and $M_{ZFC}(T)$ curves.

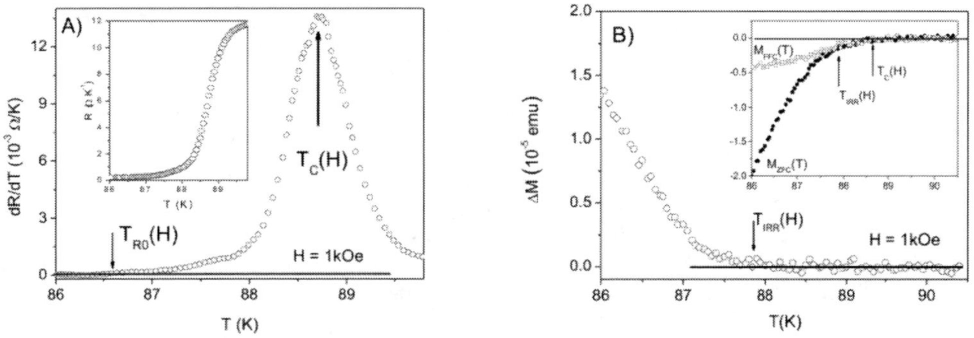

Figure 2. The determination of A) $T_{R0}(H)$ and B) the $T_{irr}(H)$ limits when the H = 1kOe is applied parallel to the c axis of the Y(Ca)BCO single crystal.

The figures 2A and 2B illustrates the method adopted to the determination of $T_{R0}(H)$ and $T_{irr}(H)$ temperatures. The $T_{R0}(H)$ is the temperature value that approximately sets the deviate from the zero base line of the $dR(T,H)/dT$ plot in the main figure 2A. The same analogous proceed is applied to the determination of the $T_{irr}(H)$ determined from the $\Delta M(T)$ plot [$\Delta M(T) = M_{FCC}(T) - M_{ZFC}(T)$] of the figure 2B.

Other aspect of the results showed in the figure 2A is the temperature asymmetric of the $dR(T,H)/dT$ data. It is possible identify in the figure 2A that the temperature resistance transition of the Y(Ca)BCO sample is characterized by a main peak with a hump at its lower temperature side. When its behavior is contrasted to the inset $dR(T,H)/dT$ data result of the figure 1B is possible to point the effect of the magnetic field. While the main peak of $dR(T,H)/dT$ data is smoothly affected for the enhancement of the applied field the hump at its lower temperature side is affected by it becoming more prominent as the applied magnetic field strength is enhanced. The classification of the superconductor magnetoresistance transition as a two stage process with the lower temperature step more affected by the magnetic field represents the signature of a granular superconductor [1-5]. In this scenario while the $dR(T,H)/dT$ peak marks the pairing critical temperature $T_C(H)$ within the superconducting grains, the hump is connected to the grain coupling and the coherence transition of the granular system. An applied field is well known to distort the phase of the GL order parameter within the junctions between the grains thereby weakening the coupling energy and the pinning strength of the Josephson flux dynamics while lets almost intact the superconducting transition within the grains governed by the Abrikosov flux dynamics [5-8].

The $T_C(H) \approx 88.8$K estimated from ac resistance and the dc magnetization, inset of figure 2B, measurements to $H = 1$kOe is in agreement. Otherwise very different were the $T_{R0}(H) \approx 86.65$K and $T_{irr}(H) \approx 87.8$K values determined to the same applied field. This is the first symptom that a coupling grain process reacts in a different way in an inhomogeneous superconductor and it does not matter if the superconductor is a single crystal. In other to disclose the correlation between magnetic irreversibility and zero resistance state of the $Y_{0.9}Ca_{0.1}Ba_2Cu_3O_{7-\delta}$ single crystal we plot the $T_{irr}(H)$ together with the corresponding $T_{R0}(H)$ data in the figure 3 where the inset stressed the low field behavior.

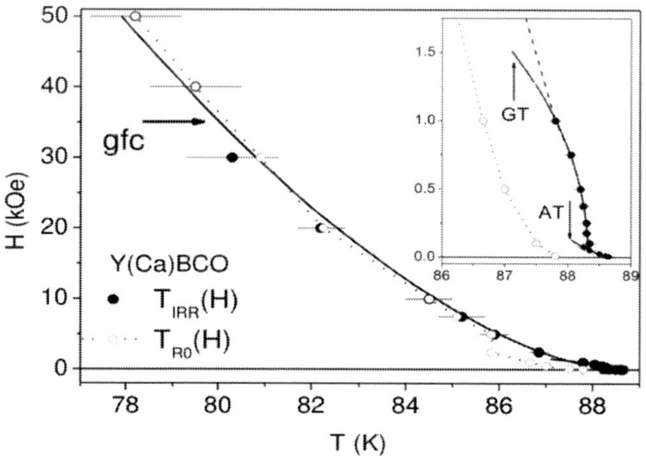

Figure 3. The correlation between the $T_{R0}(H)$ (open symbols) and $T_{irr}(H)$ (closed symbols) data of the Y(Ca)BCO sample. The inset highlight the low field details of this correlation.

In the figure 3 the solid lines identified as gfc, GT and AT are $T_{irr}(H)$ fittings with the general power law [1-5,7]:

$$H(T) = H_0\left(1 - \frac{T_{irr}(H)}{T_{irr}(0)}\right)^{\alpha} \quad (1)$$

In equation 1 the $T_{irr}(H)\ T_{irr}(0)^{-1}$ is the reduced temperature, H_0 and $T_{irr}(0)$ are fitting parameters whose are respectively the irreversibility field at zero temperature and the extrapolation of $T_{irr}(H)$ to zero field. The adjusted fitting parameters to the gfc fit were $\alpha = 1.60 \pm 0.13$, $H_0 = 1484$kOe and $T_{irr}(0) = (88.55 \pm 0.35)$K for the AT fit were $\alpha = 1.60 \pm 0.18$, $H_0 = 304$kOe and $T_{irr}(0) = (88.55 \pm 0.35)$K and for the GT fit were $\alpha = 0.48 \pm 0.01$, $H_0 = 12$kOe and $T_{irr}(0) = (88.32 \pm 0.04)$K.

The gfc continuous line through the $T_{irr}(H)$ data in the high field region ($H \geq 5$kOe) 3 is fittings with the power law [$\alpha = 3/2$ in the equation 1] predicted for the magnetic irreversibility line (MIL) by the giant-flux-creep (gfc) theory [1-5] and it is consider the signature of the intragranular Abrikosov flux dynamics [1-5,7].

The MIL in the low field region of the inset of figure 3 is adjusted by the Almeida-Thouless (AT) [1-5] [$\alpha = 3/2$ in the equation 1] and the Gabay-Toulouse (GT) [1-5] [$\alpha = 1/2$ in the equation 1] power laws. Although the AT power law has the same form as the gfc power law its physical background and the fitting parameters are quite distinct [1-5]. The description of the low field region of the MIL our sample by the AT–GT power laws is not a surprise since the grain coupling in grain aggregates of disordered superconductors under applied magnetic field are well known to be dominated by frustration, likewise the spin coupling in spin-glass systems [1,2,4]. In this scenario our Y(Ca)BCO single crystal can be described as a frustrated superconducting grain aggregates where the Josephson flux dynamics dominates [1,2,4].

The principal result of the present work is summarized when we emphasize, within our experimental precision, the correlation between the $T_{R0}(H)$ data and the MIL behaviour in the *H-T* diagram of the figure 3. It is possible to identify that at low field region ($H \leq 5$kOe) the $T_{R0}(H)$ data fall systematically below the MIL (see inset) as projected to take place to an granular superconductor. In a granular scenario when low fields are applied the starting of magnetic irreversibility is observed when the first closely packed of the aggregates of grains gets to trap vortices and blockade the magnetic flux mobility. These coherent grain clusters may remain disconnected from each other because of weaker links that are resistive down to lower temperature. A macroscopic superconducting state only stabilizes when tunneling of the Cooper pairs becomes strong enough to enforce a percolation coherence transition in the granular superconducting array justify the observation of the $T_{irr}(H) > T_{R0}(H)$ in the low field region of the Y(Ca)BCO sample *H-T* diagram.

The $T_{irr}(H)$ and $T_{R0}(H)$ correlation showed in the figure 3 to the $Y_{0.9}Ca_{0.1}Ba_2Cu_3O_{7-\delta}$ single crystal complements and magnify the preliminary results in this theme obtained to a $Y_{0.98}Ca_{0.02}Ba_2Cu_3O_{7-\delta}$ single crystal in the low field limit ($H \leq 0.5$kOe) [2]. The results displayed in the figure 3 suggests that the $T_{irr}(H)$ and $T_{R0}(H)$ correlation could be applied as a efficiently tool to separate the Josephson and Abrikosov flux dynamics ascendancy in the *H-T* diagram of an inhomogeneous superconductor.

In this work we reported ac magnetoresistance and dc magnetization measurements with the propose of the disclose the correlation between the zero resistance state and magnetic irreversibility in a hole doped $Y_{0.9}Ca_{0.1}Ba_2Cu_3O_{7-\delta}$ single crystal. The magnetoresistance transition and the magnetic irreversibility line behaviours show that the 10% partial substitution of the Ba atoms for Ca atoms introduces a superconducting granular character to the electrical and magnetic properties of the $YBa_2Cu_3O_{7-\delta}$ single crystals. The correlation between the $T_{R0}(H)$ and $T_{irr}(H)$ data shows that electrical transport remains dissipative for $T_{R0}(H) \leq T < T_{irr}(H)$ while $H \leq 5$kOe are applied parallel to the c axis of the single crystal.

Acknowledgments
The authors would like to thanks to the PRONEM PROGRAM, under contract 11/2042-7, CAPES, CNPq and FAPERGS scientific agencies for partially financing this work.

References

[1] Vieira V N, Pureur P and Schaf J 2002 *Phys Rev B* **66** 224506
[2] Vieira V N, Riegel I C and Shaf J 2007 *Phys Rev B* **76** 024518
[3] Vieira V N, Pureur P and Schaf J 2002 *Phys Rev B* **65** 144531
[4] Vieira V N, Silva J P and Schaf J 2001 *Phys Rev B* **64** 094516
[5] Dias F T, Vieira V N, Rodrigues Jr P, Pureur P and Schaf J 2008 *Phys Rev B* **77** 134503
[6] Muller K A, Takashige M and Bednorz J G 1987 *Phys Rev Lett* **58** 11 pp 1143-1146
[7] Dias F T 2012 *Magnetic irreversibility and resistive transition in YBaCuO superconductors: interpretationsand possible correlations* (Superconductors – Properties, Technology and Applications) ed by Yury Grigorashvili chapter 10
[8] Clem J 1988 *Phys C* **153-155 0**1 pp 50-55
[9] Rajivi G et al. 2005 *Phys C* **419** pp 101-108
[10] Vieira V N 2008 *J Magn Magn Mater* **320** pe500-e503
[11] Jorgensen J D et al.1990 *Phys Rev B* **41** 04
[12] Beyers R at al. 1987 *App Phys Lett* **50** 26 pp 1918-1920
[13] Hilgenkamp H and Mannhart J 2002 *Reviews of Modern Physics* **74** 485

Superconductor-insulator transition in frustrated Josephson-junction arrays on a honeycomb lattice

Enzo Granato

Laboratório Associado de Sensores e Materiais, Instituto Nacional de Pesquisas Espaciais, 12227-010 São José dos Campos, SP, Brazil

E-mail: enzo@las.inpe.br

Abstract. The zero-temperature superconductor to insulator transition is studied in a self-charging model of Josephson-junction arrays on a honeycomb lattice in an external magnetic field corresponding to f flux quantum per plaquette. Path integral Monte Carlo simulations of the equivalent (2+1)-dimensional classical model are used to study the phase transition and critical behavior. For $f = 1/3$, the transition is first order. For $f = 0$ and $f = 1/2$, the transition is second order and the corresponding correlation length exponents are estimated from finite-size scaling.

1. Introduction

The superconductor to insulator transition in Josephson junction arrays have attracted considerable interest [1, 2, 3, 4, 5, 6, 7, 8, 9, 10] as a physical realization of a quantum phase transition. Such arrays can currently be fabricated in any desired geometry both in one and two dimensions (2D) [4] as a lattice of superconducting grains coupled by the Josephson or proximity effect and with well-controlled parameters. When charging effects due to the small capacitance of the grains or junctions dominate over the Josephson coupling, strong quantum fluctuations of the phase of the superconducting order parameter may drive the system into an insulating phase at zero temperature leading to a superconductor to insulator transition as a function of the ratio of charging energy to the Josephson coupling. In presence of an external magnetic field, frustration effects lead to distinct universality classes which can depend on the geometry of the array. For a Josephson-junction array on square lattice, the universality class of these transitions have already been investigated in detail numerically [10, 11], both in relation to experiments [3, 4] and theoretical predictions [7, 8, 13]. However, for a Josephson-junction array on a honeycomb lattice the superconductor to insulator transition has not been investigated in detail. Such array is particularly interesting as a simple model [14] for ultra-thin superconducting films with a triangular pattern of nanoholes [15]. This system undergoes a superconductor to insulator transition for decreasing thickness with properties very sensitive to the applied magnetic field, which is the analog of the transition in the superconducting array for increasing ratio of the charging energy to the Josephson coupling at different frustration parameters.

In this work, we study the zero-temperature superconductor to insulator transition in a self-charging model of Josephson-junction arrays in an external magnetic field corresponding to f flux quantum per plaquette. Path integral Monte Carlo simulations of the equivalent (2+1)-dimensional classical model are used to study the phase transition and critical behavior. For

$f = 1/3$, we find that the transition is first order. For $f = 0$ and $f = 1/2$, the transition is second order and the corresponding correlation length exponents are estimated from finite-size scaling.

2. Model and simulation

We consider a Josephson-junction array on a honeycomb lattice as illustrated in Fig. 1, where charging effects are dominated by the capacitance to the ground of each grain [1, 6], and described by the Hamiltonian

$$\mathcal{H} = -\frac{E_c}{2}\sum_r \left(\frac{d}{d\theta_r}\right)^2 - E_J \sum_{<rr'>} \cos(\theta_r - \theta_{r'} - A_{rr'}). \quad (1)$$

The first term in Eq. (1) describes quantum fluctuations induced by the charging energy $E_c = 4e^2/C$ of a non-neutral superconducting grain located at site r, where e is the electronic charge and C is the effective capacitance to the ground of the grain, while the second term is the usual Josephson-junction coupling between nearest-neighbor grains described by phase variables θ_r. The effect of the applied magnetic field appears through the bond variables $A_{rr'} = (2\pi/\Phi_o)\int_r^{r'} A \cdot dr$, where $\mathbf{A}$ is the vector potential due to the external magnetic field $\mathbf{B}$ and the gauge-invariant sum around an elementary cell of the array is given by $\sum_{rr'} A_{rr'} = 2\pi f$ with $f = \Phi/\Phi_o$ defining the frustration parameter, which corresponds to the number of flux quanta per plaquette. This model is periodic in f with period $f = 1$.

In order to study the phase transition, it is convenient to use an imaginary-time path-integral formulation of the model [2]. In this formulation, the 2D quantum problem of Eq. (1) maps into a (2+1)D classical statistical mechanics problem with the extra dimension corresponding to the imaginary-time direction. The time axis τ can be discretized in slices $\Delta\tau$ and the ground state energy of the quantum model corresponds to the reduced free energy of the classical model, per unit length in the imaginary time direction. After re-scaling the time slices appropriately in order to get space-time isotropic couplings, the resulting classical partition function is given by $Z = Tr_{\{\theta\}}e^{-H}$ where the reduced classical Hamiltonian can be defined as

$$\begin{aligned} H = \quad & -\frac{1}{g}[\sum_{\tau,j} \cos(\theta_{\tau,j} - \theta_{\tau+1,j}) \\ & + \sum_{<ij>,\tau} \cos(\theta_{\tau,i} - \theta_{\tau,j} - A_{ij})]. \end{aligned} \quad (2)$$

In the above equation, j and τ label the spatial and time directions, respectively, and the ratio of the charging energy to the Josephson coupling $g = (E_c/E_J)^{1/2}$ plays the role of an effective temperature in the 3D classical model. This classical Hamiltonian can be viewed as a frustrrated XY model on a layered honeycomb lattice.

We carry out MC simulations using the 3D classical Hamiltonian in Eq. (2) regarding g as a temperature-like parameter and employing the exchange MC method (parallel tempering) [16]. In this method, many replicas of the system with different couplings g in a range above and below the critical point are simulated in parallel and the corresponding configurations are allowed to be exchanged with a probability distribution satisfying detailed balance. Simulations are performed in system sizes with equal spatial and time linear length L. This choice of the aspect ratio of the system assumes implicitly that the dynamic critical exponent z characterizing the superconductor to insulator transition is close to $z \sim 1$. In general, a quantum phase transition is characterized by intrinsic anisotropic scaling with different diverging correlation lengths ξ and ξ_τ in the spatial and time directions [2], respectively, related by the dynamic exponent z as $\xi_\tau \propto \xi^z$. Our choice is justified by the observation that the best data collapse implied by the scaling behavior discussed below is obtained for $z \sim 1$.

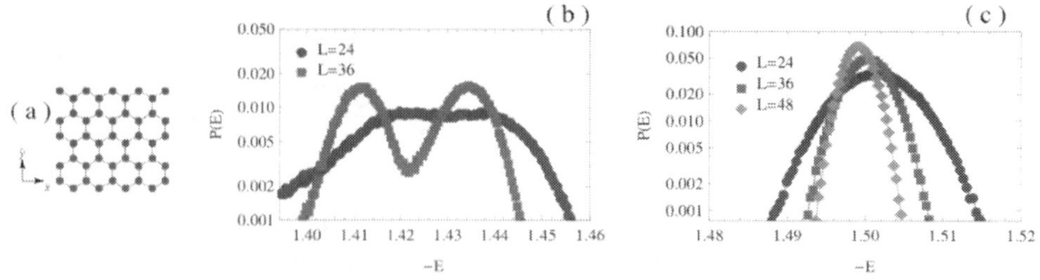

Figure 1. (a) Josephson-junction array on a honeycomb lattice. Filled circles represent superconducting grains and the lines the Josephson junctions. (b) and (c) Probability distribution of the energy $-E$ near the transition for different system sizes L in the equivalent $(2+1)$D classical model of Eq. (2), for (b) f=1/3 and (c) f=1/2.

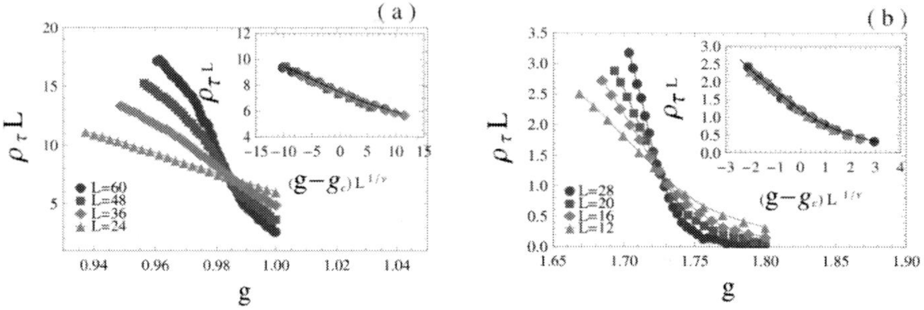

Figure 2. (a) Phase stiffness in the imaginary time direction ρ_τ for $f = 1/2$. Inset: scaling plot for data near the transition and systems sizes $L \geq 36$ with $g_c = 0.9837$ and $\nu = 0.51$. b) Phase stiffness ρ_τ for $f = 0$. Inset: corresponding scaling plot for data near the transition with $g_c = 1.7255$ and $\nu = 0.66$.

3. Numerical results and discussion

We first determine the nature of the transition by examining the histogram of the probability distribution for the energy density near the transition. As can been seen from Fig. 2, double maxima appears for $f = 1/3$, indicating that there are two coexisting phases corresponding to a first order transition. On the other hand, for $f = 1/2$, such feature is absent indicating that the transition is continuous. In the absence of the external magnetic field, $f = 0$, one expects that the superconductor-insulator transition is in the universality class of the 3D classical XY model, for which the correlation length exponent is well known. For $f = 1/2$, however, the critical behavior is unknown. Below, we verify the expected scaling behavior for the continuous transitions and obtain an estimate of the critical exponents.

To locate the superconductor-insulator transition at the critical coupling g_c and determine the critical exponents, we have performed MC calculations of the phase stiffness ρ_τ in the time direction and the finite-size correlation lengths ξ and ξ_τ measuring the spatial and time decay of the phase-correlation function for $g \geq g_c$. In the superconducting phase ρ_τ should be finite, reflecting the existence of phase coherence, while in the insulating phase it should vanish in the thermodynamic limit. For a continuous phase transition, the phase stiffness should satisfy the finite-size scaling form

$$\rho_\tau L = F(L^{1/\nu}\delta g), \tag{3}$$

where $F(x)$ is a scaling function and $\delta g = g - g_c$. This scaling form implies that data for the quantity $\rho_\tau L$ as a function of g, for different system sizes L, should cross at the critical coupling

g_c. Moreover, a scaling plot of $\rho_\tau L \times L^{1/\nu}\delta g$ sufficiently close to g_c should collapse on to the same curve if ν is chosen correctly. Fig. 2a shows the behavior of $\rho_\tau L$ as a function of g for different system sizes L, when $f = 1/2$. The transition point g_c is located where the curves cross. In the inset of Fig. 2a, we show a scaling plot of the data close to the transition for the largest system sizes ($L = 36 - 60$) according to the scaling form of Eq. (3), which provides the estimates $g_c = 0.9837(5)$ and $\nu = 0.51(6)$. For comparison, in Fig. 2b we show similar results for the unfrustrated case, $f = 0$, which gives the estimates $g_c = 1.7255(5)$ and the correlation length exponent $\nu = 0.66(4)$ in good agreement with the known result for the 3D XY model [18], $\nu = 0.671$. Note that the critical coupling $g_c(f = 1/2)$ is significantly smaller than $g_c(f = 0)$

We have also investigated the scaling behavior of the finite-size correlation lengths in the $\hat{x}$ direction ξ and in the time direction ξ_τ. The correlation length in the finite system can be obtained from a second moment calculation using the correlation function as [17]

$$\xi(L) = \frac{1}{2\sin(k_0/2)}[S(0)/S(k_0) - 1]^{1/2}, \tag{4}$$

where $S(k)$ is given by the Fourier transform of the phase correlation function $C(r)$ and k_0 is the smallest nonzero wave vector in the finite system. As in the previous work in absence of quantum fluctuations [14], it is convenient to define the correlation function in terms of the overlap order parameter $q(j) = \exp(i\theta_j^1 - i\theta_j^2)$, where 1 and 2 denote two independent copies of the system, with same parameters g and f in the model of Eq. (2). For a continuous transition, the finite-size correlation length $\xi(L, g)$ should satisfy the scaling form

$$\xi/L = G(L^{1/\nu}\delta g), \tag{5}$$

where $G(x)$ is a scaling function. Similar scaling form applies to the correlation length ξ_τ in the time direction. According to this scaling form, curves of $\xi(L, g)/L$ as a function of g, for different system sizes L, should cross at the critical coupling g_c and a scaling plot of $\xi(L, g)/L \times L^{1/\nu}\delta g$ sufficiently close to g_c should collapse on to the same curve. Figs. 3 shows the finite-size behavior of the correlation lengths ξ_τ and ξ scaled by the system size L as function of g, for $f = 1/2$. The curves for the largest systems cross at the same point, demonstrating that there is a continuous transition. In the insets of Figs. 3a and 3b, we show a scaling plot of the data according to the scaling form of Eq. (5), which provides the estimates $g_c = 0.9836(5)$ and $\nu = 0.44(5)$, and $g_c = 0.9845(5)$ and $\nu = 0.48(5)$, respectively. These alternative estimates and the value obtained from the phase-stiffness scaling are in reasonable agreement, within the errorbars, leading to the final result $\nu = 0.48(4)$.

4. Conclusions

We have studied the superconductor to insulator transition in a self-charging model of Josephson-junction arrays on a honeycomb lattice in an external magnetic field corresponding to f flux quantum per plaquette. The model can be physically as two-dimensional arrays of weakly coupled superconducting grains and ultra-thin superconducting films with a triangular pattern of nanoholes [14, 15]. From Path integral Monte Carlo simulations of the equivalent (2+1)-dimensional classical model we found that for $f = 1/3$, the superconductor to insulator transition for increasing ratio g of the charging energy to the Josephson coupling is first order. For $f = 0$ and $f = 1/2$, the transition is second order and the corresponding correlation length exponents are estimated from finite-size scaling. Since the critical coupling $g_c(f = 1/2)$ is significantly smaller than $g_c(f = 0)$, magnetoresistance measurements in the insulating side near the $f = 0$ transition should display oscillations with minima at integer values $f = n$ and maxima at $f = n + 1/2$. This is in fact observed experimentally at low temperatures in superconducting films with a triangular pattern of nanoholes [15]. The $f = 1/2$ case is of particular interest

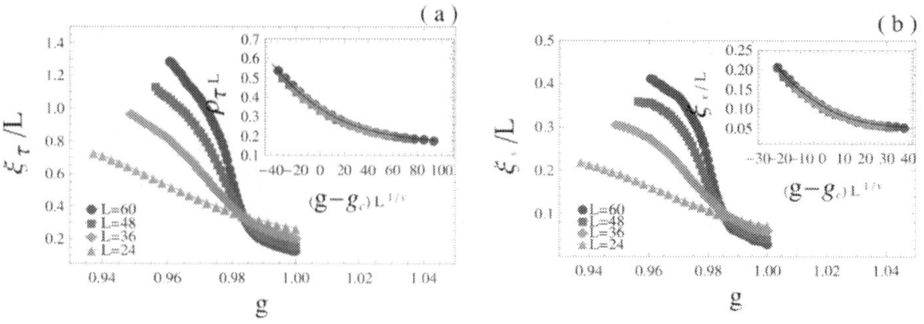

Figure 3. (a) Behavior of the scaled correlation length in the time direction ξ_τ/L for $f = 1/2$, for different systems sizes L. Inset: scaling plot for data near the transition and systems sizes $L \geq 36$ with $g_c = 0.9836$ and $\nu = 0.44$. b) Behavior of the scaled correlation length in the $\hat{x}$ spatial direction ξ_x/L for $f = 1/2$. Inset: corresponding scaling plot with $g_c = 0.9845$ and $\nu = 0.48$.

because even without charging effects, geometrical frustration and thermal fluctuations leads to an unusal phase transition as function of temperature, which is still not well understood [14, 19]. It should also be interesting to investigate the universal resistivities at the transition for $f = 1/2$ and $f = 0$ in comparison to the results obtained earlier for the same model defined on a square lattice [7, 11].

5. Acknowledgments
This work was supported by Fundação de Amparo à Pesquisa do Estado de São Paulo (FAPESP, Grant No. 2014/15372-3) and computer facilities from CENAPAD-SP.

References
[1] Fazio R and van der Zant H 2001 *Phys. Rep.* **355** 235
[2] Sondhi S L, Girvin S M, Carini J P, and Sahar D 1997 *Rev. Mod. Phys.* **69** 315
[3] Geerligs L J, Peters M, de Groot L E M, Verbruggen A, and Mooij J E 1989 *Phys. Rev. Lett.* **63** 326
[4] van der Zant H S J, Geerligs L J and Mooij J E 1992 *Europhys. Lett.* **19** 541
[5] Al-Saidi W A and Stroud D 2004 *Physica* C **402** 216
[6] Bradley R M and Doniach S 1984 *Phys. Rev.* B **30** 1138
[7] Granato E and Kosterlitz J M 1990 *Phys. Rev. Lett.* **65** 1267
[8] Fazio R and Schön G 1991 *Phys. Rev.* B **43** 5307
[9] Granato E 2005 *Phys. Rev.* B **72** 104521
[10] Lee H and Cha M -C 2002 *Phys. Rev.* B **65** 172505
[11] Cha M -C and Girvin S M 1994 *Phys. Rev.* B **49** 9794
[12] Cha M -C, Fisher M P A, Girvin S M, Wallin M, and Young A P 1991 *Phys. Rev.* B **44** 6883
[13] Fisher M P, Grinstein G and Girvin S M 1990 *Phys. Rev. Lett.* **64** 587
[14] Granato E 2013 *Phys. Rev.* B **87** 094517; Granato E 2012 *Phys. Rev.* B **85** 054508
[15] Stewart Jr. M D, Yin Aijun, Xu J M and Valles Jr. J M 2007 *Science* **318** 1273 ; Hollen S M, Fernandes G E, Xu J M and Valles J M 2013 *Phys. Rev.* B **87** 054512
[16] Hukushima K and Nemoto K 1996 *J. Phys. Soc. Jpn.* **65** 1604
[17] Ballesteros H G, Cruz A, Fenández L A, Martin-Mayor V, Pech J, Ruiz-Lorenzo J J, Tarancón A, Téllez P, Ullod C L and Ungil C 2000 *Phys. Rev.* B **62** 14237
[18] Guida R and Zinn-Justin J 1998 *J. Phys.* A **31** 8103
[19] Korshunov S E 2012 *Phys. Rev.* B **85** 134526

Journal of Physics: Conference Series **568** (2014) 022018 doi:10.1088/1742-6596/568/2/022018

Vortex Core Structure in Multilayered Rashba Superconductors

Y Higashi[1], Y Nagai[2], T Yoshida[3] and Y Yanase[4]
[1]Department of Mathematical Sciences, Osaka Prefecture University, 1-1 Gakuen-cho, Sakai 599-8531, Japan
[2]CCSE, Japan Atomic Energy Agency, 5-1-5 Kashiwanoha, Kashiwa, Chiba 277-8587, Japan
[3]Graduate School of Science and Technology, Niigata University, Niigata 950-2181, Japan
[4]Department of Physics, Niigata University, Niigata 950-2181, Japan

E-mail: higashiyoichi@ms.osakafu-u.ac.jp

Abstract. We numerically study the electronic structure of a single vortex in two dimensional superconducting bilayer systems within the range of the mean-field theory. The lack of local inversion symmetry in the system is taken into account through the layer dependent Rashba spin-orbit coupling. The spatial profiles of the pair potential and the local quasiparticle density of states are calculated in the clean spin-singlet superconductor on the basis of the quasiclassical theory. In particular, we discuss the characteristic core structure in the pair-density wave state, which is spatially modulated exotic superconducting phase in a high magnetic field.

1. Introduction

Intensive studies on locally noncentrosymmetric systems (LNCS) has been conducted [1, 2]. The LNCS are realized in various materials such as multilayered cuprates [3], the doped topological insulator $Cu_xBi_2Se_3$ [4], the pnictide superconductor SrPtAs [5] and so on. Especially, the heavy-fermion $CeCoIn_5/YbCoIn_5$ superlattice is fabricated [6] and the artificial control of the inversion symmetry breaking was realized recently [7]. Therefore, one can seek experimentally the theoretically predicted exotic superconducting phase in a magnetic field [8, 9]. The inhomogeneity due to a magnetic field is not taken into account in the existing theoretical researches on the superconductivity in LNCS [8, 9]. However, the exotic superconducting phase stabilizes in a magnetic field. So understanding the vortex states in LNCS is necessary. The observation of quasiparticle states in the vicinity of a vortex gives us a lot of information on the exotic superconducting phase. We focus on the exotic superconducting phase called the pair-density wave (PDW) phase, which is stabilized in a high magnetic field perpendicular to the layers [8]. In the PDW phase, the phase of the superconducting order parameter changes by π between the two layers, whereas no phase difference of the order parameter in the BCS phase in a low magnetic field. In this paper, we investigate the vortex core structure of the PDW and the BCS phases, and try to obtain the characteristics of the PDW phase.

2. Formulation

We consider the multilayered system characterized by the layer-dependent spin-orbit coupling (SOC) strength α_m with the layer index m. In this paper, the number of the layer is fixed to $N = 2$. The layer dependence of the SOC $(\alpha_1, \alpha_2) = (\alpha, -\alpha)$ reflects the local noncentrosymmetricity of the system [10]. We assume the simple form of the Rashba type SOC described by $\boldsymbol{g}(\boldsymbol{k}) = (-k_y, k_x, 0)/k_{\rm F}$, which satisfies the normalization condition $\langle \boldsymbol{g}(\boldsymbol{k})\rangle_{\boldsymbol{k}} = 1$ on the two dimensional Fermi surface (FS). $k_{\rm F}$ is the Fermi wave number and $\langle \cdots \rangle_{\boldsymbol{k}}$ is the average over the FS.

We consider spin-singlet s-wave pairing states within the layer. The pairing potential in the spin space is expressed as $\Delta_{s_1 s_2}(\boldsymbol{r}) = \Delta(\boldsymbol{r})(i\sigma_y)_{s_1 s_2}$. We assume that the spatial variation of the pair potential $\Delta(\boldsymbol{r})$ is the same in each layer. Then, the pair potential is expressed as $\hat{\Delta}(\boldsymbol{r}) = \Delta(\boldsymbol{r}) i\sigma_y \otimes D$, where D is the $N \times N$ diagonal matrix in the space composed of the layer (band) degree of freedom. In the case of $N = 2$, $D = \mathrm{diag}(1, s)$ with $s = 1$ for the BCS state and $s = -1$ for the PDW one. $\hat{\cdot}$ denotes the $2N \times 2N$ matrix in the spin and band space. A vortex line perpendicular to the two dimensional superconducting layer has the form, $\Delta(\boldsymbol{r}) = |\Delta(r)|\exp(i\phi_r)$ in each layer.

In this paper, we consider the superconductor in which the paramagnetic pair breaking effect is dominant. In addition, we assume the system is in the type-II limit and neglect the vector potential. Then, we incorporate only the paramagnetic pair breaking effect into the quasiclassical theory through the Zeeman term. There is three factors which split the 4-fold degenerated FS. The spin degeneracy is lifted by both the SOC and the Zeeman field $\mu_{\rm B} H$, and the band degeneracy is lifted by the interlayer hopping $t_\perp$. We assume that the strength of these three factors is sufficiently weak and incorporate them into the quasiclassical theory as perturbations. In this situation, the split of the FS is small and the difference of the Fermi velocity between the weakly split FS is negligibly small. So we can put $\boldsymbol{v}_{\rm F} = v_{\rm F}\tilde{\boldsymbol{k}}$. Then, we can expand the quasiclassical approximation described in Ref. [11] into the multilayered system.

The quasiclassical Green's function depends on the direction $\tilde{\boldsymbol{k}} = (\cos\phi_k, \sin\phi_k)$ with the azimuthal angle ϕ_k and is written as a $4N \times 4N$ matrix in the particle-hole space:

$$\check{g}(\boldsymbol{r}, \tilde{\boldsymbol{k}}, i\omega_n) = -i\pi \begin{pmatrix} \hat{g} & i\hat{f} \\ -i\hat{\bar{f}} & -\hat{\bar{g}} \end{pmatrix}, \tag{1}$$

where ω_n is the Matsubara frequency. The quasiclassical Green's function obeys the following Eilenberger equation with the spin quantization axis parallel to the z axis [12].

$$i\boldsymbol{v}_{\rm F}(\tilde{\boldsymbol{k}}) \cdot \boldsymbol{\nabla}\check{g}(\boldsymbol{r}, \tilde{\boldsymbol{k}}; i\omega_n) + \left[i\omega_n\check{\tau}_3 - \check{\Delta}(\boldsymbol{r}, \tilde{\boldsymbol{k}}) - \check{K}(\tilde{\boldsymbol{k}}), \check{g}(\boldsymbol{r}, \tilde{\boldsymbol{k}}; i\omega_n)\right] = \check{0}, \tag{2}$$

$$\check{K}(\tilde{\boldsymbol{k}}) \equiv \begin{pmatrix} t_\perp \hat{H}_{\rm inter} + \mu_{\rm B} H \hat{H}_{\rm Z} & \hat{0} \\ \hat{0} & t_\perp \hat{H}_{\rm inter} + \mu_{\rm B} H \hat{H}_{\rm Z} \end{pmatrix} + \alpha \begin{pmatrix} \hat{H}_{\rm SO}(\tilde{\boldsymbol{k}}) & \hat{0} \\ \hat{0} & \hat{H}^*_{\rm SO}(-\tilde{\boldsymbol{k}}) \end{pmatrix}, \tag{3}$$

with

$$\check{\tau}_3 = \begin{pmatrix} \sigma_0 \otimes I_{N\times N} & \hat{0} \\ \hat{0} & -\sigma_0 \otimes I_{N\times N} \end{pmatrix}, \quad \check{\Delta}(\boldsymbol{r}) = \begin{pmatrix} \hat{0} & \hat{\Delta}(\boldsymbol{r}) \\ -\hat{\Delta}^\dagger(\boldsymbol{r}) & \hat{0} \end{pmatrix}, \tag{4}$$

$$\hat{H}_{\rm Z} = -\sigma_z \otimes I_{N\times N}, \quad \hat{H}_{\rm inter} = \sigma_0 \otimes T_\perp, \quad \hat{H}_{\rm SO}(\tilde{\boldsymbol{k}}) = \boldsymbol{g}(\tilde{\boldsymbol{k}}) \cdot \boldsymbol{\sigma} \otimes S_{\rm d}. \tag{5}$$

Here, $I_{N\times N}$ is the unit matrix, $T_\perp = \mathrm{offdiag}(1,1)$ and $S_{\rm d} = \mathrm{diag}(1,-1)$. $\mathrm{offdiag}(\cdot,\cdot)$ indicates the 2×2 matrix which has the offdiagonal component only. $\hat{H}^*_{\rm SO}(-\tilde{\boldsymbol{k}}) = -\boldsymbol{g}(\tilde{\boldsymbol{k}}) \cdot \boldsymbol{\sigma}^{\rm tr} \otimes S_{\rm d}$, which is obtained from the relation $\boldsymbol{g}(-\tilde{\boldsymbol{k}}) = -\boldsymbol{g}(\tilde{\boldsymbol{k}})$. $\mu_{\rm B}$ is the Bohr magneton. We use the unit in which $\hbar = k_{\rm B} = 1$. For the numerical calculation, we employ the Riccati formalism. Regarding $\check{K}(\tilde{\boldsymbol{k}})$ as the self energy, one can use the same manner described in Ref. [13].

Starting from the pairing interaction in Ref. [14], the gap equation for the spin-singlet pair potential $\Delta(\boldsymbol{r})$ is given by [15]

$$\Delta(\boldsymbol{r}) = \lambda\pi T \frac{1}{2} \sum_{-n_{\rm c}(T)-1<n<n_{\rm c}(T)} \sum_{s'_1 s'_2} (i\sigma_y)^\dagger_{s'_2 s'_1} \left\langle f^0_{s'_1 s'_2}(\boldsymbol{r}, \tilde{\boldsymbol{k}}'; i\omega_n)(i\sigma_y)_{s'_1 s'_2} \right\rangle_{\tilde{\boldsymbol{k}}'}. \tag{6}$$

$\langle \cdots \rangle_{\tilde{k}}$ denotes the average on the circular arc with its radius 1. The coupling constant is determined from

$$\frac{1}{\lambda} = \ln\left(\frac{T}{T_{c0}}\right) + \sum_{0 \le n < n_c(T)} \frac{2}{2n+1}, \tag{7}$$

where T_{c0} is the superconducting transition temperature for $\alpha = t_\perp = \mu_B H = 0$ and $n_c(T) = (\omega_c/\pi T - 1)/2$. We fix the cut off frequency to $\omega_c = 7\pi T_{c0}$.

The local density of states per spin and layer is given by

$$N(E, \boldsymbol{r}) = -\frac{N_F}{2N}\frac{1}{\pi}\left\langle \mathrm{Im}\left[\mathrm{Tr}\hat{g}(\boldsymbol{r}, \tilde{\boldsymbol{k}}, i\omega_n \to E + i\eta)\right]\right\rangle_{\tilde{k}}, \tag{8}$$

where N_F is the density of states per spin and layer at the Fermi level in the normal state.

3. Results and discussions

In the numerical results, we fix the temperature, the SOC and the interlayer hopping to $T/T_{c0} = 0.1, \alpha/T_{c0} = 2$ and $t_\perp/T_{c0} = 1$. In Fig. 1, we show the spatial variations of the pair potential amplitude around a vortex for (a) the BCS and (b) the PDW states. The horizontal axis is normalized by the coherence length $\xi_0 = v_F/T_{c0}$. Each plot corresponds to the different magnetic field. The amplitude of the pair potential gets smaller with increasing the Zeeman field due to the paramagnetic pair breaking effect in both pairing states. One can notice that the pair potential amplitude of the PDW state is smaller than that of the BCS state. It is because the PDW state stabilizes in a high magnetic field under the influence of the paramagnetic depairing. Though the amplitude of the pair potential in the PDW state is smaller than that in the BCS state, the core size in the PDW state does not spread [compare the profiles for $\mu_B H/T_{c0} = 1.5$]. The core size in the PDW state for $\mu_B H/T_{c0} = 1.5$ and 2 is smaller than that in the BCS state for $\mu_B H/T_{c0} = 1.5$. Thus, the sudden decrease in the core radius with increasing the magnetic field is a signature of the BCS-PDW phase transition. The BCS state is completely destroyed for $\mu_B H/T_{c0} \gtrsim 2$, whereas in the PDW state, the nonzero self-consistent solutions of the pair potential exist even for the high magnetic field $\mu_B H/T_{c0} \gtrsim 2$. The profiles of the pair potential is not altered for $\alpha/T_{c0} = 50$.

In the uniform system without vortices, the stable pairing state prefers the larger bulk amplitude of the pair potential. In this paper, though there is a vortex in the system, we consider that the pair potential with the larger bulk amplitude is the more stable solution of the gap equation for a given magnetic field. The T-H phase diagram of bilayer system is displayed in Ref. [8], which is not altered qualitatively in the vortex state when the system belongs to type-II limit and has the large Maki parameter [8, 16]. According to the field dependence of the pair potential amplitude in the bulk obtained by the quasiclassical theory, the non-zero self-consistent solution of the gap equation exists only in the PDW state for $\mu_B H/T_{c0} \gtrsim 1.8$. This is consistent with the results calculated from the Bogoliubov-de Gennes equation [8].

In Fig. 2, we display the energy and spatial dependence of the local density of states (LDOS) around a vortex core. On the basis of the above consideration, the calculation of the LDOS is conducted for (a) $\mu_B H/T_{c0} = 1.5$ in the BCS state and (b) $\mu_B H/T_{c0} = 2$ in the PDW one. As shown in Fig. 2(a), the zero energy peak (ZEP) splits off due to the Zeeman field. The Zeeman split of the ZEP gets suppressed with increasing the SOC strength α for a fixed magnetic field, since the spin quantization axis is locked within the x-y plane by the Rashba SOC.

Contrastingly, though the PDW state emerges in a high magnetic field, the zero energy peak exists, which is quite different from the LDOS structure in the BCS state. If the zero energy quasiparticle states within a core are experimentally observed by scanning tunneling microscopy/spectroscopy (STM/STS), the STM/STS image of a vortex in the PDW state can be drastically different from that in the BCS state. To obtain the clue to the LDOS structure

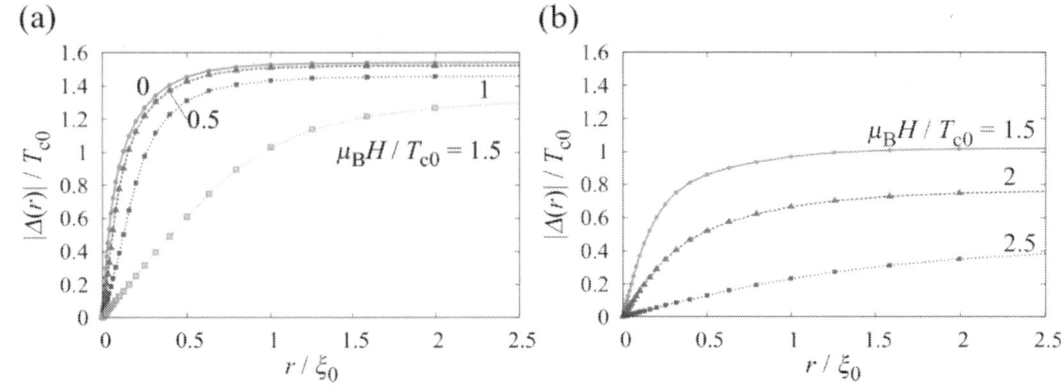

Figure 1. (color online) Spatial profiles of the pair potential amplitude $|\Delta(r)|$ for (a) the BCS state and (b) the PDW one. The horizontal axis represents the radial distance from a vortex center $r = 0$. We set $T/T_{c0} = 0.1$, $\alpha/T_{c0} = 2$ and $t_\perp/T_{c0} = 1$. The data is plotted for each Zeeman field $\mu_B H/T_{c0}$.

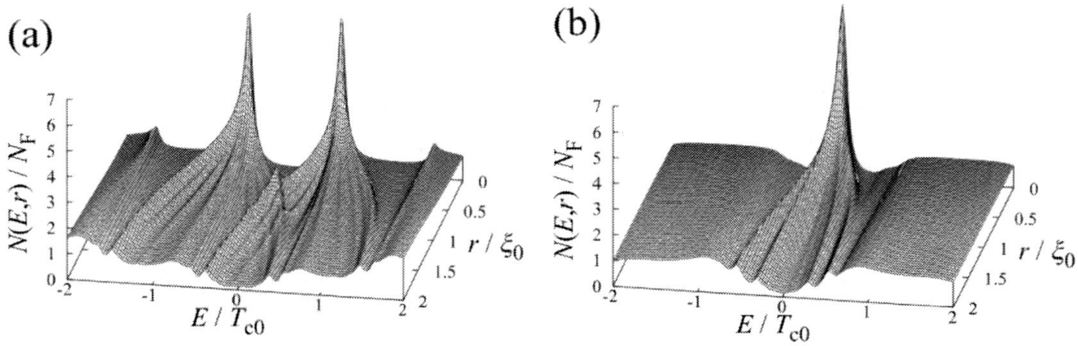

Figure 2. The local density of states $N(E, r)$ for (a) the BCS state and (b) the PDW one. The Zeeman field is set to (a) $\mu_B H/T_{c0} = 1.5$ and (b) $\mu_B H/T_{c0} = 2$. Other conditions are $T/T_{c0} = 0.1$, $\alpha/T_{c0} = 2$, $t_\perp/T_{c0} = 1$ and $\eta = 0.05T_{c0}$.

in the PDW state, we start from the case with $\alpha/T_{c0} = 0$ for simplicity. In this case, the four folding ZEP splits into four peaks. First, for $\mu_B H/T_{c0} = 0$, the band degeneracy is lifted by the interlayer hopping and the ZEP splits into the two peaks. This split of the ZEP is regarded as the level repulsion of the zero energy bound states. In the BCS state, the band degeneracy is not lifted only by the interlayer hopping. Then, the π phase difference between the layers is essential for the repulsion of the zero energy bound states. Second, each peak is split off due to the Zeeman field. Hence the four peaks appear due to the interlayer hopping and the Zeeman field. Further increasing the magnetic field, the two peaks in the low energy side shift to the lower energy side and those in the high energy side shift to the higer energy one. Eventually the two peaks in the low energy side get combined and a single zero energy peak appears.

In the real materials, for example, the $CeCoIn_5/YbCoIn_5$ superlattice, the superconducting layer is $CeCoIn_5$, which is experimentally identified as d-wave superconductors [17]. The LDOS structure obtained in this study changes considerably, since the quasiparticle states around a vortex are sensitive to the pairing symmetry. The more realistic calculation assuming the d-wave pairing symmetry is the future work.

4. Summary

We have numerically investigated the electronic structure of a vortex core in bilayer Rashba superconductors by means of the self-consistent quasiclassical calculation. We found that the local density of states (LDOS) structure in the pair-density wave (PDW) state is quite different from that in the BCS state. The zero energy vortex bound state exsists in the PDW state, whereas it is absent in the BCS state due to the Zeeman effect. This characteristic LDOS structure can be observed by scanning tunneling spectroscopy at low temperature. Another feature of the PDW state is that the core size is small compared with that in the BCS state in the vicinity of the BCS-PDW phase transition. These features in a magnetic field can be a key to identify the exotic superconducting phase.

Acknowledgments

The authors thank M. Kato for helpful discussions and reading this manuscript. The computation in this work has been done using the facilities of the Supercomputer Center of the Institute for Solid State Physics, the University of Tokyo. This study has been partially supported by JSPS KAKENHI Grant Numbers 26800197, 24740230 and 25103711.

References

[1] Yoshida T, Maruyama D and Yanase Y 2014 *Kotaibutsuri* (*Solid State Physics*) **49** 109 (in Japanese).
[2] Sigrist M, Agterberg D F, Fischer M H, Goryo J, Loder F, Rhim S H, Maruyama D, Yanase Y, Yoshida T and Youn S J 2014 *J. Phys. Soc. Jpn.* **83** 061014.
[3] Mukuda H, Shimizu S, Iyo A and Kitaoka Y 2012 *J. Phys. Soc. Jpn* **81** 011008 and references cited therein.
[4] Hor Y S, Williams A J, Checkelsky J G, Roushan P, Seo J, Xu Q, Zandbergen H. W, Yazdani A, Ong N P and Cava R J *Phys. Rev. Lett.* **104** 057001.
[5] Nishikubo Y, Kudo K and Nohara M 2011 *J. Phys. Soc. Jpn.* **80** 055002.
[6] Mizukami Y, Shishido H, Shibauchi T, Shimozawa M, Yasumoto S, Watanabe D, Yamashita M, Ikeda H, Terashima T, Kontani H and Matsuda Y 2011 *Nat. Phys.* **7** 849.
[7] Shimozawa M, Goh S K, Endo R, Kobayashi R, Watashige T, Mizukami Y, Ikeda H, Shishido H, Yanase Y, Terashima T, Shibauchi T and Matsuda Y 2014 *Phys. Rev. Lett.* **112** 156404.
[8] Yoshida T, Sigrist M, Yanase Y 2012 *Phys. Rev. B* **86** 134514.
[9] Yoshida T, Sigrist M, Yanase Y 2013 *J. Phys. Soc. Jpn.* **82** 074714.
[10] Maruyama D, Sigrist M, Yanase Y 2012 *J. Phys. Soc. Jpn.* **81** 034702.
[11] Hayashi N, Wakabayashi K, Frigeri P A and Sigrist M 2006 *Phy. Rev. B* **73** 024504.
[12] Serene J W and Rainer D 1983 *Phys. Rep.* **101** 221 and references cited therein; Schopohl N 1980 *J. Low Temp. Phys.* **41** 409; Alexander J A X, Orlando T P, Rainer D and Tedrow P M 1985 *Phys. Rev. B* **31** 5811; Rieck C T, Scharnberg K and Schopohl N 1991 *J. Low Temp. Phys.* **84** 381; Choi C H and Sauls J A 1993 *Phys. Rev. B* **48** 13684.
[13] Higashi Y, Nagai Y and Hayashi N 2014 *JPS Conf. Proc.* **3** 015003.
[14] Frigeri P A, Agterberg D F, Milat I and Sigrist M 2006 *Eur. Phys. J. B* **54** 435.
[15] Appendix B in Sigrist M and Ueda K 1991 *Rev. Mod. Phys.* **63** 239.
[16] Yoshida T, Sigrist M, Yanase Y 2014 *JPS Conf. Proc.* **3** 015026.
[17] Sakakibara T, Yamada A, Custers J, Yano K, Tayama T, Aoki H and Machida K 2007 *J. Phys. Soc. Jpn.* **76** 051004; Matsuda Y, Izawa K and Vekhter I 2006 *J. Phys.: Condens. Matter* **18** R705; An K, Sakakibara T, Settai R, Onuki Y, Hiragi M, Ichioka M and Machida K 2010 *Phys. Rev. Lett.* **104** 037202.

Development of nano and micro SQUIDs based on Al tunnel junctions

Ryosuke Ishiguro[1,2], Eiichiro Watanabe[3], Daisuke Sakuma[2], Tomoya Shinozaki[2], Shogo Tsuchiya[2], Yusuke Nago[2], Hirotaka Osato[3], Daiju Tsuya[3], Hiromi Kashiwaya[4], Satoshi Kashiwaya[4], Shintaro Nomura[5], Hideaki Takayanagi[1,6], Yoshiteru Maeno[7]

[1]Center for Emergent Matter Science, RIKEN, Wako, Saitama 351-0198, Japan

[2]Department of Applied Physics, Faculty of Science, Tokyo University of Science, Tokyo 125-8585, Japan

[3]Nanotechnology Innovation Center, NIMS, 1-2-1 Sengen, Tsukuba, Ibaraki 305-0047, Japan

[4]National Institute of Advanced Industrial Science and Technology (AIST), Tsukuba 305-8568, Japan

[5]Division of Physics, University of Tsukuba, Tsukuba, 305-8571, Japan

[6]International Center for Materials Nanoarchitectonics (MANA), National Institute for Materials Science (NIMS), Tsukuba, Ibaraki 305-0044, Japan

[7]Department of Physics, Faculty of Science, Kyoto University, Kyoto 606-8502, Japan

ryosuke.ishiguro@riken.jp

Abstract. Superconducting quantum interference devices (SQUIDs) with nano (micro)-meter dimensions are called nano (micro)-SQUIDs. The high sensitivity for flux and position of nano (micro)-SQUIDs can be applied to detect local magnetic fields induced by vortices and the magnetization of mesoscopic superconductors. Nano-SQUIDs based on carbon-nanotube junctions and niobium weak junctions are well known. However, such nano-SQUIDs are not suitable for large-scale integrated circuits and mass production. Therefore, we employ a combination of lithography using the Niemeyer-Dolan technique and the inductively coupled plasma reactive-ion etching technique to fabricate nano-SQUIDs. Here, we report the fabrication of nano (micro)-SQUIDs based on superconducting aluminum tunnel junctions and their application for vortex formation into mesoscopic chiral superconducting Sr_2RuO_4[1-3].

1. Introduction

Superconducting quantum interference devices (SQUIDs) are used as sensitive magnetometers in a wide range of fields. Recently, nanometer-sized SQUIDs (nano-SQUIDs) have attracted considerable attention because of their high resistance to external flux noise and their ability to detect magnetic fields induced by a single electron spin [4-9]. Nano-SQUIDs also play an important role in quantum computing as a quantum interface between spin qubits and SQUID-based superconducting qubits [10-15]. Nano-SQUIDs have a high sensitivity not only in the flux but also in the position. The smallest resolution of a sensor is finally determined by the size of the probe. Therefore, the nano-SQUID is applied to the scanning probe microscopy [16-18].

There are well-known nano-SQUIDs based on carbon nanotube junctions or weak-link junctions consisting of niobium [4, 7-9, 19]. However, such nano-SQUIDs are not suitable for large-scale integrated circuit fabrication, such as a quantum computing circuit or a sensor based on the SQUID array, because they must be fabricated one by one. Moreover, it is not easy to suppress the critical current of the weak-link junction. In order to realize a nano-SQUID with well-controlled Josephson junctions and high integration for such as a nano-SQUID array, we employ a combination of lithography using the Niemeyer-Dolan technique[20, 21] and the inductively coupled plasma reactive-ion etching technique (ICP-RIE)[22]. We developed a new fabrication process applicable for nano-SQUIDs comprising a superconducting aluminum loop and high-quality superconducting tunnel junctions, which are applicable to superconducting qubits[13, 14].

Dolan-bridge shadow deposition is a standard method to realize superconducting small-tunnel junctions containing superconducting qubits. However, tilted deposition undesirably broadens the width of the superconducting electrodes. That is, vortices are trapped in the wide superconducting electrodes in the magnetization measurement. By performing the RIE, it is possible to narrow the width of the electrodes independently of the shadow deposition.

In this paper, we report our work on the fabrication of micro- and nano-SQUIDs based on superconducting aluminum tunnel junctions. To fabricate the SQUIDs with aluminum tunnel junctions, we employ a combination of the Dolan bridge shadow evaporation technique [20] and the ICP-RIE method.

2. Experimental methods

Our lithography process for SQUIDs is separated into three parts. First, we fabricate the large structures: the bonding pads, Au electrodes, and registration marks for the subsequent e-beam lithography. The optical lithography for the large structures is performed in two separate processes to prevent step-disconnection between the Au electrodes and Al SQUID electrodes and to assure the sufficient film thickness for the wire bonding and registration mark for the electron beam lithography.

After suitably cleaning a silicon wafer, a common primer HMDS is used before a spin-coating of resists that serves as an adhesion promoter. Subsequently, the bilayer resist is spun on. To produce the bilayer resist, the LOR5A polydimethylglutarimide is first spin-coated onto the wafer at 3000 rpm for 60 s after the ramp up for 10 sec and soft baked to drive out the solvent at ~180 °C for 5 min. Next, the AZ5214E copolymer is spin coated in the same manner onto the coated LOR5A film and soft baked at ~110 °C for 2 min. This results in each layer being approximately 550 or 1620 nm thick for a total film thickness of approximately 2120 nm. Then, we use direct laser writing to expose the patterns for the electrical leads with a dosage of 140 mJ/cm^2 using a maskless disposer system that is manufactured by Nanosystem Solutions, Inc. The exposed wafer is developed in 2.38% TMAH (tetramethylammonium hydroxide) for 90 sec to remove the exposed AZ5214E and LOR5A, leaving windows in AZ5214E/LOR5A bilayer for the electrodes. Subsequently, the developed sample is rinsed by deionized (DI) water for 30 sec.

After the photolithography and development of the structure, a first metal layer of Ti/Au is deposited by electron beam physical evaporation in a high-vacuum chamber with a base pressure of a few 10^{-6} Pa. The first deposition of Ti 5-nm and Au 50-nm films on the surface of the sample was performed under a pressure of a few 10^{-5} Pa. After the metal deposition, liftoff is performed to remove all of the unwanted metal and resist polymers. Liftoff is achieved by soaking the metal-coated sample in NMP (N-methylpyrrolidone) for at most 40 min, at 80 °C. After the liftoff, the sample is rinsed with acetone for 30 s and then isopropanol (IPA) for 30 s and finally dried by nitrogen gas. In the 2nd process for the 2nd

Ti/Au layer, the lithography process is performed in same manner as for the 1st Ti/Au layer, but the thickness of the deposition differs between the Ti 50-nm and Au 200-nm films.

Next, smaller SQUID electrodes are fabricated by electron beam lithography. The bilayer resist is spun on to make a structure for the shadow deposition method. To produce the bilayer resist, the PMGI SF7 polydimethylglutarimide, which is a highly soluble polymer covering with a thin layer of the actual e-beam resist, is first spin coated onto the wafer at 3000 rpm for 60 sec after the ramp up for 10 sec and then soft baked to drive out the solvent at ~180 °C for 5 min. Next, the high-resolution electron beam resist ZEP520A is spin coated onto the coated PMGI SF7 film and soft baked in the same manner. This results in each layer being approximately 600 or 100 nm thick, for a total film thickness of approximately 700 nm. We then use an electron beam to expose the patterns for the SQUIDs electrodes.

Then, the wafer is cut to several chips to make tunnel junctions with different parameters. Since the degradation of the pattern starts after the development, the deposition must be done immediately. However, the exposed sample without development can be preserve for several months. The chips are developed twice first in Xylene for 30 s to dissolve the exposed ZEP520A and rinsed in IPA for 30 s, then in 2.38% TMAH for 22 s to dissolve portions of PMGI SF7 via the ZEP520A windows and rinsed by DI water for 30 s.

After e-beam lithography and development of the SQUID structure, we perform the shadow evaporation method, which was proposed by G. J. Dolan in 1977, to form the tunnel junction [20, 21]. First, a 30-nm-thick aluminum layer is deposited on the angle of plus ten degrees by electron beam physical evaporation in a high-vacuum chamber with a base pressure of a few 10^{-6} Pa.

After the first Al film deposition, the oxidation of the first deposition Al surface was performed in a sub-chamber for forming an insulating film for the tunnel junctions. The sample can be transported from the main chamber to the sub chamber without breaking the vacuum. The oxidation is controlled by the oxygen flow rate, pressure, and time. As a result, the critical current density J_c of the JJs is easily adjustable. Subsequently, a 2nd 50-nm-thick Al film is deposited in the main chamber on the angle of minus 10 degrees.

To realize a smooth surface of Al, it is important to fabricate a film that is uniform, thin, and a good insulator. When the Al is placed directly into a cold copper hearth, because of the high thermal conductivity of the aluminum and cooper, the evaporation rate is very low. Therefore, a boron nitride hearth liner is used to satisfy the deposition rate. The surface of the deposition aluminum film obtained using the boron nitride hearth liner is smoother and has a smaller grain size than that obtained using the alumina hearth liner.

Fabrication of micrometer-sized SQUIDs is finished by the previous process. However, to make nano-SQUIDs, we add an extra process using ICP-RIE. For the etching mask of the ICP-RIE, e-beam lithography is performed using a single-layer resist of the ZEP520A. The spin coating, baking, and exposing are done in the same manner as in the SQUID lithography process. After the development of the exposed ZEP520A, dry etching was performed using a Cl_2 and BCl_3 ICP-RIE process for 2 min with powers of 50 W ICP and 150 W, respectively. Cl_2 gas is suitable for etching metal such as aluminum; however, it is not suitable for etching Al_2O_3. In our structure, there are two Al_2O_3 films: one is the native oxide film on the surface, whose thickness is ordinary expected to be approximately 4 nm, and another is the film for the tunnel junction, whose thickness is expected to be 1 nm. BCl_3 gas is widely used for the etching of materials covered by native oxides owing to its effective extraction of oxygen [22, 23]. Therefore, we chose a mixture gas of Cl_2 and BCl_3 for the etching which is also widely used for the etching of thin metal films with native oxide. The etchant gases of Cl_2 and BCl_3 are introduced into the chamber at 8 and 2 sccm, respectively. The total pressure is controlled at 0.1 Pa.

Figure 1 (c) shows a scanning electron microscopic image of the fabricated nano-SQUID. The width of the electrodes of the nano-SQUID is 100 nm, which is twice the Meissner length of superconducting aluminum at zero temperature. The inner area of the nano-SQUID is 500×500 nm^2.

For setting the microscopic size samples onto the micro- and nano-SQUIDs, we typically use a nano-manipulator system that is manufactured by Omniprobe Inc., primarily for transmission electron microscope (TEM) sample preparation. During transportation, the sample was attached to the tip of a small tungsten needle. The accuracy of the setting position was approximately 100 nm, and the minimum size of the sample for the transportation was approximately 1 μm. To reduce the size to below 1 μm, the sample need to be milled by using focused ion beam on the nano-SQUIDs. Figure 1 (d) shows a scanning electron microscopic image of the micro-SQUID with a mesoscopic plate of the chiral superconductor Sr_2RuO_4[1-3].

For measurement, a dilution refrigerator with a base temperature of 30 mK or helium 3 refrigerator with a base temperature of 300 mK is used for cooling the sample to below the critical temperature of aluminum, which is T_c = 1.2 K. Because the small critical current of the Josephson junction is very sensitive to environmental noise, the supercurrent readily disappears. Therefore, a 0.1-μF chip ceramic capacitor for filtering is mounted between the measurement line and electrical ground nearby the sample, and a copper powder filter is installed at the mixing chamber stage in the dilution refrigerator[24].

3. Results and discussion

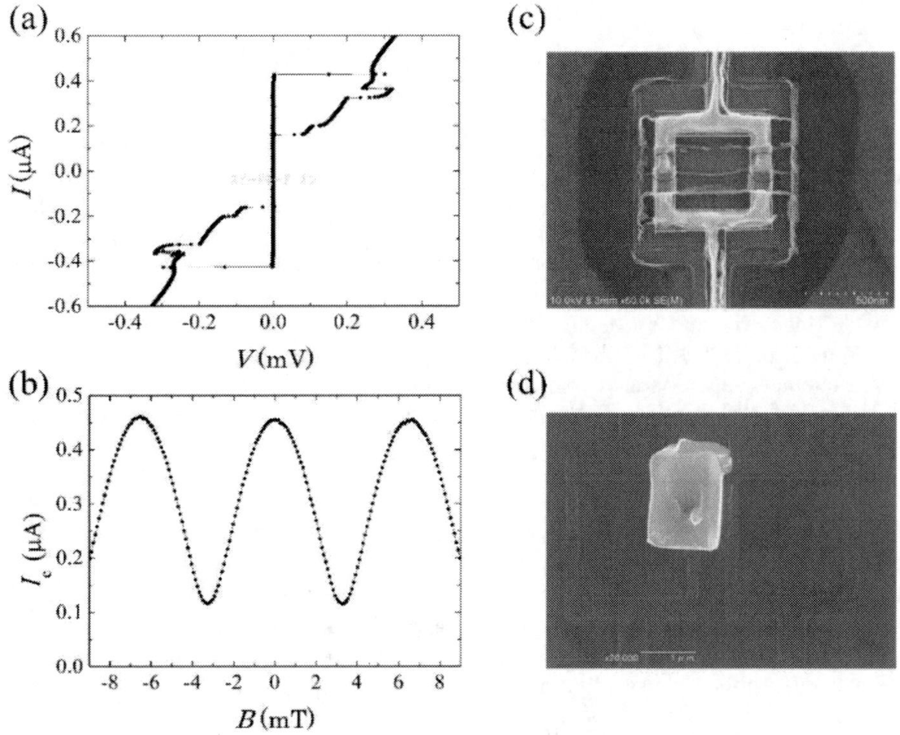

Figure 1. (a) Typical *I-V* characteristics of nano-SQUID at ~75 mK in the absence of a magnetic field. (b) Field dependence of the critical current of nano-SQUID. (c) Scanning electron microscopic image of a nano-SQUID. Red circles show the position of SIS Josephson junctions. Yellow lines show the edge of Al electrodes of SQUID. There are residual outside of the electrodes, however these are insulated. (d) Mesoscopic plate of the chiral superconductor Sr_2RuO_4 is set onto corner of the left micro SQUID using nano-manipulator system.

Figure 1 (a) shows a typical *I-V* characteristic of nano-SQUID at ~75 mK in zero magnetic field. The duration of a bias current cycle is typically 1 s. The *I-V* characteristic indicates a clear supercurrent with the critical value of I_c below 1 μA with hysteretic behavior. As a function of an external field, a periodic oscillation of the critical current I_c is observed, as shown in Fig. 1(b). The critical current is defined at switching currents between the superconducting state and the voltage state in the *I*-*V* curves. The observed period of approximately 6.6 mT is consistent with the addition of a flux quantum $\Phi_0 = h/2e$ to the effective SQUID area of 550 nm × 550 nm.

4. Conclusions

We developed a fabrication process for nano (micro)-SQUIDs based on superconducting aluminum tunnel junctions. The critical current is sufficiently small to avoid a heating at switching to the voltage state. The field dependence of the critical current shows typical behaviors of SQUIDs. In the future, we will apply this process to make SQUID arrays for the measurements of vortex formations in microscopic superconducting plates.

Acknowledgments

We thank T. Fukagawa, A. Tajima, S. Nakajima, S. Kitani, M. Matsuno, M. Yakabe, M. Kamio, H. Yabuki, M. Osugi, Y. Okada, M. Ito, Y. Tanaka, Y. Funaki, T. Sakurai, T. Koseki, S. Suyama, Y. Mochinaga and T. Yoshida for their helpful cooperation and discussions. This study was supported by the NIMS Nanofabrication Platform under the "Nanotechnology Platform Project" and by the "Topological Quantum Phenomena" (No, 25103721) MEXT KAKENHI and JSPS KAKENHI (No. 20221007, 25390046).

References

[1] Y. Maeno, S. Kittaka, T. Nomura, S. Yonezawa, and K. Ishida, J Phys Soc Jpn **81**, 011009 (2012).
[2] Y. Maeno, H. Hashimoto, K. Yoshida, S. Nishizaki, T. Fujita, J. G. Bednorz, and F. Lichtenberg, Nature **372**, 532 (1994).
[3] A. P. Mackenzie and Y. Maeno, Rev Mod Phys **75**, 657 (2003).
[4] S. K. H. Lam and D. L. Tilbrook, Applied Physics Letters **82**, 1078 (2003).
[5] V. Bouchiat, Superconductor Science & Technology **22** (2009).
[6] C. Granata, A. Vettoliere, P. Walke, C. Nappi, and M. Russo, Journal of Applied Physics **106** (2009).
[7] A. Vettoliere, C. Granata, E. Esposito, R. Russo, L. Petti, B. Ruggiero, and M. Russo, Ieee T Appl Supercon **19**, 702 (2009).
[8] W. Wernsdorfer, Superconductor Science & Technology **22** (2009).
[9] L. Bogani, R. Maurand, L. Marty, C. Sangregorio, C. Altavilla, and W. Wernsdorfer, J Mater Chem **20**, 2099 (2010).
[10] O. Buisson *et al.*, Quantum Information Processing **8**, 155 (2009).
[11] S. Kim, R. Ishiguro, M. Kamio, Y. Doda, E. Watanabe, D. Tsuya, K. Shibata, K. Hirakawa, and H. Takayanagi, Applied Physics Letters **98**, 063106 (2011).
[12] S. Kim *et al.*, AIP Conf. Proc., 383 (2011).
[13] G. Blatter, V. B. Geshkenbein, A. L. Fauchere, M. V. Feigel'man, and L. B. Ioffe, Physica C **352**, 105 (2001).
[14] H. Takayanagi, H. Tanaka, S. Saito, and H. Nakano, Physica Scripta **T102**, 95 (2002).
[15] F. Plastina and G. Falci, Physical Review B **67** (2003).
[16] D. Vasyukov *et al.*, Nature nanotechnology **8**, 639 (2013).
[17] A. Finkler *et al.*, Nano Lett **10**, 1046 (2010).
[18] J. Nagel *et al.*, Physical Review B **88** (2013).
[19] J. P. Cleuziou, W. Wernsdorfer, V. Bouchiat, T. Ondarcuhu, and M. Monthioux, Nature nanotechnology **1**, 53 (2006).
[20] G. J. Dolan, Applied Physics Letters **31**, 337 (1977).
[21] J. Niemeyer and V. Kose, Applied Physics Letters **29**, 380 (1976).

[22] J. R. Rooth, *Industrial Plasma Engineering* (Taylor & Francis Philadelphia, 1995), Vol. 1.
[23] G. M. Xue, H. F. Yu, Y. Tian, H. Deng, W. Y. Liu, Y. F. Ren, H. W. Yu, D. N. Zheng, and S. P. Zhao, Sci China Phys Mech **56**, 2377 (2013).
[24] A. Lukashenko and A. V. Ustinov, Review of Scientific Instruments **79** (2008).

Journal of Physics: Conference Series **568** (2014) 022020 doi:10.1088/1742-6596/568/2/022020

Broken time-reversal symmetry in a SQUID based on chiral superconducting Sr_2RuO_4

R. Ishiguro[1,2], T. Sakurai[2], M. Yakabe[2], T. Nakamura[3,4], S. Yonezawa[3], S. Kashiwaya[5], H. Takayanagi[2,6], and Y. Maeno[3]

[1]Center for Emergent Matter Science, RIKEN, Wako, Saitama 351-0198, Japan

[2]Department of Applied Physics, Faculty of Science, Tokyo University of Science, Tokyo 125-8585, Japan

[3]Department of Physics, Graduate School of Science, Kyoto University, Kyoto 606-8502, Japan

[4]Institute for Solid State Physics, the University of Tokyo, Chiba 277-8581, Japan

[5]National Institute of Advanced Industrial Science and Technology (AIST), Tsukuba 305-8568, Japan

[6]International Center for Materials Nanoarchitectonics (MANA), National Institute for Materials Science (NIMS), Tsukuba 305-0044, Japan

ryosuke.ishiguro@riken.jp

Abstract. Unconventional superconductors involve not only gauge-symmetry breaking but also orbital- and spin-symmetry breaking. Superconducting Sr_2RuO_4 is known as a spin-triplet chiral *p*-wave and a topological superconductor with broken time-reversal symmetry (BTRS). Kerr-effect and muon-spin-rotation (μSR) measurements have shown that the bulk superconducting state of Sr_2RuO_4 features BTRS in the orbital part; hence, it is called the chiral state. BTRS in the response of superconducting junctions or SQUIDs would appear as the shifts of magnetic interference patterns. However, it is problematic to distinguish whether the shift originates in the residual magnetic field (trapped vortex) or the effects of BTRS. Here, we show that the magnetic interference patterns of a SQUID based on Sr_2RuO_4 are explicitly asymmetric with respect to the direction of both the bias current and the applied magnetic field; namely, there is no inversion symmetry. This indicates that the superconducting state of Sr_2RuO_4 undoubtedly breaks the time-reversal symmetry of the SQUID.

Journal of Physics: Conference Series **568** (2014) 022020 doi:10.1088/1742-6596/568/2/022020

1. Introduction

The BCS superconducting transition is associated with gauge-symmetry breaking at the transition temperature T_c, above which the system is invariant under gauge transformation. Unconventional condensates, such as superfluid ^{3}He, *p*-wave superconductors, and *d*-wave superconductors, are more complicated, as they include both spin and orbital degrees of freedom. Therefore, the transition involves not only gauge-symmetry breaking but also orbital- and spin-symmetry breaking. In these superconducting states, time-reversal symmetry can be spontaneously broken.

Superconducting Sr_2RuO_4 is known as a strong candidate for a spin-triplet chiral *p*-wave and a topological superconductor with broken time-reversal symmetry (BTRS) [2-4]. A BTRS state arising from the possible magnetization of the orbital component of the superconducting wave function of the chiral *p*-wave superconducting Sr_2RuO_4 has been decisively demonstrated by Kerr-effect and muon-spin-rotation (μSR) measurements [5, 6]. Associated with the bulk BTRS of a topological superconductor, it is expected that its edge, namely its surface state, also breaks TRS as dictated by the bulk-edge correspondence, there is also a possibility of BTRS originating in the phase winding of the chiral superconducting wave function around the edge of the chiral superconductor [3, 7-10]. The wave function phase at the edge of the chiral superconductor is continuously changing with the angle of the surface normal vector [3, 11]. Therefore, superconducting hybrid junctions, or rings containing the superconducting junctions, consisting of *s*-wave and chiral superconductors have the possibility of exhibiting BTRS with a spontaneous supercurrent [3, 7-10]. The Josephson current through a superconducting junction is ordinarily proportional to $\sin\phi$, where ϕ is the phase difference between two superconductors separated by the junction. However, if BTRS exists, the Josephson current is represented by a linear combination of $\sin\phi$ and $\cos\phi$ [12].

A number of transport- and phase-sensitive measurements have been conducted on *p*- and *d*-wave superconductors to confirm the presence of BTRS [7, 9, 10, 13-16]. However, it is difficult to establish the existence of BTRS because of the subtle differences between the shifts originating in the residual magnetic field (trapped vortex) and those caused by the effects of BTRS. Here, we present the magnetic interference patterns of a SQUID consisting of a hybrid superconducting loop consisting of the *s*-wave superconductor niobium (Nb), Sr2RuO4, and Nb/Ru/Sr2RuO4 junctions. The observed interference patterns are explicitly asymmetric with respect to the direction of both the bias current and the applied magnetic field. This asymmetry clearly demonstrates the existence of BTRS.

2. Experimental methods

We fabricated hybrid Nb–Sr_2RuO_4 dc SQUIDs, which are composed of Nb, Sr_2RuO_4, and two Nb/Ru/Sr_2RuO_4 junctions, by photolithographically building a Nb bridge between the two individual Ru islands on the *ab*-plane surface of eutectic Sr_2RuO_4-Ru shown in Fig.1 [1, 17]. The Sr_2RuO_4-Ru eutectic is known to develop a filamental superconducting phase around the Ru islands for temperatures below about 3K, which is twice the transition temperature of bulk Sr_2RuO_4 ($T_{c\text{-bulk}}$ = 1.5 K) [17]. The Nb/Ru/Sr_2RuO_4 junctions, which were made by deposing an *s*-wave superconductor on Ru metal in eutectic Sr_2RuO_4-Ru, are known to behave as "topological junctions" [11, 18]. The measurements reported herein were performed at temperatures well below $T_{c\ \text{bulk}}$. Our purpose in using the eutectic material is to establish good electrical contact between the *s*-wave and chiral *p*-wave superconductors in order to exploit the proximity effect. Finally, the dc-SQUID shape was formed by the focused ion beam technique (FIB) to restrict the supercurrent path in the Sr_2RuO_4.The details of the fabrication of similar-type SQUIDs with Nb/Ru/Sr_2RuO_4 junctions and of the herein presented SQUID have been described elsewhere [19] and in a very recently submitted paper[1]. The relative angle between the two Nb/Ru/Sr_2RuO_4 junction surfaces was approximately π. In this angle condition, only π SQUID behavior is expected [8] and not BTRS. However, we observed BTRS behavior but no π SQUID features. We consider that this is related to the existence of a small number of chiral domains, each of which is a superconducting regions with the same chirality.

For the measurements, a helium-3 evaporative refrigerator with a base temperature of 0.3 K was used.

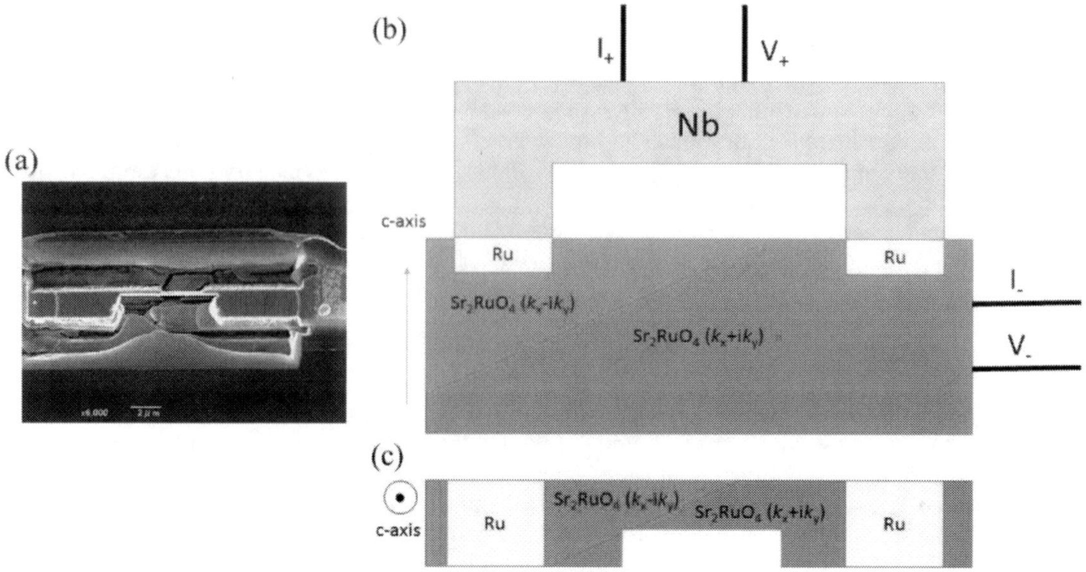

Figure 1. (a) SEM image of the SQUID. (b),(c) A possible configuration of the chiral domain in the SQUID.

3. Results and discussion

Figure 2 shows magnetic interferences pattern for the SQUID for both a positive and negative current at 0.6 K. The details of these interference patterns will be presented elsewhere[1]. It must be noted that these interference patterns are explicitly asymmetric with respect to the directions of both the bias current and the applied magnetic field; that is, there is no inversion symmetry. Usually, BTRS in hybrid SQUIDs or junctions shifts the origin (zero magnetic field) of the interference patterns [7-10]. However, it is difficult to distinguish between external residual magnetic fields and the intrinsic effects of BTRS from this shift. In the present case, the asymmetry of the observed interference pattern cannot be explained only by the shift of the origin of the external magnetic field. Therefore, this asymmetry cannot be due to the influence of the magnetic field formed by trapped flux or the test current but must be caused by the existence of $\cos\phi$ components In the Josephson current due to BTRS. Therefore, the asymmetry of the interference pattern constitutes strong evidence of the expected BTRS originating in the wave function phase winding of the chiral p-wave superconductor Sr_2RuO_4.

This reproducibility of this asymmetric interference pattern is very good; the pattern does not depend on the direction of the field sweep. In contrast, the asymmetric patterns slightly change upon thermal cycling above the superconducting transition temperature, as shown in Fig. 3. We assume that this change is influenced by the reconstruction of the chiral domains [1, 7, 18, 20]. Since it is difficult to explain BTRS only by the relative angle of two junctions, π, we believe that the origin of the displayed BTRS is related to the relative position and phase of the chiral domains present in the Sr_2RuO_4 part of the SQUID.

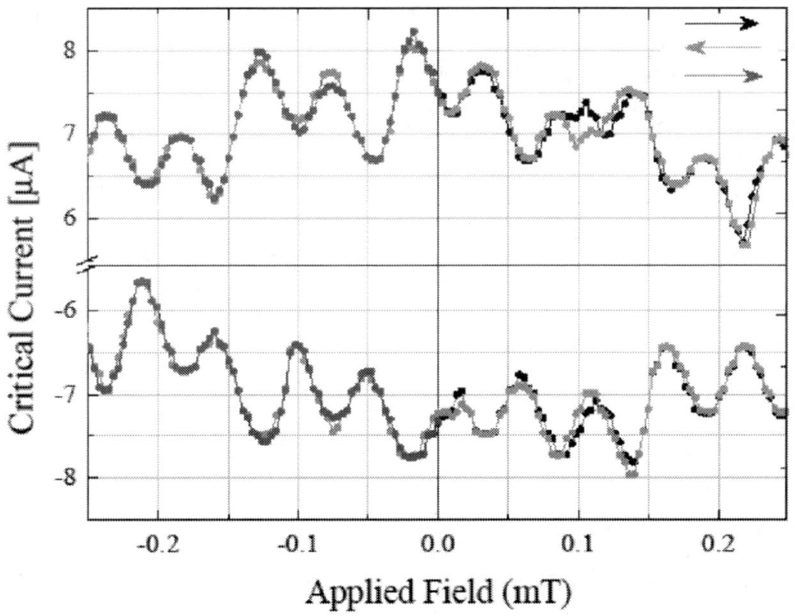

Figure 2. Eevidence of broken time-reversal symmetry. The positive and negative critical currents of SQUID are asymmetric. The asymmetries increase with decreasing temperature.The different colors indicate the field sweep directions, as shown by the arrows.

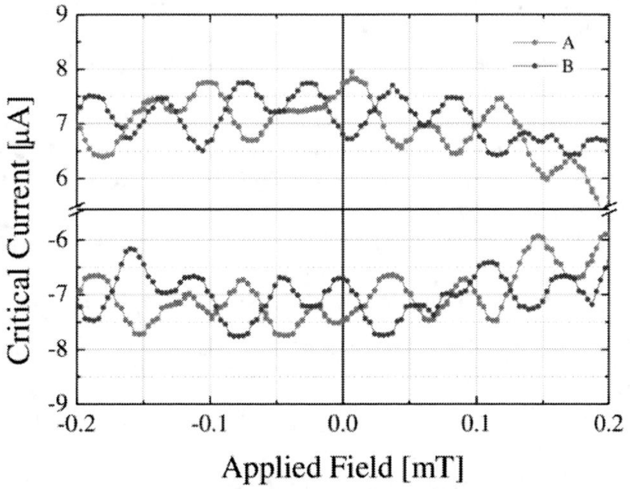

Figure 3 Thermal cycling of the magnetic interference patterns of SQUIDs at 0.6 K above the superconducting transition temperature, measured on different days[1].

4. Conclusion

We fabricated a hybrid SQUID based on the chiral p-wave superconducting Sr_2RuO_4 and an s-wave superconductor and showed that the hybrid SQUID features BTRS originating in the phase winding of the chiral superconductor and certainly breaks the time-reversal symmetry.

5. Acknowledgments

We thank M. Sigrist, M. S. Anwar, S. Nomura, H. Ikegami, and K. Kono for discussions and comments and E. Watanabe, D. Tsuya, and H. Osato for technical support. This work was supported by MEXT KAKENHI (Grants No. 22103002 and No. 25103721), as well as by JSPS KAKENHI S (Grant No. 20221007). The technical part of this study was supported by the NIMS Nanofabrication Platform of the "Nanotechnology Platform Project" sponsored by MEXT.

References

[1] R. Ishiguro, H. Sakurai, M. Yakabe, T. Nakamura, S. Yonezawa, S. Kashiwaya, H. Takayanagi, and Y. Maeno, in *submitting* 2014).
[2] Y. Maeno, H. Hashimoto, K. Yoshida, S. Nishizaki, T. Fujita, J. G. Bednorz, and F. Lichtenberg, Nature **372**, 532 (1994).
[3] Y. Maeno, S. Kittaka, T. Nomura, S. Yonezawa, and K. Ishida, J Phys Soc Jpn **81**, 011009 (2012).
[4] A. P. Mackenzie and Y. Maeno, Rev Mod Phys **75**, 657 (2003).
[5] G. M. Luke *et al.*, Nature **394**, 558 (1998).
[6] J. Xia, Y. Maeno, P. T. Beyersdorf, M. M. Fejer, and A. Kapitulnik, Phys Rev Lett **97**, 167002 (2006).
[7] F. Kidwingira, J. D. Strand, D. J. Van Harlingen, and Y. Maeno, Science **314**, 1267 (2006).
[8] K. D. Nelson, Z. Q. Mao, Y. Maeno, and Y. Liu, Science **306**, 1151 (2004).
[9] M. R. Beasley, D. Lew, and R. B. Laughlin, Phys Rev B **49**, 12330 (1994).
[10] Y. Liu, New J. Phy. **12**, 075001 (2010).
[11] T. Nakamura, R. Nakagawa, T. Yamagishi, T. Terashima, S. Yonezawa, M. Sigrist, and Y. Maeno, Phys Rev B **84**, 060512 (2011).
[12] Y. Asano, Y. Tanaka, M. Sigrist, and S. Kashiwaya, Phys Rev B **67**, 184505 (2003).
[13] W. K. Neils and D. J. Van Harlingen, Phys Rev Lett **88**, 047001 (2002).
[14] T. Lindstrom, S. A. Charlebois, A. Y. Tzalenchuk, Z. Ivanov, M. H. S. Amin, and A. M. Zagoskin, Phys Rev Lett **90**, 117002 (2003).
[15] J. R. Kirtley, C. C. Tsuei, A. Ariando, C. J. M. Verwijs, S. Harkema, and H. Hilgenkamp, Nat Phys **2**, 190 (2006).
[16] M. Covington, M. Aprili, E. Paraoanu, L. H. Greene, F. Xu, J. Zhu, and C. A. Mirkin, Phys Rev Lett **79**, 277 (1997).
[17] Y. Maeno, T. Ando, Y. Mori, E. Ohmichi, S. Ikeda, S. NishiZaki, and S. Nakatsuji, Phys Rev Lett **81**, 3765 (1998).
[18] M. S. Anwar, T. Nakamura, S. Yonezawa, M. Yakabe, R. Ishiguro, H. Takayanagi, and Y. Maeno, Sci. Rep. **3**, 2480 (2013).
[19] R. Ishiguro, M. Yakabe, T. Nakamura, E. Watanabe, D. Tsuya, H. Oosato, Y. Maeno, and H. Takayanagi, J. Phys.: Conf. Ser. **400**, 022035 (2012).
[20] A. Bouhon and M. Sigrist, New Journal of Physics **12** (2010).

Eigenvalue problem for permittivity operator of conductors with the spatial dispersion in a microwave field

M A Dresvyannikov[1], **A P Chernyaev**[2], **O M Ivanenko**[1], **A L Karuzskii**[1], **A V Perestoronin**[1], **N A Volchkov**[1]

[1] P. N. Lebedev Physical Institute of RAS, Leninsky prospekt 53, Moscow 119991, Russia
[2] Moscow Institute of Physics and Technology (State University), Institutsky Per. 9, Dolgoprudny, Moscow District 141700, Russia

E-mail: karuz@sci.lebedev.ru

Abstract. Conducting media with spatial dispersion may be described formally by the single operator - operator of dielectric permittivity, which, as it is well known, completely defines the microwave response of conductors with spatial dispersion. So the eigenvalue problem for permittivity operator of conductors and superconductors possessing strong spatial dispersion at low temperatures is of a great importance since the corresponding solutions are the stable waves for constitutive equation in a self-consistent microwave field. Here a wave problem is formulated to search the solutions, which correspond to the eigenvalues of permittivity operator, similar to the problem of wave propagation in hollow waveguides and resonators, but non-self conjugated. Dispersion relations and general solutions are obtained. Significant role of the spatial-type force resonances is considered. Conditions for the spatial resonances are derived. The obtained resonances includes particular solutions corresponding to traditional surface impedance for anomalous skin effect, surface impedance of superconductor, as well as four novel solutions, obviously related to polarization, two of which correspond to the waves with amplitude increasing into the depth of conductor, and two else describe solutions with unusual properties.

1. Introduction

An increase of the mean free path length of the current carriers at sufficiently low temperatures up to values greater than the characteristic penetration depth of electromagnetic field leads to the dominating role of the effects caused by spatial dispersion in the electromagnetic response of cooled conductors [1–6]. The effects that depend on spatial parameters, besides of the obvious interest to such effects especially in structures with small (micro and nano) sizes, open the possibility of investigation of conductivity in novel materials. Microwave effects of the spatial dispersion allow to study the properties of carriers and pairing mechanisms in superconductors, including high$-T_c$ superconductors (HTSC). To deduce this information a description of the surface impedance of superconductor should be in satisfactory agreement with experimental data. The effects of spatial dispersion and anomalous skin effect, which is a characteristic manifestation of the spatial dispersion in the microwave response of cooled conductors, are well described by the textbook kinetic theory [2–6]. Nevertheless, a calculation of the surface resistance R of superconductor for instance, which denotes the real part of complex surface impedance $\widetilde{Z} = R + iX$, in frames of the microscopic model is always complicated since the exact analytical expressions can be found only in few limiting cases. Here and below sign tilde denotes a complex value. The values of the surface resistance in the intermediate regions can be found only by the sophisticated numerical calculations [4]. These complications to a greater extent concern the HTSC since the rigorous microscopic model of microwave response of these materials is not developed yet. An anomalous skin effect is usually considered in details in many textbooks, which review the theoretical and experimental studies of the surface impedance of superconductors (see for example one of the most complete monographs [4]), because of its inherent relation to the problem of surface impedance of superconductors. Actually both the anomalous skin effect in an extremely anomalous limit and the skin effect in superconductors correspond to current carriers with the infinite length of mean free path and differ in other characteristic length parameters. Knowledge of the surface impedance in the normal state under conditions of anomalous skin effect is crucially important for calculations, e. g., of the penetration depth of electromagnetic field in superconducting state from the surface resistance experimental data [4].

Complications mentioned above in a large extent result from the fact that for an explanation of the microwave spatial effects a microscopic kinetics theory is used to obtain the constitutive equation that relates the current density and the field in a metal, and is deduced on the rigorous base of linearized Boltzmann equation [2–6]. Previously it was noted (see for instance [4]) that rigorous microscopic theory of the effects of spatial dispersion, kinetics theory of anomalous skin effect lead to sophisticated calculations and do not provide an illustrative and clear physical view of processes taking place in metal. Such sophistication of a theoretical description and absence of an illustrative and clear physical view of transport processes enhance a peril of losses of additional or novel wave solutions of the microwave kinetics problem. Really the kinetics, i. e. non-equilibrium approach leads to the electrodynamic problem with dissipation that implies the problem with non-self conjugated differential operator [7]. A theory of non-self conjugated operators is in the stage of development now and is far from that degree of completion, which corresponds to a theory of self conjugated operators [7]. Since the wave problems of the anomalous skin effect in metal and skin effect in superconductor belong to the strongly non-self conjugated electrodynamic problems, not to the so called weak disturbance of self conjugated operators [7], the mentioned peril of losses of solutions is actual. Non-self conjugated problems may possess unusual properties, for instance, may have eigenfunctions without properties of the base functions. Moreover kinetics problems with spatial dispersion are nonlinear on spatial parameters because of the self consistent field coming into the constitutive equation. So the fact that quantitative kinetics theory has given an explanation (original in [6] and replay in [2–5]) of the results obtained in certain experimental situation [8] does not ensure generally an explanation of the results obtained under slightly different experimental conditions [9–13]. To be sure it is important to analyse the possibility of existence of additional solutions of this kinetics problem or of the alternative general approach to resolve the problem.

Microwave effects of the spatial dispersion have been considered recently in [14, 15] and rather general relation has been established there between the eigenvalue of the total permittivity operator and the surface impedance. It elaborates the textbook statement [1] for the situation when the penetration depth δ becomes to be comparable with the length of mean free path l. In this situation the spatial inhomogeneity of the field makes impossible the macroscopic description of the field by use of a permittivity $\widetilde{\varepsilon}$. The $\widetilde{Z}$ surface impedance is the only value, which characterizes the metal properties. However, for its calculation the employment of the kinetics theory is required and a permittivity becomes to be an operator $\widehat{\widetilde{\varepsilon}}$. The found relation between the eigenvalue $\widetilde{\varepsilon}_a$ of the total permittivity operator $\widehat{\widetilde{\varepsilon}}_a$ and the surface impedance corresponds with the usual expression for a transverse wave propagating into a conductor without dispersion $\widetilde{Z} = (\mu_0/\widetilde{\varepsilon}_a)^{1/2}$. A substitution of the eigenvalue of permittivity operator into the

Maxwellian field-stress tensor, which determines the ponderomotive force of Abraham, results to conditions of spatial-type force resonances. The obtained resonances includes a particular solution corresponding to a traditional for anomalous skin effect value of the surface impedance $\widetilde{Z} = R + iX$ with the phase of $\pi/3$ and with the dependence of the module on the microwave frequency $|\widetilde{Z}| \sim \omega^{2/3}$. Taking into account these results and the known conclusion that the operator of dielectric permittivity alone completely defines the microwave response of conductors with spatial dispersion [1–3], the eigenvalue problem for permittivity operator is considered below because of the corresponding solutions are the stable waves for constitutive equation in a self-consistent microwave field due to the problem statement. A wave problem is formulated to search the solutions, which correspond to the eigenvalues of permittivity operator, similar to the problem of wave propagation in hollow waveguides and resonators. Dispersion relations and general solutions are obtained. A significant role of the spatial-type force resonances is considered. Due to the self-consistency of a kinetics problem, which is nonlinear on the spatial parameters, spatial-type force resonances are added to and usually dominate over an influence of boundary conditions. Conditions for the spatial resonances are derived and discussed.

2. Microwave response of conductors with spatial dispersion

When the spatial effects in a microwave response of continuous media are described by the methods of macroscopic electrodynamics, which operates the strength of electric field $\widetilde{\mathbf{E}}$ and the electric induction $\widetilde{\mathbf{D}}$, it is assumed that a characteristic scale of the correlation length between the values of $\widetilde{\mathbf{E}}$ and $\widetilde{\mathbf{D}}$ at different points of the space is much longer than the mean interparticle distance ($\sim n^{-1/3}$) [1-3]. This condition allows to apply a macroscopic description of the spatial dispersion in terms of a permittivity even in media with strong dispersion. Microscopic theories, which use, as a rule, the linearized Boltzmann kinetics equation, are applied for a deduction of a constitutive equation, which relates $\widetilde{\mathbf{E}}$ and $\widetilde{\mathbf{D}}$ in metals [2-5]. Generally this is an operator relation: an absolute permittivity operator $\widehat{\widetilde{\varepsilon}}_a$ in a constitutive equation connects the vector functions $\widetilde{\mathbf{E}}$ and $\widetilde{\mathbf{D}}$. This operator $\widehat{\widetilde{\varepsilon}}_a$, which depends on spatial variables, completely defines alone a reaction of a medium to an external microwave electromagnetic field. An inclusion of a concept of the relative magnetic permeability $\widetilde{\mu}$ in the Maxwell equations is not advantageous when the spatial dependencies are taken into account [1-3]. In media without dispersion in addition to the dielectric polarization $\widetilde{\mathbf{P}}$, which is incorporated in the definitions of $\widetilde{\mathbf{D}}$ and $\widetilde{\varepsilon}_a$, the magnetic polarization (or magnetization strength) $\widetilde{\mathbf{M}}$ is introduced as well, which comes in definitions of the magnetic field strength $\widetilde{\mathbf{H}}$ and of the absolute magnetic permeability $\widetilde{\mu}_a = \widetilde{\mu}\,\mu_0$, where μ_0 is the magnetic permeability of vacuum. It is assumed that an average microscopic current may be represented by a sum of two terms $\partial\widetilde{\mathbf{P}}/\partial t$ and rot $\widetilde{\mathbf{M}}$; in a plane wave, a combination of which is a Fourier expansion, these terms are reduced to $i\omega\widetilde{\mathbf{P}}$ and $i[\widetilde{\mathbf{k}} \times \widetilde{\mathbf{M}}]$. However, when the spatial dispersion being present, all parameters of the field depend already on the spatial parameter $\widetilde{\mathbf{k}}$ – the wave vector, the representation by two terms is excessive and the proportionality $\widetilde{\mathbf{B}} = \mu_0\widetilde{\mathbf{H}}$ must be used as a simple constitutive equation to relate the induction and the strength of magnetic field. For this choice the current and the charge density occur to be introduced in the definitions of the polarization $\widetilde{\mathbf{P}}$, the induction $\widetilde{\mathbf{D}}$ and the operator $\widehat{\widetilde{\varepsilon}}_a$, which in general case depend on both the electric $\widetilde{\mathbf{E}}$ and magnetic $\widetilde{\mathbf{B}}$ fields. However, the field $\widetilde{\mathbf{B}}$ may be expressed as a function of $\widetilde{\mathbf{E}}$ by use of the first pair of Maxwell equations [1], which include only these two values related by operations of differentiation. So the polarization $\widetilde{\mathbf{P}}$, the induction $\widetilde{\mathbf{D}}$ and the operator $\widehat{\widetilde{\varepsilon}}_a$ will be expressed as depending only on $\widetilde{\mathbf{E}}$.

The operator of absolute permittiviy $\widehat{\widetilde{\varepsilon}}_a$, which relates the vector functions $\widetilde{\mathbf{E}}$ and $\widetilde{\mathbf{D}}$, and completely defines the microwave response of the medium to the incident field, is determined in microscopic theories by an infinite system of equations of motion for individual particles together with the field equations. For description of such systems the probabilistic methods are used and kinetic equation for the distribution function with the self-consistent field should be solved [2–6]. Taking account of the dominant role of an operator of dielectric permittivity $\widehat{\widetilde{\varepsilon}}_a$, a problem of searching the wave solutions in a microwave response of conductors with spatial dispersion, which correspond to the eigenvalues of this operator is of a great importance.

3. Eigenvalue problem for permittivity operator

We shall find the wave eigensolutions compliant with the eigenvalue of permittivity operator $\widehat{\widetilde{\varepsilon}}_a$, i. e. with the numerical complex value $\widetilde{\varepsilon}_a$ in the first of constitutive equations

$$\widetilde{\mathbf{D}} = \widehat{\widetilde{\varepsilon}}_a \widetilde{\mathbf{E}} = \widetilde{\varepsilon}_a \widetilde{\mathbf{E}}\ ; \qquad \widetilde{\mathbf{B}} = \mu_0 \widetilde{\mathbf{H}}, \tag{1}$$

in the similar manner as in the problem of wave propagation in hollow waveguides and resonators [1, 4]. For monochromatic fields the Maxwell equations for vectors $\widetilde{\mathbf{E}}$ and $\widetilde{\mathbf{B}}$ in homogeneous medium with spatial dispersion [1–3] may be written, taking into account (1), as

$$\begin{aligned} &\text{rot}\ \widetilde{\mathbf{B}} = \mu_0\ \text{rot}\ \widetilde{\mathbf{H}} = \mu_0\ (\partial \widetilde{\mathbf{D}}/\partial t)\ ; \qquad \text{div}\ \widetilde{\mathbf{D}} = \text{div}\ \widetilde{\mathbf{E}} = 0\ ; \\ &\text{rot}\ \widetilde{\mathbf{E}} = -(\partial \widetilde{\mathbf{B}}/\partial t)\ ; \qquad \text{div}\ \widetilde{\mathbf{B}} = 0\ . \end{aligned} \tag{2}$$

The temporal factor $e^{i\omega t}$ is canceled out of equations below. After a differentiation with respect to time and a substitution of eqs (1) in (2), one will receive two equations

$$\text{rot}\ \widetilde{\mathbf{B}} = i\omega\mu_0\ \widetilde{\mathbf{D}} = i\omega\mu_0\widetilde{\varepsilon}_a\ \widetilde{\mathbf{E}}\ ; \qquad \text{rot}\ \widetilde{\mathbf{E}} = -i\omega\ \widetilde{\mathbf{B}} = -i\omega\mu_0\ \widetilde{\mathbf{H}}. \tag{3}$$

If we express $\widetilde{\mathbf{B}}$ from the second of eqs (3) and substitute it into the first of eqs (3), we get rot rot $\widetilde{\mathbf{E}} = \omega^2\mu_0\widetilde{\varepsilon}_a\ \widetilde{\mathbf{E}}$ or, expanding the operation of double rotation, grad div $\widetilde{\mathbf{E}}-$div grad $\widetilde{\mathbf{E}} = \omega^2\mu_0\widetilde{\varepsilon}_a\ \widetilde{\mathbf{E}}$. One can substitute in this expression the equality div $\widetilde{\mathbf{E}} = 0$ from (2) and introduce a new parameter $\widetilde{k}_d^2$ as $\omega^2\mu_0\widetilde{\varepsilon}_a = \widetilde{k}_d^2$ that result in the uniform, nonlinear in spatial parameter, vector wave equation $\nabla^2\widetilde{\mathbf{E}} + \widetilde{k}_d^2\,\widetilde{\mathbf{E}} = \nabla^2\widetilde{\mathbf{E}} + \omega^2\mu_0\widetilde{\varepsilon}_a\ \widetilde{\mathbf{E}} = 0$, where ∇^2- is the Laplace operator. If we take into account the definition of the gradient of a vector field with respect to a vector [16–18] ($\mathbf{a}$ grad) $\mathbf{V} = (\mathbf{a}\cdot\nabla)\,\mathbf{V}$, where $\mathbf{a}$ – the arbitrary constant vector and $\mathbf{V}$ – the vector field, the Laplace operator in the previous equation can be written in the form of an euclidean scalar product of two Hamilton operators div grad $\widetilde{\mathbf{E}} = (\nabla\cdot\nabla)\,\widetilde{\mathbf{E}} = \nabla^2\widetilde{\mathbf{E}}$. So we can seek a solution of the Laplace operator, which corresponds to the eigenvalue in form of the euclidean square of some vector $\widetilde{\mathbf{k}}$ in the real three-dimensional euclidean space, but generally with complex coordinates. This vector – the wave vector may be considered as a vector of the complex euclidean space [18]. Derivatives are taken on real variables here.

The equation for $\widetilde{\mathbf{B}}$ is obtained analogously. These vector equations in all cases fall into six scalar wave equations of the same type for components of vectors $\widetilde{\mathbf{E}}$ and $\widetilde{\mathbf{B}}$. Each of scalar equations can be solved by the method of variables separation. Let us assume, for example, in the rectangle coordinate system only one component of the electric field $\widetilde{E}_x$ parallel to the metal surface is presented.

Designating the three member in the Pythagoras theorem for a square of wave-vector in orthogonal coordinates by the constants $-\widetilde{k}^2$ (and $\widetilde{\varepsilon}_{ax}$ for a more illustrative physical view): $-\widetilde{k}_x^2 = -\omega^2\mu_0\widetilde{\varepsilon}_{axx}$, $-\widetilde{k}_y^2 = -\omega^2\mu_0\widetilde{\varepsilon}_{axy}$, $-\widetilde{k}_z^2 = -\omega^2\mu_0\widetilde{\varepsilon}_{axz}$, one will have $\widetilde{k}_d^2 - \widetilde{k}_x^2 - \widetilde{k}_y^2 - \widetilde{k}_z^2 = 0$ or $\widetilde{\varepsilon}_a = \widetilde{\varepsilon}_{axx} + \widetilde{\varepsilon}_{axy} + \widetilde{\varepsilon}_{axz}$. The general solution of this equations can be written in the form similar to the following $\widetilde{E}_x = \widetilde{C}_5\exp(-i\widetilde{k}_z z) + \widetilde{C}_6\exp(i\widetilde{k}_z z)$.

4. Spatial force resonance and additional waves

The strongly inhomogeneous in space field can be considered as static in a comparison with the reaction of the conducting medium, characterized by the Fermi velocity $v_F \sim 0.01\,c$, where c is the speed of light. A reaction to an external field is spatially non-local, and the state of the system is determined by the correlation of the applied forces. In the quasistatic consideration, the maximum perturbation of the conducting medium will be caused by that of possible spatial configurations of the medium, which corresponds to the influence of dominating acting force caused by the incident field. The force (per unit volume) can be calculated by the Maxwellian field-stress tensor, each component of which is the spatial density of the corresponding momentum flux $\widetilde{\mathbf{\Pi}} = [\widetilde{\mathbf{D}} \times \widetilde{\mathbf{B}}]$. Derivative $\partial\widetilde{\mathbf{\Pi}}/\partial t$ determines the Abraham force (except for the force acting on the field inside the conductor and the effects of dispersion) [1, 3].

Substituting the impedance value into the momentum flux equation one can deduce on the boundary surface

$$\widetilde{\mathbf{\Pi}} = [\widetilde{\mathbf{D}} \times \widetilde{\mathbf{B}}] = \widetilde{D}_x(0)\widetilde{B}_y(0)\mathbf{k} = \widetilde{\varepsilon}\widetilde{E}_x(0)\sqrt{\mu_0\widetilde{\varepsilon}}\widetilde{E}_x(0)\mathbf{k} = (\mu_0^{1/2}\widetilde{\varepsilon}^{3/2})\widetilde{E}_x^2(0)\mathbf{k}. \tag{4}$$

The spatial-type force resonance corresponds to a real value of the coefficient in parentheses. This condition can be expressed as the equations $\widetilde{\varepsilon} = |\widetilde{\varepsilon}|e^{i\alpha}$, $\widetilde{\varepsilon}^{3/2} = |\widetilde{\varepsilon}|^{3/2}e^{3i\alpha/2}$, $3i\alpha/2 = i\pi n$, $n = 0, 1, 2$, $\alpha_{0,1,2} = 0, 2\pi/3, 4\pi/3$. The resonance in metals corresponds to the phase $\alpha_2 = 4\pi/3$, at which the spatial force resonance takes place. Phases $\alpha_0 = 0$ and $\alpha_1 = 2\pi/3$ refer to continuous wave and wave with amplitude growing deep into conductor, correspondingly. When the phase of $\widetilde{\varepsilon}$ is $4\pi/3$, and the phase of $\widetilde{Z} = (\mu_0/\widetilde{\varepsilon})^{1/2}$ is $\pi/3$ the ratio of the absolute values of real and imaginary parts of both becomes equal to $\sqrt{3}$.

The spatial-type force resonance may also correspond to an imaginary value of a coefficient in the expression for momentum flux $\widetilde{\mathbf{\Pi}}$. This condition agrees with equalities $\widetilde{\varepsilon} = |\widetilde{\varepsilon}|e^{i\beta}$, $\widetilde{\varepsilon}^{3/2} = |\widetilde{\varepsilon}|^{3/2}e^{3i\beta/2}$, $3i\beta/2 = i(\pi n + \pi/2)$, $n = 0, 1, 2$, $\beta_{0,1,2} = \pi/3, \pi, 5\pi/3$. Phases of $\widetilde{\varepsilon}$, equal to π, and $\widetilde{Z} = (\mu_0/\widetilde{\varepsilon})^{1/2}$, equal to $\pi/2$, correspond to the surface impedance of a nearly completely superconducting component with a close to an ideal reflection. The phase $\pi/3$ corresponds to the wave with an amplitude, growing deep into a conductor, and the phase $5\pi/3$ for $\widetilde{\varepsilon}$, which would correspond to an unusual phase $\pi/6$ for the surface impedance $\widetilde{Z} = (\mu_0/\widetilde{\varepsilon})^{1/2}$, principally is possible in normal metals and its registration is of interest.

References

[1] Landau L D and Lifshiz E M 1992 *L D Landau and E M Lifshiz Theoretical Physics vol 8, Electrodynamics of Continuous Media* (Moscow: Nauka) [in Russian].
[2] Lifshiz E M and Pitaevskii L P 1979 *L D Landau and E M Lifshiz Theoretical Physics vol 10, Physical Kinetics* (Moscow: Nauka) [in Russian]
[3] Silin V P and Rukhadze A A 1961 *Electromagnetic Properties of Plasma and Plasma-Like Media* (Moscow: Gosatomizdat) [in Russian].
[4] Mende F F and Spitsyn A I 1985 *Surface Impedance of Superconductors* (Kiev: Naukova Dumka) [in Russian]
[5] Abrikosov A A 1987 *Osnovy Teorii Metallov (Fundamentals of the Theory of Metals)* (Moscow: Nauka) [in Russian. Translated into English (Amsterdam: North-Holland, 1988)]
[6] Reuter G E H and Sondheimer E H 1948 *Proc. Roy. Soc. A* **195**, 336
[7] Ilyinskii A S and Slepyan G Y 1983 *Oscillations and Waves in Dissipative Electrodynamic Systems* (Moscow: Moscow University) [in Russian]
[8] Pippard A B 1947 *Proc. Roy. Soc. A* **191** 385
[9] Shcherbakov A S *at al* 1987 *Fiz.Met.Met.* **64** 735
[10] Shcherbakov A S *at al* 1987 *Fiz.Met.Met.* **64** 742
[11] Apakina V N *at al* 1997 *Physica C* **282-287** 1585
[12] Chernyaev A P *at al* 2008 *Journal of Physics and Chemistry of Solids* **69** 3313
[13] Dresvyannikov M A *at al* 2013 *Proc. of SPIE* **8700** 87000I
[14] Volchkov N A, Karuzskii A L and Perestoronin A V 2010 *JETP* **111** 292
[15] Karuzskii A L *at al* 2012 *J. Phys.: Conf. Ser.* **400** 022048
[16] Bronshtein I N and Semendyaev K A 1995 *Teubner-Taschenbuch der Mathematik* (Leipzig-Stuttgart: Teubner)
[17] Kochin N E 1965 *Vector Analysis and Fundamentals of the Tensor Analysis* (Moscow: Nauka) [in Russian]
[18] Rashevskii P K 1967 *Riemannian Geometry and Tensor Analysis* (Moscow: Nauka) [in Russian]

Excitation spectrum of Josephson vortices on surface superconductor

Takuto Kawakami[1], Yuki Nagai[2], Shunsuke Yoshizawa[1], Howon Kim[3], Yukio Hasegawa[3], Tomonobu Nakayama[1], Takashi Uchihashi[1], and Xiao Hu[1]

[1] International Center for Material Nanoarchitectonics (WPI-MANA), National Institute for Materials Science (NIMS), 1-1 Namiki, Tsukuba 305-0044, Japan, [2]CCSE, Japan Atomic Energy Agency, 5-1-5 Kashiwanoha, Kashiwa, Chiba 277-8587 Japan, [3]The Institute for Solid State Physics, University of Tokyo, 5-1-5 Kashiwanoha, Kashiwa, Chiba 277-8587, Japan

E-mail: kawakami.takuto@nims.go.jp

Abstract. Motivated by recent discovery of superconductivity in monatomic In layer on silicon surfaces and scanning tunneling microscopy and spectroscopy (STM/STS) measurement, we study the excitation spectrum of vortex in a two-dimensional superconductor. According to the STM/STS measurements, the zero bias conductance (ZBC) for the vortices trapped at atomic steps has an anisotropic shape. In order to understand the experimental results, we study the local density of states (LDOS) as a function of strength of hopping integral at atomic steps. Our analysis is based on the self-consistent solution of the Bogoliubov-de Gennes equation coupled with a gap equation. The amplitude of the order parameter is restored at the vortex core and the phase winding becomes anisotropic with decreasing the hopping strength at atomic steps. The zero energy LDOS stretches along the step and the intensity decreases associated with the deformation of the order parameter. By comparing the theoretical results with experiments, we conclude Josephson vortices in 2D superconductivity have been observed.

1. Introduction

Experimental progress in observing the superconductivity in the monatomic layer on the silicon surface reconstruction [1, 2, 3] open new frontiers for exploring the two-dimensional superconductivity. One of the interesting points of this system is the existence of atomic steps of the silicon surface separating terraces. The atomic steps inherently interfere the electron motion between neighboring two-dimensional terraces. The observation of supercurrent thorough the atomic steps [2, 3] provides a scenario that this system behaves as Josephson junctions [2].

Very recently, the ZBC of the scanning tunneling spectroscopy (STS) measurement in the superconducting In monatomic layer on the Si(111)-($\sqrt{7}\times\sqrt{3}$) reconstruction has been reported as follows [4]. Under the magnetic field perpendicular to the surface, spots of ZBC appear both on the atomic steps and on the terrace region. The spots on the atomic steps have an elliptic shape with the longer axis parallel to the steps while those on the terrace are cylindrically symmetric. In addition, their aspect ratio differs depending on the atomic steps and the longer spots have smaller intensity.

The STS measurement images vortex shape through direct observation of quasiparticle excitation spectrum with high spatial resolution. The anisotropy of STS image should depend on the strength of interterrace Josephson coupling. Nevertheless, most of the theoretical works for the Josephson vortex [5, 6, 7] are based on the London or Ginzburg-Laudau theory, which do not include information of quasiparticles. In this paper, we study the excitation spectrum in this system by using Bogoliubov-de Gennes (BdG) theory. We investigate the spatial profile of order parameter and excitation spectrum of vortices in a self-consistent way and analyze the contribution of the quasiparticles. We conclude that the vortex image with strong anisotropic ZBC in STS measurements [4] can be considered as Josephson vortex in the surface superconductor.

2. Model and Formulation

We theoretically analyze the excitation spectrum of vortex core in terms of BdG theory. We use a two-dimensional tight-binding model with a square lattice. The BdG equation is described as

$$\sum_j \begin{pmatrix} \hat{K}_{i,j} & \hat{\Delta}_{i,j} \\ \hat{\Delta}^*_{j,i} & -\hat{K}^*_{i,j} \end{pmatrix} \begin{pmatrix} u_\gamma(\boldsymbol{r}_j) \\ v_\gamma(\boldsymbol{r}_j) \end{pmatrix} = E_\gamma \begin{pmatrix} u_\gamma(\boldsymbol{r}_i) \\ v_\gamma(\boldsymbol{r}_i) \end{pmatrix}. \tag{1}$$

The single particle part is $\hat{K}_{i,j} = -t_{ij} \exp\left[i(\pi/\Phi_0)\int_{\boldsymbol{r}_i}^{\boldsymbol{r}_j} \boldsymbol{A}(\boldsymbol{r}')\cdot d\boldsymbol{r}'\right] - \mu\delta_{ij}$, where t_{ij} is the hopping strength between the nearest neighbor atomic sites, $\boldsymbol{A}$ is the vector potential due to the external field, and $\Phi_0 = h/2e$ is the flux quantum.

We consider the vortex lattice by using the periodic boundary condition

$$\begin{pmatrix} u_\gamma(\boldsymbol{r}+\boldsymbol{R}) \\ v_\gamma(\boldsymbol{r}+\boldsymbol{R}) \end{pmatrix} = e^{i\boldsymbol{k}\cdot\boldsymbol{R}} e^{i\hat{\sigma}_z\chi(\boldsymbol{r},\boldsymbol{R})} \begin{pmatrix} u_\gamma(\boldsymbol{r}) \\ v_\gamma(\boldsymbol{r}) \end{pmatrix}, \tag{2}$$

where $\boldsymbol{R}(p_1,p_2) = \sum_\alpha p_\alpha \boldsymbol{a}_\alpha$ is the two-dimensional Bravais lattice vector and $\boldsymbol{a}_\alpha$ $(\alpha \in 1,2)$ is the primitive translation vector of the vortex lattice; $\boldsymbol{k}$ is the Bloch vector satisfying $\boldsymbol{k}\cdot\boldsymbol{R} = 2\pi\sum_\alpha p_\alpha q_\alpha/N_\alpha$ with $p_\alpha, q_\alpha \in Z$ and the system size $N_\alpha|\boldsymbol{a}_\alpha|$ for $\boldsymbol{a}_\alpha$-direction; and the condition $\chi(\boldsymbol{r},\boldsymbol{R}) = \frac{n\pi}{2}\left[\boldsymbol{g}(\boldsymbol{R})\cdot\boldsymbol{r} + p_1 p_2\right]$, where $\boldsymbol{r} = \sum_\alpha r_\alpha \hat{\boldsymbol{a}}_\alpha$, $\hat{\boldsymbol{a}}_\alpha = \boldsymbol{a}_\alpha/|\boldsymbol{a}_\alpha|$ and $\boldsymbol{g} = \sum_{\alpha\beta}\epsilon_{\alpha\beta}p_\alpha\hat{\boldsymbol{a}}_\beta$ [12, 10]. We use the symmetric gauge $\boldsymbol{A} = \frac{1}{2}\boldsymbol{H}\times\boldsymbol{r}$ which generate n vortices within the unit cell $\boldsymbol{a}_1\times\boldsymbol{a}_2$

For simplicity, we consider a straight terrace step with one atomic spacing and model it by modulating the hopping strength along the step $t_{ij}=t_s$ while keeping the hopping strength as a constant $t_{ij}=t$ elsewhere as shown in Fig. 1 (a), (b). In this work, we deal with square geometry including 192×192 atomic sites.

We obtain the following gap function and LDOS

$$\Delta(\boldsymbol{r}_i) = \delta_{ij}V\sum_\gamma u_\gamma(\boldsymbol{r}_i)v_\gamma(\boldsymbol{r}_j)f(E_\gamma) = \delta_{ij}V\int dE'\sum_\gamma u_\gamma(\boldsymbol{r}_i)v_\gamma(\boldsymbol{r}_j)f(E_\gamma)\delta(E'-E_\gamma), \tag{3}$$

$$\mathcal{N}(E,\boldsymbol{r}) = \sum_\gamma |u_\gamma(\boldsymbol{r})|^2\delta(E-E_\gamma)\,. \tag{4}$$

According to the Chebyshev-polynomial expansion scheme [8, 9, 11], we expand the Dirac delta function as

$$\delta(\omega-\omega') \sim \sum_{n=0}^{n_c} \frac{2T_n(\omega)T_n(\omega')}{\pi\sqrt{1-\omega^2}(1+\delta_{n,0})}, \tag{5}$$

where $T_n(\omega)$ is the n-th Chebyshev polynomial within the interval $\omega \in [-1,1]$. Here, we rescale the Hamiltonian as $\hat{K}'_{i,j} = \hat{K}_{i,j}/a$ and $\hat{\Delta}'_{i,j} = \hat{\Delta}_{i,j}/a$ with parameters $a=20t$ and set the cutoff constant $n_c = 1000$. In order to examine the origin of the behavior of the DOS in terms of quantized quasiparticle, we also calculate several dozen of eigenvalues and wave functions of quasiparticles by using shifted-invert Arnoldi algorithm [13].

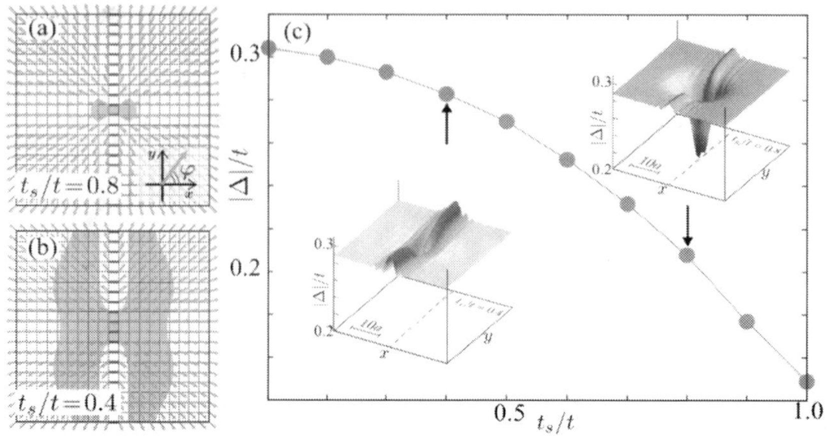

Figure 1. Spatial profile of order parameter. The direction of an arrow denotes the phase $\varphi(\boldsymbol{r})$ of order parameter on each atomic site for hopping strength $t_s/t=0.8$ (a) and 0.4 (b). The thick (blue) bonds denote the junction where the hopping strength $t_{ij}=t_s<t$. In the shaded region, the phase gradient is relatively small. The main panel in (c) shows the amplitude $|\Delta(\boldsymbol{r})|$ of order parameter at the center of vortex. The left (right) inset shows the two-dimensional profile of the amplitude $|\Delta(\boldsymbol{r})|$ for $t_s/t=0.4$ (0.8). Parameters are set as $\mu/t=-2.5$ and $V/t=3.0$.

3. Results

Figures 1 (a), (b) show profiles of the phase $\varphi(\boldsymbol{r})$ of order parameter $\Delta = |\Delta(\boldsymbol{r})|\exp[i\varphi(\boldsymbol{r})]$ for hopping strength at terrace step $t_s/t = 0.8$ and 0.4. For $t_s/t \sim 1$ (Fig. 1(a)), the profile is almost cylindrically symmetric. As the hopping strength at step t_s/t is reduced, the phase gradient along the step $\partial_y\varphi$ becomes smaller near the vortex core as shown in Fig. 1(b), which reduces the total kinetic energy. The amplitude of the order parameter for $t_s/t\sim 1$ is suppressed significantly at the vortex core $r \leq \xi$ ($\xi = v_F/\Delta_0$) (Fig. 1(c)). As the hopping strength at the step is reduced, the suppression of amplitude at vortex becomes weaker, and invisible for small hopping $t_s/t \ll 1$; simultaneously the healing length along the step becomes much larger than the vertical direction (see the insets in Fig. 1(c)). It is intriguing to observe that, as in inset of Figs. 1(c), the order parameter is enhanced at the terrace step with weak hopping strength, due to the interference of quasiparticles scattered by the step.

Figure 2 shows t_s-dependence of the quasiparticle eigenstates, which is obtained by using the self-consistent solution $\Delta(\boldsymbol{r})$ in Fig. 1. For $t_s/t\sim 1$, the vortex core bound state has mini gap $\delta E\sim\Delta^2/E_F$ [14]. With decreasing hopping t_s/t, the mini gap becomes larger as shown in Fig. 2(a). Figures 2(b) and (c) show the wave functions $u_{\gamma=1,2}(\boldsymbol{r})$ of the lowest and 2nd lowest energy states for $t_s/t = 0.8$ and 0.4. The squared absolute value of wave function $|u_\gamma(\boldsymbol{r})|^2$ is proportional to the local density of states $\mathcal{N}(\pm E_\gamma, \boldsymbol{r})$. When the hopping strength is $t_s/t=1$, $|u_{\gamma=1,2}(\boldsymbol{r})|$ exhibits four fold rotational symmetry originating from Fermi surface. As shown in Fig. 2(c), the tail of wave function approaches to the step for $t_s/t < 1$. This can be understood as follows. For $t_s/t < 1$, the phase gradient of order parameter in Fig. 2(b) becomes smaller on left and right region sandwiching the vortex core (see shaded region in Fig. 1(a) and (b)). Since the external magnetic field cannot be canceled by the phase gradient, low energy quasiparticles tend to avoid this region in order to save kinetic energy. Consequently, the low energy quasiparticle $u(\boldsymbol{r})$ is suppressed more strongly in this region with reduction of t_s/t, and the lowest energy wave function has elongated shape along y-direction. In particular, at the region close to the vortex $r \sim \xi$, the peak of the lowest energy wave function becomes elliptic and longer axis is

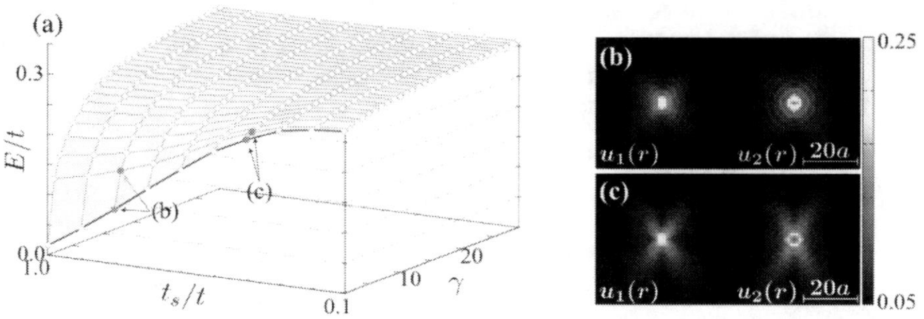

Figure 2. Eigen energy E of γ-th lowest eigenstates as a function of t_s/t (a) and the spatial profile of the wave function of lowest and second lowest eigenstates $|u_{\gamma=1,2}(\boldsymbol{r})|$ for $t_s/t=0.8$ (b) and for $t_s/t=0.4$ (c). The filled symbol in (a) denotes the eigenstate shown in (b) and (c).

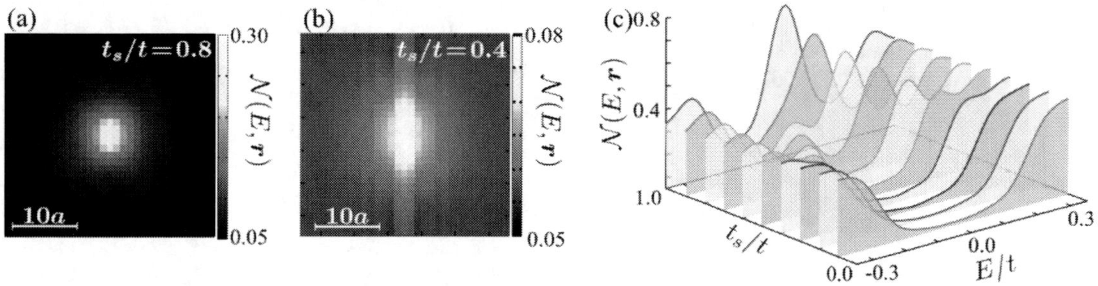

Figure 3. Two dimensional profile of the zero-energy density of states $\mathcal{N}(E=0,\boldsymbol{r})$ for hopping strength $t_s/t=0.8$ (a) and 0.4 (b), and the LDOS at the vortex center as a function of the energy and hopping t_s/t (c). We set the other parameters $\mu/t=-2.5$ and $V/t=3.0$.

y-direction.

Figures 3 (a), (b) show the zero-energy DOS (ZDOS) $\mathcal{N}(E=0,\boldsymbol{r})$ associated with the order parameter $\Delta(\boldsymbol{r})$ shown in Fig. 1. For $t_s/t \sim 1$, the ZDOS is cylindrically symmetric (Fig. 3 (a)) and becomes elliptic when the hopping strength reduces (Fig. 3 (b)) originating from the shape of elliptic peak of the lowest wave function $|u_1(\boldsymbol{r})|$. Figure 3 (c) shows the LDOS at the vortex center as a function of energy E and hopping integral t_s/t along the terrace step. The value of ZDOS decreases with reduction of t_s/t, and for small $t_s/t \ll 1$, the vortex shape cannot be distinguished as ZBC spots of STS signal. On the other hand, the mini gap at the vortex center increases with reduction of the t_s/t. This corresponds to excitation of $\gamma=1$ in Fig. 2(a). We suggest that the strength of Josephson coupling for the terrace step may be estimated by comparing the mini gap in STS at the vortex center with theoretical one shown in Fig. 3(c).

4. Summary

We have investigated the excitation spectrum at the vortex on atomic steps which has electron hopping strength t_s weaker than bulk t by using the Bogoliubov-de Gennes framework. The results are consistent with the experimental observations. We conclude that Josephson vortices have been observed at the step with weak coupling.

Acknowledgments
This work was supported by the WPI Initiative on Materials Nanoarchitectonics, and Grant-in-Aid for Scientific Research (Nos. 25103723, 25400385, 25247053 and 25286055), MEXT of Japan. The calculations have been performed using the supercomputing systems at National Institute for Materials Science and PRIMERGY BX900 at the Japan Atomic Energy Agency.

References

[1] Zhang T, *et al.*, 2010 *Nature Phys.* **6** 104
[2] Uchihashi T, Mishra P, Aono M, and Nakayama T, 2011 *Phys. Rev. Lett.* **107** 207001
[3] Uchihashi T, Mishra P, and Nakayama T 2013 *Nanoscale Res. Lett.* **8** 167
[4] Yoshizawa S, Kim H, Kawakami T, Nagai Y, Nakayama T, Hu X, Hasegawa Y, and Uchihashi T, arXiv:1405.5953
[5] Gurevich A, 1992 *Phys. Rev.* B **46** 3187
[6] Gurevich A, 1993 *Phys. Rev.* B **48** 75863
[7] Kogan V, Dobrovitski V, Clem J, Mawatari Y, and Mints R, 2001 *Phys. Rev.* B **63**, 144501
[8] Covaci L, Peeters F M, and Berciu M, 2010 *Phys. Rev. Lett.* **105** 167006
[9] Nagai Y, Ota Y, and Machida M, 2012 *J. Phys. Soc. Jpn.* **81** 024710
[10] Nagai Y, Nakai N, and Machida M, 2012 *Phys. Rev.* B **81**, 092505
[11] Y. Nagai, Y. Ota, and M. Machida, 2012 *Phys. Proc.* **27**, 72
[12] Takigawa M, Ichioka M, and Machida K, 2000 *J. Phys. Soc. Jpn* **69** 3943
[13] The subroutine libraries are available from the home page at [http://www.caam.rice.edu/software/ARPACK].
[14] Hayashi N, Isoshima T, Ichioka M, and Machida K, 1998 *Phys. Rev. Lett.* **80** 2921

Journal of Physics: Conference Series **568** (2014) 022023 doi:10.1088/1742-6596/568/2/022023

Deposition of superconducting $Ba_2Ca_{n-1}Cu_nO_{2n}(O,F)_2$ thin films by pulsed laser ablation

S KOBA[1], **A IYO**[2]

[1] Kumamoto National College of Technology, 2627 Hirayama-Shinmachi, Yatsushiro, Kumamoto 866-8501, Japan

[2] Advanced Industrial Science and Technology, Central-2, 1-1-1 Umezono, Tsukuba, Ibaraki 305-8568, Japan

E-mail: koba@kumamoto-nct.ac.jp

Abstract. The highest ever superconducting critical temperature (T_c) of 120 K has been reported for the compound $Ba_2Ca_{n-1}Cu_nO_{2n}(O,F)_2$ (F-02(n-1)n). However, only bulk samples have been synthesized to date and extremely high pressures have been required. Here, the deposition of superconducting $Ba_2Ca_{n-1}Cu_nO_{2n}(O,F)_2$ films on a $SrTiO_3(100)$ substrate was investigated using Nd-YAG pulsed laser ablation under an O_2 pressure of 16-19 Pa at a substrate temperature of 740-750 °C. The films exhibited a superconducting onset temperature (T_{co}) of 76.5 K and a zero resistance temperature (T_{ce}) of 22.5 K. The dominant diffraction peaks matched those reported for c-axis oriented $Ba_2Ca_2Cu_2O_6(O,F)_2$ (n = 3) bulk samples. However, the T_c values for the films were much lower than that for the bulk material, even though the ablation target composition was $Ba_2Ca_2Cu_3O_{6+\delta}(O,F)_2$, which yields the highest bulk T_c value.

1. Introduction

The dependence of the superconducting critical temperature (T_c) for $Ba_2Ca_{n-1}Cu_nO_{2n}(O,F)_2$ (F-02(n-1)n) multilayered superconductors on the number of CuO_2 planes (n) has been reported to be bell shaped for n=1-5 and constant for n=5-9 [1,2]. The highest T_c of 120 K was reported for F-0223 (n=3) [1]. T_c values reported for F-0212 (n=2), F-0223 (n=3), F-0234 (n=4), and F-0245 (n=5) are 90 K, 90-120 K, 70-105 K, and 90 K, respectively [1-3]. Figures 1(a) and 1(b) show crystal structures of F-0223 (n=3) and F-0234 (n=4), respectively. The F-02(n-1)n crystal structures are composed of superconducting layers (SCLs) with CuO_2 planes (n) blocked by Ca and charge reservoir layers (CRLs) with a rock salt layer of Ba-(O,F) included with substitutions of F^- at apical oxygen sites (O^{2-}) [3]. The carrier concentration in the SCL is controlled by the substitution of F^- for an apical O^{2-} in the CRL [3]. Carrier doping on the CuO_2 planes of F-02(n-1)n has been predicted to be so-called self-doping between an electron type outer-CuO_2 plane directly under the CRL included with apical (O,F) and a hole type inner-CuO_2 plane. If (O,F) substitutions are optimized, then F-02(n-1)n superconductors may have the potential to achieve a T_c that is >120 K with more optimal carrier doping on the inner CuO_2 plane by self-doping [1]. However, until now, only bulk samples have been synthesized, and extremely high pressures have been required.

We have previously proposed that deposition of F-02(n-1)n superconducting films with T_c > 100 K may be possible by optimal substitution in the CRLs, if no vacancies are required for optimal carrier doping and formation of the crystal structure [3]. This is because such high T_c values are possible for bulk samples, and pulsed laser deposition (PLD) is capable of producing films whose composition, including the number of SCLs and CRLs, and the degree of F substitution, precisely matches that of the target. In the present study, the deposition of superconducting $Ba_2Ca_2Cu_2O_{6+\delta}(O,F)_2$ (n=3) films on a $SrTiO_3(100)$ substrate was investigated using Nd-YAG PLD.

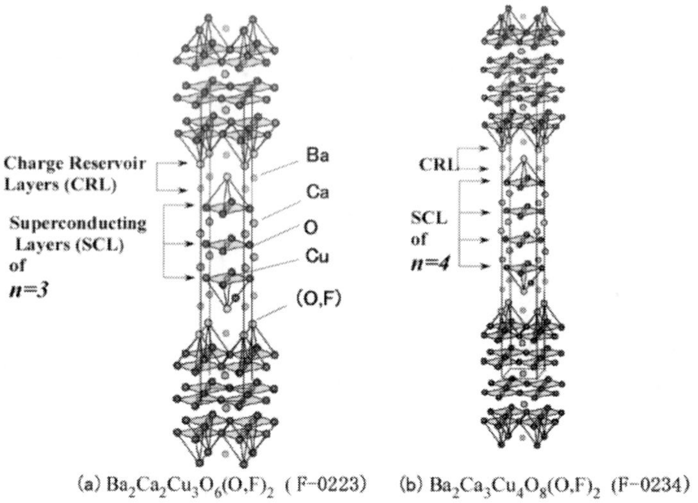

Figure 1. Crystal structures of (a) n=3, $Ba_2Ca_2Cu_3O_6(O,F)_2$ (F-0223), (b) n=4, $Ba_2Ca_3Cu_4O_8(O,F)_2$ (F-0234) on $Ba_2Ca_{n-1}Cu_nO_{2n}(O,F)_2$ (F-02(n-1)n).

2. Experimental

Targets of $Ba_2Ca_{n-1}Cu_nO_{2n}(O_{1-y},F_y)_2$ (y=0.8) were synthesized with source materials BaO_2, CaF_2, CuO, and a precursor of Ca_2CuO_3. Ca_2CuO_3 was prepared from a mixture of $CaCO_3$ and CuO by sintering at 1000 °C for 12 h under an oxygen flow with an intermediate grinding procedure. The starting materials were weighed to achieve a nominal composition of $Ba_2Ca_{n-1}Cu_nO_{2n}(O_{0.2},F_{0.8})_2$ and then ground with an agate mortar and pestle in a nitrogen-filled glovebox. The mixture was sintered at 860 °C for 12 h under oxygen flow. The resultant powder was pressed into a ca. 10 mm diameter pellet. The target was reacted in the temperature range of 880-920 °C for 12 h under an oxygen flow.

Thin film samples were deposited on $SrTiO_3(100)$ substrates (STO) using a Nd-YAG PLD system (Pascal PLD-system). F-02(n-1)n thin films were deposited at a substrate heating temperature (T_s) of 650-800 °C under pure oxygen flow at partial pressures (PO_2) of 17-19 Pa for 45-60 min, followed by a cooling process at a rate of 10 °C/min to 400 °C after the oxygen flow was stopped. Other PLD conditions were set as follows: pulsed laser energy density of 80 mJ/cm^2, laser wavelength of 266 nm, pulse frequency of 10 Hz, target-substrate distance of 38 mm. The thickness of the film was calculated based on deposition rates of 0.82, 0.76, 0.70, and 0.64 nm/min at PO_2 = 16, 17, 18, and 19 Pa, respectively. The effective surface area of the films was approximately 5×5 mm^2.

The films were characterized using inductively coupled plasma mass spectrometry (ICP-MS; Agilent 8800ICP-MS) and X-ray diffraction (XRD; Panalytical MRD) with Cu Kα radiation at 45 kV with an emission current of 40 mA. ICP-MS was conducted with a carrier gas volume of 1.03 L/min, an rf power of 1550 W, a sampling depth of 8 mm, 4 mL/min H_2 reaction gas, and 300 μL sample uptake. The temperature dependence of the resistivity of the films was measured using the four-probe method (Quantum Design Physical Property Measurement System; PPMS) with a current of 0.1-1 μA applied along the surface. The temperature was varied at a rate of 0.5 K/min.

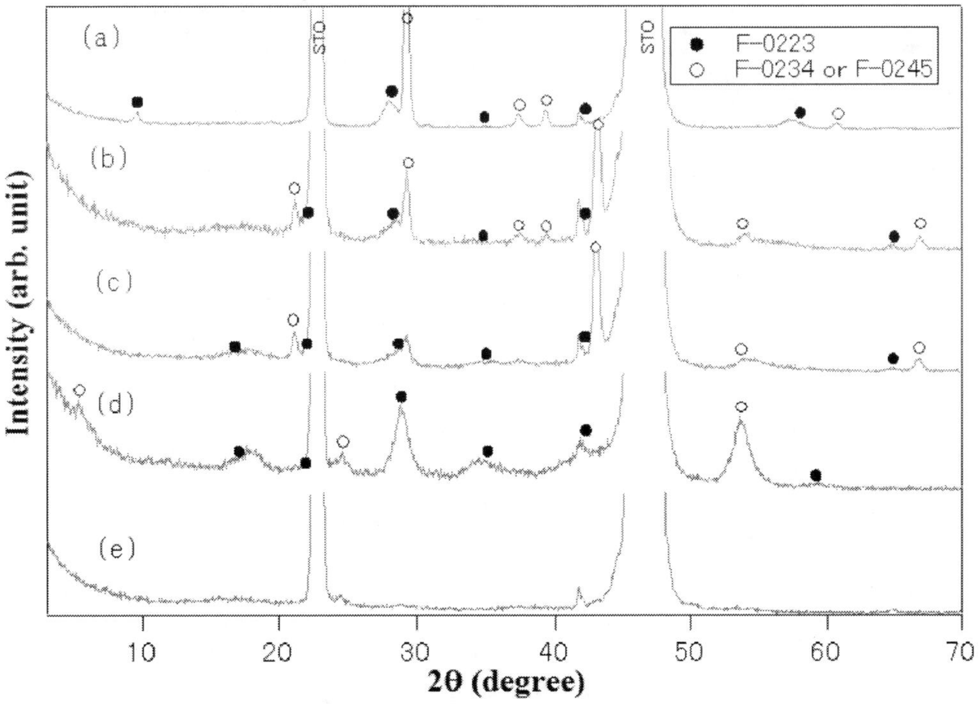

Figure 2. XRD patterns for films fabricated at substrate heating temperatures (T_s) of (a) 800, (b) 770, (c) 750, (d) 740, and (e) 650 °C.

3. Results and discussion

PLD was conducted using targets with n=2, 3, and 5 in F-02(n-1)n. Films with a superconducting transition were obtained using a target of F-0223, whereas PLD films deposited using other compound targets were insulating or semiconductor-like. However, it is considered that the PLD conditions were not optimized for F-0212 and F-0245, because other conditions, such as PO_2 much higher than 20 Pa and T_s lower than 600 °C, were not investigated in this work.

The following results are for the films prepared by PLD using the F-0223 target. PLD conditions such as T_s, PO_2, and the deposition time were examined first. Comparison with the ICDD database indicated that the XRD patterns for all samples were not consistent with those for cuprate superconductors other than F-02(n-1)n. Moreover, no microscopic metallic elements other than Ba, Ca, and Cu were detected by ICP-MS analysis of the film samples. Figure 2 shows XRD patterns for PLD films deposited at T_s of 650-800 °C. Although the film deposited at T_s=650 °C was not crystalline, all other films exhibited peaks associated with F-0223 and other phases. The resistivity of the films deposited at 800 and 770 °C was high, and the films displayed insulating and semiconducting characteristics, respectively. In contrast, a superconducting transition was observed for the films deposited at 750 and 740 °C. The F-0223 phase was evident in all samples except that prepared at 650 °C, with the strongest, sharpest peaks being obtained at 800 and 770 °C. The peaks at 2θ =21.21, 29.31, 37.31, and 39.20° are associated with the F-0234 phase, whereas that at 2θ=43.05° is due to the F-0245 phase. Figure 2 indicates that the difference in resistivity was probably due to the formation of the insulating F-0234 and F-0245 phases and the under-doped F-0223 phase, when T_s was higher than that for the optimized deposition of F-0223.

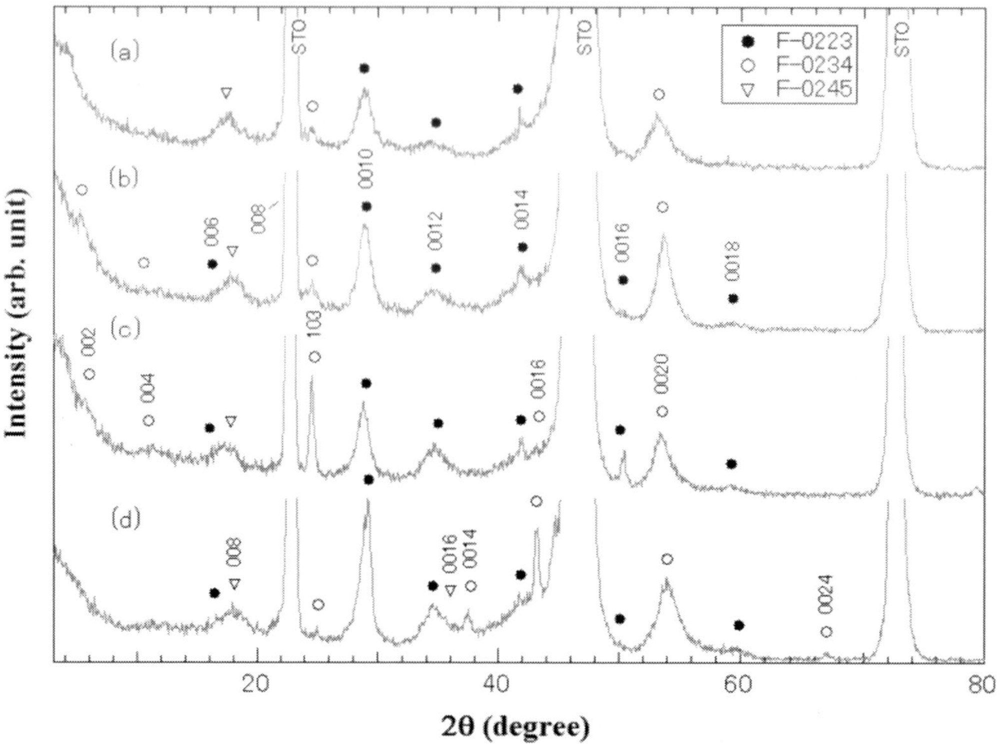

Figure 3. XRD patterns for F-0223 films fabricated at PO_2 of (a) 16, (b) 17, (c) 18, and (d) 19 Pa.

Figure 3 shows XRD patterns for superconducting samples deposited at T_s=740 °C and at PO_2=16-19 Pa, in the non-zero resistivity region at the measurable temperature limits of the PPMS. For all samples, the strongest peaks are associated with *c*-axis oriented F-0223, with some peaks due to the presence of F-0234 and F-0245 impurity phases [1,3]. The strongest peaks are consistent with (00ℓ) planes in the F-0223 phase, assuming that the lattice spacing is 0.3642 Å larger than the value of 2.7433 Å for the high-pressure phase. The sample deposited at PO_2=16 Pa has weaker F-0223 peaks than the other samples, although the F-0234 peak intensity is similar. The sample deposited at PO_2=17 Pa has XRD peaks associated with (006), (0010), (0012), (0014), (0016), and (0018) planes in F-0223, and peaks due to *c*-axis oriented F-0234 and F-0245, and the intensity of the (0010) peak is higher than that for the samples produced with a PO_2 of 16 and 18 Pa. The intensity of the (103) peak due to the F-0234 impurity phase is highest for the sample deposited at PO_2=18 Pa. The sample deposited at PO_2= 19 Pa exhibits peaks due to *c*-axis oriented F-0223, of which the (0010) peak has the highest intensity. In addition, F-0234 (0014), (0016), and (0024) peaks and an F-0245 (0014) peak are evident.

Figure 4 shows the temperature dependence of the resistivity for samples produced under various conditions. All samples exhibited a superconducting onset temperature (T_{co}), with the highest value of 81.6 K being obtained for that produced with a T_s of 750 °C and a PO_2 of 17 Pa. Figures 4(a) shows T_{co} = 37.8 K for a sample deposited at PO_2 = 16 Pa and T_s = 740 °C. In addition, the samples produced with T_s = 740 °C and PO_2 = 17 and 19 Pa exhibited T_{co} of 76.5 and 70.7 K, T_c of 47.8 and 38.7 K, and zero-resistance temperature (T_{ce}) of 22.5 and 19.0 K, respectively. The T_c values were much lower than that for the bulk material, even though the ablation target composition was $Ba_2Ca_2Cu_3O_{6+\delta}(O,F)_2$, which yields the highest T_c for the bulk [1].

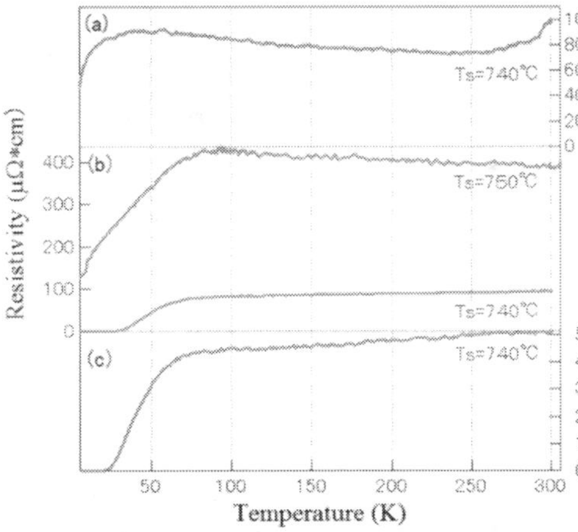

Figure 4. Temperature dependence of resistivity for samples deposited under conditions of (a) PO_2= 16 Pa and T_s= 740 °C, (b) PO_2= 17 Pa and T_s= 740 and 750 °C, and (c) PO_2= 19 Pa and T_s= 740 °C.

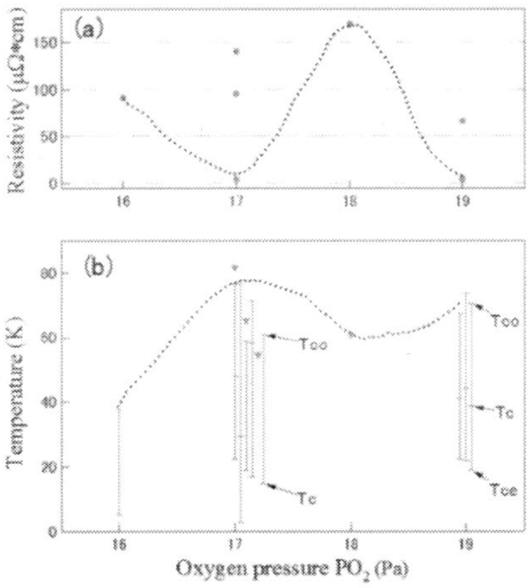

Figure 5. PO_2 dependence of (a) resistivity at 300 K and (b) T_{co} (upper or only inverted triangle), T_c (middle), and T_{ce} (lower) for superconducting films.

Figures 5(a) and 5(b) show the PO_2 dependence of the resistivity at 300 K and the T_{co}, T_c, and T_{ce} values for the superconducting films. The resistivity and T_{co} results suggest that the samples deposited at PO_2=17 and 19 Pa contain different phases. Electrical conductivity is observed for the F-0223 phase sample prepared with PO_2=17 Pa (Figure 3(b)), although there is a considerable amount of scatter in the data points for the resistivity and transition temperatures, which indicates a lack of repeatability from sample to sample. In contrast, for the F-0223, F-0234, and F-0245 multiphase samples prepared with PO_2=19 Pa (Figure 3(d)), there is a higher degree of repeatability. For samples prepared with PO_2=17-19 Pa, the molar composition was determined to be Ba:Ca:Cu = (0.4-2):2:5 from ICP-MS analysis, which indicates a lower Ba content than that typically found in F-0223 and F-0234. All of the films had T_c values that were lower than that for the bulk. This may be because, although *c*-axis oriented F-02(n-1)n films were formed, the F-0223 was the normal-pressure phase, and other phases such as F-0234 and F-0245 were also present. In addition, defects such as intergrowth in the layered structure may be present. Balestrino *et al.* [4] attributed the decrease of T_c for $(BaCuO_2)_2/(CaCuO_2)m$ superlattices, relative to the bulk T_c, to the high degree of disorder in the artificial structure and the not yet optimized chemistry of the $BaCuO_2$ in CRL block. Similarly, in this present work, the intergrowth in the layered structure and the lower T_c are probably due to a deviation from the typical growth conditions for BaF_2 layers in the CRLs with F-0223 and F-0234 phases.

4. Summary

Superconducting $Ba_2Ca_{n-1}Cu_nO_{2n}(O,F)_2$ thin films were deposited on $SrTiO_3(100)$ substrates using Nd-YAG pulsed laser ablation under O_2 partial pressures of 16-19 Pa and at substrate temperatures of 740-750 °C. The films had *c*-axis orientation and a typical sample exhibited a T_{co} of 76.5 K and a T_{ce} of 22.5 K.

References

[1] Iyo A, Tanaka Y, Tokumoto H and Ihara H 2001 *Physica C* **366** 43
[2] Iyo A, Tanaka Y, Kodama Y, Kito H, Tokiwa K and Watanabe T 2006 *J.Phys.Conf.: Series* **42** 333
[3] Iyo A, Hirai M, Tokiwa K, Watanabe T and Tanaka Y 2004 *Supercond. Sci. Technol.* **17** 143
[4] Balestrino G, Martellucci S, Medaglia P.G, Paoletti A, Petrocelli G, Varlamov A.A 1998 Physical Review B58, R8925

Emergence of superconductivity and magnetic ordering tuned by Fe-vacancy in alkali-metal Fe chalcogenides $Rb_xFe_{2-y}Se_2$

Yoshiaki Kobayashi[1,*], **Shouhei Kototani**[1], **Masayuki Itoh**[1], **Masatoshi Sato**[2]
[1] Nagoya University, Furo-cho, Chikusa-ku, Nagoya, 464-8602, Japan.
[2] Research Center for Neutron Science & Technology, Comprehensive Research Organization for Science and Society (CROSS), 162-1 Shirakata, Tokai 319-1106, Japan

*E-mail: i45323a@cc.nagoya-u.ac.jp

Abstract. Samples of $Rb_xFe_{2-y}Se_2$ exhibiting superconductivity [superconducting (SC) samples] undergo a phase-separation into two phases, a Fe-vacancy ordered phase with antiferromagnetic (AFM) transition at T_{N1}~500 K (AFM1 phase) and a phase with little Fe-vacancy and SC transition at T_c~30 K (SC phase). The samples of $Rb_xFe_{2-y}Se_2$ exhibiting no SC behaviour (non-SC samples) are phase-separated into three phases, the AFM1 phase, another AFM phase with T_{N2} ~150 K (AFM2 phase), and a paramagnetic phase with no SC transitions (paramagnetic non-SC phase). In this paper, we present the experimental results of magnetic susceptibility, electrical resistivity, and NMR measurements on single crystals of $Rb_xFe_{2-y}Se_2$ to reveal physical properties of these co-existing phases in the SC and non-SC samples. The ^{87}Rb and ^{77}Se NMR spectra show that the Fe vacancy concentration is very small in the Fe planes of the SC phase, whereas the AFM2 and paramagnetic non-SC phases in non-SC samples have larger amount of Fe vacancies. The randomness induced by the Fe vacancy in the non-SC samples makes the AFM2 and paramagnetic non-SC phases insulating/semiconducting and magnetically active, resulting in the absence of the superconductivity in $Rb_xFe_{2-y}Se_2$.

1. Introduction

In studies on the alkali metal Fe chalcogenides $Rb_xFe_{2-y}Se_2$ with various x and y, it is one of the intensively investigated issues to clarify the physical properties of the separated phases [1-5]. The Fe chalcogenides have two kinds of samples, one exhibiting the superconducting (SC) transition at ~30 K (SC sample) and the others with no SC transition (non-SC samples). Both kinds of the samples commonly have the antiferromagnetic (AFM) phase with the Neel temperature T_{N1} of ~500 K (AFM1 phase, the structure of space group I4/m) [3, 4]. In the SC sample, a phase with the SC transition temperature T_c ~30 K (SC phase, I4/mmm) exists as another phase [1, 3, 5]. On the other hand, in the non-SC sample with the nominal (x, 2-y)=(0.7, 1.9) there are two phases besides the AFM1 phase as reported in our previous papers [3, 5, 6]. One is another AFM phase with T_{N2} ~150 K (AFM2 phase, Pmna) and the other is a paramagnetic phase which has no AFM and SC transitions (paramagnetic non-SC phase, I4/mmm). However, the AFM2 phase does not always exist in the non-SC samples, suggesting that its presence sensitively depends on the concentration of the Fe-vacancy. Hereafter, non-SC samples with and without the AFM2 phase are referred as AFM2 and non-SC/non-AFM2 samples, respectively.

In this paper, we report the experimental results of magnetic susceptibility, electrical resistivity, and ^{87}Rb and ^{77}Se NMR measurements of these samples to elucidate clearly the physical properties of the SC, LT-AFM, and non-SC/non-AFM phases, and discuss the relationship between the electronic state and the local structure of the Fe planes in the three phases.

2. Experiments

$Rb_xFe_{2-y}Se_2$ single crystals were prepared by the flux method [5, 7]. The SC samples were prepared from the nominal (x, 2-y) values of (0.8, 1.7), and the AFM2 and non-SC/non-AFM2 samples were prepared from the values of (0.7, 1.9) and (1, 2), respectively. To avoid damages to the samples due to moisture or air, the samples were covered by an epoxy adhesive, Stycast 1266. In-plane electrical resistivity ρ was measured by the four terminal method with a physical property measurement system (Quantum Design, PPMS). Magnetic susceptibility χ was measured in a magnetic field H of 1 (non-SC/non-AFM2 sample) or 6.105 T (SC and AFM2 samples) using a superconducting quantum interface device (SQUID) magnetometer (Quantum Design, MPMS). ^{87}Rb NMR measurements were carried out using a coherent pulsed spectrometer and a superconducting magnet with H of 6.1050 T where the single crystal was rotated using a two-axis goniometer. Fourier-transformed (FT) spectra of spin-echo signals were measured by changing the frequency stepwise and the whole ^{87}Rb NMR spectra were obtained by superposing the FT spectra at various frequencies.

3. Results and Discussion

For the non-SC/non-AFM2 sample of $Rb_xFe_{2-y}Se_2$, the temperature T dependences of χ for $H \perp c$ and the in-plane resistivity ρ are shown in Figs. 1(a) and 1(b), respectively, for comparison, with those of the SC and AFM2 samples [3, 5]. In the non-SC/non-AFM2 sample, the χ-increase observed for the AFM2 phase with decreasing T below the AFM ordering temperature T_{N2}~150 K [3, 5] is suppressed, suggesting the absence of the AFM2 order. The zero resistivity observed at ~7 K for the non-SC/non-AFM2 sample is just due to the filamentary SC region, because the SC diamagnetism at H =10 Oe in the zero-field-cooling condition is negligibly small. The presumed suppression of the SC transition is supported by the large electrical resistivity immediately above T_c (ρ_0) of ~150 mΩ·cm corresponding to more than 300 times larger than the maximum metallic sheet resistance of $h/(4e^2)$~6.45 kΩ. Only the phase separation picture can consistently explain both the large ρ_0 and the metallic-T dependence below 200 K.

Figure 2(a) shows the ^{87}Rb NMR spectra of the SC, AFM2, and non-SC/non-AFM2 samples at 250 K for $H \parallel c$, whereas Fig. 3 presents the ^{77}Se NMR spectrum in the SC sample. All the ^{87}Rb NMR spectra split into a central line and two satellite lines due to the electric quadrupole interaction between the ^{87}Rb nuclear spin I=3/2 and the electric field gradient at the Rb site. The expanded ^{87}Rb NMR spectra of the central lines are also presented in Fig. 2(b). As in the case of the SC and AFM2 samples [3, 5], the spectrum of the non-SC/non-AFM2 sample is also considered to come from a paramagnetic phase in the sample, because the NMR spectrum with $H \parallel c$ coming from the AFM1 phase is wiped out by the spectrum broadening due to the magnetic order below T_{N1}~500 K. Thus the

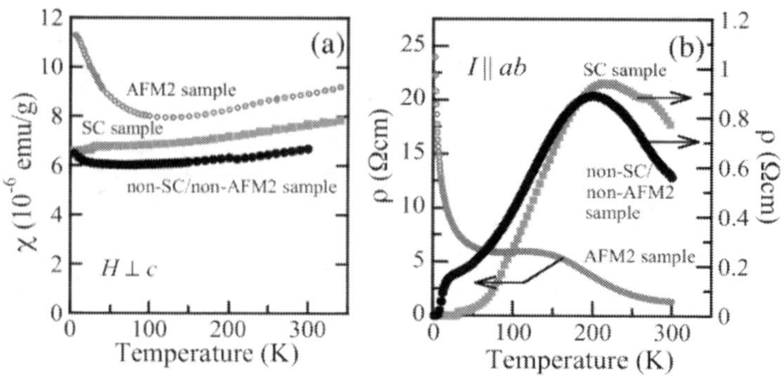

Fig. 1 Temperature dependences of (a) magnetic susceptibilities χ under a magnetic field $H \perp c$ and (b) in-plane electrical resistivities ρ of the SC, non-SC/non-AFM2, and AFM2 single crystal samples of $Rb_xFe_{2-y}Se_2$.

^{87}Rb NMR spectra can be calculated by assuming the ^{87}Rb Knight shift parallel to the c axis K_c and the nuclear quadrupole frequency ν_Q with the electric quadrupole effect up to the second-order, its distribution $\Delta\nu_Q$, the asymmetry parameter of the electric field gradient (EFG) η =0, and the Z axis parallel to the c axis. Here $\Delta\nu_Q$ is introduced, because the satellite spectra are much broadened in comparison with the central lines. By fitting the calculated spectra to the experimental ones, we obtain as K_c = 0.231%, 0.236%, and 0.270%, and ν_Q = 1.74, 1.71, and 1.65 MHz, and $\Delta\nu_Q$ = 0.06, 0.12, and 0.12 MHz for the spectra of the SC, non-SC/non-AFM2, and AFM2 samples, respectively.

The results of the fitting represented by the solid curves in Fig. 2 well reproduce the central and satellite spectra in the SC sample. This NMR spectrum monitors only the SC phase as seen in Fig. 4(a) where the ^{87}Rb Knight shift K_c shows a drop due to the SC transition below T_c. Also it should be noted that a small FWHM of the central line showing almost no distribution in the ^{87}Rb Knight shift. This is observed by the narrow ^{77}Se (I=1/2) spectrum (Fig. 3). No distribution of the ^{87}Rb and ^{77}Se Knight shifts indicates that the Fe-vacancy does not exist in the Fe plane of the SC phase, consistent with its crystal structure revealed by the neutron diffraction study on the SC phase [5], whereas $\Delta\nu_Q$ monitors a distribution of Rb. Also, although the AFM1 and SC phases coexist in the the SC sample, no effects of magnetic order in the AFM1 phase on the NMR spectra of the SC phase appear as seen in Figs. 2 and 3. This fact indicates that these phases are not microscopically but macroscopically separated in the SC sample.

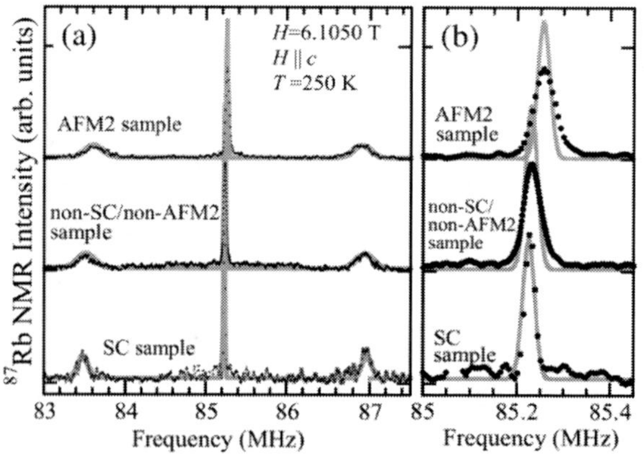

Fig. 2 (a) ^{87}Rb NMR spectra and (b) expanded spectra of the central lines in the SC, non-SC/non-AFM2, and AFM2 samples of $Rb_xFe_{2-y}Se_2$ with H (= 6.1050 T) parallel to the c axis at 250 K. The solid curves represent the calculated spectra with the parameters of the Knight shift, the electric field gradient (EFG), and its distribution of EFG presented in the text.

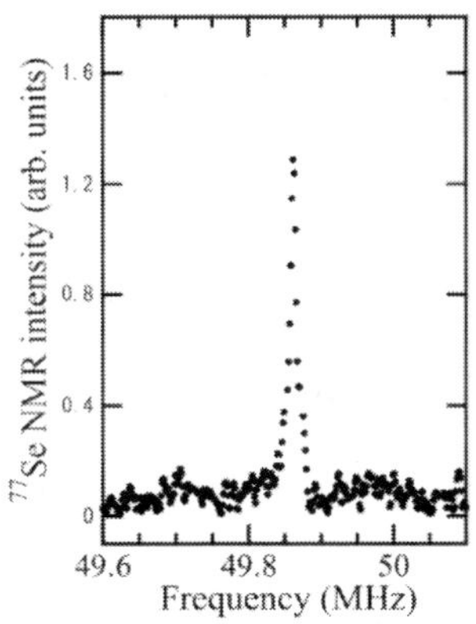

Fig. 3 ^{77}Se NMR spectrum of the SC sample of $Rb_xFe_{2-y}Se_2$ for H (= 6.1050 T) ∥ c at 250 K.

On the other hand, the fitted results for the non-SC/non-AFM2 and AFM2 samples cannot reproduce the central lines of the spectra as seen in Fig. 2(b), although they well trace the satellite lines. This discrepancy is removed by introducing a distribution of K_c, ΔK_c, of 0.020% and 0.028% which are estimated from the FWHM of the centerlines for the non-SC/non-AFM2 and AFM2 samples, respectively. Figures 4(b) and 4(c) shows the T dependences of K_c with ΔK_c denoted by the vertical bars in the non-SC/non-AFM2 and AFM2 samples, respectively. A fact that the paramagnetic NMR spectrum in the AFM2 sample disappears below T_{N2}~150 K shows that this spectrum comes from the AFM2 phase. Thus the T-dependent large K_c and ΔK_c in the AFM2 sample monitor the AFM properties of the AFM2 phase which may be mainly induced by a random distribution of Fe-vacancies in the Fe planes revealed by the neutron scattering study [5], whereas the T-independent $\Delta\nu_Q$ monitors a charge distribution due to the distributions of both Fe and Rb. This AFM2 phase is considered to have crystal and magnetic structures similar to those of a magnetic phase with T_N~300 K in a semiconducting sample reported in Ref. 2 and T_N sensitively depends on the composition of the AFM2 phase [5]. In the non-SC/non-AFM2 sample, we observed a paramagnetic NMR spectrum, which comes from the paramagnetic non-SC phase, with ΔK_c larger than that of the SC sample and no AFM transition down to the lowest temperature measured. The large ΔK_c is ascribed to a proper amount of the disordered Fe vacancies leading to no SC and AFM transitions as reported by the neutron scattering study [5].

Thus we conclude that the SC transition in $Rb_xFe_{2-y}Se_2$ emerges in the Fe plane with almost no Fe-vacancy and disperses in the Fe plane with Fe-vacancies. The presence of Fe-vacancies activates magnetically the system and consequently suppresses the superconductivity. The emergence of the superconductivity or the magnetic orders is closely related to the absence or presence of the Fe vacancies in the Fe plane.

4. Conclusion

We have made magnetic susceptibility, electrical resistivity, and ^{87}Rb and ^{77}Se NMR measurements to elucidate the physical properties of several phases separated in the alkali metal Fe chalcogenides $Rb_xFe_{2-y}Se_2$. It was found that the system has the SC samples with the AFM1 and SC phases, the non-SC samples with the AFM1, AFM2, and paramagnetic non-SC phases, and the non-SC sample without the LT-AFM. From the analyses of the NMR spectra, we found that the SC phase has almost no Fe vacancy in the Fe plane, whereas the AFM2 and paramagnetic non-SC phases have the Fe vacancies.

Journal of Physics: Conference Series **568** (2014) 022024 doi:10.1088/1742-6596/568/2/022024

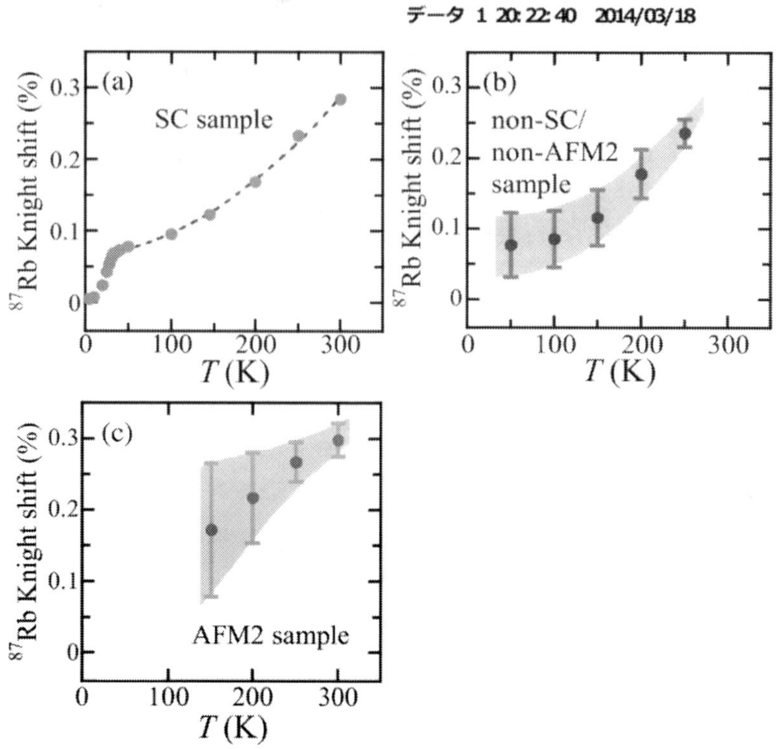

Fig. 4 Temperature dependences of the ^{87}Rb Knight shifts for $H \parallel c$, K_c, in the (a) SC, (b) paramagnetic non-SC, and (c) AFM2 phases. The vertical bars indicate the distributions of K_c estimated from the FWHM values of the central lines.

The magnetically active regions of the AFM2 and paramagnetic non-SC phases induced by the Fe vacancies were proposed to disperse the superconductivity.

This study was supported by the Grant-in-Aid (No. 2434080) for Scientific Research from the Japan Society of the Promotion of Science.

References

[1] Chen F, Xu M, Ge Q Q, Zhang Y, Ye Z R, Yang L X, Jiang Juan, Xie B P, Che R C, Zhang M, Wang A F, Chen X H, Shen D W, Hu J P, and Feng D L 2011 *Phys. Rev.* X **1** 021020

[2] Zhao J, Cao H, Bourret-Courchesne E, Lee D -H, and Birgeneau R J 2012 *Phys. Rev. Lett.* **109** 267003

[3] Kobayashi Y, Saiki S, Kototani S, Itoh M, and Sato M 2013 *J. Korean Phys. Soc.* **63** 448

[4] Bao W, Huang Q, Chen G F, Green M A, Wang D M, He J B, Wang X Q, Qiu Y 2011 *Chin. Phys. Lett.* **28** 086104

[5] Kobayashi Y, Kototani S, Kazuki Ohishi K, Itoh M, Hoshikawa A, Ishigaki T, and Sato M: submitted to *J. Phys. Soc. Jpn.*

[6] Ohishi K, Sato M, Kototani S, Saiki S, Kobayashi Y, and Itoh M 2013 *J. Korean Phys. Soc.***62** 1994

[7] Guo J, Jin S, Wang G, Wang S, Zhu K, Zhou T, He M, and Chen X 2010 *Phys. Rev.* B **82** 180520

Intrinsic Josephson properties in $Pb_{1-y}Sr_2Y_{1-x}Ca_xCu_{2+y}O_{7+\delta}$ epitaxial films

Sachio Komori and Itsuhiro Kakeya

Department of Electronic Science and Engineering, Kyoto University, Nishikyo-ku, Kyoto 615-8510, Japan

E-mail: komori@sk.kuee.kyoto-u.ac.jp, kakeya@kuee.kyoto-u.ac.jp

Abstract. We report the first observation of intrinsic Josephson junction (IJJ) characteristics in the $Pb_{1-y}Sr_2Y_{1-x}Ca_xCu_{2+y}O_{7+\delta}$ (Pb1212) epitaxial film. Pb1212 epitaxial film has been grown on $SrTiO_3$ (100) substrates by a two-step growth technique. A small mesa structure ($S = 4\mu m^2$) has been fabricated on a film surface using a standard photolithography and an Ar ion milling technique. The superconducting transition temperature of the IJJ is 43 K. At 4.2 K, the value of J_c (the critical current density) is 2.2 kA/cm^2. The depth of the hysteresis defined as $\alpha = (J_c + J_r)/\ J_c$ is 0.89, where J_r is the return current density. The temperature dependence of J_c shows good agreement with Ambegaokar-Baratoff theory.

1. Introduction

A crystal of a high-T_c cuprate is generally regarded as a stacked layer of the tunneling Josephson junctions. These junctions are called intrinsic Josephson junctions (IJJs). Since IJJs are based on the crystal structure itself and naturally formed, the surface of the junction is atomically flat and clean. For this reason, IJJs are investigated in various areas such as interlayer tunneling spectroscopy, macroscopic quantum tunneling, THz radiation, and so on. In ideal superconductor/insulator/superconductor IJJs, the current-voltage characteristics show a hysteresis and resistive branches which correspond to the resistive states of the individual IJJs. These characteristics are typically seen in IJJs of $Bi_2Sr_2CaCu_2O_8$, $Bi_2Sr_2Ca_2Cu_3O_{10}$, and $Bi_2Sr_2CuO_6$ which have crystal structures with a significantly large anisotropy around 100 or greater. $Tl_2Ba_2CaCu_2O_8$[1] and $Tl_2Ba_2Ca_2Cu_3O_{10}$[2] whose crystal structures are similar to those of Bi-family also exhibit IJJ characteristics. On the other hand, IJJ characteristics can be rarely observed in materials which have crystal structures with small anisotropy. In $YBa_2Cu_3O_7$, small hysteresis and resistive branches are observed only when the small junctions are fabricated in under-doped samples[3][4][5]. $HgBa_2CaCu_2O_7$[6] and $RuSr_2GdCu_2O_7$[7] also exhibit characteristics similar to $YBa_2Cu_3O_7$. Recently, IJJ characteristics of $Sm_{2-x}Ce_xCuO_4$[8], $La_{2-x}Sr_xCuO_4$[9], $Hg_{1-x}Re_xBa_2Ca_2Cu_3O_8$[10], and $Hg_{1-x}Re_xBa_2Ca_3Cu_4O_{10}$[11] have also been observed. As far as our knowledge, these 12 types of cuprates have been reported to exhibit IJJ characteristics. We report the first observation of IJJ characteristics in $Pb_{1-y}Sr_2Y_{1-x}Ca_xCu_{2+y}O_{7+\delta}$ (Pb1212). The crystal structure of Pb1212 is similar to that of $YBa_2Cu_3O_7$, $HgBa_2CaCu_2O_7$, and $RuSr_2GdCu_2O_7$. Due to the lack of single-crystal samples, the fundamental properties of Pb1212 have not been well investigated. We have obtained epitaxial films of Pb1212 by a two-step growth technique[12], and succeeded in the observation

Journal of Physics: Conference Series **568** (2014) 022025 doi:10.1088/1742-6596/568/2/022025

of IJJ characteristics.

2. Sample fabrication

Pb1212 epitaxial film was grown by a two-step technique consisting of a low temperature sputtering step and a high temperature *ex situ* growth step. For depositing the Pb1212 amorphous film on $SrTiO_3$ (100) substrate, we used the following sputtering conditions: composition of the sputtering target was $(Pb_{0.75}Bi_{0.25})_{1.5}Sr_2Y_{0.475}Ca_{0.525}Cu_{2.6}O_z$, the sputtering gas pressure was 100 mTorr (60 sccm Ar and 15 sccm O_2), the anode voltage was 1.4 kV, and the substrate temperature was approximately 200 °C (the substrates were not heated). The deposition time was set to 90 min. The thickness of the Pb1212 thin film measured by a stylus-based profilometer was 2600 Å.

Pb1212 containers used for *ex situ* growth were made of polycrystalline pellets with the composition of $PbSr_2Y_{0.3}Ca_{0.7}Cu_2O_z$. For epitaxial growth, the amorphous film on $SrTiO_3$ substrate was placed in a pit formed at the center of one of the pellets. The other pellet was used as a lid for the growth container. The container containing the amorphous film was heated in a muffle furnace at 970 °C for 6 h under an O_2 atmosphere, then cooled to room temperature at a rate of 200 °C/h.

The results of out-of-plane x-ray diffraction measurements indicate the absence of impurity phase and complete c-axis alignment of the film. The results of in-plane x-ray diffraction measurements indicate that the film grow epitaxially on the substrate. The concentration of the film determined by energy dispersive x-ray spectroscopy was $Pb_{0.79}Sr_s(Y_{0.65}Ca_{0.35})_{1.47}Cu_{2.21}O_z$. In Pb1212, the Ca concentration $x > 0.3$ is necessary for the occurring of the superconducting transition[13], and approximately one-fourth of Pb ions in the block layer is substituted by Cu ions.

A subsequent quenching treatment is also necessary to make the film superconducting. In this treatment, the film was placed in a quartz tube, then heated at 815 °C in air for 1 h. Within two seconds after removing the quartz tube from the furnace, it was placed in liquid nitrogen. This procedure increases the hole concentration, decreases crystallographic disorder, and consequently results in the film being superconducting[13].

Ag (50 nm) was evaporated on a film surface as a contact electrode. Then, small mesa structures ($S = 4\mu m^2$) were fabricated on a film surface using a standard photolithography and an Ar ion milling technique. The electrode configuration we employed is three-terminal type. The number of layered Josephson junctions, 21 layers, is controlled by the milling time. $I-V$ characteristics of IJJs are observed by an analog oscilloscope.

3. Results and discussion

The superconducting transition temperature of the IJJ was 43 K. Hysteretic current-voltage characteristics and multiple branches, which are peculiar features of IJJs, were observed below 39 K. $I-V$ characteristics of IJJs at 4.2 K are shown in Fig. 1. The value of J_c (the critical current density) is 2.2 kA/cm^2. The depth of the hysteresis defined as $\alpha = (J_c + J_r)/J_c$ is 0.89, where J_r is the return current density. The McCumber parameter (a measure of the damping of IJJs) is estimated to be 110 from the value of α. The fact that multiple branches can be observed near T_c and large values of α and β_c suggest that the Pb1212 is a highly anisotropic cuprate like Bi-family. However, the ab-plane resistivity measurement in magnetic fields shows that the coherent lengths of Pb1212 are approximately $\xi_{ab}(0) = 2.5$ nm and $\xi_c(0) = 0.27$ nm[13]. Therefore, the anisotropy, $\gamma = \xi_{ab}(0)/\xi_c(0)$ is 9.2, which is closer to that of $YBa_2Cu_3O_7$ ($\gamma \sim 5$) than that of Bi-family ($\gamma \sim 100$). The contradiction between characteristics of IJJs and the value of anisotropy may be peculiar to Pb1212 system. The gap voltage between the zero and the first branch is about 18 mV. This gap voltage decreases with increasing voltage or temperature.

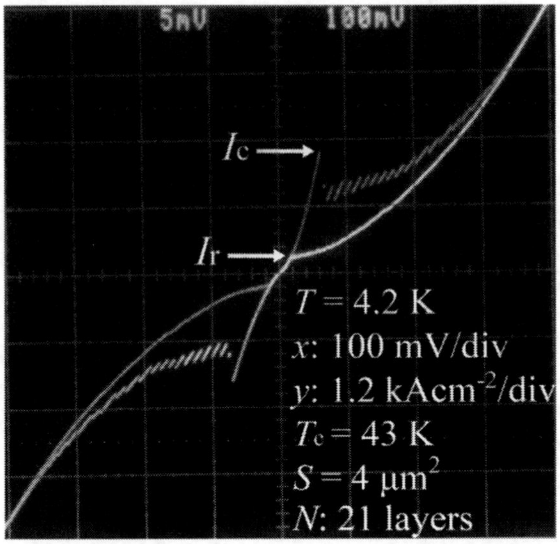

Figure 1. $I - V$ characteristics of Pb1212 intrinsic Josephson junctions

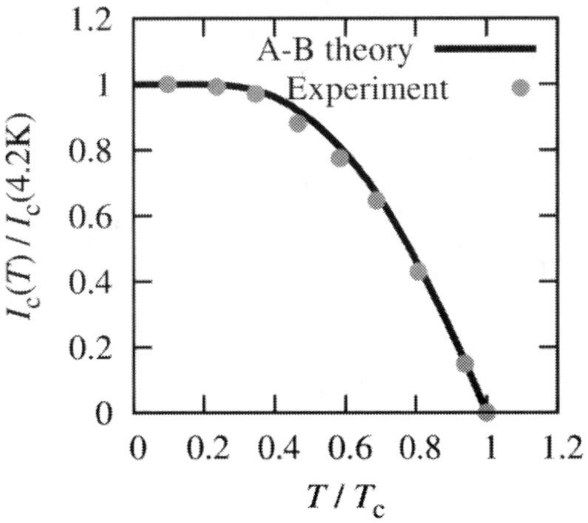

Figure 2. Temperature dependence of critical currents

The temperature dependence of the critical current of the first branch is shown in Fig. 2. The temperature and the critical current are normalized at T_c and $I_c(4.2\ \mathrm{K})$. The solid curve represents the theoretical $I_c - T$ for a superconductor-insulator-superconductor (SIS) Josephson junction based on the Ambegaokar-Baratoff (A-B) formula[14]

$$I_c(T) = \frac{\pi\Delta(T)}{2eR_N} \tanh \frac{\Delta(T)}{2k_B T} \tag{1}$$

where R_N is the normal tunneling resistance, 2Δ is the superconducting gap, and k_B is the Boltzmann constant. Experimental result has a good agreement with A-B theory. This suggests that the IJJs of Pb1212 have a character of SIS-type Josephson junction. SIS characteristics have also been observed in IJJs of under-doped $YBa_2Cu_3O_7$[5] and $RuSr_2GdCu_2O_7$[7] whose crystal structures are similar to that of Pb1212.

4. Summary

Intrinsic Josephson characteristics in the $Pb_{1-y}Sr_2Y_{1-x}Ca_xCu_{2+y}O_{7+\delta}$ (Pb1212) epitaxial film have been observed. It is found that hysteretic current-voltage characteristics and multiple branches appears below 0.9 T_c. IJJs in Pb1212 shows under-damped behavior in spite of its small anisotropy. The temperature dependence of critical currents suggests that IJJs of Pb1212 have a character of SIS-type Josephson junction.

References

[1] Warburton P A, Fenton J C, Korsah M and Grovenor C R M 2006 *Supercond. Sci. Technol.* **19** 187
[2] Schlenga K, Hechtfischer G, Kleiner R, Walkenhorst W, Müller P, Johnson H L, Veith M, Brodkorb W and Steinbeiss E 1996 *Phys. Rev. Lett.* **76** 4943
[3] Rapp M, Murk A, Semerad R and Prusseit W 1996 *Phys. Rev. Lett.* **77** 928
[4] Nishizaki T, Onodera Y and Kobayashi N 1996 *Phys. Rev.* B **53** 82
[5] Wang H B, Chen J, Nakajima K, Yamashita T, Wu P H, Nishizaki T, Shibata K and Kobayashi N 2000 *Phys. Rev.* B **61** R14948

[6] Ogawa A, Sugano T, Wakana H, Kamitani A, Adachi S, Tarutani Y and Tanabe K 2005 *J. Appl. Phys.* **97** 013903
[7] Nachtrab T, Koelle D, Kleiner R, Bernhard C and Lin C T 2004 *Phys. Rev. Lett.* **92** 117001
[8] Kawakami T and Suzuki M 2007 *Phys. Rev.* B **76** 134503
[9] Kubo Y, Takahide Y, Tanaka T, Ueda S, Ishii S, Tsuda S, Islam A T M N, Tanaka I and Takano Y 2011 *J. Appl. Phys.* **109** 033912
[10] Ueda S, Yamaguchi T, Kubo Y, Tsuda S, Takano Y, Shimoyama J and Kishio K 2009 *J. Appl. Phys.* **106** 074516
[11] Ueda S, Okutsu T, Kubo Y, Ishii S, Tsuda S, Yamaguchi T, Horii S, Shimoyama J, Kishio K and Takano Y 2008 *Physica* C **468** 1925
[12] Komori S, Inaba R, Kakeya I and Suzuki M 2014 *J. Phys. Conf. Ser.* **507** 012025
[13] Komori S, Inaba R, Kaneko K, Fujita S, Kakeya I and Suzuki M 2014 *Phys. Rev.* B **89** 174509
[14] Ambegaokar V and Baratoff A 1963 *Phys. Rev. Lett.* **10** 486

Superconductivity of the Heterofullerides Synthesized from Gallams and Amalgams

V.A. Kulbachinskii[a*], R.A. Lunin[b], B.M. Bulychev[c]
Moscow State University, Low Temperature Physics Department, Moscow, Russia
[a*]kulb@mig.phys.msu.ru, [b]lunin@mig.phys.msu.ru,
b.bulychev@highp.chem.msu.ru

Keywords: superconductivity, fullerene, fulleride, crystal structure, amalgam, gallam

Abstract. The fullerides with composition $A_2B_xD_yC_{60}$ (A=K,Rb,Cs; B,D=Ga, In, Sn, Bi, Be, Al, Mg, Hg; x,y<1) have been synthesized by using liquid amalgams (alloys of metals with mercury) or with an alloy of indium and gallium (gallams) at room temperature. It was found that the fulleride $K_2In_xGa_yC_{60}$ is a superconductor with transition temperature T_c = 24.5 K that exceed T_c=19 K for reference K_3C_{60}. The fulleride $Rb_2In_xGa_yC_{60}$ is a superconductor with T_c = 26 K. Superconductivity of heterofulleride with a composition $K_2Hg_xC_{60}$ (x<1) was observed at temperature T_c = 22 K. For the sample $Rb_2Hg_xC_{60}$ T_c=25 K, though Rb_2C_{60}, as well as K_2C_{60}, are not superconductors. Only one fulleride on the base of potassium (which does not include atoms of other alkaline metals Rb and Cs) was known before – the K_2LuC_{60} (T_c=20 K) with T_c higher than for K_3C_{60} [1]. The fullerides with composition $Rb_3In_xGa_yC_{60}$ and $Cs_3In_xGa_yC_{60}$ are not superconductors and crystallized in the orthorhombic lattice. Fullerides $K_2Ga_xBi_yC_{60}$ and $K_2Ga_xSn_yC_{60}$ (x,y<1) are superconductors.

Introduction

Fullerides of alkali and some other metals are very interesting high T_c superconductors [2-5]. The first fullerides were synthesized by a method of reactions of the metal vapor with fullerite [6-8]. Such method is suitable for alkali and other fusible metals. Also a fulleride synthesis method through an interaction of a metal solution in ammonia with fullerite or melt of alkali metals with fullerite are known [2]. Other method is a fullerides synthesis by reactions of alkali metals with a solution of fullerene in the organic solvent [4,5]. More complicated fullerides were obtained by an exchange reactions of fullerides of alkali metals with solutions of salts of other metals in organic solvents [9,10]. In all these methods one of the reagent is in a solid state that reduces a speed of the reaction and demands the execution of the synthesis at the increased temperature.

In this work we investigated superconductivity of heterofullerides with a composition $A_2B_xD_yC_{60}$ (A=K, Rb, Cs; B,D=Ga, In, Sn, Bi, Be, Al, Mg, Hg; x,y<1) obtained by a method of reactions of liquid gallams or amalgams with a solution of fullerene in organic solvent. In this method both of reagents are liquid that allows executing a reaction at room temperature. Homogeneity of the composition of obtained compounds is close to the ideal. In this case, however, mercury, indium and gallium partially remain in the obtained fulleride. We also tried to find the relations of the composition and the structure of superconducting heterofullerides with the superconducting transition temperature T_c.

Journal of Physics: Conference Series **568** (2014) 022026 doi:10.1088/1742-6596/568/2/022026

Experimental

In this paper we study the samples, synthesized by a method using liquid amalgams or gallams (alloys of metals with eutectic alloy Ga:In=70:30). We assume that Hg, Ga or heterometals can intercalate together with self-intercalated alkaline metals into the fullerite. To better deliver metal to the fullerene molecule the interaction of components (fullerite C_{60}, alkali metal and heterometal) was carried out in the organic solvent (toluene or mixture of toluene and THF in a proportion 9:1), in which fullerene is soluble. The ratio of alkali metal:fullerite was 2:1 to synthesize A_2C_{60} and to avoid the synthesis of the superconducting fulleride A_3C_{60}. The synthesis of the alkali metal fullerides and subsequent removal of toluene, gallium or gallam and mercury, drying of products of reaction, and their packaging in ampoules for measurements of temperature of the superconducting transition were carried out in vacuum in the full glass facility equipment. In more detail the technique is described in Ref's [9-10]. In the initial gallams the content of indium and gallium were in 30-70 times higher than alkali metal in order to gallams were liquid, but in the final product $A_nIn_xGa_yC_{60}$ (A=K,Rb,Cs; n=2,3) the content of indium x and gallium y were less then 1. The measurements of X-ray diffraction were carried out with Guinier G670 HUBER. Temperatures of superconducting transitions of fullerides were defined by low-frequency induction method by measuring the temperature dependence of a magnetic susceptibility in the temperature range 4.2<T<297 K [4,10,11].

Some parameters of the investigated samples are listed in table 1. We also synthesized $Rb_3In_xGa_yC_{60}$ and $Cs_3In_xGa_yC_{60}$ but they are not superconductors while $K_3In_xGa_yC_{60}$ is a superconductor with Tc=14.5 K.

Table 1. Temperature of superconducting transition T_C, type of crystal lattice and lattice parameters of investigated samples

Composition	T_C (K)	Lattice type	Lattice parameters (Å)
$K_2Ga_xIn_yC_{60}$	24.5	*fcc*	14.276(4)
$K_2Ga_xSn_yC_{60}$	15.4	*fcc*	14.282(8)
$K_2Ga_xBi_yC_{60}$	14.9	*fcc*	14.255(8)
$K_2GaHg_xC_{60}$	20	*fcc*	14,297(5)
$K_2MgHg_xC_{60}$	13.6	*fcc*	14.280(4)
$K_2AlHg_xC_{60}$	15.7	*fcc*	14,281(6)
$K_2Hg_xC_{60}$	22	*fcc*	14.279(3)
$K_3In_xGa_yC_{60}$	14.5	*fcc*	14.269(2)
$Rb_2AlHg_xC_{60}$	24.8	*fcc*+M2	14.458(4)
$Rb_2MgHg_xC_{60}$	21.8	*fcc*+~5%M2	14.447(4)
$Rb_2GaHg_xC_{60}$	25	*fcc*+ orthorhombic (10%)	14.439(3) a=9.055; b=10.057; c=14.190 Å
$Rb_2In_xGa_yC_{60}$	26	fcc+ orthorhombic	14.444(3)
Rb_2HgC_{60}	24.6	*fcc*+M2	14.440(2)
$Rb_3In_xGa_yC_{60}$	-	Orthorhombic	*a*= 9.110(13), *b*=10.105(11),*c*=14.188(15)
$Cs_3In_xGa_yC_{60}$	-	Orthorhombic	*a*=9.097(5), *b*=10.227(5), *c*=14.159(9)

M2 – monoclinic phase

Fullerides synthesized by different methods were studied in [5]. Superconducting transition temperatures for heterofullerides with similar composition but prepared from amalgams or by exchange reactions do not differ significantly [11]. It was shown that mercury get into fulleride and some charge transfer from mercury to the fulleride molecule occurs [10,11].

Results and discussion

Temperature dependence of the magnetic susceptibility for synthesized from gallams samples $K_2In_xGa_yC_{60}$, $K_3In_xGa_yC_{60}$ and $Rb_2In_xGa_yC_{60}$ is shown in **Fig. 1a**. The transitions in a superconducting state were observed at temperatures 24.5; 14.5 K and 26 K, correspondingly. The fulleride $K_2Hg_xC_{60}$ has $T_c = 22$ K and $Rb_2Hg_xC_{60}$ has $T_c = 24.6$ K (**Fig. 1b**). Thus the used methods of synthesis made superconductors $K_2In_xGa_yC_{60}$ and $K_2Hg_xC_{60}$ with T_c higher than in K_3C_{60} from not a superconductor K_2C_{60}. It indicates that mercury or components of gallams aren't inert solvents in reaction of intercalation, but intercalates itself into a fullerides lattice and change their superconducting properties. X-ray data shows that $K_2Hg_xC_{60}$ crystallized in fcc lattice (**Fig.2a**) while $K_3Hg_xC_{60}$ has a monoclinic lattice (a=16.52Å, b=10.82Å, c=10.41Å, β=108.17^0, **Fig.2b**) and not a superconductor. According to the X-ray data the heterofulleride with the composition $K_2In_xGa_yC_{60}$ was crystallized in the *fcc* lattice with the lattice parameter a=1.4276 nm (**Fig. 3a**), which is close to the value obtained for K_3C_{60} (a=1.424 nm [7]). Superconducting transitions for $K_2Ga_xBi_yC_{60}$ and $K_2Ga_xSn_yC_{60}$ are shown in **Fig. 3b**.

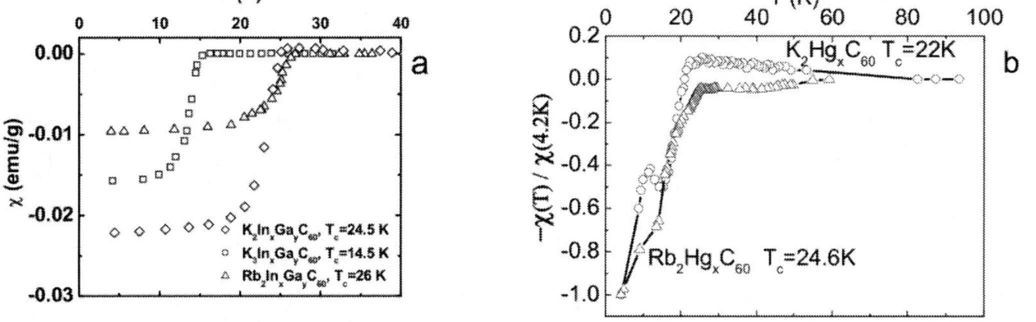

Fig. 1. Temperature dependence of magnetic susceptibility of samples $K_2In_xGa_yC_{60}$, $K_3In_xGa_yC_{60}$ and $Rb_2In_xGa_yC_{60}$ (a) and relative magnetic susceptibility of samples $K_2Hg_xC_{60}$ and $Rb_2Hg_xC_{60}$

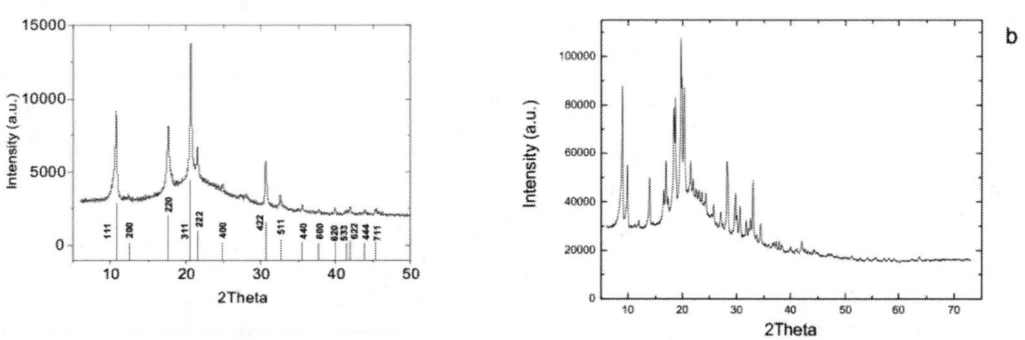

Fig. 2 X-ray diffraction of the sample (a) $K_2Hg_xC_{60}$ (indicated phase: *fcc*) and (b) $K_3Hg_xC_{60}$ (indicated phase: monoclinic)

Some temperature dependences of magnetic susceptibility for samples synthesized from amalgams are shown in **Fig. 4a**. The parameters of investigated samples are listed in table 1. The dependence of T_c on the lattice parameter a for superconducting samples synthesized from gallams and amalgams is shown in Fig. 4b by open cycles. The difference in T_c for samples $K_2In_xGa_yC_{60}$ and $K_3In_xGa_yC_{60}$ is the most probably due to the more optimal charge transfer from intercalated atoms to fullerene molecule for the sample $K_2In_xGa_yC_{60}$. The fulleride $Rb_3In_xGa_yC_{60}$ is not a superconductor. The X-ray diffraction data of the sample indicates the orthorhombic crystal lattice with the lattice parameters shown in table 1.

Transition temperature T_c increases with *fcc* lattice parameter a, as shown in **Fig. 4b.**

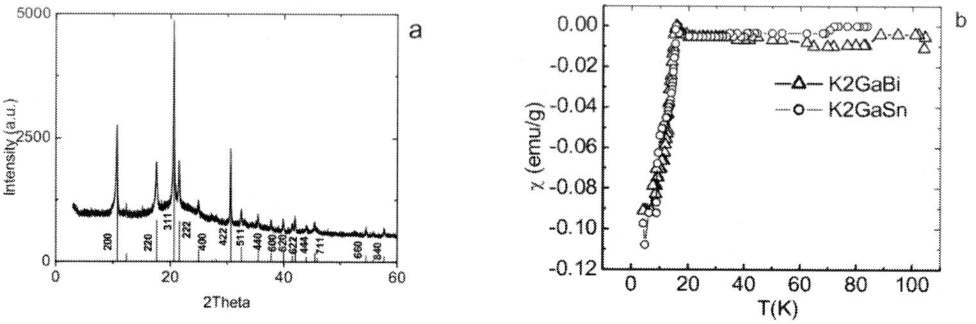

Fig. 3 X-ray diffraction of the sample (a) $K_2In_xGa_yC_{60}$ (indicated phase: *fcc*) and temperature dependence of magnetic susceptibility of samples $K_2Ga_xBi_yC_{60}$; $K_2Ga_xSn_yC_{60}$

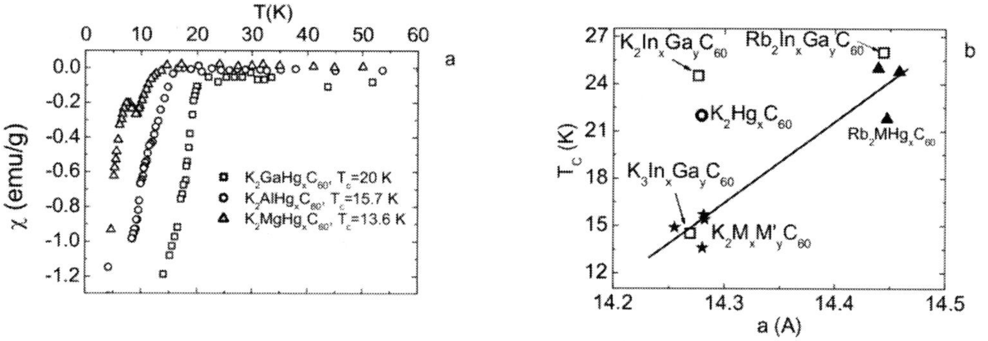

Fig. 4. (a) Temperature dependence of the magnetic susceptibility χ of the fulleride $K_2GaHg_xC_{60}$, $K_2AlHg_xC_{60}$, $K_2MgHg_xC_{60}$ (a) and dependence of the superconducting transition temperature T_c on the lattice papameter a for heterofullerides $K_2M_xM'_yC_{60}$ and $Rb_2MHg_xC_{60}$ (M = Be, Mg, Al, Ga, In, Sn, Bi)

Data shows that components of gallams were intercalated in fulleride lattice, changed the lattice type and made from superconductor Rb_3C_{60} not superconducting composition. At the same time the fulleride $Rb_2In_xGa_yC_{60}$ remain the superconductor with T_c = 26K. According to the X-ray data, this sample consists of two phases: the first phase is similar to $Rb_{2.92}C_{60}$ (*fcc* lattice) and the second phase is close to $Rb_{0.91}C_{60}$ (orthorhombic lattice).

The X-ray diffraction data for $Rb_2Hg_xC_{60}$ shows the peaks related to the *fcc* phase and also to the orthorhombic phase. The lattice parameter of the *fcc* phase equals to 1.444 nm (see table 1). This value is close to the value measured for Rb_3C_{60} (a=1.4384 nm [8]). The phase with *fcc* crystal lattice is probably superconducting, because measured T_c is relatively close to the value of superconducting transition temperature for Rb_3C_{60} (28 K [8]). Thus, under the conditions of synthesis the composition probably can decay on two phases – mono- and triple-rubidium fullerides.

The heterofulleride with the composition $Cs_3In_xGa_yC_{60}$ is not a superconductor. According to the X-ray data, this samples crystallize in the orthorhombic lattice (see table 1). The absence of a superconductivity in this fulleride is probably because its crystal lattice is not *fcc*. The absence of superconductivity in cesium fullerides, synthesized from amalgams, was also registered by us earlier [10,11]. In previous work [10] we investigated the fullerides with composition $A_nHg_xC_{60}$ (A=K,Rb; n=2,3; x<1) synthesized by method using liquid alloys of alkali metals with mercury (amalgams). It was found that the fulleride $K_2Hg_xC_{60}$ was superconductor with T_c = 22 K.

The fulleride $K_3Hg_xC_{60}$ was not a superconductor. The fulleride $Rb_2Hg_xC_{60}$ was a superconductor with T_c = 25 K. Thus the alkali metals fullerides synthesized from gallams have higher T_c than analogous fullerides synthesized from amalgams. We also synthesized from amalgams the fullerides with composition $K_2GaHg_xC_{60}$, $Rb_2GaHg_xC_{60}$ and $K_2InHg_xC_{60}$, $Rb_2InHg_xC_{60}$. The fullerides $K_2GaHg_xC_{60}$ and $Rb_2GaHg_xC_{60}$ are superconductors with T_c equal to 20 K and 25 K, correspondingly. The fullerides $K_2InHg_xC_{60}$ and $Rb_2InHg_xC_{60}$ are not superconductors though crystallize in *fcc* lattice. It means that rather gallium than indium is responsible for high T_c for the sample $K_2In_xGa_yC_{60}$.

Conclusions

The new heterofullerides with the composition $A_nB_xD_yC_{60}$ (A=K,Rb,Cs; B,D=Ga, In, Sn, Bi, Be, Al, Mg, Hg; x,y<1; n=2;3) were synthesized. Temperature dependence of the magnetic susceptibility and lattice parameters were measured in the temperature range from 4.2 K to 297 K and transitions to the superconducting state were detected at temperatures T_c ranged from 14.5 K to 26 K. For the fulleride $K_2In_xGa_yC_{60}$ the T_c is equal to 24.5 K, that is the highest T_c among fullerides on the base of potassium (which does not include atoms of other alkali metals Rb and Cs). We found that the components of gallams or mercury intercalate into a fulleride lattice and improve their superconducting properties.

References

[1] B.M. Bulychev, R.A. Lunin, V.A. Kulbachinskii, et. al. Izv. Akad. Nauk, Ser. Khim., 53(8) (2004) 1623-1628. [Russ. Chem. Bull., International Edition, 53(8) (2004) 1686-1691].

[2] O. Gunnarsson, Reviews of Modern Physics 69, (1997) 575-606.

[3] A.Y. Ganin, Y. Takjabayashi, P. Jeglič, D. et.al., Nature, 466 (2010) 221-227.

[4] V.A. Kulbachinskii, B.M. Bulychev, V.G. Kytin and R.A. Lunin, Low Temperature Physics 37 (2011) 245-260.

[5] V.A. Kulbachinskii, R.A. Lunin, I.P. Kachan, et.al., Low Temp. Phys., 39 (2013) 58-65.

[6] A.F. Hebard, M.J. Rosseinsky, R.C. Haddon, et. al., Nature, 350 (1991) 600-604.

[7] K. Tanigaki, I. Hirosawa, T.W. Ebessen, et. al., Nature 356, (1991) 419-423.

[8] M.J. Rosseinsky, A.P. Ramirez, S.H. Glarum, et. al., Phys. Rev. Lett. 66 (1991) 2830-2833.

[9] B.M. Bulychev, R.A. Lunin, A.V. Krechetov, et.al., J. of Phys. and Chem. of Solids 65 (2004) 337-342.

[10] V.G. Kytin, B.M. Bulychev, A.V. Krechetov, et. al., Russian Journal of Inorganic Chemistry 53, (2008) 30-37.

[11] V.A. Kulbachinskii1, B.M. Bulychev and R.A. Lunin, A.V. Sycheva, Journal of Physics: C.S., 507 (2014) 012026(1-4).

Journal of Physics: Conference Series **568** (2014) 022027 doi:10.1088/1742-6596/568/2/022027

Unusual Superconducting transition in Topological Insulators

Dingping Li[1,2], B. Rosenstein[3,4], B. Ya. Shapiro[5] and I. Shapiro[5]

[1]School of Physics, Peking University, Beijing 100871, China
[2]Collaborative Innovation Center of Quantum Matter, Beijing, China
[3]Department of Electrophysics, National Chiao Tung University, Hsinchu, Taiwan, R.O.C.
[4]Applied Physics Department, Ariel University Center of Samaria, Ariel 40700, Israel
[5]Department of Physics, Institute of Superconductivity, Bar-Ilan University, Ramat-Gan 52900, Israel

Abstract. Superconducting transition generally belongs to the $U(1)$ class of phase transitions. However it was pointed out long time ago that if the normal state dispersion relation is "ultra-relativistic" the transition is unusual: even the mean field critical exponents are different from the standard ones leading to a number of observable effects. Attempts to experimentally discover such a system included chiral condensate in graphene. Recently it was found that some 3D topological insulators (that possess the ultrarelativistic metal on its surface) exhibit surface superconductivity.Starting from microscopic TI Hamiltonian with local four fermions interaction, we calculated the total set of the Gor'kov equations allowing to build the Ginzburg - Landau (GL) theory including the magnetic field effects. It was shown that the GL equations reflect the novel chiral universality class, very different from original GL equations. For example the temperature dependence of the coherence length diverges at the critical temperature with critical exponent $\nu = -1$ in rather than customary $\nu = -1/2$, magnetization near the upper critical magnetic field is quadratic as a function of deviation from the upper critical field while the Superfluid density is $\psi^2 = (T_c - T)^\beta$, $\beta = 2$, not $\beta = 1$.

1. Introduction

Since best studied Topological insulator (TI) possess a quite standard phonon spectrum [1], it was predicted recently [2] that they become superconducting TI (STI) (this should be distinguished from "topological superconductors", TSC, in which superconductivity appears in the bulk[3]). The predicted critical temperature of order of $1K$ is rather low (despite a fortunate suppression of the Coulomb repulsion due to a large dielectric constant $\varepsilon \sim 50$), the nature of the "normal" state, the 2D Weyl semi-metal, might make the superconducting properties of the system unusual. Especially interesting is the case (that actually was originally predicted for the [111] surface of Bi_2Te_3 and Bi_2Se_3[4]) when the chemical potential coincides with the Weyl point. Although subsequent ARPES experiments show the location of the cone of surface states order tenths of eV off the Fermi surface; there are experimental means to shift the chemical potential, for example by the bias voltage [5].

Unlike the more customary poor 2D metals with several small pockets of electrons/holes on the Fermi surface (in semiconductor systems or even some high T_c materials), STI has two peculiarities especially important when pairing is contemplated. The first is the bipolar nature of the Weyl spectrum: there is no energy gap between the upper and lower cones. The second is that the spin degree of freedom is a major player in the quasiparticle dynamics. This degree of freedom determines the pairing channel.

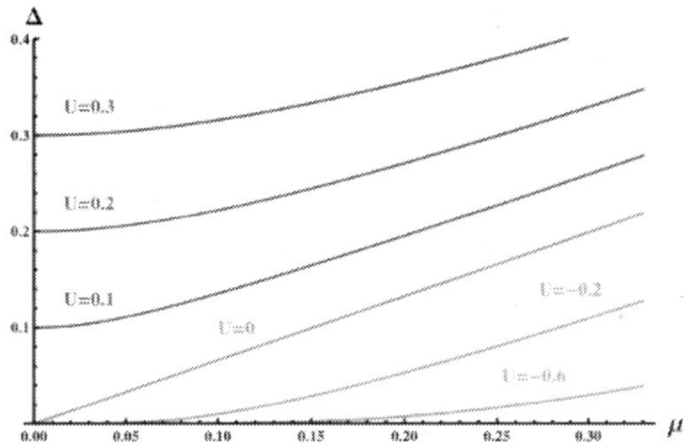

Figure 1. Order parameter at zero temperature as function of chemical potential of the TI surface Weyl semi-metal at various values of coupling parametrized by the renormalized energy U, Eq.(2). For positive U (blue lines) the superconductivity is strong and does not vanish even for zero chemical potential. There exists the critical coupling, $U = 0$ (red line), at which the second order transition occurs at quantum critical point $\mu = 0$. For negative U the superconductivity still exists at $\mu > 0$, but is exponentially weak.

In this paper we study the thermodynamic and magnetic properties of the surface superconductivity in TI with local attraction pairing Hamiltonian characterized by the coupling strength g and cutoff parameter T_D within the self-consistent approximation. The phase diagram is obtained for arbitrary temperature T and chemical potential $\mu < T_D$ We found a quantum critical point at $T = \mu = 0$ when the coupling strength g reaches a critical value g_c dependent on the cutoff parameter. We concentrate on properties of the superconducting state in a part of the phase diagram that is dominated by the QCP. Various critical exponents are obtained. In particular, the coupling strength dependence of the coherence length is $\xi \propto (g - g_c)^{-\nu}$ with $\nu = 1$, the order parameter scales as $\Delta \propto (g - g_c)^{\beta}$, $\beta = 1$. It is found that near the QCP the Ginzburg - Landau effective model is rather unconventional. The structure of the single vortex core is different from the usual Abrikosov vortex, while the magnetization curve near the upper critical magnetic field H_{c2} is quadratic: $M = (H - H_{c2})^2$, not linear.

2. TI with a local pairing interaction.

Electrons on the surface of a TI are described by a Pauli spinor $\psi_\alpha(\mathbf{r})$, where the upper plane, $\mathbf{r} = \{x, y\}$, is considered. The Hamiltonian for electrons in TI, $H = \int d^2 r \psi^+_\alpha \left(-i\hbar v_F \varepsilon_{ij} \nabla_i \sigma^j_{\alpha\beta} - \mu\delta_{\alpha\beta}\right)\psi_\beta$ where σ^j are the Pauli matrices. and interacting via four-Fermi local coupling of strength g. The effective local interaction might be generated by a phonon exchange or perhaps other mechanisms and will be assumed to be weak coupling. Therefore the BCS type approximation can be employed. In the homogeneous case, the matrix gap function can be chosen as (Δ real) $\widehat{\Delta}_{\beta\gamma} = g\langle\psi_\gamma(0)\psi_\beta(0)\rangle = i\sigma_y\Delta$. The matrix gap equation

$$\widehat{\Delta}^{*t} = -g\sum_{\omega q} D^t\widehat{\Delta}^{*t}\left(D^{-1} - \widehat{\Delta}D^t\widehat{\Delta}^{*t}\right)^{-1}. \tag{1}$$

where $D^{-1}_{\gamma\beta} = (i\omega - \mu)\delta_{\gamma\beta} - v_F\varepsilon_{ij}p_i\sigma^j_{\alpha\beta}$. The spectrum of elementary excitations, $E_p = \pm\sqrt{\Delta^2 + (v_F p - \mu)^2}$, coincides with that found within the Bogoliubov - de Gennes approach [6].

3. Zero temperature phase diagram and QCP.

At zero temperature the integrations over frequency and momentum limited by the UV cutoff Λ result in

$$U = \sqrt{\Delta^2 + \mu^2} - \frac{\mu}{2} \log \frac{\sqrt{\Delta^2 + \mu^2} + \mu}{\sqrt{\Delta^2 + \mu^2} - \mu}, \tag{2}$$

where the dependence on the cutoff is incorporated in the renormalized coupling with dimension of energy defined as $U = v_F \Lambda (1 - g_c/g)$ with $g_c = 4\pi\hbar^2 v_F/\Lambda$. Of course the superconducting solution exists only for $g > 0$. In Fig. 1 the dependence of the gap Δ as function of the chemical potential μ is presented for different values of U.

For an attractive coupling g stronger than the critical one, (when $U > 0$), blue lines in Fig. 1, there are two qualitatively different cases.

(i). When $\mu << U$ the dependence of Δ on the chemical potential is parabolic $\Delta/U \approx 1 + (\mu/U)^2$.

In particular, when $\mu = 0$, the gap equals U. As can be seen from Fig. 1, the chemical potential makes a very limited impact in the large portion of the phase diagram.

(ii) For the attraction just stronger than critical, $g > g_c$, namely for small positive U, the dependence becomes linear, see red line in Fig. 1, $\Delta = 0.663\,\mu$. So that the already weak condensate becomes sensitive to μ.

The case (i) is more interesting than (ii) since it exhibits stronger superconductivity (larger T_c, see below). Finally for $g < g_c$, namely negative U (green lines in Fig. 1), the superconductivity is very weak with exponential dependence similar to the BCS one, $\Delta \approx \mu \exp[-(|U|/\mu - 1)]$. Therefore it is possible to neglect the effect of weak doping and consider directly the $\mu = 0$ particle-hole symmetric case.

At zero temperature $\Delta = U$, while $\Delta \to 0$ as a power of the parameter $U \propto g - g_c$ describing the deviation from quantum criticality $T_c \propto U^{z\nu}$; $z\nu = 1$. Here z is the dynamical critical exponent[8]. Therefore, as expected, the renormalized coupling describing the deviation from the QCP is proportional to the temperature at which the created condensate disappears. The temperature dependence of the gap reads, $\Delta(T) = 2T \cosh^{-1}\left(\frac{1}{2}\exp\frac{U}{2T}\right)$, typical for chiral universality classes [8, 9].

4. Ginzburg - Landau effective theory and magnetic properties of the superconductor near QCP

The quadratic term of the Ginzburg-Landau energy $F_2 = \sum_{\mathbf{p}} \Delta_p^* \Gamma(p) \Delta_p$ is obtained exactly from expanding the gap equation to linear terms in Δ for arbitrary external momentum: $\Gamma(p) = -U/4\pi\hbar^2 v_F^2 + |p|/16 v_F \hbar^2$. The dependence on $\mathbf{p}$ is non-analytic and within our approximation higher powers of p do not appear. The second term is very different from the quadratic term in the GL functional for conventional phase transitions at finite temperature or even quantum phase transitions in models without Weyl fermions [8] and has a number of qualitative consequences. The coherence length as a power of parameter $U \propto g - g_c$ describing the deviation from criticality: $\xi(U) = \frac{\pi}{4} v_F \hbar U^{-\nu}$; $\nu = 1$. This is different from the dependence in non-chiral universality classes that is $\xi(T) \infty (T_c - T)^{-\nu}$, $\nu = 1/2$ in mean field. Of course in the regime of critical fluctuations this exponent is corrected in both non-chiral and chiral[9] universality classes.

Local terms in the GL energy density are calculable exactly: $f_{cond} = \propto -U|\Delta|^2 + \frac{2}{3}|\Delta|^3$. It is quite nonstandard compared to customary quartic term in conventional universality classes. The GL equations in the homogeneous case for the condensate gives $\Delta_0 = U^{\beta}$ with critical exponent $\beta = 1$, different from the mean field value $\beta = 1/2$ for the $U(1)$ universality class. The condensation energy density is $f_0 \propto U^{2-\alpha}$ with $\alpha = -1$. The free energy critical exponent at QCP therefore is also different from the classical $\alpha = 0$. In the present case the equation for the order parameter is nonlocal and nonanalytic. Near the upper critical field H_{c2} the Abrikosov hexagonal lattice is formed. The optimal Δ at external field H close to H_{c2} is $\Delta \propto (1 - H/H_{c2})^{\sigma}$, $\sigma = 1$, different from the ordinary Abrikosov lattice value of $\sigma = 1/2$. The magnetization is $M \propto -(1 - H/H_{c2})^{\tau}$, $\tau = 2$ in contrast to linear dependence, $\tau = 1$.

5. Discussion and conclusions

To estimate the pairing efficiency due to phonons, one should rely on recent studies of surface phonons in TI [2]. The coupling constant is obtained from the exchange of acoustic (Rayleigh) surface phonons $g = \lambda v_F^2 \hbar^2 / 2\pi\mu$, where λ is the dimensionless effective electron - electron interaction constant of order 0.1. It was shown in ref. [2] that at zero temperature the ratio of λ and μ is constant with well defined $\mu \to 0$ limit with value $g = 0.23\ eV\ nm^2$ for $v_F \approx 7\ \cdot 10^5 m/s$ (for Bi_2Se_3). The critical coupling constant g_c, can be estimated from the Debye cutoff $T_D = 200K$ determining the momentum cutoff $\Lambda = T_D/c_s$, where c_s is the sound velocity. Taking value to be $c_s = 2 \cdot 10^3 m/s$ (for Bi_2Se_3), one obtains $g_c = 4\pi v_F c_s \hbar^2 / T_D = 0.20\ eV\ nm^2$. Therefore the stronger superconductivity, $g > g_c$, is realized.

In this paper we focused on the qualitatively distinct case of Weyl fermions with small chemical potential. Note that a reasonable electron density of $n = 3 \cdot 10^{11} cm^{-2}$ in Bi_2Te_3 already conforms to the requirement that chemical potential $\mu = \sqrt{n}\hbar v_F / 2\pi = 100K$ is smaller that the Debye cutoff energy $T_D = 200K$. The concept of QCP at zero temperature and varying doping constitutes a very useful language for describing the microscopic origin of superconductivity in high T_c cuprates and other "unconventional" superconductors. Superconducting transitions generally belong to the $U(1)$ class of second order phase transitions, however it was pointed out a long time ago that, if the normal state dispersion relation is "ultra-relativistic", the transition at zero temperature as function of parameters like the pairing interaction strength is qualitatively distinct and belongs to chiral universality classes classified in ref. [9]. Attempts to experimentally identify second order transitions governed by QCP included quantum magnets [8], superconductor - insulator transitions and more recently chiral condensate in graphene.

Acknowledgements. We are indebted to C.W. Luo, J.J. Lin for valuable discussions. Work of D.L. and B.R. was supported by NSC of R.O.C. Grants No. 98-2112-M-009-014-MY3 and MOE ATU program. The work of D.L. also is supported by National Natural Science Foundation of China (No. 11274018).

[1] Luo C W et al 2013 *Nano Lett.* **13**, 5797.
[2] Das Sarma S and Li Qiuzi 2013 *Phys. Rev. B* **88**, 081404(R).
[3] X.-L. Qi and S.-C. Zhang 2011 *Rev. Mod. Phys.* **83**.
[4] Zhang H et al 2009 *Nat. Phys.* 5, 438.
[5] Kim D et al 2012 *Nat. Phys.* **8**, 459.
[6] Lu Chi-Ken and Herbut I F 2010 *Phys. Rev. B* **82**, 144505.
[7] Cheng M, Lutchyn R M, and Das Sarma S 2012 *Phys. Rev B* **85**, 165124.
[8] Sachdev S 2011 *Quantum Phase Transitions* (Cambridge: Cambridge University Press)
[9] Rosenstein B, Warr B J and Park S H 1989 *Phys. Rev. Lett.* **62**, 1433; Gat G, Kovner A and Rosenstein B 1992 *Nucl. Phys.[FS] B* **385**, 76.

Impurity effects on vortex core states of topological s-wave superconductor

Yusuke Masaki[1], **Yusuke Kato**[2]

[1]Department of Physics, the University of Tokyo, Tokyo 113-0033, Japan
[2]Department of Basic Science, the University of Tokyo, Tokyo 153-8902, Japan

E-mail: `masaki@vortex.c.u-tokyo.ac.jp`

Abstract. We study the impurity effects on bound states in a single vortex core of a topological s-wave superconductor (SC), which is a two-dimensional electron gas with Rashba spin-orbit coupling, Zeeman coupling and s-wave superconductivity. We calculate the Gor'kov Green's function in the presence of impurities in a way similar to Kopnin and Kravtsov (1976); We restrict the Hilbert space to that spanned by the eigenstates of the Bogoliubov-deGennes equation for a pure system, bounded near the vortex core. This topological s-wave SC has two types of vortices and each type is specified by the relative sign between the vorticity and Zeeman coupling. We find that the impurities affect bound states differently between the two types of vortices and bound states are more robust against impurities when vorticity and Zeeman coupling are opposite in signs.

1. Introduction

Topological superconductors (TSCs) have zero energy states called Majorana bound states at edges or topological defects [1, 2]. An idea has been proposed on implementation of topological quantum computation (TQC) [3], which utilizes non-Abelian statistics in the Majorana bound states in TSCs. TQC has advantage in low decoherence, compared to conventional quantum computations [4].

If we do not consider space groups, all possible TSCs are classified by Schnyder *et al.* [5, 6] into ten symmetry classes based on the random matrix classification by Altland and Zirnbauer [7]. Particularly, class D, where time reversal symmetry is broken, is one of Bogoliubov-deGennes (BdG) classes and TSCs in this class are proposed for various forms such as $\nu = 5/2$ fractional quantum Hall state [8], chiral p-wave SC [9] and cold-atoms with artificial gauge field [10]. However realization of non-Abelian statistics in solids is challenging. Sr_2RuO_4 is expected to be a chiral p-wave SC [11], but the superconducting state in this material does not have continuous rotational symmetry, because of the point group symmetry C_4 in the crystal structure. Recent theories tell us that proximity effects can lead some engineered TSCs which are equivalent to $p+ip$ SCs [12, 13]. The example is an s-wave SC on a surface of three-dimensional (3D) topological insulator with a time-reversal breaking field [12], or a heterostructure of 2D electron gas with strong Rashba spin-orbit coupling (SOC), Zeeman coupling and s-wave superconductivity [13]. In this article, we focus on the latter engineered $p + ip$ SC and call it topological s-wave SC.

TQC using non-Abelian statistics is realized through an adiabatic exchange of Majorana states in their subspace. The adiabatic time scale is characterized by the inverse of the minigap ($\Delta_{\rm mini}$), which is the energy gap between the Majorana state and the first excited state. Thus a large minigap as well as the Majorana bound state is necessary for realization of TQC. The topological s-wave SC is a promising system for TQC; the minigap for bound states in a vortex core is about $0.1 \sim 1$K, while a minigap of typical chiral p-wave SC is about $1 \sim 10$mK.

An important issue in a realistic situation is how robust this large minigap is against randomness such as impurities. In order to study the robustness of the minigap of the topological s-wave SC, the authors in the earlier work [14] introduced random potential into the BdG equation and calculate the energy spectrum. In their work, however, the effects of randomness have not been treated systematically enough in the sense that the random average has not been taken. Moreover, the difference in impurity effects on two types of vortices have not been considered seriously; In this SC, there are inequivalent vortices in the sense that one type of vortex is not transformed into the other by symmetry operations. The type of vortex is specified by the relative sign of the vorticity and the Zeeman coupling. We know that the robustness of bound states against impurities depends crucially on the type of vortex in chiral p-wave SC [15].

One of the standard approaches to treat the random impurity effects on inhomogeneous SC is to use the quasiclassical theory of superconductivity. The impurity effects on chiral p-wave SC mentioned above are studied within quasiclassical, continuous spectrum approximation. For the case of the topological s-wave SC, the large minigap implies that the spectra are discrete and the quasiclassical theory does not provide quantitative results. In addition, the quasiclassical theory itself has not been constructed for this system. Another approach is to use Green's function through the Gor'kov equation. Kopnin and Kravtsov have developed a scheme of the Green's function for an s-wave SC [16], but the form of the coherence factor is not clear in their scheme and thus it is inapplicable to other SCs, as it stands. With these backgrounds, we have derived a proper formula within the self-consistent Born Approximation, which updates the Kopnin-Kravtsov scheme. The details of this formulation are discussed in [17]. With use of this method, we can study the impurity effects on bound states systematically keeping the discreteness of the energy spectrum.

2. Formulation

Hamiltonian for a 2D electron system with s-wave pairing, Rashba SOC and Zeeman coupling is given by [13]

$$H = \int d\boldsymbol{r} \vec{\psi}^\dagger(\boldsymbol{r}) \hat{H}_0(\boldsymbol{r}) \vec{\psi}(\boldsymbol{r}) + \int d\boldsymbol{r} \left[\Delta(\boldsymbol{r}) \psi_\uparrow^\dagger(\boldsymbol{r}) \psi_\downarrow^\dagger(\boldsymbol{r}) + \text{h.c.} \right], \tag{1}$$

$$\hat{H}_0(\boldsymbol{r}) = \left(\frac{\boldsymbol{p}^2}{2m} - \mu + \alpha(\boldsymbol{p} \times \hat{\boldsymbol{\sigma}})_z + V_z \hat{\sigma}_z \right), \quad \vec{\psi}^\dagger(\boldsymbol{r}) = (\psi_\uparrow^\dagger(\boldsymbol{r}), \psi_\downarrow^\dagger(\boldsymbol{r})), \tag{2}$$

where μ is the chemical potential, α strength of Rashba type SOC, V_z Zeeman coupling. $\Delta(\boldsymbol{r})$ is s-wave pairing. We assume that there is a single vortex and the pairing function takes the form : $\Delta(\boldsymbol{r}) = \Delta(r)e^{i\kappa\theta}(\kappa = \pm 1)$, κ denotes vorticity. This system is in a topological phase when $V_z > \sqrt{\mu^2 + \Delta_0^2}$. Δ_0 denotes the magnitude of s-wave pair potential in the bulk. The BdG equation in this system can be derived as follows:

$$\check{H}_{\text{BdG}}(\boldsymbol{r}) \vec{u}_K(\boldsymbol{r}) = E_K \vec{u}_K(\boldsymbol{r}), \tag{3}$$

$$\check{H}_{\text{BdG}}(\boldsymbol{r}) = \begin{bmatrix} \hat{H}_0 & \hat{\Delta}(\boldsymbol{r}) \\ \hat{\Delta}^\dagger(\boldsymbol{r}) & -\hat{H}_0^*(\boldsymbol{r}) \end{bmatrix}, \quad \hat{\Delta}(\boldsymbol{r}) = \Delta(\boldsymbol{r}) i\hat{\sigma}_y. \tag{4}$$

The symbol ˇ denotes a 4 by 4 matrix, ˆ a 2 by 2 matrix and the basis is taken as ${}^t\vec{u}_K(\boldsymbol{r}) = (u_{K\uparrow}(\boldsymbol{r}), u_{K\downarrow}(\boldsymbol{r}), v_{K\uparrow}(\boldsymbol{r}), v_{K\downarrow}(\boldsymbol{r}))$. We take the cylindrical coordinate and each eigenstate K is labeled by (l, ν), which denotes a set of angular momentum l and quantum number ν of radial direction. In order to solve the BdG equation, we adopt Fourier-Bessel expansion [14].

We take the system size $R = 100\xi_0$ and two cut-offs: $l_c = 50, N_c = 400$ for l, ν respectively. We put $\Delta(r) = \Delta_0 \tanh(r/\xi_0)$ and $\xi_0 = v_F/\Delta_0$. This system has two kinds of bound states with energies below Δ_0; one is bounded in the vortex core known as the Caroli-deGennes-Matricon (CdGM) mode, and the other is near the edge of the system. It is known that the sign of V_z and vorticity correspond to the signs of the angular velocities of the edge mode and CdGM mode, respectively [18] and we note that the sign of V_z plays a similar role in this TSC as the sign of chirality in chiral p-wave SC. We investigate the impurity effects on the CdGM mode for both signs of V_z and the negative vorticity ($\kappa = -1$). In the following part, we call parallel (anti-parallel) vortex when the signs of V_z and vorticity are same (opposite). To study random impurity effects, we use the Gor'kov equation [17]. We take into account a mode $\nu = 0$ bounded in the vortex core only and omit this label ν hereafter. Within the self-consistent Born approximation, Green's function and self energies are

$$\check{G}(\boldsymbol{r}, \boldsymbol{r}'; \omega_n) = \sum_l \frac{\check{\tau}_z \vec{u}_l(\boldsymbol{r}) \vec{u}_l^\dagger(\boldsymbol{r})}{E_l - \sigma_l(\mathrm{i}\omega_n) - \mathrm{i}\omega_n}, \tag{5}$$

$$\sigma_l(\omega_n) = \frac{\gamma}{2\pi} \sum_{l'} \frac{M_{l,l'}}{E_{l'} - \sigma_{l'} - \mathrm{i}\omega_n}, \tag{6}$$

$$M_{l,l'} = \int r dr |\vec{u}_l^\dagger(r) \check{\tau}_z \vec{u}_{l'}(r)|^2, \tag{7}$$

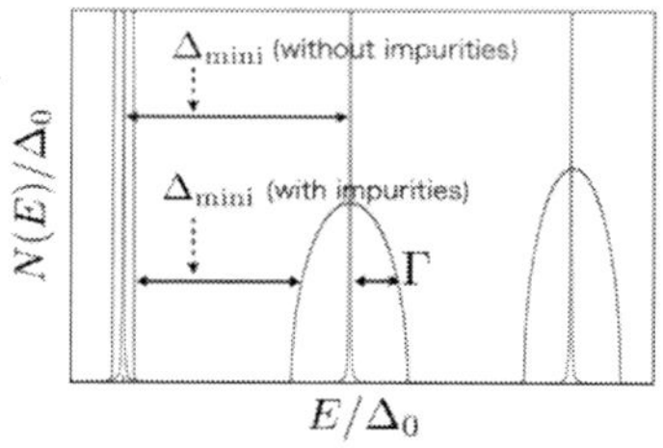

Figure 1. (Color online) Schematic picture for the relation between DOS and widths Γ.

where $\gamma = \Gamma_n/(\pi N_n) = 2v_F\Gamma_n/k_F$. Γ_n denotes the impurity scattering rate in the normal state and N_n the density of states per spin at the Fermi level in the normal state. k_F and v_F denote, respectively, the Fermi wavenumber and the Fermi velocity. $\vec{u}_l(r)$ is the radial part of $\vec{u}_l(\boldsymbol{r})$. The present scheme is similar to that derived by Kopnin and Kravtsov [16] for the s-wave SC. In contrast to their scheme, our scheme has the coherence factor (Eq. (7)) and is applicable to various kinds of SCs such as chiral p-wave SCs. We calculate the density of states from this Green's function:

$$N(\omega) = \sum_l N_l(\omega) = \int d\boldsymbol{r} \mathrm{Im} \left[\frac{1}{\pi} \mathrm{Tr}\, \check{\tau}_z \check{G}(\boldsymbol{r}, \boldsymbol{r}; \omega_n)|_{i\omega_n \to \omega + i\delta} \right]. \tag{8}$$

We evaluate the width of spectrum, denoted as Γ and defined as half the width at half maximum (HWHM) (Fig. 1). Γ corresponds to the impurity scattering rate of CdGM mode.

3. Result

We numerically calculate the density of states using Eqs. (5)-(8). In this calculation, $|V_z| = 2.0\Delta_0$, $\mu = 0.2\Delta_0$ are fixed parameters, while $m\alpha^2/\Delta_0$, $\Gamma_n/(\pi\Delta_0)$ are taken as $0.1, 0.4, 0.9, 1.6$ and $10^{-5}, 10^{-4}, 10^{-3}, 10^{-2}, 10^{-1}$ respectively.

3.1. vortex type study

Figure 2 (a) (2 (b)) shows the width calculated from the spectra of bound states scattered by impurities when the Zeeman coupling V_z and vorticity have the opposite (same) sign. The details of parameters are written in the caption. As an overall feature, we find that the impurity scattering rates in Fig. 2 (b) ($V_z < 0$) are larger than those in Fig. 2 (a) ($V_z > 0$). This reminds us of impurity effects on two types of vortices in chiral p-wave SC; the type is specified by the relative signs of the vorticity and the chirality and bound states are more robust against impurities when the vorticity and chirality have opposite sign. In this topological s-wave SC,

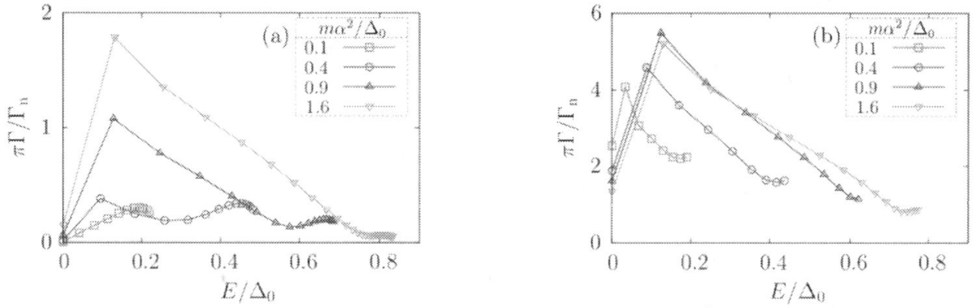

Figure 2. (Color online) The scattering rates of eigenstates. (a)$V_z/\Delta_0 = 2.0 > 0$, s.t. the anti-parallel vortex. (b)$V_z/\Delta_0 = -2.0 < 0$, s.t. the parallel vortex. Both of figures are showed for $\Gamma_n/(\pi\Delta_0) = 10^{-2}$, $m\alpha^2/\Delta_0 = 0.1$(open square), 0.4(open circle), 0.9(open triangle), 1.6(open inverted-triangle).

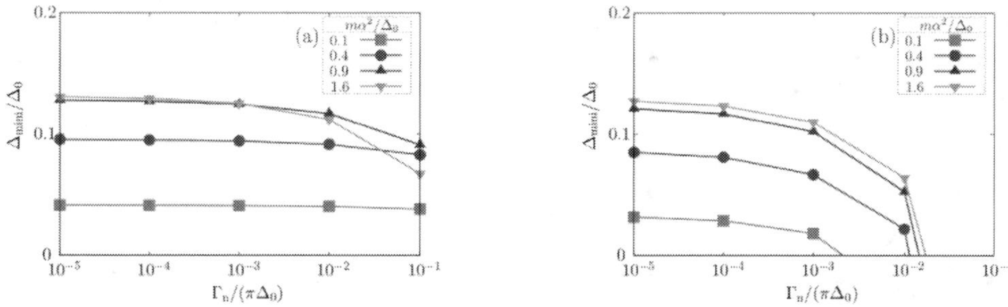

Figure 3. (Color online) Minigap in the presence of impurities. Parameters are same as those in Fig. 2, but the axes are not. $m\alpha^2/\Delta = 0.1$(solid square), 0.4(solid circle), 0.9(solid triangle), 1.6(solid inverted-triangle).

we thus understand the difference between these figures by recalling that the sign of V_z in this SC corresponds to that of the chirality in chiral p-wave SC.

We note that the robustness of bound states is crucial to realization of adiabatic manipulation of vortices in TQC. To quantify the robustness against impurities (or equivalently, realizability of adiabatic dynamics), we define "the minigaps" in the presence of impurities by subtracting the widths of the first excited state and the zero mode from the minigap without impurities. Figure 3 (a) (3 (b)) shows the dependence of minigaps on the normal scattering rates for the anti-parallel (parallel) vortex for various values of $m\alpha^2/\Delta_0$. When the signs of Zeeman coupling and vorticity are same (Fig. 3 (b)), minigaps decrease steeply as Γ_n increases. When the signs of Zeeman coupling and vorticity are opposite (Fig. 3 (a)), on the other hand, the minigaps maintain their magnitudes even for the strong Γ_n.

3.2. Rashba SOC dependence of anti-parallel vortex

Now we return to Fig. 2 and discuss the energy dependence of the scattering rate, particularly in the anti-parallel vortex. We can see that Γ for the states with finite excitation energy decreases with the energy increasing regardless of the values of $m\alpha^2/\Delta_0$ in the parallel vortex (Fig. 2 (b)). Figure 2 (a) shows that scattering rates at low energies are suppressed for small $m\alpha^2/\Delta_0$. As $m\alpha^2/\Delta_0$ increases, energy dependences become similar to those in Fig. 2 (b). Figure 2 (a) implies that the type of CdGM mode changes as $m\alpha^2/\Delta_0$ changes.

Although the relative signs of the vorticity and the Zeeman coupling are common for all data in Fig. 2 (a), different strengths of Rashba SOC cause this difference, which has not been observed in the cases of an s-wave SC and chiral p-wave SCs. We also find the $m\alpha^2/\Delta_0$ dependence of low energy scattering in Fig. 3 (a). As Γ_n increases, the minigaps for large $m\alpha^2/\Delta_0$ decrease, while the minigaps for small $m\alpha^2/\Delta_0$ are robust against impurities. For some values of Γ_n, the relative magnitude of the minigaps is reversed.

4. Discussion

We find that the scattering rates at low energies are suppressed for some values of Rashba SOC in an anti-parallel vortex. This suppression is due to small $M_{l,l}$, i.e. coherence factor. Such a suppression as this for the anti-parallel vortex is theoretically expected in chiral p-wave superconducting vortices, where the rotational symmetry is found to be crucial for the coherence factors to be small [15]. The 2D electron system in a semiconductor has the rotational symmetry, but the symmetry of chiral p-wave SCs such as Sr_2RuO_4 is affected by point groups of crystals. Therefore this suppression of the impurity scattering will be observed more clearly in the topological s-wave SC than the chiral p-wave SC.

5. Acknowledgments

This work was supported by the Program for Leading Graduate Schools, MEXT, Japan. This work was supported by KAKENHI(23244070) from JSPS.

References

[1] Hasan M Z and Kane C L 2010 *Rev. Mod. Phys.* **82**(4) 3045–3067
[2] Qi X L and Zhang S C 2011 *Rev. Mod. Phys.* **83**(4) 1057–1110
[3] Ivanov D A 2001 *Phys. Rev. Lett.* **86**(2) 268–271
[4] Nayak C, Simon S H, Stern A, Freedman M and Das Sarma S 2008 *Rev. Mod. Phys.* **80**(3) 1083–1159
[5] Schnyder A P, Ryu S, Furusaki A and Ludwig A W W 2008 *Phys. Rev. B* **78**(19) 195125
[6] Ryu S, Schnyder A P, Furusaki A and Ludwig A W W 2010 *New Journal of Physics* **12** 065010
[7] Altland A and Zirnbauer M R 1997 *Phys. Rev. B* **55**(2) 1142–1161
[8] Moore G and Read N 1991 *Nuclear Physics B* **360** 362 – 396 ISSN 0550-3213
[9] Read N and Green D 2000 *Phys. Rev. B* **61**(15) 10267–10297
[10] Sato M, Takahashi Y and Fujimoto S 2009 *Phys. Rev. Lett.* **103**(2)
[11] Maeno Y, Kittaka S, Nomura T, Yonezawa S and Ishida K 2012 *Journal of the Physical Society of Japan* **81** 011009
[12] Fu L and Kane C L 2008 *Phys. Rev. Lett.* **100**(9) 096407
[13] Sau J D, Lutchyn R M, Tewari S and Das Sarma S 2010 *Phys. Rev. Lett.* **104**(4) 040502
[14] Mao L and Zhang C 2010 *Phys. Rev. B* **82**(17) 174506
[15] Kato Y and Hayashi N 2002 *Journal of the Physical Society of Japan* **71** 1721–1727
[16] Kopnin N and Kravtsov V 1976 *JETP Lett.* **23** 578
[17] Masaki Y and Kato Y in preparation.
[18] Wu L H, Liang Q F, Wang Z and Hu X 2012 *Journal of Physics: Conference Series* **393** 012018

Lower symmetrical behaviour of electronic state in the FeAs plane of $Ba(Fe_{1-x}Co_x)_2As_2$ -^{75}As NMR study-

Masayuki Toyoda[1,*], **Yoshiaki Kobayashi**[1,**], **Masayuki Itoh**[1], **Masatoshi Sato**[2]
[1] Department of Physics, Graduate School of Science, Nagoya University, Furo-cho, Chikusa-ku, Nagoya, 464-8602, Japan.
[2] Research Center for Neutron Science & Technology, Comprehensive Research Organization for Science and Society (CORSS), 162-1 Shirakata, Tokai 319-1106, Japan.

* E-mail: toyoda.masayuki@f.mbox.nagoya-u.ac.jp
**E-mail: i45323a@cc.nagoya-u.ac.jp

Abstract. To investigate the anisotropy within the FeAs plane in the tetragonal phase of $Ba(Fe_{1-x}Co_x)_2As_2$, ^{75}As NMR measurements on the electric field gradient (EFG) at the As site have been carried out for a $Ba(Fe_{1-x}Co_x)_2As_2$ single crystal of x~0.08 with the superconducting (SC) transition temperature of T_c~23 K. We present a method how to analyse the ^{75}As NMR spectra and deduce the anisotropic parameter of the EFG, η, that shows the electric in-plane anisotropy at the As site. The EFG of the As site with no Co atoms at the nearest and next nearest Fe sites has the η value of 0.08-0.10 similar to that in the non-SC samples of x~0.02 in the tetragonal phase. The in-plane anisotropy in the x~0.08 sample remains even near T_c. We discuss the relationship between the in-plane anisotropy and local physical properties.

1. Introduction

In most Fe-based superconductors such as $LaFeAsO_{1-x}F_x$, $Ba(Fe_{1-x}Co_x)_2As_2$, and $NaFe_{1-x}Co_xAs$, an antiferromagnetic (AFM) phase with an orthorhombic structure, a paramagnetic one with a tetragonal structure, and paramagnetic and superconducting (SC) ones with a tetragonal structure appear in the phase diagram of temperature T vs. doped carrier concentration or pressure. The SC phases are located in the proximity of the AFM phases, which led on to the proposal that the AFM spin fluctuation plays a key role for making the SC pairing. Then superconducting order parameters Δ's on the different Fermi surfaces have opposite sign ($s_\pm$ symmetry of Δ) [1, 2]. However, from the early stage of studying the Fe-based superconductors, it was pointed out that Δ's with same sign (s_{++} symmetry of Δ) are realized, because the superconductivity is fairly strong against non-magnetic impurity scattering as found in $RFeAsO_{1-x}F_x$ (R =La, Nd) [3, 4], $Ca_{10}Pt_4As_8(FeAs)_{10}$ [5], and the $BaFe_2As_2$ system [6]. Kontani *et al.* suggested that the orbital fluctuation leads to the superconductivity with the s_{++} symmetry [7]. Although the orbital fluctuation is difficult to be observed directly, characteristic properties such as drastic T-dependences of elastic constants in the tetragonal phase, the impurity effects on T_c above mentioned, and so on can be explained by taking the orbital fluctuation and its ordering into account [8-10]. In addition to the superconducting mechanism, the effects of the spin and orbital fluctuations on the electronic states appear in the tetragonal phase of the samples with the structural transition or the SC transition.

Based on the in-plane anisotropy of electrical resistivities in $Ba(Fe_{1-x}Co_x)_2As_2$ single crystals, it was reported first that a lower symmetry in the FeAs plane than that of the crystal structure emerges even in the tetragonal phase [11]. This in-plane anisotropy has attracted much attention as a new phenomenon in a nematic electronic state and was discussed from theoretical aspects based on the spin [12] or orbital [7] fluctuation/order. To microscopically confirm whether there is the in-plain anisotropy or not, we studied the in-plane symmetry of the electronic field gradient (EFG) tensor at the As sites in $Ba(Fe_{1-x}Co_x)_2As_2$ with x~0.02, using the ^{75}As NMR technique [13]. We found that the EFG has the in-plane anisotropy with the same symmetrical axes as those of the orthorhombic structure even in the tetragonal phase. This anisotropy gradually decreases with increasing T and remains finite even near room temperature. However, it has not been elucidated whether the in-plain anisotropy remains in the SC region.

In this paper, we have studied the in-plane anisotropy of the EFG tensor at the As sites in a $Ba(Fe_{0.92}Co_{0.08})_2As_2$ single crystal which has the SC transition temperature T_c~23K and does not exhibit structural and AFM transitions. We present the analysis method of the ^{75}As NMR spectrum obtained for the As sites without Co impurities at the nearest and the next nearest neighbor Fe sites to obtain the in-plane anisotropy of the EFG tensor. To discuss the x dependence of the in-pláne anisotropy in the EFG tensor, we compare that in the SC sample of x~0.08 with those in the x~0.02 samples with the structural and AFM transitions. Finally, we discuss the relationship between the in-plane anisotropy and the physical properties.

2. Experiments

$Ba(Fe_{1-x}Co_x)_2As_2$ single crystals with x~0.08 were prepared by a flux method [13]. The Co concentration x was estimated by comparing the c lattice parameter ~12.9761 Å with the c-x curve reported for $Ba(Fe_{1-x}Co_x)_2As_2$ [14]. T_c was found to be 23 K from the T dependence of the superconducting diamagnetism shown in Fig. 1.

A single crystal cut in the size of 2.0 × 2.0 × 0.1 mm^3 was inserted in an NMR coil. Then we did not use an adhesive bond such as GE 7031, commonly used to fix a crystal, to avoid possible effects of different thermal contractions between the sample and the bond, because the Fe-based system is considered to be very sensitive to external stresses [15, 16]. ^{75}As NMR measurements were carried out using a coherent pulsed spectrometer and a superconducting magnet with a magnetic field H of 7.6966 T, where the single crystal was rotated using a two-axis goniometer. We measured Fourier-transformed (FT) spectra of spin-echo signals by changing the frequency stepwise and obtained the whole ^{75}As NMR spectra by superposing the FT spectra at various frequencies.

3. Results and Discussion

The tensor of the nuclear electric quadrupole frequency ν_Q^{ij}, ($i, j = x, y$, and z), which is proportional to the EFG tensor, is useful for investigating the in-plain anisotropy of the electronic state in the Fe-based superconductors. In general, ν_Q^{ij} has no in-plane anisotropy in the tetragonal symmetry, whereas it shows the twofold in-plane anisotropy in the orthogonal symmetry. The principal axes of the EFG were reported to be parallel to the symmetrical axes, the a_O, b_O, and c axes, in the orthorhombic phase of $Ba(Fe_{1-x}Co_x)_2As_2$ with x=0 [17]. Note that the a_O and b_O axes within the Fe-plane are located at ~45° from the a_T axis, one of the symmetrical axes in the tetragonal phase above T_s. The in-plane anisotropy of ν_Q^{ij} with the twofold symmetry was clearly observed in the orthorhombic phase below T_s.

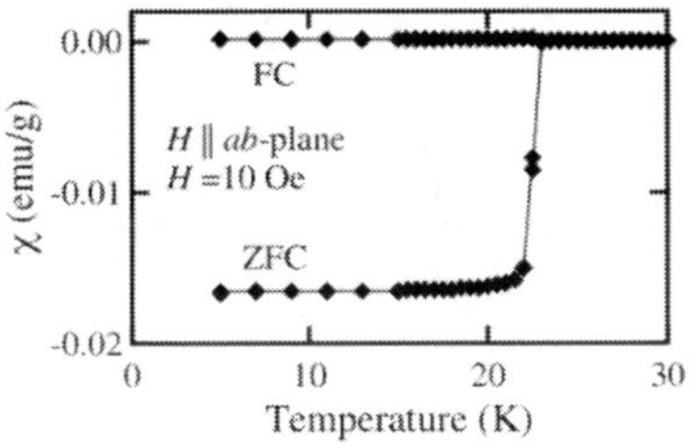

Fig. 1 Temperature dependence of superconducting diamagnetic susceptibility in $Ba(Fe_{0.92}Co_{0.08})_2As_2$.

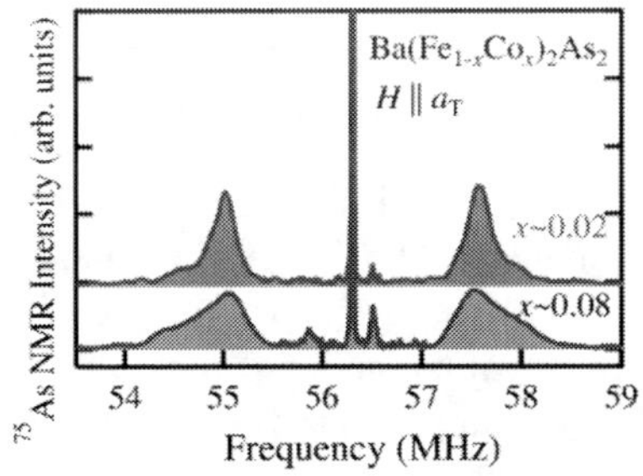

Fig. 2 ^{75}As NMR spectra at 180 K in $Ba(Fe_{1-x}Co_x)_2As_2$ of x ~0.02 and 0.08 with H parallel to the aT axis in the FeAs plane.

Figure 2 shows ^{75}As NMR spectra of $Ba(Fe_{1-x}Co_x)_2As_2$ with x~0.08 at 180K together with those of $Ba(Fe_{1-x}Co_x)_2As_2$ with x~0.02 at 180 K, for comparison. Both the samples are in the tetragonal phase at 180 K, and H is parallel to the tetragonal axis a_T. The spectra split by the nuclear electric quadrupole interaction between the ^{75}As nuclear spin I (=3/2) and EFG at the As site into the central line at ~56.3 MHz and two satellite ones at ~55.0 and ~57.5 MHz as seen in Fig. 2. Satellite spectra in the x ~0.08 sample can be decomposed into the three Gaussians as denoted by As00, As01, and As02 in Fig. 3 which shows for example the satellite spectrum at ~57.5 MHz and 30 K. The intensity ratio of the As00, As01, and As02 spectra, 17:12:4, is well reproduced by the calculation based on the binominal distribution for the Co impurities, if it is assumed that the As00, As01, and As02 spectra may come from the three As sites, which have no, one, and two Co atoms, respectively, at the eight next nearest neighbor Fe sites, without Co at the four nearest neighbor Fe sites. On the other hand, the small peaks observed at ~55.9 and ~56.5 MHz in Fig. 2 are central lines coming from the As1 site which is surrounded by one Co and three Fe at the four nearest neighbour Fe sites [13, 18]. From the analysis on the angular dependence of the NMR frequency of the central lines (Fig. 4) based on the second-order electric quarupole effect, we found that the As1 site has the ν_Q^{ij} tensor with the principal axis almost parallel to the As1-Co directions, where the tensor has the maximum value ν_Q of 14.9 MHz much larger than 2.22-3.81 MHz of the As00-As02 spectra, and the asymmetry parameter of the EFG tensor η=0.63. Almost the same result was obtained for the x~0.02 sample [13]. These facts show that the As1 satellite spectra do not overlap the As00-As02 ones in both the samples. Thus we can separately monitor the local As sites using the ^{75}As NMR technique.

We have measured the angular dependence of the ^{75}As NMR spectrum with H rotated in the FeAs plane for the x~0.08 sample at 50 K, and show the result in Fig. 4. The satellite spectra for H tilted from the a_T axis can be reproduced by assuming not the three Gaussians due to one group of the As00, As01, and As02 spectra but the four Gaussians coming from a set of two As00, one As01 and one As02 spectra with the intensity ratio of As00:As00:As01:As02=8.5:8.5:12:4 following the binominal distribution for x~0.08. The angular dependences of the Gaussians coming from the As00 site show the twofold anisotropy with the orthogonal symmetry axes, a_O or b_O, parallel to the Fe-Fe direction as denoted by the red and blue spectra in Fig. 4. This means that the As00 sites are located in two sample regions with the local symmetry axes orthogonal to each other, clearly showing the presence of the in-plane anisotropy even in the tetragonal phase. The difference between the NMR frequencies of the red

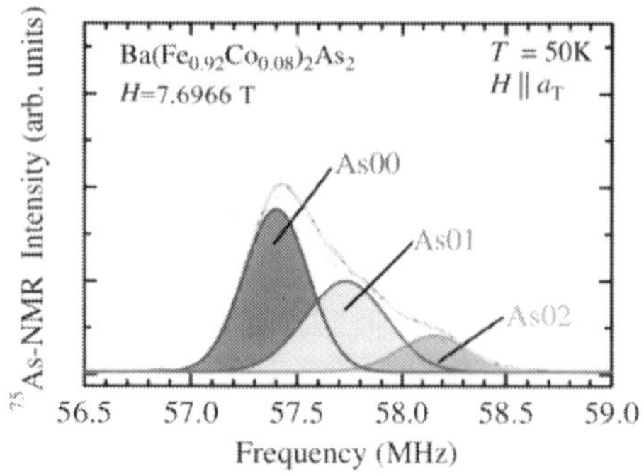

Fig.3 ^{75}As satellite spectrum at 50 K in $Ba(Fe_{0.92}Co_{0.08})_2As_2$ with H (=7.6966 T) parallel to the a_T axis in the FeAs plane. The spectrum is composed of the three Gaussians coming from the As00, As01, and As02 sites.

and blue satellite spectra parallel to the a_O and b_O axes provides the ν_Q^{ao} and ν_Q^{bo} values as 0.98 and 1.24 MHz, respectively. Using ν_Q^{ao} and ν_Q^{bo}, the in-plane anisotropy can be described by η defined as $\eta = |(\nu_Q^{ao} - \nu_Q^{bo})/\nu_Q^{c}|$ with $\nu_Q^{ao} + \nu_Q^{bo} + \nu_Q^{c} = 0$. Figure 5 shows the temperature dependence of η at the As00 site in the x~0.08 sample. It should be noted that the η value of 0.08-0.10 for x~0.08 is ~1/10 times as large as the value of η~1.2 at ~5 K in the orthorhombic phase of the x=0 sample [17]. Also the η value for x~0.08 without structural and AFM transitions is comparable to that for x~0.02 with T_s~100 K and T_N~96 K. The non-zero η value in the wide temperature range above T_s of the non-SC sample and above T_c of the SC one gradually decreases with increasing T.

Thus we observed the in-plane anisotropy of ν_Q at the As00 site in the SC sample with x~0.08 in addition to the non-SC one with x~0.02. The in-plane anisotropy, which indicates that the electronic state and/or the crystal structure are locally distorted with the orthorhombic symmetrical axes in the tetragonal phase of $Ba(Fe_{1-x}Co_x)_2As_2$, is induced by Co impurities introduced in the Fe plane [19]. Also the emergence of the in-plane anisotropy around room temperature is consistent with this picture.

4. Conclusion

We have made ^{75}As NMR measurements to study the in-plane anisotropy in the tetragonal phase of $Ba(Fe_{1-x}Co_x)_2As_2$ of x~0.08 with the superconducting transition temperature of T_c~23 K. We obtained the asymmetry parameter of the EFG tensor, η, at the As00 site with no Co atoms at the nearest and next nearest Fe sites, clearly showing the presence of the electric in-plane anisotropy above T_c. The

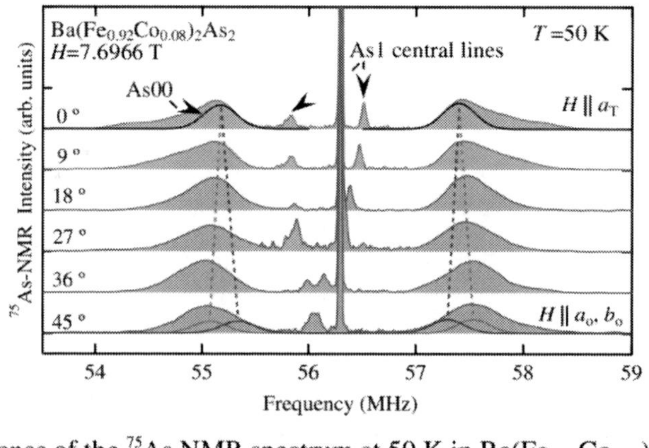

Fig. 4 Angular dependence of the ^{75}As NMR spectrum at 50 K in $Ba(Fe_{0.92}Co_{0.08})_2As_2$ with H (=7.6966 T) rotated in the FeAs plane. The red and blue spectra, whose angular dependences are presented by the dashed red and blue curves, respectively, come from the As00 sites in the two sample regions with the symmetry axes orthogonal to each other. The As1 spectra are central lines coming from the As1 site.

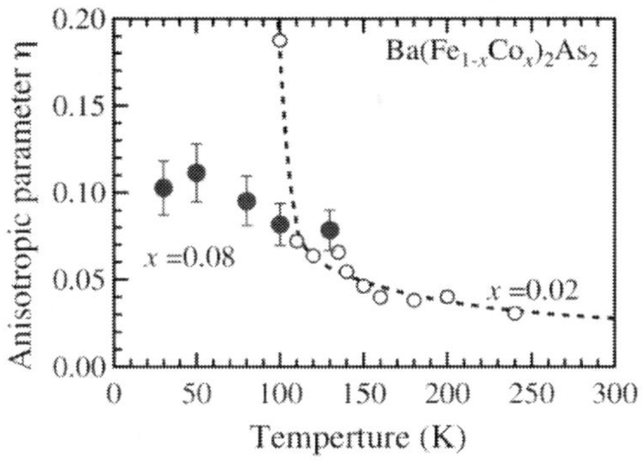

Fig.5 Temperature dependence of the asymmetry parameter of the electric field gradient tensor, η, at the As00 site in $Ba(Fe_{1-x}Co_x)_2As_2$ with x ~0.02 and 0.08.

EFG was found to have the η value of 0.08-0.10 comparable to those in the non-SC samples of x~0.02. Also the in-plane anisotropy was found to remain even near T_c.

This study was supported by the Grant-in-Aid (No. 2434080) for Scientific Research from the Japan Society of the Promotion of Science.

References

[1] Mazin I I, Singh D J, Johannes M D, and Du M H 2008 *Phys. Rev. Lett.* **101** 057003
[2] Kuroki K, Onari S, Arita R, Usui H, Tanaka Y, Kontani H, and Aoki H 2008 *Phys. Rev. Lett.* **101** 087004
[3] A. Kawabata, S. C. Lee, T. Moyoshi, Y. Kobayashi, and M. Sato 2008 *J. Phys. Soc. Jpn.* **77** 103704
[4] Sato M, Kobayashi Y, Lee S C, Takahashi H, Satomi E and Miura Y 2010 *J. Phys. Soc. Jpn.* **79** 014710
[5] Kobayashi Y, Iida T, Suzuki K, Kawamata T, Itoh M and Sato M 2014 *J. Phys. Soc. Jpn.* **83** 014712
[6] Nakajima Y, Taen T, Tsuchiya Y, Tamegai T, Kitamura H and Murakami T 2010 *Phys. Rev.* B **82** 220504(R)
[7] Onari S and Kontani H 2012 *Phys. Rev. Lett.* **109** 137001
[8] Kontani H, Saito T, Onari S 2011 *Phys. Rev.* B **84** 024528
[9] Onari S and Kontani H 2009 *Phys. Rev. Lett.* **103**, 177001
[10] Kontani H and Yamakawa Y 2014 *Phys. Rev. Lett.* **113**, 047001.
[11] Chu J -H, Analytis J G, Greve K D, McMahon P L, Islam Z, Yamamoto Y, Fisher I R 2010 *Science* **329** 824
[12] Fernandes R M, VanBebber L H, Bhattacharya, P. Chandra P, Keppens V, Mandrus D, McGuire M A, Sales B C, Sefat A S, and Schmalian J 2011 *Phys. Rev. Lett.* **105** 157003
[13] Kobayashi Y, Ichikawa A, Toyoda M, Itoh M, and Sato M 2013 *J. Korean Phys. Soc.* **63** 481
[14] Ni N, Tillman M E, Yan J -Q, Kracher A, Hannahs S H, Bud'ko S L, and Canfield P C 2008 *Phys. Rev.* B **78** 214515
[15] Goto T, Kurihara R, Araki K, Mitsumoto K, Akatsu M, Nemoto Y, Tatematsu S, Sato M 2011 *J. Phys. Soc. Jpn.* **80** 073702
[16] Yoshizawa M, Kimura D, Chiba T, Simayi S, Nakanishi Y, Kihou K, Lee C -H, Iyo A, Eisaki H, Nakajima M, and Uchida S 2012 *J. Phys. Soc. Jpn.* **81** 024604
[17] Kitagawa K, Katayama N, Ohgushi K, Yoshida M, Takigawa M 2008 *J. Phys. Soc. Jpn.* 77 114709
[18] Ning F L, Ahilan K, Imai T, Sefat A S, McGuire M A, Sales B C, Mandrus D, Cheng P, Shen B, and Wen H –H 2010 *Phys. Rev. Lett.* **104** 037001
[19] Inoue Y, Yamakawa Y, and Kontani H 2012 *Phys. Rev.* B **85** 224506

Journal of Physics: Conference Series **568** (2014) 022030 doi:10.1088/1742-6596/568/2/022030

Robustness against non-magnetic impurities in topological superconductors

Y Nagai, Y Ota and M Machida

CCSE, Japan Atomic Energy Agency, 5-1-5 Kashiwanoha, Kashiwa, Chiba 277-8587, Japan

E-mail: nagai.yuki@jaea.go.jp

Abstract. We study the robustness against non-magnetic impurities in a three-dimensional topological superconductor, focusing on an effective model (massive Dirac Bogoliubov-de Gennes (BdG) Hamiltonian with s-wave on-site pairing) of $Cu_xBi_2Se_3$ with the parameter set determined by the first-principles calculation. With the use of the self-consistent T-matrix approximation for impurity scattering, we discuss the impurity-concentration dependence of the zero-energy density of states. We show that a single material variable, measuring relativistic effects in the Dirac-BdG Hamiltonian, well characterizes the numerical results. In the nonrelativistic limit, the odd-parity fully-gapped topological superconductivity is fragile against non-magnetic impurities, since this superconductivity can be mapped onto the p-wave superconductivity. On the other hand, in the ultrarelativistic limit, the superconductivity is robust against the non-magnetic impurities, since the effective model has the s-wave superconductivity. We derive the effective Hamiltonian in the both limit.

1. Introduction

The discovery of topological insulators triggers off a number of the studies on topological aspects of matter [1]. Specifically, the interplay of topology and superconductivity leads to intriguing issues in low-temperature physics. One of the key topics is to seek a bulk topological superconductor. A strong candidate for this superconductor is $Cu_xBi_2Se_3$ [2]; the copper intercalation into a topological insulator Bi_2Se_3 induces superconductivity. Thus, identifying the characters of $Cu_xBi_2Se_3$ is highly desirable now.

An advantageous way of clarifying superconducting properties is to ask whether a superconductor is conventional or not. A distinct feature of a topological superconductor shows up on the boundaries with different topological states (e.g., vacuum); the closing of the gap at interface indicates the presence of gapless edge modes [1]. An intrinsic feature of unconventional superconductivity (e.g., p-wave) is the presence of gapless states around non-magnetic impurities, owing to the violation of Anderson's theorem. In specific models for topological superconductivity [3, 4], non-magnetic impurities induce the gapless states in the quasiparticle spectrum; one may claim that the topological superconductor is unconventional.

In this paper, we study the robustness of a topological superconductor against non-magnetic impurities, focusing on a set of the material variables for $Cu_xBi_2Se_3$. Our approach is divided into two parts. First, we build up an effective theory to intuitively understand our system. Second, we examine the presence of gapless excitations, by numerical calculations with a self-consistent T-matrix approach. We show that the system has two aspects, i.e., both conventional and unconventional properties, depending on the amount of a relativistic effect in the normal Hamiltonian. In other words, an analogy between $Cu_xBi_2Se_3$ and an unconventional superconductor is dominated by the mass-to-momentum ratio in the Dirac-type Hamiltonian.

2. Model and formulation

2.1. Dirac Bogoliubov-de Gennes Hamiltonian

The Dirac Bogoliubov-de Gennes (BdG) Hamiltonian matrix [2, 3, 5] is

$$\mathcal{H}(\boldsymbol{k}) = \begin{pmatrix} h_0(\boldsymbol{k}) & \Delta_{\rm pair}(\boldsymbol{k}) \\ \Delta^{\dagger}_{\rm pair}(\boldsymbol{k}) & -h_0^*(-\boldsymbol{k}) \end{pmatrix}, \tag{1}$$

with the normal-state Hamiltonian $h_0(\boldsymbol{k}) = \varepsilon(\boldsymbol{k}) + d_0(\boldsymbol{k})\gamma^0 + \sum_{i=1}^{3} d_i(\boldsymbol{k})\gamma^0\gamma^i$ and the 4×4 pairing potential matrix $\Delta_{\rm pair}$. The 4×4 complex matrices γ^{μ} ($\mu = 0, 1, 2, 3$) are the Gamma matrices in the Dirac basis. Using the orbital 2×2 Pauli matrices σ^i and the spin 2×2 Pauli matrices s^i, we have $\gamma^0 = \sigma^3 \otimes 1_2$ and $\gamma^i = i\sigma^2 \otimes s^i$. In the diagonal block of h_0, we have $\varepsilon = -\mu + \bar{D}_1\epsilon_c(\boldsymbol{k}) + (4/3)\bar{D}_2\epsilon_{\perp}(\boldsymbol{k})$ and $d_0 = M_0 - \bar{B}_1\epsilon_c(\boldsymbol{k}) - (4/3)\bar{B}_2\epsilon_{\perp}(\boldsymbol{k})$, with $\epsilon_c = 2 - 2\cos(k_z)$, $\epsilon_{\perp} = 3 - 2\cos(\sqrt{3}k_x/2)\cos(k_y/2) - \cos(k_y)$. The off-diagonal block has contributions from the spin-orbit couplings, $d_1 = (2/3)\bar{A}_2\sqrt{3}\sin(\sqrt{3}k_x/2)\cos(k_y/2)$, $d_2 = (2/3)\bar{A}_2[\cos(\sqrt{3}k_x/2)\sin(k_y/2) + \sin(k_y)]$, and $d_3 = \bar{A}_1\sin(k_z)$. The material variables $\bar{D}_1$, $\bar{D}_2$, $\bar{B}_1$, $\bar{B}_2$, $\bar{A}_2$, $\bar{A}_1$ are determined by the data from the first-principle calculations in [9]. The other material variable M_0 and the chemical potential μ are changed, since they are related to a doping level. In this paper, we focus on a pseudo-scalar-type order parameter in the superconducting part,

$$\Delta_{\rm pair} i\gamma^2 = i\gamma^0 f^{\rm ps}. \tag{2}$$

This pairing potential indicates the topological fully-gapped s-wave superconductivity [3]. We mention that a scalar-type gap function (a non-topological fully-gapped s-wave state) is robust against non-magnetic impurities [3]. This behavior is consistent with Anderson's theorem for s-wave superconductivity.

2.2. Indicator of relativistic effects

We show the indicator of relativistic effects [3]. The linearized normal-state Hamiltonian is

$$h_0(\boldsymbol{k}) \sim \gamma^0 \left[M_0 + \bar{A}_2(k_x\gamma^1 + k_y\gamma^2) + \bar{A}_1 k_z\gamma^3\right]. \tag{3}$$

We define the indicator of the relativistic effects as

$$\beta = \frac{\bar{A}_2 k_{\rm F}}{|M_0|} = \sqrt{\left(\frac{\mu}{M_0}\right)^2 - 1}. \tag{4}$$

The Dirac-BdG Hamiltonian has two distinct behaviors, depending on β. In the limit of $\beta \to 0$, we have a nonrelativistic limit (i.e. large-mass limit). In the limit of $\beta \to \infty$, we have an ultrarelativistic limit (i.e. massless limit).

2.3. Self-consistent T-matrix approximation

We consider the non-magnetic impurity effects with the use of the self-consistent T-matrix approximation. We assume that the impurities are randomly distributed, leading to the impurity potential

$$V(\boldsymbol{r}) = \sum_i \delta(\boldsymbol{r} - \boldsymbol{r}_i)V^{\rm NM}, \quad V^{\rm NM} = V_0 \begin{pmatrix} 1_{2\times 2} & 0 \\ 0 & -1_{2\times 2} \end{pmatrix}. \tag{5}$$

The T-matrix is written as

$$T(\Omega) = \left[1 - V^{\rm NM}\frac{1}{N}\sum_{\boldsymbol{k}} G_{\boldsymbol{k}}(\Omega)\right]^{-1} V^{\rm NM}, \tag{6}$$

where N is the number of meshes in momentum space and $G_{\boldsymbol{k}}(\Omega)$ is the Green's function with self-energy $\Sigma(\Omega)$. We have

$$G_{\boldsymbol{k}}(\Omega) = \frac{1}{\Omega - \mathcal{H}(\boldsymbol{k}) - \Sigma(\Omega)} = \begin{pmatrix} g_{\boldsymbol{k}}(\Omega) & f_{\boldsymbol{k}}(\Omega) \\ f^{\dagger}_{\boldsymbol{k}}(\Omega) & \bar{g}_{\boldsymbol{k}}(\Omega) \end{pmatrix}, \quad \Sigma(\Omega) = n_{\rm imp}T(\Omega) - n_{\rm imp}V^{\rm NM}, \tag{7}$$

with the impurity concentration $n_{\rm imp}$. Solving 6 and 7 self-consistently, we obtain the density of states (DOS) as

$$N(E) = -\frac{1}{2\pi}\frac{1}{N}\sum_{\boldsymbol{k}} {\rm tr}[{\rm Im} \lim_{\eta\to 0} g_{\boldsymbol{k}}(E + i\eta)]. \tag{8}$$

Similarly, we obtain the DOS of the normal state $N_{\rm normal}(E)$, setting $f^{\rm ps} = 0$. Note that the second term in the self energy in 7 corresponds to the renormalization of the chemical potential μ, since in our calculations μ is fixed.

3. Effective Hamiltonian

Let us examine an effective Hamiltonian, depending on β. First, we focus on the nonrelativistic limit ($\beta \to 0$). Using the Dirac basis with respect to the Gamma matrices, the eigenvalue equations of the Dirac-BdG Hamiltonian are

$$\begin{pmatrix} \mu' & \boldsymbol{k}'\cdot\boldsymbol{s} & 0 & \Delta^{12} \\ \boldsymbol{k}'\cdot\boldsymbol{s} & -2M_0 - \mu' & \Delta^{21} & 0 \\ 0 & \Delta^{21\dagger} & \mu' & \boldsymbol{k}'\cdot\boldsymbol{s}^* \\ \Delta^{12\dagger} & 0 & \boldsymbol{k}'\cdot\boldsymbol{s}^* & 2M_0 + \mu' \end{pmatrix} \begin{pmatrix} \boldsymbol{F} \\ \boldsymbol{G} \\ \boldsymbol{F}' \\ \boldsymbol{G}' \end{pmatrix} = E \begin{pmatrix} \boldsymbol{F} \\ \boldsymbol{G} \\ \boldsymbol{F}' \\ \boldsymbol{G}' \end{pmatrix}, \tag{9}$$

with $\boldsymbol{k}' = (\bar{A}_2 k_x, \bar{A}_2 k_y, \bar{A}_1 k_z)$ and the effective chemical potential $\mu' \equiv \mu - M_0$. The indicator β is zero when μ' is zero. The "large-component" ($\boldsymbol{F}$ and $\boldsymbol{F}'$) equations are

$$\mu'\boldsymbol{F} + \boldsymbol{k}'\cdot\boldsymbol{s}\boldsymbol{G} + \Delta^{12}\boldsymbol{G}' = E\boldsymbol{F}, \tag{10}$$
$$\mu'\boldsymbol{F}' + \boldsymbol{k}'\cdot\boldsymbol{s}^*\boldsymbol{G}' + \Delta^{21\dagger}\boldsymbol{G} = E\boldsymbol{F}', \tag{11}$$

whereas the "small-component" ($\boldsymbol{G}$ and $\boldsymbol{G}'$) equations are

$$\boldsymbol{G} = \frac{\boldsymbol{k}'\cdot\boldsymbol{s}}{E + 2M_0 + \mu'}\boldsymbol{F} + \frac{\Delta^{21}}{E + 2M_0 + \mu'}\boldsymbol{F}', \tag{12}$$
$$\boldsymbol{G}' = \frac{\boldsymbol{k}'\cdot\boldsymbol{s}^*}{E - 2M_0 - \mu'}\boldsymbol{F}' + \frac{\Delta^{21}}{E - 2M_0 - \mu'}\boldsymbol{F}. \tag{13}$$

Substituting the small components ($\boldsymbol{G}$ and $\boldsymbol{G}'$) into 10 and 11, we obtain the equations composed only of the large components. Imposing the conditions that $E/|M_0| \ll 1$, $\mu'/|M_0| \ll 1$, and $f^{\rm ps}/|M_0| \ll 1$, we find that

$$\begin{pmatrix} \frac{\boldsymbol{k}'^2}{2M_0} - \mu' - \frac{|f^{\rm ps}|^2}{2M_0} & \frac{f^{\rm ps}}{M_0}\boldsymbol{k}'\cdot\boldsymbol{s}(is^2) \\ \left[\frac{f^{\rm ps}}{M_0}\boldsymbol{k}'\cdot\boldsymbol{s}(is^2)\right]^{\dagger} & -\frac{\boldsymbol{k}'^2}{2M_0} + \mu' + \frac{|f^{\rm ps}|^2}{2M_0} \end{pmatrix} \begin{pmatrix} \boldsymbol{F} \\ \boldsymbol{F}' \end{pmatrix} = E \begin{pmatrix} \boldsymbol{F} \\ \boldsymbol{F}' \end{pmatrix}. \tag{14}$$

Journal of Physics: Conference Series **568** (2014) 022030 doi:10.1088/1742-6596/568/2/022030

These effective equations indicate that the spin-orbit coupling term makes the effective gap function anisotropic. This p-wave gap function is equal to that of the B-phase in superfluid ^{3}He [6, 7, 8]. Since Anderson's theorem is not valid in the p-wave superconductivity, the topological fully-gapped s-wave superconductor in the nonrelativistic limit is not robust against non-magnetic impurities. The equivalence to the B-phase also indicates that the topological nature in this limit originates from the property of the Cooper-pair wavefunction.

Next, we focus on the ultrarelativistic limit ($\beta \to \infty$). The Weyl basis with respect to the Gamma matrices is convenient. We use the over-line symbol to specify the quantities in the Weyl basis. The linearized Dirac BdG equations is

$$\begin{pmatrix} -\boldsymbol{k}'\cdot\boldsymbol{s}-\mu' & M_0 & \bar{\Delta}^{11} & 0 \\ M_0 & \boldsymbol{k}'\cdot\boldsymbol{s}-\mu & 0 & \bar{\Delta}^{22} \\ \bar{\Delta}^{11\dagger} & 0 & -\boldsymbol{k}'\cdot\boldsymbol{s}^*+\mu' & -M_0 \\ 0 & \bar{\Delta}^{22\dagger} & -M_0 & \boldsymbol{k}'\cdot\boldsymbol{s}^*+\mu \end{pmatrix} \begin{pmatrix} \bar{\boldsymbol{F}} \\ \bar{\boldsymbol{G}} \\ \bar{\boldsymbol{F}}' \\ \bar{\boldsymbol{G}}' \end{pmatrix} = E \begin{pmatrix} \bar{\boldsymbol{F}} \\ \bar{\boldsymbol{G}} \\ \bar{\boldsymbol{F}}' \\ \bar{\boldsymbol{G}}' \end{pmatrix}, \tag{15}$$

with $\bar{\Delta}^{11} = -is^2 f^{\rm ps}$ and $\bar{\Delta}^{22} = is^2 f^{\rm ps}$. Taking $M_0 \to 0$, we find that the BdG equations are divided into a left-handed sector ($\bar{G}$ and $\bar{G}'$) and a right-handed sector ($\bar{F}$ and $\bar{F}'$),

$$\begin{pmatrix} \boldsymbol{k}'\cdot\boldsymbol{s}-\mu & is^2 f^{\rm ps} \\ -is^2 f^{\rm ps *} & \boldsymbol{k}'\cdot\boldsymbol{s}^*+\mu \end{pmatrix} \begin{pmatrix} \bar{\boldsymbol{G}} \\ \bar{\boldsymbol{G}}' \end{pmatrix} = E \begin{pmatrix} \bar{\boldsymbol{G}} \\ \bar{\boldsymbol{G}}' \end{pmatrix}. \tag{16}$$

$$\begin{pmatrix} -\boldsymbol{k}'\cdot\boldsymbol{s}-\mu & -is^2 f^{\rm ps} \\ is^2 f^{\rm ps *} & -\boldsymbol{k}'\cdot\boldsymbol{s}^*+\mu \end{pmatrix} \begin{pmatrix} \bar{\boldsymbol{F}} \\ \bar{\boldsymbol{F}}' \end{pmatrix} = E \begin{pmatrix} \bar{\boldsymbol{F}} \\ \bar{\boldsymbol{F}}' \end{pmatrix}. \tag{17}$$

When the chemical potential is large and positive, one can only consider the left-handed sector ($\bar{G}$ and $\bar{G}'$) to discuss the superconducting properties, since the Fermi wave-length $k_{\rm F} = \mu$ in normal states characterizes the quasiparticle excitations. In both of the sectors, the formulae are obtained at Γ point ($\boldsymbol{k} = 0$). The effective Hamiltonian is equivalent to in an s-wave superconductor, except $\boldsymbol{k} = 0$. This result indicates that the topological fully-gapped s-wave superconductor in the ultra-relativistic limit is robust against non-magnetic impurities, owing to Anderson's theorem. We note that the difference between the topological and non-topological fully-gapped s-wave superconductors occurs at zero-momentum, where the quasiparticles with zero-momentum feel the sign change of the gap functions between the left- and right-handed sectors and induce surface bound states in the topological superconductors.

4. Numerical calculations of gapless excitations

Now, we show the numerical results. The $\boldsymbol{k}$-mesh size is $256 \times 256 \times 256$. The unitary-like scattering model with $V_0 = 10$eV is adopted. The gap amplitude is $f^{\rm ps} = 0.1$eV. The smearing factor for the DOS is $\eta = 0.5$meV. The material parameters are set by $(\bar{D}_1, \bar{D}_2, \bar{B}_1, \bar{B}_2, \bar{A}_1, \bar{A}_2) = (0.0024, 19.6/a^2, 0.216, 56.6/a^2, 0.32, 4.1/a)$, in unit of eV, with $a = 4.076$. These parameters are determined by the data from the first-principles calculations in [9]. We remark that they are different from in [2, 3]. We calculate the zero-energy DOS $N(E=0)$, changing the impurity concentration $n_{\rm imp}$. The zero-energy DOS of the normal state, $N_{\rm normal}(E=0)$ is also calculated, by setting $f^{\rm ps} = 0$.

Figure 1 shows that β characterizes well the robustness against the non-magnetic impurities. In the nonrelativistic region ($\beta = 0.88$ in figure 1), the non-magnetic impurities easily induce the zero-energy DOS. The zero-energy DOS drastically increases above the specific impurity concentration, which depends on β. We note that the $T_{\rm c}$ robustness against impurities depends on the point where the zero-energy DOS becomes finite. As shown in Fig. 1, the robustness monotonically increases with increasing β. These results indicate that the system has two aspects, p- and s- wave features, depending on the amount of relativistic effects.

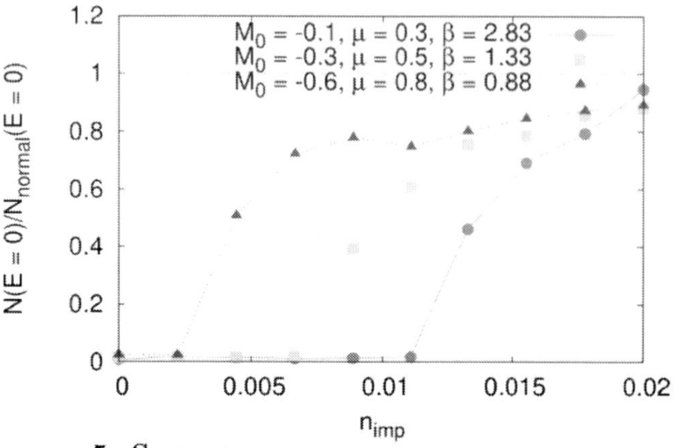

Figure 1. Impurity-concentration dependence of zero-energy density of states, for a topological s-wave superconductor with gap amplitude $\Delta = 0.1$eV. We change an indicator of relativistic effects , $\beta = \sqrt{(\mu/M_0)^2 - 1}$, where M_0 corresponds to the mass and μ is the chemical potential. Other material variables are determined by the data from the first-principles calculations in [9].

5. Summary

In conclusion, we studied the robustness against the non-magnetic impurities, in a bulk topological s-wave superconductor. The calculations were performed for the data from the first-principles calculations of $Cu_xBi_2Se_3$ [9]. We found that the results are summarized well by a single variable β, characterizing relativistic effects in the normal Hamiltonian. In the nonrelativistic region, the full-gapped topological s-wave superconductor is not robust against the non-magnetic impurities. In contrast, in the ultrarelativistic region, this superconducting state is robust against non-magnetic impurities. Consequently, the present system has two aspects, p- and s- wave features, with a typical set of the material variables of $Cu_xBi_2Se_3$.

Acknowledgments

We thank H. Nakamura and K. Kobayashi for helpful discussions and comments. The calculations were performed using the supercomputing system PRIMERGY BX900 at the Japan Atomic Energy Agency. This study has been partially supported by JSPS KAKENHI Grant Number 24340079, 26800197.

References

[1] Hasan M Z and C L Kane 2010 *Rev. Mod. Phys.* **82** 3045, and references therein
[2] Sasaki S *et al* 2011 *Phys. Rev. Lett.* **107** 217001
[3] Nagai Y, Ota Y and Machida M 2014 *Phys. Rev.* B **89** 214506
[4] Nagai Y, Ota Y and Machida M *Preprint* arXiv:1312.3065
[5] Hao L and Lee T K 2011 *Phys. Rev.* B **83** 134516
[6] Legett A 1975 *Rev. Mod. Phys.* **47** 331
[7] Murakawa S *et al* 2011 *J. Phys. Soc. Jpn.* **80** 013602
[8] Nagai Y, Nakamura H and Machida M 2014 *J. Phys. Soc. Jpn.* **83** 053705
[9] Zhang H *et al* 2009 *Nature Phys.* **5** 438

Journal of Physics: Conference Series **568** (2014) 022031 doi:10.1088/1742-6596/568/2/022031

Superconducting transition of Ru in SQUIDs with Nb/Ru/Sr_2RuO_4 junctions

Y. Nago[1], R. Ishiguro[2], T. Sakurai[1], M. Yakabe[1], T. Nakamura[3], S. Yonezawa[4], S. Kashiwaya[5], H. Takayanagi[1,6] and Y. Maeno[4]

1 Department of Applied Physics, Faculty of Science, Tokyo University of Science, Tokyo, Japan
2 Center for Emergent Matter Science, RIKEN, Wako, Saitama, Japan
3 Institute of Solid State Physics (ISSP), University of Tokyo, Kashiwa, Chiba, Japan
4 Department of Physics, Graduate School of Science, Kyoto University, Kyoto, Japan
5 National Institute of Advanced Industrial Science and Technology (AIST), Tsukuba, Ibaraki, Japan
6 International Center for Materials Nanoarchitectonics (MANA), National Institute for Materials Science (NIMS), Tsukuba, Ibaraki, Japan

E-mail: `yusukenago@rs.tus.ac.jp`

Abstract. We have measured transport properties and magnetic responses of direct current superconducting quantum interference devices(dc-SQUIDs) with Nb/Ru/Sr_2RuO_4 junctions on the same Ru inclusion of a Sr_2RuO_4-Ru eutectic crystal. We observed a superconducting transition of the Ru at 0.50 K and another transition with anomalous interference patterns of the SQUID at 0.40 K. The second transition led to change of the SQUID structure due to energy competition of alternative supercurrent paths. Phase shifts of the peak of the Fraunhofer diffraction pattern were observed below the transition temperatures, implying topological effects due to a chiral p-wave state or its competition with a s-wave state.

1. Introduction

Layered perovskite Sr_2RuO_4 is one of strong candidates of a spin-triplet chiral p-wave superconductor [1–3]. In recent ten years, Ru-Sr_2RuO_4 eutectic crystal has attracted research interest because it exhibits an inhomogeneous superconducting phase at 3 K (3-K phase), different from the bulk phase which appears at 1.5 K [4]. A theoretical research [5] shows that a 3-K phase nucleates as a state conserving time reversal symmetry at the interface with Ru inclusion and evolves into a state breaking time reversal symmetry with decreasing temperature. A number of transport measurements [6–9] revealed novel properties of 3-K phase around Ru inclusions, following the theoretical prediction. However the generation mechanism of this enhanced superconductivity has not yet been completely identified [10].

The eutectic crystal is also useful for investigation of a competition between a s-wave state of Ru and a p-wave state of Sr_2RuO_4 at sufficiently low temperatures below the superconducting transition temperature T_c^{Ru} of Ru inclusion (0.49 K); it is helpful to understand a superconducting state of a 3-K phase as well as a bulk phase [11]. A topological mismatch of s- and p-wave phases in the eutectic crystal is expected to cause Josephson vortex nucleation in Ru inclusion [12]: a frustrating coupling of the order parameters of both states induces

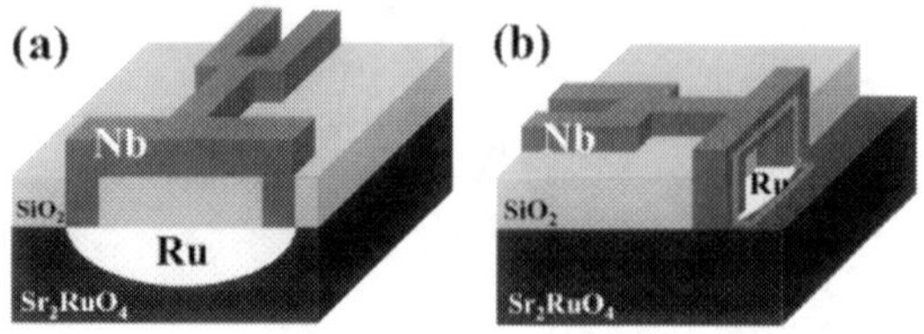

Figure 1. Configuration of a dc-SQUID with $Nb/Ru/Sr_2RuO_4$ junctions: front view (a) and side view (b). Electrodes consist of niobium. Two junctions with a distance of about 20 μm are located on the same Ru inclusion. The red line indicates one expected SQUID loop.

a spontaneous magnetic flux distribution at the interface of Ru inclusion, growing up with decreasing temperature to form a single vortex state at the Ru center.

A topological superconducting junction can be easily fabricated using the eutectic crystal and a s-wave superconductor [8,9]. We have so far fabricated dc-SQUIDs with $Nb/Ru/Sr_2RuO_4$ junctions using Ru-Sr_2RuO_4 eutectic crystal [13] and measured its transport properties [14]. In this study, a dc-SQUID whose two junctions are composed on the same Ru inclusion was prepared. We measured a superconducting transition of the Ru inclusion and magnetic responses of the SQUIDs below the transition in order to investigate the effects due to a topological mismatch of s- and p-wave phases. We discuss anomalous interference patterns of the SQUIDs observed below the transition.

2. Experimental

The dc-SQUIDs consist of Nb, Sr_2RuO_4 and two $Nb/Ru/Sr_2RuO_4$ junctions on the ab-plane surface of a Sr_2RuO_4-Ru eutectic crystal, cut from a crystalline rod grown by a floating-zone method [4]. This device was fabricated by the same technique as used in previous studies [13,14]: the eutectic crystal is coated with an SiO_2 layer of 300-nm thickness and two small windows are made on the small parts above Ru inclusion for contacting a niobium electrode using a photo lithography and a reactive ion etching. The structure of the device is similar to one in previous studies [13, 14], except for the point that two junctions are located on each edge of the same Ru inclusion as shown in Fig. 1. A size of the Ru inclusion is about 1 μm $\times$ 20 μm and hence the distance of junctions is about 20 μm. Above the Ru superconducting transition T_c^{Ru}, a SQUID loop should be formed so that a current flows on Sr_2RuO_4 in the vicinity of the Ru-Sr_2RuO_4 interface, for example, as indicated by a red line in Fig. 1 (b). Therefore it is expected that the effective SQUID-loop area consists of SiO_2 cover area surrounded by the Nb bridge, some part of the Ru inclusion area and the penetration depth λ of a magnetic flux into Nb and Sr_2RuO_4 which is order of 100 nm at sufficiently low temperatures below 1 K, as seen in Fig. 2 (a). Below T_c^{Ru}, on the other hand, the junctions change from S(s-wave)-N-S(p-wave) to S(s-wave)-S(s-wave)-S(p-wave) and a supercurrent is expected to flow on the whole Ru inclusion, forming a superconducting ring with Nb/Ru(s-s) junctions and a single Ru/Sr_2RuO_4(s-p) junction, as seen in Fig. 2 (b).

For four-terminal measurements, two of the four Nb electrodes with a thickness of 600 nm are directly linked to the SQUID line as seen in Fig. 1 and the others deposited on the other Ru inclusions (see Ref. [13, 14]). The device was placed in a dilution refrigerator and cooled below 1 K. We measured IV characteristics as shown in Fig. 3 and estimated the Josephson maximum current I_c as a function of external magnetic field at each temperature. Since I_c depending on a superconducting gap $\Delta(T)$ increases up to order of mA at sufficiently low temperatures as seen in Fig. 3, a considerably large current needs to be driven. In that case, heat generated by the current at the junctions and electrical wires made of a normal metal near the device is an inevitable problem; we actually performed continuous measurements of IV characteristics at each temperature and found a gradual decrease of I_c due to heating despite the temperature is

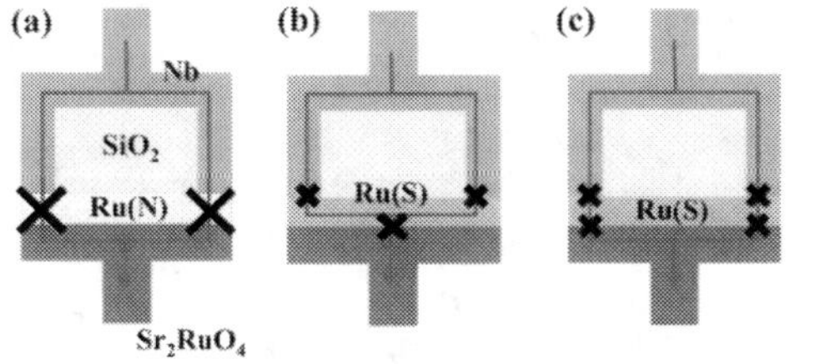

Figure 2. Schematics of flowing path of a current on the SQUID. (a) above the superconducting transition temperature of T_c^{Ru}. (b) below T_c^{Ru}, the case in which a path on superconducting Ru is more favorable. (c) below T_c^{Ru}, the case in which a path through two junctions is more favorable. Crosses indicate superconducting junction.

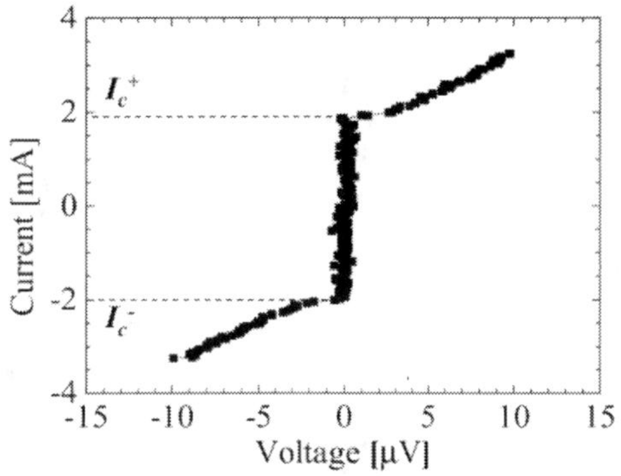

Figure 3. Current-Voltage characteristics of the SQUID at 0.55 K and zero magnetic field. Josephson maximum current I_c is determined alternatively as the current indicated in I_c^+ and I_c^- where a finite voltage suddenly appears in each current direction.

set constant. Therefore we verified effects of heating with continuous measurements of I_c and finally fixed measurement intervals of 5 min - 15 min, depending on temperature.

3. Results and discussion

Figure 4 shows the Josephson maximum current I_c^+ estimated from IV characteristics (Fig. 3) as a function of external magnetic field B at 0.55 K. Periodic patterns can be seen in a Fraunhofer diffraction pattern and its period ΔB_0 is approximated to be 0.1 mT, which corresponds to a flux quantum Φ_0 devided by the effective SQUID-loop area. Though the precise depth of the Ru inclusion is unidentified, it is speculated that only a part of the Ru near its surface contributes to the effective loop area shown in Fig. 2 (a) because the Ru is surrounded by Sr_2RuO_4 as shown in Fig. 1. Therefore, assuming λ of Nb and Sr_2RuO_4 to be 100 nm, the effective loop area is approximated to be $\sim 0.5\ \mu m \times 20\ \mu m$, consistent with the observed period ΔB_0.

The field at the maximum of I_c^+ shifts in the negative direction from $B = 0$ mT as seen in Fig. 4. We found that the field at the minimum of I_c^- in Fig. 3 is also shifted by almost the same amount of field in the positive(opposite-sign) direction. These results indicate that a self-induced flux by a bias current causes this peak shift.

Figure 5 shows temperature dependence of SQUID diffraction patterns at 0.55 K - 0.35 K. Josephson maximum current I_c^+ gradually increases with decreasing temperature, due to a temperature dependence of energy gap of both superconductors at the junction. A period ΔB_0 of the SQUID pattern also shows a very weak temperature dependence indicating gradual decrease of the effective SQUID-loop area which magnetic fluxes thread through. This can be attributed to a temperature dependence of a penetration depth λ. However we found the excess increase of ΔB_0 below 0.50 K, whose increase rate is approximately 30 % from 0.50 K to 0.40 K. When Ru undergoes a superconducting transition, the Meissner screening at the superconducting Ru area reduces the effective SQUID area. Therefore the large increase of a period evidently indicates that a superconducting transition of the Ru inclusion occurs at about 0.50 K.

A junction structure changes from S-N-S to S-S-S at T_c^{Ru} as mentioned in the section 2 and the system is likely to form series circuits with a superconducting ring with Nb/Ru(s-s) junctions and a single Ru/Sr_2RuO_4(s-p) junction as seen in Fig. 2 (b). It is not clear of which junction I_c

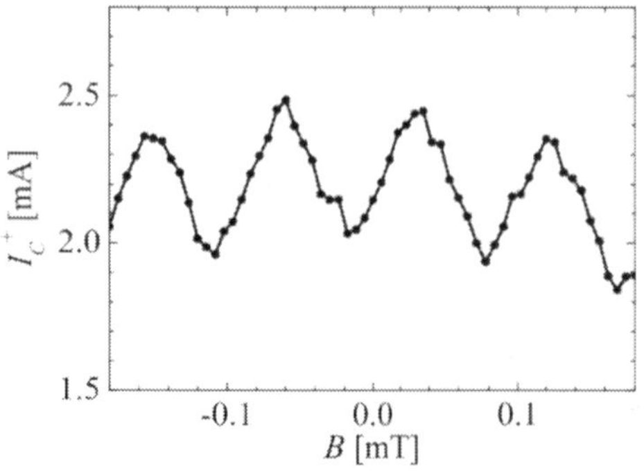

Figure 4. Josephson maximum current I_c^+ as a function of external magnetic field at 0.55 K.

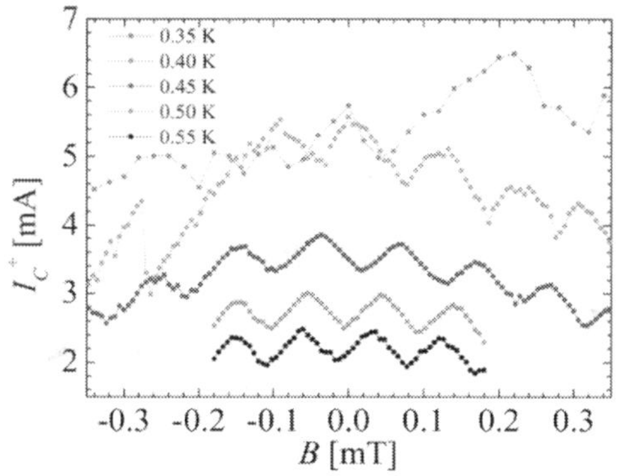

Figure 5. Josephson maximum current I_c^+ as a function of external magnetic field at 0.55 K - 0.35 K. The periodic pattern collapses below 0.40 K.

is smaller (more detectable) just below T_c^{Ru}; in either case, the proximity effect at the junction is expected to become weak compared with the case of the original S-N-S junction, because of a small gap of the Ru near T_c^{Ru}. If it is the case, I_c should decrease according to decrease of a tunneling rate through the junction. However I_c^+ seems to increase noticeably below $T_c^{Ru} =$ 0.50 K as seen in Fig. 5. This result suggests that a coupling of superconducting pair functions at the original Nb/Ru/Sr_2RuO_4 junction is still strong: a superconducting path through the Ru inclusion newly appears between two junctions at the Ru transition but, just below T_c^{Ru}, a critical current of this thin superconducting path on the Ru is too small to be energetically more favorable than the paths to bulk Sr_2RuO_4 through the two original junctions. This scenario means that the system forms a SQUID with Nb/Ru/Sr_2RuO_4(s-s-p) junctions at 0.50 K, *i.e.*, the SQUID structure changes from (a) not to (b) but to (c) shown in Fig. 2. A transition from S-N-S to S-S-S junction implies that a pair function of superconducting Ru strengthens a coupling between superconducting states of Nb and Sr_2RuO_4 and consequently I_c^+ increases, consistent with our present result.

A critical supercurrent of Ru inclusion increases with decreasing temperature and may exceed the Josephson maximum current at the Nb/Ru/Sr_2RuO_4 junctions at a certain temperature, resulting in change of the supercurrent path from Fig. 2 (c) to (b). One can find that the periodic SQUID pattern starts to collapse below about 0.40 K as seen in Fig. 5. This anomalous characteristics were not observed for the previous SQUID with two junctions connected on two different Ru inclusions [13, 14]. It is therefore plausible that the SQUID undergoes the second transition at which a SQUID structure changes into a Nb/Ru superconducting ring with a single Ru/Sr_2RuO_4 junction shown in Fig. 2 (b). Furthermore, the field value of the peak of the Fraunhofer pattern largely shifts in the positive direction below about 0.40 K as seen in Fig. 5. Such large shifts were never observed at temperatures higher than 0.40 K. Magnetic fluxes trapped in holes, defects or impurities in a superconductor cause a large Fraunhofer-peak shift [15]. In this case, I_c modulation would show hysteresis between up and down sweeps of applied field, and I_c^+ and I_c^- modulations obtained with each current direction would show the Fraunhofer peak of which the field shifts in the same-sign direction. In contrast, our results revealed no hysteresis and the negative shift of I_c^--modulation peak which is opposite to the case of I_c^+, indicating different mechanisms. The peak shift dependent on field direction implies effects of a magnetic flux induced by a bias current due to an asymmetric structure of the SQUID. The structure of our present sample seems almost symmetric in the case of Fig. 2 (c) but can become asymmetric in the case of Fig. 2 (b) if the supercurrent flows asymmetrically only through a particular part of the Ru/Sr_2RuO_4 junction. These pictures are consistent with our results that the large peak-shift was observed only below 0.40 K; in other words it can be an evidence that structure of the SQUID changed from (c) to (b) at 0.40 K. The asymmetric current flow at the Ru/Sr_2RuO_4(s-p) junction might be caused by a spontaneously generated current as a result of competition of s- and p-wave states, which possibly leads to a Josephson vortex nucleation [12]. However it should be noted that topological properties of Sr_2RuO_4 such as odd-parity pairing symmetry [16] or chiral domains [7, 9] can also cause phase shifts in the interference pattern. Further theoretical and experimental approaches are needed for clarification.

4. Conclusion

We fabricated dc-SQUID with Nb/Ru/Sr_2RuO_4 junctions on the same Ru inclusion of Ru-Sr_2RuO_4 eutectic crystals and observed two transitions in the magnetic interference patterns of the SQUID at very low temperatures. One corresponds to the superconducting transition of the Ru inclusion at 0.50 K, which changes the effective SQUID-loop area due to Meissner screening, resulting in an increase of a period of the interference patterns. It also strengthens a coupling of pairing states at the junction, resulting in an increase of a Josephson maximum current. The other is the collapse of the periodic pattern of the SQUID and its phase shifts at 0.40 K, where a supercrrent path changes due to the energy balance between a Josephson current through the junction and a critical supercurrent through the Ru inclusion. The anomalous phase shifts below 0.40 K raise a possibility of effects due to topological properties of a chiral p-wave state or topological mismatch between p- and s-wave phases.

Acknowledgments

We would like to thank S. Nomura from Univ. of Tsukuba for stimulating discussions. This work was supported by MEXT KAKENHI (Grants No. 22103002 and No. 25103721) and JSPS KAKENHI S (Grant No. 20221007). Device fabrication was supported by the NIMS Nanofabrication platform of the "Nanotechnology Platform Project" sponsored by MEXT.

References

[1] Maeno Y, Hashimoto H, Yoshida K, Nishizaki S, Fujita T, Bednorz J G and Lichtenberg F 1994 *Nature* **372** 532
[2] Mackenzie A P and Maeno Y 2003 *Rev. Mod. Phys.* **372** 657
[3] Maeno Y, Kittaka S, Nomura T, Yonezawa S and Ishida K 2012 *J. Phys. Soc. Jpn.* **81** 011009
[4] Maeno Y, Ando T, Mori Y, Ohmichi E, Ikeda S, Nishizaki S and Nakatsuji S 1998 *Phys. Rev. Lett.* **81** 3765
[5] Kaneyasu H, Hayashi N, Gut B, Makoshi K and Sigrist M 2010 *J. Phys. Soc. Jpn.* **79** 104705
[6] Hooper J, Mao Z Q, Nelson K D, Liu Y, Wada M and Maeno Y 2004 *Phys. Rev.* B **70** 014510
[7] Kambara H, Kashiwaya S, Yaguchi H, Asano Y, Tanaka Y and Maeno Y 2008 *Phys. Rev. Lett.* **101** 267003
[8] Nakamura T, Nakagawa R, Yamagishi T, Terashima T, Yonezawa S, Sigrist M and Maeno Y 2011 *Phys. Rev.* B **84** 060512
[9] Anwar M S, Nakamura T, Yonezawa S, Yakabe M, Ishiguro R, Takayanagi H and Maeno Y 2013 *Sci. Rep.* **3** 2480
[10] Kittaka S, Taniguchi H, Yonezawa S, Yaguchi H and Maeno Y 2010 *Phys. Rev.* B **81** 180510
[11] Ying Y A, Xin Y, Clouser B W, Hao E, Staley N E, Myers R J, Allard L F, Fobes D, Liu T, Mao Z Q and Liu Y 2009 *Phys. Rev. Lett.* **103** 247004
[12] Kaneyasu H and Sigrist M 2010 *J. Phys. Soc. Jpn.* **79** 053706
[13] Ishiguro R, Yakabe M, Nakamura T, Watanabe E, Tsuya D, Oosato H, Maeno Y and Takayanagi H 2012 *J. Phys.: Conf. Ser.* **400** 022035
[14] Ishiguro R, Sakurai T, Yakabe M, Nakamura T, Yonezawa S, Kashiwaya S, Maeno Y and Takayanagi H submitted
[15] Golod T, Rydh A and Krasnov V M 2010 *Phys. Rev. Lett.* **104** 227003
[16] Nelson K D, Mao Z Q, Maeno Y and Liu Y 2004 *Science* **306** 1151

Journal of Physics: Conference Series **568** (2014) 022032 doi:10.1088/1742-6596/568/2/022032

Resistive superconducting transition and effects of atmospheric exposure in the intercalation superconductor Ax(C2H8N2)yFe2-zSe2 (A = Li, Na)

T. Hatakeda, T. Noji, S. Hosono, T. Kawamata, M. Kato, and Y. Koike

Department of Applied Physics, Tohoku University, 6-6-05 Aoba, Aramaki, Aoba-ku, Sendai 980-8579, Japan

noji@teion.apph.tohoku.ac.jp

Abstract. We have succeeded in observing zero-resistivity in newly discovered intercalation superconductors Ax(C2H8N2)yFe2-zSe2 (A = Li, Na) with T_c = 45 K, using the sintered pellet samples. The electrical resistivity, ρ, in the normal state is metallic and T_c^{onset}, defined in the ρ measurements, is as high as ~ 57 K. We have also investigated effects of the atmospheric exposure in $Li_x(C_2H_8N_2)_yFe_{2-z}Se_2$. It has been found that both the crystal structure and superconductivity are maintained at least up to several days, indicating this material is comparatively resistant to the atmospheric exposure.

1. Introduction

Recently, it has been found that the superconducting transition temperature, T_c, of the iron-based superconductor FeSe increases from 8 K to 40 - 46 K via intercalation of alkaline or alkaline-earth metals and ammonia or pyridine [1-5]. Moreover, we have succeeded in synthesizing new superconductors $A_x(C_2H_8N_2)_yFe_{2-z}Se_2$ (A = Li, Na) with T_c = 45 K via intercalation of alkaline metals and ethylenediamine (EDA), $C_2H_8N_2$ [6,7]. The drastic increase in T_c via the intercalation has attracted great interest in relation to the mechanism and the application of the intercalation superconductors. For applications, zero-resistivity is important, but zero-resistivity is observed only in $Li_x(C_2H_8N_2)_yFe_{2-z}Se_2$ [6]. Moreover, T_c^{zero}, defined as the temperature where the electrical resistivity, ρ, reach zero, is much lower than T_c obtained in the measurements of the magnetic susceptibility, χ. For applications, the stability of the intercalation superconductors at the atmosphere is also important, but generally superconductors including alkaline metals are not so stable at the atmosphere [1].

In this paper, we have investigated the resistive superconducting transition of alkaline metals- and EDA-intercalated $A_x(C_2H_8N_2)_yFe_{2-z}Se_2$ (A = Li, Na) and their stability at the atmosphere.

2. Experimental

Polycrystalline samples of FeSe were prepared by the solid-state reaction method. Starting materials were powders of Fe and Se, which were weighted stoichiometrically, mixed thoroughly and pressed into pellets. The pellets were sealed in an evacuated quartz tube and heated at 800℃ for 40 h. The obtained pellets of FeSe were pulverized into powder to be used as a host of the intercalation. Dissolved alkaline metal (Li or Na) in EDA was intercalated into the powdery FeSe as follows. An appropriate amount of the powdery FeSe was placed in a beaker filled with 0.2 M solution of pure alkaline metal dissolved in EDA. The reaction was carried out at 45℃ for 7 days. The product was washed with fresh EDA. All the process was performed in an argon-filled glove box. Both FeSe and the intercalated samples were characterized by powder x-ray diffraction using CuK_α radiation. For the intercalated samples, an airtight sample-holder was used. In order to detect the superconducting transition, χ was measured using a superconducting quantum interference device (SQUID) magnetometer. Measurements of ρ were also carried out by the standard dc four-probe method. For the ρ measurements, as-intercalated powdery samples were pressed into pellets. Then, the pellets were sintered at 170 or 200℃ for 20 - 30 h in an evacuated or argon-filled glass tube. In order to investigate effects of exposure to the atmosphere, as-intercalated samples, pelletized at room temperature, and sintered pellet samples were exposed to the atmosphere up to 7 days.

3. Results and discussion

Figure 1 shows powder x-ray diffraction patterns of as-intercalated samples of $A_x(C_2H_8N_2)_yFe_{2-z}Se_2$ (A = Li, Na). The broad peak around $2\theta = 20°$ is due to the airtight sample-holder. Although there are unknown peaks, most of sharp Bragg peaks are due to the intercalation compound of $A_x(C_2H_8N_2)_yFe_{2-z}Se_2$ (A = Li, Na) and the host compound of FeSe, so that they are able to be indexed based on the $ThCr_2Si_2$-type and PbO-type structures, respectively. Therefore, it is found that alkaline metal and EDA are intercalated into FeSe, while there remains a non-intercalated region of FeSe in the samples. The lattice constants of $Li_x(C_2H_8N_2)_yFe_{2-z}Se_2$ are calculated to be a = 3.440(3) Å and c = 20.81(3) Å. The c-axis length of $Na_x(C_2H_8N_2)_yFe_{2-z}Se_2$ is 21.9(1) Å. Taking into account our previous results that the intercalation of only lithium into Fe(Se,Te) has neither effect on the superconductivity nor crystal structure[8], it is concluded that not only lithium or sodium but also EDA has been intercalated.

Figure 2 displays the temperature dependence of χ in a magnetic field of 10 Oe on zero-field cooling (ZFC) and on field cooling (FC) for as-intercalated powdery samples consisting of $A_x(C_2H_8N_2)_yFe_{2-z}Se_2$ (A = Li, Na) and FeSe. The superconducting transition is observed at 45 K. Taking into account the powder x-ray diffraction results, it is concluded that the superconducting transition is due to bulk superconductivity of $A_x(C_2H_8N_2)_yFe_{2-z}Se_2$ (A = Li, Na). The positive value of χ will be due to magnetic impurities taken into the sample. It is noted that the second superconducting transition due to FeSe observed at 8 K in our previous paper [6] is not observed clearly in the present lithium-intercalated sample. This is reasonable, because powder x-ray diffraction peaks due to FeSe are strongly suppressed in the present lithium-intercalated sample, as shown in Fig. 1.

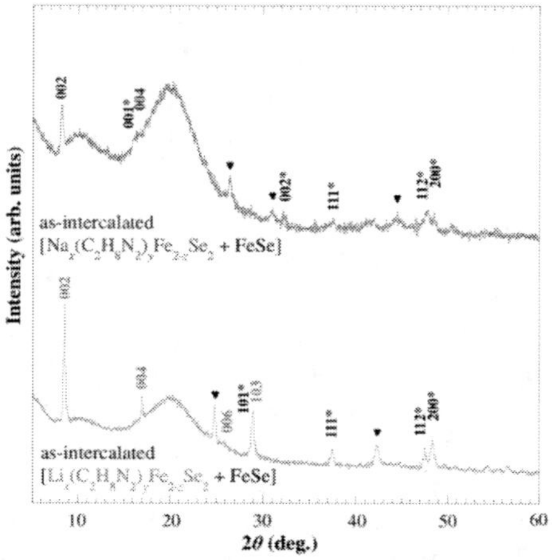

Fig. 1. Powder x-ray diffraction patterns of as-intercalated samples consisting of $A_x(C_2H_8N_2)_yFe_{2-z}Se_2$ (A = Li, Na) and FeSe, using CuK$_\alpha$ radiation. Indexes without and with asterisk are based on the $ThCr_2Si_2$-type and PbO-type structures, respectively. Peaks marked by ▼ are unknown.

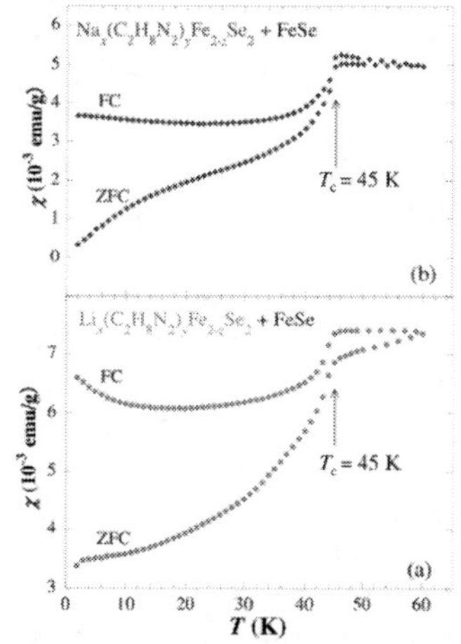

Fig. 2. Temperature dependence of the magnetic susceptibility, χ, in a magnetic field of 10 Oe on zero-field cooling (ZFC) and field cooling (FC) for as-intercalated powdery samples consisting of $A_x(C_2H_8N_2)_yFe_{2-z}Se_2$ ((a) A = Li, (b) A = Na) and FeSe.

Figure 3 displays the temperature dependence of ρ for several pellet samples. As shown in Fig. 3 (a), the as-intercalated sample consisting of $Li_x(C_2H_8N_2)_yFe_{2-z}Se_2$ and FeSe, pelletized at room temperature, exhibits the onset of the superconducting transition at 44 K, but zero-resistivity is not observed probably due to the insulating grain-boundary [6]. Sintered pellet samples consisting of $A_x(C_2H_8N_2)_yFe_{2-z}Se_2$ (A = Li, Na) and FeSe, on the other hand, exhibit zero-resistivity. It is found that the lithium-intercalated pellet sample sintered at 170℃ shows a metallic temperature-dependence of ρ and that the value of ρ in the normal state is much smaller and T_c^{zero} is much higher than those of the lithium-intercalated pellet sample sintered at 200℃, respectively. These results are understood to be due to the progress of the deintercalation of EDA with increasing sintering-temperature [7]. It is also found that T_c^{mid}, defined as the temperature where ρ shows a half of the normal-state value, is ~ 42 K for both lithium- and sodium-intercalated pellet samples sintered at 170℃ and almost the same as T_c = 45 K obtained from the χ measurements shown in Fig. 2. It is remarkable that T_c^{onset}, defined as the temperature where ρ starts to decrease with decreasing temperature due to the superconducting transition, obtained from the ρ measurements is as high as ~ 57 K and ~ 55 K for the lithium- and sodium-intercalated samples, respectively.

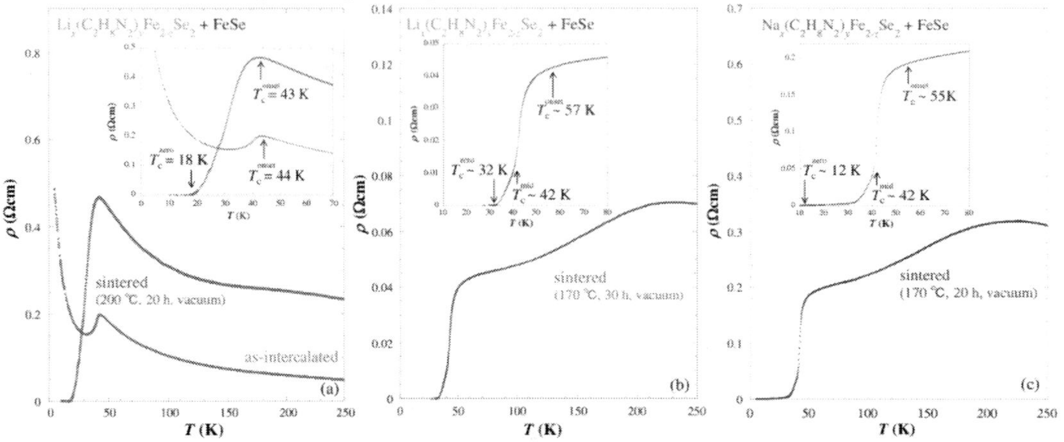

Fig. 3. Temperature dependence of the electrical resistivity, ρ, for (a) as-intercalated (pelletized at room temperature) and sintered (200℃, 20 h, vacuum) pellet samples consisting of $Li_x(C_2H_8N_2)_yFe_{2-z}Se_2$ and FeSe [6], (b) the sintered (170℃, 30 h, vacuum) pellet sample consisting of $Li_x(C_2H_8N_2)_yFe_{2-z}Se_2$ and FeSe, and (c) the sintered (170℃, 20 h, vacuum) pellet sample consisting of $Na_x(C_2H_8N_2)_yFe_{2-z}Se_2$ and FeSe. Insets show the temperature dependence of ρ around T_c.

Figure 4 (a) shows the change of the x-ray diffraction pattern of the as-intercalated sample pelletized at room temperature, consisting of $Li_x(C_2H_8N_2)_yFe_{2-z}Se_2$ and FeSe, by exposure to the atmosphere. It is found that the diffraction peaks do not change so much even for the sample exposed for 7 days, indicating that the crystal structure of $Li_x(C_2H_8N_2)_yFe_{2-z}Se_2$ is maintained at the atmosphere at least up to 7days. Observing the enlarged diffraction patterns shown in Fig. 4 (b), it is found that the c-axis lengths of the samples exposed for 2 - 7 days are a little larger than that of the as-intercalated sample, as listed in Table I. The intercalants (lithium and EDA) may be relaxed structurally in the sample or lithium in the sample may react with something at the beginning of the exposure to the atmosphere. The change of the superconductivity of the as-intercalated sample pelletized at room temperature, consisting of $Li_x(C_2H_8N_2)_yFe_{2-z}Se_2$ and FeSe, by exposure to the atmosphere has been observed in the χ measurements, as shown in Fig. 5. It is found that both T_c and the superconducting volume fraction gradually decrease by the atmospheric exposure. Furthermore, changes of ρ and the resistive superconducting transition of the sintered (170℃, 30 h, Ar) pellet sample consisting of $Li_x(C_2H_8N_2)_yFe_{2-z}Se_2$ and FeSe by the atmospheric exposure have been observed, as shown in Fig. 6. It is found that the value of ρ in the normal state gradually increases by the atmospheric exposure, while T_c^{mid} = 34 - 36 K for all the samples. These results indicate that the superconductivity of $Li_x(C_2H_8N_2)_yFe_{2-z}Se_2$ is maintained at the atmosphere at least up to several days, though the superconducting volume fraction gradually decreases and the disorder at the grain boundary is gradually enhanced. Accordingly, it is concluded that $Li_x(C_2H_8N_2)_yFe_{2-z}Se_2$ is more resistant to the atmospheric exposure than the other intercalation superconductors such as $A_x(NH_3)_yFe_{2-z}Se_2$ (A =

alkaline and alkaline-earth metals) [1], so that $Li_x(C_2H_8N_2)_yFe_{2-z}Se_2$ is more suitable for applications.

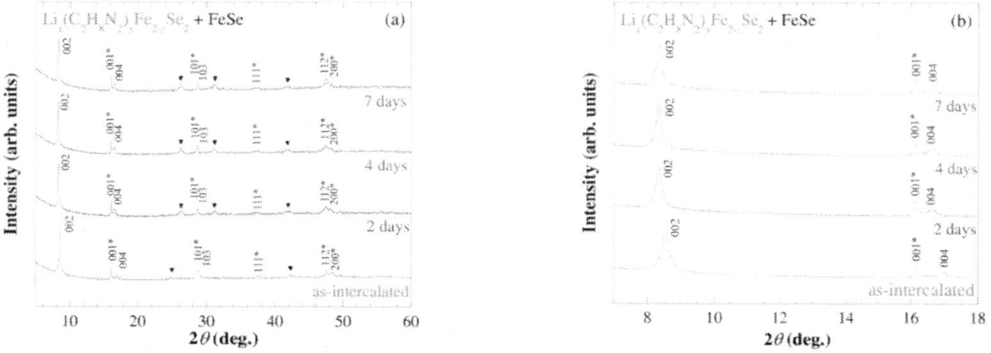

Fig. 4. (a) X-ray diffraction patterns of the as-intercalated sample pelletized at room temperature, consisting of $Li_x(C_2H_8N_2)_yFe_{2-z}Se_2$ and FeSe, and pellet samples exposed to the atmosphere for 2, 4 and 7 days using CuK_α radiation. (b) Enlarged x-ray diffraction patterns of (a) at 2θ = 7 - 18°. Indexes without and with asterisk are based on the $ThCr_2Si_2$-type and PbO-type structures, respectively. Peaks marked by ▼ are unknown.

Table 1. Lattice constants a and c of $Li_x(C_2H_8N_2)_yFe_{2-z}Se_2$ for the as-intercalated sample pelletized at room temperature, consisting of $Li_x(C_2H_8N_2)_yFe_{2-z}Se_2$ and FeSe, and pellet samples exposed to the atmosphere for 2, 4 and 7 days.

	a (Å)	c (Å)
$Li_x(C_2H_8N_2)_yFe_{2-z}Se_2$ (as-intercalated)	3.447(3)	20.85(4)
$Li_x(C_2H_8N_2)_yFe_{2-z}Se_2$ (2 days)	3.426(2)	21.33(3)
$Li_x(C_2H_8N_2)_yFe_{2-z}Se_2$ (4 days)	3.435(2)	21.32(2)
$Li_x(C_2H_8N_2)_yFe_{2-z}Se_2$ (7 days)	3.433(2)	21.30(3)

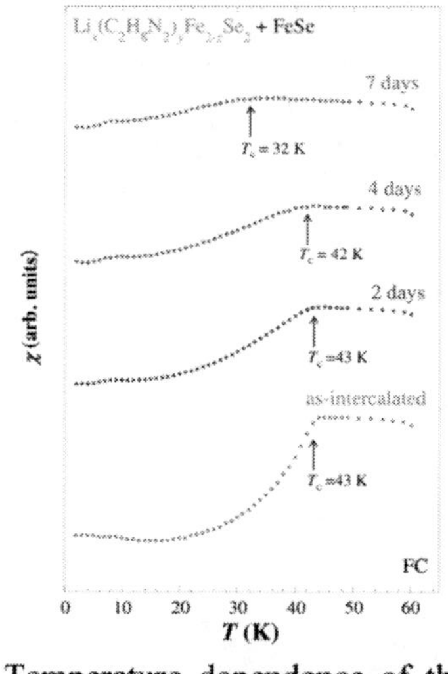

Fig. 5. Temperature dependence of the magnetic susceptibility, χ, in a magnetic field of 10 Oe on field cooling (FC) for the as-intercalated sample pelletized at room temperature, consisting of $Li_x(C_2H_8N_2)_yFe_{2-z}Se_2$ and FeSe, and pellet samples exposed to the atmosphere for 2, 4 and 7 days.

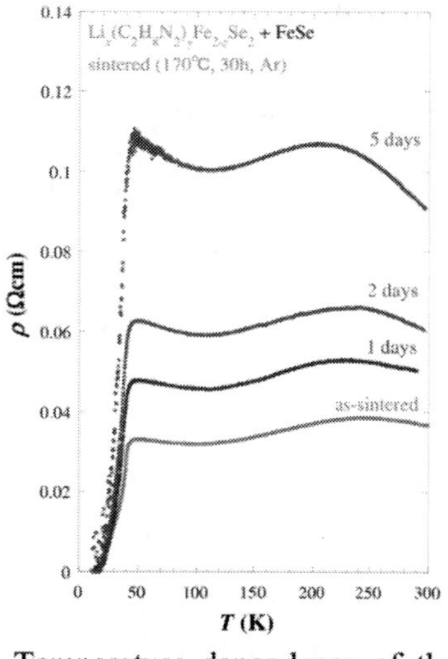

Fig. 6. Temperature dependence of the electrical resistivity, ρ, for the as-sintered (170℃, 30 h, Ar) pellet sample, consisting of $Li_x(C_2H_8N_2)_yFe_{2-z}Se_2$ and FeSe, and sintered (170℃, 30 h, Ar) pellet samples exposed to the atmosphere for 1, 2 and 5 days.

4. Summary

We have succeeded in observing zero-resistivity in our newly discovered intercalation superconductors $A_x(C_2H_8N_2)_yFe_{2-z}Se_2$ (A = Li, Na) with T_c = 45 K, using pellet samples sintered at 170℃ in an evacuated or argon-filled glass tube. It has been found that the sintered pellet samples show a metallic temperature-dependence of ρ and that T_c^{zero} is ~ 32 K and ~ 12 K for A = Li and Na, respectively. For both A = Li and Na, T_c^{mid} is ~ 42 K and almost the same as T_c obtained from the χ measurements. Moreover, T_c^{onset} is as high as ~ 57 K and ~ 55 K for A = Li and Na, respectively.

In order to investigate the stability of the lithium- and EDA-intercalated pellet samples at the atmosphere, they have been exposed to the atmosphere up to 7 days. From the x-ray diffraction analysis, it has been found that the crystal structure of $Li_x(C_2H_8N_2)_yFe_{2-z}Se_2$ is maintained at the atmosphere at least up to 7 days, though the c-axis lengths of the samples exposed for 2 - 7 days are a little larger than that of the as-intercalated sample. From the χ measurements, it has been found that the superconductivity is maintained at least up to 7 days, though both T_c and the superconducting volume fraction gradually decrease. From the ρ measurements of the sintered pellet samples, moreover, it has been found that zero-resistivity is maintained at least up to 5 days, though the value of ρ in the normal state gradually increases. In conclusion, $Li_x(C_2H_8N_2)_yFe_{2-z}Se_2$ is comparatively resistant to the atmospheric exposure and comparatively suitable for applications.

Acknowledgments

One of the authors (T. H.) is indebted to the Motizuki Foundation for the support of his travel expenses to Buenos Aires to present this work at the 27th International Conference on Low Temperature Physics.

References

[1] T. P. Ying, X. L. Chen, G. Wang, S. F. Jin, T. T. Zhou, X. F. Lai, H. Zhang, W. Y. Wang: Sci. Rep. **2** (2012) 426.
[2] E.-W. Scheidt, V. R. Hathwar, D. Schmitz, A. Dunbar, W. Scherer, F. Mayr, V. Tsurkan, J. Deisenhofer, A. Loidl: Eur. Phys. J. B **85** (2012) 279.
[3] M. Burrard-Lucas, D. G. Free, S. J. Sedlmaier, J. D. Wright, S. J. Cassidy, Y. Hara, A. J. Corkett, T. Lancaster, P. J. Baker, S. J. Blundell, S. J. Clarke: Nat. Mater. **12** (2013) 15.
[4] A. Krzton-Maziopa, E. V. Pomjakushina, V. Y. Pomjakushin, F. Rohr, A. Schilling, K. Conder: J. Phys.: Condens. Matter **24** (2012) 382202.
[5] L. Zheng, M. Izumi, Y. Sakai, R. Eguchi, H. Goto, Y. Takabayashi, T. Kambe, T. Onji, S. Araki, T. C. Kobayashi, J. Kim, A. Fujiwara and Y. Kubozono : Phys. Rev. B **88** (2013) 094521.
[6] T. Hatakeda, T. Noji, T. Kawamata, M. Kato and Y. Koike: J. Phys. Soc. Jpn. **82** (2013) 123705.
[7] T. Noji, T. Hatakeda, S. Hosono, T. Kawamata, M. Kato, Y. Koike: Physica C (in press)
[8] H. Abe, T. Noji, M. Kato, Y. Koike: Physica C **470** (2010) S487.

Journal of Physics: Conference Series **568** (2014) 022033 doi:10.1088/1742-6596/568/2/022033

Macroscopic quantum tunneling in BiPb2201

Y Nomura[1], **H Kambara**[1], **Y Nakagawa**[1] **and I Kakeya**[1]

[1]Department of Electronic Science and Engineering, Kyoto University, Japan

E-mail: `nomura@sk.kuee.kyoto-u.ac.jp`

Abstract. We measured the switching probability distributions of the first and second switches in $Bi_{2-x}Pb_xSr_{2-y}La_yCuO_{6+\delta}$ (BiPb2201) intrinsic Josephson junctions. The effective temperature of the first switch is temperature independent below 0.6 K, which is slightly higher than the crossover temperature estimated from the thermal activation model. On the other hand, the effective temperature of the second switch is temperature independent below 2.0 K, which is one-order higher than the crossover temperature. Our study shows that the temperature independent state at high temperature is measured not only in $Bi_2Sr_2CaCu_2O_{8+\delta}$ but also in $Bi_2Sr_2CuO_{6+\delta}$.

1. Introduction

In intrinsic Josephson junction (IJJ) included in $Bi_2Sr_2CaCu_2O_{8+\delta}$ (Bi2212), macroscopic quantum tunneling (MQT) is an indispensable phenomenon to realize qubits. MQT was observed up to 1 K in Bi2212 IJJs, which is one-order higher than Josephson junction made by metallic superconductors [1]. The result shows that IJJs of Bi2212 can be used as qubits at higher temperatures than Josephson junction made by metallic superconductors.

The current-biased single Josephson junction can be well described by the RCSJ (resistively and capacitively shunted junction) model. In this model, the junction is modeled by the motion of a particle in a tilted washboard potential. The switch of the Josephson junction from zero voltage state to resistive state corresponds the escape from the minima of the potential. The switch is mainly caused by thermal activation (TA) or MQT. TA is described as the escape from the minima by thermal fluctuation. The escape rate is $\sim \exp(-\Delta U/k_B T)$, where $\Delta U$is the potential barrier and k_B is the Boltzman constant [2]. MQT is described as the quantum tunneling of the particle through the potential barrier. The escape rate is $\sim \exp(-\Delta U/\hbar\omega_p)$, where ω_p is the current dependent plasma frequency [3]. At high temperatures, TA becomes dominant. The escape rate of TA suppressed with the decreasing temperature. The escape rate of TA and MQT are equal at crossover temperature given by $\sim \hbar\omega_p/2\pi k_B$. MQT is observed by measuring switching probability distributions which are histogram of currents where a Josephson junction switches from zero voltage state to resistive state. The distribution is temperature independent when MQT is dominant and temperature dependent when TA is dominant. The temperature independent region of the distribution means that MQT is observed.

Recently, several unclear phenomena have been reported in IJJs [4][5][6]. One of the unclear phenomenon is that the temperature independent behavior observed below $\sim$ 8 K for the second switch. The second switch is a phenomenon of the switch another IJJ included in the stack to resistive state subsequent to the first switch. This behavior is not explained by MQT because the temperature is one-order higher than crossover temperature [5][6]. This temperature independent behavior at high temperature is reported only in Bi2212.

In this paper, we measured the switching probability distributions of the first and second switches in the stack of $Bi_{2-x}Pb_xSr_{2-y}La_yCuO_{6+\delta}$ (BiPb2201). BiPb2201 has single CuO_2 layer so that Bi2201 has thiner superconducting layer than Bi2212 with double CuO_2 layer. The difference between Bi2201 and Bi2212 is only the number of CuO_2 layer. The feature of the first switch can be roughly explained by the single Josephson junction model. On the other hand, temperature independent behavior of the second switch is observed at higher temperature than the crossover temperature. This is not due to Pb substitution since the feature of the first switch is not changed by Pb substitution. The temperature independent behavior of the second switch is a common phenomenon observed in Bi2201 and Bi2212.

2. Experimental method

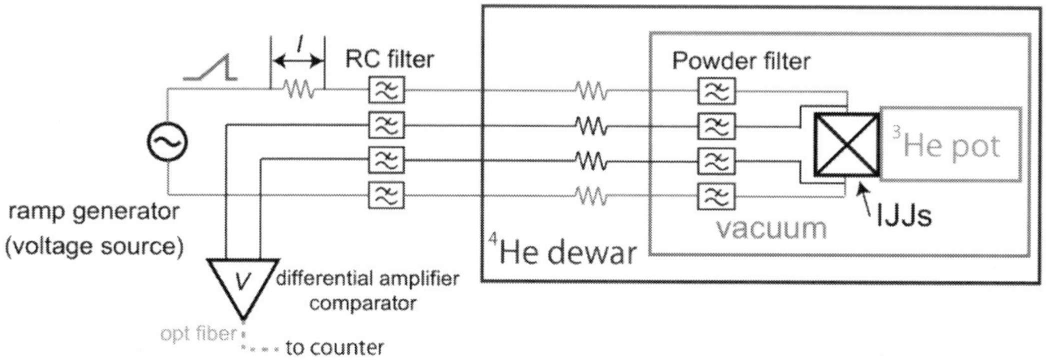

Figure 1. Schematic of the measurement system for the switching probability distribution

We used BiPb2201 single crystal grown by the traveling solvent floating zone method. The single crystal was grown as follows. Powders of Bi_2O_3, PbO, $SrCO_3$, La_2O_3 and CuO were mixed in the nominal compositions of $Bi_{1.8}Pb_{0.3}Sr_{1.5}La_{0.4}Cu_{1.0}$. Mixed powders were calcined twice at 740 °C for 24 hour. Calcined powders were put into rubber tube and pressed under 50 MPa for 1 hour. The pressed rod was sintered at 840 °C for 24 hour. The rod was pre-melted at a velocity of 10 mm/hour. After that, the crystal was grown by the rate of 0.5 mm/hour. The analyzed composition of the crystal by EDS is $Bi_{1.9}Pb_{0.1}Sr_{1.4}La_{0.6}Cu_{1.0}O_{6+\delta}$.

The mesa structure can be fabricated easily by two dimensional processing, and the number of the IJJs include in the mesa structure can be changed by the milling time. In this study, we fabricated the mesa structure on BiPb2201 single crystal with the size of 1×1 μm^2 by using electron-beam lithography and Ar-ion milling. The detail of the fabrication is described elsewhere [7].

switching probability distributions were measured down to 0.4 K by using ^{3}He cryostat. The measurement system is shown in Fig. 1. We used current source which is same as that described in Ref [8] before. However the critical current of the first switch is the largest among the branches in this mesa structure. In this case, all IJJs are switched to voltage state simultaneously if IJJs are biased by current source because the current does not decrease by the switch. We connected the home-made current-controlled voltage source to the output of the current source. The voltage source consists of three operation amplifiers (Burr-Brown OPA 277) and chip resistors. Two operation amplifiers are used for the inverting amplifier and another operation amplifier is used as the voltage follower. The stack of IJJs is biased by the output of the inverting amplifier.

The bias current is monitored by means of the voltage drops across a series of chip resistors (100 ohm). The power source of these home-made circuits is supplied by a DC stabilized voltage source. Two low-pass filters are inserted in this circuit, one is a RC filter at room temperature and another is powder filter inside the vacuum space of the cryostat. The cut-off frequency of the RC filter is 300 kHz and that of powder filter is 1 MHz. The voltage source and the IJJs are separated from digital electronics connected to the PC by the optical fiber to reduce the noise of the commercial power supply. We measured the switch current about 10,000 times to obtain its distribution.

We calculated the effective temperature ($T_{\rm eff}$) to fit the switching probability distribution by thermal activation model. $T_{\rm eff}$ is a parameter indicating the distribution. If $T_{\rm eff}$ shows temperature independent, it is considered that MQT is the dominant switching process.

3. Results and discussion

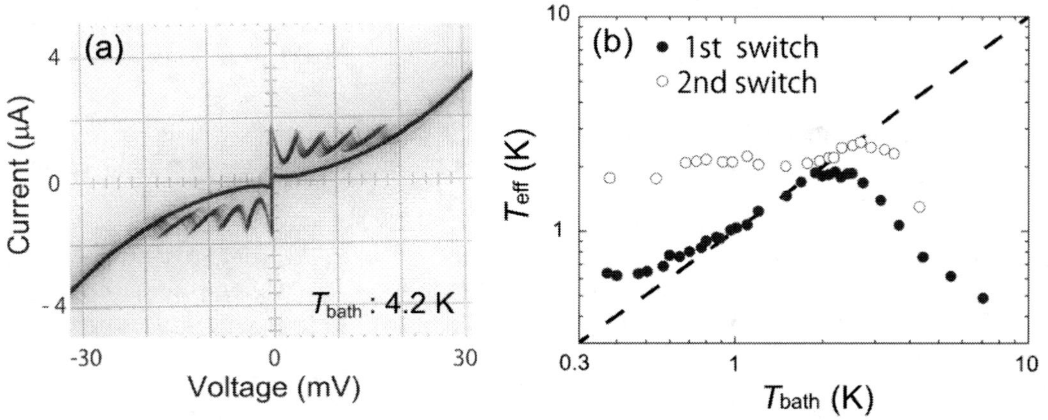

Figure 2. (a) I-V characteristics in Bi2201 at 4.2 K. (b) Effective temperature versus bath temperature in Bi2201 for the first (solid symbol) and second (open symbol) switches.

Fig. 2(a) shows an oscilloscope image of I-V characteristics for the IJJ stack, measured at 4.2 K. One can find five branches separated equally in voltage of 4 mV at $I = 1.6\ \mu$A. The critical current densities are found to be 180 A/cm^2 and the critical current of the first switch is the largest among branches of the I-V characteristics. Thus we consider that the number of the IJJs is five and these five IJJs are uniform. Fig. 2(b) shows the $T_{\rm eff}$ vs bath temperature($T_{\rm bath}$) plots. Below 0.6 K, $T_{\rm eff}$ of the first switch is temperature independent. $T_{\rm eff}$ corresponds with $T_{\rm bath}$ between 0.6 K and 2.0 K. This result can be explained that MQT is dominant below 0.6 K so that $T_{\rm eff}$ shows temperature independent behavior. The crossover temperature is $\sim$ 0.35 K. The temperature independent state is observed at a little higher temperature. The feature of $T_{\rm eff}$ between 0.6 K and 2.0 K can be explained by the TA model. We conclude that the first switch in the stack of IJJs can be roughly explained by the single Josephson junction model.

The switching probability distribution of the second switch is different from that of the first switch. $T_{\rm eff}$ shows temperature independent behavior below 2.0 K. $T_{\rm eff}$ corresponds with $T_{\rm bath}$ between 2.0 K and 3.0 K. The crossover temperature of the second switch is $\sim$ 0.30 K. The temperature independent behavior is observed one-order higher than the crossover temperature. This can not be explained by MQT of the single Josephson junction model. The behavior of $T_{\rm eff}$between 2.0 K and 3.0 K can be explained by the TA model because $T_{\rm eff}$ and $T_{\rm bath}$ shows good agreement in this region.

We consider this temperature independent state at high temperatures is not due to Pb substitution by following reasons. This peculiar behavior of $T_{\rm eff}$ is not observed at the first switch. The first switch is well described single Josephson junction model as well as in $Bi_2Sr_{2-y}La_yCuO_{6+\delta}$ (Bi2201) [9]. This means that Pb substitution does not affect the switch dynamics in Bi2201. Second, same tendency has been observed in Bi2212 despite of the no Pb substitution [5][6][10]. Therefore, we conclude that the fluctuation of the second switch is much larger than that of the first switch in IJJs consist of $Bi_2Sr_2Ca_{n-1}Cu_nO_{4+2n+\delta}$ (BSCCO) with single and double CuO_2 layer. Someone considers that the behavior of the second switch is attributed to heating effect [5]. However, we consider that the self-heating is too small to raise temperature. The mesa structure is fabricated to measure the switching probability distributions in this report. In the mesa structure, the heating in the IJJs is removed by the thick Ag electrode covering the top of the IJJs. Moreover, the previous report calculated the temperature increase due to the self-heating in Bi2212 [6]. In mesa structure of Bi2212, the temperature rise is estimated about ~ 0.14 K which is one-order smaller than the $T_{\rm eff}$ of the second switch at 0.38 K. The large fluctuation of the second switch is attributed to the charge coupling between IJJs because the $T_{\rm eff}$ of the first and second switches of $Bi_2Sr_2Ca_2Cu_3O_{10+\delta}$ IJJs are almost identical [10].

4. Conclusion

We measured switching probability distributions of the first and second switches in the stack of IJJs in BiPb2201. The first switch can be explained by the single Josephson junction model. This result is consistent with the previous report in Bi2201. Pb-substitution does not affect the switch mechanism of the first switch in Bi2201. However, the temperature independent behavior of the second switch was observed at high temperatures in comparison with the first switch. The large fluctuation of the second switch is a peculiar phenomenon in BSCCO IJJs with superconducting layer being less than 0.3 nm.

Acknowledgments

We thank Professor Osamu Sakai for allowing us to use the electron-beam lithography facility.

Reference

[1] Inomata K, Sato S, Nakajima K, Tanaka A, Takano Y, Wang H B, Nagao M, Hatano H and Kawabata S 2005 *Phys. Rev.* Lett. **95** 107005.
[2] Kramers H 1940 Physica **7** 284-304.
[3] Caldeira A O and Leggett A J 1981*Phys. Rev.* Lett. **46** 211.
[4] Jin X Y, Lisenfeld J, Koval Y, Lukashenko A, Ustinov A V and Müller P 2006 *Phys. Rev.* Lett. **96** 177003.
[5] Kashiwaya H, Matsumoto T, Shibata H, Kashiwaya S, Eisaki H, Yoshida Y, Kawabata S and Tanaka Y 2008 *Phys. Soc. Jpn* **77** 104708.
[6] Ota K, Hamada K, Takemura R, Ohmaki M, Machi T, Tanabe K, Suzuki M, Maeda A and Kitano H 2009 *Phys. Rev.* B **79** 134505.
[7] Kakeya I, Hamada K, Tachiki T, Watanabe T and Suzuki M 2009 Supercond. Sci. Technol. **22** 114014.
[8] Kubo Y, Yamaguchi T, Ueda S, Takano Y and Ootuka Y 2010 Appl. Phys. Express **3** 063104.
[9] Kashiwaya H, Matsumoto T, Shibata H, Eisaki H, Yoshida Y, Kambara H, Kawabata S and Kashiwaya S 2010 Appl. Phys. Express **3** 043101.
[10] Nomura Y, Mizuno T, Kambara H, Nakagawa Y, Watanabe T, Kakeya I and Suzuki M 2014 J. Phys.: Conf. Ser. **507** 012038.

Journal of Physics: Conference Series **568** (2014) 022034 doi:10.1088/1742-6596/568/2/022034

Transport Properties of the Layered Transition Metal Oxypnictide Sr_2ScMPO_3 with MP layers (M =Mn, Ni and $Co_{0.5}Fe_{0.5}$)

S Okada[1], Y Kamihara[2], N Ohkubo[1], S Ban[1] and M Matoba[2]

[1] Laboratory of Physics, College of Science and Technology, Nihon University, Funabashi, Japan

[2] Department of Applied Physics and Physico-Informatics, Faculty of Science and Technology, Keio University, Yokohama, Japan

E-mail: okada@phys.ge.cst.nihon-u.ac.jp

Abstract. Polycrystalline samples of novel oxypnictides, Sr_2ScMPO_3, with MP layer (M = Mn, Ni and $Co_{0.5}Fe_{0.5}$) were synthesized, and their resistivities and Seebeck coefficients were measured. Sr_2ScMPO_3 crystallizes in a stacked, layered structure comprised of a $ThCr_2Si_2$–type MP layer alternating with a K_2NiF_4–type Sr_2ScO_3 layer. $Sr_2ScMnPO_3$ is an insulator at room temperature. The resistivities (ρ) of $Sr_2ScNiPO_3$ and $Sr_2ScCo_{0.5}Fe_{0.5}PO_3$ decrease with decreasing temperature like a metal. The Seebeck coefficients (S) of these materials are negative at room temperature. For $Sr_2ScNiPO_3$, S initially decreases slightly with decreasing temperature, and increases with decreasing temperature below 50 K. However, for $Sr_2ScCo_{0.5}Fe_{0.5}PO_3$, S increases with decreasing temperature, and attains to a positive value below 270 K.

1. Introduction

Since the discovery of high T_c superconductivity in F-doped LaFePnO (Pn = P and As), numerous scientists have been interested in the transport properties of oxypnictides and oxychalcogenides [1, 2]. These oxypnictides are multi anion systems, containing O and Pn, and have a quasi-two-dimensional structure, in which an anti-fluorite transition metal (M = Fe, Co, Mn, Ni, $\cdots$) pnictide layer is separated by a carrier blocking layer [3]. The FePn series, in particular, have been intensively studied because the quasi-two-dimensional structure and the FePn layer plays an important role in their superconductivity. However, various phenomena are observed in the materials with an MP layer. For example, LaMnPO with an MnP layer is an antiferromagnetic semiconductor [4–6], LaCoPO with a CoP layer is a ferromagnetic metal having a Curie temperature of ~60 K [7], and LaNiPO with an NiP layer is a Pauli para metal with a superconducting critical temperature (T_c) of 3 K [8].

In past research, we have focused on perovskite-type oxypnictides and oxychalcogenides that have a large blocking layer [10, 11], with particular interest in $Sr_2ScFePO_3$ with a Sr_2ScO_3 layer that was reported by Ogino *et al.* [9]. $Sr_2ScFePO_3$ crystallizes in an unusual intergrowth structure consisting of an anti-fluorite FeP layer and a perovskite related K_2NiF_4-type Sr_2ScO_3 layer, as shown in Fig. 1. It was determined that $Sr_2ScFePO_3$ exhibited superconductivity below 17 K. In addition, we have successfully synthesized and evaluated the transport properties

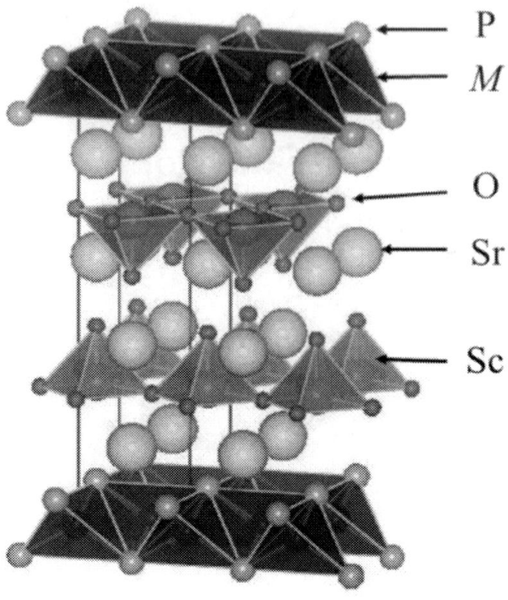

Figure 1. Crystal structure of Sr_2ScMPO_3 (M=Fe, Co, Mn, Ni and $Co_{0.5}Fe_{0.5}$). This figure was created by VESTA [13].

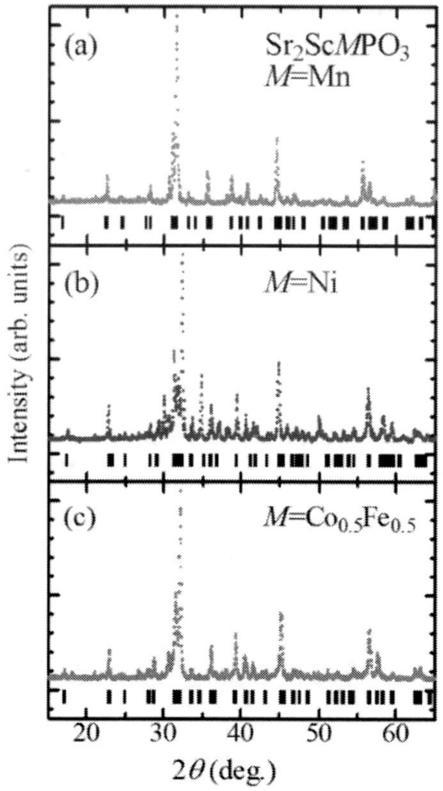

Figure 2. Powder XRD patterns of (a) $Sr_2ScMnPO_3$, (b) $Sr_2ScNiPO_3$, and (c) $Sr_2ScCo_{0.5}Fe_{0.5}PO_3$. Vertical bars denote Bragg positions.

of $Sr_2ScCoPO_3$ [12]. The resistivity (ρ) of $Sr_2ScCoPO_3$ is 4.5 mΩcm at room temperature and decreases with decreasing temperature. The Seebeck coefficient (S) of $Sr_2ScCoPO_3$ is -12 μV/K at room temperature, decreases with decreasing temperature, and rapidly increases with decreasing temperature below 50 K . To evaluate the role of the Sr_2ScO_3 layer in $Sr_2ScCoPO_3$, it is necessary to compare the transport properties of other compounds that have a Sr_2ScO_3 layer. To this end, we have, in the present study, successfully synthesized the novel oxypnictides Sr_2ScMPO_3 with MP layers (M = Mn, Ni and $Co_{0.5}Fe_{0.5}$) and evaluated their transport properties.

2. Experiment

Polycrystalline samples of Sr_2ScMPO_3 (M =Mn and Ni) were synthesized by a solid state reaction method in a quartz tube from M_2P, P, SrO, Sr and Sc_2O_3. A mixture of starting materials with a molar ratio of M_2P:P:SrO:Sr:Sc_2O_3 = 1:1:3:1:1 was pressed into pellets in a glovebox filled with Ar gas. In the case of $Sr_2ScCo_{0.5}Fe_{0.5}PO_3$ (M = $Co_{0.5}Fe_{0.5}$), the starting materials were $Co_{0.5}Fe_{0.5}P$, SrO, Sr and Sc_2O_3. The precursor $Co_{0.5}Fe_{0.5}P$ was prepared by a solid state reaction of Co, Fe_2P and P with a molar ratio of 2:1:1 at 700 °C for 40 h. These pellets were calcined at 1200 °C for 30 h in a quartz tube filled with Ar gas under atmospheric pressure

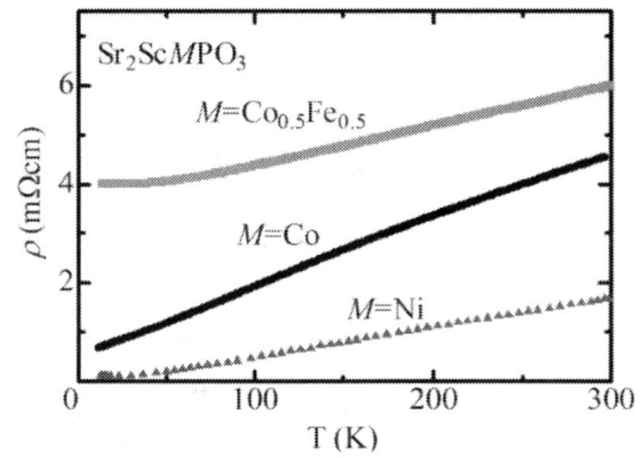

Figure 3. Temperature (T) dependence of th resistivity (ρ) for Sr_2ScMPO_3 (M = Co, Ni and $Co_{0.5}Fe_{0.5}$).

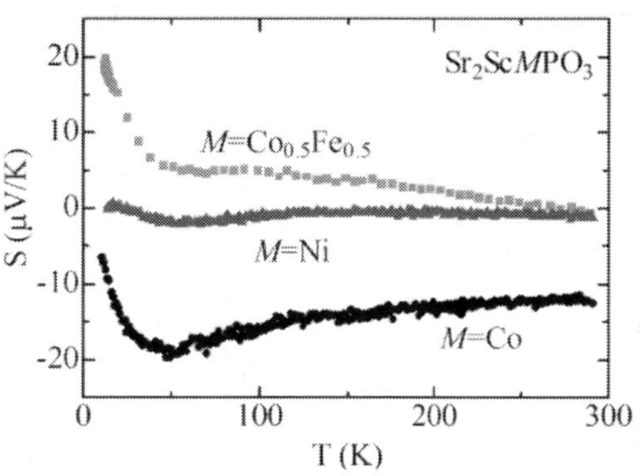

Figure 4. Temperature (T) dependence of the Seebeck coefficient (S) for Sr_2ScMPO_3 (M = Co, Ni and $Co_{0.5}Fe_{0.5}$).

of 0.1 at room temperature. Crystallographic phases were examined by powder X–ray diffraction (XRD, Rigaku, Rint2500). Lattice parameters were roughly obtained by a least–squares fitting method using the diffraction angles collected in the 2θ range from 10 to 70°.

Resistivity measurements were performed by a four–probe method at temperatures from 20 to 300 K. Seebeck coefficients were measured using a steady–state technique at temperatures from 20 to 300 K. A temperature gradient of 0.3 or 0.1 K/mm was generated by a small resistive heater placed on one edge of the sample, and was monitored with a copper–constantan differential thermocouple. The thermopower of the voltage leads was carefully subtracted.

3. Results and Discussion

Figure 2 shows the powder XRD patterns of Sr_2ScMPO_3 (M = Mn, Ni and $Co_{0.5}Fe_{0.5}$) with the accompanying Bragg positions calculated by the analysis program RIETAN-FP [14]. These patterns are similar to the pattern of $Sr_2ScFePO_3$ [9], which indicates that these compounds are isomorphic to $Sr_2ScFePO_3$. The powder XRD patterns were successfully indexed except for some peaks assigned to secondary phases. The space group of Sr_2ScMPO_3 is $P4/nmm$. The lattice constants are a = 0.4065(1) nm and c = 1.5803(6) nm for M = Mn, a = 0.4027(2) nm and c = 1.5310(11) nm for M = Ni, and a = 0.4009(1) nm and c = 1.5506(9) nm for M = $Co_{0.5}Fe_{0.5}$.

Figure 3 shows the temperature dependence of ρ for Sr_2ScMPO_3 (M = Ni and $Co_{0.5}Fe_{0.5}$) with that of $Sr_2ScCoPO_3$ previously reported [12]. In $Sr_2ScMnPO_3$, ρ at room temperature was found to be a large value like that of an insulator. In all samples except M = Mn, ρ decreases with decreasing temperature. This metallic behavior is similar to that of $Sr_2ScFePO_3$ [9]. The superconducting transition is not observed in Figure 3 for any of the materials measured in the temperature interval 20–300 K. The value of ρ for M = Ni at room temperature is 1.7 mΩ cm, smaller than that for M = Co over the entire range 20–300 K. The value of ρ for M=$Co_{0.5}Fe_{0.5}$ at room temperature is 6.0 mΩ cm, larger than that for M = Co over the entire temperature range considered.

Figure 4 shows the temperature dependence of S for Sr_2ScMPO_3 (M = Ni and $Co_{0.5}Fe_{0.5}$) with that of $Sr_2ScCoPO_3$ previously reported [12]. In $Sr_2ScNiPO_3$, S is observed to slightly decrease with decreasing temperature down to 50 K, and increase with decreasing temperature below 50 K. Although this behavior is similar to that of $Sr_2ScCoPO_3$, the absolute value of S is smaller than that of $Sr_2ScCoPO_3$ over the entire temperature range considered. In

$Sr_2ScCo_{0.5}Fe_{0.5}PO_3$, S slightly increases with decreasing temperature down to 50 K, and rapidly increase with decreasing temperature below 50 K. Figure 4 indicateds that S is negative for M = Ni over the entire temperature range 20–300 K, and that S is negative for M = $Co_{0.5}Fe_{0.5}$ only over the range 270–300 K. As such, the sign inversion of S is obtained at about 270 K for M = $Co_{0.5}Fe_{0.5}$, and S retains a positive value below 270 K. This sign inversion of S is also observed for $Sr_2ScFeAsO_3$ at 100 K [15]. The $S-T$ curves for each of the materials considered are clearly different from each other. The observed differences of the $S-T$ curves reflect the influence of multiband electronic structures.

4. Conclusion

In summary, we have successfully synthesized polycrystalline samples of novel oxypnictides, Sr_2ScMPO_3 with MP layers (M = Mn, Ni and $Co_{0.5}Fe_{0.5}$). The transport properties of these samples are demonstrated. $Sr_2ScMnPO_3$ is found to be an insulator at room temperature. In Sr_2ScMPO_3 (M = Ni and $Co_{0.5}Fe_{0.5}$), *rho* decreases with decreasing temperature like in a metal. In Sr_2ScMPO_3 (M = Ni and $Co_{0.5}Fe_{0.5}$), S exhibits negative values at room temperature. In $Sr_2ScNiPO_3$, S slightly decreases with decreasing temperature, and increases with decreasing temperature below 50 K. However, S for $Sr_2ScCo_{0.5}Fe_{0.5}PO_3$ increases with decreasing temperature, and attains a positive value around 270 K. Local structural refinement and theoretical band structure calculations are required to quantitatively analyze the transport and magnetic properties of the materials considered in the present study. Higher accuracy and precisely Rietveld refinement and application of density functional theory [16–19] are therefore chosen for future work.

Acknowledgments

M. M. and Y. K. are supported by the research grants from Keio University, the Asahi Glass Foundation, and Japan Society for Promotion of Science (JSPS) KAKENHI Grant Numbers: 26400337.

References

[1] Kamihara Y, Hiramatsu H, Hirano M, Kawamura R, Yanagi H, Kamiya T and Hosono H 2006 *J. Am. Chem. Soc.* **128** 10012.
[2] Kamihara Y, Watanabe T, Hirano M, and Hosono H 2008 *J. Am. Chem. Soc.* **130** 3296.
[3] Hosono H 2009 *Physica* **C 469** 314.
[4] Yanagi H, Watanabe T, Kodama K, Iikubo S, Shamoto S, Kamiya T, Hirano M and Hosono H. 2009 *J. Appl. Phys.* **105** 093916.
[5] Kayanuma K, Hiramatsu H, Kamiya T, Hirano M and Hosono H 2009 *J. Appl. Phys.* **105** 073903.
[6] Yanagi H, Fukuma K, Kamiya T, Hirano M and Hosono H 2010 *Mater. Sci. Eng.* **B173** 47.
[7] Yanagi H, Kawamura R, Kamiya T, Kamihara Y, Hirano M, Nakamura T, Osawa H and Hosono H 2008 *Phys. Rev.* **B 77** 224431.
[8] Watanabe T, Yanagi H, Kamiya T, Kamihara Y, Hiramatsu H, Hirano M and Hosono H 2007 *Inorg. Chem.* **46** 7719.
[9] Ogino H, Matsumura Y, Katsura Y, Ushiyama K, Horii S,Kishio K and Shimoyama J 2009 *Supercond. Sci. Technol.* **22** 075008.
[10] Okada S, Matoba M, Fukumoto S, Soyano S, Kamihara Y, Takeuchi T, Yoshida H, Ohoyama K and Yamaguchi 2002 *J. Appl. Phys.* **91** 8861.
[11] Kamihara Y, Matoba M, Kyomen T and Itoh M 2002 *J. Appl. Phys.* **91** 8864.
[12] Okada S, Kamihara Y, Ohkubo N, Ban S and Matoba M 2012 *J. Phys.: Conf. Ser.* **400** 022091.
[13] Momma K and Izumi F 2008 *J. Appl. Crystallogr.* **41** 653.
[14] Izumi F and Momma K 2007 *Solid State Phenom.* **130** 15.
[15] Xie YL, Liu RH, Wu T, Wu G, Song YA, Tan D, Wang XF, Chen H, Ying JJ, Yan YJ, Li QJ and Chen XH 2009 *Europhys. Lett.* **86** 57007.
[16] Blochl E 1994 *Phys. Rev.* **B 50** 17953.
[17] Kresse G and Furthmuller J 1996 *Comput. Mater. Sci.* **6** 15.

Journal of Physics: Conference Series **568** (2014) 022034 doi:10.1088/1742-6596/568/2/022034

[18] Kresse G and Furthmuller J 1996 *Phys. Rev.* **B 54** 11169.
[19] Kresse G and Joubert D 1999 *Phys. Rev.* **B 59** 1758.

Effects of Mn substitution on the thermoelectric properties of the electron-doped perovskite $Sr_{1-x}La_xTiO_3$

T Okuda[1], H Hata[1], T Eto[1], K Nishina[2], H Kuwahara[2], M Nakamura[3] and R Kajimoto[3]

[1]Graduate School of Science and Engineering, Kagoshima University, Kagoshima, 890-0065, Japan
[2]Department of Physics, Sophia University, Chiyoda, Tokyo 102-8554, Japan
[3]J-PARC Center, Japan Atomic Energy Agency, Ibaraki 319-1195, Japan

E-mail: okuda@eee.kagoshima-u.ac.jp

Abstract. We have tried to improve the n-type thermoelectric properties of the electron-doped Perovskite $Sr_{1-x}La_xTiO_3$ by a Mn substitution. The 1 ∼ 2 % Mn substitution enhances the Seebeck coefficient (S) and reduces the thermal conductivity (κ) by about 50 % at room temperature (RT) without largely increasing the resistivity for the 5 % electron-doped $SrTiO_3$. Consequently, the power factor at RT keeps a large value comparable to that of Bi_2Te_3 and the dimensionless figure-of-merits at RT increases twofold by the slight Mn substitution. Such a large reduction of κ at RT is perhaps due to the effect of Jahn-Teller active Mn^{3+} ions, around which dynamical local lattice distortion may occur.

1. Introduction

Since the good p-type thermoelectric (TE) property was found in Co oxides [1], transition metal oxides have been expected to be next generation TE materials, because of abundant natural resources and being nontoxic, environmental benign, and stable at high T. However, there is a problem to be solved in oxide TE materials. Good n-type oxide TE material being a counterpart of Co oxide has not been found till now, while several oxides, such as ZnO [2], $SrTiO_3$ [3, 4], $CaMnO_3$ [5], Fe pnictides [6, 7, 8], and etc., were known to show relatively good n-type TE properties.

Electron-doped perovskite $SrTiO_3$ is one of good candidates of n-type oxide TE materials. The power factor ($PF = \frac{S^2}{\rho}$) at room temperature (RT) is comparable to that of Bi_2Te_3 [3], but the dimensionless figure-of-merit ($ZT = \frac{S^2}{\rho\kappa}T$) is about 0.1 at RT [3] and about 0.3 at 1000 K [4]. There have been many attempts to improve the TE properties by various methods [9, 10].

In this study, we have tried to improve the TE properties of the electron-doped perovskite $Sr_{1-x}La_xTiO_3$ by a Mn substitution. We have found the Mn substitution enhances ZT up to about 0.15 at RT. The enhancement of ZT is due to the enhancement of the Seebeck coefficient (S) and the large reduction of the thermal conductivity (κ) without a large increase of resisitivty (ρ). In this proceeding, the origins of the changes of these physical properties will be discussed.

Journal of Physics: Conference Series **568** (2014) 022035 doi:10.1088/1742-6596/568/2/022035

2. Experimental procedures

Crystals of $Sr_{1-x}La_xTi_{1-y}Mn_zO_3$ were grown by a floating zone method. The starting materials were $SrCO_3$, La_2O_3, TiO_2, and $MnCO_3$. La_2O_3 was dried in air at 1000 C° for 12 hours before use. The stoichiometric mixture (Sr : La : Ti : Mn = $1-x$: x : 1 : 0 or $1-x$: x : 0.98 : 0.02) of starting materials were ground, calcined at 1000 C° for 12 hours, reground, calcined at 1300 C° for 12 hours, pressed into a rod, and sintered in air at 1350 C° for 12 hours. The crystal growth was performed at a rate of 30 - 35 mm/hour with rotation of the feed and seed rods in opposite directions in a flow of pure Ar gas.

Crystal structures and compositions for the obtained samples were investigated by the powder X-ray diffraction measurements and inductive coupled plasma (ICP) mass spectroscopy measurements, respectively. ρ of the obtained crystals was measured by the conventional four-probe method below 300 K by using a cryogen free cryostat. S and κ were simultaneously measured in a vacuum ($\sim 10^{-6}$ torr) by the steady method by using a handmade measurement system in a cryostat.

3. Results and discussions

3.1. Lattice parameters and compositions of $Sr_{1-x}La_xTi_{1-y}Mn_zO_3$

Powder X-ray diffraction measurements of the obtained samples indicate no impurity phase and they also indicate that the obtained crystals have a cubic structure with a space group of $Pm3m$ (No. 221). Figure 1 shows the nominal Mn concentration (y) dependences of lattice parameters of the obtained $SrTi_{1-y}Mn_zO_3$ and $Sr_{0.95}La_{0.05}Ti_{1-y}Mn_zO_3$ crystals deduced from the Rietveld analysis of the powder XRD patterns. If the Mn ions are not lost during the crystal growth, $y = z$. The lattice parameter for $Sr_{0.95}La_{0.05}Ti_{1-y}Mn_zO_3$ monotonically increases with the increase in y, while that for $SrTi_{1-y}Mn_zO_3$ monotonically decreases with the increase in y. These results indicate at least that the Mn^{3+} ions are produced at the Ti site in $Sr_{0.95}La_{0.05}Ti_{1-y}Mn_zO_3$, since the size relation of ion radii (d) in six-fold coordination (B (Ti) site) is $d(Mn^{4+}) < d(Ti^{4+}) < d(Mn^{3+})$ and that in twelve-fold coordination (A (Sr) site) is $d(Mn^{2+}) < d(Sr^{2+})$.

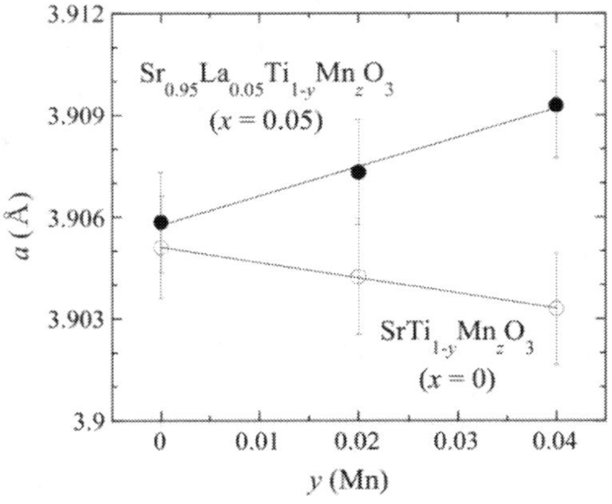

Figure 1. Nominal Mn concentration (y) dependences of lattice parameters (a) for $SrTi_{1-y}Mn_zO_3$ and $Sr_{1-x}La_xTi_{1-y}Mn_zO_3$.

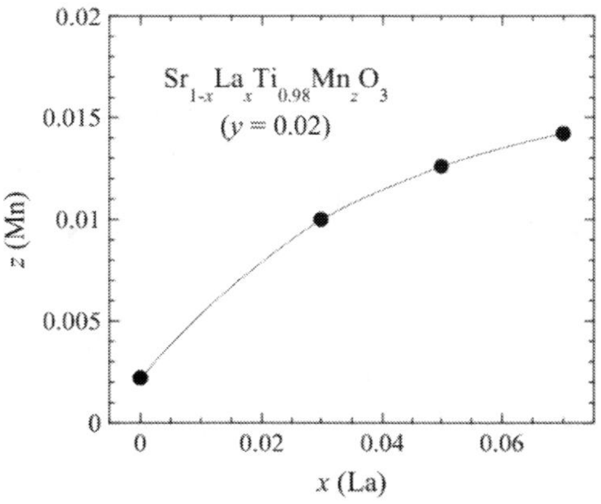

Figure 2. La substitution (x) dependence of the quantity of Mn cation (z) for $Sr_{1-x}La_xTi_{0.98}Mn_zO_3$ (y = 0.02).

However, according to ICP mass spectroscopy measurements, the quantity of Mn cation (z) for the obtained Mn substituted electron-doped $SrTiO_3$ crystal is smaller than the nominal one (y). Figure 2 shows the La substitution (x) dependence of the quantity of Mn cation (z) for $Sr_{1-x}La_xTi_{0.98}Mn_zO_3$, whose nominal value of z (= y) is 0.02. As shown in Fig.

2, the observed quantities of Mn cation are smaller than the nominal ones and the quantity monotonically increases with the increase in x. Then, the obtained Mn substituted crystal for $x = 0.05$ which shows the better TE properties (as will be discussed later) actually becomes $Sr_{0.95}La_{0.05}Ti_{0.98}Mn_{0.0126}O_3$ with lattice defects. This reduction of Mn ions is consistent with the observed absolute value of magnetization (M) smaller than the expected values of M which well obeyed the Curie-Weiss law. The spin quantum number of Mn ions estimated from the corrected M-T curve is much larger than 1.5, which also indicates that Mn^{3+} or Mn^{2+} ions are produced in $Sr_{0.95}La_{0.05}Ti_{0.98}Mn_{0.0126}O_3$.

3.2. Effects of Mn substitution on the thermoelectric properties of the electron-doped $SrTiO_3$

Figure 3 shows T dependences of ρ, S, and κ for $Sr_{0.95}La_{0.05}TiO_3$ and $Sr_{0.95}La_{0.05}Ti_{0.98}Mn_{0.0126}O_3$ crystals, the latter of which shows the best TE property in this study.

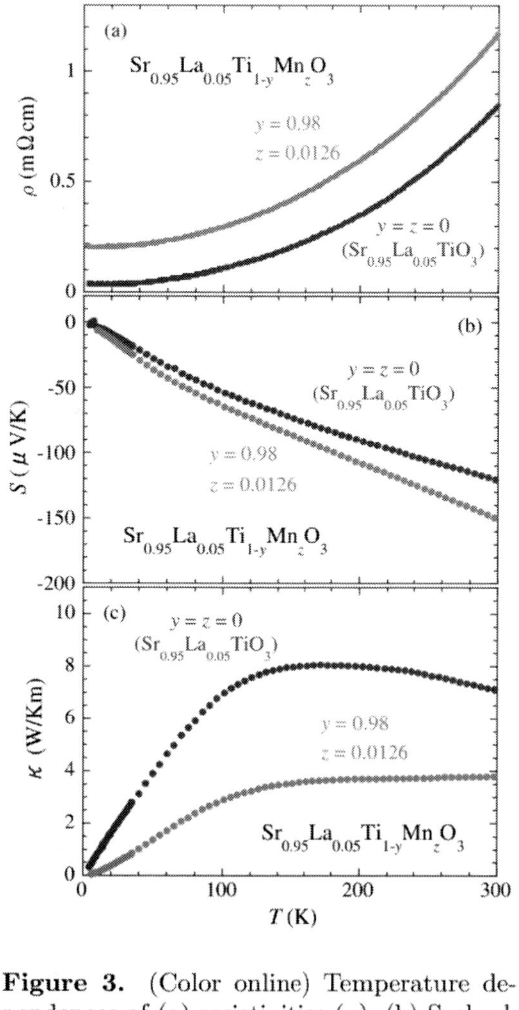

Figure 3. (Color online) Temperature dependences of (a) resistivities (ρ), (b) Seebeck coefficients (S), and (c) thermal conductivities (κ) for $Sr_{0.95}La_{0.05}Ti_{1-y}Mn_zO_3$.

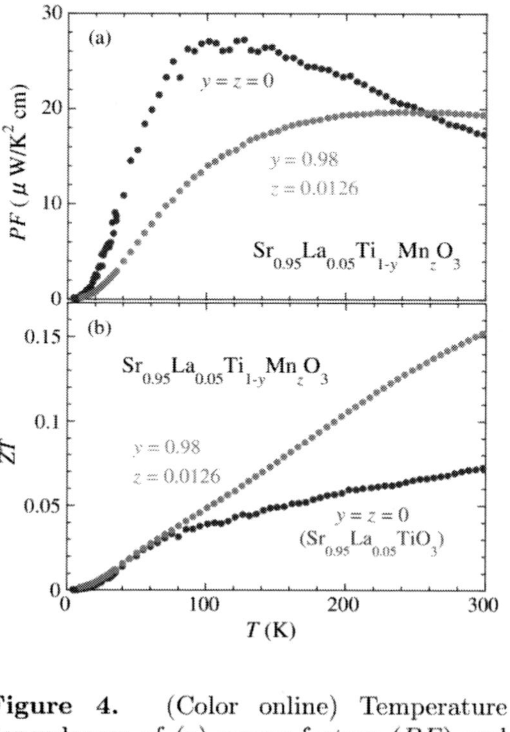

Figure 4. (Color online) Temperature dependences of (a) power factors (PF) and (b) dimensionless figure-of-merits (ZT) for $Sr_{0.95}La_{0.05}Ti_{1-y}Mn_zO_3$.

As shown in the Fig. 3(a), ρ slightly increases by the Mn substitution and the increase of ρ is almost constant below RT, which is perhaps due to the introduction of Mn ions and lattice defects. The absolute value of S ($|S|$) also increases by the Mn substitution. (Fig. 3(b)) According to the Hall measurements, the estimated carrier concentration (n) for $Sr_{0.95}La_{0.05}Ti_{0.98}Mn_{0.0126}O_3$ ($\sim$ 7 $\times$ 10^{20} cm^{-3}) is about 10 % smaller than that of $Sr_{0.95}La_{0.05}TiO_3$ ($\sim$ 8 $\times$ 10^{20} cm^{-3}), which should cause the increase of $|S|$. However, if the effective mass and relaxation time of electron is assumed to be invariant, the observed increase of $|S|$ ($\Delta S \sim 30$ μV/K at RT) cannot quantitatively be explained only by the decrease of n. Additional reason is necessary to fully explain the change. It is interesting that κ at RT decreases by about 50 % with the Mn substitution. (Fig. 3(c)) In comparison with those by the other substitutions [4, 11], the reduction by the slight Mn substitution is rather dramatic.

Figure 4 shows T dependences of PF and ZT, deduced by using the measured ρ, S, and κ. Although ρ increases with the Mn substitution, since the $|S|$ increases, the deduced PF at RT keeps the relatively large value comparable to that of Bi_2Te_3. Furthermore, since κ decrease by about 50 % at RT, the ZT value increases twofold at RT. The observed ZT value is about 0.15 at RT, which is the best value among the reported ones for bulk electron-doped $SrTiO_3$.

3.3. Origin of the reduction of thermal conductivity by the Mn substitution

The possible origins for the dramatic reduction of κ of the La and Mn cosubstituted compounds are the effects of the produced Mn^{3+} ions at the Ti sites and the lattice defects due to a loss of Mn ions. Since κ of $SrTi_{0.98}Mn_{0.0022}O_3$ having more lattice defects (Fig. 2) is as large as that of non-doped $SrTiO_3$ at RT (not shown), the reduction of κ of $Sr_{0.95}La_{0.05}Ti_{0.98}Mn_{0.0126}O_3$ is not thought to be due to the effect of lattice defects. Furthermore, taking into account the size relation of ion radii in six-fold coordination (B (Ti) site), since the lattice parameters decreases with the increase of Mn in $SrTi_{1-y}Mn_zO_3$ [12] (Fig. 1), the Mn ions in $SrTi_{0.98}Mn_{0.0022}O_3$ are thought to be Mn^{4+} at the Ti sites. Then, these results indirectly indicate that the strong candidate of the origin of the reduction of κ for the La and Mn cosubstituted $SrTiO_3$ is the effect of the produced Mn^{3+} ions at the Ti sites. The Mn^{3+} ion in six-fold coordination is a so called Jahn-Teller active ion, around which local lattice distortion occurs. Such a local lattice distortion around the Jahn-Teller active ions perhaps causes the reduction of κ. Furthermore, since, in comparison with the results of other substitutions [4, 11], the reduction is rather dramatic, the local lattice distortion may not be static but dynamic one which should effectively reduce κ. Lastly, it should be mentioned that the observed reduction of n in $Sr_{0.95}La_{0.05}TiO_3$ by the Mn substitution may originate from localization of a part of doped-electrons at the Mn ions placed at the Ti sites. Such localization of electrons at the Mn ions should change the Mn^{4+} ions at the Ti sites to the Jahn-Teller active Mn^{3+} ions.

4. Summary

We have tried to improve the thermoelectric (TE) properties of the electron-doped perovskite $SrTiO_3$ by a Mn substitution at the Ti sites and found that the dimensionless figure-of-merit for $Sr_{0.95}La_{0.05}Ti_{0.98}Mn_{0.0126}O_3$ was about 0.15 at RT. The relatively good n-type TE property for the compound was due to the enhancement of Seebeck coefficient (S) and the large reduction of thermal conductivity (κ) without a large increase of resistivity by the slight Mn substitution. The enhancement of S is partially due to the reduction of the carrier (electron) concentration perhaps due to localization of a part of doped-electrons at the Mn ions, although additional reason is necessary to fully explain the enhancement. Such localization of electrons at the Mn ions should lower the valence of Mn ions at the Ti sites to trivalent, i.e. it should produce a Jahn-Teller active Mn^{3+} ions. The Mn^{3+} ions at the Ti site may produce dynamical local lattice distortion around the ions, which may cause the dramatic reduction of κ. Further macroscopic and microscopic experimental investigations, such as specific heat measurements and neutron scattering measurements, are desired to evaluate the scenario.

Acknowledgments

This work was supported by JSPS KAKENHI Grant No. 25400378.

References

[1] Terasaki I, Sasago Y and Uchinokura K 1997 *Phys. Rev. B* **56** R12685
[2] Ohtaki M, Tsubota T, Eguchi K and Arai H 1996 *J. Appl. Phys.* **79** 1816
[3] Okuda T, Nakanishi K, Miyasaka S and Tokura Y 2001 *Phys. Rev. B* **63** 113104
[4] Muta H, Kurosaki K and Yamanaka S 2003 *J. Alloys Compd.* **350** 292
[5] Ohtaki M, Koga H, Tokunaga T, Eguchi K and Arai H 1995 *J. Solid State Chem.* **120** 105
[6] Pinsard-Gaudart L, Bérardan D, Bobroff J and Dragoe N 2008 *Phys. Status Solidi (RRL)* **2** 185
[7] Kihou K, Lee C H, Miyazawa K, Shirage P M, Iyo A and Eisaki E 2010 *J. Appl. Phys.* **108** 033703
[8] Okuda T, Hirata W, Takemori A, Suzuki S, Saijo S, Miyasaka S and Tajima S 2011 *J. Phys. Soc. Jpn.* **80** 044704
[9] Ohta H, Sungwng K, Mune Y, Mizoguchi T, Nomura K, Ohta S, Nomura T, Nakanishi Y, Ikuhara Y, Hirano M, Hosono H and Koumoto K 2007 *Nature Mat.* 129
[10] Ohta H, Masuoka Y, Asahi R, Kato T, Ikuhara Y, Nomura K and Hosono H 2009 *Appl. Phys. Lett.* **95** 113505
[11] Fukuyado J, Narikiyo K, Akaki M, Kuwahara H and Okuda T 2012 *Phys. Rev. B* **85** 075112
[12] Tkach A, Vilarinho P M and Kholkin A L 2005 *Acta Materialia* **53** 5061

Journal of Physics: Conference Series **568** (2014) 022036 doi:10.1088/1742-6596/568/2/022036

High field paramagnetic effect in YBCO single crystals with different oxygen contents

J.P. Peña[1], **O.J. Freitas**[2], **P. Pureur**[1]
[1]Instituto de Física, Universidade Federal do Rio Grande do Sul, Av. Bento Gonçalves 9500, Porto Alegre, RS, Brazil
[2] Centro Universitrio Metodista IPA, Rua Coronel Joaquim Pedro Salagado 80, Porto Alegre, RS, Brazil
E-mail: jullypaola@if.ufrgs.br

Abstract. Magnetization vs. temperature measurements were carried out on superconducting $YBa_2Cu_3O_x$ (YBCO) single crystals. Samples with oxygen concentrations corresponding to $x =$ 6.90, 6.85, 6.78 and 6.72 were studied. Zero field cooling (ZFC), field cooled cooling (FCC) and field cooled warming (FCW) prescriptions were employed. Measurements were performed in the temperature interval between 20 K and 120 K in constant magnetic fields ranged from 0.05 T to 5 T. A negative slope was observed in the FCC and FCW curves below a field-dependent minimum located in temperatures less than the irreversibility point (T_{irr}). This negative slope is identified as the high-field paramagnetic effect (HFPE). With a fixed magnetic field, the magnitude of the observed HFPE increases as the temperature decreases. The field dependence of the HFPE at fixed temperatures is not trivial. The overall magnitude of the measured HFPE depends on the oxygen content and becomes smaller when x decreases. The results are discussed on the basis of the effects of flux compression and spin moment polarization inside the vortex cores.

1. Introduction

Below some specific values of magnetic field and temperature superconductors go into Meissner state, i.e. they expel the magnetic flux from the inside. Above those field and temperature values, type II superconductors remain in the superconducting phase but rather develop a mixed state with partial penetration of the magnetic flux. In the mixed state, type II superconductors should still re-create a diamagnetic response when submitted to external magnetic fields. Surprinsingly, in the late eighties, it was experimentally shown that some samples of high temperature cuprate superconductors (HTSC) exhibit in fact, a paramagnetic response when cooled in the presence of very low magnetic fields ($B <1$ mT) [1, 2]. The effect is eventually suppressed when the field intensity surpasses a given value. This phenomenon was called the Wohlleben effect or Paramagnetic Meissner Effect (PME). Later on, it was found in textured YBCO monoliths that above a "neutral zone" of fields, where the expected diamagnetic response was observed, the paramagnetic effect reappeared at relatively higher fields and became stronger as the field was increased. This high field effect was then named as the High Field Paramagnetic Effect (HFPE) [3, 4].

When first discovered, it was thought that the PME was exclusive of unconventional granular superconductors. It was argued that the development of a paramagnetic response could be produced by the d-wave symmetry of the order parameter [5] and formation of spontaneous currents. Those currents would be caused by the occurrence of Josephson π junctions between grains [6]. However, since it was found that the effect can be observed in single crystals [7, 8] and conventional s-wave superconductors such as Nb [9, 10], the mechanism of Josephson π junctions may not be the only cause of the PME.

Since then, some attempts to explain both, the PME and the HFPE, have been proposed. These explanations include the formation of a giant vortex state [11] and compression of magnetic flux [12, 13]. The HFPE is the least studied and continues to be an open question. Here, we present some magnetization measurements as an effort to elucidate some possible causes of the appearance of the HFPE in deoxygenated YBCO single crystals.

2. Experimental details

Single crystals of YBCO were growth by the auto-flux method. The complete preparation process is published elsewhere [14]. The crystals were oxygenated by performing a thermal treatment in oxygen atmosphere. In order to obtain deoxygenated samples, some crystals were selected for new thermal processes in air atmosphere. These crystals were separately submitted to a 48 h annealing process at 500 ℃, 520 ℃ and 540 ℃. Then, the crystals were fastly cooled to room temperature. Samples with four different oxygen contents were obtained (see Table 1).

We carried out Magnetic measurements by using a SQUID magnetometer model XL5-MPMS® by Quantum Design, Inc. Magnetic fields up to 5 T were applied parallel to the c axis. Demagnetization corrections were carried out according to calculations in Ref. [15] by approaching the geometry of our samples to ellipsoids. Three sets of measurements were performed for each sample and each value of magnetic field. ZFC curves were obtained by cooling the sample through the critical temperature in zero field and registering the magnetic response while the temperature was increased. FCC curves were obtained by measuring the sample while cooling it in the presence of the field. For the FCW curves, magnetization measurements were carried out while heating the sample to high temperatures in the presence of the previously applied magnetic field.

3. Results

The critical temperatures of our samples were estimated from the intersection point of the extrapolated straight lines which fit the ZFC measurements in the superconducting and normal states performed in a magnetic field of 1 mT. The correspondence between T_c and x in the formula $YBa_2Cu_3O_x$ was estimated by using the graphs in Ref. [16]. Those data are presented in Table 1. We use both T_c and x for naming the samples.

Table 1. Equivalence of the oxygen content and the critical temperature for the studied $YBa_2Cu_3O_x$ crystals [16]

T_c	x
91.5	6.90
88.9	6.85
81.9	6.78
71.5	6.72

In Fig. 1, we present the FCC results for all samples in several applied fields. The FCW data (not shown) have a closely similar behavior to results shown in Fig. 1.

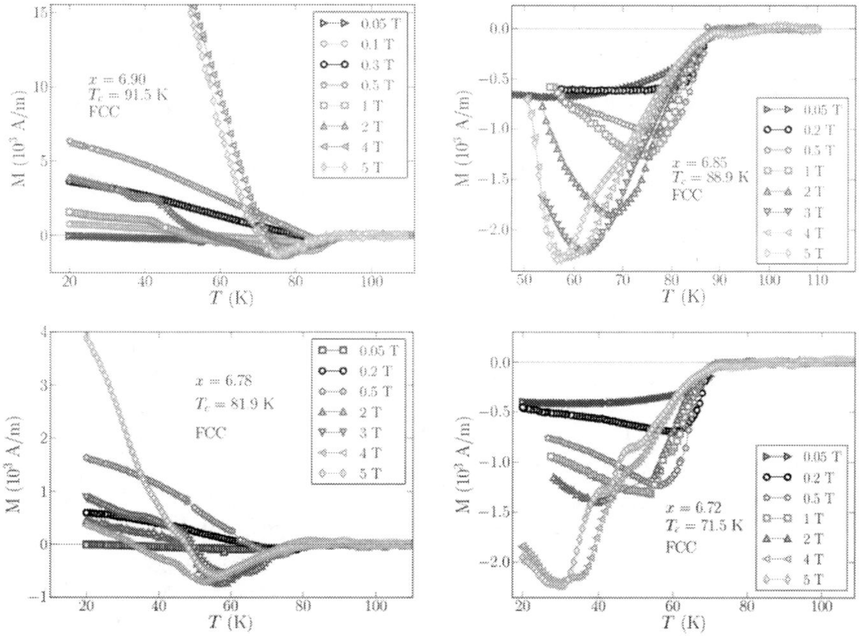

Figure 1. FCC magnetization vs. temperature results for YBCO crystals with different oxygen contents. The applied magnetic fields range between 0.05 T and 5 T.

As can be seen in Fig. 1, in all samples a negative slope occurs in the magnetization at low temperatures which we identify as a manifestation of HFPE. All curves present a minimum a few degrees below the critical temperature. This minimum is produced by the escape of the flux trapped in excess during the cooling procedure [4]. As the minimum occurs at temperatures below T_{irr}, we conclude that the HFPE is restricted to the irreversible region where the pinning forces acting on the vortices are substantial.

A real paramagnetic behavior is observed in samples with x =6.90 and x =6.78. We believe that a positive signal would be observed also in the sample with x =6.85 in fields above $B = 1$ T and temperatures below 50 K. For the most deoxygenated specimen, a linear extrapolation of the data to temperatures smaller than those measured does not reach a positive value even near the absolute zero. It is clear, however, that the HFPE occurs in this sample, though with a smaller intensity.

In the three more oxygenated samples, we observe a non-monotonic field dependence of the HFPE at fixed temperatures. As can be seen in Fig. 2, an "oscillatory behavior" is observed in the samples with x =6.90 and x =6.78.

In order to compare the HFPE in samples with different oxygen contents, in Fig. 3 we show FCC and FCW curves measured in all samples in several fields. The irreversibility between the FCC and FCW curves is observed in all crystals in low applied fields. However, only in the sample with x =6.90 the FCC-FCW irreversibility remains in the entire range of applied fields. The simultaneous ocurrence of HFPE and the FCC-FCW irreversibilities, as revealed by the results in Fig. 3, suggest that flux compression and pinning have an important role to explain the high field paramagnetic effect [3]. As the pinning becomes less efficient in deoxygenated YBCO samples, one expects that the magnitude of the HFPE progressively decreases as oxygen

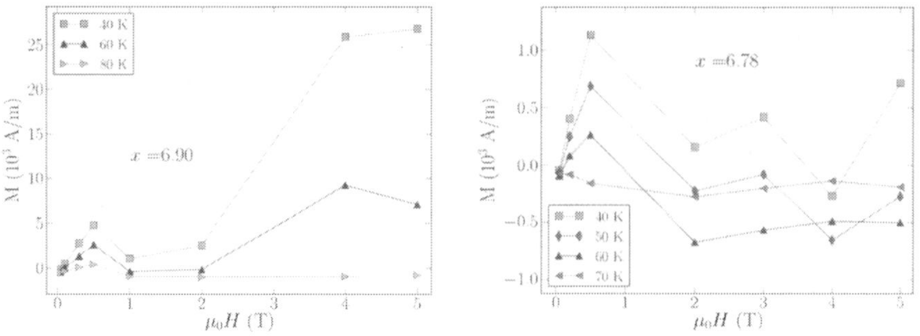

Figure 2. Magnetization as a function of the magnetic field for fixed temperatures as obtained from FCC curves for samples with x =6.90 and x =6.78.

is removed, as indeed observed in the series of crystals studied by us

In the last years, an out-of-plane spin magnetic moment in YBCO and other HTSC structures was discovered [17, 18]. This moment is observed when a moderate magnetic field is applied and is related to the canting of the antiferromagnetically ordered Cu^{2+} spins. The canting is generated by the Dzyaloshinskii-Moriya interaction within the CuO_2 planes. This interaction is reported to be non-zero even in samples with doping near the optimum value [18]. Here, we propose that the HFPE could also result from the polarization of those Cu^{2+} magnetic moments inside the vortex cores, where the magnetic induction attains locally significant magnitudes. In that case, the vortex pinning should promote the spin localization, thus contributing to generate a paramagnetic response. By using the unit cell volume ($V_c \approx 1.73 \times 10^{-22}$ cm^3), and the value for the magnetic moment in the superconducting planes of an YBCO sample with $T_c = 90$ K, $m_z = 0.019\mu_B/atom$ [18], we estimate that the contribution of the Cu moments to the magnetization of our most oxygenated sample is approximately M $\approx 2.04 \times 10^3$ A/m for an applied field of 4 T in a temperature of 9 K. This value is one order of magnitude less than the obtained by us, as can be checked by extrapolating the 4 T curve in the upper left panel in Fig. 1. So, even that the Cu atoms magnetization fills most of the HFPE size, it is clear that this is not the only contribution. Finally, as the magnetic moment developed by the Cu ions is larger in La_2CuO_4 than in YBCO, we anticipate that the HFPE should be higher in that compound if our statement is correct. Observation of the HFPE in La_2CuO_4 should be helpfull to elucidate if polarization inside the vortices cores is really a relevant mechanism to explain this unexpected though commonly occurring effect.

4. Conclusion

The high field paramagnetic effect was observed in YBCO single crystals with different oxygen contents. This fact reveals that the occurence of the HFPE is more general than previously thought. The effect is observed in low temperatures and relatively high magnetic fields. The temperature dependence of the HFPE reminds that of a superparamagnetic material. This suggests that polarization of the Cu spins inside the vortex cores might be important to explain the effect. However, as the magnetic signal of the localized Cu^{2+} spins is not directly related to the vortex lattice behavior and its contribution do not fulfill the size of the HFPE, we conclude that other mechanisms are also responsible for this effect. These mechanisms are probably pinning and flux compression, as deduced from the observation of FCC-FCW irreversibilities as well as from the progressive decrease of the effect intensity in crystals with lower oxygen contents.

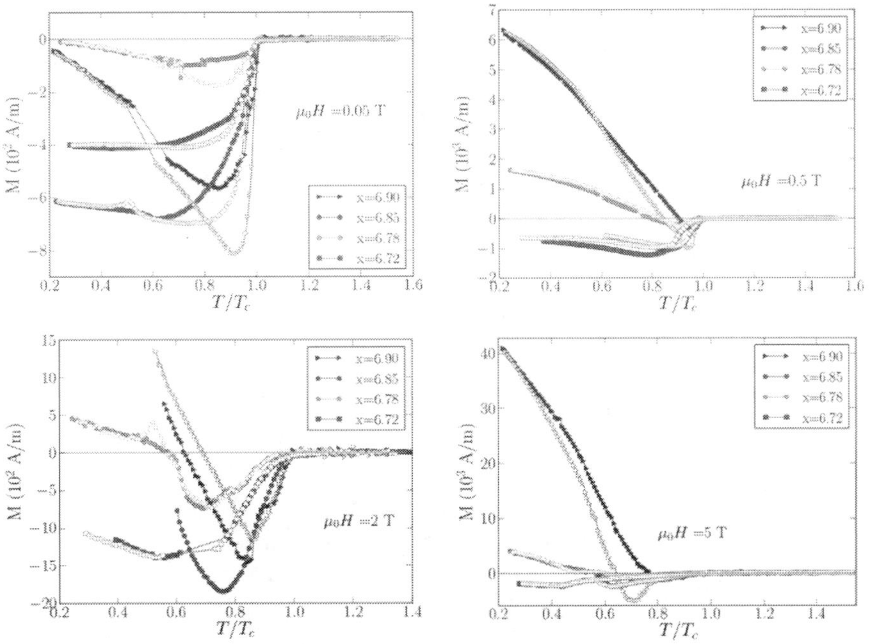

Figure 3. Magnetization of the studied YBCO crystals measured at several applied fields. Filled symbols represent FCC curves and open symbols represent FCW curves.

Acknowledgements
This work was partially financed by the Brazilian agencies FAPERGS and CNPq under grant PRONEX 10/0009-2. J.P. Peña benefits from a CNPq fellowship.

References
[1] P. Svendlindh, K. Niskanen, P. Norling, P. Nordblad, L. Lundgren, B. Lonnöberg, T. Lundstrom, Physica C 162-164, 1365-1366 (1989)
[2] W. Braunisch, N. Knauf, V. Kataev, S. Neuhausen, A. Grtz, A. Kock, B. Roden, D. Khomskii, and D. Wohlleben, Phys. Rev. Lett. 68, 19081911 (1992)
[3] C.P. Gouvea, F.T. Dias, V.N. Vieira, D.L. da Silva, J. Schaff, F. Wolff-Fabris, J.J. Roa Rovira, J. Korean Phys. Soc. 62, 1414 - 1417 (2012)
[4] F.T. Dias, P. Pureur, P. Rodrigues Jr., X, Obradors, Physica C 354, 219 - 222, (2001)
[5] M. Sigrist and T.M. Rice, J. Phys. Soc. Jpn. 61, 4283 (1992)
[6] F. V. Kusmartsev, Phys. Rev. Lett. 69, 22682271 (1992)
[7] S. Riedling, G. Brauchle, R. Lucht, G. Muller-Vogt. et al., Phys. Rev. B 49, 13283 - 13286 (1994)
[8] A.I. Rikov, S. Tajima, F.V. Kusmartsev. Phys. Rev. B 55, 8557 (1997)
[9] L. Pust, L.E. Wenger, and M.R. Koblischka, Phys. Rev. B 58, 14191 (1998)
[10] A. Terentiev, D.B. Watkins, L.E. De Long, D.J. Morgan, and J.B. Ketterson, Phys. Rev. B 60, R761 (1999)
[11] A.E. Koshelev and A.I. Larkin, Phys. Rev. B 52, 13559 (1995)
[12] V.V. Moshchalkov, X.G. Qiu, and V. Bruyndoncx, Phys. Rev. B 55, 11793 (1997)
[13] G.F. Zharkov, Phys. Rev. B 63, 214502 (2001)
[14] J.P. Peña, O.J. Freitas, P. Pureur, J. of Super. Nov. Mag. August (2014)
[15] J.A. Osborn, Demagnetizing factors of the general ellipsoid, Phys. Rev. 67, 351 (1945)
[16] R. Liang, D.A. Bonn, W.N. Hardy, Phys. Rev. B 73, 180505 (2006)
[17] B. Lake et. al., Nature Materials 4, 658 - 662 (2005)
[18] G.M. De Luca et. al., Phys. Rev. B 82, 214504 (2010)

Crystal growth and upper critical field of the noncentrosymmetric superconductor BiPd

Darren C. Peets

Max-Planck-Institut für Festkörperforschung, Stuttgart D-70569, Germany
Center for Correlated Electron Systems, Institute for Basic Science, Seoul National University, Seoul 151-747, Korea

E-mail: dpeets@snu.ac.kr

Abstract. The crystal growth of BiPd is reported and its magnetic phase diagram is clarified through measurements of the magnetization, resistivity, and specific heat. The bulk, thermodynamic upper critical field is roughly an order of magnitude lower than in several recent reports, and there are no noteworthy deviations from BCS *s*-wave expectations, again in contrast to most previous reports.

1. Introduction

Noncentrosymmetric superconductors do not possess spatial inversion symmetry, which means that parity of the superconducting wavefunction is not a meaningful concept, and singlet and triplet pairing are consequently not eigenstates. With our usual simplifying assumptions for understanding superconductivity no longer valid, a vast array of exotic physics becomes possible [1, 2]. Unfortunately, few such materials are known, many do not superconduct under ambient pressure, and single crystals have only been grown of a very few. In addition, significant spin-orbit splitting of the bands near the Fermi level is a required prerequsite for all proposed novel behaviour. As a consequence, most theoretical predictions remain unrealized.

Known to be both superconducting and noncentrosymmetric by 1953 [3] (predating BCS theory), α-BiPd was probably the first noncentrosymmetric superconductor identified as such, and crystals of the material may be readily prepared. That such materials were particularly interesting is a far more recent discovery [4], and this material is now attracting renewed attention [5–7]. Several techniques have been applied to establish whether the material exhibits nontrivial pairing, including NMR [8] and microwave susceptibility [7], turning up limited hints of unusual behaviour. The symmetry or mechanism of the pairing in this material, whether it has nodes in its gap function, and whether it exhibits mixing of singlet and triplet components have not been established. In fact, even some basic parameters of the material remain unclear. Recent resistivity and point-contact spectroscopy measurements suggest an upper critical field H_{c2} around 0.7 T [5, 6, 9], while more bulk-sensitive magnetization measurements indicate a much lower value [10]. The actual growth process, while relatively straightforward, has also not been reported in detail.

This paper details the crystal growth procedure and presents the bulk upper critical field as determined by magnetization, resistivity and specific heat measurements.

Journal of Physics: Conference Series **568** (2014) 022037 doi:10.1088/1742-6596/568/2/022037

2. Experimental

Growth progressed by a modified Bridgman-Stockbarger technique, shown schematically in Fig. 1(a). Pieces of bismuth (Aldrich, 99.999%) and palladium metal (Degussa, 99.95%) in a stoichiometric ratio were loaded into a degassed quartz tube with a cone-shaped end, and sealed under a vacuum of $\sim 10^{-4}$ torr. The sealed tube was suspended vertically in a two-zone furnace, supporting a modest temperature gradient of $\approx 1.3°C/cm$. The entire ampoule was heated above 850°C — well above the congruent melting transition at 600°C [11, 12] — and left at high temperature for four hours to fully melt and homogenize. The ampoule was cooled through the melting transition at 0.2°C/h, corresponding to a growth rate of 1.5 mm/h. Differential thermal analysis on a test sample grown under similar conditions indicated additional apparent structural transitions at 189°C at 245°C, including the transition between α-BiPd and β-BiPd [13, 14], so the ampoule was cooled through this temperature range at 0.5°C/h before being cooled more rapidly to room temperature.

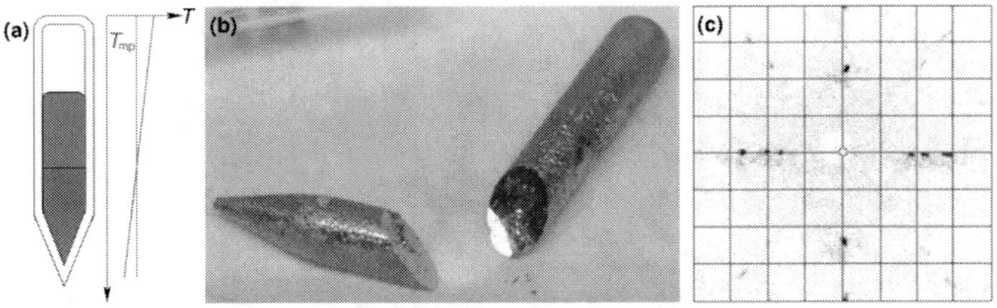

Figure 1. BiPd single crystal growth. (a) Schematic growth process: Molten BiPd was cooled through its melting point (T_{mp}) in a temperature gradient (black) while sealed in a quartz tube, leading to (b) an essentially-single-domain crystal. (c) Laue pattern of a cleaved surface.

Thanks to the cone-shaped end of the ampoule from which growth started, which rapidly selected one preferred orientation, the resulting crystal was essentially single-domain, although heavily twinned and somewhat cracked internally due to the α–β phase transition. As shown in Fig. 1, it cleaves readily to expose [010] faces (taking the b-axis as the unique monoclinic axis and $\beta > 90°$). While this has been implied previously, it is worth emphasizing that the material cleaves, as this may enable a number of surface-sensitive techniques that have not been previously applied to noncentrosymmetric superconductors due to an inability to obtain suitable cleaved surfaces.

Magnetization measurements were performed for various field orientations in Quantum Design MPMS-7 and MPMS-XL magnetometers with the RSO option, and resistivity was measured in a Quantum Design PPMS by a standard four-wire technique — the absolute value may not be intrinsic due to possible cracking in the sample or scattering at twin boundaries, and the accuracy is limited by estimates of the sample dimensions. To confirm the bulk, thermodynamic T_c and H_{c2}, specific heat was measured at low temperatures in a Quantum Design PPMS in zero field and for fields along the monoclinic b axis.

3. Results

Magnetization in a 5 mT applied field is shown as a function of temperature in Fig. 2(a) for several field orientations, and $M(H)$ loops at 1.8 K are shown in Fig. 2(b). These data closely resemble those reported previously [6, 10]. The hysteresis loops exhibit only minor differences with field orientation, with higher onset fields for in-plane applied field. These data have not been corrected for demagnetization effects, and the crystal measured was a platelet, thin in the

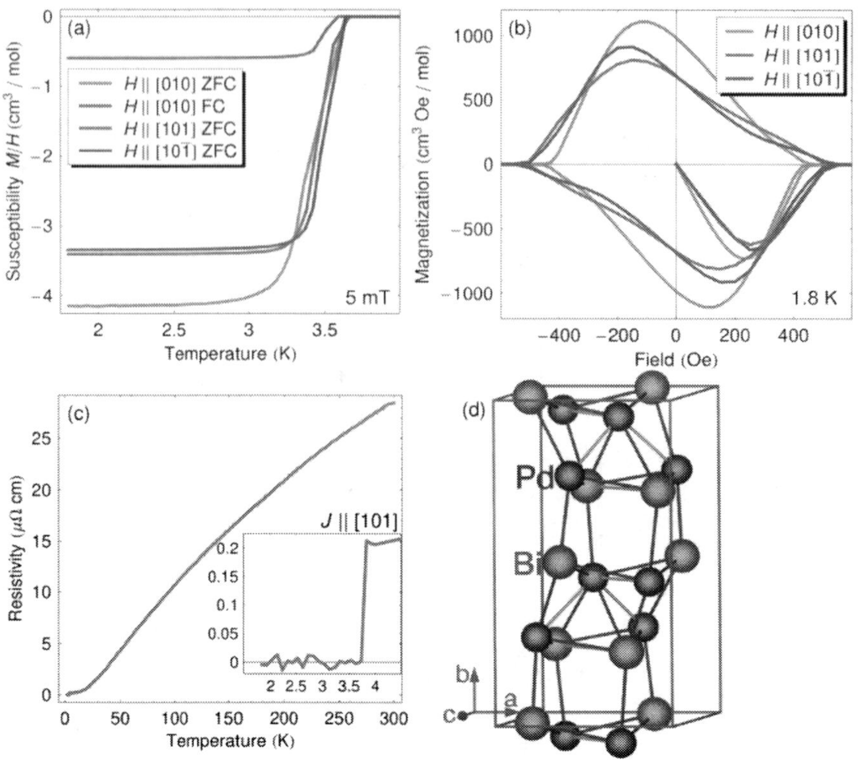

Figure 2. (a) Magnetization curves in $H = 5$ mT for three field directions; (b) $M - H$ loops for the three field orientations; (c) zero-field resistivity; and (d) Crystal structure of BiPd, based on Ref.[14].

b direction and irregularly shaped — the overall vertical scale may not be directly comparable, particularly when comparing in-plane and b-axis fields. Resistivity is presented in the same figure, but absolute values may be unreliable due to internal cracking, and the low resistivity of metallic BiPd made this measurement challenging. The residual resistivity ratio of 140 is comparable with those reported by other groups [5], as is the resistive T_c of 3.8 K.

Specific heat was measured from 0.37 to 5 K in fields up to 1.5 T, to determine the bulk superconducting transitions and gain insight on the structure of the gap function; selected results are presented in Fig. 3. The curve qualitatively matches the weak-coupling BCS expectation [15] up to at least $0.5T_c$, although falling slightly below it, with a clear exponential onset implying a full gap and suggesting a relatively isotropic gap function. The data fall significant above the BCS curve on the approach to T_c, leading to a jump height $\Delta c_{el}/\gamma T_c$ of 1.50 (BCS: 1.43). The superconducting transition is sharp, and the extracted T_c of 3.77 K agrees very well with that determined from the magnetization and resistivity measurements. Magnetic field suppresses the transition, which is completely absent above 0.4 K in a field of only 0.07 T. Fig. 3(d), depicting the data above the bulk transition $H_{c2}(T)$ for all fields, shows that there is no evidence for a second phase bulk transition which would correspond to the transition previously reported based on resistivity measurements. A polynomial fit [also shown in Fig. 3(d)] was performed to all normal-state data (up to T^7 because the data began deviating from standard phonon

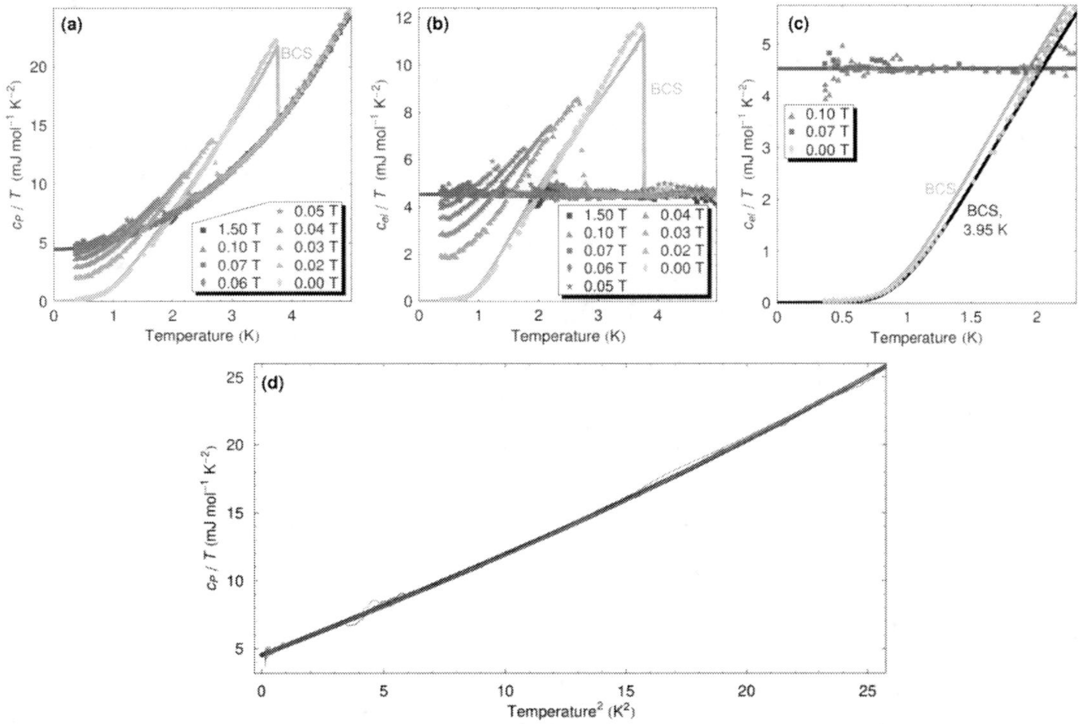

Figure 3. (a) $c_P(T)/T$ vs. T for several fields, showing the bulk superconducting transition and its suppression by field. The weak-coupling s-wave BCS form is included for comparison [15]. There is no transition visible above 0.4 K under 0.07 T or higher fields. (b) $c_{el}(T)/T$ vs. T. (c) The low-temperature region is fit better by assuming the T_c of 3.95 K that would be predicted from the STM gap [16]; this is pursued in the paper cited. (d) All data taken above the obvious (field-dependent) bulk transition, in all fields, and all data in fields of 70 mT and above — there is no evidence to suggest an additional bulk phase transition. The apparent feature near 4 K^2 was not reproduced in lower-temperature measurements.

T^3 behaviour around 3 K) to model the normal-state background. The Sommerfeld electronic specific heat coefficient γ is a modest 4.53 mJ/molK^2, and the phonon T^3 term's prefactor β is 0.710 mJ/molK^4, corresponding to a Debye temperature of 176 K. The phonon contribution's rapid departure from T^3 behaviour suggests that at least one phonon mode is rather low in energy.

Interestingly, the gap obtained from scanning tunneling spectroscopy is 0.601 meV [16], from which weak-coupling BCS theory would predict a T_c of 3.95 K. If this is used instead of the actual bulk T_c, the data are fit significantly better below ~1.8 K, as shown in Fig. 3(c). This suggests that the weak coupling approximation is not completely valid in this material. An analysis of the specific heat incorporating the known ratio of the pairing gap to T_c is pursued elsewhere [16]. The otherwise strong resemblance to the BCS form and lack of any sign of nodes are striking in light of previously published results.

The transitions obtained from magnetization, resistivity and specific heat measurements taken in a variety of applied magnetic fields may be combined into an $H-T$ phase diagram, shown in Fig. 4. The sparse data below 1.8 K do not allow a reliable estimate of $H_{c2}(0)$,

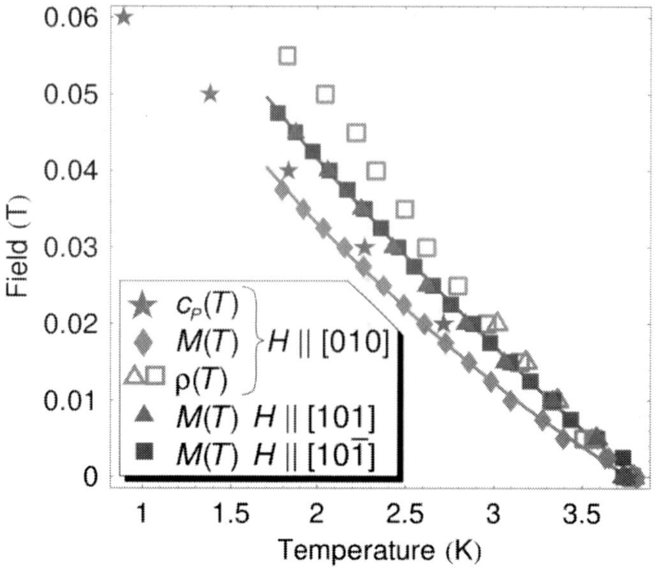

Figure 4. $H-T$ phase diagram constructed from the transitions measured by resistivity, magnetization and specific heat measurements.

but it is clearly inconsistent with the previously-reported $H_{c2}(0)$ of 0.7 T ($H \parallel b$) and 1.2 T ($H \perp b$) from resistivity and point-contact spectroscopy [5, 9]. Indeed, it would appear to be a full order of magnitude lower, and more consistent with values of 92 [$H \parallel (010)$] and 138 mT [$H \parallel (100)$] reported based on magnetization measurements [10], although even those values are likely incompatible with these data. The slight upward curvature observed near T_c and apparent linear dependence at intermediate temperatures would be unusual for an ordinary BCS superconductor, but are commonly observed in multi-band systems. BiPd's large number of valence electrons, low symmetry, and potential band splitting due to spin-orbit coupling make it exceedingly likely that this system has multiple bands crossing the Fermi level.

The zero-field electronic specific heat c_{el} may be integrated to obtain the thermodynamic critical field H_c, using

$$\frac{\mu_0 H_c^2}{2} = -\frac{\gamma T_c^2}{2} + \int_0^{T_c} c_{el} dT \tag{1}$$

in appropriately chosen volume units. The resulting thermodynamic H_c of 41 mT, combined with the $H_{c2}^b(0)$ of roughly 80 mT for $H \parallel b$ obtained by scanning tunneling microscopy [16] and consistent with the higher-temperature data reported here, would indicate a $\kappa_{GL}^b \approx 1.4$ for this field orientation. This would make the superconductivity weakly Type-II. If the ac-plane coherence length ξ_{ac} can be assumed to be isotropic, as the magnetization would suggest, then the $\xi_{ac}(0)$ extracted from $H_{c2}^b(0)$ is 64 nm. If the same assumption is extended to the penetration depth, $\lambda_{ac}(0) = 89$ nm, although such an assumption is not well justified at this stage. It is worth noting that this penetration depth is very far from the value assumed in recent microwave work [7]. If one further assumes that the factor of ~1.2 anisotropy seen in the magnetization-derived transition in Fig. 4 continues to zero temperature, $H_{c2}^{ac}(0)$ would be 96 mT, κ_{GL}^{ac} would take a slightly stronger Type-II value of 1.67, $\xi_b(0)$ would be 53 nm, and $\lambda_b(0)$ would be 107nm. Obtaining these values required several very significant assumptions, and they will require confirmation, either by extending measurements of H_{c2}^i to low temperature, or through independent measurements of the coherence length and penetration depth, perhaps from muon spin rotation.

4. Conclusion

The data presented here in characterizing BiPd show no evidence of any significant deviation from weak-coupling BCS superconductivity. $H_{c2}(T)$ exhibits upward curvature suggestive of multi-band superconductivity, which would be wholly unsurprising in this material, and the specific heat suggests that BiPd is not fully in the weak-coupling limit, but there is no evidence for nodes. The upper critical field is found to be significantly lower than reported previously based on resistivity and point-contact measurements. Possible reasons for this include differences in stoichiometry: a significant Bi deficiency has been shown to lower T_c and the residual resistivity ratio [6] and both metrics indicate the crystals in question to be of similar quality, but it's conceivable that minor deviations from the ideal stoichiometry could have significant effects on a nodal gap. Samples could also differ in twin domain size. If the twins are small compared to the coherence length, conceivably the material may no longer appear noncentrosymmetric to the superconducting carriers. It is perhaps most likely that the higher H_{c2} values benefit from thin limit physics, with filamentary superconductivity occurring along extended defects such as grain or twin boundaries. It is also possibly worth noting that in Sr_2RuO_4, widely thought to be a chiral p-wave superconductor, the superconducting onset temperature can double at boundaries with Ru metal inclusions [17], and there is some evidence that this boundary may nucleate one component of the multi-component order parameter. Twin boundaries may exhibit analogous physics in BiPd. To resolve the apparent contradiction in upper critical fields, it would be desirable to apply bulk probes to crystals known to have a high H_{c2}, and to use local probes to check for (and characterize) remnant superconductivity above H_{c2} at twin boundaries.

Acknowledgments

The author thanks C. Stahl and E. Goering of Abt. Schütz, MPI-IS, for help with magnetization measurements, and H. Jo for help measuring low-temperature specific heat. The author also is grateful for the assistance of the MPI-FKF Crystal Growth and Chemical Service Groups, and for stimulating discussions with A. Schnyder and P. Wahl. A portion of this work was supported by the Institute for Basic Science (IBS) in Korea.

References

[1] E. Bauer and M. Sigrist, eds., *Non-Centrosymmetric Superconductors: Introduction and Overview* (Springer-Verlag, 2012).

[2] S. Fujimoto, J. Phys. Soc. Japan **76**, 051008 (2007).

[3] D. M. Kheiker, G. S. Zhdanov, and N. N. Zhuravlev, Zh. Eksp. Teor. Fiz. **25**, 621 (1953).

[4] E. Bauer, G. Hilscher, H. Michor, C. Paul, E. W. Scheidt, A. Gribanov, Y. Seropegin, H. Noël, M. Sigrist, and P. Rogl, Phys. Rev. Lett. **92**, 027003 (2004).

[5] B. Joshi, A. Thamizhavel, and S. Ramakrishnan, Phys. Rev. B **84**, 064518 (2011).

[6] K. Okawa, M. Kanou, T. Katagiri, H. Kashiwaya, S. Kashiwaya, and T. Sasagawa, Physics Procedia **45**, 101 (2013).

[7] L. Jiao, J. L. Zhang, Y. Chen, Z. F. Weng, Y. M. Shao, J. Y. Feng, X. Lu, B. Joshi, A. Thamizhavel, S. Ramakrishnan, and H. Q. Yuan, Phys. Rev. B **89**, 060507(R) (2014).

[8] K. Matano, S. Maeda, H. Sawaoka, Y. Muro, T. Takabatake, B. Joshi, S. Ramakrishnan, K. Kawashima, J. Akimitsu, and G.-Q. Zheng, J. Phys. Soc. Japan **82**, 084711 (2013).

[9] M. Mondal, B. Joshi, S. Kumar, A. Kamlapure, S. C. Ganguli, A. Thamizhavel, S. S. Mandal, S. Ramakrishnan, and P. Raychaudhuri, Phys. Rev. B **86**, 094520 (2012).

[10] B. Joshi, A. Thamizhavel, and S. Ramakrishnan, JPS Conf. Proc. **3**, 015010 (2014), arXiv:1402.7232 [cond-mat.supr-con] .

[11] B. Predel, "Bi-Pd (bismuth-palladium)," in *Landolt-Börnstein – Group IV Physical Chemistry*, Vol. 5b, edited by O. Madelung (Springer-Verlag, 2014).

[12] J. Vřešt'ál, J. Pinkas, A. Watson, A. Scott, J. Houserová, and A. Kroupa, Computer Coupling of Phase Diagrams and Thermochemistry **30**, 14 (2006).

[13] Y. Bhatt and K. Schubert, Journal of the Less Common Metals **64**, 17 (1979).

[14] V. M. Ionov, N. A. Tomlin, A. E. Prozorovskii, A. N. Klimenko, Y. V. Titov, S. G. Zhukov, and G. V. Fetisov, Sov. Phys. Crystallogr. **34**, 496 (1989).

[15] B. Mühlschlegel, Zeitschrift für Physik **155**, 313 (1959).

[16] Z. Sun, M. Enayat, A. Maldonado, C. Lithgow, E. Yelland, D. C. Peets, A. Yaresko, A. P. Schnyder, and P. Wahl, (2014), arXiv:1407.5667 [cond-mat.supr-con] .

[17] Y. Maeno, T. Ando, Y. Mori, E. Ohmichi, S. Ikeda, S. Nishizaki, and S. Nakatsuji, Phys. Rev. Lett. **81**, 3765 (1998).

Torque magnetometry in $YBa_2Cu_3O_{7-\delta}$ single crystals using high sensitive micromechanical torsional oscillator

D. Perez-Morelo, H. Pastoriza

Centro Atómico Bariloche, Comisión Nacional de Energía Atómica and CONICET, R8402AGP S. C. de Bariloche, Argentina.

E-mail: djperez@cab.cnea.gov.ar

Abstract.
We have developed a high-sensitive micro-torsional oscillators and fabricated with a standard silicon process, as a platform to study the low temperatures magnetic properties of micron sized samples. When the oscillator-sample system interacts with a magnetic field, the displacement of the oscillator from its equilibrium position can be detected as a small change in the capacitance (≈ 0.001 pF).

We use this device to study the magnetic response of a thick twinned $YBa_2Cu_3O_{7-\delta}$ disk when the magnetic field is tilted away from the c axis in the temperature range $80\,K < T < 100\,K$ ($T_c \approx 92\,K$).

1. Introduction

Microelectromechanical torsional oscillators have previously been shown to be particularly useful for high sensitive measurements of small signals in a variety of nonmagnetic and magnetic experiments. These measurements include the de Haas–van Alphen oscillations [1], mesoscopic superconductors [2, 3] and Casimir effect [4, 5].

When the magnetic sample is affixed to a micromechanical oscillator and an external magnetic field is applied, the magnetic torque produces a measurable tilt of the oscillator. The magnetic properties of the attached sample can be extracted from of this change in the resonator[6].

In this work we present the response of a micro-oscillator with a $YBa_2Cu_3O_{7-\delta}$ (YBCO) disk mounted on top of it, in the presence of an external magnetic field H for varying temperatures and angles.

2. Torsional oscillator

The polysilicon mechanical micro-oscillators, based on an original design by Chan *et al.* [4], was fabricated in the MEMSCAP [7] foundry using its Polysilicon Multi-User MEMS Processes (Poly MUMPS). The device consists in a $500 \times 500\,\mu m^2$ suspended central plate that is anchored to the substrate by means two torsional rods. Two independent electrodes under the plate at each side allow the measurement of the capacitance between electrodes and the plate. The nominal separation on rest between the plate and electrodes is $2\,\mu m^2$. Photomicrograph images of the completed device are displayed in the Figure 1.

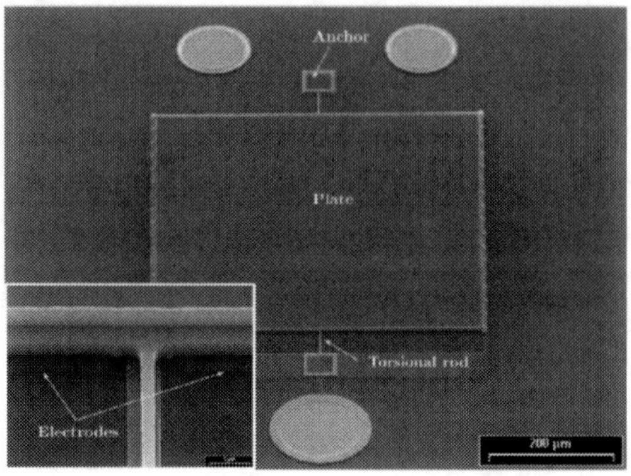

Figure 1. SEM image of the mechanical torsional oscillator. The inset shows a detail of the two electrodes located directly under the paddle. The gap between the electrodes and the paddle is fixed by the multilayer process and is 2 μm.

This structure has a variety of the possible modes of vibration. The torsional mode is the lowest in energy and will be the only one described in this paper as the device was designed for applications requiring torsional oscillations. In this normal torsion mode, i.e., not taken into account deformation of the arms, the restitution elastic constant for a torsion beam with an almost square profile $w \simeq t$ can be calculated analytically [8]:

$$k = \frac{E}{2(1+m)l}\left[\frac{w^3t + wt^3}{6}\right] \tag{1}$$

where l, w and t are respective length, width and thickness of the torsion arms. For polysilicon the Young's modulus is $E = 158\,\mathrm{GPa}$ and the the Poisson's ratio $m = 0.28$. The total torsional elastic constants for different designs are show in Table 1.

The normal torsion resonance frequency is given by

$$\nu_0 = \frac{1}{2\pi}\sqrt{\frac{k}{I}} \tag{2}$$

where I is the moment of inertia of the full structure along the rotational axis, $\approx 4.23{\cdot}10^{-17}\,\mathrm{kg}{\cdot}\mathrm{m}^2$. Using this value in (2) we obtain $\nu_0 \approx 2\,\mathrm{kHz}$. To validate this analytical solution we compare with the results from the modal analysis obtained by finite element method simulations (FEM 3D) of the proposed device.

Table 1. Elastic constants to different designs ($w = t \approx 2\,\mu$m)

Design	rods	Length(μm)	$k(\mathrm{N}\cdot\mathrm{m}\cdot\mathrm{rad}^{-1})$
A	2	$l \approx 15$	$4.38 \cdot 10^{-8}$
B	2	$l \approx 25$	$2.63 \cdot 10^{-8}$
C	4	$l \approx 200$	$5.41 \cdot 10^{-9}$

3. Capacitive Detection

When a sample is mounted and affixed to the center of polysilicon plate both rotate together. In practice, this rotation can be detected measuring the capacitance between the mobile plate and each fixed electrode (Fig 2). The plates of this capacitor are no longer parallel, however we can approximate it by a combination of N differential parallel plate capacitors with length L and infinitesimal width dr [9]:

$$dC = \frac{\epsilon_0 L \cos(\phi) dr}{g(r)} \tag{3}$$

where $g(r) = h_0 + r\tan(\phi)$. Taking the limit $N \to \infty$, the total capacitance is then given by

$$C = \frac{\epsilon_0 L}{\sin(\phi)} \ln(\frac{h_0 + r_2\tan(\phi)}{h_0 + r_1\tan(\phi))}) \tag{4}$$

This equation can be solved numerically to obtain ϕ from capacitance data.

The torsional rods provide a restoring torque $k\phi$ to counter an external torque on the top plate. In the case of anisotropic magnetic samples the existence an external magnetic field H non parallel to the sample magnetization produces a torque:

$$\tau = \mathbf{M} \times \mathbf{H} = MVH\sin(\alpha) = k\phi \tag{5}$$

where M is the magnetization, V is the sample's volume and $\alpha = \theta_H - \phi$ is the angle between **M** and **H**. In equilibrium, the magnetic torque and the restoring torque $k\phi$ are equal (stable condition).

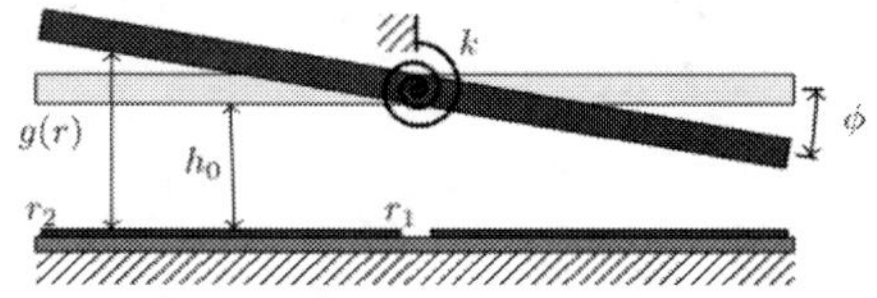

Figure 2. Schematic view of the mechanical torsional micro-oscillator. The width of the electrode is $r_2 - r_1$ with $r_1 = 4\,\mu$m and $r_2 = 249\,\mu$m.

4. Measuring YBCO samples with the micro-oscillators

The microoscilator–sample system is mounted onto a rotable sample holder with an angular resolution of $\theta \approx 0.05°$ inside a cryostat with a superconducting 18 T magnet. In this experiment have been used YBCO twinned single crystals, grown using the flux growth technique (the details of the heat treatments are described in Ref.[10]). Disks of diameter $100\,\mu$m and thickness $8\,\mu$m were prepared using Ga ion-milling (FIB) in CNN Argonne National Laboratory.

We did FC (field cooling) and FCW (field cooling warming) measurements. The changes in the capacitance were measured using a commercial high-precision capacitance bridge in a differential configuration. In the Fig.3 shows a measurement of the capacitance as a function of temperature in both electrodes with external magnetic field $H = 2\,$T. On measuring the capacitance in both electrodes allow us to subtract any time dependent drift usually found in low capacitance measurements.

The superconducting response of the YBCO disk was checked by measuring the torque as a function of temperature with the field H held constant. The result, shown in Fig. 4, provides a direct measure of T_c ($= 92\,$K). With a torsional spring constant as small as $5.41 \cdot 10^{-9}\,$N·m·rad^{-1}, the device yields a sensitivity $\Delta\tau = 2.4 \cdot 10^{-13}\,$N · m for torques acting at the plate. As **H** is rotated away from the c-axis, the torque increases. This behavior is related to the progressive rotation of the magnetic moment, which yields a zero torque when fully aligned to **H**.

From the inset Fig. 4 we can observe two different behavior on the temperature range and to a small angle θ_H. First, the torque data are softly moved to lowest temperature as the magetic field is increased, therefore T_c decreases slightly. On the other hand, as the temperature decreases (in the a FC protocol), the torque suddenly begins to decrease as from a $T < T_c$, indicating a change in the direction and magnitud of the magnetization.

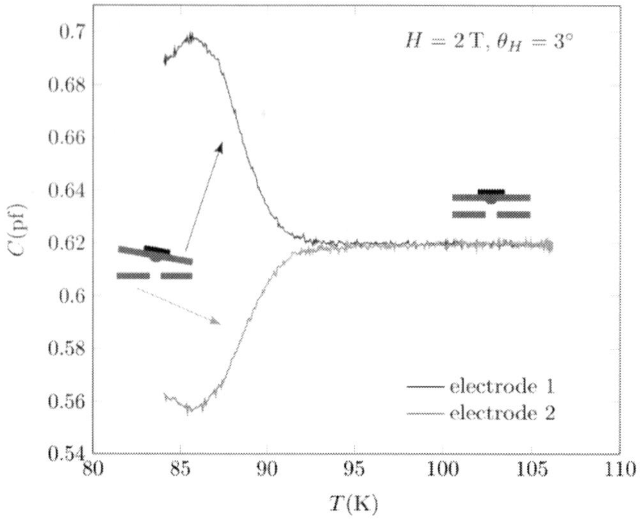

Figure 3. Temperature dependence of the capacitance in boths electrodes (differential configuration) for FC measurements with external magnetic field $H = 2\,\mathrm{T}$. The magnetic fiel is initially (high temperatures) tilted 3° away from the c axis.

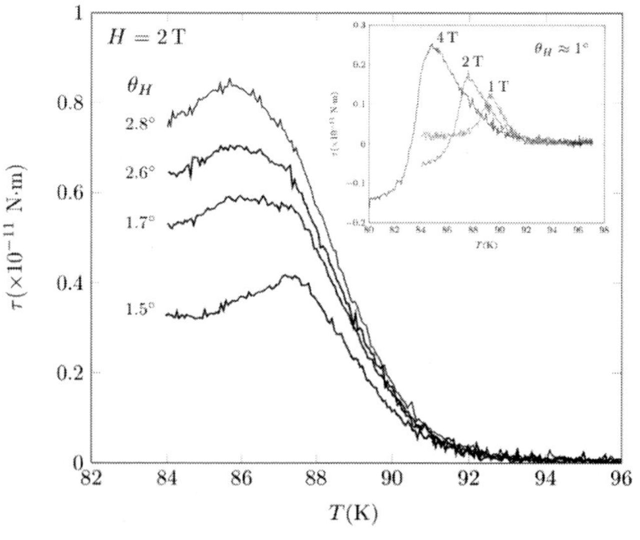

Figure 4. Temperature dependence of the torque of the twinned disk of YBCO for a succession of angles between H and c-axis. The magnetic field was held constant at 2 T. Inset: Temperature dependence of the torque for different applied magnetic fields.

5. Conclusion

The Si mechanical micro-oscillators proposed here with rectangular plate and thin torsion bars are suitable to measure high T_c samples with high sensitivity. Detection of torques of $\approx 2.4 \cdot 10^{-13}\,\mathrm{N} \cdot \mathrm{m}$, corresponding to a magnetic moment of $\approx 2.4 \cdot 10^{-13}\,\mathrm{A} \cdot \mathrm{m}^2$ at 1T has been achieved.

References

[1] Aksyuk V, Balakirev F F, Boebinger G S, Gammel P L, Haddon R C and Bishop D J 1998 *Science* **280** 720
[2] Bolle C A, Aksyuk V, Pardo F, Gammel P L, Zeldov E, Bucher E, Boie R, Bishop D J and Nelson D R 1999 *Nature* **399** 43
[3] Dolz M, Antonio D and Pastoriza H 2007 *Physica B* **398** 329
[4] Chan H B, Aksyuk V A, Kleiman R N, Bishop D J and Capasso F 2001 *Science* **291** 1941
[5] Decca R S, López D, Fischbach E and Krause D E 2003 *Phys. Rev. Lett.* **91** 050402
[6] Dolz M, Bast W, Antonio D, Pastoriza H, Curiale J, Sánchez R D and Leiva A G 2008 *J. Appl. Phys.* **103** 083 909
[7] MEMSCAP Inc., Durham, NC, USA. `http://www.memscap.comi`
[8] Rossel C, Willemin M, Gasser A, Bothuizen H, Meijer G I and Keller H 1998 *Rev. Sci. Instrum.* **69** 3199
[9] Antonio D and Pastoriza H 2009 *J. Microelectromech. Syst.* **18** 1396
[10] de la Cruz F and Nievas G 1994 *Philos. Mag.* B **70** 773

Journal of Physics: Conference Series **568** (2014) 022039 doi:10.1088/1742-6596/568/2/022039

Superconductivity in $Y_3Ru_4Ge_{13}$ and $Lu_3Os_4Ge_{13}$: A comparative study

Om Prakash, A. Thamizhavel, S. Ramakrishnan

Department of Condensed Matter Physics and Material Science, Tata Institute of Fundamental Research, Mumbai-400005, India

E-mail: op1111shukla@gmail.com

Abstract. A variety of unconventional superconductors have low carrier density as a common factor. However, the underlying mechanism of superconductivity in such low carrier density systems is not well understood. Besides, small carrier density is an unfavourable component for conventional superconductivity as described by the Bardeen-Cooper-Schrieffer (BCS) theory. Therefore, studying low carrier density systems can lead to a better understanding in such systems. In this paper, we report superconductivity property studies in low carrier density systems, $Y_3Ru_4Ge_{13}$ and $Lu_3Os_4Ge_{13}$, using various experimental techniques. Single crystals of $Y_3Ru_4Ge_{13}$ and $Lu_3Os_4Ge_{13}$ have been grown using the Czochralski crystal pulling method in a tetra-arc furnace. The x-ray diffraction experiment reveals that both compounds crystallize in cubic structure (space group $Pm\bar{3}n$, no. 223). The transport, magnetization and heat capacity measurements show that $Y_3Ru_4Ge_{13}$ single crystal undergoes a superconducting transition at 2.85 K, whereas, $Lu_3Os_4Ge_{13}$ becomes superconductor at 3.1 K.

1. Introduction

A variety of unconventional superconductors present low density of the charge carriers as a common factor, implying that it could be the basis for a unifying picture to understand the superconductivity in such exotic systems. Low density of charge carriers is one of the characteristic features which is shared by cuprates, fullerenes and MgB_2 [1, 2, 3]. This is quite surprising since low carrier density is an unfavourable element for superconductivity within the conventional framework of BCS [4] or Migdal–Eliashberg [5, 6] theories. Moreover, a small superfluid density, is unavoidably related to poor screening and strong electronic correlations, ingredients which are expected to be also detrimental for conventional superconductivity. On these grounds it is hard to understand why these low carrier materials are the best superconductors. As far as the superconductivity exhibited by inter-metallic compounds is concerned, the role of electron-phonon interaction cannot be overlooked. However, one may have to look beyond the conventional framework of BCS or Migdal–Eliashberg theories in order to understand the unconventional superconductivity in these compounds. From the experimental side, it is important to look for new superconducting materials with low carrier density.

2. Experimental Details

Single crystals of $Y_3Ru_4Ge_{13}$ and $Lu_3Os_4Ge_{13}$ have been grown using Czochralski crystal pulling method in a tetra-arc furnace under high purity argon atmosphere. Stoichiometric ammount of $Y_3Ru_4Ge_{13}$ and $Lu_3Os_4Ge_{13}$ (10 g each) was taken and melted 4-5 times in the tetra-arc furnace to make a homogeneous polycrystalline mixture. Single crystals were pulled using a tungsten seed rod at the rate of 10 mm/h for about 6 h to get 5-6 mm long and 3-4 mm thick crystals. The phase purity was characterized by powder X-ray diffraction using PANanalytical X-ray diffractometer. Single crystals were oriented along the crystallographic direction [100] using Laue back reflection using Huber Laue diffractometer and cut to desired shape and dimensions using a spark erosion cutting machine. Resistivity measurements were done in a home made setup using standard four-probe technique. Magnetization measurements were done in commercial SQUID magnetometer (MPMS5, Quantum Design, USA) and heat capacity measurements were done using PPMS.

3. Results and Discussion

The crystal structure of $Y_3Ru_4Ge_{13}$ is shown in Fig. 1. Both compounds have same crystal structure and cubic symmetry ($Pm\bar{3}n$, space group # 223). Rietveld analysis [7] of the Powder X-ray diffraction (PXRD) of $Y_3Ru_4Ge_{13}$ is shown in Fig. 2. The temperature dependence of resistivity $\rho(T)$ from 300 to 2K for $Y_3Ru_4Ge_{13}$ and $Lu_3Os_4Ge_{13}$ are shown in Fig. 3. A semi-metallic behaviour ($\frac{d\rho}{dT} < 0$) can be observed in the normal state resistivity data of both the compounds. The magnetoresistance data for $Y_3Ru_4Ge_{13}$ and $Lu_3Os_4Ge_{13}$ is shown in Fig. 4.

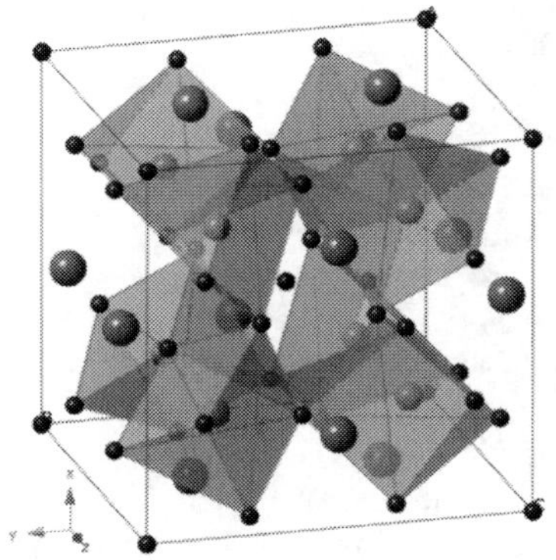

Figure 1. Crystal structure of $Y_3Ru_4Ge_{13}$. Yttrium atoms are shown in green in 6d position, Ruthenium are shown in light pink in 8e position and Germanium are shown in dark blue in 2a and 24k Wyckoff positions.

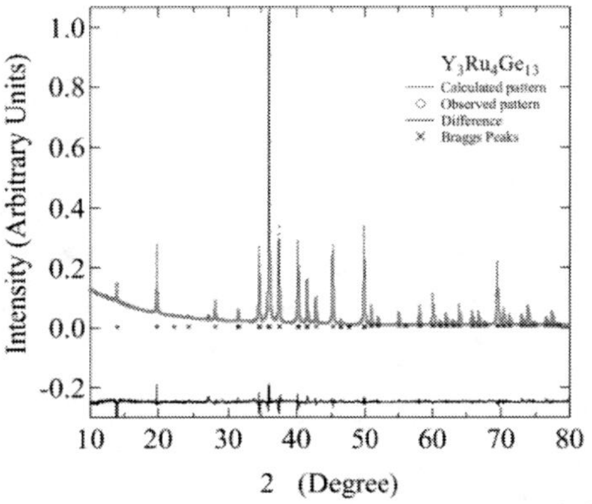

Figure 2. Rietveld analysis of the powder XRD pattern of $Y_3Ru_4Ge_{13}$. No impurity peaks are observed indicating single phase nature of the compound. Similar PXRD pattern is also observed for $Lu_3Os_4Ge_{13}$.

The width of the superconducting transition increases with increasing magnetic field. The transition temperature is taken at the point where resistivity becomes half of its normal state value. The temperature dependence of the upper critical field for $Y_3Ru_4Ge_{13}$ and $Lu_3Os_4Ge_{13}$ is shown in Fig. 5. We estimate the orbital upper critical field, $\mu_0 Hc_2(0)$, for both the compounds using Werthamer-Helfand-Hohenberg (WHH) expression[8], $\mu_0 Hc_2(0) = -0.693\, T_c \frac{dHc_2}{dT}|_{T=T_c}$ in the dirty limit for type-II superconductors. A nearly linear relationship is observed in Fig. 5 between $\mu_0 Hc_2$ and T_c in the proximity of the transition temperature (T_c at H = 0) for both the compounds but the linear trend is more prominent for $Lu_3Os_4Ge_{13}$. The slope $\frac{dHc_2}{dT}|_{T=T_c}$ is used to calculate $\mu_0 H_{c2} = 4.63 \pm 0.09$ T for $Y_3Ru_4Ge_{13}$ and $\mu_0 Hc_2 = 5.68 \pm 0.12$ T for $Lu_3Os_4Ge_{13}$ using the WHH formula in the dirty limit. The value of $\mu_0 H_{c2}$ is smaller than the weak coupling Pauli paramagnetic limit $\mu_0 H^{\mathrm{Pauli}} = 1.82 T_c = 5.09$ T for $Y_3Ru_4Ge_{13}$ and

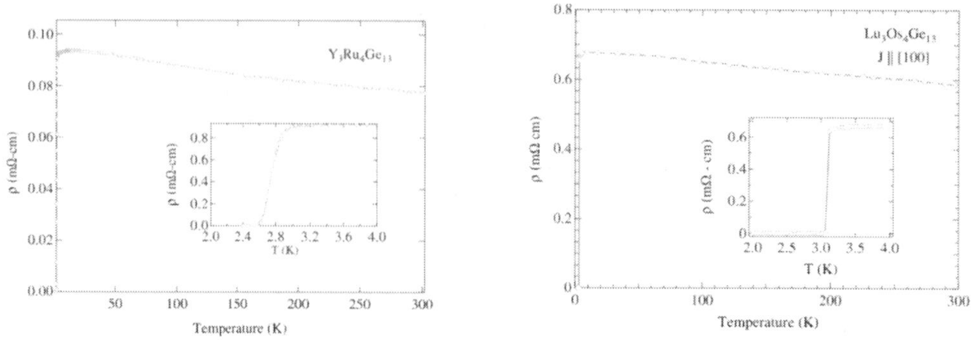

Figure 3. The temperature dependence of the electrical resistivity (ρ) along the (100) directions of cubic (Pm3n) Y3Ru4Ge13 and $Lu_3Os_4Ge_{13}$. Insets show the low temperature data indicating superconducting transition in both compounds. Resistivity data from 2 to 300K clearly shows the semi-metallic nature of both the compounds.

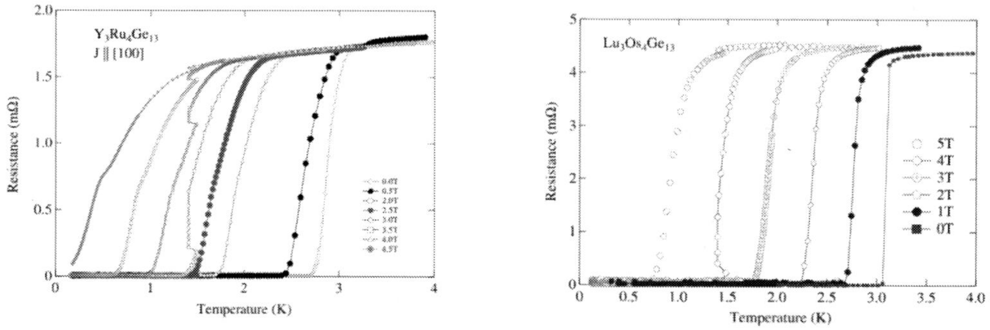

Figure 4. The resistance vs temperature data at different magnetic fields for $Y_3Ru_4Ge_{13}$ and $Lu_3Os_4Ge_{13}$. With the increase in applied magnetic field the superconducting transition becomes slightly broader for both compounds. The temperature dependence of the upper critical field ($\mu_0 Hc_2(T)$) is extracted from these measurements.

$\mu_0 H^{\text{Pauli}} = 5.80$ T for $Lu_3Os_4Ge_{13}$. The upper critical field value $\mu_0 H_{c2}(0)$ can be used to estimate the Ginzburg-Landau coherence length $\xi(0)_{GL} = \sqrt{\Phi_0/2\pi H_{c2}(0)} = 80.4 \pm 0.5\mathring{A}$ for $Y_3Ru_4Ge_{13}$ and $\xi(0)_{GL} = 76.1 \pm 0.7\mathring{A}$ for $Lu_3Os_4Ge_{13}$, where $\Phi_0 = hc/2e$ is the magnetic flux quantum.
The DC-magnetisation data of both compounds is shown in Fig. 6 indicating diamagnetic transitions of $Y_3Ru_4Ge_{13}$ at 2.8 K and $Lu_3Os_4Ge_{13}$ at 3.1 K. Very similar values of T_c from both resistivity and susceptibility data confirm that our single crystals are of very high quality. Large vortex pinning can be observed in the field cooled (FC-Meissner) data in shown in Fig. 6.

The characterisation of the superconducting transition using heat capacity measurements is shown in Fig. 7. The specific heat jump at the thermodynamic transition confirm the bulk superconductivity in both the compounds. The low temperature normal state specific heat can be well fitted with $\frac{C}{T} = \gamma + \beta T^2$, where γT represents the electronic contribution and βT^3 describe the lattice-phonon contributions to the specific heat in the normal state. Fitting the above formula give electronic specific heat coefficient $\gamma = 7.08\frac{mJ}{molK^2}$ ($\gamma = 25.4\frac{mJ}{molK^2}$) and the

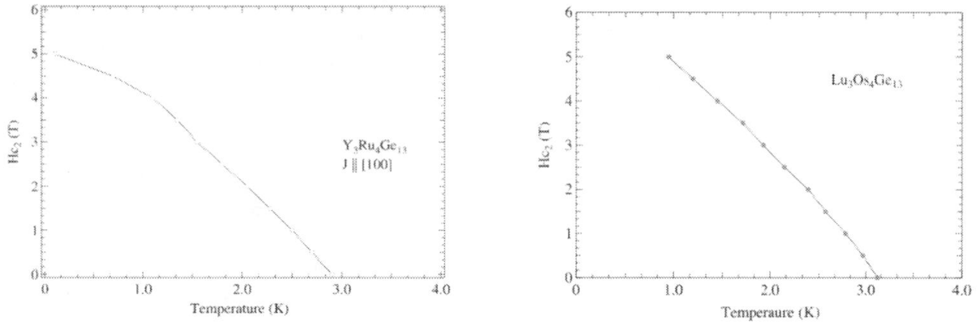

Figure 5. $\mu_0 H_{c2}$ as a function of temperature for $Y_3Ru_4Ge_{13}$ and $Lu_3Os_4Ge_{13}$. The upper critical field $\mu_0 H_{c2}$ increases linearly as the temperature is lowered in the vicinity of the transition temperature T_c for both compounds, though the linear dependence is more prominent for $Lu_3Os_4Ge_{13}$.

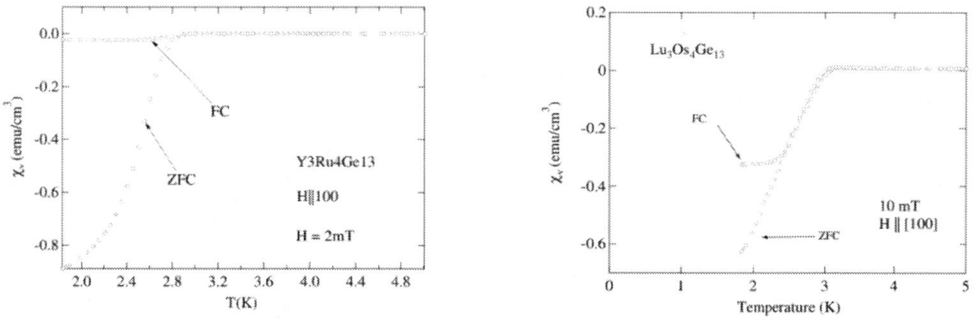

Figure 6. DC magnetic susceptibility as a function of temperature for magnetic field $H \parallel [100]$ direction for $Y_3Ru_4Ge_{13}$ and $Lu_3Os_4Ge_{13}$. The superconducting transition temperatures determined from susceptibility measurements are in excellent agreement with the resistivity data reflecting the high quality of the single crystals. The ZFC and FC susceptibility data indicate significant amount of pinning of vortices in both the compounds.

phonon/lattice contributions $\beta = 3.52\frac{mJ}{molK^4}$ ($\beta = 2.30\frac{mJ}{molK^2}$) for $Y_3Ru_4Ge_{13}$ ($Lu_3Os_4Ge_{13}$). The ratio $\frac{\Delta C}{\gamma Tc}$ can be used to measure the strength of the electron-phonon coupling. The specific heat jump $\frac{\Delta C}{Tc}$ is $6.07\frac{mJ}{molK^2}$ ($29\frac{mJ}{molK^2}$), setting the value of $\frac{\Delta C}{\gamma Tc} = 0.85$ ($\frac{\Delta C}{\gamma Tc} = 1.15$) for $Y_3Ru_4Ge_{13}$ ($Lu_3Os_4Ge_{13}$). These values are smaller than the weak-coupling limit of 1.43 for a conventional BCS superconductor, suggesting that these two compounds are moderately electron-phonon coupled superconductor.
The comparison among the normal and superconducting state parameters of the both compounds is shown in Table-1. We also notice from Table-1 that value of γ is larger in $Lu_3Os_4Ge_{13}$ suggesting stronger electronic correlations in $Lu_3Os_4Ge_{13}$ as compared to electronic correlations in $Y_3Ru_4Ge_{13}$.

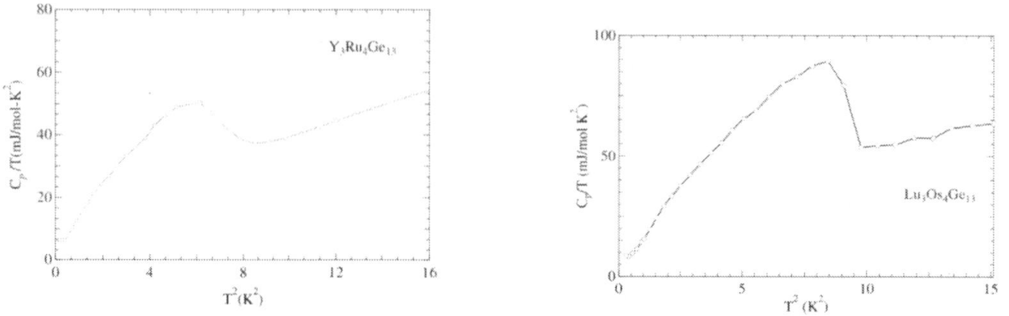

Figure 7. C/T vs T^2 data for $Y_3Ru_4Ge_{13}$ and $Lu_3Os_4Ge_{13}$ respectively. Large jump in heat capacity confirms bulk superconductivity in both the compounds.

Table 1. Normal and superconducting state parameters of $Y_3Ru_4Ge_{13}$ and $Lu_3Os_4Ge_{13}$

Parameters	$Y_3Ru_4Ge_{13}$	$Lu_3Os_4Ge_{13}$
T_c (K)	2.85	3.1
γ (mJ/mol K^2)	7.1	25.4
Θ_D (K)	223	257
$\Delta C_{el}/\gamma T_c$	0.85	1.15
$\mu_0 H_{c2}$ (T)	4.63	5.68
$\xi(0)_{GL}$ ($\mathring{A}$)	80.4	78

4. Conclusion

We have grown single crystals and characterised the superconducting properties of two semi-metallic compounds $Y_3Ru_4Ge_{13}$ and $Lu_3Os_4Ge_{13}$. A bulk superconducting transition is confirmed and characterised through electrical transport, magnetisation and heat capacity measurements on the single crystals. The magnetic susceptibility measurements show large pinning of vortices in both the compounds. The analysis of the low temperature heat capacity data suggests that both these compounds are moderately electron-phonon coupled type-*II* superconductors.

References

[1] Fleming R M, Ramirez A P, Rosseinsky M J, Murphy D W, Haddon R C, Zahurak S M and Makhija A V 1991 *Nature* **352** 787–788 URL `http://dx.doi.org/10.1038/352787a0`
[2] Holczer K, Klein O, Huang S m, Kaner R B, Fu K j, Whetten R L and Diederich F 1991 *Science* **252** 1154–1157 URL `http://www.sciencemag.org/content/252/5009/1154`
[3] Nagamatsu J, Nakagawa N, Muranaka T, Zenitani Y and Akimitsu J 2001 *Nature* **410** 63–64 ISSN 0028-0836 URL `http://dx.doi.org/10.1038/35065039`
[4] Bardeen J, Cooper L N and Schrieffer J R 1957 *Phys. Rev.* **108**(5) 1175–1204 URL `http://link.aps.org/doi/10.1103/PhysRev.108.1175`
[5] Migdal A 1958 *Soviet Physics JETP* **34**(7)
[6] Eliashberg G 1960 *Soviet Physics JETP* **11**(3)
[7] Carvajal 1993 *Physica B: Condensed Matter* **192** 55 – 69 ISSN 0921-4526 URL `http://www.sciencedirect.com/science/article/pii/092145269390108I`
[8] Werthamer N R, Helfand E and Hohenberg P C 1966 *Phys. Rev.* **147**(1) 295–302 URL `http://link.aps.org/doi/10.1103/PhysRev.147.295`

A solitary superconductivity in layered ferromagnet-superconductor heterostructures

Y Proshin, M Avdeev and M M Khusainov
Kazan Federal University, Kremlevskaya, 18, Kazan, Russia
E-mail: yurii.proshin@ksu.ru

Abstract. An electron-electron interaction in a ferromagnet (F) was neglected in the standard approach to the proximity effect theories for the layered FS structures (here S is used for a superconductor). Actually this interaction exists, but it is suppressed by strong exchange field and can reveal itself if an exchange fields of adjacent F layers compensate each other. In the clean FS structures limit the preceding consideration of this interaction could explain a surprisingly high critical temperature T_c in the short-periodic Gd/La superlattice. Here we analyze this problem for dirty case based on the Usadel equations solutions changing parameters of asymmetrical FS systems (thicknesses of layers, boundary transparencies, etc). Taking into account an electron-electron interaction leads to an appearance of hidden superconductivity of F layers which can manifest itself in the proximity effect conditions. It is especially expressed if magnetizations in F layers have opposite signs. The solitary superconductivity is also predicted for asymmetrical dirty FSF trilayers.

1. Introduction

In artificial layered structures consisting of the superconductor (S) and ferromagnet (F) layers the interplay between the S and F parameter orders can lead to a number of striking phenomena [1–3]. For example, the nonmonotonic behaviors of the critical temperature and the Josephson current as a function of the ferromagnetic layer thickness appears (see reviews [1–3] and references therein). As limit cases the re-entrant and periodically re-entrant superconductivity was predicted in works [4, 5]. Later the re-entrant superconductivity experimentally was discovered in bilayers V/Fe [6] and $Nb/Cu_{1-x}Ni_x$ [7]. Note, an appearance of superconductivity as solitary peak in dependence of critical temperature T_c versus the F layer thickness d_f was recently theoretically proposed for clean asymmetrical FS system [8, 9]. Most recently, the appearance of peculiar solitary superconductivity caused by external magnetic field is predicted for the F_1F_2S system [10].

These FS layered heterostructures can be perspective due to possible superconducting spin switch applications. Thus the spin valve device based on the three layered FS systems switched by weak external magnetic field was proposed in works [11–13]. Note, the superconducting switch based on the four-layered $F_1S_1F_2S_2$ system can have up to seven different states [14].

An electron-electron interaction in a ferromagnet (F) was neglected in the standard approach to the proximity effect theories for the layered SF structures (see for example [1–3] and references therein). In other words a superconducting order parameter Δ_f and an electron-electron interaction constant λ_f were taken as zero for a ferromagnet. Actually this interaction exists, but it is "hidden" because it is suppressed by strong exchange field I. This interaction can reveal itself if an exchange field could be "disabled". So, in this imaginary case a ferromagnet transforms to normal metal and that "unhidden" electron-electron interaction can lead to superconducting correlations and, therefore, a superconductivity onset with critical temperature T_{cf}, estimated by the standard BCS expression.

Previously we have shown that a consideration of this interaction in the clean *symmetrical* F/S structures limit [15–17] can explain a surprisingly high critical temperature T_c ($T_c \sim 5$ K) and an independence of T_c from the F layer thickness in the short-periodic Gd/La superlattice [18, 19] for case when magnetizations in adjacent F layers were antiparallel (AP). The solutions of Eilenberger equations were obtained in Cooper limit case for ideal FS boundaries transparencies. At these conditions the full compensation of pair-breaking effect of exchange field and sharing all interactions over sample are possible for thin layered symmetrical systems [15, 16]. Moreover, for clean *asymmetrical* FS structures (FSF′ [16], FSF′S′ [20], superlattice [17]) we have also predicted an appearance of solitary peak in the dependence $T_c(d_f)$ at fixed $d_{f'}$. This peak was centered at $d_f = d_{f'}$, its height and its width strongly depend from the FS structure parameters. By analogy with re-entrant superconductivity, we called this phenomenon as a *solitary* re-entrant superconductivity or more appropriately a *solitary superconductivity*. Last term will be frequently used below.

In this paper we consider a dirty limiting case and analyze solutions of boundary value problem for the Usadel function, changing different parameters of asymmetrical F_1SF_2 trilayer (thicknesses of layers, boundary transparencies, and so on). Taking into account an electron-electron interaction can lead to an appearance of "hidden" superconductivity of F layers which can be manifest itself in the proximity effect conditions. It will be especially expressed if exchange fields in both F layers have opposite signs. We discuss influence of the electron-electron interaction on the critical temperature and predict the *solitary* superconductivity for dirty asymmetrical F_1SF_2 trilayer too.

2. Theoretical background

For dirty FS system we will use Usadel-like equations [21]. The order parameter $\Delta_{s,f}$ and electron-electron interaction $\lambda_{s,f}$ were taken into account for both metals S and F, respectively.

The critical temperature T_c at the second order transition for F_1SF_2 trilayers is obtained from the set of the self-consistent equations [22] for the superconducting gaps $\Delta_{s,f}(\mathbf{r})$ in S and F layers, respectively

$$\Delta_s(\mathbf{r}) \ln t = 2\pi T_c \,\mathrm{Re} \sum_{\omega>0}^{\infty} \left(F_s(\mathbf{r},\omega) - \frac{\Delta_s(\mathbf{r})}{\omega} \right), \tag{1}$$

$$\Delta_{fi}(\mathbf{r})(\ln t + \ln \frac{T_{cs}}{T_{fi}}) = 2\pi T_c \,\mathrm{Re} \sum_{\omega>0}^{\infty} \left(F_{fi}(\mathbf{r},\omega) - \frac{\Delta_i(\mathbf{r})}{\omega} \right), \quad i = (1,2), \tag{2}$$

where $t = T_c/T_{cs}$ is the reduced critical temperature (T_{cs} is the superconducting critical temperature for the bulk S material, T_{fi} is "virtual" critical temperature T_{cf} for normal metal corresponding F_i material in which an exchange field is assumed be lacking, $I_i = 0$), ω is the Matsubara frequency.

The pair amplitude $F_{s,(i)}$ satisfies the Usadel-like equations [21, 23, 24] for S layer

$$\left[|\omega| - \frac{D_s}{2} \frac{d^2}{dx^2} \right] F_s(x,\omega) = \Delta_s(x), \tag{3}$$

and for F layers

$$\left[|\omega| - iI_i - \frac{D_{fi}(I)}{2} \frac{d^2}{dx^2} \right] F_{fi}(x,\omega) = \Delta_{fi}(x), \;\; D_{fi}(I_i) = \frac{D_{fi}}{1 - i2I_i\tau_{fi}}, \tag{4}$$

where $D_{s,fi}$ is the diffusion constant in corresponding layer, and τ_{fi} is the elastic scattering time of non-magnetic impurities in F_i layer. Note, that in the presence of exchange interaction the diffusion constant $D_{fi}(I)$ in (4) is complex [2, 25] (this form of complex diffusion constant is uniquely determined in the microscopic approach given in [10], the consistency of approach is confirmed from the fact that the complex diffusion constant enters both in the differential equations (4) and in the boundary conditions in the same manner [10]). For completeness sake we should mention variant of complex diffusion constant obtained in work [26] and corresponding discussion [27]. Unfortunately the limited size of paper does not allow to explain this interesting problem more carefully.

For brevity here we do also not describe the boundary conditions of Kupriyanov-Lukichev type [28] for pair amplitude F, microscopically derived in the works [10, 25] and a process of solution of this boundary value problem. Moreover we use the same approximations as in our recent works [10, 29].

3. Results and discussion

In this section we present and discuss the numerical results for the F_1SF_2 systems. We only consider the case of antiparallel orientation (AP state) of the magnetizations in plane of F layers. Similarly to [16], we also assume highly transparent SF interfaces ($\sigma_s = \sigma_f = 100$). All lengths related to the S and $F_{1,2}$ layers are normalized on the coherence lengths $\xi_s = \sqrt{D_s/2\pi T_{cs}}$ and $\xi_{I1,2} = \sqrt{D_{f1,2}/I_{1,2}}$, respectively.

We first discuss the symmetrical F_1SF_2 system when the thicknesses of F layers are equal ($d_{f1} = d_{f2}$). In Fig. 1 the $T_c(d_f)$ dependence is shown at various ratio T_{cs}/T_{cf}. It is clearly seen that the critical temperature T_c of the FSF system is practically equal to the critical temperature T_{cs} of isolated S layer (see solid and dashed lines) at small thicknesses of F layers $d_f \le 0.3\xi_I$, when $T_{cf} \ge T_{cs}$. For comparison, the dependence $T_c(d_f)$ has a rapid initial decline in the absence of electron-electron interaction ($T_{cs}/T_{cf} = \infty$). Both dependencies are in agreement with the results [8] for clean FSF trilayers. The reason that $T_c \simeq T_{cs}$ (for small thicknesses) is easy to understand: the effective exchange field is compensated by the

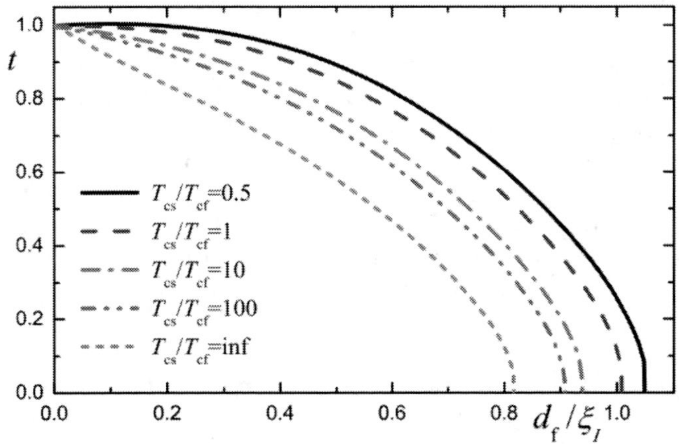

Figure 1. (Color online) The influence of the interelectronic interaction on the critical temperature T_c of the F_1SF_2 system. The critical temperature T_c vs thicknesses d_f of F layers at few values of parameter T_{cs}/T_{cf}. Other parameters of the system are $d_s/\xi_s = 1.25$, $l_s/\xi_s = 0.7$, $\sigma_s = \sigma_f = 100$, $l_{f1}/\xi_{I1} = l_{f2}/\xi_{I2} = 0.3$, $I_1/\pi T_{cs} = I_2/\pi T_{cs} = 6$.

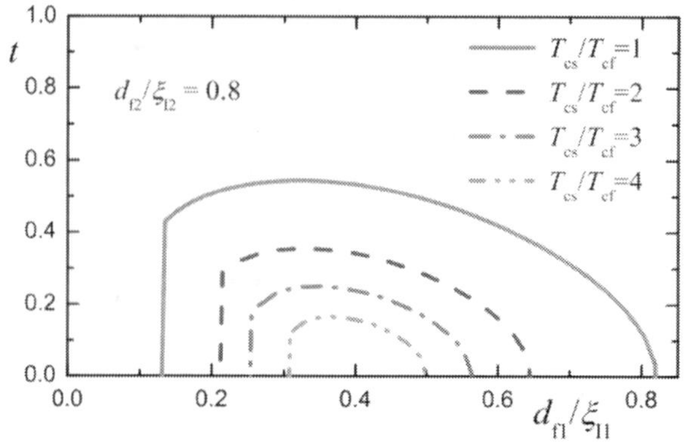

Figure 2. (Color online) The phase diagrams with the solitary superconductivity for the F_1SF_2 trilayer. The solitary peaked superconductivity $T_c(d_{f1})$ at various value of the ratio T_{cs}/T_{cf} with fixed thickness of F_2 layer $d_{f2} = 0.8\xi_{I2}$. Other parameters of the system are $d_s/\xi_s = 1.25$, $l_s/\xi_s = 0.7$, $\sigma_s = \sigma_f = 100$, $l_{f1}/\xi_{I1} = l_{f2}/\xi_{I2} = 0.5$, $I_1/\pi T_{cs} = I_2/\pi T_{cs} = 2$.

antiparallel mutual orientation of the magnetizations ($I_{eff} \simeq 0$) and it leads to contact of two "same superconductors" with $T_{cf} \simeq T_{cs}$ due to "hidden" electron-electron pairing interaction in F layers. For greater thicknesses ($d_f \gtrsim l_f$) the difference with clean case appears owing to the increasing role of the elastic scattering time on non-magnetic impurities in F layers.

Then we consider the asymmetrical F_1SF_2 system.

For investigation of the solitary superconductivity we will examine the case of very weak exchange fields $I = 2\pi T_{cs}$. So, in Fig. 2 the influence of the electron-electron interaction on the phase diagrams with the solitary superconductivity is presented for the case of soft ferromagnet. We remark that the F_1SF_2 system is asymmetrical: the thickness of the F_1 layer can be only changed but the F_2 layer has fixed thickness ($d_{f2} = 0.8\xi_{I2}$). We found parameters when solitary peak is pronounced. In this case we observe decreasing the maximum of the critical temperature with increasing the T_{cs}/T_{cf} parameter. We can find that the superconductivity completely disappears in this system when $T_{cs} > 5.4T_{cf}$. Thus we conclude that the appearance of the solitary superconductivity in dirty FSF trilayers can be say about a manifestation of "hidden superconductivity" in the F layers, but it requires the fulfilment of a number of the sufficiently specific conditions. We list them here once again: the FSF system should be in the AP state (magnetizations $\mathbf{M}_1$ and $\mathbf{M}_2$ are antiparallel); the "virtual" critical temperature T_{cf} must be comparable with the critical temperature T_{cs} of the S layer; the thickness of the S layer should be of the order of the coherence length ξ_s for a more effective compensation of the exchange fields. At the same time, the F_1S and SF_2 interfaces should be highly transparent; if the the S layer is thin, it is desirable to use the soft ferromagnets (with a low Curie temperature) such as $Cu_{1-x}Ni_x$ alloy [30].

4. Conclusions

In this work we considered the asymmetrical F_1SF_2 trilayers. Our theoretical approach takes into account of the electron-electron interaction in both F layers. We showed that in dirty limit, the asymmetry and electron-electron pairing interaction can lead to an appearance of the solitary superconductivity. In particular, we found that the kind of dependence $T_c(d_f)$ at small F layers thicknesses may indicate about possible presence of electron-electron interaction in ferromagnets. Obtained results for dirty F_1SF_2 trilayer are in qualitative agreement with ones for clean limiting case [8]. But there are important differences. So, in clean systems we should take into account the transverse Fulde-Ferrell-Larkin-Ovchinnikov states, which can be neglected in dirty case. The solitary peak of superconductivity in clean case [8] has always a maximum at $d_{f1} = d_{f2}$, but the dirty FSF system has, as rule, asymmetrical condition on position of this maximum. So, the unconventional experimental behavior of Gd/La system [18, 19] can be explained only in "clean" model for F metals with one-domen structure.

Acknowledgments
The work is partial supported by the subsidy allocated to Kazan Federal University for the state assignment in the area of scientific activities. YP is thankful to the RFBR (13-02-01202) for partial support. MA is also supported by Russian Program of Competitive Growth of Kazan Federal University.

References

[1] Izyumov Y A, Proshin Y N and Khusainov M G 2002 *Physics-Uspekhi* **45** 109–148
[2] Buzdin A 2005 *Rev. Mod. Phys.* **77** 935–976
[3] Efetov K B, Garifullin I A, Volkov A F and Westerholt K 2008 *Magnetic Heterostructures* (*Springer Tracts in Modern Physics* vol 227) (Springer Berlin / Heidelberg) chap Proximity Effects in Ferromagnet/Superconductor Heterostructures, pp 251–290
[4] Khusainov M and Proshin Y 1997 *Phys. Rev. B* **56** R14283–R14286
[5] Proshin Y N and Khusainov M 1997 *Journal of Experimental and Theoretical Physics Letters* **66** 562–568
[6] Garifullin I A, Tikhonov D A, Garif'yanov N N, Lazar L, Goryunov Y V, Khlebnikov S Y, Tagirov L R, Westerholt K and Zabel H 2002 *Phys. Rev. B* **66**(2) 020505
[7] Zdravkov V, Sidorenko A, Obermeier G, Gsell S, Schreck M, Müller C, Horn S, Tidecks R and Tagirov L R 2006 *Phys. Rev. Lett.* **97**(5) 057004
[8] Khusainov M, Khusainov M, Ivanov N and Proshin Y 2009 *JETP Letters* **90** 359–364 ISSN 0021-3640
[9] Proshin Y, Khusainov M and Khusainov M 2010 *Physica C: Superconductivity* **470** 874–876
[10] Avdeev M and Proshin Y 2014 *Superconductor Science and Technology* **27** 035006
[11] Oh S, Youm D and Beasley M 1997 *Appl. Phys. Lett* **71** 2376–2378
[12] Tagirov L R 1999 *Phys. Rev. Lett.* **83** 2058–2061
[13] Buzdin A I, Vedyayev A V and Ryzhanova N V 1999 *Europhys. Lett.* **48** 686–691
[14] Proshin Y, Zimin A, Fazleev N and Khusainov M 2006 *Phys. Rev. B* **73** 184514
[15] Khusainov M G, Khusainov M, Ivanov N and Proshin Y N 2009 *JETP letters* **89** 626–631
[16] Khusainov M G, Khusainov M, Ivanov N and Proshin Y N 2009 *JETP letters* **90** 124–129
[17] Proshin Y N, Khusainov M M and Minnullin A 2014 *Physica status solidi (c)* **11** 1080–1085
[18] Goff J, Deen P, Ward R, Wells M, Langridge S, Dalgleish R, Foster S and Gordeev S 2002 *Journal of Magnetism and Magnetic Materials* **240** 592–594
[19] Deen P P, Goff J P, Ward R C C, Wells M R, Langridge S, Dalgliesh R, Foster S and McIntyre G J 2005 *Journal of Physics: Condensed Matter* **17** 3305–3315
[20] Minnullin A N, Khusainov M M and Proshin Y N 2012 *Journal of Physics: Conference Series* **394** 012016
[21] Usadel K D 1970 *Phys. Rev. Lett.* **25**(8) 507–509
[22] Abrikosov A A and Gorkov L P 1961 *Sov. Phys. JETP* **12** 1243–1253
[23] Radović Z, Dobrosavljević-Grujić L, Buzdin A I and Clem J R 1988 *Phys. Rev. B* **38** 2388–2393
[24] Krunavakarn B and Yoksan S 2006 *Physica C: Superconductivity and its Applications* **440** 25–34
[25] Proshin Y N and Khusainov M G 1998 *Zh. Eksp. Teor. Fiz.* **113** 1708–1730
[26] Buzdin A and Baladié I 2003 *Phys. Rev. B* **67**(18) 184519
[27] Khusainov M G and Proshin Y N 2003 *Physics-Uspekhi* **46** 1311–1313
[28] Kuprianov M Y and Lukichev V F 1988 *Zhurnal Eksperimentalnoi I Teoreticheskoi Fiziki* **94** 139–149
[29] Avdeev M V and Proshin Y N 2013 *Journal of Experimental and Theoretical Physics* **117** 1101–1108
[30] Antropov E, Kalenkov M S, Kehrle J, Zdravkov V I, Morari R, Socrovisciuc A, Lenk D, Horn S, Tagirov L R, Zaikin A D, Sidorenko A S, Hahn H and Tidecks R 2013 *Superconductor Science and Technology* **26** 085003

The spin valve device based on asymmetrical ferromagnet-superconductor trilayer in an external magnetic field

Yu Proshin and M Avdeev
Kazan Federal University, 18 Kremlevskaya, Kazan, Russia
E-mail: yurii.proshin@kpfu.ru

Abstract. We theoretically explore superconducting proximity effect for the three-layered ferromagnet/superconductor (FS) heterostructures. We consider the boundary value problem for the Usadel-like equations in the case of so-called "dirty" limit. The different mutual orientations of the F layers magnetizations are examined in presence of the external magnetic field. The results of numerical calculations for critical temperature T_c at various parameters of FS structures are presented. The appearance of solitary superconductivity is predicted for the F_1F_2S system. We also discuss possible spin valve applications for F_1F_2S trilayers.

1. Introduction

In recent years the study of layered heterostructures consisting of alternating layers of ferromagnetic (F) and superconducting (S) metals is attracted a keen interest. This is primarily due to the fact that in artificial layered structures the interplay between the S and F order parameters can lead to several striking phenomena (see reviews [1–3] and references therein).

These FS layered heterostructures are also interesting due to possible spin valve applications. Thus the spin valve device based on the three layered FS systems switched by weak external magnetic field was proposed for "asymmetrical" FFS trilayer [4] and for "symmetrical" FSF case [5,6]. Changing the mutual orientation of the F layers magnetizations can control the critical temperature T_c of these systems, and, therefore, the switching between two different states, i.e. the superconducting state with antiparallel (AP) orientation of the F layers magnetizations and the resistive state with parallel (P) one. Note that the superconducting switch based on the four-layered FSFS system can have up to seven different states [7]. A more detailed implementation of the spin valve is described in the works [8–10] (see also references therein) and in reviews [1–3].

The large set of effects is connected with the nonmonotonic behavior of the critical temperature and the Josephson current as a function of the F layers thickness [1–3]. Thus, the re-entrant and periodically re-entrant superconductivity was predicted in works [11–13]. Later the re-entrant superconductivity experimentally was discovered in bilayers V/Fe [14] and $Nb/Cu_{1-x}Ni_x$ [15]. Recently the re-entrant superconductivity with increasing magnetic field, predicted in theoretical work [16], was observed in symmetrical $Cu_{41}Ni_{59}/Nb/Cu_{41}Ni_{59}$ trilayers [17]. Note, a solitary superconductivity was also recently theoretically proposed for clean FS system [18, 19]. Lately the appearance of peculiar solitary re-entrant superconductivity caused by external magnetic field is predicted for the F_1F_2S system [20].

These phenomena are closely related to one another. In this paper we present and discuss the numerical results of the phase diagrams for FSF and FFS trilayers in the presence of external magnetic field H applied parallel to the plane of the contact. In particular, we calculate the difference $\Delta T_c = T_c^{AP} - T_c^P$ as function thicknesses of the F layers. The asymmetry ($d_{f1} \neq d_{f2}$) in some cases can lead to the increase of this value ΔT_c in comparison with symmetrical samples. Note, that obtaining the higher magnitude of the difference ΔT_c is very important for stable operation of the spin valve device.

Basing on the method proposed in work [20], we solve the boundary value problem for the Usadel function and, using self-consistent equation, we calculate the critical temperature for the F_1F_2S system as function of the F layers thicknesses d_f in external magnetic field H.

2. Main equations

Let us shortly describe our approach [20]. The critical temperature T_c at the second order transition is obtained from the self-consistent equation for the superconducting gap $\Delta(\mathbf{r})$ [21]

$$\Delta(\mathbf{r}) \ln t = \pi T_c \sum_{\omega>0}^{\infty} \mathrm{Tr}\left(\hat{F}(\mathbf{r},\omega) - \frac{\Delta(\mathbf{r})}{\omega}\right), \tag{1}$$

where $t = T_c/T_{cs}$ is the reduced critical temperature (T_{cs} is the superconducting critical temperature for the bulk superconductor), ω is the Matsubara frequency, and $\hat{F}$ is the pair amplitude in spin space including singlet and triplet parts [20].

In presence of the external magnetic field $\mathbf{H} = \mathrm{rot}\mathbf{A}$ (we use the Coulomb gauge $\nabla\mathbf{A} = 0$ and $\mathbf{A} = \frac{1}{2}[\mathbf{Hr}]$) and triplet correlations the pair amplitude $\hat{F}$ satisfies the Usadel-like equations [20, 22–24]

$$\left[|\omega| - i\mathbf{I}\hat{\boldsymbol{\sigma}} - \frac{1}{2}D\hat{\mathcal{L}}^2\right]\hat{F}(\mathbf{r},\omega) = \Delta(\mathbf{r}), \quad \hat{\mathcal{L}} = \nabla - \frac{2\pi i}{\Phi_0}\mathbf{A}, \tag{2}$$

where Φ_0 is the magnetic flux quantum, $\mathbf{I}$ is the exchange field in ferromagnet, $\hat{\boldsymbol{\sigma}}$ is Pauli matrix and D is the diffusion constant. The Kupriyanov-Lukichev [25] type boundary conditions derived by approach [13] including triplet correlations [20] are

$$\frac{4D_i}{\sigma_i v_F^i}(\hat{\mathcal{L}}\mathbf{n})\hat{F}^i = \frac{4D_j}{\sigma_j v_F^j}(\hat{\mathcal{L}}\mathbf{n})\hat{F}^j = \hat{F}^i - \hat{F}^j, \quad (\hat{\mathcal{L}}\mathbf{n})\hat{F}^{i,j} = 0, \quad i,j = (s, f1, f2), \tag{3}$$

for the inner interfaces and the outer surfaces, correspondingly. For inner interfaces the detailed balance condition [1] ($\sigma_s n_{sf} = \sigma_f$ where $n_{sf} = v_F^s N_s / v_F^f N_f$) is fulfilled . Here $\mathbf{n}$ is the unit vector normal to boundary planes, the indices s and f denote S and F layers, σ is the boundary transparency, N and v_F are the density of states and electron velocity on the Fermi surface [1]. The equations (1)-(3) are sufficient to calculate the critical temperature T_c of the SF structures.

3. Results and discussion

In this paper we explore the appearance of the *solitary* superconductivity and discuss its application to spin valve problem. We use the numerical approach proposed in our recent work [20] where we obtained a good agreement with the experimental data for the symmetrical CuNi/Nb/CuNi trilayer in external magnetic field [17]. In last work the re-entrant superconductivity was observed with the magnetic field increase. In present paper we used the same fitting procedure as in [20] for symmetrical FSF trilayer and evaluated the parameters values which we will use further for asymmetrical F_1F_2S system. The obtained parameters are close to ones found in paper [20]. For brevity, we will not describe here this fitting procedure.

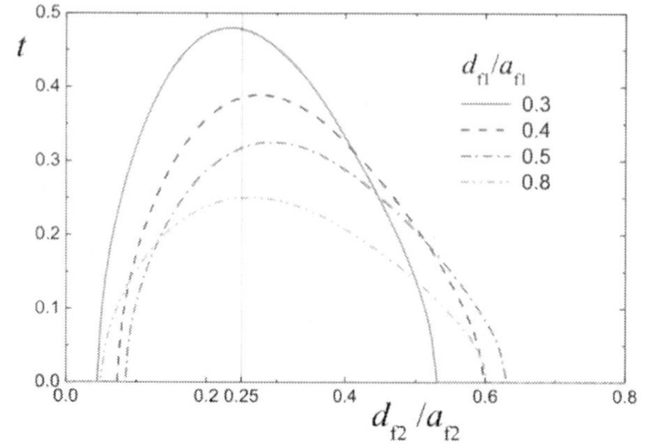

Figure 1. (Color online) The reduced critical temperature $t = T_c/T_{cs}$ (T_{cs} is critical temperature for bulk superconductor) as function of the thicknesses F_2 layer d_{f2} for the AP state. The F_1 layer thicknesses is fixed $d_{f1} = 0.3a_{f1}$;
Other parameters of the system are $d_s/\xi_{s0} = 0.6$, $l_s/\xi_{s0} = 0.2$; boundary transparencies $\sigma_s = 3$ and $\sigma_f = 100$, respectively; parameter $n_{sf} = 0.9$, $l_f/a_f = 0.5$, $I/\pi T_{cs} = 4$.

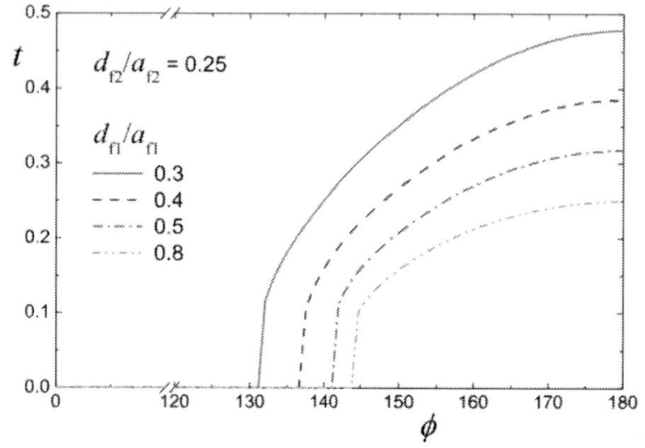

Figure 2. (Color online) The angular dependence of the reduced critical temperature t at various thicknesses of the F_1 layer and at fixed thicknesses $d_{f2} = 0.25a_{f2}$ (see thin vertical line in figure 1.
Other parameters are same as in figure 1).

Note, that the theoretical approach and fit procedure in work [17] also gave good description of the experimental points.

In figure 1 the phase diagrams of the asymmetrical F_1F_2S trilayer are shown without magnetic field. In this figure the dependencies $T_c(d_{f2})$ are plotted at various fixed values of the F_1 thicknesses. The superconductivity appears only in finite range of the F_2 layer thickness and furthermore, starting at some finite thickness of the F_2 layer. Note that the mutual orientation of the F layers magnetizations is antiparallel (AP) for this case. This condition is very important for the appearance of solitary superconductivity. This phenomenon is easy to understand due to partial compensation of the effective exchange interactions of the adjacent ferromagnets at the antiparallel mutual orientation of the magnetizations. At the same time, it is important that the F_2 interlayer has the thin thickness i.e. $d_f \sim \xi_I = \sqrt{l_f a_f/3} = \sqrt{D/2I}$ (l_f and $a_f = v_F/2I$ are the mean free path length and spin stiffness length in F layers, respectively), because the pair amplitude decays into F layer on the ξ_I length. It should also be noted that the maximum of the function $T_c(d_{f2})$ corresponds to the asymmetrical case with different F_1 and F_2 thicknesses (i.e $d_{f1} \neq d_{f2}$), that distinguishes this dirty F_1F_2S system from the clean F_1SF_2 trilayer [18], when the maximum of solitary peak was observed at $d_{f1} = d_{f2}$.

The angular dependence $T_c(\phi)$ (where ϕ is the angle between the magnetization $\mathbf{M}_1$ and $\mathbf{M}_2$) is shown in figure 2 at various thicknesses of the F_1 layer. The values $\phi = 180°$ and $\phi = 0°$ correspond to the AP and P state, respectively. It is clearly seen that the critical temperature T_c strongly depends on mutual orientation of the magnetizations. Thus, the maximum T_c

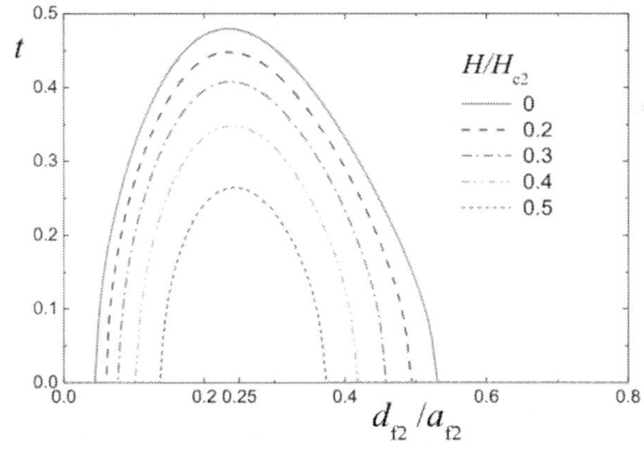

Figure 3. (Color online) The influence of the magnetic field on the critical temperature at at fixed thicknesses $d_{f1} = 0.3a_{f1}$). Other parameters are same as in figure 1.

corresponds to the AP state ($\phi = 180^\circ$) and drops sharply to zero with a further change of the mutual orientation of the magnetizations. So, at the angle $\phi \simeq 130^\circ$ superconductivity is completely absent. So, we may conclude that the solitary superconductivity appears as a result of partial compensation of effective exchange field in F layers at $\phi = 180^\circ$ (AP state).

It is important to note, that the difference $\Delta T_c = T_c^{AP} - T_c^P$ for the case of the solitary superconductivity is equal to T_c^{AP}. As discussed above, the magnitude of the difference ΔT_c is important characteristic for the spin valve devices. In this regard, we believe that the search for such exotic solitary superconductivity is very actual for possible spin valve applications.

The influence of the magnetic field on the phase diagrams of the solitary superconductivity is plotted in figure 3 at five different values of the reduced magnetic field $h = H/H_{c2}$ (where H_{c2} is upper critical field of the isolated S layer). The thickness of the F_1 layer is fixed $d_{f1} = 0.3a_{f1}$. In particular, it is seen that the magnetic field monotonically suppresses superconductivity and the critical temperature decreases, which agrees with both experiments and theory. However we note that the superconductivity is present even at comparatively high value of the magnetic fields $H \simeq 0.5H_{c2}$. Such stability of the critical temperature relative to the external magnetic field magnitude is very important, since real experimental setup for the FS spin valve involves explorations in external magnetic field.

On the basis of the phase diagrams in figure 3, we can consider the simple model of the spin valve. Actually, the external magnetic field can change the mutual orientation of the magnetizations from the AP state ($\uparrow_1\downarrow_2$S) to the P state ($\uparrow_1\uparrow_2$S). The magnetization of the F_1 layer is fixed by an additional layer of antiferromagnetic insulator (e.g., cobalt oxide) [10, 26]. For definiteness, we set the coercivity field for the F_2 layer as $H_{coer} = 0.2H_c$. As is easy to see in figure 3, the curve $T_c(d_{f2})$ has a maximum at $d_{f2} = 0.25a_{f2}$. Thus we consider a F_1F_2S system with fixed thicknesses $d_{f1} = 0.3a_{f1}$ and $d_{f2} = 0.25a_{f2}$ and at a fixed temperature $T = 0.4T_{cs}$. So, without magnetic field the system $\uparrow_1\downarrow_2$S is in the superconducting state since in this case $T < T_c$. With increasing magnetic field to the value of the coercive field $H = H_{coer}$ the system switching from superconducting to the resistive state $\uparrow_1\uparrow_2$S with $T_c^P = 0$. Finally we note once more that for the considered case of the solitary superconductivity the difference $\Delta T_c = T_c^{AP}$.

4. Conclusions

In this work we consider the asymmetrical F_1SF_2 trilayers. The values of parameters are found from comparison our theory and real experimental data for symmetrical CuNi/Nb/CuNi trilayer [17]. It is shown that asymmetry and external magnetic field can essentially influence on the critical properties of considered systems. We show that the solitary superconductivity can occur both in the presence of magnetic field and in its absence. We conclude that the states are important for spin valve applications. The simple model of the spin switch based on the FFS system with solitary superconductivity is proposed taking into account the external magnetic field.

Acknowledgments
The work is partial supported by the subsidy allocated to Kazan Federal University for the state assignment in the area of scientific activities. MA is thankful to the RFBR (13-02-01202) for partial support. YuP is also supported by Russian Program of Competitive Growth of Kazan Federal University

References
[1] Izyumov Y A, Proshin Y N and Khusainov M G 2002 *Physics-Uspekhi* **45** 109–148
[2] Buzdin A 2005 *Rev. Mod. Phys.* **77** 935–976
[3] Efetov K B, Garifullin I A, Volkov A F and Westerholt K 2008 *Magnetic Heterostructures* (*Springer Tracts in Modern Physics* vol 227) (Springer Berlin / Heidelberg) chap Proximity Effects in Ferromagnet/Superconductor Heterostructures, pp 251–290
[4] Oh S, Youm D and Beasley M 1997 *Appl. Phys. Lett* **71** 2376–2378
[5] Tagirov L R 1999 *Phys. Rev. Lett.* **83** 2058–2061
[6] Buzdin A I, Vedyayev A V and Ryzhanova N V 1999 *Europhys. Lett.* **48** 686–691
[7] Proshin Y, Zimin A, Fazleev N and Khusainov M 2006 *Phys. Rev. B* **73** 184514
[8] Kim D H and Hwang T 2007 *Physica C: Superconductivity* **455** 58–62
[9] Nowak G, Zabel H, Westerholt K, Garifullin I, Marcellini M, Liebig A and Hjörvarsson B 2008 *Phys. Rev. B* **78** 134520
[10] Leksin P V, Garif'yanov N N, Garifullin I A, Fominov Y V, Schumann J, Krupskaya Y, Kataev V, Schmidt O G and Büchner B 2012 *Phys. Rev. Lett.* **109**(5) 057005
[11] Khusainov M and Proshin Y 1997 *Phys. Rev. B* **56** R14283–R14286
[12] Proshin Y N and Khusainov M 1997 *Journal of Experimental and Theoretical Physics Letters* **66** 562–568
[13] Proshin Y N and Khusainov M G 1998 *Journal of Experimental and Theoretical Physics* **86** 930–942
[14] Garifullin I A, Tikhonov D A, Garif'yanov N N, Lazar L, Goryunov Y V, Khlebnikov S Y, Tagirov L R, Westerholt K and Zabel H 2002 *Phys. Rev. B* **66**(2) 020505
[15] Zdravkov V, Sidorenko A, Obermeier G, Gsell S, Schreck M, Müller C, Horn S, Tidecks R and Tagirov L R 2006 *Phys. Rev. Lett.* **97**(5) 057004
[16] Avdeev M, Khusainov M, Proshin Y and Tsarevskii S 2010 *Superconductor Science and Technology* **23** 105005
[17] Antropov E, Kalenkov M S, Kehrle J, Zdravkov V I, Morari R, Socrovisciuc A, Lenk D, Horn S, Tagirov L R, Zaikin A D, Sidorenko A S, Hahn H and Tidecks R 2013 *Superconductor Science and Technology* **26** 085003
[18] Khusainov M, Khusainov M, Ivanov N and Proshin Y 2009 *Journal of Experimental and Theoretical Physics Letters* **90** 359–364
[19] Proshin Y, Khusainov M and Khusainov M 2010 *Physica C: Superconductivity* **470** 874–876
[20] Avdeev M and Proshin Y 2014 *Superconductor Science and Technology* **27** 035006
[21] Abrikosov A A and Gorkov L P 1961 *Sov. Phys. JETP* **12** 1243–1253
[22] Usadel K D 1970 *Phys. Rev. Lett.* **25**(8) 507–509
[23] Radović Z, Dobrosavljević-Grujić L, Buzdin A I and Clem J R 1988 *Phys. Rev. B* **38** 2388–2393
[24] Krunavakarn B and Yoksan S 2006 *Physica C: Superconductivity and its Applications* **440** 25–34
[25] Kuprianov M Y and Lukichev V F 1988 *ZhETF* **94** 139–149
[26] Leksin P V, Garif'yanov N N, Garifullin I A, Schumann J, Vinzelberg H, Kataev V, Klingeler R, Schmidt O G and Bchner B 2010 *Appl. Phys. Lett.* **97** 102505

Critical temperature of inhomogeneous magnetic superconductor: effective tensor field approach

Yu N Proshin and V A Tumanov
Institute of Physics, Kazan Federal University, Kazan, Russian Federation
E-mail: yurii.proshin@kpfu.ru

Abstract. Superconducting state with the inhomogeneous effective exchange field background is studied. We calculate the critical temperature of magnetic superconductor on the basis of the Hamiltonian that takes into account the interaction of electrons with the effective exchange field in the direction of inhomogeneity. We use the local unitary rotation in spinor space to rewrite the Hamiltonian in the new basis, where this interaction is diagonal. In this case the exchange field becomes homogeneous but the effective tensor field appears. This method allows us to simplify the Gor'kov equations in many symmetric cases and to find the Green's functions and the critical temperature. We test our approach on the known case of magnetic superconductor with helical magnetization and focus on the critical temperature and the Fulde–Ferrell–Larkin–Ovchinnikov (FFLO) states.

1. Introduction

Interplay of superconductivity and magnetism leads to numerous significant effects in general, but many of them occur only with an inhomogeneous background magnetization [1-3]. The origin of these inhomogeneities can be various: domain walls, magnetic helical structures, artificial multilayers with different magnetization directions, etc. The effective exchange field with non-uniform orientation produces the triplet superconducting correlations in a layered ferromagnet-superconductor (FS) structure [1]. The triplet pairs penetrate much deeper into the ferromagnet compared with the singlet ones. The domain ordering affects also the critical temperature of the superconducting transition [2]. The nature of this effect can be explained by the following qualitative argument: the exchange field acting on the Cooper pairs is averaged over the region of the order of superconducting correlation length [2] Thus, the exchange is less than the average field near a domain wall, and this leads to a more intense Cooper pairing and local increase of the critical temperature (T_c) of the FS contacts in these regions. In the theoretical calculations the derivatives of the average exchange field with respect to the relative coordinates can be neglected [4]. The size of natural magnetic inhomogeneities is much larger compared with the superconducting coherence length but there are many interesting and practically significant effects in artificial systems out of this case. For example, the existence of π-phase magnetism in the FS superlattices [5,6] leads to increasing T_c. In the superconducting multilayer comprising several monolayers of iron and chromium, the critical temperature varies with large amplitude when varying the thickness of Cr layer [7]. The deposition of PdFe on the Nb film increases the critical current density [8]. The latter systems and materials are favorable for applications in spin switches like FFS, FSF and FSFS systems [9-12].

2. Effective tensor field

We use a model of collectivized electrons with Bardeen-Cooper-Schrieffer (BCS) type electron-electron interaction and ferromagnetic interaction with localized spins. Below we assume that the

Planck constant, the Boltzmann constant, and the Bohr magneton are equal to 1. Here we use the effective exchange field approximation and the Hamiltonian should be averaged over the spin operators of the localized electrons:

$$H = \int \sum_{\sigma\sigma'} \psi_{\sigma}^{\dagger} \left(\left(\frac{\hat{\mathbf{p}}^2}{2m^*} - \mu \right) \hat{\mathbf{1}} - (\mathbf{I}, \hat{\boldsymbol{\sigma}}) \right)_{\sigma\sigma'} \psi_{\sigma'} d\mathbf{r} + \lambda \int \psi_{\uparrow}^{\dagger} \psi_{\downarrow}^{\dagger} \psi_{\downarrow} \psi_{\uparrow} d\mathbf{r}, \tag{1}$$

where $\psi^{\dagger}$ and ψ are electron creation and annihilation operators; m^* is the effective mass of the electron, μ is the chemical potential, λ is the electron-electron interaction constant, σ is the electron spin projection, $\hat{\mathbf{p}}$ is the electron momentum operator. The matrix, which is responsible for the interaction between the electron spin and the mean-field localized spins, has the following form:

$$(\mathbf{I}, \hat{\boldsymbol{\sigma}}) = \begin{pmatrix} \cos(\theta) & \sin(\theta) e^{-i(\varphi)} \\ \sin(\theta) e^{i(\varphi)} & -\cos(\theta) \end{pmatrix}, \tag{2}$$

where $\varphi(\mathbf{r})$ and $\Theta(\mathbf{r})$ are the angles defining the magnetization direction in spherical coordinate system. We diagonalize the matrix $(\mathbf{I}, \hat{\boldsymbol{\sigma}})$ by rotating in the spinor space using unitary matrices. The BCS term is invariant under such transformations. The Hamiltonian H can be written in the form

$$H = \int \sum_{\sigma\sigma'} \psi_{\sigma}^{\dagger} \left(-\frac{\hat{\mathbf{D}}^2}{2m^*} - \mu \hat{1} - I \hat{\sigma}_3 \right)_{\sigma\sigma'} \psi_{\sigma'} d\mathbf{r} + \int \left(\Delta(\mathbf{r}) \psi_{\uparrow}^{\dagger} \psi_{\downarrow}^{\dagger} + \Delta^*(\mathbf{r}) \psi_{\downarrow} \psi_{\uparrow} \right) d\mathbf{r}, \tag{3}$$

where $\hat{\mathbf{D}} = \nabla + \hat{\mathbf{A}}$, $\hat{\mathbf{A}} = \hat{\mathbf{U}} \nabla \hat{\mathbf{U}}^{-1}$; $\hat{\sigma}_i$ are the Pauli matrices, Δ is the self-consistent order parameter. Here we use the standard self-consistent approximation and introduce the order parameter in the BCS term. The matrix $\hat{\mathbf{U}}$ is determined from the requirement $\hat{\mathbf{U}}(\mathbf{I}, \hat{\boldsymbol{\sigma}}) \hat{\mathbf{U}}^{-1} = I \hat{\sigma}_3$ and has the form:

$$\hat{\mathbf{U}} = \begin{pmatrix} \cos\left(\frac{\theta}{2}\right) e^{i\xi} & \sin\left(\frac{\theta}{2}\right) e^{i(\xi - \varphi)} \\ -\sin\left(\frac{\theta}{2}\right) e^{-i(\xi - \varphi)} & \cos\left(\frac{\theta}{2}\right) e^{-i\xi} \end{pmatrix}. \tag{4}$$

where $\xi(\mathbf{r})$ is arbitrary real function. Substituting $\hat{\mathbf{U}}$ into the expression for the field we obtain the following result

$$\hat{\mathbf{A}} = \frac{i}{2} (\sin(2\xi - \varphi) \hat{\sigma}_1 - \cos(2\xi - \varphi) \hat{\sigma}_2) \nabla \theta + i \hat{\sigma}_3 \left(\sin\left(\frac{\theta}{2}\right)^2 \nabla \varphi - \nabla \xi \right) + \\ + \frac{i \sin \theta}{2} (\cos(2\xi - \varphi) \hat{\sigma}_1 - \sin(2\xi - \varphi) \hat{\sigma}_2) \nabla \varphi. \tag{5}$$

The tensor field has freedom, because the real function $\xi(\mathbf{r})$ is chosen arbitrarily and the observable parameters should be independent of $\xi(\mathbf{r})$. Thus, the problem of the origin of superconducting correlations on the background of inhomogeneously directed effective exchange field is reduced to the homogeneous problem in the presence of effective tensor field. The effective tensor field has the freedom that allows us to require further invariance theory from the function $\xi(\mathbf{r})$. The critical temperature can be found by solving the equations for the Matsubara Green's functions. The matrix Green's functions are selected as follows:

$$\hat{\mathrm{G}} = \begin{pmatrix} G_{\uparrow\uparrow}(\mathbf{r}_1,\mathbf{r}_2,\omega) & G_{\uparrow\downarrow}(\mathbf{r}_1,\mathbf{r}_2,\omega) \\ G_{\downarrow\uparrow}(\mathbf{r}_1,\mathbf{r}_2,\omega) & G_{\downarrow\downarrow}(\mathbf{r}_1,\mathbf{r}_2,\omega) \end{pmatrix} \qquad \hat{\mathrm{F}} = i\hat{\sigma}_2 \begin{pmatrix} F_{\uparrow\uparrow}(\mathbf{r}_1,\mathbf{r}_2,\omega) & F_{\uparrow\downarrow}(\mathbf{r}_1,\mathbf{r}_2,\omega) \\ F_{\downarrow\uparrow}(\mathbf{r}_1,\mathbf{r}_2,\omega) & F_{\downarrow\downarrow}(\mathbf{r}_1,\mathbf{r}_2,\omega) \end{pmatrix}. \tag{6}$$

Here $\hat{G}_{\sigma\sigma'}(\mathbf{r}_1,\mathbf{r}_2,\omega)$ and $\hat{F}_{\sigma\sigma'}(\mathbf{r}_1,\mathbf{r}_2,\omega)$ are the Fourier components of the normal and anomalous Matsubara Green's functions, ω is the Matsubara frequency. In the matrix Green's functions notation the Gorkov's equations are rewritten as

$$\begin{gathered} i\omega\hat{\mathrm{G}}(\mathbf{r}_1,\mathbf{r}_2,\omega) - \hat{\mathrm{H}}_0(I)\hat{\mathrm{G}}(\mathbf{r}_1,\mathbf{r}_2,\omega) + \Delta^*(\mathbf{r}_1)\hat{\mathrm{F}}^\dagger(\mathbf{r}_1,\mathbf{r}_2,\omega) = \delta(\mathbf{r}_1-\mathbf{r}_2) \\ i\omega\hat{\mathrm{F}}^\dagger(\mathbf{r}_1,\mathbf{r}_2,\omega) + \hat{\mathrm{H}}_0(-I)\hat{\mathrm{F}}^\dagger(\mathbf{r}_1,\mathbf{r}_2,\omega) + \Delta^*(\mathbf{r}_1)\hat{\mathrm{G}}(\mathbf{r}_1,\mathbf{r}_2,\omega) = 0, \\ \Delta^*(\mathbf{r}) = \frac{\lambda T}{2}\sum_{\omega}{}' \mathrm{Sp}\hat{\mathrm{F}}^\dagger(\omega,\mathbf{r},\mathbf{r}), \qquad \hat{\mathrm{H}}_0 = -\frac{\hat{\mathbf{D}}^2}{2m^*} - \mu\hat{1} - I\hat{\sigma}_3. \end{gathered} \tag{7}$$

The prime at the summation sign means a cutoff at the Debye frequency. To calculate the critical temperature the Gor'kov equation is expanded in the linear approximation for the order parameter [13]. After that we assume the conventional self-consistency equation [13] with kernel $\hat{\mathrm{K}}^*(\omega,\mathbf{r},\mathbf{s})$

$$\begin{gathered} \hat{\mathrm{K}}^*(\omega,\mathbf{r},\mathbf{s}) = \hat{\mathrm{G}}(-\omega,-I,\mathbf{r},\mathbf{s})\hat{\mathrm{G}}(\omega,I,\mathbf{s},\mathbf{r}), \\ i\omega\hat{\mathrm{G}}(\mathbf{r}_1,\mathbf{r}_2,\omega) - \hat{\mathrm{H}}_0(\theta,\varphi,\xi,I)\hat{\mathrm{G}}(\mathbf{r}_1,\mathbf{r}_2,\omega) = \delta(\mathbf{r}_1-\mathbf{r}_2). \end{gathered} \tag{8}$$

To test our approach, we examine the case of a magnetic superconductor with helical magnetization. This case was firstly considered by Bulaevskii, Rusinov and Kulic [14] for theoretical study of superconducting substances like $ErRh_4B_4$ where helicoidal magnetization occurs. They calculated the Green's function without diagonalization of the Hamiltonian, found the expression for free energy and analysed the phase transitions in the system. Here we focus on the critical temperature of the superconducting transition and the Fulde–Ferrell–Larkin–Ovchinnikov (FFLO) states that add new information to this well-known case. Note that we use another notations than authors [14]: θ is a spatial variable only, and it is a linear function of the coordinate, and $2\xi-\varphi = 0$. The effective tensor field is coordinate independent, and the Green's function depends only on the difference $\mathbf{r}_1 - \mathbf{r}_2$. In this case the Fourier transform of the Green's function has the form

$$\hat{\mathbf{G}}(\mathbf{p},\omega) = \frac{1}{\zeta^2 - I^2 - (\mathbf{p},\nabla\theta)^2/4m^{*2}} \begin{pmatrix} \zeta - I & -i\dfrac{(\mathbf{p},\nabla\theta)}{2m^*} \\ \dfrac{i(\mathbf{p},\nabla\theta)}{2m^*} & \zeta + I \end{pmatrix}, \tag{12a}$$

$$\zeta = i\omega - \frac{p^2}{2m^*} + \mu - \frac{(\nabla\theta)^2}{8m^*}, \tag{12b}$$

After inverse Fourier transform we can obtain the well-known expression for the magnetic superconductor if grad $\theta = 0$. This kernel is an oscillating function. As was shown Fulde, Ferrell [15] and Larkin-Ovchinnikov [16], such behaviour of the kernel leads to spatial oscillations of the order parameter in a certain temperature range. In the simplest case, the order parameter can be assumed as an oscillating function $\Delta(\mathbf{r}) \sim \exp(i\mathbf{kr})$. We assume that the wave vector **k** is along the helical magnetic structure because it is the most interesting case, in our opinion. As shown in [2], under certain strict requirements for the parameters of the superconductor-ferromagnet bilayers, the problem of finding the critical temperature can be reduced to the determination of the critical temperature of the magnetic superconductor with a renormalized effective exchange field.

As usual, the equation for the critical temperature can be obtained from the self-consistent equation. The wave vector **k** of the FFLO state modulation is chosen from the condition of the critical temperature maximum [3]. In Figure 1 the calculated critical temperature of magnetic superconductor is shown for the various values of the rotation parameter $b = v_f \text{grad}\theta / T_{cs}$. Here T_{cs} is the critical temperature of bulk superconductor. In the case $b >> 1$ the influence of the effective exchange field on the critical temperature becomes much weaker. The same conclusion was drawn by Volkov [17] who considered superconductivity on the background of rotating magnetization using the Eilenberger equations. The FFLO state decays rapidly with decreasing the spatial period of magnetization rotation. Comparing the dependences like shown in figure 1 with different parameters of the magnetization rotation, we conclude that the influence of the magnetization rotation on the critical temperature can be reduced to suppression of the effective exchange field. For $b >> 1$ the suppression can be approximately represented as power law (See inset of figure 1).

Note, that in the framework of suggested approach the Usadel and Eilenberger equations can be also rewritten. Relative the known quasiclassical approach [4] we do not need to assume that the effective exchange field together with the superconducting order parameter change slowly compared

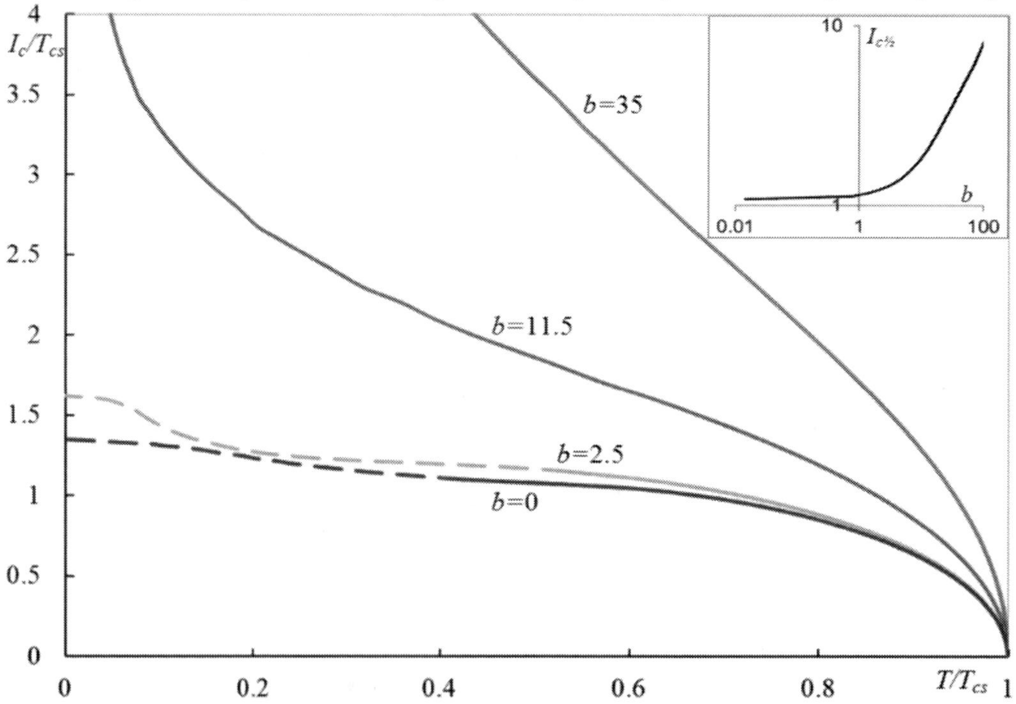

Figure 1. (Color online) The dependencies of critical effective exchange field I_c versus temperature T are presented at few values of the rotation parameter $b = v_f \text{grad}\theta / T_{cs}$. The dashed lines correspond to the FFLO states. In the inset the dependence $I_{c½}(b)$ is shown in a double logarithmic scale. $I_{c½}$ is determined from $T_c(I_{c½})/T_{cs}=1/2$ condition.

with the superconducting coherence length. In this sense our approach is more consistent, and the effect of magnetization inhomogeneity scale of correlation length is responsible to the appearance of the triplet component of the superconducting condensate.

3. Conclusions

The problem of superconducting correlations with the background of inhomogeneously directed effective exchange field can be reduced to the homogeneous problem in the presence of effective tensor field. The effective tensor field has the freedom that allows us to simplify Green's functions and quasiclassical equations. Proposed approach is confirmed in known case of superconductor with helical magnetization. The critical temperature is calculated via Gor'kov equations taken into account the FFLO states. Rotation of magnetization leads to reduction of effective exchange field and increasing the critical temperature. The approach can be expanded on the other cases of magnetic inhomogeneities.

Authors are thankful Prof. Uspenskaya L. for discussion and information about experiment [8]. The work is partially supported by the subsidy allocated to Kazan Federal University for the state assignment in the area of scientific activities. YP thanks to the RFBR (13-02-01202) for partial support. VT is also supported by Russian Program of Competitive Growth of Kazan Federal University.

References

[1] Efetov K B *et al.* 2008 *Springer Tracts in Modern Physics* **227** 251
[2] Buzdin A I 2005 *Rev. of Mod. Phys.* **77** 935
[3] Izyumov Yu A, Proshin Yu N and Khusainov M G 2002 *Physics-Uspekhi* **45** 109
[4] Demler E A, Arnold G B, Beasley M R 1997 *Phys. Rev. B* **55** 15174
[5] Khusainov M G, Izyumov Yu A and Proshin Yu N 2001 *JETP Letters* **73** 344
[6] Proshin Yu N, Izyumov Yu A and Khusainov M G 2001 *Phys. Rev. B* **64** 645221
[7] Goryunov Yu V 2003 *arXiv cond-mat* 0305428
[8] Uspenskaya L S *et al.* 2014 *Materials Research Express* – to be published
[9] Oh S, Youm D and Beasley M 1997 *Appl. Phys. Lett.* **71** 2376
[10] Tagirov L R 1999 *Phys. Rev. Lett.* **83** 2058
[11] Buzdin A I, Vedyayev A V and Ryzhanova N V 1999 *Europhys. Lett.* **48** 686
[12] Proshin Yu N *et al.* 2006 *Phys. Rev B* **73** 184514
[13] Abrikosov A A, Gor'kov L P, Dzyaloshinskii I E 1965 *Methods of Quantum Field Theory in Statistical Physics* (Pergamon Press, New York)
[14] Bulaevskii L N, Rusinov A I, Kulic M 1980 *Journal of Low Temperature Physics* **39** 255
[15] Fulde P and Ferrell R A 1964 *Phys.Rev.Lett.* **135** A550
[16] Larkin A I and Ovchinnikov Yu N, 1965 *Sov. Phys. JETP.* **20** 762
[17] Volkov A F 2008 *Phys. Rev. B* **77** 064521

Journal of Physics: Conference Series **568** (2014) 022043 doi:10.1088/1742-6596/568/2/022043

Niobium stripline resonators for microwave studies on superconductors

Markus Thiemann[1], **Daniel Bothner**[2], **Dieter Koelle**[2], **Reinhold Kleiner**[2], **Martin Dressel**[1], **Marc Scheffler**[1]

[1] 1.Physikalisches Institut, Universität Stuttgart, Pfaffenwaldring 57, D-70550 Stuttgart, Germany

[2] Physikalisches Institut and Center for Collective Quantum Phenomena in LISA$^+$, Universität Tübingen, Auf der Morgenstelle 14, D-72076 Tübingen, Germany

E-mail: `scheffl@pi1.physik.uni-stuttgart.de`

Abstract. Microwave spectroscopy is a powerful experimental tool to reveal information on the intrinsic properties of superconductors. Superconducting stripline resonators, where the material under study constitutes one of the ground planes, offer a high sensitivity to investigate superconducting bulk samples. In order to improve this measurement technique, we have studied stripline resonators made of niobium, and we compare the results to lead stripline resonators. With this technique we are able to determine the temperature dependence of the complex conductivity of niobium and the energy gap $\Delta(0) = 2.1$ meV. Finally we show measurements at the superconducting transition of a tantalum bulk sample using niobium stripline resonators.

1. Introduction

Optical spectroscopy on metals and superconductors can reveal information about charge carrier dynamics as well as electromagnetic excitations [1]. In the case of superconductors and correlated metals, these are connected to low energy scales both in temperature and in frequency and therefore have to be probed with low-energy optics, i.e. in the THz and GHz ranges [2]. Obtaining spectral information on highly conductive materials at cryogenic temperatures in the microwave range has proven technically challenging. While broadband measurements on a number of superconductors [3–7] have successfully been performed on thin films using Corbino spectrometers [8–10] or on single crystals using a bolometric approach (without obtaining phase information) [11–13], there is so far no experimental technique that allows broadband, phase-sensitive measurements on low-loss bulk samples. Therefore, in recent years new techniques have been developed that are resonant in nature, thus are very sensitive to small losses in bulk samples, and at least give certain information about the frequency-dependent response by operating at several resonant frequencies [2, 14, 15]. Our approach is based on superconducting striplines where the sample of interest replaces one of the ground planes [16]. So far, our striplines were made of Pb thin films [17, 18]. In the present work, we investigate in which regard Nb thin films can also be employed for this purpose and whether their performance can even surpass that of the Pb films.

2. Experiment

The stripline is a layered structure, consisting of the center strip sandwiched between two dielectric plates and two conducting ground planes (see Fig. 1(a)). In case of the Nb stripline

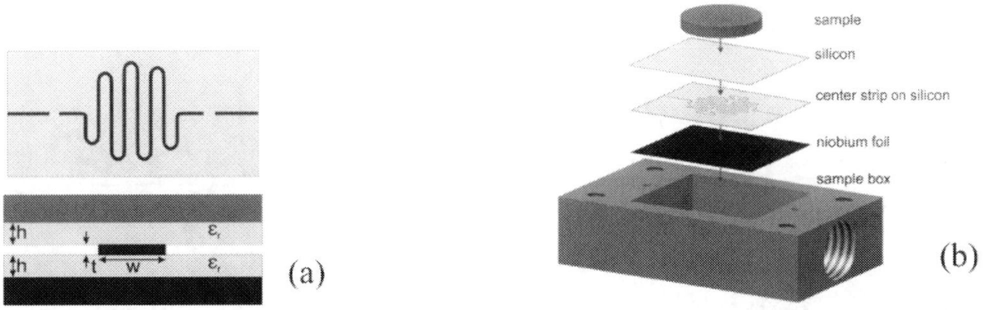

Figure 1: (a) Top: Top view on the meandered center strip with the gaps. Bottom: Cross section of a stripline. The characteristic impedance of a stripline is a function of height h of the dielectric planes, width w and height t of the center strip and the dielectric constant ϵ_r of the dielectric [19]. (b) Assembly of the stripline resonator into the brass sample box.

resonators, two 100 μm thin 12x10 mm^2 silicon plates were used as a dielectric, with a resistivity >5000 Ω/cm, due to the low dielectric losses and high dielectric constant [20] and 8 μm thick Nb foils as ground planes. Nb (T_c=9.2 K) was used as the conductor material instead of the previously used Pb (T_c=7.2 K) due to the higher transition temperature. The center strip was fabricated by sputtering Nb onto one of the silicon plates, followed by optical lithography and SF_6 etching. Due to two 80 μm gaps in the center conductor the stripline becomes a resonant structure. Since the frequency of the fundamental mode depends on the length of the stripline section between the gaps, the center strip was meander-shaped as shown in Fig. 1(a) to achieve resonance frequencies as low as possible for a given sample size. The stripline resonator was mounted inside a brass box and connected via coaxial microwave connectors to the 50 Ω measurement circuitry (see Fig. 1(b), [17]). By replacing one ground plane by any conducting sample, it is possible to determine the frequency and temperature dependence of the surface resistance R_s and the penetration depth λ of the sample [17]. In the present study all-Nb resonators and Nb resonators loaded with a tantalum sample were investigated in a temperature range between 1.6 K and 9.2 K.

3. Results

3.1. All-Nb resonator

In Fig. 2(a) the frequency dependence of R_s (obtained for 6 resonances at frequencies between 1.5 GHz and 11 GHz) of an all-Nb resonator is compared to R_s of an all-Pb resonator. Both show a quadratic frequency dependence in accord with theoretical predictions [21]. Surprisingly R_s of the Nb remains in the same order of magnitude as that of the Pb, despite the higher T_c. It has already been shown that defects on the resonator surface can increase R_s [22]. Therefore microscopic non-superconducting defects in the sputtered center strip may be an explanation for the higher R_s [23]. The measured temperature dependence of the resonance

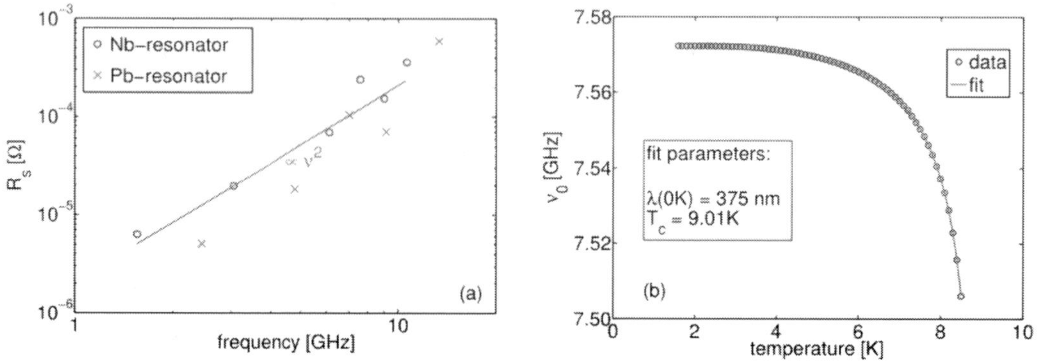

Figure 2: (a) Frequency dependence of R_s of Nb and Pb at T = 1.6 K. Despite the higher T_c, Nb shows a higher R_s. (b) Temperature dependence of the resonance frequency of the 5th mode. The green line represents a fit from which $\lambda(0\text{K})$ and T_c can be determined as described in [17].

frequency is shown in Fig. 2(b). The drop of ν_0 towards higher temperatures is due to the Nb entering the normal conducting state. The temperature dependence of ν_0 gives access to the penetration depth $\lambda(0\,\text{K}) \approx 377$ nm and from that the imaginary part of the impedance X_s can be determined [1]. In Fig. 3(a) the temperature dependence of the impedance is shown. The complex conductivity $\hat{\sigma} = \sigma_1 + i\sigma_2$ can now be calculated from the complex impedance [1].

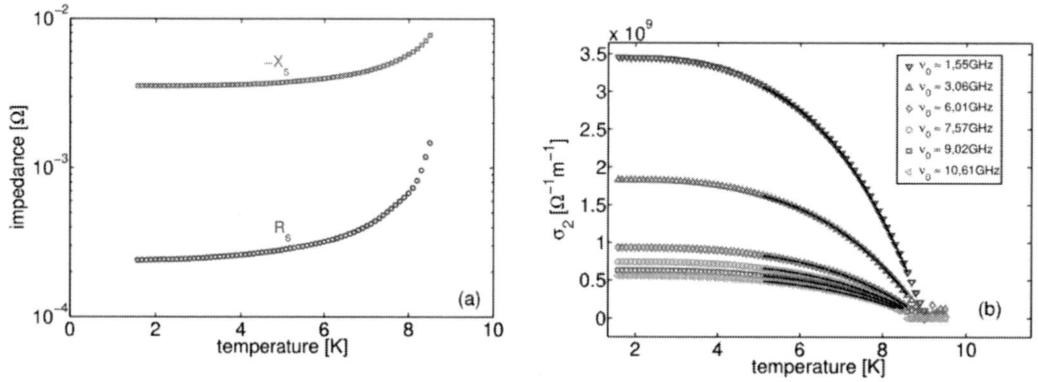

Figure 3: (a) Temperature dependence of the real and imaginary parts of the complex impedance of a pure Nb resonator obtained from the 5th mode. (b) Temperature dependence of the imaginary part σ_2 of the complex conductivity obtained from the fundamental mode and higher harmonics. The black lines represent fits of Eq. (1).

Fig. 3(b) shows the temperature dependence of the imaginary part σ_2 of the conductivity, which is connected to the superconducting energy gap Δ via:

$$\frac{\sigma_2(T)}{\sigma_n} \approx \frac{\pi\Delta(T)}{\hbar\omega} \tanh\left(\frac{\Delta(T)}{2k_{\mathrm{B}}T}\right) \tag{1}$$

By using the high temperature approximation $\Delta(T) \approx \Delta(0\mathrm{K})\sqrt{3.016\left(1-\frac{T}{T_c}\right) - 2.4\left(1-\frac{T}{T_c}\right)^2}$ [24], Eq. (1) can be fitted to the measured data with the fit parameters $\Delta(0\mathrm{K})$, T_c, and the real part σ_n of the conductivity in the normal conducting state. Averaging $\Delta(0\mathrm{K})$ obtained from all measured modes gives a value of 2.1 meV or $2\Delta(0\mathrm{K}) = 5.3k_{\mathrm{B}}T_c$. The energy gap of Nb thin films has already been investigated with THz-spectroscopy and a value of $2\Delta(0) = 4.1k_{\mathrm{B}}T_c$ was observed [25]. Both studies show deviations from the weak coupling BCS limit ($2\Delta(0\mathrm{K}) = 3.5k_{\mathrm{B}}T_c$ [26]) and may indicate a strong electron-phonon interaction.

3.2. Nb resonator loaded with tantalum sample

As a further test measurement we present data obtained from a bulk tantalum sample. Tantalum has a suitable T_c = 4.5 K which lies between the transition temperature of the Nb resonators and the lowest temperature accessible with our ^{4}He cryostat. Fig. 4(a)

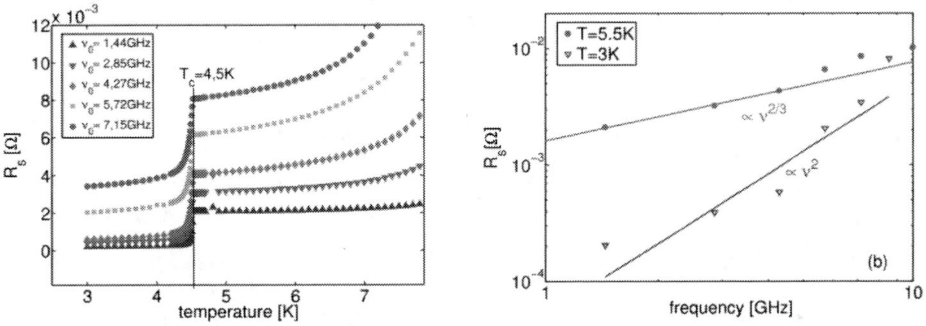

Figure 4: (a) Temperature dependence of R_s of tantalum for several modes. The drop in R_s at 4.5 K is due to the tantalum entering the superconducting state. (b) Frequency dependence of R_s of tantalum in the metallic and superconducting states. Lines are guides to the eye, representing $\nu^{2/3}$ and ν^2 dependencies.

shows the temperature dependence of R_s of the tantalum sample obtained from the different modes. Clearly a drop in R_s is visible at 4.5 K, which is due to the tantalum entering the superconducting state. At a temperature of 5.5 K tantalum is in the metallic state and R_s shows a frequency dependence of $R_s(\nu) \propto \nu^{2/3}$, which indicates that the tantalum is in the anomalous skin effect regime (see Fig. 4(b)) [17]. At 3 K the tantalum sample has entered the superconducting state, and a clear change in the frequency dependence is visible.

4. Conclusions and outlook

Using Nb as a superconducting material for stripline resonators, we increased the upper limit of the measurable temperature range from 6 K to 8 K, but a lower R_s was not achieved.

Future optimization in film growth might reduce the surface impedance and then allow even more sensitive measurements on low-loss bulk superconductors. In contrast to the previous studies using Pb resonators [17], Nb is a type-II superconductor that might allow operation in higher magnetic fields. This might be aggravated by field-dependent microwave losses due to vortices, but strategies to overcome these problems are being developed [27, 28]. The test measurements on superconducting tantalum prove the applicability of our technique. Since our approach can easily be implemented in a dilution refrigerator, microwave studies on a number of exotic superconductors, e.g. heavy-fermion superconductors [2, 29] or Sr_2RuO_4 [15] now become feasible.

Acknowledgements
We thank Gabriele Untereiner and Conrad Clauß for experimental support, and we acknowledge financial support by the DFG, including SFB/TRR21.

References
[1] Dressel M and Grüner G 2002 *Electrodynamics of Solids* (Cambridge: Cambridge University Press)
[2] Scheffler M et al 2013 *Phys. Status Solidi B* **250**, 439
[3] Wu D H, Booth J C and Anlage S M 1995 *Phys. Rev. Lett.* **75** 525
[4] Ohashi T, Kitano H, Maeda A, Akaike H and Fujimaki A 2006 *Phys. Rev. B* **73** 174522
[5] Steinberg K, Scheffler M and Dressel M 2008 *Phys. Rev. B* **77** 214517
[6] Pompeo N, Silva E, Sarti S, Attanasio C and Cirillo C 2010 *Physica C* **470** 901
[7] Liu W, Kim M, Sambandamurthy G and Armitage N P 2011 *Phys. Rev. B* **84** 024511
[8] Booth J C, Wu D H and Anlage S M 1994 *Rev. Sci. Instrum.* **65** 2082
[9] Scheffler M and Dressel M 2005 *Rev. Sci. Instrum.* **76** 074702
[10] Steinberg K, Scheffler M and Dressel M 2012 *Rev. Sci. Instrum.* **83** 024704
[11] Turner P J et al 2003 *Phys. Rev. Lett.* **90** 237005
[12] Turner P J, Broun D M, Kamal S, Hayden M E, Bobowski J S, Harris R, Morgan D C, Preston J S, Bonn D A and Hardy W N 2004 *Rev. Sci. Instrum.* **75** 124
[13] Bobowski J S et al 2010 *Phys. Rev. B* **82** 094520
[14] Huttema W A, Morgan B, Turner P J, Hardy W N, Zhou X, Bonn D A, Liang R and Broun D M 2006 *Rev. Sci. Instrum.* **77** 023901
[15] Ormeno R J, Hein M A, Barraclough T L, Sibley A, Gough C E, Mao Z Q, Nishizaki S and Maeno Y *Phys. Rev. B* **74** 092504
[16] DiIorio M S, Anderson A C and Tsaur B-Y 1988 *Phys. Rev. B.* **38** 7019
[17] Hafner D, Dressel M and Scheffler M 2014 *Rev. Sci. Instrum.* **85** 014702
[18] Scheffler M, Fella C and Dressel M 2012 *J. Phys. Conf. Ser.* **400** 052031
[19] Wheeler H A 1978 *IEEE Transactions on Microwave Theory and Techniques* **26** pp 866-76
[20] Krupka J, Breeze J, Centeno A, Alford N, Claussen T and Jenson L 2006 *IEEE Transactions on Microwave Theory and Techniques* **54** pp 3995-4001
[21] Halbritter J 1974 *Zeitschrift für Physik* **266** pp 209-17
[22] Pierce J M 1973 *Journal of Applied Physics* **44** pp 1342-7
[23] Benvenuti C, Calatroni S, Darriulat P, Peck M A, Valente A-M and Van't Hof C A 2001 *Physica C* **351** pp 421-8
[24] Ferrell R A 1964 *Zeitschrift für Physik* **182** 1
[25] Pronin A V, Dressel M, Pimenov A, Loidl A, Roshchin I V and Greene L H 1998 *Phys. Rev. B* **57** 14416
[26] Buckel W, Kleiner R 2004 *Superconductivity Fundamentals and Applications* (Wiley-VCH Verlag GmbH)
[27] Bothner D, Gaber T, Kemmler M, Koelle D, Kleiner R 2011 *Appl. Phys. Lett.* **98** 102504
[28] Bothner D et al 2012 *Appl. Phys. Lett.* **100** 012601
[29] Truncik C J S et al 2013 *Nature Commun.* **4** 2477

Quasiclassical Theory of Spin Imbalance in a Normal Metal-Superconductor Heterostructure with a Spin-Active Interface

O Shevtsov and T Löfwander

Department of Microtechnology and Nanoscience – MC2, Chalmers University of Technology, SE-412 96 Göteborg, Sweden

E-mail: shevtsov@chalmers.se, tomas.lofwander@chalmers.se

Abstract. Non-equilibrium phenomena in superconductors have attracted much attention since the first experiments on charge imbalance in the early 1970's. Nowadays a new promising line of research lies at an intersection between superconductivity and spintronics. Here we develop a quasiclassical theory of a single junction between a normal metal and a superconductor with a spin-active interface at finite bias voltages. Due to spin-mixing and spin-filtering effects of the interface a non-equilibrium magnetization (or spin imbalance) is induced at the superconducting side of the junction, which relaxes to zero in the bulk. A peculiar feature of the system is the presence of interface-induced Andreev bound states, which influence the magnitude and the decay length of spin imbalance. Recent experiments on spin and charge density separation in superconducting wires required external magnetic field for observing a spin signal via non-local measurements. Here, we propose an alternative way to observe spin imbalance without applying magnetic field.

1. Introduction

Studying non-equilibrium phenomena in mesoscopic systems is a challenging yet important task as it paves a way towards microscopic understanding of the underlying physical processes. In particular, non-equilibrium effects in superconducting hybrid structures have been an active direction of research since the pioneering experiments on charge imbalance by Clarke et al. [1]. They observed a voltage drop in the superconductor (S) close to the interface with a normal metal (N) subject to finite bias voltage. Theoretical picture proposed to explain this effect [2] was based on imbalance between the number of electron-like and hole-like quasi-particles in the superconductor when the bias was higher than the superconducting gap. This non-equilibrium quasi-particle distribution has to relax as we move towards the bulk of the superconductor, and it was shown [3] that any pair-breaking mechanism can be responsible for this.

An obvious extension of the previous effect is a situation when there is an imbalance between quasi-particles with opposite spin-projections, which can be achieved by using a ferromagnet (F) as an injector, or applying an external magnetic field. It would create an induced magnetization in the superconductor which relaxes away from the interface, or so-called spin imbalance. In recent experiments [4–6] this possibility has been investigated via non-local transport measurements in N-S-F and F-S-F structures. It is important to emphasize that for all the experiments it was necessary to apply an external magnetic field in order to see

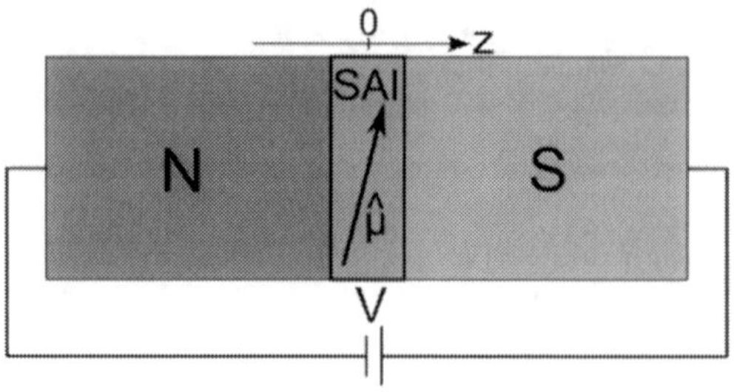

Figure 1. (Color online) Cartoon of the setup: normal metal (N) - superconductor (S) junction with a spin-active interface (SAI) characterized by an intrinsic magnetic moment directed along $\hat{\mu}$. Finite bias V is applied across the junction.

a significant spin imbalance signal. In our work we focus on a system consisting of a normal metal (N) connected to a superconductor (S) via a spin-active interface[1] (SAI) [8, and references therein], see Fig. 1. The system is kept under finite bias voltage V. One can imagine the spin-active interface as a magnetic barrier, which has different transparencies for quasi-particles with opposite spins and is able to rotate the spin of a quasi-particle via Larmor precession around its magnetic moment. Such an interface can host a pair of in-gap spin-polarized Andreev bound states, which has been observed in recent experiments [9]. We use quasiclassical Green's function method [10, 11, and references therein] to calculate the induced magnetization profile in the superconductor. It turns out that it is possible to observe significant spin imbalance signal in this system without applying external magnetic field for voltages below the superconducting gap due to presence of the bound states. Moreover the properties of the observed signal can be tuned by the parameters of the interface. For the reader interested in the physics of the processes responsible for such behavior we refer to Ref. [8]. In this paper we explain some details of calculations left out from Ref. [8]. In Sec. 2, we briefly describe the general theoretical framework we use. In Sec. 3, we describe the theoretical model of our system. In Sec. 4 we demonstrate the self-consistent procedure for calculating bulk superconducting properties in presence of magnetic and scalar impurities. Sec. 5 concludes the paper.

2. Methods

All the relevant information necessary to describe our system is contained in the quasiclassical Green's function $\check{g}(\epsilon, \mathbf{p}_{\mathrm{F}}, \mathbf{r}, t)$. Here ϵ is the quasi-particle energy, $\mathbf{p}_{\mathrm{F}}$ is the quasi-particle momentum on the Fermi surface, $\mathbf{r}$ is the spatial coordinate, and t is the time. This function satisfies the quasiclassical Eilenberger equation,

$$[\epsilon\hat{\tau}_3\check{1} - \check{h}, \check{g}]_{\otimes} + i\hbar\mathbf{v}_{\mathrm{F}} \cdot \boldsymbol{\nabla}\check{g} = \check{0}, \tag{1}$$

accompanied by the normalization condition[2] $\check{g} \otimes \check{g} = -\pi^2\check{1}$. Here and below we will suppress function arguments for brevity. Note that $\check{g}$ and the self-energy $\check{h}$ are matrices in multidimensional space: they have a 2×2 matrix structure in Keldysh space denoted by "check", a 2×2 matrix structure in particle-hole space denoted by "hat", and in general a 2×2 matrix structure in spin space[3],

$$\check{\chi} = \begin{pmatrix} \hat{\chi}^{\mathrm{R}} & \hat{\chi}^{\mathrm{K}} \\ 0 & \hat{\chi}^{\mathrm{A}} \end{pmatrix}, \ \chi = \{g, h\}, \tag{2}$$

$$\hat{h}^{\mathrm{R,A}} = \begin{pmatrix} \Sigma & \Delta \\ \tilde{\Delta} & \tilde{\Sigma} \end{pmatrix}^{\mathrm{R,A}}, \ \hat{h}^{\mathrm{K}} = \begin{pmatrix} \Sigma & \Delta \\ -\tilde{\Delta} & -\tilde{\Sigma} \end{pmatrix}^{\mathrm{K}}. \tag{3}$$

[1] Such interfaces can be engineered with the help of ferromagnetic materials, see Ref. [7].
[2] The $\otimes$ operation is defined by $\check{A} \otimes \check{B}(\epsilon, t) = e^{i\hbar(\partial_\epsilon^A \partial_t^B - \partial_t^A \partial_\epsilon^B)/2}\check{A}(\epsilon, t)\check{B}(\epsilon, t)$.
[3] The "tilde" or particle-hole conjugation operations is defined as $\tilde{Y}(\epsilon, \mathbf{p}_{\mathrm{F}}, \mathbf{r}, t) = Y(-\epsilon, -\mathbf{p}_{\mathrm{F}}, \mathbf{r}, t)^*$.

Finally, $\hat{\tau}_3$ and $\check{1}$ are third Pauli matrix in particle-hole space and unit matrix in Keldysh space, respectively. We will use $\{\sigma_i, i = 1, 2, 3\}$ Pauli matrices for spin. There is a convenient and numerically stable method for solving Eq.(1) by using the so-called Riccati parametrization [11, and references therein] for the elements of the matrix in Eq.(2),

$$\hat{g}^{\mathrm{R,A}} = \mp 2\pi i \begin{pmatrix} \mathcal{G} & \mathcal{F} \\ -\tilde{\mathcal{F}} & -\tilde{\mathcal{G}} \end{pmatrix}^{\mathrm{R,A}} \pm i\pi\hat{\tau}_3, \tag{4}$$

$$\hat{g}^{\mathrm{K}} = -2\pi i \begin{pmatrix} \mathcal{G} & \mathcal{F} \\ -\tilde{\mathcal{F}} & -\tilde{\mathcal{G}} \end{pmatrix}^{\mathrm{R}} \otimes \begin{pmatrix} x & 0 \\ 0 & \tilde{x} \end{pmatrix} \otimes \begin{pmatrix} \mathcal{G} & \mathcal{F} \\ -\tilde{\mathcal{F}} & -\tilde{\mathcal{G}} \end{pmatrix}^{\mathrm{A}}. \tag{5}$$

Here functions $\mathcal{G}$ and $\mathcal{F}$ are in turn parametrized as

$$\mathcal{G} = (1 - \gamma \otimes \tilde{\gamma})^{-1}, \; \mathcal{F} = (1 - \gamma \otimes \tilde{\gamma})^{-1} \otimes \gamma, \tag{6}$$

while their particle-hole conjugated counterparts are obtained by applying the "tilde"-operation. The $\gamma^{\mathrm{R,A}}$ and $\tilde{\gamma}^{\mathrm{R,A}}$ are *coherence functions* describing the electron-hole coherence in the superconducting state, while x and $\tilde{x}$ are *distribution functions*. So, using this parametrization the Eq.(1) simplifies to the following Riccati-type differential equations,

$$(i\hbar\mathbf{v}_{\mathrm{F}} \cdot \boldsymbol{\nabla} + 2\epsilon)\gamma^{\mathrm{R,A}} = \left[\gamma \otimes \tilde{\Delta} \otimes \gamma + \Sigma \otimes \gamma - \gamma \otimes \tilde{\Sigma} - \Delta\right]^{\mathrm{R,A}}, \tag{7}$$

$$(i\hbar\mathbf{v}_{\mathrm{F}} \cdot \boldsymbol{\nabla} + i\hbar\partial_t)x - [\gamma \otimes \tilde{\Delta} + \Sigma]^{\mathrm{R}} \otimes x - x \otimes [\Delta \otimes \tilde{\gamma} - \Sigma]^{\mathrm{A}}$$
$$= -\gamma^{\mathrm{R}} \otimes \tilde{\Sigma}^{\mathrm{K}} \otimes \tilde{\gamma}^{\mathrm{A}} + \Delta^{\mathrm{K}} \otimes \tilde{\gamma}^{\mathrm{A}} + \gamma^{\mathrm{R}} \otimes \tilde{\Delta}^{\mathrm{K}} - \Sigma^{\mathrm{K}}, \tag{8}$$

and the corresponding particle-hole conjugated equations obtained applying "tilde"-operation. They have to be supplemented by the corresponding boundary conditions in order to find unique solution. It is important to mention the symmetry relations for coherence and distribution functions,

$$\gamma^{\mathrm{A}} = \left[\tilde{\gamma}^{\mathrm{R}}\right]^{\dagger}, \; x = x^{\dagger}. \tag{9}$$

In equilibrium[4],

$$x^{\mathrm{eq}} = (1 - \gamma^{\mathrm{R}}\tilde{\gamma}^{\mathrm{A}}) \tanh\left(\frac{\epsilon}{2k_B T}\right), \tag{10}$$

and in the non-superconducting (or normal) case $\gamma^{\mathrm{R,A}}$=0. Finally, since the method we use is a mean field theory, equations (7)-(8) have to be in general solved self-consistently together with the corresponding self-consistency equations for self-energies. In particular, the order parameter for a spin-singlet s-wave superconductor $\Delta_0^{\mathrm{R}}(\mathbf{r}) = i\sigma_2\Delta_0(\mathbf{r})$ satisfies,

$$\Delta_0(\mathbf{r}) = -\frac{i\lambda N_{\mathrm{F}}}{8\pi} \int_{-\epsilon_c}^{\epsilon_c} d\epsilon \int \frac{d\Omega_{\mathbf{p}_{\mathrm{F}}}}{4\pi} \mathrm{Tr}\left[i\sigma_2(\hat{\tau}_1 - i\hat{\tau}_2)\hat{g}^{\mathrm{K}}(\epsilon, \mathbf{p}_{\mathrm{F}}, \mathbf{r})\right], \tag{11}$$

where $\lambda < 0$ is the electron-phonon coupling constant and ϵ_c is the high-energy cut-off of the order of the Debye frequency. As soon as the Riccati equations are solved we can find various physical observables as[5],

$$\text{spin imbalance: } \mathbf{M}(\mathbf{r}) = 2\mu_{\mathrm{B}}^2 N_{\mathrm{F}}\mathbf{B}(\mathbf{r}) + \frac{i\mu_{\mathrm{B}} N_{\mathrm{F}}}{8\pi} \int d\epsilon \int \frac{d\Omega_{\mathbf{p}_{\mathrm{F}}}}{4\pi} \mathrm{Tr}\left[\hat{\boldsymbol{\alpha}}\hat{g}^{\mathrm{K}}(\epsilon, \mathbf{p}_{\mathrm{F}}, \mathbf{r})\right], \tag{12}$$

$$\text{local density of states: } \mathrm{N}(\epsilon, \mathbf{r}) = -\frac{N_{\mathrm{F}}}{2\pi}\mathrm{Im}\left\{\int \frac{d\Omega_{\mathbf{p}_{\mathrm{F}}}}{4\pi} \mathrm{Tr}\left[\hat{\tau}_3\hat{g}^{\mathrm{R}}(\epsilon, \mathbf{p}_{\mathrm{F}}, \mathbf{r})\right]\right\}. \tag{13}$$

[4] Here k_B is the Boltzmann constant and T is the temperature.

[5] Here $\hat{\boldsymbol{\alpha}} = \mathrm{diag}(\boldsymbol{\sigma}, \boldsymbol{\sigma}^*)$ is a block-diagonal matrix in Nambu space, $\mathbf{B}(\mathbf{r})$ is the external magnetic field, μ_{B} is the Bohr magneton, $|e|$ is the elementary charge, and N_{F} is the density of states at Fermi level in the normal state.

3. Model for N-SAI-S junction

Let us now specify our theoretical model. We consider a conventional spin-singlet s-wave superconductor S. We assume translational invariance in the plane of the interface so that all physical quantities depend only on one spatial coordinate z. In order to achieve spin relaxation in S we consider a dilute concentration of magnetic spin-flip impurities (SFI) [12], and for completeness we also allow the presence of scalar impurities (SI). Both contributions are described in the Born approximation as

$$\check{h}_{\rm sf}(\epsilon,z)=\frac{\hbar}{2\pi\tau_{\rm sf}}\int\frac{d\Omega_{\mathbf{p}_{\rm F}}}{4\pi}(\hat{\tau}_3\check{1})\check{g}(\epsilon,\mathbf{p}_{\rm F},z)(\hat{\tau}_3\check{1}),\quad \check{h}_{\rm s}=\frac{\hbar}{2\pi\tau_{\rm s}}\int\frac{d\Omega_{\mathbf{p}_{\rm F}}}{4\pi}\check{g}(\epsilon,\mathbf{p}_{\rm F},z), \tag{14}$$

where $\tau_{\rm sf}$ and $\tau_{\rm s}$ are the spin-flip and mean free times, respectively. To simplify the calculation we further assume low transparency of the interface and therefore approximate all self-energies by their *bulk self-consistent values* constant in space. The s-wave order parameter is then defined by a real constant[6]: $\Delta_0^{\rm R}=i\sigma_2\Delta_0$, $\Delta_0=$ const. So, with the approximations mentioned above we have to solve Eqs.(7)-(8). Solutions with $\mathbf{p}_{\rm F}$ pointing towards the interface are called *incoming* and according to our approximation they are given by the bulk self-consistent values $\gamma^{\rm R}_{\rm S,in}\equiv\gamma^{\rm R}_{\rm S,bulk}$, see Sec.4. The coherence function in N is assumed $\gamma^{\rm R}_{\rm N}=0$, i.e. we neglect the proximity effect. The interface is usually described by a scattering matrix, which defines (via boundary conditions) together with the incoming solution $\gamma^{\rm R}_{\rm S,in}$ the corresponding interface value of the *outgoing* solution $\gamma^{\rm R}_{\rm S,out}(z=0)$, i.e. with $\mathbf{p}_{\rm F}$ pointing away from the interface. This value is then used as the initial condition for Eq.(7) for calculating the spatial dependence of $\gamma^{\rm R}_{\rm S,out}(z)$. The same algorithm holds for the distribution function $x_{\rm S}$ with the bulk value given by Eq.(10). We refer the reader to Refs.[8, 11, and references therein] for further details on the description of the interface and the boundary conditions. In the next section we will show how to find the self-consistent bulk value $\gamma^{\rm R}_{\rm S,bulk}$ for the coherence function, as this is the main initial ingredient one needs to know, according to our model, in order to find any physical observable.

4. Self-consistent solution in the bulk of superconductor

We now consider the bulk of the superconductor in presence of self-energies Eq.(14) and the order parameter defined in Eq.(11). Since we are looking for the bulk homogeneous solution we can omit the derivative on the left hand side of Eq.(7), which also makes the coherence function isotropic, $\gamma^{\rm R}_{\rm bulk}=\gamma^{\rm R}(\epsilon)$. Taking into account the spin-singlet pairing in the superconductor, both the order parameter and the coherence function have the same spin structure: $\Delta_0^{\rm R}=i\sigma_2\Delta_0$ and $\gamma^{\rm R}(\epsilon)=i\sigma_2\gamma(\epsilon)$. Therefore we can easily solve the Riccati equation demanding that $\gamma(\epsilon)\to 0$ if $\Delta_0\to 0$. So we find,

$$\gamma=-\frac{\Delta}{\Sigma+\sqrt{\Delta^2-\Sigma^2}},\quad \tilde{\gamma}=-\gamma \tag{15}$$

$$\Sigma=\epsilon+\left(\frac{\hbar}{\tau_{\rm sf}}+\frac{\hbar}{\tau_{\rm s}}\right)\frac{\Sigma}{2\sqrt{\Delta^2-\Sigma^2}},\quad \Delta=\Delta_0-\left(\frac{\hbar}{\tau_{\rm sf}}-\frac{\hbar}{\tau_{\rm s}}\right)\frac{\Delta}{2\sqrt{\Delta^2-\Sigma^2}}. \tag{16}$$

Finally, we have to simplify the self-consistency equation for the order parameter Eq.(11). Using Eq.(5) and Eq.(10) we obtain,

$$\Delta_0=\frac{|\lambda|N_{\rm F}}{2}\int_{-\epsilon_c}^{\epsilon_c}d\epsilon\,{\rm Im}\left[\frac{\Delta}{\sqrt{\Delta^2-\Sigma^2}}\right]\tanh\frac{\epsilon}{2k_{\rm B}T}. \tag{17}$$

Finally we have to eliminate the high-energy cut-off ϵ_c in favor of the experimentally measurable critical temperature [10] by demanding that $\Delta_0\to 0$ if $T\to T_{\rm C}$. Introducing the dimensionless quantities Z_1 and Z_2 via $\Sigma=\epsilon Z_1$ and $\Delta=\Delta_0 Z_2$, we obtain

[6] Phase of the order parameter is irrelevant because, for simplicity, we neglect bulk superflow.

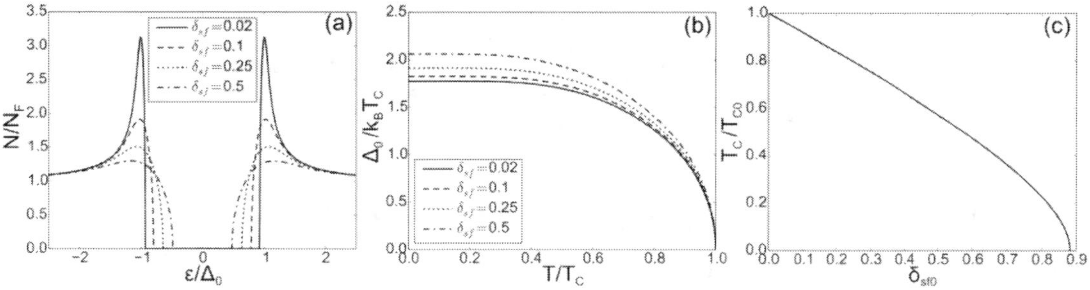

Figure 2. (a),(b),(c) Density of states $N(\epsilon)$, order parameter Δ_0, and critical temperature T_C in presence of magnetic impurities calculated self-consistently. Impurity strength is characterized by the dimensionless parameters $\delta_{sf} = \hbar/\tau_{sf} k_B T_C$ in (a)-(b) and $\delta_{sf0} = \hbar/\tau_{sf} k_B T_{C0}$ in (c). T_{C0} is the critical temperature for a clean system.

$$\ln \frac{T}{T_C} = \int_0^\infty \frac{d\epsilon}{\epsilon} \left\{ \mathrm{Im} \left[\frac{\epsilon Z_2}{\sqrt{\Delta_0^2 Z_2^2 - \epsilon^2 Z_1^2}} \right] - \frac{1}{1 + \frac{\hbar^2}{\tau_{sf}^2 \epsilon^2}} \right\} \tanh \frac{\epsilon}{2k_B T}$$
$$- \int_0^\infty \frac{d\epsilon}{\epsilon} \left[1 - \frac{1}{1 + \frac{\hbar^2}{\tau_{sf}^2 \epsilon^2}} \right] \left(\tanh \frac{\epsilon}{2k_B T} - \tanh \frac{\epsilon}{2k_B T_C} \right). \qquad (18)$$

By solving together Eq.(16) and Eq.(18) one can obtain the dependence $\Delta_0 = \Delta_0(T, T_C, \tau_{sf})$ and calculate, for example, the density of states, see Fig. 2. One can also verify with Eq.(18) that the bulk value Δ_0 is independent of the presence of scalar impurities [13].

5. Conclusions

The algorithm described in this paper has to be considered as a first iteration in a fully self-consistent solution to the problem of N-SAI-S junction. In general, all self-energies are spatially dependent as well as both incoming and outgoing solutions to Riccati equations. However, our simple analysis helps to capture qualitatively the most essential physical features of the model.

References

[1] Clarke J 1972 *Phys. Rev. Lett.* **28** 1363
[2] Tinkham M and Clarke J 1972 *Phys. Rev. Lett.* **28** 1366
[3] Lemberger T R 1984 *Phys. Rev. B* **29** 4946
[4] Hübler F, Wolf M J, Beckmann D and v Löhneysen H 2012 *Phys. Rev. Lett.* **109** 207001
[5] Wolf M, Hübler F, Kolenda S, v Löhneysen H and Beckmann D 2013 *Phys. Rev. B* **87** 024517
[6] Quay C H L, Chevallier D, Bena C and Aprili M 2013 *Nat. Phys.* **9** 84
[7] Khaire T S, Khasawneh M A, Pratt W P and Birge N O 2010 *Phys. Rev. Lett.* **104** 137002
[8] Shevtsov O and Löfwander T 2014 (*Preprint* `arXiv:1404.6172 [cond-mat.mes-hall]`)
[9] Hübler F, Wolf M J, Scherer T, Wang D, Beckmann D and v Löhneysen H 2012 *Phys. Rev. Lett.* **109** 087004
[10] Serene J W and Rainer D 1983 *Phys. Rep.* **101** 221
[11] Eschrig M 2009 *Phys. Rev. B* **80** 134511
[12] Abrikosov A A and Gor'kov L P 1961 *Sov. Phys. JETP* **12** 1243
[13] Anderson P W 1959 *J. Phys. Chem. Solids* **11** 26

Journal of Physics: Conference Series **568** (2014) 022045 doi:10.1088/1742-6596/568/2/022045

Scanning tunneling microscopy and spectroscopy at very low temperatures

I. Guillamon, J.G. Rodrigo, S. Vieira and H. Suderow

Laboratorio de Bajas Temperaturas, Departamento de Física de la Materia Condensada, Instituto de Ciencia de Materiales Nicolás Cabrera, Facultad de Ciencias
Universidad Autónoma de Madrid, E-28049 Madrid, Spain
Unidad Asociada de Bajas Temperaturas y Altos Campos Magnéticos, UAM, CSIC, Cantoblanco E-28049 Madrid, Spain

E-mail: isabel.guillamon@uam.es, hermann.suderow@uam.es

Abstract. During past decades, an increasing number of laboratories is using cryogenic scanning tunneling microscopy and spectroscopy (STM/S) to probe different kinds of electronic systems. Measurements in a dilution refrigerator are particularly useful to study superconductors, because temperatures of order of 100 mK are well below most critical temperatures and effectively reduce thermally excited quasiparticles. The local electronic density of states is then obtained at atomic level with a resolution in energy of some tens of μeV. Visualizing spatial variations of the local density of states allows characterizing vortex cores and the vortex lattice. Vortex core electronic features provide the anisotropy of the superconducting properties, and help understanding the influence of competing orders such as charge density waves. Here we will review results in dichalcogenide superconductors, in the magnetic borocarbide $TmNi_2B_2C$ and in thin films, discussing in some detail a few relevant aspects of thermal depinning and melting in thin films.

1. Introduction

Tunneling spectroscopy at zero field can be used to obtain the electronic density of states of a superconducting sample. The resulting tunneling conductance vs bias voltage is the temperature smeared density of states, integrated over the reciprocal space volume occupied by electrons participating in the tunneling process [1]. In a STM experiment, tunneling is made by using an atomically sharp tip, providing images of the spatial variation of the tunneling conductance at the surface of the sample [2–4]. STM spectroscopy of superconductors at the surface has been proven to successfully provide a good measure of the bulk electronic density of states in many materials [2–14].

In Fig.1 we plot the superconducting tunneling conductance obtained in several materials in our laboratory [5–14]. We also show its position in a plot of the critical temperature T_c vs the Fermi temperature T_F (Uemura plot [15]). Systems with nearly free electrons have very low T_c/T_F and are located on the bottom right part of the plot, whereas systems with strong electronic correlations have T_c/T_F approaching 10^{-2}-10^{-1}. It is obvious that materials showing bulk properties at the surface are spread over the whole Uemura plot. STM is thus a very powerful technique to obtain the superconducting density of states in many superconductors of topical interest.

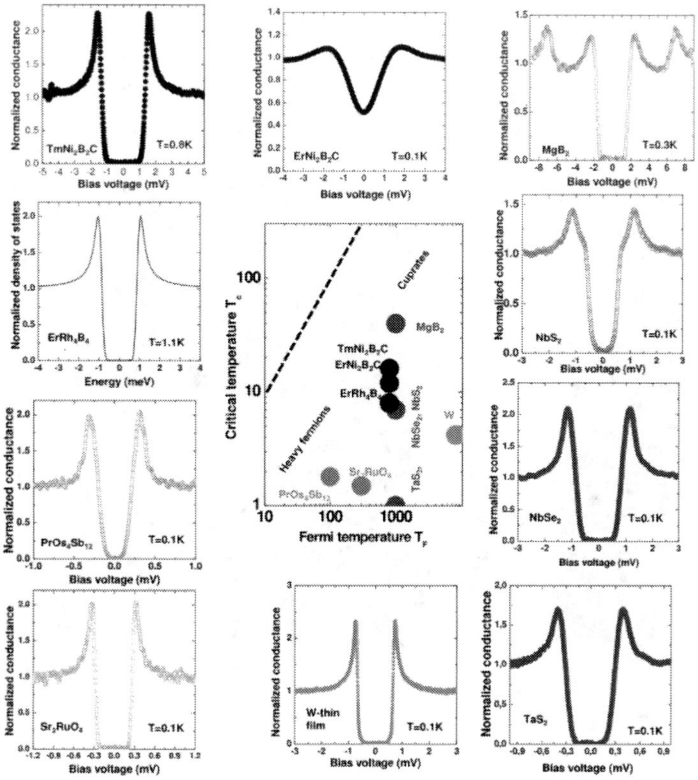

Figure 1. Tunneling spectroscopy of several superconducting materials, obtained with the STM at temperatures well below the superconducting T_c. We also show the position of each compound in a plot of the critical temperature T_c vs Fermi temperature T_F (dashed line is $T_c = T_F$). The shape of the superconducting density of states is very different in each compound, showing the variations of the superconducting gap over the Fermi surface. Finding low energy excitations does not necessarily show strong electronic correlations, but can be also due to large gap anisotropy or pair breaking. Data adapted from Refs. [5–14].

The vortex lattices of several of these compounds have been measured using STM. They are more difficult to obtain than just the superconducting density of states, because scanning conditions must be good over areas spanning many hundreds of nm. Here we will discuss results about the vortex core of the anisotropic superconductor 2H-$NbSe_2$, about vortex lattice symmetry in crystalline superconductors, and about the vortex lattice in an amorphous W-based thin film.

2. Scanning tunneling microscopy at very low temperatures

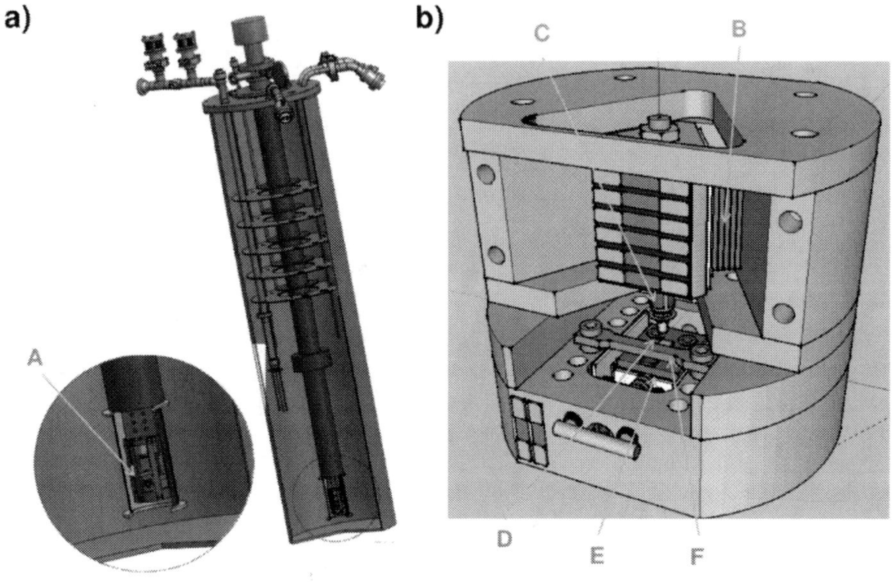

Figure 2. In a we show a simple arrangement of a cryogenic tunneling microscope. The microscope (A) is posed or suspended on the bottom of a dilution refrigerator cryostat or another cooling machine, eventually at the center of a superconducting solenoid. The microscope may be isolated from vibrations using fiber glass, as discussed in Ref. [16]. In b we show a schematic picture of one microscope. Piezostacks (B) are used to move a prism with a scanning tube inside (C). The tip (D) is located on top of a sliding sample holder (E). The sample holder is moved using a wire (not shown, see Ref. [16]) that goes up to ambient temperature. Eventually, a beam (F) is located in the way of the sample holder to break the sample. The whole microscope is made of Ti, using non-magnetic springs. We have built microscopes using similar designs with diameters ranging from 30 mm up to 50 mm.

Succesful microscopy requires an environment free of vibrations, which is inherently rather difficult to achieve at very low temperatures in a dilution refrigerator [16]. In Fig.2 we describe schematically a cryogenic system with a scanning microscope inside (Fig.2a) that is shown in more detail in Fig.2b. Useful tricks to reduce damping out mechanical noise from pumps are sandboxes, silicone tubing (even for the helium 3 pumping lines), increasing the rigidity of the cryostat as a whole, and surrounding the dewar with acoustic foam (such as for example Copopren®). Designs of a microscope often include an exchangeable sample and tip holder, which can be moved back to ambient conditions using a long stick. Eventually, a vacuum preparation chamber is attached to the top or the bottom of the cryostat to allow for sample and tip preparation [17–20]. In Ref. [16] we discuss in detail a procedure consisting of in-situ sample preparation by cryogenic cleaving or breaking a surface using a sliding sample holder. We have applied this technique succesfully in dichalcogenides, pnictides, heavy fermions and borocarbide materials. We prepare our tips using soft metals as Au, Al or Pb, and clean them in-situ by controlled repeated indentation in a sample made of the same element [21].

Temperatures of order of 100 mK and below can be achieved by using high conductivity wires between the cold plate and the microscope, and carefully thermalizing them [4]. To meaningfully measure at these temperatures, however, the resolution in energy should be below a few tens of μeV, which is difficult to achieve. The best method, in our view, to test the energy resolution of the system is to measure the tunneling spectroscopy of superconducting tip and sample of a simple s-wave superconductor as Al [22]. At low enough temperatures, the sharpness of the quasiparticle peaks should be independent of temperature if sample and tip are far enough to avoid Andreev reflection, and only shows smearing due to jitter of the voltage source [6].

3. The vortex core and superconducting anisotropies induced by charge order in 2H-NbSe$_2$

In a superconducting vortex, the superfluid density, or the square of the pair wavefunction, decreases towards the core, and vanishes exactly at the vortex center. In the clean limit, in-gap quasiparticle excitations appear throughout the core due to multiple Andreev scattering [25]. The density of states at each position and energy inside vortex core is the result of the combination of the different quantized states at the core weighted by their corresponding probability density. At the vortex center the probability density is higher for the fundamental state which gives a peak located usually very close to zero bias (actually, at the zero point energy $\Delta^2/2E_F$, with E_F being the Fermi energy and Δ the superconducting gap). At a distance of the core center, higher energy states present a larger probability density and thus contribute more to the tunneling conductance. The first observation of vortex core in-gap states was made by Hess et al in 2H-NbSe$_2$ [26–28], who carefully followed the spatial dependence of these states. In particular, they found that the tunneling conductance is highly bias voltage and position dependent. At the center of the vortex core (Fig.3a), there is a pronounced peak which splits in energy when leaving the core (Figs.3 and 4). The splitting depends on the position, and we can identify two length scales, shown by the dashed lines in Fig.3a. These length scales correspond to the vortex core extension along the red lines (insets of Fig.3a) and provide the anisotropic star shaped vortex core.

We can discuss this behavior by assuming that the extension of core states is given by an effective coherence length $\xi_{eff} = \frac{\hbar v_F}{\pi \Delta}$ where v_F is an average over all Fermi velocities of possible Fermi surface orbits at a plane perpendicular to the magnetic field, weighted by their respective contribution to the density of states, and Δ depends on the reciprocal space dependence of the superconducting gap at a plane perpendicular to the magnetic field direction [29]. The anisotropy of the vortex core observed in 2H-NbSe$_2$ shows that superconducting gap and Fermi surface have a strong six-fold modulation [30, 31]. This anisotropy is absent in 2H-NbS$_2$, a compound having a very similar T$_c$ than 2H-NbSe$_2$, but no charge density wave, suggesting that the charge density wave provides the six-fold anisotropy in ξ_{eff}.

We can understand the vortex as a perturbation of the superconductor, and the surrounding patterns in the density of states show the anisotropy of the superconducting properties in reciprocal space. The patterns are wakes of extended quasiparticle excitations and their length

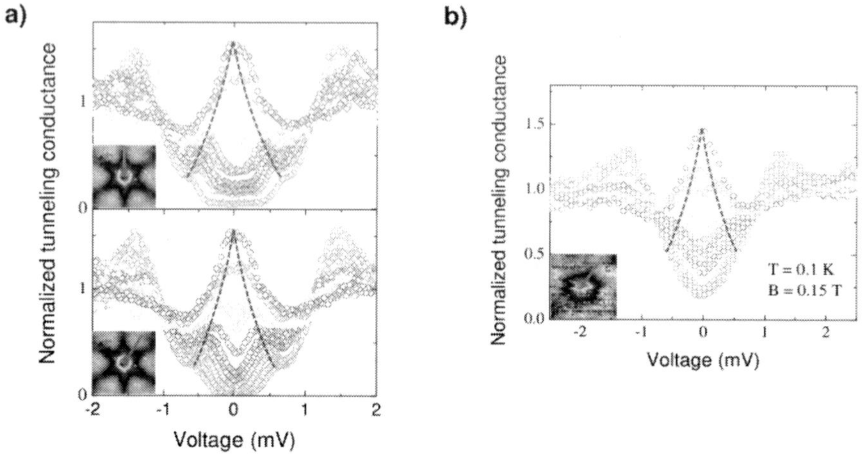

Figure 3. a) shows spatial variation of superconducting density of states in vortex cores of 2H-$NbSe_2$ at 0.03T and 0.1K in two different directions: along the ray (top panel) and in between rays (bottom panel). At the vortex center, we find a zero bias peak due to Andreev quasiparticles at the lowest quantized energy state. When moving out of the vortex core, the zero bias peak splits into two peaks that move to higher bias voltages (blue dotted lines in figure). The way the central peak evolves with energy and position is different along the two directions. b) shows Andreev localized states inside vortex cores of 2H-NbS_2 at 0.1 K and 0.15 T. Similarly to 2H-$NbSe_2$, we find a zero bias peak at the core center and higher energy states when moving out of the vortex core. But the tunneling conductance is radially symmetric. Data adapted from Refs. [23, 24].

depends on the shape of the Fermi surface and on the anisotropy of the superconducting properties [29, 32, 33]. Similarly, other point like perturbations, as a point defect in a metal, also produce anisotropic scattering processes which provide real space image of the reciprocal space electronic properties [34, 35]. By imaging the electronic properties in real space around a vortex, we obtain a picture of the anisotropies of the superconductor over the Fermi surface [23, 24].

Note that the six-fold modulation is found at all energies in 2H-$NbSe_2$ (Fig.4), but the rays present at zero bias split and rotate when approaching the bias voltage of the quasiparticle peak. In 2H-NbS_2, the cores remain radially symmetric at all energies. The splitting in 2H-$NbSe_2$ has been reproduced by introducing a six-fold cosine modulation to the superconducting gap [31], suggesting that the anisotropy in ξ_{eff} is mostly due to the superconducting gap anisotropy. Band structure calculations, on the other hand, show that there is a maximum in the non-interacting susceptibility along the directions where ξ_{eff} is largest [36].

Conductance maps for several magnetic fields are shown in Fig.5. In 2H-$NbSe_2$, the in-

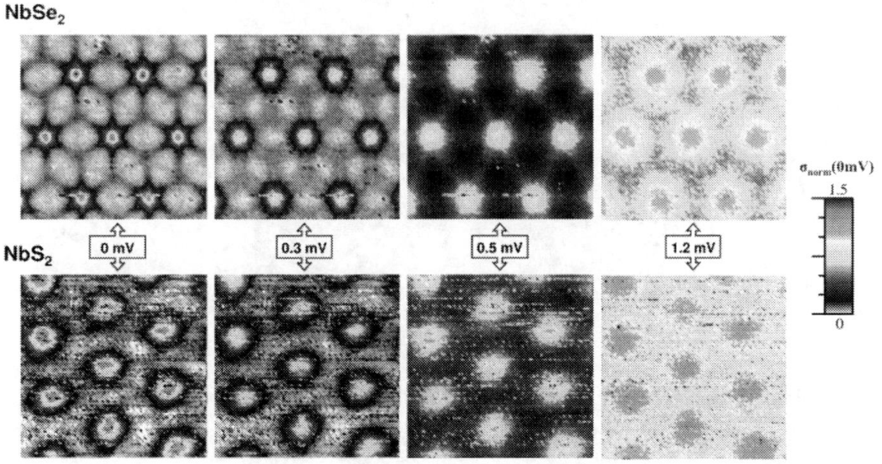

Figure 4. Conductance maps showing the spatial evolution of vortex core states at different energies in 2H-NbSe$_2$ (top panels) and 2H-NbS$_2$ (lower panels), both at 0.1K and 0.15 T. The changes in the conductance are given by the color scale at the right part of the figure. In 2H-NbSe$_2$ we find rich patterns with strong sixfold anisotropy. In 2H-NbS$_2$ we find rounded vortex cores at all energies.

plane sixfold anisotropy is more pronounced at low magnetic fields although it is also present at magnetic fields as high as 1T. This is likely due to overlapping effects between neighboring vortices which become significant at high fields where intervortex distance is within the same order of magnitude than the superconducting coherence length. For instance, the vortex image at 0.03T, where the intervortex distance is 290 nm, does not show overlapping between vortex core states of neighboring vortices. When increasing the magnetic field in 2H-NbSe$_2$ we observe overlap effects already at relatively low fields. At the middle point between three adjacent vortices we find a local enhancement of the density of states which increases (Fig.5a) due to overlap between the rays of the sixfold star shape of vortices at the point where they intersect. Note that the intersection is not located along the line of nearest neighbor vortices, but along the line where the core states have a larger extension. Similar vortex core merging effects have been predicted to occur in nanostructured superconductors [37].

At high magnetic field, we observe that the contrast rapidly decreases, due to pair breaking effects. In multiband superconductors such as 2H-NbSe$_2$ or 2H-NbS$_2$, the multigap structure is affected by the magnetic field, with a destruction of smaller sized superconducting gaps [38], and an extension of associated core states.

Thus, the spatial and magnetic field variations of superconducting vortices provide a useful probe of multigap structure and spatial anisotropies of the superconducting state. Here we have

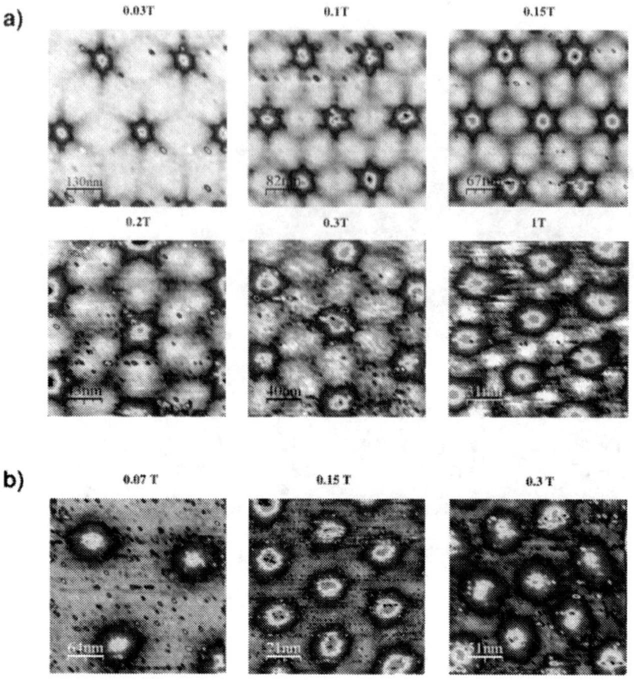

Figure 5. Zero bias conductance maps at 0.1 K and for different values of the magnetic field in 2H-$NbSe_2$ (a) and in 2H-NbS_2 (b).

illustrated the application of this technique to the case of 2H-$NbSe_2$, and shown that the charge order significantly shapes its superconducting anisotropy. The multigap features are shared with 2H-NbS_2, and are probably related to the similar bands crossing the Fermi level in these materials.

Band structure calculations indicate that the charge density wave in 2H-$NbSe_2$ is not related to Fermi surface features but rather to electron-phonon coupling [36]. The charge density wave gives spatially modulated normal state density of states along the [100] direction. Core states in 2H-$NbSe_2$ have largest extension along the [110] direction of the crystalline lattice. Further work is needed to determine the relationship between the anisotropy of the normal state density of states and of the superconducting gap with the vortex core features. Interestingly, superposed to the spatial variation of the Andreev core levels, we also find an atomically modulated superconducting density of states. The atomic size modulation follows the charge density wave. Calculations show that similar results at zero field can be explained by an anisotropic tip and an anisotropic superconducting gap [23]. It is yet unclear how to relate the atomic size modulation with the vortex core features. Similarly shaped tunneling conductance maps are found in the cuprate superconductors, where vortex cores show a checkerboard pattern [39]. A comparative study between 2H-$NbSe_2$ and 2H-NbS_2 could bring light into the actual relationship of the charge density wave in the atomic size superconducting density of states inside vortex cores.

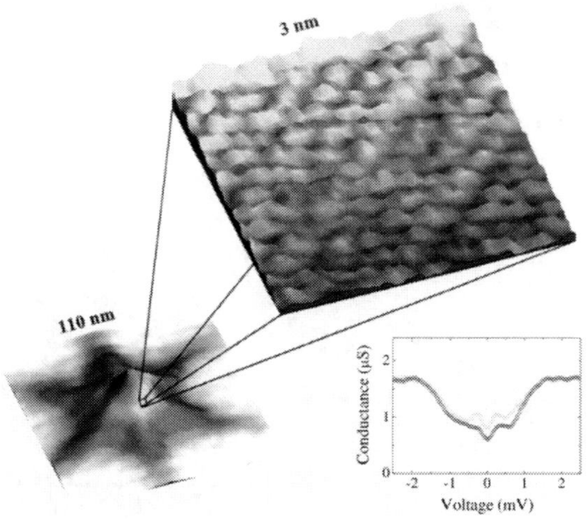

Figure 6. Zero bias conductance maps at 0.1 K and 0.03 T showing the spatial changes of the superconducting density of states at E_F on the scale of the coherence length around an isolated vortex and at atomic scale in the small region marked by a black square in the figure. Lower right panel shows the tunneling conductance measured when the tip is on top of a Se atom (green) and in between Se atoms (blue). Data adapted from [23].

4. The vortex lattice and crystalline anisotropies

Anisotropic superconducting properties also influence vortex arrangements and can modify the vortex lattice symmetry with respect to the conventional hexagonal configuration. In superconductors with hexagonal crystalline lattice as 2H-$NbSe_2$ and 2H-NbS_2, the vortex lattice remains hexagonal and is always oriented with the atomic lattice. In superconductors with different crystal symmetries such as tetragonal or orthorhombic, the vortex lattice may form a square or distorted hexagonal lattice. In the nickel borocarbide superconductors STM and neutron diffraction measurements have shown hexagonal to square transitions in the vortex lattice as a function of temperature and magnetic field [41–43]. In $TmNi_2B_2C$ such a transition has been observed by STM when increasing the magnetic field parallel to the c axis [44]. When the magnetic field is parallel to the basal plane, the anisotropy induced by the tetragonal crystal structure is weaker and the transition to the square vortex lattice occurs at higher magnetic fields [44]. The hexagonal vortex lattice is nevertheless oriented with one of its high symmetry axis parallel to the c-axis, as shown in the inset of Fig.7.

The vortex arrangement is not only influenced by the crystalline and Fermi surface symmetries but also by other existing kinds of order which break symmetries of the superconductor. A relevant case is found in magnetic superconductors where ferromagnetic or antiferromagnetic ordering have a significant influence in the superconducting properties [41, 45]. Excellent

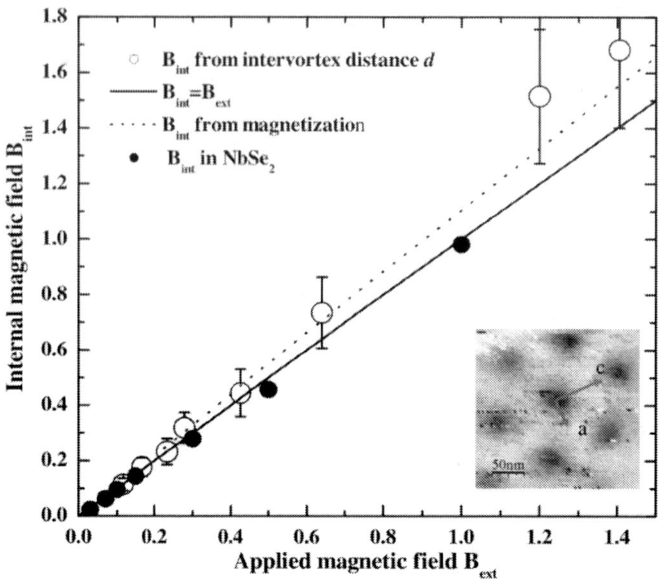

Figure 7. Internal magnetic field estimated from the intervortex distance as a function of the applied magnetic field for different superconductors: $TmNi_2B_2C$ (black circles) and $2H\text{-}NbSe_2$ (white circles). Continuous black line represents $B_{int} = B_{ext}$, which means that the intervortex distance follows the expected magnetic field dependence $d = (\frac{4}{3}^{1/4}(\frac{\Phi_0}{B})^{1/2}$. Dashed line gives B_{int} in $TmNi_2B_2C$ estimated from magnetization data from Ref. [40]. Inset shows hexagonal vortex lattice obtained at 0.1 K and 0.23 T with the field applied within the plane of the tetragonal structure of $TmNi_2B_2C$, i.e. perpendicular to the c-axis. Red arrows show the direction of the crystalline axes a and c. Data adapted from Ref. [44].

examples are the nickel borocarbides which remain the only family of magnetic superconductors where vortex lattice has been studied in depth.

In $TmNi_2B_2C$ we observe that the intervortex distance is slightly lower than expectations, indicating that the internal magnetic field is higher than the applied magnetic field. Fig.7 shows the intervortex distance in $TmNi_2B_2C$ compared with the intervortex distance found in $2H\text{-}NbSe_2$ and in $2H\text{-}NbS_2$. Clearly, magnetism of $TmNi_2B_2C$ gives slightly denser vortex lattices, possibly inducing anomalous behavior in other macroscopic properties such as the upper critical field.

5. Thermally activated vortex lattice

A generic phase diagram of type II superconductors is shown in Fig.8. When fluctuations become important, a vortex liquid phase appears at the boundary between the Shubnikov phase and the normal and Meissner phases (Fig.8b). The liquid occupies a large region of the phase diagram in many superconductors [3,46–49].

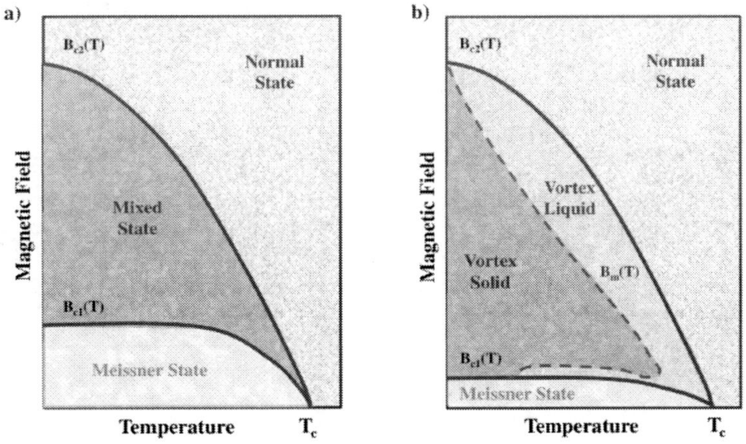

Figure 8. Generic phase diagrams of type II superconductors where thermal fluctuations close to the transition to the normal state $B_{c2}(T)$ are negligible (a) and relevant (b) giving a vortex liquid phase. In b) $B_m(T)$ is the melting line which separates the vortex solid and liquid phases.

Vortex lattice melting in amorphous W-based thin films deposited using focused ion beam [50] has been observed in Ref. [51]. A good measure of the strength of the action of fluctuations in the vortex lattice is provided by the Levanyuk-Ginzburg parameter $LG = \frac{1}{2}\left(\frac{k_B T_c(0)}{4\pi\mu_0 B_c^2(0)\xi^3(0)}\right)^2 \approx 10^{-7}\frac{\kappa^4 T_c^2(0)}{B_{c2}(0)}$ (κ being the Ginzburg-Landau parameter) [52, 53]. LG compares the condensation energy in a volume of order of the coherence length with the thermal energy at critical temperature. In conventional superconductors LG is very low, around $10^{-11} - 10^{-14}$ and the vortex liquid eventually exists in an extremely narrow region. The melting line $B_m(T)$ practically coincides with $B_{c2}(T)$. Cuprate high T_c superconductors have large values of LG, around 10^{-1} to 10^{-3}, due to their high critical temperatures, small coherence length and large anisotropy. These superconductors are characterized by the presence of a extended region in the phase diagram B-T where thermal fluctuations induce the formation of a vortex liquid (Fig.8b). Thin films of amorphous superconductors which are often extreme type II superconductors with relatively low values of T_c have values of LG rather close to those found in cuprate superconductors, with $LG \approx 10^{-4}$ and have shown a small but sizeable range close to $B_{c2}(T)$ with a vortex liquid phase [50].

Before melting takes place, the thermal de-pinning transition may occur. De-pinning happens when the vortex lattice is frozen into a distorted hexagonal lattice by interaction with defects. The depinning transition in the W thin films has been directly obtained by imaging the changes in vortex arrangements induced by temperature at different magnetic fields. Fig.9a shows the

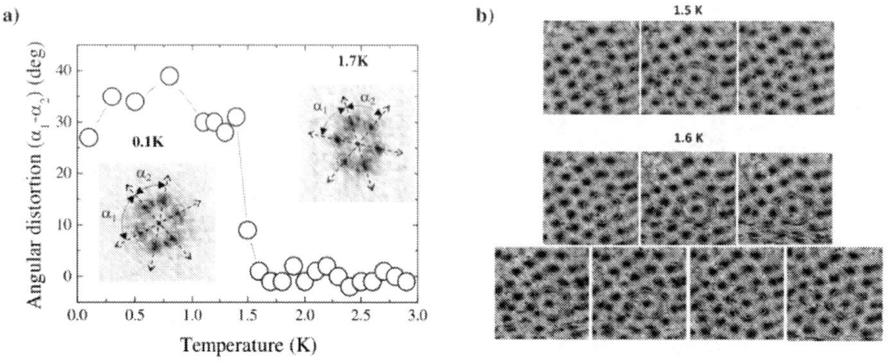

Figure 9. a) Shows the angular distortion of the hexagonal vortex lattice in W amorphous superconducting thin film across the de-pinning transition, taken at a magnetic field of 1T. The de-pinning transition occurs sharply in the temperature interval between 1.5 K and 1.6 K. b) shows a sequence of vortex lattice images taking during the de-pinning transition where blue and magenta hexagons highlight respectively the low and high temperature vortex arrangement. Data adapted from Ref. [51].

thermal de-pinning of the lattice at 1T. At low temperatures the hexagonal lattice is distorted due to pinning. When increasing temperature, thermal energy increases above pinning energy and vortices arrange forming an undistorted hexagonal lattice. The de-pinning transition is very sharp. Here it occurs in the temperature interval between 1.5 K and 1.6 K.

Below and above the depinning transition, the lattice remains stable with time. But at the depinning transition we observe time dependent rearrangements. We have followed the behavior of the lattice during one hour, taking eight consecutive vortex images at each temperature. In the temperature range from 1.5 K to 1.6 K, the vortex arrangement changes between two well defined configurations, which are highlighted in Fig.9b by blue and magenta hexagons. These correspond respectively to the vortex arrangement found at low temperature (blue hexagon) which follows the pinning potential, and the vortex configuration observed after the de-pinning transition (magenta hexagon), freed from the local pinning potential. Most dramatic changes between both configurations involve vortices between pinning centers, always moving coherently in a particular direction. On the other hand, the mobility of the vortices close to pinning features is reduced. This indicates that the de-pinning in thin films is dominated by thermal activation of vortex bundles formed between surface pinning centers at low temperatures. Images taken during the transition at 1.5K and 1.6K show an increase in the frequency (R) of the vortex rearrangement from one oscillation each hour at 1.5 K (three representative images, of the eight made at 1.5 K are shown in top panel of Fig.9b) to eight oscillations each hour at 1.6 K (images at bottom panel of Fig.9b). We can estimate the local pinning strength using an Arrhenius law given by $R = R_0 e^{-U/K_B T}$, where R_0 is the characteristic frequency for thermal fluctuations (typically $10^{-11} s^{-1}$) and U is the activation energy associated to the pinning potential barrier. Using the frequency found in the experiments, we obtain a value for U of about 50 K which is comparable to the values obtained from macroscopic measurements in superconducting thin films with similar properties. This value is also consistent with the pinning energy associated to a surface corrugation of around 10 nm, as found in the experiment.

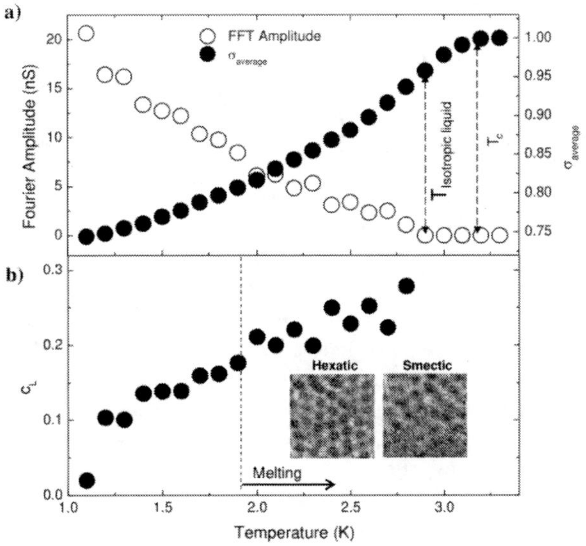

Figure 10. a) Temperature dependence of the Fourier amplitude (white circles) and average conductance $\sigma_{average}(x, y)$ (black circles) of the vortex lattice images taken during the melting process. The formation of the isotropic liquid occurs at the temperature $T_{isotropic}$ where the Fourier amplitude becomes zero whereas the transition to the normal state is found at higher temperatures, at the point T_c where $\sigma_{average} = 1$. b) Temperature dependence of the Lindemann number in the two dimensional vortex lattice. Black dotted line marks the temperature at which melting process starts with the first appearance of dislocations in the images. At the transition c_L has a value around 0.2 in good agreement with the Lindemann criterion. Insets show vortex lattice images at the hexatic and smectic phases. Data adapted from Ref. [51].

In Ref. [51], we acquire conductance maps showing the vortex lattice as a function of temperature to reach vortex melting. We identify different vortex phases, from the vortex solid to an isotropic vortex liquid. A hexatic phase with orientational order and a smectic like phase showing 1D vortex arrangements were observed in between the solid and isotropic liquid phases. Temperature dependence of the Fourier amplitude and average conductance at zero bias of the vortex maps show the temperature region at which isotropic vortex liquid forms (Fig.10a).

Lindemann criterion in the 2D melting has been discussed for different systems [54]. Due to the continuous nature of the melting transition, as opposed to first order abrupt transition, it is natural to ask how large the amplitude fluctuations are when entering the liquid phases. In Fig.10b we show that Lindemann's criterion is satisfied when entering the hexatic phase. We plot the amplitude of the positional fluctuations due to thermal smearing as a function of temperature. The melting transition is expected when the mean quadratic displacement with respect to the equilibrium positions ($< u^2 >^{1/2}$) is equal to a fraction of the lattice parameter of the ordered solid (d), $< u^2 >^{1/2} \approx c_L d$, where c_L is empirically found to be around 0.1-0.3 in most systems. We estimate c_L from the reduction of the amplitude in the Fourier transforms of the vortex lattice images due to the arrival of disorder during the melting process (A_{th}^{FFT}). We use that A_{th}^{FFT} is proportional to $e^{-2\pi c_L^2}$ [55]. To obtain A_{th}^{FFT} we renormalize the Fourier amplitude at each temperature by a term proportional to $\sigma_{averaged}[(1-\sigma_{average}(T))/(1-\sigma_{average}(0))]$. This term accounts for temperature induced reduction of the superconducting gap and increased thermal smearing in the tunneling conductance, which both deplete the Fourier amplitude. Thus, the renormalization allows us to sutract from the Fourier amplitude any temperature dependence which is not due to the increased disorder during melting. We assume that $< u^2 >^{1/2}$ is negligible at low temperatures so that $c_L = \sqrt{-ln(A_{th}^{FFT}/A_{th}^{FFT}(lowT))/2\pi}$. As shown in Fig.10b, we find that c_L is close to 0.2 when we observe the start of the melting process in our experiments. We also observe a slight increase of c_L above the melting temperature which could be due to enhanced vortex motion within the hexatic phase.

6. Conclusion and outlook

We have discussed imaging experiments with the STM which provide direct measurement of the superconducting anisotropies, of the interaction between crystalline and vortex lattices, and of thermally induced processes such as vortex lattice melting. Measurements in nanostructures, as a function of the applied current and in strongly correlated electron systems are now ongoing by several groups. They will provide vivid images of vortex behavior and of the electronic structure that will improve our understanding of superconductors.

7. Acknowledgments

This work was supported by the Spanish MINECO (FIS2011-23488), by the Comunidad de Madrid through program Nanobiomagnet and by Marie-Curie actions under the project FP7-PEOPLE-2013-CIG-618321. We also acknowledge SEGAINVEX workshop of UAM, Banco Santander and COST MP1201 action. We acknowledge discussions with A. Carrington, J.P. Brison, P.C. Canfield, V. Moshchalkov, V. Vinokur, R. Cordoba, J.M. De Teresa, J. Sese, M.A. Ramos, R. Ibarra, J. Flouquet and F. Guinea. The drawing of the STM was made by A. Fente. We acknowledge help and discussions with J.A. Galvis, A. Maldonado, V. Crespo and E. Herrera-Vasco.

8. References

[1] Wolf E 1989 *Principles of Electron Tunneling Spectroscopy* (Oxford: Oxford University Press)
[2] Fischer O, Kugler M, Maggio-Aprile I and Berthod C 2007 *Review of Modern Physics* **79** 353
[3] Hoffman J 2011 *Reports Progress Physics* **74** 124513
[4] Suderow H, Guillamon I, Rodrigo J G and Vieira S 2014 *Superconductor Science Technology* **27** 063001
[5] Martínez-Samper P, Suderow H, Vieira S, Brison J P, Luchier N, Lejay P and Canfield P C 2003 *Physical Review B* **67** 014526
[6] Rodrigo J G, Suderow H and Vieira S 2004 *European Physical Journal B* **40** 483
[7] Crespo V, Rodrigo J G, Suderow H, Vieira S, Hinks D and Schuller I 2009 *Phys. Rev. Lett.* **102** 237002
[8] Guillamon I, Suderow H, Vieira S, Cario L, Diener P and Rodiere P 2008 *Physical Review Letters* **101** 166407

[9] Guillamon I, Suderow H, Rodrigo J G, Vieira S, Rodiere P, Cario L, Navarro-Moratalla E, Marti-Gastaldo C and Coronado E 2011 *New Journal of Physics* **13** 103020
[10] Maldonado A, Guillamon I, Rodrigo J G, Suderow H, Vieira S, Aoki D and Flouquet J 2012 *Physical Review B* **85** 214512
[11] Galvis J, Rodiere P, Guillamon I, Osorio M, Rodrigo J G, Cario L, Navarro-Moratalla E, Coronado E, Vieira S and Suderow H 2013 *Physical Review B* **87** 094502
[12] Rubio-Bollinger G, Suderow H and Vieira S 2001 *Physical Review Letters* **86** 5582
[13] Suderow H, Vieira S, Strand J D, Bud'ko S and Canfield P C 2004 *Physical Review B* **69** 060504(R)
[14] Suderow H, Martínez-Samper P, Vieira S, Luchier N, Brison J P and Canfield P C 2001 *Physical Review B* **64** 020503(R)
[15] Uemura Y 2009 *Physica B: Condensed Matter* **404** 3195
[16] Suderow H, Guillamon I and Vieira S 2011 *Review of Scientific Instruments* **82** 033711
[17] Kambara H, Matsui T, Niimi Y and Fukuyama H 2007 *Review of Scientific Instruments* **78** 073703
[18] Song Y J, Otte A F, Shvarts V, Zhao Z, Kuk Y, Band S R B A, Hess F M and Stroscio J A 2010 *Review of Scientific Instruments* **81** 121101
[19] Assig M, Koch A, Stiepany W, Straer C, Ast A, Kern K and Ast C 2012 *Review of Scientific Instruments* **83** 033701
[20] Singh U, Enayat M, White S and Wahl P 2013 *Review of Scientific Instruments* **84** 013708
[21] Rodrigo J G, Suderow H, Vieira S, Bascones E and Guinea F 2004 *Journal of Physics: Condensed Matter* **16** 1151
[22] Crespo V, Maldonado A, Galvis J, Kulkarni P, Guillamon I, Rodrigo J G, Suderow H, Vieira S, Banerjee S and Rodiere P 2012 *Physica C* **479** 19
[23] Guillamón I, Suderow H, Guinea F and Vieira S 2008 *Physical Review B* **77** 134505
[24] Guillamón I, Suderow H, Vieira S, Cario L, Diener P and Rodiére P 2008 *Physical Review Letters* **101** 166407
[25] Caroli C, de Gennes P G and Matricon J 1964 *Physics Letters* **9** 307
[26] Hess H F, Robinson R B, Dynes R C, Valles J M and Waszczak J V 1989 *Physical Review Letters* **62** 214
[27] Hess H F, Robinson R B and Waszczak J V 1990 *Physical Review Letters* **64** 2711
[28] Hess H F, Robinson R B and Waszczak J V 1991 *Physica B* **169** 422
[29] Wang Y, Hirschfeld P and Vekhter I 2012 *Physical Review B* **85** 020506
[30] Gygi F and Schlüter M 1991 *Physical Review B* **43** 7609
[31] Hayashi N, Isoshima T, Ichioka M and Machida K 1998 *Physical Review Letters* **80** 2921
[32] Balatsky A V, Vekhter I and Zhu J X 2006 *Rev. Mod. Phys.* **78** 373
[33] Pan S, Hudson E, Lang K M, Eisaki H, Uchida S and Davis J C 2000 *Nature* **403** 746
[34] Weisman A, Wenderoth M, Lounis S, Zagn P, Quaas N, Ulbrich R, Dederichs P and Blügel S 2009 *Science* **323** 1190
[35] Ruby M, Heinrich B W, Pascial J I and Franke K 2014 *arXiv* 1409.6638
[36] Johannes M D, Mazin I I and Howells C A 2006 *Physical Review B* **73** 205102
[37] Melnikov A, Ryzhov D and Silaev M 2009 *Physical Review B* **79** 134521
[38] Eskildsen M R, Kugler M, Tanaka S, Jun J, Kazakov S, Karpinski J and Fischer O 2002 *Physical Review Letters* **89** 187003
[39] Hoffman J E, Hudson E W, Lang K M, Madhavan V, Eisaki H, Uchida S and Davis J 2002 *Science* **295** 466
[40] Brison J, Luchier N, Sulpice A, Suderow H, Martínez-Samper P, Vieira S, Buzdin A and Canfield P C 2004 *Journal of Magnetism and Magnetic Materials* **272** 158
[41] Canfield P C, Gammel P L and Bishop D J 1998 *Physics Today* **51** 40
[42] Gammel P L, Barber B, Ramirez A, Varma C, Bishop D, Canfield P, Kogan V, Eskildsen M, Andersen N, Mortensen K and Harada K 1999 *Physical Review Lettrs* **82** 1756
[43] Eskildsen M, Abrahamsen A B, Kogan V G, Gammel P L, Mortensen K, Andersen N H and Canfield P C 2001 *Physical Review Letters* **86** 5148
[44] Guillamon I, Crespo M, Suderow H, Vieira S, Brison J, Bud'ko S and Canfield P 2010 *Physica C* **470** 771
[45] Aoki D, Huxley A, Ressouche E, Braithwaite D, Flouquet J, Brison J, Lhotel E and Paulsen C 2001 *Nature* **413** 613–616
[46] Blatter G, Feigel'man M, Geshkenbein V, Larkin A and Vinokur V 1994 *Rev. Mod. Phys.* **66** 1125
[47] Zeldov E, Majer D, Konczykowski M, Geshkenbein V, Vinokur V and Shtrikman H 1995 *Nature* **375** 373
[48] Paltiel Y, Zeldov E, Myasoedov Y, Rappaport M, Jung G, Bhattacharya S, Higgins M, Xiao Z, Andrei E, Gammel P and Bishop D 2000 *Physical Review Letters* **85** 3712
[49] Brandt E 1993 *Physical Review B* **48** 6699
[50] Guillamón I, Suderow H, Vieira S, Fernandez-Pacheco A, Sese J, Cordoba R, Teresa J D and Ibarra M 2008 *New Journal of Physics* **10** 093005
[51] Guillamón I, Suderow H, Fernandez-Pacheco A, Sese J, Cordoba R, De Teresa J M, Ibarra M R and Vieira

Sl 2009 *Nature Physics* **5** 651
[52] Levanyuk A 1959 *Sov. Phys. JETP* **36** 571
[53] Ginzburg V 1960 *Sov. Phys. Solid State* **2** 1284
[54] Zheng X and Earnshaw J 1998 *Europhysics Letters* **41** 635
[55] Klein T, Joumard J, Marcus R and Cubbit R 1999 *Journal of Low Temperature Physics* **117** 1353

Journal of Physics: Conference Series **568** (2014) 022046 doi:10.1088/1742-6596/568/2/022046

Observation of unreconstructed square atomic square lattice in $Ca(Fe_{0.965}Co_{0.035})_2As_2$ cleaved at very low temperatures

A. Fente[1], I. Guillamón[1,2], S. Ran[3], S. Vieira[1,2], H. Suderow[1,2], S.L. Bud'ko[3] and P.C. Canfield[3]

[1] Laboratorio de Bajas Temperaturas, Departamento de Física de La materia Condensada, Instituto de Ciencia de Materiales Nicolás Cabrera and Instituto de Física de la Materia Condensada (IFIMAC), Universidad Autónoma de Madrid, Cantoblanco, E-28049 Madrid, Spain.
[2] Unidad Asociada de Bajas Temperaturas y Altos Campos Magnéticos, UAM, CSIC, Cantoblanco, E-28049 Madrid, Spain.
[3] Ames Laboratory and Department of Physics and Astronomy, Iowa State University, Ames Iowa 50011, USA.

E-mail: `anton.fente@uam.es, hermann.suderow@uam.es`

Abstract. Scanning tunneling microscopy (STM) has been by now made in many pnictide layered superconductors. In AFe_2As_2 (A=Ca, Sr, Ba), surface reconstructions dominate the STM topographic images and no surface termination with the bulk atomic lattice has been reported. Here, we present STM measurements on the surface of $Ca(Fe_{0.965}Co_{0.035})_2As_2$ cleaved in-situ below liquid helium temperatures. We observe the ubiquitous surface reconstruction consisting of 1D stripes that can adopt two degenerate orientations with respect to the crystal lattice. By searching over many different scanning windows, we find that low temperature cleaving also exposes surfaces showing the As unreconstructed lattice of the bulk tetragonal crystal, over areas of some tens of nm lateral size.

1. Introduction

Iron based compounds have phase diagrams showing magnetism, structural transitions and superconductivity. Different phases are obtained by small compositional changes, pressure or internal stress produced by annealing[1, 2, 3, 4, 5, 6]. Scanning tunneling microscopy and spectroscopy (STM) has been widely used to study normal and superconducting properties of pnictide and related superconductors[7, 8, 9, 10]. In the case of doped $CaFe_2As_2$, quasiparticle interference scattering has provided evidence for nematicity in the normal state and a detailed measurement of the bandstructure[7, 11, 12, 13]. STM topographies ubiquitously show reconstructed surfaces. However, it has been shown that quasiparticle interference is not influenced by the surface reconstruction and provides relevant bulk electronic features[7, 11, 12, 13, 14].

The tetragonal crystal structure of $CaFe_2As_2$ can show different surface terminations. Calculations show that Fe-As bonds are too strong, so only Ca or As terminations are expected[15]. Surfaces studied until now are reconstructed and Ca terminated[14]. Here, we

Journal of Physics: Conference Series **568** (2014) 022046 doi:10.1088/1742-6596/568/2/022046

have made STM experiments at 100 mK in Co doped $CaFe_2As_2$ and intensively searched for new surfaces using the in-situ positioning mechanism described in Ref.[9]. We have chosen $Ca(Fe_{1-x}Co_x)_2As_2$, with with x=0.035, where optimal T_c is obtained after annealing (Fig.1a and Ref.[5]). We indeed find unreconstructed areas, of lateral sizes up to 10 nm, showing square As surface with bulk interatomic distances.

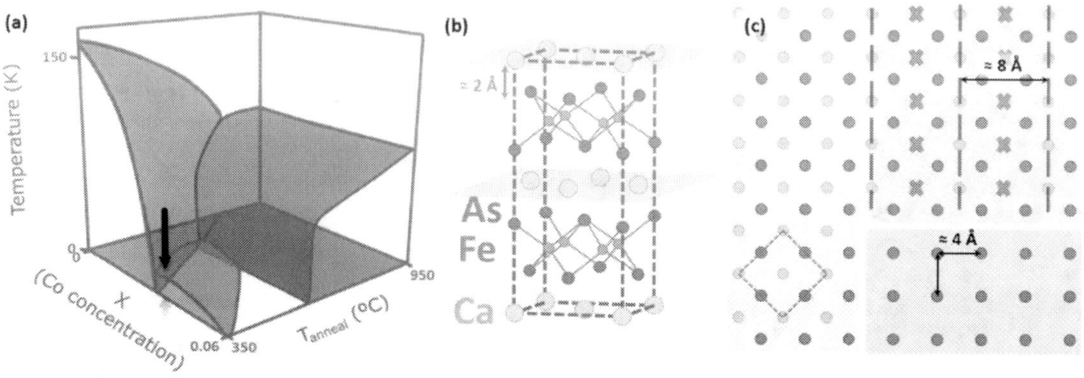

Figure 1. In a we schematically show the phase diagram of $Ca(Fe_{1-x}Co_x)_2As_2$. It highlights the different observed transition temperatures as a function of doping x and annealing temperature. The detailed phase diagram, including transition temperatures, is given in Ref. [5]. Red is the antiferromagnetic phase, blue is the collapsed tetragonal phase and green is the superconducting phase. Here we present results in the sample at the composition and annealing shown by the arrow, which is close to the optimal T_c. In b we show the crystal structure of $Ca(Fe_{1-x}Co_x)_2As_2$, with the blue arrow representing the distance between As and Ca atoms. In c we show different surface terminations. Left panel shows the structure viewed from above, with only Ca and As atoms. Ca terminations are unstable and produce a surface reconstruction consisting of rows of Ca atoms, as shown in the upper right panel by the dashed black lines. Red crosses show missing Ca atomic rows. The unreconstructed As atomic lattice observed here in some areas at the surface is shown in the bottom right panel.

2. Experimental

Our sample, $Ca(Fe_{9.65}Co_{0.35})_2As_2$, is located at the phase diagram point represented by an arrow in Fig.1 and has been grown and annealed as described in [5, 16, 17]. We have used a microscope similar to the one described in Ref. [9], and present here data taken at constant current mode (a few nA) and at 100 mK, with a bias voltage of 10 mV. We cleave our sample by first gluing a brass piece on top of it and then hitting this piece in-situ at low temperatures. To do so, we move the sample holder with a mechanism described in Ref.[9]. First we fix a copper beam on the path of the sample holder at a few mm above the surface of the sample (Fig.2). We then glue a small piece of brass on top of the sample. Once the whole system is below 4.2 K, we move the sample holder. The copper beam hits the brass piece, breaking the sample. The brass piece flies away to the bottom of the internal vacuum chamber, and a fresh surface is exposed. We have used this system already five times, obtaining fresh and large surfaces with atomic resolution.

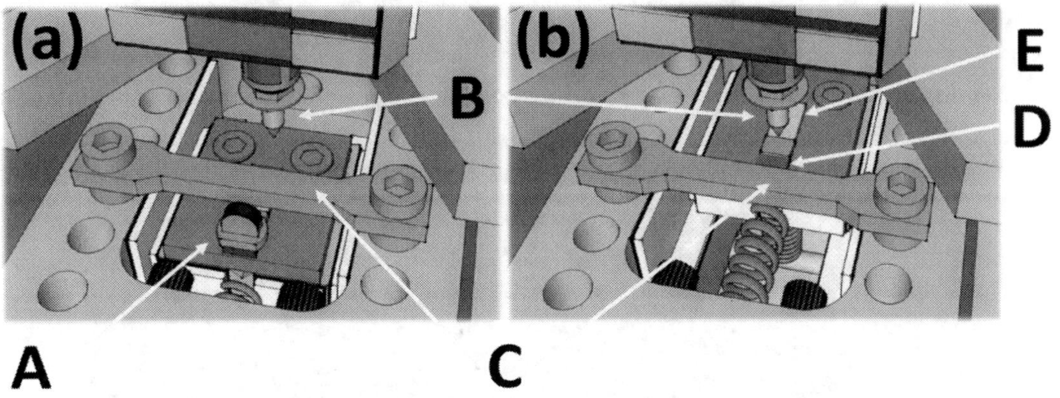

Figure 2. View of the *in-situ* cleaving system. In a we show the sample, with a stick glued on the top (A). The tip (B) lies on top of an Au sample (shown in b). A copper bar (C) is used to push the stick when moving the sliding sample holder. The holder is moved using the device described in Ref. [9] In b we show the result after moving the sample holder. The stick has gone to the bottom of the vacuum chamber and the tip (B) now lies on top of the cleaved sample (D). The sample shows now a freshly exposed surface. The results shown here have been obtained with a sample broken at temperatures below liquid helium temperature.

3. Results

Low temperature cleaving of our sample exposes large and flat surfaces with atomic resolution. A typical image is shown in Fig.3a. The surface is completely covered by the 2×1 reconstruction[7, 18, 13, 19, 20]. It consists of stripes made of one dimensional rows of atomic size, which extend over the whole surface (upper right panel of Fig.1c). We observe defects in the reconstruction due to missing rows or shifts in the wavelength of the striped modulation. The orientation of the reconstruction is degenerate with respect to the square in-plane cross section of the tetragonal lattice. Accordingly, we find reconstructions as shown in Fig.3a and along a perpendicular direction.

We have studied many different scanning windows using the positioning mechanism described in Ref.[9]. We have found mostly reconstructed surfaces along one of the two degenerate orientations of the 2×1 reconstruction. But we have also found surfaces where we observe both directions at the same time. These are separated in different domains. Close to the domain boundaries, we can find areas as in figure 3b, where the two degenerate orientations for the stripes appear randomly. The reconstruction shows disorder, and there are places unveiling a square atomic lattice below.

The lattice parameter observed in unreconstructed surfaces is of ≈ 4Å and line scans from the square atomic lattice into the reconstruction show height changes of ≈ 2Å (blue arrow in Fig.3e). This height corresponds to the distance between Ca and As atoms in the structure of $Ca(Fe_{0.965}Co_{0.035})_2As_2$ (Fig.1b).

In the unreconstructed areas we observe sometimes small groups possibly made of arrangements of Ca atoms. For instance, in Fig. 3f there are small sized stripes oriented along 45° from the atomic lattice (marked by blue ellipses).

4. Discussion

The surfaces of the AFe_2As_2 (A=Ca, Sr, Ba) compounds have been reviewed in Ref. [14]. Two relevant structures have been discussed, the $\sqrt{2}\times\sqrt{2}$ and the 2×1[15, 12]. The 2×1

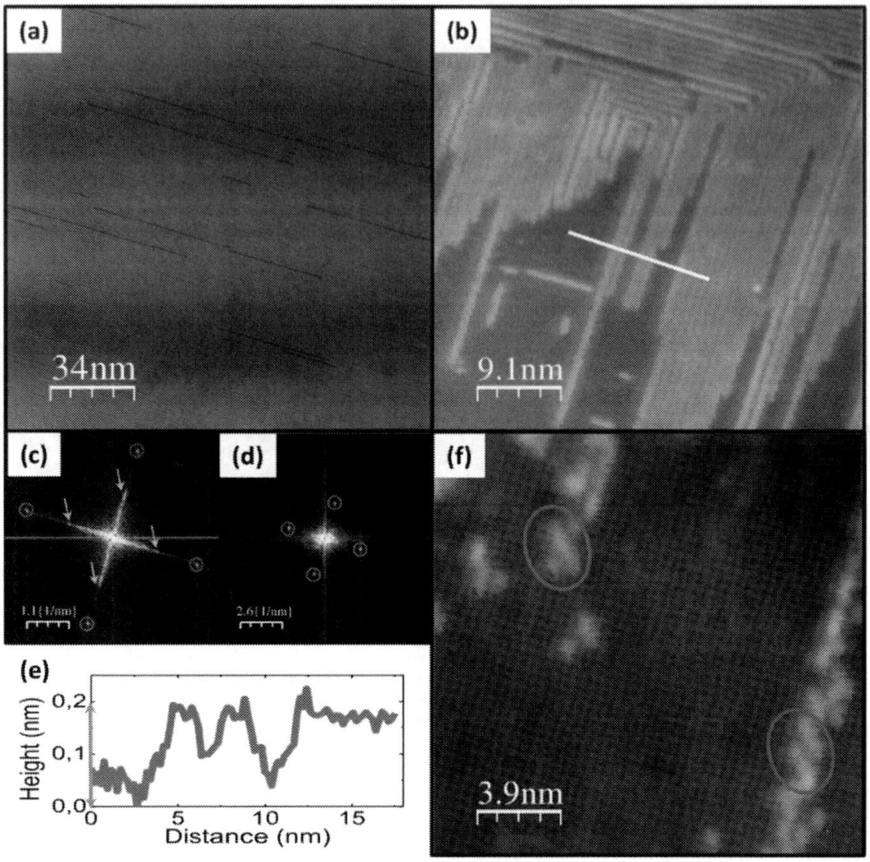

Figure 3. a) Topography of $Ca(Fe_{0.965}Co_{0.035})_2As_2$ taken with a bias voltage of 10 mV and a current of 10 nA, at a temperature of 100 mK. We use constant current mode and the gray scale corresponds to a height change of 0.3 nm. Stripes showing the surface reconstruction dominate the whole figure. b) STM topography image at a smaller sized region in another scanning window using the same tunneling conditions. We also observe stripes due to the surface reconstruction. Nevertheless, a large part of the image shows the unreconstructed square lattice expected within the tetragonal structure. The corresponding Fourier transform is shown c. Blue circles give the atomic lattice Bragg peaks. We also highlight the 2×1 surface reconstruction by orange arrows. Note that the Fourier transform does not show Bragg peaks but a cross, due to disorder. We also find regions where the atomic lattice is observed practically over the whole image, with only a few reconstructed rows. We show corresponding Fourier transform in d and topography image in f. In d, blue circles mark the atomic Bragg peak positions, and in f blue ellipses mark positions with Ca arrangements at 45° to the lattice. e shows a line scan between atomic lattice and reconstruction. Height difference matches approximately the distance between Ca (upper atomic rows) and As (lower atomic lattice) in the tetragonal crystal structure (Fig.1b).

reconstruction is schematically shown in Fig.1c (upper right panel). For the particular case of $CaFe_2As_2$, experiments until now have shown only the 2×1 reconstruction[7, 18].

Here we find unreconstructed areas with the square atomic lattice of the tetragonal structure (Fig. 3f). These are found close to reconstructed areas with strongly disordered stripe arrangements (Fig. 3b). Calculations of Ref.[15] show that the As atomic lattice over large areas is a metastable situation, but may well appear in a real cleave, because the energy difference between configurations is small[15].

It is interesting to raise the question if there are features in the bulk sample favoring the appearance of unreconstructed surfaces. Internal strain or stacking faults may favor the formation of small Ca free surfaces during the cleave at some locations. The observed disorder in the stripe arrangement is possibly also favored by defects or strain in the crystal. Our results show that low temperature cleaving can produce unreconstructed surface terminations over relatively large areas in the $CaFe_2As_2$ compound. Spectroscopic measurements will reveal further properties of surfaces showing the bulk crystalline structure.

5. Acknowledgments

This work was supported by the Spanish MINECO (FIS2011-23488), by the Comunidad de Madrid through program Nanobiomagnet, and by EU through Marie Curie Actions FP7-PEOPLE-2013-CIG-618321 and COST MP1201. We acknowledge technical support of UAM's workshops, SEGAINVEX. Work at the Ames Laboratory was supported by the US Department of Energy, Basic Energy Sciences, Division of Materials Sciences and Engineering under Contract No. DE-AC02-07CH11358.

References

[1] Stewart G R 2011 *Rev. Mod. Phys.* **83** 1589–1652
[2] Mazin I and Schmalian J 2009 *Physica C: Superconductivity* **469** 614 – 627
[3] Kamihara Y, Watanabe T, Hirano M and Hosono H 2008 *Journal of the American Chemical Society* **130** 3296–3297
[4] Ni N, Nandi S, Kreyssig A, Goldman A I, Mun E D, Bud'ko S L and Canfield P C 2008 *Phys. Rev. B* **78** 014523
[5] Ran S, Bud'ko S L, Straszheim W E, Soh J, Kim M G, Kreyssig A, Goldman A I and Canfield P C 2012 *Phys. Rev. B* **85** 224528
[6] Gati E, Köhler S, Guterding D, Wolf B, Knöner S, Ran S, Bud'ko S L, Canfield P C and Lang M 2012 *Phys. Rev. B* **86** 220511
[7] Chuang T M, Allan M P, Lee J, Xie Y, Ni N, Budko S L, Boebinger G S, Canfield P C and Davis J C 2010 *Science* **327** 181–184
[8] Guillamon I, Crespo M, Suderow H, Vieira S, Brison J, Budko S and Canfield P 2010 *Physica C: Superconductivity* **470** 771 – 775
[9] Suderow H, Guillamon I and Vieira S 2011 *Review of Scientific Instruments* **82** 033711
[10] Suderow H, Guillamn I, Rodrigo J G and Vieira S 2014 *Superconductor Science and Technology* **27** 063001
[11] Zhang H, Dai J, Zhang Y, Qu D, Ji H, Wu G, Wang X F, Chen X H, Wang B, Zeng C, Yang J and Hou J G 2010 *Phys. Rev. B* **81**(10) 104520
[12] Massee F, de Jong S, Huang Y, Kaas J, van Heumen E, Goedkoop J B and Golden M S 2009 *Phys. Rev. B* **80** 140507
[13] Yin Y, Zech M, Williams T L, Wang X F, Wu G, Chen X H and Hoffman J E 2009 *Phys. Rev. Lett.* **102**(9) 097002
[14] Hoffman J E 2011 *Rep. Prog. Phys.* **74** 124513
[15] Gao M, Ma F, Lu Z Y and Xiang T 2010 *Phys. Rev. B* **81** 193409
[16] Canfield P C 2007 2nd annual european school in material science *Solution Growth of Intermetallic Single Crystals: A Beginner's Guide*
[17] Canfield P C and Fisk Z 1992 *Phil. Mag. B* **65** 6
[18] Saparov B, Cantoni C, Pan M, Hogan T C, Ratcliff II W, Wilson S D, Fritsch K, Gaulin B and Sefat A 2014 *Scientific Reports* **4**
[19] Massee F, Huang Y, Huisman R, de Jong S, Goedkoop J B and Golden M S 2009 *Phys. Rev. B* **79**(22) 220517
[20] Boyer M C, Chatterjee K, Wise W D, Chen G F, Luo J L, Wang N L and Hudson E W 2008 *ArXiv e-prints* (*Preprint* 0806.4400)

Journal of Physics: Conference Series **568** (2014) 022047 doi:10.1088/1742-6596/568/2/022047

Pressure dependence of T_c in LnFeAsO$_{1-y}$ (Ln = La, Ce, Nd, Tb)

N. Takeshita[1], K. Miyazawa[1,2], A. Iyo[1,2], S. Furuta[3], M. Mito[3], and H. Eisaki[1]

[1]National Institute of Advanced Industrial Science and Technology, Tsukuba, Ibaraki 305-8568, Japan

[2]Department of Applied Electronics, Tokyo University of Science, Noda, Chiba 278-3510, Japan

[3]Faculty of Engineering, Kyushu Institute of Technology, Kitakyushu, Fukuoka 804-8550, Japan

E-mail: takeshita.n@aist.go.jp

Abstract. We measured the temperature dependence of electrical resistivity of non-fluorine-substituted, oxygen-deficient LnFeAsO$_{1-y}$ (Ln = La, Ce, Nd, Tb) under hydrostatic high pressure up to 18 GPa in order to observe pressure dependence of superconducting transition temperature T_c. In LaFeAsO$_{0.65}$, onset T_c initially enhances below 4 GPa, then decreases by applying further pressure. However, the zero-resistivity temperature does not show enhancement by applying pressure. In the case of NdFeAsO$_{1-y}$ and TbFeAsO$_{1-y}$, T_c shows monotonic decrease as increasing pressure. T_c is much more likely to go down faster if Ln with small ionic size was taken. Therefore, the bulk superconductivity is suppressed finally at ~7 GPa in TbFeAsO$_{0.7}$.

1. Introduction

It was a pleasant surprise for the scientific community when Kamihara *et al.* discovered iron pnictide superconductivity in LaFeAsO$_{1-x}$F$_x$ [1]. The highest superconducting transition temperature (T_c) in this series of iron pnictide compounds, LnFeAs(O$_{1-x}$F$_x$) (Ln-1111 type family where Ln is various lanthanide elements), is around 55 K, which was observed in Sm-1111 [2] or Nd-1111 [3]. Even though T_c in Ln-1111 is three times smaller than that of the record in high-T_c cuprates [4,5], Ln-1111 family of superconductors still has the second-highest T_c. It is well-known that T_c in Ln-1111 systematically changes on the changing the Ln ion [6]. The relationship between crystal structure and T_c was first pointed out by Lee et al. [7]. They predicted the dependence of T_c on the bond angles of As-Fe-As in the Fe-As tetrahedron. External pressure drastically changes lattice length and also changes T_c. The pressure dependence of T_c will help in understanding the mechanisms and important factors that vary T_c in the Ln-1111 type family.

Hydrostatic high-pressure is the best tool to investigate the relationship between crystal structure and T_c in Ln-1111 because it changes the lattice parameter without causing randomness in the sample. However, generating hydrostatic high-pressures up to 10 GPa is very difficult. Very few electrical resistivity measurements have succeeded under such hydrostatic high-pressure conditions in Ln-1111, and few apparatuses are applicable for such measurements [8,9]. As non-isotropic high-pressure provides randomness to the compounds, the transition curve of the resistivity becomes broad [10].

Journal of Physics: Conference Series **568** (2014) 022047 doi:10.1088/1742-6596/568/2/022047

Therefore hydrostaticity is necessary to observe the intrinsic behavior of superconductivity at high pressures. In order to achieve this condition, we have developed a cubic-anvil-type high-pressure apparatus [11], that enables us to measure physical properties precisely at high-pressures up to 20 GPa temperatures as low as that of liquid helium. In this study, we show an accurate pressure phase diagram of T_c in *Ln*-1111 (*Ln* = La, Ce, Nd, Tb) by measuring electrical resistivity and ac magnetic susceptibility in hydrostatic high-pressure conditions.

2. Experimental

All samples were synthesized using high-pressure techniques. The high-quality polycrystalline samples are suitable for isotropic pressure studies. Many studies on *Ln*-1111 have used fluorine-substituted samples. However, we have succeeded in demonstrating superconductivity in an oxygen deficient, non-fluorine-substituted sample (LnFeAsO_{1-y}) [3,6]. The details of the synthesis techniques are described elsewhere [6]. On examining the T_c of the samples, we concluded that they are optimally doped compounds with the exception of Tb-1111 in resistivity measurement. Judging from T_c and lattice parameters, the Tb-1111 sample in resistivity observation was relatively under-doped. We determine transition temperature T_c in this study as the temperature at which zero resistivity occurs. Zero resistivity is the most fundamental feature of superconductivity and its temperature generally matches the onset temperature of the Meissner signal. Since T_c is very sensitive to the crystal structure in the *Ln*-1111-type family, randomness in the sample will cause some parts to show a T_c higher than the bulk T_c. This will result in a possible overestimation of the onset T_c in resistivity measurement. To obtain the pressure phase diagram for bulk superconductivity, zero-resistivity temperature is appropriate for T_c. In this paper, we refer to our previous results [9,12], with redefining T_c as zero-resistivity temperature.

The resistivity at high pressure is measured in cubic-anvil-type high-pressure apparatus [11]. Samples are set in a Teflon case and homogeneously compressed with gasket material independently from six directions. Teflon case is filled with a liquid pressure medium (Daphne7474, distributed by Idemitsu Kosan Co. [13]). The temperature of the sample is monitored using a calibrated resistance-temperature sensor (Lake Shore Cryotronics, RF-100). We also measured the ac magnetic susceptibility of Tb-1111 at ac field of 10 Hz and 4 Oe with a miniature diamond anvil cell (DAC) designed for use in MPMS (Quantum Design Co.) [14].

3. Results and discussions

The temperature dependence of resistivity at high pressures in La-1111 is shown in Figure 1. The sample shows sharp transition and zero resistivity at low pressure. In La-1111, the onset T_c is enhanced at first from ~30 K to ~35 K. Above a pressure of ~4.2 GPa, T_c starts to decrease. This result is consistent with a previous study [15].

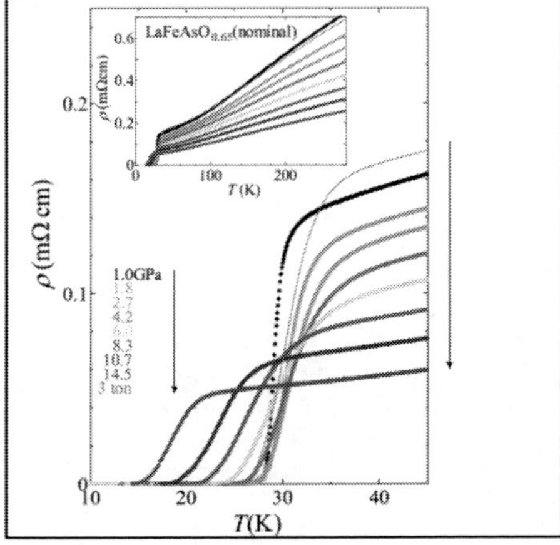

Figure 1. The electrical resistivity of polycrystalline $LaFeAO_{0.65}$ (nominal composition) under various pressures. Superconductivity survives at the highest pressure of 14.5 GPa. After the measurement at 14.5 GPa, we reduced load to check the reproducibility (3 ton).

Journal of Physics: Conference Series **568** (2014) 022047 doi:10.1088/1742-6596/568/2/022047

However, the zero resistivity temperature is seen to decrease monotonously on applying pressure. This result clearly suggests that there is no enhancement in 'bulk' T_c in La-1111 due to pressure. Furthermore, no enhancement of T_c in La-1111 is observed in magnetization measurements [16]. We also confirmed the reproducibility of our measurement. After reducing applied load, the zero resistivity temperature and resistivity value are almost completely recovered. However, onset T_c is still high and the transition curve becomes broad, as shown in Figure 1 (see the result of '3 ton').

Figure 2 shows the resistivity results in Tb-1111 up to 13 GPa. T_c is decreased by the application of pressure. Above 6.5 GPa, T_c is decreased to lower than 4.2 K. Although the resistivity drop remained in higher pressure, bulk superconductivity clearly vanished. AC magnetic susceptibility data also supports the resistivity results. The Meissner signal also disappeared above 6.5 GPa as shown in Figure 3. In the susceptibility measurements, the transition temperatures are relatively higher than that of the resistivity measurements data because the sample was improved [6]. But pressure dependence of T_c is qualitatively the same. Nd-1111 also shows negative pressure dependence of T_c [9], but it is clear that superconductivity in Tb-1111 is much more sensitive to external pressure than that in Nd-1111.

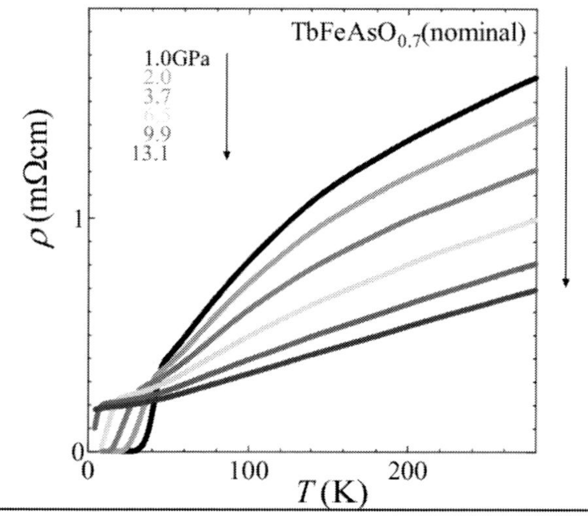

Figure 2. The electrical resistivity of polycrystalline $TbFeAO_{0.75}$ (nominal composition) at high pressures. Bulk superconductivity vanishes above ~6.5 GPa.

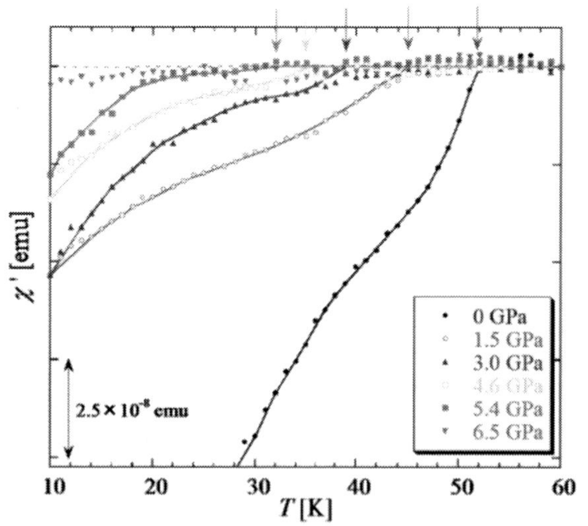

Figure 3. The ac magnetic susceptibility of Tb-1111 at various pressures. At 6.5 GPa, it becomes difficult to confirm the Meissner signal from the sample.

We obtained a T_c-Pressure phase diagram of *Ln*-1111 (*Ln* = La, Ce [12], Nd [9], Tb), as shown in Figure 4. By defining superconducting transition temperatures as the temperature at which zero resistivity occurs, we can conclude that the T_c of all the compounds show monotonic and negative pressure dependence. In La-1111, spontaneous randomness of the structure occurs when hydrostatic pressure is applied to the crystal. This randomness induces a 'pseudo' enhancement of T_c. However, it is not a bulk behavior. Negative pressure dependence of T_c is an intrinsic property in La-1111. Ce-1111 is a special case [12] in the *Ln*-1111 family. Owing to the pressure-induced enhancement of correlation to 4f electrons in the Ce ion, superconductivity suddenly vanishes at around 4 GPa. Nd-1111 and Tb-1111 shows similar behavior. Both show negative pressure dependence of T_c, and superconductivity in Nd-1111 is more robust for external pressure when compared to Tb-1111.

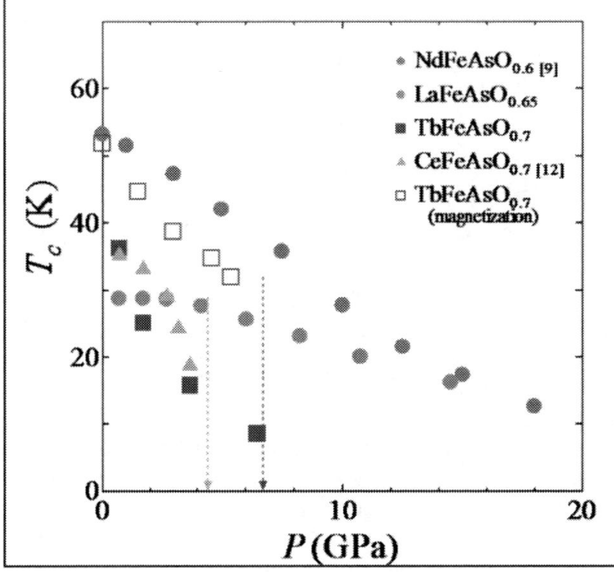

Figure 4. Obtained T_c-Pressure phase diagram of *Ln*-1111 (*Ln* = La, Ce [12], Nd [9], Tb). Because of the quality of the sample , T_c of Tb-1111 are different between the electrical resistivity results and the magnetic measurements. However, the superconductivity vanishes above ~6.5 GPa in both observations.

The pnictogen height, which is the distance between the arsenic atom and iron square lattice, has been considered as one of the potential factor [17] that determines T_c. In this model, there is an optimum value of height to explain the highest value of T_c in *Ln*-1111. Tb-1111 and Nd-1111 has nearly the same T_c and anion height [17]. However, pressure dependence of T_c is very different between these two materials. Therefore, this model would not be adequate. The bond angle [7,18] is another strong candidate as T_c determinative factor, but we cannot elucidate the effect of the bond angle from this study. It is not clear how the bond angle behaves at high pressure. Precise structural study in hydrostatic high pressure and low temperature is required. Accurate information of atom positions at high pressure is needed to determine the key parameter affecting T_c in *Ln*-1111. Since all *Ln*-1111 samples show monotonic negative pressure dependence, a very simple parameter such as the volume of iron arsenic tetrahedron could be one of the T_c determinative factors. Since La-1111, which has the largest volume, does not exhibit the highest T_c, the volume of the iron arsenic tetrahedron is not obviously a unique parameter. There would be some independent key factor for determining T_c. For example, the bond angle and the volume of iron arsenic tetrahedron are the independent parameter in *Ln*-1111.

In summary, the electrical resistivity of a high-quality *Ln*-1111 (*Ln* = La, Ce, Nd, Tb) polycrystalline sample was measured at hydrostatic pressures up to 18 GPa. All compounds show negative pressure dependence for T_c. La-1111 exhibits spontaneous randomness of structure under pressure, which causes a pseudo enhancement of the onset temperature for superconductivity. Since superconductivity in Tb-1111 is very sensitive to the applied pressure, its T_c is suppressed, whereas superconductivity in Nd-1111 survives up to a high pressure of 18 GPa.

Acknowledgments
This work was supported by JSPS KAKENHI Grant Number 26289091 and the Strategic International Collaborative Research Program (SICORP) of the Japan Science and Technology Agency (JST).

References

[1] Kamihara Y, Watanabe T, Hirano M and Hosono H 2008 *J. Am. Chem. Soc.* **130** 3296
[2] Liu R H, Wu G, Wu T, Fang D F, Chen H, Li S Y, Liu K, Xie Y L, Wang X F, Yang R L, Ding L, He C, Feng D L and Chen X H 2008 *Phys. Rev. Lett.* **101** 087001
[3] Kito H, Eisaki H and Iyo A 2008 *J. Phys. Soc. Jpn.* **77** 063707
[4] Schilling A, Cantoni M, Guo J D and Ott H R 1993 *Nature* **363** 56
[5] Takeshita N, Yamamoto A, Iyo A and Eisaki H 2013 *J. Phys. Soc. Jpn.* **82** 023711
[6] Miyazawa K, Kihou K, Shirage P M, Lee C H, Kito H, Eisaki H and Iyo A 2009 *J. Phys. Soc. Jpn.* **78** 034712
[7] Lee C H, Iyo A, Eisaki H, Kito H, Fernandez-Diaz M T, Ito T, Kihou K, Matsuhata H, Braden M and Yamada K 2008 *J. Phys. Soc. Jpn.* **77** 083704
[8] Takahashi H and Môri N 1996 *Study of High Temperature Superconductors ed. A. Narlikar* (Nova Science, New York) Vol. 16, p. 1
[9] Takeshita N, Iyo A, Eisaki H, Kito H and Ito T 2008 *J. Phys. Soc. Jpn.* **77** 075003
[10] Zocco D, Hamlin J, Baumbach R, Maple M, McGuire M, Sefat A, Sales B, Jin R, Mandrus D, Jeffries J, Weir S and Vohra Y 2008 *Physica C* **468** 2229
[11] Môri N, Takahashi H and Takeshita N 2004 *High Pressure Res.* **24** 225
[12] Takeshita N, Miyazawa K, Iyo A, Kito H and Eisaki H 2009 *J. Phys. Soc. Jpn.* **78** 065002
[13] Murata K, Yokogawa K, Yoshino H, Klotz S, Munsch P, Irizawa A, Nishiyama M, Iizuka K, Nanba T, Okada T, Shiraga Y and Aoyama S 2008 *Rev. Sci. Instrum.* **79** 085101
[14] Mito M, Hitaka M, Kawae T, Takeda K, Kitai T and Toyoshima N 2001 *Jpn. J. Appl. Phys.* **40** 6641
[15] Takahashi H, Igureawa K, Arii K, Kamihara Y, Hirano M and Hosono H 2008 *Nature* **453** 376
[16] Miyoshi K, Kojima E, Ogawa S, Shimojo Y and Takeuchi J 2013 *Phys. Rev. B* **87** 235111
[17] Mizuguchi Y, Hara Y, Deguchi K, Tsuda S, Yamaguchi T, Takeda K, Kotegawa H, Tou H and Takano Y 2010 *Supercond. Sci. Technol.* **23** 054013
[18] Lee C H, Kihou K, Iyo A, Kito H, Shirage P M, Eisaki H 2011 *Solid State Comm.* **152** 644

Journal of Physics: Conference Series **568** (2014) 022048 doi:10.1088/1742-6596/568/2/022048

Imaging of local temperature distributions in mesas of high-T_c superconducting terahertz sources

M. Tsujimoto, H. Kambara, Y. Maeda, Y. Yoshioka, Y. Nakagawa, and I. Kakeya

Department of Electronic Science and Engineering, Kyoto University, Kyoto, Japan

E-mail: tsujimoto@sk.kuee.kyoto-u.ac.jp

Abstract. Stacks of intrinsic Josephson junctions in high-T_c superconductors are a promising source of intense, continuous, and monochromatic terahertz waves. In this paer, we establish a fluorescence-based temperature imaging system to directly image the surface temperature on a $Bi_2Sr_2CaCu_2O_{8+\delta}$ mesa sample. Intense terahertz emissions are observed in both high- and low-bias regimes, where the mesa voltage satisfies the cavity resonance condition. In the high-bias regime, the temperature distributions are shown to be inhomogeneous with a considerable temperature rise. In contrast, in the low-bias regime, the distributions are rather uniform and the local temperature is close to the bath temperature over the entire sample.

1. Introduction

The terahertz wave with a frequency range of 0.3–10 THz is thought to be the most unique electromagnetic wave, since it provides numerous opportunities for a host of applications [1]. In recent years, much effort has been put in to the development of compact and solid-state terahertz sources using semiconductors [2, 3] and laser technologies [4]. Since the first demonstration of terahertz emissions from intrinsic Josephson junctions (IJJs) [5] in $Bi_2Sr_2CaCu_2O_{8+\delta}$ (Bi-2212) [6], terahertz generation using IJJ stacks has become a major focus of research, both experimental [6–26] and theoretical [27–41]. In Bi-2212, application of dc voltage V leads to an ac current and electromagnetic emission at the Josephson frequency in the form: $f_J = (2e/h)V/N$, where e is the electric charge, h is the Plank's constant, and N is the number of active junctions [42]. To generate the intense terahertz wave, the mesa structures were milled from the Bi-2212 single crystal in order to excite the cavity resonance: in the case of a long rectangular mesa, the cavity condition is expressed as $f^c = c_0/2nw$, where n is the refractive index of Bi-2212 and w is the width of the mesa [6].

The most interesting physics of the terahertz emissions is the synchronization of thousands of stacked IJJs with distributed widths, due to the trapezoidal cross-section of the mesa. This system is considered an instructive demonstration of the Kuramoto model, in which a large number of nonlinear oscillators synchronize due to weak couplings [35]: a Josephson junction array shunted by an LCR load can spontaneously synchronize to a common frequency, despite differences in bare frequencies [43]. Essentially the same phenomena can be perceived in many physical and biological systems, including relaxation-oscillator circuits, networks of neurons, and fireflies that flash in unison [44]. Meanwhile, mutual synchronization based on hot-spot formation was recently identified in a mixing experiment [40]. A complete study of increases in local temperature in the emitting IJJ stack will further our understanding of the nonlinear phase dynamics of the IJJ system. Moreover, effective prevention of overheating is required to increase the emission power, and may be essential for designing powerful sources [23].

In this study, we establish a temperature imaging system based on a florescent technique in order to visualize slocal temperature rises in the emitting stack. We observe intense emissions both in high- and low-bias regimes, and image the temperature distributions during emission.

2. Sample preparation and experimental setup

Single crystals of Bi-2212 grown by a traveling-solvent floating-zone technique were annealed at 650°C for 12 h. The temperature dependence of the c-axis resistance exhibited a behavior typical of underdoped crystals with $T_c = 78$ K. A small piece of a cleaved crystal was glued onto a sapphire substrate using epoxy resin. The mesa structure was then milled from the crystal surface by photolithography and argon-ion milling techniques. The top width of the mesa (w) was 73 μm and length (L) 400 μm. The thickness of 1.1 μm corresponds to $N \sim 720$.

To image the local temperature distribution on the sample, we use a fluorescent technique based on the strong temperature dependence of fluorescence intensity on the coordination complex [45]. Figure 1 shows a schematic view of the temperature imaging system. For the temperature marker, a film consisting of europium thenoyltrifluoroacetonate (Eu-TFC) in a polymer matrix of polybutylmethacrylate (PBMA) had been deposited on the surface of the sample using a spin-coating technique and baked in a drying oven at 175°C for 30 min. The sample was then installed into the microscopy cryostat and irradiated by 365-nm light emitted by an UV LED. A room-temperature CMOS camera, with a UV filter, was used to acquire the fluorescent image. The spatial resolution, estimated to be less than 5 μm, was determined by the diffraction limit of the lens system. The acquired data can then be directly converted to a surface local temperature (T_{local}) by calibrating the temperature dependence of the fluorescence intensity (cf. Fig. 2). Since the sensitivity of the film depends on the surface materials, *i.e.*, Bi-2212 (mesa and base crystal) or silver (electrode), a calibration curve must be determined for each surface. Nevertheless, as the inevitable edge effect at the narrow electrode stripes may degrade image quality and obscure details [21], in the present experiment we imaged T_{local}

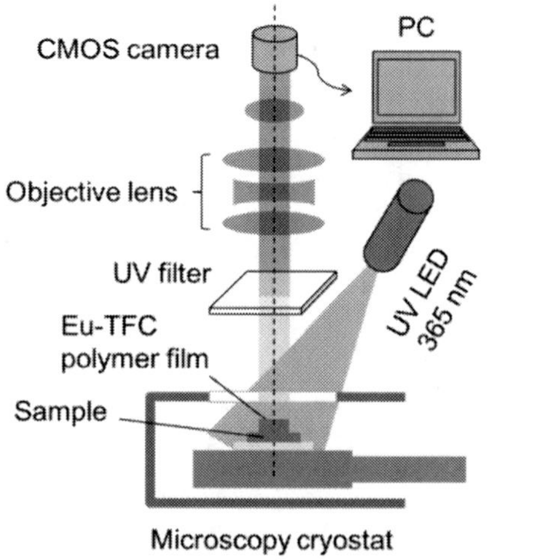

Figure 1. Schematic view of the temperature imaging system.

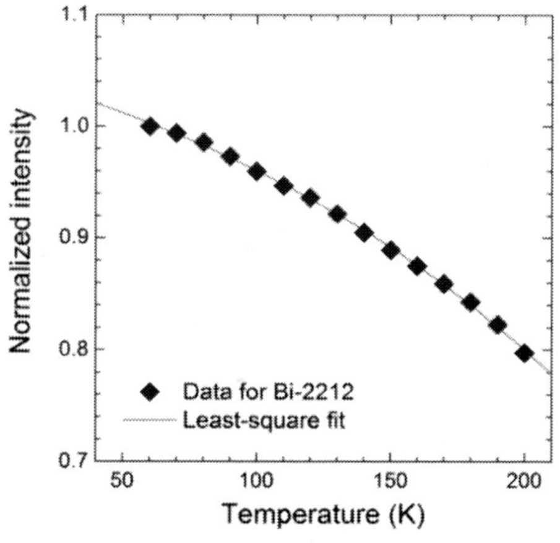

Figure 2. Temperature dependence of the fluorescence intensity for the Eu-TFC polymer film deposited on the Bi-2212 surface.

distributions only for the Bi-2212 surface, in order to facilitate the analysis.

3. Results and discussion

Figure 3(a) shows the four-terminal current-voltage (I-V) characteristics (IVCs) at bath temperatures of $T_b = 10$ K (left scale) and the emission intensity (right scale) measured using a Si-composite bolometer with a 1-THz low-pass filter. Dashed lines at $V = \pm 0.884$ V in Fig. 3(a) represent calculated emission voltages that satisfy $f_J = f^c$, namely, $V = N \cdot (c_0/2nw) \cdot (h/2e)$, where we assumed that whole N junctions in the stack were in the resistive state. In Fig. 3(b), the emission intensity was plotted versus I corresponding to Fig. 3(a). Intense emissions were indeed observed when f_J matched f^c in two characteristic bias regimes: a high-bias regime ($I = 11$–23 mA) and a low-bias regime ($I = 4$–5 mA). As discussed earlier, we suggest that the extreme temperature inhomogeneity in the mesa is key for the synchronization of the IJJ stack in the high-bias regime.

Five panels in Fig. 4(b) display the $T_{\rm local}$ distributions at various I-V points, indicated by A–E in Fig. 4(a). In the $T_{\rm local}$ image, the bounded rectangular region indicates the mesa. The images were taken while decreasing I, from 33 mA to 9 mA at a 6-mA interval. The orange parts in the mesa are no longer superconducting with $T_{\rm local} > T_c$. The exposing condition of the CMOS camera had been optimized for measuring the surface temperature of Bi-2212. The data for the silver electrodes, however, cannot be presented (cf. blacked-out stripes).

In the high-bias regime, the mesa was inevitably heated at a rate of about 30 mW, corresponding to the enormous heating power density. As I had been decreased from 33 mA to 21 mA, the hot spot that initially localized on the left side moved to the center of the mesa. The hot spot disappeared at 12 mA, resulting in a uniform distribution. Such hot-spot behavior in the high-bias regime correlates closely with previous observations [7, 11, 12, 26], which indicates that the huge heating power density cannot be removed quickly enough from the mesa to maintain equilibrium, which results in a considerable rise of the mesa temperature. Although the local heating may induce a chaotic nonequilibrium state and may adversely affect the terahertz emission, mutual synchronization based on hot-spot formation has been identified in a mixing experiment [40]. Since heat conduction became progressively worse as $T_{\rm local}$ lowered

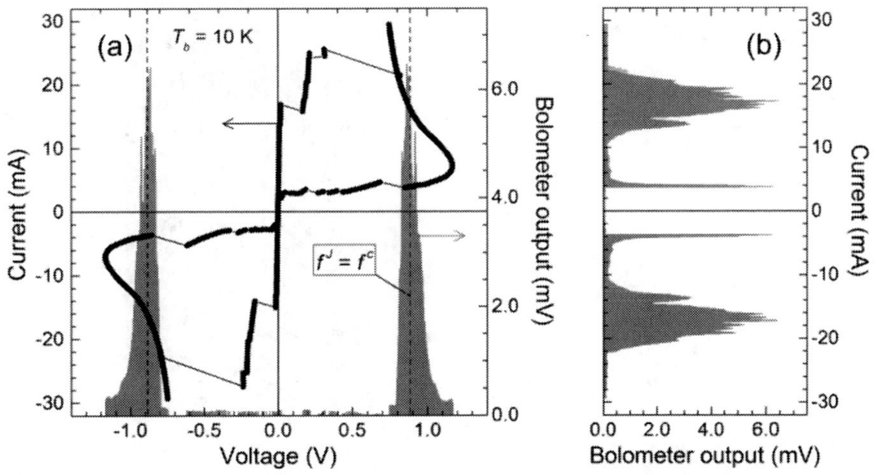

Figure 3. (a) Outermost IVC (left scale) and emission intensity detected by a Si-composite bolometer (right scale) at $T_b = 10$ K. Dashed lines at ± 0.884 V represent the calculated emission voltages that satisfy $f_J = f^c$. (b) Emission intensity versus I corresponding to (a).

and the gap vanishes as $T_{\text{local}} \to T_c$, these features may account for the peculiar temperature dependence of the emission intensity [26].

The hot-spot formation mechanism, based on the strongly negative temperature coefficient of the c-axis resistivity of Bi-2212, allows for hot spots near the center of the mesa with a maximum temperature above T_c [39]. As a contact resistance of a few ohms, separately measured using the two-terminal method, is much less than the c-axis resistivity, power dissipation at the contact is small and does not affect the temperature distribution.

From the local temperature imaging results, we can presume that emission intensity also has a strong correlation with the volume of the superconducting parts in the emitting mesa. The steady-state temperature distribution in the Bi-2212 mesa had been determined by a delicate balance between current distribution and local Joule heating. Thus, we can control the temperature distributions and the emission intensity by altering the dc current distributions. Further studies on the external control of temperature distribution could provide more information for understanding the mechanism of terahertz emission from IJJs.

4. Conclusion

In conclusion, we established a fluorescent temperature imaging system based on the strong temperature dependence of sfluorescence intensity of the coordination complex. We observed intense and continuous terahertz emissions from the IJJ stack in Bi-2212. The intense emissions take place in both high- and low-bias regimes when f_J matches f^c. The local temperature on the sample surface was directly imaged and shown to be inhomogeneously distributed, with a considerable temperature rise in the high-bias regime. In contrast, in the low-bias regime, the temperature distributions have been rather uniform and the local temperature had been close to T_b over the whole mesa. Since no regard has been given to eliminating of excess heat from mesas, or to controlling the position of the hot spot in the emitting sample, further improvements in terms of the sample structure and bias conditions may enable the development of powerful terahertz sources using IJJ stacks.

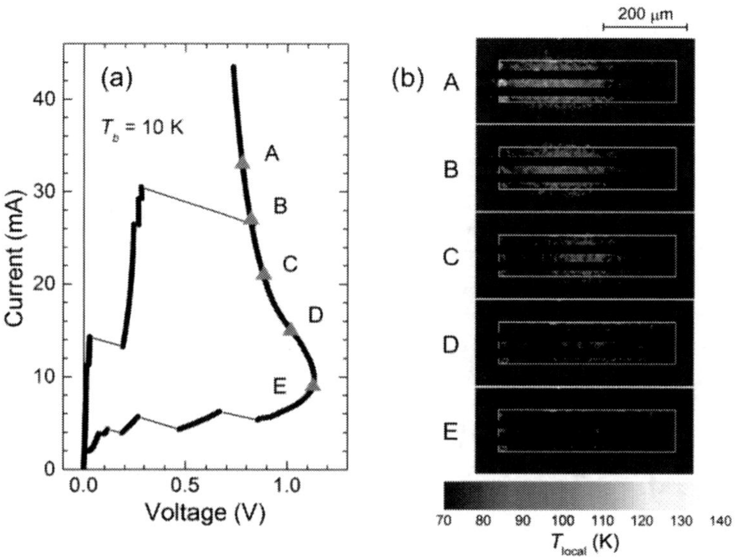

Figure 4. (a) Outermost IVC at $T_b = 10$ K. (b) T_{local} images for the temperature distribution on the Bi-2212 surface at I-V points indicated by A–E in (a).

Acknowledgments
The authors would like to thank T. Kashiwagi, H. Minami, and K. Kadowaki for valuable discussions. Comments from H. Asai are greatly appreciated. This work was supported by KAKENHI (Grants No. 23681030, No. 13J04811, and No. 26790032).

References
[1] M. Tonouchi, Nat. Photonics **1**, 97 (2007).
[2] E. R. Brown *et al.*, Appl. Phys. Lett. **58,** 2291 (1991).
[3] H. Ito *et al.*, Semicond. Sci. Tech. **20,** S191 (2005).
[4] J. Faist *et al.*, Science **264,** 553 (1994).
[5] R. Kleiner *et al.*, Phys. Rev. Lett. **68**, 2394 (1992).
[6] L. Ozyuzer *et al.*, Science **318**, 1291 (2007).
[7] H. B. Wang *et al.*, Phys. Rev. Lett. **102**, 017006 (2009).
[8] H. Minami *et al.*, Appl. Phys. Lett. **95**, 232511 (2009).
[9] K. Kadowaki *et al.*, J. Phys. Soc. Jpn. **79**, 023703 (2010).
[10] M. Tsujimoto *et al.*, Phys. Rev. Lett. **105**, 037005 (2010).
[11] H. B. Wang *et al.*, Phys. Rev. Lett. **105**, 057002 (2010).
[12] S. Guénon *et al.*, Phys. Rev. B **82**, 214506 (2010).
[13] K. Yamaki *et al.*, Opt. Express **19**, 3193 (2011).
[14] T. M. Benseman *et al.*, Phys. Rev. B **84**, 064523 (2011).
[15] M. Tsujimoto *et al.*, Phys. Rev. Lett. **108**, 107006 (2012).
[16] I. Kakeya *et al.*, Appl. Phys. Lett. **100**, 242603 (2012).
[17] M. Tsujimoto *et al.*, J. Appl. Phys. **111**, 123111 (2012).
[18] M. Li *et al.*, Phys. Rev. B **86**, 060505(R) (2012).
[19] K. Delfanazari *et al.*, Opt. Express **21**, 2171 (2013).
[20] D. Y. An *et al.*, Appl. Phys. Lett. **102**, 092601 (2013).
[21] T. M. Benseman *et al.*, J. Appl. Phys. **113**, 133902 (2013).
[22] T. M. Benseman *et al.*, Supercond. Sci. Technol. **26**, 085016 (2013).
[23] T. M. Benseman *et al.*, Appl. Phys. Lett. **103**, 022602 (2013).
[24] S. Sekimoto *et al.*, Appl. Phys. Lett. **103**, 182601 (2013).
[25] T. Kashiwagi *et al.*, Appl. Phys. Lett. **104**, 022601 (2014).
[26] H. Minami *et al.*, Phys. Rev. B **89**, 054503 (2014).
[27] A. E. Koshelev and L. N. Bulaevskii, Phys. Rev. B **77**, 014530 (2008).
[28] S. Lin and X. Hu, Phys. Rev. Lett. **100**, 247006 (2008).
[29] M. Tachiki *et al.*, Phys. Rev. Lett. **102**, 127002 (2009).
[30] N. Pedersen and S. Madsen, IEEE Trans Appl. Supercond. **19**, 726 (2009).
[31] A. E. Koshelev, Phys. Rev. B **82**, 174512 (2010).
[32] A. Yurgens, Phys. Rev. B **83**, 184501 (2011).
[33] M. Krasnov, Phys. Rev. B **83**, 174517 (2011).
[34] T. Koyama *et al.*, Supercond. Sci. Technol. **24**, 085007 (2011).
[35] S. Lin *et al.*, Phys. Rev. B **84**, 104501 (2011).
[36] R. A. Klemm *et al.*, J. Phys.: Condens. Matter **23**, 025701 (2011).
[37] S. Lin and X. Hu, Phys. Rev. B **86**, 054506 (2012).
[38] H. Asai *et al.*, Appl. Phys. Lett. **101**, 112602 (2012).
[39] B. Gross *et al.*, Phys. Rev. B **86**, 094524 (2012).
[40] B. Gross *et al.*, Phys. Rev. B **88**, 014524 (2013).
[41] F. Liu *et al.*, Supercond. Sci. Technol. **26**, 025003 (2013).
[42] B. D. Josephson, Phys. Lett. **1**, 251 (1962).
[43] K. Wiesenfeld *et al.*, Phys. Rev. E **57**, 1563 (1998).
[44] D. J. Watts and S. H. Strogatz, Nature **393**, 440 (1998).
[45] P. Kolodner and J. A. Tyson, Appl. Phys. Lett. **40**, 782 (1982).

Journal of Physics: Conference Series **568** (2014) 022049 doi:10.1088/1742-6596/568/2/022049

Effects of hole doping and chemical pressure on the average superconducting kinetic energy of $YBa_2Cu_3O_{7-\delta}$ single crystals

V N Vieira[1], A P A Mendonça[1], F T Dias[1], D L da Silva[1], P Pureur[2], J Schaf[2], M L Hneda[2], F Mesquita[2]

[1] Instituto de Física e Matemática, Universidade Federal de Pelotas, UFPEL, C.P. 354, 96010-900, Pelotas, RS, Brazil

[2] Instituto de Física, Universidade Federal do Rio Grande do Sul, UFRGS, C. P. 15051, 91501-970, Porto Alegre, RS, Brazil

E-mail: vdnvieira@gmail.com (VN Vieira)

Abstract. We reported on $M_{ZFC}(T)$ and $M_{FCC}(T)$ reversible dc magnetizations of $YBa_2Cu_3O_{7-\delta}$, $Y_{0.99}Ca_{0.01}Ba_2Cu_3O_{7-\delta}$ and $YBa_{1.75}Sr_{0.25}Cu_3O_{7-\delta}$ single crystals with a strong focus on the effects of Ca and Sr doping on the average superconducting kinetic energy density, $k(T)$ of the $YBa_2Cu_3O_{7-\delta}$. The $k(T)$ is used as a relevant tool to provide physical information about the HTSC paring mechanism. The determination of the $k(T)$ from $M_{ZFC}(T)$ and $M_{FCC}(T)$ data is supported by virial theorem of superconductivity [$k(T) = -\boldsymbol{M}.\boldsymbol{B}$]. The $M_{ZFC}(T)$ and $M_{FCC}(T)$ measurements were performed with a SQUID magnetometer to $H \leq 50$kOe applied parallel to the c axis of the samples. The results show that the samples present a common $k(T)$ behavior that is characterized by a maximum value for $T \ll T_c$ that gradually decreases as the temperature rises towards to the T_c, becoming null to $T \leq T_c$. The magnetic field affects smoothly the $k(T)$ data behavior. The $k(T)$ results contrasting of our samples shows that the Ca and Sr doping promotes a reduction of its amplitude. A possible explanation to this feature could be associated to the fact that the hole doping character promoted by Ca doping and the chemical pressure effect motivated by Sr doping affects considerably the superconducting paring mechanism of the $YBa_2Cu_3O_{7-\delta}$.

1. Introduction

According to the BCS theory the superconductor electronic state gets kinetic energy during the condensation of Cooper pairs [1-4]. Infrared experiments performed in high temperature superconductors (HTSC) detects a large spectral amount transferred to the HTSC superfluid condensate, supportive of a kinetic energy driven mechanism [1-4]. The average kinetic energy density, $k(T,B)$ can be obtained from zero field cooled, $M_{ZFC}(T)$ and field cooled cooling, $M_{FCC}(T)$ dc magnetizations measurements through the application of the virial theorem of superconductivity (VTS) given by [1-4].

$$k(T,B)=\left\langle \frac{\hbar^2}{2m}\left|\left(\nabla-\frac{2\pi}{\Phi_0}A\right)\psi\right|^2\right\rangle=-\vec{M}\cdot\vec{B} \qquad (1)$$

In the equation 1 the symbol $\langle\ldots\rangle$ represents the spatial average, $\hbar^2$, $2m$ and Φ_0 are phenomenological constants, ψ is the superconductor order parameter, M is the equilibrium magnetization and B is the vector magnetic field induction.

The $k(T)$ study can be applied as a relevant tool to get some relevant physical information about superconductor order parameter behaviour on the superconducting-normal transition [1-4]. In particular the VTS equation was applied to determination of the $k(T,B)$ behaviour of the some low and high temperatures superconductors. For instance, the $k(T)$ behaviour reported to Pb-In alloys [1] and Nb [2] samples for $T \leq T_c$ are in agreement to the Abrikosov treatment of the Ginsburg-Landau theory and to the general BCS theory predictions. Otherwise the identification of the $k(T,B) > 0$ for $T > T_c$ have been reported for underdoped $YBa_2Cu_3O_x$ [2] and optimally doped $Bi_2Sr_2CaCu_2O_8$ [2] single crystals as well as a $La_{1.9}Sr_{0.1}CuO_4$ sample[3]. The authors of these studies elect the pseudogap effect as responsible for this behaviour eliminating the thermal fluctuations as an alternative explanation. In a recent study performed with $Sm_2Ba_2Cu_3O_{7-\delta}$ ceramic samples [5] the microstructural aspects of the samples (granularity) were associated to the $k(T,B)$ behaviour.

An alternative manner of modify the superconducting state of a superconductor material is to promote the partial chemical substitution at its structure. The hole doping character or the chemical pressure mechanism promoted respectively by partial substitutions of Y for Ca and of Ba for Sr in the $YBa_2Cu_3O_{7-\delta}$ structure are potential examples of the last predictions [6-8]. The Ca doping solubility limit at Y site is almost 15% as the Sr doping solubility limit at the Ba site is almost 50% [7-8]. The Ca (+2) and Y (+3) atoms have diferent valences which justify the hole doping character introduced by this particular doping. In contrast, the Sr atom radius is smaller than the Ba atom radius which supports the chemical pressure effects on the superconducting state of $YBa_2Cu_3O_{7-\delta}$ [7-8].

Motivated by the last questions about the $k(T,B)$ behaviour we decided to plan a dc magnetization experimental study in $YBa_2Cu_3O_{7-\delta}$, $Y_{0.99}Ca_{0.01}Ba_2Cu_3O_{7-\delta}$ and $YBa_{1.75}Sr_{0.25}Cu_3O_{7-\delta}$ single crystals focusing on the effects of Ca and Sr chemical doping on the $k(T,B)$ behaviour.

2. Sample preparation and experimental procedures

The single crystals of $YBa_2Cu_3O_{7-\delta}$ (YBCO), $Y_{0.99}Ca_{0.01}Ba_2Cu_3O_{7-\delta}$ [Y(Ca)BCO] and $YBa_{1.75}Sr_{0.25}Cu_3O_{7-\delta}$ [YB(SrCO] were grown by self-flux method [6-8]. The selected single crystals were submitted to an extra oxygen process to improve their superconducting temperature transition, T_C [6-8]. The picked samples had their structure and superconducting transition characterized respectively by x-ray diffraction (XRD) and zero field cooled DC magnetization, $M_{ZFC}(T)$ [6-8]. The figure 1 displays their XRD results.

It is possible identify the characteristic YBCO-123 orthorhombic crystalline structure along their c crystallographic axis. The c lattice parameters obtained from the XRD analysis to the samples were $c = (11.65 \pm 0.01)$Å to the YBCO, $c = (11.65 \pm 0.01)$Å to the Y(Ca)BCO and $c = (11.58 \pm 0.01)$Å to the YB(Sr)CO. The estimation of the superconducting transition of the samples through of $M_{ZFC}(T)$ measurements to $H = 10$Oe applied parallel to the c axis of the single crystals resulted in the following T_c values (93.4±0.2)K to the YBCO, (91.8±0.2)K to the Y(Ca)BCO and (87.9±0.1)K to the YB(Sr)CO. The results presented to the c crystallographic axis and superconducting transition of the samples are in agreement with those reported by literature [6-8]

The isofield dc magnetization measurements were performed with a quantum design SQUID magnetometer. The usual zero field cooled, $M_{ZFC}(T)$ and field cooled cooling, $M_{FCC}(T)$ dc magnetizations were recorded while dc magnetic fields up to 50kOe were applied parallel to c axis of samples (H // c).

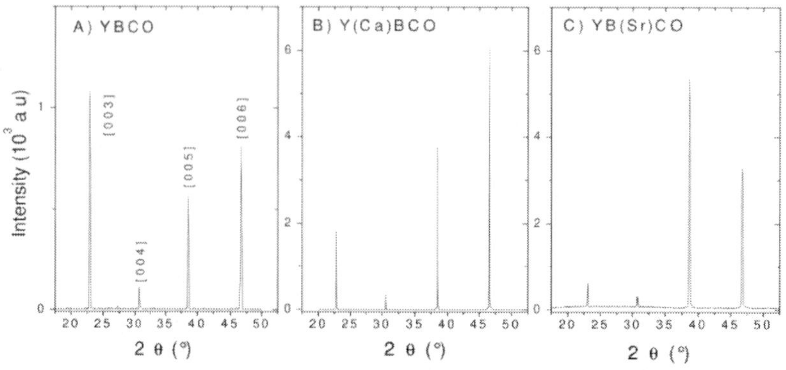

Figure 1. The XRD of the A)YBCO, B) Y(Ca)BCO and C) YB(Sr)CO samples obtained for monochromated *CuK α* radiation oriented to their c axis.

The magnetization measurement proceeds consisted in first cooling down the sample to temperatures well below T_c in zero field (ZFC). Then the zero field cooled magnetization, $M_{ZFC}(T)$ was measured under constant magnetic field while slowly warming the sample (0.4 K/min or less) up to temperatures well above T_c. Subsequently the field cooled cooling magnetization, $M_{FCC}(T)$ was measured while cooling (0.4 K/min or less) the sample back to low temperatures in the same field (FCC) [6-8].

The samples size was determined with a SEM inset of a FIB, JIB 4SOU model from JEOL Company. The possible demagnetization factor contribution to the $M(T)$ data was checked. The geometrical factors were estimated on the basis of the calculation in reference [9] and the paramagnetic signal contribution of sample holder was measured to all the applied dc magnetic fields and its magnetic contribution was subtracted from $M_{ZFC}(T)$ and $M_{FCC}(T)$ data.

3. Results and discussion

The figure 2 displays a representative result of the $\Delta M(T)$ data to the Y(Ca)BCO sample while a $H =$ 10kOe is applied parallel to c axis of the sample. The $T_{irr}(H)$ temperature is put in evidence. Otherwise its insets highlight the correspondently $M_{ZFC}(T)$ and $M_{FCC}(T)$ dc magnetizations results when the $T_c(H)$ and $T_{irr}(H)$ temperatures are indicated.

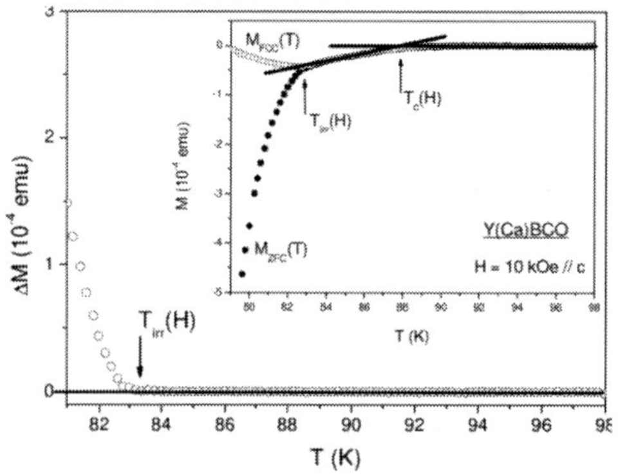

Figure 2. The determination of $T_{irr}(H)$ and $T_c(H)$ temperatures from the $M_{ZFC}(T)$ and $M_{FCC}(T)$ dc magnetizations (inset) when the $H = 10$kOe is applied parallel to the c axis of the Y(Ca)BCO sample.

In the figure 2 the $T_{irr}(H)$ is the temperature which leaves the zero base line of the $\Delta M(T)$ plot [$\Delta M(T) = M_{FCC}(T) - M_{ZFC}(T)$]. The $T_c(H)$ is determined by the intersection between the linearly extrapolated $M_{ZFC}(T)$ and $M_{FCC}(T)$ dc magnetizations in the normal and superconducting phases, as illustrated by the inset of figure 2.

The $T_{irr}(H)$ defines the limit below which pinning effects become important. We carried out our analyses in the temperature regime above $T_{irr}(H)$ in which the experimental data describe the equilibrium magnetization with no ambiguity [1-5].

We assume that GL theory describes the equilibrium magnetization adequately in most of the reversible superconducting regimes of our samples [5]. Then using the VTS [1-5] we plot in figures 3A and 3B the $k(T,B)$ versus (TT_c^{-1}) for YBCO, Y(Ca)BCO and YB(Sr)CO samples where the normalized $T_C(H)$ is indicated by arrows in the plots. In particular the figure 3A shows the $k(T,B)$ results to $0.1T \leq H \leq 1T$ as well as figure 3B to $1T \leq H \leq 5T$.

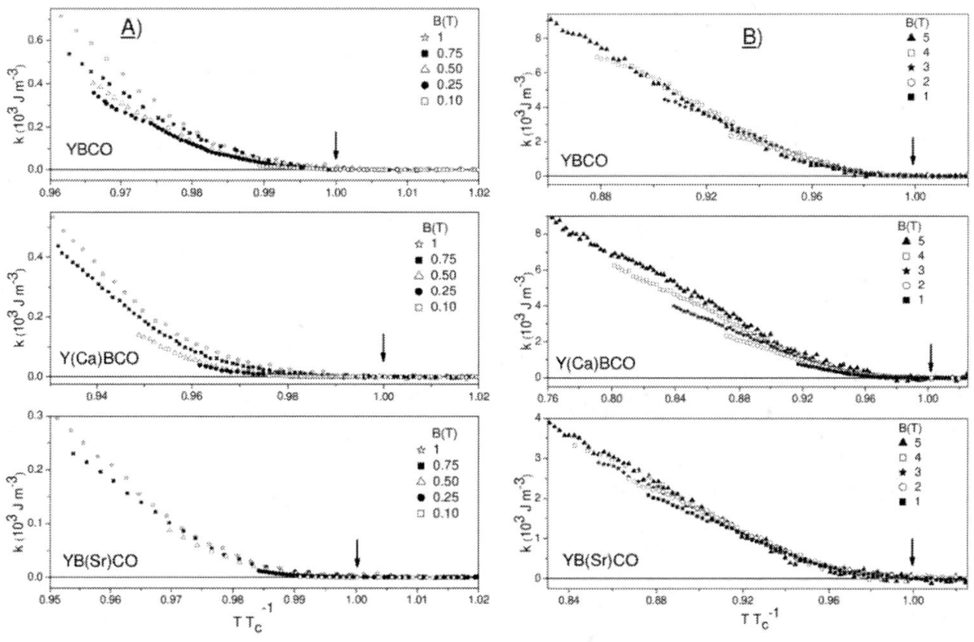

Figure 3. The $k(T,B)$ behavior to the for YBCO, Y(Ca)BCO and YB(Sr)CO samples to $0.1T \leq H \leq 5T$.

The results showed in the figure 3 are not on the same range because the reversible magnetization regime is sample dependent and according to the VTS in the light of Abrikosov theory the $k(T,B)$ data is evaluated to all the revesible magnetization regime. The results show that the samples presents a common $k(T,B)$ behavior that is characterized by a $k(T,B)$ maximum value for $T << T_c$ that gradually decreases as the temperature rises towards to the T_c, becoming null to $T \leq T_c$.

The magnetic field affects smoothly the $k(T,B)$ versus TT_c^{-1} profile of our single crystals as compared to the response observed for this quantity in microestructural modificated $Sm_2Ba_2Cu_3O_{7-\delta}$ ceramic samples [5]. In particular, the figure 3 shows that the $k(T,B)$ profile of the doped single crystals is more affected by the magnetic field application then that displayed by the pure sample especially to $B \geq 1T$. This behavior is compatible to that observed to microestructural modificated $Sm_2Ba_2Cu_3O_{7-\delta}$ ceramic samples where the $k(T,B)$ profile of the $Sm_2Ba_2Cu_3O_{7-\delta}$ sample with more microestructural defects is more affected by the applied magnetic field.

In contrast to the reported results that identify the $k(T,B) > 0$ for $T > T_c$ in underdoped $YBa_2Cu_3O_x$ [2] and optimally doped $Bi_2Sr_2CaCu_2O_8$ [2] single crystals as well as a $La_{1.9}Sr_{0.1}CuO_4$ sample [3] the results displayed in the figure 3 do not show a significant contribution of the $k(T,B)$ to $T \geq T_c$.

The contrasting between $k(T)$ results of the YBCO sample with those to Y(Ca)BCO and YB(Sr)CO samples at the same $T\ T_c^{-1}$ shows that the Ca and Sr doping promotes a reduction of its amplitude as well as produces the enlargement of the $k(T,B)$ temperature regime. These peculiarities are more pronunced to the YB(Sr)CO sample.

We realized a dc magnetization experimental study in $YBa_2Cu_3O_{7-\delta}$, $Y_{0.99}Ca_{0.01}Ba_2Cu_3O_{7-\delta}$ and $YBa_{1.75}Sr_{0.25}Cu_3O_{7-\delta}$ single crystals with a strong foucus on the effects of Ca and Sr chemical doping on the $k(T,B)$ behaviour. The Sr and Ca doping does not change significantly the $k(T,B)$ versus TT_c^{-1} profile but strongly reduces the $k(T,B)$ intensity at the same time that colaborates to the enlargement of the $k(T,B)$ temperature regime. Otherwise we could not identify a $k(T,B)$ behavior that supports the pseudgap behavior. We conclude that the chemical pressure effects introduced by 12,5 % of Sr doping [6] were more nocive to the superconducting state of $YBa_2Cu_3O_{7-\delta}$ single crystal than the 1% of Ca doping [8]. If we connect the lower T_c value of our doped single crystals to the suppression of the $YBa_2Cu_3O_{7-\delta}$ superconducting state then it will be possible to suggest that the significant reduction of the $k(T,B)$ magnitude observed to the doped single crystals is associated to the proportional breaking of the $YBa_2Cu_3O_{7-\delta}$ superconducting paring mechanism promoted more efficiently by 12,5% of Sr doping than 1% of Ca doping.

Acknowledgments
The authors would like to thanks to the PRONEM 03/2011 (under contract: 11/2042-7) scientific program for partially financing this work.

References

[1] Dória M M, Salem-Sugui Jr S, de Oliveira I G et al 2002 *Phys. Rev. B* **65** 14450902
[2] Salem-Sugui Jr S, Doria M M, Alvaraga A D et al 2007 *Phys. Rev. B* **76** 132502
[3] Dória M M, Gubernatis J E and Rainer D 1989 *Phys. Rev. B* **39** 9573
[4] Dória M M 2009 *J Supercond Nov Mag* **22** 325
[5] Pena A J P, Martinez D. B, Parra Vargas C A, et al 2013B J Phys **25** 1250
[6] Vieira V N, Pureur P, Schaf J 2002 *Phys. Rev. B* **66** 224506
[7] Vieira V N and Schaf J 2002 *Phys. Rev. B* **65** 144531-9
[8] Vieira V N, Riegel I C and Schaf J 2007 *Phys. Rev. B* **76** 024518-7
[9] Osborn J A 1945 *Physical Review* **67**, 11

Journal of Physics: Conference Series **568** (2014) 032001 doi:10.1088/1742-6596/568/3/032001

Thermal expansion and magnetostriction measurements using a high sensitive capacitive dilatometer at millikelvin temperatures

Daiki Inoue, Daisuke Kaido, Yuta Yoshikawa, Mitsuyuki Minegishi, Koichi Matsumoto and Satoshi Abe

Department of Physics, Faculty of Science, Kanazawa University, Kanazawa, Ishikawa, Japan

E-mail: `daiki010@stu.kanazawa-u.ac.jp`

Abstract. We have developed a dilatometric measuring system for thermal expansion and magnetostriction, those are more singular than speci c heat in approaching to a quantum critical point. With decreasing temperature, thermal expansion becomes small in proportional to the square of temperature, thus, high sensitivity and reproducibility are necessary for the dilatometric measurements in millikelvin temperatures. Our dilatometer composed of the sample and the reference capacitor provides the extremely high resolution of $\Delta L/L \sim 10^{-10}$ using the ratio-transformer-based capacitance bridge. The dilatometer was installed on the ^{3}He-^{4}He dilution refrigerator with the 9 T superconducting magnet, and temperature was measured by the ^{3}He melting curve thermometer. We have measured thermal expansion and magnetostriction of the typical heavy fermion compound $CeRu_2Si_2$ along *a*-axis at temperature down to 10 mK in magnetic elds up to 9 T.

1. Introduction

Dilatometry has been a powerful technique for investigating the thermodynamic properties of materials. Thermal expansion and magnetostriction, those are the strain responses of a material to temperature and an external or internal magnetic eld, have a close relationship with the partial derivative of speci c heat and magnetization with respect to pressure. Thus, these properties have been good measures of phase transitions, quantum criticality, and other interesting phenomena [1]. Among the various methods for measuring length changes induced by temperature and magnetic eld, capacitive dilatometry excels in its sensitivity reaching values of L/L 10 ^{10} at low temperatures in magnetic elds. Many types of capacitive dilatometers have been developed for various experiments, such as absolute expansion dilatometer [1], dilatometer with an open architecture sample mounting arrangement [2], and small dilatometers for magnet systems [3]. In the low millikelvin temperature range, the capacitive dilatometer has another advantage in low power dissipation. However, su cient thermal contact for a sample cooling without any heat exchange gases and liquids is important in this temperature range. The limited space in magnet systems for magnetostriction measurement also requires a small size of dilatometer. Therefore, the capacitive dilatometer has been used in only a few investigations down to low millikelvin temperatures [4–6]. We developed the compact capacitive dilatometer and the capacitance bridge, and have reported the results of the heavy fermion compound $CeRu_2Si_2$ along *c*-axis at temperature down to 1 mK in magnetic elds up to 52.6 mT [7,8]. In

Journal of Physics: Conference Series **568** (2014) 032001 doi:10.1088/1742-6596/568/3/032001

the present report, we have developed our dilatometric measurements extending the magnetic eld range up to 9 T at 10 mK. The thermal expansion and magnetostriction of $CeRu_2Si_2$ along a-axis have been measured where the accuracy of L/L is better than 1 10 ^{10} and the temperature stability T/T is 10 ^{2} in isothermal magnetostriction measurements.

2. Experiment

2.1. Dilatometer

Figure 1 shows our three-terminal capacitive dilatometer with a diameter of 12 mm and length of 30 mm, which was small enough to be used in the limited space [8]. Most of the parts were made of oxygen-free high-conductivity copper for high thermal conductivity. Capacitor plates with a diameter of 5 mm and thickness of 2 mm were glued to the upper, middle holder and sample using Stycast 2850FT. The thickness of Stycast on the upper and the movable capacitor plate were typically 0.1 mm. Cigarette papers were inserted in the Stycast for ensuring an electrical insulation. The sample was cut into a cylinder or a cube with 3 5 mm in diameter and 4 5 mm in length, and was glued to the lower holder using Arzerite VL-10 silver paste for thermal contact. The copper spacers were inserted among holders to obtain a close spacing which was less than 20 m, and three holders and plates composed two capacitors, sample and reference. Thermal expansion and magnetostriction as a linear strain is given by $L/L = d_x/L \quad C/C$ in an ideal parallel plate model, where L, C, and d_x are the sample length, capacitance, and distance between plates, respectively. However, the dilation of holder, capacitor plates and Stycast a ect the measured capacitance values [9]. Hence, change rates of C/C in the sample capacitance C_x and the reference capacitance C_{ref} are given by

$$\frac{C_x}{C_x} = \frac{L}{d_x}\left\{\frac{L}{L}_{\mathrm{sample}} \quad \frac{l}{l}_{\mathrm{Cu}}\right\} + \frac{l_{\mathrm{sty}}}{d_x}\left\{\frac{l}{l}_{\mathrm{sty}} \quad \frac{l}{l}_{\mathrm{Cu}}\right\} \quad \frac{S_p}{d_x} \quad \frac{l}{l}_{\mathrm{Cu}}, \tag{1}$$

$$\frac{C_{\mathrm{ref}}}{C_{\mathrm{ref}}} = \frac{l_{\mathrm{sty}}}{d_{\mathrm{ref}}}\left\{\frac{l}{l}_{\mathrm{sty}} \quad \frac{l}{l}_{\mathrm{Cu}}\right\} \quad \frac{S_p}{d_{\mathrm{ref}}} \quad \frac{l}{l}_{\mathrm{Cu}}, \tag{2}$$

where d_x and d_{ref} are distance between capacitor plates of C_x and C_{ref}, and l_{sty} 0.1 mm, $l_{\mathrm{Cu}} = 2$ mm, $S_p = 15$ m are the thickness of Stycast, capacitor plate, copper foil spacer, respectively. The second and last terms in Eq. (1), known as the cell e ect, are usually obtained by measurements for the copper standard sample using the same dilatometer. In our dilatometer design where the sample is glued to the holder and capacitor plate, it is di cult to replace the sample so that the cell e ect has completely the same value. From Eqs. (1) and (2), the cell e ect of the sample capacitor is expected to be same as that of the reference capacitor, if $d_x = d_{\mathrm{ref}}$. Figure 2 shows the temperature dependence of the sample capacitance C_x and the reference capacitance C_{ref} of the dilatometer for the heavy fermion compound $CeRu_2Si_2$ along the a-axis measured by the automatic capacitance meter AH 2700A [10]. Both capacitance changes are proportional to T^2, but $C_{\mathrm{ref}}/C_{\mathrm{ref}}$ is less than only 2% for C_x/C_x at 1 K.

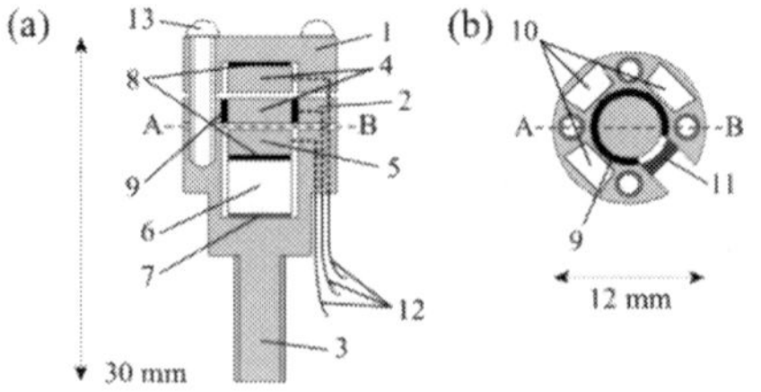

Figure 1. Schematic drawing of the capacitive dilatometer. (a) Three-terminal capacitive dilatometer. (b) Cross-sectional view of the capacitor: (1) Upper holder; (2) middle holder; (3) lower holder; (4) xed capacitor plate; (5) movable capacitor plate; (6) sample; (7) silver paste; (8) Stycast FT and cigarette paper; (9) Stycast FT; (10) copper foil spacers; (11) silver epoxy; (12) silver-coated copper coaxial cables; (13) M2 brass screws (four in total).

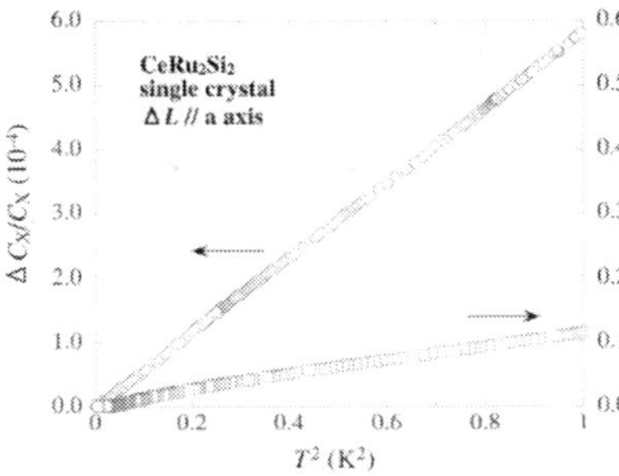

Figure 2. Capacitance change rate of the sample and the reference capacitor for $CeRu_2Si_2$ along the a-axis as a function of T^2 in a zero magnetic eld.

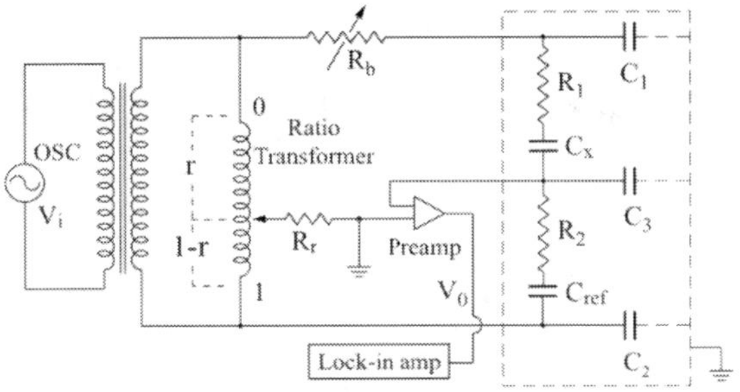

Figure 3. Diagram of the ratio-transformer-based capacitance bridge, where C_x and C_{ref} are respectively the sample and the reference capacitor, and C_1, C_2 and C_3 are the cable capacitances. R_b is adjusting resistor. The region among the dashed line is installed in the refrigerator.

2.2. Capacitance bridge

Our ratio-transformer-based capacitance bridge is shown in Fig. 3. A seven-decade ratio transformer with output impedance R_r of 5 Ω was used to compare a sample capacitance to be determined to a reference capacitance. A six-decade resistor R_b with resolution of 1 m and stability of approximately 20 ppm/K was used to balance the resistive parts R_1 and R_2. A high-sensitivity lock-in ampli er with a 40 dB preampli er was used as a null detector. The input voltage of the circuit V_i was 5 V_{rms} at frequencies between 1 and 3 kHz. If the resistive part of the bridge is adjusted by R_b, the output o -balance voltage V_0 can be given as [11]

$$\frac{V_0}{V_i} = \frac{(r \quad 1)C_x + rC_{\mathrm{ref}}}{C_x + C_{\mathrm{ref}} + C_3} \quad j\omega R_r \frac{C_x + C_{\mathrm{ref}}}{C_x + C_{\mathrm{ref}} + C_3}((r \quad 1)C_1 + rC_2), \tag{3}$$

where r is the ratio-transformer setting. When the in-phase component is balanced, that gives $C_x/C_{\mathrm{ref}} = r/(1 \quad r)$ equating to the Wheatstone bridge. Practically, the capacitance change resulting from thermal expansion and magnetostriction induces the o -valance V_0. Thus, balanced r is calculated from the o -balance voltage of the in-phase component, the ratio setting, and the ratio sensitivity. Accordingly, thermal expansion and magnetostriction as L/L using the capacitance bridge are given by

$$\frac{L}{L}_{\mathrm{sample}} \quad \frac{l}{l}_{\mathrm{Cu}} = \frac{d_x}{L} \quad \frac{C_x}{C_x} \quad \frac{C_{\mathrm{ref}}}{C_{\mathrm{ref}}} = \frac{d_x}{L}\frac{r}{r(1 \quad r)}. \tag{4}$$

Thermal expansion of $CeRu_2Si_2$ is ($L/L)_{\mathrm{sample}}$ 1 $10^{\ 6}$ at 1K, therefore, ($l/l)_{\mathrm{Cu}}$ is negligibly small because this is $10^{\ 10}$ below 1 K [12].

2.3. High magnetic field system

We developed an experimental setup in the ^3He-^4He dilution refrigerator with the 9 T superconducting magnet for measuring thermal expansion and magnetostriction in high magnetic elds. Figure 4 shows our experimental set up and magnetic eld pro le in the eld of 9 T at Z = 0 mm. Thermally-annealed oxygen-free high-conductivity copper thermal link intervening ^3He melting curve thermometer (MCT), the dilatometer, and the experiment stage had the particular structure that was the "U" shaped with a cross-section of 12 12 mm^2 and height of 256 mm. The dilatometer placed in the center of the magnetic eld can be settled in either direction, parallel and perpendicular to the magnetic eld. One extremity of the thermal link was connected to the experiment stage where there were a carbon resistance thermometer and the heater for temperature control. In contrast, the other extremity was connected to the MCT but thermally insulated to the experiment stage by Vespel. The eld dependence of the MCT is known to be $T/T = 2$ % at 25 mK and 5 T [13], hence the location of the MCT at Z = 240 mm where B -16mT at the maximum eld of 9 T ensured the accuracy of temperature measurement. Thermal link has six slits in order to reduce the eddy current heating $\dot{Q}_{eddy}(t)$ $(\partial B/\partial t)^2$ which is estimated to be 0.5 W when a rate of applied magnetic eld is $\dot{B}(t) = 2$ 10^{4} T/sec. Figure 5 shows the typical temperature stability in eld sweeps at 98 mK. It indicates that the temperature change by the eld sweep is less than about 1 mK even at the rate of 4 10^{4} T/sec. In fact, we applied the magnetic elds in the rate of 10^{5} T/sec for measuring the magnetostriction, thus, the typical temperature stability was $T/T = 5$ 10^{3} at 98 mK, and less than 10^{2} down to 14 mK.

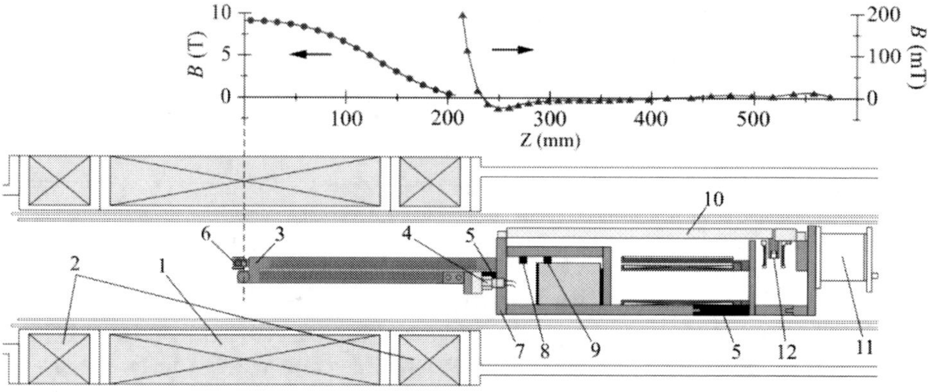

Figure 4. Schematic drawing of our experimental set up and magnetic eld pro le: (1) main magnet; (2) compensation coil; (3) copper thermal link; (4) ^{3}He melting curve thermometer; (5) Vespel; (6) dilatometer; (7) experiment stage; (8) carbon resistance thermometer; (9) heater; (10) silver thermal link; (11) mixing chamber of the ^{3}He-^{4}He dilution refrigerator; (12) indium heat switch. The eld pro le is represented by the circle in the left scale and the triangle in the right magni ed scale.

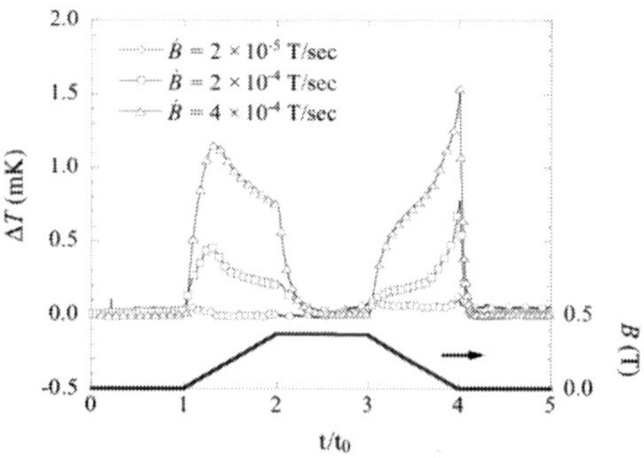

Figure 5. Temperature stability in eld sweeps at 98 mK as a function of normalized time t/t_0. t_0 are 9.0×10^2, 1.8×10^3, and 1.8×10^4 s in the rate of 4×10^{-4}, 2×10^{-4}, and 2×10^{-5} T/s, respectively. Solid line represents magnetic eld strength.

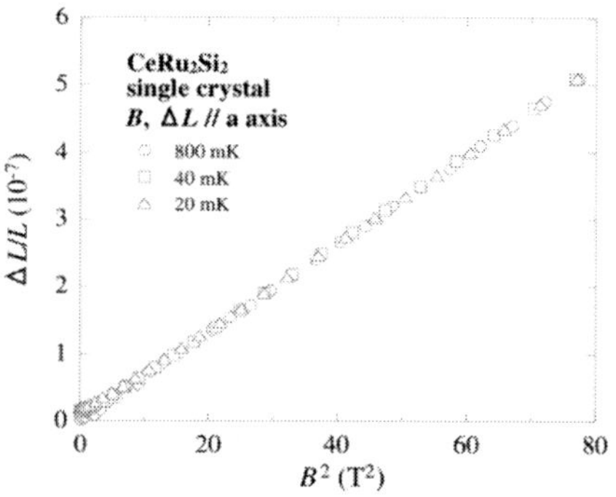

Figure 6. Isothermal magnetostriction of $CeRu_2Si_2$ along the a-axis as a function of B^2.

3. Results

We have measured magnetostriction of a typical heavy fermion compound $CeRu_2Si_2$ at temperature down to 10 mK for the magnetic field direction along the a-axis up to 9 T. Figure 6 shows measured magnetostrictions up to 9 T at 20, 40, 800 mK, where the field sweep rate were $\dot{B}(t) = 2.4 \times 10^{-5}$, 4×10^{-5}, and 2×10^{-4} T/sec, respectively. Below 200 mK, the accuracy of magnetostriction measurements being better than 1×10^{-10} were achieved by the ratio-transformer-based capacitance bridge. In contrast, the accuracy was 2×10^{-8} at 800 mK using the capacitance meter AH 2700A in the averaging time of 40 sec. The isothermal linear magnetostriction is proportional to B^2 and the metamagnetic behavior cannot be observed up to 9 T. The magnetostriction coefficient along a-axis, $\lambda_a = \partial(\Delta L/L)/\partial B$, is independent of temperature as $\lambda_a/B = 1.3 \times 10^{-8}$ T^{-1}, which is 800 times smaller than that along c-axis [14]. These differences among axes in the high field region agree with the anisotropic nature of this compound observed by other magnetic properties [15]. On the other hand, the deviation from B^2 dependence became significant around zero magnetic field with decreasing temperature. The similar behavior was observed by the previous measurements along c-axis [7]. Thus, it indicates that this compound shows an axis-independent quantum criticality below 40 mK around zero field.

4. Summary

We reported the new design of a capacitive dilatometer, a capacitance bridge system and high magnetic field system for accurate measurement of thermal expansion and magnetostriction at millikelvin temperatures in high magnetic fields. The capacitance bridge system ensured the high accuracy and high stability of $\Delta L/L$ being less than $\sim 2 \times 10^{-10}$ in the long-term experiment. The high magnetic field system enabled the isothermal magnetic measurement up to 9 T in the rate of $\dot{B}(t) \sim 10^{-5}$ T/sec. Thus, our experimental device provides sufficient performance under multi-extreme conditions.

References

[1] Barron T H K and White G K 1999 *Heat Capacity and Thermal Expansion at Low Temperatures* (New York: Plenum) and references therein
[2] Schmiedeshoff G M *et al.* 2006 *Rev. Sci. Instrum.* **77** 123907–1 and references therein
[3] Rotter M, Muller H, Gratz E, Doerr M and Loewenhaupt M 1998 *Rev. Sci. Instrum.* **69** 2742–6 and references therein
[4] Hill R W, Janssen T J B M, Meeson P J, Cuthbert M N, Springford M and Wasserman A L 1996 *Czech. J. Phys.* **46** 2077–8
[5] Sawada A, Kubo T, Fujii Y, Komatsubara T, Ōnuki Y, Kimura N, Yamamoto E and Haga Y 1996 *Czech. J. Phys.* **46** 803–4
[6] Tsujii H, Tanaka E, Ode Y, Katoh T, Mamiya T, Araki S, Settai R and Ōnuki Y 2000 *Phys. Rev. Lett.* **84** 5407–10
[7] Yoshida J, Abe S, Takahashi D, Segawa Y, Komai Y, Tsujii H, Matsumoto K, Suzuki H and Ōnuki Y 2008 *Phys. Rev. Lett.* **101** 256402
[8] Abe S, Sasaki F, Oonishi T, Inoue D, Yoshida J, Takahashi D, Tsujii H, Suzuki H and Matsumoto K 2012 *Cryogenics* **52** 452–6
[9] Pott R and Schefzyk R 1983 *J. Phys. E: Sci. Instrum.* **16** 444–9
[10] *AH 2700A* ANDEEN-HAGERLING, Inc., 31200 Bainbridge Road Cleveland, Ohio 44139-2231 U.S.A.
[11] Adams E D 1993 *Rev. Sci. Instrum.* **64** 601–11
[12] Swenson C A 1997 *Rev. Sci. Instrum.* **68** 1312–1315
[13] Fukuyama H, Yawata K, Ito D, Ikegami H and Ishimoto H 2003 *Physica* **B329** 1560–1
[14] Matsuhira K, Sakakibara T, Nomachi A, Tayama T, Tenya K, Amitsuka H, Maezawa K and Ōnuki Y 1999 *J. Phys. Soc. Jpn* **68** 3402–6
[15] Saha S R, Sugawara H, Namiki T, Aoki Y and Sato H 2002 *Phy. Rev.* B **65** 214429

Journal of Physics: Conference Series **568** (2014) 032002 doi:10.1088/1742-6596/568/3/032002

Thermal expansion and magnetostriction measurements using a Quantum Design Physical Property Measurement System

Oshi Iwakami, Naoaki Kawata, Misato Takeshita, Yusuke Yao, Satoshi Abe and Koichi Matsumoto

Department of Physics, Kanazawa University, Kanazawa, Ishikawa 920-1192, Japan

E-mail: `ou4.e-66@stu.kanazawa-u.ac.jp`

Abstract. Capacitive dilatometers that could be installed in a Quantum Design Physical Properties Measurements System (PPMS) were developed. Two types of dilatometers were constructed. One dilatometer was compact enough to be installed in small magnet bore of various cryostats and had a reference capacitor so that high sensitivity and stability were achieved. The other dilatometer with a movable capacitor supported by a flat spring enabled to measure large strain and exchange samples easily. We measured the thermal expansion and magnetostriction of single crystalline clathrate compound $Pr_3Pd_{20}Ge_6$ and polycrystalline hydrogenated $La(Fe_{0.88}Si_{0.12})_{13}$ in the temperature range between 2 and 340 K and magnetic field up to 8 T. The dilatometers allowed to detect the strain responses with a sensitivity of 10 ppb. We could measure strain from 10^{-8} to 10^{-3} using two dilatometers.

1. Introduction

Thermal expansion and magnetostriction, the strain responses of a material to temperature and a magnetic field, are thermodynamic quantities closely related to specific heat and magnetization. These quantities are very useful to study electronic and phononic properties, phase transitions, quantum criticality, and other interesting phenomena [1]. Quantum Design Physical Properties Measurements System (PPMS [2]) has been widely used for measurements of thermodynamic quantities, such as specific heat, magnetization, electrical resistance, and thermal conductivity.

In this report, we present the thermal expansion and magnetostriction measurements using homemade capacitive dilatometers installed in a PPMS in the temperature range of 2 to 340 K and magnetic fields up to 8 T. Capacitive dilatometers have been used in various thermal expansion and magnetostriction measurements because capacitive methods give very high sensitivity in detecting small length changes [3]. We developed a compact capacitive dilatometer that was installed in a dilution refrigerator [4], and quantum criticality was studied for $CeRu_2Si_2$ [5]. Similar dilatometer was made in order to measure thermal expansion and magnetostriction using a PPMS in higher temperature and magnetic field. Using this system, we observed temperature and magnetic field dependence of the linear thermal expansion of single crystalline $Pr_3Pd_{20}Ge_6$. This compound has attracted attention because of the electric quadrupole order of 4*f*-electrons and a rattling motion of a guest rare-earth ion in an oversize cage [6]. There are lots of magnetic materials that have large thermal expansion and magnetostriction around magnetic phase transition. For example, $La(Fe_{0.88}Si_{0.12})_{13}$ has been

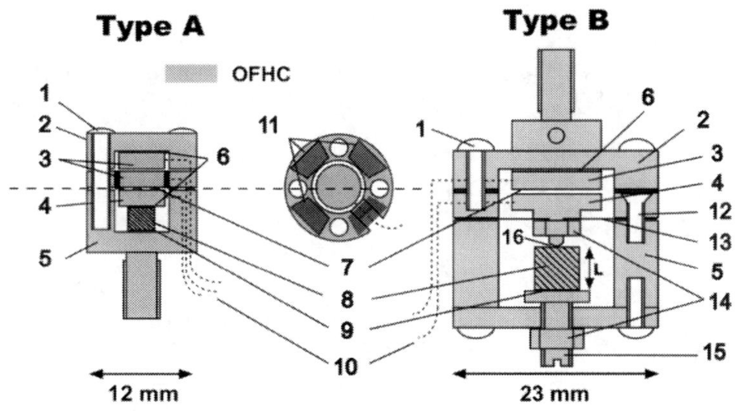

Figure 1. Schematics of the capacitive dilatometers. (1) brass screws, (2) upper holder, (3) fixed capacitor plate, (4) movable capacitor plate, (5) lower holder, (6) cigarette paper and Stycast, (7) gap D, (8) sample (length L), (9) silver paste, (10) flexible coaxial cables, (11) copper foil spacers, (12) stainless screws, (13) BeCu flat spring, (14) locknut, (15) sample platform. (16) Stycast

studied as a magnetic refrigerant for magnetic refrigeration because of the giant magnetocaloric effect around the ferromagnetic phase transition at the Curie temperature $T_c = 195$ K. The working temperature range having large magnetocaloric effects is around T_c. T_c of $La(Fe_{0.88}Si_{0.12})_{13}$ is shown to be increased up to about 330 K by hydrogen absorption [7]. Therefore, this material is expected as magnetic refrigerant for room temperature magnetic refrigerator. Volumetric change around transition is thought to contribute significantly to the giant magnetocaloric effects in this material [8, 9]. In order to measure large strain, we developed new capacitive dilatometer that had a movable capacitor supported by a flat spring. This dilatometer was used to measure linear thermal expansion of hydrogenated $La(Fe_{0.88}Si_{0.12})_{13}$. It was confirmed that these dilatometers worked with enough sensitivity and stability.

2. Dilatometers

When a capacitive dilatometer with a pair of capacitor plates is used, the dilation ΔL due to longitudinal thermal expansion and magnetostriction of a sample with length L is measured as a change in the gap D between capacitor plates. Many types of capacitive dilatometers have been developed for various experiments, such as absolute-expansion dilatometers [1], dilatometers with an open-architecture sample mounting arrangement [10], and small dilatometers for magnet systems [3]. Figure 1 schematically illustrates two different types of capacitance dilatometers used in the present study. Most of the parts were made of oxygen-free high-conductivity (OFHC) copper. Our dilatometers were small enough to be installed in a PPMS. The dilatometer of type A which has a composite structure with a reference capacitor was constructed similarly to the one used in our previous studies [4, 5]. The dilatometer of type B has a movable capacitor supported by a flat spring made of 0.1 mm thick BeCu that was annealed at 315 °C for about 3 hours. Movable capacitor plate supported by a spring is used by other researchers such as Schmiedeshoff *et al.* [10]. The spring is insulated from the movable capacitor plate using kapton films. The sample was glued to the sample platform with Arzerite VL-10 silver paste. The platform was adjusted by screw thread so that the sample was touched softly to the movable capacitor plate, then the platform was fixed with locknut. Sample and capacitor plate were insulated with thin layer of Stycast 2850GT. The fixed capacitor plate is attached to the upper holder with Stycast 2850FT using cigarette paper for insulation. The flexible, silver-coated copper coaxial cables as electrical leads were glued to the side of the capacitor plate using silver epoxy Eccobond 56C. The surfaces of the capacitor plate and holder were sandpapered to be flat, smooth, and coplanar. The gap D was adjusted by the thickness of the copper foil spacers and monitored by capacitance bridge. The parallelism and the edge effect of two capacitor plates in the dilatometer of type A has been discussed in our previous study [4]. The dilatometer of type B was checked and calibrated in the same way as Ref. [10], and was used within the range of linear response to the gap.

3. Experimental procedures

We have measured the thermal expansion and magnetostriction of $Pr_3Pd_{20}Ge_6$ using the dilatometer of type A. The single crystal of $Pr_3Pd_{20}Ge_6$ used in our investigation was from the same batch of high-quality samples, grown by a floating zone method, for ultrasound and ac-susceptibility experiments [6, 11]. The sample size of $Pr_3Pd_{20}Ge_6$ was about $4 \times 4 \times 6$ mm^3. The polycrystalline hydrogenated $La(Fe_{0.88}Si_{0.12})_{13}$ was measured using the dilatometer of type B. T_c of the present sample was determined as 313 K by magnetization and specific heat measurements. The sample size of $La(Fe_{0.88}Si_{0.12})_{13}$ was about $5 \times 5 \times 5$ mm^3.

The capacitive dilatometers were installed in a PPMS using the homemade insert that had two coaxial cables for capacitance bridge and a cernox resistance thermometer (Lake Shore) that was thermally anchored near the dilatometer. Capacitance was determined by a digital capacitance bridge (Andeen-Hagerling, 2500A [12]).

For an ideal parallel-plate capacitor with area A and gap D, the capacitance is given as $C = \epsilon_r \epsilon_0 A/D$, where $\epsilon_0 = 8.854 \times 10^{-12}$ Fm^{-1} and ϵ_r is the dielectric constant of the medium between pair of capacitor plates. In our experiments, the medium is gaseous helium for heat transfer in a sample chamber of the PPMS, and the chamber pressure varied from about 1.3 to 0.03 kPa in the temperature from 300 to 2 K. Though the dielectric constant of the gas depends on the gas density, we can assume $\epsilon_r = 1$ since even at atmospheric pressure helium has ϵ_r very close to unity ($\epsilon_r = 1.00007$ at 273 K [13]). The relative capacitance change $\Delta C/C$ due to change in ϵ_r was estimated on the order of 10^{-6} ($\Delta L/L \sim 10^{-8}$) and was sufficiently smaller than that due to the dilation of the sample.

The observed gap change ΔD is caused by difference between sample dilation $(\Delta L)_{\mathrm{Sample}}$ and dilation of a copper cell body $(\Delta L)_{\mathrm{Cu}}$ as $-\Delta D = (\Delta L)_{\mathrm{Sample}} - (\Delta L)_{\mathrm{Cu}}$. In this study, the thermal expansion of $(\Delta L)_{\mathrm{Cu}}$ was calculated using that of copper from the Ref. [14, 15, 16], and the magnetostriction of the one was negligible. As discussed by Pott and Schefzyk [17], ΔD also contains unspecified gap change that is known as "cell effect". It is thought that the cell effect is originated from epoxy, silver paste, and so on. The cell effect can be determined by measuring a copper standard sample. The cell effects of our two dilatometers were almost the same. The temperature dependence of the cell effect was confirmed to be small compared to the dilation of our sample as shown in Fig. 2. The field dependence of the cell effect was negligibly small compared to the dilation of the sample. In this paper, we show the thermal expansion and the magnetostriction including the cell effect.

4. Results and discussion

We can measure strain $\Delta L/L$ from 10^{-8} to 10^{-3} using two dilatometers. Figure 2 shows the temperature dependence of linear thermal expansion of $Pr_3Pd_{20}Ge_6$ along the [001] axis in zero magnetic field and 8 T. No hysteresis was observed in the temperature sweeps. In zero magnetic field, the thermal expansion decreases monotonically with decreasing temperature down to 2 K. The magnitude of $\Delta L/L$ from 300 to 2 K is about 2.5×10^{-3}. In 8 T, the thermal expansion was almost the same as that in zero magnetic field.

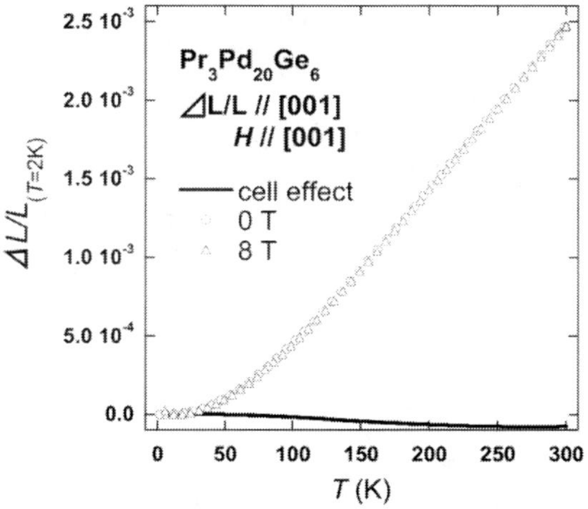

Figure 2. Temperature dependence of linear thermal expansion of $Pr_3Pd_{20}Ge_6$ along the [001] axis. Open circles ○ and open triangles △ represent the $\Delta L/L$ in zero magnetic field and 8 T, respectively. The solid line shows the cell effect.

Figure 3 shows the temperature dependence of linear thermal expansion of polycrystalline hydrogenated $La(Fe_{0.88}Si_{0.12})_{13}$. As shown in Fig. 3, large negative thermal expansion was observed in the temperature range of 300 to 320 K. Small hysteresis was also observed in the temperature sweep. In this temperature range, the cell effect of the dilatometer of type B was negligibly small compared to the sample dilation. Figure 4 shows the field-dependence of the isothermal linear magnetostriction as a function of H^2 at several temperatures. In paramagnetic phase, the magnetostriction increased approximately in proportion to the squared magnetic field. Large magnetostriction was observed near the transition temperature. In ferromagnetic phase, the magnetostriction showed saturation.

We checked a magnetic torque effect. In a magnetic field, the magnetic anisotropy and the ferromagnetic domain may lead to a torque on the sample. $Pr_3Pd_{20}Ge_6$ has magnetic anisotropy, and [001] axis is the axis of hard magnetization. The hydrogenated $La(Fe_{0.88}Si_{0.12})_{13}$ shows ferromagnetic phase transition at 313 K. By measuring magnetostriction in positive and negative directions, it is possible to distinguish the torque effect from magnetostriction. The magnetostriction of $Pr_3Pd_{20}Ge_6$ at 2 K and $La(Fe_{0.88}Si_{0.12})_{13}$ at 310 K were measured in both directions up to 8 T and 5 T, respectively. In each sample, the measured magnetostriction for both directions was identical. This result indicates that the torque effect was small for these samples. We thought that it was hard to detect of magnetic torque effect in our measurements since the magnetization of $Pr_3Pd_{20}Ge_6$ was small in paramagnetic phase and our $La(Fe_{0.88}Si_{0.12})_{13}$ sample was polycrystal and hysteresis was small.

5. Summary

We constructed two different dilatometers that could be used with a PPMS. We measured thermal expansion and magnetostriction of $Pr_3Pd_{20}Ge_6$ and hydrogenated $La(Fe_{0.88}Si_{0.12})_{13}$. Our dilatometers allow to detect the strain responses with a sensitivity of 10 ppb. We can measure strain from 10^{-8} to 10^{-3} using two dilatometers.

Acknowledgments

This work was supported in part by Grants-in-Aid for Scientific Research from MEXT and JSPS of Japan. We thank T. Goto and H. Kitazawa for providing the sample, and K. Nunomura for technical assistance.

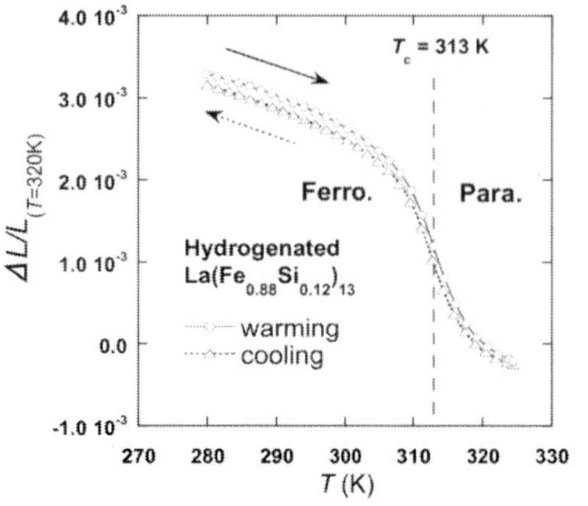

Figure 3. Temperature dependence of linear thermal expansion of hydrogenated $La(Fe_{0.88}Si_{0.12})_{13}$. Open circles ○ and open triangles △ represent the $\Delta L/L$ in warming and cooling, respectively. Broken line represents the ferromagnetic phase transition temperature.

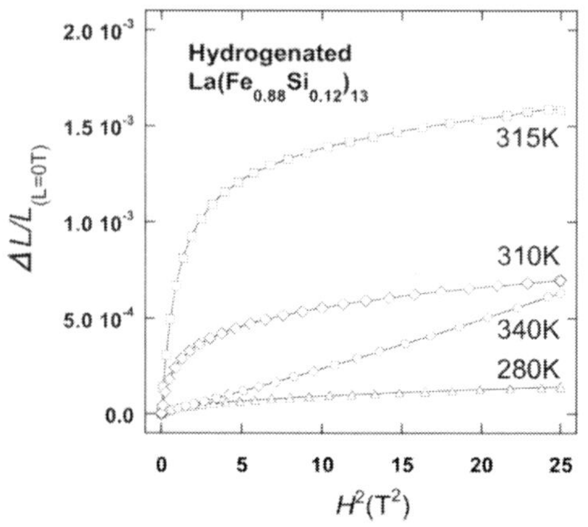

Figure 4. Field-dependence of the isothermal linear magnetostriction of polycrystalline hydrogenated $La(Fe_{0.88}Si_{0.12})_{13}$. Open circles ○ , open triangles △, open diamonds ◇, and open squares □ represent the $\Delta L/L$ at 340, 315, 310, and 280 K, respectively.

References

[1] Barron T H K and White G K 1999 *Heat Capacity and Thermal Expansion at Low Temperatures* International Cryogenics Monograph Series (New York, NY : Plenum) and references therein

[2] Quantum Design, 11578 Sorrento Valley Rd., San Diego, CA 92121.

[3] Rotter M, Müller H, Gratz E, Doerr M and Loewenhaupt M 1998 *Rev. Sci. Instrum.* **69** 2742–6 and references therein

[4] Abe S, Sasaki F, Oonishi T, Inoue D, Yoshida J, Takahashi D, Tsujii H, Suzuki H and Matsumoto K 2012 *Cryogenics* **52** 452–6

[5] Yoshida J, Abe S, Takahashi D, Segawa Y, Komai Y, Tsujii H, Matsumoto K, Suzuki H and Ōnuki Y 2008 *Phys. Rev. Lett.* **101**(25) 256402

[6] Ano G *et al.* 2012 *J. Phys. Soc. Jpn.* **81** 034710

[7] Fujita A, Fujieda S, Hasegawa Y and Fukamichi K 2003 *Phys. Rev.* B **67**(10) 104416

[8] Fujita A, Fujieda S, Fukamichi K, Mitamura H and Goto T 2001 *Phys. Rev.* B **65**(1) 014410

[9] Fujieda S, Fujita A, Fukamichi K, Yamazaki Y and Iijima Y 2001 *Appl. Phys. Lett.* **79** 653

[10] Schmiedeshoff G M *et al.* 2006 *Rev. Sci. Instrum.* **77** 123907 and references therein

[11] Iwakami O *et al.* 2014 *Phys. Rev. B* **90**(10) 100402

[12] Model 2500A digital capacitance bridge manufactured by Andeen-Hagerling, Inc., 31200 Bainbrigde Road, Cleveland, Ohio, USA.

[13] Kessom W H 1942 *Helium* (Amsterdam: Elsevier)

[14] Nix F C and MacNair D 1941 *Phys. Rev.* **60**(8) 597–605

[15] Rubin T, Altman H W and Johnston H L 1954 *J. Am. Chem. Soc.* **76** 5289–93

[16] Kroeger F and Swenson C 1977 *J. Appl. Phys.* **48** 853–64

[17] Pott R and Schefzyk R 1983 *J. Phys.* E: *Sci. Instrum.* **16** 444

Signal analysis and characterization of a micro-electro-mechanical oscillator for the study of quantum fluids

C S Barquist[1], J Bauer[1], T Edmonds[1], P Zheng[1], W G Jiang[1], M Gonzalez[1], Y Lee[1] and H B Chan[2]

[1] Department of Physics, University of Florida, Gainesville, Florida, USA
[2] Department of Physics, Hong Kong University of Science and Technology, HK

Abstract. We present the full characterization of a micro-electro-mechanical system (MEMS) based probe developed for the study of quantum fluids. The device consists of a pair of parallel plates (200 x 200 μm^2) with a well-defined gap (1.25 μm) in which a fluid film forms when immersed in liquid. The mobile plate is suspended above the fixed plate (substrate) by four serpentine springs. This geometry allows for the study of the properties of the surrounding liquid through the resonant behavior of the mobile plate. This device demonstrated its potential as a high precision probe suitable for the study of films of quantum fluids and also of quantum turbulence. In this work we present our detailed analysis of the device signal in our current measurement scheme. Based on our analysis, we determined the transduction factor of a device at room temperature and 4 K, which allows us to convert the measured quantities to physical quantities such as displacement, velocity, and force.

1. Introduction

Micro-Electro-Mechanical Systems or MEMS have a wide variety of uses in the engineering and scientific worlds. In our laboratory we have developed a MEMS-based probe for the study of quantum fluids functioning in the ultra low temperature regime [1]. The device is a comb-drive resonator designed specifically to form a thin film of superfluid ^{3}He and also to perform interface sensitive measurements. An overview of the MEMS device is shown in Fig. 1. The device is composed of the center plate which is suspended above the substrate by the four serpentine springs and the two sets of integrated capacitors on each side. The gap between the center plate and the substrate which defines a slab when immersed in liquid could be 0.75, 1.25, or 2.00 μm. The device used in this work has a 1.25 μm gap and a 200 $\times$ 200 μm^2 center plate.

Similar devices have been tested in liquid ^{3}He down to 300 μK and in liquid ^{4}He down to 50 mK and demonstrated their potential as a unique and versatile experimental probe for quantum fluids [1]. It is essential to calibrate a device to be able to convert the measured quantity (electrical voltage) to physically relevant quantities such as displacement and force to accomplish quantitative understanding of the results and phenomena. This conversion is done through the transduction factor, $\tilde{\beta}$ which is device specific and independent of the surrounding fluid. In this paper we present a full theoretical analysis of our measurement scheme. We demonstrate the validity of our analysis through various measurements and finally present the measured transduction factor at room temperature and 4.2 K in vacuum.

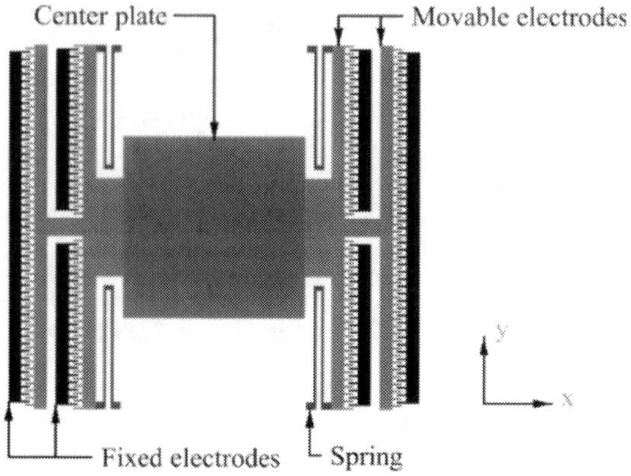

Figure 1. Top view of the MEMS device. The darker colored regions are fixed, they are where the carrier signal and actuating voltages are applied. The lighter colored regions are movable and interact with the fluid. The movable parts are actuated by the capacitive interaction between the teeth of the electrodes.

2. Measurement scheme

Investigation of fluid properties using the MEMS oscillator is carried out by observing how the fluids affect the resonant properties of the device. Because the properties of the fluid change with controllable parameters such as temperature and pressure so do those of the device in the fluid. The resonant properties of the device – resonance frequency, resonance width or quality factor, and resonant peak amplitude – can be measured by driving the device in the linear regime and sweeping the frequency through the resonance. These resonant properties can then be related back to physical quantities such as the damping coefficient and the velocity.

The MEMS device is driven by the capacitive interaction between the fixed and movable electrodes. The inter-locking teeth of the electrodes form an array of parallel plate capacitors, with the capacitance being dependent on the overlap of the teeth of the movable electrode with those of the fixed electrode as shown in Fig. 1. Because the capacitance depends on the amount of overlap between the teeth, the energy stored in the field between the teeth depends on this overlap. The overlap between the teeth depends on the displacement of the movable structure, therefore there is a position dependent energy in the system and consequently a force. It is this force that is used to actuate the device. Choosing the frequency of the oscillating force to be that of the resonant motion in a desired direction insures that the motion of the device is in the desired direction. The displacement of the device is detected by measuring the capacitance change, differential capacitance to be more specific. We adopted a capacitance bridge to perform high resolution differential capacitance measurement. Our measurement scheme is displayed in Fig. 2.

Our scheme is basically an amplitude modulation. The high frequency (hf) carrier signal labeled as V_{hf} is fed into the circuit through an isolation transformer for the bridge. The carrier signal is then split using an inductive ratio transformer. The ratio then is adjusted to balance out any asymmetries in the differential capacitance when the device is in its equilibrium position in the presence of a DC bias. The actuation of the device is preformed by the low frequency (lf) signal labeled as V_{lf} which is swept through the mechanical resonance. The lf signal is added to the hf carrier signal via a splitter used in reverse as a signal adder. The DC is added after a blocking capacitor in order to isolate the rest of the circuit from the DC voltage. A summing amplifier and blocking capacitor are also placed on the other side of the circuit for

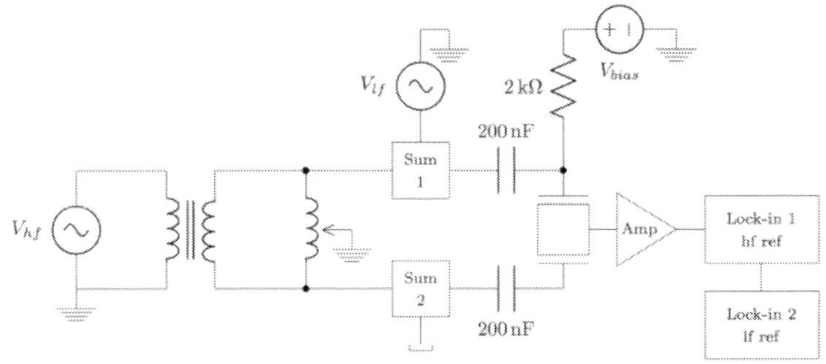

Figure 2. Measurement Scheme. The movable portion of the MEMS device is represented by the box connected to the 'Amp' with the fixed electrodes on either side of it represented by one plate of a capacitor.

symmetry, although they serve no functional purpose. When the device is actuated from the balance point, a net charge is induced on the center plate and is detected using an Amptek A250 charge sensitive amplifier. The signal from the amplifier is then demodulated by two lock-in amplifiers: first high frequency and then low frequency.

3. Analysis

Consider the voltage output A of the charge-senstive amplifier when the plate has a charge q. The amplifier has a charge-to-voltage conversion factor α which is determined by a feedback capacitor. The amplitude A in volts is given by:

$$A = \alpha q. \tag{1}$$

The total charge comes from the capacitive interaction between the fixed and the movable electrodes, and the net charge q can be split into the charge $q_{r(l)}$ coming from the interaction between electrodes on the right (left) side of the device:

$$q_l = C_l V_l, \tag{2}$$
$$q_r = C_r V_r. \tag{3}$$

C_l and C_r are the capacitances between the fixed and the movable electrodes on the left and right side, respectively. V_l and V_r are the voltages between the fixed and the center electrodes. For small displacements, x, from the equilibrium position, x_0, the capacitance changes linearly:

$$C_l = \tilde{\beta}(x_0 + x), \tag{4}$$
$$C_r = \tilde{\beta}(x_0 - x). \tag{5}$$

$\tilde{\beta}$ is a geometrical parameter that gives the capacitance between the fixed and moving electrodes as a function of the overlap of the comb teeth. The voltages applied to each side are:

$$V_L = \frac{V_{hf}}{2} + V_{lf} + V_{bias}, \tag{6}$$

$$V_R = -\frac{V_{hf}}{2}. \tag{7}$$

For brevity, the sinusoidal time dependence of V_{hf} and V_{lf} are implied. Using Eqs. (2) - (7) provides a relationship between the applied voltages, the displacement, and the charge on the plate:

$$q = \tilde{\beta}x[V_{hf} + V_{bias} + V_{lf}]. \tag{8}$$

In Eq. (8) terms that depend only on x_0 and not on x are neglected since these terms will be filtered by our detection scheme. The displacement x is the solution to the equation of motion for a damped driven harmonic oscillator with the natural frequency ω_0 and damping coefficient γ:

$$\ddot{x} + \frac{\gamma}{m}\dot{x} + \omega_0^2 x = \frac{F}{m}, \tag{9}$$

where m is the mass of the oscillator (center plate). The solution then takes the form:

$$x = Be^{-i\omega t}, \tag{10}$$

where B is the amplitude of oscillation and ω is the frequency of the driving force:

$$B = \frac{F}{m}\frac{(\omega_0^2 - \omega^2) + i(\frac{\gamma}{m}\omega)}{(\omega_0^2 - \omega^2)^2 + (\frac{\gamma}{m}\omega)^2}. \tag{11}$$

On the other hand the force on the plate is found by differentiating the capacitive energy with respect to the overlap:

$$\vec{F_j} = -\frac{d}{dx}[\frac{1}{2}C_j V_j^2]\hat{x} = -\frac{1}{2}\frac{dC_j}{dx}V_j^2\hat{x}, \tag{12}$$

where j denotes the left or right side. The net force is given by:

$$\vec{F} = \vec{F_L} + \vec{F_R} = \frac{\tilde{\beta}}{2}[V_{lf}^2 + V_{bias}^2 + V_{hf}V_{lf} + V_{hf}V_{bias} + 2V_{lf}V_{bias}]\hat{x}. \tag{13}$$

Each term in Eq. (13) contains its own respective frequency. The only relevant force term is the one that drives the oscillator through the resonance. All other terms are off resonance or filtered out during detection. Therefore, the force can then be written as an effective force:

$$F = \tilde{\beta}V_{lf}V_{bias}. \tag{14}$$

Upon substituting Eq. (14) into (11) one has a solution for the amplitude of oscillation as a function of frequency, which can be linked to the detected output voltage with the aid of Eqs. (8) & (1). It is worth mentioning that the driving force does not depend on V_{hf}.

One needs a way to determine $\tilde{\beta}$ experimentally. For this consider the signal at resonance corresponding to the imaginary (absorption) term in Eq. (11). Since $\omega = \omega_0$ at resonance and $\frac{\gamma}{m} = \Delta\omega$, where $\Delta\omega$ is the full-width-half-maximum (FWHM) of the resonance [2], the out of phase amplitude is given by:

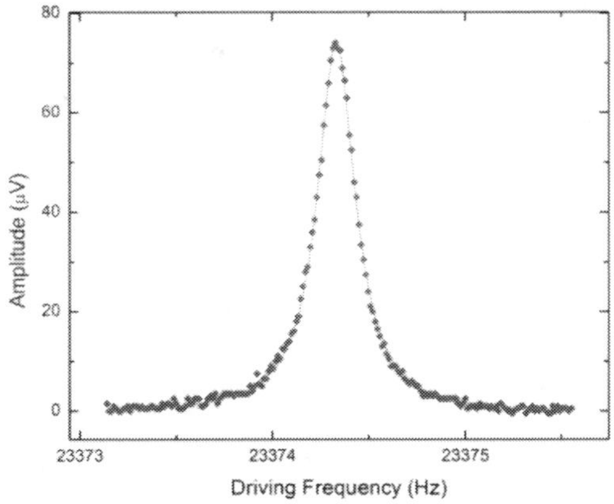

Figure 3. Absorption curve, the red line is the fitted Lorentzian curve. $V_{lf} = 274\ \mu V_{peak}$, $V_{bias} = 10\ V$, $V_{hf} = 1.41\ V_{peak}$

$$\Im m(B) = \frac{\tilde{\beta} V_{lf} V_{bias}}{m \Delta\omega \omega_0}. \tag{15}$$

An expression is obtained for the detected voltage A after the charge sensitive amplifier by combining (15), (8), and (1):

$$A = \frac{\alpha \tilde{\beta}^2 V_{bias} V_{hf} V_{lf}}{m \Delta\omega \omega_0}. \tag{16}$$

In this expression only the transduction factor is unknown. All other quantities can be measured experimentally or are determined by the known geometry of the device and material parameters. Unlike the driving force, the signal amplitude is proportional to all three excitation voltages. By measuring the signal amplitude at resonance as a function of the applied voltage one can be sure not to include any intrinsic offset not accounted for in Eq. (16). Doing this one uses the slope of the amplitude vs. applied voltage to determine the transduction factor. Experimentally, the transduction factor can be determined through:

$$\tilde{\beta} = \sqrt{\frac{dA}{dV_{lf}} \frac{m \Delta\omega \omega_0}{\alpha V_{bias} V_{hf}}}. \tag{17}$$

4. Measurement

To test the validity of Eq. (16) the device signal is measured as a function of the three voltages: V_{lf}, V_{hf}, and V_{bias}. The transduction factor is calculated using Eq. (17) and compared with the theoretical value. The amplitude, A, is measured by sweeping the frequency of V_{lf} through the resonance. Each sweep contained 200 data points over a frequency range 5 times the FWHM (see Fig. 3). All the measurements were done in the linear regime of the oscillator where the resonance curves fall on a single Lorentzian when scaled by the excitation. The data is then fitted to a Lorentzian to obtain the peak amplitude at resonance.

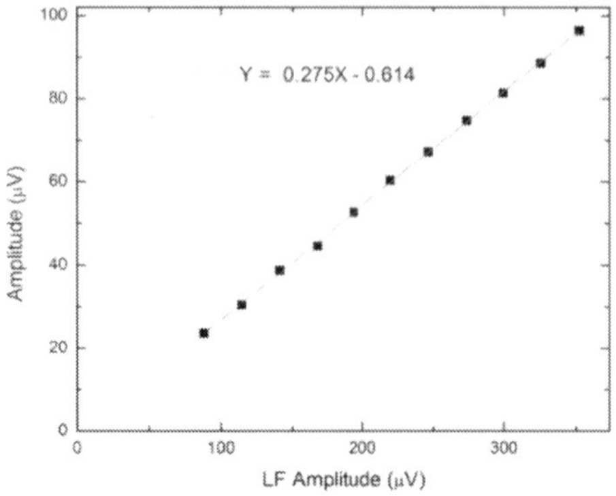

(a) Amplitude vs. V_{lf}. $V_{hf} = 1.41\ V_{peak}$ & $V_{bias} = 10\ V$

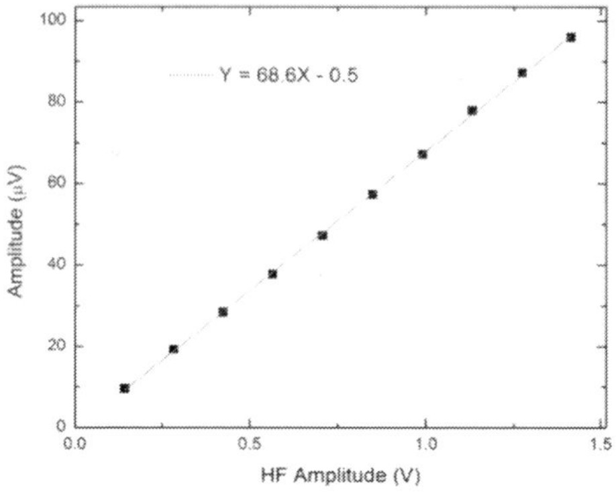

(b) Amplitude vs. V_{hf}. $V_{lf} = 353\ \mu V_{peak}$ & $V_{bias} = 10\ V$

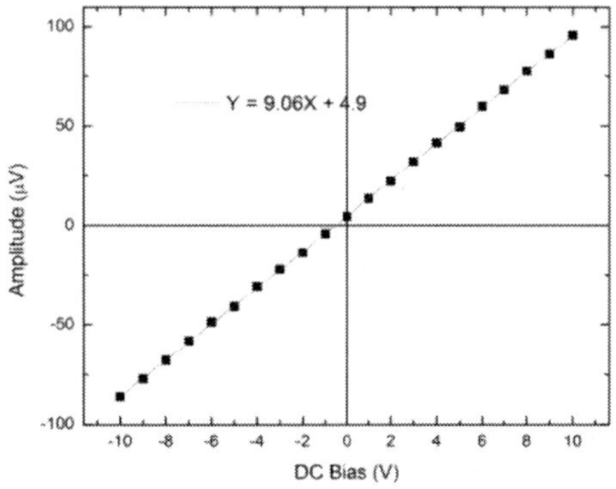

(c) Amplitude vs. DC bias. $V_{hf} = 1.41\ V_{peak}$ & $V_{lf} = 353\ \mu V$

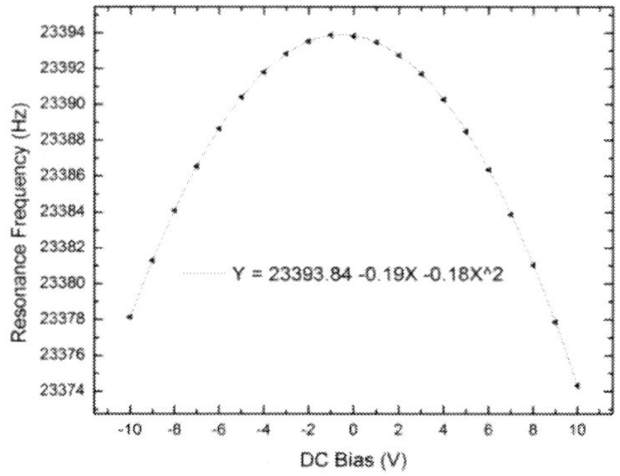

(d) Resonance frequency vs. DC bias. $V_{hf} = 1.41\ V_{peak}$ & $V_{lf} = 353\ \mu V$

Figure 4. MEMS response to excitation voltages

The signal amplitudes as a function of V_{lf}, V_{hf}, and V_{bias} are shown in Fig. 4. For each of these measurements the other two excitations were held constant. For example, V_{hf} was held at 1.41 V_{peak} and V_{bias} at 10 V while V_{lf} was varied in Fig. 4(a). These measurements confirm the linear dependences on all three excitations as expected in Eq. (16). Altering the DC bias has an interesting effect on the MEMS device. V_{bias} affects the resonance frequency of the device (see Fig. 4(d)). The shape of the bias dependence indicates softening of the device at higher bias. It might be caused by levitation or tilting of the center plate caused by asymmetric electric fields due to the presence of the substrate.

The transduction factor, $\tilde{\beta}$, is calculated from the slope in Fig. 4(a), $\tilde{\beta} = 1.36 \times 10^{-9}\ Fm^{-1}$. The transduction factor can be estimated by computing the capacitance of the overlapping comb-drive teeth in the parallel plate approximation:

$$\tilde{\beta} = \frac{N\epsilon\epsilon_0 t}{d}, \tag{18}$$

where N is the number of capacitors made by the teeth, ϵ is the relative permittivity, ϵ_0 is the

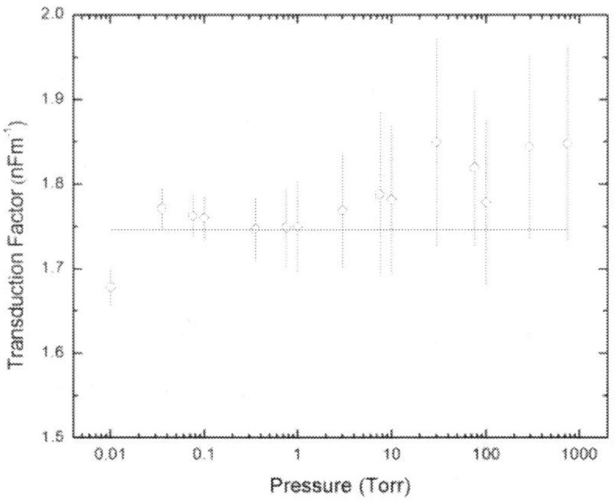

Figure 5. Pressure dependence of another MEMS device of the same type at room temperature in air. The solid line indicates the weighted mean of the transduction factor.

permittivity of free space, t is the height of the teeth perpendicular to the line pointing form one tooth to the other, and d is the distance between teeth. This formula is the simplest form of parallel plate capacitance ignoring the fringe field effect and gives $\tilde{\beta} = 1.68 \times 10^{-9}\ Fm^{-1}$, which is in good agreement with our measurement. We also performed multiple measurements of $\tilde{\beta}$ for the same device after thermal cycling and at RT. The transduction factor at room temperature is found to be $\tilde{\beta} = 1.36 \times 10^{-9}\ Fm^{-1}$ and remained the same after thermal cycling, verifying that $\tilde{\beta}$ is a device specific geometric parameter. We also measured the pressure dependance of the transduction factor for another device of the same design at room temperature in air (Fig 5). At higher pressures the uncertainties become large due to higher damping. The solid line indicates the weighted average of the transduction factor, $\tilde{\beta} = 1.75 \times 10^{-9}\ Fm^{-1}$. Within the uncertainty, no significant pressure dependance is present.

5. Conclusions & Discussion

We performed a full circuit analysis of our measurement scheme involving three excitation voltages and a capacitance bridge. The transduction factor $\tilde{\beta}$, introduced in our analysis, is a device specific geometrical parameter that links electrical quantities to physically meaningful quantities. Our analysis was experimentally verified and $\tilde{\beta}$ was determined experimentally and compared with the theoretical value.

Acknowledgments

This work is supported by NSF DMR-1205891 (YL).

References

[1] Gonzalez M, Zheng P, Moon B H, Garcell E, Lee Y and Chan H B 2014 *J. Low Temp. Phys.* **171** 200-6
[2] Landau L D and Lifshitz E M *Mechanics, Third Edition* vol 1 (Oxford: Butterworth-Heinenann) p 77
[3] Gonzalez M, Zheng P, Garcell E, Lee Y and Chan H B 2013 *Rev. Sci. Instrum.* **84** 025003

Shear piezoelectric coefficients of PZT, $LiNbO_3$ and PMN-PT at cryogenic temperatures

Syed Bukhari[1], **Md Islam**[1], **Ariel Haziot**[1,2], **John Beamish**[1]

[1]Department of Physics, University of Alberta, Edmonton, Alberta, Canada T6G 2E1

[2]Laboratoire de Physique Statistique de l'ENS, associe au CNRS et aux Universites Pierre-et-Marie-Curie & Denis-Diderot, 24, rue Lhomond, 75231 Paris cedex 05, France

E-mail: jbeamish@ualberta.ca

Abstract.
Piezoelectric transducers are used to detect stress and to generate nanometer scale displacements but their piezoelectric coefficients decrease with temperature, limiting their performance in cryogenic applications. We have developed a capacitive technique and directly measured the temperature dependence of the shear coefficient d_{15} for ceramic lead zirconium titanate (PZT), 41° X-cut lithium niobate ($LiNbO_3$) and single crystal lead magnesium niobium-lead titanate (PMN-PT). In PZT, d_{15} decreases nearly linearly with temperature, dropping by factor of about 4 by 1.3 K. $LiNbO_3$ has the smallest room temperature d_{15}, but its value decreased by only 6% at the lowest temperatures. PMN-PT had the largest value of d_{15} at room temperature (2.9×10^{-9} m/V, about 45 times larger than for $LiNbO_3$) but it decreased rapidly below 75 K; at 1.3 K, d_{15} was only about 8% of its room temperature value.

1. Introduction

A mechanical stress applied to a piezoelectric material produces a charge proportional to the force; a voltage applied across a piezoelectric produces a strain. These properties can be exploited to measure stresses and to generate displacements in mechanical actuators. Although most applications are at room temperature, scanning tunneling microscopes (STM) are often designed to operate below 4 K, and we have used PZT transducers to measure the elastic and plastic properties of solid helium [1, 2] at temperatures as low as 15 mK. For high sensitivity and/or maximum displacements, large piezoelectric coefficients d_{ij} are needed, but most materials' piezoelectric coefficients are reduced at cryogenic temperatures. The most commonly used material, PZT, has relatively large piezoelectric coefficients at room temperature. However, d_{33} and d_{13} are about 5 times smaller at 4 K [3, 4, 5, 6] and little is known about the shear coefficient d_{15} at cryogenic temperatures. The temperature dependence for lithium niobate is much weaker, but its largest coefficient, d_{15}, is about 8 times smaller than for PZT [7, 8]. New materials like single crystal PMN-PT have exceptionally large room temperature coefficients (e.g. d_{15} about

5 times larger than for PZT) but limited information is available about their behavior at low temperatures, particularly below 75 K [9, 10]. Also, the displacements in piezoelectrics often have non-linear or time-dependent behavior [12, 11, 13] which produces hysteresis when large voltages are applied and limits their performance in STM or other positioning applications.

In this paper we report measurements of d_{15} for ceramic PZT 5A[14], 41° X-cut $LiNbO_3$[15] and single crystal PMN-PT[16] transducers. Measurements were made between room temperature and 1.3 K, at voltages up to 150 V. The large room temperature values of d_{15} for PMN-PT are reduced by a factor of 12 at the lowest temperatures, compared to a 4-fold reduction for PZT. At the lowest temperatures, all three materials show linear, non-hysteretic behavior up to 150 V, but the PMN-PT transducers have significant hysteresis above 100 K.

2. Experimental design

In order to directly measure the shear piezoelectric coefficient d_{15} at low temperature, we used a capacitive displacement detection technique as shown in Fig. 1. A voltage ΔV applied across the shear transducer induces a transverse displacement $\Delta x = d_{15}\Delta V$ which then produces a capacitance change ΔC proportional to the change in spacing $\Delta L = \Delta x$ between the two blocks

$$\frac{\Delta C}{C} = \frac{\Delta L}{L} = \frac{d_{15}\Delta V}{L}. \quad (1)$$

where the gap spacing L (typically about 300 μm) is related to the initial capacitance $C = \epsilon_o A/L$, where A is the area of the gap where the blocks overlap. The piezoelectric coefficient is then

$$d_{15} = \frac{A\epsilon_o}{C^2}\frac{\Delta C}{\Delta V}. \quad (2)$$

The capacitance/transducer assembly was mounted on a copper bracket attached to the ^{4}He pot of a dilution refrigerator. Above 4 K, helium exchange gas was used for cooling.

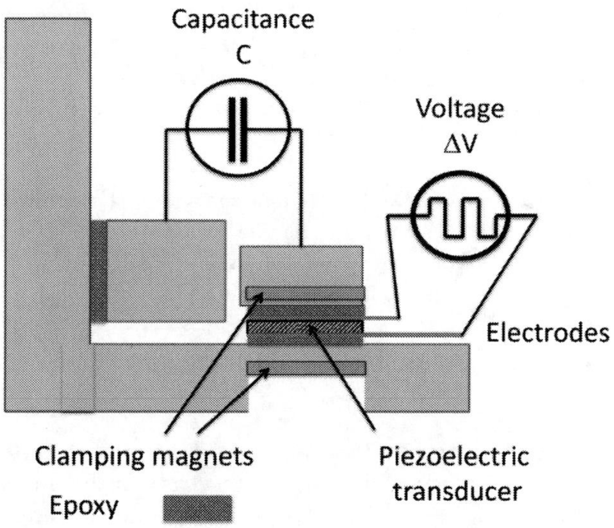

Figure 1. Displacement measurement to determine d_{15}. A Keithley 2400 voltage source applies an alternating ± 20 V potential to the transducer's electrodes. Coaxial cables connect a capacitance bridge[17] to a fixed brass block and to a block clamped on top of the shear transducer. A displacement of the transducer changes the gap between the two blocks and their capacitance.

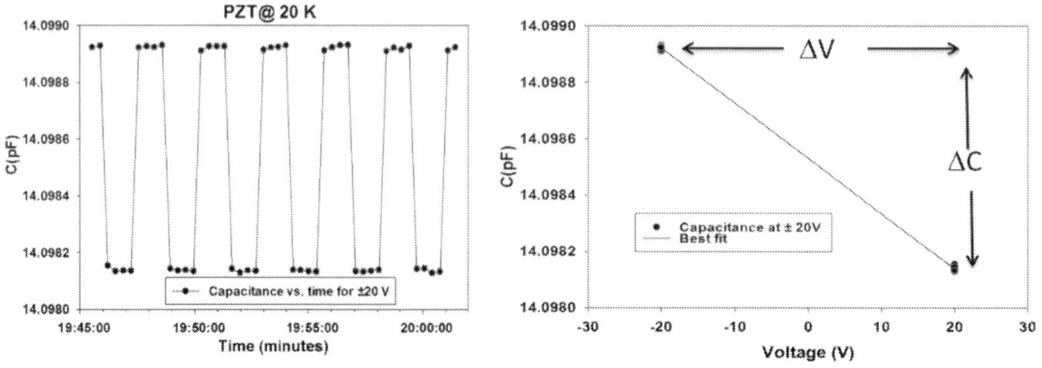

Figure 2. Capacitance changes due to voltage ±20V applied to a PZT-5A transducer at 20 K.

Figure 2 shows typical capacitance changes when an alternating DC voltage ±20 V (ΔV =40 V) is applied across the PZT-5A shear transducer at 20 K. The average capacitance $C = 14.0985$ pF and the change ΔC (about 56 ppm) give $d_{15} \approx 1.8 \times 10^{-10}$ m/V.

3. Results

The temperature dependences of the shear piezoelectric constant for the ceramic PZT-5A and 41° X-cut $LiNbO_3$ shear transducers are plotted in Fig. 3. The error bars for $LiNbO_3$ correspond to the resolution of our capacitance bridge; the errors for PZT and PMN-PT are smaller than the symbol sizes. For PZT (Fig. 3a), d_{15} decreases roughly linearly over the entire temperature range. Before taking data, the temperature was stabilized at each point for at least 50 minutes; a complete cooling or warming run took at least 4 days. The data were during cooling (green points) and during warming (red points) agree. Thermally cycling to low temperature did not affect the room temperature properties of the PZT or other transducers. The measured value of d_{15} for our PZT transducer decreased from 6.1×10^{-10} m/V at room temperature (close to the value quoted by the manufacturer: 5.85×10^{-10} m/V). At our lowest temperature (1.3 K), this dropped to 1.44×10^{-10} m/V, a reduction of more than a factor of 4. Figure 3b shows the corresponding data for $LiNbO_3$. The room temperature value of d_{15} is about ten times smaller, 6.4×10^{-11} m/V, but only drops by about 6% at low temperature.

The corresponding temperature dependence for the PMN-PT single crystal is shown in Fig. 4a. At room temperature d_{15} has a value of 2.85×10^{-9} m/V in this material, about 5 times larger than PZT. It decreases with temperature, but is nearly constant at 2.0×10^{-9} m/V between 225 and 75 K. However, below 75 K, d_{15} drops rapidly, by a factor of nearly 10 at 1.3 K.

4. Discussion

At room temperature, d_{15} of the three materials varies by a factor of 45, from 6.4×10^{-11} for $LiNbO_3$ to 2.85×10^{-9} m/V for PMN-PT. In all of the materials, d_{15} decreases with temperature, but the variation is smallest for $LiNbO_3$ and largest for PMN-PT. At 1.3 K, d_{15} ranges from

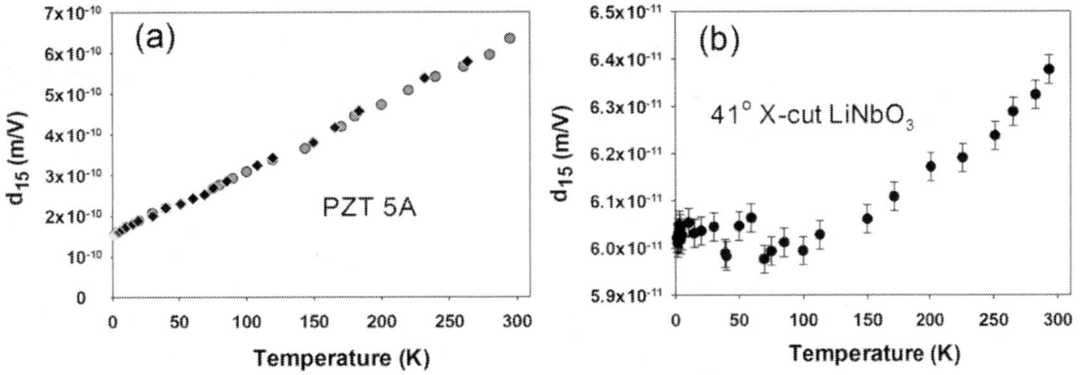

Figure 3. a) Temperature dependence of the piezoelectric coefficient d_{15} of a PZT-5A shear transducer. Green diamonds are data taken during cooling; red circular points during warming. b) Temperature dependence of d_{15} for a 41° X-cut $LiNbO_3$ crystal.

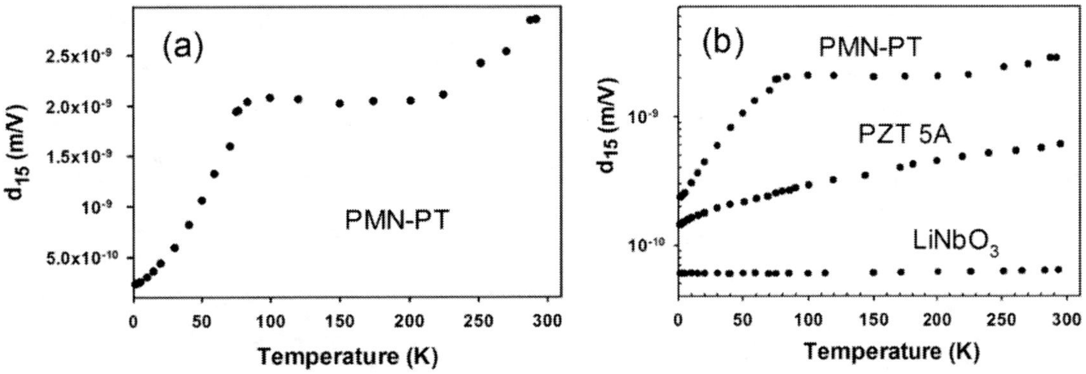

Figure 4. a) The temperature dependence of d_{15} for a PMN-PT single crystal shear transducer. b) Comparison of the shear coefficients of PZT, $LiNbO_3$ and PMN-PT transducers.

6.0×10^{-11} for $LiNbO_3$ to 2.3×10^{-10} m/V for PMN-PT, less than a factor of 4. Single crystal PMN-PT does have a significant advantage over more commonly used PZT ceramics at intermediate temperatures. At 75 K its value of d_{15} is about 8 times larger than that of PZT-5A but this advantage nearly disappears at lower temperatures (at 1.3 K it is only 60% larger). Lithium niobate has the smallest d_{15} at all temperatures, but at 1.3 K it is nearly half the value for PZT-5A.

Lithium niobate has a very high ferroelectric Curie temperature (1415 K) and its relatively small piezoelectric coefficients and weak temperature dependence reflect the intrinsic behavior of single domain crystals. The much larger room temperature coefficients of PZT-5A ceramics and PMN-PT crystals are achieved with compositions close to a transition between tetragonal and rhombohedral structures. Their piezoelectric behavior is determined by the temperature dependence of this phase boundary and by extrinsic effects associated with grain boundaries and domain walls, which we plan to study in future work.

In addition to the measurements described above, we also measured the displacements when voltages between -150 and +150 V were applied. Over this voltage range, the response was linear at all temperatures for $LiNbO_3$ and PZT. At the lowest temperatures this was also true for PMN-PT but the response was non-linear in the plateau region above 75 K. This resulted in hysteresis in the displacement when the voltage was ramped up and down between -150 and +150 V, limiting the use of PMT-PT in applications like STMs where precise, reproducible positioning is required.

Lithium niobate shear transducers have some advantages for low temperature actuators. In addition to the linear response described above, $LiNbO_3$ shear transducers can tolerate much higher voltages than PZT or PMN-PT without depoling. Also, they can be cut much thinner than PZT or PMN-PT shear transducers, which are usually at least a millimeter thick to allow them to be electrically poled parallel to the transducer surface. A compact, large displacement, linear actuator could be constructed from a stack of thin $LiNbO_3$ transducers.

In other applications, e.g. precision stress measurements, the calibration depends on d_{15} and therefore depends on temperature. For $LiNbO_3$, d_{15} is essentially constant (within the resolution of our measurements) below 100 K. For PZT and PMN-PT, on the other hand, d_{15} continues to change to the lowest temperatures, dropping by about 4% and 6%, respectively, between 4 and 1.3 K.

4.1. Acknowledgments

This work was supported by a grant from NSERC Canada. AH was also supported by a grant from EDC (AdG 247258-SUPERSOLID).

References

[1] J. Day and J. Beamish, Nature **450**, 853 (2007).
[2] A. Haziot, X. Rojas, A.D. Fefferman, J.R. Beamish and S. Balibar, Phys. Rev. Lett. **110**, 035301 (2013).
[3] X.L. Zhang, Z.X. Chen, L.E. Cross and W.A. Schulze, J. Mat. Sci. **18**, 968 (1983).
[4] B. Yurke, P.G. Kaminsky and D.M. Eigler, Cryogenics **26**, 435 (1986).
[5] M.D. Walker, NASA Report CR1998-208708 (1998).
[6] R.G. Sabat, B.K. Mukherjee, W. Ren and G. Yang, J. Appl. Phys. **101**, 064111 (2007).
[7] T. Yamada, N. Niizeki and H. Toyoda, Japan. J. Phys. **6**, 151 (1967).
[8] R.T. Smith and F.S. Welsh, **42**, 2219 (1971).
[9] F. Wang, W. Shi, S.W. Or, X. Zhao and H. Luo, Mater. Chem. Phys. **125**, 718 (2011).
[10] F. Martin, H. J. M. ter Brake, L. Lebrun, S. Zhang, and T. Shrout, J. Appl. Phys. **111**, 104108 (2012).
[11] H. Jung, J.Y. Shim and D.-G. Gweon, Rev. Sci. Instr. **71**, 3436 (2000).
[12] K.G. Vandervoort, R.K. Zasadzinski, G.G. Galicia and G. W.Crabtree, Rev. Sci. Instr. **64**, 896 (2011).
[13] M. Fouaidy, G. Martinet, M. Hammoudi, F. Chatelet and A. Olivier, Proc. 14th Intl. Conf. on Mixed Design (2007).
[14] PZT-5A ceramic shear transducers, 12.77 × 9.58 × 1.78 mm with chrome/gold electrodes, supplied by Boston Piezo-Optics, Inc. (http://www.bostonpiezooptics.com/)
[15] 41° X-cut lithium niobate shear transducers, 10.0 × 10.0 × 0.25 mm with chrome/gold electrodes, supplied by Boston Piezo-Optics, Inc.
[16] TRS X2A single crystal lead magnesium niobate-lead titanate (PMN-PT) shear plates, 10.0 × 10.0 × 1.0 mm with chrome/gold electrodes, supplied by TRS Ceramics, Inc. (http://www.trstechnologies.com/)
[17] Andeen-Hagerling 2550A 1 kHz AC capacitance bridge with a resolution of about 0.5 ppm for a typical averaging time of 9 seconds (http://www.andeen-hagerling.com/).

Journal of Physics: Conference Series **568** (2014) 032005 doi:10.1088/1742-6596/568/3/032005

Low-Temperature Scanning Capacitance Probe for Imaging Electron Motion

S Bhandari, R M Westervelt

School of Engineering and Applied Sciences, Harvard University, MA, 02138, U.S.A

Email: sbhandar@fas.harvard.edu

Abstract. Novel techniques to probe electronic properties at the nanoscale can shed light on the physics of nanoscale devices. In particular, studying the scattering of electrons from edges and apertures at the nanoscale and imaging the electron profile in a quantum dot, have been of interest [1]. In this paper, we present the design and implementation of a cooled scanning capacitance probe that operates at liquid He temperatures to image electron waves in nanodevices. The conducting tip of a scanned probe microscope is held above the nanoscale structure, and an applied sample-to-tip voltage creates an image charge that is measured by a cooled charge amplifier [2] adjacent to the tip. The circuit is based on a low-capacitance, high-electron-mobility transistor (Fujitsu FHX35X). The input is a capacitance bridge formed by a low capacitance pinched-off HEMT transistor and tip-sample capacitance. We have achieved low noise level (0.13 e/√Hz) and high spatial resolution (100 nm) for this technique, which promises to be a useful tool to study electronic behavior in nanoscale devices.

1. Introduction

Scanning probe techniques have been of great significance in studying local electronic behavior of a two dimensional electron gas (2DEG) in materials such as graphene [4] and GaAs/Al_3Ga_7As heterostructures [1][2][3]. In these techniques, images of local electron flow were obtained by having a very narrow (20 nm to 1 um) charged AFM tip scanned or positioned directly above the 2DEG while the conductance through the sample [1][4][5], or the capacitance between the sample and tip, is measured [2][3]. The later technique [2][3] uses an STM tip, which is held above the 2DEG surface and is capacitively coupled to the 2DEG immediately below. When a small AC voltage is applied between the tip and the 2DEG, the resulting flow of charge in the gas induces an oscillating image charge on the tip; that oscillation is detected by a sensitive electrometer. This technique can be very useful in mapping the local electron density. In [2][3] the technique was used to understand the Quantum Hall Effect, (QHE) which arises when 2DEG is in a strong magnetic field at low temperatures.

We plan to image the electronic density profile in graphene nanostructures and quantum dots using this technique. These nanostructures will be created on high mobility graphene-on-boron-nitride-sandwich-structures [5] and suspended graphene samples. In this paper, we present the design and implementation of a cooled scanning capacitance probe that operates at liquid He temperatures to image electron waves in nanodevices. A metal tip is held at 20 nm above the nanoscale structure, and an applied sample-to-tip voltage creates an image charge that is measured by a cooled charge amplifier [2] adjacent to the tip. The circuit is based on a low-capacitance, high-electron-mobility transistor (Fujitsu FHX35X). The input is a capacitance bridge formed by a low capacitance pinched-off HEMT transistor and tip-sample capacitance. We use a sharp metal (tungsten) tip (1um diameter) to create an image charge on a gold plate placed on a SiO_2/Si substrate. The tip is scanned at a height 20 nm above the surface of gold plate while the capacitance is measured. A model of the experimental setup was simulated in COMSOL. This model includes a metal sphere of diameter 1 um (analogue of a conducting tip) placed 20 nm above a 15 nm thick conducting metal plate that sits on top of SiO_2 substrate. The experimental result and the simulation are compared and agree quite well.

2. Method

We use a scanning probe capacitance microscopy technique to image local capacitance at the scale of nanometers. As shown in figure 1, the technique uses a metal tip scanned above the sample as the capacitance between the tip and sample is measured. The sample consists of a 15 nm thick gold electrode deposited on a Si/SiO_2 substrate. The tip was scanned across the edge of this plate so that corresponding change of capacitance can be measured. In addition, another experiment was done where a tip of radius 2.5 um was used. In this experiment, the distance between the tip and sample is varied while the change in capacitance is recorded.

The printed circuit board that holds both the sample and circuit sits on a piezotube that can be controlled electrically to move in x,y and z direction. The tip is held fixed on a support.

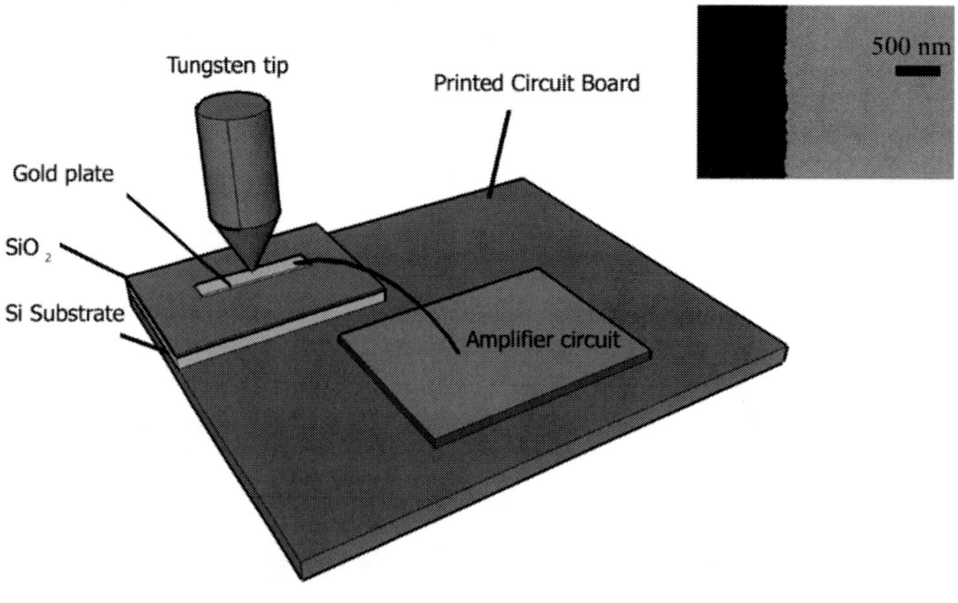

Figure 1: Schematic of experimental apparatus. Tungsten tip of diameter 1 um is scanned at 20 nm above a 15 nm gold plate deposited on top of a SiO_2 substrate. The amplifier circuit measures the change in capacitance as the tip is scanned. Both the sample and circuit are mounted on a printed circuit board. The inset on the top right shows the SEM of gold plate on Silicon Substrate.

The charge sensing circuit shown in figure 2 uses a capacitance bridge circuit followed by a two-stage common base amplifier circuit. As the experiments need to be performed at low temperatures to observe quantum mechanical effects, we use High Electron Mobility Transistors for the amplifier circuit. We chose a low noise (0.55 dB @ f = 12 GHz) pHEMT for this purpose.

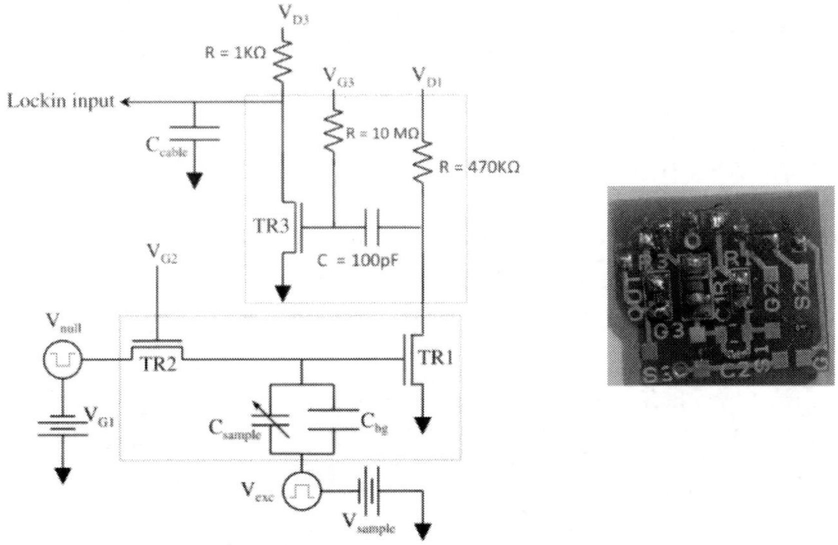

Figure 2: Left image is the schematic of pre-amplifier circuit. TR1, TR2 and TR3 are High Electron Mobility Transistors. The circuit is a capacitance bridge formed by tip-sample capacitance and a reference capacitor followed by at two stage common base amplifiers. The right image is the picture of the circuit on a printed circuit board.

As shown in figure 2, the capacitance between the tip and sample can be modeled as C_{samp}. The transistors TR1, TR2 and TR3 are HEMTs in our circuit. We use TR2 as a reference capacitor by pinching it off as a pinch voltage is applied in V_{G2}. The nominal source-drain capacitance for the transistor is 150 fF. As a typical capacitance bridge circuit, an ac voltage V_{S1} is applied to the sample while a second ac voltage V_{R1} is applied to the reference capacitor. The sample voltage V_{R1} is adjusted (both phase and amplitude) such that the voltage at junction node of reference capacitor and sample capacitor is zero. This means that the bridge is balanced and any change in the sample-tip capacitance would bring the bridge off balance and hence a change in voltage at the node.

The two-stage common base amplifier circuit amplifies the small change in voltage. The first stage consists of TR1 and a resistor R = 470 kΩ. The value of the resistor was chosen in order to achieve maximum gain at our first stage. The transistor in this stage is operated at saturation point again to maximize gain. The second stage consists of TR 3 and a resistor R = 1k ohms. The transistor TR3 is biased at triode region to allow maximum current so that it's enough to drive the long cables coming out of the cryostat to the lock-in amplifier that sits outside the cryostat. The estimated impedance in the cables is about 5 kΩ and hence we set out output impedance of the amplifier to about 2 kΩ. The 100 pF capacitor that sits between the first stage and second stage amplifier allows us to bias the TR2 as it only allow ac coupling between first stage and second stage circuits. The operating frequency of our experiment is set to 210 kHz (maximum frequency possible in lock-in amplifier) to minimize the 1/f noise in our signal. The bypass capacitor with a value of 100 pF has a lower cut-off frequency of 1 kHz which minimizes attenuation of signal going from stage 1 to stage 2.

The dc voltages that bias the amplifier circuits were generated using a homemade low-noise, battery-operated dc source, shown in figure 3. Two 6V batteries along with two low-noise (15.1 uV rms @ 100 kHz) adjustable output TPS7A30001 dc-dc converters were used. The output of the dc-dc converter was fed to a simple potential divider connected to a unity gain amplifier. Six different replicas of the circuit were made to create all the necessary biasing/gate voltages for the amplifier circuits.

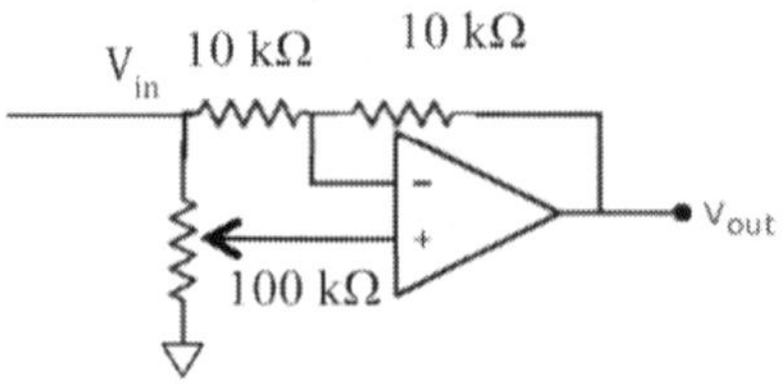

Figure 3: Schematic of the dc voltage supply to biasing transistors in the preamp. The circuit consists of dc-to-dc converter that generates 6 V, followed by a potentiometer and a unity gain amplifier.

The ac voltage source for the amplifier circuit was also homebuilt to achieve a low noise figure. It consists of two 6 V batteries which are connected to a dc-to-dc converter shown in figure 4 to maintain a constant voltage supply. A Wien Bridge Oscillator (I) is used for generating a low-distortion sinusoidal wave. The components are selected such that the frequency is set to 200 kHz ($f = 1/2\pi RC$). This is followed by a buffer amplifier (II) with unity gain to have low output impedance. A phase shifter (III) along with a potential divider (IV) is used for changing the phase and amplitude of the sine wave. Two copies of this circuit are used to generate two ac signals with independently adjustable phase and amplitude needed for the capacitance bridge described earlier.

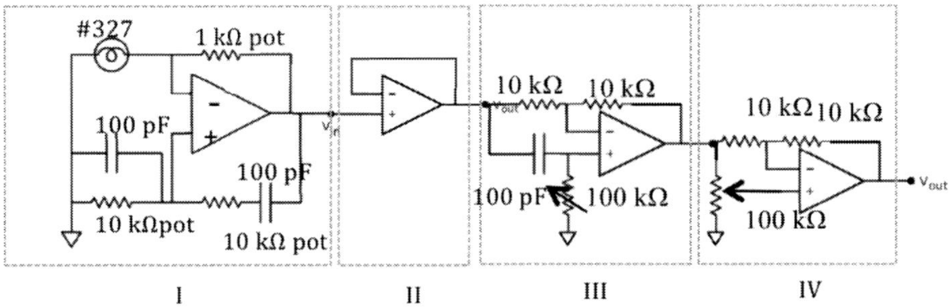

Figure 4: Schematic of variable phase and amplitude ac source. The dc-to-dc converter provides a constant power supply to the ac generator circuit. The ac generator circuit consists of four stages. Stage I is a Wien bridge oscillator, that generates sine wave of frequency 210 kHz followed by Stage II, a buffer amplifier. Stage III and IV consists of phase shifting circuit and a buffered potential divider respectively.

The dc voltage sources and ac voltage sources are put in separate enclosures for preventing ac interference into the dc source. These sources remain outside the cryostat while the preamp operates at liquid helium temperature mounted right on the scanning probe microscope. All the cables that lead to and from the pre amplifier circuit are standard coaxial cables. For the purpose of mixing the dc voltage with the ac signal, a simple capacitor resistor junction is used as a mixer.

The input voltage noise of the circuit was obtained by grounding both the input terminals and measuring the output voltage noise using a lock-in amplifier. The input noise measured at the temperature 4.2 K is 20 nV/√Hz. The stray capacitance of the preamp was measured to be 1 pF. The charge noise was obtained by the relation $\Delta Q = C\Delta V = 0.13\ e^{-}/\sqrt{Hz}$.

3. Results

3.1. Experimental results:

The experimental result of the two-dimensional capacitance scan over the edge of gold plate is shown in figure 5. The change in capacitance of 30 aF is clearly visible and a step can be noticed right along the edge of the electrode.

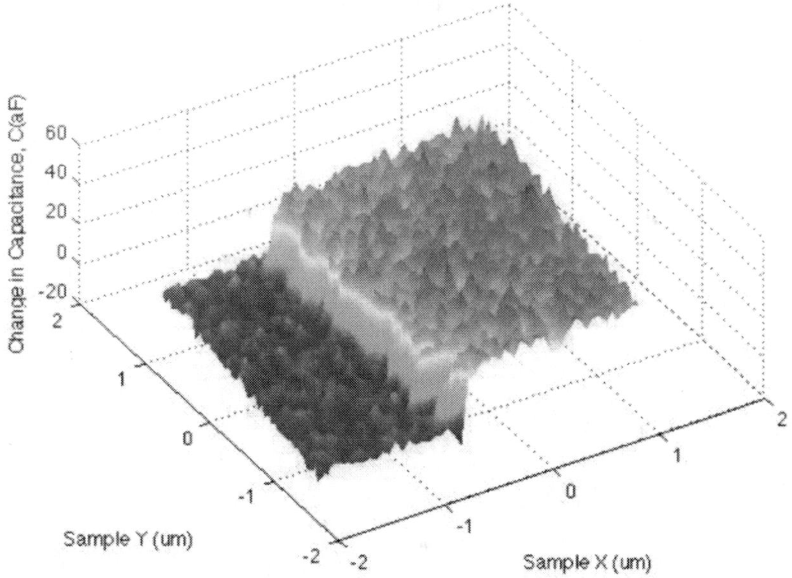

Figure 5: Capacitance-scan of the two-dimensional sample. The height of plot represents change in capacitance (aF) as the tip is scanned at a height of 20 nm on top of the edge of the 15 nm gold plate. The change in capacitance is ~ 30 aF.

To compute the spatial resolution of our scanning capacitance setup, the capacitance measurements were averaged over a range $\Delta y = 0.30$ µm of the flat edge and plotted vs. x in figure 6. The total width of the transition is $\Delta x = 100$ nm.

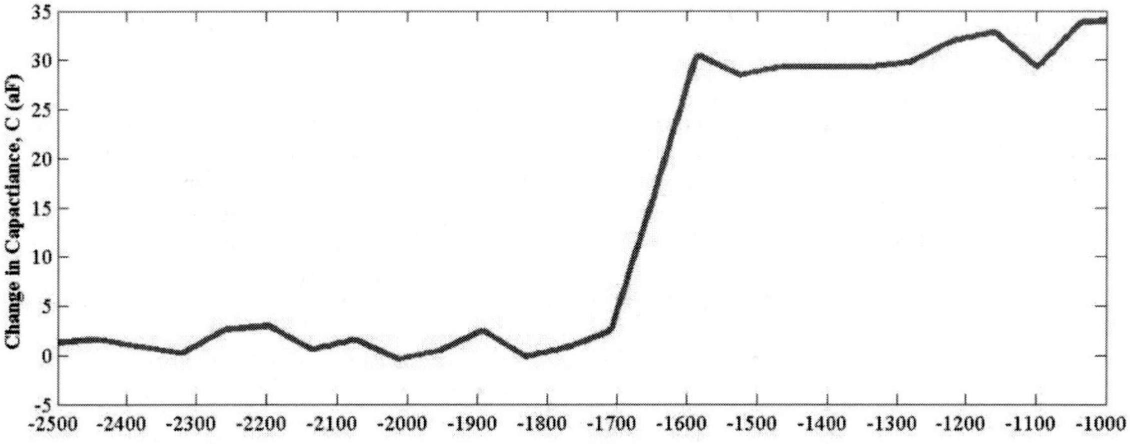

Figure 6: Capacitance change *vs.* x averaged between -0.2 µm < y < -0.5 µm. The full spatial width of the capacitance transition is 100 nm.

The change in tip-sample capacitance as the tip is swept over the edge of the conducting sheet is determined by the tip-sample distance. Figure 7 shows the measured tip-sample capacitance measured for 2.5 um radius tip as the tip-sample distance is varied, along with COMSOL simulations of the capacitance between a conducting sphere and plate. As shown, the theoretical and experimental results agree quite well for tip-sample distances less than 200 nm.

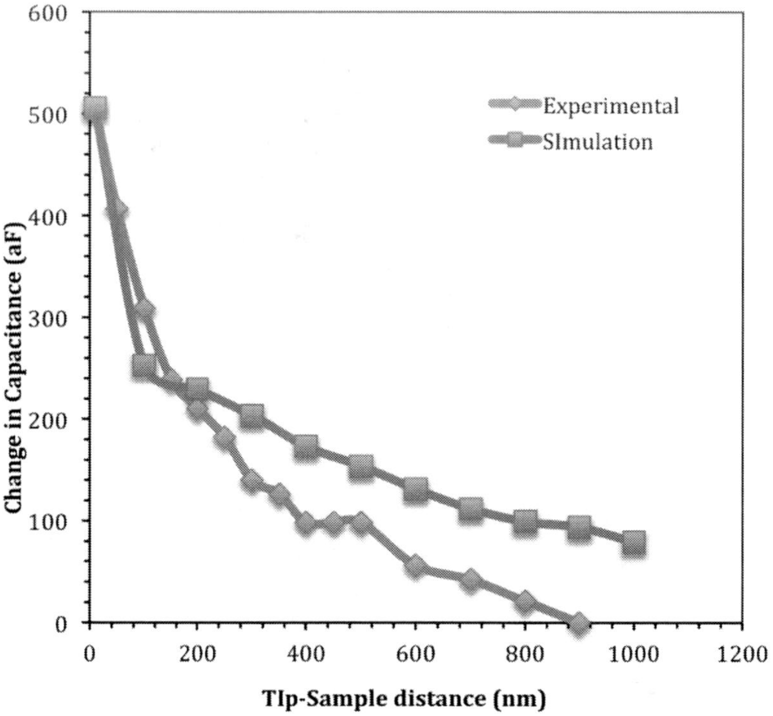

Figure 7: The tip-sample capacitance is plotted *vs.* tip-sample distance. The red curve shows COMSOL simulations while the blue is the experimental result. A tip with diameter 5 um was used.

3.2. Theory and Simulation:

To analyze the capacitive image of the conducting plate edge shown in figures 5 and 6, we used COMSOL simulations of the tip-sample conductance between a conducting sphere and a conducting plate in figure 8. The simulation geometry is shown on the left, where a metal sphere (reddish brown) is located at a distance of 20 nm above a conducting plate (reddish brown), on top of a SiO_2 layer (grey). The right side of figure 8 shows the image charge in the conducting plate, represented by the normal component of electric field at the plate's surface in the simulations. As seen in figure 8, the FWHM of the image charge induced by the voltage on the tip is 210 nm, which gives a spatial resolution of the half width at half maximum HWHM ~ 105 nm. This agrees with the experimental results of the scanning capacitance over the electrode edge.

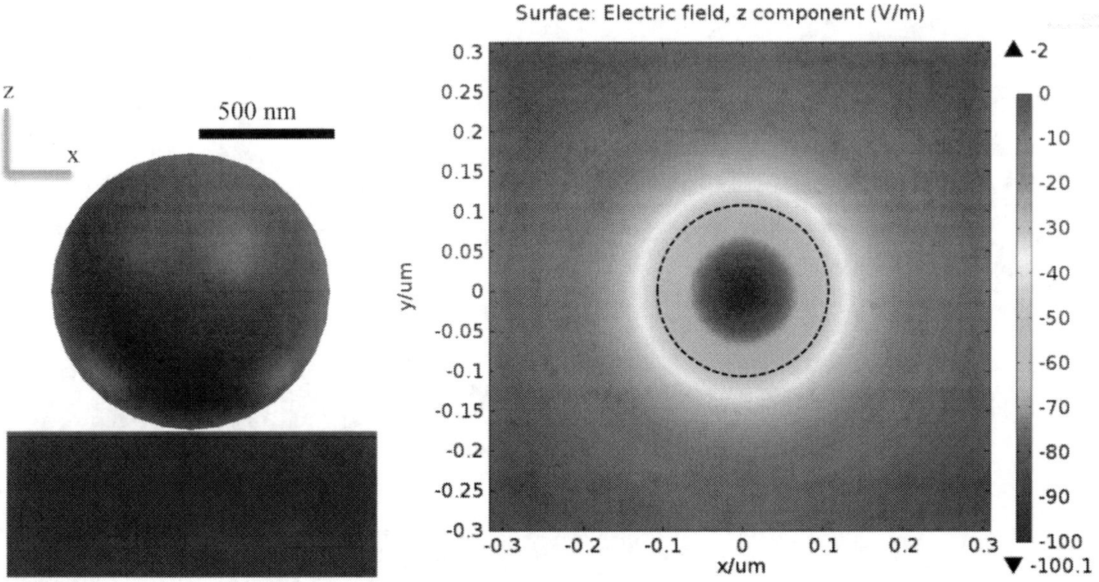

Figure 8: (left) Geometry for COMSOL simulations showing the tip (modelled as a conducting sphere) positioned at a height of 20 nm above a conducting metal plate (reddish brown) sitting on a SiO_2 substrate (grey). (right) Normal electric field on the metal plate surface, proportional to the image charge density. The FWHM from the dashed line fit is 210 nm, which predicts a spatial resolution ~ 100 nm in the experiments.

The agreement between the experimental results and simulation suggests that the technique can image local capacitance with a spatial resolution ~100 nm when scan is performed at with the tip 20 nm above the sample surface, with a tip diameter of 1 um. The noise level of the preamp circuit is $0.1e^{-}\sqrt{Hz}$. Future experiments using this setup would be to image electronic density profile in grapheme edges or quantum dots [5]. It could equally be used to obtain a high-resolution two-dimensional image of local change in capacitances in other materials. This research was supported by DoE DE-FG02-07ER46422 "Imaging Electrons in Graphene".

Acknowledgement

We would like to thank Dr. Estelle Cohen for contributing on ideas and assisting on sample fabrication. We also acknowledge Kevin Tian and John Free for providing helpful suggestions. DoE DE-FG02-07ER46422 "Imaging Electrons in Graphene" supported this research work.

References

[1] Topinka M A and Westervelt R M et al 2001 Nature **410** 183
[2] Ashoori R C 1996 Nature **379** 413
[3] Steele G A and Ashoori R C et al 2005 Phys. Rev. Lett. **95** 136804.
[4] Geim A K and Novoselov 2007 Nature Mat. **6** 183.
[4a] Berezovsky J and Westervelt R M et al 2010 Nanotech Special issue **21** 274014
[4] Boyd E E and Westervelt R M 2011 Phys. Rev. B **84** 205308
[5] Dean C R Young A F Meric I Lee C Wang L Sorgenfrei S Watanabe K Tanaguchi T Kim P Shepard K L and Hone J 2010 Nature Nanotech. **5** 10 pp 722-726.

Journal of Physics: Conference Series **568** (2014) 032006 doi:10.1088/1742-6596/568/3/032006

Vacuum properties of high quality value tuning fork in high magnetic field up to 8 Tesla and at mK temperatures

M Člověčko, M Kupka, P Skyba and F Vavrek

Centre of Low Temperature Physics, Institute of Experimental Physics SAS and P J Šafárik University Košice, Watsonova 47, Košice, Slovakia

E-mail: clovecko@saske.sk

Abstract. Tuning forks are very popular experimental tools widely applied in low and ultra low temperature physics as mechanical resonators and cantilevers in the study of quantum liquids, STM and AFM techniques, etc. As an added benefit, these forks being cooled, have very high Q-value, typically 10^6 and their properties seems to be magnetic field independent. We present preliminary vacuum measurements of a commercial tuning fork oscillating at frequency 32 kHz conducted in magnetic fields up to 8 T and at temperature $\sim$ 10 mK. We found an additional weak damping of the tuning fork motion depending on magnetic field magnitude and we discuss physical nature of the observed phenomena.

1. Introduction

Quartz tuning forks are versatile mechanical resonators with really high Q-values responsible for their superior sensitivity. No wonder that they are applied in all sub-fields of low temperature physics in the study of quantum liquids and solids, quantum turbulence, etc. [1, 2, 3, 4, 5, 6, 7, 8] and even as the important part in scanning probe techniques like AFM, STM, etc. [9, 10, 11]. In the latter case the fork measurements are often carried out in vacuum and at low temperatures including strong magnetic fields applied. Despite its obvious importance, the influence of strong magnetic fields on the vacuum resonant properties of such tuning fork at mK temperature range (to our knowledge) has not been studied yet. Therefore, the main aim of our work was to investigate the response of tuning fork under such conditions.

2. Experimental details

To perform our measurements a commercially available quartz tuning fork resonating at $\sim$ 32 kHz was used. In order to be able to measure the tuning fork in strong magnetic fields a few modifications have been made. Firstly, the metal can was removed and magnetic leads were replaced with non-magnetic ones (twisted thin copper wire pair in our case). Once out of the can, the dimensions of the tuning fork were measured by an optical microscope (see fig.1). Finally, the bare tuning fork was fixed on a specially designed copper holder and this whole setup was mounted on the cold finger of our cryogen-free dilution refrigerator Oxford Triton 200, which is capable to cool our samples down to 10 mK in the magnetic fields up to 8 T. The orientation of tuning fork's prongs was parallel to the applied magnetic field.

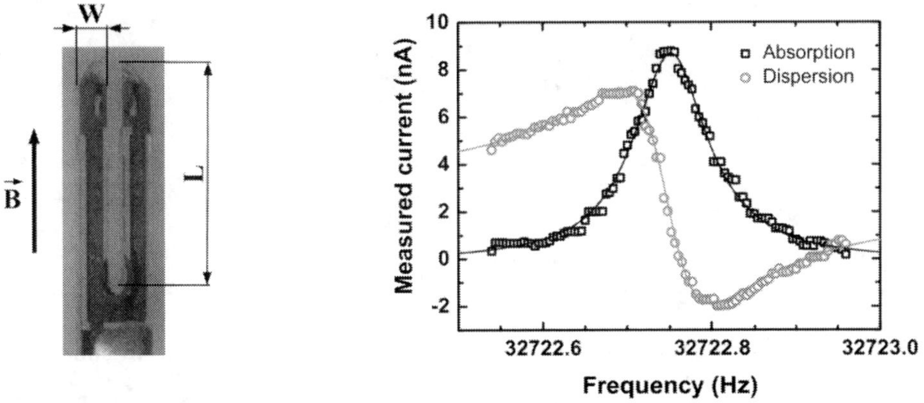

Figure 1. *Left:* 32 kHz quartz tuning fork used with the prong width $W = 400\ \mu$m and length $L = 3117\ \mu$m. The wafer thickness $T = 230\ \mu$m. *Right:* A typical frequency dependence of piezoelectric current (resonant curves) measured. Lines represents fits of experimental data with the resulting quality factor $Q \sim 5 \times 10^5$.

The response of tuning fork in the form of piezoelectric current on driving force was measured by verified technique [2, 12]. The driving force i.e. an AC-voltage slowly swept in frequency was provided by the function generator Agilent 33521A. An additional attenuator attenuated this excitation voltage by 40 dB. When driven with AC voltage at frequency close to the fork resonant frequency, the quartz crystal starts to oscillate and the piezoelectric current flows as a direct consequence of periodic changes of the crystal lattice polarization in time. This resulting piezoelectric current was detected and converted to voltage by home-made current-to-voltage (I/V) converter with gain 10^5 V/A [13]. The output voltage signal from I/V converter was measured by the phase-sensitive (lock-in) amplifier SR 830, which splits the measured signal into two phase components: in phase (absorption) component and quadrature (dispersion) component relative to the reference signal provided by above mentioned function generator.

The typical resonant curves are shown on the right side of fig. 1. Experimental data can be fitted using well known Lorentz relations

$$I_{\rm abs} = I_0 \frac{(f\Delta f)^2}{(f_0^2 - f^2)^2 + (f\Delta f)^2} \tag{1}$$

$$I_{\rm dis} = I_0 \frac{f\Delta f(f_0^2 - f^2)^2}{(f_0^2 - f^2)^2 + (f\Delta f)^2} \tag{2}$$

to obtain the resonant frequency f_0, the frequency linewidth Δf and the amplitude of piezoelectric current I_0. There are still some constant and linear backgrounds in actual measurements present, which were taken into account as well [12]. The piezoelectric current I_0 is directly proportional to the velocity v of prong tip of tuning fork. The constant of proportionality is the tuning fork constant α and can be determined experimentally using following expression [2]:

$$\alpha = \sqrt{\frac{2\, m_{\rm vac} \Delta\omega}{R}}. \tag{3}$$

Here $\Delta\omega = 2\pi\Delta f$ is the angular linewidth, R is the resistance of tuning fork at resonance, that models the damping of the fork motion and can be obtained from the fit of I_0 vs. the amplitude of driving AC voltage $U_{\rm exc}$. Finally, $m_{\rm vac} = 0.25\,\rho TWL$ is the effective mass of one

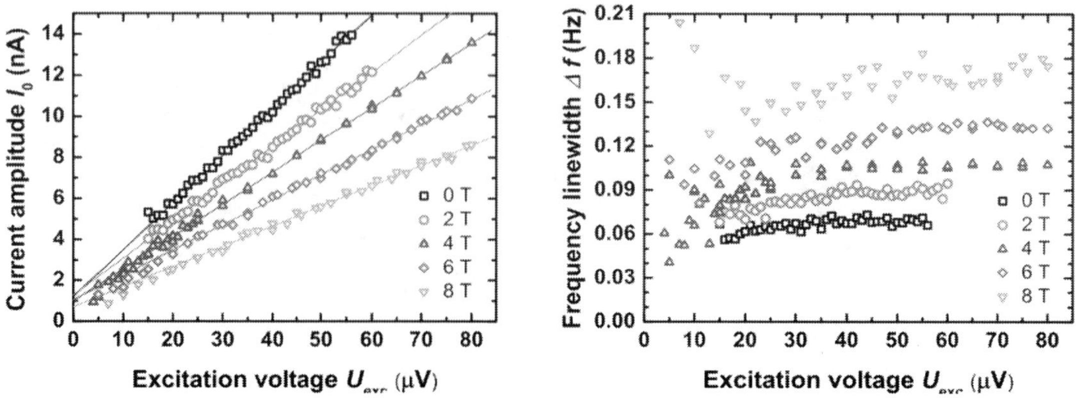

Figure 2. *Left:* Dependencies of the piezoelectric current amplitude I_0 on the excitation voltage V_{exc} and magnetic field applied. *Right:* Frequency linewidth Δf as a function of the excitation voltage V_{exc} and magnetic field applied.

fork's prong (quartz density $\rho = 2659$ kg.m^{-3}). The resulting effective mass for our tuning fork $m_{vac} = 1.90645 \times 10^{-7}$ kg.

3. Results and discussion

All measurements were performed in vacuum at $\sim$ 10 mK. Figure 2 shows dependencies of the piezoelectric current amplitude (left) and frequency linewidth Δf (right) on excitation voltage amplitude V_{exc} as measured at different values of magnetic field. The dependence of current amplitude I_0 on excitation voltage amplitude V_{exc} suggests that the tuning fork resistance R (i.e. the damping of the fork motion) is rising with the increase of magnetic field (the corresponding slope - the conductance showed in fig 2 is decreasing). Similarly, the frequency linewidth Δf is increasing as well. As measurements were carried out in vacuum, there is no influence of an external environment on the tuning fork motion and the tuning fork resistance in zero magnetic field R_0 reflects only an intrinsic damping processes. This intrinsic process of energy dissipation can be associated with shear friction as consequence of periodically bending crystal lattice of quartz.

The linear fits to the experimental data show a non-zero offset and there is a slight deviation of frequency linewidth Δf observed at small excitation voltages $U_{exc} \leq 30$ μV. Both these phenomena can be attributed to the contribution of the TTL logic (used as a reference signal for lock-in amplifier) to the excitation voltage U_{exc} due to the presence of capacitive coupling [12]. Thus, a non-zero piezoelectric current I_0 can also be detected even for zero excitation voltage amplitude U_{exc}. However, the corresponding signal measured by lock-in amplifier is continuously shifted from 0° to 90° in phase as $U_{exc} \to 0$ V. This also explains deviation of the measured linewidths Δf from constant value for low excitation voltage amplitudes.

The increase of the fork resistance R with the rising magnitude of magnetic field indicates the presence of of additional damping mechanism acting on tuning fork motion. Assuming that intrinsic damping process in zero field (R_0) i.e. the shear friction is magnetic field independent, a magnetic contribution to the tuning fork resistance R_{mag} can be simply estimated as $R_{mag} = R_B - R_0$. The fig. 3 illustrates the resulting dependence of R_{mag} on applied magnetic field. What could be an origin of the additional, the field depended damping in quartz tuning fork?

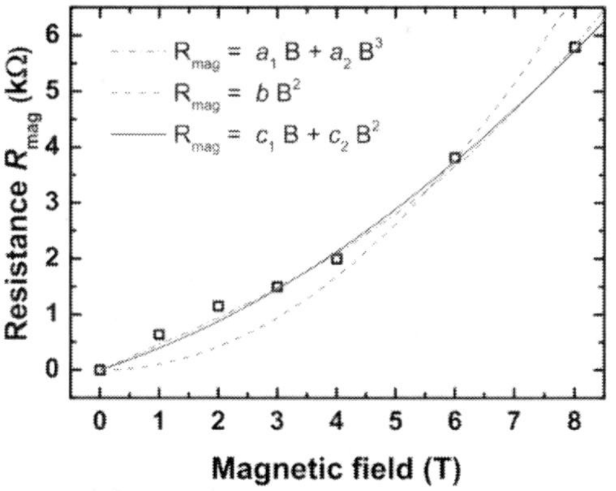

Figure 3. Magnetic contribution to the tuning fork resistance $R_{\rm mag}$ plotted against magnetic field applied. Presented fits correspond to dipole twist (— · —), pure eddy currents (– – –) and combination of both (——).

The forced motion of one prong of the tuning fork pointing in the direction of z axis in zero magnetic field is described by the Euler-Bernoulli equation

$$\rho A \frac{\partial^2 x}{\partial t^2} + 2\gamma\rho A \frac{\partial x}{\partial t} + E_m J_y \frac{\partial^4 x}{\partial z^4} = F(z)\cos(\omega t), \tag{4}$$

where A is the arm cross-section, ρ is the density, E_m is the Young's modulus of the material, J_y is the moment of inertia of the prong cross-section and γ characterizes the damping process. The external force $F(z)$ is provided by the AC-voltage applied on electrodes of the tuning fork and resulting harmonic electric field $\mathbf{E}$ creates the time dependent charge polarization $\mathbf{p}(t)$ of the crystal lattice of the tuning fork. This time dependent charge polarization $\dot{\mathbf{p}}(t)$ i.e. the piezoelectric current carries information about fork motion.

Once the magnetic field is applied on tuning fork, in general, there are two effects acting simultaneously on dynamics of the tuning fork motion. The first, in our geometry - magnetic field applied in parallel with prongs of the tuning fork, magnetic field affects the excitation force provided by electric field $\mathbf{E}$ by adding a force acting perpendicularly to the former one $(q\mathbf{E} + \dot{\mathbf{p}} \times \mathbf{B})$. This additional Lorentz force has a tendency to twist the oscillating dipole moments inside quartz crystal from the direction of electric field $\mathbf{E}$. This effect effectively reduces the magnitude of the piezoelectric current and thus increases the fork's resistance. The ratio $(\dot{\mathbf{p}} \times \mathbf{B})/q\mathbf{E}$ defines a tangent of an angle (θ), by which the oscillating dipoles are twisted. Then, using this simplest physical picture and assuming the smallness of θ, the magnetic contribution to the tuning fork resistance can be expressed in form $R_{\rm mag} = a_1 B + a_2 B^3$, where a_1 and a_2 are constants. The second effect is a damping associated with generation of the eddy currents in fork's metallic electrodes of during its oscillating motion. An EDAX analysis of the electrodes showed that electrodes consist of Cr, Ag and Sn. The damping caused by eddy currents is proportional to $b\,B^2$.

The lines in fig. 3 show the fits to the experimental data considering each mechanism (dipole twist and eddy currents) separately and acting together (without B^3 term). As follows from data fits the contribution due to the eddy currents does not fit experimental data properly, so additional contribution originating in dipole twist needs to be considered as well. Moreover, the dipole twist mechanism is capable to describe our experimental data by itself. Both above mentioned mechanisms should depend on relative orientation of the tuning fork and magnetic field, which could help to discriminate these two effects. However, the most important fact which comes out from data analysis is that in our configuration (tuning fork's prongs orientated in parallel with magnetic field), the magnitude of the α constant is almost independent on magnetic field within error of $\sim 5\%$ (see Table 1.).

Table 1. Values (and their errors) determined from experiment for frequency linewidth Δf, tuning fork resistance R and calculated values (errors) of fork constant α according to relation (3) for different magnetic fields applied.

B T	Δf Hz	Δf_{err} $\pm$ Hz	R Ω	R_{err} $\pm\,\Omega$	α A.s.m^{-1}	α_{err} $\pm$ A.s.m^{-1}
0	0.06894	0.00225	4370.59876	57.15386	6.14709×10^{-6}	1.08088×10^{-7}
1	0.07905	0.00237	5011.08829	76.42912	6.14755×10^{-6}	1.03427×10^{-7}
2	0.08885	0.00271	5523.22733	84.74155	6.20787×10^{-6}	1.05905×10^{-7}
3	0.09695	0.00261	5871.39784	39.92771	6.28943×10^{-6}	8.72780×10^{-8}
4	0.10649	0.00151	6369.90292	40.71070	6.32847×10^{-6}	4.91689×10^{-8}
6	0.13341	0.00165	8185.47428	85.27016	6.24880×10^{-6}	5.04657×10^{-8}
8	0.17520	0.00448	10164.13935	94.14669	6.42605×10^{-6}	8.73431×10^{-8}

4. Conclusions

We have presented that the fork constant α of our 32 kHz quartz tuning fork is almost independent on the magnetic field applied. This result can be of large importance for measurements performed in high magnetic fields, e.g. in various scanning probe techniques utilizing quartz tuning forks as probes. The origin of additional damping of tuning fork motion due to the presence of magnetic field is still an open question. To elucidate this problem more experiments with different relative orientation of tuning fork's prongs with respect to magnetic field are needed.

Acknowledgments

We would like to thank to V. Komanický for performing the spectral analysis of the tuning fork. CLTP is operated as the CFNT MVEP of the Slovak Academy of Sciences and P. J. Šafárik University. We acknowledge support provided by grants: APVV-0515-10, VEGA 2/0128/12, CEX-Extrem ITMS 26220120005 (SF of EU) and the Microkelvin, project of 7th FP of EU. Support provided by the U.S. Steel Košice s.r.o. is also very appreciated.

References

[1] Clubb D O, Buu O V L, Bowley R M, Nyman R, Owers-Bradley J R 2004 *J. Low Temp. Phys.* **136** 1
[2] Blaauwgeers R, Blažková M, Človečko M, Eltsov V B, de Graaf R, Hosio J, Krusius M, Schmoranzer D, Schoepe W, Skrbek L, Skyba P, Solntsev R E and Zmeev D E 2007 *J. Low Temp. Phys.* **146** 537
[3] Blažková M, Človečko M, Gažo E, Skrbek L, Skyba P 2007 *J. Low Temp. Phys.* **148** 305
[4] Blažková M, Človečko M, Eltsov V B, Gažo E, de Graaf R, Hosio J J, Krusius M, Schmoranzer D, Schoepe W, Skrbek L, Skyba P, Solntsev R E, Vinen W F 2000 *J. Low Temp. Phys.* **150** 525
[5] Pentti E M, Tuoriniemi J T, Salmela A J, Sebedash A P 2008 *Phys. Rev.* B **78** 064509
[6] Bradley D I, Fear M J, Fisher S N, Guenault A M, Haley R P, Lawson C R, McClintock P V E, Pickett G R, Schanen R, Tsepelin V, Wheatland L A 2009 *J. Low Temp. Phys.* **156** 116
[7] Schmoranzer D, La Mantia M, Sheshin G, Gritsenko I, Zadorozhko A, Rotter M, Skrbek L 2011 *J. Low Temp. Phys.* **163** 317
[8] Ahlstrom S L, Bradley D I, Človečko M, Fisher S N, Guenault A M, Guise E A, et al 2014 *J. Low Temp. Phys.* **175** 140
[9] Gunther P, Fischer U Ch, Dransfeld K 1989 *Appl. Phys.* B **48** 89
[10] Rychen J, Ihn T, Studerus P, Hermann A, Ensslin K 1999 *Rev. Sci. Instrum.* **70** 2765
[11] Rychen J, Ihn T, Studerus P, Hermann A, Ensslin K, Hug H J, van Schendel P J A and Guntherod H J 2000 *Rev. Sci. Instrum.* **71** 1695
[12] Skyba P 2010 *J. Low Temp. Phys.* **160** 219
[13] Holt S, Skyba P 2012 *Rev. Sci. Instrum.* **83** 064703

STRENGTH OF NANOSTRUCTURED AUSTENITIC STEEL 316LN AT CRYOGENIC TEMPERATURES

P. Czarkowski[#], A.T. Krawczyńska, T. Brynk, M. Nowacki, M. Lewandowska, K.J. Kurzydłowski

Warsaw University of Technology, Faculty of Materials Science and Engineering

Wołoska 141, Warsaw, 02-507, Poland

[#] e-mail: pawel.czarkowski@inmat.pw.edu.pl

Abstract. The aim of this work was to investigate the effect of nano-refinement on the properties of austenitic steel. The material with the initial grain size of 40-50μm was subjected to hydrostatic extrusion at a room temperature to the total accumulated strain exceeding 1. The microstructure developed was investigated by Transmission Electron Microscopy (TEM) and Focus Ion Beam (FIB). The strength of the extruded samples was tested at 293K, 77K and 4.2K by means of cryostat for static tensile tests. The results show that the hydrostatically extruded steel 316LN has excellent strength in cryogenic conditions, which make this material interesting for applications in cryogenic devices.

1. Introduction

Austenitic stainless steels are common structural materials for cryogenic elements such as elements of fusion devices [1, 2]. Although strength of austenitic steels increases considerably with a decrease of temperature, it is still insufficient for many applications, and other alloys, such as Ni-Cr superalloys, are also used. Components made of dissimilar materials thermally deform in less uniform way what leads to disturbances of stress field. It justifies efforts to increase strength of austenitic steels which can be substantially improved by microstructure refinement down to the nanometre scale. Studies of nanocrystalline and nanostructured austenitic steels tested at room temperature [e.g. 3-5] have shown that its strength can achieve values higher than 1300 MPa. The aim of the present study was to investigate mechanical properties of nano-austenitic stainless steel at cryogenic conditions relevant for fusion applications, following the results obtained for pure metals in [6, 7].

2. Material and methods

The material studied was 1.4429 stainless steel with chemical composition given in Tab. 1. Billets with initial diameter of 12 mm, 9 mm and 7 mm were subjected to Hydrostatic Extrusion, HE, with total accumulated strain of 0.25, 0.84 and 1.36 (denoted as HE_0.25, HE_0.84 and HE_1.36). As has already been reported in [5] hydrostatic extrusion with the total accumulated strain exceeding 1.0 transforms standard austenitic stainless steel into nano-twinned with the size of twins in the range below 100 nm. In the context of fusion applications it can be noted that the extrusion can be used to fabricate such parts as bolts, pins and shafts [8, 9] which in fusion devices transmit heavy loads [10].

Ni	Cr	Mo	Mn	Cu	N	C	Si	P	S
12.6	17.2	2.66	1.15	0.15	0.15	0.023	0.53	0.065	0.013

Table 1 Chemical composition (wt. %) of investigated steel 1.4429

The microstructures of the processed billets have been investigated by Transmission Electron Microscopy[1] and Focus Ion Beam Microscopy[2] using electrochemically polished samples. A representative TEM image of the material with total strain of 1.36 is shown in Figure 1. It can be noted that this microstructure consists of shear bands and deformation twins with thickness of tens of nanometers (Fig. 1). The volume fraction of twins was assessed by analysis of images obtained by FIB microscopy employing the channelling contrast. The results are plotted as an insert in Fig. 2 against microhardness of the HE processes samples in room temperature[3]. Assuming that the density of twins in the as-received samples is negligible, a nearly linear dependence of the resistance to the plastic flow on the density of twin is observed as already reported for fcc NiMn alloy in [11]. A profound increase in dislocations density is also observed, as recovery of processed material is limited due to very high strain rate (about 100 1/s).

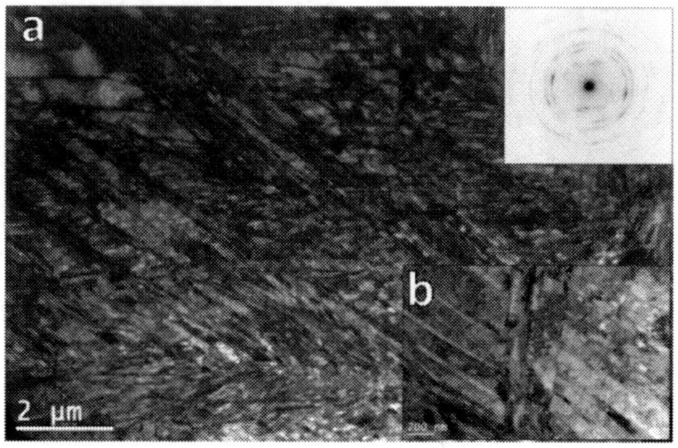

Fig. 1 Steel 1.4429 processed by hydrostatic extrusion with total strain of 1.36: typical microstructure (a) and nanotwins cut by shear band (b) - transmission electron microscopy

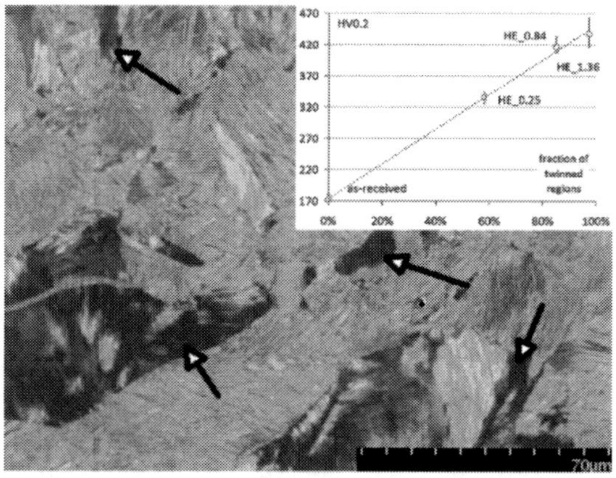

Fig. 2 Microstructure of steel 1.4429 with total strain of 0.84 revealed by Focused Ion Beam microscopy. Examples of not-twinned region indicated by arrows. The insert shows twinned regions fractions and hardness for subsequent steps of extrusion.

Since some austenitic steels undergo partial phase transformation into ferromagnetic α'-martensite at room temperature during severe plastic deformation, it should be noted that in the current case no effects of $\gamma \rightarrow \alpha'$ transformation have been revealed by means of Hall sensor. The fact that HE material remains paramagnetic is important if the material is foreseen to work in the strong magnetic field.

Strength of hydrostatically extruded steel 1.4429 was measured using standard M6 specimens with gauge diameter of 3 mm and parallel length 20 mm, submerged in liquid helium inside a vacuum insulated cryostat adapted for tensile tests. The straining temperature was about 4.2 K. The initial strain rate was 10^{-3} s^{-1}.

[1] JEOL JEM 1200X microscope, acceleration voltage 120kV,
[2] Hitachi NB5000 microscope, Ga ions, acceleration voltage 40kV
[3] Vickers microhardness determined using an Anton Paar MHT10 unit under load of 200 g

3. Results and discussion

Representative engineering stress-strain curves obtained at 4.2 K are presented in Fig. 3. All batches reveal serration effect, therefore increase of strength can be assessed using discontinuous yield strength (DYS) value, which corresponds to the stress level at the top of first relaxation tooth. Comparing the curves obtained for HE_1.36 and as-received material, one can note 80% growth of DYS up to about 2.6 GPa.

As expected, the increase in the strength is accompanied by decrease in ductility caused by early localization of plastic deformation. However, scanning electron microscopy observations of the fracture surfaces (Fig. 4) reveal that both as-received and strengthened materials have similar pattern of fracture at 4.2 K, with the main difference between these two in the ratio of the surface of the outer shear lips to inner fibrous zone. As a general rule, a shear zone becomes larger with the increase in the level of cold work and with the decrease in temperature. It can be concluded that HE processed samples in cryogenic temperatures are not inherently brittle and exhibit small strain to failure because of the strong tendency for strain localization. A strong tendency for strain localization within the neck during deformation of the HE material is caused by the limited ability of the material for further propagation of the plastic flow due to saturation of deformation mechanisms.

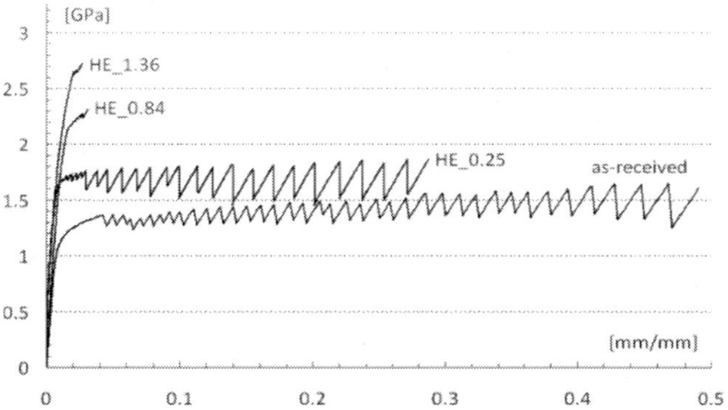

Fig. 3 Engineering stress-strain curves of steel 1.4429 at 4.2K

Bearing in mind fusion applications, the current study addressed also thermal expansion of the HE processed billets, as this parameter is of prime importance for cryogenic devices. According to the literature, the difference in thermal expansion coefficient between micro and nano-crystalline materials might be observed, mostly due to an increased volume fraction of grain boundaries [12-14]. On the other hand, in the currently studied material the population of the grain boundaries is dominated by twin boundaries, which should not change the thermal expansion. In order to measure the thermal expansion coefficient of the HE processed billets an experimental set up with attached strain gauges circuit was assembled which is described in more detail in [15]. The set-up was cooled down by immersion in liquid nitrogen, so the temperature difference employed in the measurements was 216 K. The corresponding difference in the value of the thermal expansion coefficient between as-received state and HE_1.36 material was measured as 0.24e-6 [1/K] with an experimental error 0.12e-6 [1/K]. The literature

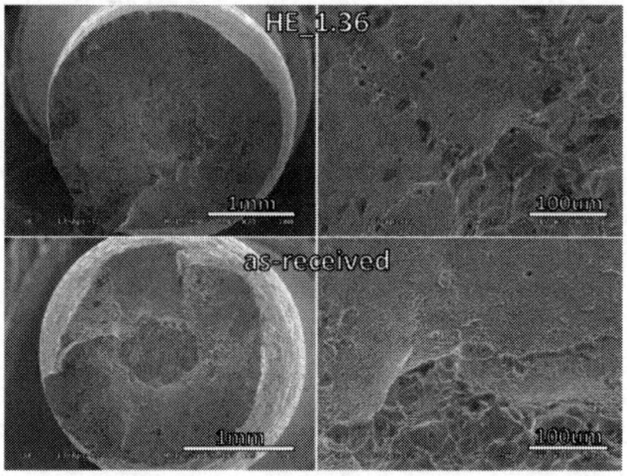

Fig. 4 SEM images of fractures of HE_1.36 and as-received material after tests at liquid helium temperature

data indicate that the thermal expansion coefficient for the investigated steel is ~12.8e-6 [1/K]. Therefore one can conclude, that hydrostatic extrusion does not change thermal expansion of steel 1.4429 considerably.

Summarizing, excellent strength and limited content of ferromagnetic martensite phase are the main advantages of hydroextruded steel 1.4429, which makes this material competitive to high strength alloys used at cryogenic conditions.

References

[1] R.L. Tobler, A. Nishimura, J. Yamamoto, Design-relevant mechanical properties of 316-type steels for superconducting magnets, Cryogenics 37 (1997) 533-550
[2] W. Daenner et al., Cryogenic materials database for the W7-X magnet system, Cryogenics 50 (2010) 72
[3] S.V. Dobatkin, O.V. Rybalchenko, G.I. Raab, Strucutre formation, phase transformations and properties in Cr-Ni austenitic steel after equal-channel angular pressing and heating, Materials Science and Engineering A 463 (2007) 41-45
[4] O.V. Rybalchenko et al., Strength of ultrafine-grained corrosion resistant steels after severe plastic deformation, Materials Science and Engineering A 387-389 (2004) 244-248
[5] P. Czarkowski et al., Low temperature mechanical properties of 316L type stainless steel after hydrostatic extrusion, Fusion Engineering and Design vol. 86 (2011) 2517-2521
[6] Y. Wang et al., Tough nanostructured metals at cryogenic temperatures, Advanced Materials 16 (2004) 328-331
[7] Y.M. Wang, E. Ma, On the origin of ultrahigh cryogenic strength of nanocrystalline metals, Applied Physics Letter 85 (2004) 2750-2752
[8] http://www.hydroextrusion.pl; 17.06.2014
[9] Klaus Mueller et al., Fundamentals of extrusion technology, Giesel Verlag GmbH 2004
[10] M. Gasparotto et al., The WENDELSTEIN 7-X mechanical structure support elements: Design and tests, Fusion Engineering and Design 74 (2005) 161-165
[11] J. Bystrzycki, K.J. Kurzydłowski, W. Przetakiewicz, On the geometry of twin boundaries and their contribution to the strengthening and recovery of FCC polycrystals, Materials Science and Engineering A225 (1997) 188-195
[12] T. Turi, U. Erb, Thermal expansion and heat capacity of porosity-free nanocrystalline materials, Materials Science and Engineering A204 (1995) 34-38
[13] Y. Tang, et al., Thermal expansion of a composite of single-walled carbon nanotubes and nanocrystalline aluminium, Carbon vol. 42 (2004) 3260-3262
[14] H.J. Klam, H. Hahn, H. Gleiter, The thermal expansion of grain boundaries, Acta Metallurgica vol. 35 (1987) 2101-2104
[15] Vishay Precision Group, Measurement of Thermal Expansion Coefficient Using Strain Gages, technical note TN-513-1

Acknowledgements

This work was carried out under the auspices of a NANOMET project financed by the European Regional Development Fund (contract No. POIG.01.03.01-00-015/08). The hydrostatic extrusion was carried out at the Institute of High Pressure Physics of the Polish Academy of Sciences within the scope of the project coordinated by the Faculty of Materials Science and Engineering of Warsaw University of Technology.
TERMET project financed by the European Regional Development Fund (contract No. POIG.01.01.02-00-09709) is gratefully acknowledged.
Authors would like to thank colleagues from Institute of Plasma Physics in Greifswald, Germany for supply of annealed state of steel 1.4429.

Top-loading small-sample calorimeters for measurements as a function of magnetic field angle

N A Fortune[1], **S T Hannahs**[2]

[1] Department of Physics, Smith College, College Lane, Northampton MA 01063, USA
[2] National High Magnetic Field Laboratory, Florida State University, 1800 E. Paul Dirac Road, Tallahassee FL 32310, USA

E-mail: nfortune@smith.edu

Abstract. In quasi-low-dimensional systems, the existence of a particular physical state and the temperature and magnetic-field-dependence of its phase boundary often strongly depends on magnetic field orientation. To investigate magnetic field orientation dependent phase transitions in these materials, we have developed rotatable miniature and sub-miniature sample-in-vacuum calorimeters that operate in dc magnetic fields up to 18 and 45 tesla. The calorimeters cover the temperature range from below 0.1 K to above 10 K; they are able rotate a full 360 degrees relative to the applied magnetic field while remaining at base temperature. Samples are typically on the order of 1 mg in mass and up to 2 mm^2 x 0.5 mm in volume.

1. Introduction

The ability to rotate a sample relative to an applied magnet field offers a number of advantages for high field, low temperature specific heat and magnetocaloric measurements, including (1) more accurate alignment of the field along one or more crystal axes [1, 2, 3], (2) the opportunity to investigate strongly field-orientation-dependent novel superconducting and magnetic phases [4, 5], and (3) a method of establishing the gap structure of anisotropic superconducting ground states [6, 7, 8]. Field-orientation-dependent calorimetric measurements are especially useful for the study of layered structure and/or quasi-low dimensional materials. Recent examples include layered-structure heavy-fermion superconductors [2, 9, 10, 11, 12] and quasi-2D molecular superconductors [13, 14, 15]. In our experiments, we use a pair of calorimeters designed to fit inside the National High Magnetic Field Laboratory (NHMFL) top-loading dilution refrigerator single-axis rotating sample probes [16]. The dilution refrigerators have a cooling power of approximately 400μW at 0.1 K and a maximum cooling power of 4 mW; the very-low friction rotating stages at the bottom of the probes allow for a full 360 degrees of rotation (with a resolution of 0.020°) at base temperature. The SCM1 dil fridge incorporates an 18 T superconducting magnet; the PDF portable dilution refrigerator fits inside the NHMFL's 32 mm bore 36 T resistive and 45T hybrid magnets. The calorimeters can be used for field-dependent magnetocaloric [17], ac calorimetric [18] and thermal-relaxation calorimetric [19] measurements at temperatures between 0.1 K and 10 K. Samples are typically on the order of 1 mg in mass and up to 2 mm^2 x 0.5 mm in volume.

Journal of Physics: Conference Series **568** (2014) 032008 doi:10.1088/1742-6596/568/3/032008

2. Calorimeter Design and Performance

2.1. Vacuum constraints

Our measurements require the sample to be in vacuum even though the housing is immersed in liquid helium (and is exposed to He vapor during the entire loading process). We accomplish this by compressing an indium gasket between matching step edges in the calorimeter base and cap [2] at room temperature, resulting in a superfluid-He leak-tight seal. A photo of the interior of a PDF calorimeter cap (with indium gasket resting on the interior step edge) is shown in figure 1, a top view of the calorimeter assembly, platform, base and step edge is shown in figure 2, and a close up view of the calorimeter platform and sample assembly is shown in figure 3. A locknut threads into the cap behind the base, drawing the pieces together as it advances.

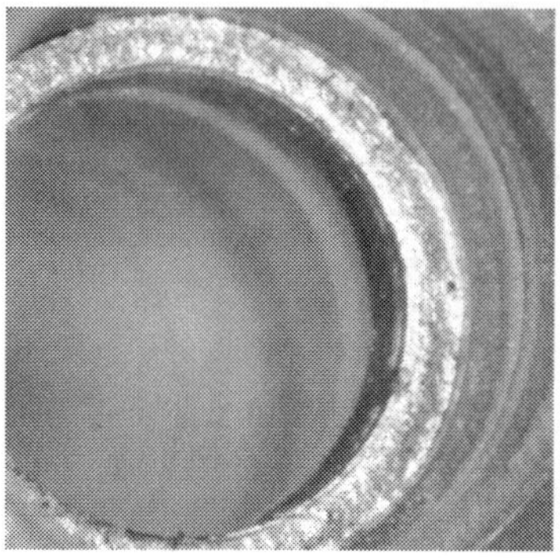

Figure 1. PDF calorimeter cap interior, showing sample chamber, step edge, indium gasket, and internal threads.

Figure 2. Top view of temperature regulated platform, and Ag sample thermometer/ spacer, and heater frames.

Figure 3. side view of PDF calorimeter platform and the heater/sample/thermometer sample assembly (& frames).

A vacuum is formed below 30 K by cryopumping of the air inside the housing [1]. The absence of a leak at low temperature is readily checked by measuring the heater power required to raise the interior temperature controlled platform well above that of the external cryogenic bath.

2.2. Space constraints

The calorimeter needs to (1) fit within the rotator body and (2) freely rotate 360° inside the mixing chamber of the dilution refrigerator. To accomplish this, we require that no part of the calorimeter (including leads) extend beyond the outer radius of the tip of the top loading probe at any point during the rotation. For a cylindrical calorimeter of diameter ϕ and length L rotating about a horizontal axis within a vertically aligned cylinder of diameter d, this requires that $L < \sqrt{d^2 - \phi^2}$. For the PDF rotator probe (with $\phi = 7.5$mm), $L \leq 12$mm (including room for leads). The alignment of the calorimeter within the rotator body can checked by sliding a tight-fitting clear plastic tube over the probe end and rotating the calorimeter 360°.

Once the external diameter is known, the largest possible inner diameter for the calorimeter cap is set by the minimum wall thicknesses of the indium gasket without fracturing of the housing or threads. For our PDF calorimeter constructed out of Stycast 1266 epoxy, the inner diameter of the cap is 4.2 mm (0.165 in). In practice, this limits the effective diameter of the PDF calorimeter sample assembly, supporting frames, and calorimeter platform (including sensors and wiring) to 3.5 mm (0.140 in), half what is available in the SCM calorimeter [2]. In that earlier design, the sample assembly is attached to a single 2.5 mm inner diameter (3.5 mm outer diameter) circular saphhire frame, surrounded by electrical leads heat sunk to the 7 mm diameter Ag platform. The reduced interior space available in the PDF calorimeter prompted us to try adifferent design: in this new calorimeter, we mount each component of the sample assembly sample heater, sample, and sample thermometer on a separate frame with integrated electrical contacts, then flip the orientation of the sample assembly by 90° relative to the calorimeter's longitudinal axis, as shown in figure 3. This gives us a 2 mm diameter sample space (compared to 2.5 mm before); the 3 frame design can accommodate samples up to 0.5 mm in thickness.

2.3. Thermal constraints

In these calorimeters, the sample heater, sample, and thermometer are weakly thermally linked through two sets of electrical leads to a temperature controlled platform (inside the vacuum space) as shown in figure 3; the phosphor bronze electrical leads and stainless steel support post passing through the base of the calorimeter serve as the thermal link between the platform and the cryogenic bath. The platform heater power P_{PH} required to raise the platform temperature T_s a specific amount above the bath temperature T_b depends on the mean thermal conductance $\bar{K}$ of the platform to bath thermal link over that temperature range:

$$P_{PH} = \int_{T_b}^{T_p} K(T)\, dT = \bar{K}\Delta T \tag{1}$$

where $K(T)$ depends on the choice of materials and T_b depends on the fridge's cooling power.

From a practical standpoint, there are 3 key constraints on P_{PH} and hence $K(T)$: (1) stable PID temperature regulation for $T_p \geq 0.1$ K, (2) quick recovery of the dilution refrigerator for $T_p \leq 4$ K and (3) the ability to reach $T_p = 10$ K when neeed. In our operating environment, this requires $P_{PH} \gtrsim 20$nW at $T_p = 0.1$ K, $P_{PH} \leq 400\mu$W at $T_p = 4$ K, and $P_{PH} \leq 4$mW at $T_p = 10$ K. As shown in figure 4, the measured dependences of T_p on P_{PH} satisfy these 3 key constraints for both calorimeters. Moreover, by scaling the cross-sectional area of the leads passing through the calorimeter base to match the reduction in calorimeter length, we are also able to closely match the qualitiative and quantitative dependence of T_p on P_{PH} for the two calorimeters. We find that in both cases, T_p is well-described by the following power law function

$$T = T_0 + A(P_0 + P)^n \tag{2}$$

with $n \lesssim 0.5$. A and n depend only on the calorimeter; P_0 and T_0 also depend on the fridge.

3. Temperature Calibration in Magnetic Field

Prior to cross-calibration, the sample and platform thermometers were repeatedly thermally cycled between room temperature and 4.2 K (after mounting inside the calorimeter). Thermal cycling was deemed complete once both the room temperature and low temperature equilibrium values remained unchanged for 10 consecutive cycles (for a total of 60 cycles). The thermometers were cross-calibrated in zero field against a LakeShore CX1010SD Cernox between 0.1 K and 10

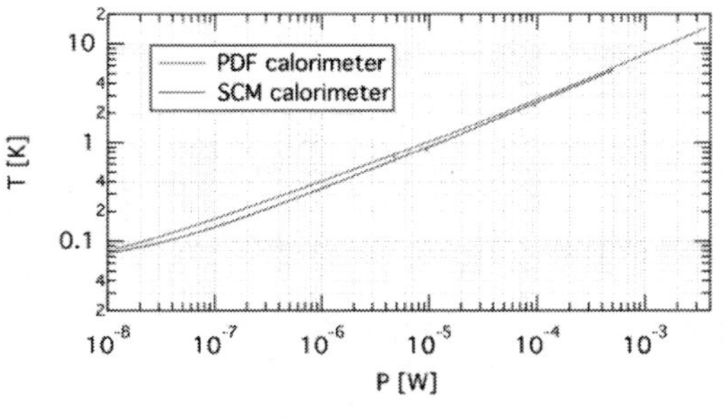

Figure 4. Platform temperature T_p as a function of platform heater power P_{PH} for the 45 tesla PDF and 18 tesla SCM field-rotatable small sample calorimeters. The calorimeters are designed for measurements between 0.1 K and 10 K. A heater power of 400 μW brings the interior to 4.2 K while the mixing chamber remains below 100 mK, allowing for easier and faster temperature regulation of the platform.

K and an Entropy Cryogenics Type D RuOx between 0.050 K and 7.5 K. This data was used to fit the temperature dependence of each resistive sensor to the Chebyshev function

$$\log R(T, B) = \sum_{n=0}^{N} c_n(B) t_n(x) \qquad (3)$$

for $B = 0$, where x is a function of $\log T$ [20]. A set of platform heater power sweeps in fixed magnetic fields (starting from the same refrigerator base temperature) then allowed us to determine the fit coefficients A, n, T_0, and P_0 in Eq. 2 and the magnetic field dependence of $c_n(B)$ in Eq. 3. Thermomolecular-corrected measurements of ^{3}He vapor pressure [21, 22, 23] and sensor resistance were used to test the calibrations in liquid ^{3}He between 0.4 K and 1.4 K in magnetic fields up to 36 T.

4. Experimental Results

Angle-dependent calorimetric measurements by our group were the first to demonstrate the existence of a field-induced phase transition into a magnetically modulated phase within the superconducting state of the layered-structure heavy fermion $CeCoIn_5$ [2, 9].

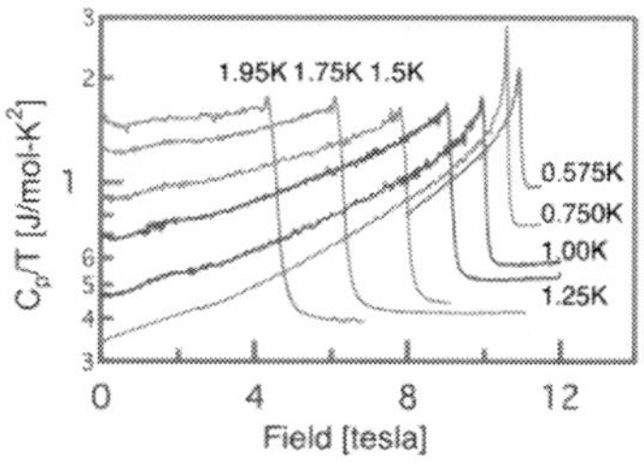

Figure 5. Field dependence of the specific heat of $CeCoIn_5$ above 0.5 K for H parallel to the highly conducting plane.

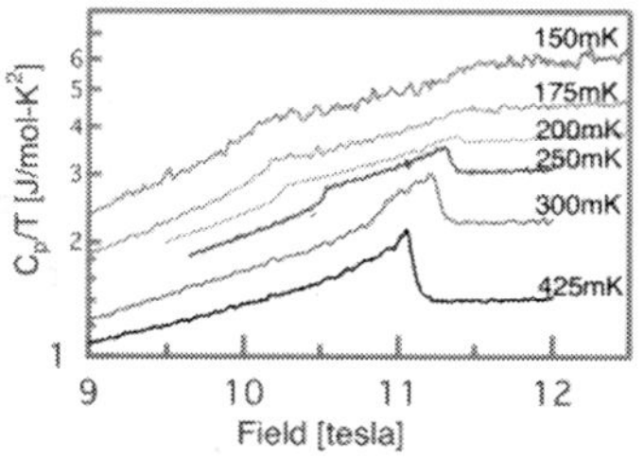

Figure 6. Lower temperature data revealing a field-induced transition to a new superconducting state below 0.35 K.

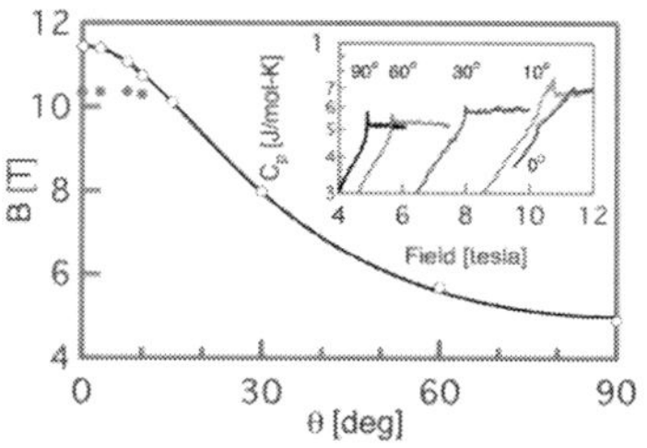

Figure 7. Magnetic-field orientation dependence of the high field phase for rotation out of the conducting plane.

As shown in Figs. 5, 6, and 7, the transition to the field-induced superconducting phase occurs for a narrow range of magnetic field at temperatures below 0.35 K when the field is directed along the [110] axis (parallel to the layers), but is quickly extinguished as the magnetic field is rotated out of the plane towards [001]. This may represent 'FFLO' enhancement of superconductivity due to the formation of electron spin domains; another interpretation attributes this field-angle-dependent phase transition to the onset of spin-density wave ordering [12].

More recently, we have investigated the field-angle dependence of magnetic-field-induced phases in a number of frustrated triangular-lattice $S = \frac{1}{2}$ quantum Heisenberg antiferromagnets with weak xy anisotropy, including $Ba_3CoSb_2O_9$, Cs_2CuCl_4 and Cs_2CuBr_4 [24]. Three magnetic phases are expected [25], each corresponding to a different spin arrangement: a low field 'Y' phase, an intermediate 'up - up - down' phase, and a high field 'V' phase. In angle-dependent measurements up to 36 T on $Ba_3CoSb_2O_9$, we have discovered (1) that the 'Y' and 'V' phases split into alternating and non-alternating co-planar subphases, (2) that contrary to previous reports, multiple field-induced phases occur for H $\|$ c as well, and (3) that the field-induced phases seen for these two orientations are in fact related to each other. The angle-dependence of these field-induced antiferromagnetic phases at 0.350 K is shown in Fig. 8. Details will be published elsewhere.

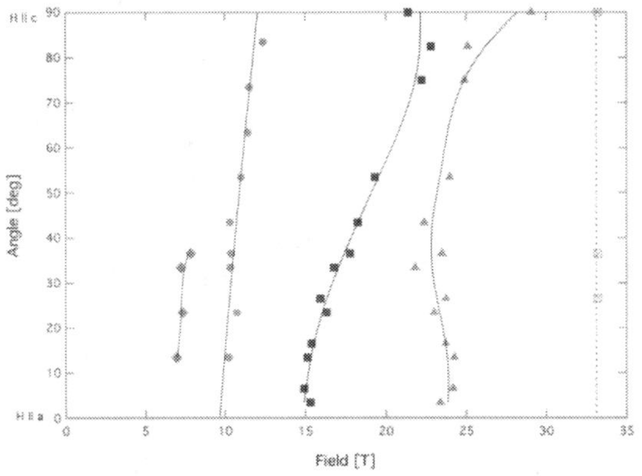

Figure 8. The angle-dependence of field-induced antiferromagnetic phases for field rotation from H||a ($\theta = 0°$) to H||c ($\theta = 90°$) for $Ba_3CoSb_2O_9$, an $S = \frac{1}{2}$ triangular-lattice Heisenberg antiferromagnet. The lowest and highest phases of the expected three phases — a 'Y' phase, an 'uud' phase, and a 'V' phase — split into alternating and non-alternating coplanar subphases in this system. The solid lines are guides to the eye indicating the evolution of each phase with field-angle; the dashed line represents the saturation field of 32.5 tesla.

Acknowledgments
A portion of this work was performed at the National High Magnetic Field Laboratory, which is supported by National Science Foundation Cooperative Agreement No. DMR-1157490, the State of Florida, and the U.S. Department of Energy. NAF was supported by NHMFL Visiting Scientist Grant 12411.

References
[1] Bondarenko V A, Tanatar M A and Kovalev A E 2000 *Review Of Scientific Instruments* **71** 3148
[2] Hannahs S T and Fortune N A 2003 *Physica B-Condensed Matter* **329-333** 1586–1587
[3] Deguchi K, Ishiguro T and Maeno Y 2004 *Review Of Scientific Instruments* **75** 1188
[4] Alicea J, Chubukov A V and Starykh O A 2009 *Physical Review Letters*
[5] Agosta C C, Jin J, Coniglio W A, Smith B E and Cho K 2012 *Physical Review B*
[6] Vorontsov A and Vekhter I 2007 *Physical Review B* **75** 224501
[7] Boyd G, Hirschfeld P, Vekhter I and Vorontsov A 2009 *Physical Review B* **79** 064525
[8] Das T, Vorontsov A B, Vekhter I and Graf M J 2013 *Physical Review B* **87** 174514
[9] Radovan H A, Fortune N A, Murphy T P, Hannahs S T, Palm E C, Tozer S W and Hall D 2003 *Nature* **425** 51–55
[10] Deguchi K, Mao Z, Yaguchi H and Maeno Y 2004 *Physical Review Letters* **92** 047002
[11] An K, Sakakibara T, Settai R, Onuki Y, Hiragi M, Ichioka M and Machida K 2010 *Physical Review Letters* **104** 037002
[12] Tokiwa Y, Bauer E D and Gegenwart P 2012 *Physical Review Letters* **109** 116402
[13] Kovalev A E, Ishiguro T, Yamada J, Takasaki S and Anzai H 2001 *Journal of Experimental and Theoretical Physics* **92** 1035–1037
[14] Malone L, Taylor O J, Schlueter J A and Carrington A 2010 *Physical Review B* **82** 014522
[15] Beyer R and Wosnitza J 2013 *Low Temperature Physics* **39** 225–231
[16] Palm E C and Murphy T P 1999 *Review Of Scientific Instruments* **70** 237–239
[17] Fortune N A, Eblen M A, Uji S, Aoki H, Yamada J, Tanaka S, Maki S, Nakatsuji S and Anzai H 1999 *Synthetic Metals* **103** 2078–2079
[18] Sullivan P and Seidel G 1968 *Physical Review* **173** 679–685
[19] Bachmann R, DiSalvo F J, Geballe T H, Greene R L, Howard R E, King C N, Kirsch H C, Lee K N, Schwall R E, Thomas H U and Zubeck R B 1972 *Review Of Scientific Instruments* **43** 205–214
[20] Fortune N A, Gossett G, Peabody L, Lehe K, Uji S and Aoki H 2000 *Review Of Scientific Instruments* **71** 3825–3830
[21] Rusby R L 1985 *Journal of Low Temperature Physics* **58** 203–205
[22] Rusby R L and Durieux M 1984 *Cryogenics* **24** 363–366
[23] Roberts T R and Sydoriak S G 1956 *Physical Review* **102** 304–308
[24] Fortune N, Hannahs S T, Yoshida Y, Sherline T, Ono T, Tanaka H and Takano Y 2009 *Physical Review Letters* **102** 257201–257204
[25] Chubukov A V and Golosov D I 1991 *Journal Of Physics-Condensed Matter* **3** 69–82

Ultra-low noise HEMTs for high-impedance and low-frequency preamplifiers: realization and characterization from 4.2 K to 77 K

Y Jin, Q Dong, YX Liang, A Cavanna, U Gennser, L Couraud, C Ulysse
CNRS/LPN, route de Nozay, 91460 Marcoussis, France

yong.jin@lpn.cnrs.fr

Abstract. We report on the experimental results of specially designed HEMTs made at CNRS/LPN. These HEMTs, with a resistance input and different capacitance inputs, have been characterized from 4.2 K to 77 K with a power consumption of 100 μW. At 4.2 K, the lowest input noise voltage, 6 $nV/Hz^{1/2}$ at 1 Hz, has been obtained with the HEMT having the largest input capacitance; the lowest input noise current of about 3 $aA/Hz^{1/2}$ at 1 Hz has been observed with the HEMT with the smallest input capacitance; and the white noise voltage in these HEMTs is of about 0.2 $nV/Hz^{1/2}$. By increasing the temperature from 4.2 K to 77 K, noise voltage and noise current increase, but their values are limited within a factor of about 3 compared with their lowest values at 4.2 K. Our results show that the HEMT can be a promising transistor to fill the gap for FETs below 100 K for high-impedance and low-frequency readout electronics.

1. Introduction

High impedance ultra sensitive sensors operate at few tens of mK in order to avoid the thermal noise perturbation. High performance low-frequency readout electronics have been based for decades on Si JFETs. However, their operating temperature is limited to above about 100 K because of the charge freeze-out. Consequently, a long cable is required between readout electronics and sensors, which degrades the sensors intrinsic performance and the readout rate. From the intrinsic point of view, the charge freeze-out can be avoided by using a degenerate electron gas. Currently, there are two types of FETs available to operate at very low-temperature: MOSFETs and HEMTs. It is well known that the MOSFET suffers from the effects of an extremely high low-frequency noise due to the oxide layer between the metal gate and the active conducting channel [1].

The HEMT is based on a 2DEG (Two Dimensional Electron Gas), which is realized in a heterostructure with a high purity material interface. In particular at cryogenic conditions, high electron mobility can be obtained and it has been widely used for mesoscopic field-effect devices operating at tens of mK for quantum coherent electron transport investigations, as well as for the demonstration of a fully ballistic FET [2]. However, for cryogenic readout electronics, commercially available HEMTs are used in a frequency range above a few hundreds of kHz, and suffer a relative high noise current and especially a large low-frequency noise or an $1/f$ noise. Based on our early investigations [3], one $1/f$ noise source in the HEMT at low temperature has been found out [4]. In this work, we show that specially designed HEMTs can reach unprecedented low noise values at low frequencies and deep cryogenic conditions [5].

2. Experiments

The HEMTs in this work are based on an AlGaAs/GaAs heterostructure grown by MBE (Molecular Beam Epitaxy). It consists of a GaAs buffer layer, a 20 nm AlGaAs spacer layer, a Si δ-doping layer, then a 15 nm undoped AlGaAs barrier layer, and finally a 6 nm undoped GaAs cap layer. At 4.2K, the 2DEG carrier concentration and mobility are 4.5×10^{15} m^{-2} and 29 $m^2V^{-1}s^{-1}$, respectively. HEMTs with various gate lengths and gate widths are fabricated and individually packaged in ceramic SOT23 as shown in figure 1. In this work, experimental results are based on the HEMTs with three gate surfaces of 6.4×10^4, 2.0×10^4 and 2.0×10^3 μm^2, respectively.

Details of the characterization method can be found in [5]. As usual, a real FET is considered as a noiseless transistor with two noise sources in its input, i.e., the noise-voltage e_n, which is the lowest noise level of a FET, and the noise-current i_n (see figure 1). With an input impedance z, i_n induces a noise voltage $e_{ni} = i_n \times |z|$. In practice, only the equivalent total input noise voltage e_{nt} can be measured directly. By supposing that e_n and i_n are uncorrelated, e_{nt} can be expressed as:

$$e_{nt} = \sqrt{e_n^2 + e_{ni}^2} = \sqrt{e_n^2 + i_n^2 z^2}\ . \qquad (1)$$

e_n can be obtained with a sufficiently small input resistor R_{input} or a large enough input capacitance C_{input}; i_n can thus be grounded, and consequently $e_{ni} \ll e_n$, and $e_{nt} = e_n$. The determination of i_n needs an $|z|$ big enough to have $e_{nt} > e_n$ and then i_n can be deduced from:

$$i_n = \sqrt{e_{nt}^2 - e_n^2}\,/|z| = e_{ni}\,/|z| \qquad (2)$$

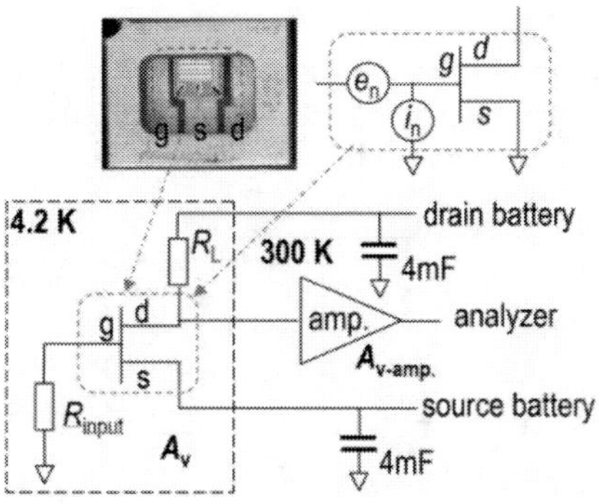

Figure 1. Fabricated HEMT, equivalent input noise voltage e_n, noise current i_n and the common-source voltage amplifier at 4.2 K (in the frame with black dashed line) based on the HEMT, R_{input} = 50 Ω and R_L = 300 Ω for e_n characterization.

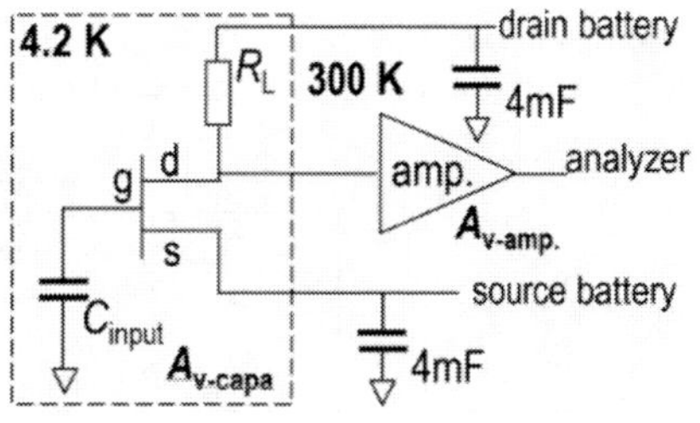

Figure 2. Experimental setup of the common-source voltage amplifier at 4.2 K (in the frame with black dashed line) based on C_{input}, R_L = 300 Ω and the under test HEMT for i_n characterization.

A HEMT based common-source amplifier can be used to determine e_n and i_n. For e_n measurement, the HEMT, R_{input} = 50 Ω and the load resistance R_L = 300 Ω, are mounted in a cryogenic insert as illustrated in figure 1. Using a lock-in amplifier, the voltage gain A_v and the output impedance or the channel impedance R_e can be measured in-situ. The signal at the drain of the HEMT is amplified once more by a low-noise amplifier with a voltage gain $A_{v\text{-amp}}$. The output voltage noise spectrum $e_{measured}$ is recorded by a vector signal analyzer. The noise voltage spectrum at the drain e_{drain} is deduced directly from $e_{measured}/A_{v\text{-amp}}$ and e_n is obtained from e_{drain}/A_v. For i_n characterization, to avoid the thermal noise perturbation, we use a C_{input} common-source voltage amplifier as shown in figure 2. In this configuration, the effective voltage gain $A_{v\text{-capa}}$ cannot be measured directly due to the feedback effect induced by C_{input}, the gate-source capacitance C_{gs}, the gate-drain capacitance C_{gd} and the Miller effect. The detail of the determination of $A_{v\text{-capa}}$ and C_{total} is described in [5]. By neglecting the real resistance due to the gate leakage current, $|z|$ can be simply expressed as $(2\pi f C_{total})^{-1}$, where f is the frequency. i_n

can be deduced from the results on the R_{input} (see figure 1) and the C_{input} (see figure 2) according to equation (2).

3. Results

We report measured e_{nt} at a chosen working point of V_{ds} = 100 mV and I_{ds} = 1 mA. This choice is in order to have sufficient large values of the transconductance g_m. Corresponded C_{gs} under the chosen working point, according to the measurement method described in [5], are of about 92, 26.5 and 5.3 pF, respectively. Our experimental results show that e_n is approximately inversely proportional to the square root of C_{gs} [6]; on the other hand, using capacitance input measurement method, i_n increases with the increase of C_{gs}. Therefore, both e_n and i_n must be taken into account for designing readout electronics.

3.1. HEMT with C_{gs} = 92 pF

We plot e_{nt} spectra in figure 3 with different input configurations at 4.2 K. At the chosen working point, g_m and the output conductance g_d are 35 mS and 0.75 mS, respectively. e_{nt} with 50 Ω input (see figure 1) and 1 nF input (see figure 2) show the lowest noise spectra. e_n reaches a value of 6 nV/Hz$^{1/2}$ at 1 Hz, and about 0.3 nV/Hz$^{1/2}$ at 1 kHz. The corner frequency (at which the 1/*f* noise value = the white noise value) is 1.2 kHz and the white noise is 0.22 nV/Hz$^{1/2}$ [5]. With the decrease of C_{input} from 300 pF to 10 pF, e_{nt} increases and reaches its maximum value of about 20 nV/Hz$^{1/2}$ at 1 Hz. At 1 Hz, the highest input impedance equals about 1 GΩ with C_{input} = 10 pF and $C_{total} \approx$ 150 pF [5]. We have $i_n \approx$ 19 aA/Hz$^{1/2}$ at 1 Hz according to equation (2). It is interesting to notice that the maximum e_{nt} is about 3 times larger than e_n at 1 Hz even C_{input} being almost zero.

At 77 K, only R_{input} = 50 Ω and C_{input} = 100 pF are used for the noise characterizations, e_{nt} spectra are plotted in figure 4, in the same figure e_{nt} at 4.2 K are reported for comparison. At 77 K and 1 Hz, e_n is around 20 nV/Hz$^{1/2}$ and i_n (with $C_{total} \approx$ 240 pF) can be estimated at 52 aA/Hz$^{1/2}$. Each e_{nt} component increases with the increase of temperature from 4.2 K to 77 K: e_n and i_n increase about three times.

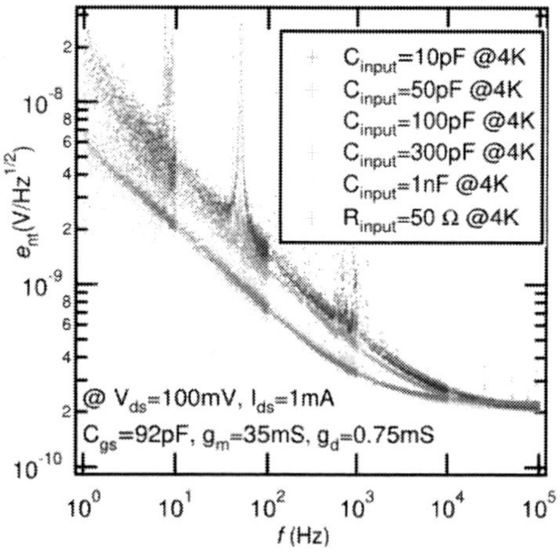

Figure 3. e_{nt} spectra of the HEMT with C_{gs} = 92 pF at 4.2 K, with different C_{input} from 10 pF to 1 nF and R_{input} = 50 Ω.

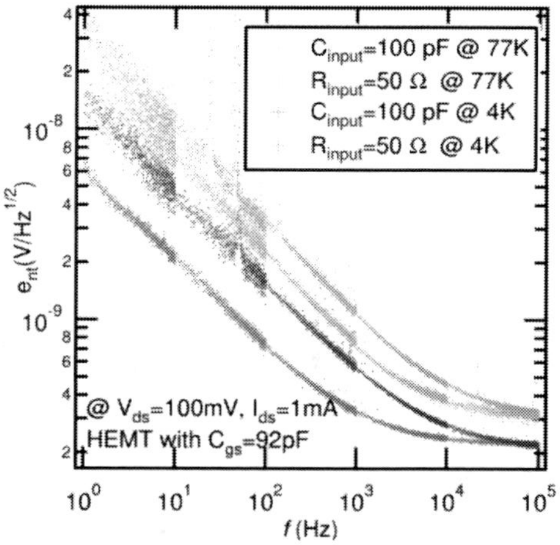

Figure 4. e_{nt} spectra of the HEMT with C_{gs} = 92 pF at 4.2 K and 77 K, with C_{input} = 100 pF and R_{input} = 50 Ω.

3.2. HEMT with C_{gs} = 26.5 pF

In Figure 5, we plot e_{nt} spectra of the HEMT with C_{gs} = 26.5 pF at 4.2 K with different C_{input} and R_{input}. At the chosen working point, g_m and g_d are 110 mS and 1.3 mS, respectively. The input noise voltage e_n is of about 15 nV/Hz$^{1/2}$ at 1 Hz and 0.54 nV/Hz$^{1/2}$ at 1 kHz. The corner frequency can be estimated at 13 kHz and the white noise is only 0.15 nV/Hz$^{1/2}$. Such a low white-noise value is owing to the high

g_m value [5]. At 1 Hz with C_{input} = 10 pF, e_{nt} is around 30 nV/Hz$^{1/2}$ and C_{total} is about 67 pF, the assessed i_n is thus of about 10 aA/Hz$^{1/2}$. At 77 K, only R_{input} = 50 Ω is chosen for the noise characterizations, e_{nt} spectrum is plotted in figure 6, in the same figure e_{nt} at 4.2 K is reported for comparison. Again, e_n increases about three times from 4.2 K to 77 K.

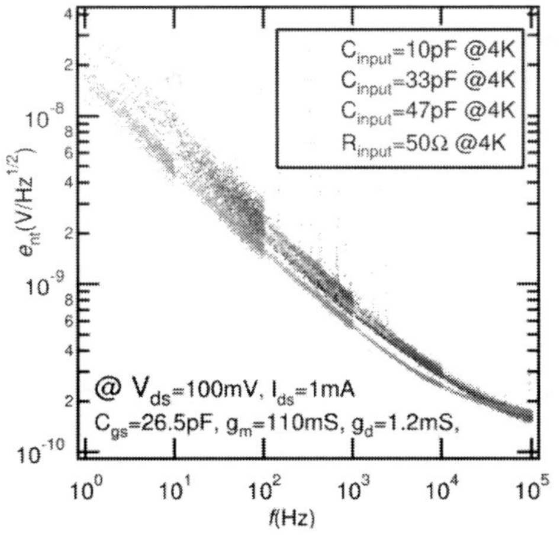

Figure 5. e_{nt} spectra of the HEMT with C_{gs} = 26.5 pF at 4.2 K, with C_{input} from 10 to 47 pF and R_{input} = 50 Ω.

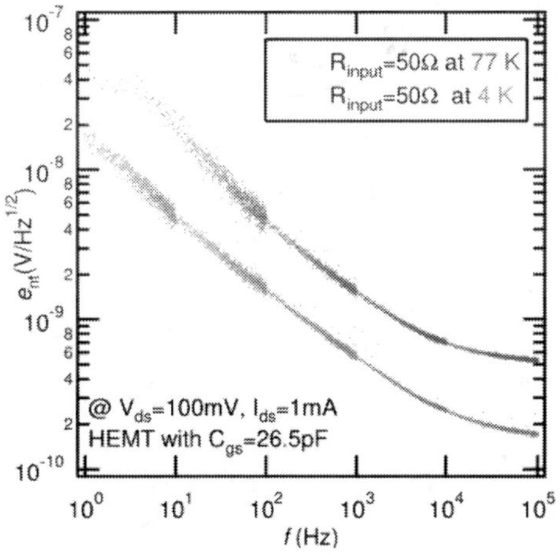

Figure 6. e_{nt} spectra of the HEMT with C_{gs} = 26.5 pF at 4.2 K and 77 K, only with R_{input} = 50 Ω.

3.3. HEMT with C_{gs} = 5.3 pF

In Figure 7, we plot e_{nt} spectra of the HEMT with C_{gs} = 5.3 pF at 4.2 K with different C_{input} and R_{input}. At the chosen working point, g_m and g_d are 44 mS and 1.3 mS, respectively. The input noise voltage e_n is of about 30 nV/Hz$^{1/2}$ at 1 Hz and 1.4 nV/Hz$^{1/2}$ at 1 kHz. The corner frequency can be estimated at 55 kHz and the white noise is 0.23 nV/Hz$^{1/2}$. By contrast, no significant increase of e_{nt} values can be found with C_{input} = 5 pF compared to e_n obtained with R_{input} = 50 Ω.

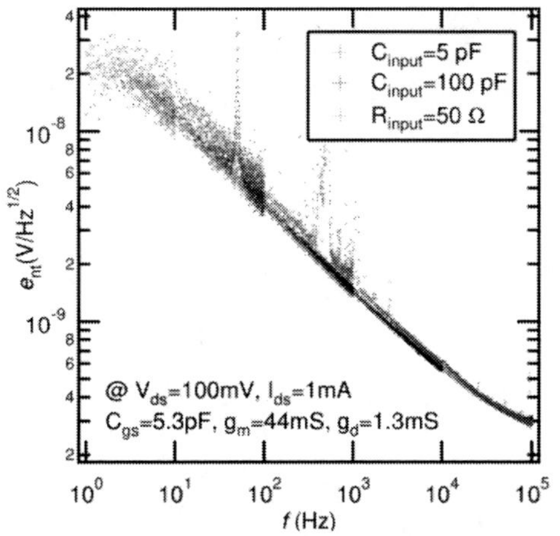

Figure 7. e_{nt} of the HEMT with C_{gs} = 5.3 pF at 4.2 K, with C_{input} of 5 and 100 pF, and R_{input} = 50 Ω.

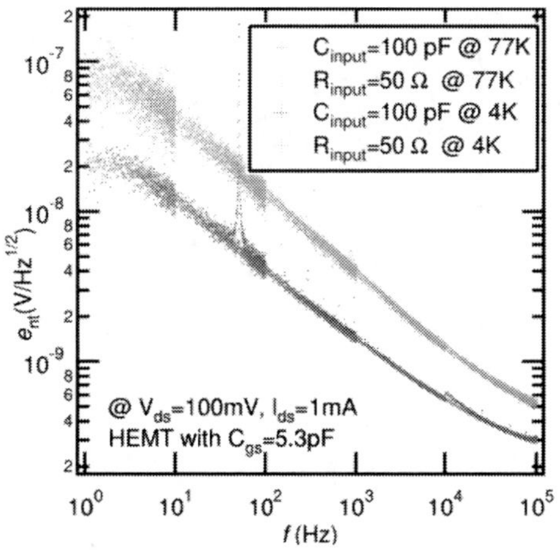

Figure 8. e_{nt} of the HEMT with C_{gs} = 5.3 pF at 4.2 K and 77 K, with C_{input} = 100 pF and R_{input} = 50 Ω.

At 1 Hz with C_{input} = 5 pF, e_{nt} can be estimated at 40 nV/Hz$^{1/2}$ and C_{total} can be found of about 17 pF (its equivalent impedance at 1 Hz is 9.4 GΩ), the assessed i_n is only 2.8 aA/Hz$^{1/2}$. Indeed, the 5.3 pF HEMT shows an extremely low i_n, however, its total input noise e_{nt} is higher than that can be obtained with the 92 pF HEMT. Obviously, 5.3 pF HEMT is more favorable compared to 92 pF HEMT for increasing the operating frequency of the readout electronics.

In Fig. 8, we report e_{nt} spectra with R_{input} = 50 Ω and C_{input} =100 pF at 77 K and 4.2 K. Again, e_{nt} increase of about three times from 4.2 K to 77 K. e_{nt} remains almost the same with C_{input} = 100 pF and with R_{input} = 50 Ω at a given temperature. It is worthy to note that the input impedance at 1 Hz with C_{input} = 100 pF (C_{total} = 112 pF) is of about 1.4 GΩ. So, e_{nt} is dominated by e_n, and the i_n contribution is negligible even the input impedance is in the giga-ohm range.

The results of this work provide a useful basis to design HEMTs for each specific application, i.e., to find the best compromise between e_n and i_n at a chosen operating frequency.

Acknowledgment
This work was supported in part by the French RENATECH network, le RTRA Triangle de la Physique grants No. 2008-015T and No. 2009-004T, European FP7 space project CESAR grant No. 263455 and DEFI Instrumentation aux limites CryoHEMTs 2013. Q. D. is funded by the BDI CNRS/CEA. We thank Drs. A. Juillard, B. Sadoulet, A. Anthore, F. Pierre, F. Parmentier and E. Cambril for stimulating discussions and help.

References
[1] A. Van der Ziel, Noise in Solid State Devices and Circuits, New York: Wiley, (1986).
[2] E. Gremion, D. Niepce, A. Cavanna, U. Gennser, and Y. Jin, "Evidence of a fully ballistic one-dimensional field-effect transistor: Experiment and simulation," Appl. Phys. Lett. **97**, 233505 (2010).
[3] E. Grémion, A. Cavanna, Y.X. Liang, U. Gennser, M.C. Cheng, M. Fesquet, G. Chardin, A. Benoît, Y. Jin, "Development of Ultra-Low Noise HEMTs for Cryoelectronics at ≤4.2 K," J. Low Temp. Phys. **151**, 971 (2008).
[4] Y. Liang, Q. Dong, M.-C. Cheng, U. Gennser, A. Cavanna, Y. Jin, "Insight into low frequency noise induced by gate leakage current in AlGaAs/GaAs high electron mobility transistors at 4.2 K," Appl. Phys. Lett. **99**, 113505 (2011).
[5] Q. Dong, Y. X. Liang, D. Ferry, A. Cavanna, U. Gennser, L. Couraud, and Y. Jin, "Ultra-low noise high electron mobility transistors for high-impedance and low-frequency deep cryogenic readout electronics," Appl. Phys. Lett. **105**, 013504 (2014).
[6] Q. Dong, Y. X. Liang, U. Gennser, A. Cavanna, and Y. Jin, "The Role of the Gate Geometry for Cryogenic HEMTs: Towards an Input Voltage Noise Below 0.5 nV/√Hz at 1 kHz and 4.2 K," J. Low Temp. Phys. **167**, 626 (2012).

Journal of Physics: Conference Series **568** (2014) 032010 doi:10.1088/1742-6596/568/3/032010

Application of cold beam of atoms and molecules for studying luminescence of oxygen atoms stimulated by metastable helium

V. V. Khmelenko[1,*], S. Mao[1], A. Meraki[1], S. C. Wilde[1], P. T. McColgan[1], A. A. Pelmenev[2,3], R. E. Boltnev[2,3], and D. M. Lee[1]

[1] Institute for Quantum Science and Engineering, Department of Physics and Astronomy, Texas A&M University, College Station, Texas 77843, USA

[2] Branch of Talroze Institute for Energy Problems of Chemical Physics, Russian Academy of Sciences, Chernogolovka 142432, Russia

[3] Joint Institute for High Temperatures, Russian Academy of Sciences, Moscow 125412, Russia

E-mail: khmel@physics.tamu.edu

Abstract. We describe a method for creating a high flux beam of cold atoms and molecules. By using this beam method, spectroscopic studies of the afterglow of oxygen-helium gas mixtures at cryogenic temperatures were performed. The cooling by helium vapor of a helium jet containing trace amounts of oxygen after passing through a radiofrequency discharge zone led to the observation of strong emissions from atomic oxygen. The effect results from the increased efficiency of energy transfer from metastable helium atoms and molecules to the atomic oxygen in the cold dense helium vapor. The effect might find application for the detection of small quantities of impurities in helium gas as well as possible laser action.

1. Introduction

The studies of inelastic collisions of cold atoms and molecules have recently attracted theoretical and experimental interest [1, 2]. Cold molecules are ideal objects for precise measurements of fundamental constants [3, 4]. Both these areas of research benefit from high flux sources of cold atoms and molecules. Recently a large variety of techniques to produce cold, low velocity samples of atoms and molecules were developed [5, 6, 7]. The brightest source of cold molecules (3×10^{12} s^{-1}) was based on the buffer gas cooling method [8].

In this work we present a method for obtaining cold atoms and molecules that uses a hydrodynamic effect to enhance the beam flux by two orders of magnitude as compared to the best existing sources. By using this method we studied processes of energy transfer from metastable He atoms and molecules to oxygen atoms. This method is based on the injection of a gaseous jet containing a small admixture of impurities which passed through a radio frequency discharge into superfluid helium [9]. When the impurity content in the helium gas is of order 1%, this method is very efficient for producing nanoclusters [10]. In the present work we found that reducing the ratio between impurity content and helium atoms in the jet from 10^{-2} to 10^{-5} allows one to avoid the impurity clustering process in the cold helium gas and to create a beam of cold atoms and molecules. As the jet descends into superfluid helium (see Fig. 1) the helium and impurity atoms and molecules in the jet are cooled efficiently by the dense helium vapor and the temperature of the atoms in the jet is decreased from 150 K near the orifice to 1.4 K near the surface of HeII. The process of cooling occurs over a distance of ~2.5 cm. The average travel time of an atom or molecule from the orifice to the surface of superfluid helium is of order 100 μs. Although we cannot obtain direct spectroscopic evidence for the temperature of the oxygen atoms, following Ref.[5], we can estimate the collision numbers needed for "warm" (~150 K) oxygen and helium atoms injected from discharge zone to be fully thermalized to a temperature of 1.4 K. These numbers are ~30 collisions for oxygen atoms and ~20 collisions for helium atoms. Near the surface of HeII in the cold dense helium gas, oxygen atoms undergo 300 collisions in 1 μs [11], which leads to full thermalization of the oxygen and helium atoms in the jet to 1.4 K. The temperature of excited atoms and molecules produced by energy transfer from metastable helium in the lower part of the jet might be higher than the temperature of thermalized atoms and molecules.

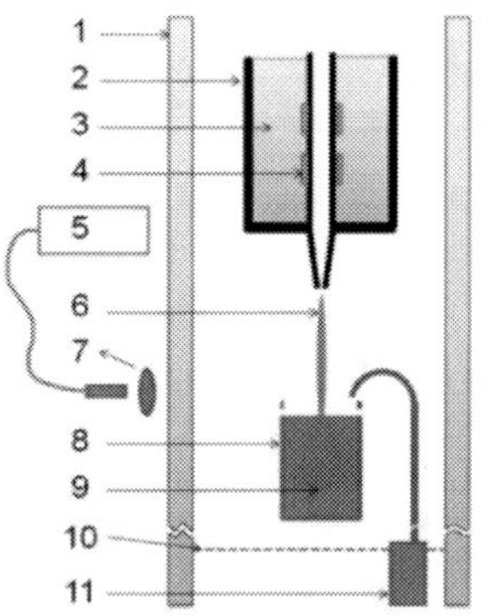

Figure 1. (Color online) Scheme of the set-up for creation of a high flux beam of cold atoms and molecules and the spectroscopic study of the afterglow in cold helium vapor. 1-helium dewar, 2-quartz tube, 3-space filled with liquid nitrogen, 4-electrodes for radiofrequency discharge, 5-system for registration of optical spectra, 6-gas jet, 7-lens, 8-quartz beaker, 9-superfluid helium in the beaker, 10-level of liquid helium in the main helium bath, 11-fountain pump.

2. Experimental setup

The overall experimental arrangement is shown in Fig. 1. The gas mixture consisting of helium and small concentrations of impurities enters the helium bath region via a quartz capillary cooled by liquid nitrogen. Electrodes placed around the quartz capillary provide a 53 MHz, 90 Watt radiofrequency discharge to form metastable helium atoms and dissociate the oxygen molecules. In the experiment, we used research grade helium from Linde Electronics&Specialty Gases with 99.9999% purity. The oxygen content in the discharge space (1 ppm) results from contamination of the helium gas. The content of impurities in the gas mixture can be adjusted by adding trace amounts of impurities to the helium gas or by purifying the helium gas in a cold trap at liquid helium temperatures. Under a small gradient of pressure (Δp ~ 2 Torr) between the discharge zone and the cryostat, a well-formed jet is created after passage of the gas mixture through a small orifice (0.75 mm dia) in the quartz capillary. The flux of $5 \cdot 10^{19}$ atoms/s of the impurity-helium mixture entering the cryostat is measured by a Brooks Instrument 5850E flow controller. A temperature of 1.4 K of the liquid helium in the cryostat is maintained by pumping the main helium bath and was measured by a germanium thermometer. A collection beaker was placed below the orifice with a distance of 2.5 cm between the orifice and the top of the beaker. The spectra of the gas jet afterglow were studied when the beaker was empty and also when the beaker was filled with superfluid helium by the fountain pump shown in Fig. 1. We studied the dependence of spectra on the pressure of helium gas in the helium dewar by varying the pumping speed on the main bath. We also studied the dependence of spectra on

Journal of Physics: Conference Series **568** (2014) 032010 doi:10.1088/1742-6596/568/3/032010

oxygen content in the oxygen-helium gas mixture. In these experiments the light from the jet afterglow was collected near the top of the collection beaker at a distance of 2.5 cm below the orifice by a lens which focused it onto the end of an optical fiber (see 7 on Fig. 1). The other end of the fiber was attached to an Andor Shamrock SR500 spectrograph. The Andor spectrograph with a Newton EMCCD camera was used to obtain high resolution (0.05 nm) spectra over a broad wavelength range (200 -1100 nm).

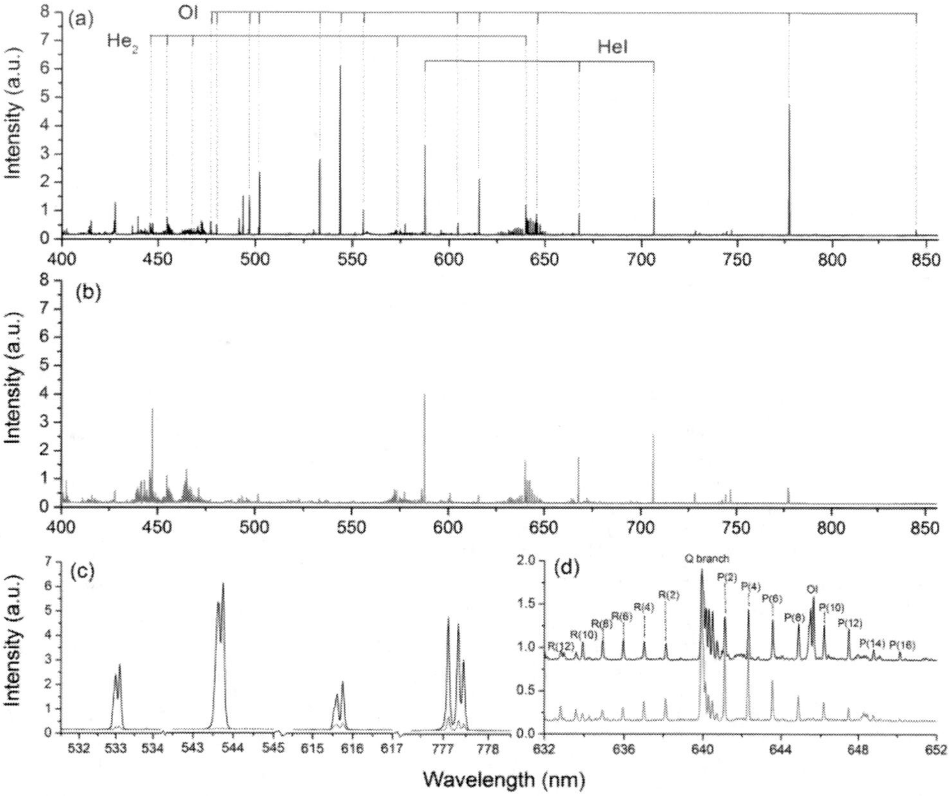

Figure 2. (Color online) Spectra of the helium jet with traces of oxygen (1 ppm): a) when the beaker is filled with HeII and b) when the beaker is empty. Light is collected at the top of the beaker c) High resolution spectra of the atomic oxygen lines. The spectra with larger intensity are from the jet when the beaker is filled with HeII while the ones with smaller intensity are from the jet when the beaker is empty. d) High resolution spectrum of the He_2 band near 639.6 nm. The He_2 band from cold jet was shifted up vertically for better comparison.

3. Experimental results

In the method we used, the impurities and helium gas in the jet are moving in the same direction in the cold dense helium gas. During the cooling process, the velocities of atoms in the jet are reduced substantially. Relative velocities between helium atoms and impurities become negligible, and it is possible to study processes of energy transfer and chemical reactions in the gas phase at low temperatures. As an example, in this work we have studied the influence of temperature on the spectra of oxygen atoms excited by metastable helium atoms and molecules [12]. Fig. 2a and 2b show the emission spectra of helium jet with 1 ppm concentration of

Table 1. Identification of observed lines and bands shown in Fig. 2.

λ, nm	Transition	λ, nm	Transition
844.6	O I ($3s^3S$-$3p^3P$)	477.3	O I ($3p^5P$-$7d^5D$)
777.4	O I ($3s^5S$-$3p^5P$)	728.3	He I ($2p^1P$-$3s^1S$)
645.4	O I ($3p^5P$-$5s^5S$)	706.5	He I ($2p^3P$-$3s^3S$)
615.7	O I ($3p^5P$-$4d^5D$)	667.9	He I ($2p^1P$-$3d^1D$)
604.6	O I ($3p^3P$-$6s^3S$)	587.6	He I ($2p^3P$-$3d^3D$)
555.6	O I ($3p^3P$-$7s^3S$)	501.6	He I ($2s^1S$-$3p^1P$)
543.6	O I ($3p^5P$-$6s^5S$)	639.6	$He_2(d^3\Sigma_u^+ \rightarrow b^3\Pi_g)$
533.0	O I ($3p^5P$-$5d^5D$)	573.2	$He_2(f^3\Delta_u \rightarrow b^3\Pi_g)$
502.0	O I ($3p^5P$-$7s^5S$)	465.0	$He_2(e^3\Pi_g \rightarrow a^3\Sigma_u^+)$
496.8	O I ($3p^5P$-$6d^5D$)	454.7	$He_2(h^3\Sigma_u^+ \rightarrow b^3\Pi_g)$
480.2	O I ($3p^5P$-$8s^5S$)	445.0	$He_2(i^3\Sigma_u^+ \rightarrow b^3\Pi_g)$

oxygen detected by the Andor spectrometer for the cases when the collection beaker contained liquid helium and when it was empty. Both spectra contained atomic oxygen lines as well as atomic helium lines and molecular helium bands listed in Table 1. When the beaker was empty, helium atomic lines and molecular bands dominate the spectrum. Only three weak oxygen lines were present in this spectrum (see Fig. 2b). When liquid helium was filled into the beaker, the oxygen lines become much stronger (see Fig. 2a and 2c) whereas the He_2 molecular bands became smaller. High resolution spectra for the oxygen lines and He_2 molecular band are shown in Fig. 2c and 2d. The fine structure obtained helped to definitely identify the spectral lines observed. The unresolved spectra of oxygen atomic lines were obtained earlier in studying helium jets with unknown contaminations of oxygen in similar conditions, but the authors did not make identifications of oxygen lines in their work [13]. In our experiment we observed a large enhancement of the oxygen line intensities during filling the collection beaker with superfluid helium, which provided conditions for efficient cooling of the oxygen-helium jet .

We studied experimental conditions for which the large enhancement of the oxygen line intensities could be observed. The spectra shown in Fig. 2 were obtained at the helium vapor pressure in the cryostat of 2 torr and for the ratio between O_2 and He in the gas jet equal to 10^{-6}. Fig. 3a shows the dependence of intensities of atomic lines and molecular bands on the pressure of helium vapor in the cryostat. It was found that as the pressure increased up to 12 torr, the intensity of all spectral lines and bands gradually decreased. Fig. 3b shows the dependence of the intensities of atomic oxygen and helium lines and He_2 molecular bands on the content of O_2 molecules in the oxygen-helium gas mixture. The dependence of intensities of different oxygen atomic lines have maxima in the range 20-200 ppm of O_2 gas mixture. The intensity of He atomic line does not change upon changing concentration of O_2 molecules. However, the intensity of the He_2 molecular band decreases when the content of O_2 molecules in the gas mixture increases. This indicates that the observed bright green luminescence of O atoms requires a very special set of conditions. For pressures above 12 torr and temperatures above 1.8 K the phenomena was not observed. At oxygen contents above 400 ppm in helium gas, the effect also became very weak.

4. Discussion

In our experiments the average travel time of an atom or a molecule from the orifice to the surface of superfluid helium is of order 100 μs. This means that all atoms and molecules excited in the

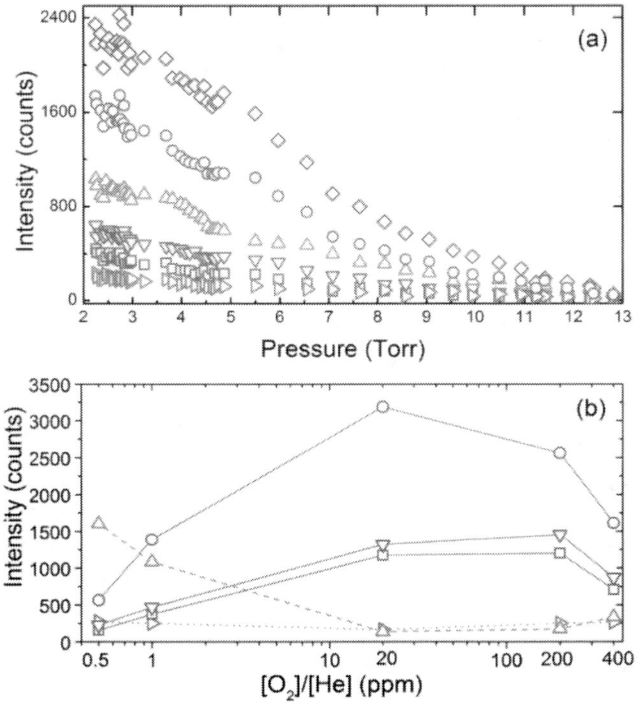

Figure 3. (Color online) a) Dependence of intensity of oxygen and helium lines and helium band on helium pressure in the dewar, b) Dependence of intensity of selected atomic oxygen and helium lines and molecular helium band on molecular oxygen content in oxygen-helium gas mixture: ◇- O atomic line ($\lambda = 543.6$ nm), ○ - O atomic line ($\lambda = 777.4$ nm), △- He_2 molecular band ($\lambda = 639.6$ nm), ▽- O atomic line ($\lambda = 615.6$ nm), □- O atomic line ($\lambda = 533.0$ nm), ▷- He atomic line ($\lambda = 706.5$ nm).

discharge zone with lifetimes in the range 14 - 1000 ns should emit light only in the discharge zone and near the orifice. Therefore the observed emission spectra well below the orifice should result from processes taking place in the jet. It is well known that metastable helium species easily excite, dissociate, and ionize any admixtures of impurity species in helium plasmas and in the afterglow [14, 15, 16]. The process of Penning ionization (PI) of molecular oxygen by helium atoms and molecules has been intensively studied [17]. In our experimental condition when we observed bright green emission, the very low content (1-25 ppm) of O_2 molecules in helium gas are completely dissociated in the discharge zone; therefore all of the oxygen should be in an atomic state in the expanded helium jet. The PI processes of excitation of oxygen atoms by helium atoms and molecules is a dominant process in the jet. Formation of highly excited states of oxygen atoms is a result of PI of oxygen atoms by helium atoms and molecules followed by electron-oxygen ion recombination. Qualitatively the effect of large enhancements of the oxygen line intensities and the appearance of new, even more intense oxygen lines in the dense cold helium vapor observed in this work might be explained by an increase in density of the cold oxygen and metastable helium atoms and molecules as the jet enters the region of dense helium vapor near the surface of liquid helium. The process of the PI of O atoms should be much more efficient in this region. Another possible factor for enhancement of the oxygen line intensities is the increasing cross sections of PI processes for oxygen atoms by metastable helium atoms and molecules at low temperatures. In our experiments the temperature of the atoms and molecules in the jet before entering bulk superfluid helium should be thermalized to the temperature close to the temperature of superfluid helium and both reacting species, helium and oxygen, are moving in the same direction, representing a situation with cold merged beams, which provides an increasing time for interaction of the reagents [18, 19]. An example of the consequences of these increased interaction times of the slow beams at low temperatures was the observation of quantum effects during the PI process [19].

Journal of Physics: Conference Series **568** (2014) 032010 doi:10.1088/1742-6596/568/3/032010

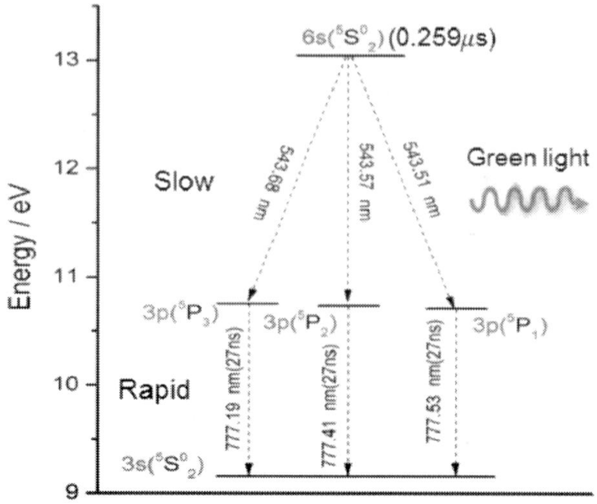

Figure 4. (Color online) Energy level diagram and transitions of oxygen atoms involved in possible lasing process.

The bright green triplet lines of 543.6 nm (see Fig. 2c) are emitted from the $6s^5S$ state which has a long decay time ($0.259\mu s$) whereas the final state lifetimes corresponding to these transitions are much shorter ($\sim$ 27ns). This provides the possibility for developing a population inversion between the $6s^5S$ and $3p^5P$ levels to study the lasing at λ=543.6 nm of atomic oxygen. An energy level diagram and the transitions of the oxygen atom involved in the lasing process are shown in Fig. 4. An atomic oxygen laser was first operated with an RF discharge in Ne-O_2 and Ar-O_2 mixtures at λ=844.67 nm [20]. The excitation mechanism involving energy transfer collisions between metastable rare gas atoms and ground state O_2 molecules to generate a population inversion was analyzed later [21]. We plan to investigate the possibility of laser action at λ=543.6 nm in a cold oxygen-helium jet excited in a RF discharge.

Conclusion

All spectral lines and bands observed in this work correspond to oxygen and helium atoms and helium molecules in the gas phase. We did not observe any sign of recombination or clusterization of oxygen atoms. This means that we had realized conditions for the creation of an intense cold gas phase oxygen-helium beam in which the flux of oxygen atoms is of order $3 \cdot 10^{14}$ atoms/s. This flux is almost two orders of magnitude larger than that obtained by using a buffer helium gas method [5, 8]. In our experiments this flux is injected into the volume of superfluid helium, but it is possible to realize a geometry in which the jet will pass the region of cold helium gas near the surface of HeII and enter a high vacuum region for further investigations and applications.

The effect of substantial enhancement of the intensity of atomic oxygen luminescence by the cooling of a helium jet with small traces of O atoms by helium vapor was observed. The effect was explained by effective energy transfer from metastable helium atoms and molecules to oxygen atoms and molecules under cryogenic conditions. This effect might provide a practical method for the detection of small traces of impurities in helium

Observation of the intense atomic oxygen emission at λ=543.6 nm in cold oxygen-helium jets after RF discharge opens possibility of creating population inversions of atomic oxygen. We plan experiments to search for laser action of atomic oxygen at λ=543.6 nm in the cold oxygen-helium beam excited by an RF discharge.

Acknowledgement
This work was supported by NSF grant No DMR 1209255. We are grateful to Marlan Scully and Suhail Zubairy for helpful conversations. We thank Trevor Dragon for participation in the experiments.

References

[1] Bethlem H L and Meijer 2003 *Int. Rev. Chem.* **22** 73128
[2] Bohn J 2001 *Phys. Rev. A* **63** 052714
[3] Hudson J, Sauer B, Tarbutt M and Hinds E A 2002 *Phys. Rev. Lett* **89** 023003
[4] Truppe S, Hendricks R J, Tokunaga S K, Lewandowski H J, Kozlov M G, Henkel C, Hinds E A and Tarbutt M R 2013 *Nat. Commun.* **4** 2600
[5] Hutzler N R, Lu H-I and Doyle J M 2012 *Chem. Rev.* **112** 4803
[6] Narevicius E and Raizen M G 2012 *Chem. Rev.* **112** 4879
[7] Chervenkov S, Wo X, Bayerl J, Rohlfes A, Gantner T, Zeppenfield M and Rempe G 2014 *Phys. Rev. Lett.* **112** 013001
[8] Patterson D and Doyle J M 2007 *J. Chem. Phys.* **126** 154307
[9] Gordon E B, Mezhov-Deglin L P, and Pugachev O F, 1974 *JETP Lett* **19** 63
[10] Khmelenko V V, Kunttu H, and Lee D M 2007 *J. Low Temp. Phys.* **148** 1
[11] Wentworth W E, Qin Y, Wiedeman S, Stearns S D, and Madabushi J 1995 *Appl. Spectraosc.* **49** 1282
[12] Khmelenko V V, Mao S, Meraki A, Wilde S C, McColgan P, Pelmenev A A, Botlnev R E and Lee D M 2013 *Phys. Rev. Lett.* **111** 183002
[13] Popov E A, Ahokas J, Eloranta J and Kunttu H 2005 *J. Low Temp. Phys.* **138** 85
[14] Collins C B and Robertson W W 1964 *J. Chem. Phys.* **40** 2208
[15] Richardson W C and Setser D W 1973 *J. Chem. Phys.* **58** 1809
[16] Tokaryk D W, Brooks R L and Hunt J L 1993 *Phys. Rev.* A **48** 364
[17] Druyvesteyn M J and Penning F M 1940 *Rev. Mod. Phys.* **12** 87
[18] Wei Q, Lyuksyutov I and Herschbach D 2012 *J. Chem. Phys.* **137** 054202
[19] Henson A B, Gersten S, Shagam Y, Narevicius J and Narevicius E 2012 *Science* **338** 234
[20] Bennett W R Jr., Faust W L, McFarlane R A and Patel C K N 1962 *Phys. Rev. Lett.* **8** 470
[21] Schaefer G and Kirkici H 1990 *IEEE J. of Quant. Elect.* **26** 1418

Journal of Physics: Conference Series **568** (2014) 032011 doi:10.1088/1742-6596/568/3/032011

Cryogen free cryostat for neutron scattering experiments

O Kirichek, R B E Down, P Manuel, J Keeping and Z A Bowden

ISIS Facility, STFC, Rutherford Appleton Laboratory, Harwell Oxford Campus, Chilton, Didcot, OX11 0QX, UK

E-mail: oleg.kirichek@stfc.ac.uk

Abstract. Most very low temperature (below 1K) experiments at advanced neutron facilities are based on dilution and ^{3}He refrigerator inserts used with Orange cryostats, or similar systems. However recent increases in the cost of liquid helium caused by global helium supply problems, has raised significant concern about the affordability of such cryostats. Here we present the design and test results of a cryogen free top-loading cryostat with a standard KelvinoxVT® dilution refrigerator insert which provides sample environment for neutron scattering experiments in the temperature range 35 mK – 300 K. The dilution refrigerator insert operates in a continuous regime. The cooling time of the insert is similar to one operated in the Orange cryostat. The main performance criteria such as base temperature, cooling power, and circulation rate are compatible with the technical specification of a standard dilution refrigerator. In fact the system offers operating parameters very similar to those of an Orange cryostat, but without the complication of cryogens. The first scientific results obtained in ultra-low temperature neutron scattering experiment with this system are also going to be discussed.

1. Introduction

Low temperature (LT) sample environment below 1K at neutron scattering facilities is usually provided by dilution and ^{3}He refrigerator inserts set in a top-loading Orange cryostat [1] or similar system. The demand for LT sample environment has dramatically increased in the last decade and in the case of the ISIS facility [2] has exceeded eighty LT experiments per year. This large scale usage of cryogenic equipment requires significant resources and can create a number of problems including health and safety issues and the considerable cost of the required cryogens. This last problem has become more significant due to increases in liquid helium costs caused by global helium supply problems [3].

The latest progress in cryo-cooler technology offers a new generation of cryogenic systems in which the cryogen consumption can be significantly reduced and in some cases completely eliminated. These systems also offer the advantage of operational simplicity, require less space than conventional cryogen-cooled systems and can significantly improve user safety. The most successful representative of the cryo-cooler family today is the pulse tube refrigerator (PTR). The unique feature of the PTR is the absence of any cold moving parts, which considerably reduces noise and vibrations generated by the cooler [4, 5]. It also increases the reliability of the cold head because no expensive high-precision seals are required and the cold head can be operated without any service inspection.

Here we discuss the operational experience and first scientific results obtained in a neutron beam line experiment where ultra-low temperature sample environment has been provided by a dilution refrigerator insert accommodated in the newly developed 50 mm diameter top-loading cryogen-free cryostat [6]. The cryostat provides sample environment in the temperature range of 1.4 – 300K with an exceptionally high cooling power of 0.23W @ 1.9K. Employment of an Oxford Instruments' KelvinoxVT® dilution refrigerator insert loaded into the cryostat's variable temperature insert (VTI) allowed us to perform neutron diffraction measurements at a base temperature of 36 mK. This system has been developed through an ISIS and Oxford Instruments collaborative project.

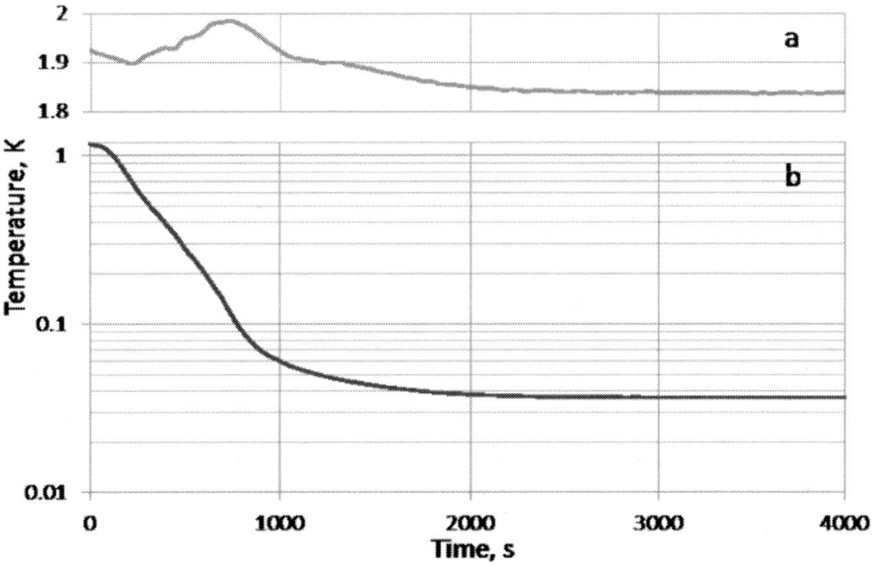

Figure 1. Cooling of the cryostat down to base temperature: (a) time dependence of temperature at the VTI exchanger; (a) time dependence of the mixing chamber temperature.

2. System design

The design of the system is based on the idea of combining a top-loading cryogen-free system described in Ref. [4] with a helium condensation loop [7], which has been successfully realized in the first prototype described in [8]. The major changes in design of the second prototype [6] used in the experiment presented in this paper are the replacement of the cryostat's top-loading insert with a standard Oxford Instruments VTI and the inclusion of a PTR second stage heat exchanger developed by ISIS to provide greater efficiency at high flow rates preferred for beam line operation. The detailed description of the system design, performance and test results can be found in Ref. [6].

Most of the modern dilution refrigerators used in neutron scattering experiments are based on the Grenoble design which incorporates sintered silver heat exchangers [9-11]. A relatively small number of neutron scattering experiments that require very low temperature in combination with high cooling power use a powerful cryogen free dilution refrigerators built around a PTR cryo-cooler [2, 12, 13]. Yet predominantly the sample environment below 1K is required for small samples. For this purpose compact dilution refrigerator inserts are usually used with the VTI of a cryostat. These systems consume cooling power produced at the VTI heat exchanger. In our case the procedure of loading and cooling of a standard KelvinoxVT® dilution insert was similar to the one employed in conventional

Journal of Physics: Conference Series **568** (2014) 032011 doi:10.1088/1742-6596/568/3/032011

cryostats. The cool down procedure consisted of three stages. Stage one was pre-cooling of the insert with heat-exchange helium gas in the refrigerator's vacuum jacket from room temperature down to ~ 2K. After that the VTI was warmed up to 5K and exchange gas was pumped out for approximately one hour. Once good vacuum in the vacuum jacket was achieved, the VTI heater was switched off and dilution refrigerator insert started helium mixture condensation and, eventually, circulation in an automated regime.

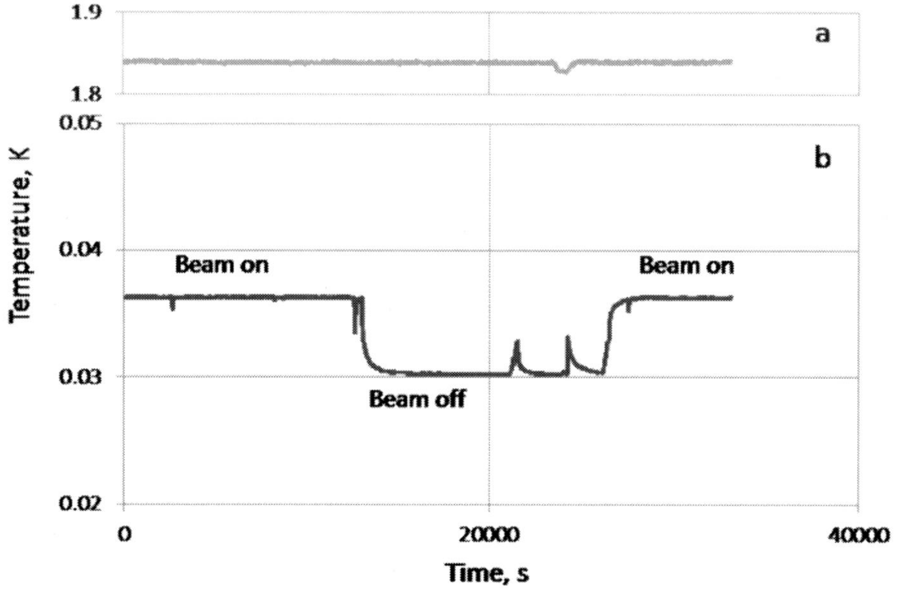

Figure 2. Time dependence of the mixing chamber temperature. (a) time dependence of temperature at the VTI exchanger; (b) time dependence of the mixing chamber temperature. Point A - neutron beam is switched off. Point B - neutron beam is back to its previous value (accelerator proton beam current 42 μA).

3. System operation

Cooling of the refrigerator down to base temperature is shown in Figure 1, where (a) represents the behavior of temperature at the VTI exchanger and (b) time dependence of the mixing chamber temperature. The refrigerator has been operated in a fully automated mode. The total cool down time of the dilution insert starting from inserting the refrigerator in the VTI was similar to the typical time taken for a conventional cryogen based top-loading cryostat.

During the first overnight base temperature run the system demonstrated remarkably stable performance. In Figure 2 we present: (a) time dependence of the VTI heat-exchanger temperature and (b) mixing chamber temperature. This high stability can be explained by the absence of time dependent processes such as changes of liquid levels in the helium and nitrogen vessels present in conventional cryostat operation. In the middle of the plot (point A) it is possible to see a step-like drop in the base temperature caused by switching the neutron beam off. After some time (point B) the neutron beam was back and the mixing chamber temperature returned to its previous value. Based on the technical specification of a standard KelvinoxVT® insert we would estimate the heating power of the neutron beam at a level of a few microwatts.

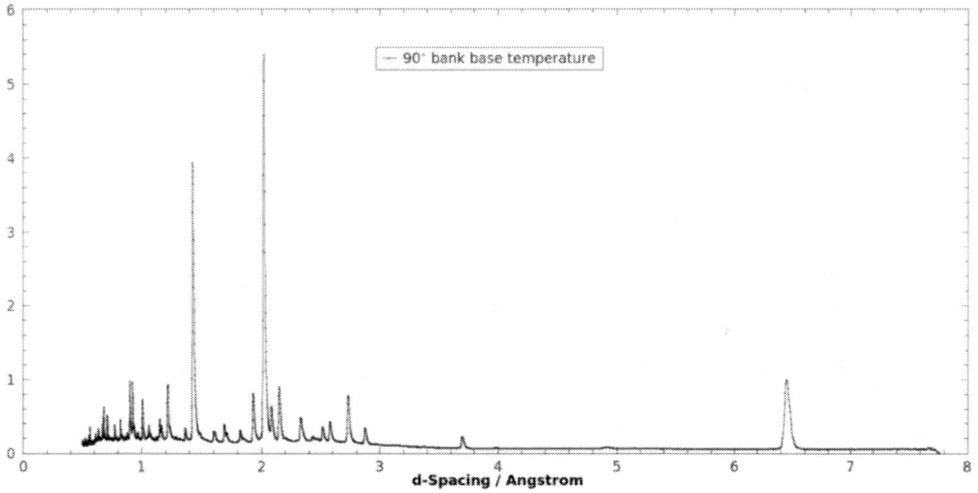

Figure 3. Powder diffraction from $Sr_2Ir_{0.5}Rh_{0.5}O_4$ sample. Measurements are taken at base temperature of the dilution refrigerator insert - 36 mK.

4. Experimental results

The first neutron scattering experiment where ultra-low temperature sample environment was provided by the cryostat described here has been conducted on WISH [9] at the ISIS Facility, STFC Rutherford Appleton Laboratory, UK. WISH is a high count rate, long-wavelength diffractometer with remarkably low neutron scattering background.

The diffraction pattern shown in Fig. 3 is obtained from a powder sample of $Sr_2Ir_{0.5}Rh_{0.5}O_4$, prepared by Dr D. Prabhakaran (Oxford University). It is an Anderson localized insulator thought to display an unusual magnetic order at 0.6K. There is an increasing interest in 5d oxides due to the richness of phases of matter they display, owing to similar energy scales of U, spin orbit coupling and other magnetic interactions in competition. The undoped Sr_2IrO_4 has been particularly studied over the past few years [10].

WISH provides excellent quality neutron powder diffraction capabilities to study magnetic ordering. As the ordered phase occurs at ~ 0.6K, a dilution fridge is required to cool the sample significantly below this temperature. The high-quality polycrystalline powder sample has been sealed in the thin-wall vanadium can with helium gas added as a heat-exchange media. This approach has been proven to be an efficient way for providing temperature equilibrium across the sample and vanadium can [16]. The diffraction data obtained in the experiment demonstrate very good angular coverage and data quality (low noise and background). Another important achievement is the high-level temperature stability at the refrigerator's base temperature which remained ± 0.1 mK during the whole measurement.

5. Conclusions

The ISIS facility carries on an internal development program intended to gradually substitute all conventional cryogenic systems with cryogen free systems preferably based on PTR. Here we discuss operational experience and first scientific results obtained in a neutron beam line experiment where very low temperature sample environment has been provided by a KelvinoxVT® dilution refrigerator insert accommodated in a newly developed top-loading cryogen-free cryostat. The system demonstrated high long time stability performance with a base temperature of 36 mK. High resolution neutron diffraction data have been obtained in the experiment. From a user perspective, the system offers operating parameters very similar to those of standard Orange cryostats but without the complication of cryogens.

Acknowledgements
The authors are grateful to Dave Bunce, Paul Frodsham and Dmitry Khalyavin for the valuable support of this experiment.

References

[1] Brochier D 1977 *ILL Tech. Report* 77/74
[2] Kirichek O 2012 *Modern Physics Letters B* **26** 1230006
[3] Feder T 2010 *Phys. Today* **3** 28
[4] Evans B, Down R, Keeping J, Kirichek O and Bowden Z 2008 *Meas. Sci. Technol.* **19** 034018
[5] Kirichek O, Carr P, Johnson C and Atrey M 2005 *Review of Scientific Instruments* **76** 055104
[6] Kirichek O, Foster J, Down R B E., Clapton D, Chapman C R, Garside J and Bowden Z A 2013 *J. of Low Temp. Phys.* **171** 737
[7] Wang C 2001 *Cryogenics* **41** 491
[8] Chapman C R, Evans B E, Dudman M P, Keeping J, Down R B E, Kirichek O and Bowden Z A 2011 *Cryogenics* **51** 146
[9] Frossati G, Godfrin H, Hebral B, Schumacher G and Thoulouze D 1977 *Proceedings of the Ultralow Temperatures Symposium*, Hakone, Japan
[10] Frossati G 1978 Thèse d'Etat, Université de Grenoble
[11] Godfrin H 1981 Thèse d'Etat, Université de Grenoble
[12] Uhlig K 2004 *Cryogenics* **44** 53
[13] Prouve T, Godfrin H, Gianèse C, Triqueneaux S and Ravex A 2007 *Journal of Low Temperautre Physics* **148** 909
[14] Chapon L C, Manuel P, Radaelli P G, Benson C, Perrott L, Ansell S, Rhodes N J, Raspino D, Duxbury D, Spill E and Norris J 2011 *Neutron News* **22** 22
[15] Kim B J, Ohsumi H, Komesu T, Sakai S, Morita T, Takagi H and Arima T 2009 *Science* **323** 1329
[16] Bailey I F 2003 *Z. Kristallogr.* **218** 84

A calculable and correlation-based magnetic field fluctuation thermometer

A Kirste, M Regin, J Engert, D Drung and T Schurig

Physikalisch-Technische Bundesanstalt (PTB), Abbestr. 2-12, 10587 Berlin, Germany

E-mail: alexander.kirste@ptb.de

Abstract. We have developed a new Magnetic Field Fluctuation Thermometer (MFFT) specifically designed for operation in primary mode, which requires the determination of the relation between thermal flux noise density and thermodynamic temperature. The noise thermometer combines a correlation-based SQUID readout and an integrated conductivity measurement on the metallic temperature sensor with an in situ flux calibration. The operation of the MFFT is modelled theoretically. First temperature measurements in secondary mode between 9 mK and 4.2 K showed excellent agreement with a copy of the PLTS-2000 within 0.5%.

1. Introduction

The magnetic field fluctuation thermometer (MFFT) is a special variation of noise thermometry, which has found application from the highest to the lowest temperatures [1]. The MFFT exploits the fact that thermally activated noise currents inside a conductor generate fluctuating magnetic fields above its surface, which can be detected by means of a Superconducting Quantum Interference Device (SQUID) and used as a measure of the conductor's temperature [2]. The noise signals are detected as thermal magnetic flux noise (TMFN) using dc SQUID based magnetometers or gradiometers. The power spectral density (PSD) S_Φ of the TMFN exhibits a low-pass like shape and depends on the geometries of both the conductor and the detection coil as well as on the electrical conductivity of the conductor. All these properties enter into the complex impedance $Z(f)$ of the detection coil that interacts with the metallic temperature sensor. The PSD of the TMFN as measured by the detection coil at temperature T is then given by [3]:

$$S_\Phi(f,T) = \frac{4k_B T \mathrm{Re}(Z(f))}{(2\pi f)^2}, \quad (1)$$

where k_B is the Boltzmann constant and f is the frequency. The application of a MFFT for temperature measurement is based on two additional assumptions: (i) the electrical conductivity of the conductor is assumed to be constant in the considered temperature range and (ii) the geometry of the conductor-detector set-up is fixed. Then, the values for the PSD are proportional to temperature for any frequency.

While MFFTs can be made up of wire-wound coils connected to SQUID current sensors [2], [4], PTB has mainly developed MFFTs with multiloop SQUID gradiometers on a single chip [5], [6], [7]. As these MFFTs basically measure ratios of temperature by measuring ratios of noise power a calibration is required at a known reference temperature, i.e. they are operated in secondary mode.

For application of a MFFT in primary mode (pMFFT) one has to relate the measured TMFN directly to thermodynamic temperature. This was achieved by designing temperature sensor and detection coils with plain geometries to allow an easy determination of all relevant dimensions and an efficient calculation of the TMFN. In addition, the pMFFT can perform in situ measurements of the electrical conductivity of the temperature sensor and of the flux-to-voltage transfer coefficient of the SQUID gradiometer detecting the TMFN. Thus, the new pMFFT permits direct measurements of temperature in a wide range without the need for calibration at a known reference temperature.

2. Design of the pMFFT and theory of operation

The new pMFFT comprises four main parts: a metallic noise temperature sensor, a system of two gradiometric detection coils read out by two separate SQUID current sensors, a room temperature electronics to read out the SQUID sensors and a calibration coil. For the particular temperature range of the PLTS-2000 we have designed a compact device, in which the SQUID sensors are placed close to the detection coils and are thus integrated in the metallic body of the temperature sensor.

2.1. Cross-correlation technique

In order to minimize the influence of non-thermal noise sources on the measurement of the TMFN and, finally, on the measured temperature a cross-correlation technique is applied in the pMFFT similar to [4]. The suppression of the non-thermal noise is possible because these noise components occur basically uncorrelated in the two SQUID channels.

Aiming at a large signal from the TMFN the correlation must be large. This is achieved if the detection coils of the two channels are as similar as possible and sense the same volume part of the temperature sensor. The design of the detection coils is optimized to obtain almost complete correlation for the TMFN while the magnetic coupling between both channels is minimized. The proper function of the correlation technique is checked by a zero measurement without temperature sensor.

If we consider two stochastic processes $x(t)$ and $y(t)$, we can define the cross-power spectral density (CPSD) $S_{xy}(f)$ as cross-covariance between the complex Fourier amplitudes $X(f)$ and $Y(f)$. The various methods for the estimation of the PSD can be applied to estimate the CPSD. For example, the multitaper spectral estimate of the CPSD is obtained similarly to that of the PSD,

$$\hat{S}_{xy}(f) \propto \frac{1}{K} \sum_k X_k(f) Y_k^*(f) \;, \qquad (2)$$

where K is the number of tapers. The CPSD is in general a complex function with amplitude and phase. The CPSD can be normalized with the square roots of the two PSDs, $\sqrt{S_x}$ and $\sqrt{S_y}$, giving the coherency $C_{xy}(f) = S_{xy}(f)/\sqrt{S_x(f)S_y(f)}$ as normalized measure of association. For all evaluations containing the CPSD or derived expressions, we will only use the real part of the corresponding quantities. This is to avoid the inclusion of (product) terms, which do not contribute to a correlation carried out in the time domain, and would not normally contribute to a correlation carried out in the frequency domain [8].

2.2. Block diagram and theory of operation

Figure 1 shows the schematic diagram of the correlation-based pMFFT. There are two detection coils located on one Si chip, which are individually read out by two SQUID current sensors. Each SQUID sensor is connected to a flux-locked loop (FLL) electronics at room temperature. The output voltages of both channels are simultaneously read out by a spectrum analyzer.

A calibration coil, located on a second Si chip, is placed on the opposite side of the temperature sensor. The shape of this coil is similar to that of the detection coils, so that a calculable flux can be produced in the detection coils by feeding a known (dc or ac) current into the calibration coil. In the dc case, this coil is used to determine the relation between magnetic flux in the detection coils and the output voltage of the SQUID electronics. In the ac case it is used to determine the electrical conductivity of the temperature sensor.

For the measurement of magnetic flux the SQUIDs are operated in FLL mode. By using PTB SQUID current sensors [9] there are two ways to form the feedback loop for FLL operation. The common feedback directly into the SQUID (gray current path in figure 1) functions independently of the properties of the input circuit but has the disadvantage that the flux transfer from the detection coils into the SQUID becomes frequency dependent. The reason is the frequency dependent inductance of the detection coil due to the skin effect in the temperature sensor, affecting the total inductance of the input circuit. As the SQUID input circuit is superconducting, feedback can also be applied to it by means of the integrated feedback transformer. This feedback scheme has the advantage that interaction between both detection coils and between detection coils and temperature sensor is suppressed within the FLL bandwidth. Hence this FLL scheme is favored for the pMFFT.

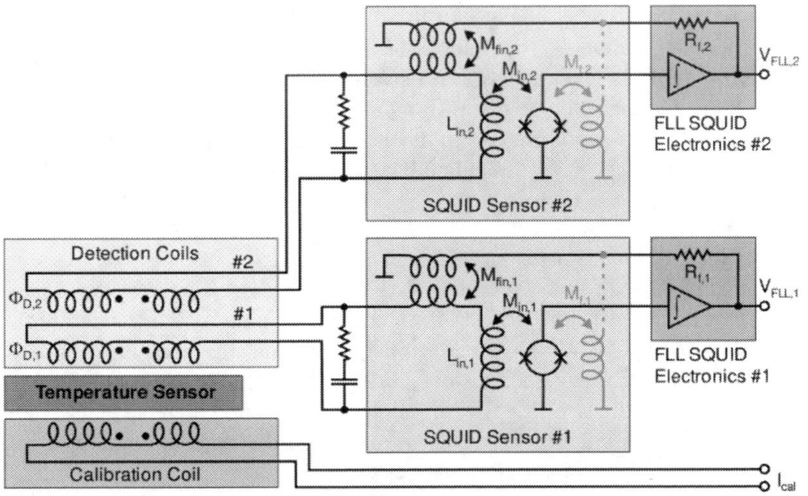

Figure 1. Schematic of the correlation-based pMFFT readout. The feedback (via M_{fin}) is into the superconducting detection coils using the feedback transformer integrated in the SQUID sensors. An alternative feedback path (gray colour) into the SQUID inductance is available for test purposes.

2.3. *Theoretical model for the pMFFT and design calculations*

In [10] the PSD S_Φ was calculated that is caused by the TMFN of an infinite conducting slab in a detection coil of contour c. Simplifications were possible for planar contours parallel to the slab, i.e. if $c = c(x, y, z)$ is at constant height z above the surface of the slab. We have extended this approach to calculate the CPSD caused by the TMFN in a system of two planar coils, c_1 and c_2. Following [10] we obtain

$$S_{12} \triangleq S_\Phi(c_1, c_2) = \mu^2 \sigma k_\mathrm{B} T \iint_{-\infty}^{\infty} I_A \mathrm{e}^{-\sqrt{a^2+b^2}(z_1+z_2)} \left(I_x(c_1) I_x^*(c_2) + I_y(c_1) I_y^*(c_2)\right) \mathrm{d}a\mathrm{d}b \qquad (3)$$

with

$$I_x(c_j) = \oint_{c_j} \mathrm{e}^{\mathrm{i}ax_j} \mathrm{e}^{\mathrm{i}by_j} \mathrm{d}x_j, \quad I_y(c_j) = \oint_{c_j} \mathrm{e}^{\mathrm{i}ax_j} \mathrm{e}^{\mathrm{i}by_j} \mathrm{d}y_j, \quad j = 1, 2 \qquad (4)$$

$$I_A = \int_0^t |A(a,b)|^2 dz', \quad A(a,b) = \mathrm{e}^{-\rho' z'} (\pi \rho')^{-1} (u + v \mathrm{e}^{-2\rho'(t-z')}) (u^2 - v^2 \mathrm{e}^{-2\rho' t})^{-1}, \qquad (5)$$

$$\rho^2 = a^2 + b^2, \quad \rho'^2 = \rho^2 + k^2, \quad u = 1 + \frac{\rho}{\rho'}, \quad v = 1 - \frac{\rho}{\rho'}, \quad k^2 = \mathrm{i}\mu\sigma\omega, \quad \omega = 2\pi f\,. \qquad (6)$$

Accordingly, z_j, ($j = 1, 2$) is the height of the contour c_j above the surface of the conducting slab, which has the electrical conductivity σ, magnetic permeability μ and a thickness of t. The integral I_A describes the eddy current distribution in the slab, whereas the geometry of the contours c_j is completely contained in the path integrals $I_x(c_j)$ and $I_y(c_j)$. For $c_1 = c_2$ the CPSD (3) is equal to the PSD derived in [10]. While S_{12} is generally complex its real part, Re(S_{12}), contains the desired CPSD.

Strictly speaking, this model is valid only for a conductor of infinite size. However, the deviations of the PSD or CPSD for real conductor geometries are made very small by proper design of the detection coils. This requires that the slab overlaps the detection coils by a few times of their characteristic size (e.g. coil radius), while this condition is further relaxed for gradiometric detection coils. The approximation of a filamentary (and superconducting) coil conductor is justified if the line width of the coil conductor is small compared to all other relevant distances (e.g. coil radius, distance to the slab, gap between coil conductors). However, for an exact calculation of the deviations a comparison with FEM calculations in 2D or 3D is necessary.

The in situ determination of the electrical conductivity σ of the temperature sensor is based on the skin effect within the conductor. In the pMFFT the frequency dependence of the penetrated flux through the temperature sensor as picked up by the detection coils is measured (figure 1), where the conductivity enters via the product $\mu\sigma f$. An analytical solution for this problem is available for systems of filamentary, concentric, circular coils parallel to an infinite slab [11], [12]. This model can only be an approximation of real geometries, but the resulting deviations for σ can be made very small, requiring the same conditions as given above for the calculation of the CPSD.

Performing the same measurement with a known dc current ($f = 0$) through the calibration coil allows to calibrate both SQUID gradiometers in terms of FLL voltage ($V_{\mathrm{FLL},j}$) per magnetic flux ($\Phi_{\mathrm{D},j}$) in the detection coils ($j = 1, 2$) using calculated mutual inductances between calibration coil and detection coils.

2.4. Mechanical construction and sensor material

Figure 2 shows an inside view of the assembled pMFFT. The basic shape of the device is a cylinder with a diameter of 18 mm and a length of 100 mm. The whole body is made from high-purity copper of nominal 99.9998% purity [13]. We have determined a residual resistivity ratio of RRR ≈ 103. The cylinder has large openings from opposite sides. From one side, it takes up the Si chip with the detection coils surrounded by the two SQUID chips glued onto chip carriers (PCBs). The Si chip with the calibration coil together with another PCB is mounted on the opposite side at a fixed distance in parallel to the detection coils with the temperature sensor in between.

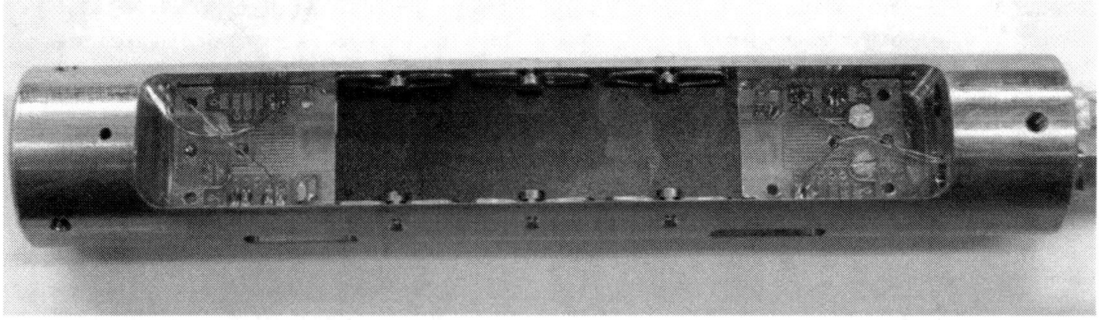

Figure 2. Inside view of the pMFFT with the magnetic shielding removed. The two SQUID current sensors are placed on PCBs directly besides the central Si chip with the detection coils. They can be seen as a shadow through the PCBs.

Only a central part near the cylinder axis serves as actual temperature sensor. This part has a thickness of about 2 mm and is virtually defined by the detection coils. In order to allow an arbitrary and minimal gap (≈50 μm or ≈100 μm depending on the side) between detection coils and temperature sensor, the Si chips are mounted face down. While the PCBs are fixed by screws, small springs (CuSn6) press the Si chips towards the edge guides providing always a proper alignment. No glue is needed to fix the coil chips. Magnetic shielding of the thermometer's inner volume is provided by an outer Nb tube, which can be removed to both sides.

2.5. Thin-film coil assemblies and SQUID sensors

Both the detection coils and the calibration coil are realized as planar thin-film coils fabricated on Si wafers using PTB's standard Nb thin-film technology. Due to the photolithography the coils can be manufactured with (sub-)micrometer precision, which is advantageous for modeling if they are mounted at distances of about 50 μm or 100 μm above the surface of the temperature sensor.

All coils are designed as series gradiometers using only a fraction of the available chip area. Despite the fact that series-parallel gradiometers with multiple degrees of freedom could couple more noise energy into the SQUID than series gradiometers with single degree of freedom, they are disadvantageous in some aspects. In particular, the calculation of the PSD of the TMFN can become more difficult and the interaction with the environment during FLL operation is not always suppressed (see section 2.2).

When designing the detection coils three properties were taken into consideration: (i) minimized capacitive coupling between both coils to avoid unnecessary interaction between the SQUIDs, (ii) maximum correlation of the TMFN from both coils to obtain a large CPSD signal or to allow an approximation of the CPSD by the single PSDs, (iii) simple coil geometries to allow simplified calculations of the TMFN and the electrical conductivity as described in section 2.3.

We have designed four different detection coils by combining different shapes (coaxial, circular or second-order gradiometers) with different arrangements formed by both coils (stripline or coplanar). For the first experiments we selected a coaxial, circular gradiometer with coplanar arrangement of both coils. The radius of the outer turns is ≈2.0 mm, eight inner turns have a radius of around 0.7 mm. In comparison with that, the distance between coil center and side walls of the temperature sensor is 6.5 mm. Line width and minimum gaps between the coils are 2.5 μm. With an estimated total coil inductance of ≈400 nH, we chose SQUID current sensors with an input inductance of 400 nH (PTB single-stage, double-transformer SQUIDs [9] of size "L") for the readout.

The SQUID sensors are glued on PCBs and mounted with sufficient distance to the conducting sensor body. Since the bond loops of the chip-to-chip connections cannot be made arbitrarily small (typical length/height 1 mm), their influence must be made sufficiently small. Depending on the detection coil this is solved either by incorporating the interconnections into a fully gradiometric design or by a large distance to the detection coils.

3. Measurements

3.1. Noise measurements with and without temperature sensor

To ensure the proper function of the cross-correlation technique, we have performed reference measurements without the temperature sensor. This was done by mounting the chip with the detection coils and the SQUID sensors on a non-metallic platform and measuring the intrinsic noise and the cross-correlation at 4.2 K. The results, referring to the flux noise in the SQUIDs, are given in figure 3. The single-channel white noise corresponds to $1.3\cdot10^{-6}\ \Phi_0/\mathrm{Hz}^{1/2}$, which is slightly increased compared to the intrinsic noise without input coil. An electronics noise contribution as well as $1/f$ noise are visible in the single-channel noise spectra below 100 Hz and 10 Hz, respectively. An inspection of the CPSD reveals that both channels are uncorrelated to a level of $\leq10^{-7}\ \Phi_0/\mathrm{Hz}^{1/2}$ (after ≈98000 averages) except at the mains frequency (50 Hz) and harmonics.

Figure 4 shows the CPSD of the TMFN at 4.23 K after integration of the temperature sensor. The calculated CPSD according to (3)-(6) is based on the parameter set $z_1 = z_2 = 51.5$ μm, $1/\sigma = (1.67\cdot10^{-8}$ / 103) Ωm, t = 2.00 mm, which is close to the independently determined values. As the originally intended calibration of the pMFFT necessary for primary thermometer mode could not be carried out, we had to measure instead the FLL parameters ($R_{\mathrm{f},j}$ and $1/M_{\mathrm{fin},j}$, j = 1, 2) and the flux transfer coefficients depending on the inductances in the SQUID input circuits. There is good agreement between experimental and calculated data.

In addition, we have examined the magnetic shielding of the MFFT, which reduces the influence of external magnetic fields onto the CPSD. Sensitivities exceed 310 $\mu\mathrm{T}/\Phi_0$ in dc fields for a 130 mm long shield. Even higher shielding can be expected for ac fields due to the additional skin effect in the end sections of the sensor body.

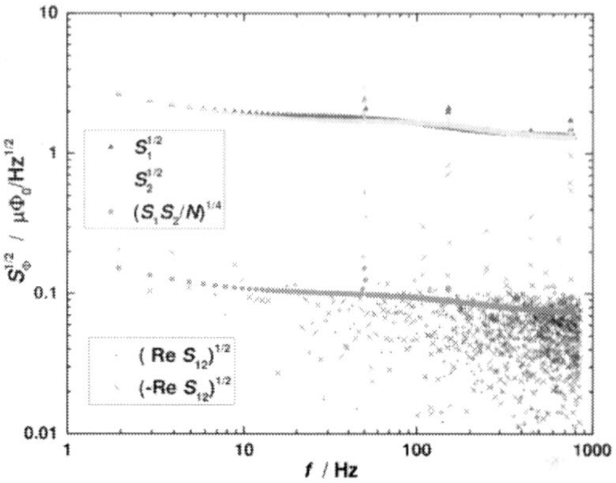

Figure 3. Flux noise referring to the SQUIDs at 4.2 K without temperature sensor. The CPSD comprises $N \approx 98000$ averages.

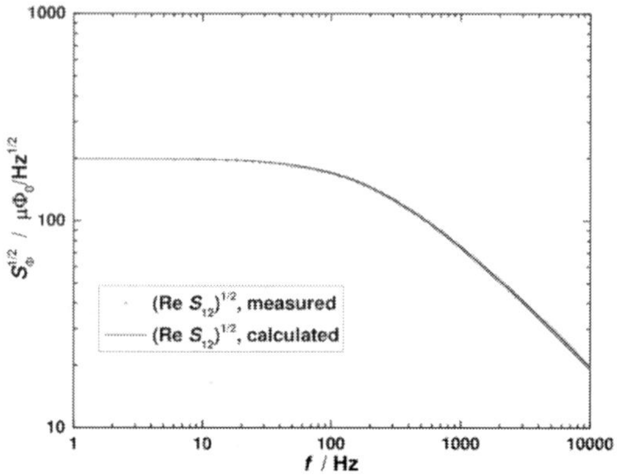

Figure 4. Flux noise referring to the SQUIDs at 4.23 K with temperature sensor. $N \approx 12200$ for experimental CPSD data.

3.2. Temperature measurements

To demonstrate the applicability of the concept, first test measurements with the pMFFT operated in secondary mode have been carried out in the temperature range from 4.2 K to 9 mK. In secondary mode the realization of only one precisely known reference temperature T_{Ref} is needed at which a so-called calibration spectrum $S_{12}(f,T_{\mathrm{Ref}})$ is acquired for the TMFN. The unknown temperature one is interested in is then obtained by averaging the ratio of the CPSDs over all frequency bins:

$$T = T_{\mathrm{Ref}} \left\langle \frac{\mathrm{Re}(S_{12}(f,T))}{\mathrm{Re}(S_{12}(f,T_{\mathrm{Ref}}))} \right\rangle . \quad (7)$$

This equation was evaluated using an improved non-parametric estimation according to [14] for frequencies up to 1 kHz, although the usable correlation bandwidth [8] was much larger. The reference temperatures T_{2000} were provided by a MFFT calibrated according to the PLTS-2000 with a relative uncertainty of $4 \cdot 10^{-4}$. The calibration of the pMFFT using the MFFT was carried out at $T_{\mathrm{Ref}} = 0.478$ K.

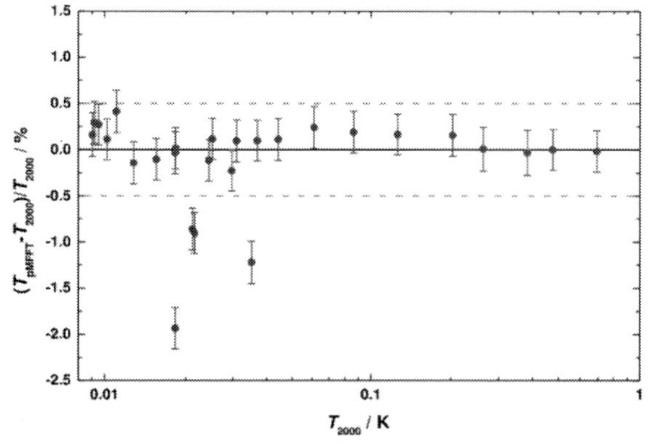

Figure 5. Relative deviations of the temperature values measured with the pMFFT in secondary mode from a copy of the PLTS-2000. Error bars indicate the relative uncertainty for a coverage factor $k = 2$. The red dashed lines mark the ±0.5% deviation range.

Figure 5 shows the relative deviations of the temperature readings of the pMFFT from the reference thermometer. Most of the data points are well within a ±0.5% interval having a typical relative uncertainty of 0.25% ($k = 2$) dominated by the uncertainty of T_{Ref} and the statistical uncertainty. From the good agreement between the pMFFT and the reference thermometer we conclude that the influence of non-thermal magnetic flux noise present in both thermometers in different forms is negligible in the investigated temperature range. This is because the pMFFT results are based on a cross-correlation measurement whereas the reference thermometer values are based on a single-channel PSD. For the temperature sensor material of both the pMFFT and the reference thermometer, high purity copper containing only negligible amount of magnetic impurities was used from different batches purchased from the same supplier [13] within a few years. As no significant deviation in temperatures is visible over almost three decades of temperature we conclude that the influence of residual magnetic impurities in the sensor materials is negligible and the electrical conductivity for the copper used can be assumed to be in residual resistance limit and constant. Four data points fall below T_{2000} by more than 0.5%. Although the reason for that deviation is not yet clear, these points were identified as outliers by an abnormal single-channel noise in both channels.

4. Conclusions

We have developed and successfully tested the prototype of a calculable MFFT designed for primary thermometry at low temperatures. The new pMFFT utilizes a cross-correlation technique for the readout of the TMFN signals. The two SQUID channels are virtually independent from each other due to a special layout of the planar thin-film detection coils that minimizes the mutual induction and because the FLL feedback is applied into the detection coils. The magnetic flux measurements agree well with analytical calculations for the set-up. First measurements in the temperature range from 9 mK to 4.2 K in secondary mode of operation showed an excellent agreement within 0.5% with a reference thermometer carrying a copy of the PLTS-2000.

As a next step the calibration coil will be integrated in the pMFFT. This will enable us to operate the pMFFT in primary mode and to calculate thermodynamic temperature values directly from the measured TMFN. The pMFFT will finally contribute to resolve the long-standing discrepancies in the background data of the PLTS-2000 [15].

Acknowledgments

The research reported here was carried out within the EURAMET project "InK – Implementing the new kelvin" [15] and funding is kindly acknowledged from the European Community's 7th Framework Programme, ERA-NET Plus, (Grant Agreement No. 217257).

References

[1] White D R, Galleano R, Actis A, Brixy H, De Groot M, Dubbeldam J, Reesink A, Edler F, Sakurai H, Shepard R and Gallop J 1996 *Metrologia* **33(4)** 325
[2] Netsch A, Hassinger E, Enss C and Fleischmann A 2005 *AIP Conf. Proc.* vol 850 p 1593
[3] Harding J T and Zimmerman E 1968 *Phys. Lett.* **27A N10** 670
[4] Rothfuß D, Reiser A, Fleischmann A and Enss C 2013 *Appl. Phys. Lett.* **103** 052605
[5] Beyer J, Drung D, Kirste A, Engert J, Netsch A, Fleischmann A, Enss C 2007 *IEEE Trans. Appl. Supercond.* vol. 17, issue 2, part 1, p 760
[6] Engert J, Beyer J, Drung D, Kirste A, Heyer D, Fleischmann A, Enss C and Barthelmess H-J 2009 *Journal of Physics: Conference Series* **150** 012012
[7] Beyer J, Schmidt M, Engert J, AliValiollahi S, Barthelmess H J, 2013 *Supercond. Sci. Technol.* **26** 065010
[8] White D R, Benz S P, Labenski J R, Nam S W, Qu J F, Rogally H and Tew W L 2008 *Metrologia* **45** 395
[9] Drung D, Aßmann C, Beyer J, Kirste A, Peters M, Ruede F and Schurig T 2007 *IEEE Trans. Appl. Supercond.* vol. 17, issue 2, part 1, p 699
[10] Kirste A, Drung D, Beyer J and Schurig T 2008 *J. Phys.: Conf. Ser.* **97** 012320
[11] Rajotte R J and Drouet M G 1975 *IEEE Trans. Power Appar. Syst.* **PAS-94** 89
[12] Sakakibara T, Goto T and Miura N 1989 *Rev. Sci. Instrum.* **60(3)** 444
[13] MaTecK GmbH, Im Langenbroich 20, D-52428 Jülich, Germany
[14] Wübbeler G and Elster C 2013 *Meas. Sci. Technol.* **24** 115004
[15] http://projects.npl.co.uk/ink/

Journal of Physics: Conference Series **568** (2014) 032013 doi:10.1088/1742-6596/568/3/032013

Effects of Large Nuclear Quadrupoles on Dielectric Properties of Glasses at Very Low Temperatures

A Luck[1], **A Fleischmann**[1], **A Reiser**[1] **and C Enss**[1]

[1] Kirchhoff-Institut für Physik, Universität Heidelberg, Im Neuenheimer Feld 227, 69120 Heidelberg, Germany

E-mail: `annina.luck@kip.uni-heidelberg.de`

Abstract.
The universal behaviour of amorphous solids at low temperatures, governed by atomic tunneling systems as described by the standard tunneling model, has long been a generally accepted fact. In the last years, however, measurements of dielectric two-pulse polarization echoes have revealed that nuclear quadrupole moments involved in atomic tunneling systems can cause specific material-dependent effects in magnetic fields. We have performed measurements of the dielectric properties of the two multicomponent glasses N-KZFS11 and HY-1, containing several percent of tantalum oxide and holmium oxide respectively. As ^{181}Ta and ^{165}Ho both carry very large nuclear quadrupole moments, these glasses are ideal candidates to study the influence of nuclear quadrupole moments on the properties of glasses at very low temperatures. Our measurements not only show unique dielectric behaviour in both glasses, but also differ significantly from various predictions of the standard tunneling model.

1. Introduction

The low temperature properties of amorphous solids are governed by atomic tunneling systems. A widely accepted phenomenological description of these tunneling systems is given by the standard tunneling model (STM) [1, 2]. The model assumes that the tunneling systems can be described as particles moving in double-well potentials. The tunneling motion results in a ground state splitting of the double-well potential given by energy splitting $E = \sqrt{\Delta_0^2 + \Delta^2}$, where $\Delta_0 = \hbar\Omega \mathrm{e}^{-\lambda}$ denotes the tunneling splitting determined by the ground state energy of the double-well potential $\hbar\Omega$ and the tunneling probability λ. The asymmetry energy Δ reflects a possible difference in depth of the two wells. In amorphous solids the parameters are broadly distributed and a central assumption of the STM is a uniform distribution of the form $P(\Delta, \lambda)\mathrm{d}\Delta\mathrm{d}\lambda = \overline{P}\,\mathrm{d}\Delta\mathrm{d}\lambda$.

For the real part of the low temperature and low frequency dielectric susceptibility of insulating glasses, the STM predicts a minimum in the dielectric constant as a function of temperature. Below this minimum, the temperature dependence of the dielectric constants is governed by a resonant interaction of the atomic tunneling systems with the electric field, leading to a logarithmic variation $\delta\epsilon/\epsilon = -C\,\ln(T/T_0)$, where C is the coupled density of states and T_0 an arbitrarily chosen reference temperature. Above the minimum, one phonon relaxational processes add to the temperature variation of the dielectric constant and lead to a logarithmic increase given by $\delta\epsilon/\epsilon = C/2\,\ln(T/T_0)$, thus the slope ratio in the two logarithmic regimes should be $-2 : 1$ according to the STM. The relaxational processes also lead to a frequency dependence of the minimum temperature, which is expected to vary as $T_{\min} \propto f^{1/3}$, where f denotes the frequencies (for a review see for example [3, 4]).

In many aspects the STM describes the thermal, acoustic and dielectric properties of glasses at low temperatures quite well. However, at the same time significant deviations have been reported that clearly show this model is incomplete. Examples are the systematic deviations from the predicted slope ratio of $-2:1$ both of the dielectric constant [5, 6, 7, 8] and the sound velocity [9, 10]. The degree of observed universality has also lead to proposed modifications of the STM [11, 12].

In some glasses one also observes an unexpected dependence of the dielectric constant on small magnetic fields [13, 14, 15]. Two models have been proposed to explain the magnetic field sensitivity of the low frequency dielectric properties by a coupling of the magnetic field to loop-like tunneling paths of mutually interacting charged tunneling particles [16, 17]. The observation of a magnetic field dependence in dielectric polarization echoes performed on multicomponent glasses at about 1 GHz [18] and the explanation of such an effect as being caused by tunneling particles carrying a nuclear quadrupole moment [17, 19, 20] has naturally brought up the question whether the deviations at low frequencies are also caused by nuclear moments. Several extensions of the STM regarding the role of nuclear quadrupoles have been proposed, in which corrections for the dielectric permittivity in the resonant regime at temperatures where the thermal energy becomes comparable to the quadrupole splitting were predicted [21, 22, 23, 24]. In particular, it was argued that the nuclear quadrupole interaction of atomic tunneling atoms increases the number of levels and thereby increases the transition probability, which in turn enhances the dielectric constant. In addition, a transition from coherent to incoherent tunneling was proposed for tunneling systems whose bare tunneling splitting matches the quadrupole splitting. Since the typical quadrupole splittings of the constituents in most glasses are of the order of 1 mK and below, such effects should be only relevant at ultralow temperatures.

In order to investigate the importance of nuclear quadrupoles at low frequencies, we have conducted measurements on both a standard probe of fused silica, Herasil (Heraeus Quartz Glass), and the two multicomponent glasses N-KZFS11 (Schott) and HY-1 (Hoya) containing several percent of tantalum oxide and holmium oxide respectively. Since both elements carry very large quadrupole moments, a possible effect of nuclear quadrupoles on the dielectric constant at low frequencies should be most pronounced.

2. Experimental Technique

In order to enable strong thermal coupling of our glass samples to the mixing chamber of our dilution refrigerator, we used a novel capacitor design instead of a classical parallel plate setup. In our case, the capacitor consists of a thin glass plate (12 mm×12 mm×0.5 mm) with

Figure 1. Photographic image of a 1mm×1mm section of our microstructured electrodes on one of our samples. Both the width and the pitch of the digits are 25 μm.

microstructured interdigital electrodes on one side, as depicted in figure 1, and a metallic film as the ground plate on the other side. The metallization layers on both sides consist of a layer of copper covered by a thin gold layer in order to prevent a degradation of the metal by oxidation, as shown schematically in figure 2. The digits have a length of about 1 cm and both a width

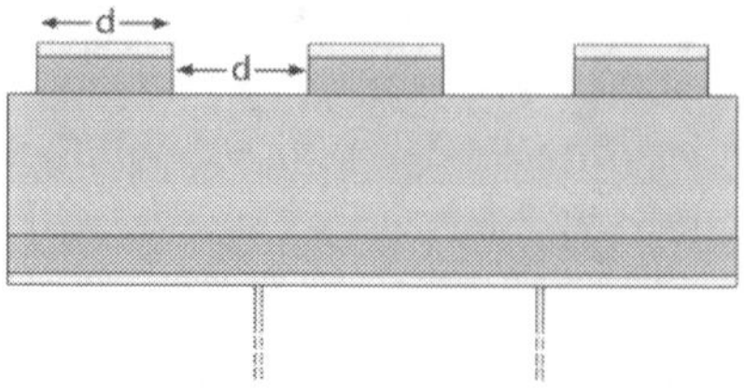

Figure 2. Schematic image of the capacitor with the microfabricated interdigital electrodes and the ground plate consisting of copper and gold layers on the glass sample. Both the width and the pitch of the digits are $d = 25\,\mu$m. Thermal contact can be improved by adding gold bonds.

and a pitch of $d = 25\,\mu$m. This very compact design has two advantages: it provides rather large absolute values of the capacitances ranging from 35 to 80 pF, depending on the dielectric susceptibility ϵ of the samples, which makes the setup somewhat less sensitive to changes of the cable capacitance or possible stray capacitances in our experiment. At the same time our setup allows for good thermal coupling of the sample via gold bonds between the ground plate and the sample holder. The electrical contacts to the interdigital electrodes were made by aluminum bonds to minimize parasitic heating of the samples through the coaxial cables that connected the experimental stage to the room temperature electronics.

The temperature dependence of the capacitance of our samples was measured at frequencies ranging from 120 Hz to 20 kHz using a commercial capacitance bridge (Andeen Hagerling AH2700A). The excitation voltage of the bridge was chosen sufficiently low in order not to exceed the range of linear response of the samples. This was studied in detail on the Herasil sample by applying voltages both high enough to cause nonlinear responses and low enough to see a voltage independent response of the sample.

3. Experimental Results

The measurements on the Herasil sample yielded results similar to those seen in previous investigations on quartz glass samples [6, 7] which proved the proper functionality of our new setup. In particular, it demonstrates that nonlinear effects due to the inhomogeneous electric fields caused by the interdigital structure are negligible, if the excitation voltage is chosen sufficiently low. In figure 3, the Herasil data are shown for frequencies between 120 Hz and 16 kHz. Note that the data points for different frequencies where shifted to be aligned at the

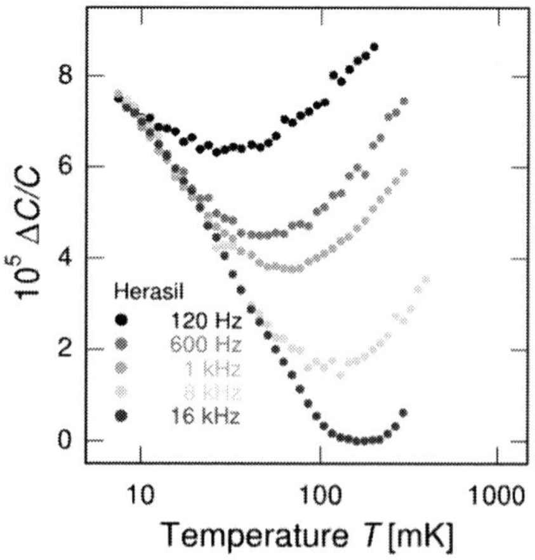

Figure 3. Dielectric constant of the quartz glass Herasil as a function of temperature at frequencies between 120 Hz and 16 kHz.

lowest temperatures, where the frequency independent resonant contribution should dominate.

It thus becomes clearly visible that all curves show the same temperature dependence in the resonant regime, while showing a clear frequency dependence at higher temperatures, where this is expected due to the relaxational contribution. Both logarithmic regimes are visible and the minimum of the dielectric constant shifts as expected for the one phonon relaxation. The data also show the typical deviation from the STM, namely the slope ratio being $-1:1$ rather than $-2:1$. We would like to point out that this is an observation that has been made virtually in all low frequency dielectric measurements on glasses so far.

Let us now turn to the results obtained for N-KZFS11, containing several percent of ^{181}Ta, which are shown in figure 4. As in the case of Herasil the data were shifted to align at the low temperature resonant regime. At first glance the overall behaviour looks very similar to that of the Herasil sample. One finds a minimum and both logarithmic regimes, as theoretically predicted. However, closer inspection reveals two striking differences, namely the fact that for all frequencies below 4 kHz only a very small frequency dependent shift of the minimum is seen and that the slope ratio is $-2:1$, as predicted by the STM but not observed in other glasses. The second remarkable observation is made clearly visible in figure 5, where the data taken at

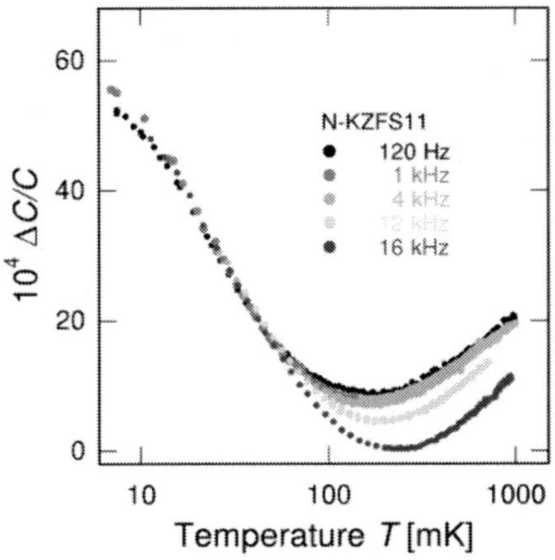

Figure 4. Dielectric constant of the multicomponent glass N-KZFS11 as a function of temperature at frequencies between 120 Hz and 16 kHz.

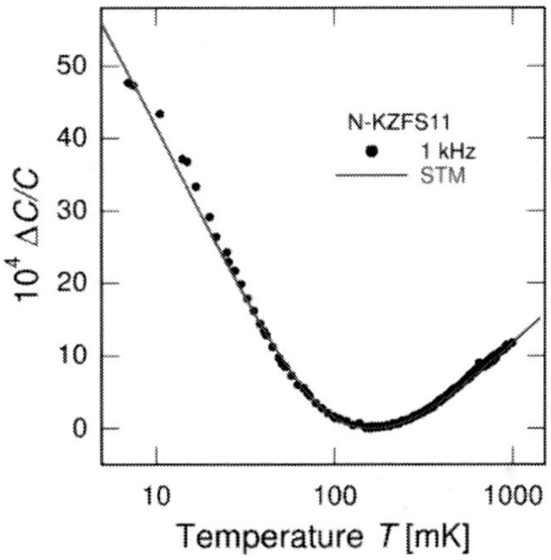

Figure 5. Dielectric constant of the multicomponent glass N-KZFS11 as a function of temperature at a frequency of 1 kHz (●). The solid line represents a numerical fit of the data based on the STM.

1 kHz are shown together with a numerical calculation based on the STM. Obviously, for this glass the data are in very good agreement with the prediction of the STM. We would like to point out, however, that there is still a small but systematic deviation between fit and data at temperatures below about 50 mK. It is also remarkable that the temperature at which the minimum appears is significantly higher than in other samples.

Figure 6 shows the results we have obtained for the third glass that we investigated with our new setup, the multicomponent glass HY-1 containing ^{165}Ho. The measurements were performed at frequencies between 120 Hz and 20 kHz. The most striking observation for this sample is that there appears to be almost no frequency dependence of the data at all. In particular, we do not see any shift of the minimum of the dielectric constant in the entire frequency range investigated. This is a behaviour not seen at any other amorphous material before. The slope ratio, in contrast, being roughly $-0.8:1$, is in line with the vast majority of measurements on glasses, except the one shown above on N-KZFS11.

In figure 7 we summarize our findings regarding the frequency dependence of the minima

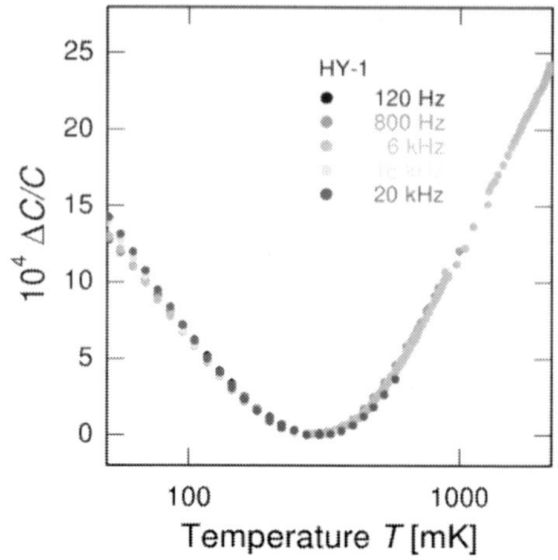

Figure 6. Dielectric constant of the multicomponent glass HY-1 as a function of temperature at frequencies between 120 Hz and 20 kHz.

for all three samples. Starting with Herasil we see that the minima appear at relatively low temperatures, indicating a strong coupling between the tunneling systems and thermal phonons. The minima shift proportional to $f^{1/3}$, as expected by the STM and indicated by the red solid line. The N-KZFS11 sample shows minima at systematically higher temperatures than Herasil. The temperature dependence of the minima only agrees with the STM prediction above 4 kHz and becomes frequency independent below this frequency. The minima observed for the ^{165}Ho containing HY-1 glass lie at temperatures well above those of both other samples and show only a very slight frequency dependence. Although we have no detailed theoretical explanation for

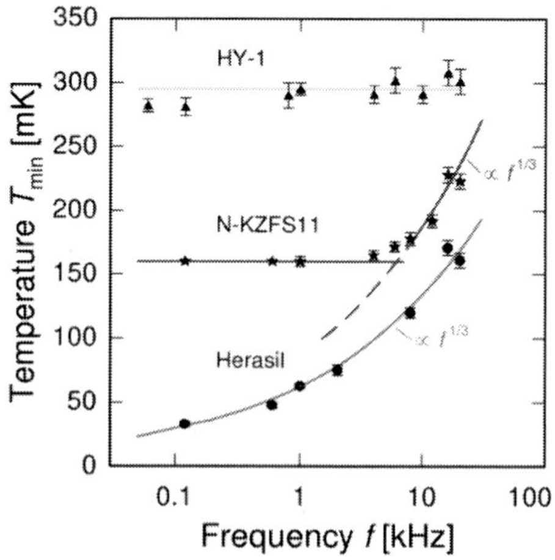

Figure 7. Frequency dependence of the minima of all three glasses being investigated in this work. Both the frequency independent regions and the regions where the minima vary proportional to $f^{1/3}$ are indicated by solid lines.

our findings at this point, it seems to be clear that the presence of nuclear quadrupole moments influences the low frequency dielectric behaviour of glasses quite dramatically. In particular, one is tempted to conclude from figure 7 that there is another relaxational mechanism acting besides the one phonon process in the two glasses containing elements with large nuclear quadrupole moments. We can only speculate at this point whether this relaxational mechanism is perhaps driven by the interaction of atomic tunneling systems carrying nuclear quadrupoles with all other surrounding nuclear moments. On the contrary, we have not seen any clear and systematic indications of the resonant part being modified by the presence of the quadrupole moments as suggested by [21, 22, 23, 24]. It is well possible that the temperatures we have reached in our experiments were not sufficiently low to observe the transition from coherent to incoherent tunneling. The only hint that the resonant part is perhaps affected by presence of large nuclear quadrupole moments is the observation, that the slope ratio is different for the N-KZFS11 sample compared to all other glasses investigated so far. One could interpret the steeper resonant part seen in this experiment as the expected enhanced resonant susceptibility, however, one would then expect a similar effect for the HY-1 glass, which is not observed.

4. Conclusions

We have investigated the frequency and temperature dependence of the dielectric constant of three different glasses at low temperatures using a novel setup. Two of those glasses contained elements with large quadrupole moments on the percent level and were chosen specifically to investigate a possible influence of these nuclear quadrupoles on the dielectric properties of glasses in this frequency and temperature range. Our findings clearly show that the dielectric properties of insulating glasses are not universal as predicted by the STM and that elements with large nuclear quadrupole moments strongly influence their low frequency dielectric behaviour. The striking dependence on specific elements may guide the way to solve several puzzles regarding the low temperature properties of glasses in the future.

5. Acknowledgements

We would like to thank M. Hempel, D. Rothfuß and S. Hunklinger for their support and many fruitful discussions. Part of this work has been supported by the European Community Research Infrastructures under the FP7 Capacities Specific Program, MICROKELVIN Project No. 228464.

References

[1] Anderson P W, Halperin B I and Varma C M 1972 *Phil. Mag.* **25** 1
[2] Phillips W A 1972 *J. Low Temp. Phys.* **7** 351–360
[3] Hunklinger S and Enss C 2000 Insulating and semiconducting glasses *Series of Directions in Condensed Matter Physics* vol 17 ed Boolchand P (World Scientific) p 499
[4] Esquinazi P (ed) 1998 *Tunneling systems in amorphous and crystalline solids* (Springer)
[5] Enss C, Bechinger C and v Schickfus M 1989 *Phonons 89, World Scientific, Singapore*
[6] Classen J, Enss C, Bechinger C, Weiss G and Hunklinger S 1994 *Annalen der Physik* **506** 315–335
[7] van Rooijen R, Marchenkov A, Akimoto H, Jochemsen R and Frossati G 1998 *J. Low. Temp. Phys.* **110**(1) 269–274
[8] Rogge S, Natelson D, Tigner B and Osheroff D D 1997 *Phys. Rev.* B **55** 11256–11262
[9] Classen J, Burkert T, Enss C and Hunklinger S 2000 *Phys. Rev. Lett.* **84** 2176–2179
[10] Fefferman A D, Pohl R O, Zehnder A T and Parpia J M 2008 *Phys. Rev. Lett.* **100** 195501
[11] Leggett A J 1991 *Phys. B Condens. Matter* **169** 322–327
[12] Leggett A J and Vural D C 2013 *J. Phys. Chem. B* **117** 12966–12971
[13] Strehlow P, Enss C and Hunklinger S 1998 *Phys. Rev. Lett.* **80** 5361–5364
[14] Strehlow P, Wohlfahrt M, Jansen A G M, Haueisen R, Weiss G, Enss C and Hunklinger S 2000 *Phys. Rev. Lett.* **84** 1938–1941
[15] Haueisen R and Weiss G 2002 *Physica* B **316317** 555 – 557
[16] Kettemann S, Fulde P and Strehlow P 1999 *Phys. Rev. Lett.* **83** 4325–4328
[17] Würger A, Fleischmann A and Enss C 2002 *Phys. Rev. Lett.* **89** 237601
[18] Ludwig S, Enss C, Strehlow P and Hunklinger S 2002 *Phys. Rev. Lett.* **88**(7) 075501
[19] Würger A 2004 *J. Low Temp. Phys.* **137** 143–166
[20] Parshin D 2004 *J. Low Temp. Phys.* **137** 233–250
[21] Burin A L, Hunklinger S, Enss C and Fleischmann A 2006 *AIP Conference Proceedings* **850** 1127–1128
[22] Burin A L, Polishchuk I Y, Fulde P and Sereda Y 2006 *Phys. Rev.* B **73** 014205
[23] Burin A L, Polishchuk I Y, Fulde P and Sereda Y 2006 *Phys. Rev. Lett.* **96**(2) 025505
[24] Sereda Y, Polishchuk I Y and Burin A L 2007 *Phys. Rev. B* **75**(2) 024207

Journal of Physics: Conference Series **568** (2014) 032014 doi:10.1088/1742-6596/568/3/032014

A new ultra-low-temperature cryogen-free experimental platform

G Batey, A J Matthews and M Patton

Oxford Instruments Omicron NanoScience
Tubney Woods, Abingdon, Oxfordshire, UK, OX13 5QX

E-mail: Anthony.Matthews@oxinst.com

Abstract. We report the introduction of a new cryogen-free dilution refrigerator experimental platform that provides significant performance enhancements, in several key areas, over the current generation of systems. In particular the ability to: install more experimental services; install higher-field experimental magnets; dissipate more power at the ~ 4 K stage; and to attain higher cooling powers and lower base-temperatures (below 3.5 mK) at the mixing chamber plate.

1. Introduction

The cryogen-free dilution refrigerator[1], often integrated with a superconducting magnet[2], has become the workhorse of many low-temperature laboratories around the world, and such systems are routinely used to attain temperatures below ~ 10 mK. In some fields of research, such as quantum information processing, these cryogen-free machines are preferred to "wet" systems using liquid ^{4}He as the installation of experimental services is more straightforward, for a discussion see[3].

There is now a drive from researchers in these fields for the ability to install more experimental hardware, particularly rigid coaxial microwave transmission lines and the associated signal conditioning elements (attenuators, circulators, bias-tees etc.) as well as the ability to run dissipative "cold" electronics at temperatures ~ 4 K. The heat loads from these experimental services need to be balanced carefully against the (limited) available cooling power from the mechanical refrigerators[4] used to generate the cold environment for these systems.

To facilitate the higher cooling powers required at 4 K we have developed a new experimental platform that is configurable with either one or two 1 or 1.5 W pulse tube coolers, that offers more access for the installation of services, and that can be integrated with large, high-field magnets. Optionally, this platform can be fitted with a powerful dilution refrigerator to enhance the cooling power available at lower temperatures. The design and performance of this platform shall be discussed in the following sections.

2. The experimental platform

The overall layout and principle of operation of this new system (single vacuum space, pre-cool circuit, Joule-Thomson condensing stage, automated operation etc.) is similar to smaller systems that have been described previously[2] and so here we focus on the enhancements this new platform offers.

2.1. Experimental access

The size of all the experimental plates have increased significantly. The mixing chamber plate on this platform is now 430 mm diameter and features 6×50 mm and 1×100 mm diameter line-of-sight ports to the system top plate.

The ~ 4 K plate (which determines the maximum dimensions of any magnet that can be integrated with the system) is 650 mm diameter.

The separations between the lower experimental plates have all been increased (to ~ 250 mm for the 4 K plate $\rightarrow$ still plate and cold-plate $\rightarrow$ mixing chamber plate, and to ~ 175 mm for the still plate $\rightarrow$ cold-plate) to provide more space to mount components, and to make fitting them more straightforward. Greater user access to the system is also facilitated by modifications to the support structure, as described in section 2.2.

The larger plates result in a longer cool down time than for smaller systems. The exact time from room temperature to base temperature will depend of the configuration of the pulse tube cooler(s) and the mass of any installed magnet, but is typically in the region of 50 hours for dual-cooler systems without magnets.

2.2. Support structure

The upper sections of the support structure (room-temperature $\rightarrow$ 70 K and 70 K $\rightarrow$ 4 K) are manufactured from stainless-steel tube and comprise three equally spaced straight legs. Between the 70 K plate and 4 K plate these tubes can be used to provide self-contained heat-pipes[5] to reduce the cool down time of large integrated magnets[6].

For the lower temperature stages `MIL-I-24768/2` G10 tubes are preferred between the 4 K plate and the still, and carbon-fibre sections are used below the still.

This approach gives an open support structure, as shown in figure 1, which is mechanically very stiff and can support heavy experimental payloads (several hundred kg at the 4 K stage and several tens of kg at the mixing chamber) whilst also minimising its thermal conductivity. A discussion of the resulting heat load to the mixing chamber shall be given in section 4.2.

Whilst not yet measured, it is anticipated that the stiff support structure and more massive plates should result is vibration levels that are as good, if not better, than those measured on similar smaller systems of $< 0.1\ \mu m^2\ Hz^{-1}$ from 0.1 Hz to 1 kHz.

3. 4 K stage cooling power

Pulse tube coolers are a source of mechanical vibration and are normally decoupled from the refrigerator plates using high thermal-conductivity, flexible copper-braids[2]. However, the conductivity of these braids is finite, so there is always some δT between the pulse tube stage and the plate.

Of interest to the experimentalist is the effective cooling power of the plate (to which their equipment is mounted) and the maximum dissipation permissible before the performance of the dilution refrigerator (and / or integrated magnet) is compromised. To quantify this a system with two Cryomech `PT415-RM`[7] pulse tube coolers, each of which is specified to have a second stage cooling power of 1.35 W at 4.2 K, was used. The temperature of both the pulse tube head and the experimental plate were determined using calibrated `cernox`[8] sensors and the power was supplied electrically *via* a resistive heater mounted near the centre of the plate. The results of these cooling power tests are summarised in figure 2. For the system described above, it was found that applying powers ~ 1 W had no measurable effect on the base temperature of the dilution refrigerator (which was below 6 mK) or the available cooling power at 100 mK. 2 W warmed the mixing chamber by around 2 mK. The operation of the integrated magnet will be discussed in section 3.1.

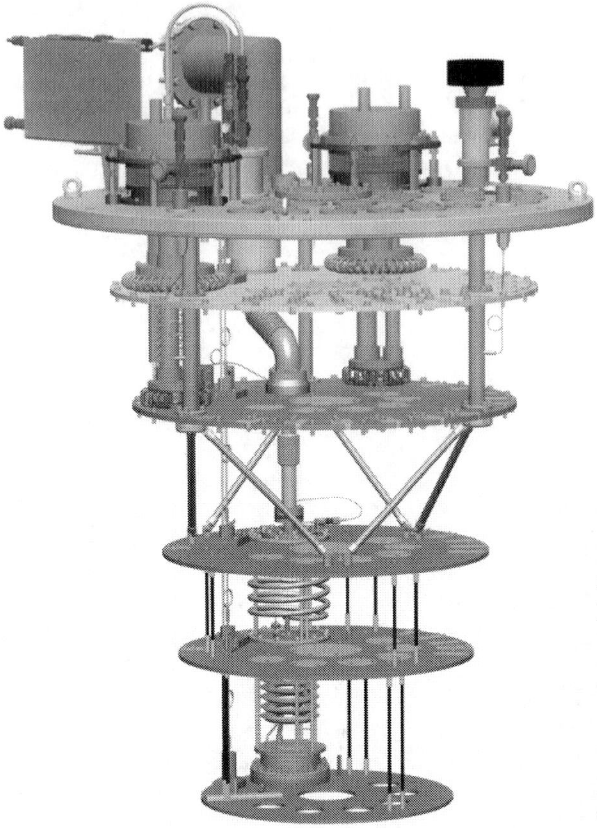

Figure 1. An image showing the general layout of the refrigerator with the vacuum and radiation shields removed. In this case the two pulse tube coolers can be seen at the top of the system and the large dilution unit at the bottom. For scale, the lowest plate (at the temperature of the mixing chamber) is 430 mm in diameter.

3.1. High-field magnets

Since the introduction of cryogen-free refrigerators with integrated superconducting magnets[2] there has been a push for ever higher fields at the sample position. Both the physical size and mass of a magnet increases rapidly as its design field goes up, meaning they can only be integrated onto a platform capable of supporting these heavy loads. The same is true for vector-rotation magnets (which allow the angle of the applied field with respect to the sample to be varied) as their formers become large to withstand the high forces between the coil sets as the fields become bigger.

The recent demonstration of the integration of nuclear demagnetisation stages with cryogen-free refrigerators to attain temperatures in the μK regime[9] has generated interest in providing a high-field experimental magnet (in addition to the demagnetisation stage magnet) in the same system.

The platform described in this article, with its larger size and upgraded support structure is suitable for all these applications.

A 16 T, 57 mm bore solenoid (which the authors believe to be the highest field magnet to be integrated with a cryogen-free refrigerator to date) that has a mass $\sim$ 130 kg, an inductance $\sim$ 60 H and an operating current $\sim$ 140 A has been installed onto this refrigerator platform.

For systems requiring high cooling powers at the 4 K stage, magnets with higher critical-temperatures can also be manufactured. A wide bore (150 mm) 5 T magnet that can be ramped to full field at a temperature of $>$ 5 K has been demonstrated. This means, that for the dual pulse tube system described in section 3, both the magnet and dilution refrigerator can operate

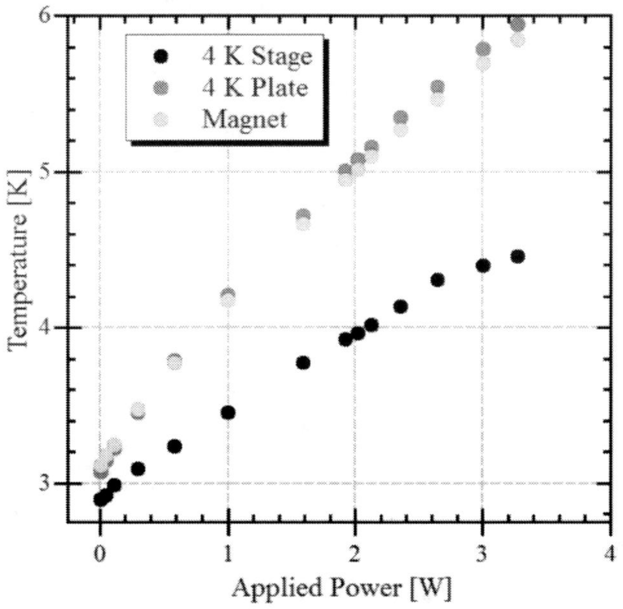

Figure 2. A graph showing the temperatures of the pulse tube cooler second stage, 4 K plate and magnet as a function of the power applied to the 4 K plate.

whilst a heat load ≈ 2 W is applied to the "4 K" plate.

4. High-performance dilution refrigerator

Dilution refrigerators with lower base temperatures and / or higher cooling powers at a given temperature are becoming a requirement for researchers working to develop general quantum computers. This is either because they hope their qubit will exploit a delicate quantum system that only manifests itself at extremely low temperatures[10], or (and) they hope to scale up to multiple qubit devices[11] and require higher cooling powers to overcome the additional dissipation per qubit whilst maintaining the system at a suitably low temperature.

The performance of a powerful dilution unit designed to be integrated with this new platform shall be described in the following sections.

4.1. Cooling Power

The cooling power of a dilution refrigerator at some elevated temperature (usually around 100 mK) is often chosen as a figure of merit. However it is not always appropriate to scale this performance to very much lower temperatures with simple relationships such as $\dot{Q} \propto T^2$ since the total heat load to the mixing chamber (including effects such as viscous heating) will be temperature dependant[12], as will be the optimal ^{3}He circulation rate.

The high-performance dilution unit designed for this experimental platform was characterised with a series of measurements. The cooling capacity was determined in vacuum, on the mixing chamber plate, as a function of temperature, figure 3. We find the cooling capacity at 100 mK is ≈ 1 mW, and is $> 25\ \mu$W at 20 mK.

4.2. Base temperature

The base temperature was determined with a commercial[13] noise thermometer[14] and verified using nuclear orientation[15]. Both thermometers were mounted in vacuum on the mixing chamber plate. The averaged temperature reported by the noise thermometer was

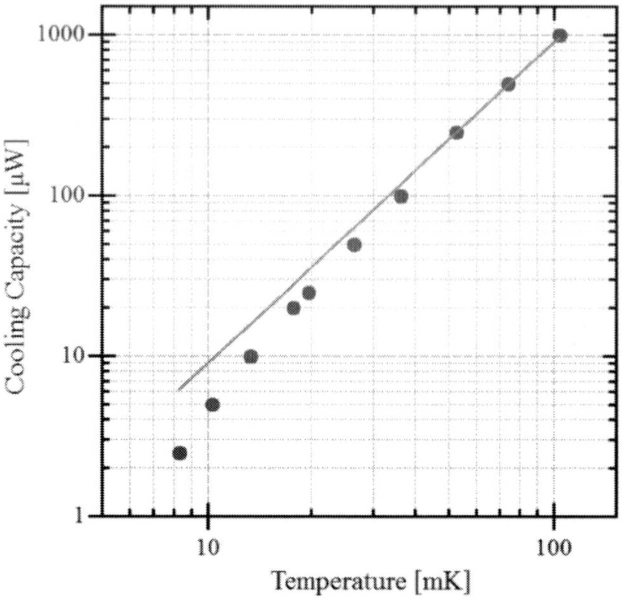

Figure 3. The cooling power available on the mixing chamber plate for a system equipped with the high-cooling power dilution refrigerator discussed in the main text. The 100 mK cooling power can be scaled as $\dot{Q} \propto T^2$ with reasonable accuracy at temperatures > 35 mK.

3.253 ± 0.016 mK (which the authors believe to be the lowest continuous temperature attained with a cryogen-free dilution refrigerator). From an analysis of the cooling power at low temperatures the total background heat load into the mixing chamber can be extracted. In this case we find a heat leak ≈ 240 nW.

References

[1] Uhlig K 2002 *Cryogenics* **42** 73–77
[2] Batey G, Buehler M, Cuthbert M, Foster T, Matthews A J, Teleberg G and Twin A 2009 *Cryogenics* **49** 727 – 734
[3] Batey G, Chappell S, Cuthbert M, Erfani M, Matthews A J and Teleberg G 2014 *Cryogenics* **60** 24 – 32
[4] Wang C and Gifford P E 2002 *AIP Conference Proceedings* **613** 641 – 648
[5] Tien C L 1975 *Annual Review of Fluid Mechanics* **7** 167 – 185
[6] Teleberg G, Matthews A J, Batey G and Elford J 2012 *Journal of Physics: Conference Series* **400** 052036
[7] Cryomech: `http://www.cryomech.com/products/cryorefrigerators/pulse-tube/pt415/`
[8] Lakeshore: `http://lakeshore.com/products/Cryogenic-Temperature-Sensors/Cernox/Models/`
[9] Batey G, Casey A, Cuthbert M N, Matthews A J, Saunders J and Shibahara A 2013 *New Journal of Physics* **15** 113034
[10] Das Sarma S, Freedman M and Nayak C 2005 *Phys. Rev. Lett.* **94**(16) 166802
[11] Fowler A G, Mariantoni M, Martinis J M and Cleland A N 2012 *Phys. Rev. A* **86**(3) 032324
[12] Pobell F 2007 *Matter and Methods at Low Temperatures* (Springer) 3rd ed
[13] Magnicon: `http://www.magnicon.com/squid-systems/noise-thermometer/`
[14] Casey A, Arnold F, Levitin L V, Lusher C P, Nyki J, Saunders J, Shibahara A, Vliet H, Yager B, Drung D, Schurig T, Batey G, Cuthbert M and Matthews A J 2014 *Journal of Low Temperature Physics* **175** 764–775
[15] Marshak H 1983 *Journal of Research of the National Bureau of Standards* **88** 175 – 217

Acknowledgments

The authors should like to thank C Wilkinson and L Otwell for their contributions to the mechanical design of the systems described in this article, and C Mitchell and N Ashdown for their assistance with the acquisition of the experimental results presented.

Journal of Physics: Conference Series **568** (2014) 032015 doi:10.1088/1742-6596/568/3/032015

Making SQUIDs a practical tool for quantum detection and material characterization in the micro- and nanoscale

Thomas Schurig

Physikalisch-Technische Bundesanstalt
Abbestr. 2-12, D-10587 Berlin, Germany

E-mail: thomas.schurig@ptb.de

Abstract. The investigation of quantum effects and materials at low and ultra-low temperature often requires very sensitive measurements of weak magnetic signals, small electric currents or voltages. Superconducting Quantum Interference Devices (SQUIDs) have been proven as very attractive tools in this field. However, well established fabrication technology and readout techniques usually fail, particularly, when going to nanoscale magnetic detection where off-the-shelf devices can rarely be used. As an alternative to elaborate nanoSQUID technology, SQUID concepts for nanoscale magnetic detection which are employing conventional, and hence, reliable technology are discussed. Magnetic coupling of nano-sized samples to conventional SQUIDs, e.g. simple gradiometers or more complex devices as fully integrated susceptometers, can be improved significantly by integrating nanoscale detection loops into these devices. Furthermore, appropriate SQUID current sensors are a prerequisite for the read-out of micro- and nanoSQUIDs and small-area detection coils. The conventionally made devices are intended for fabrication in moderate numbers to make them available for a broader community.

1. Introduction

Experiments at low and very low temperature often require sensitive measurements of magnetic properties of materials e.g. magnetic susceptibility, magnetic signals as in NMR, or detection of physical quantities which are connected with a magnetic field like weak electrical currents. Superconducting quantum interference devices (SQUIDs) are extremely sensitive cryogenic detectors for magnetic signals and have been proven as very attractive tools for these experiments. Meanwhile, a variety of SQUID based detectors from magnetometers, gradiometers to sophisticated current sensors have been made available even on a commercial basis [1]. Commonly, these devices are designed with microscale dimensions and fabricated using well-established, reliable multilayer thin-film processes with niobium being the superconducting material. Typical pickup coil dimensions for magnetometers and gradiometers range from millimeters up to centimeters, where the coils can be fabricated from superconducting wire or as thin-film components integrated on the SQUID chip. Operating these devices is not straightforward, as they are very sensitive to electromagnetic interference (EMI). Magnetic shielding using superconducting encapsulations is often used for protecting the devices or the complete measuring setup. A number of state-of-the-art SQUID current sensors are designed using gradiometric sensor configurations for suppressing environmental magnetic fields or field gradients, and hence, completely unshielded operation is enabled. Magnetic flux noise levels in the order of 1 $\mu\Phi_0/\sqrt{\text{Hz}}$ are achievable even in unshielded operation.

The characterization of material properties with ultimate sensitivity in the micro- and nano-scale (e.g. SQUID microscopy, micro- and nano-susceptometry), has undergone a growing interest recently and cannot be addressed by classical SQUID designs. For these applications novel SQUID designs have to be developed featuring micro- or nano-sized sensitive areas. So-called nanoSQUIDs, i.e., devices with overall loop dimension in the sub-μm regime (see, e.g., [2], [3] and references therein) have been developed enabling coupling improvement between sample and sensing loop. Furthermore, the inductance of the SQUID ring and the capacitance of the Josephson junctions are lowered to such low values that flux noise levels of the order of $10\ n\Phi_0/\sqrt{Hz}$ should be possible at $T = 4.2$ K. Unfortunately, the best noise levels obtained so far experimentally with thin-film SQUIDs fabricated on a common chip, are more than an order of magnitude higher, $0.2\ \mu\Phi_0/\sqrt{Hz}$ with 1/f corner frequencies above 10 Hz [4], [5]. It is worth mentioning here that the same white noise levels are achieved at 4.2 K both with conventional SQUIDs if the inductance is reduced to about 10 pH [6] and with non-dissipative SQUIDs read out in a dispersive scheme [7]. The performance and the architecture of nanoSQUIDs are mainly limited by the presently available junction technology. Sophisticated SQUID setups have been successfully tested, such as nanoSQUIDs fabricated on the apex of a sharp pulled quartz tube which are intended for nanomagnetic imaging [8]. Even nanoSQUIDs with carbon nanotube junctions have been reported [9]. Constriction junctions can be fabricated more easily by using e-beam lithography or focused ion beam techniques. Overdamped SNS and SNIS junctions have been considered as well and a number of groups have obtained promising results with these junction types as pointed out in more detail below. Furthermore, it is difficult to integrate a feedback coil on the nanoSQUID chip with a sufficiently good coupling to the SQUID loop [10]. In this case if a feedback coil is missing on chip, an external coil system is required to set the nanoSQUID working point and to operate the nanoSQUID in flux-locked-loop mode. Elaborated read-out techniques have to be applied to nanoSQUIDs in order to take full advantage of their improved thermal noise limited sensitivity.

Whereas the manufacturing and operation of nanoSQUIDs is a challenging task, miniaturized SQUIDs and complete SQUID susceptometers can be designed and manufactured employing conventional $Nb/AlO_x/Nb$ technology. In addition, these devices can be upgraded by a single nanopatterning step (e.g. e-beam-lithography, Focused Ion Beam (FIB) technique) to compete with the performance of nanoSQUIDs.

As above, the nanoSQUID still remains an object of dedicated research which has not yet reached the level of an applicable tool. So, an experimentalist not focused on the development of SQUIDs, but interested in a reasonably reliable, robust and easy-to-handle SQUID sensor as a practical instrument for his nano-scale or quantum detection experiment is facing an unsatisfactory situation. This paper is intended to shed light on the current situation and to point out what is achievable with conventional SQUID technology and what is available or will be made available in the near future. Devices manufactured utilizing a hybrid technology as mentioned above, will be introduced and compared to state-of-the-art nanoSQUIDs. Concepts of devices which have been developed mainly at PTB will be discussed in more detail.

2. State-of-the-art nanoSQUIDs

Similar to any other SQUID, a nanoSQUID consists of a superconducting ring with nanoscale dimension interrupted in the case of a dc SQUID (which will be considered in this paper) at two points by a weak link (the so-called "Josephson junction") which has a clearly reduced superconducting ampacity. Conventional SQUIDs with SIS junctions are usually employing a well established multilayer junction technology using Nb as superconductor, AlO_x as insulator and a normal metal (e.g. Pd) for the shunt resistors. However, nanoSQUIDs cannot be fabricated using this technology simply by reducing the linewidth and dimension of the device components. In order to guarantee a nonhysteretic *I-V* characteristics of the nanoSQUID, the damping parameter β_c of the nanoSQUID junctions

$$\beta_c = \frac{2\pi}{\Phi_0} I_0 R^2 C \tag{1}$$

has to be kept low, where Φ_0 is the flux quantum, R is the shunt resistance of the junction, C is the junction capacitance, and I_0 is the junction critical current. In conventional SQUIDs, β_c can be easily controlled by the shunt resistance typically consisting of a normal metal thin-film of an area of about (10-100) μm^2. Unfortunately, such resistors can hardly be implemented in a nanoSQUID, consequently another junction technology has to be considered. As mentioned above, constriction junctions are a good candidate for nanoSQUIDs and have been fabricated from various thin-films including YBCO [2, 11, 12]. Constriction junctions typically exhibit relatively large critical currents I_0 and therefore, they show hysteretic *I-V* characteristics. Due to the strong temperature dependence of the critical current when working near to the critical temperature T_c of the superconductor, β_c can be adjusted by the operating temperature of the nanoSQUID. However, this makes the operating condition and the entire nanoSQUID setup quite laborious [4]. In order to overcome this difficulty, superconductors other than Nb, with a lower critical temperature T_c may be used for nanoSQUID fabrication. If operation of the nanoSQUID at very low temperature is intended, additionally, excess heat in the junctions becomes a problem. These obstacles could be addressed by junctions based on a proximity effect bilayer. A nobel metal layer in combination with a superconducting thin-film is used, where the nobel metal acts as a thermal shunt and is used to control the T_c of the superconductor [13]. Intrinsically shunted SNS junctions could pave the way for a reproducible and mature nanoSQUID technology as it will be pointed out in more detail in section 4.3. Other approaches (such as the nanoSQUID on a capillary [8]) which are not based on a pure thin-film-on-chip technology will not be considered here.

Reading out a SQUID that is operated in a flux-locked loop, generally requires SQUID electronics which is operated at ambient temperature and consists of a low-noise amplifier and an integrator as schematically shown in Figure 1. The output signal of the integrator is fed back into the SQUID as a magnetic flux using a feedback coil in order to linearize the periodic flux-to-voltage characteristics of the sensor. This feedback coil is usually an integrated component of a common SQUID device. In nanoSQUIDs it is much more difficult to integrate such a feedback coil on chip due to technological difficulties and furthermore, the achievable mutual coupling of the coil to the SQUID loop is rather limited [10].

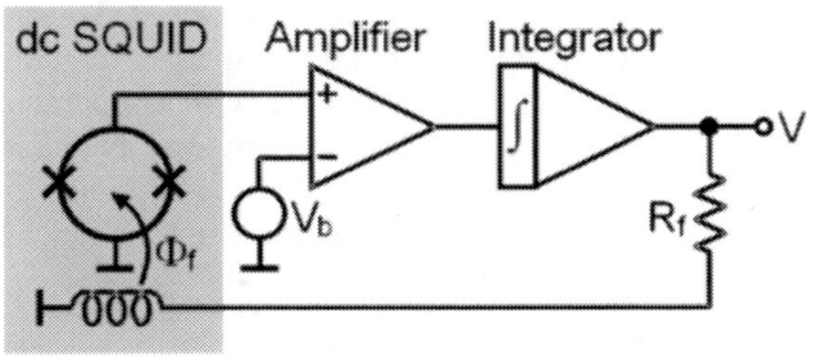

a)

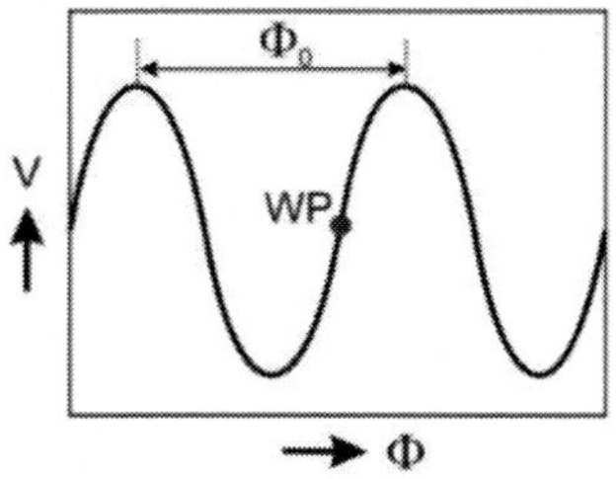

b)

Figure 1. a) Schematic circuit diagram of a SQUID system (grey area: cryogenic part). The SQUID with the two Josephson junctions is symbolized by a circle with two crosses. It is directly coupled to an amplifier in a so-called flux-locked loop (FLL). The amplified output signal of the SQUID is fed back via an integrator, a resistor R_f and a feedback coil as a magnetic flux Φ_f in such a way that the working point WP is kept stable on the periodic flux-to-voltage characteristics of the SQUID shown in b) even though the external magnetic field changes. It should be mentioned that alternatively to the readout with a directly coupled amplifier, flux modulation readout is widely used. In this case, the SQUID is coupled to the amplifier via a transformer and lock-technique is used (for more details see [14]).

It not only merits using nanoSQUIDs for the characterization of nanoscale samples, they are also attractive for achieving very low flux noise levels. Due to the small loop diameter and junction dimension, the SQUID inductance L and junction capacitance C are both low, in contrast to common SQUID sensors. This enables a low spectral flux noise density S_Φ according to the approximation by Tesche and Clarke [15]:

$$S_\Phi \approx 2L\ 16\ k_B T\ (LC)^{1/2} \tag{2}$$

where k_B is the Boltzmann constant and T is the operating temperature.

Though this approximation is not necessarily valid for nanobridge junctions as often used in nanoSQUIDs, it gives an idea of the potential possible noise level. With inductances L of a few pH and capacitances C of a few hundreds of fF, spectral flux noise of levels of $\sqrt{S_\Phi}$ < few 10 $n\Phi_0/\sqrt{Hz}$ could be achieved. However, the flux to voltage transfer coefficient V_Φ of a single nanoSQUID is not sufficient for direct readout with a room temperature preamplifier as shown in Figure 1. In order to benefit from the low intrinsic flux noise of a nanoSQUID, the equivalent flux noise $S_{\Phi,VAmp}$ generated by the voltage noise $S_{V,Amp}$ of the SQUID electronics should be well below the nanoSQUID noise:

$$S_{\Phi,Nano} >> S_{\Phi,VAmp} = S_{V,Amp} / V_\Phi^2 \tag{3}$$

Even with state of the art SQUID electronics which have a voltage noise of $\sqrt{S_{\Phi,VAmp}} = 0.33$ nV/$\sqrt{Hz}$ this is difficult to achieve. In order to overcome this obstacle, a two-stage configuration using a SQUID preamplifier is a useful approach [16]. At PTB, a SQUID series array (SSA) with an integrated bias resistor R_B and a low spectral current noise of $\sqrt{S_{I,SSA}}$ < 10 pA/$\sqrt{Hz}$ at 4.2 K has been developed and used as a cryogenic amplifier for the nanoSQUID [4, 17]. This array consists of 16 SQUIDs with a total input inductance $L_{In,SSA}$ < 3 nH. A schematic setup of the two-stage nanoSQUID readout configuration is shown in Figure 2.

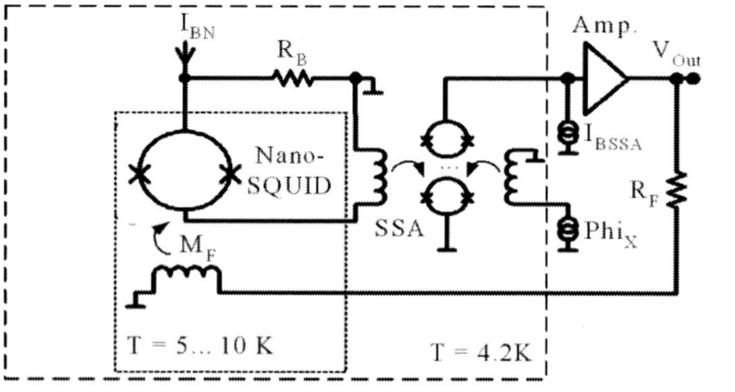

Figure 2. Scheme of two-stage nanoSQUID readout with a SSA amplifier as second stage (Amp. stands for the whole SQUID electronics including integrator). The resistor R_B for biasing the nanoSQUID is integrated on the SSA chip. The SSA is operated at liquid helium temperature whereas the nano SQUID is mounted on a variable temperature setup. For more details see [17].

This scheme has been successfully used for nanoSQUID readout, demonstrating a white flux noise of 0.23 $\mu\Phi_0/\sqrt{Hz}$ at an operating temperature of the constriction junction type nanoSQUID of 7.8 K with a readout contribution of 0.10 $\mu\Phi_0/\sqrt{Hz}$ [17].

Although appropriate SSAs are manufactured using conventional SQUID technology and commercially available together with adopted readout electronics [1], the overall setup is quite sophisticated. In order to provide optimum operating conditions for the nanoSQUID, it is operated on a variable temperature platform. The optimum working point is set up by an external superconducting wire-wound coil. In addition, a second superconducting field coil is used for exciting the sample under investigation (e.g. magnetic nanoparticle) which is attached to the nanoSQUID. Figure 3 shows how the probehead of the nanoSQUID dipstick intended for operation in liquid helium looks like.

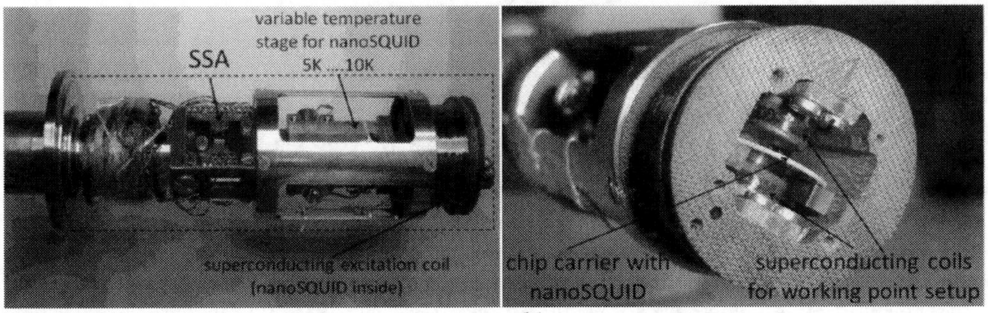

a) b)

Figure 3. Variable temperature platform for operating nanoSQUIDs in a LHe-dipstick.
a) A SSA is mounted on the 4 K platform for preamplifying the nanoSQUID signal according to the scheme in Figure 2.
b) The working point of the nanoSQUID can be set up using superconducting solenoids. Another superconducting coil providing a field perpendicular to that of the working point coil pair is intended for exciting a sample under investigation attached to the nanoSQUID [17].

Using a constriction type nanoSQUID with L ~ 1 pH, and C ~ 0.1 pF in the setup depicted in Figure 3, a white flux noise of $\sqrt{S_\Phi} = 230\ \mathrm{n}\Phi_0/\sqrt{\mathrm{Hz}}$ @ 10 Hz has been obtained at an operation temperature of $T = 7.8$ K. This is more than an order of magnitude higher than the theoretically predicted value of $\sqrt{S_\Phi} = 16\ \mathrm{n}\Phi_0/\sqrt{\mathrm{Hz}}$ [17].

3. Conventional SQUIDs with reduced dimensions

As pointed out in the previous section, nanoSQUID technology is not yet mature and requires further development. Fortunately, the main objectives pursued with nanoSQUIDs can be partly achieved by conventionally manufactured micro-sized SQUID sensors. Although the reduced SQUID dimension of a micro-sized SQUID leads only to a slightly improved noise performance according to relation (2), practical nanoSQUIDs do not perform much better than state-of-the-art microSQUIDs. A more difficult task is to couple the microSQUID to a nanoscale sample, which will be considered in more detail in this chapter.

3.1. Miniature pickup coils

A very simple way to detect magnetic signals from a sample is to use a wire wound pickup coil connected to a sensitive (commercially available) SQUID current sensor. This works even if the sample has nanoscale dimensions as it has been demonstrated in NMR experiments [18]. However, for nano-sized samples it is more appropriate to use a thin-film miniature pickup coil. This approach is in particular useful for experiments at ultralow temperature, if the SQUID can be operated remotely from the coil, and hence, parasitic heat input to the sample by the dissipative SQUID can be avoided. It has been demonstrated in [19] that such a configuration can be used for the detection of NMR signals from a ^{3}He gas sample. To obtain a high energy resolution of the NMR setup, a very sensitive two-stage SQUID current sensor consisting of a front end SQUID and a SQUID array has been used in this experiment. The SQUID current sensors fabricated at PTB are available with various inductances of the input coil. This allows one to select a sensor that provides a good match between the microcoil and the input coil of the current sensor [16]. In this particular case a sensor with an input coil inductance

Journal of Physics: Conference Series **568** (2014) 032015 doi:10.1088/1742-6596/568/3/032015

of $L_i = 29$ nH has been used in combination with the microcoil shown in Figure 4. The intention is to apply the microcoil setup for NMR experiments on nanofluidic samples.

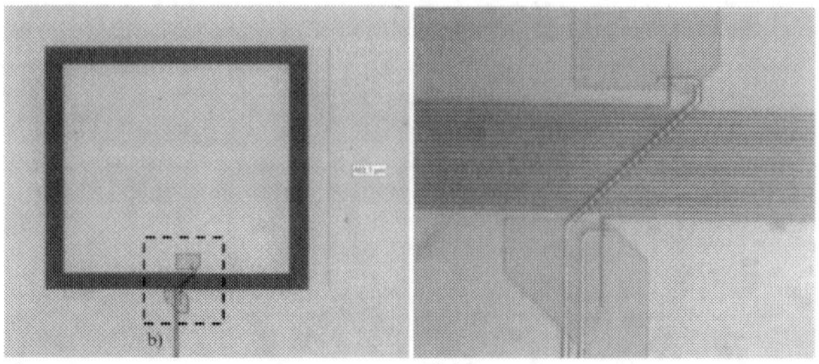

a) b)

Figure 4. Miniature superconducting pickup coil with 18 turns, an internal loop size of 432 μm × 432 μm, and an inductance of 394 nH.

a) Optical micrograph of the whole coil The bonding pads are not shown
b) Section with crossing lines [19].

Setups like this can be effectively used for other experiments as well, where microcoils are adopted to the experimental requirements and connected to an appropriate SQUID current sensor. Auxiliary components are already integrated on the current sensor chip, like filters and input current limiters, which makes it more convenient to use these sensors.

3.2. Micro-SQUID gradiometers

Conventional $Nb/AlO_x/Nb$ tunnel junction technology can be employed for manufacturing SQUIDs with a very small sensing area. The idea is to reduce the sensing area of the SQUID by covering the SQUID loop partly by a superconducting layer, while the other device components (junctions, shunt resistors) still remain their normal size. The microSQUID gradiometer presented in Figure 5 has made use of this idea. It can be clearly seen, that the areas of the Josepson junctions (4 μm × 4 μm) are much larger than the sensing area of the SQUID gradiometer. The diameter of the sensing area of about 3 μm equals approximately the linewidth of the superconducting lines. Two integrated lines enable feedback and magnetic sample excitation. With this device, a minimum intrinsic flux noise of $\sqrt{S_\Phi} = 122\ \mathrm{n}\Phi_0/\sqrt{\mathrm{Hz}}$ @ 10 kHz has been measured at an operational temperature of 4 K. With a two-stage readout an overall white noise of $\sqrt{S_\Phi} = 163\ \mathrm{n}\Phi_0/\sqrt{\mathrm{Hz}}$ @ 10 kHz was obtained which is comparable to the practically achieved noise levels of nanoSQUIDs [6].

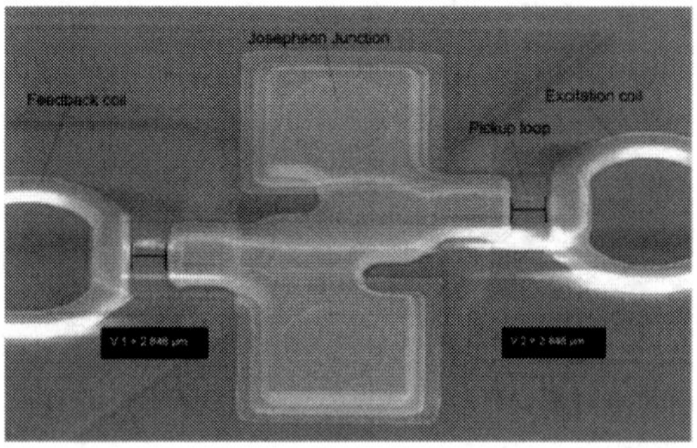

Figure 5. Electron micrograph of a micro-sized first-order SQUID series gradiometer fabricated in $Nb/AlO_x/Nb$ junction technology. Note that the width of the pickup loops (V1 and V2) is much smaller than the area of the junctions. For analysis, a sample can be attached to the pickup loop [6].

3.3. MicroSQUID susceptometers

Already 30 years ago, integrated SQUID susceptometers have been designed and successfully operated [20]. For scanning SQUID microscopy, fully integrated devices have been reported recently [21]. At PTB, we have developed integrated susceptometers with various pickup loop dimensions [6, 22]. In addition to a first-order SQUID gradiometer consisting of two circular loops, an excitation field coil is integrated on-chip as shown in Figure 6. It provides an excitation field with a sensitivity of about 100 mT/A at the sample position in the pickup loops, whereas the Josephson junctions are just exposed to a reduced field. Devices with pickup loop diameters of 30 µm, 60 µm and 100 µm are available. Compared with the microSQUID gradiometer depicted in 3.2, the micro-sized susceptometer have higher SQUID inductances which leads to higher white flux noise levels, e.g. 1.3 $\mu\Phi_0/\sqrt{Hz}$ for the 30 µm device. These susceptometers enable measurements of both the real and the imaginary part of the magnetic susceptibility in a wide frequency range up to 1 MHz at temperatures down to the mK range. Preferably, the susceptometers should be read out as well with an SSA in a two-stage configuration.

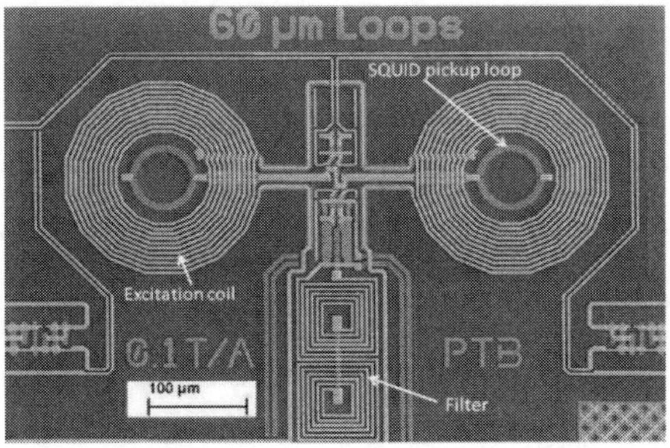

Figure 6. Fully integrated microSQUID susceptometer with gradiometric SQUID loop and excitation coils around the pickup loops (SEM picture). Dummy structures are added for maximum balance of the gradiometer type susceptometer.

4. Advanced device concepts

The devices described in section 3 are manufactured utilizing conventional optical lithography only. Their operation is quite straightforward by making use of state-of-the-art SQUID current sensors in a two-stage configuration in combination with conventional FLL electronics. Nevertheless, the coupling of the devices to real nanoscale samples can be significantly improved. This can be achieved if nano-patterning is applied to parts of the SQUID pickup loops after the basic device has been manufactured in a conventional way. Furthermore, the readout configuration can be simplified.

4.1. Micro-sized SQUIDs with integrated nanoloops

The basic idea to improve the coupling of a nanoscale sample to a micro-sized SQUID is to connect a nanoloop in series to the common SQUID pickup loop (see Figure 7). This can be done by removing material from the SQUID pickup loop which has typically a linewidth ≥ 1.5 µm. Focused Ion Beam (FIB) patterning or e-beam-lithography in combination with plasma etching are appropriate techniques for this fabrication step.

From FEM simulations a ten-fold improvement of the coupling between a magnetic dipole and the pickup loop was predicted for an optimum design. In order to verify this prediction, a microsusceptometer as shown in Figure 5 with a 30 µm pickup loop has been used to implement a nanoloop with an inner diameter of 450 nm and a linewidth of 250 nm. Test measurements with this device have been successfully performed on a Permalloy sample covering the nanoloop. Both, the in-phase signal and the out-of-phase signal from the sample have been observed in a frequency range up to 100 kHz. The sample was excited with an ac magnetic flux density of about 250 µT rms by using the integrated excitation coils of the susceptometer [22].

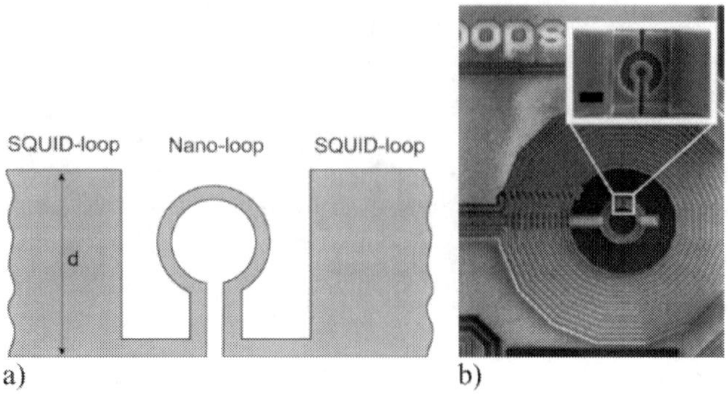

Figure 7. SQUID pickup loop with integrated nanoloop.
a) Scheme of nanoloop implementtation. The minimum pickup loop linewidth d of a conventional SQUID is typically about 1.5 μm
b) REM picture of a nanoloop with an inner loop diameter of 450 nm integrated in a pickup loop of a susceptometer depicted in Figure 6. [22].

4.2. MicroSQUIDs with on-chip preamplifier

As described in section 2, the low-noise SQUID devices have to be read out in a two-stage configuration, where the signal from the front-end SQUID is preamplified by a SQUID current sensor (in most cases a SSA). This configuration allows flexibility for the placement of the devices in a certain setup, e.g. if the SSA has to be advantageously mounted far from a magnetic excitation field. If the SSA can be operated near to the front-end SQUID, it is reasonable to integrate both devices on the same chip. It makes the entire setup much more robust and easy-to-handle and can avoid interferences due to poor wiring. Auxiliary elements like pads for heat sinking and filters are on chip as well. Figure 8 shows a chip design with the latest version of the susceptometer presented in subsection *3.3* integrated on chip with a 24-SQUID series array current sensor. Three variants with different pickup loop diameters are available [23].

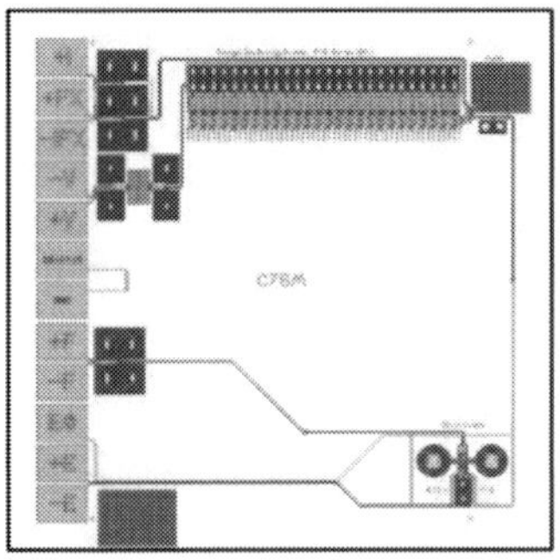

Figure 8. Chip layout of a SQUID micro-susceptometer with on-chip SSA. The chip size is 3.3 mm × 3.3 mm. The susceptometer is located in the lower right corner of the chip and the 24-SQUID series array is positioned in the upper chip part. On the left hand side are the bonding pads and filters [23].

Stimulated by the promising results obtained from the nanoloop experiments described in the previous subsection, optimized susceptometers for nanoloop integration have been developed. The large pickup loops of the susceptometer depicted in Figure 6 have shrunken to a minimum and the multi-turn excitation coils have been replaced by single lines surrounding the pickup loops. The susceptometer is integrated on a 3.3 mm × 3.3 mm chip together with an SSA in two-stage configuration as well (see Figure 9). A design with galvanically separated feedback and excitation lines by coupling transformers has also been developed and fabricated but not tested yet [22].

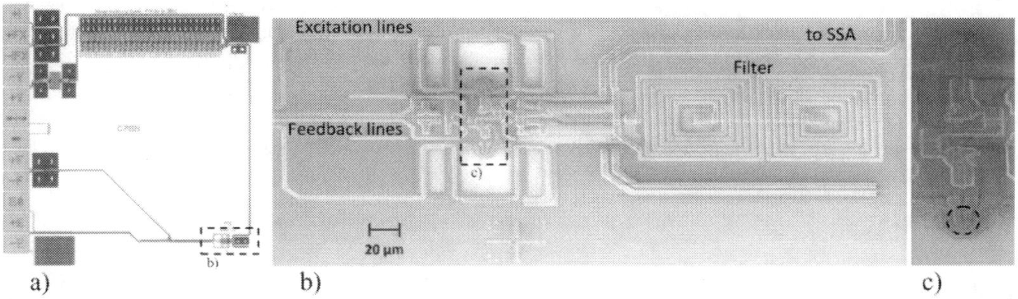

Figure 9. Novel SQUID microsusceptometer with integrated SSA. a) Chip design with the susceptometer in the lower right corner and the 24-SSA in the upper part of the 3.3 mm × 3.3 mm chip. b) Electron micrograph of the entire susceptometer and c) its center part. Both feedback and excitation lines are closely surrounding the pickup loops. The part of the pick-up loop marked with a dashed circle in c) is intended for placing the sample under investigation. It can be used for implementing a nanoloop similar to the configuration shown in Figure 7. Dummy structures are added similarly to the device in Figure 6 [22].

4.3. NanoSQUIDs with intrinsically shunted SNS junctions

As mentioned in section 2, a mature technology for nanoSQUIDs isn not available yet, but intrinsically shunted SNS junctions have a great potential to meet upcoming needs for more complex nanoSQUID based devices. Recently, SNS junctions have been used by PTB for manufacturing nanoSQUIDs employing a chemical-mechanical-polishing process (CMP) [5]. These nanoSQUIDs are based on Nb thin films and overdamped SNS Josephson junctions with HfTi as a normal conducting barrier. The SQUIDs can be operated in nonhysteretic mode at 4 K. In these devices, the SQUID loop is perpendicular to the chip plane similar to devices with SNIS junctions reported in [25] but fabricated without using FIB patterning. A line in the bottom Nb layer forms one part of the SQUID loop. The other part is formed by a line in the top Nb layer that is separated from the bottom layer by a 225 nm thick insulating SiO_2 layer. The two stacked Nb lines are connected by the HfTi Josephson junctions with lateral dimensions of 200 nm × 200 nm. The loop geometry is determined by the distance of the junctions of 1,6 µm and the SiO_2 layer thickness. These devices can be operated in magnetic fields up to $B \approx \pm 50$ mT with a white flux noise $\sqrt{S_\Phi} = 250\ \text{n}\Phi_0/\sqrt{\text{Hz}}$ for frequencies above 1 kHz. Operation in very high magnetic fields up to 0.5 T could be also demonstrated. Even at those high fields, a flux noise of $\sqrt{S_\Phi} = 680\ \text{n}\Phi_0/\sqrt{\text{Hz}}$ has been achieved, which makes this device concept very attractive for experiments where e.g. large magnetic excitation fields are required. A practical application for SQUID magnetometry has been reported in [25].

In order to make this junction technology more flexible for practical use, several designs for our standard 3.3 mm × 3.3 mm chips have been developed. The chips contain the basic Nb wiring including bond pads, and filters similar to the chip designs in Figure 8 and 9. The HfTi junctions are implemented using CMP. After the junction fabrication step is finished, the top Nb wiring is applied. Figure 10 gives an impression of the structure of a special SQUID gradiometer with HfTi junctions. A number of designs including serial and parallel nanoSQUID gradiometers and nanoSQUID current sensors are currently under investigation.

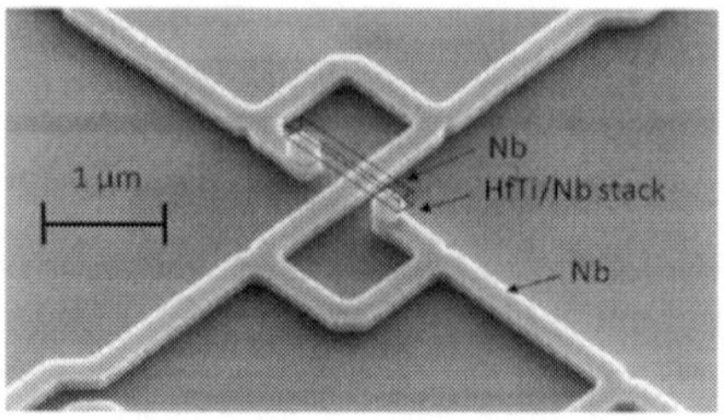

Figure 10. Serial nanoSQUID gradiometer with HfTi junctions before CMP has applied and Nb top wiring has been deposited. The position of the Nb top line is schematically shown by the dashed bar [26].

5. Summary

The aim of this paper was to deal with an overview of micro- and nano-sized SQUID devices which are manufactured using more or less conventional fabrication technology. In contrast to nanoSQUIDs based on a sophisticated junction technology, these devices are already available or can be made available for a broader community. In order to use the micro and nanoSQUIDs to full capacity, appropriate SQUID current sensors are an essential component of the readout setup. These current sensors are operated with a common SQUID readout electronics. The requirements of a number of low-temperature experiments (e.g. NMR) can be already addressed with small wire wound or thin-film coils combined with a state-of-the-art SQUID current sensor. MicroSQUIDs with a number of auxiliary components on chip are already available, even with an integrated SSA if required. The coupling of a microSQUID fabricated in conventional technology to a nano-sized sample can be significantly improved by integrating a nanoloop in the pickup coil. However, this needs additional nano-fabrication steps using FIB or e-beam lithography. Templates made in Nb technology can be used advantageously for implementing real nanoSQUIDs. This has been shown for HfTi junction based SQUIDs but could be of interest for other junction technology as well, e.g. constriction junctions. The devices fabricated in conventional technology could be further improved by shrinking the junction dimension as it is obvious from [27]. If operation in comparatively high magnetic fields is required, real nanoSQUIDs with e.g. HfTi junctions are inevitable. Figure 11 tries to summarize these issues.

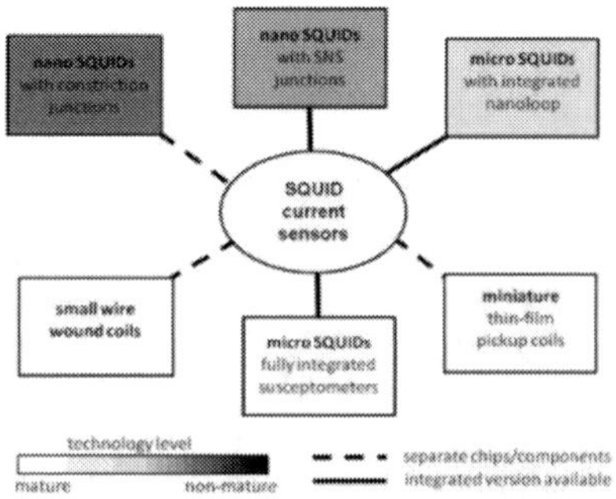

Figure 11. Overview of micro- and nanoSQUID setups with respect to technological maturity.

Acknowledgement

The author would like to thank his colleagues at PTB, especially D. Drung, J. Beyer, S. Bechstein, F. Ruede, C. Aßmann, A. Kirste, O. Kieler, J. Kohlmann and N. Kranz for helpful discussions and technical assistance. A considerable amount of the work reviewed here has been done in collaboration with groups at NPL, Royal Holloway University of London, University of Zaragoza and CSIC Zaragoza. It has been supported by the European Union in the framework of the FP7 European Microkelvin Collaboration and the European Metrology Research Programme EMRP, which is jointly funded by the participating countries within EURAMET and the EU.

References

[1] see e.g. www.magnicon.com, www.starcryo.com, www.supracon.com
[2] C. P. Foley and H. Hilgenkamp, "Why NanoSQUIDs are important: an introduction to the focus issue," Supercond. Sci. Technol., **22,** 064001, (2009).
[3] L. Hao, "Quantum Detection Application of NanoSQUIDs fabricated by Focussed Ion Beam", Journal of Physics: Conf. Series **286,** 012013 (2011)
[4] L. Hao, J. C. Macfarlane, J. C. Gallop, D. Cox, J. Beyer, D. Drung, and T. Schurig, "Measurement and noise performance of nano-superconducting-quantum-interference devices fabricated by focused ion beam," Appl. Phys. Lett., **92**, 192507, (2008).
[5] R. Wölbing, J. Nagel, T. Schwarz, O. Kieler, T. Weimann, J. Kohlmann, A. B. Zorin, M. Kemmler, R. Kleiner, and D. Koelle, "Nb nano superconducting quantum interference devices with high spin sensitivity for operation in magnetic fields up to 0.5 T," Appl. Phys. Lett., **102**, 192601, (2013).
[6] S. Bechstein, A. Kirste, D. Drung, M. Regin, O. Kazakova, J. Gallop, L. Hao, D. Cox, and T. Schurig, "Investigation of material effects with micro-sized SQUID sensors," IEEE Trans. Appl. Supercond., **23**, 1602004, (2013).
[7] J.-M. Mol, F. Foroughi, J. Arps, E. Kammerloher, P. Bethke, G. W. Gibson, Jr., Y.K.K. Fung, B. Klopfer, K. Nowack, P. A. Kratz, M. E. Huber, K. A. Moler, J. R. Kirtley, and H. Bluhm, "High-sensitivity SQUIDs with dispersive readout for scanning microscopy," http://meetings.aps.org/link/BAPS.2014.MAR.B24.11
[8] D. Vasyukov, Y. Anahory, L. Embon, D. Halbertal, J. Cuppens, L. Ne'eman, A. Finkler, Y. Segev, Y. Myasoedov, M.L. Rappaport, M.E. Huber, E. Zeldov; "Scanning nano-SQUID with single electron spin sensitivity", Nature Nanotechnology **8**, 639-644 (2013)
[9] J.-P. Cleuziou, W. Wernsdorfer, V. Bouchiat, T. Ondarcuhu, M. Monthioux, "Carbon nanotube superconducting quantum interference device", Nature Nanotechnology **1**, 53-9 (2006)
[10] C. Granata, E. Esposito, A. Vettoliere, L. Petti, M. Russo, "An integrated superconductive magnetic nanosensor for high-sensitivity nanoscale applications", Nanotechnology **19,** 275501 (2008)
[11] R. Arpia, M. Arzeo, S. Nawaz, S. Charpentier, F. Lombardi, T. Bauch, "Ultra low noise YBCO nanoSQUIDs implementing nanowires", Appl. Phys. Lett. **104**, 072603 (2014)
[12] R. Wölbing, T. Schwarz, B. Müller, J. Nagel, M. Kemmler, R. Kleiner, D. Koelle, "Optimizing the spin sensitivity of grain boundary junction nanoSQUIDs – towards detection of small spin systems with single-spin resolution", arXiv:1301.1189v2 (2014)
[13] A. Blois, S. Rozhko, L. Hao, J. C. Gallop, E. J. Romans "Proximity effect bilayer nano superconducting quantum interference devices for millikelvin magnetometry", J. Appl. Phys. **114**, 233907 (2013)
[14] D. Drung, M. Mück, "SQUID Electronics", The SQUID Handbook Vol. I, ed. by J. Clarke and A. I. Braginski, Wiley-VCH, 2004, 127-170
[15] C.D. Tesche, J. Clarke, "dcSQUID noise and optimization", J. Low Temp Phys. **27**, 301-331 (1977)
[16] D. Drung, C. Aßmann, J. Beyer, A. Kirste, M. Peters, F. Ruede, Th. Schurig, "Higly Sensitive and Easy-to-Use SQUID Sensors ", IEEE Trans. Appl. Supercond. **17**, 699-704 (2007)
[17] F. Ruede, S. Bechstein, L. Hao, C. Aßmann, Th. Schurig, J. Gallop, O. Kazakova, J. Beyer, D. Drung, "Readout System for NanoSQUID Sensors Using a SQUID Amplifier", IEEE Trans. Appl. Supercond. **21**, 408 (2011)
[18] L.V. Levitin, R.G. Bennett, A. Casey, B. Cowan, J. Saunders, D. Drung, Th. Schurig, J.M. Parpia, "Phase Diagram of the Topological Superfluid 3He Confined in a Nanoscale Slab Geometry", Science **340**, 841 (2013)
[19] A. Shibahara, A. Casey, C.P. Lusher, J. Saunders, C. Aßmann, Th. Schurig and D. Drung, "Microcoil detection of nuclear magnetic resonance for nanofluidic samples"AIP Advances **4**, 027107 (2014)

[20] M. B. Ketchen, T. Kopley, and H. Ling, "Miniature SQUID susceptometer," Appl. Phys. Lett., **44**, 1008–1010 (1984).

[21] M. E. Huber, N. C. Koshnick, H. Bluhm, L. J. Archuleta, T. Azua, P.G. Björnsson, B. W. Gardner, S. T. Halloran, E. A. Lucero, K.A. Moler, " Gradiometric micro-SQUID susceptometer for scanning measurements of mesoscopic samples", Rev. Sci. Instrum. **79**, 053704 (2008).

[22] D. Drung, J.-H. Storm, F. Ruede, A. Kirste, M. Regin, T. Schurig, A. M. Repolles, J. Sesé, F. Luis, "Thin-Film Microsusceptometer With Integrated Nanoloop", IEEE Trans. Appl. Supercond **24**., 1600206 (2014)

[23] unpublished design by D. Drung, F. Ruede, PTB

[24] M. Fretto, E. Enrico, N. De Leo, L. Boarino, V. Lacquaniti, C. Granata, R. Russo, A. Vettoliere, "Niobium nanoSQUIDs based on sub-micron tunnel junction fabricated by three-dimensional Focused Ion Beam sculpting", Journal of Physics: Conf. Series **507**, 042011 (2014)

[25] A. Buchter, J. Nagel, D. Rüffer, F. Xue, D.P. Weber, O.F. Kieler, T. weimann, J. Kohlmann, A.B. Zorin, E. Russo-Averchi, R. Huber, P. Berberich, A. Fontcuberta i Morral, M. Kemmler, R. Kleiner, D. Koelle, D. Grundler, M. Poggio, "Reversal Mechanism of an Individual Ni Nanotube Simultaneously Studied by Torque and SQUID Magnetometry", Phys. Rev. Lett. **111**, 067202 (2013)

[26] unpublished design by D. Drung, R. Ruede, O. Kieler, PTB

[27] M. Schmelz, R. Stolz, V. Zakosarenko, T. Schönau, S. Anders, L. Fritzsch, M. Mück, H.-G. Meyer, "Field-stable SQUID magnetometer with sub- fTHz-1/2 resolution based on sub-micrometer cross-type Josephson tunnel junctions", Supercond. Sci. Technol. **24**, 065009-1 (2011)

Journal of Physics: Conference Series **568** (2014) 032016 doi:10.1088/1742-6596/568/3/032016

Cryogen-free dilution refrigerator for ACTPOL polarization-sensitive receiver

V Shvarts[1], Z Zhao[1], M J Devlin[2], J Klein[2], M Lungu[2], B Schmitt[2] and R Thornton[2,3]

[1] Janis Research Company, LLC, 225 Wildwood Ave., Woburn, MA 01801, USA
[2] Department of Physics and Astronomy, University of Pennsylvania, 209 S 33rd St., Philadelphia, PA 19104, USA
[3] Department of Physics, West Chester University of Pennsylvania, 720 South Church St., West Chester, PA 19383, USA

E-mail: vshvarts@janis.com

Abstract. We present a new cryogenic receiver for the Atacama Cosmology Telescope (ACT), a six-meter diameter off-axis Gregorian telescope located at an altitude of 5,200 meters (17,000 ft.) on Cerro Toco, in Northern Chile. The focal plane contains 3000 polarization-sensitive transition edge sensor (TES) bolometers, and is cooled to below 100 mK with a removable pulse-tube based customised JDry-100 dilution refrigerator insert. The optical tubes and the rest of the receiver are cooled with a dedicated pulse tube to below 3 K. Details of the receiver-to-telescope integration and first season on-site operation are described, including detector base temperature stability in vertical and tilted position as well as remote operation via Ethernet link.

1. Introduction

The Atacama Cosmology Telescope (ACT) is a six-meter off-axis Gregorian telescope [1] (Figure 1) dedicated to investigating the formation and evolution of structure in the early universe, including the study of dark matter and dark energy. It accomplishes this by directly observing the Cosmic Microwave Background (CMB) radiation with arcminute resolution through limited atmospheric opacity from its unique high and dry location in the mountains of Chile's Atacama Desert.

ACT's new microwave receiver ACTPol (Figure 2Figure 2), designed and built in collaboration between the ACTPol team and Janis Research Company, consists of three optical tubes containing silicon reimaging optics, filters, and polarization-sensitive detector array package modules (dual 150 GHz, and one multi-chroic 90/150 GHz). Many of the optical tube components require cooling to 1 K in order to reduce their thermal emission and insure good performance of the cold detector readout electronics. The TES detector packages must be cooled to a stable temperature below 100 mK in order to maximize detector yield.

2. Design

The receiver "dewar" consists of a cylindrical outer vacuum can (OVC) of about 1 meter diameter, sealed with top and bottom flanges, all made of 6061 aluminum. The top flange has a removable hatch with three anti-reflection coated ultra-high molecular weight polypropylene windows transparent to microwaves, and serves as a mechanical support for the internal structure.

The receiver cross-section is shown in Figure 2. The 50 K and 4 K aluminum plates with corresponding removable shields are supported with two thin-wall G-10 drums. The shields are cooled by a dedicated PT-410 pulse-tube cryocooler [2], via a combination of flexible and rigid OFHC copper links. The detectors are cooled below 100 mK by a customized cryogen-free ^{3}He-^{4}He dilution refrigerator JDry-100-ACTPol insert (DR, Figure 3), equipped with a PT-407 pulse-tube cryocooler [2]. The insert pre-cooling from room temperature is done via a manually operated mechanical heat-switch connected to the PT-407 2nd stage. The automated gas handling system GHS4 is positioned next to the receiver in the temperature-controlled receiver cabin attached to the telescope and is remotely operated via Ethernet computer interface during all phases of the experiment. Its air cooling system for the electronics and solenoid valves, as well as the XDS 35i scroll and High Pace 300 turbo pumps [3] inside GHS4 allow for continuous operation at an atmospheric pressure near 500 mbar.

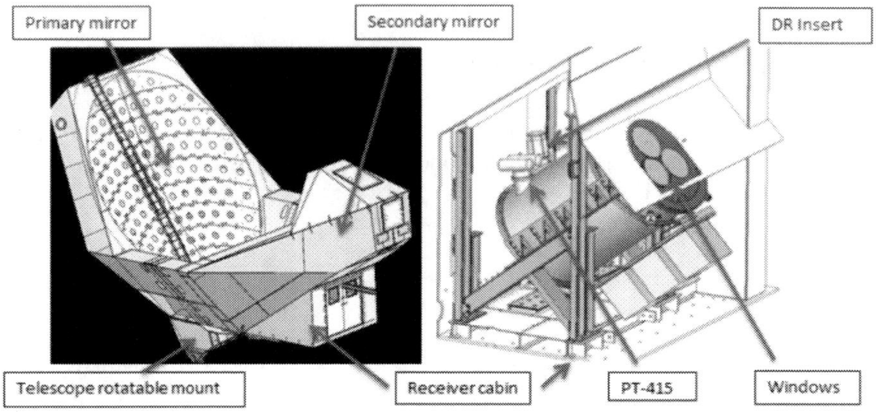

Figure 1. Telescope components arrangement on the left and receiver position inside the cabin on the right.

Pulse tube cryo-cooler, as well as DR stage performance degrades when operated in tilted position. To avoid the degradation or system shut-down in ACTPol, which scans in elevation from 32 to 60 degrees from horizon, both PT's are mounted on top ports at about 30 degree angle to its optical axis (see Figure 1).

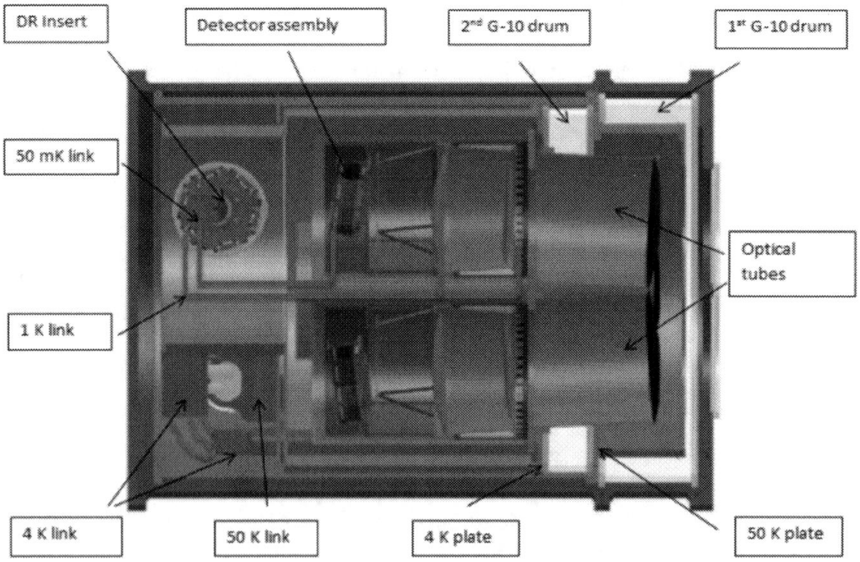

Figure 2. Receiver cross-section bottom view

3. Results

Initial tests were performed at the Janis lab with the DR insert enclosed in a small test vacuum shroud. The mixing chamber reached a base temperature near 10 mK and had 120 μW of available cooling power at 100 mK (as shown in Figure 4), exceeding the specification of 100 μW. After that, the system was integrated with the ACTPol receiver without the optical tubes installed and all windows blanked. Subsequent tests showed that although the pre-cooling time from 300 K to 6 K had increased due to additional loading from the larger "dewar", overall DR performance had not changed.

Figure 3. JDry-100 insert prior to installation; note tilted PT407 and DR stages

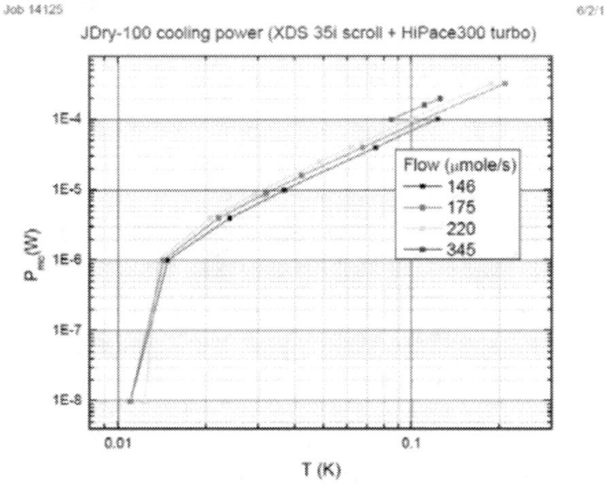

Figure 4. Cooling power of JDry-100-ACTPOL

During the spring of 2013, the first 150 GHz kilo-TES polarimeter array package (PA1) was assembled and installed in the ACTPol receiver[1], along with high-purity copper heat-straps that connect the 1K optical tubes and the sub-100 mK focal plane to the DR insert. Upon completion, pre-cooling was observed to take longer (just under 6 days – see Figure 5-Figure 8) than during prior tests due to the additional thermal-mass of the new optical tube components attached to the still and mixing chambers of the dilution refrigerator.

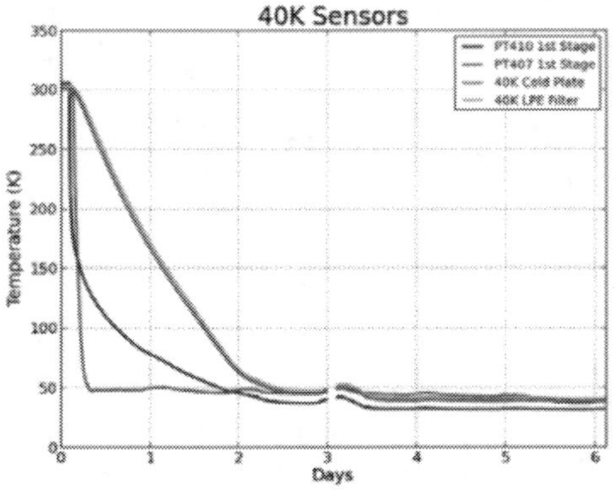

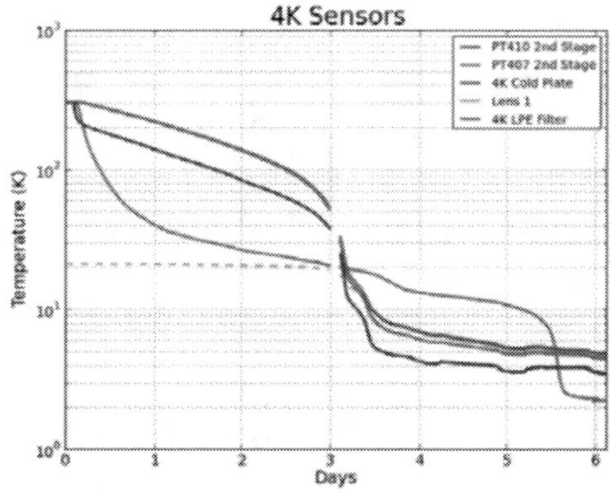

[1] One additional 150 GHz array (PA2) is currently in the process of being deployed while a third 90/150 GHz multi-chroic array (PA3) is slated for early 2015

Figure 5. Cooling of the first shield and plates

Figure 6. Cooling of the second shield and plates

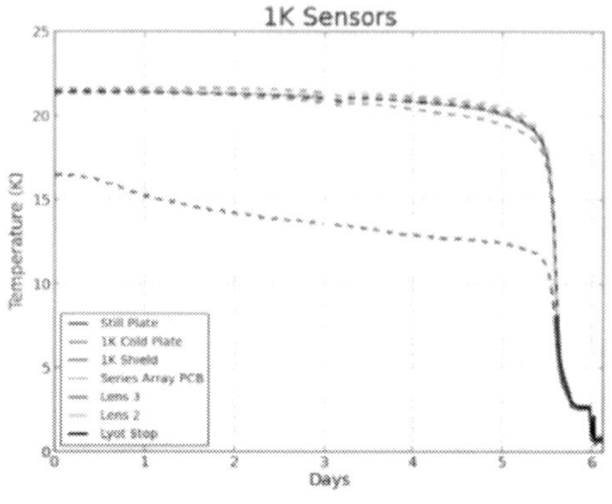

Figure 7. Cooling of components attached to the still

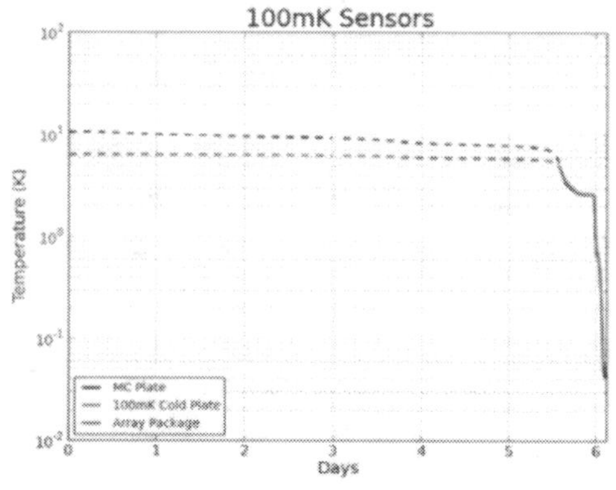

Figure 8. Cooling of components attached to the mixing chamber

Once everything had finished pre-cooling, the mechanical heat switch was manually disconnected and ^{3}He-^{4}He mixture condensation was initiated, taking approximately 2 hours to complete. This was followed by mixture phase separation, with the system reaching base temperature in just over one hour thereafter. Following several hours of biasing and tuning the TES bolometers, the instrument achieved first-light with PA1 on the night of July 18th, 2013 with a test observation of the planet Saturn.

After several weeks of testing, calibrating, and focusing of the telescope optics, first observations of the CMB began on September 11th, 2013 and lasted until December 14th, 2013, when weather conditions became unfavorable. During this period, several patches of the sky were observed while they were rising and setting by scanning the telescope back and forth in azimuth at a fixed elevation. In order to maximize the time spent on each patch (and thereby maximize the total signal-to-noise of the data), the telescope often tracked patches as they moved across the sky by changing its elevation.

Aside from a few brief interruptions due to system maintenance and testing activities (~ 3.7% of total time), the DR mixing chamber maintained a relatively stable base temperature throughout the entire season. The longest continuous stretch at base lasted just over 24 days and is shown in Figure 10 (for comparison, the temperature variation of the 4 K cold plate during the same period is also shown in Figure 9).

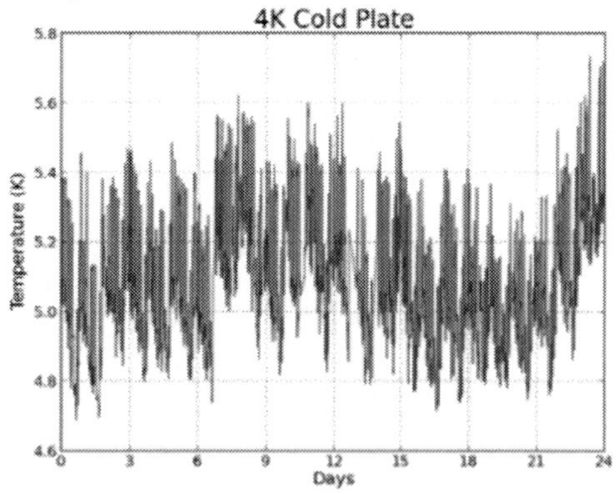

Figure 9. 4 K cold plate temperature variation during 24 day span of continuous data taking

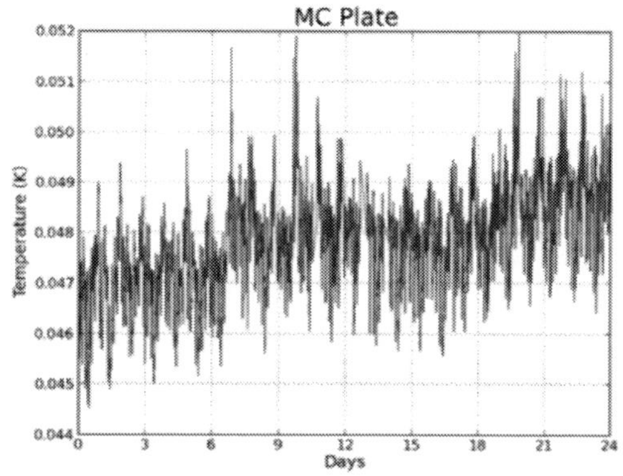

Figure 10. Base temperature variations at multiple elevations during 24 span of continuous data taking

Although diurnal oscillations caused by air-cooled PT cryo-coolers were a major source of variation, their amplitude was not large enough to adversely affect the detector data, with the array package maintaining a stable temperature of 77.3 ± 1.0 mK. Changes in the tilting angle of the DR insert that were essential to executing the aforementioned observing strategy had no significant impact on the temperature stability of the detector array (Figure 12), even though temperatures for some of the warmer components attached to the pulse-tube cooler show a clear dependence on elevation (Figure 11).

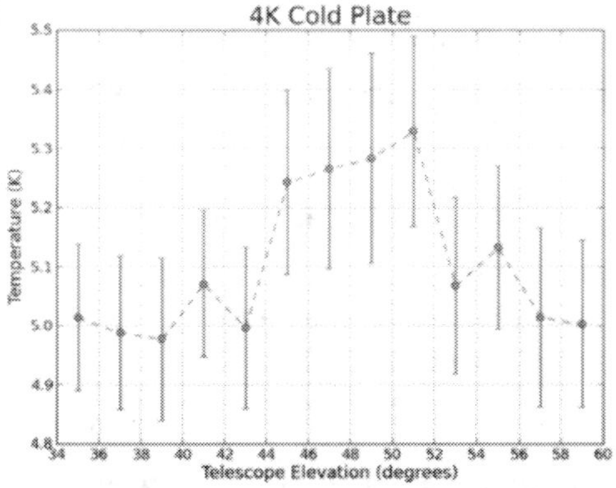

Figure 11. 4 K cold plate temperature variation with change in tilting angle

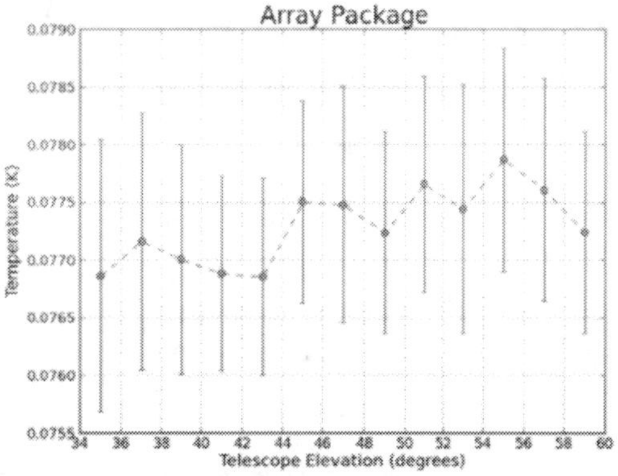

Figure 12. Array package temperature variation with change in tilting angle

4. Summary

We have designed, built and tested a new polarization-sensitive receiver using a ^3He–^4He dilution refrigerator as opposed to the more conventionally used ADR system. The DR's large cryogenic cooling capacity permits the detector arrays to reach an operating base temperature below 100 mK despite numerous sources of thermal loading from higher temperature components. This feature, in combination with the system's thermal stability over time and across a large range of tilting angles, resulted in consistent collection of high quality scientific data. Furthermore, continuous operation and remote access via Ethernet allowed for uninterrupted day and night-time observations of the CMB, making the DR an important reason for the success of the experiment.

References

[1] arXiv:astro-ph/0402234v1
[2] Cryomech, Inc., 113 Falso Drive, Syracuse, NY 13211 USA
[3] Edwards Vacuum, Inc., 88700 Expedite Way, Chicago, IL 60695 USA, and Pfeiffer Vacuum, 24 Trafalgar Square, Nashua, NH 03063-1988 USA, correspondingly.

Forced convection heat transfer from a wire inserted into a vertically-mounted pipe to liquid hydrogen flowing upward

H Tatsumoto[1], Y Shirai[2], M Shiotsu[2], Y Naruo[3], H Kobayashi[3] and Y Inatani[3]

[1]J-PARC Center, Japan Atomic Energy Agency, Tokai, Ibaraki, JP
[2]Department of Energy Science & Technology, Kyoto University, Kyoto, JP
[3]Institute of Space and Astronautical Science, JAXA, Kanagawa, JP

tatumoto@post.j-parc.jp

Abstract. Forced convection heat transfer from a PtCo wire with a length of 120 mm and a diameter of 1.2 mm that was inserted into a vertically-mounted pipe with a diameter of 8.0 mm to liquid hydrogen flowing upward was measured with a quasi-steady increase of a heat generation rate for wide ranges of flow rate under saturated conditions. The pressures were varied from 0.4 MPa to 1.1 MPa. The non-boiling heat transfer characteristic agrees with that predicted by Dittus-Boelter correlation. The critical heat fluxes are higher for higher flow rates and lower pressures. Effect of Weber number on the CHF was clarified and a CHF correlation that can describe the experimental data is derived based on our correlation for a pipe.

1. Introduction

Recently, liquid hydrogen draws attention as a coolant for large-scale HTS superconducting devices, especially MgB_2, because of lower boiling point, higher thermal conductivity, higher specific heat and lower viscosity. There has been a lack of systematic experimental data of forced flow liquid hydrogen, although some experimental data in a pool of liquid hydrogen at around atmosphere pressure have been reported [1-2].

Tatsumoto et al. [3-4] developed a thermal-hydraulics experimental system for liquid hydrogen, which implements explosion-proof measures, and carry out a systematic investigation of the forced convection heat transfer of liquid hydrogen. Tatsumoto et al. [5] measured forced convection heat transfer of saturated liquid hydrogen and clarified the effect of the heated pipe length on the critical heat flux (CHF). They derived a CHF correlation for forced flow of saturated liquid hydrogen. It is considered that a direct-cooled superconducting conductor such as a cable-in-conduit conductor (CICC) is more effective to activate the cooling characteristics of liquid hydrogen as much as possible, although there are some safety issues that need to be solved.

In this study, heat transfer from a wire inserted into a vertically-mounted pipe to forced flow of saturated liquid hydrogen are measured over wide ranges of pressure and flow velocity in order to clarify the effect of the flow rate and the pressure on the CHF. The measured CHF are compared with our CHF correlation for the vertical-mounted pipe [5].

2. Experimental apparatus and method

2.1. Explosion-proof experimental system

Details of the experimental system with the design pressure of 2.1 MPa and the measurement procedure have already been described in a previous paper [3]. The main tank (volume of 50 L), in which a test tube heater mentioned in 2.2 is vertically mounted at one end of a transfer line, is connected with a receiver tank through a transfer line with a control valve. The main tank is placed on a scale with 0.002-kg resolution. Forced flow through the test tube heater is achieved by pressurizing the main tank to a desired value with pure hydrogen gas (99.99%) and adjusting the valve opening. The mass flow rate is estimated by the weight change and the flow rate of the feed hydrogen gas, which is measured by a turbine flow meter. The flow measurement error is estimated to be within 0.1 g/s [1]. The experimental system is installed in an explosion-proof laboratory and explosion-proof devices such as a valve and a pressure transmitter are used. The power leads that supply the electrical current up to 500 A to the test heater, which is considered as an explosive source, is covered with an enclosure filled with GN2 whose pressure is kept 5kPaG slightly higher than the atmospheric pressure.

2.2. *Test heater*

A wire heater inserted into a vertical-mounted pipe is shown in Fig.1. The wire heater, which is made of Pt-Co alloy, has a diameter (d) of 1.2 mm and a heated length (L) of 120 mm. The wire heater is located on the central axis of a channel with a diameter (d_p) of 8mm and a length of 210 mm, which is made of a Fiber-Reinforced Plastic (FRP) for thermal and electrical insulation. The hydraulic equivalent diameter (D) is 6.8 mm and the aspect ratio (L/D) is 17.6. The entrance lengths of the tube heaters are set to be more than ten times longer than D. This is because the flow velocities used in this study correspond to Reynolds number, Re, higher than 1.0×10^4 and can be regarded as turbulent flow. The channel with the wire heater is vertically mounted in the main tank and the liquid hydrogen flows upward through it.

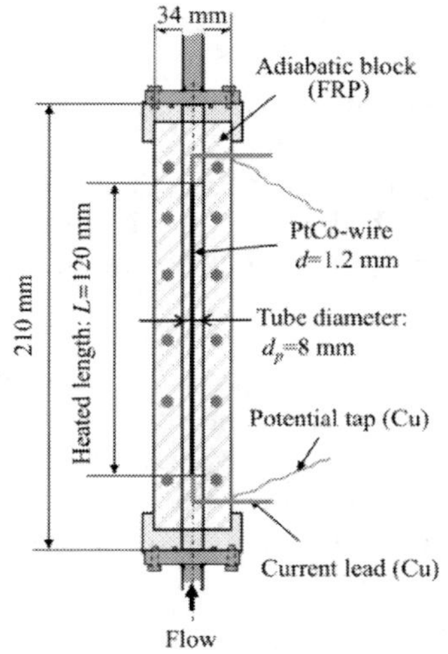

Fig. 1. Test heater

2.3. *Experimental procedure*

The wire heater is electrically heated by using a fast response direct current source (24 V, 500 A max.), which is controlled by a digital computer so as to give a desired time function for the heat input. Exponential heat generation of $Q = Q_0 \exp(t/\tau)$, for τ=10 s, where the heat transfer phenomenon can be regarded as a continuous series of steady-state, is applied to the heater. The average temperature of the heater is estimated by temperature-dependence of the electric resistance, which had been previously calibrated ranging from 20 K to ambient temperature. The electric resistance of the heater is measured using a double-bridge circuit including the heater as a branch of the bridge. The double bridge circuit is balanced at a bath temperature. The output voltage of the bridge circuit caused by the heater resistance deviation with the current heating, the voltage drop across the potential taps of the heater and that across a standard resistance, are amplified and are simultaneously sampled at a constant time interval. The heat generation rate in the heater is calculated from the measured voltage drops across the heater and the standard resistance. The surface heat flux, q, is the difference between the heat generation rate and the time rate of change of energy storage in the heater. The average surface temperature of the heater, T_w, is calculated from the average temperature and the surface heat flux by solving a conduction equation in the radius direction of the wire. The double-bridge circuit to measure the heater resistance has the accuracy of 1×10^{-4}. The temperature deviation of about 0.1 K can be measured by the bridge. The bath temperature and the inlet temperature are measured by Cernox sensors with the accuracy of 10 mK. Only the temperature increment of the heater from the inlet temperature is necessary for the analysis. Accordingly, experimental error is estimated to be within the heater surface temperature of 0.1 K and the heat flux of 2 %.

Heat transfer from the wire located at the center axis of the vertically-mounted tube to forced flow of saturated liquid hydrogen at the pressures of 0.4, 0.7 and 1.1 MPa was measured with a quasi-steady increase of the heat generation rate, $Q_0\exp(t/\tau)$, with $\tau = 10.0$ s for various flow velocities.

3. Results and Discussion

3.1. Forced convection heat transfer characteristics

Fig.2 shows forced convection heat transfer characteristics at the pressures of 0.7 MPa under saturated condition. The transverse axis is the wall superheat, ΔT_{sat} ($=T_w - T_{sat}$), that indicates the temperature difference between the heated wall temperature, T_w, and the saturated temperature, T_{sat}. With an increase in the heat input quasi-steadily, the heat flux, q, gradually increases along the curves predicted by the Dittus-Boelter correlation [6], where hydraulic equivalent diameter is used as representative length. The nucleate boiling occurs at the wall superheat slightly higher than the saturated temperature and the heat flux steeply increases up to a certain upper limit heat flux, which is called a critical heat flux (CHF), q_{cr}, with relatively little increase in the wall superheat. Above q_{cr}, it seems that the heat transfer characteristic continuously changes to film boiling regime, because the averaged heated wall temperature is used in the figure although the wire heater has a temperature distribution along a longitudinal direction. Although the heat transfer in the developed nucleate boiling regime is not affected by the flow velocity, the non-boiling heat transfer and the CHF are higher for higher flow velocities. With increase in the pressure, the heat transfers of the developed nucleate boiling get better but the CHF becomes lower, like pool boiling of saturated liquid hydrogen [7].

3.2. Critical heat flux

Fig.3 shows the effect of the flow velocity on the measured CHF at the pressures of 0.4 to 1.1 MPa. It seems that the CHFs increase in proportion to the flow for lower flow velocity. For higher flow velocity, the increasing rate is smaller. With increase in the pressure, the threshold value of the flow velocity becomes lower and the CHFs also become smaller, like that for a pool boiling [7]. For comparison, the CHF data for a tube heater with the diameter of 6 mm and the length of 100 mm

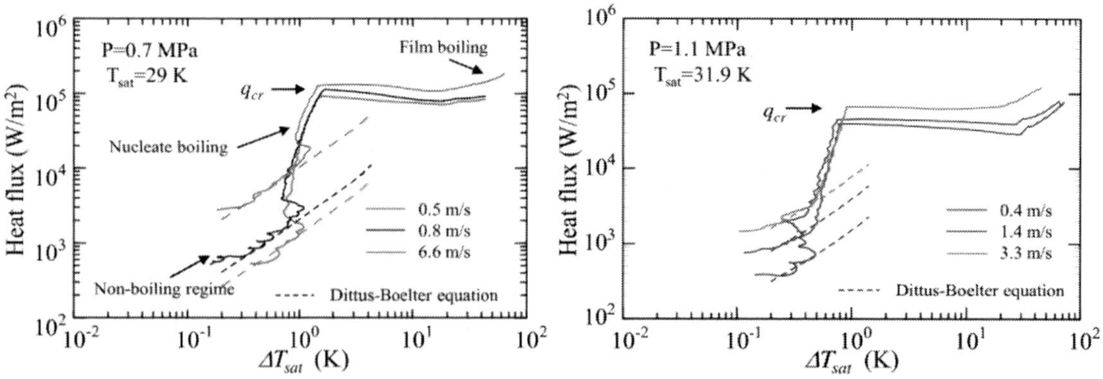

Fig. 2. Typical forced convection heat transfer characteristics of liquid hydrogen.

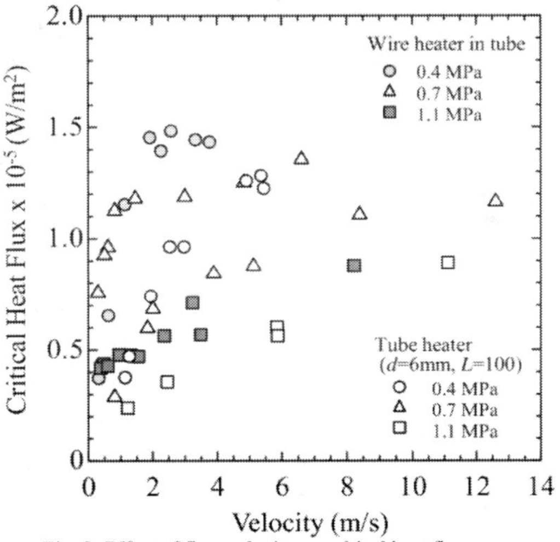

Fig. 3. Effect of flow velocity on critical heat flux.

measured by Tatsumoto et al. [5] are also plotted in the figure. The aspect ratio of *L/D* is 16.6, which is exactly similar to that used in this experiment. The CHFs for the wire heater inserted in the tube are higher than those for the heated tube, although they are affected by the flow velocity and the pressure as well as those for the heated tube.

3.3. Comparison with our CHF correlation for a vertically-mounted heated pipe

Tatsumoto et al. [5] have measured heat transfers from the inner side of vertically-mounted heated pipes with various heated lengths and diameters to forced flow of saturated liquid hydrogen at the pressures of 0.4 to 1.1 MPa and have presented the following CHF correlations for forced flow of saturated liquid hydrogen.

$$q_{cr} = Gh_{fg}\left(\frac{\rho_v}{\rho_l}\right)^{0.43}\left(\frac{L}{D}\right)^{-0.35} F_b \tag{1}$$

$$F_b = 0.038\left(\frac{L}{D}\right)^{-0.45} \quad \text{for } We < We_b \tag{2}$$

$$F_b = 0.32We^{-0.45} + 0.0017 \quad \text{for } We \geq We_b \tag{3}$$

$$We_b^{-0.45} = 0.118\left(\frac{L}{D}\right)^{-0.45} - 0.0053 \tag{4}$$

$$We = \frac{G^2 D}{\rho\sigma} \tag{5}$$

where *We* is the Weber number, *G* is the mass flux, h_{fg} is the latent heat of vaporization, σ is the surface tension, ρ is the density and *L* is the heated length. Subscripts *l* and *v* in eq. (1) indicate liquid and vapor.

As shown in Fig.3, the measured CHFs for the wire heater are different from those for the heated pipe with almost the same *L/D*. An equivalent heated diameter, D_H (= $(d_p^2-d^2)/d$), is applied to eqs.(1) to (3) instead of the equivalent hydraulic diameter, *D*. Comparison of the measured CHF for the wire inserted into the pipe with the correlations is plotted in Fig.4. It is confirmed that the correlations can also describe the measured CHFs for the wire heater as well as those for the heated tube.

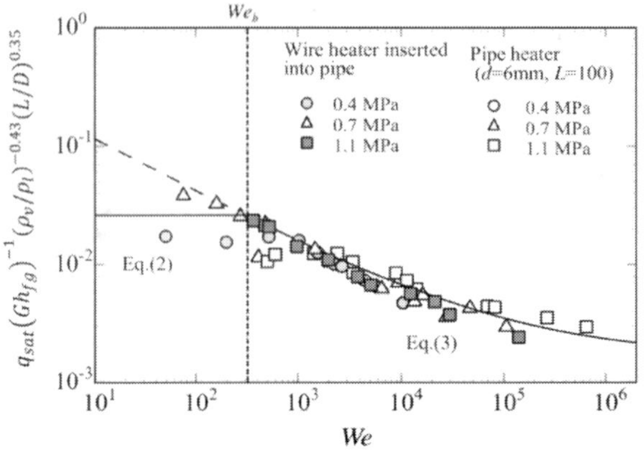

Fig. 4. Comparison of the CHF for the wire inserted into the pipe with authors' correlations.

4. Conclusion

The heat transfers from PtCo wire heater with a length of 120 mm and a diameter of 1.2 mm that was inserted into a vertically-mounted pipe with a diameter of 8.0 mm to forced flow of saturated liquid hydrogen were measured by quasi-steadily increasing the heat input for various flow rates at pressures of 0.4, 0.7 and 1.1 MPa. The experimental results lead to the following conclusions.

The non-boiling heat transfer coefficients agree with those predicted by Dittus-Boelter's equation and are higher for higher flow velocity, although the heat transfer in the developed nucleate boiling regime is not affected by the flow velocity under the same pressure.

For lower flow velocity, the CHF increases in proportion to the flow velocity. For higher flow velocity, the increasing rate of the CHF is smaller. With increase in the pressure, the CHF decreases like that for a pool boiling. The measured CHFs for the wire heater are higher than those for the heated pipe with almost the same *L/D*. It is confirmed that the measured CHFs for the wire heater can be described by our correlation for the heated tube by means of equivalent heated diameter, instead of the equivalent hydraulic diameter.

Acknowledgements

This work is supported in part by JSPS KAKENHI (20360127). The authors thank technical staffs of JAXA for technical assistance.

References

[1] Coeling K J and Merte JR H 1969 *J. Eng. Indu.* **91** 513-524
[2] Steward W G 1990 *Adv. in Cryo. Eng.* **35**, 403-412.
[3] Tatsumoto H, Shirai Y, Shiotsu M, Hata K, Kobayashi H, Naruo Y and Inatani Y 2010 *J.Phys:Conf. Seri.* 234, 032056
[4] Tatsumoto H, Shirai Y, Shiotsu M, Hata K, Kobayashi H, Naruo Y Inatani Y Narita N 2013 *Proc. Int. Cryo. Eng. Conf. 24* 157-160
[5] Tatsumoto H, Shirai Y, Shiotsu M, Hata K, Naruo Y, Kobayashi H and Inatani Y 2014 *Adv. in Cryo. Eng.* 1573 59A, 403-412
[6] Van Sciver S W 1986 *Helium Cryogenics* Plenum Press New York USA p. 251
[7] Shirai Y, Tatsumoto H, Shiotsu M, Hata K, Kobayashi H, Naruo Y, Inatani Y 2010 *Cryogenics* **50** 410-416

Experimental Investigation on Thermoresistance between AlN, Bi-2223 and OFHC in High T_C - Direct Cooling Technology *

HL Wang, RS Rao, J Wang
Huazhong University of Science and Technology, 1037 Luoyu Road, Wuhan 430074, P.R.China

E-mail: whlclf@163.com

Abstract. In the development of high temperature superconducting (HTS) direct cooling technology, the high electric insulation high heat conducting AlN has become one of the important components. The thermal contact resistance between AlN, Bi-2223 and OFHC is investigated by experiment with a G-M cryocooler as the source of cooling. The heat conductivity of AlN is measured between 29 and 160 K temperatures. When the temperature on the interface layer side of Bi-2223 is 55 K, under the action of the contact pressure of 0.5469 MPa, the thermal contact resistance between AlN and Bi-2223 is 38.86 times to the thermal conduction resistance of a 10 mm thick AlN pad. Baced on micro-nanocryogenics, it is proposed that the thermal contact resistance is one of the crucial techniques to be attacked in HTS direct cooling technology.

1. Introduction

The high heat conductance and high electric insulation aluminium nitride (AlN) ceramic material is widely used in fields such as the manufacture of electronic devices, power modules and the super-large-scale integrated circuit [1,2]. The heat conductivity of pure AlN single crystal with a hexagonal fibrous zinc ore structure at room temperature can be as high as 320 w/(m·K) [3,4]. Its high heat conductivity is nearly equivalent to that of copper. The usual heat conductivity value of actually supplied materials is between 100~170 w/(m·K) or thereabouts [3].

Despite numerous investigations on the thermal characteristics of AlN at normal and high temperatures [5], little has been done in learning about AlN material application in the

* Project supported by National Science Foundation of China, No. 51276070.

cryogenic environment. Particularly in recent years, the development of high temperature superconducting materials and the advance of cryogenic refrigerators have enhanced the high temperature superconducting direct cooling technology [2]. The high heat conductivity and electrically insulating AlN has become one of the acknowledged materials for devices using in HTS direct cooling technology.

With HTS system using direct cooling, various joining pads (washers) made of AlN possess not only have characteristics such as high heat conductivity, high electric insulation (insulation resistivity being $10^{12}\,\Omega\cdot$cm ~ $10^{14}\,\Omega\cdot$cm). They also exhibit properties such as high mechanical strength, good chemical stability and freedom from toxicity and contamination. These excellent properties make them important joining components using in the cold head of cryogenic refrigerator, binary current lead and magnet, etc. So far, not much is reported on the heat conducting characteristics of actual application of AlN ceramics in HTS projects within the 30~50 K temperature range from the angle of micro-scale cryogenic engineering and 3-D interface resistance.

2. Experimental method

It can be seen from the angle of micro- and nano-cryogenics that there exists an interfacial layer at the interface between the two solids in contact. Neither the organizing structure nor property of this interfacial layer is the same as that of the two solids. The author has observed the micro-structure interface joined by copper-aluminium cold-pressing under a scanning electron microscope (SEM) [6,7] and found that the alloy content percentages each 4 μm from the two sides of the so-called Cu-Al 2-D interface are: on the Cu side (Al: 16.37%; Cu: 83.63%); on the Al side (Cu: 1.4%, Al: 98.6%). Therefore, in high temperature superconducting direct cooling technology, research on the heat transport process and characteristics at the 3-D cryogenics interfacial layer [6] and the magnitude of the heat resistance at the 3-D cryogenics interfacial layer is of great significance.

When the heat flow passes through the interfacial layer f formed by the two objects in contact, there will be generated a temperature drop ΔT at the interfacial layer. The thermal resistance R_b at the interfacial layer is the ratio of the temperature drop to the heat flow density q,

$$R_b = \Delta T / q \quad [\mathrm{K} \cdot m^2 \cdot w^{-1}] \tag{1}$$

The principle of the experimental equipment is shown in Figure 1. In the experiment, the cooling is provided by the G-M cryorefrigerator while a controlled heat input to the sample is achieved with an electric heater. The AlN ceramics, Bi-2223 high temperature superconductor and oxygen-free copper (OFHC) specimens are all made into cylinders of nominal dimensions Ø10×30 mm. The two end-faces of a specimen are polished on a grinding machine, the end-face parallelism being < 0.008 mm. The graph of the superconducting material Bi-2223 undergoing X-ray refraction is as shown in Figure 2.

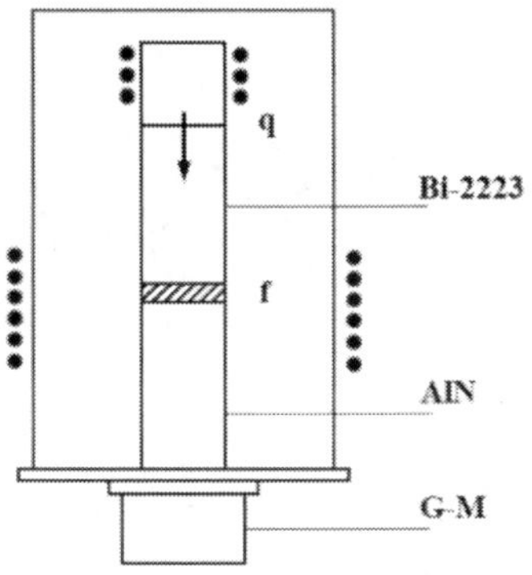

Figure 1. Schematic of the experimental principle of AlN – Bi-2223 cryogenic interfacial heat resistance.

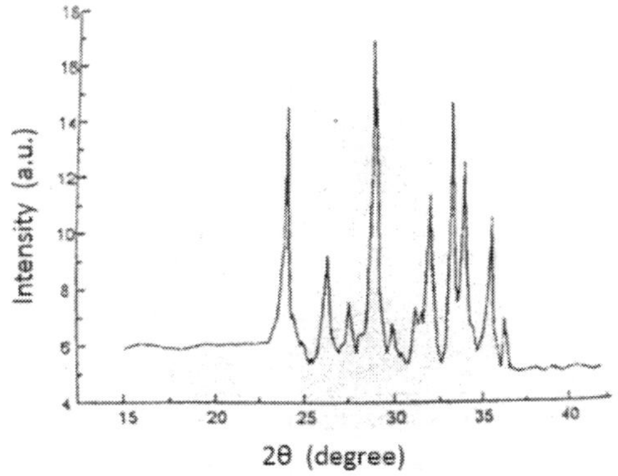

Figure 2. The x diffraction graph of the Bi-2223 specimen.

To satisfy the requirements of the 1-D axial steady-state method, a radiation screen is incorporated on the specimen shelf onto which a cryogenic rare-magnetic Cu-Fe thermocouple (NiCr-CuFe) is soldered [8,9]. A cryogenic temperature control is used for ensuring that the difference in temperature between the radiation screen and specimen is less than 1 K. A diffusion pump and G-M machine are used to ensure that the vacuum in the measuring Dewar is better than 5.0×10^{-4} Pa [10].

Self-developed rare magnet Cu-Fe cryogenic thermocouples (NiCr-Cu + 0.13% at Fe) [11] are used for experiment, with each of the thermocouples calibrated by our laboratory. The testing system is made up of US-made data collectors Keithley 2700 as the secondary instrumentation. Prior to testing, standard samples of copper and pure iron with known heat conductivity are used for measurement to verify the reliability of the experimental devices and measuring system.

3. Experimental results

3.1. The AlN cryogenic heat conductivity

Owing to the absence of free electrons in AlN, phonons play a significant role in the heat conductance of electrically insulated AlN. The measured thermal conductivity of the AlN sample is shown in Figure 3. Because of the excellent heat transmission ability of AlN at low temperature, the difference in temperature at the interface between AlN and HTS materials in contact such as oxygen free copper (OFHC) can be relatively small, which means the heat generated by the superconducting magnet is rapidly transmitted outside. In addition, the insulated resistivity of AlN is $10^{12}\Omega\cdot cm$ ~ $10^{14}\Omega\cdot cm$, the dielectric constant ε' is 9.1 [3]. The good electric insulation and low dielectric loss make AlN very conducive to highly efficient and reliable operation of the cryogenic refrigerator.

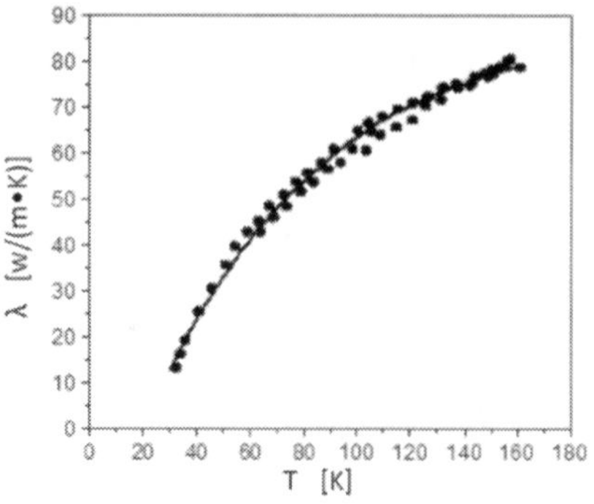

Figure 3. The relation of variation of AlN cryogenic heat conductivity with temperature.

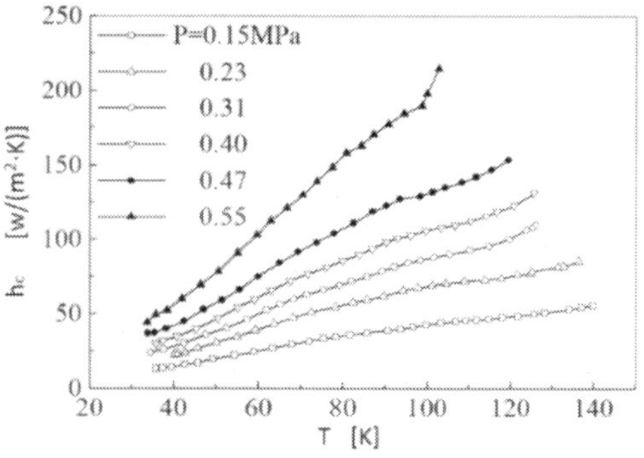

Figure 4. Variation of heat conductance at the interfacial layer between AlN and Bi-2223 with pressure.

3.2. The heat resistance at the interface between AlN and Bi-2223

It is shown by experiment that, within the range of the experimental contact pressure (0.15 ~ 0.55 MPa), with the rise of temperature and contact pressure, the heat conductance at the interfacial layer increases. Shown in Figure 4 is the curve representing the relation of variation of the contact pressure in heat conductance at the interface between AlN and Bi-2223. It can be seen from the figure that, when the temperature on the Bi-2223 side of the interfacial layer is 55K, and the contact pressure rises from 0.23 MPa to 0.55 MPa, the heat conductance rises from 34.87 w/(m^2·K) to 90.19 w/(m^2·K), equivalent to a reduction in the interfacial heat resistance by 61.3%. So, within the allowable range of material stress intensity, appropriate increase in the contact pressure can decrease the heat resistance at the contact interfacial layer.

With a contact pressure of 0.55 MPa, suppose the AlN pad has the thickness of 10 mm. When the interface temperature are 35.2 K, 55.2 K, 73.5 K and 94.5 K, the interfacial layer heat resistance is 48.6, 38.9, 33.5 and 30.5 times that of the AlN pad, respectively. Therefore, in a direct cooling HTS magnet, the interfacial layer heat resistance has an important influence on the cooling efficiency and the thermostability of the superconducting magnet.

3.3. The contact heat conductance at the interface between AlN and oxygen free copper

It is shown by experiment that the interfacial contact heat conductance between AlN and OHFC increases with the rise of contact interface pressure, as shown in Figure 5. On the interface between ceramics and metals, such as AlN and OFHC, the heat carriers are mainly phonons (on the AlN side) and electrons (on the OHFC side). Increasing the interfacial contact pressure can improve the defects on the contact interfacial layer, contract and decrease the characteristics dimensions of the superficial local micro-defects. On one hand the interfacial heat conductance and, on the other, the average free path of heat carriers reaching the micro-sub-interface is relatively increased, leading to reduction of the contact heat resistance. Under the same pressure, the relation between the interface heat conductance and temperature is approximately linear.

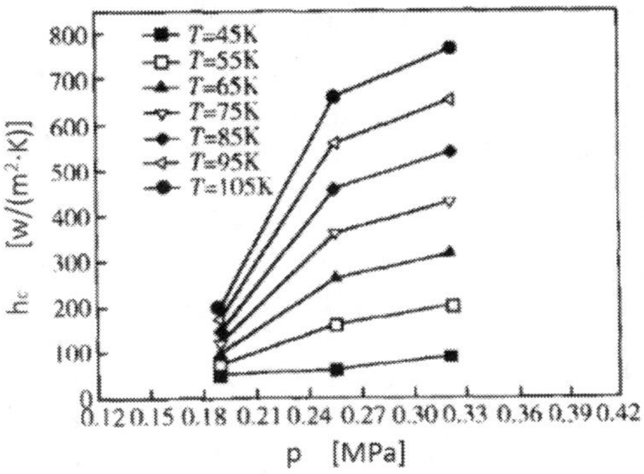

Figure 5. The variation of contact heat conductance of AlN and OFHC with pressure.

An analysis of the experimental curve variation shows that, with an increase in the interface pressure, the contact heat conductance value rapidly increases at the beginning. With further increase in the pressure, the amplitude of increase of the contact heat conductance tends to become gentle step by step as if being passivated. Meanwhile, the greater the interfacial pressure, the more abrupt the variation of the contact heat conductance with temperature.

4. Conclusions

- Heat conductivity characteristics data on AlN material for HTS application using within the 29~160 K low temperature range have been measured by experiment. The total error is $\pm 15\%$. The heat conductivity of AlN material at 29 K is 9.24 w/(m·K) and is 80.61 w/(m·K) when the temperature reaches 157 K. Its heat conductivity is dozens of times that of HTS materials.
- Within the range of contact pressure 0.15 MPa ~ 0.55 MPa, the interfacial conductance AlN – Bi-2223 increases with the rise of the contact pressure and temperature. When the temperature on the Bi-2223 side of the interfacial layer is 55 K, with the contact pressure enhanced to 0.55 MPa, the heat conductance on the interfacial layer rises from 34.87 $w/(m^2 \cdot K)$ to 90.19 $w/(m^2 \cdot K)$, equivalent to a reduction of the heat resistance on the interfacial layer by 61.3%.
- In-depth investigation on the 3-D interfacial heat resistance and the 3-D layer heat transport mechanism based on micro- and nano-cryogenics is of important scientific significance [12].

5. References

[1] Wang H L, Rao R S, Li J D, Tang Y J and Cheng S J 2001 Progress on cryogenic technology applied in superconducting electrical engineering *AEPS.* **17** 65

[2] Iwase N, Anzai K 1985 Aluminium nitride substracts having high thermal conductivity *Toshiba Rev.* (Int Ed) **153** 49

[3] Wang D F, Li W L, Zhuang H R and Guo J K 1998 Recent process in high thermal conductivity AlN ceramics *Materials Review.* **12** 29

[4] Yu Y D, Hundere A M, Høier R 2002 Microstructural characterization and microstructural effects on the thermal conductivity of AlN(Y_2O_3) ceramics *Journal of the European Ceramic Society* **22** 247

[5] Graebner J E, Jin S, Kammlott G W, Bacon B, Seibles L and Banholzer W 1992 Anisotropic thermal conductivity in chemical vapor deposition diamond *Journal of Applied Physics* **71** 5353

[6] Wang H L, Wanger T and Eska G 2000 An aluminium heat switch made from cold-pressed Cu-Al composite *Physica B: Condensed Matter* **284-288** 2024

[7] Shi L and Wang H L 2007 Investigation on thermal contact resistance by photothermal technique at low temperature *Heat and Mass Transfer* **11** 1179

[8] Rao R S and Wang H L 2003 Identification and experimental investigation on thermal interface resistance for cryocooler cooled High-Tc superconductor *Chinese Journal of Low Temperature Physics* **25** 209

[9] Chen J, Wang J R, Rao R S, Wang J, Zhang Y L, Li J D, Tang Y J and Wang H L 2003 Research on the temperature control of thermal properties experiment of superconducting material *Physica C: Superconductivity* **386** 544

[10] Rao R S, Wang J R, Zhuang H R, Wang T C, Chen J, Li J D, Tang Y J and Wang H L 2003 An investigation on interface thermal resistance at superconductor cooling by cryocooler *Physica C: Superconductivity* **386** 547

[11] Wang H L and Gao H C 1990 An investigation on the thermoelectric properties of dilute magnetic copper-iron alloys at low temperature *Physica B: Condensed Matter* **165-166** 41

[12] Xie J B, Wang H L, Wu G, Zhao Y, Li J D and Tang Y J 2006 Experimental investigation of heat transport for 35 kJ superconducting magnet energy storage system *Cryogenics* **1** 31

Construction of a 25-T cryogen-free superconducting magnet

K Watanabe[1]*, S Awaji[1], H Oguro[1], Y Tsuchiya[1], S Hanai[2], H Miyazaki[2], T Tosaka[2], M Takahashi[2] and S Ioka[2]
[1]Institute for Materials Research, Tohoku University, Sendai 980-8577, Japan
[2]Toshiba Corporation, Yokohama 230-0045, Japan

E-mail:kwata@imr.tohoku.ac.jp

Abstract. The construction of a 25-T cryogen-free superconducting magnet (25T-CSM) has started in 2013 at the High Field Laboratory for Superconducting Materials, Institute for Materials Research, Tohoku University. The 25T-CSM consists of a low-T_c superconducting (LTS) coil and a high-T_c superconducting (HTS) coil. A high-strength CuNb/Nb_3Sn Rutherford cable with the reinforcing stabilizer CuNb composite is adopted for the middle LTS section coil. The characteristic feature of the new technology using a CuNb/Nb_3Sn Rutherford cable is a react-and-wind method for the coil-winding process. The LTS coil of 300-mm winding inner diameter is fabricated, and a central magnetic field of 14 T is generated at an operation current of 851 A. The HTS insert coil wound with $GdBa_2Cu_3O_y$ (Gd123) tape has a 52-mm experimental room temperature bore, and a central magnetic field of 25.5 T will be generated at an operation current of 150 A in a background field of 14 T.

1. Introduction

A cryogen-free superconducting magnet conductively cooled by a tiny GM-cryocooler has been demonstrated in practice in 1992 [1]. At present, cryogen-free high-field superconducting magnets generating 10-15 T are widely used in new research fields such as magneto-science. The distinctive feature of the liquid-helium-free and nitrogen-free operation is the realization of an easy-to-use superconducting magnet. There is no need to supply any liquid helium or nitrogen for the magnet operation.

We have already developed an 18-T cryogen-free superconducting magnet (18T-CSM), which consists of a low-T_c superconducting (LTS) coil and a high-T_c superconducting (HTS) $Bi_2Sr_2Ca_2Cu_3O_{10}$ (Bi2223) insert coil. The 18T-CSM generated a magnetic field of 18.1 T in a 52-mm experimental room temperature bore [2]. Recently, the HTS insert coil of 18T-CSM was replaced by employing the high performance Ag/Bi2223 tape, and consequently the magnet was successfully improved to generate 20.1 T [3].

Since a high magnetic field is one of the important physical parameters in condensed matter physics and materials science, the combination of a high field and a cryogen-free superconducting magnet is expected to greatly contribute to the progress of high magnetic field research. In 2013, the construction of the 25-T cryogen-free superconducting magnet (25T-CSM) was funded by the Japanese government at the High Field Laboratory for Superconducting Materials, Institute for Materials Research, Tohoku University.

*correspondence author

In this paper, a new technology related to LTS and HTS coils for the 25T-CSM is reported.

2. Design of the magnet system for the 25T-CSM

The magnet system for the 25T-CSM, consisting of an LTS coil and an HTS coil, is shown in figure 1. The parameters of the 25T-CSM superconducting coils are summarized in table 1. Total inductance is 97 H, and stored magnetic energy is 10.7 MJ. The HTS coil, denoted as H1, consists of 68 single-pancake coils. The HTS and LTS coils are individually operated by dual power supplies. The 25T-CSM LTS coil is wound with Rutherford cable conductors. Nb_3Sn Rutherford cables are utilized for three coils (L1-L3), and NbTi Rutherford cables are used for two coils (L4-L5). A central magnetic field of 14 T is generated at an operation current of 851 A.

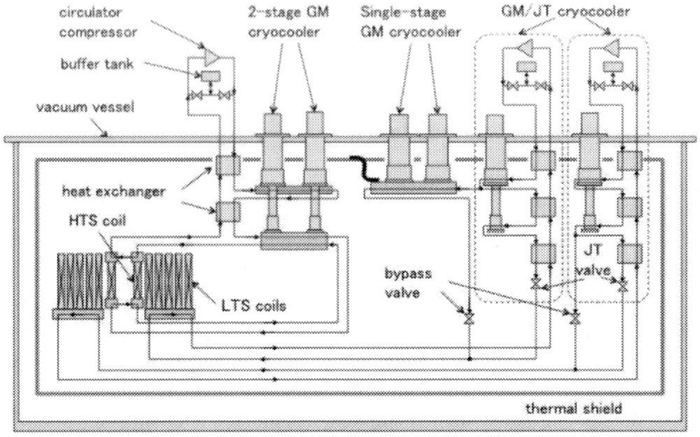

Figure 1. Schematic view of the 25-T cryogen-free superconducting magnet (25T-CSM).

Table 1. Coil parameters of the 25T-CSM (52-mm room temperature bore).

Coil ID		H1	L1	L2	L3	L4a	L4b	L5
Conductor material		Gd123	Nb_3Sn	Nb_3Sn	Nb_3Sn	NbTi	NbTi	NbTi
Operation current	A	150			851			
Inner diameter	mm	102	300	372	458	545	603	628
Outer diameter	mm	276	366	452	539	603	622	712
Height	mm	408	540	628	628	628	628	628
No. of turns/pancake, layer	-	435	80	93	93	95	107	107
No. of pancakes, layers	-	68	18	22	22	16	6	26
B_{max}	T	25.7	13.7	11.3	8.37	6.83	6.22	5.84
Magnetic field contribution	T	11.5	2.43	2.91	2.73	1.91	0.78	3.24
Width of conductor	mm	5.00	6.45	6.45	6.45	6.30	5.57	5.57
Thickness of conductor	mm	0.13	1.83	1.83	1.83	1.80	1.61	1.61
Thickness of layer insulation	mm	0.070	0.075	0.075	0.075	0.075	0.075	0.075
Space current density	A/mm^2	125	68.9	68.9	68.9	71.6	90.0	90.0
T_{cs}	K		5.87	7.28	8.58	5.92	6.12	6.32
Hoop stress	MPa	488	251	243	200	138	112	52

The current share temperature T_{cs} is 5.87 K for the L1 coil. This results in a temperature margin of over 1.3 K for all coils at an operation temperature of 4.5 K. Concerning the cooling system for the 25T-CSM, the cold mass of the 25T-CSM including the superconducting coils and the structures is 2410 kg. Since the HTS coil with a large temperature margin can be operated at higher temperatures compared to the LTS coil, the cooling structure is separated between the HTS and the LTS coil. To cool the mass, two GM-JT cryocoolers with a cooling power of 4.2 W at 4.3 K are used for the LTS coil, and two GM cryocoolers with a cooling power of 1.5 W at 4.2 K are used for the HTS coil. Even if the temperature of the HTS coil rises, the LTS coil temperatures can be kept at a low level owing to the separated cooling structure. The radiation shield is cooled by two single-stage GM cryocoolers with a cooling power of 100 W at 55 K.

The total cooling time from room temperature for the 25T-CSM is approximately 12 d. When the magnet is charged up to 25 T in 60 min, the total heat loads are estimated to be 10.3 W for HTS coils and 4.4 W for LTS coils. Further detailed cooling properties for the 25T-CSM will be reported elsewhere [4]. A large temperature rise due to AC losses limits the magnet sweep rate. We estimated the AC losses for HTS coils on the basis of the Brandt model [5] for a film shape. However, a practical HTS pancake coil will become a bulk shape. Therefore, the designed AC losses for HTS coils are considered to be overestimation, and a ramp-up time will be reduced.

3. React-and-wind method of CuNb/Nb_3Sn Rutherford cable superconducting coils

It is worth noting that we developed the high strength Nb_3Sn strand with CuNb reinforcing stabilizer (CuNb/Nb_3Sn) using the Nb-rod fabrication method [6]. Figure 2 schematically shows the Nb-rod-processed CuNb/Nb_3Sn using the drawings of Nb rods embedded into a Cu sheath. CuNb/Nb_3Sn compounds were formed by heat treatment at 670 °C for 96 h, and the characteristics of CuNb/Nb_3Sn strands were examined. The CuNb/Nb_3Sn strand and Rutherford cable parameters are listed in table 2. As shown in figure 3, the Rutherford flat cable without insulation, composed of 16 strands using a 0.8-mm-diameter CuNb/Nb_3Sn strand, is 1.53 mm in width, 6.45 mm in thickness, and 65 mm in twist pitch. A glass epoxy tape is used as an insulating material in 1/4 lap winding. In order to fabricate a coil using a react-and-wind method, the bending treatment for heat-treated Rutherford flat cables was repeated ten times at 0.5% bending strain because we found that the repeated bending treatment significantly enhances the critical current for CuNb/Nb_3Sn strands owing to reduction of the residual strain [7].

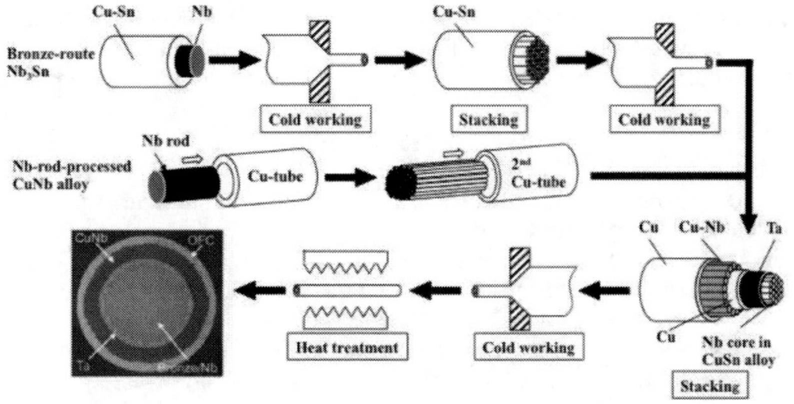

Figure 2. Fabrication process of Nb-rod-processed CuNb/Nb_3Sn strands.

Table 2. Characteristic parameters of (a) the Nb-rod-processed CuNb/Nb_3Sn strand and (b) relevant Rutherford cable.

(a) Strand	
Superconductor	Bronze-processed Nb_3Sn
Reinforcement	Nb-rod-processed Cu-17.4wt%Nb
Diameter [mm]	0.8
Cu/CuNb/non-Cu [%]	20/35/45
Filament diameter [μm]	3.3
No. of filaments	6973
Twist pitch [mm]	20

(b) Rutherford cable	
No. of strands	16
Dimensions [mm×mm]	6.4×1.5
Cabling pitch [mm]	65
I_c (4.2K, 14.5T) [A]	1600 <

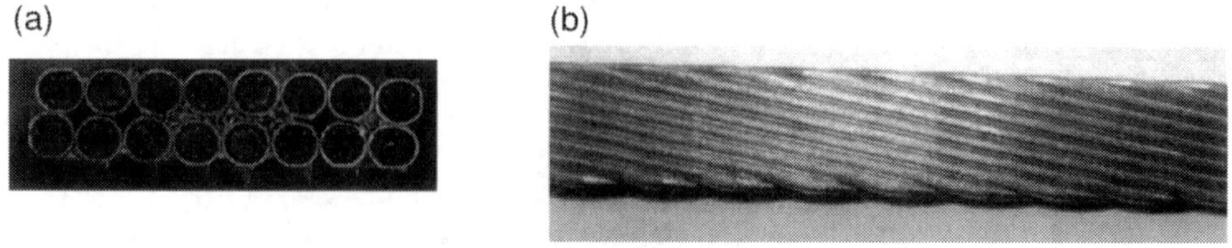

Figure 3. CuNb/Nb_3Sn Rutherford flat cable with 16 strands. (a) Cross Sectional view and (b) longitudinal flat surface view.

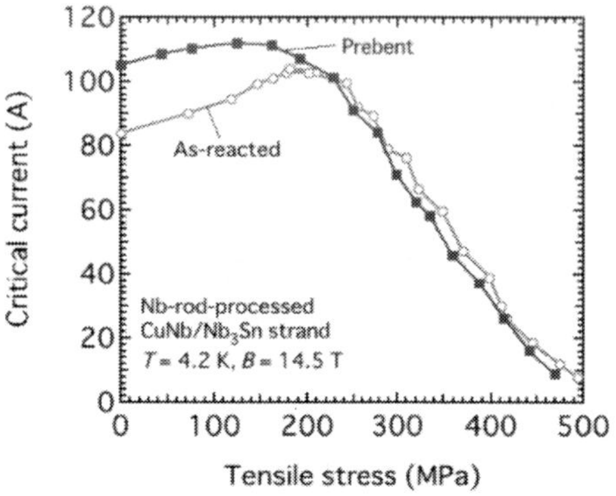

Figure 4. Critical current as a function of applied tensile stress at 4.2 K in 14.5 T for Nb-rod-processed CuNb/Nb_3Sn strands with and without prebending treatment.

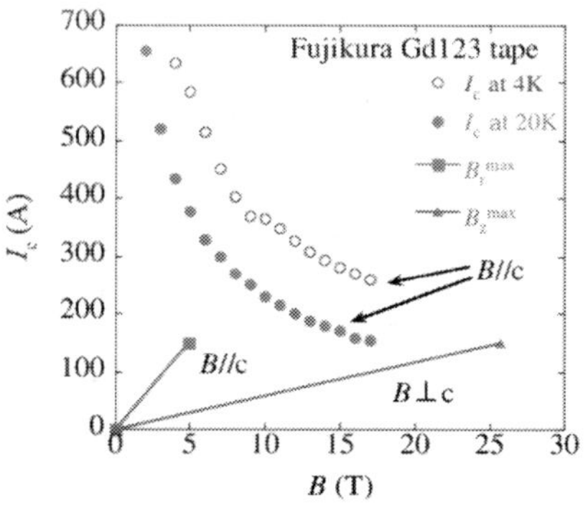

Figure 5. Critical current properties in fields for B//c at temperatures of 4.2 K and 20 K for one $GdBa_2Cu_3O_y$ (Gd123) tape. HTS coil load lines of the magnet operation are indicated in fields for B⊥c and B//c at 4.2 K.

The mechanical properties of CuNb/Nb_3Sn strands at low temperatures were measured using a tensile stress testing apparatus [8]. The critical currents I_c of CuNb/Nb_3Sn strands were determined with an electric field criterion of 1 μV/cm.

The prebending treatment changes the mechanical properties of the CuNb/Nb_3Sn strand, suggesting that the mechanical properties are enhanced because of the work hardening of the Cu stabilizer and the CuNb composite by the prebending treatment [9]. The Young's modulus of the CuNb/Nb_3Sn strand changed from 125 GPa for an as-reacted sample to 180 GPa for a prebent sample. Figure 4 shows the stress dependence of I_c for the as-reacted and 0.8% prebent Nb-rod-processed CuNb/Nb_3Sn strands. I_c values of the prebent strands under a tensile stress of 230 MPa were higher than those of the as-reacted strands. The peak positions of the I_c-stress curves were 210 MPa for the as-reacted strand and 125 MPa for the prebent strand. The hoop stress calculated by Wilson's model [10] is 251 MPa for the L1 coil. The hoop stress value lies in the mechanical tolerance of the CuNb/Nb_3Sn strand. The maximum I_c of 104 A for the as-reacted strand was increased to 112 A by the prebending treatment. These results indicate that the CuNb/Nb_3Sn Rutherford cable composed of 16 strands has enough critical current properties over 1600 A in fields up to 14.5 T at 4.2 K [11], because the designed operation current of 851 A for a 25T-CSM LTS coil is approximately half the critical current.

4. $GdBa_2Cu_3O_y$ (Gd123) superconducting coils with large hoop stress tolerance

As listed in the design parameters of the 25T-CSM HTS coil in table 1, the HTS coil can generate 11.5 T at an operation current of 150 A in a background field of 14 T. In this design, a single bundle

conductor composed of one Gd123 tape is adopted. The maximum radial component B_r of the HTS coil that corresponds to the magnetic field parallel to the c-axis of the Gd123 tape is 4.9 T. Figure 5 shows both field components, the coil-axial field B_z and the coil-radial field B_r, in comparison with I_c properties of the Gd123 tape. Considering a Gd123 single bundle conductor, the coil load lines for the operation current of 150 A per one Gd123 tape are compared with the critical current properties. Since the critical currents in fields for B⊥c are very high for the Gd123 tape, the critical currents in fields for B//c limit the coil performance. The hoop stress is estimated to be 488 MPa using Wilson's model.
When the LTS coil happens to quench, a large induced current is added to the operation current for the HTS coil. Since it is very difficult to detect a small voltage due to the hot spot for HTS, we cannot apply the traditional quench protection like a LTS coil system. We found that it is important for an HTS coil design to not have an overcurrent above the critical current. Therefore, the operation current was established below the critical current at 20 K, even in the case of a coil quench for the HTS coil. As a result, the coil performance is limited by the coil radial field B_r for the Gd123 tape.

5. Summary

We are constructing a cryogen-free 25-T superconducting magnet consisting of a low-T_c 14-T superconducting coil and a high-T_c 11-T insert coil. This magnet will generate 25.5 T in a 52-mm room temperature bore, and the magnetic field of 25.5 T would be the highest generated by a practical superconducting magnet to date.

A high-strength CuNb/Nb_3Sn Rutherford cable with the reinforcing stabilizer CuNb composite was successfully developed for the middle low-T_c section coil. A new technology of a react-and-wind method using a CuNb/Nb_3Sn Rutherford cable is adopted for the coil-winding process. The high-T_c insert coil is fabricated by stacking 68 single-pancake coils using $GdBa_2Cu_3O_y$ tape.

6. References

[1] Watanabe K, Yamada Y, Sakuraba J, Hata F, Chong C K, Hasebe T and Ishihara M 1993 *Jpn. J. Appl. Phys.* **32** L488
[2] Nishijima G, Awaji S, Hanai S and Watanabe K 2006 *Fusion Engineering and Design* **81** 2425
[3] Hanai S, Tsuchihashi T, Minemoto Y, Ioka S, Watanabe K, Awaji S and Oguro H 2014 *IEEE Trans. Appl. Supercond.* **24** 4301204
[4] Iwai S, Takahashi M, Miyazaki H, Tosaka T, Tasaki K, Hanai S, Ioka S, Watanabe K, Awaji S and Oguro H presented in Applied Superconductivity Conference 2014
[5] Brandt E H and Indenbom M 1993 *Phys. Rev. B* **48** 12893
[6] Oguro H, Awaji S, Watanabe K, Sugimoto M and Tsubouchi H 2013 *Supercond. Sci. Technol.* **26** 094002
[7] Watanabe K, Awaji S, Oguro H, Nishijima G, Miyoshi K and Meguro S 2005 *IEEE Trans. Appl. Supercond.* **15** 3564
[8] Katagiri K, Fukumoto M, Saito K, Ohgami M, Okada T, Nagata A, Noto K and Watanabe K 1990 *Adv. Cryo. Eng.* **36** 69
[9] Oguro H, Awaji S, Nishijima G, Badica P, Watanabe K, Shikanai F, Kamiyama T and Katagiri K 2007 *J. Appl. Phys.* **101** 103913
[10] Wilson M N 1982 *Superconducting Magnets* (Oxford, UK: Oxford Science) pp 41–67
[11] Watanabe K, Oguro H, Awaji S, Kumakura H, Sugimoto M and Tsubouchi H 2014 *Adv. Cryo. Eng.* **60** 186

Acknowledgments

This work was partly supported by the Grant-in-Aid for Scientific Research from JSPS, (KAKENHI 25246032), Japan. CuNb/Nb_3Sn Rutherford cables were fabricated by Furukawa Electric Co., Ltd, and Gd123 tapes by Fujikura Ltd.

Journal of Physics: Conference Series **568** (2014) 032020 doi:10.1088/1742-6596/568/3/032020

Application of low frequency SQUID NMR to the ultra-low temperature study of atomically layered ^{3}He films adsorbed on graphite

F. Arnold, B. Yager, J. Nyéki, A. J. Casey, A. Shibahara, B. P. Cowan and J. Saunders

Royal Holloway, University of London, TW20 0EX Egham, United Kingdom

E-mail: b.yager@rhul.ac.uk, j.saunders@rhul.ac.uk

Abstract. Low frequency pulsed NMR is a powerful technique for the investigation of the nuclear susceptibility and spin-dynamics of strongly correlated systems down to ultra-low temperatures. We describe a versatile broadband pulsed NMR spectrometer using a two-stage Superconducting Quantum Interference Device (SQUID). This instrument has enabled the investigation of the spin-dynamics of ^{3}He films adsorbed on exfoliated graphite into the microkelvin temperature range. In prior work we reported a SQUID NMR study of two-dimensional ferromagnetism on a triangular lattice, where ^{3}He films adsorbed on graphite provide an ideal model system. Here we describe the detection of the much weaker signals from strongly correlated fluid films in the second layer of ^{3}He on graphite, using a spectrometer with improved sensitivity. We have measured the low frequency spin-spin relaxation and spin-lattice relaxation of two-dimensional ^{3}He into the microkelvin range. These show an unusual time and frequency dependence. We also describe the use of the ^{13}C-signal from the exfoliated graphite for thermometry, and the unusual properties of the spin-lattice relaxation of that system.

1. Introduction

Helium isotopes adsorbed onto the atomically flat basal planes of exfoliated graphite provide model systems for the investigation of quantum matter in two dimensions. For an overview and a review of earlier work see [1]. These atomically layered films provide a quantum simulator to further the understanding of key contemporary problems in condensed matter physics including: frustrated magnetism [2, 3]; heavy-fermion quantum criticality [4]; the Mott-Hubbard transition [5]. NMR provides a powerful tool to study ^{3}He films, but until now the focus of measurements at ultra-low temperatures has been on studies of a static property, the ^{3}He nuclear magnetization, complementing measurements of heat capacity. While NMR studies of spin dynamics played a central role in the pioneering studies of ^{3}He submonolayer films on graphite [6, 7], their use in measurements at ultra-low temperatures have so far been limited. In this article we briefly survey our recent work to apply SQUIDs to the detection of NMR signals at relatively low frequencies. The gains are: improved precision in the measurement of magnetization; the ability to conveniently measure as a function of magnetic field; the ability to study spin-dynamics into the microkelvin range.

2. The Spectrometer

Our pulsed NMR spectrometer includes a helium adsorption cell and home-made NMR magnet assembly mounted at the demagnetization stage and mixing chamber of a dilution refrigerator respectively.

The adsorption cell contains a stack of 64 exfoliated graphite sheets (Grafoil [8]), enclosed in a Stycast-1266 body [9]. The cell is oriented such that the Grafoil sheets are normal to the static NMR field B_0. In Grafoil, graphite platelets show a preferential alignment of their c-axis normal to the foil surface. The mosaic spread of these platelets was found to be $\pm 19°$ [2] for a sample from the same batch. In the cell, each Grafoil sheet is individually heatsunk to the experimental stage through a diffusion bonded silver foil which is connected to the central thermal anchor of the cell (see Fig. 1). Helium can be admitted to the cell via a narrow fill line that extends up to a room temperature gas handling system. The total surface area of the graphite substrate is $12\,\text{m}^2$ as determined from point-B of a ^{4}He adsorption isotherm at 4.2 K [10].

The low field NMR magnet and coil assembly is contained within a 31 mm diameter, 105 mm long superconducting niobium shield and is of a similar design to those described previously [11]. The NMR magnet is a 4 layer solenoid wound using 106 µm CuNi clad NbTi wire. The transmitter coil is wound around a 25 mm diameter former using similar wire following the optimum saddle coil geometry described by Hoult and Richards [12] and has a field current ratio of $0.275\,\text{mT}\,\text{A}^{-1}$. An overlapping superconducting Hechtfischer [13] shield between the NMR magnet former and transmitter coil screens eddy-current transients in the copper induced by transmitter pulses.

The receiver coil is wound directly onto the epoxy sample cell using NbTi wire and has a 10 mm square cross-section. The NMR cell is mechanically and thermally decoupled from the magnet assembly in order to reduce magneto-acoustic resonances and heat leaks into the nuclear demagnetization stage. Orthogonality of the transmitter and receiver coils is optimized whilst assembling the setup by cross-coupling measurements.

NMR signals are coupled into the SQUID amplifier by a superconducting flux transformer

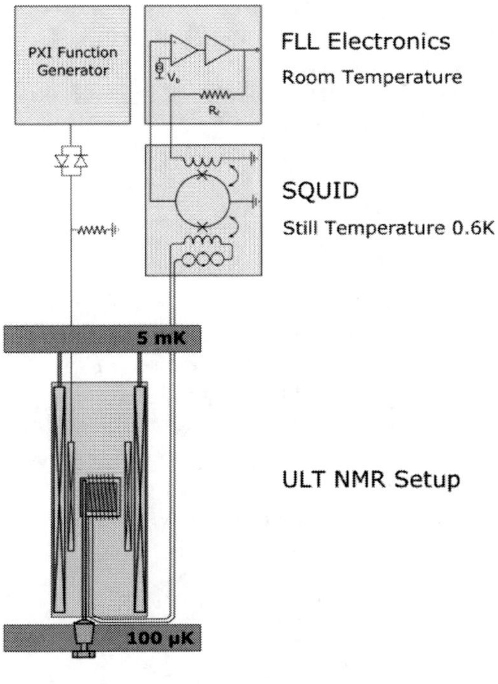

Figure 1. Schematics of the ultra-low temperature SQUID NMR setup. The NMR cell is mounted on the 100 µK-plate of a nuclear demagnetization stage whereas the magnet assembly is supported by the mixing chamber. NMR signals are coupled into a two-stage DC-SQUID, which is controlled by a room-temperature FLL electronics. Tipping pulses are generated with a PXI arbitrary waveform generator.

Journal of Physics: Conference Series **568** (2014) 032020 doi:10.1088/1742-6596/568/3/032020

consisting the receiver coil and the integrated SQUID input coil, which are connected by a superconducting twisted pair shielded by a niobium capillary. Both coils have matching inductances of $L_i = 1.8\,\mu\text{H}$ to maximize the flux transfer ratio. This set-up provides a broadband detection scheme allowing measurements at different B_0, and hence NMR frequencies.

We use a two stage DC-SQUID with integrated current limiter developed by PTB [14]. When operated in flux-locked loop mode (FLL) [15] it serves as a low noise, high bandwidth, high gain flux-to-voltage converter. Control and feedback for the the SQUID is provided by room-temperature Magnicon XXF-1 FLL electronics with 6 MHz bandwidth [16]. The coupled energy sensitivity of our spectrometer is $\varepsilon_c = L_i S_\phi / 2M_i^2 = 9.1\,h$, where M_i is the mutual inductance between the input coil and SQUID and S_ϕ the flux noise per unit bandwidth, with the SQUID cooled to 700 mK.

NMR pulse sequences and signal acquisition are controlled by a National Instruments PXI system [17]. Tipping pulses are produced by a PXI-5412 arbitrary waveform generator and the resultant NMR signals captured by a PXI-5922 digitizer with 20-bit resolution at 5MS/s [17]. A 4 kHz high-pass filter is used to attenuate low frequency eddy-current transients. The captured signals are background-subtracted with an off-resonance and zero-field transient to eliminate magneto-acoustics around the Larmor frequency and low-frequency transients respectively. The signals are analyzed in the frequency domain by applying a conventional FFT. However the exceptionally high signal-to-noise ratio also permits study directly in the time domain by using lock-in demodulation; this allows us to see the precise shape of the recorded free induction decays (FIDs).

3. 2D ^{3}He Adsorbed on Graphite at Ultra-Low Temperatures

This section describes pulsed NMR measurements on a monolayer ^{3}He film sample with areal density $5.5\,\text{nm}^{-2}$ adsorbed onto graphite preplated by a monolayer of ^{4}He, to highlight the versatility of our broadband spectrometer. We show magnetization, spin-spin and spin-lattice relaxation time measurements of this strongly correlated 2D Fermi fluid.

3.1. Nuclear Susceptibility

The nuclear susceptibility of the sample was measured at 100 kHz and temperatures from 240 µK to 10 mK during the natural warmup of the demagnetization stage and from 6 mK to 300 mK using thermally stabilized data points. The magnetization was determined from a stretched exponential fit to the envelope function of the FID in the time-domain. Figure 2 shows the temperature dependence of the total magnetization. The data is fitted by a Curie-Weiss, plus a Dyugaev Fermi fluid magnetization [18] of the form:

$$MT = \frac{N_1 C}{1 + \Theta/T} + \frac{N_2 C}{\sqrt{1 + (T_F^{**}/T)^2}}, \tag{1}$$

where C is the Curie constant per spin, Θ the Curie temperature and T_F^{**} the renormalized Fermi temperature. The first term phenomenologically describes an upturn above the Pauli susceptibility at lowest temperatures. We find an effective Fermi temperature, $T_F^{**} = 131\,\text{mK}$, consistent with previous measurements [19]. We used the Landau Fermi liquid susceptibility enhancement $\chi/\chi_0 = (m^*/m)(1 + F_0^a)^{-1}$, together with the fact that the spin-antisymmetric Landau parameter $(1 + F_0^a)^{-1} = (1 + \sqrt{1 - m/m^*})^2$ depends weakly on effective mass [5, 18, 19], to obtain an effective mass of $5.8 \pm 0.2\,m_0$. This effective mass is consistent with data obtained by Greywall [20] when density corrections due to first layer compression [21] are taken into account, and also with results of Lusher [22, 23] and Morhard [24].

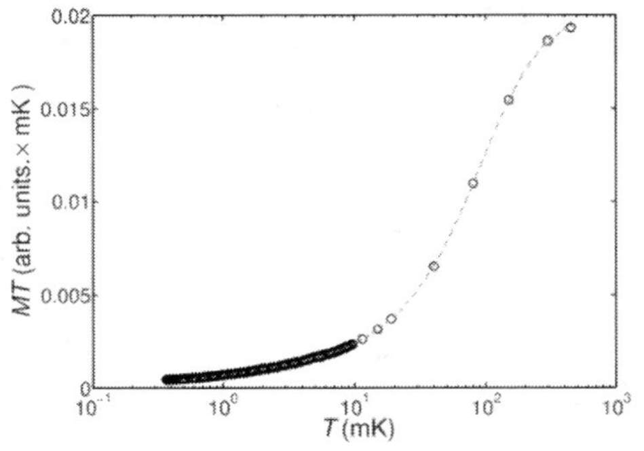

Figure 2. Sample magnetization measured over 3 orders of magnitude in temperature, obtained using small angle tipping pulses. The dashed line shows a fit to Eq. 1.

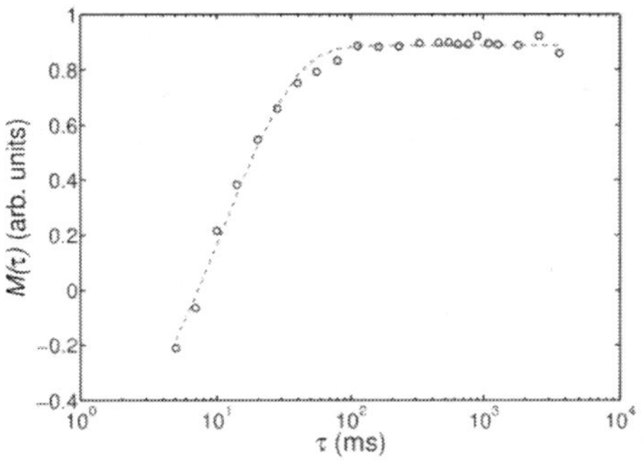

Figure 3. A 750 µK measurement of T_1 at 100 kHz made using an inversion-recovery pulse sequence. The vertical scaling is relative to the magnetization obtained using the same readout pulse, without the preparation pulse, before the measurement. The data (open circles) is fitted by an exponential decay (dashed blue line)

3.2. Spin-Spin relaxation
The transverse relaxation times T_2^* were extracted from fits to the Fourier transformed FIDs. Since T_2^* includes contributions from both intrinsic spin-spin interactions and from inhomogeneities in the static magnetic field, measurements of the intrinsic T_2 are desirable. T_2 was measured using the usual $90° - \tau - 180° - \tau$–spin-echo pulse sequence, requiring additional hardware. A Stanford DG535 delay generator provided a precise trigger source for two Agilent 33220A arbitrary waveform generators and for the PXI digitizer. An example of the quality of data obtained for a spin-echo measurement is shown in Fig. 4.

Fig. 5 shows the spin-echo heights as a function of delay, τ, for frequencies between 100 and 300 kHz. Here the decay of the magnetization was well described by a superposition of two exponential functions, as previously found for similar systems [25, 26]. We observe that the shorter component $T_{2S} = 14.8$ ms is frequency independent whilst the relaxation rate of the longer component increased non-linearly with increasing frequency (see inset Fig. 5) in contrast to previous measurements at higher frequencies [26]. Here the low frequency T_2 is governed by the intrinsic dipole-dipole interaction [27]. The magnetization of the shorter component was an order of magnitude larger than that of the longer component.

3.3. Spin-Lattice Relaxation
Measurements of the spin-lattice relaxation time of adsorbed ^{3}He have previously been used to, for example, measure exchange [28] and dimensionality at $T > 1$ K [29]. Fig. 3 shows a preliminary measurement of T_1, using a $180° - \tau -$ readout sequence, where the sample was at 750 µK before the measurement. Here the vertical scale is the magnetization normalized to its equilibrium value before each data point was taken, sampled with a 0.6° pulse. The 11% reduction of magnetization at long τ is due to an overheating of the sample during the measurement and corresponds to a temperature increase of 80 µK using the magnetization data in Fig. 2.

Journal of Physics: Conference Series **568** (2014) 032020 doi:10.1088/1742-6596/568/3/032020

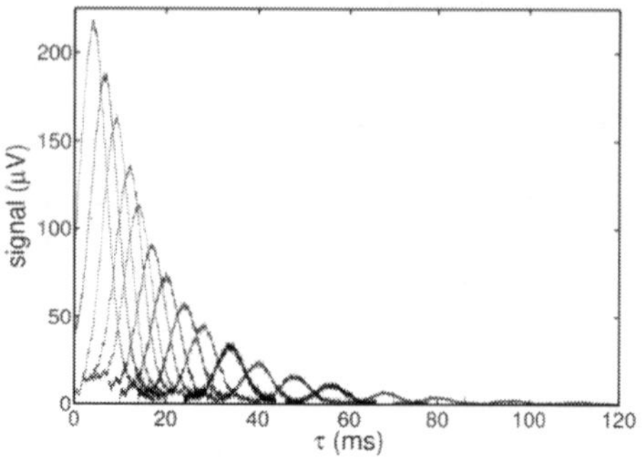

Figure 4. Spin-echoes obtained from a $90° - \tau - 180° - \tau$–spin-echo pulse sequence at 20 mK and 100 kHz. A digital lock-in technique was used to demodulate the data before fitting. The peak height of the echoes obtained from the fits, plotted in Fig. 5, was used to determine T_2.

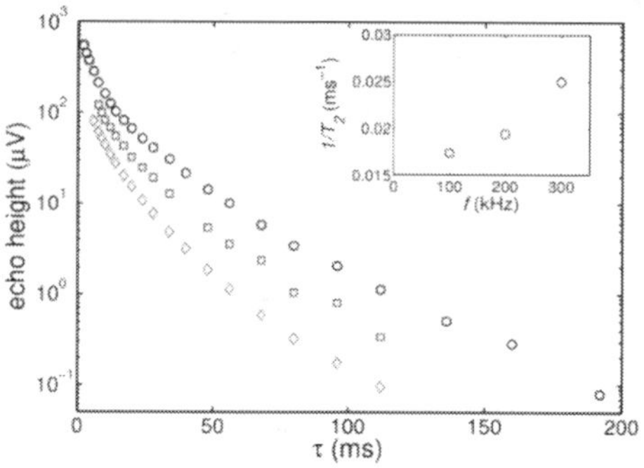

Figure 5. Results of spin-echo measurements at 20 mK. The plotted data is of the peak height of the spin echoes analysed in the time-domain using a digital lock-in technique. (○) 100 kHz, (□) 200 kHz, (◇) 300 kHz. $1/T_2$ for the long components are plotted in the inset. The gain of the amplifier was changed for different frequencies to avoid saturation.

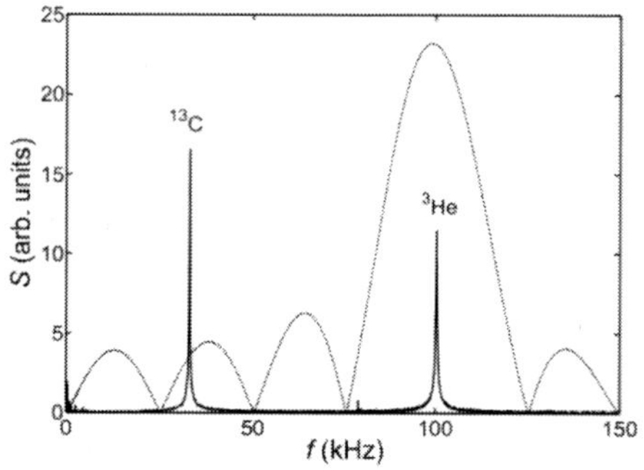

Figure 6. Frequency spectrum of a FID at 370 μK (solid line) showing the on-resonance ^{3}He signal of the 5.5 nm^{-2} sample and the ^{13}C signal of the substrate excited by the power spectrum wing of the excitation pulse (dashed line).

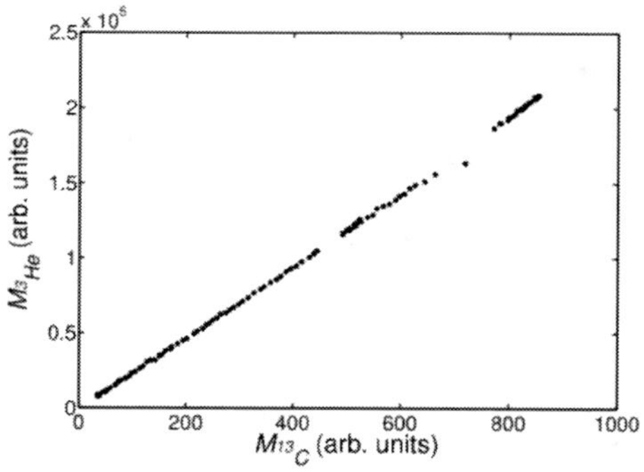

Figure 7. Carbon vs helium magnetizations of a pure ^{3}He sample, where both helium and carbon nuclei are paramagnetic. The ^{13}C is used as the main experimental thermometer below 1 mK. Data in this figure goes down to 250 μK.

4. ^{13}C NMR Thermometry

In our setup we used the spin-½ ^{13}C nuclei present in the Grafoil substrate, at a natural abundance of 1.1 %, as a thermometer below 1 mK. This enabled us to measure directly the temperature of the substrate, which should have the smallest possible temperature gradient to the helium sample. Within graphite ^{13}C forms a dilute nuclear paramagnet with weak dipole-dipole interactions and hence a narrow NMR line width ($T_2^* = 2.4$ ms).

An advantage of the broadband spectrometer is that it allows us to observe FIDs from ^{3}He

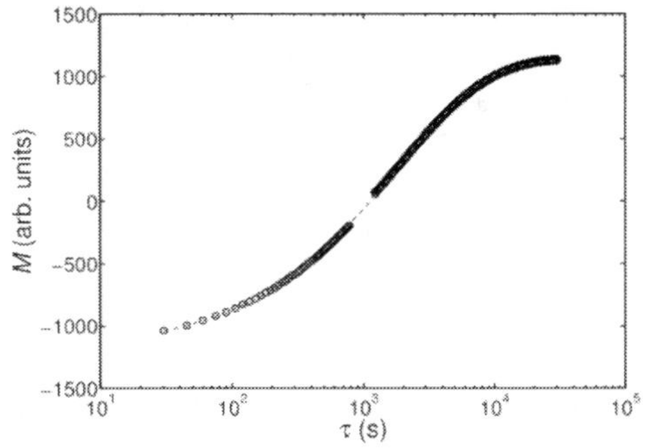

Figure 8. T_1 relaxation of ^{13}C measured at 500 μK following a 180° excitation pulse. The recovering magnetization was probed by small-angle tipping pulses. The data (open circles) is fitted to a stretched exponential relaxation with exponent $n = 0.65$ and $T_1 = 1860\,\mathrm{s}$ (dashed blue line).

and ^{13}C simultaneously. ^{13}C has a gyromagnetic ratio of $\gamma_{^{13}C} = 10.8\,\mathrm{MHz\,T^{-1}}$, compared to $\gamma_{^{3}He} = 32.4\,\mathrm{MHz\,T^{-1}}$. Fig. 6 shows the magnitude of a FFT of a 40 μs transmitter pulse at 100 kHz, demonstrating that there is significant spectral power over a broad frequency range. Fig. 6 also shows a FFT of the NMR response to this pulse where $B_0 = 3.08\,\mathrm{mT}$, so that the ^{3}He is on resonance and the ^{13}C resonance is close to the second side maximum of the pulse power spectrum.

The magnetization of the ^{13}C is obtained from a Lorentzian fit to the complex FFT of the FID. This is calibrated against the ^{3}He melting curve thermometer, MCT, over the temperature range 5.5 mK to 20 mK using thermally stabilised data points. The MCT is self-calibrated against the superfluid A-transition.

The thermal coupling of the ^{13}C and the adsorbed helium sample is demonstrated in Fig. 7 where the magnetization of the ^{13}C is compared to that of a pure bilayer ^{3}He film of areal density $13.0\,\mathrm{nm^{-2}}$, for which the first layer paramagnetism strongly dominates at low temperatures. The linearity shows that both spin species were in thermal equilibrium over the entire temperature range.

Since the pulse repetition rate of NMR thermometers at ultra-low temperatures is limited by their T_1 relaxation time, we investigated the longitudinal recovery of ^{13}C down to the lowest temperatures. The T_1 recovery of the ^{13}C, shown in Fig. 8, was measured using small angle tipping [30], where the equilibrium magnetization was initially prepared by a 180° pulse and the recovery is subsequently probed by repeated 2° tipping pulses. The recovery is fitted by a stretched exponential according to:

$$M[\tau] = M[\infty] - (M[\infty] - M[0])\,e^{-(\tau/T_1)^{\alpha}}, \tag{2}$$

where τ is the time between the preparation pulse and readout pulse and α is a fitting parameter. Non-exponential relaxation is commonly observed where a continuous range of relaxation times exist, see for example [31]. However, we find that the exponent $\alpha = 0.65$ is close to ⅔, which is the expected value for a 2D rigid lattice of spins [30], suggesting that relaxation may be due to a low dimensional effect.

5. Summary

We have installed a versatile broadband NMR spectrometer to measure samples thermalized to a nuclear demagnetization stage. The broadband nature of the spectrometer allows both frequency and temperature as tuning parameters. The nuclear magnetization of ^{13}C within the graphite substrate has been used as a thermometer below 1 mK and is shown to be well thermalized to the ^{3}He film sample. We have demonstrated the feasibility of measuring T_1, T_2,

T_2^* and the nuclear magnetism of ^{3}He films on graphite with high precision. This is currently being applied to the study of quantum criticality of the Mott-Hubbard transition and to identify a possible quantum-spin liquid phase. This work has been financially supported by EPSRC grant EP/H048375/1.

References

[1] Godfrin H and Lauter H J 1995 *Prog. Low Temp. Phys.* **14** 213 – 320
[2] Casey A J, Neumann M, Cowan B, Saunders J and Shannon N 2013 *Phys. Rev. Lett.* **111** 125302
[3] Fukuyama H 2008 *J. Phys. Soc. Jpn* **77** 111013
[4] Neumann M, Nyéki J, Cowan B and Saunders J 2007 *Science* **317** 1356
[5] Casey A J, Patel H, Nyéki J, Cowan B P and Saunders J 2003 *Phys. Rev. Lett.* **90** 115301
[6] Cowan B P 1980 *J. Phys. C: Solid State Phys.* **13** 4575
[7] Owers-Bradley J R, Thomson A L and Richards M G 1978 *J. Phys. Colloques* **39**(C6) 298
[8] Grafoil is a product of GrafTech International Advanced Electronics Technology, 11709 Madison Ave. Lakewood, OH 44107, United States, formerly UCAR - http://www.graftech.com
[9] Stycast is a product of the Henkel Corporation, One Henkel Way, Rocky Hill, CT 06067, USA
[10] Bretz M, Dash J G, Hickernell D C and Vilches O E 1973 *Phys. Rev.* B **8** 1589
[11] Lusher C, Li J, Digby M, Reed R, Cowan B, Saunders J, Drung D and Schurig T 1999 *Applied Superconductivity* **6** 591
[12] Hoult D I and Richards R E 1976 *J. Magn. Reson.* **24** 71
[13] Hechtfischer D 1987 *Cryogenics* **27**(9) 503
[14] Drung D, Assmann C, Beyer J, Kirste A, Peters M, Ruede F and Schurig T 2007 *IEEE Trans. Appl. Supercond.* **17** 699–704
[15] Drung D 1996 *Advanced Squid Read-Out Electronics* vol 329 (Springer Netherlands)
[16] Drung D, Hinnrichs C and Barthelmess H J 2006 *Supercond. Sci. Technol.* **19** S235
[17] National Instruments Corporation 11500 N Mopac Expwy Austin TX 78759-3504
[18] Dyugaev A M 1990 *Sov. Sci. Rev. A Phys.* **1**
[19] Bäuerle C, Bunkov Y M, Chen A S, Fisher S N and Godfrin H 1998 *J. Low Temp. Phys* **110** 333
[20] Greywall D S 1990 *Phys. Rev.* B **41**(4) 1842–1862
[21] Roger M, Bäuerle C, Godfrin H, Pricoupenko L and Treiner J 1998 *J. Low. Temp. Phys* **112**(5) 451
[22] Lusher C P, Saunders J and Cowan B P 1991 *Europhys. Lett.* **14** 809
[23] Lusher C P, Cowan B P and Saunders J 1991 *Phys. Rev. Lett* **67** 2497
[24] Morhard K D, Bäuerle C, Bossy J, Bunkov Y, Fisher S N and Godfrin H 1996 *Phys. Rev.* B **53** 2658
[25] Neumann M, Casey A J, Levitin L, Cowan B and Saunders J 2010 *J. Low. Temp. Phys* **158**
[26] Takayoshi S, Obata K, Sato D, Matsui T and Fukuyama H 2009 *J. Phys.: Conf. Series* **150** 032104
[27] Cowan B, El-Nasr L A, Fardis M and Richards M 1987 *Jpn. J. Appl. Phys.* **26** 309–310
[28] Cowan B, El-Nasr L A, Fardis M and Hussain A 1987 *Phys. Rev. Lett.* **58** 2308
[29] Yager B, Nyéki J, Casey A, Cowan B P, Lusher C P and Saunders J 2013 *Phys. Rev. Lett* **111**(21) 215303
[30] Cowan B P 1997 *Nuclear Magnetic Resonance and Relaxation* (Cambridge University Press)
[31] Narayanan A, Hartman J S and Bain A D 1995 *J. of Magn. Reson. Series* A **112** 58

Journal of Physics: Conference Series **568** (2014) 042001 doi:10.1088/1742-6596/568/4/042001

Bose-Einstein condensation in antiferromagnets at low temperatures

E M Alakshin[1], Yu M Bunkov[1,2], R R Gazizulin[1,2], L I Isaenko[3], A V Klochkov[1], T R Safin[1], K R Safiullin[1], M S Tagirov[1] and S.A.Zhurkov[3]
[1] Kazan Federal University, Kremlevskaya 18, 420008 Kazan, Russia
[2] Institut Neel, CNRS et Universite Joseph Fourier, F-38042 Grenoble, France
[3]V.S.Sobolev Institute of geology and mineralogy, SBranch RAS, Novosibirsk, 630090,Russia

E-mail: alakshin@gmail.com

Abstract. The Bose-Einstein condensation (BEC) was predicted by Einstein in 1925 and this effect is characterized by the formation of a collective quantum state, when macroscopic number of particles is governed by a single wave function. The BEC of magnons was discovered experimentally in superfluid phase of ^{3}He. In the present work we report our progress on the BEC of magnons investigations in solid antiferromagnets at low temperatures by magnetic resonance methods. The duration of the FID signal in two samples of easy-plane antiferromagnets $CsMnF_3$ has been studied. Obtained data confirm the formation of magnon BEC in antiferromagnet $CsMnF_3$.

1. Introduction

The Bose-Einstein condensation (BEC) corresponds to the formation of a collective quantum state in which macroscopic number of particles is governed by a single wave function. The formation of this state was predicted by Einstein in 1925 [1]. The BEC of magnons was discovered experimentally in superfluid phase of ^{3}He-B [2, 3]. It manifests itself by coherent precession of magnetization. Then 6 different states of superfluid ^{3}He with BEC formation were observed. The review of various experiments on the BEC observation can be found in [4-6]. In all cases BEC forms by excited non-equilibrium magnons. To excite it the pulse or continuous pumping at nuclear magnetic resonance (NMR) frequency was used.

In [7] it was assumed that the BEC formation is also possible in solid antiferromagnets $CsMnF_3$ and $MnCO_3$ with coupled nuclear-electron precession. The predictions were successfully confirmed. It was found that the coupled nuclear-electron precession shows all properties of coherent spin precession and magnon BEC [8]. The main experimental fact of magnon BEC evidence was independence of the nuclear-electron magnetic resonance (NEMR) signal amplitude on applied RF power [9-11]. These experiments were done by means of continuous wave NMR and "switch-off" NMR. Two regimes of radiofrequency (RF) pumping were found. In the first regime the induction signal is observed after short (about 1 μs) resonant RF pulse. This pumped state will be called as a normal one in the text. In this state spins precess with its local frequency due to the external magnetic field inhomogeneity. In the second regime the induction signal is observed after long (about hundreds of ms and longer) non-resonant RF pulse (so called "switch-off" NMR) and the signal amplitude is well described in framework of magnon BEC.

We report here the experimental investigations of the free induction decay (FID) signal behaviour in both pumping regimes in two samples of easy-plane antiferromagnet $CsMnF_3$.

2. Results and discussion

The experiments on both $CsMnF_3$ were performed at the temperature of 1.5 K. The first sample $CsMnF_3$ was grown by S.V.Petrov in the P.L. Kapitza Institute for physical problems RAS. The second one was grown by L.I.Isaenko and S.A.Zhurkov in the V.S. Sobolev Institute of geology and mineralogy SB RAS. The main difference between samples is their quality. X-ray investigations show that the second sample is more homogeneous.

The complete details of the experimental setup were described in [12]. We applied RF pulses of different amplitudes and durations with resonant NEMR frequencies or higher ones for an applied external magnetic field. The FID signal was observed in our experiments. The typical FID signals in both RF pumping regimes in the first sample are shown in figure 1.

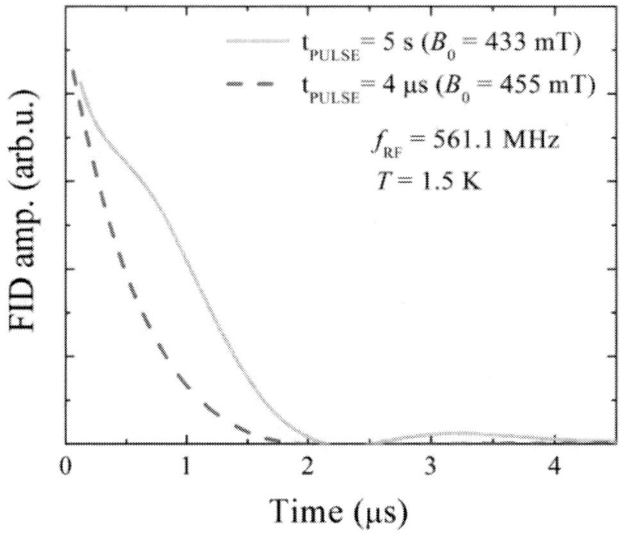

Figure 1. The FID signal $CsMnF_3$ amplitude in the first sample in two RF pumping regimes.

It can be seen that after the 5 s pulse the FID signal duration in BEC state is only about 2 times longer than in normal pumped state. Such negligible difference does not allow us to interpret data in terms of BEC. Why the FID is so short in time dimension? As an example, the process of magnon BEC state formation is very sensitive to impurities and magnetic defects in the experiments with ^{3}He. Thus, the formation of a BEC state of ^{3}He-B in aerogel (high porosity SiO_2) was observed for the first time in Grenoble as a formation of a signal localized in some region of the sample [13]. Later, a global BEC signal at the sample with better homogeneity was observed [14]. The same was observed in a pulsed NMR [15]. The FID of pure ^{3}He-B may have duration of a second, while from the ^{3}He-B in aerogel it occurs to be only about few milliseconds. The similar behavior of the FID signal was observed in superfluid ^{3}He-A in aerogel [16]. The FID duration in BEC state was only about an order of magnitude longer than in the normal state. Furthermore, the beating of a FID signals from a few local BEC states was observed in short pulse NMR. In the case of "switch-off" NMR the beats were not observed and the duration of the signal in BEC state increases (figure 2).

It is well known [17] that the relaxation times of CsMnF3 crystals can vary more than order of magnitude. The origin of the observed short FIDs in BEC state for the first sample can come from its low quality. The next experiments were performed on the second sample, that supposed to be much more homogenous The typical FID signals in both RF pumping regimes in the second sample are shown in figure 3.

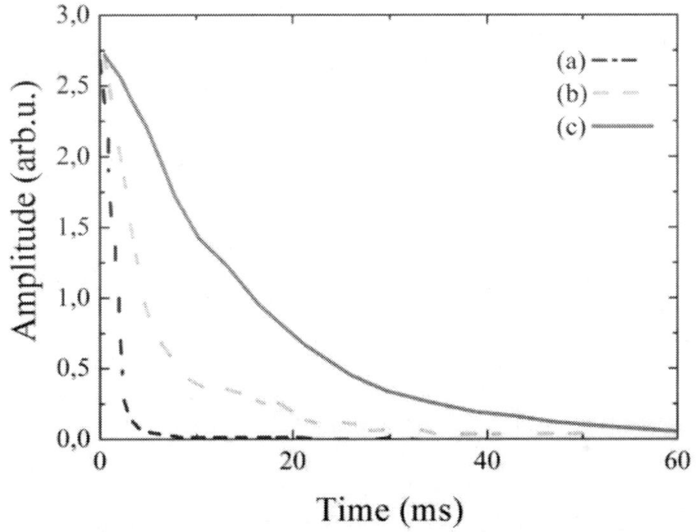

Figure 2. The FID signal amplitude in superfluid 3He-A in aerogel in two RF pumping regimes. (a) - FID from a normal ^{3}He-A. (b) - FID from a superfluid ^{3}He-A after short RF pulse . One can see the beating between different BEC signals; (c) - FID after "switch-off" NMR [16].

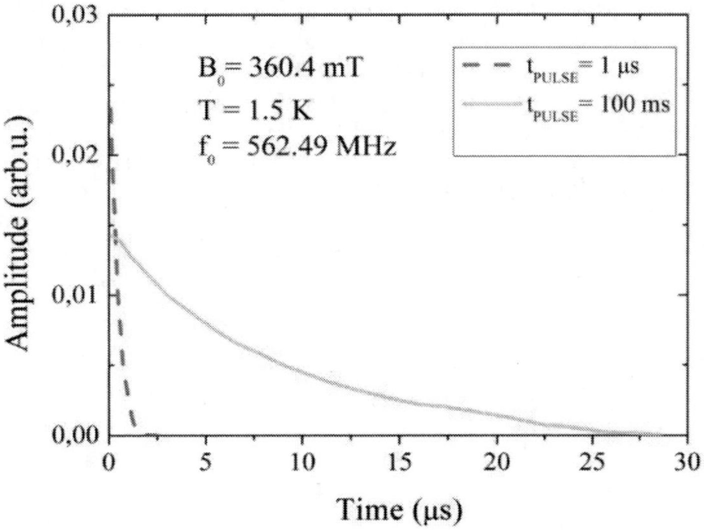

Figure 3. The FID signal $CsMnF_3$ amplitude in the second sample in two RF pumping regimes.

It is well seen in figure 3 that in both cases the FID time is well described by the exponential function. The FID time is about 0.5 μs in normal state and about 9.5 μs in magnon BEC state. The FID signal duration in BEC state is more than an order of magnitude longer than in normal pumped state for the second (more homogeneous) $CsMnF_3$ sample. Obtained results are in a full agreement with the ones observed in superfluid ^{3}He-A in aerogel (figure 2).

3. Conclusions

We have investigated the duration of the FID signal in two samples of easy-plane antiferromagnets $CsMnF_3$. In addition to our previous investigations of the signal amplitude by means of cw NMR and pulse NMR these results also confirm the formation of magnon BEC in antiferromagnet $CsMnF_3$. It is found out that the quality of samples plays significant role in BEC experiments: the long-lived free induction decay signal in the coherent precession state is observed only for higher quality sample.

Acknowledgements

The work is performed according to the Russian Government Program of Competitive Growth of Kazan Federal University, partly supported by the Russian Foundation for Basic Research (project no. 14-02-31290 mol_a) and by the Ministry of Education and Science of the Russian Federation (project no. 02.G25.31.0029).

References

[1] Giorgini S, Pitaevskii L P, Stringari S *Rev. Mod. Phys.* 2008 **80** 1215
[2] Borovik-Romanov A S et al. *JETP Letters* 1984 **40** 1033
[3] Borovik-Romanov A S, et al. *Phys. Rev. Lett.* 1989 **62** 1631
[4] Bunkov Yu M and Volovik G *J. of Low Temp. Phys.*2008 **150** 135
[5] Bunkov Yu M *J. Phys: Condens. Matter* 2009 **21** 164201
[6] Bunkov Yu M and Volovik G *J. Phys: Condens. Matter* 2010 **22** 164210
[7] Bunkov Yu M *Phys. Usp.* 2010 **53** 848
[8] Bunkov Yu M, Alakshin E M, Gazizulin R R et al. *JETP Letters* 2011 **94** 68
[9] Bunkov Yu M, Alakshin E M, Gazizulin R R et al. *J. Phys: Conf. Series* 2011 **324** 012006
[10] Bunkov Yu M, Alakshin E M, Gazizulin R R et al. *J. Phys: Conf. Series* 2012 **400** 032001
[11] Bunkov Yu M, Alakshin E M, Gazizulin R R et al. *Phys. Rev. Lett.* 2012 **108** 177002
[12] Bunkov Yu M, Alakshin E M, Gazizulin R R et al. *Appl. Magn. Reson.* 2013 **44** 595
[13] Bunkov Yu M, Collin E, Godfrin H, Harakaly R *Physica B* 2003 **329** 305
[14] Dmitriev V V, Kosarev I V, Mulders N, Zavjalov V V, Zmeev D Ye, *Physica B* 2003 **329** 324
[15] Hunger P, Bunkov Yu M, Collin E, Godfrin H *J. Phys: Conf. Series* 2012 **400** 012019
[16] Hunger P, Bunkov Yu M, Collin E et al. *J. of Low Temp. Phys.* 2010 **158** 129
[17] Welsh L B *Phys. Rev.* 1967 **156** 370

Journal of Physics: Conference Series **568** (2014) 042002 doi:10.1088/1742-6596/568/4/042002

Unusual field-driven anisotropy of the low-lying ferromagnetic state in $NdOs_4As_{12}$

Andrii Rudenko, Zygmunt Henkie, and Tomasz Cichorek

Institute of Low Temperature and Structure Research, Polish Academy of Sciences, Wroclaw, Poland

E-mail: T.Cichorek@int.pan.wroc.pl

Abstract. We present the low-temperature specific heat $C(T)$ of the filled skutterudite compound $NdOs_4As_{12}$, which undergoes a ferromagnetic transition at $T_C = 1.10$ K. In the absence of an external field B, the $C(T)$ results show a Schottky-like anomaly at 0.93 K whose origin is attributed to a lowering of the T_h cubic point symmetry of the Nd^{3+} ions rather than to an impact of the molecular field. More intriguingly, however, for $B = 0.25$ T applied parallel to the [111] direction, a sharp peak in $C(T)$ occurs at $T^* \approx 0.7$ K. This is in striking contrast to $B \parallel [100]$ where the Curie temperature is essentially unchanged. While more experimental input is needed to resolve the nature of a T^* phase, our observations point at anomalous field-driven ground-state properties of ferromagnetic $NdOs_4As_{12}$.

1. Introduction

Unique crystal structure of the filled skutterudites is the basis of their intriguing physical properties. The MT_4Pn_{12} materials are composed of rigid covalently bonded cage-forming frameworks T_4Pn_{12} enclosing differently bonded guest atoms M, where M is an electropositive element like alkali, alkaline-earth, lanthanide, light actinide or thallium metal, T a transition metal of the 8-th group, and Pn a pnictogen (P, As, or Sb). For the M atom with partially filled $4f$ orbitals, the MT_4Pn_{12} compounds exhibit a large variety of emergent quantum phenomena such as, e.g., metal-insulator transition, BCS-type and unconventional superconductivity, magnetic and quadrupole order, and heavy-fermion behavior [1].

In the case of M = Nd, all compounds exhibit ferromagnetic (FM) phase transition. Whereas $NdFe_4Sb_{12}$ orders ferromagnetically below 16.5 K, one order of magnitude lower critical temperatures have been found in other neodymium-filled phosphide and antimonide skutterudites [1]. Among them, $NdOs_4Sb_{12}$ with the Curie temperature $T_C \approx 0.9$ K has attracted considerable attention. For this low-T_C ferromagnet, a possible heavy-fermion behavior was suggested because of a large electronic specific heat coefficient $\gamma = 520\,\mathrm{mJ/molK^2}$, but a mechanism that could lead to composed quasiparticles remains unsolved [2]. Recent progress in the synthesis of arsenide skutterudites enable studies of their transport and thermodynamic properties. A salient feature of $NdFe_4As_{12}$ with $T_C = 14.6$ K is a Schottky-like anomaly deep inside the ferromagnetic state which is caused by the Zeeman splitting of a quasi-degenerate sextet ground state of the Nd^{3+} multiplet in the presence of a molecular field [3]. In contrast, a Schottky-like anomaly in $NdOs_4As_{12}$ ($T_C = 1.10$ K) occurs on the border of the FM order and shifts to the paramagnetic state upon small external fields [4]. We suppose that low-lying thermal excitations in $NdOs_4As_{12}$ originate from a structural rearrangement rather than from the splitting in a molecular field. Finally, $NdRu_4As_{12}$ shows the ferromagnetic transition at a Curie point $T_C \approx 2.3$ K. For this compound, our preliminary experiments suggest that a quartet is the Nd^{3+} ground state [5].

2. Experimental

Single crystals of $NdOs_4As_{12}$ were grown utilizing a molten Cd:As flux and an enhanced pressure as dercribed elsewhere [4]. Single crystals of the Sb-based counterpart were grown using an Sb flux method [2]. High-temperature magnetic susceptibility measurements point at an effective moment $\mu_{\rm eff} \approx 3.52\,\mu_{\rm B}$ and $3.47\ \mu_{\rm B}$ for $NdOs_4As_{12}$ and $NdOs_4Sb_{12}$, respectively. These values, being very close to the Nd^{3+} free-ion value of $\mu_{\rm eff} = 3.62\,\mu_{\rm B}$, indicate almost a complete filling of the skutterudite structure with the guest Nd atoms in both materials. Specific heat $C(T)$ was determined with the aid of the thermal-relaxation method utilizing a commercial ^{3}He microcalorimeter (PPMS). In-field measurements were performed for $B = 0.25$ T aligned parallel to the [100] and [111] directions of the same single crystal.

3. Results and discussion

A recent study showed that, similarly to the phosphide and antimonide counterparts, $NdOs_4As_{12}$ orders ferromagnetically due to a well-localized character of the $4f$ electron states [4]. However, unlike other neodymium-filled skutterudites, $NdOs_4As_{12}$ shows a Schottky-like anomaly in the vicinity of the Curie temperature. In addition, enhanced values of the specific heat are observed at liquid-helium temperatures where both phonon excitations and magnetic fluctuations were found to be negligible. With small applied fields such that $\mu_{\rm B}B \ll k_{\rm B}T_{\rm C}$, a maximum of the Schottky-like peak shifts above the ferromagnetic transition and thus, provides evidence for an anomalous origin of low-lying thermal excitations in $NdOs_4As_{12}$. We suppose that the Schottky-like effect with an energy separation $\Delta/k_B \simeq 2.2$ K is caused by a lowering of the T_h cubic point symmetry of the Nd^{3+} ion due structural disorder. A plausible possibility concerns a small distortion of the Os cubes. Indications for such a disorder have been observed in the closely related compound $NdOs_4Sb_{12}$ [6]. Because an unusual local disorder in $NdOs_4Sb_{12}$ is mainly restricted to the Nd–Os species with a minor significance of the Sb_{12} cage, it is reasonable to assume a similar Os–Os distortion in $NdOs_4As_{12}$ [4].

Figure 1 presents the specific heat of $NdOs_4As_{12}$ in the vicinity of the ferromagnetic phase transition at $T_C \approx 1.10$ K. The zero-field data are in perfect agreement with our previous results for a collection of several single crystals. In particular, we observed a hump in $C(T)$ at around $T = 0.93$ K due to the Schottky-like effect whose anomalous origin is evidenced by the measurement in a magnetic field applied along the [100] easy magnetization axis [4]. Indeed, while our findings indicate a weak $T_C(B)$ dependence in the small field range, the application of an external field rapidly shifts the Schottky-like anomaly to higher temperatures. However, for $B \parallel [111]$ we have found a *qualitatively* different behavior of the specific heat. Though the Schottky-like anomaly is substantially smaller and its maximum is shifted to somewhat higher temperatures, the FM peak is not resolved at around 1 K. Most remarkably, a sharp anomaly is observed at $T^* \approx 0.7$ K. Interesting that the zero-field and $B \parallel [111]$ dependences are essentially the same below T^*. Finally, we note that differences between various $C(T, B)$ curves tend to disappear above 3 K.

The temperature dependence of the magnetic entropy $S_{\rm mag}$ for $NdOs_4As_{12}$ is depicted in the inset of Fig. 1. The $S_{\rm mag}(T)$ behavior can be simply obtained after subtraction of the electronic term with the coefficient $\gamma = 125\,{\rm mJ/molK^2}$ (cf. the solid line in Fig.1), since the lattice contributions $C_{\rm lat}$ to the specific heat are negligible at $T \lesssim 5$ K [4]. We additionally assumed that the specific heat at $T < 0.4$ K follows the $C(T)$ dependence expected for a simple isotropic ferromagnet. At around 0.7 K, $S_{\rm mag}(T)$ reaches a value of $\approx 0.6R\ln 2$. Upon heating above T^*,

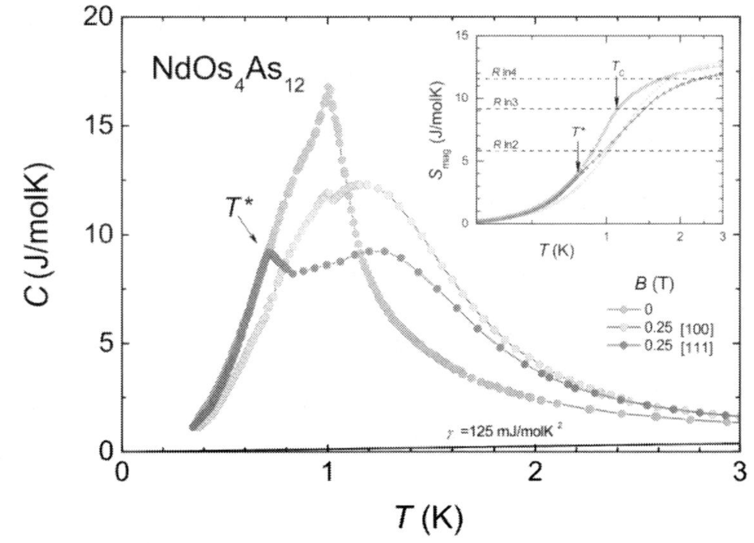

Figure 1. (Color online) Low-temperature specific heat of $NdOs_4As_{12}$ in the vicinity of the ferromagnetic phase transition at $T_C = 1.10$ K. In-field measurements were preformed for B aligned parallel to the [100] and [111] directions. $T^* \approx 0.7$ K marks a position of unknown phase transition for $B \parallel [111]$. The solid line represents the electronic specific heat with a coefficient $\gamma = 125$ mJ/molK^2. Inset: Influence of a magnetic field of 0.25 T on the temperature dependence of the magnetic entropy at $T \leq 3K$.

the S_{mag} data shows a much weaker temperature dependence for $B = 0.25$ T applied along the [111] direction then in the absence of an external magnetic filed. For $B = 0.25$ T applied along the easy direction of magnetization, the magnetic entropy starts to increase at slightly higher temperatures, but no feature at the Curie point is resolved. We also note that the zero-field magnetic entropy in $NdOs_4As_{12}$ levels off above ≈ 2.5 K at a value of ≈ 12.3 J/molK, i.e., only by a few percent larger than $R\ln 4$ and thus, suggestive of a fourfold degeneracy of the ground state.

At this stage of research, a nature of the T^* anomaly is unclear. Forthcoming experiments should uncover interactions leading to anomalous low-T physical properties of $NdOs_4As_{12}$. Obviously, a large field-induced anisotropy is driven from the partially filled $4f$ orbitals. If the magnetic dipoles are involved in the phase transition at $T^* \approx 0.7$ K in $B = 0.25$ T, then an exciting possibility concerns a FM quantum critical (end) point accessed by tuning pressure and magnetic field applied along the [111] direction. Such a scenario seems to be supported by the lack of a FM anomaly in the specific heat at around 1 K. If high-rank multipole moments of the $4f$ shell interact with the main conduction band, then a phase transition could be governed by a nondipolar order parameter. Note that phase transitions driven by electric quadrupole relatively often occur in lanthanide-filled skutterudites but, in general, a multiopolar order parameter of higher rank is very rare. A comparison with the closely related ferromagnet $NdOs_4Sb_{12}$ seems to support an existence of sizable multipolar interactions in $NdOs_4As_{12}$.

In Fig. 2 we show the specific heat of $NdOs_4Sb_{12}$ in the vicinity of the ferromagnetic phase transition at $T_C \approx 0.9$ K. The zero-field $C(T)$ results are in satisfactory agreement with the literature data for a collection of several single crystals [2]. This holds true for enhanced

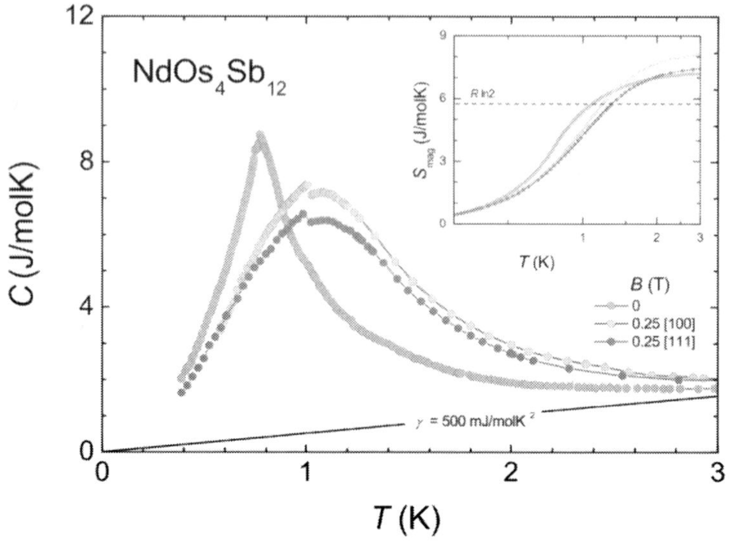

Figure 2. (Color online) Low-temperature specific heat of $NdOs_4Sb_{12}$ in the vicinity of the ferromagnetic phase transition at $T_C \approx 0.9\,K$. In-field measurements were preformed for B aligned parallel to the [100] and [111] directions. The solid line represents the electronic specific heat with a coefficient $\gamma = 500\,mJ/molK^2$. Inset: Influence of a magnetic field of 0.25 T on the temperature dependence of the magnetic entropy at $T \leq 3K$.

values of the specific heat in the low-T paramagnetic state as well as a temperature of about 0.78 K at which a FM peak occurs. Moreover, a simple model based on the relation $C(T) = \gamma T + C_{lat}$ leads to a coefficient γ as large as $\approx 500\,mJ/molK^2$. (Note that, alike for $NdOs_4As_{12}$, the phonon specific heat is negligible below 3 K.) On the other hand, the magnitude of a FM anomaly in our $NdOs_4Sb_{12}$ single crystal has been found to be smaller than in previous studies. As a consequence, there is a relatively large dissimilarity in zero-field $S_{mag}(T)$. As shown in the inset of Fig.2, the magnetic entropy saturates at a value of $0.87R$ ($\approx 1.25R\ln2$) which is substantially smaller than $1.14R$ ($\approx R\ln3$) reported previously [2]. Such a difference is puzzling, especially that a large difference in the occupancies of the Nd site is unlikely. Indeed, both $NdOs_4Sb_{12}$ specimens, being obtained in various crystal-growth processes, show the high-temperature effective magnetic moments close to the Nd^{3+} free-ion value [2, 4]. Nevertheless, magnetic entropy consideration lead to the suggestion that the $\Gamma^{(5)}$ doublet rather then the $\Gamma^{(2)}_{67}$ quartet is the ground state of the Nd^{3+} multiplet in $NdOs_4Sb_{12}$. Additionally shown in Fig. 2 are the low-temperature $C(T)$ data taken at $B = 0.25$ T applied parallel to the [100] and [111] directions. In a sharp contrast to $NdOs_4As_{12}$, the antimony-based compound exhibits an isotropic response to externally applied fields. Down to 0.4 K, both $C(T)$ dependencies are virtually the same. In particular, FM anomalies are clearly seen at around 1 K, in accord with the resistivity data [2].

4. Conclusions

We measured the specific heat of $NdOs_4As_{12}$ and $NdOs_4Sb_{12}$ in the vicinity of the ferromagnetic phase transition at $T_C = 1.10\,K$ and $T_C \approx 0.9\,K$, respectively. These closely related compounds display qualitatively different response of $C(T)$ to externally applied fields. Solely for the arsenic-based compound, we observed a pronounced anomaly at $T^{\star} \approx 0.7\,\mathrm{K}$ in $B = 0.25$ T applied parallel to the [111] direction. We tentatively attribute an anomalous field-induced anisotropy of the ferromagnetic state in $NdOs_4As_{12}$ to characteristics of fourfold degenerated ground state of the Nd^{3+} multiplet.

4.1. Acknowledgments

This work was supported by the Polish National Science Centre within Grant no. 2011/01/B/ST3/05685. Technical support by L. Bochenek is gratefully acknowledged.

References

[1] Sato H, Sugawara H, Aoki Y and Harima H 2009 *Handbook of Magnetic Materials*, edited by K.H.J. Buschow
[2] Ho P-C, Yuhasz W M, Butch N P, Frederick N A, Sayles T A, Jeffries J R, Maple M B, Betts J B, Lacerda A H, Rogl P and Giester G 2005 *Phys. Rev. B* **72** 094410
[3] Higashinaka R, Takeda K, Namiki T, Aoki Y and Sato H 2013 *J. Phys. Soc. Jpn.* **82**, 114710 (2013).
[4] Cichorek T, Rudenko A, Winiewski P, Wawryk R, Kpiski L, Pietraszko A and Henkie Z *submitted*
[5] Henkie Z et. al. *in preparation*
[6] T. Keiber T, F. Bridges F, R. E. Baumbach R E, and M. B. Maple M B 2012 *Phys. Rev. B* **86** 174106

Magnetism in α-R_2S_3 (R = Pr and Nd) single crystals

S Ebisu, T Nagata, K Fuji and Y Shibayama

College of Design and Manufacturing Technology, Muroran Institute of Technology, 27-1 Mizumoto-cho, Muroran, Hokkaido, 050-8585 Japan

ebisu@mmm.muroran-it.ac.jp

Abstract. Magnetic susceptibility, χ, and specific heat, C, of the isostructural rare-earth sesquisulfides α-R_2S_3 (R = Pr and Nd) have been investigated on single crystals. The $\chi(T)$ of α-Nd_2S_3 shows a broad peak around 3.5 K when a magnetic field of 100 Oe is applied along the b-axis of the orthorhombic crystal structure. A Neel temperature T_N above 2 K has been determined as 2.75 K by an inflection point in the $\chi(T)$ and a sharp peak in the $C(T)$. Furthermore, the $C(T)$ indicates another peak at around 0.65 K. The ground states of Nd^{3+} on both the two Nd sites are concluded as Kramers doublets. The $\chi(T)$ of α-Pr_2S_3 shows a broad peak around 3.8 K only in the case of $\boldsymbol{H}//\boldsymbol{b}$. The χ for $\boldsymbol{H}\perp\boldsymbol{b}$ is almost constant below 5 K, which suggests existence of Van Vleck paramagnetic contribution. An inflection point in the $\chi(T)$ ($\boldsymbol{H}//\boldsymbol{b}$) exists at T_N = 2.95 K which is consistent with the temperature of a sharp peak observed in the $C(T)$. A Schottky-type anomaly around 8 K in the $C(T)$ has also been found. The ground states of Pr^{3+} on two Pr sites are considered as a pseudo-doublet and a singlet.

1. Introduction

Rare-earth sesquisulfides α-R_2S_3 (R = Pr and Nd) possess an orthorhombic crystal structure having the *Pnma* symmetry [1, 2]. The crystal structure is shown in figure 1, where S atoms are omitted for clarity. There are two crystallographically inequivalent R sites labeled $R1$ and $R2$ in this structure. We have reported fascinating physical properties in the isostructural compounds of α-R_2S_3 (R = Sm, Gd, Tb and Dy) related to their magnetic transitions [2-13]. The α-Gd_2S_3 shows an antiferromagnetic transition at T_N = 9.9 K with large anisotropy [2-5]. The compounds of α-Tb_2S_3 and α-Dy_2S_3 demonstrate successive antiferromagnetic transitions at T_{N_1} = 12.5 and 11.4 K, T_{N_2} = 3.5 and 6.4 K, respectively [6, 8]. A neutron scattering measurement for α-Tb_2S_3 has clarified the Tb1 site moments order at T_{N1} and Tb2 moments do at T_{N_2} [7]. In the case of α-Sm_2S_3, weak ferromagnetic transition occurs at around 4.5 K [11]. Recent our study on the specific heat of α-Sm_2S_3 has clarified that successive weak ferromagnetic transitions occur at T_{C_1} = 3.8 K and T_{C_2} = 1.9 K, which will be reported elsewhere. Anomalies in the temperature dependence of the electrical resistivity $\rho(T)$ for α-R_2S_3 (R = Sm, Gd, Tb and Dy) have also been found at the temperatures near magnetic transition temperatures [2, 6, 9, 11], especially the extremely large changes of the resistivity in narrow temperature ranges for α-Dy_2S_3 and α-Sm_2S_3 are noticeable. These features appear as sharp peaks in the $\rho(T)$ [9, 11]. Furthermore, extremely broad hysteresis in magnetization process of α-Dy_2S_3 after cooling in magnetic field of 18 T has also been reported [12].

On the other hand, detail of the low temperature physics in α-R_2S_3 (R = Pr and Nd) has not been reported. We have successfully grown single crystals of these compounds. In this paper, we present the experimental results for magnetic susceptibility and specific heat in α-Pr_2S_3 and α-Nd_2S_3, and discuss the ground states of the R^{3+} ions on the two R sites in each compound.

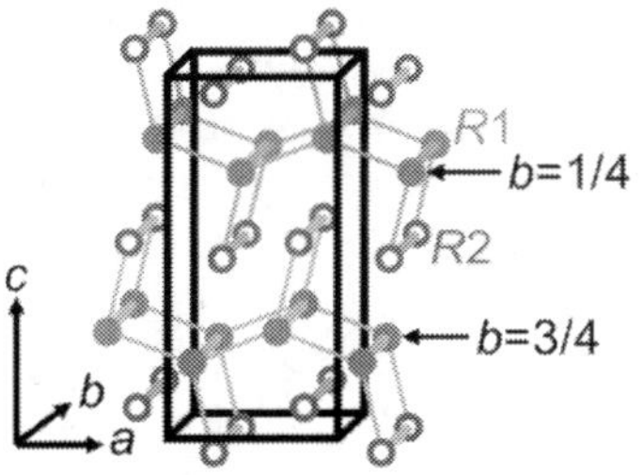

Figure 1. Crystal structure of α-R_2S_3, in which the S atoms are omitted for clarity. Two crystallographically inequivalent R sites are shown as closed circles labeled $R1$ and open circles labeled $R2$. Thin lines are drawn between neighbor atoms on the same *ac*-plane. The orthorhombic unit cell is also shown as a box.

2. Experimental

The single crystals of α-Pr_2S_3 and α-Nd_2S_3 were grown by using a chemical transport reaction method with iodine for carrier gas [2]. Identification of the powder sample and determination of the crystal orientation were performed by using X-ray (Cu K_α radiation) diffraction measurements. The magnetization measurements were made in the temperature range of 1.8-300 K using an *rf*-SQUID magnetometer with Magnetic Property Measurement System (MPMS, Quantum Design). The specific heat was measured at the temperatures of 0.4-300 K without applying magnetic field by using a relaxation method with Physical Property Measurement System (PPMS, Quantum Design).

3. Results and discussion

3.1. Magnetic susceptibility

The temperature dependence of molar magnetic susceptibility χ, magnetization M divided by a constant magnetic field H, of α-Nd_2S_3 is shown in figure 2. Magnetic field of 100 Oe was applied along the three crystal axes of orthorhombic system after cooling the sample to 1.8 K. The inset is an enlargement of the temperature region below 20 K. The $\chi(T)$ shows a broad peak around 3.5 K for $\boldsymbol{H}/\!/\boldsymbol{b}$, while it increases monotonically with decreasing temperature for $\boldsymbol{H}\perp\boldsymbol{b}$. An inflection point is clearly observed at 2.7 K below the peak temperature in $\chi(T)$ for $\boldsymbol{H}/\!/\boldsymbol{b}$. The χ values at temperatures higher than 100 K are well fitted to the Curie-Weiss law,

$$\chi = \frac{C}{T-\theta}, \tag{1}$$

where C is the molar Curie constant and θ is the Weiss constant. The fitting in the case of $\boldsymbol{H}/\!/\boldsymbol{b}$ gives

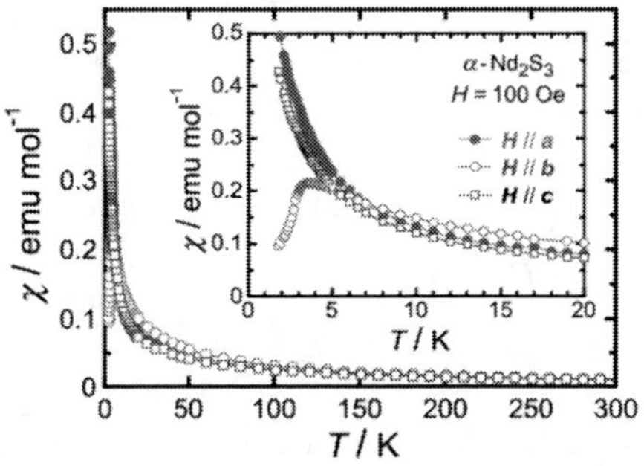

Figure 2. Molar magnetic susceptibility of α-Nd_2S_3 as a function of temperature.

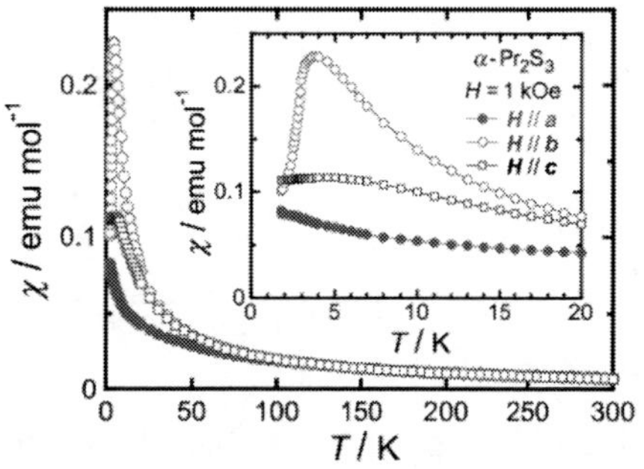

Figure 3. Molar magnetic susceptibility of α-Pr_2S_3 as a function of temperature.

$C = 3.69$ emu K mol^{-1} and $\theta = -19.9$ K. The effective magnetic moment $\mu_{\rm eff}$ is estimated as 3.84 $\mu_{\rm B}$ per one Nd atom. This value is almost the same in the case of $\boldsymbol{H}\perp\boldsymbol{b}$ and close to the theoretical value of 3.62 $\mu_{\rm B}$ for free Nd^{3+}. The $\chi^{-1}(T)$ curves for $\boldsymbol{H}\perp\boldsymbol{b}$ below 10 K are convex, which is consistent that Nd^{3+} ion has electronic configuration of $4f^3$. The Kramers double degeneracy in the system having odd number electrons cannot be lifted by crystalline electric field (CEF).

The magnetic susceptibility of α-Pr_2S_3 under the magnetic field of 1 kOe is plotted as a function of temperature in figure 3. The $\chi(T)$ indicates a broad peak around 3.5 K with an inflection point at 2.7 K in the case of $\boldsymbol{H}/\!/\boldsymbol{b}$. On the other hand, the increase of the χ for $\boldsymbol{H}\perp\boldsymbol{b}$ becomes dull at temperatures below 20 K. It is marked under $\boldsymbol{H}/\!/\boldsymbol{c}$, namely the susceptibility is almost constant below 5 K. This feature suggests existence of Van Vleck paramagnetic contribution. The ground state of Pr^{3+} ions on one Pr site might be singlet. In these series compounds α-R_2S_3, both the R sites have only the mirror symmetry of C_{1h}. According to group theory, all the J-multiplets of R^{3+} in α-R_2S_3 are splitted into singlets for the multiplets with even number $4f$ electrons or into Kramers doublets for the multiplets with odd number $4f$ electrons by CEF. Because Pr^{3+} has electronic configuration of $4f^2$, the ground singlet is reasonable. The Curie-Weiss fitting of the χ data at higher temperatures than 100 K gives the $\mu_{\rm eff}$ of 3.03 $\mu_{\rm B}$ per one Pr atom. This value is somewhat smaller than the value of 3.58 $\mu_{\rm B}$ calculated for free Pr^{3+}. Furthermore, the linearity of the χ^{-1} to T at the fitting range is not so good. These facts also support the existence of Van Vleck paramagnetic contribution.

3.2. Specific heat

Figure 4 shows the specific heat C of α-Nd_2S_3 for mole of formula-unit as a function of temperature. Although the data at higher temperatures are somewhat scattered, the C approaches the value of $3R \times 5 = 125$ J K^{-1} mol^{-1} asymptotically with increasing temperature. Here, R is gas constant and 5 means the number of atoms in formula-unit of α-Nd_2S_3. The lattice contribution $C_{\rm lat}$ evaluated by using the Debye model is shown as dashed curve in figure 4. The Debye temperature $\Theta_{\rm D}$ was estimated as 387 K by fitting the data at temperatures higher than 150 K to the Debye equation,

$$C_{\rm lat} = 9N_{\rm A} r k_{\rm B}\left(\frac{T}{\Theta_{\rm D}}\right)\int_0^{\Theta_{\rm D}/T}\frac{z^4 e^z}{\left(e^z-1\right)^2}dz\ , \qquad (2)$$

where r is the number of atoms per formula-unit, which is 5 for Nd_2S_3. $N_{\rm A}$ is the Avogadro constant, and $k_{\rm B}$ is the Boltzman's constant. The $\Theta_{\rm D}$ of 387 K is close to the value of 370 K estimated for α-Dy_2S_3 [13]. The magnetic specific heat $C_{\rm m}$ derived by subtracting $C_{\rm lat}$ from C is plotted in figure 5. The $C_{\rm m}$ demonstrates a sharp peak at 2.75 K, which is consistent to the temperature of the inflection point in the $\chi(T)$ for $\boldsymbol{H}/\!/\boldsymbol{b}$. Therefore we conclude antiferromagnetic ordering of Nd^{3+} moments occurs

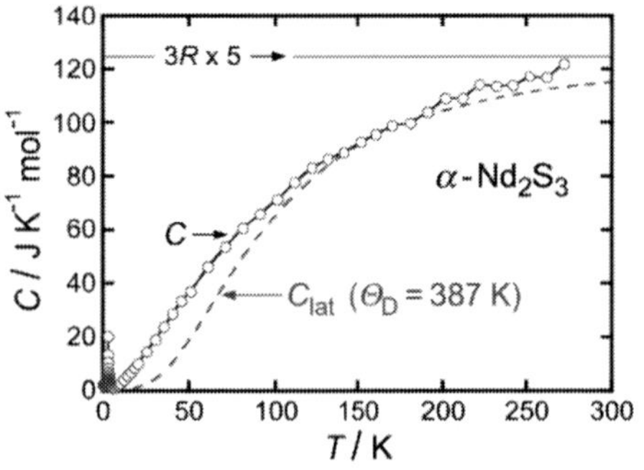

Figure 4. Temperature dependence of the molar specific heat of α-Nd_2S_3.

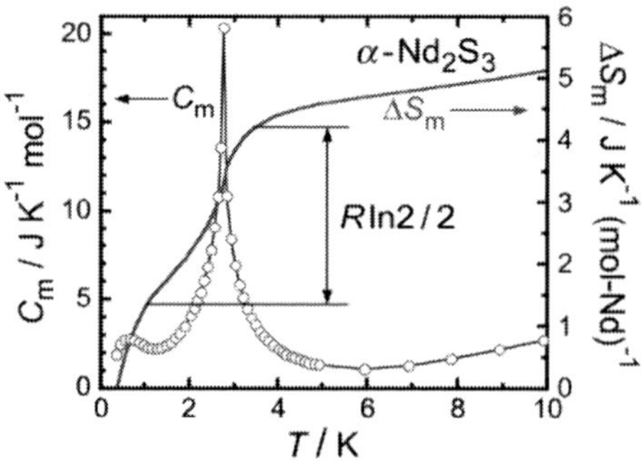

Figure 5. The magnetic specific heat and the magnetic entropy change of α-Nd_2S_3.

at the Neel temperature T_N of 2.75 K. Another peak with roundness is also observed at around 0.65 K. Magnetic entropy change ΔS_m per mol-Nd from the lowest temperature in the present measurement is also shown as a solid curve in figure 5. It is evaluated using the equation,

$$\Delta S_{\mathrm{m}} = \int_{T_0}^{T} \frac{C_{\mathrm{m}}}{2T} \mathrm{d}T \,, \tag{3}$$

where T_0 is the lowest temperature of about 0.4 K, and 2 is number of Nd atoms in the formula-unit of Nd_2S_3. The magnitude of the magnetic entropy change across the transition at T_N = 2.75 K is about $(R\ln2)/2$ as shown in figure 5. This fact means that only the magnetic moments on one Nd site order at T_N and the ground state of Nd^{3+} on this site is (Kramers) doublet. It seems that extrapolation of the ΔS_m curve to 0 K gives the entropy change with the magnitude of $(R\ln2)/2$ across another transition at around 0.65 K. It is speculated the magnetic moments on another Nd site having ground state of Kramers doublet order at around 0.65 K.

The C_m per mol-(formula unit) and the ΔS_m per mol-Pr for α-Pr_2S_3 are shown as a function of temperature in figure 6. Here the lattice contribution C_{lat} calculated by the Debye model with Θ_D = 440 K is also shown as a dashed curve. The value of Θ_D is rather larger than the value for α-Nd_2S_3, which shows difficulty of estimation of the lattice contribution by the present method. However, the lattice contribution is smaller than the magnetic one sufficiently near the magnetic transition temperature. The C_m exhibits a sharp peak at 2.95 K. Since this temperature agrees with the temperature of the inflection point in $\chi(T)$ for $\boldsymbol{H} // \boldsymbol{b}$, it is inferred that antiferromagnetic ordering of Pr^{3+} takes place at T_N = 2.95 K. At temperature below T_N, the C_m starts rising again at about 1 K with decreasing temperature. It suggests some degrees of freedom remains in this system in the temperature region below 1 K. The magnitude of the magnetic entropy change across the transition at T_N = 2.95 K is about $(R\ln2)/4$. If complete magnetic ordering occurs on one Pr site of which ground state is doublet, the expected entropy change is $(R\ln2)/2$. It is speculated that the ordering of Pr^{3+} moments on one site is not perfect but leaves some degrees of freedom. The upturn of C_m below T_N mentioned above supports this speculation. At higher temperature than T_N a hump is observed around 8 K, which is considered to be the Schottky anomaly. The Schottky specific heat C_S for mol-(magnetic atom) under taking account of only two lowest levels is expressed as

$$C_{\mathrm{S}} = R\left(\frac{\Delta}{T}\right)^2 \frac{g_0}{g_1} \frac{\exp(\Delta/T)}{\left[1+(g_0/g_1)\exp(\Delta/T)\right]^2} \,, \tag{4}$$

where Δ is energy separation measured in K between the lowest and the first excited state, and the degeneracy for each state is denoted as g_0 and g_1. Two convenient equations about T_{max} and $C_{S,max}$ representing the values of T and C_S at the maximum point of the hump in $C_S(T)$ deduced from equation (4) are as follows:

$$\frac{g_0}{g_1} \exp\left(\frac{\Delta}{T_{\max}}\right) = \frac{(\Delta/T_{\max})+2}{(\Delta/T_{\max})-2} \,, \tag{5}$$

$$C_{\mathrm{S,max}} = \frac{R}{4}\left[\left(\frac{\Delta}{T_{\max}}\right)^2 - 4\right]. \tag{6}$$

It is natural to consider the ground state of Pr^{3+} on one Pr site is singlet and the ground state on the other is pseudo-multiplet, because both the magnetic ordering and the behavior of the Van Vleck paramagnetism are observed in the $\chi(T)$. If we assume one hump exists around T_{max} = 8 K with $C_{S,max}$ = 5 J K^{-1} mol^{-1}, the values of Δ = 20.2 K and g_0/g_1 = 0.68 ~ 2/3 are deduced from equations (6) and (5). The calculated C_S using these values (g_0 = 2, g_1 = 3) is drawn as the solid curve labeled by $C_S(a)$ with the C_m. Subsequently, we assume the lowest two states on the other site are the ground singlet and the first excited doublet separated by Δ = 71.6 K. This contribution is shown as the curve labeled $C_S(b)$ in figure 7. The total of the Schottky contribution, namely $C_S(a)$ + $C_S(b)$, is also drawn. This curve reproduces the C_m up to about 20 K. Large residual part of the C_m above 20 K might be caused by the higher states. The analysis mentioned above is too rough to give precise energy level scheme; e.g. the Δ of 71.6 K might be too large. However, we consider the qualitative explanation of the ground states of Pr^{3+} on each Pr site is valid.

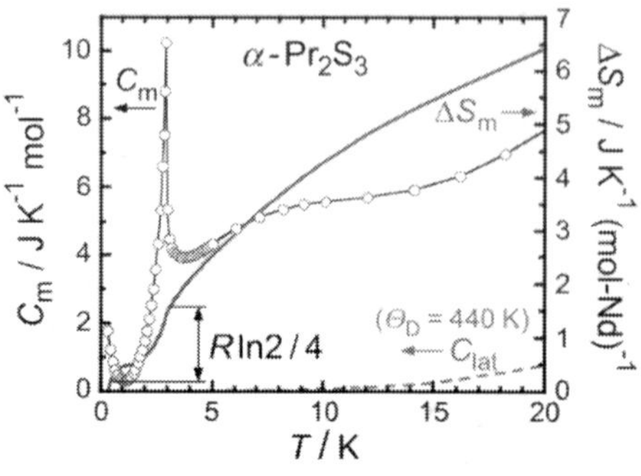

Figure 6. The magnetic specific heat and the magnetic entropy change of α-Pr_2S_3.

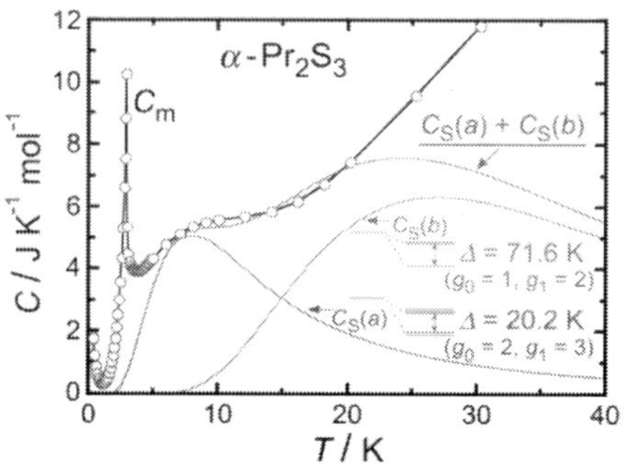

Figure 7. The Schottky contribution to the magnetic specific heat of α-Pr_2S_3.

4. Summary

The antiferromagnetic ordering of Nd^{3+} moments on one Nd site at $T_N = 2.75$ K was confirmed in α-Nd_2S_3. It was inferred the Nd^{3+} moments on the other Nd site order near 3.5 K. The degeneracies of the ground states of Nd^{3+} on both Nd sites were considered to be Kramers doublets. As for α-Pr_2S_3, it was concluded the magnetic moments of Pr^{3+} having pseudo-doublet ground state on one Pr site order at $T_N = 2.95$ K, however, the ordering at T_N is not complete but leaves a certain degree of freedom. An indication of freezing of this degree of freedom below 1 K was observed as an upturn in the $C(T)$. It was considered the Pr^{3+} on the other site has singlet ground state. Hence almost constant $\chi(T)$ at lower temperatures is due to the Van Vleck paramagnetism. The large $C_m(T)$ above T_N was analyzed by superposition of two Schottky-type contributions of Pr^{3+} on both Pr sites.

Acknowledgements

The present work was partially supported by a Grant-in-Aid for Scientific Research (C) (KAKENHI No. 23540384) from the Japan Society for the Promotion of Science.

References

[1] Picon M, Domange L, Flahaut J, Guittard M and Patrie M 1960 *Bull. Soc. Chim. France* **2** 221-8.
[2] Ebisu S , Iijima Y, Iwasa T and Nagata S 2004 *J. Phys. Chem. Solids* **65** 1113-20
[3] Kikkawa A, Katsumata K, Ebisu S and Nagata S 2004 *J. Phys. Soc. Jpn.* **73** 2955-8
[4] Matsuda M, Kikkawa A, Katsumata K, Ebisu S and Nagata S 2005 *J. Phys. Soc. Jpn.* **74** 1412-5
[5] Katsumata K, Kikkawa A, Tanaka Y, Shimomura S, Ebisu S and Nagata S 2005 *J. Phys. Soc. Jpn.* **74** 1598-601
[6] Ebisu S, Gorai M, Maekawa K and Nagata S 2006 *AIP Conf. Proc.* **850** 1237-8
[7] Matsuda M, Kakurai K, Ebisu S and Nagata S 2006 *J. Phys. Soc. Jpn.* **75** 074710
[8] Ebisu S, Narumi M, Gorai M and Nagata S 2007 *J. Magn. Magn. Mater.* **310** 1741-3
[9] Ebisu S, Narumi M and Nagata S 2006 *J. Phys. Soc. Jpn.* **75** 085002
[10] Ebisu S, Koyama K, Omote H and Nagata S 2009 *J. Phys.: Conf. Ser.* **150** 042027
[11] Ebisu S, Omote H and Nagata S 2010 *J. Phys.: Conf. Ser.* **200** 092005
[12] Ebisu S, Koyama K, Horikoshi T, Kokita M and Nagata S 2012 *J. Phys.: Conf. Ser.* **400** 032010
[13] Ebisu S, Ushiki Y and Takahashi S 2013 *J. Kor. Phys. Soc.* **63** 571-574

Magnetic-field-induced Heisenberg to XY crossover in a quasi-2D quantum antiferromagnet

N A Fortune[1], **S T Hannahs**[2], **C P Landee**[3], **M M Turnbull**[4], **F Xiao**[5]

[1] Department of Physics, Smith College, Northampton MA, USA
[2] National High Magnetic Field Laboratory, Florida State University, Tallahassee FL, USA
[3] Department of Physics, Clark University, Worcester MA, USA
[4] Department of Chemistry, Clark University, Worcester MA, USA
[5] Department of Physics, Durham University, Durham, UK

E-mail: `nfortune@smith.edu`

Abstract. The magnetic-field-dependent ordering temperature of the quasi-2D quantum Heisenberg antiferromagnet (QHAF) $Cu(pz)_2(ClO_4)_2$ was determined by calorimetric measurement in applied dc fields up to 33 tesla. The magnetic phase diagram shows a round maximum at 5.95 K and 17.5 T (at $\approx 1/3$ of its saturation field), a 40 percent enhancement of the ordering temperature above the zero field value of 4.25 K. The enhancement and reentrance are consistent with predictions of a field-induced Heisenberg to XY crossover behavior for an ideal 2D QHAF system.

1. Introduction

The copper pyrazine (pz)-based antiferromagnet $Cu(pz)_2(ClO_4)_2$ is a quasi-2D $S = \frac{1}{2}$ square-lattice quantum Heisenberg antiferromagnet with weak intraplanar anisotropy. As shown in Figure 1 the Cu ions are connected within an a-b layer via pyrazine (pz) molecules [1]; adjacent layers are offset along the crystal a and b axes, resulting in an even weaker interlayer exchange interaction (on the order of $J/J' < 10^{-4}$). With a similar structure to the square-lattice copper-oxides, but much smaller interaction strengths and saturation fields ($\mu_0 H = 48$ T), molecular magnets such as $Cu(pz)_2(ClO_4)_2$ can be used to study the 2D Quantum Heisenberg antiferromagnet (QHAF) in the strong field limit.

Quantum fluctuations act to suppress magnetic order. An ideal Heisenberg square lattice antiferromagnet orders only at $T = 0$ — with an expected magnetic moment of $0.6\mu_B$ — but weak interplanar coupling and/or small XY anisotropy can reduce the magnitude of the quantum fluctuations, stabilizing Néel-type antiferromagnetic ordering at temperatures above zero K in zero field [2, 3]. In the case of $Cu(pz)_2(ClO_4)_2$, long range 3D magnetic order occurs at $T_N = 4.25$ K in zero field. Evidence for the suppression of magnetic ordering by quantum fluctuations is seen in both the reduced magnetic moment of $0.47\mu_B$ per Cu^{2+} ion at $T \to 0$ in zero field and in the surprising increase of T_N in an applied magnetic field. The increase is attributed to field-induced suppressions of those fluctuations [4].

Interestingly, the low-field susceptibility [2] shows evidence of an exchange-anisotropy crossover from Heisenberg to XY-like behavior just above T_N, as shown in Figure 2. The crossover from Heisenberg to XY anisotropy is taken to occur at $T_{\min}$, the minimum in the χ_z

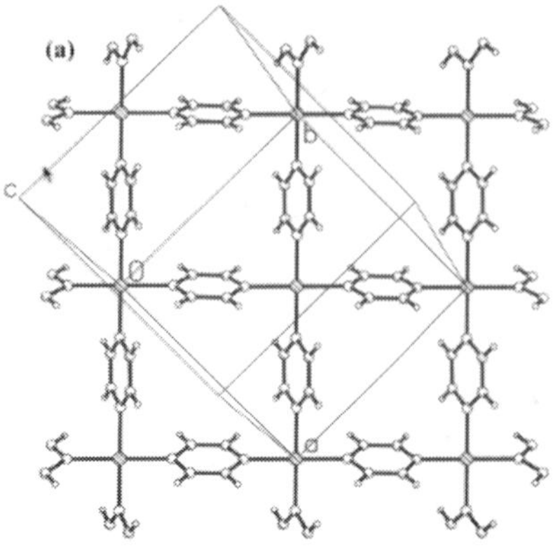

Figure 1. $Cu(Pz)_2(ClO_4)_2$ crystal structure (as viewed ⊥ ab plane). Cu ions are connected within a layer via pyrazine (pz) molecules. ClO_4 ions not shown. Adjacent layers are offset along a and b, leading to a very weak interlayer exchange interaction J′.

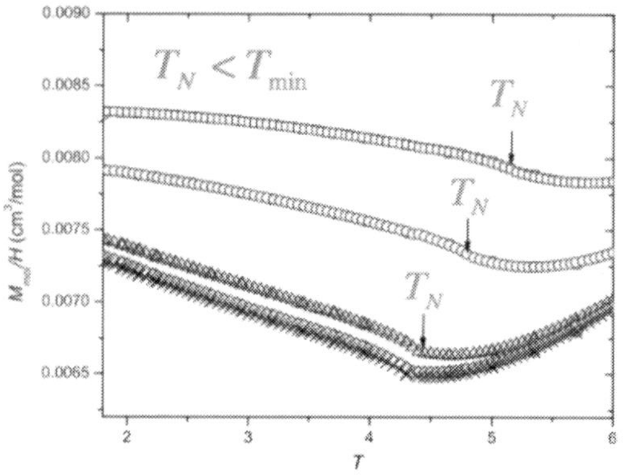

Figure 2. χ_z for weak fields of 1 to 50 kOe applied ⊥ to layers (from Ref. [2]). As temperature decreases, the out of plane susceptibility passes through a minimum at $T_{\rm min}$, followed by long range 3D order at T_N. The minimum is a signature of XY anisotropy.

versus T curve. 3D long range antiferromagnetic ordering occurs at T_N (where $T_N < T_{\rm min}$), as confirmed by μ^+SR [2]. This can be understood as follows: as the temperature decreases, the small amount of XY anisotropy in the system causes a larger proportion of spins to antialign in the xy plane, thereby increasing the fraction of antiferromagnetic coupled spins available to cant in the direction of an applied field along the z axis. The result is an increase in χ_z with decreasing temperature at sufficiently low temperatures, as observed. Like T_N, the crossover temperature $T_{\rm min}$ also increases with increasing field.

2. Model Hamiltonian

In the simplest model Hamiltonian that can capture this physics,

$$H = J\sum_{nn}[S_i^x S_j^x + S_i^y S_j^y + (1-\Delta)S_i^z S_j^z] + J'\sum_{i,i'}\mathbf{S}_i \cdot \mathbf{S}_{i'} - g\mu_B\mu_0\mathbf{H}\cdot\sum_j \mathbf{S}_j \quad (1)$$

where the first summation is over nearest neighbors, the second summation links each spin to its counterparts in adjacent layers, and the third summation includes all spins. J is the intraplanar nearest-neighbor coupling strength, J' is the effective interplanar coupling strength, and Δ is an exchange anisotropy parameter. For an ideal 2D quantum Heisenberg antiferromagnet, $J' = 0$ and $\Delta = 0$, whereas for an ideal 2D XY model, $J' = 0$ and $\Delta = 1$. The actual values of J, J', and Δ must be experimentally determined.

The addition of weak antiferromagnetic next-nearest-neighbor interactions are expected to dramatically increase quantum fluctuations, destabilizing the ordering in the antiferromagnet state [5]. Next nearest neighbor interactions are not explicitly included in Eq. 1, but in a 2D effective spin model (in which both the exchange anisotropy Δ and interplanar coupling J' terms are taken to be zero), they can be approximated by replacing J with an effective value $J_{\rm eff} = J_1 - J_2$, where J_1 and J_2 represent the strength of nearest neighbor (NN) and next nearest neighbor (NNN) calculations, respectively. Applying the effective spin model approximation to neutron inelastic scattering data [5] yields a estimate $\frac{J_2}{J_1} \simeq 0.02$ for the relative strengths of the next nearest neighbor and nearest neighbor terms. Given an experimentally inferred value of $J_1 = 1.54$meV, we then have $J_{\text{eff}} = J_1 - J_2 = (0.98)J_1 = 1.51$ meV (17.5 K).

New angle-dependent ESR data [3] suggests, however, that this choice of an XY model with single anisotropy parameter Δ overlooks a weak in-plane ansiotropy that is important for the low-field behavior, resulting in a spin-flop phase transition at 0.42 T (in the zero temperature limit) for magnetic fields applied along the easy axis. Ignoring interplanar coupling terms, a model Hamiltonian for this biaxial anisotropy becomes

$$H = J\sum_{i,j}[S_i^x S_j^x + (1-\Delta_y)S_i^y S_j^y + (1-\Delta_z)S_i^z S_j^z] - g\mu_B \mathbf{H}\cdot\sum_j \mathbf{S}_j \qquad (2)$$

where the first term is a sum over nearest neighbors and the second term is a sum over all spins. Assuming a nearest neighbor coupling strength of $J = 18.1$K (1.56 meV), a fit to the data yields exchange-anisotropy values of $\Delta_y = 3.1 \times 10^{-4}$ and $\Delta_z = 3.1 \times 10^{-3}$ [3]. Interestingly, the in-plane anisotropy changes sign above the spin-flop transition (for fields directed along x). Defining $\delta_y \equiv J(1-\Delta_y)$ and $\delta_z \equiv J(1-\Delta_z)$, the ESR biaxial model yields values of $\delta_y = 5.3$mK and $\delta_z = 53.2$mK below the spin-flop-transition, while above the spin-flop transition, $\delta y^* = -6.7$mK. In the calorimetric measurements presented here, however, the field was applied along the z axis rather than x; we plan to investigate the low-field field-angle dependent behavior in the vicinity of the spin-flop transition in a future experiment.

Combining the two models to include both the intraplanar variation in the exchange anisotropy parameter Δ and the interplanar coupling parameter J',

$$H = J\sum_{i,j}[S_i^x S_j^x + (1-\Delta_y)S_i^y S_j^y + (1-\Delta_z)S_i^z S_j^z] - g\mu_B \mathbf{H}\cdot\sum_j \mathbf{S}_j + J'\sum_{i,i'} \mathbf{S}_i \cdot \mathbf{S}_{i'} \qquad (3)$$

where $J = 18.1$K and $\Delta_z = 3.1 \times 10^{-3}$, while $\Delta_y = 3.1 \times 10^{-4}$ for fields directed along z and $\Delta_y = -3.9 \times 10^{-4}$ for fields directed along x. This model Hamiltonian does not directly include contributions next nearest neighbors; one solution to this could be to replace $J = 18.1$ K with an effective $J_{\text{eff}} = 17.5$ K as before.

Because the interplanar coupling constant J' and anisotropy parameter Δ (or Δ_y and Δ_z) and their interplay determine so much of the physical behavior in these materials — including the zero-field-ordering temperature and the initial enhancement of antiferromagnetic ordering with field at low field — it is critical to obtain an accurate value for J' as well. Unfortunately, currently available theoretical models used to estimate J' from J and T_N [6] rely on the assumption that the exchange interactions are purely Heisenberg (and therefore assume $\Delta = 0$).

One promising alternative is to experimentally determine J' and Δ in a self-consistent manner from a numerical Quantum Monte Carlo (QMC) fit [7] to the magnetic-field-dependence of the antiferromagnetic phase boundary $T_N(\mu_0 H)$. In this paper we present the experimental data needed for such a calculation for $Cu(pz)_2(ClO_4)_2$. QMC fits to the data using Eq. 1 and Eq. 3 are ongoing; the full results of these calculations will be presented in a later publication.

3. Experimental Methods and Results

For these measurements, we used a custom-built miniature sample-in-vacuum calorimeter [8] designed to fit inside the National High Magnetic Field Laboratory (NHMFL) top-loading dilution refrigerator single-axis rotating sample probes [9]. The calorimeter can be used for both ac calorimetric [10] and thermal-relaxation calorimetric [11] measurements in field.

Figure 3 shows the 1 mm x 1.3 mm x 0.3 mm thick $Cu(Pz)_2(ClO_4)_2$ sample (in blue) sandwiched between a low mass sample thermometer and sample heater. The sample thermometer and heater leads serve as the weak thermal link to a temperature-controlled silver/sapphire platform. The platform and sample assembly are in vacuum; the platform's

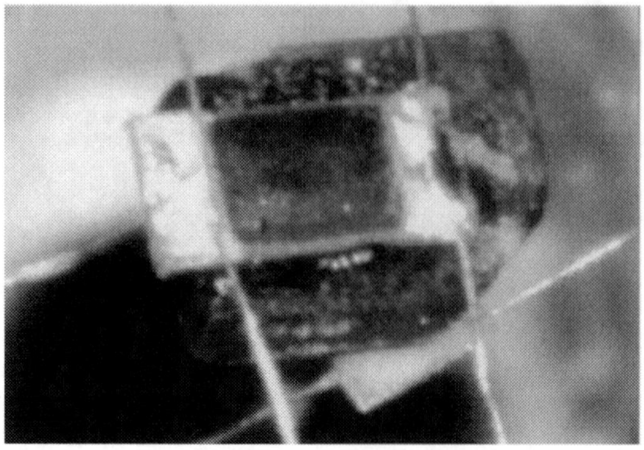

Figure 3. $Cu(Pz)_2(ClO_4)_2$ sample (blue) sandwiched between a sample thermometer and heater. The leads serve as the thermal link to a temperature-controlled platform.

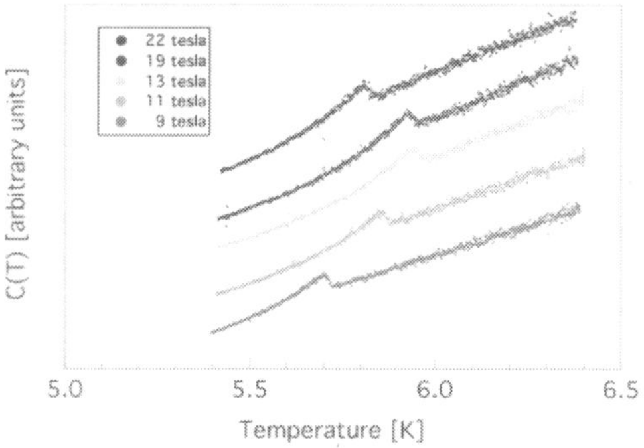

Figure 4. Heat capacity as a function of temperature for series of fixed magnetic fields near the peak of phase boundary at 5.95 K, 17.5 T. Curves offset for clarity.

supporting post and sensor leads provide a thermal link to the external cryogenic bath. The dependence of platform temperature on platform heater power is well described by the magnetic-field-independent power law function $T = T_0 + A(P_0 + P)^n$ where $n \lesssim 0.5$. A and n depend only on the properties of the calorimeter; P_0 and T_0 also depend on the properties of the fridge.

After repeated thermal cycling, the thermometers were cross-calibrated in zero field against a commercially calibrated RuOx resistive thermometer. This data was used to fit the temperature dependence of each resistive sensor to the Chebyshev polynomial $\log R(T, B) = \sum_{n=0}^{N} c_n(B)t_n(x)$ where x is a function of $\log T$ [12]. A series of platform heater power sweeps in fixed magnetic fields (starting from the same refrigerator base temperature) then allowed us to fit the magnetic field dependence of $c_n(B)$ as described in detail elsewhere [12].

In Figure 5 we present the measured ac-calorimetric response — which is inversely proportional to the heat capacity $C(T/T_N)$ — for a series of field-corrected constant temperature field sweeps between 0 and 33 tesla. The entrances into and then out of the antiferromagnetic state are visible as dips in the ac-calorimetric response (peaks in heat capacity). The crossings of the phase boundary are plotted as a function of magnetic field and temperature in Figure 6. The ordering temperature $T_N(\mu_0 H)$ increases by 40% in field, reaching a maximum of 5.95 K at 17.5 tesla. Similar re-entrant behavior has been reported for $Cu(HF_2)(pz)_2BF_4$ [13].

The experimentally observed field dependence of the phase boundary in $Cu(Pz)_2(ClO_4)_2$ is characteristic of the behavior expected for a magnetic-field-dependent Berezinksii-Kosterlitz-Thouless (BKT) transition in a 2D XY system with anisotropy Δ [14, 15]. An idea 2D Quantum Heisenberg antiferromagnet (QHAF) with small XY anisotropy can also undergo a BKT transition, if it first undergoes a crossover from Heisenberg to XY behavior [16, 17]. The application of a magnetic field can also induce a Heisenberg to XY crossover and a finite temperature BKT transition T_{BKT} [4, 15]. For small fields, the reduction in quantum fluctuations of the z components of the spins dominates, leading to an increase of T_{BKT} with increasing field. At higher fields, the reduction in the XY spin moment due to spin-canting grows in importance, leading to a broad peak and then decline in T_{BKT} with increasing field [15].

An Initial QMC calculation [13] applying the field-dependent BKT model to the anisotropic quasi-2D QHAF system $Cu(HF_2)(pz)_2BF_4$ is promising, but this simplified calculation failed to take into account the XY anisotropy Δ. A complete QMC calculation that includes both Δ and J' is needed determine self-consistent values for these parameters and confirm the applicability of the general model to these materials.

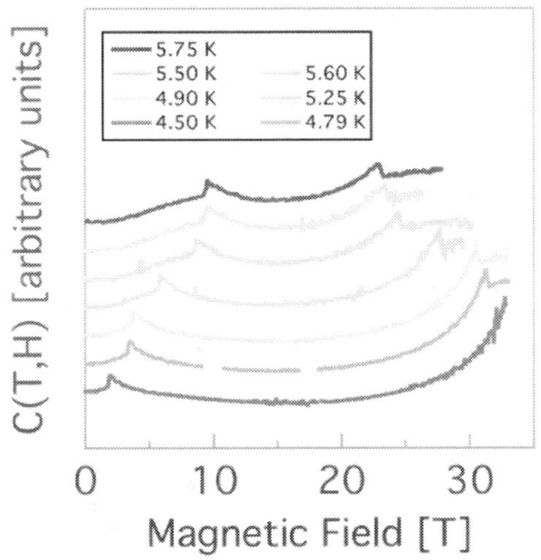

Figure 5. Heat capacity of $Cu(Pz)_2(ClO_4)_2$ as a function of magnetic field for series of fixed temperatures. Curves offset for clarity.

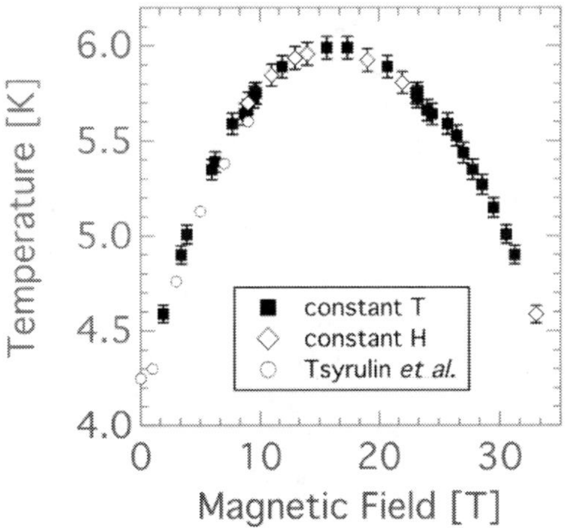

Figure 6. Field dependence of T_N from C(T,H): this work and Tsyrulin *et al.* [4].

Acknowledgments

A portion of this work was performed at the National High Magnetic Field Laboratory, which is supported by National Science Foundation Cooperative Agreement No. DMR-1157490, the State of Florida, and the U.S. Department of Energy.

References

[1] Woodward F M, Gibson P J, Jameson G B, Landee C P, Turnbull M M and Willett R D 2007 *Inorganic Chemistry* **46** 4256–4266
[2] Xiao F, Woodward F, Landee C, Turnbull M, Mielke C, Harrison N, Lancaster T, Blundell S, Baker P, Babkevich P and Pratt F 2009 *Physical Review B* **79** 134412
[3] Povarov K Y, Smirnov A I and Landee C P 2013 *Physical Review B* **87** 214402
[4] Tsyrulin N, Xiao F, Schneidewind A, Link P, Ronnow H M, Gavilano J, Landee C P, Turnbull M M and Kenzelmann M 2010 *Physical Review B* **81** 134409
[5] Tsyrulin N, Pardini T, Singh R, Xiao F, Link P, Schneidewind A, Hiess A, Landee C, Turnbull M and Kenzelmann M 2009 *Physical Review Letters* **102** 197201
[6] Yasuda C, Todo S, Hukushima K, Alet F, Keller M, Troyer M and Takayama H 2005 *Physical Review Letters* **94** 217201
[7] Bauer B *et al.* 2011 *Journal of Statistical Mechanics: Theory and Experiment* **2011** P05001
[8] Hannahs S T and Fortune N A 2003 *Physica B-Condensed Matter* **329-333** 1586–1587
[9] Palm E C and Murphy T P 1999 *Review Of Scientific Instruments* **70** 237–239
[10] Sullivan P and Seidel G 1968 *Physical Review* **173** 679–685
[11] Bachmann R *et al.* 1972 *Review Of Scientific Instruments* **43** 205–214
[12] Fortune N A, Gossett G, Peabody L, Lehe K, Uji S and Aoki H 2000 *Review Of Scientific Instruments* **71** 3825–3830
[13] Sengupta P *et al.* 2009 *Physical Review B* **79** 060409
[14] Cuccoli A, Roscilde T, Vaia R and Verrucchi P 2003 *Physical Review Letters* **90** 167205
[15] Cuccoli A, Roscilde T, Vaia R and Verrucchi P 2003 *Physical Review B* **68** 060402
[16] Ding H 1992 *Physical Review Letters* **68** 1927–1930
[17] Cuccoli A, Roscilde T, Tognetti V, Vaia R and Verrucchi P 2003 *Physical Review B* **67** 104414

High-frequency ESR measurements and ESR/NMR double resonance experiments of lightly phosphorous-doped silicon

Y Fujii[1], S Mitsudo[1], K Morimoto[2], T Mizusaki[1], M Gwak[3], SG Lee[3], A Fukuda[4], A Matsubara[5], T Ueno[6] and S Lee[7]

[1]Research Center for Development of Far-Infrared Region, University of Fukui, Fukui 910-8507, Japan
[2]Department of Applied Physics, University of Fukui, Fukui 910-8507, Japan
[3]Division of Materials Science, Korea Basic Science Institute (KBSI), 169-148 Gwahak-ro, Yuseong-gu, Daejeon 305-806, Korea
[4]Department of Physics, Hyogo College of Medicine, Nishinomiya 663-8501, Japan
[5]Research Center for Low Temperature and Materials Sciences, Kyoto University, Kyoto 606-8501, Japan
[6]Graduate School of Medicine, Kyoto University, Kyoto 606-8501, Japan
[7]Department of Physics, Korea Advanced Institute of Science and Technology, 291 Daehak-ro, Yuseong-gu, Daejeon 305-701, Korea

E-mail: yfujii@u-fukui.ac.jp

Abstract. We studied lightly doped Si:P with high-frequency (80–120 GHz) ESR and ESR/NMR double magnetic resonance techniques in the temperature range down to 1.4 K. We found dynamic nuclear polarization of ^{31}P from steady-state ESR measurements with approximately 3.6 T. We derived the nuclear spin relaxation time, T_{1N}, of ^{31}P by analysing the time-evolution of ESR spectra utilizing the dynamic nuclear polarization effect. We derive temperature and magnetic field dependence of T_{1N} and compare with experimental data. Furthermore, from our ESR measurements, we modulate the nuclear polarization of ^{31}P by applying an RF field.

1. Introduction

A quantum computer design using phosphorous-doped silicon (Si:P) proposed by B. Kane [1] has been attracting much interest, since it is one of the best practical quantum computer designs reported thus far. The model uses phosphorus (P) nuclear spins embedded regularly in silicon crystal as qubits, where for quantum computer operations, electron spins should be completely polarized by high fields (a few Tesla) and low temperatures in the range of 100 mK. Experimental demonstrations, however, have not yet been successful. In this study, we study an ensemble of ^{31}P nuclear spins in phosphorous-doped silicon (Si:P), rather than detecting the state of a single ^{31}P nuclear spin, as required in Kane's model. In Si:P, the donor electrons are well localized around the donor ions if the donor concentration is below the critical doping concentration, $n_c = 3.7 \times 10^{18}\,\mathrm{cm}^{-3}$, while they are delocalized above n_c. The P atoms in lightly doped Si:P are in a similar environment to those in Kane's model, and therefore, it is important to study ^{31}P nuclear spin state and spin dynamics by NMR especially to measure the coherence time of nuclear spins. Until now, however, it has been difficult to observe the ^{31}P NMR signal directly because the critical concentration of P atoms to keep donors in their proper separation for quantum computation is too low.

Recently, we found that ^{31}P nuclear spin polarization is enhanced by two or three orders of magnitude over the thermal equilibrium value using dynamic nuclear polarization (DNP) caused by microwave irradiation (~80 GHz) [2]. Since the NMR signal would be enhanced following enhancement of the nuclear spin polarization, it may be possible to observe the ^{31}P NMR signal from low concentration Si:P sample with the assistance of DNP. With this aim, we constructed an ESR/NMR double-resonance system.

We report here on ^{31}P nuclear spin dynamics measured by cw-ESR with the assistance of DNP. Further, we show that the magnitude of the nuclear spin polarization was changed when applying NMR pulses using an ESR/NMR double resonance system. We used Si:P samples in the insulating regime with P-concentration $n = 6.5 \times 10^{16}$ cm^{-3} where doped P atoms in Si crystal are isolated impurities.

2. Experimental – ESR/NMR double resonance system

The isolated donor electron spin system was studied using cw-ESR at the frequency of $\omega/2\pi = 80–120$ GHz and at temperatures from 20 K to 1.4 K with field modulation. The experimental setup (see figure 1) was basically the same as that in our previous report [2], except for the position of the sample. To add an NMR system to the ESR system, we made a new probe for the NMR/ESR double resonance experiment. Inside the waveguide we placed a coil for NMR, with its axis perpendicular to the magnetic field. The resonance frequency for NMR was tuneable with a capacitor on the probe and was adjustable from outside the probe. It should be noted that the microwaves are perpendicular to the magnetic field at the sample position in this system, whereas they were parallel to the magnetic field in the previous system. This different microwave propagation direction may cause a different excitation efficiency of electron spin.

A Si:P crystal sample with the size $3 \times 3 \times 0.3$ mm^3 was set inside the cylindrical waveguide. To observe the absorption signal, a simple transmission method was adopted without a cavity resonator. We optimised the microwave frequency and phase and the field modulation as described in references 2 and 3.

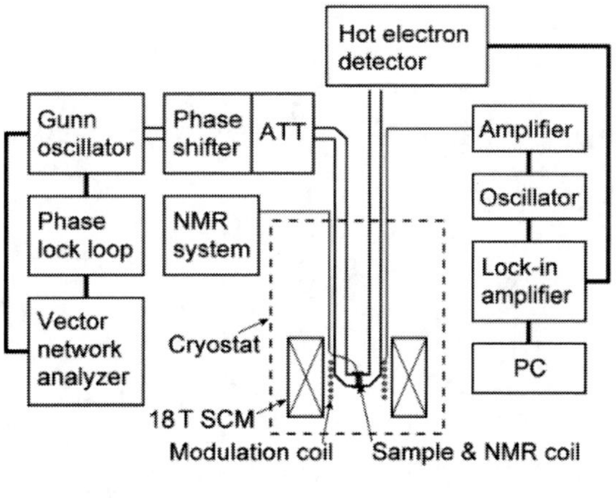

Figure 1: Block diagram of the high-frequency ESR/NMR double resonance apparatus. The microwave source is a Gunn oscillator that is phase-locked by a Millimetre-wave Vector Network Analyser (MVNA). The microwave signal is measured by an InSb hot electron detector operating at 4.2 K. The coil for the field modulation is wound on a plastic holder and fixed around the sample. The NMR coil is fixed inside the waveguide.

3. Results and discussions

3.1. DNP effect and T_{1N} measurement by cw-ESR

Electron spin **S** and nuclear spin **I** are coupled by an isotropic hyperfine constant A (A/h = 117.5 MHz) in an isolated donor in Si:P. The Hamiltonian for this two-spin system under the magnetic field B_0 along the z-axis is given by

$$\mathcal{H} = -\gamma_e \hbar S^z B_0 + A\mathbf{S}\cdot\mathbf{I} - \gamma_N \hbar I^z B_0 \,, \tag{1}$$

where γ_e and γ_N are gyromagnetic ratios for electron and nuclear spins, respectively. ESR spectra at low temperatures consist of two resonance lines separated by approximately 4 mT. We refer to an ESR absorption signal appearing at the lower (higher) field as the L-line (H-line). From equation (1), the intensity of the L-line (H-line) is almost proportional to the population of up (down) state of the ^{31}P nuclear spin. Therefore, if ^{31}P nuclear polarization is enhanced over thermal equilibrium by the DNP effect, we expect to observe an asymmetric ESR spectrum where the L-line is more intense than the H-line.

We measured the spectra at 1.44 K that is lower than our previous work [2]. After 67 minutes irradiation on the H-line, the ESR spectrum exhibited the expected asymmetry (see figure 2(a)). Note that these ESR spectra were recorded with extremely small microwave power (approximately 14dB attenuation). Approximately 20 % nuclear polarization was obtained by DNP, where the intensity of each line was evaluated from the integrated intensity of the line. We repeated the measurements by sweeping the magnetic field, and obtained the time t evolution of the ESR spectra, showing the nuclear polarization P relaxing to its thermal equilibrium (see figure 2(b)). The nuclear relaxation time, T_{1N}, was found from the best fit of

$$P = P_0 + a\exp(-t/T_{1N}) \,, \tag{2}$$

and was obtained as 24 ± 7 min at 3.65 T and 1.44 K, where P_0 and a are also fitting parameters.

The values of T_{1N} were derived similarly for above 3 K and under 8.58 T [4], and below 2.2 K and under 4.6 T [5]. In ref. 5, the relaxation times were attributed to two components: the faster one originated in the dynamic polarization of ^{29}Si nuclear spins (I=1/2) located nearby the P atom, and the slower one was observed $t \gtrsim 500$ s. In this work, we observed the longer relaxation time. It is worth comparing T_{1N} data taken across a wide range of the magnetic field. Figure 3 shows $1/(T\cdot T_{1N})$ vs. $1/T$, including data from these references. To analyse these data, we start with a general formula for T_{1N}

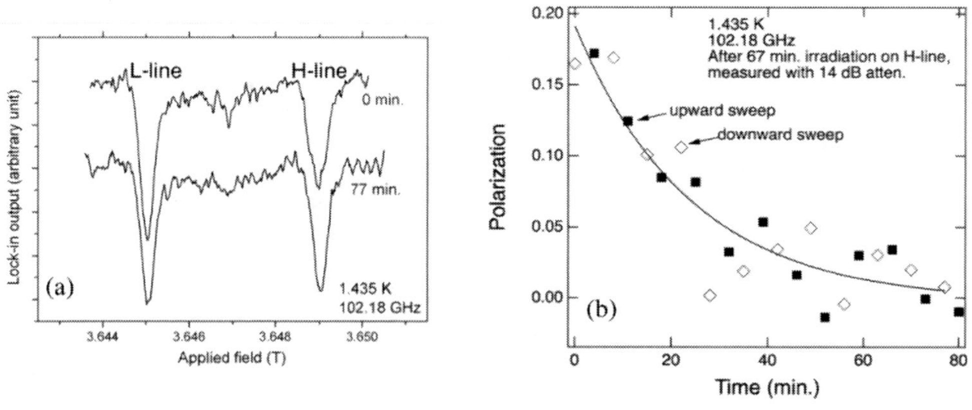

Figure 2: (a) ESR spectra recorded at less than 1 min, and 77 min after switching off the irradiation on the H-line for 67 minutes at 1.44 K. Only signals in-phase with the modulation field are shown. (b) Time dependence of ^{31}P nuclear polarization after switching off the irradiation. The solid line shows the best fit to the data of equation (2).

which is expressed in terms of fluctuating local field caused at the nucleus:

$$\frac{1}{T_{1N}} = \frac{\gamma_N^2}{2}\int_{-\infty}^{\infty}\left\langle\left\{H_{loc}^-(0), H_{loc}^+(t)\right\}\right\rangle \exp(i\omega_N t)dt\ , \tag{3}$$

where ω_N is the nuclear Larmor angular frequency, $H_{loc}^{\pm}$ is the local field caused at ^{31}P nucleus perpendicular to B_0, and $\{a,b\}= (ab+ba)/2$. Assuming an exponential correlation function with a correlation time τ, $\left\langle H_{loc}^-(0), H_{loc}^+(t)\right\rangle \sim \exp(-t/\tau)$, we obtain,

$$\frac{1}{T_{1N}} = \frac{\tau M_2}{1+\omega_N^2\tau^2}\ , \tag{4}$$

where M_2 is a relevant second moment for the hyperfine coupling between ^{31}P and donor electrons. Because the $H_{loc}^{\pm}$ originates from S^xI^z and S^yI^z terms in the Hamiltonian (1), T_{1N} is associated with the cross relaxation which involves the electron-nuclear flip-flop transition. Let us put τ as the cross relaxation time $T_x \sim 1/(B_0^2T)$ [6]. For the relevant situation, $\omega_N = \gamma_N(B_0 + B_{HF})$ where $B_{HF} = A/\gamma_N$ is the hyperfine field, and $\omega_N\tau >> 1$. Taking into account the effect of saturation of electron magnetization [7], we finally derive the following equation:

$$\frac{1}{TT_{1N}} \approx CB_0^2\left(B_0 + B_{HF}\right)^{-2}\left(1-P_e^2\right), \tag{5}$$

where $P_e = \tanh\left(\gamma_e\hbar B_0/2k_BT\right)$ is the polarization of the electron spin; and C is a constant, independent of the magnetic field and temperature. Here, the activation-type temperature dependence of T_{1N} comes from the saturation effect of the magnetization. In figure 3, the dashed lines are the best fits of equation (5) to the experimental data, where the only fitting parameter is the prefactor C. For our data at 3.6 T, we used the same value of C with that for 4.6 T. The fitted results agree reasonably well to the experimental data. The obtained values of C for different conditions of temperature ranges and magnetic fields are nearly equal with each other, suggesting that the above discussion correctly accounts for ^{31}P nuclear relaxation process in lightly doped Si:P. It should be noted here that if the hyperfine coupling has anisotropy, the correlation function $<S^zS^z>$ may contribute to T_{1N} [8]. For this case, τ is put as T_{1e} and then T_{1N} is proportional to T_{1e}. However, because of B_0^{-4} dependence for the direct process of T_{1e} relaxation [2,9], the field dependence of the experimental data of T_{1N} can not be

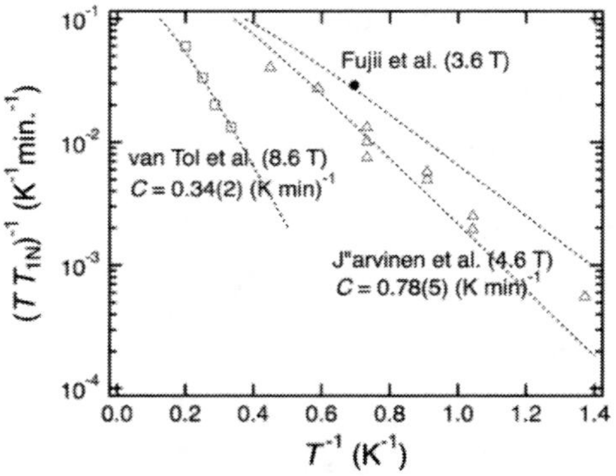

Figure 3: $1/(T{\cdot}T_{1N})$ as a function of $1/T$ for 3.6 T (solid circle, this work), 4.6 T (open triangle) [5], and 8.6 T (open square) [4]. The dashed line shows the best fit of equation (5) for each data set.

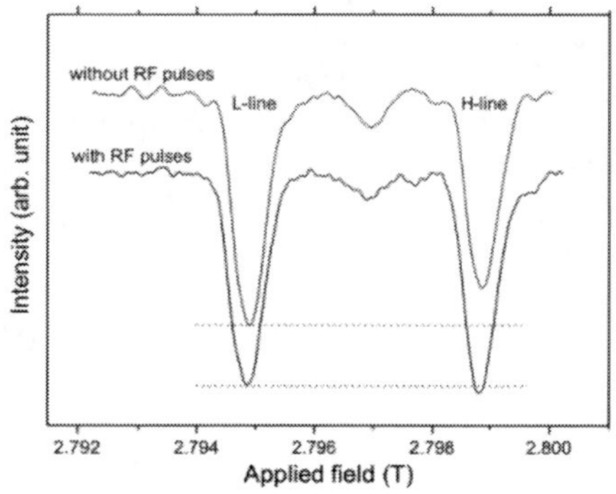

Figure 4: ESR spectra recorded after irradiating H-line for 20 minutes at 1.5 K. Only in-phase signals are shown. The lower spectrum was recorded with RF pulses of 107.1 MHz. The background signal was subtracted from the raw data for each spectrum. The dotted lines are drawn to show the intensity of L-line.

explained. We tentatively consider that the anisotropic component of the hyperfine coupling is negligible.

3.2. cw-ESR with RF pulses
To control nuclear spin polarization by NMR, we performed cw-ESR measurements with and without RF pulses using the ESR/NMR double resonance system. In preparation, microwave of about 102.1 GHz were irradiated on H-line for 20 minutes. Just after the preparation, ESR spectra were recorded *with* or *without* RF pulses at 107.1 MHz, of 3–5 μs, applied every 2 s. Without RF pulses, ESR spectrum exhibited asymmetry because of DNP (see Fig. 4). With RF pulses, the intensities of the two lines were nearly the same, suggesting that the distribution of nuclear spin states returned to nearly thermal equilibrium. This technique would be useful to obtain a thermal equilibrium state much faster than T_{1N}.

4. Summary
We studied the spin dynamics of lightly doped Si:P with the concentration $n = 6.5 \times 10^{16}$ cm^{-3}. High-frequency steady-state ESR experiments of the donor electron spin in this sample were performed at 3.6 T and 1.44 K. We observed the ^{31}P nuclear spin-lattice relaxation time T_{1N} from the DNP enhanced state to thermal equilibrium. We proposed an analysis based on that the nuclear relaxation is caused by the fluctuation of electron spins associated with nuclear-electron flip-flop transition. The equation from this analysis correctly accounted for the behaviour of T_{1N} across a wide range of temperatures and magnetic fields. We also showed that the nuclear polarization can be changed by NMR pulses. We plan to measure at further lower temperatures for large Zeeman splitting because perfect nuclear polarization is required to realize quantum computation.

Acknowledgement
This work was supported by domestic cooperative research projects of Research Center for Development of Far-Infrared Region, University of Fukui, Japan, by a Grant-in-Aid for Scientific Research (C) (26400331) from Japan Society for the Promotion of Science (JSPS), and by a JSPS bilateral joint research project under the Japan-Korea basic scientific cooperation program.

References
[1] Kane B E 1998 *Nature* **133** 393.
[2] Song M *et al.* 2010 *J. Phys.: Condens. Matter* **22** 206001.
[3] Fujii Y *et al.* 2012 *Proc. of IW-FIRT2012* (University of Fukui, Japan) 313.
[4] van Tol J *et al.* 2009 *Appl. Magn. Reson.* **36** 259.
[5] Järvinen J *et al.*, *Preprint* cond-mat/1402.4288.
[6] Pines D, Bardeen J and Slichter C P 1959 *Phys. Rev.* **106** 489.
[7] Waugh J and Slichter C P 1988 *Phys. Rev.* B **37** 4337.
[8] Abragam A and Goldman M 1982 *Nuclear Magnetism: Order and Disorder* (Claredon Press Oxford) Ch 6.
[9] Feher G and Gere E A 1959 *Phys. Rev.* **114** 1245

Magnetic properties and large magnetocaloric effect in Laves phase metallic compound

Tian Gao[*], **Ningning Qi, Yufeng Zhang, Tao Zhou**

Department of Physics, Shanghai University of Electric Power, Shanghai 201300, China

[*]Corresponding author, E-mail: gaotian@shiep.edu.com

Abstract. We investigated the magnetic properties and magnetocaloric effect of C15 Laves phase $TbMn_{1.6}Fe_{0.4}$ by magnetization and heat capacity measurements. A sharp second-order magnetic Tb-Tb ordering transition at Curie temperature $T_C \sim 120$ K and a short-range Fe-Fe ordering transition at ~230 K are observed. Around T_C, the compound shows a large magnetocaloric effect with no obvious thermal and magnetic hysteresis loss. The maximum value of magnetic entropy change reaches 8.72 J kg^{-1} K^{-1} for a magnetic field change $\Delta H = 7$ T over a wide temperature range. The obtained relative cooling power reaches 184, 560, and 803 J kg^{-1} for $\Delta H = 2$, 5 and 7 T, respectively. Large reversible magnetocaloric effect and the wide operating temperature range indicate that $TbMn_{1.6}Fe_{0.4}$ could be a promising candidate for magnetic refrigeration.

1. Introduction

The RT_2 binary compounds (where R = Y, Gd, Tb, Dy and T = Mn, Fe) crystallize in the C15 Laves phase type cubic crystal structure and exhibit distinct complex magnetic behaviors. In $TbMn_2$, the interaction between Tb and Mn ion is unstable below the Neel temperature $T_N \sim 40$ K and not all the Mn ions carry localized magnetic moments,[1] while $TbFe_2$ shows ferromagnetic (FM) ordering at Curie temperature $T_C \sim 697$ K with an effective Fe magnetic moment of 2.10 μ_B.[2] The magnetic properties in intermetallic compound $TbMn_{2-x}Fe_x$ are mainly determined by the 4f-4f interactions of Tb and 3d-3d interactions of Fe and Mn atoms. The substitution of Fe for Mn induces a smaller interatomic spacing and a rapid decay of the Mn magnetism, and therefore increase both the Curie temperature and the magnetic moment compensation of Tb,[3] which would induce a large magnetocaloric effect (MCE).

Compared with conventional gas compression refrigeration, magnetic refrigeration methods based on the MCE is considered to be advantageous for both their potential energy savings and reduced environmental impact. The MCE presents a promising environmentally friendly refrigeration method as an alternative to conventional refrigeration, which can be observed by changing the magnetic entropy of a material. Many efforts have been made to find out advanced MCE materials for improving the application of this cooling technology. For FM materials, the largest MCE is often obtained in the vicinity of a magnetic phase transition. However, the narrow temperature region and the less efficient in fast cycling refrigerators because of the considerable thermal and magnetic hysteresis accompanied by the first-order phase transition will consume the relative cooling power (RCP) of magnetic refrigerants. In contrast to a first-order phase transition, the materials showing a second-order phase transition are suitable for practical applications because of the reversible MCE with negligible thermal and magnetic loss.[4,5]

In this paper, the MCE of metallic compound $TbMn_{1.6}Fe_{0.4}$ under a modest magnetic field has been investigated. A large isothermal magnetic entropy change (8.72 J kg^{-1} K^{-1} for 0-7 T) centered at T_C ~ 120 K are observed, accompanied by a second-order phase transition with no thermal and magnetic hysteresis over a wide temperature span. The full width at half maximum δT_{fwhm} reaches 92 K at 7 T, and the corresponding RCP reaches 803 J kg^{-1}. These values, especially the huge δT_{fwhm} and RCP, which implies a larger operating temperature range, make the applicability of $TbMn_{1.6}Fe_{0.4}$ as an ideal candidate for magnetic refrigerant.

2. Experimental

The $TbMn_{1.6}Fe_{0.4}$ polycrystal was prepared by an arc-melting method using a tungsten electrode in an argon atmosphere. Firstly, stoichiometric amounts of high purity Tb, Mn, and Fe were turned over and remelted six times to ensure homogeneity in a water-cooled copper hearth. Then the obtained ingot was wrapped in a molybdenum foil, sealed in a high-vacuum quartz tube, annealed at 700 °C for one week and then quenched into cold water. The x-ray powder diffraction pattern shows a single phase of the cubic C15 Laves $MgCu_2$-type structure (space group Fd3m) in the final sample. The lattice parameter was evaluated to be 7.564 Å. The magnetization measurements were carried out by using a superconducting quantum interference device magnetometer, MPMS-7 of Quantum Design Company of America. The specific heat measurement was performed in a physical property measurement system, PPMS-9 of Quantum Design Company.

3. Results and discussion

The temperature dependence of magnetization (M) and its inverse ($1/M$) for $TbMn_{1.6}Fe_{0.4}$ in an external magnetic field H = 0.1 T are plotted in Fig. 1. The compound undergoes a clear FM transition at T_C ~ 120 K. At temperatures above ~230 K, it shows the Curie-Weiss behavior with a PM Curie temperature θ_P ~ 156 K, which is much higher than T_C. A visible anomaly of $1/M$ at T ~ 230 K is indicated by arrow, which may be induced by the short-range 3d-3d ordering of Fe atoms.[6] It is considered that the observed sharp FM transition at ~120 K and the anomaly of $1/M$ at ~230 K in $TbMn_{1.6}Fe_{0.4}$ compound is corresponding to the 4f-4f ordering of Tb and the 3d-3d ordering of Fe, respectively. The effective magnetic moment is evaluated to be 9.08 μ_B/f.u., which is slightly smaller than the theoretical PM moment of free Tb^{3+} (9.72 μ_B). Below T_C, the sample shows strong FM properties, and the magnetization increases with temperature down to the lowest temperature reached.

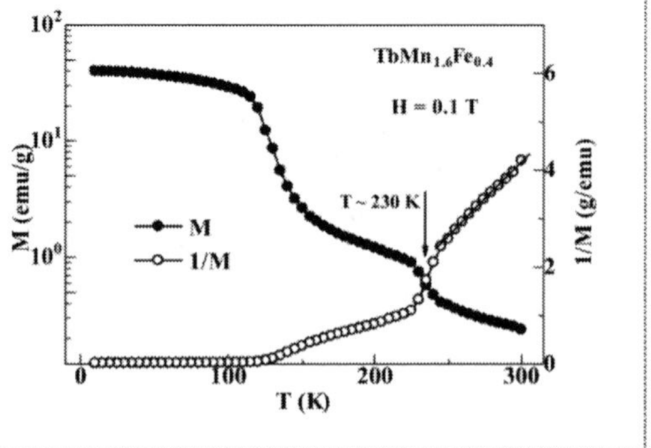

Figure 1. The temperature dependence of magnetization and its inverse in an external field H = 0.1 T. The solid line is the best fit for the Curie-Weiss law.

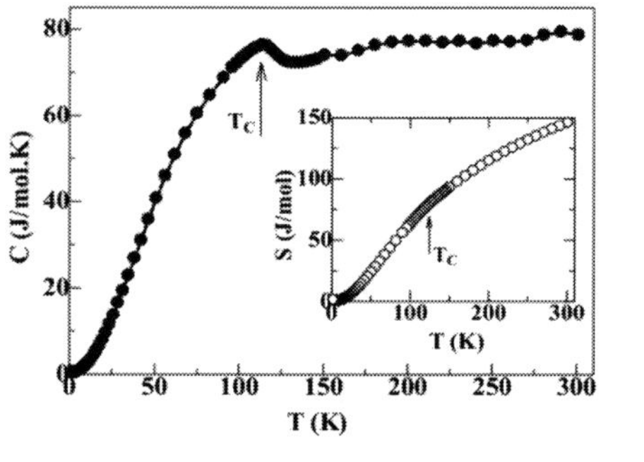

Figure 2. The specific heat $C(T)$ results for $TbMn_{1.6}Fe_{0.4}$ under zero-field. The inset shows the temperature dependence of zero-field magnetic entropy.

The temperature dependence of zero field specific heat *C(T)* for $TbMn_{1.6}Fe_{0.4}$ is displayed in Fig. 2. The clear transition at T_C ~ 120 K indicates the magnetic transition of Tb-Tb sublattice ordering. Similar to those transitions induced by Co-Co interactions in $TbCo_3B_2$[7] and $DyCo_3B_2$[8] compounds, no obvious anomaly is observed around ~230 K. In the inset of Fig. 2, the temperature dependence of zero field magnetic entropy *S(T)* for $TbMn_{1.6}Fe_{0.4}$ has been displayed. *S* increases with temperature and changes the slope at T_C.

Fig. 3 shows the behavior of the magnetization isotherms as a function of magnetic field. Below T_C ~ 120 K, it exhibits typical FM characteristics. At high temperatures far above T_C, it shows the characteristics of a paramagnet. One can find that the isothermal magnetization curves obtained well above T_C show strong curvatures at a low magnetic field. Similar results have been observed in some other intermetallic compounds,[9] which indicates the short-range Fe-Fe correlations below ~230 K. We extrapolate the high field data to zero field and get the saturation moment is ~7.56 μ_B/f.u. at 30 K, much less than the theoretical FM moment of the free ion Tb^{3+} (9.0 μ_B). It is considered that the short-range Fe-Fe ordering and the unstable 3d-3d interaction of Mn partly compensate the long-range ordered magnetic moment of Tb, in the forms of (canted) antiferromagnetism and magnetocrystalline anisotropy.[6,10]

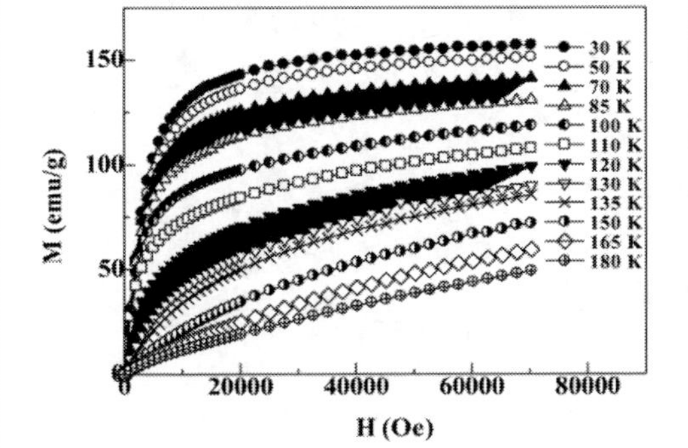

Figure 3. Initial magnetization as a function of magnetic field at selected temperatures for $TbMn_{1.6}Fe_{0.4}$.

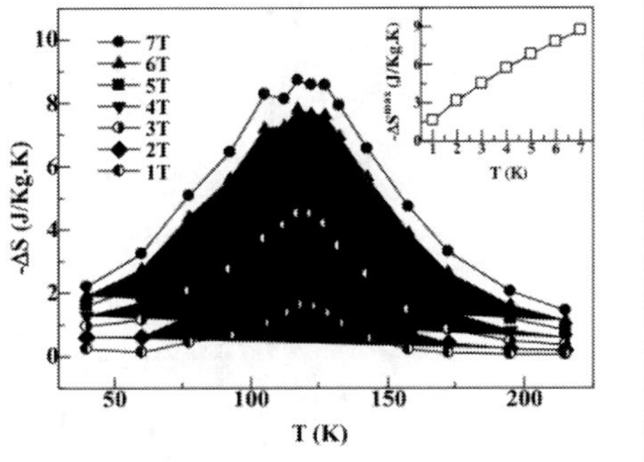

Figure 4. Isothermal magnetic entropy change ($-\Delta S_M$) of $TbMn_{1.6}Fe_{0.4}$. The inset shows the field dependence of maximum values of $-\Delta S_M$ as a function of field.

To illustrate the magnetic inhomogeneity at low temperatures, magnetic hysteresis loops of the $TbMn_{1.6}Fe_{0.4}$ compound at 30 K and 120 K are measured (not shown here). At 30 K, the compound exhibits a typical FM characteristic. The coercive field is ~0.08 T, and the remanent magnetization is ~16 emu/g. Both the coercivity and remanence may be caused by the magnetocrystalline anisotropy for the integral anisotropy of 3d atoms in transition elements.[3] At 120 K, the coercive field and the remanent magnetization cannot be detected. The M^2 versus *H/M* curves, i.e., Arrott plots (not shown here), shows positive slopes around T_C, which suggests that the magnetic Tb-Tb ordering transition at ~120 K is of second order and declares the absence of both thermal and magnetic hysteresis loss.

A large value of MCE is expected in $TbMn_{1.6}Fe_{0.4}$ from the magnetization curves, which often appears at the magnetic transition temperature and is considered to be an important requirement for application of magnetic refrigeration. The approximate temperature dependence of isothermal magnetic entropy change ΔS_M, corresponding to a magnetic field change ΔH starting from zero field, can be numerically calculated from the discrete magnetization data by means of the following expression that can be obtained from the Maxwell relationship,

$$\Delta S_M(T,H) = S(T,H) - S(T,0) = \int_0^H \left(\frac{\partial M}{\partial T}\right)_H dH$$

Fig. 4 gives the curves of -ΔS_M versus T around T_C for $TbMn_{1.6}Fe_{0.4}$ compound with ΔH up to 7 T. Here the sign of -ΔS_M is positive, which indicates the FM nature and a conventional magnetic cooling effect by adiabatic demagnetization in the present compound. No distinct temperature alteration of the peak position of -ΔS_M versus T curves is observed under variable magnetic field. Around T_C, the maximum value of magnetic entropy change which goes up with increasing field, is calculated to be 3.16, 6.81 and 8.72 J kg^{-1} K^{-1} for a field change ΔH = 2, 5 and 7 T, respectively. The large values of MCE are ascribed to the second-order Tb-Tb sublattice FM ordering. Accord well with the specific heat data in Fig. 2, no significant characteristic can be observed at ~230 K, where the short-range FM ordering is formed. One may see the wide full width at half maximum (δT_{fwhm}) of -$\Delta S_M(T)$ plots, which implies a large operating temperature range. As displayed in Fig. 5, δT_{fwhm} increases with magnetic field and reaches 58, 82 and 92 K for ΔH = 2, 5 and 7 T, respectively. The values are even larger than those in $TbMn_{1.8}Fe_{0.2}$.[6] The large temperature span may be ascribe to the long-range Tb-Tb ordering at T_C ~ 120 K and the short-range Fe-Fe ordering at T ~ 230 K.

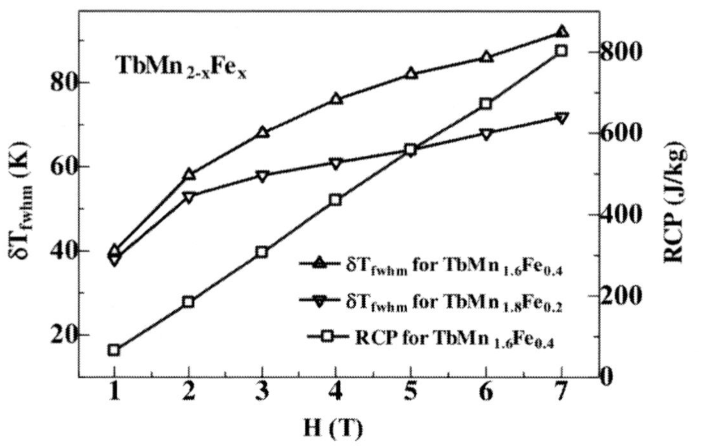

Figure 5. The full width at half maximum of the -$\Delta S_M(T)$ plots (left side) and the relative cooling power (right side) as a function of magnetic field for $TbMn_{2-x}Fe_x$ compound..

To further illustrate the MCE application of $TbMn_{1.6}Fe_{0.4}$, we have evaluated the total RCP, which is an important criterion for magnetic refrigeration materials and represents the amount of heat transferred in one thermodynamic cycle. The RCP of a reversible refrigeration cycle operating is defined as RCP = -$\Delta S_M^{max} \times \delta T_{fwhm}$, where ΔS_M^{max} is the maximum value of magnetic entropy change at the hot and cold ends of the cycle and the full width at half maximum δT_{fwhm} shows the operating temperature range in one thermodynamic cycle. For a given magnetic refrigerant, the optimum refrigeration cycle occurs when the quantity of RCP has a maximum value. In Fig. 5 for $TbMn_{1.6}Fe_{0.4}$, the RCP for ΔH = 2, 5, 7 T are evaluated to be 184, 560 and 803 J kg^{-1}, respectively. The parameters RCP, -ΔS_M^{max}, and δT_{fwhm} have similar variation tendencies with the magnetic field. Although the present value of -ΔS_M^{max} for 5 T is smaller than those of $CdCr_2S_4$ (7.04 J kg^{-1} K^{-1}),[11] TbCoAl (10.5 J kg^{-1} K^{-1}),[12] Ho_2In (11.2 J kg^{-1} K^{-1}),[13] HoGa (17.1 J kg^{-1} K^{-1}),[14] $HoCo_2$ (22 J kg^{-1} K^{-1})[15] and Gd_5Sn_4 (~36 J kg^{-1} K^{-1})[16] at adjacent magnetic transition temperatures, the present RCP value for 5 T is larger than those of $CdCr_2S_4$ (~360 J kg^{-1}), TbCoAl (~420 J kg^{-1}), Ho_2In (360 J kg^{-1}), HoGa (455 J kg^{-1}), $HoCo_2$ (216 J kg^{-1}), and Gd_5Sn_4 (~400 J kg^{-1}) due to its large value of δT_{fwhm}. In addition, the RCP of $TbMn_{1.6}Fe_{0.4}$ reaches 184 J kg^{-1} at small ΔH = 2 T with δT_{fwhm} = 58 K, both RCP and δT_{fwhm} can reach higher values in higher magnetic fields.

4. Conclusions

To summarize, the polycrystalline $TbMn_{1.6}Fe_{0.4}$ compound exhibits a large MCE over a wide effective temperature range, related to a second-order Tb-Tb ordering transition. The observed values of $-\Delta S_M$ is not so large in comparison with other potential magnetic refrigerant materials. However, the present material has a huge δT_{fwhm} around $T_C \sim 120$ K, which results in a considerable RCP (803 J kg^{-1} K^{-1} for $\Delta H = 7$ T). The above parameters increase linearly with the field, i.e., the MCE could attain higher values in higher magnetic fields. In addition, there are several additional advantages, such as no hysteresis effect, easy synthesis procedure, and high chemical and physical stabilities, which suggest that the $TbMn_{1.6}Fe_{0.4}$ compound can be a promising candidate for the magnetic refrigeration application in 70 - 162 K.

Acknowledgments

This work is supported by the National Natural Science Foundation of China (Nos. 11204171, 11374204), "Shu Guang" project of Shanghai Municipal Education Commission and Shanghai Education Development Foundation (No. 13SG52), and Science and Technology Commission of Shanghai Municipality (No. 12JC1404400).

References

[1] P J Brown, B Ouladdiaf, R Ballou, J Depoetes, A S Markosyan, J. Phys.: Condens. Matter. 4, 1103 (1992).
[2] K Chakraborty, S K Paranjpe, V Siruguri, O D Jayakumar, S K Kulshreshtha, Pramana 48, 1115 (1997).
[3] W Zhang, E M Levin, Jr K A Gschneidner, J. Magn. Magn. Mater. 250, 170 (2002).
[4] Y S Liu, J C Zhang, Y Q Wang, Y Y Zhu, Z L Yang, J Chen, and S X Cao, Appl. Phys. Lett. 94, 112507 (2009).
[5] J Chen, B G Shen, Q Y Dong, F X Hu, and J R Sun, Appl. Phys. Lett. 95, 132504 (2009).
[6] T Gao, K Nishimura, L Li, T Namiki, Y Isikawa, J. Alloy. Comp. 530, 26 (2012).
[7] M Dubman, E N Caspi, H Ettedgui, L Keller, M Melamud, and H Ahaked, Phys. Rev. B 72, 024446 (2005).
[8] L W Li, H Igawa, K Nishimura, and D X Huo, J. Appl. Phys. 109, 083901 (2011).
[9] P Arora, P Tiwari, V G Sathe, and M K Chattopadhyay, J. Magn. Magn. Mater. 321, 3278 (2009).
[10] B Ouladdiaf, J Deportes, R M Galera, A S Markosyan, J L Soubeyroux, Physica B 180–181, 85 (1992).
[11] L Q Yan, J Shen, Y X Li, F W Wang, Z W Jiang, F X Hu, J R Sun, and B G Shen, Appl. Phys. Lett. 90, 262502 (2007).
[12] X X Zhang, F W Wang, and G H Wen, J. Phys.: Condens. Matter. 13, L747 (2001).
[13] Q Zhang, J H Cho, B Li, W J Hu, and Z D Zhang, Appl. Phys. Lett. 94, 182501 (2009).
[14] J Chen, B G Shen, Q Y Dong, J R Sun, Solid State Communications 150, 157 (2010).
[15] T Tohei, and H Wada, J. Magn. Magn. Mater. 280, 101 (2004).
[16] D H Ryan, M Elouneg-Jamroz, J van Lierop, Z Altounian, and H B Wang, Phys. Rev. Lett. 90, 117202 (2003).

Spin-phonon induced magnetic order in magnetized Spin Ice systems

F A Gómez Albarracín[1], D C Cabra[1], H D Rosales[1] and G L Rossini[1]

[1] IFLP - Departamento de Física, Universidad Nacional de La Plata, C.C. 67, 1900 La Plata, Argentina

E-mail: albarrac@fisica.unlp.edu.ar

Abstract. We study the behavior of spin ice pyrochlore systems above the well known [111] 1/3 plateau, under slight deviations of the direction of the external field. We model the relevant degrees of freedom by Ising spins on the kagome lattice. We propose the inclusion of lattice deformations, which imply phononic degrees of freedom in the adiabatic limit. We use analytical calculations to estimate how these new degrees of freedom affect the short and long range spin interactions in the presence of an external magnetic field. We then obtain the magnetization curves, explore the phases and the ground states of this system in the presence of magnetic field by Monte Carlo simulations. We discuss comparisons with experimental results.

1. Introduction

Spin ice systems are one of the most studied experimental realizations of magnetic frustration. Their ground state is highly degenerate and they present an effective finite entropy. Materials such as $Ho_2Ti_2O_7$ [1, 2] and $Dy_2Ti_2O_7$ [3, 4] have the magnetic rare earth ions forming a tetrahedral corner sharing lattice, the so called pyrochlore lattice. The intense crystal fields in this materials force their magnetic moments in the direction towards the center of the tetrahedron. Therefore it is possible to model them as local Ising variables. Due to frustration, the stable configuration is two magnetic moments pointing inside the tetrahedron, and two pointing out (the ice rule "2in2out"). This is analogous to the relative position of hydrogen and oxygen in water ice [5], thus the name spin ice.

Several models have been considered for this materials: nearest neighbor interaction, inclusion of long-range dipolar interactions (which lifts the degeneracy but keeps the 2in2out configuration) and a modified dipolar model that considers exchange interactions up to third nearest neighbors [6].

The response of this materials to a magnetic field along the [111] direction is well known. The [111] direction is parallel to one of the spins of the tetrahedron, the so-called "apical" spin. Before saturation, there is an intermediate plateau at magnetization 1/3 [7–10] where the apical spins are aligned with the external magnetic field and the spins still obey the ice rules. Since the apicals are fixed and the remaining degrees of freedom are in the kagome layers, this is often referred to as kagomé ice, where in each kagome cells two magnetic moments must be *in* and one *out* (the apical spin is already *out*, fulfilling the ice rules).

Experiments [11] have studied the magnetization of $Dy_2Ti_2O_7$ upon slight deviations of the magnetic field from the [111] direction. Results show an intermediate peak in the susceptibility between the $1/3$ plateau and saturation when the field is tilted towards the [112] direction. This feature is absent when the inclination is towards the [110] direction.

We have already presented a toy model [12] with the idea of studying lattice deformations in the kagome plane in order to introduce new degrees of freedom that could reproduce the above mentioned features. In this work, we aim for a more realistic description and introduce lattice deformations in a kagome lattice where spins are in the same direction as in the pyrochlore lattice. An external magnetic field orthogonal to the kagome plane mimicks the field in the [111] direction. Deviations from this direction are obtained introducing in-plane components of the field.

2. Effective Model

We study a kagome plane with local Ising spins $\vec{S}_i = S\vec{s}_i = S\sigma_i\hat{e}_i$, where S is the spin magnitude, $\sigma_i = \pm 1$ (+1 indicating *in* and -1 *out*) and $\hat{e}_i$ are the local directions out of the plane towards the center of the tetrahedron formed by apical spins in the pyrochlore lattice. This means that $\vec{s}_i \cdot \vec{s}_j = -\frac{1}{3}\sigma_i\sigma_j$ and that for nearest neighbours $\vec{s}_i \cdot \hat{e}_{ij} = \pm\sqrt{\frac{2}{3}}$.

We consider nearest neighbor exchange interactions and long range dipolar interactions, as in the following Hamiltonian:

$$H = \sum_{\langle ij \rangle} J(r_{ij})\vec{s}_i \cdot \vec{s}_j + Da^3 \frac{1}{2} \sum_{i \neq j} \frac{\vec{s}_i \cdot \vec{s}_j}{r_{ij}^3} - 3\frac{(\vec{s}_i \cdot \hat{r}_{ij})(\vec{s}_j \cdot \hat{r}_{ij})}{r_{ij}^3} - \vec{h}\sum_i \vec{s}_i \tag{1}$$

J_0 is the nearest neighour exchange interaction, D the dipolar constant, a the distance between nearest neighbors, $\vec{h}$ the external magnetic field, r_{ij} the distance between spins at sites i and j (i.e. $r_{ij} = |\vec{r}_j - \vec{r}_i|$) and $\hat{r}_{ij}$ is the unit vector that points from site i to site j.

We follow the procedure presented in [12]: we introduce first order lattice deformations in each site as $\vec{r}_i \approx \vec{r}_i^{\,0} + \vec{u}_i$. We calculated how this changes the exchange and dipolar terms of the Hamiltonian doing a first order Taylor expansion. We treat deformations in the standard Einstein phonon model [13], taking deformations with elastic constant K, noting that bond phonon model [14] leads to trivial results in this case.

As in [12], these deformations can be integrated out as Gaussian function, provided it be sharp enough, which gives rise to effective interactions. Since we only consider deformations for nearest neighbour bonds, we get these effective interactions up to third-nearest neighbours. We include dipolar interactions up to that order and get an effective third-nearest neighbour Ising model under a magnetic field, shown below

$$H_{\text{eff}} = J_0 \left(j_1^{eff} \sum_{\langle i,j \rangle} \sigma_i\sigma_j + j_2^{eff} \sum_{\langle\langle i,j \rangle\rangle} \sigma_i\sigma_j + j_{3d}^{eff} \sum_{\langle\langle\langle i,j \rangle\rangle\rangle} \sigma_i\sigma_j + j_{3\ not\ d}^{eff} \sum_{\langle\langle\langle i,j \rangle\rangle\rangle} \sigma_i\sigma_j - \vec{h} \cdot \sum_i \vec{s}_i \right) \tag{2}$$

where

$$j_1^{eff} = \left(-\frac{1}{3} + \frac{5}{3}d - \delta\right);\ j_2^{eff} = \left(\frac{-d}{3\sqrt{3}} + \delta\right);\ j_{3d}^{eff} = \frac{d}{8};\ j_{3\ not\ d}^{eff} = \frac{-d}{8} + 2\delta \tag{3}$$

where δ is the contribution arising from the fluctuation of the phonons, $\delta = \frac{1}{4k}\left(\frac{\lambda}{3} - 5d\right)^2$, with $d = \frac{D}{J_0}$, $k = \frac{Ka^2}{J_0}$, $\lambda = \frac{\alpha}{J_0}$. Notice that as in [12] lattice deformations in nearest neighbor bonds produce only effective third neighbor couplings along the bonds and not across the hexagon diagonals. The expression for δ and the values of the couplings differ from [12] due to the fact that the spins have an out of plane component.

For our simulations, we use for $Di_2Ti_2O_7$, $d = D/J_0 = 1.44K/3.72K = 0.38$ [15] and adimensionalize choosing $J_0 = 1$. To get from (2) to Kelvin, one must simply multiply by $J_0 = 3.72K$. For example, in our simulations we chose $\delta = 0$, 0.01, 0.02..., which in Kelvin are $\delta = 0$, 0.0372K, 0.0744K.... This non zero values of delta, calculating $\lambda \approx 20$ from [16], imply $k \sim 500, 250$. This value of k fullfills the requirement of a sharp Gaussian in the integration of the phonon variables.

3. Monte Carlo Simulations and Discussion

We simulated arrays with $3L^2$ sites, where L^2 is the number of cells, $L = 8, 12, 24$. We used a simple Monte Carlo algorithm combined with the annealing technique, lowering the temperature as $T_{i+1} = 0.9T_i$, doing 40 steps from $T_0 = 0.3(1.116\text{K})$ up to $T_f = 0.00443426(0.016\text{K})$.

We chose δ from 0 to 0.1 in steps of 0.01, and introduced an external magnetic field $\vec{h}$ fixing the h_z (perpendicular to the kagome plane, [111] direction) component and calculating the h_y component in the kagome plane ([11$\bar{2}$] direction) as $h_y = \tan\theta h_z$, where θ is deviation from the [111] direction. Positive θ means that the h_y component of the field favours that the spin that is in this direction is *out*, as the [112] direction. We worked with both positive and negative θ, mimicking inclination of the magnetic field in [111] towards [112] and [110].

We calculated the magnetization per site of the tetrahedron associated to the kagome cell in the $\hat{z}$ direction, $\langle M_z \rangle$ as $\langle M_z \rangle = \frac{1+\frac{1}{3}\sum_i \sigma_i}{4}$.

Figures 1 and 2 show the magnetization $\langle M_z \rangle$ as a function of h for different values of the parameter of the deformations, δ, and three different inclinations of the magnetic field ($\theta = 0°, 5°, 10°$). In both cases, the system shows a magnetization plateau at 1/3 (where we checked that the configuration satisfies the 2in1out rule). At $\delta = 0$ (no deformations) there is no intermediate plateau. Such a plateau appears at $\langle M_z \rangle = 2/3$ for $\theta > 0$ and $\delta > 0$. This feature is not seen for negative values of θ. This behaviour holds for $\delta \leq 0.04$. For larger δ, more intermediate plateaux appear, which are not compatible with the experimental results mentioned before. We point out that at $\delta > 0.04$, $j_{3\ not\ d}^{eff}$ becomes positive.

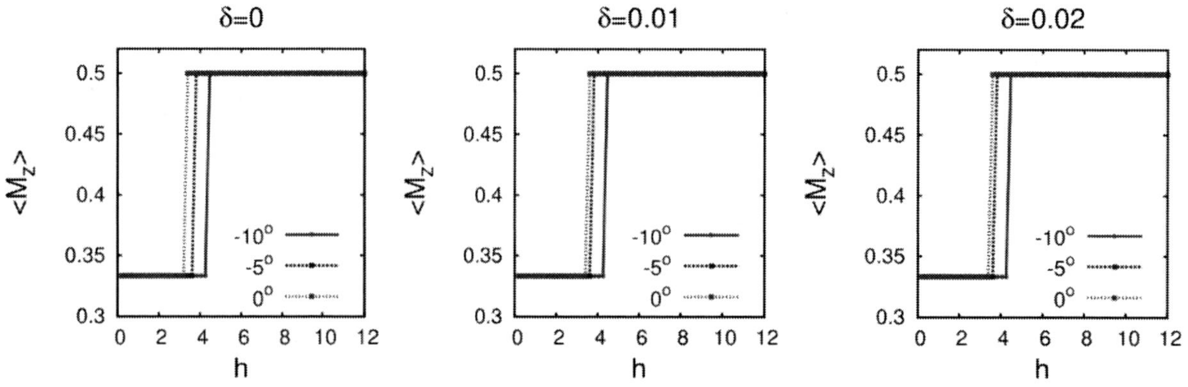

Figure 1. $\langle M_z \rangle$ as a function of h for $\delta = 0, 0.01, 0.02$ and $\theta \leq 0$ at the lowest simulated temperature ($T_f = 0.00443426 \equiv 0.016\text{K}$) for $L = 12$.

The experimental results in [11] do not include magnetization curves. However, they do include susceptibility data. For the [110] inclination, which corresponds in our work with negative θ, there is a single peak in the susceptibility presented in figure 3(a) of [11], the peak where there is a spin flip to the three-in, one-out state in pyrochlore. Our work is consistent with

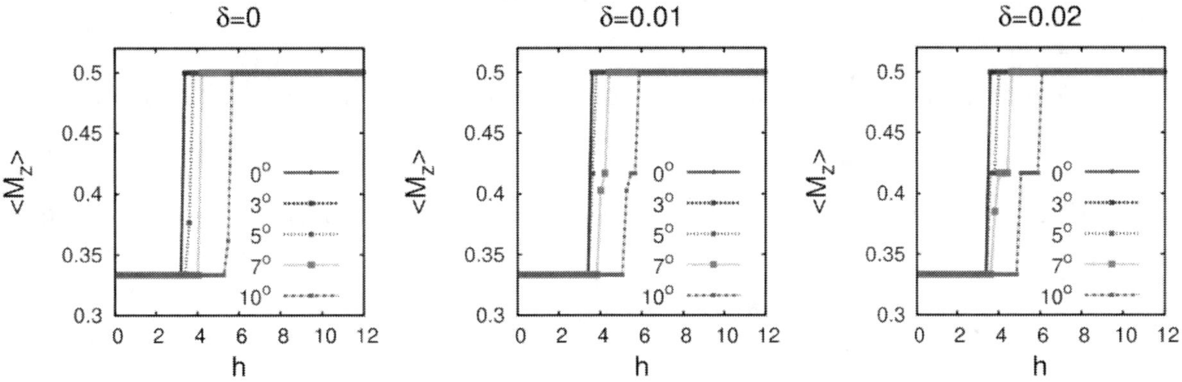

Figure 2. $\langle M_z \rangle$ as a function of h for $\delta = 0, 0.01, 0.02$ and $\theta \geq 0$ at the lowest simulated temperature ($T_f = 0.00443426 \equiv 0.016$K) for $L = 12$.

this: for negative θ there is no intermediate plateau before saturation, even when deformations are considered.

For $\theta > 0$, to compare with these experimental results, we plot the susceptibility as a function of h and three values of δ at a higher simulated temperature, $T = 0.06176734(0.23\text{K})$, in figure 3. It can be seen that without deformations ($\delta = 0$), there is only one peak, corresponding to saturation. This peak simply appears at a larger value of the magnetic field h as the inclination increases. This does not match the results from [11], where it is found that when the magnetic field is inclined in the [112] direction, the single peak is washed out, an in fact at larger inclination angles two peaks can be distinguished.

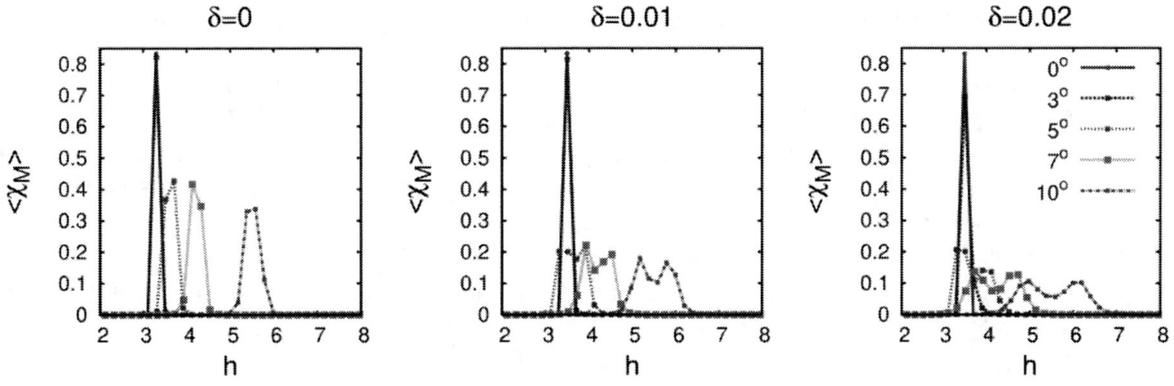

Figure 3. Susceptibility $\chi = \frac{\partial M_z}{\partial h}$ vs h for $L = 12$, $\delta = 0, 0.01, 0.02$ f, $\theta \geq 0$ at a higher temperature $T = 0.06176734 \equiv 0.23$K, to compare with figure 3 of [11]

In our work, as deformations are included, the single saturation peak becomes less defined. The middle panel of figure 3 shows that for the higher inclinations there seem to be two peaks. This is more clear in the right panel of figure 3, which correspons to a larger contribution from the deformations. Two peaks can be identified for $\theta = 7°, 10°$. Notice that the position of the first peak does not change with θ as much as that of the second peak. This is the same

Journal of Physics: Conference Series **568** (2014) 042007 doi:10.1088/1742-6596/568/4/042007

behaviour shown in figure 3(b) from [11], which only shows data up to $\theta = 7°$. It remains to be seen experimentally if these two peaks are observed at larger θ.

4. Outlook

Figure 3 suggests that the addition of elastic degrees of freedom, which change the magnetic moment couplings in a specific way, induce a second peak in the susceptibility when the magnetic field is slightly tilted from the direction perpendicular to the kagome plane. This describes quite well the experimental results in [11].

Encouraged by this fact, one should care about the role of phonons in spin ice at zero magnetization. In a possible scenario they could partially lift the large ground state degeneracy and explain the low temperature residual entropy.

Further work is being done in collaboration with Prof. Santiago Grigera and Prof. Rodolfo Borzi including simulations in pyrochlore and new experiments.

References

[1] Harris M J, Bramwell S T, McMorrow D F, Zeiske T and Godfrey K W 1997 *Phys. Rev. Lett.* **79**(13) 2554–2557 URL `http://link.aps.org/doi/10.1103/PhysRevLett.79.2554`

[2] Bramwell S T, Harris M J, den Hertog B C, Gingras M J P, Gardner J S, McMorrow D F, Wildes A R, Cornelius A L, Champion J D M, Melko R G and Fennell T 2001 *Phys. Rev. Lett.* **87**(4) 047205 URL `http://link.aps.org/doi/10.1103/PhysRevLett.87.047205`

[3] Ramirez A P, Hayashi A, Cava R J, Siddharthan R and Shastry B S 1999 *Nature* **399** 333–335 URL `http://www.nature.com/nature/journal/v399/n6734/full/399333a0.html`

[4] Fennell T, Petrenko O A, Balakrishnan G, Bramwell S T, Champion J D M, Fåk B, Harris M J and Paul D M 2002 *Applied Physics A: Materials Science & Processing* **74** 889–891 (*Preprint* `arXiv:cond-mat/0107414`)

[5] Pauling L 1935 *Journal of the American Chemical Society* **57** 2680–2684 (*Preprint* `http://pubs.acs.org/doi/pdf/10.1021/ja01315a102`) URL `http://pubs.acs.org/doi/abs/10.1021/ja01315a102`

[6] Yavors'Kii T, Fennell T, Gingras M J P and Bramwell S T 2008 *Phys. Rev. Lett.* **101** 037204 (*Preprint* `0707.3477`)

[7] Cornelius A L and Gardner J S 2001 *Phys. Rev.* B **64**(6) 060406 URL `http://link.aps.org/doi/10.1103/PhysRevB.64.060406`

[8] Fukazawa H, Melko R G, Higashinaka R, Maeno Y and Gingras M J P 2002 *Phys. Rev.* B **65**(5) 054410 URL `http://link.aps.org/doi/10.1103/PhysRevB.65.054410`

[9] Isakov S V, Raman K S, Moessner R and Sondhi S L 2004 *Phys. Rev.* B **70**(10) 104418 URL `http://link.aps.org/doi/10.1103/PhysRevB.70.104418`

[10] Moessner R and Sondhi S L 2003 *Phys. Rev.* B **68**(6) 064411 URL `http://link.aps.org/doi/10.1103/PhysRevB.68.064411`

[11] Sato H, Matsuhira K, Sakakibara T, Tayama T, Hiroi Z and Takagi S 2007 *J. Phys.: Condens. Matter* **19** 145272

[12] Gómez Albarracín F A, Cabra D C, Rosales H D and Rossini G L 2013 *Phys. Rev.* B **88**(18) 184421 URL `http://link.aps.org/doi/10.1103/PhysRevB.88.184421`

[13] Wang F and Vishwanath A 2008 *Phys. Rev. Lett.* **100**(7) 077201 URL `http://link.aps.org/doi/10.1103/PhysRevLett.100.077201`

[14] Penc K, Shannon N and Shiba H 2004 *Phys. Rev. Lett.* **93**(19) 197203 URL `http://link.aps.org/doi/10.1103/PhysRevLett.93.197203`

[15] Bramwell S T and Gingras M J P 2001 *Science* **294** 1495–1501 (*Preprint* `arXiv:cond-mat/0201427`)

[16] Zhou H, Bramwell S, Cheng J, Wiebe C, Li G, Balicas L, Bloxsom J, Silverstein H, Zhou J, Goodenough J and Gardner J 2011 *Nat. Commun.* **2** 478

NMR STUDY ON KAGOME-LATTICE ANTIFERROMAGNET $CS_2MN_3LIF_{12}$

S Moriya[1], Y Usui[2], H Tanaka[2], K Matsui[1], Y Mori[1], H Kuroe[1], T Sasaki[3] and T Goto[1]

[1] Physics division, Sophia University, Tokyo 102-8554 Japan
[2] Faculty of Science, Tokyo Institute of Technology, Tokyo 152-8550 Japan
[3] Institute for material research, Tohoku University, 980-8577 Japan

E-mail: gotoo-t@sophia.ac.jp

Abstract. ^{133}Cs/^{7}Li-NMR study was performed on a quasi-two-dimensional spin system $Cs_2Mn_3LiF_{12}$, in which $S = 2$ Mn atoms form undistorted Kagome lattice. We have observed a clear splitting in the Li-NMR resonance line for a single crystal at low temperatures below T_N = 3.6 K. The longitudinal nuclear spin relaxation rate T_1^{-1} of both Cs and Li-NMR showed a critical divergence at the same temperature. These observations are considered to be a microscopic evidence for an occurrence of long range magnetic order due to a weak inter-plane spin interactions.

1. Introduction

Kagomé lattice antiferromagnets attracted much interest for a long time among both theorists and experimentalists[1] for their strong spin frustration, caused by the geometric character of corner-sharing equilateral triangles, which does not bind the chirality of neighboring two triangles, and hence is expected to bring a novel type of ground state. So far, for quantum spin magnets with Kagomé lattice, many model compounds have been intensively studied, such as $Rb_2Cu_3SnF_{12}$ with a pin-wheel VBS state[2,3] or the vorbothite[4] with short-range magnetic ground state. On the other hand, as for

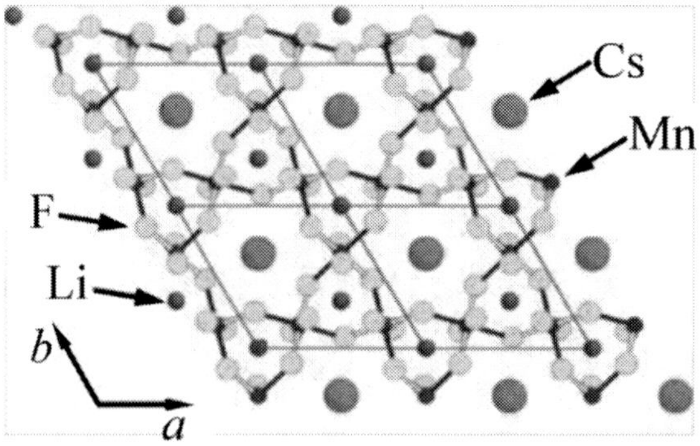

Figure 1. Schematic crystal structure of $Cs_2Mn_3LiF_{12}$. The Kagomé plane lies in ab-plane.

classical spin systems with larger quantum spin number S, although theoretical investigation predicts that the $\sqrt{3}\times\sqrt{3}$ structure is favorable[5], there has been few experimental reports except for the jarosite $KFe_3(SO_4)_2(OH)_6$, where $q = 0$ magnetic structure with positive chirality is confirmed[6].

In this article, we focus attention on $S = 2$ quasi-two-dimensional antiferromagnet $Cs_2Mn_3LiF_{12}$. This compound belongs to the space group of $R\bar{3}$ and consists of well separated non-distorted Kagomé planes. The distance between the two neighboring planes is 5.576 Å, and hence the system is considered to be suited to study the nature of Kagomé lattice[7]. We utilized the microscopic probe of NMR technique and investigated its magnetic state down to 1.6 K and found that it shows a magnetic order at $T_N = 3.6$ K and that the Néel temperature may possibly be dependent on the applied field.

2. Experimental

Single crystals with a typical size of 2×3×5 mm^3 were melt in a platinum tube from CsF, MnF_3 and LiF powder. Some of the obtained crystals were crushed into powder and tested by X-ray powder diffraction. Unfortunately, the crystal utilized for NMR and other measurements was too small to determine its axes, so that in following measurements, one has no information on the field direction. The magnetic susceptibility χ was measured by SQUID magnetometer under 4T, which is approximately the same as one of the fields for ^{7}Li-NMR measurements. With lowering temperature, χ monotonically increased down to 2 K, showing no apparent anomalies such as a kink.

NMR measurements for ^{7}Li and ^{133}Cs nuclei were performed under the applied fields of 2 and 4 T for the former and 6 T for the latter, respectively. Spectra were obtained by recording the spin-echo amplitude while sweeping the external field. The longitudinal spin relaxation rate T_1^{-1} was measured by the saturation-recovery method with a pulse train. In determining T_1, the time evolution of the nuclear magnetization was traced until its difference with the thermal equilibrium value becomes less than one percent for Li and three percent for Cs, respectively.

Before proceeding to NMR results, we briefly describe the local structure of this system, for it will be of help to understand how the NMR probes the magnetism of Mn moments. The schematic crystal structure is shown in Fig. 1. Although its overall structure is very close to the quantum spin Kagomé $Rb_2Cu_3SnF_{12}$[2], the local structure appreciably differs in that each MnF_6 octahedron is elongated along the Kagomé plane while in Rb-system along c-axis. Both Cs and Li sites locate between the two

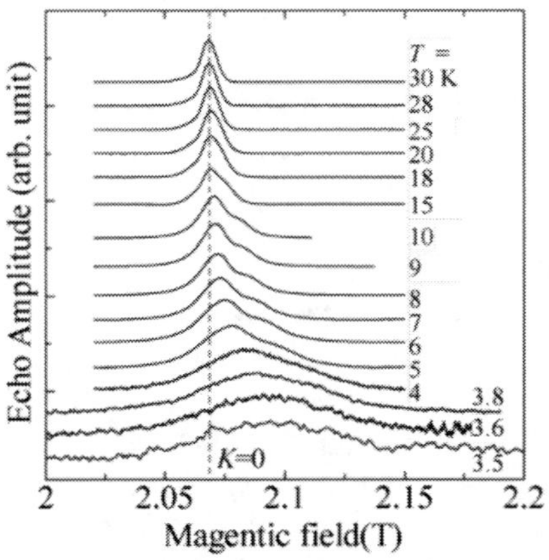

Figure 2. Typical profiles of ^{7}Li-NMR spectra at various temperatures under the field around 2 T. The vertical dashed line shows the zero shift position.

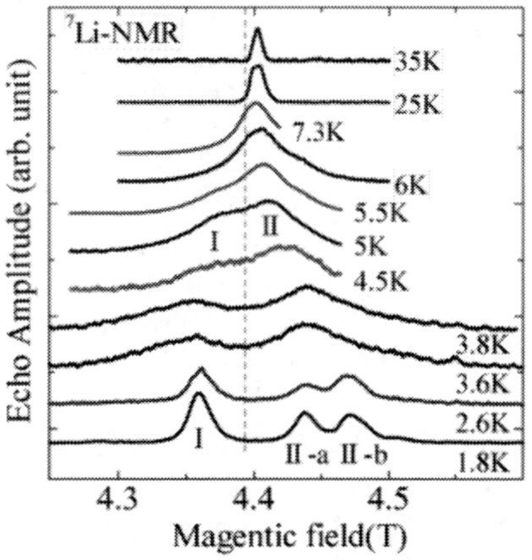

Figure 3. Typical profiles of ^{7}Li-NMR spectra at various temperatures under the field 4.4 T. The vertical dashed line shows the zero shift position. See text for indices on each peak.

neighboring Kagomé planes; the former on the vertical line from the center of hexagon formed by Mn atoms to that of triangle on the adjacent plane, while the latter on the vertical line connecting the center of two triangles on the adjacent planes. The minimum distances between Mn and NMR nuclei are 4.33Å for Cs and 3.59 Å for Li, indicating that they both well probe the magnetism of Mn moments.

3. Results

Figure 2 and 3 shows the typical profiles of ^{7}Li-NMR spectra taken at 2 T and 4 T respectively. At high temperatures around 25 K, each spectrum consists of a narrow single peak, indicating that the nuclear electrical quadruple splitting in ^{7}Li with $I = 3/2$ is negligibly small. At the low field of 2 T shown in Fig. 2, with decreasing temperature, the peak started to broaden below 10 K and continued broadening down to 3.5 K, where the width was approximately 0.05 T in FWHM. In the entire experimental temperature region, no apparent split was observed. At 4 T, in contrast, it split first at around 6 K into the two peaks denoted as I and II as shown in Fig. 3, and at still lower temperatures below 3.6 K, only the latter split again into II-a and II-b. The entire spectrum becomes broad in the vicinity of 3.6 K, and each peak becomes narrower again at the temperatures below it. The maximum splitting between II-a and b is $\delta H = 0.035$ T at 1.8 K.

In order to view the temperature dependence of each peak, their shifts are plotted against the uniform magnetization M measured under $H = 4$ T with the temperature as the implicit parameter in Fig. 4. In high temperature region above 10 K, the NMR shift shows the linear dependence on M, from which the hyperfine coupling constant was estimated to be $A = +530$ Oe/μ_B, which is moderately large, assuring that the Li-NMR is a good probe for the Mn magnetization. The inset shows the temperature dependence of the splitting distance between II-a and b. It obeyed the behavior of molecular-field approximation, that is, with the critical exponent $\beta = 0.5$ below $T_N = 3.6$ K, where the shift of the other peak I coincidentally shows a kink.

Next we show results of the longitudinal nuclear spin relaxation rate T_1^{-1} for ^{7}Li ($I = 3/2$) measured at 2 T and ^{133}Cs ($I = 7/2$) at 6 T as shown in Fig. 5. Note that in the two measurements, the same NMR resonance frequency 34.4 MHz was used. The spin relaxation of Li-nuclei was described by a

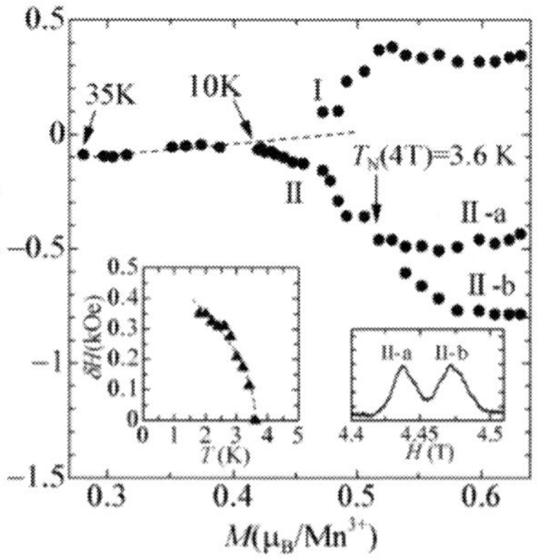

Figure 4. The plot of the shift ΔH for ^{7}Li versus uniform magnetization M under H = 4T. Two insets show the temperature dependence of δH, the splitting width of the peak II-a and II-b and its typical spectral profile. The dashed curve in the former corresponds to the critical exponent β = 0.5.

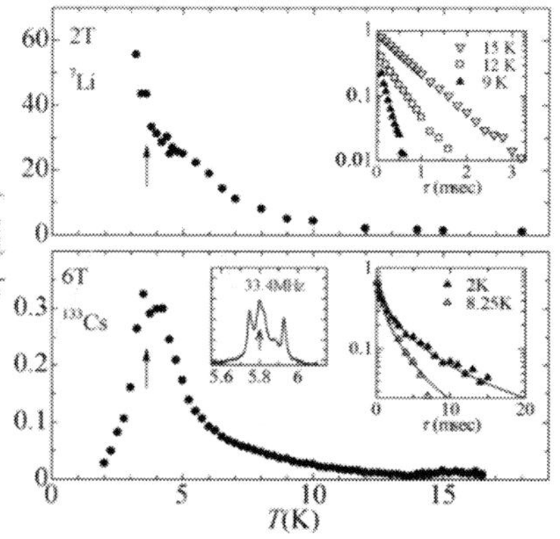

Figure 5. The temperature dependence of T_1^{-1} for the Li (upper) and Cs (lower) site. Solid arrows indicate T_N(4T) = 3.6 K obtained by spectra at 4 T as shown in Fig. 4. Insets show a typical profile of Cs spectrum with the measurement field shown by an arrow and relaxation curves (*see text*).

single exponential function as shown in the inset of the upper panel, indicating that the nuclear sub levels are nearly degenerated as was also seen in the spectrum. In ^{133}Cs measurements, the relaxation curves were well described by the stretched form of conventional equation for $I = 7/2$ nuclear spin, $8585 \cdot e^{-(28\tau/T_1)^{\beta}} + 2454 \cdot e^{-(15\tau/T_1)^{\beta}} + 819 \cdot e^{-(6\tau/T_1)^{\beta}} + 143 \cdot e^{-(\tau/T_1)^{\beta}}$, where β, the constant corresponding to the inhomogeneity in the system[8,9] stayed nearly constant at 0.7 in the entire temperature region. With decreasing temperature, both the two relaxation rates of Cs and Li increased, and the former showed a peak at around 4 K and decreased again, while the latter continued increasing until 3.2 K, indicating that the expected peak temperature of Li must be lower than Cs. There were no anomalies at around 6 K where the Li-NMR spectral peak split into I and II.

4. Discussion

First, we note the peak split observed under $H = 4$ T and at the two different temperatures 6 and 3.6 K shown in Fig.'s 3 and 4. Among them, only the anomaly at the lower temperature is associated with a critical divergence[10] in T_1 shown in Fig. 5. The coincidental appearance of the static hyperfine field and of the critical slowing down of the spin fluctuation clearly demonstrates the existence of second order phase transition and hence of the long range magnetic order. Considering the absence of the large anomaly in the uniform magnetization at this temperature, its spin structure is considered to be an antiferromagnetic like one.

Taking the value of the splitting width 0.35 kOe at the lowest temperature 1.8 K and the hyperfine coupling constant A in the paramagnetic state, and assuming the spin structure to be a simple antiferromagnetic one, one can roughly estimate the ordered moment at 1.8 K to be $\mu_{Mn} = 0.33\ \mu_B$. However, with no information of the easy axis, one must note that this value only suggests the lower limit.

Next, the peak-split of I and II at around 6 K is not derived by the magnetic order, because it is associated with no sign of anomalies in T_1. Though one cannot conclude at this stage, there are some possible origins for this split, such as a microscopic phase separation into two domains possessing appreciably different χ, or an existence of two differently oriented domains in the single crystal. In the measurement at 2 T, this split is masked by the peak broadening due to the critical slowing down toward the magnetic order at still lower temperatures. For further discussion, NMR measurements with the various direction of the field, and hence the information of the crystal axes are necessary, and they are in preparation.

Finally, we refer the difference in the peak temperature of T_1^{-1} for Cs measured at 6 T and that for Li at 2 T. For the same Larmor frequency 34.4 MHz is adopted in these two measurements, the difference in the peak temperature directly suggests that the critical temperature at 6 T is higher than that at 2 T. This argument is further supported by the peak splitting measured at 4 T is observed at 3.6 K, which is just located between the two peak temperatures of T_1^{-1} measured at 2 and 6 T. However, for the determinant argument, one requires measurements under wider field region, and these are also in progress.

Summary

In summary, we have investigated the $S = 2$ antiferromagnet with Kagomé lattice by ^{7}Li/^{133}Cs - NMR in the field range 2 and 6 T to find that it shows the long range magnetic order at $T_N = 3.6$ K.

Acknowledgements

This work was partly supported by JSPS KAKENHI Grant Number 24540350 and 26108713. This work was partly performed under the Inter-university Cooperative Research Program of the Institute for Materials Research, Tohoku University.

References

[1] I. Syôzi, Prog. 1951 Theor. Phys. **6** 306-308
[2] Ono T *et al.* 2009 Phys. Rev. **B79** 174407

[3] Matan K *et al.* 2010 Nature Physics **6** 865
[4] Bert F *et al.* 2005 Phys. Rev. Lett. **95** 087203
[5] Chubukov A 1992 Phys. Rev. Lett. **69** 832
[6] Grohol D *et al.*, 2005 Nature Materials **4** 323
[7] Englich U, Frommen C and Massa W 1997 J. Alloys and Compounds **246** 155
[8] S. Mishima et al.: J. Appl. Phys. 87 (2000) 5134
[9] T. Suzuki et al.: Phys. Rev. B79 (2009) 104409.
[10] K. Kanada et al.: J. Phys. Soc. Jpn. 76 (2007) 064706.

The two orbital Hubbard model in a square lattice: a DMFT + DMRG approach

Y. Núñez Fernández, D. García, and K. Hallberg
Instituto Balseiro and Centro Atómico Bariloche, CONICET, CNEA, Argentina
E-mail: karen@cab.cnea.gov.ar

Abstract. We develop a precise and reliable numerical method for the calculation of electronic properties of the two orbital Hubbard model in a square lattice at half filling, based on the Dynamical Mean Field Theory (DMFT). We use the Density Matrix Renormalization Group (DMRG) as the impurity solver for the DMFT 's selfconsistent equations to obtain accurate values for the Green's functions on the real axis. This way, reliable densities of states are obtained that do not need to resort to analytical continuation methods as those using quantum Monte Carlo techniques. Large system sizes can be achieved with increasing accuracy.

1. Introduction

The Dynamical Mean Field Theory (DMFT) has become one of the basic methods to calculate realistic electronic band structure in strongly correlated systems [1, 2]. A key point of the DMFT method is the solution of an associated quantum impurity model where the environment of the impurity has to be determined self-consistently. Therefore the possibility to obtain reliable DMFT solutions of lattice model Hamiltonians relies directly on the ability to solve quantum impurity models. Among the available numerical algorithms there is the Hirsch-Fye [3, 4] and Continues Time [5] Quantum Monte Carlo [3, 4] method and Wilson's Numerical Renormalization Group (NRG)[6, 7, 8]. While the former, a finite-temperature method, is very stable and accurate at the Matsubara frequencies, its main drawback is the access to real frequency quantities for the calculation of spectral functions which requires less controlled techniques for the analytic continuation of the Green functions. It also leads to sign problems for the case of two or more bands with interband hopping. The second method, the NRG, can be formulated both at T=0 and finite (small) T and provides extremely accurate results at very small frequencies, offering a less accurate description of higher energy features.

In order to overcome the difficulties encountered by these other methods, several other techniques have been proposed. We have shown that the Density Matrix Renormalization Group (DMRG)[9, 10, 11, 12] can be used very reliably to solve the related impurity problem within DMFT[13, 14]. By using the DMRG to solve the related impurity problem, no *a priori* approximations are made and the method provides equally reliable solutions for both gapless and gapfull phases. More significantly, it provides accurate estimates for the distributions of spectral intensities of high frequency features such as the Hubbard bands, that are of main relevance for analysis of x-ray photoemission and optical conductivity experiments. Other calculations using alternative methods for the calculation of dynamical properties of the impurity have been proposed [15, 16, 17] and more recenty methods using the time evolution DMRG algorithm results for the one and two orbital model where shown on [18] and a restricted active basis set based on natural orbitals [19]

2. The Model

In this paper we extend the method described above (that is, the DMRG used as the impurity-solver of the DMFT self-consistent equations), to the two-orbital Hubbard model on a square lattice. The model we are considering is [20]:

$$H = \sum_i h_i - t \sum_{<ij>m\sigma} \left(c^\dagger_{im\sigma} c_{jm\sigma} \right),$$

where i, j are the sites of a square lattice and brackets indicate nearest neighbors, m indicates each of the two orbitals and σ is the spin of the electron, whose creation and destruction operators are $c^\dagger$ and c, respectively.

Defining n_i and s_i as the on-site charge and spin operators respectively, the rotationally invariant on-site Hamiltonian is:

$$h_i = \frac{U}{2} n_i^2 + \frac{J}{2} s_i^2 - \mu n_i.$$

For two orbitals per site, we have:

$$n_i^2 = \sum_{mm'\sigma\sigma'} n_{im\sigma} n_{im'\sigma'} = n_i + 2\sum_m n_{im\uparrow} n_{im\downarrow} + 2\sum_{\sigma\sigma'} n_{i1\sigma} n_{i2\sigma'},$$

$$s_i^2 = S_{i1}^2 + S_{i2}^2 + 2\mathbf{S}_{i1} \cdot \mathbf{S}_{i2} = \frac{3}{4} n_i - \frac{3}{2} \sum_m n_{im\uparrow} n_{im\downarrow} + 2\mathbf{S}_{i1} \cdot \mathbf{S}_{i2}.$$

So the local Hamiltonian is:

$$h_i = \left(U - \frac{3}{4} J \right) \sum_m n_{im\uparrow} n_{im\downarrow} + U \sum_{\sigma\sigma'} n_{i1\sigma} n_{i2\sigma'} + J\mathbf{S}_{i1} \cdot \mathbf{S}_{i2} + \left(\frac{U}{2} + \frac{3J}{8} - \mu \right) n_i.$$

Considering a ferromagnetic J (J< 0) and a triplet ground state which leads to $\mu = 2U$, the total Hamiltonian reads:

$$H = \left(U - \frac{3}{4} J \right) \sum_{im} n_{im\uparrow} n_{im\downarrow} + U \sum_{i\sigma\sigma'} n_{1\sigma} n_{2\sigma'} + J \sum_i \mathbf{S}_{i1} \cdot \mathbf{S}_{i2} - t \sum_{<ij>m\sigma} \left(c^\dagger_{im\sigma} c_{jm\sigma} \right) + \left(\frac{-3}{2} U + \frac{3}{8} J \right) \sum_i n_i. \tag{1}$$

Applying DMFT to this model leads to a mapping of the original lattice model onto an associated quantum impurity problem in a self-consistent bath. In the particular case of the two orbital Hubbard model, the associated impurity problem is the single impurity Anderson model (SIAM) with two levels, where the hybridization function $\Delta(\omega)$, which in the usual SIAM is a flat density of states of the conduction electrons, is now to be determined self-consistently.

Starting from a guessed hybridization $\Delta(\omega + i\eta)$ for the impurity, its Green function $G(\omega + i\eta)$ is obtained using the correction vector method of DMRG [21, 22]. From this we can compute the self energy $\Sigma(\omega + i\eta) = G^{-1} - g_0^{-1}$ where g_0 is the non interacting Green function corresponding to $\Delta(\omega + i\eta)$. The self energy allows us to compute the Green function on a lattice, in this case on a square lattice (SL):

$$G_{SL}(\omega + i\eta) = \sum_{k_x, k_y} \frac{1}{\omega_{sl}(k_x, k_y) + \omega + i\eta - \Sigma(\omega + i\eta)}. \tag{2}$$

where $\omega_{sl}(k_x, k_y) = t(\cos(k_x) + \cos(k_y))$ with $t = 1/2$ to have a band of half-width $D = 1$. All energies are given in units of D. The lattice Green function G_{SL} defines a new non interacting Green function $g_0^{-1} = G_{SL}^{-1} + \Sigma$ which in turn defines a new hybridization $t^2\Delta(w) = \omega + i\eta - g_0^{-1}(\omega + i\eta)$ which is the seed to restart the cicle. The procedure is repeated until converged lattice or impurity Green functions are obtained (typically between 5 to 10 iterations). [23]

To implement the algorithm we consider [24, 25] a general representation of the hybridization function in terms of continued fractions that define a parametrization of $\Delta(\omega + i\eta)$ in terms of a set of real and positive coefficients. Since it is essentially a Green's function, $\Delta(z)$ can be decomposed into "particle" and "hole" contributions as $\Delta(z) = \Delta^>(z) + \Delta^<(z)$ with $\Delta^>(z) = \langle gs|c\frac{1}{z-(H'-E_0)}c^\dagger|gs\rangle$ and $\Delta^<(z) = \langle gs|c^\dagger\frac{1}{z+(H'-E_0)}c|gs\rangle$ for a given non interacting Hamiltonian, H' with ground-state energy E_0. By standard Lanczos technique, H' can be in principle tri-diagonalized and the functions $\Delta^>(z)$ and $\Delta^<(z)$ can be expressed in terms of respective continued fractions[26]. As first implemented in Ref.[24, 25], each continued fraction can be represented by a chain of auxiliary atomic sites whose energies and hopping amplitudes are given by the continued fraction diagonal and off-diagonal coefficients respectively.

As a result of the self-consistency condition, the two chains representing the hybridization, are "attached" to the right and left of each of both levels of the impurity to obtain a new SIAM Hamiltonian, H. In fact $G(\omega + i\eta)$ constitutes the local Green's function of the site plus the chain system.

The SIAM Hamiltonian therefore reads

$$\begin{aligned} H &= \sum_{m\sigma\alpha=-N_C;\alpha\neq 0}^{N_C} a_\alpha c^\dagger_{\alpha m\sigma}c_{\alpha m\sigma} + \sum_{m\sigma\alpha=-N_C;\alpha\neq 0,-1}^{N_C-1} b_\alpha(c^\dagger_{\alpha m\sigma}c_{\alpha+1 m\sigma} + h.c.) \\ &+ \sum_{m\sigma\alpha=\pm 1} b_0(c^\dagger_{0m\sigma}c_{\alpha m\sigma} + h.c.) + \left(U - \frac{3}{4}J\right)\sum_m n_{m\uparrow}n_{m\downarrow} + U\sum_{\sigma\sigma'} n_{1\sigma}n_{2\sigma'} + J\mathbf{S_1}\cdot\mathbf{S_2} \quad (3) \end{aligned}$$

with a two-level impurity, each of which is coupled to $2N_C$ sites (N_C to the right and similarly to the left). The set of parameters $\{a_\alpha, b_\alpha\}$ are directly obtained from the coefficients of the continued fraction representations of $\Delta(z)$ by the procedure just described.

3. Results

In Figs.1, 2 and 3 we show the DMFT+DMRG results for the density of states (DOS) and the real and imaginary parts of the self-energy, respectively, for two values of the interaction U and a finite ferromagnetic J. We have kept around 700 states in the DMRG decimation procedure and our results were benchmarked with exact diagonalization calculations for smaller systems.

These figures show the insulating (large U) and the metallic (small U) regimes and the lower and upper Hubbard bands in both cases. We are currently analyzing the whole parameter space to obtain the coexistence regime and its dependence with the Hund coupling J.

As a conclusion, we have developed an accurate real-frequency axis method for calculating electronic structure properties of the two-level Hubbard model. It relies on using the Density Matrix Renormalization Group as the impurity-solver of the Dynamical Mean Field Theory self-consistent equations. With this method, large systems can be analyzed, thus reducing finite-size effects. It also deals with all energy scales on equal footing which allows to find interesting substructure in the Hubbard bands of the correlated metallic state with the corresponding effect on other experimentally accessible quantities such as the optical conductivity and high resolution photoemission spectroscopies. In addition, realistic band-structure calculations of systems with a larger number of degrees of freedom can be handled.

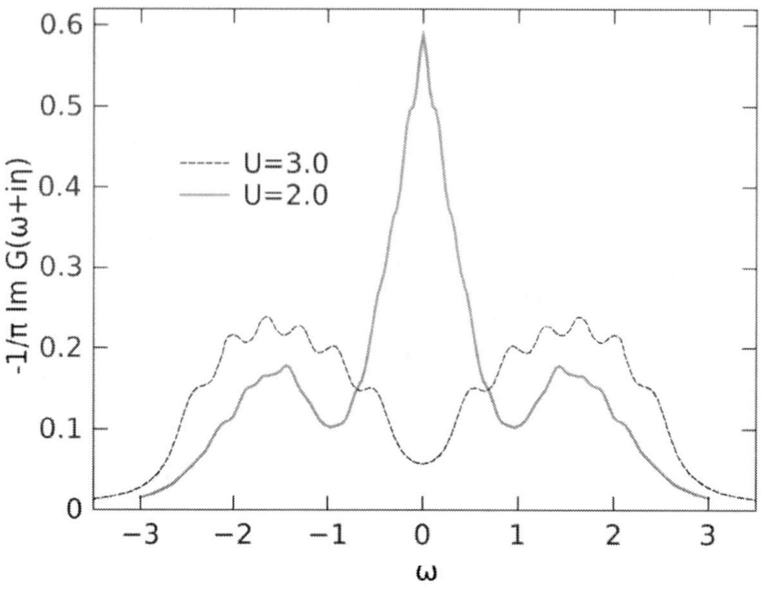

Figure 1. Converged density of states for the half-filled two-level Hubbard Model (1) for metallic (full line) and insulating (dotted line). Here $L = 42(26)$ sites, $J = -0.1$ and $\eta = 0.1(0.2)$ for the metallic (insulating) cases.

Acknowledgments
We thank Gabriel Kotliar for interesting discussions and useful feeback. We acknowledge support from CONICET, PICT2012-1069 of ANPCyT and partial support by the National Science Foundation under Grant No. PHYS-1066293 and the hospitality of the Aspen Center for Physics.

References

[1] G. Kotliar and D. Vollhardt, Physics Today **57**, 53 (2004)
[2] A. Georges, G. Kotliar, Werner Krauth and Marcelo J. Rozenberg, Rev. Mod. Phys. **68**, 13 (1996)
[3] J.E. Hirsch and R.M. Fye, Phys. Rev. Lett. **56**, 2521 (1986)
[4] M.J. Rozenberg, Phys. Rev. B **55**, R4855 (1997)
[5] E. Gull, A. J. Millis, A. I. Lichtenstein, A. N. Rubtsov, M. Troyer, and P. Werner, Rev. of Mod. Phys. **83**, 349 (2011); P. Werner, A. Comanac, L. de Medici, M. Troyer, and A. J. Millis, Phys. Rev. Lett. **97**, 076405 (2006); E. Gull, P. Werner, O. Parcollet, and M. Troyer, EPL (Eu- rophysics Letters) **82**, 57003 (2008).
[6] K. G. Wilson, Rev. Mod. Phys. **47**, 773 (1975)
[7] R. Bulla, Phys. Rev. Lett. **83**, 136 (1999)
[8] R. Bulla, T. Costi and D. Vollhardt, Phys. Rev. B **64**, 045103 (2001)
[9] S. White, Phys. Rev. Lett. **69**, 2863 (1992); S. White, Phys. Rev. B **48**, 10345 (1993)
[10] *Density Matrix Renormalization*, edited by I. Peschel, X. Wang, M. Kaulke and K. Hallberg (Series: Lecture Notes in Physics, Springer, Berlin, 1999). ISBN:3-540-66129-8
[11] U. Schollwöck, Rev. Mod. Phys. **77**, 259 (2005)
[12] K. Hallberg, Adv. Phys. **55**, 477-526 (2006)
[13] D. García, K. Hallberg and M. Rozenberg, Phys. Rev. Lett. **93**, 246403 (2004)
[14] D. J. García, E. Miranda, K. Hallberg, M. J. Rozenberg, Phys. Rev. B **75**, 121102(R) (2007); E.Miranda, D.J.Garcia, K.Hallberg, and M.J.Rozen- berg, Physica B: Cond. Mat. **403**, 1465 (2008); D.J.Garcia, E.Miranda, K.Hallberg, and M.J.Rozen- berg, Physica B: Cond. Mat. **398**, 407 (2007).
[15] F. Gebhard, E. Jeckelmann, S. Mahlert, S. Nishimoto and R. Noack, Eur. Phys. J B **36**, 491 (2003), S. Nishimoto, F. Gebhard, and E. Jeckelmann, J. Phys.: Condens. Mat. **16**, 7063 (2004).

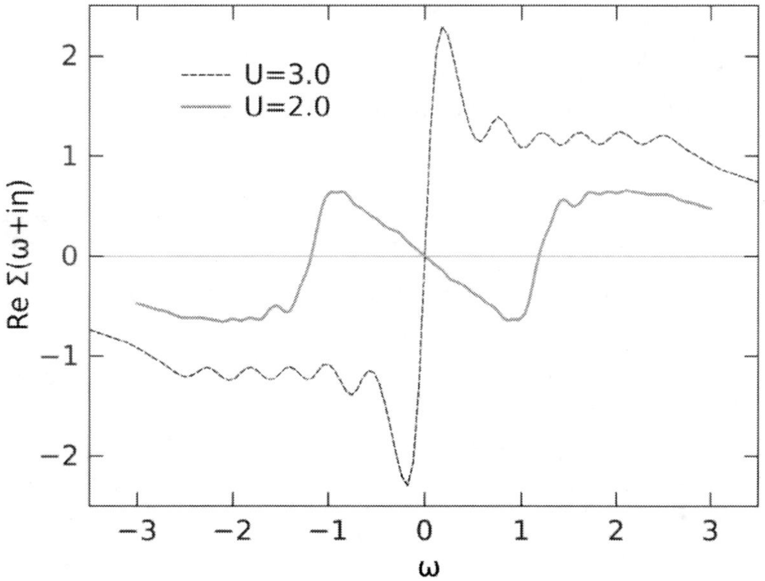

Figure 2. Real part of the self energy for the half-filled two-level Hubbard Model for metallic (full line) and insulating (dotted line). Same parameters as in Fig. 1

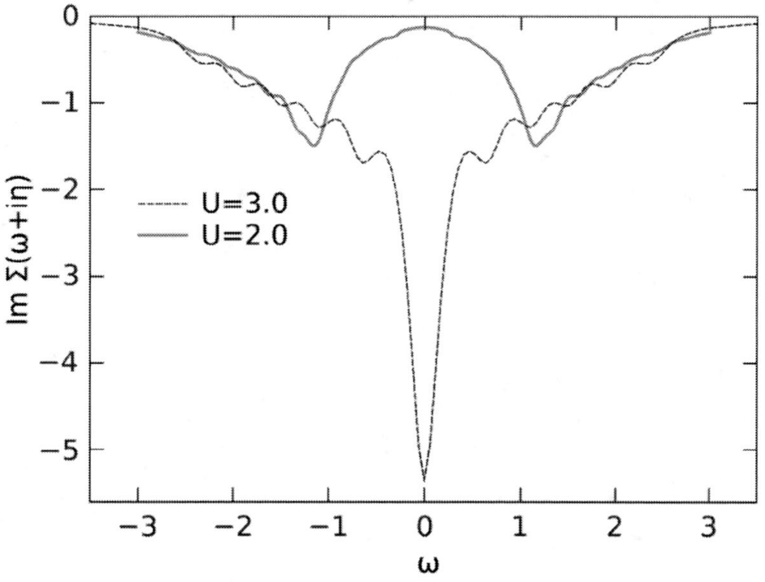

Figure 3. Imaginary part of the self energy for the half-filled two-level Hubbard Model for metallic (full line) and insulating (dotted line). Same parameters as in Fig. 1

[16] M. Karski, C. Raas, and G. Uhrig, Phys. Rev. B **72**, 113110 (2005); M. Karski, C. Raas, and G. Uhrig, Phys. Rev. B **77**, 075116 (2008); C. Raas, P. Grete, and G. Uhrig, Phys. Rev. Lett. **102**, 076406 (2009); C. Raas and G. S. Uhrig, Eur. Phys. J. B **45**, 293 (2005)
[17] R. Peters, Phys. Rev. B **84**, 075139 (2011)
[18] M. Ganahl, M. Aichhorn, H. G. Evertz, P. Thunström, K. Held and F. Verstraete, arXiv:1405.6728
[19] Y. Lu, M. Höppner, O. Gunnarsson and M. W. Haverkort, arXiv:1402.0807
[20] L. de' Medici, Phys. Rev. B **83**, 205112 (2011)
[21] T. D. Kühner and S. R. White, Phys. Rev. B **60**, 335 (1999)
[22] S. Ramasesha et al., Synthetic Metals **85**, 1019 (1997)
[23] T. Pruschke in *Computational Many-Particle Physics*, H. Fehske, R. Schneider and A. Weisse (Eds), (Series: Lecture Notes in Physics, Springer, Berlin Heidelberg, 2008). DOI 10.1007/978-3-540-74686-7, p.473
[24] Q. Si, M. J. Rozenberg, G. Kotliar, and A. E. Ruckenstein, Phys. Rev. Lett. **72**, 2761 (1994)
[25] M. J. Rozenberg, G. Moeller and G. Kotliar, Mod. Phys. Lett. B bf8, 535 (1994)
[26] K. Hallberg, Phys. Rev. B **52**, 9827 (1995)

Calorimetric determination of the angular dependent phase diagram of an S=1/2 Heisenberg triangular-lattice antiferromagnet

S. T. Hannahs[1], N. A. Fortune[2], J.-H. Park[1], Y. Takano[3], T. Ono[4] and H. Tanaka[5]

[1]National High Magnetic Field Laboratory, Tallahassee, FL, United States of America
[2]Department of Physics, Smith College, Northampton, MA, United States of America
[3]Department of Physics, University of Florida, Gainesville, FL, United States of America
[4]Department of Physics, Osaka Prefecture University, Osaka, Japan
[5]Department of Physics, Tokyo Institute of Technology, Meguro-ku, Tokyo, Japan

E-mail: sth@magnet.fsu.edu

Abstract. An antiferromagnetic system on a 2-D triangular lattice leads to geometric topological frustration. This ideal system has been the subject of theoretical investigations. One experimental realization of this system is the compound Cs_2CuCl_4. Various magnetization, heat capacity, neutron scattering and NMR studies have identified several magnetic transitions when the magnetic field is applied along one of the three principal axes. The current work investigates the evolution of these phases at intermediate angles as the crystal is rotated relative to the magnetic field. These phases were investigated using a novel rotating calorimeter allowing complete coverage of the experimental parameter space. New magnetic phases only existing at intermediate angles have been found.

1. Introduction

Geometric frustration for an antiferromagnetic system has long been studied on a 2-D triangular lattice as the simplest realization of the phenomenon. Geometric frustration even with a simple basic interaction leads to a highly degenerate ground state. Application of a magnetic field reveals the highly complex phase structure Gekht1996. If a Dzyaloshinskii-Moriya interaction is present, different phases dominate when the magnetic field is applied along different crystal lattice directions. The 2-D $S = \frac{1}{2}$ Heisenberg antiferromagnet on a triangular lattice with nearest neighbor interactions is an excellent system to investigate this phenomenon.

The ideal triangular lattice would require that $J = J'$ where J is the antiferromagnetic spin coupling along the base of the triangular lattice and J' is the coupling along the diagonal bonds. In compound Cs_2CuCl_4 the coupling constants are unequal ($J'/J \simeq 0.33$), and the adjacent layers are very weakly coupled ($\approx 0.045J$) [2]. The weak exchange energies $J \simeq 4$ K allow the phase diagram to be accessed at moderate magnetic field. It has been shown that in different orientations the phase diagram is sensitive to very small interactions introducing entirely new phases [3].

The alternating layers of Cs ensure weak coupling between adjacent layers as shown in Figure 1. The most natural start of investigation is to apply a magnetic field along the a

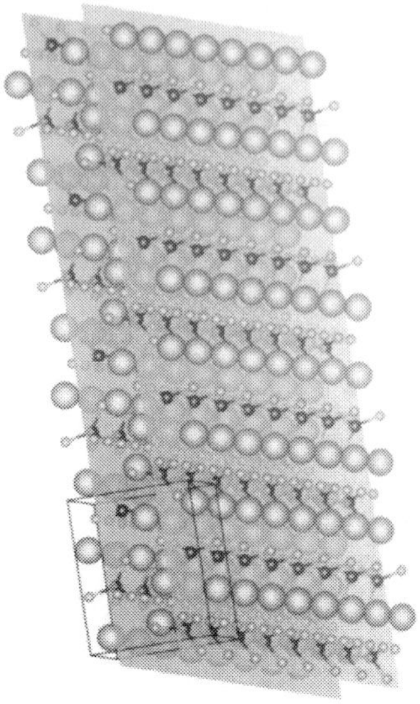

Figure 1. Cs_2CuCl_4 lattice showing inequivalent layered planes (red and blue). Cs atoms dark green and Cu atoms dark blue.

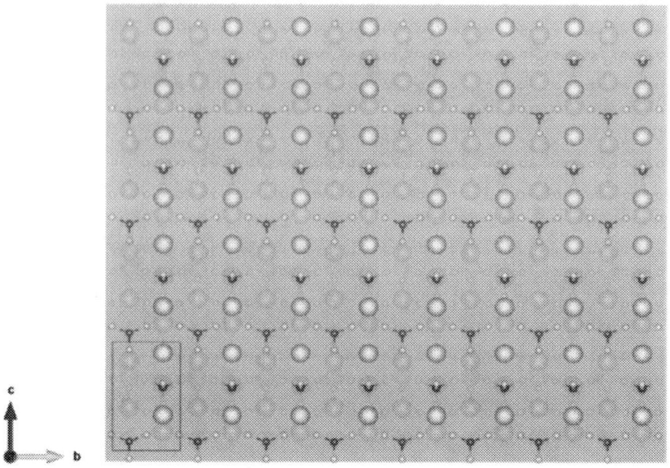

Figure 2. Cs_2CuCl_4 lattice viewed along the *a* axis showing triangular coordination of Cu atoms (dark blue).

axis perpendicular to the triangular lattice planes. This leads to a very simple phase diagram as derived by Coldea *et al* [4] and shown in Figure 3. The present work re-examines this phase diagram at the temperature profile marked by the red line.

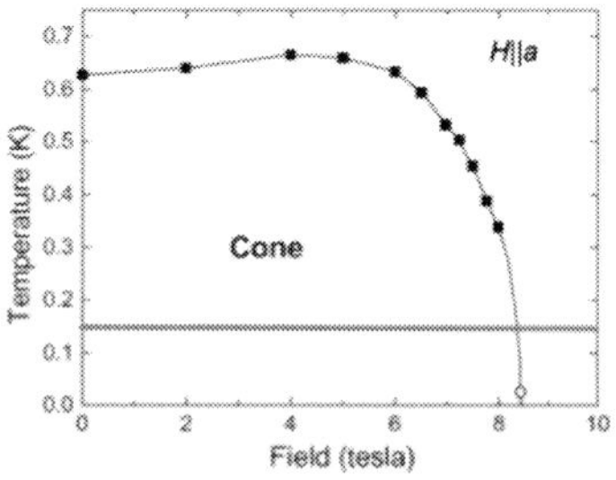

Figure 3. Phase diagram for Cs_2CuCl_4 with the magnetic field applied along the *a* axis. The cone phase persists and evolves at higher field to the saturated paramagnetic state.

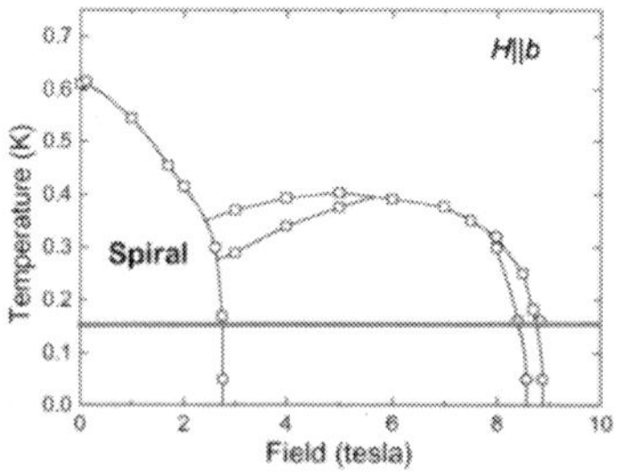

Figure 4. Phase diagram for Cs_2CuCl_4 with the magnetic field applied along the *b* axis showing more complex behavior with the spiral state and other regions of stability.

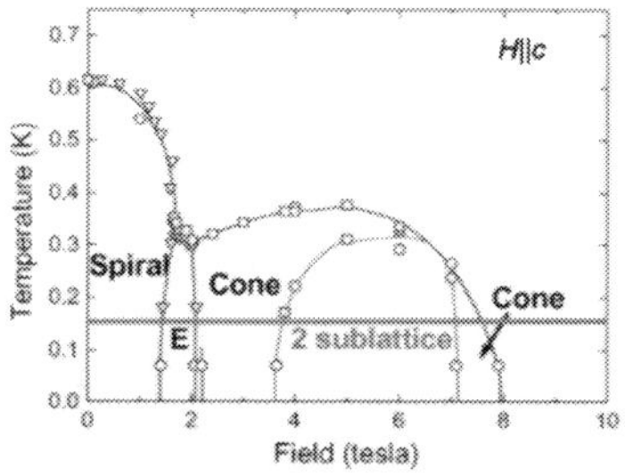

Figure 5. Phase diagram for Cs_2CuCl_4 with the magnetic field applied along the *c* axis. A complex phase diagram where the spiral state ends at even lower field.

Further work by Tokiwa *et al* [5] extended the phase diagram to fields applied along the *b* axis (along the triangle bases) and *c* axis. This reveals a much richer phase diagram with several additional phases identified by NMR measurements [6]. The *b* axis and *c* axis are similar – aside from additional phases when the magnetic field is applied along the *c* axis – but on their own provide no indication how the cone state of the *a* axis evolves to the spiral states of the *b* and *c* axes. The current work utilizes a unique low temperature rotatable calorimeter [7] using magnetocaloric effects to map the evolution of these phases as the magnetic field is rotated between the principal axes of the lattice.

2. Results

As the magnetic field is swept while holding the sample at constant temperature in the calorimeter, the transitions between phases will involve either a peak in heat capacity or a latent heat depending on the nature of the transition. Our very sensitive determination of phase boundaries can detect small (< 1 mK) changes in temperature as the field is swept. The Cs_2CuCl_4 sample of approximately 1 mg mass was thermally linked to a silver platform in a vacuum cell. The platform was regulated at a constant temperature above the base temperature of the dilution refrigerator. This temperature was regulated independent of field by correcting the magnetic field dependence of the resistance using a previously developed algorithm [8]. Small variations in the sample temperature as the field was swept give the magnitude of the magnetocaloric effect.

The phase diagram was determined at 175 mK for all angles as the calorimeter was rotated from the field along the *c* axis (most complex) to the *a* axis (simplest phase diagram). The data are shown in Figure 6, along with the phase identifications as used by Coldea *et al* [2] and Yoshida *et al* [6] as determined by NMR.

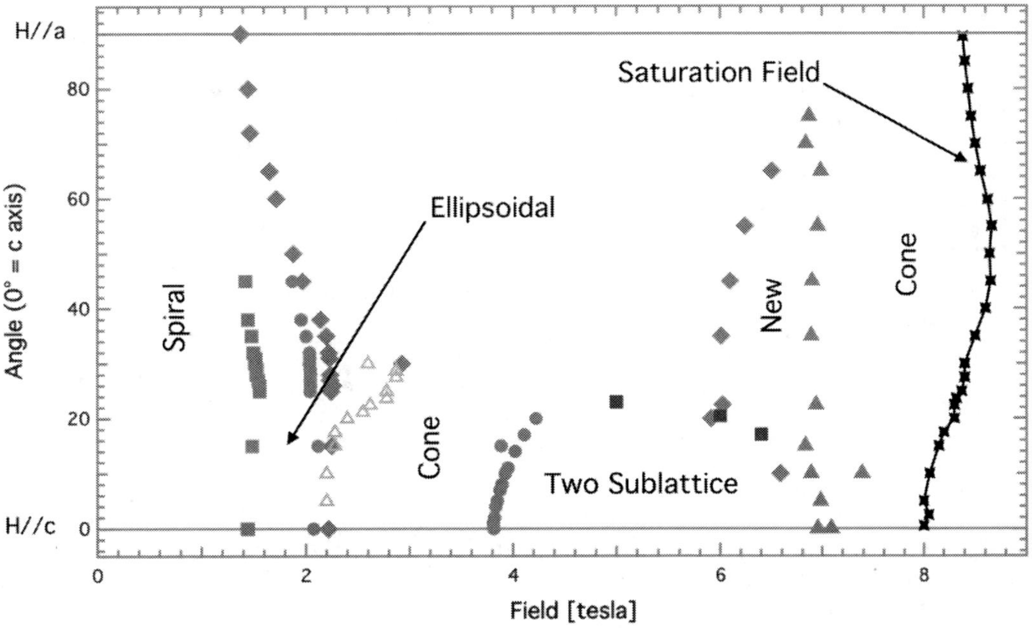

Figure 6. Phase diagram of Cs_2CuCl_4 as the sample angle to the magnetic field is changed. The black x symbols show the saturation field. The black x symbols and the open light blue triangles were measured by torque magnetometry at a temperature of 21 mK [9].

The dome between 3.8 and 7 tesla up to 20° becomes very difficult to resolve in the region where the phase boundary becomes parallel to the abscissa in the figure. In this region the ordinary magnetocaloric effect becomes very weak. In this case we make use of the gyro-magnetocaloric effect [10], where holding the temperature and magnetic field constant the sample is rotated with respect to the field while recording small changes in temperature. Clearly defined transitions are resolved using this technique. The blue square symbols show transitions determined in this manner.

It is known that Cs_2CuCl_4 has a spiral ground state due to the Dzyaloshinskii-Moriya interactions [2]. We find that the high and low field cone phases merge as the magnetic field vector approaches the a axis above 80° and that these phases must be identical. At angles above 20° where the two-sublattice dome closes a new phase appears that is stable from 20° to 80° and disappears just before the field becomes parallel with the a axis. The narrow ellipsoidal region separating the spiral and cone phases collapses at high angle and disappears resulting in the simple phase diagram along the a axis. Finally note that the phase boundary separating the spiral phase from the cone phase extends continuously to the a axis as an addition to the previous phase diagram shown in Figure 3.

If the sample is rotated from the many transitions of the c axis into the b axis there are similar regions identified but with a different angular dependence. This angular dependence of the phase diagram is shown in Figure 7.

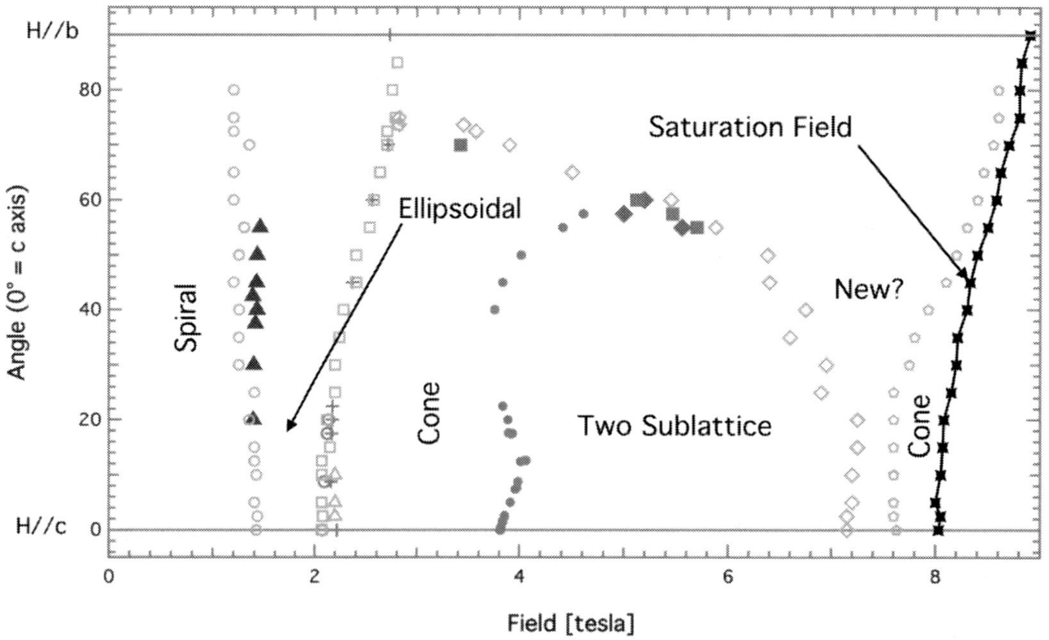

Figure 7. Phase diagram of Cs_2CuCl_4 as the sample angle to the magnetic field is changed. The black x symbols show the saturation field. The black x symbols and the open light blue symbols were measured by torque magnetometry at a temperature of 21 mK [9].

In this case the ellipsoidal region extends to the b axis; however, the transitions between this phase and the spiral phase becomes very weak as the phases merge. The two-sublattice phase extends further from the c axis but eventually closes and disappears at angles higher than 60°. In this phase diagram, the lower-field cone phase also disappears at high angles and does not merge with the higher-field phase which survives at angles up to 90°. The exact nature of the new high field phase has not been determined by these bulk thermodynamic measurements.

3. Conclusions

We have determined the phase diagram of Cs_2CuCl_4 at low temperature as a function of magnetic field and angle. The phases identified by magnetocaloric and gyro-magnetocaloric effects largely confirm previous work along the three principal crystallographic axes. However, the previous identification of phases along these axes must be modified to account for the tracking of these phase boundaries as a function of angle. The extension of the spiral-cone boundary to the a axis and the appearance of a new intermediate angle phase in this material show that the rich and complex behaviors of this spin $\frac{1}{2}$ triangular-lattice Heisenberg antiferromagnet need further investigation and modeling. Extension of this work to higher temperatures would confirm and map the intermediate angle behavior on the previously determined field-temperature phase diagrams.

Acknowledgments

The authors would like to acknowledge experimental support from the National High Magnetic Field Lab; in particular Scott Maier and Jonathan Billings provided technical assistance. The National High Magnetic Field Lab is supported by National Science Foundation, Cooperative Agreement No. DMR-1157490, the State of Florida, and the U.S. Department of Energy.

References

[1] Gekht R S and Bondarenko I N, 1997 "Triangular antiferromagnets with a layered structure in a uniform field", *Journal of Experimental and Theoretical Physics* **84** 345.
[2] Coldea R, Tennant D A, Habicht K, Smeibidl P, Wolters C and Tylczynski Z, 2002, "Direct measurement of the spin Hamiltonian and observation of condensation of magnons in the 2D frustrated quantum magnet Cs_2CuCl_4" *Phys. Rev. Lett.* **88** 137203.
[3] Starykh O A, Katsura H and Balents L, 2010 "Extreme sensitivity of a frustrated quantum magnet: Cs_2CuCl_4", *Phys. Rev. B* **82** 014421.
[4] Coldea R, Tennant D A, Tsvelik A M and Tylczynski Z 2001 "Experimental realization of a 2D fractional quantum spin liquid", *Phys. Rev. Lett.* **86** 1335.
[5] Tokiwa Y, Radu T, Coldea R, Wilhelm H, Tylczynski Z and Steglich F 2006 "Magnetic phase transitions in the two-dimensional frustrated quantum antiferromagnet Cs_2CuCl_4", *Phys. Rev. B* **73** 134414.
[6] Yoshida M, Suzuki T, Waki T and Takigawa M, 2008 Poster presentation at *"International Conference on Highly Frustrated Magnetism"* 7-12 September 2008 Braunschweig, Germany, Unpublished.
[7] Hannahs S T and Fortune N A 2003 "Heat capacity cell for angular measurements in high magnetic fields", Proc. of the 23rd International Conference on Low Temperature Physics (LT23), Aug 20-27, 2002 Hiroshima, Japan, *Physica B - Condensed Matter* **329** 1586-1587.
[8] Fortune N, Gossett G, Peabody L, Lehe K, Uji S and Aoki H 2000 "High magnetic field corrections to resistance thermometers for low temperature calorimetry", *Rev. Sci. Instrum.* **71** 3825-3830.
[9] Takano Y, Park J.-H., Ono T. and Tanaka H. 2014 Unpublished.
[10] Nikitin S A, Skokov K P, Koshkid'ko Y S, Pastushenkov Y G and Ivanova T I 2010 "Giant rotating magnetocaloric effect in the region of spin-reorientation transition in the $NdCo_5$ single crystal", *Phys. Rev. Lett.* **105** 137205.

Journal of Physics: Conference Series **568** (2014) 042011 doi:10.1088/1742-6596/568/4/042011

Fe epilayer magnetic features induced by a covalent, magnetic and metallic substrate

C. Helman[1,2], V. Ferrari[1,3], A.M. Llois[1,3]

[1]Gerencia de Investigación y Aplicaciones, CNEA, Av. Gral. Paz 1499, B1650KNA San Martín, Buenos Aires, Argentina

[2]Escuela de Ciencia y Tecnología, Universidad Nacional de San Martin, M. de Irigoyen 3100, 1650 San Martín, Buenos Aires, Argentina

[3]CONICET, Av. Rivadavia 1917, C1033AAJ Ciudad Autónoma de Buenos Aires, Argentina

E-mail: `helman@tandar.cnea.gov.ar`

Abstract.
In recent years there has been theoretical and experimental interest in the properties of Fe thin layers grown on metallic substrates. We present here a theoretical study based on *ab initio* calculations of the structure and magnetic characteristics for a very thin Fe film grown on an As-terminated MnAs substrate, which is metallic. Particular emphasis is put into the relaxation and reconstruction of the interface and on the characterization of the magnetic coupling with the substrate. The absence of Fe interdiffusion and the presence of an intermediate As layer ensure the partial magnetic decoupling between epilayer and substrate.

1. Introduction

Thin films of different kinds of materials with potential electronic, magnetic and spintronic applications, along with other functionalities, have been investigated during the last decades because of their importance and interest in the development of micro- and nanotechnological devices. Among them, manganese arsenide (MnAs) thin films have been studied quite intensively in the last years due to the good compatibility of this material with standard semiconducting substrates[1, 2, 3] and also due to its high spin polarisation[3], both characteristics being well suited for the development of spintronic devices[1, 4].

MnAs is, in fact, a metallic material which presents an interesting variety of polymorphic transformations as a function of temperature[1, 5]. In bulk, the low temperature hexagonal structure of the NiAs-type (α phase) is ferromagnetic and it has a first order transition at 40°C, where the system evolves into a paramagnetic orthorhombic structure of the MnP-type (β-phase). At 127°C, it undergoes a second order transition transforming again into a low temperature hexagonal structure, which is paramagnetic, namely the γ phase. Most of the latest works on MnAs have been done for thin films of this material grown on GaAs(001). In a temperature window around room temperature, whose spread depends on the MnAs film thickness, the contact between GaAs and MnAs gives rise to a strain-driven pattern of coexisting and alternating magnetic and non-magnetic MnAs stripes[1, 2, 4, 5].

More recently, Fe epilayers of 5 to 10 nm were grown on top of MnAs($1\bar{1}00$)/GaAs(001) in order to tune the period of the above mentioned stripes[6], or to use these heterostructures as magnetic active templates as a function of temperature[7]. In Ref. [7] it has been

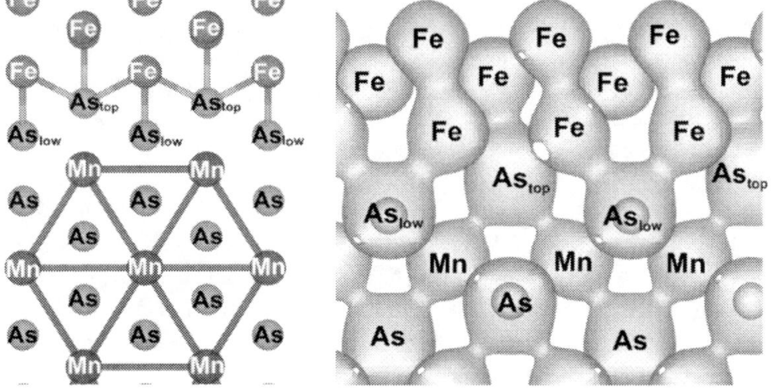

Figure 1. Three Fe layers on MnAs($1\bar{1}00$). a) Schematic lateral view of the relaxed structure. b) Charge isosurface plot corresponding to 0.05 e/Å^3, which emphasizes the Fe-As bonds.

reported the magnetic response of these Fe epilayers when the MnAs thin films go from the α to the β phase, through the phase coexistence temperature region. A magnetic decoupled behaviour of the Fe epilayers was observed, characterized by the fact that the magnetization could be either parallel or anti-parallel to the MnAs substrate. This behaviour, revealed a weak magnetic coupling between both magnetic materials. This result was supported by FMR studies, where the evolution of the magnetic anisotropies of the combined system Fe(5nm)/MnAs(100nm)/GaAs(1mm)[8] as a function of temperature was measured. It was also found that, outside the MnAs coexistence temperature range, the Fe films are also magnetically decoupled from the substrate[8].

The observed magnetic behaviour of the Fe epilayers should be traced back to interfacial effects, but the characteristics and nature of the structure at the interface of an ultrathin Fe epilayer have not yet been approached. Actually, this interface was only characterized for Fe overlayers of thickness beyond 30nm, for which a mixed MnFeAs antiferromagnetic compound in the interfacial region was found[9]. The antiferromagnetic structure of this interfacial compound made it possible to explain the magnetic decoupling between the Fe films and the MnAs substrates when the thickness of the overlayer competes with the thickness of the MnAs substrate. For thinner Fe overlayers, less than 10nm, this discussion is still open and, in this work, we try to give a partial answer to it. We theoretically study the structure of a Fe film of approximately 4Å grown on an As-terminated MnAs substrate. Particular emphasis is put into the relaxation and reconstruction of the interface, which characterize this ultrathin film regime.

2. Results and discussion

Electronic calculations are performed within the framework of density functional theory as implemented in the VASP code[10]. PAW pseudopotentials are used with the choice of exchange and correlation potential within the Generalized Gradient Approximation (GGA) as parametrized by Perdew *et al* [11, 12]. We consider a plane waves basis set with kinetic energy values of up to 350 eV, and a $20 \times 20 \times 1$ grid in k-space. The internal coordinates are fully relaxed until forces are smaller than 0.04 eV/Å.

Regarding the crystal structure of the MnAs substrate, in our calculations we consider the experimental data which states that α-MnAs is hexagonal (B_{31}) and that it grows epitaxially along the ($1\bar{1}00$) direction [6]. We simulate the ferromagnetic α-MnAs substrate using three

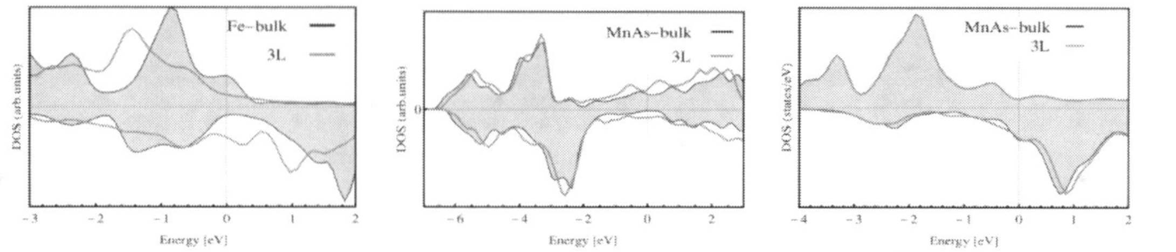

a) Fe b) As c) Mn

Figure 2. Projected densities of states (PDOS) on a) Fe, b) As and c) Mn atoms, at the interface of the 3L-Fe/MnAs heterostructure (with red lines) and the PDOS projected on the same atoms for the corresponding bulk (shown with shadow areas). The energy values are referred to the Fermi level.

MnAs unit cells separated by a vacuum region of 10 Å and perform periodic slab calculations with in-plane cell parameters $a = 3.73$ Å and $b = 5.69$ Å[3]. No inversion symmetry is imposed in order to have full formula units of MnAs in the unit cell of the slab, and as a consequence, the two free surfaces have different atomic terminations. Therefore, two possible interfaces between Fe and MnAs can be analysed depending on which of the free surfaces the Fe-film is deposited.

We find that the Mn-terminated slab is energetically preferred to the As-terminated one when a Fe overlayer is deposited on them. However, MnAs is usually grown in an As-atmosphere, which favours an As-terminated surface[2]. We focus, then, in this work on an ultrathin Fe film deposited on the As-terminated substrate.

As indicated by RHEED and TEM characterizations, Fe grows on MnAs($1\bar{1}00$) along the Fe($2\bar{1}1$) direction, perpendicular to MnAs [0001] and [$01\bar{1}0$] [8]. Following this, our ultrathin film consists of three Fe layers (3L) grown on MnAs along ($2\bar{1}1$), as shown in Fig. 1a). We assume that the interfacial Fe atoms follow the MnAs substrate occupying Mn sites, as it is expected for epitaxial growth conditions. The other two Fe layers keep a bcc-like structure, which is under strain due to epitaxy with the MnAs substrate. The epilayer, as it is usual in epitaxial thin film growth, adopts the parameters of the substrate giving rise to biaxial strain. This clamping propagates throughout the overlayer, it is tensile in one in-plane direction and compressive in the other one. This has already been analysed in Ref.[13], where it was shown that the biaxial strain contributes to a change in the magnetic anisotropy of the Fe epilayer.

After relaxation of the internal coordinates of the Fe epilayer, the resulting interface is sharp with no Fe interdiffusion, as it is shown in Fig. 1a). The interplane distances among Fe layers are reduced by approximately 20% with respect to the bulk values.

At the interface, the Fe atoms present a magnetic moment similar to the Fe bulk one ($\sim 2.26\mu B$), while the magnetic moments and charge of the Mn atoms of the MnAs slab do not change in the presence of the Fe epilayer. However, the interfacial As atoms bind covalently with the first and second Fe overlayers. In Fig. 1b) we show a charge isosurface plot of the system and the covalent As-Fe bonds can be clearly recognized.

Regarding the magnetic properties, the lowest energy configuration of the heterostructure under study is ferromagnetic. The difference in energy between the magnetic configuration with the Fe epilayer antiferromagnetically and ferromagnetically aligned with respect to the MnAs substrate is around 10 meV/Fe atom. This value is smaller than the Fe-Fe coupling reported for Fe bulk [14] and also smaller than the one among Mn atoms in the corresponding MnAs bulk[3].

In Fig.2 we show the projected densities of states (PDOS) of interfacial Fe, As and Mn atoms and also the PDOS for the same type of atoms in $\alpha-$MnAs and Fe bulk structures. We note first, that due to strain the PDOS of Fe in the epilayer presents a smaller average exchange splitting than the one in bcc Fe bulk, while the PDOS of the interfacial Mn atoms remains unchanged with respect to MnAs bulk. The Mn atoms do not sense, then, the presence of the Fe overlayer, while the interfacial As atoms build covalent bonds with the adjacent Fe atoms, as it can be seen in Fig. 2b) as well as in Fig 1b).

3. Final remarks

We have obtained that the energy difference between ferromagnetic and antiferromagnetic alignments of an ultrathin Fe film grown on As-terminated MnAs is roughly 10 meV, that is equivalent to $\sim$ 110K. This temperature is nearly 200K lower than the temperature of MnAs at phase coexistence, where a magnetic decoupled behaviour was experimentally observed. From our results, this decoupling may be attributed to the absence of Fe interdiffusion and to the presence of an intermediate As layer. The existence of an antiferromagnetic spacer is not needed to achieve the partial decoupling. This goes along with previous results which showed that the magnetic anisotropy energy of this system depends mainly on the substrate induced strain and not on its magnetic state[13].

Acknowledgments

We thank J.Milano for helpful discussions. C.H., V.F and A.M.L. belong to the Institute of Nanoscience and Nanotechnology (INN) of the Atomic Energy Agency (CNEA), Argentina. They acknowledge financial support from ANPCyT (PICT-2011-1187 and 1857) and CONICET (PIP00273).

References

[1] Das A K, Pampuch C, Ney A, Hesjedal T, Däweritz L, Koch R and Ploog K H 2003 *Phys. Rev. Lett.* **91**(8) 087203 URL `http://link.aps.org/doi/10.1103/PhysRevLett.91.087203`

[2] Breitwieser R, Vidal F, Graff I L, Marangolo M, Eddrief M, Boulliard J C and Etgens V H 2009 *Phys. Rev. B* **80**(4) 045403 URL `http://link.aps.org/doi/10.1103/PhysRevB.80.045403`

[3] Rungger I and Sanvito S 2006 *Phys. Rev. B* **74**(2) 024429 URL `http://link.aps.org/doi/10.1103/PhysRevB.74.024429`

[4] Garcia V, Jaffrès H, George J M, Marangolo M, Eddrief M and Etgens V H 2006 *Phys. Rev. Lett.* **97**(24) 246802 URL `http://link.aps.org/doi/10.1103/PhysRevLett.97.246802`

[5] Mattoso N, Eddrief M, Varalda J, Ouerghi A, Demaille D, Etgens V H and Garreau Y 2004 *Phys. Rev. B* **70**(11) 115324 URL `http://link.aps.org/doi/10.1103/PhysRevB.70.115324`

[6] Vidal F, Spezzani C, Breitwieser R, Marangolo M, Eddrief M, Sacchi M and Etgens V H 2010 *Applied Physics Letters* **97** 251914 (pages 3) URL `http://link.aip.org/link/?APL/97/251914/1`

[7] Sacchi M, Marangolo M, Spezzani C, Breitwieser R, Popescu H, Dealaunay R, Rache Salles B, Eddrief M and Etgens V H 2010 *Phys. Rev. B* **81**(22) 220401 URL `http://link.aps.org/doi/10.1103/PhysRevB.81.220401`

[8] Helman C, Milano J, Tacchi S, Madami M, Carlotti G, Gubbiotti G, Alejandro G, Marangolo M, Demaille D, Etgens V H and Pini M G 2010 *Phys. Rev. B* **82** 094423

[9] Demaille D, Patriarche G, Helman C, Eddrief M, Etgens V H, Sacchi M, Llois A M and Marangolo M 2013 *Crystal Growth & Design* **13** 4279–4284 (*Preprint* `http://pubs.acs.org/doi/pdf/10.1021/cg400576m`) URL `http://pubs.acs.org/doi/abs/10.1021/cg400576m`

[10] Kresse G and Furthmüller J 1996 *Computational Materials Science* **6** 15–50 ISSN 0927-0256 URL `http://www.sciencedirect.com/science/article/pii/0927025696000080`

[11] Kresse G and Joubert D 1999 *Phys. Rev. B* **59**(3) 1758–1775 URL `http://link.aps.org/doi/10.1103/PhysRevB.59.1758`

[12] Perdew J P, Burke K and Ernzerhof M 1996 *Phys. Rev. Lett.* **77** 3865

[13] Helman C and Llois A 2013 *IEEE Transactions on Magnetics* **49** 4675–4678 URL `http://ieeexplore.ieee.org/xpl/articleDetails.jsp?arnumber=6566111`

[14] Liechtenstein A, Katsnelson M, Antropov V and Gubanov V 1987 *Journal of Magnetism and Magnetic Materials* **67** 65–74 ISSN 0304-8853 URL `http://www.sciencedirect.com/science/article/pii/0304885387907219`

Thermal Conductivity in the Frustrated Two-Leg Spin-Ladder System $BiCu_2PO_6$

H Nagasawa[1], T Kawamata[1], K Naruse[1], M Ohno[1], Y Matsuoka[1], H Sudo[1], Y Hagiya[1], M Fujita[2], T Sasaki[2] and Y Koike[1]

[1]Department of Applied Physics, Tohoku University, Sendai, Japan
[2]Institute for Materials Research, Tohoku University, Sendai, Japan

Email : tkawamata@teion.apph.tohoku.ac.jp

Abstract. We have measured temperature and magnetic-field dependences of the thermal conductivity of single crystals of the frustrated two-leg spin-ladder system $BiCu_2PO_6$ in magnetic fields up to 14 T. It has been found that the temperature dependence of the thermal conductivity along every principal crystallographic axis shows two peaks in zero field but that the magnitude of the thermal conductivity along the *b*-axis parallel to spin ladders, κ_b, is significantly larger than those of the thermal conductivity along the *a*-axis, κ_a, and along the *c*-axis, κ_c, at high temperatures above 7 K. These results suggest that the thermal conductivity due to spins probably exists only in κ_b. Furthermore, it has been found that both magnetic-field dependences of κ_a and κ_b at 3 K show kinks at ~ 7 T and ~ 10 T, where the spin state may change.

1. Introduction

In some low-dimensional quantum spin systems, the thermal conductivity has attached great interest, because large contribution of the thermal conductivity due to spins, κ_{spin}, has been observed [1]. In the two-leg spin-ladder system $Sr_{14}Cu_{24}O_{41}$, for example, the large contribution of κ_{spin} has been observed only along the *c*-axis parallel to the leg direction of spin ladders at high temperatures around 150 K [2,3]. Since the thermal conductivity due to phonons, κ_{phonon}, has been observed markedly at low temperatures around 20 K along every principal axis, the temperature dependence of the thermal conductivity along the *c*-axis where the magnetic interaction is strong has shown two peaks due to κ_{phonon} and κ_{spin}. Similar two peaks in the temperature dependence of the thermal conductivity have been observed also in several spin-gap systems such as $SrCu_2(BO_3)_2$ [4,5] and $CuGeO_3$ [6,7]. In these systems, two peaks have been observed in the temperature dependence of the thermal conductivity along every principal axis. Therefore, both two peaks have been concluded to be due to κ_{phonon}.

Thermal conductivity is a useful probe detecting a change of the spin state in a spin system, because the scattering rate of heat carriers is sensitively affected by the change of the spin state. In fact, anomalies of the thermal conductivity due to the appearance of a new spin state have been observed in several frustrated spin systems [8].

The compound $BiCu_2PO_6$ has zigzag two-leg spin-ladders of Cu^{2+} spins with the quantum spin number $S = 1/2$ along the *b*-axis. That is, the leg direction of spin ladders is along the *b*-axis. The

Cu^{2+} spins are frustrated due to both the nearest neighbor exchange interaction along the *b*-axis and the next nearest neighbor exchange interaction along the *b*-axis. The magnitude of the spin gap, Δ, has been estimated from the specific heat, magnetic susceptibility, NMR and magnetization measurements as ~ 35 K [9,10].

In this paper, we have measured the thermal conductivity of $BiCu_2PO_6$ single crystals in magnetic fields to investigate the presence or absence of κ_{spin} and also some change of the spin state.

2. Experimental

Single crystals of $BiCu_2PO_6$ were grown by the traveling-solvent floating-zone technique. Thermal conductivity measurements were carried out by the conventional steady-state method in magnetic fields up to 14 T at High Field Laboratory for Superconducting Materials, Institute for Materials Research, Tohoku University.

3. Results and Discussion

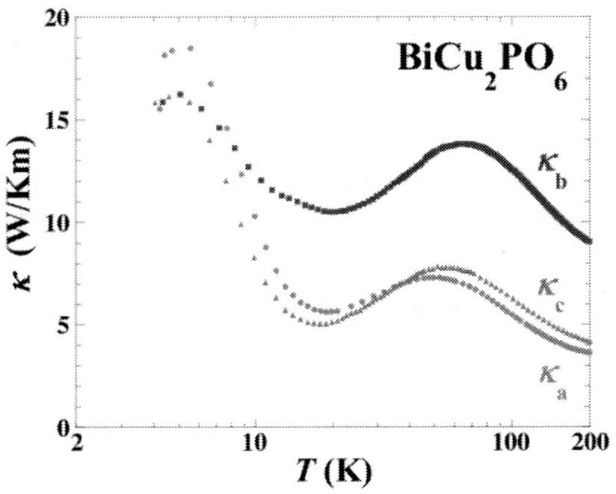

Figure 1. Temperature dependence of the thermal conductivity along the *a*-axis, κ_a, along the *b*-axis, κ_b, and along the *c*-axis, κ_c, of $BiCu_2PO_6$ single crystals in zero field.

Figure 1 shows the temperature dependence of the thermal conductivity along the principal crystallographic axes of $BiCu_2PO_6$ single crystals in zero field. It is found that the thermal conductivity along every principal direction shows two peaks, namely, a sharp peak at ~ 5 K and a broad peak at ~ 60 K. Since two peaks are observed in the thermal conductivity along every principal direction, these two peaks are inferred to be due to κ_{phonon}, as in the case of other spin-gap systems such as $SrCu_2(BO_3)_2$ [4,5]. That is, the broad peak at ~ 60 K is due to the increase of the mean free path of phonons, l_{phonon}, with decreasing temperature at high temperatures above ~ 60 K and due to the decrease of l_{phonon} on account of the enhancement of spin fluctuations owing to the frustration, namely, the increase of the phonon-magnon scattering rate with decreasing temperature at low temperatures below ~ 60 K. The sharp peak at ~ 5 K is due to the increase of l_{phonon} on account of the decrease of the phonon-magnon scattering rate owing to the decrease of the number of magnons with decreasing temperature at low temperatures below ~ 20 K where the spin gap is effective. Furthermore, it is found that the magnitude of the thermal conductivity along the *b*-axis parallel to spin ladders, κ_b, at high temperatures above 7 K is significantly larger than those of the thermal conductivity along the *a*-axis, κ_a, and along the *c*-axis, κ_c, perpendicular to spin ladders. Therefore, it is possible that κ_{spin} contributes to κ_b in a wide region of temperature above 7 K in addition to κ_{phonon}. There still remains a possibility that the anisotropy of thermal conductivity is ascribed the anisotropy of κ_{phonon}. To clarify the presence of κ_{spin}, it may be effective to perform thermal conductivity measurements of $BiCu_{2-x}Zn_xPO_6$, where κ_{spin} is expected to be suppressed more strongly than κ_{phonon} by nonmagnetic Zn^{2+} ions substituted of Cu^{2+} ions.

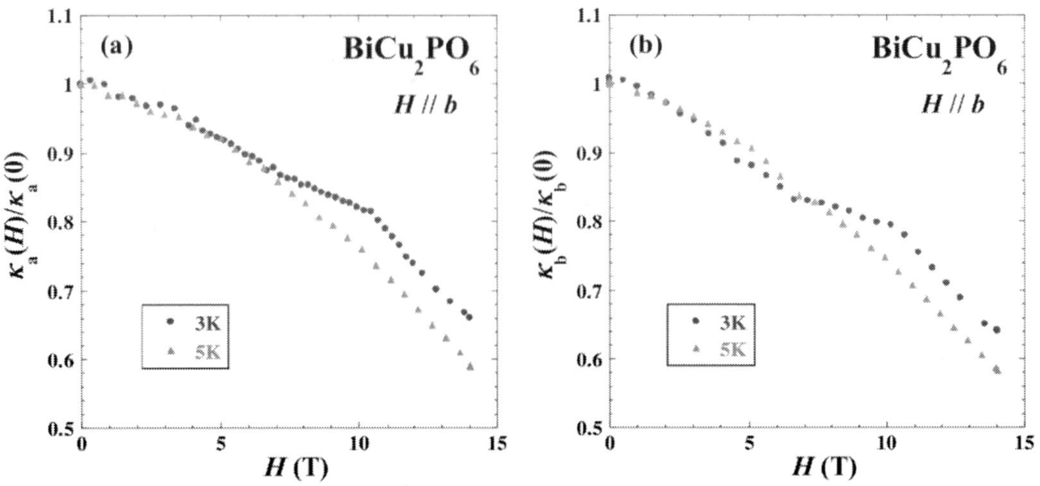

Figure 2. Magnetic-field dependence of the thermal conductivity κ_a and κ_b normalized by the value in zero field, $\kappa_a(H)/\kappa_a(0)$ and $\kappa_b(H)/\kappa_b(0)$, of $BiCu_2PO_6$ single crystals in magnetic fields parallel to the *b*-axis up to 14 T at 3 K and 5 K.

Figure 2 shows the magnetic-field dependence of κ_a and κ_b normalized by the value in zero field, $\kappa_a(H)/\kappa_a(0)$ and $\kappa_b(H)/\kappa_b(0)$, of $BiCu_2PO_6$ single crystals in magnetic fields parallel to the *b*-axis. It is found that both $\kappa_a(H)/\kappa_a(0)$ and $\kappa_b(H)/\kappa_b(0)$ are similarly suppressed by the application of magnetic field due to the reduction of the spin gap. At a low temperature of 3 K, it is remarkable that both $\kappa_a(H)/\kappa_a(0)$ and $\kappa_b(H)/\kappa_b(0)$ exhibit kinks at ~ 7 T and ~ 10 T. Since the kinks are observed in both κ_a and κ_b, they will be due to the change of κ_{phonon} via the phonon-magnon scattering caused by the change of the spin state. However, there has been no report on the anomaly at ~ 7 T and ~ 10 T at low temperatures, through field-induced magnetically ordered states have been reported to appear in magnetic fields as high as ~ 20 T and ~ 35 T [11]. Very recent inelastic neutron scattering experiments in magnetic fields parallel to the *a*-axis have revealed that there is a level crossing of magnon branches, which are split by the anisotropic interaction in zero field, around 11T [12]. Therefore, the kinks of $\kappa_a(H)/\kappa_a(0)$ and $\kappa_b(H)/\kappa_b(0)$ in magnetic fields parallel to the *b*-axis might be due to such a level crossing. In any case, this result suggests to perform detailed experiments in magnetic fields below ~ 10 T at low temperatures to clarify the possible change of the spin state.

4. Conclusions

We have measured temperature and magnetic-field dependences of the thermal conductivity of $BiCu_2PO_6$ single crystals in magnetic fields up to 14 T. In zero field, it has found that the temperature dependence of the thermal conductivity along every principal crystallographic axis shows two peaks. These peaks have been explained as being due to the temperature dependence of κ_{phonon}, namely, the temperature dependence of the phonon-magnon scattering rate, taking into account the spin fluctuations owing to the frustration and the spin gap. It has been found that it is possible that κ_{spin} contributes to κ_b in the leg direction of spin ladders, because κ_b is significantly larger than κ_a and κ_c at high temperatures above 7 K. To clarify the contribution of κ_{spin}, thermal conductivity measurements of Zn-substituted $BiCu_{2-x}Zn_xPO_6$ may be effective. It has also been found that there are two kinks in the magnetic-field dependence of $\kappa_a(H)/\kappa_a(0)$ and also $\kappa_b(H)/\kappa_b(0)$ at ~ 7 T and ~ 10 T at a low temperature of 3 K. These kinks suggest some change of the spin state or a level crossing of magnon branches. To clarify them, other detailed experiments are desired in magnetic fields below ~ 10 T at low temperatures.

References

[1] Hess C 2007 *Eur. Phys. J. Special Topics* **151** 73
[2] Kudo K , Ishikawa S, Noji T, Adachi T, Koike Y, Maki K, Tsuji S and Kumagai K 2001 *J. Phys. Soc. Jpn.* **70** 437
[3] Hess C, Baumann C, Ammerahl, Büchner B, Heidrich-Meisner F, Brenig W and Revcolecshi A 2001 *Phys. Rev.* B **64** 184305
[4] Kudo K, Noji T, Koike Y, Nishizaki T and Kobayashi N 2001 *J. Phys. Soc. Jpn.* **70** 1448
[5] Hofmann M, Lorenz T, Uhrig G S, Kierspel H, Zabara O, Freimuth A, Kageyama H, and Ueda Y 2001 *Phys. Rev. Lett.* **87** 047202
[6] Salce B, Devoille L, Calemczuk R, Buzdin A I , Dhalenne G and Revcolevschi A 1998 *Phys. Lett.* A **245** 127
[7] Ando Y, Takeya J, Sisson D L, Doettinger S G, Tanaka I, Feigelson R S and Kapitulnik A 1998 *Phys. Rev.* B **58** R2913
[8] Li Q J, Zhao Z Y, Fan C, Zhang F B, Zhou H D, Zhao X and Sun X F 2013 *Phys. Rev.* B **87** 214408
[9] Casola F, Shiroka T, Wang S, Conder K, Pomjakushina E, Mesot J and Ott H-R 2010 *Phys. Rev. Lett.* **105** 067203
[10] Tsirlin A A, Rousochatzakis I, Kasinathan D, Janson O, Nath R, Weickert F, Geibel C, Lauchli A M and Rosner H 2010 *Phys. Rev.* B **82** 144426
[11] Kohama Y, Wang S, Uchida A, Prsa K, Zvyagin S, Skourski Y, McDonald R D, Balicas L, Ronnow H M, Rüegg C and Jaime M 2012 *Phys. Rev. Lett.* **109** 167204
[12] Plumb K W, Hwang K, Qiu Y, Harriger L W., Granroth G E, Shu G J, Chou F C, Ruegg Ch, Kim Y B and Kim Y-J arXiv:cond-mat/1408.2528.

Thermal Conductivity of the $S = 1/2$ Quasi-One-Dimensional Ferromagnetic Spin System $CsCuCl_3$

T Kawamata[1], H Sudo[1], Y Matsuoka[1], K Naruse[1], M Ohno[1], T Sasaki[2], Y Koike[1]

[1] Department of Applied Physics, Tohoku University, Sendai, 980-8579 Japan
[2] Institute for Materials Research, Tohoku University, Sendai, 980-8579 Japan

E-mail: tkawamata@teion.apph.tohoku.ac.jp

Abstract. We have measured the thermal conductivity along the c-axis, $\kappa_{\parallel c}$, parallel to ferromagnetic spin-chains of single crystals of the $S = 1/2$ quasi-one-dimensional spin system $CsCuCl_3$ in magnetic fields up to 14 T, in order to investigate the thermal conductivity due to spins, κ_{spin}, and the change of thermal conductivity corresponding to the change of the spin state. In the temperature dependence of $\kappa_{\parallel c}$, no contribution of κ_{spin} has been observed, while a dip has been observed at the antiferromagnetic phase transition temperature, T_N. Furthermore, it has been found that $\kappa_{\parallel c}$ at a low temperature of 5.1 K below T_N changes with increasing field perpendicular to the c-axis in good correspondence to the field-induced change of the spin state.

1. Introduction

Thermal conductivity in low-dimensional quantum spin systems has attracted interest, owing to the large contribution of the thermal conductivity due to spins, κ_{spin} [1, 2]. Especially, a number of studies on κ_{spin} have been carried out in the spin quantum number $S = 1/2$ one-dimensional (1D) antiferromagnetic (AFM) systems [3–11]. The large contribution of κ_{spin} has been observed as a peak or a shoulder in the temperature dependence of the thermal conductivity along spin chains together with the contribution of the thermal conductivity due to phonons, κ_{phonon}. In the 1D ferromagnetic (FM) systems, on the other hand, the contribution of κ_{spin} has not been observed clearly, though several studies on κ_{spin} have been performed [12–16].

The compound $CsCuCl_3$ belongs to the so-called ABX_3 family including well-known quasi-1D spin systems with frustration. The compound contains linear spin-chains of face-sharing $CuCl_6$ octahedra running along the c-axis. The spin chains separated by Cs^+ ions form an equilateral triangular lattice in the c-plane. The intrachain exchange interaction J is FM and estimated as ~ -28 K, while the interchain exchange interaction J' is AFM and estimated as ~ 5 K [17]. Therefore, frustration is expected to exist more or less between the spin chains. The $CuCl_6$ octahedra are distorted due to the Jahn-Teller effect, so that the Dzyaloshinsky-Moriya interaction is induced between spins along the c-axis. An AF long-range order appears at the phase transition temperature, T_N, ~ 10.7 K, owing to J' [18]. The spin structure at low temperatures below T_N is helical with the pitch of $\sim 5.1^\circ$ along the c-axis owing to the combination of J and the Dzyaloshinsky-Moriya interaction. In the c-plane, the so-called 120°

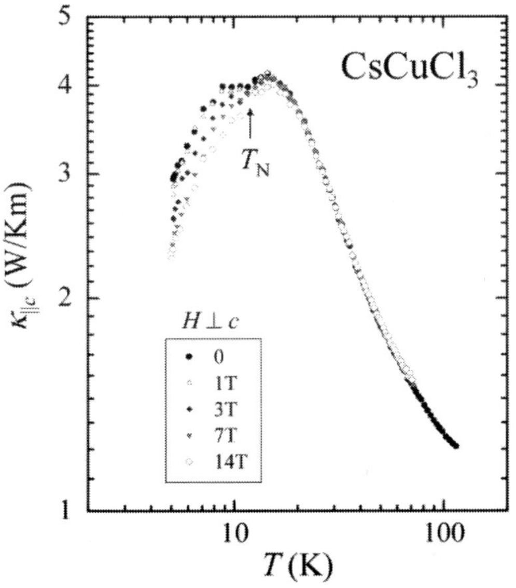

Figure 1. Temperature dependence of the thermal conductivity along the c-axis, $\kappa_{\parallel c}$, parallel to spin chains in zero field and magnetic fields perpendicular to the c-axis.

spin structure is formed owing to J' and the spin directions are confined in the c-plane [18]. Moreover, it has been known that the pitch decreases by the application of magnetic field perpendicular to spin chains [19–21].

In this paper, we have measured the thermal conductivity along the c-axis, $\kappa_{\parallel c}$, parallel to spin chains of $CsCuCl_3$ in order to investigate the contribution of κ_{spin}. Furthermore, we have also investigated the relation between the spin state and the behavior of the thermal conductivity, because thermal conductivity is a good probe detecting a change of the spin state in a spin system via the change of the scattering rate of heat carries [22–26].

2. Experimental

Single crystals of $CsCuCl_3$ were grown from solution. Thermal conductivity measurements were carried out by the conventional steady-state method in magnetic fields up to 14 T.

3. Results and Discussion

Figure 1 shows the temperature dependence of $\kappa_{\parallel c}$ parallel to spin chains in zero field and magnetic fields perpendicular to the c-axis at low temperatures below $\sim$ 100 K. It has been found that $\kappa_{\parallel c}$ show a peak around 10 K with a dip at T_N in zero field. This kind of dip at T_N has been observed in several antiferromagnets [5, 27, 28], and is interpreted as being due to the strong scattering of heat carries caused by the critical fluctuations around T_N or due to the increase of the mean free path of heat carries just below T_N on account of the marked reduction of the scattering caused by the development of the long-rage order. Therefore, this behavior of $\kappa_{\parallel c}$ in zero field is likely to be mainly due to the contribution of κ_{phonon} affected by the scattering of phonons by magnetic excitations. The contribution of κ_{spin} cannot be observed clearly in $CsCuCl_3$. Recently, single crystals of $CsCuCl_3$ of a single domain with homochirality of spins have been grown by Kousaka *et al.* [29]. On the other hand, our single crystals of the zigzag shape may be of multi-domains. Therefore, κ_{spin} might strongly be suppressed due to the strong scattering of magnons at the domain boundaries.

By the application of magnetic field perpendicular to the c-axis, it has been found that $\kappa_{\parallel c}$ at low temperatures below T_N is suppressed, as shown Fig. 1. To see the suppression in detail, the

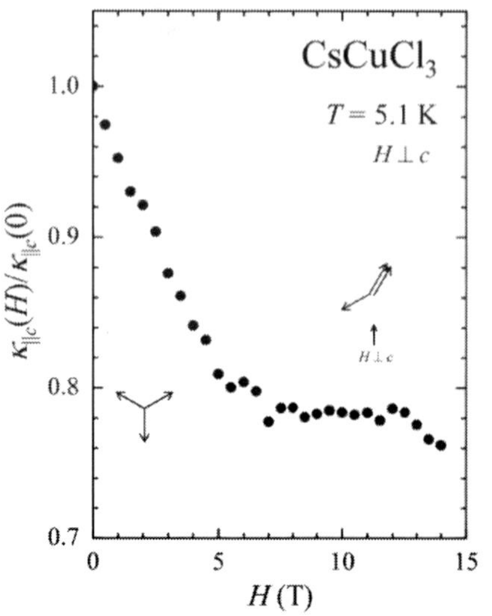

Figure 2. Magnetic-field dependence of the thermal conductivity along the c-axis normalized by the value in zero field, $\kappa_{\parallel c}(H)/\kappa_{\parallel c}(0)$, of $CsCuCl_3$ in magnetic fields perpendicular to the c-axis. Three arrows on the left and right sides mean 120° and coplanar 2-1-type spin structures, respectively.

magnetic-field dependence of $\kappa_{\parallel c}$ normalized by the value in zero field, $\kappa_{\parallel c}(H)/\kappa_{\parallel c}(0)$, at 5.1 K is shown in Fig. 2. It has been found that $\kappa_{\parallel c}(H)/\kappa_{\parallel c}(0)$ monotonically decreases with increasing field up to $\sim$ 9 T, becomes constant between $\sim$ 9 T and $\sim$ 12 T, and then decreases again above $\sim$ 12 T. This behavior is very similar to the magnetic-field dependence of the incommesurabiltiy, δ, which is proportional to the pitch of the helical spin structure along the c-axis [19–21]. It is known that the helical and 120° spin structures are maintained in low magnetic fields below $\sim$ 9 T, though the pitch of the helical structure along the c-axis decreases with increasing field [21]. The 120° spin structure in the c-axis is expected to become unstable by the application of magnetic field in the c-plane. Therefore, it is likely that $\kappa_{\parallel c}$ is suppressed with increasing field due to the enhancement of the magnetic fluctuations scattering phonons.

In a region between 9 T and 16 T, it is known that the helical and the so-called coplanar 2-1-type spin structures, where two thirds of spins are roughly parallel and the others are roughly antiparallel to the magnetic-field direction as shown in Fig.2, are formed. The constant value of $\kappa_{\parallel c}(H)/\kappa_{\parallel c}(0)$ in this region implies that the number of magnetic excitations scattering phonons does not change so much. This is consistent with the locking phenomenon of the spin structure in this region observed in the neutron scattering experiment [21] and with the plateau-like behavior in this region observed in the magnetization curve also [30].

In high magnetic fields above $\sim$ 12 T, on the other hand, the suppression of $\kappa_{\parallel c}$ can be explained by the enhancement of magnetic fluctuations scattering phonons, because the field-induced transition to the non-helical commensurate phase with $\delta = 0$ takes place at $\sim$ 16 T [20, 21]. The behavior of $\kappa_{\parallel c}$ is also roughly consistent with the neutron scattering result that the value of δ starts to decrease at $\sim$ 14 T with increasing field [20, 21].

4. Conclusions

We have measured $\kappa_{\parallel c}$ parallel to spin chains of $CsCuCl_3$, in order to investigate the presence or absence of $\kappa_{\rm spin}$ and the change of the spin state. Neither peak nor shoulder due to the contribution of $\kappa_{\rm spin}$ has been observed in the temperature dependence of $\kappa_{\parallel c}$ in zero field. This may be due to the strong scattering of magnons at boundaries of domains with homochirality of spins in the helical spin structure along the c-axis. By the application of magnetic field perpendicular to the c-axis up to 14 T, it has been found that $\kappa_{\parallel c}$ at 5.1 K below $T_{\rm N}$ decreases with increasing field up to ~ 9 T, is constant between ~ 9 T and ~ 12 T, and decreases again above ~ 12 T. These behaviors have been explained as being due to the change of κ_{phonon} caused by the change of the scattering of phonons by magnetic excitations due to the filed-induced change of the spin state.

Acknowledgments
The thermal conductivity measurements were performed at the High Field Laboratory for Superconducting Materials, Institute for Materials Research, Tohoku University. This work was supported by a Grant-in-Aid for Scientific Research from the Ministry of Education, Culture, Sports, Science and Technology, Japan.

References
[1] Hess C 2007 *Eur. Phys. J. Special Topics* **151**1 73
[2] Sologubenko A V, Lorenz T, Ott H R and Freimuth A 2007 *J. Low. Temp. Phys.* **147**3 387
[3] Miike H and Hirakawa K 1975 *J. Phys. Soc. Jpn.* **38** 1279
[4] Sologubenko A V, Giannó K, Ott H R, Vietkine A and Revcolevschi A 2001 *Phys. Rev. B* **64** 054412
[5] Sologubenko A V, Ott H R, Dhalenne G and Revcolevschi A 2003 *Europhys. Lett.* **62** 540
[6] Parfen'eva L S, Smirnov I A, Misiorek H, Mucha J, Jezowski A, Prokof'ev A V and Assmus W 2004 *Phys. Solid State* **46** 357
[7] Kawamata T, Takahashi N, Adachi T, Noji T, Kudo K, Kobayashi N and Koike Y 2008 *J. Phys. Soc. Jpn.* **77** 034607
[8] Kawamata T, Kaneko N, Uesaka M, Sato M and Koike Y 2010 *J. Phys.: Conf. Ser.* **200** 022023
[9] Hlubek N, Ribeiro P, Saint-Martin R, Revcolevschi A, Roth G, Behr G, Büchner B and Hess C 2010 *Phys. Rev. B* **81**2 020405
[10] Kawamata T, Uesaka M, Sato M, Naruse K, Kudo K, Kobayashi N and Koike Y 2014 *J. Phys. Soc. Jpn.* **83** 054601
[11] Matsuoka Y, Kawamata T, Naruse K, Ohno M, Nishiwaki Y, Kato T, Sasaki T and Koike Y 2014 *J. Phys. Soc. Jpn.* **83** 064603
[12] de Gronckel H A M, de Jonge W J M, Kopinga K and Lemmens L F 1988 *Phys. Rev. B* **37**16 9915
[13] de Gronckel H A M, de Jonge W J M, Kopinga K and Lemmens L F 1991 *Phys. Rev. B* **44**9 4654
[14] Kudo K, Koike Y, Kurogi S, Noji T, Nishizaki T and Kobayashi N 2004 *J. Magn. Magn. Mater.* **272–276** 94
[15] Choi J H, Messina T C, Yan J, Drandova G I and Markert J T 2004 *J. Magn. Magn. Mater.* **272–276** 970
[16] Cheng J G, Zhou J S and Goodenough J B 2009 *Phys. Rev. B* **79**18 184414
[17] Mekata M, Ajiro Y, Sugino T, Oohara A, Ohara K, Yasuda S, Oohara Y and Yoshizawa H 1995 *J. Magn. Magn. Mater.* **140–144** 1987
[18] Adachi K, Achiwa N and Mekata M 1980 *J. Phys. Soc. Jpn.* **49** 545
[19] Mino M, Ubukata K, Bokui T, Arai M, Tanaka H and Motokawa M 1994 *Physica B* **201** 213
[20] Nojiri H, Takahashi K, Fukuda T, Fujita M, Arai M and Motokawa M 1997 *Physica B* **241–243** 210

Journal of Physics: Conference Series **568** (2014) 042013 doi:10.1088/1742-6596/568/4/042013

[21] Stüßer N, Schotte U, Hoser A, Meschke M, Meißner M and Wosnitza J 2002 *J. Phys.: Condens. Matter* **14** 5161

[22] Kudo K, Noji T, Koike Y, Nishizaki T and Kobayashi N 2001 *J. Phys. Soc. Jpn.* **70** 1448

[23] Kudo K, Yamazaki M, Kawamata T, Noji T, Koike Y, Nishizaki T, Kobayashi N and Tanaka H 2004 *J. Magn. Magn. Mater.* **272–276** 214

[24] Sato M, Kawamata T, Sugawara N, Kaneko N, Uesaka M, Kudo K, Kobayashi N and Koike Y 2010 *J. Phys.: Conf. Ser.* **200** 022054

[25] Li Q J, Zhao Z Y, Fan C, Zhang F B, Zhou H D, Zhao X and Sun X F 2013 *Phys. Rev. B* **87**21 214408

[26] Zhang F B, Li Q J, Zhao Z Y, Fan C, Li S J, Liu X G, Zhao X and Sun X F 2014 *Phys. Rev. B* **89**9 094403

[27] Slack G A and Newman R 1958 *Phys. Rev. Lett.* **1** 359

[28] Lewis F B and Saunders N H 1973 *J. Phys. C: Solid State Phys.* **6** 2525

[29] Kousaka Y, Koyama T, Miyagawa M, Tanaka K, Akimitsu J and Inoue K 2014 *J. Phys.: Conf. Ser.* **502** 012019

[30] Nojiri, H, Tokunaga, Y and Motokawa, M 1988 *J. Phys. Colloques* **49** C8–1459

Journal of Physics: Conference Series **568** (2014) 042014 doi:10.1088/1742-6596/568/4/042014

Thermal Conductivity in the Triangular-Lattice Antiferromagnet $Ba_3CoSb_2O_9$

K Naruse[1], T Kawamata[1], M Ohno[1], Y Matsuoka[1], H Sudo[1], H Nagasawa[1], Y Hagiya[1], T Sasaki[2] and Y Koike[1]

[1] Department of Applied Physics, Tohoku University, Sendai 980-8579, Japan
[2] Institute for Materials Research, Tohoku University, Sendai 980-8577, Japan

E-mail: tkawamata@teion.apph.tohoku.ac.jp

Abstract. We have measured the thermal conductivity in the *ab*-plane, κ_{ab}, and along the *c*-axis, κ_c, of single crystals of the $S = 1/2$ triangular-lattice antiferromagnet $Ba_3CoSb_2O_9$ in zero field and magnetic fields. In zero field, it has been found that both κ_{ab} and κ_c show a similar broad peak around 40 K, suggesting that the thermal conductivity due to phonons is dominant in this compound and that the mean free path of phonons is suppressed so much by strong magnetic-fluctuations due to the magnetic frustration. It has also been found that both κ_{ab} and κ_c show a kink at the antiferromagnetic transition temperature T_N. In magnetic fields parallel to the *ab*-plane up to 14 T, magnetic-field dependences of both κ_{ab} and κ_c at temperatures below and above T_N have been found to be understood taking into account the phonon-magnon scattering in the so-called 120° structure phase and the up-up-down (UUD) phase, while the minima of κ_{ab} and κ_c observed in the middle of the UUD phase have not been understood. This suggests the occurrence of some change of the magnetic state in the middle of the UUD phase.

1. Introduction

Thermal conductivity is a useful probe to investigate the change of a magnetic state and phase transitions, because the change and phase transitions affect the thermal conductivity due to phonons, κ_{phonon}, via the phonon-magnon scattering. For example, the thermal conductivity shows various changes by the application of magnetic field in several spin-gap systems such as $SrCu_2(BO_3)_2$ [1, 2], $TlCuCl_3$ [3, 4] and $Ba_3Mn_2O_8$ [5], because the number of magnetic excitations scattering phonons is determined by the magnitude of the spin gap depending on magnetic field, leading to the change of κ_{phonon}. Recently, furthermore, interesting behaviors of thermal conductivity have been reported in various geometrically frustrated pyrochlore antiferromagnets $R_2Ti_2O_7$ (R : rare-earth elements) [6, 7]. In $Tb_2Ti_2O_7$, extremely low values of κ_{phonon} have been observed in the spin-liquid state due to the strong magnetic-fluctuations [6]. In $R_2Ti_2O_7$ (R : Tb, Gd and Er), a complicated magnetic-field dependence of the thermal conductivity has been observed in correspondence to the change of the magnetic state [6, 7].

The compound $Ba_3CoSb_2O_9$ is a spin system where Co^{2+} spins with the spin quantum number $S = 1/2$ form an uniform triangular-lattice in the *ab*-plane and the layer of the triangular-lattice is repeatedly stacked along the *c*-axis. From ESR and magnetization measurements, the exchange interaction between the nearest neighboring spins in the *ab*-plane, J, and that between the triangular-lattice layers, J', have been estimated to be 18.5 K and 0.48 K, respectively [8]. An

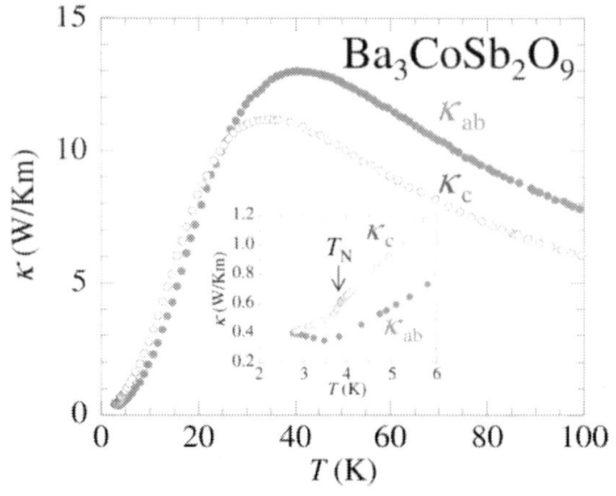

Figure 1. Temperature dependence of the thermal conductivity in the *ab*-plane, κ_{ab}, and along the *c*-axis, κ_c, of $Ba_3CoSb_2O_9$ single crystals in zero field. The inset shows the magnified plot around the antiferromagnetic transition temperature T_N = 3.8 K indicated by the arrow.

antiferromagnetic (AF) ordered state with the so-called 120° structure in the *ab*-plane appears at low temperatures below the AF transition temperature, T_N, ~ 3.8 K due to the weak J' in zero field [9]. In high magnetic fields above 9 T parallel to the *ab*-plane, the so-called up-up-down (UUD) structure, where two thirds of spins are parallel and the others are antiparallel to the magnetic-field direction, appears at low temperatures below ~5 K [8]. In the UUD state, the magnetization shows a 1/3 magnetization plateau in a field-range of 9–15 T at 1.3 K.

In this paper, we have measured the thermal conductivity in the *ab*-plane, κ_{ab}, and along the *c*-axis, κ_c, of $Ba_3CoSb_2O_9$ single crystals in zero field and magnetic fields parallel to the *ab*-plane, in order to investigate the magnetic state and the change of the magnetic state by the application of magnetic field.

2. Experimental

Single crystals of $Ba_3CoSb_2O_9$ were grown by the floating-zone method. The quality of the single crystals was checked by the x-ray back-Laue photography to be good. Thermal conductivity measurements were carried out by the conventional steady-state method. Magnetic fields up to 14 T were applied.

3. Results and discussion

Figure 1 shows the temperature dependence of κ_{ab} and κ_c of $Ba_3CoSb_2O_9$ single crystals in zero field. It is found that both κ_{ab} and κ_c show a broad peak around 40 K and then decrease with decreasing temperature. Since the magnitude of the peak and the temperature dependence of κ_{ab} are almost the same as those of κ_c, respectively, it is suggested that the contribution of phonons is dominant in the thermal conductivity. It is well known that the typical temperature dependence of κ_{phonon} in nonmagnetic insulators shows a peak around 10 K. Therefore, the appearance of the broad peak around 40 K in $Ba_3CoSb_2O_9$ may be due to strong magnetic-fluctuations, which arise from the frustration of Co^{2+} spins forming triangular-lattices and decrease the mean free path of phonons, l_{phonon}.

As shown in the inset of Fig. 1, κ_{ab} shows a kink and κ_c drops at T_N. At low temperatures below T_N, κ_{ab} increases and κ_c slowly decreases with decreasing temperature. Since κ_{phonon} is expressed as the product of the specific heat of phonons, C_{phonon}, the velocity of phonons, v_{phonon}, and l_{phonon} and v_{phonon} is little dependent on temperature, the temperature dependence of κ_{phonon} is determined by the temperature dependence of C_{phonon} and l_{phonon}. Therefore, the changes of κ_{ab} and κ_c below T_N are understood to be due to the increase of l_{phonon} owing to

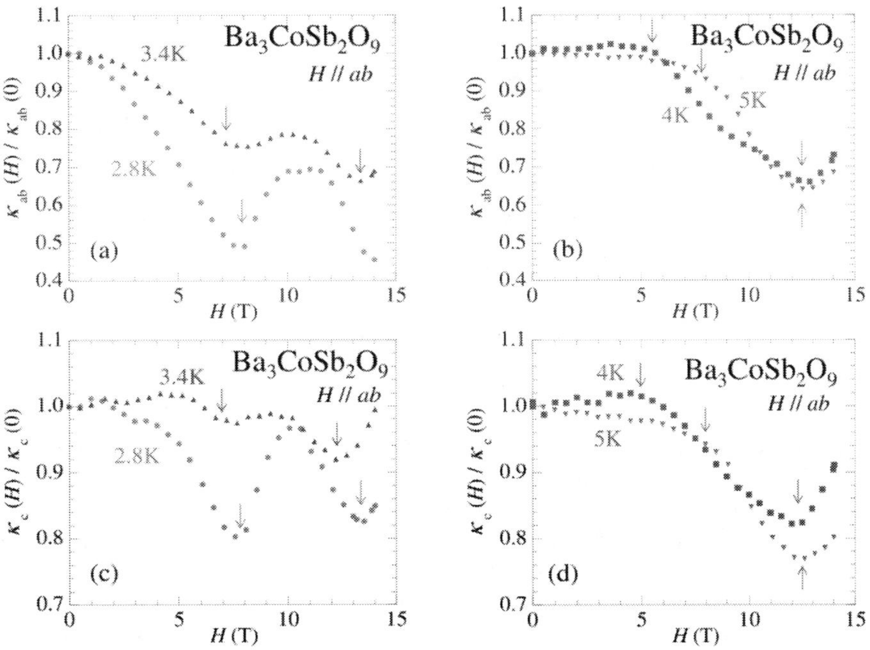

Figure 2. Magnetic-field dependence of the thermal conductivity normalized by the values in zero field for $Ba_3CoSb_2O_9$ single crystals. (a) and (b) show $\kappa_{ab}(H)/\kappa_{ab}(0)$ at low and high temperatures, respectively. (c) and (d) show $\kappa_c(H)/\kappa_c(0)$ at low and high temperatures, respectively. Magnetic fields are applied parallel to the *ab*-plane. Arrows indicate fields where the thermal conductivity drastically changes.

the suppression of the phonon-magnon scattering caused by the suppression of the magnetic fluctuations in the AF ordered state.

Figure 2 shows the magnetic-field dependence of $\kappa_{ab}(H)$ and $\kappa_c(H)$, normalized by the values in zero field, $\kappa_{ab}(0)$ and $\kappa_c(0)$, respectively, of $Ba_3CoSb_2O_9$ single crystals in magnetic fields parallel to the *ab*-plane at various temperatures. At low temperatures below T_N, it is found that both κ_{ab} and κ_c decrease with increasing field in low fields, where the 120° structure is formed, as shown in Figs. 2 (a) and (c). The decrease is interpreted as being due to the enhancement of magnetic fluctuations scattering phonons with increasing field, because magnetic fields applied parallel to the *ab*-plane disturb the ordered state with the 120° structure. With further increasing field, it is found that both κ_{ab} and κ_c increase above ~8 T indicated by arrows in Figs. 2 (a) and (c). The fields ~8 T indicated by the arrows are in good agreement with those where the transition from the 120° structure phase to the UUD phase occurs, as shown in the H–T magnetic phase diagram of Fig. 3 obtained from the magnetic susceptibility and specific heat measurements [10, 11]. The increase above ~8 T is interpreted as being due to the decrease of the number of magnetic excitations, because the UUD phase has a spin gap [12, 13]. In high magnetic fields above 8 T, it is found that both κ_{ab} and κ_c show a peak around 11 T and a dip around 13 T. The dip fields ~13 T indicated by arrows in Figs. 2 (a) and (c) are plotted by stars in Fig. 3. It is found that the dip fields are located in the middle of the UUD phase. However, the magnitude of the spin gap in the UUD phase is expected to disappear at the boundary of the UUD phase and show the maximum in the middle of the UUD phase. Therefore, the peaks around 11 T and the dips around 13 T cannot be interpreted in terms of the change of

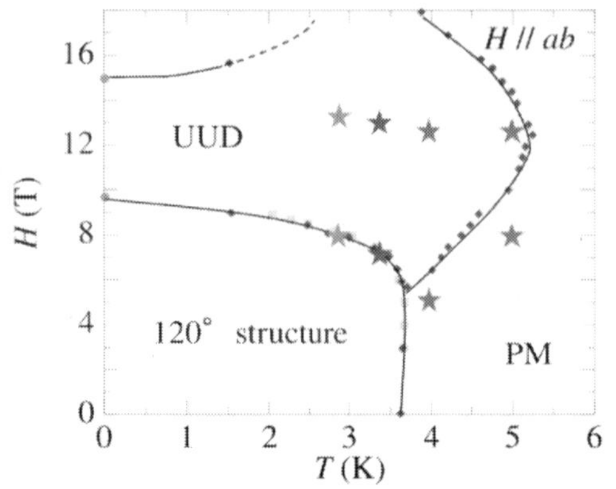

Figure 3. H–T magnetic phase diagram of $Ba_3CoSb_2O_9$ in magnetic fields parallel to the ab-plane. Data are from refs. [10, 11]. Stars correspond to arrows in Fig. 2.

the magnitude of the spin gap with changing field. These results suggest that some magnetic change takes place in the middle of the UUD phase showing the 1/3 magnetization plateau.

At high temperatures above T_N, on the other hand, it is found that both κ_{ab} and κ_c are almost constant in low magnetic fields below ~5 T, as shown in Fig. 2 (b) and (d). This means that the number of magnetic excitations does not change with increasing field in the paramagnetic (PM) state. It is found that both κ_{ab} and κ_c start to decease at ~8 T with increasing field. The fields ~8 T indicated by arrows in Figs. 2 (b) and (d) are also plotted by stars in Fig. 3. It is found that the fields are a little smaller than the boundary between the PM and UUD phases. The decrease above ~8 T implies that the increase of the number of magnetic excitations, but this cannot be explained by the increase of the spin gap in the UUD phase. The decrease above ~8 T may be due to the enhancement of magnetic fluctuations, which has been observed around the phase boundary in the specific heat measurements [10]. That is, the strong magnetic-fluctuations around the phase boundary may scatter phonons markedly, leading to the decrease in thermal conductivity. This may be the reason why the fields ~8 T indicated by arrows are a little smaller than the phase boundary. In high magnetic fields, it is found that both κ_{ab} and κ_c show a dip around 13 T. The dip fields ~13 T indicated by arrows in Figs. 2 (b) and (d) are also plotted by stars in Fig. 3. It is found that the dip fields are located in the middle of the UUD phase as well as those at low temperatures. As mentioned above, therefore, it is very likely that the magnetic state changes in the middle of the UUD phase. However, no anomaly has been observed in the field dependence of the specific heat around 13 T [10]. Accordingly, it is necessary to investigate the magnetic state using another probe such as NMR and neutron scattering measurements.

4. Summary

We have measured κ_{ab} and κ_c of $Ba_3CoSb_2O_9$ single crystals in zero field and magnetic fields. In zero field, it has been found that both κ_{ab} and κ_c show a similar broad peak around 40 K, suggesting that κ_{phonon} is dominant and that l_{phonon} is suppressed so much by strong magnetic-fluctuations due to the frustration of Co^{2+} spins. It has also been found that both κ_{ab} and κ_c show a kink at T_N, owing to the increase of l_{phonon} in the AF ordered state. By the application of magnetic field parallel to the ab-plane in the 120° structure phase at low temperatures, it has been found that both κ_{ab} and κ_c decrease with increasing field due to the enhancement of magnetic fluctuations arising from the instability of the 120° structure. With further increasing field, it has been found that both κ_{ab} and κ_c increase above ~8 T, where the transition from the 120° structure phase to the UUD phase occurs, owing to the decrease of the number of magnetic excitations in the UUD phase with a spin gap. In high magnetic fields above 8 T in the UUD phase, it has been found that both κ_{ab} and κ_c show a peak around 11 T and a dip around 13 T, which cannot be interpreted in terms of the change of the magnitude of the spin gap. Even at high temperatures above T_N, it has also been found that both κ_{ab} and κ_c show a dip around 13 T in the UUD phase. These results suggest that some magnetic change takes place in the middle of the UUD phase showing the 1/3 magnetization plateau.

Acknowledgments

The thermal conductivity measurements were performed at High Field Laboratory for Superconducting Materials, Institute for Materials Research, Tohoku University. This work was supported by a Grant-in-Aid for Scientific Research from the Ministry of Education, Culture, Sports, Science and Technology, Japan.

References

[1] Kudo K, Noji T, Koike Y, Nishizaki T and Kobayashi N 2001 *J. Phys. Soc. Jpn.* **70** 1448
[2] Hofmann M, Lorenz T, Uhrig G S, Kierspel H, Zabara O, Freimuth A, Kageyama H and Ueda Y 2001 *Phys. Rev. Lett.* **87** 047202
[3] Kudo K, Yamazaki M, Kawamata T, Noji T, Koike Y, Nishizaki T, Kobayashi N and Tanaka H 2004 *J. Phys. Soc. Jpn.* **73** 2358
[4] Kudo K, Yamazaki M, Kawamata T, Noji T, Koike Y, Nishizaki T, Kobayashi N and Tanaka H 2004 *J. Magn. Magn. Mater.* **272–276** 214
[5] Ke W P, Wang X M, Fan C, Zhao Z Y, Liu X G, Chen L M, Li Q J, Zhao X and Sun X F 2011 *Phys. Rev.* B **84** 094440
[6] Li Q J, Zhao Z Y, Fan C, Zhang F B, Zhou H D, Zhao X and Sun X F 2013 *Phys. Rev.* B **87** 214408
[7] Zhang F B, Li Q J, Zhao Z Y, Fan C, Li S J, Liu X G, Zhao X and Sun X F 2014 *Phys. Rev.* B **89** 094403
[8] Susuki T, Kurita N, Tanaka T, Nojiri H, Matsuo A, Kindo K and Tanaka H 2013 *Phys. Rev. Lett.* **110** 267201
[9] Shirata Y, Tanaka H, Matsuo A and Kindo K 2012 *Phys. Rev. Lett.* **108** 057205
[10] Zhou H D, Xu C, Hallas A M, Silverstein H J, Wiebe C R, Umegaki I, Yan J Q, Murphy T P, Park J-H, Qiu Y, Copley J R D, Gardner J S and Takano Y 2012 *Phys. Rev. Lett.* **109** 267206
[11] Koutroulakis G, Zhou T, Batista C D, Kamiya Y, Thompson J D, Brown S E and Zhou H D 2013 *Preprint* cond-mat/1308.6331
[12] Tsujii H, Rotundu C R, Ono T, Tanaka H, Andraka B, Ingersent K and Takano Y 2007 *Phys. Rev.* B **76** 060406R
[13] Fujii Y, Hashimoto H, Yasuda Y, Kikuchi H, Chiba M, Matsubara S and Takigawa M 2007 *J. Phys.: Condens. Matter* **19** 145237

High Pressure Properties for Electrical Resistivity and Ce Valence State of Heavy-Fermion Antiferromagnet Ce_2NiGa_{12}

N Kawamura[1], **R Sasaki**[2], **K Matsubayashi**[2], **N Ishimatsu**[3], **M Mizumaki**[1], **Y Uwatoko**[2], **S Ohara**[4], **and S Watanabe**[5]

[1] Japan Synchrotron Radiation Research Institute, 1-1-1 Kouto, Sayo, Hyogo 679-5198, Japan
[2] Institute for Solid State Physics, The University of Tokyo, Kashiwa, Chiba 277-8581, Japan
[3] Department of Physics, Graduate School of Science, Hiroshima University, 1-3-1 Kagamiyama, Higashi-Hiroshima, Hiroshima 739-8526, Japan
[4] Department of Engineering Physics, Electronics and Mechanics, Graduate School of Engineering, Nagoya Institute of Technology, Gokiso, Showa, Nagoya, Aichi 466-8555, Japan
[5] Quantum Physics Section, Kyushu Institute of Technology, 1-1 Sensui, Tobata, Kitakyushu, Fukuoka 804-8550, Japan

E-mail: naochan@spring8.or.jp

Abstract. A heavy-fermion antiferromagnet Ce_2NiGa_{12} has been studied at a viewpoint of pressure effects. Electrical resistivity measurements show that the antiferromagnetic ordering is suppressed at $P_C \sim 5.5$ GPa, and that Fermi liquid behaviour is observed above P_C. To investigate how Ce valence state is associated with the observed critical phenomena, X-ray absorption spectra at Ce L_3-edge have been performed. Ce valence state drastically varies at $P_v \sim 9$ GPa in the initial applying pressure process, whereas anomalous behaviour is not observed in the Ce valence under pressures around P_C. Consequently, the quantum critical phenomena in Ce_2NiGa_{12} around P_C is not strongly connected with Ce valence state.

1. Introduction

Quantum criticality phenomena have attracted much attention for exotic physical properties such as spin and orbital ordering, unconventional superconductivity, and magnetic phase transitions. Such phenomena have been observed in many Ce intermetallic compounds [1]. In particular, the unconventional superconductivity often appears near a quantum critical point (QCP), at which the magnetic order disappears. Moreover, the QCP can be controlled by external fields such as pressure and magnetic field since a competition between RKKY interaction and Kondo effect has an influence on the QCP. Recently, several theoretical approaches for the unconventional superconductivity have been proposed [2–4]. Among them, importance of valence transition near the QCP is predicted [3], and a number of experimental efforts for valence transition have been performed in Ce intermetallic compounds [5, 6]. However, an association of the QCP with the valence transition still remains as an open question.

Pressure as external fields provides large energy to the electron system compared with magnetic field because pressure directly affects band structure through the hybridization of electronic states. In fact, pressure-induced heavy-fermion (HF) superconductors have been reported for the Ce-based intermetallic compounds such as $CeCu_2Si_2$ and $CeCoIn_5$ whose crystallized symmetry is tetragonal [7, 8]. $Ce_2M Ga_{12}$ (M=Ni, Pd, and Pt) also crystallizes in the tetragonal symmetry (space group $P4/nbm$), and the Ce atom is aligned in a two-dimensional square lattice site perpendicular to the c-axis [9].

Moreover, these compounds have been reported as HF antiferromagnets with Néel temperature $T_N = 10$, 11, and 7 K the ground state at ambient pressure (AP) for M = Ni, Pd, and Pt, respectively. Therefore, it is important to investigate the properties of isostructural Ce compounds for understanding the quantum criticality behavior from a view point of Ce valence state.

To capture the valence transition, X-ray absorption spectroscopy (XAS) at Ce L_3-absorption edge is suitable for observation of pressure effect because chemical shift occurs in the spectrum due to valence transition [10]. To study the quantum criticality of Ce_2NiGa_{12} and its Ce valence state, the electrical resistivity and XAS measurements were performed under high pressure and low temperature.

2. Experimental

Single crystals of Ce_2NiGa_{12} is synthesized by a gallium self-flux method [11]. Crystal structure with tetragonal *P4/nbm* is confirmed by the X-ray powder diffraction using Cu $K\alpha$ line. Estimated lattice parameters, $a = 0.6108$ nm and $c = 1.5506$ nm, are slightly larger than those of previous report [9]. The electrical resistivity measurements at high pressure were performed by Palm Cubic Anvil Cell. Gasket of MgO and pressure medium of glycerin were used. Pressure is applied up to 8.0 GPa in a temperature range of 0.45 ~ 300 K. The residual resistivity ratio was evaluated to be ~14 at AP.

Diamond-Anvil-Cell (DAC) is often used as a high-pressure cell for XAS measurements. A compact DAC with ϕ23.5 × 17t mm made of CuBe alloy was used to achieve lower temperature below 10 K. This compact DAC can be easily inserted to the superconducting magnet or cryostat. Pressure variation between 300 K and 4 K depends on the culet size of anvils. The variation of 30~40% and 5~10% was observed by using anvils with culet diameter of 300 and 600 μm, respectively.

Nano-Polycrystalline Diamond (NPD) anvils with culet diameter of 300 μm and with height of 1.0 mm were used to avoid glitches in XAS spectra [12, 13]. A tapered hole with depth of 0.7 mm is fabricated to one anvil in pressure direction to increase X-ray intensity. X-ray transmissivity is improved 22 times higher by the fabrication for X-ray energy of 5.72 keV. Within present work, a fabricated NPD anvil withstands pressure of 15 GPa at 3.4 K.

XAS measurements at high pressure were performed at BL39XU of SPring-8 facility [14]. All measurements were carried out in transmission mode using ionization chambers. X-ray beam of 2 (vertical) × 9 (horizontal) μm^2 is focused on the sample position by Kirkpatrick and Baez mirror [15]. The compact DAC was mounted on the pulse-tube type cryostat and temperature cools down to 3.4 K. A small chip of single crystal was put into the sample room of ϕ100 μm, and mixed liquid of methanol:ethanol = 4:1 was used as a pressure-transmitting medium. Pressure at both low temperature and room temperature inside the DAC was estimated by ruby-fluorescence method [16].

3. Results and Discussion

Figure 1 shows a magnetic component of electrical resistivity (ρ_{mag}) dependent on temperature in Ce_2NiGa_{12} at selected pressures, where ρ_{mag} is obtained by subtracting the resistivity of the isostructural non-magnetic La_2NiGa_{12} compound [9]. ρ_{mag} at AP shows a typical HF feature of Ce compounds, i.e. a cusp peak due to an antiferromagnetic (AFM) ordering at $T_N = 10$ K and a broad maximum due to Kondo scattering at $T^H_{max} = 60$ K are observed in ρ_{mag} curve. As increasing pressure, the overall ρ_{mag} increases in magnitude whereas the peak position at T^H_{max} decreases. The T_N decreases with increasing pressure, and another anomaly appears at $T_{N2} < T_N$ above 2.7 GPa. With further increasing pressure, both T_N and T_{N2} are suppressed and no longer visible above 5.8 GPa, indicating the disappearance of AFM ordering at $P_C \sim 5.5$ GPa. At low temperatures near P_C, there appears an additional broad peak at T^L_{max}, which increases with pressure above 5.3 GPa.

The behavior of ρ_{mag} at low temperature region is analyzed according to Fermi-liquid (FL) character, $\rho_{mag} = \rho_0 + AT^2$, where ρ_0 represents a residual resistance for Ce_2NiGa_{12}. The pressure dependence of ρ_0 and A values are shown in figure 2(a). Both ρ_0 and A are enhanced with approaching P_C. In addition, $A \propto (T^L_{max})^{-2}$ relation is also satisfied above 5.8 GPa as shown in figure 2(b). The relationship indicates that FL state realizes in Ce_2NiGa_{12} above P_C. Therefore, T^L_{max} is associate with the Kondo temperature T_K, which increases with applying pressure as observed in Ce-based HF compounds [17]. Figure 3 shows the pressure-temperature (*P*-*T*) phase diagram of Ce_2NiGa_{12} drawn by the characteristic temperatures. The *P*-*T* phase diagram clearly shows that the FL phase appears around P_C after disappearance of AFM ordering with applying pressure. In addition, the ρ_0 value is enhanced inside the AFM phase (see figure 2(a)). Similar results have been observed in $CeAl_2$ and $CeCu_5Au$ [18, 19]. It is suggested that the ρ_0 enhancement is attributed to the energy scale of T_K comparable to T_N [4]. Another possibility is that quantum criticality of valence fluctuation induces instability of ρ_0 [3]. In particular, the latter possibility can be explored since the estimation of Ce valence state under high pressure is feasible by XAS measurements.

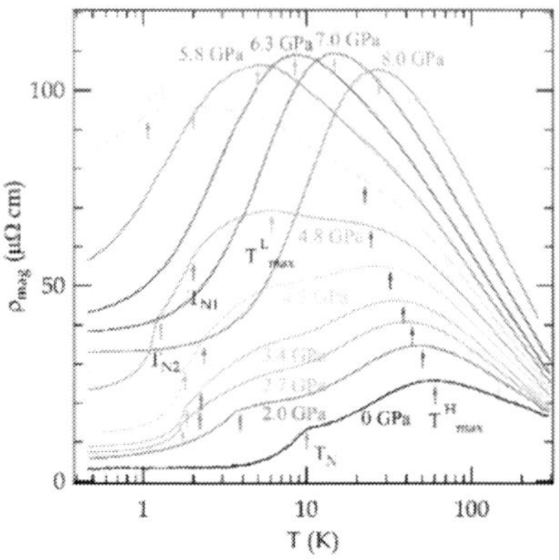

Figure 1. A magnetic component of electrical resistivity ρ_{mag} as a function of temperature in Ce_2NiGa_{12} at various pressures.

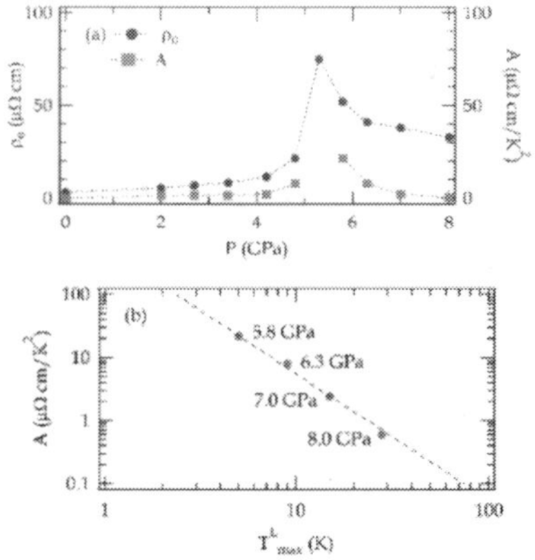

Figure 2. (a) ρ_0 and *A* coefficient dependent on pressure (see text). (b) *A* coefficient as a function of T^L_{max}. A broken line denotes $A \propto (T^L{}_{max})^{-2}$.

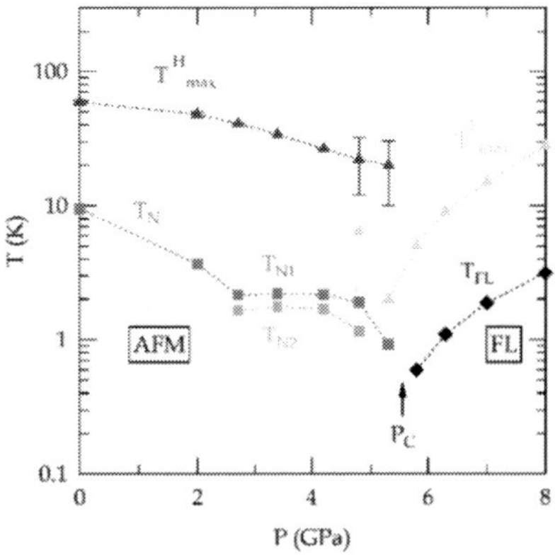

Figure 3. *P*-*T* phase diagram in Ce_2NiGa_{12} deduced by the electrical resistivity.

Figure 4(a) shows several XAS spectra at the Ce L_3-edge dependent on pressure at room temperature. The spectral profile at AP indicates Ce trivalent state, reflecting localized *f* electron feature. As increasing pressure up to 9 GPa, the profile moderately changes with pressure, i.e. a main-peak (called white-line) intensity at 5726.0 eV decreases and the intensity of a shoulder structure at 5735.7 eV increases. This variation is well known as valence variation [10]. However, no anomaly of Ce valence variation is observed around P_C. Therefore, the quantum critical phenomena in Ce_2NiGa_{12} observed by electrical resistivity may not be strongly connected with Ce valence state. On the other hand, a drastic change in XAS profile, in which the shoulder structure at 5735.7 eV becomes clear, is observed above 10 GPa. The phenomenon obviously indicates Ce valence transition.

To elucidate pressure-dependent Ce valence, XAS profiles were analyzed by the profile fitting method as shown in figure 4(b). The averaged Ce valence is obtained by the rate of Ce^{4+} content to Ce^{3+} one. The result estimated by the fitting is shown in figure 4(c). Ce valence is almost trivalent at AP. The valence moderately increases with pressure up to $P_v \sim 9$ GPa, and is suddenly enhanced above P_v. The change rate is comparable to the case in Ce γ-α transition [20]. Here, it is necessary to distinguish the spectral change due to valence variation from the effect of lattice compression according to the Natoli's rule, $k \cdot r$ = constant, where k and r represent X-ray wave number ($\propto E^{-1/2}$) and interatomic distance, respectively [21]. Natoli's rule gives energy shifts for XAS profile so that moderate variation of Ce valence below P_v seems to be the effect of Natoli's rule.

Ce valence state was investigated for pressure-decreasing process. As shown in figure 4(c), Ce valence returns to the previous state for pressure-decreasing process. However, Ce valence for reapplying process is different from that for initial applying-pressure process. On the other hand, the result in another sample is the same as the pressure variation for both increasing and decreasing processes. The difference is attributed to the strain of sample for powderization of single crystal. The situation is often observed in intermetallic compounds such as $CeCu_{6-x}Au_x$ [22]. In fact, different pressure dependence is observed in Ce valence state when pressure decreases after applying above P_v. The crystal structure of Ce_2NiGa_{12} might change above P_v. Note that the sample for electrical resistivity measurements is corresponding to the initial applying process because of no strain for shape forming. Further investigation such as X-ray powder diffraction and/or electrical resistivity measurements are necessary to discuss Ce valence state in higher pressure region above P_v.

XAS measurements at low temperature were also performed. We succeeded in measuring at 3.4 K up to 15 GPa. Unfortunately, the measurements were performed for only reapplying process. Within current results, no temperature dependence of Ce valence was observed. In addition, pressure dependence at low temperature is as same as the result at room temperature.

For Ce_2NiGa_{12}, absence of obvious valence transition near P_C provides no correlation between the quantum critical phenomena and Ce valence state. The situation might be similar to the case of $CeCu_2Si_2$, in which the different characteristic pressures P_C and P_v are located on two superconducting domes in P-T phase diagram [23]. Moreover, it seems that P_v agrees the pressure which T^{H}_{max} merges into T^{L}_{max}. To ensure the appearance of P_v, low temperature measurements are required for the initial applying-pressure process. Additionally, structural information and electrical resistivity at higher pressure are desired since the drastic variation of Ce valence above P_v is curious.

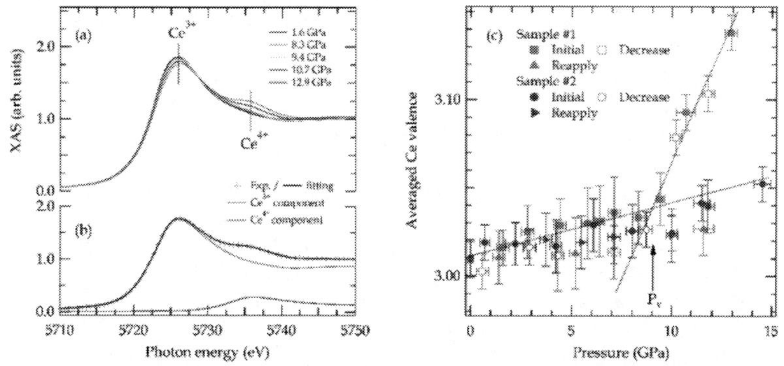

Figure 4. (a) Ce L_3-XAS spectra in Ce_2NiGa_{12} at 300 K at selected pressures. (b) Fitting result of the XAS spectrum at 12.9 GPa for Ce valence estimation. (c) Averaged Ce valence dependent (300 K) on pressure estimated by analysing Ce L_3-XAS spectra.

4. Conclusion

The electrical resistivity and XAS measurements at Ce L_3-edge in Ce_2NiGa_{12} were performed under high pressure. Quantum criticality phenomena were observed at $P_C \sim 5.5$ GPa in the electrical resistivity. To investigate a relationship between Ce valence state and the quantum criticality, XAS spectra up to 15 GPa were analyzed. Ce valence is drastically enhanced above $P_v \sim 9$ GPa only in the initial applying pressure process. The result indicates that Ce valence variation is not directly associated with the quantum criticality in Ce_2NiGa_{12} around P_C. Detailed study of electrical resistivity and crystal structure determination above P_v is necessary to discuss about the Ce valence variation.

Acknowledgements

We thanks to Drs. H. Sumiya and T. Irifune for providing the NPD anvils. This study was performed with the approval of the Japan Synchrotron Radiation Research Institute (JASRI) (Proposal Nos. 2012A1843, 2012B1976, 2013B0046, 2013B1922 and 2014A0046). This work is partially supported by a Grants-in-Aid for Scientific Research (C) (No. 24540389) from the Ministry of Education, Culture, Sports, Science, and Technology (MEXT), Japan.

References

[1] Doniach S 1977 *Valence Instability and Related Narrow Band Phenomena* ed R D Parks (New York: Plenum)
[2] Misawa T, Yamaji Y and Imada M 2009 *J. Phys. Soc. Jpn.* **78** 084707
[3] Watanabe S and Miyake K 2010 *Phys. Rev. Lett.* **105** 186403
[4] Hattori K and Miyake K 2010 *J. Phys. Soc. Jpn.* **79** 073702
[5] Rueff J –P, Raymond S, Taguchi M, Sikora M, Itié J –P, Baudelet F, Braithwaite D, Knebel G and Jaccard D 2010 *Phys. Rev. Lett.* **106** 186405
[6] Yamaoka H, Tsujii N, Yamamoto K, Oohashi H, Vlaicu A M, Kunitani K, Uotani K, Horiguchi D, Tochio T, Ito Y and Shin S 2007 *Phys. Rev.* B **76** 075130
[7] Assmus W, Herrmann M, Rauchschwalbe U, Riegel S, Lieke W, Spille H, Horn S, Weber G, Steglich F and Cordier G 1984 *Phys. Rev. Lett.* **52** 469
[8] Petrovic C, Pagliuso P G, Hundley M F, Movshovich R, Sarrao J L, Thompson J D, Fisk Z and Monthoux P 2001 *J. Phys.: Condens. Matter* **13** L337
[9] Cho J Y, Millican J N, Capan C, Sokolov D A, Moldovan M, Karki A B, Young D P, Aronson M C and Chan J Y 2008 *Chem. Mater.* **20** 6116
[10] Wohlleben D and Röhler J 1984 *J. Appl. Phys.* **55** 1904
[11] Ohara S, Yamashita T, Shiraishi T, Matsubayashi K and Uwatoko Y 2012 *J. Phys.: Conf. Ser.* **400** 042048
[12] Irifune T, Kurio A, Sakamoto S, Inoue T and Sumiya H 2003 *Nature* **421** 806
[13] Ishimatsu N, Matsumoto K, Maruyama H, Kawamura N, Mizumaki M, Sumiya H and Irifune T 2012 *J. Synchrotron Rad.* **19** 768
[14] Kawamura N, Ishimatsu N and Maruyama H, 2009 *J. Synchrotron Rad.* **16** 730
[15] Yumoto H, Hirata K, Nisawa A, Ueno G, Sato M, Son J –Y, Koganezawa T, Machida M, Muro T, Hirosawa I, Suzuki M, Kawamura N, Mizumaki M, Ohashi H, Yamamoto M, Watanabe Y and Goto S 2009 *Proc. SPIE* **7448** 74480Z
[16] Mao H K, Bell P M, Shaner J W and Steinberg D J 1978 *J. Appl. Phys.* **49** 3276
[17] Pfleiderer C 2009 *Rev. Mod. Phys.* **81** 1551
[18] Barbara B, Beille J, Cheaito B, Laurant J M, Rossignol M F, Waintal A and Zemirli S 1987 *J. Phys. France* **48** 635
[19] Wilhelm H, Raymond S, Jaccard D, Stockert O, Löhneysen H v and Rosch A 2001 *J. Phys. Condens. Matter.* **13** L329
[20] Rueff J –P, Itié J –P, Taguch M, Hague C F, Mariot J –M, Delaunay R, Kappler J –P and Jaoen N 2006 *Phys. Rev. Lett.* **96** 237403
[21] Natoli C R 1983 *EXAFS and Near Edge Structure, Springer Series in Chemical Physics* ed A Bianconi, L Incoccia and S Stipcich (Berlin: Springer)
[22] Hirose Y, *Private communications*
[23] Holmes A T, Jaccard D and Miyake K 2007 *J. Phys. Soc. Jpn.* **76** 051002

27th International Conference on Low Temperature Physics (LT27) IOP Publishing
Journal of Physics: Conference Series **568** (2014) 042016 doi:10.1088/1742-6596/568/4/042016

Magnetic excitations in the ferromagnetic superconductor UGe_2 under pressure.

M W Kepa[1], D A Sokolov[1], M Böhm[2] and A D Huxley[1]

[1]SUPA, School of Physics and Astronomy, Centre for Science at Extreme Conditions, The University of Edinburgh, Mayfield Road, Edinburgh EH9 3JZ, UK

[2]Institut Laue-Langevin, BP 156, F-38042 Grenoble Cedex 9, France

E-mail: M.Kepa@sms.ed.ac.uk

Abstract. Ferromagnetism and superconductivity coexist in UGe_2 under pressures in the range of 10-16 kbar. Here, equal spin electrons are paired to give a spin triplet state but the pairing mechanism is still not verified experimentally. The work presented here attempts to verify whether the longitudinal magnetic fluctuations associated with a magnetic transition within the ferromagnetic state might be responsible for the electron pairing. We show that the energy scale of these fluctuations must be smaller than those associated with the second order phase transition at the Curie temperature. Furthermore, there is no significant change of the energy scale of the fluctuations at the Curie temperature as a function of pressure.

1. Introduction

UGe_2 is one of the most extensively studied ferromagnetic superconductors [1-7]. The magnetic properties of UGe_2 are due to the uranium $5f$ electrons. The crystal structure of the material is orthorhombic with the space group $Cmmm$ [2]. The a-axis is the easy axis with an ordered magnetic moment $M_0 \approx 1.4\,\mu_B$ at zero temperature. The Curie temperature ($T_c = 53\,\mathrm{K}$ at ambient pressure) decreases with increasing pressure and is completely supressed at the critical pressure of $p_c = 16\,\mathrm{kbar}$ as shown on Fig. 1 [7]. This paramagnetic to ferromagnetic (PM-FM) transition is 2nd order in nature at lower pressures but at the tricritical point (TCP) at $p_{TCP} = 14.1\,\mathrm{kbar}$, it changes to 1st order [6]. Within the ferromagnetic region, there are two distinct ferromagnetic phases, the high temperature high pressure FM1 phase and the low temperature low pressure FM2 phase. At low pressures, there is a cross-over region between the two phases but at the critical end point (CEP), very close to 12 kbar, a 1st order FM1-FM2 transition emerges where the ordered magnetic moment jumps from $M_0 \approx 0.9\,\mu_B$ (FM1 phase) to $M_0 \approx 1.4\,\mu_B$ (FM2 phase) [6]. Superconductivity appears only within the ferromagnetic state and the superconducting critical temperature is maximised at pressure of $p_x = 12\,\mathrm{kbar}$ with $T_{SC} \approx 0.8\,\mathrm{K}$ [2].

The superconducting pairing mechanism in UGe_2 is not yet fully understood. Theory suggested that magnetic fluctuations in the vicinity of a ferromagnetic phase transitions could induce the Cooper pairing with odd-parity resulting in spin triplet superconductivity [8]. It was also shown experimentally with the NMR spectroscopy in another uranium based ferromagnetic superconductor, UCoGe, that longitudinal spin fluctuations might be responsible for spin triplet superconductivity [9]. The work we present here attempts to verify using inelastic neutron scattering (as a direct probe) whether the longitudinal magnetic fluctuations in UGe_2 are responsible for unconventional superconductivity.

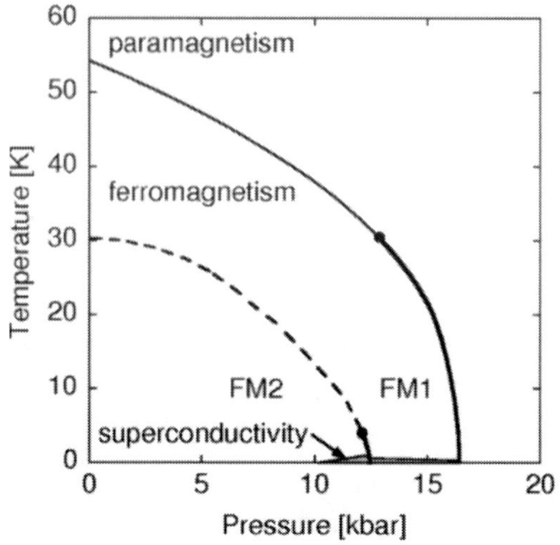

Fig. 1: The schematic space phase diagram of UGe_2 [7]. The paramagnetic to ferromagnetic phase line is second order at low pressures (thin line) and first order at higher pressures (thick line). The dashed line indicates the crossover region between FM1 and FM2 phases. Solid thick line between FM1 and FM2 is first order. The shaded region is the region of superconductivity with the maximal critical temperature of $T_{SC} \approx 0.8\,\text{K}$. Dots indciate the critical points.

2. Experiment

The sample used in the experiment was a single crystal grown by the Czochralski technique. The sample was spark cut to the dimensions of 6.1×3.4×4.0 mm. The electrical resistivity was measured at ambient pressure using the four-point method with a lock-in amplifier and is presented on Fig. 2(a). The clear drop of resistivity indicates the PM-FM transition at $T_c = 53\,\text{K}$. The value of the residual resistivity ratio (RRR) was 62.

The pressure cell used in the experiment was the large volume two-layered piston-cylinder pressure cell described in [10]. The cell combines a large sample volume with a safe operating pressure limit of 18 kbar and was optimised for inelastic neutron scattering experiments [10]. Pressure inside the cell at low temperature was determineted to be 11.8 kbar using the established phase diagram of UGe_2, Fig. 1. The measured crystal parameters at that pressure were: $b = 14.968\,\text{Å}$ and $c = 4.100\,\text{Å}$. The rocking curve, Fig. 2(b), taken through $Q = 2\pi(0,0,1)/c$ at the same pressure gave the mosaic of 1.3°. This value of Q was chosen since this Bragg peak is almost entirely magnetic due to very small nuclear contribution.

To obtain elastic and inelastic neutron scattering data we used the triple axis spectrometer IN14 at the Institut Laue-Langevin (ILL). For measurements at low temperatures, the cell was loaded in the ILL orange cryostat with a base temperature of 1.5 K. The inelastic data points were taken on $Q = 2\pi(0,0,1.04)/c$ at different temperatures at two energy transfers of 0.15 meV and 0.25 meV. A cool beryllium filter was placed on the incident momentum side. We used a neutron beam with fixed final momentum $k_f = 1.3\,\text{Å}^{-1}$. The energy resolution was determined from the incoherent scattering with $Q = 2\pi(0,0.4,1.04)/c$ and had a full width at half maximum (FWHM) of 0.10 meV. We used W-configuration as a setup geometry and a PG (002) analyser with horizontal and vertical focusing. The temperature was measured using a Cernox thermometer which was mounted at the end of the probe very close to the pressure cell. The neutron scattering intensity of the elastic scan taken on $Q = 2\pi(0, 0, 1)/c$ is proportional to the square of magnetisation and clearly showed the two transitions: first, PM-FM with $T_C = 33.5 \pm 0.5\,\mathrm{K}$; and second, FM1-FM2 with $T_x = 8.5 \pm 0.5\,\mathrm{K}$, Fig. 3. An increase of inelastic scattering intensity is observed at $T_C = 33.5 \pm 0.5\,\mathrm{K}$ at energy transfer of 0.15 meV. This increase is due fluctuations associated with the second order PM-FM transition. The intensity of inelastic scattering (relative to background) decreases at higher energy transfer, Fig. 3. The data do not show any significant increase of scattering intensity at $T_x = 8.5 \pm 0.5\,\mathrm{K}$.

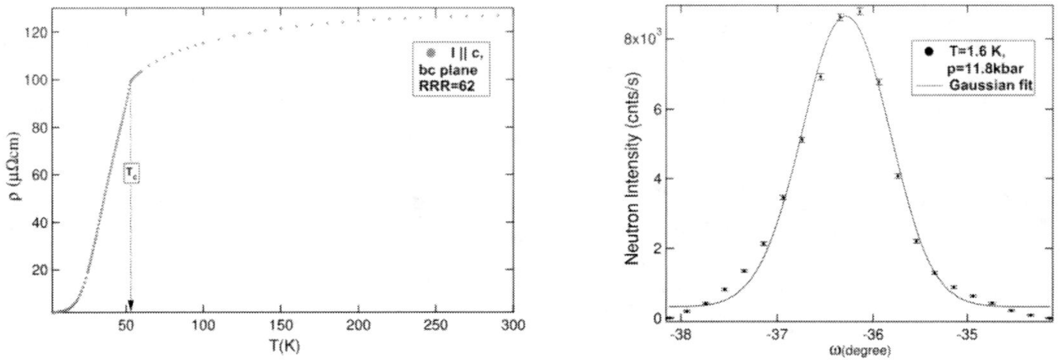

(a) **(b)**

Fig. 2: **(a)** Resistivity of UGe_2 at ambient pressure with current along c-axis within bc-plane. The PM-FM transition is observed at $T_c = 53\,\mathrm{K}$. **(b)** The rocking curve taken at $p = 11.8\,\mathrm{kbar}$ on $Q = 2\pi(0, 0, 1)/c$ at $T = 1.6\,\mathrm{K}$ (background subtracted). Solid line is the Gaussian fit to the data. The measured mosaic was 1.3° at FWHM.

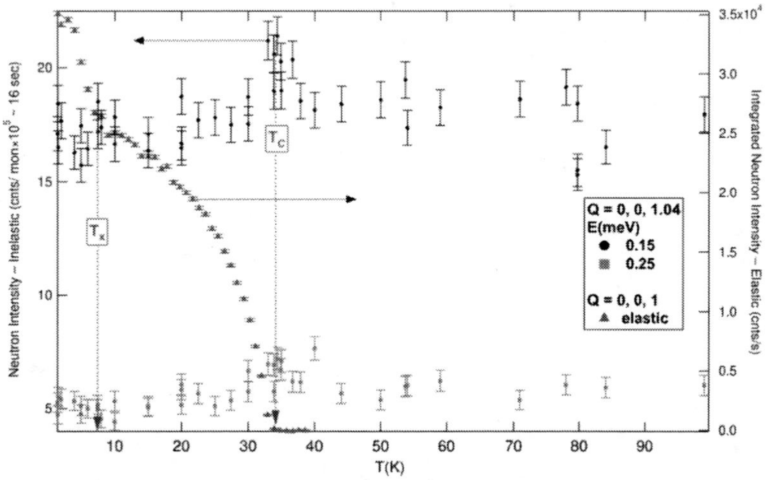

Fig. 3: Right axis shows elastic scattering points (▲) taken at $Q = 2\pi(0, 0, 1)/c$ and at $p = 11.8\,\mathrm{kbar}$. Two phase are clearly seen: PM-FM with $T_C = 33.5 \pm 0.5\,\mathrm{K}$; and FM1-FM2 with $T_x = 8.5 \pm 0.5\,\mathrm{K}$. Left axis shows inelastic scattering points at $Q = 2\pi(0, 0, 1.04)/c$ and at $p = 11.8\,\mathrm{kbar}$ with two different energy transfers of 0.15 meV (●) and 0.25 meV (■).

3. Conclusion and future outlooks

The work presented here does not reveal ferromagnetic spin fluctuations at T_x associated with the FM1-FM2 transition. Instead, we observed fluctuations at T_C associated with the PM-FM transition. Furthermore, the energy scale of the fluctuations at T_C at high pressure is similar to the energy scale at ambient pressure where there is no superconductivity [7]. This might suggest that the fluctuations at T_C are unlikely to play a significant role in inducing superconductivity. On the other hand, the previous results [7, 11-12] show that the $5f$ electron still contribute to most of the differential susceptibility and are responsible for most of bulk magnetic properties at the temperature and pressure conditions similar to those presented in our work. However, it was also suggested by the muon spin spectroscopy measurements [12] that the FM1-FM2 transition might be due to conduction electrons which would not be detected in our experiment. The inelastic scattering from the conduction electrons is strongly suppressed by the magnetic form factor and cannot be measured in the standard three-axis geometry. More work is required to fully study the magnetisation fluctuations from components due to other orbitals and to understand their role in forming superconductivity.

4. Acknowledgments

We would like to acknowledge EPSRC for financial support (Grant Number EP/E063985/1).

[1] Saxena S S *et al.* 2000 *Nature* **406** 587–592
[2] Huxley A D *et al.* 2001 *Phys. Rev.* B **63** 144519
[3] Hardy F, Meingast C, Taufour V, Flouquet J, v. Lohneysen H, Fisher R A, Phillips N E, Huxley A D and Lashley J C 2009 *Phys. Rev.* B **80** 174521
[4] Guritanu V, Armitage N P, Tediosi R, Saxena S S, Huxley A D and van der Marel D 2008 *Phys. Rev.* B **78** 172406
[5] Sokolov D A, Ritz R, Pfleiderer C, Keller T and Huxley A D 2011 *J. Phys.: Conf. Ser.* **273** 012085
[6] Taufour V, Aoki D, Knebel G and Flouquet J 2010 *Phys. Rev. Lett.* **105** 217201
[7] Huxley A D, Raymond S and Ressouche E 2003 *Phys. Rev. Lett.* **91** 207201
[8] Fay D and Appel J 1980 *Phys. Rev. Lett.* B **22** 3173
[9] Hattori T *et al.* 2012 *Phys. Rev. Lett.* **108** 066403
[10] Wang W, Sokolov D A, Huxley A D and Kamenev K V 2011 *Rev. Sci. Instrum.* **82** 073903
[11] Pfleiderer C and Huxley A D 2002 *Phys. Rev. Lett.* **89** 147005
[12] Sakarya S *et al.* 2010 *Phys. Rev.* B **81** 024429

Journal of Physics: Conference Series **568** (2014) 042017 doi:10.1088/1742-6596/568/4/042017

Magnetic properties of S=1/2 J_1-J_2 one-dimensional magnets, $VO(XO_4)$(2,2'-bpy) (X=S, Mo; bpy = bipyridine)

H. Kikuchi[1], Y. Ishikawa[1], Y. Fujii[2], A Matsuo[3], K Kindo[3]

[1]Department of Applied Physics, University of Fukui, Fukui 910-8507, Japan

[2]Research Center for Development of Far-Infrared Region, University of Fukui, Fukui 910-8507, Japan

[3]The Institute for Solid State Physics (ISSP), the University of Tokyo , 277-8581 Kashiwa, Japan

E-mail: kikuchi@u-fukui.ac.jp

Abstract. Magnetic susceptibility, high eld magnetization and speci c heat of $S = 1/2$ J_1-J_2 spin chain compounds $VO(XO_4)$(2,2'-bpy) (X=S, Mo; bpy = bipyridine) are measured. $VO(SO_4)$(2,2'-bpy) is found to be a one-dimensional Heisenberg antiferromagnet with J_1/k_B = 6.5 K and J_2 =0. A ratio $j = J_2/J_1$ of $VO(MoO_4)$(2,2'-bpy) is estimated to be about 0.2 with J_1/k_B = 51 K. Spin gap is not observed in both the compounds, which is consistent with the theoretical prediction.

1. Introduction

One dimensional Heisenberg antiferromagnets (1D HAF) with the nearest J_1 and the second-nearest J_2 interactions (J_1-J_2model or zigzag chain model) are the simplest frustrated magnets. Because several interesting magnetic phases[1], including the nematic spin phase[2], are predicted to be realized in the J_1-J_2 model with S =1/2, several experimental and theoretical works on J_1-J_2 model have been conducted. Given the fact that the quantum phase transitions are mainly determined by the relative ratio of J_1 of J_2, actual model compounds with di erent parameters are required to perform experimental studies on the ground state phase diagram.

The V^{4+} ions ($S = 1/2$) in $VO(XO_4)$(2,2'-bpy) (X=S, Mo; bpy = bipyridine) [3] form double chain such that these compounds can be regarded as new materials for the J_1-J_2 model (Fig. 1). The exchange interactions between the V^{4+} ions in each compound are assumed to be mediated through the XO_4 ionic group. We have already reported the measurement results of magnetic susceptibility and high eld magnetization of up to 35 T of $VO(MoO_4)$(2,2'-bpy) [4]. However it was not possible to accurately determine the values of exchange coupling, because of the masking by magnetic impurities and very large saturation magnetic eld.

In the present study, we report the results of magnetic susceptibility, high eld magnetization up to about 60 T and speci c heat of $VO(XO_4)$(2,2'-bpy). In addition, we discuss the determination of the exchange coupling constants of these compounds.

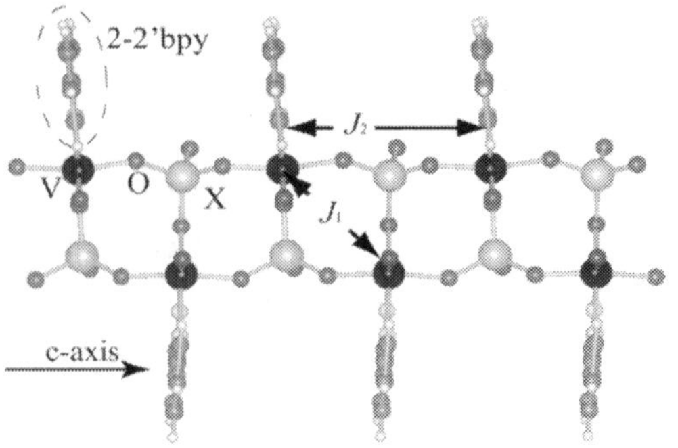

Figure 1. Crystal structure of $VO(MoO_4)$(2-2'-bpy). Here, J_1 and J_2 indicate exchange interactions between V^{4+} ions and X is Mo^{6+} or S^{6+} ions.

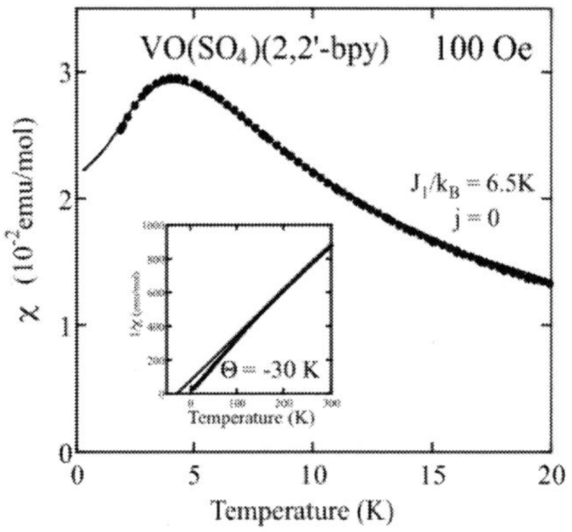

Figure 2. Temperature dependence of the magnetic susceptibility of $VO(SO_4)$(2-2'-bpy). The solid line is corresponds to the best- t result to the theoretical calculation with J_1=6.5 K and $j = 0$. Inset is the reciprocal susceptibility and the solid line is the Curie-Weiss tting result.

2. Experiments

Powder samples of $VO(XO_4)$(2,2'-bpy) were synthesized by hydrothermal method, according to the procedure reported in Ref. 3. The quality of the specimen was investigated by powder X-ray di raction. All the measurements were performed using the prepared powder samples. Magnetic susceptibility was measured using a SQUID (superconducting quantum interference device) magnetometer (MPMS, Quantum Design) in the temperature range of 2 - 300 K. High- eld magnetizations of up to 60 T were measured using a pulsed high magnet at the Institute for Solid State Physics. Speci c heat measurements were performed using a Quantum Design PPMS by the relaxation method under an applied eld of up to 7 T.

3. Results and Discussion

The following sections present the measurement results of $VO(SO_4)$(2,2'-bpy) and $VO(MoO_4)$(2,2'-bpy), and the corresponding discussion and analysis.

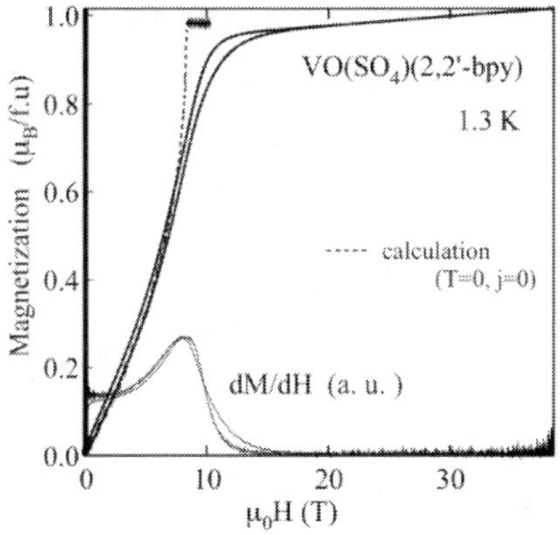

Figure 3. High eld magnetization of $VO(SO_4)$(2-2'-bpy) measured at 1.3 K. The eld derivative of the magnetization dM/dH is plotted in thin line. The dotted line is the calculated magnetization with J_1/k_B=6.5 K and $j = 0$.

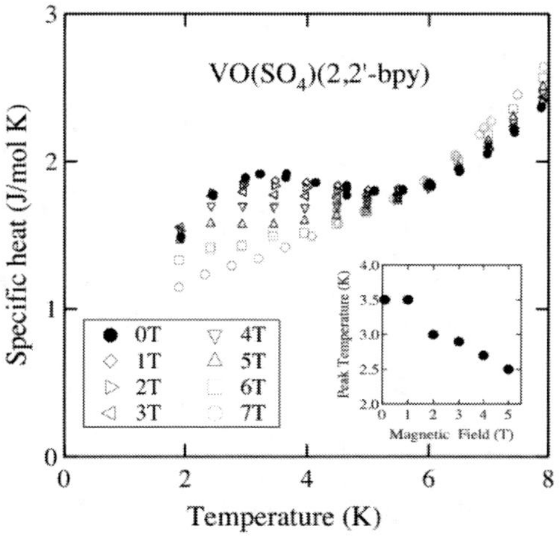

Figure 4. The temperature dependence of the speci c heat of $VO(SO_4)$(2-2'-bpy) measured at di erent magnetic elds of up to 7 T. Inset shows the magnetic eld dependence of the peak temperature of speci c heat.

3.1. VO(SO$_4$)(2,2'-bpy)

Figure 2 shows the temperature dependence of the magnetic susceptibility of $VO(SO_4)$(2,2'-bpy) measured at 100 Oe. The inset indicates the temperature dependence of the reciprocal susceptibility. A t of a high temperature region of (T) to the Curie-Weiss law yields an antiferromagnetic Weiss temperature of Θ = -30 K. A broad peak observed at 4.1 K con rms the one-dimensional nature of this compound. Magnetic susceptibility of the $S = 1/2$ J_1-J_2 model was calculated using the density-matrix renormalization-group method [5]. Best- t results of the calculation to the experimental data are obtained for J_1/k_B=6.5 K and j (=J_2/J_1) = 0, as indicated by a solid line in Fig. 2. Furthernore, the Lande's g value is estimated to be 1.85,

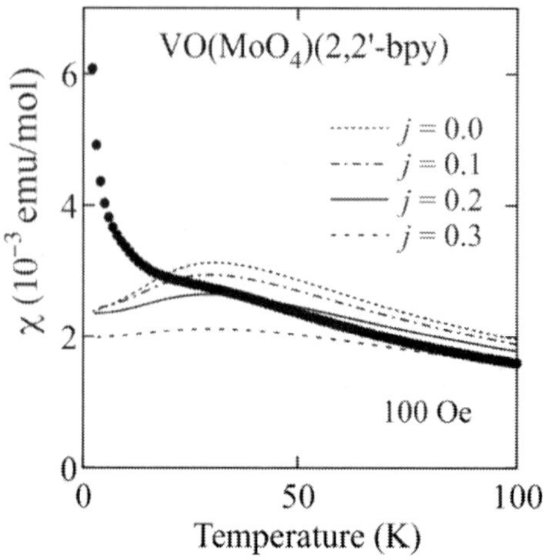

Figure 5. The temperature dependence of the magnetic susceptibility of $VO(MoO_4)$(2-2'-bpy). Theoretical calculation for J_1/k_B =51 K and di erent values of j=0-0.3 are plotted.

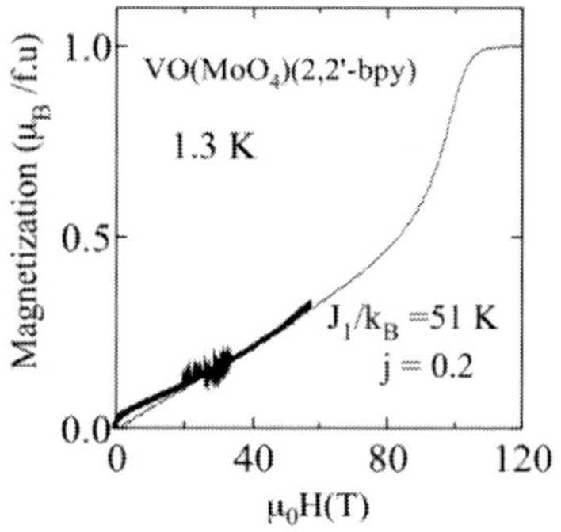

Figure 6. High eld magnetization of $VO(MoO_4)$(2-2'-bpy) measured at 1.3 K up to 60 T. The solid line is the calculated magnetization with $J_1/k_B = 51K$ and $j = 0.2$.

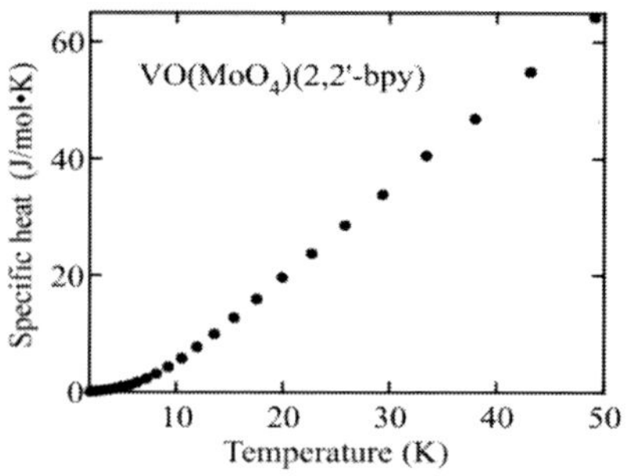

Figure 7. The temperature dependence of the speci c heat of $VO(MoO_4)$(2-2'-bpy).

which is consistent with the reported values 1.925-1.976 as determined by the ESR experiment [6]. Given the ratio parameter $j = 0$, VO (SO_4)(2,2'-bpy) is described as a simple HAF rather than the J_1-J_2 model material. Figure 3 shows the high eld magnetization curve measured at 1.3 K. The magnetization curve exhibits a down-convex form, which is commonly observed in the magnetization of 1D HAF. The eld derivative of the magnetization dM/dH has a peak at around 9 T, which is the saturation magnetic eld. The theoretical curve at T= 0 calculated for 1D HAF without J_2 [7] agrees well with the experimental data as shown in Fig. 3. Figure 4 shows the temperature dependence of speci c heat measured at di erent elds up to 7 T. The lattice contribution is not subtracted because an isostructural compound without magnetic ions is not known. A broad peak is observed in the speci c heat measured in zero eld at around

3.5 K, which corresponds to the temperature at which (*T*) is the maximum. Inset shows the magnetic eld dependence of the peak temperature of speci c heat. Upon increasing the applied eld, the peak temperature monotonically decreases.

3.2. VO(MoO$_4$)(2,2'-bpy)

Figure 5 shows the temperature dependence of (*T*) of VO(MoO$_4$)(2,2'-bpy). A broad maximum is observed at around 40 K, which is higher than that observed for VO(MoO$_4$)(2,2'-bpy). Low temperature upturn is because of a small amount of magnetic impurities. Given the fact that the simple 1D HAF model is insu cient to t the experimental data, nite J_2 is suggested to be present in this compound. We thus made an attempt to obtain the best- t for the experimental data. Although quantitative tting is not obtained, the experimental data qualitatively agree to the theoretical curve for J_1/k_B =51 K and $j = 0.2$, as plotted in Fig. 5. Figure 6 shows the high eld magnetization curve measured at 1.3 K. As expected from large exchange constant J_1, magnetization does not attain its saturation value of 1 $_B$ even at the highest eld of 60 T. Using the parameters determined from (*T*) J_1/k_B =51 K and $j = 0.2$, we calculated the theoretical magnetization curve, as shown in Fig. 6, which agrees to the experimental data. Figure 7 shows the temperature dependence of speci c heat at zero eld. Here, the lattice contribution is not subtracted.

No anomaly associated with the magnetic phase transition was observed despite the large values of the exchange constant. This implies that VO(MoO$_4$)(2,2'-bpy) has good one-dimensionality. Spin gap is not observed both in the magnetization and the speci c heat data. The gapless behavior for the J_1-J_2 spin chain with $j = 0.2$ is consistent with the theoretical expectation that the quantum phase transition occurs at $j_c = 0.2441$ from gapless spin uid to the gapped dimer phase with an increase in j from 0 [8]. It is found that the ground state of VO(MoO$_4$)(2,2'-bpy) belongs to the gapless spin uid phase.

4. Conclusion

In summary, we have determined the magnetic Magnetic susceptibility, high- eld magnetization and speci c heat of $S = 1/2$ J_1-J_2 spin chain compounds VO(XO$_4$)(2,2'-bpy) (X = S, Mo; bpy = bipyridine). VO(SO$_4$)(2,2'-bpy) is found to be a one-dimensional Heisenberg antiferromagnet with $J_1/k_B = 6.5$ K and J_2 =0. The ratio $j = J_2/J_1$ of VO(MoO$_4$)(2,2'-bpy) is estimated to be about 0.2 with $J_1/k_B = 51$ K. Spin gap is not observed in both compounds, which is consistent with the theoretical prediction.

This work was supported by a Grant-in-Aid for Scienti c Research on Priority Areas "Novel States of Matter Induced by Frustration" (19052005) and by a Grant-in-Aid for Scienti c Research(C) (23540389).

References

[1] Hikihara T, Kaburagi M and Kawamura H 2001 *Phys. Rev. B* **63** 721 and references therein
[2] Hikihara T, Kecke L, Momoi T and Furusaki A 2008 *Phys. Rev. B* **78** 144404
[3] Khan M I , Giri S, Ayesh S and Doedens R J 2002 *Inorg. Chem. Commun.* **7** 721
[4] Kikuchi H, Fujii Y, Chiba M, Mitsudo S and Idehara T 2005 *Polyhedron* **24** 2835
[5] Maeshima N and Okunishi K 2000 *Phys. Rev. B* **62** 934
[6] Takahashi N, Okubo S, Ohta H, Sakurai T, Ishikawa Y, Fujisawa M and Kikuchi H 2011 *Appl. Magn. Reson.* **40** 481
[7] Gri ths R B *Phys. Rev.* 1964 **133** A768
[8] Okamoto K and Nomura K 1992 *Phys. Lett. A* **169** 433

Journal of Physics: Conference Series **568** (2014) 042018 doi:10.1088/1742-6596/568/4/042018

Emergence of novel phase of solid oxygen in ultrahigh magnetic field

T C Kobayashi[1], T Nomura[2], Y H Matsuda[2], J L Her[3], S Takeyama[2], A Matsuo[2] and K Kindo[2]

[1] Department of Physics, Okayama University, Okayama 700-8530, Japan
[2] Institute for Solid State Physics, University of Tokyo, Kashiwa 277-8581, Japan
[3] Chang Gung University, Kwei-Shan, Tao-Yuan 333, Taiwan

E-mail: kobayashi@science.okayama-u.ac.jp

Abstract. Faraday discovered the magnetism of molecular oxygen in 1850, and since then, oxygen has attracted significant interests as a ubiquitous but exotic molecular magnet. In solid oxygen, the magnetic energy through the exchange interaction is comparable to the van der Waals cohesive energy, resulting in a strong spin-lattice correlation. We reported the discovery of a magnetic-field-induced phase transition of solid oxygen using ultrahigh magnetic fields up to 186 T. This is the first experiments demonstrating the phase control through the spin by applying a magnetic field. Magneto-optical absorption measurements of the α phase at low temperatures revealed that visible light absorption suddenly ceases and the background transparency dramatically improves above the critical field. The large hysteresis in these data indicates a first-order transition with a relaxation time on the order of microseconds. The results strongly suggest the reconstruction of the O_2 molecular geometry by the magnetic field, which causes that the antiferromagnetic phase collapses and the field-induced phase should have an isotropic crystal structure. Since all of the known phases of O_2 are induced in high pressure, the discovery of a field-induced phase is noteworthy in the long history of solid O_2 studies.

1. Introduction

In molecular solids, the electric intermolecular interaction generally determines the alignment of molecules and the resultant crystal structure. The van der Waals force forms a closed packed structure in noble gas elements, and the electrostatic quadrupolar interaction is also important in diatomic molecules such as N_2 and also in CO_2 [1].

In contrast to these molecules, molecular oxygen O_2 has a magnetic moment with the spin quantum number $S = 1$. The exchange interaction, which depends on the alignment of molecules, strongly affects the binding energy. For each spin state of an O_2–O_2 pair (total spin of $S_T = 0$ or 2), the intermolecular potential as a function of the O_2 orientation and intermolecular distance can be obtained from *ab initio* calculations of the electrostatic and exchange contributions to the binding energy [2]. The spin-dependent intermolecular potential indicates that the energy of an O_2–O_2 pair is the lowest in the rectangular parallel H geometry with the singlet spin state $S_T = 0$ (an intermolecular distance of 3.23 Å) and that the lowest energy for the quintet state $S_T = 2$ is in a crossed X geometry. For solid O_2, the molecular orientation and intermolecular distance are close to those of the H geometry, which is stable for the singlet state.

In solid oxygen γ phase (54.4–43.8 K), orientationally disordered molecules form an A15-type cubic structure [3]. Disk-like rotating sites on the cubic face show antiferromagnetic correlation mainly through the H geometry coupling. The γ–β transition occurs at 43.8 K due to the ordering of the molecular axis [3]. As a result, the crystal transforms to the rhombohedral structure with the basic arrangement of the H geometry. Geometrical frustration in the triangular lattice of the β phase suppresses the antiferromagnetic long-range ordering, but ordering takes place at the β–α transition at 23.9 K [3]. The deformation of the triangular lattice, while retaining the H geometry molecular coupling, releases the antiferromagnetic frustration. Even under high-pressure conditions, the H geometry is robust in the δ and ε phases [4-7], which are stable under pressures of 6–8 and 8–96 GPa, respectively. However, the H geometry should not be stable in a strong magnetic field in which the magnetization is saturated.

Ten years ago, the O_2-O_2 dimer was successfully synthesized by using nanoporous coordination polymers [8]. X-ray diffraction measurements at low temperatures revealed that the H geometry, as in the solid O_2, is robust in the dimer system. However, the magnetization process with a metamagnetic-like behavior could not be interpreted with $S = 1$ Heisenberg antiferromagnetic dimer model. To explain the experimental results, they invoked a scenario in which there are excited states of other geometries, e.g., the S or X type [9, 10]. This scenario indicates that a molecular rearrangement may occur in a magnetic field. It was also found that the thermally excited states lead to deviations from the H geometry at higher temperatures. These combined results, obtained on dimers, led to the prediction of a field-induced structural phase transition in solid O_2.

Recently we reported the discovery of a novel phase of solid O_2 that can be demonstrated by the magnetization measurements and magneto-optical spectroscopy in an extraordinarily strong magnetic field [11]. The experiments indicate this is the result of a first-order transition that is both a magnetic and structural transition: the antiferromagnetic phase collapses and the crystal symmetry changes. In this paper, the results of magneto-optical spectroscopy are mainly discussed.

2. Experimental

A single-turn coil technique was used to generate pulsed magnetic fields [12]. The field was pulsed for approximately 8 μs in a destructive manner. Polycrystalline α oxygen was grown with cooling in a ^{4}He-flow-type cryostat made of Bakelite [13]. In the sample cell of the cryostat, the internal gas was substituted with high-purity oxygen (99.999%). A high-speed streak camera with a polychromator was used for the magneto-optical spectroscopy. The time dependence of the transmitted light intensity through the solid oxygen was measured using a Xe arc flash lamp as the light source. Optical fibers were used for the delivery and collection of light. The temperature was monitored with a RuO_2 thermometer, which was buried in the sample space. The details of the experimental procedure are shown in the supplement materials in the previous paper [11].

3. Results and Discussion

Figure 1a shows the magnetic field waveform used for the magneto-absorption spectroscopy of solid oxygen in the α phase; absorption spectra at 21.6 K, measured continuously as a function of time, are shown in Fig. 1b. The absorption strength is represented by the optical density (O.D. = $-\log_{10}(I/I_0)$) on a color scale, where I and I_0 are the intensities of the transmitted light through the sample cell with and without oxygen, respectively. Figure 1c shows the absorption spectra obtained by taking the cross section of the results in Fig. 1b at different magnetic field strengths. The spectra are vertically shifted for better visualization. The absorption band observed at around 2.14–2.20 eV is due to the bimolecular transition of O_2 [14-17] and defined as the R-band in this paper (this absorption band is responsible for the light blue color of liquid and solid oxygen). As the field increases, the absorption peak intensity decreases accompanied by a considerable broadening, before the peak disappears at around 150 T. The broadening of the absorption peak is understood as a result of the orbital Zeeman effect in the excited state ($^1\Delta^1\Delta$) [14, 15]. With the disappearance of the peak, the transmission abruptly doubles in magnitude (Fig. 1b). At higher fields, the absorption peak is completely absent

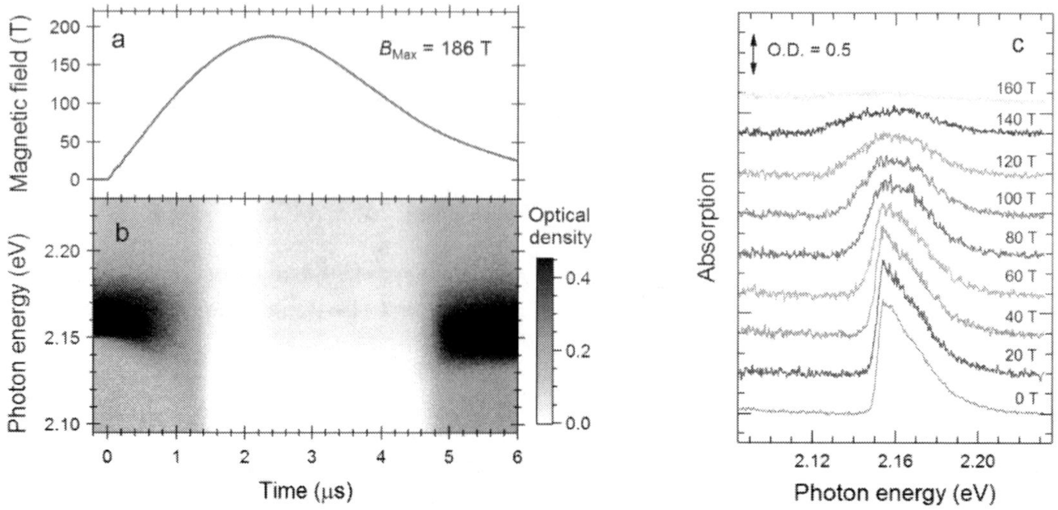

Figure 1. Magneto-absorption spectra of solid oxygen at 21.6 K. (a) Waveform of the pulsed magnetic field. (b) Time dependence of the absorption spectrum. The absorption intensity (optical density) is shown on a color scale. (c) Absorption spectra during up sweeps of the magnetic field, obtained by taking the cross section of Fig. 1(b).

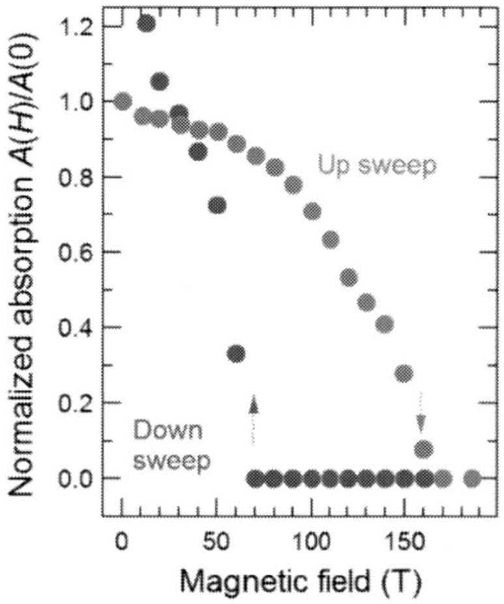

Figure 2. Magnetic field dependence of the bimolecular absorption intensity. The integrated intensity of the R-band is normalized by that at zero magnetic field. Red and blue circles correspond to up and down sweeping of the magnetic field, respectively.

within the limits of the present experimental accuracy. When the field is decreased, the absorption peak reappears at around 70 T, and simultaneously, the oxygen becomes opaque, reverting to the initial condition before the magnetic field is applied. The observed large hysteresis strongly indicates the first-order transition. Therefore, the drastic changes of the optical properties are suggested to be due to the structural phase transition.

The magnetic field dependence of the absorption intensity of the R-band is shown in Fig. 2, where the intensity $A(H)$ is normalized by that at zero field $A(0)$. In the bimolecular optical absorption process, two oxygen molecules need to simultaneously transit from magnetic initial states ($^3\Sigma^3\Sigma$) to nonmagnetic excited states ($^1\Delta^1\Delta$) via one-photon absorption, where $^3\Sigma$ and $^1\Delta$ represent the electronic states of an O_2 molecule of $S = 1$ and 0, respectively. Only antiferromagnetic pairs ($S_T = 0$) of initial states participate in the optical transition, as spin must be conserved in the dipole transition. Therefore, the disappearance of the R-band at 150–170 T strongly indicates the disappearance of singlet pairs, *i.e.*

the saturation of the magnetization [18]. The abrupt change in the absorption intensity, indicated by the arrows in Fig. 2, suggests the presence of a metamagnetic transition to the saturated magnetization state. The first-order metamagnetic transition has also been confirmed in the magnetization measurement [11].

The large increase of the transmission at 150–170 T over the whole photon energy range (Fig. 1b) cannot be explained in terms of electronic optical transitions that occur only in a specific photon energy range. It is proposed, therefore, that classical light scattering from the domain boundaries inherent to polycrystalline α oxygen is significantly reduced in this region due to the field-induced phase transition. In general, polycrystalline domains are formed during the γ–β phase transition with decreasing temperature due to the large volume contraction and the symmetry reduction from the cubic (γ) to rhombohedral (β) structure [3]. Since the crystal symmetry is lowered further to monoclinic in the α phase, it is almost impossible to obtain a single crystal of α oxygen. The α and β phases are thus optically opaque, similar to frosted glass, because of the strong light scattering from the domain boundaries with an anisotropic refractive index. The significant decrease in the light scattering is thus understood to result from the change in the crystal structure. When the crystal symmetry is isotropic, e.g. cubic, the matching conditions between the refractive indices of the domains would be improved. Moreover, domains with isotropic symmetry may form larger domains, and hence light scattering from the domain boundaries almost disappears in the high-field phase.

Figure 3a shows the field dependence of the transparency at 2.10 eV for fields with different maximum field strengths; each field has a different sweep speed, as shown by the waveforms in Fig. 3b. Around this photon energy, absorption due to the R-band is negligible and the optical transmission reflects the degree of classical light scattering. The normalized scattering $S(H)$ is defined here as

$$S(H) = \log_{10}(I_0/I) / [\{\log_{10}(I_0/I)\}_{H=0} - \{\log_{10}(I_0/I)\}_{\min}],$$

where $\log_{10}(I_0/I)$ has the same definition as the optical density but in this case represents the degree of light scattering, $\{\log_{10}(I_0/I)\}_{H=0}$ is the initial value at 0 T, and $\{\log_{10}(I_0/I)\}_{\min}$ is the minimum value for each field strength. The scattering $S(H)$ is nearly constant for a magnetic field of 117 T (green curve), while there is an abrupt decrease in $S(H)$ for a field of 128 T (yellow curve). The critical field is thus

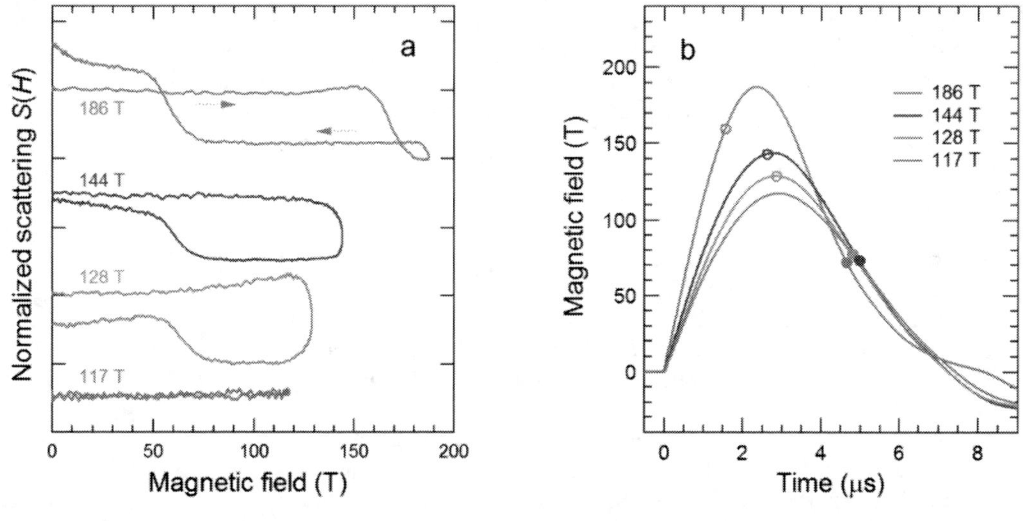

Figure 3. Field dependence of the transparency of solid oxygen. (a) Normalized light scattering $S(H)$ at 2.10 eV in pulsed magnetic fields with different maximum field strengths and initial temperatures of 186 T and 21.6 K (red), 144 T and 12.6 K (blue), 128 T and 21.6 K (yellow), and 117 T and 19.9 K (green). (b) The waveforms of the pulsed magnetic fields. Open and closed circles denote the transition field positions for the up and down sweeps of the magnetic field, respectively.

between 117 and 128 T for this time scale of the field sweep. The dependence on the sweep speed implies the transition has a long relaxation time on the microsecond scale, which supports the structural phase transition.

After applying the magnetic field, light scattering from the domain boundaries in the solid oxygen changes as shown in Fig. 3a, and Fig. 2 also showed that $A(H)/A(0)$ takes a value larger than the value before applying magnetic field, indicating different optical paths in the oxygen crystal. Thus, rearrangement of different-sized domains before and after the appearance of the high-field phase is likely to occur.

Both the metamagnetism and the structural phase transition can be explained consistently by a cooperative phase transition involving a change in the exchange interaction and a geometrical rearrangement of the oxygen molecules. Since the exchange interaction dominates the molecular arrangement in solid O_2 as described above, it is plausible that the O_2–O_2 couplings change from the antiferromagnetic H geometry to another geometry with the weak antiferromagnetic interaction or the ferromagnetic interaction. This molecular rearrangement produces an isotropic crystal structure with the saturated magnetization.

Determining the crystal structure of the high-field phase is an intriguing and important task. Standard methods such as X-ray and neutron diffraction are almost technically impossible to perform in magnetic fields of over 100 T, and so a technical breakthrough will be required to experimentally clarify the crystal structure. At present, we speculate that one possible crystal structure is a modified dry ice structure (cubic phase of solid CO_2 at atmospheric pressure), which is a stable arrangement with respect to the quadrupolar interaction. The solid N_2 α-phase and D_2 α-phase both have the dry ice structure [1]. The quadrupole moment of -0.4×10^{-26} esu in O_2 is smaller than that of N_2 (-1.4×10^{-26} esu) and CO_2 (-4.3×10^{-26} esu) but is comparable to that for D_2 (0.64×10^{-26} esu) [1]. *Ab initio* calculations of the exchange interaction in this arrangement are essential for further discussion of the crystal structure in the high-field phase.

4. Summary

We found the magnetic-field-induced phase transition of solid oxygen using ultrahigh magnetic fields up to 186 T. The magneto-optical measurements of the α phase revealed that the absorption intensity of the R-band suddenly ceases and the background transparency dramatically improves above the critical field. The disappearance of the R-band strongly indicates the saturation of the magnetization. The significant decrease in the light scattering is understood to result from the change in the crystal structure. These results strongly suggest the reconstruction of the O_2 molecular geometry by the magnetic field, which causes that the antiferromagnetic phase collapses and the field-induced phase should have an isotropic crystal structure.

References

[1] Kihara T and Koide 1975 *Advances in Chemical Physics* **33** 51
[2] Bussery B and Wormer P E S 1993 *J. Chem. Phys.* **99** 1230
[3] Freiman Y A and Jodl H J 2004 *Phys. Rep.* **401** 1 and references therein.
[4] Goncharenko I N, Makarova O L and Ulivi L 2004 *Phys. Rev. Lett.* **93** 055502
[5] Klotz S, Strässle Th, Cornelius A L, Philippe J and Hansen Th 2010 *Phys. Rev. Lett.* **104** 115501
[6] Lundegaard L S, Weck G, McMahon M I, Desgreniers S and Loubeyre P 2006 *Nature* **443** 201
[7] Fujihisa H, Akahama Y, Kawamura H, Ohishi Y, Shimomura O, Yamawaki H, Sakashita M, Gotoh Y, Takeya S and Honda K 2006 *Phys. Rev. Lett.* **97** 08553
[8] Kitaura R, Kitagawa S, Kubota Y, Kobayashi T C, Kindo K, Mita Y, Matsuo A, Kobayashi M, Chang H C, Ozawa T C, Suzuki M, Sakata M, Takata M 2002 *Science* **298** 2358
[9] Kobayashi T C, Matsuo A, Suzuki M, Kindo K, Kitaura R, Matsuda R, Kitagawa S 2005 *Prog. Theor. Phys. Suppl.* **159** 271
[10] Hori A, Kobayashi T C, Kubota Y, Matsuo A, Kindo K, Kim J, Kato K, Takata M, Sakamoto H,

Matsuda R, Kitagawa S 2013 *J. Phys. Soc. Jpn.* **82** 084703
[11] Nomura T, Matsuda Y H, Takeyama S, Matsuo A, Kindo K, Her J L and Kobayashi T C 2014 *Phys. Rev. Lett.* **112** 247201
[12] Miura N 1994 *Physica B* **201** 40.
[13] Nomura T, Matsuda Y H, Her J L, Takeyama S, Matsuo A, Kindo K, Kobayashi T C 2013 *J. Low. Temp. Phys.* **170** 372
[14] Gaididei Y B, Loktev V M, Prikhotko A F and Shanskii L I 1975 *Phys. Status Solidi B* **72** 795
[15] Gaididei Y B, Loktev V M, Prikhotko A F and Shanskii L I 1976 *Phys. Status Solidi B* **73** 415
[16] Fujiwara T 1974 *J. Phys. Soc. Jpn.* **36** 1530
[17] Litvinenko Y G, Eremenko V V and Garber T I 1968 *Phys. Status Solidi B* **30** 49
[18] Uyeda C, Yamagishi A and Date M 1988 *J. Phys. Soc. Jpn.* **57** 3954

27th International Conference on Low Temperature Physics (LT27) IOP Publishing
Journal of Physics: Conference Series **568** (2014) 042019 doi:10.1088/1742-6596/568/4/042019

Self consistent study of the quantum phases in a frustrated antiferromagnet on the bilayer honeycomb lattice.

M. Arlego
IFLP - CONICET, Departamento de Física, Universidad Nacional de La Plata, C.C. 67, 1900 La Plata, Argentina.

C. A. Lamas
IFLP - CONICET, Departamento de Física, Universidad Nacional de La Plata, C.C. 67, 1900 La Plata, Argentina.

Hao Zhang
Institute for Solid State Physics, University of Tokyo, Kashiwa, Chiba 277-8581, Japan

Abstract. We study the frustrated Heisenberg model on the bilayer honeycomb lattice. The ground-state energy and spin gap are calculated, using different bosonic representations at mean field level and numerical calculations, to explore different sectors of the phase diagram. In particular we make use of a bond operator formalism and series expansion calculations to study the extent of dimer inter-layer phase. On the other hand we use the Schwinger boson method and exact diagonalization on small systems to analyze the evolution of on-layer phases. In this case we specifically observe a phase that presents a spin gap and short range Néel correlations that survives even in the presence of non-zero next-nearest-neighbor interaction and inter-layer coupling.

1. Introduction

The study of the possible disordered ground states on two-dimensional antiferromagnets has received a great interest in the last years. In particular, the existence of quantum disordered phases has been studied in the phase diagram of antiferromagnets in a single layer honeycomb lattice [1, 2, 3, 4, 5, 6, 7, 8, 9, 10]. However, the study of the influence of possible interlayer coupling on these phases is scarce [11, 12, 13, 14]. From the experimental side, a significative progress on the study of the bismuth oxynitrate, $Bi_3Mn_4O_{12}(NO_3)$, has been made by Smirnova *et al.*[15]. In this material the Mn^{4+} ions form a honeycomb lattice and two layers of such honeycomb lattices are separated by bismuth atoms, forming a bilayer structure. The study of the magnetic susceptibility indicates two-dimensional magnetism and no long-range ordering down to 0.4 K, suggesting a nonmagnetic ground state[15, 16]. In addition, density functional studies indicate that dominant interactions are the interlayer interaction $J_\perp$ and the nearest-neighbor interaction J_1 on each layer[17].

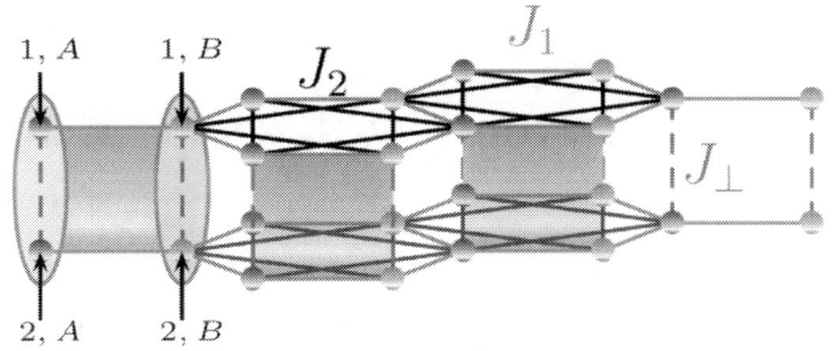

Figure 1. (Color online) Schematic representation of the coupling interactions in $Bi_3Mn_4O_{12}(NO_3)$. Colored areas correspond to the unit cells. The sites in each unit cell are labeled as $(1, A)$, $(2, A)$, $(1, B)$ and $(2, B)$.

The aim of this paper is to study the zero temperature ground state of the frustrated spin-1/2 Heisenberg model on the bilayer honeycomb lattice. We study the $S = 1/2$ case where the quantum fluctuations becomes more important in order to characterize the quantum phases in the model. On the other hand, although the material $Bi_3Mn_4O_{12}(NO_3)$ has $S = 3/2$, the substitution of Mn^{4+} in $Bi_3Mn_4O_{12}(NO_3)$ by V^{4+} may lead to the realization of the $S = 1/2$ Heisenberg model on the honeycomb lattice. We use two different mean field self-consistent approaches based on bosonic representations of the spin operators to study this system, combined with Lanczos and series expansion methods to support the mean field results.

2. Self consistent calculations on the bilayer Model

We study the following Heisenberg model on the bilayer honeycomb lattice

$$H = \sum_{\vec{r},\vec{r}',\alpha,\beta} J_{\alpha,\beta}(\vec{r},\vec{r}')\vec{\mathbf{S}}_\alpha(\vec{r}) \cdot \vec{\mathbf{S}}_\beta(\vec{r}') \qquad (1)$$

where, $\vec{\mathbf{S}}_\alpha(\vec{r})$ is the spin operator on site α corresponding to the unit cell $\vec{r}$. α takes the values $\alpha = (1, A)$, $(2, A)$, $(1, B)$, $(2, B)$ corresponding to the four sites on each unit cell as depicted in Fig. 1, together with the couplings $J_{\alpha,\beta}(\vec{r},\vec{r}')$. The coupling $J_\perp$ does not introduce frustration in the system and then, at the classical level and $T = 0$, it does not affect the classical Néel phase, present for $J_2/J_1 < 1/6$. In the quantum case the situation is much subtle, increasing $J_\perp$ Néel order is likely to melt giving rise to a non-magnetic phase.

For large values of $J_\perp$ we expect the ground state to be an interlayer valence bond crystal (IVBC) with corresponding spins from both layers forming dimers (as ilustrated in Figure 1). This limit represents an excellent starting point for the bond operators formalism and series expansion calculations.

On the other hand, starting from the magnetically ordered phase, the Néel order can be destroyed both by increasing the frustration on each layer or increasing the coupling between layers. The destruction of Néel order in a frustrated single layer honeycomb lattice has been studied by means of various approaches [3, 4, 8, 11, 14, 19, 20, 21, 22]. In the following we use two different bosonic representations of the spin operators to study the influence of the interlayer coupling in the ground state of the bilayer model.

2.1. Bond operators Mean field approach.

First, we use the well known bond-operator method to study the bilayer antiferromagnet described by Hamiltonian (1). We start by introducing a bond-operator representation of spin operators

$$\vec{\mathbf{S}}^{\alpha}_{\eta,A}(\vec{r}) = \frac{1}{2}\Big((-1)^{\eta+1}\left[\mathbf{s}^{\dagger}_{A}(\vec{r})\mathbf{a}_{\alpha}(\vec{r}) + \mathbf{a}^{\dagger}_{\alpha}(\vec{r})\mathbf{s}_{A}(\vec{r})\right] - i\epsilon_{\alpha\beta\gamma}\mathbf{a}^{\dagger}_{\beta}(\vec{r})\mathbf{a}_{\gamma}(\vec{r})\Big) \quad (2)$$

$$\vec{\mathbf{S}}^{\alpha}_{\eta,B}(\vec{r}) = \frac{1}{2}\Big((-1)^{\eta+1}\left[\mathbf{s}^{\dagger}_{B}(\vec{r})\mathbf{b}_{\alpha}(\vec{r}) + \mathbf{b}^{\dagger}_{\alpha}(\vec{r})\mathbf{s}_{B}(\vec{r})\right] - i\epsilon_{\alpha\beta\gamma}\mathbf{b}^{\dagger}_{\beta}(\vec{r})\mathbf{b}_{\gamma}(\vec{r})\Big), \quad (3)$$

where $\eta = 1, 2$ is the layer index, $\vec{\mathbf{S}}^{\alpha}_{\eta,A}(\vec{r})$ is the spin operator on the sublattice A of layer η corresponding to the unit cell located at $\vec{r}$ (See figure 1). Operators $\mathbf{s}^{\dagger}_{\mathbf{A}}(\vec{r})$ and $\mathbf{a}^{\dagger}_{\alpha}(\vec{r})$ create singlet and triplets states (out of a vacuum $|0\rangle$) in the vertical bond placed in sub-lattice A and are defined as: $\mathbf{s}^{\dagger}_{A}|0\rangle = \frac{1}{\sqrt{2}}(|\uparrow\downarrow\rangle - |\downarrow\uparrow\rangle)$, $\mathbf{a}^{\dagger}_{x}|0\rangle = -\frac{1}{\sqrt{2}}(|\uparrow\uparrow\rangle - |\downarrow\downarrow\rangle)$, $\mathbf{a}^{\dagger}_{y}|0\rangle = \frac{i}{\sqrt{2}}(|\uparrow\uparrow\rangle + |\downarrow\downarrow\rangle)$, $\mathbf{a}^{\dagger}_{z}|0\rangle = \frac{1}{\sqrt{2}}(|\uparrow\downarrow\rangle + |\downarrow\uparrow\rangle)$, and similar expressions for sublattice B. These kind of representations where proposed by Sachdev[24] in order to treat quantum phase transitions between Neél and dimerized phases. Operators belonging to the same unit cell satisfy the bosonic commutation relations whereas operators belonging to different unit cells commute. The restriction that the physical states are either singlets or triplets leads to the constraints in each unit cell, $\mathbf{s}^{\dagger}_{A}\mathbf{s}_{A} + \sum_{\alpha}\mathbf{a}^{\dagger}_{\alpha}\mathbf{a}_{\alpha} = 1$ and $\mathbf{s}^{\dagger}_{B}\mathbf{s}_{B} + \sum_{\alpha}\mathbf{b}^{\dagger}_{\alpha}\mathbf{b}_{\alpha} = 1$. Introducing the bond-operators representation of the spin operators in (1) we obtain a bosonic version of the Hamiltonian. We transform Fourier and retain terms up to second order to write $\mathbf{H} = \mathbf{H}_{\perp} + \mathbf{H_1} + \mathbf{H}_{\lambda}$, where

$$\mathbf{H}_{\perp} = -\frac{3}{2}J_{\perp}s^2N + \frac{J_{\perp}}{4}\sum_{\vec{k},\alpha}\left\{\mathbf{a}^{\dagger}_{\alpha}(\vec{k})\mathbf{a}_{\alpha}(\vec{k}) + \mathbf{b}^{\dagger}_{\alpha}(\vec{k})\mathbf{b}_{\alpha}(\vec{k})\right\}, \quad (4)$$

$$\begin{aligned}\mathbf{H_1} = \frac{J_1}{4}\sum_{\vec{k},\alpha}\Big\{&\gamma(\vec{k})\Big(\mathbf{a}_{\alpha}(\vec{k})\mathbf{b}_{\alpha}(-\vec{k}) + \mathbf{a}_{\alpha}(\vec{k})\mathbf{b}^{\dagger}_{\alpha}(\vec{k}) + \mathbf{b}^{\dagger}_{\alpha}(\vec{k})\mathbf{a}_{\alpha}(\vec{k}) + \mathbf{b}^{\dagger}_{\alpha}(\vec{k})\mathbf{a}^{\dagger}_{\alpha}(-\vec{k})\Big) \quad (5)\\ + &\gamma(\vec{k})\Big(\mathbf{b}_{\alpha}(\vec{k})\mathbf{a}_{\alpha}(-\vec{k}) + \mathbf{a}^{\dagger}_{\alpha}(\vec{k})\mathbf{b}_{\alpha}(\vec{k}) + \mathbf{a}^{\dagger}_{\alpha}(\vec{k})\mathbf{b}^{\dagger}_{\alpha}(-\vec{k}) + \mathbf{b}_{\alpha}(\vec{k})\mathbf{a}^{\dagger}_{\alpha}(\vec{k})\Big)\Big\},\end{aligned}$$

$$\mathbf{H}_{\lambda} = (2s^2 - 5)N\lambda + \lambda\sum_{\vec{k},\alpha}\left\{\mathbf{a}^{\dagger}_{\alpha}(\vec{k})\mathbf{a}^{\dagger}_{\alpha}(\vec{k}) + \mathbf{b}^{\dagger}_{\alpha}(\vec{k})\mathbf{b}^{\dagger}_{\alpha}(\vec{k})\right\}, \quad (6)$$

where, we have assumed that condensation of singlets occurs, *i.e.* $\langle \mathbf{s}_{\alpha}(\vec{r})\rangle = s$, $\gamma(\vec{k}) = s^2(1 + e^{i\vec{k}\cdot\vec{e}_1} + e^{i\vec{k}\cdot\vec{e}_2})$, $\vec{e}_1$ and $\vec{e}_2$ are the primitive vectors on a triangular lattice and λ is a Lagrange multiplier related to the constraint in the number of bosons. Diagonalization by using a Bogoliubov transformation allows us to write the following expression for the ground state energy

$$\frac{E}{N} = (2s^2 - 5)\lambda - \frac{3}{4}J_{\perp}(2s^2 + 1) + \sum_{\alpha}\int\frac{d^2k}{V}\left(\omega^{(A)}_{\alpha}(\vec{k}) + \omega^{(B)}_{\alpha}(\vec{k})\right), \quad (7)$$

where $\omega^{(A)}_{\alpha}(\vec{k})$ and $\omega^{(B)}_{\alpha}(\vec{k})$ are triplet energies. The parameters s^2 and λ are determined by solving self-consistently the saddle point conditions $\frac{\partial E}{\partial\lambda} = 0$ and $\frac{\partial E}{\partial s^2} = 0$ which are used to evaluate the energy and gap of the system and compare with numerical techniques.

In order to complement our study, we have performed series expansion (SE) calculations, starting from the limit of isolated dimers connecting spins from both layers via $J_{\perp}$. Notice that, this kind of expansions remain valid in the same limit that the bond operator approach (*i.e.* in the limit of strong inter-layer coupling.) To this end we have performed a continuous unitary transformation on the original Hamiltonian, using the flow equation method. This technique allows to obtain perturbatively an effective Hamiltonian that keeps the block diagonal structure of decoupled dimers. We refer for details of the method to ref.[25].

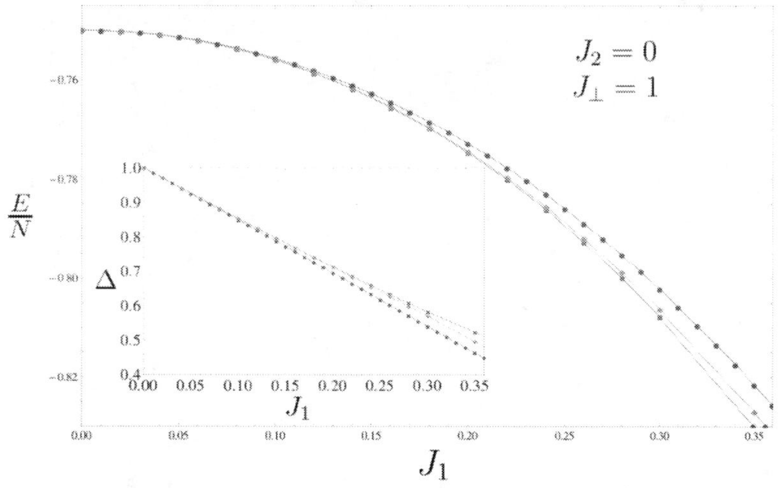

Figure 2. (Color online) Ground state energy per dimer as a function of J_1 obtained by the self consistent bond-operator approach (blue circles), Lanczos (ED) on a 24 sites system (red squares) and $O(5)$ Series Expansion (SE) (yellow rhombi). Inset: triplet gap (same set of parameters as main panel) BO (blue circles) ED (red squares) and SE (yellow rhombi).

For the present model we have performed $O(5)$ and $O(4)$ SE in $J_{1,2}$ for ground state energy and for triplet dispersion, respectively. Explicit expressions are too long to be printed explicitly but are available electronically upon request. In Fig.2 we show the ground state energy per site as a function of J_1 and $J_2 = 0$, obtained by BO (blue circles), O(5) SE (yellow rhombi) and ED (red squares) on a system of 24 sites. As it can be observed, all the techniques predict an energy decreasing with the coupling of interlayer-dimer via J_1. Furthermore, there is an excellent quantitative agreement between the three methods for small values of J_1. On the other hand, triplet gap is shown in the inset of Fig.2 for the same set of parameters as the ground state energy. Here we also observe that all the techniques predict a tendency to a closure of the gap, when J_1 is turned on. Our calculations shows that BO, ED and SE predict the same behavior.

2.2. Self consistent Schwinger Boson Mean-Field Theory

As we have seen previously, bond operator and series expansion methods are both suitable to study the interlayer-dimer phase. In order to investigate the evolution of on-layer phases as a function of inter-layer coupling we apply a representation of the spin operators in terms of Schwinger bosons [27], $\vec{\mathbf{S}}_\alpha(\vec{r}) = \frac{1}{2}\vec{\mathbf{b}}_\alpha^\dagger(\vec{r}) \cdot \vec{\sigma} \cdot \vec{\mathbf{b}}_\alpha(\vec{r})$. Here $\vec{\mathbf{b}}_\alpha(\vec{r})^\dagger = (\mathbf{b}_{\alpha,\uparrow}^\dagger(\vec{r}), \mathbf{b}_{\alpha,\downarrow}^\dagger(\vec{r}))$ is a bosonic spinor corresponding to the site α in the unit cell at position $\vec{r}$, $\vec{\sigma}$ are Pauli matrices, and the constraint in the number of bosons $\sum_\sigma \mathbf{b}_{\alpha,\sigma}^\dagger(\vec{r})\mathbf{b}_{\alpha,\sigma}(\vec{r}) = 2S$ has to be satisfied on each site. In order to perform a mean field decomposition, we define the following $SU(2)$ invariants, $\mathbf{A}_{\alpha\beta}(\vec{x},\vec{y}) = \frac{1}{2}\sum_\sigma \sigma \mathbf{b}_{\alpha,\sigma}(\vec{x})\mathbf{b}_{\beta,-\sigma}(\vec{y})$ and $\mathbf{B}_{\alpha\beta}(\vec{x},\vec{y}) = \frac{1}{2}\sum_\sigma \mathbf{b}_{\alpha,\sigma}^\dagger(\vec{x})\mathbf{b}_{\beta,-\sigma}(\vec{y})$. This decomposition allow us to treat ferromagnetism and antiferromagnetism on equal footing and has been successfully used to describe a number of quantum frustrated antiferromagnets[1, 2, 3, 14, 28, 29, 30]. We perform a Hartree-Fock decoupling where the mean field parameters are the expectation values of the SU(2) invariants **A** and **B**. The mean field equations for these parameters $A_{\alpha\beta}$ and $B_{\alpha\beta}$ and the constraints in the number of bosons must be solved self-consistently (see refs. [1], [3] and ,[14] for further details). To obtain the phase boundary between the magnetically ordered and disordered phases using the self consistent Schwinger

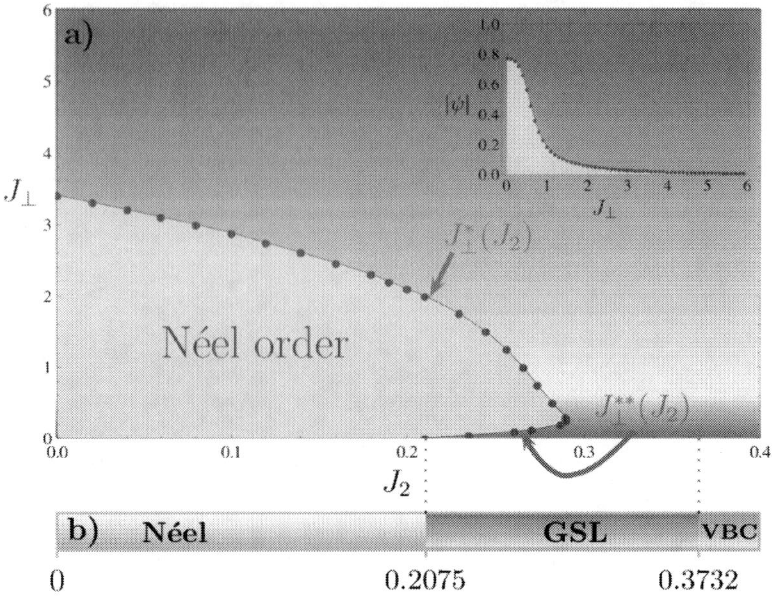

Figure 3. (Color online) a) Phase diagram in the $J_2 - J_\perp$ plane obtained with SBMFT. Gray region corresponds to the Néel phase whereas green region corresponds to magnetically disordered phases. b) Phase diagram of the single layer case corresponding to Ref. [3]. Inset: Z3 order parameter corresponding to the line $J_2 = 0.38$

boson mean field theory we study the boson spectrum. In the gapless region the excitation spectrum is zero at $\vec{k} = \vec{0}$, where the boson condensation occurs, this is characteristic of the Néel ordered phase. On the other hand, in the gapped region, the absence of Bose condensation indicates that the ground state is magnetically disordered.

In Fig. 3-a) we show the phase diagram in the $J_2 - J_\perp$ plane. For $J_\perp \gg J_2$ one can expect a IVBC ground state adiabatically connected with the limit of decoupled dimers, i.e. two singlets per unit cell, between spins $1, A(1, B)$ and $2, A(2, B)$ (see Fig. 1). In the region $0.2075 \lesssim J_2 \lesssim 0.289$ there is a reentrant effect. In this range, Néel phase separates from J_2 axis, leaving a tiny space for a magnetically disordered phase. In this way, Néel phase is here not only limited by some value $J_\perp^*(J_2)$ from above, but also by a second value $J_\perp^{**}(J_2)$ from below (See Figure 3).

On the other hand, in the range $0.3732 \lesssim J_2 \lesssim 0.398$ ($J_\perp = 0$), there is evidence of the existence of an on-layer valence bond phase [3] (see Figure 3-b). In this phase, SU(2) and translational symmetries are preserved, but Z_3 symmetry is broken. By turning on $J_\perp$ the system moves to the IVBC where the Z_3 symmetry is recovered. In the inset of Figure 3-a, we depict the Z_3 directional symmetry-breaking order parameter ρ (defined in [18]) vs $J_\perp$. The behavior of this parameter suggest a Z_3 symmetry restoring. Finally, in the region $0.289 \lesssim J_2/J_1 \lesssim 0.3732$ the ground state preserves SU(2), lattice translational and Z_3 symmetries and the spin-spin correlations are short ranged[14]. This agrees with the evidence of a spin liquid phase in the phase diagram corresponding to $J_\perp = 0$ [3, 8].

3. Conclusions

In summary, we have studied the phase diagram corresponding to a frustrated Heisenberg model on the bilayer honeycomb lattice, by means of bosonic mean field theories, complemented with exact diagonalization and series expansion, and described the behavior of the quantum phases as the interlayer coupling is increased. Using the Schwinger boson description we have determined the region where the system is Néel ordered and the lines where the Néel order is destroyed. We have determined an intermediate region where the phase diagram shows signatures of a reentrant behavior and we observe that for values of the interlayer coupling between $(0.289 \lesssim J_2/J_1 \lesssim 0.3732)$ the Néel order is absent at $J_\perp = 0$ and the system presents a nonzero spin gap, whereas in the region $(0.3732 \lesssim J_2/J_1 \lesssim 0.398)$ each layer presents a nematic disordered phase[3]. In all the range of J_2 studied, the system presents signatures of an interlayer-valence bond crystal (IVBC) phase for $J_\perp/J_1 > 4$. This phase evolves adiabatically from the limit of decoupled interlayer-dimers. This is corroborated by bond operators self-consistent calculations and series expansions starting explicitly from the limit of isolated interlayer dimers.

ACKNOWLEDGMENTS

C. A. Lamas and M. Arlego are partially supported by CONICET (PIP 1691) and ANPCyT (PICT 1426 and PICT 2013-0009).

[1] D. C. Cabra, C. A. Lamas, and H. D. Rosales, Phys. Rev. B 83, 094506 (2011).
[2] D. C. Cabra, C. A. Lamas, and H. D. Rosales, Modern Physics Letters B (MPLB) 25, 891 (2011).
[3] Hao Zhang and C. A. Lamas, Phys. Rev. B 87, 024415 (2013).
[4] A. F. Albuquerque, D. Schwandt, B. Hetényi, S. Capponi, M. Mambrini, and A. M. Läuchli, Phys. Rev. B **84**, 024406 (2011).
[5] P. H. Y. Li, R. F. Bishop, D. J. J. Farnell, and C. E. Campbell, Phys. Rev. B **86**, 144404 (2012).
[6] C. A. Lamas, A. Ralko, D. C. Cabra, D. Poilblanc, P. Pujol, Phys. Rev. Lett. **109**, 016403 (2012)
[7] C.A. Lamas, A. Ralko, M. Oshikawa, D. Poilblanc, P. Pujol, Phys. Rev. B **87**, 104512 (2013)
[8] F. Mezzacapo and M. Boninsegni, Phys. Rev. B **85**, 060402(R) (2012).
[9] R. Ganesh, J. van den Brink, and S. Nishimoto, Phys. Rev. Lett. **110**, 127203 (2013).
[10] Hong-Yu Yang, A F. Albuquerque, S. Capponi, A. M Lauchli and K. P. Schmidt, New J. Phys. **14** 115027 (2012).
[11] R. Ganesh, D.N. Sheng, Y.-J. Kim and A. Paramekanti, Phys. Rev. B **83**, 144414 (2011).
[12] R. Ganesh, S.V. Isakov, and A. Paramekanti, Phys. Rev. B **84**, 214412 (2011).
[13] J. Oitmaa and R. R. P. Singh, Phys. Rev. B **85**, 014428 (2012).
[14] Hao Zhang, M. Arlego, C. A. Lamas. Phys. Rev. B **89**, 024403 (2014).
[15] O. Smirnova, M. Azuma, N. Kumada, Y. Kusano, M. Matsuda, Y. Shimakawa, T. Takei, Y. Yonesaki, and N. Kinomura, J. Am. Chem. Soc., **131**, 8313 (2009).
[16] S. Okubo, F. Elmasry, W. Zhang, M. Fujisawa, T. Sakurai, H. Ohta, M. Azuma, O. A. Sumirnova, and N. Kumada,, J. Phys.: Conf. Series **200**, 022042 (2010).
[17] H. C. Kandpal and J. van den Brink, Phys. Rev. B **83**, 140412(R) (2011).
[18] S. Okumura, H. Kawamura, T. Okubo, and Y. Motome, J. Phys. Soc. Jpn. **79**, 114705 (2010).
[19] E. Rastelli, A. Tassi, and L. Reatto, Physica B **97**, 1 (1979).
[20] J. B. Fouet, P. Sindzingre, and C. Lhuillier, Eur. Phys. J. B **20**, 241 (2001).
[21] T. Einarsson and H. Johannesson, Phys. Rev. B **43**, 5867 (1991).
[22] A. Mattsson, P. Frojdh, and T. Einarsson Phys. Rev. B **49**, 3997 (1994)
[23] J. Oitmaa and R. R. P. Singh,Phys. Rev. B **84**, 094424 (2011)
[24] S. Sachdev and R. N. Bath; Phys. Rev: B **41** 9323 (1990).
[25] C. Knetter and G.S. Uhrig, Eur. Phys. J. B **13**, 209 (2000).
[26] M. Arlego and W. Brenig, Phys. Rev. B **84**, 134426 (2011).
[27] D. P. Arovas and A. Auerbach, Phys. Rev. B **38**, 316 (1988); A. Auerbach and D. P. Arovas, Phys. Rev. Lett. **61**, 617 (1988).
[28] H. A. Ceccato, C. J. Gazza, and A. E. Trumper, Phys. Rev. B **47**, 12329 (1993).
[29] A. E. Trumper, L. O. Manuel, C. J. Gazza, and H. A. Ceccatto, Phys. Rev. Lett. **78**, 2216 (1997).
[30] A. Mezio, C. N. Sposetti, L. O. Manuel, and A. E. Trumper, Europhys. Lett. **94**, 47001 (2011).

Journal of Physics: Conference Series **568** (2014) 042020 doi:10.1088/1742-6596/568/4/042020

Exchange bias effect in polycrystalline Sr_2IrO_4

P K Manna[*], A Thamizhavel and S Ramakrishnan

Department of Condensed Matter Physics and Materials Science, Tata Institute of Fundamental Research, Mumbai-400005, INDIA

*E-mail: pkmanna1@gmail.com

Abstract. We report the presence of an exchange bias effect in a layered iridate compound, Sr_2IrO_4, in polycrystalline form. The Rietveld refinement of the x-ray diffraction pattern shows that Sr_2IrO_4 crystallizes in a tetragonal structure ($I4_1/acd$). The temperature dependence of zero-field cooled (ZFC) and field cooled (FC) magnetization measurements show a magnetic ordering temperature of ~240 K, which is considered as its canted antiferromagnetic ordering temperature. An irreversibility between ZFC and FC curves is observed at T_{irr}, which decreases with an increasing field (H). The T_{irr} vs. $H^{2/3}$ plot follows the Almeida-Thouless (AT) line. The presence of an AT line and a significant amount of coercivity (~0.3 T) at 5 K hint a cluster-glass like behaviour in Sr_2IrO_4 polycrystals. A negative exchange bias effect has been observed at 5 K with an exchange bias field of 0.018 T, which is attributed to an interfacial exchange interaction between a canted antiferromagnetic phase and a probable cluster-glass phase. This is the first report on the observation of an exchange bias effect in Sr_2IrO_4. Further studies are underway for a detailed understanding.

1. Introduction

The 5-d electron based iridate compounds, with the general formula R_2IrO_4 (R= Sr, Ba), are in the limelight recently because of their potential to exhibit various interesting phenomena like metal to insulator transition, superconductivity [1-2]. Both these compounds crystalize in a K_2NiF_4-type layered perovskite structure, which is very similar to the parent compound of high-T_c cuprate superconductors, La_2CuO_4 [3-5]. Consequently, a significant amount of interest has been generated to search for superconductivity in such layered ididate compounds. Moreover, these compounds possess different comparable energies, *viz.* spin-orbit, Coulomb repulsion, and electronic band width, which may give rise to various interesting properties [1-5].

In this paper, we have studied polycrystalline Sr_2IrO_4, and report a rather unexplored behaviour of this compound, *i.e.* the presence of an exchange bias effect. The phenomenon of exchange bias occurs in a coupled magnetic system (ferromagnetic/antiferromagnetic, ferromagnetic/spin glass, ferrimagnetic/antiferromagnetic, *etc.* [6]) because of an interface exchange coupling between two magnetically ordered components. In the present study, we found the same in phase-pure Sr_2IrO_4 polycrystals, which was not reported even on the single crystals of this compound.

2. Experimental details

The Sr_2IrO_4 polycrystals were synthesized by the conventional solid state reaction method [7]. High purity $SrCO_3$ and iridium metal powder were heated in a furnace (in an alumina crucible) at 1000 °C for 24 hrs with intermediate grindings. The X-ray diffraction (XRD) pattern was recorded in a

Journal of Physics: Conference Series **568** (2014) 042020 doi:10.1088/1742-6596/568/4/042020

PANalytical (Xpert PRO MPD) x-ray diffractometer using the Cu K_α radiation. The elemental composition of the synthesized polycrystals was determined by energy dispersive X-ray analysis (EDX) technique using a ZEISS Ultra FESEM instrument. The dc-magnetization measurements were carried out by using a SQUID-VSM (Quantum Design Inc.).

3. Results and discussion

3.1. Structural and elemental characterization

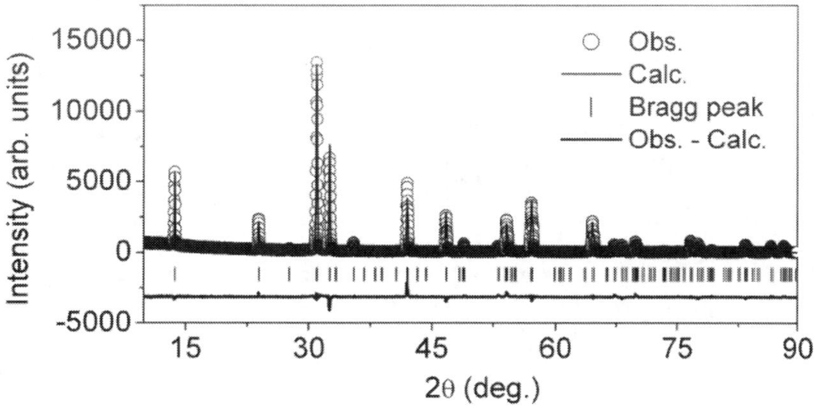

Figure 1. The Rietveld-refined XRD pattern of Sr_2IrO_4 polycrystals at room temperature.

The phase purity of the synthesized Sr_2IrO_4 polycrystals has been checked from the room temperature XRD pattern [figure 1]. The Rietveld refinement of the XRD pattern (FullProf program [8]) shows that this compound crystallizes in a tetragonal structure under the space group $I4_1/acd$ [figure 1]. The lattice parameters are determined as $a = b = 5.498(1)$ Å and $c = 25.812(1)$ Å, which agree well with earlier report on this compound [7].

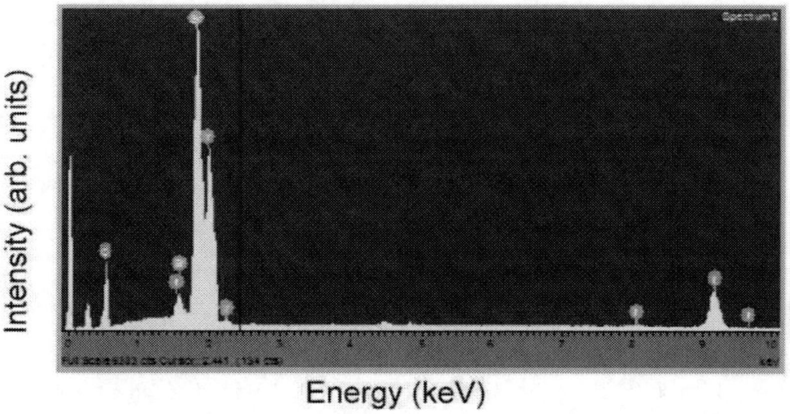

Figure 2. The EDX spectrum of the Sr_2IrO_4 polycrystals.

The EDX measurement [figure 2] shows the elemental composition of the polycrystals as: Sr = 21.42 at%, Ir = 10.66 at% and O = 67.92 at%.

3.2. DC magnetization and exchange bias

Figure 3(a) shows the temperature dependence of the zero field-cooled (ZFC) and field cooled (FC) magnetizations of the Sr_2IrO_4 polycrystals, measured in the temperature range of 2-305 K, under the magnetic fields of 0.05, 0.2 and 2 T. As the temperature decreases, the polycrystals become magnetically ordered at ~240 K, which corresponds to the canted antiferromagnetic ordering temperature of this compound [1-5, 7]. With further lowering of temperature, an irreversibility between ZFC and FC curves is observed at T_{irr} [figure 3(a)]. However, with an increasing magnetic field (H), the bifurcation between ZFC and FC curves decreases, which results in a decrease in T_{irr}. The T_{irr} vs. $H^{2/3}$ plot is linear in nature [figure 3(b)] suggesting that it follows the de Almeida–Thouless (AT) line [9-11]. Such an AT line is the characteristics of a cluster glass/spin glass or superparamagnetic system [9-11]. However, we have observed a significant amount of coercivity at 5 K (0.3 T) [figure 4], signifying that an intercluster interaction exists in the studied Sr_2IrO_4 polycrystals which drives the system to behave like a cluster-glass. It can be noted that, in Sr_2IrO_4 single crystal, Chikara *et al.* [12] observed a transition temperature, T_M (~100 K), below which the dominant exchange interaction changed from ferromagnetic to antiferromagnetic. Therefore, in order to ensure the presence of such a cluster-glass phase in our Sr_2IrO_4 polycrystals, we plan to perform other measurements, like frequency dependence of ac-susceptibility. By summarizing our results of dc-magnetization so far, we could see that these polycrystals exhibit a usual canted antiferromagnetic ordering as reported earlier in Sr_2IrO_4 in polycrystalline as well as single crystalline form. In addition, we observe a cluster-glass like behaviour in our studied polycrystals.

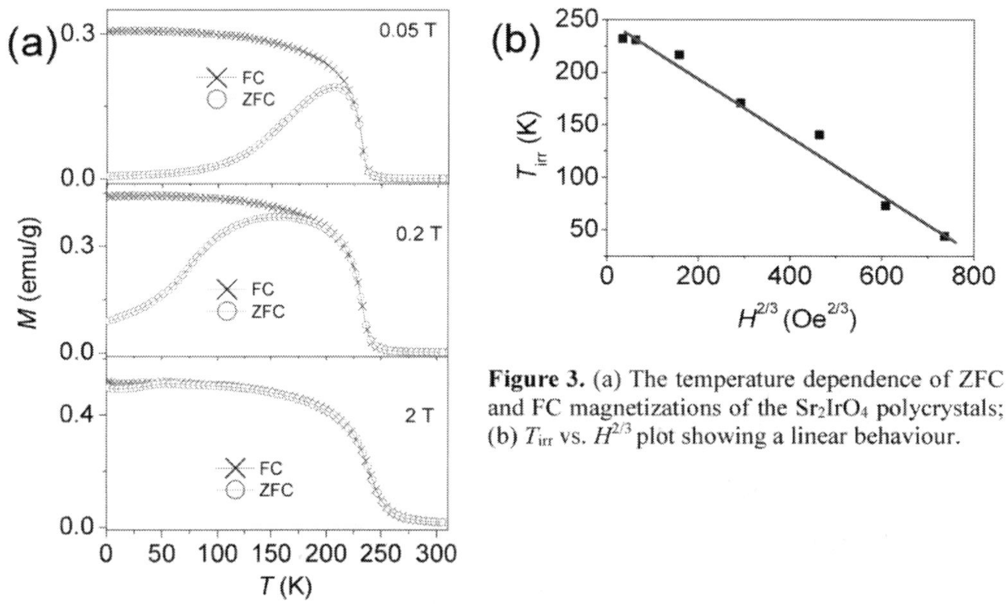

Figure 3. (a) The temperature dependence of ZFC and FC magnetizations of the Sr_2IrO_4 polycrystals; (b) T_{irr} vs. $H^{2/3}$ plot showing a linear behaviour.

Such a bi-magnetic system is very appealing for studying the exchange bias effect. In order to check the presence of an exchange bias effect, we have measured the FC (cooling field = 3T) and ZFC hysteresis loops at 5 K [figure 4]. The ZFC hysteresis loop is found to be symmetric around the origin. However, the FC hysteresis loop shifts along the negative magnetic field axis, signifying the presence of a negative exchange bias effect in the Sr_2IrO_4 polycrystals. At 5 K, the value of the exchange bias field is calculated to be 0.018 T. We believe that the origin of the exchange bias phenomenon in our studied system is nothing but an interfacial exchange coupling between the coexisting canted antiferromagnetic and cluster-glass phases. This is the first report on the observation of an exchange bias effect in Sr_2IrO_4. This effect was not reported even in the single crystalline form of this compound. Further studies are, therefore, underway to elucidate the origin of such a behaviour of the Sr_2IrO_4 polycrystals.

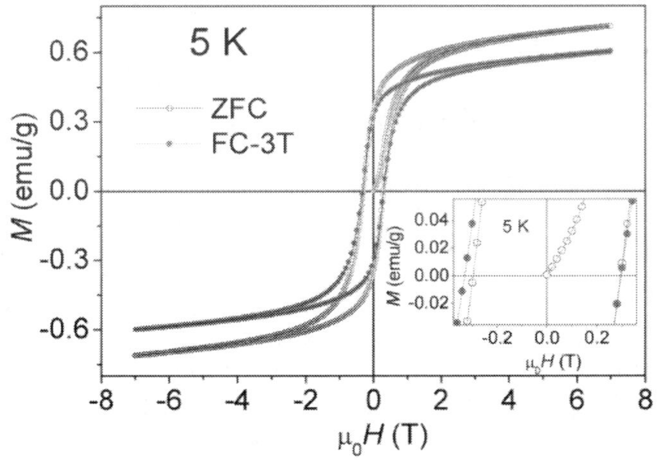

Figure 4. The ZFC and FC hysteresis loops of the Sr_2IrO_4 polycrystals recorded at 5 K. The FC loop was measured under a cooling field of 3 T. In the inset, the low field region has been enlarged, to show the presence of an exchange bias effect.

4. Summary and conclusion

We have studied Sr_2IrO_4 polycrystals by using the XRD, EDX, and dc-magnetization techniques. The Rietveld refinement of the XRD pattern confirms the single-phase nature of this compound. The EDX measurement ensures the presence of Sr, Ir, O elements in the polycrystals in the desired ratio. From the dc-magnetization measurements, we have observed the possible presence of a coexisting canted antiferromagnetic phase along with a cluster-glass phase. However, to ensure the cluster-glass nature of the second component further experiments are required. An interfacial exchange interaction between these two magnetic phases gives rise to a negative exchange bias effect in the Sr_2IrO_4 polycrystals, which has not been reported either in the polycrystalline form or in the single crystalline form of this compound.

5. References

[1] Wang F and Senthil T 2011 *Phys. Rev. Lett.* **106** 136402
[2] Kim Jungho, Casa D, Upton M H, Gog T, Kim Young-June, Mitchell J F, van Veenendaal M, Daghofer M, van den Brink J, Khaliullin G and Kim B J 2012 *Phys. Rev. Lett.* **108** 177003
[3] Watanabe Hiroshi, Shirakawa Tomonori and Yunoki Seiji 2013 *Phys. Rev. Lett.* **110** 027002
[4] Orii Daisuke, Sakata Masafumi, Miyake Atsushi, Shimizu Katsuya, Okabe Hirotaka, Isobe Masaaki, Takayama-Muromachi Eiji and Akimitsu Jun 2013 *J. Korean Phys. Soc.* **63** 349
[5] Boseggia S, Springell R, Walker H C, Rønnow H M, Rüegg Ch, Okabe H, Isobe M, Perry R S, Collins S P and McMorrow D F 2013 *Phys. Rev. Lett.* **110** 117207
[6] Manna P K and Yusuf S M 2014 *Phys. Rep.* **535** 61
[7] Kini N S, Strydom A M, Jeevan H S, Geibel C and Ramakrishnan S 2006 *J. Phys.: Condens. Matter* **18** 8205
[8] Rodriguez-Carvajal J, 2011 *Commission on Powder Diffraction (IUCr) Newsletter* **26** 12
[9] Singh M K, Katiyar R S, Prellier W and Scott J F 2009 *J. Phys.: Condens. Matter* **21** 042202
[10] Ding J F, Lebedev O I, Turner S, Tian Y F, Hu W J, Seo J W, Panagopoulos C, Prellier W, Van Tendeloo G and Wu T 2013 *Phys. Rev. B* **87** 054428
[11] Kumar Devendra and Banerjee A 2013 *J. Phys.: Condens. Matter* **25** 216005
[12] Chikara S, Korneta O, Crummett W P, DeLong L E, Schlottmann P and Cao G 2009 *Phys. Rev. B* **80** 140407(R)

NMR Studies of the Spinel-Type $Cu(Cr_{1\ x}Hf_x)_2S_4$

H Niki[1], N Higa[1], H Kuroshima[1], T Toji[1], M Oshiro[1], M Yogi[1], S Ebisu[2] and S Nagata[2]

[1] Department of Physics, Faculty of Science, University of the Ryukyus, Nishihara, Okinawa 903-0213, Japan

[2] Muroran Institute of Materials Research, Muroran, Hokkaido 980-8577, Japan

E-mail: niki@sci.u-ryukyu.ac.jp

Abstract. To investigate the magnetic properties of the Hf-rich side ($0.5 \leq x \leq 1.0$) of $Cu(Cr_{1-x}Hf_x)_2S_4$ from a microscopic point of view, ^{63}Cu NMR has been performed at temperatures between 4.2 and 300 K. Temperature dependencies of the Knight shift and spin-lattice relaxation time (T_1) have been measured. As for $0.7 \leq x \leq 1.0$, the Knight shift gradually shifts to a negative side with decreasing temperature due to the core polarization of d conduction electrons. The temperature dependence of $1/T_1T$ in this range of x indicates that the T_1T is constant in all temperature region. On the other hand, in the range of $0.5 \leq x \leq 0.7$, the Knight shift rapidly shifts to negative side below about 130 K because of the core polarization of d conduction electrons. In this range of x , the values of $1/T_1T$ at high and low temperatures take constant values. The value at low temperatures is approximately twice compared with those at high temperatures. The density of state at the Fermi level in the low temperature region would be about 1.3 times larger than that in the high temperature region.

1. Introduction

Chalcogenide thiospinel CuX_2S_4 including transition metals has a large variety of physical properties because the ions at X site have a mixed valence $X^{3.5+}$ (i.e., combination of X^{3+} and X^{4+}) [1, 2]. Cu is a monovalent cation. The cation sites, which are occupied by Cu and X ions, are located at the centers of tetrahedrons and octahedrons surrounded by $S^{2\ }$ ions, respectively. The spinel-type structure has cubic symmetry. Much research on the metal-insulator transition in $CuIr_2S_4$ has been conducted in the last decade [1, 2, 3]. A few ferromagnetic spinel compounds exist, such as $CuCr_2S_4$ [4], $CuCr_2Se_4$ [4], and $CuCr_2Te_4$ [5].

The chalcogenide tiospinel $CuCr_2S_4$ with a mixed valence state of $CuCr^{3+}Cr^{4+}S_4$ [4] is a metallic ferromagnet with a Curie temperature $T_C = 380$ K; its net magnetic moment is close to 5.0 $_B$, consisting of Cr^{3+} ion with 3 $_B$ ($S = 3/2$) and Cr^{4+} ion with 2 $_B$ ($S = 1$). Conversely, $CuHf_2S_4$ [6], which also has the spinel structure, shows metallic properties without any magnetic anomaly. Increasing values of x in the mixed spinel $Cu(Cr_{1\ x}Hf_x)_2S_4$ system [7] produce successive changes from a dominant ferromagnetic ($x\ \ 0.5$) to a re-entrant spin-glass($x \simeq 0.5$), then a spin-glass ($0.5\ \ x\ \ 0.7$), and nally to a pure paramagnetic state ($0.8\ \ x\ \ 1.0$). Hf substitutions cause the magnetic frustration, which originates from competition between ferromagnetic and antiferromagnetic interactions. A step-like anomaly in the magnetization in the range of $0.5\ \ x\ \ 0.7$ of $Cu(Cr_{1\ x}Hf_x)_2S_4$ is also found around 160 K because of a spin crossover phenomenon.

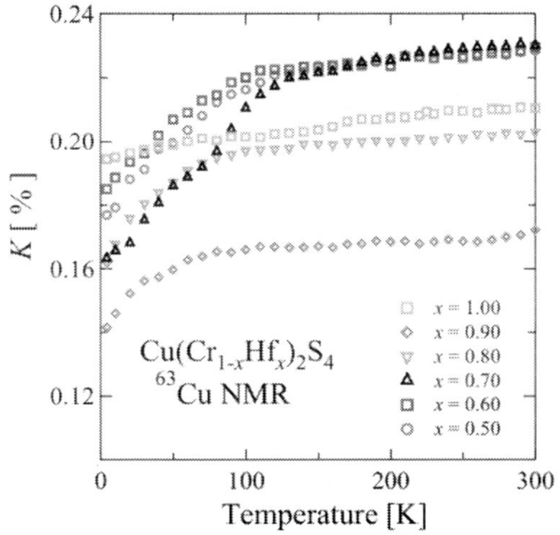

Figure 1. Knight shift vs. temeprature in ^{63}Cu NMR spectra of $Cu(Cr_{1\ x}Hf_x)_2S_4$ at temperatures between 4.2 and 300 K (x = 1.0, 0.9, 0.8, 0.7, 0.6 and 0.5).

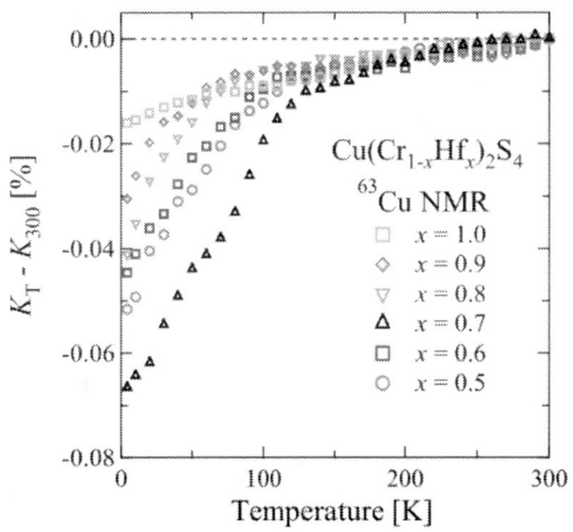

Figure 2. Deviation of the Knight shift vs. temperature in ^{63}Cu NMR of $Cu(Cr_{1\ x}Hf_x)_2S_4$ at temperatures between 4.2 and 300 K (x = 1.0, 0.9, 0.8, 0.7, 0.6 and 0.5). The deviation of the Knight shift represents the value obtained by subtracting the Knight shift at 300 K from the Knight shift at each temperature.

To investigate the magnetic properties of the Hf rich side of $Cu(Cr_{1\ x}Hf_x)_2S_4$ from a microscopic point of view, ^{63}Cu NMR has been performed at temperatures between 4.2 and 300 K. Temperature dependencies of the Knight shift and spin-lattice relaxation time (T_1) have been measured.

2. Experimental

The polycrystalline specimens were prepared by a solid state reaction method [7]. Mixtures of high-purity ne powders of Cu (purity 99.99%), Cr(99.99%), Hf(99.8%) and S(99.999%) with normal stoichiometry were calcined in sealed quartz tubes at 950 °C for 5 days. Subsequently, the specimens were reground and pressed to form pellets, and then annealed at 400 °C for 4 h. No secondary phase in the specimens was detected in the powder X-ray di raction pro les. The powdered specimens were prepared for NMR measurements. A conventional pulse NMR spectrometer was used in ^{63}Cu NMR measurements. A magnetic eld of about 6.9 T (about 78 MHz) was applied by a superconducting magnet with magnetic eld homogeneity of $10^{\ 5}$. Shifts in ^{63}Cu-NMR were obtained at a gyro-magnetic ratio of 11.285 MHz/T. The NMR spectrum was derived from a summation of the FFT spectrum as a function of frequency. The resonance frequency was determined from a peak position of the spectrum. The spin-lattice relaxation time T_1 was measured by saturation recovery method.

3. Results and Discussion

The Knight shift in ^{63}Cu NMR of $Cu(Cr_{1\ x}Hf_x)_2S_4$ for 0.5 x 1.0 has been measured at temperatures between 4.2 and 300 K. The Knight shift vs. temperature of ^{63}Cu NMR in $Cu(Cr_{1\ x}Hf_x)_2S_4$ for 0.5 x 1.0 are shown in Fig. 1. The behavior of the Knight shift in the vicinity of 300 K for 0.8 x 1.0 is di erent from that for 0.5 x 0.7. Especially the Knight

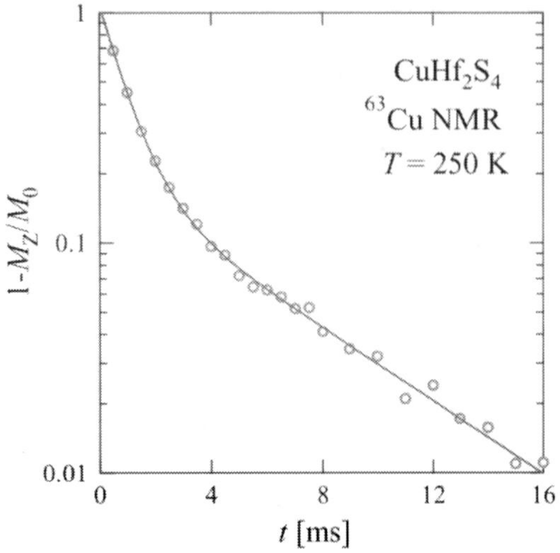

Figure 3. Nuclear longitudinal magnetization recovery $f(t) = 1 \quad (M(t)/M_0)$ in ^{63}Cu NMR of Cu(Cr$_{1\ x}$Hf$_x$)$_2$S$_4$ (x = 1.0).

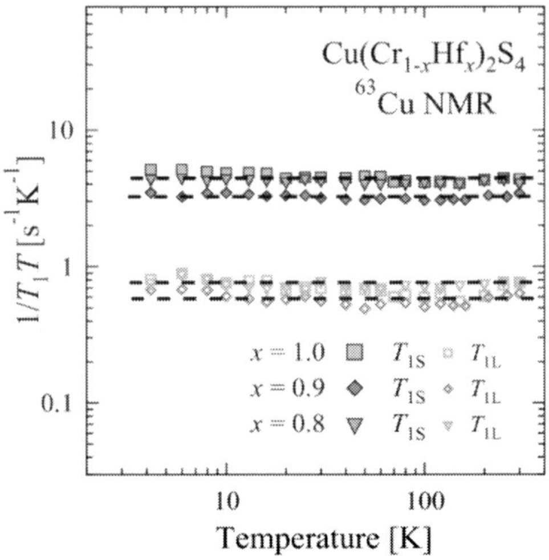

Figure 4. Temperature dependence of $1/T_1T$ in ^{63}Cu NMR of Cu(Cr$_{1\ x}$Hf$_x$)$_2$S$_4$ (x = 1.0, 0.9 and 0.8).

shift for x = 0.9 is largely deviated from others. Since the magnetization shows a diamagnetic state for x = 1.0 [2] and paramagnetic states for $x \quad 0.9$ [6], this di erence of the magnetization may a ect the large discrepancy of the Knight shift around 300 K. But the detailed reasons are not clear now.

To compare the temperature dependence of Knight shift for each concentration x, we examine the value gained by subtracting the Knight shift at 300 K from the Knight shift at each temperature. Figure 2 shows the temperature dependence of the deviation of the ^{63}Cu Knight shift. Here, K_T-K_{300} represents the value obtained by subtracting the Knight shift at 300 K from the Knight shift at each temperature. The Knight shift gradually shifts to a negative side with decreasing temperature for all samples because of the core polarization of d conduction electrons at Cu site. For x = 1.0, 0.9, and 0.8, the negative shift becomes larger with decreasing x. The Knight shift for x = 0.7 rapidly shifts to negative side below 130 K and the deviation toward negative side represents a maximum. Therefore, the ratios of negative shift for x = 0.6 and 0.5 become smaller compared with that for x = 0.7.

In the range of $0.5 \quad x \quad 0.7$, the magnetization in the spin glass phase keeps an almost constant value in eld cooling process due to the freezing of moments [7]. The Knight shift of ^{63}Cu NMR would take the almost constant value if internal eld at Cu site is mainly a ected by the Cr static moment in the spin glass phase. However, as shown in Fig. 2, the Knight shift shifts to negative side below 10 K. In the range of $0.5 \quad x \quad 0.7$, a step-like anomaly is found in the magnetization measurements around 160 K [7]. However, no anomaly can be seen in the Knight shift measurements around this temperature.

The Spin-lattice relaxation time T_1 in ^{63}Cu NMR of Cu(Cr$_{1\ x}$Hf$_x$)$_2$S$_4$ for $0.5 \quad x \quad 1.0$ has been measured at temperatures between 4.2 and 300 K. The nuclear longitudinal magnetization recovery $f(t) = 1 \quad (M(t)/M_0)$ cannot be expressed by a single exponential type. For x = 1.0, 0.9 and 0.8, the recovery $f(t)$ is explained well by $f(t) = a\exp(\ t/T_{1S}) + (1 \quad a)\exp(\ t/T_{1L})$ consisting of two components: short recovery time T_{1S} and long recovery time T_{1L}. Figure 3 shows the time dependence of the nuclear longitudinal magnetization recovery in ^{63}Cu NMR for

Journal of Physics: Conference Series **568** (2014) 042021 doi:10.1088/1742-6596/568/4/042021

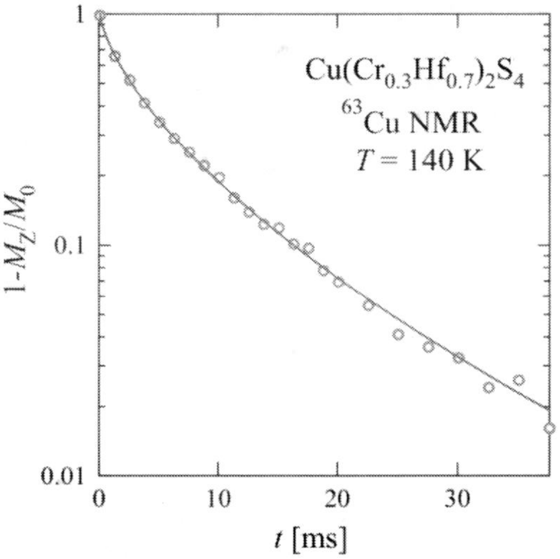

Figure 5. Nuclear longitudinal magnetization recovery $f(t) = 1 \quad (M(t)/M_0)$ in the ^{63}Cu NMR of $Cu(Cr_{1\ x}Hf_x)_2S_4$ ($x = 0.7$).

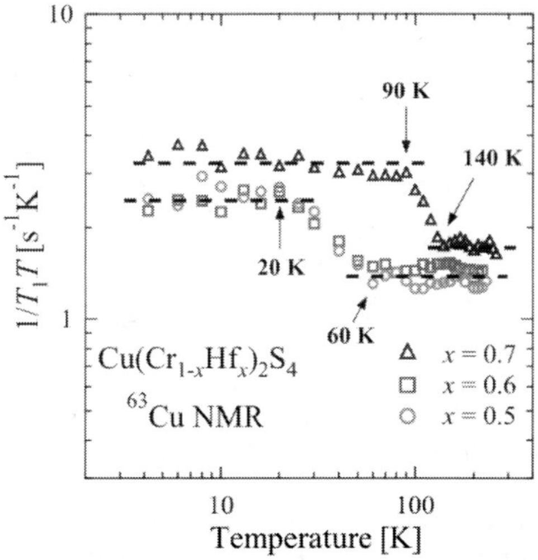

Figure 6. Temperature dependence of $1/T_1T$ in ^{63}Cu NMR of $Cu(Cr_{1\ x}Hf_x)_2S_4$ (x = 0.7, 0.6 and 0.5).

x =1.0 at 250 K. The values of a for x = 1.0, 0.9 and 0.8 are about 0.8 in all temperature region.

Since no secondary phase in cubic phase specimens of x = 1.0 was detected in the powder X-ray di raction pro le, it can be expected that the time dependence of the recovery for x = 1.0 should be expressed by a single exponential type. However, the experimental recovery curves represent two exponential types, indicating the distribution of T_1. The reason for the T_1 distribution in these materials is not clear. The same situation occurs in the T_1 measurements in ^{63}Cu NMR of $CuIr_2S_4$ [3].

The temperature dependence of $1/T_1T$ for x = 1.0, 0.9 and 0.8 at temperatures between 4.2 and 300 K is shown in Fig. 4. The $1/T_1T$ for x = 1.0, 0.9 and 0.8 indicates that the T_1T is constant. According to the experimental results of the Knight shift of ^{63}Cu NMR shown in Fig. 2, the Knight shift in the range of 0.8 x 1.0 weakly shifts to the negative side due to the core polarization of d electrons with decreasing temperatures. However, the T_1T is constant, which means that the contribution of d electrons to the uctuating local eld at Cu site is small.

On the other hand, for x = 0.7, 0.6 and 0.5, nuclear longitudinal magnetization recovery can be analyzed as a stretched exponential relaxation (SER) type as follows [8, 9]:

$$f(t) = y_0 \exp\left\{ \quad \left(\frac{t}{T_1}\right)^p \right\}. \tag{1}$$

Figure 5 shows the time dependence of the nuclear longitudinal magnetization recovery $f(t) = 1 \quad (M(t)/M_0)$ in ^{63}Cu NMR for x = 0.7 at 140 K. According to the magnetization measurements [7], the magnetization in the region of 0.5 x 0.7 is much larger than that in the region of 0.8 x 1.0. Since the local eld caused by magnetic Cr atoms in the region of 0.5 x 0.7 is very large compared with that in the region of 0.8 x 1.0, the T_1 in the region of 0.5 x 0.7 would be distributed widely. Because of this reason, the recovery would be analyzed as the SER type.

The plot of $1/T_1T$ vs. temperature for x = 0.7, 0.6 and 0.5 measured at temperatures between 4.2 and 260 K is shown in Fig. 6. The temperature dependence of $1/T_1T$ for x = 0.7 above 140 K shows a constant value. However, the value of $1/T_1T$ for x = 0.7 abruptly changes about 140

K and the $1/T_1T$ below about 90 K indicates a constant value which is about two times larger than that above 140 K. For $x = 0.6$ and 0.5, the values of $1/T_1T$ remain constant above about 60 K, increase below about 60 K and below about 20 K take constant values which are about twice compared with those above 60 K. From these results, for $x = 0.7$, 0.6 and 0.5, the density of state at the Fermi level in the low temperature region would be about 1.3 times larger than that in the high temperature region.

In summary, to investigate the magnetic properties of the Hf-rich side (0.5 x 1.0) of $Cu(Cr_{1\ x}Hf_x)_2S_4$ from a microscopic point of view, ^{63}Cu NMR has been performed at temperatures between 4.2 and 300 K. Temperature dependencies of the Knight shift and spin-lattice relaxation time (T_1) have been measured. The Knight shift for 0.7 x 1.0 gradually shifts to a negative side with decreasing temperature due to the core polarization of d conduction electrons. The temperature dependence of $1/T_1T$ in this range of x indicates that the T_1T is constant in all temperature region. On the other hand, in the range of 0.5 x 0.7, the Knight shift rapidly shifts to negative side below about 130 K because of the core polarization of d conduction electrons. In this range of x, the values of $1/T_1T$ at high and low temperatures take constant values. The value at low temperatures is approximately twice compared with those at high temperatures. The density of state at the Fermi level in the low temperature region would be about 1.3 times larger than that in the high temperature region.

References

[1] Nagata S 2005 *Chin. J. Phys.* **43** 722.
[2] Kawashima K, Horibe N, Awaka J, Yamamoto H, Ebisu S and Nagata S 2007 *Physica* B **387** 208.
[3] Niki H, Okuda H, Okada Y, Higa K, Oshiro M, Fukuyoshi N, Mahoe R, Yogi M, Nakama T, Yagasaki K, Ebisu S and Nagata S 2011 *J. Phys.: Conf. Ser.* **273** 012139.
[4] Wohlfarth E P (Ed.) 1982 *Ferro-Magnetic Materials in A Handbook on the Properties of Magnetically Ordered Substances*, vol. 3 (Amsterdam: North-Holland) p 603.
[5] Suzuyama T, Awaka J, Yamamoto H, Ebisu S, Ito M, Suzuki T, Nakama T, Yagasaki K and Nagata S 2006 *J. Solid State Chem.* **179** 140.
[6] DiSalvo F J and Waszczak J V 1982 *Phys. Rev.* B **26** 2501.
[7] Kariya F, Ebina K, Hasegawa H, Koshimizu K, Wuritunasitu B, Hondou K, Ebisu S and Nagata S, 2009 *J. Solid State Chem.* **182** 2018.
[8] Sobol W T, Cameron I G, Pintar M M and Blinc R 1987 *Phys. Rev.* B **35** 7299.
[9] Palmer R G, Stein D L, Abrahams E and Anderson P W 1984 *Phys. Rev. Lett.* **53** 958.

Magnetic properties of nickel nanoparticles embedded in amorphous Al_2O_3 matrix

Nikolina Novosel[1,2], Michael Reissner[1], Krešo Zadro[2], Damir Pajić[2], Marko Jerčinović[3], Maja Buljan[3], Nikola Radić[3]

[1]Institute of Solid State Physics, Vienna University of Technology, 1040 Vienna, Austria
[2]Faculty of Science, University of Zagreb, Bijenička 32, 10000 Zagreb, Croatia
[3]Ruđer Bošković Institute, Bijenička 54, 10000 Zagreb, Croatia

Corresponding author: Nikolina Novosel (nnovosel@phy.hr)

Abstract. Multilayer thin films consisting of 10 bilayers of $(Al_2O_3+Ni)/Al_2O_3$ were deposited by magnetron sputtering onto Si(100) substrate. Nickel nanoparticles are formed inside amorphous alumina matrix by self-assembly growth process. It was determined by GISAXS measurements that nickel particles are spheroidal with diameter < 3 nm and that they form paracrystal-like body-centered tetragonal lattice. Magnetic properties of the prepared thin films were studied. Due to the nanometer size of nickel particles, their magnetic structure is single domain and they show superparamagnetic behaviour. Anisotropy of magnetic properties was observed when magnetic field is applied parallel or perpendicular to the thin film surface and it was attributed to dipole-dipole interactions between particles. This was confirmed using simulations of the $M(H)$ curve of the 2D superlattice of the identical superparamagnetic particles, which was performed using Monte Carlo method and Metropolis algorithm.

1. Introduction

Nanosized magnetic particles have been extensively investigated in the last few decades due to the variety of their potential application, such as biomedical applications (enhancing quality of magnetic resonance imaging (MRI), hyperthermic treatment of tumor cells, drug delivery) [1,2], data storage [3] etc. Magnetic structure of nanosized magnetic particles is single-domain [4] and they can be characterised by their magnetic moment. Single domain particles exhibit superparamagnetic behaviour at high temperatures and blocking of magnetic moment due to the anisotropy barrier at low temperatures, which results in a slow relaxation of the magnetization of the system. The temperature, at which transition between blocked and superparamagnetic regime occurs, is strongly dependent on the parameters of the system, applied magnetic field and the time scale of the experiment. Besides individual properties of the particles (size, shape, crystalline structure etc.), their mutual interaction also affects magnetic behaviour of the system. This is especially important for the systems with high concentration of magnetic particles as dipole-dipole interaction between particles can lead to appearance of a collective state at low temperatures, such as super spin glass state or even long-range ferromagnetic ordering [5]. Understanding magnetic behaviour of assembly of single-domain particles is important for their possible applications, as well as for better fundamental knowledge of magnetic phenomena in nanosized structures.

Magnetron sputtering deposition is a relatively simple and efficient method for preparation of well-ordered 3D lattice of nanosized nickel particles embedded in amorphous alumina matrix. Size, shape and separation between particles can be easily tuned by changing the deposition parameters, which also allows tailoring magnetic properties of the thin film.

2. Experimental

Thin films consisting of 10 alternating Al_2O_3 and Al_2O_3+Ni bilayers were prepared by magnetron sputtering deposition (KJLC CMS-18 System) on Si(111) substrate. Pure Ni (99.995%) and Al_2O_3 (99.995%) were used as targets in the DC (50 W) and RF (300 W) magnetrons, respectively. The substrate was held at 300°C and the pressure of the process gas (Ar) was 0.4 Pa. The deposition time for each mixed layer was 30 s, and deposition time of the Al_2O_3 layer was varied from 10 to 75 s in order to obtain samples with different separation between layers containing nickel particles. Magnetic properties of the thin films were investigated using Quantum Design PPMS with VSM option and magnetic measurements were performed in temperature range 2-300 K in applied magnetic field up to 9 T.

3. Results and discussion

Structural properties

Nanosized nickel particles are formed inside amorphous Al_2O_3 matrix during the deposition of Ni+Al_2O_3 layer by the self-assembly growth process combined with the surface morphology effects. A similar effect was previously observed for nickel particles in alumina matrix with single Ni+Al_2O_3 multilayer [6], as well as for Ge particles formed in amorphous silica and alumina matrices [7,8]. Size, shape and spatial arrangement of nickel particles in the sample with larger separation between nickel containing layers were investigated using Grazing Incidence Small Angle X-ray Scattering (GISAXS) technique[1]. Numerical analysis of the measured GISAXS map was performed according to a model described in [9] which revealed spheroidal shape of the nickel particles with radius 1.4 ± 0.1 nm. Also, it was determined that nickel particles are arranged in paracrystal-like body-centered tetragonal lattice with lattice [9] parameters $a = 5.8 \pm 0.2$ nm and $c = 3.7 \pm 0.1$ nm.

Magnetic properties

In Fig. 1 zero field cooled (ZFC) and field cooled (FC) magnetization curves measured in different magnetic fields applied parallel to the thin film surface are shown. For all measuring fields ZFC magnetization increases with the increasing temperature, obtains maximum at temperature T_{max} and decreases with further increase in temperature, whereas FC magnetization decreases with increasing temperature. ZFC and FC branch of magnetization curve separate at temperature T_{irr} which is slightly higher than T_{max}. Observed behaviour of the ZFC and FC magnetization curves is in accordance with the model of superparamagnetic particles which are blocked in non-equilibrium state due to the anisotropy barrier at low temperatures and with a rather narrow distribution of sizes which causes small deviation of T_{irr} from T_{max}. But with closer inspection of the ZFC and FC curves in Fig. 1 it is found that T_{max} (and T_{irr}) remain practically constant with increasing magnetic field up to 0.1 T, which contradicts the model of noninteracting single domain particles where applied magnetic field lowers the barrier so that magnetic moments become thermally unblocked at lower temperature with expected magnetic field dependence $T_{max} \sim (1 - H/H_0)^2$ [10]. The deviation of experimental results from theoretical predictions suggests that the model of noninteracting single domain particles cannot be applied for the studied system. In samples with rather high concentrations of magnetic particles, dipole-dipole magnetic interaction between particles needs to be taken into consideration. Several

[1] GISAXS measurements were performed at Synchrotron Elettra in Trieste. Detailed results and analysis will be published elsewhere.

Journal of Physics: Conference Series **568** (2014) 042022 doi:10.1088/1742-6596/568/4/042022

models have been proposed to describe effects of the dipole-dipole interaction [11]. Although they agree with some experimental observations, they cannot fully explain magnetic behaviour of dense magnetic nanoparticle systems. It should be noted that due to the large number of parameters (size, shape, distribution of sizes of the particles, surface effects, crystalline structure, interactions) it is extremely difficult to model magnetic nanoparticle systems and some approximations must be used. A modified superparamagnetic model is often used for systems with moderate concentration of particles in which particles retain typical superparamagnetic behaviour but with barriers modified by interactions. With increasing concentration of the particles, their mutual dipole-dipole interaction becomes dominant and the system is no longer described as assembly of individual particles, as experimental results point to collective behaviour. In order to estimate the influence of dipole-dipole interactions on the macroscopic magnetic properties of our thin films, we prepared samples with smaller separation between the layers of the nickel particles. Preliminary measurements of ZFC and FC curves (not shown) indicate possible occurrence of a collective super spin glass state: in addition to splitting of ZFC and FC curves and distinct maximum of the ZFC curve, FC magnetization curve does not increase upon cooling and obtains small minimum in low temperature region, which is an indication of the super spin glass state. It seems that dipole-dipole interactions are significant in the studied systems and that they can lead to a collective state with the reduction of the separation between particles.

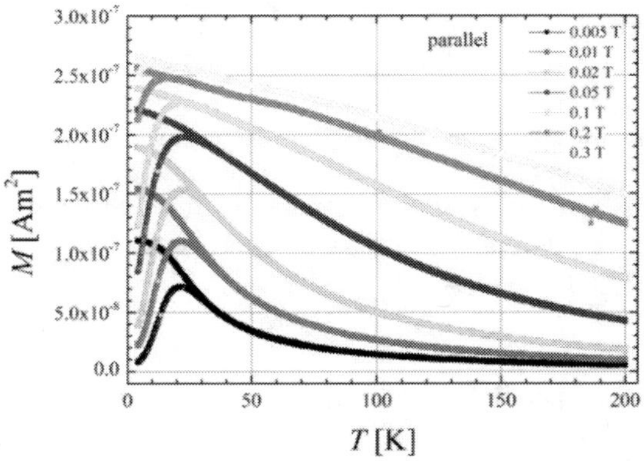

Figure 1. ZFC (open symbols) and FC (full symbols) $M(T)$ curves measured in applied field parallel to the thin film surface.

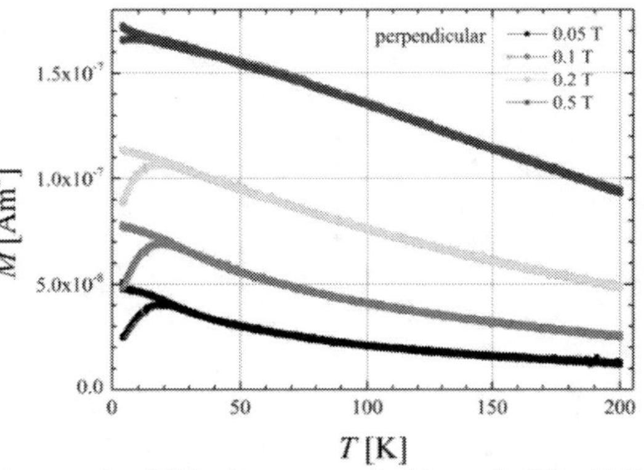

Figure 2. ZFC (open symbols) and FC (full symbols) $M(T)$ curves measured in applied field perpendicular to the thin film surface.

As our system is essentially 2-dimensional i.e. spatial arrangement of the particles is infinite in the thin film plane (xy-plane) with only 10 layers of particles in direction perpendicular to the thin film (z-direction), measurements were also performed with magnetic field applied perpendicular to the thin film to investigate possible anisotropy of the magnetic properties (Fig. 2). Comparing $M(T)$ ZFC and FC measurements with parallel and perpendicular applied field orientation (Figs. 1 and 2) striking anisotropy is observed: magnetic moment of the sample is approximately one order of magnitude smaller when measured in perpendicular orientation of the applied field with respect to the parallel one. Possible reason for the observed anisotropy of the magnetic properties are dipole-dipole interactions, as it will be discussed in the following paragraphs.

Measurements of the field dependence of magnetization of the sample for parallel and perpendicular orientation of the applied field at various constant temperatures were also performed. At temperatures ≤ 20 K $M(H)$ loops show a hysteresis which is a consequence of the slow relaxation of the magnetization at low temperatures (in Fig. 3 measurement at 2 K is shown). At high temperatures thermal energy is high enough so that magnetization of the system achieves equilibrium value during the measurement of one point and $M(H)$ loops are reversible (Fig. 4 shows measurement at 300 K). $M(H)$ curves are consistent with the $M(T)$ curves and also show anisotropy. Namely, $M(H)$ loops for magnetic field applied perpendicular to the thin film are less steep and reach saturation at higher field. This anisotropy is more pronounced at low temperature (2 K), but can also be noticed even at high temperature (300 K).

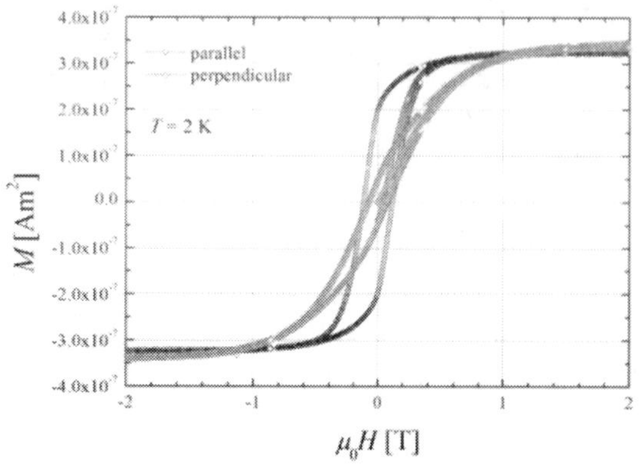

Figure 3. $M(H)$ curves for applied field in parallel and perpendicular direction at 2 K.

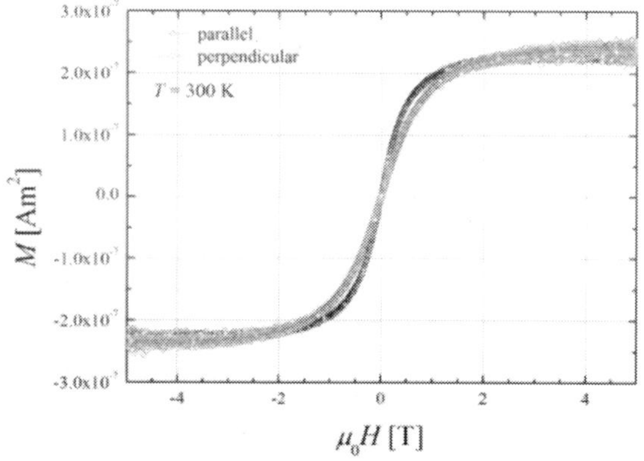

Figure 4. $M(H)$ curves for applied field in parallel and perpendicular direction at 300 K.

Both $M(T)$ and $M(H)$ measurements for both orientations of thin film with respect to the applied field suggest that particles' magnetic moments tend to lie in the thin film plane, which makes it harder to magnetize the system when magnetic field is applied perpendicular to the thin film. If we assume that anisotropy axes are oriented randomly, only the dipole-dipole magnetic interaction, which is anisotropic in its nature, can account for the observed anisotropy. To verify this assumption Monte Carlo simulations of the $M(H)$ curves were performed for the magnetic field applied in both parallel and perpendicular direction. The energy of the assembly of single domain particles is given by:

$$E = -K\sum_i\left(\frac{\vec{\mu}_i\cdot\hat{n}}{\mu_i}\right)^2 + \frac{\mu_0}{4\pi}\frac{1}{2}\sum_{i,j}\frac{1}{r_{ij}^3}\left(\vec{\mu}_i\cdot\vec{\mu}_j - 3\frac{(\vec{\mu}_i\cdot\vec{r}_{ij})(\vec{\mu}_j\cdot\vec{r}_{ij})}{r_{ij}^2}\right) - \mu_0\sum_i\vec{H}\cdot\vec{\mu}_i\,, \tag{1}$$

where K is the anisotropy constant, μ_i is the magnetic moment of the particle i, $\hat{n}$ is the unit vector of the anisotropy axis of the particle i, r_{ij} is the separation between particles i and j, H is the applied magnetic field and other symbols have their usual meaning. For convenience equation (1) was transformed to:

$$E' = -\sum_i(\hat{\mu}_i\cdot\hat{n})^2 + \frac{\mu_0\mu^2}{4\pi KVa^3}\frac{1}{2}\sum_{i,j}\frac{a^3}{r_{ij}^3}\left(\hat{\mu}_i\cdot\hat{\mu}_j - 3\frac{(\hat{\mu}_i\cdot\vec{r}_{ij})(\hat{\mu}_j\cdot\vec{r}_{ij})}{r_{ij}^2}\right) - \frac{\mu_0\mu}{KV}\sum_i\vec{H}\cdot\hat{\mu}_i\,, \tag{2}$$

where a is the lattice constant. We assumed that the particles are identical and that the anisotropy axes are oriented randomly. In the simulation constant $\mu_0\mu^2/8\pi KVa^3$ was set to 0.15 and the values of other parameters were typical for superparamagnetic systems. Simulation was performed for 100 particles arranged in 10x10 quadratic array in xy plane and the periodic boundary conditions were used for the calculation of the dipole-dipole energy. Monte Carlo simulation of the $M(H)$ curve was performed at constant temperature $T = 10$ K according to the standard Metropolis algorithm [12]. Results of the simulation are shown in Fig. 5. They confirm that for a regular 2D array of superparamagnetic particles the anisotropy of the macroscopic magnetic properties is a consequence of the dipole-dipole interactions between particles. In addition, coercive field of the simulated $M(H)$ curves is lower for the perpendicular case, which is also in accordance with the experimental results. Exact values of the saturation magnetization and coercive field of the simulated curves are different than the experimental ones due to the different choice of the parameters of the system. Main aim of the simulation was to demonstrate that dipole-dipole interactions between particles lead to anisotropic magnetic behaviour of the 2D superlattice of single domain particles.

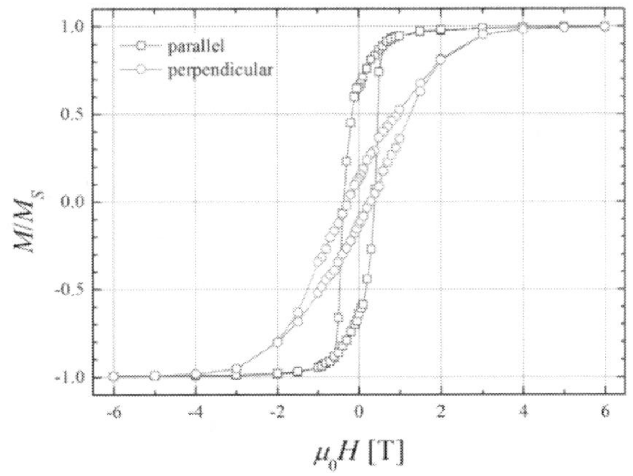

Figure 5. Simulated M(H) curves for parallel and perpendicular orientation of magnetic field.

4. Conclusion

Thin films of nanosized nickel particles embedded in amorphous alumina matrix were prepared by magnetron sputtering deposition and magnetic properties of the prepared samples were investigated. The nickel particles are single domain and they show blocking of magnetic moment at low temperatures. The dipole-dipole magnetic interaction between the particles influences the dynamics of the system and can lead to collective behaviour. Simulations performed using Monte Carlo method are in accordance with experimental results confirming that the used model is appropriate.

Acknowledgements
This work was supported by the Austrian agency for international mobility and cooperation in education, science and research (OeAD) under Ernst Mach Grant.

References

[1] Tatraj P, Morales M del P, Veintemillas-Verdaguer S, González-Carreño T and Serna C J 2003 *J. Phys. D: Appl. Phys.* **36** R182
[2] Pankhurst Q A, Connolly J, Jones S K and Dobson J 2003 *J. Phys. D: Appl. Phys.* **36** R167
[3] Moser A,Takano K, Margulies D T, Albrecht M, Sonobe Y, Ikeda Y, Sun S and Fullerton E E 2002 *J. Phys. D: Appl. Phys.* **35** R157
[4] Kittel C 1949 *Rev. Mod. Phys.* **21** 541
[5] Bedanta S and Kleemann W 2009 *J. Phys. D: Appl. Phys.* **42** 013001 (28pp)
[6] Jerčinović M, Radić N, Buljan M, Grenzer J, Delač-Marion I, Kralj M, Bogdanović-Radović I, Hübner R, Dubček P, Salamon K and Bernstorff S 2014 *J. Nanopart. Res.* **16** 2296
[7] Buljan M, Desnica U V, Ivanda M, Radić N, Dubček P, Dražić G, Salamon K, Bernstorff S and Holý V 2009 *Phys. Rev. B* **79** 035310
[8] Buljan M, Jerčinović M, Siketić Z, Bogdanović-Radović I, Delač Marion I, Kralj M, Ivanda M, Turković A, Dražić G, Bernstorff S and Radić N 2013 *J. Appl. Cryst.* **46** 1490
[9] Buljan M, Radić N, Bernstorff S, Dražić G, Bogdanović-Radović I and Holý V 2012 *Acta Crystallogr. A* **68** 124
[10] Bean C P and Livingston J D 1959 *J. Appl. Phys.* **30** 120S
[11] Knobel M, Nunes W C, Socolovsky L M, De Biasi E, Vargas J M and Denardin J C 2008 *J. Nanosci. Nanotechnol.* **8** 2836 and references therein
[12] Metropolis N, Rosenbluth A W, Teller A H and Teller E 1953 *J. Chem. Phys.* **21** 1087

Magnetization plateaux and jumps in frustrated four-leg spin tubes in magnetic fields

H D Rosales[1], **M Arlego** [1] **and F A Gómez Albarracín** [1]
[1] IFLP - Departamento de Física, Universidad Nacional de La Plata, C.C. 67, 1900 La Plata, Argentina
E-mail: rosales@fisica.unlp.edu.ar

Abstract. We study the ground state phase diagram of a frustrated spin-1/2 four-leg tube in an external magnetic field. We explore the parameter space of this model in the regime of all-antiferromagnetic exchange couplings by means of three different approaches: density matrix renormalization group (DMRG), a low-energy effective Hamiltonian (LEH) and a Hartree variational approach (HVA). We find that in the limit of weakly interacting plaquettes, singlet and triplet states play an important role in the formation of magnetization plateaux. We study the transition regions numerically and analytically, and find that they are described, at first order in a strong- coupling expansion, by an XXZ spin-1/2 chain in a magnetic field. These results are consistent with the DMRG and HVA calculations.

1. Introduction

Frustrated magnetism is a subject that has attracted much attention in the last decades and magnetic frustration is the key element in the search of exotics ground states[1]. Very often, low-energy effective models are used to get a better understanding of the physics of such frustrated systems[2]. In the last years there has been a growing interest to the phenomenon of magnetization plateaux in quasi one-dimensional spin systems, comprising chain, ladder and more involved magnetic structures spin systems. In a variety of models, the magnetization as a function of the applied magnetic field h exhibits plateaus at certain values; at those plateaux the magnetization in units of the saturation $m = M/M_s$ is locked at some rational number.

Recently, $Cu_2Cl_4{\cdot}D_8C_4SO_2$ has been established as a new spin 1/2 tube with an even number of legs [3] with diagonal coupling between adjacent legs. Recently in Ref.[4] this system has been studied at zero magnetic field revealing the existence of a gaped phase stable at a wide range of parameters. Motivated by this in this work, we study a frustrated four spin $\frac{1}{2}$ tube with a geometry depicted in Fig. 1 in a magnetic field and described by the following Hamiltonian:

$$\mathcal{H} = J_0 \sum_{n,a} \mathbf{S}_{n,a} \cdot \mathbf{S}_{n,a+1} + J_1 \sum_{n,a} \mathbf{S}_{n,a} \cdot \mathbf{S}_{n+1,a} + J_2 \sum_{n,a} \mathbf{S}_{n,a} \cdot \mathbf{S}_{n+1,a+1} - h \sum_{n,a} S^z_{n,a} \tag{1}$$

with a lattice structure and exchange antiferromagnetic couplings $J_{0,1,2}$ as shown in Fig. 1. Here $a = 1, ..., 4$ (resp. $n = 1, ..., N$) is a chain (resp. site) index, J_0 is the coupling in each plaquette, $J_{1,2}$ the couplings along the chains and the site $(n, 5)$ is identified with the site $(n, 1)$. For $J_{1,2} \ll J_0$, the quantum properties of the frustrated four spin tube can be understood in terms of weakly coupled four-spin plaquettes.

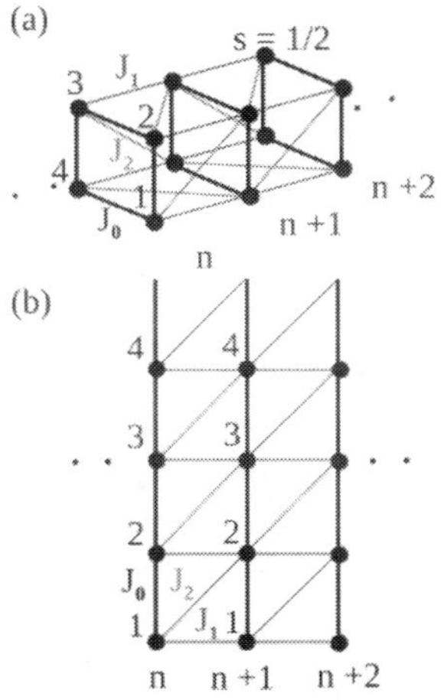

Figure 1. Diagram of the four spin tube studied. The four spins in the plaquette are indicated. Bonds mediated by J_0 are in red, by J_1 in blue and by J_2 in green. Figure taken from Ref. [4]

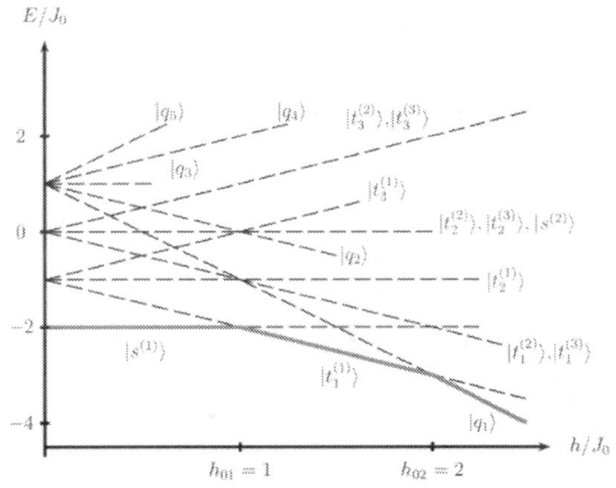

Figure 2. Energy of the states of a single plaquette of $S = 1/2$ as a function of the magnetic field. The lowest energy states and the values of the magnetic field where the energy is degenerate are indicated.

2. Low Energy Models

We treated the terms which contain interactions between plaquettes as a perturbation from the exactly solvable Hamiltonian of independent plaquettes $J_1, J_2 \ll J_0$. For $J_{1,2} = 0$, the system consists of independent plaquettes. The sixteen states of a given plaquette fall into a two spin-0 singlets ($|s^{(1)}\rangle$, $|s^{(2)}\rangle$), three spin-1 triplets ($|t_i^{(1)}\rangle$, $|t_i^{(2)}\rangle$ and $|t_i^{(3)}\rangle$ with $i = 1, ..., 3$) and one spin-2 quintuplet $|q_i\rangle$ (with $i = 1, ..., 5$). For no magnetic field, the ground state is not degenerate. When the magnetic field is switched on there are two particular values of the magnetic field $h_{01} = J_0$ and $h_{02} = 2\,J_0$ where two states cross (see Fig. 2). These two special fields define two separate regimes where we perform the perturbation expansion. For independent plaquettes, there are three magnetization plateaux, at $m = 0$, $m = 1$ and $m = 2$ corresponding to a singlet, a triplet and the quintuplet states. However the inclusion of interaction between plaquettes is expected to produce the emergence of other plateaux as well as the renormalization of existing ones.

In the first regime, when $|(h - J_0)| \ll J_0$ each plaquette can be found either in the singlet or in the triplet state. In the second regime, when $|(h - 2\,J_0)| \ll J_0$ each plaquette can be found either in the triplet or in the quintuplet state. Effectively, we can treat this as a system with only two states per site, using perturbation theory in J_1, J_2 to define a Hamiltonian that acts on this reduced Hilbert space. Defining the usual spin-1/2 matrices S^x, S^y and S^z acting on the reduced Hilbert space, one has at first order

$$\mathcal{H}_{\text{eff}} = \sum_{n=1}^{N} J_{xy}(S_n^x S_{n+1}^x + S_n^y S_{n+1}^y) + J_{zz}\, S_n^z S_{n+1}^z - h^{\text{eff}} \sum_{n=1}^{N} S_n^z \tag{2}$$

where

$$h^{\rm eff} = h - J_0 - \frac{J_1 + J_2}{4}, J_{xy} = \frac{4(J_1 - J_2)}{3}, J_{zz} = \frac{(J_1 + J_2)}{4}, \quad \text{for} \quad |(h - J_0)| \ll J_0 \qquad (3)$$

$$h^{\rm eff} = h - 2J_0 - \frac{3(J_1 + J_2)}{4}, J_{xy} = J_1 - J_2, J_{zz} = \frac{J_1 + J_2}{4}, \quad \text{for} \quad |(h - 2J_0)| \ll J_0 \qquad (4)$$

The XXZ chain is gapped in the antiferromagnetic regime $J_{zz}/|J_{xx}| > 1$ and $h^{\rm eff} = 0$, with zero magnetization in the ground state. The presence of the energy gap means that the magnetization remains exactly zero in a finite range of fields $-h_{\rm min} < h^{\rm eff} < h_{\rm min}$. At $\pm h_{\rm min}$ the energy gap vanishes and the spinons condense. As $|h^{\rm eff}|$ is increased further, the ground state becomes a sea of spinons with a growing magnetization. The system in this regime is a Luttinger liquid with continuously varying critical exponents. Finally, at $h^{\rm eff} = \pm h_{\rm max}$ the magnetization of the XXZ chain saturates. The critical fields $h_{\rm min}$ and $h_{\rm max}$ of the XXZ chain are known exactly by Bethe ansatz solution[5].
To be more specific for the first regime around h_{01}, the effective S^z represents the plaquette $S^z = 0(1)$, corresponding to the singlet(triplet) state. So the plateaux in the curve of magnetization vs $h^{\rm eff}$ in the effective model at $M = -\frac{1}{2}$, 0 and $\frac{1}{2}$ translates into plateaux at $M = 0$, $\frac{1}{2}$ and 1 in the curve of magnetization vs h. The same idea applies to the field sector around h_{02}, where effective S^z represents plaquette $S^z = 1(2)$, corresponding to triplet(quintet) involved in the high field crossing. In this case, plateaux at $M = 0$, $\frac{1}{2}$ and 1 in effective model translate into $M = 1$, $\frac{3}{2}$ and 2 in plaquette model.

3. Variational approach

The simplest way to map out the phase diagram is to use a Hartree variational approach (HVA) for the wave function allowing a linear combination of low-energy states comprising of singlet, triplet and quintuplet states on each plaquette (we keep the plaquette wave function entangled), while we retain the product form over the bonds[6]

$$|\psi\rangle = \prod_{n=1}^{N/4} |\nu_n\rangle \qquad (5)$$

where $|\nu_n\rangle = \sum_{i=1}^{Ne} \alpha_i |i\rangle$; $|i\rangle$ are the eigenstates of the plaquette and α_i are complex constants which satisfy $\sum_i |\alpha_i|^2 = 1$. The variational parameters α_i are then determined by minimizing the energy $E = \langle\psi|\mathcal{H}|\psi\rangle$. Replacing Eq. (5) in Eq. (1) we obtain

$$\begin{aligned} E &= J_0 \sum_n \vec{\alpha}_n^\dagger . M_d . \vec{\alpha}_n - h \sum_{n,a} \vec{\alpha}_n^\dagger \cdot M_a^z \cdot \vec{\alpha}_n + J_1 \sum_{n,\gamma,a} \left(\vec{\alpha}_n^\dagger \cdot M_a^\gamma \cdot \vec{\alpha}_n\right)\left(\vec{\alpha}_{n+1}^\dagger \cdot M_a^\gamma \cdot \vec{\alpha}_{n+1}\right) \\ &+ J_2 \sum_{n,\gamma,a} \left(\vec{\alpha}_n^\dagger \cdot M_a^\gamma \cdot \vec{\alpha}_n\right)\left(\vec{\alpha}_{n+1}^\dagger \cdot M_{a+1}^\gamma \cdot \vec{\alpha}_{n+1}\right) \end{aligned} \qquad (6)$$

The ground state is obtained using simulated annealing on lattices with periodic boundary condition and choosing the state with lowest energy per site. Simulations on each size were done by an exponential annealing schedule and the whole process was repeated many times to ensure stability of results. While the algorithm does not guarantee convergence to the ground state we nevertheless believe an accurate picture emerges since this method captures the main features of the low energy physics, compatible with the analytical calculations and DMRG simulations. In figures 3 and 4 we show typical magnetization curves obtained with HVA.

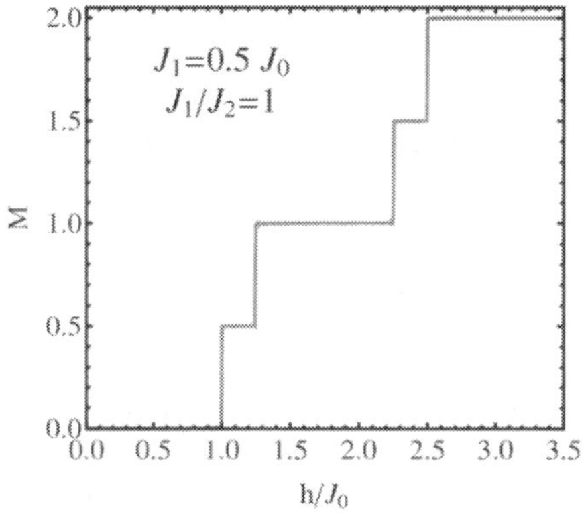

Figure 3. M vs h obtained by variational wave function approach for $J_1 + J_2 = 1$ and $J_2/J_1 = 1$.

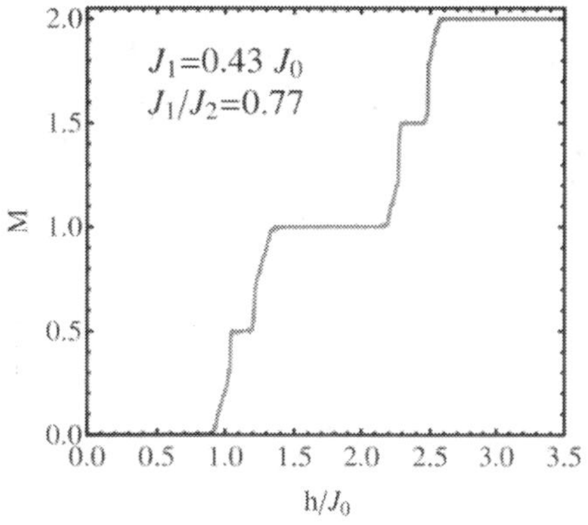

Figure 4. M vs h obtained by variational wave function approach for $J_1 + J_2 = 1$ and $J_2/J_1 = 0.77$.

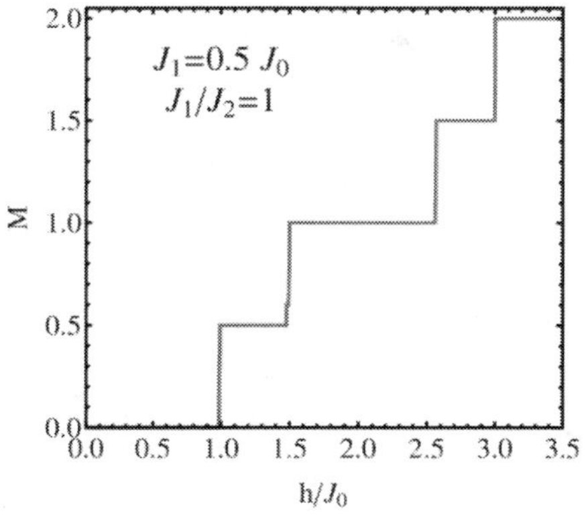

Figure 5. M vs h obtained by DMRG for $N = 4 \times 20$ sites and $J_1 + J_2 = 1$ and $J_2/J_1 = 1$.

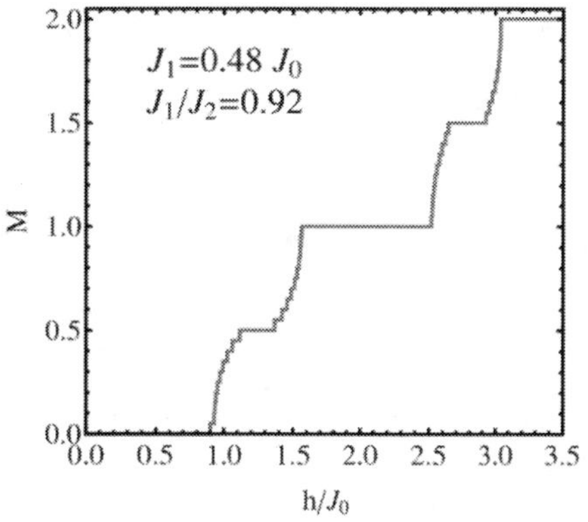

Figure 6. M vs h obtained by DMRG for $N = 4 \times 20$ sites and $J_1 + J_2 = 1$ and $J_2/J_1 = 0.48$.

4. DMRG analysis

In previous sections we provide the perturbative and qualitative picture expected when the couplings $J_1 - J_2$ connect the plaquettes. To have a complete picture we study the system using extensive DMRG computations. More specifically, we compute the ground state energy $E(S^z_{total}, h = 0)$ of Eq. (1) in the complete set of S^z_{total} subspaces using periodic boundary conditions, and keeping just 500 states was shown to be enough to assure the accuracy of the calculation. As usual, adding the Zeeman term, we solve the equation $E(S^z_{total}, h) = E(S^z_{total} + 1, h)$ to obtain the normalized magnetization where the plateaux are showing up. This procedure allows us to compute the actual width of the plateaux. From the DMRG data, we confirm the presence of two fractional plateaux at $M = 1/2$ and $M = 3/2$ in a wide range of parameter space J_1-J_2. In figures 5 and 6 we show typical magnetization curves obtained numerically which confirm those obtained by HVA and perturbation analysis.

5. Conclusion

In this paper, we have shown that frustrated interactions in a system consisting of non-frustrated plaquettes of spins-1/2 leads to the formation of rational plateaux that are not present in the non-frustrated case. With our analisys based in degenerate perturbation theory, a variational wave function approach and DMRG simulations show that the phases of plateaux are robust and hold up to large values of the coupling between plaquettes.

References

[1] Balents L 2010 *Nature* **464** 199–208 URL `http://dx.doi.org/10.1038/nature08917`

[2] Cabra D and Pujol P 2004 *Quantum Magnetism* (*Lecture Notes in Physics* vol 645) ed Schollwck U, Richter J, Farnell D and Bishop R (Springer Berlin Heidelberg) pp 253–305 ISBN 978-3-540-21422-9 URL `http://dx.doi.org/10.1007/BFb0119596`

[3] Garlea V O, Zheludev A, Regnault L P, Chung J H, Qiu Y, Boehm M, Habicht K and Meissner M 2008 *Phys. Rev. Lett.* **100**(3) 037206 URL `http://link.aps.org/doi/10.1103/PhysRevLett.100.037206`

[4] Arlego M, Brenig W, Rahnavard Y, Willenberg B, Rosales H D and Rossini G 2013 *Phys. Rev.* B **87**(1) 014412 URL `http://link.aps.org/doi/10.1103/PhysRevB.87.014412`

[5] Yang C N and Yang C P 1966 *Phys. Rev.* **151**(1) 258–264 URL `http://link.aps.org/doi/10.1103/PhysRev.151.258`

[6] Romhányi J, Totsuka K and Penc K 2011 *Phys. Rev.* B **83**(2) 024413 URL `http://link.aps.org/doi/10.1103/PhysRevB.83.024413`

Exotic Quantum Phase Transition of the Spin Nanotube

Toru Sakai, Hiroki Nakano and Kouichi Okunishi

Japan Atomic Energy Agency, SPring-8, Hyogo 679-5148, Japan
Graduate School of Material Science, University of Hyogo, Hyogo 678-1297, Japan

Department of Physics, Niigata University, Niigata 950-2181, Japan

E-mail: sakai@spring8.or.jp

Abstract. The S=1/2 three-leg spin tube is investigated using the numerical diagonalization up to the 42-spin clusters. A precise finite-size scaling analysis, so called the level spectroscopy, is carried out. It indicates a quantum phase transition between the spin-gap and gapless phases, with respect to the ratio of the leg and rung exchange interactions. The present analysis gives a precise estimate for the critical value of the ratio.

1. Introduction

The spin nanotube[1] is one of interesting nanomaterials which are expected to be developed towards some functional devices in the near future, like the carbon nanotube. Since the S=1/2 three-leg spin tube $[(CuCl_2tach \cdot H)_3Cl]Cl_2$ was synthesized[2], several theoretical works about this systems have been published[3-12]. One of the interesting point of the S=1/2 three-leg spin tube is that it has the spin gap, which is an energy gap between the singlet ground state and the triplet excitation, although the S=1/2 three-leg spin ladder is gapless. The previous density matrix renormalization group (DMRG) calculation[11] revealed that the spin gap is open for sufficiently large exchange interactions in the rung (interchain) direction of the spin tube. On the other hand, the system is gapless in the weak rung coupling limit, because it corresponds to the three independent S=1/2 antiferromagnetic spin chains. Thus a quantum phase transition should occur between the spin-gap and gapless phases, if the ratio of the rung and leg exchange interactions is varied. In our previous level spectroscopy analysis combined with the numerical diagonalization up to 30-spin clusters[8], it was difficult to determine whether a finite critical ratio exists or an infinitesimal rung coupling induces the spin gap, because the finite-size effect is quite large. Recently we performed the numerical diagonalization of the S=1/2 three-leg spin tube with system sizes up to 42 spins. In particular, calculations of the 42-spin cluster was carried out by our program running at the K Computer at RIKEN, Kobe. In the present paper, we report a result from the level spectroscopy analysis applied for the data of the numerical diagonalization up to 42-spin clusters and conclude that the gapless-gapped quantum phase transition occurs at a finite critical ratio.

2. Model

We consider the S=1/2 three-leg spin tube, shown in Fig. 1, described by the Hamiltonian

$$\hat{H} = J_1 \sum_{i=1}^{3} \sum_{j=1}^{L} \vec{S}_{i,j} \cdot \vec{S}_{i,j+1} + J_2 \sum_{i=1}^{3} \sum_{j=1}^{L} \vec{S}_{i,j} \cdot \vec{S}_{i+1,j} \qquad (1)$$

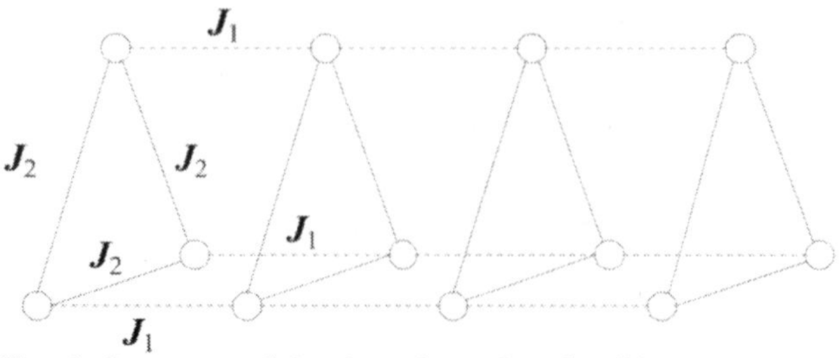

Fig. 1 Structure of the three-leg spin tube (1).

where $\vec{S}_{i,j}$ is the spin-1/2 operator and L is the length of the tube in the leg direction. The exchange interaction constant $J_1$is for the neighbouring spin pairs along the legs, while J_2 is the rung interaction constant. All the exchange interactions are supposed to be antiferromagnetic (namely, positive).

We performed the numerical diagonalization based on the Lanczos algorithm for finite-size clusters with system sizes up to 42 spins, namely L=14, under the periodic boundary condition, and obtained the single ground-sate and triplet excited-state energies. In addition we calculated the single excitation energy, namely the second eigenvalue in the S=0 sector, for the level spectroscopy analysis. The single-triplet excitation energy and single-single one are denoted as E_{01} and E_{00}, respectively.

We carried Lanczos diagonalizations using our MPI-parallelized code which was originally developed in the study of the Haldane gaps[13]. The usefulness of our program was confirmed in large-scale parallelized calculations[14,15]. Note also that this program contributed much for the progress of the studies of anomalous magnetization jump in frustrated magnets[16,17,18].

3. Quantum Phase Transition

The previous DMRG calculation[11] suggested that the system (1) has a spin gap for sufficiently large J_2. According to the Lieb-Schultz-Mattis theorem, the spin gap phase of the S=1/2 three-leg spin tube should have two-fold degeneracy in the ground state, due to the translational symmetry break down. A schematic picture of the gapped phase for the symmetric spin tube was proposed as shown in Fig. 2, where the three equivalent singlet dimer covering patterns are resonating at each two unit cell triangles.

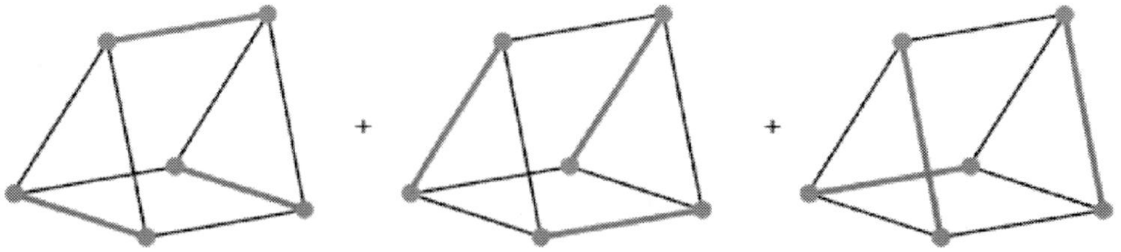

Fig. 2. A schematic picture of the spin gap formation mechanism of the S=1/2 three-leg spin tube. Circles and thick lines are spins and singlet bonds, respectively. The three equivalent patterns at each two unit cells are resonating.

On the other hand, the system is gapless in the weak J_2 limit where it corresponds to three independent spin chains. The purpose of the present work is to determine the critical point of J_2/J_1 where a quantum phase transition occurs between the gapless and gapped phases.

4. Level Spectroscopy

The quantum phase transition in the present case is expected to belong to the universality class of the Berezinkii-Kosteritz-Thouless transition. Around the transition the logarithmic system size correction proportional to 1/log L sometimes appears and as a result it is difficult to determine a precise critical point in the thermodynamic limit because of quite large finite-size effects. In such a case the level spectroscopy[19] is one of the best method to determine it. According to this method, the critical point is estimated as an intersection between the triplet and singlet excitation energies, where the logarithmic correction vanishes. The result of our previous level spectroscopy analysis up to 30-spin clusters was J_2/J_{1c}=0.5 ± 0.5. The error due to the extrapolation to the thermodynamic limit was still too large to determine whether the critical point is finite or zero. Then in the present work, we calculated the triplet and singlet excitation energies up to 42-spin clusters and applied the same method for them. The level cross points for 36- and 42-spin clusters are shown in Figs. 3 and 4, respectively.

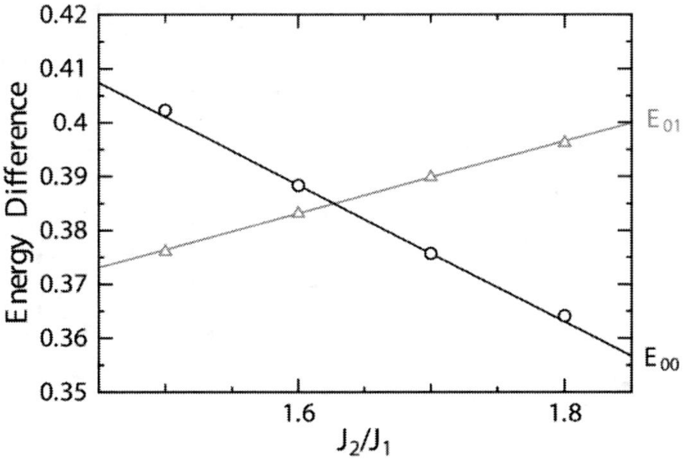

Fig.3. Singlet-triplet excitation energy E_{01}(triangles) and singlet-singlet one E_{00}(squares) for L=36 are plotted versus J_2/J_1 to estimate a level cross point.

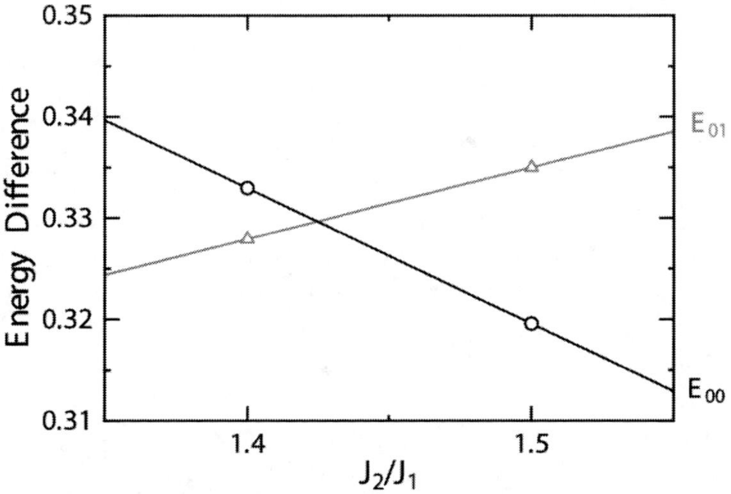

Fig.4. Singlet-triplet excitation energy E_{01}(triangles) and singlet-singlet one E_{00}(squares) for L=42 are plotted versus J_2/J_1 to estimate a level cross point.

Since the size correction of the level cross points is expected to be proportional to $1/L^2$, we estimated a quantum critical point in the thermodynamic limit by plotting them versus $1/L^2$ shown in Fig. 5. It includes the present results for L=12 (36 spins) and 14 (42 spins), as well as previous ones up to L=10. The present extrapolation indicates J_2/J_{1c}=0.71±0.21. If we estimate it only from the largest two system sizes (L=12 and 14), the result is J_2/J_{1c}=0.864. This is a strong evidence to confirm that the quantum phase transition occurs at a finite critical point of J_2/J_1. Thus the S=1/2 three-leg spin tube is gapless for weak rung exchange interactions. The present result suggests that the spin gap is not open, until the rung coupling is comparable with the leg one.

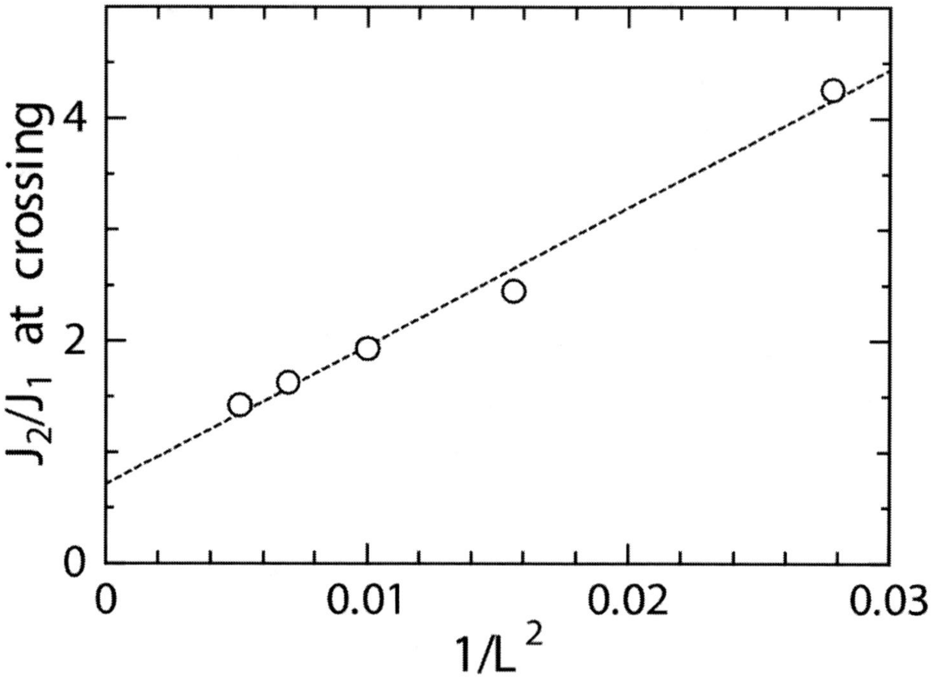

Fig.5. Extrapolation of the level cross point to the thermodynamic limit, assuming that the size correction is proportional to $1/L^2$. The estimated critical point is J_2/J_1=0.71±0.21.

5. Summary
The S=1/2 three-leg spin tube is investigated by the level spectroscopy combined with the numerical diagonalization up to the 42-spin cluster. It leads to the conclusion that a quantum phase transition between the gapless and spin-gap phases occurs at a finite critical ratio of the rung and leg coupling constants. The critical value is estimated as J_2/J_1=0.71±0.21.

Acknowledgments

This work has been partly supported by Grants-in-Aids for Scientific Research (B) (No.23340109) and (C)(Nos. 24540348 and 26400387) from the Ministry of Education, Culture, Sports, Science and Technology of Japan (MEXT). This research used computational resources of the K computer provided by the RIKEN Advanced Institute for Computational Science through the HPCI System Research projects (Project ID:hp130070 and hp130098). We further thank the Supercomputer Center, Institute for Solid State Physics, University of Tokyo; the Cyberscience Center, Tohoku University; and the Computer Room, Yukawa Institute for Theoretical Physics, Kyoto University; the Department of Simulation Science, National Institute for Fusion Science; Center for Computational Materials Science, Institute for Materials Research, Tohoku University; Supercomputing Division, Information Technology Center, The University of Tokyo for computational facilities. This work was partly supported by the Strategic Programs for Innovative Research, MEXT, and the Computational Materials Science Initiative, Japan. The authors would like to express their sincere thanks to the staff members of the Center for Computational Materials Science of the Institute for Materials Research, Tohoku University for their continuous support of the SR16000 supercomputing facilities.

References

[1] T. Sakai, M. Sato, K. Okunishi, K. Okamoto and C. Itoi, *J. Phys.: Condens. Matter* **22**, 403201 (2010).
[2] J. Schnack, H. Nojiri, P. Koegerler, G. J. T. Cooper and L. Cronin, *Phys. Rev. B*, **70**, 174420 (2004).
[3] T. Sakai, M. Matsumoto, K. Okunishi, K. Okamoto, and M. Sato, *Physica E*, **29** 633 (2005)
[4] J. -B. Fouet, A. Laeuchli, A. Pilgram, R. M. Noak and F. Mila, *Phys. Rev. B*, **73**, 014409 (2006).
[5] K. Okunishi, S. Yoshikawa, T. Sakai and S. Miyashita, *Prog. Theor. Phys. Suppl.*, **159**, 297 (2005).
[6] M. Sato and T. Sakai, *Phys. Rev. B*, **75**, 014411 (2007).
[7] S. Nishimoto and M. Arikawa, *Phys. Rev. B*, **78**, 054421 (2008).
[8] T. Sakai, M. Sato, K. Okunishi, Y. Otsuka, K. Okamoto and C. Itoi, *Phys. Rev. B*, **78**, 184415 (2008).
[9] D. C. Cabra, A. Honecker and P. Pujol, *Phys. Rev. B*, **58**, 6241 (1998).
[10] H. J. Schulz, in *Correlated Fermions and Transport in Mesoscopic Systems,* ed. T. Martin, G. Montambaux, J. Tran Than Van (1996); arXive:9605075.
[11] K. Kawano and M. Takahashi, *J. Phys. Soc. Jpn.*, **66**, 4001 (1997).
[12] K. Okunishi, M. Sato, T. Sakai, K. Okamoto and C. Itoi, *Phys. Rev. B*, **85**, 054416 (2012).
[13] H. Nakano and A. Terai, *J. Phys. Soc. Jpn.***78**, 014003 (2009).
[14] H. Nakano and T. Sakai, *J. Phys. Soc. Jpn.* **80**, 053704 (2011).
[15] H. Nakano S. Todo, and T. Sakai, *J. Phys. Soc. Jpn.* **82**, 043715 (2013).
[16] H. Nakano and T. Sakai, *J. Phys. Soc. Jpn.* **82**, 083709 (2013)
[17] H. Nakano, M. Isoda, and T. Sakai, *J. Phys. Soc. Jpn.* **83**, 053702 (2014)
[18] H. Nakano, T. Sakai, and Y. Hasegawa, arXiv:1406.4964.
[19] K. Okamoto and K. Nomura, *Phys. Lett.*, **169**,433 (1992).

Novel Field Induced Quantum Phase Transition of the Kagome Lattice Antiferromagnet

Tôru Sakai[1,2] **and Hiroki Nakano**[2]

[1]Japan Atomic Energy Agency, SPring-8, Sayo, Hyogo 679-5148, Japan
[2]Graduate School of Material Science, University of Hyogo, Kamigori, 678-1297, Japan

E-mail: sakai@spring8.or.jp

Abstract. The $S = 1/2$ kagomelattice antiferromagnet is investigated by the numerical diagonalization. We successfully obtained the ground-state full magnetization curve for the 42-spin cluster. Based on the finite-size scaling alnalysis including this new result, we consider the quatnum critical magnetization behavior around the 1/3 of the saturation magnetization. As a result, some unconventional features of the 1/3 magnetization ramp, which was proposed by our previous work, are clarified.

1. Introduction

The $S = 1/2$ Kagome-lattice antiferromagnet[1] is one of the most popular frustrated quantum spin systems. The previous theoretical studies indicated that the system is disordered in the ground state [2, 3, 4, 5, 6, 7, 8, 9, 10, 11, 12, 13]. In contrast to the triangular lattice antiferromagnet which has the 120° long-range order[14, 15, 16]. Experimental studies to observe a novel spin liquid phase in the Kagome-lattice antiferromagnet have been accelarated since discoveries of several realistic materials; the herbertsmithite[17, 18], the volborthite[19, 20] and the vesignieite[21, 22]. Since the quantum Monte Carlo simulation is useless for the system, the numerical exact diagonalization is one of the best numerical method for it, as well as the density matrix renormalization group (DMRG) calculation. The numerical diagonalization studies suggested a magnetization plateau-like behavior at 1/3 of the saturation magnetization [23, 24, 25, 26, 27], although the classical spin systems have no plateau in the ground state[28]. In our recent numerical diagonalization study on the $S = 1/2$ Kagome-lattice antiferromagnet up to $N = 36$, the calculated field derivatives revealed an anomalous behavior at 1/3 of the saturation magnetization[29]. Namely, the field derivative is divergent at the low-field side of the critical field H_c, while almost zero at the high-field side. This critical behavior is quite different from conventional magnetization plateaux in two-dimensional systems where the field derivative is finite at both sides of H_c. To distinguish such an anomalous property at the 1/3 magnetization of the Kagome-lattice from conventional plateaux, we called it a "magnetization ramp". Our recent finite-size scaling analyses up to N=39 revealed some characteristic features of the magnetization ramp quantitatively[30, 31]. In these works the critical exponent defined by

$$|m - \frac{1}{3}| \sim |H - H_c|^{\frac{1}{\delta}}, \quad (1)$$

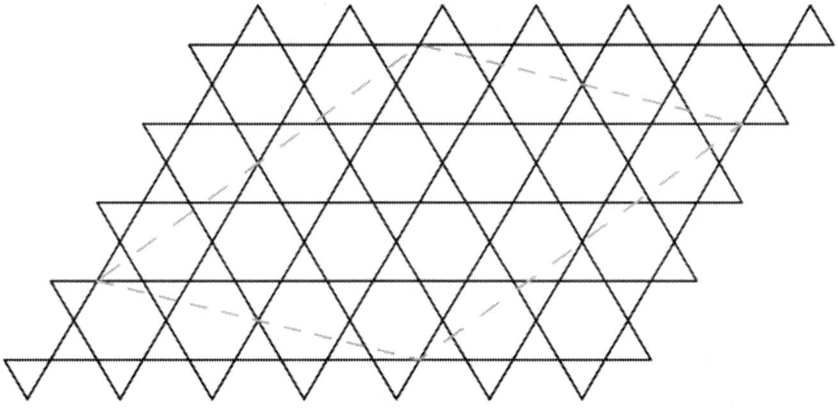

Figure 1. Shape of the kagome-lattice cluster for N=42 under the periodic boundary condition.

was revealed to be larger than one at the lower-field side, while smaller than one at the higher-field side. Namely, the divergent(zero) field derivative at the lower(higher)-field side was confirmed. However, whether a finite plateau width (flat region) appears or not at the 1/3 magnetization was not cnclusive, within the finite-size scaling analysis up to N=39, because finite-size effects are still quite large.

In the present study, in order to clarify whether a finite plateau exists or not at the 1/3 magnetization of the S=1/2 kagome-lattice antiferromagnet, we extended the same finite-size scaling analysis up to the larger system size $N = 42$, using the numerical diagonalization at K Computer Center in Kobe, Japan. The result based on the analysis up to $N = 42$ is reported. The magnetization curve for $N = 42$, which is the largest system size in present in the world obtained by the numerical diagonalization, is also presented.

2. Model

The magnetization processes of the $S = 1/2$ Kagome- and triangular-lattice antiferromagnets are described by the Hamiltonian

$$\mathcal{H} = \mathcal{H}_0 + \mathcal{H}_Z, \qquad \mathcal{H}_0 = \sum_{\langle i,j \rangle} \mathbf{S}_i \cdot \mathbf{S}_j, \qquad \mathcal{H}_Z = -H \sum_j^N S_j^z, \tag{2}$$

where $\langle i, j \rangle$ means all the nearest neighbor pairs on each lattice. Throughout we use the unit such that $g\mu_B = 1$. For N-site systems, the lowest energy of $\mathcal{H}_0$ in the subspace where $\sum_j S_j^z = M$ is denoted as $E(N, M)$. The macroscopic magnetization m is defined by $m \equiv M/M_s = 2M/N$, where $M_s = N/2$ is the saturation magnetization. In our previous work[30] we restrict us to the rhombic cluster under the periodic boundary condition to investigate the 1/3 magnetization. The used system sizes are N =9, 12, 21, 27, 36 and 39. Using the numerical exact diagonalization, we can calculate all the values of $E(N, M)$ for these clusters.

In the present work, we treat the larger system size, $N = 42$, although it is not rhombic. The shape of the 42-spin cluster used here is shown in Fig. 1.

3. Results

Using the numerical diagonalization based on the Lanczos method, we successfully obtained the full ground-state magnetization curve for $N = 42$, which is shown in Fig. 2. At least as the

Journal of Physics: Conference Series **568** (2014) 042025 doi:10.1088/1742-6596/568/4/042025

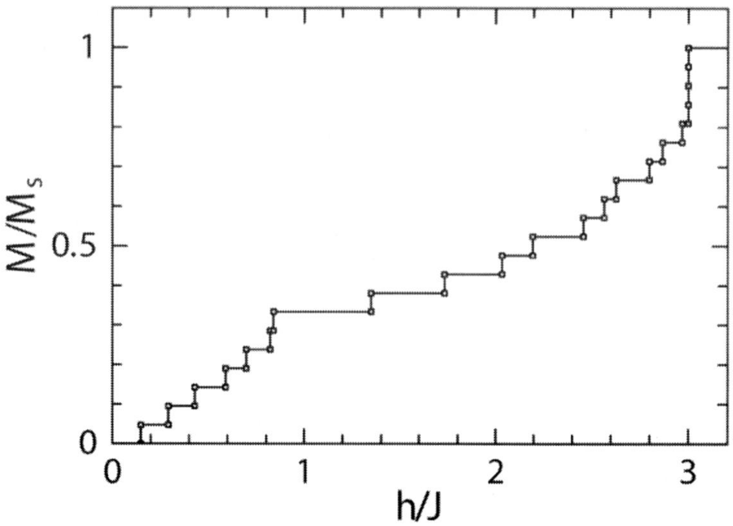

Figure 2. Ground-state magnetization curve of the S=1/2 kagome-lattice antiferromagnet calculated by the numerical diagonalization for the 42-spin cluster.

magnetization curve of the S=1/2 kagome-lattice antiferromagnet obtained by the numerical diagonalization, this system size is the present world record.

The field derivative χ estimated by the form

$$\chi^{-1} = \frac{E(N, M+1) - 2E(N, M) - E(N, M+1)}{1/M_{\rm sat}}, \qquad (3)$$

is more useful to clarify characters of the magnetization ramp, than the simple magnetization curve. The magnetization dependence of χ for N =27, 36, 39 and 42 is shown in Fig. 3 The present result for $N = 42$ qualitatively confirms the properties of the magnetization ramp; $\chi \to \infty$ at lower-field side, while $\chi = 0$ at higher-field side in the thermodynamic limit.

Finally, using the present result for $N = 42$, we try to clarify whether the 1/3 magnetization plateau exists or not. The plateau widths $H_{c+} - H_{c-}$ of the finite size systems for N =9, 12, 15, 18, 21, 24, 27, 30, 36, 39 and 42 are plotted versus $1/N^{1/2}$ in Fig. 4. If we use all the points to extrapolate them to the thermodynamic limit, the result is $H_{c+} - H_{c-} = -0.3 \pm 0.4$. It can still suggest the absence of the plateau. However, the plot in Fig. 4 obvious ly indicates a drastic change between $N = 18$ and $N = 21$. If we neglect smaller-size data than the change, the extraplation leads to the result $H_{c+} - H_{c-} = 0.5 \pm 0.3$. It possibly suggests a finite plateau, which is consistent with the recent numerical diagonalization[36] and density matrix renormalization group (DMRG)[37] studies. Even in such a case, the divergent χ before the plateau and the zero χ after the plateau are still unconventional and interesting phenomena.

4. Summary

The magnetization process of the $S = 1/2$ kagome-lattice antiferromagnet is investigated using the numerical diagonalisation up to $N = 42$. The present results support the caracters of the 1/3 magnetization ramp; $\chi \to \infty$ at the lower field side and $\chi = 0$ at the higher field side of it. On the other hand, it would be difficult to exclude the existence of the 1/3 plateau with the present result up to $N = 42$.

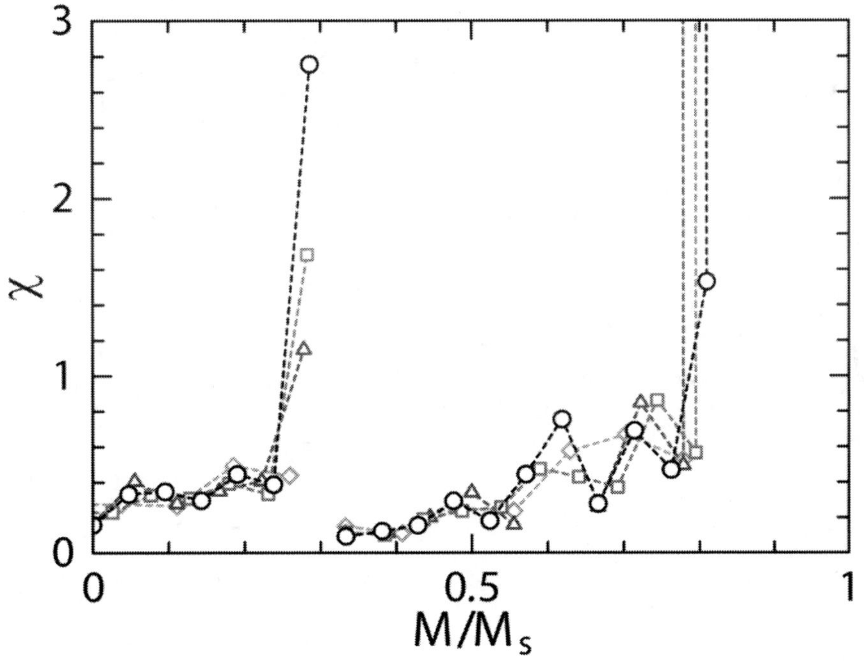

Figure 3. Magnetization dependence of the field derivative $\chi = dM/dH$ calculated by the numerical diagonalization for N =27(diamonds), 36(triangles), 39(squares) and 42(circles).

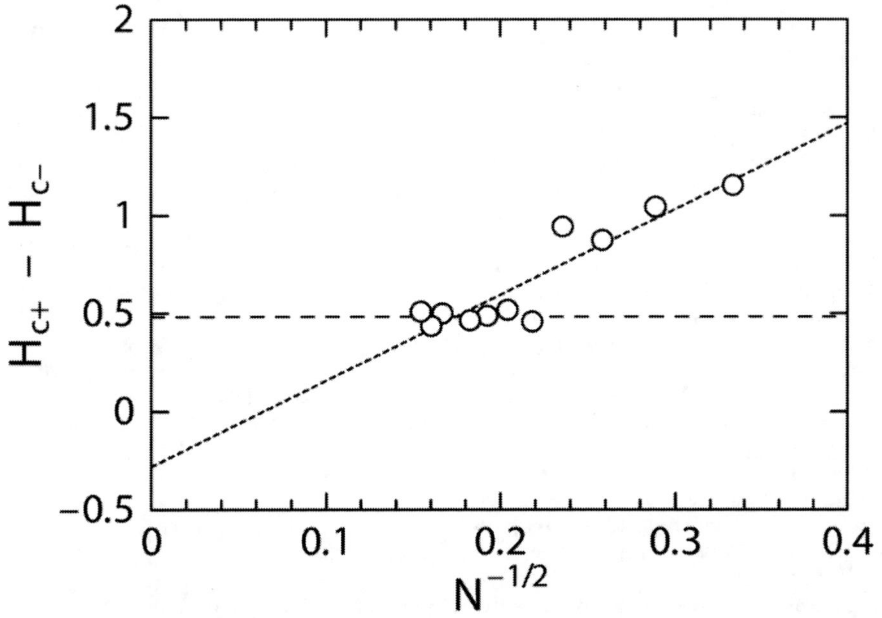

Figure 4. Magnetization plateau width $H_{c+} - H_{c-}$ plotted versus $1/N^{1/2}$.

4.1. Acknowledgments

This work has been partly supported by Grants-in-Aids for Scientific Research (B) (No.23340109) and (C)(Nos. 24540348 and 26400387) from the Ministry of Education, Culture, Sports, Science and Technology of Japan (MEXT). This research used computational resources of the K computer provided by the RIKEN Advanced Institute for Computational Science through the HPCI System Research projects (Project ID:hp130070 and hp130098). We further thank the Supercomputer Center, Institute for Solid State Physics, University of Tokyo; the Cyberscience Center, Tohoku University; and the Computer Room, Yukawa Institute for Theoretical Physics, Kyoto University; the Department of Simulation Science, National Institute for Fusion Science; Center for Computational Materials Science, Institute for Materials Research, Tohoku University; Supercomputing Division, Information Technology Center, The University of Tokyo for computational facilities. This work was partly supported by the Strategic Programs for Innovative Research, MEXT, and the Computational Materials Science Initiative, Japan. The authors would like to express their sincere thanks to the staff members of the Center for Computational Materials Science of the Institute for Materials Research, Tohoku University for their continuous support of the SR16000 supercomputing facilities.

References

[1] A brief history of the study on the Kagome-lattice magnetism was given in M. Mekata, Phys. Today Feb., 12 (2003).
[2] P. Lecheminant, B. Bernu, C. Lhuillier, L. Pierre, and P. Sindzingre, Phys. Rev. B **56**, 2521 (1997).
[3] Ch. Waldtmann, H.-U. Everts, B. Bernu, C. Lhuillier, P. Sindzingre, P. Lecheminant and L. Pierre, Eur. Phys. J. B **2**, 501 (1998).
[4] F. Mila, Phys. Rev. Lett. **81**, 2356 (1998).
[5] Y. Ran, M. Hermele, P. A. Lee and X. -G. Wen, Phys. Rev. Lett. **98**, 117205 (2007).
[6] H. C. Jiang, Z. Y. Weng and D. N. Sheng, Phys. Rev. Lett. **98**, 117203 (2008).
[7] P. Sindzingre and C. Lhuillier, Europhys. Lett. **88**, 27009 (2009).
[8] O. Cepas, C. M. Fong, P. W. Leung, and C. Lhuillier, Phys. Rev. B **78**, 140405(R) (2008).
[9] G. Evenbly and G. Vidal, Phys. Rev. Lett. **104**, 187203 (2010).
[10] A. M. Läuchli, J. Sudan and E. S. Sorensen, Phys. Rev. B **83**, 21241 (2011).
[11] Y. Iqbal, F. Becca and D. Poilblanc, Phys. Rev. B **83**, 100404 (2011); ibid **84**, 020407 (2011).
[12] S. Yan, D. A. Huse and S. R. White, Science **332**, 1173 (2011).
[13] H. Nakano and T. Sakai: J. Phys. Soc. Jpn. **80**, 053704 (2011).
[14] D. A. Huse and V. Elser, Phys. Rev. Lett. **60**, 2531 (1988).
[15] B. Bernu, P. Lecheminant, C. Lhuillier and L. Pierre, Phys. Rev. B **50**, 10048 (1994).
[16] L. Capriotti, A. E. Trumper and S. Sorella, Phys. Rev. Lett. **82**, 3899 (1999).
[17] M. P. Shores, E. A. Nytko, B. M. Barlett, and D. G. Nocera, J. Am. Chem. Soc. **127**, 13462 (2005).
[18] P. Mendels and F. Bert, J. Phys. Soc. Jpn. **79**, 011001 (2010).
[19] H. Yoshida, Y. Okamoto, T. Tayama, T.Sakakibara, M.Tokunaga, A. Matsuo, Y. Narumi, K. Kindo, M. Yoshida, M. Takigawa, and Z. Hiroi, J. Phys. Soc. Jpn. **78**, 043704 (2009).
[20] M. Yoshida, M. Takigawa, H. Yoshida, Y. Okamoto, and Z. Hiroi, Phys. Phys. Lett. **103**, 077207 (2009).
[21] Y. Okamoto, H. Yoshida, and Z. Hiroi, J. Phys. Soc. Jpn. **78**, 033701 (2009).
[22] Y. Okamoto, M. Tokunaga, H. Yoshida, A. Matsuo, K. Kindo and Z. Hiroi, Phys. Rev. B **83**, 180407 (2011).
[23] K. Hida, J. Phys. Soc. Jpn. **70**, 3673 (2001).
[24] D. C. Cabra, M. D. Grynberg, P. C. W. Holdsworth, and P. Pujol, Phys. Rev. B **65**, 094418 (2002).
[25] J. Schulenberg, A. Honecker, J. Schnack, J. Richter, and H.-J. Schmidt, Phys. Rev. Lett. **88**, 016207 (2002).
[26] A. Honecker, J. Schulenberg, and J. Richter, J. Phys.: Condens. Matter **16**, S749 (2004).
[27] D. C. Cabra, M. D. Grynberg, P. C. W. Holsworth, A. Honecker, P. Pujol, J. Schmalfus and J. Schulenburg, Phys. Rev. B **71**, 144420 (2005).
[28] M. E. Zhitomirsky, Phys. Rev. Lett. **88**, 057204 (2002).
[29] H. Nakano and T. Sakai, J. Phys. Soc. Jpn. **79**, 053707 (2010).
[30] T. Sakai and H. Nakano, Phys. Rev. B **83**, 100405(R) (2011).
[31] H. Nakano, T. Shimokawa and T. Sakai, J. Phys. Soc. Jpn. **80**, 033709 (2011).
[32] I. Affleck, Phys. Rev. B **43**, 3215 (1991).
[33] T. Sakai and M. Takahashi, Phys. Rev. B **43**, 13383 (1991); J. Phys. Soc. Jpn. **60**, 3615 (1991).
[34] N. Katoh and M. Imada, J. Phys. Soc. Jpn. **63**, 4529 (1994).

[35] T. Sakai and M. Takahashi, Phys. Rev. B **57**, R8091 (1998).
[36] S. Capponi et al., Phys. Rev. B **88**, 144416 (2013).
[37] S. Nishimoto et al., Nature Comm. **4**, 2287 (2013).

Phase transition in 2-d system of quadrupoles on square lattice with anisotropic field

A K Sallabi[1] and M Alkhttab

Department of Physics, Faculaty of Science, Misurata University, Libya

sallabiabdulwahab@yahoo.ca

Abstract. Monte Carlo method is used to study a simple model of two-dimensional interacting quadrupoles on ionic square lattice with anisotropic strength provided by the ionic lattice. Order parameter, susceptibility and correlation function data, show that this system form an ordered structure with $p(2\times1)$ symmetry at low temperature. The $p(2\times1)$ structure undergoes an order–disorder phase transition into disordered (1×1) phase at 8.3K. The two-point correlation function show exponential dependence on distance both above and below the transition temperature. At T_c the two-point correlation function shows a power law dependence on distance, e.g. $C(r) \sim r^{\eta}$. The value of the exponent η at T_c shows small deviation from the Ising value and indicates that this system falls into the same universality class as the XY model with cubic anisotropy. This model can be applied to prototypical quadrupoles physisorbed systems as N_2 on NaCl(100).

1. Introduction

The research on two-dimensional (2-d) continuous phase transition has a long history and continues to inspire basic theoretical work [1-10] as well as computational and experimental research on diverse phenomena such as magnetic reorientation in ultrathin metal films [5] and phase transitions at solid surfaces, e.g. the N_2/NaCl(001) [11,12]. Monte Carlo simulation [11], using pairwise additive potential with adjustable parameters and distributed point dipoles on atomic sites, predicted that a monolayer of nitrogen molecules adsorbed on a NaCl surface will, like the CO/NaCl system[13], undergoes an order-disorder transition at 25K from $p(2\times1)$ phase to $p(1\times1)$ phase, and that the phase transition is predicted to be fallen in the same universality class as the XY model with cubic anisotropy. This class is of considerable interest because it is predicted to exhibit nonuniversal critical exponents [3], i.e. exponents whose values depend on the relative strengths of an anisotropic external potential (provided by the NaCl substrate in this case) and the molecule-molecule interactions. The XY model with cubic anisotropy will exhibit Ising behaviour in the limit of infinite anisotropy, whereas Kosterlitz-Thouless (K-T) behaviour occurs at zero anisotropy. In recent article, Vogt [12] reported experimental work about N_2/NaCl(001) at temperatures between 20K and 40K, and provides experimental evidence that nitrogen molecules form a phase with $p(1\times1)$ symmetry at the NaCl(100) surface but found no hints for a disorder-order phase transition from a $p(1\times1)$ to $p(2\times1)$ structure above 20K (the experiment was performed at temperatures above 20K), as predicted by MC simulations [11]. It is worth noting that Vogt [12] does not exclude the existence of a phase transition from a $p(1\times1)$ to $p(2\times1)$ phases below 20K. In this article we present the results of Monte Carlo

[1] To whom any correspondence should be addressed.

methods of a simple model of two-dimensional interacting quadrupoles on square lattice with anisotropic strength provided by the ionic lattice and pure quadrupole – quadrupole potential. Order parameter, susceptibility and correlation function data results are presented. This model can be applied to prototypical quadrupoles physisorbed systems as N_2 on NaCl(100).

2. Method of simulation

The discrepancy between Monte Carlo simulation [11] and the experimental work [12] regarding the existence of phase transition of N_2 on NaCl(001) above 20K reflects the need to use a model potential with less or no adjustable parameters. In the Monte Carlo simulation [11] the potential parameters in the repulsion part of Tang-Toennies and Buckinghan atom-atom/ion potentials have been adjusted to reproduce the elastic constants and the lattice properties of N_2 and NaCl crystals. Nitrogen on graphite was well described by the electric quadrupole – quadrupole (EQQ) interaction only [14]. It is well known that most of the contribution to the interaction potential comes from the electrostatic potential. In our present work, instead of using a full pairwise potential, we use pure quadrupole – quadrupole potential between fixed point quadrupoles on square lattice and the field is provided by the lattice with no adjustable parameters.

Details of the Monte Carlo methods used in this work can be found in our previous work to simulate N_2/NaCl [11] system. The coordinate system used to describe the positions of the quadruples on the (001) face of a square lattice has an arbitrary origin in the plane of the lattice. The x axis runs along the [1,−1, 0] and the y axis runs along the [1,1,0] crystallographic directions, while the z axis was set perpendicular to the plane of the lattice. The orientation of the quadrupoles is described by a polar angle θ (tilt with respect to the surface normal) and an azimuthal angle φ (angle between the x axis and the projection of the quadrupole axis onto the plane of the lattice). These angles may be used to define an order parameter $\Phi = <(\Phi_x^2 + \Phi_y^2)^{1/2}>$ through the relations $\Phi_x = N^{-1}\sum_{i=1}^{N}[(-1)^{n_{yi}}\cos(\varphi_i)]$ and $\Phi_y = N^{-1}\sum_{i=1}^{N}[(-1)^{n_{xi}}\sin(\varphi_i)]$, where the indices n_{xi}, and n_{yi} are integers with values from 1 to L which label the x and y positions of the site i on a square lattice. The summations are with respect to index i and run over all of the localized sites (n_{xi}, n_{yi}) from 1 to $N=L^2$. The phase factors $(-1)^{n_{xi}}$ and $(-1)^{n_{yi}}$ are needed for systems with antiferro-like ground states. The angled brackets denote an ensemble average over the configurations generated from the MC simulations. The susceptibility is constructed from fluctuations in the order parameter, viz. $\chi=[<\Phi_x^2+\Phi_y^2>-\Phi 2]N/k_BT$. Similarly, the two point correlation function as a function of distance is defined as $C(\vec{r}) = \langle \vec{S}(0)\cdot\vec{S}(\vec{r})\rangle - \langle \vec{S}(0)\rangle\langle \vec{S}(\vec{r})\rangle$, where $\vec{S}$ is the transformed spins and $\vec{r}$ is the position vector of the quadrupoles.

The quadrupoles (each one is similar to the quadrupole of a nitrogen molecule and has a value of -1.41D·Å) were simulated by placing an L×L (L=20, 60) array of quadrupoles in an anisotropic strength provided by the ionic lattice similar to the lattice of NaCl(001) with lateral periodic boundary conditions. The quadrupoles are constrained to the lattice sites at a definite polar angle θ=29° to the plane lattice normal, the projections of the quadrupoles lie in the lattice plane and are allowed to azimuthally rotate within the plane of the lattice. The potentials consists of the potential U(φj) provided by square lattice of the NaCl(001), which has a fourth-order symmetry and can be approximated by the hindered-rotation potential $U(\varphi_i)=-2.4939-0.01096\cos(4.0\varphi_i)$. This equation is obtained by fitting the surface potential of N_2 molecule adsorbed on NaCl(001) surface as a function of φ. The quadrupole –quadrupole interactions are performed in the similar way as discussed in reference [15].

Figure 1. A segment from a 60×60 lattice generated by Monte Carlo simulation for quadrupoles on square lattice at 1K.

Figure 2. A segment from a 60×60 lattice generated by Monte Carlo simulation for quadrupoles on square lattice at 8.4K.

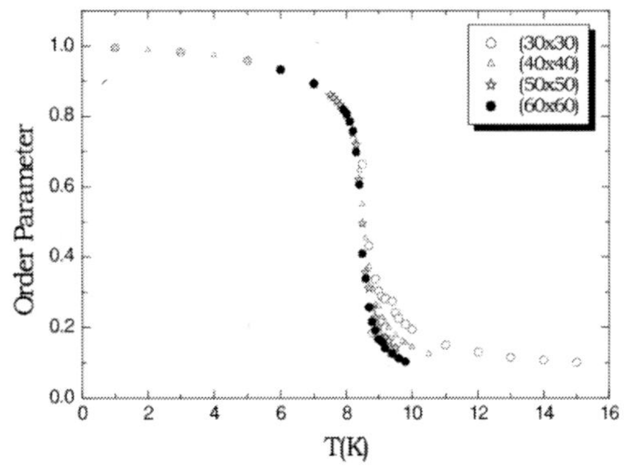

Figure 3. Monte Carlo results for the order parameter, Φ, shown as a function of temperature. Results are shown for various values of L (the linear size of the system).

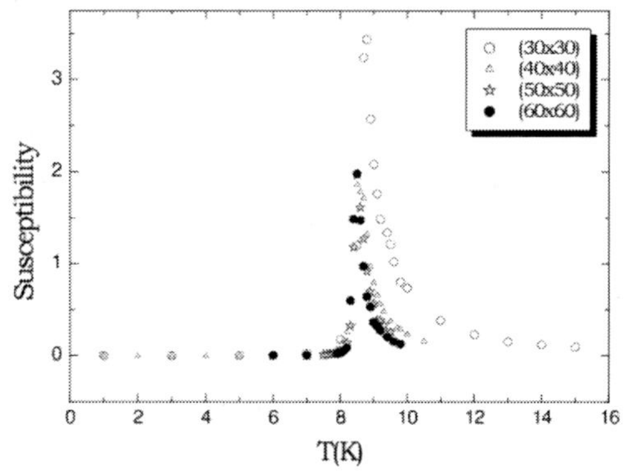

Figure 4. Monte Carlo results for the susceptibility, χ, shown as a function of temperature. Results are shown for various values of L ((the linear size of the system).

3. Results

Our focus is to check the existence of an order–disorder phase transition and to investigate whether this phase transition belongs to XY- with cubic anisotropic university class. The results of our Monte Carlo simulation show that the system of quadrupoles on ionic square lattice form an ordered structure with $p(2\times1)$ symmetry at low temperatures as shown in figure 1. As shown in figure 2, this structure undergoes an order–disorder phase transition into disorder (1×1) phase around 8.3K. In figure 3 the order parameter is plotted as a function of temperature for systems ranging from L=20 to L=60. The associated susceptibility, χ, is plotted in figure 4 as a function of temperature. The peak at 8.3K in the susceptibility coincides with a sharp drop in the order parameter and has features consistent with a continuous order- disorder phase transition at 8.3K.

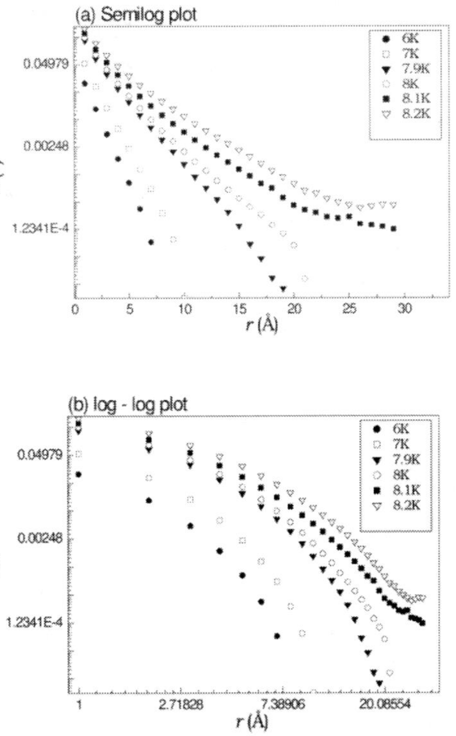

Figure 5. Dependence of the correlation function on distance r for various temperatures below T_c, in a $N = 60 \times 60$ system. Data are shown on a semilog (a) and log-log (b) plots.

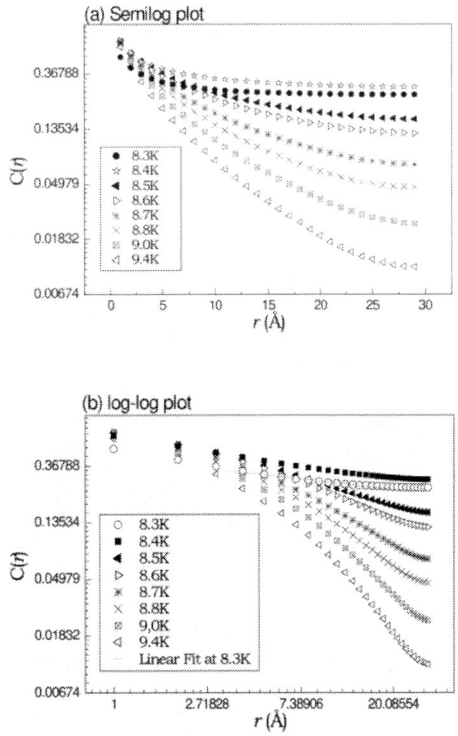

Figure 6. Dependence of the correlation function on distance r for various temperatures above T_c, in a $N = 60 \times 60$ system. Data are shown on a semilog (a) and log-log (b) plots.

In order to shade some light on the nature of the transition, we have analyzed the dependence of the correlation function on distance below T_c. If the transition is as that observed in the four-state clock model, the correlation function will have an exponential dependence on distance as described. However if the system is isotropic, the correlation function will have a power law dependence on distance. The two point correlation functions as function of distance at several values of temperature are obtained. Semi-log and log-log plots of the correlation function as a function of distance below and above the critical temperature are shown in figure 5 and figure 6 respectively. In the case of semilog plots a reasonable straight line fit can be obtained for distances greater than 5 lattice spacing and less than half the system size (15 lattice spacing). It is obvious in the log-log plots of the correlation function that the data deviates from a straight line fit. At T_c (8.3K), as shown in figure 6 the two-point correlation function show power law dependence on distance C(r) ~r^{η} where η is calculated by fitting the log-log plot to straight line and is found to have a values ranged between 0.22 (for 5 spacing to 15 lattice spacing) and 0.255 (for one lattice spacing to 20 lattice spacing) depending on the number of lattice spacings considered in the fitting. The exponent η at the critical temperature shows deviation from the Ising value of 0.25 [16] and indicate that this system may falls into the same universality class as the XY model with cubic anisotropy.

4. Conclusion

The two-point correlation functions show exponential dependence on distance both above and below the critical temperature (T_c), $C(r) \sim \exp(-r/\xi)$ where ξ is the correlation length. This result is consistent with an ordering transition in a system with rotational anisotropy. At T_c (8.3K) the exponent η shows small deviation from the Ising value [15] of 0.25 and indicate that this system may falls into the same universality class as the XY model with cubic anisotropy. This model is applied to prototypical quadrupoles physisorbed systems as N_2 on NaCl(100) [11,12].

References

[1] Kosterlit J M and Thouless D J 1973 *J. Phys. C: Solid State Phys.* **6** 1181
[2] Kosterlitz J M 1974 *J. Phys. C: Solid State Phys.* **7** 1046
[3] Jose J V, Kadanoff L P, Kirkpatrick S and Nelson D R 1977 *Phys. Rev. B* **16** 1217
[4] Abraham D B and Upton P J 2000 *Phys. Rev. Lett.* **85** 2541
[5] De'Bell, MacIsaac, and Whitehead J 2000 *Rev. Mod. Phys.* **72** 225
[6] Géza Ódor 2004 *Rev. Mod. Phys.* **76** 663
[7] Lapilli C M, Pfeifer P and Wexler C 2006 *Phys. Rev. Lett.* **96** 140603
[8] Park J H, Onoda S, Nagaos N, and Han J H 2008 *Phys. Rev. Lett.* **101** 167202
[9] Yuri M 2011 *arXiv preprint arXiv* **1102** 1085
[10] Poderoso F C, Arenzon J J and Levin Y 2011 *Phys. Rev. Lett.* **106** 067202
[11] Sallabi A K and Jack D B 2000 *Phys. Rev. B* **62** 4841
[12] Vogt J 2012 *J. Chem. Phys.* **137** 174705
[13] Vu N -T, Jakalian A, and Jack D B 1997 *J. Chem. Phys.* **106** 2551
[14] O. G. Mouritsen and A. J. Berlinsky,1982 Phys. Rev. Lett. **48**, 181
[15] O'Shea S F, Dubey G S and Rasaiah J C 1997 *J. Chem. Phys.* **107** 237
[16] Back D H, Chung J W and Han W K 1993 *Phys. Rev. B* **47** 8461

Journal of Physics: Conference Series **568** (2014) 042027 doi:10.1088/1742-6596/568/4/042027

Anomaly of Specific Heat in High Quality Single Crystal of $PrAg_2In$

Y Sato[1], H Morodomi[1], Y Inagaki[1], T Kawae[1], H S Suzuki[2]

[1]Department of Applied Quantum Physics, Kyushu university, Motooka, Fukuoka, 819-0395, Japan

[2]National Institutes for Material Science, Tsukuba, Ibaraki, 305-0047, Japan

E-mail: y.sato.483@s.kyushu-u.ac.jp

Abstract. $PrAg_2In$ has a non-Kramers doublet Γ_3 ground state, which is expected to show a quadrupolar phenomena. Although many studies have been devoted to investigate the novel features of $PrAg_2In$, there is no clear result how to remove the degeneracy of quadrupolar moment because this compound has a Heusler structure, in which the substitution between Pr and In ions is inevitable. We therefore prepare a high quality single crystal of $PrAg_2In$ with the residual resistivity ratio of 14, which is much larger than that of the previous samples. We measure the magnetic susceptibility with the SQUID magnetometer down to $T = 0.4$ K and the specific heat in the magnetic fields up to $H = 8$T along [100] crystal direction down to $T = 0.1$ K. The susceptibility is consistent with the CEF calculation. The specific heat shows a sharp peak at around $T_P = 0.33$ K. When the magnetic field is applied, the peak at T_P does not move up to $H = 2.5$ T. In contrast, it shifts to higher temperatures with broadening above $H = 2.5$ T. The broad peak above $H = 2.5$ T is well reproduced by the Schottky specific heat due to the splitting of the Γ_3 doublet in the magnetic fields. From these results, we conclude that a phase transition related to the Γ_3 doublet occurs at around T_P.

1. Introduction

In $4f$-electron systems, the spin-orbital interaction is larger than the crystalline-electric-field (CEF) effect so that the magnetic properties of $4f$-electrons are characterized by the total angular momentum $\boldsymbol{J}(= \boldsymbol{L} + \boldsymbol{S})$. For Pr-compound, a trivalent Pr ion forms $4f^2$ configuration in a crystal, indicating that the ninefold degenerate $J = 4$ multiplet is a ground state of the trivalent Pr ion. If the crystal structure of the trivalent Pr ion has a cubic symmetry, $J = 4$ multiplet splits into four levels: a singlet Γ_1, a doublet Γ_3, a triplet Γ_4 and a triplet Γ_5 by the cubic CEF effect.

A cubic compound of $PrAg_2In$ has the non-Kramers doublet Γ_3 ground state, which carries two quadrupolar moments; $O_2^0(= [3J_z^2 - \boldsymbol{J}]/2)$ and $O_2^2(= \sqrt{3}[J_x^2 - J_y^2]/2)$, and one octapolar moment $T_{xyz}(= \sqrt{15}/6\overline{J_xJ_yJ_z})$. The overline on the product denotes the operation of taking all possible permutations in terms of cartesian components, e.g., $\overline{J_xJ_y} = J_xJ_y + J_yJ_x$. The first report by Yatskar *et al.* showed no sign of quadrupolar ordering down to $T = 50$ mK [1]. Thus, it is very interesting how to dissolve the non-Kramers doublet degeneracy in $PrAg_2In$. In previous studies, a large Sommerfeld coefficient $\gamma \sim 6.5$ J/(mol K^2) and an anomalous logarithmic temperature dependence of the magnetic susceptibility are observed [1, 2]. Therefore,

Journal of Physics: Conference Series **568** (2014) 042027 doi:10.1088/1742-6596/568/4/042027

the quadrupolar Kondo effect [3] is discussed as the origin of a heavy fermion behavior with the large γ value.

The sample quality is crucial to investigate the quadrupolar phenomena. For example, we report that the antiferro-quadrupolar(AFQ) transition temperature $T_Q = 0.4$ K in in $PrPb_3$ compound is suppressed to $T_Q = 0.28$ K only by 1% nonmagnetic La substitution of Pr [4]. Moreover, $PrAg_2In$ is a Heusler alloy, implying that the displacement between Pr-ion and In-ion occurs easily. Hence, it is difficult to distinguish the essential features from the effects such as the randomness, the presence of the grain boundaries, and the site-disorder. These show that the improvement of the sample quality is essential for studying the nature of the Γ_3 doublet ground state in $PrAg_2In$.

2. Experimental

We prepare a new single crystal by the Bridgeman method using a Mo crucible and a Ar arc melting. The residual resistivity ratio of the new sample is about 14, which is much larger than that of the previous ones with about 2–3. This strongly suggests the sample quality is improved enormously.

We study the DC magnetic susceptibility and the specific heat in the magnetic fields along [100] direction with the new sample of $PrAg_2In$. The DC magnetic susceptibility is measured by the Quantum Design SQUID magnetometer with a home-made ^{3}He insert [5]. The insert is made from a thin-walled stainless steel pipe with an inner diameter of 6.2 mm and is easily attached to the SQUID magnetometer in a few minutes. ^{3}He gas is condensed in the pipe, which is liquefied by ^{4}He gas at $T = 1.8$ K generated by the magnetometer via the heat exchanger of a Cu vacuum jacket with an outer diameter of 8.6 mm soldered to the stainless steel pipe. This system enables us to measure the DC magnetization down to $T = 0.4$ K with a high resolution $\sim 10^{-7}$ emu.

The specific heat is measured by a pseudo-adiabatic heat pulse method with a ^{3}He–^{4}He dilution refrigerator from $T = 0.1$ K to 1.2 K and a superconducting magnet up to $H = 8$ T along [100] crystal direction.

3. Results and Discussion

We show the temperature dependence of the DC magnetic susceptibility of $PrAg_2In$ at $H = 0.5$ T in Fig.1. The susceptibility above 15 K follows the Curie-Weiss law and is well reproduced by the CEF level scheme with Γ_3(0K)–Γ_4(75K)–Γ_5(145K)–Γ_1(181K). In contrast with the CEF calculation which the susceptibility gradually approaches a constant value below $T = 15$ K, the magnetic susceptibility shows a logarithmic temperature dependence down to $T = 0.4$ K. It is worthy to note that the same non-Fermi-liquid behavior of the DC magnetic susceptibility M/H is observed in other Pr-based compounds with the Γ_3 doublet in the CEF ground state [7, 8].

In Fig.2, we show the temperature dependence of the specific heat in zero field together with the previous result by Yatskar *et al.* [1]. The specific heat in zero field clearly shows a peak at around $T_p = 0.33$ K. Compared with the previous result, the peak becomes sharp and the maximum value grows to ~ 3 J/(mol K) from ~ 2 J/(mol K) at around 0.4 K. These features are identical to $Pr_{1-x}La_xPb_3$ system, in which the AFQ transition peak broadens with increasing the La-concentration [4]. These facts indicate that the quality of the new sample is largely improved and the observed features are the essential nature of $PrAg_2In$.

Next, we study the magnetic field dependence of $PrAg_2In$ to clarify the origin of the peak at around T_P. The magnetic filed is applied in [100] crystal direction up to $H = 8$ T. When the magnetic field is increased, the peak at T_P does not move up to $H \sim 2.5$ T as shown in Fig.3. Above $H \sim 2.5$ T, the peak shifts to higher temperatures with broadening. To understand the origin of the broad peak at high fields, we fit the experimental results using the nuclear specific

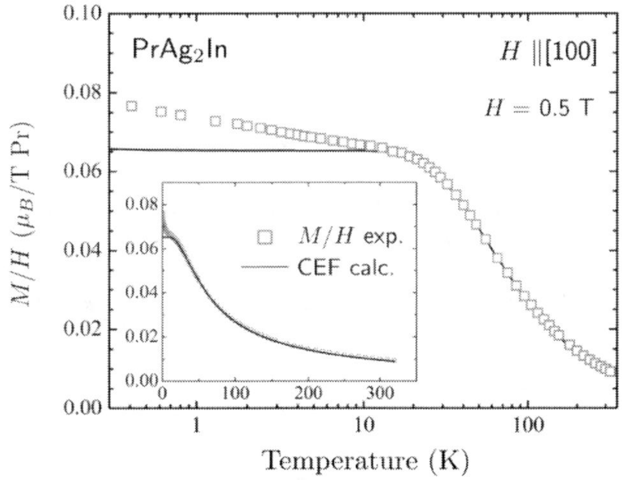

Figure 1. Temperature dependence of the magnetic susceptibility of $PrAg_2In$ at H = 0.5 T. Solid line shows the calculation result by CEF model with CEF parameter [6] of x = 0.39, W = −2.30 K.

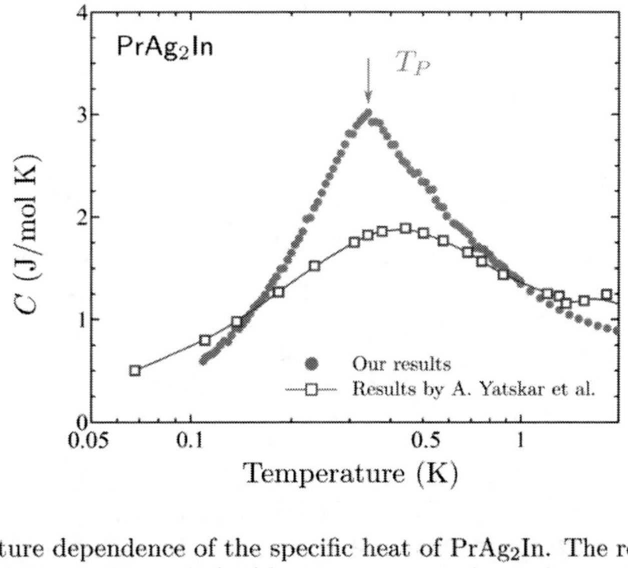

Figure 2. Temperature dependence of the specific heat of $PrAg_2In$. The red solid circle shows the results of this measurements and the blue open square shows the results by A. Yatskar *et al.* [1].

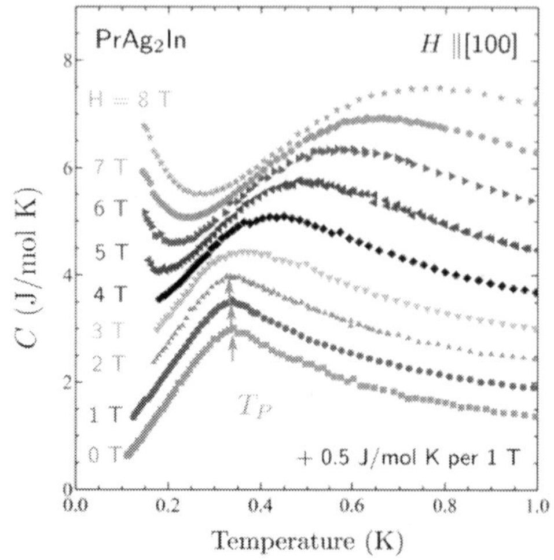

Figure 3. Temperature dependence of the specific heat of $PrAg_2In$ in the magnetic fields along [100] direction. The data from bottom to top show $H = 0$ T to 8 T per 1 T, which are shifted by 0.5 J/(mol K), respectively.

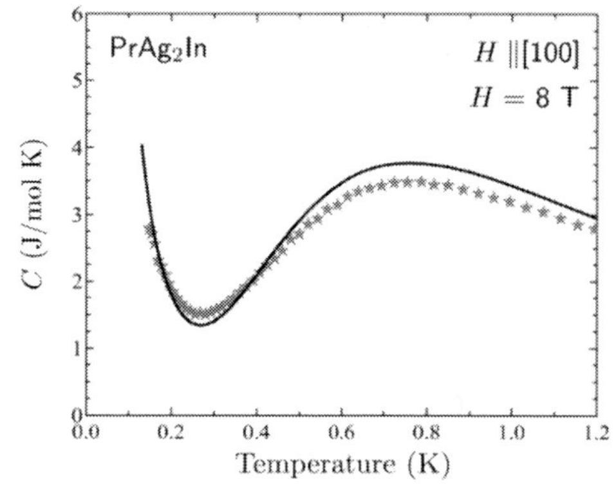

Figure 4. The Schottky peak fitting for the experimental results at $H = 8$ T(red star). The solid line is obtained using a eq.(1) with $\Delta = 1.87$ K and $A_n = 0.069$ J·K/mol .

heat and the two-level Schottky model expressed by the following equation:

$$\begin{aligned} C &= C_{\text{nuclear}} + C_{\text{schottky}}, \\ &= \frac{A_n}{T^2} + R\left(\frac{\Delta}{T}\right)\frac{\exp(\Delta/T)}{[1+\exp(\Delta/T)]^2}, \end{aligned} \tag{1}$$

where R is the gas constant with the value of 8.314 J/(mol K). We can reproduce well the experimental results at $H = 8$T with $\Delta = 1.87$ K and $A_n = 0.069$ J·K/mol. On the other hand, the splitting of the Γ_3 doublet at $H = 8$ T is calculated to be $\Delta_{\text{CEF}} = 1.58$ K by the CEF scheme obtained by the magnetic susceptibility measurements. This demonstrates that the broad peak is caused by the splitting of the Γ_3 doublet by the magnetic fields. From these results, we conclude that the sharp peak at around T_P is caused by a phase transitions, such as quadrupolar or octapolar orderings, related to the Γ_3 doublet.

4. Conclusion

We measured the temperature dependence of magnetic susceptibility at $H = 0.5$ T and the specific heat up to $H = 8$ T along [100] direction with a high quality single crystal of $PrAg_2In$. Above $T = 15$ K, the magnetic susceptibility is well reproduced by the CEF level scheme with a non-Kramers Γ_3 doublet ground state. In the specific heat, a sharp peak appears at around

$T_P = 0.33$ K. In the magnetic fields, the peak at T_P does not move up to $H = 2.5$ T. In contrast, it shifts to higher temperatures above $H = 2.5$ T with broadening. The broad peak at high magnetic fields is in good agreement with the Schottky specific heat due to the splitting of the Γ_3 doublet by the magnetic fields. From these results, we conclude that a phase transition related to the Γ_3 doublet occurs at around T_P.

References

[1] Yatskar A, Beyermann W P, Movshovich R, and Canfield P C 1996 *Phys. Rev. Lett.* **77** 3637.
[2] Tanida H, Takagi S, Suzuki H S, Satoh I and Komatsubara T 2006 *J. Phys. Soc. Jpn.* **75** 074721.
[3] Cox D L, 1987 *Phys. Rev. Lett.* **59** 1240.
[4] Kawae T, Shimogai M, Mito M, Takeda K, Ishii H, and Kitai K, 2001 *Phys. Rev.* B **65** 012409.
[5] Sato Y, Makiyama S, Sakamoto Y, Hasuo T, Inagaki Y, Fujiwara F, Suzuki H S, Matsubayashi K, Uwatoko Y, and Kawae T 2013 *Jpn. J. Appl. Phys.* **52** 106702.
[6] Lea K R, Leask M J M, and Wolf W P 1962 *J. Phys. Chem. Solids* **23** 1381.
[7] Tanida H, Suzuki H S, Takagi S, Onodera H, and Tanigaki K 2006 *J. Phys. Soc. Jpn.* **75** 073705.
[8] Kawae T, Koga M, Sato Y, Makiyama S, Inagaki Y, Tateiwa N, Fujiwara T, Suzuki H S, and Kitai T 2013 *J. Phys. Soc. Jpn.* **82** 073701.

High-field optical spectroscopy of the chromium spinel $CdCr_2O_4$

Y Sawada[1], S Kimura[1], K Watanabe[1] and H Ueda[2]

[1] Institute for Materials Research, Tohoku University, Sendai 980-8577, Japan
[2] Department of Chemistry, Kyoto University, Kitashirakawa Oiwake, Sakyo ward, Kyoto 606-8502, Japan

E-mail: yuya_sawada@imr.tohoku.ac.jp

Abstract. We have performed high-field optical spectroscopy measurements of the geometrically frustrated chromium spinel oxide $CdCr_2O_4$ in the visible light region. We succeeded in observing the pure exciton excitation peak, and identified some other absorption peaks as exciton-magnon excitation peaks from the theoretical calculations of density of states of the spin wave. Furthermore, from the high-field experiment, large variations of absorption peaks were observed at around 4 T. This result indicates that some magnetic phase transition occurs around this field.

1. Introduction

The chromium spinel oxides ACr_2O_4 (A = Zn, Mg, Cd, Hg), in which Cr^{3+} ions interact antiferromagnetically between nearest-neighbour sites and form a pyrochlore lattice, are typical examples of a three-dimensional geometrically frustrated magnet. Due to the strong geometrical frustration, these materials do not show magnetic ordering and remain a paramagnetic state down to low temperature. In such materials, it is known that a novel magnetic ground state or a peculiar phenomenon sometimes appears, and therefore, their physical properties have attracted much interest and have been studied extensively.

In this work we focus on $CdCr_2O_4$. This material remains a paramagnetic state far below its Weiss temperature $|\Theta_W|$ = 70 K. Eventually, it undergoes the magnetic phase transition from the paramagnetic state with a cubic symmetry to the antiferromagnetic ordered phase with a tetragonal symmetry at T_N = 7.8 K [1]. In this magnetic ordering accompanied by the lattice distortion, the spin-lattice coupling plays an important role to relieve the geometrical frustration. From the neutron scattering experiment [2], helical spin ordering, represented by an incommensurate wave vector $Q = (0, \delta, 1)$ with $\delta \sim 0.09$, has been suggested. Furthermore, by applying a magnetic field at low temperature, it is known that the transition to the 1/2 magnetization plateau phase accompanied by a large magnetostriction appears at 28 T [3]. A small anomaly of the magnetization was also observed around 4 T [4]. From the recent optical spectroscopy measurements of $CdCr_2O_4$, complicated absorption peaks including a spin-forbidden d-d transition of ${}^4A_{2g} \rightarrow {}^2E_g$ were reported [5]. These absorption peaks were found in the lowest energy side of the absorption peaks in the visible light region and were considered to come from the various types of multiple excitations including exiciton, magnon, and phonon excitation. In this work, we report high-field optical spectroscopy measurements of $CdCr_2O_4$ in the magnetically ordered phase. We observed an exciton absorption peak in the visible light region and large variation of absorption spectra at 4 T. In addition, from the theoretical calculation, we identified that absorption peaks located at higher energy side of an exciton line come from the simultaneous excitation of the exciton and magnon.

2. Experimental procedures

High-field optical spectroscopy measurements of $CdCr_2O_4$ were carried out by utilizing a conventional superconducting magnet. In the high-field optical spectroscopy system that was developed in the High-field Laboratory for Superconducting Materials, a deuterium and a halogen-tungsten lamp as light sources that cover from the near-infrared to the ultraviolet region, a single grating spectrometer (HORIBA iHR550) with the focal length 550 mm, and a multichannel CCD camera (ANDOR iDus DV401A) were used. Three types of grating (150, 600, and 1800 g/mm) were employed. The details of the measurement system can be found in Ref. [6]. The thickness of a polished sample is approximately 100 μm, and is mounted on a quartz substrate. The direction of the applied magnetic field is parallel to the [100] direction. The measurements were performed in the Voigt configuration. A $CdCr_2O_4$ single crystal was prepared according to the method described in Ref. [7].

3. Results and discussions

3.1 Absorption spectrum at zero magnetic field

Figure 1 shows the absorption spectrum of $CdCr_2O_4$ in the visible light region measured at 4.2 K using a 150 g/mm grating. Three broad absorption peaks with strong absorbance, indicated by solid arrows, and several sharp absorption peaks, indicated by a dotted arrow, are observed. The obserbed spectrum is compatible with that reported by the previous work [5]. The observed absorption peaks are assigned to be the d-d transitions according to the Tanabe-sugano diagram [8]. The strong absorption peaks are assigned to be the spin-allowed d-d transitions as follows: ${}^4A_{2g}\rightarrow{}^4T_{1g}(1)$ around 29000 cm^{-1}, ${}^4A_{2g}\rightarrow{}^4T_{1g}(2)$ around 22500 cm^{-1}, and $A_{2g}\rightarrow{}^4T_{2g}$ around 17000 cm^{-1}. The absorption peaks around 14500 cm^{-1} correspond to the spin-forbidden d-d transition. A number of sharp absorption peaks observed around this energy were considered to come from simultaneous excitations of exciton, magnon, or phonon [5].

Figure 2 shows the absorption spectrum at around 14500 cm^{-1} obtained by using a 600 g/mm grating. In this region, the spectrum shows complex structures. The inset shows the enlarged absorption spectrum from 14100 to 14150 cm^{-1} which corresponds to the hatched region in figure 2. This enlarged spectrum is obtained by using an 1800 g/mm grating. A weak absorption peak at 14115

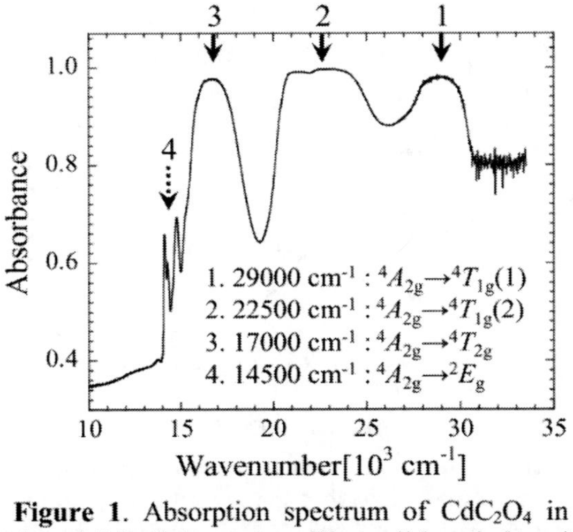

Figure 1. Absorption spectrum of CdC_2O_4 in the visible light region. The solid and dotted arrows indicate the spin-allowed and spin-forbidden d-d transition, respectively.

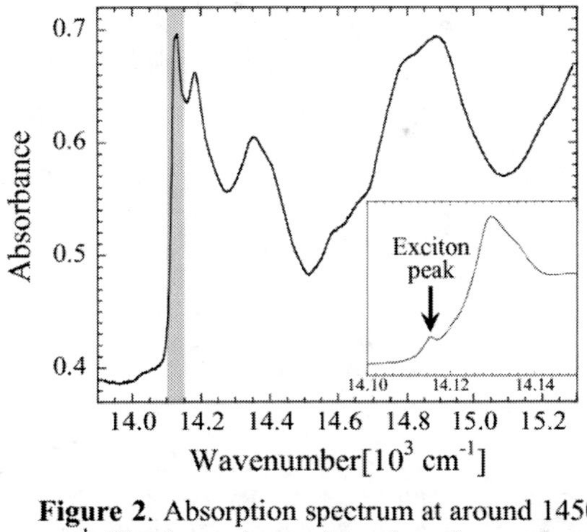

Figure 2. Absorption spectrum at around 14500 cm^{-1}. The inset shows the enlarged spectrum in the low energy region. The exciton peak was observed at 14115 cm^{-1}.

cm^{-1} was found. Because this peak is located at the lowest energy side of the complex absorption spectrum, we conclude that the absorption peak at 14115 cm^{-1} is due to a pure exciton excitation.

Next, we move on to the discussion about other absorption peaks observed around the exciton peak. From the previous study, Schmidt *et al.* indicated that the peaks at 14130 and 14149 cm^{-1} come from the simultaneous excitation of the exciton and magnon [5]. To confirm this indication, we calculate the magnon energy of $CdCr_2O_4$. The recent theories demonstrated that the spin structure of $CdCr_2O_4$ suggested by neutron scattering measurements [2] is compatible with a staggered lattice distortion with the symmetry lowering of the crystal structure to E_u symmetry, which leads to the $I4_122$ space group with lacking inversion symmetry [9]. This lattice distortion changes the direct exchange interaction between the nearest-neighbour sites. As a result, magnetic frustration is relieved and the 8 sub-lattice collinear Neel ordering with wave vector $Q = (0, 0, 1)$ depicted in figure 3 is stabilized. Then, it was indicated that the Dzyaloshinskii-Moriya (DM) interaction inherent in the pyrochlore lattice modulates the collinear structure to the helical one, and the experimentally observed $Q = (0, \delta, 1)$ structure is stabilized by the third neighbour antiferromagnetic interaction J_3. It was also suggested that the magnon spectrum, which is numerically calculated by assuming the 8 sub-lattice collinear ordering model without DM interaction, can reproduce overall features of the experimentally observed magnon dispersion [9].

Thus, we analytically calculate the magnon excitation of the Neel ordered state with the 8 sub-lattice collinear spin structure, shown in figure 3, in terms of a classical spin wave theory, and derive the density of states (DOS) of the magnon. In the spin wave calculation, we assume that the value of the nearest neighbour exchange interaction J_1 between the spins, aligning antiparallel to each other, is different from that of J_2 with a parallel spin alignment because of the lattice distortion. The third-neighbour interaction J_3 is also taken into account. We evaluated the exchange interaction parameters as $J_1 = 9.04$ K, $J_2 = 5.8$ K and $J_3 = 1.05$ K. These parameters are determined so as to reproduce the experimentally observed magnon dispersion. The details of the spin wave calculation will be published elsewhere.

The DOS of the magnon is shown in figure 4 together with the observed absorption spectrum. The wavenumber of the spectrum is shifted by 14115 cm^{-1}, which corresponds to the energy of the pure exciton. The calculated DOS shows two peaks at around 17 and 30 cm^{-1}. This result is in good agreement with the observed absorption peaks at 14130 and 14149 cm^{-1} that are 15 and 34 cm^{-1} higher

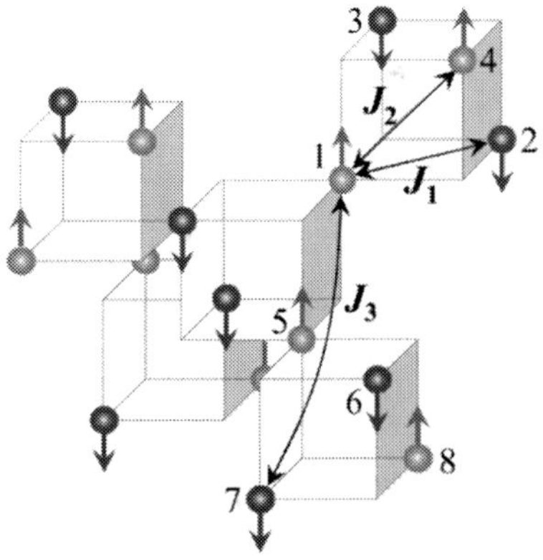

Figure 3. Collinear 8 sub-lattice magnetic structure. J_1 and J_2 indicate the nearest-neighbour interactions, and J_3 indicates the third-neighbour interaction, respectively.

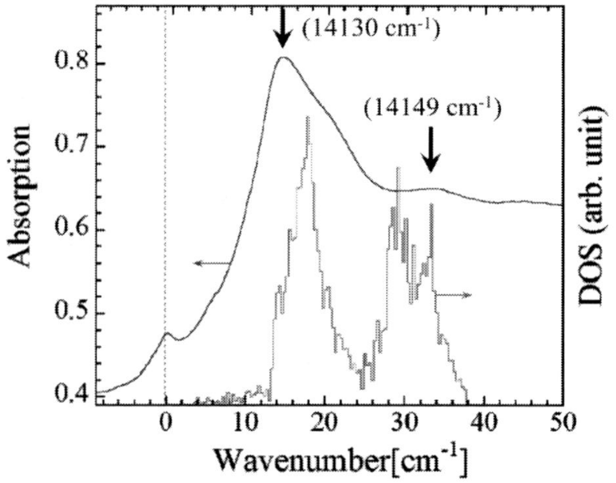

Figure 4. Density of states (DOS) of the magnon excitation and the absorption spectrum around the exciton excitation peak. The vertical arrows indicate the absorption peaks at 14130 and 14149 cm^{-1}.

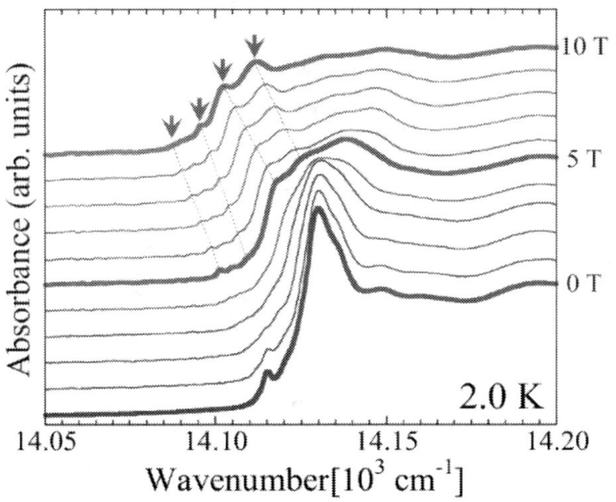

Figure 5. Magnetic field dependence of the absorption peaks measured at 2.0 K. Large variations of exciton-magnon peaks and the splitting of the exciton peak were observed.

energy than that of the exciton peak, respectively. Thus, we conclude that these two absorption peaks are exciton-magnon excitation peaks. The sharp absorption peak at 14190 cm^{-1} is likely to come from the excitation of the exciton and multiple magnons, as suggested by Schmidt *et al.* [5]

3.2 High-field experiment
Figure 5 shows the magnetic field dependence of exciton-magnon peaks measured in magnetic fields from 0 to 10 T at 2.0 K. Large variations of the absorption peaks were observed at around 4 T. Drastic decrease of the absorbance of exciton-magnon peaks above 4 T suggests the change of the spin wave dispersions due to the change of the magnetic structure. Furthermore, the number of the exciton peaks increases above 4T. At least four peaks are found above 4 T. The positions of the exciton peaks shift toward lower wavenumber as the field is increased above 4 T, as indicated by arrows in figure 5. These behaviour of the exciton peaks above 4 T is probably ascribed to the Davydov splitting [10]. The Davydov splitting occurs when the exciton migrates through the system. Such migration can take place only if the nearest neighbour spins are aligned parallel each other. Therefore, our results suggest the abrupt increase of the ferromagnetic component in the ground state above 4T. We consider that this behaviour relates to the magnetic transition from the incommensurate helical spin structure to the commensurate spin structure, suggested from the previous high-field ESR measurement [4]. For further detailed analysis, the calculations of the exciton dispersion in magnetic fields are necessary.

4. Conclusions
From the optical spectroscopy measurements of $CdCr_2O_4$, we found the exciton excitation peak in the visible light region, and succeeded in identifying the exciton-magnon peaks by the theoretical calculations. In the high-field experiment, we observed large variations of absorption peaks at 4 T, where the anomaly of the magnetization was found. This result indicates that the magnetic phase transition occurs at this field.

5. References
[1] Ueda H, Mitamura H, Goto T and Ueda Y 2006 *Phys. Rev. B* **73** 094415
[2] Chung J H, Matsuda M, Lee S H, Kakurai K, Ueda H, Sato T J, Takagi H, Hong K P and Park S 2005 *Phys. Rev. Lett.* **95** 247204
[3] Ueda G, Katori H A, Mitamura H, Goto T and Takagi H 2005 *Phys. Rev. Lett.* **94** 047202
[4] Kimura S, Hagiwara M, Ueda H, Narumi Y, Kindo K, Yashiro H, Kashiwagi T and Takagi H 2006 *Phys. Rev. Lett.* **97** 257202
[5] Schmidt M, Wang Z, Kant Ch, Mayr F, Toth S, Islam A T M N, Lake B, Tsurkan V, Loidl A

and Deisenhofer 2013 *Phys. Rev. B* **87** 224424
[6] Sawada Y, Kimura S, Watanabe K and Nakano M 2013 *J. Low. Temp. Phys.* **170** 424
[7] Dabkowska H 1981 *J. Crystal Growth* **54** 607
[8] Tanabe Y and Sugano S 1954 *J. Phys. Soc. Jpn.* **9** 766
[9] Choi E, Chern G and Perkins N B 2013 *Phys. Rev. B* **87** 054418
[10] Kojima N, Aoyagi K, Tsushima K, Tsujikawa I and Sugano S 1980 *J. Phys. Soc. Jpn.* **49** 1463

Acknowledgements
We are grateful to Dr. I Mogi at Institute for Materials Research for valuable discussions. This work was partly supported by Grant-in-Aid for Japan Society for the Promotion of Science Fellows (No. 26·8161).

Journal of Physics: Conference Series **568** (2014) 042029 doi:10.1088/1742-6596/568/4/042029

Itinerant to local transformation and critical point in Ni-rich $Ce_2(Ni_{1-y}Pd_y)_2Sn$ alloys

J.G. Sereni [1], G. Schmerber [2], M. Gómez Berisso [1] and J.P. Kappler [2]

[1] División Bajas Temperaturas (CAB - CNEA), Conicet, 8400 S.C. Bariloche, Argentina
[2] IPCMS, UMR 7504 CNRS-ULP, 23 rue de Loess, B.P. 43 Strasbourg Cedex 2, France

E-mail: jsereni@cab.cnea.gov.ar

Abstract. Structural, magnetic (M) and thermal (C_m) studies on $Ce_2(Ni_{1-y}Pd_y)_2Sn$ alloys are presented within the $0 \leq y \leq 0.55$ range of concentration, showing evidences for itinerant to local electronic transformation. At variance with RKKY type interactions between localized moments μ_{eff}, the substitution of Ni by isoelectronic Pd leads the antiferromagnetic transition to decrease from T_N 3.8 K to 1.2 K between $y = 0$ and 0.48, while $M(H)$ measured at $H = 5$ T and 1.8 K rises from 0.12 up to $0.75\mu_B$/Ce-at. Furthermore, the $C_m(T_N)$ jump increases with concentration whereas $|\theta_P|$ decreases. The magnetic entropy $S_m(T)$ grows moderately with temperature for $y = 0$ due to a significant contribution of excited levels at low energy, while at $y = 0.5$ it shows a incipient plateau around $S_m = Rln2$. All these features reflect the progressive ground state transformation of from itinerant to a local character.

Another peculiarity of this system is the nearly constant value of $C_m(T_N)$ that ends in an *entropy bottleneck* as T_N decreases. Consequently, the system shows a critical point at $y_{cr} \approx 0.48$ with signs of ferromagnetic behavior above $H_{cr} \approx 0.3$T. A splitting of the $C_m(T_N)$ maximum, tuned by field and concentration, indicates a competition between two magnetic phases, with respective peaks at $T_N \approx 1.2$ K and $T_I \approx 1.45$ K.

1. Introduction

Itinerant magnetism is mostly observed in U or Pu-based compounds because of their $5f$ orbitals [1] whereas a local character is mostly expected in $4f$ Ce and Yb ones. However, one basic question, not fully elucidated yet, concerns the possibility of a local to itinerant transformation of the Ce-$4f^1$ orbital. This possibility was investigated for 30 years within the so called *local-itinerant dilemma* [2]. Although the $4f^1$ delocalization is well documented among systems showing $4f$ and conduction band hybridization [3], that mechanism implies the screening of the magnetic moment with the consequent weakening of the magnetic order. In spite of that, there are Ce compounds showing magnetic order with record high values, being the outstanding examples $CeRh_3B_2$ ($T_C = 115$ K [4]), CeScGe ($T_N = 47$ K [5]) and $CeRh_2Si_2$ ($T_N = 36$ K [6]), which are claimed to exhibit itinerant magnetism.

One characteristic of a well localized Ce-$4f$ orbital is related to the fact that the doublet ($N = 2$) ground state (GS) is not significantly affected by Kondo screening (T_K) and the excited crystal field (CEF) levels have a comparatively large splitting (Δ_{CEF}), i.e. $T_K << \Delta_{CEF}$. The opposite scenario, with $T_K \geq \Delta_{CEF}$, is the proper situation where to search for itinerant magnetic behavior provided that the magnetic order still occurs, which only occurs in a few system. As a consequence of an eventual $N_{eff} > 2$ character, the magnetic contribution to the entropy (S_m) is expected to overcame the $RLn2$ value at relatively low temperature.

These conditions were recently observed in the ternary system $Ce_2(Pd_{1-x}Ni_x)_2Sn$ [7]. There, the substitution of Pd atoms by isoelectronic Ni leads to a change from local (Pd-rich side) to itinerant (Ni-rich side) character. Such a change in the electronic structure is associated to a crystallographic modification from tetragonal Mo_2FeB_2 to centered orthorhombic structure of W_2CoB_2 type [8] between $0.25 \leq y \leq 0.35$, where a gap of miscibility occurs.

With the aim to compare the Ce-4f orbital nature in these two phases, the magnetic properties of the $Ce_2(Ni_{1-y}Pd_y)_2Sn$ alloys on the Ni-rich side (i.e. within the orthorhombic structure) were investigated. Apart from the usual magnetic characterization, for a proper analysis of the itinerant to localized evolution of the system we focus on the study of two basic parameters: the hight of the specific heat anomaly at $T = T_N$ and the thermal dependence of the entropy $S_m(T)$. These quantities allow to evaluate the role of the Kondo effect and how the excited CEF levels contribute to the ground state properties.

2. Experimental results

2.1. Magnetic Susceptibility

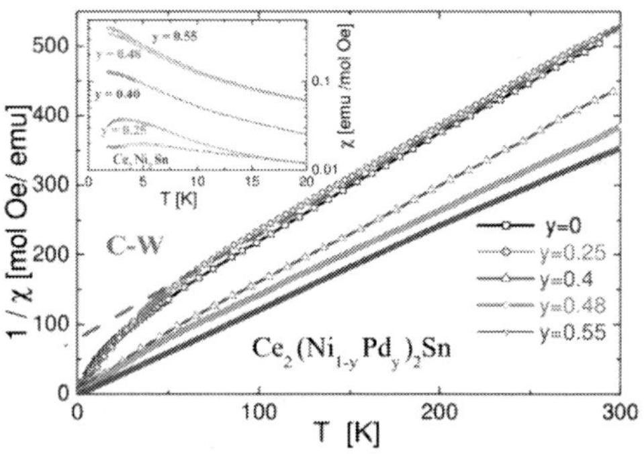

Figure 1. High temperature inverse susceptibility of Ni-rich alloys measured with $\mu_0 H = 0.1\,T$. Inset: low temperature magnetic susceptibility in a logarithmic scale.

The high temperature magnetic behavior ($T \geq 80\,K$) is well described by a Curie-Weiss type contribution ($\chi = C_C/(T - \theta_P)$) plus a minor Pauli contribution ($\chi_P = 5 \pm 2 \times 10^{-3}$ emu/mol Oe). In Fig. 1 the inverse of χ is presented to extract the respective paramagnetic temperatures $\theta_P(y)$ and effective moments $\mu_{eff}(y) \propto \sqrt{C_C}$. The $\theta_P(y)$ values range between -58 K at the Ni-rich limit and ≈ 6 at $y = 0.55$. Simultaneously, $\mu_{eff}(y)$ increases from $\mu_{eff} = 2.3\mu_B$ up to $\approx 2.54\mu_B$ between those concentration values. A detail of the $\chi(T)$ dependencies at low temperature ($T < 20\,K$) are included in the inset of Fig. 1 using a logarithmic scale to cover the large variation of $\chi(y)$ at $T = 1.8\,K$.

In order to explore this strong increase of $\chi(y)$ at low temperature the isothermal field dependence of the magnetization $M(H)$ was measured at T =1.8 K, 3 K and 4 K up to $\mu_0 H = 5$ Tesla. A clear difference in the $M(H)$ dependence can be appreciated in the results obtained at $T = 1.8\,K$ between Ni-rich samples and those with similar Pd/Ni concentration (see Fig. 2a). While $M(H)$ shows a slight positive curvature with low values of M in $y = 0$ ($0.15\mu_B$) and 0.25 alloys, those with $y \geq 0.4$ show a tendency to saturation (M_{sat}) at $H = 5$Tesla. Despite the fact that the former alloys have a $T_N \geq 1.8\,K$ while in the latter $T_N < 1.8\,K$, similar measurements performed at $T = 4\,K$ (i.e. in the paramagnetic phase) show the same variation. Notably, $M_{sat}(y)$ undergoes a maximum of $0.88\mu_B$ at $y = 0.40$ and then slightly decreases to $0.76\mu_B$ for $y = 0.55$. In coincidence with the M_{sat} decrease at these concentrations, the initial slope of $M(H)$ increases as a sign of a stronger interaction between magnetic moments. Additionally, a further slight increase in the slope is observed around $H = 1$ Tesla. To remark these features, the $\partial M/\partial H$ derivative is presented in Fig. 2b, showing that the change of slope

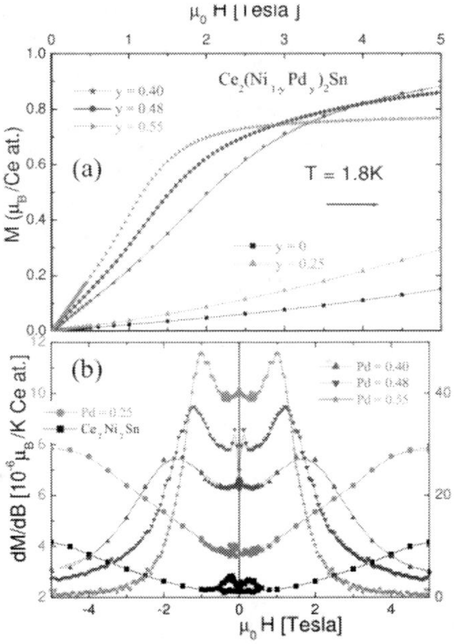

Figure 2. a) Field dependent magnetization of the 1.8 K isotherms for the Ni-rich range measured up to 5T. b) dM/dH vs H derivative showing the difference between Ni-rich samples (left axis) and those with similar Pd/Ni concentration (right axis).

only occurs in the alloys with strong magnetization (i.e. $y \geq 0.4$) and it grows up to $y = 0.55$. Notably, the $M(H)$ dependence above that critical field is properly described by a $\tanh(H)$ function which characterizes ferromagnetic systems.

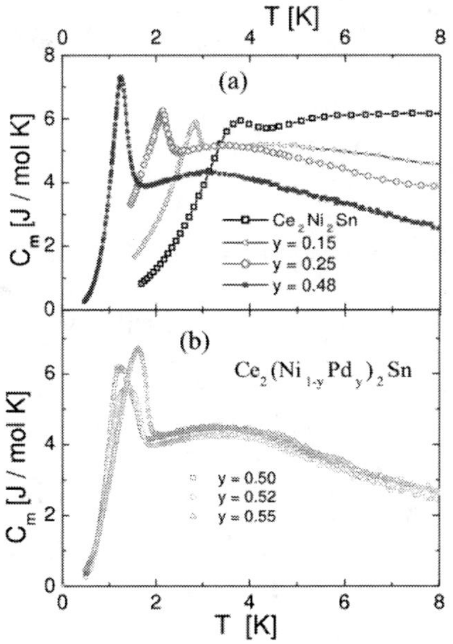

Figure 3. Thermal dependence of the magnetic contribution to the specific heat C_m for: a) Ni-rich alloys and b) critical region at similar Ni/Pd concentration.

2.2. Specific heat

The magnetic contribution (C_m) to the measured (C_P) specific heat was obtained by subtracting the phonon contribution extracted from the reference isotypic compound La_2Ni_2Sn as $C_m = C_P - C_{La}$. From the thermal dependence of C_m one can see in Fig. 3a that the ordering temperature $T_N(y)$, defined as the temperature of the maximum of $C_m(T)$, decreases with Pd concentration. This evolution extrapolates to $T_N = 0$ at $y \approx 0.70$, beyond the structural stability of this system. However, between $0.48 \leq y \leq 0.52$, $T_N(y)$ remains nearly constant at $T_N \approx 1.2\,\mathrm{K}$ while another anomaly arises for $y = 0.55$ at $T_I \approx 1.5\,\mathrm{K}$ as shown in Fig. 3b.

While the $C_m(T)$ dependence at $T < T_N$ is typical for an antiferromagnetic system, the $C_m(T)$ maximum (C_{max}) behaves quite unusual because its value slightly changes up to $y = 0.48$ and becomes sharper as Pd concentration increases, see Fig. 3a. According to the law of corresponding states, C_{max} should decrease with decreasing $T_N(y)$, to become zero at $T = 0$. As a consequence the magnetic transition with chances to reach zero temperature have to show $C_{max}/T_N \approx$ const. [9]. On the contrary, this system shows C_{max}/T rising as $\propto 1/T_N(y)$ as indicated in Fig. 4a. This divergent C_{max}/T_N dependence places the $y = 0.48$ alloy at a critical concentration point.

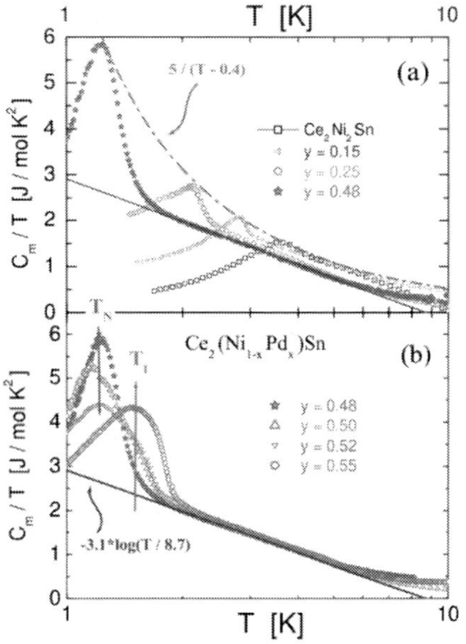

Figure 4. Logarithmic dependence of the magnetic contribution to specific heat divided by temperature C_m/T for: (a) $0 \geq y \geq 0.48$ samples. Dash-dot curve describes the $C_m/T(y)$ maxima dependence. (b) Alloys with Pd/Ni equi-atomic concentrations. Arrows at T_N and T_I indicate a two phases competition. Straight line represents a logarithmic reference fitting a universal function (see the text).

Above $T_N(y)$, $C_m(T)$ shows a very broad maximum which progressively decreases in temperature and intensity as Pd concentration increases, see Fig. 3. Such a temperature dependence is characteristic of non-Fermi-liquids (NFL) [10, 12] in the region of quantum critical behavior that affects the thermal properties because of emerging quantum fluctuations. These low energy fluctuations overcome classical thermal fluctuations below 2 or 3 K as it was observed in Ce NFL systems [11]. Within that regime, a characteristic $C_m(T)/T = -A\log(T/T^*)$ thermal dependencies occurs, like the observed in this system above T_N, see Fig. 4. The parameter T^* represents a characteristic energy scale that describes the range of the $C_m(T)/T$ tail. The obtained A and T^* parameters allow to scale this system into an universal curve $C_m/t = -7.2\log(t)$, with $t = T/T^*$, extracted for a number of Ce intermetallic compounds showing logarithmic thermal dependencies [13].

3. Discussion

3.1. Itinerant to local electronic transformation

Comparing magnetic and thermal properties one observes that, within the $0 \leq y \leq 0.48$ range, this system presents an unusual discrepancy between the decreasing $T_N(y)$ and the increasing amplitude of the $\Delta C_m(T_N)$ jump, see Fig. 3a. The latter behavior is in line with the increasing

Journal of Physics: Conference Series **568** (2014) 042029 doi:10.1088/1742-6596/568/4/042029

values of magnetization presented in Fig. 2a. In an usual scenario of competing RKKY magnetic interactions with Kondo effect, the former mechanism tends to increase magnetic order whereas the later to screen magnetic moments [14]. This classical description does not apply to this system because a $T_N(y)$ decrease would imply a weakening of the RKKY interaction due to an enhancement of the Kondo screening and the consequent decrease of $\Delta C_m(T_N)$ [7]. On the contrary, the observed growing and narrowing of $\Delta C_m(T_N)$ suggests a tendency towards the localization of Ce-$4f$ states driven by a weakening of $4f$-band hybridization (i.e. a decrease of $T_K(y)$). Furthermore, the ferromagnetic type dependence of $M(H)$ above $H_{cr} \approx 0.5$ Tesla observed in the $y \geq 0.48$ alloys clearly suggests a minor role of Kondo effect at this range of concentration. These facts make evident that different mechanisms are involved in the evolution of the ground state properties of this system.

The most likely is a decreasing contribution of the excited crystal electric field CEF levels at low energy, which leads the system into an itinerant to localized transformation. Taking into account that $|\theta_P|$ can be considered $\propto T_K$ [15], if $|\theta_P|$ is comparable to the splitting Δ_{CEF} of the first CEF level the contribution of that doublet is relevant, but it becomes irrelevant once $|\theta_P| << \Delta_{CEF}$. The reported decrease of $|\theta_P|(y)$ from 58 K (at $y = 0$) to 6 K (at y=0.55), together with the increase of μ_{eff} and $M(H)$ with concentration, support this description. From $C_m(T)$ measurements up to $T = 30$ K on sample $y = 0.55$, the first CEF splitting was evaluated as $\Delta_{CEF} = 40 \pm 5$ K.

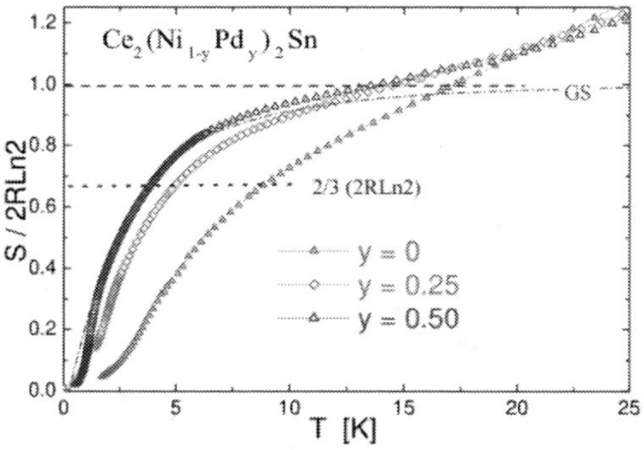

Figure 5. Thermal dependence of the entropy (normalized to one Ce at.) for $y = 0$ and the higher Pd concentrations. Dashed curve is a qualitative representation for the ground state contribution in sample $y = 0.55$.

An alternative way to test this scenario is the analysis of the entropy variation for different concentrations, $S_m(y, T)$. In Fig. 5 we show the thermal evolution of the magnetic entropy computed as $S_m = \int C_m/T \times dT$. On the Ni-rich limit, one sees that Ce_2Ni_2Sn presents a moderate and monotonous increase of $S_m(T)$ that overcomes $S_m = 2RLn2$ around 15 K. This behavior reveals a significant broadening of ground and excited CEF levels that partially overlap each other because of comparable T_K and Δ_{CEF} energies. On the contrary, the $y = 0.50$ and 0.55 alloys show an incipient plateau of $S_m(T)$ around $T \geq 10$ K. This change confirms that the doublet GS is mostly occupied at that temperature whereas the first excited CFE doublet has reduced significantly its contribution to the GS properties. Applying Desgranges-Schotte [16] criterion of $S_m(T_K) \simeq 2/3R\ln 2$ for single Kondo moments, the extracted values of $T_K(y)$ clearly show a decrease of Ce-$4f$ orbitals hybridization. This is the key to understand the unusual behavior of this system that can be explained as a combination between the decrease of the *4f-band* hybridization intensity and the consequent reduction of the CFE levels contribution to the low energy density of states.

3.2. Critical region and Phase Diagram

Another peculiarity of this system emerges from the analysis of the $S_m(T)$ dependence at low temperature since the entropy gain at T_N is remarkably low, i.e. ≈ 0.3 Rln2 per Ce-at. Such a low value is observed in Ce and Yb compounds which do not order because of e.g. magnetic frustration effects [17] but present a $C_m(T)/T$ anomaly characterized by a divergent power law tail above the temperature of the maximum T_{max}. Although in $Ce_2(Pd_{1-y}Ni_y)_2Sn$ the tail follows a $-\log(T/T^*)$ dependence, the C_m/T maximum diverges at $T_N(y)$ following a: $5/(T-0.4)\,K^{-1}$ dependence, as depicted in Fig. 4a, that should end in a critical point because of an *entropy bottleneck* [11]. This type of behavior was also observed in URu_2Si_2 under magnetic field [18] and $CeTi_ySc_{1-y}Ge$ [19]. The latter shows a very similar divergency of its C_m/T maximum also approaching a structural instability.

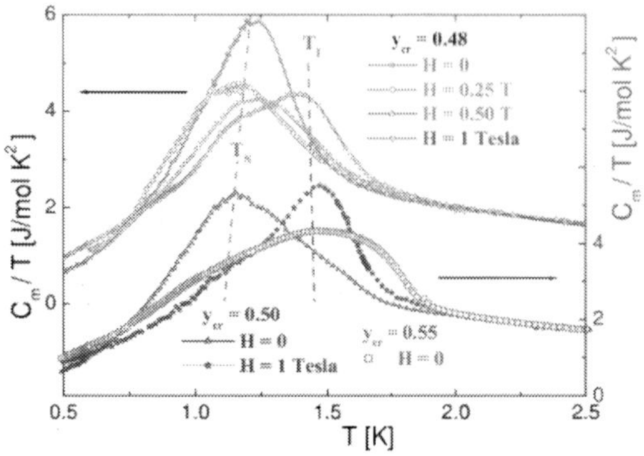

Figure 6. Left axis (shifted up): specific heat of the magnetic anomaly of sample $y_{cr} = 0.48$ at zero and three magnetic fields. Right axis: Comparison with other two concentrations. Dashed lines mark the mean temperatures of the anomalies at $T_N \approx 1.2\,K$ and $T_I \approx 1.5\,K$.

In Fig. 4a and b one can see how the temperature of the magnetic anomaly decreases down to $T_N \approx 1.2\,K$ at $y_{cr} = 0.50$ where another contribution at $T_I \approx 1.45\,K$ arises while the intensity of the former decreases. A detailed study of those specific heat anomalies around the critical region is presented in Fig. 6, performed at zero and applied magnetic field. Under magnetic field, the anomaly of the $y_{cr} = 0.48$ alloy splits into two maxima. In that figure one can clearly see how the C_m/T maximum at $T_N \approx 1.2\,K$ decreases in intensity while the one at T_I increases. The intrinsic character of this competition between a vanishing phase and an emerging one is confirmed by the same behavior observed in the $y = 0.50$ alloy, right axis in Fig. 6. Beyond that concentration, sample $y = 0.55$ shows how the relative contributions are reversed. The growing intensity of the T_I anomaly with field and the $M(H)$ dependence close to that temperature make evident a ferromagnetic character of this alloy despite of the lack of hysteresis in the $M(H)$ curves.

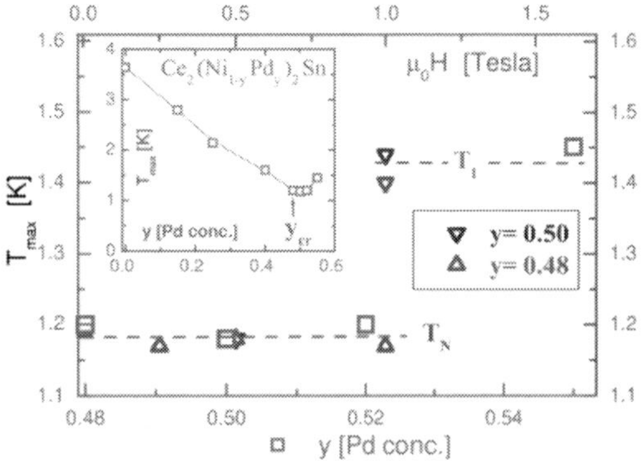

Figure 7. Magnetic phase diagram around the critical point comparing $T_N(0.48 \le y \le 0.55)$ (□) and field dependencies of $y_{cr} = 0.48$ (△) and $y = 0.50$ (▽). Inset: full Pd concentration range for $T_N(y)$.

The overall concentration dependence of $T_N(y)$ is included in the inset of Fig. 7, showing its decrease from 3.8 K down to 1.2 K at $y_{cr} = 0.48$. At that concentration the competition between two phases occurs as it shown by specific heat measurements presented in Fig. 6. Those details are summarized in the main body of Fig. 7. There, concentration and magnetic field effects on both contributions with respective maxima at $T_N \approx 1.2\,\mathrm{K}$ and $T_I \approx 1.45\,\mathrm{K}$ are compared. Notice that thermal ($k_B T_{max}$) and magnetic ($\mu_B H$) energy scales used in the phase diagram are comparable.

4. Conclusions

The outstanding characteristics observed in the $Ce_2(Ni_{1-y}Pd_y)_2Sn$ family are the transformation of the magnetic properties from itinerant to mostly localized and the presence of a critical concentration with signatures of ferromagnetic contribution.

The former characteristic is recognized from the Pd concentration dependence of the magnetic properties like: i) the decrease of $|\theta_P|$, reflecting a decrease of T_K, ii) the increase of the effective magnetic moment and iii) the sharpening of the specific heat transition. All these properties behave at variance to the decrease of the ordering temperature T_N in a usual RKKY scenario of localized moments. In this system, the $T_N(y)$ decrease is driven by the reduction of the CEF excited levels contribution to the ground state properties as revealed by the thermal evolution of the magnetic entropy with growing Pd concentration.

The second peculiarity of this system concerns the fact that the $C_m(T)$ anomaly associated to $T_N(y)$ shows a nearly constant maximum which ends into a *bottleneck* of the magnetic entropy at a critical concentration $y_{cr} \approx 0.48$. At that point $M(H)$ reveals a ferromagnetic type dependence above $H_{cr} \approx 0.4$T. The splitting of the $C_m(T)$ maximum, tuned by field and concentration indicates a competition between two magnetic phases within the $0.48 \leq y \leq 0.55$ range. Their respective characteristic energies are $T_N \approx 1.2\,\mathrm{K}$ and $T_I \approx 1.45\,\mathrm{K}$, with the former contribution vanishing and the latter arising as a function of concentration and/or field.

Lower temperature magnetic measurements (i.e. $T < 1.8\,\mathrm{K}$) are required to better characterize the magnetic ground state within the critical region. Spectroscopic measurements may shed light into the dynamical nature of these competing phases and band calculations are highly desirable to trace the modification of the Fermi surface.

References

[1] L.E. De Long, J.G. Huber and K.S. Bedell, J. Magn. Magn. Mat 99 (1991) 171.
[2] A.R. Mackintosh; Physica 130B (1985) 112.
[3] See e.g. J.G. Sereni; Physica B 215 (1995) 273.
[4] S K Dhar, S K Malik and R Vijayaraghavan; J. Phys. C: Solid State Phys., 14 (1981) L321-L324.
[5] S. Singh, S.K. Dahr, C. Mitra, P. Paulose, P. Manfrinetti, A. Palenzona; J. Phys. Cond. Mat. 13 (2001) 3753.
[6] See e.g. S. Kawarazaki, M. sato, Y. Miyako, N. Chigusa, K. Watamabe; Phys. Rev. B 61 (2000) 4167.
[7] J.G. Sereni, G. Schmerber, A. Braghta, B. Chevalier, J.P. Kappler; arXiv 1103.0190 cond-mat, 1 March 2011.
[8] F.J. DiSalvo, R.A. Gordon, C.M. Spencer, Y. Ijiri; J. Alloys and Compd. **224** (1995) 101.
[9] J.G. Sereni; J. Low Temp. Phys. **147** 179 (2007).
[10] H.v. Löhneysen, A. Rosch, M. Vojta, P. Wölfle, Rew. Mod. Phys. **79** (2007) 1015.
[11] J.G. Sereni; Phil. Mag. 93 (2013) 409.
[12] G.R. Stewart; Rew. Mod. Phys. **73** 797 (2001).
[13] J.G. Sereni, C. Geibel, M. G.-Berisso, P. Hellmann, O. Trovarelli and F. Steglich, Physica B 230 (1997) 580.
[14] M. Lavagna, C. Lacroix, M.J. Cyrot, J. Phys. F 12 (1982) 745.
[15] H.R. Krishna-murthy and C. Jayaprakash, Phys. Rev. B 30 (1984) 2806.
[16] H.-U. Desgranges and K.D. Schotte, Physics Letters **91A** 240 (1982).
[17] J.G. Sereni, 2014, to be published.
[18] M. Jaime, K.H. Kim, J. Guillermo, S. McCall, J.A. Mydosh; Phys. Rev. Lett., 89 (2002) 287201.
[19] J.G. Sereni, P. Pedrazzini, M. Gmez Berisso, A. Chacoma, S. Encina, T. Gruner, N. Caroca-Canales, C. Geibel; ArXiv:cond.mat 1403.4490, 18 March 2014.

Novel multiferroic phase of $CsCuCl_3$ in High Magnetic Fields

J. Shibuya, M. Akaki, Y. Kohama, A. Miyake, M. Tokunaga, and H. Tanaka [1]

The Institute for Solid State Physics, The University of Tokyo, Kashiwa, Chiba 277-8581, Japan

[1]Department of Physics, Tokyo Institute of Technology, Ookayama, Kanagawa, 152-8551, Japan

E-mail: jshibu@issp.u-tokyo.ac.jp

Abstract. We measured the magnetization and the electric polarization under pulsed magnetic field up to 44 T. In magnetization measurements for $H//b^*$ (perpendicular to the *ca*-plane), we found a new phase transition through the differential magnetization at 17 T, which is slightly higher than the field range of the magnetization plateau. This finding was supported by the electronic polarization measurements, with the observation of a bump structure in dP/dB signal. These results suggest the new phase transition.

1. Introduction

$CsCuCl_3$ is a family of a hexagonal perovskite, which belongs to the $P6_122$ space group. The magnetic Cu^{2+} ions ($S = 1/2$) forms a stacked triangular-lattice and exhibits exotic magnetic ground states with and without an external magnetic field. In this material, a ferromagnetic interaction along the chain direction (*c*-axis) is dominant, and a frustrated antiferromagnetic interaction in the *ab*-plane is as small as one-sixth of the main ferromagnetic interaction.

In the absence of an external magnetic field, $CsCuCl_3$ undergoes a transition to a spin-helix structure at 10.65 K (T_N), in which the spins are arranged into the well-known 120° structure in the *ab*-plane with an incommensurate (IC) modulation along the *c*-axis owing to the Dzyaloshinkii-Moriya interaction. [1] Nojiri *et al.* reported that when the magnetic field is applied parallel to the *c*-axis, the magnetization curves exhibit a jump at $\mu_0 H_J \sim 12.5$ T and then saturate at $\mu_0 H_{sat} \sim 30$ T. When an external magnetic field is applied within the *ab*-plane, a plateau-like anomaly was observed between $\mu_0 H_{p1} \sim 11$ T and $\mu_0 H_{p2} \sim 13$ T.

These field-induced anomalies were theoretically investigated by Nikumi and Shiba with taking into account a quantum effect [2,3]. They proposed that the magnetization jump at $\mu_0 H_J$ can be attributed to the transition from an in-plane 120° structure to a coplanar structure. In a high-field phase (coplanar structure), the spins gradually tilt to the field orientation by further increasing the magnetic field towards $\mu_0 H_{sat}$. On the other hand, application of $H \perp c$ is expected to change the in-plane IC solution to the out-of-plane one at a certain field, and then cause a transition to the commensurate (C) state at larger fields. This model is widely believed to explain these field-induced phenomena.

In this paper, we report the experimental observation of an additional field-induced phase transition in $CsCuCl_3$ through magnetization measurements in magnetic fields applied normal to the *c*-axis. Owing to the low crystallographic symmetry, some kinds of symmetry breaking caused by magnetic order result in the emergence of polarization as observed in a related material of $TlCoCl_3$ [4]. Measurement of electric polarization in pulsed high magnetic field on $CsCuCl_3$ revealed that this novel phase has spontaneous electric polarization.

2. Experiment

In this study, we investigated magnetic and dielectric properties in single crystals of $CsCuCl_3$ in high magnetic fields. Measurements in high magnetic fields were performed using non-destructive pulse magnets installed at The Institute for Solid State Physics at The University of Tokyo. For magnetization and electric polarization measurements, magnetic fields up to 44 T (pulse duration of 8 ms), were generated using unipolar and bipolar pulse magnets, respectively.

Single crystals of $CsCuCl_3$ were synthesized by dissolving stoichiometric amounts of CsCl and $CuCl_2$•H_2O in water. And the solution evaporates slowly. The single crystals grew as hexagonal bipyramids and were about 4 cm long along the *c*-axis. By the back Laue method, we determined the principal axes of the sample and cut them along perpendicular or parallel to the *c*-axis with using a wiresaw.

Magnetization (*M*) measurements were performed by the induction method using two pickup coils placed coaxially. In electric polarization (*P*) measurement, we measured polarization currents caused by the field-induced changes in electric polarization. To eliminate the effect of induced voltage in wiring, the data in positive and negative field scans were averaged [5], and then integrated.

3. Results and discussion

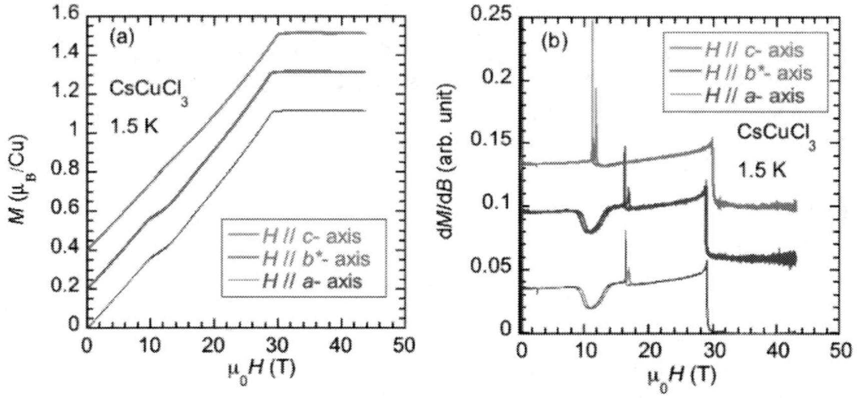

Figure 1. (colors online) (a) The field dependence of magnetization up to 44 T at 1.5 K for fields along *c* (red), b^* (green), *a* (blue), respectively. The data were vertically offset for clarity. (b) The field dependence of differential magnetization for each field direction.

Figure 1(a) shows the field dependence of magnetization. We applied the magnetic fields along *c*, *a*, and b^*-axis (perpendicular to the *ca*-plane). The overall features of the magnetization curves are consistent with the results in the previous report [1]. However, close look into the present data reveals the existence of novel features as exhibited in the differential magnetization curves in Fig. 1(b). For H // c, a sharp peak was observed at around 11 T corresponding to the magnetization jump. On the other hand, the differential magnetization for H // b^*, which is nearly identical with that for H // a, decreases between 9 T and 14 T, corresponding to the plateau-like structure. In addition to these features reported earlier, the present data show clear peaks at ~ 17 T indicating abrupt jumps in magnetization. Since we observed difference in the peak fields between up- and down-sweeps, we regard these anomalies as the transition of first order.

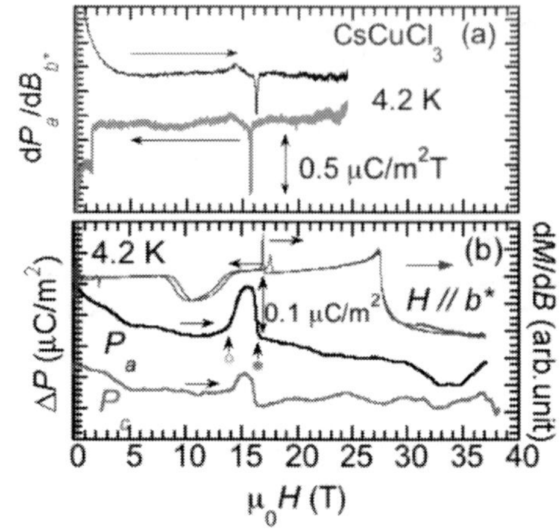

Figure 2. (colors online) (a) Differential electric polarization parallel to the a-axis as a function of applied magnetic fields along the b^*-axis in the field increasing (black) and decreasing (red) processes. The data were vertically offset for clarity. (b) The field dependence of differential magnetization (blue) and electric polarization along the a (P_a) and c (P_c) -axes for H // b^*.

Figure 2(a) shows differential electric polarization (dP/dB) parallel to the a-axis as a function of applied magnetic fields along the b^*-axis. A bump and successive opposite sharp peak structures are observed at about 14 T and 17 T, respectively. The profiles in both up- and down-sweeps show negative peaks in dP/dB_{b*} at the fields of 16.5 T and 15.5 T, respectively. The integrated field dependence of the electric polarization is shown in Fig. 2(b) together with the differential magnetization at the same temperature. Significant ΔP along the a- and c-axes are observed between 14 T and 16 T. The ΔP along the a-axis is roughly two times larger than that along the c-axis. The observed ΔP ~ 0.5 μC/m^2 is less than 10^{-3} of P in the related material, $TlCoCl_3$, and can be discernible only in highly sensitive measurements in pulsed-fields.

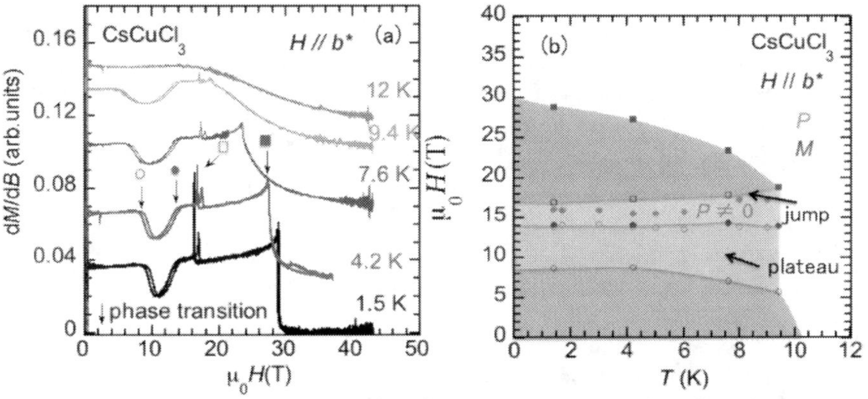

Figure 3. (colors online) (a) The field dependence of differential magnetization at various temperatures for H // b-axis. The data were vertically offset for clarity. (b) Magnetoelectric phase diagram in the H_{b*}-T plane. Red (Blue) circles represent transition fields determined by the inflection points of the P–H (M–H) curves.

Figure 3(a) shows the field dependence of differential magnetization at various temperatures between 1.5 K and 12 K. For H // b^*, a sharp jump is observed at ~17 T below T_N (= 10.65 K). The cusp-like structure that emerges at the saturation field moves to lower field with increasing temperature, whereas the other features for plateau-like anomalies and peaks are less sensitive to the temperature. The arrows in Fig. 3(a) represent our definition of the transition fields. Figure 3(b) shows magnetoelectric phase diagrams in the H_{b*}-T plane determined from these results. The novel polar phase emerges in the orange area between 14 T and 17 T. Multiple phase transitions of $CsCuCl_3$ for in-plane fields have been reported through specific heat measurements and neutron diffraction [6,7,8]. Our present results revealed the emergence of a novel multiferroic phase in between the IC3 and commensurate phases shown in Fig. 8 of Ref. 7. Careful reader might notice the discrepancy of the transition field at ~17 T determined by magnetization and electric polarization, which can be ascribed to the finite misalignment of the field angle as shown in the following.

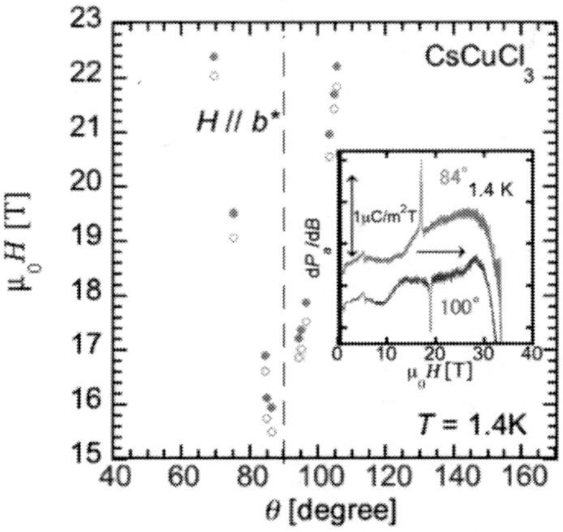

Figure 4. (colors online) Angular dependence of the transition field determined by polarization measurements at 1.4 K. The θ denotes polar angle from the c to b^* axis. The insets show differential polarization along the a-axis as a function of magnetic field at θ = 84° and 100°.

Figure 4 shows angular dependence of the transition fields determined by measurements of electric polarization along the *a*-axis. The θ denotes the polar angle of the applied field from the *c*- to the b^*-axis. The inset shows field dependence of differential polarization along the *a*-axis at $\theta = 84°$ and $100°$. The transition field steeply increases with tilting away from the b^* axis. The dP_a/dB signal changed its sign between $\theta < 90°$ and $\theta > 90°$, and was hardly discernible at $\theta = 90°$.

Here, we will discuss the possible origin of the phase transitions at ~17 T for $H \perp c$-axis. Early neutron experiments reported deformation of the helical structure up to the plateau-like region [8], at which field dependence of the index δ of the incommensurate magnetic peak at (1/3, 1/3, δ) also shows a plateau at $\delta \sim 0.047$ [9]. With further increasing the field, the δ started to desrease again toward zero at ~17.5 T [9]. The present polar phase emerges in the region with this small δ, and hence, the period of the incommensurate spin moduration becomes sufficiently long in this field region.

According to the theoretical study by Nikuni and Jacobs [3], application of the transverse fields realizes "out-of-plane" incommensurate spin order in the intermediate field region. Although the details of the "out-of-plane" state was not well described, simple cycloidal structure cannot explain the coexistence of P_a and P_c in the polar state within the framework of the spin-current model [10] while the modulation vector pointing to the *c*-axis. On the other hand, the same authors also mentioned the solitonic phase as one of the in-plane solutions at large fields. Recently, a solitonic lattice state was found in a chiral heli-magnet $Cr_{1/3}NbS_2$ in fields applied normal to the ferromagnetic spin chains [11]. Although it is not clear whether such solitonic state can accommodate with electric polarization in $CsCuCl_3$ even if it exists, the observed polar phase will be of interest as a novel phase of chiral magnets in the transverse fields.

Recent X-ray study on $CsCuCl_3$ revealed that usual crystals of this material contain mixed crystallographic domains with right- and left-hand chirality [12], which may diminish the net electric polarization detectable in the whole sample. Future experiments on crystals having single chirality will clarify the intrinsic nature of the novel multiferroic phase in this material.

4. Conclution

We measured magnetization and electric polarization of single crystals of $CsCuCl_3$ in pulsed magnetic field up to 44 T. For fields applied normal to the *c*-axis, we found novel multiferroic phase between 14 – 17 T.

References

[1] Nojiri H. *et al. Journal De Physique* **C8**, 1459 (1988).
[2] Nikuni T. and Shiba. H *J. Phys. Soc. Jpn.* **62**, 3268 (1993).
[3] Nikuni T. and Jacobs.A.E *Phy. Rev. B* **57**, 5205 (1998).
[4] Nishiwaki Y. *et al. J. Phys. Soc. Jpn.* **72,** 2608 (2003).
[5] Mitamura H. *et. al. J. Phys. Soc. Jpn.* **76,** 094709 (2007).
[6] Bügel R. *et al. Phys. Rev. B* **65**, 052402 (2001).
[7] Stüβer N. *et al. J. Phys.: Condens. Matter* **14,** 5161 (2002).
[8] Stüβer N. *et al. Physica B* **350**, e47 (1998).
[9] Nojiri H. *et al. Physica B* **241-243,** 210 (1998).
[10] Katsura H., Nagaosa N., and Balatsky A.V. *Phys. Rev. Lett* **95,** 057205 (2005).
[11] Togawa Y. *et al. Phys. Rev. Lett.* **108,** 107202 (2012).
[12] Ohsumi H. *et al. Angew. Chem. Int. Ed.* **52,** 8718 (2013).

Journal of Physics: Conference Series **568** (2014) 042031 doi:10.1088/1742-6596/568/4/042031

Multiple First Order Transitions in Cr Thin Films

R. H. Rodriguez[1], **L. Tosi**[1], **M. Menghini**[2] **H. Pastoriza**[1], **C. A. Balseiro**[1]

[1] Centro Atómico Bariloche and Instituto Balseiro, Comisión Nacional de Energía Atómica, 8400 Bariloche, and CONICET, Argentina

[2] Dept. of Physics and Astronomy, KU Leuven, Leuven, Belgium.

E-mail: tosil@ib.cnea.gov.ar

Abstract. The incommensurate spin density wave (SDW) in metalic Cr forms below $T_N = 311\,\mathrm{K}$ and is characterized by a temperature dependent wave-vector **Q** along the <100> direction. The wavelength varies from 78 Å at 300 K to 60 Å at 150 K. In thin films, the Q-vector orients perpendicular to the film surface and the SDW is forced to satisfy boundary conditions. As a consequence of the confinement, the SDW wavelength is quantized in a set of available modes, each one having n nodes. As the temperature decreases the wavelength does not follow a continuous bulk-type variation but jumps between different allowed values. These jumps occur at certain determined temperatures T_n as first order type transitions which are accompanied with a decrease (n to $n+1$ nodes) or an increase ($n+1$ to n nodes) of the resistance. In this work, we present transport measurements on Cr thin films showing hysteretic behavior in the resistance as a function of temperature in cooling-warming cycles. The hysteresis loops are thickness dependent and present structures which we associate with the possible multiple transitions that may occur as temperature is swept.

1. Introduction

In recent experiments attention was drawn back to the antiferromagnetic ground state of Cr system [1, 2]. The study of the spin density wave (SDW) was very important in the understanding of giant magnetoresistance and spin dependent transport (spintronics) in CrFe superlattices [3, 4, 5, 6] and the SDW ground state in Cr films [7, 8] became an active research area with experimental and theoretical studies involving different magnetic materials [9, 10, 11]. It was realized that confined geometries could induce non-trivial modifications in the magnetic and transport properties of these multilayered systems [12, 13]. Kummamuru's recent work [1] reopened the interest in this system showing how transport properties are modified by the antiferromagnetic state of Cr when the SDW is confined in very thin films. As further explored in other works [2], confinement leads to quantization of the wavelength and to the formation of antiferromagnetic domains. Transport properties are strongly influenced by the presence of these domains and their evolution with temperature, and since antiferromagnetic spintronics is a well considered candidate for future applications [14, 15], different ideas about the control and manipulation of these domains and domain-walls are now being rehearsed [16]. In Cr thin films domain structures emerge in a very natural way and it is possible to dessign basic experiments in several directions: *i)* it is clear than the domain structure depends on the films roughness, but how much this affects transport properties has not yet been addressed. *ii)* The average dimension and nature of domains is still an open question and also *iii)* it is interesting to study how sensitive are transport properties to individual domain nucleation, growing or switching.

In this work we present experimental results on transport properties in thin (~400 Å) Cr films grown by molecular beam epitaxy (MBE). In thin films, the Q-vector of the SDW has been found to be perpendicular to the films surface in the whole range of studied temperatures (from 330 K to 30 K) [7, 10] and it has been observed that the strong boundary conditions emerging from the confinement within the film thickness produce a thickness-dependent hysteretic behavior of the transport properties [2]. In previous reports epitaxial thin films grown by sputtering have been investigated. In those cases, the observed phenomena are governed by the confinement of the SDW within the film thickness L with an incertitude given mainly by the mean roughness ~ 20 Å. The results of the present article may give some answers to the three points stated above. In first place, our films grown by MBE have a mean roughness of ~ 5 Å and the effect of the roughness reduction can be directly analyzed. As shown in Fig. 1A), our experimental setup allows us to measure simultaneously different parts of the sample (with different dimensions) and thus to gain some insight into the size (and number) of domains, although this can be inferred only roughly. Finally, the hysteresis in resistance is an evidence of the motion of domains walls, i.e. growing or shrinking of domains and for that reason, a detailed study of its structure gives some information of the sensitivity to individual processes.

2. Experimental Method

Cr thin films were grown on MgO <100> substrates by molecular beam epitaxy following a similar procedure as reported for sputtering deposition. In the latter, the resulting films showed a very good degree of epitaxy [1, 2]. We characterized our films with X-ray diffraction and as shown in Fig. 1B), the spectrum shows a well defined peak at $2\theta = 64.6°$ corresponding to the <002> Cr characteristic peak and an angular dispersion of 0.5° (FWHM from rocking curve). The other peaks correspond to the substrate. From small angle reflectometry, shown in Fig. 1C), we obtained the thickness of the sample L =370 Å. The films surface were characterized by AFM (see Fig. 1D)) and the mean roughness found was ~5 Å. This is approximately four times better than what was found in epitaxial films grown by sputtering (~20 Å).

In order to perform transport measurements we patterned the films by means of optical lithography and chemical etching in a hallbar-like configuration with several terminals (see Fig. 1A)). We contacted the samples with 25 microns gold wires glued with Epo-tek epoxy to different pads being thus able to measure fours regions of the hallbar (four independent resistances) simultaneously (i.e. during the same cooling-warming cycle). Samples were thermally anchored to the cold plate of a commercial close cycle cryocooler and temperature cycles were performed from 330 K to 20 K with a step of 1 K and a temperature stabilization protocol that assures fluctuation of the temperature lower than 20 mK.

3. Results and discussion

Resistivity as a function of temperature $\rho(T)$ (see Fig. 2) shows an hysteretic behavior in cooling-warming cycles. This effect is rather small (0.1 %) and is consistent with the findings of Kummamuru *et al.* [1] and Osquiguil *et al.* [2] for thin Cr films grown by sputtering. A better look at the effect of the SDW in transport properties can be obtained from the hysteresis loop: the difference $\Delta\rho(T) = \rho_c(T) - \rho_w(T)$ between the resistivity when cooling $\rho_c(T)$ and warming $\rho_w(T)$ is shown in Fig. 2B). The hysteresis loop presents some remarkable features: *i)* first of all, hysteresis is different from zero only in the temperature window ranging from 60 to 260 K. *ii)* It is recognizable the presence of two regions where the hysteresis is maximum but also *iii)* a very sharp decrease is observed around 180 K.

The origin of the hysteresis in thin Cr films has been recently discussed and so far understood in terms of the confinement of the SDW within the film thickness [1, 2, 14, 17]. Assuming that

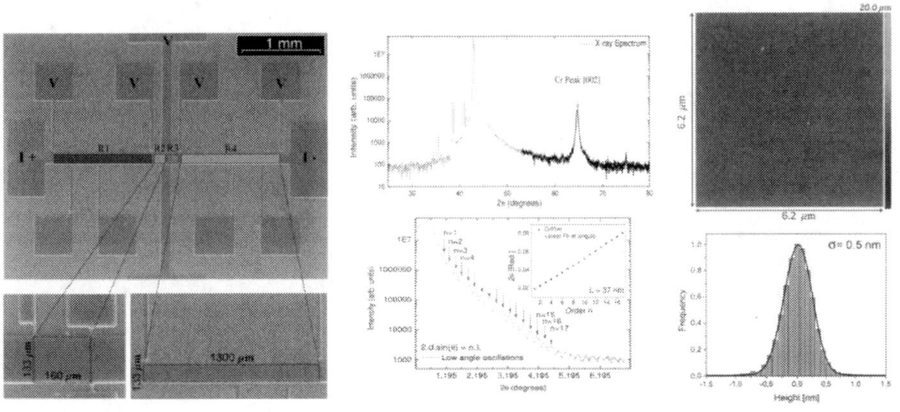

Figure 1. Sample fabrication and characterization: A) Cr thin film patterned as a hallbar with several terminals allowing different configuration measurements. The low panels show a zoom to two sample regions. B) X-ray difraction spectrum showing the <002> characteristic Cr peak. The other peaks correspond to the substrate. C) Small angle X-ray reflectometry spectrum from which the film thickness can be obtained. D) AFM topography map and height histogram (inferior panel) showing a mean roughness of ~5 Å.

both, vacuum at the surface and non-magnetic MgO at the bottom interface define strong boundary conditions for the magnetic state of Cr, a discrete set of wavelengths Λ_n would be stable. As in bulk samples the SDW wavelength changes with the temperature, from 78 Å at 300 K to 60 Å at 150 K [18], it is reasonable to assume that the hysteresis in thin films is the result of the discontinuous jumps between a high-temperature wavelength Λ_n to a smaller low-temperature one Λ_{n+1}. Here n indicates the number of nodes inside the film and, as shown by the hysteresis loops in Hall measurements [1], in the low-temperature phase a larger number of effective carriers is present (thus a lower resistance) than in the high temperature phase. Indeed, as shown in the inset of Fig. 2A), resistivity when warming ρ_w is always lower than when cooling ρ_c and hysteresis loop is positive. The jumps that lead to the hysteretic behavior occur as a result of a competition between the energy cost of deforming the SDW (a different wavelength with respect to that expected from bulk, continuously decreasing with temperature, $\Lambda_{SDW}(T)$) and the surface energy related to the boundary conditions (imposing $\Lambda_n = 2L/n$). The change from n to $n+1$ occurs as a first order type transition at a temperature that is determined by this energetic balance, close to a commensurability condition $L \sim (n+1)\Lambda_{SDW}(T)/2$, here we recall, L is the film thickness [2].

From the derivative of the resistivity $d\rho(T)/dT$ some other features of the hysteretic behavior become apparent. As shown in Fig. 2C), there is a slope change at 300 K, then a wide minimum and a maximum. The nucleation of the SDW, due to the nesting of electron and hole sheets, is accompanied with a partial destruction of the Fermi surface (through the opening of a gap). As a consequence, resistivity increases at the transition temperature T_N and a change in the slope is thus expected. The combined effect with electron-phonon scattering makes the resistivity decrease once again with temperature and this is the observed minimum. In general, the Néel temperature is defined at the minimum of $d\rho/dT$, however in the present case it is more appropiate define T_N as the temperature at which the slope changes. The former case would be 240 K and the latter, more consistent with the literature, 300 K. The reason why there is so much difference between the position of the minimum in MBE vs sputtering grown Cr thin films is not clear to us at the moment. What is nevertheless more notorious is the fact that cooling

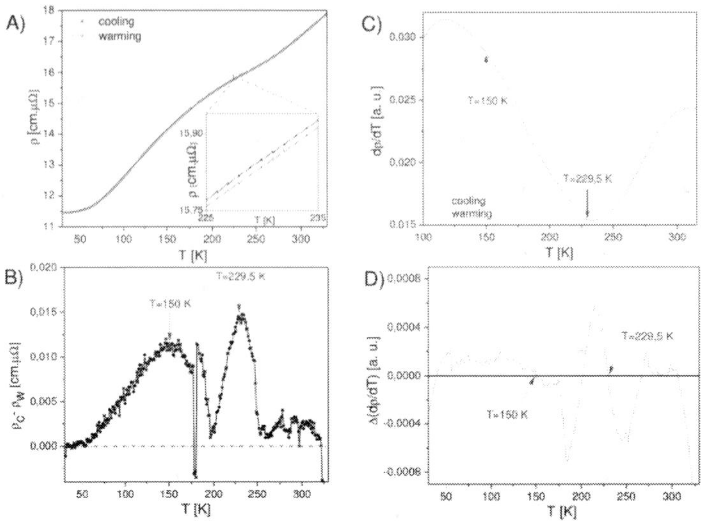

Figure 2. Hysteretic Behavior: A) Resistivity as a function of temperature when cooling (black curve) and warming (red curve). A zoom is shown in the inset to better apreciate the hysteresis. B) Hysteresis loop: difference between resistivity when cooling and warming. The maximum of hysteresis is observed at 150 K and 230 K (indicated with arrows). C) Derivative of the resistivity as a function of temperature showing crossing at the same tempeartures where the hysteresis maximum occur. D) Difference between the derivatives of the resistivity when cooling and warming.

and warming derivatives curves have their minimum occurring at different temperatures and as shown in Fig. 2C), they cross at the temperatures at which the maxima of $\Delta\rho$ in the hysteresis loop are found, $T = 230\,\mathrm{K}$ and $T = 150\,\mathrm{K}$ as indicated in the figure.

The hysteresis loop has been shown to be thickness dependent, being this related to the fact the commensurability condition $L \sim (n+1)\Lambda_{SDW}(T)/2$ involves the thickness L [2, 19]. For different values of L, it is possible to obtain several corresponding sets of available modes with $\Lambda_n \in [60, 78]\,\text{Å}$ and for that reason, depending on the case, more than one transition may be expected. For a 370 Å film, two transitions, the first one $n = 10 \rightarrow 11$ and the second one $n = 11 \rightarrow 12$, may be predicted simply from the strong boundary conditions. Whether these transitions can be resolved is another problem. Each of the first order transitions does not occur in the whole sample all at once, at a given temperature. On the contrary, as temperature is swept there is region of coexistence. An antiferromagnetic domains structure develops consisting on regions of SDW's with different n (i.e. different wavelength, neither orientation nor polarization). The hysteresis loop is a direct consequence of this coexistence and the temperature at the maximum can be taken as the mean (or more probable) transition temperature. The width of the hysteresis loop and the temperature separation between consecutive commensurability conditions determine whether more than one transition can be resolved as in the case presented in Fig. 2. The interesting point to be addressed is that up to now it had not been reported so clearly the observation of multiple phase transitions in Cr thin films. We believe that the reduction of the roughness by growing our films by MBE was crucial for our finding.

We have observed that the hysteresis loops all together with the noise patterns vary little

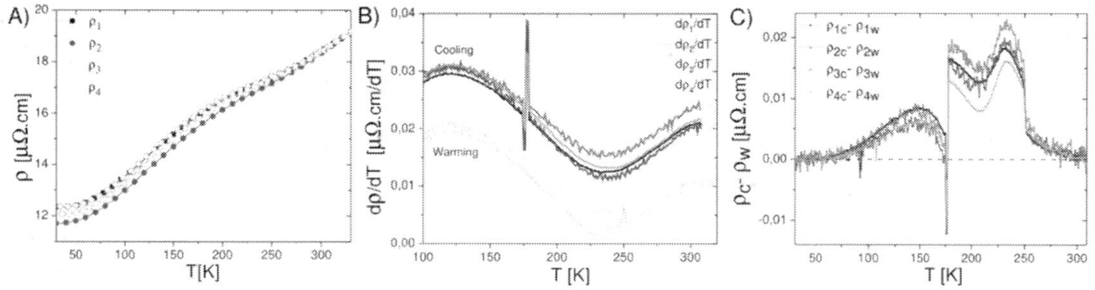

Figure 3. Size effects: A) Resistivities as a function of temperature when cooling for four different regions of the sample identified as 1-4. B) Derivatives of the resistivities as a function of temperature when cooling and warming with an arbitrary offset. C) Hysteresis loops for 1-4.

from run to run but also, rescaling appropriately with the respective dimensions, the hysteresis loops from the different parts of the sample (see Fig. 1A)) can be reproduced with the same parameters. In Fig. 3A) we show the resistivities of these parts, designated 1-4, measured in a cooling-warming cycle with a better than 5 mK tempeature stabilization. They coincide at high temperatures above T_N and then follow a similar temperature dependence. The derivatives are shown in Fig. 3B). In all the sample parts there is a strong peak at 180 K when cooling and another peak at 250 K when warming. From a careful look at the resistivities we observe that nothing happens during the cooling at 250 K and similarly, no change occurs in the resistivity at 180 K when warming. The sudden decrease of the resistivity at 180 K when cooling and the sudden increase at 250 K when warming generate a big distortion of the hysteresis loop as shown in Fig. 3C). We do not have an explanation for the origin of these jumps which we think are not directly related to the hysteresis. We believe that an hysteretic transition from a transversal to a longitudinal SDW [7] may be the cause of such behavior. The fact that it seems to occur in the whole sample at once (and considering our AFM characterization) makes us disregard a possible step pinning the SDW and thus creating a big hysteron.

It is also remarkable that the loops are almost the same independently from the size of the sample (there is a factor ten between 1-4 and 2-3, see Fig. 1A)). We believe this is a consequence of the domains being always much smaller than the sample dimensions even for the parts 2 and 3 (whose length is 130 μm). It also tells us that the number of hysterons (related indirectly to the number of domains) scales proportionally to the length of the sample.

L.T. has a scholarship from CONICET and R.H.R has a scholarship from CNEA. Work partially supported by PICT 1060 ANPCYT.

References

[1] Kummamuru R K and Soh Yeong-Ah 2008 *Nature (London)* **452** 859.
[2] Osquiguil E, Kaul E E, Tosi L and Balseiro C A 2012 *Phys. Rev. B* **85** 104410.
[3] Fawcett E 1988 *Rev. Mod. Phys.* **60** 209 and references therein.
[4] Shi Z-P and Fishman R S 1997 *Phys. Rev. Lett.* **78** 1351.
[5] Fishman R S and Shi Z-P 1999 *Phys. Rev. B* **59** 13849.
[6] Fishman R S 2001 *J. Phys. Condens. Matter* **13** R235.
[7] Sonntag P *et al.* 1998 *J. Magn. Magn. Mater.* **183** 5.
[8] Zabel H 1999 *J. Phys.: Condens. Matter* **11** 9303.
[9] Fullerton E E *et al.* 1996 *Phys. Rev. Lett.* **77** 1382.
[10] Bödeker P *et al.* 1998 *Phys. Rev. Lett.* **81** 914.
[11] Niklasson A M N *et al.* 1999 *Phys. Rev. Lett.* **82** 4544.
[12] Mibu K *et al.* 2002 *Phys. Rev. Lett.* **89** 287202-1.

[13] Fullerton E E *et al.* 2003 *Phys. Rev. Lett.* **91** 237201-1.
[14] Soh Y-A and Kummamuru R K 2011 *Phil. Trans. R. Soc. A* **369** 3646.
[15] Gomonay E V and Loktev V M 2014 *Low Temperature Physics* **40** 17.
[16] Logan J M *et al.* 2014 arXiv:1202.4401 [cond-mat.str-el].
[17] Tosi L, Osquiguil E, Kaul E E and Balseiro C A 2012 *Europhys. Lett.* **100** 67005.
[18] Werner S A *et al.* 1967 *Phys. Rev.* **155** 528.
[19] A complete discussion of the thickness dependence of the hysteresis loop in sputtering and MBE grown samples is currently under preparation.
[20] Barkhausen H 1919 *Z. Phys.* **20** 401.
[21] Preisach F 1935 *Z. Phys.* **94** 277.
[22] Michel R P, Israeloff N E, Weissman M B, Dura J A and Flynn C P 1991 *Phys. Rev. B* **44** 7413.
[23] Evans P G *et al.* 2002 *Science* **295** 1042.
[24] Shpyrko O G *et al.* 2007 *Nature* **447** 68.

Magnetic and transport properties of $EuNi(Si_{1-x}Ge_x)_3$ compounds

K Uchima[1], Y Takaesu[1], H Akamine[2], M Kakihana[2], K Tomori[2], T Uejo[2], A Teruya[2], A Nakamura[2], M Hedo[3], T Nakama[3], K Yagasaki[3], K Matsubayashi[4] and Y Uwatoko[4]

[1]General Education, Okinawa Christian Junior College, Nishihara, Okinawa, 903-0207, Japan
[2]Graduate School of Science and Engineering, University of the Ryukyus, Nishihara, Okinawa 903-0213, Japan
[3]Faculty of Science, University of the Ryukyus, Nishihara, Okinawa, 903-0213, Japan
[4]Institute for Solid State Physics, University of Tokyo, Kashiwa, Chiba 277-8581, Japan

E-mail: uchima@ocjc.ac.jp

Abstract. The magnetization M, electrical resistivity ρ, thermopower S and specific heat C of $EuNi(Si_{1-x}Ge_x)_3$ compounds have been measured at temperatures from 2 to 300 K. For the compounds of $EuNi(Si_{1-x}Ge_x)_3$, we obtained an effective magnetic moment of $\mu_{eff} \approx 7.7\ \mu_B$, which is close to the divalent Eu value of $\mu_{eff}=7.94\mu_B$. All compounds of $EuNi(Si_{1-x}Ge_x)_3$ order antiferromagnetically. The Néel temperature T_N decreases monotonously with increasing the Ge concentration x from T_N=49 K for $EuNiSi_3$ to T_N=14 K for $EuNiGe_3$. In the low temperature region below T_N, anomalies corresponding to an additional magnetic phase transition into ferromagnetic state for compounds with $x < 0.3$, and into another antiferromagnetic for $x > 0.3$ were observed. The Curie temperature T_C rapidly decreases with increasing x and vanishes at $x \approx 0.3$. It is found that the magnetic phase transition temperatures of T_N and T_C in $EuNi(Si_{1-x}Ge_x)_3$ are strongly connected with the change of volume induced by the atomic substitution of Si by Ge.

1. Introduction

Ternary intermetallic RT_2X_2 (122) and RTX_3 (113) systems (R= rare earth, T=transition metal and X=Ge or Si) have been studied from scientific and practical viewpoints. RT_2X_2 compounds with $ThCr_2Si_2$-type crystal structure have attracted much attention because of the large variety of interesting phenomena [1, 2]. Especially, the compounds with R=Ce and Yb show the characteristic phenomena such as superconductivity, heavy fermion, mixed valence and Kondo behaviors. It is well known that some Eu compounds reveal valence instability, depending on temperature, magnetic field, pressure and atomic composition. The accumulated data show that the stable states of Eu^{2+} ($4f^76s^2$), Eu^{3+}($4f^65d^16s^2$) and dynamic intermediate valence state are possible for Eu. The electronic state of Eu^{2+} is magnetic with $J = 7/2$, while the Eu^{3+} state is non-magnetic with $J = 0$. In RT_2X_2 series, $EuNi_2Ge_2$, in which the Eu atoms are in the Eu^{2+} ($J = 7/2$) state, manifests an antiferromagnetic order at the $T_N \approx 30$ K. On the other hand, $EuNi_2Si_2$, in which Eu atoms are in the Eu^{3+} ($J = 0$) state, is the Pauli-paramagnetism with temperature-independent magnetic susceptibility [3, 4, 5]. RTX_3 (R= rare earth, T=transition metal, and X=Si, Ge) compounds, which crystallize in the $ThCr_2Si_2$-type or $BaNiSi_3$-type tetragonal structures [6], have been studied extensively from scientific and practical viewpoints. However, few studies on $EuNiX_3$ (X=Si, Ge), as far as we know, have been reported in the literature.

In our previous work [7, 8], the effective magnetic moments of $EuNiSi_3$ and $EuNiGe_3$ were obtained as $\mu_{eff} \approx 7.5\mu_B$/Eu and $\mu_{eff} \approx 7.7\ \mu_B$/Eu, respectively. These values of effective magnetic moments are close to a divalent value of 7.94 μ_B/Eu. For $EuNiGe_3$ compound, antiferromagnetic transition was observed at Néel temperature T_N = 14 K, and another antiferromagnetic transition at T_a = 11.4 K. On the other hand, $EuNiSi_3$ is reported to exhibit antiferromagnetic transition at $T_N \approx$ 49 K and ferromagnetic transition at $T_C \approx$ 35 K, which are confirmed by the magnetization, specific heat, electrical resistivity and thermopower measurements. An additional anomaly corresponding to phase transition into the spin freezing state at $T^* \approx 5$ K is observed in the magnetization and thermopower measurements. The magnetic transition temperatures of T_N and T_C of $EuNiSi_3$ compound increase with increasing pressure. In this study, in order to investigate the effect of atomic substitution on magnetism and transport properties of $EuNi(Si_{1-x}Ge_x)_3$ compounds, we have carried out the measurements of the magnetization M, specific heat C, electrical resistivity ρ and thermopower S in the temperature range from 2 to 300 K.

2. Experimental procedures

Polycrystalline samples of $EuNi(Si_{1-x}Ge_x)_3$ were prepared from the pure components by melting in an arc furnace under a protective Ar atmosphere and was subsequently annealed in vacuum at 800 °C for five days. The crystal structures were verified by the powder X-ray diffraction measurement over the whole range of the $EuNi(Si_{1-x}Ge_x)_3$ compounds. $EuNiSi_3$ compound is a $ThCr_2Si_2$-type tetragonal structure with lattice parameters of a = 4.158 Å and c = 9.636 Å [6], and $EuNiGe_3$ is $BaNiSn_3$-type tetragonal structure with lattice parameters of a = 4.339 Å and c = 9.887 Å [8]. The crystal structures of $EuNi(Si_{1-x}Ge_x)_3$ compounds of $0 < x < 1.0$ were verified by the powder X-ray diffraction measurement as tetragonal structure. However, these compounds were not confirmed as to whether they are $ThCr_2Si_2$-type tetragonal structure or $BaNiSn_3$-type. The electrical resistivity ρ measurements were carried out by means of the standard four-probe dc method. The differential method with the seesaw-heating procedure was used for the thermopower S measurement [9]. The electrical resistivity ρ and thermopower S were measured simultaneously in the temperature range from 2 to 300 K. The magnetization M was measured at temperatures between 2 and 300 K by using a commercial SQUID magnetometer (MPMS, Quantum Design), and the specific heat C was measured from 2 to 80 K by using PPMS (Quantum Design).

3. Results and Discussion

Figure 1 shows the temperature dependence of the magnetic susceptibility χ of $EuNi(Si_{1-x}Ge_x)_3$ in the temperature region from 2 to 300 K in a magnetic field of 0.5 T. In the pure $EuNiSi_3$(x = 0.0) compound, χ increases with decreasing temperature, having a small peak at $T_N \approx$ 49 K, and shows a drastic increase at $T_C \approx$ 35 K. In addition, a sudden decrease is observed in $\chi(T)$ at $T^* \approx$ 5 K. From the magnetization measurement, it was found that these anomalies at T_N, T_C and T^* correspond to the magnetic phase transitions into antiferromagnetic, ferromagnetic and spin freezing states, as reported in our previous paper [7]. At higher temperatures than T_N, χ decreases with increasing temperature, following the Curie-Weiss law. The effective magnetic moments, obtained from the temperature dependence of the reciprocal susceptibility for compounds of $EuNi(Si_{1-x}Ge_x)_3$, are almost the same value of $\mu_{eff} \approx 7.7\ \mu_B$/Eu, indicating divalent state of Eu with $4f^7$ electronic configuration. The Curie temperature T_C and the spin freezing temperature T^* are confirmed in compounds with $x < 0.3$.

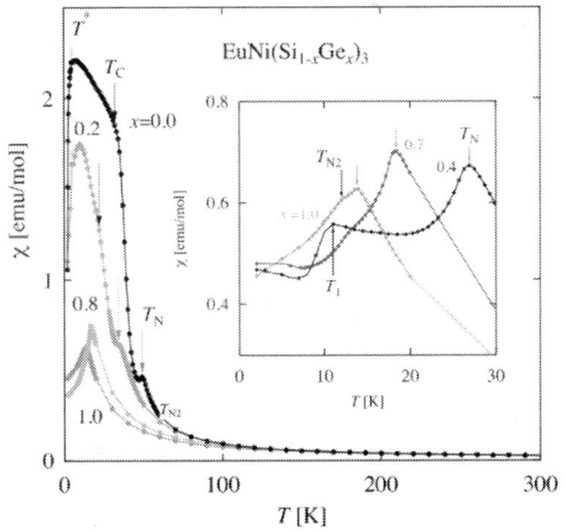

Figure 1. Temperature dependence of the magnetic susceptibility χ of $EuNi(Si_{1-x}Ge_x)_3$ in the temperature region from 2 to 300 K at a magnetic field of 0.5 T. The inset depicts the enlarged portion of $\chi(T)$ at temperatures below 30 K.

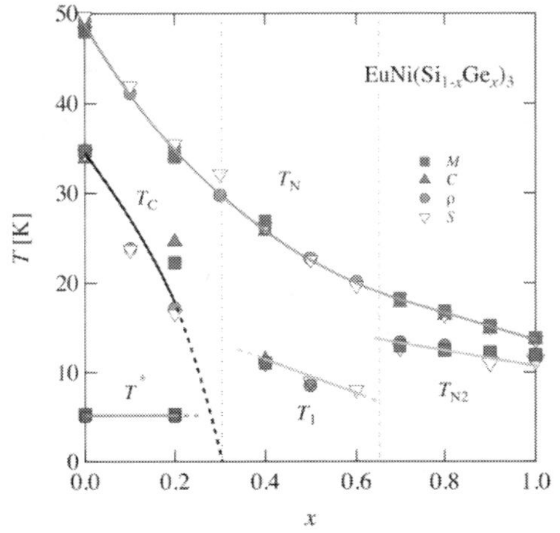

Figure 2. Ge concentration x dependences of the magnetic transition temperatures of T_N, T_C, T_{N2}, T_1 and T^* in $EuNi(Si_{1-x}Ge_x)_3$ compounds. The open and closed symbols of square, triangle, circle and reverse triangle are obtained from the measurements of magnetization, specific heat, electrical resistivity and thermopower, respectively.

The concentration x dependences of T_N, T_C and T^* obtained from $\chi(T)$ are plotted using square symbols (□ and ■) in Fig. 2. As shown in Fig. 2, the Néel temperature T_N decreases monotonously with increasing the Ge concentration x from $T_N = 49$ K for $EuNiSi_3$ to $T_N = 15$ K for $EuNiGe_3$. The Curie temperature T_C decreases rapidly with increasing x, and vanishes around $x = 0.3$. The inset depicts the enlarged view of the temperature dependence of χ at temperature region below 30 K. An additional anomaly in $\chi(T)$ for the compounds of $x \geq 0.3$ has been observed at low temperatures below T_N, which is considered to be another antiferromagnetic transition. As seen in Fig. 2, the x dependence of the phase transition temperature below T_N in compounds of $x > 0.3$ shows discontinuous change at $x \approx 0.7$, implying the different ground states in the phase diagram in the regions of $0 < x < 0.3$, $0.3 < x < 0.7$ and $0.7 < x < 1.0$. We denote the magnetic phase transition temperatures as T_1 for the compounds with $0.3 < x < 0.7$ and as T_{N2} for $0.7 < x < 1.0$.

Figure 3 indicates the temperature dependence of the specific heat in the form of C/T for selected compounds of $EuNi(Si_{1-x}Ge_x)_3$ in the temperature region below 80 K. In $EuNiSi_3$ compound, two appreciable peaks at T_N and T_C are observed, and a small anomaly at $T^* \approx 5$ K. With increasing x, a peak at T_N becomes sharply, and T_N shifts toward the low temperature region. For the compounds of $x > 0.3$, an anomaly corresponding to the magnetic phase transitions at T_1 for $0.3 < x < 0.7$ and T_{N2} for $0.7 < x < 1.0$ below T_N is observed. The x dependences of T_N, T^*, T_1 and T_{N2} obtained from the specific heat are also shown in Fig. 2 using open and closed triangular symbols.

Figure 4 and 5 show the temperature dependences of the electrical resistivity ρ and thermopower S of compounds of $EuNi(Si_{1-x}Ge_x)_3$, respectively. The insets in Fig. 4 and 5 are the enlarged views of $\rho(T)$ and $S(T)$, respectively, at low temperatures below 40 K. No indication for phase transition into the spin freezing state in compounds with $x < 0.3$ is confirmed in $\rho(T)$

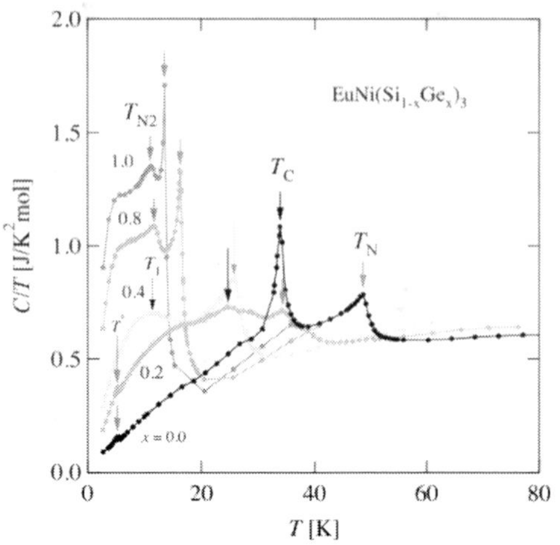

Figure 3. Temperature dependence of the specific heat in the form of C/T of $EuNi(Si_{1-x}Ge_x)_3$ in the temperature region below 80 K.

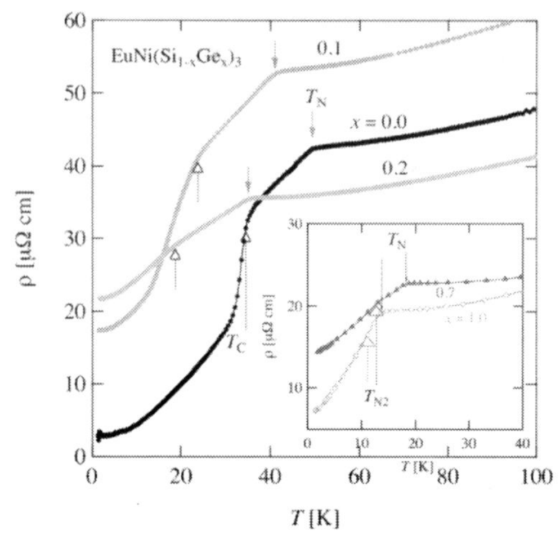

Figure 4. Temperature dependence of the electrical resistivity ρ of the compounds with $x \leq 0.2$ at temperatures below 100 K. The inset depicts $\rho(T)$ for $x = 0.7$ and 1.0 at temperatures below 50 K.

and $S(T)$. Anomalies corresponding to magnetic phase transitions into antiferromagnetic and ferromagnetic states are observed in the both $\rho(T)$ and $S(T)$ curves, except for an anomaly at T_1 in $\rho(T)$. The magnetic phase transitions obtained from $\rho(T)$ and $S(T)$ are shown in Fig. 2. As seen in Fig. 2, these results are in good agreement with those obtained from the magnetic susceptibility χ and specific heat C.

In our previous paper [7], we reported the pressure effect on the magnetic phase transition temperatures of T_N and T_C in $EuNiSi_3$. Both T_N and T_C increase with increasing pressure P, and T_N merges with T_C at $P \approx 2.5$ GPa. This result implies that the increase of the magnetic phase transition temperatures is strongly connected with the atomic volume contraction, *i.e.* the distance between the magnetic moments of Eu. It is plausible that the effect of the atomic substitution of Si by Ge on T_N and T_C is the same as that of negative pressure, resulting in a decrease of T_N and T_C. The volume dependences of the magnetic phase transition temperatures of T_N and T_C can be expressed by using the following Grüneisen parameters,

$$\Omega_A = -\frac{d \ln T_N}{d \ln V},\ \Omega_F = -\frac{d \ln T_C}{d \ln V}. \tag{1}$$

The inset of Fig. 6 indicates the x dependence of the atomic volume V of $EuNi(Si_{1-x}Ge_x)_3$, obtained from the lattice parameters of a and c, $V = a^2c$. V shows the linear dependence on x because of the larger ion radius of Ge than that of Si. The plots of T_N and T_C against V in double logarithmic scale are shown in Fig. 6. Both plots lie on straight lines, indicating the validity of eq. (1). These results reveal that the changes of T_N and T_C strongly depend on the change of the atomic volume of crystal induced by the atomic substitution or application of high pressure.

In summary, we have measured the magnetic susceptibility and specific heat, together with the electrical resistivity and thermopower of the compounds of $EuNi(Si_{1-x}Ge_x)_3$. For all compounds of $EuNi(Si_{1-x}Ge_x)_3$, we obtained the effective magnetic moment $\mu_{eff} \approx 7.7\ \mu_B$, which is close to a divalent Eu value of μ_{eff}=7.94μ_B. The Néel temperature T_N decreases monotonously with

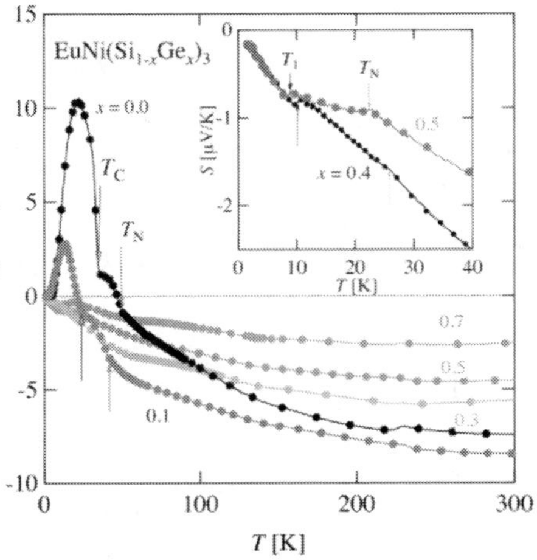

Figure 5. Temperature dependence of the thermopower S of $EuNi(Si_{1-x}Ge_x)_3$ at temperatures from 2 to 300 K. The inset shows the enlarged view of $S(T)$ for $x = 0.4$ and 0.5 at low temperatures below 40 K.

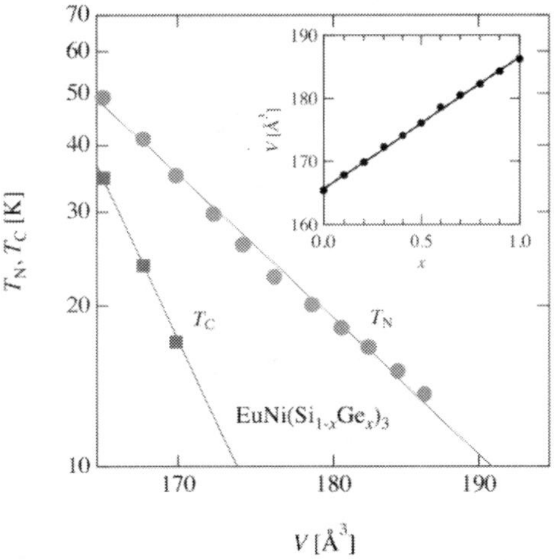

Figure 6. Magnetic transition temperatures of T_N and T_C against the atomic volume V in double logarithmic scale. The inset depicts the Ge concentration x dependence of the atomic volume V of $EuNi(Si_{1-x}Ge_x)_3$.

increasing the Ge concentration x from T_N=49 K for $EuNiSi_3$ to T_N=14 K for $EuNiGe_3$. At low temperatures below T_N, anomalies corresponding to additional magnetic phase transitions into ferromagnetic and spin freezing states for the compounds with $x < 0.3$, and into another antiferromagnetic for $x > 0.3$ were observed. It is found that the magnetic phase transition temperatures of T_N and T_C in $EuNi(Si_{1-x}Ge_x)_3$ are strongly connected with the change of the atomic volume induced by the atomic substitution of Si by Ge.

References

[1] Szytula A and Leciejewicz J. *Handbook on the physics and chemistry of rare earths*, **12** (1989) 133.
[2] Bud'ko S L, Islam Z, Wierner T A, Fisher I R, Lacerda A H and Canfield P C. *J. Magn. Magn. Mater.* **205** (1999) 53.
[3] Hesse H J, Lubers R, Winzenick M, Neuling H W and Wortmann G J. *Alloys Compd.* **246** (1997) 220.
[4] Wada H, Nakamura A, Mitsuda A, Shiga M, Tanaka T, Mitamura H and Goto T. *J. Phys.:Condes. Matter* **9** (1997) 7913.
[5] Nakamura A, Nakama T, Uchima K, Arakaki N, Zukeran C, Komesu S, Takeda M, Takaesu Y, Nakamura D, Hedo M and Yagasaki K. *J. Phys.: Conf. Ser.* **400** (2012) 032106.
[6] Nowik I , Felner I and Wertheim G K. *Hyp.1987 Int.* **33** 145.
[7] Uchima K, Arakaki N, Hirakawa S, Hiranaka Y, Uejo T, Nakamura A, Takeda M, Takaesu Y, Hedo M, Yagasaki K, Sakai E, Matsubayashi K and Uwatoko Y. *JPS Conf. Proc.* **3** 011016 (2014)
[8] Uchima K, Arakaki N, Hirakawa S, Hiranaka Y, Uejo T, Teruya A, Nakamura A, Takeda M, Takaesu Y, Hedo M, Nakama T, Yagasaki K, and Uwatoko Y. *JPS Conf. Proc.* **1** 012015 (2014).
[9] Resel R, Gratz E, Burkov A T, Nakama T, Higa M and Yagasaki K. *Rev. Sci. Instrum.* **67** (1996) 1970.
[10] Morzkin A V and Seropegin Yu D. *J. Alloys Compd.* **237** (1996) 124.

Transport properties of Heusler Compound Mn_3Si under high pressure

Y Takaesu[1], **K Uchima**[1], **S Nakamura**[2], **M Yogi**[3], **H Niki**[3], **M Hedo**[3], **T Nakama**[3], **and S Tomiyoshi**[4]

[1]General Education, Okinawa Christian Junior College, Nishihara, Okinawa 903-0204, Japan
[2]Graduate School of Engineering and Science, University of the Ryukyus, Nishihara, 903-0213, Japan
[3]Faculty of Science, University of the Ryukyus, Nishihara, Okinawa 903-0213, Japan
[4]Faculty of Engineering, Ehime University, Matsuyama 790-8577, Japan

Abstract. The electrical resistivity ρ and the thermopower S of the single crystalline sample of the Heusler compound Mn_3Si have been measured in the temperature range between 2 and 300 K under high pressures up to 2.2 GPa. The temperature variations of ρ and S indicate the characteristic features at the Néel temperature T_N. An additional anomaly in $S(T)$, related to the **3Q** satellites in SDW, is observed at $P < 1$ GPa, and it disappears at $P > 1$ GPa. The Néel temperature, obtained from $\rho(T)$ and $S(T)$ curves, increases with increasing pressure. The pressure dependences of the residual resistivity and the thermopower at $T = 2$ K show the discontinuous changes at $P \approx 1$ GPa, indicating a pressure induced phase transition.

1. Introduction

The Heusler compound of Mn_3Si with the DO_3-type cubic crystal structure orders antiferromagnetically at the Néel temperature $T_N = 23$ K with incommensurate spin structure of a propagation vector $\mathbf{Q} = 0.425\mathbf{G}_{111}$, where $\mathbf{G}_{111}$ is the reciprocal lattice vector along the [111] direction [1, 2]. The Mn atoms have two crystallographic sites of Mn(I) and Mn(II). The Mn(I) and Mn(II) atoms are surrounded by eight Mn nearest neighbors, and four Mn and four Si nearest neighbors, respectively. Due to this difference of the nearest neighbor atomic configurations, the atoms of Mn(I) and Mn(II) sites have different magnetic moments in magnitude. This crystal structure can be seen as a layered structure stacking the (111) planes in a sequence of planes of Mn(I), Mn(II), and Si atoms, respectively.

From the neutron diffraction measurements, the spin structure in the SDW of Mn_3Si has been found to be either a helical or a transversal sinusoidal structures [2, 3, 4]. However, the correct spin structure is not determined yet. The estimated values of magnetic moments of Mn(I) and Mn(II) from the analysis of the fundamental satellites by assuming the helical spin structure are 1.7 μ_B and 0.2 μ_B, respectively, and for the transversal sinusoidal structure the maximum amplitude of their magnetic moments are 2.4 μ_B and 0.28 μ_B, respectively [1, 2].

Tomiyoshi *et al.* have investigated higher harmonics SDW in Mn_3Si by the neutron diffraction [2]. At low temperatures below $T_a \approx 8$ K, the third-order harmonics satellites have been observed at the **3Q** positions expected by the propagation wave vector **Q** of the fundamental satellite. The anomalies in the temperature dependence of the specific heat C were

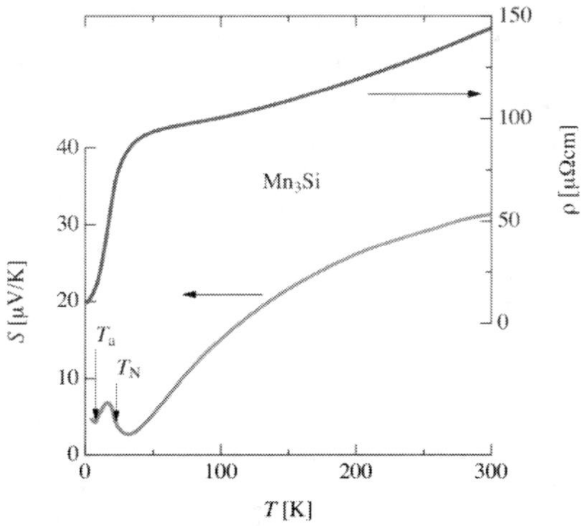

Figure 1. Temperature dependences of the electrical resistivity ρ and the thermopower S at ambient pressure.

observed at T_N and T_a. It was also reported that T_a is independent of an external magnetic field up to 14 T [5].

The results of the band structure calculations for Mn_3Si reveal that the large electronic density of states (DOS) due to mainly Mn 3d electrons lies near the Fermi level [6]. Since the electron scattering is sensitive to the electronic state in the vicinity of the Fermi level, the measurements of the transport properties, such as the electrical resistivity and the thermopower, are instructive in investigating the magnetism and the electronic state around the Fermi level. In order to clarify the pressure effect on the magnetic and the transport properties of Mn_3Si, we have measured the electrical resistivity ρ and the thermopower S at temperatures from 2 to 300 K under hydrostatic pressures up to 2.2 GPa.

2. Experimental

Single crystalline sample of Mn_3Si was prepared by the Bridgemann method from the stoichiometric ratio of the pure constituents of 99.9 % Mn and 99.999 % Si. The detailed procedure for the sample preparation has been described elsewhere [2]. No impurity phase in the specimen was detected in the powder X-ray diffraction measurement. The measurements of the electrical resistivity ρ and the thermopower S were carried out by the standard four-probe dc method and the differential method with seesaw heating procedure [7], respectively. The piston-cylinder-type high-pressure cell with Daphne 7373 as the pressure transmitting medium was used for the simultaneous measurements of ρ and S at temperatures from 2 to 300 K under hydrostatic pressures up to 2.2 GPa [8]. The electrical current and the heat flow were applied in the same direction of the [111] axis.

3. Results and Discussion

Figure 1 shows the temperature dependences of the electrical resistivity ρ and thermopower S of Mn_3Si in the temperature range between 2 and 300 K at ambient pressure. ρ decreases with decreasing temperature, and shows a sharp drop below T_N. S also decreases monotonously with decreasing temperature, and shows a characteristic feature at low temperatures below about 50 K, as seen in Fig. 1. S shows upturn below $T \approx 30$ K, having a maximum at $T \approx 17$ K, and

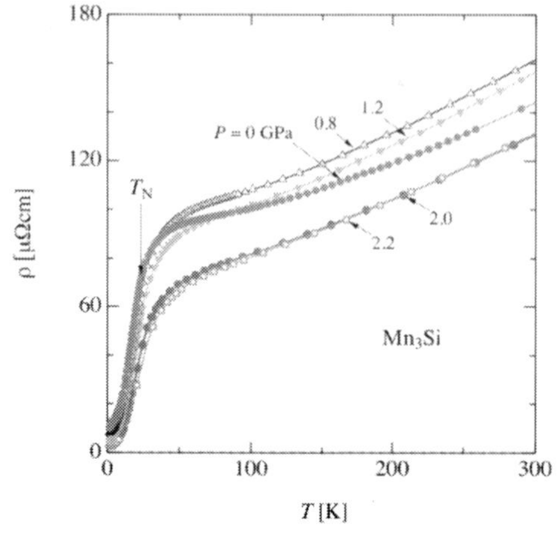

Figure 2. Temperature dependence of the electrical resistivity ρ of Mn_3Si under pressures up to 2.2 GPa.

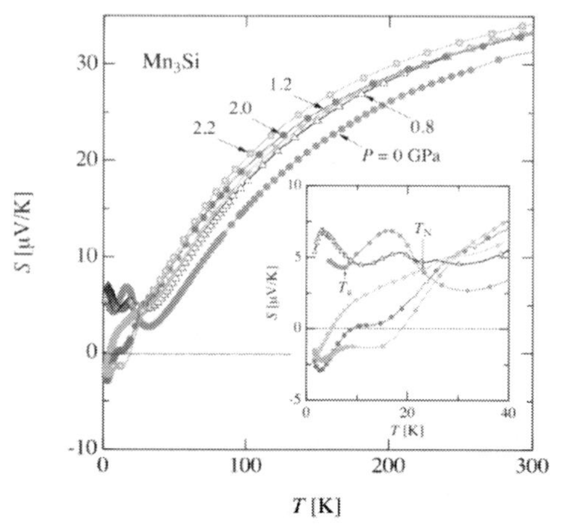

Figure 3. Temperature dependence of the thermopower S of Mn_3Si under pressures up to 2.2 GPa. The inset shows an enlarged view of $S(T)$ at low temperatures.

an additional anomaly in the form of a local minimum, as indicated in Fig. 1, was observed at $T_a \approx 7.8$ K, where the 3**Q** satellites in SDW were confirmed by the neutron diffraction [2]. No distinct change around T_a in $\rho(T)$ curve, however, was detected.

Figure 2 and 3 show the temperature dependences of ρ and S under pressures up to 2.2 GPa, respectively. The temperature variation of ρ under pressure is almost the same as that in ambient pressure in the whole temperature range. ρ decreases with decreasing temperature in the paramagnetic phase, and shows a steep decrease in the antiferromagnetic phase. The residual resistivity ρ_0, which is determined as the resistivity at the lowest temperature of 2 K, increases with increasing pressure P. The electrical resistivity at higher temperatures above T_N increases with increasing pressure up to $P \approx 1$ GPa, and decreases with further increase of P. In the whole pressure range, S shows qualitatively the same temperature dependence at temperatures above $T \approx 50$ K. On the other hand, the low-temperature thermopower shows the large pressure variation, as can be seen in the inset of Fig. 3. The pressure dependences of T_N and T_a, obtained from $\rho(T)$ and $S(T)$ curves, are shown in Fig. 4. T_N increases linearly with increasing pressure P, $dT_N/dP \approx 3.5$ K/GPa. T_a also increases with increasing P at $P < 1$ GPa, and most likely disappears at the critical pressure $P_c \approx 1$ GPa.

In low temperature limit, the diffusion thermopower S_d for metallic conductor is expressed by the Mott's formula:

$$S_d = \frac{\pi^2 k_B^2}{3e} T \left(\frac{1}{\sigma} \frac{\partial \sigma}{\partial \varepsilon} \right).$$

Here, k_B is the Boltzmann's constant, e is the electronic charge and σ is the energy dependent electrical conductivity [9]. This formula predicts a linear temperature dependence for S at the low temperature region. As shown in the inset of Fig. 3, no linear temperature variations of $S(T)$ in Mn_3Si, however, are observed at low temperatures, indicating a large contribution of the magnetic scattering of conduction electrons to the low temperature transport properties. Then, we use the thermopower at $T = 4$ K, S_{4K}, instead of the temperature gradient of thermopower S/T, to investigate the effect of pressure on S. In Fig. 5, we show the pressure dependences of

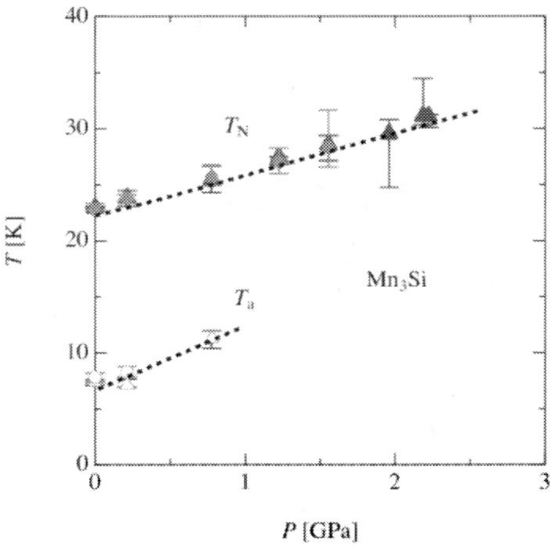

Figure 4. Pressure dependences of T_N and T_a of Mn_3Si. The symbols of the circles and triangles are obtained from $\rho(T)$ and $S(T)$, respectively.

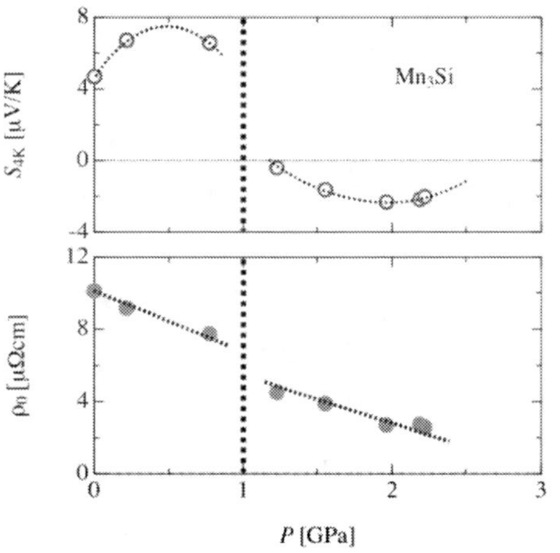

Figure 5. Pressure dependences of the thermopower at $T = 4$ K, S_{4K}, and the residual resistivity ρ_0.

S_{4K} and the residual resistivity ρ_0 by the closed and open symbols, respectively. As seen in Fig. 5, both S_{4K} and ρ_0 show the discontinuous changes at $P_c \approx 1$ GPa, where an anomaly in $S(T)$ related to the **3Q** satellites in SDW disappears. The sign of S_{4K} is positive for $P < 1$ GPa and is negative for $P > 1$ GPa. The variation of ρ_0 also indicate a discontinuous change around P_c. These behaviors imply that the electronic state around the Fermi level is modified by applying pressure, namely a pressure induced phase transition takes place.

In summary, the electrical resistivity and the thermopower have been measured at temperatures from 2 to 300 K under hydrostatic pressures up to 2.2 GPa. The Néel temperature T_N, obtained form $\rho(T)$ and $S(T)$ curves, increases linearly with increasing pressure P, $dT_N/dP \approx 3.5$ K/GPa. An additional anomaly in $S(T)$ related to the **3Q** satellites in SDW is confirmed at $P < 1$ GPa, and it most likely disappears at $P_c \approx 1$ GPa. The pressure dependence of the residual resistivity and the thermopower at $T = 4$ K show the discontinuous changes at $P_c \approx 1$ GPa, indicating a pressure induced phase transition.

References

[1] Kalishevich G I, Vereshagin Y A and Gel'd P V 1974 *Sov. Phys. Solid State* **16** 1151
[2] Tomiyoshi S, Cowley E R and Onodera H 2006 *Phys. Rev. B* **73** 024416
[3] Niki H, Okada Y, Oshiro M, Higa K, Yogi M and Tomiyoshi S 2012 *J. Phys.: Conf. Ser.* **400** 032067
[4] Niki H, Yogi M, Nakamura S, Uechi A and Tomiyoshi S 2012 *Hyperfine Interact.* **212** 7
[5] Pfleiderer C, Boeuf J and v Lohneysen H 2002 *Phys. Rev. B* **65** 172404
[6] Jeong T 2012 *Physica B* **407** 888
[7] Resel R, Gratz E, Burkov A T, Nakama T, Higa M and Yagasaki K 1996 *Rev. Sci. Instrum.* **67** 1970
[8] Hedo M, Nakamura D, Takaesu Y, Fujiwara T, Uchima K, Yagasaki K and Nakama T 2010 *J. Phys.: Conf. Ser.* **215** 012186
[9] Barnard R D 1972 *Thermoelectricity in Metals and Alloys* (London: Taylor and Francis)

Journal of Physics: Conference Series **568** (2014) 042034 doi:10.1088/1742-6596/568/4/042034

Dynamic equilibrium of collective degrees of freedom in strongly correlated quantum matter

Peter Wölfle

Institute for Theory of Condensed Matter, KIT, 76021 Karlsruhe, Germany

E-mail: peter.woelfle@kit.edu

Abstract. Strongly correlated quantum systems may assume states stabilized by dynamic equilibrium of competing collective degrees of freedom. As a first example we consider the Kondo effect, which may be viewed as governed by a subtle balance of an infinite growth of the exchange coupling constant controled by an increasing local spin relaxation rate, leading to a Fermi liquid stable fixed point at low temperatures. As a second example the quantum critical behavior in antiferromagnetic metals will be considered. It is often found experimentally that the quasiparticle effective mass appears to diverge at the quantum critical point in spite of the fact that the critical spin fluctuations seem to be in the weak coupling domain - a contradiction. It will be shown that the feedback of the diverging mass into the spin fluctuation spectrum allows to establish a balance of the fermionic and bosonic singular degrees of freedom, leading to strong-coupling type quantum critical behavior. The resulting self-consistent theory obeys hyper-scaling of the dominant fermionic degrees of freedom, while the bosonic interaction is still in the weak coupling domain. Excellent agreement is found with experiment for the two candidate cases $CeCu_{6-x}Au_x$ and $YbRh_2Si_2$.

1. Introduction

Over the last four decades strongly correlated quantum systems have attracted considerable attention owing to the rich variety of novel behavior found in these systems. Starting with the discovery of the superfluid phases of 3He (for a comprehensive treatise see [1]), the Kondo effect and the advent of heavy fermion metallic compounds [2] , the fractional quantum Hall effect, the cuprate high temperature superconductors and most recently the pnictide superconductors [3] it has become clear that correlations play a dominant role in all of these systems. The theoretical advances in attempting to describe these systems have been impressive: renormalized perturbation theory, renormalization group theories, slave particle theories, dynamical mean field theory, to name just a few. Yet there are problems that appear to defy our theoretical capabilities, like the cuprate superconductor problem. What is lacking is a systematic approach to tackle many-body problems at strong coupling. A first simple but powerful approximation scheme working at weak and strong coupling has been established more than a century ago: mean field theory. First developed to describe the liquid-gas system or magnetic systems, later generalized to describe superconductivity, charge density waves, spin density waves, or indeed any type of order, mean field theory appears to work resonably well if there is a mean field order parameter of some sort. Generally speaking the mean field is an expectation value of an operator, which involves averaging over a number of neighbors of a given entity (charge, spin, Cooper pair,..). The larger the number of neighbors the better defined the mean field is.

However, fluctuations may destroy the mean field. In this case the fluctuations are the object of interest, and one may ask whether it is possible to devise a kind of mean field theory for the fluctuations. This is the subject of the present paper. I will argue in the following that there are examples of mean field theories of dynamical quantities, which give at least a good qualitative description of a situation in which fluctuations dominate the behavior. A first example is the Kondo effect, which is the formation of a quantum many-body resonance state of conduction electrons at a magnetic impurity in a metal. The famous perturbative renormalization group treatment of this model, Anderson's "poor man's scaling" [4], suggests infinite growth of the spin exchange coupling. We will see how this growth is controled by the increasing inelastic scattering rate of the inpurity spin such that a stable Fermi liquid state is attained. The second example is the quantum critical behavior at an antiferromagnetic instability point of a metal. Here the conventional Hertz-Millis theory (a perturbative renormalization group theory of the bosonic degrees of freedom) is found to often be in conflict with experiment, whenever the fermionic quasiparticles are characterized by a diverging effective mass (for an overview see [5]*)*. It will be shown that in this case the fermions rather than the bosons are the dominant critical entities. The critical fermions modify the (bosonic) spin fluctuation in such a way that a self-consistent theory of the fermionic mass emerges which has a strong coupling solution in addition to the usual weak coupling solution, provided that the initial effective mass enhancement is sufficiently large.

2. Kondo effect

The Kondo effect is arguably the best studied quantum many-body problem. It is defined by a local spin $\mathbf{S}$, representing a quantum impurity with internal degrees of freedom, coupled by isotropic spin exchange interaction J to a bath of conduction electrons [2]. At the heart of the Kondo problem is the excitation of low energy particle-hole pairs occurring after each spin-flip process, when the conduction electron Fermi sea has to adjust to the suddenly switched potential provided by the spin, and undergoes an *orthogonality catastrophe* [6]. The quantum coherent readjustment of the Fermi sea is interrupted by any phase breaking process, which may happen at a rate Γ, the local spin relaxation rate. A perturbative treatment of this problem requires a systematic resummation of infinitely many relevant contributions, which may be achieved by the renormalization group (RG) method. While an analytic formulation in the form of Anderson's *poor man's scaling* [4] provides a correct and simple description in the perturbative regime, i.e. for dimensionless coupling $g_0 = JN(0) \ll 1$, or equivalently, temperatures much larger than the Kondo temperature, $T \gg T_K = D\exp(-1/2g_0)$, the strong coupling regime has so far only been accessible by Wilson's Numerical RG (NRG) [7]. The latter method has established that the low energy behavior of a Kondo system ($T \ll T_K$) is governed by local Fermi liquid theory, with far-reaching consequences as noted by Nozieres [8] . These results, at least as far as thermodynamic properties are concerned, have been confirmed by the Bethe-Ansatz solution [9, 10] .

2.1. Kondo model

We consider a local spin $S = 1/2$ exchange coupled to the local conduction electron spin density, as described by the Hamiltonian [12]

$$H = \sum_{\mathbf{k},\sigma} \epsilon_{\mathbf{k}} c^\dagger_{\mathbf{k}\sigma} c_{\mathbf{k}\sigma} + J \sum_{\mathbf{k},\mathbf{k}',\sigma,\sigma',i} \frac{1}{2} c^\dagger_{\mathbf{k}\sigma} \tau^i_{\sigma\sigma'} c_{\mathbf{k}'\sigma'} S^i \quad (1)$$

Here τ is the vector of Pauli matrices, and $c^\dagger_{\mathbf{k}\sigma}$ creates an electron of momentum $\mathbf{k}$ and spin σ . In order to make use of standard many-body techniques we represent the spin operator by $\mathbf{S} = \frac{1}{2}\sum_{\sigma,\sigma'} f^\dagger_\sigma \tau_{\sigma\sigma'} f_{\sigma'}$, where $f^\dagger_\sigma, f_\sigma$ are pseudo-fermion (pf) operators ($\sigma = \uparrow, \downarrow$). We take the pf chemical potential to be $\lambda = 0$ so that the pf spectral function is particle-hole symmetric, $A(\omega) = A(-\omega) \simeq \frac{\Gamma}{\omega^2+\Gamma^2}$. The projection onto the physical sector of Hilbert space, $Q = \sum_\sigma f_\sigma^\dagger f_\sigma = 1$, is done by omitting all higher pf-loop diagrams. In case of the pf-selfenergy this amounts to taking only diagrams without any pf-loop, whereas in case of the conductivity/conductance and of the local spin susceptibility only diagrams with a single loop are kept. In addition, the latter quantities are corrected by a factor [11]

$$Y = \frac{2}{2 - \frac{Z_0}{Z}}, \quad \frac{Z_0}{Z} = \frac{1}{4}\exp[\frac{1}{T}\int_0^\infty d\lambda \, \langle Q \rangle_\lambda], \tag{2}$$

where $Z_0 = \mathrm{Tr}[e^{-H_0/T}]$ and $Z = \mathrm{Tr}[e^{-H/T}]$ are the partition functions of, respectively, the noninteracting and interacting systems. A useful approximation is provided by $Z_0/Z = \frac{1}{4}\exp[2\int d\omega \ln(1+e^{-\omega/T})A(\omega)]$.

2.2. RG-equation.
The coupling g as a function of the temperature T, three frequencies (which we put to zero) and the running UV-cutoff D obeys the RG-equation [13, 14]

$$\frac{dg}{dD} = \beta(g)\frac{D}{D^2+\Gamma^2}. \tag{3}$$

At weak coupling ($\Gamma \ll D$), $\beta(g) = -c_2 g^2 + O(g^3)$, while at strong coupling ($D \ll \Gamma$) $\beta(g) = -\frac{1}{2}c_3 g^3 + O(g^2)$. Assuming the interpolation formula $\beta = -\frac{[g(c_2+c_3 g)]^2}{c_2+2c_3 g}$, the RG-equation may be integrated to give $\frac{1}{g(c_2+c_3 g)} - \frac{1}{g_0 c_2} = \ln(\sqrt{D^2+\Gamma^2}/D_0)$ ($g_0 \ll 1$). The leading g^3-dependence of $\beta(g)$ at $g \gg 1$, which is reached at scales $D \ll T_K$, follows from an analysis of the contributions to $\beta(g)$ in arbitrary order of perturbation theory. It turns out that a third order (in g) contribution is dominant, as all higher order contributions vanish as $(D/\Gamma)^n$, $n > 4$, in the limit $D \ll \Gamma$. The two-loop order vertex correction diagram at strong coupling, $D \ll \Gamma$, which is the dominant contribution since self-energy corrections are smaller by a factor $(D/\Gamma)^2$, gives $c_3 = 1$. We also use the weak coupling result $c_2 = 2$. The temperature dependence of g (at zero energy) is found by solving the RG equation, setting $D = T$ as [13, 14]

$$g(T) = -1 + \sqrt{1 + 1/\ln(\sqrt{(T^2+\Gamma^2)}/T_K)}. \tag{4}$$

, where $T_K = D_0 e^{-1/2g_0}$ is the Kondo temperature and $g_0 = N(0)J$ is the bare coupling. We observe that the growth of $g(T)$ as $T \to 0$ is controled by the relaxation rate Γ, which is again a function of temperature and depends on $g(T)$ in turn. In the limit of $T \gg T_K$ we recover the weak coupling result $g(T) = 1/[2\ln(\sqrt{(T^2+\Gamma^2)}/T_K)]$, and $\Gamma \ll T_K$. In the limit $T \ll T_K$, considering that $\Gamma \to T_K + O(T^2)$, and consequently $\ln(\sqrt{(T^2+\Gamma^2)}/T_K) \propto T^2$, as will be shown below, we find $g(T) \propto T_K/T$, diverging in the limit $T \to 0$.

It is worthwhile to note that the coefficient of the g^3-term of $\beta(g)$ at $g \ll 1$ has the opposite sign, $c_3 = -4$. We may improve the approximate form of the β-function by requiring that the g^3-term at weak coupling be reproduced, in the form $\beta(g) = -2g^2\frac{1+g^2/4}{1+g}$.

2.3. Selfenergy
The spin relaxation rate Γ is given by the imaginary part of the pf self-energy $\mathrm{Im}\Sigma$ plus vertex corrections, which we omit at present, as we do not expect those to have a major effect. It is

best to consider the diagrams of the components $\Sigma^{>,<}$, which may be classified according to the number n of particle-hole (ph) excitations in the intermediate state:

$$\mathrm{Im}\Sigma(\omega) = \Gamma(\omega) = \frac{i}{2}\sum_{n=1}^{\infty}\left[\Sigma_n^{>}(\omega) - \Sigma_n^{<}(\omega)\right] \tag{5}$$

$$\begin{aligned}\Sigma_n^{>}(\omega) &= \sum_{n=1}^{\infty}\int d\omega\prime A(\omega\prime)f(-\omega\prime)\int d\epsilon_1 d\bar{\epsilon}_1....d\epsilon_n \times \\ &\times\ d\bar{\epsilon}_n f(\epsilon_1)f(-\bar{\epsilon}_1)...f(\epsilon_n)f(-\bar{\epsilon}_n)\times|U_n|^2\times \\ &\times\ \delta(\omega\prime-\omega+\bar{\epsilon}_1+\epsilon_1......+\bar{\epsilon}_n+\epsilon_n)\end{aligned} \tag{6}$$

and a similar expression for $\Sigma_n^{<}(\omega)$. Here $f(\omega) = 1/[e^{\omega/T}+1]$, the pf spectral function is given by $A(\omega) = \frac{1}{\pi}\Gamma(\omega)/[(\omega-\mathrm{Re}\Sigma)^2+\Gamma^2(\omega)]$, U_n is the n-particle-hole vertex function and the local conduction electron spectral functions $N(\epsilon) = N(0)\Theta(D-|\epsilon|)$ can be taken as constant and are combined with the (renormalized) exchange coupling J to give g . We will first analyze the single particle-hole contribution ($n = 1$), for which $U_1(\epsilon_1) = g(\epsilon_1)$. Since the Fermi functions confine the integrations to $|\omega'|, |\epsilon_1| \lesssim T$, we may take $g(\epsilon_1) \approx g(T)$ out of the integral. Further we replace $\Gamma(\omega')$ by $\Gamma(T)$ under the integral, resulting in the self-consistency equation

$$\Gamma(T) = 3g^2(T)\int d\omega\omega\, n(\omega)f(-\omega)\frac{\Gamma(T)}{\omega^2+\Gamma^2(T)}$$

with $n(\omega)$ being the Bose function. Here we neglected $\mathrm{Re}\Sigma(\omega')$, which is justified at all T. At $T \ll T_K$, we may neglect ω^2 in the denominator of $A(\omega)$. Then one finds $\Gamma^2(T) = c_{T,2}T^2g^2(T)$, and further, substituting g, we find

$$\Gamma/T_K = 1 + c_\Gamma\,(T/T_K)^2 + \mathcal{O}\left((T/T_K)^3\right), \qquad g = c_g(T_K/T) \tag{7}$$

with $c_{T,2} = \frac{3}{4}\pi^2$, $c_\Gamma = (c_{T,2}-\frac{1}{2})$ and $c_g = 1/\sqrt{c_{T,2}}$. We observe that g is indeed scaling to infinite strength as $T \to 0$, while the rate Γ assumes a finite value T_K with Fermi liquid corrections $\propto T^2$.

Let us now take a brief look at the higher order contributions. Analyzing the diagrams of U_n we note that only the n vertices g with external legs grow large. The internal g's are integrated over and enter only the prefactor. We therefore get as generalized self-consistent equation $\Gamma^2 = \sum_{n=1}^{\infty} c_{T,n}g^{2n}(T^{2n}/\Gamma^{2n-2})$, which has the low T solution $\Gamma = T_K + c_\Gamma T^2$, and $g = c_g(T_K/T)$, with modified coefficients c_Γ, c_g , where $\sum_{n=1}^{\infty} c_{T,n}c_g^{2n} = 1$.

The limiting value of $\Gamma = T_K$ has been obtained before by Mattuck and collaborators [15], using the reasoning applied above. However, these authors employed the weak coupling form of the RG equation, which leads to finite temperature corrections linear in T, in disagreement with Fermi liquid theory. It is the form of the RG β-function $\propto g^3$, which is responsible for recovering the correct T^2-dependence, the hallmark of Fermi liquid theory.

2.4. Conductance and spin susceptibility

With the aid of the above results one may now proceed to calculate the conductance G of a Kondo quantum dot (in units of $2e^2/\hbar$)

$$\frac{G}{G_0} = 3\pi^2g^2Y\int d\epsilon\, d\epsilon' f(\epsilon)f(-\epsilon')f(\epsilon'-\epsilon)A(\epsilon)A(\epsilon').$$

Here Y is the normalization factor introduced above. We may calculate Y analytically at high and low temperatures to find $Y = 2$ and $Y = 1$, respectively. The conductance at $T \gg T_K$ is found as $G/G_0 = 3\pi^2/[16 \ln^2(T/T_K)]$ in agreement with perturbation theory. At low temperatures $T \ll T_K$ we may drop ϵ^2 and $(\epsilon')^2$ in the denominators of A compared to Γ^2. The remaining integral is exactly the same as in the selfconsistency equation for Γ, and therefore $G/G_0 = c_{T,2}g^2\frac{T^2}{\Gamma^2} = 1$ at $T = 0$. In other words, the requirement of unitarity of the scattering amplitude is met. The finite temperature corrections are $O(T^2)$ and are obtained, e.g. by keeping the ϵ^2 in the denominator, as well as similar terms in the equation for the selfenergy and results in

$$G/G_0 = 1 - c_G\left(\frac{T}{T_K}\right)^2 \qquad \text{for} \quad \frac{T}{T_K} \ll 1 \tag{8}$$

The spin susceptibility is given by

$$\chi^R_{imp}(T) = (g\mu_B)^2[\chi^R_{SS}(T) + 2\chi^R_{sS}(T) + \chi^R_{ss}(T) - \chi^R_{ss,0}(T)], \tag{9}$$

where $\chi^R_{\alpha\beta}(T)$, $\{\alpha, \beta\} = s, S$ are the spin-spin-response functions of impurity spin and conduction electron spins, respectively. Neglecting vertex corrections, $\chi^R_{SS}(T)$ is given by the pf-bubble diagram,

$$\chi^R_{SS}(T) = \pi Y \int d\omega \tanh(\frac{\omega}{2T})A^2(\omega)\frac{\omega}{\Gamma} \tag{10}$$

The mixed response function $\chi^R_{sS}(T)$ is given by a pf-bubble and a local conduction electron bubble, joined by an interaction vertex g as $\chi^R_{sS}(T) = -8\Pi^R_g\chi^R_{SS}(T)$ with $\Pi^R_g(T) = \frac{1}{N(0)}\int d\omega g(\omega)\omega \tanh(\frac{\omega}{2T})N^2(\omega)$. The impurity induced c-electron contribution is given by two c-bubbles at the ends and a pf-bubble in the middle $\chi^R_{ss}(T) - \chi^R_{ss,0}(T) = c_{ss}[\Pi^R_g]^2\chi^R_{SS}(T)$. In the limit $T \to 0$ the susceptibility is found to tend to a finite value, with quadratic temperature correction, as required by Fermi liquid theory, $\chi^R(T) = \alpha\frac{1}{T_K} - O(T^2)$. For high temperatures we find the perturbative result $\chi^R(T) = 1/(4T)\,(1 - 1/\ln(T/T_K) \pm \ldots)$.

2.5. Summary of the Kondo problem part

In the above we demonstrated that the Kondo problem may be solved in the framework of renormalized perturbation theory, employing a self-consistent relation for the central quantity, the spin relaxation rate Γ. As to be expected, it is necessary to sum infinite classes of contributions in perturbation theory, which is done with the aid of the renormalization group method. The relaxation rate Γ of the local spin acts to separate the weak coupling (scale parameter $D >> \Gamma$) from the strong coupling regime ($D << \Gamma$) of the renormalization group flow. Surprisingly, the renormalization group β-function in the strong coupling regime is dominated by a two-loop (cubic in g) contribution, all higher contributions being small in the parameter D/Γ. The coupling function acquires a strong dependence on energy or temperature, $g(T) \propto T_K/T$, and thus diverges in the limit $T \to 0$. In this work we determine Γ from renormalized perturbation theory, keeping all orders, in principle. As noticed in the 1970s by Mattuck and collaborators [15], the divergence of the coupling function $g(T)$ is exactly cancelled by the phase space factors of the energy integrations, so that Γ tends to the finite value T_K as $T \to 0$. This statement is correct even in the case of multiple particle-hole excitations in the intermediate state (in this case the numerical coefficients have not yet been calculated). The finite temperature corrections are $\propto T^2$, as required by Fermi liquid theory. Given the relatively simple structure of the theory, one may calculate thermodynamic, transport or dynamic properties. We presented results on the linear conductance through a Kondo quantum-dot. We observe that the conductance reaches the unitarity limit at $T \to 0$ and decreases $\propto T^2$ in the regime $T << T_K$. We further calculated the temperature dependent spin susceptibility, again in good agreement with known results.

3. Critical quasiparticles at an antiferromagnetic quantum critical point

The theory of quantum criticality in metals goes back to the 1960s, when the first RPA-type theories of metals with strong spin fluctuations were formulated. A field-theoretical formulation was pioneered by J. Hertz [16] and later completed by A. Millis [17]. The latter theories are based on the assumption that the critical excitations are the bosonic spin fluctuations, while the fermionic quasiparticle excitations may be integrated out. The resulting ϕ^4 - field theory has an upper critical dimension $d = 4$. Observing that the temporal dimension is given by the dynamical exponent z, the bare value of which is $z = 3$ for ferromagnets and $z = 2$ for antiferromagnets, one finds that the effective dimension $d_{eff} = d + z$ is frequently above the upper critical dimension, meaning that the theory is Gaussian. Nonetheless the results of the theory are often in conflict with experimental observation. The reason is that in those cases the fermionic quasiparticles show critical behavior, as deduced from a specfic heat coefficient $\gamma(T) = C(T)/T$ diverging as $T \to 0$. Provided that the single-particle self-energy $\Sigma(\omega, k)$ is a weak function of momentum k, which appears to be the case in these systems, it follows that the quasiparticle effective mass ratio $m^*/m \propto \gamma(T)$ diverges, implying that the quasiparticles are critical.

3.1. Collective excitations in antiferromagnetic metals

The clue to resolving this puzzle lies in a careful study of vertex corrections [18, 21]. The (three-point) vertices coupling to spin density, $\Lambda(\mathbf{k}, \omega; \mathbf{q}, \Omega)$, and spin current density, $\mathbf{\Lambda}(\mathbf{k}, \omega; \mathbf{q}, \Omega)$ satisfy a Ward identity based on the spin conservation law: $\Omega\Lambda - \mathbf{q}\mathbf{\Lambda} = G^{-1}(\mathbf{k} + \mathbf{q}/\mathbf{2}, \omega + \Omega/2) - G^{-1}(\mathbf{k} - \mathbf{q}/\mathbf{2}, \omega - \Omega/2)$. It follows that in the case that the effective mass ratio $m^*/m = 1 - \partial Re\Sigma/\partial\omega$ is singularly enhanced in the limit $\omega \to 0$, the vertices Λ will be enhanced, too, in particular for momenta at the antiferromagnetic ordering wave vector $\mathbf{q} = \mathbf{Q}$, $\Lambda(\mathbf{k_F}, 0; \mathbf{Q}, 0) = \lambda_Q \propto m^*/m$.The latter vertex corrections enter the evaluation of diagrammatic contributions in several places. First in the Landau damping term, accounting for the decay of collective spin excitations into particle-hole pairs, and second at the vertices coupling the spin fluctuations to quasiparticles. The imaginary part of the spin susceptibility (the spectrum of spin fluctuations) for wave vectors near the AFM ordering wave vector $\mathbf{Q}$ thus takes the form

$$\mathrm{Im}\chi(\mathbf{q}, \omega) = \frac{N_0(\omega\lambda_Q^2/v_F Q)}{[(r + (\mathbf{q} - \mathbf{Q})^2\xi_0^2]^2 + (\omega\lambda_Q^2/v_F Q)^2}, \tag{11}$$

Here, N_0 is the bare density of states at the Fermi surface, v_F is the bare Fermi velocity and $\xi_0 = k_F^{-1}$ is the microscopic AFM correlation length. The control parameter r is a function of both, the tuning field and the temperature. Outside the critical regime we have $r = r_0^{2\nu}$, where $\nu = 1/(1 + z/2)$, whereas in the critical regime $r \propto T^{2/z}$.

The above spin fluctuation propagator may be viewed as mediating a retarded interaction transferring momentum $\mathbf{q} \approx \mathbf{Q}$. It follows that a quasiparticle with momentum $\mathbf{k}$ on the Fermi surface scattering off a spin fluctuation will end up in a state of momentum $\mathbf{k} + \mathbf{Q}$, in general far from the Fermi surface, except if $\mathbf{k}$ and $\mathbf{k} + \mathbf{Q}$ are located in so-called "hot spots" . There are, however, ways to get singular scattering even on the cold parts of the Fermi surface. One possibility is that impurity scattering helps to distribute the singular scattering at the hot spots all over the Fermi surface [19, 20]. Another mechanism operative even in clean systems is provided by the simultaneous scattering off two spin fluctuations carrying opposite momenta, such that the net transferred momentum is small [18, 22]. An excitation combined of two spin fluctuations may be considered as an (exchange) energy fluctuation χ_E . Its spectral weight is given by the expression

$$\mathrm{Im}\chi_E(q,\omega) \approx N_0^3\lambda_Q^{-2}\frac{(\omega\lambda_Q^2/v_FQ)^{d-1/2}}{[\frac{|\omega|\lambda_Q^2}{v_FQ}+q^2\xi_0^2+r]^{(d+1)/2}} \tag{12}$$

3.2. Selfconsistent theory of fermion effective mass

The effective mass m^* is calculated from the fermion selfenergy Σ. The leading scale-dependent contribution to the latter is obtained from the single boson exchange diagram of the energy fluctuation propagator, the imaginary part of which is given by

$$\begin{aligned}\mathrm{Im}\Sigma(\mathrm{k},\omega) &= -\lambda_{\mathrm{E}}^2\int\frac{\mathrm{d}\nu}{\pi}\sum_{\mathbf{q}}\mathrm{ImG}(\mathbf{k}+\mathbf{q},\omega+\nu)\mathrm{Im}\chi_E(\mathbf{q},\nu)[b(\nu)+f(\omega-\nu)]\\ &\approx v_FQ(m^*/m)^2(\omega\lambda_Q^2/v_FQ)^{d-1/2}\\ &\propto |\omega|^{d-1/2-\eta(2d+1)},\end{aligned} \tag{13}$$

The interaction vertex $\lambda_E = \lambda_Q^2\lambda_v$, where λ_v is $\propto (m^*/m)$, as it arises through a Ward identity connected to energy conservation. We used $\lambda_Q \propto (m^*/m)$, as discussed above. The Fermi and Bose functions $f(\omega), b(\omega)$ confine the ν-integration at low T to the interval $[0,\omega]$. In Eq. (4), we used the power law $m^*(\omega) \propto |\omega|^{-\eta}$. The scale dependent contribution to $\mathrm{Re}\Sigma(\omega)$ follows from analyticity as $\mathrm{Re}\Sigma(\omega) \propto (\omega/v_FQ)^{d-1/2}(m^*/m)^{2d+1})$. This leads to the self-consistency relation for $m^*(\omega)$ [18]

$$\begin{aligned}m^*(\omega)/m &= 1-\partial\mathrm{Re}\Sigma(\omega)/\partial\omega\\ &\approx 1+(m^*/m)^{(2d+1)}(\omega/v_FQ)^{d-3/2}\end{aligned} \tag{14}$$

In dimensions $d > 3/2$ the scale dependent correction term $\propto (\omega/v_FQ)^{d-3/2}$ is seen to vanish in the limit $\omega \to 0$. This would suggest that m^* does not diverge in this limit, in agreement with the result of Hertz-Millis theory in that case. If, however, the effective mass is already sufficiently enhanced, e.g. by AFM fluctuations around the hot spots or by additional fluctuations, the scale-dependent term on the right-hand-side will dominate and a second solution becomes available, which is of a strong coupling type.

Before solving the self-consistent equation for $m^*(\omega)$ in the strong coupling regime we first note that in general, the ω and T dependence of m^* is obtained by substituting $\sqrt{\omega^2+a^2T^2}$ for ω, where a is a constant of order unity. For frequencies less than the temperature, we may replace ω by T. We then find

$$m^*(T) \propto (T/v_FQ)^\eta, \tag{15}$$

where the exponent η characterizing the scaling of m^* is found to be

$$\eta = (2d-3)/4d \tag{16}$$

which gives $\eta = 1/4$ in the case of threedimensional AFM fluctuations and $\eta = 1/8$ in the case of quasi-twodimensional spin fluctuations ($d = 2$) in a threedimensional metal.

3.3. Critical exponents and dynamical scaling

The critical behavior of the spin-excitation spectrum as obtained above is given by power laws in frequency and in the bare control parameter r_0 (magnetic field H, pressure P, ..)

$$\omega \;\propto\; q^z\,,\quad z=\frac{2}{1-2\eta}=\frac{4d}{3} \tag{17}$$

$$\xi \;\propto\; r_0^{-\nu}\,,\quad \nu=\frac{3}{3+d} \tag{18}$$

The spin correlation length is a function of T inside the critical regime, and a function of H (for example) outside

$$1/\xi(H,T) \propto T^{1/z}, \qquad T > |H - H_c|^{z\nu} \tag{19}$$
$$1/\xi(H,T) \propto |H - H_c|^{\nu}, \qquad |H - H_c|^{z\nu} > T \tag{20}$$

The scaling properties of the fermions may be derived from the single particle Green's function: $\Sigma(\omega) \propto v_F|k-k_F|$, i.e. the dynamical critical exponent of the fermions is given by $z_f = 1/(1-\eta)$ and the effective dimension of the fermions is $d_f = 1$ (the dimension perpendicular to the Fermi surface). The correlation length of the critical quasiparticles ξ_f follows power laws in temperature and the control parameter, respectively, $\xi_f \propto T^{-1/z_f}$ and $\xi_f \propto |H - H_c|^{-1/\nu_f}$ The exponent ν_f follows from the relation of the momentum to the (spindependent) shift in chemical potential $v_F|k - k_F| \propto \delta\mu_\sigma$ as $\nu_f = 1$. It is interesting to note that the Harris criterion $\nu_f > 2/3$ is satisfied so that the solution found here is stable with respect to spatial fluctuations induced by impurities. The correlation lengths of fermions and bosons are therefore related by $\xi \propto \xi_f^{z_f/z}$ inside the critical regime and $\xi \propto \xi_f^{\nu_f/\nu}$ outside.

Inside the critical regime and at $\mathbf{q} = \mathbf{Q}$ one finds E/T-scaling of the dynamical structure factor [18]

$$S(\mathbf{Q}, E) \sim T^{-2/z} \frac{\phi(E/T)}{1 + (\phi(E/T))^2}, \tag{21}$$

where $\phi(x) = x(x^2 + a^2)^{-\eta}$ and a is a constant.

3.4. Free energy

The true critical excitations of the above strong coupling scenario are the fermionic quasiparticles. Their contribution to the free energy may be derived from the expression for the entropy density in terms of the self-energy:

$$\frac{S}{V} = \frac{1}{2\pi} N(0) \int \frac{d\omega\omega}{T^2 \cosh^2 \frac{\omega}{2T}} [\omega - \mathrm{Re}\Sigma(\omega)] \tag{22}$$

Substituting the critical self energy found above and integrating over temperature we find the scaling form of the free energy density

$$f(H,T) = \xi_f^{-(d_f+z_f)} \Phi_f(r_0 \xi_f^{1/\nu_f}, T\xi_f^{z_f}) \tag{23}$$

which obeys hyperscaling. This expression translates into

$$f(H,T) = \xi^{-(2d+1)} \Phi_f(r_0 \xi^{1/\nu}, T\xi^z) \tag{24}$$

in terms of the bosonic variables. This is not of the hyperscaling form. Indeed, in the cases considered the effective dimension of the bosons is $d_{eff} = d + 4d/3 = 7d/3 > 4$ for spatial dimension $d > 12/7$, implying that the boson-boson interaction scales to zero.

The consequences of the free energy expression have been analyzed in detail in [19, 20, 18]. The power laws in temperature (inside the critical regime) and in the control parameter (in the quantum disordered regime) have been determined. The exponents found agree very well with the available extensive experimental data. There is not a single observable where a distinct discrepancy of theory and experiment is found.

3.5. Summary of the self-consistent theory of quantum criticality

The theory of quantum criticality in metallic systems has traditionally been formulated in terms of bosonic incoherent excitations such as spin fluctuations, charge fluctuations, pair fluctuations. While it is true that these fluctuations are harbingers of the incipient order, one should keep in mind that the more fundamental excitations of an itinerant interacting Fermi system are the fermionic quasiparticles. The concept of quasiparticles is not confined to the realm of Fermi liquids, where the Landau quasiparticles are stable in the sense that their line width is $\propto \omega^2$, which is much less than the quasiparticle energy ω in the limit $\omega \to 0$. In fact quasiparticles may be well defined even in so-called non-Fermi liquid states, with power law self-energy $\Sigma(\omega) \propto \omega^{1-\eta}$, provided the exponent $\eta < 1/2$. In that case the quasiparticle decay rate is still less than the energy and the quasiparticle weight is non-zero at any finite energy, a necessary condition for the existence of a well-defined quasiparticle peak in the single particle spectral function. Consequently, one may still use the familiar tools of renormalized perturbation theory, namely the expansion of response functions in terms of quasiparticle-quasihole pairs and the infinite summation in the spirit of RPA. There is one important new element here: if the quasiparticles turn critical, i.e. if their effective mass diverges at the QCP, this has fundamental consequences for the interaction vertices. Namely, the Ward identities deriving from conservation laws or from additional symmetries cause the vertices (e.g. the spin vertex at finite momentum $\mathbf{Q}$ in the antiferromagnetic case) to scale with the divergent effective mass. As a result the usual single boson exchange diagram for the self-energy, which may be proportional to a positive power of energy ω^s and thus irrelevant in the limit $\omega \to 0$, will be multiplied by a factor originating from vertex corrections, which may diverge and may offset the vanishing of ω^s. This is the scenario considered in the above. It leads to a self-consistent relation for the effective mass, which allows for two entirely different solutions, depending on the initial condition, i.e. the system parameters at the far end of the critical regime. One of the solutions is the familiar weak coupling solution, which is experimentally found to be realized in a number of systems. The second and new solution is activated at strong coupling, if the effective mass is already enhanced by additional critical fluctuations. This case is apparently met in at least two of the most intensively studied systems. It turns out that the theory agrees in detail with experimental observation.

4. Conclusion

Dynamic equilibrium of collective degrees of freedom in strongly correlated quantum matter may characterize certain well-defined states at strong coupling. In particular, the dynamic equilibrium of renormalized single particle quantities such as the fermionic quasiparticle relaxation rate or/and the effective mass and of two-particle quantities such as dynamical interaction vertices has been shown here to lead to novel equilibrium states. In the cases considered here there are reasons to expect that the correction terms to the self-consistent problems describing the dynamic equilibrium do not change the character of the solution. It is hoped that the approach sketched here will be useful in solving more strong-coupling many-body problems of this type in the future.

5. Acknowledgments

I acknowledge the fruitful collaboration with Holger Schmidt and Jens Paaske on the Kondo problem. It is a pleasure to thank Elihu Abrahams and Joerg Schmalian for collaboration on quantum criticality in antiferromagnetic metals. This work has been supported by the research unit "Quantum Phase Transitions" of the Deutsche Forschungsgemeinschaft. Partial support by the University of Wisconsin-Madison and by an ICAM senior fellowship during a stay at UW and of several stays at the Aspen Center for Physics by the National Science Foundation under Grant No. PHYS-1066293 is gratefully acknowledged.

6. References

[1] Vollhardt D and Wölfle P 2013 *The Superfluid Phases of Helium 3* (New York: Dover)
[2] Hewson A C 1993 *The Kondo Problem to Heavy Fermions* (Cambridge University Press)
[3] Chubukov A V 2012 *Annu. Rev. Condens. Matter Phys.* **3** 57
[4] Anderson P W 1970 *J. Phys. C* **3** 2439
[5] v. Lohneysen H, Rosch A, Vojta M and Wölfle P 2007 *Rev. Mod. Phys.* **79** 1015
[6] Anderson P W 1967 *Phys. Rev.* **164** 352
[7] Wilson K G 1975 *Rev. Mod. Phys.* **47** 773
[8] Nozieres P 1974 *J. Low Temp. Phys.* **17** 31
[9] Andrei N 1980 *Phys. Rev. Lett.* **45** 379
[10] Wiegmann P B 1980 Sov. Phys. JETP Lett. **31** 392
[11] Larsen U *Z. Phys.* **256** 65
[12] Kondo J 1964 *Prog. Theor. Phys.* **32** 37
[13] Paaske J, Schmidt H and Wölfle P 2012 unpublished
[14] Fresard R, Kroha J and Wolfle P 2011 in *Theoretical Methods for Strongly Correlated Systems* (Berlin: Springer) p 111
[15] Mattuck R 1967 *A Guide to Feynman Diagrams in the Many-Body Problem* (New York: McGraw-Hill); Larsen U 1974 *Proceedings of the Int. Conf. on Magnetism ICM-73* vol.V 88
[16] Hertz J A 1976 *Phys. Rev. B* **14** 1165
[17] Millis A J 1993 *Phys. Rev. B* **48** 7183
[18] Abrahams E, Schmalian J and Wölfle P 2014 *Phys. Rev. B, in press*
[19] Wölfle P and Abrahams E 2011 *Phys. Rev. B* **84** 041101
[20] Abrahams E and Wölfle P 2012 *Proc. Nat. Acad. Sciences* **109** 3238
[21] Varma C M and Wölfle P 2014 unpublished
[22] Hartnoll S A, Hofman D M, Metlitski M A and Sachdev S 2011 Phys. Rev. B **84** 125115

Journal of Physics: Conference Series **568** (2014) 042035 doi:10.1088/1742-6596/568/4/042035

Spin and Hole Dynamics in Carrier-Doped Quantum Haldane Chain

T Yokoo[1,2], S Itoh[1,2], S Ibuka[1], H Yoshizawa[3], and J Akimitsu[4]

[1] Institute of Materials Structure Science, High Energy Accelerator Research Organization (KEK), Tsukuba, Ibaraki 305-0801, Japan
[2] Materials and Life Science Experimental Facility, J-PARC Center, Tokai, Ibaraki 319-1106, Japan
[3] Institute for Solid State Physics, the University of Tokyo, Tokai, Ibaraki 319-1106, Japan
[4] Department of Physics and Mathematics, Aoyama-Gakuin University, Sagamihara, Kanagawa 229-8558, Japan

E-mail: tetsuya.yokoo@kek.jp

Abstract. Quantum spins in one-dimensional (1D) chains exhibit characteristic features such as the Haldane gap ($S = 1$) and the spin-Peierls state ($S = 1/2$). Multiple degrees of freedom associated with quantum spins have recently become a focus of attention. We present here the spin dynamics in lightly hole-doped $Nd_{2-x}Ca_xBaNiO_5$ ($x = 0.1$), a 1D Haldane system, using pulsed neutron inelastic scattering. The entire dispersion of the 1D Haldane chain was clearly observed with a ~14 meV spin gap (Haldane gap) at the magnetic zone center. Below the gap, much spectral weight was observed, indicating an in-gap state possibly originating from the doped holes in Haldane chains.

1. Introduction

In a low-dimensional system, the quantum nature of spins appears in macroscopic phenomena. A Haldane gap is one such *unconventional* property of one-dimensional (1D) integer spin systems in which a spin gap exists between the ground state and the first excited state [1, 2]. Inorganic nickelate Y_2BaNiO_5 was found to be a good candidate for a Haldane system with a spin gap of 10 meV at the magnetic zone center (MZC) [3]-[6]. In contrast, R_2BaNiO_5 (where R is a magnetic rare earth) is known as a classical antiferromagnet showing three-dimensional (3D) long-range magnetic ordering at low temperature [7]. However, the saturation moment of the sublattice magnetization of the Ni^{2+} spins is somewhat suppressed, indicating a remaining quantum spin fluctuation. Indeed, inelastic neutron scattering experiments revealed the existence of a spin gap at the MZC even in R_2BaNiO_5 with magnetic rare earth ions [8]-[11]. In R_2BaNiO_5 (R = Nd, Pr), the observed excitation energy is quite similar to the Haldane gap in Y_2BaNiO_5 and shows a structure factor reflecting a 1D Ni chain correlation. In addition, the magnetization and gap properties have been investigated by synthesizing a solid solution of Y^{3+} and Nd^{3+}, namely, $(Y_{1-x}Nd_x)_2BaNiO_5$, to ascertain how the magnetic system (R = Nd) and the nonmagnetic system (R = Y) are connected within the context of the Haldane system [10, 11]. Surprisingly, for all values of x at various ordered temperatures, the gap properties can be scaled by the temperature. Above the Néel temperature T_N for each value of x, spin gap excitations at around $\Delta \approx 10$ meV are clearly observed. Once the temperature drops below

T_N, however, the gap energy abruptly increases with decreasing temperature. Moreover, the gap intensity starts decreasing below T_N. All these experimental results indicate that the new Haldane system Nd_2BaNiO_5 can be interpreted as a "pure 1D Haldane chain" only above T_N without any magnetic perturbation from magnetic rare earth ions. Near T_N, Nd^{3+} sublattices begin ordering and applying an internal effective field to the Ni^{2+} sublattice (1D chains). Thus, in this temperature region, the Ni chain can be explained well as a "Haldane chain in a staggered field" [12]. A theoretical evaluation of this scenario also supports the temperature dependence of the observed gap energy [13, 14]. Thus, Nd_2BaNiO_5 is found to *not* be a simple Haldane system.

Another topic of interest regarding transition metal chain oxides is carrier doping, which yields phenomena ranging from strongly renormalized Fermi liquid behavior to high-temperature superconductivity. In view of the enhanced quantum fluctuations in 1D systems, doping of transition metal chains could well lead to equally surprising discoveries. R_2BaNiO_5 is a charge-transfer insulator [15, 16], and some quantity of hole carriers is successfully introduced primarily on oxygen sites by replacing part of the off-chain R^{3+} with Ca^{2+} [4]. The spin dynamics of hole-doped Ni 1D chains have also been addressed using a theoretical approach [17]-[19] and through neutron experiments [20, 21]. In this article, we present the spin and hole dynamics of hole-doped Nd_2BaNiO_5 measured by means of a pulsed neutron scattering technique to survey the entire area of the Brillouin zone in momentum and energy space.

2. Experimental procedure

The carrier content can be controlled by replacing a portion of the Nd^{3+} ions with Ca^{2+}. Lightly doped ($x = 0.1$) $Nd_{2-x}Ca_xBaNiO_5$ was chosen for the neutron experiments. T_N is evaluated as 30 K. Two cylindrical single crystals were assembled with a total mass of 13 g in an aluminum can. Inelastic neutron scattering experiments were conducted at the High-Resolution Chopper spectrometer (HRC) installed at the Materials and Life Science Experimental Facility in J-PARC [22, 23]. The HRC is a direct geometry time-of-flight neutron spectrometer. The measurement temperatures were 4, 20, 30, 40, and 150 K, which are above and below T_N. An incident neutron energy E_i of 102.0 meV was chosen to measure a wide Q–ω space. A Fermi chopper [23] was used to monochromatize the incoming neutron energy; its rotational frequency of 200 Hz yielded an energy resolution of $\frac{dE}{E_i} \sim 6\%$.

3. Results and discussion

In the neutron scattering experiment, the $[H\ 0\ L]$ zone was into the horizontal scattering plane, where H corresponds to the chain direction (crystallographic a axis). Figure 1 shows the obtained dynamical structures $S(Q, \omega)$ at $T = 4.0$ K (well below $T_N = 30$ K) and $T = 40$ K (above T_N). Both spectra are normalized by the measurement time (proton current measured just before the neutron target). Intense crystal electric field (CEF) excitations of Nd^{3+} ions lie at $E = 24$ and 38 meV. The site symmetry of Nd^{3+} in Nd_2BaNiO_5 is sufficiently low to split the tenfold degenerate $J = 9/2$ multiplet into five Kramers doublets. In addition to the CEF excitations, excellent dispersion relations for the Ni 1D chain are observed at both temperatures for the entire first and second Brillouin zones. The minima of the dispersion observed at the MZC, $H = 0.5$, 1.5, and even 2.5, occur where the magnetic scattering intensity becomes weak because of the magnetic form factor, and phonon scattering is dominant in this high-Q region. On the other hand, at $H = 1.0$ and 2.0, the weight of the magnon dispersion is known to be zero at the zone boundary. At low temperature [Fig. 1 (a)], the magnon bandwidth reaches 60 meV, with a somewhat large gap energy of $\Delta \sim 20$ meV compared to the bare (unperturbed) 1D Haldane gap [5]. In addition, a flat band lying at $E = 65$ meV and corresponding to a phonon or magnon optical branch was observed. At $T = 40$ K, the Haldane gap is lowered, and

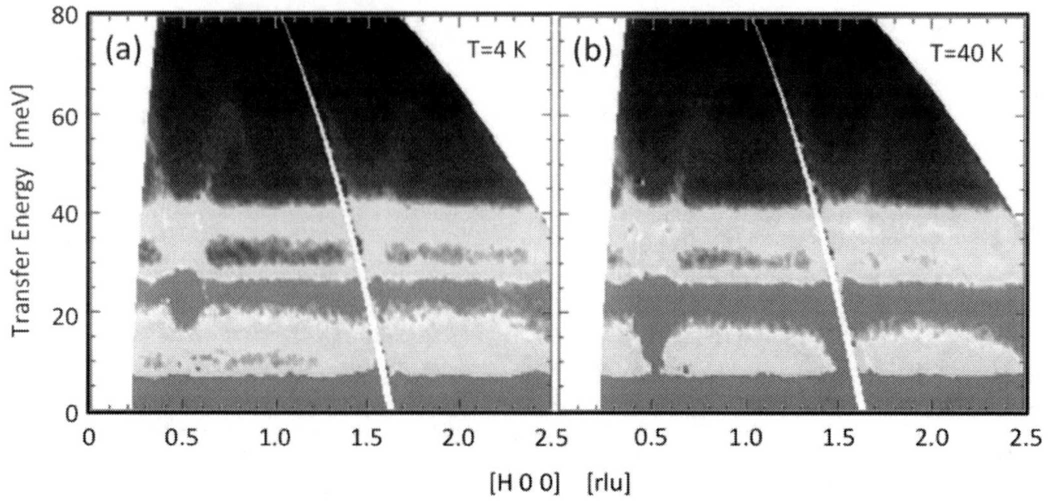

Figure 1. Observed dispersion relation of $x = 0.1$ at (a) $T = 4.0$ K ($<T_N$) and (b) $T = 40$ K ($>T_N$). Flat bands at $E = 24$ and 38 meV are contributions from crystal field excitation of Nd^{3+} ions. Oscillating signals showing minima at magnetic zone center (H=0.5 and 1.5) are from 1D chain.

the spectral weight is slightly shifted with respect to the MZC. The boundary energy is almost the same at both temperatures.

To survey the spectra near the gap, constant-Q cuts are shown in Fig. 2 for both temperatures. Because the transfer energy in Figs. 1 and 2 is strongly related to the wave vector parallel to the incoming neutron beam k_i (crystallographic c axis in our experiment), the 1D spectra are also a function of L. However, in the 1D system, only the chain direction can be varied, and the wave numbers in the other direction are totally unchanged. Thus, we can safely accumulate the spectra with a finite index L. The spectra at $T = 40$ K, where the 1D Ni chain can be bare without any magnetic perturbation, shows the original gap value of ~14 meV. In contrast, the gap peak merged into the CEF peaks at 24 meV, which shows a much narrower profile. Although the energy resolution is not sufficient to show the detailed structure below the Haldane gap, considerable weight was observed, indicating a hole dynamical structure as an *in-gap state*. By analogy with the slightly doped ($x = 0.035$) compound [24], the weight at the MZC ($H =$ 0.5) appeared as a result of hole doping. We now plan to perform further experiments at a high energy resolution to observe the dynamical structure below the gap.

Thus, the entire dynamical set of properties of a hole-doped 1D Ni chain was successfully measured by pulsed neutron inelastic scattering. At T_N, the dynamical structure changes dramatically, presenting new features—a local excitation at $E \sim 10$ meV and doped hole dynamics below the gap energy—in addition to a view of the entire 1D chain dispersion.

Acknowledgments

TY is supported by JSPS KAKENHI Grants [Scientific Research (C) No. 26400376 and Scientific Research (B) No. 25287081]. In addition, TY and SI are supported by a JSPS KAKENHI Grant [Scientific Research (C) No. 24540352]. The neutron scattering experiment was approved by the Neutron Scattering Program Advisory Committee of IMSS, KEK (Proposal No. 2014S01).

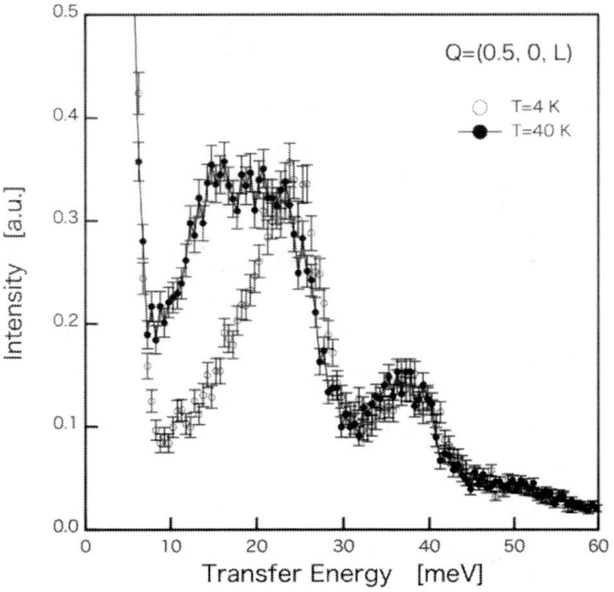

Figure 2. Energy spectra at $Q = (0.5, 0, L)$. Open symbol is the measurement at $T = 4$ K, which is well below T_N. Solid symbol is at $T = 40$ K (above T_N). The gap value was evaluated at Δ=14 meV from 1D profiles.

References

[1] Haldane F D M 1983 *Phys. Lett.* **93** 464
[2] Haldane F D M 1983 *Phys. Rev. Lett.* **50** 1153
[3] Darriet J and Regnault L P 1993 *Solid State Commun.* **86** 409
[4] DiTusa J F, Cheong S W, Park J H, Aeppli G, Broholm C and Chen C T 1994 *Phys. Rev. Lett.* **73** 1857
[5] Yokoo T, Sakaguchi T, Kakurai K and Akimitsu J 1995 *J. Phys. Soc. Jpn.* **64** 3651
[6] Sakaguchi T, Kakurai K, Yokoo T and Akimitsu J 1996 *J. Phys. Soc. Jpn.* **65** 3025
[7] Sachan V, Buttrey D J, Tranquada J M and Shirane G 1994 *Phys. Rev. B* **49** 9658
[8] Zheludev A, Tranquada J M, Vogt T and Buttrey D J 1996 *Phys. Rev. B* **54** 7210
[9] Zheludev A, Tranquada J M, Vogt T and Buttrey D J 1996 *Phys. Rev. B* **54** 6437
[10] Yokoo T, Zheludev A, Nakamura M and Akimitsu J 1997 *Phys. Rev. B* **55** 11516
[11] Yokoo T, Raymond S, Zheludev A, Maslov S, Ressouche E, Zaliznyak I, Erwin R, Nakamura M and Akimitsu J 1998 *Phys. Rev. B* **58** 14424
[12] Raymond S, Yokoo T, Zheludev A, Nagler S E, Wildes A and Akimitsu J 1999 *Phys. Rev. Lett.* **82** 2382
[13] Maslov S and Zheludev A 1998 *Phys. Rev. B* **57** 68
[14] Maslov S and Zheludev A 1998 *Phys. Rev. Lett.* **80** 5786
[15] Mattheiss L F 1993 *Phys. Rev. B* **48** 4532
[16] Eisaki H and Uchida S 1995 *J. Phys. Chem. Solids* **56** 1811
[17] Dagotto E, Riera J, Sandvik A and Moreo A 1996 *Phys. Rev. Lett.* **76** 1731
[18] Batista C D, Aligia A A and Eroles J 1998 cond-mat/9810387
[19] Penc K and Shiba H 1995 *Phys. Rev. B* **52** R715
[20] Xu G, Aeppli G, Bisher M E, Broholm C, DiTusa J F, Frost C D, Ito T, Oka K, Paul R L, Takagi H and Treacy M M J 2000 *Science* **289** 419
[21] Sasaki T, Yokoo T, Katano S and Akimitsu J 2005 *J. Phys. Soc. Jpn.* **74** 267
[22] Itoh S, Yokoo T, Satoh S, Yano S, Kawana D, Suzuki J and Sato T J 2011 *Nucl. Instrum. Methods A* **631** 90
[23] Itoh S, Ueno K and Yokoo T 2012 *Nucl. Instrum. Methods A* **661** 58
[24] Yokoo T, Itoh S, Kawana D, Yoshizawa H and Akimitsu J 2014 *J. Phys. Conf. Ser.* **502** 012045

Journal of Physics: Conference Series **568** (2014) 052001 doi:10.1088/1742-6596/568/5/052001

Weak antilocalization in (111) thin films of a topological crystalline insulator SnTe

R Akiyama, K Fujisawa, R Sakurai, and S Kuroda

Institute of Materials Science, University of Tsukuba, 1-1-1 Tennoudai, Tsukuba, Ibaraki 305-8573, Japan

E-mail: akiyama@ims.tsukuba.ac.jp

Abstract.
We grew single-crystal thin films of a topological crystalline insulator (TCI) SnTe with a smooth surface at the atomic scale by molecular beam epitaxy (MBE). In the magnetoresistance (MR) measurement, we observed both positive and negative components near zero magnetic field at lowest temperatures of 2 - 3 K, while we observed only a negative MR at elevated temperatures of 6 - 10 K. The positive MR is attributed to the weak antilocalization (WAL) in the transport through the topological surface state (SS), demonstrating π berry phase which is essential to the topological SS, while the negative MR to the weak localization (WL) in the transport through the bulk state (two-dimensional bulk subbbands). The absolute value of the prefactor α deduced from the fitting of the observed positive MR to the Hikami-Larkin-Nagaoka equation was much smaller than expected from the number of transport channel of the SS, suggesting the coupling of the SS to the bulk state.

1. Introduction

Topological insulator (TI) is a new class of materials. The surface state (SS) of TIs has a band structure of Dirac-cone shape where the spin of carriers is locked with their momentum except for k=0[1]. The SS of TIs is metallic protected by time-reversal symmetry and exceedingly high mobility is expected in the transport of mass-less Dirac fermions in the SS. The topological SS has so far been investigated intensively, for example, in $Bi_{1-x}Sb_x$[2], Bi_2Se_3[3], Bi_2Te_3[4], and $TlBiSe_2$[5] through electrical transport measurements and the angle-resolved photoemission spectroscopy (ARPES). Recent explorations of TIs have opened the door of a new type of TIs, that is, topological crystalline insulators (TCIs), in which the topological SS is protected by the mirror symmetry (reflection symmetry) of the crystal[6][7], instead of the time-reversal symmetry. SnTe was predicted to be one of the TCIs whose SS is protected by mirror symmetry with respect to the {111}, {001}, and {110} planes of the rock-salt crystal structure. The APRES measurement confirmed the topological phase in the (001) and (111) surfaces of SnTe by observing Dirac-cone band dispersion derived from the SS[8][9] as well as $Pb_{1-x}Sn_xTe$($x <$ 0.25)[10] and in the (001) surface of $Pb_{0.73}Sn_{0.27}Se$[11]. In addition, the electrical transport properties which are characteristics of two-dimensional (2D) SS have been reported, such as the Shubnikov-de Haas (SdH) oscillation[12], and the Altshuler-Aronov-Spivak (AAS)/ the Aharonov-Bohm (AB) effect[13].

In the present study, we investigated the magneto-transport properties in a SnTe(111) thin film, which was grown by molecular beam epitaxy (MBE). In the transverse magnetoresistance

(MR), we observed a positive MR, which is attributed to the weak antilocalization (WAL) in the transport through the SS. The WAL in the SS of TIs is considered to arise from a different mechanism from the case of the conventional transport; the prohibition of backscattering due to the accumulated π Berry phase of Dirac fermions is essential for the WAL instead of the spin-orbit interaction (SOI)[14]. Therefore, the investigation of the WAL in TIs or TCIs is useful for gaining an insight into the transport properties of the 2D SS[15][16]. In the case of SnTe, the SS has not been confirmed sufficiently from the viewpoint of the electrical transport because of the recent recognition as a TCI. In this article, we report the observation of a positive MR originating from the WAL in the SS in a (111) thin film of single-crystal SnTe. This result suggests that the surface transport of 2D SS of SnTe can be detected even in the existence of a large contribution to the electrical transport of high-density bulk carriers due to Sn vacancies.

2. Experimental

Thin films of the (111) plane of single-crystal SnTe were epitaxially grown on BaF_2(111) substrates by molecular beam epitaxy (MBE) equipped with elementary sources of Sn and Te. The substrate temperature was kept at 220°C during the growth of SnTe. The molecular beam flux ratio of Sn/Te was 1/20 (Te-rich condition). The thickness of the grown SnTe film was 115 nm. Figure 1 (a) and (b) show a reflection high energy electron diffraction (RHEED) image along [1-10] direction and an atomic force microscope (AFM) image of the surface of the grown SnTe film, respectively. As shown in these images, the surface of the SnTe film is smooth in the range of the atomic scale although the macroscopic morphology shows a characteristic structure. The surface morphology was found to be quite sensitive to the molecular beam flux ratio Sn/Te and the substrate temperature during growth. The electrical transport measurements under magnetic fields were performed in the standard four-probe and the Van der Pauw configurations using Quantum Design Physical Properties Measurement System (PPMS).

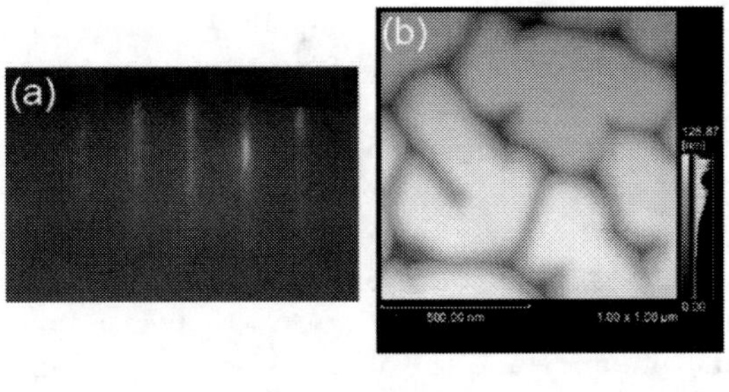

Figure 1. (a) The RHEED image along [1-10] direction of the grown SnTe film. (b) The atomic force microscope (AFM) image of the surface of the grown SnTe film.

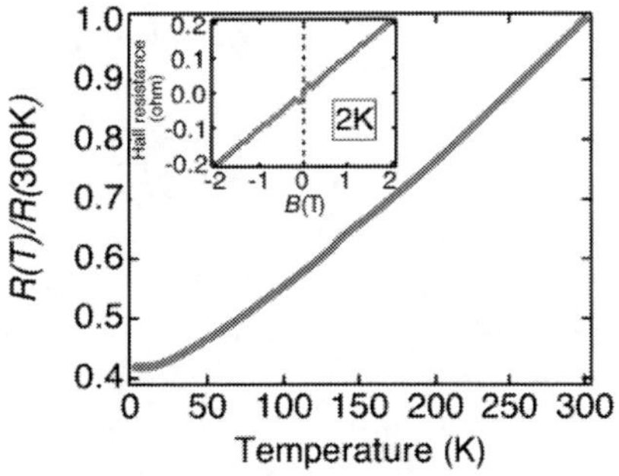

Figure 2. The temperature dependence of the resistance of the SnTe film. Inset: The magnetic field dependence of the Hall resistance measured in the Van der Pauw configuration at 2K.

3. Results and discussion

Figure 2 shows the temperature dependence of the resistance of the grown SnTe film. In the figure, the value of resistance is normalized by the value at 300 K. The temperature dependence shows a metallic behavior with decrease in resistance as the temperature decreases. The inset of Fig. 2 shows the Hall resistance, which was obtained by extracting a component of the odd function of magnetic field from the raw data. From the Hall coefficient, it was deduced that the carrier type is a hole and the hole density is 5.0×10^{20}cm^{-3}. This value is approximately consistent with the previous report[8]. Figure 3 shows the magnetic field dependence of the longitudinal resistance at 2 K under a magnetic field perpendicular to the film plane (transverse MR). In the MR curve, a component of the odd function of magnetic field, which originates from the Hall voltage due to a small deviation of the alignment of the electrodes, was subtracted. As shown in the figure, the MR curve depends on the magnetic field B in a quadratic way in the magnetic field range of $|B| \gtrsim 2.5$ T, and there appears a small peak near zero field. The quadratic dependence on B originates from the semiclassical motion of carriers deflected by the Lorentz force. From this dependence, the mobility μ is estimated at 273 cm^2/V·s using the Kohler's rule[17], $R(B)/R(B=0) \approx 1 + (\mu B)^2$.

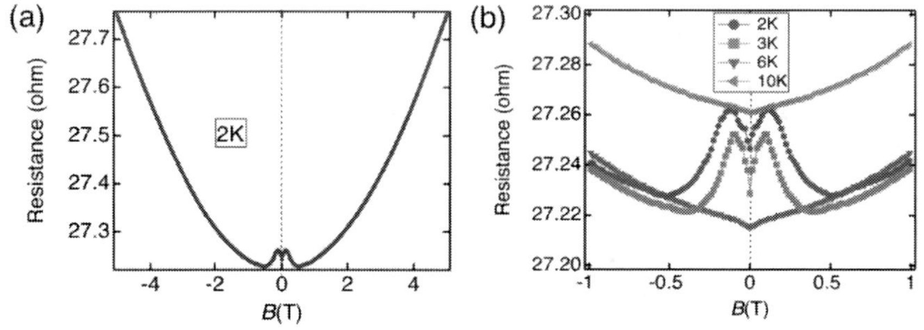

Figure 3. (a)The magnetic field dependence of the longitudinal resistance measured in the four-probe configuration at 2 K. The component of the odd function of magnetic field was subtracted from the raw data. (b)The magnetoresistance (MR) in the range of $|B| \leq 1$ T at 2 - 10 K.

In order to take a closer look of the peak near zero field, the MR in a small field range within ±1 T is plotted in Fig. 3(b), together with the data at temperatures of 3 - 10 K. As shown in the figure, the MR curves at 2 K and 3 K consist of two components; a negative MR forming a gentle hill in the range of ±0.5 T and a positive component forming a sharp dip in the range of ±0.2 T. At higher temperatures of 6 K and 10 K, only a positive MR appears. The positive MR is attributed to the WAL in the surface transport through the SS. On the other hand, the negative MR could be interpreted as a result of the weak localization (WL) in the transport through the bulk subbands; due to a small thickness of the film in the present measurement, the bulk state inside the film is quantized into 2D subbands and the WL can be expected in the transport through these quantized subbands[18]. The MR curves shown in Fig. 3(b) are plotted in the form of the magnetoconductance (MC) $\Delta G(B) = 1/R(B) - 1/R(B=0)$ in Fig. 4.

According to the theory of quantum correction in the 2D diffusive transport, the MC due to the WAL or WL is given by the Hikami-Larkin-Nagaoka (HLN) equation as follows[19],

$$\Delta\sigma = \sigma(B) - \sigma(0) = -\frac{\alpha e^2}{2\pi^2\hbar}\left[\ln\frac{\hbar}{4Bel_\phi^2} - \psi\left(\frac{1}{2} + \frac{\hbar}{4Bel_\phi^2}\right)\right], \tag{1}$$

where e is the charge of an electron, $\hbar$ is the Plank's constant, l_ϕ is the phase coherent length, and $\psi(x)$ is the digamma function. α is a coefficient which should be -0.5 per single transport channel for the symplectic case of the WAL, and 1 for the case of the WL, respectively. $\Delta G(B)$ at 2 K and 3 K, represented by blue and green marks respectively in Fig. 4(a), is fitted with

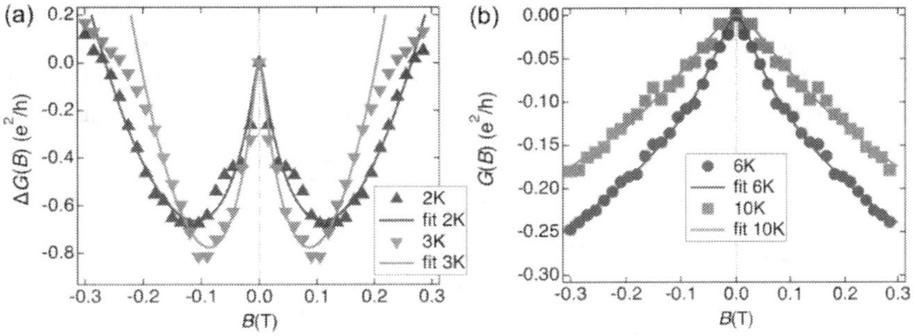

Figure 4. The magnetoconductance (MC) $\Delta G(B)$ in the range of $|B| \leq 0.3$ T. (a) Blue (green) marks represent the experimental data at 2 K (3 K). Lines of the respective colors represent the result of fitting to the HLN equation. (b) Purple (red) marks represent the experimental data at 6 K (10 K) and lines represent the result of fitting.

Eq. (2), expressing the sum of the contributions of the WAL and WL.

$$\Delta\sigma = \sigma(B) - \sigma(0) = -\frac{\alpha e^2}{2\pi^2\hbar}\left[\ln\frac{\hbar}{4Bel_\phi^2} - \psi\left(\frac{1}{2}+\frac{\hbar}{4Bel_\phi^2}\right)\right] - \frac{\beta e^2}{2\pi^2\hbar}\left[\ln\frac{\hbar}{4Bel_\xi^2} - \psi\left(\frac{1}{2}+\frac{\hbar}{4Bel_\xi^2}\right)\right], \quad (2)$$

where, the first and second terms correspond to the MCs due to the WAL and WL, with the respective fitting parameters α, l_ϕ for the WAL and β, l_ξ for the WL. The parameters obtained from the fitting at 2 K (3 K) are $\alpha = -1.67$ (-2.13), $l_\phi = 224$ nm (235 nm), and $\beta = 276$ (291), $l_\xi = 15$ nm (17 nm), respectively.

$\Delta G(B)$ at temperatures of 6 K and 10 K are plotted in Fig. 4(b). Since a negative MR due to the WL is missing, $\Delta G(B)$ is fitted with Eq. (1). The parameters obtained from the fitting at 6 K (10 K) are $\alpha = -0.45$ (-0.52), and $l_\phi = 131$ nm (86 nm), respectively. In Fig. 5, the fitting parameters α and l_ϕ for the WAL are plotted as a function of temperature. With the increase of temperature, α approaches -0.5 and the phase coherent length l_ϕ decreases. The decrease of the absolute value of α with temperature suggests that the number of the transport channel through the SS decreases.

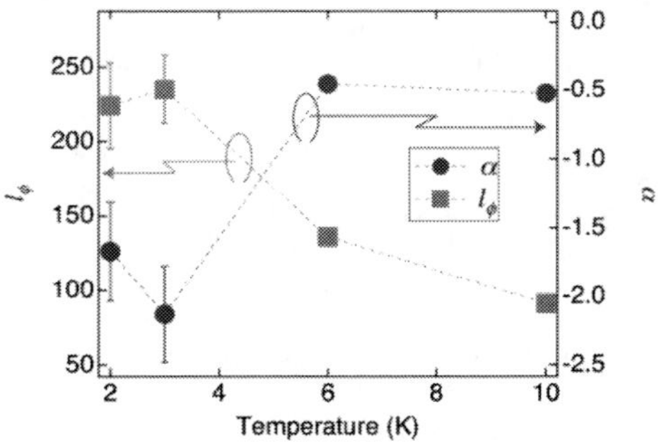

Figure 5. The fitting parameters α and l_ϕ obtained from the fitting are plotted as a function of temperature.

In the surface transport through the SS of 3D TIs, which is classified into the symplectic group, the value of α should be -0.5 per single transport channel. In the (111) surface of SnTe, the total number of Dirac cones is four, three on the $\overline{\mathrm{M}}$ points which are projections of L points, and one on the $\overline{\Gamma}$ point which are projections of Γ or L points[9]. In addition, considering the surface transport through both the top and bottom surfaces of the film, the total number of the channel should be 8, resulting in the value of $|\alpha| = 4$. However, $|\alpha|$ around 2 was obtained from the measurement at 2K, giving the number of channels in the SS around 4 . A possible explanation for this deviation in the value of α would be given by the coupling between the SS and the bulk state or the coupling between the different SSs, as recently suggested for a small value of $|\alpha|$ [20] [21]; if there is a coupling between different transport channels due to the scattering of carriers from one transport channel to another with keeping the phase coherence, they contribute to the conductance as a single phase-coherent channel.

4. Summary

We investigated the MR in the SnTe(111) thin film at low temperatures in order to clarify the properties of the SS of TCIs from the viewpoint of electrical transport. We observed that positive and negative components coexist in the MR curve near zero field at lowest temperatures of 2 - 3 K while only the positive MR appears at elevated temperatures of 6 - 10 K. The positive MR is attributed to the WAL originating from the surface transport in the SS, while the negative MR to the weak localization (WL) effect originating from the transport in the quantized 2D subbands of the bulk state. We deduced the prefactor α and the phase coherence length l_ϕ by fitting the observed MC due to the WAL with the HLN equation. The value of $|\alpha|$ was much smaller than expected from the nominal total number of the transport channels of the SS in the SnTe (111) surface, which suggests the coupling between the SS and bulk state.

Acknowledgments

The authors would like to thank T. Koyano in the Cryogenics Division of Research Facility Center for Science and Technology, University of Tsukuba for PPMS measurements, and T. Suemasu and K. Ito for AFM measurements. This work is partially supported by Grant-in-Aid for Scientific Research (KAKENHI) from the Japan Society for the Promotion of Science, CASIO Science Promotion Foundation, and Sumitomo Electric Industries Group CSR Foundation.

References

[1] Fu L, Kane C L, and Mele E J, *Phys. Rev. Lett.* **98**, 106803 (2007).
[2] Hsieh D, Qian D, Wray L, Xia Y, Hor Y S, Cava R J, and Hasan M Z, *Nature* **452**, 970 (2008).
[3] Moore J E, and Balents L, *Phys. Rev. B* **75**, 121306 (2007).
[4] Chen Y L *et al.*, *Science* **325**, 178 (2009).
[5] Kuroda K *et al.*, *Phys. Rev. Lett.* **105**, 146801 (2010).
[6] Hsieh T H, Lin H, Liu J, Duan W, Bansil A, and Fu L, *Nat. Commun.* **3**, 982 (2012).
[7] Fu L, *Phys.Rev.Lett.* **106**, 106802 (2012).
[8] Tanaka Y *et al.*, *Nat. Phys.* **8**, 800 (2012).
[9] Tanaka Y *et al.*, *Phys. Rev. B* **88**, 235126 (2013).
[10] Xiu S -Y *et al.*, *Nat. Commun.* **3**, 1192 (2012).
[11] Dziawa P *et al.*, *Nat. Mater.* **11**, 1023 (2012).
[12] Taskin A A, Yang F, Sasaki S, Segawa K, and Ando Y, *Phys.Rev. B* **89**, 121302(R) (2014).
[13] Safdar M, Wang Q, Mirza M, Wang Z, Xu K, and He J, *Nanoletters* **13**, 5344 (2013).
[14] Nomura K, Koshino M, and Ryu S, *Phys. Rev. Lett.* **99**, 146806 (2007).
[15] He H -T *et al.*, *Phys. Rev. Lett.* **106**, 166805 (2011).
[16] Ning W, Du H, Kong F, Yang J, Han Y, Tian M, and Zhang Y, *Scientific Reports* **3**, 1654 (2013).
[17] Ziman J M, *Electrons and Phonons* (Oxford University Press, 1960) p 490.
[18] Lu H -Z, and Shen S-Q, *Phys. Rev. B* **84**, 125138 (2011).
[19] Hikami S, Larkin A, and Nagaoka Y, *Prog. Theor. Phys.* **63**, 707 (1980).
[20] Assaf B A *et al.*, *ArXiv* 1403.1810 (2014).
[21] Steinberg H *et al.*, *Phys. Rev. B* **84**, 233101 (2011).

Magnetic and orbital instabilities in a lattice of SU(4) organometallic Kondo complexes

A. M. Lobos[1] **and A. A. Aligia**[2]

[1]Joint Quantum Institute and Condensed Matter Theory Center, Department of Physics, University of Maryland, College Park, Maryland 20742, USA
[2]Centro Atómico Bariloche and Instituto Balseiro, Comisión Nacional de Energía Atómica, 8400 Bariloche, Argentina

E-mail: aligia@cab.cnea.gov.ar

Abstract. Motivated by experiments of scanning tunneling spectroscopy (STS) on self-assembled networks of iron(II)-phtalocyanine (FePc) molecules deposited on a clean Au(111) surface [FePc/Au(111)] and its explanation in terms of the extension of the impurity SU(4) Anderson model to the lattice in the Kondo regime, we study the competition between the Kondo effect and the magneto-orbital interactions occurring in FePc/Au(111). We explore the quantum phases and critical points of the model using a large-N slave-boson method in the mean-field approximation. The SU(4) symmetry in the impurity appears as a combination of the usual spin and an orbital pseudospin arising from the degenerate $3d_{xz}$ and $3d_{yz}$ orbitals in the Fe atom. In the case of the lattice, our results show that the additional orbital degrees of freedom crucially modify the low-temperature phase diagram, and induce new types of orbital interactions among the Fe atoms, which can potentially stabilize exotic quantum phases with magnetic and orbital order. The dominant instability corresponds to spin ferromagnetic and orbital antiferromagnetic order.

1. Introduction

Organometallic complexes containing magnetic centers are currently under intense investigation for their potential uses as building blocks for nanotechnologies. Low-dimensional magnetic nanostructures have a high potential for applications in spintronics, magnetic recording and sensing devices [1, 2] These systems also offer a unique platform to study exotic phases of matter. In particular, Minamitani *et al.* [3] have shown that the Kondo effect observed in isolated iron(II) phtalocyanine (FePc) molecules deposited on top of clean Au(111) [FePc/Au(111)] (in the most usual on-top configuration) is a new realization of the SU(4) Kondo model, in which not only the spin degeneracy but also the orbital degeneracy between $3d_{xz}$ and $3d_{yz}$ orbitals of Fe play a role (z is the direction normal to the surface). The SU(4) Kondo effect manifests itself as a dip at the Fermi energy in the differential conductance observed in scanning tunneling spectroscopy (STS). The half width at half maximum of the dip is about $T_K \sim 0.4$ meV, with T_K the Kondo temperature.

In another set of experiments, a self-organized square lattice of FePc/Au(111) as well as small clusters were studied by STS [4]. It was found that as for the isolated molecule, a single dip in the differential conductance remains for the molecules that lie at the corners of the clusters. However, increasing the coordination, the peak tends to split and for the lattice, a clear splitting of approximately 2 meV becomes apparent. Romero and the present authors were able to explain these experimental results, using a natural generalization of the single-impurity SU(4) Anderson model to the lattice, including hoppings between effective orbitals of nearest-neighbor molecules respecting the symmetry of these orbitals [5]. The relevant effective orbitals (i.e., the closest to the Fermi level), are essentially the $3d_{xz}$ and $3d_{yz}$ of Fe, with some admixture of other orbitals of the molecule. The splitting of the Kondo dip is a consequence of the anisotropy in the hopping for a given effective orbital.

Preliminary results suggests that the square lattice of FePc/Au(111)FePc is not far from a magnetic instability [5]. Ferromagnetic order was observed in a two-dimensional layer of organic molecules absorbed on graphene [6] and in metal-organic networks on Au surfaces [7, 8]. In addition, magnetic and orbital ordering are intertwined [9, 10].

In this work, starting from the orbitally degenerate Hubbard-Anderson model that describes the observed STS in different arrays of FePc molecules [5], we perform a Schrieffer-Wolff transformation to obtain an effective model which includes spin and orbital interactions. Solving this model in a slave-boson mean-field approximation (SBMFA), we obtain the critical values of the interactions leading to different symmetry-breaking magnetic and orbital instabilities. The dominant one turns out to be a spin ferromagnetic and orbital antiferromagnetic order. We show that the effect of the Ruderman-Kittel-Kasuya-Yosida (RKKY) interactions is small and can be neglected.

2. The Hubbard-Anderson model

Our starting model was derived Ref. [5] and discussed in detail in the supplemental material of that paper. The Hamiltonian $H = H_1 + H_2$ can be separated into one-body (H_1) and two-body (H_2) parts. The first one can be written as

$$
\begin{aligned}
H_1 &= \sum_{ij}^{N}[E_h n_{\mathbf{r}_{ij}} - \sum_{\sigma\nu}\left(t_2 h^{\nu\dagger}_{\mathbf{r}_{ij},\sigma} h^{\nu}_{\mathbf{r}_{ij}\pm\mathbf{a}_\nu,\sigma} + t_1 h^{\bar{\nu}\dagger}_{\mathbf{r}_{ij},\sigma} h^{\bar{\nu}}_{\mathbf{r}_{ij}\pm\mathbf{a}_\nu,\sigma}\right) + \sum_{\xi\sigma\nu}\epsilon_\xi c^{\nu\dagger}_{\mathbf{r}_{ij},\xi,\sigma} c^{\nu}_{\mathbf{r}_{ij},\xi,\sigma} \\
&+ V\sum_{\xi\sigma\nu}\left(h^{\nu\dagger}_{\mathbf{r}_{ij},\sigma} c^{\nu}_{\mathbf{r}_{ij},\xi,\sigma} + \text{H.c.}\right)].
\end{aligned} \tag{1}
$$

Here the operators $h^{\nu}_{\mathbf{r}_{ij},\sigma}$ annihilate a hole (create an electron) in the state $|\nu_{\mathbf{r}_{ij},\sigma}\rangle$, where $\nu = (x, y)$ denotes one of the two orbitally degenerate molecular states with spin σ at site with position $\mathbf{r}_{ij}$ of the square lattice. $n_{\mathbf{r}_{ij}} = \sum_{\sigma\nu} n^{\nu}_{\mathbf{r}_{ij},\sigma}$, with $n^{\nu}_{\mathbf{r}_{ij},\sigma} = h^{\nu\dagger}_{\mathbf{r}_{ij},\sigma} h^{\nu}_{\mathbf{r}_{ij},\sigma}$ is the total number of holes at the molecule lying at site ij. The lattice vectors $\mathbf{a}_\nu$ connect nearest-neighbor sites, $\bar{x} = y$, and $\bar{y} = x$. The operator $c_{\mathbf{r}_{ij},\xi,\sigma}$ annihilates a conduction hole with spin σ and quantum number ξ at position $\mathbf{r}_{ij}$. The first and second terms of H_1 describe the molecular states and the hopping between them. The hopping t_2 between x (y) orbitals in the x (y) direction is larger than the hopping t_1 between x (y) orbitals in the y (x) direction. The third term of H_1 corresponds to a band of bulk and surface conduction electrons of the substrate and the last term is the hybridization between molecular and conduction states.

The interaction term can be written as [10, 11]

$$
\begin{aligned}
H_2 &= \sum_{ij}^{N}\left[U\sum_{\nu} n^{\nu}_{\mathbf{r}_{ij},\uparrow} n^{\nu}_{\mathbf{r}_{ij},\downarrow} + \sum_{\sigma\sigma'}[(U-2J_H) n^{x}_{\mathbf{r}_{ij},\sigma} n^{y}_{\mathbf{r}_{ij},\sigma'} + J_H h^{x\dagger}_{\mathbf{r}_{ij},\sigma} h^{y\dagger}_{\mathbf{r}_{ij},\sigma'} h^{x}_{\mathbf{r}_{ij},\sigma'} h^{y}_{\mathbf{r}_{ij},\sigma}]\right. \\
&+ \left. J_H\sum_{\nu} h^{\nu\dagger}_{\mathbf{r}_{ij},\uparrow} h^{\nu\dagger}_{\mathbf{r}_{ij},\downarrow} h^{\bar{\nu}}_{\mathbf{r}_{ij},\downarrow} h^{\bar{\nu}}_{\mathbf{r}_{ij},\uparrow}\right].
\end{aligned} \tag{2}
$$

For the system we are considering, the occupation of the molecular states is nearly one-hole per site ($\langle n_{\mathbf{r}_{ij}} \rangle \sim 1$) and the main effect of H_2 is to inhibit double occupancy. Therefore as a first approximation one can take $U \longrightarrow +\infty$ and therefore neglect the Hund rules exchange J_H in comparison with U. In this case, the interaction takes the simpler form $H_2^s = \frac{U}{2}\sum_{ij}^N n_{\mathbf{r}_{ij}}\left(n_{\mathbf{r}_{ij}} - 1\right)$.

3. Symmetry of H_i

For the case of one molecule only, the resulting impurity Anderson Hamiltonian has SU(4) symmetry, which in simple terms means that permutations of the four states $|\nu_\sigma\rangle$ leave the Hamiltonian invariant. The fifteen generators of the SU(4) symmetry are three trivial diagonal matrices, six permutations of two states and other six permutations with a change of phases for the permuted states [12]. The twelve non trivial generators can also be written as a generalization of the raising and lowering operators for SU(2) [13]. Specifically, for the impurity Anderson model they are $S_{\nu\sigma}^{\nu'\sigma'} = \left(h_\sigma^{\nu\dagger} h_{\sigma'}^{\nu'} + \sum_\xi c_{\xi,\sigma}^{\nu\dagger} c_{\xi,\sigma'}^{\nu'}\right)$ for $\nu\sigma \neq \nu'\sigma'$ (note that the conduction electron degrees of freedom must be taken into account to keep the SU(4) symmetry of the total system.)

For the lattice, the simplest generalization of these generators leads to $S_{\nu\sigma}^{\nu'\sigma'} = \sum_{ij}^N \left(h_{\mathbf{r}_{ij},\sigma}^{\nu\dagger} h_{\mathbf{r}_{ij},\sigma'}^{\nu'} + \sum_\xi c_{\mathbf{r}_{ij},\xi,\sigma}^{\nu\dagger} c_{\mathbf{r}_{ij},\xi,\sigma'}^{\nu'}\right)$. All these generators commute with H_2^s, but those with $\nu \neq \nu'$ commute with H_1 only in the particular case $t_2 = t_1$, which seems incompatible with the observed STS in the square lattice of FePc/Au(111) [5]. In the general case, $t_2 \neq t_1$, H_1 has however SU(4) symmetry with the following non-trivial generators $S_{\nu\sigma}^{\nu\bar{\sigma}} = \sum_{ij}^N \left(h_{\mathbf{r}_{ij},\sigma}^{\nu\dagger} h_{\mathbf{r}_{ij},\bar{\sigma}}^{\nu} + \sum_\xi c_{\mathbf{r}_{ij},\xi,\sigma}^{\nu\dagger} c_{\mathbf{r}_{ij},\xi,\bar{\sigma}}^{\nu}\right)$ and $S_{\nu\sigma}^{\bar{\nu}\sigma'} = \sum_{ij}^N \left(h_{\mathbf{r}_{ij},\sigma}^{\nu\dagger} h_{R\mathbf{r}_{ij},\sigma'}^{\bar{\nu}} + \sum_\xi c_{\mathbf{r}_{ij},\xi,\sigma}^{\nu\dagger} c_{R\mathbf{r}_{ij},\xi,\sigma'}^{\bar{\nu}}\right)$, where R is the reflection that permutes x and y (it is an element of the point group C_{4v} of the system). It can be verified easily that these generators commute with H_1. However, inclusion of H_2 reduces the symmetry to spin SU(2) times orbital $Z_2\times$U(1). Only the term of H_2^s with $\mathbf{r}_{ij} = 0$ commutes with the $S_{\nu\sigma}^{\bar{\nu}\sigma'}$ generators which contain R. Nevertheless, in a Fermi liquid, the interaction becomes irrelevant at the Fermi energy and we expect that SU(4) is an emergent symmetry at low energies [14] if there is not a symmetry breaking (a magnetic or orbital instability). In fact, in the SBMFA, where the action is reduced to an effective non-interacting one near the Fermi energy [5] or in a dynamical-mean field approximation in which the interaction is treated exactly at one site in an effective medium, the effective model has SU(4) symmetry if the symmetric form H_2^s is taken.

4. Effective generalized Heisenberg interactions

When two nearest-neighbor sites are singly occupied and if $U \gg t_i$ (as it seems to be the case for FePc/Au(111) [5], the hopping terms connecting these sites can be eliminated by means of a canonical transformation, in a similar fashion as the $t-J$ model is derived from the Hubbard model. This leads to an effective exchange model for spins and orbitals, as in the Kugel-Khomskii model [9, 10]. For simplicity we write first the result using the SU(4) symmetric form of the interaction H_2^s. After a lengthy but straightforward algebra we obtain

$$\begin{aligned} H_H &= \sum_i \sum_\nu \left[\tfrac{4t_2^2}{U}\mathbf{S}_i^\nu \cdot \mathbf{S}_{i+\nu}^\nu + \tfrac{4t_1^2}{U}\mathbf{S}_i^{\bar{\nu}}.\mathbf{S}_{i+\nu}^{\bar{\nu}}\right] + \sum_i \sum_\nu \left[\tfrac{t_2^2+t_1^2}{4U}\left(4T_i^z T_{i+\nu}^z - 3\right)\right] \\ &\quad + \sum_i \sum_\nu \tfrac{2t_1t_2}{U}\left(T_i^- T_{i+\nu}^+ + T_i^+ T_{i+\nu}^-\right)\left(\tfrac{1}{2} + 2\mathbf{S}_i \cdot \mathbf{S}_{i+\nu}\right), \end{aligned} \tag{3}$$

where $\mathbf{S}_i^\nu = \sum_{\alpha\beta} h_{i\alpha}^{\nu\dagger}\sigma_{\alpha\beta}h_{i\beta}^\nu/2$ is the spin of the orbital ν at site i, and $\mathbf{S}_i = \sum_\nu \mathbf{S}_i^\nu$. The operator $\mathbf{T}_i$ denotes the orbital SU(2) pseudospin (with the identification of x for pseudospin up). The subscript $i+\nu$ denotes the nearest neighbor of i in the $+\nu$ direction. Note that in the case $t_1 = t_2 = t$, this Hamiltonian reduces to

$$H_H^{SU(4)} = \frac{2t^2}{U}\sum_i\sum_\nu\left(2\mathbf{S}_i\cdot\mathbf{S}_{i+\nu}+\frac{1}{2}\right)\left(2\mathbf{T}_i\cdot\mathbf{T}_{i+\nu}+\frac{1}{2}\right)+\text{cte}. \tag{4}$$

where we have used that $\mathbf{S}_i^\nu\cdot\mathbf{S}_{i+\nu}^\nu+\mathbf{S}_i^{\bar\nu}\cdot\mathbf{S}_{i+\nu}^{\bar\nu}=\mathbf{S}_i\cdot\mathbf{S}_{i+\nu}\left(2T_i^zT_{i+\nu}^z+1/2\right)$. This Hamiltonian is a sum of products of a spin SU(2) invariant form (first factor) times a pseudospin SU(2) invariant (last factor). Thus, it is explicitly SU(2)×SU(2) invariant. However, it has been shown [13] that the symmetry of $H_H^{SU(4)}$ is actually SU(4), which is larger than SU(2)×SU(2).

When t_1 and t_2 are very different, as in the realistic case for for FePc/Au(111) [5], the first two terms of H_H Eq. (3) are the most important ones. The first one is optimized for orbital ferromagnetic and spin antiferromagnetic order, while the second one favors orbital antiferromagnetic order. In a classical picture, the energy of both orders would be the same, $-(t_1^2+t_2^2)/U$ per site. However when Hund rules are included [the form H_2, Eq. (2) is used for the interaction] the spin ferromagnetic order is favored. Projecting over intermediate double occupied triplet states, the dominant term of H_H takes the form

$$H_H^d = \frac{J}{2}\sum_{\mathbf{r}_{ij},\mathbf{a}}\left(-1+4T_{\mathbf{r}_{ij}}^zT_{\mathbf{r}_{ij}+\mathbf{a}}^z\right)\left(\frac{3}{4}+\mathbf{S}_{\mathbf{r}_{ij}}.\mathbf{S}_{\mathbf{r}_{ij}+\mathbf{a}}\right),\qquad J=\frac{t_1^2+t_2^2}{2(U-3J_H)} \tag{5}$$

5. Instabilities due to H_H

The SBMFA is valid at temperatures $T \ll T_K$ and when the degeneracy N is sufficiently large, although good results are obtained already for $N = 2$ [15]. In the present case, where the experimental temperature $T_{\text{exp}} = 0.4$ K $\approx 0.1T_K$ and $N = 4$, we expect it to be quantitatively valid. However, the simplest SBMFA of H [5] is not enough to treat the magnetic and orbital instabilities. As it is usually done in mean-field treatments of the Kondo lattice, in which the RKKY interaction should be included explicitly [16], in this section we consider $H + H_H^d$ within the SBMFA to study the instabilities induced by H_H^d. In this approximation, the limit $U \to +\infty$ is taken with a constraint of forbidden double occupancy. However, the effects of a finite U are considered explicitly in H_H.

As before [5], the hole operators are written in terms of auxiliary particles as $h_{\mathbf{r}_{ij},\sigma}^\nu = b_{\mathbf{r}_{ij}}^\dagger f_{\mathbf{r}_{ij},\sigma}^\nu$. The spin and pseudospin operators take the form

$$\mathbf{S}_{\mathbf{r}_{ij}} \equiv -\frac{1}{2}\sum_{\sigma,\nu} f_{\mathbf{r}_{ij},\sigma}^{\nu\dagger}\hat\sigma_{\sigma,\sigma'}f_{\mathbf{r}_{ij},\sigma'}^{\nu},\qquad \mathbf{T}_{\mathbf{r}_{ij}} \equiv -\frac{1}{2}\sum_{\sigma,\nu} f_{\mathbf{r}_{ij},\sigma}^{\nu\dagger}\hat\tau_{\nu,\nu'}f_{\mathbf{r}_{ij},\sigma}^{\nu'}, \tag{6}$$

We also define the mixed operator $U_{\mathbf{r}_{ij}}^z \equiv S_{\mathbf{r}_{ij}}^zT_{\mathbf{r}_{ij}}^z = \frac{1}{4}\sum_{\sigma,\nu}\eta(\sigma)\eta(\nu)f_{\mathbf{r}_{ij},\sigma}^{\nu\dagger}f_{\mathbf{r}_{ij},\sigma}^{\nu}$, where $\eta(\uparrow)=\eta(x)=+1$ and $\eta(\downarrow)=\eta(y)=-1$, and where the constraint of no double occupation has been used.

Without loss of generality, we can assume that the ferromagnetic spin order occurs along the $\hat z$ axis. Then H_H^d can be written as

$$H_H^d = \frac{J}{2}\sum_{\mathbf{r}_{ij},\mathbf{a}}\left[-S_{\mathbf{r}_{ij}}^zS_{\mathbf{r}_{ij}+\mathbf{a}}^z+3T_{\mathbf{r}_{ij}}^zT_{\mathbf{r}_{ij}+\mathbf{a}}^z+4U_{\mathbf{r}_{ij}}^zU_{\mathbf{r}_{ij}+\mathbf{a}}^z\right]+\text{cst}, \tag{7}$$

We now introduce the following Hubbard-Stratonovich (HS) decouplings

$$-\frac{J}{2}\sum_{\mathbf{r}_{ij},\mathbf{a}} S^z_{\mathbf{r}_{ij}}.S^z_{\mathbf{r}_{ij}+\mathbf{a}} \rightarrow \sum_{\mathbf{r}_{ij},\mathbf{a}}\left[\tfrac{1}{2J}m_{\mathbf{r}_{ij}}m_{\mathbf{r}_{ij}+\mathbf{a}} + \tfrac{1}{2}m_{\mathbf{r}_{ij}+\mathbf{a}}S^z_{\mathbf{r}_{ij}} + \tfrac{1}{2}m_{\mathbf{r}_{ij}}S^z_{\mathbf{r}_{ij}+\mathbf{a}}\right], \tag{8}$$

$$\frac{3J}{2}\sum_{\mathbf{r}_{ij},\mathbf{a}} T^z_{\mathbf{r}_{ij}}T^z_{\mathbf{r}_{ij}+\mathbf{a}} \rightarrow \sum_{\mathbf{r}_{ij},\mathbf{a}}\left[-\tfrac{1}{6J}n_{\mathbf{r}_{ij}}n_{\mathbf{r}_{ij}+\mathbf{a}} + \tfrac{1}{2}n_{\mathbf{r}_{ij}+\mathbf{a}}T^z_{\mathbf{r}_{ij}} + \tfrac{1}{2}n_{\mathbf{r}_{ij}}T^z_{\mathbf{r}_{ij}+\mathbf{a}}\right], \tag{9}$$

$$2J\sum_{\mathbf{r}_{ij},\mathbf{a}} U^z_{\mathbf{r}_{ij}}U^z_{\mathbf{r}_{ij}+\mathbf{a}} \rightarrow \sum_{\mathbf{r}_{ij},\mathbf{a}}\left[-\tfrac{8}{J}p_{\mathbf{r}_{ij}}p_{\mathbf{r}_{ij}+\mathbf{a}} + 4p_{\mathbf{r}_{ij}}U^z_{\mathbf{r}_{ij}+\mathbf{a}} + 4p_{\mathbf{r}_{ij}+\mathbf{a}}U^z_{\mathbf{r}_{ij}}\right], \tag{10}$$

where $m_{\mathbf{r}_{ij}}$ is related to the usual magnetization order parameter, $n_{\mathbf{r}_{ij}}$ is the orbital order parameter and $p_{\mathbf{r}_{ij}}$ is a magneto-orbital order parameter. In order for Eqs. (9) and (10) to be well-defined HS decouplings, the fields $n_{\mathbf{r}_{ij}}$ and $p_{\mathbf{r}_{ij}}$ should have antiferromagnetic order. Introducing the Fourier decompositions

$$m_{\mathbf{r}_{ij}} = \sum_{\mathbf{q}}\frac{e^{i(\mathbf{Q}_m+\mathbf{q})\mathbf{r}_{ij}}}{\sqrt{N}}m_{\mathbf{q}},\quad n_{\mathbf{r}_{ij}} = \sum_{\mathbf{q}}\frac{e^{i(\mathbf{Q}_n+\mathbf{q})\mathbf{r}_{ij}}}{\sqrt{N}}n_{\mathbf{q}},\quad p_{\mathbf{r}_{ij}} = \sum_{\mathbf{q}}\frac{e^{i(\mathbf{Q}_p+\mathbf{q})\mathbf{r}_{ij}}}{\sqrt{N}}p_{\mathbf{q}}, \tag{11}$$

where $\mathbf{Q}_m = (0,0)$ and $\mathbf{Q}_n = \mathbf{Q}_p = \mathbf{Q} = \left(\frac{\pi}{a_0},\frac{\pi}{a_0}\right)$, H^d_H becomes

$$\begin{aligned} H^d_H &= \sum_{\mathbf{q},\mathbf{a}} e^{-i\mathbf{q}\mathbf{a}}\left[\frac{\beta}{2J}m_{\mathbf{q}}m_{-\mathbf{q}} - m_{\mathbf{q}}\sum_{\nu}\frac{1}{\sqrt{N}}\sum_{\mathbf{k},\omega_n}\frac{1}{2}\left(\bar{f}^{\nu}_{\mathbf{k},\uparrow}f^{\nu}_{\mathbf{k}-\mathbf{q},\uparrow} - \bar{f}^{\nu}_{\mathbf{k},\downarrow}f^{\nu}_{\mathbf{k}-\mathbf{q},\downarrow}\right)_{\omega_n}\right] \\ &+ \sum_{\mathbf{q},\mathbf{a}} e^{-i\mathbf{q}\mathbf{a}}\left[\frac{\beta}{6J}n_{\mathbf{q}}n_{-\mathbf{q}} - n_{\mathbf{q}}\sum_{\sigma}\frac{1}{\sqrt{N}}\sum_{\mathbf{k},\omega_n}\frac{1}{2}\left(\bar{f}^{x}_{\mathbf{k},\sigma}f^{x}_{\mathbf{k}-(\mathbf{Q}+\mathbf{q}),\sigma} - \bar{f}^{y}_{\mathbf{k},\sigma}f^{y}_{\mathbf{k}-(\mathbf{Q}+\mathbf{q}),\sigma}\right)_{\omega_n}\right] \\ &+ \sum_{\mathbf{q},\mathbf{a}} e^{-i\mathbf{q}\mathbf{a}}\left[\frac{\beta 8}{J}p_{\mathbf{q}}p_{-\mathbf{q}} - p_{\mathbf{q}}\sum_{\sigma\nu}\frac{1}{\sqrt{N}}\sum_{\mathbf{k},\omega_n}2\eta(\sigma)\eta(\nu)\left(\bar{f}^{\nu}_{\mathbf{k},\sigma}f^{\nu}_{\mathbf{k}-(\mathbf{Q}+\mathbf{q}),\sigma}\right)_{\omega_n}\right], \end{aligned} \tag{12}$$

The minimum of the free energy with respect to the order parameters defines the saddle-point equations

$$m_{\mathbf{q}} = J\sum_{\nu}\frac{1}{\beta N}\sum_{\mathbf{k},\omega_n}\frac{1}{2}\left\langle \bar{f}^{\nu}_{\mathbf{k},\uparrow}(i\omega_n)f^{\nu}_{\mathbf{k}+\mathbf{q},\uparrow}(i\omega_n) - \bar{f}^{\nu}_{\mathbf{k},\downarrow}(i\omega_n)f^{\nu}_{\mathbf{k}+\mathbf{q},\downarrow}(i\omega_n)\right\rangle, \tag{13}$$

$$n_{\mathbf{q}} = 3J\sum_{\sigma}\frac{1}{\beta N}\sum_{\mathbf{k},\omega_n}\frac{1}{2}\left\langle \bar{f}^{x}_{\mathbf{k},\sigma}(i\omega_n)f^{x}_{\mathbf{k}-(\mathbf{Q}-\mathbf{q}),\sigma}(i\omega_n) - \bar{f}^{y}_{\mathbf{k},\sigma}(i\omega_n)f^{y}_{\mathbf{k}-(\mathbf{Q}-\mathbf{q}),\sigma}(i\omega_n)\right\rangle, \tag{14}$$

$$p_{\mathbf{q}} = \frac{J}{4}\sum_{\sigma\nu}\frac{1}{\beta N}\sum_{\mathbf{k},\omega_n}2\eta(\sigma)\eta(\nu)\left\langle \bar{f}^{\nu}_{\mathbf{k},\sigma}(i\omega_n)f^{\nu}_{\mathbf{k}-(\mathbf{Q}-\mathbf{q}),\sigma}(i\omega_n)\right\rangle. \tag{15}$$

In order to determine which of these order parameters is the first to develop, we compute the change of sign in the second derivative of the free energy in the symmetric phase. This leads to the following conditions for the critical value of J for each instability (similar to Stoner criteria)

$$\frac{1}{J_{c,m}} = \chi(\mathbf{0},0),\quad \frac{1}{J_{c,n}} = 3\chi(\mathbf{Q},0),\quad \frac{1}{J_{c,p}} = 4\chi(\mathbf{Q},0), \tag{16}$$

where we have defined the static susceptibility of the electronic system,

$$\chi(\mathbf{q},0) \equiv \chi(\mathbf{q},\omega=0) = -\frac{4}{\beta N}\sum_{\mathbf{k},\omega_n} \frac{G^{ff}_{\nu,\sigma}(\mathbf{k},i\omega_n) - G^{ff}_{\nu,\sigma}(\mathbf{k}+\mathbf{q},i\omega_n)}{\epsilon_{\mathbf{k}} - \epsilon_{\mathbf{k}+\mathbf{q}}}, \tag{17}$$

and the pseudofermion Green function is given in [5]

The problem now reduces to computing Eq. (17) for a generic $\mathbf{q}$. Performing the Matsubara sum at $T = 0$ we obtain

$$\chi(\mathbf{q},0) = \frac{4}{N}\sum_{\mathbf{k}} \frac{1}{\pi} \frac{\arctan\left(\frac{E_h+\lambda+\epsilon_{\mathbf{k}}}{z^2\Gamma}\right) - \arctan\left(\frac{E_h+\lambda+\epsilon_{\mathbf{k}+\mathbf{q}}}{z^2\Gamma}\right)}{\epsilon_{\mathbf{k}} - \epsilon_{\mathbf{k}+\mathbf{q}}}, \tag{18}$$

where $z = \left\langle b^{\dagger}_{\mathbf{r}_{ij}} \right\rangle$, λ is a Lagrange multiplier used to impose the constraint $z^2 + \sum_{\sigma,\nu=x,y} f^{\nu\dagger}_{\mathbf{r}_{ij},\sigma} f^{\nu}_{\mathbf{r}_{ij},\sigma} = 1$, W is half the band width and Γ the resonant level width of the isolated impurity. This function is represented in Fig. 1

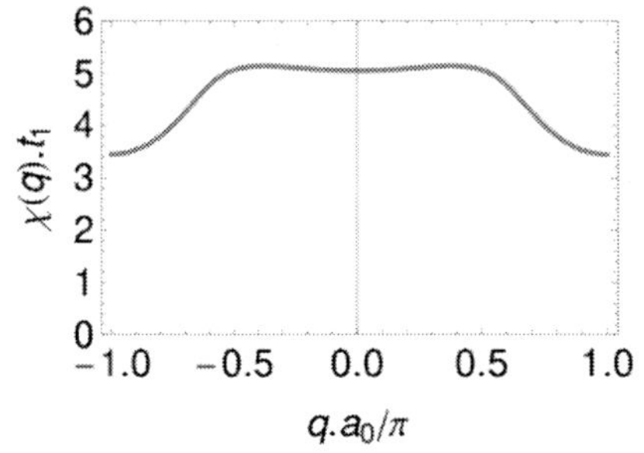

Figure 1. Susceptibility $\chi(\mathbf{q},0)$ vs $\mathbf{q}$ in the (1,1) direction.

For $\mathbf{q} \to 0$ we recover the usual expression $\chi(\mathbf{0},0) = \rho^t_f(0)$, where $\rho^t_f(0)$ is the total spectral density (summing over spin and orbital) at the Fermi level. For the parameters λ, z that minimize the SBMFA action, we obtain $\chi(0,0) = 5.05/t_1$, $\chi(\mathbf{Q},0) = 3.45/t_1$ which replaced in Eq. (16) lead to the following critical values ($t_1 = 0.007$ eV)

$$J_{c,m} \approx 16.1\text{K}, \quad J_{c,n} \approx 7.9\text{K}, \quad J_{c,p} \approx 5.9\text{ K}. \tag{19}$$

We conclude that the first instability occurs in the magneto-orbital channel.

For FePc/Au(111) one can estimate $t_1 = 0.007$ eV, $t_2 = 3t_1$ and $U = 1.6$ eV [5]. The exchange constant J_H is difficult to estimate for effective molecular orbitals. For pure Fe orbitals it is of the order of 0.7 eV. From these estimates and the second Eq. (5), one expects $J \geq 1.8$ K. Thus, one might infer that the system is not too far from an instability against spin ferro- and orbital antiferro-magnetic order.

6. RKKY interactions

Another possible source of magnetic instabilities is the RKKY interaction, which consists in the indirect interaction I between spins mediated by conduction electrons generated by the effective Kondo coupling between spins and conduction electrons $J_K \mathbf{S}_i \cdot \mathbf{s}_i$ at second order in J_K, where $\mathbf{s}_i$ is the spin of the conduction electrons at site i.

An advantage of Au and its (111) surface is that both the bulk sates and the surface Shockley states near the Fermi energy can be described as free electrons and therefore the calculations in Ref. [17] for three dimensions (3D) and Ref. [18] for the 2D case are valid. Following these works one can write for dimension N=2 or 3 for two spins $\mathbf{S}_1$ and $\mathbf{S}_2$ at a distance R (we will consider nearest neighbors only)

$$H_{RKKY}(R) = I_{\rm ND}(R)\,\mathbf{S}_1.\mathbf{S}_2, \qquad I_{\rm ND}(R) = -\frac{1}{4}\tilde{J}^2_{\rm ND}\chi_{\rm ND}(R), \tag{20}$$

where $\tilde{J}_{\rm 3D} = J_K V_{\rm at}$, $\tilde{J}_{\rm 2D} = JS_{\rm at}$, $V_{\rm at}$ ($S_{\rm at}$) is the volume (surface) per Au atom in the bulk (surface) and $\chi_{\rm ND}(R)$ is the spin susceptibility, given by Eqs. (14) and (15) of Ref. [18]:

$$\chi_{\rm 3D}(R) = -\rho_{\rm 3D}(\epsilon_F)\frac{4k_F^3}{\pi}F_{\rm 3D}(2k_FR), \chi_{\rm 2D}(R) = -\rho_{\rm 2D}(\epsilon_F)k_F^2F_{\rm 2D}(k_FR), \tag{21}$$

where $\rho_{\rm 3D}(\epsilon_F) = mk_F/2\pi^2\hbar^2$ ($\rho_{\rm 2D}(\epsilon_F) = m^*/2\pi\hbar^2$) is the density of states per spin and per unit volume (surface), and

$$F_{\rm 3D}(x) \equiv \frac{x\cos - \sin x}{x^4}, \quad F_{\rm 2D}(x) \equiv J_0(x)Y_0(x) + J_1(x)Y_1(x) \xrightarrow[x\to\infty]{-} \frac{\sin(2x)}{\pi x^2}, \tag{22}$$

with $J_\nu(x)(Y_\nu(x))$ the Bessel function of the first (second) kind.

For the more realistic 3D case, using the value $k_F = 1.21$ Å^{-1} for Au [19], one obtains $\rho_{\rm 3D}(\epsilon_F) = 0.00805/$ (eV Å^3). The density per atom and spin projection is $\rho = \rho_{\rm 3D}(\epsilon_F)V_{\rm at} = 0.137/\text{eV}$, where we have used $V_{\rm at} = 17.0$ Å^3 (the lattice parameter of f.c.c. Au is $a = 4.08$ Åand $V_{\rm at} = a^3/4$). Keeping the product $\tilde{J}_{\rm 3D}\rho_{\rm 3D}(\epsilon_F) = J_K\rho$ that leads to the observed $T_K \simeq W\exp[-1/(2J_K\rho)] = 4.5$ K, with $W = 1/(2\rho)$ for the SU(4) impurity Kondo model, one obtains $J_K \simeq 0.4$ eV.

Using the above equations with $|F_{\rm 3D}(x)| \leq 1/x^3$ for large x and $R = 14.7$ Åfor the intermolecular distance [4] we obtain

$$|I_{\rm 3D}| \leq (JV_{\rm at})^2\rho_{\rm 3D}(\epsilon_F)\frac{1}{\pi(2R)^3} = 0.05\text{K}. \tag{23}$$

For 2D, the effective mass of the surface Shockley states is $m^* = 0.28m$ [19]. This leads to $\rho_{\rm 2D}(\epsilon_F) = 0.00589/\left(\text{eVÅ}^2\right)$. Using $S_{\rm at} = \sqrt{3}a^2/4 = 7.21$ Å^2 one obtains $\rho = \rho_{\rm 2D}(\epsilon_F)S_{\rm at} = 0.0425/\text{eV}$. Knorr *et al.* have shown that the bulk states dominate the hybridization with the impurity [20]. Assuming (as an overestimation) that half of the contribution to $J_K\rho$ is due to surface states leads to $J_K \simeq 0.65$ eV. From Eq. (22) for large x, $|F_{\rm 2D}(x)| \leq 1/\left(\pi x^2\right)$, and using Eqs. (20) and (21) for $R = 14.7$ Åwe obtain

$$|I_{\rm 2D}| \leq (JS_{\rm at})^2\rho_{\rm 2D}(\epsilon_F)\frac{1}{4\pi R^2} = 0.54\text{K}. \tag{24}$$

A calculation that follows the same steps as done in the previous section shows that to have a ferromagnetic instability in the system $I_{\rm ND} < 0$, with $|I_{\rm ND}| > I_c = J_{cm} \approx 16.1$ K. Therefore, magnetic instabilities driven by the RKKY interaction are unlikely for FePc/Au(111).

7. Summary and discussion

We have studied the magnetic and orbital instabilities of a model used before to explain the scanning tunneling spectroscopy (STS) of a system of FePc molecules on Au(111). The model generalizes to the lattice the SU(4) Anderson model and is expected to have emergent SU(4) symmetry at low energies.

We find that due to effective generalized exchange interactions originated by the hopping terms, the system is close to a combined instability of spin ferromagnetic and orbital antiferromagnetic character. Due to this combined character it is possible that the application of a magnetic field induces not only a finite magnetization but also a checkerboard orbital ordering, which might be observed by STS if the tip is not radially symmetric.

Acknowledgments

AML acknowledges support form JQI-NSF-PFC. AAA is partially supported by CONICET, Argentina. This work was sponsored by PICT 2010-1060 and 2013-1045 of the ANPCyT-Argentina and PIP 112-201101-00832 of CONICET.

References

[1] Bogani L and Wernsdorfer W, 2008 *Nat. Mater.* **7**, 179
[2] Bader S D, 2006 *Rev. Mod. Phys.* **78**, 1
[3] Minamitani E *et al* 2012 *Phys. Rev. Lett.* **109**, 086602
[4] Tsukahara N, *et al* 2011 *Phys. Rev. Lett.* **106**, 187201
[5] Lobos A M, Romero M A, and Aligia A A 2014 *Phys. Rev. B* **89**, 121406(R)
[6] Garnica M *et al* 2013 *Nat. Phys.* **9**, 368
[7] Umbach T R *et al* 2012 *Phys. Rev. Lett.* **109**, 267207
[8] Abdurakhmanova N *et al* 2013 *Phys. Rev. Lett.* **110**, 027202
[9] Khomskii D I and Kugel K I 1973 *Solid State Commun.* **13**, 763
[10] Aligia A A and Gusmão M A, 2004 *Phys. Rev. B* **70**, 054403
[11] Aligia A A and Kroll T, 2010 *Phys. Rev. B* **81**, 195113
[12] Sbaih M A A *et al* 2013 *EJTP* **10**, 9
[13] Li Y Q *et al* 1998 *Phys. Rev. Lett.* **80**, 3527
[14] Batista C D and Ortiz G 2004 *Adv. in Phys.* **53**, 1
[15] Hewson A C, in *The Kondo Problem to Heavy Fermions* (Cambridge, University Press, 1993).
[16] Coqblin B, *et al* 2003 *Phys. Rev. B* **67**, 064417
[17] Kittel C, *Quantum Theory of Solids* (Wiley, New York, 1987).
[18] Béal-Monod M T, 1987 *Phys. Rev. B* **36**, 064417
[19] Kevan S D and Gaylord R H, 1987 *Phys. Rev. B* **36**, 5809
[20] Knorr N *et al* 2002 *Phys. Rev. Lett.* **88**, 096804

27th International Conference on Low Temperature Physics (LT27) IOP Publishing
Journal of Physics: Conference Series **568** (2014) 052003 doi:10.1088/1742-6596/568/5/052003

Impurity states and Localization in Bilayer Graphene: the Low Impurity Concentration Regime

H. P. Ojeda Collado[1], **Gonzalo Usaj**[1], **C. A. Balseiro**[1]

[1] Centro Atómico Bariloche and Instituto Balseiro, Comisión Nacional de Energía Atómica, 8400 Bariloche, and CONICET, Argentina

E-mail: hpablo1988@gmail.com

Abstract. We study the problem of non-magnetic impurities adsorbed on bilayer graphene in the diluted regime. We analyze the impurity spectral densities for various concentrations and gate fields. We also analyze the effect of the adsorbate on the local density of states (LDOS) of the different C atoms in the structure and present some evidence of strong localization for the electronic states with energies close to the Dirac point.

The problem of adatoms in graphene has been the subject of an intense activity for they could be used to modify and control the electronic properties of the material. Diluted adatoms or molecules necessarily generates disorder [1] and in some regimes may lead to strong localization of the electronic states [2, 3, 4, 5, 6, 7]. Electron localization in graphene is quite peculiar, Dirac fermions tend to elude localization in systems with Anderson-type disorder. However impurities leading to short range disorder at the atomic scale generate inter-valley mixing and break the symplectic symmetry opening the route to strong localization. The problem of adatoms and electron localization in bilayer graphene (BLG), although considered by several groups [11, 12], received much less attention [10, 13]. In the most common structure of BLG, known as the Bernal stacking, only one of the two non-equivalent sites (A, B) of the honeycomb lattice of the top layer lies on top of a site of the bottom layer. The resulting structure, shown in Fig. 1, induces a weak coupling of the two layers. The unit cell has four carbon atoms leading to four bands, two of them having a parabolic dispersion relation around the K and K' points of the Brillouin Zone, touch each other at the Fermi energy.

In most of the experimental setups, BLG lies on top of a substrate and the impurities are adsorbed on the top layer only. Due to the difference of the A and B sites of the layer, there is a small difference in the absorption energy on the two inequivalent sublattices. This difference favors absorption on the B sites and in what follows we assume that all impurities are on the B sublattice. A very interesting aspect of BLG is its response to a gate field [14, 15, 16, 17, 18]. An electric field perpendicular to the layers opens a gap at the Fermi level an effect that can be used to modify the impurity states. Here we study the problem of non-magnetic impurities adsorbed on BLG in the diluted regime. We analyze the impurity spectral densities for various concentrations and gate fields. The effect of the adsorbate on the local density of states (LDOS) of the different C atoms in the structure is analyzed and we present some evidence of strong localization for the electronic states with energies close to the Dirac point.

The Hamiltonian of the system includes the bilayer Hamiltonian H_{BLG}, the impurity contribution H_{imp} and the hybridization term H_{hyb}. In what follows, as there are no spin effects we ignore the spin index.

$$H_{BLG} = -\sum_{j,\boldsymbol{k}}[V(-1)^j(a^\dagger_{j\boldsymbol{k}}a_{j\boldsymbol{k}} + b^\dagger_{j\boldsymbol{k}}b_{j\boldsymbol{k}}) + t(\phi(\boldsymbol{k})\,a^\dagger_{j\boldsymbol{k}}b_{j\boldsymbol{k}} + \phi^*(\boldsymbol{k})\,b^\dagger_{j\boldsymbol{k}}a_{j\boldsymbol{k}})] - \sum_{\boldsymbol{k}} t_\perp(a^\dagger_{1\boldsymbol{k}}b_{2\boldsymbol{k}} + b^\dagger_{2\boldsymbol{k}}a_{1\boldsymbol{k}}) \quad (1)$$

here $a_{j\boldsymbol{k}}$ and $b_{j\boldsymbol{k}}$ destroy electrons with wavevector $\boldsymbol{k}$ in sublattices A and B respectively, the subindex $j = 1$ ($j = 2$) refers to the top (bottom) plane. V is the potential induced by the gate voltage, t and $t_\perp$ are the intra and inter-plane hoppings, respectively and $\phi(\boldsymbol{k}) = \sum_{\boldsymbol{\delta}} e^{i\boldsymbol{k}\cdot\boldsymbol{\delta}}$ where $\{\boldsymbol{\delta}\}$ are the three vectors connecting one site with its neighbors in the same plane. We consider adatoms that are bounded to a single C atom, $H_{imp} = \sum_l' \varepsilon_0 f_l^\dagger f_l$ with f_l the destruction operator of an electron on the impurity orbital of the adatom at site l, ε_0 is the energy of the orbital and the sum runs over the sites of the carbon lattice having an impurity on top. Finally $H_{hyb} = \gamma \sum_l'(f_l^\dagger b_{1,l} + b^\dagger_{1,l} f_l)$ here $b_{1,l} = 1/\sqrt{N}\sum_{\boldsymbol{k}} e^{i\boldsymbol{k}\cdot\boldsymbol{R}_l} b_{1,\boldsymbol{k}}$ where $\boldsymbol{R}_l$ is the coordinate of site l. The parameters used to describe the BLG are $t = 0.25$ eV, $t_\perp = t/9$ and we take $\gamma = 2t$ [19, 20]. Three values of the bias voltage V are taken; $V = -0.02t$, $V = 0$eV, $V = +0.02t$ and without any loss of generality we take $\varepsilon_0 \geq 0$.

We first present results for the impurity contribution $\rho_{imp}(\omega)$ to the total density of states (DOS). We use the Chebyshev polynomials method which has proven to be very efficient to deal with realistic impurity concentrations [10, 21, 22].

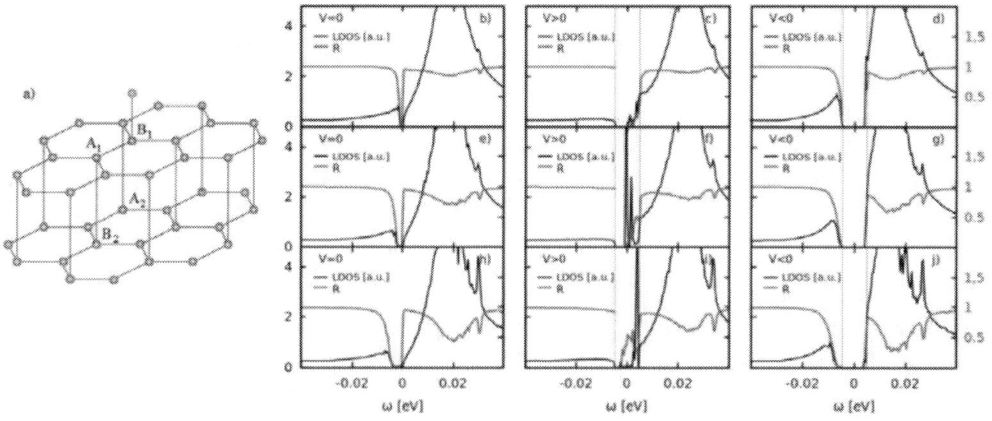

Figure 1. Structure of the bilayer graphene (a) and impurity spectral densities (b-j), left scale. Upper, central and lower panels correspond to impurity concentrations 1/1000, 1/500 and 1/250 respectively. The three columns correspond to unbiased BLG (left), positive (centre) and negative (right) bias. The red thick line is the function $R(\omega)$, right scale, see text.

The average impurity spectral density is then given by $\rho_{imp}(\omega) = -\frac{1}{\pi}\langle \mathrm{Im}\mathcal{G}_{ll}\rangle_{\mathrm{avg}}$ where $\mathcal{G}_{ll}$ is the retarded impurity propagator and $\langle \ldots \rangle_{\mathrm{avg}}$ indicates the configurational average over the impurities. The results for three different concentrations are shown in Fig 1. There, the three columns correspond to different values of the gate voltage V, the rows to different impurity concentrations. Interestingly, for $V = 0$ the impurity spectral density shows a gap close to the Dirac point that increases with increasing impurity concentration. This effect is reminiscent of the gap induced in monolayer graphene when impurities are adsorbed in a single sublattice [23]. For a non-zero gate voltage V, the pristine BLG develops a gap at the Dirac point (indicated by vertical lines in the figure). For positive V the gap is partially filled by impurity states. For large impurity concentrations the gap closes, while for small concentrations a reduced gap remains. In the thermodynamic limit and for all cases discussed above, we expect the gaps to be just pseudogaps with exponentially small DOS [24]. The case of negative V is completely different: the gap of the pristine BLG remains unaltered for small or moderate values of $|V|$. This effect can be understood by looking at the response of a single impurity to the gate field [25]. In the single impurity case and for $\varepsilon_0 \neq 0$ and small values of V a bound state appears in the gap only for one polarity of the field.

To better understand the nature of the electronic states and the effect of the gates, we calculate the LDOS at the different C atoms, in the upper and lower layer. Results for different concentrations and polarities of the bias field are shown in Fig 2. In all cases the impurity contribution to the LDOS for small energies is large at the $A1$ sublattice, an effect that is also characteristic of impurities adsorbed on top of a single $B1$-carbon atom in monolayers [20].

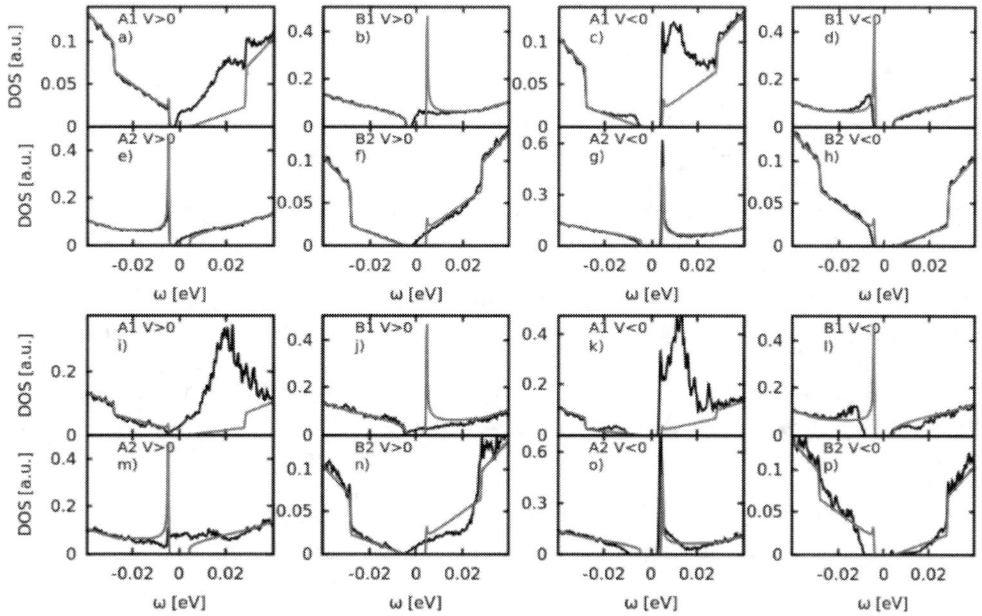

Figure 2. LDOS of the four sublattices for impurity concentrations 1/1000 (upper panels) and 1/250 (lower panels) and different polarities of the bias field. Smooth curves correspond to the pristine sample. Note the different scales of the panels

For low concentrations the LDOS of the other sublattices—namely $B1$, $A2$ and $B2$—present only small modifications. In particular for $V > 0$ the valence band remains essentially unaltered with its $1D$-like van Hove singularity in the $A2$ sublattice. This strongly suggests that at least for this polarity there is no strong localization of the electronic states in the valence band. We could draw similarly conclusions from the structure of the LDOS of the $A2$ sublattice for $V < 0$ as illustrated in the upper panels of Fig 2. As the concentration increases (lower panel of the figure), the modifications of the LDOS become more important and it is necessary to look for a better and more qualitative criterion for localization. To this end we evaluate the function

$R(\omega) = \rho_{typ}(\omega)/\rho(\omega)$ where [26]

$$\rho_{typ}(\omega) = \left[\prod_{l=1}^{N} \rho_l(\omega)\right]^{\frac{1}{N}} \quad \text{and} \quad \rho(\omega) = \frac{1}{N}\sum_{l=1}^{N} \rho_l(\omega) \,. \tag{2}$$

Here N is the number of impurities and $\rho_l(\omega) = -\frac{1}{\pi}\mathrm{Im}\mathcal{G}_{ll}$ is the spectral density of the lth impurity. When, for a given energy ω, the states are extended $\rho_l(\omega)$ has small fluctuations and $\rho_{typ}(\omega) \approx \rho(\omega)$ resulting in $R(\omega) \approx 1$. In contrast, if states are localized $\rho_{typ}(\omega)$ is dominated by small values of $\rho_l(\omega)$ and $R(\omega) \ll 1$, with $R(\omega) \to 0$ as $N \to \infty$. We stress that our analysis is based only on the properties of the impurity spectral densities and gives a very qualitative estimation of the energy range where we may expect strong localization effects. While, as mentioned above, the Chebyshev polynomials method is very efficient to evaluate the average spectral densities, it becomes numerically costly to evaluate a large number of $\rho_l(\omega)$. To estimate $R(\omega)$ we then resort to the method described in Ref. [6]. The impurity propagator matrix $\boldsymbol{\mathcal{G}}$ with matrix elements $\mathcal{G}_{ij}(\omega)$ satisfies the Dyson equation

$$\left[(\omega + i0^+ - \varepsilon_0)\boldsymbol{I} - \gamma^2\tilde{\boldsymbol{g}}\right]\boldsymbol{\mathcal{G}} = \boldsymbol{I} \,, \tag{3}$$

where $\boldsymbol{I}$ is the unit matrix and $\tilde{\boldsymbol{g}}$ is a matrix whose elements are the propagators of pristine biased BLG, $g_{B1i,B1j}(\omega)$, between impurity sites adsorbed on the $B1$ sublattice. For large distances and low frequencies we evaluate $g_{B1i,B1j}(\omega)$ in the continuous limit [27]. The required spectral densities are the imaginary part of the diagonal terms of the matrix $\boldsymbol{\mathcal{G}}$.

Results for $R(\omega)$ are presented in Fig. 1. They show that strong localization effects are to be expected for energies very close to the Dirac point. In the case of positive bias, when the gate induced gap is filled by impurity states, these states in the gap are strongly localized. Our results suggest that, independently of the bias and for large impurity concentration, localization effects are also to be expected in the energy window close to the maximum of the impurity LDOS. This last effect is also observed in monolayer graphene although the localization length may be quite different in the two systems. Away from these energies, only weak localization effects are likely. A more quantitative estimation of the localization phenomena requires evaluation of the localization length.

As a final remark we mention that for the parameters used in the present work, that are suitable to describe fluorine on graphene, and the small bias voltages of the figures, a single impurity on an $A1$ sites does not generate a bound state inside the gap for positive V, for negative V there is a bound state exponentially close to the gap edge. As a consequence, a small amount of impurities added to the $A1$ sublattice would not change the results obtained for small energies (within the field induced gap).

In summary, we have presented results for diluted impurities on BLG with and without gate voltages that open a gap in the pristine sample. We have shown the existence of drastic effects of the polarity of the electric field. For impurity parameters appropriate to describe fluorine on BLG, the gap induced in the pristine sample for positive polarity is filled with strongly localized states. Conversely, for negative polarity the gap remains. In all cases, strong localization occurs only at low energies.

We acknowledge useful discussions with J. Sofo and financial support from PICT Bicentenario 2010-1060 from ANPCyT, PIP 11220080101821 and 11220110100832 from CONICET and 06/C400 and 06/C415 SeCyT-UNC. HPOC acknowledges scholarship from ICTP.

References

[1] S. Das Sarma, Shaffique Adam, E. H. Hwang, and Enrico Rossi. *Rev. Mod. Phys.*, 83(2), 470 (2011).
[2] P. M. Ostrovsky, I. V. Gornyi, and A. D. Mirlin. *Phys. Rev. B*, 74, 235443 (2006).

[3] S Gattenloehner, W R Hannes, P M Ostrovsky, I V Gornyi, A D Mirlin, and M Titov. *arXiv.org*, June 2013.
[4] S. Roche, N. Leconte, F. Ortmann, A. Lherbier, D. Soriano, and J. C. Charlier. *Solid State Comm.*, 152, 1404 (2012).
[5] X. Hong, S. H. Cheng, C Herding, and J. Zhu. *Phys. Rev. B*, 83(8), 085410 (2011).
[6] G. Usaj, P. S. Cornaglia, and C. A. Balseiro. *Phys. Rev. B*, 89, 085405 (2014).
[7] A. Cresti, F. Ortmann, T. Louvet, D. Van Tuan, and S. Roche. *Phys. Rev. Lett.*, 110, 196601 (2013).
[8] A. Weiße, G. Wellein, A. Alvermann, and H. Fehske. *Rev. Mod. Phys.*, 78, 275 (2006).
[9] L. Covaci, F. Peeters, and M. Berciu. *Phys. Rev. Lett.*, 105, 167006 (2010).
[10] S. Yuan, H. De Raedt, and M. I. Katsnelson. *Phys. Rev. B*, 82, 115448 (2010).
[11] E. McCann, and M. Koshino, *Rep. Prog. Phys.*, 76, 056503 (2013)
[12] J. Nilsson and A.H. Castro Neto, *Phys. Rev. Lett.*, 98, 126801 (2007)
[13] A.H. Castro Neto, F. Guinea, N. M. R. Peres, K. S. Novoselov, and A. K. Geim, *Rev. Mod. Phys.*, 81, 109 (2009).
[14] E. McCann and V.I. Falko, *Phys. Rev. Lett.*, 96, 086805 (2006).
[15] F. Guinea, A. H. Castro Neto, and N. M. R. Peres, *Phys. Rev. B.*, 73, 245426 (2006).
[16] E. McCann, *Phys. Rev. B.*, 74, 161403(R) (2006).
[17] T. Ohta, A. Bostwick, T. Seyller, K. Horn, and E. Rotenberg, *Science*, 313, 951 (2006).
[18] E. V. Castro, K. S. Novoselov, S. V. Morozov, N. M. R. Peres, J. M. B. Lopes dos Santos, J. Nilsson, F. Guinea, A. K. Geim, and A. H. Castro Neto, *Phys. Rev. Lett.*, 99, 216802 (2007).
[19] J. O. Sofo, G. Usaj, P. S. Cornaglia, A. M. Suarez, A. D. Hernndez-Nieves, C. A. Balseiro, *Phys. Rev. B.*, 85, 115405 (2012).
[20] J. O. Sofo, A. M. Suarez, G. Usaj, P. S. Cornaglia, A. D. Hernndez-Nieves, C. A. Balseiro, *Phys. Rev. B.*, 83, 081411(R) (2011).
[21] A. Weisse, G. Wellein, A. Alvermann, and H. Fehske, *Rev. Mod. Phys.*, 78, 275 (2006)
[22] L. Covaci, F. Peeters, and M. Berciu, *Phys. Rev. Lett.*, 105, 167006 (2010).
[23] Vitor M. Pereira, J. M. B. Lopes dos Santos and A. H. Castro Neto, *Phys. Rev. B.*, 77, 115109 (2008).
[24] V. V. Mkhitaryan, and M. E. Raikh, *Phys. Rev. B.*, 78, 195409 (2008)
[25] V. V. Mkhitaryan, E. G. Mishchenko, *Phys. Rev. Lett.*, 110, 086805 (2013).
[26] J. H. Garcia, B. Uchoa, L. Covaci, and T. G. Rappoport, *arXiv*:1309.2306
[27] F. Parhizgar, M. Sherafati, R. Asgari, and S. Satpathy, *Phys. Rev. B.*, 87, 165429 (2013)

Journal of Physics: Conference Series **568** (2014) 052004 doi:10.1088/1742-6596/568/5/052004

Low temperature transport measurements on atomically smooth metallic and oxygen deficient strontium titanate

C S Barquist, I H Kwak, J Bauer, T Edmonds, A Biswas and Y Lee

Department of Physics, University of Florida, Gainesville, Florida, USA

Abstract. Atomically smooth, TiO_2 terminated $SrTiO_3$ (STO) substrates were prepared using a combination of chemical and thermal annealing treatments. The TiO_2 terminated surface was obtained by etching with aqua regia solution and thermal annealing at 1000 °C for 30 min. The subsequent vacuum annealing at 830 °C for 10 min generated an atomically smooth and metallic surface of STO. In this paper, we report low temperature transport measurements down to 50 mK on these samples which clearly exhibit a metallic temperature dependence in the resistance. The samples show no sign of superconductivity down to the lowest temperatures.The $R_\square(T)$ data provide information on the physical origin of metallic behavior in STO, which might also be relevant to the current research interest in oxide interfaces.

1. Introduction

For the past decade $SrTiO_3$ (STO) has been on the vanguard of complex oxide physics. Perhaps the most intriguing discovery is the existence of a metallic state at the heterointerface of $LaAlO_3/SrTiO_3$, which also displays Shubnikov-de Hass oscillations [1]. This junction was later shown to undergo a superconducting transition around 200 mK [2]. The authors demonstrate that the superconducting transitions is Kosterlitz-Thouless-like indicating the two dimensional nature of the electron gas at the interface. Interestingly, a two dimensional electron gas has been observed on the surface of cleaved STO in vacuum, using angle resolved photo emission spectroscopy [3]. Metal-insulator transitions have also been observed in bulk and thin film STO [4, 5]. A superconducting ground state has also been observed in bulk, oxygen deficient STO for carrier concentrations as low as $5.5 \times 10^{17} cm^{-3}$ [4]. Apart from the interesting physics of intrinsic STO, it is also valuable as a substrate for growing materials such as thin films of perovskite oxides. A prerequisite for epitaxial crystal growth is an atomically smooth substrate. There are several methods for making atomically smooth STO [6, 7, 8]. However, producing atomically smooth metallic STO has proven to be a formidable task. Doping STO by vacuum annealing for long periods of time ($\sim$ 1-10 hr) has led to rough surfaces [9], unacceptable for epitaxy.

We have produced atomically smooth TiO_2 terminated samples that exhibit metallic behavior down to 50 mK. The procedure is relatively simple and the results are reproducible. The annealing time used is relatively short, so it is expected that only a thin film on the surface of the substrate is conducting. In this paper we present transport measurements on two samples prepared in the same manner and an estimation of the conducting layer thickness.

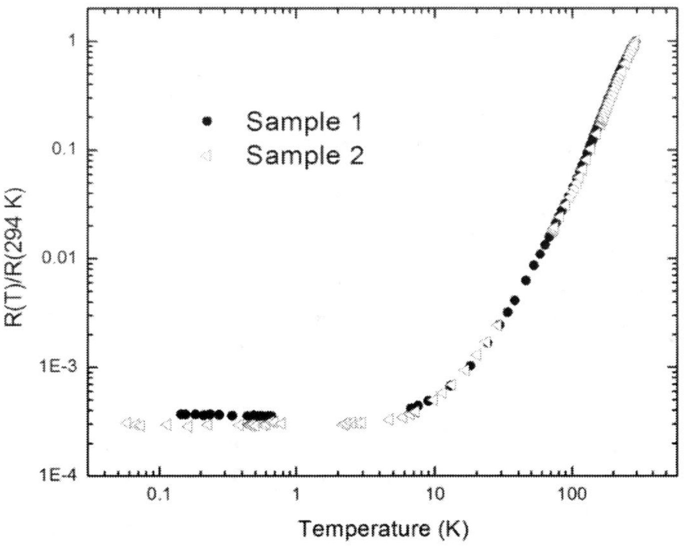

Figure 1. Temperature dependence of the normalized resistance for two samples by the resistance at 294 K. $R(294\ K)$ is 34.8 Ω and 19.8 Ω for sample 1 and 2, respectively.

2. Sample Preparation

The samples are prepared from commercially available STO substrates from the CrysTec company. The substrates used in this study are 0.5 mm thick and oriented in the (100) direction. The samples are treated further to obtain the atomically smooth TiO_2 terminated surface with a conducting layer. The atomically smooth TiO_2 layer was obtained using a wet-etch technique [6]. The untreated samples are first cleaned with ethanol for 5 min. They are then soaked in 50 °C DI water for 30 min. After the soak the samples are etched with an aqua regia solution (3:1 HCl to HNO_3) for 12 min. The soaking and etching take place in a sonicator. After etching, the samples are rinsed with DI water. Additional cleaning with acetone and ethanol is preformed before annealing the samples in air at 1000 °C for 30 min. In order to obtain the metallic surface the samples are annealed at 830 °C for 10 min. in vacuum (10^{-7} Torr). Vacuum annealing is known to cause oxygen deficiency in STO, and the oxygen vacancies introduce charge carriers [9]. Removal of an oxygen atom from STO leaves behind a doubly charged (+) oxygen vacancy and introduces two free electrons. Therefore, it is expected that the charge carriers from oxygen vacancy doping are electrons. This has been confirmed by preliminary Hall effect measurements which indicate negative charge carriers. This result will be published elsewhere. Samples made using this process have been characterized by AFM and LFM. The samples have an rms surface roughness of about 1.7 Å with wide terraces of unit cell height (3.9 Å).

3. Metallic Layer Thickness

One quantity of particular interest is the thickness of the conducting layer. Currently no experiments have been done to directly determine this quantity. However, the thickness can be estimated from the diffusion constant of oxygen in STO and the annealing time:

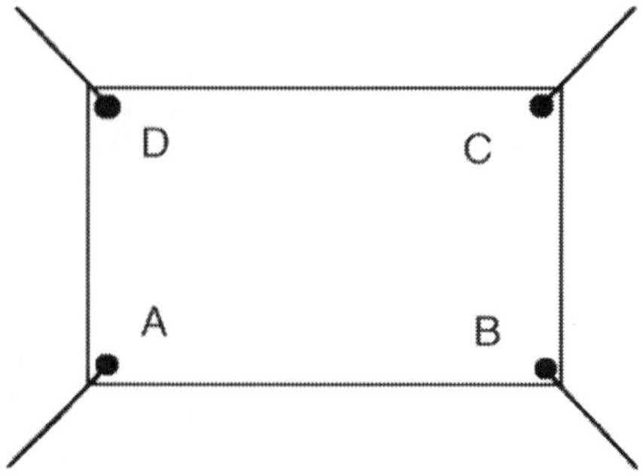

Figure 2. Resistance measurement scheme: $R = V_{CD}/I_{AB}$ and $R' = V_{AD}/I_{BC}$. As an example, V_{CD} indicates the voltage measured across contacts C and D. Sheet resistance is plotted against temperature in Fig. 3.

$$\ell = \sqrt{Dt}, \tag{1}$$

where ℓ is the thickness of the metallic layer, D is the diffusion constant, and t is the annealing time. The diffusion constant of oxygen ions in single crystal STO has been measured by Paldino *et al.* [10]. For the temperature used in annealing the samples, one can estimate the diffusion constant to be 5×10^{-11} cm^2 s^{-1}, which combined with an annealing time of 10 min gives a thickness of roughly 2 μm.

4. Resistance Measurements

Resistance measurements were made on two samples. Both samples were prepared using the same method described above. The resistance was measured down to 145 mK for one sample and 58 mK for the other. The resistance was measured using a standard four wire technique. The measurements were made with a 13 Hz ac driving current and the voltage across the sample was measured using a lock-in referenced at that frequency. At low temperature, the excitation current was 9.7 μA. This current was chosen to minimize heating to the sample. The resistance of each sample was measured while cooling and warming the cryostat. Each sample was measured on a different run of the cryostat. The whole range of R vs. T for both samples is shown in Fig. 1. The samples show metallic behavior down to the lowest temperatures measured. The RRR of both samples, inferred form Fig. 1, is about 3000, indicating high purity. In Fig. 1 the resistance of each sample is normalized by the resistance value at room temperature: $R(294\ K)$ is 34.8 Ω for sample 1 and 19.8 Ω for sample 2. The dependence on sample size and lead placement is removed by taking the ratio of the resistance to a value at a reference temperature. The two curves lie on top of each other for most of the temperature range and only begin to deviate below ~ 1 K, implying that this process produces reproducible metallic samples.

It was observed that in a thin film of vacuum annealed STO with a thickness ~ 2.5 μm a metal-insulator transition, attributed to carrier freeze-out, occurred at 78 K [5]. The thin film was annealed at 950 °C at ~ 10^{-7} Torr for one hour, where it is expected that the distribution of carriers in the film is homogenous. The expected conducting film thickness of our samples is similar to that of the aforementioned thin film. However, we do not observe a clear metal-insulator transition in our samples down to the lowest temperature, although sample 1 exhibits a slight increase in resistance below 0.7 K of about 3% indicating the presence of weak localization.

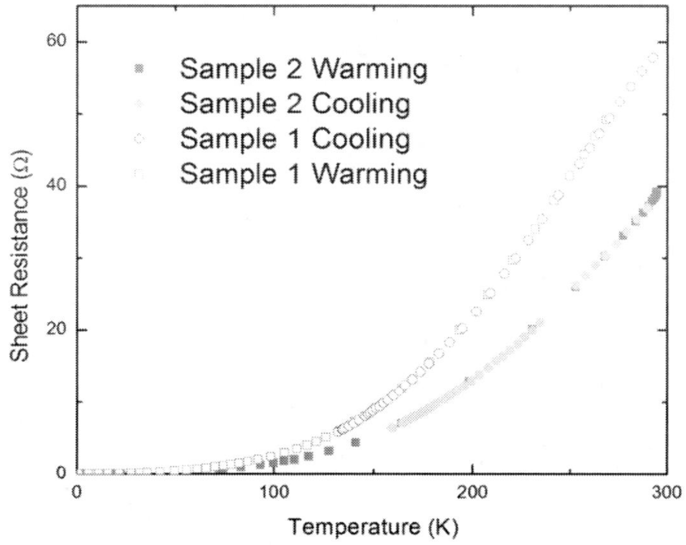

Figure 3. Sheet resistance $R_\square$ of both samples. The upper curve is for sample 1, and the lower curve is for sample 2. Warming and cooling data are included for both graphs.

This behavior might be caused by a large gradient in charge carrier density where the metallic behavior is confined to a small layer on the surface.

The sheet resistance of both samples was calculated, at room temperature, using the van der Pauw method [11]:

$$R_\square = \frac{\pi}{\ln(2)} \frac{1 + R'/R}{2} f \left[\frac{R}{R'} \right] R, \tag{2}$$

R and R' are defined in Fig. 2 and f is a slowly varying function given in [11]. The sheet resistance at temperatures lower than room temperature is inferred from the ratio of sheet resistance to sample resistance at room temperature. The value of the ratio $R_\square/R$ is 1.69 and 1.98 for sample 1 and 2, respectively. It is assumed that the change of this ratio with temperature is negligible. The sheet resistance for the whole temperature range is shown in Fig. 3.

5. Conclusion

It was shown that samples of TiO_2 terminated STO which were subsequently vacuum annealed at 830 °C for 10 min show metallic behavior down to 58 mK. After normalization the resistances curves of both samples fall directly onto of one another, indicating the reproducibility of the procedure. It was also estimated, based on the diffusion of oxygen atoms in STO, that the thickness of the conducting layer is approximately 2 μm. Although, it is more likely the metallic behavior is contained in a thinner layer near the surface

Acknowledgments
This work is supported by NSF DMR-1205891 (YL).

References

[1] Ohtomo A and Hwang H Y 2004 *Nature* **427** 423-6.
[2] Reyren N *et al.* 2007 *Science* **317** 1196-9.
[3] Santander-Syro *et al.* 2011 *Nature* **469** 189-93.
[4] Lin X, Zhu X, Fauqué B and Behnia K 2013 *Phys. Rev. X* **3** 021002.
[5] Liu Z Q *et al* 2011 *Phys. Rev. Lett.* **107** 146802.
[6] Kareev M, Prosandeev S, Liu J, Gan C, Kareev A, Freeland J W, Xian M and Chakhalian J 2008 *Appl. Phys. Lett.* **93** 061909.
[7] Connell J G, Isaac B J, Ekanayake G B, Strachan D R and Seo S S A 2012 *Appl. Phys Lett.* **101** 251607.
[8] Kawasaki M, Takahashi K, Maeda T, Tsuchiya R, Shinohara M, Ishiyama O, Yonezawa T, Yoshimoto M and Koinuma H 1994 *Science* **226** 1540-2.
[9] Hanzig J *et al.* 2011 *J. Appl. Phys.* **110** 064107.
[10] Paladino A E, Rubin L G and Waugh J S 1965 *J. Phys. Chem. Solids* **26** 391-7.
[11] van der Pauw L J 1958 *Philips Res. Repts.* **13** 1-9.

Journal of Physics: Conference Series **568** (2014) 052005 doi:10.1088/1742-6596/568/5/052005

Finite frequency noise for Laughlin state investigated by a resonant circuit

M Carrega[1], D Ferraro[2,3], A Braggio[1] and M. Sassetti[4,1]

[1] SPIN-CNR, Via Dodecaneso 33, 16146 Genova, Italy

[2] Aix Marseille Université, CNRS, CPT, UMR 7332, 13288 Marseille, France

[3] Université de Toulon, CNRS, CPT, UMR 7332, 83957 La Garde, France

[4] Dipartimento di Fisica, Università di Genova, Via Dodecaneso 33, 16146 Genova, Italy

E-mail: `alessandro.braggio@spin.cnr.it`

Abstract. We study the finite frequency (F.F.) noise properties of edge states in the Laughlin state. We investigate the model of a resonant detector coupled to a quantum point contact in the weak-backscattering limit. In particular we discuss the impact of possible renormalization of the Luttinger exponent, due to environmental effects, on the measured quantities and we propose a simple way to extract such non-universal parameters from noise measurements.

1. Introduction

Strong correlations in low dimensional systems can lead to very intriguing effects such as charge fractionalization and non-Fermi liquid behaviour. A remarkable example is the fractional quantum Hall effect [1], where quasiparticles carry fractional charges and have non-trivial statistics [2, 3, 4, 5, 6]. Transport measurements in a quantum point contact (QPC) geometry are among the simplest ways to probe these peculiar properties. In particular, by studying current fluctuations in the shot noise limit, one can directly measure the charge of the excitations involved in the tunnelling through the QPC [7, 8, 9]. Indeed, the zero frequency current-current correlation in the weak-backscattering regime, is predicted to be proportional to the induced backscattering current via the fractional charge associated to the tunnelling excitation between the opposite edges of the Hall bar [2, 3, 4, 10]. Clear experimental signatures of this fact have been reported for the Laughlin sequence [5], with filling factor $\nu = 1/(2n+1)$ ($n \in \mathbb{N}$). Here the measured fractional charge is $e^* = e/(2n+1)$ (e the electron charge), in agreement with theoretical predictions [7, 8, 9].

In the case of composite edges, such as in the Jain sequence [6] with $\nu = p/(2np+1)$ ($p \in \mathbb{Z}$), the situation is more involved and, at low energies, various excitations with different fractional charges can contribute to the tunnelling. Indeed, different experiments have reported the observation of a crossover in the value of the effective charge (as a function of temperature) which has been explained in terms of competition between different excitations, namely the single quasiparticle (QP) (with fundamental charge $e^* = \nu e/|p|$) and p- agglomerates (with charge νe) [11, 12, 13, 14, 15, 16, 17]. Furthermore, renormalizations of the chiral Luttinger liquid exponents are usually needed to fully explain this phenomenology [11, 12, 13, 14, 15, 16, 17, 18], and many mechanisms responsible for these renormalizations were proposed in literature [18, 19, 20, 21, 22]. Even for the simple Laughlin sequence renormalization parameter for the Luttinger

exponent are often introduced to fully explain anomalous current-voltage characteristics reported at very low temperature [23, 24].
An important tool in order to gain more information about the properties of fractional excitations is represented by finite frequency current correlations [25, 26, 27]. For example, for quantum Hall QPC transport, the F.F. noise is predicted to show resonances in correspondence of Josephson frequencies, which are proportional to the fractional charges [26, 27, 28, 29, 30, 31, 32]. In this case the presence of fractionally charged tunnelling excitations may be revealed, at extremely low temperatures, by the presence of peaks or dips in the noise spectrum occurring at frequency $\omega \backsim e^*V/\hbar$. The study of F.F current fluctuations [31, 32, 33] can also serve as an alternative tool to probe the power-law behaviour expected for these chiral Luttinger liquids.
However, at these energy scales the detection scheme has to be considered with care, in order to properly identify which quantity is effectively probed in real experiments.
In this context Lesovik and Loosen [34] introduced a model based on a resonant LC circuit as prototypical scheme for F.F. noise measurement. It has been shown that the measured quantity for the LC detector setup can be expressed in terms of the non-symmetrized F.F. noise which reflects the emission and adsorption contributions of the active system under investigation, *i.e.* the QPC [33, 34, 35, 36]. The non-symmetrized noise has been considered in literature for different systems as the ultimate source of information of quantum noise properties [37, 38, 39, 40, 41, 42].
Here we consider the F.F. detector output power of a resonant circuit coupled to a QPC in the fractional quantum Hall regime, as discussed in Ref. [33]. We analize symmetric and measured noise as a function of bias and temperature at a fixed frequency ω. For sake of simplicity we limit the discussion to the case of the Laughlin sequence with $\nu = 1/(2n+1)$. In particular we will consider possible renormalizations of the Luttinger exponent, due to the interaction with the external environment, extending the previous work of Ref. [33], to investigate the robustness of the relevant features associated to F.F. current fluctuations and clarify the range of validity of the results discussed in Ref. [33].

2. Theoretical model

The field theoretical description of the edge states in the Laughlin sequence [5], with filling factor $\nu = 1/(2n+1)$ is given in terms of a single bosonic mode through the Lagrangian density

$$\mathcal{L} = -\frac{1}{4\pi\nu}\partial_x\varphi(\partial_t + v_c\partial_x)\varphi, \tag{1}$$

where $\varphi(x)$ is a right-moving field with commutation relation

$$\left[\varphi(x), \varphi(x')\right] = i\pi\nu\text{sign}(x - x') \ . \tag{2}$$

This bosonic mode propagates along the edge at velocity v_c and is related to the electron particle density by

$$\rho(x) = \frac{1}{2\pi}\partial_x\varphi(x). \tag{3}$$

It is possible to show that the single QP is the most relevant excitation in the tunnelling process [4] and, using bosonization techniques [43], we can express the single QP operator as

$$\Psi(x) = \frac{\mathcal{F}}{\sqrt{2\pi\alpha}}e^{i\nu\varphi(x)} \tag{4}$$

being α a finite length cut-off and $\mathcal{F}$ the so-called Klein factor [10]. The fundamental charge associated to this excitation is $e^* = \nu e$, with scaling dimension [4]

$$\Delta = g\frac{\nu}{2} \tag{5}$$

being $g \geq 1$ a possible renormalization factor associated to external interactions like, among the others, the coupling with one dimensional phonon mode, the Coulomb interaction at the QPC or the interplay between $1/f$ noise and dissipation generated by the circuit [18, 19, 20, 21, 22].

2.1. Measured finite frequency noise

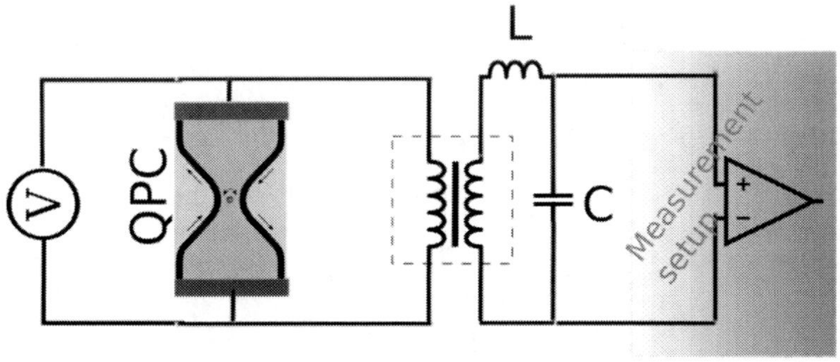

Figure 1. Schematic view of an Hall bar (blue) with a QPC coupled with LC detector. A bias voltage V is applied to the QPC and QP excitations can tunnel between the edges. The two circuits are impedance matched via a coupling circuit (inside dashed line).

We consider the F.F. backscattering current fluctuations in a single QPC geometry coupled to the detector setup shown in Fig. 1. Here the QPC is subjected to a bias voltage V and coupled to a resonant LC circuit, the detector (with frequency $\omega = \sqrt{1/LC}$), via an impedance matching circuit (inside the dashed line in the figure) [44, 45]. We assume a small electrical coupling $K \ll 1$ between the QPC and the detector [33, 34], with a very high quality factor and we keep constant the resonant frequency ω as usually done in experiments. The point-like tunnelling process of single QP between the upper and lower edge of the Hall bar through the QPC can be described by the tunnelling Hamiltonian

$$\hat{\mathcal{H}}_T = t\ \Psi(0)\Psi^{\dagger}(0) + h.c.\ \ , \tag{6}$$

where t represents the tunnelling amplitude, which is assumed energy-independent for sake of simplicity. A finite bias voltage V is included in the tunnelling amplitude through the usual gauge transformation $t \to te^{i\omega_0 t}$, where $\omega_0 = e^*V/\hbar$ is the Josephson resonance associated to the fundamental charge e^* [10].
The non-symmetrized noise [34, 35, 37] is defined in terms of the backscattering current fluctuations

$$S_+(\omega) = \frac{1}{2}\int_{-\infty}^{+\infty} dt\, e^{i\omega t}\langle \delta I_B(0)\delta I_B(t)\rangle, \tag{7}$$

and represents, for $\omega > 0$, the noise emission of the system into the detector. In the above expression we considered the back-scattering current fluctuation $\delta I_B = I_B - \langle I_B\rangle$ with the average $\langle ...\rangle$ taken over the quantum statistical ensemble. Using the time-reversal symmetry properties of Eq. (7) we can deduce the absorptive part of the spectrum as

$$S_-(\omega) = S_+(-\omega)\ . \tag{8}$$

Usually the theoretical study of F.F. properties relies on the so-called symmetrized noise, *i.e.*

$$S_{sym}(\omega) = S_+(\omega) + S_-(\omega) \ , \tag{9}$$

however the measurable quantity in the scheme of Fig. 1 is the output power proportional to the variation of the energy stored in the LC before and after the switching on of the LC-QPC coupling. In the following we refer to it as measured noise $S_{meas}(\omega)$ and, in terms of the non-symmetrized noise, it reads

$$S_{meas}(\omega) = K\left\{S_+(\omega) + n_B(\omega)\left[S_+(\omega) - S_-(\omega)\right]\right\} \ , \tag{10}$$

with $n_{\mathrm{B}}(\omega) = \left[e^{\beta_c\omega} - 1\right]^{-1}$ the boson distribution and $\beta_c = 1/k_B T_c$ the detector inverse temperature. We recall that, at the lowest order in the tunnelling Hamiltonian, the non-symmetrized emission noise can be written in terms of the single QP tunnelling rate as [10, 35]

$$S_+(\omega, \omega_0) = \frac{(e^*)^2}{2}\left[\mathbf{\Gamma}\left(-\omega + \omega_0\right) + \mathbf{\Gamma}\left(-\omega - \omega_0\right)\right], \tag{11}$$

where

$$\mathbf{\Gamma}(E) = |t|^2 \int_{-\infty}^{+\infty} d\tau \ e^{iEt} \mathcal{G}_-^<(-t)\mathcal{G}_+^>(t), \tag{12}$$

with $\mathcal{G}_\pm^\gtrless(t) = \langle \Psi_{\nu,\pm}(t)\Psi_{\nu,\pm}^\dagger(0)\rangle = (\mathcal{G}_\pm^\lessgtr(-t))^*$ the greater/lesser correlation functions of the single QP for the edge $j = \pm$. For the Laughlin sequence these correlation functions are [14, 15]

$$\mathcal{G}_\pm^>(t) = \frac{1}{2\pi\alpha}\left[\frac{|\Gamma(1 + (\beta\omega_c)^{-1} - it/\beta)|^2}{\Gamma^2(1 + (\beta\omega_c)^{-1})(1 \pm i\omega_c t)}\right]^{2\Delta}, \tag{13}$$

being $\Gamma(x)$ the Euler gamma function and $\omega_c = v_c/a$ the high frequency cut-off. This leads, in the zero temperature limit, to the simple expression [46, 47]

$$\mathbf{\Gamma}(E) = \frac{|t|^2}{2\pi\alpha^2}\left(\frac{E}{\omega_c}\right)^{4\Delta} \frac{E^{-1}}{\Gamma(2g\nu)}\Theta(E) \propto \Theta(E)E^{4\Delta - 1} \tag{14}$$

with $\Theta(x)$ the Heaviside step function.

It is worth to note that the exponent depends on the scaling dimension of the QP operator and hence is affected by the renormalization parameter $g \geq 1$.

3. Results and discussion

In the Ref. [33] one can find a complete discussion of the proposed detection scheme in the general contest of the fractional quantum Hall states. Here we limit only to discuss F.F. symmetrized and measured noise in the Laughlin sequence. For sake of clarity we focus on the state at $\nu = 1/3$ which, as stated before, carries a tunnelling charge $e^* = e/3$. For this state we investigate the effects of the renormalization of the Luttinger parameter on the behaviour of the physical quantities. We consider the very low temperature regime for both the QPC and the detector ($T = T_c \sim 15$ mK)[1] and we keep fixed the frequency ω scanning on the bias ω_0. This will allow us to be closer to realistic experimental conditions.

In Fig. 2 a) we show the behaviour of S_{sym} as a function of ω_0/ω. In absence of renormalization

[1] Note that for $T = T_c$ the measured noise coincides with the excess noise power $S_{ex}(\omega, \omega_0) = S_{meas}(\omega, \omega_0) - S_{meas}(\omega, \omega_0 = 0)$, a quantity that is extracted keeping fixed the coupling K. This is experimentally easier since the critical step is the frequency dependence of the fine tuning in the impedance matching circuit.

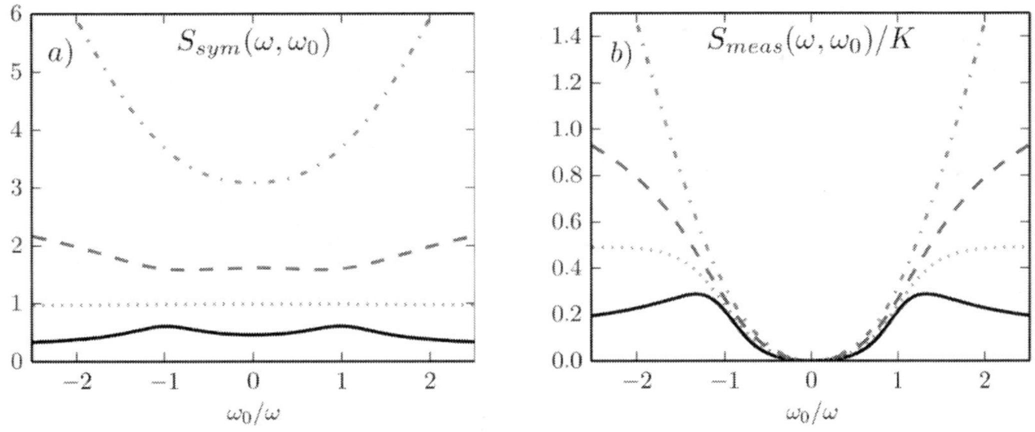

Figure 2. (Color online) a) Symmetrized noise $S_{sym}(\omega, \omega_0)$ (in units of $S_0 = e^2|t|^2/((2\pi\alpha)^2\omega_c)$) as a function of ω_0/ω and b) measured noise $S_{meas}(\omega, \omega_0)/K$ (in units of S_0) as a function of ω_0/ω. The figures are for $\nu = 1/3$ and for different renormalizations of the Luttinger parameter: $g = 1$ black full, $g = 1.5$ red dotted, $g = 2$ blue dashed and $g = 3$ green dashed dotted. Other parameters are: $T = T_c = 15$ mK, $\omega = 60$ mK.

($g = 1$, black full curve) we observe a two peaks structure associated to the resonances at the Josephson frequency ($\omega_0/\omega = \pm 1$) [31, 33]. By increasing the renormalization we first find a completely flat regime ($g = 1.5$, red dotted curve) that turns into a double-dipped structure ($g = 2$, blue dashed curve). At even higher renormalization ($g = 3$, green dashed dotted curve) every feature associated to the Josephson resonances is washed out and the signal increase at high voltage with a progressively increasing power-law. For not so strong renormalization this phenomenology can be easily explained by looking at the behaviour of the noise near the resonances. Indeed, according to Eq. (14), around these peaks ($\omega_0/\omega \approx \pm 1$), up to a smearing of the singularities associated to thermal effects, the noise is symmetric and scale as

$$S_{sym}(\omega, \omega_0) \approx |\omega_0 - \omega|^{2\nu g - 1} \tag{15}$$

in agreement with what shown in Fig. 2 a). However, when the power-law growth is too strong, all the relevant features near Josephson resonances are washed away.

Another remarkable characteristic of S_{sym} is its finite value at $\omega_0/\omega \approx 0$. This contribution, which scale as a power law with temperature, is essentially due to the so-called zero point fluctuations [34] which are taken into account by the symmetric combination of the tunnelling rates in Eq. (9).

Different is the situation for what it concerns S_{meas} shown in Fig. 2 b). In this case all the curves start from zero at zero voltage due to the absence of contribution from zero point fluctuations. All curves are very similar for $|\omega_0/\omega| \leq 1$, where the physics is essentially dominated by thermal effects. Then, for $\omega_0 - \omega \gg k_B T$, they follow a similar power-law behaviour around the resonance

$$S_{meas}(\omega, \omega_0) \approx (\omega_0 - \omega)^{2\nu g - 1}. \tag{16}$$

as already discussed for the symmetrized noise.

In order to better understand the main differences between S_{meas} and S_{sym} it is useful to consider the ratio

$$\mathcal{R}(\omega, \omega_0) = \frac{S_{meas}(\omega, \omega_0)}{K S_{sym}(\omega, \omega_0)}. \tag{17}$$

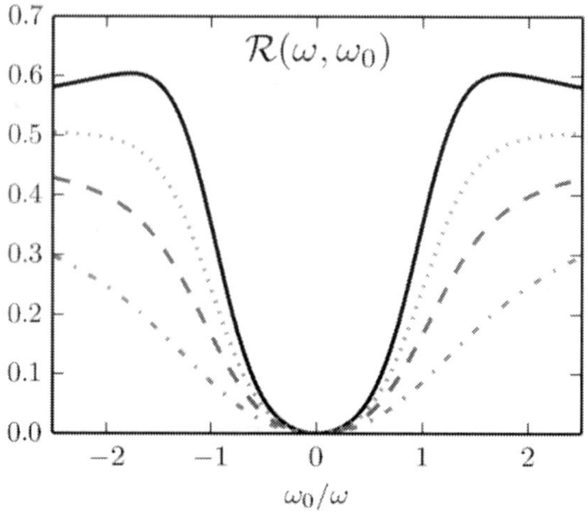

Figure 3. (Color online) Ratio $\mathcal{R}(\omega, \omega_0)$ between $S_{meas}(\omega, \omega_0)/K$ and $S_{sym}(\omega, \omega_0)$ for $\nu = 1/3$ and different values of the renormalization: $g = 1$ black full, $g = 1.5$ red dotted, $g = 2$ blue dashed and $g = 3$ green dashed dotted. Other parameters are: $T = T_c = 15$ mK.

In the zero temperature limit ($T = T_c \to 0$) this quantity reduces to

$$\mathcal{R}(\omega, \omega_0) \approx \frac{S_+(\omega, \omega_0)}{[S_+(\omega, \omega_0) + S_-(\omega, \omega_0)]}. \tag{18}$$

Using the expression for the tunnelling rate at zero temperature in Eq. (14), one obtains

$$\mathcal{R}(\omega, \omega_0) \approx \begin{cases} 0 & |\omega_0/\omega| < 1 \\ \left[\frac{1}{2} + \left(\frac{1}{2} - g\nu\right) \left|\frac{\omega}{\omega_0}\right|\right] & |\omega_0/\omega| \gg 1 \end{cases}. \tag{19}$$

At finite temperature, keeping $T = T_c$ these two asymptotic behaviours remains valid respectively for $\omega_0/\omega \approx 0$ and $\omega_0/\omega \gg 1$ as clearly shown is Fig. (3), while the other regions of the curves are rounded due to thermal effects. According to the previous considerations, at extremely high voltage with respect to the measurement frequency the ratio reaches the asymptotic value $\mathcal{R} \approx 1/2$ and the way to approach it, namely from above or from below, crucially depends on the renormalization parameter. In particular for $\nu < 1/2g$ (black full curve) this ratio presents two maxima symmetric with respect to $\omega_0/\omega = 0$, while for $\nu > 1/2g$ the curves increase (decrease) monotonically for positive (negative) voltage. Expression in Eq. (18) therefore suggests that measurement of S_{sym} and S_{meas} in the same setup could provide an estimation of the non universal renormalization of the Luttinger parameters.[2]

4. Conclusions

We have studied the finite frequency (F.F.) noise properties of edge states in the Laughlin state. We have considered a realistic measuring setup based on a resonant circuit coupled to a quantum point contact in the fractional Hall regime. Symmetric and measured F.F. noise have been investigated, properly discussing the effects of possible renormalization of the Luttinger exponents on these observables, mainly focusing on the robustness of the relevant peaked or dipped structures. We finally considere their ratio as a useful way to access the value of the non universal renormalization parameter.

[2] Note that, in general, the measurement of S_{sym} requires a different detection scheme.[25, 29, 36]

Acknowledgements
We acknowledge the support of the MIUR-FIRB2012 - Project HybridNanoDev (Grant No.RBFR1236VV) and the EU FP7/2007-2013 under REA grant agreement no 630925 – COHEAT.

References

[1] Das Sarma S and Pinczuk A 1997 *Perspective in Quantum Hall Effects: Novel Quantum Liquid in Low-Dimensional Semiconductor Structures* (New York: Wiley)
[2] Wen X G 1995 *Adv. Phys.* **44** 405
[3] Kane C L, Fisher M P A and Polchinski J 1994 *Phys. Rev. Lett.* **72** 4129
[4] Kane C L and Fisher M P A 1995 *Phys. Rev. B* **51** 13449
[5] Laughlin R B 1983 *Phys. Rev. Lett.* **50** 1395
[6] Jain J K 1989 *Phys. Rev. Lett.* **63** 199
[7] de Picciotto R, Reznikov M, Heiblum M, Umansky V, Bunin G and Mahalu D 1997 *Nature* 389 162
[8] Saminadayar L, Glattli D C, Jin Y and Etienne B 1997 *Phys. Rev. Lett.* 79 2526
[9] Reznikov M, de Picciotto R, Griffiths T G, Heiblum M and Umansky V 1999 *Nature* **399** 238
[10] Martin T 2005 *Course 5 Noise in mesoscopic physics* Les Houches, France
[11] Chung Y C, Heiblum M, and Umansky V 2003 *Phys. Rev. Lett.* **91** 216804
[12] Bid A, Ofek N, Heiblum M, Umansky V, and Mahalu D 2009 *Phys. Rev. Lett.* **103** 236802
[13] Dolev M, Gross Y, Chung Y C, Heiblum M, Umansky V, and Mahalu D 2010 *Phys. Rev. B* **81** 161303
[14] Ferraro D, Braggio A, Merlo M, Magnoli N and Sassetti M 2008 *Phys. Rev. Lett.* **101** 166805
[15] Ferraro D, Braggio A, Magnoli N and Sassetti M 2010 *Phys. Rev. B* **82** 085323
[16] Ferraro D, Braggio A, Magnoli N and Sassetti M 2010 *New J. Phys.* **12** 013012
[17] Carrega M, Ferraro D, Braggio A, Magnoli N and Sassetti M 2011 *Phys. Rev. Lett.* **107** 146404
[18] Braggio A, Ferraro D, Carrega M, Magnoli N and Sassetti M 2012 *New J. Phys.* **14** 093032
[19] Rosenow B and Halperin B I 2002 *Phys. Rev. Lett.* **88** 096404
[20] Papa E and MacDonald A H 2004 *Phys. Rev. Lett.* **93** 126801
[21] Mandal S S and Jain J K 2002 *Phys. Rev. Lett.* **89** 096801
[22] Yang K 2003 *Phys. Rev. Lett.* **91** 036802
[23] Roddaro S, Pellegrini V, Beltram F, Biasiol G, Sorba L, Raimondi R and Vignale G 2003 *Phys. Rev. Lett.* **90** 046805
[24] Roddaro S, Pellegrini V, Beltram F, Biasiol G and Sorba L 2004 *Phys. Rev. Lett.* **93** 046801
[25] Blanter Ya M and Büttiker M 2000 *Phys. Rep.* **336** 1
[26] Rogovin D and Scalapino D J 1974 *Ann. Phys.* **86** 1
[27] Chamon C, Freed D E and Wen X G 1995 *Phys. Rev.* B **51** 2363
[28] Chamon C, Freed D E and Wen X G 1996 *Phys. Rev.* B **53** 4033
[29] Dolcini F, Trauzettel B, Safi I and Grabert H 2005 *Phys. Rev. B* **71** 165309
[30] Bena C and Safi I 2007 *Phys. Rev. B* **76** 125317
[31] Carrega M, Ferraro D, Braggio A, Magnoli N and Sassetti M 2012 *New J. Phys.* **14** 023017
[32] Ferraro D, Braggio A, Magnoli N and Sassetti 2012 *Phys. Scr.* 014025
[33] Ferraro D, Carrega M, Braggio A, and Sassetti M 2014 *New J. Phys.* **16** 043018
[34] Lesovik G B and Loosen R 1997 *JETP Lett.* **65** 295
[35] Gavish U, Levinson Y, and Imry Y *Phys. Rev. B* **67** 10637 R
[36] Bednorz A, Bruder C, Reulet B and Belzig W 2013 *Phys. Rev. Lett.* **110** 250404
[37] Aguado R and L. P. Kouwenhoven L P 2000 *Phys. Rev. Lett.* **84**, 1986
[38] Sukhorukov E V, Burkard G and Loss D 2001 *Phys. Rev. B* **63** 125315
[39] Johansson G, Kack A and Wendin G 2002 *Phys. Rev. Lett.* **88** 046802
[40] Deblock R, Onac E, Gurevich L and Kouwenhoven L P 2003 *Science* **301** 203
[41] Billangeon P M, Pierre F, Bouchiat H and Deblock R 2006 *Phys. Rev. Lett.* **96** 136804
[42] Safi I, Bena C, Crepieux A 2008 *Phys. Rev. B* **78** 205422
[43] Miranda E, 2003*Braz. J. Phys* **33** 3
[44] C. Altimiras, O. Parlavecchio, P. Joyez, D. Vion, P. Roche, D. Esteve, F. Portier *Phys. Rev. Lett.* in press, arXiv:1403.5999
[45] C. Altimiras, O. Parlavecchio, P. Joyez, D. Vion, P. Roche, D. Esteve, F. Portier 2013 *Appl. Phys. Lett.* **103** 212601
[46] Guinea F, Gomez Santos G, Sassetti M and Ueda M 1995 *Europhys. Lett.* **30** 561
[47] Cuniberti G, Sassetti M and Kramer B 1997 *J. Phys.: Condens. Matter* **8** L21

Journal of Physics: Conference Series **568** (2014) 052006 doi:10.1088/1742-6596/568/5/052006

D.C. Josephson transport by quartets and other Andreev resonances in superconducting bijunctions.

R. Mélin[1], D. Feinberg[1], H. Courtois[1], C. Padurariu[1], A. Pfeffer[2], J.E. Duvauchelle[2], F. Lefloch[2], T. Jonckheere[3], J. Rech[3], T. Martin[3], B. Douçot[4]

[1]Université Grenoble-Alpes and CNRS, Institut Néel, Grenoble, France.
[2]SPSMS/LaTEQS, UMR-E 9001, CEA-INAC and Université Grenoble-Alpes, Grenoble, France
[3]Aix-Marseille Université, Université de Toulon, CNRS, CPT, UMR 7332, 13288 Marseille, France
[4]Laboratoire de Physique Théorique et des Hautes Energies, CNRS UMR 7589, Universités Paris 6 et 7, 4 Place Jussieu, 75252 Paris Cedex 05

E-mail: `denis.feinberg@neel.cnrs.fr`

Abstract. Bijunctions are three-terminal Josephson junctions where three superconductors are connected by a single weak link made of a metallic region or of quantum dots. Biasing two of the superconductors with commensurate voltages yields Andreev resonances that produce d.c. Josephson currents made of correlated Cooper pairs. For instance with applied voltages $(0, V, -V)$, quartets formed by two entangled Cooper pairs are emitted by one reservoir towards the two others. Theory involving non-equilibrium Green's functions reveal the microsopic mechanism at play, e.g multiple coherent Andreev reflections that provide an energy-conserving and fully coherent channel. Recent experiments on diffusive Aluminum-Copper bijunctions show transport anomalies that are interpreted in terms of quartet resonances.

1. Introduction

Josephson junctions are weak links connecting two superconductors [1]. Cooper pairs crossing between the two condensates establish phase coherence [2]. The transport properties depend on the phase difference φ between the two reservoirs. Zero voltage transport involves the d.c. Josephson current made of a steady flow of Cooper pairs, while at nonzero voltage, due to the relation $\frac{d\varphi}{dt} = \frac{2eV}{\hbar}$, pair transport manifests through the a.c. Josephson effect with oscillating Cooper pairs. In transparent enough SNS junctions, transport occurs through Andreev reflections at the SN interfaces, forming Andreev electron-hole bound states at zero voltage. At nonzero voltage below the superconducting gap Δ, quasiparticle transport can occur through multiple Andreev reflections (MAR) [3]. Thus, dissipationless d.c. transport at $V = 0$ and dissipative pair-assisted quasiparticle transport at $V \neq 0$ are well-separated phenomena.

Recently, set-ups involving three superconductors coupled by a common link have been proposed theoretically [4, 5, 6, 7, 8, 9] and recently achieved experimentally [10, 11, 12]. Transport is then governed by two (instead of one) phase or voltage variables. Coherent mechanisms coupling the three superconductors altogether become possible and provide new d.c. channels which involve correlated motion of Cooper pairs. The lowest order processes correspond to quartets, e.g, two pairs exiting simultaneously from a reservoir and crossing the bijunction, one towards each of the other reservoirs [8, 9]. This process is deeply related to so-called Cooper pair splitting that occurs in an hybrid NSN bijunction. Cooper pairs are split from S by so-called Crossed Andreev Reflection, and two energy- and spin- entangled electrons are sent in the two separated normal metal reservoirs [13, 14, 15, 16, 17, 18, 19, 20, 21, 22, 23, 24, 25]. When those reservoirs are instead superconducting, such a process should be duplicated in order to produce two quasiparticles in each of the side superconductors, that recombine into Cooper pairs [8, 9, 12]. The basic process thus involves *four* fermions, and it originates from splitting *two* Cooper pairs from a given superconductor. It involves four (instead of two) Andreev reflections within the weak link N, and it exchanges opposite-spin electrons of the two split Cooper pairs.

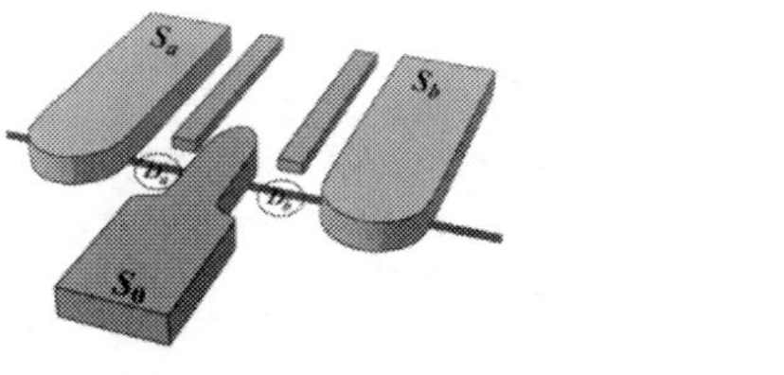

Figure 1. A Josephson bijunction. Superconductors $S_{a,b}$ are biased at voltages $V_{a,b}$, S_0 is grounded. The distance between the two quantum dots is comparable to the coherence length. Gates (in grey online) are figured.

Here we review the present theory of quartet and multipair coherent transport, and summarize a recent experiment where transport signatures of quartets have been obtained in a triterminal $Al-Cu$ bijunction[12]. Section 2 recalls the adiabatic argument for multipair transport. Section 3 presents theoretical results for a double dot bijunction. Section 4 summarizes the recent experiment and the quartet interpretation.

2. The adiabatic limit of multipair transport

Let us consider three superconductors $S_{0,a,b}$ with voltages $V_0 = 0, V_a$ and V_b, connected by a short metallic link or a quantum (double) dot (Figure 1). Defining the phases $\varphi_0 = 0, \varphi_a$, φ_b, the total free energy at equilibrium ($V_{a,b} = 0$) can be written as a doubly periodic function $F(\varphi_a, \varphi_b)$. Fourier transforming and expressing the current in one lead (say, S_a), one obtains:

$$I_a = \frac{2e}{\hbar}\frac{\partial F}{\partial \varphi_a} = \sum_{n,m} I_a(n,m)\,\sin(n\varphi_a + m\varphi_b) \tag{1}$$

At nonzero voltages, one writes:

$$\varphi_a = \varphi_{0a} + \frac{2e}{\hbar}V_a t\,, \quad \varphi_b = \varphi_{0b} + \frac{2e}{\hbar}V_b t \tag{2}$$

where the phases at the origin of time should not be omitted. Within the adiabatic approximation, one plugs the time-dependent phases into the $I_a(\varphi_a, \varphi_b)$ relation [1]. Thus, if $V_{a,b}$ satisfy $nV_a + mV_b = 0$, only the harmonics $\sin p(n\varphi_a + m\varphi_b)$ of I_a have a nonzero static value, and a d.c. current flows in S_a, as well as in S_0 and S_b. This current involves n (m) pairs crossing altogether to $S_a (S_b)$, and $n+m$ pairs coming from S_0. The simplest of these multipair processes involves quartets, e.g. $n = m = 1$. It is important to notice that the lowest-order Josephson processes taking one single pair from one to the other superconductor average to zero.

From Equations (1,2), one finds that a d.c. multipair current in S_a appears:

$$I_a^{d.c.}(\varphi_{0a}, \varphi_{0b}) = \sum_p I_{a,max}(p)\,\sin p(n\varphi_{0a} + m\varphi_{0b}) \tag{3}$$

This shows the relevance of the phase combination $(n\varphi_{0a} + m\varphi_{0b})$ which controls coherent d.c. transport by multipair processes. This phase can be emulated either by imposing suitable fluxes in a circuit containing the bijunction, or by sending a d.c. current in the bijunction. The first method is demanding as it implies the control of two phase variables. The second one is somewhat similar to driving a current into a Josephson junction and exploring its zero-voltage branch. An equivalent in the bijunction of this zero-voltage branch is a $nV_a + mV_b = 0$ line in the (V_a, V_b) plane.

The above simple calculation is valid only for tunneling barriers and voltages well below the gap. Yet, in this case, the quartet process which is a second order process might be weak. On the contrary, if the bijunction is metallic and has a high transparency, quartet transport can be strong and even resonant. In this case, a full nonequilibrium calculation should be performed. In what follows, we explore two situations, first, a bijunction made of a double quantum dot, achievable with a carbon nanotoube or a nanowire, and second, with disordered long metallic junctions.

3. Quartet and multipair transport in a double quantum dot bijunction

Recent progress on Cooper pair splitting has arisen with the fabrication of hybrid $N_aD_aS_0D_bN_b$ bijunctions where the dots D are created within a single wall carbon nanotube [23] or a semiconducting nanowire [24, 25]. Evidence for Cooper pair splitting comes out from measurement of the nonlocal conductance. Consider now a bijunction $S_aD_aS_0D_bS_b$ with a similar geometry, but with all leads being superconducting. For sake of simplicity, one considers one single electronic level $\varepsilon_{a,b}$ in each dot. The coupling between the dots and the superconductors is supposed to be symmetric and described by the level broadening $\Gamma = \pi\nu(0)t^2$ where ν is the normal metal density of states in the superconductors and t is a dot-superconductor hopping parameter. The adiabatic argument of Section 2 suggests that the multipair current depends on the phase combination $n\varphi_{0a} + m\varphi_{0b}$. The calculation of the current-phase characteristics is performed using nonequilibrium Keldysh Green's functions [9]. This yields the solution as an effective partition function for the dot "dressed" by tunneling into the superconductors. This approach is nonperturbative and provides exact results. Details about the formalism can be found in Ref. [26]. The parameters of the calculation (in units of the gap) are : the dot energies $\varepsilon_{a,b}$, the coupling Γ, and the voltage V. Notice that the calculation gives the full current, which for a given commensurate combination of voltages comprises not only the coherent d.c. multipair current, but also MAR quasiparticle currents.

3.1. Open dot regime

One first considers a regime in which each dot mimics a metallic junction, achieved by placing energy levels out of resonance $|\varepsilon_{a,b}| > \Delta$, with large couplings $\Gamma > \Delta$. The results for the largest resonances are represented on Figure 2 and compared to the phase-independent part of the quasiparticle current. One sees that the multipair features appear as strong resonances. The quartet dispersion has a negative slope at the origin and is thus a π-junction in terms of the "quartet phase" $\varphi_a + \varphi_b$. This is due to the exchange process occurring as two split Cooper pairs are emitted by S_0 and recombine into one pair in S_a and one in S_b.

3.2. Closed and nonresonant dot regime

Consider now small coupling and nonresonant dots, e.g. $\Gamma < |\varepsilon_{a,b}| < \Delta$. Figure 3 shows the current-phase profiles for various values of V, increasing from 0.09Δ to 0.8Δ. The maximum current increases by several orders of magnitude and is maximum when $V \sim |\varepsilon_{a,b}|$. Moreover, above the maximum, currents I_a and I_b start to depart from each other, signalling the onset of a strong MAR quasiparticle current. Remarkably, this MAR current is phase-dependent, a

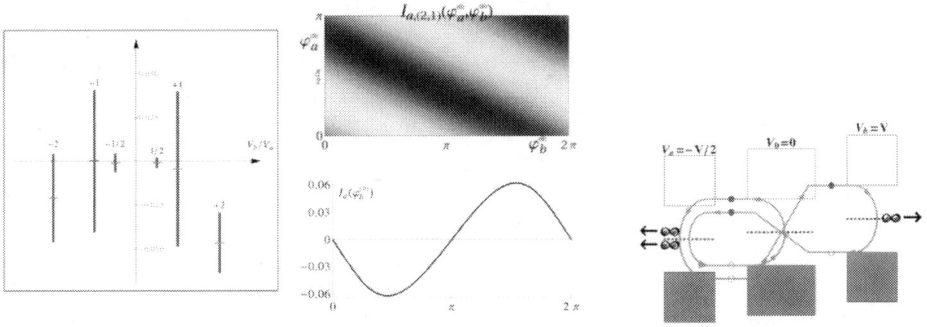

Figure 2. (color online)"Broad" dots regime: $|\varepsilon_{a,b}| = 6\Delta$, $\Gamma_{a,b} = 4\Delta$. Left: phase-dispersion (in blue online) of the d.c. current for the main resonances (with $|n| + |m| \leq 3$), centered around the values of the phase-independent quasiparticle current (small horizontal bars). Horizontal axis is V_a/V_b, with $V_b/\Delta = 0.3$. Upper middle: Current $\langle I_a \rangle$, for the resonance $2V_a + V_b = 0$, as a function of the phases $\varphi_a^{(0)}$ and $\varphi_b^{(0)}$, showing the dependence in $2\varphi_a^{(0)} + \varphi_b^{(0)}$. Lower middle: Current-phase relation $I_a(\varphi_b^{(0)})$ at $\varphi_a^{(0)} = 0$ for the quartet resonance $V_a + V_b = 0$, which shows the π-phase behavior. Right: Energy diagram for the $S_aD_aS_0D_bS_b$ bijunction, with a higher order diagram associated with a "sextet" current with 3 pairs emitted from S_0, 2 into S_a and 1 into S_b, with $V_a = -V_b/2$.

specific feature of the resonant quartet behaviour, which can be ascribed to interferences between various amplitudes within the bijunction (Figure 4).

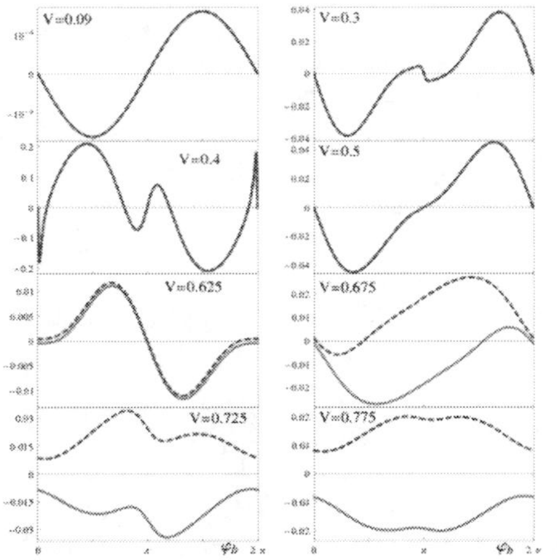

Figure 3. (color online) Current-phase relations $I_a(\varphi_b^{(0)})$ (red, full curve) and $I_b(\varphi_b^{(0)})$ (blue, dashed curve) in the quartet configuration $V_a = -V_b$, for the resonant dots regime: $\varepsilon_a = -\varepsilon_b = 0.4\Delta$, $\Gamma = 0.1\Delta$, $V = 0.09 - 0.775\Delta$. Note the different y scales in the different panels.

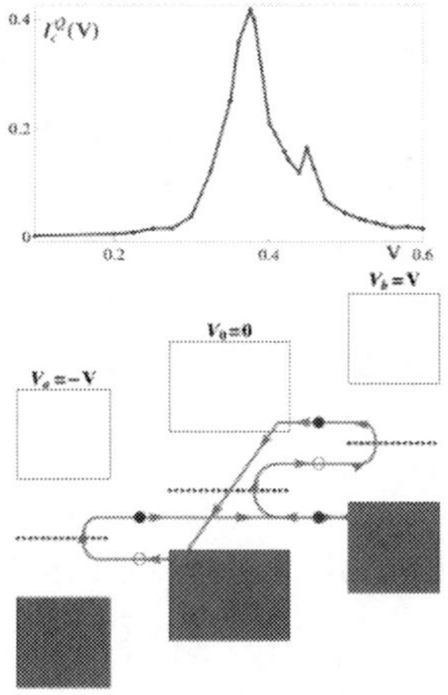

Figure 4. (color online) Upper panel: critical current I_c^Q a as function of the voltage in the quartet configuration, for resonant dots (same parameters as in Figure 3. Lower panel: Lowest order diagram contributing to the phase-dependent MAR current.

This calculation validates the existence of quartet and higher-order resonances featuring a d.c. current of correlated Cooper pairs. These currents are functions of a well-defined combination of the phases, that becomes a relevant and observable variable. Phase coherence is thus established between the three superconductors, despite the presence of voltage biases. One should underline that instead of Andreev bound states forming in a junction at equilibrium, here the d.c. Josephson-like multipair resonances are due to Andreev *resonances*, and they coexist with dissipative MAR channels. This coexistence is a novel feature in superconductivity. As a corollary, the maximum "critical" quartet current depends on the applied voltage, in a dramatic way in the example shown above (Section 3.2).

4. Quartet anomalies in metallic bijunctions

Copper and Aluminum bijunctions have been fabricated [12] by shadow mask evaporation technique [27], leading to highly transparent and uniform SN interfaces. They form three-terminal junctions with a T-shape normal conductor connecting three superconducting electrodes S_0, S_a and S_b (Figure 5). The width and length of the normal metal are about $0.6\,\mu m$ and $1\,\mu m$. Using a diffusion constant for copper $D = 100\,cm^2/s$, one gets a Thouless energy $E_{Th} = \hbar D/L^2 = 7\,\mu eV$. The superconducting Al energy gap is $\Delta = 170\,\mu eV$. These Josephson junctions are thus in the long junction regime defined as the Thouless energy E_{Th} being much smaller than the gap Δ, or equivalently a junction length L larger than the superconductor coherence length $\xi_s = \sqrt{\hbar D/\Delta}$. The diffusion time is $\tau_D = L^2/D \simeq 0.1ns$ is much smaller than the inelastic time $\tau_{in} \simeq 1\,ns$ at $100\,mK$.

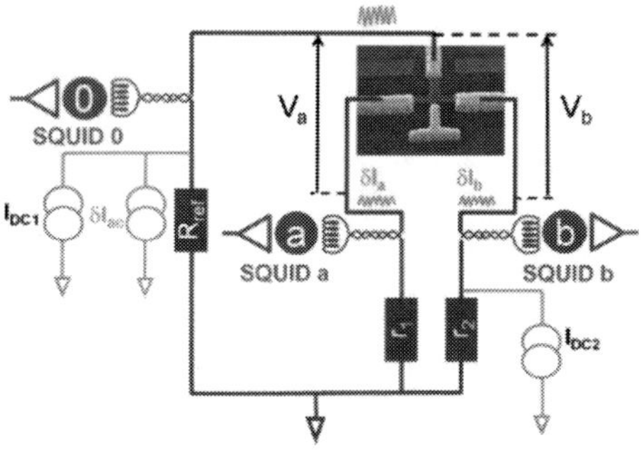

Figure 5. Experimental set-up [28]. The three macroscopic resistors have low resistance values ($\simeq$ 0.1 Ω) allowing voltage-biasing of the samples. The SEM image shows a bijunction sample with a T-shape geometry.

Three-terminal differential resistances were measured using a specific experimental set-up including 3 SQUIDs as current amplifiers (Figure 5) [28]. The differential resistance $\frac{\partial V_a}{\partial I_a}$ is plotted as a 2D map in the (V_a, V_b) plane (Figure 6). First, it displays anomalies along the axis, that correspond to the "direct" Josephson effect between S_0 and S_a or S_b (the one between S_a and S_b is not visible due to the location of the ac lock-in). Second, clear features appear at voltage values such as $V_i + V_j - 2V_k = 0$, where $(i, j, k) = (a, b, 0), (b, 0, a), (0, a, b)$, corresponding to quartets emitted respectively by S_0, S_a, S_b. The transversal $I(V)$ traces shown in Figure 7 display comparable shape and amplitude with respect to the Josephson features obtained at $V_a = 0$ or $V_b = 0$.

An explanation in terms of the synchronization of ac Josephson oscillations in junctions S_0S_a and S_0S_b [29, 30] is unlikely, since in the experimental regime where eV is as large as $8E_{Th}$, ac Josephson oscillations should be strongly decreased. In addition, the amplitude of the anomalies is nearly independent of the voltage value, pointing towards a fully coherent phenomenon. Figure 8 indeed shows why the quartet mechanism is immune against the decoherence brought by such voltage in a single Josephson junction. Two Andreev reflections in S_0 produce two splitted Cooper pairs, materializing in electrons and holes at energies $eV \pm \varepsilon$ in the $a-$ branch, and

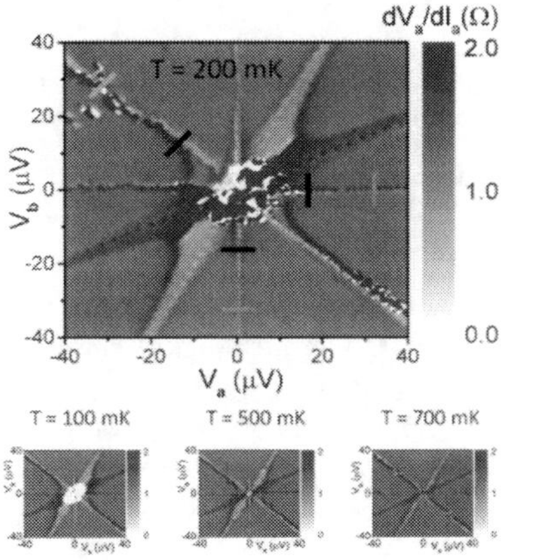

Figure 6. Differential resistance $R_{diff,a}$ of a T-shape junction in the (V_a, V_b) plane for various temperatures.

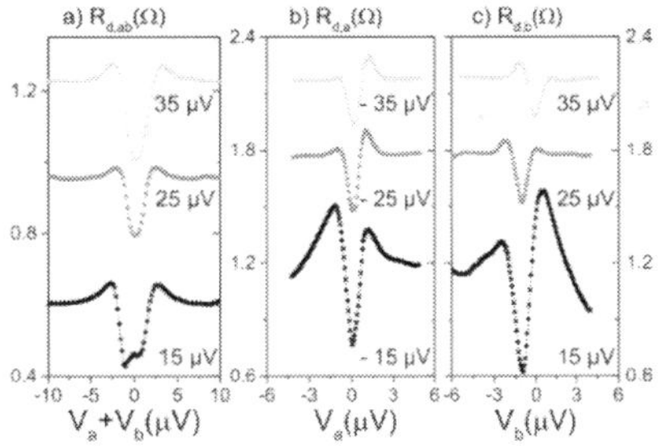

Figure 7. Line-traces at various values of the applied voltage of a) the differential resistance of the full sample as a function of the voltage $V_a + V_b$, for several values of $(V_a - V_b)/2$, b) c) the differential resistance of branch a vs V_a (V_b) for various values of V_b (V_a).

at energies $-eV \pm \varepsilon$ in the $b-$ branch. Provided $\varepsilon < E_{Th}$, such energies allow recombination by Andreev reflection at S_a and S_b, without any loss of coherence introduced by the voltage. This coherent process has the structure of a MAR process, but in the present case it is energy conserving and it provides a Josephson-like quartet channel.

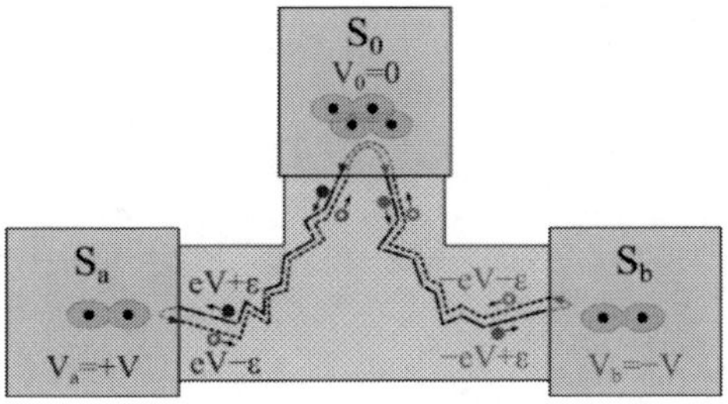

Figure 8. Schematic diagram for Q_0 quartet production, from S_0 to S_a and S_b. Two Cooper pairs are split simultaneously at S_0 with one electron of each pair diffusing towards S_a and S_b where, under the appropriate energy condition ($V_a = -V_b$), they recombine to create two separated Cooper pairs.

A perturbative non-equilibrium Green's function [12] calculations gives the following estimate for the maximum quartet current:

$$eI_Q \sim -G_{CAR}E_{Th}\sin(\varphi_{a0} + \varphi_{b0}), \tag{4}$$

where the characteristic phase dependence of the quartet mode stems from the four involved Andreev reflections, one at S_a, one at S_b and two at S_0. The conductance G_{CAR} is the Crossed Andreev conductance of a $N_aNS_0NN_b$ structure in which the electrodes S_a and S_b are in the normal state and at voltages $\pm V$. By "CAR conductance" we mean the conductance formally associated to a process in which an electron from N_a, at energy eV is transmitted as a hole in N_b at energy $-eV$, and a Cooper pair is therefore split from S_0 [19]. This formula is similar to the one for a SNS junction where the *pair* critical current $eI_c \sim G_N E_{Th}$ is related to the single particle normal conductance and the Thouless energy sets the coherence of Andreev reflection.

The CAR conductance can be evaluated as $G_{CAR} \sim \frac{G_{Na}G_{Nb}}{G_0(\xi_s)}$, where $G_{Na,b}$ is the conductance within each normal branch of the bijunction, and $G_0(\xi_s) = (2e^2/h)\mathcal{N}(l_e/\xi)$ is the normal-state conductance of a region of size ξ of the superconductor S_0 ($\mathcal{N}$ is the channel number). This shows that the ratio between the quartet maximum current at a bias V and the single junction critical current at zero bias is $I_{Qmax}(V)/I_c(0) \sim G_{CAR}/G_N \sim G_N/G_0(\xi_s)$, which is not necessarily small. Based on measured sample parameters, we estimate this ratio to $\sim 0.1 - 0.5$, in fair agreement with the experiment. Notice that if $eV \ll \Delta$, then G_{CAR} as well as $I_{Q,max}$ do not decrease with V, in agreement with the present experiment.

5. Conclusion

The theory and the experiment reported here open the way to a new kind of Josephson devices. Phase coherence is possible between three superconductors biased at suitable voltages, and open new channels for transport, made of correlated Cooper pairs. Such phase coherence remains to be demonstrated by interference experiments. Beyond the understanding and probing of the microscopic properties of such bijunctions (made of quantum dots or metals), future prospects include the exploration of new entanglement properties, and the production of phase-correlated microwave fields.

Acknowledgments
This work has been partially funded by ANR-NanoQuartet (ANR12BS1000701). We acknowledge the Nanoscience Foundation for the PhD grant of A. H. Pfeffer and the NanoFab facility at Institut Néel-CNRS for sample fabrication.

References

[1] Tinkham M 1996, Introduction to Superconductivity (Mc Graw-Hill, Singapore).
[2] Josephson BD 1962 Physics Letters **1**, 251.
[3] Octavio M, Tinkham M, Blonder GE and Klapwijk TM 1983, Phys. Rev. B **27** 6739.
[4] Cuevas JC and Pothier H 2007, Phys. Rev. B**75**, 174513.
[5] Duhot S, Lefloch F and Houzet M 2009, Phys. Rev. Lett. **102**, 086804.
[6] Houzet M and Samuelsson P 2010, Phys. Rev. B**82**, 060517.
[7] Chtchelkatchev NM, Baturina TI, Glatz A and Vinokur VM 2010, Phys. Rev. B **82**, 024526.
[8] Freyn A, Douçot B, Feinberg D and Mélin M 2011, Phys. Rev. Lett. **106**, 257005.
[9] Jonckheere T, Rech J, Martin T, Douçot B, Feinberg F and Mélin R 2013, Phys. Rev. **87**, 214501.
[10] Cleuziou JP, Wernsdorfer W, Bouchiat V, Ondarçuhu T and Monthioux M 2006, Nature Nanotech. **1**, 53.
[11] Kaviraj B, Coupiac O, Courtois C and Lefloch F 2011, Phys. Rev. Lett. **107**, 077005.
[12] Pfeffer AH, Duvauchelle JE, Courtois H, Mélin R, Feinberg D and Lefloch F 2013, arXiv:1307.4862, Phys. Rev. B in press.
[13] Byers JM and Flatté ME 1995, Phys. Rev. Lett. **74**, 306.
[14] Martin T and Torrès J 1999, Eur. Phys. J. B **12**, 319.
[15] Deutscher G and Feinberg D 2000, Appl. Phys. Lett. **76**, 487.
[16] Falci G, Feinberg D and Hekking FWJ 2001, Europhys. Lett. **54**.
[17] Recher P, Sukhorukov EV and Loss D 2001, Phys. Rev. B **63**, 165314.
[18] Lesovik GB, Martin T and Blatter G 2001, Eur. Phys. J. B **24**, 287.
[19] Mélin R 2005, Phys. Rev. B **72**, 054503.
[20] Beckmann D, Weber HB and v. Löhneysen H 2004, Phys. Rev. Lett. **93**, 197003.
[21] Russo S, Kroug M, Klapwijk TM and Morpurgo AF 2005, Phys. Rev. Lett. **95**, 027002.
[22] Zimansky PC and Chandrasekhar V 2006, Phys. Rev. Lett. **97**, 237003.
[23] Hermann LG, Portier F, Roche P, Levy Yeyati A, Kontos T and Strunk C 2010, Phys. Rev. Lett. **104**, 026801.
[24] Hofstetter L, Csonka S, Nygård J and C. Schönenberger 2009, Nature (London) **461**, 960.
[25] Das A, Ronen Y, Heiblum M, Mahalu D, Kretinin AV and Shtrikman H 2012, Nature Comm. **3**, 1165.
[26] Chevallier D, Rech J, Jonckheere T and Martin T 2011, Phys. Rev. B **83** 125421.
[27] Dubos P, Courtois H, Buisson O and Pannetier B 2001, Phys. Rev. B **87**, 206801.
[28] Pfeffer A, Kaviraj B, Coupiac O and Lefloch F 2012, Review of Scientific Instruments **83**, 115107.
[29] Nerenberg NAH, Blackburn JA and Jillie DW 1980, Phys. Rev. B **21**, 118.
[30] Argaman N 1999, Superlattices and Microstructures **25**, 861.

X-boson cumulant approach to the topological Kondo insulators

E Ramos[1], R Franco[1], J Silva-Valencia[1], M E Foglio[2] and M S Figueira[3]

[1] Departamento de Física, Universidad Nacional de Colombia, A. A. 5997, Bogotá, Colombia
[2] Instituto de Física "Gleb Wataghin" Universidade Estadual de Campinas, 13083-970 Campinas, São Paulo, Brazil
[3] Instituto de Física, Universidade Federal Fluminense, 24210-340, Niterói, RJ, Brazil

E-mail: `figueira@if.uff.br`

Abstract.
In this work we present a generalization of our previous work of the $X-$boson approach to the periodic Anderson model (PAM), adequate to study a novel class of intermetallic $4f$ and $5f$ orbitals materials: the topological Kondo insulators, whose paradigmatic material is the compound SmB_6.

For simplicity, we consider a version of the PAM on a $2D$ square lattice, adequate to describe $Ce-$based compounds in two dimensions. The starting point of the model is the $4f-Ce$ ions orbitals, with $J = 5/2$ multiplet, in the presence of spin-orbit coupling. Our technique works well for all of the parameters of the model and avoids the unwanted phase transitions of the slave boson mean field theory. We present a critical comparison of our results with those of the usual slave boson method, that has been intensively used to describe this class of materials. We also obtain a new valence first order transition which we attribute to the $\vec{k}$ dependence of the hybridization.

1. Introduction

In this work we study a novel class of intermetallic $4f$ and $5f$ orbitals materials: the topological Kondo insulators; in which strong interactions between itinerant and predominately localized degrees of freedom give rise to a bulk insulating state at low temperatures, while the surface remains metallic. This effect arises due to inversion of even parity conduction bands and odd parity very narrow f electron bands. For an odd number of band inversions, the metallic surface states are chiral and therefore remain robust against disorder and time reversal invariant perturbations.

The topological Kondo insulators have been studied employing the slave boson mean field theory (SBMFT) [1], both in the limit of the Coulomb repulsion $U \to \infty$ [2, 3], and for finite correlation U [4]. This approximation is attractive because with a small numerical effort, it is capable of qualitatively describe the Kondo regime, but as the temperature is increased or when $(\mu >> E_{f,\tau} = E_f)$ at low temperatures, and also for all parameters at intermediate temperatures, the SBMFT presents an unphysical second order phase transition with the conduction and the localized electrons decoupling from each other. In the impurity case this transition occurs when $T > T_K$, and defines the Kondo temperature.

To circumvent these problems, maintaining the simplicity of the calculation and the ideas involved, we have developed the X-boson method [6], inspired by the slave boson formalism. To solve the problem of non-conservation of probability (completeness) that appears when we use the X Hubbard operators, we use the chain cumulant Green's functions.

The SBMFT continues to be an attractive approach to treat the extreme Kondo limit of the Anderson model, where the localized occupation number is equal to $n_f = 1$ at very low temperatures, but it is less adequate to describe intermediate valence (IV) systems like the new topological Kondo insulator SmB_6, where the localized occupation number associated with the opening a gap at low temperatures, is $n_f \simeq 0.7 - 0.8$ [8] in this compound. On the other hand, the X-boson approach does not capture the Kondo peak physics in the limit of $n_f = 1$, as the SBMFT does, but it is more adequate to describe IV systems because this formalism does not present spurious phase transitions for any of the system parameters.

We generalize our previous work of the $X-$boson approach to the periodic Anderson model (PAM) for f electrons states with a total angular momentum J and z-axis component M, while the conduction electron states are described by a momentum $\mathbf{k}$ and spin σ; in our previous calculation we considered $J = 1/2$. The spin-orbit coupled Wannier states of the conduction electrons are then decomposed in terms of plane-wave states, and this gives rise to momentum-dependent form factors with symmetries that are uniquely determined by the local symmetry of the f states.

2. The $2D$ periodic Anderson model

The present paper deals with the periodic Anderson model (PAM) in the limit of infinite Coulomb repulsion ($U = \infty$), adequate to treat Ce-based compounds in two dimensions (2D) [5], and we employ the Hubbard X operators to write its Hamiltonian in the form

$$H = \sum_{\mathbf{k}\sigma} E_{\mathbf{k},\sigma} c^{\dagger}_{\mathbf{k},\sigma} c_{\mathbf{k},\sigma} + \sum_{j\tau} E_{f,j\tau} X_{j,\tau} + \sum_{j\tau,\mathbf{k}\sigma} \left(V_{\sigma\tau}(\vec{k}) X^{\dagger}_{j,0\tau} c_{\mathbf{k},\sigma} + V^{*}_{\sigma\tau}(\vec{k}) c^{\dagger}_{\mathbf{k},\sigma} X_{j,0\tau} \right), \quad (1)$$

where we follow the notation of the paper in the reference [3]. The first term is the Hamiltonian of the conduction electrons (c-electrons), with momentum $\vec{k}$ and spin σ, the second term describes independent localized electrons (f-electrons), with pseudospin τ belonging to one representation γ of some multiplet state at the site j. The last term is the hybridization Hamiltonian giving the interaction between the c-electrons and the f-electrons. As there is no local hybridization process between s and f electrons in rare earth ions, the hybridization process results from the nearest-neighbor hopping from the f electrons at a site j to the s electron at the vicinity of this site. To determine the structure of the $V_{\sigma\tau}(\vec{k})$ we must specify the representation γ of the localized electron and the corresponding conduction electrons. Considering the $j = 5/2$ multiplet structure the hopping hybridization matrix can be written as $V_{\sigma\tau}(\vec{k}) = V[\mathbf{\Phi}]_{\sigma\tau}(\vec{k})$, where $[\mathbf{\Phi}]_{\sigma\tau}(\vec{k})$ is the form factor, which is associated with the $\vec{k}$ dependence and the non trivial orbital structure of the hybridization. Following the derivation presented in reference [3], the form factor can be written as [5] $\mathbf{\Phi}(k) = \vec{d}(k) \circ \vec{\sigma}$, where for a $2D$ square lattice, $\vec{d}(k) = 2[sin(k_x), sin(k_y)]$ and $\vec{\sigma}$ are the Pauli spin matrices.

The X operators are very convenient to work with local states associated to the sites j of a lattice, and are defined in general by $X_{j,ab}=|j,a\rangle\langle j,b|$, where the set $\{|j,a\rangle\}$ is an orthonormal basis in the space of local states of interest. When $U \to \infty$ the identity I_j at site j should satisfy the completeness relation: $X_{j,00} + X_{j,\tau\tau} + X_{j,\overline{\tau}\overline{\tau}} = I_j$, where $\overline{\tau}$ is the pseudospin component opposite to τ, and the $X_{j,aa}$ are the projectors into $|\, j,a\rangle$.

3. The X-boson approach

The Hamiltonian can be treated by a generalization of the X-boson approach [6] for the lattice case, and the cumulant Green's function (GF) are then given by

$$G^{f}_{\mathbf{k}\sigma}(z) = \frac{-D_{\tau}\left(z-\varepsilon_{\mathbf{k}\sigma}\right)}{\mathcal{E}(z)}, \quad G^{c}_{\mathbf{k}\sigma}(z) = \frac{-\left(z-\widetilde{E}_{f,\tau}\right)}{\mathcal{E}(z)}, \quad G^{fc}_{\mathbf{k}\sigma}(z) = \frac{-|V|D_{\tau}\Delta(\vec{k})}{\mathcal{E}(z)}, \tag{2}$$

with $\mathcal{E}(z) = \left(z-\widetilde{E}_{f,\tau}\right)(z-\varepsilon_{\mathbf{k}\sigma}) - |V|^2 D_{\tau}\mathbf{\Delta}^2(\vec{k})$, where $z=\omega+i\eta$, with $\eta\to 0^+$ and the roots of $\mathcal{E}(z)$ are given by

$$\omega_{\mathbf{k},\sigma}(\pm) = \frac{1}{2}\left(\varepsilon_{\mathbf{k},\sigma}+\widetilde{E}_f\right) \pm \frac{1}{2}\sqrt{\left(\varepsilon_{\mathbf{k},\sigma}-\widetilde{E}_f\right)^2 + 4|V|^2 D_{\tau}\mathbf{\Delta}^2(\vec{k})}, \tag{3}$$

where $\mathbf{\Delta}^2(\vec{k}) = \frac{1}{2}Tr[\mathbf{\Phi}(\vec{k}).\mathbf{\Phi}^*(\vec{k})]$. The correlations appear in the X-boson approach through the quantity $D_{\tau} = R + n_{f,\tau}$, with $R = \langle X_{0,0}\rangle$ and $n_{f,\tau} = \langle X_{\tau,\tau}\rangle$. In the X-boson approach the quantity D_{τ} must be calculated self-consistently through the minimization of the corresponding thermodynamic potential with respect to the parameter R, at the same time $\widetilde{E}_{f,\tau} = E_{f,\tau} + \Lambda$, where Λ is a Lagrange multiplier. We shall consider a spin independent tight-binding conduction band on a $2D$ square lattice

$$\varepsilon_{\mathbf{k}} = -2t[cos(k_x) + cos(k_y)], \tag{4}$$

where we employ the hopping between neighboring sites of the $2D$ square lattice, $t=1$ as the energy unit. In the X-boson method [6], after the minimization of the thermodynamic potential $\Omega = -k_B T\ln(\mathcal{Q})$, (where $\mathcal{Q}$ is the grand partition function) with respect to the R parameter we obtain the X-boson parameter Λ

$$\Lambda = \frac{-V^2}{\pi^2}\int_{-\pi}^{\pi} dk_x \int_{-\pi}^{\pi} dk_y \mathbf{\Delta}^2(\vec{k}) \frac{n_F\left(\omega_{\mathbf{k}}(+)\right) - n_F\left(\omega_{\mathbf{k}}(-)\right)}{\sqrt{\left(\varepsilon_{\mathbf{k}}-\widetilde{E}_f\right)^2 + 4V^2 D_{\tau}\mathbf{\Delta}^2(\vec{k})}}, \tag{5}$$

where $n_F(x)$ is the Fermi-Dirac distribution $n_F(z) = [1+\exp(\beta\, z)]^{-1}$, N_s is the number of sites and $\mathbf{\Delta}^2(\vec{k}) = 4[sin^2(k_x) + sin^2(k_y)]$. After the numerical calculation of the parameter Λ, we calculate the occupation numbers employing the Green's functions and we use the completeness relation $\langle X_{j,00}\rangle + \langle X_{j,\tau\tau}\rangle + \langle X_{j,\overline{\tau\tau}}\rangle = I_j$, to calculate the parameter $R = D_{\tau} - n_{f,\tau}$. All the calculations are repeated again until the convergence of the X-boson parameters Λ and R is attained.

The density of states is obtained numerically through the relation

$$\rho(\omega) = \frac{-1}{\pi} Im \int_{-\pi}^{\pi} dk_x \int_{-\pi}^{\pi} dk_y \frac{1}{\omega - \omega_{\mathbf{k}}(\pm) + i\eta}, \tag{6}$$

where $\omega_{\mathbf{k}}$ is given by the Eq. 3.

In a similar way as it is done with the slave boson Hamiltonian [1], we have shown [6] that the X-boson Hamiltonian Eq. 1 can be written in an uncorrelated form H^u as

$$H^u = \sum_{\mathbf{k},\sigma} \varepsilon_{\mathbf{k}}\, c^{\dagger}_{\mathbf{k},\sigma} c_{\mathbf{k},\sigma} + \sum_{k,\tau} \widetilde{\varepsilon}_f\, f^{\dagger}_{\mathbf{k},\tau} f_{\mathbf{k},\tau} + N_s\Lambda(R-1) + \sum_{\mathbf{k},\sigma\tau}\left(\overline{V}_{\sigma,\tau}(\mathbf{k}) f^{\dagger}_{\mathbf{k},\tau}\, c_{\mathbf{k},\sigma} + \overline{V}_{\sigma,\tau}(\mathbf{k})^* c^{\dagger}_{\mathbf{k},\sigma}\, f_{\mathbf{k},\tau}\right), \tag{7}$$

with $\overline{V}_{\sigma\tau}(\mathbf{k}) = \sqrt{D_{\tau}}V_{\sigma\tau}(\mathbf{k})$. This Hamiltonian can be easily diagonalized, and following Tran's work [3], we can write

$$\mathcal{H}^u = \sum_{\mathbf{k}} \Psi^{\dagger}(\mathbf{k}) H^u(\mathbf{k}) \Psi(\mathbf{k}) + N_s\Lambda(R-1), \tag{8}$$

with

$$H^u(\mathbf{k}) = \begin{pmatrix} \varepsilon_{\mathbf{k}} & V\Phi_{\mathbf{k}} & 0 & 0 \\ V\Phi^*_{\mathbf{k}} & \widetilde{\varepsilon}_f & 0 & 0 \\ 0 & 0 & \varepsilon_{\mathbf{k}} & -V\Phi_{\mathbf{k}} \\ 0 & 0 & -V\Phi^*_{\mathbf{k}} & \widetilde{\varepsilon}_f \end{pmatrix}, \tag{9}$$

where $\Psi^\dagger(\mathbf{k}) = (c^\dagger_{\mathbf{k},\uparrow} f^\dagger_{\mathbf{k},(-)} c^\dagger_{\mathbf{k},\downarrow} f^\dagger_{\mathbf{k},(+)})$ is a four component Dirac spinor, with $\varepsilon_{\mathbf{k}} = E_{\mathbf{k}} - \mu$, $\widetilde{\varepsilon}_f = \widetilde{E}_f + \Lambda - \mu$ and $\Phi_{\mathbf{k}} = 2[sin(\mathbf{k_x}) - isin(\mathbf{k_y})]$. Expanding $\Phi_{\mathbf{k}}$ for small values of $\mathbf{k_x}$ and $\mathbf{k_y}$, we obtain an effective Dirac theory given by the Hamiltonian

$$H^u(\mathbf{k}) = \begin{pmatrix} \varepsilon_{\mathbf{k}} & 2V(\mathbf{k_x} - i\mathbf{k_y}) & 0 & 0 \\ 2V(\mathbf{k_x} + i\mathbf{k_y}) & \widetilde{\varepsilon}_f & 0 & 0 \\ 0 & 0 & \varepsilon_{\mathbf{k}} & -2V(\mathbf{k_x} + i\mathbf{k_y}) \\ 0 & 0 & -2V(\mathbf{k_x} - i\mathbf{k_y}) & \widetilde{\varepsilon}_f \end{pmatrix}, \tag{10}$$

whose spectrum can be written in a Dirac form $E_\pm(p) = \varepsilon \pm \sqrt{A^2(p_x^2 + p_y^2) + M^2}$, with $\varepsilon = (\varepsilon_{\mathbf{k}} + \widetilde{\varepsilon}_f)/2$, $M = (\varepsilon_{\mathbf{k}} - \widetilde{\varepsilon}_f)/2$ and $A = 4V$. This discussion shows that the X-boson captures the behavior of the Dirac cones in the spectral density at around $\mathbf{k} \approx \mathbf{0}$, as indicated in Fig. 3.

4. Results and discussion

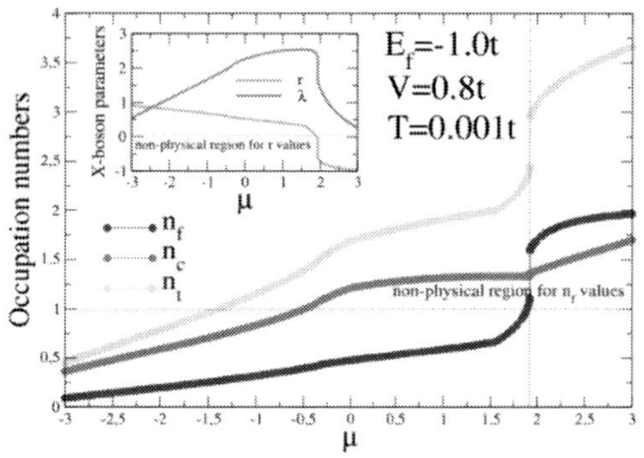

Figure 1. Slave boson occupation numbers.

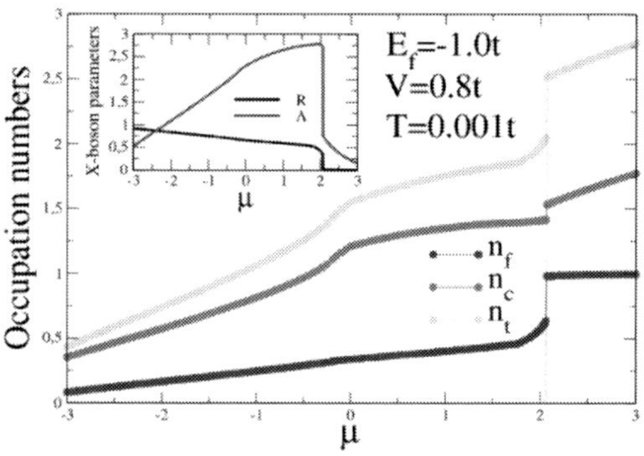

Figure 2. X-boson occupation numbers.

In Figs. 1 and 2 we compare the occupations numbers calculated employing both the SBMFT [1] and the X-boson approach [6]. We choose the following set of parameter: $E_f = -1.0t$, $V = 0.8t$ and $T = 0.001t$. The new interesting result, is that both methods show a first order transition in the occupation numbers in the local magnetic moment regime (LMM)[7]. In the SBMFT the transition is not physical, because it occurs in the region where $n_f > 1$, but in the X-boson case the corresponding transition, presented in Fig. 2, is physical because it occurs in a region where $n_f < 1$. It is possible to observe in Fig. 1 that the proximity of the spurious transition of the SBMFT distorts all the occupation numbers in the intermediate valence (IV) region, indicating that the SBMFT is not completely adequate do describe the topological Kondo insulators. This X-boson first order transition is not observed when we consider a constant hybridization V (cf. Fig. 12 of our paper [6]), and in this case we obtain a crossover from the IV to the Kondo region. We can associate this first order transition to the $j = 5/2$ multiplet structure of the hopping hybridization matrix employed to describe the Ce ions, which introduces a $\vec{k}$ dependence and a non trivial orbital structure in the hybridization.

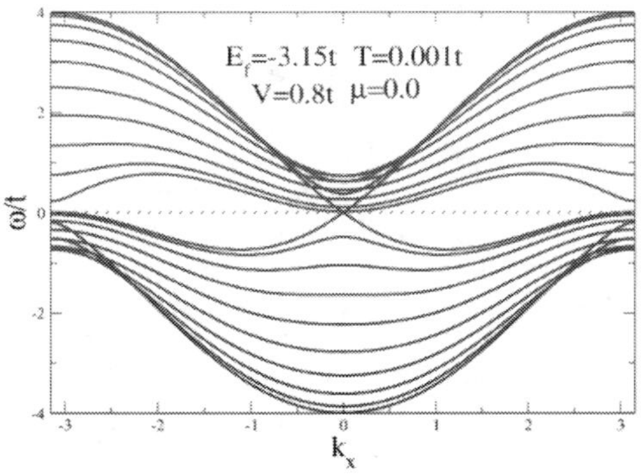

Figure 3. Single-particle band spectrum for the X-boson.

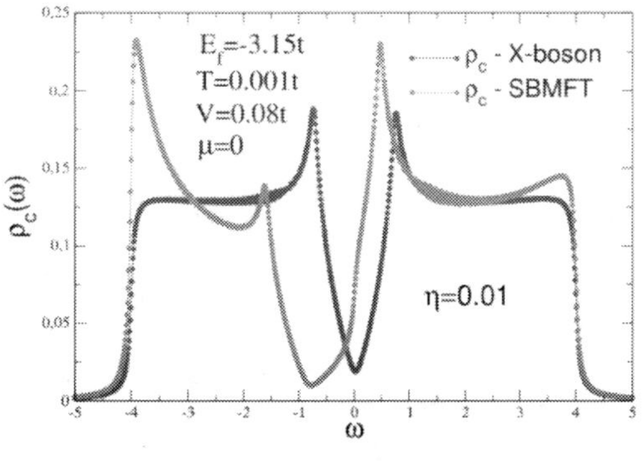

Figure 4. Density of states of the conduction electrons for the X-boson and slave boson.

In Figs. 3 we represent the X-boson single-particle band spectrum for $E_f = -3.15t$, $V = 0.8t$, $T = 0.001t$, along the x axis, $k = k_x$, and several k_y values for $\mu = 0.0$ in the IV regime, corresponding to an insulator situation. The Fig. 3 represents a strong topological Kondo insulator, because the spectral function presents a small gap at the border of the zone boundary and the Dirac cones crosses the chemical potential at $\mu = 0.0$ [2]. In Fig. 4 we plot the density of states of the conduction electrons for the X-boson and the slave boson methods, calculated with the same parameters of Fig. 3. The X-boson presents a typical insulator situation, with the localized occupation number being $n_f = 0.66$ whereas the slave boson breaks down in this region producing an unphysical density of states of $n_f = 1.37$. Due to numerical instabilities [9] associated with the analytic continuation of the Green's functions to the real axis, we represent the density of states for $\eta = 0.01$, but it is clear from the figure that it represents an insulator.

Acknowledgments
We are thankful for the financial support of DINAIN and DIB (Colombia National University), COLCIENCIAS (Colombia), CNPq (Brazil) and ICTP Trieste.

[1] Coleman P 1984 *Phys. Rev.* B **29** 3035
[2] Dzero M, Sun K, Galitski V and Coleman P, 2010 *Phys. Rev. Lett.* **104** 106408; Dzero M, Sun K, Coleman P, and Galitski V, 2012 *Phys. Rev.* B **85** 045130; Alexandrov V, Dzero M, and Coleman P, 2013 *Phys. Rev. Lett.* **111** 226403
[3] Tran M T, Takimoto T and Kim K S 2012 *Phys. Rev. B* **85** 125128
[4] Legner M, Rueg A and Sigrist M *Phys. Rev. B* **89** 085110
[5] Werner J and Assaad F F 2013 *Phys. Rev. B* **88** 035113
[6] Franco R, Figueira M S and Foglio M E 2003 *Phys. Rev.* B **66** 045112
[7] Steglich F, Geibel C, Gloss K, Olesch G, Schank C, Wassilew C, Loidl A, Krimmel A, and Stewart G F, 1994 *Journal of Low Temperature Physics* **95**, 3
[8] Derr J, Knebel G, Lapertot G, Salce B, Méasson M-A and Flouquet J 2006 *Journal of Physics: Cond. Matter* **18** 2089
[9] Natoli V D, Cohen H M and Fornberg B 1996 *Journal of Computacional Physics* **126** 99

Correlations between charge and energy current in ac-driven coherent conductors

Francesca Battista[1], **Federica Haupt**[2] **and Janine Splettstoesser**[3]

[1]Departamento de Física, FCEyN, Universidad de Buenos Aires and IFIBA, Pabellón I, Ciudad Universitaria, 1428 CABA Argentina

[2]JARA Institute for Quantum Information, RWTH Aachen University, D-52056 Aachen, Germany

[3]Department of Microtechnology and Nanoscience (MC2), Chalmers University of Technology, SE-41298 Göteborg, Sweden

E-mail: battista@df.uba.ar

Abstract. We study transport in coherent conductors driven by a time-periodic bias voltage. We present results of the charge and energy noise and complement them by a study of the mixed noise, namely the zero-frequency correlator between charge and energy current. The mixed noise presents interference contributions and transport contributions, showing features different from those of charge and energy noise. The mixed noise can be accessed by measuring the correlator between the fluctuations of the power provided to the system and the charge current.

1. Introduction

In the last years, big effort has been put in the theoretical study and in the experimental investigation of charge transport in mesoscopic structures. Single particle sources, allowing for the control in time and space of the flow of a small number of electrons or holes, have been realized with time-periodically driven mesoscopic systems [1]. These setups are usually based on the emission of particles from a confined region.

Levitov and coworkers [2] proposed an alternative way to achieve single-particle emission without using quantum confinements. Special Lorentzian-shaped voltage pulses applied to a conductor give rise only to single-particle excitations from a Fermi sea, free from electron-hole pairs excitations. Such Lorentzian pulse $V(t)$ satisfies the condition $\int_0^{\mathcal{T}} dteV(t)/h = l$ where l is an integer number. Experimentally, Lorentzian pulses carrying an integer number of particles can approximately be realized by superposing several harmonic driving potentials. In this case a reduction of the charge current noise was measured in Refs. [3, 4].

Charge current noise in ac-driven systems has been widely studied. However the charge carriers, electrons and holes, transport energy as well. It is thus important for future applications to complement the existing literature (see e.g. [3, 5, 6, 7]) with an accurate study of energy noise in these systems [8]. Recently, energy noise in normal conducting [9] and superconducting systems [10] has been studied in the stationary regime; also fluctuation relations in systems with a time-dependent driving have been considered [11, 12]. In this paper, we model the setups of the latest experiments [3, 4] with a two-terminal conductor with a central scatterer. After reporting results [8] on the correlator of charge current fluctuations (charge noise) and the correlator of energy current fluctuations (energy noise) we calculate and discuss the behavior of the correlator between the charge and the energy current fluctuations (mixed noise) [9].

2. Formalism and model

Our considerations are based on the scattering theory for photon-assisted transport [13]. We consider a coherent mesoscopic conductor connected to metallic contacts (reservoirs) by ballistic leads. The reservoirs are subjected to time-periodic voltages $V_\alpha(t)$. The charge current operator in contact α is [14]

$$\hat{I}^{\mathcal{C}}_\alpha(t) = \frac{e}{h}\int_{-\infty}^{\infty}\int_{-\infty}^{\infty} dEdE' e^{i(E-E')t/\hbar}\hat{\imath}_\alpha(E,E'), \quad (1)$$

while the corresponding energy current operator is [15]

$$\hat{I}^{\mathcal{E}}_\alpha(t) = \frac{1}{h}\int_{-\infty}^{\infty}\int_{-\infty}^{\infty} dEdE' \frac{(E+E')}{2} e^{i(E-E')t/\hbar}\hat{\imath}_\alpha(E,E'). \quad (2)$$

Here, $e<0$ is the electron charge and $\hat{\imath}_\alpha(E,E') = \hat{\mathbf{b}}^\dagger_\alpha(E)\hat{\mathbf{b}}_\alpha(E') - \hat{\mathbf{a}}^\dagger_\alpha(E)\hat{\mathbf{a}}_\alpha(E')$. The vectors $\hat{\mathbf{a}}_\alpha$ and $\hat{\mathbf{b}}_\alpha$ have the operators $\hat{a}_{\alpha n}$ and $\hat{b}_{\alpha n}$ as their components, annihilating an electron in channel n in lead α moving towards or away from the scatterer, respectively. Of our interest are the zero-frequency auto- and cross-correlators of charge current (charge noise), $S_{\alpha\beta}$, of energy current (energy noise), $S^{\mathcal{E}}_{\alpha\beta}$, as well as the correlators between charge and energy current (mixed noise), $S^X_{\alpha\beta}$, which are expected to be nonzero because electrons and holes carry both charge and energy. In the case of time periodic driving considered here, we can write in energy space [16]

$$S^{xy}_{\alpha\beta} = h\int_{-\infty}^{\infty}\int_{-\infty}^{\infty} dEdE' \langle \Delta\hat{I}^x_\alpha(E)\Delta\hat{I}^y_\beta(E')\rangle, \quad (3)$$

with $\Delta\hat{A} = \hat{A} - \langle\hat{A}\rangle$, and the charge current density, $\hat{I}^{\mathcal{C}}_\alpha(E) = e/h\cdot\hat{\imath}_\alpha(E,E)$, and energy current density, $\hat{I}^{\mathcal{E}}_\alpha(E) = E/h\cdot\hat{\imath}_\alpha(E,E)$. This equation yields the charge noise for $x=y=\mathcal{C}$, the energy noise for $x=y=\mathcal{E}$, as well as the mixed noise for $x=\mathcal{C}, y=\mathcal{E}$. The quantum statistical averages, $\langle\hat{A}\rangle$, in Eq. (3) can be evaluated substituting $\hat{\mathbf{b}}_\alpha(E) = \sum_\beta \mathbf{s}_{\alpha\beta}(E)\hat{\mathbf{a}}_\beta(E)$, where $\mathbf{s}_{\alpha\beta}(E)$ is the scattering matrix relating $\hat{\mathbf{a}}_\alpha$ to $\hat{\mathbf{b}}_\alpha$, and for leads with N_α and N_β channels has dimensions $N_\alpha \times N_\beta$. Interacting with $V_\alpha(t)$, the electrons in the reservoir α can absorb k energy quanta $\hbar\Omega$ with probability amplitude $c_{\alpha k} = \int_0^{\mathcal{T}} \frac{dt}{\mathcal{T}} e^{-i\frac{e}{\hbar}\int_0^t dt'[V_\alpha(t')-\overline{V}_\alpha]} e^{ik\Omega t}$, where $\Omega = 2\pi/\mathcal{T}$ is the frequency of the driving and $\overline{V}_\alpha$ is the dc component of $V_\alpha(t)$. A state with energy E in the leads corresponds to a superposition of reservoir states with energy $E_{-k} = E - k\hbar\Omega$ [13]. The statistics of the operators $\hat{a}_{\alpha n}(E)$ and $\hat{a}^\dagger_{\alpha n}(E)$ is thus

$$\langle \hat{a}^\dagger_{\alpha n}(E)\hat{a}_{\beta m}(E')\rangle = \delta_{\alpha\beta}\delta_{mn}\sum_{k,\ell=-\infty}^{+\infty} c^*_{\alpha k}c_{\beta k+\ell} f_\alpha(E_{-k})\delta(E_\ell - E'), \quad (4)$$

with the Fermi function $f_\alpha(E) = [1+\exp\{(E-eV_\alpha)/k_{\mathrm{B}}T\}]^{-1}$, the temperature T and the Boltzmann constant k_{B}. As a consequence of the unitarity of the scattering matrix we have that $S^{xy}_{LL} = S^{xy}_{RR} = -S^{xy}_{LR} = -S^{xy}_{RL}$. For simplicity, from now on we will always refer to the auto-correlators in the right reservoir $S \equiv S_{\mathrm{RR}}$, $S^{\mathcal{E}} = S^{\mathcal{E}}_{\mathrm{RR}}$ and $S^X \equiv S^X_{\mathrm{RR}}$.

Inspired by the experiments in Refs. [3, 4], we consider a two-terminal conductor and we assume an energy-independent transmission D of the scatterer [3]. In order to simplify the notation, we focus on a spinless single-channel conductor. The left contact is subject to a periodic time-dependent potential $V_{\mathrm{L}}(t) = V_{\mathrm{ac}}(t) + \bar{V}$, where $\bar{V}$ is a dc voltage offset and $V_{\mathrm{ac}}(t)$ is a pure ac component. The right contact is grounded $V_{\mathrm{R}}(t) = 0$, and energies are taken with respect to the electrochemical potential $\mu = 0$ of this reservoir.[1] The time-dependent driving excites the Fermi sea of the left reservoir. Because of Pauli exclusion principle, only occupied states above μ participate to transport and thus we define the excitations above μ as electron-like. Empty states below μ contribute to transport as well. We refer to the excitations below μ as hole-like excitations. Correspondingly, we write the energy resolved current operators as the sum of two contributions, one carried by electron-like and the other by hole-like excitations, $\hat{I}^{\mathcal{C}}_{\alpha}(E) = \sum_{i=\mathrm{e,h}} \hat{I}^{\mathcal{C}(i)}_{\alpha}(E)$, and $\hat{I}^{\mathcal{E}}_{\alpha}(E) = \sum_{i=\mathrm{e,h}} \hat{I}^{\mathcal{E}(i)}_{\alpha}(E)$ [16].

3. Results

We want to investigate the correlations between charge and energy currents carried by the electron and hole excitations participating to transport. It is instructive to decompose the charge, energy and mixed noise in terms of contributions that account for the correlations between excitations of the same or of different kind [8]

$$S^{xy(ij)} = h \int_{-\infty}^{\infty} \int_{-\infty}^{\infty} dE dE' \langle \Delta \hat{I}^{x(i)}(E) \Delta \hat{I}^{y(j)}(E') \rangle. \tag{5}$$

These correlators show two contributions, $S^{xy(ij)} = S^{xy(ij)}_{\mathrm{tr}} + S^{xy(ij)}_{\mathrm{int}}$, see Ref. [8], where

$$S^{xy(ij)}_{\mathrm{tr}} = \frac{D(1-D)}{h} \int dE\, \tilde{x}\tilde{y} \sum_{\ell=-\infty}^{\infty} |c_\ell|^2 \left\{ f_{\mathrm{L}}(E_{-\ell}) \left[1 - f_{\mathrm{R}}(E)\right] + f_{\mathrm{R}}(E)[1 - f_{\mathrm{L}}(E_{-\ell})] \right\} \theta_i(E)\theta_j(E), \tag{6}$$

$$S^{xy(ij)}_{\mathrm{int}} = \frac{D^2}{h} \sum_{\alpha=\mathrm{L,R}} \sum_{k,\ell,q=-\infty}^{\infty} c^*_{\alpha\ell} c_{\alpha(\ell+q)} c^*_{\alpha(k+q)} c_{\alpha k} \int dE\, \tilde{x}\tilde{y}_q f_\alpha(E_{-\ell}) \left[1 - f_\alpha(E_{-k})\right] \theta_i(E)\theta_j(E_q), \tag{7}$$

with distinct physical origins. The expressions for the charge, energy, and mixed noise, are obtained by replacing $\tilde{x} = e$ ($\tilde{x} = E$) when $x = \mathcal{C}$ ($x = \mathcal{E}$) and $\tilde{y} = e$ and $\tilde{y}_q = e$ ($\tilde{y} = E$ and $\tilde{y}_q = E_q$) when $y = \mathcal{C}$ ($y = \mathcal{E}$). We introduced $\theta_{\mathrm{e/h}}(E) = \theta(\pm E)$ with the Heaviside step function $\theta(E)$. The first term, Eq.(6), depends on the Fermi distribution of the two reservoirs and it is related to correlations due to the exchange of particles between the two contacts. We call it transport part of the noise [8], and it is nonzero only if one considers correlations between the same type of excitations, i.e. $S^{xy(\mathrm{eh})} = S^{xy(\mathrm{he})} = 0$. This indicates that electrons and holes contribute independently to the transport terms. The scatterer randomly transmits and reflects charge and energy carriers and this is reflected in the factor $D(1-D)$. The second term, Eq.(7) originates ultimately from correlations due to the exchange of particles between states with different energies in the same reservoir. Without periodic driving, only elastic exchange processes (i.e. thermal fluctuations) contribute to the noise, since $c_{\mathrm{L}k} = \delta_{k0}$ if $V_{\mathrm{ac}} = 0$. In the presence of a periodic driving, correlations between states with different energies E_ℓ, E_k do in general contribute to the charge, energy and mixed noise. However, how large this contribution is depends on the interference between different "paths" in energy space, that electrons can take when being promoted from one energy state to another by interacting with the ac-field. Therefore we refer to $S^{xy(ij)}_{\mathrm{int}}$ as the interference part of the noise [8].

In the following we summarize the main features of the interference and the transport part of the charge and energy noise studied in Ref. [8], and then focus on the mixed noise. In all cases, we make examples taking the drivings used in Ref. [3]: a simple harmonic driving $V^{\mathrm{h}}_{\mathrm{L}}(t) = \bar{V} + V_0 \cos(\Omega t)$ and a bi-harmonic driving $V^{\mathrm{bh}}_{\mathrm{L}}(t) = \bar{V} + V_0 \cos(\Omega t) + \frac{V_0}{2} \cos(2\Omega t)$.

4. Charge noise

We first analyse the interference part of the charge noise, $S_{\mathrm{int}} = \sum_{ij} S^{(ij)}_{\mathrm{int}}$. It can be shown that this simply corresponds to thermal fluctuations $S_{\mathrm{int}} = 2\frac{e^2}{h} D^2 k_{\mathrm{B}} T$, indicating that, even in

[1] Note that with this choice the energy current in the right reservoir coincides with the heat current.

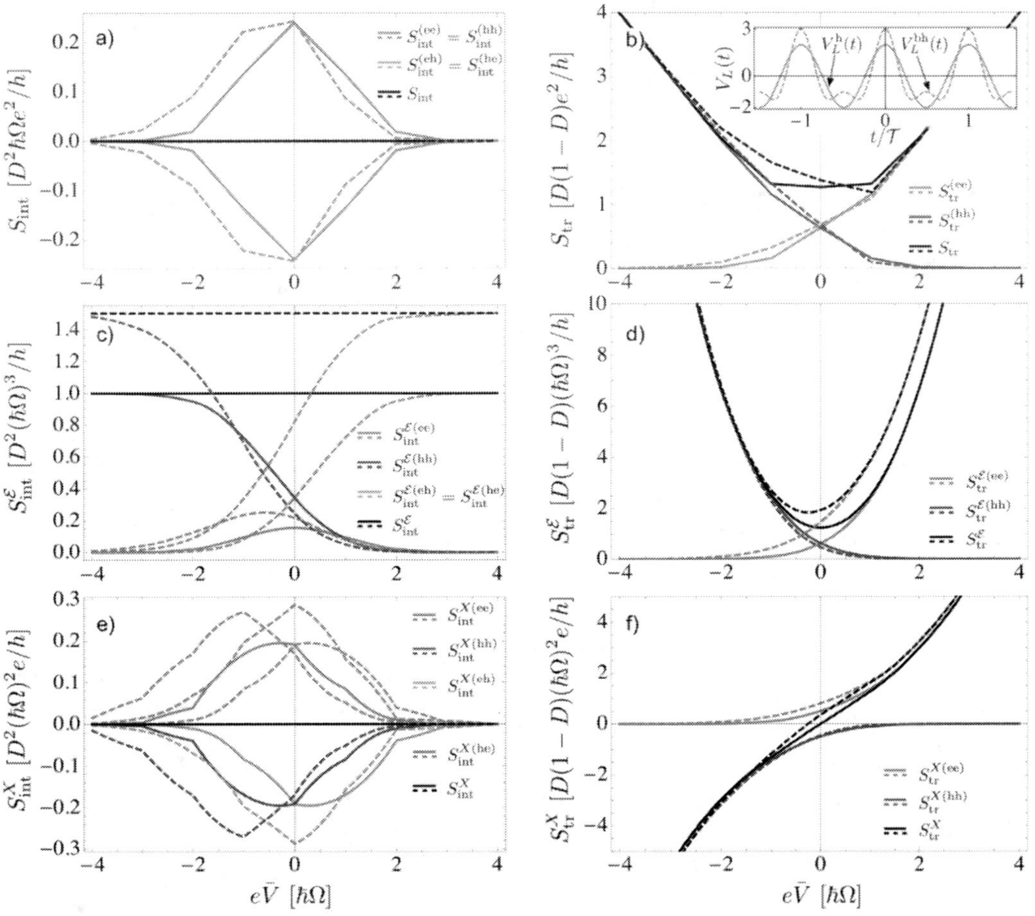

Figure 1. Plots of the interference and transport contributions to the charge ($S_{\rm int}^{(ij)}, S_{\rm tr}^{(ij)}$), energy ($S_{\rm int}^{\mathcal{E}(ij)}, S_{\rm tr}^{\mathcal{E}(ij)}$) and mixed ($S_{\rm int}^{X(ij)}, S_{\rm tr}^{X(ij)}$) noise as a function of the dc offset of the driving. Inset in panel b) Line-shape of the applied voltages: harmonic driving, $V_{\rm L}^{\rm h}(t) = \bar{V} + V_0\cos(\Omega t)$, and bi-harmonic driving of the form $V_{\rm L}^{\rm bh}(t) = \bar{V} + V_0\cos(\Omega t) + \frac{V_0}{2}\cos(2\Omega t)$. In all panels $k_{\rm B}T = 0$ and $eV_0 = 2\hbar\Omega$. Full lines correspond to the case of harmonic driving while dashed lines represent the case of bi-harmonic driving.

the presence of an ac-field, inelastic exchange processes in one reservoir interfere destructively when they all contribute to a certain observable with the same weight (e^2, for the case of the charge noise) [8]. In Fig. 1a) we plot the interference parts of the different contributions $S^{(ij)}$ as a function of the dc component of the bias at zero temperature. We find $S_{\rm int}^{(\rm ee)} = S_{\rm int}^{(\rm hh)} = -S_{\rm int}^{(\rm eh)} = -S_{\rm int}^{(\rm he)}$ giving indeed $S_{\rm int}(k_{\rm B}T = 0) = 0$.

We now consider the transport contribution to the charge noise, $S_{\rm tr} = \sum_i S_{\rm tr}^{(ii)}$, given by

$$S_{\rm tr} = \frac{e^2 D(1-D)}{h} \sum_{\ell=-\infty}^{+\infty} |c_{{\rm L}\ell}|^2 (\ell\hbar\Omega + e\bar{V}) \coth\left(\frac{\ell\hbar\Omega + e\bar{V}}{2k_{\rm B}T}\right). \quad (8)$$

At zero temperature this is the total charge noise $S_{\rm tr}(k_{\rm B}T = 0) = S(k_{\rm B}T = 0)$ and it can be written as the sum of a contribution due to electron-like excitations and a contribution due to hole-like excitations [7],

$$S(k_{\rm B}T=0)=\sum_{i={\rm e,h}}\frac{e^2\Omega}{2\pi}D(1-D)N_i\ . \tag{9}$$

Here, $N_i=\sigma_i\sum_{\ell=-\infty}^{+\infty}|c_{{\rm L}\ell}|^2(\ell+e\bar{V}/\hbar\Omega)\theta_i(\ell\hbar\Omega+e\bar{V})$ is the number of electrons ($i={\rm e}$) or holes ($i={\rm h}$) that arrive on the scatterer during one period, with $\sigma_{\rm e/h}=\pm$. Gabelli and Reulet [3] observed a reduction of the shot noise in a tunnel junction driven by a bi-harmonic signal with respect to the shot noise due to a simple harmonic driving. This is a consequence of the inhibition of electron-hole pair creation [6]. In Fig. 1b) we plot the transport contributions to the charge correlator $S_{\rm tr}^{(ii)}$ at zero temperature. It indeed shows a minimum in the noise when a biharmonic signal is applied, which corresponds to a reduction of $S_{\rm tr}^{\rm (ee)}$ and $S_{\rm tr}^{\rm (hh)}$ with respect to the case of harmonic driving.

5. Energy noise

The interference part of the energy noise, $S_{\rm int}^{\mathcal{E}}=\sum_{ij}S_{\rm int}^{\mathcal{E}(ij)}$ reads

$$S_{\rm int}^{\mathcal{E}}=\frac{D^2}{h}\left[\frac{2\pi^2\,(k_{\rm B}T)^3}{3}+\sum_{k=-\infty}^{+\infty}\frac{|e\,v_{{\rm L}k}|^2}{2}k\hbar\Omega\coth\left(\frac{k\hbar\Omega}{2k_{\rm B}T}\right)\right], \tag{10}$$

with $v_{{\rm L}k}=\int_0^{\mathcal{T}}\frac{dt}{\mathcal{T}}V_{\rm L}(t)e^{ik\Omega t}$. While the first term is again only given by thermal fluctuations, the second one is given by all the inelastic processes due to the ac driving field. This second term corresponds to the variance, $|ev_{{\rm L}k}|^2/2$, of the energy of a classical, charged particle in the oscillating potential $v_{{\rm L}k}\cos(k\Omega t)$ (i.e. the kth Fourier component of $V_{\rm ac}(t)$), multiplied by the characteristic rate at which electrons exchange the energy $k\hbar\Omega$ with the ac field by fluctuating between states in the same reservoir [8]. In Fig. 1c) we plot the interference contributions $S_{\rm int}^{\mathcal{E}(ij)}$ to the energy noise as a function of the dc offset of the driving. The correlators between different kinds of particles, $S_{\rm int}^{\mathcal{E}\rm (eh)}=S_{\rm int}^{\mathcal{E}\rm (he)}$, have the same order of magnitude as the ones for equal kinds of particles, as long as the dc component of the driving signal is very small, but they decay rapidly for increasing $\bar{V}$. Depending on the sign of $\bar{V}$ and on the shape of $V_{\rm ac}(t)$ we find $S_{\rm int}^{\mathcal{E}\rm (ee)}>S_{\rm int}^{\mathcal{E}\rm (hh)}$ or $S_{\rm int}^{\mathcal{E}\rm (ee)}<S_{\rm int}^{\mathcal{E}\rm (hh)}$.

The shape of the ac driving also determines the features of the transport contribution to the energy noise, $S_{\rm tr}^{\mathcal{E}}=\sum_i S_{\rm tr}^{\mathcal{E}(ii)}$, given by

$$S_{\rm tr}^{\mathcal{E}}=\frac{D(1-D)}{3h}\sum_{\ell=-\infty}^{+\infty}|c_{{\rm L}\ell}|^2\coth\left(\frac{\ell\hbar\Omega+e\bar{V}}{2k_{\rm B}T}\right)\left[(\ell\hbar\Omega+e\bar{V})^3+(\ell\hbar\Omega+e\bar{V})(\pi k_{\rm B}T)^2\right]. \tag{11}$$

This quantity depends not only on the number of electron and hole excitations but also on their energy distribution, determined by $V(t)$: for example, if $\max V_{\rm ac}(t)>|\min V_{\rm ac}(t)|$, the spread in energy of the electron-like excitations is larger than the one of the corresponding holes and this leads to $S_{\rm tr}^{\mathcal{E}\rm (ee)}>S_{\rm tr}^{\mathcal{E}\rm (hh)}$, as shown in the case of biharmonic driving in Fig. 1d). Differently from what observed for the charge noise, the suppression of the creation of charge-neutral excitations (electron-hole pairs) is therefore not enough to lead to a minimum of $S_{\rm tr}^{\mathcal{E}}$. It has been shown in Ref. [8] that a trade-off between the number of the excitations and the energy range they span has to be reached in order to minimize the transport part of the energy noise, see Fig. 1d).

6. Mixed noise

We now turn to the mixed correlator [9]. The total interference contribution, $S^X_{\rm int}=\sum_{ij}S^{X(ij)}_{\rm int}$ is

$$S^X_{\rm int} \;=\; \frac{e}{h}D^2 e\bar{V}k_{\rm B}T. \tag{12}$$

The interference part of the mixed noise shows different features from $S_{\rm int}$ and $S^{\mathcal{E}}_{\rm int}$. It is proportional to the interference part of the charge noise $S_{\rm int}$ times $e\bar{V}/2$, which is the average energy of the electrons injected into the system at the left contact, with respect to $V_{\rm R}=\mu=0$. It is hence non-vanishing only if the system is driven out of equilibrium and temperature is finite.

In Fig. 1e) we plot the contributions to $S^X_{\rm int}$ from excitations of the same or of different kind, $S^{X(ij)}_{\rm int}$, as a function of $e\bar{V}$ at zero temperature. We find $S^{X({\rm ee})}_{\rm int}=-S^{X({\rm he})}_{\rm int}$ and $S^{X({\rm hh})}_{\rm int}=-S^{X({\rm eh})}_{\rm int}$ giving as a consequence that the interference contribution to the correlator between fluctuations of the energy carried by the electrons (holes) and the total charge current is zero. Moreover, we have in general $S^{X({\rm ee})}_{\rm int}\neq S^{X({\rm hh})}_{\rm int}$, apart for the case when $e\bar{V}=0$ and $V_{\rm ac}(t)$ is symmetric with respect to one of its nodes, so that the holes and the electrons are symmetrically distributed in energy with respect to $\mu=0$. In that case they just have opposite sign. If an asymmetry in the energy distribution of the electron- and hole-like excitations is introduced *only* as an effect of the dc-bias, see e.g. the case of the harmonic signal shown in Fig. 1e), then we have $S^{X({\rm ee})}_{\rm int}(e\bar{V})=-S^{X({\rm hh})}_{\rm int}(-e\bar{V})$.

The transport contribution, $S^X_{\rm tr}=\sum_{ij}S^{X(ii)}_{\rm tr}$, is given by

$$S^X_{\rm tr}=\frac{eD(1-D)}{2h}\sum_{\ell=-\infty}^{+\infty}|c_{{\rm L}\ell}|^2(\ell\hbar\Omega+e\bar{V})^2\coth\left(\frac{\ell\hbar\Omega+e\bar{V}}{2k_{\rm B}T}\right) \tag{13}$$

At zero temperature, it is the only non-vanishing contribution to the mixed noise, $S^X(k_{\rm B}T=0)=S^X_{\rm tr}(k_{\rm B}T=0)$. We can separate a contribution due to electron-like excitations from the one due to hole-like excitations

$$S^X(k_{\rm B}T=0)=\sum_{i={\rm e,h}}\frac{\Omega}{2\pi}D(1-D)\Xi_i. \tag{14}$$

The quantity Ξ_i is formally similar to the definition of N_i,

$$\Xi_i=\sigma_i e\sum_{\ell=-\infty}^{+\infty}|c_{{\rm L}\ell}|^2\frac{(\ell\hbar\Omega+e\bar{V})(\ell+e\bar{V}/\hbar\Omega)}{2}\theta_i(\ell\hbar\Omega+e\bar{V}), \tag{15}$$

but each contribution coming from a different subband l is weighted by its corresponding energy. It can thus be understood as the total average energy transported by i-like excitations multiplied by $\sigma_{\rm e/h}e=\pm e$ the corresponding charge. Indeed, $(\Xi_{\rm e}-\Xi_{\rm h})/(e\mathcal{T})=\overline{eV_{\rm L}(t)^2}/2h$ corresponds to the power P dissipated to one of the leads per period. In a two-terminal conductor characterized by energy-independent transmission probability at the central scatterer, the energy is equally dissipated in the two reservoirs resulting in the factor 1/2.

In Fig. 1f) we plot $S^{X(ii)}_{\rm tr}$ at zero temperature as a function of the dc part of the driving signal. The correlators $S^{X({\rm ee})}_{\rm tr}$ and $S^{X({\rm hh})}_{\rm tr}$ have opposite signs, as a consequence of the fact that electron- and hole- excitations have different charges but both give a contribution to the energy current of the same sign. Moreover, they become negligible when $e\bar{V}\ll 0$ or $e\bar{V}\gg 0$, respectively, since $\Xi_{\rm e}$ and $\Xi_{\rm h}$ are suppressed in these voltage ranges. This leads to a change of sign of the total transport contribution $S^X_{\rm tr}$ as a function of $e\bar{V}$, in contrast to $S_{\rm tr}$ and $S^{\mathcal{E}}_{\rm tr}$.

We now relate the mixed noise to the zero frequency correlator between the power fluctuations and the charge current fluctuations, $\overline{\langle \Delta P(t)\Delta \hat{I}_{\rm R}(t+\tau)\rangle} = \int_0^{\mathcal{T}} dt \int_{-\infty}^{\infty} d\tau \langle\{\Delta\hat{P}(t), \Delta\hat{I}_{\rm R}(t+\tau)\}/(2\mathcal{T})$, a measurable quantity. Here, we define $\hat{P}(t) = -V_{\rm L}(t)\hat{I}_{\rm L}(t)$, the operator for the power provided by the time-dependent voltage source. We find $\langle\Delta P(t)\Delta\hat{I}_{\rm L}(t+\tau)\rangle + \langle\Delta P(t)\Delta\hat{I}_{\rm R}\ (t+\tau)\ \rangle = 0$, since neither charge nor energy can be accumulated in the conductor. Explicitly, we have

$$\overline{\langle \Delta P(t)\Delta I_{\rm R}(t+\tau)\rangle} = 2S_{\rm int}^X + 2S_{\rm tr}^X\,, \tag{16}$$

with the mixed correlator contributions discussed in Eqs. (12) and (13). Importantly, for a completely transparent scatterer, $D = 1$, $2S_{\rm int}^X$ is the only contribution to the zero-frequency correlator of power and charge current fluctuations; if the transparency of the scatterer is reduced, the second term which corresponds to $2S_{\rm tr}^X$, is the dominant one. Both terms should thus be well observable.

7. Conclusions

We considered a two-terminal conductor subjected to a time-periodic bias voltage. We discussed the correlator of charge current fluctuations and energy current fluctuations and the correlator between charge and energy current fluctuations. All the considered correlators show a transport contribution, due to correlations between particles exchanged between the two leads, and an interference contribution, due to the exchange of particles between different energy states in the same lead. The mixed noise is nonzero since electrons and holes are carriers both of charge and energy. At zero temperature, it is formally analogous to the charge noise, but now what is partitioned at the scatterer is not the number of impinging electrons or holes, but their energy, averaged over the different subbands. In contrast with charge and energy noise, it does not show a minimum as a function of an applied dc bias, but it changes sign when the bias is reversed. All discussed effects are expected to be observable by looking at the correlations between power and charge current fluctuations, under given conditions.

References

[1] Pekola J P, Saira O-P, Maisi V F, Kemppinen A, Möttönen M, Pashkin Y A and Averin D V 2013 *Rev. Mod. Phys.* **85** 1421
[2] Keeling J, Klich I and Levitov L S 2006 *Phys. Rev. Lett.* **97** 116403
[3] Gabelli J and Reulet B 2013 *Phys. Rev. B* **87** 075403
[4] Dubois J, Jullien T, Grenier C, Degiovanni P, Roulleau P and Glattli D C 2013 *Nature* **502** 659
[5] Moskalets M and Büttiker M 2004 *Phys. Rev. B* **70** 245305
Moskalets M and Büttiker M 2006 *Phys. Rev. B* **73** 125315
Gasse G, Spietz L, Lupien C and Reulet B 2013 *Phys. Rev. B* **88** 241402
Dubois J, Jullien T, Grenier C, Degiovanni P, Roulleau P and Glattli D C 2013 *Phys. Rev. B* **88** 085301
[6] Vanević M, Nazarov Y V and Belzig W 2007 *Phys. Rev. Lett.* **99** 076601
Vanević M, Nazarov Y V and Belzig W 2008 *Phys. Rev. B* **78** 245308
Vanević M and Belzig W 2012 *Phys. Rev. B* **86** 241306
[7] Reydellet L-H, Roche P, Glattli D C, Etienne B and Jin Y 2003 *Phys. Rev. Lett.* **90** 176803
[8] Battista F, Haupt F and Splettstoesser J 2014 *Preprint* arXiv:1405.4326
[9] Crépieux A and Michelini F 2014 *Preprint* arXiv:1403.8035
[10] Virtanen P and Giazotto F 2014 *Preprint* arXiv:1405.1362
[11] Moskalets M 2014 *Phys. Rev. Lett.* **112** 206801
[12] Virtanen P and Giazotto F 2014 *Preprint* arXiv:1407.0493
[13] Prêtre A, Thomas H and Büttiker M 1996 *Phys. Rev. B* **54** 8130
Pedersen M H and Büttiker M 1998 *Phys. Rev. B* **58** 12993
[14] Büttiker M 1992 *Phys. Rev. B* **46** 12485
[15] Ludovico M F, Lim J S, Moskalets M, Arrachea L and Sánchez D 2014 *Phys. Rev. B* **89** 161306
[16] Rychkov V S, Polianski M L and Büttiker M 2005 *Phys. Rev. B* **72** 155326

Journal of Physics: Conference Series **568** (2014) 052009 doi:10.1088/1742-6596/568/5/052009

Effect of the Inert Gas Adsorption on the Bilayer Graphene to the Localized Electron Magnetotransport

A Fukuda[1], D Terasawa[1], Y Ohno[2], K Matsumoto[2]

[1] Department of Physics, Hyogo College of Medicine, 1-1 Mukogawacho, Nishinomiya, Hyogo 663-8501, Japan

[2] The Institute of Scientific and Industrial Research, Osaka University, 8-1 Mihogaoka, Ibaraki, Osaka 567-0047, Japan

E-mail: fuku@hyo-med.ac.jp

Abstract. Graphene has a fascinating property that the two-dimensional electron gas is easily accessible externally and it is challenging to investigate the effects of the adsorption of inert gases on graphene, which may be the least effective chemically and physically. We carry out the magnetotransport measurements of ^{4}He-adsorbed bilayer graphene at low temperatures and the magnetic field B ranging from 0 to 4 T. The magnetoresistance ΔR_{xx} change from the pristine graphene is measured as a function of gate voltage V_g and B for partial coverage of 1/10 (= 0.1) layers and one layer ^{4}He-adsorbed graphene. The overall magnitudes of ΔR_{xx} for one layer are larger than the one for 1/10 layers. Signs of ΔR_{xx} depend on the V_g for the entire range of B, associated with the magnetoresistance oscillation owing to the weak localization in the pristine graphene.

1. Introduction

The experimental and theoretical studies of graphene, a two-dimensional (2D) monolayer of carbon atoms tightly packed into a honeycomb lattice, have attracted a major interest of modern condensed-matter physics, since an unusual quantum Hall effect (QHE) in high magnetic field was discovered in 2005 by two different groups[1, 2]. The most salient features of graphene contains the pseudo-relativistic physics performed by the massless Dirac fermions. The pseudo-relativistic QHE is understood in terms of Landau quantization for massless Dirac fermions, which is on the theoretical basis on the electron-electron interactions. In the case of bilayer graphene, unconventional QHE was found with massive chiral fermions with Berry's phase 2π[3]. On the contrary, in low magnetic field, weak localization (WL) dominates the transport properties of graphene. The interference of electron waves that are scattered by disorders or impurities, and form closed trajectories, usually decreases the conductivity, resulting in the WL of electrons. A negative magnetoresistance as a signature of WL was reported including the electroneutrality region in the bilayer graphene[4]. The WL in the bilayer graphene was also theoretically investigated[5].

Another remarkable feature of graphene is the accessibility to the two-dimensional electron gas (2DEG) from outside. Though the 2DEG in the conventional semiconductor heterostructures such as GaAs/AlGaAs is known to have an extremely high mobility, it has been rather difficult

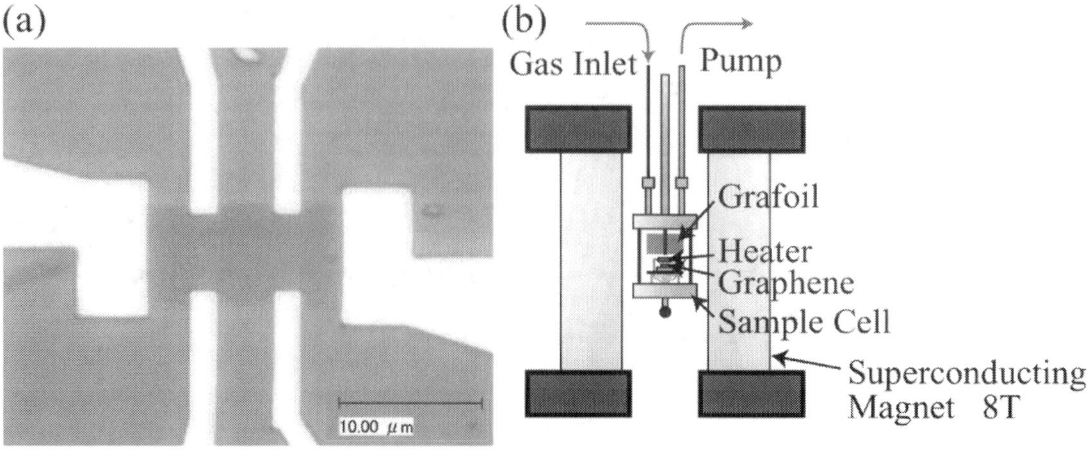

Figure 1. (a) Optical microscope image of our exfoliated graphene device. (b) Design of the cryostat for transport measurements of graphene device with ^{4}He atom adsorption.

to access directly from the outside probe such as a scanning tunnel microscope. Graphene is a promising material not only to solve this problem but also to study the effects of the adsorption or chemical adjunction of various kinds of molecules on graphene. Therefore, to study the transport properties with an adsorption or chemical modification by external molecules on graphene is intriguing. Shedin *et al.* demonstrated that micrometre-size sensors made from graphene are capable of detecting individual events when a gas molecule such as CO, NH_3, H_2O, NO_2 attaches to or detaches from graphene's surface[6]. Another prominent achievement is findings of the evidence of hydrogenated graphene called "graphane"[7], where the appearance and the disappearance of the QHE are fully controllable by reversible hydrogenation. Applied techniques are utilized for such as an electrical detecting pH and protein adsorptions[8] or a selective permeation of water[9]. These works envision us the higher feasibility of graphene-based gas sensors with an ultimate sensitivity. Therefore, it is most challenging but has been less explored to investigate the effects of the adsorption of an inert gas on graphene, which may be the least effective to the transport properties.

In this report, we observe the magnetotransport of ^{4}He-adsorbed bilayer graphene at low temperatures and under the magnetic field B ranging from 0 to 4 T. We measure the ΔR_{xx}, the difference of the magnetoresistance R_{xx} between the pristine graphene and helium adsorbed graphene, where the layer number of adsorbed atoms is precisely controlled. We discuss the relationship between ΔR_{xx} and the weak localization effect of 2DEG transport. We also remark the possibility to utilize the graphene device as a highly sensitive sensor for adsorbed gas as a powerful tool of the condensed matter physics.

2. Experiments

The bilayer graphene is fabricated by the mechanical exfoliation method[10] onto the surface of SiO_2 300 nm apart from the n^+ doped Si substrate. Optical microscope image of our exfoliated graphene device is shown in Fig. 1(a). We choose a natural bilayer graphene flake that has almost a rectangular shape. We fabricate a Hall bar with the dimension of 13 μm-long for the source - drain electrodes, 5 μm-long for the longitudinal voltage electrodes, and 7 μm-wide for the Hall voltage electrodes. The contact pattern is depicted by electron beam lithography. The ohmic contact materials (10 nm of Ti followed by 100 nm of Au) are deposited by thermal evaporation then followed by liftoff in warm 1-methyl-2-pyrrolidone. The number of the layers

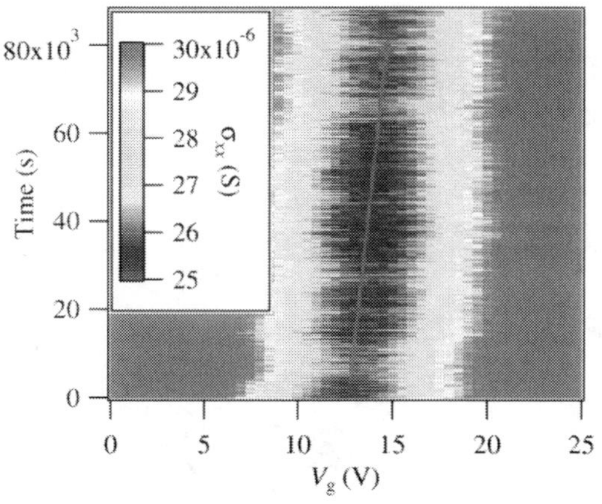

Figure 2. Displacement of the charge neutral point (CNP). Time evolution of the conductivity σ_{xx} as a function of gate voltage $V_{\rm g}$ is demonstrated.

is confirmed by the Raman shift spectra[11]. The sample is first annealed at 700 K in the H_2 atmosphere for 30 minutes.

Overview of our cryostat for the transport measurements of graphene with gas adsorption is illustrated in Fig. 1(b). The sample is placed in the vacuum-tight sample cell with an external heater and a grafoil sheet having large effective areas of 1.2 m^2. The grafoil sheet is utilized to regulate and determine the precise number of layers of adsorbed molecules. The heater inside the sample cell enables us *in situ* annealing even at low temperatures. The sample is annealed again by using this heater for two hours at 410 K before it is cooled down. Resistances are measured by a standard ac lock-in technique of 4-terminal measurements with a frequency of 37 Hz. Carrier densities are controlled by the back gate voltage $V_{\rm g}$ applied between the graphene sheet and the substrate. The helium-free Gifford-MacMahone (GM) refrigerator is used to cool down the sample as well as the superconductor magnet with the maximum magnetic field of 8 T. The magnetic field is applied up to 4 T in this experiment.

Using the relationship between the carrier density n and the gate voltage $n = (C/e) \cdot V_{\rm g} = 7.2 \times 10^{10}\, V_{\rm g}\, \mathrm{cm}^{-2}$, where C denotes the capacitance of 300-nm thick SiO_2 and e denotes the elementary charge, the carrier density at 18V apart from the charge neutral point (CNP) is estimated to be $1.3 \times 10^{16}\, \mathrm{cm}^{-2}$. Thus the mobility of the sample at room temperature is estimated to be $2500\, \mathrm{cm}^2/\mathrm{Vs}$.

3. Results and Discussions

First, to check the sensitivity of our graphene device for an adsorption of the external gas molecules at room temperature, we investigate the conductivity σ_{xx} change induced by the air adsorption at the field B of 0 T, followed by the well-annealed process mentioned above, while we monitor that the CNP moves to the lower voltage. Figure 2 demonstrates the time t evolution of σ_{xx} for gate voltage $V_{\rm g}$ sweep, where the graphene is exposed to air at the constant pressure $P = 815$ Pa. At $t = 0$ sec, the minimum of σ_{xx} is located at $V_{\rm g} = 12.5\,\mathrm{V}$, which is located at the CNP. At $t = 8.78 \times 10^4$ sec, the CNP moves to $V_{\rm g} = 14.5\,\mathrm{V}$, indicated by a red arrow in Fig. 2. Therefore, the velocity of the CNP displacement $dV_{\rm g}^{\rm CNP}/dt = 2.3 \times 10^{-5}\mathrm{V/s}$. Subsequent experiments reveal that the quantity of $dV_{\rm g}^{\rm CNP}/dt$ sensitively depends on the gas pressure P, where $dV_{\rm g}^{\rm CNP}/dt$ increases as P increases.

Next, we cool down the device of a graphene to about 7 K and observe the temperature dependence of the resistance R_{xx}. Figure 3 presents the R_{xx} as a function of $V_{\rm g}$ for various temperature T. The position of the CNP $V_{\rm g}^{\rm CNP}$ is almost temperature independent. Above

Journal of Physics: Conference Series **568** (2014) 052009 doi:10.1088/1742-6596/568/5/052009

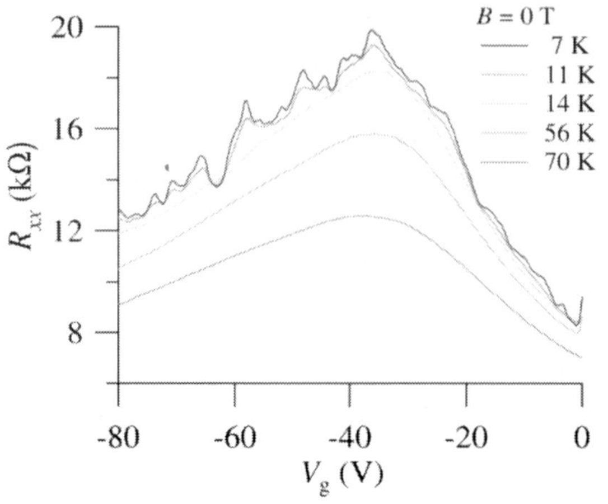

Figure 3. Observed resistance R_{xx} of a pristine graphene as a function of V_g for various temperatures.

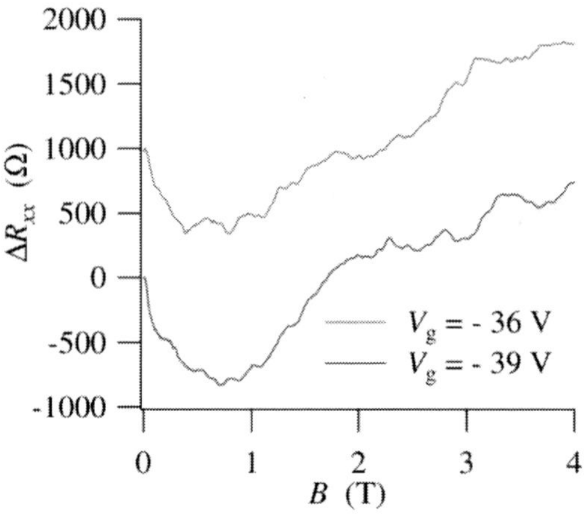

Figure 4. Magnetoresistance R_{xx} of a pristine graphene as a function of magnetic field B for two gate voltages V_g. The temperature is 7 K.

$T = 14$ K, the curve of R_{xx} is apparently smooth for the entire range of V_g. At $T = 11$ K, a subtle oscillation of the R_{xx} appears. As T decreases to 7 K, the amplitude of the oscillation grows more prominently. The pattern of the oscillation is fully reproducible, which is a typical manifestation of the universal conductance fluctuation (UCF) in the mesoscopic system. It should be also noted that the derivative of dR_{xx}/dT is negative everywhere in the whole range of V_g, which indicate that the graphene device is slightly insulative (not metallic), but here we emphasize that the monotonous dependence of R_{xx} on T is a key property.

To investigate the physical origin of such oscillations in detail, we measured the magnetoresistance R_{xx} for two different gate voltages at $V_g = -36\,\text{V}$ and $V_g = -39\,\text{V}$ at a low temperature of 7 K. In both cases, R_{xx} decreases as B increases in the weak magnetic field. As B increases more, R_{xx} turns to increase via the minima of R_{xx} that is located at $B_{\text{min}} = 0.55\,\text{T}$ at $V_g = -36\,\text{V}$ and $B_{\text{min}} = 0.70\,\text{T}$ at $V_g = -39\,\text{V}$. Here two gate voltages are chosen in the vicinity of the CNP, but these behaviors about B dependence of R_{xx} are similarly observed for the V_g far from the CNP.

The properties of R_{xx} in the low magnetic field limit are well explained by the weak localization (WL) theory in the bilayer graphene[4]. The interference of the electron waves that are scattered by disorders or impurities in the closed trajectories is weakened by the magnetic

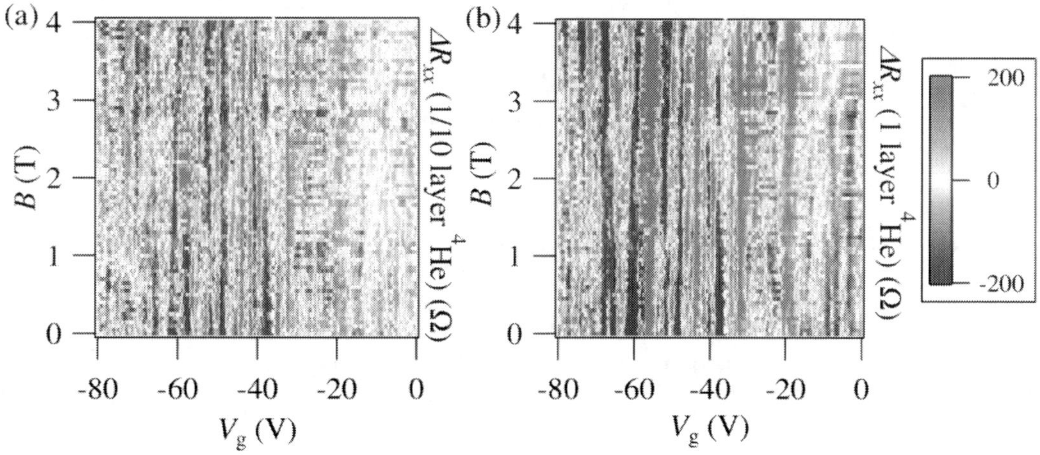

Figure 5. The magnetoresistance difference ΔR_{xx} between ^{4}He adsorbed graphene and pristine one as a function of B and V_g at $T = 7$ K. The adsorbed layer of ^{4}He is (a) 1/10 layers and (b) 1 layer.

field, resulting in a negative magnetoresistance. Since the interference of the electron waves in two different trajectories also causes the UCF of the order of e^2/h, where e is the elementary charge and h is the Plank constant, it concurs to the results of the magnetoresistance oscillation for both V_g and B in Fig. 3.

When $B > B_{\min}$, the conventional property of the observed enhancement in R_{xx} for increment of B implies that the crossover from WL to the strong Anderson localization. The R_{xx} behaviors are well analyzed by the model slightly modified from the equations for monolayer graphene suggested by McCann *et al.*[12] associated with various possible ways of breaking a 'hidden' valley symmetry of the graphene[13].

Figure 5 contains the most central part of this study. First, we pre-measure the magnetoresistance of the well-annealed pristine graphene R_{xx}^{pristine} as a function of B and V_g at $T = 7$ K. Afterward, we pour the ^{4}He gas to the sample cell through thin capillary from the room temperature, monitoring the decrement of the gas pressure in the constant volume. The adsorption energy of ^{4}He atom on graphite is about 12.2 K[14]. Though the binding energy of ^{4}He atom on graphene has not been experimentally observed yet, theory calculated it on the monolayer graphene about 13.4 K[15]. However, since we employ the bilayer graphene that is attached on the SiO_2 substrate, actual adsorption energy of ^{4}He on our sample may approach to the one on graphite. Thus we assume that adsorption energy of ^{4}He on graphene is nearly equivalent to the one on graphite, almost all ^{4}He atoms are homogenously adsorbed to mainly grafoil at $T = 7$ K, and we estimate the layer number of adsorbed ^{4}He atoms. We also measure the magnetoresistance with ^{4}He atoms, R_{xx}^{He}, as a function of B and V_g at $T = 7$ K. Then the difference $\Delta R_{xx} = R_{xx}^{\text{He}} - R_{xx}^{\text{pristine}}$ is demonstrated as an image plot for B and V_g in Fig. 5 for partial coverage of 1/10 (= 0.1) layers (a) and 1 layer (b) ^{4}He. The red and blue areas indicate the positive and negative change of magnetoresistance, respectively. The deeper colors represent the larger amplitude of ΔR_{xx}. The partial coverage of 1/10 layers is selected to attain the higher sensitivity on behalf of further investigation for 2D ^{4}He properties.

Firstly, at a glance of Fig. 5, overall colors of graph in Fig. 5(b) are darker than ones in Fig. 5(a). This fact suggests that the magnetoresistance is more influenced by the adsorption of larger amount of ^{4}He atoms. Secondly, the total amount of red areas and blue ones are almost equal. The equality of the proportion of each area with different sign in ΔR_{xx} implies the changes of the resistance are NOT owing to the temperature variation, because the R_{xx} at 0 field monotonically depend on T. Finally, the shape of red and blue areas seems stripe extended to the vertical direction. It means that though ΔR_{xx} are strongly influenced by V_g, it depends on B very weakly.

Here we discuss the effects of adsorbed inert atoms to electron transport in graphene and their mechanisms. As mentioned above, we could perfectly eliminate the possibility of sample heating by the thermal influx from the hot part of the cell to the sample. Therefore the R_{xx} change reflects intrinsic 2DEG properties. Now we focus on the relationship between the sign of ΔR_{xx} and the UCF oscillation. When we compare the ΔR_{xx} in Fig. 5 to R_{xx}^{pristine} carefully, we find the tendencies that the positive sign of ΔR_{xx} appears around the R_{xx} local maxima and negative one does around the R_{xx} local minima. These tendencies suggest that the ΔR_{xx} is significantly affected by the UCF pattern. The weak dependence of ΔR_{xx} on B at low magnetic field enlightens that the electron paths responsible for the interference does not change so much by the adsorption of ^{4}He atoms. However, the fact that 1 ^{4}He layer is more effectively change R_{xx} than 1/10 ^{4}He layers implies that the amplitude of the Coulomb potential that electrons feel caused by disorders or impurities of graphene varies by the adatoms, though the allocations of such defects, and thus electron paths, does not change drastically. Because the ^{4}He atoms are fully isotropic and Coulomb interactions between the ^{4}He atom and substrate are relatively small, the magnitude of ΔR_{xx} is also small. However, such subtle change of magnetoresistance may open the feasibility for detecting the intrinsic ^{4}He properties such as two dimensional Kosterlitz-Thouless type superfluid transition[16] or phase transition of two dimensional ^{4}He solid related to the commensurability[17]. Additionally, it should be noted that the phase diagram of ^{4}He on graphene is theoretically proposed[18], and thus the experimental investigation is invaluable. ΔR_{xx} is highly associated with the UCF, we infer that the more "dirty" graphene may be suitable for more sensitive detectors. Therefore the relationship between the density of the defects or impurities in artificially flawed graphene and the sensitivity as gas detectors should be further investigated.

In conclusion, we have carrid out the magnetotransport measurements of ^{4}He-adsorbed bilayer graphene at low temperature and the magnetic field B ranging from 0 to 4 T. The magnetoresistance difference ΔR_{xx} from the pristine graphene is observed as a function of gate voltage V_g and B for partial coverage of 1/10 (= 0.1) layers and one layer ^{4}He-adsorbed graphene. The overall magnitudes of ΔR_{xx} for one layer are larger than the one for 1/10 layer. Signs of ΔR_{xx} depend on the V_g for the entire range of B, associated with the magnetoresistance oscillation owing to the weak localization in the pristine graphene.

4. Acknowledgments

We acknowledge to A. Sawada and H. Yayama for the experimental setup at low temperature and high magnetic field with helium-free cryostat. We thank T. Nakajima for the machining the cryostat parts and the electronics of experiments. We are thankful to T. Terashima and S. Sasaki for utilizing the clean room facilities in the LTM center of Kyoto university. We are also thankful to A. Fujimoto for fruitful discussions about the weak localization of graphene. This research was performed under the Cooperative Research Program of "Network Joint Research Center for Materials and Devices". This work was supported by the "Topological Quantum Phenomena" (No. 25103722) Grant-in Aid for Scientific Research on Innovative Areas from the Ministry of Education, Culture, Sports, Science and Technology (MEXT) of Japan, JSPS KAKENHI Grant Number 24540331 and 25870966, Grants-in-Aid for Researchers from Hyogo College of Medicine (2010 for D. T. and 2011 for A. F.) and Grant for Basic Science Research Projects from the Sumitomo Foundation.

References

[1] Y. Zhang, Y. W. Tan, H. L. Stormer, and P. Kim, *Nature (London)* **438**, 201 (2005).
[2] K. S. Novoselov, A. K. Geim, S. V. Morozov, D. Jiang, M. I. Katsnelson, I. V. Grigorieva, S. V. Dubonos and A. A. Firsov, *Nature (London)* **438**, 197 (2005).
[3] K. S. Novoselov, E. McCann, S. V. Morozov, V. I. Fal'ko, M. I. Katsnelson, U. Zeitler, D. Jiang, F. Schedin and A. K. Geim, *Nat. Phys.* **2**, 177 (2006).
[4] R.V. Gorbachev, F.V. Tikhonenko, A. S. Mayorov, D.W. Horsell, and A. K. Savchenko, *Phys. Rev. Lett.* **98**, 176805 (2007)
[5] K. Kechedzhi, V. I. Fal'ko, E. McCann, and B. L. Altshuler, *Phys. Rev. Lett.* **98**, 176806 (2007)
[6] F. Schedin, A. K. Geim, S. V. Morozov, E. W. Hill, P. Blake, M. I. Katsnelson and K. S. Novoselov, *Nat. Mat.* **6**, 652 (2007).
[7] D. C. Elias, R. R. Nair, T. M. G. Mohiuddin, S. V. Morozov, P. Blake, M. P. Halsall, A. C. Ferrari, D. W. Boukhvalov, M. I. Katsnelson, A. K. Geim and K. S. Novoselov, *Science* **323**, 30 (2009). o, *Phys. Rev. Lett.* **98**, 176805 (2007)
[8] Y. Ohno, K. Maehashi, Y. Yamashiro, and K. Matsumoto, *Nano. Lett.* **9**, 3318 (2009).
[9] R. R. Nair, H. A. Wu, P. N. Jayaram, I. V. Grigorieva and A. K. Geim, *Science* **335**, 442 (2012).
[10] K. S. Novoselov, A. K. Geim, S. V. Morozov, D. Jiang, Y. Zhang, S. V. Dubonos, I. V. Grigorieva, A. A. Firsov, *Science* **306**, 666 (2004).
[11] A. C. Ferrari, J. C. Meyer, V. Scardaci, C. Casiraghi, M. Lazzeri, F. Mauri, S. Piscanec, D. Jiang, K. S. Novoselov, S. Roth, and A. K. Geim, *Phys. Rev. Lett.* **97**, 187401 (2006).
[12] E. McCann, K. Kechedzhi, V. I. Fal'ko, H. Suzuura, T. Ando, B. L. Altshuler, *Phys. Rev. Lett.* **97**, 146805 (2006).
[13] D. Terasawa, A Fukuda, Y. Ohno and K. Matsumoto, *to be published.*
[14] M. W. Cole and D. R. Frankl, *Rev. Mod. Phys.* **53**, 199 (1981).
[15] M. C. Gordillo and J. Boronat, *Phys. Rev. Lett.* **102**, 085303 (2009).
[16] D. J. Bishop and J. D. Reppy, *Phys. Rev. Lett.* **40**, 1727 (1978).
[17] D. S. Greywall and P. A. Busch, *Phys. Rev. Lett.* **67**, 3535 (1991).
[18] J. Happacher, P. Corboz, M. Boninsegni and L. Pollet, *Phys. Rev. B* **87**, 094514 (2013)

Journal of Physics: Conference Series **568** (2014) 052010 doi:10.1088/1742-6596/568/5/052010

Room-temperature quantum Hall effect in graphene: the role of the two-dimensional nature of phonons

A A Greshnov[1,2]

[1] Ioffe Physical-Technical Institute RAS, Polytechnicheskaya 26, St Petersburg, Russian Federation
[2] Saint Petersburg Electrotechnical University, Professora Popova 5, St Petersburg, Russian Federation

E-mail: a_greshnov@hotmail.com

Abstract. We consider two-dimensional nature of the electron-phonon coupling in graphene as a source for the room-temperature quantum Hall effect discovered in 2007. It is shown that magnetic field introduces strong cut-off for coupling with the two-dimensional acoustic phonons, *viz.* the processes with energy tranfer exceeding $\hbar s l_B^{-1}$ are exponentially suppressed, while for three-dimensional phonons the cut-off is set by a temperature T (here s is the sound velocity and $l_B \propto B^{-1/2}$ is the magnetic length). Consequently, at sufficiently high temperatures and magnetic fields only a small part ($\sim \hbar s l_B^{-1}/T$) of the electron states is involved in coupling with a given electron state in comparison with the case of three-dimensional phonons. Hence, the percolation threshold is postponed, and the quantum Hall effect survives up to $T = 300\,\text{K}$.

1. Introduction

The integer quantum Hall effect is known to be irrefutable proof of atomistically two-dimensional nature of the graphene sheets [1], produced originally by the exfoliation technique, and now by the epitaxial methods suitable for massive production of graphene. Though it was speculated to be "half-integer", in fact the Hall conductivity of graphene σ_{xy} is quantized in the units of e^2/h, while really intriguing is consequence of the plateaus, described by the redoubled odd numbers, so that $\sigma_{xy} = 2(2n+1)$. This particular sequence is readily understood from the Landau level spectrum of graphene,

$$E_n = \mathrm{sgn}(n)\sqrt{2|n|}\hbar v_0 l_B^{-1}, \tag{1}$$

taking into account fourthfold degeneracy due to valleys and spin, and symmetry of the ground Landau level $n = 0$ with respect to zero energy point $E = 0$ (here $v_0 \approx 10^8\,\text{cm/s}$ is the graphene effective electron velocity and $l_B = \sqrt{\hbar c/eB}$ is the magnetic length). In fact, the "half-integer" rule of quantization is not really universal, since $\sigma_{xy} = 0, \pm 4$, and even $\sigma_{xy} = \pm 1$ plateaus have been observed experimentally [2] at sufficiently low temperatures, and the latter set of plateaus ($\sigma_{xy} = \pm 1$) raises questions to a theory of Ref. [3], employing the non-linear σ-model with artificially added topological term to explain the "half-integer" effect.

However, the most striking feature of the quantum Hall effect in graphene is, in our opinion, its immunity to elevated temperatures, as high as 300 K [4], while 20 K is the best of what the traditional two-dimensional systems can boast. Although this fact has been considered by many researches just as manifestation of the large gaps between the ground ($n = 0$) and the first excited ($n = \pm 1$) Landau levels, described by Eq. (1), physics behind the room-temperature quantum Hall effect in graphene is indeed much more exciting. Using Eq. (1) together with the usual expression for the cyclotron energy suitable for convenient two-dimensional structures, $\hbar\omega_c = \hbar eB/m^*c$, it is easy to check that the distances between $n = 0$ and $n = 1$ Landau levels of graphene and $InAs$ quantum wells ($m^* = 0.023m_0$) differ less than by a factor of 1.5 at $B = 30\,\mathrm{T}$, while the limiting temperatures differ by more than one order of magnitude. This means it is not just $T/\hbar\omega_c$ parameter that restricts observation of the quantum Hall effect at $T > 20\,\mathrm{K}$ in the convenient two-dimensional structures, but there are relevant inelastic scatterings, which destroy coherence in the electron subsystem, namely electron-electron and electron-phonon interactions. Though it is not easy to describe influence of the interactions on the quantum Hall effect explicitly, they are proved experimentally to limit maximum temperatures for observation of the effect, and the transition widths between adjacent plateaus are shown to depend on temperature as $\Delta B \propto T^{\kappa}$ with $\kappa \approx 0.42$, reflecting ideas of the so-called scaling theory of the integer quantum Hall effect [5]. In other words, dephasing processes at strong magnetic fields are rather tricky (compared to the case of zero magnetic field [6]) since almost all the electron states are localized and require self-consistent treatment at essentially non-zero temperature T, introducing heat bath of either photons or phonons [7].

In this paper we focus on the electron-phonon interaction in graphene as being more relevant at not too low temperatures [6], considering the two-dimensional acoustic phonons of graphene. Some time ago, the two-dimensional nature of the graphene phonons was shown theoretically [8] to change appreciably the temperature dependence of resistivity in the Bloch-Grüneisen regime, from $\rho \sim T^5$ to $\rho \sim T^4$, and this prediction was confirmed experimentally by Evefov and Kim [9]. And though graphene physics is far from being settled down, for example reported magnitude of the deformation potential Ξ varies from 3 to 30 meV (see discussion in Ref. [10]), we believe that the two-dimensional intrinsic graphene phonons dominate over the three-dimensional substrate phonons. Our results show that reduced dimensionality of the phonons is really dramatic at a strong magnetic field, which introduces strong cut-off for energy of a phonon to be absorbed or radiated. Notably, a crucial thing here is not the total rate of transitions from a given electron state (*i.e.* the inverse lifetime), but number of the electron states coupled with this state by phonons, which is reduced by factor $\hbar s l_B^{-1}/T \sim 10^{-1}$ in comparison with the case of three-dimensional phonons. This feature explains robustness of the quantum Hall in graphene to the temperatures, which are an order of magnitude higher than allowed for traditional two-dimensional structures with the essentially three-dimensional phonons.

2. Theory

It is well known that lateral dynamics of the electrons set on the Landau levels of graphene does not differ cardinally from the case of convenient two-dimensional electron gas, except for the multi-component (chiral) structure of the wave function [11]. However, the room-temperature quantum Hall effect in graphene exhibits only two plateaus, both belonging to the ground Landau level $n = 0$, and this particular case is reduced to the scalar one [11]. Therefore, we describe graphene electrons at a strong magnetic field using the model of Ref. [12], so the one-particle wavefunctions can be presented in basis of the "Landau stripes" Ψ_{nk} with $n = 0$, $\Psi_\alpha(\mathbf{r}) = \sum_k C_{\alpha k}\Psi_{0k}(\mathbf{r})$. To calculate the electron lifetime due to electron-phonon coupling, we apply the Fermi golden rule, and it is convenient to represent the total rate W_α for a given electron state α as a sum of the rates due to spontaneous emission $W_\alpha^{(1)}$, stimulated emission $W_\alpha^{(2)}$, and absorption $W_\alpha^{(3)}$ of the acoustic phonons, and each of them can be written as

$$W_\alpha^{(i)} = \frac{2\pi}{\hbar}\sum_{\beta\mathbf{Q}}(1 - f_\beta)|V_{\beta\alpha}|^2[n_Q]^{a_i}\delta(E_\alpha - E_\beta - b_i\hbar sQ), \tag{2}$$

where $\beta \neq \alpha$ runs over all the electron states of the ground Landau level $n = 0$, $f_\beta = f_0(E_\beta)$ and $n_Q = n_0(\hbar s Q)$ are the Fermi-Dirac and Bose-Einstein distributions of the electrons and phonons, respectively, $a_1 = 0$, $a_2 = a_3 = 1$, $b_1 = b_2 = 1$, $b_3 = -1$, and $V_{\beta\alpha}$ represents matrix element of the electron-phonon interaction,

$$\hat{V} = \sum_{\alpha\beta\mathbf{Q}} i\Xi \sqrt{\frac{\hbar Q}{2s\rho_d \mathcal{V}_d}} P^{\mathbf{Q}}_{\beta\alpha} \hat{a}^+_\beta \hat{a}_\alpha, \tag{3}$$

$$P^{\mathbf{Q}}_{\beta\alpha} = \int d^3 r \Psi^*_\beta(\mathbf{r}) e^{i\mathbf{Q}\mathbf{r}} \Psi_\alpha(\mathbf{r}). \tag{4}$$

Here Ξ is the deformation potential, s is the sound velocity, ρ_d and $\mathcal{V}_d$ are graphene/graphite density and normalization area/volume, depending on the dimensionality d (either 2 or 3). Essential difference between the cases of two-dimensional and three-dimensional phonons is hidden inside the sum in Eq. (2), which turns into the integral over either two-dimensional or three-dimensional phonon momentum $\mathbf{Q}$. To estimate the electron lifetime, we employ a model of uncorrelated random C-coefficients, which actually neglects the effects of localization, but has been found suitable for approximate treatment of the density of states and even σ_{xx} peaks [12]. Of course, this approximation makes $W^{(i)}_\alpha$ independent of the particular form of the eigenstate α, so only the energy dependence remains, and it is convenient to introduce the averaged rates

$$W^{(i)}(\varepsilon) = \overline{W^{(i)}_\alpha} = \frac{\sum_\alpha W^{(i)}_\alpha \delta(\varepsilon - E_\alpha)}{\sum_\alpha \delta(\varepsilon - E_\alpha)}, \tag{5}$$

where summation over all the electron states of the infinite system is implied, or, equivalently, disorder averaging is carried out. Under this approach only the averaged value of $|P^{\mathbf{Q}}_{\alpha\beta}|^2$ matters, and it is readily calculated using the relationships [12]

$$\int d^3 r \Psi^*_{0k_1}(\mathbf{r}) e^{i\mathbf{Q}\mathbf{r}} \Psi_{0k_2}(\mathbf{r}) = \delta_{q_x, k_1 - k_2} e^{i\frac{k_1+k_2}{2} q_y l_B^2} e^{-\frac{q^2 l_B^2}{4}}, \tag{6}$$

$$\sum_{k_1,k_2} \overline{C^*_{\alpha,k_1-k} C_{\alpha,k_2-k} C_{\beta,k_1} C^*_{\beta,k_2}} e^{i\gamma(k_1-k_2)} = N_L^{-1} \tag{7}$$

to be

$$\overline{|P^{\mathbf{Q}}_{\alpha\beta}|^2} = N_L^{-1} e^{-\frac{q^2 l_B^2}{2}}, \tag{8}$$

where $N_L = S/2\pi l_B^2$ is degeneracy of the Landau levels on the area S, and $\mathbf{q}$ is the in-plane component of the momentum $\mathbf{Q}$ (being generally three-dimensional). Eq. (8) shows that phonons with the in-plane momentum q exceeding l_B^{-1} couple electron states of the ground Landau level $n = 0$ exponentially weak. Thus, in the approximation neglecting possible corrections to W due to carrier localization, which are believed to be small, $W^{(i)}(\varepsilon)$ can be represented as

$$W^{(i)}(\varepsilon) = \frac{\pi\Xi^2}{\hbar s^2 \rho_d} \int \frac{d^d Q}{(2\pi)^d} [1 - f_0(\varepsilon - b_i \hbar s Q)] \mathcal{D}(\varepsilon - b_i \hbar s Q) [n_0(\hbar s Q)]^{a_i} \hbar s Q \exp\left(-\frac{q^2 l_B^2}{2}\right), \tag{9}$$

where $\mathcal{D}(\varepsilon)$ is the density of electron states, given for the ground Landau level $n = 0$ in case of the short-range disorder by exact solution due to Wegner [13]. Let us focus further on the regime of relatively high temperatures and magnetic fields, so that $\hbar s l_B^{-1}/T \ll 1$, assuming $T \ll \sqrt{2}\hbar v_0 l_B^{-1}, \Gamma$ at the same time (here Γ is the Landau level width). Typical value of $\hbar s l_B^{-1}/T$ at $T = 300\,\mathrm{K}$ and really strong magnetic field $B = 30\,\mathrm{T}$, required for the effect being observed at this temperature, is 0.1 (using $s = 2 \cdot 10^6$ cm/s), so this case is quite relevant experimentally. Since we are really interested in electron states lying near the point of transition, with the energies $\varepsilon \sim T^{\kappa}$, the density of states in Eq. (9) can be taken outside the integral, reducing with good precision to the constant $\mathcal{D}_0 = \mathcal{D}(E=0) \approx 0.36\Gamma^{-1}$ [13].

For two-dimensional phonons, the lifetime can be calculated under given conditions explicitly. The last factor in Eq. (9) introduces rather sharp cut-off of the integral at the inverse magnetic length l_B^{-1}, therefore the function $f_0(E)$, slowly varying on this scale, can be taken exactly at $Q = 0$, and T/E should be substituted for $n_0(E)$. The rest is easily calculated, yielding

$$W_{d=2}^{(i)} = (1 - f_0(\varepsilon))\frac{\Xi^2 \mathcal{D}_0 T}{2\rho_2 \hbar s^2 l_B^2}\left(\frac{\sqrt{\pi}\hbar s l_B^{-1}}{\sqrt{2}T}\right)^{1-a_i}, \tag{10}$$

so stimulated emission and absorption of phonons give equal contributions to the total rate, while a contribution of spontaneous emission is negligible. Such allotment is direct result of the magnetic length cut-off, since only "cold" phonons, with $\hbar sQ \ll T$ and, consequently, $n_0(\hbar sQ) \gg 1$, contribute to the process in the two-dimensional case.

The case of three-dimensional phonons is more complicated, since there is no magnetic length cut-off of the full phonon momentum Q, but only suppression at large Q, which competes with the three-dimensional phonon density of states. Technically, to evaluate the rates it is convenient to go to spherical coordinates, so that $q = Q\sin\vartheta$, and to integrate over the angles φ and ϑ first. The integral over ϑ reduces to the Dawson function $\mathcal{F}(x)$ with the well-known asymptotics,

$$\frac{1}{2}\int_0^{\pi} \exp\left(-\frac{t^2\sin^2\vartheta}{2}\right)\sin\vartheta d\vartheta = \mathcal{G}(t) = \frac{\sqrt{2}\mathcal{F}(t/\sqrt{2})}{t} \sim \begin{cases} 1 - \frac{1}{3}t^2, & t \ll 1 \\ t^{-2}, & t \gg 1. \end{cases} \tag{11}$$

Although $\mathcal{G}(t)$ decays quadratically at large $t = Ql_B \gg 1$, this drop is completely compensated by three-dimensional density of the phonon states, therefore cut-off in Q is set by a temperature. This implies that the contributions of small $Q \ll l_B^{-1}$ can be neglected, and t^{-2} can be substituted for $\mathcal{G}(t)$ in further calculations. Hence, the only dimensionless parameter remains, namely $\delta = (\varepsilon - \mu)/T$, and all the contributions can be expressed as

$$W_{d=3}^{(i)} = C_i(\delta)\frac{\Xi^2 \mathcal{D}_0 T^2}{2\pi\rho_3 \hbar^2 s^3 l_B^2}, \tag{12}$$

where the coefficients C_i are integrated combinations of f_0 and n_0. For our purposes δ dependencies are not very important, and we choose the most natural case of $\delta = 0$, where the coefficients can be evaluated analytically, yielding $C_1 = \frac{\pi^2}{12}$, $C_2 = \frac{\pi^2}{24}$, $C_3 = \frac{\pi^2}{8}$. Finally, we obtain the ratio between the full rates W at $\varepsilon = \mu$ for two different phonon dimensionalities,

$$w = \frac{W_{d=3}}{W_{d=2}} = \frac{\pi}{4}\frac{Td}{\hbar s}, \tag{13}$$

where $d = \rho_2/\rho_3 \approx 0.34$ nm is the interlayer distance of graphite. Though w grows with a temperature, at $T = 300$ K it is equal to 0.5, so the lifetime due to the two-dimensional phonons is half of that of the three-dimensional case. Hence, the electron-phonon interaction turns out to be even stronger in the two-dimensional case in the sense of the electron lifetimes. However, it more important how much electron states the interaction mixes with a given one than how intensive the overall mixing is (like in any other percolation problem), and this aspect, vital for the room-temperature quantum Hall effect, is discussed in the final section.

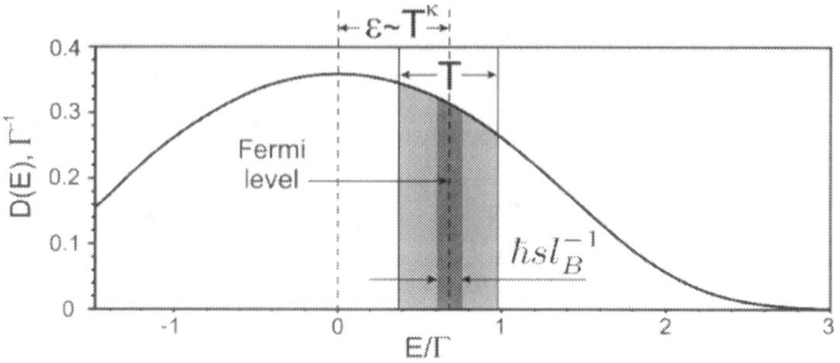

Figure 1. Electron states on the ground Landau level of graphene, producing fundamental plateaus of the integer quantum Hall effect ($\sigma_{xy} = \pm 2$). Active mixing of the electron states by the two-dimensional phonons takes place inside $\hbar s l_B^{-1}$ wide stripe, which is substantially narrower than the temperature stripe, germane to the case of three-dimensional phonons.

3. Discussion and conclusions

Let us recall qualitative picture of the quantum Hall transitions (occurring between the plateaus), using density of the electron states on the ground Landau level of graphene, as sketched in Fig. 1. In order to ascertain magnitude of the conductivity σ_{xx} at given position of the Fermi level μ, we have to consider a kind of percolation problem [7], where localized electron states (α, β, *etc.*) are treated as nodes of the resistor network, connected by the links (conductivities) with strengths, proportional to the rates $W_{\alpha\beta}$ due to electron-phonon interaction, which are given by Eq. (2) with omitted summation over the final states ("β"). Indeed, if we put electron into a certain state α, after the interval $\tau_\alpha = W_\alpha^{-1}$ it will be smeared over all the states β with non-exponentially small transition rates $W_{\alpha\beta}$, allowing transport within a cluster, formed from these states. Usually, number of the states involved in percolation is restricted by a temperature, so all of them lie inside the temperature stripe (shown by dim filling in Fig. 1), and the answer for percolation threshold in this case is well known, $\varepsilon \sim T^\kappa$ [5]. However, two-dimensional phonons do not allow mixing of the electron states outside narrower stripe of the width $\hbar s l_B^{-1}$ (dark area in Fig. 1), hence the threshold is postponed to $\varepsilon \sim (\hbar s l_B^{-1})^\kappa$. This means that the integer quantum Hall effect at high magnetic fields is not restricted by temperature at all, provided that phonons are strictly two-dimensional. Of course, three-dimensional substrate phonons must spoil this picture, but they seem to be suppressed by the two factors, namely weakness of the interlayer chemical links and strong magnetic field, as high as 45 T [4].

Acknowledgments

This work has been supported by Russian President Grant for Young Scientists MK-718.2014.2.

References

[1] Novoselov K S *et al* 2005 *Nature* **438** 197
[2] Jiang Z, Zhang Y, Tan Y-W, Stormer H L and Kim P 2007 *Solid State Comm.* **143** 14
[3] Ostrovsky P M, Gornyi I V and Mirlin A D 2008 *Phys. Rev. B* **77** 195430
[4] Novoselov K S *et al* 2007 *Science* **315** 1379
[5] Huckestein B 1995 *Rev. Mod. Phys.* **67** 357
[6] Altshuler B L, Aronov A G and Khmelnitsky D E 1982 *J. Phys. C* **15** 7367
[7] Greshnov A A and Beltukov Y M 2014 *Semiconductors* **48** 228
[8] Hwang E H and Das Sarma S 2008 *Phys. Rev. B* **77** 115449
[9] Efetov D K and Kim P 2010 *Phys. Rev. Lett.* **105** 256805
[10] Kaasbjerg K, Thygesen K S and Jacobsen K W 2012 *Phys. Rev. B* **85** 165440
[11] Castro Neto A H, Guinea F, Peres N M R, Novoselov K S and Geim A K 2009 *Rev. Mod. Phys.* **81** 109
[12] Greshnov A A, Zegrya G G and Kolesnikova E N 2008 *JETP* **107** 491
[13] Wegner F 1983 *Z. Phys.* **51** 279

Journal of Physics: Conference Series **568** (2014) 052011 doi:10.1088/1742-6596/568/5/052011

The quantum Hall effect in quantum dot systems

Y M Beltukov[1] **and A A Greshnov**[1,2]
[1] Ioffe Physical-Technical Institute RAS,
Polytechnicheskaya 26, St Petersburg, Russian Federation
[2] Saint Petersburg Electrotechnical University,
Professora Popova 5, St Petersburg, Russian Federation

E-mail: a_greshnov@hotmail.com

Abstract. It is proposed to use quantum dots in order to increase the temperatures suitable for observation of the integer quantum Hall effect. A simple estimation using Fock-Darwin spectrum of a quantum dot shows that good part of carriers localized in quantum dots generate the intervals of plateaus robust against elevated temperatures. Numerical calculations employing local trigonometric basis and highly efficient kernel polynomial method adopted for computing the Hall conductivity reveal that quantum dots may enhance peak temperature for the effect by an order of magnitude, possibly above 77 K. Requirements to potentials, quality and arrangement of the quantum dots essential for practical realization of such enhancement are indicated. Comparison of our theoretical results with the quantum Hall measurements in $InAs$ quantum dot systems from two experimental groups is also given.

1. Introduction

The integer quantum Hall effect has been a subject for intense research in the two-dimensional systems, quantum wells and heterojunctions, since the moment of its discovery in 1980 [1], while the quantum dot systems were undeservedly overlooked, except for a few papers [2],[3]. It is traditional that the quantum Hall effect is a part of the low-temperature physics since it demands both strong magnetic fields B and liquid helium temperatures T to become visible in the conventional two-dimensional systems. So hard experimental conditions reflect fundamental physics of the effect, which is interplay between localization due to random disorder potential and dephasing processes due to electron-phonon interaction [4],[5]. Experiments show that the plateau widths scale with a temperature like $\Delta B \propto T^{\kappa}$ with $\kappa \approx 0.4$, and this non-trivial value of the scaling index κ (substantially different from unity) proves that influence of electron-phonon interaction does not reduce to temperature broadening of the distribution function. Strictly speaking, all the electron states in a quantum Hall system are localized at $T = 0$ except for a null set [4], and the electron-phonon interaction cannot be treated in the same simple manner as for zero magnetic field. Therefore, a resulting "degree of delocalization", described quantitatively by the plateau widths, is governed by characteristics of disorder potential (its amplitude and correlation length), providing that the parameters of electron-phonon interaction are fixed. Of course, it is desirable to have a deep and smooth disorder potential for better localization of carriers, which guarantees higher temperatures for observation of the quantum Hall effect. Using modulation doping, it is possible to make disorder potential really smooth (as compared to a scale of the magnetic length) and observe the quantum Hall plateaus for high Landau levels (up to $n \sim 10$), however, amplitude of the potential produced by the remote dopants is not enough to overcome $T \sim 10K$, even for $n = 0$. Evidently, the quantum dots produce a potential which is both deep and smooth, and it is worth to apply them for the quantum Hall effect.

We develop a theory of the integer quantum Hall effect in the quantum dot systems at finite temperatures, where electron-phonon interaction is taken into account by artificial broadening of the spectral functions $\mathcal{D}(\varepsilon)$ in Eq. (6) for the quantum correlator of velocities $w_{xy}(\varepsilon_1, \varepsilon_2)$. This approach allows us to reproduce low temperature behaviour of the transition width ($\Delta\varepsilon \propto T^\kappa$ with $\kappa \approx 0.4$) and carry out numerical calculations of the Hall conductivity efficiently. The latter is achieved by using a compact form of the one-particle hamiltonian (local trigonometric basis [6]) and the kernel polynomial method [7]. We have studied temperature dependencies of the magnetotransport curves between 4 K and 80 K as well as influence of the quantum dot parameters (size and potential depth), their density and arrangement, and our calculations show that the temperatures higher than $40K$ can be achieved for observation of the quantum Hall effect with a quantum dot array with a density of $1.4 \cdot 10^{10}\,\mathrm{cm}^{-2}$ at magnetic field $B = 10\,\mathrm{T}$. However, it is vital to preserve the short-range component of disorder small enough, the condition hardly satisfied in the available experimental papers [2],[3]. So, finally we give specific requirements essential for practical observation of the integer quantum Hall effect at elevated temperatures.

2. Theory: the model, spectrum and Hall conductivity

Basically, we consider rather simple model of the two-dimensional electron gas at strong magnetic field and an external potential, described in detail in Ref. [8]. To take into account both the quantum dots and disorder, we present the potential in the form $V(\mathbf{r}) = V_{\mathrm{QD}}(\mathbf{r}) + V_{\mathrm{dis}}(\mathbf{r})$, where potentials of the individual quantum dots are modelled by the gaussians,

$$V_{\mathrm{QD}}(\mathbf{r}) = -\sum_i V_i \exp\left(-\frac{(\mathbf{r}-\mathbf{r}_i)^2}{2d_i^2}\right), \tag{1}$$

which gives parabolic expansion near potental minima, a simple and practical approximation used in many papers, including Refs. [9],[10] (here, we have projected the task on plane of a structure by integrating out the third dimension). Physically, it is a very common case of the quantum dots composed due to variation of the alloy composition, for example $InGaAs$ quantum dots [11]. Though some researches call those quantum dots $InAs$, it is very unlikely that the effects of size quantization and strain alone are so strong and homogeneous to produce $\sim$ 1eV blueshift of the optical spectra, almost equal to difference between the bandgaps of $InAs$ and $GaAs$, and reproduced in all the papers on this issue. Thus, we estimate quantum dot potential amplitude as tens of meV instead of hundreds of meV, and their lateral dimensions as tens of nm. Such magnitudes are comparable with the cyclotron energy and magnetic length at magnetic field $B = 10\,\mathrm{T}$, which are $\hbar\omega_c = \hbar eB/m^*c \approx 30\,\mathrm{meV}$, $l_B = \sqrt{\hbar c/eB} \approx 8\,\mathrm{nm}$ when considering electrons with the effective mass $m^* = 0.04m_0$ (something between $InAs$ and $GaAs$), meaning that magnetoconfined states are formed [12], and spectrum of carriers can be described by the Fock-Darwin solution [13]:

$$\varepsilon_{n,m} = \frac{\hbar\omega_c}{2}\left[2n + 1 + (n+m+1)(\sqrt{1+(2\omega_0/\omega_c)^2} - 1)\right], \tag{2}$$

where $\hbar\omega_0 = \sqrt{V_0\hbar^2/m^* d_0^2} \approx 20\,\mathrm{meV}$ for $V_0 = 45\,\mathrm{meV}$, $d_0 = 14\,\mathrm{nm}$, $m^* = 0.04m_0$, and distance between the successive magnetoconfined levels of a given Landau level n is about 10 meV, thus compared to the cyclotron energy of about 30 meV. As a result, we get $N = 3$ magnetoconfined levels from each quantum dot for each Landau level, these states are strongly localized and form special intervals of the quantum Hall plateaus robust to elevated temperatures. It is easy to estimate ratio of these states for a given density of the quantum dots (say, $n_{QD} = 1.4 \cdot 10^{10}\,\mathrm{cm}^{-2}$),

$$\mathcal{R} = 2\pi l_B^2 n_{QD} N \approx 20\%. \tag{3}$$

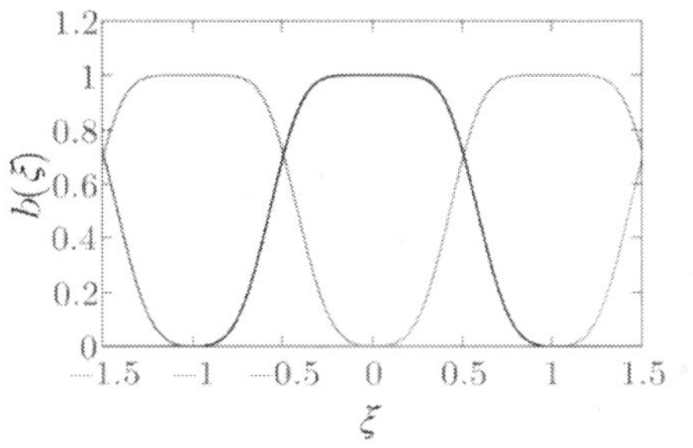

Figure 1. Optimal window function $b(x-q)$ is shown for odd and even q.

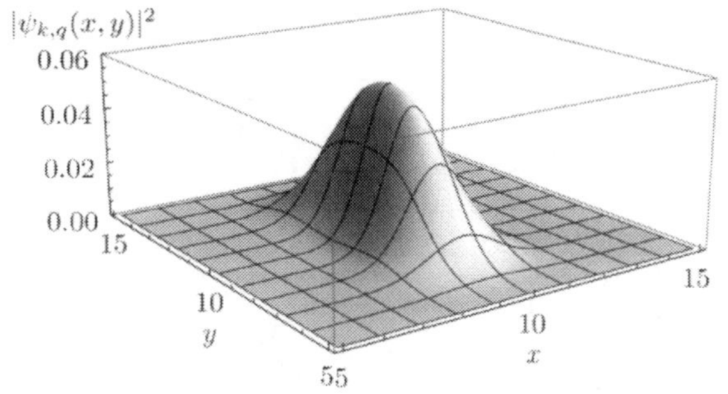

Figure 2. One of two peaks of the wave function $|\psi_{k,q}(x,y)|^2$ at $q=2$, $k=16$, $N=16^2$, $M=2\sqrt{N}$.

However, the estimation given by Eq. (3) does not imply that these special parts of plateaus are *absolutely* robust against temperature, and to find the maximum temperature maintaining the quantum Hall effect, we have carried out numerical calculations of the Hall conductivity treating electron-phonon interaction in an approximate manner, by smearing delta-functions in definition of the quantum correlator of velocities w_{xy}, so that

$$\sigma_{xy}(\varepsilon) = \frac{2e^2\hbar}{S} \int\limits_{\varepsilon_1=-\infty}^{\varepsilon} \int\limits_{\varepsilon_2=\varepsilon}^{\infty} \frac{w_{xy}(\varepsilon_1,\varepsilon_2)}{(\varepsilon_1-\varepsilon_2)^2} d\varepsilon_1 d\varepsilon_2, \tag{4}$$

$$w_{xy}(\varepsilon_1,\varepsilon_2) = \sum_{\alpha,\beta} \mathrm{Im}\{(v_x)_{\alpha\beta} v_y)_{\beta\alpha}\} \mathcal{D}(\varepsilon_1-\varepsilon_\alpha)\mathcal{D}(\varepsilon_2-\varepsilon_\beta), \tag{5}$$

$$\mathcal{D}(\varepsilon) = \frac{1}{\pi}\frac{\delta}{\delta^2+\varepsilon^2}, \tag{6}$$

where $S = L_x \times L_y$ is a sample area, α and β count eigenstates of the one-particle hamiltonian [8], and δ is a temperature induced broadening, which describes the low temperature behaviour $\Delta\varepsilon \propto T^\kappa$ ($\kappa \approx 0.4$) best when defined as $\delta(T) = \mathcal{C}T^2$. Since to reveal the effects of a temperature we need to model rather extended samples, more than $100 l_B \sim 1\,\mu$m long, it is important to optimize computing by means of compact representation of the one-particle hamiltonian as described in Sec. 3 and avoiding explicit hamiltonian diagonalization by use of the kernel polynomial method [7], whose implementation to current problem will be published elsewhere.

3. Local trigonometric basis

In order to write the one-particle hamiltonian in a compact form, we start from its representation in basis of the "Landau stripes" [8], which we denote as $\Psi_k(x,y)$ (let us fix the Landau level number here, say $n=0$, and set $l_B=1$). These elements are extended over x and localized near $y = L_y k/N_L$, where $N_L = S/2\pi$ is number of the electron states on each Landau level. Alternatively, we can apply fourier transform with respect to the quantum number k, resulting in the same set of functions but of the exchanged coordinates $x \leftrightarrow y$. To construct the *localized* basis instead, it is needed to appy the *short-time* Fourier transform,

$$\varphi_{k,q}(x,y) = \frac{1}{\sqrt{M}} \sum_{l=1}^{N_L} w\,(l/M-q)\exp(-2\pi i l k/M)\Psi_l(x,y), \tag{7}$$

where $w(\xi)$ is a window function, vanishing outside the interval $[-\frac{1}{2},\frac{1}{2}]$, and M is the integer of the order of $\sqrt{N}$. Though it is expected that the functions $\varphi_{k,q}(x,y)$ are localized near the

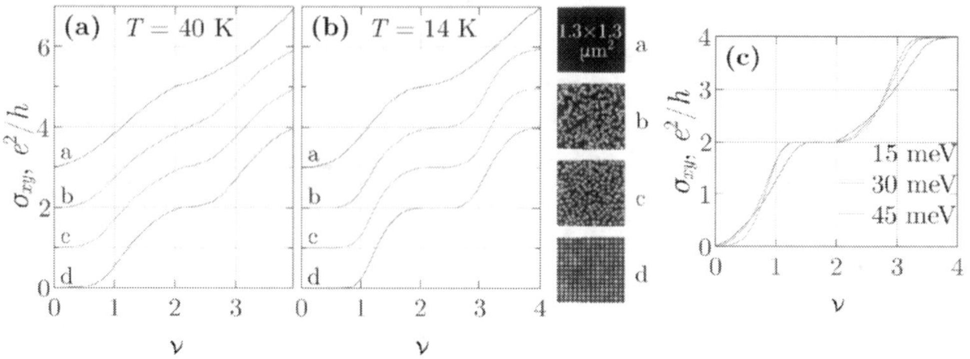

Figure 3. Picture of the quantum Hall effect in the quantum dot systems with different degree of arrangement presented for the temperatures (a) 40 K and (b) 14 K. Plot (c) demonstrates influence of the quantum dot potential amplitude for the model d ("crystal" of quantum dots). Filling factor $\nu = 2\pi l_B^2 n_e$ is proportional to electron concentration n_e and inversely proportional to magnetic field B (which was fixed during this set of calculations).

point $(L_x k/M, L_y qM/N_L)$, the Balian-Low theorem [14] tells us that $(\Delta x)^2 = \infty$ or $(\Delta y)^2 = \infty$ due to restrictions imposed on the window function $w(\xi)$ in order to keep orthogonality of the transformation. However, this recipe can be happily improved [6], and the functions

$$\psi_{k,q}(x,y) = \sqrt{\frac{2}{M}} \sum_{l=1}^{N_L} b\,(l/M - q) \cos[\pi(k - 1/2)(l/M - q + 1/2)]\Psi_l(x,y) \tag{8}$$

form the *local trigonometric basis*. Speaking formally, the function given by Eq. (8) are not localized at all, however, each of them consists of the two well localized functions, one situated at $(L_x k/2M, L_y qM/N_L)$, and another one at $(L_x(1 - k/2M), L_y qM/N_L)$, but this is enough to form *sparse* representation of the one-particle hamiltonian. Surely, the window function $b(\xi)$ still obeys some relations protecting orthogonality of the transformation [6], and it is practical to find specific shape of $b(\xi)$ supporting best localization within these restrictions numerically. Shape of the optimized window function and a hump of one of the functions (8) are shown in Figs. 1, 2. As a result, the hamiltonian matrix has fixed number (~ 100) of non-vanishing elements in each row, providing for efficient modelling of the systems larger than $1\,\mu\text{m}^2$.

4. Numerical results and discussion

We have carried out numerical calculations of the Hall conductivity following the approach described in previous sections, and our results are shown in Fig. 3 for three types of the quantum dot arrangement in comparison with case of a quantum well, as illustrated between the plots (b) and (c) of Fig. 3 (the potentials produced by quantum dot systems are shown there). Let us indicate these models as quantum dot "vacuum" (a), "gas" (b), "liquid" (c) and "crystal" (d). We used parameters typical for strong magnetic field measurements in *InGaAs* quantum dot systems, as expained in Sec. 2: $B = 10\,\text{T}$ ($l_B \approx 8\,\text{nm}$), $m^* = 0.04 m_0$ ($\hbar\omega_c \approx 30\,\text{meV}$), $g^* = 0$ [15], $n_{QD} = 1.4 \cdot 10^{10}\,\text{cm}^{-2}$, $d_0 = 14\,\text{nm}$, $V_0 = 45\,\text{meV}$ (except for the two curves in Fig. 3(c) given for V_0 of 15 and 30 meV). Relative scattering of the quantum dot potential depths V_i and radiuses d_i was taken as $\pm 20\%$, and the short-range disorder potential producing $2\Gamma = 4\,\text{meV}$ broadening of the Landau levels was included. Figs. 3(a, b) demonstrate developing picture of the integer quantum Hall effect in the quantum dot systems at $T = 40\,\text{K}$ (for the model "d"), while only a hint of quantization is seen for the quantum well model "a" at $T = 14\,\mathrm{K}$ (where all the quantum dot models exhibit really wide plateaus). Thus, quantum dots are shown theoretically to increase maximum temperature allowed for observation of the quantum Hall effect by almost an order of magnitude, especially when they are properly arranged, a property expected from the Stranski-Krastanov quantum dots. However, it is vital to protect quantum dots from the pernicious short-range potential, its effective amplitude Γ must be much smaller than the quantum dot potential depth V_0 and cyclotron energy $\hbar\omega_c$. Once this condition is not entirely satisfied, like for one of the curves in Fig. 3 corresponding to $V = 15\,\mathrm{meV}$, we notice considerable narrowing of the plateaus, and at $2\Gamma/V_0 \sim 0.5$ they fully disappear.

Experimentally, the quantum Hall effect in quantum dot systems has been studied in just a few papers [2],[3], and experimental results are not so optimistic as our theoretical predictions. The phenomenon has been observed only at liquid helium temperatures ($T \lesssim 4\,\mathrm{K}$ in Ref. [2] and $T \lesssim 0.3\,\mathrm{K}$ in Ref. [3]), only the fundamental plateau (corresponding to filling factor $\nu = 2$) was pronounced (as expected for the short-range disorder [8]), and quantization precision was not high (about 1%). This should not be perceived as a surprise since quantum dots enhance the quantum Hall effect only if they have enough power for this, in particular, their potential amplitude is much larger than the effective amplitude of the short-range disorder potential. We attribute unsatisfactory experimental results to *insufficient quality* of the investigated samples in the sense that they contain enough short-range (as compared to l_B) variations of composition and *insufficient electron concentration*, which drives the fundamental quantum Hall plateau ($\nu = 2$) at the moderate magnetic fields, corresponding to (relatively) large values of magnetic length ($B = 3\,\mathrm{T}$, $l_B = 15\,\mathrm{nm}$ in Ref. [2] and $B = 1.5\,\mathrm{T}$, $l_B = 21\,\mathrm{nm}$ in Ref. [3]). Taking into account $\sim 1\,\mathrm{eV}$ difference between $InAs$ and $GaAs$ band gaps, it is easy to estimate that short-range compositional fluctuations of the order of 1% produce random potential amplitude and Landau level width 2Γ of the order of $10\,\mathrm{meV}$, which is by no means smaller than the cyclotron energy corresponding to magnetic fields mentioned above ($\hbar\omega_c \approx 10\,\mathrm{meV}$ in Ref. [2] and $\hbar\omega_c \approx 5\,\mathrm{meV}$ in Ref. [3]). However, since nanotechnology is developing quite fast, we expect observation of the predicted effect of the *quantum Hall enhancement by quantum dots* in future quantum dot structures with smoothed compositional fluctuations and enhanced electron concentration.

Acknowledgments

The authors are grateful to Dynasty Foundation and the Federal Programme on Support of Leading Scientific Schools (NSH-5062.2014.2) for financial support of this work.

References

[1] von Klitzing K, Dorda G and Pepper M 1980 *Phys. Rev. Lett.* **45** 494
[2] Kulbachinskii V A, Lunin R A, Rogozin V A, Golikov A V, Kytin V G, Zvonkov B N, Nekorkin S M, Filatov D O and de Visser A 2003 *Phys. Solid State* **45** 762
[3] Kim G H, Liang C.-T., Huang C F, Nicholls J T, Ritchie D A, Kim P S, Oh C H, Juang J R and Chang Y H 2004 *Phys. Rev.* B **69** 073311; Kim G H, Nicholls J T, Khondaker S I, Farrer I and Ritchie D A 2000 *Phys. Rev.* B **61** 10910
[4] Huckestein B 1995 *Rev. Mod. Phys.* **67** 357
[5] Greshnov A A and Beltukov Y M 2014 *Semiconductors* **48** 228
[6] Coifman R R and Meyer Y 1991 *Comptes rendus de l'Académie des sciences. Série 1, Mathmatique* **312** 259
[7] Weiße A, Wellein G, Alvermann A and Fehske H 2006 *Rev. Mod. Phys.* **78** 275
[8] Greshnov A A, Zegrya G G and Kolesnikova E N 2008 *JETP* **107** 491
[9] Glazov M M 2010 *J. Phys.: Cond. Matt.* **22** 025301
[10] Wojs A, Hawrylak P, Fafard S and Jacak L 1996 *Phys. Rev. B* **54** 5604
[11] Oshinowo J, Nishioka M, Ishida S and Arakawa Y 1994 *Appl. Phys. Lett.* **65** 1421
[12] Henriques A B and Efros Al L 2009 *Phys. Rev.* B **80** 153302
[13] Fock V 1928 *Z. Phys.* **47** 446
[14] Benedetto J J, Heil C and Walnut D F 1994 *Jour. Fourier Anal. Appl.* **1** 355
[15] Medeiros-Ribeiro G, Pinheiro M V B, Pimentel V L, Marega E 2002 *Appl. Phys. Lett.* **80**, 4229

Superconductor-Metal-Insulator Transitions in two dimensional amorphous Nb_xSi_{1-x}

Vincent Humbert, François Couëdo, Olivier Crauste, Laurent Bergé, Anne-Aelle Drillien, Claire Akiko Marrache-Kikuchi and Louis Dumoulin

CNRS-CSNSM UMR8609, Université Paris Sud, 91405 Orsay Cedex

E-mail: vincent.humbert@u-psud.fr

Abstract. We report on the study of the two-dimensional Disorder-induced Superconductor-Insulator Transition (D-SIT) in Nb_xSi_{1-x} thin films. In this proceeding, we present new results on the emergence of an insulating state from a 2d metallic state.

1. Introduction

Amorphous Nb_xSi_{1-x} is an alloy presenting disorder-driven Superconductor-Metal-Insulator Transitions in three dimension. These transitions can usually be tuned either by changing the composition x or the thickness of the sample t. An alternative solution - specific to a-Nb_xSi_{1-x} - is to apply a heat treatment which permits to finely tune the transitions.

These transitions have been studied in different materials for several decades now. Nevertheless, it still is a hot topic of condensed matter physics since the competition between a superconducting behavior and disorder effects - inducing localization and enhanced Coulomb interactions - is known to give rise to new phenomena which are far from being understood.

In two dimensions, the lowered dimensionality makes the picture more complex because of enhanced fluctuations. A direct Superconductor-to-Insulator Transition (SIT) is then expected [1][2]. One subject of interest resides in the understanding of the system's behavior in the vicinity of the SIT. Notably what are the mechanisms for the destruction of superconductivity - phase or amplitude fluctuations [3][4] - and what is the nature of the insulating state - fermionic or bosonic [5] - ?

To address these questions, the present work focuses on the emergence of the insulating phase at low temperature by increasing the disorder in a-Nb_xSi_{1-x} thin films.

2. Experimental details

The samples are deposited onto a sapphire substrate by a co-deposition process in a dedicated e-beam evaporator under ultra-high vacuum. Si and Nb are simultaneously evaporated. The deposition rate of each source is separately controlled by a set of dedicated piezoelectric quartz. The films are then protected by a thin SiO layer. The composition is later cross-checked ex-situ by Rutherford Back-scattering Spectroscopy. Previous measurements [6] have shown that the composition and thickness of a given sample are well controlled and highly reproducible.

Each sample has undergone successive heat treatments that were performed following a systematic procedure. The sample is put into an oven for 1 hour at the considered heat treatment temperature θ, with a constant flux of nitrogen to avoid any spurious oxidation. TEM measurements have shown that the heat treatment leaves the film amorphous and homogeneous up to 250°C [7].

The transport measurements were performed from 300 K to 10 mK using conventional four probes techniques in a dilution fridge. The electrical lines are filtered at room temperature and well thermally anchored at low temperature. The measurements were performed through a low temperature resistance measurement bridge and completed by conventional DC and low frequency AC techniques. All measurements - down to the lowest temperatures - have been checked to be in the Ohmic regime. Moreover, the experimental setup has been carefully designed to prevent any incoming electromagnetic radiation.

3. The SIT in a-Nb_xSi_{1-x}

Nb_xSi_{1-x} is an alloy in which the ground state can be tuned by modifying the disorder of the film. In the three-dimensional limit it has indeed been shown, for instance by tuning the composition x, that a Superconductor-to-Metal Transition occurs at $x = 12\%$ and a Metal-to-Insulator Transition occurs at $x = 10\%$ [8]. It has also been shown that a heat treatment of the films allows to finely tune the disorder in our system [7]. Indeed the heat treatment in such binary alloys is - counter-intuitively - increasing the disorder without modifying the morphology of the films [9].

In two-dimensional systems, where the film thickness t is smaller than the characteristic lengths, and particularly smaller than the superconducting coherence length ξ, one should expect a direct SIT since metallic states are theoretically forbidden [1]. It has then been widely accepted [2] that the superconductor is characterized by a positive TCR - Temperature Coefficient of Resistance, defined as $\frac{dR}{dT}$ - and the insulator by a negative TCR at low enough temperature.

Previous studies have shown that, in a-Nb_xSi_{1-x} 2d films, the situation is more complex. As partially shown figure 1, the transport properties near the disorder-driven SIT can present 4 different regimes [10].

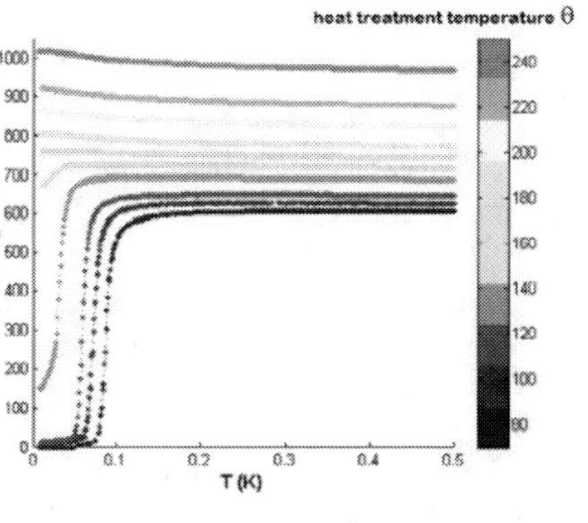

Figure 1. Transport characteristics of a single sample of $Nb_{13.5}Si_{86.5}$ with d=23 nm which sustained several thermal treatments from $\theta = 70$°C to $\theta = 250$°C.

The first regime is a superconducting one, with no dissipation below T_C. By increasing the disorder, we observe a first metallic-like regime where the conductivity saturates at low temperature and the TCR is positive. These saturations of the resistance are intrinsic and cannot be explained by experimental artifacts, such as heating for example [11]. By increasing the disorder even more, a second metallic-like regime is observed with a negative TCR and a saturation of the conductivity at the lowest measured temperature. Finally this metallic state evolves into an insulating one - with an infinite resistivity at zero temperature.

In this work we will focus on the low temperature transport properties of the two latter states.

4. Results

When the dimensionality is reduced, previous studies have shown that the critical composition x_c for the transition to occur is dependent on the sample thickness in a-Nb_xSi_{1-x}[12]. In particular, for $t_c = 15$ nm, the composition corresponding to a change of the TCR sign is $x_c = 13.5\%$. We have studied a set of 6 samples of a-$Nb_{13.5}Si_{86.5}$ with thicknesses ranging from 2 to 5 nm lower than $t_c = 15$ nm, that is to say close to the SIT but with a negative TCR. Initially, the four samples with t between 2 and 4 nm were on the insulating side of the transition and ones with $t = 4,5$ and 5 nm were in the second metallic state described above. Each sample was then progressively driven towards a more resistive state by successive heat treatments. A typical set of data representing the electronic transport properties evolution with the temperature of the heat treatment θ is given figure 2 for the 3,5 nm thick sample. Let us now study the evolution of the laws describing the transport properties in the different regimes that we will identify.

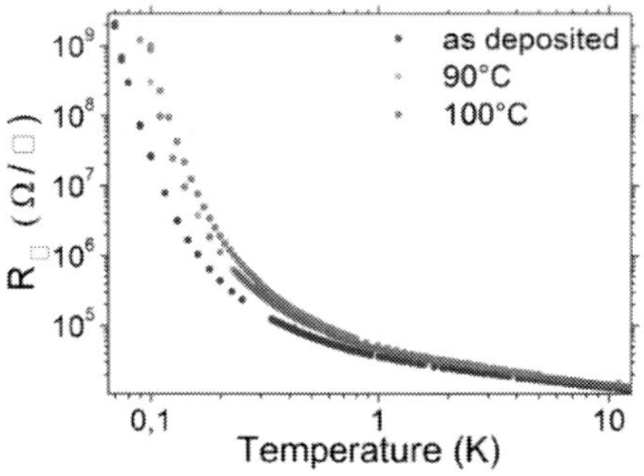

Figure 2. R□ as a function of temperature for a single sample of $Nb_{13.5}Si_{86.5}$ with $t = 3,5$ nm before and after sustaining heat treatments at $\theta = 90°C$ and $\theta = 100°C$

For each sample, at a given θ, we have identified different behaviors depending on the temperature range. For the least disordered sample ($t = 5$ nm), which is on the second metallic state of the transition, we have found three different low temperature regimes as presented figure 3 where we have plotted the sheet resistance $R_\square$ of the sample as a function of temperature. We observe that the electronic transport for temperature higher than 500 mK can be described by a $R_\square \propto lnT$ law, corresponding to a two dimensional weak localization process. From 30 to 500 mK, we observe an insulating law such as :

$$R = R_0 e^{(T_0/T)^{1/n}} \tag{1}$$

The precise nature of the insulating law is given by the value of the parameter n. For instance, n=4 and n=3 correspond respectively to a three dimensional and two dimensional Mott hopping law which are regimes resulting from the disorder-induced localization in the material. $n = 2$ corresponds to an Efros-Shklovskii hopping law, resulting from localization taking into account electron-electron interactions. Finally, $n = 1$ corresponds to an Arrhenius regime, reflecting a hard gap insulator. The parameter n can be determined by using a method described by Zabrodskii et al[13]. We have found that a 2 dimensional Mott regime well fits the intermediate temperature regime on the data presented figure 3. However, at the lowest temperatures, we observed the data systematically deviate from this insulating law. Indeed, in the third regime, the resistance saturates, translating the above-described metallic state. This behavior is intrinsic to the sample, as mentioned above, and the study of this feature is beyond the scope of this paper. However it is noteworthy that - in our system - a metallic ground state emerges from an insulating behavior at higher temperatures.

For more disordered films, on the insulating state, we have identified two low temperature regimes shown figure 4. At the higher presented temperatures we observe a Efros-Shklovskii law, meaning that the transport properties correspond to a hopping law sensitive to Coulomb

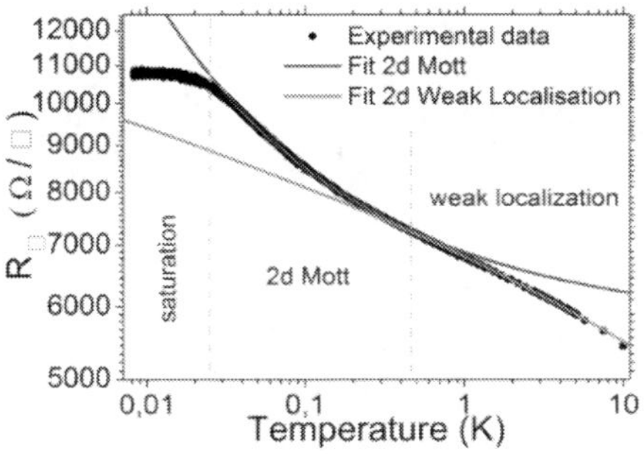

Figure 3. $R_\square$ as a function of temperature for a sample of $Nb_{13.5}Si_{86.5}$ with d=5 nm, as deposited. The sample is on the metallic side of the transition. The gray dashed line corresponds to a crossover between the different identified regimes.

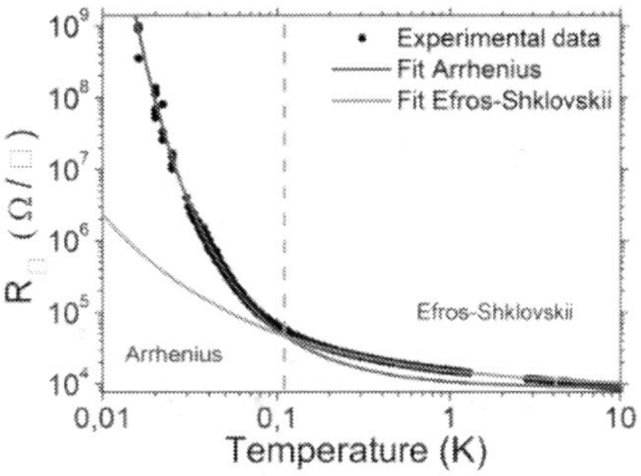

Figure 4. $R_\square$ as a function of temperature for a sample of $Nb_{13.5}Si_{86.5}$ with d=4 nm, as deposited. The sample is on the insulating side of the transition. The gray dashed line corresponds to a crossover between the Arrhenius and the Efros-Shklovskii regimes.

interactions. At the lowest temperatures, the sample evolves towards an Arrhenius-like law corresponding to a hard gap. The existence of this low temperature Arrhenius law seems to be inseparable from the Efros-Shklovskii law at higher temperature.

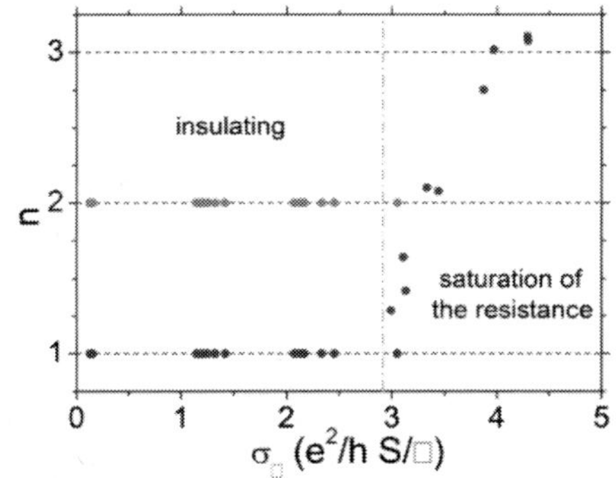

Figure 5. n as a function of $\sigma_\square$ taken at 4 K. The blue value corresponds to the insulating law recorded at the lowest temperature. The red value corresponds to the insulating law at higher temperature. The gray dashed line corresponds to the limit between the samples which have a saturating resistivity at the lowest temperature and the truly insulating ones. Other dashed lines are guides to the eye.

We can study the transition from the two extreme situations presented above by plotting the evolution of the parameter n - extracted from the Zabrodskii method and readjusted to perform the best fit of the data - as a function of $\sigma_\square = \frac{1}{R_\square} = k_f l \frac{e^2}{h}$, taken at 4 K, which can be viewed as measure of the disorder. The corresponding plot is displayed figure 5. For metallic samples, the value of n reported is the one corresponding to the intermediate temperature regime described above. For the insulating samples, two values of n, corresponding to the low temperature (blue) and the high temperature (red) regimes are plotted. In addition, the vertical line corresponds to the transition between the metallic and the insulating states, defined by the existence (or non existence) of a saturation of the resistance at the lowest measured temperatures. At the transition between the metal and the insulator - as defined section 3, when the disorder is increased, the value of the low temperature n evolves quickly, but continuously, from 3 to 1. This is consistent with a system evolving towards a more insulating state. This strikingly continuous evolution tends to indicate that this still unexplained metal can be seen as a precursor state for the insulator.

5. Conclusion

In this paper, progressive heat treatments enabled us to finely tune the disorder in thin a-Nb_xSi_{1-x} films between a 2d metallic state and an insulating one. Studying the low temperature properties of this system, we have shown that the insulating state is characterized by an Arrhenius law at low temperature. Metallic samples - in the sense of the second metallic state described section 3 - exhibit an activated 2d Mott law at intermediate temperatures before abruptly saturating at the lowest temperature of the measurements. Furthermore, this hoping law continuously evolves towards a hard gap at the Metal-to-Insulator transition.

References

[1] Abrahams E, Anderson P, Licciardello D and Ramakrishnan T 1979 *Phys. Rev. Lett.* **42** 673
[2] Goldman A and Markovic N November 1998 *Physics Today* 39
[3] Finkel'stein A 1994 *Physica B* **197** 636
[4] Fisher M and Lee D 1989 *Phys. Rev. B* **39** 2756
[5] Dobrosavlejevic V, Trivedi N and Valles J 2012 *Conductor Insulator Quantum Phase Transitions* (OUP Oxford)
[6] Crauste O, Marrache-Kikuchi C, Bergé L, Collin S, Dolgorouky Y, Marnieros S, Nones C and Dumoulin L 2011 *J. Low Temp. Phys.* **163** 60
[7] Crauste O, Gentils A, Couedo F, Dolgorouky Y, Bergé L, Collin S, Marrache-Kikuchi C and Dumoulin L 2013 *Phys. Rev. B* **87** 144514
[8] Marnieros S 1998 *Couches minces d'isolant d'Anderson. Application à la bolométrie à très basse température* Ph.D. thesis Paris XI, Orsay
[9] Béal M and Friedel J 1964 *Phys. Rev.* **135** A466
[10] Couëdo F, Crauste O, Bergé L, Marrache-Kikuchi C and Dumoulin L paper in preparation
[11] Couëdo F 2014 *Transitions de phase quantiques dans les systèmes désordonnés de basses dimensions* Ph.D. thesis Paris-Sud, Orsay
[12] Crauste O, Couëdo F, Bergé L, Marrache-Kikuchi C and Dumoulin L paper in preparation
[13] Zabrodskii A and Zinov'eva K 1984 *Zh. Eksp. Teor. Fiz* **86** 742

Shubnikov de Haas Effect and Electron Mobilities in the Isomorphic InGaAs Quantum Well With the InAs Insert on the InP Substrate

V.A. Kulbachinskii[1,a*], L.N. Oveshnikov[1,b], R.A. Lunin[1,c], N.A. Yuzeeva[2,d], G.B. Galiev[2,e], E.A. Klimov[2,f]

[1]M.V. Lomonosov Moscow State University, Physics Department, Moscow, Russia

[2]Institute of Ultra-High-Frequency Semiconductor Electronics, RAS, Moscow, Russia

[a*]kulb@mig.phys.msu.ru, [b]ragnos@list.ru, [c]lunin@mig.phys.msu.ru, [d]yuzeeva.nataliya@gmail.com, [e]galiev_galib@mail.ru, [f]klimov_evgenyi@mail.ru

Keywords: Shubnikov - de Haas effect; Quantum well, InGaAs; electron mobility, electron effective mass; InP

Abstract. The Shubnikov – de Haas (SdH) effect at T=4.2 and 8.4 K and the Hall effect have been investigated in isomorphic $In_{0.52}Al_{0.48}As/In_{0.53}Ga_{0.47}As/In_{0.52}Al_{0.48}As/InP$ quantum well with InAs inserts to the center of the quantum well. The structures were both-sides delta-doped by silicon. The effective mass m* was measured by SdH effect by a new method which allowed experimentally to determine m* in the every dimensionally quantized subband using the temperature dependence of the SdH effect amplitude. Central InAs inserts in a quantum well lead to a decreasing of m* by about 20% as compared with the uniform $In_{0.53}Ga_{0.47}As$ lattice-matched quantum well. We also calculated the transport and quantum mobility of electrons in dimensionally quantized subbands using the SdH effect. The calculated and experimentally determined values are in a good agreement.

Introduction

One of the most important problems of ultra-high-frequency (UHF) electronics is the optimization of base material of the transistor. In the InAlAs/InGaAs/InAlAs quantum well on the InP substrate, two-dimensional electron gas with a high electron concentration up to ns=3*1012 cm−2, has high mobility μ=10 000 cm^2 (Vs)$^{-1}$ at room temperature [1,2] and highly saturated electron drift velocity. These structures are extensively used for making UHF devices for millimeter and submillimeter wave applications. The lattice-matched $In_yAl_{1-y}As/In_xGa_{1-x}As/In_yAl_{1-y}As$ (y = 0.52) structure on the InP substrate has a high InAs content in a quantum well $In_xGa_{1-x}As$ (x = 0.53). It leads to a decrease of the effective electron mass m* as compared with GaAs, the stronger capture of electrons in a deep $In_xGa_{1-x}As$ quantum well and a significant increase of the electron mobility μ. However, with increasing x the thickness of strained layer $In_xGa_{1-x}As$ has to decrease to avoid dislocations and hence the electron mobility drops. According to results [3-6] the thin InAs layer in the quantum well InGaAs leads to the electron mobility enhancement as compared with the uniform quantum well. Thus on the one hand InAs insert results in the growth of μ due to an increase of the energy gap between dimensional quantization subbands and decreasing of m* in a quantum well. On the other hand, the InAs layer thickness d is limited by a critical value [7]. Exceeding the critical thickness results in a degradation of the insert layer and the formation of lattice misfit dislocations.

Here we report the results of the experimental determination of the electron effective mass separately in different dimensionally quantized subbands by exploring the Shubnikov–de Haas (SdH) effect at different T. We developed a new method of a digital separation of SdH frequencies at different temperatures in the case of several filled subbbands. It gave the possibility experimentally to measure the electron effective mass separately in the every subband using the temperature dependence of the SdH oscillations amplitude. The SdH effect also was used to evaluate the subband quantum and transport mobilities. In this study for reducing the electron effective mass and mobility enhancement we introduce the InAs insert to the quantum well center.

Experimental

A schematic diagram of the $In_{0.52}Al_{0.48}As/In_xGa_{1-x}As/In_{0.52}Al_{0.48}As$ structure with a central InAs insert on InP substrate is shown in Fig. 1. The samples were grown by the molecular beam epitaxy on InP (100) substrates. All samples have $In_{0.52}Al_{0.48}As$ buffer matched (isomorphic) to InP. Samples have a double-side delta-doping by Si. The spacer thickness was 4.7 nm (for a sample #4 – 7.3 nm). A cap layer was undoped $In_{0.53}Ga_{0.47}As$. Temperature dependence of the resistance and Hall effect were measured in the temperature interval 4.2 K<T<300 K.

$In_{0.53}Ga_{0.47}As$	54 Å
$In_{0.52}Al_{0.48}As$	173 Å
δ_2-Si	2.9·10^{12} cm^{-2}
$In_{0.52}Al_{0.48}As$	47 Å
$In_{0.53}Ga_{0.47}As$	(168 Å - d)/2
InAs	D
$In_{0.53}Ga_{0.47}As$	(168 Å - d)/2
$In_{0.52}Al_{0.48}As$	47 Å
δ_1-Si	0.88·10^{12} cm^{-2}
$In_{0.52}Al_{0.48}As$	2440 Å
InP (100)	

Fig. 1 Sample structure. A quantum well is shown by a grey colour.

At T=4.2 K the SdH effect was investigated in magnetic field provided by a superconducting solenoid up to 6 T. We found that the Hall mobility depends on the width of the InAs insert to a quantum well. Some experimental parameters of samples at T=4.2 K are listed in table 1. The sample 4 was taken as the reference sample with d=L, that is this sample has the InAs strained quantum well. The maximal value of mobility was observed in the sample #3 with wide InAs insert. The mobility in the sample 1 without the insert and the reference sample #4 with InAs quantum well is less and close each to other.

Table 1. The InAs insert width *d*, the quantum well width *L*, the Hall electron concentration N_{Hall} and the mobility μ_{Hall} at T= 4,2 K, SdH electron concentration in 2 filled subbands n_{SdH}

Sample #	*L*, Å	*d*, Å	N_{Hall}, 10^{12} cm^{-2}	μ_{Hall}, cm^2/(V·s)	n_{SdH}, 10^{12} cm^{-2}
1	168	0	4.03	27600	2.76; 1.29
2	168	20	2.99	26000	2.3; 0.75
3	168	34	3.03	29500	2.44; 0.6
4	97	97	1.30	28400	1.29

As we see in table 1 the summary of electron concentration in both filled subbands are in a very good agreement with the Hall concentration of electrons.

Results and discussion

Amplitude of magnetoresistance oscillations (SdH effect) increases with temperature decrease. This gives the opportunity to determine the electron effective mass m* at extremal sections of Fermi surface. The relation between amplitudes of oscillations $A(T_1)$ and $A(T_2)$ determined at temperatures T_1 and T_2 ($T_1 > T_2$) for the same value of magnetic field $B = B_n$ in case of Dingle temperature T_D independence on temperature is equal to [8]

$$\frac{A(T_2)}{A(T_1)} = \frac{T_2}{T_1}\frac{sh(2\pi^2 k_B T_1 / \hbar\omega_c^*)}{sh(2\pi^2 k_B T_2 / \hbar\omega_c^*)}. \qquad (1)$$

The formula (1) is a transcendental equation for the determination m_c^*. However, at $T_1 = 2T_2$ this equation simplifies and it's possible to get an expression for m_c^*:

$$m_c^* = \frac{e\hbar B_n}{2\pi^2 k_B T_2} Arch\left[A(T_2, B_n)/A(T_1\ B_n)\right] \qquad (2)$$

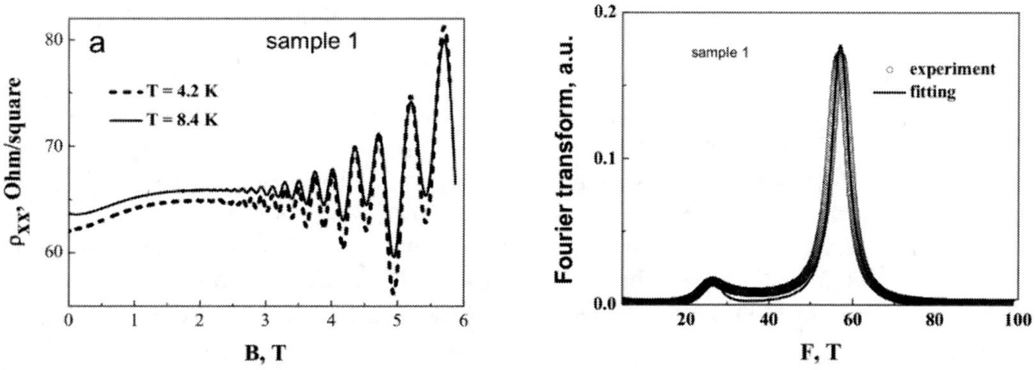

Fig.2. Shubnikov – de Haas oscillations at two temperatures for sample #1 (a) ant its Fourier spectra (b) at T=4.2 K (points – experiment, solid line – fitting)

This formula may be used only in the case of a single oscillation frequency. We observed two frequencies in all samples except sample #4 in which only one subband is filled by electrons. As an example the SdH oscillations at two temperatures for sample #1 are given in Fig. 2a. Two frequencies in the SdH effect correspond to two filled subbands. We use conventional Fourier transform (FT) of the SdH oscillations to obtain the electron subband concentrations n_i from the frequencies of oscillations. Fast Fourier transform spectra are shown in Fig. 2b.

In order to measure accurately the amplitude of the SdH oscillation corresponding to every frequency we digitally separated observed frequencies by an original program using an idea developed in Ref. [9]. In fig. 3 we plotted SdH oscillation at two temperatures separately for high (Fig.3a) and low (Fig.3b) frequencies. We used separated high and low frequencies of the SdH oscillations at two temperatures using formula 2 for the experimental determination of the electron effective masses listed in table 2.

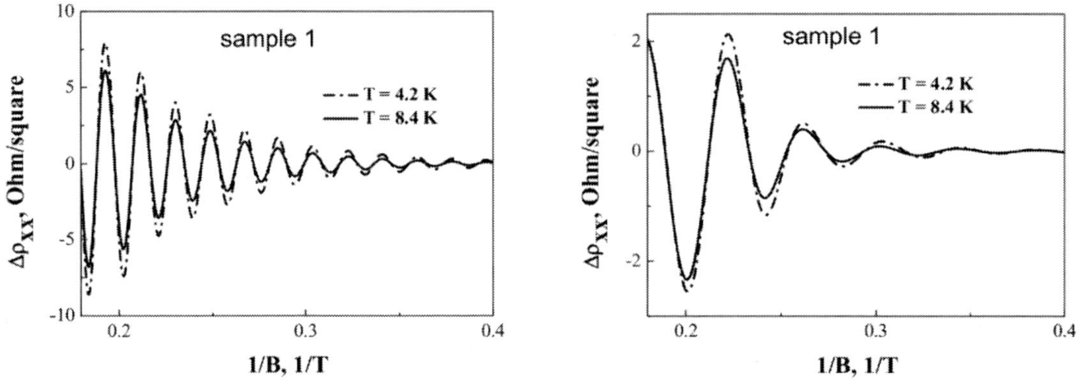

Fig. 3 The SdH oscillations in sample 1 at two temperatures separately for high (a) and low (b) frequencies

Table 2. Electron effective masses in the first (m^*_1) and the second (m^*_2) subbands

sample #	m^*_1	m^*_2
1	0.059	0.060
2	0.050	0.057
3	0.048	0.049
4	0.042	–

As we see in table 2 the effective mass in the second upper subband higher than in the first subband. This fact pointed to the some nonparabolicity of the electron energy spectrum.

Reliable electron densities and mobilities can be derived from Shubnikov-de Haas oscillations, for which each subband has an oscillation with its own period. The part of the density of states oscillating in magnetic field, Δg, can be expressed as [10,11]

$$\frac{\Delta g(\varepsilon_F)}{g_0} = 2\sum_{s=1}^{\infty} \exp\left(-\frac{\pi s}{\mu_q B}\right) \cos\left[\frac{2\pi s(E_F - E_i)}{\hbar\omega_c} - s\pi\right] \frac{\left(2\pi^2 s k_B T/\hbar\omega_c\right)}{\sinh\left(2\pi^2 s k_B T/\hbar\omega_c\right)}, \quad (3)$$

which leads to the following expressions for the conductivity tensor components in the 2-D case (the Landau level width is assumed to be independent of energy and magnetic field):

$$\sigma_{xx} = \frac{en_S\mu_t}{1+\mu_t^2B^2}\left[1+\frac{2\mu_t^2B^2}{1+\mu_t^2B^2}\frac{\Delta g(\varepsilon_F)}{g_0}\right], \tag{4}$$

$$\sigma_{xy} = -\frac{en_S\mu_t^2B}{1+\mu_t^2B^2}\left[1-\frac{3\mu_t^2B^2+1}{\mu_t^2B^2(1+\mu_t^2B^2)}\frac{\Delta g(\varepsilon_F)}{g_0}\right], \tag{5}$$

where μ_t is the transport mobility at B=0, g_0 is the density of states at zero magnetic field B, μ_q is the quantum mobility, n_s is the electron density, and e is the absolute value of the electron charge. The oscillation frequency F in the reciprocal magnetic field determines the two- dimensional electron density: $n_s=eF/\pi\hbar$, and Fermi-energy is $E_F=(\pi\hbar^2/m^*)n_s$. Scattering through small angles makes a negligible contribution to transport relaxation time. If scattering is large at small angles (in the case of Coulomb scattering), the transport relaxation time may be an order of magnitude larger than quantum one. Hence the transport mobility is larger than quantum one. Analysis of the oscillation amplitude as a function of magnetic field and temperature yields the transport and quantum mobility's of 2D electrons in each dimensional subband. By varying μ_t and μ_q in each subband, one can fit the theoretical dependence (3-5) of oscillation on magnetic field and FT to experimental data (see Fig. 2b). As an example the fit to the experimental data for the sample 1 is shown in Fig. 2b by a solid line. The resulting transport μ_t and quantum μ_q mobility's are listed in table 3. Scattering through small angles makes a negligible contribution to transport relaxation time.

Table 3 Quantum μ_q transport μ_t electron mobility in the first and the second subbands and Hall mobility μ_{Hall}, at 4.2 K for all samples

sample #		μ_q, cm^2/(V·s)	μ_t, cm^2/(V·s)	μ_{Hall}, cm^2/(V·s)
1	2	2100	24900	27600
	1	2200	27600	
2	2	2200	26800	26000
	1	3000	28700	
3	2	1900	30700	29500
	1	3100	36200	
4	1	2300	27600	28400

For all structures, the electron wavefunctions $\psi_i(z)$ and the energies E_i were determined in the effective mass approximation by the self-consistent solution of the one-dimensional Schrödinger and Poisson equations [10,12]. We used the following values of the effective masses at the bottom of the conduction band in Γ-valley: in InAs – m* = 0.03 m_0, in $In_{0.53}Ga_{0.47}As$ – m* = 0.043 m_0, in GaAs – m* = 0.067 m_0 and in InAlAs – m* = 0.075 m_0. The conduction band discontinuities were equal to δE_c=–0.7 eV between InGaAs and InAlAs and δE_c = –0.45,–0.3 eV between InGaAs and InAs and InGaAs and GaAs [3,13-15]. The value of a surface potential for all heterostructures was equal to φs = 0.5 eV. Fig. 4 shows the band diagrams of the sample #1 and #3 with the wide InAs insert. By inserting a wide InAs layer the lowest electron energy level E_1 falls down to the bottom of the quantum well and forms level bounded mainly with insert. Effective width of the quantum well determined as a half-width of electron wave function decreases. The distance between E_1 and the next level E_2 increases. On the one hand, it leads to an increase of the electron mobility μ due to the weakening of the inter-subband scattering and decreasing of electron effective mass in the quantum well. On the other hand, lateral roughness scattering increases in a narrow quantum well. Inserted InAs layers are strained as compared to the isomorphic $In_{0.53}Ga_{0.47}As$ layer and electron mobility can increase only up to some critical width of the InAs insert. Thus it is necessary to choose the optimal width of an insert. As it is seen in table 1 the maximal electron mobility was observed in sample 3 with the wide InAs insert.

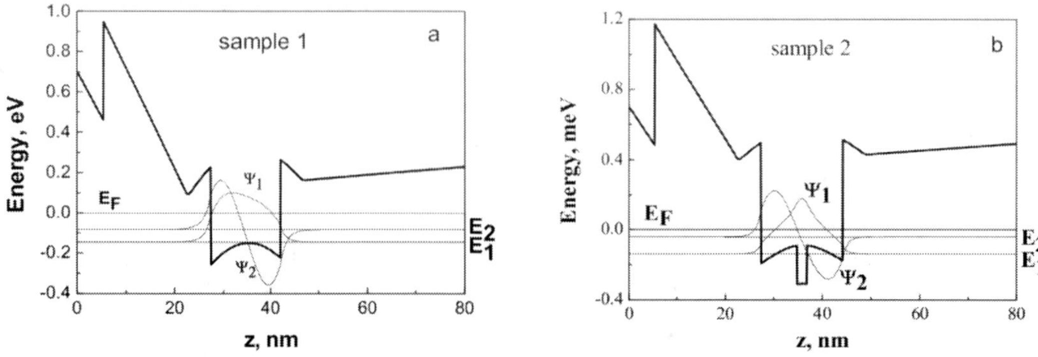

Fig. 3. Band diagram and electron wavefunctions ψ_i for samples 1 (a) and 2 (b). Fermi energy E_F is at the zero level; E_1 and E_2 are the electron levels

Conclusion

We have measured the electron effective masses m* separately in the every subbands and found that m* decreases due to InAs insert to the center of the quantum well. We observed the electron mobility enhancement due to the InAs insert of the center of the quantum well. Our numerical calculations indicate that the electron mobility decreases with the subband number. The mobility in the 2nd subband is less because the wave function spread to the delta-layer where charged impurity scattering is higher. The calculated transport mobility μ_t is much higher than the quantum mobility μ_q. This fact indicates the essential role of the small-angle scattering which is a characteristic of the Column scattering in the investigated structures.

Literature

[1] Y .Yamashita, A. Endoh, K. Shinohara, et al, IEEE Electron Device Lett. 23 (2002) 573-575.
[2] X. Wallart, B. Pinsard and F. Mollot, J. Appl. Phys. 97, 053706 (2005).
[3] V.A. Kulbachinskii, N.A. Yuzeeva, et.al., Semicond. Sci. Technol. 27, (2012) 035021(1-5).
[4] D.S. Ponomarev, I.S. Vasil'evskii, et.al., Semiconductors, 46 (2012) 484–490.
[5] T. Akazaki, K. Arai, T. Enoki, Y. Ishiil, IEEE Electron Device Lett. 13 (1992) 325-327.
[6] T. Akazaki, J. Nitta, H. Takayanagi, Appl. Phys. Lett. 65 (1994) 1263-1265.
[7] J.W. Matthews, and A.E. Blakeslee, J. Cryst. Growth 27, (1974) 118-125.
[8] D. Shoenberg, Magnetic Oscillations in Metals, Cambridge Univ. Press, Cambridge, 1984.
[9] J.F. Kaiser and W.A. Reed, Review of Scientific Instruments, 49, 1103 (1978).
[10] V.A. Kulbachinskii, R.A. Lunin, V.G. Kytin, et.al., JETP, 83 (1996) 841-848.
[11] V.A. Kulbachinskii, R.A. Lunin, V.A. Rogozin, V.G. Mokerov, Yu.V. Fedorov, Yu. V.Khabarov and A. de Visser, Semicond. Sci. Technol. 17 (2002) 947–951.
[12] V.A. Kul'bachinski, V.G. Kytin, R.A. Lunin, et.al., Semiconductors, 33 (1999) 771-778.
[13] 12. Averkiev N S, Berezovets V A, Mikhailova, et.al., Phys. Solid State, 46 (2004) 2153.
[14] R.F. Korf, H.P. Wei, A.P. Perley, G. Livescu, Appl. Phys. Lett. 60(19), (1992) 2386-2388.
[15] W. Chen, M. Fritze, W. Walecki, A.V. Nurmikko, D. Ackley, J.M. Hong, and L.L. Chang, Phys.Rev. B, 45 (1992) 8464-8477.

Influence of Thallium on the Shubnikov – de Haas effect and Thermoelectric Properties of Sb_2Te_3 and Bi_2Se_3

V.A. Kulbachinskii[1], **A.A. Kudryashov, V.G. Kytin**
M.V. Lomonosov Moscow State University, Low Temperature Physics Department, 119991, Moscow, Russia

E-mail: kulb@mig.phys.msu.ru

Abstract. Influence of Tl-doping on the Shubnikov de Haas effect (SdH) at T=4.2 K in magnetic field up to 38 T of p-$Sb_{2-x}Tl_xTe_3$ (x=0; 0.005; 0.015; 0.05) and n-$Bi_{2-x}Tl_xSe_3$ (x=0, 0.01; 0.02; 0.04; 0.06) single crystals has been investigated. By increasing the Tl content, the frequency of the SdH effect and hence the extremal cross-sections of the Fermi-surface decreases in both materials. The hole concentration decreases in $Sb_{2-x}Tl_xTe_3$ due to a donor effect of Tl and the electron concentration decreases in n-$Bi_{2-x}Tl_xSe_3$ due to an acceptor effect of Tl. Temperature dependence of the Seebeck coefficient S, electrical conductivity σ, thermal conductivity k and the figure of merit ZT single crystals were measured in the temperature range 77 K – 300 K. The values of k and σ decrease due to Tl doping in $Sb_{2-x}Tl_xTe_3$ and n-$Bi_{2-x}Tl_xSe_3$ and the Seebeck coefficient S for all compositions increases in the whole temperature range. The figure of merit ZT increases in both materials. The preferential scattering mechanism in Tl-doped samples changes from the acoustic phonon scattering to the ionized impurity scattering.

1. Introduction

The efficiency of thermoelectric generators is limited to a fraction of their Carnot efficiency, determined by the dimensionless thermoelectric material figure of merit $ZT=S^2\sigma T/k$, where S is the thermoelectric power or the Seebeck coefficient of the thermoelectric material, σ and k are the electrical and thermal conductivities, respectively, and T is the absolute temperature [1]. Because ZT also depends on the carrier's group velocity via the electrical conductivity, the value of E_F that maximizes ZT is somewhat different from the value that maximizes the Seebeck coefficient S. Crystals of Sb_2Te_3 always exhibit p-type conductivity due to the high concentration of charged point defects of predominantly antistructural type (i.e., Sb atoms occupy the positions of Te atoms). The formation of such defects is due to weak polarity of Sb–Te bonds. A change in the bond polarity upon doping changes the concentration of point defects and, hence, the hole concentration. The single crystals of Bi_2Se_3 prepared from stoichiometric melt show an excess of bismuth over the stoichiometric composition. This is the reason why the crystal matrix contains negatively charged defects of bismuth atoms on selenium sites Bi_{Se}^{-1} and positively charged vacancies in the selenium sublattice V_{Se}^{+2} [2]. The concentration of vacancies exceeds the concentration of Bi_{Se}^{-1} defects and the $Bi_{2+\delta}Se_3$ crystals exhibit n-type conductivity. Despite a considerable research on this material, there are

[1] To whom any correspondence should be addressed.

few papers concerning the doping with Tl of Bi_2Se_3 [3], Sb_2Te_3 [4] and $Sb_{1.5}Bi_{0.5}Te_3$[5]. Here we investigated Shubnikov-de Haas effect of Tl-doped single crystals Sb_2Te_3 and n-Bi_2Se_3 in a magnetic field up to 30-38 T. We report also the influence of Tl doping on the thermoelectric properties of p-$Sb_{2-x}Tl_xTe_3$ and n-$Bi_{2-x}Tl_xSe_3$ single crystals in the temperature range 77K–300 K.

2. Experimental

In this work, we investigated pristine and Tl-doped p-$Sb_{2-x}Tl_xTe_3$ (x=0; 0.005; 0.015; 0.05) and n-$Bi_{2-x}Tl_xSe_3$ (x=0, 0.01; 0.02; 0.04; 0.06) single crystals grown by the Bridgman method. In the measurements of the SdH effect, the current flowed along the C_2 axis and the magnetic field was directed along the C_3 axis of the crystals. In the measurements of the thermopower S and heat conductivity k, the temperature gradient and a heat flux were directed along the C_2 axis. The concentration of Tl in the studied samples is shown as loaded for the growth.

In order to determine the concentration of light holes (or electrons) and the Fermi energy E_F, we used the SdH effect at $T = 4.2$ K in high pulsed magnetic fields [6]. Some parameters of samples are listed in table 1 for p-$Sb_{2-x}Tl_xTe_3$ and table 2 for n-$Bi_{2-x}Tl_xSe_3$.

Table 1. SdH oscillation frequency F, Fermi-energy E_F and concentration of light holes p in p-$Sb_{2-x}Tl_xTe_3$

№	composition	F, T	E_F, meV	p, cm^{-3}
1	Sb_2Te_3	54	97,1	$2,8*10^{19}$
2	$Sb_{1.995}Tl_{0.005}Te_3$	52,1	93,7	$2,7*10^{19}$
3	$Sb_{1.985}Tl_{0.015}Te_3$	51,3	92,2	$2,6*10^{19}$
4	$Sb_{0.95}Tl_{0.05}Te_3$	34	61,1	$1,4*10^{19}$

Table 2. SdH oscillation frequency F, Fermi-energy E_F and concentration of electrons in n-$Bi_{2-x}Tl_xSe_3$

№	composition	F, T	E_F, meV	n, cm^{-3}
1	Bi_2Se_3	167,6	161,7	$2,2*10^{19}$
2	$Bi_{1.99}Tl_{0.01}Se_3$	166	160,1	$2,1*10^{19}$
3	$Bi_{1.98}Tl_{0.02}Se_3$	159	153,4	$1,9*10^{19}$
4	$Bi_{1.96}Tl_{0.04}Se_3$	145,4	140,3	$1,6*10^{19}$
5	$Bi_{1.94}Tl_{0.06}Se_3$	134,8	130	$1,4*10^{19}$

As can be seen from tables 1 and 2, the carrier concentration decreases with Tl doping in the both p-Sb_2Te_3 (holes) and n-Bi_2Se_3 (electrons) single crystals. The simultaneous measurements of the temperature dependence of the Seebeck coefficient, a heat conductivity, and the electrical conductivity were performed using an original setup including a low temperature cryostat [7].

3. The Shubnikov-de Haas effect data

In high magnetic field there are SdH oscillations in p-$Sb_{2-x}Tl_xTe_3$ and n-$Bi_{2-x}Tl_xSe_3$. In figure 1a oscillations of the transverse magnetoresistance are shown for different samples p-$Sb_{2-x}Tl_xTe_3$ in magnetic field parallel to the C_3 axis of the crystal. In this orientation the extremal cross-sections of all six ellipsoids of the Fermi surface of the upper valence band in p-$Sb_{2-x}Tl_xTe_3$ coincide. We observed a single frequency F in $Sb_{2-x}Tl_xTe_3$ single crystals as it is seen in the Fourier transform in figure 1b. Tl-doping decreases the value of F in p-$Sb_{2-x}Tl_xTe_3$. The six-ellipsoidal non parabolic model satisfactory describe the energy spectrum of the light holes in the highest valence band of Sb_2Te_3 [6,8]. Using this model and the SdH effect data we calculated the carriers concentrations and Fermi energies, listed in Table 1. The method of calculation is described elsewhere [6,8].

In n-$Bi_{2-x}Tl_xSe_3$ we observed a clear SdH oscillations with a single frequency as it is shown in figure 2 for different samples in magnetic field parallel to the C_3 axis of the crystals. As it is seen in figure 2 Tl-doping decreases the value of F in n-$Bi_{2\text{-}x}Tl_xSe_3$. From the frequency F of the SdH oscillation the extremal cross section S_c of the Fermi surface in the momentum space perpendicular to the direction of B may be evaluated by a relation $S_c = 2\pi e\hbar F$. According to experimental and theoretical investigations, [9,10] there are two conduction bands in Bi_2Se_3. However, almost nothing is known about the upper conduction band. The lower conduction band is located at the Γ point of the Brillouin zone and is represented by an ellipsoid elongated in the C_3 direction with cross-sections, $S_a = S_b = \eta S_c$ with a volume V. From the known dependence of the anisotropy η of the

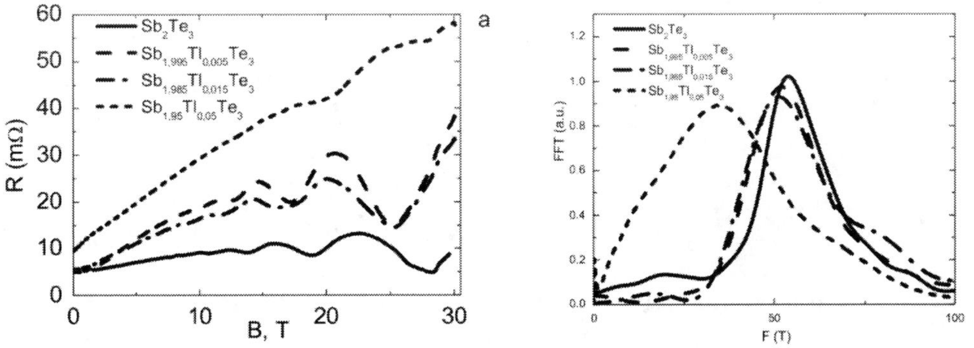

Figure 1. The SdH oscillations on magnetic field B at T=4.2 K of p-$Sb_{2\text{-}x}Tl_xTe_3$ single crystals with different x (a); the Fourier transforms of the SdH oscillations (b).

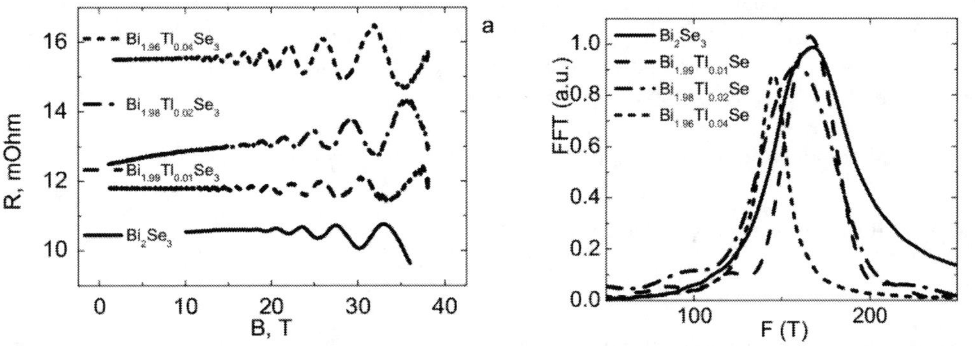

Figure 2. Oscillating part of the SdH oscillations on magnetic field B at T=4.2 K in n-$Bi_{2\text{-}x}Tl_xSe_3$ single crystals with different x (a) and the Fast Fourier Transforms of the SdH oscillations (b)

ellipsoidal Fermi surface on the electron concentration or the frequency F, we can calculate the volume V of the ellipsoid and the value of the electron concentration n [10]. Evaluated electron concentrations are listed in Table 2 for various samples. For the last sample the frequency of oscillations was extrapolated using experimental data for all other samples. As the effective masse does not depend on the carrier concentration [10] it is possible to use simple parabolic dispersion relation to calculate the Fermi energy from the SdH effect data $S_c = \pi p_{\perp F}^2 = 2\pi m^* E_F$, where $p_{\perp F}$ is the Fermi momentum in the direction perpendicular to the magnetic field. From the known value of m* [10] and experimentally determined the extremal cross section S_c of the Fermi surface, E_F was calculated for the lower conduction band, and listed in Table 2.

4. Thermoelectric properties

The value of the Seebeck coefficient S is positive in $Sb_{2-x}Tl_xTe_3$. Tl-doping increases the thermoelectric power of Sb_2Te_3 single crystal in the whole temperature interval. The heat conductivity k and the conductivity σ of $Sb_{2-x}Tl_xTe_3$ crystals decreases due to Tl doping as compared with pristine Sb_2Te_3. A decrease in the heat conductivity and an increase of the Seebeck coefficient lead to the increase of the value of the dimensionless thermoelectric figure of merit (figure 3a) from ZT=0.20 in Sb_2Te_3 up to 0.37 at T=292 K in the Tl-doped sample $Sb_{0.95}Tl_{0.05}Te_3$.

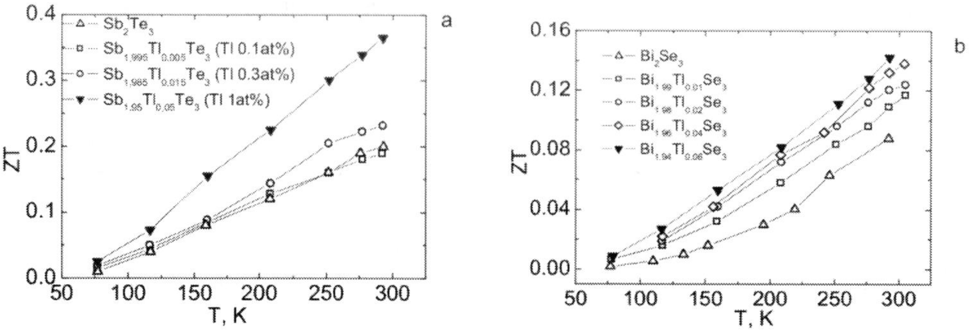

Figure 3. Temperature dependence of t the dimensionless figure of merit *ZT* for (a) of $Sb_{2-x}Tl_xTe_3$ and $Bi_{2-x}Tl_xSe_3$ single crystals

The Seebeck coefficient S increases under Tl-doping while the heat conductivity k decreases in $Bi_{2-x}Tl_xSe_3$. When temperature decreases the conductivity always increases in $Bi_{2-x}Tl_xSe_3$ The conductivity has a tendency to decrease under Tl-doping. An increase of the Seebeck coefficient and a decrease of the heat conductivity increase the value ZT in the whole temperature range: ZT=0.02 in pristine Bi_2Se_3 and ZT=0.14 in $Bi_{1.94}Tl_{0.06}Se_3$ at T=290 K (Figure 3b). Note that it is very important to subtract the thermoelectric voltage during the measurements of the conductivity σ in order to get a proper value of σ. Otherwise it can lead to a higher value of the conductivity in doped samples as compared with undoped.

The simple model with square dispersion law and isotropic relaxation time τ given by

$$\tau=\tau_0 E^r \quad (1)$$

yields in the following expression for the Seebeck coefficient:

$$S(T)=\frac{k_B}{e}\left(\frac{(2r+5)F_{r+3/2}(\eta)}{(2r+3)F_{r+1/2}(\eta)}-\eta\right), \quad (2)$$

where k_B being the Boltzmann constant, e being electron charge, E_F being the Fermi energy, $\eta=E_F/k_BT$ is the reduced Fermi energy, and r is a parameter characterizing the scattering mechanism ($r=-1/2$ for acoustic phonon scattering, $r=1/2$ for polar optical scattering and $r=3/2$ for ionized impurity scattering); $F_s(\eta)$ is the Fermi integral. Using simple model with a square dispersion law and isotropic relaxation time τ given by (1) we evaluate the temperature dependence of the scattering parameter r using formula (2). In Tl-free crystals parameter r is close to -1/2 at high temperatures indicating the major role of the acoustical phonon scattering. Tl doping leads to a pronounced increase of r. This points to the change of the scattering mechanism from the phonon scattering to the ionized impurity scattering in doped samples. Thus in all samples the reason in a increase of ZT is a decrease of the carrier concentration (holes in $Sb_{2-x}Tl_xTe_3$) (electrons in n-$Bi_{2-x}Tl_xSe_3$) and the Fermi energy and change of the scattering mechanism due to the Tl-doping.

Incorporating of the Tl atoms into the Sb_2Te_3 lattice decreases the free hole concentration. The bonds between Sb and Te atoms are very weakly polarized, Sb atoms may enter a Te sublattice forming antistructural (AS) defects $Sb_{Te}^{'}$ (Sb in the Te site). The substitution of Sb atoms by Tl atoms leads to the formation of uncharged defects Tl_{Sb}^{x}. Tl atoms have a lower value of the electronegativity with respect to Sb atoms. Hence, the bonds between the defects Tl_{Sb}^{x} and surrounding Te atoms are more polar than the Sb-Te bonds. Positive charges arise at the points defects whose increase the bond ionicity. The higher ionicity of the Tl_{Sb}^{x} defects results in the suppression of the formation of the antistructural defects $Sb_{Te}^{'}$, responsible for the hole concentration.

As mentioned above, doping with Tl decreases the free electron concentration in *n*-Bi_2Se_3 single crystals. The crystal matrix contains negatively charged AS defects of Bi atoms on Se sites Bi_{Se}^{-1} and positively charged vacancies in the selenium sublattice V_{Se}^{+2} [2,3]. The concentration of vacancies exceeds the concentration of AS defects and the $Bi_{2+\delta}Se_3$ crystals exhibit *n*-type conductivity; the concentration of free electrons *n* is given by the difference of the concentration of both native defects. Tl atoms enter the sublattice of Bi forming Tl_{Bi} substitutional defects. Let us assume that Tl forms uncharged defects Tl_{Bi}^{x} with the valence state +3. In this way Tl produces no free carriers. From the chemical point of view the more stable form of Tl is the valence +1, i.e. it can capture two electrons from conduction band and reduce the free electron concentration. However the nonlinear concentration dependence of the lattice parameter on the Tl content indicates that this is not the only process running upon a Tl doping. The actual concentration of Tl is smaller than the nominal one [3]. A single crystal with Tl grew from the melt containing an overstoichiometric amount of Se leads to a crystal with suppressed concentration of selenium vacancies. This fact can account for the observed decrease in free electron concentration. Thus both the formation of Tl_{Bi}^{-2} and the descending vacancies concentration V_{Se}^{+2} can account for the decrease in the free electron concentration. The experimental results do not allow us to specify the dominant mechanism.

5. Conclusions

We investigated SdH effect and thermoelectric properties of Tl-doped *p*-$Sb_{2-x}Tl_xTe_3$ (x=0; 0.005; 0.015; 0.05) and *n*-$Bi_{2-x}Tl_xSe_3$ (x=0, 0.01; 0.02; 0.04; 0.06) single crystals. We found that Tl has a donor behavior in the solid solutions *p*-$Sb_{2-x}Tl_xTe_3$ while in *n*-$Bi_{2-x}Tl_xSe_3$ Tl has an acceptor behavior. The figure of merit ZT increases up to 0.37 at T=292 K in *p*-$Sb_{2-x}Tl_xTe_3$ with x=0.05 and increases up to ZT=0.14 in $Bi_{2-x}Tl_xSe_3$ with x=0.6 at T=290 K.

6. References

[1] G. J. Snyder, E. S. Toberer, Nat. Mater. **7**, 105 (2008).
[2] A. Sklenář, Č. Drašar, A. Krejčová, and P. Lošťák, Cryst. Res. Technol. **35**, 1069 (2000).
[3] P. Janíček, Č. Drašar, L. Beneš, and P. Lošťák, Cryst. Res. Technol. **44**, 505 (2009).
[4] P. Lošťák, R. Novotný, J. Horák and J. Klikorka, physica status solidi (a), **89**, K55 (1985).
[5] A. Sher, M. Shilon, and L. Ben-dor, J. Electron. Mater., **12**, 983 (1983).
[6] V.A. Kul'bachinskii, A.Yu.Kaminski, V.G. Kytin, et. al., JETP, **90**, No.6, 2000, 1081 (2000). [Translated from ZÉTF, **117**, No. 6, 1242 (2000)].
[7] E.A. Kelm, A.V. Olenev, M.A. Bykov, et. al., Z.Anorg.Allg.Chem.**637**, 2059 (2011).
[8] V.A. Kulbachinskiia, N. Miura, H. Najagawa, et. al., J. Phys. Condens. Matter, **11**, 5273 (1999).
[9] P. Pecheur and G. Toussaint, J. Phys. Chem. Solids, **55**, 327 (1994).
[10] V.A. Kulbachinskii, N. Miura, H. Nakagawa, H. Arimoto, T. Ikaida, P. Lostak and C. Drasar, Phys. Rev. B., **59**, 15733 (1999).

Journal of Physics: Conference Series **568** (2014) 052015 doi:10.1088/1742-6596/568/5/052015

Features of the hopping conductivity in gallium and cobalt doped ZnO thin films

V G Kytin[1,a], O V Reukova[1,b], V A Kulbachinskii[1,c], L I Burova[2], A R Kaul[2], A G Ulyashin[3]

[1]Physics Faculty, M.V. Lomonosov Moscow State University, 119991, Moscow, Russia
[2]Chemistry Department, M.V. Lomonosov Moscow State University, 119991, Moscow, Russia
[3]SINTEF Materials and Chemistry, P.O. Box 124, Forskingsveien 1, Blindern, 0314 Oslo, Norway

E-mail: [a]kytin@mig.phys.msu.ru; [b]reukova@physics.msu.ru; [c]kulb@mig.phys.msu.ru;

Abstract. The effect of Ga and Co doping on the hopping transport of electrons in ZnO films was studied. All investigated films were deposited on monocrystalline substrates by water assisted metal organic chemical vapour deposition method. Temperature dependencies of the resistivity of all films obeyed the Mott's law at low temperatures. Both positive and negative magnetoresistance were observed in the temperature range of the Mott's law. The values of the localization length obtained from magneto transport data for ZnO and ZnO:Ga films are larger than the effective Bohr radii of a shallow donor state and are independent on the Ga content. These observations point to the localization of electrons in the random potential originating from the structural disorder. In Co doped ZnO films large positive magnetoresistance was observed in low magnetic fields at low temperatures. The value of positive magnetoresistance increases with an increase of Co content and decrease of temperature. Observed magneto transport properties of Co-doped films were interpreted by the decrease of electron hopping probability due to decrease of the density of localized states in magnetic field.

Introduction

Conducting properties of ZnO films are the subject of intensive investigation due to their importance for applications and understanding of electron transport mechanisms in disordered semiconductors. Electrophysical properties of doped and undoped ZnO films depend strongly on the structure of the films and can be essentially modified by doping with different impurities [1].

Gallium is group III element. Therefore it behaves as donor impurity substituting Zn in ZnO lattice. In interstitial position Ga can also supply electrons in conduction band. Cobalt can be solved in ZnO up to very high concentration [2-3]. According to X-ray diffraction and X-ray absorption data it preferably substitutes zinc in the lattice in the same charge state. Therefore it should be electrically inactive. However there are observations that doping of ZnO films with Co essentially changes the resistivity, galvanomagnetic and magnetic properties of the films [2-3]. At low temperatures large positive magnetoresistance was reported in ZnO films doped with Co and Mn [1-3]. The value of this magnetoresistance increases with an increase of magnetic impurity content. Therefore recent theoretical models consider this magnetoresistance as spin-related and suggest exchange interaction between d-electrons of impurity and conduction electrons. The presence of this interaction is supported by an increase of band gap with an increase of Co content [4]. The resent models for spin related magnetoresistance were developed for hopping transport of electrons. Therefore investigation of hopping conductivity in ZnO films with and without magnetic impurity is important for understanding the effect of magnetic impurity on electron transport.

This work reports the investigation of hoping conductivity in thin ZnO films doped with Co and Ga in the wide range of temperatures and magnetic fields. Temperature dependence of resistivity and magnetoresistance of the films were compared and analyzed in frame of available models of hopping transport.

Experimental

Investigated films were synthesized by water-assisted chemical vapor deposition (MOCVD). The details of this modification of MOCVD deposition method are presented in [5-6]. Water assisted MOCVD allows the deposition at lower temperatures than oxygen-assited MOCVD. All films were deposited on monocrystalline substrates of R- and C-sapphire and Yttrium stabilized zirconia (YSZ). Gallium and cobalt content were determined by energy dispersive X-ray microanalysis. Structure of the films was studied by X-ray diffraction. It was found that all films were polycrystalline with chaotic orientation of crystallites. Surface morphology of the films was investigated by scanning electron microscopy (SEM). According to SEM images surface of the films was nearly smooth although some crystallites can be seen. This is consistent with X-ray diffraction data. Some parameters of investigated films are listed in Table.1.

Table 1. Deposition parameters of studied films

Film	Substrate	$T_{deposition}$ (^0C)	Thickness (nm)	Impurity content at%
GaC	C-sapphire	350	900	7 (Gab)
GaR	R-sapphire	350	750	21 at% (Ga)
GaY	$Zr_2O_3(Y_2O_3)$	350	900	25 wt% (Ga)
Co1	R-sapphire	500	70	0.4 (Co)
Co2	R-sapphire	500	80	1.9 (Co)

Resistivity and magnetoresistance of the films were measured by the conventional four probe method. Samples for measurements were prepared in rectangular shape with typical dimensions 5×3 mm. Contact plates were made by vacuum deposition of gold and copper wires were soldered to the plates. For the measurements samples were mounted in the cryostat where the temperature was varied from 4.2 to 293 K. Magnetic fields up to 6 T for magnetoresistance measurements was created by superconducting solenoid. The presence of CoO or other phases was not detected

Results and discussion

For all investigated films significant increase of resistivity with the decrease of temperature was observed in the in all studied temperature range. Observed temperature dependencies of resistivity can be well described by Mott's law [7] at sufficiently low temperatures:

$$\rho=\rho_0\exp\left[\left(\frac{T_0}{T}\right)^{1/4}\right] \qquad (1)$$

where ρ and T is the resistivity and the temperature, ρ_0 and T_0 are parameters. This is illustrated in figure 1. Mott's law was derived for variable range hopping conduction in disordered semiconductors. Parameter T_0 depends on the density of localized states at Fermi energy $g(E_F)$ and localization length r_{loc} of electron wave function [7]:

$$T_0 = \frac{\beta}{k_B g(E_F) r_{loc}^3}, \quad (2)$$

where β=21,2 is the digital coefficient, k_B is the Boltzmann constant.

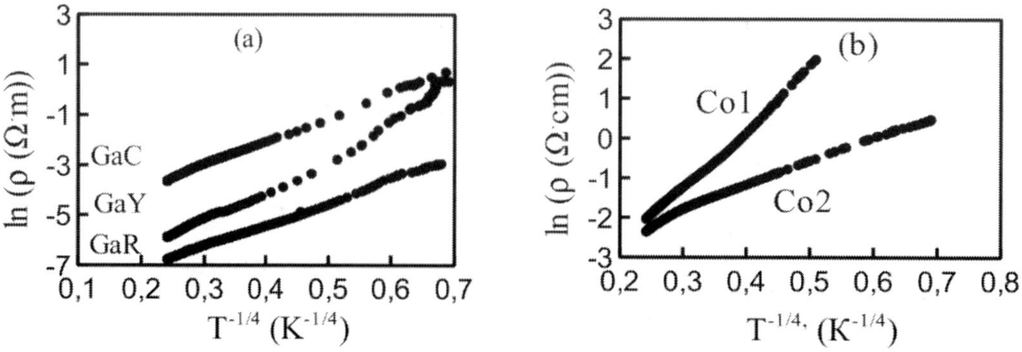

Figure 1.Dependencies of ln(ρ) on $T^{-1/4}$ for ZnO:Ga (a) and ZnO:Co (b) films

All investigated films exhibited positive and negative magnetoresistance depending on temperature and magnetic field range. Magnetoresistance of some Ga and Co-doped film is shown in figure 2. In low magnetic fields magnetoresistance of ZnO:Ga films is negative while in higher magnetic field it is positive. At low temperatures magnetoresistance of ZnO:Co films was strong and positive. In higher magnetic fields the saturation or change the sign of magnetoresistance from positive to negative was observed for these films. These observations suggest different origin of magnetoresistance for Ga and Co-doped films.

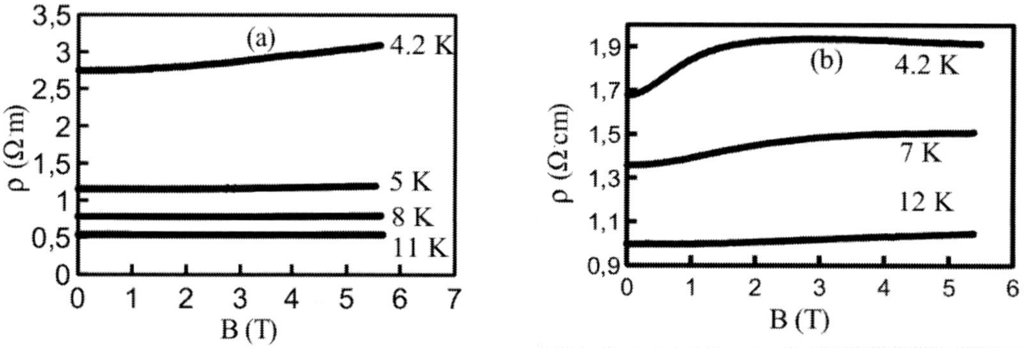

Figure 2.Magnetoresistance of GaC (a) and Co2 (b) films

Several mechanisms of magnetoresistance were proposed for hopping transport of electrons. Some of them consider the effect of magnetic field on the localization length of electron wave function. In sufficiently strong magnetic field the localization length of electrons decreases in the directions perpendicular to magnetic field. This is the origin of positive magnetoresistance for hopping conduction [7]. The value of this positive magnetoresistance is significant when magnetic length (l_B $l_B = \sqrt{\hbar / eB}$) is comparable or smaller than localization length in zero magnetic field. Expressions for this type of positive magnetoresistance were presented in [7]. For example in low magnetic field ($l_B >> r_{loc}(B=0)$):

$$\rho = \rho_0 \exp\left[\alpha \frac{e^2 r_{loc}^4 B^2}{\hbar^2}\left(\frac{T_0}{T}\right)^{3/4}\right] \quad (3)$$

In this expression α=5/2016 is the digital coefficient, B is the magnetic field, e is the elementary charge, $\hbar$ is the Plank constant. In low magnetic field logarithm of magnetoresistance is proportional to the square of magnetic field while in high magnetic field it is proportional to the square root of magnetic field. In figure 3 (a) logarithm of resistivity is plotted versus square of magnetic field for ZnO:Ga films. Taking a slope of linear parts of the plots and using expressions (2) and (3) we calculated the localization length and density of electronic states at Fermi energy. Obtained values are listed in table 2.

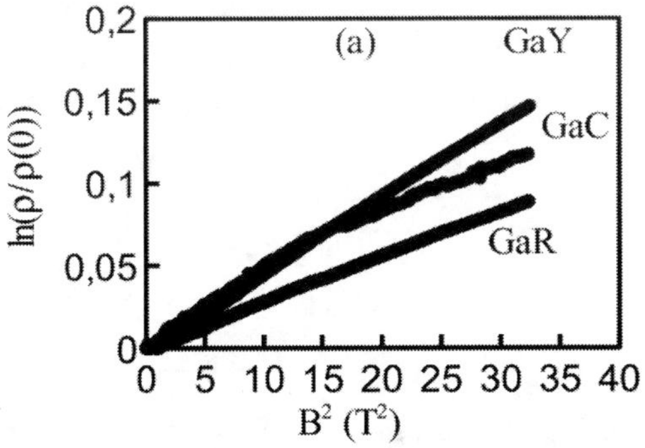

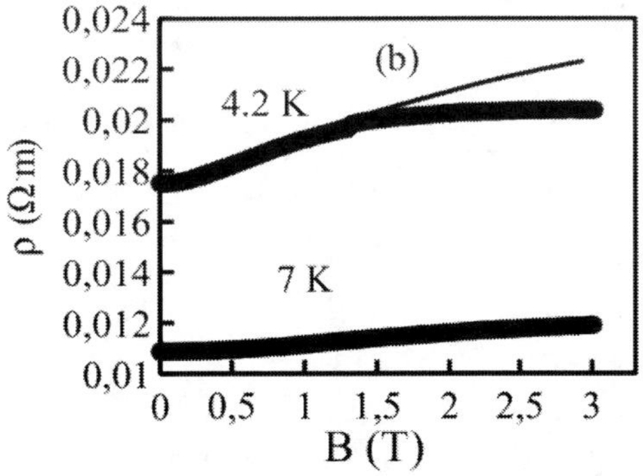

Figure 3.Dependencies of ln(ρ) on B^2 for ZnO:Ga films at 4.2 K (a) Magnetoresistance fitting for Co2 film by expressions (4-5) (b) Points are experimental data, Solid lines are fitting by expressions (4-5)

The value of localization length is significantly larger than the effective Bohr radii in ZnO (1.3 nm). Thus the localization of electrons most probably occurs due to polycrystalline structure of structure and random potential caused by disorder and random distribution of donors. Then localization can be partially caused by interference of electron waves scattered by separated donors and/or grain boundaries. Magnetic filed destroys the interference introducing additional phase shift between scattered electron waves. This mechanism of negative magnetoressistance for variable range hopping conduction was proposed in many publications [8-10]. Therefore the observation of negative magnetoresistance is consistent with the large value of localization length listed in table 2.

Large positive magnetoresistance observed in Co-doped films should be somehow related to the presence of magnetic impurity. There are several models for spin-related magnetoresistance and hopping conduction of electrons. The resent and detailed interpretation of positive magnetoresistance in magnetically doped semiconductors was presented in [11]. The authors of this work showed that the density of electronic states at Fermi energy can be reduced in magnetic field. They proposed the method for determination of the relative variation of width σ of the density of localized states assuming Gaussian form of density of states. The method is based on the following expressions:

$$\frac{\rho}{T}=f\left(\frac{\sigma}{\sigma_0 T}\right) \quad (4)$$

$$\frac{\sigma(B)}{\sigma_0}=\sqrt{1-\gamma_1^2+\left(\gamma_1+\gamma_2 BF\left(\frac{g\mu_B B}{k_B T}\right)\right)^2} \quad (5)$$

where σ – is width of the energy distribution of donor levels, f is the universal function, BF is the Brillouin function, μ_B and g is Bohr magneton and factor Lande. The fitting of magnetoresistance of Co2 film by expressions (4-5) is shown in figure 3 (b). Fiiting parameters were γ_1 and γ_2. Expressions (4-5) describes the experimental magnetoresistance quite well for intermediate magnetic fields. For low and especially for high magnetic fields theoretical dependence is stronger than the experimental one. The deviation can be due the difference between BF and real magnetization. Negative magnetoresistence (NMR) observed at 4.2 K can originate from bound magnetic polarons [12]. This type of NMR can also decrease the total MR in smaller magnetic field.. Parameters γ_1 and γ_2 are related to the localization length of donor states via expressions [11]:

$$r_{loc}=\left(\frac{d\,E_c/dx}{y\,x^{1/2}\,N^{3/2}}\right)^{2/9} \qquad (6)$$

$$y=\alpha J\frac{\gamma_1}{2\gamma_2} \qquad (7)$$

where N is the Zn concentration, x is the Co content,α is the exchange energy, E_c is the bottom of conduction band, J is angular momentum of Co ion.

Table 2. Density of states at Fermi energy and localization length for ZnO:Ga films

Film	$g(E_F)$, (10^{19} eV^{-1} cm^{-3})	r_{loc} (nm)
GaC	7.1	7.3
GaR2	4.2	5.5
GaY1	13	6.9

Taking α and dE_c/dx from [4] we got an estimate of 1.2 nm for localization length in Co2. This value is close to effective Bohr radii of shallow donors in ZnO (1.3 nm).

Summary

Thus the hopping transport of electrons in ZnO films doped with Ga and Co was investigated and compared. Density of localized states at Fermi energy and localization length was estimated for ZnO:Ga films by analysis of temperature dependencies of resistivity and magnetoresistance data. Obtained values of localization length are essentially larger than effective radii of shallow donors and nearly independent on Ga content. Magnetoresistance of ZnO:Co films was interpreted by reduction of electron density of states at Fermi energy in magnetic field due to Zeeman effect and exchange interaction between conduction electrons and magnetic impurity.

References

[1] Özgür Ü, Alivov Ya I, Liu C, Teke A, Reshchikov M A, Do an S, Avrutin V, Cho S-J, Morkoç ğ H 2005 J. Appl. Phys. 96 041301
[2] Yan Shi-shen, Liu J P, Mei L M, Tian Y F, Song H Q, Chen Y X, Liu G L 2007 J. Phys.:Condens. Matter 88 10469-10480
[3] Pearton S J, Norton D P, Ivill M P, Hebard A F, Zavada J M, Chen W M, Buyanova I A 2007 IEEE Transaction on electron devices 54 1040-1048
[4] Kim J H, Kim H, Kim, Yoon S G, Choo W K 2004 Solid State Communications 131677-680
[5] Burova L I, Samollenkov S V, Fonin M, Biegger E, Dedkov Y, Ganshina E A, Gorbenko O Yu,Ruediger U, Kaul A R 2007 Thin Solid Films 515 8490-8494
[6] Kaul A R, Gorbenko O Yu, Botev A N, Burova L I 2005 Superlattices and Microstructures 38272-282
[7] Shklovskii B I, Efros A L 1984 Electronic properties of doped semiconductors (SpringerVerlag. Berlin Heidelberg NewYork Tokyo)
[8] Altshuler B.L., Aronov A.G., Khmelnitskii D.E, 1982 JETP Letters 36 195-198
[9] Choy T.C., Stoneham A.M., Ortuño M., Somoza A.M. 2008 Applied Physics Letters 92, 012120
[10] Zhang Y., Sarachik M.P. 1991 Phys. Rev. B, 43, 7212-7215
[11] Nenashev A V, Jansson F, Wiemer M, Petznick S, Klar P J,Hetterich M 2013 Phys. Rev. B 88115210
[12] Petukhov A.G., Foygel M. 2000 Phys. Rev. B 62, 520-531.

Thermoelectric effects in quantum Hall systems beyond linear response

Rosa López[1], **Sun-Yong Hwang**[1], **David Sánchez**[1]
[1] Institut de Física Interdisciplinària i de Sistemes Complexos IFISC (CSIC-UIB), E-07122 Palma de Mallorca, Spain

E-mail: rosa.lopez-gonzalo@uib.es

Abstract. We consider a quantum Hall system with an antidot acting as a energy dependent scatterer. In the purely charge case, we find deviations from the Wiedemann-Franz law that take place in the nonlinear regime of transport. We also discuss Peltier effects beyond linear response and describe both effects using magnetic-field asymmetric transport coefficients. For the spin case such as that arising along the helical edge states of a two-dimensional topological insulator, we investigate the generation of spin currents as a result of applied voltage and temperature differences in samples attached to ferromagnetic leads. We find that in the parallel configuration the spin current can be tuned with the leads' polarization even in the linear regime of transport. In contrast, for antiparallel magnetizations the spin currents has a strict nonlinear dependence on the the applied fields.

1. Introduction

Under the application of strong fields, electron transport becomes nonlinear and new effects arise: rectification [1, 2, 3, 4, 5, 6, 7], magnetic-field asymmetries[8, 9, 10, 11, 12, 13, 14, 15, 16, 17, 18, 19, 20, 21, 22, 23, 24], and the generation of higher harmonics [25, 26, 27]. Quantum coherent conductors are excellent platforms to observe and study nonlinearities since small voltage biases cause a sizeable effect over rather short lengths. In the ballistic regime of transport, current response is determined by the transmission probability $t(E)$. When t depends weakly on the carrier's energy E, current is always a linear function of V independently of the background temperature T. Quite generally, however, transmission across nanostructures shows sharp features arising from strongly energy dependent $t(E)$. Furthermore, in the nonlinear regime of transport, the transmission becomes a function of the applied voltage V because the applied field modifies the potential landscape which in turn affects the scattering properties of the sample [28, 29]. The self-consistent procedure must then include electron-electron interactions that restore current conservation and gauge invariance beyond linear response. In fact, it is the nontrivial dependence of the screening potential on voltage and magnetic fields that explains both rectification effects and magnetic-asymmetries, respectively.

Electron motion can also be induced with the application of external thermal gradients. The two main thermoelectric effects are the Seebeck and Peltier effects. The former leads to the creation of a voltage drop in response to a temperature difference θ in the open circuit case. The latter is based on the fact that electrons carry energy in addition to charge, and a heat current then flows in the presence of an electric current. Both phenomena have been observed in mesoscopic systems [30, 31, 32], and the agreement with the scattering approach [33] is remarkable. In these nanodevices, the thermodynamic efficiency can be tuned with an external magnetic field [34, 36, 37, 38, 39, 40].

Beyond linear response, one must also take into account that the transmission is not only a function of E and V but also depends on θ: $t(E, V, \theta)$ [41]. As a consequence, the injected charge that builds up in the vicinity of the sample is determined from both particle and entropic injectivities [41, 42, 43]. Upper bounds to the performance of heat engines and coolers are thus to be carefully evaluated [44, 45]. Surprisingly, the thermovoltage can vanish for a finite value of θ, as recently demonstrated in experiments with nanowire quantum dots [46]. Furthermore, reciprocal relations are shown to break down for sufficiently high heating currents applied to multiterminal setups [47]. Interestingly, departures from the Onsager-Kelvin relations differ from auto or cross-terminal measurements [41, 48].

In this work, we discuss thermoelectric effects beyond linear response for quantum Hall bars with an inserted antidot. This system allows for an investigation of the underlying symmetries while providing, at the same time, a strong interaction and energy dependent scattering [49, 50]. We find large deviations of the Wiedemann-Franz law and interesting nonlinear behavior of the Peltier effect. Additionally, we discuss the spin transport in quantum spin Hall systems [51] coupled to ferromagnetic leads. Whereas the normal case already leads to the generation of pure spin currents [52], we here find a competition with the polarized currents injected from the magnetic contacts.

2. Theoretical formalism

When a mesoscopic conductor is coupled to multiple terminals $\alpha, \beta, \ldots$, each terminal can be characterized by voltage bias $eV_\alpha = \mu_\alpha - E_F$ (μ_α is the electrochemical potential and E_F is the Fermi energy) and also by the temperature gradient $\theta_\alpha = T_\alpha - T$ (T_α and T are the reservoir and the background temperature, respectively). In the presence of these two driving fields, the electronic and heat transport is entirely described by the scattering matrix $s_{\alpha\beta} = s_{\alpha\beta}(E, eU)$, which is a function of the carrier energy E and the potential landscape U inside the conductor. The potential $U = U(\vec{r}, \{V_\gamma\}, \{\theta_\gamma\})$ is, in general, a function of the position $\vec{r}$ and the set of applied voltages $\{V_\gamma\}$ and thermal gradients $\{\theta_\gamma\}$. The charge and heat currents at lead α are respectively given by [41, 43]

$$I_\alpha = \frac{2e}{h} \sum_\beta \int dE A_{\alpha\beta}(E, eU) f_\beta(E), \tag{1}$$

$$\mathcal{J}_\alpha = \frac{2}{h} \sum_\beta \int dE (E - \mu_\alpha) A_{\alpha\beta}(E, eU) f_\beta(E), \tag{2}$$

where $A_{\alpha\beta} = \text{Tr}[\delta_{\alpha\beta} - s^\dagger_{\alpha\beta} s_{\alpha\beta}]$ and $f_\beta(E) = (1 + \exp[(E - \mu_\beta)/k_B T_\beta])^{-1}$ is the Fermi distribution function in the reservoir β. In the weakly nonlinear regime of transport, one can expand these currents around the equilibrium state, i.e., $\mu_\alpha = E_F$ and $T_\alpha = T$, up to second order of V_α and θ_α [41, 42, 43]:

$$I_\alpha = \sum_\beta \Big(G_{\alpha\beta} V_\beta + L_{\alpha\beta} \theta_\beta \Big) + \sum_{\beta\gamma} \Big(G_{\alpha\beta\gamma} V_\beta V_\gamma + L_{\alpha\beta\gamma} \theta_\beta \theta_\gamma + 2M_{\alpha\beta\gamma} V_\beta \theta_\gamma \Big), \tag{3}$$

$$\mathcal{J}_\alpha = \sum_\beta \Big(R_{\alpha\beta} V_\beta + K_{\alpha\beta} \theta_\beta \Big) + \sum_{\beta\gamma} \Big(R_{\alpha\beta\gamma} V_\beta V_\gamma + K_{\alpha\beta\gamma} \theta_\beta \theta_\gamma + 2H_{\alpha\beta\gamma} V_\beta \theta_\gamma \Big). \tag{4}$$

The linear conductance coefficients are given by [33] $G_{\alpha\beta} = (2e^2/h) \int dE A_{\alpha\beta}(E)(-\partial_E f) \approx (2e^2/h) A_{\alpha\beta}(E_F)$, $L_{\alpha\beta} = (2e/hT) \int dE (E - E_F) A_{\alpha\beta}(E)(-\partial_E f) \approx (2e\pi^2 k_B^2 T/3h) \partial_E A_{\alpha\beta}(E)|_{E=E_F}$, $R_{\alpha\beta} = (2e/h) \int dE (E - E_F) A_{\alpha\beta}(E)(-\partial_E f) \approx (2e\pi^2 k_B^2 T^2/3h) \partial_E A_{\alpha\beta}(E)|_{E=E_F}$, and $K_{\alpha\beta} =$

Journal of Physics: Conference Series **568** (2014) 052016 doi:10.1088/1742-6596/568/5/052016

$(2/h)\int dE\frac{(E-E_F)^2}{T}A_{\alpha\beta}(E)(-\partial_E f) \approx (2\pi^2 k_B^2 T/3h)A_{\alpha\beta}(E_F)$, where we have resorted to the Sommerfeld expansion for the approximate expressions. We emphasize here that the linear coefficients $G_{\alpha\beta}$, $L_{\alpha\beta}$, $R_{\alpha\beta}$, and $K_{\alpha\beta}$ are equilibrium quantities and hence are independent of the screening potential U. In contrast, the nonlinear coefficients $G_{\alpha\beta\gamma}$, $L_{\alpha\beta\gamma}$, $M_{\alpha\beta\gamma}$, $R_{\alpha\beta\gamma}$, $K_{\alpha\beta\gamma}$, and $H_{\alpha\beta\gamma}$ manifestly depend on U in response to the applied voltage and temperature biases. Explicitly, they are written by [41, 42, 43] $G_{\alpha\beta\gamma} = (-e^2/h)\int dE\Big(\frac{\partial A_{\alpha\beta}}{\partial V_\gamma} + \frac{\partial A_{\alpha\gamma}}{\partial V_\beta} + e\delta_{\beta\gamma}\frac{\partial A_{\alpha\beta}}{\partial E}\Big)\partial_E f$, $L_{\alpha\beta\gamma} = (e/h)\int dE\frac{E_F-E}{T}\Big(\frac{\partial A_{\alpha\beta}}{\partial\theta_\gamma} + \frac{\partial A_{\alpha\gamma}}{\partial\theta_\beta} + \delta_{\beta\gamma}\frac{E-E_F}{T}\frac{\partial A_{\alpha\beta}}{\partial E}\Big)\partial_E f$, $M_{\alpha\beta\gamma} = (e^2/h)\int dE\Big(\frac{E_F-E}{eT}\frac{\partial A_{\alpha\gamma}}{\partial V_\beta} - \frac{\partial A_{\alpha\beta}}{\partial\theta_\gamma} - \delta_{\beta\gamma}\frac{E-E_F}{T}\frac{\partial A_{\alpha\beta}}{\partial E}\Big)\partial_E f$, $R_{\alpha\beta\gamma} = (e^2/h)\int dE\Big\{\delta_{\alpha\gamma}A_{\alpha\beta} + \delta_{\alpha\beta}A_{\alpha\beta} - (E-E_F)\Big(\frac{\partial A_{\alpha\beta}}{\partial eV_\gamma} + \frac{\partial A_{\alpha\gamma}}{\partial eV_\beta}\Big) - \delta_{\beta\gamma}\Big[(E-E_F)\frac{\partial A_{\alpha\beta}}{\partial E} + A_{\alpha\beta}\Big]\Big\}\partial_E f$, $K_{\alpha\beta\gamma} = (-1/h)\int dE\frac{(E-E_F)^2}{T}\Big\{\Big(\frac{\partial A_{\alpha\beta}}{\partial\theta_\gamma} + \frac{\partial A_{\alpha\gamma}}{\partial\theta_\beta}\Big) + \delta_{\beta\gamma}\Big[\frac{(E-E_F)}{T}\frac{\partial A_{\alpha\beta}}{\partial E} + \frac{A_{\alpha\beta}}{T}\Big]\Big\}\partial_E f$, and $H_{\alpha\beta\gamma} = (-e/h)\int dE(E-E_F)\Big\{\Big(\frac{\partial A_{\alpha\gamma}}{\partial\theta_\beta} + \frac{(E-E_F)}{T}\frac{\partial A_{\alpha\beta}}{\partial eV_\gamma} - \delta_{\alpha\gamma}\frac{A_{\alpha\beta}}{T}\Big) + \delta_{\beta\gamma}\Big[\frac{(E-E_F)}{T}\frac{\partial A_{\alpha\beta}}{\partial E} + \frac{A_{\alpha\beta}}{T}\Big]\Big\}\partial_E f$.

Leading order interaction effects can be incorporated into the nonequilibrium potential U via characteristic potentials (CPs) $u_\alpha = (\partial U/\partial V_\alpha)_{\rm eq}$ and $z_\alpha = (\partial U/\partial\theta_\alpha)_{\rm eq}$:

$$U = U_{\rm eq} + \sum_\alpha u_\alpha V_\alpha + \sum_\alpha z_\alpha\theta_\alpha. \tag{5}$$

Note that CPs u_α and z_α relate the variation of U to voltage and temperature shifts, respectively. We determine U self-consistently by considering the net charge of the system $q = q_{\rm bare} + q_{\rm scr}$. The bare charge $q_{\rm bare} = e\sum_\alpha(D_\alpha^p eV_\alpha + D_\alpha^e\theta_\alpha)$ has two contributions due to voltage bias and temperature shift, described by the particle injectivity [28, 29] $\nu_\alpha^p(E) = (2\pi i)^{-1}\sum_\beta {\rm Tr}\big[s_{\beta\alpha}^\dagger\frac{ds_{\beta\alpha}}{dE}\big]$ and the entropic injectivity [41] $\nu_\alpha^e(E) = (2\pi i)^{-1}\sum_\beta {\rm Tr}\big[\frac{E-E_F}{T}s_{\beta\alpha}^\dagger\frac{ds_{\beta\alpha}}{dE}\big]$ with $D_\alpha^{p,e} = -\int dE\nu_\alpha^{p,e}(E)\partial_E f$. The screening charge $q_{\rm scr}$ building up inside the sample can be obtained from the response of the screening potential, $\Delta U = U - U_{\rm eq}$, away from the equilibrium state $U_{\rm eq}$. One can write $q_{\rm scr} = e^2\Pi\Delta U$ where Π is the Lindhard function which becomes $\Pi = \int dED(E)\partial_E f$ in the long wavelength approximation, with $D(E)$ being the density of states. Then, the set of equations for the CPs is closed via the Poisson's equation $\nabla^2\Delta U = -4\pi q$.

Finally, for a practical calculation, one can resort to the WKB approximation valid in the long wavelength limit and make the replacement $\delta/\delta U \to -e\partial/\partial E$, i.e., $\frac{\partial A_{\alpha\beta}}{\partial V_\gamma} = \frac{\partial U}{\partial V_\gamma}\frac{\delta A_{\alpha\beta}}{\delta U} \approx -eu_\gamma\frac{\partial A_{\alpha\beta}}{\partial E}$ and $\frac{\partial A_{\alpha\beta}}{\partial\theta_\gamma} = \frac{\partial U}{\partial\theta_\gamma}\frac{\delta A_{\alpha\beta}}{\delta U} \approx -ez_\gamma\frac{\partial A_{\alpha\beta}}{\partial E}$. For a two-terminal setup which we consider below, the matrix elements of $A_{\alpha\beta}$ are given by $A_{11} = A_{22} = -A_{12} = -A_{21} = t(E)$, where $t(E)$ is the transmission probability through the system.

2.1. Magnetic-field asymmetry

At equilibrium, the screening potential $U = U_{\rm eq}$ is symmetric with respect to the reversal of an applied magnetic field, i.e., $U_{\rm eq}(B) = U_{\rm eq}(-B)$, due to the fundamental microscopic reversibility principle. In Ref. [33], magnetic-field *symmetry* of *linear* thermoelectric and heat transport has been shown in the framework of scattering theory. When the system is driven into the nonequilibrium regime, however, this magnetic-field symmetry can be broken since the CPs in Eq. (5) are in general magnetic-field asymmetric, i.e., $u_\alpha(B) \neq u_\alpha(-B)$ and $z_\alpha(B) \neq z_\alpha(-B)$. In Ref. [8], the nonlinear conductance $G_{\alpha\beta\gamma}$ in the isothermal case has been shown to be magnetic-field asymmetric since u_α, the voltage response of U, is not an even function of B. Here, we generalize this result to all thermoelectric and heat transport coefficients.

We quantify the magnetic-field asymmetry in the nonlinear transport regime by defining the symmetry(Σ) and the asymmetry(A) parameters [48] for G, L, R, and K coefficients in Eqs. (3)

and (4):

$$\Sigma^X_{\alpha\beta,\gamma\delta} \equiv \frac{X_{\alpha\beta}(B)X_{\gamma\delta}(-B)}{X^{\rm linear}_{\alpha\beta}(B)X^{\rm linear}_{\gamma\delta}(-B)}, \qquad A^X_{\alpha\beta,\gamma\delta} \equiv \frac{X_{\alpha\beta}(B)}{X_{\gamma\delta}(-B)}, \tag{6}$$

where $X_{\alpha\beta}$ indicates the differential transport coefficients $\mathcal{G}_{\alpha\beta}$, $\mathcal{L}_{\alpha\beta}$, $\mathcal{R}_{\alpha\beta}$, and $\mathcal{K}_{\alpha\beta}$ defined by

$$\mathcal{G}_{\alpha\beta} \equiv \left.\frac{\partial I_\alpha}{\partial V_\beta}\right|_{\{\theta\}=0} = G_{\alpha\beta} + 2G_{\alpha\beta\beta}V_\beta + \sum_{\epsilon\neq\beta}(G_{\alpha\beta\epsilon} + G_{\alpha\epsilon\beta})V_\epsilon, \tag{7}$$

$$\mathcal{L}_{\alpha\beta} \equiv \left.\frac{\partial I_\alpha}{\partial \theta_\beta}\right|_{\{V\}=0} = L_{\alpha\beta} + 2L_{\alpha\beta\beta}\theta_\beta + \sum_{\epsilon\neq\beta}(L_{\alpha\beta\epsilon} + L_{\alpha\epsilon\beta})\theta_\epsilon, \tag{8}$$

$$\mathcal{R}_{\alpha\beta} \equiv \left.\frac{\partial \mathcal{J}_\alpha}{\partial V_\beta}\right|_{\{\theta\}=0} = R_{\alpha\beta} + 2R_{\alpha\beta\beta}V_\beta + \sum_{\epsilon\neq\beta}(R_{\alpha\beta\epsilon} + R_{\alpha\epsilon\beta})V_\epsilon, \tag{9}$$

$$\mathcal{K}_{\alpha\beta} \equiv \left.\frac{\partial \mathcal{J}_\alpha}{\partial \theta_\beta}\right|_{\{V\}=0} = K_{\alpha\beta} + 2K_{\alpha\beta\beta}\theta_\beta + \sum_{\epsilon\neq\beta}(K_{\alpha\beta\epsilon} + K_{\alpha\epsilon\beta})\theta_\epsilon, \tag{10}$$

and $X^{\rm linear}_{\alpha\beta}$ refers to the linear terms $G_{\alpha\beta}$, $L_{\alpha\beta}$, $R_{\alpha\beta}$, and $K_{\alpha\beta}$. We here consider either an isothermal, i.e., $\{\theta\}=0$, or an isoelectric case, i.e., $\{V\}=0$, hence the terms $M_{\alpha\beta\gamma}$ and $H_{\alpha\beta\gamma}$ in Eqs. (3) and (4) are not considered. Notice that $X_{\alpha\beta}$ contains both linear and nonlinear terms and in the linear response regime it satisfies $\Sigma^X_{\alpha\beta,\beta\alpha} = A^X_{\alpha\beta,\beta\alpha} = 1$, due to the microscopic reversibility $X^{\rm linear}_{\alpha\beta}(B) = X^{\rm linear}_{\beta\alpha}(-B)$. Hence, a deviation from 1 indicates the magnetic-field asymmetry in the nonlinear regime. Moreover, these definitions have direct relevance to experiments [47] and also are related to the efficiency of the thermoelectric power generation or the refrigeration [34, 37, 38].

3. Quantum Hall bar wih an antidot

As a model system for the general formalism described above, we consider a conductor in the quantum Hall regime in a two-terminal setup, see Fig. 1. We suppose that B is strong enough so that only the lowest Landau level is occupied and consider the reversal of its direction $B \to -B$. A gate-controllable antidot [49, 50] can connect two counter-propagating edge states, which we regard as a quantum impurity with a Breit-Wigner resonance at $E_d + eU_d(B)$ where $U_d(B)$ is the potential shift at the antidot in the presence of magnetic field B. The upper and lower edge states are tunnel-coupled to the antidot level via coupling strengths Γ_1 and Γ_2, respectively. If the direction of the magnetic field is reversed, the resonant level at the antidot is located at $E_d + eU_d(-B)$. It should be noted that in general $U_d(B) \neq U_d(-B)$ beyond the linear response regime [8]. The B-asymmetry can appear either via scattering asymmetry, $\Gamma_1 \neq \Gamma_2$, or electrical asymmetry if the Coulomb interaction between the upper edge and the antidot charges is stronger (or weaker) than that between the lower edge and antidot.

In Fig. 1, we divide the conductor potential into five regions Ω_i with $i = 1, \dots, 5$, with $\Omega_5 \equiv \Omega_d$ indicating the antidot region. We assume that the potential U_i in each region is constant and the Coulomb interactions between different regions are described by a capacitance matrix C_{ij} [28], capturing the essential physics of our interest [8, 28, 29]. For definiteness, we also assume the equal density of states for all regions, i.e., $D_i = D$ and the symmetric injectivities between two terminals, i.e., $D^{p,e}_{i\alpha} = D^{p,e}$ and $\Pi_i = \Pi$. Finally, we self-consistently determine the internal potential and therefore the CPs by solving the Poisson's equation explained previously.

We here consider the two cases: (i) the scattering asymmetry, i.e., $C_i = C$, $\Gamma_1 = (1+\eta)\Gamma/2$, $\Gamma_2 = (1-\eta)\Gamma/2$, (ii) the electrical asymmetry, i.e., $\Gamma_1 = \Gamma_2$, $C_1 = C_2 = (1+\xi)C$, $C_3 = C_4 = (1-\xi)C$. In the above two cases, the inherent asymmetry is described with a

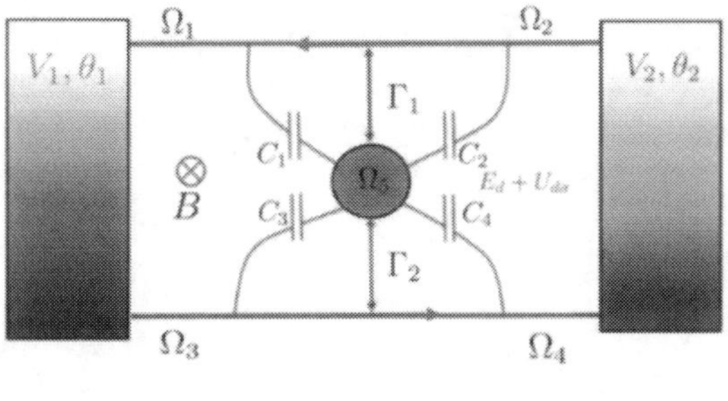

Figure 1. A two-terminal quantum Hall bar-antidot system, where voltages V_1, V_2 and temperature gradients θ_1, θ_2 are applied. An antidot (Ω_5) is coupled to the edge states with the hybridization widths Γ_1 and Γ_2 and the Coulomb interactions are described by capacitances C_1, C_2, C_3, C_4.

parameter η or ξ, respectively. We evaluate the CPs as

$$u_1(B) = u_2(-B) = \begin{cases} \frac{1}{2} + \eta c_{sc} \\ \frac{1}{2} + \xi c_{el} \end{cases}, \qquad u_1(-B) = u_2(B) = \begin{cases} \frac{1}{2} - \eta c_{sc} \\ \frac{1}{2} - \xi c_{el} \end{cases}, \tag{11}$$

$$z_1(B) = z_2(-B) = \frac{D^e}{eD^p} u_1(B), \qquad z_1(-B) = z_2(B) = \frac{D^e}{eD^p} u_1(-B), \tag{12}$$

with

$$c_{sc} = \left(2 + \frac{4\pi C D^p \Gamma}{r(C - e^2\Pi)}\right)^{-1}, \quad c_{el} = \frac{-\pi e^2 \Pi D^p C \Gamma t}{(C - e^2\Pi)[2\pi C D^p \Gamma + r(C - e^2\Pi)]}, \tag{13}$$

where $r = 1 - t = \Gamma_1\Gamma_2/|\Lambda|^2$ is the Breit-Wigner reflection (t: transmission) through the antidot at equilibrium, and $\Lambda = E_F - E_d - eU_d^{eq} + i\Gamma/2$.

In Eq. (11), one can easily prove the fundamental sum rule $u_1(B) + u_2(B) = u_1(-B) + u_2(-B) = 1$ due to the gauge invariance. Importantly, the CPs are generally magnetic-field asymmetric, i.e., $u_\alpha(B) \neq u_\alpha(-B)$ and $z_\alpha(B) \neq z_\alpha(-B)$. We also point out the properties $u_1(\pm B) = u_2(\mp B)$ and $z_1(\pm B) = z_2(\mp B)$ in Eqs. (11) and (12), due to the chiral nature of the quantum Hall system.

The symmetry(Σ) and the asymmetry(A) parameters defined in Eq. (6) can be evaluated straightforwardly with the CPs in Eqs. (11) and (12). The general expressions of these parameters for a two-terminal quantum conductor can be found in Ref. [48], to which we also refer the readers for the detailed analysis of our quantum Hall system. We here briefly summarize some of the interesting results.

Firstly, we find all the off-diagonal asymmetry parameters $A^X_{\alpha\beta,\beta\alpha}$ as well as the symmetry parameter $\Sigma^{\mathcal{G}}_{11,11}$ for electric conductance always exhibit B-symmetry even in the nonlinear regime:

$$\Sigma^{\mathcal{G}}_{11,11} = A^{\mathcal{G}}_{12,21} = A^{\mathcal{L}}_{12,21} = A^{\mathcal{R}}_{12,21} = A^{\mathcal{K}}_{12,21} = 1. \tag{14}$$

This originates from the chiral property $u_1(\pm B) = u_2(\mp B)$ and $z_1(\pm B) = z_2(\mp B)$. In addition to this, the gauge invariance condition $\sum_\alpha u_\alpha(\pm B) = 1$ gives a contribution for the derivation of $\Sigma^{\mathcal{G}}_{11,11} = 1$ because $u_1(B) + u_1(-B) = u_1(B) + u_2(B) = 1$ where the latter equality comes from the gauge invariance (see Ref. [48] for the details).

Secondly, we find that the symmetry parameters for $\mathcal{L}$(thermoelectric) and $\mathcal{K}$(thermal) coefficients depend on the lead indices:

$$\Sigma^{\mathcal{L}}_{11,11} = 1 + c^{\mathcal{L}}_1(2\theta/T), \qquad \Sigma^{\mathcal{L}}_{12,21} = 1 + c^{\mathcal{L}}_2(2\theta/T), \tag{15a}$$

$$\Sigma^{\mathcal{K}}_{11,11} = 1 + c^{\mathcal{K}}_1(2\theta/T), \qquad \Sigma^{\mathcal{K}}_{12,21} = 1 + c^{\mathcal{K}}_2(2\theta/T). \tag{15b}$$

This different tendency between diagonal and off-diagonal parameters have a relevance to recent experiments [47]. Intriguingly, we note that this is a high-temperature effect since in the limit $k_BT \to 0$ we find $\Sigma^{\mathcal{L}}_{11,11} = \Sigma^{\mathcal{L}}_{12,21} = \Sigma^{\mathcal{K}}_{11,11} = \Sigma^{\mathcal{K}}_{12,21} = 1 + 2\theta/T$, irrespective of the system parameters (see Ref. [48] for the details).

In this quantum Hall system, eight parameters $\Sigma^X_{11,11}$ and $A^X_{12,21}$ for all $X = \mathcal{G}, \mathcal{L}, \mathcal{R}, \mathcal{K}$, are independent of the scattering asymmetry(η) and the electrical asymmetry(ξ) factors due to the chiral nature, including five B-symmetric parameters in Eq. (14). However, the off-diagonal elements $\Sigma^{\mathcal{L}}_{12,21}$ and $\Sigma^{\mathcal{K}}_{12,21}$ are dependent on η or ξ. Consequently, the distinction between the diagonal and the off-diagonal elements disappears for $\eta = \xi = 0$. Thus, an asymmetry inherent in the system is crucial to observe this distinction, which is consistent with an asymmetric scattering in a recent experiment [47].

Even with a nonzero η or ξ, our results show that we can tune the antidot level to make $c^{\mathcal{L}}_1 = c^{\mathcal{L}}_2 = 1$ in Eq. (15a) and recover the universality of the thermoelectric coefficients, i.e., $\Sigma^{\mathcal{L}}_{11,11} = \Sigma^{\mathcal{L}}_{12,21} = 1 + 2\theta/T$. But, this is not the case for the heat current counterparts $\Sigma^{\mathcal{K}}_{11,11}$ and $\Sigma^{\mathcal{K}}_{12,21}$.

In addition to the above parameters, one can also write $\Sigma^{\mathcal{G}}_{12,21} = 1 - c^{\mathcal{G}}(eV/\Gamma)$, $A^{\mathcal{G}}_{11,11} = 1 + c^{\mathcal{G}}(eV/\Gamma)$, $A^{\mathcal{L}}_{11,11} = 1 + c^{\mathcal{L}}_A(2\theta/T)$, $A^{\mathcal{R}}_{11,11} = 1 + c^{\mathcal{R}}_A(eV/\Gamma)$, and $A^{\mathcal{K}}_{11,11} = 1 + c^{\mathcal{K}}_A(2\theta/T)$, where $c^{\mathcal{G}} = c^{\mathcal{L}}_A = c^{\mathcal{R}}_A = c^{\mathcal{K}}_A = 0$ if $\eta = \xi = 0$. Hence B-asymmetry of these parameters is due only to the underlying asymmetry in the system. We note here that $\Sigma^{\mathcal{G}}_{12,21}$ and $A^{\mathcal{G}}_{11,11}$ are described by a single constant $c^{\mathcal{G}}$.

Finally, we write the electrothermal symmetry parameters $\Sigma^{\mathcal{R}}_{11,11} = 1 + c^{\mathcal{R}}_1(eV/\Gamma)$ and $\Sigma^{\mathcal{R}}_{12,21} = 1 - c^{\mathcal{R}}_2(eV/\Gamma)$ in heat current measurements. Here, $\Sigma^{\mathcal{R}}_{12,21}$ is in general a function of η or ξ, whereas $\Sigma^{\mathcal{R}}_{11,11}$ is not.

3.1. Nonlinear Peltier effect

When an electrical current (I) flows through a conductor in a isothermal configuration a heat current (J) is generated. The linear Peltier coefficient is defined as the ratio between the heat and electrical currents,

$$\Pi_0 = \frac{R_{11}}{G_{11}}, \tag{16}$$

In Ref. [43] the Π coefficient was generalized to the nonlinear regime. By keeping only the leading order corrections in powers of the electrical current one has

$$\delta\Pi = \Pi - \Pi_0 = \frac{I}{G_{11}}\left[\frac{R_{111}}{R_{11}} - \frac{G_{111}}{G_{11}} + \cdots\right], \tag{17}$$

The conversion factor for electric currents into heat flow in the nonlinear regime is given by the relative strength of the nonlinear conductances, R_{111}, and G_{111} to the linear ones (R_{11}, and G_{11}). In terms of the symmetries of the nonlinear coefficients we can rewrite Eq. (17) as

$$\delta\Pi = \Pi - \Pi_0 = \frac{I}{2VG_{11}}\left[\Sigma^{\mathcal{R}}_{11,11} + A^{\mathcal{R}}_{11,11} - \left(\Sigma^{\mathcal{G}}_{11,11} + A^{\mathcal{G}}_{11,11}\right) + \cdots\right], \tag{18}$$

In Fig. 2 we plot $\delta\Pi$ normalized to a given value of the electric current I. We observe large

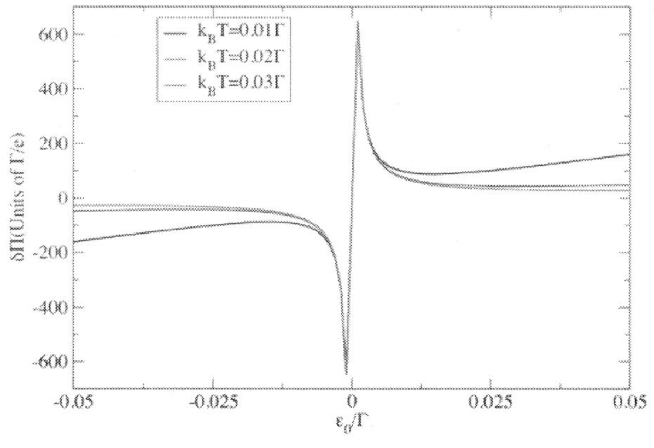

Figure 2. Plots of $\delta\Pi$ versus antidot level ε_0/Γ for several background temperatures k_BT. We used $k_B\theta/\Gamma = eV/\Gamma = 0.1$ and $\eta = 0.1$.

deviations of the Peltier effect arising around resonance (i.e., when $\epsilon_0 \approx E_F$). Otherwise, the nonlinear Peltier coefficient attains its linear value. Besides, we find that deviations occurring for $\epsilon_0 \approx E_F$ are almost independent of the background temperature. Only when ϵ_0 lies off resonance, the largest deviations take place for the highest temperature.

3.2. Wiedemann-Franz law

We now explore the degree of fulfilment of the Wiedemann-Franz law beyond linear response. For a metallic system and at very low temperatures the ratio between the thermal and electric conductances normalized to the background temperatue T equals a universal value,

$$\Lambda_0 = \frac{K_{11}}{TG_{11}}, \tag{19}$$

where $\Lambda_0 = \frac{\pi^2}{3}(\frac{k_B}{e})^2$ is the Lorentz number. In the nonlinear regime, Ref. [43] suggests a similar quantity as the ratio between the heat flow (normalized to the temperature shift) in the isoelectrical configuration and the electrical current in the isothermal case (normalized to the applied voltage),

$$\Lambda = \frac{(J/\theta)|_{V=0}}{T(I/V)|_{\theta=0}}, \tag{20}$$

The normalized deviation of the Wiedemann-Franz law from the Lorentz number can be expressed in terms of the nonlinear transport coefficients,

$$\frac{\Lambda - \Lambda_0}{\Lambda_0} = \frac{K_{111}}{K_{11}}V - \frac{G_{111}}{G_{11}}\theta + \cdots \tag{21}$$

which, in turn, can be recast using the symmetry and asymmetry coefficients defined above,

$$\frac{\Lambda - \Lambda_0}{\Lambda_0} = \left(\frac{\Sigma^{\mathcal{K}}_{11,11} + A^{\mathcal{K}}_{11,11}}{2} - 1\right)\frac{V}{2\theta} - \left(\frac{\Sigma^{\mathcal{G}}_{11,11} + A^{\mathcal{G}}_{11,11}}{2} - 1\right)\frac{\theta}{2V} + \cdots \tag{22}$$

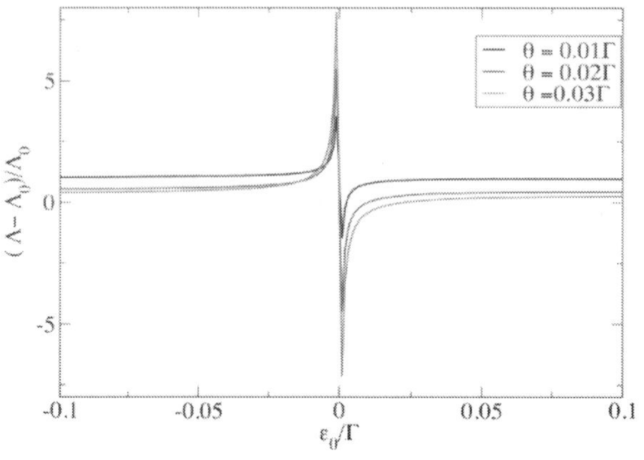

Figure 3. Plots of $(\Lambda - \Lambda_0)/\Lambda_0$ versus antidot level ε_0/Γ for several temperatures $k_B\theta$. We use $k_BT/\Gamma = 0.1$, $eV/\Gamma = 0.05$ and $\eta = 0.5$.

We find that departures become stronger for level resonances around the Fermi energy within a energy scale given by Γ. This is so because the Γ determines the energy variation of the transmission function. In addition, the deviations increase for higher θ since the heat transport is more linear with increasing θ.

4. Thermoelectric transport in quantum spin Hall insulators

4.1. Spin-generalized screening potential

In the previous section, a quantum Hall antidot in a two-terminal setup has been investigated, where we determine the CPs by solving the Poisson's equation. Here, we consider a simple extension to the quantum spin Hall (QSH) system, see Fig. 4. In order to deal with the spintronic case with the same methodology, we generalize the scattering formalism in a spin-dependent manner. The potential $U_\sigma = U(\vec{r}, \{V_\gamma\}, \{\theta_\gamma\}, \sigma)$ is now a function of the spin index $\sigma = \uparrow, \downarrow$ as well. This σ-dependence is crucial in our QSH system due to the underlying helicity, i.e., the spin-momentum correlation. Indeed, the property $U_\uparrow \neq U_\downarrow$ through the antidot filter is the operational principle for the spin-polarized currents even with normal metallic contacts.

We generalize Eqs. (1) and (2) into the spin-resolved form [52]

$$I_\alpha^\sigma = \frac{e}{h} \sum_\beta \int dE A_{\alpha\beta}^\sigma(E, eU) f_\beta(E), \tag{23}$$

$$\mathcal{J}_\alpha^\sigma = \frac{1}{h} \sum_\beta \int dE (E - \mu_\alpha) A_{\alpha\beta}^\sigma(E, eU) f_\beta(E), \tag{24}$$

for which we divide $2A_{\alpha\beta}$ in Eqs. (1) and (2) into $A_{\alpha\beta}^\uparrow = A_{\alpha\beta}(U_\uparrow)$ and $A_{\alpha\beta}^\downarrow = A_{\alpha\beta}(U_\downarrow)$, explicitly incorporating the spin-dependent screening effect. In a two-terminal setup, we have $A_{11}^\sigma = A_{22}^\sigma = -A_{12}^\sigma = -A_{21}^\sigma = t^\sigma(E)$ where $t^\sigma(E)$ is the spin-resolved transmission probability. We disregard the spin-flip scattering, hence the current conservation condition is satisfied for each spin component as $\sum_\alpha I_\alpha^\sigma = 0$ and $\sum_\alpha(\mathcal{J}_\alpha^\sigma + I_\alpha^\sigma V_\alpha) = 0$. One can define the direction

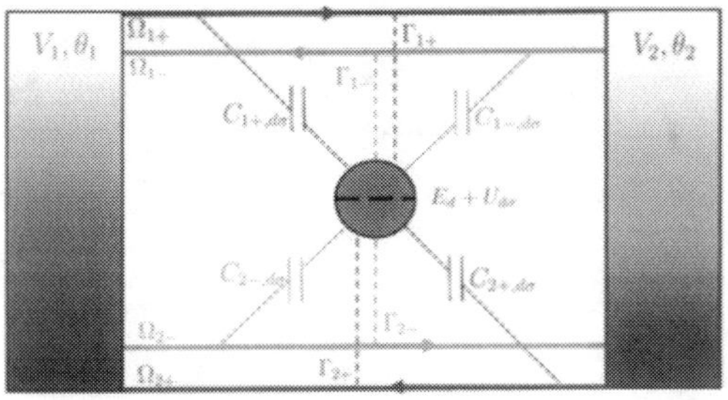

Figure 4. Schematics of quantum spin Hall bar antidot setup. Spin-generalized interactions are described by spin-dependent capacitance $C_{is,d\sigma}$, with $i = 1, 2$ being the edge label, $s = \pm$ the helicity, d standing for the dot, and $\sigma = \uparrow, \downarrow$ the electronic spin. Spin-dependent couplings between the helical edges and the single level antidot are denoted with Γ_{is}.

of spin-resolved currents $I_\sigma \equiv I_1^\sigma = -I_2^\sigma$ and $\mathcal{J}_\sigma \equiv \mathcal{J}_1^\sigma = -\mathcal{J}_2^\sigma - I_\sigma(V_1 - V_2)$, and hence the spin-polarized currents as well:

$$I_s = I_\uparrow - I_\downarrow \tag{25}$$

$$\mathcal{J}_s = \mathcal{J}_\uparrow - \mathcal{J}_\downarrow \tag{26}$$

Total charge and heat fluxes are given by $I_c \equiv I_\uparrow + I_\downarrow$ and $\mathcal{J}_c \equiv \mathcal{J}_\uparrow + \mathcal{J}_\downarrow$, respectively.

The screening potential $U = \sum_\sigma U_\sigma$ in Eq. (5) can now be generalized as

$$U = U_{\rm eq} + \sum_{\alpha,\sigma} u_{\alpha\sigma} V_\alpha + \sum_{\alpha,\sigma} z_{\alpha\sigma}\theta_\alpha, \tag{27}$$

where $u_{\alpha\sigma} = (\partial U_\sigma/\partial V_\alpha)_{\rm eq}$ and $z_{\alpha\sigma} = (\partial U_\sigma/\partial\theta_\alpha)_{\rm eq}$ are spin-dependent CPs which relate the variation of spin-resolved potential U_σ to voltage and temperature shifts at each terminal.

As described earlier, the self-consistent determination of U can be accomplished via Poisson's equation $\nabla^2\Delta U = -4\pi q$, with $\Delta U = U - U_{\rm eq} = \sum_\sigma \Delta U_\sigma$ and

$$q = \sum_\sigma q_\sigma = e\sum_{\alpha,\sigma}\left[D_\alpha^p(\sigma)eV_\alpha + D_\alpha^e(\sigma)\theta_\alpha\right] + e^2\sum_\sigma \Pi_\sigma\Delta U_\sigma\,. \tag{28}$$

In our model, σ-dependences of $D_\alpha^{p,e}(\sigma)$ and Π_σ appear for unequal spin populations arising, e.g., from ferromagnetic leads. In order to understand this, note that the first two terms in a square bracket in Eq. (28) are the contributions from the lead injection of charges by means of voltage and thermal driving, respectively. Only when the spin population is unequal ($p \neq 0$, p: polarization) via using ferromagnetic contacts, one expects $D_\alpha^{p,e}(\uparrow) \neq D_\alpha^{p,e}(\downarrow)$ and $\Pi_\uparrow \neq \Pi_\downarrow$, where the latter comes directly from $D_\uparrow \neq D_\downarrow$. For normal metallic contacts, the only term in Eq. (28) which can give rise to a spin imbalance inside the system is the screening potential ΔU_σ.

The solution procedure is quite analogous to the quantum Hall case (see Ref. [52] for details). It should be noted that for QSH case the edge-antidot couplings Γ_{1s} and Γ_{2s} in general depend on the helicity $s = \pm$ corresponding to spin channels $\uparrow(+)$ and $\downarrow(-)$, when coupled to spin-polarized ferromagnetic contacts. This is also the case for the reflection and transmission probabilities since $r_\sigma = 1 - t_\sigma = \Gamma_{1s}\Gamma_{2s}/|\Lambda_s|^2$, where $\Lambda_s = E_F - E_d + i\Gamma_s/2$ with $\Gamma_s = \Gamma_{1s} + \Gamma_{2s}$. But, for normal contacts with $p = 0$, there is no spin imbalance at the edge states, which leads to

Journal of Physics: Conference Series **568** (2014) 052016 doi:10.1088/1742-6596/568/5/052016

$t_\uparrow = t_\downarrow$ via antidot scattering. In this case, the linear conductances are spin-independent and the spin-polarization appears only in the nonlinear transport regime. In contrast, for general case with $p \neq 0$, unequal spin density leads to $t_\uparrow(p) \neq t_\downarrow(p)$ giving rise to spin-polarized electric and heat currents already in the linear regime (see Ref. [52] for details). However, even for $p \neq 0$ case, if the two contacts are magnetized in an antiparallel configuration, one finds $t_\uparrow(p) = t_\downarrow(p)$ and hence the linear spin-polarization vanishes as in normal contacts, due to the helical nature of the QSH system.

The effective Poisson's equation reads

$$q_{is} = e\sum_\alpha (D^p_{is,\alpha} eV_\alpha + D^e_{is,\alpha}\theta_\alpha) + e^2\Pi_{is}\Delta U_{is} = \sum_\sigma C_{is,d\sigma}(\Delta U_{is} - \Delta U_{d\sigma}). \tag{29}$$

Note here that the charge with spin $\sigma =\uparrow(\downarrow)$ at the antidot is supplied from the edge with helicity $s = +(-)$ since in our model we neglect the spin-flip scattering, by which one can maximize spin-polarization effects. The density of states for all regions are given by $D_{is} = D_s = (1 + sp)D/2$, and the symmetric injectivities from the two terminals give $D^{p,e}_{is,\alpha} = D^{p,e}_s = (1 + sp)D^{p,e}/2$ and $\Pi_{is} = \Pi_s = (1 + sp)\Pi/2$.

4.2. Normal contacts

Here, we focus on the result for normal metallic contacts ($p = 0$) since this is the most attractive case of an all-electrical setup, in which our main physics of interest resides. For $p = 0$, it should be mentioned that the spin-channel density of states does not depend on the helicity, i.e., $D_+ = D_- = D/2$ with $D = D_+ + D_-$.

We consider the scattering asymmetric case, i.e., equal interaction strength $C_{is,d\sigma} = C_{is} = C_s = C/2$ with $C = C_+ + C_-$, but asymmetric hybridizations $\Gamma_{1s} = (1+\eta)\Gamma/4$, $\Gamma_{2s} = (1-\eta)\Gamma/4$ with $\Gamma = \Gamma_+ + \Gamma_-$ ($\Gamma_s = \Gamma_{1s} + \Gamma_{2s} = \Gamma/2$). The coupling asymmetry is thus quantified with a nonzero $\eta = (\Gamma_1 - \Gamma_2)/\Gamma$ where $\Gamma_i = \sum_s \Gamma_{is}$. In an experiment, this can be the general case where the antidot is closer to one of the edge states. Also, this asymmetry can be generated by tuning the width and the height of the tunnel barriers between the antidot level and the edge channels. One can find $\Delta U_{d\sigma} = u_{1\sigma}V_1 + u_{2\sigma}V_2 + z_{1\sigma}\theta_1 + z_{2\sigma}\theta_2$, where the corresponding CPs are given by

$$u_{1\uparrow} = u_{2\downarrow} = \frac{1}{2} + \eta c_{\rm sc}, \quad u_{1\downarrow} = u_{2\uparrow} = \frac{1}{2} - \eta c_{\rm sc}, \tag{30}$$

$$z_{1\uparrow} = z_{2\downarrow} = \frac{D^e}{eD^p}u_{1\uparrow}, \quad z_{1\downarrow} = z_{2\uparrow} = \frac{D^e}{eD^p}u_{1\downarrow}. \tag{31}$$

Here, $c_{\rm sc} = [2 - 2C/e^2\Pi]^{-1} = C_\mu/2C$ with $1/C_\mu = 1/C + 1/e^2D$. Importantly, we can expect electronic transport to be spin polarized for asymmetric couplings due to the spin-dependent screening. Intriguingly, our result is a purely interaction-driven effect and this effect disappears in the noninteracting limit $C \to \infty$.

Up to the second order expansion of $V_1 = V$ and $\theta_1 = \theta$ ($V_2 = \theta_2 = 0$), we have the current expressions

$$I_s = -\eta c_{\rm sc}\left(\frac{2e^3}{h}t'V^2 + \frac{2e\pi^2k_B^2T}{3h}\frac{D^e}{D^p}t''\theta^2 + \frac{2e^2}{h}\left[\frac{\pi^2k_B^2T}{3}t'' + \frac{D^e}{D^p}t'\right]V\theta\right), \tag{32}$$

$$\mathcal{J}_s = -\eta c_{\rm sc}\left(\frac{2e^2\pi^2(k_BT)^2}{3h}t''V^2 + \frac{2\pi^2k_B^2T}{3h}\frac{D^e}{D^p}t'\theta^2 + \frac{2e\pi^2(k_BT)^2}{3h}\left[\frac{1}{T}t' + \frac{D^e}{D^p}t''\right]V\theta\right), \tag{33}$$

$$I_c = \frac{2e^2}{h}tV + \frac{2e\pi^2k_B^2T}{3h}t'\theta + \frac{e\pi^2k_B^2}{3h}\left(t' - T\frac{D^e}{D^p}t''\right)\theta^2 + \frac{e^2}{h}\left(\frac{\pi^2k_B^2T}{3}t'' - \frac{D^e}{D^p}t'\right)V\theta, \tag{34}$$

$$\begin{aligned}\mathcal{J}_c = &\frac{2e\pi^2(k_BT)^2}{3h}t'V + \frac{2\pi^2k_B^2T}{3h}t\theta \\ &-\frac{e^2}{h}\left(t+\frac{\pi^2(k_BT)^2}{6}t''\right)V^2 + \frac{\pi^2k_B^2}{3h}\left(t-T\frac{D^e}{D^p}t'\right)\theta^2 + \frac{e\pi^2k_B^2T}{3h}\left(t'-T\frac{D^e}{D^p}t''\right)V\theta,\end{aligned} \tag{35}$$

where $t \equiv t(E_F)$, $t' \equiv \partial_E t(E)|_{E=E_F}$, and $t'' \equiv \partial_E^2 t(E)|_{E=E_F}$. These are central to our analytic results. The spin-polarized electronic and heat currents can be generated if $\eta c_{sc} \neq 0$. Note also that spin-polarized currents appear in the nonlinear regime only. Isothermal ($\theta = 0$) and

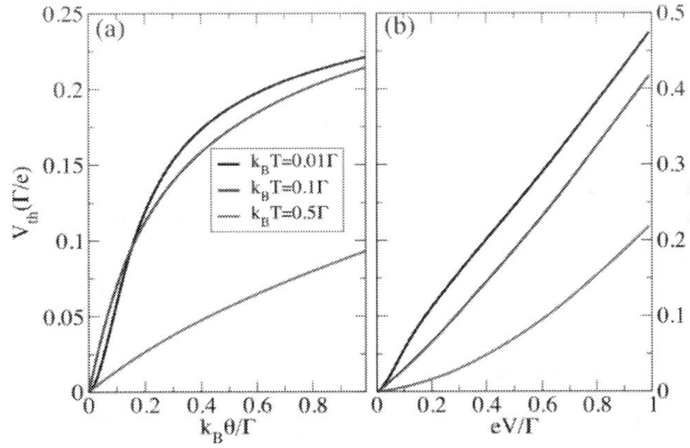

Figure 5. (a) $V_{\rm th}$ versus $k_B\theta/\Gamma$ and (b) $\theta_{\rm ad}$ versus eV/Γ, at $E_d = 0.25\Gamma$ for several k_BT. Used parameters are $\eta = c_{\rm sc} = 0.5$ and $E_F = 0$.

isoelectric ($V = 0$) cases are treated in Ref. [52]. Here, we focus on the pure spin currents, i.e., $I_c = 0$, $\mathcal{J}_c = 0$, by means of thermoelectric Seebeck and Peltier effects.

In open-circuit conditions, a thermovoltage $V_{\rm th}$ can be generated in response to a temperature bias θ, making $I_c = 0$. Figure 5(a) shows the numerically evaluated set of biases $\{\theta, V\}$ which satisfies $I_c(V_{\rm th}, \theta) = 0$, where the slope can be identified as the Seebeck coefficient. Note that the thermovoltage acquires a nonlinear component with increasing θ [41, 46]. One finds the pure spin current expression as a function of thermal gradient only

$$I_s = \eta c_{\rm sc}\frac{2e\pi^2k_B^2T}{3h}\left(\frac{\pi^2k_B^2T}{3}\left[\frac{t't''}{t}-\frac{(t')^3}{t^2}\right]+\frac{D^e}{D^p}\left[\frac{(t')^2}{t}-t''\right]\right)\theta^2, \tag{36}$$

up to leading order in θ. Figure 6(a) shows the numerical plots of pure I_s versus θ beyond the quadratic regime (a comparison with the analytical results can be found in Ref. [52]). We observe that the amplitude of I_s shows a nonmonotonic behavior with T, providing another way to maximize the effect.

In our setup, also the *pure spin heat flows* can be generated using electrical means only. By adiabatically isolating the sample, thermal bias $\theta_{\rm ad}$ is generated in response to the applied voltage V, making $\mathcal{J}_c(V, \theta_{\rm ad}) = 0$, see Fig. 5(b). We find the pure spin heat current expression

$$\mathcal{J}_s = \eta c_{\rm sc}\frac{2e^2\pi^2(k_BT)^2}{3h}\left(\left[\frac{(t')^2}{t}-t''\right]+T\frac{D^e}{D^p}\left[\frac{t't''}{t}-\frac{(t')^3}{t^2}\right]\right)V^2. \tag{37}$$

up to leading order in V. Corresponding plots are shown in Fig. 6(b) as a function of V.

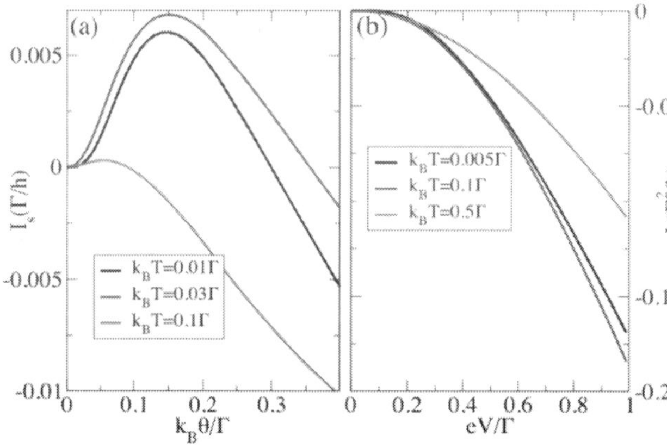

Figure 6. (a) Pure I_s versus $k_B\theta/\Gamma$ and (b) pure $\mathcal{J}_s$ versus eV/Γ, at $E_d/\Gamma = 0.2$ for several k_BT. Here, $\eta = c_{\rm sc} = 0.5$ and $E_F = 0$.

4.3. Polarized contacts

For polarized contacts with $p \neq 0$, an unequal spin injection from the reservoirs with a majority spin density $(1+p)D/2$ and a minority one $(1-p)D/2$ would generate spin-polarized currents. In this sense, the results are to some extent anticipated and hence less interesting than those of the unpolarized case. For an equal and antiparallel magnetization, however, it is interesting to note that linear conductances are spin-independent giving rise to vanishing linear spin-polarization. This will be explained below.

Although the QSH bar itself preserves the time-reversal invariance, we can also investigate the reciprocity relation by considering ferromagnetic contacts with a polarization p, where the role of a magnetic-field reversal ($B \to -B$) can be achieved by simultaneous reversal of polarization $p \to -p$ and spin index $\sigma \to \bar{\sigma}$ ($\sigma =\uparrow, \downarrow$) [53]. As we will explain below, in any case the Onsager-Casimir relation is always satisfied for linear coefficients, i.e., $X^{\uparrow}_{\alpha\beta}(p) = X^{\downarrow}_{\beta\alpha}(-p)$ for $X = G, L, R, K$, which holds even for $p = 0$. Beyond the above argument, we here discuss about the breakdown of this fundamental symmetry relation in nonlinear regime of transport for parallel and antiparallel magnetization respectively, and view the normal contacts as a limiting case of these two.

4.3.1. Parallel magnetization For an equal and parallel magnetization between two reservoirs with a polarization p, the density of majority spin-component (say, $\uparrow$) is given by $D_{\ell\uparrow} = (1+p)D_\ell/2$ and the minority one by $D_{\ell\downarrow} = (1-p)D_\ell/2$ where $\ell = L, R$ denotes the left or right leads. Then, including the helicity s-dependence, we have $\Gamma_{\ell s} = (1+sp)\Gamma_\ell/2$ where $\ell = 1, 2$ corresponds to the upper or lower counterpart for edge-antidot coupling [see Fig. 4]. This gives polarization- and spin-dependent transmission as $t^\sigma(p, E) = 16(E-E_d)^2/[16(E-E_d)^2 + (1+sp)^2\Gamma^2]$ with $\Gamma = \Gamma_1 + \Gamma_2$. Hence, as expected, spin-polarized currents appear already in the linear regime since $t_\uparrow(p) - t_\downarrow(p) \neq 0$. Moreover, one can notice here that $t_\uparrow(p) = t_\downarrow(-p)$ because the relation $sp = \bar{s}(-p)$ is always satisfied in the expression of transmission, with $s = +(-)$ corresponding to $\sigma =\uparrow (\downarrow)$. Thus, we have $G^{\uparrow}_{\alpha\beta}(p) = G^{\downarrow}_{\beta\alpha}(-p)$, $L^{\uparrow}_{\alpha\beta}(p) = L^{\downarrow}_{\beta\alpha}(-p)$, $R^{\uparrow}_{\alpha\beta}(p) = R^{\downarrow}_{\beta\alpha}(-p)$, and $K^{\uparrow}_{\alpha\beta}(p) = K^{\downarrow}_{\beta\alpha}(-p)$ for $\alpha, \beta = 1, 2$ (we refer the reader to Ref. [52] for explicit expressions of all coefficients). This is a manifestation of the fundamental symmetry relations near equilibrium [33].

We calculate the screening potential $\Delta U^{\sigma}(p) = u_1^{\sigma}(p)V_1 + u_2^{\sigma}(p)V_2 + z_1^{\sigma}(p)\theta_1 + z_2^{\sigma}(p)\theta_2$, where the corresponding CPs in the presence of (i) scattering and (ii) electrical asymmetries are given by

$$u_1^{\sigma}(p) = \left\{ \begin{array}{l} \frac{1}{2} + \eta c_{\rm sc}^{\sigma}(p) \\ \frac{1}{2} + \xi c_{\rm el}^{\sigma}(p) \end{array} \right. , \quad u_2^{\sigma}(p) = \left\{ \begin{array}{l} \frac{1}{2} - \eta c_{\rm sc}^{\sigma}(p) \\ \frac{1}{2} - \xi c_{\rm el}^{\sigma}(p) \end{array} \right. , \tag{38}$$

and $z_{\ell=1,2}^{\sigma}(p) = (D^e/eD^p)u_{\ell=1,2}^{\sigma}(p)$, with

$$c_{\rm sc}^{\sigma}(p) = s\left[2 - \frac{2C(1+sp)}{e^2\Pi}\right]^{-1}, \tag{39}$$

$$c_{\rm el}^{\sigma}(p) = \frac{-(s)e^4\pi\Gamma C\Pi D^p(1+sp)\left(\frac{(1+sp)^2(1-r_\sigma^2)}{[2C-e^2\Pi(1+sp)(1+r_\sigma)]^2} - \frac{(1+\bar{s}p)^2(1-r_{\bar\sigma}^2)}{[2C-e^2\Pi(1+\bar{s}p)(1+r_{\bar\sigma})]^2}\right)}{4\pi\Gamma C(1+sp) + 8e^2 r_\sigma}. \tag{40}$$

Here, $t_\sigma(p) = 1 - r_\sigma(p) = 16(E_F - E_d)^2/[16(E_F - E_d)^2 + (1+sp)^2\Gamma^2]$ and $s = \pm$ refers to $\sigma = \uparrow, \downarrow$.

We note that $c_{\rm sc}^{\uparrow}(p) = -c_{\rm sc}^{\downarrow}(-p)$ and $c_{\rm el}^{\uparrow}(p) = -c_{\rm el}^{\downarrow}(-p)$, and hence $u_1^{\uparrow}(p) = u_2^{\downarrow}(-p)$. Furthermore, one can show that $u_{\ell=1,2}^{\uparrow}(p) - u_{\ell=1,2}^{\downarrow}(-p) \propto \eta, \xi$ and $u_{\ell=1,2}^{\uparrow}(p) - u_{\ell=1,2}^{\downarrow}(p) = \pm\eta[c_{\rm sc}^{\uparrow}(p) - c_{\rm sc}^{\downarrow}(p)]$ or $\pm\xi[c_{\rm el}^{\uparrow}(p) - c_{\rm el}^{\downarrow}(p)]$. Importantly, the former characterizes the breakdown of Onsager-Casimir symmetry in nonlinear regime [48, 8], e.g., $G_{111}^{\uparrow}(p) - G_{111}^{\downarrow}(-p) \propto \eta, \xi$, while the latter does the nonlinear contribution of the spin-polarized currents, e.g., $G_{111}^{\uparrow}(p) - G_{111}^{\downarrow}(p) = g(\eta, \xi; p)$, with $g(\eta, \xi; p)$ being a function of η or ξ and p. For $p = 0$, these two relations merge into a single one $u_1^{\uparrow} - u_1^{\downarrow} = 2\eta c_{\rm sc}, 2\xi c_{\rm el}$, leading to $G_{111}^{\uparrow} - G_{111}^{\downarrow} \propto \eta, \xi$. Previously, this has been shown to generate spin-polarized currents in the nonlinear transport regime with normal metallic contacts.

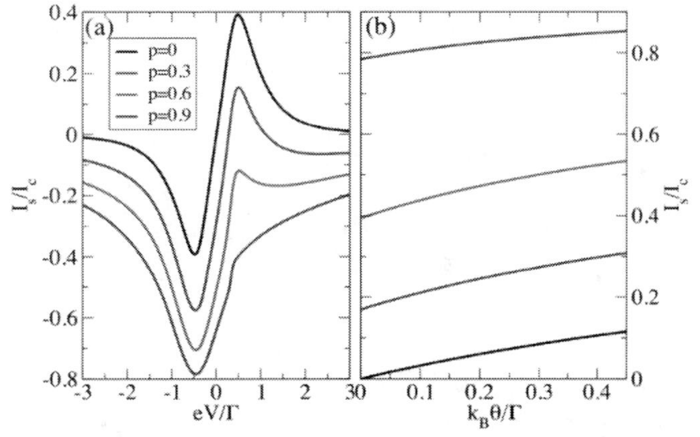

Figure 7. Plots of I_s/I_c versus (a) eV/Γ at $k_BT/\Gamma = 0.01$ and (b) $k_B\theta/\Gamma$ at $k_BT/\Gamma = 0.5$, for several (parallel) polarization p, with $\eta = c_{\rm sc} = 0.5$, $E_d/\Gamma = 0.25$, and $E_F = 0$. With $p \neq 0$, spin-polarization occurs even at zero biases, i.e., $V = 0$ or $\theta = 0$.

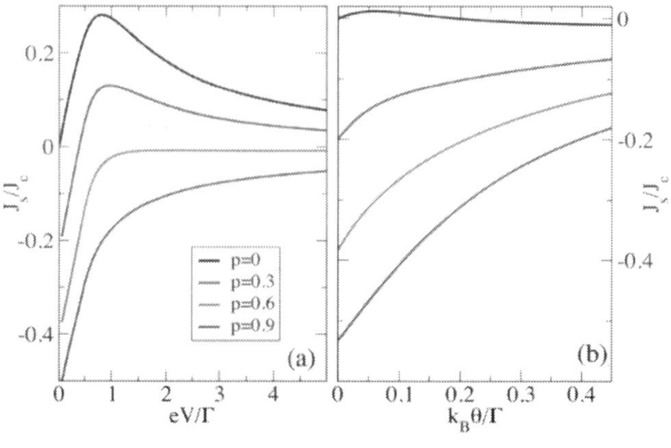

Figure 8. Plots of $\mathcal{J}_s/\mathcal{J}_c$ versus (a) eV/Γ at $k_BT/\Gamma = 0.01$ and (b) $k_B\theta/\Gamma$ at $k_BT/\Gamma = 0.05$, for several (parallel) polarization p, with $\eta = c_{\rm sc} = 0.5$, $E_d/\Gamma = 0.3$, and $E_F = 0$. With $p \neq 0$, spin-polarization occurs even at zero biases, i.e., $V = 0$ or $\theta = 0$.

4.3.2. Antiparallel magnetization For an equal but antiparallel magnetization between two reservoirs, however, the upper edge is occupied by right-moving ↑- and left-moving ↓-electrons both with an equal portion $(1+p)/2$ of the total density of states in each lead, because in this configuration the majority spin component is up (down) for the left (right) reservoir. In contrast, the lower helical edge is filled only with a minority portion $(1-p)/2$ for both left-moving $[+(\uparrow)]$ and right-moving $[-(\downarrow)]$ channels. The net effect of this antiparallel configuration gives spin-independent coupling and transmission, i.e., $\Gamma_{1s} = (1+p)\Gamma_1/2$, $\Gamma_{2s} = (1-p)\Gamma_2/2$, and $t^\sigma(p,E) = [16(E-E_d)^2 + p^2\Gamma^2]/[16(E-E_d)^2+\Gamma^2]$. We then have $t_\uparrow(p) = t_\downarrow(p)$, hence linear conductance coefficients make no contribution to spin-polarized currents, i.e., $G^\uparrow_{\alpha\beta}(p) = G^\downarrow_{\alpha\beta}(p)$, $L^\uparrow_{\alpha\beta}(p) = L^\downarrow_{\alpha\beta}(p)$, $R^\uparrow_{\alpha\beta}(p) = R^\downarrow_{\alpha\beta}(p)$, and $K^\uparrow_{\alpha\beta}(p) = K^\downarrow_{\alpha\beta}(p)$. This vanishing linear order contribution even with a nonzero p but with a specific antiparallel configuration is due to the helical nature of the QSH system. Nevertheless, the symmetry relation still holds, i.e., $t_\uparrow(p) = t_\downarrow(-p)$, hence $G^\uparrow_{\alpha\beta}(p) = G^\downarrow_{\beta\alpha}(-p)$, $L^\uparrow_{\alpha\beta}(p) = L^\downarrow_{\beta\alpha}(-p)$, $R^\uparrow_{\alpha\beta}(p) = R^\downarrow_{\beta\alpha}(-p)$, and $K^\uparrow_{\alpha\beta}(p) = K^\downarrow_{\beta\alpha}(-p)$

We analogously evaluate $\Delta U^\sigma(p) = u^\sigma_1(p)V_1 + u^\sigma_2(p)V_2 + z^\sigma_1(p)\theta_1 + z^\sigma_2(p)\theta_2$, where

$$u^\sigma_1(p) = \begin{cases} \frac{1}{2} + c^\sigma_{\rm sc}(\eta,p) \\ \frac{1}{2} + c^\sigma_{\rm el}(\xi,p)\, , \end{cases} \quad u^\sigma_2(p) = \begin{cases} \frac{1}{2} - c^\sigma_{\rm sc}(\eta,p) \\ \frac{1}{2} - c^\sigma_{\rm el}(\xi,p)\, , \end{cases} \tag{41}$$

and $z^\sigma_{\ell=1,2}(p) = (D^e/eD^p)u^\sigma_{\ell=1,2}(p)$, with

$$c^\sigma_{\rm sc}(\eta,p) = s(\eta+p)\left[2(1+\eta p) - \frac{2C(1-p^2)}{e^2\Pi}\right]^{-1}, \tag{42}$$

$$c^\sigma_{\rm el}(\xi,p) = sp\left[2 - \frac{2C(1-p^2)(1+\xi p)}{e^2\Pi}\right]^{-1}. \tag{43}$$

Here, $t(p) = 1 - r(p) = [16(E_F-E_d)^2 + p^2\Gamma^2]/[16(E_F-E_d)^2+\Gamma^2]$ and $s = \pm$ corresponds to $\sigma =\uparrow,\downarrow$.

It is easy to show that $c_{\rm sc}^{\uparrow}(\eta,p) = -c_{\rm sc}^{\downarrow}(\eta,p)$, $c_{\rm el}^{\uparrow}(\xi,p) = -c_{\rm el}^{\downarrow}(\xi,p)$, $c_{\rm sc}^{\uparrow}(\eta=0,p) = c_{\rm sc}^{\downarrow}(\eta=0,-p)$, $c_{\rm el}^{\uparrow}(\xi=0,p) = c_{\rm el}^{\downarrow}(\xi=0,-p)$, $c_{\rm sc}^{\sigma}(\eta=0,p) \propto sp$, $c_{\rm el}^{\sigma}(\xi,p) \propto sp$, $c_{\rm sc}^{\sigma}(\eta,p=0) \propto \eta s$, and $u_1^{\sigma}(p) = u_2^{\bar{\sigma}}(p)$. Thus, we have, e.g., $u_1^{\uparrow}(p) - u_1^{\downarrow}(-p) = c_{\rm sc}^{\uparrow}(\eta,p) - c_{\rm sc}^{\downarrow}(\eta,-p) = c_{\rm sc}^{\uparrow}(\eta,p) + c_{\rm sc}^{\uparrow}(\eta,-p)$ as well as $u_1^{\uparrow}(p) - u_1^{\downarrow}(p) = c_{\rm sc}^{\uparrow}(\eta,p) - c_{\rm sc}^{\downarrow}(\eta,p) = 2c_{\rm sc}^{\uparrow}(\eta,p)$.

The relation such as $u_1^{\uparrow}(p) - u_1^{\downarrow}(-p) = c_{\rm sc}^{\uparrow}(\eta,p) + c_{\rm sc}^{\uparrow}(\eta,-p)$ indicates the Onsager-Casimir symmetry breaking in nonlinear regime, which vanishes for $\eta = 0$ but survives for $p = 0$, $\eta \neq 0$ with the latter corresponding to the normal metal leads. The nonlinear spin-polarization term, e.g., $u_1^{\uparrow}(p) - u_1^{\downarrow}(p) = 2c_{\rm sc}^{\uparrow}(\eta,p)$ remains finite provided that either of η or p is nonzero. For $p = 0$, as in the parallel configuration, these two relations merge into a single one, e.g., $u_1^{\uparrow} - u_1^{\downarrow} = 2\eta c_{\rm sc}, 2\xi c_{\rm el}$. Only when $\eta, \xi = p = 0$, all spin-polarized currents vanish, which again explains the spin-filter effect in an unpolarized case [Eqs. (32) and (33)], as a limiting case $p \to 0$ of the antiparallel configuration.

From the symmetry arguments in this section, one can notice that the Onsager-Casimir symmetry breakdown in nonlinear regime for $p \to 0$ with a nonzero asymmetry factor η or ξ suggests the underlying principle for the observed spin-polarization for QSH antidot system.

5. Conclusions

We have analysed the nonlinear transport of quantum Hall setups subjected to the voltage and temperature biases. Our theory is based on the scattering transport formalism that incorporates electron-electron interaction within a mean-field description. The potential landscape of the nanostructure depends on the injected charges due to both, the voltage and temperature gradients. Using this theory we find the symmetry relations of the nonlinear transport coefficients for a quantum Hall bar with an inserted antidot. We find large deviations for the Wiedemann-Franz law and Peltier coefficient from their linear values. We also have reported results for the spin Hall bar including the antidot system in a spin-dependent capacitative model. We have found pure spin current generation for which the spin energy and spin current values depend on the magnitude of the lead polarization. Our results are relevant in view of recent works that emphasize nonlinear properties of thermopower in nanostructures [54, 55, 56, 57, 58, 59, 60, 61, 62, 63, 64, 65]. In general, the nonlinear regime of transport has not been explored to a large extend. Therefore, we expect the coming years to be full of exciting discoveries and developments in this field.

Acknowledgments

Work supported by MINECO Grant No. FIS2011-23526, the Conselleria d'Educació, Cultura i Universitats (CAIB) and FEDER.

References

[1] Song A M, Lorke A, Kriele A, Kotthaus J P, Wegscheider W and Bichler M 1998 *Phys. Rev. Lett.* **80** 3831
[2] Linke H, Sheng W D, Svensson A, Lofgren A, Christensson L, Xu H Q, Omling P and Lindelof P E 2000 *Phys. Rev.* B **61** 15914
[3] Shorubalko I, Xu H Q, Maximov I, Omling P, Samuelson L and Seifert W 2001 *Appl. Phys. Lett.* **79** 1384
[4] Fleischmann R and Geisel T 2002 *Phys. Rev. Lett.* **89** 016804
[5] Büttiker M and Sánchez D 2003 *Phys. Rev. Lett.* **90** 119701
[6] González T, Vasallo B G, Pardo D and Mateos J 2004 *Semicond. Sci. Technol.* **19** S125
[7] Hackens B, Gence L, Gustin C, Wallart X, Bollaert S, Cappy A and Bayot V 2004 *Appl. Phys. Lett.* **85** 4508
[8] Sánchez D and Büttiker M 2004 *Phys. Rev. Lett.* **93** 106802; Sánchez D and Büttiker M 2005 *Phys. Rev.* B **72** 201308
[9] Spivak B and Zyuzin A 2004 *Phys. Rev. Lett.* **93** 226801
[10] De Martino A, Egger R and Tsvelik A M 2006 *Phys. Rev. Lett.* **97** 076402
[11] Andreev A V and Glazman L I 2006 *Phys. Rev. Lett.* **97** 266806
[12] Kalina R, Szafran B, Bednarek S and Peeters F M 2009 *Phys. Rev. Lett.* **102** 066807

[13] Hernández A R and Lewenkopf C H 2009 *Phys. Rev. Lett.* **103** 166801
[14] Lim J S, Sánchez D and López R 2010 *Phys. Rev.* B **81** 155323
[15] Kubo T, Ichigo Y and Tokura Y 2011 *Phys. Rev.* B **83** 235310
[16] Ludovico M F and Arrachea L 2013 *Phys. Rev.* B **87** 115408
[17] Wei J, Schimogawa M, Whang Z, Radu I, Dormaier R and Cobden D H 2005 *Phys. Rev. Lett.* **95** 256601
[18] Marlow C A, Taylor R P, Fairbanks M, Shorubalko I and Linke H 2006 *Phys. Rev. Lett.* **96** 116801
[19] Leturcq R, Sánchez D, Götz G, Ihn T, Ensslin K, Driscoll D C and Gossard A C 2006 *Phys. Rev. Lett.* **96** 126801
[20] Zumbühl D M, Marcus C M, Hanson M P and Gossard A C 2006 *Phys. Rev. Lett.* **96** 206802
[21] Angers L, Zakka-Bajjanni E, Deblock R, Guéron S, Bouchiat H, Cavanna A, Gennser U and Polianksi M 2007 *Phys. Rev.* B **75** 115309
[22] Hartmann D, Worschech L and Forchel A 2008 *Phys. Rev.* B **78** 113306
[23] Chepelianskii A D and Bouchiat H 2009 *Phys. Rev. Lett.* **102** 086810
[24] Brandenstein-Köth B, Worschech L and Forchel A 2009 *Appl. Phys. Lett.* **95** 062106
[25] Vegvar P G N de 1993 *Phys. Rev. Lett.* **70** 837
[26] Ma Z-S, Guo H, Schülke L, Yuan Z-Q, and Li H-Z 2000 em Phys. Rev. B **61** 317
[27] Safi I, and Joyez P 2011 *Phys. Rev. B* **84** 205129
[28] Büttiker M 1993 *J. Phys. Condens. Matter* **5** 9361
[29] Christen T and Büttiker M 1996 *Europhys. Lett.* **35** 523
[30] Molenkamp L W,Gravier Th, Houten H van, Buijk O J A,Mabesoone M A A , and Foxon C T 1992 *Phys. Rev. Lett.* **68** 3765
[31] Dzurak A S, Smith C G, Barnes C H W, Pepper M , Martn-Moreno L, Liang C T, Ritchie D A, and Jones G A C 1997 *Phys. Rev. B* **55** R10197
[32] Godijn S F, Mller S, Buhmann H, Molenkamp L W, and Langen S A van 1999 *Phys. Rev. Lett.* **82** 2927
[33] Butcher P N 1990 *J. Phys. Condens. Matter* **2** 4869
[34] Saito K, Benenti G, Casati G and Prosen T 2011 *Phys. Rev.* B **84** 201306(R)
[35] Sánchez D and Serra L 2011 *Phys. Rev.* B **84** 201307(R)
[36] Entin-Wohlman O and Aharony A 2012 *Phys. Rev.* B **85** 085401
[37] Benenti G, Saito K and Casati G 2011 *Phys. Rev. Lett.* **106** 230602
[38] Brandner K, Saito K and Seifert U 2013 *Phys. Rev. Lett.* **110** 070603
[39] Balachandran V, Benenti G and Casati G 2013 *Phys. Rev.* B **87** 165419
[40] Apertet Y, Ouerdane H, Goupil C and Lecoeur P 2013 *Physical Review E* **88** 022137
[41] Sánchez D, and López 2013 *Phys. Rev. Lett* **110** 026804
[42] Meair J and Jacquod P 2013 *J. Phys. Condens. Matter* **25** 082201
[43] López R and Sánchez D 2013 *Phys. Rev.* B **88** 045129
[44] Whitney R S 2012 arXiv:1208.6130 (preprint)
[45] Whitney R S 2013 *Phys. Rev.* B **87** 115404
[46] Svensson S F, Hoffmann E A, Nakpathomkun N, Wu P M, Xu H Q, Nilsson H A, Sánchez D, Kashcheyevs V and Linke H 2013 *New J. Phys.* **15** 105011
[47] Matthews J, Battista F, Sánchez D, Samuelsson P and Linke H 2013 *arXiv:1306.3694* (preprint)
[48] Hwang S-Y, Sánchez D, Lee M and López R 2013 *New J. Phys.* **15** 105012
[49] Ford C J B, Simpson P J, Zailer I, Mace D R, Yosefin M, Pepper M, Ritchie D A, Frost J E F, Grimshaw M P and Jones G A C 1994 *Phys. Rev.* B **49** 17456
[50] Kirczenow G, Sachrajda A S, Feng Y, Taylor R P, Henning L, Wang J, Zawadzki P and Coleridge P T 1994 *Phys. Rev. Lett.* **72** 2069
[51] Qi X-L and Zhang S-C 2010 *Phys. Today* **63** 33
[52] Hwang S-Y, López R, Lee M and Sánchez D 2014 *Preprint* cont-mat/1406.3187
[53] López R, Lim J S and Sánchez D 2012 *Phys. Rev. Lett.* **108** 246603
[54] Boese D and Fazio R 2001 *Europhys. Lett.* **56** 576
[55] Dong B and Lei X L 2002 *J. Phys.: Condens. Matter* **14** 11747
[56] Azema J, Daré A-M, Schäfer S and Lombardo P 2012 *Phys. Rev.* B **86** 075303
[57] Kulik I O 1994 *J. Phys.: Condens. Matter* **6** 9737
[58] Bogachek E N, Scherbakov A G and Landman U 1999 *Phys. Rev.* B **60** 11678
[59] Segal D and Nitzan A 2005 *J. Chem. Phys.* **122** 194704
[60] Chang C W, Okawa D, Majumdar A and Zettl A 2006 *Science* **314** 1121
[61] Ruokola T and Ojanen T 2011 *Phys. Rev.* B **83** 241404(R)
[62] Kuo D M T and Chang Y 2001 *Phys. Rev. B* **81** 205321
[63] Krawiec M, and Wysokinski K I 2007 *Phys. Rev. B* **75** 155330
[64] Hershfield S, Muttalib S K A, and Nartwot B J 2013 *Phys. Rev. B* **88** 085426

[65] Zimbovskaya N A 2014 *arXiv:1405.6968* (preprint)

Journal of Physics: Conference Series **568** (2014) 052017 doi:10.1088/1742-6596/568/5/052017

Time resolved heat exchange in driven quantum systems

María Florencia Ludovico

Departamento de Física, FCEyN, Universidad de Buenos Aires and IFIBA, Pabellón I, Ciudad Universitaria, 1428 CABA Argentina

Jong Soo Lim

School of Physics, Korea Institute for Advanced Study, Seoul 130-722, Korea

Michael Moskalets

Department of Metal and Semiconductor Physics, NTU "Kharkiv Polytechnic Institute", 61002 Kharkiv, Ukraine

Liliana Arrachea

Departamento de Física, FCEyN, Universidad de Buenos Aires and IFIBA, Pabellón I, Ciudad Universitaria, 1428 CABA Argentina

David Sánchez

Instituto de Física Interdisciplinar y Sistemas Complejos IFISC (UIB-CSIC), E-07122 Palma de Mallorca, Spain

Abstract. We study time-dependent heat transport in systems composed of a resonant level periodically forced with an external power source and coupled to a fermionic continuum. This simple model contains the basic ingredients to understand time resolved energy exchange in quantum capacitors that behave as single particle emitters. We analyse the behaviour of the dynamic heat current for driving frequencies within the non-adiabatic regime, showing that it does not obey a Joule dissipation law.

1. Introduction

Recent progress in the miniaturisation of electronic circuits puts on agenda the development of new theoretical methods. The energy transport in systems on the nano- and already molecular and atomic scales cannot be treated classically but requires a quantum-mechanical treatment [1, 2, 3]. In particular, time-dependent quantum transport in systems that act as quantum capacitors have recently received a lot of attention both experimentally and theoretically [4, 5, 6, 7]. Recent experiments show that a quantum dot tunnel-coupled to a reservoir can be used for on-demand single-electron injection [8]. Electron emission and absorption are periodically generated by applying an external time-periodic potential. Harmonic potentials are also crucial for the creation of directed transport in asymmetric systems, such as charge [9] and spin [10] ratchets. Furthermore, ac fields are shown to control matter tunneling in Bose-Einstein condensates [11].

Electrons in periodically driven coherent conductors carry energy in addition to current. Thus, the role of electron-like and hole-like excitations created by ac driving in the energy current noise is investigated in Ref. [12]. Heat production in nanoscale engines is analyzed in Refs. [13, 14] while Ref. [15] finds a universal thermal resistance for the low-temperature dynamical transport.

In a recent work [16], we studied the time-resolved energy production and redistribution in ac-driven quantum coherent electron systems. We showed that the coupling between the different parts of the system contributes to the energy transport, and that this contribution is of ac nature. We also presented an appropriate definition for the time resolved heat current in accordance with the fundamental principles of thermodynamics. Interestingly, we showed that for low frequencies of the driving potential (adiabatic regime) we found that the time-dependent heat flux is instantaneously given by the Joule law with a universal resistance. The purpose of this work is to explore if this behaviour of the time resolved heat current remains also valid for higher driving frequencies beyond the adiabatic regime. To this end, we consider a single dot connected to a fermionic band of continuous density of states (a reservoir) and driven with a harmonically time-dependent potential with a frequency within the non-adiabatic regime.

2. Model and theoretical treatment

The system under consideration is sketched in Fig. 1, in which a resonant level is driven harmonically by a power source. The Hamiltonian of the setup is $H = H_C + H_T + H_D$, where the first term, $H_C = \sum_k \varepsilon_k c_k^\dagger c_k$, represents the fermionic continuum (reservoir) with the energy band ε_k. The Hamiltonian of a tunneling region reads $H_T = \sum_k (w_k d^\dagger c_k + h.c)$, where w_k is the coupling amplitude. The term $H_D(t) = \varepsilon_d(t) d^\dagger d$ denotes a driven system, in which the energy level, $\varepsilon_d(t) = \varepsilon_0 + V_{ac}\sin(\Omega t)$, is varied in time by the power source. For simplicity, we consider noninteracting, spinless electrons.

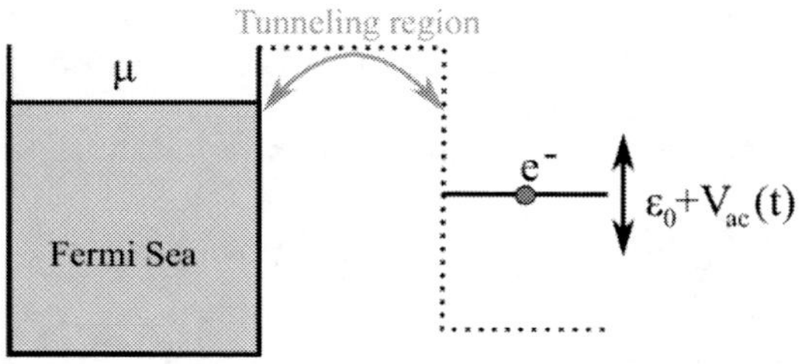

Figure 1. Sketch of the setup under consideration. A single electron level is connected to a fermionic band (reservoir with chemical potential μ) via a tunnel barrier. Energy is supplied to the system by a time periodic power source $V_{ac}(t)$ with characteristic frequency Ω.

In order to define the energy fluxes entering each part of the system, we analyze the evolution in time of the total energy,

$$\frac{d\langle H\rangle}{dt} = J_C^E(t) + J_T^E(t) + J_D^E(t) + P(t), \tag{1}$$

where we identify the energy flux entering the reservoir $J_C^E = i\langle [H, H_C]\rangle/\hbar$, the change per unit time of the energy stored in the tunneling region $J_T^E = i\langle [H, H_T]\rangle/\hbar$ and the energy flux entering the resonant level $J_D^E = i\langle[H, H_D]\rangle/\hbar$. The total Hamiltonian conserves the number of particles but not the total energy due to the power developed by the ac forces $P(t) = d\langle H\rangle/dt = \langle\partial H_D/\partial t\rangle = n_d(t)d\varepsilon_d(t)/dt$, where $n_d(t) = \langle d^\dagger(t)d(t)\rangle$ is a resonant level occupation probability, i.e., the time-average of the number of electrons present on the resonant level.

We solve the problem with the non-equilibrium Green function procedure of Refs. [17, 18]. The different energy fluxes can be computed in terms of the retarded Green function $\mathcal{G}^r(t,t') = -i\theta(t-t')\langle\{d(t), d^\dagger(t')\}\rangle$, and we use the Floquet-Fourier representation

$$\mathcal{G}^r(t,t') = \sum_{n=-\infty}^{\infty} \int \frac{d\omega}{2\pi} e^{-in\Omega t} e^{-i\omega(t-t')} \mathcal{G}^r(n,\omega), \tag{2}$$

with Ω being the fundamental driving frequency. The Floquet components $\mathcal{G}^r(n,\omega)$ can be computed in term of the equilibrium Green function $\mathcal{G}^0(\omega) = [\omega - \varepsilon_0 + i\Gamma/2]^{-1}$, where Γ represent the hybridization between the electron level and the reservoir. We find that the energy current entering the reservoir at time t reads

$$\begin{aligned} J_C^E(t) &= -\sum_l e^{-il\Omega t} \int \frac{d\varepsilon}{h} \{ i\mathcal{G}^{r*}(-l,\varepsilon)\Gamma[(\varepsilon - l\hbar\Omega)f(\varepsilon - l\hbar\Omega) - \varepsilon f(\varepsilon)] - \\ & \sum_n [(\varepsilon + \frac{l\hbar\Omega}{2})f(\varepsilon - n\hbar\Omega) - \varepsilon f(\varepsilon)]\mathcal{G}^r(l+n, \varepsilon - n\hbar\Omega)\Gamma^2 \mathcal{G}^{r*}(n, \varepsilon - n\hbar\Omega)\}, \end{aligned} \tag{3}$$

where $f(\varepsilon) = 1/[1 + e^{(\varepsilon-\mu)/k_B T}]$ is the Fermi- Dirac distribution. Following the same procedure, we can compute the other fluxes entering Eq. (1). The variation of the energy stored in the tunneling region reads

$$J_T^E(t) = \int \frac{d\varepsilon}{h} \Omega f(\varepsilon) \sum_n n\, 2\mathrm{Im}\{e^{-in\Omega t}\mathcal{G}^r(n,\varepsilon)\Gamma\}. \tag{4}$$

Finally, the energy flux entering the dot with a single resonant level can be expressed as $J_D^E = \varepsilon_d(t)dn_d(t)/dt = -\varepsilon_d(t)J_C(t)/e$. In this expression, we have used charge conservation to rewrite $dn_d(t)/dt$ in terms of the reservoir current ($J_C(t) = -edn_d(t)/dt$). Applying the above method, the charge current can be expressed as

$$\begin{aligned} J_C(t) &= -\frac{e}{h}\sum_l e^{-il\Omega t} \int d\varepsilon \{ i\mathcal{G}^{r*}(-l,\varepsilon)\Gamma[f(\varepsilon - l\hbar\Omega) - f(\varepsilon)] - \\ & \sum_n [f(\varepsilon - n\hbar\Omega) - f(\varepsilon)]\mathcal{G}^r(l+n, \varepsilon - n\hbar\Omega)\Gamma^2 \mathcal{G}^{r*}(n, \varepsilon - n\hbar\Omega)\}. \end{aligned} \tag{5}$$

We stress that Eq. (3), (4) and (5) are general and valid up to any order in Ω and the ac amplitude V_{ac}.

The dc components of the currents $\overline{J_C^E}$ and $\overline{J_D^E}$ satisfy $\overline{J_C^E} = -\overline{J_D^E}$. The energy conservation expressed by Eq. (1) implies that knowledge on how energy is absorbed or emitted in the contact region, $J_T^E(t)$, plays an important role in the definition of heat in the time domain, even though this term satisfies $\overline{J_T^E} = 0$. The dc component of heat flowing into the reservoir is related to energy and charge currents as follows, $\overline{J_C^Q} = \overline{J_C^E} - \mu\overline{J_C}$. In Ref. [16] we focused on the adiabatic regime and adopted a thermodynamical analysis to define heat, based on the fact that the reservoir is a macroscopic system, which experiences little changes under slow variations due to the driving at the quantum dot. Thus, the rate of change of the internal energy in the reservoir leads to the appropriate definition of heat exchange between the reservoir and the driven part of the system,

$$J^Q(t) = J_C^E(t) + J_T^E(t)/2 - \mu J_C(t), \tag{6}$$

where we can interpret the quantity $J_T^E(t)$ as the chemical work developed by the contact when an electron is flowing through it.

The above definition of the heat flow in the time domain within the adiabatic regime is fully consistent with the treatment based on the scattering matrix formalism, as shown in [16]. Furthermore, it was also shown that such definition leads to a generalized Joule law in the time domain within the adiabatic regime

$$J^Q(t) = R_q[J_C(t)]^2, \tag{7}$$

where $R_q = h/2e^2$ is the universal resistance of the contact. This law reflects the fact that within the adiabatic regime the heat generated by an ac driving flows into the reservoir and increases its entropy at *every time*.

3. Results

The question arises now to what extent the definition of both a time-resolved heat current (6) and a corresponding time-dependent Joule law (7) remain valid when the driving frequency is high and the transport is non-adiabatic. In order to explore this issue we have explicitly evaluated time-dependent heat and charge currents $J^Q(t)$ and $J_C(t)$ for different driving frequencies Ω. Results are shown in Fig. 2, where, for simplicity, we have considered zero temperature ($T = 0$). We present results for slow driving frequencies (solid circles), where the heat current is linear as a function of $J_C^2(t)$, with the universal slope R_q. On the other hand, we show (open triangles) that the induced heat current for higher frequencies (non-adiabatic regime) departs from the Joule law. Furthermore it may attain negatives values for some times, which is in seeming contradiction with the second law of the thermodynamics. This puzzling result, a consequence of system's non-equilibrium dynamics, raises a question about the correct formulation of the second law of the thermodynamics for strongly nonequilibrium systems. At slow driving frequencies the heat current behaves as expected for stationary systems (remains positive and obeys a Joule dissipation law), for which the classical laws of thermodynamics remain valid. However, at high driving frequencies, the non-equilibrium and quantum effects may become relevant. In particular, the uncertainty relation between time and energy may play a role. This feature deserves a more detailed analysis.

4. Summary and conclusions

In conclusion, we have discussed dynamical heat generation in a resonant level system due to coupling to an external time-dependent potential with a typical frequency within the non-adiabatic regime. Unlike a slow driving frequency regime, where the time-dependent heat flux is instantaneously given by the Joule law with universal resistance, for higher driving frequencies the heat flux can attain negative values, which may arise from a non-equilibrium dynamics. Further analysis is necessary to better understand this phenomenon.

5. Acknowledgment

We acknowledge support from CONICET, MINCYT and UBACYT, Argentina (MFL and LA), MINECO Grant No. FIS2011-23526, CAIB and FEDER.

References

[1] S. Jezouin, F. D. Parmentier, A. Anthore, U. Gennser, A. Cavanna, Y. Jin, and F. Pierre, Science **343**, 601 (2013).

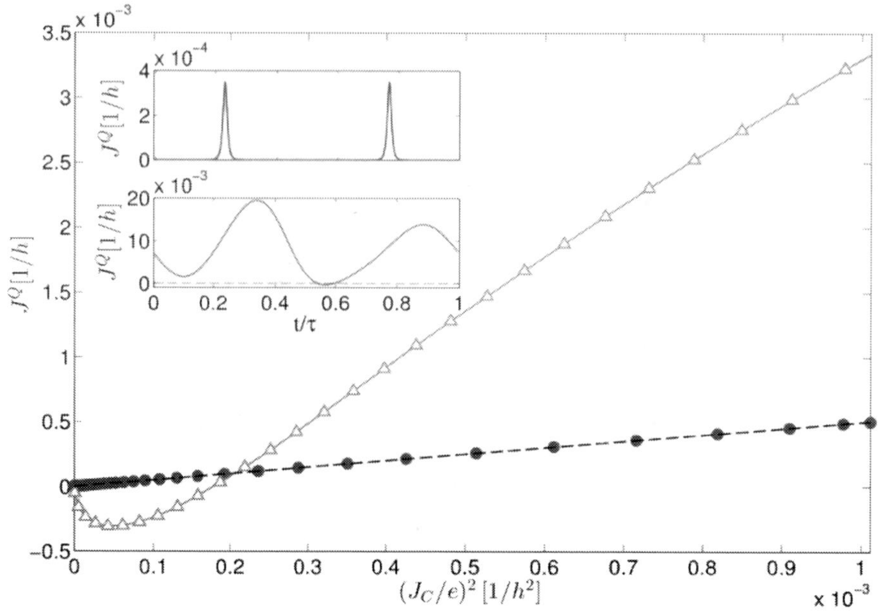

Figure 2. Heat fluxes $J^Q(t)$ as a function of the charge current $J_C^2(t)$ for two different driving frequencies Ω. Open triangles corresponds to a frequency within the non-adiabatic regime $\hbar\Omega = 0.3$, $\mu = 0.2$, $\varepsilon = 0$, $V_{ac} = 0.6$, and $T = 0$. Solid circles corresponds to a slow driving frequency $\hbar\Omega = 10^{-3}$, $\mu = 0$, $\varepsilon = -1.2$, $V_{ac} = 10$, and $T = 0$. Energies are expressed in units of Γ. Dashed line represents the Joule dissipation relation with a resistance $R = h/2e^2$. Inset: Upper panel shows $J^Q(t)$ within the slow driving regime; lower panel depicts $J^Q(t)$ for the non-adiabatic regime.

[2] M. Büttiker, A. Prêtre, and H. Thomas, Phys. Lett. A **180**, 364 (1993).
[3] M. Büttiker, A. Prêtre, and H. Thomas, Phys. Rev. Lett. **70**, 4114 (1993).
[4] J. Gabelli, G. Fève, J.-M. Berroir, B. Plaçais, A. Cavanna, B. Etienne, Y. Jin, and D.C. Glattli, Science **313**, 49 (2006).
[5] J. Splettstoesser, S. Ol'khovskaya, M. Moskalets and M. Büttiker, Phys. Rev. B **78**, 205110 (2008)
[6] M. Moskalets and M. Büttiker, Phys. Rev. B **80**, 081302(R) (2009)
[7] M. Moskalets, P. Samuelsson and M. Büttiker, Phys. Rev. Lett. **100**, 086601 (2008).
[8] G. Fève, A. Mahé A, J.M. Berroirm, T. Kontos, B. Plaçais, D.C. Glattli, A. Cavanna, B. Etienne and Y. Jin, Science **316** 1169 (2007).
[9] H. Linke, T.E. Humphrey, A. Löfgren, A. Sushkov, R. Newbury, R.P. Taylor, and P.Omling, Science **286**, 2314 (1999).
[10] M.V. Costache and S.O. Valenzuela, Science **330**, 1645 (2010).
[11] H. Lignier, C. Sias, D. Ciampini, Y. Singh, A. Zenesini, O. Morsch, and E. Arimondo, Phys. Rev. Lett. **99**, 220403 (2007).
[12] F. Battista, F. Haupt and J. Splettstoesser, arXiv:1405.4326 (preprint).
[13] L. Arrachea, M. Moskalets, and L. Martin-Moreno, Phys. Rev. B **75**, 245420 (2007).
[14] M. Moskalets and M. Büttiker, Phys. Rev. B **80**, 081302(R) (2009).
[15] J.S. Lim, R. López and D. Sánchez, Phys. Rev. B **88**, 201304(R) (2013).
[16] M.F. Ludovico, J.S. Lim, M. Moskalets, L. Arrachea, and D. Sánchez, Phys. Rev B **89**, 161306(R) (2014).
[17] L. Arrachea, Phys. Rev. B **66** (2002) 045315; L. Arrachea, Phys. Rev. B **70** (2004) 155407
[18] L. Arrachea, Phys. Rev. B **72** (2005) 125349; L. Arrachea, M. Moskalets, Phys. Rev. B **74** (2006) 245322; L. Arrachea, Phys. Rev. B **75** (2007) 035319.
[19] R. Balian, *From Microphysics to Macrophysics*, Vol. I, Springer, Berlin-Heidelberg, 1991-2007.

Anomalous low-temperature specific heat of $Fe_{2-x}V_{1+x}Al$ (x=0; 0.1) alloys

A.T. Lonchakov[1], V.V. Marchenkov[1], V.I. Okulov[1], K.A. Okulova[1], T.E. Govorkova[1], S.M. Podgornykh[1], H.W. Weber[2]

[1] Institute of Metal Physics, 620990 Ekaterinburg, Russia
[2] Atominstitut, Vienna University of Technology, 1020 Vienna, Austria

march@imp.uran.ru

Abstract. The low-temperature specific heat of stoichiometric and off-stoichiometric $Fe_{2-x}V_{1+x}Al$ (x = 0; 0.1) Heusler alloys was measured in the temperature interval from 1.8 to 12 K. We report on specific features in the low-temperature heat capacity, which reflect the scattering of the current carriers by magnetic defects in both alloys and the existence of a narrow pseudogap in the V-enriched alloy.

1. Introduction

Systems of transition elements combining semiconducting behavior and magnetism have become of great interest in solid state physics in recent years. Among such systems are the iron-vanadium-aluminum alloys with near stoichiometric composition. In the stoichiometric Fe_2VAl alloy with the $L2_1$ structure the Fe atoms are in a tetrahedral and the V atoms are in an octahedral environment. The magnetic properties of $Fe_{2-x}V_{1+x}Al$ near the stoichiometric composition are largely due to magnetic defects related to the redistribution of the iron atoms in the crystalline lattice [1]. The $Fe_{2-x}V_{1+x}Al$ alloys with x close to 0 are of great interest, because these alloys show semiconducting behavior of the electoresistivity $\rho(T)$ up to 1000 K [2]. It was shown in Ref. [3] that in the case of the $Fe_{2-x}V_{1+x}Al$ alloy with x=0.1 the electroresistivity, the Hall coefficient and the magnetic susceptibility change much more sharply in the low-temperature interval 2 K $\leq T \leq$ 20 K than at high temperatures. A peak in the temperature dependence of the thermoelectric power was found in this alloy in the same temperature range [4]. In Refs. [3, 4] we succeeded to consistently describe the low-temperature features of a number of phenomena based on the ideas of the Fermi energy being located in a narrow pseudogap (width $\approx$ 4 meV) in the electron density of states. The narrow pseudogap, i.e. a dip in the density of states, can be related to the displacement of the energy bands in the $Fe_{2-x}V_{1+x}Al$ alloys enriched by V.

The main purpose of this paper is to report on new data on the existence of the pseudogap state and on physical effects due to the presence of magnetic moments in the $Fe_{2-x}V_{1+x}Al$ alloys. We analyzed the low-temperature heat capacity in the temperature range (1.8 – 12) K in the stoichiometric Fe_2VAl alloy and in the above mentioned $Fe_{1.9}V_{1.1}Al$ alloy enriched by V. Our study confirms anomalies in the temperature dependence of the heat capacity, which reflect the scattering of the current carriers by magnetic defects in both alloys and the existence of a narrow pseudogap in the $Fe_{1.9}V_{1.1}Al$ alloy).

2. Experimental results and discussion

2.1. The role of electron scattering by magnetic defects

The procedure of preparing the alloys and of controlling their chemical composition and homogeneity were described in detail in Refs. [4, 5]. The chemical composition was determined with an accuracy of $\pm 1\%$ by energy-dispersive X-ray spectroscopy (EDX). The composition analysis was carried out at the central part of the samples and at their edges. The conductivity of the $Fe_{2-x}V_{1+x}Al$ (x=0; 0.1) alloys is of the hole type according to our measurements of the Hall effect and the thermoelectric power at T=4.2 K. The heat capacity was measured by a PPMS-9 (Quantum Design).

The temperature dependence of the heat capacity $C(T)$ of the Fe_2VAl and the $Fe_{1.9}V_{1.1}Al$ alloys are shown in Figs. 1 and 2, respectively. The insets confirm a C/T versus T^2 dependence. According to Refs. [1, 2] the contribution of the current carriers to the heat capacity at temperatures below 10 K noticeably exceeds that of the lattice, which can, therefore, be neglected. As seen from the insets in Figs. 1 and 2, C/T increases significantly with decreasing T, i.e. the γ coefficient for the hole heat capacity $c_h = \gamma T$ grows, which was already observed in Refs. [1, 2, 6], but not fully interpreted. According to the evaluation in Ref. [2], the increase of γ can lead to a growth of the effective mass of the holes at the Fermi level to approximately 20 times the free electron mass, which means that the alloy could be considered as a heavy fermion system. Other authors believe that the growth of C/T can be related to Schottky contributions from the magnetic defects [1]. We give here another explanation for this effect.

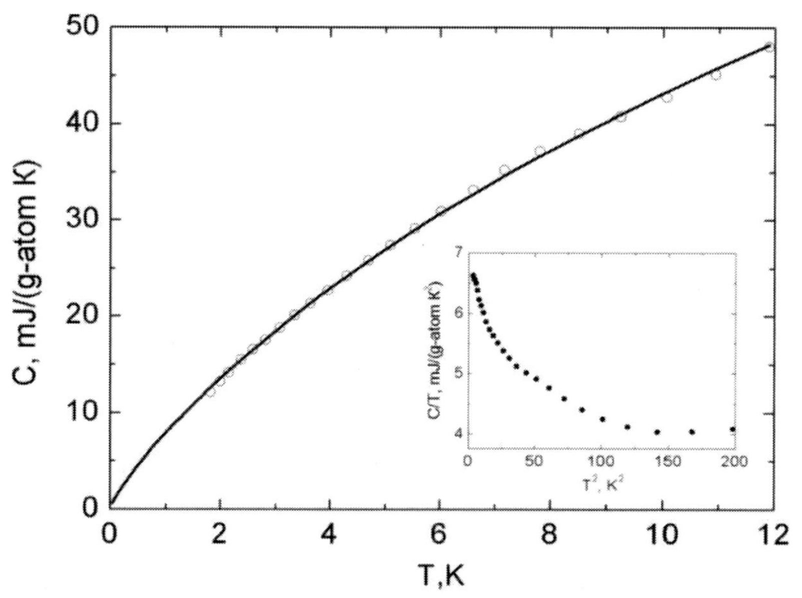

Figure 1. The heat capacity C as a function of temperature for the Fe_2VAl sample. The symbols represent the experimental data; the solid line is a fit to Eq. (1). Inset: C/T vs. T^2.

In the $Fe_{2-x}V_{1+x}Al$ alloys with $x \geq 0$, the Fe atoms, which are placed on the V sites, can be considered as magnetic anti-site defects. The paramagnetic character of the magnetic susceptibility described in Refs. [3, 6] for the polycrystalline $Fe_{1.9}V_{1.1}Al$ and Fe_2VAl alloys provides evidence for the presence of local magnetic moments at low concentration. Ishikawa et al. [7] have also inferred a low concentration of magnetic anti-site defects in $Fe_{1.98}V_{1.02}Al$, $Fe_{1.95}V_{1.05}Al$ and Fe_2VAl single crystals, based on their relatively low magnetization at temperatures below 50K. Therefore, the model of dilute magnetic systems can be applied to describe the low-temperature heat capacity of these alloys with a small deviation from stoichiometry. To describe the observed dependence of $C(T)$, besides the usual electron term γT, an additional contribution to the heat capacity $c_{AK} = bTln(T_K/T)$, according to the Appelbaum-Kondo theory [8, 9] for paramagnetic centers in metals, should be taken into account. Here T_K is the Kondo temperature and b is a fit parameter of the order of γ. This contribution is a consequence of changing the energy by forming a singlet ground state in the "magnetic impurity - conduction electron" bound system. Hence the low-temperature heat capacity of the $Fe_{2-x}V_{1+x}Al$ alloys can be presented in form

$$c_h(T) = \gamma T + bT\,ln(T_K/T). \tag{1}$$

The additional term $b\,ln(T_K/T)$ leads to the sharp increase of C/T at low temperatures.

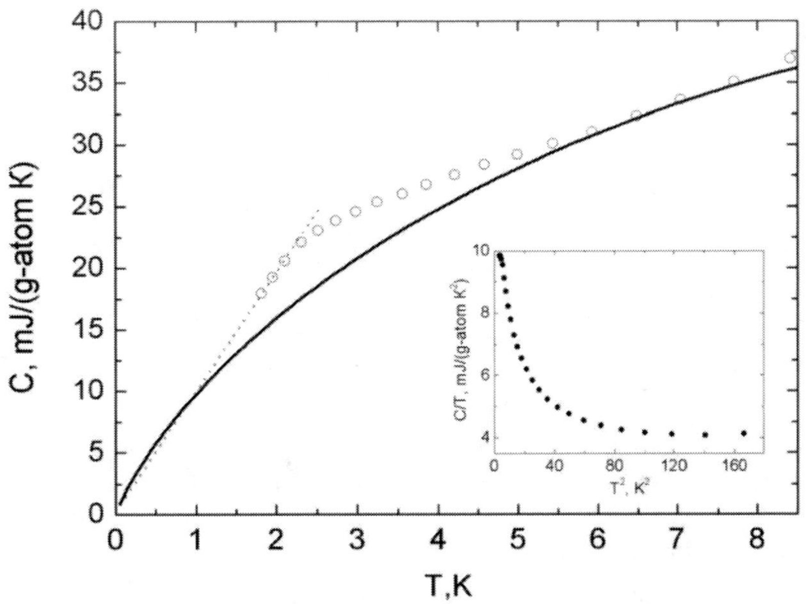

Figure 2. The heat capacity C as a function of temperature for the $Fe_{1.9}V_{1.1}Al$ sample. The symbols represent the experimental data; the solid line is a fit to Eq. (1). The dashed line is an extrapolation of the $C(T)$ dependence to $T = 0$. Inset: C/T vs. T^2.

The results of a calculation of the heat capacity using Eq. (1) are presented by the solid lines in Figs. 1 and 2, respectively, where the following fit parameters were used: $\gamma = 2.73$ and 2.07 mJ/g-atom·K^2, b = 1.54 and 2.56 mJ/g-atom·K^2, $T_K = 28$ and 20 K. The close values of the fit parameters for $Fe_{1.9}V_{1.1}Al$ and Fe_2VAl can be associated with a rather low concentration of magnetic anti-site defects, inherent in the V-enriched samples and Fe_2VAl. [6, 7]. As seen from Fig. 1, the experimental dependence of $C(T)$ for Fe_2VAl is in a good agreement with the theoretical curve. However, for the $Fe_{1.9}V_{1.1}Al$ sample enriched by V the theoretical curve lies below the experimental values of the heat capacity in low-temperature range. This disagreement can be attributed to the presence of an additional contribution to the heat capacity $c_{add} = C(T) - (\gamma T + c_{AK})$, which peaks at $T = 2.5$ K (Fig.3).

2.2. Contribution of the pseudogap state in the V-enriched sample

It should be noted that a maximum in the temperature dependence of the thermoelectric power was observed in the $Fe_{1.9}V_{1.1}Al$ sample in the same temperature range [4]. It was shown that the corresponding contribution to the thermoelectric power of $\alpha_{add}(T)$ in $Fe_{1.9}V_{1.1}Al$ (inset in Fig. 3) could be explained by the location of the Fermi level in the pseudogap with a width Δ of about 4 meV. In

Ref. [3] based on the same hypothesis and with approximately the same value of Δ we achieved a consistent theoretical description of the anomalous low-temperature dependence of the electroresistivity and the magnetic susceptibility, observed in the alloy with x=0.1. Therefore, there is good reason to believe that the maximum of the heat capacity in the above alloy is also caused by the existence of a narrow pseudogap in the density of states at the Fermi edge. For calculating the electron specific heat $c_0(T)$, taking the effect of a pseudogap into account, we used the simple approach based

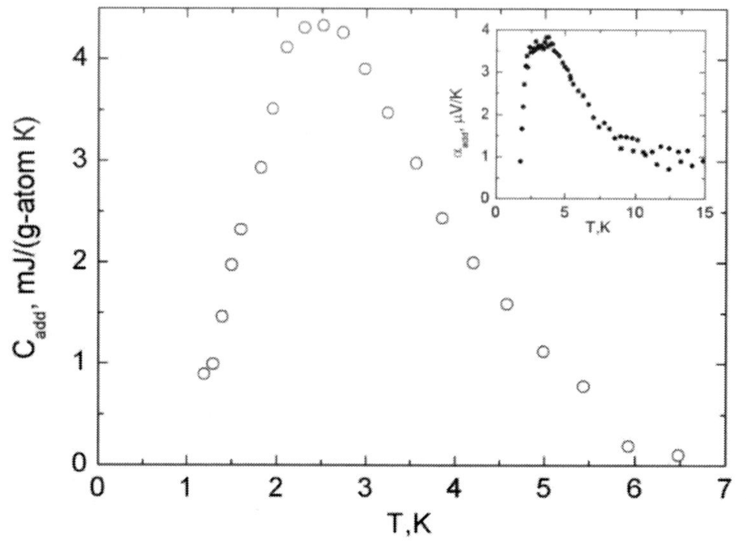

Figure 3. Temperature dependence of an additional contribution to the heat capacity c_{add} for the $Fe_{1.9}V_{1.1}Al$ sample. The inset shows the temperature dependence of the pseudogap contribution to the thermoelectric power α_{add} for this sample [4].

on approximations as employed in Refs. [3, 4]. In the framework of this approach the pseudogap contribution is characterized by three parameters, which are the width of the pseudogap Δ and the values of the electron density of states at the bottom of and outside the pseudogap. The corresponding equation has the form

$$c_0(T) = \gamma T + \gamma_0 T(3/2\pi^2)\int_0^{\Delta/2T} x^2[ch(x/2)]^{-2}dx. \quad (2)$$

The term γT reflects the contribution of the density of states outside the pseudogap and, in essence, coincides with the first term of Eq. (1) in the vicinity of the pseudogap. The value of γ_0 is equal to the density of states at the bottom of the pseudogap, where the Fermi energy is located. The second term in Eq. (2) has a maximum and describes the temperature dependence of $c_{add}(T)$ shown in Fig. 3. Figure 4 shows the result of fitting the theoretical dependence of the heat capacity, presented as the sum of $c_0(T)$ and the second term in Eq. (1), reflecting the contribution of scattering by magnetic defects, to the experimental data in the vicinity of the maximum with the following fit parameters: $\gamma_0 = 3$ mJ/g-at K^2, $\Delta = 20$ K, $\gamma = 2.04$ mJ/g-at K^2, $b = 2.3$ mJ/g-at K^2, $T_K = 20$ K. Note that the temperature dependence of the heat capacity is in good agreement with the theoretical ideas stated here. This is consistent with Refs. [3, 4], where the thermoelectric power and other physical parameters were studied.

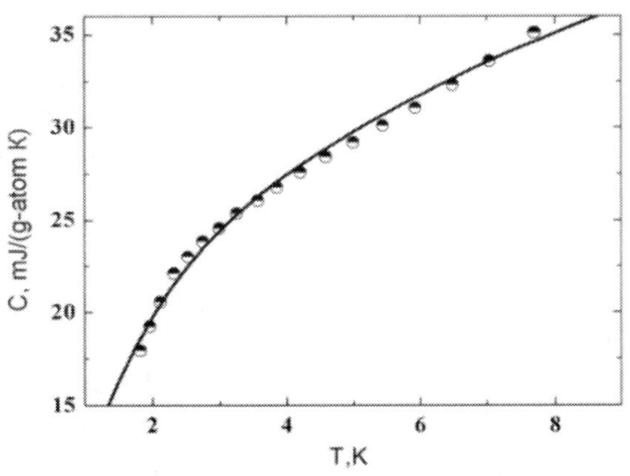

Figure 4. Fit of the theoretical dependence (Eq. (2)) to the experimental data for $Fe_{1.9}V_{1.1}Al$ in the vicinity of the pseudogap anomaly.

3. Conclusions
We conclude that the pseudogap state and the scattering of the current carriers by magnetic defects play a significant role in the electronic properties of $Fe_{2-x}V_{1+x}Al$ (x=0; 0.1). The anomalous monotonic temperature dependence of *C(T)* is caused by strong exchange scattering of electrons. In the case of the V-enriched alloy the non-monotonical contribution to the specific heat can be explained by the existence of a narrow pseudogap in the electron density of states.

4. Acknowledgements
This work was partly supported by RFBR (Grant No 14-02-01238), by the Program of RAS (No 12-02-T-1016) and by the Scientific School (Grant No SS-1540.2014.2).

References
[1] C.S. Lue, Joseph H. Ross, C.F. Chang, H.D. Yang, Phys. Rev. B **60**, R13941 (1999).
[2] Y. Nishino, M. Kato, S. Asano, K. Soda, M. Hayasaki, U. Mizutani, Phys. Rev. Lett. **79**, 1909 (1997).
[3] V.I. Okulov, V.E. Arkhipov, T.E. Govorkova, A.V. Korolyov, V.V. Marchenkov, K.A. Okulova, E.I. Shreder, H.W. Weber, Low Temp. Phys. **33**, 692 (2007).
[4] A.T. Lonchakov, V.V. Marchenkov, V.I. Okulov, K.A. Okulova, Low Temp. Phys. **39**, 84 (2013).
[5] S.M. Podgornykh, A.D. Svyazhin, E.I. Shreder, V.V. Marchenkov, V.P. Dyakina, JETP **105**, 42 (2007).
[6] Y. Nishino, H. Sumi, and U. Mizutani, Phys. Rev. B **71**, 094425 (2005).
[7] F. Ishikawa, K. Fukuda, N. Kawase, M. Mochigi, Yuh Yamada, A. Matsushita, J. Magn. Magn. Mater. **310**, e616 (2007).
[8] J.A. Appelbaum, J. Kondo, Phys. Rev. Lett. **19**, 906 (1967).
[9] J.A. Appelbaum, J. Kondo, Phys. Rev. **170**, 524 (1968).

Electronic transport in Co-based half-metallic ferromagnetic Heusler alloys

V.V. Marchenkov[1], N.I. Kourov[1], K.A. Belozerova[1], S.M. Emelyanova[1], V.P. Dyakina[1], E.B. Marchenkova[1], M. Eisterer[2], H.W. Weber[2]

[1] Institute of Metal Physics, 620990 Ekaterinburg, Russia
[2] Atominstitut, Vienna University of Technology, 1020 Vienna, Austria

march@imp.uran.ru

Abstract. The electroresistivity of the Co-based half-metallic ferromagnetic Co_2CrGa, Co_2CrAl, Co_2VAl, Co_2MnAl Heusler alloys and the Ni_2MnGa ferromagnets were measured in the temperature range from 4.2 to 900 K. Specific features in the electrical resistivity of the half-metallic ferromagnetic Heusler alloys were observed, which can be explained in terms of the two-current conduction model, taking into account the existence of an energy gap in the electron spectrum near the Fermi level.

1. Introduction

The creation of a spin-polarized electron current and its control in conducting materials represent a promising trend in solid state physics. 100% spin-polarization can be realized in Heusler alloys, i.e. intermetallic X_2YZ compounds with the $L2_1$ structure (X and Y are transition metals, and Z is an element from the III-V groups), which are being actively studied because of their possible application in spintronics. Band calculations performed for some Heusler alloys demonstrate the presence of an energy gap at the Fermi level in one of the spin subbands and a metallic type of electron states in the other subband. This peculiarity of the electron structure explains the term «half-metallic ferromagnets» (HMF) used for these alloys [1]. This feature of the electron spectrum, detected as a result of theoretical "ab initio" band calculations, is usually observed experimentally in the optical properties (see, e.g., [2-3]). Due to significant changes in the spectral parameters at the Fermi level, it should manifest itself also in other electronic properties, particularly, in the electronic transport. The purpose of this work is to study the effect of "gap" features in the electron energy spectrum near the Fermi level on the electrical resistivity of HMF Heusler alloys.

2. Experimental

We investigated the electrical properties of the Co-based HMF Co_2CrGa, Co_2CrAl, Co_2VAl, Co_2MnAl Heusler alloys as well as the ferromagnetic Ni_2MnGa compound. The alloys were prepared by arc melting in an argon atmosphere. The melted ingots were annealed at 1123 K for 24 hours in an argon atmosphere, followed by furnace cooling. The chemical composition was determined by energy-dispersive x-ray spectroscopy with an accuracy of $\pm 1\%$.
The structure analysis was performed by XRD investigations. The diffraction patterns were obtained at room temperature in a range of 2θ angles from 35° to 135° in steps of 0.02°. Monochromatic Cr-K_α

radiation with the wavelength $\lambda = 0.229$ nm was used. The analysis, carried out using the PowderCell program [4], shows that the Heusler alloys X_2YZ (X = Co; Y = Cr, V, Mn; Z = Ga, Al) have the structure $L2_1$, which contains eight bcc cells. The corners of each bcc cell, i.e. the 4a (0, 0, 0) and 4b (0.5, 0.5, 0.5) positions, are occupied by the X atoms, whereas the body-centered components, i.e. the 4c (0.25, 0.25, 0.25) and 4d (0.75, 0.75, 0.75) positions, are occupied by the Y and Z atoms, respectively [5]. The electrical resistivity was measured from 4.2 to 900 K by a conventional four-probe method.

3. Results and discussion

Figure 1 shows the temperature dependence of the resistivity $\rho(T)$ for the ferromagnetic Ni_2MnGa and the half-metallic ferromagnetic Co_2CrGa and Co_2CrAl Heusler alloys [6-8]. The magnitude of ρ and the form of the $\rho(T)$ dependence are significantly different. The residual resistivity (at T = 4.2 K) is $\rho_0 = 12$ μΩ·cm in the ferromagnetic Ni_2MnGa alloy, $\rho_0 = 151$ μΩ·cm and $\rho_0 = 128$ μΩ·cm in the HMF Co_2CrAl and Co_2CrGa alloys, respectively. Furthermore, $\rho(T)$ in Ni_2MnGa shows a typical "metallic" behavior [9] with a positive temperature coefficient of the resistance (TCR), i.e. it increases with increasing temperature and has a kink near the Curie temperature T_C. The compound Ni_2MnGa undergoes a martensitic transformation near 200 K, which was not detected by resistance measurements, but manifested itself in the thermopower (see inset in Fig.1). In case of the HMF Co_2CrGa and Co_2CrAl alloys $\rho(T)$ shows a more complicated dependence with typical minima and maxima and temperature ranges with a negative TCR.

Table 1 lists the Curie temperatures, the HMF-state and the gap value for the Ni_2MnGa, Co_2CrGa and Co_2CrAl Heusler alloys. The Ni_2MnGa alloy is a typical ferromagnetic metal with no gap near the Fermi level. The other two, Co_2CrGa and Co_2CrAl, are half-metallic ferromagnets, which have an energy gap in the electronic spectrum for one spin subband (with spin down). The energy gap value is about 1 eV and 0.3 eV for Co_2CrGa and Co_2CrAl, respectively. Apparently, these features of the electronic band structure manifest themselves in the electrical properties.

Table 1. The Curie temperatures T_C, the HMF-state (HMF - half-metallic ferromagnets, FM - ferromagnet), and the values of the energy gap ΔE_F for the Ni_2MnGa, Co_2CrGa, Co_2CrAl, Co_2VAl and Co_2MnAl Heusler alloys.

Alloy	T_C (K)	HMF/FM	ΔE_F (eV)
Ni_2MnGa	390 [8]	FM	0 [10]
Co_2CrGa	495 [8]	HMF	1 [8]
Co_2CrAl	305 [6]	HMF	0.3 [7]
Co_2VAl	310	HMF	0.08 [11]
Co_2MnAl	697 [12]	near HMF	-

At temperatures below the gap temperature, there are two channels of conductivity in the HMF alloys: one for the electrons with spin up (↑) and another for the electrons with spin down (↓). The first channel of conductivity $\sigma_\uparrow$ has the usual form for ferromagnetic metals, i.e.

$$\sigma_\uparrow = \frac{1}{\rho_\uparrow}, \tag{1}$$

where

$$\rho_\uparrow = \rho_0 + \rho_{ee} + \rho_{ph} + \rho_m . \tag{2}$$

Here ρ_0 is the residual resistivity, ρ_{ee}, ρ_{ph} and ρ_m are the contributions to the total electroresistivity due to the electron-electron, electron-phonon and electron-magnon scattering, respectively.

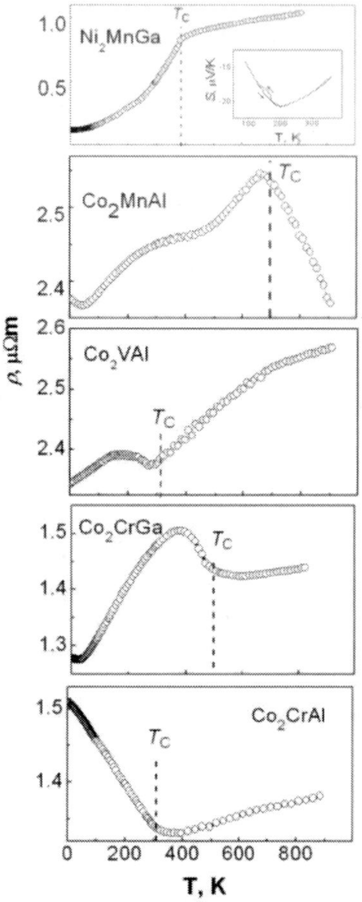

Figure 1. Temperature dependence of the resistivity: the "usual" metallic behavior for the ferromagnetic Ni_2MnGa and the "anomalous" behavior for the half-metallic ferromagnetic Co_2MnAl, Co_2VaAl, Co_2CrGa, Co_2CrAl alloys. The martensitic transformation in Ni_2MnGa near 200 K is detected by thermopower measurements (inset).

The second channel of conductivity $\sigma_\downarrow$, for electrons with spin down, should strongly depend on the parameters of the energy gap in the electronic spectrum near the Fermi level. $\sigma_\downarrow$ should have either an exponential dependence, i.e., $\sigma_\downarrow \sim \exp(-T_0/T)$, where T_0 is the gap temperature, or a power dependence $\sigma_\downarrow \sim T^n$, where n is a power index. Since these two channels have a completely different temperature dependence, i.e. $\sigma_\uparrow$ decreases and $\sigma_\downarrow$ increases with T, the peculiarities of $\rho(T)$ ($\sigma(T)$), particularly the negative TCR, the extremes near T_C and the high resistivity value, can easily appear. Moreover a temperature dependence of the conductivity (resistivity) proportional to $\exp(-B/T^{1/4})$ (B is a coefficient) was observed in such high-resistivity alloys (see, e.g., Ref. [13]). This looks very similar to the Mott law "$T^{1/4}$" [14], i.e. like the hopping conductivity with a variable hopping length. We assumed that the conductivity of the Heusler alloys can also be proportional to $\exp(-B/T^{1/4})$. To verify this assumption, the resistivity (conductivity) of the Co_2MeAl (Me = Cr, Mn, V) alloys was studied.

For the data analysis the concept of coexistence of elastic and inelastic electron scattering was used [13]. According to Ref. [13], the temperature dependence of the conductivity in the high-resistance alloys can be written as

$$\sigma = \sigma(0) + \sigma_{in} = \sigma(0) + A \cdot \exp(-B / T^{1/4}), \quad (3)$$

where A and B are coefficients, $\sigma(0)$ is the usual metallic conductivity due to the elastic scattering processes, and σ_{in} is the conductivity associated with the inelastic scattering of charge carriers. The conductivity $\sigma(0)$ includes the residual conductivity and the conductivity due to the electron-electron and electron-phonon interactions. It is assumed that $\sigma_{in} >> \sigma(0)$.

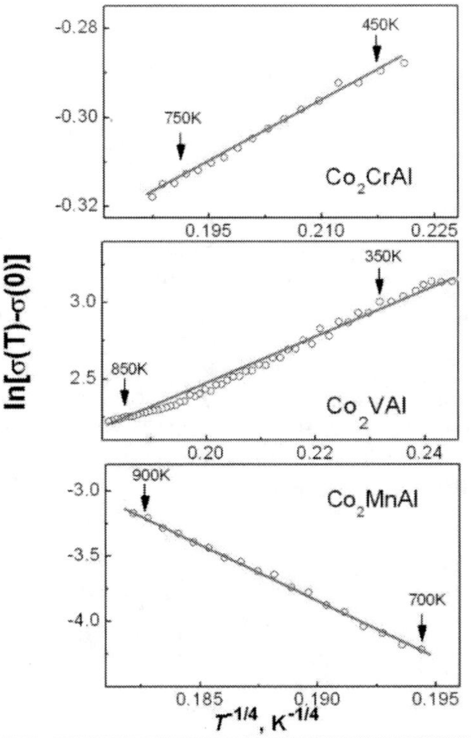

Figure 2. Dependence of $\ln[\sigma(T) - \sigma(0)]$ on $T^{-1/4}$ for Co_2CrAl Co_2VAl and Co_2MnAl.

Fig. 2 presents the temperature dependence of the conductivity in the coordinates $\ln[\sigma(T) - \sigma(0)]$ versus $T^{-1/4}$ for the high-resistivity Co_2MeAl (Me = Cr, V, Mn) alloys. The conductivity follows $\sigma \sim A \cdot \exp(-B / T^{1/4})$ over a wide temperature range, which might also be a manifestation of peculiarities in the electron band structure near the Fermi level E_F of these alloys [8, 13]. Apparently, the sign of the coefficient B, and, hence, the type of the temperature dependence of the conductivity, can also be determined by features of the electron structure near E_F as well as by the relative contributions from different scattering mechanisms of current carriers to the total conductivity.

4. Conclusions

We demonstrated that the temperature dependence of the electrical resistivity in the half-metallic ferromagnetic Co-based Heusler alloys is predominantly determined by specific features of their band structure. The presence of an energy gap at the Fermi level E_F in one of the spin subbands of the conduction electrons with spin down can lead to anomalies in the resistivity, i.e. to a high residual resistivity, to extrema near the Curie temperature, and to a negative coefficient of resistivity. The conductivity is found to be proportional to $\exp(-B/T^{1/4})$ in these high-resistivity alloys at higher temperatures.

5. Acknowledgments

This work was partly supported by the Program of UB RAS (project No 12-T-2-1011), by RFBR (grant No 12-02-00271), by the Program of UB RAS for young researchers (project No 14-2-NP-223), and by the Scientific School (Project No SS-1540.2014.2).

References

[1] V.Yu. Irkhin and M.I. Katsnelson, Phys. - Usp. **37**, 659 (1994).
[2] K.A. Fomina, V.V. Marchenkov, E.I. Shreder, H.W. Weber, Solid State Phenomena **168-169**, 545 (2011).
[3] E.I. Shreder, A.D. Svyazhin, K.A. Belozerova, The Physics of Metals and Metallography **114**, 904 (2013).
[4] W.Kraus, G.Nolze, J. Appl. Cryst., **29**, 301 (1996).
[5] P.J.Webster, J.Phys.Chem.Sol. **32**, 1221 (1971).
[6] N.I. Kourov, A.V. Korolev, V.V. Marchenkov et al., Physics of the Solid State **55**, 977 (2013).
[7] N.I. Kourov, A.V. Lukoyanov, V.V. Marchenkov, Physics of the Solid State **55**, 2487 (2013).
[8] N.I. Kourov, V.V. Marchenkov, V.G. Pushin, K.A. Belozerova, Journal of Experimental and Theoretical Physics 117, 121 (2013).
[9] S.V. Vonsovsky, Magnetism, Wiley, New York (1974).
[10] O.I. Velikokhatnyi, I.I. Naumov, Physics of the Solid State **41**, 617 (1999).
[11] D.P. Rai, A. Shankar, M.P. Ghimire and R K Thapa, Phys. Scr. 86 045702(2012).
[12] I. Galanakis and P.H. Dederichs N. Papanikolaou Phys. Rev. B 66, 174429 (2002).
[13] A.S. Shcherbakov, A.F. Prekul, R.V. Pomortsev, JETP Letters **32**, 401 (1980).
[14] N.F. Mott, Philosophical Magazine **19**, 835 (1969).

Simultaneous study by scanning tunneling spectroscopy and transport measurements in adsorbate-induced two-dimensional systems

Ryuichi Masutomi and Tohru Okamoto

Department of Physics, University of Tokyo, 7-3-1 Hongo, Bunkyo-ku, Tokyo 113-0033, Japan

E-mail: masutomi@phys.s.u-tokyo.ac.jp

Abstract.
An adsorbate-induced two-dimensional electron system at cleaved surfaces of InSb is investigated by low-temperature scanning tunneling microscope and spectroscopy combined with transport measurements in magnetic fields up to 10 T. The magnitude of the potential disorder obtained from the spatially averaged density of states agrees with that deduced from the analysis of the Shubnikov-de Haas oscillations.

1. Introduction

It is well-known that conduction electrons can be induced by submonolayer deposition of other materials at surfaces of narrow band-gap III-V semiconductors. Recently, the present authors have successfully performed in-plane magnetotransport measurements on adsorbate-induced two-dimensional (2D) electron systems (2DESs) at *in situ* cleaved surfaces of *p*-type InAs [1, 3, 2, 4] and InSb [5]. For an Ag-deposited InSb surface, perfect two dimensionality of the adsorbate-induced 2DES was established from the observation of the integer quantum Hall (QH) effect [5]. Moreover, results of low-temperature annealing experiments at the InSb surface suggest that surface morphology of the deposited materials strongly affects the transport properties of the induced 2DES [5]. In this paper, we report not only magnetotransport measurements but also scanning tunneling microscopy (STM) images of surface morphology of adatoms in an adsorbate-induced 2DES at InSb surfaces. In addition, we compare the magnitude of the potential disorder obtained from the spatially averaged density of states (DOS) directly with that estimated from the Shubnikov-de Haas (SdH) oscillations.

2. Experiment

The InSb samples used were cut from a Ge-doped single crystal with an acceptor concentration of (1-2) $\times$ 10^{21} m^{-3}. In the case of *p*-type crystals, surface electrons are separated from three-dimensional carriers in the substrate by a depletion layer. Sample preparation is similar to that used in the previous work on InSb surfaces [5]. After sample cleaving in an ultrahigh vacuum chamber, deposition of Fe atoms was done at the surface of the cold substrate to avoid inhomogeneous distribution of adsorbates. A home-made low-temperature STM was operated with a PtIr probe tip in magnetic fields up to 10 T. Sample holder has six electrodes to measure the diagonal resistivity ρ_{xx} and the Hall resistance R_H of the adsorbate-induced 2DES

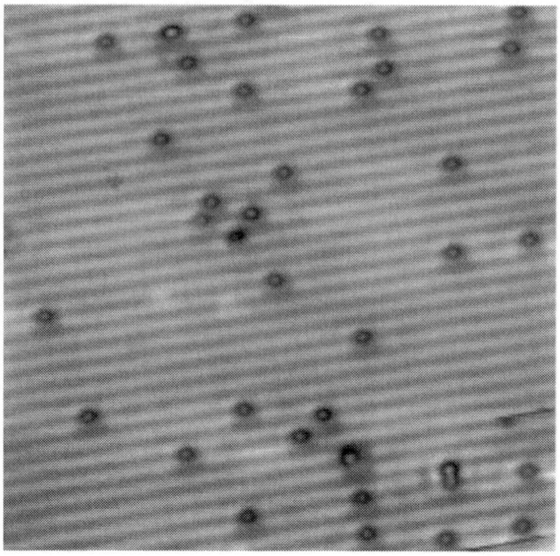

Figure 1. STM topograph of Fe atoms on the (110) cleaved surface of InSb; tunneling current $I = 100$ pA, sample bias voltage $V = 100$ mV, 20×20 nm^2. The Fe atoms appear as a protrusion with a dark rim. The bright rows represent Sb lattice.

and to apply a sample bias voltage (V) for STM and spectroscopy (STM/S) measurements. Magnetotransport data on the (110) cleaved surface were taken in a Hall bar geometry (4×0.4 mm^2) using the standard four-probe lock-in technique with two current electrodes and four voltage electrodes prepared by deposition of gold films onto noncleaved surfaces at room temperature. All the data shown here were taken on optically flat InSb surfaces at 4.2 K.

3. Results and discussion

Fig. 1 shows a typical STM image of Fe atoms on the (110) cleaved surface of InSb at zero magnetic field. The Fe atoms act as donors, which bend the bands downwards and induce a 2D electron gas (2DEG) at the InSb surface. Therefore, tunneling current can flow through the induced 2DEG. Sb lattice appears as bright rows because we use a positive sample bias voltage ($V = 100$ mV), while the Fe atoms are visible as a protrusion with a dark rim between rows of the Sb lattice. The Fe atoms are positioned slightly lower than the surface Sb atoms, which is consistent with Ref. [6]. Thirty-five Fe atoms are counted in an area of 20×20 nm^2, which is equivalent to the Fe atomic density N_{Fe} of 8.8×10^{16} m^{-2}. At this Fe atomic density, each Fe atom is isolated and the formation of chain structures, which is expected for deposition on the substrate at room temperature, is less than five percent.

Fig. 2 shows ρ_{xx} and R_H at 4.2 K in magnetic fields up to 9 T at the same sample used in Fig.1. Due to low DOS of InSb 2DESs, more than one subband is occupied even at moderate total electron density [7, 8]. From the analysis of the SdH oscillations, we deduced the electron density of the first (second) subband $N_1 = 5.8 \times 10^{15}$ m^{-2} ($N_2 = 1.7 \times 10^{15}$ m^{-2}), while the total electron density $N_s = 7.5 \times 10^{15}$ m^{-2} was obtained from the Hall coefficient in low magnetic fields. From the ratio N_s/N_{Fe}, almost every 12th Fe atom donates one electron into the InSb surface. As a result of the doping, the 2DEG is formed at the cleaved surface of InSb.

The transport mobility μ_t obtained from the zero-field conductivity σ_0 ($\mu_t = \sigma_0/eN_s$) is estimated to be 14 m^2/Vs, which is the highest μ_t of the adsorbate-induced 2DESs [1, 5]. The

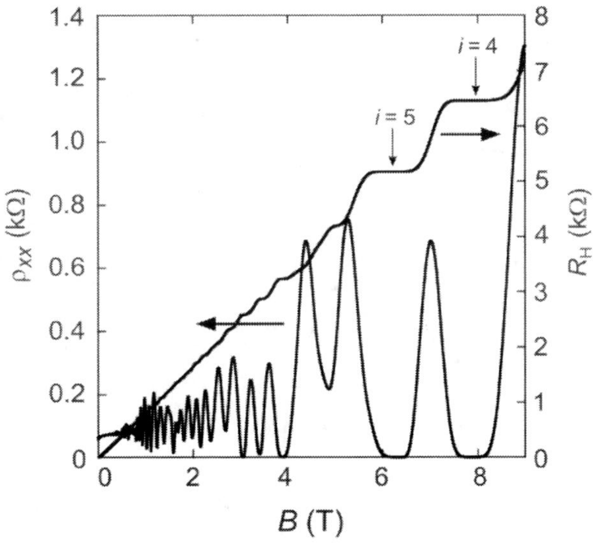

Figure 2. Diagonal resistivity ρ_{xx} (left axis) and Hall resistance R_H (right axis) at 4.2 K as a function of magnetic field.

quantum mobility μ_q, in contrast, obtained from the damping of the SdH oscillations is found to be 1.0 ± 0.2 m^2/Vs, which is similar to that of the 2DES induced by submonolayer Ag deposition at cleaved surfaces of InSb [5]. This result suggests that the charged acceptors within the inversion layer play a major role in the potential disorder.

It is well-known that the Dingle ratio is large ($\mu_t/\mu_q \gg 1$) for long-range potential fluctuations and small ($\mu_t/\mu_q \approx 1$) for short-range potential fluctuations [9]. The obtained ratio of two electron mobilites ($\mu_t/\mu_q \simeq 14$) is sufficiently larger than unity, suggesting that long-range potential fluctuations are dominant in the Fe-induced 2DES at InSb surfaces. In general, μ_t/μ_q can be large in high-mobility modulation-doped GaAs/AlGaAs heterostructures, where the potential disorder is mainly caused by remote ionized impurities [10, 11, 12]. This discrepancy can be explained as follows. In our Fe-induced 2DES, almost every 12th Fe atom donates one electron into the InSb surface, i.e., only 8.5% of the adsorbate Fe atoms contribute to form the 2DEG. Therefore, low ionization probability may be able to make the potential disorder caused by the ionized Fe atoms moderate. Additionally, the adsorbate-induced 2DES is not covered with a dielectric layer that causes interface roughness, such as a Si metal-oxide-semiconductor field-effect transistor. Hence, μ_t/μ_q can be large in the Fe-induced 2DES at InSb surfaces.

Both the first and second subbands form the Landau level (LL) structures and contribute to the magnetoconductance. The QH effect occurs only when the Fermi level E_F is in the localized states between the LLs for both subbands, in which ρ_{xx} vanishes and R_H is quantized as h/ie^2 for an integer i [13, 14]. The QH effect with $i = 5$ ($i = 4$), which consists of a combination of the first subband $i_1 = 4$ ($i_1 = 3$) with the second subband $i_2 = 1$, is observed at $B = 6.2$ T (7.9 T). The odd-integer QH states correspond to the spin splitting.

Fig. 3 displays spatially averaged differential conductance (dI/dV) spectra at $B = 0$ and 10 T perpendicular to the (110) cleaved surface of InSb. We obtain the spatially averaged dI/dV spectra, which coincide with the DOS of the Fe-induced 2DES, as an accumulation of typically 10×10 dI/dV spectra in an area of 100×100 nm^2. The tip-induced quantum dot states as described in Ref. [15] are not observed in our dI/dV spectra. In Fig. 3(a), the spatially averaged

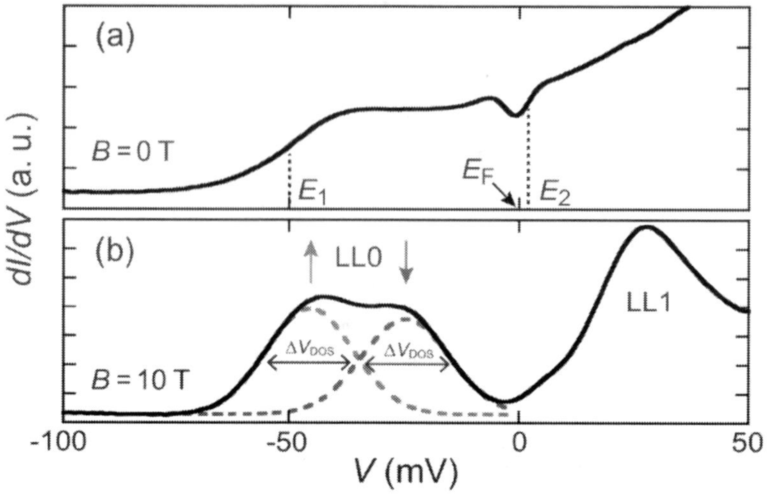

Figure 3. Spatially averaged dI/dV spectra at $B = 0$ (a) and 10 T (b); $I = 100$ pA, $V = 100$ mV and a modulation voltage $V_{\rm mod} = 2.0$ mV. The lowest LL (LL0) and the first excited LL (LL1) of the first subband are marked. Red and blue arrows in Fig. 3(b) represent spin direction. Dashed red and blue curves are result of a double Gaussian fitting for the LL0 at 10 T.

dI/dV spectrum at $B = 0$ T shows an apparent step at $V = -50 \pm 5$ meV, representing the first subband edge E_1. Moreover, the second subband edge is found to be $E_2 = 2 \pm 2$ meV from the spatially averaged dI/dV spectrum at $B = 14$ T (not shown). A small dip at $E_{\rm F}$ (V=0) of Fig. 3(a) shows a Coulomb gap, which is discussed elsewhere. Based on N_1 and N_2 obtained from the analysis of the SdH oscillations, we estimated the subband edges E_{N1} and E_{N2} [3, 16, 17]. The values of $E_{N1} = -80$ and $E_{N2} = -29$ meV slightly differ from those of E_1 and E_2, respectively. This difference in the subband energies between $E_1(E_2)$ and $E_{N1}(E_{N2})$ is attributed to the influence of a tip-induced band bending, which is caused by a work function mismatch between the STM tip and the sample [18]. However, it is known that a small difference between the surface potentials does not significantly affect the dI/dV spectra except for a slight energetic shift [19].

The spatially averaged dI/dV spectrum at $B = 10$ T shown in Fig. 3(b) exhibits pronounced peak structures corresponding to spin-split LLs as a function of V. The effective mass m^* and the Landé g-factor g can be deduced from the separation of LLs $\hbar eB/m^*$ and of spin levels $|g|\mu_{\rm B}B$ with the Bohr magneton $\mu_{\rm B}$. From the lowest peak distances of the first subband at $B = 10$ T, we obtained $m^* = 0.018m_e$ (m_e: free electron mass) and $|g| = 36$. These values slightly differ from those of InSb at the conduction band minimum $m^* = 0.0135m_e$ and $|g| = 51$ [20] due to the well-known nonparabolicity of the conduction band of InSb [21].

The potential disorder can be obtained not only from the SdH oscillations but also from the dI/dV spectra. The full width at half maximum of the lowest LL in the spatially averaged dI/dV spectrum at $B = 10$ T is $\Delta V_{\rm DOS} \simeq 22.3$ mV, analyzed by a double Gaussian fit with a single width parameter $\Delta V_{\rm DOS}$ [22, 23]. In contrast, the average width of the single pot dI/dV is $\Delta V_s = 20.7 \pm 1.2$ mV at $B = 10$ T. We evaluate the width of the potential disorder to be $\Delta_{\rm dis} = \sqrt{\Delta V_{\rm DOS}{}^2 - \Delta V_s{}^2} \simeq 8.4$ meV, which is in the same way in Ref. [23]. On the other hand, from μ_q obtained from the damping of the SdH oscillations, the level broadening of the LLs is found to be $\hbar/\tau_q = \hbar e/m^*\mu_q \simeq 8.6$ meV, where τ_q is the single-particle scattering time. The magnitude of the potential disorder obtained from the spatially averaged dI/dV spectra is in good agreement with that estimated from the damping of the SdH oscillations in the Fe-induced 2DES at InSb surfaces.

4. Summary

An Fe-induced 2DES at cleaved surfaces of InSb is investigated by low-temperature STM/S combined with transport measurements in magnetic fields up to 10 T. The magnitude of the potential disorder obtained from the spatially averaged DOS is estimated to be approximately 8.4 meV, which is comparable to that deduced from the damping of the SdH oscillations. Our STM/S combined with magnetotransport measurements will be applied to further studies on magnetism [2] and superconductivity [24] in one-atomic-layer metal films.

Acknowledgments

The authors thank M. Morgenstern, T. Eguchi, Y. Hasegawa, K. Hashimoto, H. Shinaoka and T. Matsui for helpful discussions. This work has partly supported by Grant-in-Aid for Scientific Research (A) (No. 21244047), Young Scientists (B) (No. 23740233) and Scientific Research (C) (No. 25400316) from MEXT, Japan.

[1] Y. Tsuji, T. Mochizuki, and T. Okamoto, Appl. Phys. Lett. **87**, 062103 (2005).
[2] T. Mochizuki, R. Masutomi, and T. Okamoto, Phys. Rev. Lett. **101**, 267204 (2008).
[3] M. Minowa, R. Masutomi, T. Mochizuki, and T. Okamoto, Phys. Rev. B **77**, 233301 (2008).
[4] T. Okamoto, T. Mochizuki, M. Minowa, K. Komatsuzaki, and R. Masutomi, J. Appl. Phys. **109**, 102416 (2011).
[5] R. Masutomi, M. Hio, T. Mochizuki, and T. Okamoto, Appl. Phys. Lett. **90**, 202104 (2007).
[6] A. A. Khajetoorians, B. Chilian, J. Wiebe, S. Schuwalow, F. Lechermann, and R. Wiesendanger, Nature **467**, 1084 (2010).
[7] N. Kotera, Y. Katayama, and Kiichi F. Komatsubara, Phys. Rev. B **5**, 3065 (1972).
[8] A. Därr and J. P. Kotthaus, Surf. Sci. **73**, 549 (1978).
[9] S. Das Sarma and F. Stern, Phys. Rev. B **32**, 8442 (1985).
[10] M. A. Paalanen, D. C. Tsui, and J. C. M. Hwang, Phys. Rev. Lett. **51**, 2226 (1983); **52**, 484(E) (1984).
[11] J. P. Harrang, R. J. Higgins, R. K. Goodall, P. R. Jay, M. Laviron, P. Delescluse, Phys. Rev. B **32**, 8126 (1985).
[12] P. T. Coleridge, Phys. Rev. B **44**, 3793 (1991).
[13] S. Kawaji and J. Wakabayashi, in *Physics in High Magnetic Fields*, proceedings of the Oji International Seminar, Hakone, Japan, 1980, edited by S. Chikazumi and N. Miura (Springer-Verlag, Berlin, 1981), p. 284.
[14] K. von Klitzing, G. Dorda, and M. Pepper, Phys. Rev. Lett. **45**, 494 (1980).
[15] M. Morgenstern, J. Klijn, Chr. Meryer, M. Getzlaff, R. Adelung, R. A. Römer, K. Rossnagel, L. Kipp, M. Skibowski, and R. Wiesendanger, Phys. Rev. Lett. **89**, 136806 (2002).
[16] Y. Takada, K. Arai, N. Uchimura, and Y. Uemura, J. Phys. Soc. Jpn. **49**, 1851 (1980).
[17] T. Ando, J. Phys. Soc. Jpn. **54**, 2676 (1985).
[18] R. Dombrowski, Chr. Steinebach, Chr. Wittneven, M. Morgenstern, and R. Wiesendanger, Phys. Rev. B **59**, 8043 (1999).
[19] K. Hashimoto, C. Sohrmann, J. Wiebe, T. Inaoka, F. Meier, Y. Hirayama, R. A. Römer, R. Wiesendanger, and M. Morgenstern, Phys. Rev. Lett. **101**, 256802 (2008).
[20] I. Vurgaftman, J. R. Meyer, and L. R. Ram-Mohan, J. Appl. Phys. **89**, 5815 (2001).
[21] U. Merkt, M. Horst, T. Evelbauer, and J. P. Kotthaus, Phys. Rev. B **34**, 7234 (1986).
[22] A. Endo and Y. Iye, J. Phys. Soc. Jpn. **77**, 064713 (2008).
[23] See Supplemental Material for Phys. Rev. Lett. **106**, 156805 (2011).
[24] T. Sekihara, R. Masutomi, and T. Okamoto, Phys. Rev. Lett. **111**, 057005 (2013).

Direct spectral analysis of environmental noise in SQUID and SQUBIT type circuits using a Josephson bifurcation amplifier

G Ithier, G Tancredi and P J Meeson

Department of Physics, Royal Holloway, University of London, TW20 0EX, Egham, Surrey, UK.

E-mail: Phil.Meeson@rhul.ac.uk

Abstract. The Josephson Bifurcation Amplifier is a threshold detector that may be viewed as a 1-bit analog-to-digital converter. Such a device is subject to two random processes, firstly the bifurcation itself is a dynamical process that is subject to quantum or thermal fluctuations that broaden the threshold, secondly the device is immersed in an environment that induces low frequency parametric noise. We have developed a new measurement technique that quantitatively and directly links the environmental noise spectrum to the spectrum calculated from the repetitively acquired binary output of the Josephson Bifurcation Amplifier. The technique has considerable advantages over the NMR techniques presently in common use for studies of qubit decoherence and the techniques used for SQUIDs. It is non-dissipative, enabling operation in the milliKelvin range, it has a wide bandwidth, it is operated at low photon numbers, and its sensitivity approaches the shot noise limit of a weak continuous measurement.

1. Introduction

Superconducting circuits provide a promising framework for Quantum Information Processing. However, as solid state systems, they suffer from strong environmental noise sources that limit their quantum coherence. Despite a great improvement in coherence times over the recent years they are not yet sufficient for the construction of a quantum computer. As a result, the careful characterisation of environmental noise sources, so as to provide information that might lead to their elimination, is a key issue. Until recently, all characterisation techniques (except notably [1, 2]) of noise sources in superconducting quantum bit circuits used the decay of coherence functions borrowed from NMR [3, 4, 5]. In contrast, the repetitive sampling technique we present here provides a direct access to the noise spectrum with a frequency resolution given by the inverse of the total acquisition time, similar to any discrete Fourier transform. In addition, we make use of the high sensitivity of the "Josephson Bifurcation Amplifier". The JBA provides, in effect, a 1-bit analog-to-digital converter. We hence trade dynamic range for sensitivity.

Our device is located on a dilution refrigerator at a temperature of 11 mK ± 2 mK. It consists of a section of superconducting niobium coplanar waveguide enclosed between input and output capacitors which provide coupling to this resonant Fabry-Perot like structure (see Fig.1, more details are given in [6, 7]). An array of Superconducting QUantum Interference Devices (SQUIDs) is located in the middle of this structure at anti-nodes of the electric current

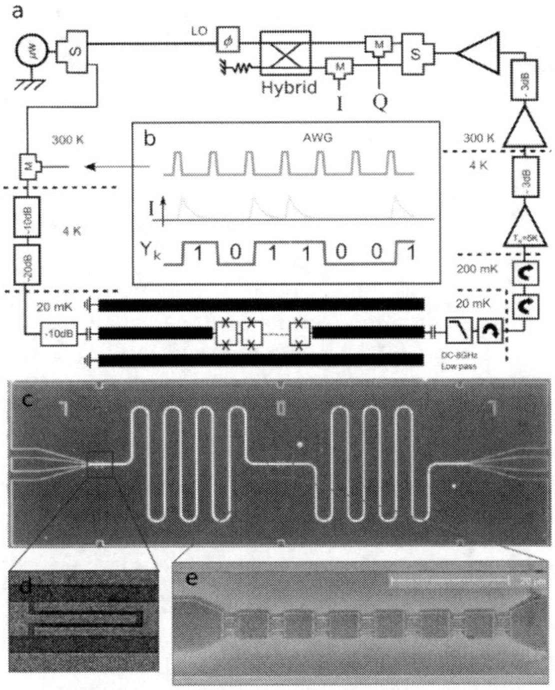

Figure 1. Experimental Setup. a) Homodyne detection: A continuous wave radiofrequency source is split at S providing the drive signal and a reference signal. The drive is mixed, M, with a gate to provide pulses of microwaves and sent to the sample via attenuators and lossy coaxial cables. The signal returns via a low pass filter and circulators designed to protect the sample from amplifier noise. The signal is amplified and homodyne detected. "Hybrid" is a microwave hybrid coupler providing the two microwave quadratures of the phase shifted, ϕ, Local Oscillator LO. b) Waveforms: trapezoidal gate pulses shape the microwaves applied to the sample (red). The system is tuned so that one quadrature of the demodulated signal shows switching events (as random jumps in green). The switching signal is recorded as a binary signal (blue). c) Micrograph of the sample chip, 200 nm thick niobium deposited on an oxidised silicon wafer to form a meandering coplanar waveguide. d) Interdigitated input capacitor. e) Array of 7 aluminium SQUIDs located at an anti-node of the electric current distribution of odd harmonics.

distribution of odd numbered harmonic modes. This setup provides a strong and tunable non-linearity for these modes. Due to the Jospehson non-linearity the system exhibits a bifurcation phenomenon above a critical number n_c of photons populating the resonator. Here $n_c \approx 250$ for the third harmonic mode we consider in the following. The bifurcation is a transition between two dynamical states of oscillation: one of small and one of large amplitude, which can be easily detected using commercially available cryogenic amplifiers. The transition rate between these two states depends on experimental parameters which are for instance the resonant frequency of the mode, its quality factor, or the frequency or amplitude of the microwave drive (see [8, 9, 10] for calculations of this rate).

We first acquire a set of cumulative switching probability curves (or S-curves) at two different pulse repetition rates of 500 Hz and 5 kHz. See Fig. 1, panel **b**. The result shows no dependence of the switching curves on the repetition rate, which allows us to exclude heating effects as a source of correlations between elements of the array. We next set the working point of our experiment to $p_{exp} \approx 1/2$, meaning that the probability of recording a low amplitude state or a high amplitude state as the result of any one probing pulse is approximately equal. We then repetitively apply pulses and record the resulting state of the JBA as a binary array over a time (≈ 3 min) and for the same two different repetition rates: 500 Hz and 5 kHz.

What information does this array convey? Ordinarily a threshold detector or 1-bit A-D convertor possessing a sharp threshold would provide only crude information about any input signal. It would, for example, distort a sine wave input to a square wave output provided the sampling theorem is satisfied and the sine wave crosses the trigger level of the threshold detector. However, in analog to digital conversion systems it is common practice to add a small amount of noise to the input signal prior to digitisation in order to control or "shape" the associated distortion. For instance, this technique is implemented in fast oscilloscopes where a noise is generated by the pre-amplifier stage of the scope before the input voltage is sampled. A careful engineering of this input noise can increase the effective resolution of the converter at the expense of the sampling rate, this is the so-called "oversampling" technique [11]. We may consider the Josephson bifurcation amplifier with thermal and quantum fluctuations taken into account as analogous to a 1−bit analog to digital conversion with an additive random noise. It is possible to make use of the thermal or quantum fluctuations that broaden the JBA bifurcation threshold to achieve the same aim of increasing the effective resolution of the converter and to use this to sample quantitatively the environmental noise affecting the device. Surprisingly, the effect is to linearise the transfer function between the spectral density of the environmental noise source and the spectral density of the binary array. The effects of thermal and quantum noises can be modelled as the addition of a gaussian random variable prior to thresholding and the random variables can be assumed to be independent since the sampling interval is much larger than the reset time of the detection process. The probability distribution of the additive noise is directly related to the cumulative probability switching curve (or 'S-curve') and can be obtained from the Dykman model in both the thermal and quantum regimes.

The relation between the spectral densities of the acquired binary array, Y_k, and the 'input signal' from the environmental noise X_k (respectively S_Y and S_X) may be written as [12];

$$S_Y(\nu) \approx \frac{2}{\pi} \frac{\sigma_Y^2}{\sigma_D^2 + \sigma_X^2} S_X(\nu) + 2\frac{\sigma_Y^2}{\nu_{rep}} \quad (1)$$

where σ_D is the standard deviation of the additive (thermal or quantum) noise. This equation shows that S_Y directly provides S_X and that the digitisation noise σ_Y^2 is spread as a white background over the acquisition bandwidth $\nu_{rep}/2$. The constant background gives the sensitivity at which S_X can be measured. It is important to note that this noise level can be squeezed down just by increasing the sampling rate. The coefficient of S_X may be termed a 'gain' factor and it can be shown [12] that this is subject to some subtle constraints.

Fig 2 shows the spectral density of the switching signal acquired at two different repetition rates. The digitisation background is indicated for both repetition rates. It is interesting to compare this sensitivity to a fundamental scale, namely the standard quantum limit of a weak continuous measurement of the frequency of a resonator [13] under comparable experimental conditions. We consider the average photon number in the cavity (here $\overline{n} \approx 2n_c \approx 500$) leaking at rate $2\pi\gamma$ (here ≈ 2 MHz). The frequency fluctuations of the resonator equivalent to the shot noise of the driven coherent state are given by $S^{sn}_{\delta\nu/\nu} = 2\nu_0^2/(\gamma^2\dot{n})$ where $\dot{n} = 2\pi\gamma\overline{n}$. Remarkably, for the maximal theoretical repetition rate of this detector ($\nu_{rep} \approx \gamma/5 \approx 50$ kHz) the theoretical prediction for the sensitivity of the bifurcation as a noise spectrum analyser would be comparable to the standard quantum limit. Experimentally, we used a maximum repetition rate of 5 kHz, giving a sensitivity within an order of magnitude of the standard quantum limit.

In addition to the digitisation noise, a significant A/f frequency noise is present in our

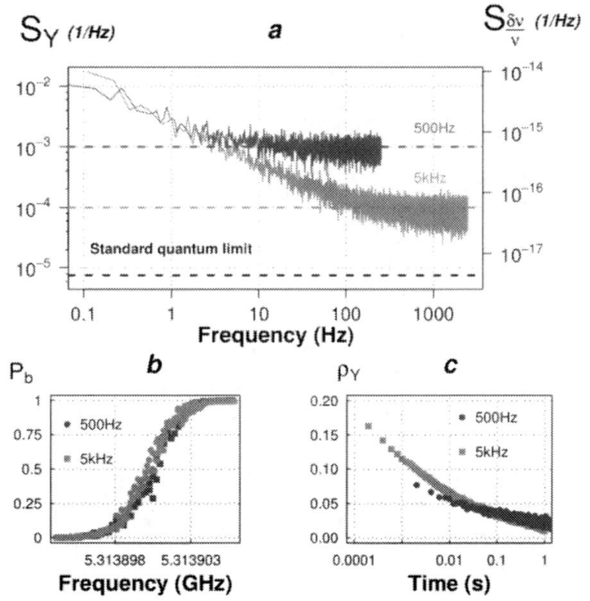

Figure 2. a) The spectral density of the switching signal for two repetition rates: 5 kHz (red) and 500 Hz (blue). Right scale: equivalent relative frequency jitter calculated using Eq.(1). The white background is indicated with dashed lines for both repetition rates (red: 5 kHz, blue: 500 Hz) and is consistent with the expected digitization noise. The standard quantum limit is displayed (black dashed line). b) Switching probability curves at 5 kHz (red) and 500 Hz (blue) repetition rates as a function of the microwave driving frequency. Each point is calculated over 1000 events (thus 0.2 or 2 s of acquisition). The average 10% – 90% width is 4.5 kHz or 1 ppm of the resonance frequency of the resonator. c) Autocorrelation of the switching signal. Note the log scale on the time axis.

sample with $A \approx 10^{-15}$. From flux modulation measurements [6], we can put an upper bound on the contribution of flux noise at the optimal working point ($\phi = 0$ where sensitivity to flux noise is only second order), and show that it has negligible contribution. In addition, because of the small value of the participation ratio $L_{SQUIDs}/L_{tot} \approx 2.5\%$ (where L_{SQUIDs} is the total inductance of the SQUID array, and L_{tot} the total inductance of the cavity), critical current noise has also negligible contribution. Finally, we note that the noise amplitude and frequency dependence observed is comparable to previous observations made in Kinetic Inductance Detectors [14, 15, 16]. We conclude that dielectric noise is probably the source for the observed $1/f$ noise in this device.

2. Conclusion

We have demonstrated that a repetitive sampling of the binary outcome of a bifurcation detector provides a quantitative measure of the environmental noise sources affecting a superconducting non-linear resonator. The technique combines the advantages of state of the art noise measurement techniques in superconducting circuits [14, 17, 18, 19, 20] with the advantages of non dissipative quantum bit readout setups, achieving together a high bandwidth given by half the repetition rate of the measurement; a high sensitivity which may approach the standard quantum limit of a weak continuous measurement and a low temperature of operation (< 15 mK) due to the absence of on-chip dissipation and a low probing energy (in our case $\approx 10^3$ photons in the JBA). The detection bandwidth of the method is half the repetition rate which is, in our case, limited by the reset time of the bifurcation detector. As the sensitivity of the bifurcation depends weakly on the quality factor, a lower Q than that used in our experiment could allow repetition rates of order of several hundreds of MHz. Obtaining the noise spectrum over this frequency range with a lower digitization noise would be of great interest. The technique can be applied, in principle, to any detector involving a threshold effect and in particular qubit state measurement setups, it may also have important applications for the measurement of the full counting statistics of a quantum conductor.

Acknowledgments
We wish to thank D. Estève and all at the Quantronics Group at CEA Saclay for their support over many years, especially P. Bertet and A. Palacios-Laloy who fabricated the sample, A. Tzalenchuk and T. Lindström (NPL) for helpful discussions and the loan of equipment. We wish to thank the EPSRC (EP/D001048/1 and EP/F041128/1), the European MicroKelvin consortium and the EXL03 Microphoton project of the European Metrology Research Program for financial support. The EMRP is jointly funded by the EMRP participating countries within EURAMET and the European Union.

References

[1] Ithier G, Tancredi G, and Meeson PJ *"Noise spectral analysis of environmental noise in SQUID and superconducting qubit type circuits using a Cavity Bifurcation Amplifier"* 2011 Poster 81 presented at the Conference on Condensed Matter and Materials Physics (CMMP11), December 13-15, Manchester University, UK.
[2] Yan F, Bylander J, Gustavsson S, Yoshihara F, Harrabi K, Cory DG, Orlando TP, Nakamura Y, Tsai J-S, and Oliver WD 2012 *Phys. Rev.* B, **85** 174521
[3] Yoshihara F, Nakamura Y, and Tsai JS 2010 *Phys. Rev.* B, **81**, 132502
[4] Ithier G, Collin E, Joyez P, Meeson PJ, Vion D, Esteve D, Chiarello F, Shnirman A, Makhlin Y, Schriefl J and Schön G, *Phys. Rev.* B, **72**, 134519
[5] Bylander J, Gustavsson S, Yan F, Yoshihara F, Harrabi K, Fitch G, Cory DG, Nakamura Y, Tsai JS, and Oliver WD 2011 *Nature Physics*, **7** 565
[6] Tancredi G, Ithier G, Meeson PJ 2013 *Appl. Phys. Lett.* **103** 063504
[7] Palacios-Laloy A, Nguyen F, Mallet F, Bertet P, Vion D, and Esteve D 2008 *Journal of Low Temperature Physics* **151** 1034
[8] Vijay R 2008 Ph.D.thesis, Yale University
[9] Dykman MI and Krivoglaz MA 1980 *Physica* A **104** 480
[10] Dykman MI and Smelyansky VN 1988 *Zh. Eksp. Teor. Fiz. Pis'ma Red.* **94** 61
[11] Hauser MW, 1991 *J. Audio Eng. Soc.* **39** 3
[12] Ithier G, Tancredi G, Meeson PJ 2014 *New Journal of Physics* **16** 055010
[13] Clerk AA, Devoret MH, Girvin SM, Marquardt F and Schoelkopf RJ 2010 *Rev. Mod. Phys.* **82** 1155
[14] Gao J, Zmuidzinas J, Mazin BA, LeDuc HG and Day PK 2007 *Applied Physics Letters* **90** 102507
[15] Gao J, Daal M, Vayonakis A, Kumar S, Zmuidzinas J, Sadoulet B, Mazin BA, Day PK, and Leduc HG 2008 *Applied Physics Letters* **92** 152505
[16] Gao J, Daal M, Martinis JM, Vayonakis A, Zmuidzinas J, Sadoulet B, Mazin BA, Day PK, and Leduc HG 2008 *Applied Physics Letters* **92** 212504
[17] Lindström T , Burnett J, Oxborrow M, and Tzalenchuk, AYu 2011 *Review of Scientific Instruments* **82** 104706
[18] Burnett J, Lindström T, Oxborrow M, Harada Y, Sekine Y, Meeson PJ and Tzalenchuk AYu, 2013 *Phys. Rev.* B **87** 140501
[19] Burnett J, Faoro L, Wisby I, Gurtovoi VL, Chernykh AV, Mikhailov GM, Tulin VA, Shaikhaidarov R, Antonov V, Meeson PJ , Tzalenchuk AYu and Lindström T 2014 *Nature Communications* **5** 4119
[20] The SQUID Handbook: Applications of SQUIDs and SQUID Systems, Volume II (2004) edited by John Clarke and A.I. Braginski, Wiley-VCH, Weinheim, Germany.

Journal of Physics: Conference Series **568** (2014) 052022 doi:10.1088/1742-6596/568/5/052022

Spectroscopy of the modes of a non-linear superconducting microwave resonator via inter-mode coupling and bifurcation amplification

G Tancredi, G Ithier and P J Meeson

Department of Physics, Royal Holloway, University of London, TW20 0EX, Egham, Surrey, UK.

E-mail: `Phil.Meeson@rhul.ac.uk`

Abstract. The Josephson non-linearity provides a route to high sensitivity measurements through the bifurcation of a resonant mode into two metastable dynamical states. We probe the bifurcation regime of a superconducting microwave co-planar resonator incorporating a controlled non-linearity made of an array of SQUIDs. The switching probability between metastable states may be determined by repeated measurements under nominally identical conditions. We show that measurements of this probability and the inter-mode coupling introduced by the SQUIDs may be used for spectroscopy of other resonant modes. We also find the width of the switching-curve is within a factor two of quantitative agreement with Dykman's model, but that the switching probability is affected by a low frequency noise that has its source on-chip.

1. Introduction

The phenomena associated with the Josephson effect may be regarded as a method of introducing non-linearity into an electrical circuit. The concept is particularly important as non-linearity is a fundamental necessity in the field of superconducting quantum circuits and the Josephson junction is the only known method of introducing non-linearity without dissipation. In the present discussion the non-linearity is based on the fact that the inductance of a Josephson junction is a function of the current through the junction. Including Josephson elements in an electrical resonator provides a method for controlling by design the degree of non-linearity introduced and considerably enhancing it so that it is the dominant non-linearity. In our case the resonator is constructed from a superconducting co-planar waveguide and the non-linearity is introduced by placing an array of SQUIDs at the centre of the resonator. See Figure 1.

The consequences of adding non-linearity are multiple [1, 2, 3, 4, 5, 6, 7, 8, 9, 10]. In the first instance, the resonant modes are no longer harmonic multiples of each other, this follows because the non-linear inductance (which controls the frequency of the resonant modes) depends on the current flowing in the circuit and this is different for the different resonant modes. Secondly each mode becomes non-linear such that at high oscillator energies the mode is first 'pulled' in frequency and then 'bifurcates'. In the present electrical circuit this is a consequence of the inductance changing with the power level applied or equivalently the magnitude of the electrical current through the Josephson junction. The bifurcation phenomenon occurs when (descriptively speaking) the peak of the resonance curve is pulled further than the resonant

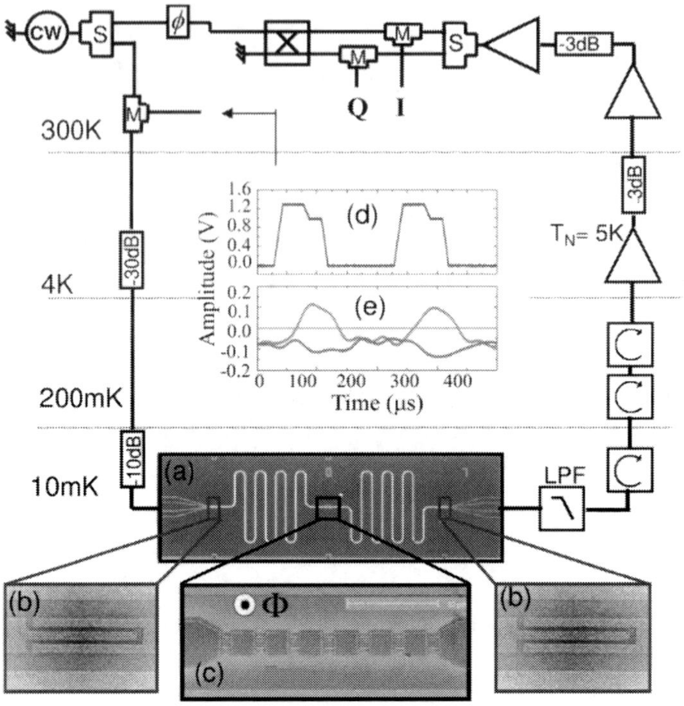

Figure 1. The experimental arrangement for pulsed radiofrequency measurements of a non-linear superconducting resonator. Radiofrequency pulses with the envelope shown in insert **d** are fed to the sample (insert **a**, located at T ≈ 20 mK) through attenuators and lossy coaxial lines. The power and the frequency of the pulses are tuned so that the bifurcation regime of the non-linear resonator is probed. The transmitted signal is fed via circulators, superconducting coax, a cold amplifier and Cu coax to room temperature and then homodyne detected. Insert **e** shows the signal obtained when the sample does or does not exhibit bifurcation. Inserts **b** and **c** show the input and output capacitance and the SQUID array respectively.

linewidth, resulting in three possible amplitudes at the same drive frequency, one of which is unstable. Under these conditions at a given drive frequency the system must choose between a high amplitude state and a low amplitude state. The transition between the two states is narrower than the natural linewidth of the resonant mode and hysteretic. By analogy with the Kerr effect in the field of quantum optics, the bifurcation phenomenon may be viewed as being a consequence of the 'self-Kerr effect'. Thirdly, the resonant modes become coupled. This follows because the current flowing in the Josephson junction and attributed to one mode alters the value of the inductance that must be considered for the evaluation of a different mode. This coupling is called the 'cross-Kerr effect'. The dispersive coupling described by the cross-Kerr effect has not been widely explored but has, for example, been discussed in the context of quantum non-demolition measurements [11, 12, 13].

It is usual to regard resonant phenomena as 'amplifiers' that enhance measurements by a factor of order the Quality-factor of the resonator, at the expense of being required to wait long enough that the oscillator may follow the parameter being measured. The bifurcation phenomena may also be used to measure experimental parameters but it is, in contrast, a threshold measurement. The 'bifurcation amplifier' may be biased close to the transition between

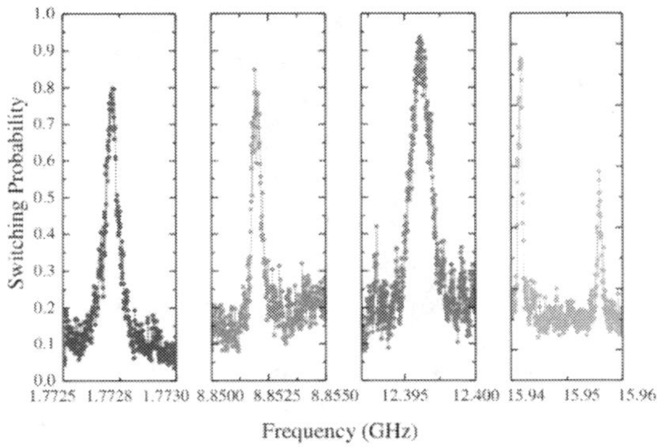

Figure 2. Spectroscopy of Kerr-coupled microwave modes. The probability of switching of the 3rd mode of a superconducting non-linear resonator versus frequency, measured while simultaneously applying a continuous swept microwave excitation in the vicinity of (from left to right) the 1st, 5th, 7th, and 9th modes.

the two states and the external parameter to be measured may be used to influence whether or not the bifurcation threshold is actually crossed. The output is therefore binary but the threshold is sharper than the natural linewidth of the resonant mode, hence there is a gain in sensitivity at the expense of dynamic range. Through repetitive measurements of whether the system crosses the threshold a switching probability may be determined. Measurements of the switching probability as a function of drive frequency (or power) result in a cumulative probability curve or 'S-curve', whose width is a measure of the sensitivity of the technique. In the experiments described below the natural linewidth of the ≈ 2 GHz resonant mode is ≈ 212 kHz, while the width of the S-curve is ≈ 4.5 kHz.

Our device [14] consists of a 200 nm thick sputter-deposited Nb film fabricated into a meandering half-wavelength coplanar waveguide resonator through optical lithography, with end capacitors of design value $C = 7\,\mathrm{pF}$. The fundamental frequency is $\simeq 1.77\,\mathrm{GHz}$. An array of $N = 7$ Al/AlO$_x$/Al SQUIDs was formed with standard e-beam lithography and double-angle shadow evaporation. For a fixed inductance, using an array rather than a single junction allows the critical current (and hence the magnitude of the non-linearities) to be varied at the design stage. In this design only the odd numbered modes couple to the SQUID array, the even numbered modes possess nodes of current at the location of the array.

In the present experiments we drive the third mode to the bifurcation regime and tune it to a switching probability of 10%. We then inject a low-power continuous-wave signal in the vicinity of the coupled mode to be detected and sweep its frequency across the mode to measure the response. As shown in Fig. 2, we were able to detect the change in photon occupation of the 1st, 5th, 7th and 9th modes via a variation in the switching probability of the 3rd mode, thus demonstrating simultaneously both the self-Kerr bifurcation of the 3rd mode and the cross-Kerr coupling between the modes. With this technique, the cross-Kerr effect enables spectroscopy of modes that lie outside our measurement bandwidth (because of the limited bandwidth of the circulators) and thus are undetectable via the usual transmission detection. An important figure of merit is obtained by comparing the S-curve width to the magnitude of the relevant inter-mode coupling strength, allowing us to conclude that our device is sensitive to $\simeq 10$ photons in the 1st mode. The resonant frequencies of the coupled modes are obtained via a Lorentzian fit providing a value for the participation ratio at zero magnetic flux of $\beta = (2.55 \pm 0.1)\%$, which is compatible with the value obtained from flux modulated vector network analyser measurements. The 9th

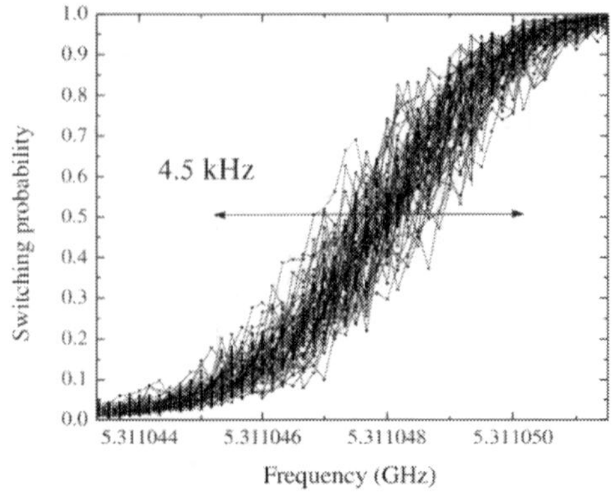

Figure 3. Measurements of the switching probability of the 3rd mode of a superconducting non-linear resonator as a function of the driving frequency at 8mK and $\Phi/\Phi_0 = 0$. Each point represents the probability of switching of the mode as determined from 1000 pulses. 50 separate consecutively measured curves are shown. The white line is the average of the 50 curves. The expected statistical error of the black lines is approximately the thickness of the white line.

mode is split for reasons that we do not understand; possibly the mode is bifurcating.

Figure 3 shows 50 switching probability curves acquired consecutively under nominally identical conditions, shown as a function of the driving frequency. The 10%–90% width of the S-shaped averaged probability curve (white line) is 4.5 ± 0.5 kHz or 0.9 ppm of the frequency of the third mode. This value can be compared with a theoretical value calculated from Dykman's model [15, 16] which predicts an S-curve width due to quantum fluctuations of $\simeq 0.5$ ppm. Hence our experimental value is about twice the theoretical value. In attempting to reconcile this, we note that the bifurcation curves of Fig. 3 show that the switching probability is affected by low frequency fluctuations that significantly exceed the expected statistical fluctuations. A noise source, whose low frequency components appear in these measurements, may also be present at higher frequencies where it may increase the S-curve width and may be the source of the discrepancy between our experimental values and the theory of Dykman. Such a noise source would be parametrically coupled to our CJBA, either through the frequency of the cavity or through the amplitude and frequency of the biasing pulses. Estimates and measurements of the extrinsic sources of noises, such as temperature fluctuations and power (frequency) fluctuations of the biasing pulses, suggest that they cannot account for the observed noise fluctuations and we are led to conclude that they originate on-chip due to the parametric coupling of our oscillator to local fluctuators embedded in the chip.

In summary [17], we have created a device to assess the usefulness of on-chip superconducting microwave optical components. By inserting a controllable, non-linear element into a superconducting microwave resonator, we can simultaneously exploit the self- and cross-Kerr effects. We have shown that high-sensitivity spectroscopy of nearby modes may be performed through measurements of the probability of switching of another mode via non-linear coupling. Analysis of the device sensitivity suggests that there is a source of noise that manifests as fluctuations in the cavity resonant frequency, it has a low frequency component and it arises on-chip. The noise may be related to the ubiquitous noise seen in similar SQUID and qubit systems [18, 19]; we have excluded microscopic flux noise and the remaining sources include two-level fluctuators in the environment and critical current noise in the tunnel barriers. However, we note that the intrinsic sensitivity of the device is close to the limits given by the theory of Dykman. Further analysis of repetitively acquired switching data from Josephson bifurcation phenomena can be further explored ([20] and poster No. 785, this conference).

Acknowledgments
We wish to thank D. Esteve and the Quantronics Group at CEA Saclay for their support over many years, especially P. Bertet and A. Palacios-Laloy for fabricating the sample [14] and A. Tzalenchuk, T. Lindstrom (NPL) for helpful discussions. Part of this work was done within the project EXL03 MICROPHOTON of the European Metrology Research Programme (EMRP). The EMRP is jointly funded by the EMRP participating countries within EURAMET and the European Union. We thank also EPSRC for their financial support.

References
[1] Wallraff A *et al.* 2004 *Nature* **431** 162-167
[2] I. Chiorescu *et al.* 2004, *Nature* **431** 159-162
[3] Blais A, Huang RS, Wallraff A, Girvin SM and Schoelkopf RJ 2004 *Phys. Rev.* A **69** 062320
[4] Castellano-Beltran MA and Lehnert KW 2007 *Appl. Phys. Lett.* **91** 083509
[5] E. Boaknin *et al.*, arXiv:cond-mat/0702445
[6] Castellano-Beltran MA *et al.* 2008 *Nature Phys.* **4** 929
[7] Vijay R, Slichter DH and Siddiqi I 2011 *Phys. Rev. Lett.* **106** 110502
[8] Metcalfe M *et al.* 2007 *Phys. Rev.* B, **76** 174516
[9] Mallet F *et al.* 2009 *Nature Phys.* **5** 791
[10] Ong FR *et al.* 2011 *Phys. Rev. Lett.* **106** 167002
[11] Buks E and Yurke B 2006 *Phys. Rev.* A **73** 023815
[12] Kumar S and DiVincenzo DP 2010 *Phys. Rev.* B **82** 014512
[13] Suchoi S *et al.* 2010 *Phys. Rev.* B **81** 174525
[14] A. Palacios-Laloy *et al.* 2008 *J. Low Temp. Phys.* **151** 1034
[15] Dykman MI and Smelyanskii VN 1988 *Zh. Eksp. Teor. Fiz.* **94** 61-74
[16] Dykman MI and Krivoglaz MA, 1979 *Zh. Eksp. Teor. Fiz.* **77** 60-73
[17] Tancredi G, Ithier G, Meeson PJ 2013 *Appl. Phys. Lett.* **103** 063504
[18] Burnett J, Lindström T, Oxborrow M, Harada Y, Sekine Y, Meeson PJ and Tzalenchuk AYu, 2013 *Phys. Rev.* B **87** 140501
[19] Burnett J, Faoro L, Wisby I, Gurtovoi VL, Chernykh AV, Mikhailov GM, Tulin VA, Shaikhaidarov R, Antonov V, Meeson PJ , Tzalenchuk AYu and Lindström T 2014 *Nature Communications* **5** 4119
[20] Ithier G, Tancredi G, Meeson PJ 2014 *New Journal of Physics* **16** 055010

Thermal resistance at constrictions in 2D mesoscopic ribbons at low temperatures

A Ramiere[1], J Amrit[1] and S Volz[2]

[1]Laboratoire d'Informatique pour la Mécanique et les Sciences de l'Ingénieur, UPR CNRS 3251, Université Paris-Sud, Rue John von Neumann, 91403 Orsay, France

[2]Energétique Moléculaire et Macroscopique, Combustion, UPR CNRS 288, Ecole Centrale Paris, Grande Voie des Vignes, 92295 Chatenay-Malabry, France

E-mail: aymeric.ramiere@limsi.fr

Abstract. Phonon transport in junctions formed between suspended membranes of the same material is studied using Monte Carlo method. The investigation is conducted at 1K for systematically chosen ratios of *(d/D)* where *d* is the junction width and *D* the membrane length. The influence of surface roughness is taken into account. The dependency of the constriction thermal resistance (in µmK/W) is established: $R_J \approx [2.4 + 2.0 \times (d/D)^{-1/2}] \times 10^9$. The average and frequency dependent transmission coefficients are determined as a function of *(d/D)*. These structural configurations are inherent in thermoelectric; and graphene and silicene related applications.

1. Introduction

In this paper phonon transport in a uniform material composed of two rigid suspended mesoscopic membranes connected with an abrupt junction/bridge is studied at low temperatures (see figure 1). In particular we show the apparition of a thermal boundary resistance R_J at the membrane/junction interface. The latter is due to the mismatching of vibrational modes because of the sharp change in the geometry. The junction diameter *d* and the root mean square surface roughness σ of the structure are also taken into account. We recall that thermal characterization of these structural configurations are vital for thermoelectric; and graphene and silicene related nanotechnology applications at room temperatures.

2. Sample set-up and Monte Carlo technique

Our study is conducted on a 2D material with a surface roughness of 2.4 nm. The membranes are 100 x100 µm squares and the junction length is fixed to 10µm. The junction diameter is varied from 1µm to 10µm. The sample is divided into cells of constant length $\Delta = 0.5\,\mu$m. Our approach in implementing the Monte Carlo method is similar to that of Peterson[1] and to that described in ref. [2]. Phonons (of different polarizations *L* or *T* and frequencies) are randomly selected in a Planck-like distribution to populate the hot membrane until the total energy corresponding to the set temperature of the membrane is attained. Flexural modes, which are fully diffusely scattered, are neglected as the main objective is to highlight the geometrical dependencies of R_J. The trajectories of the phonons (defined by randomly chosen angles) are ballistic and the nature of the boundary scattering at the walls is determined by the specularity factor $p(\sigma,\lambda) = exp(-16\pi^2 cos(\theta)\sigma^2/\lambda^2)$ where θ is the angle between the phonon wave vector and the normal to the ideal smooth surface. Under the physical conditions of this simulation, phonon-phonon and Umklapp processes are inoperative.

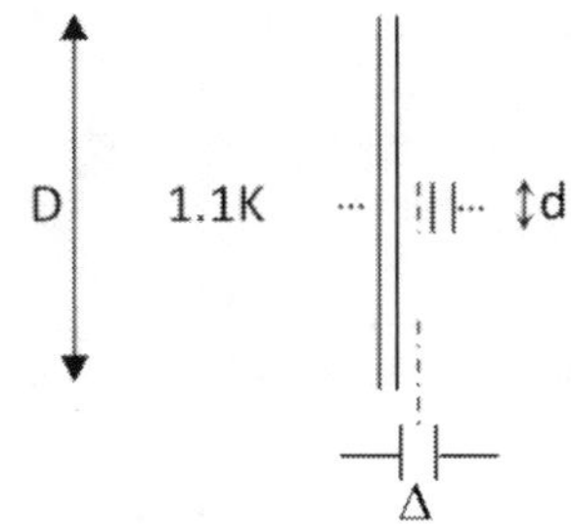

Figure 1. Junction/Bridge of width *d* between two suspended membranes of sides *D*. The abrupt form is chosen to enhance the effect of the junction/membrane thermal resistance. The latter is determined over the length Δ, defined in text.

3. Thermal resistance and transmission coefficients at membrane-junction interface

The thermal conductance h_J at the junction/membrane interface is defined as: $h_J = (Q^+ - Q^-)/\Delta T$ where Q^+ and Q^- are respectively the energy associated with the phonon flux leaving and entering the membrane to and from the junction, and ΔT is the temperature jump over the cell length Δ at the interface. The junction/membrane thermal boundary resistance is now defined as: $R_J = d/h_J$. In figure 2, R_J is plotted as a function of $(d/D)^{-1/2}$ to highlight the following linear behaviour in µm.K/W: $R_J \approx [2.4 + 2.0 \times (d/D)^{-1/2}] \times 10^9$. The numerical coefficient is simply the thermal resistance due to the conductance over Δ.

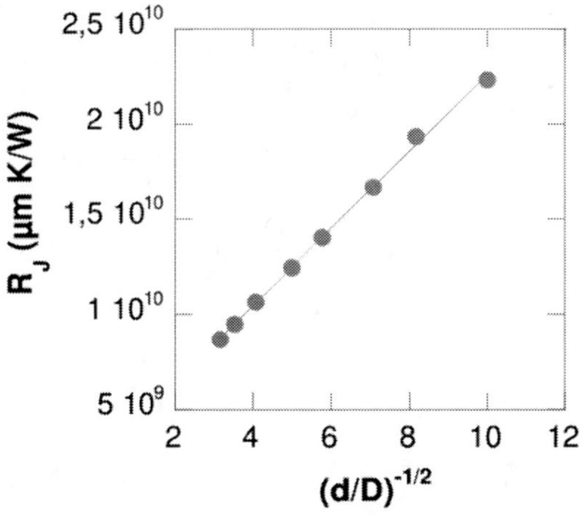

Figure 2 Thermal resistance at membrane/ junction interface in Silicene material.

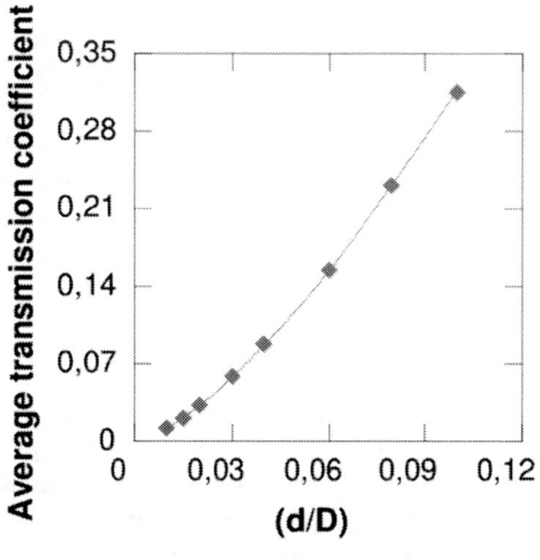

Figure 3. Power law evolution of the average transmission coefficient as a function of *(d/D)*.

The average transmission $\bar{\tau}$ is determined using the definition[3] $h_J = (1/4)C_2 \upsilon_D \bar{\tau}$ where the specific heat $C_2 = (\partial E_{2D} / \partial T)$, with E_{2D} being the 2D energy, and υ_D is the Debye velocity defined as $\upsilon_D^{-2} = \sum_i (1/\upsilon_i^2)$, with $i = L, T$. All numerical values used are that of Silicene material[5]. Figure 3 shows that $\bar{\tau}$ follows a power law with (d/D) given by: $\bar{\tau} = 8.23(d/D)^{\sqrt{2}}$. The frequency dependant transmission for each polarisation entering the junction undergoes a change as a function of the junction width as shown in figure 4 (a) and (b). Also, the transmission of *L* modes dominates over that of *T* modes.

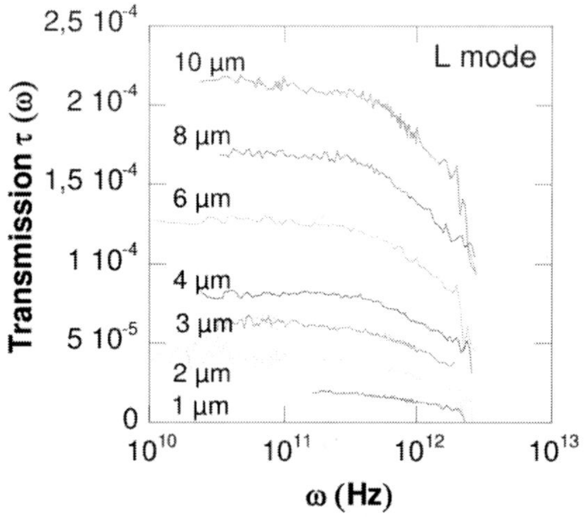

Figure 4a. Transmission of mode L membrane/ junction interface.

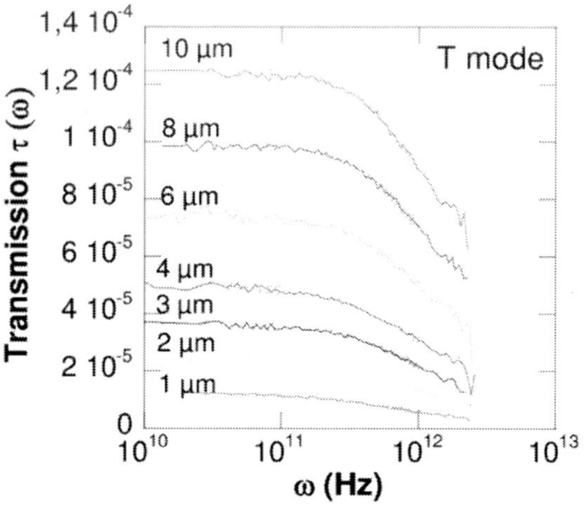

Figure 4b. Transmission of mode T at membrane/junction interface.

Finally, we note that an analytical solution of the thermal transport in this 2D configuration is discussed by Cross and Lifshitz[4]. The novelty in our study is that we have taken into account the surface roughness. Details of our full study shall be given in ref. [5] in preparation.

Conclusions

Using the Monte Carlo method at ~1K, we have demonstrated that the thermal resistance at the membrane/bridge interface varies as $(d/D)^{-1/2}$. The average transmission coefficient $\bar{\tau} \propto (d/D)^{\sqrt{2}}$. The transmission coefficient of modes *L* and *T* are independent of frequency for small ω, but it varies with the junction width *d*.

References

1. R. B. Peterson, Journal of Heat Transfer, **116**, 815 (1994)
2. A. Ramiere, J. Amrit and S. Volz, Nanoenergy Letters, **6**, 13 (2013)
3. L. J. Challis, Journal of Physics C: Solid State Phys. **7**, 481 (1974)
4. M. C. Cross and R. Lifshitz, Phys. Rev. B. **64**, 085324 (2001)
5. M. Hu, X. Zhang and D. Poulikakos, Phys. Rev. B. **87**, 195417 (2013)
6. A. Ramiere, J. Amrit and S. Volz, *in preparation*

Acknowledgements

This work is supported by the Labex LaSips. One of us (AR) benefited from a grant from ED 534 MIPGE, Paris-Sud University.

Single-electron transitions in one-dimensional native nanostructures

M Reiche[1], M Kittler[2], M Schmelz[3], R Stolz[3], E Pippel[1], H Uebensee[4], M Kermann[4], and T Ortlepp[4]

[1] Max Planck Institute of Microstructure Physics, Halle, Germany
[2] BTU CS, JointLab IHP/BTU, Cottbus, Germany
[3] Leibniz Institute of Photonic Technology, Jena, Germany
[4] CIS Research Institute of Microsensorics and Photovoltaics, Erfurt, Germany

E-mail: reiche@mpi-halle.de

Abstract. Low-temperature measurements proved the existence of a two-dimensional electron gas at defined dislocation arrays in silicon. As a consequence, single-electron transitions (Coulomb blockades) are observed. It is shown that the high strain at dislocation cores modifies the band structure and results in the formation of quantum wells along dislocation lines. This causes quantization of energy levels inducing the formation of Coulomb blockades.

1. Introduction

Nanomaterials are low-dimensional materials, such as zero-dimensional quantum dots, one-dimensional nanowires, or two-dimensional sheets. The shrinkage of dimensions is expected to result in new exceeding properties related to size and quantum effects suggesting numerous new applications in electronics, optoelectronics, and other areas (e.g. [1, 2]). In recent years, silicon nanomaterials and nanostructures have been attracted extensive interest, not only for technology and device development, but also for fundamental research [3]. Here, studies of Si nanowires produced by different growth techniques were main issues [4, 5]. Preparation, handling, and packaging of individual or arrays of nanowires, however, are difficult. An alternative are dislocations. Dislocations are basic one-dimensional crystal defects. Their dimensions of a few nanometers (diameter) and lengths of more than 1 µm characterize these defects as native nanowires embedded in a perfect crystalline matrix.

The present paper deals with analysis of electronic properties of dislocations. Room temperature measurements proved already some exceptional properties such as a significant increase of the drain current of metal-oxide-semiconductor field-effect transistors (MOSFETs) if defined numbers and types of dislocations are placed in the device channel [6]. A supermetallic conductivity of dislocations was stated which is about eight orders of magnitude higher than of the surrounding silicon matrix [7]. Here, we present results of low-temperature measurements showing the presence of a two-dimensional electron gas (2DEG) at dislocation arrays. Effects related to the presence of the 2DEG (single electron transitions) are explained.

2. Sample preparation and measurement technique

A smart technique to realize reproducible arrays of dislocations is semiconductor wafer direct bonding [6, 8]. Hydrophobic bonding process initiate the formation of a two-dimensional dislocation network in the interface between both wafers. The dislocation distance in the network is well-controlled by the

Journal of Physics: Conference Series **568** (2014) 052024 doi:10.1088/1742-6596/568/5/052024

misfit between both wafers. Because dislocations are crystal defects, their type depends on the crystal symmetry. Bonding of two {100}-oriented silicon wafers results, for instance, in a network of square-like meshes of screw dislocations (Burgers vector $\mathbf{b}$ = ½<110>) running parallel to both orthogonal <110> directions. To avoid interactions of dislocations with other possible defects in the bulk of wafers (point defects, etc.) silicon-on-insulator (SOI) wafers were applied. This results in dislocation networks in the center of an about 60 nm thick Si layer (device layer) electrically isolated by a buried oxide (BOX) layer from the bulk of the wafer [7].

For measurements n-channel MOSFETs (nMOSFETs) and diodes were prepared using conventional CMOS processing. The channel direction was in all cases parallel to <110>-directions, i.e. parallel to the dislocation lines (Fig.1) [6, 7]. The channel length of all devices was constant (L = 1 μm), while the channel width varied between 30 nm and 1 μm. This allows us to vary the number of dislocations in the device channel if the same dislocation network (i.e. constant dislocation spacing) is applied.

Low-temperature measurements were carried out using a cryogenic probe station ($T_{min} \cong 4$ K, Lake Shore Cryotronics) and a Physical Property Measurement System (PPMA, Model 6000, $T_{min} \cong 2$ K Quantum Design, Inc.). The latter enables also measurements in magnetic fields up to 8 Tesla. A modified Heliox VL refrigator system (Oxford Instruments) was applied for measurements down to T ≅ 300 mK. Device measurements (output and transfer characteristics of MOSFETs) were carried out using either the 4156C or B1500A Precision Semiconductor Parameter Analyzer (Agilent).

3. Results and discussion

The typical structure of a dislocation network formed by bonding of {100}-oriented Si wafers is shown in Fig. 1. The network is formed by two sets of screw dislocations parallel to orthogonal <011>-directions. Analysis at higher magnification revealed splitting of the screw dislocations into Shockley partials (Burgers vector of the type $\mathbf{b}$ = 1/6<121>) producing a stacking fault (SF) between both. The distance between both partials (or SF size) is about 6 nm [7].

Low-temperature measurements refer to Shubnikov-de Haas (SdH) oscillations if magnetic fields are applied (Fig. 2). Because SdH oscillations are proved also by setting the gate-source voltage (V_{GS}) to zero, they are induced by dislocations and not by confinement of carriers in the space charge region of

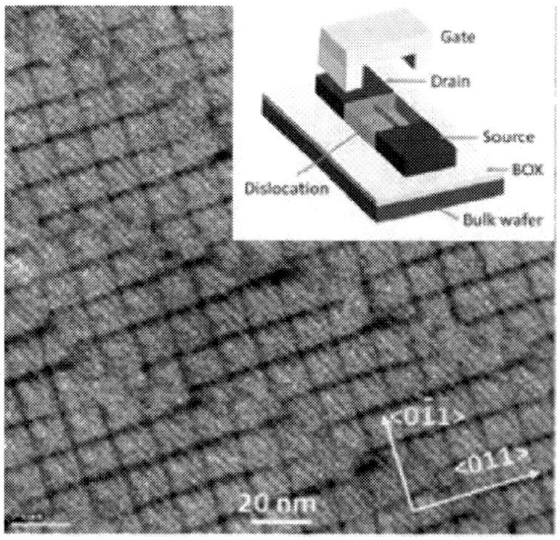

Figure 1. Transmission electron microscope (STEM) plan-view image of a screw dislocation network formed by wafer bonding of Si(100) wafers. The inset shows a schema of a MOSFET with dislocation in the channel. The gate was lifted for clarity.

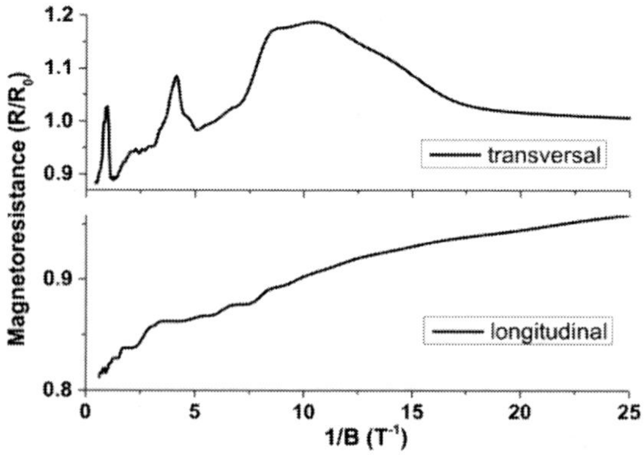

Figure 2. Transversal and longitudinal magnetoresistance vs. inverse magnetic field B. Measurement on an nMOSFET prepared on the network shown in Fig. 1 at T = 3.1 K (V_{DS} = 3V, V_{GS} = 0V).

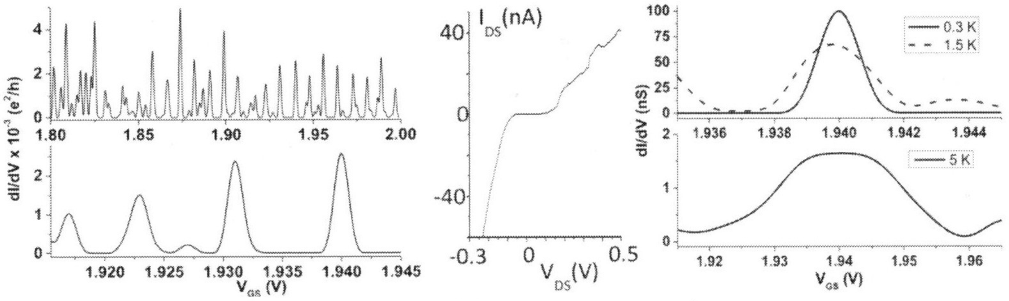

a) b) c)

Figure 3. Coulomb blockade oscillations as a function of gate-source voltage (V_{GS}) measured at T = 0.3 K and V_{DS} = 10 mV (a). The upper part shows that oscillations appear over large ranges of V_{GS}, while the lower part presents details. I_{DS}-V_{DS} characteristics at V_{GS} = 0 V and T = 0.3 K (b). Temperature dependence of the oscillations (c). Measurements at T = 0.3 K, 1.5 K, and 5 K.

MOSFETs [9, 10]. Furthermore, SdH oscillations appear also by applying a reverse bias (source-drain voltage V_{DS}). The asymmetry of the oscillations is frequently observed but different reasons are discussed (for instance [10, 11]). While dislocations represent nanowires, the asymmetry may be attributed to the fact that field-dependent oscillations are extremely sensitive to the alignment between the field and wire axis [12]. All the data suggest the presence of a two-dimensional electron gas (2DEG) on dislocations in the channel of nMOSFETs

As a consequence of the 2DEG, single electron transitions (Coulomb blockades oscillations) are observed for nMOSFETs containing dislocations in the channel. Fig. 3a shows an example of equidistant oscillations in V_{GS}, which is typical for a single-island system where each peak corresponds to the addition of one extra electron onto the island. The current-voltage (I_{DS}-V_{DS}) characteristics illustrates Fig. 3b. At T = 0.3 K a nonohmic semiconductor-like behavior with zero conductivity in the low voltage limit exists. Devices prepared on a dislocation network as in Fig. 1 were used for measurement. The period (ΔV_G) of the oscillations is about 9 mV (Fig. 3a) resulting in a gate capacitance $C_G = e/\Delta V_G = 18$ aF. Variations of the amplitude indicate the quantum regime which can be also deduced from temperature-dependent measurements (Fig. 3c). Plots of full-width-half-maximum (FWHM) of the peak width vs. increasing temperature result in a linear dependence with a slope of the curve equal to 3.5 kT [13]. Furthermore, a conversion factor [14]

$$\alpha = \frac{5kT}{\Delta V_{FWHM}} \operatorname{arc\,cosh} \sqrt{2} \cong 0.22 \tag{1}$$

is extracted from temperature-dependent measurements, resulting in an energy level spacing between discrete states along the wire axis $\Delta E = \alpha \cdot \Delta V_{GS} = 2$ meV which is close to data reported from grown silicon nanowires [15].

Further information about Coulomb blockades are obtained by applying external magnetic fields. Fig. 4 shows the evolution of Coulomb blockade oscillations in dependence on the strength of magnetic fields (B) oriented perpendicular (top) and parallel to the dislocation plane (bottom). A screw dislocation network with dislocation spacing of about 10 nm was used. Without a magnetic field (B = 0 T), periodic oscillations are observed (ΔV = 54 mV). Applying a magnetic field perpendicular to the dislocation plane does not change the oscillation distance and peak width for both sets of orthogonal dislocations. Higher bias conditions, however, decrease the amplitude of the oscillations. On the other hand, an increase of the amplitude is found for Coulomb blockade oscillations proved by measurements of devices containing the orthogonal set of screw dislocations of the network. This cannot be explained by an increase of diffusion-controlled transport to the dominant tunneling at low bias conditions. Instead, a different modification of the spatial extent of the ground state wave function by the magnetic field is assumed which causes changes of the level width and density, resp. [16].

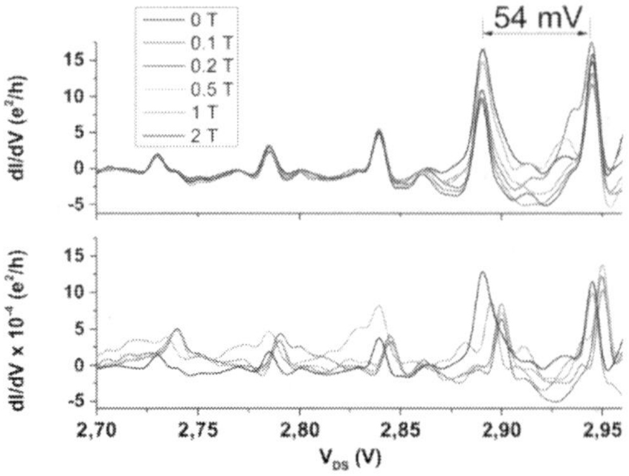

Figure 4. Evolution of Coulomb blockade oscillations with magnetic field perpendicular (top) and parallel to the dislocation plane (bottom).The strength of the magnetic field is indicated in the figure. Measurements at T = 3.1 K.

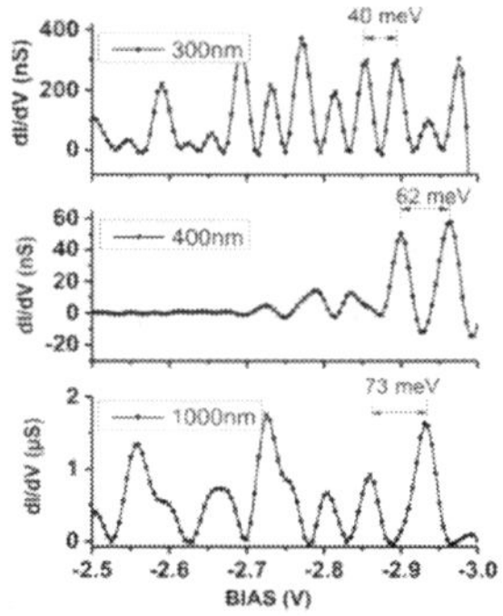

Figure 5. Dependence of the Coulomb blockade oscillations on the number of parallel dislocations. Increasing the channel width (indicated in figures) increases the number of dislocations. The same dislocation network is used. Measurements at T = 4.3 K.

Characteristic changes of the oscillations appear by applying a magnetic field parallel to the dislocation plane (Fig. 4, bottom). Increasing B results not only in shifts of the oscillation peaks, but also in their splitting. The shifts of the oscillation peaks increase for $B \leq 0.5$ T and are constant for higher values of B. The maximum shift is about 10 mV but is not constant for all peaks. Similar values of the shifts of oscillation peaks have been also measured for the orthogonal set of dislocations. Furthermore, peak splitting is observed with varying magnetic field. Both, the peak shift and peak splitting are caused by the fact that the chemical potential of a quantum dot in a parallel magnetic field has two additional terms [17], the diamagnetic shift ($\gamma_N B_{\parallel}^2$) describing the shift of the energy levels due to a squeezing of the wave functions, and the Zeeman splitting

$$\Delta E_{Zeeman} = \pm \frac{1}{2} g^* \mu_B B_{\parallel} \quad . \tag{2}$$

Here , γ_N denotes the effective diamagnetic factor, g^* the effective g factor, and μ_B the Bohr magneton. While the diamagnetic shift is proportional to $B_{\parallel}^2$, the Zeeman splitting is linear in $B_{\parallel}$.

The periodicity of the Coulomb blockade oscillations depends not only on the dislocation type but also on the number of parallel dislocations in the device channel. Fig. 5 shows the oscillations of three devices prepared on the same dislocation network. The width of the device channel was varied from 300 nm to 1000 nm, which means an increase of the number of dislocations by more than a factor of three. Note that the dislocation spacing is the same in all cases. At the smallest channel width (300 nm, corresponding to about 15 dislocations), the period of the oscillations is 40 mV and increases by a factor of 1.5 to 62 mV if the channel width increases to 400 nm (i.e. about 26 dislocations). A further increase of the channel width to 1000 nm (corresponding to about 67 dislocations) increases the period to 73 mV. Increasing the number of dislocations increases not only the period of the Coulomb blockade oscillations but also their amplitude. Both refer to interactions between the quantum dots on different dislocations.

The formation of a 2DEG along individual dislocations lines is a direct consequence of the high strain level on the dislocation core exceeding 10 %, or more [7]. In the case of a screw dislocation, discussed here, the strain is tensile. Such high strain levels result in dramatic changes of the electronic band structure. Band structure calculations using an 8-band-**k**·**p**-model and an empirical $sp^3d^5s^*$ tight binding model clearly showed that the effective electron mass shrinks and the band gap at the Γ-point is reduced at strain levels of $\varepsilon \geq 10$ % [18]. The band gap at the Γ-point reaches the same value as for the indirect band gap between the Γ- and Δ-point. This means that Si changes into a direct semiconductor and the dislocation forms a quantum wire structure where electrons are confined in the direction of the wire axis.

4. Conclusions

Dislocations are one-dimensional crystal defects in a perfect crystalline matrix. Their properties characterize the defects as one-dimensional nanowires. In particular, the strain in the dislocation core causes changes of the band structure resulting in the formation of a quantum well along the dislocation line. Within the well, quantization of energy levels exists where the energy levels depend on the width (or diameter) of a dislocation. A consequence of the quantum well is the formation of a 2DEG for two-dimensional dislocation networks and the quantization of energy levels induces the formation of Coulomb blockades. The measured distances of Coulomb blockades corresponds to the lowest energy levels if widths of the blockades of about 5 nm are assumed in calculations [18].

References

[1.] Cao G and Wang Y 2011 *Nanostructures and nanomaterials - synthesis, properties, and applications*. 2nd ed (Singapore: World Scientific Publ.).
[2.] Zhang R-Q 2014 *Growth mechanisms and novel properties of silicon nanostructures from quantum-mechanical calculations* (Heidelberg: Springer).
[3.] Koshida N (editor) 2009 *Device applications of silicon nanocrystals and nanostructures* (New York: Springer).
[4.] Huang Y and Tu K-N (editors) 2013 *Silicon and silicide nanowires* (Boca Raton: CRC Press).
[5.] Schmidt V, Wittemann JV, Senz S and Gösele U 2009 Adv Mater **21** 2681.
[6.] Reiche M, Kittler M, Scholz R, Hähnel A and Arguirov T 2011 J Phys Conf Ser **281** 012017.
[7.] Reiche M, Kittler M, Erfurth W, Pippel E, Sklarek K, Blumtritt H, Haehnel A and Uebensee H 2014 J Appl Phys **115** 194303.
[8.] Reiche M and Gösele U 2012 Direct Wafer Bonding *Handbook of Wafer Bonding* eds P Ramm, JJ-Q Lu, MMV Taklo (Weinheim: Wiley-VCH) pp 81-100.
[9.] Kravchenko SV, Shashkin AA, Bloore DA and Klapwijk TM 2000 Solid State Commun **116** 495.
[10.] Haug RJ, von Klitzing K and Ploog K 1987 Phys Rev B **35** 5933.
[11.] Griffin N, Dunford RB, Pepper M, Robbins DJ, Churchill AC and Leong WY 2000 J Phys: Condens Matter **12** 1811-8.
[12.] Shim W, Ham J, Kim J and Lee W 2009 Appl Phys Lett **95** 232107.
[13.] Kouwenhoven LP, Marcus CM, McEuen PL, Tarucha S, Westervelt RM and Wingreen NS 1997 Electron transport in quantum dots *Mesoscopic electron transport* eds LL Sohn, LP Kouwenhoven, G Schön (Dordrecht: Kluwer).
[14.] Tilke AT, Simmel FC, Blick RH, Lorenz H and Kotthaus JP 2001 Progr Quant Electron **25** 97.
[15.] Zhong Z, Fang Y, Lu W and Lieber CM 2005 Nano Letters **5** 1143.
[16.] Aravind K, Lin MC, Ho IL, Wu CSK, W., Kuan CH, Chang-Liao KS and Chen CD 2012 J Nanosc Nanotechnol **12** 2509.
[17.] Rahimi-Iman A, Schneider C, Fischer J, Holzinger S, Amthor M, Höfling S, Reitzenstein S, Worschech L, Kamp M and Forchel A 2011 Phys Rev B **34** 165325.
[18.] Reiche M, Kittler M, Uebensee H, Pippel E, Haehnel A and Birner S 2014 submitted to Phys Rev B.

Journal of Physics: Conference Series **568** (2014) 052025 doi:10.1088/1742-6596/568/5/052025

Weak localization in ZnO:Ga and ZnO:Al thin films

O V Reukova[1,a], V G Kytin[1,b], V A Kulbachinskii[1,c], L I Burova[2], A R Kaul[2], A G Ulyashin[3]

[1]Physics Faculty, M.V. Lomonosov Moscow State University, 119991, Moscow, Russia
[2]Chemistry Department, M.V. Lomonosov Moscow State University, 119991, Moscow, Russia
[3]SINTEF Materials and Chemistry, P.O. Box 124, Forskingsveien 1, Blindern, 0314 Oslo, Norway

E-mail: [a]reukova@physics.msu.ru;[b]kytin@mig.phys.msu.ru; [c]kulb@mig.phys.msu.ru;

Abstract. The temperature dependences of the resistivity, magnetoresistance and the Hall effect of ZnO thin films doped with Ga and Al were investigated at low temperatures. According to obtained experimental data the dimensionality of the investigated films with respect to the weak localization theory changes from 2D to 3D with an increase of magnetic field and temperature. To describe the observed negative magnetoresistance under these conditions we derived a new expression for the weak localization correction to the conductivity. It was found that the obtained expression describes magnetoresistance of investigated films much better than all known expressions for negative magnetoresistence related to weak localization. The values of the electron diffusion length during the phase relaxation time of wave function were obtained by fitting of experimental magnetoresistance with derived expression. Obtained values are consistent with the applied approach.

Introduction

The theory of quantum correction to conductivity (QC) was developed to explain experimentally observed peculiarities in the low temperature behavior of the resistivity and negative magnetoresistance (NMR) in disordered conductors [1-3]. Typically NMR in such systems is described by correction to the conductivity due to weak localization (WL). The expressions for corrections to conductivity due to WL for two-dimensional (2D), quasi two-dimensional (quasi-2D) and three-dimensional (3D) systems and corresponding magnetoresistance are known and widely used [4-5]. But there are only few publications devoted to the analysis of weak localization phenomena for the systems with dimensionality intermediate between purely 3D and quasi 2D [6].

The dimensionality of electron system with respect to WL theory in absence of magnetic field is determined by the ratio of diffusion length for the phase relaxation time of wave function ($L_\varphi = \sqrt{D\tau_\varphi}$) to the least size of the sample (*d*) [4-5]. In the presence of magnetic field the value of magnetic length ($l_B = \sqrt{\hbar/eB}$) also should be taken into account [4-5]. The magnetic length decreases with an increase of magnetic field. If $L_\varphi >> d$ but $l_B << d$ the electron system will behave as 3D system with respect to the WL theory. Therefore expressions for quasi-2D or 3D electron system will not describe

[a] To whom correspondence should be addressed

the magnetoconductivity of the system in the range of magnetic fields where dimensionality transition caused by the decrease of l_B takes place. The accurate consideration of weak localization in the film should account the boundary effects as shown in Ref. [6]. The author of this ref. showed that the WL correction to the conductivity of semi-infinite samples and films always comprises 2D and 3D contribution. The formula for the magnetoconductivity of thick films related to WL and accounting both contributions was derived in Ref. [6]. However the expression for magnetoconductivity related to weak localization for the case of 2D-3D crossover is absent in available literature.
This research is devoted to the experimental and theoretical study of the weak localization phenomenon in heavily doped ZnO films with a crossover from quasi 2D to 3D induced by temperature or magnetic field.

Experimental
Investigated films were synthesized by the oxygen-assisted chemical vapor deposition method (MOCVD) and magnetron sputtering technique (MS). Films were deposited on monocrystalline substrates of R-saphire, Yittrium stabilized zirconia (YSZ) and on glass. The films differed in thickness and donor impurity concentration. Some characteristics and deposition parameters of studied films are listed in Table 1.

Table 1. Deposition parameters of studied films

Film	Deposition technique	Substrate	$T_{deposition}$ (^{0}C)	Thickness (nm)	Concentration impurity
R1	MOCVD	R-saphire	600	60	7.2 at% (Ga)
Y1	MOCVD	$Zr_2O_3(Y_2O_3)$	600	90	6.8 at% (Ga)
Al	MS	Glass	70	100[b]	2 wt% (Al)

According to the X-ray diffraction data all films deposited by oxygen-assisted MOCVD were epitaxial with orientation determined by substrate. The details on synthesis and structure analysis of films could be found in Ref. [7-8].
Resistivity and Hall effect of the films have been measured by the standard four probe method. Samples were made rectangular form with typical dimensions 8×3 mm. Plates of electrical contacts were made by evaporation of Au and copper wires were soldered to the plates. For the measurements samples were mounted in the cryostat where the temperature was varied from 4.2 to 293 K. The Hall effect and magnetoresistance were measured in magnetic fields up to 6 T using a superconducting solenoid.

Results and discussion
Temperature dependencies of resistivity for studied films are shown in figure 1. Resistivity of ZnO:Ga films as well as ZnO:Al film is a nearly temperature independent except a slight increase at helium temperatures. Such behavior of resistivity is specific for band electron transport in degenerated semiconductors and disordered metals.

Hall resistivity of all investigated films depends linearly on magnetic field in the measured range of the magnetic fields. The Hall effect data at 4.2 K enabled us to determine concentration (n) and mobility (μ) of electrons in films. The results are presented in table 2.

All films exhibit NMR. We used the values of n and μ to verify the fulfillment of Ioffe-Regel criterion in studied films: $k_F l >> 1$, where k_F is the Fermi wave vector of electrons, l is the electron mean free path. The criterion is satisfied for all investigated films therefore we can use the theory of QC for analysis of observed NMR.

[b] The thickness was estimated from the conditions of film deposition

We tried to approximate observed magnetoconductivity by the expressions for WL correction to the conductivity for quasi-2D, 3D electron system and by expression (1) for thick films derived in Ref. [6]

$$\Delta\sigma(B) = d \cdot \Delta\sigma_{3D}(B) + \frac{1}{2}\Delta\sigma_{2D}(B) \qquad (1)$$

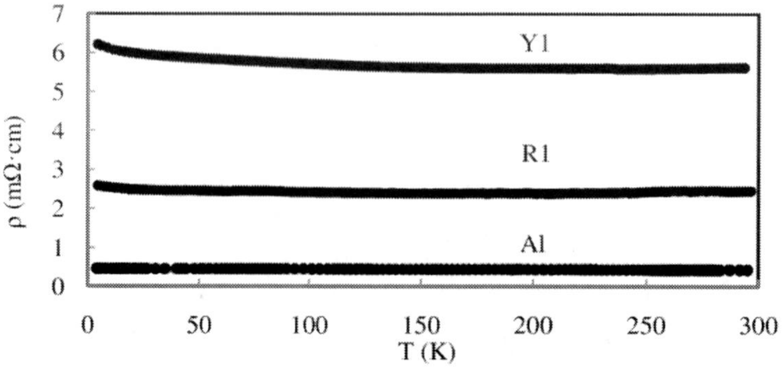

Figure 1. Temperature dependences of resistivity of ZnO:Ga and ZnO:Al films

Table 2. Calculated characteristics of investigated films at 4.2K

Film	n, 10^{19} (cm^{-3})	μ (cm^2/(V·sec))	E_F (meV)	l, nm	$k_F l$
R1	6.7	38	240	3.3	3.9
Y1	5.6	18	210	1.6	2.1
Al	140	9.8	1650	2.2	7.7

$\Delta\sigma_{3D}$,$\Delta\sigma_{2D}$ are WL correction to conductivity for quasi-2D, 3D electron system correspondingly. The expressions for σ_{3D}, σ_{2D} were taken from Ref. [9]. The ratio of parameter L_φ to the least size of the sample (d) determines the dimensionality of the system with respect to the WL theory. If $d<\lambda<<L_\varphi$, (λ is de Broglie wavelength), then electron system is considered to be 2D with respect to the WL theory, if $\lambda<d<<L_\varphi$, then it is quasi-2D and in case of $d>>L_\varphi$ it is 3D [4-5]. The parameter L_φ was determined by fitting of experimental MR data. The example of approximation of magnetoconductivity of the investigated film by expressions for WL correction to conductivity of quasi-2D, 3D electron system and expression for thick films (1) is shown in figure 2. The expression for quasi-2D system satisfactory approximates experimental data in low magnetic fields up to 0.7 T. Resulted value of L_φ=275 nm exceeds the thickness of the film d =100nm and thus satisfied to the applied quasi-2D WL theory. The calculated magnetoconductivity for 3D system goes lower than the experimental curve for all reasonable values of L_φ. Expression (1) satisfactory fits the experimental data, however resulted value of L_φ=467 nm is not consistent with applied theory, because obtained value of L_φ exceeds the thickness of the film (100nm), while the expression (1) is valid for $d>>L_\varphi$ [6].

The calculated magnetoconductivity of quasi-2D electron system deviates from the experimental dependence in high magnetic fields. The deviation of the theoretical dependence from the experimental one points to the magnetic field induced 2D-3D crossover of the dimensionality of electron system with respect to WL theory. To describe observed NMR in the whole range of magnetic fields we derived an expression for magnetoresistance which is valid for the arbitrary film thickness within diffusion limit.

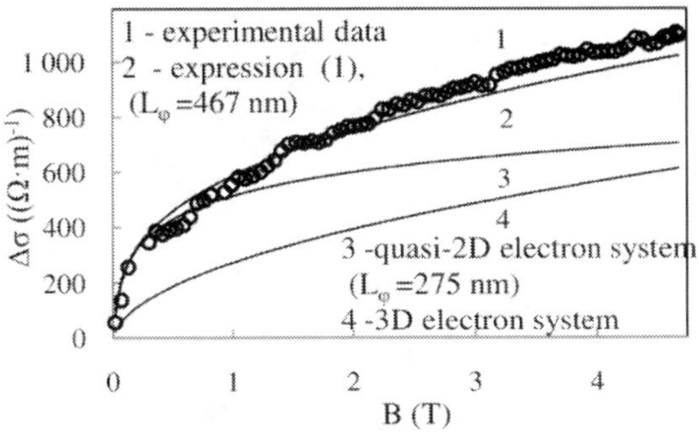

Figure 2. Experimental data (symbols) and calculated curves (line) for Al film at 4.2K

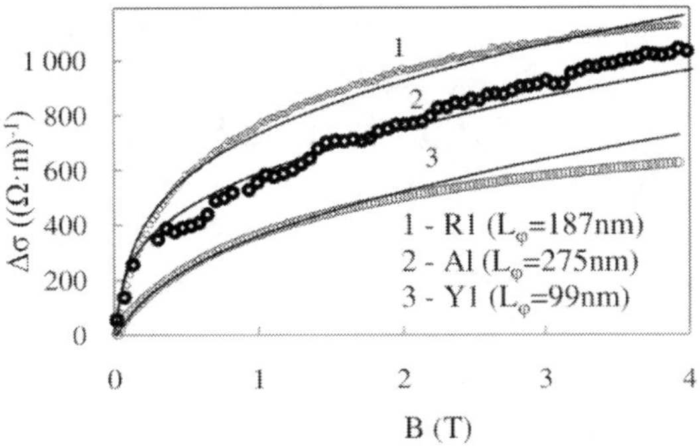

Figure 3. Magnetoconductivity of films experimental data (symbols) and fitting by (5) (line)

During this procedure we followed conventional approach developed in Ref. [4]. Namely the equation (2) for Cooperon *C(r,r,ω)* was solved with boundary conditions (3) :

$$(-i\omega + D(-i\vec{\nabla} - 2e/\hbar\,\vec{A})^2 + \tau_\varphi^{-1})C(\boldsymbol{r},\boldsymbol{r}',\omega) = \frac{\delta(\boldsymbol{r}-\boldsymbol{r}')}{\tau} \qquad (2)$$

$$(\nabla_z - 2ie/\hbar\,A_z)|_{z=0}C = 0$$

$$(\nabla_z - 2ie/\hbar\,A_z)|_{z=d}C = 0 \qquad (3)$$

The expression for QC and Cooperon are related as following [4]

$$\delta\sigma(r,\omega) = \frac{-2e^2D}{\pi^2\hbar}\tau C(r,r,\omega)$$

where $\vec{A}$ is a magnetic vector potential, ω is an external field frequency, D is an electron diffusion coefficient, τ is electron momentum relaxation time. Using mathematical technique described in [10] we found the expression for magnetoconductivity at $\omega=0$:

$$\Delta\sigma(B) = \frac{e^2}{2\pi^2\hbar}\sum_m\left[\ln x_m + \Psi\left(\frac{1}{x_m}+\frac{1}{2}\right)\right] \qquad (5)$$

where $x_m = \frac{\omega_c}{\tau_\varphi^{-1}+D(\frac{\pi m}{d})^2}$, $\omega_c = \frac{4eBD}{\hbar}$, m=0,1,...

Approximation of experimental magnetoconductivity measured at 4.2 K by formula (5) for investigated films and corresponding values of the fitting parameter L_φ are shown in figure 3. The expression (5) describes the experimental data better than all other expressions for magnetoconductivity related to WL. In contrast to the expression (1) the formula (5) could be applied for the analysis of films of arbitrary thickness and resulted values of L_φ are consistent with applied theory. We utilized expression (5) to analyze observed NMR of films R1 and Y1 at different temperatures (figure 4).

Values of L_φ are listed in figure 4. Formula (5) satisfactory approximates magnetoconductivity of the films at helium temperatures as well as at higher temperatures. With increase of the temperature the value of diffusion length L_φ decreases. This is consistent with the majority of the experimental data and theoretical predictions [3, 4].

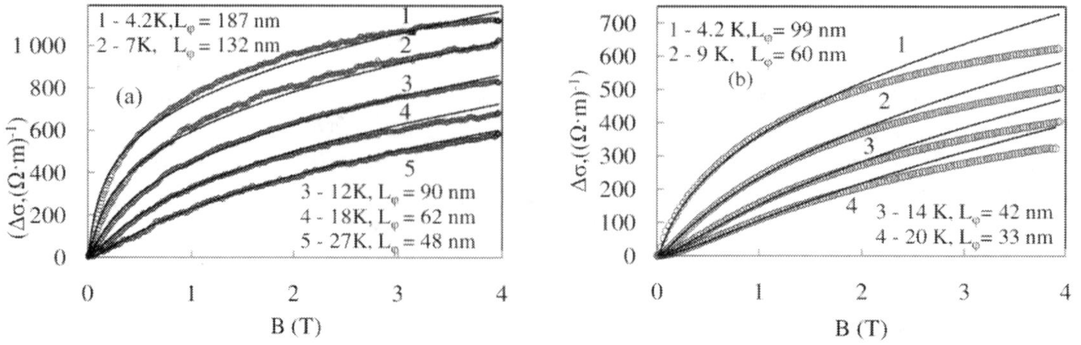

Figure 4. Magnetoconductivity at different temperatures (a) – R1 film, $d_{film}=90nm$, (b) – Y1 film, $d_{film}=60nm$ experimental data – scatter, approximation by (5) – solid line

The decrease of the value of L_φ is accompanied by the change of the dimensionality of electron system and at about T=20K the electron system becomes effectively 3D in respect to the WL theory. The deviation of calculated curve from experimental data for Y1 film at high magnetic field in figure 4 (b) has systematical character and most probably associated with the electron-electron interaction.

Summary

Temperature dependence of resistivity, magnetoresistance and Hall effect were studied for thin ZnO:Ga and ZnO:Al films. All investigated films exhibited negative magnetoresistane at low temperatures explained by quantum corrections to the conductivity. Analysis of the experimental results points to the magnetic field and temperature induced 2D-3D crossover of the dimensionality of electron system with respect to the WL theory. An expression for magnetoresistance caused by weak localization valid for arbitrary film thickness was derived. The obtained expression satisfactory describes observed NMR. Obtained values of the electron diffusion length during the phase relaxation time of electron wave function are consistent with the theory.

References

[1] Kawabata A 1980 *J Phys Soc Japan* **49** 628-637
[2] Altshuler B L, Khmel'nitzkii D, Larkin A I, Lee P A 1980 *Phys. Rev. B* **22** 5142
[3] Lee P A, Ramakrishnan T V 1985 *Rev Mod Phys* **57** 287-337
[4] Al`tshuler B L, Aronov A G 1985 *Electron-electron interaction in disordered systems* (*Modern Problems in Condensed Matter Sciences* vol 10) ed Efros A L, Pollak M (Amsterdam:Elsevier Science Publishers B V) p 1
[5] Polyanskaya T A, Shmartsev Yu V 1989 *Semiconductors* **23** 3-32
[6] Volkov V A 1982 *JETP Lett* **36** 394-396
[7] Kaul A R, Gorbenko O Yu, Botev A N, Burova L I 2005 *Superlattices and Microstructures* **38** 272
[8] Breivik T H, Diplas S, Ulyashin A G 2007 *Thin Solid Films* **515** 8479–8483
[9] Al`tshuler B L, Aronov A G, Larkin A I, Khmel`nitskii D E 1981 *Sov. Phys. JETP* **54** 411-419
[10] Rammer J 2007 *Quantum Field Theory of Non-Equilibrium States* (New York : Cambridge University Press) p 423

Ferromagnetic Kondo lattice model with Rashba coupling: effects of antiferromagnetic coupling between localized spins.

Giovany A. Meza and José A. Riera
Instituto de Física Rosario y Departamento de Física,
Universidad Nacional de Rosario, Rosario, Argentina

E-mail: riera@ifir-conicet.gov.ar

Abstract. Motivated by emergent phenomena at oxide surfaces and interfaces involving transition metal oxides, we examine the ferromagnetic Kondo lattice model (FKLM) in the presence of a Rashba spin-orbit coupling (RSOC) at zero temperature using numerical techniques. We find that the main effect of the RSOC is the destruction of the ferromagnetic state present in the FKLM at low electron fillings, with the consequent suppression of conductivity. In addition, near half-filling the RSOC leads to a departure of the antiferromagnetic state of the FKLM with a consequent reduction to the intrinsic tendency to electronic phase separation. A similar behavior is observed at very low electron density in the presence of an antiferromagnetic coupling between localized spins.

1. Introduction

So-called emergent electrodynamics phenomena at the interface between strongly correlated materials, and particularly transition metal oxides (TMO), or at the surface of such materials, have been revealed by a number of theoretical and experimental studies [1]. In essence, this exciting new physical phenomena is induced by the breaking of the inversion symmetry at the interface (or surface) itself. As a result of this broken symmetry, and due to relativistic arguments, it appears the so-called Rashba effect [2, 3, 4], which describes various momentum-dependent spin-splitting processes, including spin currents and the spin-Hall effect [5, 6]. As emphasized in a rather extensive literature, the RSOC leads to new applications in spintronics [7] since in an appropriate device this coupling can also be tuned by an external gate voltage.

Various types of interfaces between TMOs where RSOC is present have been studied, mostly involving $SrTiO_3$ [8, 9], but also $LaMnO_3/SrMnO_3$ interfaces [10] have been considered. The contribution of gate tunable RSOC in addition to the in-plane RSOC of broken symmetry origin, has been measured in $LaAlO_3/SrTiO_3$ interfaces [9]. Experimental indications of RSOC have also been reported at surfaces in $SrTiO_3$ [11].

The microscopic description of several transition metal oxides and heavy fermions is achieved through generalizations of the ferromagnetic Kondo lattice model [12, 13, 14, 15, 16] (FKLM) also called the double exchange model particularly when the Hund coupling is much larger than the hopping integral. For a large Hund coupling, it is known that there is a metallic ferromagnetic (FM) phase up to a filling $\nu \sim 0.8$, followed by a tendency towards an antiferromagnetic (AFM) state with semiconducting or insulating properties [17, 18, 19, 20]. Another important issue that has been thoroughly discussed in the context of manganites is the presence of an instability towards phase separation (PS) [16, 19], which will be also examined here.

In this work we will continue the study of the RSOC effects on the magnetic and transport properties in the single-orbital 2D FKLM, which was started in Ref. [21]. Of course, a more realistic model would take into account the fact that in TMOs with perovskite structure, the originally five-fold degenerate 3d orbitals are split into three-fold degenerate t_{2g} orbitals xy, yz, zx and two-fold degenerate e_g orbitals $x^2 - y^2, 3z^2 - r^2$. There is additional complexity at the interface such as dynamical transfer of electrons from the bulk and location of the interface layer [22, 23, 24]. By starting with the study of a single orbital model, we follow the program which has been pursued for example in the study of manganites and on the competition between the Rashba and Hubbard couplings [25]. In the present work we take a step towards a more realistic model by including an antiferromagnetic coupling between localized spins.

2. Model and method

The Rashba-FKLM for a single delocalized orbital coupled to classical localized spins is defined by the Hamiltonian $H_{1o} = H_0 + H_{int}$, where:

$$\begin{aligned} H_0 &= -t_0 \sum_{<l,m>,\sigma} (c^\dagger_{l\sigma} c_{m\sigma} + H.c.) + \lambda_{SO} \sum_l [c^\dagger_{l+x\downarrow} c_{l\uparrow} \\ &- c^\dagger_{l+x\uparrow} c_{l\downarrow} + i(c^\dagger_{l+y\downarrow} c_{l\uparrow} + c^\dagger_{l+y\uparrow} c_{l\downarrow}) + H.c.] \\ H_{int} &= -J_H \sum_l \mathbf{S}_l \cdot \mathbf{s}_l + U \sum_l n_{l\uparrow} n_{l\downarrow} + J \sum_{<l,m>} \mathbf{S}_l \cdot \mathbf{S}_m \end{aligned} \quad (1)$$

The first term in the noninteracting part H_0, is the usual hopping term, $H_{0,hop}$, and the second one corresponds to the RSOC, $H_{0,SO}$, assuming a square lattice in the xy plane (z is the spin quantization axis) [26]. The first term in the interacting part of the Hamiltonian H_{int} is the ferromagnetic Hund term, H_H, between localized $\mathbf{S}_l$ and conduction electron $\mathbf{s}_l$ spins. The second term is the Hubbard repulsion between conduction electrons, H_U, and the last one corresponds to the AFM exchange Hamiltonian between localized spins, H_J. We set $U = 0$, which is a reasonable assumption at least for $SrTiO_3$ [22, 23]. We normalize the hopping and RSOC parameters as $t_0^2 + \lambda_{SO}^2 = 1$ whose square root will be henceforth adopted as the unit of energy. The solely parameters left are then the ratio λ_{SO}/t_0, J, and the electron filling $\nu \equiv N_e/N$ ($N \equiv L \times L$).

In this work, we will employ a Monte Carlo technique that is based on the assumption that the localized spins are described by classical continuum spins $\mathbf{S}_l = (S, \theta_l, \phi_l)$ in spherical coordinates [16, 17]. In the present study, we resort to the so-called perturbative Monte Carlo (PMC) [27, 28], in which the full Hamiltonian is only diagonalized after a sweep on the whole lattice. Since in this work we limit ourselves to study zero temperature properties, the Monte Carlo simulation is reduced to a simple optimization procedure, the so-called simulated annealed optimization, in which a Boltzmann weight is used with a parameter that plays the role of "temperature". Further details can be found in Ref. [21].

3. Results

3.1. Magnetic properties

In the pure FKLM, various types of magnetic order have been detected in the J_H-density phase diagram in 2D, and this variety is even richer in the presence of an AFM interaction between localized spins J [17, 18]. As we show in this Subsection, this is also the case when λ_{SO} is turned on, even at a fixed $J_H = 10$. To understand the magnetic behavior, the spin-spin correlations between localized spins, $C(\mathbf{r}) = \langle \mathbf{S_r} \cdot \mathbf{S_0} \rangle$ ($\mathbf{0}$ is the reference site), and their Fourier transform leading to the static magnetic structure function $\chi(\mathbf{k})$, have been computed.

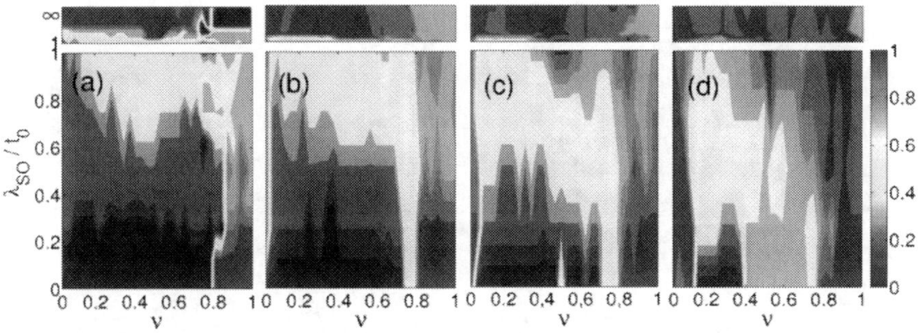

Figure 1. Magnetic phase diagram in the electron density-λ_{SO}/t_0 plane determined from the modulus of the peak of the magnetic structure factor, where dark blue corresponds to FM order ($|\mathbf{k}_{peak}| = 0$) and dark red to AFM order ($|\mathbf{k}_{peak}| = 1$). From left to right, $J = 0$, 0.01, 0.025, and 0.05. Results for the 8×8 cluster with open BC.

The RSOC leads to a very rich magnetic landscape. Let us start by examining the results for $\chi(\mathbf{k})$, in the density-λ_{SO}/t_0 plane, and various values of J. In figure 1 we show the modulus of the peak of $\chi(\mathbf{k})$, $|\mathbf{k}_{peak}|$, which essentially describes the proximity to the FM state ($|\mathbf{k}_{peak}| = 0$), or to the AFM state ($|\mathbf{k}_{peak}| = 1$), its maximum value (in units of $\sqrt{2}\pi$) for the 8×8 cluster with open boundary conditions (BC).

For $J = 0$ (leftmost panel) and $\lambda_{SO} = 0$ there is a neat crossover from $\mathbf{k}_{peak} = (0, 0)$ to $\mathbf{k}_{peak} = (\pi, \pi)$ states at $\nu \approx 0.8$, consistently with previous studies. In the $\nu \gtrsim 0.8$ region, the behavior has been explained as a phase separated AFM-FM state [16] which we discuss below. As λ_{SO}/t_0 is increased, the peak of $\chi(\mathbf{k})$ departs from both FM and AFM states but remain close to them in the low and high electron density regions respectively, up to $\lambda_{SO}/t_0 \sim 2$. It is interesting to note that for very large relative values of the RSOC, $\lambda_{SO}/t_0 \gtrsim 4$ the situation is reversed, that is, the low (high) density region holds now a AFM (FM) state. As J is turned on, the region of FM is further shrunk, and the region with AFM order is extended to lower ν. Moreover, a new AFM region appears near $\nu = 0$. The region of FM order is suppressed by increasing λ_{SO}/t_0 similarly to what happens at $J = 0$.

3.2. Transport properties

The Drude weight D is calculated from the optical conductivity and the f-sum rule as:

$$\frac{D}{2\pi} = -\frac{\langle H_{0,x}\rangle}{2L} - \frac{1}{L}\sum_{n\neq 0}\frac{\langle \Psi_n|j|\Psi_0\rangle|^2}{E_n - E_0} \tag{2}$$

where $K_x \equiv -\langle H_{0,x}\rangle$ is the total kinetic energy of electrons along the x-direction [29].

The kinetic energy associated with the RSOC along the x direction, $K_{SO,x} = - < H_{0,SO,x} >$ per site, and the total Drude peak, D, are shown in figure 2 for various values of λ_{SO}/t_0 and J. These results were obtained after averaging over periodic BC and "strip" BC (open along the y-direction and periodic along the x-direction), and in the later case we impose a phase in the electron wave-function or "twist", $\Phi = 0, \pi/2$ and π in the appropriate units [21].

$K_{SO,x}/N$ shows an expected increase with both λ_{SO}/t_0 and J, in the later case due to suppression of FM order, which is unfavourable for RSOC (figure 2(a)). On the other hand, the Drude peak, within the dispersion of the data corresponding to different BC, is suppressed by increasing RSOC in a monotonous way, for $J \geq 0$ (figure 2(b,c)). For $J = 0$ and densities larger than $\nu \sim 0.75$, the Drude peak starts to decrease in a rather abrupt fashion (figure 2(b)). It has been noticed [16, 18, 21] that this suppression could be related to the Fermi level moving from

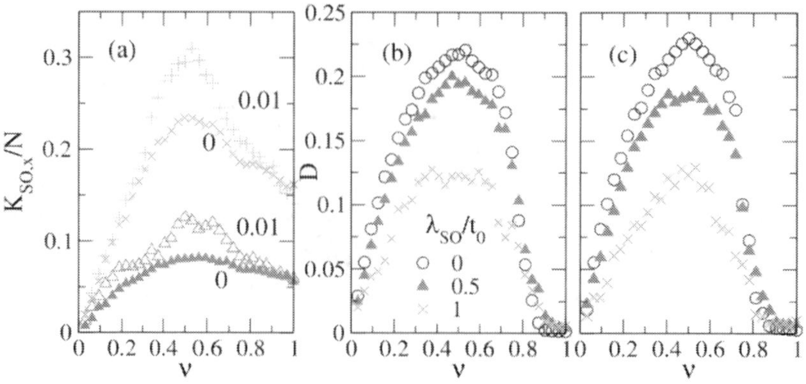

Figure 2. (a) Kinetic energy associated with the RSOC along the x direction, for two values of J as indicated in the plot. Total Drude peak for various values of λ_{SO}/t_0 and $J = 0$ (b) and $J = 0.01$ (c). Results for the 8×8 cluster averaged over periodic and twisted BC.

the bulk of the conduction band towards its high energy edge as ν is increased from $\nu \sim 0.75$, forming a pseudogap. Now, as J is increased, D starts to be suppressed at lower densities, as illustrated in figure 2(c) for $J = 0.01$, which is consistent with the reduction of the FM region shown in figure 1. At the same time another suppression of the Drude peak can be observed close to $\nu = 0$ where the FM order is also replaced by an AFM one by a positive J.

3.3. Phase separation

As emphasized in many previous studies on FKLM [16, 19], the behavior of magnetic and transport properties in the high density region can be understood by the presence of a phase separated state between AFM and FM orders. These two orders correspond to different stable electron fillings, one smaller than ~ 0.75 and the other equal to 1. The electron filling stability is determined by computing the so-called Maxwell construction.

Results for the 8×8 cluster with open BC are shown in figure 3(a) for $J = 0$ and various values of λ_{SO}/t_0. For $\lambda_{SO} = 0$, that is, for the pure FKLM, we recover the well-known PS [16, 19] which extends between densities $\nu = 0.72$ and 1. As λ_{SO}/t_0 is increased such PS state is gradually suppressed, for example for $\lambda_{SO}/t_0 = 1$, the largest PS state extends between $\nu = 0.82$ and 0.94. This PS region is further reduced by increasing λ_{SO}/t_0. As J is increased, first to 0.025 and then to 0.05 (figure 3(b) and (c) respectively) it can be seen that the PS region close to half-filling, for $\lambda_{SO} = 0$, extends to lower densities reaching $\nu = 0.6$ for $J = 0.05$. As λ_{SO}/t_0 is increased, this tendency to PS is suppressed. This behavior can be understood by the fact that the AFM effective exchange, which provides the attractive force leading to PS, is weakened and replaced by an effective stripe order with momentum close to (π, π) [21]. In addition, it should be noticed that close to $\nu = 0$, $\lambda_{SO} = 0$, another PS state appears for finite J, as originally reported in Ref. [17] for the FKLM. Similarly to what happens for high densities, the RSOC suppresses this PS state by the same mechanism discussed above. The PS region between $\nu = 0.14$ and $\nu = 0.42$ for $J = 0.05$ is explained by the fact that the two striped quasi-AFM states at the ends of this interval have lower energies than the FM states inside it. Then, the PS in this region is suppressed by the RSOC by raising the energies of the end AFM states.

4. Conclusions

In this work we have analyzed the interplay between the Rashba coupling on one side, and the hopping and Hund couplings that characterize the FKLM on the other. Near quarter-filling, the RSOC moves the system away from the ferromagnetic metallic state that is present in the pure

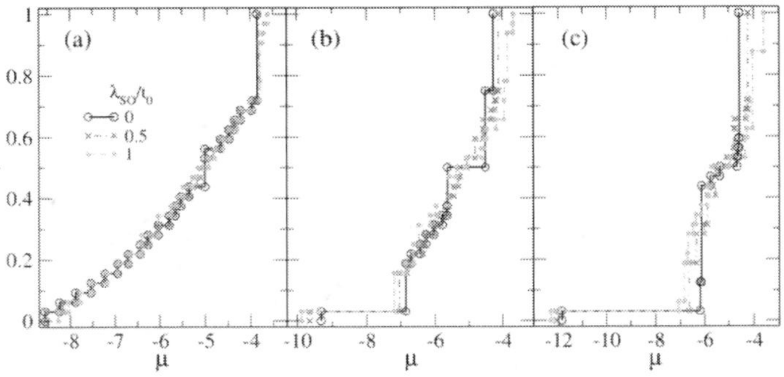

Figure 3. Density-chemical potential stability diagram for various values of λ_{SO}/t_0, and $J = 0$ (a), 0.025 (b), and 0.05 (c). Results for the 8×8 cluster with open BC.

FKLM, leading to a rich variety of magnetic states and to a loss of conductivity. These effects are enhanced in the presence of a $J > 0$. Near half-filling, on the other hand, the mechanism that favours an AFM order in the pure FKLM is no longer fully acting in the presence of the RSOC, and the system moves away from this order. As a consequence the presence of phase separation between AFM and FM regions in the FKLM is suppressed by the RSOC. A similar effect is observed near the empty band where in the presence of $J > 0$ a new AFM phase appears.

5. References

[1] Hwang H Y, Iwasa Y, Kawasaki M, Keimer B, Nagaosa N and Tokura Y 2012 *Nature Mater.* **11** 103
[2] Rashba E I 1960 *Sov. Phys. Solid State* **2** 1109
[3] Bychkov Y A and Rashba E I 1984 *JETP Lett.* **39** 78
[4] R. Winkler 2003 *Spin-orbit Coupling Effects in Two-Dimensional Electron and Hole Systems* (Berlin: Springer)
[5] D'yakonov M I and Perel V I 1971 *Sov. Phys. JETP* **33** 1053
[6] Hirsch J E 1999 *Phys. Rev. Lett.* **83** 1834
[7] Wolf S A, Awschalom D D, Buhrman R A, Daughton J M, von Molnar S, Roukes M L, Chtchelkanova A Y and Treger D M 2001 *Science* **294** 1488
[8] Caviglia A D, Gabay M, Gariglio S, Reyren N, Cancellieri C and Triscone J M 2010 *Phys. Rev. Lett.* **104** 126803
[9] Caprara S, Peronaci F and Grilli M 2012 *Phys. Rev. Lett.* **109** 196401
[10] Solovyev I V 2011 *Phys. Rev.* B **83** 054404
[11] Nakamura H, Koga T and Kimura T 2012 *Phys. Rev. Lett.* **108** 206601
[12] Zener C 1951 *Phys. Rev.* **81** 440
[13] Anderson P W and Hasegawa H 1955 *Phys. Rev.* **100** 675
[14] de Gennes P G 1960 *Phys. Rev.* **118** 141
[15] Riera J, Hallberg K and Dagotto E 1997 *Phys. Rev. Lett.* **79** 713
[16] Dagotto E 2002 *Nanoscale Phase Separation in Manganites* (Heidelberg: Springer-Verlag)
[17] Yunoki S, Hu J, Malvezzi A L, Moreo A, Furukawa N and Dagotto E 1998 *Phys. Rev. Lett.* **80** 845
[18] Moreo A, Yunoki S and Dagotto E 1999 *Phys. Rev. Lett.* **83** 2773
[19] Dagotto E, Hotta T and Moreo A 2001 *Phys. Rep.* **344** 153
[20] Santos C and Nolting W 2002 *Phys. Rev.* B **65** 144419
[21] Meza G A and Riera J A 2014 Magnetic and transport signatures of Rashba spin-orbit coupling on the ferromagnetic Kondo lattice model in two dimensions *Preprint* arXiv:1405.1240
[22] Khalsa G, Lee B and MacDonald A H 2013 *Phys. Rev.* B **88** 041302
[23] Bucheli D, Grilli M, Peronaci F, Seibold G and S. Caprara 2014 *Phys. Rev.* B **89** 195448
[24] Shanavas K V and Satpathy S 2014 *Phys. Rev. Lett.* **112** 086802
[25] Riera J A 2013 *Phys. Rev.* B **88** 045102
[26] Pareek T P and Bruno P 2002 *Phys. Rev.* B **65** 241305
[27] Duane S, Kennedy A D, Pendleton B J and Roweth D 1987 *Phys. Lett.* B **195** 216
[28] Kennett M P, Berciu M and Bhatt R N 2002 *Phys. Rev.* B **66** 045207
[29] Fye R M, Martins M J, Scalapino D J, Wagner J and Hanke W 1991 *Phys. Rev.* B **44** 6909

Effective tunneling processes in an interferometer of helical edge states with an antidot

Bruno Rizzo
Departamento de Fisica and IFIBA, Facultad de Ciencias Exactas y Naturales, Universidad de Buenos Aires, Pab. I, Ciudad Universitaria, 1428 Buenos Aires, Argentina

Alberto Camjayi
Departamento de Fisica and IFIBA, Facultad de Ciencias Exactas y Naturales, Universidad de Buenos Aires, Pab. I, Ciudad Universitaria, 1428 Buenos Aires, Argentina

Liliana Arrachea
Departamento de Fisica and IFIBA, Facultad de Ciencias Exactas y Naturales, Universidad de Buenos Aires, Pab. I, Ciudad Universitaria, 1428 Buenos Aires, Argentina

Abstract. We consider an interferometer of edge states of a two-dimensional topological insulator with an antidot. We analyze the mechanisms leading to an effective tunneling with spin flip between different helical states.

1. Introduction

Recent experiments in quantum wells of the two-dimensional topological insulator HgTe demonstrated a dissipationless charge transport through its helical edge states (HES) [1, 2, 3]. Due to the time reversal symmetry that holds in this material, the edge states come in Kramer's pairs [4, 5, 6], through which the electrons move with a defined spin associated with the direction of propagation [7].

The nature of the electron propagation in HES along with the topological protection against backscattering, make these materials very appealing to be employed for spintronics and quantum computation [8, 9]. This motivates the study of configurations of helical interferometers in setups including voltages and constrictions akin to those fabricated with quantum Hall edge states [10, 11, 12, 13, 14, 15, 16, 17, 18, 19]. So far, most of the studies were based on scattering at quantum point contacts connecting two HES [10, 11, 12, 13, 14, 16, 17, 19], while the effect of a wide tunnelling contact was also recently analyzed [15, 18]. Quite generally, the (local) time-reversal invariance only allows for the direct scattering between edge states situated at different sides of the sample, while the direct (back) scattering within the same Kramer's pair is forbidden even in the presence of a constriction. Therefore, the scattering at the point contact is characterized by two parameters, one describing a spin-preserving tunneling and one describing a spin-flip tunneling between two Kramer's pairs. However it is still unclear whether these effects are present in real setups. For this reason, in a previous work we have proposed a simple test

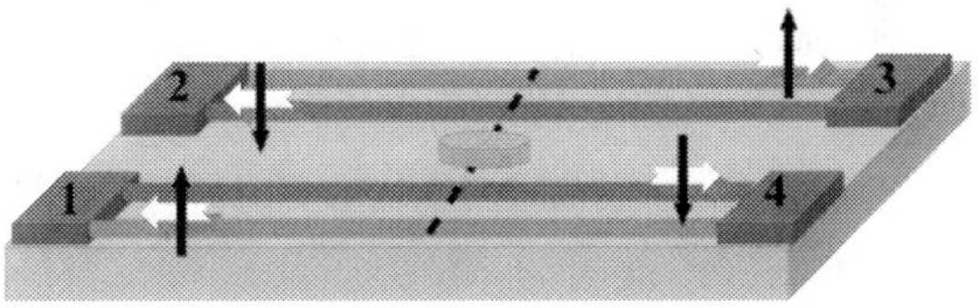

Figure 1. (Color online) Sketch of the helical edge states of a topological insulator with an antidot in the center. The pair of edge states in contact to the voltages $V_1, \ldots, V_4$ define the terminals $l = 1, \ldots, 4$.

based on the measurement of noise, as a concrete procedure to check the presence or absence of spin-flip tunneling between HES at the quantum point contact [19].

In interferometers fabricated with edge states of a two dimensional electron gas in the quantum Hall regime, the effects introduced by antidots in the bulk of the bar have been widely studied [20]. Antidots in quantum Hall systems are generated by applying a negative voltage on a surface gate, which generates an artificial repulsive impurity and a center for many-body interactions in local contact to the edge states. A similar type of structure can be also generated in a topological insulator, as sketched in Fig.(1). The aim of this work is to analyze the effect of an antidot in generating backscattering in a Kramer's pair of helical edge states as well as providing channels for inter-pair tunneling processes.

2. Model and theoretical framework

We model the topological insulator (TI) as a bar with infinite length sketched in Fig. 1. Each longitudinal edge of the TI hosts a Kramer's pair of HES with a given helicity. This is described by a Hamiltonian of one-dimensional (1D) free electrons with a definite spin and momentum. The TI is assumed to have an antidot inserted in the center of the bar, which hybridizes with the upper and lower edge states by means of spin-preserving tunneling processes. The electron levels of the dot also correspond to a pair of HES circulating around the small region of the sample where the surface gate is applied. For simplicity we model it by a single spin-degenerate level with a Coulomb repulsion energy for double occupancy. We also consider a local spin-flip process at the dot, representing a possible local tunneling within the pair induced by the surface voltage. Thus, the electrons propagating through a given edge channel of the TI could tunnel into the dot and tunnel again into another edge channel, with a different helicity. The full Hamiltonian reads $H = H_0 + H_d + H_t$. The Hamiltonian for the HESs is

$$H_0 = -i\hbar v_F \sum_{\sigma=\uparrow,\downarrow} \int dx [: \Psi^{\dagger}_{R,\sigma}(x)\partial_x \Psi_{R,\sigma}(x) : - : \Psi^{\dagger}_{L,\overline{\sigma}}(x)\partial_x \Psi_{L,\overline{\sigma}}(x) :], \quad (1)$$

where $\overline{\sigma}$ stands for the spin opposite to σ. We have assumed that a Kramer's pair of right-moving(R) with $\uparrow$ spin electron states and left-moving (L) with $\downarrow$ spin ones lie along the top edge of the bar, while another pair $L, \uparrow$ and $R, \downarrow$ lie on the bottom, as shown in Fig 1. The Fermi velocity v_F is assumed to be the same for the two pairs and $: O :$ denotes normal ordering.

The antidot (impurity) is described by the following Hamiltonian

$$H_d = \sum_{\sigma=\uparrow,\downarrow} \left[\varepsilon_0 d_\sigma^\dagger d_\sigma + \frac{U}{2} n_\sigma n_{\bar\sigma} + R d_\sigma^\dagger d_{\bar\sigma} \right], \tag{2}$$

with ε_0 is the single energy level, $n_\sigma = d_\sigma^\dagger d_\sigma$ and U represents the Coulomb interaction. We also consider a spin-flip scattering process inside the dot represented by the last term of Eq.(2) [21]. Finally the tunneling between the quantum dot located at x_0 and the HESs is,

$$H_t = \sum_{\alpha=R,L} \sum_{\sigma=\uparrow,\downarrow} \int dx \Gamma(x) [\Psi_{\alpha,\sigma}^\dagger(x) d_\sigma + H.c.], \tag{3}$$

where $\Gamma(x) = \sqrt{a}\gamma\delta(x - x_0)$ and a is a characteristic length of the coupling region between the antidot and the HESs.

The current operator for electrons flowing into the terminal l reads $\hat{I}^l(x) = ev_F \left[\hat\rho_+^l(x) - \hat\rho_-^l(x)\right]$, where $\hat\rho_\pm^l(x)$ is the density operator for particles incoming (outgoing) $l+$ ($l-$) the terminal l. The associated indices α, σ, with $\alpha = L, R$ and $\sigma =\uparrow, \downarrow$, depend on the terminal under consideration. For example, for $l = 3$, the $l+$ indices are $R, \uparrow$. The mean value of this current can be evaluated by recourse to the non-equilibrium Green's function formalism presented in Ref. [19], by introducing suitable generalizations to treat the Coulomb interaction U at the dot. We find for small bias voltage difference V at the terminals [22],

$$I^l(x) = I_0^l - I_{sp}^l - I_{sf}^l - I_b^l. \tag{4}$$

The first term, $I_0^l = \frac{e}{\hbar}\int_{-\infty}^{+\infty} \frac{d\omega}{2\pi} \left[f_{\alpha\sigma}(\omega) - f_{\bar\alpha\bar\sigma}(\omega)\right]$ is the perfect ballistic current induced in the terminal l in the absence of the tunneling to the dot. The other terms account for the contact between the HESs and the dot and tend to decrease of the current I^l with respect to the ballistic limit. We can distinguish different contributions, which describe, respectively, the tunneling to the wires in the opposite side of the bar preserving the spin (sp) and flipping the spin (sf). In addition, there is a backscattering contribution I_b^l which represents an effective resistance in the wire l. Explicitly, these terms read

$$\begin{aligned} I_{sp}^l &= \frac{e}{\hbar}\int_{-\infty}^{+\infty} \frac{d\omega}{2\pi} T_{\sigma\sigma}(\omega) \left[f_{\alpha\sigma}(\omega) - f_{\bar\alpha\sigma}(\omega)\right], \quad I_{sf}^l = \frac{e}{\hbar}\int_{-\infty}^{+\infty} \frac{d\omega}{2\pi} T_{\sigma\bar\sigma}(\omega) \left[f_{\alpha\sigma}(\omega) - f_{\alpha\bar\sigma}(\omega)\right], \\ I_b^l &= \frac{e}{\hbar}\int_{-\infty}^{+\infty} \frac{d\omega}{2\pi} T_{\sigma\bar\sigma}(\omega) \left[f_{\alpha\sigma}(\omega) - f_{\bar\alpha\bar\sigma}(\omega)\right], \end{aligned} \tag{5}$$

being

$$T_{\sigma\sigma'}(\omega) = (\sqrt{a}\gamma)^4 |g_{\alpha\sigma}^{0r}(\omega, x, x_0)|^2 |G_{\sigma,\sigma'}^r(\omega)|^2 \tag{6}$$

In the above expressions, $g_{\alpha\sigma}^{0,r}(x, x', \omega) = -\frac{i}{v_F\hbar}\Theta\left(s_\alpha(x - x')\right)\exp\left[i\frac{\omega}{v_F\hbar}(x - x')\right]$, with $s_\alpha = +1(-1)$ if $\alpha = R(L)$, are the retarded Green's functions of the isolated helical state, while $G_{\sigma,\sigma'}^r(\omega)$ is the Fourier transform of the retarded Green's function of the interacting dot in contact to the helical states $G_{\sigma,\sigma'}^r(t - t') = -i\Theta(t - t')\langle\{d_\sigma(t), d_{\sigma'}^\dagger(t')\}\rangle$.

3. Results

The retarded Green's function of the interacting quantum dot is calculated exactly in the non-interacting limit $U = 0$ and with quantum Monte Carlo by following the procedure of Refs. [23] in the symmetric Anderson model for finite U. In the latter case, the corresponding density of states $\rho_\sigma(\omega) = -2\text{Im}[G_{\sigma\sigma}^r(\omega)]$ is obtained by analytic continuation from the Matsubara to the

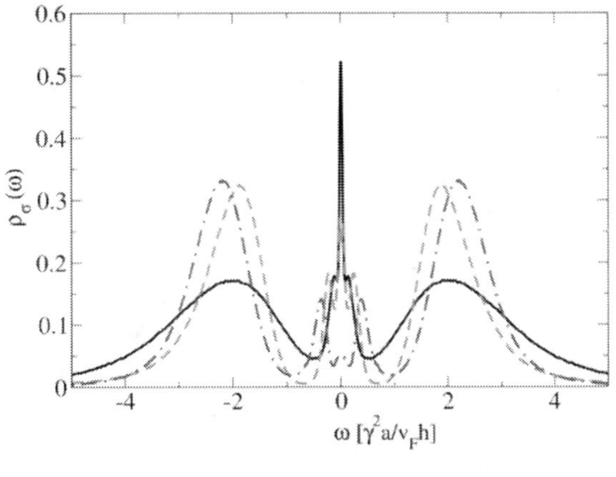

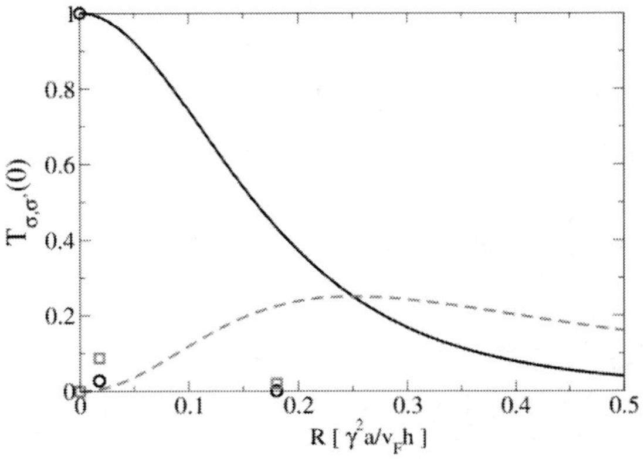

Figure 2. (Color online) a) The density of states ρ_σ ($\rho_\downarrow = \rho_\uparrow$) in the dot for three different values of spin-flip: $R = 0$ (black solid line),0.018 (red dashed line) and 0.18 (blue dashed-dotted line). $\varepsilon_0 = -U/2$, $T = 0.0025$, $U = 4$ and $\sqrt{a}\gamma = 0.5$. b) Transmission coefficient $T_{\sigma\sigma}(0)$ as functions of the amplitude of the spin-flip parameter R for $U = 0$ (black solid line) and for $U = 4$ (black open circles) and $T_{\sigma\overline{\sigma}}(0)$ as functions of R for $U = 0$ (red dashed line) and for $U = 4$ (red open squares) .

real frequency axis and typical results are shown in Fig. 2a. We see that for these parameters the Kondo peak at $\omega = 0$ is clearly distinguished for $R = 0$. The effect of R is to introduce strong spin fluctuations at the dot, which conspire against the constitution of the Kondo resonance. For this reason, the peak melts down for increasing R.

In order to study the effect of the different interactions at the dot on the transport properties of the interferometer wires, we analyze the transmission functions at the Fermi level of the wires, assumed at $\omega = 0$. Results are shown in Fig. 2b. The energy of the bare level is assumed to be $\varepsilon_0 = -U/2$, in order to have the impurity at half filling and the resonance aligned with the Fermi energy. We see that in the non-interacting case ($U = 0$), the transmission $T_{\sigma,\sigma}(0) = 1$ for $R = 0$ and decreases as this parameter increases. Instead, the transmission with spin flip shows the opposite behavior. It vanishes at $R = 0$ and increases with this parameter. We see that for $R = 0$ the behavior is the same for $U = 0$ and for $U = 4$. This is because for this value of the Coulomb interaction the dot is in the Kondo regime (see Fig. 2a) and the Kondo resonance opens a perfect transmission channel with $T_{\sigma,\sigma}(0) = 1$ and $T_{\sigma,\overline{\sigma}}(0) = 0$. For $R \neq 0$ the behavior is, however, clearly different for both values of U. In the non-interacting case, for increasing R, the spin preserving transmission decreases as the spin-flip one increases. This is also the case for $U = 4$ and small values of R, where the Coulomb interaction seems to enhance the spin fluctuations induced by R, in comparison to the non-interacting case. However, for large enough R, the Kondo resonance decreases and tends to disappear, which means a vanishing small spin preserving transmission and a spin-flip transmission which is much smaller than the non-interacting one.

4. Summary and Conclusions

We have studied scattering effects of an antidot coupled to helical edge states of a topological insulator. We have considered a local spin-flip mechanism as well as a Coulomb interaction. We have shown that these two interactions generate relevant effects in the transmission functions that define the electron transport through the helical edge states. These processes open new possibilities in the manipulation of electron currents in topological insulator interferometers, which may be useful for generating controlled orbital entanglement similar to the one proposed in Ref. [24].

References

[1] König M, Wiedmann S, Brüne C, Roth A, Buhmann H, Molenkamp L W, Qi X L and Zhang S C 2007 *Science* **318** 766
[2] König M, Buhmann H, Molenkamp L W, T Hughes, Liu C X, Qi X L, and Zhang S-C 2008 *J. Phys. Soc. Jpn.* **77** 031007
[3] Roth A, Brüne C, Buhmann H, Molenkamp L W, Maciejko J, Qi X L, and Zhang SC 2009 *Science* **325** 294
[4] Kane C L and Mele E J 2005 *Phys. Rev. Lett.* **95** 226801
[5] Wu C, Bernevig B A and Zhang S C 2006 *Phys. Rev. Lett.* **96** 106401
[6] Xu C and Moore J 2006 *Phys. Rev. B* **73** 045322
[7] Kane C L and Mele J E 2005 *Phys. Rev. Lett.* **95** 146802
[8] Prinz G A 1998 *Science* **282** 1660
[9] Wolf S A, Awschalom D D, Buhrman R A, Daughton J M, von Molnár S, Roukes M L, Chtchelkanova A Y and Treger D M 2001 *Science* **294** 1488
[10] Hou C Y Y, Kim E A, and Chamon C 2009 *Phys. Rev. Lett.* **102** 076602
[11] Ström A and Johannesson H 2009 *Phys. Rev. Lett.* **102** 096806
[12] Teo J and Kane C L 2009 *Phys. Rev. B* **79** 235321
[13] Schmidt T L 2011 *Phys. Rev. Lett.* **107** 096602
[14] Ferraro D, Wahl C, Rech J, Jonckheere T and Martin T 2014 *Phys. Rev. B* **89** 075407
[15] Dolcetto G, Barbarino S, Ferraro D, Magnoli N and Sassetti M 2012 *Phys. Rev. B* **85** 195138
[16] Dolcini F 2011 *Phys. Rev. B* **83** 165304
[17] Romeo F, Citro R, Ferraro D, Sassetti M 2012 *Phys. Rev. B* **86** 165418
[18] Sternativo P and Dolcini F 2014 *Phys. Rev. B* **89** 035415
[19] Rizzo B, Arrachea L and Moskalets M 2013 *Phys. Rev. B* **88** 155433
[20] Sim H S, Kataoka M and Ford C J B 2007 *Phys. Rep.* **456** 127; references therein
[21] Lopez R and Sanchez D 2003 *Phys. Rev. Lett.* **90** 116602
[22] Rizzo B, Camjayi A and Arrachea L, in preparation
[23] Camjayi A and Arrachea L 2012 *Phys. Rev. B* **86** 235143 ; ibid 2014 it J. Phys.: Condens. Matter **26** 035602
[24] Rizzo B, Arrachea L and Paz J P 2012 *Phys. Rev. B* **85** 045442

Weak localization and conductance fluctuations-like effects in Qubits driven by biharmonic signals

Alejandro Ferrón[1,2] **Daniel Domínguez**[3] **and María José Sánchez**[3]
[1]Instituto de Modelado e Innovación Tecnológica (CONICET-UNNE), 3400 Corrientes
[2]International Iberian Nanotechnology Laboratory (INL), Av. Mestre Jos Veiga, 4715-330, Braga, Portugal.
[3]Centro Atómico Bariloche and Instituto Balseiro, 8400 San Carlos de Bariloche, Argentina.

Abstract.
We investigate the effect of broken time reversal symmetry in flux qubits driven by a biharmonic magnetic flux signal with a phase lag. In the regime of large relaxation times, we explicitly compute the transition rate between the ground and the excited state, accounting for decoherence as a classical noise. Through a direct analogy between interference effects at the avoided level crossing and scattering events in weakly disordered electronic mesoscopic systems, the transition rate plays the role of an effective transmittance while the phase lag acts as a time reversal control parameter. Clear signatures of both weak localization and conductance fluctuations-like effects are predicted. Their behavior is studied as a function of the coherence rate, and a comparison with recent experimental results is performed. Our study shows that it is decoherence, and not the driving protocol, what limits the experimental detection of weak localization effects.

1. Introduction

Flux qubits (FQ) can be considered as artificial atoms whose energy levels are sensitive to an external magnetic flux. [1, 2] Its energy spectrum exhibits, as a function of the static magnetic flux, a rich structure of avoided crossings that can be explored by driving the FQ with a strong ac magnetic flux. This driving protocol -known as amplitude spectroscopy (AS) [3]- was successfully applied to reconstruct the FQ energy level spectrum [4, 5] and to estimate the qubit coherence times. [6] In the AS, the FQ is prepared in the ground state for a given value of the dc flux, and evolves quasi adiabatically under the driving, until at the first avoided crossing the state undergoes a transition and splits into a coherent superposition of transmitted and reflected amplitudes. In this way, each avoided crossing acts as an effective beam splitter in where a scattering event takes place. This analogy has been discussed in extent for FQ under weak driving, when the lowest two energy levels are explored and a single avoided crossing is attained by the amplitude of the ac flux.[3, 7]

Besides flux qubits, AS has been also employed in other systems, like charge qubits [12], ultracold molecular gases [13] and single electron spins. [14]

In addition, other phenomena relevant to quantum control- like population inversion- [15] have been tested with AS taking into account the coupling of the driven system to an environmental bath. [16]

One of the most efficient ways to probe coherent effects is by the phase sensitivity of the wave function to an external parameter that breaks time reversal symmetry. [18] In this way, fundamental effects like weak localization (WL) and universal conductance fluctuations (UCF) have been tested in a wide variety of electronic mesoscopic system for decades. [19, 20, 21] UCF describe the sample specific fluctuations of the quantum conductance due to the sensitivity of the scattering configuration to a change in a parameter, like an external magnetic field or the electron density.[19] The WL effect originates from the constructive interference of pairs of time-reversed trajectories that survives disorder average. The enhanced return probability results in a negative quantum correction to the classical conductance, characterized by a peak in the resistance (dip in the conductance) at zero magnetic field. To suppress the WL effect, a critical magnetic field $B_c \sim 1/(D\tau_\phi)$ -being D the diffusion coefficient and τ_ϕ the dephasing time- is required. This critical field washes out the constructive interference and thus the weak localization correction. [20]

While a priori there is not an obvious connection between driven two level systems (TLS) and the physics of disordered electronic systems, the analogy between transitions at the avoided crossing and scattering events, suggests a way to study fluctuation effects in FQ. Along this line Gustavsson and collaborators [7] investigated, using as a test system the lowest two levels of a FQ driven by a biharmonic ac flux, fluctuations in the transition rate W, similar to UCF. The driving was implemented by a biharmonic signal $f(t) = A_1 \cos(\omega t + \alpha) - A_2 \cos(2\omega t)$ of fundamental frequency $\omega = 2\pi/T$, that could be turned asymmetric in time by means of a phase lag α. In addition, as the driving period $T \sim 8$ ns was comparable to the qubit dephasing time $T_2 = 1/\Gamma_2 \sim 10$ ns, coherence was preserved within one period, although multiphoton resonances were not clearly resolved because the resonance width was comparable to the resonances spacing.[9] In the experiments of Ref.[7] the energy relaxation time $T_1 = 1/\Gamma_1 \sim 20\mu s$ was much longer than the driving period T. Under this regime population could build up in the excited state as a function of time, decaying exponentially to the stationary state with a characteristic rate Γ, which was estimated from experimental results by a fitting procedure.[7] Although the restriction to the lowest two levels in the FQ constrains the mesoscopic analogy to the few-scatterers limit, the scattering events could be modified by either tuning the external dc flux or by changing the phase lag parameter α. In this way, the qubit could be driving up to four times through a given avoided crossing in one period T, given rise to non trivial interference patterns (see Fig. 1 (a)).

In this work we will show that, besides the detection of fluctuations analogous to UCF [7], weak localization (WL) effects could also be experimentally observable for the biharmonic driving protocol, in the regime of large coherence times, $T_2 \gg T$.

2. FQ under biharmonic driving

2.1. *Transition rate and decoherence effects*

In this section we derive an explicit expression for the decay rate W for a TLS under biharmonic driving in the presence of decoherence sources. The analysis is restricted to the lowest two levels of the FQ, and disregards additional energy levels not explored in the experiment of Ref.[7]. We start by writing the two level hamiltonian H for a FQ subjected to a biharmonic driving in the flux. In the TLS the diabatic states carry a persistent current $\mp I_p$ and are coupled via the tunneling matrix element Δ. [3, 4, 9] The detuning from the avoided crossing is defined as $\varepsilon = 2I_p f$, with f the flux difference respect to half the superconducting flux quantum. [9] Under the biharmonic driving we define $\varepsilon(t) = \epsilon + A_1 \cos(\omega t + \alpha) - A_2 \cos(2\omega t)$.

Figure 1 sketches the energy level diagram of the TLS and the biharmonic drive waveform chosen to traverse the avoided crossing four times in one period T. To account for decoherence we follow the approach of Berns *et. al.* [9] and include classical diagonal noise, neglecting energy relaxation processes (consistent with the experimental regime $T \lesssim T_2 \ll T_1$). For $\hbar = 1$, the

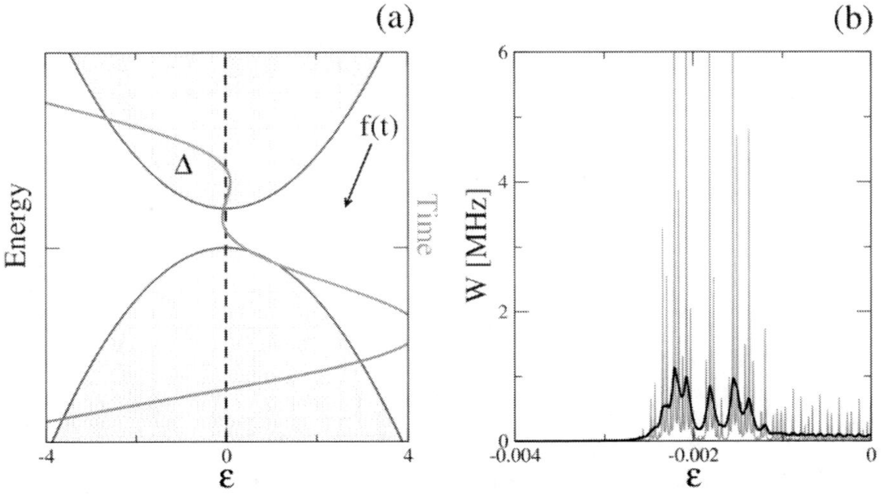

Figure 1. (color online) (a) Qubit energy diagram and biharmonic pulse for $\alpha = 0.4$ (b) Simulated transition rates W using modified Eq.(4) for qubit parameters $\Delta/h = 19\,MHz$, $\omega/(2\pi) = 125\,MHz$, $A_1 = 3\,m\Phi_0$, $A_2 = 1.65\,m\Phi_0$ and $\alpha = 0$ for $\Gamma_2 = 200\,MHz$ (black line) and $\Gamma_2 = 20\,MHz$ (red line).

hamiltonian H reads

$$H = -\frac{h(t)}{2}\hat{\sigma}_z - \frac{\Delta}{2}\hat{\sigma}_x$$

where $h(t) = \varepsilon(t) + \delta\varepsilon$ takes into account a classical fluctuating noise $\delta\varepsilon$ in the flux, and $\hat{\sigma}_z$, $\hat{\sigma}_x$ are Pauli matrices. After an unitary transformation, the hamiltonian can be written as

$$\tilde{H} = -\frac{1}{2}\begin{pmatrix} 0 & \Delta \\ \Delta^* & 0 \end{pmatrix} ,$$

where $\Delta(t) = \Delta e^{-i\phi(t)}$, and $\phi(t) = \int_0^t h(\tau)d\tau$. In the fast driving regime, where the driving frequency $\omega > \Delta$, the qubit population changes slowly on the time scale of T_2. In this way the time evolution operator can be expanded as

$$U(t, t_0) = 1 - i\int_{t_0}^{t} \tilde{H}(\tau)d\tau + \mathcal{O}(\Delta^2) ,$$

and the transition rate W between the ground and the excited state can be computed using perturbation theory,

$$W = \lim_{\delta t \to \infty} \frac{|A_{t,t'}|^2}{\delta t} \quad , A_{t,t'} = \frac{1}{2}\int_{t'}^{t} \Delta(\tau)d\tau . \tag{1}$$

In order to evaluate W we must solve

$$W = \frac{1}{4}\int_{t'}^{t}\int_{t'}^{t} \Delta(\tau_1)\Delta^*(\tau_2)d\tau_1 d\tau_2 . \tag{2}$$

We expand the oscillating exponentials in Bessel functions and average $\delta\phi(t) = \int_0^t \delta\varepsilon(\tau)d\tau$ for a white noise model $\langle e^{i\delta\phi(t)}e^{-i\delta\phi(t')}\rangle = e^{-\Gamma_2|t-t'|}$. In this way, we get after integration of Eq.(1)

$$W = \frac{\Delta^2}{2}\sum_{nn'mm'} J_n(x_1)J_{n'}(x_1)J_m(x_2)J_{m'}(x_2) \tag{3}$$

Journal of Physics: Conference Series **568** (2014) 052028 doi:10.1088/1742-6596/568/5/052028

$$e^{i(n-n')\alpha}e^{i\omega(n-n')t}e^{2i\omega(m-m')t}\frac{\Gamma_2}{(\varepsilon - n\omega - 2m\omega)^2 + \Gamma_2^2},$$

where we define $x_1 = A_1/\omega$ and $x_2 = -A_2/(2\omega)$. In the secular approximation, we neglect the fast oscillating terms, keeping only the $\omega(n-n') + 2\omega(m-m') \approx 0$ terms, obtaining

$$W = \frac{\Delta^2}{2}\Gamma_2 \sum_{nn'mm'} \lambda_{nn'mm'} \frac{\cos((n-n')\alpha)}{(\varepsilon - n\omega - 2m\omega)^2 + \Gamma_2^2}, \quad (4)$$

where $\lambda_{nn'mm'} = J_n(x_1)J_{n'}(x_1)J_m(x_2)J_{m'}(x_2)\,\delta_{n-n',2(m-m')}$. The last equation shows the explicit dependence of the transition rate W on the phase lag α. In addition, as will be shown in the next section, W will display strong fluctuations both with α and the dc flux detuning ϵ.

2.2. Transition and decay rates. Computing fluctuations

In the scattering approach the transition rate W is identified with an effective transmission amplitude -the essential ingredient to compute the electronic conductance within the Landauer formalism. [11] In addition, to complete the analogy with electronic transport in disordered mesoscopic systems, the parameter α plays the role of an effective magnetic (flux) field. In Ref.[7], the experimentally accessible quantity is the decay rate Γ. From phenomenological rate equations that describe the TLS dynamics, it is straightforward to show that $\Gamma = 2W + 2\Gamma_1$.[9] For large relaxation times T_1, it is satisfied that $\Gamma_1 \ll W$ and then $W \approx \Gamma/2$. Larger values of Γ_1 would require the explicit inclusion of relaxation processes in the analysis to avoid important differences between W and the experimentally measured Γ. [22] In the following we will keep our analysis consistent with the hypothesis $W \approx \Gamma/2$ and in this way, Eq.(4) provides an explicit expression to compute Γ. In Ref. [7], the specific value of the amplitudes ratio $A_2/A_1 = 0.55$ has been selected. In this way, the FQ has been driven up to four times through the avoided crossing in one period T, giving three different interference phases (one phase for two successive passages) and eight possible superposition states.

Figure 1 (b) shows the value of W obtained from Eq.(4) for the qubit parameters used in Ref.[7], i.e. $\Delta/h = 19\,MHz$, $\omega/(2\pi) = 125\,MHz$, $A_1 = 3\,m\Phi_0$ and $A_2 = 1.65\,m\Phi_0$ ($A_2/A_1 = 0.55$). The experimentally reported values of $T_{2_{exp}} = 10 - 20\,ns$ give $\Gamma_{2_{exp}} = 50 - 100\,MHz$. We explicitly set the phase lag $\alpha = 0$. W exhibits a clear pattern of oscillations as a function to the static flux detuning ϵ. These strong fluctuations are a manifestation of the sensitivity of the total phase accumulated during one period of the driving with the value of ϵ, as a consequence of the strong asymmetry in the driven signal. When analyzing the behavior of W as a function of the decoherence rate, we find that for small values of Γ_2, the transition rate exhibits high and sharp peaks that turn smoother and wider as the decoherence rate is increased. This behavior is fully consistent with the transition from the non overlapping to the overlapping resonances limit, also observed experimentally for the single harmonic driving protocols. [3, 4] In Fig. 2 we plot W from Eq. (4) for the same parameters used in Fig 1 (b) but for $\alpha \neq 0$. We can appreciate the results for $\alpha = 0.2$ (left panel, Fig. 2 (a)) and $\alpha = 0.4$ (right panel, Fig. 2 (b)), in both cases the structure of the peaks described in Fig.1 (b) is also obtained. Indeed, the asymmetry of the peaks as a function of the detuning ϵ is clearly observed.

For the chosen amplitudes ratio and depending on ϵ the wave form may drive the qubit through the avoided crossing zero, two or four times per cycle, producing different phase accumulations and interference conditions. In addition, the total phase accumulated- besides its dependence on ϵ- strongly depends on the biharmonic waveform, which is controlled by the asymmetry parameter α. To explore these effects, in Fig 3 we construct a map of the transition rate W as a function of the detuning and the phase lag. W exhibits in its interference patterns strong fluctuations, both as function of ϵ and α, in fully agreement with the fluctuations detected

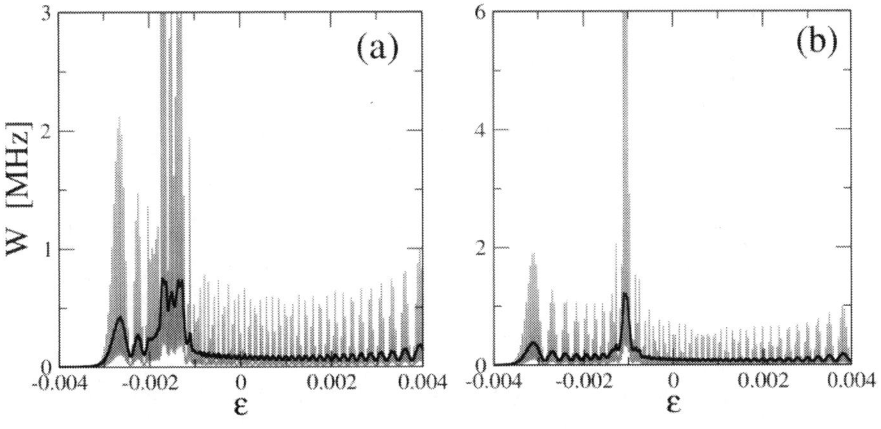

Figure 2. (color online) Simulated transition rates W using Eq.(4) for qubit parameters $\Delta/h = 19\,MHz$, $\omega/(2\pi) = 125\,MHz$, $A_1 = 3\,m\Phi_0$ and $A_2 = 1.65\,m\Phi_0$ for $\Gamma_2 = 200\,MHz$ (black line) and $\Gamma_2 = 20\,MHz$ (red line). (a) $\alpha = 0.2$ and (b) $\alpha = 0.4$.

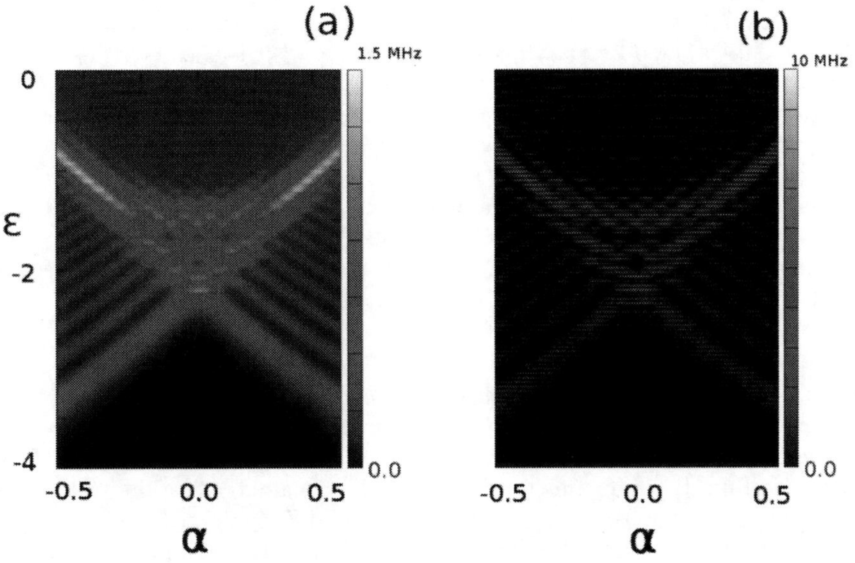

Figure 3. (color online) Contour plot of transition rate W using Eq.(4) for qubit parameters $\Delta/h = 19\,MHz$, $\omega/(2\pi) = 125\,MHz$, $A_1 = 3\,m\Phi_0$ and $A_2 = 1.65\,m\Phi_0$. (a)$\Gamma_2 = 200\,MHz$ and (b)$\Gamma_2 = 20\,MHz$.

experimentally in the FQ population interference patterns.[7] Another interesting feature is the strong effect of the decoherence rate on the results, as we already mentioned. We can appreciate important qualitative differences in the results for $\Gamma_2 = 200\,MHz$ (Fig. 3 (a)) and $\Gamma_2 = 20\,MHz$ (Fig. 3 (b).

As we show below, even after averaging over different values of the dc flux, strong fluctuations are still visible in W as a function of α. In order to evaluate them, we can compute the average $\langle W\rangle = \frac{1}{2\varepsilon^{max}}\int_{-\varepsilon^{max}}^{\varepsilon^{max}} W\, d\varepsilon$ and using Eq. (4) we can evaluate the integral, obtaining

$$\langle W\rangle = \frac{\Delta^2}{4\varepsilon^{max}} \sum_{nn'mm'} \lambda_{nn'mm'} \cos\left((n-n')\alpha\right) \tag{5}$$
$$\left[\arctan\left(\frac{n\omega+2m\omega+\varepsilon^{max}}{\Gamma_2}\right) - \arctan\left(\frac{n\omega+2m\omega-\varepsilon^{max}}{\Gamma_2}\right)\right] .$$

In the same way we can compute

$$\langle W^2\rangle = \frac{1}{2\varepsilon^{max}}\int_{-\varepsilon^{max}}^{\varepsilon^{max}} W^2\, d\varepsilon \; . \tag{6}$$

Although $< W^2 >$ does not have a simple analytic expression, its numerical calculation is straightforward. In Fig 4 we plot both the average transition rate $< W >$ and its fluctuations $\sigma = \sqrt{< W^2 > - < W >^2}$ for the experimental FQ parameters but for different values of the decoherence rate Γ_2. We can see that, for values of the decoherence rate similar to the experimental ones $\Gamma_2 \sim 20 - 200MHz$, $\langle W\rangle$ is almost independent of α and σ- characterized as the UCF- presents a well defined peak at $\alpha = 0$, in full agreement with Ref.[7].

The fingerprint of the WL effect should be a dip in $< W >$ at $\alpha = 0$ - a dip in the average transmission probability- following the electronic transport analogy. In the experiments of Ref.[7] this WL effect wast not observed. However, there is not an experimental constrain in the driving protocol that precludes the observation of WL. As we show, it is the decoherence time T_2 which limits the width of the WL peak, and its experimental detection. From the theory of disordered electronic systems [20] the value of the critical magnetic field needed to destroy the WL peak is $B_c \sim 1/(D\tau_\phi)$, being D the diffusion coefficient and $\tau_\phi \propto T_2$. In the present analysis, the critical value α_c needed to destroy the WL correction, should scale as $\alpha_c \sim 1/(T_2)$. Thus it is expected, that large decoherence times should be needed in order to experimentally detect the whole WL peak with accessible values of α_c. Consistent with this analysis, in fig. 4 we show that for $\Gamma_2 = 6\,MHz$ there is a well defined dip present in $\langle W\rangle$ at $\alpha = 0$. As a consequence, we can conclude that for long decoherence time ($T_2 \simeq 150\,ns$) WL should emerge in the experiments.

3. Conclusions and perspectives

In this work we have tested fluctuation effects associated to broken time reversal symmetry in FQ driven by a biharmonic ac magnetic flux. Employing a simplified model that accounts only for decoherence, we computed the transition rate W which displays strong fluctuations as a function of the dc flux detuning and the time reversal parameter α. The results presented are derived under the assumptions that $W \ll \Gamma_2 = {T_2}^{-1}$ and $T \lesssim T_2 \ll T_1$, and in agreement with the experimentally reported regime of Ref. [7]. However it is not difficult to attain experimental regimes in which some of the above conditions are not satisfied. In particular, for small driving frequencies or larger decoherence times, T_2, important difference could emerge between the transition rate W, Eq.(4), and the measured decay rate Γ. Besides conductance fluctuations-like effects, we also show that WL effect could be detected for the biharmonic driving protocol. However to observe this effect, the experiments should be performed in a more coherent regime, in which larger values of T_2 could be attained. By increasing the driving amplitude, more avoided crossings of the FQ can be reached, enabling the computation of averages and fluctuations with reasonable accuracy. This regime is experimentally attainable as the amplitude spectroscopy experiments have been proven. In addition the extension of the calculations for a realistic model that includes, besides decoherence, relaxation processes due to the interaction of the FQ with the circuitry environment is also under study. [22]

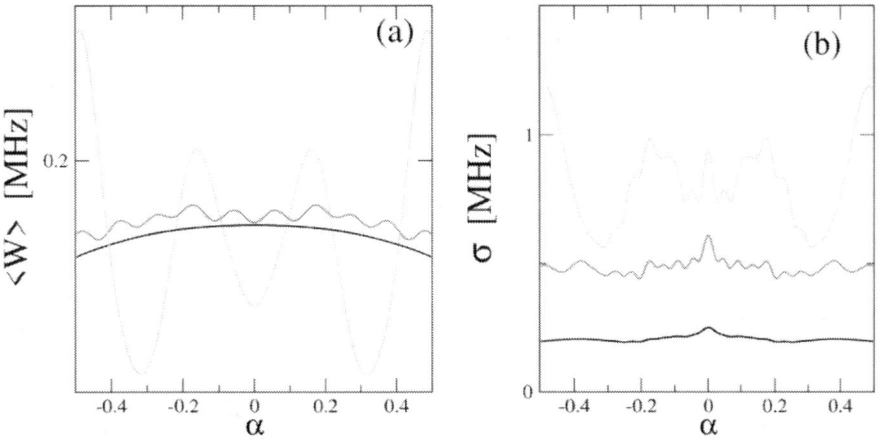

Figure 4. (color online) (a) Averaged transition rate $< W >$ and (b) σ for qubit parameters $\Delta/h = 19\,MHz$, $\omega/(2\pi) = 125\,MHz$, $A_1 = 3\,m\Phi_0$ and $A_2 = 1.65\,m\Phi_0$ calculated using Eq.(4), (5) and (6) for $\Gamma_2 = 200\,MHz$ (black line), $\Gamma_2 = 20\,MHz$ (red line) and $\Gamma_2 = 6\,MHz$ (green line). The averages were performed in the range $-4m\Phi_0$ to $4m\Phi_0$.

Acknowledgments
We acknowledge support from CNEA, CONICET and ANPCyT PICT2011-1537.

References

[1] Orlando T P *et al* 1999 *Phys. Rev.* B **60** 15398
[2] Chiorescu I *et al* 2003 *Science* **299** 1869
[3] Oliver W D *et al* 2005 *Science* **310** 1653
[4] Berns D M *et al* 2008 *Nature* **455** 51; Oliver W D and Valenzuela S O 2009 *Quantum Inf. Process* **8** 261 (2009)
[5] Ferrón A, Domínguez D and Sánchez M J 2010 *Phys. Rev.* B **82** 134522
[6] Shevchenko S N, Ashhab S and Nori F 2010 *Phys. Rep.* **492** 1
[7] Gustavsson S, Bylander J and Oliver W D 2013 *Phys. Rev. Lett.* **110** 016603
[8] You J Q and Nori F 2011 *Nature* **474** 589
[9] Berns D M *et al* 2006 *Phys. Rev. Lett.* **97** 150502
[10] Izmalkov A *et al* 2008 *Phys. Rev. Lett.* **101** 017003
[11] Datta S 1995 *Electronic transport in Mesoscopic Systems* (Cambridge: Cambridge University Press)
[12] Nakamura Y, Pashkin Y A and Tsai J S 2001 *Phys. Rev. Lett.* **87** 246601; Sillanpaa M *et al* 2006 *Phys. Rev. Lett.* **96** 187002; Wilson CM *et al* 2007 *Phys. Rev. Lett.* **98** 257003
[13] Mark M *et al* 2007 *Phys. Rev. Lett.* **99** 113201
[14] Huang P *et al* 2011 *Phys. Rev.* X **1** 011003
[15] Weber J 1959 *Rev. Mod. Phys.* **31** 681; Astafiev O *et al* 2007 *Nature* **449** 588; You J Q *et al* 2007 *Phys. Rev.* B **75** 104516
[16] Ferrón A, Domínguez D and Sánchez M J 2012 *Phys. Rev. Lett.* **82**,134522
[17] Ferrón A and Domínguez D 2010 *Phys. Rev. B* **81** 104505
[18] For a review, see Althsuler B L, Lee P A and Webb R A 1991 *Mesoscopic Phenomena in Solids* (Amsterdam: Elsevier)
[19] Benoit A *et. al.* 1986 *Phys. Rev. Lett.* **58** 2343; Washburn S and Webb R A 1992 *Rep. Prog. Phys.* **55** 1311 (1992)
[20] Bergmann G 1982 *Phys. Rev.* B **25** 2937 (1982); Bishop D J *et. al.* 1982 *Phys. Rev.* B **26** 773
[21] Akkermans E, Montambaux C, Pichard J L and Zinn-Justin J 1994 *Mesoscopic quantum physics* (Amsterdam: Elsevier)
[22] Ferrón A, Domínguez D and Sánchez M J 2014 in preparation

Journal of Physics: Conference Series **568** (2014) 052029 doi:10.1088/1742-6596/568/5/052029

Witten Effect and Fractional Electric Charge on the Domain Wall between Topological Insulators and Spin Ice Compounds

T Sasaki, E Imai, I Kanazawa

Department of Physics, Tokyo Gakugei University, Koganeishi, Tokyo 184-8501, Japan

E-mail: f107143f@st.u-gakugei.ac.jp

Abstract. We propose that there might be emergent quasiparticles with fractional electronic charge such dyons on the domain wall between topological insulators and spin ice compounds through the Witten effect and interaction between the Dirac fermions and excited magnetic monopoles.

1. Introduction

The Quantum Spin Hall state has been theoretically predicted [1] and experimentally observed in HgTe quantum wells [2]. Time-Reversal-Invariant topological insulators have also been classified in (3+1) dimension [3, 4]. These three-dimensional states carry spin Hall current in the insulating state [5]. Recently the three-dimensional topological insulator $Bi_{1-x}Sb_x$ in a certain range of compositions [6] has been predicted. In addition, Dirac fermions on the surface of three-dimensional topological insulator have been observed by angle-resolved photoemission spectroscopy [7]. Also Castelnovo *et al.* [8] have proposed that magnetic monopoles emerge in a class of exotic magnets known collectively as spin ice compounds $Dy_2Ti_2O_7$ and $Ho_2Ti_2O_7$. Spin ice is identified as a very unusual magnet when it is noted that it does not order to the lowest temperatures T even though it appeared to have ferromagnetic interaction. Indeed, spin ice is found to have a residual entropy at low temperature, which is well-approximated by the Pauling entropy for water ice. Recently, one of us [9] has proposed exotic quasiparticles with fractional charges in semiconductor-dot from collectively induced-charge effect on a domain wall shell. In this study, we propose that there might be emergent quasiparticles with fractionalized electric charge on the domain wall between topological insulator and spin ice compounds, through the Witten effect [10] and interaction between the Dirac fermions and excited magnetic monopoles.

2. Exotic particle on the domain wall between the topological insulator and spin ice

To study the response properties of the (3+1) dimension system, the effective action $S_{3D}(A, \theta)$ with an adiabatic parameter θ can be defined as

$$e^{iS_{3D}(A,\theta)} = \int D\psi D\overline{\psi} \exp\left[i \int dt \sum_{\vec{x}} \overline{\psi}_{\vec{x}}(i\partial_\tau - A_{\vec{x}0})\psi_{\vec{x}} - \mathcal{H}[A, \theta]\right].$$

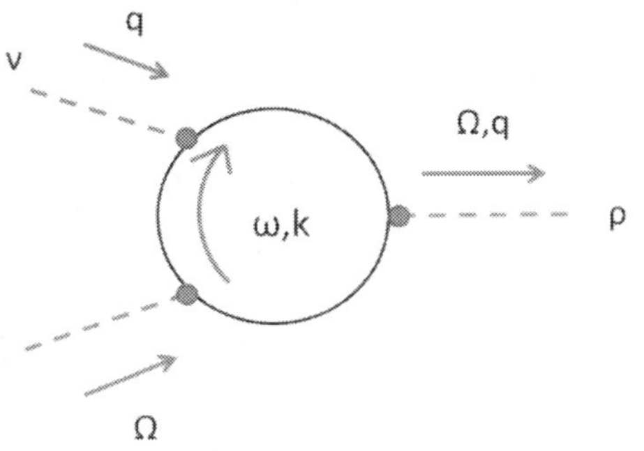

Figure 1. The Feynman diagram that contributes to topological term. The loop is a fermion propagetor and the dashed lines are external legs corresponding to the gauge field.

$\mathcal{H}(A, \theta)$ is the (3+1) dimensional Hamiltonian of Dirac model coupled to an external $U(1)$ gauge field A [11]. ψ is the wavefunction of the Dirac electron. Where $\vec{x}$ stands for the three-dimension coordinates. A Taylor expansion of S_{3D} can be carried out around the gauge field configuration $A_s(\vec{x}, t)$, whose $s = 1, 2, 3$ stands for the x,y,z directions, $\theta(\vec{x}, t) \equiv \theta_0$, which contains a nonlinear-response term derived from the (4+1) dimensional Chern-Simon action,

$$S_{3D}(A, \theta) = \frac{G_3(\theta_0)}{4\pi} \int d^3x\, dt\, \varepsilon^{\mu\nu\sigma\tau} \delta\theta \partial_\mu A_\nu \partial_\sigma A_\tau \ .$$

The field $\delta\theta(\vec{x}, t) = \theta(\vec{x}, t) - \theta_0$ plays the role of A_4 in the (4+1) dimension, and the coefficient $G_3(\theta_0)$ is determined by Goldstone and Wilczek-type Feynman diagram [12] in Fig. 1. Consequently, $G_3(\theta_0)$ can be calculated as follows.

$$G_3(\theta_0) = -\frac{\pi}{6} \int \frac{d^3k\, d\omega}{(2\pi)^4} Tr \varepsilon^{\mu\nu\sigma\tau} \left[\left(G\frac{\partial G^{-1}}{\partial q^\mu} \right) \left(G\frac{\partial G^{-1}}{\partial q^\nu} \right) \left(G\frac{\partial G^{-1}}{\partial q^\sigma} \right) \left(G\frac{\partial G^{-1}}{\partial q^\tau} \right) \left(G\frac{\partial G^{-1}}{\partial \theta_0} \right) \right]$$

where $q^\mu = (\omega, k_x, k_y, k_z)$ and $G(q_\mu) = [\omega + i\delta - h(k_i)]^{-1}$ is the single-particle Green's function. In addition, $G_3(\theta_0)$ is determined from the Berry phase curvature as

$$G_3(\theta_0) = \frac{1}{8\pi^2} \int d^3k\, \varepsilon^{ijh} Tr[f_{\theta i} f_{jh}]$$

in which the Berry phase gause field is defined in the four-dimensional space $(k_x, k_y, k_z, \theta_0)$, ie.,

$$a_i^{\alpha\beta} = -i \left\langle \vec{k}, \theta_0; \alpha | \left(\frac{\partial}{\partial k_i} \right) | \vec{k}, \theta_0; \beta \right\rangle$$

and

$$a_\theta^{\alpha\beta} = -i \left\langle \vec{k}, \theta_0; \alpha | \left(\frac{\partial}{\partial \theta_0} \right) | \vec{k}, \theta_0; \beta \right\rangle.$$

A generalized polarization $P_3(\theta_0)$ can also be defined in (3+1) dimensions so that $G_3(\theta_0) = \frac{\partial P_3(\theta_0)}{\partial \theta_0} = \int d^3k \partial_\theta K^\theta$, with

$$P_3(\theta_0) = \int d^3k\, K_\theta = \frac{1}{16\pi^2} \int d^3k\, \varepsilon^{\theta ijk} Tr \ \left(f_{ij} - \frac{1}{3}[a_i, a_j] \right) \cdot a_k \ .$$

The effective action for the (3+1)-dimensional system is finally written as

$$S_{3D}(A, \theta) = \frac{1}{16\pi^2} \int d^3x\, dt\, \theta_0(\vec{x}, t) \varepsilon^{\mu\nu\sigma\tau} \partial_\mu A_\nu \partial_\sigma A_\tau$$

In the case of the (3+1) dimensional topological insulator, $\theta_0(\vec{x},t)$ correspond to π. Here we shall consider the domain wall between the topological insulator and spin ice as shown in Fig. 2. We must consider the (3+1) dimensional Dirac fermion with a domain wall configuration of the $\theta_0(z)$ field given by

$$\theta_0(z) = \frac{\pi}{2}\left[1 - \tanh\left(\frac{z}{4\xi}\right)\right]$$

which has the asymptotic behavior $\theta_0(z \to -\infty) = \pi$, $\theta_0(z \to \infty) = 0$. The domain wall width is ξ.

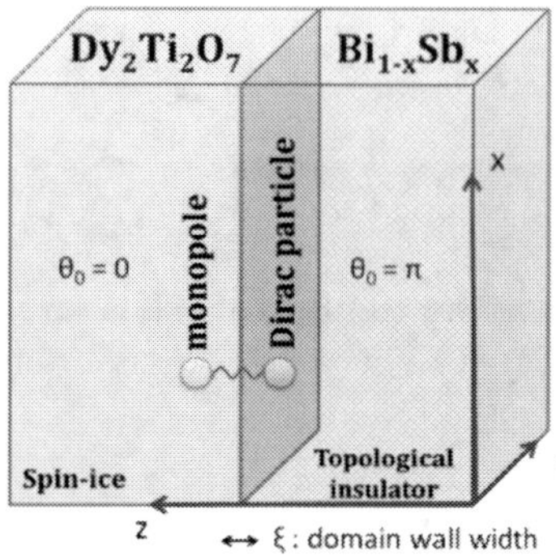

Figure 2. The bilayer of spin ice $Dy_2Ti_2O_7$ and topological insulator $Bi_{1-x}Sb_x$

For the spin ice compound $Dy_2Ti_2O_7$, the magnitude μ of the magnetic moments equals approximately ten Bohr magnetrons $\mu \sim 10\ \mu_B$. The thermodynamic properties of this compound are known to be described by an energy term that accounts for the nearest-neighbor exchange and the long-range dipolar interactions [13, 14],

$$\mathcal{H} = \frac{J}{3}\sum_{<i,j>} S_i \cdot S_j + Da^3 \sum_{<i,j>}\left[\frac{\hat{e}_i\hat{e}_j}{|r_{ij}|^3} - \frac{3(\hat{e}_i r_{ij})(\hat{e}_j r_{ij})}{|r_{ij}|^5}\right] S_i S_j.$$

The distance between spins is r_{ij} and $a \sim 3.54\ \mathring{A}$ is the pyrochlore nearest-neighbor distance,

$$D = \frac{\mu_0\mu^2}{4\pi a^3} = 1.41\ \mathrm{K}$$

is the coupling constant of the dipolar interaction. Here we shall consider the excited states. The excited inverted-dumbbell corresponds to two adjacent sites with net magnetic charge $\pm q_m = \pm 2\frac{\mu}{a_d}$ $-a$ nearest neighbor monopole-antimonopole pair. Where $a_d = \sqrt{\frac{3}{2}}a$ is the diamond lattice bond length. The monopoles can be separated from one another without further violation of local neutrality by flipping a chain of adjacent dumbbell. As shown in Fig. 2, when excited magnetic monopoles in the spin ice inject into the domain wall with the $\theta_0(z)$ as follows,

$$\theta_0(z) \sim \frac{\pi}{2}\left[1 - \tanh\left(\frac{z}{4\xi}\right)\right].$$

The magnetic monopole transfers to the dyon with the fractional electric charge

$$Q \sim \pm\frac{\theta_0(z)}{2\pi} q_m = \pm\frac{\theta_0(z)}{2\pi}\frac{2\mu}{a_d}$$

due to the Witten effect [10]. Now we shall consider the relaxation of the fermion on the domain wall due to the dyons, which are derived from magnetic monopoles in the spin ice. The Green-function of fermion is represented approximatly as follow,

$$G(k, \varepsilon + i\delta) = \frac{1}{\varepsilon + i\delta - \varepsilon_k - \sum(k, \varepsilon + i\delta)}$$

$$\sum(k, \varepsilon + i\delta) \propto i\pi N_{\mathrm{dyon}} \left(\frac{e}{4\pi}\theta_0 \frac{2\mu}{a_d}\right)^2 n \equiv \frac{i}{2\tau}$$

where N_{dyon} is the density of dyons and n is the fermion density per spin on the domain wall. τ is relaxation time. N_{dyon} is proportional to the magnetic monopole density ρ_{mono}. Taking into accounts the monopole density in the spin ice [15], it has been deduced that N_{dyon} at 10 K is approximately 10 times of one at 1 K in the case of the domain wall between the spin ice $Dy_2Ti_2O_7$ and the topological insulator $Bi_{1-x}Sb_x$. That is, it is expected that the relaxation time τ at 10 K is reduced 1/10 in comparison with the τ at 1 K.

3. Conclusion

It has been proposed that there might be emergent quasiparticle with fractional electronic charge such as dyons on the domain wall between topological insulator and spin ice components, through the Witten-effect and interaction between the Dirac fermions and excited magnetic monopoles.

References

[1] Bernevig B A, Hughes T L and Zhang S C 2006 *Science* **314** 1757
[2] König M *et al.* 2007 *Science* **318** 766
[3] Fu L, Kane C L and Mele E J 2007 *Phys. Rev. Lett.* **98** 106803
[4] Hasan M Z and Kane C L 2010 *Rev. Mod. Phys.* B **82** 3045
[5] Murakami S, Nagaosa N and Zhang S C 2004 *Phys. Rev. Lett.* **93** 156804
[6] Fu L and Kane C L 2007 *Phys. Rev.* B **76** 045302
[7] Hsieh D *et al.* 2008 *Nature* **452** 970
[8] Castelnovo C, Moessner R and Sondhi S L 2008 *Nature* **451** 42
[9] Kanazawa I 2012 *J. Phys.: Conf. Series* **400** 042028
[10] Witten E 1979 *Phys. Lett.* B **86** 283
[11] Qi X L, Hughes T L and Zhang S C 2008 *Phys. Rev.* B **78** 195424
[12] Goldstone J and Wilczek F 1981 *Phys. Rev. Lett.* **47** 986
[13] Siddharthan R *et al.* 1999 *Phys. Rev. Lett.* **83** 1854
[14] Ruff J P C, Melko R G and Gingras M J P 2005 *Phys. Rev. Lett.* **95** 097202
[15] Castelnovo C, Moessner R and Sondhi S L 2011 *Phys. Rev.* B **84** 144435

Observation and calculation of higher-order sideband transitions in a flux qubit coupled to a SQUID-based resonator

Y Shimazu[1]**, R Shirasaki, S Toda and H Yamanashi**

Yokohama National University, 79-5 Tokiwadai, Hodogaya, Yokohama 240-8501, Japan

[1]E-mail: yshimazu@ynu.ac.jp

Abstract. Qubit–resonator coupled systems have been intensively studied because of their potential usefulness in quantum information processing. We investigated a superconducting flux qubit coupled to a resonator mode (SQUID plasma mode) associated with a dc SQUID used to read out the qubit state. Higher-order red and blue sideband transitions with $|M|$ = 1, 2, and 3, where M represents the change in the quantum number of the resonator state, were clearly observed away from the symmetry point. We calculated the transition matrix elements between the dressed states and examined their dependence on both the coupling strength and the flux bias for the qubit. The observation of the sidebands up to $|M|$ = 3 is consistent with the calculation results.

1. Introduction

Circuit quantum electrodynamics (QED) systems [1], in which superconducting qubits are coupled to a resonator, are considered to be important hybrid quantum systems for applications in quantum information processing. Recently, we have shown experimental results of higher-order sideband transitions with $|M|$ =1 and 2 and sideband Rabi oscillations on a flux qubit coupled to a SQUID-based resonator deep in the dispersive regime [2], where M represents the change in the quantum number of the resonator state. Higher-order sideband transitions in a superconducting qubit coupled to a resonator have also been reported with the use of a linear LC resonator [3] and a micromechanical resonator [4]. In quantum information processing, these sideband transitions may be useful for building quantum gates [5].

In this paper, we present experimental results from a flux qubit coupled to a SQUID-based resonator with enhanced coupling strength compared with that of our previous study. We observed higher-order sideband transitions with $|M|$ up to 3. To elucidate the origin of these transitions, we calculated the relevant transition matrix elements. The calculated results are consistent with our experimental results.

2. Experimental

Figure 1 shows a schematic of the sample, which was fabricated using electron-beam lithography and angled aluminum deposition. The three-Josephson-junction flux qubit [6] is galvanically connected to the readout dc SQUID. The effective area of the Al shunt capacitor is ~2000 μm^2. The qubit is

[1] To whom any correspondence should be addressed.

dominantly coupled to a single arm of the dc SQUID [7], in contrast to the conventionally studied flux qubits with symmetric geometry [8–10]. In our geometry, the qubit–SQUID coupling is considerably larger at zero bias current at the expense of higher susceptibility to bias-current noise. This large coupling enabled us to observe higher-order sideband transitions [2]. The measurement was performed at 20 mK. For spectroscopy and measurement of Rabi oscillations, the switching probability P_{sw} of the SQUID was recorded using a short bias-current pulse applied to the SQUID immediately after a microwave pulse [8].

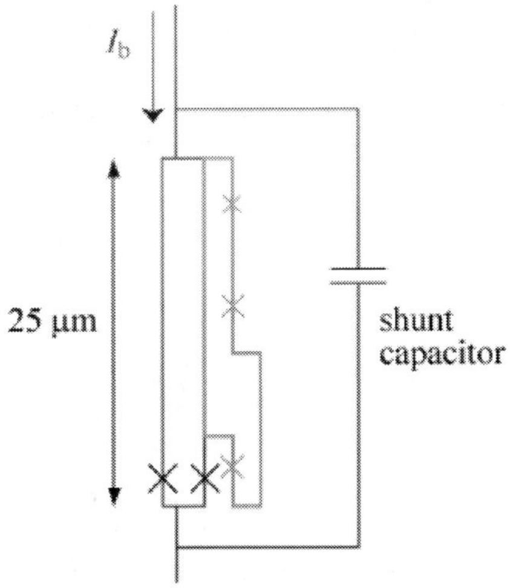

Figure 1. (color online) Schematic diagram of the sample. The crosses represent small Josephson junctions (JJ). The blue loop with three JJ represents a three-JJ flux qubit, which is coupled to the single arm of the dc SQUID. The frequency of the SQUID plasma mode can be tuned via the bias current I_b and the magnetic flux in the SQUID loop.

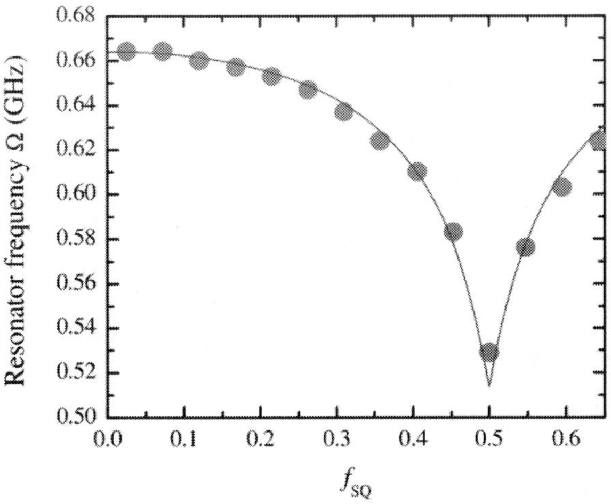

Figure 2. (color online) The resonator frequency, that is, the frequency of the SQUID plasma mode, as a function of $f_{SQ} = \Phi_{SQ}/\Phi_0$, where Φ_{SQ} is the magnetic flux threading the SQUID loop. The parameters used for the calculated result shown by the line are as follows: minimum Josephson inductance of the dc SQUID L_{SQ} = 280 pH, stray inductance L_s = 1560 pH, shunt capacitance C = 31 pF, and normalized SQUID-loop inductance $\beta = 2LI_c/\Phi_0 = 0.12$, where L and I_c are the SQUID-loop inductance and the critical current of each SQUID junction, respectively.

3. Results and discussion

As shown later, sideband transitions were observed at frequencies separated by ~0.6 GHz. The resonator mode of ~0.6 GHz was indeed observed as a sharp resonant peak in P_{sw}, which was apparent at temperatures as high as 0.5 K. The dependence of the resonant frequency on $f_{SQ} = \Phi_{SQ}/\Phi_0$, where Φ_{SQ} is the magnetic flux threading the SQUID loop and $\Phi_0 = h/2e$, is shown in figure 2. This resonance is attributed to the SQUID plasma mode [9,10] that is an LC resonance involving the Josephson inductance of the SQUID, the stray inductance, and the shunt capacitance C. The observed resonant frequency Ω is in good agreement with the calculated frequency of the SQUID plasma mode as shown in figure 2. In the calculation, a loop inductance for the SQUID of reasonable magnitude was required to attain good agreement with the experimental data.

In the spectroscopic measurement of the qubit, we observed higher-order red- and blue-sideband transitions, as shown in figure 3, under strong microwave drive and away from the symmetry point. The peak frequency is given by $\nu_Q + M\Omega$, where $\nu_Q = 9.9$ GHz is the qubit transition frequency. These sidebands are attributed to the transitions among energy levels of the dressed states of the qubit–resonator coupled system. The resonator energy $\Omega \sim 0.6$ GHz corresponding to 30 mK should be below the thermal energy; the effective temperature of the system is estimated to be ~200 mK on the basis of the width of the qubit step [11]. Therefore, a significant amount of photons is thermally excited, in contrast to the case of the previous study with $\Omega \sim 3$ GHz [9]. These thermally excited photons seed the red-sideband generation process. In figure 3, the transitions with $|M| = 1$, 2, and 3 are clearly shown. This result is compared with that of our previous measurement on a similar sample [2], where only the transitions with $|M| \leqq 2$ were clearly observed. We ascribe this difference to the enhanced coupling strength between the qubit and the resonator. The critical current of the SQUID in the present sample is higher by a factor of approximately four compared with the previous sample, which should lead to enhanced coupling strength.

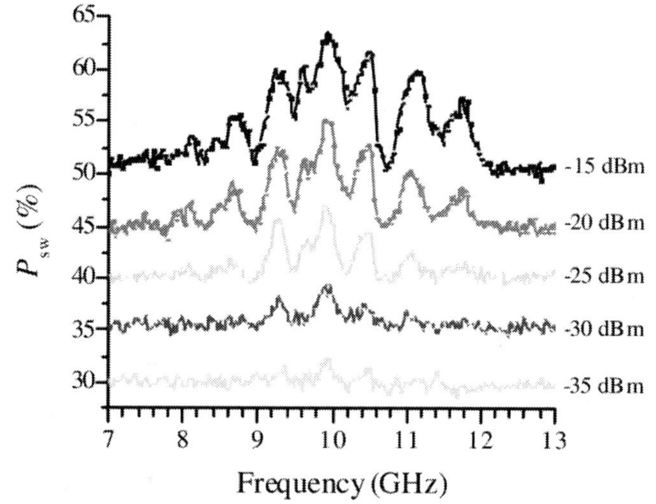

Figure 3. (color online) P_{sw} as a function of microwave frequency for various values of microwave power, measured at $I_b = 0$ and $\cos\theta = 0.84$. The duration of the microwave pulse was 100 ns. The traces are offset vertically for clarity. With increasing power, sideband transitions of higher orders appear beside the carrier transition.

We calculated the matrix elements that are relevant to the observed sideband transitions. The energy eigenvectors of the qubit–resonator system can be derived by diagonalizing the coupled Hamiltonian [2,12],

$$H_{QR} = \frac{h}{2}\nu_Q\sigma_z + h\Omega(a^\dagger a + \frac{1}{2}) + hg(\sigma_z\cos\theta - \sigma_x\sin\theta)(a^\dagger + a),$$

where g is the qubit–resonator coupling constant, $\cos\theta = \sqrt{1-(\Delta/\nu_Q)^2}$, and Δ is the qubit energy gap [6]. Deep in the dispersive regime, which is the case for the sample under investigation, the eigenvectors can be labeled as $|\mathrm{g}\ n\rangle$ and $|\mathrm{e}\ n\rangle$ using the labels of the dominant bare states; the bare states are the eigenstates of $\frac{h}{2}\nu_Q\sigma_z + h\Omega(a^\dagger a + \frac{1}{2})$. Here, g and e denote the ground and excited states of the qubit, respectively, and n is the photon number. The transition matrix elements for the sideband transition from the initial state $|\mathrm{g}\ n\rangle$ changing the photon number by M is given by

$$T(n, M) = |\langle \mathrm{e}\ n{+}M\ |(\sigma_z\cos\theta - \sigma_x\sin\theta)|\ \mathrm{g}\ n\rangle|.$$

In order to compare these calculated results with the data shown in figure 3, $T(n, M)$ was calculated by fixing the parameters as $\nu_Q = 9.9$ GHz, $\Omega = 0.6$ GHz, and $\cos\theta = 0.84$. Since the coupling constant g is unknown from the experimental data obtained thus far, g was varied from 0.05 to 0.2 GHz. We believe that these values are in the relevant range, in view of $g = 0.1$ GHz estimated for a separate sample [2]. In a future experiment, g will be estimated for the present sample by using a spectroscopic measurement in which the bias current is varied. Because several photons are thermally excited, $T(n, M)$ for n ranging from 0 to 5 is shown here. With increasing g, $T(n, M)$ with higher M increases, as expected. For g higher than 0.15 GHz, $T(n, M)$ with $|M| = 3$ has appreciable magnitude, which is consistent with our observation of transitions with $|M| = 3$. Interestingly, for the higher values of g and n, $T(n, M)$ with nonzero M is larger than $T(n, 0)$. The appearance of the higher-order sideband transitions reflects the fact that the eigenstate consists of several photon-number states. If we assume that the coupled Hamiltonian is of Jaynes-Cummings type [1], the eigenstate consists of two bare states and only transitions with $M = 0$, -1, and -2 are possible. Experimentally, we observed that the sideband transitions are gradually suppressed as the operating point approaches the symmetry point ($\theta = \pi/2$). This behavior was also described in the calculation of $T(n, M)$.

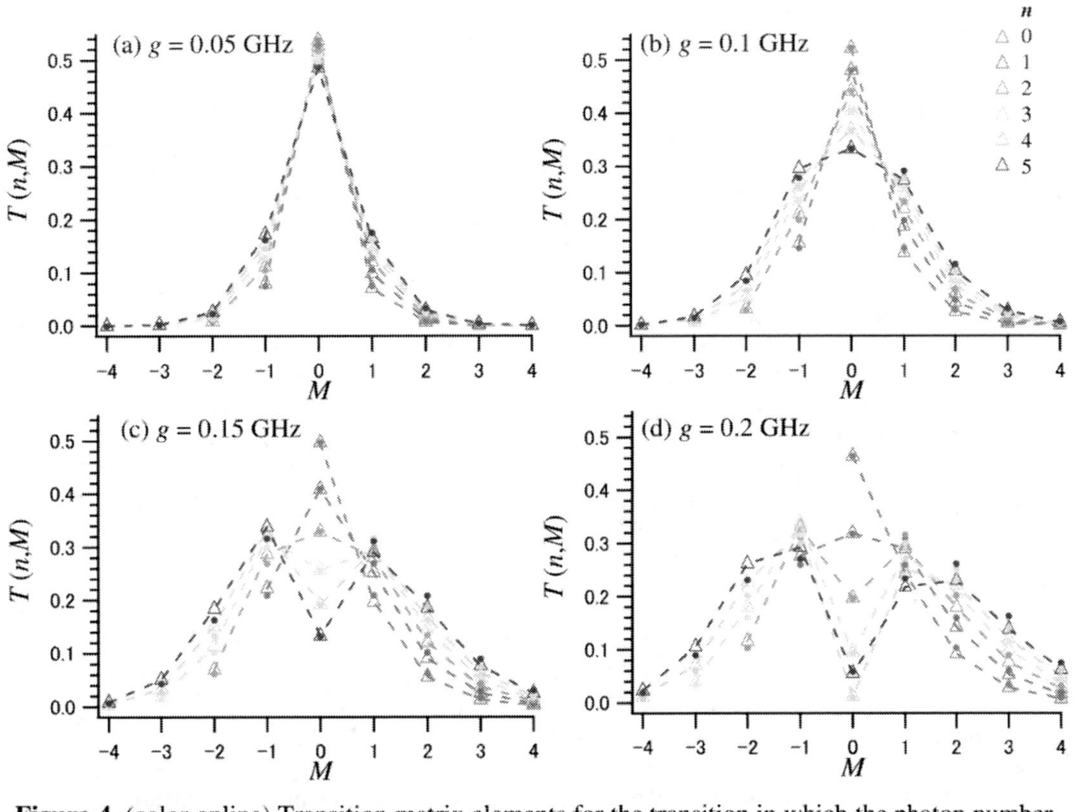

Figure 4. (color online) Transition matrix elements for the transition in which the photon number is changed by M; $T(n, M) = |\langle e\ n{+}M\ |(\sigma_z \cos\theta - \sigma_x \sin\theta)|\ g\ n\rangle|$ for g = 0.05, 0.1, 0.15, and 0.2 GHz. The parameters used for the calculation are ν_Q = 9.9 GHz, Ω = 0.6 GHz, and $\cos\theta$ = 0.84. The values denoted with triangles are obtained using the eigenstates of H_{QR}, while the values denoted with dots are obtained using those of H_{QR}'.

If we neglect the term $-hg\,\sigma_x \sin\theta(a^+ + a)$ in H_{QR}, the eigenstates of the resulting Hamiltonian

$$H_{QR}' = \frac{h}{2}\nu_Q\sigma_z + h\Omega(a^\dagger a + \frac{1}{2}) + hg\sigma_z \cos\theta(a^\dagger + a)$$

can be analytically derived using a displacement operator for the harmonic oscillator states [13]. We found that the transition matrix elements between the eigenstates of H_{QR}' nearly agree with those between the eigenstates of H_{QR} for the parameters used in the calculation, as indicated by the dots and triangles, respectively, in figure 4.

To understand this agreement, we consider the eigenstates of H_{QR} and H_{QR}'. In general, the eigenstates of H_{QR} can be written as

$$|g\ n\rangle = \sum_k G_k^{g,n} |g\ k\rangle^{(0)} + \sum_k E_k^{g,n} |e\ k\rangle^{(0)} \text{ and } |e\ n\rangle = \sum_k G_k^{e,n} |g\ k\rangle^{(0)} + \sum_k E_k^{e,n} |e\ k\rangle^{(0)},$$

where $|g\ k\rangle^{(0)}$ and $|e\ k\rangle^{(0)}$ are the bare states of the system. Meanwhile, the eigenstates of H_{QR}' can be written as

$$|g\ n\rangle' = \sum_k \tilde{G}_k^{g,n} |g\ k\rangle^{(0)} \text{ and } |e\ n\rangle' = \sum_k \tilde{E}_k^{e,n} |e\ k\rangle^{(0)},$$

because H_{QR}' does not couple the qubit ground state $|g\rangle$ with the qubit excited state $|e\rangle$. Deep in the dispersive regime, which is the case considered in this study, we observe that the $E_k^{g,n}$ terms in $|g\ n\rangle$ and the $G_k^{e,n}$ terms in $|e\ n\rangle$ are very small. This is because these terms are proportional to g sin θ/ν_Q, as derived from the first-order perturbation theory. Subsequently, we note that the matrix elements of H_{QR} and H_{QR}' in the subspace spanned by $|g\ k\rangle^{(0)}$ (or $|e\ k\rangle^{(0)}$) are the same:
$^{(0)}\langle g\ k | H_{QR} | g\ k'\rangle^{(0)} = {}^{(0)}\langle g\ k | H_{QR}' | g\ k'\rangle^{(0)}$ and $^{(0)}\langle e\ k | H_{QR} | e\ k'\rangle^{(0)} = {}^{(0)}\langle e\ k | H_{QR}' | e\ k'\rangle^{(0)}$.
Given the aforementioned considerations, we conclude that the eigenstates of H_{QR} are approximately equal to those of H_{QR}'. Therefore, the transition matrix elements between the eigenstates of H_{QR}' nearly agree with those between the eigenstates of H_{QR}. The substitution of H_{QR}' for H_{QR} may be advantageous in some analyses because the eigenstates of H_{QR}' can be analytically derived.

The transition matrix element is closely related to the period of Rabi oscillations associated with the transition. We have shown that the coupling constant can be estimated from the ratio between the Rabi periods for the carrier transition and the sideband transitions [2]. For the sample under investigation, although we have observed Rabi oscillations for the sideband transition, quantitative analysis of the Rabi periods on the basis of the calculated $T(n, M)$ has not been performed, partly because of the unknown coupling constant and the significant frequency dependence of the microwave attenuation, which strongly affects the ratio between the Rabi periods for different transitions. Further measurement and analysis will be performed in a future experiment. In the sample studied here with a relatively small value of Ω, the significant amount of thermal excitation of photons causes difficulty in analyzing the data and photon noise [10], leading to a short coherence time. However, this system may become a feasible platform for quantum operations involving many dressed states with easily accessible energies, including sideband cooling of the resonator.

4. Conclusion

In conclusion, we observed higher-order red- and blue-sideband transitions with $|M| = 1, 2$, and 3 in a coupled system of a flux qubit and a SQUID-based resonator, the coupling strength of which is larger than that of a similar sample we studied previously. Higher-order red sidebands were observed in a simple scheme because the resonator was strongly excited by thermal energy. The significant qubit–resonator coupling at $I_b = 0$ in these observations was due to the sample geometry; the qubit was coupled to a single arm of the dc SQUID. We calculated transition matrix elements between the dressed states obtained by diagonalizing the coupled Hamiltonian. The appearance of the sidebands up to $|M| = 3$ is consistent with the results of calculations in which reasonable coupling strength is assumed. The observed suppression of the sideband transition near the symmetry point is also consistent with the calculated transition matrix elements.

Acknowledgments
This work was supported by JSPS KAKENHI Grant Number 24540317.

References

[1] Blais A, Huang R S, Wallraff A, Girvin S M and Schoelkopf R J 2004 *Phys. Rev.* A **69** 062320
[2] Shimazu Y, Takahashi M and Okamura N 2013 *J. Phys. Soc. Jpn.* **82** 074710
[3] Plantenberg J H 2007 Ph. D. Thesis (Delft University of Technology)
[4] Pirkkalainen J M, Cho S U, Li J, Paraoanu G S, Hakonen P J and Sillanpaa M A 2013 *Nature* **494** 211
[5] Blais A, Gambetta J, Wallraff A, Schuster D I, Girvin S M, Devoret M H and Schoelkopf R J 2007 *Phys. Rev.* A **75** 032329
[6] Mooij J E, Orlando T P, Levitov L S, Tian L, van der Wal C H and Lloyd S 1999 *Science* **285** 1036, Orlando T P, Mooij J E, Tian L, van der Wal C H, Levitov L S, Lloyd S, and Mazo J J 1999 *Phys. Rev.* B **60** 15398
[7] Shimazu Y, Nakajima M, Takahashi M, Okamura N and Yoshiyama K 2012 *Physics Procedia* **27** 360
[8] Chiorescu I, Nakamura Y, Harmans C J P M and Mooij J E 2003 *Science* **299** 1869
[9] Chiorescu I, Bertet P, Semba K, Nakamura Y, Harmans C J P M and Mooij J E 2004 *Nature* **431** 159
[10] Bertet P, Chiorescu I, Burkard G, Semba K, Harmans C J P M, DiVincenzo D P and Mooij 2005 *Phys. Rev. Lett.* **95** 257002
[11] van der Wal C H, ter Haar A C J, Wilhelm F K, Schouten R N, Harmans C J P M, Orlando T P, Lloyd S and Mooij J E 2000 Science **290** 773
[12] Fedorov A, Feofanov A K, Macha P, Forn-Díaz P, Harmans C J P M and Mooij J E 2010 *Phys. Rev. Lett.* **105** 060503
[13] Scully M O and Zubairy M S 1997 *Quantum Optics* (Cambridge University Press)

Surface Acoustic Waves Probe of the Spin Phase Transition at ν=2/3 in n-GaAs/AlGaAs structure

I L Drichko[1], I Yu Smirnov[1,4], A V Suslov[2], L N Pfeiffer[3] and K W West[3]
[1] A.F. Ioffe Physico-Technical Institute, St. Petersburg, 194021, Russia
[2] National High Magnetic Field Laboratory, Tallahassee, FL 32310, USA
[3] Princeton University, Princeton, NJ 08544, USA

E-mail: ivan.smirnov@mail.ioffe.ru

Abstract. High frequency (ac) conductivity in the single quantum well AlGaAs/GaAs/AlGaAs with high mobility was investigated by contactless acoustic methods in the fractional quantum Hall effect regime in perpendicular and tilted magnetic fields. We studied the dependence of ac conductivity $\sigma^{ac}=\sigma_1-i\sigma_2$ on both the temperature and magnetic field tilt angle. Tilting the magnetic field relative to the sample surface enabled us to change the position of the conductivity oscillation minimum at ν= 2/3. We measured the temperature dependence of ac conductivity for each tilt angle and for the 2/3 state we calculated the activation energy ΔE which was derived by constructing the Arrhenius plot ln σ_1 against 1/T. Analyzing behavior of the activation energy in total magnetic field for the filling factor 2/3 we observed a distinct minimum which can be interpreted as the spin unpolarized-polarized phase transition.

1. Introduction

In high mobility GaAs/AlGaAs structures the fractional quantum Hall effect (FQHE) is observed in strong magnetic fields at low temperatures. In this regime at the filling factor ν=2/3 the spin unpolarized-polarized transition occurs with change of the magnetic field or electron density. In accordance with the composite fermions (CF) picture the ν=2/3 state becomes the CF state with the filling factor $\nu^{CF}=2$. The energy spacing between CFs Landau levels (LLs) is determined by a little portion of the Coulomb energy $\alpha_c E_c=\alpha_c e^2/\varepsilon l_B \sim\sqrt{B}$, where $l_B=\sqrt{\hbar/eB}$ is the magnetic length, $\alpha_c \ll 1$ is the critical parameter [1]. Each CF LL could in turn be spin-split into two levels separated by the Zeeman energy $E_Z= g^{CF}\mu_B B \propto B$, where μ_B is the Bohr magneton, and $g^{CF}=g=-0.44$. The energy gap between nearest CF LLs with different spin orientation at $\nu^{CF}=2$ can be expressed as $\Delta E=\alpha_c E_c - g\mu_B B$. Because of different scaling of the energies E_Z and E_C with B, CF LLs could cross at some critical magnetic field B_c, resulting in $\Delta E\to 0$. Thus, the two CF LLs with the same spin orientation could turn to be under the Fermi level, i.e. the system becomes spin-polarized.

This spin phenomenon has been investigated in numerous papers by different methods [1-16]. In Fig.1 we have summarized the available data on the spin transition critical magnetic field B_c. It is seen from the figure that B_c increases with the electron density in the sample [2-5,13,14].

[4] To whom any correspondence should be addressed.

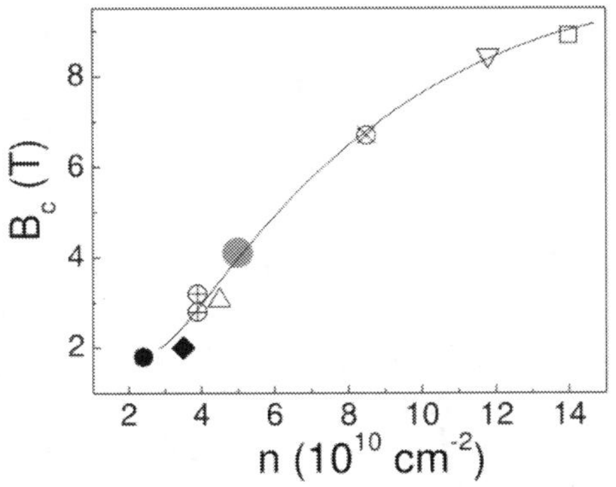

Figure 1. The transition critical field vs the electron density following the results of Ref.2 (⊕), Ref.3 (●), Ref.4 (∇), Ref.5 (Δ), Ref.13 (⊗, □), Ref.14 (♦), and the current paper (●).The line is a guide for the eye.

A study of the FQHE using acoustic methods at ν=2/3 was implemented in Ref. [4], where the acoustic velocity shift was measured, but no transition effects were observed. Earlier we also used the acoustic methods [12] for study the ac conductivity of the fully spin-polarized state of composite fermions at ν=2/3 in the GaAs/AlGaAs with electron density $n \cong 2\times10^{11}$ cm^{-2} and mobility $\mu \cong 1.5\times10^6$ cm^2/Vs. Rather high concentration of our sample did not allowed us to explore this transition directly, since the ν=2/3 state in our experiment was at B=12 T, while the spin transition occurs at lower magnetic fields. That is why we have determined just the dependence of the energy gap ΔE on magnetic field in the fully spin polarized state, where it turns out that $\Delta E \propto B$.

In this paper, we report the activation energy study of the 2/3 state using the Surface Acoustic Wave (SAW) contactless technique in a high mobility heterostructure GaAlAs/GaAs in tilted magnetic fields. The concentration of electrons $n=5.5\times10^{10}$ cm^{-2} provides position of the oscillation corresponding to ν=2/3 at $B_\perp$~3.2 T, which is rather low. Accordingly with Fig. 1, we shall be possibly able to determine the energy gap in the transition region by tilting the magnetic field. Thus, for the first time we would apply the acoustic methods to observe the spin unpolarized-polarized phase transition at ν=2/3.

2. Experiment

In the experiments we used the "hybrid" configuration of the contactless acoustic method: the surface acoustic wave, excited by interdigital transducers, propagates along the surface of the piezodielectric lithium niobate, and the studied structure is pressed onto the surface of the $LiNbO_3$ (see Fig. 2a). The electric field produced by the SAW interacts with the carriers in 2D channel. Thus, the attenuation Γ and velocity $\Delta v/v_0$ of the acoustic wave are affected by the conductivity of the 2DEG. In this "hybrid" setup no deformation is transmitted into the sample. The technique was first employed for GaAs/AlGaAs structures in [17]. A detailed acoustic study of the FQHE has been performed in [18].

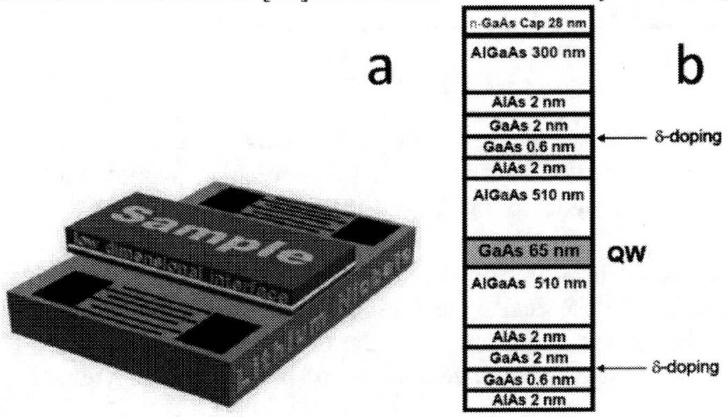

Figure 2. A sketch of the acoustic experimental setup (a) and a cross-section of the studied sample (b).

The samples grown by molecular beam epitaxy has a 65 nm wide symmetrically doped GaAs/AlGaAs quantum well (QW) with the density of n=5.5×10^{10} cm^{-2} and mobility of μ=8.5×10^{6} cm^{2}/Vs and contained undoped $Al_{0.24}Ga_{0.76}As$ spacer layers and Si δ-doped layers (see Fig. 2b). The low temperature measurements in the perpendicular field were done at a dilution refrigerator while tilting the sample was carried out in a ^{3}He system equipped with a one-axis rotator.

3. Experimental results and discussion

The measurements of the SAW attenuation Γ and velocity change $\Delta v/v_0$ in this system were done at the frequency of 86 MHz in magnetic fields B up to 18 T and in the temperature range of 0.1 – 1.6 K. Figure 3 shows the dependences of the SAW attenuation change Γ (top panel) and the change of the SAW velocity $\Delta v/v_0$ (middle panel) on the magnetic field B applied along the QW normal z for different temperatures in the QHE regime. One can see that both the absorption coefficient and the velocity change exhibit IQHE and FQHE type oscillations in the magnetic field.

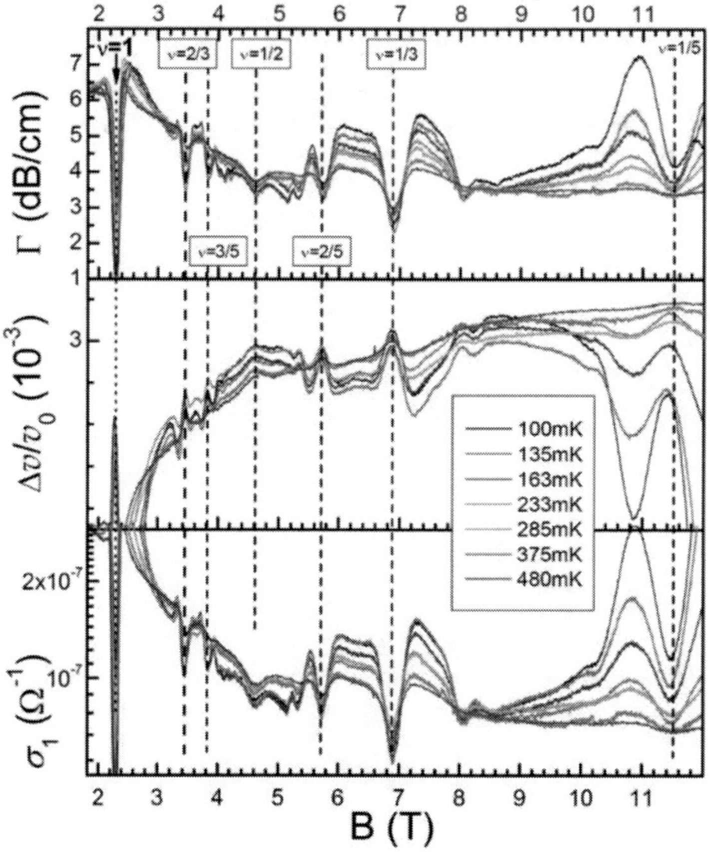

Figure 3. Temperature evolution of the SAW attenuation and velocity shift as well as the real part of the ac conductivity σ_1 in the QHE regime in the perpendicular magnetic field, f=86 MHz. The values of the filling factors ν are shown by vertical dash lines.

From the experimentally measured values of the SAW absorption Γ and the relative change of the SAW velocity $\Delta v/v_0$, one can calculate the real σ_1 and imaginary σ_2 components of the high-frequency conductivity $\sigma^{ac}=\sigma_1-i\sigma_2$ in the electron channel using the Eqs. (1) and (2) of Ref. 12. Below, we focus only on the real part of the ac conductivity since analysis of this conductivity enables us to determine the parameters of the energy spectrum. The corresponding dependence of σ_1 on the magnetic field for different temperatures at f=86 MHz is presented in the bottom panel of Fig. 3. The magnetic field dependence of the ac conductivity also contains a rich oscillations pattern. At the fields higher than 3 T σ_1(B) manifests pronounced FQHE oscillations, which are caused by the formation of CF Landau levels in the two-dimensional electron gas and are similar to the oscillations of the magnetoconductivity in the quantum Hall effect observed in dc transport measurements.

In this paper we aimed at the activation gap studies. To conduct such measurements properly one indeed needs to find at first the temperature domain where the conductivity is activated. Compassed in this way we studied the behavior of ac magnetoconductivity as a function of temperature down to 100 mK. We found that the ac conductivity at ν=2/3 is reliably activated at T>0.3 K.

We also carried out measurements of the acoustoelectric effects and calculated σ_1 in a tilted magnetic field, being focused mainly on the 2/3 state which corresponds to the normal component of total magnetic field $B_\perp = B_{TOTAL}\cos(\Theta) \approx 3.2$ T. We measured the angle Θ by tracing position of the most pronounced oscillations, namely the one of ν=1. Tilting the magnetic field relative to the sample surface enabled us to change the position of the conductivity oscillation minimum at ν=2/3.

We measured the temperature dependence of Γ and $\Delta v/v_0$ in the range of 0.3 to 1 K for each tilt angle and derived σ_1. The activation energy $\Delta E_{2/3}$ was derived by constructing the Arrhenius plot ln σ_1 against 1/T assuming that the conductivity $\sigma_1 \propto \exp(-\Delta E_{2/3}/2k_BT)$. Dependence of the activation energy for ν=2/3 $\Delta E_{2/3}$ on total magnetic field is illustrated in Fig. 4. As one can see, the energy gap in the magnetic field for filling factor 2/3 passes through a distinct minimum, which can be interpreted as the unpolarized-polarized phase transition when a crossing of the composite fermion's Landau levels with different spin directions at ν^{CF}=2 occurs. At the crossing between two CF LLs the energy gap should disappear at the transition point. However, as seen in Fig. 4 $\Delta E_{2/3}$ does not approach the zero. This anticrossing could evolve as a result of the electron exchange interaction as discussed in Ref.[4].

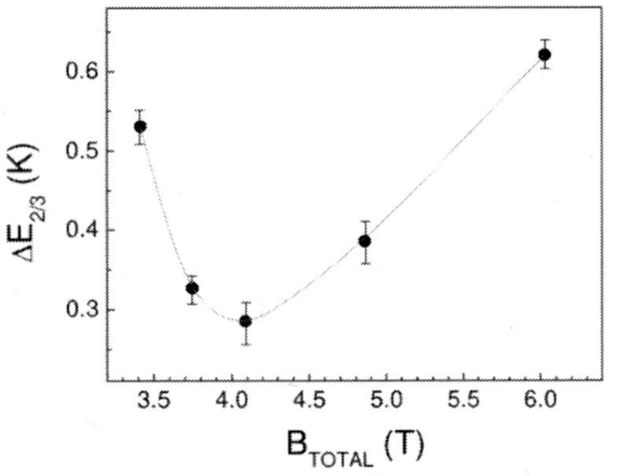

Figure 4. Dependence of the activation gap for ν=2/3 on total magnetic field. The line is a guide to the eye.

It is known that if a quantum well is wide the Coulomb interaction could be weakened. As a result, at the layer thickness exceeding $3l_b$ the FQHE quickly collapses (see [19] and refs therein). In our sample the QW width is 65 nm, which several times exceeds the magnetic length at B≈4T. However, using the acoustic technique we were able not only to observe a rich FQHE pattern, but also to measure the temperature dependence of conductivity at ν=2/3, as well as to determine the dependence of the energy gap for ν^{CF}=2 on the magnetic field. These measurements allowed us to observe the transition from unpolarized to polarized FQHE states at the CF filling factor ν^{CF}=2.

Critical field of this transition B=4.2 T is indicated in Figure 1 by the red dot. If we assume that CF g-factor $|g^{CF}|= |g|=0.44$, then in the spin transition point at B_c we can estimate the value of the critical parameter $\alpha_c=E_Z/E_c=0.012$, which is close to the value of $\alpha_c\approx0.008$, determined in the Ref. [1] for the 65 nm wide quantum well for $\nu= 2/3$.

In conclusion, we performed Surface Acoustic Waves contactless measurements of the activation energy of the 2/3 state in tilted magnetic fields, thus, having found the critical field of the spin transition in dilute GaAs/AlGaAs structure with wide quantum well.

Acknowledgments
The authors would like to thank E. Palm, T. Murphy, J.-H. Park, and G. Jones for technical assistance. NHMFL is supported by NSF Cooperative Agreement No. DMR-1157490, the State of Florida, and the U.S. Department of Energy. The work at Princeton was partially funded by the Gordon and Betty Moore Foundation through Grant GBMF2719, and by the National Science Foundation MRSEC-DMR-0819860 at the Princeton Center for Complex Materials.

4. References

[1] Liu Y Hasdemir S Wojs A Jain J K Pfeiffer L N West K W Baldwin K W and Shayegan M 2014 *Arxiv.org* 1406.2387
[2] Eisenstein J P, Stormer H L, Pfeiffer L N and West K W 1990 *Phys. Rev.* B **41** 7910
[3] Engel L W, Hwang S W, Sajoto T, Tsui D C and Shayegan M 1992 *Phys. Rev.* B **45** 3418
[4] Stern O 2004 *Ph.D. thesis* (Stuttgart: Max Planck Institute)
[5] Clark R G Haynes S R Branch J V Suckling A M Wright P A Oswald P M W Harris J J and Foxon C T 1990 *Surf. Sci.* **229** 25
[6] Nicholas R J Leadley D R Daly M S van der Burgt M van der Gee P Singleton J Maude D K Portal J C Harris J J and Foxon C T 1996 *Semicond. Sci. and Technol.* **11** 1477
[7] Leadley D R Nicholas R J Maude D K Utjuzh A N Portal J C Harris J J and Foxon C T 1997 *Phys. Rev. Lett.* **79** 4246
[8] Schulze-Wischeler F Mariani E Hohls F and Haug R J 2004 Phys. Rev. Lett. **92** 156401
[9] Haug R J von Klitzing K Nicholas R J Maan J C and Weimann G 1987 *Phys. Rev.* B **36**, 4528
[10] Boebinger G S Stormer H L Tsui D C Chang A M Hwang J C M Cho A Y Tu C W and Weimann G 1987 *Phys. Rev.* B **36** 7919
[11] Khrapai V S Shashkin A A Trokina M G Dolgopolov V T Pellegrini V Beltram F Biasiol G and Sorba L 2007 *Phys. Rev. Lett.* **99** 086802
[12] Drichko I L Smirnov I Yu Suslov A V and Leadley D R 2011 *Phys. Rev.* B **83** 235318
[13] Freytag N Tokunaga Y Horvatic M Berthier C Shayegan M and Lévy L P 2001 *Phys. Rev. Lett.* **87** 136801
[14] Kukushkin I V von Klitzing K and Eberl K 1999 *Phys. Rev. Lett.* **82** 3665
[15] Vanovsky V V Khrapai V S Shashkin A A Pellegrini V Sorba L and Biasiol G 2013 *Phys. Rev.* B **87** 081306(R)
[16] Verdene B Martin J Gamez G Smet J von Klitzing K Mahalu D Schuh D Abstreiter G and Yacoby A 2007 *Nature Physics* **3** 392
[17] Wixforth A Scriba J Wassermeier M Kotthaus J P Weimann G and Schlapp W 1989 *Phys. Rev.* B **40** 7874; Wixforth A Kotthaus J P and Weimann G 1986 *Phys. Rev. Lett.* **56** 2104
[18] Willett R L 1997 *Adv. Phys.* **46** 447
[19] Shayegan M 2006 Flatland Electrons in High Magnetic Fields *High Magnetic Fields: Science and Technology, Theory and Experiment* vol 3 ed F Herlach and N Miura (Singapore: World Scientific) pp 31-60

Journal of Physics: Conference Series **568** (2014) 052032 doi:10.1088/1742-6596/568/5/052032

Equilibrium axial current due to a static localized spin in Weyl semimetals

Katsuhisa Taguchi

Department of Applied Physics, Nagoya University, Nagoya, 464-8603, Japan

E-mail: `taguchi@rover.nuap.nagoya-u.ac.jp`

Abstract. We theoretically study the equilibrium axial current $\boldsymbol{j}_5^{\mathrm{eq}}$, which is the di erence between the charge current of left-handed and right-handed helicity of the Weyl fermions, in the junction of a magnetic insulator attached to Weyl semimetals with time-reversal breaking and inversion symmetry. The charge current of each helicity $\boldsymbol{j}_{\gamma=\pm}^{\mathrm{eq}}$ is induced by the localized spin $\boldsymbol{S}$ of the magnetic insulator. $\boldsymbol{j}_{\gamma=\pm}^{\mathrm{eq}}$ is proportional to $\nabla \times \boldsymbol{S}$ and is independent of the sign of γ. As a result, the di erence between $\boldsymbol{j}_{\gamma=+}^{\mathrm{eq}}$ and $\boldsymbol{j}_{\gamma=-}^{\mathrm{eq}}$ are canceled out, and $\boldsymbol{j}_5^{\mathrm{eq}}$ is zero.

1. Introduction

The axial current is de ned by the di erence between the charge current of right- and left-handed helicity in Dirac electron systems [1, 2, 3, 4, 5, 6, 7]. Although this concept was initially devised in elementary particle physics, currently, it strongly a ects condensed matter physics as well because candidate materials of massless Dirac material, such as Weyl semimetal (WS) has been reported [8, 9, 10, 11, 12, 13, 14, 15]. It is noted that the axial current can ow without an accompanying charge current [1, 2, 3, 4, 5, 6, 7]. The property of the axial current is similar to pure spin current in spintronics[16]. Therefore, one can expect that the axial current can be applied to a method of low-consumption electronics.

The axial current is induced when we apply a static magnetic eld $\boldsymbol{H}$[1, 2, 3, 4, 5, 6, 7]. This phenomenon is called as the chiral separation e ect. Its origin lies in the di erence of helicity between right-handed and left-handed fermions in the WS. The helicity $= \hat{\boldsymbol{\sigma}} \ \hat{\boldsymbol{p}}$ indicates the relative angle between the direction of the spin $\hat{\boldsymbol{\sigma}}$ and that of the momentum $\hat{\boldsymbol{p}}$ of Weyl fermions. The helicity of right-handed fermions is $= +1$, whereas that of left-handed ones is $= \ 1$, but both spins are parallel to each other along the applied magnetic eld (Figure. 1(a)). Here the axial current can be regard as the equilibrium ow. Recently, the nonequilibrium axial current has been also studied in the WS / a magnetic insulator (MI) junction. The nonequilibrium axial current is driven by the spin transfer from the spin angular momentum of the the localized spin into the conduction electrons spin in the WS[16]. The preexisting work considered only the nonequilibrium axial current due to the spin transfer. The equilibrium axial current due to the spin transfer has not been discussed, so far.

In this paper, we study the equilibrium axial current $\boldsymbol{j}_5^{\mathrm{eq}}$ due to the spin transfer in the WS / MI junction, where WS is time reversal breaking and inversion symmetry within a di usive regime. Using the Green's functions techniques, we calculate the charge current of each helicity, $\boldsymbol{j}_{=\pm}^{\mathrm{eq}}$ and obtain the axial current $\boldsymbol{j}_5^{\mathrm{eq}} \quad \boldsymbol{j}_+^{\mathrm{eq}} \quad \boldsymbol{j}^{\mathrm{eq}}$ in the linear response of the localized spin$\boldsymbol{S}$ in the MI.

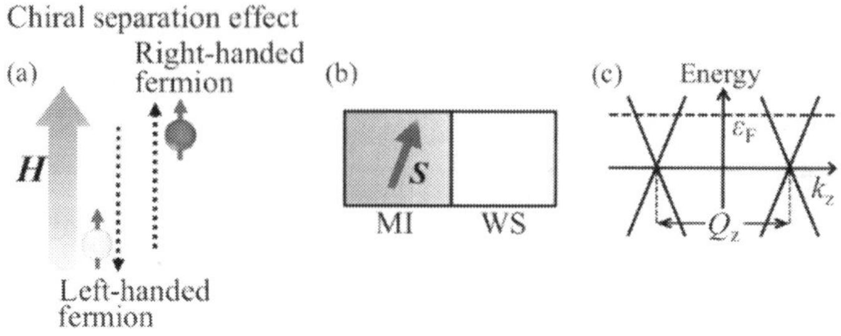

Figure 1. (a) Schematic illustration of the chiral separation e ect. When a magnetic eld $\boldsymbol{H}$ is applied, right-handed and left-handed fermions are separated along the $\boldsymbol{H}$ direction. (b) MI/WS junction with the chiral separation e ect resulting from the static localized spin $\boldsymbol{S}$. (c) Schematic illustration of the energy dispersion of the WS with time-reversal symmetry breaking and inversion symmetry.

2. Charge current of each helicity due to the localized spin

We will consider how the localized spin contributes to $\boldsymbol{j}_5^{\mathrm{eq}}$ in the WS / MI junction (represented in the Figure. 1 (b)). The total Hamiltonian we consider is given by

$$\mathcal{H} = \mathcal{H}_{\mathrm{W}} + \mathcal{H}_{\mathrm{ex}} + V_{\mathrm{i}}, \tag{1}$$

where the rst term,

$$\mathcal{H}_{\mathrm{W}} = \sum_{=\pm} \mathcal{H}_{\mathrm{W},} = \sum_{=\pm} \left\{ \sum_{\boldsymbol{k}} {}^{\dagger}_{\boldsymbol{k},} \, [\hbar v_{\mathrm{F},} \, (\boldsymbol{k} \quad \boldsymbol{Q}/2) \; \hat{\boldsymbol{\sigma}} \quad {}_{\mathrm{F}}] \, {}_{\boldsymbol{k},} \right\}, \tag{2}$$

is the Hamiltonian describing the doped WS. Here ${}^{\dagger}_{\boldsymbol{k},} = ({}^{\dagger}_{\boldsymbol{k},\,,\uparrow} \; {}^{\dagger}_{\boldsymbol{k},\,,\downarrow})$, and ${}_{\boldsymbol{k},}$ are the creation and annihilation operators of the Weyl fermions of each helicity sector , respectively (where indices $\uparrow$ and $\downarrow$ represent spin), ${}_{\mathrm{F}}$ is the Fermi energy [Figure. 1(c)], and $v_{\mathrm{F},} \; = \; v_{\mathrm{F}}$ is the Fermi velocity. We assume that a single pair of Dirac cones exists in the WS with inversion-symmetry and time-reversal-symmetry breaking with nonzero $\boldsymbol{Q}$. The parameter $\boldsymbol{Q}$ of Eq. (2) denotes the position of the Weyl node with $\boldsymbol{Q}/2$ and its magnitude $|\boldsymbol{Q}|$ is the distance between two Dirac cones. The second term in Eq. (1),

$$\mathcal{H}_{\mathrm{ex}} = \sum_{=\pm} \mathcal{H}_{\mathrm{ex},} \; = \; \sum_{=\pm} J_{\mathrm{ex}} \int d\boldsymbol{x} \boldsymbol{S} \; (\; {}^{\dagger} \hat{\boldsymbol{\sigma}} \quad), \tag{3}$$

describes the interaction between the localized spin in the MI and the spin in the WS, where $J_{\mathrm{ex}} > 0$ is the exchange coupling constant, $\boldsymbol{S} = S\boldsymbol{n}(\boldsymbol{x})$ is the classical vector representing the spin structure, S is its magnitude, and $\boldsymbol{n}$ is the unit vector representing the direction. Finally, the last term in Eq. (1), V_{i}, represents nonmagnetic impurity scattering, which causes a relaxation time of the transport of conduction electrons in the WS.

In the following calculation, $\boldsymbol{Q}$ is chosen to be parallel to the quantization axis of the localized spin (z axis) as $\boldsymbol{Q} = Q_z \boldsymbol{z}$ and Q_z is independent of time. In addition, we incorporate the term proportional to $\boldsymbol{Q}$ in $H_{\mathrm{W},}$ into $\mathcal{H}_{\mathrm{ex},}$ by using the following transformation: $\boldsymbol{S} \to \boldsymbol{S}' = (S_x, S_y, S_z \quad \frac{\hbar v_{\mathrm{F}}}{2J_{\mathrm{ex}}} Q_z)$. This transformation enables us to calculate the axial current rather easily. Then, we assume that the e ect of $\mathcal{H}_{\mathrm{ex},}$ is weak and can be treated as a perturbation. This condition is satis ed by $J_{\mathrm{ex}}|\boldsymbol{S}'| \;\; /\hbar \quad 1$ within the di usive transport regime.

To consider the axial current, we will calculate the current $\boldsymbol{j}$ using the above assumptions. We de ne the charge current of each helicity sector as $\boldsymbol{j} \; = \; ev_{\mathrm{F},} \langle \; {}^{\dagger}\boldsymbol{\sigma} \quad \rangle$ from the

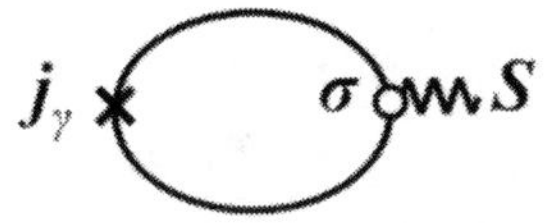

Figure 2. Feynman diagram of j . The bold line shows the Green's function of $\mathcal{H}_{W,}$ and the wavy line represents the exchange interaction $\mathcal{H}_{ex,}$.

conservation law $\dot{} = \nabla \; j$, where $e\langle \; ^{\dagger} \; \rangle$ is the charge density of helicity . The current is represented by using the same space and time of lesser Green's functions $\hat{G}^{<} = \langle \; ^{\dagger} \; \rangle / (\; i\hbar)$ as

$$j_{i,} (\boldsymbol{x}, t) = i\hbar e v_{F,} \ \mathrm{tr}[\hat{}_i \hat{G}^{<}(\boldsymbol{x}, t : \mathbf{0}, 0)]. \quad (4)$$

$j_{i,}$ due to the linear response of the localized spin is described diagrammatically in Figure. 2. We note that in this paper, we consider the equilibrium charge current $j_{i,}^{eq} \quad j_{i,}^{eq}(\boldsymbol{x}, 0)$. Here $j_{i,}^{eq}$ is derived from the spin-spin response function $\Pi_{ij,} (\boldsymbol{q}, \quad = 0) \quad \Pi_{ij,} (\boldsymbol{q}, 0)$ and is given by

$$j_{i,}^{eq}(\boldsymbol{x}, 0) = \frac{i\hbar J_{ex} e v_{F,}}{V} \sum_{\boldsymbol{q}} e^{\ i\boldsymbol{q}\ \boldsymbol{x}} \Pi_{ij,} (\boldsymbol{q}, \quad = 0) S'^{j}_{\boldsymbol{q},\Omega=0}, \quad (5)$$

$$\Pi_{ij,} (\boldsymbol{q}, 0) = \frac{1}{V} \sum_{\boldsymbol{k},\omega} \mathrm{tr}[\hat{}_i \hat{g}_{\boldsymbol{k} \ \frac{\boldsymbol{q}}{2},\omega,} \ \hat{}_{j,} \ \hat{g}_{\boldsymbol{k}+\frac{\boldsymbol{q}}{2},\omega,} \ ]^{<}, \quad (6)$$

where $\hat{}_{j,}$ is the vertex function of V_i and V is the system volume in the WS. The vertex function $\hat{}_{j,}$ is represented from the 4 4 matrix $\hat{}$ and the Pauli matrix $\boldsymbol{\sigma}$ as $\hat{}_{j,} = [\hat{} \ ]_j \ \hat{}$. Here the matrix $\hat{}$ is de ned by $\hat{} = \sum_{n=0}^{\infty} (\hat{} \ )^n$, where 4 4 matrix $\hat{}$ satis es $\hat{}_{\ ,} = [\hat{} \ ] \ \hat{}$. This $\hat{}_{\ ,}$ is represented as

$$\hat{}_{\ ,} (\boldsymbol{q}, 0) = \frac{1}{V} \sum_{\boldsymbol{k}} n_i u_i^2 \hat{g}_{\boldsymbol{k} \ \frac{\boldsymbol{q}}{2},\omega,} \ \hat{} \ \hat{g}_{\boldsymbol{k}+\frac{\boldsymbol{q}}{2},\omega,} \ . \quad (7)$$

In the Eqs. (6) and (7), $\hat{g}_{\boldsymbol{k},\omega,}$ is the Green's function of $\mathcal{H}_{W,}$ including V_i. The retarded (advanced) Green's function $g^r(g^a)$ in $\mathcal{H}_{W,}$ is given by

$$\hat{g}^r_{\boldsymbol{k},\omega,} = [\hbar\omega + \ _F \quad \hbar v_{F,} \ \boldsymbol{k} \ \hat{\boldsymbol{\sigma}} + i\eta]^{\ 1}, \quad (8)$$

where $\eta \quad \hbar/(2 \ ) = n_i u_i^2 \ _e/4$ is the self-energy of V_i and is obtained in the Born approximation. Here n_i, u_i, and $_e$ are the concentration of impurities, the potential energy of impurities, and the density of states at $_F$, respectively. We calculate $\Pi_{ij,}$ by using $\hat{g}^{<}_{\boldsymbol{k},\omega,} = f_\omega(\hat{g}^a_{\boldsymbol{k},\omega,} \quad \hat{g}^r_{\boldsymbol{k},\omega,} \ )$ [17], where f_ω is the Fermi distribution function. Then, the response function is represented by

$$\Pi_{ij,} (\boldsymbol{q}, 0) = \sum_{\boldsymbol{k},\omega} f_\omega \mathrm{tr}[\hat{}_i \hat{g}^a_{\boldsymbol{k} \ \frac{\boldsymbol{q}}{2},\omega,} \ \hat{}^{aa}_{j,} \ \hat{g}^a_{\boldsymbol{k}+\frac{\boldsymbol{q}}{2},\omega,} \ ] \quad \mathrm{h.c}, \quad (9)$$

where $\hat{}^{nm}_{j,} = [\hat{}^{nm}]_j \ \hat{}$ is constructed by the advanced Green's function $(n, m = a)$ or the retarded Green's function $(n, m = r)$. $\hat{}^{aa}_{j,} = \hat{}_j + \hat{}^{aa}_{j,} +$ is described by $\hat{}^{aa}_{j,} = \frac{1}{V} \sum_{\boldsymbol{k}} n_i u_i^2 \hat{g}^a_{\boldsymbol{k} \ \frac{\boldsymbol{q}}{2},\omega,} \ \hat{} \ \hat{g}^a_{\boldsymbol{k}+\frac{\boldsymbol{q}}{2},\omega,}$. Here $\hat{}^{aa}_{j,}$ can be approximately estimated by $\hat{}^{aa}_{j,} \simeq \hat{}_j + o(q)$. Then, $\Pi_{ij,}$ is obtained by expanding with $q/k_F \quad 1$ within $\hbar/(\ _F \ ) \quad 1$ as

$$\Pi_{ij,} (\boldsymbol{q}, 0) = \frac{_e}{4 \ _F} v_{F,} \ q \ \ _j \ _a \mathrm{tr}[\hat{}_i \hat{}_a] + o(q^2). \quad (10)$$

Substituting Eq. (10) into Eq. (5), we nally obtain $j_{i,}^{\mathrm{eq}}$ in the lowest order of $\boldsymbol{q}$ as

$$j_{i,}^{\mathrm{eq}}(\boldsymbol{x},0) = \frac{i\hbar \ {}_e J_{\mathrm{ex}} e v_{\mathrm{F},}^2}{2 \ {}_{\mathrm{F}} V} \sum_{\boldsymbol{q}} e^{\ i\boldsymbol{q}\cdot\boldsymbol{x}} q \ {}_j \ {}_i S'^j_{\boldsymbol{q},\Omega=0} + o(q^2) \quad (11)$$

$$= \frac{\hbar \ {}_e J_{\mathrm{ex}} e v_{\mathrm{F}}^2}{2 \ {}_{\mathrm{F}}} \ {}_i \ {}_j \nabla \ S'^j(\boldsymbol{x}). \quad (12)$$

In the above equation, we use $v_{\mathrm{F},}^2 = v_{\mathrm{F}}^2$. From Eq. (12), $j_{i,}^{\mathrm{eq}}$ is induced by $\boldsymbol{\nabla} \ \boldsymbol{S}$ and is independent of .

3. Discussion of axial current, charge current, and spin polarization density

Using Eq. (12), we discuss the axial current $\boldsymbol{j}_5^{\mathrm{eq}} \ \boldsymbol{j}_+^{\mathrm{eq}} \ \boldsymbol{j}^{\mathrm{eq}}$. Because $j_{i,}^{\mathrm{eq}}$ in the Eq. (12) is independent of , the di erence between $\boldsymbol{j}_+^{\mathrm{eq}}$ and $\boldsymbol{j}^{\mathrm{eq}}$ are canceled out each other. Therefore, the axial current $\boldsymbol{j}_5^{\mathrm{eq}}$ becomes

$$\boldsymbol{j}_5^{\mathrm{eq}} = 0. \quad (13)$$

However, the total charge current, which is de ned by $j_i^{\mathrm{eq}} = j_{i,+}^{\mathrm{eq}} + j_{i,}^{\mathrm{eq}}$, is given by

$$\boldsymbol{j}^{\mathrm{eq}}(\boldsymbol{x},0) = \frac{\hbar \ {}_e J_{\mathrm{ex}} e v_{\mathrm{F}}^2}{{}_{\mathrm{F}}} \boldsymbol{\nabla} \ \boldsymbol{S}'. \quad (14)$$

From the above result, $\boldsymbol{j}^{\mathrm{eq}}$ is proportional to $\boldsymbol{\nabla} \ \boldsymbol{S}' = \boldsymbol{\nabla} \ (\boldsymbol{S} \ \frac{\hbar v_{\mathrm{F}}}{2J_{\mathrm{ex}}}\boldsymbol{Q})$. When $\boldsymbol{Q}$ is also independent of the space in the WS, the charge current becomes $\boldsymbol{j}^{\mathrm{eq}} = \frac{\hbar \ {}_e J_{\mathrm{ex}} e v_{\mathrm{F}}^2}{\epsilon_{\mathrm{F}}} \boldsymbol{\nabla} \ \boldsymbol{S}$. Here $\boldsymbol{j}^{\mathrm{eq}}$ can be regard as the magnetization current density as $\boldsymbol{j}^{\mathrm{eq}} \propto \boldsymbol{\nabla} \ \boldsymbol{M}$, because the localized spin $\boldsymbol{S}$ can be replaced with the magnetization $\boldsymbol{M} = \ g \ {}_B \boldsymbol{S}/a^3$, where g is the Lande factor, ${}_B$ is the Bohr magneton, a is the lattice constant.

We next discuss the spin density $\boldsymbol{s}^{\mathrm{eq}}$, which is de ned by $\boldsymbol{s}^{\mathrm{eq}} = \boldsymbol{s}_+^{\mathrm{eq}} + \boldsymbol{s}^{\mathrm{eq}}$, in the WS from Eq. (12). Here $\boldsymbol{s}^{\mathrm{eq}}_{=\pm} = \frac{1}{2}\langle \ ^{\dagger} \ \hat{} \ \rangle$ is the spin density in each helicity. Because the direction of the spin and the momentum of right (left)-handed are parallel (antiparallel) to each other in the WS, $\boldsymbol{s}^{\mathrm{eq}}$ is represented by $\boldsymbol{s}^{\mathrm{eq}} = \boldsymbol{j}^{\mathrm{eq}}/(\ 2ev_{\mathrm{F},} \ )$. From Eq. (12), $\boldsymbol{s}^{\mathrm{eq}}$ is given by

$$\boldsymbol{s}^{\mathrm{eq}}(\boldsymbol{x},0) = \frac{\hbar \ {}_e J_{\mathrm{ex}} v_{\mathrm{F}_\gamma}}{2 \ {}_{\mathrm{F}}} \boldsymbol{\nabla} \ \boldsymbol{S}'. \quad (15)$$

From the above equation, $\boldsymbol{s}^{\mathrm{eq}}$ is induced by $\boldsymbol{\nabla} \ \boldsymbol{S}'$ and is proportional to the sign of $=$ as $\boldsymbol{s}_+^{\mathrm{eq}} = \ \boldsymbol{s}^{\mathrm{eq}}$. Therefore, total spin density $\boldsymbol{s}^{\mathrm{eq}}$ vanishes as

$$\boldsymbol{s}^{\mathrm{eq}} = 0. \quad (16)$$

From the above equation, there is no spin polarization due to the static localized spin in the WS. Equations (13), (14), and (16) are the main results of this section.

Next, we will compare the property of $\boldsymbol{j}_5^{\mathrm{eq}}$ and the nonequilibrium axial current $\boldsymbol{j}_5^{\mathrm{neq}}$. $\boldsymbol{j}_5^{\mathrm{eq}}$ is zero, but $\boldsymbol{j}_5^{\mathrm{neq}}$ is induced by the dynamical localized spin [16]. The summary of the di erence is shown in Table I. In a similar way, there is the di erence between $\boldsymbol{j}^{\mathrm{eq}}$ and the nonequilibrium charge current $\boldsymbol{j}^{\mathrm{neq}}$. $\boldsymbol{j}^{\mathrm{eq}}$ is generated by $\boldsymbol{\nabla} \ \boldsymbol{S}$ as $\boldsymbol{j}^{\mathrm{eq}} \propto \boldsymbol{\nabla} \ \boldsymbol{S}$, but $\boldsymbol{j}^{\mathrm{neq}}$ is zero. The property of $\boldsymbol{s}^{\mathrm{eq}}$ and the nonequilibrium spin density $\boldsymbol{s}^{\mathrm{neq}}$ are di erence. $\boldsymbol{s}^{\mathrm{eq}}$ is zero and $\boldsymbol{s}^{\mathrm{neq}}$ is driven by the dynamics of the localized spin $\partial_t \boldsymbol{S}$. Therefore, there is no spin polarization in the WS unless the localized spin depends on time.

Table 1. Summary of the axial current, the charge current, and the spin density due to the localized spin.

	Axial current	Charge current	Spin density
Static $\boldsymbol{S}$	$j_5^{\mathrm{eq}} = 0$	$\boldsymbol{j}^{\mathrm{eq}} \propto \boldsymbol{\nabla} \quad \boldsymbol{S}$	$\boldsymbol{s}^{\mathrm{eq}} = 0$
Dynamical $\boldsymbol{S}$	$j_5^{\mathrm{neq}} \propto \partial_t \boldsymbol{S}$	$\boldsymbol{j}^{\mathrm{neq}} = 0$	$\boldsymbol{s}^{\mathrm{neq}} \propto \partial_t \boldsymbol{S}$

Finally, we discuss the gauge invariance in the presence of the localized spin in the WS. Owing to spin-momentum locking, $\boldsymbol{S}$ plays a role like the electromagnetic vector potential as $\mathcal{H}_{\mathrm{W},} + \mathcal{H}_{\mathrm{ex},} \propto \boldsymbol{\sigma} \quad (\boldsymbol{k} \quad \frac{e}{\hbar}\boldsymbol{\mathcal{A}} \quad)$, where the vector potential $\boldsymbol{\mathcal{A}} \quad = J_{\mathrm{ex}}\boldsymbol{S}/(ev_{\mathrm{F},} \quad)$ is conjugate to $\boldsymbol{j}^{\mathrm{eq}}$. Therefore, the observable quantity should be proportional to the gauge invariant form as $\partial_t \boldsymbol{\mathcal{A}} \quad \boldsymbol{\mathcal{E}}$ or $\boldsymbol{\nabla} \quad \boldsymbol{\mathcal{A}} \quad \boldsymbol{\mathcal{B}}$. From viewpoint of the e ective electromagnetic eld, $\boldsymbol{j}^{\mathrm{eq}}$ can be driven by the e ective magnetic eld $\boldsymbol{\mathcal{B}}$ from Eq. (14).

4. Conclusion

We study the equilibrium axial current $\boldsymbol{j}_5^{\mathrm{eq}}$ due to $\boldsymbol{S}$ in the WS / MI junction. Using the Green's function, we calculate $j_{i,}^{\mathrm{eq}}$ in the linear response of $\boldsymbol{S}$. From the results, $j_{i,}^{\mathrm{eq}}$ is proportional to $\boldsymbol{\nabla} \quad \boldsymbol{S}'$ and $v_{\mathrm{F},}$ as you can see Eq. (12). Then, $\boldsymbol{j}_5^{\mathrm{eq}}$ becomes zero because of the cancelation between $j_{i,+}^{\mathrm{eq}}$ and $j_{i,}^{\mathrm{eq}}$. In addition, the total spin density $\boldsymbol{s}^{\mathrm{eq}}$ is also zero, but the total charge current is induced by the localized spin from Eq. (14).

Acknowledgments

This work was supported by Grant-in-Aid for JSPS Fellows.

References

[1] Vilenkin A 1980 *Phys. Rev.* D **22**, 3080.
[2] Metlitski A M and Zhitnitsky R A 2005 *Phys. Rev.* D **72**, 045011.
[3] Newman M G and Son T D 2006 *Phys. Rev.* D **73**, 045006.
[4] Kharzeev D and Zhitnitsky A 2007 *Nucl. Phys.* A **797**, 67.
[5] Kharzeev E D, McLerran D L, and H. J. Warringa 2008 *Nucl. Phys.* A **803**, 227.
[6] Gorbar V E, Miransky A V, Shovkovy A I, and Wang X 2013 *Phys. Rev.* D **88**, 025025.
[7] Kharzeev D, Landsteiner K, Schmitt A, Yee H 2013 *Lect. Notes Phys.* **871**, 241.
[8] Wan X, Turner M A, Vishwanath A, and Savrasov Y S 2011 *Phys. Rev.* B **83** 205101.
[9] Balents L 2011 *Physics* **4**, 36.
[10] Burkov A A and Balents L 2011 *Phys. Rev. Lett.* **107**, 127205.
[11] Halasz B G and Balents L 2012 *Phys. Rev.* B **85**, 035103.
[12] Xu G, Weng H, Wang Z, Dai X, and Fang Z 2011 *Phys. Rev. Lett.* **107**, 186806.
[13] Liu C, Ye P, and Qi X 2013 *Phys. Rev.* B **87**, 235306.
[14] Hosur P and Qi X 2013 *C. R. Physique* **14**, 857.
[15] Tominaga J, Kolobov V A, Fons P, Nakano T, and Murakami S 2014 *Adv. Mater. Interfaces* **1**, 1300027.
[16] Taguchi K and Tanaka Y, 2014 Axial Current driven by Magnetization Dynamics in Weyl Semimetals *Preprint* 1406.4636.
[17] Haug H and Jauho P A 2007 *Quantum Kinetics in Transport and Optics of Semiconductors* (Springer, New York, 2nd ed.), pp. 45–46.

Journal of Physics: Conference Series **568** (2014) 052033 doi:10.1088/1742-6596/568/5/052033

Electronic and Thermodynamic Properties of Sn and SbI_3 Doped Single Crystals $p - Bi_2Te_3$

M Z Tahar[1], **D I Popov**[2] **and S A Nemov**[3]

[1]Department of Physics, State University of New York at Brockport, NY 14420, USA
[2]ZAO NevInPat, per. Pirogova 7, St.-Petersburg 190000 Russia
[3]State Polytechnical University, Politechnicheskaya ul.29, St.-Petersburg 195251 Russia

E-mail: `mtahar@brockport.edu`

Abstract. $Bi_2Te_3 - Sb_2Te_3$ solid solutions, the most commonly used thermoelectrics (~300K), are characterized by inhomogeneities and defects of various types, and doping contributes to spatial inhomogeneity. These impurity energy states fall into the band gap with small high-field SdH oscillations. However, it is known that Sn impurity pins the Fermi level and tremendously improves the spatial homogeneity, leading to observation of high amplitude SdH oscillations in lower magnetic field. The iodine (I) impurity was chosen to shift the Fermi level and affect the filling factor of impurity states. Numerous experimental results on magnetoresistivity, specific heat and magnetic susceptibility indicate that the most probable model for one-electron states in Bi_2Te_3 solid solutions doped with Sn is the presence of two impurity bands one filled and one empty- with the Fermi level pinned in-between. Earlier we reported on SdH oscillations at $T = 4.2$K and in magnetic Field up to 10T. Here, we report on investigations of the magnetoresistivity ($2 < T < 295$K) and its oscillations ($T < 12$K). We deduce the carriers mobility and concentration and their temperature dependence along with that of the dampening of SdH amplitude. We also report on specific heat measurements in the range ($2 < T < 295$K), with a low T power law of exponent p=3.5 behavior.

1. Introduction

Bismuth telluride (Bi_2Te_3) is a narrow gap semiconductor with Eg ~ 0.1 eV and is of practical importance as a foundation material for developing high efficiency thermoelectric materials for use in cooling devices operating at ~ 300 K. Bi_2Te_3 belongs to the class of layered R_{3mm} structures with anisotropic physical properties. Further, due to peculiarities in its phase diagram, material's synthesis can result in the fact that even its most perfect crystals are plagued by a large quantity of various point defects and inhomogeneities (~ 10^{19}cm^{-3}). These defects lead to considerable fluctuations of the electro-physical properties of crystals and their technical characteristics.

Usually, the doping of a semiconductor perturbs the material's homogeneity and the introduced impurity states fall in the forbidden zone. However, in narrow gap semiconductors IV-VI (A^4B^6) and V-VI (A^5B^6), some species of impurities create quasi-local states located on the background of the allowed zone(s). Sn is such an unusual acceptor-like impurity in Bi_2Te_3. According to Zhitinskaya *et. al.*[1], and Kulbachinskii *et. al.*[2], doping of Bi_2Te_3 with Sn results in the formation of a band of the quasi-local states on the background of the valence band (VB) of bismuth telluride. Additionally, the impurity states of Sn stabilize the concentration of holes in the VB by pinning down the Fermi level. This leads to considerable improvement of the spatial homogeneity of electro-physical properties of crystals $Bi_2Te_3 : Sn$.

The Sn doped Bi_2Te_3 crystals studied by Zhitinskaya[1] , Kulbachinskii[2] , and Laiho[3] had different content of Sn and location of E_F. For meaningful crystal properties comparison, it is necessary to fix the content of Sn and to vary the Fermi level by doping with an additional impurity. In view of this, we pursue work on samples of Bi_2Te_3 doped with 0.5% Sn, which were additionally doped with iodine (I), as introduced in the form of SbI_3 (0.05% in sample #97 and 0.1% in #98). We investigate the nuances between the two compositions by obtaining magnetoresistance at various temperatures the range of 300 K down to 2 K.

The textbook [4] magnetoconductivity is given by

$$\sigma_{xx}(H) = \frac{\sigma_0}{1 + (\omega_c\tau)^2} \tag{1}$$

with $\omega_c = \frac{eH}{m}$. Rewriting this as the magnetoresistivity, $\rho_{xx}(H) = \rho_0(1 + (\omega_c\tau)^2)$ and further as:

$$\frac{\rho_{xx}(H) - \rho_0}{\rho_0} = \frac{R_{xx}(H) - R_0}{R_0} = (\omega_c\tau)^2 = (\frac{e\tau H}{m})^2. \tag{2}$$

Thus, for low to moderate field, the coefficient of the quadratic term in H is $(\frac{e\tau}{m})^2 = \mu^2$, just the squared mobility. The conductivity $\sigma_0 = \frac{1}{\rho_0} = pe\mu$, therefore resistivity measurements, in zero field along with scans to high fields, allow for the mobility to be obtained, and for concentration of carriers p to be deduced, using the standard four-contact method instead of five or six contacts. This is especially convenient for highly magnetoresistive materials where Hall measurements may require more steps due to alignment difficulty of transverse contacts, as compared to longitudinal. This is also in the reverse order of the traditional method of measuring the resistivity and the Hall coefficient, which depend on the thickness of the sample, and then deducing the carriers mobility.

Previously, we reported on the resistivity temperature dependence, the low temperature SdH[5, 6], and specific heat[6]. Here, we expand on those results of the two compounds in light of the above and report on the carriers mobility and concentration and their temperature dependence, as well as the temperature dependence of SdH oscillations and estimate the carrier effective mass.

2. Experimental Details

The resistance measurements were carried out on thin samples using the standard four-point method, as has been described in previous reports [5, 6] and references therein. The applied magnetic field is perpendicular to the basal the layers of the material and the probing current is parallel to them. All measurements were done in a Quantum Design PPMS cryostat with a superconducting magnet capable of ramping the field at 0.6 T/min. up to 9 T and sample's temperature control for setting, holding and reproducibility $< 1\%$.

The heat capacity measurements were carried out in the same PPMS cryostat using the pulsed method, whereby a heat pulse is applied while monitoring the sample's temperature during the heating and drifting without heat. Then, standard routines were used in fitting temperature time response to obtain the heat capacity. This was done twice at the same temperature in the range 2 K - 295 K, first for determining the heat capacity of the addenda and then for the sample and addenda. The difference yields the sample's heat capacity which can be converted into the molar specific heat.

3. Results and Discussion

First, we present the magnetoresistance in an applied magnetic field perpendicular to basal planes ($0 < H < 9$T) for the two compounds at a multitude of temperatures (12 K- 280 K).

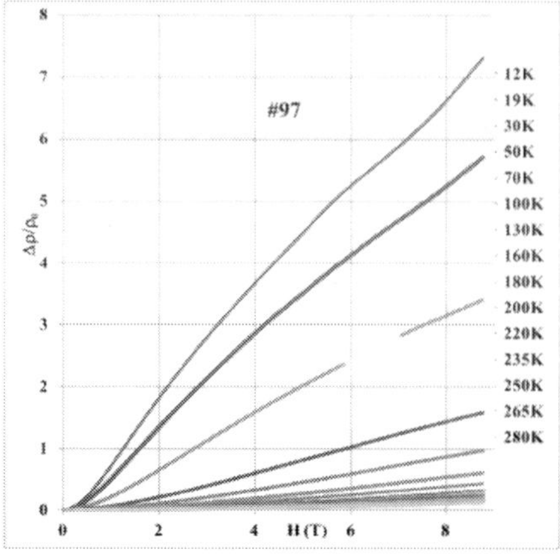

Figure 1. Magnetoresistivity, $\frac{\rho(H)-\rho_0}{\rho_0}$ vs. H of Be_2Te_3 doped with 0.5% Sn and 0.05% SbI_3 (#97) at various temperatures.

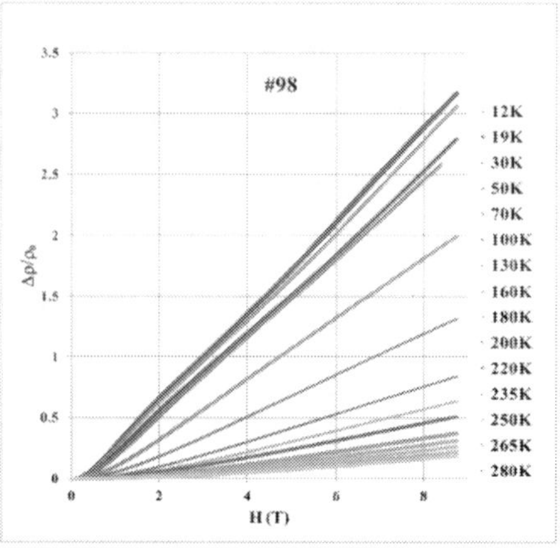

Figure 2. Magnetoresistivity, $\frac{\rho(H)-\rho_0}{\rho_0}$ vs. H of Be_2Te_3 doped with 0.5% Sn and 0.1% SbI_3 (#98) at various temperatures.

The data are shown in Fig. 1 for #97 and Fig. 2 for #98. Both sets of data show a positive magnetoresistance with a quadratic field dependence ($H < 2$T), which deviates at high fields and decreases with temperature. Note that the magnetoresistance for #97 is $\sim 2\times$ higher than for #98 and the spacings between the curves from one temperature to the next are different. Additionally, the magnetoresistance starts exhibiting quantum oscillations for $H \sim 2.5$ T and at temperature as high as 12 K.

For every temperature $\frac{\rho(H)-\rho_0}{\rho_0}$ vs. H, for $H < 1$ T, was least squares fit to a second order polynomial and the resulting coefficient of H^2 was used to obtain the mobility, μ. Fig. 3 shows a log-log plot of μ as a function of T; as highlighted by the straight lines, μ decreases with temperature as T^{-1} for #97 sample and slightly steeper then that for sample #98, indicative of additional scattering besides phonons, due to the added impurities. Below 10 K, both compounds have constant mobilities but of lower value for #98 due to higher impurities and the other way around above 30 K due to phonons.

Further, using μ and $\rho(0)$ at that same temperature, one can deduce the carrier density, which is graphed in Fig. 4. Note the carrier density for #97 is constant ($\sim 2 \times 10^{18}\text{cm}^{-3}$) up to 50 K then increases with temperature, and is higher than for #98, which is constant ($\sim 1.5 \times 10^{18}\text{cm}^{-3}$) up to 10 K at most, and decreases to minimum between 150 and 200K and then increases with temperature. This behavior is consistent with the additional SbI_3 impurities, as they populate the valence band and create defects whose scattering effects become more prominent at intermediate temperatures.

The magnetoresistance for #97 for various low temperatures is shown in Fig. 5 and exhibits only one period of oscillations, which indicates that only one type of holes (light holes) effectively contribute to all transport properties in this system. The single period of oscillation is confirmed by extraction of the oscillations from the signal. This was accomplished by fitting the original data to a power law in H^{-1}, judged better than a polynomial, and subtracting it to obtain the

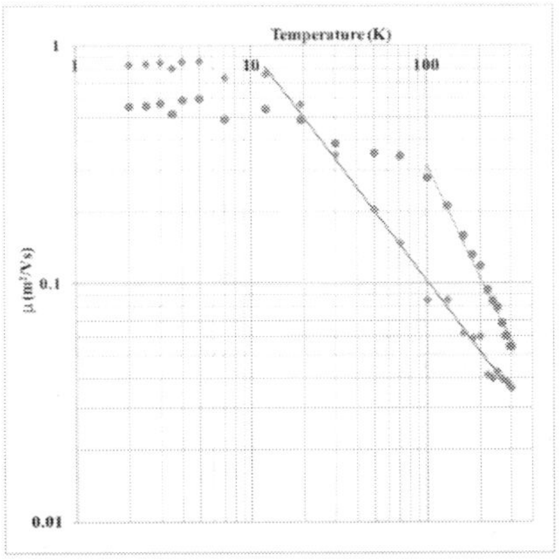

Figure 3. μ vs. T for #97 (blue) and #98 (red), as obtained from the coefficient of H^2 in $\frac{\rho(H)-\rho_0}{\rho_0}$ vs. H at the relevant temperature.

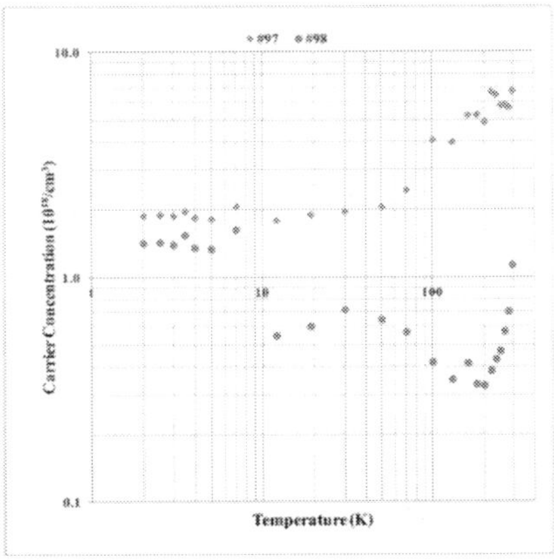

Figure 4. The Carrier density, $p(T)$, as deduced from the mobility and resistivity at the relevant temperature.

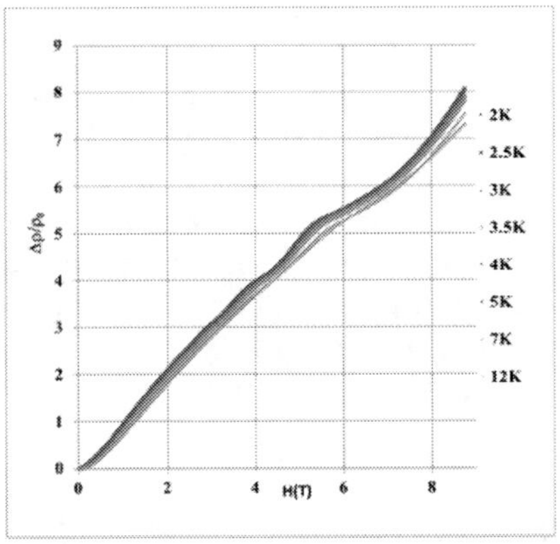

Figure 5. SdH Oscillations of #97 for field scans at several low temperatures (2 K to 12 K).

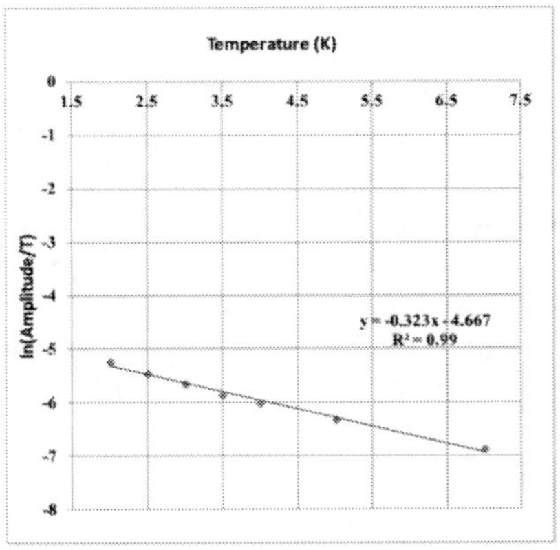

Figure 6. $ln(A/T)$ vs. $ln(T)$ (#97) for $2K < T < 7K$, showing the linear fit used to estimate m^*.

oscillating part. Then, using sine and cosine Fourier Transforms to assert the single frequency (11 T) and obtain the amplitude and its weak temperature dependence, as shown in Fig. 6. From the linear fit of $ln(A/T)$ vs. $ln(T)$ one can estimate [7] the carriers effective mass, $m^* \sim 0.2m_e$.

Second, we show the specific heat results in a log-log plot of the molar specific heat vs. temperature in Fig. 7. Note the linear behavior in low T and the saturation at high. The low

Journal of Physics: Conference Series **568** (2014) 052033 doi:10.1088/1742-6596/568/5/052033

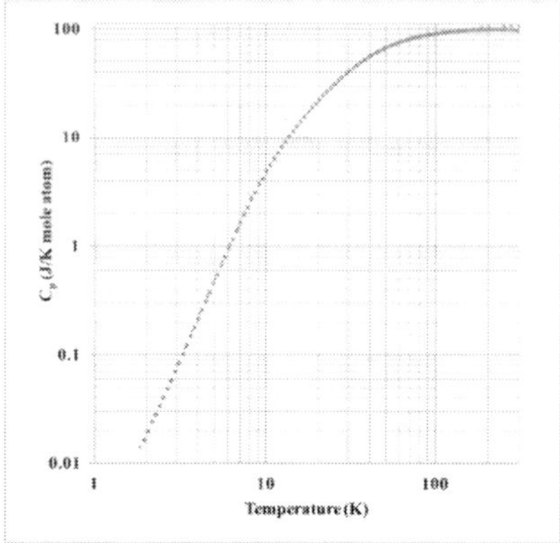

Figure 7. Molar specific heat, $C_P(T)$ for both compounds in the range 2K - 300K.

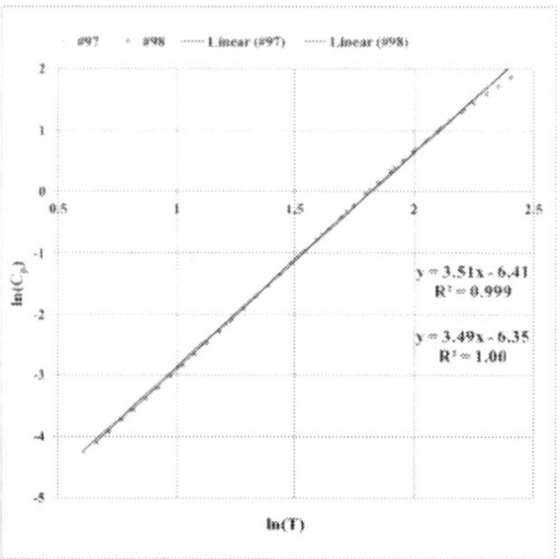

Figure 8. $ln(C_P)$ vs. $ln(T)$ in the range 2K - 10K, with least squares linear fits.

T behavior is further shown to exhibit a power law with an exponent, p= 3.5, as seen in Fig. 8. The high T saturation has a value of $\sim 12R$ indicative of 24 degrees of freedom for atomic vibrations.

4. Conclusion

The introduction of donor impurity (iodine) with the same order of concentration as Sn impurity, maintains the Sn impurity zone but affects the filling factor of these states and shifts the Fermi level. The observation of quantum oscillations is a demonstration of excellent spatial homogeneity of electrical properties of quasi-local states of Sn. The observed frequency is 11 Tesla and the temperature variation of the amplitude of oscillations allows for an estimate of an effective mass of $0.2m_e$. Both results are an indication of a light effective mass and skinny cross-section ellipsoidal pocket. The specific heat temperature dependence is typical except for the low temperature $\sim T^{3.5}$, which warrants even lower temperature heat capacity measurements.

4.1. Acknowledgments

MZT acknowledges the support by the Department of Physics, The College at Brockport, State University of New York where all the measurements were carried out using the newly acquired PPMS as a result of facilities renovation.

References

[1] M K Zhitinskaya, S A Nemov and T E Svechnikova 1997 *Semiconductors* **31**, 375
[2] V A Kulbachinskii, A Y Kaminskii, N Miyajima, M Sasaki, H Negish, and H Kadomatsu 1999 *JETP Letters* **70 (11)**, 767
[3] R Laiho, S A Nemov, A V Lashkul, E Lahderanta, T E Svechnikova and D S Dvornik 2007 *Semiconductors* **41 (5)**, 546
[4] Kittel, 2005 *Introduction to Solid State Physics* J. Wiley
[5] M Z Tahar, S A Nemov, D I Popov and T E Svechnikova *J. Phys.:Conf. Ser.* **150** (2009) 022082
[6] M Z Tahar, D I Popov and S A Nemov *J. Phys.:Conf. Ser.* **400** (2012) 042056
[7] D Shoenberg, *Magnetic Oscillations in Metals* Cambridge University Press 1984

Isolating electrons on superfluid helium

Maika Takita and S.A. Lyon
Department of Electrical Engineering, Princeton University, Princeton, NJ, USA
E-mail: mtakita@princeton.edu

Abstract. Electrons floating on the surface of superfluid helium have been suggested as promising mobile spin quantum bits (qubits). Transferring electrons extremely efficiently in a narrow channel structure with underlying gates has been demonstrated, showing no transfer error while clocking 10^9 pixels in a 3-phase charge coupled device (CCD). While on average, one electron per channel was clocked, it is desirable to reliably obtain a single electron per channel. We have designed an electron turnstile consisting of a narrow (0.8μm) channel and narrow underlying gates (0.5μm) operating across seventy-eight parallel channels. Initially, we find that more than one electron can be held above the small gates. Underlying gates in the turnstile region allow us to repeatedly split these electron packets. Results show a plateau in the electron signal as a function of the applied gate voltages, indicating quantization of the number of electrons per pixel, simultaneously across the seventy-eight parallel channels.

1. Introduction

Electrons are bound above the surface of liquid helium by their weak image potential as well as potentials from underlying electrostatic gates [1–4]. These underlying gates control the position of electrons on helium and can move them across the sample into various regions [5, 6]. The electrons reside in vacuum, about 11nm above the surface, forming a very clean classical two dimensional electron system. Electron spins are expected to have little spin decoherence, due to the small spin-orbit interaction [7], allowing for the possibility of mobile spin qubits in a condensed matter system. To realize mobile spin qubits, control of individual electrons is necessary. With a single-electron transistor (SET) placed under the helium surface, individual electrons have been counted as they come in and out of a trap [8]. Electron transport of a single line of charge has been observed with a point constriction formed in a microchannel using a split gate [9]. We show that in addition to the confinement from a channel, narrow underlying gates perpendicular to the channel can be used to trap electrons over one electrostatic gate. The narrow channel and gates for isolating single electrons and wider channels with CCD's are integrated, moving towards single electron control.

2. Device Structure

The complementary metal-oxide-silicon (CMOS) process is a standard foundry process that consists of multiple metal layers above a silicon substrate, with the metal layers separated by insulators. The helium channel device [6, 9–12] with turnstiles was fabricated utilizing this technology. Seventy-eight parallel channels were designed using one of the metal layers as the top ground plane, with the metal layer below defining the gates that underlay all the channels as shown in Fig.1. The device consists of an electron collecting region to the right of the

Journal of Physics: Conference Series **568** (2014) 052034 doi:10.1088/1742-6596/568/5/052034

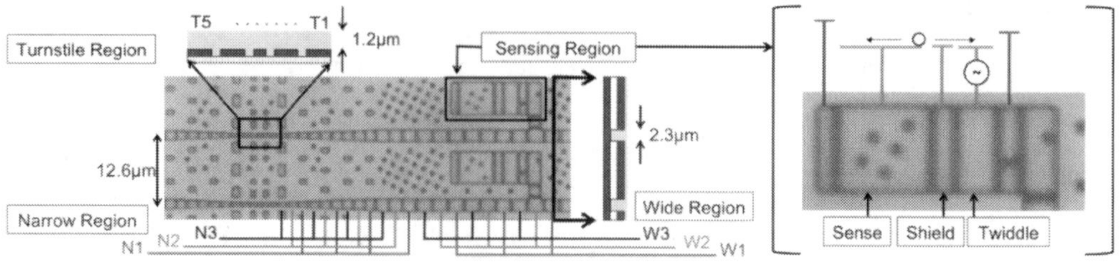

Figure 1. Microscope image of the device with cartoon of the channel profile. Two out of seventy-eight channels are shown with gate labels. There are two sets of three-phase CCD's in the narrow region and the wide region. Dark spots on the images are holes etched through the metal to improve the yield of the fabrication processes by having a similar metal density over large areas. An enlarged cartoon of the turnstile cross-section and the side-view of the channels show their dimensions. The cartoon of electron detecting scheme; confining the electrons over sense, shield, and twiddle gate, by keeping neighboring gates repulsive and modulating twiddle gate to move electrons on and off the sense gate; is shown over magnified image of sensing region.

image in Fig.1, sensing region, wide CCD (WCCD) region, narrow CCD (NCCD) region, and a turnstile region. Electrons are collected over a wide underlying gate, termed *plate*, and clocked into 8.7μm long pixels (note: three gates comprise a pixel, with a 2.9μm gate period), both of which have 2.3μm wide channels. The WCCD region is followed by a NCCD region with four three-phase pixels. The first NCCD has a 7.5μm pixel length while it remains in a 2.3μm wide channel. It then necks down to a 0.8μm wide channel over three 6μm period pixels. The NCCD section is followed by the turnstile region which contains five individual gates. The first two gates, designated *T1* and *T2*, are 1μm wide, with 0.4μm spacing. The center gate, *T3*, is the narrowest gate (0.5μm), with 0.4μm spacing to the neighboring gates. The fourth and fifth gates, *T4* and *T5*, have the same dimensions as the first two, which forms a symmetric turnstile region. Electrons are detected in the sensing region which branches out from the channel, creating a larger area for electrons to be detected through a capacitive coupling to a wide *sense* gate [5]. A small ac-voltage (V_{ac}) is applied to the *twiddle* gate to push electrons on and off the *sense*, while a narrow gate between these two gates shields the direct capacitive coupling between them. The periodically induced voltage on the *sense* gate induced by the electrons at the frequency of V_{ac} (111kHz) is buffered by a high electron mobility transistor (HEMT) and detected using a lock-in amplifier. The HEMT is mounted on the backside of the printed circuit board, on which the sample is wirebonded.

3. Electron Detection

The large *sense* gate in the wider channels allows for improved electron detection due to a decreased coupling to the top ground plane. Figure 2 (a) shows the potential simulation of the sensing region when electrons are pushed on and off the *sense* gate. The star is placed at the potential minimum in each case. When electrons are pushed away from the *twiddle* gate the potential minimum is on the *sense* gate. When they are pulled towards the *twiddle*, the potential minimum is shifted away from the *sense*. The signal will increase as electrons are brought closer to the center of the *sense* and then pushed completely off the *sense* onto the *twiddle*. This is accomplished by applying a larger V_{ac} on *twiddle*. Figure 2 (b) is a plot of measured electron signals from a range of applied voltages on the *shield* gate for different V_{ac} on the *twiddle* gate. The signal increases as the amplitude of V_{ac} increases, confirming that electrons are getting closer to the middle of the *sense* where the induced voltage will be maximized.

Journal of Physics: Conference Series **568** (2014) 052034 doi:10.1088/1742-6596/568/5/052034

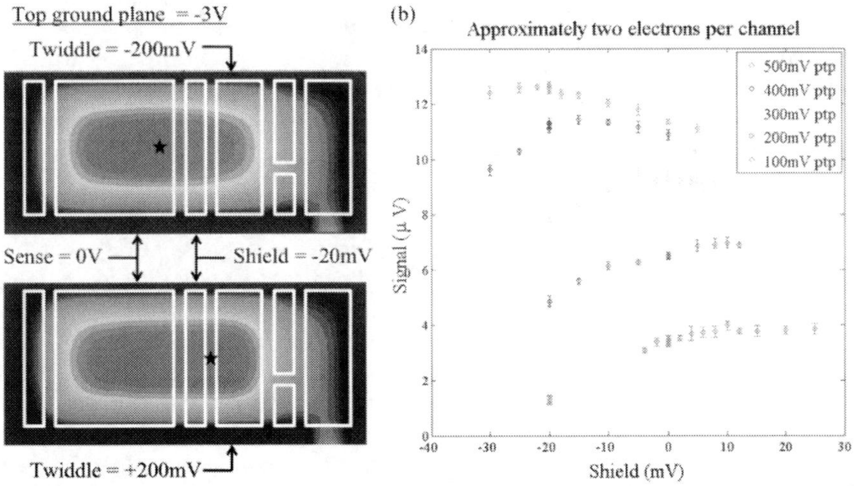

Figure 2. Maximizing electron signal. (a) Potential simulation of the sensing region when electrons are on the *sense* being pushed away from the *twiddle* when it is at the most repulsive potential (-200mV) and when electrons are off the *sense* being pulled over to the *twiddle* when it is at the most attractive potential (+200mV). The gates to the right of the *twiddle* are kept at -2V and the gate to the left of the *sense* is kept at -3V. (b) The plot of electron signal while varying *shield* gate voltage with different V_{ac} on *twiddle*.

The potential minimum when electrons are pushed onto the *sense* gate could also be shifted by applying a small negative bias on the *shield* gate placed in between *twiddle* and *sense*. This also shifts the minimum away from the *sense* gate when electrons are pulled onto the *twiddle* gate. The experiments were performed with -20mV on the *shield* gate, and applying a 400mV *twiddle* V_{ac}.

4. Experiment and Results

Electrons are photoemitted into the vacuum above the *plate* from a zinc film placed a few millimeters above the device [13]. Electrons collected on the *plate* are subsequently clocked into the sensing region using the WCCD gates. Starting with an arbitrary number of electrons per channel, these packets of electrons are brought into the NCCD gates. As the charge packet goes through the NCCD gates, the number of electrons that are allowed on each gate decreases as the channel narrows. Approximately 20 electrons per channel remain above turnstile gate, *T1*, when appropriate voltages are applied. These electrons are our initial charge packet. Other electrons that are not held on *T1* are left behind above other gates, and these are clocked back into the *plate*. The initial packet of electrons are brought onto *T2*, *T3*, and *T4* with +2V bias on those gates, while *T1* and *T5* are made repulsive at -3V, which is the same as the top ground plane bias. While voltage applied to *T4* (V_{T4}) is always held fixed at +2V, we chose a *T2* voltage (V_{T2}) to set a voltage difference, $\Delta V = V_{T2} - V_{T4}$. For each measurement, ΔV is chosen to make the potential well on the turnstile gates somewhat asymmetric. The voltage applied to *T3* is gradually changed from +2V to -3V over 5 seconds, splitting the electron packet onto gates *T2* and *T4*. The electrons above *T2* are then clocked out from the turnstile, measured in the sensing region and brought into a gate in the wide channel for storage. The remaining electrons above *T4* are again distributed onto gates *T2*, *T3*, and *T4* by bringing the bias on

T2 and *T3* back to +2V with *T1* and *T5* at -3V. *T2* is set to the same V_{T2}, and the splitting sequence is repeated until no more electrons are found above *T2* after splitting. Electrons above *T4* are then brought back to the sensing region for measurement. These electrons are returned to those stored during the splitting and we perform the experiment again with the same group of electrons but a new ΔV. Typical clock gate periods are 160 milliseconds. Given the rather low electron mobility at the operation temperature of 1.6K, hot electron effects are expected to be negligible at these time scales.

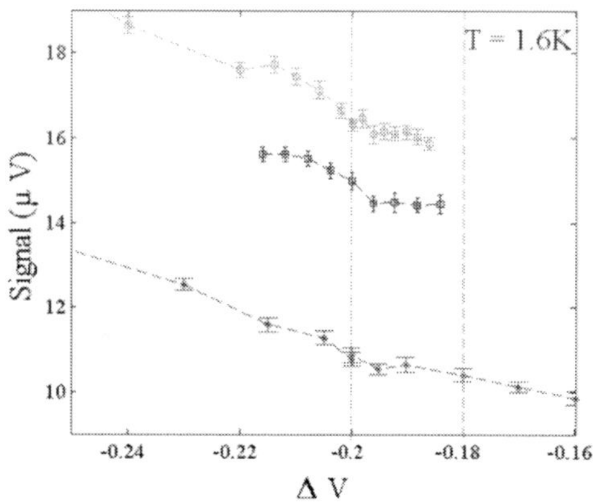

Figure 3. Plot of electron signal left above *T4* after multiple splitting of electron packets, versus ΔV, the voltage difference between *T2* and *T4*. A plateau is seen in the region between ΔV = -0.20V to -0.18V. 1V change on the *T2* gate corresponds to changing the potential on the helium surface by approximately 20meV. Each dataset is from a different experiment. Dotted lines connecting data points are guide to the eye.

The signal, proportional to the number of electrons which were found above gate *T4* at the end of the splitting process integrated across all seventy-eight channels, is shown in Fig.3 as a function of ΔV. The asymmetry while splitting the electron packets with varying ΔV determines the number of electrons left in *T4*. When there are many electrons in the turnstile approximately half of the electrons go onto *T2* and the other half onto *T4*. After multiple "splits" there will be two electrons left, and the Coulomb repulsion assists the splitting of the electrons onto *T2* and *T4*, as long as ΔV is small (in magnitude) so the two electrons will prefer to split onto two gates, rather than both electrons residing above *T4*. This sets the lower bound of V_{T2} which would allow one electron on *T4*. When there is one electron in a channel, the electron will move onto *T4* as long as ΔV is large enough for the last electron to fall onto *T4* instead of *T2*. This sets the upper bound of V_{T2}, which would allow the final electron to go onto *T4* after splitting. A plateau is seen in Fig.3 over multiple experiments between ΔV = -0.20 and -0.18V. From finite element calculations, we find that applying 1V on either *T2* or *T4* changes the potential on the helium surface by 20meV and 1V on *T3* changes the potential by 15meV. For most experiments, 4mV steps were taken on *T2* gate, which corresponds to about 0.08mV (0.9K) on the surface. All the experiments are done at 1.6K, so the plateaus shown in Fig.3 are a few kT wide.

5. Conclusion

The existence of a plateau suggests a quantized electron signal, where the number of electrons per channel is fixed. Signal calibration using the measured capacitance (room temperature) and an estimate of the induced voltage on the *sense* gate with the parameters used in the experiments gives a signal of approximately 8.5μV for one electron per channel when all seventy-eight channels are filled. The measured signal at the plateau region is between 10μV to 16μV. These values are larger than our estimate for a single electron per channel signal. If we take into account that not all the channels are filled with electrons, the measured signal approximately corresponds to two electrons per channel. The uniformity of electron occupation across the channels has been measured using a vertical channel as previously demonstrated [14]. We find that the number of channels filled with electrons varies from cooldown to cooldown with some empty channels at the top and bottom of the array. The smaller signal for the blue set of data points in Fig.3, as compared to the other two experiments probably arises from a larger number of empty channels in that experiment. From these considerations we conclude that the plateaus probably correspond to two electrons in a channel.

To investigate whether the plateaus in this region result from having one or two electrons per channel, detailed measurements at smaller ΔV are necessary. Nonetheless, the plateaus seen in the measurements are a clear indication of the existence of a regime in which the number of electrons in a channel is always fixed, and that all the channels are operating simultaneously to quantize the charge with one set of gate voltages.

Acknowledgments
We would like to thank J. Donnal for the electronics to apply clock voltages to the CCD. Work at Princeton was supported by the NSF under Grant No. DMR-1005476.

References
[1] Sommer W T 1964 *Phys. Rev. Lett.* **12**(11) 271
[2] Williams R and Crandall R S 1971 *Phys. Rev. A* **4**(5) 2024
[3] Grimes C C 1978 *Surface Science* **73** 379–395
[4] Andrei E Y (ed) 1997 *Two-dimensional electron systems on helium and other cryogenic substrates* (Kluwer Academic Publishers)
[5] Bradbury F R, Takita M, Gurrieri T M, Wilkel K J, Eng K, Carroll M S and Lyon S A 2011 *Phys. Rev. Lett.* **107** 266803
[6] Sabouret G, Bradbury F R, Shankar S, Bert J A and Lyon S A 2008 *Applied Physics Letters* **92** 082104
[7] Lyon S A 2006 *Phys. Rev. A* **74** 052338
[8] Papageorgiou G, Glasson P, Harrabi K, Antonov V, Collin E, Fozooni P, Frayne P G, Lea M J, Rees D G and Mukharsky Y 2005 *Applied Physics Letters* **86** 153106
[9] Rees D G, Kuroda I, Marrache-Kikuchi C A, Höfer M, Leiderer P and Kono K 2011 *Phys. Rev. Lett.* **106** 026803
[10] Marty D 1986 *Journal of Physics C: Solid State Physics* **19** 6097
[11] Glasson P, Andresen S, Ensell G, Dotsenko V, Bailey W, Fozooni P, Kristensen A and Lea M 2000 *Physica B: Condensed Matter* **284-288** 1916–1917
[12] Ikegami H, Akimoto H and Kono K 2009 *Phys. Rev. Lett.* **102** 046807
[13] Shankar S, Sabouret G and Lyon S A 2010 *Journal of Low Temperature Physics* **161** 410–416
[14] Takita M, Bradbury F R, Gurrieri T M, Wilkel K J, Eng K, Carroll M S and Lyon S A 2012 *Journal of Physics: Coference Series* **400** 042059

Critical current in semiconductor nanowire Josephson junctions in the presence of magnetic field

Tomohiro Yokoyama[1,2], Mikio Eto[3], and Yuli V. Nazarov[1]

[1] Kavli Institute of Nanoscience, Delft University of Technology, Delft, The Netherlands
[2] Center for Emergent Matter Science, RIKEN Institute, Wako, Japan
[3] Faculty of Science and Technology, Keio University, Yokohama, Japan

E-mail: `T.Yokoyama@tudelft.nl`

Abstract. We study theoretically the critical current in semiconductor nanowire Josephson junction with strong spin-orbit interaction. The critical current oscillates with an external magnetic field. We reveal that the oscillation of critical current depends on the orientation of magnetic field in the presence of spin-orbit interaction. We perform a numerical simulation using a tight-binding model. The Andreev levels are calculated as a function of phase difference φ between two superconductors. The DC Josephson current is evaluated from the Andreev levels in the case of short junctions. The spin-orbit interaction induces the effective magnetic field. When the external field is parallel with the effective one, the critical current oscillates accompanying the 0-π like transition at the cusp of critical current. The distance of cusps increases gradually with increasing of the angle between the external and effective fields. The magnetic anisotropy of critical current is attributed to the spin precession due to the spin-orbit interaction.

1. Introduction

The spin-orbit (SO) interaction has attracted a lot of interest. In narrow-gap semiconductors, such as InAs and InSb, the strong SO interaction has been reported and many phenomena based on the SO interaction are investigated intensively, e.g., spin Hall effect [1]. The SO interaction has a great advantage also for application to the spintronic devices and to quantum information processing. InAs and InSb nanowires are interesting nanostructure for the application and studied in recent experiments, e.g., the electrical manipulation of single electron spin in quantum dots fabricated on the nanowires [2]. Nanowire-superconductor hybrid systems have been also examined to investigate the Majorana fermions induced by the SO interaction and the Zeeman effect [3].

In Josephson junctions, the supercurrent flows when the phase difference φ between two superconductors is present. In this paper, we investigate theoretically the Josephson junction of semiconductor nanowires with strong SO interaction. The supercurrent in semiconductor nanowires has been reported by experiment groups [4, 5, 6, 7]. The Josephson effect with SO interaction has been studied theoretically for some materials, e.g., magnetic normal metals [8], where the combination of SO interaction and exchange interaction results in an unconventional current-phase relation, $I(\varphi) = I_0 \sin(\varphi - \varphi_0)$. The phase shift φ_0 deviates the ground state of junction from $\varphi = 0$ or π, which is so-called φ_0-state. The anomalous supercurrent is obtained at $\varphi = 0$. In previous studies, we have pointed out that the anomalous effect is attributed

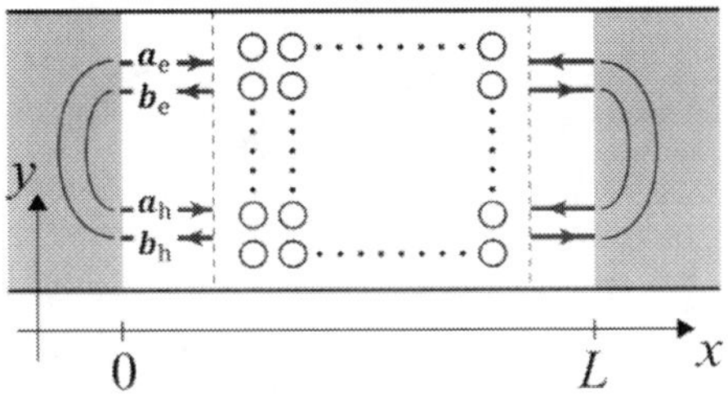

Figure 1. Model for a semiconductor nanowire Josephson junction. A tight-binding model is applied to the normal region with hard-wall potentials forming the nanowire. The nanowire is represented by a quasi-one dimensional system along the x direction. The normal region is $0 < x < L$ and the superconducting region induced by the proximity effect is $x < 0$, $x > L$.

to the spin-dependent channel mixing due to the SO interaction [9, 10]. In the present study, we focus on the critical current oscillation when an external magnetic field is applied. The DC Josephson current is evaluated from the Andreev levels in the case of short junction. We examine a numerical calculation using a tight-binding model for the nanowire. In this model, a particular form of SO interaction can be considered. When the angle between the external field and an effective magnetic field due to the SO interaction is smaller, the oscillation period of critical current is shorter.

2. Model

The nanowire along the x direction is connected to two superconductors (Fig. 1). At $x < 0$ and $x > L$, the superconducting pair potential is induced into the nanowire by the proximity effect. We assume that the pair potential is $\Delta(x) = \Delta_0 e^{i\varphi/2}$ at $x < 0$ and $\Delta(x) = \Delta_0 e^{-i\varphi/2}$ at $x > L$, where φ is the phase difference between the two superconductors. In the normal region at $0 < x < L$, $\Delta(x) = 0$. When a magnetic field is applied to the junction, the Zeeman effect is taken into account in the nanowire. The magnetic field is not too large to break the superconductivity and screened in the superconducting regions. The Hamiltonian is given by $H = H_0 + H_{\rm SO} + H_{\rm Z}$ with $H_0 = \boldsymbol{p}^2/2m + V_{\rm conf} + V_{\rm imp}$, the Rashba interaction $H_{\rm SO} = (\alpha/\hbar)(p_y\hat{\sigma}_x - p_x\hat{\sigma}_y)$, and the Zeeman term $H_{\rm Z} = g\mu_{\rm B}\boldsymbol{B}\cdot\hat{\boldsymbol{\sigma}}/2$, using effective mass m, g-factor g (< 0 for InSb), Bohr magneton $\mu_{\rm B}$, and Pauli matrices $\hat{\boldsymbol{\sigma}}$. We neglect the orbital magnetization effect in the nanowire. $V_{\rm conf}$ describes the confining potential forming the nanowire. $V_{\rm imp}$ represents the impurity potentials. We consider short junction, where the spacing between two superconductors is much smaller than the coherent length in the normal region, $L \ll \xi$. There is no potential barrier at $x = 0, L$. The Zeeman energy $E_{\rm Z} = |g\mu_B B|$ and the pair potential Δ_0 are much smaller than the Fermi energy $E_{\rm F}$.

The Bogoliubov-de Gennes (BdG) equation is written as

$$\begin{pmatrix} H - E_{\rm F} & \hat{\Delta} \\ \hat{\Delta}^\dagger & -(H^* - E_{\rm F}) \end{pmatrix} \begin{pmatrix} \boldsymbol{\psi}_{\rm e} \\ \boldsymbol{\psi}_{\rm h} \end{pmatrix} = E \begin{pmatrix} \boldsymbol{\psi}_{\rm e} \\ \boldsymbol{\psi}_{\rm h} \end{pmatrix} \tag{1}$$

with $\hat{\Delta} = \Delta(x)\hat{g}$. $\boldsymbol{\psi}_{\rm e} = (\psi_{\rm e+}, \psi_{\rm e-})^{\rm T}$ and $\boldsymbol{\psi}_{\rm h} = (\psi_{\rm h+}, \psi_{\rm h-})^{\rm T}$ are the spinors for electron and hole, respectively. $\hat{g} = -i\hat{\sigma}_y$. The energy E is measured from the Fermi level $E_{\rm F}$. The BdG equation determines the Andreev levels E_n ($|E_n| < \Delta_0$) as a function of φ.

The ground state energy of junction is given by $E_{\rm gs}(\varphi) = -(1/2){\sum_n}' E_n(\varphi)$, where the summation is taken over all the positive Andreev levels, $E_n(\varphi) > 0$. The contribution from continuous levels ($|E| > \Delta_0$) can be disregarded in the case of short junctions [11]. At zero temperature, the supercurrent is calculated as $I(\varphi) = (2e/\hbar)(dE_{\rm gs}/d\varphi)$. The current is a periodic function for $-\pi \le \varphi < \pi$. The maximum (or absolute value of minimum) of $I(\varphi)$ yields the critical current $I_{\rm c}$.

The BdG equation in eq. (1) is expressed in terms of the scattering matrix [11]. The scattering matrix of electrons (holes) transport in the normal region is given by $\hat{S}_{\rm e}$ ($\hat{S}_{\rm h}$). $\hat{S}_{\rm e}$ and $\hat{S}_{\rm h}$ are

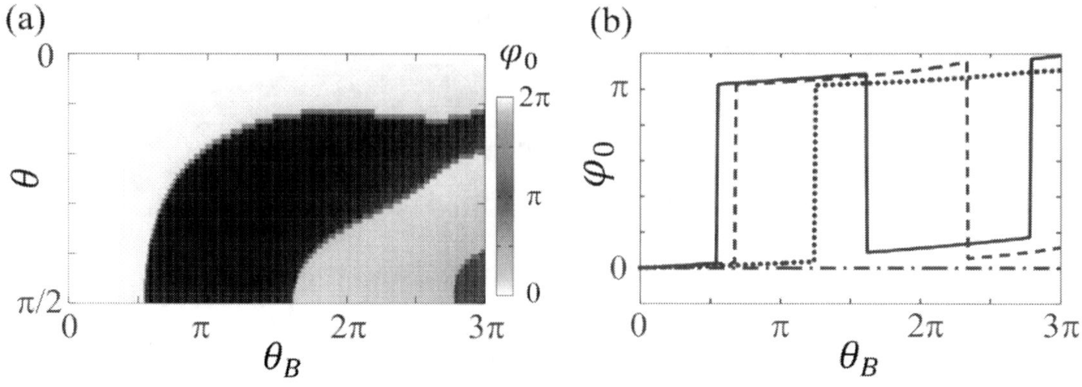

Figure 2. Numerical results of phase difference φ_0 at the minimum of ground-state energy when the number of conduction channel is unity and $l_{\text{mfp}}/L = 1$. The SO interaction is $l_{\text{SO}}/L = 0.2$. The results are for a sample. (a) Grayscale plot of φ_0 in the plane of magnetic field $\theta_B = E_Z L/(\hbar v_F)$ and its direction θ. (b) Cross section of panel (a) at $\theta = 0.5\pi$ (solid), 0.3π (broken), 0.15π (dotted), and 0 (dotted broken lines).

related to each other by $\hat{S}_e = \hat{S}_h^*$ for the short junctions. We denote $\hat{S}_e = \hat{S}$ and $\hat{S}_h = \hat{S}^*$. The Andreev reflection at $x = 0$ and L is described by the scattering matrix $\hat{r}_{\text{he}}$ for the conversion from electron to hole and $\hat{r}_{\text{eh}}$ for that from hole to electron. The normal reflection can be neglected. The matrix coefficients of $\hat{r}_{\text{he}}$ and $\hat{r}_{\text{eh}}$, e.g., $\exp\{-i \arccos(E/\Delta_0) - i\varphi/2\}$ for $\hat{r}_{\text{he}}$ at $x = 0$, are calculated from the boundary condition at $x = 0$ and L. The SO interaction does not affect the Andreev reflection coefficients. The Andreev levels, $E_n(\varphi)$, are obtained from the product of $\hat{S}$, $\hat{r}_{\text{he}}$, and $\hat{r}_{\text{eh}}$,

$$\det\left(\hat{1} - \hat{r}_{\text{eh}}\hat{S}^*\hat{r}_{\text{he}}\hat{S}\right) = 0. \tag{2}$$

Equation (2) is equivalent to the BdG equation (1).

To calculate the scattering matrix $\hat{S}$, we adopt the tight-binding model which discretizes a two-dimensional space (xy plane). The edges of nanowire are represented by a hard-wall potential. The width of nanowire is $W = 12a$ with the lattice constant $a = 10$nm. The Fermi wavelength is fixed at $\lambda_F = 18a$, where the number of conduction channels is unity. The length of normal region is $L = 50a$. The on-site random potential by impurities is taken into account, the distribution of which potential is uniform. We set that the mean free path due to the impurity scattering is $l_{\text{mfp}}/L = 1$. The SO length is $l_{\text{SO}}/L = 0.2$ with $l_{\text{SO}} = k_\alpha^{-1} = \hbar^2/(m\alpha)$. The magnetic field is $\boldsymbol{B} = B\boldsymbol{e}_\theta$ with the angle θ from the x axis in the xy plane.

3. Results

We consider a sample for the nanowire. For the magnetic field, we introduce a parameter, $\theta_B = E_Z L/(\hbar v_F)$, which means an additional phase due to the Zeeman effect in the propagation of electron and hole. Here, v_F is the Fermi velocity in the absence of SO interaction.

Figure 2 shows the phase difference φ_0 at the minimum of E_{gs} when the magnetic field is increased and rotated. In the absence of SO interaction, φ_0 takes only 0 or π exactly and clear 0-π transition happens (see Ref. [10]). In the presence of SO interaction, φ_0 is deviated from 0 and π, where the anomalous Josephson current is obtained. When the magnetic field is in the y direction ($\theta = \pi/2$), the transition between $\varphi_0 \approx 0$ and $\varphi_0 \approx \pi$ takes place around $\theta_B = \pi/2, 3\pi/2, \cdots$. The transition points are shifted gradually to large θ_B with decreasing of

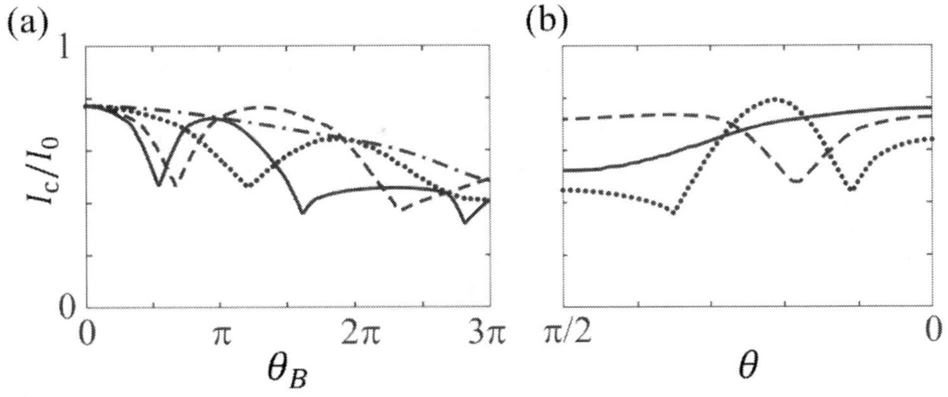

Figure 3. Numerical results of critical current I_c when $N = 1$ and $l_{mfp}/L = 1$. The SO interaction is $l_{SO}/L = 0.2$. $I_0 \equiv e\Delta_0/\hbar$. The sample is the same as that in Fig. 2. (a) I_c as a function of magnetic field $\theta_B = E_Z L/(\hbar v_F)$ when $\theta = 0.5\pi$ (solid), 0.3π (broken), 0.15π (dotted), and 0 (dotted broken lines). (b) I_c as a function of magnetic field orientation θ when $\theta_B = \pi/2$ (solid), π (broken), and 2π (dotted lines).

angle θ. At $\theta \approx 0$, the transition is not observed in Fig. 2.

The transition points correspond to the positions of cusps of critical current. Figure 3(a) exhibits the critical current when the magnetic field increases. The critical current I_c oscillates as a function of θ_B. The distance of cusps becomes long when the direction of magnetic field is tilted from the y axis. In the case of the parallel magnetic field to the nanowire ($\theta = 0$), I_c has no cusp in accordance with no transition. The critical current at $\theta = 0$ decreases with increase of θ_B although φ_0 is almost fixed at zero in Fig. 2(b). Figure 3(b) shows I_c when the magnetic field is rotated in the xy plane. The strength of magnetic field is fixed. We find the transition with the angle θ in Fig. 2(a). The critical current also oscillates as a function of θ. At small magnetic field ($\theta_B < \pi/2$), I_c changes monotonically. The oscillation with θ can be obtained when the magnetic field is $\theta_B > \pi/2$. If the state of junction at $\theta = \pi/2$ is $\varphi_0 \approx \pi$ (0), the critical current shows one cusp (two cusps) in Fig. 3(b). Therefore we can estimate the state at $\theta = \pi/2$ from the magnetic anisotropy of critical current.

4. Conclusions and Discussion

We have studied the DC Josephson effect in the semiconductor nanowire with strong SO interaction. We have examined a numerical simulation using the tight-binding model in the case of short junction. The combination of SO interaction and Zeeman effect in the nanowire results in the anomalous Josephson effect. The critical current oscillates as a function of magnetic field. In the presence of SO interaction, the oscillation of critical current depends on the magnetic field orientation. The oscillation period is the shortest when the magnetic field is perpendicular to the nanowire. For a parallel magnetic field to the nanowire, the 0-π (like) transition and the cusp of critical current are not found.

In this numerical model, we have considered the Rashba interaction. In the quasi-one-dimensional nanowire, the effective magnetic field induced by the Rashba interaction is in the y direction. The magnetic anisotropy of critical current is understood intuitively by the spin precession in the propagation of electron and hole. When the external magnetic field is parallel to the effective SO field (the y direction), the spin quantization axis is fixed in that direction. The electron and hole receive the additional phase in the propagation. On the other hand, when the external field is in the x direction, the spin quantization axes for electron and hole are not parallel with each other since the effective fields for electron and hole are antiparallel. The spin of electron and hole forming the Andreev bound state is rotated, which rotation cancels out the phase θ_B due to the Zeeman splitting. As a result, the critical current oscillation disappears. In the case of general SO interaction, the effective field would be deviated from the y axis. By measuring the magnetic anisotropy of critical current oscillation, we can evaluate the direction of effective field due to the SO interaction.

Acknowledgments

We acknowledge financial support by the Motizuki Fund of Yukawa Memorial Foundation. We acknowledge fruitful discussions about experiments with Professor L. P. Kouwenhoven, A. Geresdi, V. Mourik, K. Zuo of Delft University of Technology. T.Y. is a JSPS Postdoctoral Fellow for Research Abroad.

References

[1] Kato Y K, Myers R C, Gossard A C, and Awschalom D D 2004 *Science* **306** 1910
[2] Nadj-Perge S, Pribiag V S, van den Berg J W G, Zuo K, Plissard S R, Bakkers E P A M, Frolov S M, and Kouwenhoven L P 2012 *Phys. Rev. Lett.* **108** 166801
[3] Mourik V, Zuo K, Frolov S M, Plissard S R, Bakkers E P A M, and Kouwenhoven L P 2012 *Science* **336** 1003
[4] Doh Y-J, van Dam J A, Roest A L, Bakkers E P A M, Kouwenhoven L P, and Franceschi S De 2005 *Science* **309** 272
[5] Gharavi K, Holloway G W, Haapamaki C M, Ansari M H, Muhammad M, LaPierre R R, and Baugh J arXiv: 1405.7455
[6] Li C, Kasumov A, Murani A, Sengupta S, Fortuna F, Napolskii K, Koshkodaev D, Tsirlina G, Kasumov Y, Khodos I, Deblock R, Ferrier M, Guéron S, and Bouchiat H arXiv: 1406.4280
[7] Kouwenhoven L P, Geresdi A, Mourik V, and Zuo K (private communications)
[8] Buzdin A 2008 *Phys. Rev. Lett.* **101** 107005
[9] Yokoyama T, Eto M, and Nazarov Yu V 2013 *J. Phys. Soc. Jpn.* **82** 054703
[10] Yokoyama T, Eto M, and Nazarov Yu V 2014 *Phys. Rev.* B **89** 195407
[11] Beenakker C W J 1991 *Phys. Rev. Lett* **67** 3836; 1992 *Phys. Rev. Lett* **68** 1442(E)